Metallic Materials Specification Handbook

Metallic Materials Specification Handbook

Metallic Materials Specification Handbook

Robert B. Ross

Managing Director
W. H. Herdsman Ltd
Chemical Analysts and Metallurgical Consultants

THIRD EDITION

London   New York

E. & F. N. Spon

First published 1968 as
Metallic Materials
by Chapman and Hall Ltd
Second edition published 1972
by E. & F. N. Spon Ltd
11 New Fetter Lane, London EC4P 4EE
Third edition 1980

Published in the U.S.A.
by E. & F. N. Spon Ltd
in association with Methuen, Inc.
733 Third Avenue, New York, N.Y. 10017

© 1968, 1972, 1980 Robert B. Ross

Typeset by Elanders Ltd, Inverness and Kungsbacka, Sweden
Printed in Great Britain by the Fakenham Press Ltd, Fakenham

ISBN 0 419 11360 6

British Library Cataloguing in Publication Data

Ross, Robert Ballantyne
   Metallic materials specification handbook.
   – 3rd ed.
   1. Metals – Handbooks, manuals, etc.
   I. Title
   669'.0021'2   TA459   79-40761
   ISBN 0-419-11360-6

Contents

Preface

How to use this book

Preface

This third edition of **Metallic Materials Specification Handbook** is basically identical to the first two editions with the addition of approximately 10 000 new entries, giving a total of somewhere in the region of 50 000 trade names specifications or symbols relating to metals. As before, the materials are classified according to chemical analysis with a short note at the start of each group for the use of technicians. The opportunity has been taken to correct some errors and to rationalize the layout of the information, and the tabulated sections have been given a new format in order to conserve space. Abbreviations have been eliminated as far as possible, with the exception of chemical symbols used in the tabulated data section and the use of conventional abbreviations where mechanical properties are involved.

Under no circumstances can this book be taken as giving authority to use a material in lieu of the designer's specification. It will, however, greatly assist in the preparation of a list of alternative materials for consideration. The addresses of manufacturers, of the various trade names and specifications are given in appendixes, so that further detailed information can be obtained where necessary. In this edition information on specifications and symbols of unknown origin have also been included, as it was felt in many instances readers would find this of some value.

The size of the volume clearly indicates that there is something wrong with the present method of identifying materials. There is no obvious solution, but as a first step users and suppliers should call for the appropriate national specification which already exists. National bodies should be prepared to be more flexible and produce workable specifications before materials enter general use and become known by proprietary names.

The reader will note that the nature and size of this work make it impossible to be completely up-to-date. Considerable efforts have been made to ensure accuracy but the author can accept no responsibility for any mistakes. He would be grateful if any errors could be brought to his attention for future correction.

Many of the materials listed are the subject of patent rights and nothing in this book should be taken as permission to infringe these rights.

Robert B. Ross
May 1979

How to use this book

There will be two main reasons for using this book:

(a) To find the ingredients and properties of a symbol representing a specification or trade name.

(b) Knowing the desired properties, to find the specification and trade names which fulfil these requirements.

Taking (a), the reader will have a symbol representing the material in question. Reference should first be made to the index at the rear, which covers all the specifications listed, giving the full nomenclature, as well as the commonly known abbreviations; for example BS 970, En 24 appears under BS 970 En 24, and also under En 24. This list is in numero-alphabetical order, with numbers always taking precedence over letters, and refers the symbol to the section containing all the relevant information.

While every attempt has been made to include all known variations of trade names and specifications it is possible that some have been missed. It is therefore suggested that the reader use his initiative by adding to, or subtracting from, the symbols in his possession. Before each of the sections on the commonly used materials appears a short note on the methods of classification used by the various national bodies. Having identified the section in which the desired specification appears, this will be found to contain, again in numero-alphabetical order, all the specifications and trade symbols similar to the wanted specification. Against each is shown the nominal composition, mechanical properties, condition, maker's name if appropriate and any relevant remarks. The reader can thus find the composition and properties of his unknown symbols.

Prior to each group of specifications is listed certain physical properties, with a note on the metallurgy, thermal treatment, weldability, method of flaw detection, corrosion protection, machinability, and uses of that range of materials. This information is of necessity brief and as it covers more than one specification must be to some extent general. Every care has however been taken to ensure that the specifications are in groups small enough to make the information as meaningful as possible.

At the end of the book, under Appendix I are listed the names and addresses of the companies supplying the material, or national body responsible for producing the specification. Thus the reader has at his finger tips technical information on the actual specification or symbol being queried and also all known similar specifications and trade symbols with the name and address of the material supplier, or the body producing the specification. It should be emphasized, however, that the information in this book is general rather than specific and reference must be made to the actual specifications involved prior to any important decision being made.

Buyers, store-keepers, planners and production engineers should all find that being able rapidly and easily to obtain information on alternative materials can save time and money.

Turning now to (b) where the reader knows his design requirements and wants to give it a name, the list of contents gives the sections devoted to each group of materials. Listed with each group are certain physical properties such as specific gravity, conductivity and expansion. In each group are all known specifications and trade symbols of alloys with the chemical composition to give these properties and against each specification are the nominal composition, mechanical properties and condition. Thus the designer can see at a glance whether or not his requirements are covered by existing specifications, and has the means of rapidly checking on the availability of the material.

While on the whole, abbreviations have been avoided it has been found convenient to use some for the sake of clarity or where the additional space otherwise required did not appear to be justified. Physical property tabulated matter uses the standard abbreviations and element names appear as the chemical symbol. Full translations are given in Appendix III. The headings in the tabulated matter have been abbreviated for space saving reasons and translations appear in Appendix III and immediately prior to the table in each section.

Conversion tables and factors for all common physical properties are given in Appendix III and for some, conversion factors are given at the head of each section.

Appendix I lists in alphabetical order the suppliers and associations which appear throughout the text, and Appendix II lists trade names with their originator.

1. Aluminium Al

Physical properties

Atomic number	13	
Atomic weight	26.97	
Crystal structure	Face centred cubic	
Colour	Tin-white	
Specific gravity	2.7	
Density	2700 kg/m^3	(0.098 lb/in.^{3})
Melting point	658 °C	
Boiling point	2270 °C	
Specific heat	0.90 J/g °C	(0.214 cal/g °C)
Thermal conductivity	210 W/m °C	(50 cal/m s °C)
Coefficient of linear expansion (20–100° C)	24 × 10^{-6}/ °C	
Latent heat of fusion	386.9 J/g	(92.4 cal/g)
Latent heat of vaporization	9462 J/g	(2260 cal/g)
Thermal neutron absorption cross-section	0.215 barns/atom	
Electrical conductivity	62–62.9% IACS (copper 100%)	
Specific resistance	27.8–27.4 microhm mm	
Temperature coefficient of electrical resistance	0.0041/ °C	
Electrochemical equivalent	0.3354 g/A/h	
Electrode potential	-1.69 V	
Magnetic susceptibility	0.6 × 10^{-6}	
Young's modulus of elasticity	68.3 × 10^9 N/m^2	(9.9 × 10^6 lbf/in.^{2})
Tensile strength	annealed 45 N/mm^2	(3 tonf/in.^{2})
Hardness	annealed 15 DPN	

1.1 General notes on aluminium

Aluminium is available in commercial grades from 99.0 per cent to 99.9 per cent purity, with a well-defined series of alloys based on the metal, which are found in all the normal forms of castings and wrought products, and also as a fine powder used in paints and metal spraying.

The metal and its alloys are used because of their lightness, corrosion resistance and strength.

Aluminium alloys weigh approximately one.third that of steel, have a much better corrosion resistance to atmospheric conditions and depending on the alloys can be of a comparable strength to low alloy steel.

Following are listed all known specifications and trade names covering aluminium and its alloys. These are grouped in such a manner that all specifications for each well-defined alloy appear under the one heading, with a general note covering their properties, processing and uses.

1.2 Notes on specifications and trade names

An explanation of the different classifications and systems used by the various bodies and companies whose alloys are listed in the following pages may be of help in rapidly finding the desired material and in understanding the symbols.

British Standards Institution (BS)

This body issues individual specifications covering a particular alloy.

The present policy however, is to have a master specification related to the form of the material and in the main these replace individual specifications. Thus:

BS 1470 Sheet strip and plate – all alloys
BS 1471 Drawn tube – all alloys
BS 1472 Forgings and forging stock – all alloys
BS 1473 Rivet, bolt and screw stock for forging – all alloys
BS 1474 Bar, extruded tube and sections – all alloys
BS 1475 Wire – all alloys
BS 1490 Castings – all alloys

BS 1476 and BS 1477 have now been withdrawn and included in BS 1470 and BS 1474 respectively.

The specification number is followed by the letters N or H, denoting non-heat-treatable (N) alloys, or heat-treatable (H) alloys. Commercially pure aluminiums have no letter.

The form is next identified by the following code letters:

B   Bolt and screw stock for forging
C   Clad sheet and strip
E   Bars and sections
F   Forgings and forging stock

G   Wire
LM   Casting
P   Plate
PC   Clad plate
R   Rivet stock
S   Sheet up to and including 0.252 in. thick,
         and strip up to and including 0.192 in. thick
T   Drawn tube
V   Extruded round tube and hollow section

This is followed by a number with which the actual alloy is coded. Thus 1 is always 99.99%; aluminium 1A is always 99.98%; aluminium 6 is always 5% magnesium-aluminium alloy.

In the index alloys are listed under their full title, and also in the abbreviated form but only the full form is shown in the classified sections. Thus high purity aluminium is shown under 1A 99.98% aluminium, and under BS 1470 S1A 99.98% aluminium sheet, BS 1471 T1A 99.98% aluminium tube, etc.; and 1.5% copper 1% magnesium 1% silicon aluminium alloy is shown under H 11, and under BS 1472 HF11–1.5% copper 1% magnesium 1% silicon aluminium alloy forging, etc.

The abbreviated alloy forms will be encountered under other specification numbers than those listed above. They are also sometimes incoporated into trade names.

The condition in which the material is supplied or used is often coded and then added to the specification as a suffix.

This practice has not been adhered to in the following pages, as it is felt that the code is too arbitrary to be easily memorized. Also the BS code is similar to, but differs in detail from the American code. Thus the condition has been added to the description wherever possible.

The code used by the BS range is as follows:

H8   Fully hardened by cold working (previously H)
H2   Cold worked by approximately one quarter of possible amount, or fully cold worked then partially annealed to remove approximately three quarters of the cold work (previously H4)
H4   As H2 but cold worked to half possible (previously H$\frac{1}{2}$)
H6   As H2 but cold worked to three-quarters possible (previously H$\frac{3}{4}$)
M   As cast; as rolled; as extruded
O   Annealed
TE   Precipitation treated (aged only) (previously P)
TP   Solution treated only (may room age, depending on the alloy) (previously W)
TH   Drawn after solution treatment, then aged (previously WDP)
TF   Solution treated and aged (previously WP)

In the above and in the following pages the term 'aged' indicates artificial ageing. Room ageing is always described as such.

In addition to the normal BS range, a series of aircraft alloys are listed. These are individual specifications which have the prefix L. The range also includes magnesium alloys. As these are commonly known as BSL and L alloys

both forms of identification are listed in the index the L alloy being referred to the full BSL specification to avoid confusion.

American Aluminium Association (AA)

This body does not issue mandatory specifications, but has organized a four figure system which is partly descriptive. It is often referred to as 'Commercial Designation'. In the following pages the system is referred to as 'designation used by AA'. This alloy classification is now used by ASTM and SAE but not under AMS. There are five groups of alloys, identified by the first digit thus:

1xxx   Pure aluminium
2xxx   Copper aluminium alloys
3xxx   Manganese aluminium alloys
4xxx   Silicon aluminium alloys
5xxx   Magnesium aluminium alloys
6xxx   Silicon, magnesium aluminium alloys
7xxx   Zinc aluminium alloys

The remaining three digits signified by x are used to describe the alloying elements.

To date this system applies only to wrought products and not to castings.

American Society for Testing & Materials (ASTM)

This body issues several volumes of specifications at frequent intervals. The non-ferrous alloys are pre-fixed by the letter B followed by a number which identifies the specification. In some instances this will cover one material only, but more commonly it will cover all alloys in one form. Thus B209 is the specification for sheet aluminium alloys and must be followed by the four figures (designation used by AA), identifying the actual alloys.

The full specification then has the date of issue and in many instances the letter T signifying that the specification is still tentative. An example of such a specification would be:

ASTM B209 3003/61. This is for manganese aluminium alloy sheet, the specification being issued in 1961. The full specification appears only in the index, the year of issue being omitted in the classified section of this book.

Previous to 1961 ASTM used their own alloy designation. This code is still used for casting alloys. Letters are given for up to two principal alloying elements, then up to three figures code the amount of these elements. Letters following the figures denote slight variations of the same basic specification. The alphabetical code used is:

A	Al	H	Th	Q	Ag
B	Bi	K	Zr	R	Cr
C	Cu	L	Li	S	Si
D	Cd	M	Mn	T	Sn
E	Rare earths	N	Ni	Y	Sb
F	Fe	P	Pb	Z	Zn
G	Mg				

Thus ASTM B209 M1A58 is for manganese aluminium alloy sheet, now ASTM B209 3003/61. In the following pages only the alloy designation (i.e. M1A) is listed, and cross-referred to the present designation (i.e. 3003). The

ASTM casting alloys still employ this system of classification.

American Society of Mechanical Engineers (ASME)

This body issues specifications which are identical as far as material content is concerned to ASTM. They use the prefix S in lieu of ASTM (e.g. SB 125 would be identical to ASTM B125).

Society of Automotive Engineers (SAE) of America

This body issues general purpose wrought and cast alloys which were identified by two or three figures, prefixed by SAE. The wrought alloys now use the AAA four figure code prefixed AA. There is no attempt to code the casting alloys.

This society also issue a series of specifications for aircraft materials. Aerospace Material Specifications have four figures prefixed by the letters AMS. The suffix letter indicates the issue number.

Thus AMS 4006 is a manganese aluminium alloy. These four figures bear no relationship to the AA code.

Temper designations used by American societies

The condition in which the material is supplied or used has been coded and this is in general use as a suffix to specifications in America. This has not been used in the following pages, as the condition is described in full.

The code is as follows:

Type 1   Alloys strengthened only by cold work
Type 2   Alloys which respond to heat treatment
F   As fabricated
H   Strain hardened – (cold worked),
     sub-divided as follows:
H1   Strain hardened (cold worked) only. A second number defines the degree of cold work, up to 8, which is fully hard. Thus H14 is half hard, H16 three quarters hard.
H2   Strain hardened (cold worked) and partly annealed. A second number indicates the degree of cold work remaining after softening, up to 8, which is fully hard. Thus, for example, H24 is half hard.
H3   Strain hardened (cold worked) and stabilized. A second number indicates the degree of cold work in the stabilized alloy, up to 8, which is fully hard. Thus H34 is half hard.
O   Annealed
T2   Annealed – castings only
T3   Solution treated and cold worked

T4   Solution treated – room aged
T5   Aged
T6   Solution treated and aged
T7   Solution treated and stabilized
T8   Solution treated, cold worked and aged
T9   Solution treated, aged and cold worked
T10   Aged and cold worked
W   Solution treated – unstable condition

In the above, and in the following pages the term 'aged' indicates artificial ageing: room ageing is always described as such.

Trade names and symbols

These generally include some means of identifying the company, and often attempt to describe the alloy, either by percentage or using one of the national alloy designations.

Wherever possible all the forms under which the alloy might be used have been included in the following pages. It is possible, however, that some have been omitted. If, therefore, the desired alloy cannot be found in the index, attempts should be made to add or subtract the firm's name as a prefix to the symbol.

German National Standards (DIN)

This body is analogous to the British Standards Institution and like them has issued individual specifications covering certain alloys used under particular conditions.

There are now two distinct systems of designations, one using a descriptive code with chemical symbols and numbers, the second using numbers only, known as 'Werkstoff number'. Both systems of designation are contained in master specifications of which there are four principal ones relating to aluminium.

These are DIN 1712, high purity aluminium; DIN 1725, aluminium alloys; DIN 1732, welding wire and DIN 8512, brazing and soldering fillers.

Examples of the methods of designation are, first the descriptive code:

Al Mg 7   7% magnesium aluminium alloy
Al Cu Mg 2   4.2% copper 1.6% Magnesium
              0.6% manganese aluminium alloy

These have largely been replaced by the numeral code which is designed to suit modern computing machines. All aluminium alloys are included in the range 3.000 to 3.499; examples are:

3.3557   7% magnesium aluminium alloy
3.1355   4.2% copper 1.6% magnesium 0.7%
         manganese aluminium alloy

1A Aluminium – commercially pure – wrought and cast

Specific gravity	2.7	
Density	2700 kg/m^3	(0.098 lb/in.^{3})
Solidus/liquidus	wrought 646–660 °C	
	cast 645–655 °C	
Thermal conductivity	230 W/m °C	(55.0 cal/m s °C)
Coefficient of linear expansion	24 × 10^{-6}/ °C	
Electrical conductivity	wrought, annealed 59–61% IACS (copper 100%)	
	cast, annealed 57% IACS	
Specific resistance	wrought, annealed 28 microhm mm	
	cast, annealed 30 microhm mm	
Young's modulus of elasticity	68.3 × 10^9 N/m^2	(9.9 × 10^6 lbf/in.^{2})
Impact	wrought 27 J	(20 ft lb) Izod
Fatigue strength (50 × 10^6 cycles)	wrought, annealed ± 40 N/mm^2	(± 2.5 tonf/in.^{2})
	wrought, ½-hard ± 55 N/mm^2	(± 3.5 tonf/in.^{2})
	wrought, hard ± 75 N/mm^2	(± 5.0 tonf/in.^{2})
	cast, annealed ± 30 N/mm^2	(± 2.0 tonf/in.^{2})

Hot strength

Temperature	Annealed		Half-hard	
°C	Tensile strength N/mm^2	Elongation %	Tensile strength N/mm^2	Elongation %
150	50	65	110	16
260	22	85	23	85
315	17	90	17	90
370	9.0	95	9.0	95

The properties on this page are typical of the following group, and may not apply exactly to any one specification. It is possible that with certain specifications some of the values may not be applicable.

General metallurgical characteristics

The listed specifications cover aluminium in its various forms from 99% to 99.99% purity. The materials have low strength and high ductility with excellent corrosion resistance, and are capable of being deep drawn or rolled to very thin foils.

The materials have poor casting properties, having low fluidity with high solidification shrinkage. This combined with a low strength after solidification, causes the defect known as 'hot shortness', resulting in cracking. Most of the cast materials listed are primary metals.

Thermal treatment

The only heat treatment possible is annealing at a temperature of 350–400 °C for 1 hour to remove all traces of cold work. Lower temperature will remove a proportion of the cold work but it is important to note that the hardness cannot be regained by any form of thermal treatment once the material has been annealed.

An anneal may be necessary between deep drawing operations, but only if an exceptional amount of cold work is being carried out.

Annealing can be carried out in an air furnace for all normal purposes, although the surface achieved will not be a good reflector.

Welding and brazing

Pre-treatment. No thermal treatment. The weld areas must be thoroughly cleaned and scratch brushed.

Welding & brazing. Normal gas and electric welding using fluxes, which must be completely removed after welding since they are corrosive.

Inert gas shielded welding is the most satisfactory method, as no flux is required. Resistance welding is possible using special equipment.

Brazing requires fluxes and uses silicon alloy rods which reduce the corrosion resistance.

Post treatment. None required. It should be noted however that welding will have locally removed the effect of any cold work, leaving a soft area.

Flaw detection methods

Crack test. Oil penetrant chalk method, but seldom required.

X-ray. Not normally required on these materials.

Ultrasonic test. Not required.

Chemical etch. Only required for grain size or directional properties on wrought alloys. A 10% caustic solution is satisfactory.

Corrosion protection

Temporary. None required.

Permanent. None required except to preserve appearance. Where necessary the parts can be anodized using either the chromic, oxalic, or sulphuric acid process. Dyeing after anodizing is also possible. Where a measure of corrosion protection is required which does not justify anodizing, there are various forms of chemical oxidation available. Special grades of high purity aluminium are available and should be specified when components are to be electropolished and colour anodized.

Plating. Electroplating is possible, but difficult and requires special techniques.

Painting. This is not normally required as the metal can be colour anodized. Where necessary, either air drying or stoving enamels can be used, but in either case an etch primer, or chemical pre-treatment prior to painting is necessary. This is much cheaper than colour anodizing.
Lead based primers must not be used.

Machinability

Difficulty is often experienced in obtaining a good finish with this material in the soft conditions. Where it is essential and the material must be annealed, it may be necessary to machine in the hard condition and anneal the finished part. High speeds and low tool pressures give best results.

Uses

Wrought. For cladding age hardening alloys. Foils for food wrapping and packaging (silver paper). In the colour anodized condition it is used for decorative purposes, jewellery, ash trays and toys.
Certain grades are used for electrical bus bars and conductors.
Some architectural uses where high strength is not important.

Cast. As primary metal for alloying. As low strength corrosion resistant structural castings. Ornamental castings where surface defects are unimportant, and accuracy of reproduction does not matter.

Symbol	Nominal analysis, supplier, condition and remarks.
1S	99.5% Al: Sheet and strip; Alcan
2S	99.0% Al: Sheet; Alcan
2S	99.0% Al: Sheet; Alcan for BS alloy S1C
2S	99.0% Al: Previous AA designation; now 1100
3.0185	98% Al with Cu and Cr: German Standard DIN 1712
3.0200	99.0% Al: German Standard DIN 1712
3.0205	99.0% Al: German Standard DIN 1712
3.0250	99.5% Al: German Standard DIN 1712
3.0255	99.5% Al: Designation used in German Standard DIN 1712
3.0256	99.5% Al: Wrought; German Standard DIN 1712
3.0257	99.5% Al: German Standard DIN 1712
3.0270	99.7% Al: Wrought; German Standard DIN 1712
3.0275	99.7% Al: Designation used in German Standard DIN 1712
3.0280	99.8% Al: Wrought; German Standard DIN 1712
3.0285	99.8% Al: German Standard DIN 1712
3.0305	99.85% Al: German Standard DIN 1712
3.0385	99.99% Al: German Standard DIN 1712
3.040	99.99% Al: Wrought; German Standard DIN 1712
3.4415	1% Zn Al alloy: Used for cladding; German Standard DIN 1725
72S	1% Zn Al alloy: Used for cladding; previous AA designation; now 7072
99.8	99.8% Al: Bar, tube, etc; British Al Co for BS alloy 1A
990A	99.0% Al: Previous ASTM designation; now 1100
1130	99.0% Al: Wrought; Designations used by AA; no ASTM specification

Symbol	Nominal analysis, supplier, condition and remarks.
1145	0.55% Si + Fe 99.45% Al: Foil for cladding; AA designation; no ASTM alloy
1175	99.75% Al: Wrought; designation used by AA; no ASTM specification
1230	0.1% Cu 0.1% Zn Al: Used for cladding; designation used by AA
1235	99.35% Al: Foil, AA designation; no ASTM alloy
1345	99.45% Al: Wire; AA designation; no ASTM alloy
1350	99.5% Al: For electrical conductivity; designation used by ASTM and AAA
A 2	99.8% Al: Obsolete; Southern Forge; replaced by Impalco 030
A 3	99.5% Al: Obsolete; Southern Forge; replaced by Impalco 050
A 4	99% Al: Sheet, tube, bar, etc; Southern Forge; replaced by Impalco 102
A 4	99.00% Al: Ingot, billet, etc; L'Aluminium Francais
A 4	99.00% Al: Ingot; French Standard; part of NFA S7.101
A 4.5	0.1% Cu Al: Wrought; French Standard
A 5	99.5% Al: Ingot; French Standard; part of NFA S7.101
A 5	99.5% Al: Wrought; French Standard
A 5	99.5% Al: Ingot, billet, etc; L'Aluminium Francais
A 5B	99.5% Al: Low Si & Fe for foil production; L'Aluminium Francais
A 5L	99.5% Al: For electrical conductors; L'Aluminium Francais
A 7	99.70% Al: Ingot, billet, etc.; L'Aluminium Francais
A 7	99.70% Al: Ingot; French Standard; part of NFA S7.101
A 8	99.80% Al: Ingot; French Standard; part of NFA S7.101
A 8	99.8% Al: Ingot, billet, etc.; L'Aluminium Francais
A 9	99.99% Al: Super purity, ingot, billet, etc.; L'Aluminium Francais
AA 1100	0.2% Cu 99.0% Al: Bar, sheet, tube etc.; SAE designation; formerly SAE 25

Note. The following abbreviations and units are used in the tables:

DPN	Hardness, diamond pyramid number
UTS	Ultimate tensile strength, N/mm^2
Elon	Elongation, %
Proof	0.1 % proof strength, N/mm^2

$1 \ N/mm^2 = 0.1 \ hbar = 0.102 \ kgf/mm^2 = 0.06475 \ tonf/in^2 = 145.04 \ lbf/in^2$

Symbol	Nominal analysis, supplier, condition and remarks.
AA 1230	0.1% Cu 99.3% Al: Used only for cladding; SAE designation; formerly SAE 28
AA 5005	0.7% Mg Al alloy: Wrought
AGG 1	99.5% Al: Wrought; further information from Aluminium Zentrale
AGG 2	99.0% Al: Wrought; further information from Aluminium Zentrale
AL 3	99.0% Al: Sheet, tube, forgings etc.; James Booth for BS alloy 1C
AL 5	99.5% Al: Sheet, tube, forgings & plate; James Booth for BS alloy 1B
AL 99/99R	99.99% Al: For high reflectivity; previous German Standard designation
ALCAN GB 1S	99.5% Al: Sheet, tube, red forgings, etc.; Alcan for BS alloy 1B
ALCAN GB 2S	99.25% Al: Wire & bar; Alcan for BS alloy 1C
ALCAN GB 99.7	99.7% Al: Sheet Alcan; annealed — UTS: 80 Elon: 50%
ALCAN GB 99.8	99.8% Al: Alcan for BS alloy S1A
ALCAN GB 99.99	99.99% Al: Alcan for BS alloy S1
ALCAN GB 100	99.5% Al: Casting; Alcan; as cast — DPN: 24 UTS: 8 Elon: 55% Proof: 30
ALCAN GB B1S	99.5% Al: Wire for electrical purposes; Alcan for BS alloy 1E
ALCAN GB C1S	99.5% Al: Wire for electrical purposes; Alcan for BS alloy 1E
ALCOA 050	99.5% Al: Wrought; Alcoa; annealed — UTS: 75 Elon: 30% Proof: 30
ALCOA 051	99.5% Al Cu+Si+Fe 0.5% (max): Wrought; Alcoa; annealed — UTS: 60 Elon: 25%
ALCOA 102	99.0% Al: Wrought; Alcoa; annealed — UTS: 80 Elon: 28%
ALMINAL 3	0.1% Cu 0.5% Si 0.1% Mn Al: Wrought alloy; origin unknown
ALMINTRODE 96.10	99.6% Al 0.2% Fe 0.2% Si: For electrodes; ESAB
ALOLINE	Hg free Al for anodes: Wilson Walton
ALUM T	99.0% Al: Bar; E Kaye for BS alloy 1E
ALUMOWELD	Mild steel core wire coated with high conductivity aluminium; Copperweld Co
Al Zn 1	1% Zn Al alloy: Previous German Standard designation
AMS 4000 B	99.7% Al: Sheet; annealed — UTS: 80
AMS 4001 B	99.0% Al: Sheet; annealed — UTS: 80 Elon: 20%
AMS 4003 B	99.0% Al: Sheet; ½-hard — UTS: 110 Elon: 2%
AMS 4062 C	99.0% Al: Seamless tube; ½-hard; AMS for AA alloy 1100
AMS 4102 A	99.9% Al: Bar as rolled; AMS for AA alloy 1100
AMS 4180 B	99.0% Al: Wire for spraying; cold drawn; AMS for AA alloy 1100
AMTROD 18.01	0.15% Fe 0.06% Zn Al: Welding wire; ESAB; for inert gas; as welded — UTS: 70 Elon: 25%

Symbol	Nominal analysis, supplier, condition and remarks.
ASTM B37	Aluminium for use in steel manufacture: Purity denoted by grade symbol
ASTM B179/995 A	99.5% Al: Ingot; primary metal
ASTM B184/A12	99.5% Al (min): For electrode filler wire; may be flux coated
ASTM B209/1060	99.6% Al: Sheet; annealed — UTS: 45 Elon: 25% Proof: 7
ASTM B209/1060	99.6% Al: Sheet; cold rolled ½-hard — UTS: 60 Elon: 10% Proof: 45
ASTM B209/1100	99.0% Al: Sheet; annealed — UTS: 80 Elon: 30% Proof: 7
ASTM B209/1100	99.0% Al: Sheet; hard — UTS: 140 Elon: 4%
ASTM B209/1230	0.7% Si+Fe 99.3% Al: Sheet & plate
ASTM B209/5257	0.4% Mg Al alloy: Sheet & plate
ASTM B209/5557	0.2% Mn 0.6% Mg Al alloy: Sheet & plate
ASTM B209/5657	0.8% Mg Al alloy: Sheet & plate
ASTM B209/7072	0.7% Si+Fe 1.0% Zn Al alloy: Sheet & plate
ASTM B210/1060	99.6% Al: Tube; annealed — UTS: 45 Proof: 7
ASTM B210/1060	99.6% Al: Tube; cold drawn — UTS: 90 Proof: 70
ASTM B210/1100	99.0% Al: Tube; annealed — UTS: 90
ASTM B210/1100	99.0% Al: Tube; cold drawn — UTS: 140
ASTM B210/7072	0.7% Si+Fe 1.0% Zn Al alloy: Tube
ASTM B211/1060	99.6% Al: Bar; annealed — UTS: 45 Elon: 25% Proof: 7
ASTM B211/1060	99.6% Al: Bar; cold drawn — UTS: 90 Proof: 70
ASTM B211/1100	99.0% Al: Bar; annealed — UTS: 60 Elon: 25% Proof: 7
ASTM B211/1100	99.0% Al: Bar; cold drawn — UTS: 140
ASTM B221/1060	99.6% Al: Extruded bar tube etc.
ASTM B221/1100	99.0% Al: Extruded bar; annealed — UTS: 60 Elon: 25% Proof: 7
ASTM B221/7072	0.7% Si+Fe 1.0% Zn Al alloy: Extruded bar etc.
ASTM B233 1350	99.5% Al: Rod for electrical purposes; 61.5% 1ACS
ASTM B234/1060	99.6% Al: Tube; cold drawn — UTS: 60 Proof: 45
ASTM B234/7072	1.0% Zn Al alloy: Seamless drawn tube
ASTM B235/1060	99.6% Al: Tube; extruded annealed — UTS: 60 Proof: 15
ASTM B235/1100	99.0% Al: Tube; extruded annealed — UTS: 90
ASTM B241/1060	99.6% Al: Pipe & tube
ASTM B241/1100	1.0% Si+Fe Al: Pipe & tube
ASTM B241/7072	0.7% Si+Fe 1.0% Zn Al: Pipe & tube
ASTM B247/1100	99.0% Al: Forging; as forged — DPN: 20 UTS: 60 Elon: 25% Proof: 15
ASTM B285 ER1060	99.6% Al: Electric welding rod; obsolete
ASTM B285 ER1100	99.0% Al: Welding wire
ASTM B285 ER1260	99.6% Al: Welding wire
ASTM B307/1100	99.0% Al: Coiled tube; annealed — UTS: 60
ASTM B307/5005	0.8% Mn Al alloy: Drawn tube
ASTM B313/1100	99.0% Al: Welded tube; cold rolled — UTS: 140 Elon: 4%
ASTM B316/1100	99.0% Al: For rivets; annealed — UTS: 90
ASTM B327 G1C	95.0% (min) Al 0.9% Mg Al alloy: Hardener for Zn based die castings
ASTM B345/1060	99.6% Al: Pipe; annealed — UTS: 45 Proof: 7
ASTM B361	Weld fitments on Al and Al alloys; graded by AA code (see also other A sections)
ASTM B373/1145	99.45% (min) Al alloy: Foil for capacitors

Note. The following abbreviations and units are used in the tables:

DPN	Hardness, diamond pyramid number
UTS	Ultimate tensile strength, N/mm^2
Elon	Elongation, %
Proof	0.1 % proof strength, N/mm^2

$1\ N/mm^2 = 0.1\ hbar = 0.102\ kgf/mm^2 = 0.06475\ tonf/in^2 = 145.04\ lbf/in^2$

Symbol	Nominal analysis, supplier, condition and remarks.
ASTM B373/1235	99.35% (min) Al alloy: Foil for capacitors
ASTM B396/5805	0.8% Mg Al alloy: Wire for electrical purposes
ASTM B404	99.6% (min) Al and Al alloy: For condensor tubes; graded by AA system (see also other A sections)
ASTM B544	99.5% Al: Rod for electrical conductivity; 61.5% IACS
ASTM B547	Al alloy: Drawn and welded tube; alloys designated by AAA number
ASTM B566	Cu clad Al wire
ASTM B606	Zn coated steel-cored Al alloy: Conductors
AWCO 99.5	99.5% Al: Wire; Aluminium Wire Co
AWCO 99.8	99.8% Al: Wire; Aluminium Wire Co
AWCO EP	0.05% Cu 0.5% Al: For electrical purposes; Aluminium Wire & Cable Co for BS alloy 1E
AWCO SP	99.9% Al: Wire; Aluminium Wire & Cable Co for BS alloy 1A
BA	Hg free Al: For anodes; British Al
BA 99	99.0% Al: Sheet, bar, forging, etc.; British Al Co for BS Alloy 1C
BA 99.5	99.5% Al: Bar, tube, etc.; British Al Co for BS alloy 1B
BA 99.8	99.8% Al: Sheet, bar, forging, etc.; British Al Co for BS alloy 1A
BA 177	0.7% Si 0.7% Fe Al alloy: Sheet and strip; British Al Co; $\frac{1}{2}$-hard
BA COMMERCIAL QUALITY	99.0% Al: Bar, tube, etc.; British Al Co for BS alloy 1C
BA ELECTRICAL PURITY	0.05% Cu 99.5% Al: For electrical purposes; British Al Co for BS alloy E1E DPN: 22 UTS: 75 Elon: 47% Proof: 30
DA SUPER PURITY	99.99% Al: Bar, tube, etc.; British Al Co. DPN: 15 UTS: 45 Proof: 30
BIRMETAL 1	99.5% Al: In wrought form; Birmetal
BIRMETAL 2M	99.0% Al: In wrought form; Birmetal
BS 215	99.5% Al: Wire for conductors; material conforms to BS2627 G1E
BS 359	98.0% Al: Ingot for remelting; replaced by BS 1490
BS 360	99.0% Al: Ingot for remelting; replaced by BS 1490
BS 385	Pure Al tubes. Replaced by BS 1471
BS 386	Pure Al bar & sections. Replaced by BS 1476
BS 1453 G1	99.9% Al: For filler rod for welding pure Al; BS alloy 1
BS 1453 G1A	99.8% Al: For filler rod for welding pure Al; BS alloy 1A
BS 1453 G1B	99.5% Al: For filler rod for welding pure Al; BS alloy 1B
BS 1453 G1C	99.0% Al: For filler rod for welding Al; BS alloy 1C
BS 1470 S1	99.99% Al: Sheet & strip; annealed UTS: 60 Elon: 45%
BS 1470 S1	99.99% Al: Sheet & strip; $\frac{1}{2}$-hard UTS: 90 Elon: 12%
BS 1470 S1	99.99% Al: Sheet & strip; hard UTS: 90 Elon: 6%
BS 1470 S1A	99.8% Al: Sheet & strip; annealed UTS: 70 Elon: 35%
BS 1470 S1A	99.8% Al: Sheet & strip; $\frac{1}{2}$-hard UTS: 110 Elon: 8%
BS 1470 S1A	99.8% Al: Sheet & strip; hard UTS: 120 Elon: 5%
BS 1470 S1B	99.5% Al: Sheet & strip; annealed UTS: 90 Elon: 30%
BS 1470 S1B	99.5% Al: Sheet & strip; $\frac{1}{2}$-hard UTS: 110 Elon: 8%
BS 1470 S1B	99.5% Al: Sheet & strip; hard UTS: 120 Elon: 5%

Symbol	Nominal analysis, supplier, condition and remarks.
BS 1470 S1C	99.0% Al: Sheet & strip; annealed UTS: 90 Elon: 30%
BS 1470 S1C	99.0% Al: Sheet & strip; $\frac{1}{2}$-hard UTS: 110 Elon: 12%
BS 1470 S1C	99.0% Al: Sheet & strip; $\frac{1}{2}$-hard UTS: 120 Elon: 7%
BS 1470 S1C	99.0% Al: Sheet & strip; $\frac{3}{4}$-hard UTS: 140 Elon: 5%
BS 1470 S1C	99.0% Al: Sheet & strip; hard UTS: 140 Elon: 3%
BS 1471 T1A	99.8% Al: Tube; annealed UTS: 70
BS 1471 T1A	99.8% Al: Tube; hard UTS: 90
BS 1471 T1B	99.5% Al: Tube; annealed
BS 1471 T1B	99.5% Al: Tube; hard
BS 1471 T1C	99.0% Al: Tube; annealed
BS 1471 T1C	99.0% Al: Tube; hard
BS 1471 F1A	99.8% Al: For forging & bar; annealed UTS: 45 Elon: 30%
BS 1472 F1B	99.5% Al: For forging & bar; annealed UTS: 60 Elon: 25%
BS 1472 F1C	99.0% Al: For forging & bar; annealed UTS: 60 Elon: 20%
BS 1473 R1B	0.5% Cu+Si+Fe+Mn+Zn (max) 99.5% Al: Roll and screw stock; cold rolled UTS: 110
BS 1473 R1C	99.0% Al: For rivets; $\frac{1}{2}$ hard UTS: 110
BS 1474 V1A	99.8% Al: For tube; as drawn UTS: 45 Elon: 30%
BS 1474 V1B	99.5% Al: For tube; as drawn UTS: 60 Elon: 25%
BS 1474 V1C	99.0% Al: For tube; as drawn UTS: 60 Elon: 20%
BS 1475 G1	99.99% Al: Wire; hard drawn UTS: 90
BS 1475 G1A	99.8% Al: Wire; annealed UTS: 70
BS 1475 G1A	99.8% Al: Wire; hard drawn UTS: 120
BS 1475 G1B	99.5% Al: Wire; annealed UTS: 90
BS 1475 G1C	99.0% Al: Wire; annealed UTS: 90
BS 1475 G1C	99.0% Al: Wire; hard drawn UTS: 140
BS 1476 E1A	99.8% Al: Bar; as rolled UTS: 45 Elon: 30%
BS 1476 E1B	99.5% Al: Bar; as rolled UTS: 60 Elon: 25%
BS 1476 E1C	99.0% Al: Bar; as rolled UTS: 60 Elon: 20%
BS 1477 P1A	99.8% Al: Plate; $\frac{1}{2}$-hard UTS: 110 Elon: 8%
BS 1477 P1B	99.5% Al: Plate; $\frac{1}{2}$-hard UTS: 110 Elon: 8%
BS 1477 P1C	99.0% Al: Plate; $\frac{1}{2}$-hard UTS: 110 Elon: 7%
BS 1490 LMO	99.5% Al: Ingot
BS 1490 LMOM	99.5% Al: Castings; as cast
BS 1616 A	99.5% Al: Welding rod; flux coated; wire to BS alloy 1B
BS 1683	99.0% Al: Sheet or strip for wrapping cheese with a protective coating
BS 2627 G1E	0.05% Cu (max) 99.5% Al: Wire for electrical purposes; annealed; electrical conductivity 58% IACS
BS 2627 G1E	0.05% Cu (max) 99.5% Al: Wire for electrical purposes; $\frac{3}{4}$-hard; electrical conductivity 56% IACS

Symbol	Nominal analysis, supplier, condition and remarks.
BS 2627 G1E	0.05% Cu (max) 99.5% Al: Wire for electrical purposes; hard; electrical conductivity 56% IACS
BS 2791	99.5% Al: Wire for electrical conductors as BS 2627 G1E
BS 2897 D1E	0.05% Cu (max) 99.5% Al: Strip for electrical purposes; annealed **UTS: 90 Elon: 25%**
BS 2897 D1E	0.05% Cu (max) 99.5% Al: Strip for electrical purposes; ½-hard **UTS: 110 Elon: 38%**
BS 2897 D1E	0.05% Cu (max) 99.5% Al: Strip for electrical purposes; hard **UTS: 140 Elon: 3%**
BS 2898 E1E	0.05% Cu (max) 99.5% Al: Bar for electrical purposes; ½-hard **UTS: 70 Elon: 15%**
BS 2901 G1	99.99% Al: Rod; for all types of welding
BS 2901 G1A	99.8% Al: Rod; for all types of welding
BS 2901 G1B	99.5% Al: Rod; for all types of welding
BS 2901 G1C	99.0% Al: Rod; for all types of welding
BS 3313 S1C	99.0% Al: Foil for dairy product containers; ½-hard **UTS: 120**
BS 3988 C1E	0.5% Cu (max) Cu+Si+Fe 0.5% (max) Al: Wrought; for electrical purposes **UTS: 80 Elon: 25%**
BS 4300/2 BTRS1	0.5% Mg Al alloy: For bright trim; annealed **UTS: 120**
BS 4300/7 NS41	0.9% Mg Al alloy: Sheet & strip; cold rolled **UTS: 180 Elon: 3% Proof: 160**
BS 4300/9 NG41	0.9% Mg Al alloy: Wire; cold drawn **UTS: 180**
BS 6791	0.05% Cu (max) 99.5% Al: For insulated cables order as BS2627 - N68
BS L4	High purity Al: Sheet; obsolete
BS L16	99.0% Al: Sheet; ½-hard **UTS: 120 Proof: 110**
BS L17	99.0% Al: Sheet; soft **UTS: 90**
BS L30	98.0% Al: Casting
BS L31	99.0% Al: Ingots; virgin metal for re-melting
BS L32	Aluminium bars; obsolete
BS L34	99.0% Al: Bar or section **UTS: 70**
BS L36	99.0% Al: Wire or tube for rivets **UTS: 110**
BS L48	99.7% Al: Ingots; virgin metal for re-melting
BS L49	99.0% Al: Ingots; secondary metal for re-melting
BS L54	99.0% Al: Tube; hydraulically tested cold drawn
BS L67	99.0% Al: Tube; hydraulically tested cold drawn **UTS: 110**
BE T9	Aluminium tube; obsolete **UTS: 110**
CINDAL	0.3% Cr 0.12% Zn 0.2% Mg Al alloy: Origin unknown
DIN 1712	Chemical composition of high purity, aluminium products; German Standard
DIN 1725	Chemical composition of aluminium cast and wrought alloys; German Standard

Note. The following abbreviations and units are used in the tables:

DPN	Hardness, diamond pyramid number
UTS	Ultimate tensile strength, N/mm^2
Elon	Elongation, %
Proof	0.1 % proof strength, N/mm^2

1 N/mm^2=0.1 hbar=0.102 kgf/mm^2=0.06475 tonf/in^2=145.04 lbf/in^2

Symbol	Nominal analysis, supplier, condition and remarks.
DURALBRITE 1	99.8% (min) Al extrusions: J Booth; hard drawn **UTS: 120**
DURALBRITE 2	1.0% Mg 0.3% Mn Al extrusion: J Booth; hard drawn **UTS: 210**
DURALBRITE 3	0.7% Mg Al extrusion: J Booth; hard drawn **UTS: 210**
DURALCOTE	Pre-painted aluminium sheet: Alcan Booth
DURCILIUM T	0.05% Cu 99.5% Al: For electrical purposes; E Kaye for BS alloy 1E
E Al	99.9% Al: Wrought; previous German Standard designation
ERMAL 99.0	99.0% Al: Sheet & strip; Enfield spec for BS 1470 S1C
ERMAL 99.5	99.5% Al: Sheet & strip; Enfield spec for BS 1470 S1B
ERMAL 99.7	99.7% Al: Sheet & strip; Enfield; annealed **UTS: 75 Elon: 50%**
HA 2.9900	99.00% Al: Ingot for re-melting; Canadian Standard
HA 2.9950	99.50% Al: Ingot for re-melting; Canadian Standard
HA 2.9970	99.70% Al: Ingot for re-melting; Canadian Standard
HA 2.9980	99.80% Al: Ingot for re-melting; Canadian Standard
HA 2.9990	99.90% Al: Ingot for re-melting; Canadian Standard
HA 2.9999	99.99% Al: Ingot for re-melting; Canadian Standard
HA 4.990	99.0% Al: Plate or sheet; Canadian Standard
HA 5.990C	99.0% Al: Bar etc.; Canadian Standard
HA 6.990	99.0% Al: For rivets and brazing wire; Canadian Standard
HA 7.995	99.5% Al: Tube; Canadian Standard
H Al 1	99.9999% Al: Ingot; Koch Light Ltd; high purity metal
H Al 6A	99.999% Al: Ingot; Koch Light Ltd.; high purity metal
H Al 9	99.999% Al: Sheet; 2 mm thick; Koch Light Ltd.; high purity metal
HIDUMINIUM 1A	99.8% Al: Sheet, bar, forgings, etc.; High Duty Alloy for BS alloy 1A
HIDUMINIUM 1B	99.5% Al: Sheet, bar, forgings, etc.; High Duty Alloy for BS alloy 1B
HIDUMINIUM 1C	99.0% Al: Sheet, bar, forgings, etc.; High Duty Alloy for BS alloy 1C
HIDUMINIUM 100	Commercially pure Al: Sintered powder, bar & sheet; High Duty Alloys **UTS: 350 Elon: 7% Proof: 210**
IMPALCO 030	99.8% Al: Sheet, tube, bar, etc.; Imperial Al Co for BS alloy 1A
IMPALCO 050	99.5% Al: Sheet, tube, wire bar, etc.; Imperial Al Co for BS alloy 1B
IMPALCO 051	99.5% Al: Wire & bar for electrical purposes; Imperial Al Co for BS alloy 1E
IMPALCO 102	99.0% Al: Sheet, tube bar, etc.; Imperial Al Co for BS alloy 1C
IMPALCO 700	0.7% Zn Al alloy: Imperial Al Co
IMPALCO P3	99.8% Al: Sheet strip and plate; Imperial Al Co for BS alloy 1A
IMPALCO P5	99.5% Al: Sheet, strip, tube, bar, forgings & plate; Imperial Al Co for BS alloy 1B
IMPALCO P5E	99.5% Al: For wire & rod; high conductivity aluminium; Imperial Al Co for BS alloy 1E
IMPALCO P10	99.0% Al: For sheet, strip, tube, bar, forgings & plate; Imperial Al Co for BS alloy 1C
KYNAL P5	99.5% Al: ICI obsolete; now Imperial Al Co
KYNAL P10	99.0% Al: ICI obsolete; now Imperial Al Co

Symbol	Nominal analysis, supplier, condition and remarks.
MG11A	1% Zn Al alloy: Previous ASTM designation now 7072
NFA57.101	Classification of pure Al: French Standard
NF A57-101-A4	99.0% Al: Ingot; French National Standard
NF A57-101-A5	99.5% Al: Ingot; French National Standard
NF A57-101-A7	99.7% Al: Ingot
NF A57-101-A8	99.8% Al: Ingot
NORAL S	99.0% Al: Sheet & tube; Northern Al Co for BS alloy 1C
QQ A 411b	1.0% Fe+Si 99% Al: Bar, US Federal; annealed **UTS: 90** **Elon: 25%**
QQ A 561	99.0% Al: Sheet & strip; US Federal; annealed **UTS: 70** **Elon: 20%**
RAFFINAL 990	99.99% Al: Wrought; Anglo-Swiss Al Co
REINALUMINIUM 99.0	99.0% Al: Wrought; Anglo-Swiss Al Co
REINALUMINIUM 99.5	99.5% Al: Wrought; Anglo-Swiss Al Co
ROTOR ALUMINIUM 04	99.5% Al: Ingot; Anglo-Swiss Al Co
SAE 25	99.0% Al: Wrought annealed; now AA 1100 **UTS: 90** **Elon: 30%**
SAE 25	99.0% Al: Wrought hard; now AA 1100 **UTS: 140** **Elon: 4%**
SAE 28	0.1% Cu 99.3% Al: Used only for cladding; analysis applies before cladding; now AA 1230
SAE 214	1% Zn Al alloy: Used for cladding; now 7072
SIS 40-20	99.8% Al: Billet ingot or casting; Swedish Standard
SIS 40-21	99.70% Al: Billet ingot or casting; Swedish Standard
SIS 40-22	99.50% Al: Billet ingot or casting; Swedish Standard
SIS 40-24	99.00% Al: Billet ingot or casting; Swedish Standard
SIS 4005	99.7% Al: Sheet & strip; Swedish Standard; annealed **DPN: 25** **UTS: 70** **Elon: 35%** **Proof: 30**
SIS 4005	99.7% Al: Sheet & strip; Swedish Standard; cold worked **DPN: 45** **UTS: 150** **Elon: 3%** **Proof: 110**

Symbol	Nominal analysis, supplier, condition and remarks.
SIS 4007	99.5% Al: Sheet; strip & tube; Swedish Standard; annealed **DPN: 25** **UTS: 90** **Elon: 30%** **Proof: 30**
SIS 4007	99.5% Al: Sheet, strip & tube; Swedish Standard; cold worked **DPN: 45** **UTS: 150** **Elon: 3%** **Proof: 120**
SIS 4008	99.5% Al: Bar & tube; for electrical purposes; Swedish Standard; annealed **DPN: 25** **UTS: 70** **Elon: 30%** **Proof: 30**
SIS 4008	99.5% Al: Bar and tube; for electrical purposes; Swedish Standard; cold worked **DPN: 35** **UTS: 120** **Elon: 5%** **Proof: 70**
SIS 4020	99.8% Al: Ingot; Swedish Standard; primary metal
SIS 4021	99.7% Al: Ingot; Swedish Standard; primary metal
SIS 4022	99.5% Al: Ingot; Swedish Standard; primary metal
SIS 4023	99.3% Al: Ingot; Swedish Standard; primary metal
SIS 4024	99.0% Al: Ingot; Swedish Standard; primary metal
STA 7 A2	99.8% Al: Sheet; replaced by BS 1470 S1A
STA 7 A3	99.5% Al: Bar; replaced by BS 1476 E1B
STA 7 A4B	99.0% Al: Bar; replaced by BS 1476 E1C
STA 7 A4C	99.0% Al: Tube; BS 1471 T1C
STA 7 A4D	99.0% Al: Sheet; replaced by BS 1470 S1C
STA 7 A4E	99.5% Al alloy: Wire; replaced by BS 1473/5 R1C
T 9	Al: Tube; obsolete; as BS T9
UNI 3021 APR04	0.4% Cr Al alloy: Ingot; primary metal; Italian Standard
UNI 3021 APT04	0.4% Ti Al alloy: Ingot; primary metal; Italian Standard
UNI 3567/66	99.0% Al: Italian Standard
UNI 3950 AP0	99.00% Al: Ingot; Italian Standard
UNI 3950 AP3	99.3% Al: Ingot; Italian Standard
UNI 3950 AP5	99.5% Al: Ingot; Italian Standard
UNI 3950 AP7	99.70% Al: Ingot; Italian Standard
UNI 3950 AP8	99.80% Al: Ingot; Italian Standard
UNI 4507	99.5% Al: Italian Standard
UNI 4508	99.7% Al: Italian Standard
UNI 4509	99.8% Al: Italian Standard
UNI 820	98-99.5% Al: Ingot; Italian Standard
W Al 8	99.999% Al: Powder; Light Ltd; high purity metal
WW T 783a	1.0% Fe+Si(max) 99.0% Al: Tube; US Federal

1B Aluminium–manganese wrought alloys

Specific gravity	2.73	
Density	2730 kg/m^3	(0.099 lb/in.^{3})
Solidus/liquidus	643–654 °C	
Thermal conductivity	annealed 193 W/m °C	(46 cal/m s °C)
	$\frac{1}{2}$-hard 176W/m °C	(42 cal/m s °C)
	hard 155 W/m °C	(37 cal/m s °C)
Coefficient of linear expansion	23.2 × 10^{-6}/ °C	
Electrical conductivity	annealed 51% IACS (copper 100%)	
	$\frac{3}{4}$-hard 41% IACS	
	hard 40% IACS	
Specific resistance	annealed 33.5 microhm mm	
	hard 43.1 microhm mm	
Young's modulus of elasticity	66.9 × 10^9 N/m^2	(9.7 × 10^6 lbf/in.^{2})
Impact	annealed 35 J	(25 ft lb) Izod
	$\frac{1}{2}$-hard 28 J	(21 ft lb) Izod
Fatigue strength (50 × 10^6 cycles)	annealed ± 55 N/mm^2	(±3.5 tonf/in.^{2})
	$\frac{1}{2}$-hard ±70 N/mm^2	(±4.5 tonf/in.^{2})
	hard ±85 N/mm^2	(±5.5 tonf/in.^{2})

Hot strength

Temperature	Annealed		Hard	
°C	Tensile strength N/mm^2	Elongation %	Tensile strength N/mm^2	Elongation %
150	75	47	156	12
200	55	50	114	15
260	36	60	69	25
315	25	60	35	55

The above properties are typical of the following group, and may not apply exactly to any one specification. It is possible that with certain specfications some of the values may not be applicable.

General metallurgical characteristics

The addition of 1–2% manganese to aluminium slightly increases the strength. This increase is more marked when the alloys are cold worked and is achieved with only a very slight decrease in the corrosion resistance.

The addition of manganese does not make the alloy hardenable by heat treatment, but the specifications are capable of accepting a considerable amount of cold work without intermediate anneals.

Thermal treatment

The only heat treatment possible on the alloys is an anneal at 370–450 °C. For best results this should be a salt bath treatment to prevent grain growth. In most instances normal air furnaces will give satisfactory results, but will remove the reflective surface and may give rise to some increase in grain size, owing to the slower rate of heating.

It should be noted that the strength achieved by cold work is removed by annealing and can only be regained by further cold work.

An anneal may be required between deep drawing operations, but this will only be necessary when the work involved is excessive.

Welding and brazing

Pre-treatment. No thermal treatment required. The weld areas must be thoroughly cleaned and scratch brushed.

Welding and brazing. Normal gas and electric welding techniques are satisfactory using fluxes which are corrosive and must be completely removed after welding. Inert gas shielded arc welding gives best results. Resistance welding requires special equipment.

Brazing requires fluxes and high silicon braze filler rods.

Post treatment. None required. It should be noted however that welding and brazing will remove any cold work, leaving a local weak area, very often with a low corrosion resistance.

Flaw detection methods

Crack test. Oil penetrant chalk method.

X-ray. Not normally required.

Ultrasonic test. Not required.

Chemical etch. Only required for grain size or directional estimation, when a 10 per cent caustic solution is satisfactory.

Corrosion protection

Temporary. None required.

Permanent. None required except to preserve appearance. Chromic, oxalic or sulphuric acid anodizing can be applied when necessary. Various chemical oxidation processes are available which give a lesser degree of protection than anodizing, but are generally satisfactory and considerably cheaper.

Painting. This is not normally required as these specifications can be colour anodized. Air drying and stoving enamels can be used if required, but in both instances an etch primer or chemical pre-treatment prior to painting is necessary. This is much cheaper than colour anodizing. Lead based primers must not be used.

Plating. Electroplating is possible using special techniques, but has very few applications.

Machinability

Difficulty may be countered in achieving a good finish when machining these materials in the soft condition.

If a good finish is necessary and the part must be soft it is recommended that it is machined in the hard or ¾-hard condition and finally annealed.

Uses

Structural work; where low strength with good corrosion resistance is required, for example, corrugated sheeting. Hollow-ware; where deep drawing quality and corrosion resistance are more important than strength. Vehicle panelling, where strength and corrosion resistance are required.

Symbol	Nominal analysis, supplier, condition and remarks.
3S 1	1.2% Mn Al alloy: Previous AA designation; now 3003
3S	1.25% Mn Al alloy: Alcan for BS alloy N3
3.0515	1.25% Mn Al alloy: German Standard DIN 1725
3.0570	1% Mn 0.5% Si 0.8% Fe Al alloy; German Standard
4S	1.2% Mn 1% Mg Al alloy: Previous AA designation now 3004
AA 3003	1.2% Mn Al alloy: Wrought; SAE designation; formerly SAE 29
AA 3004	1.2% Mn 1% Mg Al alloy: Wrought; SAE designation; formerly SAE 20
ALCAN GB 3S	1.2% Mn Al alloy: Alcan for BS alloy N3
ALCOA 190	1.2% Mn Al alloy: Wrought; Alcoa; ½-hard H4 condition — **UTS: 150 Elon: 6%**
ALM	1% Mn Al alloy: Sheet; J Booth for BS alloy N3
ALMINTRODE 96.20	1.30% Mn 98.0% Al 0.30% Fe: For electrodes; ESAB
Al Mn	1.2% Mn 0.2% Mg Al alloy: Wrought; previous German Standard designation
ALUMAL	1.25% Mn Al alloy: Origin unknown
ALUMAN 100	1.2% Mn Al alloy: Wrought; Anglo-Swiss Al Co
AM 1	1.2% Mn Al alloy: Wrought; L'Aluminium Francais
AMGO5	0.4% Mg 1.2% Mn Al alloy: Sheet; French Standard
AMI	1.2% Mn Al alloy: Wrought; French Standard
AMIG	1.0% Mg 1.2% Mn Al alloy: Sheet, French Standard
AMS 4006 B	1.2% Mn Al: Sheet annealed; AMS for AA alloy 3003 — **UTS: 120 Elon: 17%**

Symbol	Nominal analysis, supplier, condition and remarks.
AMS 4008 C	1.2% Mn Al: Sheet ½ hard; AMS for AA alloy 3003 — **UTS: 160 Elon: 3%**
AMS 4010	1.2% Mn Al: Foil; hard rolled; AMS for AA alloy 3003 — **UTS: 180**
AMS 4065 B	1.2% Mn Al alloy: Tube; annealed; AMS for AA alloy 3003 — **UTS: 110**
AMS 4067 B	1.2% Mn Al alloy: Tube; ½-hard; AMS for AA alloy 3003 — **UTS: 120**
ASTM B209/3003	1.2% Mn Al alloy: Sheet; annealed — **UTS: 110 Elon: 23% Proof: 7**
ASTM B209/3003	1.2% Mn Al alloy: Sheet; cold rolled; ½-hard — **UTS: 180 Elon: 12% Proof: 30**
ASTM B209/3004	1.2% Mn 1% Mg Al alloy: Sheet & plate
ASTM B209/3005	1.2% Mn 0.4% Mg Al alloy: Sheet & plate
ASTM B210/3003	1.2% Mn Al alloy: Tube; annealed — **UTS: 110 Proof: 25**
ASTM B210/3003	1.2% Mn Al alloy: Tube; cold drawn — **UTS: 180 Proof: 150**
ASTM B210/3004	1.2% Mn 1% Mg Al alloy: Tube; annealed — **UTS: 190 Proof: 45**
ASTM B210/3004	1.2% Mn 1% Mg Al alloy: Tube; cold drawn — **UTS: 250 Proof: 190**
ASTM B211/3003	1.2% Mn Al alloy: Bar; cold drawn — **UTS: 180**
ASTM B221/3003	1.2% Mn Al alloy: Extruded bar; annealed — **UTS: 70 Elon: 25% Proof: 15**
ASTM B221/3004	1.2% Mn 1% Mg Al alloy: Extruded bar, etc.
ASTM B234/3003	1.2% Mn Al alloy: Tube; cold drawn — **UTS: 150 Proof: 120**
ASTM B235/3003	1.2% Mn Al alloy: Extruded tube; annealed — **UTS: 120 Elon: 25% Proof: 30**
ASTM B235/3004	1.2% Mn 1% Mg Al alloy: Extruded tube; annealed — **UTS: 190 Proof: 45**
ASTM B241/3003	1.2% Mn Al alloy: Pipe; cold drawn — **UTS: 180 Proof: 150**
ASTM B247/3003	1.2% Mn Al alloy: Forging; as forged — **DPN: 25 UTS: 70 Elon: 25% Proof: 15**
ASTM B285 ER3004	1.2% Mn 1% Mg Al alloy: Welding rod; obsolete

Note. The following abbreviations and units are used in the tables:

DPN	Hardness, diamond pyramid number
UTS	Ultimate tensile strength, N/mm^2
Elon	Elongation, %
Proof	0.1 % proof strength, N/mm^2

1 N/mm^2=0.1 hbar=0.102 kgf/mm^2=0.06475 tonf/in^2=145.04 lbf/in^2

Symbol	Nominal analysis, supplier, condition and remarks.
ASTM B307/3003	1.2% Mn Al alloy: Coiled tube; annealed **UTS: 110**
ASTM B307/3004	1.3% Mn 1.0% Mg Al alloy: Drawn tube
ASTM B313/3003	1.2% Mn Al alloy: Welded tube; cold rolled **UTS: 180 Elon: 3%**
ASTM B318/3003	1.2% Mn Al alloy: Drawn coiled tubes; annealed **UTS: 120**
ASTM B318/3004	1.2% Mn 1% Mg Al alloy: Drawn coiled tube; annealed **UTS: 180**
ASTM B345/3003	1.2% Mn Al alloy: Pipe; annealed **UTS: 70 Proof: 15**
ASTM B345/3004	1.2% Mn 1% Mg Al alloy: Pipe; annealed **UTS: 150 Proof: 30**
ASTM B404/3003	1.25% Mn Al alloy: Condenser tubes
AWCO 60	1.25% Mn Al alloy: Wire, Al Wire Co for BS alloy N3
BA 60	1.25% Mn Al: Sheet; British Al Co for BS alloy N3
BIRMETAL 3	1.0% Mn Al alloy: In wrought form; Birmetal
BS 1453 N3	1.2% Mn Al alloy: Welding rod for welding Al alloy N3 type
BS 1470 NS3	1% Mn Al alloy: For sheet & strip; ¾-hard **UTS: 170 Elon: 5%**
BS 1470 NS3	1% Mn Al alloy: For sheet & strip; hard **UTS: 170 Elon: 3%**
BS 1470 NS3	1% Mn Al alloy: For sheet & strip; annealed **UTS: 110 Elon: 30%**
BS 1470 NS3	1% Mn Al alloy: For sheet & strip; ¼-hard **UTS: 120 Elon: 12%**
BS 1470 NS3	1% Mn Al alloy: For sheet & strip; ½-hard **UTS: 150 Elon: 7%**
BS 1472 NF3	1% Mn Al alloy: For forgings & bar; annealed **UTS: 90 Elon: 20% Proof: 45**
BS 1475 NG3	1% Mn Al alloy: Wire; annealed **UTS: 120**
BS 1475 NG3	1% Mn Al alloy: Wire; hard drawn **UTS: 170**
BS 1477 NP3	1% Mn Al: Plate; ½-hard **UTS: 150 Elon: 7%**
BS 2901 NG3	1.2% Mn Al alloy: Rod for inert gas shielded arc welding
BS 3313 NS3	1% Mn Al alloy: Strip for dairy product containers; ½-hard **UTS: 150**
BS 4300/1 NJ3	1.2% Mn Al alloy: Welded tube; as manufactured **UTS: 180 Elon: 3%**
BS 4300/6 NS31	0.7% Mn 0.4% Mg Al alloy: Sheet & strip; cold rolled **UTS: 210 Elon: 2% Proof: 190**
BS L59	1% Mn Al alloy: For sheet; ¾-hard **UTS: 170 Elon: 5%**
BS L60	1% Mn Al alloy: Sheet & strip; ¼-hard **UTS: 120 Elon: 12%**
BS L61	1% Mn Al alloy: For sheet & strip; annealed **UTS: 90 Elon: 30%**
DTD 213	Mn Al alloy: Replaced by BSL59
DTD 653	Mn Al alloy: Replaced by BSL60
ELM 15	1% Mn Al alloy: Strip; Elm Engineering
HA4 MC10	0.15% Cu 1.2% Mn Al alloy: Plate; Canadian Standard
HA7 MC10	0.1% Cu 1.2% Mn Al alloy: Tube; Canadian Standard

Symbol	Nominal analysis, supplier, condition and remarks.
HIDUMINIUM 11	1.25% Mn Al: For extrusions & tube; as rolled; High Duty Alloy **DPN: 30 UTS: 110 Elon: 30% Proof: 70**
HIDUMINIUM 11	1.25% Mn Al: For extrusions & tube; ½ hard; High Duty Alloy **DPN: 45 UTS: 150 Elon: 7% Proof: 110**
IMPALCO 190	0.15% Cu 1.2% Mn 0.6% Si Al alloy: Sheet & wire; Imperial Al Co for BS alloy N3
IMPALCO PA19	1% Mn Al alloy: For sheet, strip & wire; Imperial Al Co for BS alloy N3
KYNAL PA19	1.25% Mn Al alloy: ICI; obsolete; now Imperial Al Co
M1A	1.2% Mn Al alloy: Previous ASTM designation; now 3003
MANGALAL	1.2% Mn Al alloy: Sheet; J Booth for BS alloy N3
MG 11A	1.2% Mn 1% Mg Al alloy: Previous ASTM designation; now 3004
NORAL 3S	1.25% Mn Al: Sheet; North Al Co for BS 1470 NS3
QQ A 356b	1.2% Mn Al alloy: Bar; US Federal; annealed **UTS: 110 Elon: 25% Proof: 30**
QQ A 359a	1.2% Mn Al alloy: Sheet; US Federal; annealed **UTS: 110 Elon: 22%**
SAE 20	1.2% Mn 1% Mg Al alloy: Wrought hard; now AA 3004 **UTS: 250 Elon: 3%**
SAE 20	1.2% Mn 1% Mg Al alloy: Wrought annealed; now AA 3004
SAE 29	1.3% Mn Al alloy: Sheet; annealed; now AA 3003 **UTS: 120 Elon: 25%**
SAE 29	1.3% Mn Al alloy: Sheet; hard; now AA 3003 **UTS: 180 Elon: 4%**
SIS 4054	1.25% Mn Al alloy: Sheet & strip; Swedish Standard; annealed **DPN: 30 UTS: 120 Elon: 25% Proof: 45**
SIS 4054	1.25% Mn Al alloy: Sheet & strip; Swedish Standard; cold rolled **DPN: 60 UTS: 170 Elon: 3% Proof: 140**
STA7 AW 3C	1.2% Mn Al alloy: Sheet; replaced by BS 1470 NS3
TI 11	1.25% Mn Al: Sheet; TI Al Co. spec for BS 1470 NS3
UNI 3568	1.2% Mn Al alloy: For forging; Italian Standard
UNI 6361	1.0% Mg 1.2% Mn Al alloy: For forging; Italian Standard
V Al Mn 1	1% Mn 0.5% Si Al alloy: Previous German Standard designation
W 3	1.2% Mn Al alloy: Sheet; replaced by IMPALCO 190; Southern Forge
WW T 788A	1.2% Mn Al alloy: Tube; US Federal

Note. The following abbreviations and units are used in the tables:

DPN	Hardness, diamond pyramid number
UTS	Ultimate tensile strength, N/mm^2
Elon	Elongation, %
Proof	0.1 % proof strength, N/mm^2

$1 \ N/mm^2 = 0.1 \ hbar = 0.102 \ kgf/mm^2 = 0.06475 \ tonf/in^2 = 145.04 \ lbf/in^2$

1C Aluminium–silicon and wrought cast alloys

Specific gravity	2.6–2.7	
Density	2600–2700 kg/m³	(0.096–0.098 lb/in.³)
Solidus/liquidus	525–625 °C	
Thermal conductivity	101–126 W/m57 C	(24–30 cal/m s °C)
Coefficient of linear expansion	20–23 × 10⁻⁶/ °C	
Electrical conductivity	annealed 26–39% IACS (copper 100%)	
Specific resistance	40–50 microhm mm	
Young's modulus of elasticity	71.0×10^9 N/m²	$(10.3 \times 10^6$ lbf/in.²)
Impact	7–8.5 J	(5–6.5 foot/lb) Izod
Fatigue strength 50×10^6 cycles	sand cast ±45 N/mm²	(±3–5 tonf/in.²)
Hot strength		

Temperature °C	Tensile strength N/mm²
100	240
150	210
200	180
250	90
300	60
350	37
400	34

The above properties are chosen to show the typical values for the alloys listed, and may not apply exactly to any one specification. It is possible that with certain specifications some of the values may not be applicable.

General metallurgical characteristics

Silicon reduces the melting point progressively to the eutectic at approximately 11%, while small amounts of other alloying elements can vary this between 10–13% silicon.

These alloys are characterized by excellent fluidity at casting temperatures, allowing thin sections to be cast and giving excellent reproducibility. The defect known as 'hot shortness' is virtually absent. This means that the strength of the casting immediately after solidification is sufficient to withstand shrinkage stresses, allowing the casting of intricate shapes with rapid change in section.

The alloys are not heat treatable and their corrosion resistance is not particularly high.

Additions of small amounts of copper, iron and nickel strengthen the alloys, without reducing too much the casting properties.

Drawn wire is used as brazing filler rod.

Thermal treatment

Most of these alloys cannot have their mechanical properties improved by any form of thermal treatment. There is however, an alloy with 1% copper, magnesium, nickel and 11% silicon specially developed for pistons which can be solution treated and aged. This should receive the treatment outlined in Section 1F.

If a casting is of an intricate shape, or if dimensional stability is essential, a stress release at 250–300 °C for several hours is recommended.

Welding and brazing

Pre-treatment. No thermal treatment required. The weld area must be thoroughly cleaned and scratch brushed.

Welding and brazing. The alloys with 1% copper should not be welded or brazed. The straight silicon aluminium alloys can be readily welded using fluxes, or inert gas shielded arc welding. Brazing is not recommended as the Al/Si brazing rods have melting points close to the alloys being brazed.

Post treatment. None required, except for thorough cleaning to remove all traces of flux.

Flaw detection methods

Crack test. Penetrant oil chalk test is recommended for all castings which are to be stressed.

X-ray. This is not often specified on these alloys, as they are used for their excellent castability rather than high strength.

Ultra-sonic test. Not normally required.

Pressure test. An air in water pressure test is recommended to prove the absence of leaks. If a little wetting agent, such as soap or detergent is added to the water the sensitivity of the test is increased. High gas pressures should not be used without taking safety precautions against rupture.

Hydraulic pressure testing is recommended when high test pressures are necessary. Again soap or detergent should be added to the water.

Chemical etch. As these are good casting alloys it should seldom be necessary to etch the parts, except as a quality check. A hot 10% caustic soda solution is satisfactory. The alloys with copper and nickel may require a concentrated nitric acid dip subsequently to remove the black deposit.

Corrosion protection

Temporary. None required.

Permanent. Under many circumstances no protection is required. Only when appearance is important, or with marine or humid conditions will these alloys require anodizing or chemical oxidation. Sulphuric acid anodizing gives the best results, but it is doubtful if the increase in corrosion protection is justified by the additional cost.

Painting. This requires either an etch primer or to be preceded by chemical oxidation or anodizing. Air drying lacquers, or stoved paints are satisfactory, but lead based primers must not be used.

Plating. Electroplating is possible using special techniques. Only with excellent castings, showing no surface defects will good results be obtained. The foundry should always be informed that their castings are to be plated.

Machinability

These alloys all exhibit a tendency to tear and flow owing to the soft matrix, this being lessened when copper and nickel are present. Also, owing to the presence of silicon, tool wear is high.

The use of tungsten carbide tipped tools with high surface speed and low pressure overcomes both of the above defects to some extent.

Uses

For intricate castings, with low or medium strength, covers, electric motor casings, camera cases and portable tools. Thin walled castings such as fuse boxes, and brackets where strength and corrosion resistance are not required. Drawn wire is used as brazing filler rod and is the only known use for the wrought product.

Symbol	Nominal analysis, supplier, condition and remarks.
3.2131	1.2% Cu 5.5% Si 0.4% Mg Al alloy: Casting; German Standard; solution treated and aged **DPN: 95 UTS: 230 Elon: 1% Proof: 200**
3.2152	4% Cu 7% Si Al alloy: German Standard DIN 1725
3.2153	3% Cu 7% Si 0.2% Mg 0.4% Mn 1.2% Zn Al alloy: Casting; German Standard DIN 1725
3.2245	0.2% Mg 5.25% Si Al alloy: German Standard DIN 1725
3.2285	7.5% Si Al alloy: German Standard DIN 1725
3.2341	5% Si Al alloy: German Standard DIN 1725
3.2381	0.5% Mg 9.5% Si Al alloy: German Standard DIN 1725
3.2383	10% Si 0.3% Mg 0.2% Al alloy: Casting; German Standard
3.2385	0.3% Cu 12% Si Al alloy: German Standard DIN 1725
3.2572	0.6% Cu 12% Si 0.3% Mn Al alloy: German Standard DIN 1725
3.2573	3.7% Cu 9% Si Al alloy: German Standard DIN 1725
3.2581	11.5% Si Al alloy: German Standard DIN 1725
3.2582	0.1% Cu 12% Si Al alloy: German Standard DIN 1725
3.2583	2% Cu 12.5% Si 0.3% Mn Al alloy: Casting; German Standard DIN number not known
3.2585	0.3% Cu 10% Si Al alloy: German Standard DIN 1725
13	11% Si Al alloy: Casting; Alcoa for BS alloy LM 20
32S	1% Cu 12% Si 1% Mg Al alloy: Previous AA designation; now 4032 for pistons
43	5% Si Al alloy: Casting; AA designation for ASTM Alloy 85A or 85B

Symbol	Nominal analysis, supplier, condition and remarks.
43	5% Si Al alloy: Casting; Alcoa for BS alloy LM 18
43S	5% Si Al alloy: Wire; Alcoa for BS alloy NW 21
43S	5% Si Al alloy: Previous AA designation; now 4043
123	5% Si Al alloy: Casting; Comalco for BS alloy LM 18
160	12% Si Al alloy: Casting; Comalco for BS alloy LM 6
160S	12% Si 0.3% Mg Al alloy: Casting; Comalco for BS alloy LM 6
220	7% Si 0.3% Mg Al alloy: Casting; AA designation for ASTM alloy G 10A
443.0	5.2% Si Al alloy: Investment casting; American National Standard; ANSI
4047	12% Si Al: Wire for brazing; AA designation; no ASTM specification
A 443.0	5.2% Si Al alloy: Investment casting; American National Standard; ANSI
A 04430	Unified number for alloy 443.0; ASTM system
A 14430	Unified number for alloy A443.0; ASTM system
A 24430	Unified number for alloy B443.0; ASTM system
AA 4032	1% Cu 12% Si 1% Mg 1% Ni Al alloy: Wrought; SAE designation; formerly SAE 290
AA 4043	5% Si Al alloy: Wrought; SAE designation; formerly SAE 205
AA 4045	9.5% Si Al alloy: Used only for cladding; SAE designation
AA 4343	7% Si Al alloy: Used only for cladding; SAE designation; formerly SAE 206
AC 5	5% Si 0.7% Mg 0.2% Mn 0.1% Ti Al alloy: Anglo-Swiss
ALAR 005	5% Si Al alloy: Casting; American proprietary alloy for BS alloy LM 18
ALAR 0012	12% Si Al alloy: Casting; American proprietary alloy for BS alloy LM 20
ALCAN GB 38S	1% Cu 11% Si 1% Mg 1% Ni Al alloy: Forging; solution treated and aged; Alcan **DPN: 105 UTS: 350 Elon: 9% Proof: 270**
ALCAN GB 160	11.5% Si Al alloy: Casting; Alcan for BS alloy LM 6
ALCAN GB 33S	5% Si Al alloy: Wire; Alcan; annealed **UTS: 110 Elon: 40% Proof: 30**

Note. The following abbreviations and units are used in the tables:

DPN	Hardness, diamond pyramid number
UTS	Ultimate tensile strength, N/mm^2
Elon	Elongation, %
Proof	0.1 % proof strength, N/mm^2

$1 \ N/mm^2 = 0.1 \ hbar = 0.102 \ kgf/mm^2 = 0.06475 \ tonf/in^2 = 145.04 \ lbf/in^2$

Symbol	Nominal analysis, supplier, condition and remarks.
ALCAN GB 33S	5% Si Al alloy: Wire; Alcan; hard
	UTS: 210 Elon: 5% Proof: 170
ALCAN GB B33S	5% Si 0.2% Mg Al alloy: Wire; Alcan; solution treated & aged
	UTS: 180 Elon: 10% Proof: 150
ALMINAL 2	1% Cu 10% Si Al alloy: Casting; origin unknown
ALMINAL 6	12% Si Al alloy: Casting; origin unknown
ALMINAL 8	5% Si 0.4% Mg 0.5% Mn Al alloy: Casting; origin unknown
ALMINAL 9	12.5% Si 0.4% Mg 0.5% Mn Al alloy: Casting; origin unknown
ALMINAL 10	10.5% Si Al alloy: Casting; origin unknown
ALMINAL 20	3% Cu 9% Si 0.2% Mg 1% F 0.5% Zn Al alloy: Casting; origin unknown
ALMINTRODE 96.0	0.5% Fe 11.0% Si Al: For electrodes; ESAB
	UTS: 170 Elon: 13
ALPAX	13% Si Al alloy: Casting; Light Alloys Ltd for BS alloy LM 6
AMS 4145 E	1% Cu 11% Si 1% Mg 1% Ni Al alloy: Forging; solution treated & aged; AMS for AA 4032
AMS 4184 B	4% Cu 10% Si Al alloy: Wire for brazing; AMS for AA alloy 4145
AMS 4185 A	12% Si Al alloy: Wire for brazing; AMS for AA alloy 4047
AMS 4190 A	5% Si Al alloy: Wire for brazing; AMS for AA alloy 4043
ANQQ A-405	5.0% Si Al alloy: Casting; US Service; as cast
	UTS: 90 Elon: 3%
ANTICORODAL 5S1	5% Si 0.7% Mg 0.2% Mn 0.1% Ti Al alloy: Casting; Anglo-Swiss; solution treated & aged; obsolete
	DPN: 95 UTS: 280 Elon: 2% Proof: 250
ANTICORODAL 34	0.4% Mg 3.2% Si 0.2% Mn 0.15% Ti Al: Primary casting metal; Anglo-Swiss Al Co
A-S4G	4.2% Si 0.6% Mg 45% Mn Al alloy: Casting; L'Aluminium Francais
A-S7G	7.0% Si 0.3% Mg Al alloy: Casting; L'Aluminium Francais
A-S10G	10% Si 0.25% Mg Al alloy: Casting; L'Aluminium Francais
A-S10UG	2.3% Cu 10% Si 1.0% Fe 1.2% Mg 0.5% Mn Al alloy: Casting; L'Aluminium Francais
A-S12N2G	1.0% Cu 12% Si 1.0% Mg 2.2% Ni Al alloy: Casting; L'Aluminium Francais
A-S12UN	1.0% Cu 12% Si 1.0% Mg 1.0% Ni Al alloy: Wrought & cast; L'Aluminium Francais
A-S13	12.5% Si Al alloy: Casting; L'Aluminium Francais
A-S22	22% Si Al master alloy: L'Aluminium Francais
A-S22UNK	1.5% Cu 21% Si 1.0% Mg 0.5% Mn 1.1% Ni 1.0% Co Al alloy: Casting; L'Aluminium Francais
ASTM B26 S5A	5% Si Al alloy: Sand casting; as cast
	UTS: 110 Elon: 3% Proof: 22
ASTM B26 S5B	5% Si 0.25% Cr Al alloy: Sand casting; as cast
	UTS: 110 Elon: 3% Proof: 30
ASTM B85 S5C	5% Si Al alloy: Die casting; as cast
	UTS: 220 Elon: 9% Proof: 70
ASTM B85 S12A	12% Si Al alloy: Die casting; as cast
	UTS: 280 Elon: 3.5% Proof: 120
ASTM B85 S12B	12% Si 2% Fe Al alloy: Die casting; as cast
	UTS: 300 Elon: 2.5% Proof: 140
ASTM B108 S5A	5% Si Al alloy: Casting; as cast
	UTS: 140 Elon: 2.5% Proof: 22
ASTM B108 S5B	5% Si 0.2% Cr Al alloy: Casting; as cast
	UTS: 140 Elon: 3% Proof: 30
ASTM B108 SC41	1.5% Cu 0.7% Mg 12% Si Al alloy: Chill cast; solution treated & aged; obsolete
	DPN: 100 UTS: 300 Elon: 1% Proof: 250
ASTM B108 SC9IAI	9.0% Si 0.6% Mg Al alloy: Die casting
ASTM B179 57A	7.0% Si Al alloy: Ingot
ASTM B179 S5A	5% Si Al alloy: Ingot; primary metal

Symbol	Nominal analysis, supplier, condition and remarks.
ASTM B179 S5B	5% Si Al alloy: Ingot; primary metal
ASTM B179 S5C	5% Si Al alloy: Ingot; primary metal
ASTM B179 S12A–B	12% Si Al alloy: Ingot; primary metal
ASTM B179 SC94A	3.5% Cu 9% Si 0.3% Mg Al alloy: Ingot
ASTM B179 SC71A	7.0% Si 0.5% Mg Al alloy: Ingot
ASTM B179 SC91A	9% Si 0.6% Mg Al alloy: Ingot
ASTM B184 A143	5% Si Al alloy: For electrode filler wire; may be flux coated
ASTM B247 4032	1% Cu 12% Si 1% Mg 1% Ni Al alloy: Forging; solution treated & aged
	DPN: 115 UTS: 370 Elon: 5% Proof: 300
ASTM B260 B Al Si 1	5% Si Al alloy: Brazing filler; melting range 577–630 °C
ASTM B260 B Al Si 2	7% Si Al alloy: Brazing filler; melting range 577–613 °C
ASTM B260 B Al Si 3	10% Si 4% Cu Al alloy: Brazing filler; melting range 521–584 °C
ASTM B260 B Al Si 4	12% Si Al alloy: Brazing filler; melting range 577–582 °C
ASTM B260 B Al Si 5	7.5% Si Al alloy: Braze metal; brazing range 588–604° C
ASTM B285 ER4043	5.2% Si Al alloy: Welding wire
ASTM B285 RS5B	5% Si Al alloy: Welding rod; obsolete
ASTM B285 RSC51A	5% Si 1.2% Cu 0.5% Mg Al alloy: Welding wire
ASTM B285 RSG70A	7% Si 0.3% Mg Al alloy: Welding wire
AWCO 40	12% Si Al alloy: For welding wire; Aluminium Wire Co for BS alloy N2
AWCO 45	5% Si Al alloy: For welding wire; Aluminium Wire Co for BS alloy N21
AWCO 46	3.5% Cu 12% Si Al alloy: Wire; Aluminium Wire Co; for welding
BA 40	11% Si Al alloy: Casting; British Al Co for BS LM 6
BA 45	5% Si Al alloy: Casting; British Al Co for BS alloy LM 18
BIRMASIL SPEC	11% Si 3% Ni Al alloy: Casting; Birmingham Al Co for BS alloy LM17
BIRMASTIC	12% Si 3% Ni Al alloy: Origin unknown
BIRMETAL 005	5% Si Al alloy: Wire; Birmabright for BS alloy N21
BS 1354 NG2	12% Si Al alloy: Welding rod; for welding BS alloys H9, H10, H20, H30 and cast alloys
BS 1453 NG21	5% Si Al alloy: Welding rod; for welding BS alloys H9, H10, H20, H30 and cast alloys
BS 1475 NG2	12% Si Al alloy: Wire
BS 1475 NG21	5% Si Al alloy: Wire
BS 1490 LM2	1.5% Cu 10% Si Al alloy: Ingot; chill cast
	UTS: 150
BS 1490 LM2M	1.5% Cu 10% Si Al alloy: For die castings; chill cast
	UTS: 150
BS 1490 LM6	11% Si Al alloy: Ingots; chill cast
	UTS: 180 Elon: 7%
BS 1490 LM6M	11% Si Al alloy: For sand and die castings; chill cast
	UTS: 180 Elon: 7%
BS 1490 LM17	11% Si 3% Ni Al alloy: Casting; obsolete
BS 1490 LM18	5% Si Al alloy: Ingot; chill cast
	UTS: 140 Elon: 4%
BS 1490 LM18M	5% Si Al alloy: Sand or die casting; chill cast
	UTS: 140 Elon: 4%
BS 1490 LM20	12% Si Al alloy: Ingot; chill cast
	UTS: 180 Elon: 5%
BS 1490 LM20M	12% Si Al alloy: Die casting; chill cast
	UTS: 180 Elon: 5%
BS 1490 LM25	0.3% Mg 7.0% Si Al alloy: Casting; sand cast
	UTS: 230 Proof: 200

Symbol	Nominal analysis, supplier, condition and remarks.
BS 1616 B	5% Si Al alloy: Welding rod flux coated wire to BS alloy NG 21
BS 1942/1	3% Cu 12% Si Al alloy: For brazing; melting range 550–570 °C
BS 1942/2	12% Si Al alloy: Brazing filler BS alloy N2; melting range 565–575 °C
BS 1942/3	8% Si Al alloy: Brazing filler; melting range 565–600 °C
BS 1942/4	5% Si Al alloy: Brazing filler BS alloy N21; melting range 565–625 °C
BS 2901 NG2	12% Si Al alloy: Rod for inert gas shielded arc welding
BS 2901 NG21	5% Si Al alloy: Rod for inert gas shielded arc welding
BS L8	12% Cu Al alloy: Casting; obsolete
BS L33	11% Si Al alloy: Casting; chill cast
	UTS: 180 Elon: 7% Proof: 70
BS L99	0.3% Mg 7.0% Si Al alloy: Casting; sand cast
	UTS: 230 Elon: 2% Proof: 180
CHROMET	10% Si Al alloy: Bearings for hard shafts; origin unknown
COMALCO 160	12% Si 0.3% Fe Al alloy: Casting; Comalco for BS alloy LM6
COMALCO 160S	12% Si 0.2% Fe 0.03% Na Al alloy: Casting; Comalco for BS alloy LM6
DEF 30	Si Al alloy: For die casting; covers BS alloys LM2 to LM6 and LM24
DS 15000 Al501	11.5% Si Al alloy: Casting; Danish specification
DS 15000 Al502	11.5% Si Al alloy: Casting; Danish specification
DTD 135	Si Al casting alloy: Cancelled; obsolete
DTD 324A	1% Cu 11% Si 1% Mg 1% Ni Al alloy: Forging; solution treated & aged
	UTS: 300 Elon: 3%
DTD 5028	7.0% Si 0.3% Mg Al alloy: Sand casting
	UTS: 220 Elon: 3% Proof: 180
G Al Si 5 Cu 1	5% Si 0.4% Mg 1% Cu Al alloy: Casting; previous German Standard designation
G Al Si 5 Mg	5% Si 0.6% Mg Al alloy: Casting; previous German Standard designation
G Al Si 6 Cu 4	4% Cu 6% Si 0.2% Mg Al alloy: Casting; previous German Standard designation
G Al Si 6 Cu 6	3% Cu 5% Si 0.5% Mn Al alloy: Casting; previous German Standard designation
G Al Si 7 Cu 3	3% Cu 7% Si 0.2% Mg Al alloy: Casting; previous German Standard designation
G Al Si 9 (Cu)	9% Si 1.5% Cu Al alloy: Casting; previous German Standard designation
G Al Si 10 Mg	9.5% Si 0.2% Mg Al alloy: Casting; previous German Standard designation
G Al Si 10 Mg (Cu)	10% Si 0.3% Mg 0.2% Cu Al alloy: Casting; previous German Standard designation
G Al Si 12	11% Si Al alloy: Casting; previous German Standard designation
G Al Si 12 (Cu)	2% Cu 12.5% Si 0.3% Mn Al alloy: Casting; previous German Standard designation
GD Al Si 6 Cu 3	3% Cu 6% Si 0.4% Mn Al alloy: Casting; previous German Standard designation
Gd Al Si 10 (Cu)	0.8% Cu 10% Si 0.2% Mg Al alloy: Casting; previous German Standard designation

Symbol	Nominal analysis, supplier, condition and remarks.
GD Al Si 12 .	12% Si Al alloy: Casting; previous German Standard designation
GD Al Si 12 (Cu)	0.6% Cu 12% Si 0.3% Mn Al alloy: Casting; previous German Standard designation
GOST 2685-63 Al2	11.5% Si Al alloy: Casting, Russian specification
GOST 2685-63 Al4	0.4% Mg 11.5% Si 0.5% Mn Al alloy: Casting; Russian specification
GOST 2685-63 Al4B	0.4% Mg 11.5% Si 0.5% Mn Al alloy: Casting; Russian specification
GOST 2685-63 Al8	10% Si Al alloy: Casting; Russian specification
GOST 2685-63 Al9	0.5% Mg 5% Si Al alloy: Casting; Russian specification
GOST 2685-63 Al9B	0.5% Mg 5% Si Al alloy: Casting; Russian specification
GOST 2685-63 Al25	0.9% Cu 1.2% Mg 12% Si 1.6% Ni Al alloy: Casting; Russian specification
GOST 2685-63 Al27	10% Si Al alloy: Casting; Russian specification
GOST 2685-63 Al 30	0.9% Cu 1.2% Mg 12% Si 1.6% Ni Al alloy: Casting; Russian specification
GRINATAL	0.3% Mg 4.5% Si Al alloy: Wrought; Anglo-Swiss Al Co
GRINATAL 350	0.3% Mg 4.0% Si Al alloy: Wrought; Anglo-Swiss Al Co
H 49-2	1.6% Cu 10% Si Al casting alloy: Australian specification
H 49-5	12% Si Al casting alloy: Australian specification
H 49-6	0.5% Mg 4.7% Si Al casting alloy: Australian specification
H 49-7	0.4% Mg 11.5% Si 0.5% Mn Al alloy: Casting; Australian specification
H 49-11	0.8% Cu 1.2% Mg 12% Si 1.5% Ni Al alloy: Casting; Australian specification
H 49-14	5.2% Si Al alloy: Casting; Australian specification
H 49-15	11.5% Si Al alloy: Casting; Australian specification
HA 3 S5	5.2% Si Al alloy: Ingot; Canadian Standard
HA 3 S12N	12.5% Si Al alloy: Ingot; Canadian Standard
HA 3 S12P	12.0% Si 0.8% Fe Al alloy: Ingot; Canadian Standard
HA 3 SG70N	0.2% Fe 0.3% Mg 7.0% Si Al alloy: Ingot; Canadian Standard
HA 3 SG70P	0.3% Mg 7.0% Si Al alloy: Ingot; Canadian Standard
HA 3 SG71	0.5% Mg 7.0% Si Al alloy: Ingot; Canadian Standard
HA 3 SN122	1.0% Mg 2.5% Ni 12% Si Al alloy: Ingot; Canadian Standard
HA 8 SG121	0.9% Cu 1.0% Mg 0.9% Ni 12% Si Al alloy: Forging; Canadian Standard
HA 9 S5	5.2% Si Al alloy: Sand casting; Canadian Standard
HA 9 S12N	12.0% Si Al alloy: Sand casting; Canadian Standard
HA 9 SG70N	0.3% Mg 7.0% Si Al alloy: Sand casting; Canadian Standard
HA 9 SG71	0.5% Mg 7.0% Si Al alloy: Sand casting; Canadian Standard
HA 10 S5	5.2% Si Al alloy: Die casting; Canadian Standard
HIDUMINIUM 00	1.5% Cu 10% Si Al alloy: Die castings; High Duty Alloys for BS alloy LM2
HIDUMINIUM 08	1% Cu 11% Si 1% Mg 1% Ng Al alloy: Forging; solution treated & aged; High Duty Alloy
HIDUMINIUM 10	11% Si Al alloy: For casting thin sections; High Duty Alloys for BS alloy LM6
HIDUMINIUM 512	1% Cu 11% Si 1% Mg 1% Ni Al alloy: Forging; solution treated & aged; High Duty Alloy
IMPALCO 450	12% Si Al alloy: Wire for brazing; Imperial Al Co for BS alloy N2
IMPALCO 460	5% Si Al alloy: Wire for brazing; Imperial Al Co for BS alloy N21
IMPALCO 470	7.5% Si Al alloy: Wire for furnace brazing; Imperial Al Co

Note. The following abbreviations and units are used in the tables:

DPN	Hardness, diamond pyramid number
UTS	Ultimate tensile strength, N/mm^2
Elon	Elongation, %
Proof	0.1 % proof strength, N/mm^2

$1\ N/mm^2 = 0.1\ hbar = 0.102\ kgf/mm^2 = 0.06475\ tonf/in^2 = 145.04\ lbf/in^2$

Symbol	Nominal analysis, supplier, condition and remarks.
IMPALCO PA15	12% Si Al alloy: Wire for brazing; Imperial Al Co for BS alloy N2
IMPALCO PA16	5% Si Al alloy: Wire for brazing; Imperial Al Co for BS alloy N21
IMPALCO PA17	7.5% Si Al alloy: For brazing; Imperial Al Co
KYNAL PA15	12% Si Al alloy: Wire for brazing; ICI; obsolete; now Imperial Al Co
KYNAL PA16	5% Si Al alloy: Wire for brazing; ICI; obsolete; now Imperial Al Co
KYNAL PA17	5% Si Al alloy: Wire for brazing; ICI; obsolete; now Imperial Al Co
L 252	11.5% Si Al alloy: Casting; Spanish specification
L 253	11.5% Si Al alloy: Casting; Spanish specification
L 254	0.4% Mg 11.5% Si 0.5% Mn Al alloy: Casting; Spanish specification
L 255	0.9% Cu 1.2% Mg 12% Si 1.6% Ni Al alloy: Casting; Spanish specification
L 256	0.4% Mg 11.5% Si 0.5% Mn Al alloy: Casting; Spanish specification
L 257	0.55% Mg 5% Si Al alloy: Casting; Spanish specification
LAC 112 A	1.5% Cu 10% Si Al alloy: Casting; obsolete; Air Ministry specification similar to BS alloy LM2
L Al Si 12	0.3% Cu 12% Si Al alloy: German Standard; DIN number not known
LM 2	1.5% Cu 10% Si Al alloy: Casting; Intal for BS alloy LM2
LM 6	11% Si Al alloy: Casting; Intal for BS alloy LM6
LM 18	5% Si Al alloy: Casting; Intal for BS alloys LM18
MVC	11% Si Al: Casting; AEI; chill cast **UTS: 180**
NBN436 Al Si 5 Mg	0.6% Mg 5% Si Al alloy: Casting; Belgium specification
NBN436 Al Si 10 Mg	0.4% Mg 11.5% Si 0.5% Mn Al alloy: Casting; Belgium specification
NBN 436 Al Si 12	12% Si Al alloy: Casting; Belgium specification
NORAL 33S	5% Si Al alloy: Wire; Northern Al for BS alloy N21
NORAL 38S	1% Cu 11% Si 1% Mg 1% Ni Al alloy: Forging; solution treated & aged; Northern Al Co **UTS: 300 Elon: 3%**
NORAL 123	5% Si Al alloy: Casting; Northern Al Co for BS alloy LM 18
NORAL 158	11% Si 3% Ni Al alloy: Casting; Northern Al Co for BS alloy LM 17
NORAL 160	11% Si Al alloy: Casting; Northern Al Co for BS alloy LM20
NURAL 1761	1.0% Cu 17% Si 3.4% Ni 1.0% Mg 0.5% Cr Al alloy: Casting; Alcan (Germany) for pistons
NURAL 1761P	1.0% Cu 16.8% Si 1.0% Ni 1.0% Mg Al alloy: Forgings; Alcan (Germany) for pistons
NURAL 1762	17.2% Si 1.0% Mg 1.0% Cu Al alloy: Alcan (Germany)
NURAL 2361	1.0% Cu 23% Si 1.0% Ni 1.0% Mg 0.5% Cr Al alloy: Castings; Alcan (Germany) for pistons
NURAL 3210	1.1% Cu 12% Si 1.0% Ni 1.0% Mg 0.1% Ti Al alloy: Casting; Alcan (Germany) for pistons
QQ A 591/1	12% Si Al alloy: Casting; US Federal
QQ A 591/2	12.0% Si Al alloy: Casting; US Federal
QQ A 591/3	5.0% Si Al alloy: Casting; US Federal
QQ A 591/8	5.0% Si Al alloy: Casting; US Federal
QQ A 596/7	5.0% Si Al alloy: Casting; US Federal; as cast **UTS: 120 Elon: 2.5%**
QQ A 601/2	5.0% Si Al alloy: Casting; US Federal; as cast **UTS: 90 Elon: 3%**
SAE 35	5% Si Al alloy: Casting; sand cast **UTS: 110 Elon: 3% Proof: 30**
SAE 205	5% Si Al alloy: Welding wire; now AA4043
SAE 206	7% Si Al alloy: Used only for cladding; analysis applies before cladding; now AA4343

Symbol	Nominal analysis, supplier, condition and remarks.
SAE 290	1% Cu 12% Si 1% Mg Al alloy: Solution treated and aged; now AA 4043 **DPN: 115 UTS: 360 Elon: 5% Proof: 280**
SAE 304	5% Si Al alloy: Casting; die cast; as cast **UTS: 200 Elon: 5%**
SAE 305	12% Si Al alloy: Casting; die cast; as cast **UTS: 240 Elon: 3.5%**
S Al Si 5	5% Si Al alloy: Wrought; German Standard
S Al Si 12	8% Si Al alloy: Wrought; German Standard
Sf 1	13% Si 0.35% Mn Al alloy: Casting; Anglo Swiss code for Silafont 1
Sf 2	11% Si 0.45% Mg 0.3% Mn Al alloy: Casting; Anglo Swiss code for Silafont 2
Sf 3	9% Si 0.4% Mg 0.3% Mg Al alloy: Casting; Anglo Swiss code for Silafont 3
Sf 4	1.2% Cu 11% Si 0.65% Mg 0.3% Mn Al alloy: Casting; Anglo Swiss code for Silafont 4
Sf 5	9.5% Si 0.3% Mg 0.5% Co Al alloy: Casting; Anglo Swiss code for Silafont 5
Sf 6	10% Si 0.25% Mg Al alloy: Casting; Anglo Swiss code for Silafont 6
Sf 7	1.2% Cu 13% Si 1% Mg 1% Ni Al alloy: Casting; Anglo Swiss code for Silafont 7
SF Aluminium	6% Al Si: Wire for metal spraying; as sprayed; Metco **UTS: 240**
SILAFONT 1	13% Si 0.35% Mn Al alloy: Casting; Anglo Swiss; as cast; obsolete **DPN: 55 UTS: 210 Elon: 7% Proof: 110**
SILAFONT 2	11% Si 0.45% Mg 0.3% Mn Al alloy: Casting; Anglo Swiss; solution treated and aged; obsolete **DPN: 90 UTS: 300 Elon: 2% Proof: 240**
SILAFONT 3	9% Si 0.4% Mg 0.3% Mn Al alloy: Casting; Anglo Swiss; solution treated and aged; obsolete **DPN: 90 UTS: 270 Elon: 2% Proof: 220**
SILAFONT 4	1.2% Cu 11% Si 0.65% Mg 0.3% Mn Al alloy: Casting; Anglo Swiss: as cast; obsolete **DPN: 90 UTS: 210 Elon: 2% Proof: 150**
SILAFONT 5	9.4% Si 0.3% Mg 0.5% Co Al alloy: Casting; Anglo Swiss; solution treated and aged; obsolete **DPN: 70 UTS: 270 Elon: 4% Proof: 220**
SILAFONT 6	10% Si 0.25% Mg Al alloy: Casting; Anglo Swiss as cast; obsolete **DPN: 60 UTS: 210 Elon: 2% Proof: 150**
SILAFONT 7	1.2% Cu 13% Si 1% Mg 1% Ni Al alloy: Casting; Anglo Swiss; solution treated and aged; obsolete **DPN: 120 UTS: 250 Elon: 1% Proof: 210**
SILAFONT 14	13% Si 0.3% Mn Al alloy: Casting; Anglo-Swiss Al Co
SILAFONT 74	1.2% Cu 1.0% Mg 12% Si 1.0% Ni Al alloy: Ingot; Anglo-Swiss Al Co
SILAFONT 84	10% Si 0.3% Mn Al alloy: Casting; Anglo-Swiss Al Co
SIS 4244	7% Si 0.3% Mg Al alloy: Casting; Swedish Standard; as cast **DPN: 90 UTS: 220 Elon: 2% Proof: 200**
SIS 4247	9% Si 1% Fe 0.5% Mn Al alloy: Casting; Swedish Standard; as cast **DPN: 70**
SIS 4253	10% Si 0.3% Mg Al alloy: Casting; Swedish Standard; as cast **DPN: 90 UTS: 220 Elon: 2% Proof: 200**
SIS 4255	10% Si Al alloy: Casting; Swedish Standard; as cast **DPN: 60**
SIS 4260	12% Si 0.4% Mn Al alloy: Casting; Swedish Standard; as cast **DPN: 60**

Symbol	Nominal analysis, supplier, condition and remarks.
SIS 4261	12% Si Al alloy: Casting; Swedish Standard; as cast **DPN: 60**
STA 7 AC2	1.5% Cu 10% Si Al alloy: Casting; replaced by BS 1490 LM2
STA 7 AC6	11% Si Al alloy: Casting; replaced by BS 1490; LM6
STA 7 AW1	5.0% Si Al alloy: Wire; replaced by BS 1473/5 NW21
STA 7 AW2C	12% Si Al alloy: Wire; replaced by BS 1473/5 NW2
UNI 3022 MAS6	6% Si Al alloy: Ingot; primary metal; Italian Standard
UNI 3022 MA625	25% Si Al alloy: Ingot; primary metal; Italian Standard
UNI 3048	13% Si 0.8% Cu 0.3% Mn Al alloy: Italian Standard
UNI 3049	12% Si 0.5% Mn 0.3% Mg Al alloy: Italian Standard
UNI 3572	1.0% Mg 1.0% Cu 12% Si 0.8% Ni Al alloy: For forging; Italian Standard
UNI 4513	12% Si 2.0% Cu 0.3% Mn Al alloy: Casting; Italian Standard
UNI 4514	13% Si Al alloy: Casting; Italian Standard
UNI 5074	9.0% Si 0.4% Mg 0.6% Fe Al alloy: Casting; Italian Standard
UNI 5076	12% Si 2.0% Cu 0.7% Fe Al alloy: Casting; Italian Standard

Symbol	Nominal analysis, supplier, condition and remarks.
UNI 5077	·5.0% Si 0.8% Fe Al alloy: Casting; Italian Standard
UNI 5078	9.0% Si 0.7% Fe Al alloy: Casting; Italian Standard
UNI 5079	13% Si 0.7% Fe Al alloy: Casting; Italian Standard
UNI 6250	12.7% Si 2.2% Ni 1.0% Mg 0.8% Cu Al alloy: Casting; Italian Standard
UNI 6251	21% Si 1.6% Cu 1.5% Ni 0.7% Mn 0.6% Mg 0.8% Co Al alloy: Casting; Italian Standard
V 1036 A/B BVI	5.5% Si Al alloy: Casting; Dutch specification
V 1036 A/B BVII	11.5% Si Al alloy: Casting; Dutch specification
V 1036 A/B BIX	0.5% Mg 5% Si Al alloy: Casting; Dutch specification
V 1036 A/B BXI	0.4% Cu 11.5% Si 0.5% Mn Al Alloy: Casting; Dutch specification
WILMIL	13% Si Al alloy: Casting; W Mills Ltd for BS alloy LM6

Note. The following abbreviations and units are used in the tables:

DPN	Hardness, diamond pyramid number
UTS	Ultimate tensile strength, N/mm^2
Elon	Elongation, %
Proof	0.1 % proof strength, N/mm^2

$1 \ N/mm^2 = 0.1 \ hbar = 0.102 \ kgf/mm^2 = 0.06475 \ tonf/in^2 = 145.04 \ lbf/in^2$

1D Aluminium–magnesium wrought alloys

Specific gravity	2.65–2.7	
Density	2650–2700 kg/m^3	(0.97 lbsf/in.^{3})
Solidus/liquidus	580–650 °C	
Thermal conductivity	annealed 120–190 W/m °C	(28–46 cal/m s °C)
	½–hard 138–147 W/m °C	(0.33–0.35 cal/m s °C)
Coefficient of linear expansion	23.8–24.3 × 10^{-6}/ °C	
Electrical conductivity	annealed 30 × 50% IACS (copper 100%)	
Specific resistance	annealed 34–59 microhm mm	
	½–hard 40–55 microhm mm	
	hard 64 microhm mm	
Young's modulus of elasticity	66.9–72.4 × 10^9 N/m^2	(9.7 × 10.5 × 10^6 lbf/in.^{2})
Impact	annealed 38–49 J	(28–36 foot lb) Izod
	½–hard 28 J	(21 foot/lb)
Fatigue strength (50 × 10^6 cycles)	annealed ± 90–140 N/mm^2	(± 6–9.5 tonf/in.^{2})
Hot strength		

Temperature °C	Tensile strength N/mm^2	Elongation %
150	150–180	40–45
200	145–160	50–55
250	110–125	60–65
300	48–54	

The above properties are included as a guide to the group of specification as a whole. There is considerable variation within the group, and as would be expected these properties vary in general with the magnesium percentage.

General metallurgical properties

Magnesium, in small quantities of less than 1% gives alloys with excellent hot working properties, and is therefore used for complicated pressings and extrusions. These alloys are not heat-treatable. By the addition of silicon a range of heat-treatable alloys are obtained. These form the separate group listed in Section 1E.

Additions of magnesium from 2–7% give alloys with excellent corrosion resistance, including seawater with an increase in mechanical properties. The increase in mechanical properties is even more marked when the alloys are cold worked. Addition of copper, manganese and chrome strengthen the matrix with little effect on the corrosion resistance.

Thermal treatment

Annealing is the only heat treatment possible on all of these alloys, and is carried out in an air furnace at 360–400 °C followed by a slow cool.

The increased hardness from cold work is removed, and can only be recovered by further cold work.

Welding and brazing

Pre-treatment. No thermal treatment is required but the weld area must be thoroughly cleaned and scratch brushed.

Welding and brazing. Normal gas or electric welding using flux is difficult but possible. As the fluxes are generally corrosive they must be completely removed on completion of the weld. The affinity of magnesium for oxygen gives rise to welding difficulties, resulting in oxide trapping and brittle welds. Inert gas shielded arc welding gives the best results as no flux is required. The rate of flow of the inert gas must be sufficient to supply a blanket until the weld has solidified. Resistance welding is possible, but requires special equipment.

Brazing, using fluxes and aluminium–silicon alloy filler rods is readily accomplished, but affects the corrosion resistance locally.

Post treatment. None required.

It should be noted however that welding will remove any cold work, leaving a local weak area.

Flaw detection methods

Crack test. Oil penetrant chalk method.

X-ray. Not normally required, except perhaps for a complicated high duty weld.

Ultra-sonic test. Not normally required.

Chemical etch. Normally only required for grain size and directional estimation. Can be used for weld or braze examination. A 10% caustic solution is satisfactory, but may require to be followed by a concentrated nitric acid dip to remove the black deposit.

Corrosion protection

Temporary. None required.

Permanent. None will be required when used under normal atmospheric conditions. Sulphuric, oxalic and chromic acid anodizing and chemical oxidzing are all suitable means of increasing the corrosion resistance for more severe conditions. Colour anodizing is often applied to these alloys.

Painting. Air drying or stove enamel paints are satisfactory, but an etch primer or chemical oxidze, prior to painting is necessary. Lead based primers must not be used.

Plating. This is possible using special etching techniques, but will seldom be required on these alloys.

Machinability

Generally very good, but some trouble with finish can be encountered in the annealed condition. If it is necessary to have the part annealed with a good finish, then it should be machined in the ¾-hard or hard condition and finally annealed.

Uses

Architectural uses where moderate strength and excellent corrosion resistance are required.

Marine use – superstructure, lift boats, spars, etc.

Hollow-ware where moderate strength and good ductility are required.

Motor car and commercial vehicle trim and panelling.

Some alloys are used in the chemical industry.

Symbol	Nominal analysis, supplier, condition and remarks
3.3308	0.5% Mg Al alloy: German Standard DIN 1725
3.3309	0.5% Mg Al alloy: German Standard DIN 1725
3.3318	0.8% Mg Al alloy: German Standard DIN 1725
3.3319	1% Mg Al alloy: German Standard DIN 1725

Note. The following abbreviations and units are used in the tables:

DPN	Hardness, diamond pyramid number
UTS	Ultimate tensile strength, N/mm²
Elon	Elongation, %
Proof	0.1 % proof strength, N/mm²

$1 \text{ N/mm}^2 = 0.1 \text{ hbar} = 0.102 \text{ kgf/mm}^2 = 0.06475 \text{ tonf/in}^2 = 145.04 \text{ lbf/in}^2$

Symbol	Nominal analysis, supplier, condition and remarks
3.3329	2% Mg Al alloy: German Standard DIN 1725
3.3350	10% Mg Al alloy: Wrought; German Standard DIN 1725
3.3315	0.9% Mg 0.2% Mn 0.2% Cr Al alloy: Wrought; German Standard DIN 1725; annealed **DPN: 30 UTS: 90 Elon: 20%**
3.3525	2.0% Mg 0.3% Mn 0.2% Cr Al alloy: Wrought; German Standard DIN 1725
3.3527	2% Mg 1% Mn 0.2% Cr Al alloy: Wrought; German Standard DIN 1725
3.3535	3% Mg 0.2% Cr Al alloy: Wrought; German Standard DIN 1725
3.3537	7% Mg 0.45% Mn Al alloy: Wrought; German Standard DIN 1725

Symbol	Nominal analysis, supplier, condition and remarks.
3.3555	5% Mg 0.2% Cr Al alloy: Wrought; German Standard DIN 1725; annealed **DPN: 55 UTS: 220 Elon: 17% Proof: 90**
3.3557	7% Mg Al alloy: German Standard DIN 1725
3.3575	6.8% Mg 0.2% Cr Al alloy: Wrought; German Standard DIN 1725
50S	1.2% Mg Al alloy: Previous AA designation now 5050
52S	2% Mg Al: Sheet & tube; Alcoa for BS alloy N4
52S	2.5% Mg 0.2% Cr Al alloy: Previous AA designation now 5052
56S	5% Mg Al alloy: Sheet & strip; Alcoa for BS alloy N6
56S	5% Mg 0.15% Cr Al alloy: Previous AA designation now 5056
AA 5050	1.5% Mg Al alloy: Wrought; SAE designation; formerly SAE 207
AA 5052	2.5% Mg 0.2% Cr Al alloy: Wrought; SAE designation; formerly SAE 201
AA 5083	0.7% Mn 4.5% Mg 0.2% Cr Al alloy: Wrought; SAE designation; annealed; no previous SAE alloy **UTS: 330 Elon: 16% Proof: 110**
AA 5086	0.5% Mn 4% Mg 0.2% Cr Al alloy: Wrought; SAE designation; annealed; no previous SAE alloy **UTS: 240 Elon: 14% Proof: 70**
AA 5154	3.5% Mg 0.2% Cr Al alloy: Wrought; SAE designation; formerly SAE 208
AA 5252	2.5% Mg Al alloy: Wrought; SAE designation; annealed; no previous SAE alloy **UTS: 160 Elon: 20%**
AA 5257	0.4% Mg Al alloy: Wrought; SAE designation; annealed; no previous SAE alloy **UTS: 70 Elon: 22%**
AA 5357	0.3% Mn 1% Mg Al alloy: Wrought; SAE designation; formerly SAE 209
AA 5454	0.7% 2.6% Mg 0.1% Cr Al alloy: Wrought; SAE designation; formerly SAE 251
AA 5456	0.7% Mn 5.1% Mg 0.1% Cr Al alloy: Wrought; SAE designation; no previous SAE alloy **UTS: 330 Elon: 16% Proof: 110**
AA 5457	0.2% Mn 1% Mg Al alloy: Wrought; SAE designation; formerly SAE 252
AA 5557	0.2% Mn 0.6% Mg Al alloy: Wrought; SAE designation; no previous SAE alloy **UTS: 110 Elon: 20%**
AA 5657	0.8% Mg Al alloy: Wrought; SAE designation; no previous SAE alloy **UTS: 110 Elon: 20%**
AG 0.5	0.5% Mg Al alloy: Wrought; L'Aluminium Francais
AG 0.6	0.8% Mg Al alloy: Wrought; French Standard
AG 1	1.0% Mg Al alloy: Wrought; L'Aluminium Francais
AG 1	1.4% Mg Al alloy: Wrought; French Standard
AG 2	2.2% Mg Al alloy: Wrought; French Standard
AG 2	2.1% Mg Al alloy: Wrought; L'Aluminium Francais
AG 3	3.1% Mg 0.8% Mn Al alloy: Wrought; L'Aluminium Francais
AG 3	3.2% Mg Al alloy: Wrought; French Standard

Note. The following abbreviations and units are used in the tables:

DPN	Hardness, diamond pyramid number
UTS	Ultimate tensile strength, N/mm^2
Elon	Elongation, %
Proof	0.1 % proof strength, N/mm^2

1 N/mm^2=0.1 hbar=0.102 kgf/mm^2=0.06475 tonf/in^2=145.04 lbf/in^2

Symbol	Nominal analysis, supplier, condition and remarks.
AG 4	4.2% Mg 0.8% Mn Al alloy: Wrought; L'Aluminium Francais
AG 4 MC	4.2% Mg Al alloy: Wrought; French Standard
AG 5	5.0% Mg 0.6% Mn Al alloy: Sheet; French Standard
AG 5	5.2% Mg 0.8% Mn Al alloy: Wrought; L'Aluminium Francais
AG 7	7.0% Mg 0.8% Mn Al alloy: Wrought; L'Aluminium Francais
AGG 4	Mg Mn Al alloy: Wrought; further information from Aluminium Zentrale
AGG 50	1% Mg 0.2% Mn 0.2% Cr Al alloy: Wrought; further information from Aluminium Zentrale
AGG 54	4% Mg Al alloy: Wrought; further information from Aluminium Zentrale
AGG 57	3% Mg 0.2% Cr Al alloy: Wrought; further information from Aluminium Zentrale
ALCAN GB 54S	3.5% Mg 0.3% Mn Al alloy: Alcan for BS alloy N5
ALCAN GB 58S	7.2% Mg 0.3% Mn Al alloy: Alcan for BS alloy N7
ALCAN GBA 56S	5% Mg 0.2% Mn Al alloy: Alcan for BS alloy N6
ALCAN GBB 53S	2.7% Mg 0.7% Mn 0.1% Cr Al alloy: Alcan **UTS: 250 Elon: 6% Proof: 200**
ALCAN GBD 54S	4.2% Mg 0.7% Mn 0.12% Cr Al alloy: Alcan for BS alloy N5
ALCAN GBL 57S	1% Mg Al alloy: Sheet and strip; Alcan; for bright anodizing trim
ALCAN GBM 57S	2% Mg 0.3% Mn Al alloy: Alcan for BS alloy N4
ALCAN GBS 57S	0.6% Mg Al alloy: Bar & forging for bright anodizing; Alcan for BT 1
ALCOA 510	2.0% Mg Al alloy: Wrought; Alcoa; ½-hard; H4 condition **UTS: 225 Elon: 5% Proof: 170**
ALCOA 520	3.5% Mg Al alloy: Wrought; Alcoa; ½-hard; H4 condition **UTS: 300 Elon: 6% Proof: 220**
ALCOA 540	1.0% Mg Al alloy: Wrought; Alcoa; ½-hard; H4 condition **UTS: 155**
ALCOA 550	0.5% Mg Al alloy: Wrought; Alcoa; ½-hard; H4 condition; for anodizing **UTS: 150**
Al Mg 1	1% Mg 0.2% Cr 0.2% Mn Al alloy: Wrought; previous German Standard designation
Al Mg 2	2% Mg 0.2% Mn 0.2% Cr Al alloy: Wrought; previous German Standard designation
Al Mg 3	3% Mg 0.2% Mn 0.2% Cr Al alloy: Wrought; previous German Standard designation
Al Mg 3 Si	3% Mg 0.6% Si 0.5% Mn 0.2% Cr Al alloy: Wrought; previous German Standard designation
Al Mg 5	5% Mg 0.4% Mn 0.2% Cr Al alloy: Wrought; previous German Standard designation
Al Mg 7	7% Mg 0.4% Mn 0.2% Cr Al alloy: Wrought; previous German Standard designation
Al Mg Mn	2% Mg 1% Mn 0.2% Cr Al alloy: Wrought; previous German Standard designation
ALMINAL 4	0.15% Cu 0.6% Si 1.8% Mg 0.5% Mn 0.5% Cr Al alloy: Wrought; origin unknown
ALMINAL 5	0.15% Cu 0.6% Si 3.0% Mg 1.0% Mn 0.5% Cr Al: Wrought; origin unknown
ALMINAL 6	0.15% Cu 0.6% Si 4.5% Mg 1.0% Mn Al alloy: Wrought; origin unknown
Al R Mg 0.5	0.5% Mg Al alloy: For reflective uses; previous German Standard designation
Al R Mg 1	1% Mg Al alloy: For reflective uses; previous German Standard designation
Al R Mg 2	2% Mg Al alloy: For reflective uses; previous German Standard designation

Symbol	Nominal analysis, supplier, condition and remarks.
ALUMAGNESE 10 C	1% Mg Al alloy: E Kaye; hard drawn **UTS: 180 Elon: 5%**
ALUMAGNESE 20 V	2% Mg Al alloy: E Kaye for BS alloy N4
ALUMAGNESE 35 B	3.5% Mg Al alloy: E Kaye for BS alloy N5
ALUMAGNESE 45	4.5% Mg 0.8% Mn Al alloy: E Kaye for BS alloy N8
ALUMAGNESE 50J	5% Mg Al alloy: E Kaye for BS alloy H6
ALUMAGNESE CS	1% Mg Al alloy: Tube; E Kaye; ½-hard **UTS: 210 Elon: 9%**
AMS 4004	2.5% Mg 0.25% Cr Al alloy: Foil
AMS 4015 E	2.5% Mg 0.25% Cr Al alloy: Sheet & plate; annealed; AMS for AA alloy 5052 **UTS: 180 Elon: 17%**
AMS 4016 E	2.5% Mg 0.25% Cr Al alloy: Sheet & plate; ½-hard; AMS for AA alloy 5052 **UTS: 220 Elon: 6%**
AMS 4017 E	2.5% Mg 0.25% Cr Al alloy: Sheet & plate; ½-hard; AMS for AA alloy 5052 **UTS: 250 Elon: 4%**
AMS 4018	3.5% Mg 0.25% Cr Al alloy: Sheet & plate; annealed; AMS for AA alloy 5154 **UTS: 240 Elon: 14% Proof: 70(0.2%)**
AMS 4019	3.5% Mg 0.25% Cr Al alloy: Sheet plate; cold rolled; AMS for AA alloy 5154 **UTS: 240 Elon: 8% Proof: 170**
AMS 4056 B	4.5% Mg 0.6% Mn 0.15% Cr Al alloy: Sheet; annealed; AMS for AA alloy 5083
AMS 4057 A	4.5% Mg 0.6% Mn 0.15% Cr Al alloy: Sheet; ½-hard; AMS for AA alloy 5083
AMS 4058 A	4.5% Mg 0.6% Mn 0.15% Cr Al alloy: Sheet; ½-hard; AMS for AA alloy 5083
AMS 4059 B	4.5% Mg 0.6% Mn 0.15% Cr Al alloy: Sheet; 1/8-hard; AMS for AA alloy 5083
AMS 4069	2.5% Mg 0.25% Cr Al alloy: Seamless tube; annealed; AMS for AA alloy 5052
AMS 4070 F	2.5% Mg 0.25% Cr Al alloy: Seamless tube; annealed; AMS for AA alloy 5052
AMS 4071 F	2.5% Mg 0.25% Cr Al alloy: Seamless tube; annealed & drawn; AMS for AA alloy 5052
AMS 4114 B	2.5% Mg 0.25% Cr Al alloy: Bar as rolled; AMS for AA alloy 5052
AMS 4182 A	5% Mg 0.12% Mn 0.12% Cr Al alloy: Wire; annealed; AMS for AA alloy 5056
AN WW C 561a/1	5.2% Mg Al alloy: Wire; US Service
ASTM B199 AZ81A	7.5% Mg 0.7% Zn Al alloy: Die casting
ASTM B209/5005	1% Mg Al alloy: Sheet; annealed **UTS: 110 Elon: 22%**
ASTM B209/5005	1% Mg Al alloy: Sheet; cold rolled **UTS: 170 Elon: 3%**
ASTM B209/5050	1.5% Mg Al alloy: Sheet; annealed **UTS: 120 Elon: 20% Proof: 22**
ASTM B209/5050	1.5% Mg Al alloy: Sheet; cold rolled **UTS: 200 Elon: 3%**
ASTM B209/5052	2.5% Mg 0.2% Cr Al alloy: Sheet; annealed **UTS: 180 Elon: 18% Proof: 45**
ASTM B209/5052	2.5% Mg 0.2% Cr Al alloy: Sheet; cold rolled **UTS: 280 Elon: 4%**
ASTM B209/5083	4.5% Mg 0.7% Mn 0.2% Cr Al alloy: Sheet; annealed **UTS: 310 Elon: 16% Proof: 110**
ASTM B209/5083	4.5% Mg 0.7% Mn 0.2% Cr Al alloy: Sheet; cold rolled **UTS: 370 Elon: 6% Proof: 270**
ASTM B209/5086	4% Mg 0.5% Mn 0.2% Cr Al alloy: Sheet; annealed **UTS: 250 Elon: 18% Proof: 70**

Symbol	Nominal analysis, supplier, condition and remarks.
ASTM B209/5086	4% Mg 0.5% Mn 0.2% Cr Al alloy: Sheet; cold rolled **UTS: 340 Elon: 4% Proof: 250**
ASTM B209/5154	3.5% Mg 0.25% Cr 0.2% Ti Al alloy: Sheet; annealed **UTS: 240 Elon: 18% Proof: 45**
ASTM B209/5154	3.5% Mg 0.25% Cr 0.2% Ti Al alloy: Sheet; cold rolled **UTS: 310 Elon: 5% Proof: 240**
ASTM B209/5155	0.4% Mn 4.2% Mg 0.15% Cr Al alloy: Sheet & plate
ASTM B209/5252	2.5% Mg Al alloy: Sheet & plate
ASTM B209/5254	3.5% Mg 0.25% Cr Al alloy: Sheet; annealed **UTS: 240 Elon: 18% Proof: 45**
ASTM B209/5254	3.5% Mg 0.25% Cr Al alloy: Sheet; cold rolled **UTS: 330 Elon: 5% Proof: 240**
ASTM B209/5454	2.7% Mg 0.7% Mn 0.15% Cr 0.2% Ti Al alloy: Sheet; annealed **UTS: 240 Elon: 18% Proof: 60**
ASTM B209/5454	2.7% Mg 0.7% Mn 0.15% Cr 0.2% Ti Al alloy: Sheet; cold rolled **UTS: 310 Elon: 10% Proof: 180**
ASTM B209/5456	5% Mg 0.7% Mn 0.15% Cr 0.2% Ti Al alloy: Sheet; annealed **UTS: 340 Elon: 16% Proof: 180**
ASTM B209/5456	5% Mg 0.7% Mn 0.15% Cr 0.2% Ti Al alloy: Sheet; ½-hard **UTS: 400 Elon: 9% Proof: 270**
ASTM B209/5457	0.3% Mn 1.0% Mg Al alloy: Sheet & plate
ASTM B209/5652	2.5% Mg 0.25% Cr Al alloy: Sheet; annealed **UTS: 200 Elon: 20%**
ASTM B209/5652	2.5% Mg 0.25% Cr Al alloy: Sheet; cold rolled **UTS: 280 Elon: 4%**
ASTM B210/5050	1.5% Mg Al alloy: Tube; annealed **UTS: 140 Proof: 15**
ASTM B210/5050	1.5% Mg Al alloy: Tube; cold drawn **UTS: 210 Proof: 150**
ASTM B210/5052	2.5% Mg 0.2% Cr Al alloy: Tube; annealed **UTS: 170 Proof: 45**
ASTM B210/5052	2.5% Mg 0.2% Cr Al alloy: Tube; cold drawn **UTS: 270 Proof: 210**
ASTM B210/5083	0.7% Mn 4.4% Mg 0.2% Cr Al alloy: Tube
ASTM B210/5086	0.5% Mn 4% Mg 0.15% Cr Al alloy: Tube
ASTM B210/5154	3.5% Mg 0.2% Cr 0.2% Ti Al alloy: Tube; annealed **UTS: 200 Proof: 60**
ASTM B210/5154	3.5% Mg 0.2% Cr 0.2% Ti Al alloy: Tube; cold drawn **UTS: 310 Proof: 220**
ASTM B210/5254	3.5% Mg 0.2% Cr Al alloy: Tube; annealed **UTS: 200 Proof: 60**
ASTM B210/5254	3.5% Mg 0.2% Cr Al alloy: Tube; cold drawn **UTS: 310 Proof: 220**
ASTM B210/5652	2.5% Mg 0.2% Cr Al alloy: Tube; annealed **UTS: 200 Proof: 45**
ASTM B210/5652	2.5% Mg 0.2 °Cr Al alloy: Tube; cold drawn **UTS: 280 Proof: 210**
ASTM B211/5052	2.5% Mg 0.2% Cr Al alloy: Bar; cold drawn **UTS: 270**
ASTM B211/5056	5% Mg 0.1% Cr Al alloy: Bar; cold rolled **UTS: 390**
ASTM B211/5154	3.5% Mg 0.2% Cr 0.2% Ti Al alloy: Bar; cold drawn **UTS: 310**
ASTM B211/5254	3.5% Mg 0.2% Cr Al alloy: Bar; cold drawn **UTS: 310**
ASTM B211/5652	2.5% Mg 0.2% Cr Al alloy: Bar; cold drawn **UTS: 270**
ASTM B221/5052	2.4% Mg 0.2% Cr Al alloy: Extruded bar, etc.

Symbol	Nominal analysis, supplier, condition and remarks.
ASTM B221/5083	0.8% Mn 4.5% Mg 0.2% Cr Al alloy: Extruded bar; cold drawn
	UTS: 270 **Elon: 12%** **Proof: 150**
ASTM B221/5086	0.6% Mn 4% Mg 0.2% Cr Al alloy: Extruded bar; cold drawn
	UTS: 240 **Elon: 12%** **Proof: 110**
ASTM B221/5154	3.5% Mg 0.2% Cr Al alloy: Extruded bar; cold drawn
	UTS: 200 **Proof: 75**
ASTM B221/5454	0.8% Mn 2.8% Mg 0.1% Cr Al alloy: Extruded bar; cold drawn
	UTS: 210 **Elon: 12%** **Proof: 60**
ASTM B221/5456	0.8% Mn 5% Mg 0.1% Cr Al alloy: Extruded bar; cold drawn
	UTS: 280 **Elon: 12%** **Proof: 170**
ASTM B221/5652	2.4% Mg 0.2% Cr Al alloy: Extruded bar, etc.
ASTM B234/5052	2.5% Mg 0.2% Cr Al alloy: Tube; cold drawn
	UTS: 220 **Proof: 170**
ASTM B234/5454	2.8% Mg 0.7% Mn 0.1% Cr Al alloy: Tube; cold drawn
	UTS: 270 **Proof: 200**
ASTM B235/5052	2.5% Mg 0.2% Cr Al alloy: Tube; extruded
	UTS: 220 **Proof: 60**
ASTM B235/5154	3.5% Mg 0.2% Cr Al alloy: Tube; annealed
	UTS: 220 **Proof: 75**
ASTM B235/5254	3.5% Mg 0.2% Cr Al alloy: Extruded tube; annealed; free of copper
	UTS: 220 **Proof: 75**
ASTM B235/5454	0.7% Mn 2.6% Mg 0.1% Cr Al alloy: Extruded tube; annealed
	UTS: 240 **Elon: 14%** **Proof: 75**
ASTM B235/5456	0.7% Mn 5% Mg 0.1% Cr Al alloy: Extruded tube; hard drawn
	UTS: 310 **Elon: 12%** **Proof: 120**
ASTM B235/5652	2.5% Mg 0.2% Cr Al alloy: Extruded tube; annealed
	UTS: 220 **Proof: 60**
ASTM B241/5052	0.45% Si+Fe 2.5% Mg 0.25% Cr Al: Pipe & tube
ASTM B241/5083	0.7% Mn 4.5% Mg 0.15% Cr Al: Pipe & tube
ASTM B241/5154	3.5% Mg 0.2% Cr Al alloy: Pipe; cold drawn
	UTS: 310 **Proof: 220**
ASTM B241/5254	3.5% Mg 0.2% Cr Al alloy: Pipe; low copper; cold drawn
	UTS: 310 **Proof: 220**
ASTM B241/5454	0.7% Mn 2.7% Mg 0.1% Cr Al alloy: Pipe; annealed
	UTS: 210 **Elon: 12%** **Proof: 60**
ASTM B241/5456	0.7% Mn 5% Mg 0.1% Cr Al alloy: Pipe; annealed
	UTS: 280 **Elon: 12%** **Proof: 120**
ASTM B241/5652	0.40% Si+Fe 2.5% Mg 0.25% Cr Al: Pipe & tube
ASTM B247/5083	0.7% Mn 4.5% Mg 0.15% Cr Al: Die and hand forgings
ASTM B247/5456	0.8% Mn 5.1% Mg 0.15% Cr Al: Die and hand forgings
ASTM B285 ER5050	1.5% Mg Al alloy: Electric welding rod; obsolete
ASTM B285 ER5052	2.5% Mg 0.2% Cr Al alloy: Welding rod; obsolete
ASTM B285 ER5154	3.5% Mg 0.2% Cr Al alloy: Welding wire
ASTM B285 ER5183	0.7% Mn 0.15% Cr 4.8% Mg Al alloy: Welding wire

Symbol	Nominal analysis, supplier, condition and remarks.
ASTM B285 ER5254	3.5% Mg 0.2% Cr Al alloy: Welding wire; low Cu
ASTM B285 ER5356	0.1% Mn 5% Mg 0.1% Cr 0.1% Ti Al alloy: Welding wire
ASTM B285 ER5554	0.8% Mn 2.8% Mg 0.1% Cr 0.1% Ti Al alloy: Welding wire
ASTM B285 ER5556	0.7% Mn 5.1% Mg 0.1% Cr 0.1% Ti Al alloy: Welding wire
ASTM B285 ER5652	2.5% Mg 0.2% Cr Al alloy: Welding wire
ASTM B307/5050	1.5% Mg Al alloy: Coiled tube; annealed
	UTS: 120
ASTM B307/5052	2.5% Mg 0.25% Cr Al alloy: Drawn tube
ASTM B308/5083	0.7% Mn 4.5% Mg 0.1% Cr Al alloy: Section; annealed
	UTS: 250 **Elon: 16%** **Proof: 90**
ASTM B308/5086	0.5% Mn 4% Mg 0.1% Cr Al alloy: Section; as rolled or extruded
	UTS: 240 **Elon: 12%** **Proof: 110**
ASTM B308/5454	0.7% Mn 2.7% Mg 0.1% Cr Al alloy: Section; annealed
	UTS: 210 **Elon: 14%** **Proof: 60**
ASTM B308/5456	0.7% Mn 5% Mg 0.1% Cr Al alloy: Section; annealed
	UTS: 280 **Elon: 16%** **Proof: 120**
ASTM B313/5050	1.5% Mg Al alloy: Welded tube; cold rolled
	UTS: 270 **Elon: 4%**
ASTM B313/5052	2.5% Mg 0.2% Cr Al alloy: Welded tube; cold rolled
	UTS: 270 **Elon: 4%**
ASTM B313/5086	0.5% Mn 4% Mg Al alloy: Welded tube; cold rolled
	UTS: 330 **Elon: 5%** **Proof: 250**
ASTM B313/5154	3.5% Mg 0.2% Cr Al alloy: Welded tube; cold rolled
	UTS: 4%0 **Proof: 240**
ASTM B316/5052	2.5% Mg 0.2% Cr Al alloy: Rivets; annealed
	UTS: 210
ASTM B316/5056	0.1% Mn 5% Mg 0.1% Cr Al alloy: Rivets; annealed
	UTS: 310
ASTM B316/5652	2.5% Mg 0.2% Cr Al alloy: Rivets; annealed
	UTS: 210
ASTM B318/5005	0.8% Mg Al alloy: Drawn coiled tubes; annealed
	UTS: 120
ASTM B318/5050	1.5% Mg Al alloy: Drawn coiled tube; annealed
	UTS: 140
ASTM B318/5052	2.5% Mg 0.2% Cr Al alloy: Drawn coiled tube; annealed
	UTS: 220
ASTM B345/5050	1.4% Mg Al alloy: Pipe; annealed
	UTS: 110 **Proof: 15**
ASTM B345/5052	2.5% Mg 0.2% Cr Al alloy: Pipe; annealed
	UTS: 170 **Proof: 45**
ASTM B345/5083	0.8% Mn 4.5% Mg 0.1% Cr Al alloy: Pipe; as drawn
	UTS: 270 **Elon: 12%**
ASTM B345/5086	0.5% Mn 4% Mg Al alloy: Pipe; annealed
	UTS: 240 **Elon: 14%** **Proof: 75**
ASTM B345/5154	3.5% Mg 0.2% Cr Al alloy: Pipe; annealed
	UTS: 200 **Elon: 12%** **Proof: 60**
ASTM B345/5456	0.8% Mn 5.1% Mg 0.1% Cr Al alloy: Pipe; annealed
	UTS: 280 **Elon: 16%** **Proof: 120**
ASTM B404/5052	2.5% Mg 0.25% Cr Al alloy: Condenser tubes
ASTM B404/5454	0.75% Mn 2.7% Mg 0.12% Cr Al alloy: Condenser tubes
ASTM B531	0.7% Mg Al alloy: Rod for electrical purposes; 54% IACS conductivity
AWCO 07	7% Mg Al alloy: Wire; Aluminium Wire Co for BS alloy N7

Note. The following abbreviations and units are used in the tables:

DPN	Hardness, diamond pyramid number
UTS	Ultimate tensile strength, N/mm^2
Elon	Elongation, %
Proof	0.1 % proof strength, N/mm^2

$1 \text{ N/mm}^2 = 0.1 \text{ hbar} = 0.102 \text{ kgf/mm}^2 = 0.06475 \text{ tonf/in}^2 = 145.04 \text{ lbf/in}^2$

Symbol	Nominal analysis, supplier, condition and remarks.
AWCO 21	2% Mg Al alloy: Wire; Aluminium Wire Co for BS alloy N4
AWCO 27	3% Mg Al alloy: For wire; Aluminium Wire Co for BS N5
AWCO 28	5% Mg Al alloy: Wire; Aluminium Wire Co for BS alloy N6
AWCO 282	4.8% Mg 0.8% Mn 0.15% Cr Al alloy: Wire; Aluminium Wire Co; for welding
AWCO 283	5.2% Mg 0.8% Mn 0.15% Cr 0.15% Ti Al alloy: Wire; Aluminium Wire Co; for welding
AWCO 284	2.7% Mg 0.8% Mn 0.15% Cr 0.15% Ti Al alloy: Wire; Aluminium Wire Co; for welding
AWCO 285	5.0% Mg 1.0% Mn 0.15% Cr 0.15% Ti Al alloy: Aluminium Wire Co; for welding
BA 21	2% Mg Al alloy: Sheet & tube; British Al Co for BS alloy N4
BA 27	3% Mg Al alloy: Sheet, strip & tube; British Al Co for BS alloy N5
BA 28	5% Mg Al alloy: Sheet, strip & tube; British Al Co for BS alloy N6
BA 211	1.1% Mg Al alloy: Sheet & strip for bright anodizing; British Al Co for BT2
BA 212	0.6% Mg Al alloy: Sheet & strip for bright anodizing; British Al Co for BT1
BA 213	1.1% Mg Al alloy: Sheet & strip for bright anodizing; British Al Co for BT 2
BA 218	2.0% Mg Al alloy: Sheet; British Al Co
BA 226	1.0% Mg Al alloy: Tube; British Al Co
BA 227	7% Mg Al alloy: Forgings; British Al Co for BS alloy N7
BA 271	2.7% Mg 0.8% Mn 0.1% Cr Al alloy: Sheet & strip; British Al Co
BA 281	4.5% Mg 0.7% Mn Al alloy: British Al Co for BS alloy N8
BA 284	4.0% Mg 0.5% Mn 0.1% Cr Al alloy: Sheet & strip; British Al Co; ½-hard **UTS: 375**
BA 5052	2.5% Mg 0.25% Cr Al alloy: Sheet; British Al Co; ¼-hard; for transportation containers **Elon: 5.5% Proof: 240**
BA SP11	0.5% Mg Al alloy: Sheet & strip for bright anodizing; British Al Co for BT4
BA SP12	1% Mg Al alloy: Sheet & strip for bright anodizing; British Al Co for BT5
BB 1	1% Mg Al alloy: Tube; Birmabright; annealed **DPN: 40 UTS: 150 Elon: 25% Proof: 60**
BB 1	1% Mg Al alloy: Tube; Birmabright; hard **DPN: 60 UTS: 220 Elon: 7% Proof: 180**
BB 1 - X	1% Mg Al alloy: Sheet; Birmabright; annealed **DPN: 30 UTS: 110 Elon: 27% Proof: 50**
BB 1 - X	1% Mg Al alloy: Sheet; Birmabright; ¾-hard **DPN: 45 UTS: 200 Elon: 4% Proof: 180**
BB 2	2% Mg Al alloy: Sheet & tube; Birmabright for BS alloy N4
BB 3	3% Mg Al alloy: Sheet, strip & tube; Birmabright for BS alloy N5
BB 4	4.5% Mg 0.75% Mn Al alloy: Wrought form; Birmetal for BS alloy N8
BB 5	5% Mg Al alloy: Sheet, strip & tube; Birmabright for BS alloy N6
BB 5-X	4.5% Mg 0.7% Mn Al alloy: Birmabright for BS alloy N8
BB 7	7% Mg Al alloy: Sheet, strip, tube & forging; Birmabright for BS alloy N7
BB 17	0.6% Mg Al alloy: Sheet & strip for bright anodizing; Birmabright for BT 1
BB 127	1.1% Mg Al alloy: Sheet & strip for bright anodizing; Birmabright for BT2
BS 477	al alloy: Bar; replaced by BS 1476
BS 532	Al alloy: Forgings; replaced by BS 1472
BS 1453 NG5	3.5% Mg Al alloy: Welding rod; for welding Al alloys N4 and N5 type
BS 1453 NG6	5% Mg Al alloy: Welding rod; for welding Al alloys, cast & wrought with up to 5% Mg
BS 1470 NS4	2% Mg Al alloy: For sheet & strip; ½-hard **UTS: 220 Elon: 5% Proof: 170**
BS 1470 NS4	2% Mg Al alloy: For sheet & strip; ¼-hard **UTS: 200 Elon: 8% Proof: 120**
BS 1470 NS4	2% Mg Al alloy: For sheet & strip; annealed **UTS: 180 Elon: 18%**
BS 1470 NS5	3% Mg Al alloy: For sheet & strip; annealed **UTS: 210 Elon: 18%**
BS 1470 NS5	3% Mg Al alloy: For sheet & strip; ¼ hard **UTS: 240 Elon: 8% Proof: 150**
BS 1470 NS5	3% Mg Al alloy: For sheet & strip; ½-hard **UTS: 270 Elon: 5% Proof: 210**
BS 1470 NS6	5% Mg Al alloy: For sheet & strip; annealed **UTS: 250 Elon: 18% Proof: 120**
BS 1470 NS6	5% Mg Al alloy: For sheet & strip; ¼-hard **UTS: 280 Elon: 8% Proof: 210**
BS 1471 NT4	2% Mg Al alloy: Tube; ½-hard **UTS: 220 Elon: 5% Proof: 170**
BS 1471 NT4	2% Mg Al alloy: Tube; annealed **UTS: 170 Elon: 18%**
BS 1471 NT5	3% Mg Al alloy: Tube; annealed **UTS: 210 Elon: 18% Proof: 110**
BS 1471 NT5	3% Mg Al alloy: Tube; ½-hard **UTS: 240 Elon: 5% Proof: 180**
BS 1471 NT6	5% Mg Al alloy: Tube; annealed **UTS: 250 Elon: 18% Proof: 120**
BS 1471 NT6	5% Mg Al alloy: Tube; ½-hard **UTS: 260 Elon: 5% Proof: 210**
BS 1471 NT8	4.5% Mg 0.75% Mn Al alloy: Tube; ½-hard H4 condition **UTS: 310 Elon: 5% Proof: 230**
BS 1472 NF4	2% Mg Al alloy: For forgings & bar; annealed **UTS: 170 Elon: 18% Proof: 75**
BS 1472 NF5	3% Mg Al alloy: For forgings & bar; annealed **UTS: 210 Elon: 18% Proof: 90**
BS 1472 NF6	5% Mg Al alloy: For forgings & bar; annealed **UTS: 250 Elon: 18% Proof: 120**
BS 1472 NF7	7% Mg Al alloy: For forgings & bar; annealed **UTS: 300 Elon: 18% Proof: 120**
BS 1472 NF8	4.5% Mg 0.75% Mn Al alloy: For forgings & bar; annealed **UTS: 270 Elon: 15% Proof: 120**
BS 1473 NB6	5.0% Mg 0.3% Mn+Cr Al alloy: Rivet and screw stock; cold rolled **UTS: 350 Proof: 240**
BS 1473 NG7	7% Mg Al alloy: Welding rod; for welding Al cast & wrought alloys with high Mg
BS 1473 NR5	3% Mg Al alloy: For rivets; annealed **UTS: 210**
BS 1473 NR6	5% Mg Al alloy: For rivets; annealed **UTS: 240**
BS 1473 NR6	5% Mg Al alloy: For bolt stock; ½-hard **UTS: 310 Proof: 220**
BS 1474 NV4	2% Mg Al alloy: For tube; as drawn **UTS: 170 Elon: 18%**
BS 1474 NV5	3% Mg Al alloy: For tube; as drawn **UTS: 210 Elon: 18% Proof: 90**
BS 1474 NV6	5% Mg Al alloy: For tube; as drawn **UTS: 250 Elon: 18% Proof: 120**
BS 1474 NV8	4.5% Mg 0.75% Mn Al alloy: As drawn **UTS: 260 Elon: 16% Proof: 120**
BS 1475 NG4	2% Mg Al alloy: Wire; annealed **UTS: 180**
BS 1475 NG4	2% Mg Al alloy: Wire; hard drawn **UTS: 250**
BS 1475 NG5	3.5% Mg Al alloy: Wire

Symbol	Nominal analysis, supplier, condition and remarks.
BS 1475 NG6	5% Mg Al alloy: Wire; annealed **UTS: 260**
BS 1475 NG6	5% Mg Al alloy: Wire; ½-hard **UTS: 330**
BS 1475 NG6	5% Mg Al alloy: Wire; hard **UTS: 370**
BS 1476 NE4	2% Mg Al alloy: Bar; as rolled **UTS: 170 Elon: 18%**
BS 1476 NE5	3.5% Mg Al alloy: Bar; as rolled **UTS: 200 Elon: 18% Proof: 90**
BS 1476 NE6	5% Mg Al alloy: Bar; as rolled **UTS: 250 Elon: 18% Proof: 120**
BS 1476 NE8	4.5% Mg 0.75% Mn Al alloy: Bar; annealed **UTS: 250 Elon: 16% Proof: 120**
BS 1477 NP4	2% Mg Al alloy: Plate; as rolled **UTS: 180 Elon: 12%**
BS 1477 NP6	5% Mg Al alloy: Plate; annealed **UTS: 250 Elon: 20% Proof: 90**
BS 1477 NP8	4.5% Mg 0.75% Mn Al alloy: Plate; annealed **UTS: 160 Elon: 16% Proof: 120**
BS 1616 C	5% Mg Al alloy: Welding rod flux coated wire to BS alloy N6
BS 2901 NG4	2.2% Mg Al alloy: Rod for inert gas shielded arc welding
BS 2901 NG5	3.5% Mg Al alloy: Rod for all welding
BS 2901 NG6	5.0% Mg Al alloy: Rod for all welding
BS 2901 NG7	7.0% Mg Al alloy: Rod for inert gas shielded arc welding
BS 4300/1 NJ4	2.2% Mg Al alloy: Longitudinal welded tube; as welded **UTS: 240 Elon: 3% Proof: 220**
BS 4300/1 NJ5	3.4% Mg Al alloy: Longitudinal welded tube; as welded **UTS: 290 Elon: 5% Proof: 250**
BS 4300/2 BTRS1	0.5% Mg Al alloy: For bright trim; ½-hard condition **UTS: 150**
BS 4300/2 BTRS2	0.09% Mg Al alloy: For bright trim; ½-hard condition **UTS: 170**
BS 4300/2 BTRS2	1.0% Mg Al alloy: For bright trim; annealed **UTS: 150**
BS 4300/7 NS41	0.8% Mg Al alloy: Sheet & strip; ½-hard H4 condition **UTS: 160 Elon: 4% Proof: 100**
BS 4300/8 NS51	2.7% Mg 0.7% Mn 0.12% Cr Al alloy: Plate & sheet; ½-hard H4 condition **UTS: 300 Elon: 5% Proof: 200**
BS 4300/9 NG41	0.8% Mg Al alloy: Wire; hard H8 condition **UTS: 180**
BS 4300/10 NT51	2.7% Mg 0.7% Mn 0.12% Cr Al alloy: Drawn tube; ½-hard H4 condition **UTS: 300 Elon: 4% Proof: 220**
BS 4300/11 NF51	2.7% Mg 0.7% Mn 0.12% Cr Al alloy: Forging; as forged **UTS: 270 Elon: 18% Proof: 80**
BS 4300/12 NE51	2.8% Mg 0.7% Mn 0.12% Cr Al alloy: Bar, tube, etc.; as drawn **UTS: 250 Elon: 17% Proof: 90**

Note. The following abbreviations and units are used in the tables:

DPN	Hardness, diamond pyramid number
UTS	Ultimate tensile strength, N/mm^2
Elon	Elongation, %
Proof	0.1 % proof strength, N/mm^2

1 N/mm^2=0.1 hbar=0.102 kgf/mm^2=0.06475 tonf/in^2=145.04 lbf/in^2

Symbol	Nominal analysis, supplier, condition and remarks.
BS 4300/13 NG52	2.8% Mg 0.7% Mn 0.12% Cr Al alloy: Welding wire
BS 4300/13 NG52	2.7% Mg 0.7% Mn 0.7% Cr 0.1% Ti Al alloy: Wire; for welding
BS L44	2% Mg Al alloy: Bar; as rolled **UTS: 170 Elon: 18%**
BS L46	Mg Al alloy: Obsolete
BS L55	2% Mg Al alloy: Tube; ½-hard; cold drawn **UTS: 220**
BS L56	2% Mg Al alloy: Tube; annealed **UTS: 210**
BS L58	5% Mg Al alloy: Wire; as drawn **UTS: 260**
BS L80	2.25% Mg Al alloy: Sheet; annealed **UTS: 180 Elon: 18%**
BS L81	2.25% Mg Al alloy: Sheet; ½-hard **UTS: 220 Elon: 5% Proof: 170**
BS L82	3.5% Mg Al alloy: Sheet; annealed **UTS: 210 Elon: 18%**
BT 1	0.6% Mg Al alloy: For bright anodizing; annealed; designation given by Aluminium Federation **UTS: 120**
BT 1	0.6% Mg Al alloy: For bright anodizing; ¼-hard; designation given by Aluminium Federation **UTS: 140**
BT 1	0.6% Mg Al alloy: For bright anodizing; ½-hard; designation given by Aluminium Federation **UTS: 150**
BT 1	0.6% Mg Al alloy: For bright anodizing; hard; designation given by Aluminium Federation **UTS: 170**
BT 2	1.1% Mg Al alloy: For bright anodizing; annealed; designation given by Aluminium Federation **UTS: 170**
BT 2	1.1% Mg Al alloy: For bright anodizing; ½-hard; designation given by Aluminium Federation **UTS: 180**
BT 2	1.1% Mg Al alloy: For bright anodizing; hard; designation given by Aluminium Federation **UTS: 200**
BT 4	0.5% Mg Al alloy: For bright anodizing; annealed; designation given by Aluminium Federation **UTS: 90**
BT 4	0.5% Mg Al alloy: For bright anodizing; ¼-hard; designation given by Aluminium Federation **UTS: 110**
BT 4	0.5% Mg Al alloy: For bright anodizing; ½-hard; designation given by Aluminium Federation **UTS: 120**
BT 4	0.5% Mg Al alloy: For bright anodizing; hard; designation given by Aluminium Federation **UTS: 150**
BT 5	1% Mg Al alloy: For bright anodizing; annealed; designation given by Aluminium Federation **UTS: 120**
BT 5	1% Mg Al alloy: For bright anodizing; ¼-hard; designation given by Aluminium Federation **UTS: 140**
BT 5	1% Mg Al alloy: For bright anodizing; ½-hard; designation given by Aluminium Federation **UTS: 170**
BT 5	1% Mg Al alloy: For bright anodizing; hard; designation given by Aluminium Federation **UTS: 180**
DTD 175	Mg Al alloy: Obsolete
DTD 180B	3% Mg Al alloy: Sheet; ½-hard **UTS: 250 Elon: 8% Proof: 170**
DTD 182B	7% Mg Al alloy: Sheet & strip; annealed **UTS: 330 Elon: 18%**

Symbol	Nominal analysis, supplier, condition and remarks.
DTD 186B	7% Mg Al alloy: Tube; ½-hard **UTS: 370 Elon: 5% Proof: 260**
DTD 190	Obsolete. No information available
DTD 297A	7% Mg Al alloy: Bar & forging; annealed **UTS: 300 Elon: 18% Proof: 140**
DTD 303	5% Mg Al alloy: Wire as drawn; replaced by BSL 58 **UTS: 260**
DTD 310	2% Mg Al alloy: Replaced by BSL 56
DTD 440	2% Mg Al alloy: Tube; ½-hard; cold drawn **UTS: 220**
DTD 606A	2% Mg Al alloy: Sheet; ½-hard; cold rolled **UTS: 220 Elon: 15% Proof: 180**
DTD 634A	2% Mg Al alloy: Sheet & strip; annealed **UTS: 180 Elon: 18%**
DURALBRITE 4	2.5% Mg Al alloy: Extrusion; Alcan; hard drawn **UTS: 255**
ELM 8	2% Mg Al alloy: Strip and welded tube; Elm Engineering **UTS: 220 Elon: 5% Proof: 170**
ELM 10	2% Mg Al alloy: Strip and welded tube; Elm Engineering **UTS: 260 Elon: 4% Proof: 220**
ERMAL NS4	2% Mg Al alloy: Sheet & strip; Enfield specification for BS 1470 NS4
G 1A	1.5% Mg Al alloy: Previous ASTM designation, now 5050
G 1B	0.8% Mg Al alloy: Previous ASTM designation, now 5005
GM 31A	2.8% Mg 0.8% Mn 0.1% Cr Al alloy: Previous ASTM designation, now 5454
GM 40A	4% Mg 0.5% Mn 0.2% Cr Al alloy: Previous ASTM designation, now 5086
GM 41A	4.5% Mg 0.7% Mn 0.15% Cr Al alloy: Previous ASTM designation, now 5083
GM 50A	5% Mg 0.15% Mn 0.1% Cr Al alloy: Previous ASTM designation, now 5056
GM 51A	5% Mg 0.7% Mn 0.15% Cr Al alloy: Previous ASTM designation, now 5456
GR 20A	0.1% Cu 2.5% Mg 0.3% Cr Al alloy: Previous ASTM designation, now 5052
GR 20B	2.5% Mg 0.3% Cr Al alloy: Previous ASTM designation, now 5652
GR 40A	3.5% Mg 0.2% Cr Al alloy: Previous ASTM designation, now 5154
GR 40B	3.5% Mg 0.2% Cr Al alloy: Previous ASTM designation, now 5254
HA 4 GM 31N	2.7% Mn 0.8% Mn 0.1% Cr Al alloy: Plate; Canadian Standard
HA 4 GM41	4.5% Mg 0.7% Mn 0.1% Cr Al alloy: Plate; Canadian Standard
HA 4 GR20	2.6% Mg 0.25% Cr Al alloy: Plate; Canadian Standard
HA 5 GM41	4.5% Mg 0.7% Mn 0.1% Cr Al alloy: Bar, etc.; Canadian Standard
HA 5 GR20	2.6% Mg 0.2% Cr Al alloy: Bar, etc.; Canadian Standard
HA 6 GM50R	5.0% Mg 0.1% Mn 0.1% Cr Al alloy: For rivets and brazing wire; Canadian Standard
HA 6 GS11P	1.2% Mg 0.2% Cr Al alloy: For rivets and brazing wire; Canadian Standard
HA 7 GR20	2.5% Mg Al alloy: Tube; Canadian Standard
HIDUMINIUM 05	5% Mg Al alloy: For tubes & extrusions; High Duty Alloys for BS alloy N6
HIDUMINIUM 07	7% Mg Al alloy: For forgings & tubes; High Duty Alloys for BS alloy N7
HIDUMINIUM 12	1.3% Mg Al alloy: Sheet; High Duty Alloys; annealed **UTS: 150 Elon: 25% Proof: 50**

Symbol	Nominal analysis, supplier, condition and remarks.
HIDUMINIUM 14	1% Mg 0.5% Mn Al alloy: Tube; High Duty Alloys; hard drawn **UTS: 210 Elon: 6% Proof: 170**
HIDUMINIUM 16	0.6% Mg Al alloy: Sheet & strip for bright anodizing; High Duty Alloys for BT 1
HIDUMINIUM 17	1.1% Mg Al alloy: Sheet & strip for bright anodizing; High Duty Alloys for BT 2
HIDUMINIUM 22	2% Mg Al alloy: For extrusions & tubes; High Duty Alloys for BS alloy N4
HIDUMINIUM 24	2.5% Mg Al alloy: Sheet; High Duty Alloys; annealed **UTS: 200 Elon: 18%**
HIDUMINIUM 33	3% Mg Al alloy: For extrusions & tubes; High Duty Alloys for BS alloy N5
HIDUMINIUM 35	4.5% Mg 0.8% Mn Al alloy: High Duty Alloys for BS alloy N8
IMPALCO 510	2% Mg Al alloy: Sheet, tube, bar, etc.; Imperial Al Co for BS alloy N4
IMPALCO 520	3.5% Mg 1% Mn Al alloy: Sheet, tube, bar, etc.; Imperial Al Co for BS alloy N5
IMPALCO 530	4.5% Mg 1% Mn Al alloy: Plate; Imperial Al Co for BS alloy N5/6
IMPALCO 531	4.4% Mg Al alloy: Plate, bar & forgings; Imperial Al Co for BS alloy N8
IMPALCO 540	1% Mg Al alloy: Sheet & bar; Imperial Al Co; annealed **UTS: 150**
IMPALCO 540	1% Mg Al alloy: Sheet & bar; Imperial Al Co; ½-hard **UTS: 180**
IMPALCO 540	1% Mg Al alloy: Sheet & bar; Imperial Al Co; ½-hard **UTS: 200**
IMPALCO 550	0.5% Mg Al alloy: Sheet; Imperial Al Co; annealed **UTS: 120**
IMPALCO 550	0.5% Mg Al alloy: Sheet; Imperial Al Co; ½-hard **UTS: 140**
IMPALCO 560	5% Mg 0.5% Cr Al alloy: Sheet & bar; Imperial Al Co for BS alloy N6
IMPALCO 570	7% Mg Al alloy: Sheet & bar; Imperial Al Co for BS alloy N7
IMPALCO 900	1% Mg 0.2% Mn Al alloy: Tube; Imperial Al Co; annealed **UTS: 120 Elon: 20%**
IMPALCO 900	1% Mg 0.2% Mn Al alloy: Tube; Imperial Al Co; ½-hard **UTS: 200 Elon: 6% Proof: 150**
IMPALCO 900	1% Mg 0.2% Mn Al alloy: Tube; Imperial Al Co; hard **UTS: 220 Elon: 4% Proof: 200**
IMPALCO 901	1% Mg Al alloy: Tube; Imperial Al Co; as drawn **UTS: 140 Elon: 25% Proof: 75**
IMPALCO M31	0.6% Mg Al alloy: Sheet & strip for bright anodizing; Imperial Al Co for BT1
IMPALCO M32	10% Mg Al alloy: For sheet & strip; ½-hard; Imperial Al Co **UTS: 140 Elon: 12% Proof: 110**
IMPALCO M32	1% Mg Al alloy: Sheet & strip; annealed; Imperial Al Co for high reflectivity **UTS: 140 Elon: 20%**
IMPALCO M32	1% Mg Al alloy: For sheet & strip; hard; Imperial Al Co **UTS: 180 Elon: 4% Proof: 140**
IMPALCO M32X	1% Mg Al alloy: For sheet & strip; annealed; Imperial Al Co. Similar to M32 **UTS: 140 Elon: 20%**
IMPALCO M32X	1% Mg Al alloy: For sheet & strip; ½-hard; Imperial Al Co **UTS: 140 Elon: 12% Proof: 120**

Symbol	Nominal analysis, supplier, condition and remarks.
IMPALCO M32X	1% Mg Al alloy: For sheet & strip; hard; Imperial Al Co
	UTS: 180 **Elon: 4%** **Proof: 140**
IMPALCO M34	1% Mg 0.05% Cu Al alloy: For tube; hard; Imperial Al Co
	DPN: 70 **UTS: 210** **Elon: 6%** **Proof: 180**
IMPALCO M34	1% Mg 0.05% Cu Al alloy: For tube; annealed; Imperial Al Co
	DPN: 36 **UTS: 130** **Elon: 22%**
IMPALCO M35/1	2% Mg Al alloy: For sheet strip tube wire plate & sections; Imperial Al Co for BS alloy N4
IMPALCO M35/2	3.5% Mg 0.1% Cu Al alloy: For sheet, strip, tube, wire, bar & sections; Imperial Al Co for BS alloy N5
IMPALCO M35/2	4% Mg 1% Mn Al alloy: For plate; Imperial Al Co for BS alloy N5/6
IMPALCO M36	5% Mg 1% Mn Al alloy: For sheet, strip, wire, bar, rod & sections; Imperial Al Co for BS alloy N6
KYNAL M35/1	2% Mg Al alloy: ICI; obsolete; now Imperial Al Co
KYNAL M35/2	3% Mg Al alloy: ICI; obsolete; now Imperial Al Co
KYNAL M36	5% Mg Al alloy: ICI; obsolete; now Imperial Al Co
KYNAL M37	7% Mg Al alloy: ICI; obsolete; now Imperial Al Co
MG 2	2% Mg Al alloy: Sheet & tube; J Booth for BS alloy N4
MG 3	3% Mg Al alloy: Sheet, strip & tube; J Booth for BS alloy N5
MG 5	5% Mg Al alloy: Sheet, strip & tube; J Booth for BS alloy N6
MG 5S	4.5% Mg 0.7% Mn 0.15% Cr Al alloy: Tube forgings; J Booth; annealed
	UTS: 260 **Elon: 16%** **Proof: 120**
MG 7	7% Mg Al alloy: Sheet, strip & tube forging; J Booth for BS alloy N7
NORAL 54S	3% Mg Al alloy: Sheet, strip & tube; Northern Al Co for BS alloy N5
NORAL 58S	7% Mg Al alloy: Sheet, strip & tube forging; Northern Al Co for BS alloy N7
NORAL A56S	5% Mg Al alloy: Sheet, strip & tube; Northern Al Co for BS alloy N6
NORAL M57S	2% Mg Al alloy: Sheet & tube; Northern Al Co for BS alloy N4
P 3G	0.3% Si 3.1% Mg 0.2% Mn 0.1% Ti Al alloy: Casting; Anglo-Swiss code for Peraluman 3G
P 5G	0.9% Si 5.2% Mg 0.3% Mn 0.1% Ti Al alloy: Casting; Anglo-Swiss code for Peraluman 5G
P 9G	0.2% Mg 0.3% Mn 0.025% Be Al alloy: Casting; Anglo-Swiss code for Peraluman 9G
Pe 15	1.5% Mg 0.2% Mn Al alloy: Wrought; Anglo-Swiss code for Peraluman 15
Pe 30	2.8% Mg 0.4% Mn Al alloy: Wrought; Anglo-Swiss code for Peraluman 30
Pe 40	4% Mg 0.4% Mn Al alloy: Wrought; Anglo-Swiss code for Peraluman 40
Pe 50	0.4% Cu 5.2% Mg 0.4% Mn Al alloy: Wrought; Anglo-Swiss code for Peraluman 50

Note. The following abbreviations and units are used in the tables:

DPN	Hardness, diamond pyramid number
UTS	Ultimate tensile strength, N/mm^2
Elon	Elongation, %
Proof	0.1 % proof strength, N/mm^2

$1\ N/mm^2 = 0.1\ hbar = 0.102\ kgf/mm^2 = 0.06475\ tonf/in^2 = 145.04\ lbf/in^2$

Symbol	Nominal analysis, supplier, condition and remarks.
PERALUMAN 3G	0.3% Si 3.1% Mg 0.2% Mn 0.1% Ti Al alloy: Casting; Anglo-Swiss solution treated & aged; obsolete
	DPN: 60 **UTS: 210** **Elon: 6%** **Proof: 140**
PERALUMAN 5G	0.9% Si 5.2% Mg 0.3% Mn 0.1% Ti Al alloy: Casting; Anglo-Swiss as die cast; obsolete
	DPN: 70 **UTS: 210** **Elon: 5%** **Proof: 110**
PERALUMAN 9G	9.2% Mg 0.3% Mn 0.025% Be Al alloy: Casting; Anglo-Swiss as die cast; obsolete
	DPN: 70 **UTS: 220** **Elon: 2%** **Proof: 140**
PERALUMAN 15	1.5% Mg 0.2% Mn Al alloy: Wrought; Anglo-Swiss; cold drawn; obsolete
	DPN: 70 **UTS: 220** **Elon: 6%** **Proof: 210**
PERALUMAN 30	2.8% Mg 0.4% Mn Al alloy: Wrought; Anglo-Swiss; hard drawn; obsolete
	DPN: 85 **UTS: 280** **Elon: 9%** **Proof: 250**
PERALUMAN 40	4% Mg 0.4% Mn Al alloy: Wrought; Anglo-Swiss; cold drawn; obsolete
	DPN: 90 **UTS: 310** **Elon: 9%** **Proof: 280**
PERALUMAN 50	0.4% Cu 5.2% Mg 0.4% Mn Al alloy: Wrought; Anglo-Swiss; hard drawn; obsolete
	DPN: 100 **UTS: 360** **Elon: 9%** **Proof: 300**
PERALUMAN 100	0.2% Mn 0.8% Mg 0.2% Cr Al alloy: Wrought; Anglo-Swiss Al Co
PERALUMAN 150	1.6% Mg 0.2% Mn Al alloy: Wrought; Anglo-Swiss Al Co
PERALUMAN 260	2.6% Mg 0.8% Mn 0.15% Cr Al alloy: Wrought; Anglo-Swiss Co
PERALUMAN 300	2.9% Mg 0.3% Mn Al alloy: Wrought; Anglo-Swiss Al Co
PERALUMAN 400	4.0% Mg 0.3% Mn Al alloy: Wrought; Anglo-Swiss Al Co
PERALUMAN 460	4.5% Mg 0.8% Mn 0.2% Cr Al alloy: Wrought; Anglo-Swiss Al Co
PERALUMAN 500	5.2% Mg 0.3% Mn Al alloy: Wrought; Anglo-Swiss Al Co
QQ A 315	2.4% Mg 0.2% Cr Al alloy: Bar; US Federal; annealed
	UTS: 180 **Elon: 25%** **Proof: 75**
QQ A 318	2.5% Mg 0.3% Cr Al alloy: Plate; US Federal; cold rolled
	UTS: 250 **Elon: 2%** **Proof: 180**
REFLECTAL 050	0.5% Mg Al alloy: For anodizing; Anglo-Swiss Al Co
REFLECTAL 100	1.0% Mg Al alloy: For anodizing; Anglo-Swiss Al Co
SAE 201	2.5% Mg 0.25% Cr Al alloy: Wrought; annealed; now AA 5652
	UTS: 210 **Elon: 19%**
SAE 201	2.5% Mg 0.25% Cr Al alloy: Wrought; hard; now AA 5052
	UTS: 250 **Elon: 4%**
SAE 207	1.5% Mg Al alloy: Cold rolled; now AA 5050
	DPN: 63 **UTS: 210** **Elon: 6%** **Proof: 180**
SAE 208	3.5% Mg 0.2% Cr Al alloy: Cold rolled; now AA 5154
	DPN: 80 **UTS: 330** **Elon: 10%** **Proof: 280**
SAE 209	0.3% Mn 1% Mg Al alloy: Wrought; annealed; now AA 5357
	UTS: 110 **Elon: 20%**
SAE 251	0.7% Mn 2.7% Mg 0.1% Cr Al alloy: Wrought; annealed; now AA 5454
	UTS: 260 **Elon: 18%** **Proof: 60**
SAE 252	0.2% Mn 1% Mg Al alloy: Wrought; annealed; now AA 5457
	UTS: 140 **Elon: 20%**
SIS 4106	0.9% Mg Al alloy: Sheet & strip; Swedish Standard; annealed
	DPN: 30 **UTS: 110** **Elon: 25%** **Proof: 45**

Symbol	Nominal analysis, supplier, condition and remarks.
SIS 4106	0.9% Mg Al alloy: Sheet & strip; Swedish Standard; cold rolled **DPN: 60 UTS: 150 Elon: 3% Proof: 120**
SIS 4120	2.3% Mg Al alloy: Bar, sheet & strip; Swedish Standard; annealed **DPN: 55 UTS: 180 Elon: 18% Proof: 45**
SIS 4120	2.3% Mg Al alloy: Bar, sheet & strip; Swedish Standard; cold worked **DPN: 85 UTS: 250 Elon: 3% Proof: 210**
SIS 4163	5% Mg 1% Si Al alloy: Casting; Swedish Standard; as cast **DPN: 60**
STA 7 AW4A	2% Mg Al alloy: Bar; replaced by BS 1476 NE4
STA 7 AW4B	2% Mg Al alloy: Tube, replaced by BS 1471 NT4
STA 7 AW4C	2% Mg Al alloy: Sheet; replaced by BS 1470 NS4
STA 7 AW5A	3% Mg Al alloy: Bar; replaced by BS 1476 NE5
STA 7 AW5B	3% Mg Al alloy: Tube; replaced by BS 1471 NT5
STA 7 AW5C	3% Mg Al alloy: Sheet; replaced by BS 1470 NS5
STA 7 AW6A	5% Mg Al alloy: Bar; replaced by BS 1476 NE6
STA 7 AW6B	5% Mg Al alloy: Tube; replaced by BS 1471 NT6
STA 7 AW6C	5% Mg Al alloy: Sheet; replaced by BS 1470 NS6
STA 7 AW6D	5% Mg Al alloy: Wire; replaced by BS 1473-5 NR6 NW6 NG6
STA 7 AW7A	7% Mg Al alloy: Forging & bar; replaced by BS 1472 NF7 BS 1476 NE7
STA 7 AW7B	7% Mg Al alloy: Tube; replaced by BS 1471 NT7
STA 7 AW7C	7% Mg Al alloy: Replaced by BS 1470 NS7
TI 05	5% Mg Al alloy: For tube; Ti Al for BS alloy N6
TI 07	7% Mg Al alloy: Tube & forging; Ti Al for BS alloy N7
TI 22	2% Mg Al alloy: Sheet & tube; Ti Al for BS alloy N4
TI 33	3% Mg Al alloy: Sheet, strip & tube, Ti Al for BS alloy N5
UNI 3573	1.5% Mg Al alloy: For forging; Italian Standard
UNI 3574	2.5% Mg Al alloy: For forging; Italian Standard
UNI 3575	3.5% Mg Al alloy: For forging; Italian Standard
UNI 3576	5.0% Mg Al alloy: For forging; Italian Standard

Symbol	Nominal analysis, supplier, condition and remarks.
UNI 4510	0.9% Mg Al alloy: Sheet for decorative use; Italian Standard
UNI 4511	2.0% Mg Al alloy: For forging; Italian Standard
UNI 4512	0.5% Mg Al alloy: For decorative use; Italian Standard
UNI 5452	4.4% Mg Al alloy: For forging; Italian Standard
UNI 5784	0.8% Mg Al alloy: For forging; Italian Standard
UNI 6360	0.1% Cu 0.9% Mg Al alloy: For decorative purposes; Italian Standard
UV Al Mg	10% Mg Al alloy: Wrought; previous German Standard designation
W 4	2.2% Mg Al alloy: Sheet, bar, wire, etc.; Southern Forge; replaced by IMPALCO 510
W 5	3.5% Mg Al alloy: Sheet, bar, etc.; Southern Forge; replaced by IMPALCO 520
W 6	5% Mg 1% Mn Al alloy: Sheet, bar & wire; Southern Forge; replaced by IMPALCO 560
W 010	1% Mg Al alloy: Sheet, bar & forging; Southern Forge; replaced by IMPALCO 540
W 011	1% Mg Al alloy: Southern Forge; replaced by IMPALCO 900
W 012	1% Mg Al alloy: Southern Forge; replaced by IMPALCO 901
WWT 787	2.4% Mg 0.2% Cr Al alloy: Tube; US Federal; annealed **UTS: 220 Proof: 120**

Note. The following abbreviations and units are used in the tables:

DPN	Hardness, diamond pyramid number
UTS	Ultimate tensile strength, N/mm^2
Elon	Elongation, %
Proof	0.1 % proof strength, N/mm^2

$1 \ N/mm^2 = 0.1 \ hbar = 0.102 \ kgf/mm^2 = 0.06475 \ tonf/in^2 = 145.04 \ lbf/in^2$

1E Aluminium–magnesium–silicon wrought alloys

Specific gravity		2.7	
Density		2700 kg/m^3	(0.096 lbf/m^3)
Solidus/liquidus		552–652 °C	
Thermal conductivity	annealed	190–210 W/m °C	(45–50 cal/m s °C)
	solution treated and aged	170–200 W/m °C	(40–48 cal/m s °C)
Coefficient of linear expansion		23–25 × 10^{-6}/ °C	
Electrical conductivity	solution treated and aged	40–55% IACS (copper 100%)	
Specific resistance	annealed	32–38 microhm mm	
	solution treated and aged	36–45 microhm mm	
Young's modulus of elasticity		65.5–68.9 × 10^9 N/m^2	(9.5–10 × 10^6 lbf/in.^{2})
Impact	solution treated and aged	27 – 41 J	(20–30 ft/lb)
Fatigue strength (10^8 cycles)	solution treated and aged	±80–120 N/mm^2	(±5.5–7.5 tonf/in.^{2})

Hot strength

Temperature °C	Tensile strength* N/mm^2	Elongation* %
100	300	21
150	250	24
200	205	21
250	120	23
315	47	38
350	37	48
400	30	57

*The values shown apply to the 1% Mg 1% Si 0.7% Mn alloys. Other alloys have lower values above 150 °C.

The above properties are typical of the following specifications, and may not apply exactly to any one. It is possible that some of the values are inapplicable.

General metallurgical characteristics

By the addition of 0.5–1% magnesium, and 0.5–1% silicon, with and without 0.5% copper, 0.5% chromium and small amounts of manganese, a range of heat-treatable aluminium alloys is obtained. These are characterized by excellent cold working properties, making them suitable for extrusions and good corrosion resistant properties, with reasonably good mechanical strength.

The alloys show a less marked drop in corrosion resistance for a proportional increase in mechanical properties than other heat-treatable aluminium alloys.

The addition of small amounts of copper increases the strength, with a reduction in ductility and corrosion resistance. Addition of chromium acts as a grain refiner and improves the corrosion resistance.

Thermal treatment

If considerable cold work is carried out during manufacture, for example deep drawing or extruding, an anneal at approximately 400 °C for 2 hours, followed by a slow cool may be required.

The alloys are almost always used in the solution treated and aged condition, being water quenched from about 530 °C and then aged at 175–190 °C, the exact time and temperature varying for the different alloys.

All these treatments can be carried out in normal air furnaces resulting only in a reduction of surface brightness.

Welding and brazing

Pre-treatment. No thermal treatment required. Thorough cleaning and scratch brushing of the weld area are essential.

Welding and brazing. Normal gas or electric arc welding techniques with fluxes are satisfactory, but the fluxes are generally corrosive and must be completely removed.

Inert gas shielded arc welding is recommended as no flux is necessary. Resistance welding using special equipment is satisfactory.

Brazing, using special fluxes and aluminium/silicon filler rod is readily accomplished.

Post treatment. Wherever possible the parts should be solution treated and aged after welding. This will not be possible however, in the majority of cases owing to the danger of distortion, or in some cases incipient re-melting of the weld area where the melting point may have been reduced.

In all cases the parts must be re-aged after welding and brazing, otherwise the physical properties in the weld area will be seriously affected.

Flaw detection methods

Crack test. Penetrant oil chalk test method.

X-ray. May be required on high duty forgings, or to prove important welds.

Ultra-sonic test. Not normally required.

Chemical etching. Can be used to increase the sensitivity of chalk testing, by preparing the surface. Will also show grain size, forging laps and gross inclusions and is a useful check on welding and brazing. A 10% caustic solution is satisfactory.

Corrosion protection

Temporary. None required.

Permanent. In many instances none will be required, although it is doubtful if these alloys should be subjected to outdoor conditions in this country without at least some form of artificial oxidation. This will certainly apply if they are to be used near the coast. Sulphuric, oxalic or chromic-acid anodizing is recommended for these conditions. Where less severe conditions are to be encountered, e.g. garden furniture (not for coastal districts), chemical oxidation will generally suffice.

Painting. Normal air drying lacquers or stove enamels can be used. These should be preceded by an etch primer or chemical oxidation, if the part is not anodized. Lead based primers must not be used.

Plating. These alloys can be electroplated, using special techniques.

Machinability

These alloys have excellent machinability characteristics

in the heat treated condition. Owing to the presence of hard intermetallics the use of high speed or tipped tools should be considered, especially for long run production.

Uses

Architectural purposes, such as window glazing bars, cur-tain wall supports and similar items. Railway and road transport vehicle body structural members and containers used where lightness, corrosion resistance and stength are all essential. The alloys should be anodized when the above uses are considered.

Tubular furniture where only medium high strength is required.

Symbol	Nominal analysis, supplier, condition and remarks.
2EC	0.5% Mg 0.5% Si Al alloy: For electrical purposes; AA designation
3.0615	Mg Si Pb alloy: German Standard DIN 1725
3.2305	0.5% Si 0.6% Mg Al alloy: German Standard DIN 1725
3.2315	1% Si 0.7% Mg 0.6% Mn Al alloy: German Standard DIN 1725
3.3205	0.7% Mg 0.4% Si Al alloy: German Standard DIN 1725
3.3206	0.7% Mg 0.6% Si Al alloy: Wrought; German Standard DIN 1725
3.3241	1% Si 3% Mg 0.3% Mn 0.1% Ti Al alloy: Casting; German Standard DIN 1725
3.3243	0.7% Si 3% Mg 0.3% Mn 0.1% Ti Al alloy: Casting; German Standard DIN 1725 **DPN: 60 UTS: 170 Elon: 5% Proof: 75**
3.3245	3% Mg 0.6% Si 0.6% Mn 0.2% Cr Al alloy: Wrought; German Standard DIN 1725 **DPN: 45 UTS: 170 Elon: 17% Proof: 75**
62S	0.2% Cu 1% Mg 0.6% Si 0.1% Cr Al alloy: Previous AA designation now 6062
63S	0.75% Mg 0.5% Si Al alloy: Wire; Alcoa for BS alloy H9
63S	0.8% Mg 0.4% Si Al alloy: Previous AA designation now 6063
6011	1% Mg 1% Si Al alloy: Sheet; designation used AA no ASTM alloy
6101	0.5% Mg 0.5% Si Al alloy: Wrought; designation used by AA; no ASTM alloy
6201	0.8% Mg 0.7% Si Al alloy: Bar; designation used by AA; ASTM alloy
6262	1% Mg 0.6% Si 0.1% Cr Al alloy: Tube & bar; designation used by AA; no ASTM alloy
6563	0.2% Cu 0.7% Mg 0.4% Si Al alloy: Designation used by AA; no ASTM alloy listed
A 51S	0.75% Mg 1% Si Al alloy: Forgings; Alcoa
AA 6003	0.8% Si 1.2% Mg Al alloy: Wrought; used only for cladding; SAE designation; formerly SAE 210
AA 6061	0.25% Cu 0.6% Si 1% Mg 0.2% Cr Al alloy: Wrought; SAE designation; formerly SAE 281
AA 6062	0.2% Cu 0.6% Si 1% Mg 0.1% Cr Al alloy: Wrought; SAE designation; formerly SAE 212
AA 6063	0.40% Si 0.65% Mg Al alloy: Wrought; solution treated and aged; SAE designation; formerly SAE 212
AA 6151	1% Si 0.6% Mg 0.2% Cr Al alloy: Wrought; SAE designation; formerly SAE 280

Note. The following abbreviations and units are used in the tables:

DPN	Hardness, diamond pyramid number
UTS	Ultimate tensile strength, N/mm^2
Elon	Elongation, %
Proof	0.1 % proof strength, N/mm^2

1 N/mm^2=0.1 hbar=0.102 kgf/mm^2=0.06475 tonf/in^2=145.04 lbf/in^2

Symbol	Nominal analysis, supplier, condition and remarks.
AA 6262	0.3% Cu 0.6% Si 1.0% Mg 0.1% Cr 0.5% Pb 0.5% Bi Al alloy: Wrought; SAE designation; free machining
AA 6463	0.4% Si 0.7% Mg Al alloy: Wrought; SAE designation; formerly SAE 253
AA 6951	0.4% Si 0.6% Mg Al alloy: Wrought; SAE designation; formerly SAE 213
AC	1% Si 0.8% Mg Al alloy: Anglo-Swiss code for Anticordal
AGG 51	0.4% Mg 0.55% Si Al alloy: Wrought; further information from Aluminium Zentrale
AGS	0.8% Mg 0.8% Si Al alloy: Wrought; French Standard; part of NFA. 57. 350
AGS 50	0.4% Si 0.5% Mg Al alloy: Wrought; L'Aluminium Francais
AGS 65	0.5% Si 0.65% Mg Al alloy: Wrought; L'Aluminium Francais
AGS/L	0.6% Si 0.7% Mg Al alloy: Wrought; L'Aluminium Francais
ALCAN GB 50S	0.7% Mg 0.4% Si Al alloy: Alcan for BS alloy HE9
ALCAN GB 51S	0.7% Mg 1% Si Al alloy: Alcan for BS alloy HE19
ALCAN GB 65S	1% Mg 0.7% Si 0.2% Cr Al alloy: Alcan for BS alloy H20
ALCAN GB B51S	0.7% Mg 1% Si 0.5% Mn Al alloy: Alcan for BS alloy H30
ALCAN GB C50S	0.7% Mg 0.4% Si Al alloy: Bar & forging; Alcan for alloy BT3
ALCAN GB D50S	0.7% Mg 0.5% Si Al alloy: Alcan for BS 2898. E91E
ALCAN GB D57S	0.3% Cu 0.7% Mg 0.4% Si Al alloy: Bar & forging; Alcan for BT6
ALCOA 910	0.6% Mg 0.5% Si Al alloy: Wrought; Alcoa; solution treated & aged **UTS: 250 Elon: 13% Proof: 230**
ALCOA 912	0.6% Mg 0.4% Si Al alloy: Wrought; Alcoa; solution treated & aged **UTS: 240 Elon: 15% Proof: 210**
ALCOA 918	0.05% Cu max 0.6% Mg 0.5% Si Al alloy: Wrought; Alcoa; solution treated & aged 55/IACS; for electrical purposes **UTS: 240 Elon: 16% Proof: 210**
ALCOA 920	0.8% Mg 1.0% Si Al alloy: Wrought; Alcoa; solution treated & aged **UTS: 340 Elon: 11% Proof: 300**
ALCOA 940	0.25% Cu 1.0% Mg 0.6% Si Al alloy: Wrought; Alcoa; solution treated & aged **UTS: 310 Elon: 12% Proof: 270**
ALCOA 945	0.25% Cu 1.0% Mg 0.6% Si Al alloy: Wrought; Alcoa; solution treated & aged **UTS: 310 Elon: 12% Proof: 280**
ALCOA 946	0.25% Cu 1.0% Mg 0.6% Si 0.5% Ph 0.5% Bi Al alloy: Wrought; Alcoa; solution treated & aged; free machining **UTS: 310 Elon: 14% Proof: 280**
A..DREY	0.5% Si 0.4% Mg Al alloy: Anglo-Swiss, for high electrical conductivity

Symbol	Nominal analysis, supplier, condition and remarks.
ALDREY 051	0.4% Mg 0.6% Si Al alloy: Wrought; Anglo-Swiss Al Co
ALMASILIUM	1% Mg 2% Si Al alloy: Heat treatable; origin unknown
Al Mg Si 0.5	0.7% Mg 0.5% Si Al alloy: Wrought; previous German Standard designation
Al Mg Si 1	1% Mg 1.4% Si 0.4% Mn 0.2% Cr Al alloy: Wrought; previous German Standard designation
Al Mg Si Pb	1% Mg 1.2% Si 0.8% Mn 0.2% Cr + Pb Al alloy: Wrought; previous German Standard designation; free machining
ALMINAL 9	0.15% Cu 0.3% Si 0.4% Mg Al alloy: Wrought; origin unknown
ALMINAL 10	0.15% Cu 0.3% Si 0.4% Mg 1.0% Mn 0.5% Cr Al alloy: Wrought; origin unknown
ALMINAL 11	1% Mg 0.8% Si 0.5% Mg 1% Mn Al alloy: Wrought; origin unknown
AMS 4025 D	0.25% Cu 1% Mg 0.6% Si 0.25% Cr Al alloy: Sheet; annealed. AMS for AA alloy 6061
AMS 4026 D	0.25% Cu 1% Mg 0.6% Si 0.25% Cr Al alloy: Sheet; AMS for AA alloy 6061; solution treated
AMS 4027 E	0.25% Cu 1% Mg 0.6% Si 0.25% Cr Al alloy: Sheet; AMS for AA alloy 6061; solution treated & aged
AMS 4043	0.25% Cu 1% Mg 0.6% Si 0.25% Cr Al alloy: Plate; AMS for AA alloy 6061
AMS 4053	0.25% Cu 1% Mg 0.6% Si 0.25% Cu 0.25% Cr Al alloy: Solution treated stretched & aged; AMS for AA alloy 6061
AMS 4079	0.25% Cu 1% Mg 0.6% Si 0.25% Cr Al alloy: Seamless tube; annealed; AMS for AA alloy 6061
AMS 4080 E	0.25% Cu 1% Mg 0.6% Si 0.25% Cr Al alloy: Seamless tube; annealed; AMS for AA alloy 6061
AMS 4081 A	0.25% Cu 1% Mg 0.5% Si 0.25% Cr Al alloy: Seamless tube; solution treated; AMS for AA alloy 6061
AMS 4082 E	0.25% Cu 1% Mg 0.6% Si 0.25% Cr Al alloy: Seamless tube; solution treated & aged; AMS for AA alloy 6061
AMS 4083 D	0.25% Cu 1% Mg 0.6% Si 0.25% Cr Al alloy: Seamless tube; solution treated & aged; AMS for AA alloy 6061
AMS 4091	0.25% Cu 1% Mg 0.6% Si 0.1% Cr Al alloy: Tube; solution treated; AMS for AA alloy 6062
AMS 4092	0.25% Cu 1% Mg 0.6% Si 0.1% Cr Al alloy: Tube; solution treated & aged; AMS for AA alloy 6062
AMS 4093	0.25% Cu 1% Mg 0.6% Si 0.1% Cr Al alloy: Tube; solution treated & aged; AMS for AA alloy 6062
AMS 4115	0.25% Cu 1% Mg 0.6% Si 0.25% Cr Al alloy: Bar; as rolled; AMS for AA alloy 6061
AMS 4116 A	0.3% Cu 1% Mg 0.6% Si 0.25% Cr Al alloy: Bar; solution treated; AMS for AA alloy 6061
AMS 4117 A	0.3% Cu 1% Mg 0.6% Si 0.25% Cr Al alloy: Bar; solution treated & aged; AMS for AA alloy 6061
AMS 4125 E	0.6% Mg 1% Si 0.25% Cr Al alloy: Forgings; solution treated & aged; AMS for AA alloy 6151

Symbol	Nominal analysis, supplier, condition and remarks.
AMS 4127 B	0.3% Cu 1% Mg 0.6% Si 0.25% Cr Al alloy: Forging; solution treated & aged; AMS for AA alloy 6061
AMS 4146	0.3% Cu 1% Mg 0.6% Si 0.25% Cr Al alloy: Forging; solution treated; AMS for AA alloy 6061
AMS 4150 C	0.25% Cu 1% Mg 0.6% Si 0.25% Cr Al alloy: Extrusion; solution treated & aged; AMS for AA alloy 6061
AMS 4155 A	0.25% Cu 1% Mg 0.6% Si 0.1% Cr Al alloy: Extrusion; solution treated & aged; AMS for AA alloy 6062
AMS 4156 C	0.6% Mg 0.4% Si Al alloy: Extrusion; solution treated & aged; AMS for AA alloy 6063
AMS 4160	0.25% Cu 1% Mg 0.6% Si 0.25% Cr Al alloy: Extrusion; annealed; AMS for AA alloy 6061
AMS 4161	0.25% Cu 1% Mg 0.6% Si 0.25% Cr Al alloy: Extrusion; solution treated; AMS for AA alloy 6061
ANTICORODAL	1% Si 0.8% Mg Al alloy: Anglo-Swiss; cold rolled; solution treated and aged; obsolete
	DPN: 110 UTS: 340 Elon: 16% Proof: 300
ANTICORODAL	0.8% Mg 1.0% Si 0.2% Mn Al alloy: Wrought; Anglo Swiss Al Co
ANTICORODAL 041	0.35% Mg 0.4% Si Al alloy: Wrought; Anglo-Swiss Al Co
ANTICORODAL Pb 108	0.8% Mg 1.0% Si 1.2% Pb 0.3% Co Al alloy: Wrought; Anglo-Swiss Al Co; free machining
ANTICORODAL 110	0.8% Mg 1.0% Si 0.7% Mn Al alloy: Wrought; Anglo-Swiss Al Co
ANTICORODAL 990	0.45% Mg 0.55% Si Al alloy: Wrought; Anglo-Swiss Al Co
A SG	1.1% Si 1.0% Mg Al alloy: Wrought; L'Aluminium Francais
A SG	1.0% Mg 1.2% Si Al alloy: Wrought; French Standard
A SGM	1.0% Mg 1.1% Si 0.5% Mn 0.2% Cr Al alloy: Wrought; French Standard
ASTM B209/6003	0.5% Si 1.2% Mg Al alloy: Sheet & plate
ASTM B209/6061	0.2% Cu 0.6% Si 1% Mg 0.2% Cr Al alloy: Sheet; annealed
	UTS: 140 Elon: 18% Proof: 60
ASTM B209/6061	0.2% Cu 0.6% Si 1% Mg 0.2% Cr Al alloy: Sheet; solution treated & aged
	UTS: 280 Elon: 9% Proof: 220
ASTM B210/6061	0.2% Cu 0.6% Si 1% Mg 0.2% Cr Al alloy: Tube; solution treated & aged
	UTS: 280 Proof: 240
ASTM B210/5062	0.2% Cu 0.6% Si 1% Mg 0.1% Cr Al alloy: Tube; solution treated & aged
	UTS: 280 Proof: 240
ASTM B210/6063	0.5% Si 0.7% Mg Al alloy: Tube; solution treated & aged
	UTS: 220 Proof: 180
ASTM B211/6061	0.2% Cu 0.6% Si 1% Mg 0.2% Cr Al alloy: Bar; solution treated & aged
	UTS: 280
ASTM B211/6262	0.6% Si 1.0% Mg 0.1% Cr 0.5% Bi 0.5% Pb Al alloy: Bar; free machining
ASTM B221/6061	0.2% Cu 0.6% Si 1% Mg 0.2% Cr Al alloy: Extruded bar; solution treated & aged
	UTS: 250 Elon: 10% Proof: 240
ASTM B221/6062	0.2% Cu 0.6% Si 1% Mg 0.1% Cr Al alloy: Extruded bar; solution treated & aged
	UTS: 250 Elon: 10% Proof: 240
ASTM B221/6063	0.4% Si 0.5% Mg Al alloy: Extruded bar; solution treated & aged
	UTS: 200 Elon: 10% Proof: 170
ASTM B221/6351	0.6% Mg 1% Si 0.6% Mn Al alloy: Extruded bar; solution treated & aged
	UTS: 280 Elon: 10% Proof: 250

Note. The following abbreviations and units are used in the tables:

DPN	Hardness, diamond pyramid number
UTS	Ultimate tensile strength, N/mm²
Elon	Elongation, %
Proof	0.1 % proof strength, N/mm²

$1 \text{ N/mm}^2 = 0.1 \text{ hbar} = 0.102 \text{ kgf/mm}^2 = 0.06475 \text{ tonf/in}^2 = 145.04 \text{ lbf/in}^2$

Symbol	Nominal analysis, supplier, condition and remarks.
ASTM B234/6061	1% Mg 0.6% Si 0.2% Cr Al alloy: Tube; solution treated & aged **UTS: 280 Elon: 11% Proof: 240**
ASTM B234/6062	0.2% Cu 0.6% Si 1% Mg 0.1% Cr Al alloy: Tube; solution treated & aged **UTS: 280 Elon: 11% Proof: 240**
ASTM B235/6061	0.2% Cu 0.6% Si 1% Mg 0.2% Cr Al alloy: Extruded tube; solution treated & aged **UTS: 250 Elon: 10% Proof: 240**
ASTM B235/6062	0.2% Cu 0.6% Si 1% Mg 0.1% Cr Al alloy: Extruded tube; solution treated & aged **UTS: 250 Elon: 10% Proof: 240**
ASTM B235/6063	0.3% Si 0.7% Mg Al alloy: Extruded tube; solution treated & aged **UTS: 200 Elon: 10% Proof: 170**
ASTM B235/6351	1% Cu 0.6% Mn 0.6% Mg Al alloy: Extruded tube; solution treated & aged **UTS: 280 Elon: 10% Proof: 250**
ASTM B241/6061	0.6% Si 0.2% Cu 1% Mg 0.2% Cr Al alloy: Pipe; solution treated & aged **UTS: 260 Elon: 10% Proof: 240**
ASTM B241/6062	0.6% Si 0.2% Cu 1% Mg 0.1% Cr Al alloy: Pipe; solution treated & aged **UTS: 260 Elon: 10% Proof: 240**
ASTM B241/6063	0.4% Si 0.7% Mg Al alloy: Pipe; solution treated & aged **UTS: 200 Elon: 8% Proof: 170**
ASTM B241/6351	1% Si 0.6% Mn 0.6% Mg Al alloy: Pipe; solution treated & aged **UTS: 280 Elon: 10% Proof: 250**
ASTM B247/6053	1.2% Mg 0.5% Si 0.2% Cr Al alloy: Forging; solution treated & aged **DPN: 75 UTS: 240 Elon: 16% Proof: 200**
ASTM B247/6061	0.2% Cu 0.6% Si 1% Mg 0.2% Cr Al alloy: Forging; solution treated & aged **DPN: 80 UTS: 260 Elon: 10% Proof: 240**
ASTM B247/6151	0.9% Si 0.6% Mg 0.2% Cr Al alloy: Forging; solution treated & aged **DPN: 90 UTS: 300 Elon: 14% Proof: 250**
ASTM B308/6061	0.2% Cu 0.6% Si 1% Mg 0.2% Cr Al alloy: Section; solution treated & aged **UTS: 250 Elon: 10% Proof: 240**
ASTM B308/6062	0.2% Cu 0.6% Si 1% Mg 0.1% Cr Al alloy: Section; solution treated & aged **UTS: 250 Elon: 10% Proof: 240**
ASTM B308/6066	1% Cu 1.5% Si 0.8% Mn 1.2% Mg Al alloy: Section; solution treated & aged **UTS: 350 Elon: 8% Proof: 310**
ASTM B308/6351	1% Si 0.6% Mn 0.6% Mg Al alloy: Section; solution treated & aged **UTS: 280 Elon: 10% Proof: 250**
ASTM B313/6061	0.2% Cu 0.6% Si 1% Mg 0.2% Cr Al alloy: Welded tube; solution treated & aged **UTS: 280 Elon: 10% Proof: 240**
ASTM B316/6053	1.2% Mg 0.7% Si 0.2% Cr Al alloy: Rivets; annealed
ASTM B316/6061	0.2% Cu 0.6% Si 1% Mg 0.2% Cr Al alloy: Rivets; annealed
ASTM B317	0.5% Si 0.6% Mg Al alloy: For electrical purposes; solution treated & aged 55% IACS **UTS: 200 Proof: 170**
ASTM B317/6101	0.5% Si 0.55% Mg Al alloy: Bar
ASTM B345/6061	0.2% Cu 0.6% Si 1% Mg 0.2% Cr Al alloy: Pipe; solution treated & aged **UTS: 250 Elon: 8% Proof: 220**
ASTM B345/6062	0.2% Cu 0.6% Si 1% Mg 0.1% Cr Al alloy: Pipe; solution treated & aged **UTS: 250 Elon: 8% Proof: 220**
ASTM B345/6063	0.4% Si 0.7% Mg Al alloy: Pipe; solution treated & aged **UTS: 200 Elon: 8% Proof: 170**
ASTM B345/6351	1% Si 0.6% Mn 0.6% Mg Al alloy: Pipe; solution treated & aged **UTS: 280 Elon: 10% Proof: 250**
ASTM B398/6201	0.7% Si 0.75% Mg Al alloy: Wire
ASTM B404/6061	0.6% Si 0.30% Cu 1.0% Mg 0.2% Cr Al alloy: Condenser tube
ASTM B429/6061	0.6% Si 0.30% Cu 1.0% Mg 0.2% Cr Al alloy: Extruded pipe & tube
ASTM B429/6063	0.4% Si 0.7% Mg Al alloy: Extruded pipe and tube
AWCO 22	0.3% Cu 1% Mg 0.6% Si 0.6% Mn Al alloy: Aluminium Wire Co for BS alloy H 20
AWCO 24	0.75% Mg 0.5% Si Al alloy: Wire; Aluminium Wire Co for BS alloy H9
AWCO 25	0.75% Mg 1% Si Al alloy: Wire; Aluminium Wire Co for BS alloy H10
AWCO SILMALEC	0.04% Cu 0.6% Mg 0.5% Si Al alloy: For electrical purposes; Aluminium Wire Co for BS alloy E91E
BA 22	0.25% Cu 1% Mg 0.6% Si 0.6% Mn Al alloy: British Al Co for BS alloy H20
BA 24	0.75% Mg 0.5% Si Al alloy: Extrusion; British Al Co for BS H9
BA 25	0.75% Mg 1% Si Al alloy: Sheet, strip tube & extrusion; British Al Co
BA 241	0.7% Mg 0.4% Si Al alloy: Bar & forging for bright anodizing; British Al Co for BT 3
BA 251	1% Mg 1% Si Al Alloy: British Al Co for BS alloy H19
BA SILMALEC	0.04% Cu 0.6% Mg 0.5% Si Al alloy: For electrical purposes; British Al Co for BS alloy E91E
BA SP 16	0.3% Cu 0.7% Mg 0.4% Si Al alloy: Bar & forging; British Al Co for BT 6
BB 019	0.7% Mg 1% Si Al alloy: Sheet, strip & tube; Birmabright
BIRMETAL 016	0.3% Cu 1% Mg 0.7% Mn Al alloy: Birmabright for BS alloy H20
BIRMETAL 055	0.75% Mg 0.5% Si Al alloy: Birmabright for BS alloy H9
BIRMETAL 065	0.7% Mg 0.4% Si Al alloy: Bar & forging for bright anodizing; Birmabright for BT3
BIRMETAL 069	1% Mg 1% Si alloy: Birmabright for BS alloy H19
BIRMETAL 071	1% Mg 1% Si 0.7% Mn Al alloy: Birmabright for BS alloy H30
BS 414	Al alloy: Sheet & strip; heat treated; replaced by BS 1470
BS 1470 HS 20	0.3% Cu 1% Mg 0.7% Si 0.6% Mn Al alloy: Sheet & strip; solution treated & aged **UTS: 260 Elon: 8% Proof: 220**
BS 1470 HS 30	1% Mg 1% Si 0.75% Mn Al alloy: Sheet & strip; solution treated & aged **UTS: 280 Elon: 8% Proof: 240**
BS 1471 HT 9	0.7% Mg 0.5% Si Al alloy: Tube; solution treated & aged **UTS: 200 Elon: 8% Proof: 170**
BS 1471 HT19	1% Mg 1% Si Al alloy: For tube; solution treated & aged **UTS: 280 Elon: 8% Proof: 210**
BS 1471 HT20	0.3% Cu 1% Mg 0.5% Si 0.6% Mn Al alloy: Tube; solution treated & aged **UTS: 280 Elon: 8% Proof: 210**
BS 1471 HT30	1% Mg 1% Si 0.75% Mn Al alloy: Tube; solution treated & aged **UTS: 300 Elon: 8% Proof: 220**
BS 1472 HF9	0.75% Mg 0.5% Si Al alloy: For forgings & bar; solution treated & aged **UTS: 180 Elon: 10% Proof: 140**

Symbol	Nominal analysis, supplier, condition and remarks.
BS 1472 HF30	1% Mg 1% Si 0.75% Mn Al alloy: For forging & bar; solution treated & aged **UTS: 280 Elon: 10% Proof: 240**
BS 1473 HB30	1% Mg 1% Si 0.75% Mn Al alloy: For bolt stock; solution treated & aged **UTS: 280 Proof: 240**
BS 1473 HR30	1% Mg 1% Si 0.75% Mn Al alloy: For rivets; solution treated room aged **UTS: 200**
BS 1474 HV9	0.75% Mg 0.7% Si Al alloy: For tube; solution treated & aged **UTS: 200 Elon: 12% Proof: 150**
BS 1474 HV19	1% Mg 1% Si Al alloy: For tube; solution treated & aged **UTS: 260 Elon: 10% Proof: 220**
BS 1474 HV30	1% Mg 1% Si 0.75% Mn Al alloy: For tube; solution treated & aged **UTS: 280 Elon: 10% Proof: 240**
BS 1475 HG9	0.75% Mg 0.5% Si Al alloy: Wire; solution treated & drawn **UTS: 260**
BS 1475 HG20	0.3% Cu 1% Mg 0.7% Si 0.5% Mn Al alloy: Wire; solution treated & drawn **UTS: 350**
BS 1475 HG30	1% Mg 1% Si 0.75% Mn Al alloy: Wire; solution treated & aged **UTS: 280**
BS 1476 HE9	0.75% Mg 0.5% Si Al alloy: Bar; solution treated & aged **UTS: 180 Elon: 12% Proof: 150**
BS 1476 HE19	1% Mg 1% Si Al alloy: Bar; solution treated & aged **UTS: 260 Elon: 10% Proof: 220**
BS 1476 HE20	0.3% Cu 1% Mg 0.7% Si 0.6% Mn Al alloy: Bar; solution treated & aged **UTS: 260 Elon: 10% Proof: 220**
BS 1476 HE30	1% Mg 1% Si 0.7% Mn Al alloy: Bar; solution treated & aged **UTS: 280 Elon: 10% Proof: 240**
BS 1477 HP20	0.3% Cu 1% Mg 0.7% Si 0.7% Mn Al alloy: Plate; solution treated & aged **UTS: 260 Elon: 8% Proof: 210**
BS 1477 HP30	1% Mg 1% Si 0.7% Mn Al alloy: Plate; solution treated & aged **UTS: 280 Elon: 8% Proof: 220**
BS 2898 E91E	0.5% Mg 0.5% Si 0.04% Cu max Al alloy: For electrical purposes; solution treated & aged **UTS: 200 Elon: 10% Proof: 150**
BS 3242	Mg Si Al wrought alloy: For electrical conductors; electrical conductivity 52% IACS
BS 4300/4	0.6% Mg 0.3% Si Al alloy: For bright trim; solution treated & aged; bars & section **UTS: 185 Elon: 10% Proof: 160**
BS 4300/3	0.6% Mg 0.4% Si Al alloy: For bright trim; solution treated & aged; forging **UTS: 185 Elon: 10% Proof: 160**
BT 3	0.7% Mg 0.4% Si Al alloy: For bright anodizing; solution treated & aged; designation given by Al industry **UTS: 180 Elon: 12% Proof: 150**

Symbol	Nominal analysis, supplier, condition and remarks.
BT 6	0.3% Cu 0.7% Mg 0.4% Si Al alloy: For bright anodizing; solution treated & aged; designation given by Al industry **UTS: 220 Elon: 10%**
DTD 346 A	1% Mg 1% Si 0.7% Mn Al alloy: Sheet & strip; annealed **UTS: 170 Elon: 18%**
DTD 372 B	0.6% Mg 0.5% Si Al alloy: Bar; solution treated; suitable for welding **UTS: 140 Elon: 17%**
DTD 5080	1% Mg 1% Si 0.7% Mn Al alloy: Sheet; solution treated & aged; weldable **UTS: 280 Elon: 8% Proof: 240**
DURALBRITE 6	0.7% Mg 0.4% Si Al alloy: Extrusion; Alcan; solution treated & aged **UTS: 190 Elon: 8%**
DURAL F	0.25% Cu 1% Mg 0.6% Si 0.25% Cr Al alloy: Sheet & tube forging; J Booth for BS alloy H20
DURAL H	0.75% Mg 1% Si Al alloy: Sheet & tube forging; Alcan for BS alloy H30
DURAL X	0.8% Mn 1% Si 0.15% Mn Al alloy: Tube; Alcan for BS alloy H19
DURCILIUM E	0.6% Mg 0.5% Si Al alloy: For electrical purposes; E Kaye; solution treated & aged; electrical conductivity 54% IACS
DURCILIUM Q	0.3% Cu 1% Mg 0.7% Si 0.6% Mn Al alloy: E Kaye for BS alloy H20
DURCILIUM R	1% Cu 1% Mg 1% Si Al alloy: E Kaye for BS alloy H19
DURCILIUM S	1% Mg 1% Si 0.7% Mn Al alloy: E Kaye for BS alloy H30
DURCILIUM W	0.75% Mg 0.5% Si Al alloy: E Kaye for BS alloy H9
E Al Mg Si	0.4% Mg 0.5% Si Al alloy: Wrought; previous German Standard designation
ED	0.7% Mg 0.7% Si Al alloy: Anglo-Swiss code for Extrudal
EXTRUDAL	0.7% Si 0.7% Mg Al alloy: Anglo-Swiss; cold rolled; solution treated & aged **DPN: 85 UTS: 260 Elon: 14% Proof: 210**
EXTRUDAL 050	0.6% Mg 0.6% Si Al alloy: Wrought; Anglo Swiss Al Co
GS 10A	0.7% Mg 0.4% Si Al alloy: Previous ASTM designation now 6063
GS 11A	0.2% Cu 0.5% Si 1% Mg 0.2% Cr Al alloy: Previous ASTM designation; now 6061
GS 11B	1.2% Mg 0.7% Si 0.2% Cr Al alloy: Previous ASTM designation; now 6053
GS 11C	0.2% Cu 1% Mg 0.6% Si 0.1% Cr Al alloy: Previous ASTM designation; now 6062
HA 4 GS11N	0.25% Cu 1.0% Mg 0.6% Si 0.2% Cr Al alloy: Plate; Canadian Standard
HA 5 GS10	0.7% Mg 0.5% Si Al alloy: Bar, etc.; Canadian Standard
HA 5 GS11N	0.3% Cu 1.0% Mg 0.6% Si 0.2% Cr Al alloy: Bar, etc.; Canadian Standard
HA 5 GS11T	0.2% Cu 0.9% Mg 0.25% Mn 0.5% Si Al Alloy: Bar, etc.; Canadian Standard
HA 5 GS11R	0.6% Mg 0.6% Mn 1.0% Si Al alloy: Bar, etc.; Canadian Standard
HA 6 GS11N	0.3% Cu 1.0% Mg 0.6% Si 0.2% Cr Al alloy: For rivets and brazing wire; Canadian Standard
HA 7 GS10	0.6% Mg 0.4% Si Al alloy: Tube; Canadian Standard
HA 7 GS11N	0.2% Cu 1.0% Mg 0.6% Si 0.2% Cr Al alloy: Tube; Canadian Standard
HA 7 GS11T	0.2% Cu 0.8% Mg 0.3% Mn 0.5% Si Al alloy: Tube; Canadian Standard
HA 7 GS11R	0.6% Mg 0.6% Mn 1.0% Si Al alloy: Tube; Canadian Standard

Note. The following abbreviations and units are used in the tables:

DPN	Hardness, diamond pyramid number
UTS	Ultimate tensile strength, N/mm^2
Elon	Elongation, %
Proof	0.1 % proof strength, N/mm^2

1 N/mm^2=0.1 hbar=0.102 kgf/mm^2=0.06475 tonf/in^2=145.04 lbf/in^2

Symbol	Nominal analysis, supplier, condition and remarks.
HA 8 GS11N	0.2% Cu 1.0% Mg 0.6% Si 0.2% Cr Al alloy: Forging; Canadian Standard
HA 8 GS11P	0.7% Mg 0.8% Si 0.2% Cr Al alloy: Forging; Canadian Standard
HIDUMINIUM 18	0.7% Mg 0.4% Si Al alloy: Bar & forging for bright anodizing; High Duty Alloys for BT 3
HIDUMINIUM 42	1% Mg 1% Si Al alloy: High Duty Alloy for BS alloy H19
HIDUMINIUM 43	0.2% Cu 1% Mg 0.6% Si 0.6% Mn 0.2% Cr Al alloy: High Duty Alloy for BS alloy H20
HIDUMINIUM 44	0.75% Mg 1% Si Al alloy: For forgings extrusions and tubes; High Duty Alloy
HIDUMINIUM 46	0.6% Mg 0.5% Si Al alloy: High Duty Alloy for BS alloy H9
IMPALCO 910	0.5% Mg 0.5% Si Al alloy: Tube bar & forging; Imperial Al Co for BS alloy H9
IMPALCO 912	0.6% Mg 0.5% Si Al alloy: Bar & forgings; high purity 910; Imperial Al Co; solution treated & aged UTS: 180 Elon: 12% Proof: 150
IMPALCO 918	0.5% Mg 0.5% Si Al alloy: Bar; Imperial Al Co for BS 2898 E91E
IMPALCO 920	1% Mg 1% Si 0.7% Mn 0.2% Cr Al alloy: Sheet & bar, etc.; Imperial Al Co for BS alloy H30
IMPALCO 940	1% Mg 0.6% Si 0.2% Cr Al alloy: Sheet & bar, etc.; Imperial Al Co for BS alloy H20
IMPALCO 945	1% Mg 0.6% Si 0.1% Cr max. Al alloy: Bar; Imperial Al Co; solution treated & aged UTS: 280 Elon: 10% Proof: 240
IMPALCO 946	1% Mg 0.6% Si 0.5% Pb 0.5% Bi Al alloy: Bars & forgings; Imperial Al Co; solution treated & aged DPN: 110 UTS: 330 Proof: 300
IMPALCO 950	0.9% Mg 1% Si 0.1% Cr (max) Al: Bar & tube; Imperial Al Co for BS alloy H19
IMPALCO M38	0.7% Mg 0.5% Si Al alloy: Bar for electrical purposes; Imperial Al Co for BS 2898 E91E
IMPALCO M39/1	0.7% Mg 0.5% Si Al alloy: Imperial Al Co for BS alloy H9
IMPALCO M39/2	1% Mg 1% Si 0.7% Mn 0.3% Cr Al alloy: Imperial Al Co for BS alloy H30
IMPALCO M40	0.2% Cu 1% Mg 0.7% Si 0.6% Mn Al alloy: Imperial Al Co for BS alloy H20
IMPALCO M41	1% Mg 1% Si 0.4% Cr Al alloy: Strip, tube & bar; Imperial Al Co; solution treated & aged UTS: 280 Elon: 9% Proof: 220
IMPALCO M42	1% Mg 1% Si Al alloy: Imperial Al Co for BS alloy H19
IMPALCO SUPERSPEED 946	0.25% Cu 1.0% Mg 0.6% Si 0.5% Pb 0.5% Bi Al alloy: Imperial Al Co; free machining bar UTS: 270 Elon: 10% Proof: 225
KYNAL C69	1% Cu 1% Si 1% Mg Al alloy: ICI; obsolete; now Imperial Al Co
KYNAL M39/1	0.7% Mg 0.5% Si Al alloy: ICI; obsolete; now Imperial Al Co
KYNAL M39/2	0.7% Mg 1% Si Al alloy: ICI; obsolete; now Imperial Al Co
NORAL 50S	0.75% Mg 0.5% Si Al alloy: Wire; Northern Al Co for BS alloy H9
NORAL 51S	0.75% Mg 1% Si Al alloy: Sheet & tube forging; Northern Al Co
NORAL 62S	1% Cu 1% Si 1% Mg Al alloy: For tube forging; Northern Al Co; solution treated & aged UTS: 370 Elon: 6% Proof: 310
QQ A 327	0.3% Cu 1.0% Mg 0.6% Si 0.2% Cr Al alloy: Sheet; US Federal; solution treated & aged UTS: 260 Elon: 9% Proof: 200
QQ A331B	1.3% Mg 0.7% Si 0.2% Cr Al alloy: Bar; US Federal; solution treated & aged UTS: 210 Elon: 14% Proof: 150

Symbol	Nominal analysis, supplier, condition and remarks.
QQ A 334	1.3% Mg 0.7% Si 0.25% Cr Al alloy: Sheet; US Federal; solution treated & aged UTS: 220 Elon: 9% Proof: 170
SAE 210	0.8% Si 1.2% Mg Al alloy: Wrought; used only for cladding; now AA 6003
SAE 211	0.2% Cu 1% Mg 0.6% Si 0.1% Cr Al alloy: Solution treated & aged; AA alloy 6062 DPN: 95 UTS: 300 Elon: 17% Proof: 270
SAE 212	0.7% Mg 0.4% Si Al alloy: Solution treated & aged; now AA 6062 DPN: 73 UTS: 220 Elon: 12% Proof: 200
SAE 213	0.2% Cu 0.6% Mg 0.3% Si Al alloy: Now AA 6951
SAE 253	0.4% Si 0.7% Mg Al alloy: Wrought; solution treated & aged; now AA 6463 UTS: 150 Elon: 10% Proof: 110
SAE 280	1% Si 0.6% Mg 0.25% Cr Al alloy: Forgings; solution treated & aged; now AA 6151 UTS: 300 Elon: 14% Proof: 250
SAE 281	0.25% Cu 0.6% Si 1% Mg 0.25% Cr Al alloy: Tube & sheet forging; solution treated & aged; now AA 6061 UTS: 8%0 Proof: 240
SAE 282	0.7% Si 1.5% Mg 0.25% Cr Al alloy: Wrought; solution treated & aged UTS: 210 Elon: 10% Proof: 170
SG 11 B	1% Cu 1.1% Mg 1.5% Si 0.8% Mn Al alloy: Previous ASTM designation; now 6066
SIMALEC	0.04% Cu 0.5% Mg 0.5% Si Al alloy: For electrical purposes; Aluminium Wire Co & British Al Co for BS alloy E91E
SIMGAL	0.75% Mg 0.5% Si Al alloy: For wire; Alcan for BS alloy H9
SIS 4104	0.7% Mg 0.5% Si Al alloy: Bar & tube; Swedish Standard; as rolled DPN: 40 UTS: 140 Elon: 17% Proof: 110
SIS 4212	0.9% Mg 1% Si 0.8% Mn Al alloy: Bar & forging; Swedish Standard; as forged DPN: 70 UTS: 200 Elon: 15% Proof: 90
STA 7 AW9A	0.7% Mg 0.5% Si Al alloy: Bar; replaced by BS 1476 HE9
STA 7 AW9B	0.7% Mg 0.5% Si Al alloy: Bar; replaced by BS 1476 HE9
STA 7 AW10A	0.7% Mg 1% Si Al alloy: Forging & bar; replaced by BS 1472
STA 7 AW10B	0.7% Mg 1% Si Al alloy: Forging & bar; replaced by BS 1472
STA 7 AW10C	0.7% Mg 1% Si Al alloy: Tube; replaced by BS 1471
STA 7 AW10D	0.7% Mg 1% Si Al alloy: Sheet; replaced by BS 1470
STA 7 AW10E	0.7% Mg 1% Si Al alloy: Sheet; replaced by BS 1470
STA 7 AW10F	0.7% Mg 1% Si Al alloy: Wire; replaced by BS 1473-5
TI 03	1% Cu 1% Mg 1% Si Al alloy: Forging; TI Aluminium; solution treated & aged UTS: 370 Elon: 10% Proof: 300
TI 40	0.75% Mg 0.5% Si Al alloy: TI Aluminium for BS alloy H9
TI 44	0.75% Mg 1% Si Al alloy: Sheet, strip & tube; TI Aluminium
UNI 3569	0.7% Mg 0.4% Si Al alloy: For forging; Italian Standard
UNI 3570	0.5% Mg 0.5% Si Al alloy: For forging; Italian Standard
UNI 3571	0.6% Mg 1.0% Si 0.3% Mn Al alloy: For forging; Italian Standard
UNI 6170	0.25% Cu 1.0% Mg 0.6% Si 0.25% Cr Al alloy: For forging; Italian Standard

Symbol	Nominal analysis, supplier, condition and remarks.
UNI 6359	0.1% Cu 0.5% Mg 0.45% Si Al alloy: For forging; Italian Standard
VIVAL	1.0% Mg 0.5% Si 0.5% Mn Al rolling alloy: Origin unknown
W 9	0.6% Mg 0.5% Si Al alloy: Southern Forge; replacedby IMPALCO 910
W 10	1% Mg 1% Si Al alloy: Southern Forge; replaced by IMPALCO 950
W 20	0.2% Cu 1% Mg 0.6% Si 0.2% Cr Al alloy: Southern Forge; replaced by IMPALCO 940
W 30	1% Mg 1% Si 0.8% Mn Al alloy: Southern Forge; replaced by IMPALCO 920
WO 90	0.5% Mg 0.4% Si Al alloy: Southern Forge; replaced by IMPALCO 912

Symbol	Nominal analysis, supplier, condition and remarks.
WWT 789	1.0% Mg 0.6% Si 0.2% Cr 0.1% Ti Al alloy: Tube; US Federal; solution treated & aged **UTS: 270** **Elon: 9%** **Proof: 220**
WWT 790	1.2% Mg 0.7% Si 0.2% Cr Al alloy: Tube; US Federal; solution treated & aged

Note. The following abbreviations and units are used in the tables:

DPN	Hardness, diamond pyramid number
UTS	Ultimate tensile strength, N/mm^2
Elon	Elongation, %
Proof	0.1 % proof strength, N/mm^2

$1\ N/mm^2 = 0.1\ hbar = 0.102\ kgf/mm^2 = 0.06475\ tonf/in^2 = 145.04\ lbf/in^2$

1F Aluminium–copper wrought alloys
With magnesium silicon, titanium, nickel and iron additions

Specific gravity	2.7–2.75	
Density	2700–2750 kg/m³	(0.098–0.099 lb/in.³)
Solidus/liquidus	530–640 °C	
Thermal conductivity	147–180 W/m °C	(35–43 cal/m s °C)
Coefficient of linear expansion	$22–24 \times 10^{-6}/$ °C	

Electrical conductivity	solution treated and aged	34–41% IACS (copper 100%)	
Specific resistance			
	solution treated and aged	41–51 microhm mm	
Young's modulus of elasticity		$68.95 \times 10^9\ N/m^2$	$(10 \times 10^6\ lbf/in.^2)$
Impact	solution treated and aged	8–11 J	(6–8 ft/lb) Izod
Fatigue strength (50×10^6 cycles)			
	solution treated and aged	$\pm$ 120–150 N/mm^2	$(\pm$ 8–10 $tonf/in.^2)$

Hot strength

Temperature °C	Tensile strength* N/mm²	Elongation* %
100	410	18
150	370	20
200	330	20
250	270	17
300	150	18
350	75	45
400	45	60

*These are approximate values which might be expected from the medium alloyed specifications listed. Some of the 4% copper highly alloyed specifications give better results.

The properties on the previous page are chosen to be typical of the group of specifications as a whole. There is considerable variation within the group and in general the properties vary in proportion to the copper content.

General metallurgical characteristics

This range of specifications has up to 4.5% copper which forms intermetallic compounds with aluminium, giving age hardening properties. The addition of further elements, notably silicon and manganese strengthen this property.

These intermetallic compounds are dissolved in the aluminium at solution treatment, and after quenching remain in solution giving a very soft ductile material. Then, depending on the other alloying elements, the compounds will start to precipitate, causing an increase in hardness and reduction in ductility. With some specifications this takes place within hours at room temperature (room ageing). With others it takes much longer and still others require to be heated (artificially aged). All the alloys can be artificially aged and ageing can be prevented in all the alloys by holding at 0 °C or lower.

The addition of iron, nickel and titanium act as grain refiners and stabilizers and it is these elements which give certain of the alloys their high temperature properties.

None of the alloys exhibit good corrosion resistance, but are the general first choice when a strong material of low weight is required, with the alloys in Section 1E being economical alternatives.

Thermal treatment

These alloys in the main cannot accept a high degree of cold work without intermediate anneals at 400 °C.

The specifications are almost invariably supplied and used in the fully heat treated condition, that is, solution treated and aged. This consists of solution treatment at 500–530° C depending on the alloy and water quenching. Where it is necessary to keep distortion to a minimum, the parts should be boiling water quenched, which results in a slight reduction in tensile strength.

The alloys are soft and ductile when solution treated and any forming should be carried out in this condition.

All of the specifications listed will age harden to some extent at room temperature, but the full hardness is only achieved by furnace ageing at a temperature of 100–175 °C, depending on the alloy. It is always necessary to consult the full specification for the exact temperature, as the solution treatment temperature is close to the melting range, with the consequent danger of incipient melting at grain boundaries if the wrong temperature is used. Also if the incorrect ageing temperature is chosen the part will be either over or under aged, and full hardness will not be achieved.

All the treatments can be carried out in air furnaces, or neutral salt baths.

Welding and brazing

Pre-treatment. Pre-heating may be required on certain parts to reduce distortion. This will soften the components if above the age temperature. The areas being welded must be thoroughly cleaned and scratch brushed, particularly if they have been anodized.

Welding and brazing. These alloys are not easy to weld with gas or electric arc, although satisfactory welds can be obtained with experienced welders using the correct filler rods and fluxes. The weld area is much weaker, being cast, and the heat affected zone will be over aged at some point. The fluxes used are corrosive and must be removed completely. Inert gas shielded arc welding, requiring no flux, and giving a localized heat source gives the best welds. Resistance welding is possible, but requires special equipment.

Brazing, using special fluxes and Al/Si filler rods is readily accomplished, but will not have the strength of the parent metal.

Post-treatment. Where possible the parts should be re-solution treated and re-aged. This will seldom be possible owing to the danger of distortion and incipient melting of the weld area where the melting point has been reduced. The parts must always be re-aged at the correct temperature after welding or brazing. This re-age, as well as restoring the mechanical properties as much as possible, reduces the stresses caused by welding.

Flaw detection methods

Crack test. Penetrant oil chalk test method.

X-ray. This may be required on high duty pieces and to prove important welds, but is an unusual requirement on wrought products.

Ultra-sonic test. This is sometimes used on high duty complicated forgings at the forging stage or after rough machining to prove that subcutaneous cracks and folds are absent.

Chemical etch. This can be used before the oil/chalk test to increase the sensitivity, as it is a useful way to prepare and clean the surface. It is also a useful means of assessing grain size, and directional grain flow and will indicate any seams or laps. A 10% caustic etch is satisfactory, and will require to be followed by a concentrated nitric acid dip. Large weld and braze defects can be readily identified after etching, as can the heat affected zone.

Corrosion protection

Temporary. None normally required.

Permanent. These alloys have relatively poor resistance to atmospheric corrosion compared to the majority of aluminium alloys and should always be anodized, or chemically oxidized. Sulphuric and oxalic acid anodizing is suitable for all the alloys, and chromic acid anodizing for all except the highly alloyed specifications. Chemical oxidation is much cheaper, but gives a lesser degree of corrosion protection.

Painting. Air drying or stove enamel paints are suitable, and should be applied on top of anodizing or chemical pre-treatment for all outdoor uses. Lead based primer paints must not be used.

Plating. Electroplating can be used to improve the corrosion or wear resistance. Special etching techniques are necessary.

Cladding with pure or very low alloyed aluminium is commonly used (see Section 1H).

Machinability

On the whole these alloys exhibit excellent machinability when fully heat treated. Under certain conditions of variation in grain size, tearing and poor finish may be encountered; one reason for ensuring a consistent grain size.

The hard intermetallics are abrasive, and high speed or tipped tools should be used, especially for long production runs.

Some of the alloys have elements such as lead added to increase their machinability for high speed automatic lathe work.

Uses

Where high strength, low weight, is required for structural members and load bearing parts. Not used much for architectural or transport purposes, owing to low corrosion resistance.

It should be noted that the hard intermetallics in certain of the alloys can act as a lap, and cause severe wear if in sliding contact. Cases are known where hardened steel parts have been destroyed.

Seizing will be encountered if the alloys are in moving contact with each other.

Symbol	Nominal analysis, supplier, condition and remarks.
3.1190	5% Cu Al alloy: German Standard DIN number not known
3.1255	4.5% Cu 0.6% Mg 0.8% Si 0.9% Mn Al alloy: German Standard DIN 1725
3.1305	2.5% Cu 0.4% Mg Al alloy: Wrought; German Standard DIN 1725
3.1325	4% Cu 0.7% Mg 0.6% Si 0.7% Mn Al alloy: Wrought; German Standard DIN 1725
3.1335	4% Cu 0.7% Mg 0.7% Si 0.5% Mn Al alloy: Wrought; German Standard DIN 1725
3.1355	4.3% Cu 1.3% Mg 0.7% Mn Al alloy: Wrought; German Standard DIN 1725
3.1365	4.3% Cu 1.5% Mg Al alloy: German Standard DIN 1725
3.1590	3.3% Cu Al alloy: German Standard; DIN number not known
3.1645	4.2% Cu 1% Mg 0.7% Mn 2% (Pb+Sn+Bi+Sb+Cd) Al alloy: Wrought; German Standard DIN 1725
3.2131	2% Cu 3% Si Al alloy: German Standard DIN 1725
11S	5.5% Cu Al alloy: Previous AA designation; now 2011
14S	4.4% Cu 0.6% Mg 0.7% Si 0.6% Mn Al alloy: Sheet & forging; Alcoa for BS alloy H15
14S	4.5% Cu 1% Si 0.6% Mg 1% Mn Al alloy: Previous AA designation; now 2014
17S	4% Cu 0.6% Mg 0.5% Mn Al alloy: Forgings; Alcoa for BS alloy H14
17S	4% Cu 0.6% Mg 0.6% Mn Al alloy: Previous AA designation; now 2017
18S	4% Cu 1.5% Mg 2% Ni Al alloy: Forgings; Alcoa for BS alloy H17
18S	4% Cu 0.8% Mg 2% Ni Al alloy: Previous AA designation; now 2018
24S	4.5% Cu 1.5% Mg 0.6% Mn Al alloy: Previous AA designation; now 2024
25S	4.5% Cu 0.9% Mn 0.8% Si Al alloy: Previous AA designation; now 2025
A 17S	2.8% Cu 0.3% Mn Al alloy: Previous AA designation; now 2117
A 17S	2.2% Cu 0.3% Mg Al alloy: Wire for rivets; Alcoa for BS alloy H13
AA 2011	5.5% Cu 0.4% Pb 0.4% Bi Al alloy: Free machining; SAE designation; formerly SAE 202
AA 2014	4.5% Cu 0.8% Si 0.4% Mg Al alloy: Wrought; SAE designation; formerly SAE 260
AA 2017	4% Cu 0.8% Mn Al alloy: Wrought; SAE designation; formerly SAE 26
AA 2018	4% Cu 0.5% Mg 2% Ni Al alloy: Wrought; SAE designation; formerly SAE 270
AA 2024	4.5% Cu 0.6% Mn 1.5% Mg Al alloy: Wrought; SAE designation; formerly SAE 24

Note. The following abbreviations and units are used in the tables:

DPN	Hardness, diamond pyramid number
UTS	Ultimate tensile strength, N/mm^2
Elon	Elongation, %
Proof	0.1 % proof strength, N/mm^2

1 N/mm^2=0.1 hbar=0.102 kgf/mm^2=0.06475 tonf/in^2=145.04 lbf/in^2

Symbol	Nominal analysis, supplier, condition and remarks.
AA 2117	2.6% Cu 0.4% Mg Al alloy: Wrought; SAE designation; formerly SAE 204
AA 2218	4% Cu 1.5% Mg 2% Ni Al alloy: Wrought; SAE designation; formerly SAE 203
AA 2219	6.2% Cu 0.3% Mn 0.8% Ti Al alloy: Wrought; SAE designation
AA 2618	2.5% Cu 1% Fe 1.5% Mg 1% Ni Al alloy: Wrought; SAE designation
ALCAN GB 16S	2.2% Cu 0.3% Mg Al alloy: Alcan; solution treated room aged UTS: 300 Elon: 20% Proof: 150
ALCAN GB 17S	4.2% Cu 0.7% Mg 0.7% Mn Al alloy: Alcan for BS alloy H14
ALCAN GB 19S	4.25% Cu 1.5% Mg 2% Ni Al alloy: Forging; Alcan; solution treated; room aged UTS: 360 Elon: 15% Proof: 210
ALCAN GB 24S	4.5% Cu 1.5% Mg 0.7% Mn Al alloy: Alcan; solution treated; room aged DPN: 120 UTS: 450 Elon: 19% Proof: 300
ALCAN GB 26S	4% Cu 0.7% Mg 0.7% Si 0.7% Mn Al alloy: Sheet and plate; Alcan
ALCAN GB 28S	5.2% Cu 0.5% Bi 0.5% Pb Al alloy: Free machining; Alcan; solution treated & aged UTS: 330 Elon: 17% Proof: 220
ALCAN GB 42S	2.2% Cu 1.5% Mg 1.2% Ni 1% Fe Al alloy: Forging; Alcan for BS alloy H18
ALCAN GB 62S	1.5% Cu 1% Si 1% Mg 0.7% Mn Al alloy: Alcan for BS alloy H11
ALCAN GB B19S	2.5% Cu 1.5% Mg 1.2% Ni 1.1% Fe 0.1% Ti Al alloy: Alcan; solution treated & aged UTS: 400 Elon: 11% Proof: 390
ALCAN GB B26S	4.2% Cu 0.7% Mg 0.7% Si 0.7% Mn Al alloy: Alcan; solution treated & aged DPN: 145 UTS: 460 Elon: 10% Proof: 400
ALCAN GB D19S	2.5% Cu 1.5% Mg 1.3% Ni Al alloy: Plate; Alcan
ALCOA 660	4.5% Cu 0.5% Mg 0.7% Si 0.8% Mn Al alloy: Wrought; Alcoa; solution treated and aged UTS: 480 Elon: 7% Proof: 400
ALCOA 800	5.5% Cu 0.5% Pb 0.5% Bi Al alloy: Wrought; Alcoa; solution treated & aged; free machining UTS: 380 Elon: 13% Proof: 300
Al Cu Mg 0.5	2.5% Cu 0.4% Mg Al alloy: Wrought; previous German Standard designation
Al Cu Mg 1	4% Cu 0.6% Mg 0.6% Si 0.8% Mn Al alloy: Wrought; previous German Standard designation
Al Cu Mg 2	4.2% Cu 1.6% Mg 0.8% Mn Al alloy: Wrought; previous German Standard designation
Al Cu Mg Pb	4.5% Cu 1% Mg 0.7% Mn + Pb Al alloy: Wrought; previous German Standard designation; free machining
Al Cu Si Mn	4.5% Cu 0.6% Mg 1% Si 0.8% Mn Al alloy: Wrought; previous German Standard designation
ALDURAL Q	4.4% Cu 0.6% Mg 0.7% Si 0.6% Mn Al alloy: Sheet & strip; J Booth for BS alloy H15
ALMINAL 11	4.5% Cu Al alloy: Origin unknown
ALMINAL 15	4.0% Cu 1.5% Si 0.6% Mg 1.2% Mn 1.0% Fe Al alloy: Wrought; origin unknown
AMS 4014	4.5% Cu 0.8% Si 0.8% Mn 0.5% Mn Al alloy: Plate; solution treated; cold rolled & aged UTS: 450 Elon: 5%
AMS 4028 A	4.5% Cu 0.5% Mg 0.8% Si 0.8% Mn Al alloy: Sheet; AMS for AA alloy 2014; annealed

Symbol	Nominal analysis, supplier, condition and remarks.
AMS 4029 A	4.5% Cu 0.5% Mg 0.8% Si 0.8% Mn Al alloy: Sheet; AMS for AA alloy 2014; solution treated & aged
AMS 4031	6.3% Cu 0.3% Mn 0.18% Zr 0.1% V 0.06% Ti Al alloy: Sheet; AMS for AA alloy 2219; annealed **UTS: 200 Elon: 12% Proof: 120**
AMS 4033 A	4.5% Cu 1.5% Mg 0.6% Mn Al alloy: Plate; AMS for AA alloy 2024; solution treated & stretched
AMS 4035 E	4.5% Cu 1.5% Mg 0.6% Mn Al alloy: Sheet; AMS for AA alloy 2024; annealed
AMS 4037 F	4.5% Cu 1.5% Mg 0.6% Mn Al alloy: Sheet; AMS for AA 2024; solution treated
AMS 4086 F	4.5% Cu 1.5% Mg 0.6% Mn Al alloy: Seamless tube; solution treated; AMS for AA alloy 2024
AMS 4087 C	4.5% Cu 1.5% Mg 0.6% Mn Al alloy: Seamless tube; annealed; AMS for AA alloy 2024
AMS 4088 E	4.5% Cu 1.5% Mg 0.6% Mn Al alloy: Seamless tube; solution treated; AMS for AA alloy 2024
AMS 4118 C	4% Cu 0.5% Mg 0.7% Mn Al alloy: Bar; solution treated; AMS for AA alloy 2017
AMS 4119 A	4.5% Cu 1.5% Mg 0.6% Mn Al alloy: Bar; solution treated & cold rolled; AMS for AA alloy 2024
AMS 4120 F	4.5% Cu 1.5% Mg 0.6% Mn Al alloy: Bar; cold rolled; solution treated; AMS for AA alloy 2024
AMS 4121 C	4.5% Cu 0.5% Mg 0.9% Si 0.8% Mn Al alloy: Bar; cold rolled; solution treated & aged; AMS for AA alloy 2014
AMS 4130 G	4.5% Cu 0.8% Si 0.8% Mn Al alloy: Forging; solution treated & aged; AMS for AA alloy 2025
AMS 4132 A	2.3% Cu 1.6% Mg 1% Fe 1% Ni 1% Ti Al alloy: Forging; solution treated & aged; AMS for AA alloy 2618
AMS 4134 A	4.4% Cu 0.4% Mg 0.8% Si 0.8% Mn Al alloy: Forging; solution treated; AMS for AA alloy 2014
AMS 4135 J	4.5% Cu 0.5% Mg 0.9% Si 0.8% Mn Al alloy: Forging; solution treated & aged; AMS for AA alloy 2014
AMS 4140 D	4% Cu 0.7% Mg 2% Ni Al alloy: Forging; solution treated & aged; AMS for AA alloy 2018
AMS 4142 D	4% Cu 1.5% Mg 0.7% Si 2% Ni Al alloy: Forging; as forged; AMS for AA alloy 2218
AMS 4152 G	4.5% Cu 1.5% Mg 0.6% Mn Al alloy: Extrusions; solution treated; AMS for AA alloy 2024
AMS 4153 B	4.5% Cu 0.5% Mg 1% Si 0.8% Mn Al alloy: Extrusion; solution treated & aged; AMS for AA alloy 2014
AMS 4164 C	4.4% Cu 1.5% Mg 0.6% Mn Al alloy: Extrusion; AMS for AA alloy 2024
AMS 4165 C	4.5% Cu 1.5% Mg 0.6% Mn Al alloy: Extrusion; AMS for AA alloy 2024
AMS 4191 A	6.3% Cu 0.3% Mn 0.2% Zn 0.15% Ti 0.1% V Al alloy: Welding rod; AMS for AA alloy 2319
ANA 8	4.2% Cu 1.0% Mn 1.0% Si Al alloy: Forging; US Service; solution treated; room aged **UTS: 330 Elon: 12% Proof: 210**
ANA 12.1	4.2% Cu 1.5% Mg 0.2% Cr Al alloy: Sheet; US Service; solution treated & aged **UTS: 440 Elon: 14% Proof: 250**
ANA 13	4.3% Cu 1.5% Mg 0.2% Cr Al alloy: Sheet; US Service; solution treated & aged **UTS: 420 Elon: 12% Proof: 240**
AN QQ W 298/2	4.2% Cu 0.5% Mn 1.5% Mg Al alloy: Wire; US Service; solution treated & aged **UTS: 420**
ARDAL	2% Cu 1.5% Fe 0.6% Ni Al alloy: Origin unknown
ASTM B209/2024	4% Cu 1.5% Mg Al alloy: Sheet; annealed **UTS: 210 Elon: 10% Proof: 75**

Symbol	Nominal analysis, supplier, condition and remarks.
ASTM B209/2024	4% Cu 1.5% Mg Al alloy: Sheet; solution treated room aged **UTS: 420 Elon: 6% Proof: 270**
ASTM B209/2219	0.3% Mn 6.3% Cu 0.06% Ti Al alloy: Sheet
ASTM B210/2024	4% Cu 0.5% Mn 1.5% Mg Al alloy: Tube; solution treated & cold worked **UTS: 450 Proof: 280**
ASTM B211/2011	5.5% Cu 0.4% Bi 0.4% Pb Al alloy: Bar; free machining
ASTM B211/2014	4.5% Cu 1% Si 1% Mn 0.6% Mg Al alloy: Bar; solution treated & aged **UTS: 460**
ASTM B211/2017	4% Cu 0.7% Mn 0.6% Mg Al alloy: Bar; solution treated, room aged **UTS: 390**
ASTM B211/2024	4% Cu 0.6% Mn 1.5% Mg Al alloy: Bar; solution treated, room aged **UTS: 440**
ASTM B211/2219	6.2% Cu 0.3% Mn 0.05% Ti Al alloy: Bar
ASTM B221/2014	4.5% Cu 1% Si 1% Mn 0.6% Mg Al alloy: Extruded bar; solution treated & aged **UTS: 420 Elon: 7% Proof: 370**
ASTM B221/2024	4% Cu 0.6% Mn 1.5% Mg Al alloy: Extruded bar; solution treated, room aged **UTS: 400 Elon: 12% Proof: 280**
ASTM B221/2219	6.2% Cu 0.3% Mn 0.05% Ti Al alloy: Extruded bar, etc.
ASTM B235/2014	4.5% Cu 1% Si 1% Mn 0.6% Mg Al alloy: Extruded tube; solution treated & aged **UTS: 480 Elon: 7% Proof: 420**
ASTM B235/2024	4.5% Cu 0.7% Mn 1.5% Mg Al alloy: Extruded tube; solution treated, room aged **UTS: 450 Elon: 10% Proof: 350**
ASTM B241/2014	0.8% Si 4.5% Cu 0.8% Mn 0.5% Mg Al alloy: Pipe and tube
ASTM B241/2024	4.3% Cu 0.6% Mn 1.5% Mg Al alloy: Pipe and tube
ASTM B241/2219	6.3% Cu 0.3% Mn 0.6% Ti Al alloy: Pipe and tube
ASTM B247/2014	4.5% Cu 1% Si 0.8% Mn 0.6% Mg Al alloy: Forging; solution treated & aged **DPN: 125 UTS: 460 Elon: 10% Proof: 390**
ASTM B247/2018	4% Cu 0.8% Mg 2% Ni Al alloy: Forging; solution treated & aged **DPN: 100 UTS: 390 Elon: 10% Proof: 270**
ASTM B247/2025	4.5% Cu 1% Si 0.7% Mn 0.6% Mg Al alloy: Forging; solution treated & aged **DPN: 100 UTS: 420 Elon: 16% Proof: 220**
ASTM B247/2218	4% Cu 1.5% Mg 2% Ni Al alloy: Forging; solution treated & aged **DPN: 100 UTS: 390 Elon: 10% Proof: 270**
ASTM B285 ER2014	4.5% Cu 1% Mn 0.6% Mg Al alloy: Electric welding rod; obsolete
ASTM B285 RC4A	4.5% Cu Al alloy: Welding wire
ASTM B285 RCN42A	4% Cu 1.5% Mg 2% Ni Al alloy: Welding wire
ASTM B308/2014	4.5% Cu 0.9% Si 0.7% Mn 0.6% Mg Al alloy: Section; solution treated & aged **UTS: 420 Elon: 7% Proof: 370**
ASTM B316/2017	4% Cu 0.8% Mn 0.6% Mg Al alloy: Rivets; annealed **UTS: 240**
ASTM B316/2024	4.5% Cu 0.6% Mn 1.5% Mg Al alloy: Rivets; annealed **UTS: 240**
ASTM B316/2117	2.7% Cu 0.4% Mg Al alloy: Rivet; annealed **UTS: 170**
A U2G	2.4% Cu 0.8% Si 0.4% Mg Al alloy: Wrought; L'Aluminium Francais

Symbol	Nominal analysis, supplier, condition and remarks.
A U2G	2.5% Cu 0.3% Mg Al alloy: Wrought; French Standard
A U2GN	2.2% Cu 1.6% Mg 1.2% Fe 1.1% Ni Al alloy: Wrought; L'Aluminium Francais
A U2N	2.1% Cu 0.8% Si 1.0% Fe 1.0% Mg 0.1% Ti 1.0% Ni Al alloy: Wrought; L'Aluminium Francais
A U4G	4.0% Cu 0.5% Si 0.7% Mg 0.5% Mn Al alloy: Wrought; L'Aluminium Francais
A U4G	4.1% Cu 0.7% Mg 0.6% Si 0.6% Mn Al alloy: Wrought; French Standard
A U4G1	4.2% Cu 1.6% Mg 0.6% Mn Al alloy: Wrought; French Standard
A U4G1	4.2% Cu 1.4% Mg 0.7% Mn Al alloy: Wrought; L'Aluminium Francais
A U4N	4.0% Cu 1.5% Mg 2.0% Ni Al alloy: Wrought; L'Aluminium Francais
A U4PB	4.3% Cu 1.0% Mg 1.2% Pb Al alloy: Wrought; free machining; French Standard
A U4SG	4.4% Cu 0.5% Mg 0.9% Si 0.8% Mn Al alloy: Wrought; French Standard
A U4SG	4.4% Cu 0.6% Mg 0.8% Si 0.8% Mn Al alloy: Wrought; L'Aluminium Francais
Av 22	4% Cu 0.7% Mg 0.5% Mn Al alloy: Anglo-Swiss code for Avional 22
Av 24	4.3% Cu 1.1% Mg 0.8% Mn 0.1% Cr Al alloy: Anglo-Swiss code for Avional 24
AVIONAL 22	4% Cu 0.7% Mg 0.5% Mn Al alloy: Anglo-Swiss; solution treated & aged; obsolete
	DPN: 110 **UTS: 390** **Elon: 23%** **Proof: 270**
AVIONAL 24	4.3% Cu 1.1% Mg 0.8% Mn 0.1% Cr Al alloy: Anglo-Swiss; solution treated & aged; obsolete
	DPN: 130 **UTS: 480** **Elon: 16%** **Proof: 330**
AVIONAL 050	2.5% Cu 0.3% Mg Al alloy: Wrought; Anglo-Swiss Al Co
AVIONAL 100	4.0% Cu 0.6% Mg 0.5% Mn Al alloy: Wrought; Anglo-Swiss Al Co
AVIONAL Pb 118	4.7% Cu 1.0% Mg 0.7% Mn 1.0% Pb Al alloy: Wrought; Anglo-Swiss Al Co; free machining
AVIONAL 150	4.5% Cu 1.5% Mg 0.8% Mn 1.0% Cr Al alloy: Wrought; Anglo-Swiss Al Co
AWCO 31	5% Cu Al alloy: Welding wire; Aluminium Wire Co; as drawn
AWCO 301	4% Cu 0.6% Mg 0.5% Mn Al alloy: Wire; Aluminium Wire Co for BS alloy N14
AWCO 304	2.2% Cu 0.3% Mg Al alloy: Wire for rivets; Aluminium Wire Co for BS alloy N13
AWCO 305	4.4% Cu 0.6% Mg 0.7% Si 0.6% Mn Al alloy: Wire; Aluminium Wire Co for BS alloy N15
AWCO 308	5.5% Cu 0.5% Pb 0.5% Bi Al alloy: Bar etc.; Aluminium Wire Co; free machining
AWCO 309	4.5% Cu 1.6% Mg 0.6% Mn Al alloy: Bar; Aluminium Wire Co
B 18S	4% Cu 1.5% Mg 2% Ni Al alloy: Previous AA designation; now 2218
BA 35	3.5% Cu 0.7% Mg 0.5% Sb 0.2% Sn Al alloy: Extrusion; British Al Co; as rolled
	UTS: 240 **Elon: 10%** **Proof: 110**
BA 301	4% Cu 0.6% Mg 0.5% Mn Al alloy: Sheet, strip & tube; British Al Co for BS alloy N14

Symbol	Nominal analysis, supplier, condition and remarks.
BA 303	4.4% Cu 0.6% Mg 0.7% Si 0.6% Mn Al alloy: Sheet & tube; British Al Co for BS alloy N15
BA 305	4.6% Cu 0.7% Mg 0.8% Si 0.7% Mn Al alloy: British Al Co for BS alloy N15
BA 306	1.7% Cu 0.7% Mg 1% Si 0.6% Mn Al alloy: British Al Co for BS alloy N11
BA 307	2% Cu 0.9% Mg 0.8% Si 1% Ni 0.1% Ti Al alloy: British Al Co; solution treated & aged
	UTS: 370 **Elon: 8%** **Proof: 280**
BA 308	5.5% Cu 0.5% Pb 0.5% Bi Al alloy: Bar; British Al Co; solution treated & aged; free machining
	UTS: 300 **Elon: 8%** **Proof: 240**
BA 309	4.3% Cu 1.6% Mg 0.5% Mn 0.05% Pb Al alloy: Plate; British Al Co; solution treated & aged
	UTS: 420 **Elon: 10%** **Proof: 250**
BA 352	4% Cu 0.6% Mg 0.5% Mn Al alloy: Sheet & strip; British Al Co for BS alloy N14
BA 353	4.4% Cu 0.6% Mg 0.7% Si 0.6% Mn Al alloy: Sheet & strip; British Al Co for BS alloy N15
BIRMETAL 161	1.5% Cu 1% Mg 1% Si Al alloy: Birmabright for BS alloy H11
BIRMETAL 230	Cu Mg Al alloy: For rivets; Birmabright
	UTS: 270
BIRMETAL 477	4% Cu 1% Mg 0.5% Si 0.7% Mn Al alloy: Birmabright for BS alloy H14
BIRMETAL 478	4% Cu 0.5% Mg 0.5% Si 1% Mn Al alloy: Birmabright for BS alloy H15
BMB 473	4% Cu 0.6% Mg 0.5% Mn Al alloy: Sheet & strip; Birmabright for BS alloy H14
BMB 551	4.4% Cu 0.6% Mg 0.7% Si 0.6% Mn Al alloy: Forging; Birmabright for BS alloy H15
BMB 761	1.5% Cu 1% Mg 1% Si Al alloy: Forging; Birmabright; solution treated & aged
	UTS: 370 **Elon: 10%** **Proof: 300**
BRILLUM	1.5% Cu 2% Ni Al alloy: ICI; obsolete; now Imperial Al Co
BS 395	Al alloy (Duralumin): Sheet & strip; replaced by BS 1470
BS 396	Al alloy (Duralumin): Tube; replaced by BS 1471
BS 478	Al alloy bar (Y alloy): Replaced by BS 1476
BS 533	Al alloy forging (Y alloy): Replaced by BS 1472
BS 1080	3% Cu 0.5% Mg Al alloy: Bar
	UTS: 240 **Elon: 10%** **Proof: 110**
BS 1470 HS14	4% Cu 1% Mg 0.75% Mn Al alloy: For sheet & strip; solution treated, room aged
	UTS: 370 **Elon: 15%** **Proof: 220**
BS 1470 HS15	4% Cu 0.5% Mg 0.7% Si 0.75% Mn Al alloy: For sheet & strip; solution treated, room aged
	UTS: 370 **Elon: 15%** **Proof: 220**
BS 1471 HT14	4% Cu 1% Mg 0.75% Mn Al alloy: Tube
	UTS: 390 **Elon: 10%** **Proof: 270**
BS 1471 HT15	4% Cu 0.5% Mg 0.7% Si 0.75% Mn Al alloy: For tube
	UTS: 420 **Elon: 8%** **Proof: 350**
BS 1472 HF11	1.5% Cu 1% Mg 1% Si Al alloy: For forging & bar; solution treated & aged
	UTS: 360 **Elon: 8%** **Proof: 280**
BS 1472 HF12	2% Cu 1% Mg 1% Si 1% Fe 1% Ni Al alloy: For forging & bar; solution treated & aged
	UTS: 370 **Elon: 8%** **Proof: 280**
BS 1472 HF14	4% Cu 0.6% Mg 0.5% Mn Al alloy: For forging; solution treated, room aged
	UTS: 360 **Elon: 15%** **Proof: 200**
BS 1472 HF15	4% Cu 0.75% Mg 0.75% Si 1% Mn Al alloy: For forging & bar; solution treated & aged
	UTS: 420 **Elon: 8%** **Proof: 370**
BS 1472 HF16	2.2% Cu 1.5% Mg 1.2% Si 1.1% Ni Al alloy: Forging; solution treated & aged
	UTS: 430 **Elon: 5%** **Proof: 340**

Note. The following abbreviations and units are used in the tables:

DPN	Hardness, diamond pyramid number
UTS	Ultimate tensile strength, N/mm^2
Elon	Elongation, %
Proof	0.1 % proof strength, N/mm^2

1 N/mm^2=0.1 hbar=0.102 kgf/mm^2=0.06475 tonf/in^2=145.04 lbf/in^2

Symbol	Nominal analysis, supplier, condition and remarks.
BS 1472 HF17	4% Cu 1.5% Mg 2% Ni Al alloy: For forging; solution treated & aged **UTS: 330 Elon: 8%**
BS 1472 HF18	2% Cu 1.5% Mg 1% Ni 1% Fe Al alloy: For forging & bar; solution treated & aged **UTS: 360 Elon: 6% Proof: 250**
BS 1473 HB15	4% Cu 0.75% Mg 0.75% Si 1% Mn Al alloy: For bolt stock; solution treated & aged **UTS: 420 Proof: 360**
BS 1473 HR15	4% Cu 0.7% Mg 0.7% Si 1% Mn Al alloy: For rivets; solution treated, room aged **UTS: 370**
BS 1474 HV11	1.5% Cu 1% Mg 1% Si Al alloy: For tubes; solution treated & aged **UTS: 330 Elon: 8% Proof: 270**
BS 1474 HV14	4% Cu 1% Mg 0.5% Si 0.75% Mn Al alloy: For tubes; solution treated, room aged **UTS: 360 Elon: 12% Proof: 210**
BS 1474 HV15	4% Cu 0.75% Mg 0.75% Si 1% Mn Al alloy: For tubes; solution treated & aged **UTS: 420 Elon: 8% Proof: 360**
BS 1475 HG15	4% Cu 0.6% Mg 0.7% Si 1% Mn Al alloy: Wire; solution treated & aged **UTS: 420**
BS 1476 HE11	1.5% Cu 1% Mg 1% Si Al alloy: Bar; solution treated & aged **UTS: 370 Elon: 8% Proof: 300**
BS 1476 HE14	4% Cu 1% Mg 0.5% Si 0.7% Mn Al alloy: Bar; solution treated, room aged **UTS: 370 Elon: 10% Proof: 220**
BS 1476 HE15	4% Cu 0.5% Mg 0.5% Si 1% Mn Al alloy: Bar; solution treated & aged **UTS: 450 Elon: 8% Proof: 390**
BS 1477 HP14	4% Cu 1% Mg 0.5% Si 0.7% Mn Al alloy: Plate; solution treated & aged **UTS: 370 Elon: 10% Proof: 220**
BS 1477 HP15	4% Cu 0.5% Mg 0.7% Si 1% Mn Al alloy: Plate solution treated & aged **UTS: 410 Elon: 8% Proof: 310**
BS 4300/5 FC 1	5.5% Cu 0.5% Bi 0.5% Pb Al alloy: Bar; solution treated & aged; free machining **UTS: 300 Elon: 6% Proof: 200**
BS L1	4% Cu Mn Mg Al alloy: For forgings; obsolete; replaced by BSL 64
BS L3	Al alloy: Sheet & coils; obsolete; no information available
BS L24	4% Cu 1.5% Mg 2% Ni Al alloy: Y alloy; obsolete
BS L25	4% Cu 1.5% Mg 2% Ni Al alloy: Forging; solution treated & aged **UTS: 360 Elon: 15% Proof: 210**
BS L37	4% Cu 0.75% Si 1% Mn Al alloy: For wires & rivets; cold drawn; can be solution treated & aged **UTS: 370**
BS L39	4.4% Cu 0.6% Mg 0.7% Si 0.7% Mn Al alloy: Forging; solution treated, room aged **UTS: 370 Elon: 15% Proof: 220**
BS L40	Cu Mg Si Al alloy: Obsolete; replaced by BSL 65
BS L42	2.2% Cu 0.8% Si 1.2% Ni 1.5% Mg 1% Fe 0.1% Ti Al alloy: Forging; solution treated & aged **UTS: 370 Elon: 8% Proof: 250**
BS L43	Cu Mg Al alloy: Forgings; obsolete; replaced by BSL 65
BS L45	Cu Al alloy: Forgings; obsolete; replaced by BSL 65
BS L57	2.2% Cu 0.3% Mg Al alloy: Wire; solution treated **UTS: 250**
BS L62	4% Cu 0.75% Mg 0.75% Si 0.8% Mn Al alloy: For tubes; solution treated; room aged **UTS: 390 Proof: 270**

Symbol	Nominal analysis, supplier, condition and remarks.
BS L63	4% Cu 0.75% Mg 0.75% Si 0.8% Mn Al alloy: For tubes; solution treated & aged **UTS: 440 Proof: 340**
BS L64	4% Cu 0.75% Mg 0.75% Si 0.8% Mn Al alloy: For bars & forgings; solution treated, room aged **UTS: 340 Elon: 13% Proof: 210**
BS L65	4% Cu 0.75% Mg 0.75% Si 0.8% Mn Al alloy: For bars & forgings; solution treated & aged **UTS: 440 Elon: 8% Proof: 370**
BS L69	Cu Mg Al alloy: Wire; obsolete; replaced by BSL 86
BS L70	4% Cu 0.75% Mg 0.75% Si 0.8% Mn Al alloy: For sheet & strip; solution treated, room aged **UTS: 370 Elon: 15% Proof: 240**
BS L71	4% Cu 0.75% Mg 0.75% Si 0.8% Mn Al alloy: For sheet & strip; solution treated & aged **UTS: 420 Elon: 8% Proof: 340**
BS L76	4.4% Cu 0.7% Mg 0.7% Si 0.6% Mn Al alloy: Bar and forging; solution treated, room aged **UTS: 360 Elon: 15% Proof: 200**
BS L77	4.4% Cu 0.7% Mg 0.7% Si 0.6% Mn Al alloy: Bar and forging; solution treated & aged **UTS: 440 Elon: 8% Proof: 370**
BS L83	2% Cu 1% Mg 0.9% Si 1% Ni 0.9% Fe Al alloy: Bar & forging; solution treated & aged **UTS: 390 Elon: 10% Proof: 270**
BS L84	1.5% Cu 0.8% Mg 1% Si Al alloy: Bar; solution treated & aged **UTS: 300 Elon: 15% Proof: 180**
BS L85	1.5% Cu 0.8% Mg 1% Si Al alloy: Bar & forging; solution treated & aged **UTS: 360 Elon: 10% Proof: 270**
BS L86	2.5% Cu 0.4% Mg Al alloy: For gold forged rivets; cold drawn; can be solution treated & aged **UTS: 250**
BS L87	4.4% Cu 0.7% Mg 0.7% Si 0.6% Mn Al alloy: Bar; solution treated & aged **UTS: 420 Elon: 8% Proof: 360**
BS T4	Aluminium alloy: Tube; obsolete; see BS/L 62
CB 60A	5.5% Cu Al alloy: Previous ASTM designation; now 2011
CG 30A	2.5% Cu 0.3% Mg Al alloy: Previous ASTM designation; now 2117
CG 42A	4.2% Cu 0.7% Mn 1% Mg Al alloy: Previous ASTM designation; now 2024
CM 41A	4% Cu 0.8% Mn 0.6% Mg Al alloy: Previous ASTM designation; now 2017
CS 41A	4.5% Cu 1% Si 0.8% Mn 0.6% Mg Al alloy: Previous ASTM designation; now 2014
CONLOY	4.5% Cu Al alloy: Tube; Constrictor Ltd; obsolete **DPN: 75 UTS: 220 Proof: 170**
DECOLTAL 500	5.5% Cu 0.4% Ph 0.4% Si Al alloy: Wrought; Anglo-Swiss Al Co; free machining
DTD 130 B	2% Cu 1% Mg 0.15% Ti Al alloy: Bar & forgings; solution treated & aged **UTS: 370 Elon: 9% Proof: 300**
DTD 147	4% Cu 0.6% Mg 0.5% Mn Al alloy: Forgings; solution treated **UTS: 360 Elon: 15% Proof: 200**
DTD 150A	4% Cu 0.6% Mg 0.5% Mn Al alloy: Forgings; solution treated & room aged **UTS: 360 Elon: 15% Proof: 200**
DTD 246 B	2% Cu 1% Mg 1% Si 1% Ni 0.1% Ti Al alloy: Forgings; solution treated & aged **DPN: 85 UTS: 240 Elon: 16% Proof: 105**
DTD 273	4% Cu 1.4% Mg 0.4% Mn Al alloy: Tube; solution treated **UTS: 440 Elon: 7% Proof: 330**
DTD 327	2.2% Cu 0.3% Mg Al alloy: Wire; solution treated **UTS: 250**

Symbol	Nominal analysis, supplier, condition and remarks.
DTD 364	4% Cu 0.7% Si 0.7% Mg Al alloy: Forging
DTD 410	2% Cu Si Mg Fe Ni Al alloy: Forging
	DPN: 100 UTS: 340 Elon: 6% Proof: 220
DTD 423 C	1.5% Cu 1% Mg 1% Si Al alloy: Bar & forging; solution treated & aged
	UTS: 370 Elon: 10% Proof: 280
DTD 443 A	1.5% Cu 1% Mg 1% Si Al alloy: Bar & forging; solution treated, room aged
	UTS: 270 Elon: 15% Proof: 170
DTD 450 A	1.5% Cu 1% Mg 1% Si Al alloy: Tube; solution treated, room aged
	UTS: 250 Elon: 12% Proof: 170
DTD 460 A	1.5% Cu 1% Mg 1% Si Al alloy: Tube; solution treated & aged
	UTS: 370 Elon: 6% Proof: 310
DTD 464	Cu Mn Al alloy: Replaced by BSL 63
DTD 520	Cu Si Al alloy: Obsolete
DTD 546 B	4.4% Cu 0.6% Mg 0.7% Si 0.6% Mn Al alloy: Sheet; solution treated & aged
	UTS: 400 Elon: 8% Proof: 310
DTD 603 B	4.4% Cu 0.6% Mg 0.7% Si 0.6% Mn Al alloy: Sheet; solution treated, room aged
	UTS: 390 Elon: 15% Proof: 240
DTD 646 B	4.4% Cu 0.6% Mg 0.7% Si 0.6% Mn Al alloy: Sheet; solution treated & aged
	UTS: 420 Elon: 8% Proof: 340
DTD 706	4.4% Cu 0.7% Mg 0.7% Si 0.6% Mn Al alloy: Sheet; solution treated & aged
	UTS: 420 Elon: 8% Proof: 340
DTD 717	2% Cu 1.5% Mg 1% Fe 1% Ni Al alloy: Forgings; solution treated & aged; for high temperature use
	UTS: 390 Elon: 12% Proof: 240
DTD 724	2.2% Cu 1.5% Mg 1.0% Fe 1.2% Ni 0.1% Ti Al alloy: Forging; solution treated & aged
	UTS: 390 Elon: 10% Proof: 250
DTD 731A	2.2% Cu 1.5% Mg 1% Ni 1% Fe Al alloy: Forging; solution treated & aged
	UTS: 420 Elon: 6% Proof: 310
DTD 745	2.2% Cu 1.5% Mg 1% Fe 1% Ni Al alloy: Forging; solution treated & aged
	UTS: 400 Elon: 6% Proof: 300
DTD 746	4.2% Cu 0.7% Mg 0.7% Si 1% Mn Al alloy: Sheet; solution treated & aged
	UTS: 420 Elon: 8% Proof: 330
DTD 5004	6% Cu 0.3% Mn Al alloy: Forging; solution treated & aged; for high temperature use
	UTS: 360 Elon: 6% Proof: 210
DTD 5010	4.2% Cu 0.7% Mg 0.7% Si 1% Mn Al alloy: Plate; solution treated & aged
	UTS: 390 Elon: 12% Proof: 240
DTD 5014	2.2% Cu 1.5% Mg 1% Fe 1% Ni Al alloy: Bar; solution treated & aged
	UTS: 390 Elon: 7% Proof: 300
DTD 5020 A	4.4% Cu 0.7% Mg 0.7% Si 0.6% Mn Al alloy: Plate; solution treated & aged
	UTS: 390 Elon: 7% Proof: 330
DTD 5030	4.4% Cu 0.7% Mg 0.7% Si 0.6% Mn Al alloy: Plate; solution treated, & room aged
	UTS: 370 Elon: 12% Proof: 220

Note. The following abbreviations and units are used in the tables:

DPN	Hardness, diamond pyramid number
UTS	Ultimate tensile strength, N/mm²
Elon	Elongation, %
Proof	0.1 % proof strength, N/mm²

$1 \text{ N/mm}^2 = 0.1 \text{ hbar} = 0.102 \text{ kgf/mm}^2 = 0.06475 \text{ tonf/in}^2 = 145.04 \text{ lbf/in}^2$

Symbol	Nominal analysis, supplier, condition and remarks.
DTD 5084	2.2% Cu 1.6% Mg 1.2% Fe 1.2% Ni Al alloy: Forging; solution treated & aged; for high temperature use
	UTS: 370 Elon: 8% Proof: 280
DTD 5090	4.4% Cu 1.5% Mg 0.6% Mn Al alloy: Plate; solution treated, room aged
	UTS: 400 Elon: 12% Proof: 270
DU BRAND	4% Cu 0.6% Mg 0.5% Mn Al alloy: Forging & tube; High Duty Alloys; see Hiduminium 01
DURAL B	4% Cu 0.6% Mg 0.5% Mn Al alloy: Sheet, tube & forging; Alcan for BS alloy H14
DURAL E	1.8% Cu 1% Si 0.8% Mg 0.7% Mn Al alloy: Tube & forging; Alcan for BS alloy H11
DURAL G	4.2% Cu 1.5% Mg 0.2% Si 0.6% Mn Al alloy: Sheet; tube & forging; Alcan; solution treated, room aged
	UTS: 420 Elon: 12% Proof: 270
DURAL J	2.2% Cu 1.5% Mg 0.8% Si 1% Ni 0.1% Ti Al alloy: Forging; Alcan for BS alloy H18
DURAL JJ	2.2% Cu 1.5% Mg 0.15% Si 1% Fe 1.1% Ni Al alloy: Tube & forging; Alcan; solution treated & aged
	UTS: 390 Elon: 6% Proof: 300
DURAL M	2.2% Cu 0.3% Mg Al alloy: Wire for rivets; Alcan
DURAL Q	4% Cu 1.4% Mg 0.4% Mn Al alloy: Tube; Alcan; solution treated
	UTS: 440 Elon: 7% Proof: 330
DURAL S	4.4% Cu 0.6% Mg 0.7% Si 0.6% Mn Al alloy: Sheet, tube & forging; Alcan for BS alloy H15
DURAL T	2% Cu 1% Mg 0.8% Si 1% Ni 0.1% Ti Al alloy: Forging; Alcan; solution treated & aged
	UTS: 370 Elon: 8% Proof: 280
DURICILIUM F	4% Cu 0.7% Mg 0.7% Si 1% Mn Al alloy: Tube; E Kaye for BS alloy H14/15
DURICILIUM H	4% Cu 1% Mg 0.5% Si 0.7% Mn Al alloy: E Kaye for BS alloy H14
DURICILIUM K	4% Cu 0.5% Mg 0.5% Si 1% Mn Al alloy: E Kaye for BS alloy H15
DURCILILIUM M	1.5% Cu 1% Mg 1% Si Al alloy: E Kaye for BS alloy H11
FMA	5.5% Cu 0.5% Bi 0.5% Pb Al alloy: Extrusions; free machining. McKechnie; solution treated & aged
	DPN: 100 UTS: 330 Elon: 10% Proof: 250
HA 4 CG42	4.3% Cu 1.6% Mg 0.6% Mn Al alloy: Sheet & plate; Canadian Standard
HA 5 CB60	5.5% Cu Al alloy: Bar, etc.; Canadian Standard
HA 5 CG42	4.3% Cu 1.6% Mg 0.6% Mn Al alloy: Bar etc.; Canadian Standard
HA 5 CS41N	4.5% Cu 0.6% Mg 0.8% Mn 0.8% Si Al alloy: Bar etc; Canadian Standard
HA 6 CG30	2.6% Cu 0.4% Mg Al alloy: For rivets and brazing wire; Canadian Standard
HA 6 CM41	4.0% Cu 0.6% Mg 0.7% Mn 0.6% Si Al alloy: For rivets and brazing wire; Canadian Standard
HA 7 CG42	4.3% Cu 1.6% Mg 0.6% Mn Al alloy: Tube; Canadian Standard
HA 8 CN42	4.0% Cu 0.7% Mg 2.0% Ni Al alloy: Forging; Canadian Standard
HA 8 CS41N	4.5% Cu 0.6% Mg 0.8% Mn 0.8% Si Al alloy: Forging; Canadian Standard
HA 8 CS41P	4.5% Cu 0.6% Mg 0.8% Mn Al alloy: Forging; Canadian Standard
HA 8 ZG62	1.6% Cu 2.5% Mg 5.5% Zn 0.2% Cr Al alloy: Forging; Canadian Standard
HIDUMINIUM 01	4% Cu 0.6% Mg 0.5% Mn Al alloy: For forgings & tube. High Duty alloys for BS alloy H14 (DU BRAND)

Symbol	Nominal analysis, supplier, condition and remarks.
HIDUMINIUM 02	4% Cu 1.5% Mg 2% Ni Al alloy: For pistons & cylinder heads; High Duty Alloys for BS alloy H17 (Y alloy)
HIDUMINIUM 03	1% Cu 1% Mg 1% Si Al alloy: For forgings & extrusions; High Duty alloys for BS alloy H11
HIDUMINIUM 55	2.5% Cu 1% Mg 1% Si 0.7% Ni 0.3% Mn Al alloy: For forgings; High Duty alloys for BS alloy H12
HIDUMINIUM 66	4.5% Cu 0.6% Mg 0.7% Si 0.6% Mn Al alloy: For forgings; High Duty alloys for BS alloy H15
HIDUMINIUM 72	4.5% Cu 1.6% Mg 0.6% Mn Al alloy: Plate; High Duty alloys; solution treated & aged **UTS: 420 Elon: 10% Proof: 250**
IMPALCO 660	4.1% Cu 1% Mn Al alloy: Bar & forgings etc.; Imperial Al Co for BS alloy H15
IMPALCO 690	1.5% Cu 1% Mg Al alloy: Bar & forging; Imperial Al Co for BS alloy H11
IMPALCO 800	5.5% Cu 0.4% Pb 0.4% Bi Al alloy: Bar & tube; Imperial Al Co; solution treated & aged; free machining **UTS: 220 Elon: 11% Proof: 300**
IMPALCO 810	1.5% Cu 0.7% Mg 1% Si 1% Pb Al alloy: Bar & forgings; Imperial Al Co; solution treated & aged; free machining **UTS: 270 Elon: 10% Proof: 220**
IMPALCO 830	2.3% Cu 1% Mg 1% Si 1% Ni Al alloy: Bar & forging; Imperial Al Co for BS alloy H12
IMPALCO 840	2.2% Cu 1.5% Mg 1.3% Si 1% Ni Al alloy: Forging; Imperial Al Co for BS alloy H18
IMPALCO 860	4% Cu 1.5% Mg 2% Ni Al alloy: Forging; Imperial Al Co; solution treated & aged **UTS: 340 Elon: 15% Proof: 210**
IMPALCO 2024	4.2% Cu 1.6% Mn 0.6% Mn 0.2% Cr Al alloy: Plate; Imperial Al Co; solution treated & aged; temporary designation **UTS: 420 Elon: 10% Proof: 250**
IMPALCO C66	4% Cu 1% Mn 1% Fe 0.7% Mg Al alloy: Sheet strip rod; Imperial Al Co for BS alloy H15
IMPALCO C69	1.5% Cu 1% Mg 1% Mn 1% Si Al alloy: Imperial Al Co for BS alloy H11
IMPALCO C80	5.5% Cu 0.5% Pb Al alloy: For bar & section; Imperial Al Co; free machining **UTS: 280 Elon: 12% Proof: 210**
KYNAL 90	2.2% Cu 0.3% Mg Al alloy: ICI; obsolete; now Imperial Al Co
KYNAL C65	4% Cu 0.6% Mg 0.5% Mn Al alloy: ICI; obsolete; now Imperial Al Co
KYNAL C66	4.4% Cu 0.6% Mg 0.7% Si 0.6% Mn Al alloy: ICI; obsolete; now Imperial Al Co
KYNAL C67	4.4% Cu 0.6% Mg 0.7% Si 0.6% Mn Al alloy: ICI; Obsolete; now Imperial Al Co
KYNAL Y88	2.6% Cu 1% Mg 1% Si 1% Ni 0.1% Ti Al alloy: ICI; obsolete; now Imperial Al Co
KYNAL Y92	4% Cu 1.5% Mg 2% Ni Al alloy: ICI; obsolete; now Imperial Al Co
ML 2	0.4% Si 5.5% Cu 0.4% Pb 0.4% Bi Al alloy: Bar; AALCO; free machining; American equivalent 2011T3 **DPN: 95 UTS: 300 Elon: 10% Proof: 240**
NORAL 16S	2.2% Cu 0.3% Mg Al alloy: Wire for rivets; Northern Al Co
NORAL 17S	4% Cu 0.6% Mg 0.5% Mn Al alloy: Sheet, strip & tube; Northern Al Co for BS alloy H14
NORAL 19S	4% Cu 1.5% Mg 2% Ni Al alloy: Forging; Northern Al Co for BS alloy H17
NORAL 24S	4% Cu 1.4% Mg 0.4% Mn Al alloy: Tube; Northern Al Co; solution treated **UTS: 440 Elon: 7% Proof: 330**
NORAL 26S	4.4% Cu 0.6% Mg 0.7% Si 0.6% Mn Al alloy: Sheet tube & forging; Northern Al Co for BS alloy H15
NORAL 42S	2.2% Cu 1.5% Mg 0.8% Si 1% Ni 0.1% Ti Al alloy: Forgings; Northern Al Co for BS alloy H18
QQ A 351b	4.0% Cu 0.4% Mg Al alloy: Bar & forging; US Federal; solution treated, room aged **UTS: 360 Elon: 12% Proof: 210**
QQ A 353a	4.0% Cu 0.6% Mn 0.6% Mg Al alloy: Sheet; US Federal **UTS: 370 Elon: 12% Proof: 200**
QQ A 354a	4.1% Cu 0.5% Mn 1.2% Mg Al alloy: Forging; US Federal; solution treated & aged **UTS: 420 Elon: 12% Proof: 270**
QQ A 355a	4.2% Cu 1.5% Mg 0.2% Cr Al alloy: Sheet; US Federal; solution treated & aged **UTS: 440 Elon: 15% Proof: 250**
QQ A 361	4.0% Cu 0.6% Mg 0.2% Cr Al alloy: Sheet; US Federal; solution treated & aged **UTS: 340 Elon: 14% Proof: 180**
QQ A 362/1	4.2% Cu 1.5% Mg 0.2% Cr Al alloy: Sheet; US Federal **UTS: 370 Elon: 13% Proof: 180**
QQ A 367b/1	4.5% Cu 0.5% Mg 1.0% Mn Al alloy: Forging; US Federal; solution treated & aged **DPN: 125 UTS: 450 Elon: 10% Proof: 300**
RR 56	2% Cu 0.9% Mg 0.8% Si 1% Ni 0.1% Ti Al alloy: For forgings & extrusions; High Duty alloy specification for DTD 130A DTD 410
RR 57	6% Cu 0.25% Mn 0.1% Ti Al alloy: For forging; High Duty alloys; solution treated & aged **DPN: 110 UTS: 370 Elon: 8% Proof: 220**
RR 58	2% Cu 1.5% Mg 1.2% Ni 0.1% Ti Al alloy: For forging; High Duty alloy for DTD 717 DTD 724
RR 59	2% Cu 1.5% Mg 0.8% Si 1.2% Ni 0.1% Ti Al alloy: For forging; High Duty alloys for BS alloy H18
RR 257	6% Cu 1% Ni 0.25% Co 0.25% Sb 0.2% Ti Al alloy: High Duty alloys; solution treated & aged **UTS: 340 Elon: 8% Proof: 220**
RR S R	2.5% Cu 0.1% Mg 0.75% Ni 0.1% Ti Al alloy: For forgings; High Duty alloys; used for sealing rings **DPN: 70**
SAE 24	4.5% Cu 1.5% Mg 0.6% Mn Al alloy: Wrought; solution treated; now AA 2024 **UTS: 440 Elon: 12% Proof: 270**
SAE 26	4% Cu 0.5% Mg 0.5% Mn Al alloy: Bar, wire, etc.; solution treated; now AA 2017 **UTS: 390 Elon: 16% Proof: 210**
SAE 202	5.5% Cu 0.4% Pb 0.4% Bi Al alloy: For free machining; solution treated, cold worked & aged; now AA 2011 **DPN: 100 UTS: 420 Elon: 12% Proof: 300**
SAE 203	4% Cu 1.5% Mg 1% Ni Al alloy: Solution treated & stabilized, now AA 2218 **DPN: 95 UTS: 330 Elon: 11% Proof: 240**
SAE 204	2.6% Cu 0.4% Mg Al alloy: Wrought; now AA 2117
SAE 260	4.5% Cu 0.8% Si 0.4% Mg 0.8% Mn Al alloy: Bar & forging; solution treated & aged; now AA 2014 **UTS: 460 Elon: 10% Proof: 390**
SAE 270	4% Cu 0.5% Mg 2% Ni Al alloy: Forgings; solution treated & aged; now AA 2018 **UTS: 390 Elon: 10% Proof: 270**
SIS 4335	4.2% Cu 1% Mg 0.8% Mn 1.2% Pb Al alloy: Bar; Swedish Standard **DPN: 120 UTS: 420 Elon: 6% Proof: 300**

Symbol	Nominal analysis, supplier, condition and remarks.
SIS 4338	4.5% Cu 0.9% Si 0.6% Mg 0.8% Mn Al alloy: Bar, forging, etc.; Swedish Standard; as rolled **DPN: 130 UTS: 450 Elon: 8% Proof: 340**
STA 7 AW8	3.4% Cu 0.7% Mg 0.5% Sb 0.2% Sn Al alloy: Bar (free cutting); as rolled; free cutting; obsolete **UTS: 240 Elon: 10% Proof: 105**
STA 7 AW11A	1% Cu 1% Mg 1% Si Al alloy: Bar; replaced by BS 1476 HE11
STA AW11B	1% Cu 1% Mg 1% Si Al alloy: Bar; replaced by BS 1476 HE11
STA 7 AW12	2.6% Cu 1% Mg 1% Si 1% Fe 0.3% Mn 0.7% Ni Al alloy: Forging; replaced by BS 1472 HF 12
STA 7 AW13	2.2% Cu 0.3% Mg Al alloy: Wire; replaced by BS 1473-5
STA 7 AW14	4% Cu 0.6% Mg 0.5% Mn Al alloy: Wire; replaced by BS 1473-5 HR14
STA 7 AW15A	4% Cu 0.6% Mg 0.5% Mn Al alloy: Forging & bar; replaced by BS alloy H14
STA 7 AW15B	4.4% Cu 0.6% Mg 0.7% Si 0.6% Mn Al alloy: Forging; replaced by BS alloy H15
STA 7 AW15C	4.4% Cu 0.6% Mg 0.7% Si 0.6% Mn Al alloy: Tube; replaced by BS 1471 HT15
STA 7 AW15D	4.4% Cu 0.6% Mg 0.7% Si 0.6% Mn Al alloy: Tube; replaced by BS 1471 HT15
STA 7 AW15E	4% Cu 0.6% Mg 0.5% Mn Al alloy: Sheet; replaced by BS 1470 HS14
STA 7 AW15G	4.4% Cu 0.6% Mg 0.7% Si 0.6% Mn Al alloy: Sheet; replaced by BS 1470 HC15
STA 7 AW15H	4.4% Cu 0.6% Mg 0.7% Si 0.6% Mn Al alloy: Sheet; replaced by BS 1470 HC15
STA 7 AW17B	4% Cu 1.5% Mg 2% Ni Al alloy: Forgings; replaced by BS 1472 HF17
STA 7 AW18	2.2% Cu 1.5% Mg 0.8% Si 1% Fe 1.2% Ni 0.1% Ti Al alloy: Forging; replaced by BS 1472 HF18
TI 01	4% Cu 0.6% Mg 0.5% Mn Al alloy: Tube; TI Al Co for BS alloy H14
TI 04	4% Cu 0.6% Mg 0.5% Mn Al alloy: Sheet & strip; TI Al Co for BS alloy H14
TI 53	3.5% Cu 0.7% Mg 0.5% Sb 0.2% Sn Al alloy: Extrusions; TI Al Co; as rolled **UTS: 240 Elon: 10% Proof: 105**
TI 55	2.6% Cu 1% Mg 1% Si 0.3% Mn 0.7% Ni Al alloy: Forgings; TI Al Co for BS alloy H12
TI 56	2% Cu 1% Mg 0.8% Si 1% Ni 0.1% Ti Al alloy: Forging; TI Al Co; solution treated & aged **UTS: 400 Elon: 10% Proof: 320**
TI 66	4.4% Cu 0.6% Mg 0.7% Si 0.6% Mn Al alloy: Sheet & tube; TI Al Co for BS alloy H15
UNI 3022 MAC33	33% Cu Al alloy: Ingot; primary metal; Italian Standard
UNI 3577	2.5% Cu 0.3% Mg 0.3% Si Al alloy: For forging; Italian Standard

Symbol	Nominal analysis, supplier, condition and remarks.
UNI 3578	3.5% Cu 0.6% Mg 1.5% Fe 0.6% Si 0.6% Ni Al alloy: For forging; Italian Standard
UNI 3579	4.0% Cu 0.5% Mg 0.5% Mn Al alloy: For forging; Italian Standard
UNI 3581	4.4% Cu 0.4% Mg 0.8% Si 0.8% Mn Al alloy: For forging; Italian Standard
UNI 3583	4.5% Cu 1.5% Mg 0.6% Mn Al alloy: For forging; Italian Standard
UNI 6362	5.5% Cu 0.5% Pb 0.5% Bi Al alloy: For forging; Italian Standard; free machining
V Al Cu	5% Cu Al alloy: Previous German Standard designation
V Al Cu Mn 1	3.3% Cu Al alloy: Previous German Standard designation
W 11	1.5% Cu 0.8% Mg 1% Si 1% Mn Al alloy: Southern Forge; replaced by IMPALCO 690
W 12	2.2% Cu 1% Mg 1% Si 1% Fe 1% Ni Al alloy: Southern Forge; replaced by IMPALCO 830
W 17	4% Cu 1.5% Mg 2% Ni Al alloy: Southern forge; replaced by IMPALCO 860
W 18	2.2% Cu 1.5% Mg 1% Si 1% Fe 1% Ni Al alloy: Southern Forge; replaced by IMPALCO 840
W 150	4% Cu 0.8% Mg 1% Si 1% Mn Al alloy: Southern Forge; replaced by IMPALCO 660
W 152	1.5% Cu 1% Mg 1% Si 1% Pb Al alloy: Free machining; Southern Forge; replaced by IMPALCO 810
WW T 785	4.2% Cu 0.6% Mn 1.5% Mg 0.2% Cr Al alloy: Tube; US Federal; solution treated & aged **UTS: 420 Elon: 13% Proof: 270**
WW T 786a	4.0% Cu 0.6% Mn 0.6% Mg Al alloy: Tube; US Federal; solution treated & aged **UTS: 370 Elon: 13% Proof: 270**
Y ALLOY	4% Cu 5% Mg 2% Ni Al alloy: For pistons & cylinder heads; High Duty alloys; HIDUMINIUM 02
ZIRKONAL	0.5% Si 0.8% Mn 15% Cu Al alloy: Origin unknown
ZISIUM	15% Zn 2.0% Cu 0.5% Sn Al alloy: Forging; origin unknown

Note. The following abbreviations and units are used in the tables:

DPN	Hardness, diamond pyramid number
UTS	Ultimate tensile strength, N/mm^2
Elon	Elongation, %
Proof	0.1 % proof strength, N/mm^2

$1 \ N/mm^2 = 0.1 \ hbar = 0.102 \ kgf/mm^2 = 0.06475 \ tonf/in^2 = 145.04 \ lbf/in^2$

1G Aluminium–zinc wrought alloys
With copper, manganese, magnesium and chromium additions

Specific gravity		2.01–2.84	
Density		2010–2840 kg/in.^{3}	(0.102–0.103 lbf/in.^{3})
Solidus/liquidus		477–629 °C	
Thermal conductivity	solution treated & aged	134 W/m °C	(32 cal/m s °C)
Coefficient of linear expansion		20–23.2 × 10^{-6}/ °C	
Electrical conductivity		30% IACS (copper 100%)	
Specific resistance	annealed	30–37 microhm mm	
	solution treated & aged	50–57 microhm mm	
Young's modulus of elasticity		71.02 × 10^9 N/m^2	(10.3 × 10^6 lbf/in.^{2})
Impact	solution treated & aged	7 J	(5.0 ft lb) Izod
Fatigue strength	solution treated & aged	± 150 N/mm^2	(± 10 tonf/in.^{2})
Hot strength			

Temperature °C	Tensile strength N/mm^2
100	480–550
150	330–370
200	150–200
250	90
300	60

The above properties are chosen to be typical of the group of specifications as a whole. There is some variation within the group, and certain properties may not apply to some specifications.

General metallurgical properties

This range of alloys all have from 4–7% zinc, 2–4% magnesium, 0.5–1.7% copper, with or without titanium, manganese and chromium, and are the strongest aluminium alloys, the strength being obtained by thermal treatment.

The alloying elements form intermetallic compounds which are dissolved at the solution treatment and remain in solution after quenching. Age hardening then takes place if the parts are heated for a specified time, the intermetallic compounds starting to precipitate out of solution, increasing the hardness and decreasing the ductility and corrosion resistance.

The addition of manganese, titanium and chromium act as grain refiners and stabilize the structure.

None of the alloys have good corrosion resistance compared to steel, but are sometimes used in the clad condition.

Thermal treatment

These alloys cannot be subjected to any great degree of cold work without an intermediate anneal which is best carried out at 380 °C for 2–3 h then cooled very slowly to 180° C, before cooling to room temperature. The specifications are always used in the solution treated and aged condition.

The solution treatment temperature is 440–460 °C for 30 min to 9 h, depending on the alloy and ruling section, followed by quenching in oil or water. Cold water gives the best mechanical properties, but may result in considerable distortion, which can be reduced by using water at 70–90 °C.

When solution treated these materials are in their softest, most ductile condition, and any cold forming should be carried out in this state, but it should be noted that the age operation will then have a tendency to cause distortion.

After the solution treatment the parts must be aged at 130–140 °C for approximately 12 hours. It is important that the full specification is always consulted for the exact heat treatment temperatures.

Welding and brazing

Pre-treatment. No thermal treatment is necessary before welding or brazing. The area must be thoroughly cleaned, and scratch brushed.

Welding and brazing. Fusion welding of all types is generally unsuitable for these alloys owing to the presence of zinc, although specially formulated materials have been devised which are weldable.

Resistance welding is difficult but can be accomplished using special equipment and techniques. Joining by brazing is possible, but again is not recommended. These alloys should not be specified when welding or brazing is a desirable method of fabrication, without detailed discussion with suppliers.

Post-treatment. Fluxes must be completely removed, and an ageing treatment is necessary to recover some of the mechanical properties.

Flaw detection methods

Crack test. Penetrant oil chalk test methods.

X-ray. May be required on high duty parts, but this is unusual on wrought items.

Ultra-sonic test. This can be applied after rough machining, when it will identify subcutaneous defects.

Chemical etch. This can be used before chalk testing to clean out cracks and thus increase the sensitivity of the process. It is also a useful means of assessing grain size, directional grain flow, and will indicate any seams or laps.

A hot 10% caustic etch is satisfactory, followed by a concentrated nitric acid dip.

Corrosion protection

Temporary. None required.

Permanent. These alloys have comparatively poor corrosion resistance compared with the majority of aluminium alloys. Although capable of being anodized, using the chromic, sulphuric or oxalic acid process, the degree of additional protection is not as great as with other aluminium alloys. Some protection is afforded using chemical oxidation treatments.

Painting. This is probably the best method of protecting these alloys, using an etch primer or chemical oxidation prior to a stoved top coat.

Air drying paints can be used but in general give less protection. Lead based primers must not be used. These alloys in the sheet form may be coated with pure aluminium, or an aluminium alloy. They are then re-classified and are described briefly in Section 1H.

Plating. Electroplating is possible using special etching

techniques, but is more difficult than with other aluminium forging alloys.

Machinability

These alloys are readily machined in the fully heat treated condition. Certain conditions of grain growth can give rise to tearing, but the majority of the specifications listed have grain controlling elements added.

The hard intermetallics which give these alloys their strength are abrasive in nature, thus high speed or tipped tools should be used, particularly for long run production.

Uses

At present these alloys are mainly used in the aircraft industry for such items as main spars, landing gear and structural items.

The relativly poor corrosion resistance and high price prevents their more widespread use.

It should be noted that the hard intermetallic particles can act as a lap and cause severe wear even on case hardened surfaces. The parts will seize if two of these alloys are in moving contact with each other unless special precautions are taken regarding lubrication and loading conditions.

Symbol	Nominal analysis, supplier, condition and remarks.
3.0001	5% Cu 2.5% Mg 5.5% Zn Al alloy: German Standard DIN 1725
3.0002	4.5% Cu 2% Mg 3% Zn Al alloy: German Standard DIN 1725
3.0120	4% Cu 1% Mg 1.5% Zn Al alloy: German Standard DIN 1725
3.4345	0.7% Cu 3.1% Mg 0.4% Mn 4.5% Zn 0.2% Cr Al alloy: German Standard DIN 1725
3.4355	2.7% Mg 0.5% Mn 4.8% Zn 0.2% Cr Al alloy: German Standard DIN 1725
3.4365	1.6% Cu 2.5% Mg 5.6% Zn 0.2% Cr Al alloy: German Standard DIN 1725
3.4375	1% Cu 2.7% Mg 5% Zn 0.5% Cr Al alloy: German Standard DIN 1725
75S	0.4% Cu 2.7% Mg 0.5% Mn 5% Zn Al alloy: Tube; Alcoa; solution treated & aged **UTS: 450 Elon: 8% Proof: 400**
75S	1.5% Cu 2.5% Mg 5.5% Zn 0.3% Cr Al alloy: Previous AA designation - now 7075
76S	1% Cu 2.7% Mg 5% Zn Al alloy: Bar; Alcoa; solution treated & aged **UTS: 500 Elon: 460**
7001	2% Cu 3% Mg 0.3% Cr 7% Zn Al alloy: Wrought; designation used by AA; no ASTM alloy
7039	0.3% Mn 3% Mg 0.2% Cr 4% Zn Al alloy: Sheet & plate; designation used by AA; no ASTM alloy

Symbol	Nominal analysis, supplier, condition and remarks.
7076	0.8% Cu 0.5% Mn .8% Mg 7.5% Zn Al alloy: Forging; designation used by AA; no ASTM alloy
7277	1.2% Cu 2% Mg 0.3% Cr 4% Zn Al alloy: Rivets; designation used by AA; no ASTM alloy
AA 7039	0.3% Mn 3.0% Mg 4.0% Zn 0.2% Cr Al alloy: Wrought
AA 7072	1% Zn Al alloy: Used only for cladding; SAE designation; formerly SAE 214
AA 7075	1.8% Cu 2.5% Mg 5.5% Zn 0.3% Cr Al alloy: Wrought; SAE designation; formerly SAE 215
AA 7076	0.7% Cu 1.6% Mg 0.6% Mn 7.3% Zn Al alloy: Wrought
AA 7079	0.6% Cu 0.2% Mn 3% Mg 4% Zn 0.2% Cr Al alloy: Wrought
AA 7178	2.0% Cu 2.8% Mg 6.8% Zn 0.3% Cr Al alloy: Wrought; SAE designation
ALCOA 740	0.5% Cu 2.7% Mg 0.5% Mn 5.7% Zn Al alloy: Wrought; Alcoa; solution treated & aged **UTS: 550 Elon: 8% Proof: 510**
ALCOA 760	1.0% Cu 2.7% Mg 0.5% Mn 5.5% Zn Al alloy: Wrought; Alcoa; solution treated & aged **UTS: 630 Elon: 7% Proof: 580**
ALCAN 74S	1.8% Mg 4.3% Zn Al alloy: Sheet, tube, bar, etc.; Alcan; solution treated & aged weldable **UTS: 370 Elon: 15% Proof: 300**
ALCAN GB B75S	0.7% Cu 2.5% Mg 0.2% Mn 0.12% Cr 5.7% Zn Al alloy: Alcan; solution treated & aged **DPN: 170 UTS: 600 Elon: 10% Proof: 550**
ALCAN GB C75S	0.7% Cu 2.7% Mg 0.7% Mn 5.5% Zn Al alloy: Alcan; solution treated & aged **UTS: 590 Elon: 11% Proof: 540**
ALCAN GB C77S	1.7% Cu 2.2% Mg 0.12% Mn 0.12% Cr 7.2% Zn Al alloy: Alcan; solution treated & aged **DPN: 175 UTS: 610 Elon: 10% Proof: 570**
ALCAN GB M75S	1.3% Cu 2.5% Mg 0.2% Mn 0.12% Cr 6% Zn Al alloy: Alcan **DPN: 110 UTS: 250 Elon: 12% Proof: 150**

Note. The following abbreviations and units are used in the tables:

DPN	Hardness, diamond pyramid number
UTS	Ultimate tensile strength, N/mm^2
Elon	Elongation, %
Proof	0.1 % proof strength, N/mm^2

1 N/mm^2=0.1 hbar=0.102 kgf/mm^2=0.06475 tonf/in^2=145.04 lbf/in^2

Symbol	Nominal analysis, supplier, condition and remarks.
ALZEN 305K	30% Zn 5% Cu Al alloy: Extrusion; Fry
Al Zn Mg 3	2.7% Mg 5% Zn 0.2% Cr Al alloy: Wrought; previous German Standard designation
Al Zn Mg Cu 0.5	0.8% Cu 3% Mg 4.5% Zn 0.2% Cr Al alloy: Wrought; previous German Standard designation
Al Zn Mg Cu 1.5	1.8% Cu 2.7% Mg 5.5% Zn 0.3% Cr Al alloy: Wrought; previous German Standard designation
AMS 4024 A	0.6% Cu 0.2% Mn 3% Mg 4.3% Zn 0.2% Cr Al alloy: Sheet; AMS for AA alloy 7079; solution treated & aged
AMS 4038	1.6% Cu 2.5% Mg 5.6% Zn 0.3% Cr Al alloy: Plate; AMS for AA alloy 7075; solution treated; cold worked & aged
AMS 4044 C	1.6% Cu 2% Mg 5.6% Zn 0.25% Cr Al alloy: Plate; AMS for AA alloy 7075; annealed
AMS 4045 C	1.6% Cu 2.5% Mg 5.6% Zn 0.25% Cr Al alloy: Sheet; AMS for AA alloy 7075; solution treated & aged
AMS 4122 C	1.6% Cu 2.5% Mg 5.6% Zn 0.3% Cr Al alloy: Bar; AMS for AA alloy 7075; solution treated & aged
AMS 4123 A	1.5% Cu 2.5% Mg 5.6% Zn 0.3% Cr Al alloy: Bar; AMS for AA alloy 7075; cold drawn
AMS 4136	0.6% Cu 3.3% Mg 4.3% Zn 0.2% Cr Al alloy: Forging; AMS for AA alloy 7079; solution treated & aged
AMS 4137 A	0.7% Cu 1.6% Mg 0.5% Mn 7.3% Zn Al alloy: Forging; AMS for AA alloy 7076; solution treated & aged
AMS 4138	0.6% Cu 3.3% Mg 0.2% Mn 4.3% Zn 0.2% Cr Al alloy: Forging; AMS for AA alloy 7079; solution treated & aged
AMS 4139 F	1.6% Cu 2.5% Mg 5.6% Zn 0.25% Cr Al alloy: Forging; AMS for AA alloy 7075; solution treated & aged
AMS 4154 F	1.6% Cu 2.5% Mg 5.6% Zn 0.3% Cr Al alloy: Extrusion; AMS for AA alloy 7075; solution treated & aged
AMS 4158 A	2% Cu 2.7% Mg 6.8% Zn 0.3% Cr Al alloy: Extrusion; AMS for AA alloy 7178; solution treated & aged
AMS 4168 A	1.6% Cu 2.5% Mg 5.6% Zn 0.3% Cr Al alloy: Extrusion; AMS for AA alloy 7075; solution treated & aged
AMS 4169 B	1.6% Cu 2.5% Mg 5.6% Zn 0.3% Cr Al alloy: Extrusion; AMS for AA alloy 7075; solution treated & aged
AMS 4170	1.6% Cu 2.5% Mg 5.6% Zn 0.25% Cr Al alloy: Extrusion; AMS for AA alloy 7075; solution treated & aged
AMS 4171 A	0.6% Cu 3.3% Mg 4.3% Zn 0.2% Mn 0.2% Cr Al alloy: Extrusion; AMS for AA alloy 7079; solution treated & aged
ASTM B209/7039	0.3% Mn 2.8% Mg 0.2% Cr 4% Zn Al alloy: Plate & sheet
ASTM B209/7075	1.7% Cu 2.5% Mg 5.5% Zn 0.3% Cr Al alloy: Sheet; annealed UTS: 270 Elon: 10% Proof: 140
ASTM B209/7075	1.7% Cu 2.5% Mg 5.5% Zn 0.3% Cr Al alloy: Sheet; solution treated & aged UTS: 530 Elon: 6% Proof: 470
ASTM B210/7075	1.6% Cu 2.4% Mg 0.3% Cr 5.6% Zn Al alloy: Tube
ASTM B211/7075	1.5% Cu 2.5% Mg 5.5% Zn 0.3% Cr Al alloy: Bar; solution treated & aged UTS: 530
ASTM B221/7075	1.5% Cu 2.5% Mg 5.5% Zn 0.3% Cr Al alloy: Extruded bar; solution treated & aged UTS: 590 Elon: 7% Proof: 510

Symbol	Nominal analysis, supplier, condition and remarks.
ASTM B221/7079	0.6% Cu 0.2% Mn 3% Mg 4% Zn 0.2% Cr Al alloy: Extruded bar; solution treated & aged UTS: 540 Elon: 7% Proof: 480
ASTM B221/7178	2% Cu 2.8% Mg 7% Zn 0.3% Cr Al alloy: Extruded bar; solution treated & aged UTS: 600 Elon: 5% Proof: 540
ASTM B235/7075	1.7% Cu 2.5% Mg 5.5% Zn 0.3% Cr Al alloy: Extruded tube; solution treated & aged UTS: 590 Elon: 7% Proof: 510
ASTM B241/7075	1.6% Cu 2.5% Mg 0.3% Cr 5.6% Zn Al alloy: Pipe & tube
ASTM B241/7079	0.6% Cu 0.2% Mn 3.3% Mg 0.15% Cr 4.3% Zn Al alloy: Pipe & tube
ASTM B241/7178	2.0% Cu 2.8% Mg 0.3% Cr 6.8% Zn Al alloy: Pipe & tube
ASTM B247/7075	1.6% Cu 2.5% Mg 5.5% Zn 0.2% Cr Al alloy: Forging; solution treated & aged DPN: 135 UTS: 540 Elon: 10% Proof: 460
ASTM B285 R ZG61A	0.6% Mg 0.5% Cr 5.7% Zn 0.2% Ti Al alloy: Welding wire
ASTM B316/7075	1.7% Cu 2.5% Mg 5.5% Zn 0.2% Cr Al alloy: Rivets; annealed UTS: 270
ASTM B327 ZG71A	0.85% Mg 0.7% Zn Al alloy: Hardener for Zn base die casting
AWCO 701	1% Cu 3% Mg 6% Zn Al alloy: Wire; Aluminium Wire Co; solution treated & aged UTS: 490 Elon: 7% Proof: 420
A Z3G2	2.2% Mg 3.5% Zn 0.2% Cr Al alloy: Wrought; French Standard
A Z4G	2.0% Mg 0.5% Mn 3.0% Zn 0.3% Cr Al alloy: Wrought; L'Aluminium Francais
A Z5G	1.2% Mg 4.5% Zn 0.2% Cr Al alloy: Wrought; French Standard
A Z5GU	1.0% Cu 2.8% Mg 5.7% Zn Al alloy: Wrought; L'Aluminium Francais
A Z6G	2.0% Mg 0.5% Mn 6.0% Zn 0.3% Cr Al alloy: Wrought; L'Aluminium Francais
A Z8GU	2.0% Cu 2.0% Mg 7.2% Zn 0.3% Cr Al alloy: Wrought; L'Aluminium Francais
BA 703	0.3% Cu 2.5% Mg 0.2% Mn 5.7% Zn 0.1% Cr Al alloy: British Al Co; solution treated & aged UTS: 490 Elon: 6% Proof: 420
BA 704	1.5% Cu 2.2% Mg 0.2% Mn 7% Zn 0.1% Cr Al alloy: British Al Co; solution treated & aged UTS: 540 Elon: 5% Proof: 480
BA 705	0.5% Cu 2.5% Mg 0.5% Mn 5.7% Zn Al alloy: British Al Co; solution treated & aged UTS: 510 Elon: 6% Proof: 440
BA 706	1% Cu 2.79% Mg 0.5% Mn 5.7% Zn Al alloy: British Al Co; solution treated & aged UTS: 540 Elon: 5% Proof: 460
BA 707	1.3% Cu 2.9% Mg 0.4% Mn 5.7% Zn 0.1% Cr Al alloy: British Al Co; solution treated & aged UTS: 480 Elon: 6% Proof: 400
BBZ 36	1% Cu 2.7% Mg 5% Zn Al alloy: Bar; Birmabright; solution treated & aged UTS: 510 Elon: 5% Proof: 460
BIRMETAL 212	1.6% Cu 2.5% Mg 0.2% Zn 0.15% Cr Al alloy: In wrought form. Birmetal for DTD 5074
BIRMETAL Z36	3% Cu 5% Zn Mg Al alloy: Birmabright DPN: 170 UTS: 600 Proof: 520
BMB 1306	0.4% Cu 2.7% Mg 0.5% Mn 5% Zn Al alloy: Birmabright; solution treated & aged UTS: 480 Elon: 5% Proof: 400
BMB 2308	1% Cu 2.7% Mg 5% Zn Al alloy: Bar; Birmabright; solution treated & aged UTS: 530 Elon: 5% Proof: 460
BS 918	Cu Zn Al alloy: Bars; obsolete

Symbol	Nominal analysis, supplier, condition and remarks.
DTD 363 A	3% Cu 5% Zn Mg Al alloy: Extrusions **UTS: 510 Elon: 5% Proof: 450**
DTD 683 A	Cu Mg Zn Mn Al alloy: For forgings; replaced by DTD 5024
DTD 693	0.4% Cu 2.7% Mg 0.5% Mn 5.3% Zn Al alloy: Tube; solution treated & aged **UTS: 450 Elon: 8% Proof: 400**
DTD 5014	5.5% Zn 2.5% Cu 0.5% Mn Al alloy: Forging; similar to DTD 5024 with improved stress corrosion resistance
DTD 5024	0.5% Cu 2.5% Mg 0.5% Mn 5.7% Zn Al alloy: Forgings; solution treated & aged **UTS: 470 Elon: 7% Proof: 400**
DTD 5034	0.5% Cu 2.7% Mg 5.5% Zn 0.2% Cr Al alloy: Forgings; solution treated & aged **UTS: 470 Elon: 7% Proof: 400**
DTD 5044	0.5% Cu 2.7% Mg 0.5% Mn 5.5% Zn Al alloy: Bar; solution treated & aged **UTS: 480 Elon: 7% Proof: 450**
DTD 5050 B	0.9% Cu 2.7% Mg 5.7% Zn 0.16% Cr Al alloy: Plate; solution treated & aged **UTS: 480 Elon: 7% Proof: 450**
DTD 5054	0.5% Cu 2.7% Mg 5.7% Zn 0.16% Cr Al alloy: Bar; solution treated & aged **UTS: 520 Elon: 7% Proof: 450**
DTD 5064	1.1% Cu 2.8% Mg 0.5% Mn 5.5% Zn Al alloy: Bar; solution treated & aged **UTS: 540 Elon: 5% Proof: 450**
DTD 5074	1.6% Cu 2.5% Mg 6.2% Zn 0.16% Cr Al alloy: Bar; solution treated & aged **UTS: 540 Elon: 5% Proof: 470**
DTD 5094	0.5% Cu 2.5% Mg 0.5% Mn 5.5% Zn Al alloy: Forgings; solution treated & aged **UTS: 460 Elon: 8% Proof: 400**
DURALUMIN KC	1.8% Cu 2.5% Mg 6% Zn 0.05% Pb Al alloy: Bar and forging; Alcan; solution treated & aged **UTS: 530 Elon: 5% Proof: 450**
DURALUMIN LC	0.5% Cu 2.8% Mg 5.5% Zn 0.05% Pb 0.1% Cr Al alloy: Forgings; Alcan; solution treated & aged **UTS: 520 Elon: 7% Proof: 440**
DURAL K	0.4% Cu 2.7% Mg 0.5% Mn 5% Zn Al alloy: Forgings & tube; Alcan; solution treated & aged **UTS: 530 Elon: 5% Proof: 450**
DURAL L	0.6% Cu 3% Mg 0.4% Mn 5.8% Zn Al alloy: Tube, forging & plate; Alcan; solution treated & aged **UTS: 480 Elon: 5% Proof: 400**
DURICILIUM N	Zn containing Al alloy: For welded structure; E Kaye; solution treated **UTS: 330 Elon: 14% Proof: 220**
DURCILIUM P	1.6% Cu 2.5% Mg 6.2% Zn 0.2% Cr Al alloy: E Kaye; solution treated & aged **UTS: 600 Elon: 6% Proof: 530**
DURCILIUM X	1% Cu 2.8% Mg 0.5% Mn 5.5% Zn Al alloy: E Kaye; solution treated & aged **UTS: 600 Elon: 6% Proof: 530**
DURCILIUM Z	0.5% Cu 2.7% Mg 5.7% Zn 0.16% Cr Al alloy: E Kaye; solution treated & aged **UTS: 540 Elon: 8% Proof: 480**

Symbol	Nominal analysis, supplier, condition and remarks.
HA 4 ZG62	1.6% Cu 2.5% Mg 5.6% Zn 0.2% Cr Al alloy: Plate; Canadian Standard
HA4 ZG62 (ALCLAD)	1.6% Cu 2.5% Mg 5.6% Zn 0.2% Cr Al alloy: Plate clad with 1.0% Zn Al; Canadian Standard
HA 5 ZG62	1.6% Cu 2.5% Mg 5.6% Zn 0.2% Cr Al alloy: Bar, etc.; Canadian Standard
HA 7 ZG62	1.6% Cu 2.5% Mg 5.5% Zn 0.2% Cr Al alloy: Tube; Canadian Standard
HIDUMINIUM 48	Zn Mg Al alloy: Wrought; High Duty alloy **UTS: 495 Elon: 12% Proof: 450**
HIDUMINIUM 78	0.5% Cu 2.8% Mg 0.1% Mn 0.15% Cr 6% Zn Al alloy: Bar & plate; High Duty alloys; solution treated & aged **UTS: 400 Elon: 6% Proof: 440**
HIDUMINIUM 89	1.4% Cu 2.5% Mg 0.1% Mn 0.2% Cr 5.8% Zn Al alloy: Bar; High Duty alloys; solution treated & aged **UTS: 550 Elon: 5% Proof: 460**
IMPALCO 740	0.4% Cu 3% Mg 5.7% Zn Al alloy: Bar & forgings; Imperial Al Co; solution treated & aged **UTS: 500 Elon: 7% Proof: 440**
IMPALCO 750	0.5% Cu 2.8% Mg 5.8% Zn 0.2% Cr Al alloy: Bar & forgings; Imperial Al Co; solution treated & aged **UTS: 520 Elon: 7% Proof: 440**
IMPALCO 760	1% Cu 2.8% Mg 5.5% Zn Al alloy: Bar; Imperial Al Co; solution treated & aged **UTS: 540 Elon: 5% Proof: 460**
IMPALCO 770	1.7% Cu 2.5% Mg 6% Zn Al alloy: Bar; Imperial Al Co; solution treated& aged **UTS: 540 Elon: 5% Proof: 460**
KYNAL Z93	0.4% Cu 2.7% Mg 5.3% Zn Al alloy: ICI; obsolete; now Imperial Al Co
NORAL C77S	1% Cu 2.7% Mg 5% Zn Al alloy: Bar; Northern Al Co; solution treated & aged **UTS: 560 Elon: 5% Proof: 460**
NORAL M75S	0.4% Cu 2.7% Mg 0.5% Mn 5% Zn Al alloy: Forgings; Northern Al Co; solution treated & aged **UTS: 490 Elon: 7% Proof: 420**
PERUNAL	1.5% Cu 2.3% Mg 0.2% Mn 6.2% Zn 0.2% Cr 0.1% Ti Al alloy: Anglo-Swiss; solution treated & aged **DPN: 180 UTS: 600 Proof: 530**
PERUNAL 215	1.5% Cu 2.2% Mg 0.2% Mn 5.7% Zn 0.1% Ti 0.2% Cr Al alloy: Wrought; Anglo-Swiss Al Co
Pu	1.5% Cu 2.3% Mg 0.2% Mn 6.2% Zn 0.2% Cr 0.1% Ti Al alloy: Anglo-Swiss; code for Perunal
RR 77	0.4% Cu 2.7% Mg 0.5% Mn 5% Zn Al alloy: For forgings & tubes; High Duty alloys for DTD 683, DTD 693
RR 88	1% Cu 2.7% Mg 0.5% Mn 5% Zn Al alloy: For extrusions; High Duty alloys for DTD 363A
SAE 215	1.8% Cu 2.5% Mg 5.5% Zn 0.2% Cr Al alloy: Solution treated & aged; now AA 7075 **UTS: 570 Elon: 7% Proof: 510**
STA 7 AW16A	1.0% Cu 2.7% Mg 0.5% Mn 5.3% Zn Al alloy: Bar; solution treated & aged; obsolete **UTS: 540 Elon: 5% Proof: 460**
TI 77	0.4% Cu 2.7% Mg 0.5% Mn 5% Zn Al alloy: Tube; Ti Al Co; solution treated & aged **UTS: 450 Elon: 8% Proof: 400**
TI 88	1% Cu 2.7% Mg 5% Zn Al alloy: Bar; TI Al Co; solution treated & aged **UTS: 560 Elon: 5% Proof: 460**
UNI 3735	1.6% Cu 2.5% Mg 5.8% Zn Al alloy: For forging; Italian Standard
UNI 3737	1.6% Cu 2.5% Mg 7.8% Zn Al alloy: For forging; Italian Standard

Note. The following abbreviations and units are used in the tables:

DPN	Hardness, diamond pyramid number
UTS	Ultimate tensile strength, N/mm^2
Elon	Elongation, %
Proof	0.1 % proof strength, N/mm^2

1 N/mm^2=0.1 hbar=0.102 kgf/mm^2=0.06475 tonf/in^2=145.04 lbf/in^2

Symbol	Nominal analysis, supplier, condition and remarks.
UNI 3738	1.6% Cu 2.5% Mg 7.8% Zn Al alloy: For forging; Italian Standard
UNIDOR 080	0.8% Mg 4.3% Zn Al alloy: Wrought; Anglo-Swiss Al Co
UNIDOR 100	1.2% Mg 0.4% Mn 0.2% Cr 4.2% Zn Al alloy: Wrought; Anglo-Swiss Al Co
URAL 1	4% Cu 1% Mg 1.5% Zn Al alloy: German Standard
URAL III	4.5% Cu 2% Mg 3% Zn Al alloy: German Standard
URAL IV	5% Cu 2.5% Mg 5.5% Zn Al alloy: German Standard
W 16	1.8% Cu 2.5% Mg 6% Zn 0.2% Cr Al alloy: Southern Forge; replaced by IMPALCO 770
W 26	1% Cu 2.8% Mg 0.5% Mn 5.5% Zn Al alloy: Southern Forge; replaced by IMPALCO 760

Symbol	Nominal analysis, supplier, condition and remarks.
W 160	0.5% Cu 2.8% Mg 5.8% Zn 0.2% Cr Al alloy: Southern Forge; replaced by IMPALCO 750
W 260	0.5% Cu 2.8% Mg 5.8% Zn Al alloy: Southern Forge; replaced by IMPALCO 740
ZG 62A	1.5% Cu 2.5% Mg 5.5% Zn 0.3% Cr Al alloy: Previous ASTM designation; now 7075

Note. The following abbreviations and units are used in the tables:

DPN	Hardness, diamond pyramid number
UTS	Ultimate tensile strength, N/mm^2
Elon	Elongation, %
Proof	0.1 % proof strength, N/mm^2

$1 \ N/mm^2 = 0.1 \ hbar = 0.102 \ kgf/mm^2 = 0.06475 \ tonf/in^2 = 145.04 \ lbf/in^2$

1H Aluminium alloys clad with aluminium

Physical properties

These will be approximately the same as the unclad alloys, always remembering that one, or both sides of the material has a thin coating of pure, or low alloyed aluminium metal. This will influence the properties to a greater or lesser extent depending on the extent of the difference between the alloy and the coating, and the thickness of the coating. For example, pure aluminium has an electrical conductivity of approximately 60% IACS whereas the conductivity of certain of the zinc bearing materials is as low as 30% IACS.

General metallurgical properties

These are in the main the same as the unclad alloy, apart from the considerably enhanced corrosion resistance and slightly reduced physical properties. This is caused by the thin layer of pure, or low alloyed aluminium, which has excellent corrosion resistance but low mechanical strength. The bond of this thin layer to the base metal is excellent and will withstand considerably deformation. It must, however, be noted that if the coating is damaged or broken the corrosion which would ensue could be dangerous, being local and thus concentrated. It is also difficult, if not impossible, to identify visually local damage to the cladding. For these reasons considerable care should be exercised in working, and fabricating the alloys.

The only forms found at present are sheet and plate, clad one or both sides, bars or simple sections and some tubing. Also included in this group are some specifications which have been clad with braze metal for mass production brazing of components.

Thermal treatment

This will be identical to the base metal, where applicable.

Welding and brazing

Pre-treatment. No thermal treatment required. Thorough cleaning, and local scratch brushing is essential. This will remove the clad layer and thus must be confined to as small an area as possible.

Welding and brazing. As for the base metal.

Post-treatment. The weld or braze area must be thoroughly cleaned to remove all traces of flux, if used. This area will have had the cladding removed, and will require local corrosion protection.

Flaw detection methods

Crack test. Oil penetrant chalk test can be used, but this will show surface defects produced in the base metal prior to cladding, which will be filled with cladding. As this is of low strength these defects can be potentially dangerous. It is possible to specify that the base metal is crack tested prior to cladding, and this is recommended for high integrity components.

X-ray. Not normally required.

Ultra-sonic test. Not normally required. The coating join can interfere with the ultra-sonic beam giving spurious indications of defects.

Chemical etch. This must not be carried out as it will remove or damage the coating.

Corrosion protection

Temporary. None required.

Permanent. None required except to preserve appearance. These materials are not suitable for anodizing, but can be painted, using an etch primer in preference to chemical oxidation, as every effort should be made to protect the cladding metal.

Where welding or brazing has been carried out, that area must be thoroughly protected, and in this case local chemical oxidation is desirable. For complete protection the area should be metal sprayed after local blasting, using pure aluminium.

Painting. This is seldom required apart from the above. As stated it is preferable to use an etch primer rather than chemical oxidation when painting is necessary. This will be for decorative rather than corrosion protection purposes. Lead based primers must not be used.

Plating. Electroplating requires chemical etching and will remove the clad layer, thus the process will not normally be used.

Machinability

These materials are not often machined, as this will remove the clad surface. Their machining characteristics will be identical to that of the base metal.

Uses

For structural purposes where strength and first class corrosion resistance is required. Architectural purposes. Stressed members on transport, protected from mechanical damage.

Symbol	Nominal analysis, supplier, condition and remarks
AA 11	1.2% Mn Al sheet clad one side with 7.5% Al alloy: For brazing; SAE designation; formerly SAE 247
AA 12	1.2% Mn Al sheet clad both sides with 7.5% Si Al alloy: For brazing; SAE designation; formerly SAE 248
AA 21	0.4% Si 0.6% Mg Al alloy sheet clad one side with 7.5% Si Al alloy: SAE designation; formerly SAE 249 for brazing
AA 22	0.4% Si 0.6% Mg Al alloy clad both sides with 7.5% Si Al alloy: For brazing; SAE designation; formerly SAE 250
AA 23	0.2% Cu 0.3% Si 0.6% Mg Al alloy sheet clad one side with 9% Si Al alloy: For brazing; SAE designation
AA 24	0.2% Cu 0.3% Si 0.6% Mg Al alloy sheet clad both sides with 9% Si Al alloy: For brazing; SAE designation
ALCAN GB24S ALCLAD	4.5% Cu 1.5% Mg 0.7% Mn Al alloy clad with Al: Alcan; solution treated & room aged **UTS: 420 Elon: 18% Proof: 280**
ALCAN GB26S ALCLAD	4.2% Cu 0.7% Mg 0.7% Si 0.7% Mn Al alloy clad with Al: Alcan; solution treated & aged **UTS: 440 Elon: 10% Proof: 360**
ALCAN GBM75S ALCLAD	1.3% Cu 2.5% Mg 0.12% Cr 6% Zn Al alloy clad with Al: Alcan; solution treated & aged **UTS: 510 Elon: 11% Proof: 420**
ALCOA 460	5.0% Si Al alloy: Clad on Alcoa 190 sheet; for brazing
ALCOA 470	7.5% Si Al alloy: Clad on Alcoa 190 sheet; for brazing
ALDURAL B	4.2% Cu 0.7% Mg 0.6% Mn Al alloy clad with Al: J Booth; clad version of DURAL B
ALDURAL G	4.2% Cu 1.5% Mg 0.2% Ci 0.6% Mn Al alloy clad with Al: Sheet; J Booth; clad version of DURAL G
ALDURAL JJ	2.3% Cu 1.5% Mg 0.15% Si 1% Fe 1.1% Ni Al alloy clad with Al: Sheet; J Booth; clad version of DURAL JJ
ALDURAL K	1.1% Cu 3% Mg 5.7% Zn Al alloy clad with 1% Zn Al alloy: J Booth; clad version of DURAL K clad with 1% Zn Al

Note. The following abbreviations and units are used in the tables:

DPN	Hardness, diamond pyramid number
UTS	Ultimate tensile strength, N/mm^2
Elon	Elongation, %
Proof	0.1 % proof strength, N/mm^2

1 N/mm^2=0.1 hbar=0.102 kgf/mm^2=0.06475 tonf/in^2=145.04 lbf/in^2

Symbol	Nominal analysis, supplier, condition and remarks
ALDURAL L	0.6% Cu 3% Mg 5.8% Zn Al alloy sheet clad with 1% Zn Al alloy: J Booth; clad version of DURAL L clad with 1% Zn Al alloy
ALDURAL S	4.2% Cu 0.7% Mg 0.7% Si 0.7% Mn Al alloy clad with Al: Sheet; J Booth; clad version of DURAL S
AMS 4020	0.25% Cu 1% Mg 0.6% Si 0.25% Cr Al alloy clad with Al/Zn: AMS for alloy 6061 clad with 7072
AMS 4021 B	0.25% Cu 1% Mg 0.6% Si 0.25% Cr Al alloy clad with Al/Zn: AMS for alloy 6061 clad with 7072 annealed
AMS 4022 C	0.25% Cu 1% Mg 0.6% Si 0.25% Cr Al alloy clad with Al/Zn: AMS for alloy 6061 clad with 7072 solution treated
AMS 4023 C	0.25% Cu 1% Mg 0.6% Si 0.25% Cr Al alloy clad with Al/Zn: AMS for alloy 6061 clad with 7072 solution treated & aged
AMS 4034 A	4.5% Cu 1.5% Mg 0.6% Mn Al alloy clad with Al: AMS for AA alloy 2024 clad with 1230
AMS 4036	4.5% Cu 1.5% Mg 0.6% Mn Al alloy clad with Al one side: AMS for AA alloy 2024 clad with Al
AMS 4039	1.6% Cu 2.5% Mg 5.6% Zn 0.3% Cr Al alloy clad with Al/Zn: AMS for AA alloy 7075 clad with Al/Zn
AMS 4040 E	4.5% Cu 1.5% Mg 0.6% Mn Al alloy clad with Al: AMS for AA alloy 2024 clad with 1230; annealed
AMS 4041 G	4.5% Cu 1.5% Mg 0.6% Mn Al alloy clad with Al: AMS for AA alloy 2024 clad with 1230
AMS 4042 E	4.5% Cu 1.5% Mg 0.6% Mn Al alloy clad with Al: AMS for AA alloy 2024 clad with Al
AMS 4046	1.6% Cu 2.5% Mg 5.6% Zn 0.25% Cr Al alloy clad with Al one side: AMS for AA 7075 clad with Al/Zn
AMS 4047 B	1.6% Cu 2.5% Mg 5.6% Zn 0.25% Cr Al alloy Al clad Al/Zn roll tapered: AMS for AA alloy 7075 clad with Al/Zn
AMS 4048 C	1.6% Cu 2.5% Mg 5.6% Zn 0.25% Cr Al alloy clad with Al, annealed with Al/Zn: AMS for AA alloy 7075 clad with Al/Zn
AMS 4049 C	1.6% Cu 2.5% Mg 5.6% Zn 0.25% Cr Al alloy clad with Al/Zn: AMS for AA alloy 7075 clad with Al/Zn
AMS 4051 A	2% Cu 2.7% Mg 6.8% Zn 0.3% Cr Al alloy clad with Al/Zn: Annealed **UTS: 240 Elon: 10%**
AMS 4052 A	2% Cu 2.7% Mg 6.8% Zn 0.3% Cr Al alloy clad with Al/Zn: Solution treated & aged **UTS: 570 Elon: 6% Proof: 490**
AMS 4054	0.3% Cu 0.6% Mg 0.35% Si Al alloy clad with braze metal: Annealed & clad one side with 7% Si Al alloy braze metal
AMS 4055	0.3% Cu 0.6% Mg 0.3% Si Al alloy clad with braze metal: Annealed; clad both sides with 7% Si Al alloy braze metal

Symbol	Nominal analysis, supplier, condition and remarks.
ASTM B209/2014 CLAD	4.5% Cu 1% Si 0.5% Mg Al alloy sheet clad with Al: Annealed and clad with ASTM 6003 **UTS: 210 Elon: 16% Proof: 75**
ASTM B209/2014 CLAD	4.5% Cu 1% Si 0.5% Mg Al alloy sheet clad with Al: Clad with ASTM 6003; solution treated & aged **UTS: 450 Elon: 4% Proof: 400**
ASTM B209/2024 CLAD	4% Cu 1.5% Mg Al alloy sheet clad with Al: Annealed, clad with ASTM 1230 **UTS: 210 Elon: 10% Proof: 75**
ASTM B209/2024 CLAD	4% Cu 1.5% Mg Al alloy sheet clad with Al: Solution treated; room aged; clad with ASTM 1230 **UTS: 420 Elon: 6% Proof: 270**
ASTM B209/3003 CLAD	1.2% Mn Al alloy sheet clad with Al/Zn: Annealed, clad with ASTM 7072 **UTS: 110 Elon: 23% Proof: 75**
ASTM B209/3003 CLAD	1.2% Mn Al alloy sheet clad with Al/Zn: Cold rolled; ½-hard **UTS: 150 Elon: 8% Proof: 90**
ASTM B209/3004 CLAD	3004 alloy sheet clad with 7072 alloy
ASTM B209/5155 CLAD	5155 alloy clad with alloy 7072
ASTM B209/6061 CLAD	0.2% Cu 0.6% Si 1% Mg 0.2% Cr Al alloy sheet clad with Al/Zn: Annealed; clad with ASTM 7072 **UTS: 120 Elon: 18% Proof: 60**
ASTM B209/6061 CLAD	0.2% Cu 0.6% Si 1% Mg 0.2% Cr Al alloy sheet clad with Al/Zn: Solution treated & aged; clad with ASTM 7072 **UTS: 280 Elon: 8% Proof: 240**
ASTM B209/7075	1.7% Cu 2.5% Mg 5.5% Zn 0.3% Cr Al alloy sheet clad with Al/Zn
ASTM B210/3003 CLAD	1.2% Mn Al alloy tube clad with Al/Zn: Annealed; clad with ASTM 7072 **UTS: 75 Proof: 15**
ASTM B210/3003 CLAD	1.2% Mn Al alloy tube clad with Al/Zn: Cold drawn; clad with ASTM 7072 **UTS: 170 Proof: 150**
ASTM B211/5056 CLAD	5% Mg 0.1% Cr Al alloy bar clad with Al: Cold drawn; clad with ASTM 6253 **UTS: 340**
ASTM B234/3003 ALCLAD	1.2% Mn Al tube coated with Al/Zn: Cold drawn; clad with 7072 **UTS: 140 Proof: 110**
ASTM B235/3003 ALCLAD	1.2% Mn Al alloy tube clad with Al/Zn: Annealed; clad with 7072 and extruded **UTS: 100 Proof: 22**
ASTM B241/3003 ALCLAD	1.3% Mn Al pipe and tube clad with ASTM B241/7072
ASTM B307/3003 ALCLAD	1.2% Mn Al alloy coiled tube, clad with Al/Zn: Annealed; clad with 7072 **UTS: 90**
ASTM B307/5050 ALCLAD	1.4% Mg Al alloy clad with 1.2% Mg 1.6% Zn Al alloy: Drawn tube
ASTM B313/3004 ALCLAD	1.2% Mn 1% Mg Al alloy welded tube clad with Al: Cold rolled; clad with 7072 **UTS: 250 Elon: 4%**
ASTM B345/3003 ALCLAD	1.2% Mn Al alloy pipe, clad with Al/Zn: As drawn; clad with 7072 **UTS: 750 Proof: 30**
ASTM B345/6061 ALCLAD	0.2% Cu 0.6% Si 1% Mg 0.2% Cr Al alloy clad with Al/Zn: Pipe; solution treated & aged; clad with 7072 **UTS: 220 Proof: 200**
ASTM B404/3003 ALCLAD	1.25% Mn Al alloy clad with 1.0% Zn Al alloy: Condenser tubes
BA 351	4.1% Cu 0.8% Mg 0.5% Si 0.7% Mn Al alloy clad with Al: Plate; British Al Co for BS alloy HC14
BA 355	4.6% Cu 0.7% Mg 0.8% Si 0.7% Mn Al alloy clad with Al: British Al Co for BS alloy HC15

Symbol	Nominal analysis, supplier, condition and remarks.
BA 359	4.2% Cu 1.6% Mg 0.6% Mn 0.05% Pb Al alloy clad with Al: Plate; British Al Co; solution treated & aged **UTS: 400 Elon: 10% Proof: 240**
BA 751	1.3% Cu 2.9% Mg 0.4% Mn 5.7% Zn 0.1% Cr Al alloy: Clad; British Al Co; solution treated & aged; clad with 1% Zn alloy **UTS: 460 Elon: 8% Proof: 390**
BA 757	1.3% Cu 2.9% Mg 0.4% Mn 5.7% Zn 0.1% Cr Al alloy: Clad; British Al Co; solution treated & aged; clad with 1% Zn alloy **UTS: 460 Elon: 6% Proof: 390**
BIRMETAL 477 CLAD	4% Cu 1% Mg 0.5% Si 0.7% Mn Al alloy clad with Al: Birmabright for BS alloy HC 14
BIRMETAL 478 CLAD	4% Cu 0.5% Mg 0.5% Si 1% Mn Al alloy clad with Al: Birmabright for BS alloy HC15
BIRMETAL Z36 CLAD	3% Cu 5% Zn Mg Al alloy clad with Al: Birmabright **UTS: 480 Elon: 11% Proof: 420**
BS 1470 HC14	4% Cu 1% Mg 0.75% Mn Al alloy: Al coated sheet & strip; solution treated; room aged **UTS: 360 Elon: 15% Proof: 210**
BS 1470 HC15	4% Cu 0.7% Mg 0.6% Si 0.8% Mn Al alloy: Sheet and strip coated with Al; solution treated & aged **UTS: 390 Elon: 8% Proof: 300**
BS 1477 HPC14	4% Cu 1% Mg 0.5% Si 0.7% Mn Al alloy plate clad with Al: Solution treated; room aged; coated with Al **UTS: 360 Elon: 10% Proof: 210**
BS 1477 HPC15	4% Cu 0.5% Mg 0.7% Si 1% Mn Al alloy: Al clad plate; solution treated & aged; coated with Al **UTS: 390 Elon: 8% Proof: 300**
BS L38	Cu Al alloy sheet clad with Al: Obsolete
BS L47	Cu Al alloy sheet clad with Al: Obsolete; replaced by BS L73
BS L72	4% Cu 0.75% Mg 0.75% Si 0.8% Mn Al alloy sheet coated with Al: Solution treated; room aged **UTS: 370 Elon: 15% Proof: 220**
BS L73	4% Cu 0.75% Mg 0.75% Si 0.8% Mn Al alloy sheet coated with Al: Solution treated & aged **UTS: 400 Elon: 8% Proof: 310**
CLAD CG 42A	4.2% Cu 1.5% Mg Al alloy clad with Al: Previous ASTM designation; now 2024; clad with 1230
CLAD CS 41A	4.5% Cu 1% Si 0.6% Mg Al alloy clad with Si/Mg Al: Previous ASTM designation; now 2014; clad with 6003
CLAD GM 50A	5% Mg 0.1% Cr Al alloy clad with Al/Zn alloy: Previous ASTM designation; now 5056; clad with 6253
CLAD GS 11A	0.2% Cu 1% Mg 0.6% Si 0.2% Cr Al alloy clad with Al/Zn: Previous ASTM designation; now 6061; clad with 7072
CLAD M1A	1.2% Mn Al alloy clad with Al/Zn alloy: Previous ASTM designation; now 3003; clad with 7072
CLAD MG 11A	1.2% Mn 1% Mg Al alloy clad with Al/Zn alloy: Previous ASTM designation; now 3004; clad with 7072
CLAD ZG 62A	1.7% Cu 2.5% Mg 5.5% Zn 0.3% Cr Al alloy clad with Al/Zn alloy: Previous ASTM designation; now 7075; clad with 7072
DTD 390	Al clad sheet: Replaced by DTD 610
DTD 610	4% Cu 1.5% Si 1% Fe Al alloy clad with Al
DTD 687B	1% Cu 3% Mg 5% Zn 0.2% Cr Al alloy sheet coated with Al: Solution treated & aged; coated with 1% Zn Al alloy **UTS: 450 Elon: 8% Proof: 390**

Symbol	Nominal analysis, supplier, condition and remarks.
DTD 710B	4% Cu 0.7% Mg 0.7% Si 1% Mn Al sheet coated with Al: Solution treated; room aged; coated with Al **UTS: 370 Elon: 15% Proof: 220**
DTD 5040	4.4% Cu 0.7% Mg 0.7% Si 0.6% Mn Al alloy plate clad with Al: Solution treated & aged; coated with Al **UTS: 420 Elon: 8% Proof: 330**
DTD 5060	1.5% Cu 2.7% Mg 0.8% Mn 5% Zn Al alloy coated with Al: Solution treated & aged; coated with 1% Zn Al alloy **UTS: 480 Elon: 6% Proof: 400**
DTD 5070A	2.5% Cu 1.5% Mg 1.2% Ni 1% Fe Al alloy coated with Al: Sheet; solution treated & aged; coated with 1% Zn Al alloy **UTS: 370 Elon: 6% Proof: 300**
DTD 5100	4.4% Cu 1.5% Mg 0.6% Mn Al alloy clad with Al: Plate; solution treated; room aged; coated with Al **UTS: 420 Elon: 10% Proof: 250**
HA 4 CG42 ALCLAD	4.3% Cu 1.6% Mg 0.6% Mn Al plate clad with 99.7% Al: Canadian Standard
HA 4 CS41N ALCLAD	4.5% Cu 0.6% Mg 0.8% Mn 0.8% Si Al alloy plate clad with 1% Mg Al: Canadian Standard
HIDUMINIUM 04	4.2% Cu 0.8% Mg 0.5% Si 0.8% Mn Al alloy clad with Al: Plate; High Duty Alloys for BS alloy HC14
IMPALCO C66A	4% Cu 0.7% Mg 1% Mn Al alloy clad with Al: Sheet; Imperial Al Co for BS alloy HC15
KYNAL-CORE C65A	4% Cu 0.6% Mg 0.5% Mn Al alloy clad with Al: ICI; obsolete; now Imperial Al Co
KYNAL-CORE C66A	Al clad form of KYNAL C66: ICI; obsolete; now Imperial Al Co
KYNAL-CORE C67A	Al clad version of KYNAL C67: ICI; obsolete; now Imperial Al Co
KYNAL-CORE C68A	Al clad version of KYNAL C67: ICI; obsolete; now Imperial Al Co
KYNAL-CORE Z93A	Al clad form of KYNAL Z93: ICI; obsolete; now Imperial Al Co
SAE 240	4.5% Cu 1.5% Mg 0.6% Mn Al alloy clad with Al: Sheet; solution treated; SAE 24 clad with Al **UTS: 420 Elon: 12% Proof: 250**

Symbol	Nominal analysis, supplier, condition and remarks.
SAE 242	1.2% Mn 1% Mg Al alloy clad with Al/Zn alloy: Cold rolled; AA alloy 3004 clad with 7072 **UTS: 270 Elon: 5% Proof: 240**
SAE 244	0.6% Si 1% Mg Al alloy clad with Al/Zn alloy: AA alloy 6061; clad with 7072
SAE 245	1.2% Mn Al alloy clad with Al/Zn alloy: Cold rolled; AA alloy 3003 clad with 7072 **UTS: 200 Elon: 4% Proof: 170**
SAE 246	4.5% Cu 1% Si 1% Mn 0.6% Mg 0.1% Cr Al alloy clad with Al: AA alloy 2014 clad with 6003
SAE 247	1.2% Mn Al sheet clad one side with 7.5% Si Al: For brazing; alloy 3003 clad with 4343; now AA11
SAE 248	1.2% Mn Al sheet clad both sides with 7.5% Si Al alloy: For brazing; alloy 3003 clad with 4343; now AA12
SAE 249	0.4% Si 0.6% Mg Al alloy clad one side with 7.5% Si Al alloy: Alloy 6951 clad with 4343; now AA21
SAE 250	0.4% Si 0.6% Mg Al alloy sheet clad both sides with 7.5% Si Al alloy: Alloy 6951 clad with 4343; now AA22
UNI 3736	1.6% Cu 2.5% Mg 0.2% Mn 5.8% Zn 0.15% Cr 0.1% Ti Al alloy clad with 1.0% Zn Al: Italian Standard
UNI 3580	4.0% Cu 0.5% Mg 0.5% Mn Al alloy clad with Al: Italian Standard
UNI 3582	4.4% Cu 0.8% Si 0.4% Mg 0.8% Mn Al alloy clad with 1% Si Mg Mn alloy: UNI 3571; Italian Standard

Note. The following abbreviations and units are used in the tables:

DPN	Hardness, diamond pyramid number
UTS	Ultimate tensile strength, N/mm^2
Elon	Elongation, %
Proof	0.1 % proof strength, N/mm^2

$1 \ N/mm^2 = 0.1 \ hbar = 0.102 \ kgf/mm^2 = 0.06475 \ tonf/in^2 = 145.04 \ lbf/in^2$

1J Aluminium–magnesium cast alloys

Specific gravity	2.55–2.65	
Density	2550–2650 kg/m³	(0.92–0.96 lbf/in.³)
Solidus/liquidus	450–642 °C	
Thermal conductivity	86–138 W/m °C	(21–33 cal/m s °C)
Coefficient of linear expansion	24–$25 \times 10^{-6}/$ °C	
Electrical conductivity	20–30% IACS (copper 100%)	
Specific resistance	60–80 microhm mm	
Young's modulus of elasticity	$71.0 \times 10^9 \ N/m^2$	$(10.3 \times 10^6 \ lbf/in.^2)$
Impact	sand cast 8 J	(6 ft lb) Izod
	chill cast 12–15 J	(9–11 ft lb) Izod
Fatigue strength (10^6 cycles)	sand cast ± 60–75 N/mm^2	(± 4–5 $tonf/in.^2$)
	chill cast ± 90–120 N/mm^2	(± 6–8 $tonf/in.^2$)
Hot strength	No detailed figures available. Strength falls off rapidly above 100–125 °C.	

The above properties are typical of the following group, and may not apply exactly to any one specification. It is possible that with certain specifications some of the values may not be applicable.

General metallurgical characteristics

These specifications cover a range of alloys from 4% to 10% magnesium, some of which have up to 2% silicon added.

The strength and corrosion resistance increases with magnesium content, as also does the difficulty in casting. This difficulty is mainly caused by the great affinity of magnesium for oxygen. Special casting techniques and inhibitors for the mould sand are required. Even with these it is not recommended that these alloys are chosen where thin sections or pressure tightness are essential.

As the alloys are also prone to be 'hot short' (that is to have low strength just below freezing point) rapid change in section should be avoided in order to equate, as far as possible, cooling stresses.

Thermal treatment

The low magnesium (up to 8%) alloys are not heat treatable except for stress releasing at 250–300 °C to remove casting or machining stresses.

The 10% magnesium alloys should be solution treated at 420–440 °C for about 8 hours, then quenched in hot water or oil.

The actual specification involved should be consulted for exact details of times and temperatures.

Welding and brazing

Pre-treatment. No thermal treatment required. The weld area must be thoroughly cleaned and scratch brushed.

Welding and brazing. The low magnesium alloys are not suitable for welding, but can be brazed using special fluxes and Al/Si brazing rods.

The high magnesium alloys can be welded using the inert gas shielded arc process as it is essential that oxygen is prevented from contacting the weld pool. Adequate flow of the inert gas must be maintained during welding to ensure that the weld is covered until solid, otherwise oxide trapping will result.

These alloys can also be brazed using Al/Si rod and flux.

Post-treatment. If possible the high magnesium alloys should be re-solution treated after welding or brazing to regain their mechanical properties.

No other thermal treatment is necessary unless with complicated assemblies which may be stressed at welding. These should be stress relieved at 250–300 °C for 2 hours.

Al fluxes must be completely removed as soon as possible after welding, as they are corrosive and must be removed before any stress relieving treatment.

Flaw detection methods

Crack detection. Penetrant oil chalk test.

Pressure test. These alloys should not be used where high pressure tightness is essential. Pressure testing at 344 740 N/m^{-2} (50 lbf/in.$^{-2}$) air in water will find most defects. Higher air pressures should not be used without taking precautions against a casting bursting under pressure.

X-ray. High duty castings should be X-rayed. As X-raying is expensive it is recommended that castings are first chalk tested to find the obvious defects.

Ultra-sonic test. Not normally required.

Chemical etch. A hot 10% caustic soda etch is useful for finding casting defects. This can be used as a percentage quality check in the foundry as it increases the sensitivity of any subsequent chalk test.

Corrosion protection

Temporary. None required.

Permanent. None required under most circumstances. These alloys have the best corrosion resistance of the cast aluminium alloys, this resistance being proportional to the magnesium content. The alloys can be anodized, using the sulphuric, chromic or oxalic acid processes, sulphuric acid giving the best result.

It is also possible to improve the resistance to a certain extent using chemical oxidation processes.

Painting. This is not normally carried out except for decorative purposes. It will always require either etch priming, anodizing or chemical oxidation.

Lead based primers must not be used.

Plating. Electroplating requires special etch techniques which accentuate any casting defects. It is therefore not recommended that these alloys are plated. Should this be necessary, it is essential that the foundry are informed that the castings are to be plated.

Uses

The sole reason for using these difficult casting alloys is for corrosion resistance, coupled with reasonable strength.

Marine castings, decorative and utilitarian. Architectural castings. Builder's hardware, rainwater castings.

Symbol	Nominal analysis, supplier, condition and remarks.
3.3261	1% Si 4.8% Mg 0.1% Ti Al alloy: Casting; German Standard DIN 1725
	DPN: 70 UTS: 220 Elon: 5% Proof: 75
3.3282	0.7% Si 7.5% Mg 0.4% Mn Al alloy: Casting; German Standard DIN 1725
	DPN: 70 UTS: 200 Elon: 2% Proof: 140
3.3292	0.5% Si 9% Mg 0.4% Mn Al alloy: Casting; German Standard DIN 1725
	DPN: 70 UTS: 220 Elon: 2% Proof: 140
3.3591	10% Mg 0.2% Mn 0.1% Ti Al alloy: Casting; German Standard DIN 1725
220	10% Mg Al alloy: Casting; Alcoa for BS alloy LM 10
320	4% Mg Al alloy: Casting; Comalco; sand cast
	DPN: 60 UTS: 150 Elon: 4% Proof: 60
340	8% Mg Al alloy: Casting; Comalco; sand cast
	DPN: 70 UTS: 250 Elon: 7% Proof: 110
350	10% Mg Al alloy: Casting; Comalco for BS alloy LM 10
514.0	4.0% Mg Al alloy: Investment casting; American National Standard; ANSI
520.0	10.0% Mg Al alloy: Investment casting; American National Standard; ANSI
535.0	7.0% Mg 0.2% Mn Al alloy: Investment casting; American National Standard; ANSI
A 05140	Unified number for alloy 514.0; ASTM system
A 10	10% Mg Al master alloy: L'Aluminium Francais
A 05200	Unified number for alloy 520.0; ASTM system
A 05350	Unified number for alloy 535.0; ASTM system
A 15140	Unified number for alloy B514.0; ASTM system
A G3T	3.0% Mg Al alloy: Casting; L'Aluminium Francais
A G4Z	4% Mg 1.2% Zn Al alloy: Casting; L'Aluminium Francais
A G6	6.0% Mg Al alloy: Casting; L'Aluminium Francais
ALCAN GB 350	10% Mg Al alloy: Casting; Alcan for BS alloy LM 10
ALCAN GB B320	4% Mg 0.5% Mn Al alloy: Casting; Alcan for BS alloy LM 5
AMS 4238 A	6.8% Mg 0.2% Mn 0.2% Ti Al alloy: Casting; as cast
AMS 4239	6.8% Mg 0.2% Mn 0.2% Ti Al alloy: Casting; annealed
AMS 4240 C	10% Mg Al alloy: Casting; solution treated
AN QQ A 366	8.0% Mg Al alloy: Die casting; US Service; as cast
	UTS: 210 Elon: 5% Proof: 120
AN QQ A 392	10% Mg Al alloy: Casting; US Service; aged
	UTS: 280 Elon: 12% Proof: 140
AN QQ A 402	4.0% Mg Al alloy: Casting; US Service; as cast
	UTS: 110
ASTM B26 G4A	4% Mg 0.35% Si Al alloy: Sand casting; as cast
	UTS: 150 Elon: 6% Proof: 45
ASTM B26 G10A	10% Mg Al alloy: Sand casting; solution treated
	UTS: 280 Elon: 12% Proof: 140
ASTM B26 GM70B	7% Mg Al alloy: Sand casting
	UTS: 240 Elon: 9% Proof: 110
ASTM B26 GS42A	2% Si 4% Mg Al alloy: Sand casting; as cast
	UTS: 110 Proof: 50
ASTM B85 G8A	8% Mg Al alloy: Die casting; as cast
	UTS: 320 Elon: 5% Proof: 180

Note. The following abbreviations and units are used in the tables:

DPN	Hardness, diamond pyramid number
UTS	Ultimate tensile strength, N/mm^2
Elon	Elongation, %
Proof	0.1 % proof strength, N/mm^2

$1 \text{ N/mm}^2 = 0.1 \text{ hbar} = 0.102 \text{ kgf/mm}^2 = 0.06475 \text{ tonf/in}^2 = 145.04 \text{ lbf/in}^2$

Symbol	Nominal analysis, supplier, condition and remarks.
ASTM B108 GM42A	4% Mg 0.2% Cr Al alloy: Casting; as cast
	UTS: 120 Elon: 1.5%
ASTM B108 GM70B	0.2% Mn 7% Mg 0.2% Ti Al alloy: Casting; as cast
	UTS: 240 Elon: 8% Proof: 110
ASTM B108 GZ42A	4% Mg 2% Zn Al alloy: Casting; as cast
	UTS: 140 Elon: 2.5% Proof: 75
ASTM B179 G4A	4% Mg Al alloy: Ingot
ASTM B179 G8A	8% Mg Al alloy: Ingot
ASTM B179 G10A	10% Mg Al alloy: Ingot
ASTM B179 GM70B	7% Mg 0.2% Ti Al alloy: Ingot
ASTM B179 GS42A	2% Si 4% Mg Al alloy: Ingot
B 514.0	1.8% Si 4.0% Mg Al alloy: Investment casting; American National Standard
BA 28	5% Mg 0.5% Mn Al alloy: Casting; British Al Co for BS alloy LM 5
BA 29	10% Mg Al alloy: Casting; British Al Co for BS alloy LM 10
BB 5	5% Mg 0.5% Mn Al alloy: Casting; Birmabright for BS alloy LM 5
BIRMABRIGHT	5% Mg 0.5% Mn Al alloy: Casting; Birmabright for BS alloy LM 5
BS 1490 LM5	4.5% Mg 0.5% Mn Al alloy: Ingot; chill cast
	UTS: 170 Elon: 5%
BS 1490 LM5M	4.5% Mg 0.5% Mn Al alloy: For sand & die castings; chill cast
	UTS: 170 Elon: 5%
BS 1490 LM10	10% Mg Al alloy: Ingots; chill cast
	UTS: 300 Elon: 12%
BS 1490 LM10W	10% Mg Al alloy: Sand or die castings; chill cast; solution treated
	UTS: 300 Elon: 12%
BS L53	10% Mg Al alloy: Casting; chill cast; solution treated
	UTS: 300 Elon: 12% Proof: 170
DSI 5000 A1252	4.5% Mg 0.5% Mn Al alloy: Casting; Danish specification
DTD 165 A	4.5% Mg 0.5% Mn Al alloy: Castings; chill cast; as cast
	UTS: 170 Elon: 5% Proof: 75
DTD 300 A	10.5% Mg Al alloy: Casting; chill cast; solution treated
	UTS: 300 Elon: 12% Proof: 170
DTD 5018	7.5% Mg 0.2% Mn 1.2% Zn Al alloy: Sand casting
	UTS: 280 Elon: 5% Proof: 170
G Al Mg 3	3% Mg Al alloy: Casting; previous German Standard designation
G Al Mg 3(Cu)	3% Mg 0.4% Cu Al alloy: Casting; previous German Standard designation
G Al Mg 5	5% Mg 1% Si 0.2% Cu Al alloy: Casting; previous German Standard designation
G Al Mg 10	10% Mg 0.2% Cu Al alloy: Casting; previous German Standard designation
G Al Mg(Cu)	0.4% Cu 4% Mg Al alloy: Casting; previous German Standard designation
GD Al Mg 8(Cu)	0.2% Cu 8% Mg 1% Si 0.4% Mn Al alloy: Casting; previous German Standard designation
GD Al Mg 9	9% Mg Al alloy: Casting; previous German Standard designation
GOST 2685-63 AL13	4.5% Mg 0.5% Mn Al alloy: Casting; Russian specification
H 49 4	4.5% Mg 0.5% Mn Al alloy: Casting; Australian specification
H 49-8	10% Mg Al alloy: Casting; Australian specification
HA 3 G8	0.1% Cu 8% Mg Al alloy: Ingot; Canadian Standard
HA 3 G10	10% Mg Al alloy: Ingot; Canadian Standard
HA 3 GS40	4% Mg 0.5% Si Al alloy: Ingot; Canadian Standard

Symbol	Nominal analysis, supplier, condition and remarks.
HA 9 G10	10% Mg Al alloy: Sand casting; Canadian Standard
HA 9 GS40	4.0% Mg 0.5% Si Al alloy: Sand casting; Canadian Standard
HIDUMINIUM 90	10.5% Mg Al alloy: Casting; High Duty Alloys for BS alloy LM10
L231	10% Mg Al alloy: Casting; Spanish specification
LM 10	10% Mg Al alloy: Casting; INTAL for BS alloy LM 10
MG 7	7% Mg 1.5% Fe Al alloy: For die casting; Birmabright; as cast **UTS: 220 Elon: 3.5%**
NBN 436 Al Mg 6	5% Mg 0.5% Mn Al alloy: Casting; Belgium specification
NBN 436 Al Mg 11	10.2% Mg Al alloy: Casting; Belgium specification
NORAL 350	10% Mg Al alloy: Casting; Northern Al Co for BS alloy LM 10
PERALUMAN 34	3.4% Mg 0.2% Mn 0.15% Ti Al alloy: Casting; Anglo-Swiss Al Co
PERALUMAN 75	7.0% Mg 1.7% Si 0.4% Fe 0.4% Mn 0.15% Ti Al alloy: Ingot; Anglo-Swiss Al Co
QQ A 591/7	8.0% Mg Al alloy: Die casting; US Federal; as cast **UTS: 210 Elon: 4% Proof: 120**
QQ A 591/9	3.2% Mg Al alloy: Die casting; US Federal; as cast **UTS: 270 Elon: 10% Proof: 120**

Symbol	Nominal analysis, supplier, condition and remarks.
QQ A 601/5	3.8% Mg Al alloy: Casting; US Federal; as cast **UTS: 140**
SAE 320	4% Mg Al alloy: Casting; as cast; sand cast **UTS: 150 Elon: 6% Proof: 45**
SAE 324	10% Mg Al alloy: Casting; sand cast; solution treated **UTS: 280 Elon: 12% Proof: 150**
STA 7 AC5	5.0% Mg 0.5% Mn Al alloy: Casting; replaced by BS 1490 LM 5
STA 7 AC10	10.5% Mg Al alloy: Casting; replaced by BS 1490 LM 10
UNI 3056	10% Mg Al alloy: Casting; Italian Standard
UNI 3057	7.0% Mg Al alloy: Casting; Italian Standard
UNI 3058	5.0% Mg Al alloy: Casting; Italian Standard
UNI 3059	3.0% Mg Al alloy: Casting; Italian Standard
UNI 5080	7.5% Mg 0.8% Fe Al alloy: Casting; Italian Standard
V1036 A/B BXIV	10% Mg Al alloy: Casting; Dutch specification

Note. The following abbreviations and units are used in the tables:

DPN	Hardness, diamond pyramid number
UTS	Ultimate tensile strength, N/mm^2
Elon	Elongation, %
Proof	0.1 % proof strength, N/mm^2

$1 N/mm^2 = 0.1 hbar = 0.102 kgf/mm^2 = 0.06475 tonf/in^2 = 145.04 lbf/in^2$

1K Aluminium–silicon–magnesium cast alloys

Specific gravity	2.65–2.7	
Density	2650–2700 kg/m³	(0.096–0.098 lbf/in.³)
Solidus/liquidus	525–650 °C	
Thermal conductivity	117–170 W/m °C	(28–40 cal/m s °C)
Coefficient of linear expansion	$19–23 \times 10^{-6}/$ °C	
Electrical conductivity	29–43% IACS (copper 100%) The electrical resistance increases with silicon content and in the heat treated condition.	
Specific resistance	41–64 microhm mm	
Young's modulus of elasticity	$71.0 \times 10^9 N/m^2$	$(10.3 \times 10^6 lbf/in.^2)$
Impact	1–2 J	(1–1.5 ft/lb) Izod
Fatigue strength (5×10^6 cycles)		
sand cast, solution treated & aged	50–80 N/mm^2	($\pm$ 3.5–5.5 tonf/in.²)
chill cast, solution treated & aged	80–100 N/mm^2	($\pm$ 5.3–6.5 tonf/in.²)

Hot strength

Temperature °C	Tensile strength N/mm^2
100	320
150	300
200	260
250	150
300	90
350	50
400	37

The above values are chosen to show the typical properties of this group of alloys. It may be that none of the values applies to certain of the specifications listed.

General metallurgical characteristics

The addition of small amounts of magnesium to these silicon–aluminium alloys causes the formation of complex intermetallics. These can be dissolved and then precipitated by the correct heat treatment, giving enhanced mechanical properties.

The fact that a reasonably large quantity of silicon is present gives excellent fluidity at casting with little tendency for the alloys to be 'hot short', that is of low strength immediately after solidification. Care must however be exercised to prevent shrinkage cavities which tend to form readily if feeders are not carefully designed.

The additions of copper and nickel in small quantities act as grain refiners and stabilizers giving some increase in strength, with a reduction in corrosion resistance.

Magnesium is very readily and preferentially oxidized unless considerable care is taken during casting. As it is the magnesium which forms the intermetallic compounds to cause hardening, the loss can be serious. Any dross or oxide formed can also be readily trapped to give porosity and large inclusions.

Thermal treatment

These alloys can all be solution treated and aged. To solution treat the part is raised to 515–545 °C and held there for some hours, then quenched in water, which should be hot if distortion must be prevented. The solution treatment dissolves the intermetallics in the aluminium silicon matrix and puts the material into its softest possible condition. The parts are then aged for some hours at a temperature between 120–175 °C. This causes the intermetallic compounds to precipitate from solution, giving an increase in hardness and reduction in ductility.

It is imperative that the actual specification is consulted for the correct times and temperatures. The solution treatment temperatures and the melting points of the alloys in this group overlap, thus if the wrong temperature is chosen incipient melting around the grain boundaries can result and this cannot be recovered by any subsequent treatment.

Welding and brazing

Pre-treatment. No thermal treatment required. The weld area must be thoroughly cleaned and scratch brushed.

Welding and brazing. Welding uses fluxes or inert gas shielded arc is satisfactory, but some skill and experience are required. The low silicon alloys can be brazed, but the high silicon series have melting points close to that of the brazing rods.

Resistance welding is not possible with these alloys.

Post-treatment. The parts must be re-aged after welding or brazing to restore some of the mechanical properties. It is not usually possible to re-solution treat after welding, owing to the danger of partially re-melting the weld area. There will always be an area adjacent to the weld which is weaker than the remainder.

All traces of fluxes used must be removed, as these are highly corrosive.

Flaw detection methods

Crack test. Penetrant oil, chalk test should always be carried out on stressed parts.

Pressure test. A pressure test of 350 N/mm^2 (50 lbf/in.^{2}) air in water will find most defects. Higher air pressures should not be used without taking precautions against rupture of the casting under pressure. Hydraulic pressure testing should be used to test castings which require proving above 700 N/mm^2 (100 lbf/in.^{2}).

X-ray. Important high duty castings should be subjected to X-ray testing at the casting stage, after chalk testing.

Ultra-sonic test. This can be used to find subcutaneous defects.

Chemical etch. This is recommended as a quality check at the foundry stage on intricate or important castings. It shows up most casting defects and is an excellent method of surface preparation for chalk testing.

A hot 10% caustic solution is satisfactory. This will produce a black deposit with certain of the alloys and must be removed by a dip in concentrated nitric acid.

Corrosion protection

Temporary. None required.

Permanent. Under most conditions no protection will be required. Where necessary sulphuric acid anodizing will give good results, but leaves a cloudy grey surface. Chemical oxidation is recommended before painting or can be used alone where corrosive conditions are not severe.

Painting. This must always be preceded by an etch primer or preferably by chemical oxidation or anodizing.

Lead based paints must not be used as primers.

The final coat can be an air drying lacquer or stove enamel.

Plating. Electroplating requires special etches which accentuate casting defects. Thus plating should be called for only when there is no alternative and this information must be supplied to the foundry.

Machinability

These alloys should always be machined in the solution treated and aged condition, when very good surface finishes free of all tearing will result.

The presence of silicon and the hard intermetallics gives rise to tool wear. Tungsten carbide tipped tools should therefore be used, whenever possible, with adequate lubrication.

Uses

Pump castings, crank cases, gear boxes, brackets, stressed engine castings where reasonably good mechanical strength, corrosion resistance and good castability are required.

The LM13 type alloy is used for pistons as it has good hot strength and low expansion.

Symbol	Nominal analysis, supplier, condition and remarks.
135	7% Si 0.3% Mg 0.1% Ti Al alloy: Casting; Comalco; solution treated & aged
	DPN: 65 **UTS: 210** **Elon: 5%** **Proof: 110**
162	1% Cu 1.1% Mg 12% Si 2.5% Ni Al alloy: Casting; Comalco for BS alloy LM13
355	1.2% Cu 5% Si 0.5% Mg Al alloy: Casting; AA designation for ASTM alloy SC51A
355	1.1% Cu 0.5% Mg 5% Si Al alloy: Casting; Alcoa for BS alloy LM16
356	0.3% Mg 5.5% Si 0.4% Mn Al alloy: Casting; Alcoa for BS alloy LM8
356	0.3% Mg 7% Si 0.2% Fe Al alloy: Casting; Birmabright; chill cast; solution treated & aged
	DPN: 80 **UTS: 220** **Elon: 5%** **Proof: 150**
356	7% Si 0.3% Mg Al alloy: Casting; AA designation for ASTM alloy SG 70A
356.0	7.0% Si 0.3% Mg Al alloy: Investment casting; American National Standard; ANSI
360	0.5% Mg 9.5% Si Al alloy: Casting; Comalco; as cast
	DPN: 85 **UTS: 270** **Elon: 3%** **Proof: 150**
A 132	1% Cu 12% Si 1% Mg 2.5% Ni Al alloy: Casting; AA designation for ASTM alloy SN 122A
A 132	1% Cu 1% Mg 12% Si 2.5% Ni Al alloy: Casting; Alcoa for BS alloy LM 13
A 356	0.3% Mg 7% Si 0.1% Fe (max) Al alloy: Casting; Birmabright; chill cast; solution treated & aged
	DPN: 85 **UTS: 250** **Elon: 9%** **Proof: 180**
A 356.0	7.0% Al 0.3% Mg Al alloy: Investment casting; American National Standard; ANSI
A 360	0.5% Mg 9.5% Si Al alloy: Casting; AA designation for ASTM alloy SG 100A
A 360	0.5% Mg 9.5% Si Al alloy: Casting; Comalco for BS alloy LM 9
A 03560	Unified number for alloy 356.0; ASTM system
A 13560	Unified number for alloy A356.0; ASTM system
ALCAN GB 125	1.2% Cu 5% Si 0.5% Mg Al alloy: Casting; Alcan for BS alloy LM 16
ALCAN GB 162	1% Cu 1% Mg 2% Ni 11.5% Si Al alloy: Casting; Alcan for BS alloy LM 13
ALCAN GB B116	4.5% Si 0.5% Mg Al alloy: Casting; Alcan for BS alloy LM 8
ALCAN GB C125	5% Si 1.5% Cu 0.5% Mg Al alloy: Casting; Alcan for BS alloy LM 16
ALCAN GB D135	7% Si Mg Al alloy: Casting; Alcan for BS alloy LM 25
ALPAX BETA	0.4% Mg 11% Si 0.5% Mn Al alloy: Casting; Light Alloys Ltd for BS alloy LM 9
ALPAX GAMMA	0.4% Mg 11% Si 0.5% Mn Al alloy: Casting; Light Alloys Ltd for BS alloy LM 9
ALMELEC	0.7% Mg 0.5% Si Al alloy: Source unknown
	UTS: 300
AMS 4210 F	1.2% Cu 1.5% Mg 5% Si Al alloy: Casting; AMS for AA alloy 355
AMS 4212 E	1.2% Cu 0.5% Mg 5% Si Al alloy: Casting; solution treated & aged; AMS for AA alloy 355
AMS 4214 D	1.2% Cu 0.5% Mg 5% Si Al alloy: Casting; solution treated & stabilized; AMS for AA alloy 355

Symbol	Nominal analysis, supplier, condition and remarks.
AMS 4215 B	1.2% Cu 0.5% Mg 5% Si Al alloy: Casting; premium grade; AMS for AA alloy 355
AMS 4217 D	0.3% Mg 7% Si Al alloy: Casting; solution treated & aged; AMS for AA alloy 356
AMS 4218 B	0.3% Mg 7% Si Al alloy: Casting; premium grade AMS for AA alloy 356
AMS 4260	0.3% Mg 7% Si Al alloy: Casting; investment; solution treated & aged; AMS for AA alloy 356
AMS 4280 E	1.2% Cu 0.5% Mg 5% Si Al alloy: Casting; solution treated & over aged; AMS for AA alloy 355
AMS 4281 C	1.2% Cu 0.5% Mg 5% Si Al alloy: Casting; solution treated & aged; AMS for AA alloy 355
AMS 4284 D	0.3% Mg 7% Si Al alloy: Casting; solution treated & aged; AMS for AA alloy 356
AMS 4285	0.3% Mg 7% Si Al alloy: Centri-cast; solution treated & aged; AMS for AA alloy 356
AMS 4286 A	0.3% Mg 7% Si Al alloy: Casting; aged; AMS for AA alloy 356
AMS 4290	0.5% Mg 9.5% Si Al alloy: Casting; die cast
	UTS: 280 **Elon: 2%** **Proof: 150**
AN QQ A 376	1.2% Cu 5.0% Si 0.5% Mg Al alloy: Casting; US Service
	UTS: 210 **Elon: 2%** **Proof: 120**
AN QQ A 386	1.0% Cu 12% Si 1.0% Mg 2.5% Ni Al alloy: Casting; US Service; as cast
	UTS: 210
AN QQ A 394	7.0% Si 0.3% Mg Al alloy: Casting; US Service; solution treated & aged
	UTS: 170 **Elon: 3%** **Proof: 110**
ANTICORODAL	0.7% Mg 6.5% Si 0.3% Mn 0.15% Ti Al alloy: Casting; Anglo-Swiss Al Co
ANTICORODAL 70	0.3% Mg 7.0% Si 0.1% Ti Al alloy: Casting; Anglo-Swiss Al Co
ANTICORODAL 71	0.45% Mg 7.2% Si Al alloy: Primary casting ingot; Anglo-Swiss Al Co
ASTM B26 SC82A	1.5% Cu 8% Si 0.5% Mg Al alloy: Sand casting; solution treated & aged
	UTS: 220 **Elon: 1%** **Proof: 140**
ASTM B26 SG70A	7% Si 0.3% Mg Al alloy: Sand casting; solution treated & aged
	UTS: 200 **Elon: 3%** **Proof: 120**
ASTM B85 SG100A	10% Si 0.5% Mg Al alloy: Die casting; as cast
	UTS: 320 **Elon: 3.5%** **Proof: 150**
ASTM B85 SG100B	10% Si 0.5% Mg 2% Fe Al alloy: Die casting; as cast
	UTS: 2.5% **Proof: 170**
ASTM B108 SC51B	1.2% Cu 5% Si 0.5% Mg 0.2% Ti Al alloy: Casting; solution treated & aged
	UTS: 270 **Elon: 3%** **Proof: 200**
ASTM B108 SG70A	7% Si 0.3% Mg 0.2% Ti Al alloy: Casting; solution treated & aged
	UTS: 220 **Elon: 3%** **Proof: 140**
ASTM B108 SG70B	7% Si 0.3% Mg 0.2% Ti Al alloy: Casting; solution treated & aged
	UTS: 250 **Elon: 5%** **Proof: 170**
ASTM B108 SN122A	1% Cu 12% Si 1% Mg 0.2% Ti Al alloy: Casting; solution treated & aged
	UTS: 270
ASTM B179 SC51B	1% Cu 5% Si 0.5% Mg Al alloy: Ingot; primary metal
ASTM B179 SC82A	1.5% Cu 8% Si 0.4% Mg 0.5% Mn Al alloy: Ingot; primary metal
ASTM B179 SG70A	7% Si 0.3% Mg Al alloy: Ingot; primary metal
ASTM B179 SG70B	7% Si 0.3% Mg Al alloy: Ingot; primary metal
ASTM B179 SG100A-B	9.5% Si 0.5% Mg Al alloy: Ingot; primary metal
ASTM B179 SN122A	1% Cu 12% Si 1% Mg Al alloy: Ingot; primary metal

Note. The following abbreviations and units are used in the tables:

DPN	Hardness, diamond pyramid number
UTS	Ultimate tensile strength, N/mm^2
Elon	Elongation, %
Proof	0.1 % proof strength, N/mm^2

$1 \text{ N/mm}^2 = 0.1 \text{ hbar} = 0.102 \text{ kgf/mm}^2 = 0.06475 \text{ tonf/in}^2 = 145.04 \text{ lbf/in}^2$

Symbol	Nominal analysis, supplier, condition and remarks.
BA 41	0.4% Mg 11% Si 0.5% Mn Al alloy: Casting; British Al Co for BS alloy LM 9
BA 42	1% Cu 1% Mg 12% Si 2.5% Ni Al alloy: Casting; British Al Co for BS alloy LM 13
BA 451	0.5% Mg 5.0% Si Al alloy: Bar; British Al Co; solution treated & aged
	UTS: 180 Elon: 12% Proof: 150
BIRMIDAL	0.3% Mg 5.5% Si 0.4% Mn Al alloy: Casting; Birmabright for BS alloy LM 8
BS 1490 LM8	0.5% Mg 4.5% Si Al alloy: Ingot; chill cast
	UTS: 270 Elon: 2%
BS 1490 LM8M	0.5% Mg 4.5% Si Al alloy: Sand & die casting; chill cast
	UTS: 220 Elon: 3%
BS 1490 LM8P	0.5% Mg 4.5% Si Al alloy: Sand & die casting; chill cast & aged
	UTS: 180 Elon: 2%
BS 1490 LM8W	0.5% Mg 4.5% Si Al alloy: Sand & die casting; chill cast; solution treated & room aged
	UTS: 220 Elon: 5%
BS 1490 LM8WP	0.5% Mg 4.5% Si Al alloy: Sand & die casting; chill cast; solution treated & aged
	UTS: 270 Elon: 2%
BS 1490 LM9	0.4% Mg 12% Si 0.5% Mn Al alloy: Ingot; chill cast
	UTS: 280
BS 1490 LM9WP	0.4% Mg 12% Si 0.5% Mn Al alloy: Sand & die casting; chill cast; solution treated & aged
	UTS: 280
BS 1490 LM13	1% Cu 1% Mg 12% Si 2% Ni Al alloy: Ingot; chill cast
	UTS: 270
BS 1490 LM13WP	1% Cu 1% Mg 12% Si 2% Ni Al alloy: Sand or die piston casting; chill cast; solution treated & aged
	UTS: 270
BS 1490 LM16	1% Cu 0.5% Mg 5% Si Al alloy: Ingot; chill cast
	UTS: 270
BS 1490 LM16WP	1% Cu 0.5% Mg 5% Si Al alloy: Sand & die casting; chill cast; solution treated & aged
	UTS: 270
BS L75	0.4% Mg 12% Si 0.5% Mn Al alloy: Casting; chill cast; solution treated & aged
	UTS: 280 Proof: 240
COMALCO 135	7% Si 0.3% Mg 0.2% Fe 0.1% Ti Al alloy: Casting; Comalco alloy for SG 70A
COMALCO 162	1% Cu 12% Si 1.2% Mg 0.3% Fe 2.5% Ni Al alloy: Casting; Comalco alloy for LM 13
COMALCO 360	9.5% Si 0.5% Mg 0.9% Fe Al alloy: Casting; Comalco alloy for ZG 100B
D 12	7% Cu 0.3% Mg 5.5% Si Al alloy: Casting; Birmabright; solution treated & aged; pistons
	DPN: 132 UTS: 330 Proof: 300
DTD 245 A	0.5% Mg 11.5% Si 0.5% Mn Al alloy: Casting; chill cast; replaced by BS L75
	UTS: 280 Proof: 240
DTD 716 A	0.5% Mg 5% Si Al alloy: Casting; chill cast
	UTS: 160 Elon: 3% Proof: 80

Symbol	Nominal analysis, supplier, condition and remarks.
DTD 722 A	0.5% Mg 5% Si Al alloy: Casting; chill cast & aged
	UTS: 180 Elon: 2% Proof: 130
DTD 727 A	0.5% Mg 5% Si Al alloy: Casting; chill cast; solution treated
	UTS: 220 Elon: 5% Proof: 100
DTD 735 A	0.5% Mg 5% Si Al alloy: Casting; chill cast; solution treated & aged
	UTS: 270 Elon: 2% Proof: 210
DTD 5028	7.0% Si 0.3% Mg Al alloy: Casting
HA 9 SC51N	1.2% Cu 0.5% Mg 5.0% Si Al alloy: Sand casting; Canadian Standard
HA 9 SC51P	1.2% Cu 0.5% Mg 5.0% Si Al alloy: Sand casting; Canadian Standard
HA 10 SC51N	1.2% Cu 0.5% Mg 5.0% Si Al alloy: For die casting; Canadian Standard
HA 10 SC51P	1.2% Cu 0.5% Mg 5.0% Si Al alloy: For die casting; Canadian Standard
HA 10 SG70N	0.3% Mg 7.0% Si Al alloy: For die casting; Canadian Standard
HA 10 SG70P	0.3% Mg 7.0% Si Al alloy: For die casting; Canadian Standard
HA 10 SG71	0.5% Mg 7.0% Si Al alloy: For die casting; Canadian Standard
HA 10 SN122	1.0% Cu 1.0% Mg 2.5% Ni 12% Si Al alloy: For die casting; Canadian Standard
HIDUMINIUM 40	0.3% Mg 5% Si 0.4% Mn Al alloy: For sand & die casting; High Duty Alloys for BS alloy LM 8
LM 8	0.5% Mg 4.5% Si Al alloy: Casting; INTAL for BS alloy LM 8
LM 13	1% Cu 1% Mg 12% Si 2% Ni Al alloy: Casting; INTAL for BS alloy LM 13
LM 16	1% Cu 0.5% Mg 5% Si Al alloy: Casting; INTAL for BS alloy LM 16
LO-EX	1% Cu 1% Mg 12% Si 2% Ni Al alloy: Casting; BKL for BS alloy LM 13
NORAL 125	1.1% Cu 0.5% Mg 5% Si Al alloy: Casting; Northern Al Co for BS alloy LM 16
NORAL 161	0.4% Mg 11% Si 0.5% Mn Al alloy: Casting; Northern Al Co for BS alloy LM 9
NORAL 162	1% Cu 1% Mg 12% Si 2.5% Ni Al alloy: Casting; Northern Al Co for BS alloy LM 13
NURAL 25	5.5% Si 1.5% Mg 0.2% Mn Al alloy: Casting; Alcan (Germany); general purpose alloy
QQ A 596/6	1.2% Cu 5.0% Si 0.5% Mg Al alloy: Casting; US Federal; aged
	UTS: 210 Elon: 1.5%
QQ A 596/8	7.0% Si 0.3% Mg Al alloy: Casting; US Federal; solution treated & aged
	UTS: 200 Elon: 3%
QQ A 596/9	1.0% Cu 12% Si 1.0% Mg 2.5% Ni Al alloy: Casting; US Federal; aged
	UTS: 200
QQ A 596/10	1.5% Cu 12% Si 0.5% Mg Al alloy: Casting; US Federal; aged
	UTS: 180
QQ A 601/3	7.0% Si 0.3% Mg Al alloy: Casting; US Federal; solution treated & aged
	UTS: 200 Elon: 3%
QQ A 601/10	12% Cu 5.0% Si 0.5% Mg Al alloy: Casting; US Federal; aged
	UTS: 200 Elon: 2%
QQ A 601/11	1.5% Cu 5.0% Si 0.5% Mg 0.8% Ni Al alloy: Casting; US Federal; aged
	UTS: 150
QQ A 601/12	0.7% Cu 12% Si 1.0% Mg 2.5% Ni Al alloy: Casting; US Federal; aged
	UTS: 200

Note. The following abbreviations and units are used in the tables:

DPN	Hardness, diamond pyramid number
UTS	Ultimate tensile strength, N/mm²
Elon	Elongation, %
Proof	0.1 % proof strength, N/mm²

$1 \text{ N/mm}^2 = 0.1 \text{ hbar} = 0.102 \text{ kgf/mm}^2 = 0.06475 \text{ tonf/in}^2 = 145.04 \text{ lbf/in}^2$

Symbol	Nominal analysis, supplier, condition and remarks.
SAE 309	10% Si 0.5% Mg Al alloy: Casting; die cast; as cast
	UTS: 300 Elon: 5%
SAE 321	1% Cu 12% Si 1% Mg 2.5% Ni Al alloy: Casting; solution treated & aged
	UTS: 270
SAE 322	1.25% Cu 5% Si 0.5% Mg Al alloy: Casting; sand cast; solution treated & aged
	UTS: 210 Elon: 2% Proof: 130
SAE 323	7% Si 0.3% Mg Al alloy: Casting; sand cast; solution treated & aged
	UTS: 200 Elon: 3% Proof: 130
SAE 327	1.5% Cu 8% Si 0.5% Mn 0.5% Mg Al alloy: Casting; solution treated & aged
	UTS: 220 Elon: 1% Proof: 140
SAE 328	1.5% Cu 12% Si 0.8% Mn 0.8% Mg Al alloy: Casting; solution treated & aged
	UTS: 280 Proof: 250
SAE 336	7% Si 0.3% Mg Al alloy: Casting; solution treated & aged
	UTS: 250 Elon: 5% Proof: 180
SILAFONT 30	0.3% Mg 9.5% Si Al alloy: Casting; Anglo-Swiss Al Co
SILAFONT 54	0.3% Mg 9.5% Si Al alloy: Primary ingot; Anglo-Swiss Al Co
SILAFONT 64	0.2% Mg 10% Si Al alloy: Casting; Anglo-Swiss Al Co
SILAFONT 66	0.3% Mg 10% Si Al alloy: Casting; Anglo-Swiss Al Co

Symbol	Nominal analysis, supplier, condition and remarks.
STA 7 AC8	0.3% Mg 5.5% Si 0.4% Mn Al alloy: Casting; replaced by BS 1490 LM8
STA 7 AC9	0.4% Mg 11% Si 0.5% Mn Al alloy: Casting; replaced by BS 1490 LM9
STA 7 AC13	1.0% Cu 1.0% Mg 12% Si 2.5% Ni Al alloy: Casting; replaced by BS 1490 LM 13
UNI 3051	9% Si 0.5% Mn 0.35% Mg Al alloy: Casting; Italian Standard
UNI 3054	4.5% Si 0.7% Mn 0.65% Mg Al alloy: Casting; Italian Standard
UNI 3055	2.0% Si 0.7% Mn 0.65% Mg Al alloy: Casting; Italian Standard
UNI 3599	7.0% Si 0.3% Mg 0.5% Mn Al alloy: Casting; Italian Standard
UNI 3600	5.0% Si 1.3% Cu 0.5% Mg Al alloy: Casting; Italian Standard
WILMIL M	0.4% Mg 11% Si 0.5% Mn Al alloy: Casting; W Mills Ltd for BS alloy LM 9; obsolete

Note. The following abbreviations and units are used in the tables:

DPN	Hardness, diamond pyramid number
UTS	Ultimate tensile strength, N/mm²
Elon	Elongation, %
Proof	0.1 % proof strength, N/mm²

1 N/mm²=0.1 hbar=0.102 kgf/mm²=0.06475 tonf/in²=145.04 lbf/in²

1L Aluminium–copper cast alloys
With silicon, zinc, nickel etc.

Specific gravity	2.7–2.9	
Density	2700–2900 kg/m³	(0.1–0.105 lbf/in.³)
Solidus/liquidus	510–650 °C	
Thermal conductivity*	121–176 W/m °C	(29–42 cal/m s °C)
Coefficient of linear expansion	21–23 × 10⁻⁶/ °C	
Electrical conductivity*	26–35% IACS (copper 100%)	
Specific resistance	36–49 microhm mm	
Young's modulus of elasticity	69.6–71.0 N/m²	(10.1–10.3 × 10⁶ lbf/in.²)
Impact	0.5–4 J	(0.5–4.0 ft/lb) Izod
Fatigue strength (10⁶ cycles)	± 50–140 N/mm²	(± 3.5–9.0 tonf/in.²)
Hot strength		

Temperature °C	Tensile strength N/mm²	Tensile strength † N/mm²
100	300	
150	280	220
200	240	200
250	170	180
300	90	150
350	50	80

* Alloys with low percentage alloying elements have higher values.
† The values shown on the right apply to the alloys developed for hot strength by addition of titanium and nickel. Alloys with low percentage alloying elements have higher values.

The above properties are chosen to be typical of the group as a whole. Certain alloys listed may have specific properties which do not apply.

General metallurgical characteristics

With castings, the addition of alloying elements has the dual purpose of improving castability and enhancing certain physical or mechanical properties. Thus silicon enhances fluidity for thin walled or pressure tight castings and for intricate shapes, but gives no improvement in other properties. Copper also gives some improvement to fluidity, but not to the same extent as silicon. The mechanical properties can be considerably improved with little copper but corrosion resistance is reduced.

Zinc additions do little to improve castability, in fact they tend to reduce the hot strength giving a condition known as 'hot shortness' where cooling stresses cause rupture soon after solidification. Zinc, however, improves the machinability, and is used for low priced, low duty castings. The high zinc alloys do not retain their mechanical properties at high temperature as well as other alloys in this group.

Magnesium is added to improve corrosion resistance, but it does little to improve castability, as the melt tends to be sluggish, and a stable oxide is formed which is easily trapped in the molten metal.

Nickel, iron, titanium, cobalt, niobium, zirconium and antimony are added in small amounts as grain refiners and stabilizers, which give some of the alloys their high temperature properties. Most of these elements also form hard intermetallics which strengthen the solid solutions.

Some of the alloys have lead added to improve the machinability.

The alloys listed are all heat treatable to a greater or lesser extent, but are very often used in the cast condition.

Thermal treatment

Most of these alloys are of the type where the intermetallic compounds are taken into solution at a temperature of 500–550 °C, and remain there after quenching. This is the softest possible condition. The parts must be held at temperature for several hours.

These compounds can then be made to precipitate by ageing at 120–170 °C for several hours. This increases the strength and reduces the ductility.

With many of the alloys listed, the advantages to be gained by the above are negligible. With some of the alloys it is unnecessary to solution treat (casting can act as a form of solution treatment) but some precipitation is obtained by ageing.

Where no heat treatment is carried out, the casting is an intricate shape, an anneal, or stress release at 200–350 °C is recommended. All of the temperatures quoted above are listed for guidance only. It is imperative that the actual specification is consulted before heat treatment, as the melting points and solution temperatures overlap with these alloys. Too high a solution temperature can cause incipient melting at grain boundaries, and the part be rendered useless.

Welding and brazing

Pre-treatment. Pre-heating can be an advantage where possible, the part being welded while hot. The weld area must be thoroughly cleaned and scratch brushed.

Welding and brazing. Fusion welding is possible on most of the alloys listed, but requires either flux or inert gas shielded electric arc. Considerable skill and experience are required. These alloys should not be the first choice when welding is involved.

Brazing is not recommended, and resistance welding is not practicable.

Post-treatment. All traces of flux must be removed, as these are very corrosive.

The alloys should be re-aged after welding, as this recovers most of the mechanical properties and also acts as a stress release.

Flaw detection methods

Crack test. Penetrant oil chalk test detection should be carried out on all important castings.

Pressure test. Air pressure test at 350 N/mm^2 (50 lbf/in.^{2}) in water is a very searching test for porosity. A small addition of a wetting agent such as soap or soluble oil to the water increases the sensitivity by enlarging the bubbles which appear at any defect. Air at pressures above 350 N/mm^2 (50 lbf/in.^{2}) should not be used without adequate safety precautions against bursting.

Where higher pressures are required, hydraulic pressure test should be employed. Again the liquid should have a surface tension reducer such as soap added.

X-ray. This should be carried out when necessary after initial etching and crack testing on the rough casting.

Ultra-sonic test. Not normally used on castings of this type but can be used as an alternative to X-ray.

Chemical etch. This can be used on the rough casting to show any surface imperfections and as a pre-treatment to chalk test. 10% hot caustic soda solution is satisfactory. The parts will then need to be brightened using concentrated nitric acid.

Corrosion protection

Temporary. None required.

Permanent. Anodizing will increase the corrosion resistance, but not to the same extent as with the wrought alloys. The sulphuric acid process in general gives the best results, but it is doubtful if the increase in corrosion resistance warrants the cost of the process. Some of the alloys should not be chromic acid anodized, as this can reduce the fatigue strength. Chemical oxidation gives some increase in corrosion resistance, but often results in a patchy or spotted appearance.

Painting. Stoving, or air drying lacquers are recommended, provided that an etch primer or chemical oxidation is used before the top coat. Lead based primers should not be used.

Plating. The pre-plating etches required are usually too vicious, giving rise to surface imperfections. Electroplating should therefore not be used on these materials when a good finish is essential. It is important that the foundry are informed that the castings are to be plated.

should always be machined in the heat treated condition.

Machinability

The presence of silicon and the hard intermetallics in these alloys necessitates the use of either high speed or tipped tools. Provided these are used with adequate coolant, good surface finishes can be achieved. The alloys

Uses

Gear boxes, sumps, low-stressed leak proof items. Electrical fittings, brackets. The heat treatment alloys withstand shock loads. The highly alloyed specifications have good high temperature properties.

Symbol	Nominal analysis, supplier, condition and remarks.
3.1263	4.5% Cu 3% Si 2.5% Zn 0.5% Ni Al alloy: Casting; German Standard DIN 1725
3.1371	4.5% Cu 0.2% Mg 0.2% Ti Al alloy: Casting; German Standard DIN 1725
3.1841	4.5% Cu 0.2% Ti Al alloy: Casting; German Standard DIN 1725
3.2151	4% Cu 0.2% Mg 6% Si 2% Zn Al alloy: Casting; German Standard DIN 1725
113	7% Cu 3% Si 3% Zn Al alloy: Casting; Alcoa for BS alloy LM 1
113	7% Cu 3% Si Al alloy: Casting; AA designation for ASTM alloy CS 72A
117	3% Cu 5% Si 0.5% Mn Al alloy: Casting; Comalco for BS alloy LM 4
122	10% Cu 0.3% Mg Al alloy: Casting; Alcoa for BS alloy LM 12
122	10% Cu 0.2% Mg Al alloy: Casting; AA designation for ASTM alloy CG 100A
127	4.5% Cu 5% Si 0.3% Fe Al alloy: Casting, Comalco for BS alloy LM 22
138	10% Cu 4% Si 0.3% Fe Al alloy: Castings; as cast **DPN: 100 UTS: 200 Elon: 1.5% Proof: 170**
142	4% Cu 1.5% Mg 2% Ni Al alloy: Casting; Alcoa for BS alloy LM 14
142	4% Cu 1.5% Mg 2% Ni Al alloy: Casting; AA designation for ASTM alloy CN 42A
143	3.5% Cu 8.5% Si 0.8% Fe Al alloy: Casting; Comalco for BS alloy LM 24
152	6% Cu 5.5% Si 0.4% Mg Al alloy: Casting; AA designation for ASTM alloy CS 66A
195	4.5% Cu Al alloy: Casting; Alcoa for BS alloy LM 11
201.0	4.6% Cu 0.35% Mn 0.3% Mg Al alloy: Investment casting; American National Standard; ANSI
204.0	4.6% Cu 0.2% Mg Al alloy: Investment casting; American National Standard; ANSI
208.0	4.0% Cu 3.0% Si Al alloy: Investment casting; American National Standard; ANSI
222.0	10.0% Cu 2.0% Si (max) 0.2% Mg Al alloy: Investment casting; American National Standard; ANSI
226	4.8% Cu 0.2% Ti Al alloy: Casting; Comalco for BS alloy LM 11
242.0	4.0% Cu 1.6% Mg 2.0% Ni Al alloy: For investment casting; American National Standard; ANSI

Symbol	Nominal analysis, supplier, condition and remarks.
295.0	4.5% Cu 1.0% Si Al alloy: Investment casting; American National Standard; ANSI
319	3% Cu 4.5% Si 0.5% Mn Al alloy: Casting; Alcoa for BS alloy LM 4
319	4% Cu 6% Si Al alloy: Casting; AA designation for ASTM alloy SC 64B or SC 64C
319.0	3.5% Cu 6.0% Si Al alloy: Investment casting; American National Standard; ANSI
328.0	1.5% Cu 8.0% Si 0.4% Mn 0.4% Mg Al alloy: Investment casting; American National Standard; ANSI
355.0	1.2% Cu 5.0% 0.5% Mg Al alloy: Investment casting; American National Standard; ANSI
380	3.5% Cu 8% Si Al alloy: Casting; AA designation for ASTM alloy SC 84B
384	3.5% Cu 11% Si Al alloy: Casting; AA designation for ASTM alloy SC 114A
645	3.5% Cu 11% Zn Al alloy: Casting; Alcoa for BS alloy LM 3
A 02010	Unified number for alloy 201.0; ASTM system
A 02040	Unified number for alloy 204.0; ASTM system
A 02080	Unified number for alloy 208.0; ASTM system
A 02220	Unified number for alloy 222.0; ASTM system
A 02420	Unified number for alloy 242.0; ASTM system
A 02950	Unified number for alloy 295.0; ASTM system
A 03190	Unified number for alloy 319.0; ASTM system
A 03550	Unified number for alloy 355.0; ASTM system
A 108	4.5% Cu 5.5% Si Al alloy: Casting; AA designation for ASTM alloy SC 64A
A 143	3% Cu 10% Si 1% Mg 0.4% Fe 1% Ni Al alloy: Casting; Comalco; aged **DPN: 105 UTS: 240 Elon: 1% Proof: 180**
A 380	3.5% Cu 8% Si Al alloy: Casting; AA designation for ASTM alloy SC 84A
A 33550	Unified number for alloy C355.0; ASTM system
AERAL	3.5% Cu 1.8% Mg 0.6% Si 2.25% Cd Al alloy: Source unknown **UTS: 450**
AEROLITE	1.15% Cu 0.4% Mg 0.5% Si 0.1% Zn 1.0% Fe Al alloy: Source unknown
AERON	4% Cu 1.0% Si Al alloy: Solution treated & aged; source unknown **UTS: 370**
Af 3	4.5% Cu 0.2% Ti Al alloy: Casting; Anglo-Swiss code for Alufont 3
AL 75	3% Cu 5% Si 0.5% Mn Al alloy: Casting; American name for BS alloy LM 22
ALAR 308	3.5% Cu 8% Si Al alloy: Casting; American name for BS alloy LM 24
ALCAN GB 117	3% Cu 5% Si 0.5% Mn Al alloy: Casting; Alcan for BS alloy LM 4
ALCAN GB 218	4% Cu 1.5% Mg 2% Ni Al alloy: Casting; Alcan for BS alloy LM 14
ALCAN GB 226	4.5% Cu Al alloy: Casting; Alcan for BS alloy LM 11
ALCAN GB A143	9% Si 3% Cu Mg alloy: Casting for pistons; Alcan for BS alloy LM 26

Note. The following abbreviations and units are used in the tables:

DPN	Hardness, diamond pyramid number
UTS	Ultimate tensile strength, N/mm^2
Elon	Elongation, %
Proof	0.1 % proof strength, N/mm^2

1 N/mm^2=0.1 hbar=0.102 kgf/mm^2=0.06475 tonf/in^2=145.04 lbf/in^2

Symbol	Nominal analysis, supplier, condition and remarks.
ALCAN GB B117	3% Cu 5% Si 0.5% Mn Al alloy: Casting; Alcan for BS alloy LM 21
ALCAN GB C117	3.5% Cu 5% Si 0.5% Mn Al alloy: Casting; Alcan for BS alloy LM 22
ALDAL	4% Cu 0.5% Mg 0.5% Mn 0.6% Si Al alloy: Source unknown **UTS: 450**
ALFERIUM	2.5% Cu 0.6% Mg 0.5% Mn 0.3% Si Al alloy: Source unknown **UTS: 450**
ALUFONT 3	4.5% Cu 0.2% Ti Al alloy: Casting; Anglo-Swiss; solution treated & aged **DPN: 105 UTS: 340 Elon: 6% Proof: 220**
ALUFONT 42	4.7% Cu 0.25% Ti Al alloy: Casting; Anglo-Swiss Al Co.
ALUFONT 47	4.7% Cu 0.2% Mg 0.25% Ti Al alloy: Casting; Anglo-Swiss Al Co.
ALUMINAL 7	2.0% Cu 2.5% Mg 0.2% Mn 1.0% Ni Al alloy: Casting; origin unknown
AMS 4220 D	4% Cu 1.5% Mg 2% Ni 0.2% Cr Al alloy: Casting; solution treated & over aged; AMS for AA alloy 142
AMS 4222 D	4% Cu 1.5% Mg Al alloy: Casting; solution treated & over aged; AMS for AA alloy 142
AMS 4224	4% Cu 2% Mg 2% Ni 0.3% Cr 0.3% Mn 0.1% Ti 0.1% V Al alloy: Casting; stabilized
AMS 4227 A	8% Cu 6% Mg 0.5% Mn 0.5% Ni Al alloy: Casting
AMS 4230 C	4.5% Cu Al alloy: Casting; solution treated; AMS for AA alloy 195
AMS 4231 C	4.5% Cu Al alloy: Casting; solution treated & aged; AMS for AA alloy 195
AMS 4282 E	4.5% Cu 2.5% Si Al alloy: Casting; solution treated & aged
AMS 4283 D	4.5% Cu 2.5% Si Al alloy: Casting; solution treated
AMS 4291 B	3.5% Cu 5% Si or 8.5% Si Al alloy: Die casting; as cast
AN A4	3.5% Cu 3.5% Si 1.0% Fe Al alloy: Casting; US Federal **UTS: 110 Elon: 1.5%**
AN A5	4.0% Cu 2.5% Si Al alloy: Casting; US Service; aged **UTS: 170**
AN QQ A 379	4.0% Cu 1.5% Mg 2.0% Ni Al alloy: Casting; US Service; solution treated & aged **UTS: 200**
AN QQ A383	4.5% Cu 2.5% Si Al alloy: Casting; US Service; aged **UTS: 200**
AN QQ A390	4.5% Cu 1.5% Si Al alloy: Casting; US Service **UTS: 180**
AN QQ A397	4.0% Cu 3.0% Si Al alloy: Casting; US Service **UTS: 110 Elon: 1.5%**
AN QQ A399	8.0% Cu 1.2% Si 1.5% Fe Al alloy: Casting; US Service; as cast **UTS: 110**
A S2U	1.3% Cu 2.3% Si 1.0% Fe 0.1% Mg Al alloy: Casting; L'Aluminium Francais

Symbol	Nominal analysis, supplier, condition and remarks.
A S5U3	3.2% Cu 5.0% Si 0.1% Mg Al alloy: Casting; L'Aluminium Francais
ASTM B26 C4A	4.5% Cu 1.5% Si 1% Fe 0.25% Ti Al alloy: Sand casting; solution treated & aged **UTS: 210 Elon: 3% Proof: 120**
ASTM B26 CG100A	10% Cu 1.5% Fe 2% Si Al alloy: Sand casting; solution treated & aged **UTS: 200**
ASTM B26 CN42A	4% Cu 1% Fe 1.5% Mg 2% Ni Al alloy: Sand casting; solution treated & aged **UTS: 210 Proof: 120**
ASTM B26 CS43A	4% Cu 1.2% Fe 3% Si Al alloy: Sand casting; as cast **UTS: 110 Elon: 1.5% Proof: 60**
ASTM B26 CS72A	7% Cu 1.4% Fe 3% Si 2.5% Zn Al alloy: Sand casting; as cast; obsolete **UTS: 120 Elon: 1% Proof: 60**
ASTM B26 SC8	3% Cu 5% Si Al alloy: Casting; chill cast; solution treated & aged **UTS: 300 Elon: 3% Proof: 150**
ASTM B26 SC51A	1.2% Cu 5% Si 0.5% Mg Al alloy: Sand casting; solution treated & aged **UTS: 210 Elon: 2% Proof: 120**
ASTM B26 SC64D	4% Cu 6% Si Al alloy: Sand casting; solution treated & aged **UTS: 210 Elon: 2.5% Proof: 120**
ASTM B26 CS74A	7.0% Cu 1.4% Fe 3.5% Si 0.2% Ti Al alloy: Casting; replaces CS72A
ASTM B85 SC84A	3.5% Cu 8% Si Al alloy: Die casting; as cast **UTS: 330 Elon: 3.5% Proof: 150**
ASTM B85 SC84B	3.5% Cu 8% Si 2% Fe Al alloy: Die casting; as cast **UTS: 320 Elon: 2.5% Proof: 150**
ASTM B85 SC114A	3.5% Cu 11% Si Al alloy: Die casting; as cast **UTS: 330 Elon: 2.5% Proof: 150**
ASTM B108 CG100A	10% Cu 2% Si 0.2% Mg Al alloy: Casting; solution treated & aged **UTS: 270**
ASTM B108 CN42A	4% Cu 0.7% Si 1.5% Mg 2% Ni Al alloy: Casting; solution treated & aged **UTS: 270**
ASTM B108 CS42A	4.5% Cu 2.5% Si Al alloy: Casting; solution treated & aged **UTS: 240 Elon: 2% Proof: 140**
ASTM B108 CS66A	6.5% Cu 5.5% Si 0.4% Mg Al alloy: Casting; aged **UTS: 210 Proof: 170**
ASTM B108 CS72A	7% Cu 2.5% Si Al alloy: Casting; as cast; obsolete **UTS: 150 Proof: 90**
ASTM B108 CS104A	10% Cu 4% Si 0.2% Mg Al alloy: Casting; as cast
ASTM B108 SC51A	1.2% Cu 5% Si 0.8% Mg Al alloy: Casting; solution treated & aged **UTS: 250 Elon: 1.5% Proof: 150**
ASTM B108 SC64D	4% Cu 6% Si 0.2% Ti 0.3% Ni Al alloy: Casting; solution treated & aged **UTS: 270 Elon: 2% Proof: 150**
ASTM B108 SC92A	2.0% Cu 9% Si 0.5% Mg Al alloy: Die casting
ASTM B108 SC94A	3.5% Cu 9% Si 0.25% Mg Al alloy: Die casting
ASTM B108 SC103A	3% Cu 9% Si 1% Mg Al alloy: Casting; aged **UTS: 210**
ASTM B108 SC122A	1.5% Cu 12% Si 0.7% Mg 0.2% Ti Al alloy: Casting; solution treated & aged **UTS: 280 Proof: 250**
ASTM B179 C4A	4.5% Cu 1% Si Al alloy: Ingot; primary metal
ASTM B179 CG100A	10% Cu 2% Si 0.3% Mg Al alloy: Ingot; primary metal
ASTM B179 CN42A	4% Cu 1.5% Mg 2% Ni Al alloy: Ingot; primary metal
ASTM B179 CS42A	4.5% Cu 2.5% Si Al alloy: Ingot; primary metal
ASTM B179 CS43A	4% Cu 3% Si Al alloy: Ingot; primary metal

Note. The following abbreviations and units are used in the tables:

DPN	Hardness, diamond pyramid number
UTS	Ultimate tensile strength, N/mm^2
Elon	Elongation, %
Proof	0.1 % proof strength, N/mm^2

1 N/mm^2=0.1 hbar=0.102 kgf/mm^2=0.06475 tonf/in^2=145.04 lbf/in^2

Symbol	Nominal analysis, supplier, condition and remarks.
ASTM B179 CS66A	6% Cu 5.5% Si 0.3% Mg Al alloy: Ingot; primary metal
ASTM B179 CS72A	7% Cu 3% Si Al alloy: Ingot; primary metal
ASTM B179 CS76A	7% Cu 35% Si Al alloy: Ingot
ASTM B179 CS104A	10% Cu 4% Si 0.3% Mg Al alloy: Ingot; primary metal
ASTM B179 SC51A	1% Cu 5% Si 0.5% Mg 0.25% Cr Al alloy: Ingot; primary metal
ASTM B179 SC64C	4% Cu 6% Si Al alloy: Ingot; primary metal
ASTM B179 SC84A-B	3.5% Cu 8% Si 2.9% Max Zn Al alloy: Ingot; primary metal
ASTM B179 SC103A	2.5% Cu 9.5% Si 1% Mg Al alloy: Ingot; primary metal
ASTM B179 SC114A	4% Cu 11% Si Al alloy: Ingot; primary metal
ASTM B179 SC122A	1.5% Cu 12% Si 0.7% Mg Al alloy: Ingot; primary metal
ASTM B179 SC649	4% Cu 6% Si Al alloy: Ingot; primary metal
A U4NT	4.0% Cu 1.4% Mg Al alloy: Casting; L'Aluminium Francais
A U5GT	4.8% Cu 0.25% Mg Al alloy: Casting; L'Aluminium Francais
A U10G	10% Cu 0.25% Mg Al alloy: Casting; L'Aluminium Francais
A U40T5	55% Cu 5% Ti Al master alloy: L'Aluminium Francais
A U50	49% Cu Al master alloy: L'Aluminium Francais
A U1054	10% Cu 4.2% Si 0.25% Mg Al alloy: Casting; L'Aluminium Francais
B 132	1.5% Cu 12% Si 0.7% Mn 1.0% Mg Al alloy: Casting; AA designation for ASTM alloy SC122A
B 143	3% Cu 9.5% Si 1.2% Mg Al alloy: Casting; Comalco; aged **DPN: 103 UTS: 240 Elon: 1% Proof: 170**
B 195	4.5% Cu 2.5% Si Al alloy: Casting; AA designation for ASTM alloy CS42A
BA 32	4.5% Cu Al alloy: Casting; British Al Co. for BS alloy LM 11
BA 33	4% Cu 1.5% Mg 2% Ni Al alloy: Casting; British Al Co. for BS alloy LM 14
BIRMALITE	10% Cu 0.3% Mg Al alloy: Castings; Birmabright for BS alloy LM 12
BKL 305	3% Cu 5% Si 0.5% Mn Al alloy: Castings; BKL chill cast similar to BS alloy LM 4 with wider limits **DPN: 85 UTS: 170 Elon: 1.5% Proof: 75**
BKL 308	3.5% Cu 8.5% Si Al alloy: Casting; BKL for BS alloy LM 24
Bohn L4	3% Cu 1% Mg 9% Si 1% Ni Al alloy: Casting; American proprietary alloy similar to Birmabright **DPN: 105 UTS: 210 Elon: 0% Proof: 200**
BS 361	7% Cu Al alloy: Casting; replaced by BS 1490
BS 362	12% Cu Al alloy: Casting; replaced by BS alloy LM12
BS 363	Zn Cu Al alloy: Casting for crankcases; replaced by BS alloy LM 3
BS 702	Si Cu Al alloy: Casting; replaced by BS 1490
BS 703	4% Cu 1.5% Mg 2% Ni 0.2% Ti Al alloy: Castings; Y alloy; replaced
BS 1490 LM1	7% Cu 3% Si 3% Zn Al alloy: Ingot; chill cast **UTS: 150**
BS 1490 LM1M	7% Cu 3% Si 3% Zn Al alloy: Casting; as cast; chill cast **UTS: 150**
BS 1490 LM3	3% Cu 10% Zn Al alloy: Ingot; chill cast **UTS: 140**
BS 1490 LM3M	3% Cu 10% Zn Al alloy: For sand casting; chill cast **UTS: 140**
BS 1490 LM4	3% Cu 5% Si 0.5% Mn Al alloy: Ingot; chill cast **UTS: 150**
BS 1490 LM4M	3% Cu 5% Si 0.5% Mn Al alloy: For sand & die casting; chill cast **UTS: 150**
BS 1490 LM4WP	3% Cu 5% Si 0.5% Mn Al alloy: Sand & die casting; solution treated & aged **UTS: 220**
BS 1490 LM7	1.75% Cu 2% Si 1% Ni 1% Fe Al alloy: Casting; chill cast **UTS: 180 Elon: 3%**
BS 1490 LM11	4.5% Cu 0.3% Ti Al alloy: Ingot; chill cast **UTS: 300 Elon: 9%**
BS 1490 LM11WP	4.5% Cu 0.3% Ti Al alloy: For sand casting; chill cast; solution treated & aged **UTS: 300 Elon: 9%**
BS 1490 LM2	10% Cu 0.3% Mg Al alloy: Ingot; chill cast **UTS: 170**
BS 1490 LM12WP	10% Cu 0.3% Mg Al alloy: Die castings; chill cast; solution treated & aged **UTS: 270**
BS 1490 LM14	4% Cu 1.5% Mg 2% Ni Al alloy: Ingot; chill cast **UTS: 270**
BS 1490 LM14WP	4% Cu 1.5% Mg 2% Ni Al alloy: Sand or die castings; chill cast; solution treated & aged **DPN: 120 UTS: 270**
BS 1490 LM15	2% Cu 1% Mg 1.5% Si 1.5% Ni 1% Fe Al alloy: Casting; chill cast; solution treated & aged **DPN: 125 UTS: 330**
BS 1490 LM21	4% Cu 6% Si 0.5% Mn Al alloy: Ingot; chill cast **UTS: 170 Elon: 1%**
BS 1490 LM21M	4% Cu 6% Si 0.5% Mn Al alloy. Sand or die casting; chill cast **UTS: 170 Elon: 1%**
BS 1490 LM22	3% Cu 5% Si 0.5% Mn Al alloy: Ingot; chill cast **UTS: 240 Elon: 8%**
BS 1490 LM22W	3% Cu 5% Si 0.5% Mn Al alloy: Die casting; chill cast; solution treated, room aged **UTS: 240 Elon: 8%**
BS 1490 LM23	1.5% Cu 2% Si 1% Fe 1% Ni 0.2% Ti Al alloy: Ingot; chill cast **UTS: 180 Elon: 3%**
BS 1490 LM23P	1.5% Cu 2% Si 1% Fe 1% Ni 0.2% Ti Al alloy: Sand or die casting; chill cast & aged **UTS: 180 Elon: 3%**
BS 1490 LM24	3.5% Cu 8% Si Al alloy: Ingots; chill cast **UTS: 170 Elon: 1.5%**
BS 1490 LM24M	3.5% Cu 8% Si Al alloy: Die castings; as cast; chill cast **UTS: 170 Elon: 1.5%**
BS 1490 LM26	3.0% Cu 1.0% Mg 9.5% Si Al alloy: Casting; chill cast **DPN: 100 UTS: 210 Proof: 160**
BS 1490 LM27	2.0% Cu 7.0% Si 0.4% Mn Al alloy: Casting; chill cast **UTS: 160 Elon: 2% Proof: 90**
BS 1490 LM28	1.5% Cu 1.2% Mg 19% Si 1.2% Ni Al alloy: Casting; chill cast **UTS: 190 Proof: 160**
BS 1490 LM29	1.1% Cu 1.1% Mg 23% Si 1.1% Ni Al alloy: Casting; chill cast **UTS: 190 Proof: 170**
BS 1490 LM30	4.5% Cu 0.5% Mg 17% Si Al alloy: Casting; chill cast **UTS: 160 Proof: 160**
BS L35	4% Cu 1.5% Mg 2% Ni 0.2% Ti Al alloy: Ingot & casting; chill cast; Y alloy **UTS: 270 Proof: 210**
BS L50	4% Cu 0.75% Mg 0.6% Mn Al alloy: For ingots; secondary metal for re-melting

Symbol	Nominal analysis, supplier, condition and remarks.
BS L51	1% Cu 2.5% Si 0.9% Ni 1% Fe 0.2% Ti Al alloy: Casting; chill cast & aged **UTS: 190 Elon: 3% Proof: 120**
BS L52	2.5% Cu 1% Mg Ni Si Fe Al alloy: For sand or die castings; chill cast **UTS: 300 Proof: 250**
BS L78	1.25% Cu 0.5% Mg 5% Si Al alloy: Casting; chill cast **UTS: 290 Proof: 170**
BS L79	Cu Si Al alloy: Casting; obsolete
BS L91	4.5% Cu Al alloy: Casting; sand cast **UTS: 220 Elon: 7% Proof: 160**
BS L92	4.5% Cu Al alloy: Casting; sand cast **UTS: 280 Elon: 4% Proof: 200**
CERALUMIN ASM	1.5% Cu 0.6% Mg 1% Si 0.8% Ni 0.1% Nb Al alloy: Casting; J Stone; chill cast; solution treated & aged **UTS: 300 Elon: 12% Proof: 170**
C 355.0	1.25% Cu 5.0% Si 0.5% Mg Al alloy: Investment casting; American National Standard; ANSI
CERALUMIN B	1% Cu 0.1% Mg 2.5% Si 1% Ni 0.2% Ti Al alloy: Castings; J Stone for BS alloy LM 7
CERALUMIN C	1.5% Cu 0.8% Mg 0.75% Si 1% Ni 0.2% Ti Al alloy: Casting; J Stone for BS alloy LM 15
COMALCO 117	3% Cu 5% Si 0.3% Fe 0.5% Mn Al alloy: Casting; Comalco for BS alloy LM 4
COMALCO 127	4.5% Cu 5% Si 0.3% Fe Al alloy: Casting; Comalco for BS alloy LM 22
COMALCO 138	10% Cu 4% Si 0.3% Fe Al alloy: Casting; Comalco for ASTM alloy CS 104A
COMALCO 143	3.5% Cu 8% Si 0.8% Fe Al alloy: Casting; Comalco for ASTM alloy LM 24
COMALCO A143	3% Cu 9.5% Si 1% Mg 0.5% Fe 1% Ni Al alloy: Casting; Comalco
COMALCO B143	3% Cu 9.5% Si 1% Mg 0.3% Fe Al alloy: Casting; Comalco for ASTM alloy SC 103A
CQ 51A	4.6% Cu 0.35% Mn 0.3% Mg Al alloy: Investment casting; designation used by ASTM
D 7	3% Cu 4.5% Si 0.5% Mn Al alloy: Casting; Birmabright for BS alloy LM4
D 8	3% Cu 4.5% Si 0.5% Mn Al alloy: Casting; Birmabright for BS alloy LM 4
DS 15000 Al 101	10% Cu 0.3% Mg Al alloy: Casting; Danish specification
DS 15000 Al 511	3% Cu 5% Si Al alloy: Casting; Danish specification
DS 15000 Al 512	3.5% Cu 8.5% Si Al alloy: Casting; Danish specification
DTD 131	Cu Al alloy: Casting; obsolete; replaced by BS L52
DTD 133C	1% Cu 2.5% Si 1% Ni 1% Fe 0.2% Ti Al alloy: Casting; chill cast & aged **UTS: 190 Elon: 3% Proof: 120**
DTD 276A	1.2% Cu 0.5% Mg 5% Si Al alloy: Casting; solution treated & aged; chill cast; replaced by BS L78 **UTS: 250 Proof: 180**
DTD 287	1% Cu .5% Si 1% Ni 1% Fe 0.2% Ti Al alloy: Casting, chill cast & aged **UTS: 190 Elon: 3% Proof: 120**

Note. The following abbreviations and units are used in the tables:

DPN	Hardness, diamond pyramid number
UTS	Ultimate tensile strength, N/mm^2
Elon	Elongation, %
Proof	0.1 % proof strength, N/mm^2

1 N/mm^2=0.1 hbar=0.102 kgf/mm^2=0.06475 tonf/in^2=145.04 lbf/in^2

Symbol	Nominal analysis, supplier, condition and remarks.
DTD 298B	4.5% Cu 0.2% Ti Al alloy: Casting; solution treated; chill cast **UTS: 250 Elon: 13% Proof: 130**
DTD 304B	4.5% Cu 0.2% Ti Al alloy: Casting; solution treated & aged; chill cast **UTS: 300 Elon: 9% Proof: 180**
DTD 361B	4.5% Cu Al alloy: Casting; solution treated & aged; chill cast **UTS: 390 Elon: 4% Proof: 340**
DTD 424A	3% Cu 5% Si 0.5% Mn Al alloy: Casting; as cast; replaced by BS L79 **UTS: 150 Elon: 2%**
DTD 741A	4% Cu 2% Mg 0.7% Co 0.2% Nb Al alloy: Casting; chill cast; solution treated & aged **UTS: 330 Proof: 240**
DTD 5018	7.7% Mg 1.2% Zn Al alloy: Casting
E QQ A 601/13	4.0% Cu 2.5% Si Al alloy: Casting; US Federal; aged **UTS: 170**
E QQ A 601/14	3.5% Cu 3.0% Si 1.0% Fe Al alloy: Casting; US Federal **UTS: 110 Elon: 1.5%**
G Al Cu 4 Ti	4.5% Cu 0.2% Ti Al alloy: Casting; previous German Standard designation
G Al Cu 4 Ti Mg	4.5% Cu 0.2% Mg 0.2% Ti Al alloy: Casting; previous German Standard designation
G Al Cu 5 Si 3	5% Cu 3% Si Al alloy: Casting; previous German Standard designation
GMNA 100	6% Cu 6% Si Al alloy: Casting; Comalco; as cast **DPN: 65 UTS: 150 Elon: 3% Proof: 75**
GOST 2685/63 Al 1	4% Cu 1.5% Mg 2.0% Ni Al alloy: Casting; Russian specification
GOST 2685/63 Al 3	1.3% Cu 0.5% Mg 5.0% Si Al alloy: Casting; Russian specification
GOST 2685/63 Al 3B	1.3% Cu 0.5% Mg 5.0% Si Al alloy: Casting; Russian specification
GOST 2685/63 Al 6	Cu Si Mg Al alloy: Casting; Russian specification
GOST 2685/63 Al 7	4.5% Cu Al alloy: Casting; Russian specification
GOST 2685/63 Al 7 B	4.5% Cu Al alloy: Casting; Russian specification
GOST 2685/63 Al 10B	7% Cu 3% Si 3% Zn Al alloy: Casting; Russian specification
GOST 2685/63 Al 12	10% Cu 0.3% Mg Al alloy: Casting; Russian specification
GOST 2685/63 Al 14B	3.5% Cu 8.5% Si Al alloy: Casting; Russian specification
GOST 2685/63 Al 15	7% Cu 3% Si 3% Zn Al alloy: Casting; Russian specification
GOST 2685/63 Al 16B	3% Cu 5% Si 0.5% Mn Al alloy: Casting; Russian specification
GOST 2685/63 Al 18B	Cu Si Zn Al alloy: Casting; Russian specification
GOST 2685/63 Al 19	4.5% Cu Al alloy: Casting; Russian specification
GOST 2685/63 Al 20B	7% Cu 3% Si 3% Zn Al alloy: Casting; Russian specification
GOST 2685/63 Al 21	4% Cu 1.5% Mg 2% Ni Al alloy: Casting; Russian specification
H 49-1	7% Cu 3% Si 3% Zn Al alloy: Casting; Australian specification
H 49-3	3% Cu 5% Si Al alloy: Casting; Australian specification
H 49-9	4.5% Cu Al aloy: Casting; Australian specification
H 49-10	10% Cu 0.3% Mg Al alloy: Casting; Australian specification
H 49-12	4.0% Cu 1.5% Mg 2.0% Ni Al alloy: Casting; Australian specification
H 49-13	1.2% Cu 0.5% Mg 5.0% Si Al alloy: Casting; Australian specification
H 49-16	3.5% Cu 8.5% Si Al alloy: Casting; Australian specification

Symbol	Nominal analysis, supplier, condition and remarks.
HA 3 C4	4.5% Cu 0.25% Si 0.1% Ti Al alloy: Ingot; Canadian Standard
HA 3 CG50	4.8% Cu 0.2% Mg 0.2% Ti Al alloy: Ingot; Canadian Standard
HA 3 CS42	4.5% Cu 2.5% Si 0.1% Ti Al alloy: Ingot; Canadian Standard
HA 3 CS72	7.2% Cu 1% Fe 2.0% Si Al alloy: Ingot; Canadian Standard
HA 3 SC51N	1.2% Cu 0.2% Fe 0.5% Mg 5.0% Si Al alloy: Ingot; Canadian Standard
HA 3 SC51P	1.2% Cu 0.5% Mg 5.0% Si Al alloy: Ingot; Canadian Standard
HA 3 SC53	3.0% Cu 0.5% Mn 5.0% Si Al alloy: Ingot; Canadian Standard
HA 3 SC84N	3.5% Cu 8.5% Si Al alloy: Ingot; Canadian Standard
HA 3 SC84P	3.5% Cu 0.8% Fe 8.5% Si Al alloy: Ingot; Canadian Standard
HA 3 SC84R	3.5% Cu 1.0% Fe 0.6% Mg 8.5% Si 1.0% Zn Al alloy: Ingot; Canadian Standard
HA 9	4.5% Cu Al alloy: Sand casting; Canadian Standard
HA 9 CG50	4.8% Cu 0.2% Mg Al alloy: Sand casting; Canadian Standard
HA 9 CS42	4.5% Cu 2.5% Si Al alloy: Sand casting; Canadian Standard
HA 9 CS72	7.2% Cu 1.2% Fe 2.0% Si Al alloy: Sand casting; Canadian Standard
HA 9 SC53	3.0% Cu 0.5% Mn 5.0% Si Al alloy: Sand casting; Canadian Standard
HA 9 ZG61N	0.6% Mg 0.2% Ti 5.8% Zn 0.5% Cr Al alloy: Sand casting; Canadian Standard
HA 9 ZG61P	0.5% Mg 0.7% Mg 6.5% Zn Al alloy: Sand casting; Canadian Standard
HA 10 CS42	4.5% Cu 2.5% Si Al alloy: Die casting; Canadian Standard
HA 10 SC53	3.0% Cu 0.5% Mn 5.0% Si Al alloy: For die casting; Canadian Standard
HIDUMINIUM 20	3% Cu 4.5% Si 0.5% Mn Al alloy: For sand & die casting; High Duty Alloys for BS alloy LM 4
HIDUMINIUM 80	4.5% Cu 0.15% Ti Al alloy: High strength sand casting; High Duty Alloys for BS alloy LM 11
INTAL 7Q5	7.2% Cu 5.5% Si 5.0% Si Al alloy: For die casting; Canadian Standard
L 4	3% Cu 1% Mg 9% Si 1% Ni Al alloy: Casting; Birmabright; chill cast for pistons **DPN: 105 UTS: 210 Proof: 200**
L 211	10% Cu 0.3% Mg Al alloy: Casting; Spanish specification
L 212	10% Cu 0.3% Mg Al alloy: Casting; Spanish specification
L 213	7% Cu 3% Si 3% Zn Al alloy: Casting; Spanish specification
L 214	4.5% Cu Al alloy: Casting; Spanish specification
L 215	4% Cu 1.5% Mg 2% Ni Al alloy: Casting; Spanish specification
L 216	4% Cu 1.5% Mg 2% Ni Al alloy: Casting; Spanish specification
L 232	4.5% Cu 0.5% Mn Al alloy: Casting; Spanish specification
L 258	1.4% Cu 0.13% Mg 2.2% Si 1.2% Fe 1.3% Ni 0.12% Ti Al alloy: Casting; Spanish specification
L 411	7% Cu 3% Si 3% Zn Al alloy: Casting; Spanish specification
L 451	3% Cu 5% Si 0.5% Mn Al alloy: Casting; Spanish specification
L 452	4% Cu 6% Si 0.35% Mn Al alloy: Casting; Spanish specification
L 453	3.5% Cu 8.5% Si Al alloy: Casting; Spanish specification

Symbol	Nominal analysis, supplier, condition and remarks.
LAC 10	10% Cu 0.3% Mg Al alloy: Casting; obsolete; Air Ministry for BS 1490 LM 12
LAC 113B	3% Cu 10% Zn Al alloy: Casting; obsolete; Air Ministry for BS 1490 LM 3
LM 1	7% Cu 3% Si 3% Zn Al alloy: Casting; INTAL for BS alloy LM 1
LM 3	3% Cu 10% Zn Al alloy: Casting; INTAL for BS alloy LM 3
LM 4	3% Cu 5% Si 0.5% Mn Al alloy: Casting; INTAL for BS alloy LM 4
LM 11	4.5% Cu 0.3% Ti Al alloy: Casting; INTAL for BS alloy LM 11
LM 14	4% Cu 1.5% Mg 2% Ni Al alloy: Casting; INTAL for BS alloy LM 14
LM 23	1.5% Cu 2% Si 1% Fe 1% Ni 0.2% Ti Al alloy: Casting; INTAL for BS alloy LM 23
LM 24	3.5% Cu 8% Si Al alloy: Casting; INTAL for BS alloy LM 24
MIRALITE	4% Ni 0.3% Si 0.4% Fe 0.04% Na 0.05% Pb Al alloy: Casting and wire; MIRALITE
NBN 436 AlCu2NiMg	2% Cu 1.5% Mg 2% Ni Al alloy: Casting; Belgium specification
NBN 436 AlCu4Ni2Mg	4% Cu 1.5% Mg 2.0% Ni Al alloy: Casting; Belgium specification
NBN 436 AlCu5MgTi	4.5% Cu 0.3% Ti Al alloy: Casting; Belgium specification
NBN 436 AlSi2Cu	1.4% Cu 0.1% Mg 2.2% Si 1.1% Fe 1.3% Ni Al alloy: Casting; Belgium specification
NBN 436 AlSi5Cu	3% Cu 5% Si Al alloy: Casting; Belgium specification
NBN 436 AlSi9Cu3	3.5% Cu 8.5% Si Al alloy: Casting; Belgium specification
NORAL 117	3% Cu 4.5% Si 0.5% Mn Al alloy: Casting; Northern Al Co. for BS alloy LM 4
NORAL 218	4% Cu 1.5% Mg 2% Ni Al alloy: Casting; Northern Al Co. for BS alloy LM 14
NORAL 226	4.5% Cu Al alloy: Casting; Northern Al Co. for BS alloy LM 11
NORAL 237	7% Cu 3% Si 3% Zn Al alloy: Casting; Northern Al Co. for BS alloy LM 1
NORAL 252	10% Cu 0.3% Mg Al alloy: Casting; Northern Al Co. for BS alloy LM 12
NORAL A111	1% Cu 0.1% Mg 2.5% Si 1% Ni 0.2% Ti Al alloy: Casting; Northern Al Co. for BS alloy LM 7
QQ A 591/4	7.0% Cu 3.5% Si 2.0% Fe 1.5% Zn Al alloy: Casting; US Federal
QQ A 591/5	4.0% Cu 5.0% Si 2.0% Fe 1.0% Zn Al alloy: Casting; US Federal
QQ A 591/10	3.5% Cu 6.5% Si 1.2% Fe Al alloy: Casting; US Federal
QQ A 591/11	3.5% Cu 6.5% Si 2.0% Fe 1.0% Zn Al alloy: Casting; US Federal
QQ A 596/1	7.0% Cu 4.0% Si 1.2% Fe 2.5% Zn Al alloy: Casting; US Federal **UTS: 140**
QQ A 596/3	4.0% Cu 1.5% Mg 2.0% Ni Al alloy: Casting; US Federal; aged **UTS: 200**
QQ A 596/4	4.5% Cu 2.5% Si Al alloy: Casting; US Federal; aged **UTS: 200**
QQ A 596/5	4.5% Cu 5.5% Si Al alloy: Casting; US Federal; as cast **UTS: 150**
QQ A 601/4	4.5% Cu 1.2% Si Al alloy: Casting; US Federal; aged **UTS: 180**
QQ A 601/6	4.0% Cu 1.5% Mg 2.0% Ni Al alloy: Casting; US Federal; as cast **UTS: 150**

Symbol	Nominal analysis, supplier, condition and remarks.
QQ A 601/8	4.0% Cu 3.0% Si Al alloy: Casting; US Federal **UTS: 110 Elon: 1.5%**
QQ A 601/9	8.0% Cu 1.2% Si 1.5% Fe Al alloy: Casting; US Federal; as cast **UTS: 110**
RR 50	1% Cu 2.5% Si 1% Ni 0.2% Ti Al alloy: Sand & die casting; High Duty Alloys for BS alloy LM 7
RR 53B	1.5% Cu 0.8% Mg 0.75% Si 1% Ni 0.2% Ti Al alloy: Sand & die casting; High Duty Alloys for BS alloy LM 15
RR 250	5% Cu 0.25% Mn 1% Ni 0.2% Ti 0.25% Co 0.25% Sb Al alloy: For sand casting; High Duty Alloys; for use at high temperature use **DPN: 90 UTS: 210 Elon: 1% Proof: 150**
RR 350	5% Cu 1.5% Ni 0.25% Co 0.25% Zr 0.25% 0.2% Ti Al alloy: Casting; High Duty Alloys; solution treated & aged; for high temperature use **UTS: 240 Elon: 2% Proof: 150**
RRAC 9 A	1.2% Cu 0.5% Mg 1.3% Si 1.6% Ni 5% Sn Al alloy: Casting; aged; High Duty Alloys; used as a bearing alloy **DPN: 80 Proof: 150**
SAE 33	7% Cu 3% Si Al alloy: Casting; sand cast **UTS: 120 Proof: 70**
SAE 34	10% Cu 0.3% Mg Al alloy: Casting; sand cast **UTS: 200**
SAE 38	4.5% Cu 0.8% Si Al alloy: Casting; solution treated & aged **UTS: 240 Proof: 180**
SAE 39	4% Cu 1.5% Mg 2% Ni Al alloy: Casting; solution treated & aged **UTS: 210 Proof: 120**
SAE 300	6% Cu 6% Si 0.5% Mg Al alloy: Casting; die cast; aged **DPN: 95 UTS: 210 Proof: 170**
SAE 303	3.5% Cu 11% Si Al alloy: Casting; die cast; as cast **UTS: 280 Elon: 2%**
SAE 306	3.5% Cu 8% Si Al alloy: Casting; die cast; as cast **UTS: 280 Elon: 3%**
SAE 307	3.5% Cu 5% Si Al alloy: Casting; die cast; as cast **UTS: 250 Elon: 2.5%**
SAE 308	3.5% Cu 8% Si Al alloy: Casting; die cast; as cast **UTS: 300 Elon: 2%**
SAE 326	4% Cu 6% Si Al alloy: Casting; solution treated & aged **UTS: 200 Elon: 2.5% Proof: 130**
SAE 329	4% Cu 6% Si Al alloy: Casting; solution treated & aged **UTS: 210 Elon: 1.5% Proof: 130**
SAE 330	4.5% Cu 5.5% Si Al alloy: Casting; die cast; as cast **UTS: 150 Proof: 70**
SAE 331	4% Cu 9% Si Al alloy: Casting; solution treated & aged **UTS: 220**
SAE 332	3% Cu 9% Si 1% Mg 1% Ni Al alloy: Casting; aged **UTS: 210**
SAE 334	2.3% Cu 12% Si 1% Mg Al alloy: Casting; aged **UTS: 210**
SAE 335	1.2% Cu 5% Si 0.4% Mg Al alloy: Casting; solution treated & aged **UTS: 270 Elon: 3% Proof: 200**
SAE 380	4.5% Cu 2.5% Si Al alloy: Casting; solution treated & aged **UTS: 220 Elon: 2% Proof: 140**
SIS 4230	3% Cu 6% Si 0.4% Mn Al alloy: Casting; Swedish Standard; as cast **DPN: 70**
SIS 4231	3.2% Cu 5% Si 0.5% Mn Al alloy: Casting; Swedish Standard; as cast **DPN: 70**
SIS 4251	3.5% Cu 8% Si 0.5% Mn Al alloy: Casting; Swedish Standard; as cast **DPN: 70**
SIS 4252	3.5% Cu 8% Si 1% Fe 0.5% Mn Al alloy: Casting; Swedish Standard; as cast **DPN: 80**
SIS 4254	3.0% Cu 9.0% Si 2.0% Zn Al alloy: Casting; Swedish Standard **DPN: 85 UTS: 250 Elon: 1% Proof: 180**
STA 7 AC1	7% Cu 3% Si 3% Zn Al alloy: Casting; replaced by BS 1490 LM 1
STA 7 AC3	3.5% Cu 11% Zn Al alloy: Casting; replaced by BS 1490 LM 3
STA 7 AC4	3% Cu 4.5% Si 0.5% Mn Al alloy: Casting; replaced by BS 1490 LM 4
STA 7 AC7	1% Cu 2.5% Si 0.1% Mg 0.9% Ni 0.2% Ti Al alloy: Casting; replaced by BS 1490 LM 7
STA 7 AC11	4.5% Cu Al alloy: Casting; replaced by BS 1490 LM 11
STA 7 AC12	10% Cu 0.3% Mg 1% Fe Al alloy: Casting; replaced by BS 1490 LM 12
STA 7 AC14	4% Cu 1.5% Mg 2% Ni Al alloy: Casting; replaced by BS 1490 LM 14
STA 7 AC15	1.5% Cu 0.8% Mg 0.7% Si 1% Fe 1% Ni 0.2% Ti Al alloy: Casting; replaced by BS 1490 LM 15
UNI 3040	12% Cu Al alloy: Primary metal; Italian Standard
UNI 3041	10% Cu 1.0% Fe 0.3% Mg Al alloy: Primary metal; Italian Standard
UNI 3042	10% Cu 1.5% Ni 1.0% Si 0.25% Mg 0.15% Ti Al alloy: Casting; Italian Standard
UNI 3043	8% Cu Al alloy: Casting; Italian Standard
UNI 3044	4.5% Cu Al alloy: Casting; Italian Standard
UNI 3045	4% Cu 2.0% Ni 1.5% Mg Al alloy: Casting; Italian Standard
UNI 3046	3% Cu 1.5% Fe 0.7% Si 0.6% Mg 0.6% Ni 0.15% Ti Al alloy: Casting; Italian Standard
UNI 3050	10% Si 2.2% Cu 1.4% Mg 1.0% Ni Al alloy: Italian Standard
UNI 3052	5.5% Si 4.0% Cu Al alloy: Casting; Italian Standard
UNI 3601	8.5% Si 3.5% Cu Al alloy: Casting; Italian Standard
UNI 3602	0.6% Mg 1.0% Fe 0.2% Ti 5.0% Zn Al alloy: Casting; Italian Standard
UNI 5075	8.5% Si 3.5% Cu 0.7% Fe Al alloy: Casting; Italian Standard
V 1036 A/B BXIII	4.5% Cu 0.5% Mn Al alloy: Casting; Dutch specification
V 1036 A/B BXV	1.3% Cu 0.5% Mg 5% Si Al alloy: Casting; Dutch specification
V 1036 A/B BXVI	3% Cu 5% Si 0.5% Mn Al alloy: Casting; Dutch specification
VANALUIM	14% Zn 5% Cu 0.75% Fe 0.25% V Al alloy: Origin unknown
VITAL	1.2% Zn 0.9% Si 1% Cu Al alloy: Origin unknown

Note. The following abbreviations and units are used in the tables:

DPN	Hardness, diamond pyramid number
UTS	Ultimate tensile strength, N/mm^2
Elon	Elongation, %
Proof	0.1 % proof strength, N/mm^2

$1 \ N/mm^2 = 0.1 \ hbar = 0.102 \ kgf/mm\ 22 = 0.06475 \ tonf/in^2 = 145.04 \ lbf/in^2$

Symbol	Nominal analysis, supplier, condition and remarks.
X-ALLOY	3.5% Cu 1.25% Fe 0.6% Si 0.6% Ni 0.6% Mg Al alloy: Origin unknown
Y-ALLOY	4% Cu 1.5% Mg 2% Ni Al alloy: Sand & die castings; High Duty Alloys for BS alloy LM 14
Z 3	7% Cu 3% Si 3% Zn Al alloy: Casting; Birmabright for BS alloy LM 1

Note. The following abbreviations and units are used in the tables:

DPN	Hardness, diamond pyramid number
UTS	Ultimate tensile strength, N/mm^2
Elon	Elongation, %
Proof	0.1 % proof strength, N/mm^2

$1 \ N/mm^2 = 0.1 \ hbar = 0.102 \ kgf/mm^2 = 0.06475 \ tonf/in^2 = 145.04 \ lbf/in^2$

1M Aluminium–magnesium–zinc cast alloys
With and without titanium, copper and chromium

Specific gravity	2.76–2.8	
Density	2760–2800 kg/m^3	(0.1–0.101 $lbf/in.^3$)
Solidus/liquidus	570–760 °C	
Thermal conductivity	138 W/m °C	(33 cal/m s °C)
Coefficient of linear expansion	21 × $10^{-6}/$ °C	
Electrical conductivity	35% IACS (copper 100%)	
Specific resistance	49 microhm mm	
Young's modulus of elasticity	71.0 × 10^9 N/m^2	(10.3 × 10^6 $lbf/in.^2$)
Impact	– J	
Fatigue strength (5 × 10^6 cycles)	sand cast ± 50 N/mm^2	(± 3.5 $tonf/in.^2$)
	chill cast ± 75 N/mm^2	(± 5 $tonf/in.^2$)

The above properties are typical of the following group, and may not apply exactly to any one specification. It is possible that with certain specifications some of the values may not be applicable.

General metallurgical properties

This is a comparatively small range of alloys based on a 5–7% zinc alloy, always with some magnesium to improve the corrosion resistance and generally small quantities of titanium and chromium for grain refinement and corrosion resistance. They are medium strength, medium castable, low cost alloys, with fairly good corrosion resistance, which can be hardened to some extent by the precipitation of intermetallic compounds from solution. The addition of small quantities of copper to certain of the alloys improves this ability to harden.

Thermal treatment

These alloys are all capable of being solution treated and aged. By heating to a temperature of 450–500 °C for several hours, all of the intermetallic compounds are dissolved and remain in solution, after quenching in water or oil.

The alloys are then in their softest, most ductile condition. Depending on the size of the casting, this can be obtained, to a greater or lesser extent by the casting process and use is made of this with most of the specifications listed.

After the solution treatment, the parts are precipitation hardened (aged) by heating to a temperature of 100–160 °C when the intermetallic compounds are precipitated.

Ageing of most of these alloys can take place at room temperature, but this does not result in full hardening, and requires considerable time–up to three weeks in some instances.

Some of the alloys are used in the chill cast condition. This, if the casting is of the correct size, is a form of solution treatment.

Welding and brazing

Pre-weld treatment. No thermal treatment is required. The area to be welded or brazed must be thoroughly cleaned and scratch brushed.

Welding and brazing. These alloys are not suitable for welding of any sort, owing to the presence of zinc. It is possible, using fluxes, or an inert gas shielded arc and a skilled operator, to obtain welds of a sort, but these are not generally satisfactory. It should be noted that zinc fumes will be given off and that these can be dangerous if inhaled.

Brazing, using Al/Si filler rods and fluxes is possible, but not recommended.

Post-weld treatment. The parts should be re-aged, if possible, after welding or brazing. Any flux used must be completely removed before ageing.

Flaw detection methods

Crack test. Penetrant oil chalk test.

X-ray. Not normally required on this type of material, except as a percentage quality check at the foundry operation.

Ultra-sonic test. Not normally required.

Chemical etch. Again not normally required on this type of material. It is however, a useful means of checking a percentage of castings at the foundry stage to maintain quality. It is much more economical than X-raying. A 10% hot caustic soda solution shows up most surface defects and is an excellent surface preparation for crack testing.

Corrosion protection

Temporary. None required.

Permanent. These alloys have better than average corrosion resistance, compared to other aluminium alloys. Anodizing increases this to some extent, but the cost of the process is not generally warranted by the slight improvement obtained. It is however, used on polished articles to retain the attractive appearance.

The alloys should not be used in a corrosive atmosphere such as maritime conditions.

Painting. This is the normal method of protecting these alloys. An etch primer or chemical oxidizer myst always be used before either an air drying lacquer or stove enamel is applied. Lead based primer paints must not be used.

Plating. This will seldom or never be required on these alloys. When necessary the foundry must be informed and care taken at the pre-plating etching.

Machinability

Some difficulty may be found in obtaining good surface finishes with the softer of these alloys. High surface speeds and light but firm contact pressure give the best results. There are hard intermetallic particles present which will cause some tool wear, but only the longest mass production runs should require tipped tools. In general these alloys have good machinability and are capable of taking a very good polish.

Uses

These alloys are not used to any great extent in Great Britain. They have medium properties and price, and are used in the USA as general purpose aluminium casting alloys.

Symbol	Nominal analysis, supplier, condition and remarks.
3.1263	4.5% Cu 3% Si 2.5% Zn 0.5% Ni Al alloy: Casting; German Standard DIN 1725
40 E	0.6% Mg 5.5% Zn 0.2% Ti 0.5% Cr Al alloy: Casting; AA designation for ASTM alloy ZG 61A
705.0	1.6% Mg 0.5% Mn 0.3% Cr 0.3% Zn Al alloy: Investment casting; American National Standard; ANSI
707.0	2.0% Mg 0.3% Cr 4.2% Zn Al alloy: Investment casting; American National Standard; ANSI
713.0	0.7% Cu 0.35% Mg 7.5% Zn Al alloy: Investment casting; American National Standard; ANSI
771.0	0.9% Mg 0.1% Cr 7.0% Zn Al alloy: Investment casting; American National Standard; ANSI
A 612	0.5% Cu 0.7% Mg 6.5% Zn Al alloy: Casting; AA designation for ASTM alloy ZG 61B
A 712.0	0.4% Cu 0.7% Mg 6.5% Zn Al alloy: Investment casting; American National Standard; ANSI
A 07050	Unified number for alloy 705.0; ASTM system
A 07070	Unified number for alloy 707.0; ASTM system
A 07130	Unified number for alloy 713.0; ASTM system
A 07710	Unified number for alloy 771.0; ASTM system
Al 100 (79)	1% Cu 3% Mg 5% Zn Al alloy: Wrought; further information from Aluminium Zentrale
A 17120	Unified number for alloy A712.0; ASTM system
Al B42	1% Cu 3% Mg 5% Zn Al alloy: Wrought; further information from Aluminium Zentrale
Al B133	3% Mg 1% Si 1% Mn 5% Zn 0.2% Cr Al alloy: Wrought; further information from Aluminium Zentrale
AN A17	0.5% Mg 5.0% Zn 0.5% Cr 0.2% Ti Al alloy: Casting; US Service; as cast **UTS: 210 Elon: 3% Proof: 120**

Symbol	Nominal analysis, supplier, condition and remarks.
ASTM B26 ZC81A	0.7% Cu 0.3% Mg 7.5% Zn Al alloy: Sand casting; aged **UTS: 200 Elon: 3% Proof: 150**
ASTM B26 ZC32A	0.5% Mn 1.5% Mg 3% Zn 0.3% Cr Al alloy: Casting; room aged **UTS: 200 Elon: 5% Proof: 110**
ASTM B26 ZG42A	0.5% Mn 2% Mg 4% Zn 0.3% Cr Al alloy: Casting; solution treated & aged **UTS: 250 Elon: 1% Proof: 200**
ASTM B26 ZG61A	0.6% Mg 5.5% Zn 0.5% Cr Al alloy: Casting; room aged **UTS: 210 Elon: 3% Proof: 150**
ASTM B26 ZG61B	0.5% Cu 0.7% Mg 6.5% Zn Al alloy: Casting; room aged **UTS: 210 Elon: 2% Proof: 120**
ASTM B108 ZC60A	0.5% Cu 0.3% Mg 6.5% Zn Al alloy: Casting; aged **UTS: 7%0 Proof: 110**
ASTM B108 ZC81B	0.6% Cu 0.4% Mg 7.5% Zn 0.3% Cr Al alloy: Casting; aged **UTS: 210 Elon: 4% Proof: 140**
ASTM B108 ZG32A	0.5% Mn 1.5% Mg 3% Zn 0.3% Cr Al alloy: Casting; aged **UTS: 250 Elon: 10% Proof: 110**
ASTM B108 ZG42A	0.5% Mn 2% Mg 4.2% Zn 0.3% Cr Al alloy: Casting; aged **UTS: 300 Elon: 4% Proof: 180**
ASTM B179 ZC60A	0.5% Cu 0.4% Mg 6.5% Zn Al alloy: Ingot; primary metal
ASTM B179 ZC81A	0.7% Cu 0.4% Mg 7.5% Zn Al alloy: Ingot; primary metal
ASTM B179 ZC81B	0.6% Cu 0.4% Mg 7.5% Zn Al alloy: Ingot; primary metal
ASTM B179 ZG32A	1.6% Mg 3% Zn 0.5% Mn Al alloy: Ingot; primary metal
ASTM B179 ZG42A	2% Mg 4% Zn 0.3% Mn Al alloy: Ingot; primary metal
ASTM B179 ZG61A	0.6% Mg 5.5% Zn 0.5% Cr Al alloy: Ingot; primary metal
ASTM B179 ZG61B	0.5% Cu 0.7% Mg 6.5% Zn Al alloy: Ingot; primary metal
AZ 5G	0.2% Cu 0.5% Mg 0.2% Ti 0.2% Cr 5.0% Zn Al alloy: Casting; L'Aluminium Francais

Note. The following abbreviations and units are used in the tables:

DPN	Hardness, diamond pyramid number
UTS	Ultimate tensile strength, N/mm^2
Elon	Elongation, %
Proof	0.1 % proof strength, N/mm^2

$1 \ N/mm^2 = 0.1 \ hbar = 0.102 \ kgf/mm^2 = 0.06475 \ tonf/in^2 = 145.04 \ lbf/in^2$

Symbol	Nominal analysis, supplier, condition and remarks.
BS L5	Cn Zn Al alloy: Casting; obsolete
C 612	0.5% Cu 0.3% Mg 6.5% Zn Al alloy: Casting; AA designation for ASTM alloy ZC60A
D 712.0	0.6% Mg 0.5% Cr 5.7% Zn Al alloy: Investment casting; American National Standard; ANSI
DSI 5000 A1621	0.6% Mg 5.2% Zn Al alloy: Casting; Danish specification
DTD 5008A	0.6% Mg 5% Zn 0.2% Ti 0.5% Cr Al alloy: Casting; chill cast & aged
	UTS: 220 **Elon: 5%** **Proof: 180**
FRONTIER 40E	0.6% Mg 5% Zn 0.2% Ti 0.5% Cr Al alloy: Casting; chill cast; also known as 40E; American alloy
	UTS: 240 **Elon: 7%** **Proof: 170**
GOST 2685/63 A124	0.6% Cu 5.3% Zn 0.2% Ti 0.5% Cr Al alloy: Casting; Russian specification
H 49-17	0.6% Mg 5.2% Zn Al alloy: Casting; Australian specification
HA 3 ZG61N	0.55% Mg 0.2% Ti 5.7% Zn 0.5% Cr Al alloy: Ingot; Canadian Standard
HA 3 ZG61P	0.5% Cu 0.75% Mg 6.5% Zn Al alloy: Ingot; Canadian Standard
SAE 310	0.5% Mg 5.5% Zn 0.5% Cr Al alloy: Casting; sand cast & aged
	UTS: 210 **Elon: 3%** **Proof: 150**
SAE 311	1.5% Mg 0.5% Mn 3% Zn 0.3% Cr Al alloy: Casting; sand cast & aged
	UTS: 200 **Elon: 5%** **Proof: 110**
SAE 312	2% Mg 0.5% Mn 4% Zn 0.3% Cr Al alloy: Casting; solution treated & aged
	UTS: 220 **Elon: 1%**
SAE 313	0.5% Cu 0.7% Mg 6.5% Zn 0.2% Ti Al alloy: Casting; sand cast & aged
	UTS: 210 **Elon: 2%** **Proof: 130**
SAE 314	0.5% Cu 0.3% Mg 6.5% Zn 0.2% Ti Al alloy: Casting; die cast; aged
	UTS: 180 **Elon: 7%**

Symbol	Nominal analysis, supplier, condition and remarks.
SAE 315	0.8% Cu 0.3% Mg 7.5% Zn 0.3% Cr Al alloy: Casting; aged
	UTS: 200 **Elon: 5%** **Proof: 150**
SIS 4438	0.6% Mg 5% Zn 0.2% Ti 0.4% Cr Al alloy: Casting; Swedish Standard
	DPN: 80
TENZALOY	0.8% Cu 0.3% Mg 7.5% Zn Al alloy: Casting; AA designation for ASTM alloy ZC81A
TERNALLOY 5	1.5% Mg 3% Zn 0.3% Cr Al alloy: Casting; AA designation for ASTM alloy ZG 32A
TERNALLOY 7	2% Mg 0.5% Mn 4.2% Zn 0.3% Cr Al alloy: Casting; AA designation for ASTM alloy ZG 42A
U15	1% Mg 5% Zn Al alloy: Casting; Anglo-Swiss Al Co.
UNIFONT 5	1% Mg 5% Zn Al alloy: Casting; Anglo-Swiss Al Co. room aged; obsolete
	DPN: 75 **UTS: 240** **Elon: 6%** **Proof: 170**
UNIFONT 54	0.9% Mg 0.2% Cr 5.2% Zn Al alloy: Casting; Anglo-Swiss Al Co.
ZG 71B	0.9% Mg 0.1% Cr 7.0% Zn Al alloy: Investment casting; designation used by ASTM
ZG 81A	0.7% Cu 0.35% Mg 7.5% Zn Al alloy: Investment casting; designation used by ASTM
ZIMALUIM	7.5% Mg 11.5% Zn Al alloy: German origin

Note. The following abbreviations and units are used in the tables:

DPN	Hardness, diamond pyramid number
UTS	Ultimate tensile strength, N/mm^2
Elon	Elongation, %
Proof	0.1 % proof strength, N/mm^2

1 N/mm^2=0.1 hbar=0.102 kgf/mm^2=0.06475 tonf/in^2=145.04 lbf/in^2

1N Aluminium – miscellaneous alloys

Following is listed a small number of aluminium alloys which have special purposes and do not fit into any of the preceding groups.

Very little information is available on these materials apart from that appearing with each alloy.

Symbol	Nominal analysis, supplier, condition and remarks.
3.2685	11% Si 10% Sn + Cd 3% Cu + Ni Al alloy: German Standard
850.0	0.9% Cu 1.0% Ni 6.2% Sn Al alloy: Investment casting; American National Standard
A 850.0	2.5% Si 1.0% Cu 0.5% Ni 6.2% Sn Al alloy: Investment casting; American National Standard; ANSI
A 08500	Unified number for alloy 850.0; ASTM system
A 18500	Unified number for alloy A 850.0; ASTM system

Note. The following abbreviations and units are used in the tables:

DPN	Hardness, diamond pyramid number
UTS	Ultimate tensile strength, N/mm^2
Elon	Elongation, %
Proof	0.1 % proof strength, N/mm^2

1 N/mm^2=0.1 hbar=0.102 kgf/mm^2=0.06475 tonf/in^2=145.04 lbf/in^2

Symbol	Nominal analysis, supplier, condition and remarks.
A 28500	Unified number for alloy B850.0; ASTM system
A C4	3.7% Cr Al master alloy: L'Aluminium Francais
A Fe 10	10% Fe Al master alloy: L'Aluminium Francais
ALCAN DURALCOTE	Trade name for pre-coated Al sheet; Alcan
ALCAN GB 730	1% Cu 2.5% Si 0.5% Ni 6.5% Sn Al alloy: Casting; Alcan; aged; for bearings
	DPN: 50 **UTS: 150** **Elon: 5%** **Proof: 75**
Al-Zr	47.5% Zr Al alloy: Origin unknown
A M4	4.0% Mn Al alloy: Casting; L'Aluminium Francais
A M10	10% Mn Al master alloy: L'Aluminium Francais
AMS 4275B	1% Cu 1% Ni 6% Sn Al alloy: Casting for bearings; AMS for SAE 770
A N20	20% Ni Al master alloy: L'Aluminium Francais
A S9K7	9.5% Si 7% Co Al master alloy: L'Aluminium Francais
ASTM B199 HK31A	3.2% Th 0.7% Zr Al alloy: Die casting
ASTM B199 QE22A	2.2% Rare earth 0.7% Zr Al alloy: Die casting
ASTM B327	18.0% Cu 0.8% Mg Al alloy: Hardener for Zn base; die casting
CG181A	
A T4	4% Ti Al master alloy: L'Aluminium Francais
A Zr5	5.5% Zr Al master alloy: L'Aluminium Francais
B 850.0	2.0% Cu 0.7% Mg 1.2% Ni 6.2% Sn Al alloy: Investment casting; American National Standard; ANSI
BS L11	Al alloy: Casting; obsolete; no information available
BS L27	Al alloy: Obsolete; no information available

Symbol	Nominal analysis, supplier, condition and remarks.
BS L28	Aluminium alloy: Obsolete; no information available
BS L29	Aluminium alloy: Obsolete; no information available
FERRO ALUMINIUM	0.04% C 0.2% Si 0.1% Cu 50% Fe Al: Metal Alloys; primary metal
HIDUMINIUM 29	6.5% Sn 1% Cu 0.8% Ni Al alloy: Casting; High Duty Alloys; chill cast; aged for bearings **DPN: 40 UTS: 140 Elon: 15% Proof: 50**
L Al Si Sn	11% Si 10% Sn + Cd 3% Cu + Ni Al alloy: Designation used by German Standard
MB 7	1% Cu 1% Mg 0.6% Bi 1.7% Ni 7% Sn Al alloy: Casting; Birmabright; aged for bearings **UTS: 210 Elon: 1.5% Proof: 170**
NF A57 350	Classification of Al alloys: Wrought; French Standard
NF A57 650	Classification of Al alloy: Sheets; French Standard
SAE 770	1% Cu 1% Ni 6% Sn Al alloy: For bearings
SAE 780	6% Sn 1% Cu 1.5% Si Al alloy: Bearing metal; cast on steel backing
SAE 781	4% Si 1% Cd Al alloy: Bearing metal; cast on steel backing
SAE 782	1.0% Cu 1.0% Ni 3.0% Col Al alloy: For bearings bonded to steel backing
UNI 3022 MAF5	5% Fe Al alloy: Ingot; primary metal; Italian Standard
UNI 3022 MAF10	10% Fe Al alloy: Ingot; primary metal; Italian Standard
UNI 3022 MAK10	10% Co Al alloy: Ingot; primary metal; Italian Standard
UNI 3022 MAM10	10% Mn Al alloy: Ingot; primary metal; Italian Standard
UNI 3022 MAN25	25% Ni Al alloy: Ingot; primary metal; Italian Standard

Symbol	Nominal analysis, supplier, condition and remarks.
UNI 3022 MAR4	4% Cr Al alloy: Ingot; Primary metal; Italian Standard
UNI 3022 MAT3	3% Ti Al alloy: Ingot; primary metal; Italian Standard
UNI 6252	6% Si 1.2% Si 1.0% Cu 1.0% Ni 0.12% Ti Al alloy: Casting; Italian Standard
UNI 6253	2% Ni 2% Mn 0.15% Ti Al alloy: Casting; Italian Standard
WOLFRAMIUM	1.4% Sb 0.3% Cu 0.2% Fe 0.1% Sn 0.4% W Al alloy: Good corrosion resistance; origin unknown
X 8081	1.0% Cu 20% Sn Al alloy: For bearings; Alcoa; available in coil form
Z - ALLOY	6.5% Ni 0.5% Ti Al alloy: Used for hard shafts and medium bearings; origin unknown
ZICRAL	8.0% Zn 2.25% Mg 1.5% Cu 0.4% Cr 0.4% Mn 0.7% Si Al alloy: Wrought; origin unknown

Note. The following abbreviations and units are used in the tables:

DPN	Hardness, diamond pyramid number
UTS	Ultimate tensile strength, N/mm^2
Elon	Elongation, %
Proof	0.1 % proof strength, N/mm^2

$1 \ N/mm^2 = 0.1 \ hbar = 0.102 \ kgf/mm^2 = 0.06475 \ tonf/in^2 = 145.04 \ lbf/in^2$

2. Antimony Sb

Physical properties

Atomic number	51	
Atomic weight	121.76	
Crystal structure	Hexagonal	
Colour	Silver-white with bluish tinge	
Specific gravity	6.62	
Density	6620 kg/m^3	(0.239 lbf/in.^{3})
Melting point	630.5 °C	
Boiling point	1380 °C	
Specific heat	0.206 J/g °C	(0.0493 cal/g °C)
Thermal conductivity	18.6 W/m °C	(4.44 cal/m s °C)
Coefficient of linear expansion (20–100° C)	10.9 × 10^{-6}/ °C	
Latent heat of fusion	168.3 J/g	(40.2 cal/g)
Latent heat of vaporization	1604 J/g	(383 cal/g)
Thermal neutron absorption	5.7 barns/atom	
Electrical conductivity	4.5% IACS (copper 100%)	
Specific resistance	410 microhm mm	
Temperature coefficient of electrical resistance	0.0039/ °C	
Electrochemical equivalent	1.494 g/A/h	
Electrode potential	+ 0.1 V	
Magnetic susceptibility	0.82 × 10^{-6}	
Young's modulus of elasticity	77.9 × 10^9 N/m^2	(11.3 × 10^6 lbf/in.^{2})
Tensile strength	108 N/mm^2	(1560 lbf/in.^{2})
Hardness	30–60 DPN	

2.1 General notes on antimony

Antimony does not occur in the metallic state. It is generally purified from the sulphide ore (Stibnite) by reduction with wrought iron, using common salt as the flux in a reverberatory furnace.

The pure metal is brittle with little strength and is a poor conductor of heat and electricity but a good light reflector.

Antimony, like arsenic and bismuth, is a metalloid having many characteristics of the non-metals although not to quite the same degree as arsenic.

No commercial use is made of the primary metal and there are no alloys with antimony as the basic ingredient. It is however purified from its ore and is available under several names as a primary metal.

As an alloying element, antimony finds use as follows:

Lead alloys. 0.25–0.5% antimony has little effect on the physical and mechanical properties of lead but improves its castability.

5–12% antimony added to lead gives a range of alloys with increased strength, corrosion resistance and hardness. These are used in electrical storage batteries, as electroplating anodes, and vat linings. The name 'Regulus Metal' is applied to these alloys, as well as to the commercially pure form of antimony. The alloys have a better creep strength than pure lead.

Bearings. From 3–20% antimony is added to both lead based and tin based bearing alloys.

Antimony forms intermetallic compounds with tin, which are cubic in shape and give hardness and strength to an otherwise soft matrix.

Type metal. Antimony in small quantities added to certain lead based alloys lowers the melting point, and reduces the solidifation contraction. This increases the casting fluidity ensuring complete filling of intricate moulds while the lack of contraction gives excellent reproduction of the pattern being cast. Considerable quantities of these alloys are continuously used for newspaper type.

Brittanic metal. This is basically 80% tin, 18% antimony, 1.5% copper. It has good castability, is sonorous and retains its polish, hence it is used for bell castings and household ornaments.

Pewter. Previously this was 80% tin, 20% lead, with antimony additions. Now it resembles Brittanic metal.

Ammunition. Up to 15% antimony has been added to lead for bullets. This hardens the lead, making it brittle.

Following are listed the specifications, trade names and symbols used for the primary metal.

Symbol	Nominal analysis, supplier, condition and remarks.
ANTIMONY	High Purity SB metal: Impuritives less than 1 part per million; Johnson Matthey; available as cast bar for semi conductors
ANTIMONY	Sb metal: Blackwells; pure and commercial quality metal
ARSENIC-ANTIMO NY	10% As Sb alloy: Blackwells; primary metal
ASTM B237 A	0.05% As 0.15% Pb Sb metal: Primary metal
ASTM B237 B	0.1% As 0.2% Pb Sb metal: Primary metal
BRIOUDE	1.3% As 0.25% Pb 98.3% Sb metal: Brand name for antimony regulus; further information from The Lead Development Association
COOKSON'S C	99.6% Sb metal: Brand name for antimony metal; further information from The Lead Development Association
HALLETTS	98.5% Sb metal: Brand name for antimony regulus; further information from The Lead Development Association
h Sb 1	99.9999% Sb metal: Ingot; Light Ltd; high purity metal
h Sb 96	95.99% Sb 121: Stable isotope 121; Light Ltd
h Sb 97	95.99% Sb 123: Stable isotope 123; Light Ltd
LA LUCETTE	99.35% Sb metal: Brand name for antimony metal; further information from The Lead Development Association

Symbol	Nominal analysis, supplier, condition and remarks.
LONESTAR	99.7% Sb metal: US brand name for antimony regulus
REGULUS	Name given to commercially pure antimony: The same name is used for Pb/Sb alloys
RMM	99.3% Sb metal: US brand name for antimony regulus
TYNE	99.0% Mn Sb metal: Brand name for antimony regulus; further information from The Lead Development Association
WCC	99.848% Sb metal: Brand name for antimony regulus; further information from The Lead Development Association

Note. The following abbreviations and units are used in the tables:

DPN	Hardness, diamond pyramid number
UTS	Ultimate tensile strength, N/mm^2
Elon	Elongation, %
Proof	0.1 % proof strength, N/mm^2

$1 \ N/mm^2 = 0.1 \ hbar = 0.102 \ kgf/mm^2 = 0.06475 \ tonf/in^2 = 145.04 \ lbf/in^2$

3. Arsenic As

Physical properties

Atomic number		33
Atomic weight		74.9
Crystal structure		Rhombohedral
Colour		Steel-grey
Specific gravity	stable form	5.7
	yellow	3.7
	black	4.7
Density	stable	5700 kg/m³	(0.205 lbf/in.³)
	yellow	3700 kg/m³	(0.134 lb/in.³)
	black	4700 kg/m³	(0.17 lbf/in.³)

Melting point	under pressure	814 °C
Boiling point		Sublimes above 450 °C at normal pressure
Specific heat		0.33 J/g °C	(0.08 cal/g °C)
Thermal conductivity		– W/m °C	(– cal/m s °C)
Coefficient of linear expansion		4.7×10^{-6}/ °C
Latent heat of fusion		370.5 J/g	(88.5 cal/g)
Latent heat of vaporization		416.9 J/g	(102 cal/g)
Thermal neutron absorption cross-section		4.2 barns/atom

Electrical conductivity	3.8% IACS (copper 100%)
Specific resistance	460 microhm mm
Temperature coefficient of electrical resistance	0.0043/ °C
Electrochemical equivalent	0.932 g/A/h
Electrode potential	– V
Magnetic susceptibility	-0.31×10^{-6}

Young's modulus of elasticity	– N/m²
Tensile strength	– N/mm²
Hardness	– DPN

3.1 General notes on arsenic

Arsenic is found with the ores of copper, lead, tin, zinc and gold and recovered as a by-product when these ores are treated.

The metal is obtained by reduction with carbon when crude arsenic is obtained. It is purified by subliming over charcoal under vacuum.

The stable form is steel grey, brittle and crystalline. Arsenic exists in two other less stable forms, (allotropes) yellow arsenic, specific gravity 3.7, and black arsenic, specific gravity 4.7.

Arsenic forms no useful alloys where it is the main constituent but is used as an alloying element to improve the heat resistance of copper, to harden lead and also to harden certain white metal bearing alloys. In small amounts it is used to de-oxidize copper.

The metal itself is not toxic but many of the soluble compounds, notably the oxides, are poisonous. They are used in insecticides, sheep dips, and fungicides. Arsenic is a metalloid, in that it has the characteristics of non-metal and a metal, and along with boron and silicon has the least metallic nature of the elements listed here as metals.

Symbol	Nominal analysis, supplier, condition and remarks.
ARSENIC COPPER	50% Cu As alloy: Blackwells; primary metal
ARSENIC	High purity metal impurities less than 4 p.p.m.: Johnson Matthey; supplied sealed under vacuum
ARSENIC	98–99% commercially pure metal: Blackwells; primary metal
ARSENIC-IRON	45% Fe As alloy: Blackwells; primary metal
ORPIMENT	$As_2 S_3$ ore, generally associated with lead and copper ore
P ASLA	99.9999% As metallic lumps, each 100 g: Light Ltd; high purity metal
MISPICKEL	Fe S As ore: The most common arsenic ore
REALGAR	$As_2 S_2$ ore: Generally associated with lead and copper ores
WHITE ARSENIC	$As_2 O_3$: Arsenious oxide

Note. The following abbreviations and units are used in the tables:

DPN	Hardness, diamond pyramid number
UTS	Ultimate tensile strength, N/mm²
Elon	Elongation, %
Proof	0.1 % proof strength, N/mm²

1 N/mm²=0.1 hbar=0.102 kgf/mm²=0.06475 tonf/in²=145.04 lbf/in²

4. Barium Ba

Physical properties

Atomic number	56	
Atomic weight	137.36	
Crystal structure	–	
Colour	Yellowish white	
Specific gravity	3.66	
Density	3660 kg/m³	(0.126 lbf/in.³)
Melting point	704 °C	
Boiling point	1620 °C	
Specific heat	0.285 J/g °C	(0.068 cal/g °C)
Thermal conductivity	– W/m °C	(– cal/m s °C)
Coefficient of linear expansion (20–100° C)	18 × 10⁻⁶/ °C	
Latent heat of fusion	55.7 J/g	(13.3 cal/g)
Latent heat of vaporization	1290 J/g	(308 cal/g)
Thermal neutron absorption cross-section	1.17 barns/atom	
Electrical conductivity	3.5% IACS (copper 100%)	
Specific resistance	500 microhm mm	
Temperature coefficient of electrical resistance	–/ °C	
Electrochemical equivalent	2.56 g/A/h	
Electrode potential	-2.9 V	
Magnetic susceptibility	0.9 × 10⁻⁶	
Young's modulus of elasticity	– N/m²	
Tensile strength	– N/mm²	
Hardness	– DPN	

4.1 General notes on barium

Barium metal is still a laboratory curiosity, as it is very similar to calcium which is much cheaper. One limited use of barium is as an alloy with magnesium and aluminium as a 'getter' for high vacuum valves, when it is allowed to combine with the last traces of oxygen inside the finished valve. It is claimed that the thin film of barium which is produced in the process, acts as a high temperature lubricant to any moving electrode systems.

There are records of barium used in conjunction with lead and calcium in a bearing alloy, but no current information is available.

Symbol	Nominal analysis, supplier, condition and remarks.
BARIUM	Commercially pure metal: Blackwells; primary metal
w Ba 21	99.5% Ba rod: 20 mm in diameter; Light Ltd; high purity metal

Note. The following abbreviations and units are used in the tables:

DPN	Hardness, diamond pyramid number
UTS	Ultimate tensile strength, N/mm²
Elon	Elongation, %
Proof	0.1 % proof strength, N/mm²

$1 \text{ N/mm}^2 = 0.1 \text{ hbar} = 0.102 \text{ kgf/mm}^2 = 0.06475 \text{ tonf/in}^2 = 145.04 \text{ lbf/in}^2$

5. Beryllium Be

Physical properties

Atomic number	4	
Atomic weight	9.02	
Crystal structure	–	
Colour	Steel grey	
Specific gravity	1.844	
Density	1844 kg/m^3	(0.065 lbf/in.^{3})
Melting point	1285 °C	
Boiling point	2507 °C	
Specific heat	1.78 J/g °C	(0.425 cal/g °C)
Thermal conductivity	147 1.78 J/g °C	(35 cal/m s °C)
Coefficient of linear expansion (20–100° C)	13 × 10^{-6}/ °C	
Latent heat of fusion	1047–1130 J/g	(250–270 cal/g)
Latent heat of vaporization	24773 J/g	(5917 cal/g)
Thermal neutron absorption cross-section	0.009 barns/atom	
Electrical conductivity	38.9–43.1% IACS (copper 100%)	
Specific resistance	122 microhm mm	
Temperature coefficient of electrical resistance	0.006/ °C	
Electrochemical equivalent	– g/A/h	
Electrode potential	-1.7 V	
Magnetic susceptibility	-1.0 × 10^{-6}	
Young's modulus of elasticity	276 × 10^9 N/m^2	(40 × 10^6 lbf/in.^{2})
Tensile strength	370 N/mm^2	(25 tonf/in.^{2})
Hardness	55–60 DPN	

5.1 General notes on beryllium

Pure beryllium metal can be worked at about 400 °C and is available as extrusions, castings (vacuum cast) and powder compacts, but it is not possible to cold work beryllium which is relatively brittle.

The metal has a low neutron capture cross-section which renders it useful in atomic energy equipment, where the high melting point, chemical stability and relatively high strength enhance this usefulness. The metal has a tensile strength of 300–450 N/mm^2 at room temperature and about 150 N/mm^2 at 600° C. The elongation of the wrought material is about 20% in the direction of working, but is very low in the transverse direction. The pure metal finds use as electrodes in neon tubes, as a component in neutron sources, and the powder has been used to surface harden copper, nickel and iron alloys.

The metal and its oxide are dangerous to health and very great care must be taken when handling. In addition many soluble compounds can cause severe dermatitis. Because of this, the use of the metal and many of its compounds is restricted to those establishments which comply with the necessary precautions to safeguard operators' health. These do not apply however, to metallic alloys where beryllium is present in relatively small amounts in a stable form.

The largest amount of beryllium is used as an alloy with copper, when it is present up to a maximum of 2.8% and does not present any health hazards. These are the strongest copper alloys known and form the subject of a separate group, 'Copper–beryllium', Section 14D.

Alloys of beryllium and nickel, or iron, have age hardening properties, but to date have found no commercial application. When traces of beryllium are added to platinum there is a measurable increase in hardness.

Some use has been found for beryllium as an alloy with aluminium for coating steel, when there is a reduction in the interface steel/aluminium layer, with a consequent increase in the coating ductility.

There has also been a report that traces added to magnesium improve the castability of the metal and refine the grain structure.

Apart from the uses in nuclear engineering, where all concerned are familiar with the discipline necessary for handling potentially dangerous materials, no practical use is made of beryllium because of the health hazard. As stated above, this does not exist with the beryllium–copper alloys.

Symbol	Nominal analysis, supplier, condition and remarks.
AMS 7900	Be metal: Bar; rod & shapes; primary metal
AMS 7901	Be metal: Bar; rod & shapes
AMS 7902	Be metal: Plate & sheet
BERYLLIUM	Commercially pure metal: Blackwells; primary metal
I 400	4.2% BeO 92% (min) Be billet: QMV grade; Brush Beryllium Co.; for instruments
	UTS: 330
N 50A	1.0% BeO Be block: Brush Beryllium Co.
N 100A	1.2% BeO Be block: Brush Beryllium Co.
N 200A	2.0% BeO Be block: Brush Beryllium Co.
P Be 12	99.99% Be flakes: Light Ltd; high purity metal
QMV	Trade name for commercial beryllium: Brush Beryllium Co.; available in various forms and grades
S 100C	1.2% BeO 98.5% (min) Be billet: QMV grade; Brush Beryllium Co.
	UTS: 220 Elon: 1% Proof: 180
S 200C	2.0% BeO 98.0% (min) Be billet: QMV grade; Brush Beryllium Co.
	UTS: 270 Elon: 1% Proof: 200
S 300C	3.0% BeO 97.4% (min) Be billet: QMV grade; Brush Beryllium Co.
	UTS: 270 Elon: 1% Proof: 200

Symbol	Nominal analysis, supplier, condition and remarks.
SP 100C	1.2% BeO Be powder: Brush Beryllium Co.
SP 200C	2.0% BeO Be powder: Brush Beryllium Co.
SP 300C	3.0% BeO Be powder: Brush Beryllium Co.
SR 200	2.0% BeO Be sheet: QMV grade; Brush Beryllium Co.; hot rolled
	UTS: 480 Elon: 5% Proof: 330
UNS R10001–R19999	Be metals and alloys. American Standards system; unified number

Note. The following abbreviations and units are used in the tables:

DPN	Hardness, diamond pyramid number
UTS	Ultimate tensile strength, N/mm^2
Elon	Elongation, %
Proof	0.1 % proof strength, N/mm^2

$1 \ N/mm^2 = 0.1 \ hbar = 0.102 \ kgf/mm^2 = 0.06475 \ tonf/in^2 = 145.04 \ lbf/in^2$

6. Bismuth Bi

Physical properties

Atomic number	83	
Atomic weight	209.0	
Crystal structure	Rhombohedral	
Colour	White with reddish tinge	
Specific gravity	9.8	
Density	9800 kg/m³	(0.34 lbf/in.³)
Melting point	271.3 °C	
Boiling point	1420 °C	
Specific heat	0.13 J/g °C	(0.030 cal/g °C)
Thermal conductivity	10 W/m °C	(2.4 cal/m s °C)
Coefficient of linear expansion (20–100° C)	13.3×10^{-6}/ °C	
Latent heat of fusion	54.4 J/g	(13 cal/g)
Latent heat of vaporization	854 J/g	(204 cal/g)
Thermal neutron absorption cross-section	0.032 barns/atom	
Electrical conductivity	1.7% IACS (copper 100%)	
Specific resistance	1120 microhm mm	
Temperature coefficient of electrical resistance	0.0043/ °C	
Electrochemical equivalent	2.59 g/A.h	
Electrode potential	+0.2 V	
Magnetic susceptibility	-1.35×10^{-6}	
Young's modulus of elasticity	31.7×10^{9} N/m²	$(4.6 \times 10^{6}$ lbf/in.²$)$
Tensile strength	– N/mm²	
Hardness	– DPN	

6.1 General notes on bismuth

Bismuth is soft but quite brittle, being similar in many respects to antimony, arsenic and zinc. It has considerably more metallic properties than either antimony or arsenic but is a very poor heat conductor and, like antimony, expands on solidification, this property being transferred to its alloys.

The pure metal can be extruded as a very fine wire which is used in high quality pyrometer and galvanometer suspensions, as well as hair lines in optical instruments.

The bismuth alloys are characterized by their low melting point, low contraction rate and high fluidity. These low melting point alloys are used as fusible plugs or links which melt when heat is applied, and are found in fire sprinkler systems, holding open fire doors, and other safety devices which must automatically operate when the temperature rises above a certain limit. Alloys melting from 60 °C upwards are available.

Bismuth alloys exist which expand on solidification and remain dimensionally constant or contract very slightly. All of these are used in engineering to fill pipes at bending, to hold intricate parts during machining, or as patterns to simulate certain shrinkage characteristics.

Molten bismuth does not readily oxidize and it is this property which enhances the fluidity of some alloys. This, with the low contraction rate, gives castings of a very high definition, and is thus used as type metal in good quality printing.

Small quantities of bismuth added to austenitic stainless steel are reported to increase the machinability without detracting from the corrosion resisting or mechanical properties, while some zinc alloys have also been produced, notably in Germany, with bismuth added for the same purpose.

Note. The following abbreviations and units are used in the tables:

DPN	Hardness, diamond pyramid number
UTS	Ultimate tensile strength, N/mm²
Elon	Elongation, %
Proof	0.1 % proof strength, N/mm²

1 N/mm²=0.1 hbar=0.102 kgf/mm²=0.06475 tonf/in²=145.04 lbf/in²

Symbol	Nominal analysis, supplier, condition and remarks.
AAF 11078A/1	27% Pb 13% Sn 11% Cd Bi alloy: US Service; melting point 71 °C
AAF 11078A/2	42% Pb 7.0% Cd Bi alloy: US Service; melting range 80–91 °C
ANATOMICAL ALLOY	17% Pb 19% Sn 10% Hg Bi alloy: Melting point 60 °C; used for making bone replicas
BISMANOL	20% Mn Bi alloy: Pressed powder for magnets; consult PMA for further information

Symbol	Nominal analysis, supplier, condition and remarks.
BISMUTH	Commercially pure Bi metal: Blackwells; primary metal
BISMUTH	High purity Bi metal impurities 10 p.p.m.: Johnson Matthey; supplied as ingot or wire
CERROBASE	44.5% Pb Bi alloy: Mining and Chemical Co.; melting point 124 °C
CERRO-CAST	Bi base alloy: Melting range 140–170 °C; Mining & Chemical Co.
CERRO-LOW 117	Bi base alloy: Melting point 47.5 °C; Mining & Chemical Co.
CERRO-LOW 136	Bi base alloy: Melting point 58 °C; Mining & Chemical Co.
CERRO-LOW 147	Bi base alloy: Melting range 61–65 °C; Mining & Chemical Co.
CERROSAFE	37.5% Pb 11.3% Sn 8.5% Cd Bi alloy: Melting range 70–90 °C; Mining & Chemical Products Ltd
CERROTRIC	42% Sn Bi alloy: Melting point 138.5 °C; Mining & Chemical Products Ltd
CERROTRU	42% Sn 1% Pb Bi alloy: Mining & Chemical Co; melting range 134–135 °C; zero contraction on solidification
D'ARCET	25% Sn 25% Pb Bi alloy: Information from Tin Research Institute; melting range 96–98 °C
GLANCE	Bismuth sulphide ore
hBi1	99.9999% Bi rod, 12 mm diameter: Light Ltd; high purity metal
LICHTENBERG	20% Sn 30% Pb Bi alloy: Information from Tin Research Institute; melting range 96–100 °C

Symbol	Nominal analysis, supplier, condition and remarks.
LIPOWITZ	13.3% Sn 26.7% Pb 10% Cd Bi alloy: Information from Tin Research Institute; melting range 70–73 °C
MALOTTE	34% Sn 20% Pb Bi alloy: Information from Tin Research Institute; melting range 96–123 °C
NEWTON	18.8% Sn 31.2% Pb Bi alloy: Information from Tin Research Institute; melting range 96–97 °C
ONION	20% Sn 30% Pb Bi alloy: Information from Tin Research Institute; melting range 96–100 °C
ROSE	22% Sn 28% Pb Bi alloy: Information from Tin Research Institute; melting range 96–100 °C
WOODS	12.5% Sn 25% Pb 12.5% Cd Bi alloy: Information from Tin Research Institute; melting range 70–72 °C

Note. The following abbreviations and units are used in the tables:

DPN	Hardness, diamond pyramid number
UTS	Ultimate tensile strength, N/mm^2
Elon	Elongation, %
Proof	0.1 % proof strength, N/mm^2

$1\ N/mm^2 = 0.1\ hbar = 0.102\ kgf/mm^2 = 0.06475\ tonf/in^2 = 145.04\ lbf/in^2$

7. Boron B

Physical properties

Atomic number	5	
Atomic weight	10.82	
Crystal structure	Monoclinic	
Colour	Yellow	
Specific gravity	2.6	
Density	2600 kg/m³	(0.091 lbf/in.³)
Melting point	2100 °C	
Boiling point	3550 °C	
Specific heat	1.30 J/g °C	(0.31 cal/g °C)
Thermal conductivity	– W/m °C	(– cal/m s °C)
Coefficient of linear expansion (20–100° C)	8.3×10^{-6}/ °C	
Latent heat of fusion	– J/g	(– cal/g)
Latent heat of vaporization	3140 J/g	(750 cal/g)
Thermal neutron absorption cross-section	– barns/atom	
Electrical conductivity	– % IACS (copper 100%)	
Specific resistance	– microhm mm	
Temperature coefficient of electrical resistance	–/ °C	
Electrochemical equivalent	– g/A/h	
Electrode potential	– V	
Magnetic susceptibility	0.69×10^{-6}	
Young's modulus of elasticity	– N/m²	
Tensile strength	– hbar	
Hardness	– DPN	

7.1 General notes on boron

At present pure boron metal is a laboratory curiosity, although there are suggestions that it may find use in the nuclear industry as a neutron absorber.

Traces of boron in steel (less than 0.003%) increase the hardenability by an amount equal to 1% nickel 0.5% chromium or 0.25% molybdenum. These boron steels are now commonly used in the US, and to an increasing extent in this country.

Small amounts of boron, up to 0.1%, added to cast iron inhibit graphite formation, thus increasing the possibility of obtaining local surface hardening when desired.

Boron is added in amounts up to 3.0% to certain hard facing alloys as a fluxing agent. This melts above a certain temperature and acts as a bonding agent for the hard infusible particles.

Like arsenic and silicon, boron is a metalloid with only feeble metallic properties. There are no alloys based on boron.

Note. The following abbreviations and units are used in the tables:

DPN	Hardness, diamond pyramid number
UTS	Ultimate tensile strength, N/mm²
Elon	Elongation, %
Proof	0.1 % proof strength, N/mm²

1 N/mm²=0.1 hbar=0.102 kgf/mm²=0.06475 tonf/in²=145.04 lbf/in²

Symbol	Nominal analysis, supplier, condition and remarks.
BORON	Pure amorphous powder: Blackwells; primary metal
BORON CARBIDE	Blackwells
HB 6	99.9995% B rod 5 mm diameter: Light Ltd; high purity metal
UNS R01001–R01999	B and alloys: American Standards system; unified number

8. Cadmium Cd

Physical properties

Atomic number	48	
Atomic weight	112.4	
Crystal structure	Hexagonal	
Colour	White with blue tinge	
Specific gravity	8.64	
Density	8640 kg/m³	(0.305 lb/in.³)
Melting point	321 °C	
Boiling point	767 °C	
Specific heat	0.230 J/g °C	(0.055 cal/g °C)
Thermal conductivity	92 W/m °C	(22 cal/m s °C)
Coefficient of linear expansion (20–100° C)	$30 \times 10^{-6}/$ °C	
Latent heat of fusion	55.3 J/g	(13.2 cal/g)
Latent heat of vaporization	1197 J/g	(286 cal/g)
Thermal neutron absorption cross-section	2.400 barns/atom	
Electrical conductivity	22.5% IACS (copper 100%)	
Specific resistance	77 microhm mm	
Temperature coefficient of electrical resistance	0.0042/ °C	
Electrochemical equivalent	2.09 g/A/h	
Electrode potential	-0.403 V	
Magnetic susceptibility	0.18×10^{-6}	
Young's modulus of elasticity	55.2×10^9 N/m²	$(8 \times 10^6$ lbf/in.²)
Tensile strength	75 N/mm²	(5 tonf/in.²)
Hardness	22 DPN	

8.1 General notes on cadmium

Approximately 60% of the world's supply of cadmium is used as electroplating anodes for deposition of a corrosion protection coating. Although much dearer than zinc, cadmium plating is commonly used and is specified for aircraft use wherever steel has to be protected by plating. It must not be used in contact with food as it can be poisonous if ingested. Cadmium oxide fumes are toxic and are a hazard if allowed to accumulate during melting, welding or brazing.

Metallic cadmium is a neutron absorber and is used in rod form, or in electroplated graphite rods as fission moderators in atomic piles.

There are very few alloys based on cadmium, but a few brazing materials have been developed which have a high shear strength. These are zinc/cadmium or silver/cadmium, and are used for tipping high speed tools. There are also a few cadmium based alloys used for bearing materials.

Cadmium is used as an alloying element with copper, when up to 1% cadmium is added to increase the tensile strength with very little reduction in electrical conductivity. These alloys are used for traction conductor wires and for long span transmission conductor wires.

With lead, cadmium forms an alloy which does not work harden, and has been used for cable sheathing purposes, where a harder coating than pure lead is required.

Low melting point alloys used for fusible links generally have some cadmium content. Where alloys with cadmium are tested above their melting point for any reason, great care is necessary to prevent ingestion of cadmium fumes, which are now known to be extremely toxic.

Note. The following abbreviations and units are used in the tables:

DPN	Hardness, diamond pyramid number
UTS	Ultimate tensile strength, N/mm²
Elon	Elongation, %
Proof	0.1 % proof strength, N/mm²

1 N/mm²=0.1 hbar=0.102 kgf/mm²=0.06475 tonf/in²=145.04 lbf/in²

Symbol	Nominal analysis, supplier, condition and remarks.
2.2480	20% Zn Cd brazing alloy: Working temperature 280 °C; German Standard
BS 2686/1	99.95% Cd (Sb + As + Th 0.01% (max)) anodes: Cast; rolled or extruded
CADMIUM	Cd metal: In form of powder, stick, sheet, etc.; Blackwell; primary metal
CAZIN	17.4% Zn Cd alloy: For soldering steel cables; origin unknown; melting point 263 °C

Symbol	Nominal analysis, supplier, condition and remarks.
DIN 8512 L Cd Zn 20	20% Zn Cd brazing alloy: Working temperature 280 °C
ELMET SILNO	Ag + Ni with Ag and Cd oxides: Sintered material; Metro Cutanit Ltd; range of materials
h Cd 1	99.9999% Cd rod 13 mm diameter: Light Ltd; high purity metal
L Cd Zn 20	20% Zn Cd brazing alloy: Designation used by German Standard
LM 5	Ag Cd base soft solder: Melting range 338–390 °C; Johnson Matthey; electrical conductivity 22.4% IACS UTS: 130 Elon: 25%
HT 5	16.5% Zn 5% Ag Cd alloy: Braze metal; melting range 270–285 °C; Sheffield Smelting Co. DPN: 60 UTS: 200 Elon: 8% Proof: 150

Symbol	Nominal analysis, supplier, condition and remarks.
SAE 18	1.3% Ni Cd alloy: For bearings
SAE 180	0.6% Cu 0.7% Ag Cd alloy: For bearings

Note. The following abbreviations and units are used in the tables:

DPN	Hardness, diamond pyramid number
UTS	Ultimate tensile strength, N/mm^2
Elon	Elongation, %
Proof	0.1 % proof strength, N/mm^2

$1\ N/mm^2 = 0.1\ hbar = 0.102\ kgf/mm^2 = 0.06475\ tonf/in^2 = 145.04\ lbf/in^2$

9. Caesium Cs (cesium)

Physical properties

Atomic number	55	
Atomic weight	132.91	
Crystal structure	Body-centred cubic	
Colour	Silvery white	
Specific gravity	1.89	
Density	1890 kg/m^3	(0.066 lb/in.^{3})
Melting point	28.4 °C	
Boiling point	705 °C	
Specific heat	0.209 J/g °C	(0.050 cal/g °C
Thermal conductivity	– W/cm °C	(– cal/m s °C)
Coefficient of linear expansion (20–100° C)	97 × 10^{-6}/ °C	
Latent heat of fusion	15.78 J/g	(3.77 cal/g)
Latent heat of vaporization	611 J/g	(146 cal/g)
Thermal neutron absorption cross-section	29 barns/atom	
Electrical conductivity	9% IACS (copper 100%)	
Specific resistance	190 microhm mm	
Temperature coefficient of electrical resistance	0.0044/ °C	
Electrochemical equivalent	– g/A/h	
Electrode potential	-3.0 V	
Magnetic susceptibility	-0.22 × 10^{-6}	
Young's modulus of elasticity	– N/m^2	
Tensile strength	– N/mm^2	
Hardness	– DPN	

9.1 General notes on caesium

Caesium salts used in the manufacture of electronic high vacuum valves result in the reduction of caesium from the chloride salt by molten barium metal. This is a by-product of the high vacuum process required by modern electronic equipment and is not a real source of the metal.

Caesium is of interest in that it is the most basic element known, and could be termed the most metallic element. The metal reacts explosively with oxygen or water and thus finds very little commercial use.

When subjected to light, gamma or X-rays it emits electrons, in proportion to the amount falling on it. The sensitivity to light rays is very similar to the human eye. Thus, in spite of the difficulty of using the metal, photoelectric cell mechanisms are manufactured with caesium.

The metal is not radio-active.

As would be expected, there are no alloys based on caesium. To date no use has been made of the metal as an alloying element, but it has found some use as a catalyst in the resin industry.

No specifications or proprietary names could be found for caesium metal.

10. Calcium Ca

Physical properties

Atomic number	20	
Atomic weight	40.08	
Crystal structure	Face-centred cubic	
Colour	Yellowish white	
Specific gravity	1.54	
Density	1540 kg/m³	(0.056 lb/in.³)
Melting point	851 °C	
Boiling point	1482 °C	
Specific heat	0.63 J/g °C	(0.150 cal/g °C)
Thermal conductivity	126 W/m °C	(30 cal/m s °C)
Coefficient of linear expansion (20–100° C)	23 × 10⁻⁶/ °C	
Latent heat of fusion	216.4 J/g	(51.7 cal/g)
Latent heat of vaporization	4660 J/g	(1113 cal/g)
Thermal neutron absorption cross-section	0.43 barns/atom	
Electrical conductivity	48.7% IACS (copper 100%)	
Specific resistance	46 microhm mm	
Temperature coefficient of electrical resistance	0.00457/ °C	
Electrochemical equivalent	0.747 g/A/h	
Electrode potential	-2.8 V	
Magnetic susceptibility	+ 1.1 × 10⁻⁶	
Young's modulus of elasticity	23.4 × 10⁹ N/m²	(3.4 × 10⁻⁶ lbf/in.²)
Tensile strength	rolled 110 N/mm²	(7 tonf/in.²)
	annealed 40 N/mm²	(2.5 tonf/in.²)
Hardness	17 DPN	

10.1 General notes on calcium

Calcium resembles aluminium and magnesium in many of its properties but is more difficult to cast as it oxidizes much more rapidly. It is also much softer and more ductile, being readily cut with a knife and bent by hand. Although it burns readily in air it is not subject to spontaneous combustion and can be safely handled.

Calcium, either as the metal or as a compound, finds several uses as de-oxidizers in the preparation of some of the rarer metals such as beryllium, titanium, vanadium, uranium and zirconium.

There are no commercial calcium based alloys, but the metal is used as an addition to aluminium and magnesium alloys, where it acts as a grain refiner. There is reported to be an improvement in surface finish when machining these alloys.

Lead/calcium alloys have improved fatigue strength and have been used for bearing alloys and cable sheathing, as well as electrical storage battery grids.

Calcium silicide is used as a graphite controller in 'inoculated' cast irons.

The metal in particle form is now finding considerable use as a catalyst in chemical engineering.

Symbol	Nominal analysis, supplier, condition and remarks.
CALCIUM	Ca metal: In form of lumps and filings; Blackwell; primary metal
DOMAL 2	99.9% Ca metal: Dominion Magnesium
DOMAL 3	0.3% Al 0.75% Mg Ca metal: Dominion Magnesium
DOMAL 4	0.35% Al 1% Mg 98.5% Ca metal: Dominion Magnesium
NELCO	0.25% Al 0.2% Mg 99% Ca metal: Nelco Metal Corp
W Ca 17	99.9% Ca metal: In grain form; Light Ltd; high purity metal

Note. The following abbreviations and units are used in the tables:

DPN	Hardness, diamond pyramid number
UTS	Ultimate tensile strength, N/mm²
Elon	Elongation, %
Proof	0.1 % proof strength, N/mm²

1 N/mm²=0.1 hbar=0.102 kgf/mm²=0.06475 tonf/in²=145.04 lbf/in²

11. Cerium Ce

Physical properties

Atomic number	58	
Atomic weight	140.13	
Crystal structure	Face-centred cubic	
Colour	Steel grey	
Specific gravity	6.7	
Density	6700 kg/m^3	(0.238 lb/in.^{3})
Melting point	795 °C	
Boiling point	3470 °C	
Specific heat	0.188 J/g °C	(0.045 cal/g °C)
Thermal conductivity	109 W/m °C	(2.6 cal/m s °C)
Coefficient of linear expansion (20–100° C)	8.4 × 10^{-6}/ °C	
Latent heat of fusion	35.6 J/g	(8.5 cal/g)
Latent heat of vaporization	– J/g	(– cal/g)
Thermal neutron absorption cross-section	0.73 barns/atom	
Electrical conductivity	2.2% IACS (copper 100%)	
Specific resistance	780 microhm mm	
Temperature coefficient of electrical resistance	0.00087/ °C	
Electrochemical equivalent	– g/A/h	
Electrode potential	– V	
Magnetic susceptibility	15 × 10^{-6}	
Young's modulus of elasticity	41.4 × 10^9 N/m^2	(6 × 10^6 lbf/in.^{2})
Tensile strength	150 N/mm^2	(10 tonf/in.^{2})
Hardness	24 DPN	

11.1 General notes on cerium

This is one of the elements forming the group known as 'rare earth metals', which are not as rare as was at one time thought, but are by no means plentiful. In general they are found together, and great difficulty is experienced in obtaining any one element free from the remainder. The history of the group gives some indication of the difficulty in purification, in that what at one time was thought to be one element turned out in fact to be a mixture of four of the rare earth metals.

Cerium is one of the less rare metals in the group and probably the most widely used, and for this reason has been separated from the other 'rare earths' which are treated as one group in Section 35 of this book.

Cerium metal is soft, being readily cut with a knife, oxidizes rapidly in air, but does not burn spontaneously. It reacts with water, releasing hydrogen but not violently. It burns more intensely in air than does magnesium and this affinity for oxygen is utilized as a 'getter' when cerium combines with the last traces of oxygen in vacuum electronic equipment.

Cerium alloyed with other rare earth metals is used as a powerful reducing agent and catalyst in chemical engineering.

An interesting use of cerium is as an alloy with iron in proportions of approximately 50 per cent each, when it has peculiar pyrophoric qualities, giving off copious sparks when rubbed with a file. These alloys are used as 'flints' in cigarette lighters and similar equipment. This use is now diminishing with the advent of the 'catalytic reaction' type equipment.

Cerium is used as an alloying element with aluminium and magnesium, when it enhances the high temperature properties, fatigue and creep strength.

The alloy mischmetall, 50% cerium, is used as a deoxidizer in the steel industry, and has been found to give some grain refinement. In small amounts cerium gives an improvement in the hot working properties of austenitic steels and nickel alloys.

Symbol	Nominal analysis, supplier, condition and remarks.
AUER	35% Fe 24% La 4% Yb 2% Er Ce alloy: Common name; pyrophoric material
CERIUM	99.5% Ce metal: Ingot, sheet, rod & wire; Johnson Matthey
CERIUM	Ce metal: Blackwell, primary metal
h Ce 16	99.9% Ce metal: Ingot, Light Ltd; high purity metal
h Ce 19	99.9% Ce metal: Foil; Light ltd; high purity metal
h Ce 20a	99.9% Ce metal; Wire; 1mm in diameter; Light Ltd; high purity metal
h Ce 96	99.5% Ce 140 isotope: Light Ltd; stable isotopic form
HUBER	15% Mg Ce alloy: Common name; pyrophoric material
MISCHMETALL	40% La + Sa + Nd + Pr 10% Fe + Yb Ce alloy: Obtained by electrolysis; common name

Symbol	Nominal analysis, supplier, condition and remarks.
SENSITIVE	25% Pt Ce alloy: Common name; pyrophoric material
WELSBACH	30% Fe 10% rare earth Ce alloy: Common name; pyrophoric material

Note. The following abbreviations and units are used in the tables:

DPN	Hardness, diamond pyramid number
UTS	Ultimate tensile strength, N/mm^2
Elon	Elongation, %
Proof	0.1 % proof strength, N/mm^2

$1 \ N/mm^2 = 0.1 \ hbar = 0.102 \ kgf/mm^2 = 0.06475 \ tonf/in^2 = 145.04 \ lbf/in^2$

12. Chromium Cr

Physical properties

Atomic number	24	
Atomic weight	52.01	
Crystal structure	Rhombohedral	
Colour	Greyish white	
Specific gravity	7.19	
Density	7190 kg/m^3	(0.255 lb/in.^{3})
Melting point	1860 °C	
Boiling point	2200 °C	
Specific heat	0.461 J/g °C	(0.110 cal/g °C)
Thermal conductivity	69.1 W/m °C	(16.5 cal/m s °C)
Coefficient of linear expansion (20–100° C)	6.2 × 10^{-6}/ °C	
Latent heat of fusion	293.1 J/g	(70 cal/g)
Latent heat of vaporization	5945 J/g	(1420 cal/g)
Thermal neutron absorption cross-section	2.9 barns/atom	
Electrical conductivity	14% IACS (copper 100%)	
Specific resistance	130 microhm mm	
Temperature coefficient of electrical resistance	0.00588/ °C	
Electrochemical equivalent	0.971 g/A/h	
Electrode potential	-0.56 V	
Magnetic susceptibility	3.08 × 10^{-6}	
Young's modulus of elasticity	248.2 × 10^9 N/m^2	(36 × 10^6 lbf/in.^{2})
Tensile strength	890 N/mm^2	(60 tonf/in.^{2})
Hardness	annealed 90 DPN	

12.1 General notes on chromium

Metallic chromium of 99% purity is available but is too brittle for any practical purpose. Higher purity grades are generally only laboratory materials, but there are indications that a more ductile and therefore useful metal may be available in the future, which could have many potential uses. At present there is no alloy system based on chromium but the metal is used for chromium plating and as an alloy addition, often with startling results.

In electroplating, chromium is used for two distinct purposes. First, as a decorative finish when it is always plated on top of nickel in the form of a very thin cracked deposit. As this is known to be porous, recent work would appear to indicate that the microcracking produces the necessary small anode to large cathode ratio known to result in galvanic protection.

The coating is so thin that unless the undercoat is highly polished, the chromium will appear dull. The hard blue-white colour of the deposited chromium is retained under all normal atmospheric conditions. The chrome is seldom or never the source of failure, which is caused by corrosion of the base metal from a too thin, badly prepared, or porous nickel deposit.

Deposits of hard chrome for wear resistant purposes are now common. Chromium has a low coefficient of friction and this is aided by the electroplated deposit having a microcracked surface which retains a film of oil even under adverse conditions. The deposit is plated directly on the base metal, and should not be less than 0.1 mm thick after grinding. Electroplated chromium is very brittle and must not be subjected to shock loads. Considerable care must also be taken when grinding the deposit as this is very readily cracked and softened. Because of this, hard chromium plating is being replaced by metal deposit techhniques using other alloyed materials which have good wear resistance and can better withstand stock loading. Chromium deposits have however, certain high temperature wear resisting properties which appear to be unique.

Details of the use of chromium as an alloying element appear with the appropriate section on the base material, briefly these are:

Aluminium. Additions of chromium in the order of 0.3% are made as grain refiners and to improve the corrosion resistance of many high duty aluminium wrought alloys.

Copper. A chromium content of 0.5% in copper produces an age hardening alloy used for welding electrodes.

Cobalt. Up to 25% chromium is found in the cobalt based alloys, where it forms a solid solution. If carbon is present some of the chromium will form hard carbides, and there is evidence to show that it is the surface layer of chromium oxide which gives these alloys their excellent high temperature oxidation and corrosion resistance.

Steel. Chromium has two attributes when alloyed in steel. First, it combines with carbon to form very hard chromium carbides. These considerably increase the hardenability of the steel. The chromium content varies from as low as 0.5% to about 5%, and is often found in conjunction with other carbide forming elements such as vanadium, molybdenum and tungsten. Most high speed tools have some chromium added.

The second virtue of chromium is that it enhances corrosion resistance and resistance to oxidation at temperature. Chromium alone in quantities from 10% up to 20% gives steels with very good corrosion resistance. When over 7% nickel is added to steel with 17% and more chromium an austenic stainless steel is formed. Detailed information on the affects of chromium on steels appear in the relative sections concerned.

Nickel. Additions of up to 20% chromium to nickel gives a range of materials with excellent high temperature properties. The basic 80/20 nickel/chromium alloys now include many complex elements. Details of these are given in the specific sections.

Symbol	Nominal analysis, supplier, condition and remarks.
608	15% Ni 2.0% W chromium carbide: General Electric Co
	UTS: 240 Elon: 1.1% Proof: 60
ASTM A 560/50/50	0.1% C 50% Ni Cr alloy
ASTM A 560/60/40	0.1% C 40% Ni Cr alloy
CHROMIUM 99	0.01% C 0.2% Fe 0.1% Al 0.1% Si 0.01% Cu Cr metal: Metal Alloys Ltd
CHROMIUM 99 LOW GAS	0.08% C 0.03% N 0.001% H Cr metal: Metal Alloys Ltd
CHROMIUM	Cr metal: Fused powder; Blackwell; pure metal
h Cr 7a	99.999% Cr flake: Light Ltd; high purity metal
h Cr 7b	99.999% Cr pellets: Light Ltd; high purity metal
h Cr 8	99.999% Cr crystals (ex iodide): Light Ltd; high purity metal
h Cr 12	99.99% Cr flake: Light Ltd; high purity metal
p Cr 23	99.0% Cr crystalline powder: Light Ltd
CRONITE 50/50	0.1% C 48% Ni 2.0% W Cr alloy: Casting; Cronite Ltd; as cast; heat resistant

Symbol	Nominal analysis, supplier, condition and remarks.
CRONITE 60/40	0.1% C 40% Ni 2.0% W Cr alloy: Casting; Cronite Ltd; as cast; heat resistant
	DPN: 350 UTS: 520 Elon: 1.0% Proof: 400
TOLOY 60/40	0.1% C 40% Ni Cr alloy: Casting; Wellman;
UNS R20001–R29999	Cr metal and alloys. American Standards system; unified number

Note. The following abbreviations and units are used in the tables:

DPN	Hardness, diamond pyramid number
UTS	Ultimate tensile strength, N/mm^2
Elon	Elongation, %
Proof	0.1 % proof strength, N/mm^2

$1\ N/mm^2 = 0.1\ hbar = 0.102\ kgf/mm^2 = 0.06475\ tonf/in^2 = 145.04\ lbf/in^2$

13. Cobalt Co

Physical properties

Atomic number	27	
Atomic weight	58.94	
Crystal structure	Close-packed hexagonal up to 417 °C; face-centred cubic up to melting point	
Colour	Silver-white with bluish tinge	
Specific gravity	8.8	
Density	8800 kg/m³	(0.322 lb/in.³)
Melting point	1493 °C	
Boiling point	3100 °C	
Specific heat	0.440 J/g °C	(0.105 cal/g °C)
Thermal conductivity	69.21 W/cm °C	(16.53 cal/m s °C)
Coefficient of linear expansion (20–100° C)	12.5×10^{-6}/ °C	
Latent heat of fusion	259.6 J/g	(62 cal/g)
Latent heat of vaporization	3280 J/g	(1.500 cal/g)
Thermal neutron absorption cross-section	34.8 barns/atom	
Electrical conductivity	27.6% IACS (copper 100%)	
Specific resistance	62.4 microhm mm	
Temperature coefficient of electrical resistance	0.0053/ °C	
Electrochemical equivalent	1.1 g/A/h	
Electrode potential	-0.28 V	
Magnetic susceptibility	Ferromagnetic	
Young's modulus of elasticity	211×10^9 N/m²	$(30.6 \times 10^6$ lbf/in.²)
Tensile strength	240 N/mm²	(16 tonf/in.²)
Hardness	230 DPN	

13.1 General notes on cobalt

Cobalt produced by present processes usually has a purity of 99.5% but recent work has shown that with careful refining conditions malleable cobalt with a purity level of 99.8–99.9% can be produced.

Cobalt is a silver-white metal with a slight bluish tinge normally stable in the atmosphere but tending to oxidize more easily than nickel.

The pure metal has few uses at present but because of two recent developments is now commercially available as bar sheet and expanded metal, and is therefore likely to find more widespread applications. At present cobalt is used because of its good magnetic properties which are retained up to 1120 °C, considerably higher than any iron based alloy. It is also used as an electroplated deposit, and as polished sheet for reflective purposes under arduous conditions.

Cobalt base alloys and cobalt containing alloys, of which there are a great number, find important commercial applications as magnetic alloys, creep and heat resistant alloys, ultra-high tensile steels, dental and surgical alloys and in a wide range of non-ferrous alloys where its addition contributes either directly or indirectly to the required properties.

The advances made in engineering materials to meet the demands of rapidly expanding technologies have resulted in a dramatic growth in the consumption of cobalt to today's high level.

Symbol	Nominal analysis, supplier, condition and remarks.
COBALT	High purity metal: Impurities 10 p.p.m.; Johnson Matthey; supplied as sponge, rod, sheet and wire
COBALT	Co metal: In form of powder, briquettes, strip and wire; British Metal Corporation Ltd
COBALT	Co metal: In all forms; Brandhurst Co. Ltd
COBALT	Co metal: In form of powder, sponge, bar, sheet; New Metals & Chemicals Ltd
COBALT	Co metal: In form of powder, cubes, sheets, etc.; Blackwell; pure metal
L Co 8	99.999% Co metal: In form of sponge; Light Ltd; high purity metal
L Co 11	99.999% Co metal: In form of rod 5 mm diameter; Light Ltd; high purity metal
L Co 73	99.99% Co metal: In form of single crystals 6.3 × 12 mm; Light Ltd
W Co 23	99.5% Co metal: In form of powder; Light Ltd

Note. The following abbreviations and units are used in the tables:

DPN	Hardness, diamond pyramid number
UTS	Ultimate tensile strength, N/mm²
Elon	Elongation, %
Proof	0.1 % proof strength, N/mm²

1 N/mm²=0.1 hbar=0.102 kgf/mm²=0.06475 tonf/in²=145.04 lbf/in²

13A Cobalt alloys – wrought and cast

Specific gravity	8.2–9	
Density	8200–9000 kg/m³	(0.29–0.33 lb/in.³)
Solidus/liquidus	14.7 W/m °C	(3.5 cal/m s °C)
Coefficient of linear expansion (20–100° C)	9.5–11 × 10⁻⁶/ °C	
Electrical conductivity	1–2% IACS (copper 100%)	
Specific resistance	850–17000 microhm mm	
Young's modulus of elasticity	179–248 × 10⁹ N/m²	(26–36 × 10⁶ lbf/in.²)
Impact	– J	
Fatigue strength (10⁶ cycles)	± 250 N/mm²	(± 17 tonf/in.²)
	(applies to Stellite X40)	

Hot strength

Temperature °C	Tensile strength N/mm²	Elongation %
650	300 (after 1500 h)	5.6 (20 tonf/in.²)
750	200 (after 1000 h)	6.5 (13 tonf/in.²)
815	120 (after 750 h)	1.5 (8 tonf/in.²)

The above properties are typical of the following group, and may not apply exactly to any one specification. It is possible that with certain specifications some of the values may not be applicable.

General metallurgical characteristics

These alloys are based on the solid solution of cobalt and chromium. They have additions of nickel to improve the ductility, while carbon and carbide formers such as tungsten and molybdenum are added to increase the hardness.

Some of the more complex modern alloys are capable of being age hardened to a limited extent in a manner analogous to the nickel chromium alloys. To date none of the cobalt materials produced has as marked age hardening properties as those in the nickel series.

The ductility of all the alloys is very low, hence little or no cold work can be carried out without causing undue brittleness often resulting in crumbling.

Cobalt alloys all have the ability to retain their hardness at elevated temperatures and this with their corrosion resistance is the principal reason for their existence. This extreme difficulty in hot working results in few wrought products, although there are notable exceptions. The alloys are generally used as castings, or as welded or sprayed coatings on more ductile materials.

The cobalt alloys have excellent damping characteristics and this property can be of considerable value in high duty rotating assemblies operating at elevated temperatures. The magnetic properties of some cobalt alloys arise from the high anisotropy constant of cobalt which leads to magnetic properties such as coercivity and magnetic energy. Cobalt also has the highest curie point. Thus, the alloys retain their magnetism at a higher temperature than any other magnetic materials. Alloyed with iron, cobalt increases the saturation induction. Only a few of the magnetic alloys appear in the following pages as cobalt is seldom the principal constituent. These alloys appear under 20C and 44L.

The effect of alloying elements are briefly as follows:

Chromium. This is the principal element, forming a stable solid solution, analogous in many ways to the nickel chrome series. There is evidence that it is the chromium which imparts the hot oxidation resistance.

Nickel. The presence of nickel increases the malleability with the minimum effect on the remaining properties. Nickel also helps to take some of the refractory metals into solution, thus aiding the hot strength, and is an essential element in certain precipitation hardening alloys as it is a constituent in the necessary intermetallic compounds.

Molybdenum. This strengthens the solid solution matrix, increasing the hot strength and also taking part in precipitation age hardening.

Tungsten, niobium and tantalum. These act in a similar manner to molybdenum, generally being more active and in some cases more refractory.

Titanium. This is one of the elements which forms intermetallic compounds with nickel, allowing age hardening to take place.

Aluminium. This acts in a similar manner to titanium, again forming intermetallic compounds with nickel.

Zirconium and other refractory materials are added to increase the hot strength or corrosion resistance under specific conditions.

Carbon is present with carbide forming elements to give the very hard carbides which give wear resistance. Depending on the metal used, these can have a very high hardness even at elevated temperatures. With the correct thermal treatment, some age hardening can be obtained with these carbides but this is not as marked as with the intermetallic compounds.

Boron. This is used in small amounts to improve the hot strength, and in large amounts (above 1%) to act as a fluxing agent in the sprayed or welded alloys to improve the bond to the base metal.

Thermal treatment

Cold working of these alloys is generally impossible, or only possible on a very limited scale without causing undue brittleness. A small number of alloys have been developed capable of accepting cold or hot work and considerable development is being carried out in this field. Annealing after working or heavy machining, particularly before welding, is recommended and requires a temperature between 1050 and 1175 °C depending on the alloy.

This is the only thermal treatment possible with the majority of cobalt alloys.

With the age hardening alloys it is first necessary to ensure that the intermetallic compounds are in solution. This can be achieved by chill casting or heating to above about 1150 °C then cooling as rapidly as possible. Ageing requires a temperature in the region of 700–900 °C for several hours and results often in very little increase in room temperature properties, but an improvement in creep strength and other hot strength properties. The actual temperatures and times can be rather critical and it is essential that the originator of the specification is consulted for details.

Certain sprayed or welded coatings require a subsequent heat treatment to fuse the coating and base metal, and this generally requires a salt bath or controlled atmosphere.

Other heat treatments require no special equipment apart from the high temperatures necessary. Special atmospheres are not normally required and when used must be free from sulphur contamination. This generally prohibits the use of oil fired furnaces and requires careful control of the furnace atmosphere with gas furnaces.

Welding and brazing

The parts should ideally be in the annealed condition and the weld area must be thoroughly cleaned and free from all trace of oxide. This is best achieved by abrasive blasting followed by electrolytic etching immediately before welding, or by sodium hydride treatment. Careful weld design is essential to minimize cooling stresses which can cause restraint.

Welding and brazing. Electric arc and oxyacetylene welding using fluxes are possible with many of the alloys listed, but best results are generally obtained with inert gas shielded arc welding. Resistance welding is only possible with certain alloys and special equipment. Brazing generally requires a very active flux or a furnace technique using vacuum or hydrogen, as the braze filler metals require high temperature properties which match those of the base material.

It is generally advisable to obtain specialist advice before attempting to fusion join these alloys if the join is to be subjected to any degree of stress.

Post-treatment. None normally required. If there is any doubt regarding cooling stresses causing restraint, the part should be annealed then re-aged if necessary.

Specialist advice should be sought for all critical components.

Flaw detection methods

Crack test. Penetrant oil and chalk test methods are used, with fluorescent penetrants preferred for critical components. Although magnetic, none of these alloys should be magnetically crack tested as there is insufficient magnetism to give sensitive results.

X-ray. This should be carried out on all important castings and high duty welds.

Ultra-sonic test. This can be used after rough machining to identify sub-cutaneous defects. Under certain circumstances it can be used in lieu of X-ray and is generally cheaper.

Chemical etch. Ferric chloride in hydrochloric acid can be used to identify weld defects. It is also a very useful etch for removing any flowed surface layer prior to crack testing critical parts.

Corrosion protection

Temporary. None required.

Permanent. None required. These alloys have been developed for their resistance to oxidation at elevated temperatures and resist attack from a wide range of chemicals and molten metals.

Machinability

These alloys generally have an inherently high hardness which is increased by cold working. This makes them difficult to machine unless the correct techniques and equipment are used. It is important that the tool is cutting at all times. Thus, light cuts or interrupted cuts must be avoided. Best results are obtained with single point tools and heavy duty vibration free machine tools. Tools must be kept sharp and should be of the carbide tipped or high speed steel type.

The modern techniques of spark erosion, electrolytic grinding and electro-chemical machining are now being used on many of the more intractable cobalt alloys with considerable success.

Uses

High temperature creep resistant purposes such as gas turbine components, particularly where hot corrosion is present, turbine blades, nozzle guide vanes. The deposited alloys are also used in this field. Industrial furnaces, chemical equipment, oil cracking plant and other chemical engineering purposes.

Hard facing alloys are also used to resist abrasion and wear and there are certain alloys with excellent hot frettage resistance.

The use of cobalt alloys will undoubtedly increase if some of the problems of workability can be overcome.

Symbol	Nominal analysis, supplier, condition and remarks.
19	0.45% C 26% Cr 15% Ni 6% Mo 1% Ti Co alloy: Casting; American proprietary alloy listed in SAE year book
25 Ni	0.17% C 19% Cr 25% Ni 10% W 1% Fe 1.5% Nb 42.5% Co alloy: General Electric Co.
25 Ni & V	0.2% C 19% Cr 25% Ni 11% W 1.2% Nb 3% V 2% Fe Co alloy: General Electric Co.
422	0.45% C 26% Cr 15% Ni 6% Mo 1% Fe Co alloy: American proprietary alloy listed in SAE year book
422/19	0.45% C 26% Cr 15% Ni 6% Mo 1% Fe Co alloy: Information from Cobalt Information Centre
AF 94	0.12% C 15% Cr 10% Ni 5% Mo 10% W 1% Nb 2% Fe Co alloy: Allegheny Ludlum
ALLOY 6	1% C 30% Cr 1% Si 5% W Co alloy: Casting; Dewrance **DPN: 394 UTS: 670**
ALLOY 9	1.35% C 31% Cr 1% Si 8% W Co alloy: Casting; Dewrance **DPN: 521 UTS: 470**
ALTEMP S 816	0.4% C 20% Cr 20% Ni 4% Mo 4% W 4% Nb Co alloy: Allegheny Ludlum
AIR RESIST 13	0.45% C 1.0% Ni 11.0% W 2.0% Nb 3.5% Al 2.5% Fe Co alloy: Origin unknown
AIR RESIST 213	Modified Air Resist 13
AIR RESIST 215	Modified Air Resist 13
AISI 670	0.12% C 19.8% Cr 9.9% W 15.2% Ni 1.6% Fe Co alloy: Annealed **UTS: 1050 Elon: 64% Proof: 450**
AISI 671	0.4% C 19.6% Cr 20.3% Ni 10% Fe 4% Mo 4% W 4% Nb Co alloy: Solution treated & aged **UTS: 1000 Elon: 35% Proof: 370**
AMS 5373 A	28% Cr 5% W Co alloy: Sand casting
AMS 5375 B	25% Cr 1.8% Ni 5% W Co alloy: Investment casting
AMS 5378 B	25% Cr 32% Ni 5.5% Mo Co alloy: Investment casting
AMS 5380 C	26% Cr 15% Ni 6% Mo Co alloy: Investment casting
AMS 5382 B	25% Cr 10% Ni 7.5% W Co alloy: Investment casting
AMS 5385 C	27% Cr 2.8% Ni 5.5% Mo Co base: Investment casting
AMS 5387	28% Cr 5% W Co alloy: Investment casting
AMS 5530 C	15.5% Cr 16% Mo 5.5% Fe 3.8% W Co alloy: Sheet
AMS 5534	20% Cr 20% Ni 4% Mo 4% Nb Co alloy: As S816
AMS 5537 B	20% Cr 10% Ni 15% W Co alloy: Sheet
AMS 5759 B	20% Cr 10% Ni 15% W Co alloy: Bar & forging
AMS 5765 A	20% Cr 20% Ni 4% Mo 4% W 4% Nb Co alloy: Bar & forging
AMS 5788	28% Cr 4.5% W Co alloy: For coating
AMS 5796	20% Cr 10% Ni 15% W Co alloy: Wire
AMS 5797	20% Cr 10% Ni 15% W Co alloy: Coated; electrode
ANC 13	0.5% C 25% Cr 10% Ni 7% W Co alloy: Investment casting; designation used by BS

Symbol	Nominal analysis, supplier, condition and remarks.
ANC 14	0.2% C 27% Cr 2.5% Ni 5.5% Mo Co alloy: Investment casting; designation used by BS
ASTM A461/671	0.37% C 20% Cr 20% Ni 4.0% Mo 4.0% W 4.0% Nb + Ta 5.0% Fe Co alloy: Bar
ASTM A567/1	0.25% C 27% Cr 2.7% Ni 5.5% Mo 3% Fe Co alloy: Casting
ASTM A567/2	0.5% C 25.5% Cr 10.5% Ni 7.5% W 2% Fe Co alloy: Casting
ASTM A639/671	0.37% C 20.0% Cr 20% Ni 4.0% Mo 4.0% W 4.0% Nb + Ta 5.0% Fe (max) Co alloy
ATG Z⁴	0.4% C 20% Cr 20% Ni 4.0% Mo 5.0% Fe 4.0% W 4.0% Nb Co alloy: Imphy
BS 3146 ANC13	0.5% C 25% Cr 10% Ni 7% W Co alloy: Investment casting; as cast **UTS: 630 Elon: 5% Proof: 400**
BS 3146 ANC14	0.2% C 27% Cr 2.5% Ni 5.5% Mo Co alloy: Investment casting; as cast **UTS: 600 Elon: 5% Proof: 420**
BS 3531/13	29% Cr 6.0% Mo Co alloy: For implant and surgery tools; cast **UTS: 650 Elon: 8% Proof: 450**
BS 3531/1-4	0.1% C 20% Cr 15% W 10% Ni Co alloy: For implant and surgery tools; wrought **UTS: 880 Elon: 30% 4390**
CM 6	1% C 30% Cr 1% Si 5% W Co alloy: For hard facing; Dewrance; gas welding
CM 9	1.3% C 31% Cr 1% Si 8% W Co alloy: For hard facing; Dewrance; gas welding
CM 10	4.2% C 15% Cr 2% Ni 8% Mo 1.5% Si 1% V Co alloy: For hard facing; Dewrance; gas welding
CM 63	2% C 28% Cr 20% W 1% V Co alloy: For hard facing; Dewrance; gas welding
CM 106	1% C 30% Cr 1% Si 5% W Co alloy: For hard facing; Dewrance; electric welding
CM 110	4.25% C 16% Cr 2% Ni 8% Mo 1.5% Si 1% V Co alloy: For hard facing; Dewrance; electric welding
CO-ELINVAR	10% Cr 30% Fe Co alloy: Details from Cobalt Information Centre
CROFORM	27% Cr 5% Mo Co alloy: For dental uses; details from Cobalt Information Centre **DPN: 390 UTS: 700 Elon: 4% Proof: 510**
CUNICO 11	24% Ni 35% Cu Co alloy: For magnets; US material listed by PMA
DELCRO 1300K	0.1% C 28% Cr 20% Fe 0.7% Si Co alloy: Deloro; for furnace equipment; origin unknown
DIN 17410	German Standard covering magnetic alloy
DYNAMET	20% Cr 13% Ni 2.5% W 17% Fe Co alloy: Wrought; American proprietary alloy for springs
DYNAVAR	20% Cr 13% Ni 2.5% W 17% Fe Co alloy: Wrought; American proprietary alloy for springs
ELGILOY	0.15% C 20% Cr 15% Ni 7% Mo 16% Fe 0.04% Be Co alloy: American proprietary alloy listed in SAE year book **DPN: 702 UTS: 2460 Proof: 1510**
ELGILOY	0.15% C 20% Cr 15% Ni 15% Fe 7.0% Mo 0.04% Be Co alloy: Strip; Gilby-Brunton; aged **DPN: 700 UTS: 2550 Proof: 1600**
ESCO ALLOY 72	0.1% C 28% Cr 20% Fe 0.7% Si Co alloy: Eastern Stainless Steel; for furnace equipment
ESSCALLOY L 605	0.1% C 1.5% Mn 20% Cr 10% Ni 15% W 3% Fe Co alloy: Eastern Stainless Steel Co
F 484	0.4% C 25% Cr 5% Ni 4% Mo 4% W 4% Nb 1% B 2% Fe Co alloy: Allegheny Ludlum
G 32	0.3% C 19% Cr 12% Ni 2% Mo 1.2% Nb 2.8% V 13% Fe Co alloy: Bar & sheet; Jessop Saville; solution treated & aged **DPN: 320 UTS: 1000 Elon: 8% Proof: 720**

Note. The following abbreviations and units are used in the tables:

DPN	Hardness, diamond pyramid number
UTS	Ultimate tensile strength, N/mm^2
Elon	Elongation, %
Proof	0.1 % proof strength, N/mm^2

$1 \text{ N/mm}^2 = 0.1 \text{ hbar} = 0.102 \text{ kgf/mm}^2 = 0.06475 \text{ tonf/in}^2 = 145.04 \text{ lbf/in}^2$

Symbol	Nominal analysis, supplier, condition and remarks.
G 34	0.8% C 19% Cr 12.5% Ni 2% Mo 1.3% Nb 2.8% V Co alloy: Casting; Jessop Saville; as cast **DPN: 305 UTS: 600 Elon: 2.5% Proof: 400**
G 34C	0.8% C 19% Cr 12% Ni 15% Fe 2% Mo 2.5% V 1.3% Nb Co alloy: Casting; Jessop Saville
G 60	0.25% C 27% Cr 2.8% Ni 5.5% Mo Co alloy: Casting; Jessop Saville
G 65	0.45% C 25% Cr 10% Ni 7% W 2% Fe Cobalt alloy: Casting; Jessop Saville; as cast **UTS: 700 Elon: 17% Proof: 400**
G 75	0.45% C 21% Cr 11% W 2% Nb Co alloy: Casting; Jessop Saville
G 87	0.10% C 21.0% Cr 10.0% Ni 15.0% W Co alloy: Jessop Saville; scale or creep resistant
G 88	0.10% C 16.0% Cr 11.0% Ni 1.5% Mo 1.5% Nb Co alloy: Jessop Saville; scale or creep resistant
GAUSSIT 180	14% V Co alloy: Rolled strip; German alloy listed by PMA
HA 3	2.5% C 30% Cr 3% Ni 12% W 3% Fe Co alloy: Code for Haynes 3
HA 4	0.6% C 30% Cr 14% W 3% Fe Co alloy: Code for Haynes 4
HA 6	1.0% C 30% Cr 3% Ni 4% W 3% Fe Co alloy: Code for Haynes 6
HA 12	1.3% C 30% Cr 8% Ni 8% W 3% Fe Co alloy: Code for Haynes 12
HA 19	1.7% C 31% Cr 10% W 3% Fe Co alloy: Code for Haynes 19
HA 25	0.1% C 20% Cr 10% Ni 15% W 3% Fe Co alloy: Code for Haynes 25
HA 36	0.4% C 18.5% Cr 10% Ni 15% W 0.03% B Co alloy: Code for Haynes 36
HA 98 M2	2.0% C 30% Cr 3.5% Ni 18% W 1% Mo 2% Fe Co alloy: Code for Haynes 98 M2
HA 150	0.1% C 28% Cr 20% Fe 0.7% Si Co alloy: Code for Haynes 150
HA 151	0.5% C 20% Cr 12.5% W 0.05% B Co alloy: Code for Haynes 151
HA 152	0.4% C 20% Cr 1% Ni 11% W 2% Nb & Ta Co alloy: Code for Haynes 152
HA 302	0.85% C 21.5% Cr 10% W 9% Ta 0.2% Zr Co alloy: Code for Haynes 302
HAVAR	.2% C 20% Cr 13% Ni 2% Mo 2.8% W 0.04% Be Co alloy: American proprietary alloy; consult SAE **DPN: 650 UTS: 310 Proof: 1620**
HAYNES 3	2.5% C 30% Cr 3% Ni 12% W 3% Fe Co alloy: Casting; S Osborn; cutting tools; wear resistant parts **DPN: 670 UTS: 540 Elon: 1%**
HAYNES 4	0.8% C 30% Cr 14% W 8% Fe Co alloy: Casting; S Osborn; abrasion & corrosion resistant **DPN: 460**
HAYNES 6	1% C 30% Cr 3% Ni 4% W 3% Fe Co alloy: Casting; S Osborn; tough grades abrasion resistant **DPN: 460 UTS: 820 Elon: 1%**
HAYNES 6B	1% C 30% Cr 3% Ni 4% W 3% Fe Co alloy: Bar; S Osborn; abrasion and corrosion resistant **DPN: 450 UTS: 1040 Elon: 7% Proof: 600**

Note. The following abbreviations and units are used in the tables:

DPN	Hardness, diamond pyramid number
UTS	Ultimate tensile strength, N/mm^2
Elon	Elongation, %
Proof	0.1 % proof strength, N/mm^2

$1 \ N/mm^2 = 0.1 \ hbar = 0.102 \ kgf/mm^2 = 0.06475 \ tonf/in^2 = 145.04 \ lbf/in^2$

Symbol	Nominal analysis, supplier, condition and remarks.
HAYNES 6K	1.6% C 31% Cr 3% Ni 4% W 3% Fe Co alloy: Bar; S Osborn; abrasion and corrosion resistant **DPN: 500 UTS: 1280 Elon: 3.5%**
HAYNES 12	1.3% C 30% Cr 3% Ni 8% W 3% Fe Co alloy: Casting; S Osborn; tough grade; abrasion resistant **DPN: 520 UTS: 740**
HAYNES 19	1.7% C 31% Cr 10% W 3% Fe Co alloy: Casting; S Osborn; cutting alloy; wear resistant **DPN: 570 UTS: 750 Elon: 1%**
HAYNES 25	0.1% C 20% Cr 10% Ni 15% W 3% Fe Co alloy: Wrought; S Osborn; solution treated; heat resistant; wear resistant **UTS: 1000 Elon: 64% Proof: 450**
HAYNES 36	0.4% C 18.5% Cr 10% Ni 15% W 0.03% B Co alloy: Union Carbide casting
HAYNES 98 M2	2% C 30% Cr 3.5% Ni 18% W 1% Mo 2% Fe Co alloy: Casting; S. Osborn; cutting tools; wear resistant parts **DPN: 760 UTS: 660**
HAYNES 150	0.1% C 28% Cr 20% Fe 0.7% Si Co alloy: Union Carbide; for furnace equipment **DPN: 250 UTS: 520 Elon: 8% Proof: 300**
HAYNES 151	0.5% C 20% Cr 12.5% W 0.05% B Co alloy: Casting; Union Carbide
HAYNES 152	0.4% C 20% Cr 1% Ni 11% W 2% Nb & Ta Co alloy: Union Carbide **DPN: 375 UTS: 820 Elon: 5% Proof: 600**
HAYNES 302	0.85% C 21.5% Cr 10% W 9% Ta 0.2% Zr Co alloy: Union Carbide; turbine blade alloy
HAYNES STAR J	2.5% C 32% Cr 2.5% Ni 17% W 2% Fe Co alloy: Union Carbide for metal cutting **DPN: 750 UTS: 480**
HDA 8151	0.5% C 20% Cr 12.5% W 0.05% B Co alloy: Casting; Union Carbide
HE 1049	0.4% C 26% Cr 10% Ni 15% W 3% Fe 0.4% B Co alloy: Casting; American proprietary alloy listed in SAE yearbook
HEV 4	0.1% C 20% Cr 10% Ni 15% W Co alloy: For valves; annealed; Designation used by SAE **UTS: 1100 Elon: 65% Proof: 470**
HONIAL 2	Al Ni Co alloy; For magnets; French alloy listed by PMA
HONIAL 5	Al Ni Co alloy; For magnets; French alloy listed by PMA
HONIAL 12	Al Ni Co alloy; For magnets; French alloy listed by PMA
HS 21	0.25% C 27% Cr 3% Ni 5% Mo 1% Fe Co alloy: Casting; American proprietary alloy; listed in SAE yearbook
HS 23	0.4% C 24% Cr 2% Ni 6% W Co alloy: Casting; listed by Cobalt Information Centre
HS 23	0.4% C 24.0% Cr 2.0% Ni 5.0% W 1.0% Fe Co alloy: Origin unknown
HS 25	0.1% C 20% Cr 10% Ni 15% W Co alloy: American proprietary alloy listed in SAE yearbook **DPN: 220 UTS: 1020 Elon: 65% Proof: 470**
HS 27	0.4% C 25% Cr 3.2% Ni 6% W Co alloy: Casting; listed by Cobalt Information Centre
HS 27	0.4% C 25.0% Cr 32.0% Ni 5.5% Mo 1.0% Fe Co alloy: Origin unknown
HS 30	0.45% C 26% Cr 15% Ni 6% Mo 1% Fe Co alloy: American proprietary alloy listed in SAE yearbook
HS 31	0.5% C 25% Cr 10% Ni 7.5% Mo 1.5% Fe Co alloy: Casting; American proprietary alloy listed in SAE yearbook **UTS: 700 Elon: 11% Proof: 500**
HS 36	0.4% C 19% Cr 10% Ni 15% W 1% Fe 0.03% B Co alloy: American proprietary alloy listed in SAE yearbook **DPN: 300 UTS: 700 Elon: 5% Proof: 600**

Symbol	Nominal analysis, supplier, condition and remarks.
I 336	0.2% C 20% Cr 15% Ni 12% W 1% Nb Co alloy: Listed by Cobalt Information Centre
ILLIUM D	0.2% C 27% Cr 4.5% Mo 1% W 1% Fe Co alloy: American proprietary alloy listed in SAE yearbook
ILLIUM H	0.1% C 28% Cr 20% Fe 0.7% Si Co alloy: Listed in SAE yearbook
ILLIUM X	0.8% C 28% Cr 1% Ni 15% W 2% Fe Co alloy: American proprietary alloy listed in SAE yearbook
J 1570	0.2% C 20% Cr 28% Ni 6% W 4% Ti Co alloy: Universal Cyclop **DPN: 385**
J 1650	0.2% C 19% Cr 27% Ni 12% W 3.8% Ti 2% Ta 0.02% B Co alloy: Universal Cyclop
JETALLOY 209	0.02% C 20% Cr 10% Ni 15% W 2% Ti Co alloy: Quebec Metallurgical
JETALLOY 249	0.03% C 25% Cr 10% Ni 7% W Co alloy: Quebec Metallurgical
JETALLOY 1570	0.2% C 20% Cr 30% Ni 6.5% W 4% Ti Co alloy: General Electric
KOERFLEX 30	8% V 39% Fe Co alloy: Strip for magnets; German alloy listed by PMA
KOERFLEX 200	8% V 39% Fe Co alloy: Wire & strip for magnets; German alloy listed by PMA
KOERFLEX 300	8% V 4% Cr 36% Fe Co alloy: Strip & wire for magnets; German alloy listed by PMA
KOERZIT H	8% V 39% Fe Co alloy: Strip & wire for magnets; German alloy listed by PMA
KOERZIT T	4% Cr 36% Fe 0.8% V Co alloy: Permanent alloy; details from Cobalt Information Centre
L 251	0.4% C 19% Cr 10% Ni 15% W 1% Fe 0.03% B Co alloy: Casting; American proprietary alloy listed in SAE yearbook
L 605	0.1% C 20% Cr 10% Ni 15% W 3% Fe Co alloy: Used by American Companies; consult SAE yearbook
LODEX	Fe Co sintered alloy: With lead for magnets; further information from PMA
M 203	0.07% C 20% Cr 25% Ni 12% W 2% Ti 0.7% Al 1.5% Nb 1.6% Fe Co alloy: General Electric
M 204	0.07% C 18% Cr 25% Ni 12% W 1.6% Fe 1.2% Nb Co alloy: General Electric
M 205	0.07% C 18% Cr 25% Ni 12% W 2.75% Al 1.2% Nb 1.6% Fe Co alloy: General Electric
MAGNETOFLEX 20	20% Ni Co alloy: Rolled strip; German alloy listed by PMA
MALCALLOY	13.5% Al Co alloy: For magnets; Japanese alloy listed by PMA
MAR - M 918	0.05% C 20.0% Ni 7.5% Ta 0.1% Zr Co alloy: Martin **UTS: 880 Elon: 48% Proof: 380**
ML 1700	0.2% C 25% Cr 15% W 0.4% B Co alloy: Casting; American proprietary alloy; listed in SAE yearbook
NFW 1	Al Ni Co alloy: Casting for magnets; Japanese alloy listed by PMA
NFW 2	Al Ni Co alloy: Casting for magnets; Japanese alloy listed by PMA
NFW 3	Al Ni Co alloy: Casting for magnets; Japanese alloy listed by PMA
NFW 4	Al Ni Co alloy: Casting for magnets; Japanese alloy listed by PMA
NFW 5	Al Ni Co alloy: Casting for magnets; Japanese alloy listed by PMA
NFW 6	Al Ni Co alloy: Casting for magnets; Japanese alloy listed by PMA
NICKELVAC L. 605	0.12% C 20% Cr 10% Ni 15% W 1.7% Mn Co alloy: Vanadium alloy; double vacuum cast
NICROBRAZ 210	0.4% C 19.0% Cr 0.8% B 17% Ni 8.0% Si 4.0% W Co alloy: For brazing; Wall Colmonoy; melting range 1110–1150 °C

Symbol	Nominal analysis, supplier, condition and remarks.
NICROBRAZ 300	0.8% C 21.0% Cr 3.25% B 10.0% W 17.0% Ni Co alloy: For brazing; Wall Colmonoy; melting range 1040–1120 °C
NIVAFLEX	0.03% C 18% Cr 21% Ni 4% Mo 4% W 1% Ti 0.3% Be Co alloy: Information from Cobalt Information Centre
NIVCO 10	0.05% C 22.5% Ni 1.8% Ti 0.2% Al 1% Fe 1% Zr Co alloy: American proprietary alloy listed in SAE yearbook
ONERAL M 47	1% C 28% Cr 6% Ni 10% Mo 0.3% Ti 3% Fe Co alloy: Casting; listed by Cobalt Information Centre
ONERAL S 90	0.3% C 27% Cr 17% Ni 5% Mo Co alloy: Information from Cobalt Information Centre
P 6	6.0% Ni 4.0% V 45% Fe Co alloy: Rolled strip; US alloy listed by PMA for hysteresis motors
PERMENDUR 24	49% Fe 2% V Co alloy: More ductile than Permendur; Telcon
PERMEDUR V	2% V 49% Fe Co alloy: Telcon; high permeability for high flux densities
R 30816	UNS designation for ASTM A639/671
REFRACTALOY 70	0.05% C 20% Cr 20% Ni 8% Mo 4% W 15% Fe Co alloy: American proprietary alloy; consult SAE
REMENDUR	2.5% V Fe Co alloy: For magnets; American alloy listed by PMA
REXWELD 33W	2.2% C 3% Cr 18% W Co alloy: For hard facing: Crucible Steel Co. **DPN: 700**
REXWELD A	1.0% C 28% Cr 4.5% W Co alloy: For hard facing; Crucible Steel Co. **DPN: 400**
REXWELD B	1.5% C 28% Cr 7.5% W Co alloy: For hard facing; Crucible Steel Co. **DPN: 420**
REXWELD C	2.5% C 30% Cr 12.5% W Co alloy: For hard facing; Crucible Steel Co. **DPN: 550**
REXWELD VT	0.3% C 27% Cr 3% Ni 5.5% Mo Co alloy: For hard facing; Crucible Steel Co. **DPN: 350**
SAE VF 5	1.7% C 25% Cr 22% Ni 12% W 2% Fe Co alloy: For hard facing
S 816	0.38% C 20% Cr 20% Ni 3% Fe 4% Mo 4% W 4% Nb Co alloy: Allegheny Ludlum
S 844	0.25% C 25% Cr 20% Ni 2% Mo 2% W 2% Nb 3% Fe Co alloy: Allegheny Ludlum
SL 959	High C High Cr Co alloy: Casting; Swift Levrick **DPN: 450**
SL 970	0.4% C 24.5% Cr 5% W Co alloy: Swift Levrick
SM 302	0.85% C 21.5% Cr 10% W 9% Ta 0.2% Ti 0.005% B 1% Fe Co alloy: Martin Metal Co.
SM 322	1.0% C 21.5% Cr 9% W 4.5% Ta 0.7% Ti 2.2% Zr 0.7% Fe Co alloy: Martin Metal Co.
STAR J	2.5% C 32% Cr 2% Ni 17% W 3% Fe Co alloy: Casting; S Osborn; cutting tools; wear resistant parts **DPN: 760**
STELLITE 1	2.5% C 33% Cr 13% W Co alloy: Casting for hard facing rods (oxyacetylene) and electrodes; Deloro **DPN: 620 UTS: 600 Elon: 1%**
STELLITE 3	2.4% C 30% Cr 13% W Co alloy: Casting; Deloro **DPN: 620 UTS: 600 Elon: 1%**
STELLITE 4	1% C 31% Cr 14% W Co alloy: Casting; Deloro **DPN: 490 UTS: 1000 Elon: 1% Proof: 600**
STELLITE 6	1% C 26% Cr 5% W Co alloy: Casting for hard facing rods (oxyacetylene) and electrodes; Deloro **DPN: 400 UTS: 880 Elon: 1% Proof: 530**
STELLITE 7	0.4% C 26% Cr 6% W Co alloy: Casting; Deloro **DPN: 325 UTS: 900 Elon: 8% Proof: 450**

Symbol	Nominal analysis, supplier, condition and remarks.
STELLITE 8	0.2% C 30% Cr 6% Mo Co alloy: Casting; Deloro for BS 3146/ANC14 **DPN: 325 UTS: 900 Elon: 9% Proof: 470**
STELLITE 12	1.8% C 29% Cr 9% W Co alloy: Casting for hard facing rods (oxacetylene) and electrodes; Deloro **DPN: 510 UTS: 820 Elon: 1% Proof: 630**
STELLITE 12P	1.4% C 31.0% Cr 9.0% W Co alloy: Melting range 1260–1300 °C; Deloro Stellite **DPN: 470 UTS: 880 Elon: 1%**
STELLITE 20	2.5% C 3% Cr 18% W Co alloy: Casting for hard facing rods (oxyacetylene) and electrodes; Deloro **DPN: 660 UTS: 600 Elon: 1%**
STELLITE 21	0.25% C 27% Cr 2.5% Ni 4.5% Mo 2% Fe 0.007% B Co alloy: Casting; Deloro
STELLITE 31	0.5% C 25.5% Cr 10.5% Ni 7.5% W Co alloy: Casting; Deloro
STELLITE 100	2% C 34% Cr 19% W Co alloy: Casting; Deloro **DPN: 850 UTS: 690 Elon: 1%**
STELLITE 208	26.0% Cr 3.0% Mo 20% Fe alloy: Melting range 1400–1415 °C; Deloro Stellite **DPN: 280 UTS: 571 Elon: 12%**
STELLITE 306	1.4% C 25.0% Cr 5.0% Ni 2.0% W 6.0% Nb Co alloy: Melting range 1225–1305 °C; Deloro Stellite **DPN: 360**
STELLITE 506	1.6% C 35.0% Cr 7.5% W 3.0% Fe Co alloy: Deloro Stellite **DPN: 400**
STELLITE X 40	0.3% C 25% Cr 10% Ni 7% W Co alloy: Casting; Deloro for BS 3146/ANC 13 **DPN: 325 UTS: 1120 Elon: 10% Proof: 430**
SUPERMENDUR	49% Fe 2% V Co alloy: For magnets; Telcon; rectangular hysteresis loop
TANTUNG	3% C 30% Cr 16% W 4.5% Nb or Ta 2% Mn 3% Fe Co alloy: Vasoloy; harder than Tantung G **DPN: 830**
TANTUNG G	3.0% C 30% Cr 16% W 4.5% Nb or Ta 2% Mn 3% Fe Co alloy: Vasoloy; Ramet; cutting tools **DPN: 790**
THERMELAST 2602	0.03% C 12% Cr 26% Ni 13% Fe 4% W 4% Mo 1% Ti 0.2% Be Co alloy: Details from Cobalt Information Centre
TRIBALOY 100	35% Mo 10% Si Co alloy: Intermetallic material for wear resistant purposes; Du Pont
TRIBALOY 400	28% Mo 8% Cr 2% Si Co alloy: Intermetallic material for wear resistant purposes; suitable for hand facing; Du Pont
TRIBALOY 800	28% Mo 17% Cr 3% Si Co alloy: Intermetallic material for wear resistant purposes; suitable for hand facing; Du Pont
UCAR 25	0.07% C 20.0% Cr 10.0% Ni 0.5% Mo 1.5% Fe Co alloy: Casting; Union Carbide
UCAR FSX414	0.25% C 29.0% Cr 10.0% Ni 7.0% W Co alloy: Union Carbide
UCAR M509	0.6% C 21.5% Cr 10.0% Ni 7.0% W Co alloy: Union Carbide
UCAR X45	0.25% C 25.5% Cr 10.5% Ni 7.0% W Co alloy: Union Carbide

Symbol	Nominal analysis, supplier, condition and remarks.
UM Co 50	0.1% C 28% Cr 20% Fe 0.7% Si Co alloy: American proprietary alloy; consult SAE yearbook **UTS: 540 Elon: 8% 4310**
UM Co 51	0.27% C 28% Cr 20% Fe 2% Nb 0.7% Si Co alloy: American proprietary alloy; consult SAE yearbook **UTS: 630 Elon: 4% Proof: 460**
UNITEMP L 605	0.1% C 1.5% Mn 20% Cr 10% Ni 15% W 3% Fe Co alloy: Universal Cyclops
UNITEMP S 816	0.38% C 1.5% Mn 20% Cr 20% Ni 4% Mo 4% W 4% Nb Co alloy: Universal Cyclops
UNS R30001–R39999	Co metal and alloys: American Standard system; unified number
V 36	0.3% C 25% Cr 20% Ni 4% Mo 2% W 2.2% Nb 3% Fe Co alloy: Allegheny Ludlum
VELINVAR	31% Fe 8.5% V Co alloy: Details from Cobalt Information Centre
VF 2	1.2% C 28% Cr 3% Ni 4.5% W 0.5% Mo 3% Fe Co alloy: For hard facing; Designation used by SAE
VF 5	1.7% C 25% Cr 22% Ni 12% W 2% Fe Co alloy: For hard facing; Designation used by SAE
VICALLOY	38% Fe 10% V Co alloy: Magnet alloy strip; Telcon; annealed **DPN: 200 UTS: 760**
VICALLOY 1	38% Fe 10% V Co alloy: Permanent magnet; Telcon
VICALLOY 11	35% Fe 13% V Co alloy: Permanent magnet; Telcon
VIMETAL	0.05% C 9% Cr 17% Ni 5.5% Fe 4% W 4% Mo 3% Nb 1.5% Ti Co alloy: Details from Cobalt Information Centre
VITALLIUM	35% Cr 5% Mo Co alloy: American proprietary alloy **DPN: 388 UTS: 280 Elon: 4% Proof: 420**
WAI-MET 50	0.1% C 28% Cr 20% Fe 0.7% Si Co alloy: Wai-Met alloys for furnace equipment
WALLEX 1	2.2% C 30% Cr 12.5% W Co alloy: Castings & rods for facing; Wall Colomony; melting point 1270 °C; good carrosion resistance low friction **DPN: 570**
WALLEX 1NE	2.5% C 3.0% Fe 30% Cr 12.0% W Co alloy: Bronze alloy metal; Wall Colmonoy; melting point 1255 °C
WALLEX 4	1.0% C 1.5% Si (max) 30.5% Cr 14.0% W Co alloy: Bronze metal; Wall Colmonoy; melting point 1260 °C
WALLEX 6	1% C 29% Cr 4.5% W Co alloy: Castings & rods for spraying; Wall Colomony; melting point 1275 °C; heat resistant; low friction **DPN: 430 UTS: 810**
WALLEX 6NE	1.0% C 2.0% Fe (max) 31% Cr 5.5% N Co alloy: Bronze metal; Wall Colmonoy; melting point 1285 °C
WALLEX 7NE	0.3% C 3.0% Fe (max) 30% Cr 3.0% Ni 5.5% Mo Co alloy: Bronze metal; Wall Colmonoy; melting point 1260 °C
WALLEX 12	1.0% C 1.25% Si 9.0% W 29% Cr Co alloy: Bronze metal; Wall Colmonoy; melting point 1280 °C
WALLEX 12NE	1.5% C 3.0% Fe (max) 30% Cr 8.5% W Co alloy: Bronze metal; Wall Colmonoy; melting point 1280 °C
WALLEX 50	0.7% C 25% Cr 11% Ni 1% Fe 10% W 3% B Co alloy: Powder; Wall Colomony; melting point 1110 °C; similar to Wallex 1 **DPN: 550**
WF 11	0.1% C 1.5% Mn 20% Cr 10% Ni 15% W 3% Fe Co alloy: American proprietary alloy; consult SAE yearbook

Note. The following abbreviations and units are used in the tables:

DPN	Hardness, diamond pyramid number
UTS	Ultimate tensile strength, N/mm^2
Elon	Elongation, %
Proof	0.1 % proof strength, N/mm^2

$1 \text{ N/mm}^2 = 0.1 \text{ hbar} = 0.102 \text{ kgf/mm}^2 = 0.06475 \text{ tonf/in}^2 = 145.04 \text{ lbf/in}^2$

Symbol	Nominal analysis, supplier, condition and remarks.
WF 31	0.15% C 20% Cr 10% Ni 2.6% Mo 10% W 1% Ti Co alloy: American proprietary alloy listed in SAE yearbook
WI 52	0.4% C 20% Cr 1% Ni 11% W 2% Nb + Ta Co alloy: Wai-Met Alloy Co.
WIPTAM	0.1% C 28.3% Cr 24.4% Ni Co alloy: For dental use; details from Cobalt Information Centre **UTS: 620 Elon: 52% Proof: 830**
X 40	0.5% C 25% Cr 10% Ni 7.5% W 1.5% Fe Co alloy: American proprietary alloy listed in SAE yearbook
X 40	0.45% C 25% Cr 10% Ni 7% W Cobalt alloy: Casting; Jessop Saville as G 65 **UTS: 730 Elon: 17% Proof: 400**
X 50	0.75% C 22% Cr 20% Ni 12% W 2.5% Fe Co alloy: Casting; listed by Cobalt Information Centre

Symbol	Nominal analysis, supplier, condition and remarks.
X 63	0.4% C 23% Cr 10% Ni 6% Mo 1% Fe Co alloy: Casting; American proprietary alloy listed in SAE yearbook
XSH	20.0% Cr 10.0% Ni 15.0% W Co alloy: Origin unknown

Note. The following abbreviations and units are used in the tables:

DPN	Hardness, diamond pyramid number
UTS	Ultimate tensile strength, N/mm^2
Elon	Elongation, %
Proof	0.1 % proof strength, N/mm^2

$1 \ N/mm^2 = 0.1 \ hbar = 0.102 \ kgf/mm^2 = 0.06475 \ tonf/in^2 = 145.04 \ lbf/in^2$

14. Copper Cu

Physical properties

Atomic number	27	
Atomic weight	63.54	
Crystal structure	Face-centred cubic	
Colour	Reddish orange	
Specific gravity	8.96	
Density	8960 kg/m³	(0.318 lb/in.³)
Melting point	1083 °C	
Boiling point	2582 °C	
Specific heat	0.385 J/g °C	(0.092 cal/g °C)
Thermal conductivity	0.385 W/m °C	(92 cal/m s °C)
Coefficient of linear expansion (20–100° C)	16.4 × 10⁻⁶/ °C	
Latent heat of fusion	204.8 J/g	(48.9 cal/g)
Latent heat of vaporization	5234 J/g	(1250 cal/g)
Thermal neutron absorption cross-section	3.8 barns/atom	

Electrical conductivity		100% IACS (copper 100%)	
Specific resistance		16.73 microhm mm	
Temperature coefficient of electrical resistance		0.0039/ °C	
Electrochemical equivalent	monovalent	2.380 g/A/h	
	divalent	1.185 g/A/h	
Electrode potential	monovalent	-0.52 V	
	divalent	-0.34 V	
Magnetic susceptibility		-0.080 × 10⁻⁶	

Young's modulus of elasticity		110 × 10⁹ N/m²	(16 × 10⁶ lbf/in.²)
Tensile strength	annealed	210 N/mm²	(14 tonf/in.²)
Hardness	annealed	50 DPN	

14.1 General notes on copper

Copper and its alloys have been in the service of mankind for longer than any other metal, and until the second half of this century it had the largest use of any metal after iron. Aluminium has now taken second place and copper is third in the table for usefulness.

Articles made from bronze – a tin copper alloy – have been found which were made by man in his very early days as a creative animal. The history of copper illustrates the technical development and ability of men through the ages. These attributes continue today, with highly complex copper alloys soaring into space.

Relatively large amounts of native copper metal have been found widely dispersed throughout the world. The largest deposit is probably that on the shores of Lake Superior, USA. These deposits probably account for the early use of copper, but the ease with which it can be reduced from its ore must also be taken into account.

Copper is used as high purity metal and has a range of alloys all of which are described and listed in the following sections.

Copper is also used as an alloying element, the largest proportion being with nickel where it forms a series of heat and corrosion resistant alloys.

It is also commonly alloyed with zinc to form the brasses.

In small quantities of up to 5% it forms intermetallic compounds with aluminium, giving the well-known age hardening series of aluminium alloys.

Many tin and lead based bearing metals have up to 10% additions. These strengthen the matrix and also in some cases help in the formation of intermetallic compounds.

In quantities up to 0.5%, copper is sometimes added to steel where it gives a degree of atmospheric corrosion resistance and some increase in yield. It is also a constituent in the very new maraging series of steels.

Between 1 and 2% copper added to some types of cast iron ensures uniformity of properties in individual castings and can enhance the mechanical properties.

Copper has had many years in the service of man and recently has met with competition from the younger metals. There is no danger whatever of copper being completely supplanted by these materials as the versatility of copper ensures that new uses are found to replace those more economically carried out by aluminium and stainless steels.

14.2 Notes on specifications and trade names

An explanation of the different methods of classification and designation used with copper base materials may be of assistance in understanding and rapidly finding the different specifications and symbols.

British Standards Institution (BS)

This body issues specifications covering individual alloys, and until recently their policy was to have an individual specification for each material, and often for each form of the material.

The present policy is to have a master specification number covering the form, with alloy designations covering the chemical composition. These on the whole replace the individual specification. Copper base materials however tend to be used in traditional industries where the old specifications die hard.

The current specifications are:

BS 1400   Copper and alloy casting
BS 2870   Copper and alloy sheet and strip
BS 2871   Copper and alloy tube
BS 2872   Copper and alloy forgings and forging stock
BS 2873   Copper and alloy wire
BS 2874   Copper and alloy bar and section
BS 2875   Copper and alloy plate

These are followed by the alloy designation which always has a prefix letter or letters denoting the main alloying constituent.

Thus CA designates aluminium–copper, CB beryllium–copper, CN nickel–copper, CZ zinc–copper (brass), etc. A three figure code then identifies the actual copper or alloy, for example CZ101 is 10% zinc–copper, CB101 1.8% beryllium–copper.

Suffix letters and figures can be used to indicate the heat treatment condition, or degree of cold work carried out. These are as follows:

$H\frac{1}{4}$ Cold worked to approximately one quarter the possible amount, or fully cold worked, then partially annealed to remove about three-quarters the cold work.
$H\frac{1}{2}$ As $H\frac{1}{4}$ but cold worked half possible.
$H\frac{3}{4}$ As $H\frac{1}{4}$ but cold worked to three-quarters possible.
M   As cast – as rolled – as extruded
O   Annealed
W   Solution treated
P   Precipitation treated – aged.

Examples of these specifications would be:

BS 2874 CZ101   10% zinc–copper alloy in bar form
BS 2870 CB101   1.8% beryllium–copper alloy in sheet form

Full specifications covering the mechanical and physical properties are not available for the complete range of alloy designations of all the forms. In the following pages the nominal analysis has been supplied for the full range

for identification purposes, but only where the British Standards Institution supply mechanical or other properties are these quoted, and only the latter specifications are available. The alloy designations are often quoted without reference to the BS number and have been indexed under both the full specification and the alloy designation, thus BS 2874 CZ101 and CZ101 appear as separate entries.

In addition there are a limited number of aircraft specifications with the prefix BS B or T followed by a number. These are individual specifications covering one analysis and generally in one form.

American Society for Testing and Materials (ASTM)

This body issues several volumes of specifications at intervals. All the non-ferrous alloys are prefixed with the letter B followed by a number for the actual alloy. The year of issue follows, with the letter T for tentative issues of a specification. Typical examples would be:

ASTM B36/4   Copper alloy sheet containing 20% zinc
ASTM B291   Copper alloy sheet containing 1% manganese and 29% zinc

Although some of these specifications are subdivided to group together similar alloys, there is no real classification system at present with ASTM, and no attempt to code the form or condition of the material.

This body devised a designation code for the various forms of pure copper and these are listed in ASTM B224. In general the code is the initial letter of the description, thus OFC is oxygen-free copper. This method is now finding general use including some proprietary alloys in this country.

Society of Automotive Engineers (SAE) (American)

This body issues a handbook at intervals included in which are their own alloys with the prefix SAE. Previously these had a two or three digit number prefixed by SAE some of which had a letter suffix indicating the issue number. The present system uses an alloy designation agreed by the American Copper Association, with the prefix CA followed by three figures. This alloy designation is used with and without the SAE prefix and without any prefix, thus all forms are listed.

This society also issues a series of aircraft material specifications with the prefix AMS followed by four figures for the actual material condition and form. The suffix letter indicates the issue number.

Trade names and symbols

These generally incorporate some code or symbol indicating the supplier followed by letters or figures to identify the alloy. Sometimes the national designations are incorporated in the trade names.

Whenever possible all forms of the names and specifications have been included in the following lists, but it is possible that some have been omitted and it is suggested that if the name is not found at the first attempt then attempts are made to add or subtract the makers' name.

14A Copper – commercially pure – wrought and cast

Specific gravity	8.9	
Density	8900 kg/m^3	(0.3215 lb/in.^{3})
Solidus/liquidus	1083 °C	
Thermal conductivity	314–402 W/m °C	(75–96 cal/m s °C)
Coefficient of linear expansion	17.3 × 10^{-6}/ °C	
Electrical conductivity	annealed 80–101% IACS (copper 100%)	
Specific resistance	annealed 21.5–17.241 microhm mm	
Young's modulus of elasticity	124 × 10^9 N/m^2	(18 × 10^6 lbf/in.^{2})
Impact	– J	(– ft/lb)
Fatigue strength	annealed ± 60 N/mm^2	(± 4 tonf/in.^{2}) 10^8 cycles
	cadmium copper ± 150 N/mm^2	(± 10 tonf/in.^{2}) 10^8 cycles

Hot strength

Temperature °C	Tensile strength N/mm^2	Elongation %
100	205	50
200	172	48
300	150–151.5	46
400	108–130	42–48

The above are applicable to the specifications listed, but may not be exactly applicable to any one material.

General metallurgical characteristics

Depending on its method of manufacture and in some instances the source of the ore, there are several varieties of commercially pure copper. Briefly these are:

Cathode copper. This is the direct product of electrolytic refining and is the raw material for many of the other forms listed below. It can be re-melted and cast, but apart from this finds little use. It has relatively poor conductivity owing to the presence of minute quantities of impurities.

Tough pitch copper. This is copper with a controlled amount of oxygen present which improves the workability and conductivity of the metal. There are various forms of tough pitch copper.

Electrolytic tough pitch copper. This is cathode copper re-melted and allowed to pick up oxygen in the process. The product is used for all forms of electrical equipment, and also where good thermal conductivity is important.

With oxygen controlled to 0.04% annealed wrought product has a conductivity of 100% IACS.

Fire refined tough pitch copper. This is copper refined in a furnace and not electrolytically. As this is generally more difficult and does not recover any precious metals which may be present in the raw material, the type of ore and local conditions largely control the use of this method.

The final product is comparable to electrolytic tough pitch copper with a conductivity in the wrought annealed condition of 100% IACS or better.

Ordinary tough pitch copper. The oxygen content and

other impurities are not as carefully controlled as with the above two types. The electrical and thermal conductivities are lower, thus it is not used for electrical purposes, but finds many general engineering uses. This copper is also cast into ingots and used as the primary metal for alloys.

Tough pitch arsenical copper. This has approximately 0.4% arsenic in solid solution. The electrical and thermal conductivities are reduced, but the strength and corrosion resistance are improved, and these improvements are maintained up to approximately 300 °C.

Oxygen-free copper. There are various forms of copper either produced oxygen free, or subsequently de-oxidized. These are:-

Oxygen-free high conductivity copper. This is difficult and expensive to produce. In America and Finland it is made by remelting and pouring cathode copper under an inert atmosphere resulting in high conductivity copper which can be brazed and welded. This material will accept a higher degree of cold work than any other form of copper.

Phosphorus de-oxidized copper. Phosphorus up to 0.05% but not generally above 0.04% is used to de-oxidize copper. This eases casting, working, brazing and welding problems, but reduces the thermal and electrical conductivity, which falls to about 70% IACS. Higher phosphorus contents are used in brazing filler rods.

De-oxidized arsenical copper. This is phosphorus de-oxidized copper with small quantities of arsenic added which has little effect on the other properties but increases the corrosion resistance and hot strength.

Pure copper will always be one of the above when found in industry and commerce. Also classified as copper are several alloys where the alloying element is less than 1%.

These elements are:

Silver. This has no effect on the conductivity but does raise the annealing temperature required to remove cold work. Thus cold worked copper with approximately 0.1% silver has to be heated to above 300 °C before it softens, whereas ordinary copper tends to soften at 200 °C. This silver copper is used where cold worked material is to be soldered. Silver also improves the creep strength of copper.

Cadmium. This has a similar effect to silver and in addition toughens the material and improves the fatigue strength. Approximately 0.8% cadmium reduces the conductivity in the annealed state to 95% IACS. This material is used for overhead contact wires for traction and for long span conductors.

Tellurium and selenium. These also increase the softening temperature, in a similar manner to silver and cadmium. Their main use in copper however, is to improve machinability. A good surface finish and dimensional stability is obtained with annealed tellurium or selenium copper which has a conductivity above 90% IACS.

Sulphur. This has a similar effect to tellurium and is now preferred by many users.

All the above coppers are capable of accepting a considerable amount of cold work, with the de-oxidized varieties superior in this context to the tough pitch varieties.

Thermal treatment

It is not possible to harden the materials listed by any form of heat treatment. Cold work can be removed by annealing at 450–650 °C for approximately 30 minutes and air cooling or water quenching. No special equipment is necessary and the temperature required is not critical. It is not advisable however, to heat tough pitch copper above 450 °C unless the atmosphere is oxidizing, otherwise the oxide content can be reduced causing brittleness. No advantage is to be gained by heating above 650 °C as all traces of cold work are removed below this figure.

Welding and brazing

Pre-treatment. No thermal treatment necessary. The weld area must be thoroughly cleaned, and parts will often require pre-heating.

Welding and brazing. Electric arc welding is difficult owing to the high electrical and thermal conductivity of copper, while gas welding is comparatively easy with de-oxidized coppers. Tough pitch coppers however are difficult to weld as the molten metal reacts with the atmosphere to cause brittleness.

Resistance welding is again difficult owing to the high conductivity.

Brazing is readily carried out on all forms of copper.

Post treatment. None required, except to remove fluxes if used. Large complicated parts may require stress relieving at 250 °C.

Flaw detection methods

None generally required. Where necessary, penetrant oil chalk can be carried out.

Corrosion protection

Temporary. None required.

Permanent. None required except for appearance purposes. Copper will form the familiar green surface patina under atmospheric conditions, and this inhibits further corrosion. The Copper Development Association is now promoting lacquers which prevent the formation of this green product. These are air drying, but require thorough cleaning before application. These retain the natural copper colour.

Machinability

In the annealed condition copper has very poor machinability owing to its softness, and it is also difficult to hold size accurately. There is an improvement when the copper is cold worked, or with additions of tellurium or selenium, which considerably increases the machinability with very little drop in conductivity or mechanical properties. Sulphur has the same effect on machinability and is now preferred to either of the above.

Uses

Cathode copper. Primary metal.

Tough pitch copper. For all electrical purposes, where conductivity is required. Conductor wires, switch gear etc. Also sheet products for roofing, chemical purposes, etc. Not easily welded.

Oxygen-free high conductivity. As tough pitch, but this can be welded using gas heat. The high cost limits use of this material.

De-oxidized copper. Chemical purposes, water pipes, domestic, industrial and decorative uses, sheet tube and extrusions where excellent corrosion resistance is required. Food containers. Architectural uses where appearance is important. These can be welded.

Symbol	Nominal analysis, supplier, condition and remarks.
1C	High conductivity Cu: Langley Alloys **UTS: 180 Elon: 40% Proof: 30**
2.0050	99.9% Cu: Electrolytic; German Standard
2.0060	99.9% Cu: German Standard
2.0060	0.03% Ag 99.95% Cu: German Werkstoff number
2.0070	99.9% Cu: German Standard
2.0080	99.9% Cu: German Standard
2.0090	99.9% Cu: German Standard
2.0100	99.75% Cu: German Standard
2.0110	99.8% Cu: German Standard
2.0120	99.5% Cu: German Standard
2.0150	0.4% As 99.25% Cu: German Standard
2.0170	0.4% As 99.0% Cu: German Standard
2.0181	99.0% Cu: German Standard
2.1006	0.8% Sn Cu: German Standard
2.1211	1.0% Ag Cu: German Standard
101	99.96% Cu: Oxygen-free; designation agreed by CDAA; not used
104	99.95% Cu: Ag always present; oxygen-free; designation agreed by CDAA; not used
105	99.95% Cu: Ag always present; oxygen-free; designation agreed by CDAA; not used
107	99.95% Cu: Ag always present; oxygen-free; designation agreed by CDAA; not used
121	0.01% P 99.9% Cu: With Ag; designation agreed by CDAA; not used
123	0.02% P 99.9% Cu: With Ag; designation agreed by CDAA; not used
125	99.88% Cu: Fire refined; tough pitch; designation agreed by CDAA; not used
127	99.88% Cu: Fire refined; tough pitch with Ag; designation agreed by CDAA; not used
128	99.88% Cu: Fire refined; tough pitch; with Ag; designation agreed by CDAA; not used
130	99.88% Cu: Fire refined; tough pitch; with Ag; designation agreed by CDAA; not used
141	0.4% As 99.4% Cu: Arsenical tough pitch; designation agreed by CDAA; not used
142	0.4% As 0.03% P 99.4% Cu: Designation agreed by CDAA; not used
164	0.7% Cd 0.3% Sn Cu alloy: Wrought; designation agreed by CDAA; not used
165	0.8% Cd 0.6% Sn Cu alloy: Wrought; designation agreed by CDAA; not used
189	0.7% Sn 0.3% Si Cu alloy: Wrought; designation agreed by CDAA; not used
ABS/B42	99.9% Cu 0.04% P (max) Cu: Pipe; American Bureau of Shipping
AL 0 1	Low Zn Cu: Thomas Bolton; for singe plate
AMPCOLAY 90	Cu: 92% IACS; Ampco
AMPCOLAY 900	High conductivity 99.9% Cu: Ampco
AMPCOLAY 901	99.9% Cu with Ag: 93% IACS; Ampco
AMPCOLAY 910	Cu: 90% IACS; Ampco
AMS 4500 D	Cu: Sheet & strip; annealed
AMS 4701 B	Cu: Wire; annealed
ANACOS	99.90% Cu: Wire; may be tinned; Fred Smith; hard drawn or annealed
ANACOS CADMIUM	1% Cd trace P Cu: Wire; may be tinned; Fred Smith; hard drawn

Symbol	Nominal analysis, supplier, condition and remarks.
ANOCOS SILVER	0.1% Ag trace P Cu: Wire; Fred Smith; hard drawn
ASTM B1	Cu: Wire; hard drawn; mechanical properties vary with diameter
ASTM B2	Cu: Wire; medium hard drawn; mechanical properties vary with diameter
ASTM B3	Cu: Wire; annealed; 100% IACS **Elon: 25%**
ASTM B4	99.9% Cu: Wire, bar, ingots, etc.; Lake Copper; covers copper of low resistivity & high resistivity
ASTM B5	99.5% Cu: Wire, bar, ingots, etc.; electrolytic refined high conductivity copper
ASTM B11/110	99.88% Cu: Plate for fireboxes
ASTM B11/122	0.02% P 99.90% Cu: Plate for fireboxes
ASTM B11/125	99.88% Cu: Plate for fireboxes
ASTM B11/141	0.35% As 99.4% Cu: Plate for fireboxes
ASTM B11/142	0.3% As 0.03% P Cu: Plate for fireboxes
ASTM B11 ATP	0.2% As 99.4% Cu: Plate; tough pitch arsenical copper; hot rolled **UTS: 210 Elon: 35%**
ASTM B11 DHP	0.02% P 99.9% Cu: Phosphorized; hot rolled **UTS: 210 Elon: 30%**
ASTM B11 DPA	0.3% As 0.02% P 99.4% Cu: Phosphorized arsenical copper; hot rolled **UTS: 210 Elon: 35%**
ASTM B11 ETP	99.88% Cu: Plate; tough pitch copper **UTS: 200 Elon: 30%**
ASTM B11 FRTP	99.88% Cu: Plate; tough pitch copper **UTS: 200 Elon: 30%**
ASTM B12	Cu: Rods for locomotive staybolts
ASTM B12/102	99.9% Cu: Rods for staybolts; oxygen-free; previously OF
ASTM B12/110	99.9% Cu: Rods for staybolts; electrolytic tough pitch; previously ETP
ASTM B12/120	99.9% Cu: Rods for staybolts; phosphorus de-oxodized; previously DLP
ASTM B12/122	99.9% Cu: Rods for staybolts; phosphorus de-oxidized; previously DHP
ASTM B12/125	99.9% Cu: Rods for staybolts; fire refined tough pitch; previously FRTP
ASTM B12/141	0.3% As 99.88% Cu: Rods for staybolts; previously ATP
ASTM B12/142	0.3% As 99.88% Cu: Rods for staybolts; previously DPA
ASTM B42	0.04% P 99.9% Cu: Seamless pipe; as drawn **UTS: 260 Proof: 240**
ASTM B42/102	99.9% Cu: Seamless pipe; previously OF
ASTM B42/120	99.9% Cu: Seamless pipe; phosphorus de-oxidized; previously DLP
ASTM B42/122	99.9% Cu: Seamless pipe; phosphorus de-oxidized; previously DHP
ASTM B47	Cu: Wire for overhead trolleys **UTS: 310 Proof: 4%**
ASTM B48	Cu: Rod; rectangular section for conductors **Elon: 30%**
ASTM B49	Cu: Rod; hot rolled for electrical purposes; conductivity 100% IACS
ASTM B52A	14.0% P (min) Cu: Ingot
ASTM B52B	10% P (min): Ingot
ASTM B68 DHP	0.02% P 99.9% Cu: Seamless tube; annealed **UTS: 200 Elon: 40%**
ASTM B68 DLP	0.01% P 99.9% Cu: Seamless tube; annealed **UTS: 200 Elon: 40%**
ASTM B68 OF	99.92% Cu: Seamless tube; annealed **UTS: 200 Elon: 40%**
ASTM B75/102	99.95% Cu: Seamless tube; oxygen-free; previously OF
ASTM B75/120	0.01% P 99.90% Cu: Seamless tube; previously DLP

Note. The following abbreviations and units are used in the tables:

DPN	Hardness, diamond pyramid number
UTS	Ultimate tensile strength, N/mm^2
Elon	Elongation, %
Proof	0.1 % proof strength, N/mm^2

$1 \ N/mm^2 = 0.1 \ hbar = 0.102 \ kgf/mm^2 = 0.06475 \ tonf/in^2 = 145.04 \ lbf/in^2$

Symbol	Nominal analysis, supplier, condition and remarks.
ASTM B75/122	0.02% P 99.9% Cu: Seamless tube; previously DHP
ASTM B75/142	0.02% P 99.4% Cu: Seamless tube; previously DPA
ASTM B72 A	99.75% Cu: For castings; High purity copper; fire refined
ASTM B72 B	0.1% As 0.3% Pb Cu: For casting; fire refined copper
ASTM B75 DHP	0.02% P 99.9% Cu: Seamless tube; as drawn **UTS: 240**
ASTM B75 DHP	0.02% P 99.9% Cu: Seamless tube; cold drawn **UTS: 320**
ASTM B75 DLP	0.01% P 99.9% Cu: Seamless tube; as drawn **UTS: 240**
ASTM B75 DLP	0.01% P 99.9% Cu: Seamless tube; cold drawn **UTS: 330**
ASTM B75 DPA	0.02% P 0.3% As 99.4% Cu: Seamless tube; as drawn **UTS: 240**
ASTM B75 DPA	0.02% P 0.3% As 99.4% Cu: Seamless tube; cold drawn **UTS: 330**
ASTM B75 OF	99.2% Cu: Oxygen-free seamless tube; as drawn **UTS: 240**
ASTM B75 OF	99.2% Cu: Oxygen-free seamless tube; cold drawn **UTS: 330**
ASTM B88	0.04% P 99.9% Cu: Seamless tube; annealed **UTS: 200**
ASTM B111	99.4% Cu 0.4% As Cu: Tube
ASTM B111	99.9% Cu 0.04% P (max): Tube
ASTM B111/102	99.5% Cu: Seamless tube; oxygen-free; previously OF
ASTM B111/120	0.01% P 99.90% Cu: Seamless tube; previously DLP
ASTM B111/122	99.9% Cu: Seamless tube; phosphorus de-oxidized; previously DHP
ASTM B111/142	0.2% As 99.4% Cu: Seamless tube; previously DPA
ASTM B115	99.9% Cu: Electrolytic cathode copper
ASTM B124/12	99.9% Cu: Forging & bar; mechanical properties not quoted
ASTM B152/102	99.95% Cu: Sheet & bar; oxygen-free
ASTM B152/104	99.95% Cu: Sheet & bar; oxygen-free; silver bearing
ASTM B152/105	99.95% Cu: Sheet & bar; oxygen-free; silver bearing
ASTM B152/107	99.95% Cu: Sheet & bar; oxygen-free
ASTM B152/122	99.90% Cu 0.03% P: Sheet & bar; phosphorized; high residual phosphorus
ASTM B152/123	99.90% Cu: Sheet & bar; phosphorized silver bearing
ASTM B152 ATP	0.25% As 99.4% Cu: Sheet, strip, plate & bar; annealed; arsenical tough pitch **UTS: 270**
ASTM B152 DHP	0.02% P 99.9% Cu: Strip, sheet, plate & bar; annealed; phosphorized copper **UTS: 270**
ASTM B152 DPS	0.04% P (max) 99.9% Cu: Sheet, strip, plate & bar; silver bearing; phosphorized **UTS: 270**
ASTM B152 ETP	99.9% Cu: Sheet, strip, plate & bar; annealed; electrolytic; tough pitch **UTS: 270**
ASTM B152 FRTP	99.88% Cu: Sheet, plate, strip & bar; annealed; fire refined; tough pitch **UTS: 270**
ASTM B152 OF	99.92% Cu: Sheet, strip, plate & bar; annealed; oxygen-free **UTS: 270**

Symbol	Nominal analysis, supplier, condition and remarks.
ASTM B152 OFS	99.9% Cu: Sheet, strip, plate & bar; annealed; silver bearing; oxygen-free **UTS: 270**
ASTM B152 STP	99.9% Cu: Sheet, strip, plate & bar; annealed; silver bearing copper; tough pitch **UTS: 270**
ASTM B187	Cu: Bus bar rods & shapes for electrical use; conductivity IACS 97.5% (min)
ASTM B188	Cu: Pipe & tube; seamless for electrical use; conductivity IACS 97.5% (min)
ASTM B224 ATP	Arsenical tough pitch Cu: Classification of Cu
ASTM B224 CAST	Casting Cu: Classification of Cu
ASTM B224 CATH	Electrolytic cathode Cu: Classification of Cu
ASTM B224 DHP	Phosphorized Cu: High residual phosphorus classification of Cu
ASTM B224 DLP	Phosphorized Cu: Low residual phosphorus classification of Cu
ASTM B224 DPA	Phosphorized arsenical Cu: Classification of Cu
ASTM B224 DPS	Phosphorized silver bearing Cu: Classification of Cu
ASTM B224 DPTE	Phosphorized tellurium bearing Cu: Classification of Cu
ASTM B224 ETP	Electrolytic tough pitch Cu: Classification of Cu
ASTM B224 FRHC	Fire refined high conductivity Cu: Classification of Cu
ASTM B224 FRTP	Fire refined tough pitch Cu: Classification of Cu
ASTM B224 OF	Oxygen-free Cu without residual de-oxidants: Classification of Cu
ASTM B224 OFP	Oxygen-free phosphorus bearing Cu: Classification of Cu
ASTM B224 OFPTE	Oxygen free phosphorus & tellurium bearing Cu: Classification of Cu
ASTM B224 OFS	Oxygen-free silver bearing Cu: Classification of Cu
ASTM B224 OFTE	Oxygen-free tellurium bearing Cu: Classification of Cu
ASTM B224 SATP	Silver bearing arsenical tough pitch Cu: Classification of Cu
ASTM B224 STP	Silver bearing tough pitch Cu: Classification of Cu
ASTM B229	Standard copper and copper-clad steel wire composite conductors
ASTM B246	Cu: Wire; hard and medium; hard drawn; tin coated **Elon: 1%**
ASTM B260 B Cu	0.07% P 99.9% Cu: Braze filler; melting point 1080 °C
ASTM B260 B Cu 1	99.90% (min) Cu: Brazing filler metal; braze range 1093–1149 °C
ASTM B260 B Cu 1a	99.0% (min) Cu: Brazing filler metal; braze range 1093–1149 °C
ASTM B280	99.92% Cu: Seamless tube
ASTM B280 DHP	0.02% P 99.2% Cu: Seamless tube
ASTM B280 DLP	0.01% P 99.9% Cu: Seamless tube
ASTM B301	0.5% Te Cu alloy: Bar; ½-hard **UTS: 260 Elon: 12% Proof: 200**
ASTM B301	0.5% Te Cu alloy: Bar; hard **UTS: 300 Elon: 10% Proof: 260**
ASTM B359/102	99.95% Cu: Firmed tube
ASTM B359/120	99.9% Cu: Firmed tube
ASTM B359/122	99.9% Cu: Firmed tube
ASTM B359/142	99.4% Cu: Firmed tube
ASTM B360	0.02% P 99.9% Cu: Capillary tube **UTS: 450**
ASTM B370	99.88% (min) Cu: Sheet & strip for buildings
ASTM B379	Cu billets: P de-oxidized; classified by P content
ASTM B442	99.90% Cu: Chemically refined; conductivity 100% IACS
ASTM B447	99.4% Cu (min): Welded tube; graded by Cu content

Symbol	Nominal analysis, supplier, condition and remarks.
ASTM B451	99.50% Cu (min): Foil strip, etc. for printed circuits
ASTM B465	2.4% Fe 0.03% P 0.01% Zn Cu alloy: Plate & bar; conductivity 62% IACS
ASTM B469	0.02% P 1.0% Fe 98.7% Cu (min): Tube for pressure application
	UTS: 280 Elon: 18% Proof: 180
B 37	Copper casting: High conductivity; Anti-attrition Co Ltd
BOLTOMET 103	Oxygen-free high conductivity copper: Thomas Bolton
BOLTOMET 105	Electrolytically refined copper: Thomas Bolton for BS 1036
BOLTOMET 107	Fire refined copper: Thomas Bolton for BS 1037
BOLTOMET 108	Anode copper: Thomas Bolton for BS 1038; withdrawn
BOLTOMET 112	0.1% Ag oxygen-free high conductivity copper: Thomas Bolton
BOLTOMET 113	0.1% Ag Cu: For electrical purposes; Thomas Bolton for BS 1434
BOLTOMET 115	0.15% Ag Cu: For electrical conductors; hollow section; Thomas Bolton
BOLTOMET 117	Calcium boride de-oxidized copper: Thomas Bolton; suitable for brazing; withdrawn
BOLTOMET 121	Phosphorus de-oxidized electrolytically refined copper: Thomas Bolton for BS 1172; low bismuth
BOLTOMET 123	Phosphorus de-oxidized fire refined copper: Thomas Bolton for BS 1172
BOLTOMET 152	Phosphorus de-oxidized arsenical copper: Thomas Bolton for BS 1174
BOLTOMET 154	Low arsenical copper: For printing rollers; Thomas Bolton
BOLTOMET 156	Arsenical Cu: Thomas Bolton for BS 1173; withdrawn
BOLTOMET 160	High phosphorus de-oxidized non-arsenical copper: Wire; Thomas Bolton
BOLTOMET 162	0.03% P de-oxidized copper: Thomas Bolton
BOLTOMET 170	Low Zn Cu: Thomas Bolton for singe plates; withdrawn
BOLTOMET 175	9.05% P de-oxidized copper: Thomas Bolton
BOLTOMET 206	0.6% Cd Cu: Thomas Bolton for BS 672
BOLTOMET 208	0.8% Cd Cu: Thomas Bolton for BS 23
BOLTOMET 210	1% Cd Cu: For welding electrodes; Thomas Bolton
BOLTOMET 302	0.4% Sn Cu: Wire; Thomas Bolton for armature wires
BOLTOMET 304	1% Sn Cu: Wire; Thomas Bolton for armature wires
BOLTOMET 366	Si Mn de-oxidized Cu: Thomas Bolton for BS 2901. C7
BOLTOMET 820	Zr Cu alloy: For commutator bar; Thomas Bolton; withdrawn
BOLTOMET 909	Te de-oxidized high conductivity Cu: Thomas Bolton; withdrawn
BOLTOMET 952	Te tough pitch high conductivity Cu: Thomas Bolton; free machining
BOROFIL	Boron Cu alloy: Welding rod; IMI melting point 1083 °C

Symbol	Nominal analysis, supplier, condition and remarks.
BS 23	Cu of high conductivity for trolley wires can be 99.9% Cu or 0.7% Cd Cu: As BS 2873 C108
BS 24/5	99.2% Cu: For railway fireboxes, tubes, etc.; BS 1173 or BS 1174
BS 125	Hard drawn Cu 0.7% Cd Cu: Wire for overhead transmission
BS 198	Electrolytic copper: Wire, bar, cakes, slabs, etc.; replaced by BS 1035-40
BS 198	Electrolytic copper: Ingots & ingot bars; replaced by BS 1035-40
BS 200	Tough copper: Cakes & billets; replaced by BS 1035-40
BS 201	Fire copper: Cakes; replaced by BS 1035-40
BS 202	Electrolytic cathode copper: Replaced by BS 1035-40
BS 203	'Best Select' copper: Replaced by BS 1035-40
BS 378 C106	0.05% P Cu: Phosphorus de-oxidized copper tube; as drawn for condenson tubes, etc.
BS 378 C107	0.05% P 0.4% As Cu: Phosphorous de-oxidized arsenical copper tube; as drawn for condenson tubes, etc.
	DPN: 100
BS 518	Medium hard copper: Strip, bar & rod for electrical purposes; replaced by BS 1432-3
BS 659	P or P and As de-oxidized Cu: Tube; light drawn
BS 672	Cd Cu alloy: For electrical conductor wires; hard drawn
	UTS: 530 Elon: 2%
BS 899 C101	Electrolytic tough pitch high conductivity copper: Sheet & strip; annealed; electrical conductivity 99.25% IACS
	UTS: 210 Elon: 35%
BS 899 C101	Electrolytic tough pitch high conductivity copper: Sheet & strip; hard; electrical conductivity 97% IACS
	UTS: 300
BS 899 C102	Fire refined tough pitch high conductivity copper: Sheet & strip; annealed; electrical conductivity 99.25% IACS
	UTS: 210 Elon: 35%
BS 899 C102	Fire refined tough pitch high conductivity copper: Sheet & strip; hard; electrical conductivity 97% IACS
	UTS: 300
BS 899 C103	Oxygen free high conductivity copper: Sheet & strip; annealed; electrical conductivity 99.25% IACS
	UTS: 210 Elon: 35%
BS 899 C103	Oxygen free high conductivity copper: Sheet & strip; hard; electrical conductivity 97% IACS
	UTS: 300
BS 899 C104	99.85% Cu: Non-arsenical tough pitch; sheet & strip; annealed
	UTS: 210 Elon: 35%
BS 899 C104	99.85% Cu: Non-arsenical tough pitch; sheet & strip; hard
	UTS: 300
BS 899 C105	Tough pitch arsenical copper: Annealed
	UTS: 210 Elon: 35%
BS 899 C105	Tough pitch arsenical copper: Hard
	UTS: 300
BS 899 C106	Non-arsenical copper: Phosphorous de-oxidized; sheet; annealed
	UTS: 210 Elon: 35%
BS 899 C106	Non-arsenical copper: Phosphorous de-oxidized; sheet; hard
	UTS: 300
BS 899 C107	Arsenical copper: Phosphorous de-oxidized; sheet & strip; annealed
	UTS: 210 Elon: 35%

Note. The following abbreviations and units are used in the tables:

DPN	Hardness, diamond pyramid number
UTS	Ultimate tensile strength, N/mm^2
Elon	Elongation, %
Proof	0.1 % proof strength, N/mm^2

$1 \text{ N/mm}^2 = 0.1 \text{ hbar} = 0.102 \text{ kgf/mm}^2 = 0.06475 \text{ tonf/in}^2 = 145.04 \text{ lbf/in}^2$

Symbol	Nominal analysis, supplier, condition and remarks.
BS 899 C107	Arsenical copper: Phosphorus de-oxidized; sheet & strip; hard **UTS: 300**
BS 1035	99.9% Cu: Produced as cathodes by electrolysis; Ag is counted as copper
BS 1036	99.9% Cu: Electrolytic refined; tough pitch high conductivity copper
BS 1037	99.9% Cu: Fire refined tough pitch high conductivity copper
BS 1038	99.85% Cu: Tough pitch copper; electrical conductivity not specified
BS 1039	99.75% Cu: Tough pitch copper; electrical conductivity not specified
BS 1040	99.5% Cu: Tough pitch copper; electrical conductivity not specified
BS 1110	Cu: Sheet & strip; electrical; hard; replaced by BS 1432
BS 1172	99.85% Cu: Phosphorus de-oxidized non-arsenical; electrical conductivity not specified
BS 1173	0.5% As 99.2% Cu: Tough pitch arsenical copper
BS 1174	0.5% As 99.2% Cu: Phosphorus de-oxidized arsenical copper
BS 1400 HCC1C	99.9% Cu: Castings; high conductivity; electrical conductivity 95% IACS **UTS: 150 Elon: 25%**
BS 1400 HCC2C	Cu: Castings; analysis not specified; electrical conductivity 75% IACS **UTS: 150 Elon: 25%**
BS 1401	99.9% Cu: Phosphorus or arsenical de-oxidized; tube; solid drawn tube; analysis to BS 1172 or BS 1174
BS 1432 C101	99.9% Cu: Electrolytic refined tough pitch copper; sheet; annealed; electrical conductivity 99.25% IACS **UTS: 210 Elon: 35%**
BS 1432 C101	99.9% Cu: Electrolytic refined tough pitch copper; sheet; ½-hard; electrical conductivity 97% IACS **UTS: 240 Elon: 12%**
BS 1432 C101	99.9% Cu: Electrolytic refined tough pitch copper; Sheet; hard; electrical conductivity 97% IACS **UTS: 300**
BS 1432 C102	99.9% Cu: Fire refined tough pitch copper; sheet; annealed; electrical conductivity 99.25% IACS **UTS: 210 Elon: 35%**
BS 1432 C102	99.9% Cu: Fire refined tough pitch copper; sheet; ½-hard; electrical conductivity 99.25% IACS **UTS: 240 Elon: 12%**
BS 1432 C102	99.9% Cu: Fire refined tough pitch copper; sheet; hard; electrical conductivity 97% IACS **UTS: 300**
BS 1432 C103	99.9% Cu: Oxygen-free copper; sheet; annealed; 99.25% IACS **UTS: 210 Elon: 35%**
BS 1432 C103	99.9% Cu: Oxygen-free copper; sheet; ½-hard; 97% IACS **UTS: 240 Elon: 12%**
BS 1432 C103	99.9% Cu: Oxygen-free copper; sheet; hard; 97% IACS **UTS: 300**
BS 1433	99.9% Cu: Tough pitch copper; bar; annealed **UTS: 210 Elon: 50%**
BS 1433	99.9% Cu: Tough pitch copper; bar; medium hard **UTS: 240 Elon: 22%**
BS 1433	99.9% Cu: Tough pitch copper; bar; hard **UTS: 280 Elon: 9%**
BS 1434	99.9% Cu: Tough pitch copper; for electrical purposes; hard drawn; 97% IACS; for commutator bars **DPN: 82 UTS: 280**
BS 1453 C1	1% Ag Cu alloy: Weld filler rod; for welding copper
BS 1541 C106	0.06% P 99.85% Cu: Plate; annealed **UTS: 210 Elon: 35%**
BS 1845 CU1	99.9% Cu: For brazing; total impurities 0.03%; melting point 1085 °C
BS 1845 CU2	99.9% Cu: For brazing; total impurities 0.04%; melting point 1085 °C
BS 1845 CU3	99.95% Cu: For brazing; total impurities 0.03%; melting point 1085 °C
BS 1845 CU4	99.85% Cu: For brazing; total impurities 0.05%; melting point 1085 °C
BS 1845 CU5	99.5% Cu: For brazing; total impurities 0.1%; melting point 1085 °C
BS 1845 CU6	99.85% Cu: For brazing; melting point 1085 °C
BS 1861	99.95% Cu: Oxygen-free high conductivity copper
BS 1977 C101	99.9% Cu: Electrolytic tough pitch Cu; tube; annealed Cu to BS 1036 99.25% IACS **UTS: 200 Elon: 40%**
BS 1977 C102	99.9% Cu: Fire refined tough pitch Cu; tube; annealed Cu to BS 1037 99.25% IACS **UTS: 200 Elon: 40%**
BS 1977 C102	99.9% Cu: Oxygen-free Cu; tube; annealed Cu to BS 1861 99.25% IACS **UTS: 200 Elon: 40%**
BS 2027	99.8% Cu: Plate for general purposes; annealed; analysis to BS 1038, BS 1172-3-4 **UTS: 200 Elon: 35%**
BS 2755 a	High conductivity Cu: For overhead wires
BS 2755/2	Cd Cu alloy: For overhead wires
BS 2870 C101	99.9% Cu: Electrolytic tough pitch Cu; sheet; annealed **UTS: 200 Elon: 35%**
BS 2870 C101	99.9% Cu: Electrolytic tough pitch Cu; sheet; hard **UTS: 280**
BS 2870 C102	99.9% Cu: Fire refined tough pitch Cu; sheet; annealed **UTS: 200 Elon: 35%**
BS 2870 C102	99.9% Cu: Fire refined tough pitch Cu; sheet; hard **UTS: 280**
BS 2870 C103	99.95% Cu: Oxygen-free Cu; sheet; annealed **UTS: 200 Elon: 35%**
BS 2870 C103	99.95% Cu: Oxygen-free Cu; sheet; hard **UTS: 270**
BS 2870 C104	99.85% Cu: Tough pitch non-arsenical; sheet; annealed **UTS: 200 Elon: 35%**
BS 2870 C104	99.85% Cu: Tough pitch non-arsenical; sheet; ½-hard **UTS: 240 Elon: 10%**
BS 2870 C104	99.85% Cu: Tough pitch non-arsenical; sheet; hard **UTS: 280**
BS 2870 C105	0.4% As 99.2% Cu: Tough pitch arsenical; sheet; annealed **UTS: 210 Elon: 35%**
BS 2870 C105	0.4% As 99.2% Cu: Tough pitch arsenical; sheet; ½-hard **UTS: 240 Elon: 10%**
BS 2870 C105	0.4% As 99.2% Cu: Tough pitch arsenical; sheet; hard **UTS: 280**
BS 2870 C106	0.04% P 99.85% Cu: Phosphorus de-oxidized Cu; sheet; annealed **UTS: 200 Elon: 35%**
BS 2870 C106	0.04% P 99.85% Cu: Phosphorus de-oxidized Cu; sheet; hard **UTS: 280**

Symbol	Nominal analysis, supplier, condition and remarks.
BS 2870 C107	0.1% P 0.4% As 99.2% Cu: Phosphorus de-oxidized arsenical Cu; sheet; annealed **UTS: 200 Elon: 35%**
BS 2870 C107	0.1% P 0.4% As 99.2% Cu: Phosphorus de-oxidized Cu; sheet; hard **UTS: 280**
BS 2870 C108	1% Cd Cu alloy: Sheet; specification not issued
BS 2870 C109	0.5% Te Cu alloy: Sheet; specification not issued
BS 2871 C101	99.9% Cu: Electrolytic refined tough pitch; tube; annealed **UTS: 210**
BS 2871 C102	99.9% Cu: Fire refined tough pitch; tube; annealed **UTS: 210**
BS 2871 C103	99.95% Cu: Oxygen-free copper; tube; annealed **UTS: 210**
BS 2871 C104	99.8% Cu: Non-arsenical tough pitch; tube; specification not issued
BS 2871 C105	0.4% As 99.2% Cu: Tough pitch copper; tube; specification not issued
BS 2871 C106	0.04% P 99.85% Cu: Phosphorus de-oxidized; tube; annealed **UTS: 220**
BS 2871 C107	0.05% P 0.4% As 99.2% Cu: Phosphorus arsenical; tube; annealed **UTS: 220**
BS 2871 C108	Cd Cu: Tube; specification not issued
BS 2871 C109	0.5% Te Cu alloy: Tube; specification not issued
BS 2872 C101	99.9% Cu: Forging; electrolytic tough pitch Cu; specification not issued
BS 2872 C102	99.9% Cu: Fire refined tough pitch; forging; specification not issued
BS 2872 C103	99.95% Cu: Forging; oxygen-free; high conductivity; specification not issued
BS 2872 C104	99.85% Cu: Forging; tough pitch; specification not issued
BS 2872 C105	0.4% As 99.2% Cu: Forging; tough pitch; specification not issued
BS 2872 C106	0.04% P 99.85% Cu: Forging; specification not issued
BS 2872 C107	0.4% As 0.06% P 99.2% Cu: Forging; specification not issued
BS 2872 C108	1% Cd Cu alloy: Forging; specification not issued
BS 2872 C109	0.5% Te Cu alloy: Forging; specification not issued
BS 2873 C101	99.9% Cu: Electrolytic refined tough pitch; wire; annealed; electrical conductivity 97% IACS **Elon: 15%**
BS 2873 C102	99.9% Cu: Fire refined tough pitch; wire; annealed; electrical conductivity 100% IACS **Elon: 25%**
BS 2873 C103	99.95% Cu: Oxygen-free high conductivity; wire; mechanical properties not quoted; annealed; electrical conductivity 100% IACS
BS 2873 C104	99.85% Cu: Tough pitch; wire; specification not issued
BS 2873 C105	99.2% Cu: Tough pitch; wire; specification not issued
BS 2873 C106	99.85% Cu: Phosphorus de-oxidized; wire; specification not issued

Symbol	Nominal analysis, supplier, condition and remarks.
BS 2873 C107	99.2% Cu: Phosphorus de-oxidized; wire; specification not issued
BS 2873 C108	1% Cd Cu alloy: Wire; mechanical properties not quoted; hard; electrical conductivity 80% IACS
BS 2873 C109	0.5% Te Cu alloy: Wire; specification not issued
BS 2874 C101	99.9% Cu: Electrolytic tough pitch high conductivity; bar; annealed **UTS: 220 Elon: 40%**
BS 2874 C101	99.9% Cu: Electrolytic tough pitch high conductivity; bar; hard **UTS: 320 Elon: 9%**
BS 2874 C102	99.9% Cu: Fire refined tough pitch high conductivity; bar; annealed **UTS: 220 Elon: 40%**
BS 2874 C102	99.9% Cu: Fire refined tough pitch high conductivity; bar; hard **UTS: 320 Elon: 9%**
BS 2874 C103	99.95% Cu: Oxygen-free high conductivity; bar; annealed **UTS: 220 Elon: 40%**
BS 2874 C103	99.95% Cu: Oxygen-free high conductivity; bar; hard **UTS: 320 Elon: 9%**
BS 2874 C104	99.85% Cu: Tough pitch non-arsenical; bar; specification not issued
BS 2874 C105	0.4% As 99.2% Cu: Tough pitch arsenical; bar; annealed **UTS: 210 Elon: 40%**
BS 2874 C106	0.04% P 99.85% Cu: Bar; annealed **UTS: 210 Elon: 40%**
BS 2874 C106	0.04% P 99.85% Cu: Bar; hard **UTS: 220 Elon: 15%**
BS 2874 C107	0.4% As 99.2% Cu: Bar; annealed **UTS: 210 Elon: 40%**
BS 2874 C108	1% Cd Cu alloy: Bar; specification not issued
BS 2874 C109	0.4% Te Cu alloy: Bar; annealed **UTS: 210 Elon: 35%**
BS 2875 C101	99.9% Cu: Electrolytic tough pitch; high conductivity; plate; annealed **UTS: 210 Elon: 35%**
BS 2875 C101	99.9% Cu: Electrolytic tough pitch; high conductivity; plate; hard **UTS: 250 Elon: 17%**
BS 2875 C102	99.9% Cu: Fire refined tough pitch; high conductivity; plate; annealed **UTS: 210 Elon: 35%**
BS 2875 C102	99.9% Cu: Fire refined tough pitch; high conductivity; plate; hard **UTS: 250 Elon: 17%**
BS 2875 C103	99.95% Cu: Oxygen-free high conductivity; plate; annealed **UTS: 210 Elon: 35%**
BS 2875 C103	99.95% Cu: Oxygen-free high conductivity; plate; hard **UTS: 250 Elon: 17%**
BS 2875 C104	99.85% Cu: Tough pitch non-arsenical; plate; annealed **UTS: 210 Elon: 35%**
BS 2875 C104	99.85% Cu: Tough pitch non-arsenical; plate; hard **UTS: 250 Elon: 17%**
BS 2875 C105	0.4% As 99.2% Cu: Tough pitch arsenical; plate; annealed **UTS: 210 Elon: 35%**
BS 2875 C105	0.4% As 99.2% Cu: Tough pitch arsenical; plate; hard **UTS: 250 Elon: 17%**
BS 2875 C106	0.04% P 99.85% Cu: Phosphorus de-oxidized non-arsenical; plate; annealed **UTS: 210 Elon: 35%**

Note. The following abbreviations and units are used in the tables:

DPN	Hardness, diamond pyramid number
UTS	Ultimate tensile strength, N/mm^2
Elon	Elongation, %
Proof	0.1 % proof strength, N/mm^2

1 N/mm^2=0.1 hbar=0.102 kgf/mm^2=0.06475 tonf/in^2=145.04 lbf/in^2

Symbol	Nominal analysis, supplier, condition and remarks.
BS 2875 C106	0.04% P 99.85% Cu: Phosphorus de-oxidized non-arsenical; plate; annealed UTS: 250 Elon: 17%
BS 2875 C107	0.4% As 0.04% P 99.2% Cu: Arsenical; plate; annealed UTS: 210 Elon: 35%
BS 2875 C107	0.4% As 0.04% P 99.2% Cu: Arsenical; plate; hard UTS: 270 Elon: 17%
BS 2875 C108	1% Cd Cu alloy: Plate; electrical conductivity 80% IACS UTS: 280 Elon: 17%
BS 2875 C109	0.5% Te Cu: Plate; specification not issued
BS 2901 C7	0.3% Si 0.2% Mn 98.5% Cu: Rod for all welding types
BS 2901 C8	0.2% Ti 0.2% Al 99.4% Cu: Rod for all welding types
BS 4109 C101	High purity copper: Electrolytic tough pitch as BS 1036 C101
BS 4109 C102	High purity copper: Fire refined tough pitch as BS 1036 C102
BS 4393	High purity copper: Tin or tin/lead coated
BS 4577 A1/1	99.93% Cu: For resistance welding electrodes; as drawn; conductivity 98% IACS DPN: 90 UTS: 260 Elon: 25%
BS 4577 A1/2	1% Te 0.1% Ni Cu: For resistance welding electrodes; as drawn; conductivity 85% IACS DPN: 90 UTS: 250 Elon: 14%
BS 4577 A1/3	1% Cd Cu: For resistance welding electrodes; as drawn; conductivity 80% IACS DPN: 90 UTS: 300 Elon: 14%
BS 4577 A4/3	6% Ag Cu: For resistance welding electrodes; wrought; conductivity 80% IACS DPN: 150 UTS: 490 Elon: 9%
BS 4608	99.9% Cu: For electrical purposes; sheet & strip
BS 6360	99.9% Cu: Tough pitch high conductivity copper; as BS 4109; as BS 1036 and BS 1037
BS T7	0.4% As 99.2% Cu: Seamless tube; ½-hard UTS: 300
BS T51	0.5% As 99.2% Cu: Seamless tube; as drawn UTS: 240
C 80	Copper sintered material: Sintered Products Ltd; 8–11% porosity UTS: 750 Elon: 3%
C 80	Sintered pure Cu: 9% porosity; specific gravity 8.1; Durasint DPN: 40 UTS: 100 Elon: 3%
CADMIUM COPPER	0.75% Cd Cu alloy: High tensile strength; used for telephone wires; origin unknown
CHILI BAR	1% S + usual impurities Cu: Crude; origin unknown
COMBARLOY	0.1% Ag Cu: For electrical purposes; Thomas Bolton
COPPER	Cu metal: Ingots; granulated, powder, foil etc.; Blackwell; primary metal
COPPER	High purity metal: Impurities 10 p.p.m.; Johnson Matthey; supplied as ingot, powder, wire & sheet
COPPERWELD HM	Mild steel copper-coated wire: Hard drawn steel; 40% of wire is copper for electrical leads; Copperweld Steel Co
COPPERWELD LC	Mild Steel copper-coated wire: Annealed steel; 40% of wire is copper for electrical leads; Copperweld Steel Co
Cu/a1	99.9% Cu: Oxygen-free; French Standard; part of NFA 53-100
Cu/a2	99.9% Cu: French Standard; part of NFA 53-100
Cu/a3	99.75% Cu: French Standard; part of NFA 53-100
Cu/b	99.9% Cu 0.03% O: French Standard; part of NFA 53-100
Cu/c1	99.92% Cu: For conductors; French Standard; part of NFA 53-100
Cu/c2	99.96% Cu 0.0003% O (max): For conductors; French Standard; part of NFA 53-100
Cu/d	99.9% Cu cathode copper: French Standard; part of NFA 53-100
Cu–Zr MASTER ALLOY	12.5% Zr Cu alloy: Origin unknown
CUPRALITH	Series of copper–lithium alloys containing 1–10% Li
DIN 1708	Covers high purity coppers
DIN 1733 S Cu Ag	1.0% Ag Cu
DIN 1733 S Cu Sn	0.6% Sn Cu
DIN 1785 S B Cu	0.4% As 0.03% P Cu: Wrought
DIN 1785 S D Cu	0.03% P Cu: Wrought
DO	0.03% P 99.95% Cu: Extrusions; phosphorus de-oxidized; McKechnie Bros for BS alloy C106
DONA	0.04% P 0.02% As 99.9% Cu: Non-arsenical; de-oxidized; Birmingham Battery for BS 1172
DONA	Phosphorus de-oxidized non-arsenical copper: IMI 131
DOXA	0.04% P 0.35% As 99.5% Cu: Arsenical de-oxidized; Birmingham Battery for BS 1174
DTD 607	99.9% Cu: Strip; annealed; 0.05% Ag is optional DPN: 55
DTD 607	99.9% Cu: Strip; cold rolled & tempered; 0.05% Ag is optional DPN: 95
DTD 631	99.8% Cu: Silver coated strip suitable for brazing
EC	99.9% Cu: Extrusions; electrolytic tough pitch; McKechnie Bros for BS alloy C101
ELKALOY A	Cd Cu alloy: For resistance welding electrode; Johnson Matthey; electrical conductivity 85% IACS DPN: 90 UTS: 450 Elon: 20% Proof: 370
ERM ALW	0.1% Ni 0.6% Te Cu alloy: Drawn bar; Enfield Rolling Mills Ltd; electrode material for resistance welding DPN: 100 UTS: 270 Elon: 16% Proof: 240
ERM HSM	0.6% Te 0.03% O Cu alloy: Drawn bar; Enfield Rolling Mills Ltd; commutator bars etc. DPN: 90 UTS: 260 Elon: 20% Proof: 220
HC	99.95% Cu: Extrusions; fire refined tough pitch; McKechnie Bros for BS alloy C102
h Cu 6 a	99.998% Cu: Bars; Light Ltd; high purity metal
h Cu 9 a	99.999% Cu: Sheet 1 mm thick; Light Ltd; high purity metal
h Cu 13	99.998% Cu: Powder; Light Ltd; high purity metal
h Cu 73	99.99% Cu metal single crystals; 25 mm diameter: Light Ltd
h Cu 96	99.5% Cu 63 isotope: Light Ltd; stable isotopic form
h Cu 97	99.4% Cu 65 isotope: Light Ltd; stable isotopic form
HCOKOF	Oxygen-free high conductivity copper: Thomas Bolton; withdrawn
HCOKOF	High purity, high conductivity, oxygen-free Cu: Outokumpu
HSM/S	0.4% S Cu alloy: Rod for free machining; Enfield specification for BS 2874 C111
IMI 100	Oxygen-free high conductivity copper: IMI
IMI 103	Tough pitch high conductivity copper: Electrolytic; IMI for BS 125, BS 128, BS 1432, BS 1433
IMI 121	99.8% Cu: Fire refined for rivets; IMI; withdrawn
IMI 131	Phosphorus de-oxidized non-arsenical copper: IMI for BS 1172
IMI 134	99.85% Cu: Tough pitch non-arsenical; IMI for BS 1038, BS 1039
IMI 138	0.4% As Cu: Tough pitch; IMI
IMI 146	0.4% As Cu: De-oxidized arsenical; IMI
IMI 153	Ag Cu alloy: Brazing rod; IMI for BS alloy C1
IMI 161	Boron Cu: Welding rod; IMI Borofil

Symbol	Nominal analysis, supplier, condition and remarks.
IMI 161	0.2% Al 0.2% Ti Cu alloy: Welding rod; IMI Nitrofil
IMI 166	1% Cd Cu alloy: Wire & bar; IMI for BS alloy C108
IMI 181	0.5% Te Cu: For free machining; IMI KUTERN; electrical conductivity 90% IACS
IMI 651	1% Sn Cu alloy: Wire & bar; IMI for PO bronze **DPN: 170 UTS: 540 Elon: 2% Proof: 480**
IMI 981	High conductivity copper wire: Tinned; IMI wire for either IMI 103 or IMI 100
KUFIL	1% Ag Cu alloy: Welding rod; IMI BS alloy C1; melting range 1073–1078 °C
KUTERN	0.5% Te Cu alloy: Rod & sections; IMI free machining copper; IMI 181
LAKE COPPER	Copper which originates on North peninsula of Michigan, USA: See ASTM B4
MANGANIN	13% Mn Cu alloy: For resistors etc; Driver
NFA 53-100	Classification of copper: French Standard
NITROFIL	0.3% Al 0.3% Ti Cu alloy: Filler rod; IMI; melting point 1080 °C; for arcon arc welding Cu
PO Bronze	1% Sn Cu alloy: Traditional name; originally used for telephone line wire
QQ C 501	99.9% Cu: Plate or sheet; US Federal **Elon: 20%**
QQ W 336	High conductivity Cu: Wire; US Federal; hard drawn **UTS: 430 Elon: 3%**
QQ W 341	High conductivity Cu: Wire; US Federal; annealed **UTS: 240 Elon: 28%**
RWMA Class 1	Cd Cu alloy: Mallory Metallurgical code for Elkaloy A
SAE 71	99.9% Cu: Sheet & strip; ½-hard; now SAE CA 116 **UTS: 280**
SAE 71	99.9% Cu: Sheet & strip; extra hard; now SAE CA 116 **UTS: 370**
SAE 75	99.9% Cu: Tube; light drawn; now SAE CA 122 **UTS: 240**
SAE 75	99.9% Cu: Tube; hard drawn; now SAE CA 122 **UTS: 330**
SAE 83	99.9% Cu: Wire; annealed; electrical conductivity 100% IACS; now SAE CA 110 **UTS: 240 Elon: 30%**
SAE CA 102	99.95% Cu: Oxygen-free; suitable for welding
SAE CA 110	99.9% Cu: Electrolytic tough pitch; formerly SAE 71 and 83
SAE CA 111	99.9% Cu: Electrolytic tough pitch with trace Cd; formerly SAE 71
SAE CA 113	99.9% Cu trace Ag: Tough pitch; formerly SAE 71; ASTM alloy STP
SAE CA 114	99.9% Cu trace Ag: Tough pitch; formerly SAE 71; ASTM alloy STP; higher Ag than CA 113
SAE CA 116	99.9% Cu trace Ag: Tough pitch; formerly SAE 71; ASTM alloy STP; higher Ag than CA 113 & CA 114
SAE CA 120	99.9% Cu 0.008% P: De-oxidized; formerly SAE 75; ASTM alloy DLP

Symbol	Nominal analysis, supplier, condition and remarks.
SAE CA 122	99.9% Cu 0.02% P: De-oxidized; formerly SAE 75; ASTM alloy DHP
SAE CA 145	0.5% Te 0.008% P 99.5% Cu: Free machining; electrical conductivity IACS 93%
SAE CA 147	0.3% S 99.7% Cu: Free machining; IACS 93%
SAE CA 150	0.12% Zr 99.88% Cu: For welding electrode
SAE CA 162	1% Cd 99% Cu alloy
SAE CA 187	1% Pb Cu alloy: Free machining
SC	0.3% S 99.7% Cu: Extrusions; free machining; McKechnie; as drawn; electrical conductivity IACS 95% **DPN: 90 UTS: 240 Elon: 20% Proof: 150**
SIS 5010	99.9% Cu: Bar, sheet, forging, etc.; Swedish Standard; annealed; electrical conductivity 100% IACS **DPN: 50 UTS: 240 Elon: 50% Proof: 45**
SIS 5010	99.9% Cu: Bar, sheet, forging, etc.; Swedish Standard; cold worked; electrical conductivity 100% IACS **DPN: 70 UTS: 420 Elon: 30% Proof: 210**
SIS 5011	99.5% Cu: Bar & sheet; oxygen-free high conductivity; Swedish Standard; annealed; electrical conductivity 100% IACS **DPN: 50 UTS: 200 Elon: 50% Proof: 45**
SIS 5011	99.5% Cu: Bar & sheet; oxygen-free high conductivity; Swedish Standard; cold rolled; electrical conductivity 100% IACS **DPN: 100 UTS: 300 Elon: 12% Proof: 240**
SIS 5013	99.8% Cu: Plate; fire refined; Swedish Standard; annealed **DPN: 65 UTS: 210 Elon: 50% Proof: 45**
SIS 5013	99.8% Cu: Plate; fire refined; Swedish Standard; cold rolled **DPN: 100 UTS: 330 Elon: 12% Proof: 300**
SIS 5015	0.03% P 99.8% Cu: Plate & bar; fire refined; Swedish Standard; annealed **DPN: 50 UTS: 200 Elon: 45% Proof: 45**
SIS 5015	0.03% P 99.8% Cu: Plate & bar; fire refined; Swedish Standard; cold rolled **DPN: 85 UTS: 300 Elon: 30% Proof: 200**
SIS 5030	0.08% Ag 99.9% Cu: Extrusions; Swedish Standard; cold drawn **DPN: 100 UTS: 280 Elon: 10% Proof: 240**
SIS 5053	Cd bearing Cu: Bar; Swedish Standard; cold drawn **UTS: 580 Proof: 420**
STA 7 C1	99.75% Cu: Tough pitch for driving bands; annealed **DPN: 50 UTS: 180 Elon: 40%**
STA 7 C2	0.1% P 99.75% Cu: Phosphorous de-oxidized for driving bands; annealed **DPN: 50 UTS: 180 Elon: 40%**
STA 7 C3	99.9% Cu: Tough pitch high conductivity; sheet; annealed BS 899 **UTS: 200 Elon: 35%**
STA 7 C3	99.9% Cu: Tough pitch high conductivity; sheet; ½-hard **UTS: 270**
STA 7 C4	99.9% Cu: Tough pitch; tubes; annealed **UTS: 210 Elon: 40%**
STA 7 C5	99.9% Cu: Tough pitch; bars; annealed **UTS: 210 Elon: 40%**
STA 7 C5	99.9% Cu: Touch pitch; bars; ½-hard **DPN: 65 UTS: 270 Elon: 12%**
STA 7 C6	99.9% Cu cathode copper: BS 1035
STA 7 C7A	99.9% Cu: Electrolytic high conductivity; BS 1036
STA 7 C7B	99.9% Cu: Fire refined high conductivity; BS 1037
STA 7 C8A	99.85% Cu: Tough pitch non-arsenical; BS 1038
STA 7 C8B	99.75% Cu: Tough pitch non-arsenical; BS 1039
STA 7 C8C	99.5% Cu: Tough pitch non-arsenical; BS 1040

Note. The following abbreviations and units are used in the tables:

DPN	Hardness, diamond pyramid number
UTS	Ultimate tensile strength, N/mm^2
Elon	Elongation, %
Proof	0.1 % proof strength, N/mm^2

1 N/mm^2=0.1 hbar=0.102 kgf/mm^2=0.06475 tonf/in^2=145.04 lbf/in^2

Symbol	Nominal analysis, supplier, condition and remarks.
STA 7 C9	0.4% As 99.2% Cu: Tough pitch; now BS 1173
STA 7 C10	0.1% P 99.85% Cu: Phosphorous de-oxidized; now BS 1172
STA 7 C11	0.1% P 0.4% As 99.2% Cu: Now BS 1174
STA 13 C1	99.75% Cu: Tough pitch for projectile driving bands
STA 13 C2	0.05% P 99.75% Cu: Phosphorous de-oxidized for projectile driving bands

Symbol	Nominal analysis, supplier, condition and remarks.
TC	0.5% Te 99.5% Cu: Extrusions; free machining; McKechnie; electrical conductivity IACS 90% **DPN: 100 UTS: 280 Elon: 15% Proof: 220**
WHITE PINE LAKE COPPER	Ag 99.9% Cu: American trade name
WW T 799a	99.9% Cu: Seamless tube; US Federal
WW T 799a/N	99.9% Cu: Seamless tube; US Federal

14B Copper–zinc wrought alloys
Less than 38 per cent zinc – alpha brass

Specific gravity	8.4–8.8	
Density	8400–8800 kg/m^3	(0.303–0.318 lb/in.^{3})
Solidus/liquidus	5% Zn 1044 °C	
	30% Zn 900 °C	
Thermal conductivity	5% Zn 188 W/m °C	(45 cal/m s °C)
	30% Zn 109 W/m °C	(26 cal/m s °C)
Coefficient of linear expansion	18.2–20.5 × 10^{-6}/ °C	
Electrical conductivity	5% Zn alloys 44% IACS (copper 100%)	
	30% Zn alloys 27% IACS (copper 100%)	
Specific resistance	5% Zn 50 microhm mm	
	30% Zn 65 microhm mm	
Young's modulus of elasticity	105–117 × 10^9 N/m^2	(15–17 × 10^6 lhf/in.^{2})
Impact	– J	(– ft/lb)
Fatigue strength (10^8 cycles)	annealed 80–110 N/mm^2	(± 5.4–7.5 tonf/in.^{2})
	cold worked 120–140 N/mm^2	(± 8–9.8 tonf/in.^{2})
	Fatigue strength varies directly with the zinc content.	

Hot strength

Temperature °C	Tensile strength N/mm^2		Elongation %	
	5 % Zn	30 % Zn	5 % Zn	30 % Zn
100	660	570	8	7
200	670	580	8	7
300	450	570	28	8
400	420	380	50	42
500	340	300	60	45
600	320	270	65	47

The above properties are typical of the following group, and may not apply exactly to any one specification. It is possible that with certain specifications some of the values may not be applicable.

General metallurgical characteristics

The majority of copper alloys are based on the copper/zinc series. There are three groups, alpha, alpha plus beta, and beta depending on the zinc content. Each of these groups has its own characteristics, which are modified first by the zinc content and then by the addition of alloying elements to give increased strength and corrosion resistance. The alpha brasses are listed in this section 14B, while the alpha + beta and beta alloys are found in 14C.

The alpha group has up to 38% zinc present. At the higher zinc figure depending on the heat treatment and other alloying elements, some beta phase may be present.

These alpha brasses are similar in many respects to copper and are often used in lieu of copper where economy is necessary, or corrosion resistance and electrical properties are not of paramount importance, for example, decorative articles and household goods. There are several groups of these alloys, as follows:

Cap copper. Zinc content 2–5%. This, in effect, is copper de-oxidized with zinc. The result is a ductile readily worked material of low mechanical strength. The electrical properties are not much affected, the conductivity being 75–95% IACS. The alloy can be strengthened by cold work to a limited extent.

Gilding metal. Zinc content 5–20%. These alloys are used for their pleasing reddish copper colour which becomes less coppery and more brassy as the zinc content increases. They are all capable of being cold worked, the hardness increasing with zinc content.

These alloys are not prone to the defect known as 'season cracking' which affects most of the higher zinc brasses. This cracking is caused by local corrosion attack at grain boundaries when cold worked brasses are stored in certain atmospheres.

The electrical conductivity of 10% zinc gilding metal is as low as 43% IACS.

Cartridge brass. Zinc content 28–32%. This is the most ductile and easily worked of the zinc/copper alloys, finding use where deep drawing operations are required. Interstage annealing is necessary between heavy deep draws and time and temperature must be carefully controlled to prevent grain growth resulting in an orange peel surface finish. This material is subject to season cracking and should always be stress relieved after the final forming operation. The electrical conductivity of these alloys is 27% IACS.

Basis or common brass. Zinc content is 35–38%. This is the cheapest zinc copper alloy, and is generally found only as sheet. The ductility is reduced as the zinc content rises, owing to the formation of some beta phase. The material is generally used for shallow pressings for low duty purposes. The corrosion resistance and electrical conductivity are relatively low.

Effect of alloying elements

All of the above materials can have certain properties modified by the addition of alloying elements with little effect on the remaining properties.

Tin. The addition of up to 1% improves the corrosion resistance. 'Admiralty Brass' is 'Cartridge Brass' with 1% tin added. This differs from 'Naval Brass' in the zinc content.

Lead. Small amounts of lead (up to 2%) are added to the brasses to improve their machinability. The lead does not go into solution, but acts as a discontinuity giving built in lubrication and smaller chips. There is a slight reduction in corrosion resistance and ductility.

Aluminium. This improves the resistance of brasses to sea water up to 2% aluminium being used. One part by weight of aluminium has the equivalent effect of six parts zinc, as far as other properties are concerned. Thus aluminium brasses have approximately 22% zinc, 2% aluminium, but are equivalent for all practical purposes to the 30% zinc series.

In general it can be said that corrosion resistance, electrical and heat conductivity and price all fall as the zinc content rises. The ability to accept cold work rises to the 30% zinc alloy, then drops with further increase in zinc.

Thermal treatment

None of the alloys listed can be hardened by heat treatment. They are all capable of accepting cold work to a relatively high degree but must often be given an anneal between drawing operations. This should be at approximately 500 °C for the lower zinc contents and 600 °C for the 30–35% zinc alloys. Where surface finish is important it is imperative that the exact temperature is found, otherwise grain growth and an 'orange peel' surface will result. As there is no phase change involved there is no need to hold the part at the correct temperature for any longer than it takes to reach the desired temperature. Too high a temperature, apart from causing grain growth, can result in 'de-zincification' at grain boundaries. This in addition to giving an unsightly appearance seriously weakens the part. It is always advisable to stress relieve these brasses at about 250 °C for 30 minutes after working. This may remove some of the cold work but considerably reduces the danger of season cracking.

Welding and brazing

Pre-treatment. The area to be joined must be thoroughly cleaned and all traces of grease and oil removed. Owing to the good heat conductivity of these alloys it is advisable with large components to pre-heat and weld or braze while hot. Stress relieving at 250 °C can be advantageous with large complicated parts which have been cold worked as this reduces the danger of distortion.

Welding and brazing. Welding is easier with these alloys than with coppers as the heat conductivity is reduced. Considerable local heat is still required, often with one torch to pre-heat while a second operator welds. There is also a health hazard from zinc oxide fumes. Alloys containing lead cannot normally be welded. Where necessary inert gas shielded arc welding can be used, when fume extraction is recommended. The advantages of welding over brazing are often not sufficient to warrant the extra precautions required. Grain growth can be experienced above 400 °C and when this is not desirable a brazing alloy melting below this should be chosen. Brazing and soldering of all these alloys is very easy and joints of a relatively high strength can be obtained. Fluxes are required and can be corrosive, thus considerable care must be taken to ensure that all traces of flux are removed.

Post-treatment. None required except to remove all traces of flux. It should be noted that if the material has been cold worked this effect will be locally removed where the parts are heated. Stress relieving at 250 °C is recommended on complicated parts where distortion could cause stressing with season cracking following.

Flaw detection methods

Crack test. This will seldom be required. Oil penetrant chalk test can be used when necessary.

X-ray. Not required.

Ultra-sonic test. Not normally required.

Chemical etch. This will be needed only for special duty parts. Ferric chloride in dilute hydrochloric acid shows up the grain size and surface defects. The resultant surface has poor fatigue and corrosion resistance and should be polished to remove all trace of the etch. It is also possible to show a tendency for season cracking with a laboratory etch, which scraps the part.

Corrosion protection

Temporary. None required.

Permanent. None generally required except for appearance purposes. The Copper Development Association is promoting a lacquer which prevents the formation of the normal green colouration. There are also various chemical treatments which darken the surface and give a bronze effect. Neither the lacquered or the bronzed surfaces should be cleaned using abrasive materials.

Plating. The alloys can be easily electroplated with normal precautions.

Painting. Painting with air drying or stoving enamels is possible. An etch primer is recommended and thorough cleanliness is essential.

Machinability

In the annealed condition all these materials have a tendency to tear and it is sometimes difficult to hold sizes ac-

curately. These defects are reduced to some extent with higher zinc content and to a marked extent by cold working.

The inclusion of small amounts of lead gives greatly improved machinability.

Uses

Cap copper. Deep drawn pressings of low strength. Some electrical and ornamental copper uses.

Gilding metal. Architectural where strength and appearance are not important. Ornamental where a golden appearance is required. This use is now being superseded by anodized aluminium. Architectural ornamental purposes where parts can be readily cleaned, for example, gates, post boxes and grilles, medals and some coins.

Cartridge brass. Deep pressings for ornamental trim in architectural, transport and household use and also as gauze for filters. When lead is added the excellent machinability of these alloys makes them ideal for auto-machining. With tin or aluminium added they find considerable marine use for condenser tubes, handrails, non-magnetic fittings and miscellaneous items of a low duty nature.

Common brass. This is cheaper than the 30% zinc alloys and has numerous uses, but cannot be as deeply drawn and has reduced corrosion resistance.

Symbol	Nominal analysis, supplier, condition and remarks.
2.0290	35% Zn 2% Pb Cu alloy: Casting; German Standard
2.0291	35% Zn 2% Pb Cu alloy: Casting; German Standard
2.0293	35% Zn 2% Pb Cu alloy: Casting; German Standard
2.0295	35% Zn 2% Pb Cu alloy: Casting; German Standard
2.0470	1.0% Sn 28% Zn Cu alloy: German Werkstoff number
8M	35% Zn 1% Sn 0.2% Pb 1% Fe 4% Al Cu alloy: Wrought; Langley Alloys **UTS: 540 Elon: 15% Proof: 270**
20	35.5% Zn 2% Pb Cu alloy: Extrusion; McKechnie Bros; as drawn **DPN: 120 UTS: 370 Elon: 30% Proof: 200**
21	35% Zn Cu alloy: Extrusion; McKechnie Bros; annealed **DPN: 90 UTS: 330 Elon: 45% Proof: 100**
65	29% Zn 0.75% Fe 4.25% Al 0.75% Mn Cu alloy: Extrusion; McKechnie Bros; as extruded **DPN: 130 UTS: 600 Elon: 15% Proof: 300**

Note. The following abbreviations and units are used in the tables:

DPN	Hardness, diamond pyramid number
UTS	Ultimate tensile strength, N/mm^2
Elon	Elongation, %
Proof	0.1 % proof strength, N/mm^2

1 N/mm^2=0.1 hbar=0.102 kgf/mm^2=0.06475 tonf/in^2=145.04 lbf/in^2

Symbol	Nominal analysis, supplier, condition and remarks.
193	5% Zn 2.2% Fe Cu alloy: Wrought; designation agreed by CDAA; not used
205	2% Zn Cu alloy: Wrought; designation agreed by CDAA; not used
226	12.5% Zn Cu alloy: Wrought; designation used by CDAA
234	15% Zn Cu alloy: Wrought; designation agreed by CDAA; not used
250	25% Zn Cu alloy: Wrought; designation agreed by CDAA; not used
261	0.04% P 30% Zn Cu alloy: Wrought; designation agreed by CDAA; not used
262	32% Zn Cu alloy: Wrought; designation agreed by CDAA; not used
274	37% Zn Cu alloy: Wrought; designation agreed by CDAA; not used
310	8% Zn 0.5% Pb Cu alloy: Wrought; designation agreed by CDAA; not used
310	20% Zn 0.2% Pb Cu alloy: Extrusions; McKechnie; Bros; annealed **DPN: 90 UTS: 280 Elon: 50% Proof: 100**
311	30% Zn Cu alloy: Extrusion; McKechnie Bros; annealed **DPN: 90 UTS: 280 Elon: 60% Proof: 100**
314	9.25% Zn 1.7% Pb Cu alloy: Wrought; designation used by CDAA
316	9% Zn 1.8% Pb 1.0% Ni Cu alloy: Wrought; designation agreed by CDAA; not used
320	12% Zn 2% Pb Cu alloy: Wrought; designation agreed by CDAA; not used

Symbol	Nominal analysis, supplier, condition and remarks.
325	25% Zn 2.8% Pb Cu alloy: Wrought; designation agreed by CDAA; not used
332	35% Zn 1.7% Pb Cu alloy: Tube; designation used by CDAA
335	37% Zn 0.5% Pb Cu alloy: Wrought; designation used by CDAA
340	35% Zn 1% Pb Cu alloy: Wrought; designation used by CDAA
344	35% Zn 0.7% Pb Cu alloy: Wrought; designation agreed by CDAA; not used
347	35% Zn 1.5% Pb Cu alloy: Wrought; designation agreed by CDAA; not used
348	38% Zn 0.6% Pb Cu alloy: Wrought; designation agreed by CDAA; not used
353	38% Zn 2% Pb Cu alloy: Wrought; designation used by CDAA
356	38% Zn 2.5% Pb Cu alloy: Wrought; designation used used by CDAA
362	37% Zn 4.0% Pb Cu alloy: Wrought; designation agreed by CDAA; not used
365	38% Zn 0.8% Pb Cu alloy: Wrought; designation used by CDAA
366	38% Zn 0.7% Pb 0.07% As Cu alloy: Wrought; designation used by CDAA
367	38% Zn 0.7% Pb 0.07% Sb Cu alloy: Wrought; designation used by CDAA
368	38% Zn 0.7% Pb 0.06% P Cu alloy: Wrought; designation used by CDAA
370	38% Zn 1.1% Pb Cu alloy: Wrought; designation used by CDAA
371	38% Zn 1.0% Pb Cu alloy: Wrought; designation agreed by CDAA; not used
377	38% Zn 2.0% Pb Cu alloy: For forging; designation used by CDAA
405	5% Zn 1.0% Sn Cu alloy: Wrought; designation agreed by CDAA; not used
408	5% Zn 2.0% Sn Cu alloy: Wrought; designation agreed by CDAA; not used
409	5% Zn 0.7% Sn Cu alloy: Wrought; designation agreed by CDAA; not used
410	5% Zn 2.5% Sn Cu alloy: Wrought; designation agreed by CDAA; not used
411	7% Zn 0.5% Sn Cu alloy: Wrought; designation agreed by CDAA; not used
413	7% Zn 1.0% Sn Cu alloy: Wrought; designation agreed by CDAA; not used
415	6% Zn 2% Sn Cu alloy: Wrought; designation agreed by CDAA; not used
420	8% Zn 1.7% Sn 0.2% P Cu alloy: Wrought; designation agreed by CDAA; not used
421	8% Zn 2.6% Sn 0.3% P Cu alloy: Wrought; designation agreed by CDAA; not used
422	10% Zn 1% Sn 0.3% P Cu alloy: Wrought; designation agreed by CDAA; not used
425	10% Zn 2% Sn 0.3% P Cu alloy: Wrought; designation agreed by CDAA; not used
430	12% Zn 2.2% Sn Cu alloy: Wrought; designation agreed by CDAA; not used

Symbol	Nominal analysis, supplier, condition and remarks.
432	14% Zn 0.5% Sn 0.3% P Cu alloy: Wrought; designation agreed by CDAA; not used
434	15% Zn 0.8% Sn Cu alloy: Wrought; designation agreed by CDAA; not used
435	18% Zn 1.0% Sn Cu alloy: Wrought; designation agreed by CDAA; not used
436	18% Zn 0.4% Sn Cu alloy: Wrought; designation agreed by CDAA; not used
438	18% Zn 1.2% Sn Cu alloy: Wrought; designation agreed by CDAA; not used
442	28% Zn 1.0% Sn Cu alloy: Wrought; designation agreed by CDAA; not used
443	30% Zn 1.0% Sn 0.07% As Cu alloy: Wrought; designation used by CDAA
444	30% Zn 1.0% Sn 0.07% Sb Cu alloy: Wrought; designation used by CDAA
445	30% Zn 1.0% Sn 0.06% P Cu alloy: Wrought; designation used by CDAA
462	38% Zn 0.7% Sn Cu alloy: Wrought; designation agreed by CDAA; not used
476	10% Zn 2% Pb 2% Sn 0.05% P Cu alloy: Wrought; designation agreed by CDAA; not used
482	38% Zn 0.8% Pb 0.7% Sn Cu alloy: Wrought; designation agreed by CDAA; not used
485	38% Zn 1.8% Pb 0.7% Sn Cu alloy: Wrought; designation used by CDAA
665	20% Zn 1.0% Mn Cu alloy: Wrought; designation agreed by CDAA; not used
667	30% Zn 1.2% Mn Cu alloy: Wrought; designation agreed by CDAA; not used
674	38% Zn 0.4% Pb 0.2% Sn 1.5% Al 3.0% Mn 1.0% Si Cu alloy: Wrought; designation agreed by CDAA; not used
675	38% Zn 1.0% Sn 0.3% Mn Cu alloy: Wrought; designation used by CDAA
676	38% Zn 0.7% Pb 1.0% Sn 0.4% Mn Cu alloy: Wrought; designation agreed by CDAA; not used
685	12% Zn 4.0% Al 2.0% Fe Cu alloy: Wrought; designation agreed by CDAA; not used
687	20% Zn 2.1% Al 0.7% As Cu alloy: Wrought; designation used by CDAA
862	23% Zn 5% Al 3% Fe 3% Mn Cu alloy: Casting; designation used by CDAA
863	26% Zn 6% Al 3% Fe 3% Mn Cu alloy: Casting; designation used by CDAA
864	39% Zn 1% Al 1% Fe 1% Mn Cu alloy: Casting; designation used by CDAA
ABS/B43	15% Zn Cu alloy: Piping; American Bureau of Shipping
ABS TYPE 2	39% Zn 0% Sn (max) 0.4% Pb (max) 1.2% Fe 0.7% Al Cu alloy: Casting; American Bureau of Shipping **UTS: 460 Elon: 20% Proof: 175**
ADMIRALTY BRASS	28% Zn 1% Sn Cu alloy: Common name; further information from Copper Development Association
AICHS METAL	38% Zn 1.8% Fe Cu alloy: Casting; source unknown
ALDURBRA	22% Zn 2% Al Cu alloy: Wrought; C Clifford
ALDURBRA	22% Zn 2% Al Cu alloy: Casting; C Clifford **DPN: 130 UTS: 590 Elon: 16% Proof: 300**
ALUMBRO	22% Zn 2% Al 0.04% As Cu alloy: Strip & sheet; IMI; annealed **DPN: 100 UTS: 400 Elon: 60% Proof: 100**
ALUMBRO	22% Zn 2% Al 0.04% As Cu alloy: Strip & sheet; IMI; ½-hard **DPN: 160 UTS: 500 Elon: 25% Proof: 420**
ALUMINIUM BRASS	22% Zn 2% Al Cu alloy: Common name; further information from the Copper Development Association

Note. The following abbreviations and units are used in the tables:

DPN	Hardness, diamond pyramid number
UTS	Ultimate tensile strength, N/mm^2
Elon	Elongation, %
Proof	0.1 % proof strength, N/mm^2

$1 \ N/mm^2 = 0.1 \ hbar = 0.102 \ kgf/mm^2 = 0.06475 \ tonf/in^2 = 145.04 \ lbf/in^2$

Symbol	Nominal analysis, supplier, condition and remarks.
AMERICAN BRASS	35% Zn 3% Pb Cu alloy: Bar; free cutting; common name for BS alloy CZ124 type; further information from the Copper Development Association
AMS 4505 D	30% Zn Cu alloy: Sheet & strip; annealed
AMS 4507 C	30% Zn Cu alloy: Sheet & strip; ½-hard
AMS 4508 B	30% Zn Cu alloy: Laminated sheet
AMS 4555 C	31% Zn Cu alloy: Seamless tube; light annealed; similar to SAE 74C
AMS 4558 B	32% Zn 1.6% Pb Cu alloy: Seamless tube; as drawn
AMS 4610 H	35.5% Zn 3% Pb Cu alloy: Bar; free cutting; ½-hard
AMS 4611 C	38.5% Zn 0.8% Sn Cu alloy: Bar; naval brass; ½-hard
AMS 4612 D	38.5% Zn 0.8% Sn Cu alloy: Bar; naval brass; hard drawn
AMS 4614 D	38% Zn 2% Pb Cu alloy: Forging; free machining
AMS 4710	35% Zn Cu alloy: Wire; tinned; annealed
AMS 4712 A	35% Zn Cu alloy: Wire; annealed
AMS 4713 A	35% Zn Cu alloy: Wire; 1/8-hard
AMS 4862 B	24% Zn 5.2% Al 3% Fe 4% Mn Cu alloy: Casting
ASTM B16	35% Zn 3% Pb Cu alloy: Rod; annealed **UTS: 330 Elon: 20% Proof: 10**
ASTM B16	35% Zn 3% Pb Cu alloy: Rod; hard drawn **UTS: 500 Elon: 4% Proof: 24**
ASTM B19	30% Zn Cu alloy: Cartridge brass; annealed **UTS: 330 Elon: 45%**
ASTM B19	30% Zn Cu alloy: Cartridge brass; hard **UTS: 530**
ASTM B19	30% Zn Cu alloy: Cartridge brass; extra spring hard **UTS: 760**
ASTM B22 E	25% Zn 3% Fe 5% Al 4% Mn Cu alloy: Casting; as cast **DPN: 223 UTS: 800 Elon: 12% Proof: 42**
ASTM B36/1	5% Zn Cu alloy: Sheet & bar; ¼-hard **UTS: 300**
ASTM B36/1	5% Zn Cu alloy: Sheet & bar; extra spring hard **UTS: 460**
ASTM B36/2	10% Zn Cu alloy: Sheet & bar; ¼-hard **UTS: 340**
ASTM B36/2	10% Zn Cu alloy: Sheet & bar; extra spring hard **UTS: 530**
ASTM B36/3	15% Zn Cu alloy: Sheet & bar; ¼-hard **UTS: 340**
ASTM B36/3	15% Zn Cu alloy: Sheet & bar; extra spring hard **UTS: 620**
ASTM B36/4	20% Zn Cu alloy: Sheet & bar; ¼-hard **UTS: 370**
ASTM B36/4	20% Zn Cu alloy: Sheet & bar; extra spring hard **UTS: 680**
ASTM B36/6	30% Zn Cu alloy: Sheet & bar; ¼-hard **UTS: 370**
ASTM B36/6	30% Zn Cu alloy: Sheet & bar; extra spring hard **UTS: 730**
ASTM B36/8	35% Zn Cu alloy: Sheet & bar; ¼-hard **UTS: 380**
ASTM B36/8	35% Zn Cu alloy: Sheet & bar; extra spring hard **UTS: 660**
ASTM B43	15% Zn Cu alloy: Seamless pipe; annealed
ASTM B105	10% Zn 1.5% Cd (max) 5% Sn 3.5% Al 3% Si Cu: Wire; mechanical and electrical properties depend on size etc
ASTM B111/230	15% Zn Cu alloy: For condenser tubes
ASTM B111/442	1% Sn 28% Zn Cu alloy: Seamless tube
ASTM B111/443	1% Sn 28% Zn 0.05% As Cu alloy: Seamless tube
ASTM B111/444	1% Sn 28% Zn 0.05% Sb alloy: Seamless tube
ASTM B111/445	1% Sn 28% Zn 0.05% P alloy: Seamless tube
ASTM B111/687	2.1% Al 0.05% As 20% Zn Cu alloy: For condenser tubes
ASTM B111 A	30% Zn 1% Sn Cu alloy: Tube
ASTM B111 B	30% Zn 1% Sn 0.1% As Cu alloy: Tube
ASTM B111 C	30% Zn 1% Sn 0.1% Sb Cu alloy: Tube
ASTM B111 D	30% Zn 1% Sn 0.1% P Cu alloy: Tube
ASTM B111 A1 BRASS	22% Zn 2% Al Cu alloy: Tube
ASTM B111 MUNTZ	38% Zn 0.3% Pb Cu alloy: Tube
ASTM B111 RED BRASS	15% Zn Cu alloy: Tube
ASTM B119	Cast Cu based alloys: Replaced by individual specifications
ASTM B121	35% Zn 2.5% Pb Cu alloy: Sheet & bar; extra spring hard; available in all degrees of hardness **UTS: 620**
ASTM B121/1	10% Zn 0.5% Pb Cu alloy: Sheet & bar; extra hard; available in all degrees of hardness **UTS: 490**
ASTM B121/2	35% Zn 0.5% Pb Cu alloy: Sheet & bar; extra spring hard; available in all degrees of hardness **UTS: 620**
ASTM B121/3	35% Zn 1.1% Pb Cu alloy: Sheet & bar; extra spring hard; available in all degrees of hardness **UTS: 620**
ASTM B121/4	37% Zn 1.8% Pb Cu alloy: Sheet & bar; extra spring hard; available in all degrees of hardness **UTS: 620**
ASTM B121/5	35% Zn 2% Pb Cu alloy: Sheet & bar; extra spring hard; available in all degrees of hardness **UTS: 620**
ASTM B124/13	37% Zn 1.5% Pb 0.7% Sn Cu alloy: Forging & bar; mechanical properties not quoted
ASTM B129	30% Zn Cu alloy: For cartridge cups; mechanical properties not quoted
ASTM B130	10% Zn Cu alloy: Strip; ¼-hard **UTS: 320**
ASTM B130	10% Zn Cu alloy: Strip; extra spring hard **UTS: 530**
ASTM B131	10% Zn Cu alloy: For pressings; for bullet jacket cups
ASTM B134/1	5% Zn Cu alloy: Wire; ¼-hard **UTS: 300**
ASTM B134/1	5% Zn Cu alloy: Wire; spring hard **UTS: 510**
ASTM B134/2	10% Zn Cu alloy: Wire; ¼-hard **UTS: 370**
ASTM B134/2	10% Zn Cu alloy: Wire; spring hard **UTS: 600**
ASTM B134/3	15% Zn Cu alloy: Wire; ¼-hard **UTS: 420**
ASTM B134/3	15% Zn Cu alloy: Wire; spring hard **UTS: 730**
ASTM B134/4	20% Zn Cu alloy: Wire; ¼-hard **UTS: 500**
ASTM B134/4	20% Zn Cu alloy: Wire; spring hard **UTS: 870**
ASTM B134/6	30% Zn Cu alloy: Wire; ¼-hard **UTS: 530**
ASTM B134/6	30% Zn Cu alloy: Wire; spring hard **UTS: 890**
ASTM B134/7	35% Zn Cu alloy: Wire; ¼-hard **UTS: 530**
ASTM B134/7	35% Zn Cu alloy: Wire; spring hard **UTS: 890**
ASTM B134/8	38% Zn Cu alloy: Wire; ¼-hard **UTS: 530**
ASTM B134/8	38% Zn Cu alloy: Wire; spring hard **UTS: 890**

Symbol	Nominal analysis, supplier, condition and remarks.
ASTM B135/1	15% Zn Cu alloy: Tube; hard drawn
	UTS: 400
ASTM B135/2	30% Zn Cu alloy: Tube; hard drawn
	UTS: 460
ASTM B135/3	33% Zn 0.5% Pb Cu alloy: Tube; hard drawn
	UTS: 460
ASTM B135/4	32% Zn 1.5% Pb Cu alloy: Tube; hard drawn
	UTS: 460
ASTM B135/7	10% Zn Cu alloy: Tube; hard drawn
	UTS: 360
ASTM B138 A	39% Zn 1% Sn 1.5% Fe 0.1% Mn Cu alloy: Bar; annealed
	UTS: 390 Elon: 20% Proof: 140
ASTM B138 a	39% Zn 1% Sn 1.5% Fe 0.1% Mn Cu alloy: Bar; hard
	UTS: 530 Elon: 10% Proof: 360
ASTM B138 B	23% Zn 4.5% Al 3.5% Mn 3% Fe Cu alloy: Bar; annealed; manganese bronze
	UTS: 620 Elon: 10% Proof: 330
ASTM B138 B	23% Zn 4.5% Al 3.5% Mn 3% Fe Cu alloy: Bar; hard; manganese bronze
	UTS: 820 Elon: 5% Proof: 480
ASTM B140 A	15% Zn 2% Pb Cu alloy: Bar; annealed
	UTS: 220 Elon: 15% Proof: 450
ASTM B140 B	10% Zn 2% Pb 1% Ni Cu alloy: Bar; annealed
	UTS: 220 Elon: 15% Proof: 450
ASTM B171	Condenser tube plates in copper alloys; covers various types see also Section 14G
ASTM B176 ZS331A	35% Zn 1% Si Cu alloy: Casting; die cast
	UTS: 400 Elon: 15% Proof: 130
ASTM B283	Cu alloy: Die forgings; various types; see also Section 14G
ASTM B 291	29% Zn 1% Mn Cu alloy: Sheet & strip; ½-hard
	UTS: 420
ASTM B 291	29% Zn 1% Mn Cu alloy: Sheet & strip; extra spring hard
	UTS: 750
ASTM B 359/230	15% Zn Cu alloy: Fenned tube
ASTM B 359/442	1.0% Sn 30% Zn Cu alloy: Fenned tube
ASTM B 359/443	1.0% Sn 30% Zn Cu alloy: Fenned tube
ASTM B 359/444	1.0% Sn 30% Zn Cu alloy: Fenned tube
ASTM B 359/445	1.0% Sn 30% Zn Cu alloy: Fenned tube
ASTM B453	Zn Pb Cu alloy: Rods; graded by CDAA system
ASTM B505	Cu alloys: For continuous casting; graded by CDAA alloys; see also Section 14G and 14K
ASTM B508/411	9.5% Zn 0.5% Sn Cu alloy: Strip for flexible piping
B 42 HC	35% Zn 3% Pb Cu alloy: Bar & forging; Manganese Bronze Ltd for BS alloy CZ124
B 76	35% Zn 1.5% Pb Cu alloy: Extrusion; Delta Metal Ltd
BARRONIA	12.5% Zn 0.5% Pb 4.0% Sn Cu alloy: Common name
BASIS BRASS	27% Zn Cu alloy: Sheet; origin unknown
BASIS QUALITY BRASS	23.5% Zn Cu alloy: Origin unknown
BATALBRA	22% Zn 2% Al 0.03% As Cu alloy: Tube; Birmingham Battery for BS alloy CZ110

Symbol	Nominal analysis, supplier, condition and remarks.
BATNAVAL	38% Zn 1.2% Sn Cu alloy: Sheet & plate; Birmingham Battery for BS alloy CZ112
BATURNAL	35% Zn 2% Pb Cu alloy: Tube; Birmingham Battery for BS alloy CZ119
BINDING BRASS	35% Zn 1.5% Pb Cu alloy: Origin unknown
BOBIERE'S METAL	37% Zn Cu alloy: Origin unknown
BOLTOMET 506	3% Zn Cu alloy: T. Bolton; cap copper
BOLTOMET 510	10% Zn Cu alloy: T. Bolton for BS 713
BOLTOMET 514	15% Zn Cu alloy: T. Bolton for BS 712
BOLTOMET 516	20% Zn Cu alloy: T. Bolton for BS 711
BOLTOMET 518	30% Zn Cu alloy: T. Bolton for BS 267
BOLTOMET 520	35% Zn Cu alloy: T. Bolton for BS 266 & BS 2786
BOLTOMET 522	37% Zn Cu alloy: T. Bolton for BS 265
BOLTOMET 570	30% Zn Cu alloy: Wire; T. Bolton
BOLTOMET 607	27% Zn 1.25% Pb Cu alloy: Rod; T. Bolton
BOLTOMET 710	35% Zn 1.25% Sn Cu alloy: Naval brass; T. Bolton for BS 251, BS 409, BS 1541
BOLTOMET 731	10% Zn Cu alloy: T. Bolton; commercial bronze
BROWN METAL	15% Zn Cu alloy: Common name
BS 251 CZ112	35% Zn 1.25% Sn Cu alloy: Naval brass & forgings; as rolled
	UTS: 360 Elon: 20%
BS 265 CZ108	37% Zn Cu alloy: Strip; cold rolled
	UTS: 360 Elon: 15%
BS 266 CZ107	35% Zn Cu alloy: Sheet & strip; annealed
	DPN: 80 UTS: 270 Elon: 45%
BS 266 CZ107	35% Zn Cu alloy: Sheet & strip; ¼-hard
	DPN: 75 (min) UTS: 330 Elon: 35%
BS 266 CZ107	35% Zn Cu alloy: Sheet & strip; ½-hard
	DPN: 110 UTS: 360 Elon: 20%
BS 266 CZ107	35% Zn Cu alloy: Sheet & strip; hard
	DPN: 135 UTS: 450 Elon: 3%
BS 267 CZ106	30% Zn Cu alloy: Sheet & strip; cold rolled
	DPN: 75 UTS: 270 Elon: 50%
BS 378 CZ105	30% Zn 0.04% As Cu alloy: Arsenical brass tube; as drawn; for condenser tubes, etc.
	DPN: 150
BS 378 CZ110	20% Zn 2% Al 0.04% As Cu alloy: Aluminium brass; as drawn; for condenser tubes, etc.
	DPN: 150
BS 378 CZ111	28% Zn 1.2% Sn 0.04% As Cu alloy: Tube; as drawn; Admiralty Brass for condenser tubes, etc.
	DPN: 150
BS 711 CZ103	20% Zn Cu alloy: Sheet & strip; annealed 80/20 brass
	DPN: 80 UTS: 250 Elon: 40%
BS 711 CZ103	20% Zn Cu alloy: Sheet & strip; ½-hard 80/20 brass
	DPN: 95 UTS: 330 Elon: 10%
BS 711 CZ103	20% Zn Cu alloy: Sheet & strip; hard 80/20 brass
	DPN: 120 UTS: 390 Elon: 5%
BS 712 CZ102	15% Zn Cu alloy: Sheet & strip; annealed 85/15 brass
	DPN: 80 UTS: 240 Elon: 35%
BS 712 CZ102	15% Zn Cu alloy: Sheet & strip; ½-hard 85/15 brass
	DPN: 95 UTS: 330 Elon: 7%
BS 712 CZ102	15% Zn Cu alloy: Sheet & strip; hard 85/15 brass
	DPN: 110 UTS: 360 Elon: 3%
BS 713 CZ101	10% Zn Cu alloy: Sheet & strip; annealed 90/10 brass
	DPN: 80 UTS: 240 Elon: 35%
BS 713 CZ101	10% Zn Cu alloy: Sheet & strip; ½-hard 90/10 brass
	DPN: 95 UTS: 300 Elon: 7%
BS 713 CZ101	10% Zn Cu alloy: Sheet & strip; hard 90/10 brass
	DPN: 110 UTS: 350 Elon: 3%

Note. The following abbreviations and units are used in the tables:

DPN	Hardness, diamond pyramid number
UTS	Ultimate tensile strength, N/mm^2
Elon	Elongation, %
Proof	0.1 % proof strength, N/mm^2

1 N/mm^2=0.1 hbar=0.102 kgf/mm^2=0.06475 tonf/in^2=145.04 lbf/in^2

Symbol	Nominal analysis, supplier, condition and remarks.
BS 885 CZ105	30% Zn 0.04% As Cu alloy: Tube; annealed 70/30 arsenical brass **UTS: 270**
BS 885 CZ110	20% Zn 2% Al 0.04% As Cu alloy: Tube; annealed aluminium brass **UTS: 330**
BS 920	Naval brass die casting; replaced by BS 1400
BS 1400 SCB1-C	21% Zn 2% Sn 3% Pb Cu alloy: Casting; sand cast **DPN: 50 UTS: 180 Elon: 30% Proof: 60**
BS 1400 SCB2-1	30% Zn 3.5% Pb Cu alloy: Brass ingots; sand cast **UTS: 170 Elon: 12%**
BS 1400 SCB2-C	30% Zn 3.5% Pb Cu alloy: Brass castings; sand cast **UTS: 170 Elon: 12%**
BS 1400 SCB3-1	30% Zn 2.5% Pb Cu alloy: Brass ingots; sand cast **UTS: 180 Elon: 12%**
BS 1400 SCB3-C	30% Zn 2.5% Pb Cu alloy: Brass castings; sand cast **UTS: 180 Elon: 12%**
BS 1400 SCB5-C	12% Zn 1% Sn Cu alloy: Casting; brazeable sand cast **DPN: 50 UTS: 200 Elon: 30% Proof: 60**
BS 1400 SCB6-1	15% Zn Cu alloy: Brass ingots; sand cast; brazeable **UTS: 170 Elon: 20%**
BS 1400 SCB6-C	15% Zn Cu alloy: Brass castings; sand cast; brazeable **UTS: 170 Elon: 20%**
BS 1402	30% Zn Cu alloy: Tube; solid drawn & annealed; for gas installation work
BS 1403	30% Zn Cu alloy: Tube; annealed & cold drawn; for gas lighting
BS 1464 CZ105	30% Zn 0.04% As Cu alloy: Tube; arsenical brass **DPN: 95**
BS 1464 CZ110	20% Zn 2% Al Cu alloy: Tube; aluminium brass **DPN: 95**
BS 1464 CZ111	28% Zn 1.2% Sn Cu alloy: Tube; Admiralty brass **DPN: 95**
BS 1541 CZ110	20% Zn 1.8% Al 0.04% As Cu alloy: Plate; aluminium brass; as rolled **UTS: 270 Elon: 40%**
BS 1541 CZ112	35% Zn 1% Sn Cu alloy: Plate; Naval brass; as rolled **UTS: 340 Elon: 20%**
BS 1845 CZ1	49% Zn Cu alloy: For brazing; melting range 860–870 °C
BS 1845 CZ2	46% Zn Cu alloy: For brazing; melting range 870–880 °C
BS 1845 CZ3	40% Zn Cu alloy: For brazing; melting range 885–890 °C
BS 1845 CZ4	1.0% Sn 46% Zn Cu alloy: For brazing; melting range 860–870 °C
BS 1845 CZ5	1.0% Sn 40% Zn Cu alloy: For brazing; melting range 880–890 °C
BS 1845 CZ6	0.3% Si 40% Zn Cu alloy: For brazing; melting range 875–895 °C
BS 1845 CZ7	0.2% Si 0.1% Mn 40% Zn Cu alloy: For brazing; melting range 870–900 °C
BS 2785 CZ118	35% Zn 1% Pb Cu alloy: Strip; ½-hard; for clocks and watches **DPN: 130 UTS: 370 Elon: 10%**
BS 2785 CZ118	35% Zn 1% Pb Cu alloy: Strip; extra hard; for clocks and watches **DPN: 180 UTS: 510 Elon: 3%**
BS 2785 CZ119	35% Zn 2% Pb Cu alloy: Strip; ½-hard; for clocks and watches **DPN: 130 UTS: 390 Elon: 10%**
BS 2785 CZ119	35% Zn 2% Pb Cu alloy: Strip; extra hard; for clocks and watches **DPN: 180 UTS: 510 Elon: 3%**
BS 2785 CZ120	38% Zn 2% Pb Cu alloy: Strip; ½-hard; for clocks and watches **DPN: 130 UTS: 510 Elon: 10%**
BS 2785 CZ120	38% Zn 2% Pb Cu alloy: Strip; extra hard; for clocks and watches **DPN: 180 UTS: 560 Elon: 5%**
BS 2786 CZ107	35% Zn Cu alloy: Wire; hard drawn **UTS: 750**
BS 2870 CZ101	10% Zn Cu alloy: Sheet; annealed **DPN: 80 UTS: 240 Elon: 35%**
BS 2870 CZ101	10% Zn Cu alloy: Sheet; ½-hard **UTS: 900 Elon: 7%**
BS 2870 CZ101	10% Zn Cu alloy: Sheet; hard **DPN: 105 UTS: 330 Elon: 3%**
BS 2870 CZ102	15% Zn Cu alloy: Sheet; annealed **DPN: 80 UTS: 240 Elon: 35%**
BS 2870 CZ102	15% Zn Cu alloy: Sheet; ½-hard **DPN: 90 UTS: 300 Elon: 7%**
BS 2870 CZ102	15% Zn Cu alloy: Sheet; hard **DPN: 105 UTS: 340 Elon: 3%**
BS 2870 CZ103	20% Zn Cu alloy: Sheet; annealed **DPN: 80 UTS: 250 Elon: 40%**
BS 2870 CZ103	20% Zn Cu alloy: Sheet; ½-hard **DPN: 90 UTS: 330 Elon: 10%**
BS 2870 CZ103	20% Zn Cu alloy: Sheet; hard **DPN: 115 UTS: 370 Elon: 5%**
BS 2870 CZ104	20% Zn 1% Pb Cu alloy: Sheet; specification not issued
BS 2870 CZ105	30% Zn 0.05% As Cu alloy: Sheet; specification not issued
BS 2870 CZ106	30% Zn Cu alloy: Sheet; annealed **DPN: 80 UTS: 270 Elon: 50%**
BS 2870 CZ106	30% Zn Cu alloy: Sheet; ¼-hard **DPN: 75 UTS: 330 Elon: 35%**
BS 2870 CZ106	30% Zn Cu alloy: Sheet; ½-hard **DPN: 100 UTS: 340 Elon: 20%**
BS 2870 CZ106	30% Zn Cu alloy: Sheet; hard **DPN: 120 UTS: 390 Elon: 3%**
BS 2870 CZ107	35% Zn Cu alloy: Sheet; annealed **DPN: 80 UTS: 270 Elon: 45%**
BS 2870 CZ107	35% Zn Cu alloy: Sheet; ¼-hard **DPN: 75 UTS: 320 Elon: 35%**
BS 2870 CZ107	35% Zn Cu alloy: Sheet; ½-hard **DPN: 100 UTS: 360 Elon: 20%**
BS 2870 CZ107	35% Zn Cu alloy: Sheet; hard **DPN: 130 UTS: 420 Elon: 3%**
BS 2870 CZ107	35% Zn Cu alloy: Sheet; extra hard **DPN: 165 UTS: 510**
BS 2870 CZ108	38% Zn Cu alloy: Sheet; common brass; annealed **DPN: 80 UTS: 270 Elon: 40%**
BS 2870 CZ108	38% Zn Cu alloy: Sheet; ¼-hard **DPN: 75 (min) UTS: 330 Elon: 30%**
BS 2870 CZ108	38% Zn Cu alloy: Sheet; ½-hard **DPN: 100 UTS: 360 Elon: 15%**
BS 2870 CZ108	38% Zn Cu alloy: Sheet; hard **DPN: 130 UTS: 420 Elon: 3%**
BS 2870 CZ108	38% Zn Cu alloy: Sheet; common brass; extra hard **DPN: 165 UTS: 510**
BS 2870 CZ110	20% Zn 2% Al 0.05% As Cu alloy: Sheet; aluminium brass
BS 2870 CZ111	28% Zn 1.2% Sn 0.05% As Cu alloy: Sheet; Admiralty brass
BS 2870 CZ112	35% Zn 1.2% Sn Cu alloy: Sheet; annealed Naval brass **UTS: 330 Elon: 25%**

Symbol	Nominal analysis, supplier, condition and remarks.
BS 2870 CZ112	35% Zn 1.2% Sn Cu alloy: Sheet; hard; Naval brass
	UTS: 390 Elon: 20%
BS 2870 CZ116	30% Zn 1% Fe 4.5% Al 1.5% Mn Cu alloy: Sheet; high tensile brass; specification not issued
BS 2870 CZ118	35% Zn 1% Pb Cu alloy: Sheet; ½-hard
	DPN: 120 UTS: 360 Elon: 10%
BS 2870 CZ118	35% Zn 1% Pb Cu alloy: Sheet; hard
	DPN: 150 UTS: 420 Elon: 5%
BS 2870 CZ118	35% Zn 1% Pb Cu alloy: Sheet; extra hard
	DPN: 180 UTS: 510 Elon: 3%
BS 2870 CZ119	35% Zn 2% Pb Cu alloy: Sheet; ½-hard
	DPN: 120 UTS: 360 Elon: 10%
BS 2870 CZ119	35% Zn 2% Pb Cu alloy: Sheet; hard
	DPN: 150 UTS: 420 Elon: 5%
BS 2870 CZ119	35% Zn 2% Pb Cu alloy: Sheet; extra hard
	DPN: 180 UTS: 510 Elon: 3%
BS 2870 CZ124	35% Zn 3% Pb Cu alloy: Sheet; specification not issued
BS 2870 CZ125	4.0% Zn Cu alloy: Sheet, etc.; cap copper
	DPN: 75 UTS: 100
BS 2871 CZ101	10% Zn Cu alloy: Tube; specification not issued
BS 2871 CZ102	15% Zn Cu alloy: Tube; specification not issued
BS 2871 CZ103	20% Zn Cu alloy: Tube; specification not issued
BS 2871 CZ104	20% Zn Cu alloy: Tube; specification not issued
BS 2871 CZ105	30% Zn 0.04% As Cu alloy: Tube; arsenical brass; annealed
	UTS: 300
BS 2871 CZ106	30% Zn Cu alloy: Tube; specification not issued
BS 2871 CZ107	35% Zn Cu alloy: Tube; specification not issued
BS 2871 CZ108	38% Zn Cu alloy: Tube; common brass; specification not issued
BS 2871 CZ110	20% Zn 2% Al 0.03% As Cu alloy: Tube; aluminium brass; annealed
	UTS: 370
BS 2871 CZ111	28% Zn 1% Sn 0.03% As Cu alloy: Tube; Admiralty brass
BS 2871 CZ116	35% Zn 4.5% Al 1% Mn Cu alloy: Tube; specification not issued
BS 2871 CZ118	35% Zn 1% Pb Cu alloy: Tube; specification not issued
BS 2871 CZ119	35% Zn 2% Pb Cu alloy: Tube; annealed
	UTS: 330
BS 2871 CZ124	35% Zn 3% Pb Cu alloy: Tube; specification not issued
BS 2872 CZ101	10% Zn Cu alloy: Forging; specification not issued
BS 2872 CZ102	15% Zn Cu alloy: Forging; specification not issued
BS 2872 CZ103	20% Zn Cu alloy: Forging; specification not issued
BS 2872 CZ104	20% Zn 1% Pb Cu alloy: Forging; specification not issued
BS 2872 CZ105	30% Zn 0.04% As Cu alloy: Forging; specification not issued
BS 2872 CZ106	30% Zn Cu alloy: Forging; specification not issued
BS 2872 CZ107	35% Zn Cu alloy: Forging; specification not issued
BS 2872 CZ108	37% Zn Cu alloy: Forging; specification not issued
BS 2872 CZ110	20% Zn 2% Al 0.04% As Cu alloy: Forging; specification not issued
BS 2872 CZ111	28% Zn 1.2% Sn 0.04% As Cu alloy: Forging; specification not issued

Symbol	Nominal analysis, supplier, condition and remarks.
BS 2872 CZ112	35% Zn 1.2% Sn Cu alloy: Forging; as forged
	UTS: 330 Elon: 20%
BS 2872 CZ116	30% Zn 1% Pb 4.5% Al 1% Mn Cu alloy: Forging; as forged
	UTS: 450 Elon: 15% Proof: 270
BS 2872 CZ118	35% Zn 1% Pb Cu alloy: Forging; specification not issued
BS 2872 CZ119	35% Zn 2% Pb Cu alloy: Forging; specification not issued
BS 2872 CZ124	35% Zn 3% Pb Cu alloy: Forging; specification not issued
BS 2873 CZ101	10% Zn Cu alloy: Wire; specification not issued
BS 2873 CZ102	15% Zn Cu alloy: Wire; annealed
	UTS: 280 Elon: 30%
BS 2873 CZ102	15% Zn Cu alloy: Wire; hard
	UTS: 580
BS 2873 CZ103	20% Zn Cu alloy: Wire; annealed
	UTS: 300 Elon: 35%
BS 2873 CZ103	20% Zn Cu alloy: Wire; hard
	UTS: 600
BS 2873 CZ104	20% Zn 0.8% Pb Cu alloy: Wire; specification not issued
BS 2873 CZ105	30% Zn 0.04% As Cu alloy: Wire; specification not issued
BS 2873 CZ106	30% Zn Cu alloy: Wire; annealed
	UTS: 300 Elon: 45%
BS 2873 CZ106	30% Zn Cu alloy: Wire; hard
	UTS: 600
BS 2873 CZ107	35% Zn Cu alloy: Wire; annealed
	UTS: 320 Elon: 40%
BS 2873 CZ107	35% Zn Cu alloy: Wire; extra hard
	UTS: 700
BS 2873 CZ108	38% Zn Cu alloy: Wire; annealed
	UTS: 320 Elon: 40%
BS 2873 CZ108	38% Zn Cu alloy: Wire; extra hard
	UTS: 680
BS 2873 CZ110	20% Zn 2% Al 0.04% As Cu alloy: Wire; specification not issued
BS 2873 CZ111	28% Zn 1.2% Sn 0.04% As Cu alloy: Wire; specification not issued
BS 2873 CZ112	35% Zn 1.2% Sn Cu alloy: Wire; specification not issued
BS 2873 CZ116	30% Zn 4.5% Al 1% Fe 1% Mn Cu alloy: Wire; specification not issued
BS 2873 CZ118	35% Zn 1% Pb Cu alloy: Wire; specification not issued
BS 2873 CZ119	35% Zn 2% Pb Cu alloy: Wire; specification not issued
BS 2873 CZ124	35% Zn 3% Pb Cu alloy: Wire; specification not issued
BS 2874 CZ101	10% Zn Cu alloy: Bar; specification not issued
BS 2874 CZ102	15% Zn Cu alloy: Bar; specification not issued
BS 2874 CZ103	20% Zn Cu alloy: Bar; as rolled
	UTS: 300 Elon: 25%
BS 2874 CZ104	20% Zn 0.7% Pb Cu alloy: Bar; as rolled
	UTS: 300 Elon: 25%
BS 2874 CZ105	30% Zn 0.04% As Cu alloy: Bar; specification not issued
BS 2874 CZ106	30% Zn Cu alloy: Bar; annealed
	UTS: 270 Elon: 50%
BS 2874 CZ107	35% Zn Cu alloy: Bar; specification not issued
BS 2874 CZ108	38% Zn 0.3% Pb Cu alloy: Bar; specification not issued
BS 2874 CZ110	20% Zn 2% Al 0.04% As Cu alloy: Bar; specification not issued
BS 2874 CZ111	28% Zn 1.2% Sn 0.04% As Cu alloy: Bar
BS 2874 CZ116	30% Zn 1% Fe 4.5% Al 1% Mn Cu alloy: Bar; as rolled
	UTS: 530 Elon: 15% Proof: 270

Note. The following abbreviations and units are used in the tables:

DPN	Hardness, diamond pyramid number
UTS	Ultimate tensile strength, N/mm^2
Elon	Elongation, %
Proof	0.1 % proof strength, N/mm^2

1 N/mm^2=0.1 hbar=0.102 kgf/mm^2=0.06475 tonf/in^2=145.04 lbf/in^2

Symbol	Nominal analysis, supplier, condition and remarks.
BS 2874 CZ118	35% Zn 1% Pb Cu alloy: Bar; specification not issued
BS 2874 CZ119	35% Zn 2% Pb Cu alloy: Bar; as rolled **UTS: 330 Elon: 25%**
BS 2874 CZ124	35% Zn 3% Pb Cu alloy: Bar; annealed **UTS: 320 Elon: 15% Proof: 120**
BS 2874 CZ124	35% Zn 3% Pb Cu alloy: Bar; hard drawn **UTS: 530 Elon: 4% Proof: 220**
BS 2875 CZ101	10% Zn Cu alloy: Plate; specification not issued
BS 2875 CZ102	15% Zn Cu alloy: Plate; specification not issued
BS 2875 CZ103	20% Zn Cu alloy: Plate; specification not issued
BS 2875 CZ104	20% Zn 1% Pb Cu alloy: Plate; specification not issued
BS 2875 CZ105	30% Zn 0.04% As Cu alloy: Plate; annealed **UTS: 270 Elon: 45%**
BS 2875 CZ105	30% Zn 0.04% As Cu alloy: Plate; hard **UTS: 340 Elon: 22%**
BS 2875 CZ106	30% Zn Cu alloy: Plate; annealed **UTS: 270 Elon: 45%**
BS 2875 CZ106	30% Zn Cu alloy: Plate; hard **UTS: 340 Elon: 22%**
BS 2875 CZ107	35% Zn Cu alloy: Plate; specification not issued
BS 2875 CZ108	38% Zn Cu alloy: Plate; hard **UTS: 370 Elon: 20%**
BS 2875 CZ110	20% Zn 2% Al 0.04% As Cu alloy: Plate; annealed **UTS: 340 Elon: 20%**
BS 2875 CZ110	20% Zn 2% Al 0.04% As Cu alloy: Plate; hard **UTS: 390 Elon: 20%**
BS 2875 CZ111	28% Zn 1.2% Sn 0.04% As Cu alloy: Plate; specification not issued
BS 2875 CZ112	35% Zn 1.2% Sn Cu alloy; Plate
BS 2875 CZ116	30% Zn 4.5% Al 1% Fe 1% Mn Cu alloy: Plate; specification not issued
BS 2875 CZ124	35% Zn 3% Pb Cu alloy: Plate; specification not issued
BS 2901 C14	30% Zn 1.2% Sn 0.04% As Cu alloy: Rod for inert gas shielded arc welding
BS 2901 C15	30% Zn 2% Al 0.04% As Cu alloy: Rod for inert gas shielded arc welding
BS 3127/3	35% Zn Cu alloy: Tubing; ½-hard drawn for bourdon tubes **DPN: 120**
BS B11	20% Zn Cu alloy: Bar; used as brazing filler wire **UTS: 300 Elon: 25% Proof: 60**
BS T8	Seamless brass tubes; annealed
BS T18	Brass tube; hard drawn, seamless; obsolete see BS 885
BS T47	30% Zn Cu alloy: Tube for radiators; as drawn **UTS: 290**
BS T48	Brass tube for radiators; obsolete
C 14	30% Zn 1.2% Sn 0.04% As Cu alloy: Welding rod; designation used by BS
C 15	30% Zn 2% Al 0.04% As Cu alloy: Welding rod; designation used by BS
CAP COPPER	3.5% Zn Cu alloy: Strip; common name; further information from Copper Development Association
CARO - BUHLER	Zn Pb P Cu alloy: For bar & tube; Carobronze; free machining
CARTRIDGE BRASS	30% Zn Cu alloy: Sheet; common name
CHINESE ART METAL	10% Zn 1% Sn 17.5% Pb Cu alloy: Origin unknown
CLOCK BRASS	37% Zn 2% Pb Cu alloy: Bar & sheet; American trade name
COMMERCIAL BRONZE	10% Zn Cu alloy: Common American name
CON	34% Zn 0.5% Sn 0.5% Pb 1% Fe 1.5% Al 2.75% Mn 1.75% Ni Cu alloy: Extrusions; McKechnie Bros; as drawn **DPN: 150 UTS: 550 Elon: 20% Proof: 270**

Symbol	Nominal analysis, supplier, condition and remarks.
CORRONIUM	15% Zn 5% Sn Cu alloy: Origin unknown
CSC STAR	37.5% Zn Cu alloy: Extrusion; McKechnie Bros; annealed **DPN: 90 UTS: 320 Elon: 45% Proof: 100**
DIN 1709 G Ms 65	35% Zn 2% Pb Cu alloy **DPN: 60 UTS: 200 Elon: 20% Proof: 75**
DIN 1785 K Ms 63	35% Zn Cu alloy
DIN 1785 K Ms 72	30% Zn Cu alloy: Wrought
DIN 1785 So Ms 71	1% Sn 30% Zn Cu alloy: Wrought
DIN 1785 So Ms 76	20% Zn 2% Al Cu alloy: Wrought
DIN 17656	35% Zn 2% Pb Cu alloy: Casting
GBMs 65 A	
DIN 17660	Specification covering a series of copper zinc alloys; designations prefixed Ms are included
DIN 17661	Copper zinc alloys; with additions; designations with prefix So Ms
DTD 253 A	10% Zn 1% Al 1% Ni 1% Si Cu alloy: Tube **UTS: 450**
DTD 283 A	15% Zn 1% Al 1% Ni 1% Si Cu alloy: Sheet; annealed **UTS: 370**
DTD 307	10% Zn 2% Si 1.5% Mn Cu alloy: Annealed; solid drawn tube **UTS: 370**
DTD 318 A	15% Zn 4% Sn 1% Fe Cu alloy: Tube; annealed; solid drawn **UTS: 450**
DTD 319	12% Zn 1% Al 1% Ni 1% Si Cu alloy: Bars; as rolled **UTS: 450 Elon: 20% Proof: 250**
DTD 323 D	12% Zn 1% Al 1% Ni 1% Si Cu alloy: Tube; annealed **UTS: 400**
DTD 337	4.0% Zn Cu alloy: Tube for radiators
DTD 387	4.0% Zn Cu alloy: Tube for radiators
DTD 604	30% Zn 1% Sn Cu alloy: Tubes (Sn optional); annealed; solid drawn **UTS: 300 Proof: 60**
DTD 627	35% Zn 1.8% Pb Cu alloy: Bar; not suitable for rivets
DTD 5009	15% Zn 1% Al 1% Ni 1% Si Cu alloy: Bar & wire; cold drawn for springs
DTD 5019	15% Zn 1.0% Al 1.2% Ni 1.0% Si Cu alloy: Tube; annealed **UTS: 450 Elon: 40%**
E 48	30% Zn Pb Cu alloy: Extrusions; McKechnie Bros; leaded variety of 311
E 56	30% Zn As (trace) Cu alloy: Extrusions; McKechnie Bros; resists de-zincification
E 78	10% Zn Cu alloy: Extrusions; McKechnie Bros; annealed **DPN: 90 UTS: 250 Elon: 50% Proof: 90**
F 7S	37% Zn 1.2% Sn 0.5% Pb Cu alloy: Extrusions; McKechnie Bros; as drawn **DPN: 115 UTS: 39 Elon: 30% Proof: 14**
F 7S STAR	37% Zn 1.2% Sn Cu alloy: Extrusions; McKechnie Bros; lead free variety of F 7S
G Cu 65 Zn	35% Zn 2% Pb Cu alloy: Casting; designation used by German Standard
G Ms 65	35% Zn 2% Pb Cu alloy: Casting; designation used by German Standard
GB Cu 65 Zn	35% Zn 2% Pb Cu alloy: Casting; designation used by German Standard
GB Ms 65 A	35% Zn 2% Pb Cu alloy: Casting; designation used by German Standard
GB Ms 65 B	35% Zn 2% Pb Cu alloy: Casting; designation used by German Standard
GB Ms 65 C	35% Zn 2% Pb Cu alloy: Casting; designation used by German Standard

Symbol	Nominal analysis, supplier, condition and remarks.
GILDING METAL	10% Zn Cu alloy: Common name; further information from Copper Development Association
HARDWARE BRONZE	13% Zn 2% Pb Cu alloy: Rod; American trade name
IMI 210	10% Zn Cu alloy: Gilding metal; IMI for BS alloy CZ101
IMI 212	12% Zn Cu alloy: Gilding metal; IMI
IMI 215	15% Zn Cu alloy: Gilding metal; IMI for BS alloy CZ102
IMI 220	20% Zn Cu alloy: Gilding metal; IMI for BS alloy CZ103
IMI 230	30% Zn Cu alloy: Bar; IMI for BS alloy CZ106
IMI 237	37% Zn Cu alloy: Bar & wire; IMI for BS alloy CZ108
IMI 239	37% Zn Pb Cu alloy: Wire; IMI
IMI 276	30% Zn As Cu alloy: Bar
IMI 303	22% Zn 2% Al Cu alloy: Welding wire; IMI 'Alumbro'; withdrawn
IMI 345	11% Zn 1% Sn Cu alloy: Wire; IMI
IMI 432	35% Zn 2% Pb Cu alloy: Bar; IMI for BS alloy CZ119; nipple wire
IMI 433	35% Zn 2% Pb Cu alloy: Wire; IMI for BS alloy CZ119
IMI 443	37% Zn 0.25% Pb Cu alloy: Free machining; IMI
IMMADIUM II	30% Zn Fe Mn Al Cu alloy: Manganese Bronze Ltd
IMMADIUM V	30% Zn 1% Pb 4.5% Al 1% Mn Cu alloy: Bar & forging; Manganese Bronze Ltd for BS alloy CZ116
IMMADIUM VI	28% Zn Fe Mn Al Cu alloy: Manganese Bronze Ltd
JEWELRY BRONZE	12.5% Zn Cu alloy: Wrought; used for Au plated parts; traditional name
K Ms 63	35% Zn Cu alloy: Wrought; designation used by German Standard
K Ms 72	30% Zn Cu alloy: Wrought; designation used by German Standard
KUNIFORM	Zn / Cu strip; fine grained; IMI; covers 60/40 to 90/10 Cu/Zn alloys
LBZ	35% Zn 1% Sn 2.4% Pb 0.5% Fe 0.4% Al 1.75% Mn Cu alloy: Extrusions; McKechnie Bros; as drawn **DPN: 130 UTS: 480 Elon: 20% Proof: 210**
LED-O-LOY	9% Zn 2% Sn 2% Pb 0.05% P Cu alloy: Strip; American trade name
LOW BRASS	20% Zn Cu alloy: Common American name
MANGANESE BRONZE	30% Zn 4% Al 3% Mn 3% Fe Cu alloy: Common name; analysis varies considerably
METER BRONZE	7% Zn 0.6% Sn Cu alloy: American trade name
Ms 56	40% Zn 2% Pb Cu alloy: Wrought; designation used by German Standard
Ms 58	40% Zn 2% Pb Cu alloy: Wrought; designation used by German Standard
Ms 60	40% Zn Cu alloy: Wrought; designation used by German Standard
Ms 60 Pb	38% Zn 2.0% Pb Cu alloy: Wrought; designation used by German Standard
Ms 63	37% Zn Cu alloy: Wrought; designation used by German Standard
Ms 63 Pb	3.6% Zn 2.0% Pb Cu alloy: Wrought; designation used by German Standard
Ms 67	33% Zn Cu alloy: Wrought; designation used by German Standard
Ms 72	30% Zn Cu alloy: Wrought; designation used by German Standard
Ms 80	20% Zn Cu alloy: Wrought; designation used by German Standard
Ms 85	15% Zn Cu alloy: Wrought; designation used by German Standard
Ms 90	10% Zn Cu alloy: Wrought; designation used by German Standard
MUNTZ	38% Zn 0.3% Pb Cu alloy: Common name; lead not always present; further information from the Copper Development Association
NEBALOY	37% Zn Cu alloy: Strip; American trade name
NIPPLE	35% Zn 2% Pb Cu alloy: Wire; common name
QQ B 601/A	15% Zn 0.3% Pb Cu alloy: Casting; US Federal
QQ B 601/B	15% Zn 0.5% Pb Cu alloy: Casting; US Federal
QQ B 611a/E	30% Zn Cu alloy: Plate or sheet; US Federal
QQ B 621/B	30% Zn 0.5% Sn 2.0% Pb Cu alloy: Casting; US Federal **UTS: 200 Elon: 20%**
QQ B 621/C	20% Zn 1.5% Sn 3.0% Pb Cu alloy: Casting; US Federal **UTS: 220 Elon: 25%**
QQ B 721a/B	30% Zn 4.0% Al 4% Mn 3.0% Fe Cu alloy: Bar & forging; US Federal **UTS: 600 Elon: 20% Proof: 270**
QQ B 726/B	30% Zn 0.2% Pb 3% Fe 5% Al 3% Mn Cu alloy: Casting; US Federal 'manganese bronze' **UTS: 620 Elon: 20% Proof: 300**
QQ B 726/C	30% Zn 0.2% Pb 3% Fe 5% Al 3% Mn Cu alloy: Casting; US Federal **UTS: 750 Elon: 12% Proof: 420**
QQ C 591/A	4.2% Zn 1.5% Sn 3.0% Si 1.5% Mn Cu alloy: Sheet; US Federal **UTS: 370 Elon: 50% Proof: 120**
QQ C 591a/D	4.0% Zn 2.2% Sn 2.0% Si Cu alloy: Bar; US Federal **UTS: 600 Elon: 20% Proof: 300**
QQ S 551/C	30% Zn 0.3% Pb Cu alloy: Braze filler; US Federal; melting range 899–960 °C
QQ S 551/D	20% Zn 0.2% Pb Cu alloy: Braze filler; US Federal; melting range 968–996 °C
QQ W 321/A	35% Zn 0.1% Pb Cu alloy: Wire; US Federal; annealed **UTS: 300**
QQ W 321/C	28% Zn Cu alloy: Wire; US Federal; hard drawn **UTS: 820**
RED BRASS	15% Zn Cu alloy: Common name
S1B	35% Zn 1.75% Al 3% Mn 1% Si Cu alloy: Extrusions; McKechnie Bros; as extruded **DPN: 130 UTS: 580 Elon: 15% Proof: 270**
SAE 70A	30% Zn Cu alloy: Sheet & strip; ½-hard **UTS: 370**
SAE 70A	30% Zn Cu alloy: Sheet & strip; extra hard **UTS: 620**
SAE 70B	30% Zn Cu alloy: Sheet & strip; ½-hard **UTS: 370**
SAE 70B	30% Zn Cu alloy: Sheet & strip; extra hard **UTS: 620**
SAE 70C	35% Zn Cu alloy: Sheet & strip; ½-hard **UTS: 370**
SAE 70C	35% Zn Cu alloy: Sheet & strip; extra hard **UTS: 600**
SAE 72	35% Zn 3% Pb Cu alloy: Bar; annealed **UTS: 300 Elon: 25% Proof: 100**
SAE 72	35% Zn 3% Pb Cu alloy: Bar; ½-hard **UTS: 360 Elon: 20% Proof: 270**

Note. The following abbreviations and units are used in the tables:

DPN	Hardness, diamond pyramid number
UTS	Ultimate tensile strength, N/mm^2
Elon	Elongation, %
Proof	0.1 % proof strength, N/mm^2

1 N/mm^2=0.1 hbar=0.102 kgf/mm^2=0.06475 tonf/in^2=145.04 lbf/in^2

Symbol	Nominal analysis, supplier, condition and remarks.
SAE 74B	35% Zn 0.8% Pb Cu alloy: Tube; as drawn **UTS: 370**
SAE 74B	35% Zn 0.8% Pb Cu alloy: Tube; hard drawn **UTS: 460**
SAE 74C	30% Zn Cu alloy: Tube; as drawn **UTS: 370**
SAE 74C	30% Zn Cu alloy: Tube; hard drawn **UTS: 460**
SAE 74D	15% Zn Cu alloy: Tube; light drawn **UTS: 370**
SAE 74D	15% Zn Cu alloy: Tube; hard drawn **UTS: 400**
SAE 79A	15% Zn Cu alloy: Sheet & strip; ½-hard **UTS: 340**
SAE 79A	15% Zn Cu alloy: Sheet & strip; extra hard **UTS: 530**
SAE 79B	20% Zn Cu alloy: Sheet & strip; ½-hard **UTS: 370**
SAE 79B	20% Zn Cu alloy: Sheet & strip; extra hard **UTS: 600**
SAE 80A	30% Zn Cu alloy: Wire; ¼-hard **UTS: 460**
SAE 80A	30% Zn Cu alloy: Wire; extra hard **UTS: 800**
SAE 80B	35% Zn Cu alloy: Wire; ½-hard **UTS: 460**
SAE 80B	35% Zn Cu alloy: Wire; extra hard **UTS: 800**
SAE 795	10% Zn 0.5% Sn Cu alloy: For bearings; wrought
SAE 890	20% Zn 1.5% Pb Cu alloy: Sintering; density 750 kg/m³
SAE 891	20% Zn 1.5% Pb Cu alloy: Sintering; density 770 kg/m³
SAE CA 210	5% Zn Cu alloy: Gilding metal
SAE CA 220	10% Zn Cu alloy: Commercial bronze
SAE CA 230	15% Zn Cu alloy: Red brass; formerly SAE 74D and SAE 79A
SAE CA 240	20% Zn Cu alloy: Low brass; formerly SAE 79B
SAE CA 260	30% Zn Cu alloy: Cartridge brass; formerly SAE 70A, SAE 74C, SAE 80A
SAE CA 268	33.7% Zn Cu alloy: Yellow brass sheet; formerly SAE 70C
SAE CA 270	35% Zn Cu alloy: Rod & wire; formerly SAE 80B
SAE CA 330	33.5% Zn 0.5% Pb Cu alloy: Tube; formerly SAE 74B
SAE CA 331	32.5% Zn 1% Pb Cu alloy: Wrought
SAE CA 342	33.5% Zn 2% Pb Cu alloy: Wrought
SAE CA 345	35% Zn 2% Pb Cu alloy: Wrought
SAE CA 350	36.5% Zn 1% Pb Cu alloy: Wrought
SAE CA 360	35.5% Zn 3% Pb Cu alloy: Formerly SAE 72
SAE CA 670	24.2% Zn 4.5% Al 3% Fe 3.7% Mn Cu alloy: Wrought
SAE CA 673	33.7% Zn 1% Si 2% Pb 2.7% Mn Cu alloy: Wrought
SEVA	0.04% As 30% Zn Cu alloy: For tube; Yorkshire Imp Metals
SIS 5112	15% Zn Cu alloy: Plate & bar; Swedish Standard; annealed **DPN: 70 UTS: 300 Elon: 50% Proof: 75**
SIS 5112	15% Zn Cu alloy: Plate & bar; Swedish Standard; cold rolled **DPN: 130 UTS: 400 Elon: 12% Proof: 300**
SIS 5114	20% Zn Cu alloy: Plate; Swedish Standard; annealed **DPN: 60 UTS: 300 Elon: 50% Proof: 75**
SIS 5114	20% Zn Cu alloy: Plate; Swedish Standard; cold rolled **DPN: 110 UTS: 420 Elon: 20% Proof: 340**

Symbol	Nominal analysis, supplier, condition and remarks.
SIS 5122	30% Zn Cu alloy: Plate; Swedish Standard; annealed **DPN: 70 UTS: 340 Elon: 55% Proof: 100**
SIS 5122	30% Zn Cu alloy: Plate; Swedish Standard; cold rolled **DPN: 140 UTS: 450 Elon: 15% Proof: 360**
SIS 5124	35% Zn Cu alloy: Bar; Swedish Standard; annealed **DPN: 75 UTS: 300 Elon: 45% Proof: 90**
SIS 5124	35% Zn Cu alloy: Bar; Swedish Standard; cold rolled **DPN: 140 UTS: 420 Elon: 10% Proof: 340**
SIS 5140	35% Zn 1% Pb Cu alloy: Bar; Swedish Standard; annealed **DPN: 75 UTS: 320 Elon: 45% Proof: 90**
SIS 5140	35% Zn 1% Pb Cu alloy: Bar; Swedish Standard; cold rolled **DPN: 150 UTS: 440 Elon: 10% Proof: 360**
SIS 5144	30% Zn 1% Sn 2% Pb Cu alloy: Casting; Swedish Standard; as cast **DPN: 50 UTS: 200 Elon: 20% Proof: 90**
SIS 5150	36% Zn Cu alloy: Plate & bar; Swedish Standard; annealed **DPN: 75 UTS: 360 Elon: 45% Proof: 100**
SIS 5150	36% Zn Cu alloy: Plate & bar; Swedish Standard; cold rolled **DPN: 160 UTS: 600 Elon: 10% Proof: 580**
SIS 5202	9% Zn 3.5% Sn 3.5% Pb Cu alloy: Casting; Swedish Standard; as cast **DPN: 70 UTS: 180 Elon: 15% Proof: 100**
SIS 5217	31% Zn 2% Al 0.04% As Cu alloy: Bar; Swedish Standard; annealed **DPN: 90 UTS: 390 Elon: 45% Proof: 170**
SIS 5220	28% Zn 1.1% Sn 0.04% As Cu alloy: Bar; Swedish Standard; annealed **DPN: 100 UTS: 360 Elon: 45% Proof: 170**
SIS 5234	25% Zn 4.5% Al 3.5% Mn 2% Fe Cu alloy: Bar & forging; Swedish Standard; hot rolled **DPN: 200 UTS: 650 Elon: 10% Proof: 330**
SIS 5236	30% Zn 2% Pb 2% Al 0.8% Mn Cu alloy: Bar; Swedish Standard; cold rolled **DPN: 140 UTS: 480 Elon: 20% Proof: 330**
SIS 5238	34% Zn 1% Sn 0.8% Pb 0.8% Al 2% Mn 1% Fe Cu alloy: Bar & forging; Swedish Standard; cold rolled **DPN: 160 UTS: 510 Elon: 10% Proof: 340**
So Ms 57 Al 2	37% Zn 2.0% Al 2.0% Mn 1.0% Ni Cu alloy: Wrought; designation used by German Standard
So Ms 58	37% Zn 1.5% Ni 0.2% Mn 1% Fe 0.2% Sn Cu alloy: Wrought; designation used by German Standard
So Ms 58 Al 1	38% Zn 1.0% Al 1.0% Ni 2.0% Mn Cu alloy: Wrought; designation used by German Standard
So Ms 58 Pb	37% Zn 1.5% Pb 0.2% Ni 1.0% Mn 0.4% Fe 0.5% Al Cu alloy: Wrought; designation used by German Standard
So Ms 59	37% Zn 2.5% Ni 2.0% Mn 1.0% Al Cu alloy: Wrought; designation used by German Standard
So Ms 60	39% Zn 0.8% Sn Cu alloy: Wrought
So Ms 71	1% Sn 30% Zn Cu alloy: Wrought; designation used by German Standard
So Ms 76	20% Zn 2% Al Cu alloy: Wrought; designation used by German Standard
SPINNING BRASS	37% Zn 1% Pb Cu alloy: Bar & sheet; American trade name
SPRABRASSY	34% Zn Cu alloy: Wire for spraying; Metco Ltd
SPRABRONZE C	10% Zn Cu alloy: Wire for spraying; Metco Ltd
STA 7 CM4C	35% Zn 1% Sn 0.2% Pb 1% Fe 4% Al Cu alloy: Bar & forging **UTS: 530 Elon: 15% Proof: 270**

Symbol	Nominal analysis, supplier, condition and remarks.
STA 7 CM4B	35% Zn 0.7% Fe 4.5% Al Cu alloy: Bar & forging
	UTS: 530 Elon: 15% Proof: 270
STA 7 CS1A	5% Zn 2% Fe 1% Mn 4% Si Cu alloy: Bar; annealed
	UTS: 340 Elon: 45%
STA 7 CS1A	5% Zn 2% Fe 1% Mn 4% Si Cu alloy: Bar; cold worked
	UTS: 530 Elon: 13%
STA 7 CS1B	5% Zn 2% Fe 1% Mn 4% Si Cu alloy: Sheet; cold worked
	UTS: 530 Elon: 20%
STA 7 CS1D	5% Zn 2% Fe 1% Mn 4% Si Cu alloy: Sheet; annealed
	UTS: 340 Elon: 45%
STA 7 CS2	5% Zn 1% Mn 2.5% Si Cu alloy: Tube; hard drawn
	UTS: 530
STA 7 CX1	15% Zn Cu alloy: Casting; gilding metal
STA 7 CX2	20% Zn Cu alloy: Casting
STA 7 CZ1	3.5% Zn Cu alloy: Strip; cap copper
	DPN: 70
STA 7 CZ2	10% Zn Cu alloy: Sheet; gilding metal
	DPN: 70
STA 7 CZ3	22% Zn 2% Al Cu alloy: Tube; for condenser tubes
STA 7 CZ4A	30% Zn Cu alloy: Tube
STA 7 CZ4B	29% Zn 1.2% Sn Cu alloy: Tube; annealed
	UTS: 340
STA 7 CZ4B	29% Zn 1.2% Sn Cu alloy: Tube; hard drawn
	UTS: 450
STA 7 CZ5	30% Zn Cu alloy: Sheet; annealed; cartridge brass
	DPN: 70
STA 7 CZ5	30% Zn Cu alloy: Sheet; cold rolled; cartridge brass
	DPN: 130
STA 7 CZ6	35% Zn Cu alloy: Sheet; annealed
	DPN: 75 UTS: 270 Elon: 45%
STA 7 CZ6	35% Zn Cu alloy: Sheet; cold rolled
	DPN: 110 UTS: 420 Elon: 17%
STA 7 CZ7	37% Zn Cu alloy: Sheet; annealed
	DPN: 75 UTS: 270 Elon: 40%
STA 7 CZ7	37% Zn Cu alloy: Sheet; cold rolled
	DPN: 110 UTS: 420 Elon: 17%
STA 7 CZ8A	38% Zn 1.2% Sn Cu alloy: Bar
	UTS: 360 Elon: 20%
STA 7 CZ8B	38% Zn 1.2% Sn Cu alloy: Plate; annealed
	UTS: 340 Elon: 25%
STA 7 CZ8B	38% Zn 1.2% Sn Cu alloy: Plate; cold rolled
	UTS: 390 Elon: 20%
STA 7 CZ11A	35% Zn 2.5% Pb Cu alloy: Bar; annealed
	UTS: 300 Elon: 25% Proof: 120
STA 17	3.5% Zn Cu alloy: Strip for shells & caps; cap copper alloy
	DPN: 70

Symbol	Nominal analysis, supplier, condition and remarks.
STA 19	10% Zn Cu alloy: Strip for bullet envelopes; gilding metal
	DPN: 70
STA 51	30% Zn Cu alloy: Bar for vent tubes
TAURUS XXII	28% Zn 1% Mn 1.5% Fe 5% Al Cu alloy: Casting; D Brown; sand cast
	DPN: 150 UTS: 600 Elon: 17% Proof: 300
TAURUS XXIII	30% Zn 1.5% Mn 1.5% Fe 5.5% Al Cu alloy: Casting; D Brown; sand cast
	DPN: 190 UTS: 760 Elon: 13% Proof: 420
TAURUS XXIV	5% Sn 17% Zn 5% Pb Cu alloy: Casting; D Brown; centri-cast
	DPN: 120 UTS: 340 Elon: 2% Proof: 270
TAURUS XXX	10% Zn 2% Sn Cu alloy: Casting; D Brown; sand cast
	DPN: 65 UTS: 200 Elon: 32%
TUNGUM	10% Zn 1% Al 1.2% Ni 1% Si Cu alloy: Casting; Tungum Co
TUNGUM	10% Zn 1% Al 1.2% Ni 1% Si Cu alloy: Sheet, bar & tube; Tungum Co; as rolled
	UTS: 600 Elon: 20% Proof: 330
UZ 29E1	1.0% Sn 28.0% Zn Cu alloy: Origin unknown
VIALBRA	22% Zn 2% Al Cu alloy: Origin unknown
WAS	Cu alloy: Welding rod; Delta metal for BS 1724A
WHEEL BRASS	30% Zn 2% Pb Cu alloy: Origin unknown
WW T 791/1	15% Zn Cu alloy: Seamless tube; US Federal
WW T 791/2	35% Zn 0.8% Pb Cu alloy: Seamless tube; US Federal
YALE BRONZE	7.75% Zn 1% Sn 1% Pb Cu bronze; Used for nuts and bolts; origin unknown
YELLOW BRASS	40% Zn Cu alloy: Common name; further information from Copper Development Association
YELLOW INGOT METAL	1% Sn 2% Pb 32% Zn Cu alloy: Used for plumber's fitting; common name
YORCALBRO	22% Zn 2% Al 0.04% As Cu alloy: Tube; Yorkshire Imperial Metals Ltd
YR	37% Zn 1.5% Pb Cu alloy: Extrusions; McKechnie Bros; as drawn
	DPN: 125 UTS: 390 Elon: 27% Proof: 170
YS	36% Zn 2.75% Pb Cu alloy: Extrusions; McKechnie Bros; as BS alloy CZ 124

Note. The following abbreviations and units are used in the tables:

DPN	Hardness, diamond pyramid number
UTS	Ultimate tensile strength, N/mm^2
Elon	Elongation, %
Proof	0.1 % proof strength, N/mm^2

1 N/mm^2=0.1 hbar=0.102 kgf/mm^2=0.06475 tonf/in^2=145.04 lbf/in^2

14C Copper–zinc wrought and cast alloys
37–50 per cent zinc – alpha + beta and beta alloys

Specific gravity	8.4	
Density	8400 kg/m³	(0.303 lb/in.³)
Solidus/liquidus	890–920 °C	
Thermal conductivity	109 W/m °C	(26 cal/m s °C)
Coefficient of linear expansion	21 × 10⁻⁶/ °C	
Electrical conductivity	25–28% IACS (copper 100%)	
Specific resistance	62–68 microhm mm	
Young's modulus of elasticity	103×10^9 N/m²	$(15 \times 10^6$ lbf/in.²)
Impact	– J	(– ft/lb)
Fatigue strength (10^7 cycles)	± 150 N/mm²	(± 10 tonf/in.²)
Hot strength	–	

The above properties are typical of the following group, and may not apply exactly to any one specification. It is possible that with certain specifications some of the values may not be applicable.

General metallurgical characteristics

The alloys listed all have sufficient zinc – above 37% to ensure the presence of some beta phase in the alpha matrix. This causes the alloys to be stronger than the plain alpha brass (up to 38% zinc) but means that ductility is reduced to such an extent that little or no cold forming or drawing can be carried out. The strength and lack of ductility increases proportionally up to about 46% zinc, at which beta only is present and then a third phase – gamma – makes its appearance.

There will inevitably be some overlap between what are classed as true alpha brasses and the alpha–beta group. The alpha brasses are listed in section 14B.

With copper at 55–60% and alloying elements in addition to zinc, there is often sufficient overlap in the individual analysis for the one specification to be alpha or alpha–beta depending on the actual analysis.

At 50% only gamma is present and the alloys are so brittle as to be virtually useless. There is also a considerable drop in the corrosion resistance.

By far the largest and most important group is based on the 60% copper, 40% zinc analysis – the 60/40 brasses. This is a true alpha–beta brass, in that it has poor cold working properties but excellent hot working characteristics. Thus good quality forgings, stampings, plates, bars and hot extrusions can be produced from this relatively inexpensive material which has good corrosion resistance with acceptable strength, ductility, thermal and electrical properties.

Considerable modifications can be made to the basic alloy by additions of lead, tin, aluminium, silicon, manganese and nickel. In small quantities, generally replacing zinc, these metals do not affect the general metallurgy of the alloys but have the following effects:

Lead. Up to 3.5% lead additions are made to improve machinability. Bars with this amount of lead find considerable use in automachining when the lead acts as a lubricant and also produces small chips which do not clog the tool. There is a reduction in the impact value and corrosion resistance proportional to the lead content.

Tin. Up to 1% is added to improve the corrosion resistance. This gives a slight increase in tensile strength with no reduction in ductility. The alloy is termed Naval brass and is distinct from Admiralty, which is an alpha brass.

Iron. This strengthens the alpha phase, giving a higher tensile strength with little reduction in ductility. It is generally added with aluminium and manganese.

Aluminium. 1% aluminium has the same effect as about 6% zinc on the structure of copper–zinc alloys. In addition it considerably enhances the corrosion resistance, particularly to sea water, and high temperature oxidation.

Silicon. Small amounts of silicon are sometimes added to 60/40 brass to improve the initial castability prior to forging. This also enhances the corrosion resistance. Other alloying elements are always present.

These alloys must not be confused with the silicon bronze series which have different and distinct properties.

Manganese. The term 'manganese bronze' was for many years applied to 60/40 brasses with any other alloying elements, if manganese was present. This is a misnomer as they are not true bronzes and manganese probably has the least effect of any of the alloying elements. It does give some strength to the alpha matrix but its main contribution is to form an oxide during forging which in addition to enhancing the corrosion resistance has a pleasing brown colour.

Nickel. This strengthens the alpha phase and considerably improves the corrosion resistance.

High tensile brasses. These are essentially 60/40 brasses with up to 2% of aluminium, tin and nickel, and about 1% of iron and manganese. Some have small amounts of lead added to improve machinability. Apart from the increase in tensile strength and corrosion resistance these materials have all the characteristics of alpha beta brasses and are often incorrectly termed 'manganese bronze'.

High zinc alloys. These are beta alloys and are generally too brittle for most practical purposes. They can be readily hot worked but are prone to de-zincification and corrosion. Some use is made of them as brazing alloys owing to their relatively low melting range.

Thermal treatment

None of these materials can be hardened by any form of heat treatment. As they cannot accept any appreciable amount of cold work without becoming brittle there is no way of increasing the strength to any great extent. They are generally used in the annealed condition, the annealing temperature being 450–500° C. There may be a tendency for grain growth to take place if too high a temperature is used. Loss of zinc from the grain boundaries (dezincification) can also take place, resulting in a poor appearance and reduction in mechanical properties, in particular fatigue and impact strength.

Welding and brazing

Pre-treatment. No thermal treatment is required. In many cases it is however, advisable to pre-heat and weld or braze while hot. The parts must be thoroughly cleaned before heating.

Welding and brazing. Oxy-acetylene, electric arc and inert gas shielded are methods possible. Because of the high heat conductivity it is advisable to employ a second operator with a gas torch to pre-heat the weld area. Fluxes are required with gas or electric welding and fumes of zinc oxide will always be formed. These are poisonous and fume extraction equipment is recommended. The alloys are all readily brazed, but care must be taken when soldering as there is a tendency for the solder to cause severe embrittlement if the alloys are under stress when soldered. There is no danger when the material is in the annealed condition.

Post-treatment. None required, except a stress relieve if the assembled structure is intricate and liable to distortion from cooling stresses.

Corrosion protection

Temporary. None required.

Permanent. None normally required. These alloys will discolour under atmospheric conditions. The Copper Development Association has information on a lacquer which prevents this discoloration. For marine conditions the materials should always contain tin or aluminium, when no further protection is generally necessary.

Plating. These alloys are all readily electroplated using normal techniques.

Painting. Air drying or stoving paints can be applied. An etch primer is recommended.

Machinability

These alloys exhibit good machinability and with the leaded alloys this becomes excellent. The lead acts as a built-in lubricant and chip breaker. The 3.5% lead 60/40 brass is first choice for automachining of non-ferrous materials.

Uses

The alpha plus beta alloys. General purpose hot stampings, forgings and extrusions for all engineering uses. Nuts, bolts, pipe fittings machined from bar.

Beta alloys. Brazing filler rods.

Symbol	Nominal analysis, supplier, condition and remarks.
2.0340	40% Zn 0.8% Al Cu alloy: Casting; German Standard
2.0341	40% Zn 0.3% Al Cu alloy: Casting; German Standard
2.0343	40% Zn 0.3% Al Cu alloy: Casting; German Standard
2.0366	40% Zn 0.2% Sn 0.3% Si Cu alloy: Wrought; German Standard
2.0531	40% Zn Cu alloy: Wrought; German Standard
2.0590	40% Zn 1% Fe 2% Mn 1% Sn Cu alloy: Casting; German Standard
2.0592	40% Zn 2% Al 1.5% Fe 2.5% Mn 1.5% Ni Cu alloy: Casting; German Standard

Symbol	Nominal analysis, supplier, condition and remarks.
2.0596	40% Zn 4% Al 2% Fe 3% Mn 1.5% Ni Cu alloy: Casting; German Standard
2.0598	40% Zn 7% Al 3% Fe 4% Mn 1.5% Ni Cu alloy: Casting; designation used by German Standard
4	39% Zn 3% Pb Cu alloy: Extrusions; McKechnie Bros; as drawn. DPN: 140 UTS: 400 Elon: 20% Proof: 200
4M	40% Zn 5% Al 1% Fe 2% Ni 3% Mn Cu alloy: Casting; Langley Alloys
4VH	39% Zn 3% Pb Cu alloy: Extrusion code for Hidurit 11; McKechnie Bros; variant of 4
5	38% Zn 1% Pb Cu alloy: Extrusion; Delta Metal Ltd
5V	38% Zn 1% Pb Cu alloy: Extrusion; Delta Metal Ltd
7M	40% Zn 0.7% Sn 1% Fe 1.5% Mn 1% Pb 1% Al Cu alloy: Wrought; Langley Alloys; as Hidurit 15
9M	40% Zn 2.5% Al 1% Ni 5% Sn 3% Mn Cu alloy: Casting; Langley Alloys for Hidurit 12
10	37% Zn 1% Sn 1% Pb Cu alloy: Extrusion; Delta Metal Ltd
11M	40% Zn 5% Al 4% Mn 1% Ni Cu alloy: Casting; Langley Alloys; Hidurit 10

Note. The following abbreviations and units are used in the tables:

DPN	Hardness, diamond pyramid number
UTS	Ultimate tensile strength, N/mm^2
Elon	Elongation, %
Proof	0.1 % proof strength, N/mm^2

1 N/mm^2=0.1 hbar=0.102 kgf/mm^2=0.06475 tonf/in^2=145.04 lbf/in^2

Symbol	Nominal analysis, supplier, condition and remarks.
31	40% Zn Sn Pb Fe Al Ni Cu alloy: Extrusion; Delta Metal Ltd; alloying elements varied to suit requirements
37	40% Zn Sn Pb Fe Al Mn Ni Cu alloy: Extrusion; Delta Metal Ltd; alloying elements varied to suit requirements
51A	50% Zn Cu alloy: Bronze filler; melting range 860–870 °C; Langley Alloys
51A	50% Zn Cu alloy: Brazing spelter; Delta Metal for BS 1845C
60L	38% Zn 1% Pb Cu alloy: Extrusion; Delta Metal Ltd
115	41% Zn 3% Pb Cu alloy: Extrusion; McKechnie Bros; as drawn **DPN: 100 UTS: 360 Elon: 22% Proof: 170**
210	40% Zn 2.5% Pb Cu alloy: Extrusion; McKechnie Bros; as drawn **DPN: 130 UTS: 400 Elon: 22% Proof: 200**
218	40% Zn Sn Pb Fe Al Mn Ni Cu alloy: Extrusion; Delta Metal Ltd; alloying elements varied to suit requirement
280	40% Zn Cu alloy: Wrought; designation used by CDAA
298	50% Zn Cu alloy: For brazing; designation agreed by CDAA; not used
301	40% Zn 1% Pb Cu alloy: Extrusion; Delta Metal
312	40% Zn Cu alloy: Extrusion; McKechnie Bros; annealed **DPN: 100 UTS: 330 Elon: 40% Proof: 140**
380	40% Zn 2.0% Pb Cu alloy: Wrought; designation agreed by CDAA; not used
385	40% Zn 3% Pb Cu alloy: Wrought; designation used by CDAA
465	40% Zn 0.7% Sn 0.07% As Cu alloy: Wrought; designation used by CDAA
466	40% Zn 0.7% Sn 0.07% Sb Cu alloy: Wrought; designation used by CDAA
467	40% Zn 0.7% Sn 0.06% P Cu alloy: Wrought; designation used by CDAA
470	40% Zn 0.8% Sn 0.01% Al Cu alloy: Brazing rod; designation agreed by CDAA; not used
472	50% Zn 3.5% Sn Cu brazing alloy: Designation agreed by CDAA; not used
540	38.5% Zn 0.75% Sn 1.25% Pb 0.75% Fe 1.25% Mn Cu alloy: Extrusions; McKechnie Bros; as drawn **DPN: 130 UTS: 480 Elon: 20% Proof: 210**
671	40% Zn 0.2% Pb 0.7% Sn 0.2% Mn Cu alloy: Wrought; designation agreed by CDAA; not used
677	40% Zn 0.8% Pb 1.0% Fe 2.0% Ni 0.4% As 0.2% Mn Cu alloy: Wrought; designation agreed by CDAA; not used
680	40% Zn 0.9% Sn 0.6% Ni 0.4% Mn Cu alloy: Wrought; designation agreed by CDAA; not used
681	40% Zn 0.9% Sn 0.4% Mn Cu alloy: Wrought; designation agreed by CDAA; not used
796	40% Zn 1.0% Pb 10% Ni 2.0% Mn Cu alloy: Wrought; designation agreed by CDAA; not used
798	40% Zn 2.0% Pb 10% Ni 2.0% Mn Cu alloy: Wrought; designation agreed by CDAA; not used
850	41% Zn 0.2% Si Cu alloy: Braze filler rod; McKechnie Bros; melting range 890–900 °C
912	2.5% Mn 1.5% Al 0.7% Si 0.5% Pb 40% Zn Cu alloy: Bar & tube; Manganese Bronze Co
A 0	40% Zn Cu alloy: Bar & forging; Manganese Bronze Co; for BS alloy CZ109
A 71 NB	40% Zn 1.2% Sn Cu alloy: Bar & forging; Manganese Bronze Co for BS alloy CZ112

Symbol	Nominal analysis, supplier, condition and remarks.
ABS TYPE 3	43% Zn 1.0% Sn (max) 1.2% Fe 2.0% (max) Al 3.2% Ni Cu alloy: Casting; American Bureau of Shipping **UTS: 530 Elon: 18% Proof: 225**
AITCH METAL	38% Zn 1.5% Fe Cu alloy: Obsolete; Blackwell
AM	40% Zn Sn Pb Fe Al Mn Ni Cu alloy: Extrusion; Delta Metal Ltd; alloying elements varied to suit requirement
AMPCOLOY 62	40% Zn 1% Fe 1% Al 0.2% Mn Cu alloy: Casting; Ampco
AMPCOLOY 64	30% Zn 0.2% Pb 3% Fe 5% Al 4% Mn Cu alloy: Casting; Ampco
AMPCOLOY 66	30% Zn 0.2% Pb 3% Fe 5% Al 4% Mn Cu alloy: Casting; Ampco
AMPCOLOY 666	Zn Mn Cu alloy: Manganese bronze; Ampco
AMS 4619 B	40% Zn 1% Al 0.2% Mn 1% Fe Cu alloy: Forging
AMS 4860 A	39% Zn 1% Al 1% Fe 0.3% Mn Cu alloy: Casting
AN QQ B 646	40% Zn 1.0% Sn Cu alloy: Bar; US Service **UTS: 420 Elon: 30% Proof: 200**
ARCHITECTURAL BRONZE C 94	44% Zn Fe Mn Cu alloy: Manganese Bronze Co
ASTM B21 A	40% Zn 0.7% Sn Cu alloy: Bar; Naval brass; annealed **UTS: 370 Elon: 30% Proof: 120**
ASTM B21 A	40% Zn 0.7% Sn Cu alloy: Bar; Naval brass; hard drawn **UTS: 460 Elon: 13% Proof: 330**
ASTM B21 B	40% Zn 0.7% Sn 0.7% Pb Cu alloy: Bar; Naval brass; annealed **UTS: 370 Elon: 25% Proof: 120**
ASTM B21 B	40% Zn 0.7% Sn 0.7% Pb Cu alloy: Bar; Naval brass; hard drawn **UTS: 460 Elon: 11% Proof: 330**
ASTM B21 C	40% Zn 0.7% Sn 1.8% Pb Cu alloy: Bar; Naval brass; annealed **UTS: 370 Elon: 20% Proof: 120**
ASTM B21 C	40% Zn 0.7% Sn 1.8% Pb Cu alloy: Bar; Naval brass; hard drawn **UTS: 460 Elon: 10% Proof: 330**
ASTM B30/6 C	1% Sn 1% Pb 36.5% Zn 0.3% Al Cu alloy: Ingot; as cast **UTS: 270 Elon: 15%**
ASTM B30/7 A	0.7% Sn 0.7% Pb 37% Zn 1.2% Fe 0.5% Mn 0.7% Al Cu alloy: Ingot; as cast **UTS: 420 Elon: 15%**
ASTM B30/8 A	38% Zn 1.25% Fe 1.25% Al 1% Mn Cu alloy: Ingot; as cast **UTS: 460 Elon: 20%**
ASTM B30/8 B	25% Zn 3% Fe 5% Al 4% Mn Cu alloy: Ingot; as cast **UTS: 450 Elon: 18%**
ASTM B30/8 C	25% Zn 3% Fe 5% Al 4% Mn Cu alloy: Ingot; as cast **UTS: 800 Elon: 12%**
ASTM B111/280	39% Zn Cu alloy: For condenser tubes
ASTM B124/2	39% Zn 2% Pb Cu alloy: Forging & bar; mechanical properties not quoted
ASTM B 124/3	39% Zn 0.8% Sn Cu alloy: Forging & bar; mechanical properties not quoted
ASTM B 124/4	39% Zn 1% Sn 0.1% Mn Cu alloy: Forging & bar; mechanical properties not quoted
ASTM B132 A	40% Zn 1.5% Al 1.5% Sn 1% Mn Cu alloy: Casting; as cast **UTS: 420 Elon: 15% Proof: 120**
ASTM B132 B	38% Zn 3% Al 3% Fe 3% Mn 1.5% Sn Cu alloy: Casting; as cast **UTS: 580 Elon: 15% Proof: 210**
ASTM B135/5	40% Zn Cu alloy: Tube; as drawn **UTS: 370**

Symbol	Nominal analysis, supplier, condition and remarks.
ASTM B135/6	39% Zn 1% Pb Cu alloy: Tube; as drawn **UTS: 370**
ASTM B147/8 A	40% Zn 1% Fe 1% Al 0.2% Mn Cu alloy: Casting **UTS: 460 Elon: 20% Proof: 160**
ASTM B147/8 B	30% Zn 0.2% Pb 3% Fe 5% Al 4% Mn Cu alloy: Casting **UTS: 660 Elon: 18% Proof: 320**
ASTM B147/8 C	30% Zn 0.2% Pb 3% Fe 5% Al 4% Mn Cu alloy: Casting **UTS: 810 Elon: 12% Proof: 420**
ASTM B259 RB Cu Zn A	40% Zn 0.8% Sn Cu alloy: Welding rod **UTS: 340**
ASTM B259 RB Cu Zn B	40% Zn 1% Sn 0.5% Ni Cu alloy: Welding rod **UTS: 400**
ASTM B259 RB Cu Zn C	40% Zn 1% Sn Cu alloy: Welding rod **UTS: 400**
ASTM B259 RB Cu Zn D	40% Zn 10% Ni Cu alloy: Welding rod
ASTM B260 RB Cu Zn A	40% Zn 0.8% Sn Cu alloy: Braze filler; melting range 887–898 °C
ASTM B455/380	40.5% Zn 2.0% Pb Cu alloy: Extrusion
ASTM B455/385	39.6% Zn 2.9% Pb Cu alloy: Extrusion
B 31	40% Zn Sn Pb Fe Al Mn Ni Cu alloy: Extrusion; Delta Metal Ltd; alloying elements varied to suit requirement
B 41	38% Zn 1% Sn Cu alloy: Casting; Anti-Attrition Ltd for BS alloy SCB4; Naval brass
B 42F	40% Zn 2% Pb Cu alloy: Bar; Manganese Bronze Co; hard drawn for BS alloy CZ122
B 42HS	40% Zn 3% Pb Cu alloy: Bar; Manganese Bronze Co for BS alloy CZ121
B 50	40% Zn 0.5% Pb Cu alloy: Bar & forging; Manganese Bronze Co for BS alloy CZ123
BATH METAL	45% Zn Cu alloy: Origin unknown
BETA BRASS	Zn Al Fr Ni Sn Mn Cu alloy: Casting; common name for BS alloy HTB
BOLTOMET 524	39% Zn Cu alloy: T Bolton for BS 1949/A; withdrawn
BOLTOMET 611	39% Zn 1.75% Pb Cu alloy: T Bolton for BS 218
BOLTOMET 613	40% Zn 2.25% Pb Cu alloy: T Bolton for BS 249
BOLTOMET 619	40% Zn 0.4% Pb Cu alloy: T Bolton for BS 1949/B
BOLTOMET 715	35% Zn 0.7% Sn 1% Pb 1% Fe 1% Al 1% Mn Cu alloy: T. Bolton for BS 250 CZ 114; high tensile brass
BOLTOMET 721	40% Zn 1% Sn 1% Fe 1% Pb Cu alloy: T Bolton for BS 1001; high tensile brass
BS 31	Copper alloy bars for automatic machinery; obsolete
BS 207	Brass ingot & casting; replaced by BS 1400
BS 208	Brass castings; replaced by BS 1400
BS 218	40% Zn 2% Pb Cu alloy: Bars and forgings **UTS: 300 Elon: 25%**
BS 249 CZ121	40% Zn 2.5% Pb Cu alloy: Rods & sections; free machining **UTS: 370 Elon: 15%**
BS 250 CZ114	40% Zn 0.7% Sn 1% Fe 1.5% Mn 1% Pb 1% Al Cu alloy: Bar; as drawn **UTS: 450 Elon: 20% Proof: 220**

Symbol	Nominal analysis, supplier, condition and remarks.
BS 250 CZ122	40% Zn 2% Pb Cu alloy: Bar; hard drawn **UTS: 450 Elon: 20% Proof: 220**
BS 252 CZ113	40% Zn 1% Sn Cu alloy: Naval brass bars; as rolled **UTS: 400 Elon: 20%**
BS 264	Hot rolled yellow metal plate & sheet; obsolete
BS 409	40% Zn 1% Sn Cu alloy: Plate, sheet & strip; annealed; Naval brass **UTS: 330 Elon: 25%**
BS 409	40% Zn 1% Sn Cu alloy: Plate, sheet & strip; cold rolled; Naval brass **UTS: 400 Elon: 20%**
BS 932	Brass gravity die casting; replaced by BS 1400
BS 944/1	39% Zn 2% Pb Cu alloy: Bars & forgings **UTS: 300 Elon: 25%**
BS 1001 CZ115	40% Zn 1% Sn 1% Fe 1% Pb 1.5% Mn Cu alloy: Forgings; soldering quality; high tensile brass; as forged **UTS: 450 Elon: 20% Proof: 180**
BS 1002	High tensile brass; replaced by BS 1001
BS 1400 DCB1/1	40% Zn 0.4% Al Cu alloy: Brass ingots; chill cast; for die casting **UTS: 270 Elon: 25%**
BS 1400 DCB1 C	40% Zn Cu alloy: Brass castings; chill cast; for die casting **UTS: 270 Elon: 25%**
BS 1400 DCB2/1	40% Zn 1% Sn 0.4% Al Cu alloy: Brass ingots; chill cast; for die casting **UTS: 300 Elon: 20%**
BS 1400 DCB2 C	40% Zn 1% Sn Cu alloy: Naval brass casting; chill cast; for die casting **UTS: 300 Elon: 20% Proof: 100**
BS 1400 DCB3/1	40% Zn 1.5% Pb 0.7% Al Cu alloy: Brass ingots; chill cast; for die casting **UTS: 290 Elon: 15%**
BS 1400 DCB3 C	40% Zn 1.5% Pb Cu alloy: Brass casting; chill cast; for die casting **UTS: 290 Elon: 15%**
BS 1400 HTB1/1	40% Zn 2.5% Al 1.2% Fe 1% Ni 1.5% Sn 3% Mn Cu alloy: Ingots; chill cast **UTS: 480 Elon: 20% Proof: 210**
BS 1400 HTB1 C	40% Zn 2.5% Al 1.2% Fe 1% Ni 1.5% Sn 3% Mn Cu alloy: Castings; chill cast **UTS: 480 Elon: 20% Proof: 210**
BS 1400 HTB2/1	40% Zn 5% Al 1.2% Fe 2% Ni 0.5% Sn 3% Mn Cu alloy: Ingots; sand cast **UTS: 530 Elon: 15% Proof: 270**
BS 1400 HTB2 C	40% Zn 5% Al 1.2% Fe 2% Ni 0.5% Sn 3% Mn Cu alloy: Castings; sand cast **UTS: 530 Elon: 15% Proof: 270**
BS 1400 HTB3/1	40% Zn 5% Al 1.2% Fe 1% Ni 0.2% Sn 4% Mn Cu alloy: Ingots; sand cast; beta brass **UTS: 730 Elon: 12% Proof: 390**
BS 1400 HTB3 C	40% Zn 5% Al 1.2% Fe 1% Ni 0.2% Sn 4% Mn Cu alloy: Castings; sand cast; beta brass **UTS: 730 Elon: 12% Proof: 390**
BS 1453 C2	40% Zn 0.3% Si Cu alloy: Weld filler rod for brazing steel to copper or welding brasses
BS 1453 C4	40% Zn 0.2% Si 0.2% Mn Cu alloy: Weld filler rod for brazing of copper to cast iron steel and for welding brasses
BS 1541 CZ123	40% Zn 0.5% Pb Cu alloy: Plate; as rolled **UTS: 330 Elon: 20%**
BS 1845/8	50% Zn Cu alloy: Braze filler; melting range 860–870 °C
BS 1845/9	45% Zn Cu alloy: Braze filler; melting range 870–880 °C
BS 1845/10	40% Zn Cu alloy: Braze filler; melting range 885–890 °C

Note. The following abbreviations and units are used in the tables:

DPN	Hardness, diamond pyramid number
UTS	Ultimate tensile strength, N/mm^2
Elon	Elongation, %
Proof	0.1 % proof strength, N/mm^2

1 N/mm^2=0.1 hbar=0.102 kgf/mm^2=0.06475 tonf/in^2=145.04 lbf/in^2

Symbol	Nominal analysis, supplier, condition and remarks.
BS 1845/11	44% Zn 1% Sn Cu alloy: Braze filler; melting range 860–870 °C
BS 1845/12	39% Zn 1% Sn Cu alloy: Braze filler; melting range 880–890 °C
BS 1949 CZ109	40% Zn Cu alloy: Brass rods & forgings; as rolled **UTS: 310 Elon: 30% Proof: 150**
BS 1949 CZ123	40% Zn 0.5% Pb Cu alloy: Brass rod & forgings; as rolled **UTS: 310 Elon: 30%**
BS 2870 CZ109	40% Zn Cu alloy: Sheet; specification not issued
BS 2870 CZ113	40% Zn 1.0% Sn Cu alloy: Sheet; specification not issued
BS 2870 CZ114	40% Zn 0.7% Sn 1% Pb 1% Fe 1.5% Mn 1% Al Cu alloy: Sheet; high tensile brass; specification not issued
BS 2870 CZ115	40% Zn 1% Sn 1.2% Pb 1% Fe 1.5% Mn Cu alloy: Sheet; high tensile brass soldering quality; specification not issued
BS 2870 CZ120	40% Zn 2% Pb Cu alloy: Sheet; ½-hard **DPN: 120 Elon: 10%**
BS 2870 CZ120	40% Zn 2% Pb Cu alloy: Sheet; hard **DPN: 150 UTS: 490 Elon: 5%**
BS 2870 CZ120	40% Zn 2% Pb Cu alloy: Sheet; extra hard **DPN: 180 UTS: 560 Elon: 3%**
BS 2870 CZ121	40% Zn 3% Pb Cu alloy: Sheet; specification not issued
BS 2870 CZ122	40% Zn 2% Pb Cu alloy: Sheet; specification not issued
BS 2870 CZ123	40% Zn 0.5% Pb Cu alloy: Sheet; as rolled **UTS: 360 Elon: 20%**
BS 2871 CZ109	40% Zn Cu alloy: Lead free brass; specification not issued
BS 2871 CZ112	38% Zn 1.2% Sn Cu alloy: Tube; Naval brass; specification not issued
BS 2871 CZ113	40% Zn 1% Sn Cu alloy: Tube; Naval brass; specification not issued
BS 2871 CZ114	40% Zn 0.7% Sn 1% Pb 1% Fe 1.0% Al 1.5% Mn Cu alloy: Tube; specification not issued
BS 2871 CZ115	40% Zn 1% Sn 1% Pb 1% Fe 1.5% Mn Cu alloy: Tube; specification not issued
BS 2871 CZ117	40% Zn 0.7% Sn 1% Fe 2.5% Al 1% Mn Cu alloy: Tube; specification not issued
BS 2871 CZ120	40% Zn 2% Pb Cu alloy: Tube; specification not issued
BS 2871 CZ121	40% Zn 3% Pb Cu alloy: Tube; specification not issued
BS 2871 CZ122	40% Zn 2% Pb Cu alloy: Tube; specification not issued
BS 2871 CZ123	40% Zn 0.5% Pb Cu alloy: Tube; specification not issued
BS 2872 CZ109	40% Zn Cu alloy: Forging; as forged **UTS: 300 Elon: 30%**
BS 2872 CZ113	40% Zn 1% Sn Cu alloy: Forging; specification not issued
BS 2872 CZ114	40% Zn 0.7% Sn 1% Pb 1% Fe 1.5% Mn Cu alloy: Forging **UTS: 450 Elon: 20%**
BS 2872 CZ115	35% Zn 1.0% Sn 1.0% Pb 1.0% Fe 1.0% Mn Cu alloy: Forging; as forged **UTS: 450 Elon: 20% Proof: 180**
BS 2872 CZ120	40% Zn 2% Pb Cu alloy: Forging; specification not issued
BS 2872 CZ121	40% Zn 3% Pb Cu alloy: Forging; specification not issued
BS 2872 CZ122	40% Zn 2% Pb Cu alloy: Forging; as forged **UTS: 300 Elon: 25%**
BS 2872 CZ123	40% Zn 0.5% Pb Cu alloy: Forging; as forged **UTS: 300 Elon: 30%**
BS 2873 CZ109	40% Zn Cu alloy: Wire; specification not issued
BS 2873 CZ113	40% Zn 1% Sn Cu alloy: Wire; specification not issued
BS 2873 CZ114	40% Zn 0.8% Sn 1% Pb 1% Fe 1.5% Mn Cu alloy: Wire; specification not issued
BS 2873 CZ115	35% Zn 1.0% Sn 1.0% Pb 1.0% Mn 1.0% Fe Cu alloy: Wire; specification not issued
BS 2873 CZ120	40% Zn 2% Pb Cu alloy: Specification not issued
BS 2873 CZ121	40% Zn 3% Pb Cu alloy: Wire; specification not issued
BS 2873 CZ122	40% Zn 2% Pb Cu alloy: Wire; specification not issued
BS 2873 CZ123	40% Zn 0.5% Pb Cu alloy: Wire; specification not issued
BS 2874 CZ109	40% Zn Cu alloy: Bar; as rolled **UTS: 310 Elon: 30% Proof: 150**
BS 2874 CZ112	38% Zn 1.2% Sn Cu alloy: Bar; as rolled **UTS: 360 Elon: 20%**
BS 2874 CZ113	40% Zn 1% Sn Cu alloy: Bar; as rolled **UTS: 360 Elon: 20%**
BS 2874 CZ114	40% Zn 1% Sn 1% Pb 1% Fe 1.5% Mn 1% Al Cu alloy: Bar; as rolled **UTS: 480 Elon: 18% Proof: 240**
BS 2874 CZ115	40% Zn 1% Sn 1% Pb 1% Fe 1.5% Mn Cu alloy: Bar; cold worked & stress relieved **UTS: 480 Elon: 15% Proof: 240**
BS 2874 CZ120	40% Zn 2% Pb Cu alloy: Bar; specification not issued
BS 2874 CZ121	40% Zn 3% Pb Cu alloy: Bar; as rolled **UTS: 340 Elon: 16%**
BS 2874 CZ122	40% Zn 2% Pb Cu alloy: Bar; hard drawn **UTS: 400 Elon: 22% Proof: 180**
BS 2874 CZ123	40% Zn 0.5% Pb Cu alloy: Bar; as rolled **UTS: 320 Elon: 30%**
BS 2875 CZ109	40% Zn Cu alloy: Plate; specification not issued
BS 2875 CZ113	40% Zn 1% Sn Cu alloy: Plate; specification not issued
BS 2875 CZ114	40% Zn 0.7% Sn 1% Pb 1% Fe 1.0% Al 1.5% Mn Cu alloy: Plate; specification not issued
BS 2875 CZ115	40% Zn 1% Sn 1% Pb 1% Fe 1.5% Mn Cu alloy: Plate; specification not issued
BS 2875 CZ121	40% Zn 3% Pb Cu alloy: Plate; specification not issued
BS 2875 CZ122	40% Zn 2% Pb Cu alloy: Plate; specification not issued
BS 2875 CZ123	40% Zn 0.5% Pb Cu alloy: Plate; as rolled **UTS: 320 Elon: 20%**
BS 2901 C4	40% Zn 0.2% Mn 0.2% Si Cu alloy: Welding rod for inert gas shielded arc welding
BS B1	High tensile brass bar; obsolete
BS B12	40% Zn Cu alloy: Sheet; annealed
BS B13	40% Zn Cu alloy: Bar; free machining
BS B20	Brass bar for hot stamping; obsolete
B STAR H	41.5% Zn Cu alloy: Extrusion; McKechnie Bros; annealed **DPN: 100 UTS: 330 Elon: 40% Proof: 140**
C 4	40% Zn 0.2% Mn 0.2% Si Cu alloy: Welding rod; designation used by BS
C 50	40% Zn 2% Pb Cu alloy: Extrusion; Delta Metal for BS 249
C 90	40% Zn 1% Sn 1% Pb 1% Fe 1% Mn Cu alloy: Bar; Manganese Bronze Co; Parsons C 90
C 92	40% Zn Fe Mn Al Sn Cu alloy: Manganese Bronze Co; Parsons C 92
C 94	44% Zn Mn Fe Cu alloy: Manganese Bronze Co; architectural bronze
CHANNEL BRONZE	1% Sn 39% Zn Cu alloy: Origin unknown
CM 2	40% Zn Sn Pb Fe Al Mn Ni Cu alloy: Extrusion; Delta Metal for BS 1002
CX 1	40% Zn 2% Pb Cu alloy: Extrusion; Delta Metal

Symbol	Nominal analysis, supplier, condition and remarks.
D 1	40% Zn 1% Sn 1% Pb 1% Fe 1% Mn Cu alloy: Bar; Manganese Bronze Co; for BS alloy CZ114
D 7	40% Zn Cu alloy: Lead free bars; Manganese Bronze Co; annealed
	UTS: 330 Elon: 50%
D 9	40% Zn Cu alloy: Free machining bar; Manganese Bronze Co; for small sections
	UTS: 370 Elon: 25%
DIL	40% Zn 1% Sn 1% Pb 1% Fe 1% Mn Cu alloy: Bar; Manganese Bronze Co; for BS alloy CZ 114
DIN 1709 G So Ms F 30	40% Zn 1% Fe 2% Mn 1% Sn Cu alloy: Casting
	DPN: 85 UTS: 330 Elon: 25% Proof: 150
DIN 1709 G So Ms F 45	40% Zn 2% Al 1.5% Fe 2.5% Mn 1.5% Ni Cu alloy: Casting
	DPN: 130 UTS: 530 Elon: 25% Proof: 180
DIN 1709 G So Ms F 60	40% Zn 4% Al 2% Fe 3% Mn 1.5% Ni Cu alloy: Casting
	DPN: 160 UTS: 600 Elon: 20% Proof: 300
DIN 1709 G So Ms F 75	40% Zn 7% Al 3% Fe 4% Mn 1.5% Ni Cu alloy: Casting
	DPN: 100 UTS: 360 Elon: 35% Proof: 90
DIN 1709 GD Ms 60	40% Zn 0.8% Al Cu alloy: Casting; die cast
	DPN: 100 UTS: 330 Elon: 4% Proof: 150
DIN 1709 GK Ms 60	40% Zn 0.8% Al Cu alloy: Casting
DIN 1733 Ms 60	40% Zn 0.2% Sn 0.3% Si Cu alloy: Wrought
DIN 1733 S So Ms	40% Zn Cu alloy: Wrought
DIN 17656 GB Ms 60 A	40% Zn 0.3% Cu alloy: Casting
E	39% Zn 1.5% Pb Cu alloy: Extrusion; Delta Metal Ltd
E 22	40% Zn 1% Sn 1% Fe 1% Pb Cu alloy: T Bolton for BS 1001
E D	40% Zn Sn Pb Fe Al Mn Ni Cu alloy: Extrusion; Delta Metal Ltd; alloying elements varied to suit requirement
EMBEESH	40% Zn Fe Mn Al Sn Cu alloy: Manganese Bronze Co
ETS	39% Zn 0.75% Sn 0.1% Pb Cu alloy: Extrusions; McKechnie Bros; as drawn
	DPN: 120 UTS: 370 Elon: 25% Proof: 140
ETW	40% Zn 0.3% Sn 0.2% Si Cu alloy: Brazing filler rod; McKechnie Bros; melting range 890–900 °C
F 58	40% Zn 1% Sn 0.5% Pb Cu alloy: Extrusion; McKechnie Bros; as drawn
	DPN: 120 UTS: 370 Elon: 25% Proof: 140
FXY	38% Zn 1% Sn 2.25% Pb Cu alloy: Extrusion; McKechnie Bros; as drawn
	DPN: 120 UTS: 370 Elon: 22% Proof: 140
G Cu 55 Zn Al	40% Zn 2% Al 1.5% Fe 2.5% Mn 1.5% Ni Cu alloy: Casting; designation used by German Standard
G Cu 55 Zn Al 2	40% Zn 4% Al 2% Fe 3% Mn 1.5% Ni Cu alloy: Casting; designation used by German Standard
G Cu 55 Zn Al 4	40% Zn 7% Al 3% Fe 4% Mn 1.5% Ni Cu alloy: Casting; designation used by German Standard
G Cu 55 Zn Mn	40% Zn 1% Fe 2% Mn 1% Sn Cu alloy: Casting; designation used by German Standard
G So Ms F 30	40% Zn 1% Fe 2% Mn 1% Sn Cu alloy: Casting; designation used by German Standard
G So Ms F 45	40% Zn 2% Al 1.5% Fe 2.5% Mn 1.5% Ni Cu alloy: Casting; designation used by German Standard
G So Ms F 60	40% Zn 4% Al 2% Fe 3% Mn 1.5% Ni Cu alloy: Casting; designation used by German Standard
G So Ms F 75	40% Zn 7% Al 3% Fe 4% Mn 1.5% Ni Cu alloy: Casting; designation used by German Standard
GB	38% Zn 3.75% Pb Cu alloy: Extrusion; McKechnie Bros; as drawn
	DPN: 110 UTS: 370 Elon: 20% Proof: 170
GB Cu 60 Zn	40% Zn 0.3% Al Cu alloy: Casting; designation used by German Standard
GB Ms 60A	40% Zn 0.3% Al Cu alloy: Casting; designation used by German Standard
GB Ms 60B	40% Zn 0.3% Al Cu alloy: Casting; designation used by German Standard
GD Cu 60 Zn	40% Zn 0.8% Al Cu alloy: Casting; die cast
GD Ms 60	40% Zn 0.8% Al Cu alloy: Casting; die cast
GK Cu 60 Zn	40% Zn 0.8% Al Cu alloy: Casting; designation used by German Standard
GK Ms 60	40% Zn 0.8% Al Cu alloy: Casting; designation used by German Standard
H 15	39% Zn 1.5% Pb Cu alloy: Extrusion; Delta Metal Co
HIDURIT 10	40% Zn 5% Al 4% Mn 1% Fe 1% Ni Cu alloy: Casting; Langley Alloys for BS alloy HTB3
HIDURIT 11	40% Zn 5% Al 1% Fe 2% Ni 3% Mn Cu alloy: Casting; Langley Alloys for BS alloy HTB2
HIDURIT 12	40% Zn 2.5% Al 1% Fe 1% Ni 1.5% Sn 3% Mn Cu alloy: Casting; Langley Alloys for BS alloy HTB1
HIDURIT 15	40% Zn 0.7% Sn 1% Fe 1.5% Mn 1% Pb 1% Al Cu alloy: Wrought; Langley Alloys for BS alloy CZ114
HIGH TENSILE BRASS	40% Zn with Sn Fe Mn & Al Cu alloy: BS alloy CZ114; further information from Copper Development Association
No. 1 HIGH TENSILE BRASS	35% Zn 1% Ni 2% Fe 4% Mn 4.5% Al Cu alloy: Casting; Stone Manganese Marine for BS alloy HTB3
No. 3 HIGH TENSILE BRASS	35% Zn 2% Ni 2% Fe 3% Mn 5% Al Cu alloy: Casting; Stone Manganese Marine for BS alloy HTB2
No. 4 HIGH TENSILE BRASS	35% Zn 2% Al 1.5% Mn 1% Fe 0.5% Sn Cu alloy: Casting; Stone Manganese Marine
No. 6 HIGH TENSILE BRASS	1.5% Sn 2.5% Al 3% Mn 1.5% Fe Cu alloy: Casting; Stone Manganese Marine for BS alloy HTB1
IAT	40% Zn 1% Sn 0.5% Pb 1% Fe 1% Mn Cu alloy: Bar; Manganese Bronze Co for BS alloy CZ114 low Pb
IMI 246	45% Zn Cu alloy: Brazing rod; IMI for BS 1845-9; melting range 870–880 °C
IMI 312	37% Zn 3% Al 2% Mn 1% Fe Cu alloy: Bar; IMI
	DPN: 180 UTS: 630 Elon: 20% Proof: 330
IMI 330	39% Zn 0.5% Al 1.5% Mn 1.2% Sn 1% Fe 0.3% Pb Cu alloy: Bar; IMI
	DPN: 160 UTS: 540 Elon: 30% Proof: 240
IMI 333	40% Zn 1.5% Mn 1.2% Pb 0.7% Sn 1.5% Fe Cu alloy: Bar; IMI and Manganese Bronze Co
	DPN: 140 UTS: 480 Elon: 25% Proof: 240
IMI 334	38% Zn 1.2% Al 1.5% Mn 1.2% Pb 1% Fe Cu alloy: Bar; IMI
	DPN: 170 UTS: 560 Elon: 20% Proof: 300
IMI 360	37% Zn 1% Sn Cu alloy: Bar; IMI Naval brass; annealed
	DPN: 100 UTS: 330 Elon: 56% Proof: 100
IMI 365	39% Zn 1% Sn Cu alloy: Brazing rod; IMI for BS 1845-12; melting range 880–890 °C

Note. The following abbreviations and units are used in the tables:

DPN	Hardness, diamond pyramid number
UTS	Ultimate tensile strength, N/mm^2
Elon	Elongation, %
Proof	0.1 % proof strength, N/mm^2

1 N/mm^2=0.1 hbar=0.102 kgf/mm^2=0.06475 tonf/in^2=145.04 lbf/in^2

Symbol	Nominal analysis, supplier, condition and remarks.
IMI 366	40% Zn Cu alloy: Small amount Sn; IMI 'Tobin Brass' for brazing high Cu brasses
IMI 388	40% Zn 0.3% Si Cu alloy: Brazing metal; silicon brass for brazing steel & Cu alloys; melting range 880–890 °C; IMI
IMI 442	39% Zn 2.6% Pb Cu alloy: Bar; IMI
IMI 452	40% Zn 0.5% Pb Cu alloy: IMI for BS 2873 CZ123; melting range 883–890 °C
IMI 453	40% Zn 0.7% Pb (max) Cu alloy: Bar; IMI
IMI 456	40% Zn 0.7% Pb Cu alloy: Bar; IMI
IMI 467	40% Zn 2.2% Pb Cu alloy: Free machining bar for hot stamping; IMI
IMI 469	40% Zn 3% Pb Cu alloy: Free machining bar; IMI for BS alloy CZ121
IMMADIUM IV	40% Zn 1% Sn 1% Pb 1% Fe 1% Mn Cu alloy: Bar; Manganese Bronze Co for BS alloy CZ115; soldering quality
ITA	40% Zn Fe Mn Al & Sn Cu alloy: Manganese Bronze Co; Parsons ITA
K	38% Zn 0.75% Sn 1.25% Pb 0.75% Fe 0.25% Al 1.25% Mn Cu alloy: Extrusion; McKechnie Bros; as drawn
	DPN: 140 UTS: 500 Elon: 20% Proof: 240
KP	38% Zn 0.75% Sn 0.75% Pb 0.25% Fe 0.25% Al 1.25% Mn Cu alloy: Extrusion; McKechnie Bros; as drawn
	DPN: 130 UTS: 450 Elon: 25% Proof: 210
KS	40% Zn 0.75% Pb 0.25% Fe 1.75% Mn Cu alloy; Extrusion; McKechnie Bros; as drawn
	DPN: 100 UTS: 420 Elon: 30% Proof: 180
KT	40% Zn 2% Pb Cu alloy: Extrusion; Delta Metal Ltd
KZ	38.5% Zn 0.5% Sn 0.75% Pb 0.75% Fe 0.75% Al 0.75% Mn Cu alloy: Extrusion; McKechnie Bros; as drawn
	DPN: 140 UTS: 530 Elon: 15% Proof: 270
M 1	40% Zn Sn Pb Fe Al Mn Ni Cu alloy: Extrusion; Delta Metal Ltd for BS 1001
MANGANESE BRONZE IX	40% Zn 1% Sn 1% Pb 1% Fe 1% Mn Cu alloy: Bar; Manganese Bronze Co for BS alloy CZ114
MANGANESE BRONZE D1	40% Zn 1% Sn 1% Pb 1% Fe 1% Mn Cu alloy: Bar; Manganese Bronze Co for BS alloy CZ114
MANGANESE BRONZE DIL	40% Zn 1% Sn 1% Pb 1% Fe 1% Mn Cu alloy: Bar; Manganese Bronze Co for BS alloy CZ114
MANGANESE BRONZE IAT	40% Zn 1% Sn 0.5% Pb 1% Fe 1% Mn Cu alloy: Bar; Manganese Bronze Co for BS alloy CZ114; low Pb
MKDC 1	37% Zn 0.5% Sn 2.5% Pb Cu alloy: Casting; McKechnie Bros; chill cast
	DPN: 105 UTS: 340 Elon: 17% Proof: 150
NAVAL BRASS	38% Zn 1.25% Sn Cu alloy: Common name; further information from Copper Development Association
NF 2	40% Zn Sn Pb Fe Al Mn Ni Cu alloy: Extrusion; Delta Metal Ltd; alloying elements varied to suit requirements
NIL	37% Zn 1% Sn 1% Pb Cu alloy: Extrusion; Delta Metal Ltd
NS	44% Zn 10% Ni 0.2% Si Cu alloy: Brazing filler rod; McKechnie Bros; melting range 905–915 °C
ORM	40% Zn Sn Pb Fe Al Mn Ni Cu alloy: Extrusion; Delta Metal Ltd; alloying elements varied to suit requirements
PARSONS C 90	40% Zn 1% Sn 1% Pb 1% Fe 1% Mn Cu alloy: Bar; Manganese Bronze Co for BS alloy CZ115
PARSONS C 92	40% Zn Fe Mn Al Sn Cu alloy: Manganese Bronze Co
PARSONS ITA	40% Zn Fe Mn Al Sn Cu alloy: Manganese Bronze Co
PARSONS No 1	40% Zn 1% Sn 1% Pb 1% Fe 1% Mn Cu alloy:

Symbol	Nominal analysis, supplier, condition and remarks.
STANDARD	Bar; Manganese Bronze Co for BS alloy CZ114; made from virgin metal
PARSONS SB 6 PL	46% Zn 6% Ni Cu alloy: Manganese Bronze Co 40% Zn 2% Pb Cu alloy: Extrusion; Delta Metal Ltd
PROPELLER BRASS	40% Zn 1% Sn 1% Fe 0.2% Mn 0.2% Al Cu alloy: Casting; further information from Copper Development Association
PW	40% Zn 0.75% Sn 0.3% Fe Cu alloy: Brazing filler rod; McKechnie Bros; melting range 885–895 °C
QQ B 611a/A	40% Zn 2.0% Pb Cu alloy: Bar & forging; US Federal
QQ B 611a/B	40% Zn 3.0% Pb Cu alloy: Bar & forging; US Federal
	UTS: 340 Elon: 15% Proof: 120
QQ B 636	40% Zn 0.8% Sn Cu alloy: Bar; US Federal
	UTS: 400 Elon: 28% Proof: 200
QQ B 721a/A	40% Zn 1.0% Sn Cu alloy: Plate; US Federal
	UTS: 370 Elon: 22% Proof: 140
QQ S 551/A	48% Zn 0.5% Pb Cu alloy: Brazing filler; US Federal
QQ S 551/B	45% Zn 3.5% Sn 0.5% Pb Cu alloy: Brazing filler; US Federal; melting point 871 °C
S	40% Zn 3% Pb 0.3% Al Cu alloy: Extrusion; McKechnie Bros; as drawn
	DPN: 120 UTS: 400 Elon: 22% Proof: 180
SAE 43	40% Zn 1% Al 1.5% Mn Cu alloy: Casting; as cast
	UTS: 460 Elon: 20% Proof: 170
SAE 73	40% Zn 1% Sn Cu alloy: Bar & forging; annealed; ASTM B 21-A
	UTS: 370 Elon: 40% Proof: 120
SAE 73	40% Zn 1% Sn Cu alloy: Bar & forging; hard; ASTM B 21-A
	UTS: 450 Elon: 25% Proof: 240
SAE 74 A	40% Zn 0.8% Pb Cu alloy: Tube; as drawn; ASTM B 135-5
	UTS: 370
SAE 88	40% Zn 2% Pb Cu alloy: Bar & forging; as forged; ASTM B 124-2
	UTS: 370 Elon: 30%
SAE 430 A	30% Zn 5% Al 3% Mn 3% Fe Cu alloy: Casting
	UTS: 650 Elon: 20% Proof: 330
SAE 430 B	30% Zn 5% Al 3% Mn 3% Fe Cu alloy: Casting
	UTS: 810 Elon: 12% Proof: 420
SAE CA 377	39% Zn 2% Pb Cu alloy: Formerly SAE 88
SAE CA 464	39.25% Zn 0.75% Sn Cu alloy: Naval brass; formerly SAE 73
SAE CA 674	36% Zn 1.2% Al 1% Si 2.7% Mn Cu alloy: Wrought
SAE CA 675	38.8% Zn 1% Sn 0.2% Mn 1.4% Fe Cu alloy: Wrought
SB	40% Zn Sn Pb Fe Al Mn Ni Cu alloy: Extrusion; Delta Metal Ltd; alloying elements varied to suit requirement
SB 6	46% Zn 6% Ni Cu alloy: Manganese Bronze Co
SCB 4	40% Zn 1.2% Sn Cu alloy: Designation used by BS
SIFBRONZE	39% Zn 0.25% Si Cu alloy: Brazing rod; Suffolk Iron Foundry; melting point 850 °C
SILICON BRASS	40% Zn 0.3% Si Cu alloy: Braze metal; IMI 388
SIS 5163	40% Zn 0.5% Pb Cu alloy: Plate; Swedish Standard; annealed
	DPN: 90 UTS: 330 Elon: 40% Proof: 120
SIS 5163	40% Zn 0.5% Pb Cu alloy: Plate; Swedish Standard; cold rolled
	DPN: 125 UTS: 370 Elon: 20% Proof: 240
SIS 5165	40% Zn 1.2% Pb Cu alloy: Plate & bar; Swedish Standard; annealed
	DPN: 80 UTS: 330 Elon: 35% Proof: 120

Symbol	Nominal analysis, supplier, condition and remarks.
SIS 5165	40% Zn 1.2% Pb Cu alloy: Plate & bar; Swedish Standard; cold rolled **DPN: 135 UTS: 370 Elon: 15% Proof: 200**
SIS 5168	40% Zn 2% Pb Cu alloy: Plate; Swedish Standard; cold rolled **DPN: 140 UTS: 390 Elon: 12% Proof: 280**
SIS 5170	38% Zn 3% Pb Cu alloy: Bar & forging; Swedish Standard; annealed **DPN: 95 UTS: 390 Elon: 35% Proof: 140**
SIS 5170	38% Zn 3% Pb Cu alloy: Bar & forging; Swedish Standard; cold rolled **DPN: 140 UTS: 450 Elon: 15% Proof: 330**
SIS 5173	40% Zn 1% Pb Cu alloy: Extrusion; Swedish Standard; as rolled **DPN: 95 UTS: 400 Elon: 30% Proof: 150**
SIS 5272	40% Zn 0.8% Pb 0.8% Al Cu alloy: Extrusion; Swedish Standard; as rolled **DPN: 130 UTS: 480 Elon: 20% Proof: 170**
S Ms 60	40% Zn 0.2% Sn 0.3% Si Cu alloy: Wrought; designation used by German Standard
SPRABRONZE TM	40% Zn 0.8% Sn 0.2% Mn 0.7% Fe Cu alloy: Wire for spraying; Metco Ltd; as sprayed **UTS: 180**
S So Ms	40% Zn Cu alloy: Wrought; designation used by German Standard
SS	40% Zn 1.2% Pb Cu alloy: Extrusion; McKechnie Bros; as drawn **DPN: 135 UTS: 390 Elon: 30% Proof: 180**
SS 1	39.5% Zn 0.5% Pb Cu alloy: Extrusion; McKechnie Bros; annealed **DPN: 100 UTS: 370 Elon: 35% Proof: 220**
STA 7 CM1	38% Zn 1% Sn 1% Fe 2% Mn Cu alloy: Bar; as rolled BS 250, BS 1001 **UTS: 450 Elon: 20% Proof: 220**
STA 7 CM1	38% Zn 1% Sn 1% Fe 2% Mn Cu alloy: Bar; cold drawn; stress relieved **UTS: 530 Elon: 15% Proof: 240**
STA 7 CM2	40% Zn 0.5% Sn 1% Pb 1% Fe 1% Mn 1% Al Cu alloy: Bar & forging **UTS: 450 Elon: 20% Proof: 220**
STA 7 CM3	40% Zn 0.5% Sn 1% Pb 1% Fe 1.5% Mn 2.5% Al Cu alloy: Bar & forging **UTS: 530 Elon: 15% Proof: 240**
STA 7 CM4A	38% Zn 1% Sn 0.5% Pb 1% Fe 1% Mn 1.5% Al Cu alloy: Bar & forging **UTS: 530 Elon: 15% Proof: 240**
STA 7 CM5	40% Zn 0.5% Sn 2% Pb 0.7% Fe 1% Mn 0.2% Al Cu alloy: Casting; as cast **UTS: 300 Elon: 12% Proof: 120**
STA 7 CM6 A & B	40% Zn 0.5% Sn 1% Pb 0.7% Fe 2% Mn 1% Al Cu alloy: Casting **UTS: 440 Elon: 17% Proof: 200**
STA 7 CM7 A & B	40% Zn 0.2% Sn 0.2% Pb 1.5% Fe 2% Mn 4% Al Cu alloy: Casting **UTS: 600 Elon: 14% Proof: 330**
STA 7 CX3	40% Zn 1% Sn 1% Fe 0.2% Mn 0.2% Al Cu alloy: Casting; Admiralty Propeller brass **UTS: 500 Elon: 15% Proof: 240**
STA 7 CX4	40% Zn 1% Sn 1% Fe 1.5% Mn 0.8% Al Cu alloy: Casting **UTS: 500 Elon: 15% Proof: 240**
STA 7 CX4	40% Zn 1% Sn 1% Fe 1.5% Mn 0.8% Al Cu alloy: Bar & forging **UTS: 500 Elon: 25% Proof: 240**
STA 7 CX5	40% Zn 1% Fe 2% Mn 3% Al Cu alloy: Section; base fuse alloy **UTS: 600 Elon: 15% Proof: 300**
STA 7 CX6	High tensile brass; no specific composition; as drawn **UTS: 450 Elon: 28% Proof: 220**

Symbol	Nominal analysis, supplier, condition and remarks.
STA 7 CX13	40% Zn 1% Pb 2% Mn 3% Al Cu alloy: Section; leaded base fuse metal **UTS: 600 Elon: 15% Proof: 300**
STA 7 CZ9A	40% Zn 2% Pb Cu alloy: Forging **UTS: 300 Elon: 25%**
STA 7 CZ9B	40% Zn 2% Pb Cu alloy: Bar **UTS: 300 Elon: 25%**
STA 7 CZ9B1	40% Zn 1.5% Pb (max) Cu alloy: Bar & forging **UTS: 300 Elon: 25%**
STA 7 CZ9C	40% Zn 2% Pb Cu alloy: Bar; drawn & reeled **UTS: 300 Elon: 30% Proof: 180**
STA 7 CZ9D	40% Zn 2% Pb Cu alloy: Bar; drawn & reeled up to 40 mm diameter **UTS: 400 Elon: 20% Proof: 220**
STA 7 CZ9E	40% Zn 2% Pb Cu alloy: Bar; drawn & reeled up to 25 mm diameter **UTS: 450 Elon: 15% Proof: 260**
STA 7 CZ10	40% Zn 3% Pb Cu alloy: Bar; free machining **UTS: 370 Elon: 20%**
STA 7 CZ11B	40% Zn 2% Pb Cu alloy: Sheet & strip **UTS: 370**
STA 7 CZ14	40% Zn Cu alloy: Die casting; as cast **UTS: 260 Elon: 25%**
STA 7 CZ15	38% Zn 1% Sn Cu alloy: Die casting; as cast **UTS: 300 Elon: 20%**
STEREO	38% Zn 2% Fe Cu alloy: Blackwells; obsolete
SUPERSTON 50	30% Zn 5% Al 1.5% Fe 2.2% Mn 0.2% Pb Cu alloy: Casting; Stone Manganese for BS 207
TAURUS XXI	40% Zn 1% Sn 1% Mn 1% Fe 0.5% Al Cu alloy: Casting; D Brown; sand cast **DPN: 110 UTS: 480 Elon: 24% Proof: 200**
TAURUS XXXI	40% Zn 1% Pb Cu alloy: Casting; D Brown; sand cast **DPN: 70 UTS: 290 Elon: 15%**
TOBIN BRASS	40% Zn Cu alloy with small addition Sn: IMI 366
TOBIN BRONZE	39% Zn 0.75% Sn 0.1% Pb Cu alloy: Extrusions; common name for this alloy; further information from Copper Development Association
TONUM	35% Zn 1.7% Al 1% Fe 1.5% Mn Cu alloy: Casting; Stone Manganese **UTS: 520 Elon: 20% Proof: 240**
TURBISTON 3	41% Zn 0.25% Al 1% Sn 1% Fe 0.2% Mn Cu alloy: Casting; Stone Manganese **DPN: 150 UTS: 510 Elon: 18% Proof: 240**
TURBISTON M	35% Zn 1.7% Al 1% Fe 1.5% Mn Cu alloy: Casting; Stone Manganese **UTS: 52 Elon: 20% Proof: 24(0.15%)**
UNI 6896	39% Zn 1.0% Al 1.0% Fe 1.0% Mn Cu alloy: Wrought; Italian standard; high tensile brass **DPN: 140 UTS: 480 Elon: 15% Proof: 200**
VICTOR BRONZE	39% Zn 1.5% Al 1.0% Fe 0.05% V Cu alloy: Origin unknown
W	40% Zn 2% Pb Cu alloy: Extrusion; McKechnie Bros; as drawn **DPN: 130 UTS: 40 Elon: 22% Proof: 20**
WS	39% Zn 1.75% Pb Cu alloy: Extrusion; McKechnie Bros; as drawn **DPN: 125 UTS: 39 Elon: 30% Proof: 18**
XX	50% Zn Cu alloy: Brazing filler rod; McKechnie Bros; melting range 875–885 °C

Note. The following abbreviations and units are used in the tables:

DPN	Hardness, diamond pyramid number
UTS	Ultimate tensile strength, N/mm^2
Elon	Elongation, %
Proof	0.1 % proof strength, N/mm^2

1 N/mm^2=0.1 hbar=0.102 kgf/mm^2=0.06475 tonf/in^2=145.04 lbf/in^2

14D Copper–beryllium wrought and cast alloys

Specific gravity		8.25	
Density		8250 kg/m³	(0.298 lb/in.³)
Solidus/liquidus		–	
Thermal conductivity	solution treated	84 W/m °C	(20 cal/m s °C)
	solution treated and aged	105 W/m °C	(25 cal/m s °C)
Coefficient of linear expansion		17×10^{-6}/ °C	
Electrical conductivity	solution treated	16–18% IACS	
	aged maximum hardness	20–25% IACS (copper 100%)	
	aged to maximum conductivity	32–38% IACS	
Specific resistance	solution treated	96–108 microhm mm	
	aged to maximum hardness	69–86 microhm mm	
	aged to maximum conductivity	46–54 microhm mm	
Young's modulus of elasticity	–	$124–131 \times 10^9$ N/m²	$(18–19 \times 10^6$ lbf/in.²)
Impact		–	
Fatigue strength (10^8 cycles)		$\pm$ 220–270 N/mm²	($\pm$ 14.5–18.5 tonf/in.²)
Hot strength		–	

The above properties are typical of the following group, and may not apply exactly to any one specification. It is possible that with certain specifications some of the values may not be applicable.

General metallurgical characteristics

The alloys listed have up to 2% beryllium, with additions of cobalt or nickel. They all exhibit age hardening characteristics and give the best mechanical properties obtainable from any copper alloy.

In the annealed condition two phases, alpha and gamma, are present and the alloy is reasonably ductile. On heating to above about 750 °C, depending on the beryllium content, the gamma phase dissolves in the alpha and if quenched in water will remain in solution – 'solution treated'. In this condition it resembles pure copper being soft, ductile and capable of accepting a considerable amount of cold work. If however, the alloy is then heated to 315–470 °C, the gamma phase agglomerates and precipitates out of solution. This causes disturbance of the crystal lattice making the material less ductile but giving a considerable increase in hardness. Moreover this increase in hardness can be additive to that obtained from cold work. The electrical and thermal conductivity is reduced as the hardness increases. Corrosion resistance is also affected but not to such a marked degree.

There are two basic ranges of alloys, those with less than 1% beryllium, and those with about 2% beryllium. The cobalt or nickel content generally varies inversely with the beryllium content, their presence increasing the response to heat treatment.

The mechanical properties of beryllium coppers improve at sub-zero temperatures as low as -200 °C. This improvement is about 10% and includes ductility and notch sensitivity figures.

Thermal treatment

All of the alloys can be solution treated and aged. It is not economically advisable to use these materials unless advantage is taken of this ability to be hardened by heat treatment.

The mechanical properties are considerably enhanced by cold working the solution treated material, then ageing.

The approximate heat treatment temperature for each family of alloys is as follows:

Low beryllium alloys. Solution treat 850–920 °C – water quench. This puts the material into the softest possible condition.

Age 420–470 °C – air cool.

High beryllium alloys. Solution treat 760–820 °C – water quench.

Age 315–350 °C – air cool.

The time at solution treatment should be as short as possible and rapid, efficient water quenching is essential. Time at ageing varies from 30 minutes at top temperature to two hours at lower temperatures. Annealing and solution treating are the same for these alloys, but adequate softening for most purposes can be achieved by treating at 600–650 °C and air cooling.

All beryllium–copper alloys require to be carefully heat treated as the temperature and time at temperature are very critical. It is thus essential to consult the individual specification involved or carry out pilot tests to ascertain the correct temperatures and times.

If the alloys are held at the ageing temperature for a slightly longer time than that required to give the maximum hardness, an improvement in thermal and electrical conductivity is obtained at the cost of some reduction in hardness. The same effect is achieved by increasing the ageing temperature for a shorter time.

The solution treatment should ideally be carried out in an inert, or reducing atmosphere. Salt baths should not be used. If air furnaces are used there will be a drop in the fatigue strength unless the oxidized surface layer can be subsequently removed by mechanical or chemical means.

Ageing can be carried out using air furnaces or salt baths.

The oxide can be removed with a sulphuric acid pickle after solution-treatment or annealing, while no scale removal is necessary after ageing.

Welding and brazing

Pre-treatment. The weld area must be thoroughly cleaned, and should have any oxide coating mechanically removed. The material should be in the annealed or softened condition for best results but can be joined while in the solution treated condition provided certain precautions are taken. The parts must not be in the aged condition.

Welding and brazing. Beryllium oxide is very refractory and its presence prevents satisfactory joints. It is therefore not generally possible to use any gas fusion process with beryllium copper.

Electric arc using flux coated rods gives satisfactory results in skilled hands, the best welds being obtained with inert gas shielded arc welding.

These alloys are poor heat conductors and do not require a second operator pre-heating the weld area, but it is advisable to use the maximum possible heat and the minimum time to prevent grain growth.

When the parts to be joined are relatively small the welding can be carried out in the solution treated condition and immediately quenched into water. This acts as a solution treat. Wherever possible however, welding or brazing should be carried out in the soft condition, and solution treated after welding.

Resistance welding can be readily carried out, again preferably in the soft condition.

Post-treatment. All traces of flux must be removed, followed by solution-treatment and ageing. Parts which have been welded in the solution-treated condition and quenched from the weld, only require to be aged.

It is possible to improve the properties of the weld area by hammering to cause controlled cold work prior to ageing. By this means, with careful design and skilled welding, fabricated beryllium copper parts can have a weld zone which has as high a strength and ductility as the parent metal.

Flaw detection methods

Crack test. The oil penetrant chalk test method is recommended where necessary. This must always be a final operation after ageing and finish machining.

X-ray. Seldom required, except perhaps on intricate castings.

Ultra-sonic test. Not normally required.

Chemical etch. Not required.

Corrosion protection

Temporary. None required.

Permanent. None required, except to preserve appearance. Beryllium–copper has similar atmospheric corrosion resistance to pure copper and a lesser tendency to form the green corrosion product associated with copper.

High sulphur bearing atmospheres do have an adverse affect and result in a drop in the fatigue strength.

Painting. This will very seldom be required. If necessary the surface should be prepared by light grit blasting and an etch primer used. A stove enamel top coat will probably be required as beryllium coppers, properly used, are generally subjected to hard usage, which would cause failure of any but the best quality paints correctly applied.

Plating. Where necessary parts can be electroplated using normal plating techniques which may require special etching solutions when adhesion is of prime importance.

Machinability

In the annealed and solution treated condition, beryllium copper has a tendency to tear and it is difficult to hold accurate sizes.

In the finally aged condition the alloys can be too hard for normal machining techniques, although carbide tips can be used, or grinding.

Best results are obtained by machining in the solution treated, quarter or half hard cold work condition. There will be little or no distortion at the final age carried out after machining.

It is essential that adequate cooling is used during machining, otherwise the parts may age from the frictional heat.

Uses

Springs of all kinds, where high strength and corrosion resistance are important. Resistance welding electrodes. Non-spark tooling, including chisels, spanners, shovels, etc. Spring contacts where high strength, corrosion resistance and electrical conductivity are important. Dies for deep drawing of metals and moulding plastics. High strength, corrosion-resistant tubing. This can be bent to complicated shapes, then aged. While beryllium metal is very toxic, there is no health hazard involved with copper–beryllium alloys.

Symbol	Nominal analysis, supplier, condition and remarks.
10	0.6% Be 2.5% Co Cu alloy: Wrought; Brush Beryllium Co
10C	0.6% Be 2.5% Co Cu alloy: Casting; Brush Beryllium Co
10CR	0.6% Be 2.5% Co Cu alloy: Casting; Beryllium Corporation
11CR	0.6% Be 2% Co 1% Ni Cu alloy: Casting; Beryllium Corporation
20C	2.1% Be 0.5% Co Cu alloy: Casting; Brush Beryllium Co
20CR	2.2% Be 0.5% Co Cu alloy: Casting; Beryllium Corporation
25	2% Be 0.25% Co Cu alloy: Wrought; Brush Beryllium Co
30CR	0.6% Be 1.8% Ni Cu alloy: Casting; Beryllium Corporation
35	0.3% Be 1.5% Ni Cu alloy: Wrought; Brush Beryllium Co
35C	0.35% Be 1.5% Ni Cu alloy: Casting; Brush Beryllium Co
50	0.4% Be 1.5% Co 1% Ag Cu alloy: Wrought; Beryllium Corporation
50C	0.5% Be 1.1% Ag Cu alloy: Casting; Brush Beryllium Co
50CR	0.5% Be 1.5% Co 1% Ag Cu alloy: Casting; Beryllium Corporation
50P	0.5% Be 2.5% Co Cu alloy: Powder; Brush Beryllium Co; for sintering
70CR	0.15% Be 0.9% Cr Cu alloy: Casting; Beryllium Corporation; for welding electrodes
150P	1.5% Be 0.25% Co Cu alloy: Powder; Brush Beryllium Co; for sintering
165	1.7% Be 0.25% Co Cu alloy: Wrought; Brush Beryllium Co
165C	0.6% Be 2% Co 1% Ni Cu alloy: Casting; Brush Beryllium Co
165CR	1.7% Be 0.2% Co Cu alloy: Casting; Brush Beryllium Co
200P	2.0% Be Cu alloy: Powder; Brush Beryllium Co; for sintering
245C	2.5% Be 0.5% Co Cu alloy: Casting; Brush Beryllium Co
245Cr	2.5% Be 0.5% Co Cu alloy: Casting; Beryllium Corporation
275C	2.75% Be 0.5% Co Cu alloy: Casting; Brush Beryllium Co
275Cr	2.7% Be 0.5% Co Cu alloy: Casting; Beryllium Corporation
AMPCOLOY 83	1.7% Be 0.25% Co Cu alloy: Wrought; Ampco Metal
AMPCOLOY 84	Be Cu alloy: Ampco
AMPCOLOY 96	0.3% Be 1.5% Ni Cu alloy: Casting; Ampco Metal
AMS 4530 B	1.9% Be Cu alloy: Sheet & strip; solution treated
AMS 4532 A	1.9% Be Cu alloy: Sheet & strip; ½-hard
AMS 4650 D	1.9% Be Cu alloy: Bar & strip; solution treated
AMS 4725 A	1.9% Be Cu alloy: Wire; solution treated
AMS 4890 A	2% Be 0.4% Co 0.3% Si Cu alloy: Investment casting; heat treated

Note. The following abbreviations and units are used in the tables:

DPN	Hardness, diamond pyramid number
UTS	Ultimate tensile strength, N/mm^2
Elon	Elongation, %
Proof	0.1 % proof strength, N/mm^2

1 N/mm^2=0.1 hbar=0.102 kgf/mm^2=0.06475 tonf/in^2=145.04 lbf/in^2

Symbol	Nominal analysis, supplier, condition and remarks.
ASTM B194	1.9% Be 0.6% Ni & Co & Fe (max) Cu alloy: Sheet & bar; annealed **UTS: 460**
ASTM B194	1.9% Be 0.6% Ni & Co & Fe (max) Cu alloy: Sheet & bar; hard **UTS: 820**
ASTM B196	1.9% Be 0.6% Ni & Co & Fe (max) Cu alloy: Bar; annealed **UTS: 460**
ASTM B196	1.9% Be 0.6% Ni & Co & Fe (max) Cu alloy: Bar; solution treated; hard drawn & aged **UTS: 1370**
ASTM B197	1.9% Be 0.6% Ni & Co & Fe (max) Cu alloy: Wire; annealed **UTS: 460**
ASTM B197	1.9% Be 0.6% Ni & Co & Fe (max) Cu alloy: Wire; cold drawn; solution treated; cold drawn & aged **UTS: 1540**
ASTM B441	0.55% Be 2.5% Co Cu alloy: Bar; solution treated & aged; conductivity 20% IACS **UTS: 800**
BERALOY A	2% Be 0.5% Co Cu alloy: Driver
BERYDUR	1% Be 0.25% Co Cu alloy: Wrought; Beryllium Corporation
BERYLCO	4% Be Cu alloy: Ingot; Beryllium Corporation
BERYLCO	0.55% Be 2.5% Co Cu alloy: Wrought; Beryllium Corporation; solution treated & aged **UTS: 1070** **Proof: 750**
BERYLCO 25	1.9% Be 0.2% Co Cu alloy: Wrought; Beryllium Corporation; cold worked; solution treated & aged **UTS: 1540** **Proof: 1370**
BERYLCO 33/25	2% Be 0.25% Co Cu alloy: Beryllium Corporation; same as alloy 25; free machining
BERYLCO 50	0.4% Be 1.5% Co 1.0% Ag Cu alloy: Bar & forging; Beryllium Corporation; solution treated & aged **UTS: 1070** **Proof: 750**
BERYLCO 165	1.7% Be 0.2% Co Cu alloy: Strip & bar; Beryllium Corporation; cold worked; solution treated & aged **UTS: 1370** **Proof: 1140**
BERYLLIUM BRONZE	2.5% Be Cu alloy: Common name
BERYLLOY	2.7% Be 0.5% Co Cu alloy: Telcon Metals
BRUSH 190	1.9% Be 0.2% Co or Ni Cu alloy: Wrought; Brush Beryllium Co
BS 2870 CB101	1.8% Be 0.3% Co & Ni Cu alloy: Sheet; solution treated & aged **DPN: 350**
BS 2870 CB101	1.8% Be 0.3% Co & Ni Cu alloy: Sheet; solution treated & cold rolled; hard **DPN: 225 UTS: 680 Elon: 4%**
BS 2870 CB101	1.8% Be 0.3% Co & Ni Cu alloy: Sheet; solution treated; cold rolled & aged **DPN: 350**
BS 2870 CB101	1.8% Be 0.3% Co & Ni Cu alloy: Sheet; solution treated; cold rolled & aged **DPN: 350**
BS 2872 CB101	1.8% Be 0.2% Ni or Co Cu alloy: Forging; specification not issued
BS 2873 CB 101	1.8% Be 0.2% Ni & Co Cu alloy: Wire; solution treated & aged **UTS: 1050**
BS 2873 CB101	1.8% Be 0.2% Ni & Co Cu alloy: Wire; solution treated; cold drawn & aged **UTS: 1200**
BS 2874 CB101	1.8% Be 0.2% Co & Ni Cu alloy: Bar; specification not issued

Symbol	Nominal analysis, supplier, condition and remarks.
BS 2875 CB101	1.8% Be 0.2% Co & Ni Cu alloy: Plate; specification not issued
BS 3127/2	1.8% Be 0.3% Co Cu alloy: Tube; solution treated for bourdon tube **DPN: 90**
BS 3127/2	1.8% Be 0.3% Co Cu alloy: Tube; solution treated; ½-hard **DPN: 160**
BS 4577 A3/1	2.5% Co 0.4% Be Cu alloy: For resistance welding electrodes; conductivity 45% IACS; wrought **DPN: 180 UTS: 620 Elon: 9%**
BS 4577 A4/2	2.0% Be Cu alloy: For resistance welding electrodes; wrought; conductivity 23% IACS **DPN: 350 UTS: 1050 Elon: 2%**
CABRA 170	1.7% Be 0.2% Co Cu alloy: Casting; tentative SAE designation
CDA 170	1.7% Be 0.25% Co Cu alloy: Wrought; Copper Development Association of America; tentative
CDA 172	2% Be 0.25% Co Cu alloy: Wrought; Copper Development Association of America
CDA 175	0.4% Be 1.5% Co 1% Ag Cu alloy: Wrought; Copper Development Association of America; tentative
CDA 176	0.5% Be 2.5% Co Cu alloy: Wrought; Copper Development Association of America; tentative
CU BE 50	0.5% Be Cu alloy: Strip & sheet; Telcon Metals; solution treated & aged; electrical conductivity 50% IACS **DPN: 200 UTS: 760 Elon: 10%**
CU BE 50	0.5% Be Cu alloy: Strip & sheet; Telcon Metals; solution treated; cold rolled & aged; electrical conductivity 50% IACS **DPN: 250 UTS: 800 Elon: 8%**
CU BE 250	1.8% Be 0.25% Co Cu alloy: Sheet & strip; Telcon Metals; annealed **UTS: 460 Elon: 45% Proof: 170**
CU BE 250	1.8% Be 0.25% Co Cu alloy: Sheet & strip; Telcon Metals; solution treated & cold rolled **UTS: 900 Elon: 5% Proof: 720**
CU BE 250	1.8% Be 0.25% Co Cu alloy: Sheet & strip; Telcon Metals; solution treated; cold rolled & aged **UTS: 1300 Elon: 2% Proof: 1100**
CU BE 275	2.5% Be 0.5% Co Cu alloy: Casting; Telcon; for high strength moulds and dies
CUBELLOY	4.0% Be Cu master alloy: Imperial Smelting Corporation; primary material
ERM N S	0.4% Be 2.25% Co Cu alloy: Bar or forging; Enfield Rolling Mills Ltd; resistance weld electrodes; electrical conductivity 50% IACS **DPN: 250 UTS: 810 Elon: 10% Proof: 700**
IMI 185	1.8% Be 0.3% Co & Ni Cu alloy: Strip & wire; IMI previously KYCUBE 25
KYCUBE 25	1.8% Be 0.3% Co & Ni Cu alloy: Strip & wire; IMI for BS alloy CB101
KYCUBE 25/1	2% Be 0.6% Co & Fe & Ni Cu alloy: Strip; IMI for BS alloy CB101; withdrawn
MALLORY 53B	1.7% Be 0.25% Co Cu alloy: Casting; Mallory Metallurgical; American alloy
MALLORY 73	1.8% Be 0.1% Co Cu alloy: Castings; Mallory Metallurgical; electrical conductivity 20% IACS **DPN: 370 UTS: 750 Elon: 2% Proof: 670**
MALLORY 73	Be Co Cu alloy: For flash welding dies; Mallory Metallurgical for BS alloy CB101; electrical conductivity 23% IACS **DPN: 370 UTS: 1120 Elon: 4% Proof: 1020**
MALLORY 100	Be Co Cu alloy: Casting; Mallory Metallurgical; electrical conductivity 45% IACS **DPN: 210 UTS: 600 Elon: 6% Proof: 530**
MALLORY 100	Be Co Cu alloy: For resistance welding electrodes for stainless steel; Mallory Metallurgical; electrical conductivity **DPN: 200 UTS: 680 Elon: 10% Proof: 600**
MALLORY 150	0.5% Be 1.5% Co 1.1% Ag Cu alloy; Casting; Mallory Metallurgical; welding electrodes; American alloy
RWMA Class 3	Be Co Cu alloy: Mallory metallurgical code for Mallory 100
RWMA Class 4	1.8% Be 0.1% Cu alloy: Mallory metallurgical code for Mallory 73
SAE CA 170	1.7% Be 0.3% Co Cu alloy: Strip
SAE CA 172	1.9% Be Cu alloy: Strip
SAE CA 175	0.5% Be 2.5% Co Cu alloy: Strip
SAE CA 176	0.35% Be 1.5% Co Cu alloy: Bar

Note. The following abbreviations and units are used in the tables:

DPN	Hardness, diamond pyramid number
UTS	Ultimate tensile strength, N/mm^2
Elon	Elongation, %
Proof	0.1 % proof strength, N/mm^2

$1 \text{ N/mm}^2 = 0.1 \text{ hbar} = 0.102 \text{ kgf/mm}^2 = 0.06475 \text{ tonf/in}^2 = 145.04 \text{ lbf/in}^2$

14E Copper–nickel wrought and cast alloys
Up to 50 per cent nickel with additions of iron and manganese

Specific gravity	8.9–8.97	
Density	8900–8970 kg/m³	(0.323 lb/in.³)
Solidus/liquidus	1199 °C	
Thermal conductivity	25–59 W/m °C	(6–14 cal/m s °C)
Coefficient of linear expansion	14.5-17 × 10⁻⁶/ °C	
Electrical conductivity	6.5% IACS (copper 100%)	
Specific resistance	265 microhm mm	
Young's modulus of elasticity	124–159 × 10⁹ N/m²	(18–23 × 10⁶ lbf/in.²)
Impact	107 J	(79 ft/lb)
Fatigue strength	cold drawn ± 230 N/mm²	(± 15 tonf/in.²)
	annealed ± 150 N/mm²	(± 10 tonf/in.²)

Hot strength

Temperature °C	Tensile strength N/mm²
205	320
425	240
645	140
875	37

The above properties are typical of the following group, and may not apply exactly to any one specification. It is possible that with certain specifications some of the values may not be applicable.

General metallurgical characteristics

Copper and nickel form a continuous series of solid solutions. That is, copper and nickel are soluble in each other without forming intermetallic compounds, the characteristics of copper gradually changing to those of nickel from 0% nickel 100% copper, to 100% nickel, 0% copper.

Nickel increases the strength, electrical and corrosion resistance (particularly marine corrosion) in direct proportion to the alloy content. Iron is added to increase the mechanical strength and to provide resistance to impingement attack.

Manganese is added to de-sulphurize thus preventing the brittleness and reduction in corrosion resistance which results when sulphur is present.

Titanium is added to welding rods and alloys commonly welded to reduce porosity at the weld.

Silicon additions made to some of the alloys with about 2.5% nickel results in a degree of age hardening caused by the precipitation of intermetallic compounds from solution. The effect is not very marked and the alloys are as often used in the annealed or cold worked solution as the heat treated condition.

The ability to accept cold work without requiring an anneal varies inversely with the nickel content. Alloys with less than 10% nickel can be subjected to considerable cold work without trouble and all the alloys can be readily hot worked. When more than 50% is present the alloys are listed in Section 29D as nickel–copper alloys and include 'Monel' metals which accept very little cold work without becoming brittle. The copper–nickel alloys are seldom used as castings. Some copper–nickel alloys have high electrical resistance, with very low temperature coefficient of resistance.

Thermal treatment

The alloys listed are all examples of solid solutions on which the only form of thermal treatment possible is annealing to remove cold work. This should be carried out at approximately 700 °C for as short a time as possible to prevent undesirable grain growth. It must be appreciated that once annealed these alloys cannot be re-hardened except by cold working. As the nickel content increases, the amount of cold work possible between anneals decreases.

Normal air furnaces or salt baths can be used for annealing. Where the hardness achieved by cold working is desired it is recommended that stress relieving at 250–300 °C is given, as this increases the fatigue strength with very little decrease in hardness. Scale removal is not necessary with these treatments. Where necessary sulphuric acid pickling or blasting gives best results when de-scaling is required.

Welding and brazing

Pre-treatment. The part should be freshly prepared and thoroughly cleaned.

Welding and brazing. The addition of nickel progressively raises the melting point above that of copper, thus more heat is required when welding these materials than is normal for other copper alloys. Considerable care must be taken to prevent formation of manganese and nickel oxides and sulphur pick-up. Using the correct fluxes and welding rods with a high manganese content helps to prevent this.

Phosphorus bearing welding rods must be avoided as these will cause brittle welds.

All the normal welding and brazing methods are possible, with inert gas shielded arc welding the most satisfactory.

The thermal conductivity of the copper–nickel alloys drops with the nickel content thus pre-heating will probably be required to a greater or lesser extent with the low nickel materials.

Post-treatment. None required. It is advisable to remove all traces of flux as this is generally corrosive and hygroscopic. If the structure is of a complex shape stress relieving at 300 °C may be desirable.

Flaw detection

Crack test. Penetrant oil chalk test method.

X-ray. This will seldom be required, except to prove important welds.

Ultra-sonic test. This may be required on large expensive forgings which have a considerable amount of subsequent machining. This finds subcutaneous defects but cannot identify surface flaws.

Chemical etch. Not required.

Corrosion protection

Temporary. None required.

Permanent. None required under normal circumstances. These alloys are used because of their excellent corrosion resistance particularly in marine conditions. This improves with increase in nickel content.

It is of interest that the lower nickel content alloys exhibit less fouling from marine growth owing to the presence of minute quantities of copper salts (corrosion products) which are toxic to marine organisms.

Painting. Seldom required. An etch primer or a blasted or mechanically roughened surface should be used, followed by stove enamel top coat for best results.

Plating. These alloys are readily plated using normal electroplating techniques, but will very seldom require to be plated.

Machinability

The lower nickel content alloys are soft and ductile, thus have a tendency to tear and should be machined in the cold worked condition if a good finish is required.

The higher nickel alloys above 15% nickel are readily machined. The alloys approaching 50% nickel cold work readily, thus care should be taken to maintain sharp cutting edges at all times and very light cuts should be avoided. Where dimensional stability is essential interstage stress relieving at about 250–300 °C is recommended. This will remove machine stresses and some at least of the cold work caused by machining.

Uses

For miscellaneous uses, where strength and corrosion resistance are required, condensers and condenser tubes, rigging fittings, valves handling salt water, propeller gear, decorative fittings, etc.

The higher nickel alloys are used where a constant electrical resistance over a range of temperatures is required.

Symbol	Nominal analysis, supplier, condition and remarks.
2.0837	30% Ni 1% Fe 0.2% Ti Cu alloy: Wrought; German Standard
2.0882	30.0% Ni Cu alloy: German Werkstoff number
4F	2.7% Ni 0.5% Si Cu alloy: Bar & forging; Langley Alloys
30 ALLOY	2.25% Ni Cu alloy: British Driver-Harris; electrical conductivity 30% IACS
	UTS: 300
60 ALLOY	6% Ni Cu alloy: For resistors; Driver
90 ALLOY	12% Ni Cu alloy: For resistors; Driver
95 ALLOY	11% Ni Cu alloy: Driver-Harris
	UTS: 360
111 ALLOY	Ni Cu alloy: Driver-Harris; electrical conductivity 55% IACS
	UTS: 300
180 ALLOY	22% Ni Cu alloy: For resistors; Driver
190	1.1% Ni 0.2% P Cu alloy: Wrought; designation agreed by CDAA; not used
191	1.1% Ni 0.2% P Cu alloy: Wrought; designation agreed by CDAA; not used

Note. The following abbreviations and units are used in the tables:

DPN	Hardness, diamond pyramid number
UTS	Ultimate tensile strength, N/mm^2
Elon	Elongation, %
Proof	0.1 % proof strength, N/mm^2

1 N/mm^2=0.1 hbar=0.102 kgf/mm^2=0.06475 tonf/in^2=145.04 lbf/in^2

Symbol	Nominal analysis, supplier, condition and remarks.
647	2.0% Ni 0.6% Si Cu alloy: Wrought; designation agreed by CDAA; not used
702	2.5% Ni 0.4% Mn Cu alloy: Wrought; designation agreed by CDAA; not used
703	5.2% Ni 0.5% Mn Cu alloy: Wrought; designation agreed by CDAA; not used
704	1.0% Zn 5.5% Ni 1.5% Fe 0.5% Mn Cu alloy: Wrought; designation agreed by CDAA; not used
707	10% Ni 0.5% Mn Cu alloy: Wrought; designation agreed by CDAA; not used
708	11.0% Ni 0.15% Mn Cu alloy: Wrought; designation agreed by CDAA; not used
709	1.0% Zn 15% Ni 0.6% Mn Cu alloy: Wrought; designation agreed by CDAA; not used
720	42% Ni 2.0% Fe 1.2% Mn Cu alloy: Wrought; designation agreed by CDAA; not used
964	30% Ni Cu alloy: Designation used by CDAA
ADNIC	29% Ni 1% Sn 1.3% Mn Cu alloy: Origin unknown
ADVANCE	44% Ni 1.5% Mn Cu alloy: British Driver-Harris
	UTS: 450
AMPCOLOY 521	30% Ni Cu alloy: Casting; Ampco
AMPCOLOY 522	30% Ni Cu alloy: Ampco
AMPCOLOY 525	10% Ni 1.2% Fe Cu alloy: Casting; Ampco
AMPCOLOY 526	10% Ni 1% Fe Cu alloy: Ampco
ANTIMONY BRONZE	7.5% Sb 2% Ni Sn free Cu alloy: Common name
ASTM B111 70/30 Cu Ni	31% Ni Cu alloy: Tube
ASTM B111 80/20 Cu Ni	21% Ni Cu alloy: Tube

Symbol	Nominal analysis, supplier, condition and remarks.
ASTM B111 90/10 Cu Ni	10% Ni Cu alloy: Tube
ASTM B111/704	5.2% Ni 1.5% Fe 0.5% Mn Cu alloy: Seamless tube
ASTM B111/706	10% Ni 1.4% Fe Cu alloy: Seamless tube
ASTM B111/710	21% Ni 0.8% Fe Cu alloy: Seamless tube
ASTM B111/715	31% Ni 0.5% Fe Cu alloy: Seamless tube
ASTM B111/716	31% Ni 5.2% Fe Cu alloy: Seamless tube
ASTM B111/720	41% Ni 2.0% Fe 1.2% Mn alloy: Seamless tube
ASTM B122/5	30% Ni Cu alloy: Sheet; $\frac{1}{2}$-hard **UTS: 390**
ASTM B122/5	30% Ni Cu alloy: Sheet; spring hard **UTS: 500**
ASTM B122/6	20% Ni Cu alloy: Sheet; $\frac{1}{2}$-hard **UTS: 390**
ASTM B122/6	20% Ni Cu alloy: Sheet; spring hard **UTS: 590**
ASTM B171/706	10% Ni 1.2% Fe Cu alloy: Rolled plate
ASTM B171/715	31% Ni 0.55% Fe Cu alloy: Rolled plate
ASTM B225 E Cu Ni	29% Ni Cu alloy: Welding electrode **UTS: 340**
ASTM B259 R Cu Ni	1% Zn 1% Sn 30% Ni Cu alloy: Welding rod **UTS: 340**
ASTM B358 A	30% Ni Cu alloy: Ingot; primary metal
ASTM B358 B	50% Ni Cu alloy: Ingot; primary metal
ASTM B360 B	30% Ni 0.9% Fe Cu alloy: Casting; as cast **UTS: 420 Elon: 20% Proof: 210**
ASTM B369 A	10% Ni 1.2% Fe Cu alloy: Casting; as cast **UTS: 300 Elon: 20% Proof: 170**
ASTM B402/70/30	31% Ni 0.6% Fe Cu alloy: Plate & sheet
ASTM B402/90/10	10% Ni 1.5% Fe Cu alloy: Plate & sheet
ASTM B411	1.9% Ni 0.6% Si Cu alloy: Rod & bar
ASTM B412	1.9% Ni 0.6% Si Cu alloy: Wire
ASTM B422	1.9% Ni 0.6% Si Cu alloy: Sheet & strip
ASTM B433/70/30	30% Ni 1.2% Mn 0.7% Fe 0.5% Si 1.0% Nb Cu alloy: Casting
ASTM B433/90/10	10% Ni 1.2% Mn 1.3% Fe Cu alloy: Casting
ASTM B466	Ni Cu alloy: Pipe & tube; graded by CDAA system
ASTM B467	Ni Cu alloy: Welded pipe & tube; graded by CDAA system
ASTM B492	20% Ni 0.4% Si 0.8% Mn 1.0% Fe Cu alloy: Casting for tailshaft sleeves
B 47	Ni Cu alloy: Casting; Anti Attrition Ltd for electrical slip rings **UTS: 240 Elon: 25%**
B 68	Ni Cu alloy: Casting for food containers; Anti Attrition Ltd **DPN: 60 UTS: 210 Elon: 10%**
BATNICKON 5	5% Ni Fe Mn alloy: Tube, sheet, etc.; Birmingham Battery
BATNICKON 10	10% Ni Fe Cu alloy: Tube, sheet, etc.; Birmingham Battery
BATNICKON 30	30% Ni Fe Cu alloy: Sheet, tube, etc.; Birmingham Battery
BS 374 CN103	15% Ni 0.2% Mn Cu alloy: Sheet; annealed **UTS: 260 Elon: 36%**
BS 374 CN104	20% Ni 0.2% Mn Cu alloy: Sheet; annealed **UTS: 300 Elon: 36%**
BS 374 CN105	25.0% Ni 0.25% Mn Cu alloy: Sheet; annealed **UTS: 330 Elon: 33%**
BS 374 CN106	30.0% Ni Cu alloy: Sheet; annealed **UTS: 360 Elon: 33%**
BS 378 CN102	10% Ni 1.5% Fe 0.75% Mn Cu alloy: Tube; as drawn for condenser tubes, etc. **DPN: 150**
BS 378 CN107	30% Ni 0.75% Fe 1% Mn Cu alloy: Tube; as drawn for condenser tubes, etc. **DPN: 150**

Symbol	Nominal analysis, supplier, condition and remarks.
BS 1464 CN107	30% Ni 0.7% Fe 1% Mn Cu alloy: Tube **DPN: 120**
BS 1541 CN101	5.5% Ni 1.2% Fe 0.6% Mn Cu alloy: Plate; as rolled **UTS: 220 Elon: 30%**
BS 1541 CN102	10.5% Ni 1.5% Fe 0.75% Mn Cu alloy: Plate; as rolled **UTS: 260 Elon: 30%**
BS 1541 CN104	20% Ni 0.2% Mn Cu alloy: Plate; as rolled **UTS: 250 Elon: 30%**
BS 1541 CN106	30% Ni 0.5% Mn Cu alloy: Plate; as rolled **UTS: 300 Elon: 30%**
BS 1541 CN107	30% Ni 0.6% Fe 1% Mn Cu alloy: Plate; as rolled **UTS: 300 Elon: 30%**
BS 1845 C28	0.3% Si 9.5% Ni Cu alloy: For brazing; melting range 920–980 °C
BS 2870 CN101	5.5% Ni 1.2% Fe 0.5% Mn Cu alloy: Sheet; as rolled **UTS: 220 Elon: 35%**
BS 2870 CN102	10% Ni 1.5% Fe 0.7% Mn Cu alloy: Sheet
BS 2870 CN103	15% Ni 0.2% Mn Cu alloy: Sheet; annealed **UTS: 260 Elon: 35%**
BS 2870 CN104	20% Ni 0.2% Mn Cu alloy: Sheet; annealed **UTS: 300 Elon: 35%**
BS 2870 CN105	25% Ni 0.3% Mn Cu alloy: Sheet; annealed **UTS: 330 Elon: 30%**
BS 2870 CN106	30% Ni Cu alloy: Sheet; annealed **UTS: 360 Elon: 30%**
BS 2870 CN107	30% Ni 0.7% Fe 1% Mn Cu alloy: Sheet; specification not issued
BS 2871 CN101	5.5% Ni 0.5% Mn 1.2% Fe Cu alloy: Tube; annealed **UTS: 300**
BS 2871 CN102	10% Ni 0.7% Mn 1.5% Fe Cu alloy: Tube; annealed **UTS: 330**
BS 2871 CN103	15% Ni 0.2% Mn Cu alloy: Tube; specification not issued
BS 2871 CN104	20% Ni 0.2% Mn Cu alloy: Tube; specification not issued
BS 2871 CN105	25% Ni 0.4% Mn Cu alloy: Tube; specification not issued
BS 2871 CN106	30% Ni Cu alloy: Tube; specification not issued
BS 2871 CN107	30% Ni 1% Mn 0.75% Fe Cu alloy: Tube; annealed **UTS: 420**
BS 2872 CN101	5.5% Ni 1.2% Fe 0.5% Mn Cu alloy: Forging; specification not issued
BS 2872 CN102	10% Ni 1.5% Fe 0.7% Mn Cu alloy: Forging; specification not issued
BS 2872 CN103	15% Ni 0.3% Mn Cu alloy: Forging; specification not issued
BS 2872 CN104	20% Ni 0.3% Mn Cu alloy: Forging; specification not issued
BS 2872 CN105	25% Ni 0.4% Mn Cu alloy: Forging; specification not issued
BS 2872 CN106	30% Ni Cu alloy: Forging; specification not issued
BS 2872 CN107	30% Ni 1% Mn Cu alloy: Forging; specification not issued
BS 2873 CN101	5% Ni 1.2% Fe 0.6% Mn Cu alloy: Wire; specification not issued
BS 2873 CN102	10% Ni 1.5% Fe 0.8% Mn Cu alloy: Wire; specification not issued
BS 2873 CN103	15% Ni 0.3% Mn Cu alloy: Wire; specification not issued
BS 2873 CN104	20% Ni 0.2% Mn Cu alloy: Wire; specification not issued
BS 2873 CN105	25% Ni 0.4% Mn Cu alloy: Wire; specification not issued
BS 2873 CN106	30% Ni Cu alloy: Wire; specification not issued

Symbol	Nominal analysis, supplier, condition and remarks.
BS 2873 CN107	31% Ni 1% Mn 0.8% Fe Cu alloy: Wire; specification not issued
BS 2874 CN101	5.5% Ni 1.2% Fe 0.6% Mn Cu alloy: Bar; specification not issued
BS 2874 CN102	10% Ni 1.5% Fe 0.7% Mn Cu alloy: Bar; specification not issued
BS 2874 CN103	15% Ni 0.3% Mn Cu alloy: Bar; specification not issued
BS 2874 CN104	20% Ni 0.3% Mn Cu alloy: Bar; specification not issued
BS2874 CN105	25% Ni 0.4% Mn Cu alloy: Bar; specification not issued
BS 2874 CN106	30% Ni Cu alloy: Bar; specification not issued
BS 2874 CN107	30% Ni 0.7% Fe 1% Mn Cu alloy: Bar; specification not issued
BS 2875 CN101	5.5% Ni 1% Fe 0.5% Mn Cu alloy: Plate; as rolled **UTS: 220　Elon: 30%**
BS 2875 CN102	10% Ni 1.5% Fe 0.7% Mn Cu alloy: Plate; as rolled **UTS: 260　Elon: 30%**
BS 2875 CN103	15% Ni 0.2% Mn Cu alloy: Plate; specification not issued
BS 2875 CN104	20% Ni 0.3% Mn Cu alloy: Sheet; as rolled **DPN: 250　Elon: 30%**
BS 2875 CN105	25% Ni 0.25% Mn Cu alloy: Plate; specification not issued
BS 2875 CN106	30% Ni Cu alloy: Plate; as rolled **UTS: 300　Elon: 30%**
BS 2875 CN107	30% Ni 0.8% Fe 1% Mn Cu alloy: Plate; as rolled **UTS: 300　Elon: 30%**
BS 2901 C16	10% Ni 0.3% Al Cu alloy: Rod for inert gas shielded arc welding; obsolete
BS 2901 C17	20% Ni 0.3% Ti Cu alloy: Rod for all welding
BS 2901 C18	30% Ni 0.3% Al Cu alloy: Rod for all welding
BS 2901 C19	5.5% Ni 1.2% Fe 0.3% Al 0.5% Cu alloy: Rod for all welding
BS 4577 A3/2	2.5% Ni 0.5% Si Cu alloy: For resistance welding electrodes; wrought; conductivity 38% IACS **DPN: 200　UTS: 650　Elon: 14%**
BS 4577 A4/1	0.2% P 0.1% Ni Cu alloy: For resistance welding electrodes; wrought; conductivity 50% IACS **DPN: 130　UTS: 400　Elon: 20%**
CONSTANTAN	45% Ni 55% Cu alloy: Electrical conductivity 3.5% IACS; A Heckford Ltd
Cu Ni 10 Fe	10% Ni Cu alloy: Wrought & cast; Inco; annealed **DPN: 90　UTS: 310　Elon: 40%　Proof: 140**
Cu Ni 10 Fe Mn	10% Ni with Fe Mn Cu alloy: Wrought & cast; Inco; annealed **DPN: 100　UTS: 310　Elon: 30%　Proof: 140**
Cu Ni 30 Fe	30% Ni with Fe Cu alloy: Wrought & cast; Inco; annealed **DPN: 100　UTS: 300　Elon: 45%　Proof: 150**
Cu Ni 30 Fe Mn	30% Ni with Fe and Mn Cu alloy: Wrought & cast; Inco; annealed **DPN: 100　UTS: 430　Elon: 35%　Proof: 220**
CuNi30Fe2Mn2	30% Ni with Fe and Mn Cu alloy: Wrought & cast; Inco; annealed **DPN: 110　UTS: 460　Elon: 45%　Proof: 190**

Note. The following abbreviations and units are used in the tables:

DPN	Hardness, diamond pyramid number
UTS	Ultimate tensile strength, N/mm^2
Elon	Elongation, %
Proof	0.1 % proof strength, N/mm^2

1 N/mm^2=0.1 hbar=0.102 kgf/mm^2=0.06475 tonf/in^2=145.04 lbf/in^2

Symbol	Nominal analysis, supplier, condition and remarks.
CUNICO 1	21% Ni 29% Co Cu alloy: For magnets; American material listed by PMA
CUNIFE 1	20% Ni 20% Fe Cu alloy: For magnets; American material listed by PMA
CUNIFE 11	20% Ni 27.5% Fe 2.5% Co Cu alloy: For magnets; American material listed by PMA
CUPRON	45% Ni Cu alloy: Brunton for use up to 500 °C; electrical conductivity 4% IACS
CUPRO NICKEL	Series of copper–nickel alloys containing 15–17% Ni: Common name
DIN 1733 S Cu Ni 30 Fe	30% Ni 1% Fe 0.2% Ti Cu alloy: Wrought
DIN 1785 Cu Ni 10 Fe	16% Ni 0.5% Mn 1% Fe Cu alloy: Wrought
DIN 1785 Cu Ni 20 Fe	21% Ni 1% Mn 0.8% Fe Cu alloy: Wrought
DIN 1785 Cu Ni 30 Fe	31% Ni 1% Mn 0.8% Fe Cu alloy: Wrought
DTD 498	2.7% Ni 0.5% Si Cu alloy: Bar & forging; as rolled or forged **UTS: 570　Elon: 15%　Proof: 400**
DTD 504	2.7% Ni 0.5% Si Cu alloy: Bar; cold rolled; solution treated & aged **DPN: 200　UTS: 630　Elon: 15%　Proof: 560**
ERM 3A	1% Ni 0.2% P Cu alloy: Bar, forging & casting; Enfield; as drawn; electrical conductivity 60% IACS **DPN: 160　UTS: 320　Elon: 22%　Proof: 270**
FERRY	45% Ni Cu alloy: For electrical resistance; H Wiggin; annealed; electrical resistance 49 microhm/cm; non-magnetic **DPN: 156　UTS: 480　Elon: 47%　Proof: 370**
FIVEOHM	Ni Mn Cu alloy: Specific electrical resistance 5 microhm/cm/cm^2
HECNUM	45% Ni Cu alloy: Electrical conductivity 5.5% IACS; A Heckford Ltd; low temperature coefficient of electrical resistance **DPN: 155　UTS: 480　Elon: 38%　Proof: 220**
HIDURAL 5	2.7% Ni 0.5% Si Cu alloy: Bar & forging; Langley Alloys; annealed **UTS: 600　Elon: 20%　Proof: 450**
HIDURAL 5	2.7% Ni 0.5% Si Cu alloy: Forging, Langley Alloys; cold drawn **UTS: 730　Elon: 20%　Proof: 640**
HIDURAX SPECIAL 13A	3.0% Al 1.7% Fe 14.5% Ni 0.3% Mn Cu alloy: Langley; aged **DPN: 250　UTS: 860　Elon: 14%　Proof: 700**
HIDURAX SPECIAL 19A	1.7% Al 17.5% Ni 4.5% Mn Cu alloy: Langley; aged; improved properties can be obtained by cold working **DPN: 240　UTS: 790　Elon: 25%　Proof: 490**
HIDURON 501	12% Ni Al Fe Mn Cu alloy: Cast; Langley Alloys **UTS: 450　Elon: 35%　Proof: 220**
IMI 842	20% Ni Cu alloy: Strip & wire; IMI; annealed **DPN: 90　UTS: 330　Elon: 50%　Proof: 110**
IMI 842	20% Ni Cu alloy: Strip & wire; IMI; cold drawn **DPN: 160　UTS: 530　Elon: 8%　Proof: 480**
IMI 849	30% Ni Cu alloy: Sheet & wire; IMI
IS 2283NS10	10% Ni 1.5% Fe 0.7% Mn Cu alloy: Sheet; Indian Standard; annealed **DPN: 85　UTS: 320　Elon: 42%　Proof: 120**
KUNIFER 5	5.5% Ni 1.2% Fe 0.5% Mn Cu alloy: Tube, plate & sheet; ICI and Yorkshire Imperial Metals; resistant to corrosion & erosion **DPN: 70　UTS: 290　Elon: 47%　Proof: 110**
KUNIFER 5T	5% Ni 1.2% Fe 0.5% Mn 0.3% Ti Cu alloy: Welding rod; IMI

Symbol	Nominal analysis, supplier, condition and remarks.
KUNIFER 10	10% Ni 2% Fe 1% Mn Cu alloy: Sheet & tube; ICI and Yorkshire Imperial Metals; corrosion resistant **DPN: 100 UTS: 320 Elon: 47% Proof: 150**
KUNIFER 10T	10% Ni 1.5% Fe 0.7% Mn 0.3% Ti Cu alloy: Welding rod; IMI
KUNIFER 30	31% Ni 0.6% Fe 0.8% Mn Cu alloy: Tube; Yorkshire Imperial Metals
KUNIFER 30	30% Ni 0.7% Fe 0.8% Mn Cu alloy: Wrought; IMI alloy obtained from Yorkshire Imperial Metals **DPN: 100 UTS: 440 Elon: 45% Proof: 140**
KUNIFER 30A	30% Ni 2% Mn 2% Fe Cu alloy: Wrought; IMI **DPN: 110 UTS: 650 Elon: 45% Proof: 180**
KUNIFER 30T	30% Ni 0.5% Fe 0.7% Mn 0.4% Ti Cu alloy: Welding rod; IMI
LOHM	6% Ni Cu alloy: British Driver-Harris; electrical conductivity 18% IACS **UTS: 520**
MAGNETOFLEX 12	20% Ni Cu alloy: Strip for magnets; German alloy listed by PMA
MANGANIN	4% Ni 12% Mn Cu alloy: Wire; British Driver-Harris **UTS: 450**
MIDOHM	23% Ni Cu alloy: British Driver-Harris **UTS: 520**
NEN 6033/Cu Ni 10	10% Ni 1.5% Fe 0.7% Mn Cu alloy: Sheet; Dutch specification; annealed **DPN: 85 UTS: 320 Elon: 42% Proof: 120**
NEWLOY	35% Ni 1% Sn Cu alloy: Harrison Fischer
NF A51 102 Cu Ni 10 Fe 1 Mn	10% Ni 1.5% Fe 0.7% Mn Cu alloy: Tube; French Standard; annealed **DPN: 95 UTS: 320 Elon: 40% Proof: 150**
NF A51 102 Cu Ni 30 Mn 1 Fe	30% Ni 1% Mn 0.7% Fe Cu alloy: Tube; French Standard; annealed **DPN: 105 UTS: 420 Elon: 42% Proof: 180**
PERMET	21% Ni 29% Co Cu alloy: For magnets; American alloy listed by PMA
REGENT	20% Ni Cu alloy: A Heckford; electrical conductivity 6.5% IACS **DPN: 128 UTS: 330 Elon: 39%**
SAE CA 706	10% Ni 1.3% Fe Cu alloy: Wrought
SAE CA 710	20% Ni 1% Fe 1% Mn Cu alloy: Wrought
SAE CA 715	30% Ni 1% Zn (max) 1% Mn (max) Cu alloy: Wrought
SIS 5667	10% Ni 1% Mn 1.5% Fe Cu alloy: Bar; Swedish Standard; annealed **DPN: 80 UTS: 300 Elon: 35% Proof: 150**
SIS 5682	30% Ni 1% Mn 1% Fe Cu alloy: Bar; Swedish Standard; annealed **DPN: 100 UTS: 320 Elon: 32% Proof: 170**
SPECIAL ADVANCE	42% Ni Cu alloy: British Driver-Harris; thermocouples
STA CX11	0.5% Si 2.5% Ni Cu alloy: Bar; solution treated & aged **UTS: 570 Elon: 15% Proof: 400**
STA 7 CX11	0.5% Si 2.5% Ni Cu alloy: Bar; solution treated; cold worked & aged **UTS: 630 Elon: 15% Proof: 560**
STABILOHM 43	Cu Mn Ni alloy: Wire for resistors; Johnson Matthey

Symbol	Nominal analysis, supplier, condition and remarks.
TAURUS	1.5% Sn 4% Ni Cu alloy: Casting; D Brown; centri-cast **DPN: 80 UTS: 290 Elon: 35% Proof: 150**
TAURUS XXVI	11% Sn 38% Ni 7% Pb Cu alloy: Casting; D Brown; centri-cast **DPN: 170 UTS: 340 Elon: 3% Proof: 180**
TELCALLOY 1	5% Ni Cu alloy: For electrical resistance; Telcon; annealed; electrical conductivity 18% IACS **UTS: 240**
TELCALLOY 1.5	10% Ni Cu alloy: For electrical resistance; Telcon; annealed; electrical conductivity 12% IACS **UTS: 260**
TELCALLOY 2	15% Ni Cu alloy: For electrical resistance; Telcon; annealed; electrical conductivity 8.5% IACS **UTS: 300**
TELCALLOY 2	Ni Cu alloy: Wire for heaters; specific resistance 20 microhm/cm; Telcon Ltd
TELCALLOY 3	25% Ni Cu alloy: For electrical resistance; Telcon; annealed; electrical conductivity 5.5% IACS **UTS: 370**
TELCALLOY 3	Ni Cu alloy: For resistance wire; Telcon; specific resistance 30 microhm/cm
TELCALLOY 4	Ni Cu alloy: For electrical resistance; Telcon
TELCONSTAN	Ni Cu alloy: For electrical resistance; Telcon; electrical conductivity 3.5% IACS; constant electrical resistivity over normal temperature range; annealed **UTS: 400**
TENOHM	Ni Mn Cu alloy: Specific electrical resistance 10 microhm/cm/cm²; origin unknown
UN 30	31.0% Ni 0.5% Fe Cu alloy: Origin unknown
UNI 6785	10% Ni 1.5% Fe 0.7% Mn Cu alloy: Tube; Italian Standard; annealed **DPN: 85 UTS: 320 Elon: 40% Proof: 150**
UNI 6786	30% Ni 1% Mn 0.7% Fe Cu alloy: Tube; Italian specification; annealed **DPN: 105 UTS: 420 Elon: 42% Proof: 180**
VSM 10803/Cu Ni 10 Fe Mn	10% Ni 1.5% Fe 0.7% Mn Cu alloy: Sheet; Swiss Standard; annealed **DPN: 85 UTS: 320 Elon: 42% Proof: 120**
VSM 10803/Cu Ni 30 Fe Mn	30% Ni 1% Mn 0.7% Fe Cu alloy: Tube; Swiss Standard; annealed **DPN: 105 UTS: 420 Elon: 42% Proof: 180**
WYMDALOY	20% Mn 20% Ni Cu alloy: Good corrosion resistance and high resistance to wear; origin unknown
YORCORON	30% Ni 2% Fe 2% Mn Cu alloy: Wrought; Yorkshire Imperial Metals **DPN: 70 UTS: 290 Elon: 47% Proof: 110**

Note. The following abbreviations and units are used in the tables:

DPN	Hardness, diamond pyramid number
UTS	Ultimate tensile strength, N/mm²
Elon	Elongation, %
Proof	0.1 % proof strength, N/mm²

$1 \text{ N/mm}^2 = 0.1 \text{ hbar} = 0.102 \text{ kgf/mm}^2 = 0.06475 \text{ tonf/in}^2 = 145.04 \text{ lbf/in}^2$

14F Copper–nickel–zinc wrought and cast alloys
With additions of lead, tin, manganese and silicon – 'nickel silver' and 'nickel brass'

Specific gravity	8.46–8.60	
Density	8460–8600 kg/m^3	(0.303–0.311 lb/in.^{3})
Solidus/liquidus	930–1020 °C	
Thermal conductivity	37.3–46 W/m °C	(8.9–11 cal/m s °C)
Coefficient of linear expansion	15.0–16.7 $\times$ 10^{-6}/ °C	
Electrical conductivity	6–15% IACS (copper 100%)	
Specific resistance	110–120 microhm mm	
Young's modulus of elasticity	120 $\times$ 10^9 N/m^2	(17.4 $\times$ 10^6 lbf/in.^{2})
Impact on 30% Ni 15% Zn annealed alloy	108 J	(80 ft/lb) at 20 °C Izod
	118 J	(87 ft/lb) at - 190 °C
Fatigue strength (10 $\times$ 10^7 cycles) 20% Ni casting	$\pm$ 100 N/mm^2	$\pm$ 7 tonf/in.^{2}

Hot strength

Temperature °C	Tensile strength N/mm^2	Elongation %
180	420	16.5
290	390	10.5
340	320	1
400	320	2
480	290	3.5

The above properties are typical of the following group, and may not apply exactly to any one specification. It is possible that with certain specifications some of the values may not be applicable.

General metallurgical characteristics

These alloys all have 40–60% copper with from 10–30% nickel, the remainder zinc. The copper content generally remains constant, with the nickel and the zinc varying in inverse proportion. Lead is added to increase machinability, tin to improve corrosion resistance, manganese acts as a de-sulphurizer, while silicon strengthens the alpha matrix.

These alloys are all white in colour, with reasonably good corrosion resistance and the ability to take and retain a high polish.

They can all be cold worked and only require interstage annealing when very deep drawing has been carried out, although this is less marked with the high zinc alloys – nickel brass.

All the lead free alloys can be hot worked but best results are obtained with low nickel, high zinc which produces an alpha–beta structure.

The structure resembles that of the alpha brasses, except for the higher zinc materials where some beta phase appears. This reduces somewhat their ability to accept cold work.

All the alloys can be used to make castings, but these have no significant use except for ornament.

Thermal treatment

None of these alloys can have their mechanical properties improved by any form of heat treatment.

Cold working can increase the tensile and proof strengths with a reduction in the ductility, and this cold work can be removed by annealing at 700–800 °C. The time at temperature should be kept to a minimum to prevent grain growth.

It must be noted that once the parts have been annealed they can only be re-hardened by further cold work.

Annealing can be carried out using normal air furnaces or salt baths and will result in some surface discolouration which can be removed by blasting or polishing. Acid pickling can only be used with care as there is always a danger of preferential attack leaving a pitted surface. specialist advice should be obtained before pickling any of these alloys.

Welding and brazing

Pre-treatment. Joint faces should be freshly prepared immediately prior to welding and the area must be thoroughly cleaned.

Corrosion protection

Temporary. None required.

Permanent. None required under most conditions. These alloys can be highly polished, then the surfaces protected against tarnishing with clear lacquer, details of which can be obtained from the Copper Development Association.

Painting. These alloys are used for their attractive appearance, thus should not require painting.

Plating. Normal electroplating techniques give satisfaction and are much used for cutlery and household fittings, the well-known EPNS is 'electroplated nickel silver'.

Machinability

The majority of these alloys have a tendency to tear when machining. High surface speeds and light cuts give best results. All the alloys can be readily polished to give a high lustre.

Some alloys have up to 2% lead added and these exhibit excellent machinability. The alloys with silicon additions may give severe tool wear under certain conditions, but the use of tipped or high speed tools will give satisfaction.

Uses

These alloys are used in the main for their attractive white 'silver' finish. A considerable amount is used as cutlery, when much of it is subsequently silver plated EPNS.

The alloys are also used to produce 'silver' household ornaments, again generally polished then silver plated.

Architectural purposes now use considerable quantities, often as decorative castings. Taps and valves where appearance is important.

Industry finds some use for these alloys as resistance wires – high nickel – and as electrical spring contacts where constant contact resistance is important.

Symbol	Nominal analysis, supplier, condition and remarks.
1N 732X	30% Ni 2.8% Cr Cu alloy: International Nickel Co; annealed
	DPN: 150 **Proof: 340**
2A	Ni Zu Cu alloy: Nickel silver; welding rod; Delta metal for BS1724C
732	4.5% Zn 21% Ni 1.0% Mn Cu alloy: Wrought; designation agreed by CDAA; not used
735	10% Zn 18% Ni 0.5% Mn Cu alloy: Wrought; designation agreed by CDAA; not used
736	12% Zn 15% Ni 0.5% Mn Cu alloy: Wrought; designation agreed by CDAA; not used
740	18% Zn 10% Ni 0.5% Mn Cu alloy: Wrought; designation agreed by CDAA; not used
745	25% Zn 10% Ni 0.5% Mn Cu alloy: Wrought; designation agreed by CDAA
754	21% Zn 15% Ni 0.5% Mn Cu alloy: Wrought; designation used by CDAA
757	23% Zn 12% Ni 0.5% Mn Cu alloy: Wrought; designation used by CDAA
762	18% Zn 12% Ni 0.5% Mn Cu alloy: Wrought; designation agreed by CDAA; not used
764	12% Zn 17% Ni 0.5% Mn Cu alloy: Wrought; designation agreed by CDAA; not used
766	30% Zn 12% Ni 0.5% Mn Cu alloy: Wrought; designation agreed by CDAA; not used
767	28% Zn 15% Ni 0.5% Mn Cu alloy: Wrought; designation agreed by CDAA; not used
773	40% Zn 10% Ni 0.2% P 0.2% Si Cu alloy: Wrought; designation agreed by CDAA; not used
774	40% Zn 10% Ni Cu alloy: Wrought; designation agreed by CDAA; not used
776	42% Zn 13% Ni 0.2% Mn 0.2% Pb Cu alloy: Wrought; designation agreed by CDAA; not used
782	23% Zn 2.0% Pb 8% Ni 0.5% Mn Cu alloy: Wrought; designation agreed by CDAA; not used
784	30% Zn 1.2% Pb 10% Ni 0.5% Mn Cu alloy: Wrought; designation agreed by CDAA; not used
786	30% Zn 1.5% Pb 10% Ni 0.5% Mn Cu alloy: Wrought; designation agreed by CDAA; not used

Note. The following abbreviations and units are used in the tables:

DPN	Hardness, diamond pyramid number
UTS	Ultimate tensile strength, N/mm^2
Elon	Elongation, %
Proof	0.1 % proof strength, N/mm^2

$1\ N/mm^2 = 0.1\ hbar = 0.102\ kgf/mm^2 = 0.06475\ tonf/in^2 = 145.04\ lbf/in^2$

Symbol	Nominal analysis, supplier, condition and remarks.
788	24% Zn 1.7% Pb 10% Ni 0.5% Mn Cu alloy: Wrought; designation agreed by CDAA; not used
790	20% Zn 2.0% Pb 12% Ni 0.5% Mn Cu alloy: Wrought; designation agreed by CDAA; not used
792	22% Zn 1.0% Pb 12% Ni 0.5% Mn Cu alloy: Wrought; designation agreed by CDAA; not used
794	15% Zn 1.0% Pb 18.5% Ni 0.5% Mn Cu alloy: Wrought; designation agreed by CDAA; not used
ALBATRA METAL	20% Zn 20% Ni Cu alloy: May contain 1.25% Pb
ALFENIDE METAL	30% Zn 10% Ni Cu alloy: Origin unknown
AMBRAC (1)	Corrosion resistant alloy 20% Ni 5% Zn Cu alloy: Origin unknown
AMBRAC (1)	20% Ni 5% Zn Cu alloy: For corrosion resistance; Origin unknown
AMBRAC (2)	30% Ni 5% Zn Cu alloy: Tube; origin unknown
ASTM B30/10 A	2% Sn 9% Pb 20% Zn 12% Ni Cu alloy: Ingot; as cast; leaded nickel silver
	UTS: 200 Elon: 8%
ASTM B30/11 A	4% Sn 4% Pb 8% Zn 20% Ni Cu alloy: Ingot; as cast; leaded nickel silver
	UTS: 260 Elon: 10%
ASTM B30/11 B	5% Sn 1.5% Pb 2% Zn 25% Ni Cu alloy: Ingot; as cast; leaded nickel silver
	UTS: 340 Elon: 12%
ASTM B122/1	10% Zn 18% Ni Cu alloy: Sheet; ¼-hard
	UTS: 450
ASTM B122/1	10% Zn 18% Ni Cu alloy: Sheet; extra hard
	UTS: 600
ASTM B122/2	7% Zn 18% Ni Cu alloy: Sheet; ¼-hard
	UTS: 450
ASTM B122/2	17% Zn 18% Ni Cu alloy: Sheet; spring hard
	UTS: 720
ASTM B122/3	24% Zn 10% Ni Cu alloy: Sheet; ¼-hard
	UTS: 460
ASTM B122/3	24% Zn 10% Ni Cu alloy: Sheet; spring hard
	UTS: 740
ASTM B122/4	27% Zn 18% Ni Cu alloy: Sheet; ¼-hard
	UTS: 560
ASTM B122/4	27% Zn 18% Ni Cu alloy: Sheet; spring hard
	UTS: 830
ASTM B122/7	5% Zn 20% Ni Cu alloy: Sheet; ¼-hard
	UTS: 420
ASTM B122/7	5% Zn 20% Ni Cu alloy: Sheet; spring hard
	UTS: 630
ASTM B122/8	29% Zn 12% Ni Cu alloy: Sheet; ¼-hard
	UTS: 530
ASTM B122/8	29% Zn 12% Ni Cu alloy: Sheet; spring hard
	UTS: 840

Symbol	Nominal analysis, supplier, condition and remarks.
ASTM B122/9	20% Zn 10% Ni Cu alloy: Sheet; ½-hard **UTS: 440**
ASTM B122/9	20% Zn 10% Ni Cu alloy: Sheet; extra hard **UTS: 620**
ASTM B124/14	45% Zn 10% Ni Cu alloy: Forging & bar; mechanical properties not part of specification
ASTM B149/10 A	20% Zn 12% Ni 9% Pb 2% Sn Cu alloy: Casting **UTS: 200 Elon: 8% Proof: 75**
ASTM B149/11 A	8% Zn 20% Ni 4% Pb 4% Sn Cu alloy: Casting **UTS: 200 Elon: 8% Proof: 110**
ASTM B149/11 B	2% Zn 25% Ni 1.5% Pb 5% Sn Cu alloy: Casting **UTS: 330 Elon: 15% Proof: 140**
ASTM B151 A	17% Zn 18% Ni Cu alloy: Bar; hard **UTS: 560**
ASTM B151 B	27% Zn 18% Ni Cu alloy: Bar; hard **UTS: 600**
ASTM B151 B1	22% Zn 18% Ni Cu alloy: Bar; hard **UTS: 600**
ASTM B151 C	19% Zn 18% Ni 1% Pb Cu alloy: Bar; hard **UTS: 560**
ASTM B151 D	23% Zn 12% Ni Cu alloy: Bar; hard **UTS: 600**
ASTM B151 E	25% Zn 10% Ni Cu alloy: Bar; hard **UTS: 600**
ASTM B206 A	17% Zn 18% Ni Cu alloy: Wire; hard **UTS: 750**
ASTM B206 B	27% Zn 18% Ni Cu alloy: Wire; spring hard **UTS: 950**
ASTM B206 B1	22% Zn 18% Ni Cu alloy: Wire; spring hard **UTS: 950**
ASTM B206 C	19% Zn 18% Ni Cu alloy: Wire; ½-hard **UTS: 570**
ASTM B206 D	23% Zn 12% Ni Cu alloy: Wire; spring hard **UTS: 950**
ASTM B206 E	25% Zn 10% Ni Cu alloy: Wire; spring hard **UTS: 950**
ASTM B260 RB Cu Zn D	40% Zn 10% Ni 0.2% P 0.2% Si Cu alloy: Braze filler; melting range 920–931 °C
ASTM B292 A	5% Sn 5% Ni 2% Zn Cu alloy: Casting; as cast **UTS: 320 Elon: 25% Proof: 120**
ASTM B292 A	5% Sn 5% Ni 2% Zn Cu alloy: Casting; heat treated **UTS: 540 Elon: 5% Proof: 340**
ASTM B292 B	5% Sn 5% Ni 1% Pb 2% Zn Cu alloy: Casting; as cast **UTS: 180 Elon: 20% Proof: 120**
AWA	Ni Zn Cu alloy: Nickel silver welding rod; Delta
BS 790 NS103	28% Zn 10% Ni Cu alloy: Sheet; annealed **DPN: 100**
BS 790 NS103	28% Zn 10% Ni Cu alloy: Sheet; extra hard **DPN: 185**
BS 790 NS104	26% Zn 12.0% Ni Cu alloy: Sheet; annealed **DPN: 100**
BS 790 NS104	26% Zn 12.0% Ni Cu alloy: Sheet; extra hard **DPN: 190**
BS 790 NS105	25% Zn 15% Ni Cu alloy: Sheet; annealed **DPN: 105**
BS 790 NS105	25% Zn 15% Ni Cu alloy: Sheet; extra hard rolled **DPN: 195**

Symbol	Nominal analysis, supplier, condition and remarks.
BS 790 NS106	20% Zn 18% Ni Cu alloy: Sheet; annealed **DPN: 110**
BS 790 NS106	20% Zn 18% Ni Cu alloy: Sheet; extra hard rolled **DPN: 200**
BS 790 NS108	20% Zn 20% Ni Cu alloy: Sheet; annealed **DPN: 110**
BS 790 NS108	20% Zn 20% Ni Cu alloy: Sheet; extra hard rolled **DPN: 205**
BS 790 NS109	17% Zn 25% Ni Cu alloy: Sheet; annealed **DPN: 115**
BS 790 NS109	17% Zn 25% Ni Cu alloy: Sheet; extra hard rolled **DPN: 210**
BS 790 NS110	12% Zn 30% Ni Cu alloy: Strip; annealed **DPN: 115**
BS 790 NS110	12% Zn 30% Ni Cu alloy: Strip; extra hard rolled **DPN: 210**
BS 1453 C5	40% Zn 10% Ni 0.4% Si Cu alloy: Filler rod; for brazing mild steel cast iron
BS 1453 C6	40% Zn 15% Ni 0.3% Si Cu alloy: Filler rod; for brazing cast iron
BS 1824 NS104	25% Zn 12% Ni Cu alloy: Strip; annealed **DPN: 115**
BS 1824 NS104	25% Zn 12% Ni Cu alloy: Strip; cold rolled **DPN: 220**
BS 1824 NS107	25% Zn 18% Ni Cu alloy: Strip; annealed **DPN: 115**
BS 1824 NS107	25% Zn 18% Ni Cu alloy: Strip; cold rolled **DPN: 220**
BS 2870 NS101	45% Zn 10% Ni 2% Pb 0.5% Mn Cu alloy: Sheet; specification not issued
BS 2870 NS102	42% Zn 14% Ni 2% Pb 2% Mn Cu alloy: Sheet; specification not issued
BS 2870 NS103	25% Zn 10% Ni Cu alloy: Sheet; annealed **DPN: 100**
BS 2870 NS103	25% Zn 10% Ni Cu alloy: Sheet; extra hard **DPN: 185**
BS 2870 NS104	26% Zn 12% Ni Cu alloy: Sheet; annealed **DPN: 100**
BS 2870 NS104	26% Zn 12% Ni Cu alloy: Sheet; extra hard **DPN: 190**
BS 2870 NS105	25% Zn 15% Ni Cu alloy: Sheet; annealed **DPN: 105**
BS 2870 NS105	25% Zn 15% Ni Cu alloy: Sheet; extra hard **DPN: 195**
BS 2870 NS106	20% Zn 18% Ni Cu alloy: Sheet; annealed **DPN: 110**
BS 2870 NS106	20% Zn 18% Ni Cu alloy: Sheet; extra hard **DPN: 200**
BS 2870 NS107	25% Zn 18% Ni Cu alloy: Sheet; soft **DPN: 100**
BS 2870 NS107	24% Zn 18% Ni Cu alloy: Sheet; extra hard **DPN: 220**
BS 2870 NS108	20% Zn 20% Ni Cu alloy: Sheet; annealed **DPN: 110**
BS 2870 NS108	16% Zn 20% Ni Cu alloy: Sheet; ½-hard **DPN: 140**
BS 2870 NS108	16% Zn 20% Ni Cu alloy: Sheet; hard **DPN: 175**
BS 2870 NS108	16% Zn 20% Ni Cu alloy: Sheet; extra hard **DPN: 205**
BS 2870 NS109	17% Zn 25% Ni Cu alloy: Sheet; annealed **DPN: 115**
BS 2870 NS109	17% Zn 25% Ni Cu alloy: Sheet; ½-hard **DPN: 150**
BS 2870 NS109	17% Zn 25% Ni Cu alloy: Sheet; hard **DPN: 180**
BS 2870 NS109	17% Zn 25% Ni Cu alloy: Sheet; extra hard **DPN: 210**
BS 2870 NS110	12% Zn 30% Ni Cu alloy: Sheet

Note. The following abbreviations and units are used in the tables:

DPN	Hardness, diamond pyramid number
UTS	Ultimate tensile strength, N/mm^2
Elon	Elongation, %
Proof	0.1 % proof strength, N/mm^2

1 N/mm^2=0.1 hbar=0.102 kgf/mm^2=0.06475 tonf/in^2=145.04 lbf/in^2

Symbol	Nominal analysis, supplier, condition and remarks.
BS 2870 NS111	30% Zn 10% Ni 1.5% Pb Cu alloy: Sheet; specification not issued
BS 2870 NS112	24% Zn 15% Ni 0.7% Pb Cu alloy: Sheet
BS 2870 NS113	20% Zn 18% Ni 0.6% Pb Cu alloy: Sheet
BS 2871 NS101	40% Zn 10% Ni 2.0% Pb 0.5% Mn Cu alloy: Tube; specification not issued
BS 2871 NS102	40% Zn 14% Ni 2.0% Pb 2% Mn Cu alloy: Tube; specification not issued
BS 2871 NS103	28% Zn 10% Ni Cu alloy: Tube; specification not issued
BS 2871 NS 104	26% Zn 12% Ni Cu alloy: Tube; specification not issued
BS 2871 NS105	25% Zn 15% Ni Cu alloy: Tube; specification not issued
BS 2871 NS106	20% Zn 18% Ni Cu alloy: Tube; specification not issued
BS 2871 NS107	25% Zn 18% Ni Cu alloy: Tube; specification not issued
BS 2871 NS108	20% Zn 20% Ni Cu alloy: Tube; specification not issued
BS 2871 NS109	17% Zn 25% Ni Cu alloy: Tube; specification not issued
BS 2871 NS110	12% Zn 30% Ni Cu alloy: Tube; specification not issued
BS 2872 NS101	40% Zn 10% Ni 2.0% Pb Cu alloy: Forging; as forged **UTS: 450 Elon: 10%**
BS 2872 NS102	40% Zn 14% Ni 2.0% Pb 2% Mn Cu alloy: Forging; specification not issued
BS 2872 NS103	28% Zn 10% Ni Cu alloy: Forging; specification not issued
BS 2872 NS104	26% Zn 12% Ni Cu alloy: Forging; specification not issued
BS 2872 NS105	25% Zn 15% Ni Cu alloy: Forging; specification not issued
BS 2872 NS106	20% Zn 18% Ni Cu alloy: Forging; specification not issued
BS 2872 NS107	25% Zn 18% Ni Cu alloy: Forging; specification not issued
BS 2872 NS108	20% Zn 20% Ni Cu alloy: Forging; specification not issued
BS 2872 NS109	17% Zn 25% Ni Cu alloy: Forging; specification not issued
BS 2872 NS110	12% Zn 30% Ni Cu alloy: Forging; specification not issued
BS 2872 NS111	30% Zn 10% Ni 1.5% Pb Cu alloy: Forging; specification not issued
BS 2872 NS112	24% Zn 15% Ni 0.7% Pb Cu alloy: Forging; specification not issued
BS 2872 NS113	20% Zn 18% Ni 0.5% Pb Cu alloy: Forging; specification not issued
BS 2873 NS101	40% Zn 10% Ni 2% Pb 0.5% Mn Cu alloy: Wire; specification not issued
BS 2873 NS102	40% Zn 14% Ni 2% Pb 2% Mn Cu alloy: Wire; specification not issued
BS 2873 NS103	28% Zn 10% Ni Cu alloy: Wire; mechanical properties not quoted
BS 2873 NS104	26% Zn 12% Ni Cu alloy: Wire; mechanical properties not quoted
BS 2873 NS105	25% Zn 15% Ni Cu alloy: Wire; mechanical properties not quoted
BS 2873 NS106	20% Zn 18% Ni Cu alloy: Wire; specification not issued
BS 2873 NS107	25% Zn 18% Ni Cu alloy: Wire; mechanical properties nor quoted
BS 2873 NS108	20% Zn 20% Ni Cu alloy: Wire; mechanical properties not quoted
BS 2873 NS109	17% Zn 25% Ni Cu alloy: Wire; mechanical properties not quoted
BS 2873 NS110	12% Zn 30% Ni Cu alloy: Wire; mechanical properties not quoted
BS 2873 NS111	30% Zn 10% Ni 1.5% Pb Cu alloy: Wire; specification not issued
BS 2873 NS112	24% Zn 15% Ni 0.8% Pb Cu alloy: Wire; mechanical properties not quoted
BS 2873 NS113	20% Zn 18% Ni 0.6% Pb Cu alloy: Wire; mechanical properties not quoted
BS 2874 NS101	40% Zn 10% Ni 2% Pb 0.5% Mn Cu alloy: Bar; as rolled **UTS: 450 Elon: 10%**
BS 2874 NS102	40% Zn 14% Ni 2% Pb 2% Mn Cu alloy: Bar; as rolled **UTS: 490 Elon: 10%**
BS 2874 NS103	38% Zn 10% Ni Cu alloy: Bar; specification not issued
BS 2874 NS104	26% Zn 12% Ni Cu alloy: Bar; specification not issued
BS 2874 NS105	25% Zn 15% Ni Cu alloy: Bar; specification not issued
BS 2874 NS106	20% Zn 18% Ni Cu alloy: Bar; specification not issued
BS 2874 NS107	25% Zn 18% Ni Cu alloy: Bar; specification not issued
BS 2874 NS108	20% Zn 20% Ni Cu alloy: Bar; specification not issued
BS 2874 NS109	17% Zn 30% Ni Cu alloy: Bar; specification not issued
BS 2874 NS110	12% Zn 30% Ni Cu alloy: Bar; specification not issued
BS 2874 NS111	30% Zn 10% Ni 1.5% Pb Cu alloy: Bar
BS 2874 NS112	24% Zn 15% Ni 0.8% Pb Cu alloy: Bar
BS 2874 NS113	20% Zn 18% Ni 0.6% Pb Cu alloy: Bar
BS 2875 NS101	40% Zn 10% Ni 2% Pb 0.5% Mn Cu alloy: Plate; specification not issued
BS 2875 NS102	40% Zn 14% Ni 2% Pb 2% Mn Cu alloy: Plate; specification not issued
BS 2875 NS103	28% Zn 10% Ni Cu alloy: Plate; specification not issued
BS 2875 NS104	26% Zn 12% Ni Cu alloy: Plate; specification not issued
BS 2875 NS105	25% Zn 15% Ni Cu alloy: Plate; specification not issued
BS 2875 NS106	20% Zn 18% Ni Cu alloy: Plate; specification not issued
BS 2875 NS107	25% Zn 18% Ni Cu alloy: Plate; specification not issued
BS 2875 NS108	20% Zn 20% Ni Cu alloy: Plate; specification not issued
BS 2875 NS109	17% Zn 25% Ni Cu alloy: Plate; specification not issued
BS 2875 NS110	12% Zn 30% Ni Cu alloy: Plate; specification not issued
BS 2875 NS111	30% Zn 10% Ni 1.5% Pb 0.2% Mn Cu alloy: Plate; specification not issued
BS 2875 NS112	24% Zn 15% Ni 0.7% Pb Cu alloy: Plate; specification not issued
BS 2875 NS113	20% Zn 18% Ni 0.6% Pb Cu alloy: Plate; specification not issued
BS T3	Zn Ni Cu alloy: Seamless Ni brass tube; obsolete
C 63	40% Zn 10% Ni 2% Pb Cu alloy: Bar & forging; Manganese Bronze for BS alloy NS 101; Parsons C 63
CHROMAX BRONZE	15.2% Ni 12% Zn 3% Al 3% Cr Cu alloy: Origin unknown
IMI 511	28% Zn 10% Ni Cu alloy: Bar & wire; IMI
IMI 512	26% Zn 12% Ni Cu alloy: Bar & wire; IMI
IMI 513	23% Zn 15% Ni Cu alloy: Bar & wire; IMI
IMI 514	20% Zn 18% Ni Cu alloy: Bar & wire; IMI
IMI 515	18% Zn 20% Ni Cu alloy: Bar & wire; IMI

Symbol	Nominal analysis, supplier, condition and remarks.
IMI 525	40% Zn 10% Ni Cu alloy: Bar; IMI; annealed **DPN: 120 UTS: 450 Elon: 25% Proof: 200**
IMI 530	20% Zn 6% Ni 1.7% Al Cu alloy: Bar; IMI; heat treated **DPN: 200 UTS: 560 Elon: 20% Proof: 330**
IMI 551	38% Zn 10% Ni 2% Pb Cu alloy: Bar; IMI; nickel brass; free machining **DPN: 140 UTS: 480 Elon: 20% Proof: 220**
KUNIAL BRASS	20% Zn 6% Ni 2% Al Cu alloy: Bar; sheet & wire; IMI; solution treated; cold rolled & tempered for springs **DPN: 270 UTS: 810**
NICKEL BRASS	38% Zn 10% Ni Cu alloy: Common name; further information from Copper Development Association
NICKELOID	Trade name for nickel silver: Barker & Allen ltd
NICKEL SILVER	25% Zn 10% Ni Cu alloy: Common name for BS alloys NS 103 to NS 110; further information from Copper Development Association
PARSONS C 63	40% Zn 10% Ni 2% Pb Cu alloy: Bar & forging; Manganese Bronze for BS alloy NS 101
PARSONS SS 15	40% Zn 14% Ni 2.2% Pb Cu alloy: Bar & forging; Manganese Bronze for BS alloy NS 102
QQ N 321/A	20% Zn 18% Ni Cu alloy: Bar & plate; US Federal
QQ N 321/B	20% Zn 18% Ni 1.0% Pb Cu alloy: Bar & sheet; US Federal
SAE CA 752	17% Zn 18% Ni Cu alloy: Wrought
SAE CA 770	27% Zn 18% Ni Cu alloy: Wrought
SCIMITARS	18% Ni Zn Cu alloy: Barker & Allen
SILMET	Trade name for nickel silver: Barker & Allen Ltd
SPEDEX	Trade name for nickel silver: Barker & Allen Ltd
SIS 5243	24% Zn 12% Ni 0.2% Mn Cu alloy: Plate & bar; Swedish Standard; annealed **DPN: 90 UTS: 370 Elon: 40% Proof: 150**
SIS 5243	24% Zn 12% Ni 0.2% Mn Cu alloy: Plate & bar; Swedish Standard; cold rolled **DPN: 190 UTS: 690 Elon: 5% Proof: 480**
SIS 5246	16% Zn 18% Ni 0.5% Mn Cu alloy: Plate; Swedish Standard; annealed **DPN: 95 UTS: 400 Elon: 35% Proof: 170**

Symbol	Nominal analysis, supplier, condition and remarks.
SIS 5246	16% Zn 18% Ni 0.5% Mn Cu alloy: Plate; Swedish Standard; cold rolled **DPN: 150 UTS: 630 Elon: 17% Proof: 550**
SLIDEX	12% Zn Cu alloy: Wire & strip; Jones & Rooke
SS 15	40% Zn 14% Ni 2.2% Pb Cu alloy: Bar & forging; Manganese Bronze for BS alloy NS 102; Parsons SS15
SWM	44% Zn 8.5% Ni 1.5% Pb Cu alloy: Extrusions; McKechnie; as extruded **DPN: 130 UTS: 530 Elon: 20% Proof: 240**
VICTOR METAL	15% Ni 35% Zn Cu alloy: Good machinability; origin unknown
WESSELS ALLOY	1% Ag 15% Zn 25% Ni Cu alloy: Origin unknown
WHITE BENEDICT METAL	16.5% Ni 18% Zn 45% Pb 1% Sn Cu alloy: Origin unknown
WINNS BRONZE	32% Zn 2.2% Ni 0.75% Pb 0.25% Fe Cu alloy: Origin unknown
WM	41% Zn 9.5% Ni 2% Pb Cu alloy: Extrusion; McKechnie; as drawn **DPN: 150 UTS: 530 Elon: 20% Proof: 260**
WM 11	37% Zn 11% Ni 2% Pb Cu alloy: Extrusion; McKechnie; as extruded **DPN: 130 UTS: 530 Elon: 20% Proof: 240**
WMW	43% Zn 8% Ni 1.5% Pb Cu alloy: Extrusion; McKechnie; as drawn **DPN: 150 UTS: 530 Elon: 20% Proof: 260**
WOLFRAM BRASS	4% W 14% Ni 22% Zn Cu alloy: Origin unknown
ZODIAC	20% Ni 16% Zn Cu alloy: Electrical resistance alloy; origin unknown

Note. The following abbreviations and units are used in the tables:

DPN	Hardness, diamond pyramid number
UTS	Ultimate tensile strength, N/mm^2
Elon	Elongation, %
Proof	0.1 % proof strength, N/mm^2

$1 \ N/mm^2 = 0.1 \ hbar = 0.102 \ kgf/mm^2 = 0.06475 \ tonf/in^2 = 145.04 \ lbf/in^2$

14G Copper–aluminium wrought and cast alloys
With iron, manganese, nickel, tin and lead – aluminium bronze

Specific gravity	8.34	
Density	8340 kg/m^3	(0.270–0.298 lb/in.^{3})
Solidus/liquidus	990–1100 °C	
Thermal conductivity	42–88 W/m °C	(10–21 cal/m s °C)
Coefficient of linear expansion	16–19 × 10^{-6}/ °C	
Electrical conductivity	6–15% IACS (copper 100%)	
Specific resistance	120–240 microhm mm	
Young's modulus of elasticity	103–110 × 10^9 N/m^2	(15–16 × 10^6 lbf/in.^{2})
Impact	annealed 34 J	(25 ft/lb)
	heat treated 20 J	(15 ft/lb)
Fatigue strength (30 × 10^7 cycles)	hard drawn ± 150 N/mm^2	(± 10 tonf/in.^{2})

Hot strength

Temperature °C	Tensile strength N/mm^2
95	600
205	500
315	500
425	260
540	140

The above properties are typical of the following group, and may not apply exactly to any one specification. It is possible that with certain specifications some of the values may not be applicable.

General metallurgical characteristics

These alloys with up to 11% aluminium all have excellent corrosion resistance, including good oxidation resistance at elevated temperature particularly when hot, caused by the formation of a thin surface layer of aluminium oxide.

There are two basic series of alloys:

Up to 5% aluminium. These have an alpha structure.

With 7-11% aluminium. These have an alpha structure, with a second phase, beta, present at room temperature. This latter increases the strength but decreases the ductility. A third phase, 'gamma two' may be present if the alloys are cooled very slowly and this further decreases the ductility.

All the alloys generally have one or more of the following elements:

Iron. This strengthens the alpha phase with little effect on the corrosion resistance up to 2% iron. It also helps to refine the structure of high aluminium alloys.

Manganese. Small quantities are used to desulphurize during casting. Larger quantities help to strengthen the alloy, and also may improve the corrosion resistance.

Nickel. This strengthens the alpha phase, improves the corrosion resistance and like iron inhibits the formation of the undesirable 'gamma two' phase.

Lead. This is added to improve machinability. It reduces to some extent the ductility and corrosion resistance.

All these alloys maintain their properties at relatively high temperatures – up to 300–400 °C – being the best copper alloys for this purpose.

The corrosion resistance is also excellent at these temperatures.

The lower aluminium alloys can be readily cold and hot worked, while alloys with 9% or more of aluminium are not always suitable for cold working but are readily hot worked. The structure of these high aluminium alloys above approximately 550 °C is alpha plus beta. The beta however, may break down to the brittle 'gamma two' phase below 550 °C. This can be prevented by rapid cooling from above 550 °C, when the beta phase will remain.

Thermal treatment

Alpha alloys – up to 5% aluminium – cannot be thermal treated, except to anneal after cold work. This should be carried out at about 650 °C for as short a time as possible to prevent grain growth and will soften the material, which can then only be re-hardened by further cold work. The duplex alloys, those with 7–11% aluminium, cannot be readily cold worked, but can have their mechanical properties improved to a slight extent by heat treatment. This consists of quenching from 850 °C and ensures the presence of alpha plus beta phase, and the absence of the undesirable 'gamma two' phase. The alloys are used in this condition when they have better mechanical properties than the alpha alloys. A marginal improvement in strength at the expense of ductility can be obtained by reheating to above 400 °C when the beta phase breaks down to 'gamma two'. Provided this is present as a very fine dispersion the increased strength will be obtained with a minimum reduction in ductility.

Considerable skill and the correct alloys are necessary to obtain the best results.

Welding and brazing

Pre-treatment. It is important that the joint preparation is carefully carried out as soon as possible before welding or brazing. Absolute cleanliness is essential.

Welding and brazing. The ready formation of the aluminium oxide film makes brazing and soldering very difficult without the use of very corrosive fluxes and considerable skill.

Welding, using fluxes or inert gas shielded arc can be carried out. Care in the choice of rod is important to keep the properties of the weld area as high as possible. Preheating of the lower aluminium alloys may be necessary, but as the aluminium content increases the thermal conductivity drops.

Post-treatment. All traces of flux should be removed. The high aluminium alloys should be heated to above 580 °C and quenched to eliminate any brittle 'gamma two' phase which will have been formed around the weld or braze area. Stress release at 300 °C is recommended with the alpha alloys, if the assembly is of a complex shape.

Flaw detection

Crack test. Penetrant oil chalk test method.

X-ray. This is advisable on large or expensive castings, as this material is prone to form 'hot shuts' where turbulence during casting causes oxide formation preventing two streams of metal intermingling.

Ultra-sonic test. This may be required on large forgings before expensive machining takes place.

Chemical etch. May be required to identify large grain. Ferric chloride in 10% hydrochloric acid used hot is satisfactory. This will etch the surface and result in a reduction in fatigue strength unless removed by polishing or machining.

Corrosion protection

Temporary. None required.

Permanent. None required normally. Where required the Copper Development Association has information on lacquers for most conditions.

Painting. Not required, except as above. These alloys are used for their corrosion and temperature resistance.

Plating. The alloys can all be readily electroplated using appropriate modified preparation and plating techniques, but it is seldom that this will be required.

Machinability

The alpha alloys are readily machined to give a good surface finish.

The duplex alloys with more than 7% aluminium are harder and may have some undesirable 'gamma two' phase present in addition to the beta phase. Tool wear can be high, and tipped or high speed tools are recommended.

Uses

These alloys are often used when conditions are not quite severe enough to warrant austenitic stainless steels, or nickel copper alloys.

The alloys have good strength and excellent corrosion, oxidation, wear and abrasion resistance.

Gears for pumps, impellers, control rods and shafts, nuts and bolts, turbine and compressor blading, acid handling equipment.

Symbol	Nominal analysis, supplier, condition and remarks.
1A	9% Al 3% Fe + Ni (max) Cu alloy: Bar & forging; Langley Alloy for BS alloy CA103
2.0917	6% Al 0.4% Ni 1.0% Mn Cu alloy: Wrought; German Standard
2.0928	9.5% Al Cu alloy: Casting; German Standard
2.0929	9% Al Cu alloy: Casting; German Standard
2.0940	10% Al 3% Fe Cu alloy: Casting; German Standard
2.0941	10% Al 3.0% Fe Cu alloy: Casting; German Standard
2.0962	8% Al 6% Mn 1.5% Ni Cu alloy: Casting; German Standard

Symbol	Nominal analysis, supplier, condition and remarks.
2.0966	5.5% Ni 9.5% Al 2.0% Fe 1.0% Mn Cu alloy: German Werkstoff number
2.0970	8% Al 5% Ni 5% Fe Cu alloy: Casting; German Standard
2.0971	9% Al 5% Ni 5% Fe Cu alloy: Casting; German Standard
2.0975	9% Al 5% Fe 5% Ni Cu alloy: Casting; German Standard
2.0980	10% Al 6% Fe 6% Ni Cu alloy: Casting; German Standard
2A	9% Al Cu alloy: Bar & forging; Langley Alloys code for Hidurax 5
3A	9.5% Al 2% Ni 2% Fe Cu alloy: Forging; Langley Alloys code for Hidurax 2
4A	9.5% Al 2% Fe Cu alloy: Casting; Langley Alloys code for Hidurax 2
6A	10% Al 5% Fe 5% Ni Cu alloy: Bar & forging; Langley Alloys code for Hidurax 1
7A	10% Al 4.5% Ni 4.5% Fe Cu alloy: Casting; Langley Alloys code for Hidurax 1
10A	10% Al 5% Ni 5% Fe 2% Mn Cu alloy: Bar & forging; Langley Alloys
	UTS: 750 Elon: 20% Proof: 450

Note. The following abbreviations and units are used in the tables:

DPN	Hardness, diamond pyramid number
UTS	Ultimate tensile strength, N/mm^2
Elon	Elongation, %
Proof	0.1 % proof strength, N/mm^2

1 N/mm^2=0.1 hbar=0.102 kgf/mm^2=0.06475 tonf/in^2=145.04 lbf/in^2

Symbol	Nominal analysis, supplier, condition and remarks.
18A	11% Al 4% Fe 4% Ni Cu alloy: Casting; Langley Alloys UTS: 740 Elon: 15% Proof: 400
21A	10% Al 3% Fe 5% Ni 1% Mn Cu alloy: Bar; Langley Alloys UTS: 600 Elon: 10% Proof: 300
24A	5% Al Cu alloy: Bar & forging; Langley Alloys code for Hidurax 6
26A	10% Al 2% Fe 5% Ni 1% Mn Cu alloy: Wrought; Langley Alloys
41A	7% Al 2% Si Cu alloy: Wrought; Langley Alloys code for Hidurax 7
75	11.5% Al 5% Fe 1% Mn 5% Ni Cu alloy: Extrusion; McKechnie Bros; as extruded DPN: 250 UTS: 860 Elon: 8% Proof: 450
160	9.5% Al Cu alloy: Extrusions; McKechnie Bros; as drawn DPN: 150 UTS: 530 Elon: 25% Proof: 240
164	9.3% Al 1.5% Fe 1.5% Ni Cu alloy: Extrusions; McKechnie Bros; as drawn DPN: 170 UTS: 580 Elon: 18% Proof: 300
197	10% Al 4.5% Fe 4.5% Ni Cu alloy: Extrusion; McKechnie Bros; as drawn DPN: 220 UTS: 750 Elon: 15% Proof: 420
606	5.5% Al Cu alloy: Wrought; designation agreed by CDAA; not used
607	1.8% Zn 2.6% Al Cu alloy: Wrought; designation agreed by CDAA; not used
608	5.8% Al 0.3% As Cu alloy: Wrought; designation agreed by CDAA; not used
610	8% Al Cu alloy: Wrought; designation agreed by CDAA; not used
612	8% Al Cu alloy: Wrought; designation agreed by CDAA; not used
613	7% Al 0.4% Sn 3.5% Fe Cu alloy: Wrought; designation agreed by CDAA; not used
614	7% Al 3.0% Fe 1.0% Mn Cu alloy: Wrought; designation used by CDAA
616	9% Al 4.0% Fe 1.0% Ni 1.0% Zn 1.5% Mn Cu alloy: Wrought; designation agreed by CDAA; not used
618	10% Al 1.5% Fe Cu alloy: Wrought; designation agreed by CDAA; not used
620	10% Al 3.5% Fe Cu alloy: Wrought; designation agreed by CDAA; not used
622	11% Al 4.0% Fe Cu alloy: Wrought; designation agreed by CDAA; not used
626	10.3% Al 3.8% Ni 3.0% Fe 1.5% Mn Cu alloy: Wrought; designation agreed by CDAA; not used
628	9.5% Al 5.5% Ni 2.2% Fe Cu alloy: Wrought; designation agreed by CDAA; not used
634	8.8% Al Cu alloy: Extrusion; McKechnie Bros; as drawn DPN: 150 UTS: 530 Elon: 40% Proof: 220
639	7.2% Al 0.2% Ni 2.2% Si Cu alloy: Wrought; designation agreed by CDAA; not used
642	8% Al 4.0% Fe 1.0% Zn 2% Si Cu alloy: Wrought; designation agreed by CDAA; not used
705	7.0% Al 0.15% Mn Cu alloy: Wrought; designation agreed by CDAA; not used
952	9% Al 3% Fe Cu alloy: Casting; designation used by CDAA
953	10% Al 1% Fe Cu alloy: Casting; designation used by CDAA
954	11% Al 4% Fe Cu alloy: Casting; designation used by CDAA
955	4% Ni 11% Al 4% Fe Cu alloy: Casting; designation used by CDAA

Symbol	Nominal analysis, supplier, condition and remarks.
AB	9.8% Al 1.5% Pb Cu alloy: Extrusion; McKechnie Bros; as drawn DPN: 200 UTS: 550 Elon: 15% Proof: 260
AB 164	9.5% Al 2% Ni 2% Fe Cu alloy: Bar & forging; Manganese Bronze Ltd; as Al Bronze 164
AB 197	9.5% Al 5% Ni 5% Fe Cu alloy: Bar & forging; Manganese Bronze Ltd; as Al Bronze 197
ABS TYPE 4	10% Al 4.0% Fe 4.2% Ni Cu alloy: Casting; American Bureau of Shipping UTS: 600 Elon: 15% Proof: 245
ABS TYPE 5	12.5% Mn 7.8% Al 2.2% Ni 3.0% Fe Cu alloy: Casting; American Bureau of Shipping UTS: 635 Elon: 20% Proof: 280
AL 1	9.4% Al Cu alloy: Wrought; Delta Metal
AL BRONZE 184	9.5% Al 2% Ni 2% Fe Cu alloy: Bar & forging; Manganese Bronze Ltd; as DTD 164 DPN: 170 UTS: 550 Elon: 17% Proof: 260
AL BRONZE 197	9.5% Al 5% Ni 5% Fe Cu alloy: Bar & forging; Manganese Bronze Ltd; as DTD 197 DPN: 220 UTS: 720 Elon: 15% Proof: 400
Al Bz 5	5% Al 0.3% As Cu alloy: Wrought; designation used by German Standard
AMPCO 8	6.5% Al 2.5% Fe 0.25% Sn Cu alloy: Wrought; Ampco UTS: 570 Elon: 35% Proof: 310
AMPCO 12	8.9% Al 2.9% Fe Cu alloy: Casting; Ampco DPN: 130 UTS: 550 Elon: 37% Proof: 220
AMPCO 15	9.3% Al 3.1% Fe Cu alloy: Extrusion; Ampco DPN: 180 UTS: 670 Elon: 25% Proof: 330
AMPCO 16	10.1% Al 3.3% Fe Cu alloy: Casting; Ampco DPN: 160 UTS: 650 Elon: 23% Proof: 230
AMPCO 18	10.5% Al 3.5% Fe Cu alloy: Wrought & cast; Ampco DPN: 190 UTS: 720 Elon: 15% Proof: 270
AMPCO 18/13	10.5% Al 3.5% Fe Cu alloy: Casting; Ampco DPN: 175 UTS: 700 Elon: 20% Proof: 250
AMPCO 18/22	10.5% Al 3.5% Fe Cu alloy: Casting; Ampco DPN: 225 UTS: 750 Elon: 9% Proof: 380
AMPCO 18/23	10.5% Al 3.5% Fe Cu alloy: Casting; Ampco DPN: 205 UTS: 750 Elon: 15% Proof: 380
AMPCO 20	11.3% Al 3.8% Fe Cu alloy: Casting; Ampco DPN: 220 UTS: 630 Elon: 6% Proof: 270
AMPCO 20/13	11.3% Al 3.8% Fe Cu alloy: Casting; Ampco DPN: 205 UTS: 610 Elon: 8% Proof: 240
AMPCO 21	13.1% Al 4.4% Fe Cu alloy: Wrought & cast; Ampco DPN: 290 UTS: 610 Elon: 2% Proof: 390
AMPCO 22	14.1% Al 4.7% Fe Cu alloy: Casting; Ampco DPN: 330 UTS: 580 Elon: 0.5% Proof: 490
AMPCO 24	Al Cu alloy: For dies; Ampco DPN: 350
AMPCO 25	Al Cu alloy: For dies; Ampco DPN: 360
AMPCOLAY 45	10% Al 3% Fe 5% Ni 1% Mn Cu alloy: Wrought; Ampco
AMPCOLAY 405	8% Al 4% Fe 1% Ni 2% Si 1% Mn Cu alloy: Wrought; Ampco
AMPCOLAY 483	Al Ni Fe Cu alloy: Ampco
AMPCOLAY 495	8% Al 12% Mn 3% Fe 2% Ni Cu alloy: Forging; Ampco
AMPCOLAY 570	10% Al 15% Ni 1.5% Co 0.7% Fe Cu alloy: Casting; Ampco
AMPCOLAY A1	9% Al 3% Fe Cu alloy: Casting; Ampco
AMPCOLAY B2	10% Al 1% Fe Cu alloy: Casting; Ampco
AMPCOLAY B2 (wrought)	8% Al 4% Fe 1% Ni 2% Si 1% Mn Cu alloy: Wrought; Ampco
AMPCOLAY C3	11% Al 4% Fe Cu alloy: Casting; Ampco
AMPCOLAY D4	11% Al 4% Fe 4% Ni Cu alloy: Casting; Ampco
AMPCOLAY E5	8% Al 4% Fe 1% Ni 2% Si 1% Mn Cu alloy: Wrought; Ampco

Symbol	Nominal analysis, supplier, condition and remarks.
AMS 4630 E	8.5% Al Cu alloy: Tube; annealed
AMS 4631 C	7.5% Al 2% Si Cu alloy: Bar & forging
AMS 4632 C	8.5% Al Cu alloy: Bar; hard drawn
AMS 4635 B	10% Al 3% Fe Cu alloy: Bar & forging
AMS 4640 C	10.5% Al 5% Ni 2.5% Fe Cu alloy: Bar & forging
AMS 4870 B	11% Al 3.6% Fe Cu alloy: Casting; chill cast; as cast
AMS 4871 B	11% Al 3.6% Fe Cu alloy: Casting; chill cast; heat treated
AMS 4872 B	11% Al 3.6% Fe Cu alloy: Sand casting; as cast
AMS 4873 A	11% Al 3.6% Fe Cu alloy: Sand casting; heat treated
AMS 4880	10.5% Al 5% Ni 2.5% Fe Cu alloy: Centri-casting; heat treated
AN B 16	8.0% Al 2.2% Si 5.0% Ni Cu alloy: Bar; US Service
	UTS: 600 Elon: 17% Proof: 220
AN QQ B 672	11% Al 5.0% Mn 5.0% Ni Cu alloy: Casting; US Service
	UTS: 580 Elon: 3%
ASTM B30/9 A	9% Al 3% Fe Cu alloy: Ingot; as cast
	UTS: 460 Elon: 20%
ASTM B30/9 B	10% Al 1% Fe Cu alloy: Ingot; heat treated
	UTS: 580 Elon: 12%
ASTM B30/9 C	11% Al 4% Fe Cu alloy: Ingot; heat treated
	UTS: 660 Elon: 6%
ASTM B30/9 D	11% Al 4% Fe 4% Ni Cu alloy: Ingot; heat treated
	UTS: 800 Elon: 5%
ASTM B111/608	5.5% Al Cu alloy: For condenser tubes
ASTM B111/608	5.5% Al Cu alloy: Seamless tube
ASTM B111/687	2.2% Al Cu alloy: Seamless tube
ASTM B111 Al BRONZE	6% Al Cu alloy: Tube
ASTM B124/11 A	8% Al 1.5% Mn 1% Ni Cu alloy: Forging & bar; mechanical properties not quoted
ASTM B124/11 B	10% Al 1.5% Mn 5% Ni 3% Fe Cu alloy: Forging & bar; mechanical properties not quoted
ASTM B148/9 A	9% Al 3% Fe Cu alloy: Casting; as cast
	DPN: 110 UTS: 460 Elon: 20% Proof: 170
ASTM B148/9 B	10% Al 1% Fe Cu alloy: Casting; heat treated
	DPN: 160 UTS: 580 Elon: 12% Proof: 260
ASTM B148/9 C	11% Al 4% Fe Cu alloy: Casting; heat treated
	DPN: 190 UTS: 660 Elon: 6% Proof: 330
ASTM B148/9 D	11% Al 4% Fe 4% Ni Cu alloy: Casting; heat treated
	DPN: 200 UTS: 800 Elon: 5% Proof: 420
ASTM B150/1	8% Al 4% Fe 1% Ni 2% Si 1% Mn Cu alloy: Bar
	UTS: 530 Elon: 10% Proof: 240
ASTM B150/2	10% Al 3% Fe 5% Ni 1% Mn Cu alloy: Bar
	UTS: 690 Elon: 6% Proof: 330
ASTM B150/3	7% Al 2% Fe Cu alloy: Bar
	UTS: 530 Elon: 30% Proof: 240
ASTM B150/614	7.0% Al 2.5% Fe Cu alloy: Rod; previously alloy No 3
ASTM B150/630	10.0% Al 3.0% Fe 4.7% Ni Cu alloy: Rod; previously alloy No 2
ASTM B150/642	8.5% Al 4.0% Fe (max) 1.0% Ni (max) Cu alloy: Rod; previous alloy No 1

Note. The following abbreviations and units are used in the tables:

DPN	Hardness, diamond pyramid number
UTS	Ultimate tensile strength, N/mm^2
Elon	Elongation, %
Proof	0.1 % proof strength, N/mm^2

1 N/mm^2=0.1 hbar=0.102 kgf/mm^2=0.06475 tonf/in^2=145.04 lbf/in^2

Symbol	Nominal analysis, supplier, condition and remarks.
ASTM B169/612	8% Al Cu alloy: Al bronze; sheet & bar
ASTM B169/612	7% Al 2.5% Fe Cu alloy: Al bronze; sheet & bar
ASTM B169 A	5.5% Al Cu alloy: Sheet & bar; annealed
	UTS: 330 Elon: 40% Proof: 110
ASTM B169 A	5.5% Al Cu alloy: Sheet & bar; hard drawn
	UTS: 360 Elon: 30% Proof: 120
ASTM B169 C	8% Al Cu alloy: Sheet & bar; annealed
	UTS: 390 Elon: 25% Proof: 150
ASTM B169 C	8% Al Cu alloy: Sheet & bar; hard drawn
	UTS: 400 Elon: 22% Proof: 170
ASTM B169 D	7% Al 2% Fe Cu alloy: Sheet & bar; annealed
	UTS: 510 Elon: 33% Proof: 240
ASTM B169 D	7% Al 2% Fe Cu alloy: Sheet & bar; hard drawn
	UTS: 580 Elon: 32% Proof: 330
ASTM B171/464	0.75% Sn 40% Zn Cu alloy: Rolled plate
ASTM B171/614	2.5% Sn 7% Al 2.5% Fe Cu alloy: Rolled plate
ASTM B171/628	5.5% Ni 1.2% Mn 2.5% Fe 10.5% Al Cu alloy: Plate
ASTM B225 E Cu Al A 1	7% Al Cu alloy: Welding electrode
	UTS: 390
ASTM B225 E Cu Al A 2	10% Al Cu alloy: Welding electrode
	UTS: 420
ASTM B225 E Cu Al B	11% Al 4% Fe Cu alloy: Welding electrode
	UTS: 460
ASTM B259 R Cu Al A 2	10% Al Cu alloy: Welding rod
	UTS: 450
ASTM B259 R Cu Al B	11% Al 4% Fe Cu alloy: Welding rod
	UTS: 470
ASTM B359/608	5.5% Al Cu alloy: Finned tube
ASTM B359/687	2.2% Al Cu alloy: Finned tube
B 29	9.5% Al 2.5% Fe 1% Mn Cu alloy: Casting; Anti-Attrition for BS alloy AB1
B 31	0.5% Zn 9.5% Al 5.5% Ni 4.5% Fe 1.5% Mn Cu alloy: Casting; Anti-Attrition for BS alloy AB2
BATTERIUM	9% Al 1% Ni Cu alloy: Batterium Metal Ltd
	DPN: 168 UTS: 730 Elon: 48%
BOLTOMET 803	9.25% Al Cu alloy: T Bolton for BS alloy CA103
BOLTOMET 807	9.5% Al 5% Ni 5% Fe Cu alloy: T Bolton for BS alloy CA104
BS 378 CA102	7% Al Fe & Mn 2.5% Cu alloy: Tubes; as drawn for condenser tubes 150
BS 1031	Aluminium bronze ingots & castings; replaced by BS 1400
BS 1032	Aluminium bronze ingots & castings; replaced by BS 1400
BS 1072	Al bronze ingots & castings; high tensile; replaced by BS 1400
BS 1073	Al bronze ingots & castings; high tensile; replaced by BS 1400
BS 1400 AB1 C	9.5% Al 2% Fe Cu alloy: Aluminium bronze castings; chill cast
	UTS: 530 Elon: 20% Proof: 200
BS 1400 AB2 1	9.5% Al 4.5% Fe 5% Ni Cu alloy: Aluminium bronze ingots; chill cast
	UTS: 630 Elon: 15% Proof: 240
BS 1400 AB2 C	9.5% Al 4.5% Fe 5% Ni Cu alloy: Aluminium bronze castings; chill cast
	UTS: 630 Elon: 15% Proof: 240
BS 1400 CMA1	8% Al 3% Fe 3% Ni 1% Sn 12% Mn Cu alloy: Ingots; chill cast
	UTS: 650 Elon: 30% Proof: 300
BS 1400 CMA1C	8% Al 3% Fe 3% Ni 1% Sn 12% Mn Cu alloy: Castings; chill cast
	UTS: 650 Elon: 30% Proof: 300
BS 1400 CMA2 1	8.75% Al 3% Fe 3% Ni 1% Sn 12% Mn Cu alloy: Ingots; chill cast
	UTS: 730 Elon: 10% Proof: 370
BS 1400 CMA2C	8.75% Al 3% Fe 3% Ni 1% Sn 12% Mn Cu alloy: Castings; sand cast
	UTS: 730 Elon: 10% Proof: 370

Symbol	Nominal analysis, supplier, condition and remarks.
BS 1464 CA102	7% Al Cu tube; aluminium bronze; Mn Ni & Fe may be present to a total of 2.5% **DPN: 120**
BS 1541 CA102	7% Al Cu alloy: Plate; aluminium bronze; as rolled; Fe, Mn & Ni may be present up to 2.5% **UTS: 450 Elon: 35%**
BS 1541 CA105	10.0% Al 2.5% Fe 5.0% Ni 1.0% Mn Cu alloy: Plate; as rolled **UTS: 580 Elon: 12%**
BS 1541 CA106	7% Al 3% Fe Cu alloy: Plate; as rolled **UTS: 450 Elon: 35%**
BS 1867	7% Al 2% Ni + Fe & Mn Cu alloy: Tube; annealed **UTS: 400 Elon: 50%**
BS 2032 CA103	9% Al 3% Fe + Ni max Cu alloy: Bar & forging; annealed **UTS: 510 Elon: 15% Proof: 180**
BS 2033 CA104	10% Al 5% Fe 5% Ni Cu alloy: Bar & forging; as rolled or forged **UTS: 690 Elon: 15% Proof: 370**
BS 2870 CA101	5% Al Cu alloy: Sheet; as rolled; aluminium bronze **UTS: 330 Elon: 40%**
BS 2870 CA102	7% Al Cu alloy: Sheet; Ni, Fe & Mn up to 2.5%; aluminium bronze; specification not issued
BS 2870 CA103	9% Al 3% Fe & Ni Cu alloy: Sheet; specification not issued
BS 2870 CA104	10% Al 5% Ni 5% Fe Cu alloy: Sheet; specification not issued
BS 2870 CA105	10% Al 5% Ni 2.5% Fe 1.0% Mn Cu alloy: Sheet; specification not issued
BS 2870 CA106	7% Al 3% Fe Cu alloy: Sheet; specification not issued
BS 2871 CA101	5% Al Cu alloy: Tube; aluminium bronze; annealed **UTS: 390**
BS 2871 CA102	7% Al Cu alloy: Tube; aluminium bronze; annealed; Ni & Fe & Mn 2.5% **UTS: 420**
BS 2871 CA103	9% Al 3% Fe & Ni Cu alloy: Tube; aluminium bronze; specification not issued
BS 2871 CA104	10% Al 5% Fe 5% Ni Cu alloy: Tube; aluminium bronze; specification not issued
BS 2872 CA101	5% Al Cu alloy: Forging; specification not issued
BS 2872 CA102	7% Al 2.5% Ni & Fe & Mn Cu alloy: Forging; specification not issued
BS 2872 CA103	9% Al 3% Fe & Ni Cu alloy: Forging; as forged **UTS: 510 Elon: 25% Proof: 220**
BS 2872 CA104	10% Al 5% Fe 5% Ni Cu alloy: Forging; as forged **UTS: 690 Elon: 15% Proof: 370**
BS 2872 CA105	10% Al 2.5% Fe 5% Ni 1% Mn Cu alloy: Forging; specification not issued
BS 2872 CA106	7% Al 3.0% Fe Cu alloy: Forging; as forged **UTS: 520 Elon: 35% Proof: 210**
BS 2873 CA101	5% Al Cu alloy: Wire; specification not issued
BS 2873 CA102	7% Al 2.5% Ni + Fe & Mn Cu alloy: Wire; specification not issued
BS 2873 CA103	9% Al 3% Fe + Ni Cu alloy: Wire; specification not issued
BS 2873 CA104	10% Al 5% Ni 5% Fe Cu alloy: Wire; specification not issued
BS 2873 CA105	10% Al 1.0% Mn 5.0% Ni 2.5% Fe Cu alloy: Wire; specification not issued
BS 2873 CA106	7% Al 3% Fe Cu alloy: Wire; specification not issued
BS 2874 CA101	5% Al Cu alloy: Bar; specification not issued
BS 2874 CA102	7% Al 2.5% Ni + Fe & Mn Cu alloy: Bar; specification not issued
BS 2874 CA103	9.0% Al 3% Fe + Ni Cu alloy: Bar; annealed **UTS: 510 Elon: 15% Proof: 180**
BS 2874 CA104	10% Al 5% Ni 5% Fe Cu alloy: Bar; as rolled **UTS: 660 Elon: 15% Proof: 330**
BS 2874 CA105	10% Al 5% Ni 2.5% Fe 1% Mn Cu alloy: Bar; specification not issued
BS 2874 CA106	7% Al 3% Fe Cu alloy: Bar; annealed **UTS: 510 Elon: 35% Proof: 210**
BS2875 CA101	5% Al Cu alloy: Plate; specification not issued
BS 2875 CA102	7% Al 2.5% Fe + Mn + Ni Cu alloy: Plate; as rolled **UTS: 450 Elon: 35%**
BS 2875 CA103	9% Al 3% Fe + Ni Cu alloy: Plate; specification not issued
BS 2875 CA104	10% Al 5% Fe 5% Ni Cu alloy: Plate; specification not issued
BS 2875 CA105	10% Al 5% Ni 2.5% Fe 1% Mn Cu alloy: Plate; as rolled **UTS: 580 Elon: 12%**
BS 2875 CA106	7% Al 3% Fe Cu alloy: Plate; as rolled **UTS: 450 Elon: 35%**
BS 2901 C12	7% Al 2% Fe + Ni + Mn Cu alloy: Rod for all welding
BS 2901 C13	10% Al Cu alloy: Rod for all welding
BS 2901 C20	9% Al 2% Fe 5% Ni 1% Mn Cu alloy: Rod for gas & electric welding
BS 4577 A4/4	10% Al 5% Fe 5% Ni Cu alloy: For resistance welding electrodes; wrought; conductivity 10% IACS **DPN: 190 UTS: 700 Elon: 13%**
CROTORITE	Al bronze casting; Manganese Bronze Ltd for DTD 160 DTD 174
CROTORITE IV	10% Al + Ni + Mn Cu alloy: Manganese Bronze Ltd
CROTORITE V	9% Al 3% Fe + Ni Cu alloy: Bar & forging; Manganese Bronze Ltd for BS alloy CA103; low Fe
CROTORITE Z	9% Al 3% Fe + Ni Cu alloy: Bar & forging; Manganese Bronze Ltd for BS alloy CZ103
DGS 8451	9% Al 3% Fe + Ni Cu alloy: Bar & forging; Admiralty for BS alloy CA103; heat treated; further information from the Copper Development Association **DPN: 150 UTS: 550 Elon: 35 Proof: 280**
DIN 1714	9.5% Al Cu alloy: Casting
G Al Bz 9	**DPN: 110 UTS: 420 Elon: 25% Proof: 170**
DIN 1714	10% Al 3% Fe Cu alloy: Casting
G Fe Al Bz F 50	**DPN: 135 UTS: 530 Elon: 20% Proof: 210**
DIN 1714	8% Al 6% Mn 1.5% Cu alloy: Casting
G Mn Al Bz F 42	**DPN: 120 UTS: 510 Elon: 26% Proof: 220**
DIN 1714	8% Al 5% Ni 5% Fe Cu alloy: Casting
G Nl Al Bz F 50	**DPN: 150 UTS: 580 Elon: 25% Proof: 240**
DIN 1714	9% Al 5% Fe 5% Ni Cu alloy: Casting
G Ni Al Bz F 60	**DPN: 170 UTS: 630 Elon: 18% Proof: 330**
DIN 1714	10% Al 6% Fe 6% Ni Cu alloy: Casting
G Ni Al Bz F 68	**DPN: 190 UTS: 750 Elon: 8% Proof: 370**
DIN 1714	9% Al 5% Fe 5% Ni Cu alloy: Casting
GZ Ni Al Bz F 70	**DPN: 180 UTS: 750 Elon: 16% Proof: 360**
DIN 1733 S Al Bz 6	6% Al 0.4% Ni 1.0% Mn Cu alloy: Wrought
DIN 1733 S Al Bz 8	8% Al 0.4% Ni 1.0% Mn Cu alloy: Wrought
DIN 1785 Al Bz 5	5% Al 0.3% As Cu alloy: Wrought
DIN 17656	9% Al Cu alloy: Casting
GB Al Bz 9	
DIN 17656	10% Al 3.0% Fe Cu alloy: Casting
GB Fe Al Bz	
DIN 17656	9% Al 5% Ni 5% Fe Cu alloy: Casting
GB Ni Al Bz	

Symbol	Nominal analysis, supplier, condition and remarks.
DNC M 38	9% Al 4.5% Fe 1.5% Mn 5% Ni Cu alloy: Casting; Admiralty specification for BS alloy AB2.C; further information from the Copper Development Association
DTD 160	9.5% Al Cu alloy: Bar; tempered & quenched; for valve seats **DPN: 150 UTS: 530**
DTD 164 A	9.5% Al 2% Ni 2% Fe Cu alloy: Forgings; tempered **UTS: 560 Elon: 17% Proof: 270**
DTD 174 A	8% Al 2% Fe 3% Mn 3% Ni Cu alloy: Casting **UTS: 480 Elon: 20%**
DTD 197 A	9.5% Al 5% Ni 5% Fe Cu alloy: Bar & forging; as rolled or forged **UTS: 650 Elon: 15% Proof: 330**
DTD 412	10% Al 4.5% Ni 4.5% Fe Cu alloy: Casting; as cast **UTS: 600 Elon: 12% Proof: 240**
E 86	7% Al 2% Si Cu alloy: Extrusion; McKechnie Bros; as drawn **DPN: 150 UTS: 580 Elon: 35% Proof: 300**
E in C No. C 106	9% Al 4.5% Fe 1.5% Mn 5% Ni Cu alloy: Casting; Admiralty for BS alloy AB 2 C; further information from the Copper Development Association
G Cu Al 8 Mn	8% Al 6% Mn 1.5% Ni Cu alloy: Casting; designation used by German Standard
G Cu Al 9	9.5% Al Cu alloy: Casting; designation used by German Standard
G Cu Al 9 Ni	8% Al 5% Ni 5% Fe Cu alloy: Casting; designation used by German Standard
G Cu Al 10 Fe	10% Al 3% Fe Cu alloy: Casting; designation used by German Standard
G Cu Al 10 Ni	9% Al 5% Fe 5% Ni Cu alloy: Casting; designation used by German Standard
G Cu Al 11 Ni	10% Al 6% Fe 6% Ni Cu alloy: Casting; designation used by German Standard
G Fe Al Bz F 50	10% Al 3% Fe Cu alloy: Casting; designation used by German Standard
GB Cu Al 9	9% Al Cu alloy: Casting; designation used by German Standard
GB Cu Al 9 Ni	9% Al 5% Ni 5% Fe Cu alloy: Casting; designation used by German Standard
GB Cu Al 10 Fe	10% Al 3.0% Fe Cu alloy: Casting; designation used by German Standard
GZ Cu Al 10 Ni	9% Al 5% Fe 5% Ni Cu alloy: Casting; designation used by German Standard
HEUSLER ALLOY	15% Al 30% Mn Cu alloy: Magnetic alloy; further information from Permanent Magnet Association
HIDURAL 7	Al Zn Cu alloy: Wrought; composition not supplied; Langley **DPN: 145 UTS: 570 Elon: 20% Proof: 260**
HIDURAX 1	10% Al 4.5% Ni 4.5% Fe Cu alloy: Casting; Langley Alloys for BS alloy AB 2
HIDURAX 1	10% Al 5% Fe 5% Ni Cu alloy: Bar & forging; Langley Alloys for BS alloy CA104
HIDURAX 2	9.5% Al 2% Fe Cu alloy: Casting; Langley Alloys for BS alloy AB 1

Symbol	Nominal analysis, supplier, condition and remarks.
HIDURAX 2	9.5% Al 2% Ni 2% Fe Cu alloy: Forging; Langley Alloys **UTS: 690 Elon: 22% Proof: 330**
HIDURAX 3	10% Al Cu alloy: For casting; Langley; for corrosion resistance **DPN: 130 UTS: 540 Elon: 25% Proof: 230**
HIDURAX 4	Al Cu alloy: Composition not supplied; Langley; cast **DPN: 250 UTS: 820 Elon: 8% Proof: 540**
HIDURAX 4	Al Cu alloy: Composition not supplied; Langley; wrought **DPN: 270 UTS: 800 Elon: 7% Proof: 490**
HIDURAX 5	9% Al Cu alloy: Bar & forging; Langley Alloys **UTS: 530 Elon: 20% Proof: 250**
HIDURAX 6	5% Al Cu alloy: Bar & forging; Langley Alloys **UTS: 390 Elon: 60% Proof: 150**
HIDURAX 7	7% Al 2% Si Cu alloy: Wrought; Langley Alloys
HIDURON 191	Al Fe Cu alloy: Wrought; Langley Alloys; similar to BS 2033 **UTS: 490 Elon: 32% Proof: 300**
HOLFOS AB1	9.5% Al 2.5% Fe Cu alloy: Casting; J Holroyd; as cast **DPN: 130 UTS: 550 Elon: 20% Proof: 200**
HOLFOS AB2	10% Al 4.5% Fe 5.5% Ni Cu alloy: Casting; J Holroyd; as cast **DPN: 165 UTS: 660 Elon: 15% Proof: 250**
HOLFOS HTB1	1% Al 2% Mn 1% Fe 36% Zn Cu alloy: Casting; J Holroyd; as cast **DPN: 115 UTS: 500 Elon: 20% Proof: 220**
HOLFOS HTB2	3.5% Al 2% Mn 1.5% Fe 33% Zn Cu alloy: Casting; J Holroyd; as cast **DPN: 135 UTS: 590 Elon: 15% Proof: 280**
HOLFOS HTB3	5% Al 3% Mn 2% Fe 30% Zn Cu alloy: Casting; J Holroyd; as cast **DPN: 185 UTS: 750 Elon: 12% Proof: 420**
IMI 581	Al Mn Cu alloy: Resistance wire; IMI; Kuthern 41
IMI 756	5% Al Cu alloy: Welding rod & wire; IMI
IMI 757	7% Al Cu alloy: Welding rod & wire; IMI
IMI 764	10% Al Cu alloy: Bar; as drawn; IMI **DPN: 160 UTS: 560 Elon: 20% Proof: 330**
INCRAMET 800	10% Al 15% Ni 1.5% Co 0.7% Fe Cu alloy: Casting; Copper Development Association; for bottle moulds **UTS: 660**
INOXYDA	10% Al 4% Fe 4.5% Ni Cu alloy: Casting; information from International Nickel
KUMANAL	10% Mn 2% Al Cu alloy: Hard drawn wire & strip; IMI; electrical conductivity 4.5% IACS; heating elements **UTS: 670 Elon: 10%**
KUTHERM 41	Al Mn Cu alloy: Wire; IMI; electrical conductivity 4.5% IACS; withdrawn
MEIGH METAL	9% Al 4% Fe 5% Ni Cu alloy: Casting; Meigh Casting Co for BS alloy AB2C
NARITE	14% Al 1% Ni 5% Fe Cu alloy: Casting for deep drawing dies; N C Ashton; 120 tons/in^2 compression strength
NIKALIUM	9% Al 4.5% Fe 5% Ni Cu alloy: Casting for marine turbines; Manganese Bronze Ltd for BS alloy AB2C
No. 4 ALUMINIUM BRONZE	9.5% Al 4.5% Fe 5% Ni Cu alloy: Casting; Stone Manganese Marine for BS alloy AB 2
No. 6 ALUMINIUM BRONZE	9% Al 2% Fe 1% Mn 1% Ni Cu alloy: Casting; Stone Manganese Marine for BS alloy AB 1
NOVOSTON	8% Al 12% Mn 3% Fe 2% Ni Cu alloy: Forging; Stone Manganese Marine; as forged **DPN: 200 UTS: 750 Elon: 28% Proof: 400**

Note. The following abbreviations and units are used in the tables:

DPN	Hardness, diamond pyramid number
UTS	Ultimate tensile strength, N/mm^2
Elon	Elongation, %
Proof	0.1 % proof strength, N/mm^2

1 N/mm^2=0.1 hbar=0.102 kgf/mm^2=0.06475 tonf/in^2=145.04 lbf/in^2

Symbol	Nominal analysis, supplier, condition and remarks.
QQ B 666 A	5.0% Al Cu alloy: Sheet; US Federal UTS: 370 Elon: 27% Proof: 140
QQ B 666 B	8% Al 2.2% Si 5.0% Ni Cu alloy: Bar & forging; US Federal UTS: 530 Elon: 17% Proof: 210
QQ B 671 A	10% Al 3% Mn 3.5% Fe 0.5% Sn Cu alloy: Casting; US Federal UTS: 450 Elon: 22% Proof: 150
QQ B 671 B	10% Al 1% Fe Cu alloy: Casting; US Federal DPN: 170 UTS: 530 Elon: 12% Proof: 240
QQ B 671 C	10% Al 3% Mn 0.5% Sn 5% Ni Cu alloy: Casting; US Federal DPN: 180 UTS: 530 Elon: 12% Proof: 220
QQ B 671 D	10% 3% Mn 5% Ni Cu alloy: Casting; US Federal UTS: 600 Elon: 6% Proof: 300
RESISCO	7% Al Cu alloy: Tube; Yorkshire Imperial Metals Ltd
SAE 68 A	3% Fe 9% Al Cu alloy: Casting; as cast UTS: 460 Elon: 20% Proof: 170
SAE 68 B	1% Fe 10% Al Cu alloy: Casting; as cast UTS: 460 Elon: 20% Proof: 170
SAE 68 B	1% Fe 10% Al Cu alloy: Casting; solution treated & aged UTS: 580 Elon: 12% Proof: 260
SAE 701 A	5% Al Cu alloy: Bar & forging; properties not specified
SAE 701 B	10% Al 1% Ni 1.5% Mn 4% Fe Cu alloy: Bar & forging UTS: 530 Elon: 15% Proof: 240
SAE 701 C	10% Al 5% Ni 1.5% Mn 3% Fe Cu alloy: Bar & forging UTS: 690 Elon: 12% Proof: 330
SAE CA 614	7% Al 2% Fe Cu alloy: Wrought; formerly SAE 701 D
SAE CA 617	8.5% Al 1.5% Fe Cu alloy: Wrought; formerly SAE 701 B
SAE CA 623	9% Al 3% Fe Cu alloy: Wrought; formerly SAE 701 B
SAE CA 624	10.7% Al 3% Fe Cu alloy: Wrought; formerly SAE 701 B
SAE CA 630	10% Al 3% Fe 5% Ni Cu alloy: Wrought; formerly SAE 701 C
SAE CA 637	7.5% Al 1.7% Si Cu alloy: Wrought; formerly SAE 701 B
S Al Bz 6	6% Al 0.4% Ni 1.0% Mn Cu alloy: Wrought; designation used by German Standard
S Al Bz 8	8% Al 0.4% Ni 1.0% Mn Cu alloy: Wrought; designation used by German Standard
SILVA BRONZE	Al Cu wrought alloy: IMI; for fire back boilers DPN: 100 UTS: 400 Elon: 75% Proof: 90
SPRABRONZE AA	9% Al 1% Fe Cu alloy: Wire for spraying; Metco Ltd; as sprayed DPN: 146 UTS: 200
STA 7 CA1	5% Al Cu alloy: Bar & sheet UTS: 300 Elon: 25%
STA 7 CA2	9.5% Al 0.1% Pb Cu alloy: Bar DPN: 150 UTS: 530

Symbol	Nominal analysis, supplier, condition and remarks.
STA 7 CA3	10% Al 2% Fe 1% Mn 1% Fe Cu alloy: Casting; as cast UTS: 480 Elon: 20%
STA 7 CA4	10% Al 4.5% Ni 4% Fe 3% Mn Cu alloy: Casting; sand & die cast UTS: 600 Elon: 12%
STA 7 CA5	10% Al 5% Ni 5% Fe 2% Mn Cu alloy: Bar & forging DPN: 200 UTS: 660 Elon: 15%
STA 7 CX10	10% Al 3% Fe 3% Ni Cu alloy: Casting; die cast DPN: 190
SUPERSTON	1% Sn 3% Ni 12% Mn 8% Al 3% Fe Cu alloy: Casting; Phosphor Bronze Ltd for BS alloys CMA 1 C & CMA 2 C
SUPERSTON 10	4% Al 12% Mn 3% Fe 2% Ni Cu alloy: Sheet & plate; Stone Manganese Marine; as rolled UTS: 420 Elon: 62% Proof: 200
SUPERSTON 40	8% Al 12% Mn 3% Fe 2% Ni Cu alloy: Casting; Stone Manganese Marine; as cast DPN: 180 UTS: 690 Elon: 27% Proof: 300
SUPERSTON 40	8% Al 12% Mn 3% Fe 2% Ni Cu alloy: Forging; Stone Manganese Marine; as forged DPN: 200 UTS: 750 Elon: 28% Proof: 400
SUPERSTON 60	9% Al 12% Mn 3% Fe 2% Ni Cu alloy: Casting; Stone Manganese Marine; as cast DPN: 220 UTS: 750 Elon: 14% Proof: 400
SUPERSTON 60	9% Al 12% Mn 3% Fe 2% Ni Cu alloy: Casting; Stone Manganese Marine; as forged DPN: 250 UTS: 830 Elon: 15% Proof: 450
SUPERSTON 70	7.5% Al 15% Mn 3% Fe 2.5% Ni Cu alloy: Casting; Stone Manganese Marine; as cast DPN: 200 UTS: 660 Elon: 20% Proof: 300
SUPERTRODE	8% Al 3% Fe 13% Mn 2% Ni Cu alloy: Welding rod; Stone Manganese Marine; as welded DPN: 180 UTS: 680 Elon: 25% Proof: 370
TAURUS XX	9.5% Al 5% Ni 1% Mn 4% Fe Cu alloy: Casting; D Brown; centri-cast DPN: 155 UTS: 630 Elon: 14% Proof: 330
U-A 10NM	5.5% Ni 2.2% Fe 9.5% Al 1.2% Mn Cu alloy: Origin unknown
VULCAN METAL	1% Al 0.7% Cr 1.5% Ni 4.4% Fe 1% Si 0.4% Sn Cu alloy: Origin unknown
XANTAL	9.5% Al 0.2% Fe 0.2% Ni 0.05% Zn Cu alloy: Cast & wrought; origin unknown

Note. The following abbreviations and units are used in the tables:

DPN	Hardness, diamond pyramid number
UTS	Ultimate tensile strength, N/mm^2
Elon	Elongation, %
Proof	0.1 % proof strength, N/mm^2

$1\ N/mm^2 = 0.1\ hbar = 0.102\ kgf/mm^2 = 0.06475\ tonf/in^2 = 145.04\ lbf/in^2$

14H Copper alloys for brazing fillers and fuse materials
Containing silver, gold, palladium, phosphorous, cadmium or zinc
General metallurgical characteristics

These alloys have been developed in the first instance for the temperature range over which they are molten, as this very often is the deciding factor regarding their use. It is also necessary that a braze metal is capable of wetting the materials being joined and gives off no noxious fumes.

Probably the most important aspect however, is the workability of the molten metal. This must be as fluid as possible for joints relying on capillary action, but should be less so for manipulating hand brazes. The casting qualities of the braze metal should be of the first order.

The various alloying elements added to copper firstly lower the melting point then, depending on the element, have one of several effects.

Silver. Lowers the melting point, increases the fluidity, improves the strength and soundness of the braze including the fatigue properties and has good electrical conductivity and corrosion resistance. It is sometimes present alone, or with any of the listed alloying elements except gold.

Zinc. Lowers the melting point and acts as a de-oxidizer. It is usually added as the most economical method of achieving the above but has adverse effects on the strength and corrosion resistance. It is never present alone in the listed alloys but brazing alloys will be found in the copper zinc groups 14B and 14C.

Cadmium. Reduces the melting point to the lowest commonly used when present with silver and some zinc, without seriously affecting the electrical or corrosion properties. It is never found as the sole alloying element in brazing materials. Cadmium coppers for electrical purposes are found in the pure copper group 14A. These cadmium alloys present a health hazard when used for brazing or casting.

Phosporus. This is a powerful de-oxidizer, but can have serious deleterious effects on the electrical resistance and ductility. It also increases the fluidity of the molten metal. The straight copper phosphorus alloys are generally used only when brazing copper.

Gold. This is used on a limited scale where excellent corrosion resistance brazing is necessary, with high melting points. These alloys are also oxidation resistant up to their melting range, which is fairly narrow. Gold is always the sole alloying element.

Palladium. This usually, but not always, has silver also present. The alloys melt over a fairly narrow range and result in a hard, very corrosion resistant braze with stable electrical properties. The higher palladium silver alloys are white in colour.

Thermal treatment

None of these materials can have their properties improved by thermal treatment.

In many instances it is desirable to carry out some form of heat treatment on the brazed assembly. Considerable care must then be taken that the lower figure of the braze filler melting range is higher than the highest heat treatment temperature.

Welding and brazing

Pre-treatment. The joint area must be thoroughly cleaned to remove all traces of dirt and adhered oxide.

Welding and brazing. The materials listed are almost always used as filler rods for joining other materials, either copper or ferrous alloys. The remarks found in the appropriate section against the materials being joined then apply.

There are however, cases when the listed materials will be used as components which require joining. Gas welding or brazing can be readily carried out on all the materials generally with pre-heating and fluxes. The alloys containing cadmium present a serious health hazard to brazers and welders.

Post-treatment. All traces of flux must be removed as this is generally corrosive. No thermal treatment is normally required, although if the final assembly is complicated some form of stress relieving may be advisable. This should be at a temperature of 50–100 °C below the lowest melting point if possible.

Flaw detection

Crack test. Penetrant oil chalk test method.

X-ray. Seldom required as it is very unlikely these materials will ever be used to manufacture high duty parts. When used as filler rods for joining other materials X-ray may be required to prove the join.

Ultra-sonic test. Not normally required.

Chemical etch. Not required.

Corrosion protection

Temporary. None required.

Permanent. Most of these alloys have been developed in part at least for their corrosion resistance. When used for joining ferrous materials they will have superior corrosion resistance to the base metals and will normally be treated at the same time and in the same manner.

Provided all traces of flux are completely removed, other materials will seldom require protection. There should be no trouble electroplating a braze, provided simple precautions are taken regarding choice of pre-plating cleaners and etches to ensure that they are compatible with all the materials involved.

Machinability

When used as filler metals in correctly designed joints, there should be no significant effect on the machinability of the materials which are brazed.

When made into components most of the alloys listed are too soft to exhibit good machining characteristics. High surface speeds, low tool pressure with single point cutting tools, probably give the best results.

The palladium alloys are harder and therefore give better surface finishes.

Uses

Silver alloys. Principally as braze metals, the final choice of alloy depending on melting point, strength, economy and corrosion resistance. Some of these alloys have been used as bearing metals and for metal to metal revolving seals.

Phosphorus alloys. For brazing copper when electrical resistance is of no importance.

Palladium alloys. High strength, high temperature joints with good corrosion resistance, also for jewellery work where the 'silver' colour is necessary.

Gold alloys. Certain electrical joints subjected to high temperature and corrosive atmospheres. The largest use is as heat fuses to protect electrical furnaces. The electrical resistance of the fuse is unaffected by surface corrosion or deterioration and thus the melting range remains stable. The narrow melting range is itself an advantage.

Symbol	Nominal analysis, supplier, condition and remarks.
279	Pd Ni Mn Cu brazing alloy: Melting range 1060–1105 °C; Engelhard Industries; electrical conductivity 3% IACS **DPN: 135 UTS: 600 Elon: 35%**
ARGO-BOND	23% Ag Zn Cd Cu alloy: Braze metal; Johnson Matthey; melting range 616–735 °C
ARGO-SWIFT	29% Ag Zn Cd Cu alloy: Braze metal; Johnson Matthey; melting range 607–700 °C
ASTM B260 B Au 1	37% Au Cu alloy: Braze metal; temperature range 1016–1093 °C
ASTM B260 B Au 3	34.5% Au 3% Ni Cu alloy: Braze metal; temperature range 1030–1090 °C
ASTM B260 B Cu 1	99.9% (min) Cu: Braze metal; temperature range 1093–1149 °C
ASTM B260 B Cu 1 A	99.0% (min) Cu: Braze metal; temperature range 1093–1149 °C
ASTM B260 B Cu 2	86.5% (min) Cu: Braze metal; brazing range 1090–1150 °C
ASTM B260 B Cu Au 1	37.5% Au Cu alloy: Braze filler; melting range 955–990 °C
ASTM B260 B Cu P 1	5% P Cu alloy: Brazing filler; melting range 706–890 °C
ASTM B260 B Cu P 2	7% P Cu alloy: Brazing filler; melting range 706–805 °C
ASTM B260 B Cu P 3	6% P 6% Ag Cu alloy: Brazing filler; melting range 640–800 °C
ASTM B260 B Cu P 4	7% P 6% Ag Cu alloy: Brazing filler; melting range 640–719 °C
ASTM B260 B Cu P 5	5% P 15% Ag Cu alloy: Brazing filler; melting range 640–815 °C
B 6	16% Ag Zn Cu alloy: Braze metal; Johnson Matthey; melting range 790–830 °C
BS 1845/6	14% Ag 5% P Cu alloy: Braze filler; melting range 625–780 °C
BS 1845/7	7% P Cu alloy: Braze filler; melting range 705–800 °C
BS 1845 Au 3	37.5% Au Cu alloy: For brazing metal; melting range 980–1000 °C

Note. The following abbreviations and units are used in the tables:

DPN	Hardness, diamond pyramid number
UTS	Ultimate tensile strength, N/mm^2
Elon	Elongation, %
Proof	0.1 % proof strength, N/mm^2

1 N/mm^2 = 0.1 hbar = 0.102 kgf/mm^2 = 0.06475 tonf/in^2 = 145.04 lbf/in^2

Symbol	Nominal analysis, supplier, condition and remarks.
BS 1845 Au 4	30% Au Cu alloy: For braze metal; melting range 995–1020 °C
BS 1845 Cp 1	4.6% P 14.5% Ag Cu alloy: Braze metal; melting range 645–700 °C
BS 1845 Cp 2	6.5% P 2.0% Ag Cu alloy: Braze metal; melting range 645–740 °C
BS 1845 Cp 3	7.7% P Cu alloy: Braze metal; melting range 705–800 °C
BS 1845 Pd 8	18% Pd Cu alloy: For brazing metal; melting range 1080–1090 °C
BS 1845 Pd 12	20% Pd 10% Mn 15% Ni Cu alloy: For brazing metal; melting range 1060–1110 °C
BS 1845 Pd 13	30% Pd 15% Mn 20% Ni Cu alloy: For brazing metal; melting range 1070–1090 °C
C 4	24% Ag Zn Cu alloy: Braze metal; Johnson Matthey; melting range 740–780 °C
COLMONOY 15	7.0% P Cu alloy: Bronze metal; Wall Colmonoy; melting point 795 °C
CPNM 2	Pd Ni Mn Cu brazing alloy: Melting range 1060–1105 °C; Engelhard Industries; electrical conductivity 3% IACS **DPN: 135 UTS: 600 Elon: 35%**
D 3	33% Ag Zn Cu alloy: Braze metal; Johnson Matthey; melting range 700–740 °C
DIN 1734 L Ag 8	37% Zn 8% Ag Cu alloy: Braze metal; brazing temperature 860 °C
DIN 1734 L Ag 12	35% Zn 12% Ag Cu alloy: Brazing metal; working temperature 830 °C
DIN 1734 L Ag 12 Cd	30% Zn 12% Ag 7% Cd Cu alloy: Braze metal; working temperature 800 °C
DIN 1734 L Ag 15	35% Zn 15% Ag 10% Cd Cu alloy: Braze metal; working temperature 770 °C
DIN 1734 L Ag 15 P	3% P 15% Ag Cu alloy: Braze metal; working temperature 710 °C
DIN 1734 L Ag 20	30% Zn 20% Ag 15% Cd Cu alloy: Braze metal; working temperature 750 °C
DIN 1734 L Ag 25	30% Zn 25% Ag Cu alloy: Braze metal; working temperature 780 °C
DIN 1734 L Ag 25 Cd	20% Zn 25% Ag 15% Cd Cu alloy: Braze metal; working temperature 730 °C
DIN 1734 L Ag 27	20% Zn 27% Ag 8% Mn 5% Ni Cu alloy: Braze metal; working temperature 840 °C
DIN 1734 L Ag 30 Cd 5	20% Zn 30% Ag 5% Cd Cu alloy: Braze metal; working temperature 770 °C
DIN 1734 L Ag 30 Cd 12	20% Zn 30% Ag 12% Cd Cu alloy: Braze metal; working temperature 700 °C
DIN 1734 L Ag 38	25% Zn 38% Ag 3% Sn Cu alloy: Braze metal; working temperature 800 °C

Symbol	Nominal analysis, supplier, condition and remarks.
ERM SC 65	6% Ag 0.03% O Cu alloy: Forgings; Enfield Rolling Mills Ltd; resistance weld electrodes
	DPN: 140 UTS: 440 Elon: 12% Proof: 340
H 12	Ag Zn Cu alloy: For brazing; silver solder; Sheffield Smelting Co; melting point 780 °C; electrical conductivity 8.9% IACS
	DPN: 106 UTS: 260 Elon: 30% Proof: 190
JMMB BRONZE	Cu alloy: For brazing stainless steel; molybdenum or tungsten; Johnson Matthey
L 3	38% Zn 10% Ag Cu alloy: Braze metal; Sheffield Smelting; melting range 840–855 °C
	DPN: 178 UTS: 530 Elon: 7% Proof: 480
L 7	35% Zn 14% Ag Cu alloy: Braze metal; Sheffield Smelting; melting range 810–835 °C
	DPN: 174 UTS: 550 Elon: 3% Proof: 500
LX 13	30% Zn 20% Ag 10% Cd Cu alloy: Braze metal; Sheffield Smelting; melting range 718–773 °C
	DPN: 111 UTS: 370 Elon: 9% Proof: 300
LX 18	27.5% Zn 25% Ag 17.5% Cd Cu alloy: Braze metal: Sheffield Smelting; melting range 605–710 °C
	DPN: 116 UTS: 370 Elon: 14% Proof: 300
M 1	34% Zn 31% Ag Cu alloy: Braze metal; Sheffield Smelting; melting range 714–755 °C
	DPN: 112 UTS: 480 Elon: 5% Proof: 450
MX 4	Ag Zn Cu alloy: For brazing; silver solder; Sheffield Smelting; melting range 608–665 °C; electrical conductivity 26% IACS
	DPN: 103 UTS: 240 Elon: 15% Proof: 150
NICORO	35% Au 3.0% Ni Cu alloy: For brazing; melting range 1000–1030 °C; Wesgo
OROBRAZE 998	37.5% Ag Cu alloy: Braze metal; Johnson Matthey; melting range 980–998 °C
OROBRAZE 1018	30% Au Cu alloy: Braze metal; Johnson Matthey; melting range 996–1018 °C
PALLABRAZE	15% Pb Ag Cu alloy: Braze metal; Johnson Matthey; melting range 856–880 °C
PALLABRAZE 810	5% Pd Ag Cu alloy: Braze metal; Johnson Matthey; melting range 807–810 °C
PALLABRAZE 840	10% Pd Ag Cu alloy: Braze metal; Johnson Matthey; melting range 830–840 °C
PALLABRAZE 850	10% Pd Ag Cu alloy: Braze metal; Johnson Matthey; melting range 824–850 °C
PALLABRAZE 880	15% Pd Ag Cu alloy: For brazing; Johnson Matthey; melting range 856–880 °C
PALLABRAZE 900	20% Pd Ag Cu alloy: Braze metal; Johnson Matthey; melting range 876–900 °C
PALLABRAZE 950	25% Pd Ag Cu alloy: Braze metal; Johnson Matthey; melting range 901–950 °C
PALLABRAZE 1090	18% Pd Cu alloy: Braze metal; Johnson Matthey; melting range 1080–1090 °C
PHOSMALLOY No 1	15% Ag 5% P Cu alloy: Braze metal; Sheffield Smelting; melting range 625–780 °C
PHOSPHALLOY No 2	2% Ag 6.5% P Cu alloy: Braze metal; Sheffield Smelting; melting range 690–835 °C
	DPN: 130 UTS: 420 Elon: 5% Proof: 240
QQ S 561/3	15% Ag 5.0% P Cu alloy: Solder; US Federal
SB 260 B Cu 2	See ASTM B 260 BCu 2

Symbol	Nominal analysis, supplier, condition and remarks.
SILBERLOT 8	37% Zn 8% Ag Cu alloy: Braze metal; designation used by German Standard
SILBERLOT 12	35% Zn 12% Ag Cu alloy: Braze metal; designation used by German Standard
SILBERLOT 12 Cd	30% Zn 12% Ag 7% Cd Cu alloy: Braze metal; designation used by German Standard
SILBERLOT 15	35% Zn 15% Ag 10% Cd Cu alloy: Braze metal; designation used by German Standard
SILBERLOT 15 P	3% P 15% Ag Cu alloy: Braze metal; designation used by German Standard
SILBERLOT 20	30% Zn 20% Ag 15% Cd Cu alloy: Braze metal; designation used by German Standard
SILBERLOT 25	30% Zn 25% Ag Cu alloy: Braze metal; designation used by German Standard
SILBERLOT 25 Cd	20% Zn 25% Ag 15% Cd Cu alloy: Braze metal; designation used by German Standard
SILBERLOT 27	20% Zn 27% Ag 8% Mn 5% Ni Cu alloy: Braze metal; designation used by German Standard
SILBERLOT 30 Cd 5	20% Zn 30% Ag 5% Cd Cu alloy: Braze metal; designation used by German Standard
SILBERLOT 30 Cd 12	20% Zn 30% Ag 12% Cd Cu alloy: Braze metal; designation used by German Standard
SILBERLOT 38	25% Zn 38% Ag 3% Sn Cu alloy: Braze metal; designation used by German Standard
SILBRALLOY	2% Ag P Cu alloy: Braze metal; Johnson Matthey; melting range 638–694 °C; for brazing copper & alloys
	DPN: 195 UTS: 530 Elon: 5%
SILBRAZE	41% Zn 1% Ag 0.3% Si Cu alloy: Braze metal; Sheffield Smelting; melting range 886–893 °C
	DPN: 115 UTS: 480 Elon: 43% Proof: 210
SILFLO 0	7.0% P Cu alloy: Brazing rod for high electrical conductivity joints; 98% IACS; All-State; brazing range 710–790 °C
SILFLO 5	5.0% Ag 6.0% P Cu alloy: Brazing rod; All-State; high electrical conductivity; 98% IACS; brazing range 700–850 °C
SILFLO 15	15.0% Ag 5.0% P Cu alloy: Brazing rod; All-State; brazing range 700–830 °C; high electrical conductivity; 98% IACS
SIL-FOS	15% Ag P Cu alloy: Braze metal for copper & its alloys; Johnson Matthey; melting range 625–780 °C
	DPN: 187 UTS: 690 Elon: 10%
THESSCO	Trade name for brazing and soldering material; Sheffield Trade name for brazing and soldering material; Sheffield

Note. The following abbreviations and units are used in the tables:

DPN	Hardness, diamond pyramid number
UTS	Ultimate tensile strength, N/mm^2
Elon	Elongation, %
Proof	0.1 % proof strength, N/mm^2

$1 N/mm^2 = 0.1 hbar = 0.102 kgf/mm^2 = 0.06475 tonf/in^2 = 145.04 lbf/in^2$

14J Copper sintered alloys
With iron, tungsten and tungsten carbide

General metallurgical characteristics

These alloys are produced for two basic reasons, firstly as porous metals and secondly to give artificial hardness. In both cases the excellent properties of copper regarding conductivity, corrosion resistance and as a bearing metal are combined with other properties by physical, not metallurgical means.

The series of iron–copper sintered alloys are harder than most other copper alloys and their porosity can be controlled.

The tungsten–copper and tungsten carbide sintered alloys have these very hard particles set in the copper matrix with its excellent electrical conductivity.

The sintering process employs very fine particles, the various constituents being throroughly mixed then compacted to the desired shape. The pressure employed at this stage decides the percentage porosity of the finished article. The shapes are then heated when the copper melts and surrounds each particle of the alloying material. With some processes pressure and heat are applied at the same time.

Thermal treatment

It could be said that the second stage of manufacture of these alloys is a form of thermal treatment. Apart from this no form of hardening is possible and it is recommended that advice from the manufacturer is sought before any of these alloys, but particularly the iron–copper sinters, are subjected to heating for any reason. Many are oil or grease impregnated and this is burned out at a comparatively low temperature.

Welding and brazing

This should not normally be attempted on any of the materials listed without first obtaining the advice of the manufacturer. Where sintering is employed for its inherent porosity it will seldom be desirable to weld or braze as this results in a cast structure. Also the heating will drive off any impregnated oil or grease.

Some of the tungsten copper sinters can be joined when the same techniques as for pure copper can be used.

Flaw detection

Crack test. Penetrant oil chalk test can be used on non-porous sintered parts. Porous parts will absorb the penetrant oil all over the surface masking the fluid absorbed into any crack. With careful examination of the background density it may be possible to identify cracking, but fine cracks cannot be positively located.

X-ray. Never required.

Ultra-sonic test. Not normally required and very difficult to evaluate.

Chemical etch. The normal porosity of these materials makes etching difficult or impossible.

Corrosion protection

Temporary. None required.

Permanent. Long term storage should be in plastic or other non-porous containers otherwise no protection is necessary. Electroplating should not be attempted on these materials without first contacting the supplier for advice.

Machinability

The manufacturing process using fine powders ensures that these materials all have excellent machining characteristics. The porous bearings must be very carefully machined to prevent the surface pores being smeared over. This can be caused by the use of blunt tools or heavy feeds. Wherever possible single point tooling, high speeds, light cuts and copious coolant should be used.

Uses

The iron–copper sintered alloys are used as bearings.

The tungsten–copper and tungsten carbide–copper sinters are used as resistance welding electrodes.

Symbol	Nominal analysis, supplier, condition and remarks.
AMS 4805 B	10% Sn Cu alloy sinter: Oil impregnated
B 10	Iron–copper sintered material: Sintered Products Ltd; 17–22% porosity; withdrawn
	UTS: 220 Elon: 1.5%

Note. The following abbreviations and units are used in the tables:

DPN	Hardness, diamond pyramid number
UTS	Ultimate tensile strength, N/mm^2
Elon	Elongation, %
Proof	0.1 % proof strength, N/mm^2

1 N/mm^2=0.1 hbar=0.102 kgf/mm^2=0.06475 tonf/in^2=145.04 lbf/in^2

Symbol	Nominal analysis, supplier, condition and remarks.
B 10	Bronze sinter: Firth Cleveland; specific gravity 6.8–7.2; as sintered
	DPN: 115 UTS: 130 Elon: 4%
B 15	Iron–copper sintered material: Sintered Products Ltd; 17–22% porosity; withdrawn
	UTS: 240 Elon: 1.5%
B 20	Bronze sinter: Firth Cleveland; specific gravity 8.0–8.4; as sintered
	DPN: 90 UTS: 320 Elon: 40%
BM 78	Iron–copper carbon sintered material: Sintered Products Ltd; 18–22% porosity
	UTS: 250 Elon: 1.0%
C 10	Cu sinter: Firth Cleveland; specific gravity 7.0–7.4; as sintered
	DPN: 115 UTS: 160 Elon: 13%

Symbol	Nominal analysis, supplier, condition and remarks.
ELKONITE 1 W 3	W Cu alloy: For arc resistant contacts; Mallory Metallurgical; conductivity 41% IACS **DPN: 140**
ELKONITE 3 W 3	W Cu alloy: For arc resistant contacts; Mallory Metallurgical; conductivity 33% IACS **DPN: 160**
ELKONITE 10 W 3	W Cu alloy: For welding electrodes; Mallory Metallurgical
ELKONITE 20 K 3	W C Cu alloy: For riveting & welding electrodes; Mallory Metallurgical; electrical conductivity 22% IACS **DPN: 300 UTS: 530**
ELKONITE 20 W 3	W Cu alloy: For resistance welding electrodes; Mallory Metallurgical; electrical conductivity 28% IACS **DPN: 230 UTS: 630**
ELKONITE 30 W 3	W Cu alloy: For resistance welding electrodes; Mallory Metallurgical; electrical conductivity 28% IACS **DPN: 240 UTS: 65**
ELKONITE 100 M	W Cu sintered alloy: For hot riveting electrodes; Mallory Metallurgical; conductivity 30% IACS **DPN: 170 UTS: 35**
ELKONITE 100 W	W Cu sintered alloy: For resistance brazing electrodes; Mallory Metallurgical; conductivity 30% IACS **DPN: 550 UTS: 22**
FC 3/75	Iron copper sintered material: Sintered Products Ltd; 22–28% porosity **UTS: 20 Elon: 1.5%**
FC 75	Iron copper sintered material: Sintered Products Ltd; 22–28% porosity **UTS: 21 Elon: 1.5%**
FC 80	Iron copper sintered material: Sintered Products Ltd; 18–22% porosity **UTS: 24 Elon: 1.5%**
FEC 90	Iron copper sintered material: Sintered Products Ltd; 4–8% porosity; withdrawn **UTS: 34 Elon: 5.5%**
FGC 75	Iron copper carbon sintered material: Sintered Products Ltd; 22–23% porosity; now FC75/1 **UTS: 24 Elon: 1.5%**
FGC 85	Iron copper carbon sintered material: Sintered Products Ltd; 15–20% porosity; now FC85/1 **UTS: 28 Elon: 1.5%**
FX 10	Fe Cu sinter: Firth Cleveland; specific gravity 7.1–7.6; as sintered **DPN: 140 UTS: 400 Elon: 4%**
H 75	Iron copper carbon sintered material: Sintered Products Ltd; 15–20% porosity; withdrawn **UTS: 22 Elon: 1%**
MALLORY 1000	W Cu alloy: Sintered; Mallory Metallurgical; electrical conductivity 14% IACS; 'No Chat' **DPN: 270 UTS: 75 Elon: 2.5%**

Symbol	Nominal analysis, supplier, condition and remarks.
MALLORY NO-CHAT	W Cu alloy: Sintered; Mallory Metallurgical; electrical conductivity 14% IACS; Mallory 1000 **DPN: 270 UTS: 75 Elon: 2.5%**
NO-CHAT	W Cu alloy: Sintered; Mallory Metallurgical; electrical conductivity 14% IACS **DPN: 270 UTS: 75 Elon: 2.5%**
POROSINT	Bronze sintered metal filters: Sintered Products Ltd
RWMA CLASS 12	W Cu alloy: For riveting and welding electrodes; Mallory Metallurgical code for Elkonite 20 W3
RWMA CLASS 13	Mallory Metallurgical code for Elkonite 100 W
S 60	Iron carbon sintered material: Bearings; Sintered Products Ltd; 25–30% porosity; withdrawn **UTS: 9**
SAE 890	0.1% Sn 1.5% Pb 0.2% Fe Cu sintered material: Density 7.2–7.7 **UTS: 9**
SAE 891	0.1% Sn 1.5% Pb 0.2% Fe Cu sintered material: Density 7.7 (min) **UTS: 10**
SPARKOMITE 10	W Cu sintered tube: For spark erosion electrodes; as Elkonite 10W3; Mallory Metallurgical
SX 20	Fe Cu with C sinter: Firth Cleveland; specific gravity 7.1–7.6; hardened and tempered; file hard **UTS: 700 Elon: 5%**
U 3	W Cu contact material: Sheffield Smelting; electrical conductivity 28% IACS **DPN: 240**
U 4	W Cu contact material: Sheffield Smelting; electrical conductivity 35% IACS **DPN: 220**
U 8	W Cu contact material: Sheffield Smelting; electrical conductivity 40% IACS **DPN: 180**
WOLFRAM BRONZE	10% W Cu alloy: Origin unknown
Z 10	Brass sinter: Firth Cleveland; specific gravity 7.2–7.6; as sintered **DPN: 115 UTS: 160 Elon: 11%**

Note. The following abbreviations and units are used in the tables:

DPN	Hardness, diamond pyramid number
UTS	Ultimate tensile strength, N/mm^2
Elon	Elongation, %
Proof	0.1 % proof strength, N/mm^2

1 N/mm^2=0.1 hbar=0.102 kgf/mm^2=0.06475 tonf/in^2=145.04 lbf/in^2

14K Copper–tin wrought and cast alloys
With additions of phosphorus, lead, and zinc – 'bronze' and 'phosphor bronze'

Specific gravity	8.92	
Density	–	(0.322 lb/in.^3)
Solidus/liquidus	1050 °C	
Thermal conductivity	75.4 W/m °C	$(18 \text{ cal/m s } °C)$
Coefficient of linear expansion	$18 \times 10^{-6}/ °C$	
Electrical conductivity	16–17% IACS (copper 100%)	
Specific resistance	110–115 microhm mm	
Young's modulus of elasticity	$110 \times 10^9 \text{ N/m}^2$	$(16 \times 10^6 \text{ lbf/in.}^2)$
Impact 6% Sn 10% Pb cast alloy	26 J	(19 ft/lb) Izod
Fatigue strength $(10 \times 10^8 \text{ cycles})$ 6% Sn 10% Pb alloy	$\pm 90 \text{ N/mm}^2$	$(\pm 6 \text{ tonf/in.}^2)$

Hot strength

Temperature °C	Tensile strength N/mm²
95	240
150	210
205	200

The above properties are typical of the following group and may not apply exactly to any one specification. It is possible that with certain specifications some of the values may not be applicable.

General metallurgical characteristics

The term 'bronze' was originally confined to the series of copper–tin alloys, but has gradually been widened to include additions of other elements, such as phosphorus and lead. The range of alloys covered by the terms 'phosphor bronze' and 'lead bronze' all contain up to approximately 10% tin and retain the characteristics of the true 'bronze' with additional properties supplied by these supplementary elements. This is not the case with aluminium, manganese or silicon bronzes, which never have appreciable additions of tin and have properties entirely different from tin bronze. Each of these groups appears under its own heading. The term 'bronze' is now very often loosely applied to any copper alloy not based on the copper–zinc series.

In the true wrought bronzes, that is, copper–tin alloys, the tin is usually present between 5 and 8% and is soluble in the copper matrix, giving an alpha solid solution. This results in a stronger material than copper, with the ability to accept a considerable degree of cold work, thus enhancing the mechanical properties. Corrosion resistance is improved but there is a considerable drop in electrical and thermal conductivities.

Phosphorus acts as a de-oxidizer during casting and is invariably present up to about 0.5% when it increases the tensile strength and improves the cold working properties, with little effect on the ductility. With the higher phosphorous contents there is a hard constituent present which gives cast phosphor bronze, containing up to about 10% tin, good bearing properties. Lead may be added in small amounts to improve the machinability of the wrought alloys and there is a series of copper–tin lead casting materials used as high-duty bearings, particularly for reciprocating loads.

Additions of up to 10% zinc are made as an economical de-oxidizer, particularly in the cast alloys, where it has no effect on the mechanical properties. As zinc and phosphorus both act as de-oxidizers they are not found together.

The copper–tin–zinc series are known as 'gunmetals', and find considerable use for their excellent casting properties and corrosion resistance, particularly in marine conditions. When lead up to 5% is added, a range of leaded gunmetals is formed with better machining characteristics than the straight gunmentals and are cheaper to cast.

Alloys with more than 10% tin are hard and brittle, and find little industrial use apart from 'bell' and 'speculum' metal.

Bell metal has 20% tin, generally with antimony present. It has good casting but poor mechanical properties, with the advantage of a very sonorous note when struck. Copper alloys containing about 45% tin have a white lustrous colour and are capable of taking and retaining a high polish. These are the speculum metals, which have very poor mechanical properties but find some use because of their good reflectivity and are now electroplated successfully.

Thermal treatment

None of the materials listed can have their mechanical properties improved by any form of thermal treatment.

Annealing at 450–650 °C for 30 minutes will remove all traces of cold work from wrought materials and may be required between severe cold drawing operations.

Flaw detection

Crack test. Penetrant oil chalk test method.

X-ray. Not normally required, but it has been used to identify lead segregation in the cast lead bronzes for high duty bearings.

Ultra-sonic test. Not normally required.

Chemical etch. No known requirement.

Pressure test. Castings for high pressure use should be pressure tested at 350 N/mm² (50 lb/in.²) using air, and warm water. Higher air pressures should not be used without taking precautions against bursting. Hydraulic pressure testing, following an air test, is recommended where this is required, care being taken to bleed off all air before increasing the pressure.

Corrosion protection

Temporary. None required.

Permanent. None normally required. The Copper Development Association has information on a lacquer which preserves the attractive bronze appearance when this is of importance.

Painting. This is not a usual requirement. Where necessary the surface must be thoroughly cleaned to remove oxide, and if possible roughened by grit blasting or other mechanical means.

Plating. Electroplating of these alloys is straightforward. The bearing metals are sometimes silver or lead plated to reduce friction, when care must be taken at pre-plating not to destroy the surface by preferentially etching out any of the alloy constituents. The etching preparation will also highlight surface casting defects. The foundry should always be notified when a casting is to be electroplated.

Machinability

The wrought materials have excellent machining charac-

teristics, giving good surface finish with better than average tool life for copper alloys.

The cast lead bronzes have a soft matrix and exhibit a tendency to smear rather than cut particularly with high lead contents. Single point tooling used at a relatively high speed and low pressure gives best results.

Uses

Wrought. Strip and wire products for springs, clips, gauzes, filters, etc., where corrosion resistance and good mechanical properties are required. These are cheaper than stainless steel, but do not have equal corrosion resistance. Household goods where the attractive bronze colour is required. Architectural uses include gates, grilles, window frames, balustrades, doors and fittings.

Castings. The straight copper–tin alloys are seldom used as castings, apart from 'bell' metal which is still used for casting bells. The gunmetals, with and without lead, are used for general purpose castings where good corrosion resistance, economy and castability are important. These materials find more use as castings than any other series of alloys, apart from cast iron.

Lead bronze is used as a bearing metal, often being cast onto a steel shell.

Phosphor bronzes are also used as bearing metals and being harder than lead bronze do not have the same self lubricating properties, but support heavier loads.

Speculum metal is used as an electrodeposited coating on household goods and is sometimes used on reflectors.

Symbol	Nominal analysis, supplier, condition and remarks.
2.0921	8% Sn 0.4% Ni 1.0% Mn Cu alloy: Wrought; German Standard
2.1021	7% Sn 0.2% Pb Cu alloy: German Standard
2.1050	10% Sn 1.0% Pb 0.5% Zn Cu alloy: Casting; German Standard
2.1051	10% Sn Cu alloy: Casting; German Standard
2.1052	12% Sn 1.0% Pb 0.5% Zn Cu alloy: Casting; German Standard
2.1053	0.2% P 12% Sn Cu alloy: Wrought; German Standard
2.1056	14% Sn 1.0% Pb 0.5% Zn Cu alloy: Casting; German Standard
2.1057	14% Sn Cu alloy: Casting; German Standard
2.1086	2% Zn 10.5% Sn 1.0% Pb Cu alloy: Casting; German Standard
2.1087	2% Zn 10% Sn Cu alloy: Casting; German Standard

Symbol	Nominal analysis, supplier, condition and remarks.
2.1090	4% Zn 7% Sn 6% Pb Cu alloy: Casting; German Standard
	DPN: 80 UTS: 260 Elon: 19% Proof: 120
2.1091	4.5% Zn 7% Sn 6% Pb Cu alloy: Casting; German Standard
2.1096	5% Zn 5% Sn 5% Pb Cu alloy: Casting; German Standard
	DPN: 70 UTS: 250 Elon: 18% Proof: 110
2.1097	5% Zn 5% Sn 5% Pb Cu alloy: Casting; German Standard
2.1166	26% Pb 3% Sn 2% Ni Cu alloy: Casting; German Standard
2.1170	10% Sn 5% Pb 1% Ni Cu alloy: Casting; German Standard
2.1171	10% Sn 5% Pb Cu alloy: Casting; Germand Standard
2.1176	10% Sn 10% Pb 1% Ni Cu alloy: Casting; German Standard
2.1177	10% Sn 10% Pb Cu alloy: Casting; German Standard
2.1182	8% Sn 15% Pb 1.5% Ni Cu alloy: Casting; German Standard
2.1183	8% Sn 15% Pb Cu alloy: Casting; German Standard
2.1188	5% Sn 20% Pb 2% Ni Cu alloy: Casting; German Standard
2.1189	4.5% Sn 21% Pb Cu alloy: Casting; German Standard

Note. The following abbreviations and units are used in the tables:

DPN	Hardness, diamond pyramid number
UTS	Ultimate tensile strength, N/mm²
Elon	Elongation, %
Proof	0.1 % proof strength, N/mm²

1 N/mm²=0.1 hbar=0.102 kgf/mm²=0.06475 tonf/in²=145.04 lbf/in²

Symbol	Nominal analysis, supplier, condition and remarks.
419	4% Zn 5% Sn Cu alloy: Wrought; designation agreed by CDAA; not used
502	1.2% Sn 0.04% P Cu alloy: Wrought; designation used by CDAA
505	1.5% Sn 0.3% P Cu alloy: Wrought; designation agreed by CDAA; not used
507	1.8% Sn 0.04% P Cu alloy: Wrought; designation agreed by CDAA; not used
508	3.0% Sn 0.04% P Cu alloy: Wrought; designation agreed by CDAA; not used
509	3.1% Sn 0.2% P Cu alloy: Wrought; designation agreed by CDAA; not used
518	5% Sn 0.3% P 0.01% Al Cu alloy: Wrought; designation agreed by CDAA; not used
524	10% Sn 0.2% P Cu alloy: Wrought; designation used by CDAA
532	3.2% Pb 4.7% Sn 0.3% P Cu alloy: Wrought; designation agreed by CDAA; not used
534	1.0% Pb 5% Sn 0.2% P Cu alloy: Wrought; designation agreed by CDAA; not used
546	3.0% Zn 4% Pb 4.0% Sn 0.4% P Cu alloy: Wrought; designation agreed by CDAA; not used
836	5% Sn 5% Pb 5% Zn Cu alloy: Casting; designation used by CDAA
838	4% Sn 6% Pb 7% Zn Cu alloy: Casting; designation used by CDAA
842	5% Sn 2% Pb 13% Zn Cu alloy: Casting; designation used by CDAA
844	3% Sn 7% Pb 9% Zn Cu alloy: Casting; designation used by CDAA
848	3% Sn 6% Pb 15% Zn Cu alloy: Casting; designation used by CDAA
903	8% Sn 4% Zn Cu alloy: Casting; designation used by CDAA
905	10% Sn 2% Zn Cu alloy: Casting; designation used by CDAA
907	11% Sn 1.5% P Cu alloy: Casting; designation used by CDAA
910	15% Sn Cu alloy: Casting; designation used by CDAA
913	19% Sn Cu alloy: Casting; designation used by CDAA
915	10% Sn 2.5% Pb 3.5% Ni Cu alloy: Casting; designation used by CDAA
922	6% Sn 2% Pb 4% Zn Cu alloy: Casting; designation used by CDAA
923	8% Sn 1% Pb 4% Zn Cu alloy: Casting; designation used by CDAA
925	11% Sn 1% Pb 1.5% Ni 1.5% P Cu alloy: Casting; designation used by CDAA
927	10% Sn 2% Pb Cu alloy: Casting; designation used by CDAA
928	16% Sn 5% Pb Cu alloy: Casting; designation used by CDAA
932	7% Sn 7% Pb 3% Zn Cu alloy: Casting; designation used by CDAA
934	8% Sn 8% Pb Cu alloy: Casting; designation used by CDAA
935	5% Sn 9% Pb 1% Zn Cu alloy: Casting; designation used by CDAA
937	10% Sn 10% Pb Cu alloy: Casting; designation used by CDAA
938	7% Sn 15% Pb Cu alloy: Casting; designation used by CDAA
939	6% Sn 16% Pb Cu alloy: Casting; designation used by CDAA
940	13% Sn 15% Pb Cu alloy: Casting; designation used by CDAA
941	5% Sn 20% Pb Cu alloy: Casting; designation used by CDAA

Symbol	Nominal analysis, supplier, condition and remarks.
943	5% Sn 25% Pb Cu alloy: Casting; designation used by CDAA
947	5% Sn 2% Zn 5% Ni Cu alloy: Casting; designation used by CDAA
948	5% Sn 1% Pb 2% Zn 5% Ni Cu alloy: Casting; designation used by CDAA
A 25	Bronze welding rod; high tensile; low fuming; Delta Metals
AAPHOS	10% Sn 0.25% Pb 0.5% P Cu alloy: Casting; Anti Attrition for BS alloy PB 1
ACID BRONZE	Range of leaded nickel–copper–tin bronzes 9.5% Pb 9% Sn 0.75% Ni Cu alloy: Common name
ADMIRALTY GUNMETAL	2% Zn 10% Sn Cu alloy: Common name
ALLEN'S METAL	40% Pb 5% Sn Cu alloy
AMOLCTC 2	10% Sn 0.25% Pb 0.5% P Cu alloy: Continuous cast; Eyre Smelting for BS alloy PB 1; free machining DPN: 75 UTS: 280 Elon: 25%
AMPCOLOY 31	4% Zn 8% Sn Cu alloy: Casting; Ampco
AMPCOLOY 32	10% Sn 10% Pb Cu alloy: Casting; Ampco
AMPCOLOY 35	7% Sn 7% Pb 3% Zn Cu alloy: Casting; Ampco
AMPCOLOY 38	4% Zn 8% Sn Cu alloy: Casting; Ampco
AMPCOLOY 54	Sn P Cu alloy: For gears; Ampco
AMPCOLOY 71	4.5% Zn 1.5% Pb 6% Sn Cu alloy: Casting
AMPCOLOY 72	4% Zn 8% Sn Cu alloy: Casting; Ampco
AMPCOLOY 74	5% Zn 5% Pb 5% Sn Cu alloy: Casting; Ampco
AMPCOLOY 79	2% Zn 10% Sn Cu alloy: Casting; Ampco
AMPCOLOY 711	11% Sn 0.5% Pb 0.2% P Cu alloy: Casting; Ampco
AMPCOLOY 712	Sn P Cu alloy: For gears; Ampco
AMPCOLOY 715	Sn P Cu alloy: For gears; Ampco
ARTIC BRONZE	Series of leaded bearing bronzes; grain structure produced by rapid cooling in (chill cast) metal moulds; origin unknown
AMS 4510 C	5% Sn P Cu alloy: Sheet & strip; hard drawn
AMS 4520 E	3% Zn 4% Pb 4% Sn Cu alloy: Strip
AMS 4625 D	5% Sn P Cu alloy: Bar & tube; ½-hard
AMS 4720 B	5% Sn P Cu alloy: Wire; spring hard
AMS 4840 A	5.5% Sn 24.5% Pb Cu alloy: Casting
AMS 4842 A	10% Sn 10% Pb Cu alloy: Casting
AMS 4845 D	2% Zn 10% Sn Cu alloy: Casting
AMS 4846 A	2% Zn 11% Sn Cu alloy: Casting
AMS 4855 B	5% Zn 5% Sn 5% Pb Cu alloy: Casting
ARIEL BRONZE	10% Sn 0.25% Pb 0.5% P Cu alloy: Casting; Eyre Smelting for BS alloy PB 1
ASTM B22 A	19% Sn 0.25% Pb Cu alloy: Casting; deformation limit of 170 N/mm²; minimum specified
ASTM B22 B	16% Sn 0.25% Pb 1% P Cu alloy: Casting; deformation limit of 120 N/mm²; minimum specified
ASTM B22 C	10% Sn 9.5% Pb 1% Ni 0.1% P Cu alloy: Casting
ASTM B22 D	10% Sn 0.3% Pb 2.5% Zn 1% Ni Cu alloy: Casting; as cast UTS: 260 Elon: 20% Proof: 100
ASTM B30/1 A	10% Sn 2% Zn Cu alloy: Ingot 'G' Bronze; as cast UTS: 260 Elon: 20%
ASTM B30/1 B	8% Sn 4% Zn Cu alloy: Ingot 'G' Bronze; as cast UTS: 260 Elon: 20%
ASTM B30/2 A	6% Sn 1.5% Pb 4.5% Zn Cu alloy: Ingot Navy 'M'; as cast UTS: 240 Elon: 30%
ASTM B30/2 B	8.5% Sn 0.5% Pb 4% Zn Cu alloy: Ingot 'Navy PC'; as cast UTS: 240 Elon: 18%
ASTM B30/3 A	10% Sn 10% Pb Cu alloy: Ingot; as cast UTS: 220 Elon: 22%

Symbol	Nominal analysis, supplier, condition and remarks.
ASTM B30/3 B	7% Zn 7% Pb 3% Zn Cu alloy: Ingot; as cast **UTS: 200 Elon: 12%**
ASTM B30/3 C	5% Sn 9% Pb 1% Zn Cu alloy: Ingot; as cast **UTS: 170 Elon: 8%**
ASTM B30/3 D	7% Sn 15% Pb Cu alloy: Ingot; as cast **UTS: 170 Elon: 10%**
ASTM B30/4 A	5% Sn 5% Pb 5% Zn Cu alloy: Ingot; as cast **UTS: 210 Elon: 22%**
ASTM B30/4 B	4% Sn 6% Pb 7% Zn Cu alloy: Ingot; as cast **UTS: 200 Elon: 15%**
ASTM B30/5 A	3% Sn 7% Pb 9% Zn Cu alloy: Ingot; as cast **UTS: 200 Elon: 18%**
ASTM B30/5 B	2.5% Sn 6.5% Pb 15% Zn Cu alloy: Ingot; as cast **UTS: 170 Elon: 15%**
ASTM B30/6 A	1% Sn 3% Pb 24% Zn Cu alloy: Ingot; as cast **UTS: 240 Elon: 25%**
ASTM B30/6 B	1% Sn 3% Pb 29% Zn Cu alloy: Ingot; as cast **UTS: 200 Elon: 20%**
ASTM B61	6% Sn 1.5% Pb 4% Zn 1% Ni (max) Cu alloy: Casting; as cast **UTS: 220 Elon: 22% Proof: 90**
ASTM B62	5% Sn 5% Pb 5% Zn 1% Ni (max) Cu alloy: Casting; as cast **UTS: 200 Elon: 20% Proof: 75**
ASTM B66/PHOS BRONZE	8% Sn 10.5% Pb 0.2% P Cu alloy: Casting for locomotives
ASTM B66/HARD BRONZE	7.5% Sn 13% Pb Cu alloy: Casting for locomotives
ASTM B66/ MEDIUM BRONZE	7% Sn 19% Pb Cu alloy: Casting for locomotives
ASTM B66/SOFT BRONZE	5% Sn 25% Pb Cu alloy: Casting for locomotives
ASTM B100/1	4.6% Sn 0.2% P Cu alloy: Plate & sheet; as rolled or annealed **DPN: 130 UTS: 420 Elon: 10%**
ASTM B103/A	4.5% Sn 0.2% P Cu alloy: Sheet & bar; annealed **UTS: 360**
ASTM B103 A	4.5% Sn 0.2% P Cu alloy: Sheet & bar; extra spring hard **UTS: 760**
ASTM B103 B	4.5% Sn 0.2% P 3% Pb Cu alloy: Sheet & bar; annealed **UTS: 370**
ASTM B103 B	4.5% Sn 0.2% P 3% Pb Cu alloy: Sheet & bar; extra spring hard **UTS: 750**
ASTM B103 B1	4.2% Sn 0.2% P 1% Pb Cu alloy: Sheet & bar; annealed **UTS: 370**
ASTM B103 B1	4.2% Sn 0.2% P 1% Pb Cu alloy: Sheet & bar; extra spring hard **UTS: 750**
ASTM B103 C	8% Sn 0.2% P Cu alloy: Sheet & bar; annealed **UTS: 420**
ASTM B103 C	8% Sn 0.2% P Cu alloy: Sheet & bar; extra spring hard **UTS: 810**
ASTM B103 D	10% Sn 0.2% P Cu alloy: Sheet & bar; annealed **UTS: 390**

Symbol	Nominal analysis, supplier, condition and remarks.
ASTM B103 D	10% Sn 0.2% P Cu alloy: Sheet & bar; extra spring hard **UTS: 930**
ASTM B139 A	4% Sn 0.3% P Cu alloy: Bar; annealed **UTS: 370**
ASTM B139 A	4% Sn 0.3% P Cu alloy: Bar; spring hard **UTS: 730 Elon: 5%**
ASTM B139 B1	4% Sn 0.3% P 1% Pb Cu alloy: Bar; hard **UTS: 370 Elon: 12%**
ASTM B139 B2	4% Sn 0.4% P 4% Pb 3% Zn Cu alloy: Bar; hard **UTS: 370 Elon: 12%**
ASTM B139 C	8% Sn 0.3% P Cu alloy: Bar; annealed **UTS: 420**
ASTM B139 C	8% Sn 0.3% P Cu alloy: Bar; hard **UTS: 460 Elon: 10%**
ASTM B139 D	10% Sn 0.3% P Cu alloy: Bar; annealed **UTS: 480**
ASTM B139 D	10% Sn 0.3% P Cu alloy: Bar; hard **UTS: 560 Elon: 10%**
ASTM B143/1 A	2% Zn 10% Sn Cu alloy: Casting; as cast **UTS: 260 Elon: 20% Proof: 100**
ASTM B143/1 B	4% Zn 8% Sn Cu alloy: Casting; as cast **UTS: 260 Elon: 20% Proof: 100**
ASTM B143/2 A	4.5% Zn 1.5% Pb 6% Sn Cu alloy: Casting; as cast **UTS: 220 Elon: 22% Proof: 90**
ASTM B143/2 B	4% Zn 1% Pb 8% Sn Cu alloy: Casting; as cast **UTS: 240 Elon: 18% Proof: 90**
ASTM B144/3A	10% Sn 10% Pb Cu alloy: Casting; as cast **UTS: 170 Elon: 8% Proof: 60**
ASTM B144/3 B	7% Sn 7% Pb 3% Zn Cu alloy: Casting; as cast **UTS: 200 Elon: 12% Proof: 75**
ASTM B144/3 C	5% Sn 9% Pb 1% Zn Cu alloy: Casting; as cast **UTS: 170 Elon: 8% Proof: 60**
ASTM B144/3 D	7% Sn 15% Pb Cu alloy: Casting; as cast **UTS: 170 Elon: 10% Proof: 75**
ASTM B144/3 E	5% Sn 25% Pb Cu alloy: Casting; as cast **UTS: 140 Elon: 7%**
ASTM B145/4 A	5% Zn 5% Pb 5% Sn Cu alloy: Casting **UTS: 200 Elon: 20% Proof: 75**
ASTM B145/4 B	7% Zn 6% Pb 4% Sn Cu alloy: Casting **UTS: 200 Elon: 15% Proof: 60**
ASTM B145/5 A	9% Zn 7% Pb 3% Sn Cu alloy: Casting **UTS: 200 Elon: 18% Proof: 60**
ASTM B145/5 B	15% Zn 6% Pb 3% Sn Cu alloy: Casting **UTS: 170 Elon: 15% Proof: 60**
ASTM B146/6 A	24% Zn 3% Pb 1% Sn Cu alloy: Casting **UTS: 240 Elon: 25% Proof: 60**
ASTM B146/6 B	29% Zn 3% Pb 1% Sn Cu alloy: Casting **UTS: 200 Elon: 20% Proof: 60**
ASTM B146/6 C	37% Zn 1% Pb 1% Sn Cu alloy: Casting **UTS: 260 Elon: 15% Proof: 75**
ASTM B147/7 A	35% Zn 0.7% Sn 0.7% Pb 1% Fe 0.7% Al 0.2% Mn Cu alloy: Casting **UTS: 420 Elon: 15% Proof: 120**
ASTM B159 A	5% Sn 0.3% P Cu alloy: Wire; annealed **UTS: 330**
ASTM B159 A	5% Sn 0.3% P Cu alloy: Wire; hard **UTS: 880**
ASTM B159 C	8% Sn 0.3% P Cu alloy: Wire; annealed **UTS: 400**
ASTM B159 C	8% Sn 0.3% P Cu alloy: Wire; hard **UTS: 1000**
ASTM B159 D	10% Sn 0.3% P Cu alloy: Wire; annealed **UTS: 480**
ASTM B159 D	10% Sn 0.3% P Cu alloy: Wire; hard **UTS: 1080**
ASTM B176 Z 30 A	30% Zn 1.5% Sn 1.5% Pb Cu alloy: Casting; die cast **UTS: 330 Elon: 10% Proof: 150**

Note. The following abbreviations and units are used in the tables:

DPN	Hardness, diamond pyramid number
UTS	Ultimate tensile strength, N/mm^2
Elon	Elongation, %
Proof	0.1 % proof strength, N/mm^2

1 N/mm^2=0.1 hbar=0.102 kgf/mm^2=0.06475 tonf/in^2=145.04 lbf/in^2

Symbol	Nominal analysis, supplier, condition and remarks.
ASTM B225 E Cu	1% Sn 98% Cu alloy: Welding electrode **UTS: 170**
ASTM B225 E Cu Si	1.5% Sn 3% Si Cu alloy: Welding electrode **UTS: 340**
ASTM B225 E Cu Sn A	5% Sn 0.3% P Cu alloy: Welding electrode **UTS: 240**
ASTM B225 E Cu Sn C	8% Sn 0.3% P Cu alloy: Welding electrode **UTS: 260**
ASTM B259 R Cu	1% Sn 98% Cu alloy: Welding rod **UTS: 150**
ASTM B259 R Cu Sn A	5% Sn 0.3% P Cu alloy: Welding rod **UTS: 220**
ASTM B427/908	12% Sn Cu alloy: Casting for gears **DPN: 95 UTS: 350 Elon: 12%**
ASTM B427/915	10% Sn 3.2% Ni 2.7% Pb Cu alloy: Castings for gears **DPN: 75 UTS: 310 Elon: 8% Proof: 170**
ASTM B427/916	10.2% Sn 1.75% Ni Cu alloy: Castings for gears **DPN: 65 UTS: 320 Elon: 10% Proof: 180**
ASTM B427/917	12% Sn 1.75% Ni Cu alloy: Casting for gears **DPN: 85 UTS: 320 Elon: 12% Proof: 280**
ASTM B508/505	1.3% Sn Cu alloy: Strip for flexible piping
ATLAS CTC 1	10% Sn 0.25% Pb 0.5% P Cu alloy: Continuous cast; Eyre Smelting for BS alloy PB 1 **DPN: 95 UTS: 360 Elon: 17%**
B 1	High Sn Cu alloy: Casting for high duty bearings; Anti-Attrition **DPN: 103 UTS: 240 Elon: 2%**
B 2	Sn Cu alloy: Casting for thrust bearings; Anti-Attrition **DPN: 120 UTS: 240 Elon: 1%**
B 6	Sn Cu alloy: Casting for unlined bearings; Anti-Attrition **DPN: 95 UTS: 220 Elon: 10%**
B 9	10% Sn 0.25% Pb 0.5% P Cu alloy: Casting; Anti-Attrition for BS alloy PB 1
B 13	Sn P Cu alloy: Casting for rolling mill bearings; Anti-Attrition **DPN: 92 UTS: 220 Elon: 6%**
B 14	12% Sn 0.15% P Cu alloy: Casting for worm wheels; Anti-Attrition for BS alloy PB 2
B 18	Sn Cu alloy: Casting; electrical bronze; Anti-Attrition; for good conductivity fittings **DPN: 75 UTS: 260 Elon: 30%**
B 20	5% Sn 5% Pb 5% Zn 2% Ni Cu alloy: Casting; Anti-Attrition for BS alloy LG 2
B 32	5% Sn 20% Pb 1% Zn 2% Ni 0.1% P Cu alloy: Casting; Anti-Attrition for BS alloy LB 5
B 38	Sn Cu alloy: Brazing metal; Anti-Attrition
B 39	10% Sn 1.5% Pb 2% Zn Cu alloy: Casting; Anti-Attrition for BS alloy G 1
B 40	8% Sn 1.5% Pb 4% Zn Cu alloy: Casting; Anti-Attrition for BS alloy G2
B 45	Sn Pb Cu alloy: Casting; gunmetal; Anti-Attrition **DPN: 60 UTS: 120 Elon: 14%**
B 60	Copper–tin sintered material: Sintered Products Ltd; 30–33% porosity **UTS: 70 Elon: 2.5%**
B 65	Copper–tin sintered material: Sintered Products Ltd; 25–30% porosity **UTS: 750 Elon: 2.5%**
B 70	Copper–tin sintered material: Sintered Products Ltd; 22–25% porosity **UTS: 90 Elon: 3%**
B 80	Copper–tin sintered material: Sintered Products Ltd; 10–15% porosity **UTS: 100 Elon: 4%**
B 80	Sintered bronze: 22% porosity; SG 7.1; Durasint **DPN: 50 UTS: 140 Elon: 5%**

Symbol	Nominal analysis, supplier, condition and remarks.
B 80/1	Sintered graphite bronze: 18% porosity; SG 710; Durasint **DPN: 60 UTS: 140 Elon: 6%**
B 85	Sintered bronze: 7% porosity; SG 8.2; Durasint **DPN: 70 UTS: 350 Elon: 8%**
B 85/P	Copper–tin sintered material: Sintered Products Ltd; 4–8% porosity; withdrawn **UTS: 180 Elon: 20%**
B 90/P	Copper–tin sintered material: Sintered Products Ltd; 1–3% porosity; withdrawn **UTS: 300 Elon: 35%**
BEARING BRONZE	11% Sn 0.2% Pb Cu alloy: Cast; origin unknown
BELL METAL	22.5% Sn Cu alloy: Used for bells or as bearing bronze; origin unknown
BO BRONZE	11.5% Sn 0.7% P Cu alloy: Casting; spuncast; I Holroyd for BS alloy PB 1 & PB 2; as cast **DPN: 100 UTS: 290 Elon: 3% Proof: 210**
B O H T BRONZE	11.5% Sn 0.7% P Cu alloy: Casting; spuncast; I Holroyd; heat treated version of B O Bronze **DPN: 95 UTS: 320 Elon: 20% Proof: 170**
BOLTOMET 10	Previous name for Bolt 317; 5% Sn Cu alloy; Thomas Bolton; withdrawn
BOLTOMET 11	Previous name for Bolt 320; 6% Sn Cu alloy; Thomas Bolton; withdrawn
BOLTOMET 12	Previous name for Bolt 312; 3% Sn Cu alloy; Thomas Bolton; withdrawn
BOLTOMET 15	Previous name for Bolt 338; 8% Sn Cu alloy; Thomas Bolton; withdrawn
BOLTOMET 16	Previous name for Bolt 327; 7% Sn Cu alloy; Thomas Bolton; withdrawn
BOLTOMET 58	Previous name for Bolt 445 Pb Sn Cu alloy; Thomas Bolton; withdrawn
BOLTOMET 305	1% Sn Cu alloy: Strip; hard; for flexible tubing; Thomas Bolton **DPN: 145 UTS: 48 Elon: 8%**
BOLTOMET 307	1.2% Sn 0.3% P Cu alloy: Rod & strip; annealed; Thomas Bolton **DPN: 65 UTS: 260 Elon: 50% Proof: 75**
BOLTOMET 307	1.2% Sn 0.3% P Cu alloy: Rod; hard; Thomas Bolton **DPN: 140 UTS: 400 Elon: 22% Proof: 360**
BOLTOMET 307	1.2% Sn 0.03% P Cu alloy: Wire & strip; Thomas Bolton **DPN: 182 UTS: 560 Elon: 3% Proof: 480**
BOLTOMET 309	2.5% Sn 0.04% P Cu alloy: Rod & strip: annealed; Thomas Bolton **DPN: 75 UTS: 300 Elon: 50% Proof: 90**
BOLTOMET 309	2.5% Sn 0.04% P Cu alloy: Hard; Thomas Bolton **DPN: 157 UTS: 450 Elon: 21% Proof: 390**
BOLTOMET 309	2.5% Sn 0.04% P Cu alloy: Strip; hard; Thomas Bolton **DPN: 200 UTS: 600 Elon: 4% Proof: 550**
BOLTOMET 312	3.5% Sn 0.07% P Cu alloy: Rod & strip; annealed; Thomas Bolton **DPN: 70 UTS: 300 Elon: 55% Proof: 90**
BOLTOMET 312	3.5% Sn 0.07% P Cu alloy: Rod; hard; Thomas Bolton **DPN: 180 UTS: 530 Elon: 18% Proof: 460**
BOLTOMET 312	3.5% Sn 0.07% P Cu alloy: Strip; hard; Thomas Bolton **DPN: 220 UTS: 700 Elon: 3% Proof: 600**
BOLTOMET 317	5% Sn 0.15% P Cu alloy: Rod & strip; annealed; Thomas Bolton **DPN: 80 UTS: 330 Elon: 60% Proof: 120**
BOLTOMET 317	5% Sn 0.15% P Cu alloy: Rod; hard; Thomas Bolton **DPN: 210 UTS: 590 Elon: 20% Proof: 460**

Symbol	Nominal analysis, supplier, condition and remarks.
BOLTOMET 317	5% Sn 0.15% P Cu alloy: Strip; hard; Thomas Bolton **DPN: 225 UTS: 730 Elon: 3% Proof: 630**
BOLTOMET 319	0.1% P 6.25% Sn Cu alloy: Bar; hard drawn; Thomas Bolton **DPN: 170 UTS: 550 Elon: 19% Proof: 470**
BOLTOMET 320	6.3% Sn 0.25% P Cu alloy: Strip; annealed; Thomas Bolton **DPN: 82 UTS: 340 Elon: 68% Proof: 120**
BOLTOMET 320	6.3% Sn 0.25% P Cu alloy: Strip; hard; used for springs; Thomas Bolton **DPN: 241 UTS: 780 Elon: 3% Proof: 690**
BOLTOMET 327	6.8% Sn 0.25% P Cu alloy: Wire; hard; Thomas Bolton for BS 384 **UTS: 830**
BOLTOMET 338	7.8% Sn 0.3% P Cu alloy: For rod & strip; annealed; Thomas Bolton **DPN: 90 UTS: 370 Elon: 75% Proof: 140**
BOLTOMET 338	7.8% Sn 0.3% P Cu alloy: For rod; hard; Thomas Bolton **DPN: 232 UTS: 660 Elon: 20% Proof: 510**
BOLTOMET 338	7.8% Sn 0.3% P Cu alloy: For strip; hard; Thomas Bolton **DPN: 243 UTS: 800 Elon: 4% Proof: 700**
BOLTOMET 349	10% Sn 0.15% P Cu alloy: Strip; annealed; Thomas Bolton **DPN: 97 UTS: 420 Elon: 70% Proof: 150**
BOLTOMET 349	10% Sn 0.15% P Cu alloy: Strip; hard; Thomas Bolton **DPN: 248 UTS: 820 Elon: 4% Proof: 730**
BOLTOMET 445	5% Sn 1% Pb 0.15% P Cu alloy: Rod; annealed; free machining; Thomas Bolton **DPN: 85 UTS: 330 Elon: 55% Proof: 140**
BOLTOMET 445	5% Sn 1% Pb 0.15% P Cu alloy: Rod; hard; free machining; Thomas Bolton **DPN: 160 UTS: 500 Elon: 17% Proof: 360**
BOLTOMET 356	0.05% P 2.0% Sn Cu alloy: Strip; hard; Thomas Bolton **DPN: 167 UTS: 500 Elon: 4% Proof: 470**
BRONZE 44	8% Sn 15% P Cu alloy: Welding electrode; Murex Ltd
BRONZE 66	8% Sn 0.15% P Cu alloy: Welding electrode; Murex Ltd
BRONZETRODE 94.15	0.50% Mn 7.40% Sn 0.20% P Cu alloy: Electrodes; ESAB **UTS: 463 Elon: 28%**
BS 99	8% Zn 5% Sn 2% Pb Cu alloy: Covers all forms of screwed fittings **UTS: 180 Elon: 15%**
BS 352	Phosphor bronze turbine blading; obsolete
BS 369 PB102	5% Sn 0.25% P Cu alloy: Phosphor bronze bar; as rolled **UTS: 450 Elon: 20% Proof: 330**
BS 382/3	Bronze; gunmetal; ingots & castings; replaced by BS 1400
BS 384	5.7% Sn 0.2% P 0.02% Pb (max) Cu alloy: Phosphor bronze wire; cold drawn for general engineering **UTS: 830**

Symbol	Nominal analysis, supplier, condition and remarks.
BS 407 PB101	4% Sn 0.3% P Cu alloy: Sheet & strip; annealed **DPN: 80 UTS: 290 Elon: 40%**
BS 407 PB101	4% Sn 0.2% P Cu alloy: Sheet & strip; ½-hard **DPN: 100 UTS: 330 Elon: 30% Proof: 100**
BS 407 PB101	4% Sn 0.2% P Cu alloy: Sheet & strip; ½-hard **DPN: 150 UTS: 450 Elon: 8% Proof: 290**
BS 407 PB101	4% Sn 0.3% P Cu alloy: Sheet & strip; cold rolled **DPN: 180 UTS: 530 Elon: 4% Proof: 370**
BS 407 PB101	4% Sn 0.2% P Cu alloy: Sheet & strip; extra hard **DPN: 190 UTS: 600 Proof: 450**
BS 407 PB102	5% Sn 0.2% P Cu alloy: Sheet & strip; annealed **DPN: 85 UTS: 300 Elon: 45%**
BS 407 PB102	5% Sn 0.2% P Cu alloy: Sheet & strip; ½-hard **DPN: 110 UTS: 340 Elon: 35% Proof: 120**
BS 407 PB102	5% Sn 0.2% P Cu alloy: Sheet & strip; ½-hard **DPN: 160 UTS: 480 Elon: 10% Proof: 330**
BS 407 PB102	5% Sn 0.2% P Cu alloy: Sheet & strip; hard **DPN: 180 UTS: 580 Elon: 4% Proof: 420**
BS 407 PB102	5% Sn 0.2% P Cu alloy: Sheet & strip; extra hard **DPN: 210 UTS: 640 Proof: 480**
BS 407 PB103	7% Sn 0.2% P Cu alloy: Sheet & strip; spring hard; up to 15 cm wide 0.056 cm thick **DPN: 220**
BS 407 PB103	7% Sn 0.2% P Cu alloy: Sheet & strip; extra spring hard; up to 15 cm wide 0.056 cm thick **DPN: 240**
BS 407 PB103	7% Sn 0.2% P Cu alloy: Sheet & strip; hard **DPN: 200 UTS: 600 Elon: 6% Proof: 450**
BS 407 PB103	7% Sn 0.2% P Cu alloy: Sheet & strip; extra hard **DPN: 215 UTS: 690 Proof: 530**
BS 407 PB103	7% Sn 0.2% P Cu alloy: Sheet & strip; annealed **DPN: 90 UTS: 330 Elon: 50%**
BS 407 PB103	7% Sn 0.2% P Cu alloy: Sheet & strip; ½-hard **DPN: 115 UTS: 370 Elon: 40% Proof: 180**
BS 407 PB103	7% Sn 0.2% P Cu alloy: Sheet & strip; ½-hard **DPN: 170 UTS: 520 Elon: 12% Proof: 360**
BS 421	Phosphor bronze castings; replaced by BS 1400
BS 897	Leaded gunmetal casting; replaced by BS 1400
BS 898	Leaded gunmetal casting; replaced by BS 1400
BS 960	Leaded bronze casting & ingot; replaced by BS 1400
BS 961	Leaded bronze casting & ingot; replaced by BS 1400
BS 962	Leaded bronze casting & ingot; replaced by BS 1400
BS 963	Leaded bronze casting & ingot; replaced by BS 1400
BS 964	Leaded bronze casting & ingot; replaced by BS 1400
BS 965	Leaded bronze casting & ingot; replaced by BS 1400
BS 1021	Copper alloy: Ingot & casting; replaced by BS 1400
BS 1022	Copper alloy: Ingot & casting; replaced by BS 1400
BS 1023	Copper alloy: Ingot & casting; replaced by BS 1400
BS 1024	Copper alloy: Ingot & casting; replaced by BS 1400
BS 1025	Copper alloy: Ingot & casting; replaced by BS 1400
BS 1026	Copper alloy: Ingot & casting; replaced by BS 1400
BS 1027	Copper alloy: Ingot & casting; replaced by BS 1400
BS 1028	Copper alloy: Ingot & casting; replaced by BS 1400
BS 1058	Phosphor bronze ingot & casting; replaced by BS 1400

Note. The following abbreviations and units are used in the tables:

DPN	Hardness, diamond pyramid number
UTS	Ultimate tensile strength, N/mm^2
Elon	Elongation, %
Proof	0.1 % proof strength, N/mm^2

1 N/mm^2=0.1 hbar=0.102 kgf/mm^2=0.06475 tonf/in^2=145.04 lbf/in^2

Symbol	Nominal analysis, supplier, condition and remarks.
BS 1059	Phosphor bronze ingot & casting; replaced by BS 1400
BS 1060	Phosphor bronze ingot & casting; replaced by BS 1400
BS 1061	Phosphor bronze ingot & casting; replaced by BS 1400
BS 1158	Copper alloy: Ingot & casting; replaced by BS 1400
BS 1159	Copper alloy: Ingot & casting; replaced by BS 1400
BS 1400 G1/1	10% Sn 2.25% Zn Cu alloy: Gunmetal ingot; sand cast **UTS: 240 Elon: 15%**
BS 1400 G1 C	10% Sn 2% Zn Cu alloy: Gunmetal casting; chill cast **UTS: 220 Elon: 3% Proof: 120**
BS 1400 G2/1	8% Sn 4.25% Zn Cu alloy: Gunmetal ingot; sand cast **UTS: 240 Elon: 15%**
BS 1400 G2 C	8% Sn 4% Zn Cu alloy: Gunmetal casting; chill cast **UTS: 210 Elon: 3% Proof: 120**
BS 1400 G3/1	7% Sn 2.5% Zn 5.5% Ni Cu alloy: Ni gunmetal ingot; sand cast **UTS: 270 Elon: 18%**
BS 1400 G3 C	7% Sn 2.2% Zn 5.5% Ni 0.4% Pb Cu alloy: Casting; sand cast; as cast **UTS: 270 Elon: 18% Proof: 140**
BS 1400 G3 WP	7% Sn 2.2% Zn 5.5% Ni 0.4% Pb Cu alloy: Casting **DPN: 160 UTS: 420 Elon: 3% Proof: 270**
BS 1400 LB1/1	9% Sn 15% Pb Cu alloy: Lead bronze ingot; sand cast **UTS: 170 Elon: 4%**
BS 1400 LB1 C	9% Sn 15% Pb Cu alloy: Lead bronze casting; chill cast **UTS: 200 Elon: 3% Proof: 120**
BS 1400 LB2/1	10% Sn 10% Pb Cu alloy: Lead bronze ingot; sand cast **UTS: 180 Elon: 5%**
BS 1400 LB2 C	10% Sn 10% Pb Cu alloy: Lead bronze casting; chill cast **UTS: 210 Elon: 3% Proof: 140**
BS 1400 LB3/1	10% Sn 5% Pb Cu alloy: Lead bronze ingot; sand cast **UTS: 180 Elon: 5%**
BS 1400 LB3 C	10% Sn 5% Pb Cu alloy: Lead bronze casting; chill cast **UTS: 200 Elon: 3% Proof: 140**
BS 1400 LB4/1	5% Sn 10% Pb Cu alloy: Lead bronze ingot; sand cast **UTS: 150 Elon: 8%**
BS 1400 LB4 C	5% Sn 10% Pb Cu alloy: Lead bronze casting; chill cast **UTS: 200 Elon: 5% Proof: 75**
BS 1400 LB5/1	5% Sn 21% Pb Cu alloy: Lead bronze ingot; sand cast **UTS: 150 Elon: 6%**
BS 1400 LB5 C	5% Sn 21% Pb Cu alloy: Lead bronze casting; chill cast **UTS: 170 Elon: 6% Proof: 75**
BS 1400 LG1/1	3% Sn 9% Zn 5% Pb Cu alloy: Lead gunmetal ingot; sand cast **UTS: 170 Elon: 12%**
BS 1400 LG1 C	3% Sn 8.5% Zn 4.5% Pb Cu alloy: Lead gunmetal casting; sand cast **UTS: 170 Elon: 12%**
BS 1400 LG2/1	5% Sn 5.5% Zn 5% Pb Cu alloy: Lead gunmetal ingot; sand cast **UTS: 200 Elon: 15%**

Symbol	Nominal analysis, supplier, condition and remarks.
BS 1400 LG2 C	5% Sn 5% Zn 5% Pb Cu alloy: Lead gunmetal casting; chill cast **UTS: 200 Elon: 7% Proof: 100**
BS 1400 LG3/1	7% Sn 4.5% Zn 2% Pb Cu alloy: Lead gunmetal bronze ingot; sand cast **UTS: 210 Elon: 12%**
BS 1400 LG3 C	7% Sn 4% Zn 2% Pb Cu alloy: Lead gunmetal bronze casting; chill cast **UTS: 220 Elon: 5% Proof: 110**
BS 1400 LG4/1	7% Sn 2% Zn 3% Pb Cu alloy: Lead gunmetal bronze ingot; sand cast **UTS: 240 Elon: 18%**
BS 1400 LPB1/1	7.5% Sn 0.4% P 3.5% Pb Cu alloy: Ingot; sand cast; leaded phosphor bronze **UTS: 180 Elon: 3%**
BS 1400 LPB1 C	7.5% Sn 0.3% P 3.5% Pb Cu alloy: Casting; chill cast; leaded phosphor bronze **UTS: 210 Elon: 2% Proof: 120**
BS 1400 PB1/1	10% Sn 0.6% P Cu alloy: Phosphor bronze casting; sand cast **UTS: 210 Elon: 3%**
BS 1400 PB1 C	10% Sn 0.5% P Cu alloy: Phosphor bronze casting; chill cast **UTS: 300 Elon: 2% Proof: 170**
BS 1400 PB2/1	12% Sn 0.25% P Cu alloy: Phosphor bronze ingot; sand cast **UTS: 210 Elon: 5%**
BS 1400 PB2 C	12% Sn 0.15% P Cu alloy: Phosphor bronze casting; chill cast **UTS: 240 Elon: 3% Proof: 170**
BS 1400 PB3/1	9.5% Sn 0.25% P Cu alloy: Phosphor bronze ingot; sand cast **UTS: 220 Elon: 7%**
BS 1400 PB3 C	9.5% Sn 0.25% P Cu alloy: Phosphor bronze casting; chill cast **UTS: 220 Elon: 7% Proof: 120**
BS 1400 PB4/1	9.5% Sn 0.6% P Cu alloy: Phosphor bronze ingot; sand cast **UTS: 180 Elon: 3%**
BS 1400 PB4 C	9.5% Sn 0.5% P Cu alloy: Phosphor bronze casting; chill cast **UTS: 240 Elon: 2% Proof: 140**
BS 1400 SCB1/1	25% Zn 2.5% Sn 3.5% Pb Cu alloy: Brass ingot; sand cast **UTS: 170 Elon: 20%**
BS 1400 SCB1 C	25% Zn 2.5% Sn 3.5% Pb Cu alloy: Brass casting; sand cast **UTS: 170 Elon: 20%**
BS 1400 SCB4/1	40% Zn 1.2% Sn Cu alloy: Naval brass ingot; sand cast **UTS: 160 Elon: 20%**
BS 1400 SCB4 C	40% Zn 1.2% Sn Cu alloy: Naval brass casting; sand cast **UTS: 160 Elon: 20% Proof: 450**
BS 1400 SCB5/1	10% Zn 1.5% Sn Cu alloy: Brass ingot; sand cast for brazed parts in contact with salt water **UTS: 180 Elon: 20%**
BS 1400 SCB5 C	10% Zn 1.5% Sn Cu alloy: Brass casting; sand cast for brazed parts in contact with salt water **UTS: 180 Elon: 20%**
BS 2061	5% Sn 0.3% P Cu alloy: Spring washers; specification is for spring washers **DPN: 200**
BS 2870 PB101	0.3% P 4% Sn Cu alloy: Sheet; hard **DPN: 170 UTS: 520 Elon: 4% Proof: 340**
BS 2870 PB101	0.3% P 4% Sn Cu alloy: Sheet; extra hard **DPN: 190 UTS: 600 Proof: 450**
BS 2870 PB101	0.3% P 4% Sn Cu alloy: Sheet; annealed **DPN: 80 UTS: 290 Elon: 40%**

Symbol	Nominal analysis, supplier, condition and remarks.
BS 2870 PB101	0.3% P 4% Sn Cu alloy: Sheet; ¼-hard **DPN: 100 UTS: 330 Elon: 30% Proof: 100**
BS 2870 PB101	0.3% P 4% Sn Cu alloy: Sheet; ½-hard **DPN: 140 UTS: 440 Elon: 8% Proof: 240**
BS 2870 PB102	0.2% P 5% Sn Cu alloy: Sheet; annealed **DPN: 85 UTS: 300 Elon: 45%**
BS 2870 PB102	0.2% P 5% Sn Cu alloy: Sheet; ¼-hard **DPN: 110 UTS: 340 Elon: 35% Proof: 120**
BS 2870 PB102	0.2% P 5% Sn Cu alloy: Sheet; ½-hard **DPN: 150 UTS: 460 Elon: 10% Proof: 300**
BS 2870 PB102	0.2% P 5% Sn Cu alloy: Sheet; hard **DPN: 170 UTS: 530 Elon: 4% Proof: 370**
BS 2870 PB102	0.2% P 5% Sn Cu alloy: Sheet; extra hard **DPN: 210 UTS: 630**
BS 2870 PB103	0.2% P 7% Sn Cu alloy: Sheet; extra hard **DPN: 215 UTS: 680 Proof: 530**
BS 2870 PB103	0.2% P 7% Sn Cu alloy: Sheet; spring hard **DPN: 230**
BS 2870 PB103	0.2% P 7% Sn Cu alloy: Sheet; extra hard **DPN: 240 (min)**
BS 2870 PB103	0.2% P 7% Sn Cu alloy: Sheet; annealed **DPN: 90 UTS: 340 Elon: 50%**
BS 2870 PB103	0.2% P 7% Sn Cu alloy: Sheet; ¼-hard **DPN: 115 UTS: 370 Elon: 40% Proof: 180**
BS 2870 PB103	0.2% P 7% Sn Cu alloy: Sheet; ½-hard **DPN: 160 UTS: 480 Elon: 12% Proof: 300**
BS 2870 PB103	0.2% P 7% Sn Cu alloy: Sheet; hard **DPN: 180 UTS: 580 Elon: 6% Proof: 400**
BS 2870 PB104	0.3% P 8% Sn Cu alloy: Sheet; specification not issued
BS 2871 PB101	3.5% Sn 0.3% P Cu alloy: Tube; specification not issued
BS 2871 PB102	5% Sn 0.2% P Cu alloy: Tube; annealed **UTS: 320**
BS 2871 PB103	7% Sn 0.2% P Cu alloy: Tube; specification not issued
BS 2871 PB104	8% Sn 0.3% P Cu alloy: Tube; annealed **UTS: 360**
BS 2872 PB101	3.5% Sn 0.3% P Cu alloy: Forging; specification not issued
BS 2872 PB102	5% Sn 0.2% P Cu alloy: Forging; specification not issued
BS 2872 PB103	7% Sn 0.2% P Cu alloy: Forging; specification not issued
BS 2872 PB104	8% Sn 0.3% P Cu alloy: Forging; specification not issued
BS 2873 PB101	3.5% Sn 0.3% P Cu alloy: Wire; specification not issued
BS 2873 PB102	5% Sn 0.2% P Cu alloy: Wire; annealed **UTS: 330 Elon: 40%**
BS 2873 PB102	5% Sn 0.2% P Cu alloy: Wire; extra hard **UTS: 830**
BS 2873 PB103	7% Sn 0.2% P Cu alloy: Wire; annealed **UTS: 260 Elon: 45%**
BS 2873 PB103	7% Sn 0.2% P Cu alloy: Wire; extra hard **UTS: 880**
BS 2873 PB104	8% Sn 0.3% P Cu alloy: Wire; specification not issued

Note. The following abbreviations and units are used in the tables:

DPN	Hardness, diamond pyramid number
UTS	Ultimate tensile strength, N/mm^2
Elon	Elongation, %
Proof	0.1 % proof strength, N/mm^2

$1 N/mm^2 = 0.1 \text{ hbar} = 0.102 \text{ kgf}/mm^2 = 0.06475 \text{ tonf}/in^2 = 145.04 \text{ lbf}/in^2$

Symbol	Nominal analysis, supplier, condition and remarks.
BS 2874 PB101	3.5% Sn 0.3% P Cu alloy: Bar; specification not issued
BS 2874 PB102	5% Sn 0.2% P Cu alloy: Bar; as rolled **UTS: 370 Elon: 20% Proof: 300**
BS 2874 PB103	7% Sn 0.2% P Cu alloy: Bar; specification not issued
BS 2874 PB104	8% Sn 0.3% P Cu alloy: Bar; as rolled **UTS: 450 Elon: 20%**
BS 2875 PB101	3.5% Sn 0.3% P Cu alloy: Plate; as rolled **UTS: 290 Elon: 35%**
BS 2875 PB101	3.5% Sn 0.3% P Cu alloy: Plate; hard **UTS: 400 Elon: 15%**
BS 2875 PB102	5% Sn 0.2% P Cu alloy: Plate; annealed **UTS: 300 Elon: 40%**
BS 2875 PB102	5% Sn 0.2% P Cu alloy: Plate; hard **UTS: 400 Elon: 15%**
BS 2875 PB103	7% Sn 0.2% P Cu alloy: Plate; specification not issued
BS 2875 PB104	8% Sn 0.3% P Cu alloy: Plate; specification not issued
BS 2901 C10	5% Sn 0.3% P Cu alloy: Rod for all forms of welding
BS 2901 C11	7% Sn 0.3% P Cu alloy: Rod for all forms of welding
BS 3127/5	5% Sn 0.15% P Cu alloy: Tube; ½-hard; drawn for bourdon tubes **DPN: 130**
BS B8	10% Sn 0.5% P Cu alloy: Casting; chill cast for bearings **UTS: 240 Elon: 3%**
BS T52	5.5% Sn 0.2% P Cu alloy: Tube; hard drawn **UTS: 360**
CAROBRONZE	8.5% Sn 0.3% P Cu alloy: Bar; Carobronze; cold drawn **DPN: 145 UTS: 530 Elon: 25% Proof: 370**
CHINESE BRONZE	22% Sn Cu alloy: Origin unknown
COGWHEEL 11	Sn P Cu alloy: Casting; Phosphor Bronze Ltd; chill cast **DPN: 100 UTS: 290 Elon: 7% Proof: 150**
COGWHEEL VII	Sn 1.5% P Cu alloy: Casting; Phosphor Bronze Ltd; chill cast **DPN: 125 UTS: 340 Elon: 3% Proof: 180**
COGWHEEL VIII	High Sn P Cu alloy: Casting; Phosphor Bronze Ltd; chill cast; for bushes **DPN: 140 UTS: 340 Elon: 2% Proof: 220**
COGWHEEL XI	Sn P 8% Pb Cu alloy: Casting; Phosphor Bronze Ltd; chill cast; bearings **DPN: 100 UTS: 270 Elon: 1.5% Proof: 170**
COINAGE BRONZE	4% Sn 1% Zn Cu alloy: Wrought tin bronze used for coins; origin unknown
CUSILAY	1.5% Sn 2% Si 0.85% Fe Cu alloy: Origin unknown
DIN 1705 G Sn Bz 10	10% Sn 1.0% Pb 0.5% Zn Cu alloy: Casting
DIN 1705 G Sn Bz 14	14% Sn 1.0% Pb 0.5% Zn Cu alloy: Casting
DIN 1705 Rg 5	5% Zn 5% Sn 5% Pb Cu alloy: Casting
DIN 1705 Rg 7	4% Zn 7% Sn 6% Pb Cu alloy: Casting
DIN 1705 Rg 10	2% Zn 10.5% Sn 1.0% Pb Cu alloy: Casting
DIN 1716 G Pb Bz 25	26% Pb 3% Sn 2% Ni Cu alloy: Casting **DPN: 30**
DIN 1716 G Sn Pb Bz 5	10% Sn 5% Pb 1% Ni Cu alloy: Casting **DPN: 85 UTS: 220 Elon: 18% Proof: 120**
DIN 1716 G Sn Pb Bz 10	10% Sn 10% Pb 1% Ni Cu alloy: Casting **DPN: 75 UTS: 220 Elon: 14% Proof: 100**
DIN 1716 G Sn Pb Bz 15	8% Sn 15% Pb 1.5% Ni Cu alloy: Casting **DPN: 70 UTS: 210 Elon: 12% Proof: 100**
DIN 1716 G Sn Pb Bz 20	5% Sn 20% Pb 2% Ni Cu alloy: Casting **DPN: 55 UTS: 200 Elon: 10% Proof: 90**
DIN 1733 S Sn Bz6	7% Sn 0.2% Pb Cu alloy

Symbol	Nominal analysis, supplier, condition and remarks.
DIN 1733 S Sn Bz 12	0.2% P 12% Sn Cu alloy: Wrought
DIN 1785 K Sn Bz 2	1.7% Sn Cu alloy: Wrought
DIN 17656 GB Rg 10	2% Zn 10% Sn Cu alloy: Casting
DIN 17656 GB Sn Bz 14	14% Sn Cu alloy: Casting
DIN 17656 GB Sn Pb Bz 5	10% Sn 5% Pb Cu alloy: Casting
DIN 17656 GB Sn Pb Bz 10	10% Sn 10% Pb Cu alloy: Casting
DTD 265 A	8% Sn 0.3% P Cu alloy: Bar & tube; hard drawn; for bushes **DPN: 120 UTS: 450 Elon: 20%**
DTD 318 A	15% Zn 4.0% Sn 1.0% Fe Cu alloy: Tube; annealed **UTS: 450**
DTD 459	15% Sn 5% Pb Cu alloy: Casting lead bronze; sand or chill cast; for bearings
ENCON	Sn Cu alloy: Continuously cast; Enfield alloys for BS 1400
ESCO TCC1	10% Sn 0.25% Pb 0.5% P Cu alloy: Casting; Eyre Smelting for BS alloy PB 1
ESCO TCC2	7% Sn 3.5% Pb 2% Zn 1% Ni 0.3% P Cu alloy: Casting; Eyre Smelting Co for BS alloy LPB 1
G I BRONZE	10% Sn 2% Zn Cu alloy: Casting continuous cast; J Holroyd for BS alloy G1 **DPN: 105 UTS: 360 Elon: 25% Proof: 170**
GB Cu Pb 5 Sn	10% Sn 5% Pb Cu alloy: Casting; designation used by German Standard
GB Cu Pb 10 Sn	10% Sn 10% Pb Cu alloy: Casting; designation used by German Standard
GB Cu Pb 15 Sn	8% Sn 15% Pb Cu alloy: Casting; designation used by German Standard
GB Cu Pb 20 Sn	4.5% Sn 21% Pb Cu alloy: Casting; designation used by German Standard
GB Cu Sn 5 Zn Pb	5% Zn 5% Sn 5% Pb Cu alloy: Casting; designation used by German Standard
GB Cu Sn 10	10% Sn Cu alloy: Casting; designation used by German Standard
GB Cu Sn 12	12% Sn Cu alloy: Casting; designation used by German Standard
GB Cu Sn 14	14% Sn Cu alloy: Casting; designation used by German Standard
GB Rg 5	5% Zn 5% Sn 5% Pb Cu alloy: Casting; designation used by German Standard
GB Rg 7	4.5% Zn 7% Sn 6% Pb Cu alloy: Casting; designation used by German Standard
GB Rg 10	2% Zn 10% Sn Cu alloy: Casting; designation used by German Standard
GB Sn Bz 10	10% Sn Cu alloy: Casting; designation used by German Standard
GB SN BZ 12	12% Sn Cu alloy: Casting; designation used by German Standard
GB Sn Bz 14	14% Sn Cu alloy: Casting; designation used by German Standard
GB Sn Pb Bz 5	10% Sn 5% Pb Cu alloy: Casting; designation used by German Standard
GB Sn Pb Bz 10	10% Sn 10% Pb Cu alloy: Casting; designation used by German Standard
GB Sn Pb Bz 15	8% Sn 15% Pb Cu alloy: Casting; designation used by German Standard
GB Sn Pb Bz 20	4.5% Sn 21% Pb Cu alloy: Casting; designation used by German Standard
GB Cu Sn7 Zn Pb	4.5% Zn 7% Sn 6% Pb Cu alloy: Casting; designation used by German Standard
GB Cu Sn10 Zn	2% Zn 10% Sn Cu alloy: Casting; designation used by German Standard

Symbol	Nominal analysis, supplier, condition and remarks.
GC Cu Sn5 Zn Pb	5% Zn 5% Sn 5% Pb Cu alloy: Casting; designation used by German Standard
GC Cu Sn7 Zn Pb	4% Zn 7% Sn 6% Pb Cu alloy: Casting; designation used by German Standard
GC Cu Sn10 Zn	2.0% Zn 10.5% Sn 1.0% Pb Cu alloy: Casting; designation used by German Standard
GC Cu Sn 12	12% Sn 1.0% Pb 0.5% Zn Cu alloy: Casting; designation used by German Standard
GC Rg 5	5% Zn 5% Sn 5% Pb Cu alloy: Casting; designation used by German Standard
GC Rg 7	4% Zn 7% Sn 6% Pb Cu alloy: Casting; designation used by German Standard
GC Rg 10	2.0% Zn 10.5% Sn 1.0% Pb Cu alloy: Casting; designation used by German Standard
GC Sn Bz 12	12% Sn 1.0% Pb 0.5% Zn Cu alloy: Casting; designation used by German Standard
G Cu Pb 5 Sn	10% Sn 5% Pb 1% Ni Cu alloy: Casting; designation used by German Standard
G Cu Pb 10 Sn	10% Sn 10% Pb 1% Ni Cu alloy: Casting; designation used by German Standard
G Cu Pb 15 Sn	8% Sn 15% Pb 1.5% Ni Cu alloy: Casting; designation used by German Standard
G Cu Pb 20 Sn	5% Sn 20% Pb 2% Ni Cu alloy: Casting; designation used by German Standard
G Cu Pb 25	26% Pb 3% Sn 2% Ni Cu alloy: Casting; designation used by German Standard
G Cu Sn5 Zn Pb	5% Zn 5% Sn 5% Pb Cu alloy: Casting; designation used by German Standard
G Cu Sn7 Zn Pb	4% Zn 7% Sn 6% Pb Cu alloy: Casting; designation used by German Standard
G Cu Sn 10	10% Sn 1.0% Pb 0.5% Zn Cu alloy: Casting; designation used by German Standard
G Cu Sn 10 Zn	2% Zn 10.5% Sn 1.0% Pb Cu alloy: Casting; designation used by German Standard
G Cu Sn 12	12% Sn 1.0% Pb 0.5% Zn Cu alloy: Casting; designation used by German Standard
G Cu Sn 14	14% Sn 1.0% Pb 0.5% Zn Cu alloy: Casting; designation used by German Standard
G Pb Bz 25	26% Pb 3% Sn 2% Ni Cu alloy: Casting; designation used by German Standard
G Sn Bz 10	10% Sn 1.0% Pb 0.5% Zn Cu alloy: Casting; designation used by German Standard
G Sn Bz 12	12% Sn 1.0% Pb 0.5% Zn Cu alloy: Casting; designation used by German Standard
G Sn Bz 14	14% Sn 1.0% Pb 0.5% Zn Cu alloy: Casting; designation used by German Standard
G Sn Pb Bz 5	10% Sn 5% Pb 1% Ni Cu alloy: Casting; designation used by German Standard
G Sn Pb Bz 10	10% Sn 10% Pb 1% Ni Cu alloy: Casting; designation used by German Standard
G Sn Pb Bz 15	8% Sn 15% Pb 1.5% Ni Cu alloy: Casting; designation used by German Standard
G Sn Pb Bz 20	5% Sn 20% Pb 2% Ni Cu alloy: Casting; designation used by German Standard
GZ Cu Sn5 Zn Pb	5% Zn 5% Sn 5% Pb Cu alloy: Casting; designation used by German Standard
GZ Cu Sn10 Zn	2% Zn 10.5% Sn 1.0% Pb Cu alloy: Casting; designation used by German Standard
GZ Cu Sn 12	12% Sn 1.0% Pb 0.5% Zn Cu alloy: Casting; designation used by German Standard
GZ Rg 5	5% Zn 5% Sn 5% Pb Cu alloy: Casting; designation used by German Standard
GZ Rg 7	4% Zn 7% Sn 6% Pb Cu alloy: Casting; designation used by German Standard
GZ Rg 10	2% Zn 10.5% Sn 1.0% Pb Cu alloy: Casting; designation used by German Standard
GZ Sn Bz 12	12% Sn 1.0% Pb 0.5% Zn Cu alloy: Casting; designation used by German Standard
HOLFOS BO	11.5% Sn 0.7% P Cu alloy: Casting; Holroyd **DPN: 100 UTS: 300 Elon: 5% Proof: 210**

Symbol	Nominal analysis, supplier, condition and remarks.
HOLFOS BOHT	0.7% P 11.5% Sn Cu alloy: Casting; Holroyd; heat treated
	DPN: 90 UTS: 330 Elon: 20% Proof: 170
HOLFOS G1	10% Sn 2% Zn Cu alloy: Casting; Holroyd; centrifugally cast
	DPN: 95 UTS: 300 Elon: 10% Proof: 170
HOLFOS G2	8% Sn 4% Zn Cu alloy: Casting; Holroyd; centrifugally cast
	DPN: 90 UTS: 240 Elon: 9% Proof: 140
HOLFOS G3	7% Ni 2.2% Zn 0.5% P (max) 5.5% Ni Cu alloy: Casting; Holroyd; centrifugally cast; as cast
	DPN: 95 UTS: 300 Elon: 14% Proof: 150
HOLFOS G3WP	7% Ni 2.2% Zn 0.5% P (max) 5.5% Ni Cu alloy: Holroyd; centrifugally cast; heat treated
	DPN: 180 UTS: 420 Elon: 3% Proof: 270
HOLFOS JH17	1.3% P 14% Sn Cu alloy: Casting; Holroyd
	DPN: 140 UTS: 340 Elon: 1.0% Proof: 270
HOLFOS LB1	9% Sn 15% Pb Cu alloy: Casting; Holroyd; centrifugally cast
	DPN: 70 UTS: 210 Elon: 7% Proof: 140
HOLFOS LB2	10% Sn 10% Pb Cu alloy: Casting; Holroyd; centrifugally cast
	DPN: 90 UTS: 240 Elon: 8% Proof: 170
HOLFOS LB3	10% Sn 5% Pb Cu alloy: Casting; Holroyd; centrifugally cast
	DPN: 95 UTS: 250 Elon: 12% Proof: 210
HOLFOS LB4	5% Sn 10% Pb Cu alloy: Casting; Holroyd; centrifugally cast
	DPN: 70 UTS: 220 Elon: 10% Proof: 140
HOLFOS LB5	5% Sn 20% Pb Cu alloy: Casting; Holroyd; centrifugally cast
	DPN: 70 UTS: 170 Elon: 7% Proof: 140
HOLFOS LG1	3% Sn 9% Zn 5% Pb Cu alloy: Casting; Holroyd; centrifugally cast
	DPN: 70 UTS: 220 Elon: 35% Proof: 120
HOLFOS LG2	5% Sn 5% Zn 5% Pb Cu alloy: Casting; Holroyd; centrifugally cast
	DPN: 75 UTS: 210 Elon: 15% Proof: 140
HOLFOS LG3	7% Sn 5% Zn 2% Pb Cu alloy: Casting; Holroyd; centrifugally cast
	DPN: 80 UTS: 240 Elon: 12% Proof: 170
HOLFOS LG4	7% Sn 3% Zn 3% Pb Cu alloy: Casting; Holroyd; centrifugally cast
	DPN: 70 UTS: 250 Elon: 15% Proof: 140
HOLFOS LG773	7% Sn 3% Zn 7% Pb Cu alloy: Casting; Holroyd; centrifugally cast
	DPN: 85 UTS: 300 Elon: 15% Proof: 140
HOLFOS LPB1	0.3% (min) 4% Pb 7.5% Sn Cu alloy: J Holroyd; centrifugally cast
	DPN: 85 UTS: 270 Elon: 20% Proof: 170
HOLFOS PB1	0.5% P (min) 10% Sn Cu alloy: Casting; Holroyd; centrifugally cast
	DPN: 100 UTS: 300 Elon: 10% Proof: 210
HOLFOS PB2	0.15% P (min) 11.5% Sn Cu alloy: Casting; Holroyd; centrifugally cast
	DPN: 100 UTS: 300 Elon: 8% Proof: 180
HOLFOS PB3	0.3% P 9.5% Sn Cu alloy: Casting; Holroyd; centrifugally cast
	DPN: 95 UTS: 270 Elon: 9% Proof: 180

Symbol	Nominal analysis, supplier, condition and remarks.
HOLFOS SPUNCAST	11.5% Sn 0.7% P Cu alloy: Casting; spuncast; J Holroyd; alternative name for B O BRONZE
HOLFOS WW	11.5% Sn 0.2% P Cu alloy: Casting; chill cast; J Holroyd for BS alloy PB 2
	DPN: 90 UTS: 270 Elon: 5% Proof: 170
IMI 657	5% Sn 0.2% P Cu alloy: Strip & wire; IMI for BS alloy PB 102
IMI 659	7% Sn 0.2% P Cu alloy: Strip & wire; IMI for BS alloy PB 103
IMI 660	8% Sn 0.3% P Cu alloy: Bar & wire; IMI
IMI 661	8% Sn 0.3% P Cu alloy: Strip & wire; IMI for BS alloy PB 104; withdrawn
JH 17	14% Sn 1.3% P Cu alloy: Casting; spuncast; J Holroyd
	DPN: 140 UTS: 320 Elon: 1% Proof: 270
K Sn Bz 2	1.7% Sn Cu alloy: Wrought; designation used by German Standard
NBN 267/21/22 Br Sn 10	0.7% P 11.5% Sn Cu alloy: Casting; Belgian specification
NBN 267/21/22 Br Sn 10	0.5% P 10% Sn Cu alloy: Casting; Belgian specification
NBN 267/21/22 Br Sn 12	0.15% P 11.5% Sn Cu alloy: Casting; Belgian specification
NBN 267/21/22 Br Sn 5 Zn 5 Pb 5	5% Sn 5% Zn 5% Pb Cu alloy: Casting; Belgian specification
NBN 267/21/22 Br Sn 7 Zn 3 Pb 4	7% Sn 3% Zn 4% Pb Cu alloy: Casting; Belgian specification
NBN 267/21/22 Br Sn 10 Zn 2	10% Sn 2% Zn Cu alloy: Casting; Belgian specification
NBN 267/21/22 Br Sn 12	11.5% Sn 0.2% P Cu alloy: Casting; Belgian specification
NBN 267/21/23 Br Sn 9 P	0.3% P 9.5% Sn Cu alloy: Casting; Belgian specification
NBN 267/23 Br Pb 15 Sn 8	9% Sn 1.5% Pb Cu alloy: Casting; Belgian specification
NBN 267/23 Br Pb 24 Sn 5	5% Sn 20% Pb Cu alloy: Casting; Belgian specification
NBN 267/23 Br Sn 10 Pb 10	10% Sn 10% Pb Cu alloy: Casting; Belgian specification
NF A53 012	Classification of bronze alloy; French Standard
NF A53 607	General specification covering range of bronze alloys; French Standard
NF A53 707	Classification of bronze alloys; French Standard
No. 1 GUNMETAL	10% Sn 1.5% Pb 2% Zn 1% Ni Cu alloy: Casting; Stone Manganese Marine for BS alloy G1
No. 1 LEADED BRONZE	10% Sn 10% Pb Cu alloy: Casting; Stone Manganese Marine; as AMS 4842
No. 1 PHOSPHOR BRONZE	12% Sn 0.5% Pb 0.15% P Cu alloy: Casting; Stone Manganese Marine for BS alloy PB2
No. 2 GUNMETAL	7% Sn 2% Pb 4% Zn 2% Ni Cu alloy: Casting; Stone Manganese Marine for BS alloy LG3
No. 2 LEADED BRONZE	9% Sn 15% Pb 1% Zn 2% Ni 0.1% P Cu alloy: Casting; Stone Manganese Marine for BS alloy LB1
No. 2 PHOSPHOR BRONZE	10% Sn 0.25% Pb 0.5% P Cu alloy: Casting; Stone Manganese Marine for BS alloy PB1
No. 3 GUNMETAL	5% Sn 5% Pb 5% Zn 2% Ni Cu alloy: Casting; Stone Manganese Marine for BS alloy LG2
No. 3 PHOSPHOR BRONZE	9% Sn 0.25% Pb 0.3% P Cu alloy: Casting; Stone Manganese Marine for BS alloy PB 3
No. 4 LEADED BRONZE	5% Sn 9% Pb 0.5% Ni Cu alloy: Casting; Stone Manganese Marine; as SAE 66
No. 4 PHOSPHOR BRONZE	7% Sn 3% Pb 2% Zn 1% Ni 0.3% P Cu alloy: Casting; Stone Manganese Marine for BS alloy LPB1
NOIL	20% Sn Cu alloy: Obsolete; consult the Copper Development Association
	DPN: 158

Note. The following abbreviations and units are used in the tables:

DPN	Hardness, diamond pyramid number
UTS	Ultimate tensile strength, N/mm^2
Elon	Elongation, %
Proof	0.1 % proof strength, N/mm^2

1 N/mm^2=0.1 hbar=0.102 kgf/mm^2=0.06475 tonf/in^2=145.04 lbf/in^2

Symbol	Nominal analysis, supplier, condition and remarks.
OUNCE METAL	5% Sn 5% Pb 5% Zn 1% Ni (max) Cu alloy: Casting; common name; refer to Copper Development Association for further information
PHOSPHOR BRONZE	10% Sn with up to 0.5% P Cu alloy: Covered by BS alloys PB 1-4
QQ B 691/1	4.0% Zn 6.0% Sn 1.5% Pb Cu alloy: Casting; US Federal **UTS: 220 Elon: 22%**
QQ B 691/2	5% Zn 5% Sn 5% Pb Cu alloy: Casting; US Federal **UTS: 200 Elon: 20%**
QQ B 691/3	13% Zn 5% Sn 2.5% Pb 0.7% Ni Cu alloy: Casting; US Federal **UTS: 170 Elon: 15%**
QQ B 691/6	4.0% Zn 8% Sn Cu alloy: Casting; US Federal **UTS: 220 Elon: 22%**
QQ B 691/11	8% Zn 3.0% Sn 7% Pb Cu alloy: Casting; US Federal **UTS: 200 Elon: 13%**
QQ B 691b/5	4.0% Zn 8% Sn Cu alloy: Casting; US Federal **UTS: 270 Elon: 20%**
QQ B 746/A	4.0% Sn Cu alloy: Bar & forging; US Federal **UTS: 450 Elon: 15% Proof: 270**
QQ C 591a/B	1.2% Zn 2.2% Sn 1.5% Si Cu alloy: Rod; US Federal **UTS: 520 Elon: 10% Proof: 370**
QQ W 401	3.5% Sn 0.3% P Cu alloy: Wire; US Federal; hard drawn for springs **UTS: 750 Elon: 5%**
SAE 40	5% Zn 5% Sn 5% Pb Cu alloy: Casting; as cast **UTS: 200 Elon: 20%**
SAE 41	30% Zn 2.5% Pb 1.5% Sn Cu alloy: Casting; as cast **UTS: 200 Elon: 20%**
SAE 48	30% Pb Cu alloy: For bearings; as cast
SAE 49	24% Pb Cu alloy: For bearings; as cast
SAE 62	10% Sn 0.3% Pb 2% Zn 1% Ni Cu alloy: Casting; as cast **UTS: 270 Elon: 20%**
SAE 63	10% Sn 2% Pb 1% Ni 0.2% P Cu alloy: Casting; as cast **UTS: 240 Elon: 10%**
SAE 64	10% Sn 9% Pb 0.5% Ni 0.25% P Cu alloy: Casting; as cast **UTS: 170 Elon: 8%**
SAE 65	11% Sn 0.5% Pb 0.2% P Cu alloy: Casting; as cast **UTS: 240 Elon: 10%**
SAE 66	5% Sn 9% Pb 0.5% Ni Cu alloy: Casting; as cast **UTS: 170 Elon: 8%**
SAE 67	6% Sn 16% Pb 0.7% Ni Cu alloy: Casting; semi-plastic bronze; as cast **UTS: 120 Elon: 10%**
SAE 77 A	4% Sn 0.3% P Cu alloy: Sheet & strip; annealed **UTS: 340**
SAE 77 A	4% Sn 0.3% P Cu alloy: Sheet & strip; extra hard **UTS: 600**
SAE 77 C	8% Sn 0.2% P Cu alloy: Sheet & strip; annealed **UTS: 420**
SAE 77 C	8% Sn 0.2% P Cu alloy: Sheet & strip; extra hard **UTS: 750**
SAE 81	4% Sn 0.25% P Cu alloy: Bar & wire; hard drawn **UTS: 830 Elon: 3.5%**
SAE 480	35% Pb Cu alloy: For bearings; as cast
SAE 481	40% Pb Cu alloy: For bearings; as cast
SAE 482	5.0% Sn 28% Pb Cu alloy: For bearings
SAE 484	3.0% Sn 42% Pb Cu alloy: For bearings
SAE 620	8% Sn 0.3% Pb 4% Zn 1% Ni Cu alloy: Casting; as cast **UTS: 270 Elon: 20%**
SAE 621	8% Sn 1% Pb 4% Zn 1% Ni Cu alloy: Casting; as cast **UTS: 240 Elon: 18%**
SAE 622	6% Sn 1.5% Pb 4% Zn 1% Ni Cu alloy: Casting; as cast **UTS: 240 Elon: 22%**
SAE 640	11% Sn 1.2% Pb 1% Ni 0.25% P Cu alloy: Casting; as cast **UTS: 240 Elon: 10%**
SAE 660	7% Sn 7% Pb 3% Zn 0.5% Ni Cu alloy: Casting; as cast **UTS: 200 Elon: 12%**
SAE 791	4% Zn 4% Sn 4% Pb Cu alloy: For bearings; wrought
SAE 792	10% Sn 10% Pb Cu alloy: For bearings; as cast
SAE 793	4% Zn 4% Sn 8% Pb Cu alloy: For bearings; cast
SAE 794	3.5% Sn 23% Pb Cu alloy: For bearings; cast
SAE 797	10% Sn 10% Pb Cu alloy: For bearings; sintered
SAE 798	4% Zn 4% Sn 8% Pb Cu alloy: For bearings; sintered
SAE 799	3.5% Sn 23% Pb Cu alloy: For bearings; sintered
SAE 840	1.7% C 10% Sn Cu alloy: Sinter for bearings; density 6.0 g/cm³
SAE 841	1.7% C 10% Sn Cu alloy: Sinter for bearings; density 6.6 g/cm³
SAE 842	1.7% C 10% Sn Cu alloy: Sinter for bearings; density 7000 kg/m³
SAE 843	1.5% C 2.5% Pb Cu alloy: Sinter for bearings; density 6700 kg/m³
SAE CA 510	5% Sn 0.2% P Cu alloy: Wrought; formerly SAE 77A and SAE 81
SAE CA 521	8% Sn 0.2% P Cu alloy: Wrought; formerly SAE 77C
SAE CA 544	4% Zn 4% Sn 4% Pb Cu alloy: Wrought
SFG	0.2% P 5% or 7% Sn Cu alloy: Strip; IMI grade with improved properties over BS alloys PB 102-103
SIS 5203	7% Zn 5.5% Sn 3% Pb Cu alloy: Casting; Swedish Standard; as cast **DPN: 70 UTS: 200 Elon: 20% Proof: 100**
SIS 5204	5% Zn 5% Sn 5% Pb Cu alloy: Casting; Swedish Standard; as cast **DPN: 70 UTS: 200 Elon: 20% Proof: 100**
SIS 5420	1% Sn 0.2% P Cu alloy: Plate; Swedish Standard; annealed **DPN: 65 UTS: 270 Elon: 50% Proof: 60**
SIS 5420	1% Sn 0.2% P Cu alloy: Plate; Swedish Standard; cold drawn **DPN: 100 UTS: 340 Elon: 15% Proof: 290**
SIS 5428	7% Sn 0.3% P Cu alloy: Plate & bar; Swedish Standard; annealed **DPN: 85 UTS: 390 Elon: 55% Proof: 150**
SIS 5428	7% Sn 0.3% P Cu alloy: Plate & bar; Swedish Standard; cold rolled **DPN: 210 UTS: 720 Elon: 10% Proof: 630**
SIS 5431	9% Sn 0.3% P Cu alloy: Plate & bar; Swedish Standard; cold rolled **DPN: 200 UTS: 530 Elon: 25% Proof: 390**
SIS 5443	10% Sn Cu alloy: Casting; Swedish Standard; as cast **DPN: 75 UTS: 250 Elon: 18% Proof: 150**
SIS 5444	2% Zn 9% Sn 3% Pb Cu alloy: Casting; Swedish Standard; as cast **DPN: 75 UTS: 240 Elon: 16% Proof: 150**
SIS 5465	12% Sn Cu alloy: Casting; Swedish Standard; as cast **DPN: 90 UTS: 240 Elon: 12% Proof: 150**
SIS 5475	14% Sn Cu alloy: Casting; Swedish Standard; as cast **DPN: 100 UTS: 210 Elon: 4% Proof: 170**

Symbol	Nominal analysis, supplier, condition and remarks.
SIS 5640	10% Sn 10% Pb Cu alloy: Casting; Swedish Standard; as cast
	DPN: 70 UTS: 180 Elon: 12% Proof: 90
SIS 8300	0.1% C (max) sintered Cu
	UTS: 463 Elon: 28%
SIS 8320	0.5% C 10% Sn Cu alloy: Sinter
SIS 8321	1.5% C 10% Sn Cu alloy: Sinter
SIS 8330	0.5% C 10% Sn 10% P & Cu alloy: Sinter
SPECULUM	35% Sn Cu alloy: For electroplating; used for plating cutlery and household goods
SPOOLARC GRADE C	8.0% Sn 0.2% P Cu alloy: Brazing rod; All-State for welding phosphorus bronze
	DPN: 90 UTS: 420 Elon: 55% Proof: 200
SPRABRONZE P	5% Sn Cu alloy: Wire for spraying; Metco Ltd
	UTS: 100
STA 7 CG1	2% Zn 10% Sn Cu alloy: Bar; hard rolled
	UTS: 530 Elon: 18%
STA 7 CG2	2.0% Zn 10% Sn 0.5% Pb Cu alloy: Casting; sand cast
	UTS: 240 Elon: 10%
STA 7 CG3	4.0% Zn 8.0% Sn 0.5% Pb Cu alloy: Casting; sand cast
	UTS: 240 Elon: 10%
STA 7 CG4	5.0% Zn 7% Sn 2.0% Pb Cu alloy: Casting; sand cast
	UTS: 210 Elon: 10%
STA 7 CG5	5.0% Zn 5.0% Sn 5.0% Pb Cu alloy: Casting; sand cast
	UTS: 180 Elon: 10%
STA 7 CG6	9.0% Zn 3.0% Sn 5.0% Pb Cu alloy: Casting; sand cast
	UTS: 170 Elon: 10%
STA 7 CP1	5.0% Sn 0.3% P Cu alloy: Sheet, tube & wire; annealed
	DPN: 80 UTS: 300 Elon: 45%
STA 7 CP1	5.0% Sn 0.3% P Cu alloy: Sheet, tube & wire; hard
	UTS: 860
STA 7 CP2	4.0% Sn 0.3% P Cu alloy: Bar; as drawn
	UTS: 370 Elon: 20%
STA 7 CP3	2.0% Zn 7.0% Sn 3.0% Pb 1.0% Ni 0.3% P Cu alloy: Casting; chill cast
	UTS: 210 Elon: 1.5%
STA 7 CP4	8.0% Sn 0.3% P Cu alloy: Casting; as cast; for bearings
	UTS: 220 Elon: 10%
STA 7 CP5	10% Sn 0.2% Pb 0.5% P Cu alloy: Casting; chill cast
	UTS: 240 Elon: 1.5%
STA 7 CP6	12% Sn 0.5% Pb 0.5% Ni 0.15% P Cu alloy: Casting; chill cast
	DPN: 90 UTS: 250 Elon: 3%
STA 7 CX7	8.0% Sn 0.3% P Cu alloy: Bar & tube; hard drawn
	DPN: 120 UTS: 450 Elon: 20%
STA 7 CX8	2.0% Zn 9.0% Sn Cu alloy
	UTS: 210 Elon: 3.5%
STA 7 CX9A	1.0% Zn 15% Sn 0.2% Pb Cu alloy
STA 7 CX9B	0.7% Zn 18% Sn 0.2% Pb Cu alloy

Symbol	Nominal analysis, supplier, condition and remarks.
STA 7 CX12	2.0% Zn 6.0% Sn 2.0% Ni 0.3% P Cu alloy: Casting; as cast
	UTS: 300 Elon: 20% Proof: 150
STA 7 CZ12	35% Zn 2.0% Sn 3.0% Pb 1.0% Ni Cu alloy: Casting; as cast
	UTS: 170 Elon: 20%
STA 7 CZ13	35% Zn 1.0% Sn 3.0% Pb 1.0% Fe 1.0% Ni alloy: Casting; as cast
	UTS: 180 Elon: 12%
SUPER-HOLFOS WW	11.5% Sn 0.2% P Cu alloy: Casting; centri cast; J Holroyd for BS alloy PB 2
	DPN: 100 UTS: 300 Elon: 6% Proof: 180
T 52	5.5% Sn 0.2% P Cu alloy: Tube; see BS/T52
TAURUS	10.5% Sn 0.7% Ni 0.25% P Cu alloy: Casting; D Brown; centri cast
	DPN: 75 UTS: 270 Elon: 22% Proof: 170
TAURUS I	12% Sn 0.3% P Cu alloy: Casting; D Brown; centri cast
	DPN: 90 UTS: 280 Elon: 5% Proof: 210
TAURUS II	10% Sn 0.2% P Cu alloy: Casting; D Brown; sand cast
	DPN: 70 UTS: 270 Elon: 15% Proof: 120
TAURUS III	14% Sn 0.05% P Cu alloy: Casting; D Brown; sand cast
	DPN: 100 UTS: 210 Elon: 2% Proof: 180
TAURUS IV	10.5% Sn 1% Ni 0.25% P Cu alloy: Casting; D Brown; sand cast
	DPN: 80 UTS: 270 Elon: 20% Proof: 170
TAURUS VI	10% Sn 2% Zn Cu alloy: Casting; D Brown; sand cast
	DPN: 60 UTS: 270 Elon: 15% Proof: 140
TAURUS VII	12% Sn 1% Ni 5% Pb 0.3% P Cu alloy: Casting; D Brown; sand cast
	DPN: 75 UTS: 220 Elon: 6% Proof: 180
TAURUS VIII	10% Sn 1% Ni 3% Pb 0.25% P Cu alloy: Casting; D Brown; sand cast
	DPN: 70 UTS: 240 Elon: 9% Proof: 170
TAURUS IX	10% Sn 1% Ni 11.5% Pb 0.25% P Cu alloy: Casting; D Brown; sand cast
	DPN: 70 UTS: 220 Elon: 8% Proof: 170
TAURUS XI	5% Sn 5% Zn 5% Pb Cu alloy: Casting; D Brown; sand cast
	DPN: 50 UTS: 220 Elon: 20% Proof: 90
TAURUS XIV	7% Sn 5% Zn 2% Pb Cu alloy: Casting; D Brown; sand cast
	DPN: 60 UTS: 240 Elon: 10% Proof: 100
TAURUS XVI	9% Sn 0.5% Ni 0.25% P Cu alloy: Casting; D Brown; sand cast
	DPN: 75 UTS: 330 Elon: 25% Proof: 180
TAURUS XXXIII	2% Zn 9.5% Sn 1.5% Ni Cu alloy: Casting; D Brown; sand cast
	DPN: 85 UTS: 300 Elon: 12% Proof: 180
TELFOS	5% Sn Cu alloy: Bar; free machining quality; C Clifford
UE3Z9	3.0% Sn 8.2% Zn 0.2% P (max) Cu alloy: Designation for French Standard
UE 5P	5.0% Sn 0.3% P (max) Cu alloy: Designation used by French Standard
UE 3Z9	3% Sn 8.5% Zn 0.1% P Cu alloy: French Standard; part of NF A 53-607
UE 5Z4	4% Sn 4% Zn 0.1% P Cu alloy: French Standard; part of NF A 53-607
UE 5P	5% Sn 0.2% P Cu alloy: French Standard; part of NF A 53-607
UE 5Pb 5Z5	5.0% Sn 5.0% Zn 5.0% Pb Cu alloy: Casting; designation used by French Standard
UE 5Pb 5Z5	5% Sn 5% Zn 5% Pb Cu alloy: French Standard; part of NF A 53-707
UE 5Z4	4.0% Sn 4.0% Zn 0.2% P (max) Cu alloy: Designation used by French Standard

Note. The following abbreviations and units are used in the tables:

DPN	Hardness, diamond pyramid number
UTS	Ultimate tensile strength, N/mm^2
Elon	Elongation, %
Proof	0.1 % proof strength, N/mm^2

1 N/mm^2=0.1 hbar=0.102 kgf/mm^2=0.06475 tonf/in^2=145.04 lbf/in^2

Symbol	Nominal analysis, supplier, condition and remarks.
UE 7Z5Pb4	7.2% Sn 5.0% Zn 3.9% Pb Cu alloy: Casting; designation used by French Standard
UE 7Z5Pb4	7.5% Sn 6% Zn 4% Pb Cu alloy: French Standard; part of NF A 53-707
UE 8Z2	8.5% Sn 2.5% Zn 1% Pb Cu alloy: French Standard; part of NF A 53-707
UE 8Z2	8.5% Sn 2.5% Zn 2.0% Zn (max) Cu alloy: Casting; designation for French Standard
UE 9P	8.2% Sn 0.35% P (max) Cu alloy: Designation used by French Standard
UE 9P	9% Sn 0.2% P Cu alloy: French Standard; part of NF A 607
UE 10Z1	9% Sn 1.0% Zn 1% Pb Cu alloy: French Standard; part of NF A 53-707
UE 10Z1	9.5% Sn 0.2% Zn (max) 0.2% Pb (max) Cu alloy: Casting; designation used by French Standard
UE 12P	11.5% Sn 0.2% Zn (max) 1.0% Pb (max) Cu alloy: Casting; designation used by French Standard
UE 12P	11.5% Sn 0.1% Zn 0.5% Pb Cu alloy: French Standard; part of NF A 53-707
UE 12Z1	11.5% Sn 0.2% Zn (max) 1.5% Pb (max) Cu alloy: Casting; designation used by French Standard
UE 12Z1	11.5% Sn 0.1% Zn 0.7% Pb Cu alloy: French Standard; part of NF A 53-707
UE 14	13.5% Sn 1.0% Zn (max) 1.0% Pb (max) Cu alloy: French Standard; part of NF A 53-707
UE 14	13.5% Sn 1.0% Zn (max) 1.0% Pb (max) Cu alloy: Casting; designation used by French Standard

Symbol	Nominal analysis, supplier, condition and remarks.
UE 16	16.0% Sn 1.0% Zn (max) 1.0% Pb (max) Cu alloy: Casting; designation used by French Standard
UE 16	16% Sn 1.0% Zn (max) 1.0% Sn (max) Cu alloy: French Standard; part of NF A 53-707
UE 18	18% Sn 0.5% Zn (max) 1.0% Pb (max) Cu alloy: French Standard; part of NF A 53-707
UE 18	18.0% Sn 0.5% Zn (max) 1.0% Pb (max) Cu alloy: Casting; designation used by French Standard
UE Pb Z	4.5% Sn (max) 5.0% Zn 5.0% Pb Cu alloy: Casting; designation used by French Standard
UEPbZ	4.5% Sn 5% Zn 5% Pb Cu alloy: French Standard; part of NFA 53-707
VALVE BRONZE	6% Sn 4.5% Pb 6% Zn Cu alloy: Origin unknown
VALVE METAL	3% Sn 7% Pb 9% Zn Cu alloy: Origin unknown
VALVIT	9% Sn Cu alloy: For bearings; origin unknown
YORCASTON	12% Sn Cu alloy: Tube; Yorkshire Imperial Metals Ltd

Note. The following abbreviations and units are used in the tables:

DPN	Hardness, diamond pyramid number
UTS	Ultimate tensile strength, N/mm^2
Elon	Elongation, %
Proof	0.1 % proof strength, N/mm^2

$1 \ N/mm^2 = 0.1 \ hbar = 0.102 \ kgf/mm^2 = 0.06475 \ tonf/in^2 = 145.04 \ lbf/in^2$

14L Copper–chromium wrought and cast alloys

Specific gravity	8.94	
Density	8940 kg/m³	(0.322 lb/in.³)
Solidus/liquidus	1073–1081 °C	
Thermal conductivity solution treated and aged	306 W/m °C	(73 cal/m s °C)
Coefficient of linear expansion	17.7 × 10⁻⁶/ °C	
Electrical conductivity aged	80% IACS (copper 100%)	
solution treated	45% IACS (copper 100%)	
Specific resistance aged	22 microhm mm	
solution treated	39 microhm mm	
Young's modulus of elasticity	110 × 10⁹ N/m²	(16 × 10⁶ lbf/in.²)
Impact	– J	
Fatigue strength (10⁸ cycles)	± 190 N/mm²	(± 12.5 tonf/in.²)
Hot strength	The alloys retain their strength up to 350 °C	

The above properties are typical of the following group, and may not apply exactly to any one specification. It is possible that with certain specifications some of the values may not be applicable.

General metallurgical characteristics

The specifications listed have approximately 0.5% chromium added to pure copper. The resulting intermetallics can be dissolved in the copper by heating to the solution treatment temperature then quenching when the material is very soft and ductile, but suffers a drop in conductivity. After ageing at a lower temperature, there is an increase in hardness and conductivity with a decrease in ductility. These properties are maintained in service up to 350 °C at least, which gives chromium–copper an advantage over work hardened copper alloys when hardness at high temperatures is necessary.

Thermal treatment

Copper chromium is not appreciably hardened by cold working, thus annealing during forming is not normally required. Annealing when necessary will be at about 600 °C when the hardness achieved by heat treatment will be removed in addition to any cold working stresses. The parts may also be solution treated to remove cold work, but this may give rise to grain growth if repeated too often.

When heated to 1000 °C the chromium as a copper compound is taken into solution in the copper. By water quenching this compound is held in solution giving the softest most ductile condition. If slowly cooled, the material will be soft with a superior conductivity, but will not be hardenable. By ageing at 450 °C for 4 hours after solution treatment, the chromium compounds will precipitate from solution, causing strain in the lattice structure and an increase in hardness and electrical conductivity.

Welding and brazing

Pre-treatment. For welding the parts should be solution treated and not in the aged condition. The weld area must be mechanically cleaned and pre-heated.

Parts to be brazed should also be solution treated, but may be aged depending on the melting point of the braze metal.

Welding and brazing. These materials have a high thermal conductivity and welding is difficult unless adequate pre-heating is employed. Wherever possible, this should take the form of a second operator heating the immediate adjacent areas during the actual welding. Electric arc and inert gas shielded arc welding are difficult under these conditions.

Good results can be obtained by brazing, again pre-heating. Fluxes are necessary for welding and brazing, unless the inert gas shielded method is possible.

Great care must be taken when choosing the braze metal that its melting range does not coincide with the ageing temperature. If the braze metal melts below the age temperature the copper chrome alloy must be in the aged condition prior to brazing and considerable care exercised during pre-heating. It is recommended that braze metals melting above the ageing temperature are used, and the parts aged after brazing.

Welding should not be the first choice by designers for joining copper chromium materials.

Post-treatment. All traces of flux must be removed prior to heat treatment otherwise severe corrosion can occur.

After welding it is recommended that parts are solution treated and aged. In cases where distortion is a potential danger the parts should be welded in the solution treated condition, and aged after welding, but this will seldom result in the weld area being fully hardened.

The treatment for brazed parts will depend on the method employed.

Flaw detection

Crack test. Penetrant oil chalk test.

X-ray. Not normally required.

Ultra-sonic test. Nor normally required.

Chemical etch. Not required.

Corrosion protection

Temporary. None required.

Permanent. These materials have excellent corrosion resistance with a tendency to discolour. The chromium content ensures that this tendency is less than with pure copper and for the uses to which these materials are put additional corrosion protection is unusual. Lacquers are now available which protect copper alloys when necessary.

Machinability

These materials should always be machined in the aged condition, when a reasonably good finish is obtained, using high speed tools. In the annealed or solution treated condition, tearing and poor finish analogous to pure soft copper results.

Uses

Principally resistance welding electrodes.

Other uses are confined to those where conductivity and strength at high temperature are required, such as cylinder head and switch gear parts, and rotors.

Symbol	Nominal analysis, supplier, condition and remarks.
182	1.0% Cr Cu alloy: Wrought; designation agreed by CDAA; not used
185	0.7% Cr Cu alloy: Wrought; designation agreed by CDAA; not used
BOLTOMET 814	Cr Cu alloy: For commutator bars; T Bolton
BOLTOMET 818	Cr Mg Cu alloy: For machined products; Thomas Bolton
BS 4577 A2/1	1% Cr Cu alloy: For resistance welding electrodes; wrought; conductivity 78% IACS **DPN: 120 UTS: 380 Elon: 15%**

Note. The following abbreviations and units are used in the tables:

DPN	Hardness, diamond pyramid number
UTS	Ultimate tensile strength, N/mm^2
Elon	Elongation, %
Proof	0.1 % proof strength, N/mm^2

$1 \text{ N/mm}^2 = 0.1 \text{ hbar} = 0.102 \text{ kgf/mm}^2 = 0.06475 \text{ tonf/in}^2 = 145.04 \text{ lbf/in}^2$

Symbol	Nominal analysis, supplier, condition and remarks.
BS 4577 A2/2	1% Cr 0.1% Zr Cu alloy: Resistance welding electrodes; wrought; conductivity 75% IACS **DPN: 130 UTS: 380 Elon: 15%**
CHROMIUM COPPER	0.5% Cr Cu alloy: Solution treated; used for cylinder heads and resistance welding electrodes; common name
CUPALOY	0.5% Cr 0.1% Ag Cu alloy: Origin unknown
CUPROCHROM	1.8% Cr Cu alloy: For electronic grids. Driver
DTD 354	2% Sn 1.5% Cr 2% Fe Cu alloy: Bar & tube; chromium bronze; as drawn for valve guides **UTS: 330 Elon: 25%**
ERM CCS	0.6% Cr Cu alloy: Drawn bar; Enfield Rolling Mills Ltd; welding electrodes, switchgear, etc.; electrical conductivity 85% IACS **DPN: 150 UTS: 480 Elon: 25% Proof: 400**
ERM CCS/Mg	0.6% Cr 0.03% Mg Cu alloy: Bar & forging; Enfield; 80% IACS; as drawn **DPN: 150 UTS: 480 Elon: 25% Proof: 400**
ERM CCS/Z	0.6% Cr 0.15% Zr Cu alloy: Bar & forging; Enfield for switchgear; IACS 80%; as drawn

Symbol	Nominal analysis, supplier, condition and remarks.
HIDURAL 6	Cr Cu alloy: For resistance electrodes; Langley; 80% IACS
	DPN: 100 UTS: 490 Proof: 330
HIDURAL 640	Cr Zr Cu alloy: For resistance electrodes; Langley; 80% IACS
	DPN: 100 UTS: 490 Proof: 330
IMI 171	0.5% Cr Cu alloy: For electrode tips; IMI; KUMIUM
KUMIUM	0.7% Cr Cu alloy: Bar; solution treated & cold rolled; IMI; resistance weld electrodes switch contacts
	UTS: 490 Elon: 12% Proof: 400
MALLORY 3	Cr Cu alloy: Casting; Mallory Metallurgical; conductivity 75% IACS
	DPN: 120 UTS: 330 Elon: 15% Proof: 270
MALLORY 3	Cr Cu alloy: For resistance welding electrodes for mild steel; Mallory Metallurgical; electrical conductivity 80% IACS
	DPN: 140 UTS: 510 Elon: 15% Proof: 460
MALLORY 328	Cr Zr Cu alloy: For resistance welding electrodes for light alloys; Mallory Metallurgical; electrical conductivity 80% IACS
	DPN: 140 UTS: 510 Elon: 15% Proof: 460

Symbol	Nominal analysis, supplier, condition and remarks.
MALLORY 328	Cr Zr Cu alloy: Casting; Mallory Metallurgical; electrical conductivity 75% IACS
	DPN: 120 UTS: 330 Elon: 15% Proof: 240
RWMA Class 2	Cr Cu alloy: Mallory Metallurgical code for Mallory 3 and 328
SAE CA184	0.8% Cr Cu alloy: For welding electrodes
VERILITE	1.0% Cu 1.5% Cr 1.5% Ni Cu alloy: Origin unknown

Note. The following abbreviations and units are used in the tables:

DPN	Hardness, diamond pyramid number
UTS	Ultimate tensile strength, N/mm^2
Elon	Elongation, %
Proof	0.1 % proof strength, N/mm^2

$1\ N/mm^2 = 0.1\ hbar = 0.102\ kgf/mm^2 = 0.06475\ tonf/in^2 = 145.04\ lbf/in^2$

14M Copper–silicon wrought and cast alloys
With manganese, zinc and iron additions – silicon brass and bronze

Specific gravity	8.52	
Density	8520 kg/m³	
Solidus/liquidus	1020 °C	
Thermal conductivity	32.7 W/m °C	(7.8 cal/m s °C)
Coefficient of linear expansion	$18 \times 10^{-6}/$ °C	
Electrical conductivity	annealed 6.6% IACS (copper 100%)	
Specific resistance	annealed 257 microhm mm	
Young's modulus of elasticity	103×10^9 N/m²	$(15 \times 10^6$ lbf/in.²)
Impact	20°C 89 J	(66.4 ft/lb)
	30°C 100 J	(74.6 ft/lb)
	50°C 98 J	(73.1 ft/lb)
	80°C 94 J	(69.2 ft/lb)
	115°C 87 J	(64.5 ft/lb)
Fatigue strength (30×10^7 cycles)	cold drawn ± 220 N/mm²	(± 15 tonf/in.²)
	annealed ± 120 N/mm²	(± 8 tonf/in.²)

Hot strength

Temperature °C	Tensile strength N/mm²	Elongation %
20	360	66
100	330	59
200	300	54
300	270	52

The above properties are typical of the following group and may not apply exactly to any one specification. It is possible that with certain specifications some of the values may not be applicable.

General metallurgical characteristics

Silicon added to copper acts as a de-oxidizer and also enhances the strength and corrosion resistance, especially to acids. The fluidity of the molten metal is increased, which helps to give sound castings and makes welding easier. There is a large increase in the electrical resistance, allowing the use of these alloys for resistance welding purposes where good corrosion resistance is an advantage.

Manganese up to about 1% is generally present with all the straight silicon bronzes acting as a de-sulphurizer and aiding the corrosion resistance. When zinc is present up to about 15% the alloys are known as silicon brass. These alloys have excellent casting characteristics and can be readily cold worked, and are not affected by season cracking which can be troublesome with high copper–zinc alloys, the actual cracking taking place during storage. The zinc-free alloys maintain their mechanical properties including ductility at temperatures down to -100 °C.

Thermal treatment

These alloys cannot be hardened by any form of heat treatment. The effects of cold working or casting stresses can be removed by annealing at a temperature of approximately 550 °C. Time at temperature should be as short as possible to prevent grain growth. It should be noted that once annealed, parts can only be re-hardened by further cold work. Normal air atmosphere furnaces are satisfactory.

Welding and brazing

Pre-treatment. The area must be thoroughly cleaned, and should be prepared immediately before joining.

Welding and brazing. These alloys can be readily welded using gas or electric welding with flux, or inert gas shielded arc, the presence of silicon maintaining the weld pool in a de-oxidized condition. The thermal conductivity of these alloys is generally high enough not to require any pre-heating. The silicon brasses contain zinc, thus zinc oxide fumes which are toxic are given off during welding.

Brazing of these alloys is slightly more difficult than most copper alloys owing to the rather tenacious oxide film. By the use of correct fluxes no trouble should be encountered.

Post-treatment. None required except to remove all traces of flux. The weld areas of these alloys can be successfully cold worked by hammering to improve the mechanical properties.

Flaw detection

Crack detection. Penetrant oil chalk test method.

X-ray. This may be required to prove expensive castings or welds. Although the good castability and weldability reduces this requirement to a minimum.

Ultra-sonic test. None normally required.

Chemical etch. Not required.

Corrosion resistance

Temporary. None required.

Permanent. None normally required. Where appearance is important parts can be lacquered.

Painting. This will seldom be required. When necessary, the surface should be prepared by blasting or etching and an etch primer applied. Generally a stove enamel top coat will be required as these alloys are only used for their corrosion resistant properties.

Plating. The tenacious oxide requires special etching techniques to ensure good adhesion of the electroplated metal. Otherwise normal plating procedure gives good results.

Machinability

These alloys all give a good surface finish and are not susceptible to cold work to any great degree.

The presence of silicon, however, can give severe tool wear because of its abrasive nature and may necessitate the use of tipped tools.

Uses

The straight silicon bronzes find uses in chemical engineering for acid storage tanks piping, pickling crates and handling equipment.

These alloys are also used for marine purposes, such as nails, bolts and nuts, etc. Paper making equipment also uses the silicon bronze alloys. The silicon brasses have valuable bearing properties, and are used in place of tin bronze for economic reasons. Bells are made from these alloys and brazing rods are often of the silicon brass type.

Symbol	Nominal analysis, supplier, condition and remarks.
2.1461	3.5% Si Cu alloy: German Standard
651	1.5% Zn 0.7% Mn 1.5% Si Cu alloy: Wrought; designation used by CDAA
653	2.3% Si Cu alloy: Wrought; designation agreed by CDAA; not used

Note. The following abbreviations and units are used in the tables:

DPN	Hardness, diamond pyramid number
UTS	Ultimate tensile strength, N/mm^2
Elon	Elongation, %
Proof	0.1 % proof strength, N/mm^2

1 N/mm^2=0.1 hbar=0.102 kgf/mm^2=0.06475 tonf/in^2=145.04 lbf/in^2

Symbol	Nominal analysis, supplier, condition and remarks.
656	1.5% Zn 1.5% Sn 1.5% Mn 3.4% Si Cu alloy: Wrought; designation agreed by CDAA; not used
658	1.5% Mn 3.2% Si Cu alloy: Wrought; designation agreed by CDAA; not used
661	1.5% Zn 0.6% Pb 1.5% Mn 3.2% Si Cu alloy: Wrought; designation agreed by CDAA; not used
692	10% Zn 1.2% Si Cu alloy: Wrought; designation agreed by CDAA; not used
694	20% Zn 0.2% Pb 4.0% Si Cu alloy: Wrought; designation agreed by CDAA; not used
697	20% Zn 1.0% Pb 3.0% Si Cu alloy: Wrought; designation agreed by CDAA; not used
AMS 4615 C	3.2% Si Cu alloy: Bar; hard drawn
AMS 4616 B	28% Zn 3.2% Si 1.5% Fe Cu alloy: Bar, forging & tube

Symbol	Nominal analysis, supplier, condition and remarks.
AMS 4665 A	3.2% Si Cu alloy: Seamless tube; annealed
ARGOFIL	0.25% Mn 0.25% Si Cu alloy: Wire; IMI; filler rod for welding copper
ASTM B 30/12 A	0.5% Sn 3% Zn 1.5% Fe 1% Al 1% Mn 3% Si Cu alloy: Ingot; as cast **UTS: 320 Elon: 20%**
ASTM B 30/13 A	0.5% Pb 14% Zn 3.5% Si Cu alloy: Ingot; as cast **UTS: 340 Elon: 18%**
ASTM B 30/13 B	14% Zn 4% Si Cu alloy: Ingot; as cast **UTS: 420 Elon: 16%**
ASTM B53 A	11% Si Cu alloy: Ingot
ASTM B53 B	20% Si Cu alloy: Ingot
ASTM B53 C	30% Si Cu alloy: Ingot
ASTM B96 A	3.2% Si Cu alloy: Plate & sheet; annealed **UTS: 400 Elon: 40% Proof: 100**
ASTM B96 C	2.5% Si Cu alloy: Plate & sheet; annealed **UTS: 400 Elon: 40% Proof: 100**
ASTM B97	3.2% Si Cu alloy: Sheet, strip & bar; annealed **UTS: 400**
ASTM B97	3.2% Si Cu alloy: Sheet, strip & bar; spring hard **UTS: 810**
ASTM B97 B	1.4% Si Cu alloy: Sheet & bar; annealed **UTS: 290**
ASTM B97 B	1.4% Si Cu alloy: Sheet & bar; spring hard **UTS: 550**
ASTM B97 C	2.5% Si Cu alloy: Sheet & bar; annealed **UTS: 400**
ASTM B97 C	2.5% Si Cu alloy: Sheet & bar; spring hard **UTS: 810**
ASTM B98/651	1.4% Si 1.5% Zn (max) Cu alloy: Rod; previously alloy B
ASTM B98/655	3.3% Si 1.5% Mn (max) Cu alloy: Rod; previously alloy A
ASTM B98/661	3.1% Si 0.4% Pb 1.5% Mn (max) Cu alloy: Rod; previously alloy D
ASTM B98 A	3.2% Si Cu alloy: Bar; annealed **UTS: 360 Elon: 35% Proof: 90**
ASTM B98 A	3.2% Si Cu alloy: Bar; extra hard **UTS: 730 Elon: 7% Proof: 390**
ASTM B98 B	1.4% Si Cu alloy: Bar; annealed **UTS: 270 Elon: 30% Proof: 75**
ASTM B98 B	1.4% Si Cu alloy: Bar; extra hard **UTS: 580 Elon: 8% Proof: 330**
ASTM B98 D	3.1% Si 0.6% Pb Cu alloy: Bar; annealed **UTS: 360 Elon: 35% Proof: 90**
ASTM B98 D	3.1% Si 0.6% Pb Cu alloy: Bar; extra hard **UTS: 730 Elon: 7% Proof: 390**
ASTM B99 A	3.2% Si Cu alloy: Wire; annealed **UTS: 440 Elon: 47%**
ASTM B99 A	3.2% Si Cu alloy: Wire; spring hard **UTS: 950 Elon: 4%**
ASTM B99 B	1.4% Si Cu alloy: Wire; annealed **UTS: 290 Elon: 40%**
ASTM B99 B	1.4% Si Cu alloy: Wire; spring hard **UTS: 750 Elon: 6%**
ASTM B124/7	3% Si Cu alloy: Forging & bar; mechanical properties not quoted
ASTM B176 ZS 144 A	15% Zn 4% Si Cu alloy: Casting; die cast **UTS: 630 Elon: 25% Proof: 340**
ASTM B198/12 A	4% Si 1.5% Al 5% Zn 1% Sn Cu alloy: Casting; sand cast **UTS: 300 Elon: 20% Proof: 100**
ASTM B198/13 A	14% Zn 3% Si 1% Pb Cu alloy: Casting; sand cast **UTS: 330 Elon: 18% Proof: 120**
ASTM B198/13 B	14% Zn 4% Si Cu alloy: Casting; sand cast **UTS: 420 Elon: 16% Proof: 150**
ASTM B259 R Cu Si A	1% Zn 1% Sn 3% Si Cu alloy: Welding rod **UTS: 340**
ASTM B259 R Cu Si B	1% Zn 1% Sn 1.5% Si Cu alloy: Welding rod **UTS: 220**

Symbol	Nominal analysis, supplier, condition and remarks.
ASTM B315/615	1.4% Si Cu alloy: Pipe
ASTM B315/655	3.2% Si Cu alloy: Pipe
ASTM B315/658	3.2% Si 0.9% Mn Cu alloy: Pipe
ASTM B315 A	3.2% Si 1% Mn Cu alloy: Pipe
ASTM B315 A7	3.1% Si 1.2% Mn Cu alloy: Pipe
ASTM B371 A	4% Si 15% Zn Cu alloy: Rod; annealed **UTS: 560 Elon: 15% Proof: 270**
ASTM B371 B	20% Zn 3% Si Cu alloy: Rod; annealed **UTS: 420 Elon: 22% Proof: 210**
BOLTOMET 968	3% Si Cu alloy: T Bolton
BS 1029	Silicon bronze ingots & casting; replaced within BS 1400
BS 1030	Silicon bronze ingots & casting; replaced within BS 1400
BS 1541 CS101	3% Si 1% Mn Cu alloy: Plate; as rolled **UTS: 340 Elon: 45%**
BS 1866	3% Si 1% Mn Cu alloy: Tube; solid drawn; annealed **UTS: 370 Elon: 50%**
BS 1948 CS101	3% Si 1% Mn Cu alloy: Rods & sections; annealed **UTS: 360 Elon: 50%**
BS 2870 CS101	3% Si 1% Mn Cu alloy: Sheet; as rolled; silicon bronze **UTS: 360 Elon: 50%**
BS 2871 CS101	3% Si 1% Mn Cu alloy: Tube; annealed **UTS: 420**
BS 2872 CS101	3% Si 1% Mn Cu alloy: Forging; as forged **UTS: 330 Elon: 50%**
BS 2873 CS101	3% Si 1% Mn Cu alloy: Wire; mechanical properties not quoted
BS 2874 CS101	3% Si 1% Mn Cu alloy: Bar; annealed **UTS: 360 Elon: 50%**
BS 2875 CS101	3% Si 1% Mn Cu alloy: Plate; as rolled **UTS: 340 Elon: 45%**
BS 2901 C9	3% Si 1% Mn Cu alloy: Rod for all forms of welding
DIN 1733 S Cu Si	3.5% Si Cu alloy: Wrought
DTD 263	10% Zn 1.5% Si Cu alloy: Silicon brass; annealed **UTS: 420 Elon: 30%**
DTD 267	10% Zn 1.5% Si Cu alloy: Sheet; ½-hard; cold rolled **UTS: 360 Elon: 20%**
DTD 355	2% Zn 3.5% Si 1.7% Fe Cu alloy: Casting **UTS: 300 Elon: 12%**
EVERDUR	4% Si 1% Mn Cu alloy: Casting; IMI; corrosion resistant; withdrawn
EVERDUR A	3% Si 1% Mn Cu alloy: Bar, sheet, wire & rod; IMI; corrosion resistant with high ductility **UTS: 340 Elon: 70%**
EVERDUR D	4% Si 1% Mn Cu alloy: Ingot; IMI
HOLFOS JHR42	0.5% Mn 1.5% Fe 3.5% Si 2% Zn Cu alloy: Casting; J Holroyd; as cast **DPN: 110 UTS: 340 Elon: 18% Proof: 200**
IMI 176	0.25% Mn 0.25% Si Cu alloy: Welding rod; IMI; Argofil
IMI 705	3% Si 1% Mn Cu alloy: Sheet & wire; IMI; Everdur A
JB 33 A	10% Zn 1.5% Si Cu alloy: Proprietary name; consult Copper Development Association for further details
JHR 42	2% Zn 3.5% Si 1.7% Fe Cu alloy: Casting; J Holroyd for DTD 355 **DPN: 105 UTS: 340 Elon: 15% Proof: 200**
MALLORY 53	Ni Si Cu alloy: For welding electrodes; Mallory Metallurgical
PMG	2% Zn 3.5% Si Cu alloy: Casting; Vickers Armstrong; DTD 355

Symbol	Nominal analysis, supplier, condition and remarks.
QQ B 746/A	3.5% Si 0.3% P Cu alloy: Plate & sheet; US Federal
	UTS: 690 Elon: 2.5%
QQ C 593	3.0% Si 2.5% Fe 0.5% Zn Cu alloy: Casting; US Federal
	UTS: 300 Elon: 15%
SA 4	8% (Si + Al) Cu alloy: Manganese Bronze Ltd
SAE CA 655	3% Si 1.5% Fe + Mn + Zn (max) Cu alloy: Wrought
S Cu Si	3.5% Si Cu alloy: Wrought; designation used by German Standard
SILICON BRONZE SA 4	8% (Si + Al) Cu alloy: Manganese Bronze Ltd
STA 7 CS1C	5.0% Zn 2.0% Fe 1.0% Mn 4.0% Si Cu alloy: Casting; as cast
	UTS: 300 Elon: 15%

Note. The following abbreviations and units are used in the tables:

DPN	Hardness, diamond pyramid number
UTS	Ultimate tensile strength, N/mm^2
Elon	Elongation, %
Proof	0.1 % proof strength, N/mm^2

$1\ N/mm^2 = 0.1\ hbar = 0.102\ kgf/mm^2 = 0.06475\ tonf/in^2 = 145.04\ lbf/in^2$

15. Gallium Ga

Physical properties

Atomic number	31	
Atomic weight	69.72	
Crystal structure	Orthorhombic	
Colour	Silver white	
Specific gravity	5.91	
Density	5910 kg/m³	(0.21 lb/in.³)
Melting point	29.8 °C	
Boiling point	2403 °C	
Specific heat	0.33 J/g °C	(0.08 cal/g °C)
Thermal conductivity	– W/m °C	(– cal/m s °C)
Coefficient of linear expansion (20–30 °C)	58×10^{-6}/ °C	
	(thermal expansion varies in different directions)	
Latent heat of fusion	80.4 J/g	(19.2 cal/g)
Latent heat of vaporization	3894 J/g	(930 cal/g)
Thermal neutron absorption cross-section	2.8 barns/atom	
Electrical conductivity (a-axis)	10% IACS (copper 100%)	
Specific resistance (resistivity) (a-axis)	174 microhm mm	
Temperature coefficient of electrical resistance	0.0040/ °C	
Electrochemical equivalent	2.45 g/A/h	
Electrode potential	-0.52 V	
Magnetic susceptibility	-0.24×10^{-6}	
Young's modulus of elasticity	– N/m²	
Tensile strength	– N/mm²	
Hardness	50 DPN	

15.1 General notes on gallium

Gallium is one of those elements which appear very sparingly in nature, and yet whose occurrence is widespread. Most coal has a minute percentage of the element which can be recovered from the flue dust when large quantities of coal are burned.

The metal is soft being readily cut, but is also reasonably brittle. It has a brilliant lustre when first cut, then oxidizes rapidly. It resembles aluminium and indium in many respects.

The long temperature range of the liquid phase, from 30–2000 °C, makes this a potentially interesting material. Unfortunately, liquid gallium, even at reasonably low temperatures, is corrosive to most other metals. The metals tantalum and tungsten resist this attack, as do quartz, graphite and alumina. Using quartz capillary tubes gallium thermometers have been produced capable of measuring accurately temperatures up to 1200 °C.

Gallium alloys are available with very low melting points. Two eutectic alloys exist with sharp melting points.

82% Ga 12% Sn 6% Zn Melting point 17 °C
76% Ga 24% In   Melting point 15.7 °C

Again, these materials attack other metals at elevated temperatures, otherwise they would probably find considerable use as heat transfer liquids.

Like antimony and bismuth, gallium expands on solidi-fication. With the low melting points of the element and some of its alloys, this can be a serious disadvantage, causing strain on the container in the same manner as water freezing and fracturing its pipe.

Some use has been made of gallium in vapour arc lamps, usually in conjuction with cadmium, but also with mercury.

There are reports that additions of gallium to magnesium and some magnesium–tin alloys markedly increase the corrosion resistance.

The high cost of extraction and purification and the corrosive attack on other metals will probably prevent any wider use of this metal and its alloys.

Symbol	Nominal analysis, supplier, condition and remarks.
GALLIUM	High purity metal: Impurities less than 4 p.p.m.; Mallory Metallurgical; supplied as rod for semiconductors
h Ga 1	99.9999% Ga rod: Light Ltd; high purity gallium

Note. The following abbreviations and units are used in the tables:

DPN	Hardness, diamond pyramid number
UTS	Ultimate tensile strength, N/mm²
Elon	Elongation, %
Proof	0.1 % proof strength, N/mm²

1 N/mm²=0.1 hbar=0.102 kgf/mm²=0.06475 tonf/in²=145.04 lbf/in²

16. Germanium Ge

Physical properties

Atomic number	32	
Atomic weight	72.6	
Crystal structure	Cubic	
Colour	Greyish white	
Specific gravity	5.32	
Density	5320 kg/m³	(0.19 lb/in.³)
Melting point	958 °C	
Boiling point	2700 °C	
Specific heat	0.301 J/g °C	(0.072 cal/g °C)
Thermal conductivity	6.28 W/m °C	(1.5 cal/m s °C)
Coefficient of linear expansion (20–100° C)	6×10^{-6}/ °C	
Latent heat of fusion	427 J/g	(102 cal/g)
Latent heat of vaporization	3345 J/g	(799 cal/g)
Thermal neutron absorption cross-section	2.5 barns/atom	
Electrical conductivity	3% IACS (copper 100%)	
Specific resistance	500 microhm mm	
Temperature coefficient of electrical resistance	–/ °C	
Electrochemical equivalent	– g/A/h	
Electrode potential	– V	
Magnetic susceptibility	-0.12×10^{-6}	
Young's modulus of elasticity	– N/m²	
Tensile strength	– N/mm²	
Hardness	– DPN	

16.1 General notes on germanium

Germanium is a metalloid, having some non-metallic characteristics, but not as marked as in silicon, or boron. It is a very brittle material, quite stable in air up to red heat.

Until the unusual electrical properties became known germanium had very little use. There are no commercial alloys based on the metal, but additions to aluminium copper alloys ('Duralumin') increases the strength and hot working properties.

When added to tin an alloy with considerably greater hardness and some reduction in ductility is obtained. Gold with 13% germanium forms a eutectic alloy melting at 356 °C which finds some use as a special purpose solder.

The high cost and scarcity of germanium tend to persuade potential users to alternative materials.

By far the largest use of germanium is in the electronic industry where its electrical rectification properties have been found to be extremely useful.

The voltage range of germanium covers the gap between silicon and selenium rectifiers, but by far the largest use of the metal is as 'transistors'. These are germanium crystals, which rectify radio signals using considerably less power in the process than conventional valves. In addition they are very much smaller and more robust, and constitute the core of all modern electronic devices which can operatre for long periods on minute power supplies, the transistor radio set being a good example.

Note. The following abbreviations and units are used in the tables:

DPN	Hardness, diamond pyramid number
UTS	Ultimate tensile strength, N/mm²
Elon	Elongation, %
Proof	0.1 % proof strength, N/mm²

1 N/mm²=0.1 hbar=0.102 kgf/mm²=0.06475 tonf/in²=145.04 lbf/in²

Symbol	Nominal analysis, supplier, condition and remarks.
GERMANIUM	99.999% Ge ingot: Koch Light
GERMANIUM I	Ge metal: Impurities below spectrographic detection; Mallory Metallurgical; ingot; specific resistance 0.02 microhm mm
GERMANIUM II	High purity metal Ge: Impurities less than 1 p.p.m.; Mallory Metallurgical; ingot; specific resistance 0.5 microhm mm

17. Gold Au

Physical properties

Atomic number	79	
Atomic weight	197.2	
Crystal structure	Face-centred cubic	
Colour	Yellow	
Specific gravity	19.32	
Density	19320 kg/m^3	(0.69 lb/in.^{3})
Melting point	1063 °C	
Boiling point	2960 °C	
Specific heat	0.1323 J/g °C	(0.0316 cal/g °C)
Thermal conductivity	301 W/m °C	(72 cal/m s °C)
Coefficient of linear expansion (20–100° C)	14.4 × 10^{-6}/ °C	
Latent heat of fusion	66.2 J/g	(15.8 cal/g)
Latent heat of vaporization	1738 J/g	(415 cal/g)
Thermal neutron absorption cross-section	99 barns/atom	
Electrical conductivity	80% IACS (copper 100%)	
Specific resistance	22 microhm mm	
Temperature coefficient of electrical resistance	0.0034/ °C	
Electrochemical equivalent	2.45 g/A/h	
Electrode potential	+ 1.68 V	
Magnetic susceptibility	-0.15 × 10^{-6}	
Young's modulus of elasticity	80 × 10^9 N/m^2	(11.6 × 10^6 lbf/in.^{2})
Tensile strength	annealed 100 N/mm^2	(7 tonf/in.^{2})
Hardness	annealed 30 DPN	

17.1 General notes on gold

Along with copper, gold is the only truly coloured metal, being pale yellow in the massive state, and red or even purple when finely divided.

The ability of gold to resist oxidation and chemical attack, together with its attractive colour, has made it popular as a means of decoration and a standard of value since the dawn of history.

The largest use of gold to-day is as the medium of international currency and as such most of it is securely locked in underground vaults in the financial centres of the world.

Considerable quantities are still used for ornamental purposes, often being added as fine powder to pottery glazes and glass ware.

Industrially, gold has only limited uses owing to its artificially high cost and low strength.

The ability to resist oxidation over a large range of temperatures, and its high electrical conductivity are now finding use in the electrical and electronic industries. For these purposes very thin electroplated deposits are used. A thin deposit of nickel between the gold and copper base metal prevents diffusion of the gold into copper. Gold plating for purely decorative purposes is also increasing. Modern techniques are increasing the hardness of the plated deposit, which, as it does not oxidize or readily corrode has no need for abrasive cleaning and can therefore be very thin.

There are also a limited number of chemical engineering and laboratory uses for gold, notably in the man-made fibre industry. Some gold alloy brazing materials are used where excellent joint conductivity under oxidizing conditions is necessary. Gold leaf is still used on articles such as presentation leather books and for outdoor decorations, such as name signs. The high malleability and almost complete absence of cold working, coupled with the high density allows gold to be rolled or beaten into intricate shapes, or very thin foils – leaf gold.

Addition of platinum and palladium to gold can give the alloys age hardening characteristics. The temperatures required are in the region of 300–400 °C, the most popular uses being for personal ornaments and dental fillings.

The 'carat' applied to gold is not the weight measure used for precious stones, but a proportional measure giving the percentage of gold present. Twenty-four carat gold is pure metal, 12 carat gold is 50% gold alloyed with other metals, generally copper with palladium nickel or silver to give some increase in hardness.

Following are listed the known specifications and trade names with gold as the main constituent.

Symbol	Nominal analysis, supplier, condition and remarks.
9ct No. 90	Gold alloy: Reddish yellow colour; Sheffield Smelting Co; melting range 880–895 °C **DPN: 110**
9ct No. 91	Gold alloy: Yellow colour; Sheffield Smelting Co; melting range 855–870 °C **DPN: 110**
9ct No. 92	Gold alloy: Reddish yellow colour; Sheffield Smelting Co; melting range 920–940 °C **DPN: 110**
9ct No. 93	Gold alloy: Yellow colour; Sheffield Smelting Co; melting range 870–890 °C **DPN: 190**
9ct No. 94	Gold alloy: Yellow colour; Sheffield Smelting Co; melting range 835–860 °C **DPN: 110**
9ct No. 95	Gold alloy: Yellow colour; Sheffield Smelting Co; melting range 860–880 °C; age hardenable **DPN: 110**
9ct No. 97	Gold alloy: Red colour; Sheffield Smelting Co; melting range 940–955 °C **DPN: 110**
9ct No. 98	Gold alloy: Low zinc; red colour; Sheffield Smelting Co; melting range 925–945 °C **DPN: 90**
9ct No. 99	Gold alloy: Yellow colour; Sheffield Smelting Co; melting range 820–840 °C **DPN: 190**
9ct No. 910	Gold alloy: Greenish yellow colour; Sheffield Smelting Co; melting range 895–920 °C; age hardenable **DPN: 90**
9ct No. 911	Gold alloy: Reddish yellow colour; Sheffield Smelting Co; melting range 845–870 °C; age hardenable **DPN: 110**
9ct No. 912	Gold alloy: Reddish yellow colour; Sheffield Smelting Co; melting range 910–935 °C **DPN: 110**
9ct S 91 (easy)	37.5% Au alloy: For soldering; Sheffield Smelting Co; melting range 610–660 °C
9ct S 92 (medium)	37.5% Au alloy: For soldering; Sheffield Smelting Co; melting range 700–710 °C
9ct S93 (hard)	37.5% Au alloy: For soldering; Sheffield Smelting Co; melting range 740–760 °C
9ct W 91	Gold alloy: White; Sheffield Smelting Co **DPN: 140**
9ct W 93	Gold alloy: White; Sheffield Smelting Co **DPN: 85**
9ct W 94	Gold alloy: White; Sheffield Smelting Co **DPN: 60**
9ct W95	Gold alloy: White; Sheffield Smelting Co **DPN: 85**
9ct W 96	Gold alloy: White; Sheffield Smelting Co **DPN: 85**
9ct W 97	Gold alloy: White; Sheffield Smelting Co **DPN: 85**
9ct W 98	Gold alloy: White; Sheffield Smelting Co **DPN: 60**

Note. The following abbreviations and units are used in the tables:

DPN	Hardness, diamond pyramid number
UTS	Ultimate tensile strength, N/mm^2
Elon	Elongation, %
Proof	0.1 % proof strength, N/mm^2

$1 \ N/mm^2 = 0.1 \ hbar = 0.102 \ kgf/mm^2 = 0.06475 \ tonf/in^2 = 145.04 \ lbf/in^2$

Symbol	Nominal analysis, supplier, condition and remarks.
9ct W 99	Gold alloy: White; Sheffield Smelting Co **DPN: 85**
14/450	Au based solder for dental use: Engelhard Industries; melting range 620–755 °C
14ct No. 144	Gold alloy: Green colour; Sheffield Smelting Co; melting range 950–975 °C **DPN: 90**
14ct No. 145	Gold alloy: Red colour; Sheffield Smelting Co; melting range 885–895 °C **DPN: 110**
14ct No. 146	Gold alloy: Reddish yellow colour; Sheffield Smelting Co; melting range 840–880 °C **DPN: 90**
14ct No. 147	Gold alloy: Green colour; Sheffield Smelting Co; melting range 920–935 °C **DPN: 110**
14ct No. 148	Gold alloy: Yellow colour; Sheffield Smelting Co; melting range 830–855 °C; age hardenable **DPN: 190**
14ct No. 149	Gold alloy: Reddish yellow colour; Sheffield Smelting Co; melting range 810–838 °C **DPN: 90**
14ct S 143 (medium)	58.5% Au alloy: For soldering; Sheffield Smelting Co; melting range 830–835 °C
14ct S 145 (hard)	58.5% Au alloy: For soldering; Sheffield Smelting Co; melting range 835–870 °C
14ct S 146 (easy)	58.5% Au alloy: For soldering; Sheffield Smelting Co; melting range 720–740 °C
14ct S 147 (easy)	58.5% Au alloy: For soldering; Sheffield Smelting Co; melting range 700–720 °C
14ct W 142	Gold alloy: White; Sheffield Smelting Co **DPN: 140**
14ct W 143	Gold alloy: White; Sheffield Smelting Co **DPN: 60**
15ct S 151 (easy)	62.5% Au alloy: For soldering; Sheffield Smelting Co; melting range 720–750 °C
16/550	Au based solder for dental use: Engelhard Industries; melting range 767–783 °C
18/650	Au based solder for dental use: Engelhard Industries; melting range 785–798 °C
18ct S 181 (hard)	75% Au alloy: For soldering; Sheffield Smelting Co; melting range 885–905 °C
18ct W 181	Gold alloy: White; Sheffield Smelting Co **DPN: 85**
18ct W 182	Gold alloy: White; Sheffield Smelting Co **DPN: 226**
18ct W 183	Gold alloy: White; Sheffield Smelting Co **DPN: 140**
18ct No. 184	Gold alloy: Reddish yellow; Sheffield Smelting Co; melting range 920–935 °C **DPN: 150**
18ct W 184	Gold alloy: White; Sheffield Smelting Co **DPN: 140**
18ct W 185	Gold alloy: White; Sheffield Smelting Co **DPN: 60**
18ct S 185 (easy)	75% Au alloy: For soldering; Sheffield Smelting Co; melting range 660–765 °C
18ct No. 185	Gold alloy: Reddish yellow colour; Sheffield Smelting Co; melting range 890–915 °C; age hardenable **DPN: 150**
18ct No. 186	Gold alloy: Reddish yellow colour; Sheffield Smelting Co; melting range 900–920 °C **DPN: 150**
18ct S 186 (medium)	75% Au alloy: For soldering; Sheffield Smelting Co; melting range 650–800 °C
18ct No. 187	Gold alloy: Yellow colour; Sheffield Smelting Co; melting range 885–905 °C **DPN: 110**

Symbol	Nominal analysis, supplier, condition and remarks.
18ct No. 188	Gold alloy: Reddish yellow colour; Sheffield Smelting Co; melting range 890–910 °C **DPN: 150**
20/750	Au based solder for dental use: Engelhard Industries; melting range 707–822 °C
22/800	Au based solder for dental use: Engelhard Industries; melting range 722–845 °C
22ct No. 223	Gold alloy: Deep yellow: Sheffield Smelting Co; melting range 890–910 °C **DPN: 90**
424	Cu Au brazing alloy: Melting point 889 °C; Engelhard Industries; electrical conductivity 15% IACS **DPN: 250 UTS: 450 Elon: 17%**
429	Cu Au brazing alloy: Melting range 905–915 °C; Engelhard Industries; electrical conductivity 15% IACS **DPN: 112 UTS: 400 Elon: 38%**
441	Ni Au brazing alloy: Melting point 950 °C; Engelhard Industries **DPN: 245 UTS: 690 Elon: 5%**
450 (fine)	Au based solder for dental use: Engelhard Industries; melting range 620–755 °C
550 (fine)	Au based solder for dental use: Engelhard Industries; melting range 767–783 °C
625R	Cu Ag Au alloy: For sliding contacts; conductivity 14% 1ACS; Johnson Matthey **DPN: 95**
650 (fine)	Au based solder for dental use: Engelhard Industries; melting range 785–798 °C
750 (fine)	Au based solder for dental use: Engelhard Industries; melting range 707–822 °C
800 (fine)	Au based solder for dental use: Engelhard Industries; melting range 722–845 °C
AMERICAN GOLD	90% Au 10% Cu standard coinage alloy of the USA
ANORMAT	78.3% Au + Pt for dental purposes: Mallory Metallurgical; melting range 915–917 °C; hardened **DPN: 165 Elon: 26%**
ASTM B260/B Au2	20% Cu Au brazing filler metal: Brazing temperature range 890–982 °C
ASTM B260/B Au4	18.5% Ni Au brazing filler metal: Brazing temperature range 919–1004 °C
ASTM B260 B Cu Au2	20% Cu Au alloy: Braze filler; melting range 880–887 °C
AUBEL	92% (min) Au for dental purposes: Mallory Metallurgical; melting range 970–995 °C **DPN: 67 Elon: 50%**
AUDENWIRE	75% Au + Pt alloy: Wire for dental purposes; Mallory Metallurgical; melting range 965–1000 °C; hardened **DPN: 339**
AUMET	83.3% Au + Pt for dental purposes: Mallory Metallurgical; melting range 1000–1085 °C **DPN: 71 Elon: 47%**
BAKER 4	Gold alloy used in dentistry: With Pt and Pd; Engelhard Industries; annealed; melting range 850–910 °C **DPN: 130 UTS: 480 Elon: 25% Proof: 370**
BAKER 4	Gold alloy used in dentistry: With Pt and Pd; Engelhard Industries; age hardened **DPN: 228 UTS: 850 Elon: 4% Proof: 600**
BINORMAT	78.3% Au + Pt for dental purposes: Mallory Metallurgical; melting range 925–945 °C **DPN: 100 Elon: 42%**
BORAWIRE	61% Au + Pt alloy: Wire for dental purposes; Mallory Metallurgical; melting range 1080–1180 °C; hardened **DPN: 296**

Symbol	Nominal analysis, supplier, condition and remarks.
BS 1845 Au 1	19% Cu 1% Fe Au alloy: For braze metal; melting range 905–910 °C
BS 1845 Au 2	37.5% Cu Au alloy: For braze metal; melting range 930–940 °C
BS 1845 Au 5	17.5% Ni Au alloy: For braze metal; melting point 950 °C
BS 1845 Au 6	25% Ni Au alloy: For braze metal; melting point 950 °C
BS 3384 A	65% Au (min) dental solder
BS 3384 B	53% Au (min) dental solder
BS 4425	Au alloy used for dental casting graded by mechanical properties
CHICAGO 4	Gold alloy used in dentistry: High Pt; Engelhard Industries; annealed; melting range 890–940 °C **DPN: 150 UTS: 550 Elon: 30% Proof: 340**
CHICAGO 4	Gold alloy used in denstistry: High Pt; Engelhard Industries; age hardened; melting range 890–940 °C **DPN: 235 UTS: 780 Elon: 5% Proof: 660**
COINAGE GOLD	0.15% Ag Au alloy: 22 carat; balance Cu
CORODENT	71% Au + Pt for dental purposes: Mallory Metallurgical; melting range 935–995 °C; hardened **DPN: 192 Elon: 17%**
DENTECON	63% Au + Pt for dental purposes: Mallory Metallurgical; melting range 885–945 °C; hardened **DPN: 266 Elon: 12%**
DENTORMAT	76% Au + Pt for dental purposes: Mallory Metallurgical; melting range 880–925 °C; hardened **DPN: 252 Elon: 8%**
DORDENT	72.5% Au + Pt for dental purposes: Mallory Metallurgical; melting range 870–920 °C; hardened **DPN: 298 Elon: 2.5%**
GOLD	High purity metal: Impurities less than 5 p.p.m.; Mallory Metallurgical; supplied as sponge rod sheet and wire
h Au 3	99.9999% Au sponge: Light Lab Ltd; high purity metal
h Au 6	99.999% Au rod 3 mm diameter: Light Lab Ltd; high purity metal
h Au 8	99.999% Au powder; Light Lab Ltd; high purity metal
h Au 72	99.999% Au single crystals 6.35 mm diameter, 100 mm long: Light Lab Ltd
JMC 625	Cu Ag Au alloy: For wiping contacts; Mallory Metallurgical
JMC 625 R	Cu Ag alloy: For wiping contacts; Mallory Metallurgical
JMM 625	Cu Ag Au alloy: For sliding contacts; conductivity 12% 1ACS; Johnson Matthey **DPN: 175**
NIORO	18% Ni Au alloy: For brazing; melting point 950 °C; Wesgo
NICORO 80	2.0% Ni 16% Cu Au alloy: For brazing; melting range 910–925 °C; Wesgo
ORO BRAZE	Cu Fe 80% Au alloy: Brazing metal; Mallory Metallurgical; melting range 908–910 °C
ORO BRAZE 910	Cu Fe 80% Au alloy: For brazing; Mallory Metallurgical; melting range 908–910 °C
ORO BRAZE 940	Cu 62.5% Au alloy: Brazing metal; Mallory Metallurgical; melting range 930–940 °C
ORO BRAZE 950	17.5% Ni Au alloy: Bearing metal; Mallory Metallurgical; melting point 950 °C
ORO BRAZE 990	25% Ni Au alloy: Bearing metal; Mallory Metallurgical; melting range 950–990 °C
ORO BRAZE 1040	30% Ag Au alloy: For brazing; Mallory Metallurgical; melting range 1030–1040 °C
ORO CAST	Gold alloy: Used in dentistry; with Pt metals; Engelhard Industries; annealed; melting range 850–910 °C **DPN: 142 UTS: 510 Elon: 25% Proof: 330**

Symbol	Nominal analysis, supplier, condition and remarks.
ORO CAST	Gold alloy: Used in dentistry; with Pt metals; Engelhard Industries; age hardened; melting range 850–910 °C
	DPN: 229 UTS: 800 Elon: 6% Proof: 770
PALNIRO 1	25% Pd 25% Ni Au alloy: For brazing; melting range 1102–1121 °C; Wesgo
PALORO	8.0% Pd Au alloy: For brazing; melting range 1200–1240 °C; Wesgo
PLATINUM COLOUR SOLDER	Au based solder with Pd and Pt for dental use: Engelhard Industries; melting range 832–860 °C
QA Wire	Gold based Pt and Pd alloy: Engelhard Industries
SILCORO 60	20% Cu 20% Ag Au alloy: For brazing; melting range 835–845 °C; Wesgo
SILCORO 75	20% Cu 5.0% Ag Au alloy: For brazing; melting range 885–895 °C; Wesgo
SUPER ORALIUM	Gold alloy used in dentistry: With Pt and Pd; white colour; Engelhard Industries; annealed; melting range 923–975 °C
	DPN: 175 UTS: 530 Elon: 27% Proof: 390
SUPER ORALIUM	Gold alloy used in dentistry: With Pt and Pd; white colour; Engelhard Industries; age hardened; melting range 923–975 °C
	DPN: 220 UTS: 830 Elon: 2% Proof: 790
TRUCAST HARD	Gold alloy used in industry: Contains Pt; Engelhard Industries; annealed; melting range 900–950 °C
	DPN: 127 UTS: 400 Elon: 24% Proof: 210
TRUCAST HARD	Gold alloy used in dentistry: Contains Pt; Engelhard Industries; age hardened; melting range 900–950 °C
	DPN: 159 UTS: 480 Elon: 8% Proof: 320

Symbol	Nominal analysis, supplier, condition and remarks.
TRUCAST MEDIUM	Gold alloy used in dentistry: Pt free; Engelhard Industries; melting range 920–955 °C
	DPN: 90 UTS: 340 Elon: 31% Proof: 170
TRUCAST SOFT	Gold alloy used in dentistry: Pt free; Engelhard Industries; melting range 940–970 °C
	DPN: 59 UTS: 250 Elon: 33% Proof: 100
UNIGOLD	Gold alloy used in dentistry: With Pt and Pd; Engelhard Industries; annealed; melting range 850–890 °C
	DPN: 142 UTS: 460 Elon: 8% Proof: 320
UNIGOLD	Gold alloy used in dentistry: With Pt and Pd; Engelhard Industries; age hardened
	DPN: 229 UTS: 890 Elon: 1% Proof: 780
UNS P00001-P00999	Au metal and alloys: American Standards system; unified numbers
WHITE GOLD SOLDER	12% Ni 15% Zn Au alloy: Origin unknown

Note. The following abbreviations and units are used in the tables:

DPN	Hardness, diamond pyramid number
UTS	Ultimate tensile strength, N/mm^2
Elon	Elongation, %
Proof	0.1 % proof strength, N/mm^2

$1\ N/mm^2 = 0.1\ hbar = 0.102\ kgf/mm^2 = 0.06475\ tonf/in^2 = 145.04\ lbf/in^2$

18. Indium In

Physical properties

Atomic number	49	
Atomic weight	114.82	
Crystal structure	Face-centred tetragonal	
Colour	Silver-white	
Specific gravity	7.31	
Density	7310 kg/m^3	(0.264 lb/in.^{3})
Melting point	156.4 °C	
Boiling point	2100 °C	
Specific heat	0.239 J/g °C	(0.057 cal/g °C)
Thermal conductivity	24.7 W/m °C	(5.9 cal/m s °C)
Coefficient of linear expansion (20–100° C)	33×10^{-6}/ °C	
Latent heat of fusion	28.5 J/g	(6.81 cal/g)
Latent heat of vaporization	2022 J/g	(483 cal/g)
Thermal neutron absorption cross-section	196 barns/atom	
Electrical conductivity	20% IACS (copper 100%)	
Specific resistance	90 microhm mm	
Temperature coefficient of electrical resistance	0.0045/ °C	
Electrochemical equivalent	1.4 g/A/h	
Electrode potential	-0.34 V	
Magnetic susceptibility	0.11×10^{-6}	
Young's modulus of elasticity	10.8×10^9 N/m^2	(1.57×10^6 lbf/in.^{2})
Tensile strength	annealed 4.5 N/mm^2	(0.3 tonf/in.^{2})
Hardness	less than 10 DPN	

18.1 General notes on Indium

The compounds of indium occur in small quantities widely scattered throughout the world. To date no deposits large enough to warrant extraction of the metal are known, but certain tin, lead and zinc ores contain appreciable quantities.

The purified metal is very soft and ductile, being readily cut with a knife. The surface does not oxidize in air at room temperatures and indium can be easily welded to itself without heat and with very little pressure.

The only known use for the metal is as a sealing material for high vacuum equipment under sterile conditions and no alloys are known with indium as the base metal.

The largest quantity is electrodeposited, some for its attractive lustre and good wearing properties. It is however, more generally applied as a very thin top coat on lead, silver or cadmium bearings. This is then diffused into the substrate coating by a heat treatment and imparts corrosion resistance and improved frictional properties. Some modern cars use thin steel shell bearings coated with lead bronze, indium plated.

When indium is added to the lead-tin–bismuth alloy – Woods Metal – the melting point can be reduced to as low as 48 °C. This alloy is used in dentistry and plastic surgery. Small additions of indium strengthen gold and platinum alloys without detracting too much from their other properties.

Indium and its compounds are now finding some use as a semiconductor in the electronics industry.

Following are listed the known suppliers.

Note. The following abbreviations and units are used in the tables:

DPN	Hardness, diamond pyramid number
UTS	Ultimate tensile strength, N/mm^2
Elon	Elongation, %
Proof	0.1 % proof strength, N/mm^2

1 N/mm^2=0.1 hbar=0.102 kgf/mm^2=0.06475 tonf/in^2=145.04 lbf/in^2

Symbol	Nominal analysis, supplier, condition and remarks.
h In 1	99.9999% In rod 12 mm diameter: Light Ltd; high purity metal
h In 2	99.9999% In shot: Light Ltd; high purity metal
h In 6	99.999% In ingot: Light Ltd; high purity metal
h In 96	99.8% In 115 stable isotope: Light Ltd; isotope 115
INDIUM	High purity metal: Impurities 1 p.p.m.; Mallory Metallurgical - supplied as ingot sheet wire
INDIUM	In metal: Blackwells; pure metal

19. Iridium Ir

Physical properties

Atomic number	77	
Atomic weight	192.1	
Crystal structure	Face-centred cubic	
Colour	Grey	
Specific gravity	22.65	
Density	22 650 kg/m^3	(0.81 lb/in.^{3})
Melting point	2443 °C	
Boiling point	4500 °C	
Specific heat	0.135 J/g °C	(0.0323 cal/g °C)
Thermal conductivity	147 W/m °C	(35 cal/m s °C)
Coefficient of linear expansion (20–100° C)	6.5 × 10^{-6}/ °C	
Latent heat of fusion	27 600 J/g	(6600 cal/g)
Latent heat of vaporization	636 000 J/g	(152 000 cal/g)
Thermal neutron absorption cross-section	440 barns/atom	
Electrical conductivity	32% IACS (copper 100%)	
Specific resistance	47 microhm mm	
Temperature coefficient of electrical resistance	0.0041/ °C	
Electrochemical equivalent	– g/A/h	
Electrode potential	+ 1.0 V	
Magnetic susceptibility	0.15 × 10^{-6}/gm	
Young's modulus of elasticity	524 × 10^9 N/m^2	(76 × 10^6 lbf/in.^{2})
Tensile strength	cold drawn 2000 N/mm^2	(130 tonf/in.^{2})
	annealed 1000 N/mm^2	(65 tonf/in.^{2})
Hardness	cold drawn 650 DPN	
	annealed 370 DPN	

19.1 General notes on iridium

Iridium invariably occurs in association with platinum, often being found as the native platinum/osmium/iridium alloy 'osmiridium'.

Metal of 99.9% purity is obtained as a by-product in the preparation of platinum. This can be hot worked at white heat to bars or hollow shapes required for laboratory use. At room temperature it is hard and brittle, readily crumbling to a coarse powder if cold worked to any great extent. Iridium is the densest known element.

The pure metal has excellent chemical resistance to the extent of not being attacked by aquaregia. This property accounts for the limited use made of the metal which would be increased but for the extreme difficulty of working and shaping.

Platinum is hardened by the addition of iridium, with little reduction in ductility up to 20% alloy content. Above this figure the material loses its ductility. These platinum–iridium alloys are used for laboratory ware and electrical contacts where hammering action at high speeds occur. Under certain conditions of alternating current these materials have superior electrical performance to tungsten contacts. The natively occurring alloy 'osmiridium' is still used for tipping pen nibs. Until recently this alloy was used exclusively for this purpose, a typical analysis being 35% osmium, 30% iridium, remainder platinum, rhodium and ruthenium. It is now largely superseded by ruthenium sintered powder. The alloys of iridium and platinum, with and without osmium and rhodium, are used to a great extent in jewellery as mounts for precious stones. The 10% iridium/platinum alloy was used to manufacture the standard metre, the standard pound weight and the standard kilogram; the majority of sub-standards use a similar type alloy. This is indicative of the high degree of stability with which these alloys are viewed.

One of the isotopes of iridium – Ir 192 – is radioactive and increasing use is being made of this as a source of non-destructive testing.

In general this is a scarcely used, expensive material with a number of potentially useful applications limited by price and the difficulty of working the metal.

Following are listed the known iridium suppliers with their identifying symbols.

Symbol	Nominal analysis, supplier, condition and remarks.
h Ir 8	99.999% Ir sponge: Light Ltd; high purity metal
h Ir 10a	99.99% Ir wire 0.5 mm diameter: Light Ltd
h Ir 73	99.995% Ir single crystal 3 mm diameter: Light Ltd
IRIDIUM	99.9% Ir wire: International Nickel; cold drawn **DPN: 625 UTS: 2000 Elon: 2%**
IRIDIUM	99.9% Ir wire and rod: International Nickel; annealed at 1000 °C for 30 min **DPN: 500 UTS: 1020 Elon: 15%**
IRIDIUM	99.9% Ir wire and rod: International Nickel; annealed at 1200 °C for 30 min **DPN: 350**
IRIDIUM	High purity metal: Impurities less than 10 p.p.m.; Mallory Metallurgical; supplied as sponge
IRIDOSIUM	10% Pt 1.5% Rh 27% Os 6% Ru Ir alloy: Natively occurring

Symbol	Nominal analysis, supplier, condition and remarks.
OSMIRIDIUM	30% Os Ir alloy: Naturally occurring; details from International Nickel; for pen nib tips, etc.
PLATINIRIDIUM	Pt Os Rh Pd Ir alloy: Up to 75% Ir; rare natively occurring alloy
UNS P01001–P01999	Ir metal and alloys: American Standards system; unified numbers

Note. The following abbreviations and units are used in the tables:

DPN	Hardness, diamond pyramid number
UTS	Ultimate tensile strength, N/mm^2
Elon	Elongation, %
Proof	0.1 % proof strength, N/mm^2

$1\ N/mm^2 = 0.1\ hbar = 0.102\ kgf/mm^2 = 0.06475\ tonf/in^2 = 145.04\ lbf/in^2$

20. Iron Fe

Physical properties

Atomic number	26	
Atomic weight	55.85	
Crystal structure	Body-centred cubic up to 950 °C	
	Face-centred cubic up to 1425 °C	
Colour	Greyish white	
Specific gravity	7.9	
Density	7900 kg/m³	(0.202 lb/in.³)
Melting point	1535 °C	
Boiling point	2800 °C	
Specific heat	0.44 J/g °C	(0.105 cal/g °C)
Thermal conductivity	76.2 W/m °C	(18.2 cal/m s °C)
Coefficient of linear expansion (20–100° C)	12.2×10^6/ °C	
Latent heat of fusion	272 J/g	(65 cal/g)
Latent heat of vaporization	6615 J/g	(1580 cal/g)
Thermal neutron absorption cross-section	2.53 barns/atom	
Electrical conductivity	19% IACS (copper 100%)	
Specific resistance	89 microhm mm	
Temperature coefficient of electrical resistance	0.0062/ °C	
Electrochemical equivalent	1.042 g/A/h – divalent	
	0.695 g/A/h – trivalent	
Electrode potential	-0.04 V	
Magnetic susceptibility	Ferro–magnetic	
Young's modulus of elasticity	$193–200 \times 10^9$ N/m²	$(28–29 \times 10^6$ lbf/in.²)
Tensile strength	540 N/mm²	(35 tonf/in.²)
Hardness	– DPN	

20.1 General notes on iron

In this book iron, iron alloys and cast iron appear in this separate section, divorced from the carbon/iron alloys – steel – which appear later in Section 44.

Iron followed copper in the service of man and in certain parts of the world, notably India, it is possible that iron tools and ornaments were in fact in use before bronze.

The ores of iron are found in many areas, often in highly concentrated pockets, and there is little doubt that the first iron was accidentally produced when ore was used as firebricks or fireplace material. It is this ease of reduction from the ore coupled with large ore deposits which has made iron the principal tool in man's struggle to civilize and improve his lot.

All industrial countries have or had large iron ore deposits, many of which are becoming exhausted. Large deposits are available in some of the world's hinterlands, such as Alaska and North Canada, which can be mined and shipped economically using modern machinery and techniques.

A limited amount of native iron is found, notably one twenty-five ton lump in Greenland, but no commercial quantities have been discovered. Meteorites almost always contain native iron alloyed with nickel, cobalt and other metals.

Until the end of the last century a considerable quantity of wrought iron was made. Iron of a very high purity, albeit in contact with stringers of slag, was produced by this method. This material has now almost all been replaced by mild steel which is made using considerably less human sweat and muscle. For some uses, such as chains, wrought iron is still specified.

High purity iron is very much a laboratory material and has no real commercial applications. When highly purified, iron has considerable resistance to corrosion. A metallurgical enigma is the 'Delhi Pillar' which is a large structure of pure iron standing unprotected for centuries in the open air with no visible evidence of corrosion.

The 'pure' iron normally used tarnishes in the atmosphere, rapidly forming a scale of complex iron oxides which is only loosely adherent and is then readily removed, allowing the fresh surface to form new scale and repeat the process.

Industrially iron in the cast form finds considerable use.

Further details of wrought and cast iron are given in the subsequent sections.

Steel is an alloy of iron and carbon, often with other alloying elements. These materials are dealt with in Section 44.

178

There is also a range of highly alloyed heat resistant iron alloys. These are dealt with under nickel–iron, nickel, chrome–iron, etc., in Section 44, but Section 20C is devoted to the iron–nickel alloys which have been developed for their specific magnetic properties.

20A Iron – commercially pure

Specific gravity	7.5–7.8	
Density	7500–7800 kg/m^3	(0.28 lb/in.^{3})
Solidus/liquidus	1520–1530 °C	
Thermal conductivity	17–38 W/m °C	(4–9 cal/m s °C
Coefficient of linear expansion (20–100° C)	10.2–11.9 × 10^{-6}/ °C	
Electrical conductivity	3.5–9% IACS (copper 100%)	
Specific resistance	180–520 microhm mm	
Young's modulus of elasticity	211.7 × 10^9 N/m^2	(30.7 × 10^6 lbf/in.^{2})
Impact	– J	
Fatigue strength	–	
Hot strength	–	

The above properties are typical of the following group, and may not apply exactly to any one specification. It is possible that with certain specifications some of the values may not be applicable.

General metallurgical characteristcs

Iron is a relatively soft ductile material which has few industrial applications as it cannot compete economically with materials such as mild steel.

Very high purity iron, which is exceedingly difficult to produce and of academic interest only, has excellent corrosion resistance. All other grades tarnish and corrode rapidly. Apart from wrought iron, the specifications listed are generally for primary materials and laboratory use.

Thermal treatment

Pure iron cannot be hardened by any form of thermal treatment. The material is capable of accepting a considerable amount of cold work with no interstage anneals. When required annealing should be at 500–600 °C in a protective atmosphere.

Welding and brazing

Pre-weld treatment. No thermal treatment prior to welding is necessary. The areas to be welded must be thoroughly cleaned and mechanically abraded.

Welding and brazing. Probably the most satisfactory welds are produced using the old fashioned Smith's coke fire for heating, and hammer forge welding or the modern equivalent. Modern electric, gas and inert gas shield arc welding all give satisfactory welds using fluxes where necessary, as care must be taken to prevent oxide trapping.

Brazing is readily accomplished using copper alloy rods and fluxes.

Post-weld treatment. No thermal treatment is required. All traces of flux must be removed and the parts protected from corrosion.

Flaw detection methods

Crack detection. Magnetic crack test can be used, but will seldom be required as parts are not liable to be stressed. Oil penetrant chalk test is not quite as sensitive but will generally give adequate results.

X-ray. Not generally required.

Chemical etch. Not required.

Corrosion protection

Temporary. Parts should be oiled or greased at all times and protected against dampness.

Permanent. This is difficult unless great care is taken to remove all trace of moisture and oxide and immediately apply the protective coating, which can be paint or plating. There are various forms of oxide which can be produced artificially to give some measure of protection and a pleasant appearance, but these will not generally be applied to the materials listed.

Painting. Paint must be applied immediately after cleaning. A metal base primer is essential which can be zinc, lead oxide or chromate. Materials are now available which chemically convert to the metal, in effect coating the article with an adherent layer of metallic zinc or lead, both of which sacrifice themselves to protect the base iron.

The top coat should seal the undercoat. In general stoved paints applied to a mechanically roughened surface give the best results.

Plating. Electroplating will not be generally applied, but when necessary presents no problems. Galvanizing – zinc coating – is sometimes applied to wrought iron and parts may also be dip tinned.

Machinability

These materials will seldom be machined. Iron being soft and ductile, difficulty will be experienced if a good surface finish is required. High speeds with low tool pressure and a copious flow of coolant probably give best results.

Uses

Principal uses of the material listed are as primary metals, or for laboratory purposes.

Some wrought iron is still used in hand forged chains where the almost complete absence of cold working properties is an advantage. This obviates or considerably reduces the need to carry out periodic stress releases to regain the ductility.

Symbol	Nominal analysis, supplier, condition and remarks.
00 Iron	0.025% C 0.1% Si 0.12% Mn steel: Casting; Edgar Allen; Code 687
A GRADE	Low carbon iron: Low Moor for DTD 5092; DTD 5102
ALFENOL	Fe alloy containing 16% Al
ALFER	Fe alloy containing 11/13% Al
ALSIFER	Master alloy: 20% Al 40% Fe 40% Si
ALSIMIN	Ferrosilicon aluminium alloy: 45% Si 15% Al iron
AMS 7706	Fe: Commercially pure plate, sheet, bar, etc.; annealed
AMS 7707	Fe: Commercially pure plate, sheet, bar, etc.; as rolled
ARMICO	0.03% C 0.005% Si 0.15% Ni 0.2% Cu steel: Strip; Bairds
ARMCO IRON	Purest form of commercial iron: Soft iron containing less than 0.1% impurities; trade name
ASTM A42	0.09% (max) Mn wrought iron plate
ASTM A84	0.06% Mn wrought iron for staybolts **UTS: 340 Elon: 30%**
ASTM A207	Wrought iron: Bars & shapes; 0.09% Mn (max)
ASTM A382	Wrought iron: For heat exchanger and condenser tubes
BEST YORKSHIRE	0.06% Mn 0.16% P wrought iron: General name for BS 858
B GRADE	Low carbon iron: Low Moor; for DTD 5102
BS 48	Wrought iron: Replaced by BS 51
BS 51 A	0.1% Mn (max) wrought iron: Bar & plate; as rolled **UTS: 330 Elon: 25%**
BS 51 B	0.15% Mn (max) wrought iron: Bar & plate; as rolled **UTS: 330 Elon: 23%**
BS 51 C	Wrought iron: Analysis not specified; as rolled **UTS: 330 Elon: 19%**
BS 601/1	Low C steel: Strip for magnetic purposes; non-oriented; graded by magnetic properties **UTS: 330 Elon: 25% Proof: 170**
BS 601/2	Low C steel: Strip for magnetic purposes; oriented; graded by magnetic properties
BS 762	0.075% Mn wrought iron bars **UTS: 330 Elon: 25% Proof: 170**
BS 858	0.06% Mn 0.16% P wrought iron: 'Best Yorkshire' wrought iron **UTS: 330 Elon: 21% Proof: 170**
CODE 687	0.025% C 0.1% Si 0.12% Mn steel: Casting; Edgar Allen; 00 Iron

Note. The following abbreviations and units are used in the tables:

DPN	Hardness, diamond pyramid number
UTS	Ultimate tensile strength, N/mm^2
Elon	Elongation, %
Proof	0.1 % proof strength, N/mm^2

$1 \text{ N/mm}^2 = 0.1 \text{ hbar} = 0.102 \text{ kgf/mm}^2 = 0.06475 \text{ tonf/in}^2 = 145.04 \text{ lbf/in}^2$

Symbol	Nominal analysis, supplier, condition and remarks.
DTD 330	Soft iron: Replaced by DTD 5092 and DTD 5102
DTD 5092	Soft iron: Sheet, bar & forging; annealed; magnetic properties quoted in specification
DTD 5102	Soft iron: Sheet, bar & forging; annealed; magnetic properties included in specification
E 10	Fe high purity sinter: Firth Cleveland; specific gravity 6.6 – 7.0; as sintered **DPN: 100 UTS: 190 Elon: 7%**
E 20	Fe high purity sinter: Firth Cleveland; specific gravity 7.1 – 7.5; as sintered **DPN: 130 ' UTS: 250 Elon: 12%**
F 10	Fe sintered products: Specific gravity 5.7 – 6.1; Firth Cleveland; as sintered **UTS: 120 Elon: 2%**
F 20	Fe sintered products: Specific gravity 6.1 – 6.5; Firth Cleveland; as sintered **UTS: 160 Elon: 6%**
F 30	3% Cu Fe sintered product: Specific gravity 6.1 – 6.5; Firth Cleveland; as sintered **UTS: 240 Elon: 1%**
F 40	5.0% Cu Fe sintered products: Specific gravity 6.1 – 6.5; Firth Cleveland; as sintered **UTS: 250 Elon: 1%**
F 50	10% Cu Fe sintered products: Specific gravity 6.1 – 6.5; Firth Cleveland; as sintered **UTS: 270 Elon: 1%**
F 70	Iron: Sintered material; bearings; Sintered Products Ltd; 23–28% porosity **UTS: 90 Elon: 1.5%**
F 80	Iron: Sintered material; bearings; Sintered Products Ltd; 15–22% porosity **UTS: 100 Elon: 2.5%**
FE 85	High purity iron: Sintered material; for high magnetic permeability; Sintered Products Ltd; 12–14% porosity **UTS: 200 Elon: 14%**
FE 90	High purity iron: Sintered material for high magnetic permeability; Sintered Products Ltd; 4–8% porosity **UTS: 220 Elon: 26%**
FERROCOR 216	Electrical steel: For fractional horse power motors; Richard Thomas & Baldwins
FERROCOR 253	Electrical steel: For fractional horse power motors; Richard Thomas & Baldwins
FERROCOR 320	Electrical steel: For fractional horse power motors; Richard Thomas & Baldwins
FERROSIL 100	Low C iron: Strip for magnetic purposes; Richard Thomas & Baldwins for BS 601/1
FERROSIL 107	Low C iron: Strip for magnetic purposes; Richard Thomas & Baldwins for BS 601/1
FERROSIL 146	Low C iron: Strip for magnetic purposes; Richard Thomas & Baldwins for BS 601/1
FERROSIL 170	Low C iron: Strip for magnetic purposes; Richard Thomas & Baldwins for BS 601/1
FERROSIL 187	Low C iron: Strip for magnetic purposes; Richard Thomas & Baldwins for BS 601/1

Symbol	Nominal analysis, supplier, condition and remarks.
FERROSIL 216	Low C iron: Strip for magnetic purposes; Richard Thomas & Baldwins for BS 601/1
FERROSIL CR253	Low C iron: Strip for magnetic purposes; Richard Thomas & Baldwins for BS 601/1
FERROVAC E	0.007% C Fe high purity iron: Crucible Steel Co; for magnetic purposes
h Fe 8	99.999% Fe sponge: Light Ltd; high purity metal
h Fe 11a	99.998% Fe rod 5 mm diameter: Light Ltd; high purity metal
h Fe 73	99.995% Fe single crystals 2.25 mm diameter: Light Ltd
HIPERM (SUPER)	Low carbon iron: Obsolete; Low Moor for DTD 5092 & DTD 5201
IRON	Fe powder: Electrolytic; Blackwells; pure metal
IRON	High purity metal impurities less than 10 p.p.m.: Mallory Metallurgical; supplied as sponge, rod, wire & sheet
KB 90	Sintered Fe: High impact strength; specific gravity 7.25; Durasint DPN: 120 UTS: 430 Elon: 12%
KB 90/3/4	Sintered Fe: High impact strength; specific gravity 7.25; Durasint DPN: 240 UTS: 620 Elon: 2%
PERMET PF1	100% Fe: For magnets; origin unknown
S 10	5% Cu Fe with C sintered products: Firth Cleveland; hardened and tempered UTS: 500 Elon: 0.5%
S 20	Fe sinter with C: Firth Cleveland; hardened and tempered UTS: 470 Elon: 0.5%
S 30	3.0% Cu Fe with C sinter: Firth Cleveland; hardened and tempered UTS: 500 Elon: 0.5%
S 40	Cu P Fe with C sinter: Firth Cleveland; hardened and tempered; file hard UTS: 500 Elon: 0.8%
SKF REMKO	0.02% C 0.12% Mn iron: Origin unknown
TRAN-COR A 5	Low carbon steel: Strip for audio transformers; Armco; hot rolled DPN: 180 UTS: 480 Elon: 10% Proof: 370

Symbol	Nominal analysis, supplier, condition and remarks.
TRAN-COR A 6	Low carbon steel: Strip for audio transformers; Armco; hot rolled DPN: 180 UTS: 480 Elon: 10% Proof: 370
TRAN-COR M 14	Low carbon steel: Strip for transformers; Armco; hot rolled DPN: 200 UTS: 490 Elon: 2% Proof: 450
TRAN-COR M 15	Low carbon steel: Strip for transformers; Armco; hot rolled DPN: 200 UTS: 490 Elon: 2% Proof: 450
TRAN-COR M 17	Low carbon steel: Strip for transformers; Armco; hot rolled DPN: 190 UTS: 490 Elon: 3% Proof: 440
TRAN-COR M 19	Low carbon steel: Strip for transformers; Armco; hot rolled DPN: 175 UTS: 460 Elon: 11% Proof: 390
TRAN-COR M 22	Low carbon steel: Strip for transformers; Armco; hot rolled DPN: 171 UTS: 460 Elon: 12% Proof: 370
TRAN-COR M 27	Low carbon steel: Strip for transformers; Armco; hot rolled DPN: 160 UTS: 420 Elon: 18% Proof: 330
TRAN-COR M 36	Low carbon steel: Strip for magnetos; Armco; hot rolled UTS: 330 Elon: 26% Proof: 210
TRAN-COR M 43	Low carbon steel: Strip for electric motors; Armco; hot rolled UTS: 270 Elon: 29% Proof: 150
W Fe 18	99.9% Fe powder: Light Ltd; made from iron carbonyl

Note. The following abbreviations and units are used in the tables:

DPN	Hardness, diamond pyramid number
UTS	Ultimate tensile strength, N/mm^2
Elon	Elongation, %
Proof	0.1 % proof strength, N/mm^2

$1 \ N/mm^2 = 0.1 \ hbar = 0.102 \ kgf/mm^2 = 0.06475 \ tonf/in^2 = 145.04 \ lbf/in^2$

20B Iron – cast

Specific gravity	7.2–7.4	
Density	7200–7400 kg/m^3	(0.25–0.26 lb/in.^{3})
Solidus/liquidus	1150–1450 °C	
Thermal conductivity	46–63 W/m °C	(11–15 cal/m s °C)
Coefficient of linear expansion	$10–17 \times 10^{-6}/$ °C	
Electrical conductivity	5–6.5% IACS (copper 100%)	
Specific resistance	270–370 microhm cm	
Young's modulus of elasticity	$138–152 \times 10^9$ N/m^2	($20–22 \times 10^6$ lbf/in.^{2})
Impact	5.2–22 J	(7–30 ft/lb) Izod
Fatigue strength (10×10^6 cycles)	± 15	(± 10–12 tonf/in.^{2})
Hot strength	Not applicable owing to extremely wide variations	

The above figures are intended to show typical values and do not apply exactly to any one specification. As with all casting materials wide variations will be found, many of which are considerably out with the above listed properties.

General metallurgical characteristics

Cast iron can be described as an alloy of iron and carbon.

The normal accepted minimum carbon content is 1.5%, however most cast irons have carbon in the range 2.5–3.5%. Commercial cast irons are not simply alloys of iron and carbon, but invariably have some silicon, manganese, phosphorous and sulphur present. These elements affect the carbon solubility and give rise to various types of cast iron dependent on the form in which the carbon is present.

Alloying elements may be added to give special effects and various thermal treatments can be applied. The types, alloying contents and thermal treatments overlap to such an extent that it is not considered possible or desirable to split the various cast irons into separate groups.

The fact that these materials are cast in small batches, varying in some degree from each other, makes cast iron rather unique among metallic materials. Control of the low alloy irons is usually by mechanical properties only, the founder using his skill to vary the composition according to size, sectional thickness and surface required on the finished casting.

Briefly the types of cast iron are as follows:

Grey cast iron. This is the most common type, being used for general purpose castings where low cost is the prime factor. This material is easily produced, gives excellent reproduction and allows considerable flexibility in casting design. It is the cheapest metallic material used. The structure is usually a mixture of pearlite and ferrite with flakes of carbon in the form of graphite. The proportion of pearlite and ferrite present is controlled by alloying elements, notably silicon, and also by the rate of cooling from liquid to solid state, a fully pearlitic structure generally giving the best tensile and transverse properties of the unalloyed cast irons.

White cast irons. These contain no free carbon as graphite in the 'as cast' condition and are therefore very hard and brittle with limited use. When alloyed with nickel, chromium or molybdenum a tough, hard abrasion resistant iron is obtained.

Whiteheart malleable cast iron. In the Whiteheart process white iron castings are first produced and then annealed in an oxidizing atmosphere. This results in a gradual migration of the carbon from the centre of the casting to the skin where it is oxidized and removed from the metal. In this process therefore the annealed casting has a low carbon content – sometimes less than 1% with thin sections. The rest of the carbon is present partly in nodular form as free graphite and partly combined, the latter appearing as pearlite in the matrix. The final structure and properties depend on the time and temperature of the heat treatment and also on the original composition of the metal. This material is seldom used now as the long term anneal makes it expensive.

Blackheart malleable cast iron. The Blackheart process differs from the Whiteheart process in that the anneal is carried out in a neutral or slightly carburizing atmposphere; consequently the carbon content of the metal is not appreciably reduced. The carbon in the final casting approaches that of grey cast iron. In the annealing process however, it is deposited in a finely divided form (temper carbon) giving a product with properties more resembling steel than normal grey cast iron.

Modern malleabilizing techniques have substantially improved these processes. Malleable cast iron is used for many engineering components and is particularly suitable when something stronger and tougher than grey cast iron is required, but where cheapness is essential. Compared with other irons malleable cast iron shows less corrosion resistance but its other properties make it suitable for components subject to shock.

There is a certain restriction in the size of malleable castings – generally above 100 lb the process becomes uneconomical as the time at temperature to convert the carbon is excessive. The relationship between surface area and mass is the important factor in the economics of the process.

It should be noted that the term 'malleable' is now being used for spheroidal graphite irons, particularly in America.

Nodular or spheroidal graphite cast irons

These cover a range of cast irons in which the graphite present is in spheroidal form instead of the flake in ordinary cast iron. The change can be caused by various treatments (usually patented) of molten cast iron in the ladle prior to pouring. This controls the manner in which the carbon forms during solidification. The brittleness of normal grey cast iron is partly due to the edges of the graphite flakes forming easy cleavage paths, thus giving areas of weakness. When the graphite is in the nodular form, this weakness is eliminated and a material is obtained of greater strength plus some ductility. The structure of nodular iron can be pearlitic or ferritic or combinations of both. The pearlitic type as cast has high strength with relatively low elongation but this can be heat treated to produce a ferritic type with somewhat lower strength but much greater elongation. Under certain conditions the ferritic type can be produced 'as cast' but this is not usual.

Acicular cast irons

The term 'acicular' means 'needle-like' and describes the ferritic structure which forms in grey cast iron during the change from austenite into pearlite and graphite on cooling. This can only be produced on unalloyed iron by quenching from between 480 and 260 °C. This is impracticable with castings made in sand and it is necessary to retard the rate of transformation by the addition of alloying elements such as nickel and molybdenum.

Tensile strengths in the range 380–540 N/mm^2 are obtained from acicular iron with flake graphite. By over-alloying and tempering, tensiles of 630 N/mm^2 can be obtained. Impact resistance of acicular iron is about twice that of high duty iron with a pearlite matrix and resistance to wear is also better due to higher initial hardness and the fact that acicular irons have work hardening properties.

Most cast irons are covered by the preceding types. The properties of each are dependent on various elements, some of which are normally present due to their presence in the pig iron used to produce the metal. Other alloying elements may be added to produce specific results. Briefly the effect of these elements are as follows:

Silicon. This is a graphite former, helping to convert the cementite in white cast iron to graphite or aiding the formation of ferrite in grey cast iron. It also helps the fluidity at casting and very few irons are produced with less than 2% silicon. Irons with up to 5% silicon generally with nickel and chromium have good oxidation resistance, while higher silicon irons – 15–20% – although brittle and virtually unmachinably hard, have excellent resistance to sulphuric and nitric acids of all concentrations and temperatures.

Chromium. This is a carbide former, thus tends to reduce the amount of graphite present. The carbides formed are very hard and brittle, increasing the hardness and hardenability of cast iron. Chromium also forms an adherent oxide film and when present above about 5% has an appreciable effect on the corrosion resistance.

As with steel, 18% chromium when present with 8% nickel, gives an austenite structure which has excellent corrosion resistance. Cast irons never have a truly single phase structure owing to their high carbon content, thus these austenitic irons are not stainless in the sense that applies to 18/8 austenitic steels.

Nickel. This is a ferrite strengthener which encourages the formation of graphite and more than any other single element improves the toughness of cast irons. As stated above, when present in sufficient quantity with chromium, austenitic corrosion resistant irons are formed.

Molybdenum. This is a carbide former which considerably increases the hardenability of cast iron. All hardenable irons with any appreciable sectional thickness will have some molybdenum present, generally with nickel to improve the ductility.

Phosphorus. This forms a hard brittle phase which for most purposes is highly undesirable. It does however, aid the fluidity of the molten metal. Phosphorus is present in most pig irons and remains as a residual element unless special fluxing techniques are used for its removal.

Thermal treatment

Generally there are three reasons for thermal treatment of cast iron:

(a) A high temperature treatment to alter the constitution of the metal at about 800 °C. This is used mainly for the production of malleable castings and can be used for obtaining machinability where castings are hard or where heavy cuts are required and strength is not important. Slow cooling from the chosen temperature is necessary.

(b) Stress relief annealing at 450–600 °C. This lower temperature is used to remove casting stresses and prevent the possibility of distortion after machining. There is no metallurgical change.

(c) Hardening at 750–850 °C followed by tempering. In the past attempts to improve the toughness and strength of cast iron were generally unsuccessful owing to the poor quality of the material and the severe stresses set up in the quenching treatment. Modern treatments have produced superior material with the ability to withstand these stresses, and with many parts, especially those of light section, it is now relatively simple to increase the strength and hardness of the finished component by hardening and tempering. As with steel it has been found that the use of alloying elements increases the scope of thermal treatment and this applies particularly to nickel additions. Up to 5% nickel develops the martensitic structure in castings 'as cast' which is similar to that desired after the heat-treatment cycle. One of the effects of nickel is to improve the heat treatment properties of cast iron so that with increasing amounts of nickel an appreciably milder type of treatment can be used to produce a given result. Thus with an ordinary grey iron, water quenching is required for a given result, with a 1.5% nickel iron similar properties can be obtained by quenching in oil, while at 3.0% nickel the result is obtained by cooling the casting in an air blast. The use of nickel alloy cast irons therefore, facilitates the heat treatment of complex castings which would not normally withstand the more drastic forms of quenching.

By selection of the tempering temperature, various combinations, such as high hardness with low ductility, hardness with intermediate strength, or low hardness with good ductility, can be developed.

Induction and flame hardening can also be supplied for the local hardening of iron castings, such as gears, cams, or lathe-beds, where hard wearing surfaces combined with a tough strong machineable base are required. Induction hardening requires special equipment and control, while the flame hardening uses normal gas welding plant but is very difficult to control.

Some surface hardness can be obtained by the use of chills in the mould when hard carbides are formed in the vicinity of the chill instead of graphite and ferrite.

Other surface treatments concentrate on improving the corrosion resistance of the metal using either silicon for acid resistance or chromium for hot oxidation resistance. These processes require special compounds but the equipment is usually quite simple.

Welding and brazing

Pre-treatment. Careful joint preparation is important and all evidence of dirt oil and grease must be removed.

Welding and brazing. Welding should only be carried out after obtaining advice from a laboratory. Chilled castings, acicular and spheroidal graphite castings should not be welded.

Large parts should be pre-heated and cooled slowly after welding to prevent quench cracking. Special fluxes and filler rods are available for welding, using electric or gas techniques.

Brazing with copper alloy filler rods and flux is the most common method of joining cast irons and this is very readily accomplished with a minimum of skill. Wherever possible this should be used instead of welding.

Post-treatment. Fluxes should be removed. Hardenable alloy cast irons should be re-hardened and tempered if possible, or at least tempered after welding. As an absolute minimum the weld area must be stress relieved using gas torches or portable electric blankets.

Flaw detection methods

Crack test. Although these materials, with the exception of the austenitic irons, are magnetic, the use of penetrant oil chalk test rather than magnetic crack test is recommended. The chalk test method is more flexible and requires less equipment and training, as the relatively rough 'as cast' surface can lead to difficulty in interpretation.

X-ray. Most high duty castings will require to be X-rayed and use of this technique is recommended as a control of casting quality for all important or expensive parts.

Ultra-sonic test. With modern techniques and skilled personnel this can be used to find minor defects, and for information on the graphite form. Specialist advice is essential as considerable expertise is necessary in operation and interpretation. It cannot be used for surface defects.

Chemical etch. Not required.

Corrosion protection

Temporary. None generally required, the 'as cast' skin being a very satisfactory protection.

Permanent. This will depend on the use for which the castings are designed. It is often cheaper to make the casting large enough to allow for some rusting and corrosion, rather than protect the surface. The 'as cast' surface has a built-in corrosion protection in the adherent oxides.

Painting. This is the most common method of corrosion protection. It is essential that all trace of existing corrosion, dirt and grease are removed. This is best accomplished by grit blasting which should be immediately followed by zinc or lead oxide, or chromate primer.

Chemical treatments are not normally necessary before priming, but if carried out must be well swilled and oven dried, as cast iron has some absorption ability.

A top coat of stoving enamel or other high duty paint should be applied to seal the primer as soon after application as possible.

Plating. This is not often necessary for corrosion protection. When called for it requires special care at the cleaning and etch owing to the surface absorption properties of many cast irons, which tend to carry over the pre-plating liquids into the plating solution with adverse effects on adhesion.

Cast iron is often galvanized and this zinc coating can be very successful if correctly applied.

Metal spraying of zinc or aluminium gives excellent results provided a good key is made by grit blasting and the subsequent coating is sealed by wire brushing, then chemical treatment or painting.

Vitreous enamelling using special frits which are fused after application give excellent results if correctly applied and are applied to many household articles such as cookers sinks and baths.

Machinability

The grey and nodular pearlite cast irons have excellent machining characteristics. White cast irons and chilled areas are very hard, being often unmachinable. This applies also to the high silicon irons and many of the higher chromium castings. Pearlite irons are tougher to machine than ferritic irons.

Uses

Grey cast iron and blackheart malleable. General purpose engineering components where fatigue and tensile stress is secondary to compressive strength. Machine tool castings, housings of all types where weight is not important, wheels, brackets, gears, shafts. These irons are sometimes used for their excellent damping characteristics which are superior to those of wrought steel.

Household fittings, such as rainwater and drainage pipes, underground pipe lines, fireplaces, cookers.

White cast iron. Wear resistant parts, dredger buckets, excavator grabs, crusher jaws.

Nodular cast irons. In general the uses are similar to grey iron but of a higher order. For example diesel and petrol engine crankshafts could be in this iron, in addition to gears, crankcases, gearboxes, etc.

Austenitic cast iron. For marine use, pump bodies for water engineering use where good corrosion resistance is necessary and economy important.

Symbol	Nominal analysis, supplier, condition and remarks.
0707	Nodular graphite cast iron: Pearlitic/ferritic structure; designation used by Danish Standards **DPN: 230 UTS: 600 Elon: 3% Proof: 350**

Note. The following abbreviations and units are used in the tables:

DPN	Hardness, diamond pyramid number
UTS	Ultimate tensile strength, N/mm^2
Elon	Elongation, %
Proof	0.1 % proof strength, N/mm^2

1 N/mm^2=0.1 hbar=0.102 kgf/mm^2=0.06475 tonf/in^2=145.04 lbf/in^2

Symbol	Nominal analysis, supplier, condition and remarks.
0708	Nodular graphite cast iron: Pearlitic structure; designation used by Danish Standards **DPN: 265 UTS: 700 Elon: 2% Proof: 400**
0715	Nodular graphite cast iron: Pearlitic/ferritic matrix; designation used by Danish Standards **DPN: 165 UTS: 400 Elon: 17% Proof: 250**
0716	Nodular graphite cast iron: Pearlitic/ferritic matrix; designation used by Danish Standards **DPN: 165 UTS: 400 Elon: 17% Proof: 250**
0727	Nodular graphite cast iron: Pearlitic/ferritic structure; designation used by Danish Standards **DPN: 205 UTS: 500 Elon: 7% Proof: 310**
A 32/101	General standard for flake cast iron: Grey; French Standard; see designation for details

Symbol	Nominal analysis, supplier, condition and remarks.
A 32/101 F40D	Fine ground flake graphite cast iron: French Standard **DPN: 250 UTS: 400**
A 32/201	General standard for nodular graphite cast iron: French Standard; see designation for details
AMS 5310 B	Pearlitic malleable cast iron: Hardened & tempered
AMS 5315	Nodular ductile cast iron **UTS: 420**
AMS 5316	Nodular ductile cast iron **UTS: 580**
AMS 5328	0.3% C 0.8% Cr 1.8% Ni 0.35% Mo: Investment cast iron
AMS 5329	0.3% C 0.8% Cr 1.8% Ni 0.35% Mo: Sand cast iron
AMS 5330	0.42% C 0.8% Cr 1.8% Ni 0.35% Mo: Investment cast iron
AMS 5331	0.42% C 0.8% Cr 1.8% Ni 0.35% Mo: Sand cast iron
AMS 5333	0.15% C 0.5% Cr 0.5% Ni 0.2% Mo: Investment cast iron
AMS 5334A	0.3% C 0.5% Cr 0.5% Ni 0.2% Mo: Investment cast iron
AMS 5335A	0.3% C 0.5% Cr 0.5% Ni 0.2% Mo: Sand cast iron
AMS 5336	0.3% C 0.95% Cr 0.2% Mo: Investment cast iron
AMS 5338	0.4% C 0.95% Cr 0.2% Mo: Investment cast iron
ASTM A43	Pig iron for foundry use; various grades by analysis
ASTM A47	Malleable cast iron: Composition not specified **UTS: 370 Elon: 18% Proof: 240**
ASTM A48	Grey iron castings: Suffix number indicates tensile strength in p.s.i. × 1000
ASTM A53 T/T 60/40/18	Spheroidal cast iron with ferritic matrix **DPN: 160 UTS: 36 Elon: 17% Proof: 22**
ASTM A53 T/T 65/45/12	Spheroidal cast iron with mainly ferritic matrix **DPN: 170 UTS: 40 Elon: 12% Proof: 27**
ASTM A53 T/T 80/55/06	Spheroidal cast iron with ferritic/pearlitic matrix **UTS: 48 Elon: 7% Proof: 33**
ASTM A53 T/T 100/70/03	Spheroidal cast iron with pearlitic structure **UTS: 65 Elon: 2% Proof: 42**
ASTM A53 T/T 120/90/02	Spheroidal cast iron with pearlitic matrix **UTS: 72 Elon: 2% Proof: 45**
ASTM A159	Grey cast iron: For automotive use; graded in same manner as SAE alloys
ASTM A197	Malleable iron: Free of primary graphite **UTS: 280 Elon: 5% Proof: 200**
ASTM A220	Malleable pearlitic cast iron: Graded in same manner as SAE alloys
ASTM A276	Grey iron castings: For temperatures up to 300 °C; classified by tensile properties
ASTM A319/1	4.1% C grey cast iron
ASTM A319/11	3.8% C grey cast iron
ASTM A319/111	3.5% grey cast iron
ASTM A398	Welding rods for cast iron: Classified according to analysis; ASTM system
ASTM A436/1	3.0% C (max) 15% Ni 6.5% Cu 2.2% Cr cast iron: Austenitic grey
ASTM A436/1b	3.0% C (max) 15% Ni 6.5% Cu 3.0% Cr cast iron: Austenitic grey
ASTM A436/2	3.0% C (max) 20% Ni 32.2% Cr cast iron: Austenitic grey
ASTM A436/2b	3.0% C 20% Ni 4.5% Cr cast iron: Austenitic grey
ASTM A436/3	2.6% C (max) 30% Ni 3.0% Cr cast iron: Austenitic grey
ASTM A436/4	2.6% C (max) 5.5% Si 30.5% Ni 5.0% Cr cast iron: Austenitic grey
ASTM A436/5	2.4% C (max) 35% Ni cast iron: Austenitic grey
ASTM A436/6	3.0% C (max) 20% Ni 4.5% Cu 1.5% Cr cast iron: Austenitic grey

Symbol	Nominal analysis, supplier, condition and remarks.
ASTM A439 D2	3.0% C (max) 20% Ni 2.2% Cr iron: Casting austenitic; ductile
ASTM A439 D2B	3.0% C (max) 20% Ni 3.3% Cr iron: Casting austenitic; ductile
ASTM A439 D2C	2.9% C (max) 22.5% Ni iron: Casting austenitic; ductile
ASTM A439 D3	2.6% C (max) 30% Ni 3.0% Cr iron: Casting austenitic; ductile
ASTM A439 D3A	2.6% C (max) 30% Ni 1.2% Cr iron: Casting austenitic; ductile
ASTM A439 D4	2.6% C (max) 30% Ni 5.0% Cr iron: Casting austenitic; ductile
ASTM A439 D5	2.4% C (max) 35% Ni iron: Casting austenitic; ductile
ASTM A439 D5B	2.4% C (max) 35% Ni 2.5% Cr iron: Casting austenitic; ductile
ASTM A445	3.0% C (min) 2.5% Si 0.08% P cast iron: Ferritic; ductile
ASTM A447	0.2% C 12% Ni 25.5% Cr iron: Casting for high temperature use
ASTM A448	0.55% C 35% Ni 15.5% Cr iron: Casting for high temperature use; previously ASTM B207
ASTM A476	3.0% C (min) 3.0% Si 0.08% P 0.05% S cast iron: Ductile
ASTM A518	0.9% C 14.5% Si cast iron: Corrosion resistant
ASTM A532/1	3% C (total) 4% Ni 1.5% Cr white cast iron: Abrasion resistant; three grades varying slightly in analysis
ASTM A532/11	3% C (total) 16% Cr 3% Mo white cast iron: Abrasion resistant; two grades varying slightly in analysis
ASTM A532/111	2.8% C (total) 26% Cr white cast iron: Abrasion resistant; two grades varying slightly in analysis
ASTM A536 60/40/18	Nodular graphite cast iron **UTS: 414 Elon: 18% Proof: 276**
ASTM A536 65/45/12	Nodular graphite cast iron **UTS: 450 Elon: 12% Proof: 310**
ASTM A536 80/55/06	Nodular graphite cast iron **UTS: 550 Elon: 6% Proof: 380**
ASTM A536 80/60/03	Nodular graphite cast iron **UTS: 550 Elon: 3% Proof: 410**
ASTM A536 100/70/03	Nodular graphite cast iron **UTS: 690 Elon: 3% Proof: 480**
ASTM A536 120/90/02	Nodular graphite cast iron **UTS: 825 Elon: 2% Proof: 620**
ASTM A567 HH90	0.9% C 26% Cr 12.5% Ni Fe alloy: Casting
ASTM A567 HI50C	0.5% C 28% Cr 16% Ni 1.0% Nb + Ta 0.1% Fe alloy: Casting
ASTM A567 HT50C	0.5% C 14% Cr 35% Ni 1.0% Nb + Ta Fe alloy: Casting
ASTM A571	2.5% C 2.0% Si 4.2% Mn 23% Ni Fe alloy: Casting; ductile iron
ASTM A602 M3210	Spheroidal graphite malleable iron: Casting; annealed **DPN: 156 UTS: 35 Elon: 10% Proof: 22**
ASTM A602 M4504	Spheroidal graphite malleable iron: Casting; hardened and tempered **DPN: 185 UTS: 45 Elon: 4% Proof: 32**
ASTM A602 M5003	Spheroidal graphite malleable iron: Casting; hardened and tempered **DPN: 210 UTS: 53 Elon: 3% Proof: 38**
ASTM A602 M5503	Spheroidal graphite malleable iron: Casting; hardened and tempered **DPN: 210 UTS: 53 Elon: 3% Proof: 39**
ASTM A602 M7002	Spheroidal graphite malleable iron: Casting; hardened and tempered **DPN: 250 UTS: 63 Elon: 2% Proof: 50**
ASTM A602 M8501	Spheroidal graphite malleable iron: Casting; hardened and tempered **DPN: 285 UTS: 74 Elon: 1% Proof: 60**

Symbol	Nominal analysis, supplier, condition and remarks.
ASTM A667	Specification covering dual metal grey and white cast iron cylinder
ASTM A716	Ductile iron: Pipe for culverts **UTS: 410** **Elon: 10%** **Proof: 290**
BS 309 W22/4	Whiteheart malleable cast iron **UTS: 300** **Elon: 4%**
BS 309 W24/8	Whiteheart malleable cast iron **UTS: 360** **Elon: 8%**
BS 310 B18/6	Malleable Blackheart iron: Casting **UTS: 270** **Elon: 6%** **Proof: 170**
BS 310 B20/10	Malleable Blackheart iron: Casting **UTS: 300** **Elon: 10%** **Proof: 80**
BS 310 B22/14	Malleable Blackheart iron: Casting **UTS: 330** **Elon: 14%** **Proof: 200**
BS 321 A	Grey cast iron: General purpose; as cast **UTS: 320**
BS 321 C	Grey cast iron: General purpose; as cast **UTS: 140**
BS 821 (High)	Iron castings: For gears & gear blanks; as cast **DPN: 220** **UTS: 300**
BS 821 (Medium)	Iron castings: For gears & gear blanks; stress relieved **DPN: 200** **UTS: 220**
BS 821 (Ordinary)	Iron castings: For gears & gear blanks; as cast **DPN: 160** **UTS: 170**
BS 1452/10	Grey iron castings **DPN: 200** **UTS: 150**
BS 1452/12	Grey iron castings **DPN: 200** **UTS: 180**
BS 1452/14	Grey iron castings **DPN: 200** **UTS: 210**
BS 1452/17	Grey iron castings **DPN: 210** **UTS: 240**
BS 1452/20	Grey iron castings **DPN: 210** **UTS: 300**
BS 1452/23	Grey iron castings **DPN: 210** **UTS: 340**
BS 1452/26	Grey iron castings **DPN: 260** **UTS: 390**
BS 1453 B1	3.3% C 3.2% Si 1.5% P (max) low S: Weld filler rod; for welding cast iron
BS 1453 B2	3.3% C 2.2% Si 1.5% P low S: Weld filler rod; for welding cast iron; harder deposit than B1
BS 1453 B3	3.2% C 2.2% Si 1.5% Ni: Weld filler rod; for welding Ni bearing & high strength cast irons
BS 1591	0.8% C 15% Si 1.0% Mn 1.0% P 0.1% S cast iron: Annealed; for acid resistant purposes
BS 2789/1	Cast iron: Replaced by BS alloys SNG 32/2 and SNG 32/7
BS 2789 2 A	Cast iron: Replaced by BS alloy SNG 27/12
BS 2789/2 B	Cast iron: Replaced by BS alloy SNG 24/17
BS 2789 SNG24/17	Spheroidal cast iron: With ferritic matrix **DPN: 160** **UTS: 360** **Elon: 17%** **Proof: 220**
BS 2789 SNG27/12	Spheroidal cast iron: With mainly ferritic matrix **DPN: 170** **UTS: 400** **Elon: 12%** **Proof: 270**
BS 2789 SNG32/7	Spheroidal cast iron with ferritic/pearlitic matrix **UTS: 480** **Elon: 7%** **Proof: 330**
BS 2789 SNG37/2	Spheroidal cast iron with pearlitic matrix **UTS: 560** **Elon: 2%** **Proof: 370**

Symbol	Nominal analysis, supplier, condition and remarks.
BS 2789 SNG42/2	Spheroidal cast iron with pearlitic structure **UTS: 650** **Elon: 2%** **Proof: 420**
BS 2789 SNG47/2	Spheroidal cast iron with pearlitic matrix **UTS: 720** **Elon: 2%** **Proof: 450**
BS 3333 P28/6	Pearlitic malleable cast iron **UTS: 420** **Elon: 6%** **Proof: 270**
BS 3333 P33/4	Pearlitic malleable cast iron **UTS: 500** **Elon: 4%** **Proof: 300**
BS 3468 AUS101 A	3% C 2% Si 2% Cr 15% Ni 6% Cu cast iron: Austenitic structure; high expansion **DPN: 212** **UTS: 140** **Elon: 2%**
BS 3468 AUS101 B	3% C 2% Si 3% Cr 15% Ni 6% Cu cast iron: Austenitic structure; high expansion **DPN: 248** **UTS: 180**
BS 3648 AUS 102 A	3% C 2% Si 2% Cr 20% Ni cast iron: Austenitic structure; high expansion **DPN: 212** **UTS: 140** **Elon: 2%**
BS 3468 AUS102 B	3% C 2% Si 3% Cr 20% Ni cast iron: Austenitic structure; high expansion **DPN: 248** **UTS: 180**
BS 3468 AUS104	2% C 5% Si 3.5% Cr 10% Ni cast iron: Austenitic structure; for heat resistance **DPN: 248** **UTS: 180** **Elon: 2%**
BS 3468 AUS105	2.5% C 1.5% Si 3% Cr 30% Ni cast iron: Austenitic structure; corrosion resistance **DPN: 212** **UTS: 170**
BS 3468 AUS202 A	3% C 2.5% Si 2% Cr 20% Ni cast iron: Austenitic structure with spheroidized graphite **DPN: 201** **UTS: 360** **Elon: 8%** **Proof: 220**
BS 3468 AUS202 B	3% C 2.5% Si 3% Cr 20%Ni cast iron: Austenitic structure with spheroidized graphite **DPN: 255** **UTS: 360** **Elon: 6%** **Proof: 220**
BS 3468 AUS203	3% C 2.5% Si 0.5% Cr 22% Ni cast iron: Austenitic structure with spheroidized graphite; good impact strength **DPN: 170** **UTS: 360** **Elon: 20%** **Proof: 220**
BS 3468 AUS204	3% C 5% Si 2% Cr 20% Ni cast iron: Austenitic structure; spheroidized graphite **DPN: 230** **UTS: 360** **Elon: 10%** **Proof: 220**
BS 3468 AUS205	2.5% C 2.0% Si 3% Cr 30% Ni cast iron: Austenitic structure; spheroidized graphite **DPN: 201** **UTS: 360** **Elon: 7%** **Proof: 220**
BS 5001	3.5% C 1.8% Si 1.0% Mn 1.0% P 0.12% S cast iron: Piston rings **UTS: 240**
BS 5022	Malleable iron: Castings for automobiles; withdrawn
BS 5024	Cast iron: For air cooled & jacketed cylinders for automobiles; withdrawn
BS 5025	Cast iron: For sand cast pistons & valve guides for automobiles; withdrawn
BS 5026	Cast iron: For flywheels for automobiles; withdrawn
BS K6	3.5% C (max) 1.8% Si 0.12% S 1.0% P cast iron: Combined carbon 0.7%; for piston rings
BS K11	3.5% C 2.0% Si 1.0% Mn 0.12% S 1.0% P cast iron: Combined carbon 0.7% **UTS: 200**
CAUSAL METAL	Austenitic grey cast iron: Corrosion resistance comparable to nickel; origin unknown
CB	Cast iron: For chemical corrosion resistance; Meehanite; as cast for salt water and acid resistance **DPN: 190** **UTS: 300**
CB 3	Cast iron: For use with concentrated sulphuric acid; Meehanite; as cast; suitable for up to 100% at 95 °C **DPN: 200** **UTS: 300**

Note. The following abbreviations and units are used in the tables:

DPN	Hardness, diamond pyramid number
UTS	Ultimate tensile strength, N/mm^2
Elon	Elongation, %
Proof	0.1 % proof strength, N/mm^2

$1 \ N/mm^2 = 0.1 \ hbar = 0.102 \ kgf/mm^2 = 0.06475 \ tonf/in^2 = 145.04 \ lbf/in^2$

Symbol	Nominal analysis, supplier, condition and remarks.
CC	Cast iron: Excellent corrosion resistance; Meehanite; as cast for salt water resistance, etc. **DPN: 190 UTS: 300**
CRS	Austenitic nodular graphite cast iron: Corrosion resistance; Meehanite **DPN: 250 UTS: 380 Elon: 20%**
D	14.5% Si Fe casting: Duriron Co; symbol for Duriron
D 51	14.5% Si 4.5% Cr Fe casting: Duriron Co; symbol for Durichlor 51
DIN 1691	General standard for flake cast iron: Grey; German Standard; see designation for details
DIN 1691 GG40	Fine ground flake graphite cast iron: German Standard **DPN: 250 UTS: 400**
DIN 1693	General standard for nodular cast irons: German Standard; see designation for details
DIN 1693 GGG38	Spheroidal cast iron with ferritic matrix **DPN: 160 UTS: 360 Elon: 17% Proof: 220**
DIN 1693 GGG42	Spheroidal cast iron with mainly ferritic matrix **DPN: 170 UTS: 400 Elon: 12% Proof: 270**
DIN 1693 GGG45	Spheroidal cast iron with ferritic/pearlitic matrix **UTS: 480 Elon: 7% Proof: 330**
DIN 1693 GGG50	Spheroidal cast iron with ferritic/pearlitic matrix **UTS: 480 Elon: 7% Proof: 330**
DIN 1693 GGG60	Spheroidal cast iron with pearlitic matrix **UTS: 560 Elon: 2% Proof: 370**
DIN 1693 GGG70	Spheroidal cast iron with pearlitic matrix **UTS: 270 Elon: 2% Proof: 450**
DRICO	3.2% C 1.5% Si cast iron: Origin unknown
DS 11/301	General standard for flake cast iron: Grey; Danish Standard; see designation for details
DS 11/301 GG40	Fine ground flake graphite cast iron: Danish Standard **DPN: 250 UTS: 400**
DS 11/303	General Standard for nodular graphite cast iron: Danish Standard; see designation for details
DTD 233 A	3.5% C total 0.7% C combined 2% Si 0.5% Cr 0.7% Mo cast iron: For piston rings; centri-cast pots **DPN: 275**
DTD 413	3.4% C total 0.7% C combined 2% Si 1% Mo cast iron: For piston rings
DTD 462	1.7% C total 30% Cr 1.5% Si 1% Mn cast iron: Centri-cast; for piston rings; annealed **DPN: 300**
DTD 485 A	2.9% C total 2% C combined 1% Cr 0.9% Mn 8% Mo cast iron: Centri-cast; for piston rings; annealed **DPN: 280**
DTD 614	1.8% C total 1.8% Si 16% Cr 0.1% S & P cast iron **DPN: 300**
DTD 719	1.7% C total 14% Cr 0.4% Mo cast iron: Centri-cast; for rings & liners; annealed
DURATLAS GM	3.3% C 0.5% Si 0.5% Mn 2% Cr 4.8% Ni 0.5% Mo cast iron: Firth Brown **DPN: 500**
DURICHLOR 51	0.9% C 14.5% Si 4.5% Mn iron alloy: Duriron Co; excellent corrosion resistance
DURIRON	0.9% C 1.5% Mn 14.5% Si iron alloy: Duriron Co; excellent corrosion resistance
FC 10 GRADE 1	Fine grained grey cast iron: Designation used for Japanese Standards **UTS: 150**
FC 15	Fine grained flake graphite cast iron: Designation used by Portugese Standards **DPN: 170 UTS: 150**

Symbol	Nominal analysis, supplier, condition and remarks.
FC 15 GRADE 2	Fine grained grey cast iron: Designation used for Japanese Standards **UTS: 150**
FC 20	Fine grained flake graphite cast iron: Designation used by Portugese Standard **DPN: 195 UTS: 225**
FC 20 GRADE 3	Fine grained grey cast iron: Designation used for Japanese Standards **UTS: 200**
FC 25	Fine grained flake graphite cast iron: Designation used by Portugese Standard **DPN: 200 UTS: 250**
FC 25 GRADE 4	Fine grained grey cast iron: Designation used by Japanese Standard **UTS: 250**
FC 30	Fine grained flake graphite cast iron: Designation used by Portugese Standards **DPN: 210 UTS: 300**
FC 30 GRADE 5	Fine grained grey cast iron: Designation used by Japanese Standard **UTS: 300**
FC 35	Fine grained flake graphite cast iron: Designation used by Portugese Standards **DPN: 220 UTS: 350**
FC 35 GRADE 6	Fine grained grey cast iron: Designation used by Japanese Standards **UTS: 350**
FC 40	Fine grained flake graphhite cast iron: Designation used by Portugese Standards **DPN: 250 UTS: 400**
FG 10	Fine grained flake graphite cast iron: Designation used by Spanish Standards **DPN: 170 UTS: 150**
FG 15	Fine grained flake graphite cast iron: Designation used by Spanish Standards **DPN: 170 UTS: 150**
FG 15D	Fine grained flake graphite cast iron: Designation used by French Standards **DPN: 170 UTS: 150**
FG 20	Fine grained flake graphite cast iron: Designation used by Spanish Standards **DPN: 195 UTS: 225**
FG 20D	Fine grained flake graphite cast iron: Designation used by French Standards **DPN: 190 UTS: 200**
FG 20D	Fine grained flake graphite cast iron: Designation used by French Standards **DPN: 195 UTS: 225**
FG 25	Fine grained flake graphite cast iron: Designation used by Spanish Standards **DPN: 200 UTS: 250**
FG 25D	Fine grained flake graphite cast iron: Designation used by French Standards **DPN: 205 UTS: 275**
FG 30	Fine grained flake graphite cast iron: Designation used by Spanish Standards **DPN: 210 UTS: 300**
FG 30D	Fine grained flake graphite cast iron: Designation used by French Standards **DPN: 210 UTS: 300**
FG 35	Fine grained flake graphite cast iron: Designation used by Spanish Standards **DPN: 220 UTS: 350**
FG 35	Fine grained flake graphite cast iron: Designation used by Spanish Standards **DPN: 250 UTS: 400**
FG 35D	Fine grained flake graphite cast iron: Designation used by French Standards **DPN: 220 UTS: 350**

Symbol	Nominal analysis, supplier, condition and remarks.
FG 40D	Fine grained flake graphite cast iron: French Standards **DPN: 250 UTS: 400**
FGE 38.17	Nodular graphite cast iron: Mainly ferritic structure; designation used by Spanish Standards **DPN: 180 UTS: 380 Elon: 17% Proof: 230**
FGE 42.12	Nodular graphite cast iron: Ferritic structure; designation used by Spanish Standards **DPN: 200 UTS: 420 Elon: 12% Proof: 250**
FGE 50.7	Nodular graphite cast iron: Pearlitic/ferritic structure; designation used by Spanish Standards **DPN: 205 UTS: 500 Elon: 7% Proof: 310**
FGE 60.2	Nodular graphite cast iron: Pearlitic/ferritic structure; designation used by Spanish Standards **DPN: 235 UTS: 600 Elon: 3% Proof: 350**
FGE 70.2	Nodular graphite cast iron: Pearlitic structure; designation used by Spanish Standards **DPN: 265 UTS: 700 Elon: 2% Proof: 400**
FGE 80.2	Nodular graphite cast iron: Pearlitic structure; designation used by Spanish Standards **DPN: 300 UTS: 800 Elon: 2% Proof: 460**
FGG	Fine grained flake graphite cast iron: Belgian Standard **DPN: 250 UTS: 400**
FGG 10	Fine grained flake graphite cast iron: Designation used by Belgian Standards
FGG 15	Fine grained flake graphite cast iron: Designation used by Belgian Standards **DPN: 170 UTS: 150**
FGG 20	Fine grained flake graphite cast iron: Designation used by Belgian Standards **DPN: 190 UTS: 200**
FGG 20	Fine grained flake graphite cast iron: Designation used by Belgian Standards **DPN: 195 UTS: 225**
FGG 25	Fine grained flake graphite cast iron: Designation used by Belgian Standards **DPN: 205 UTS: 275**
FGG 30	Fine grained flake graphite cast iron: Designation used by Belgian Standards **DPN: 210 UTS: 300**
FGG 35	Fine grained flake graphite cast iron: Designation used by Belgian Standards **DPN: 220 UTS: 350**
FGS 38.15	Nodular graphite cast iron: Pearlitic/ferritic matrix; designation used by French Standards **DPN: 165 UTS: 400 Elon: 17% Proof: 250**
FGS 42.12	Nodular graphite cast iron: Mainly ferritic matrix; designation used by French Standards **DPN: 200 UTS: 420 Elon: 12% Proof: 250**
FGS 50.7	Nodular graphite cast iron: Pearlitic/ferritic structure; designation used by French Standards **DPN: 205 UTS: 500 Elon: 7% Proof: 310**
FGS 60.2	Nodular graphite cast iron: Pearlitic/ferritic structure; designation used by French Standards **DPN: 230 UTS: 600 Elon: 3% Proof: 350**
FGS 70.2	Nodular graphite cast iron: Pearlitic structure; designation used by French Standards **DPN: 265 UTS: 700 Elon: 2% Proof: 400**

Symbol	Nominal analysis, supplier, condition and remarks.
FNG 1	Nodular graphite cast iron: Pearlitic/ferritic structure; designation used by Belgian Standards **DPN: 205 UTS: 500 Elon: 7% Proof: 310**
FNG 38.17	Nodular graphite cast iron: Pearlitic/ferritic matrix; designation used by Belgian Standards **DPN: 165 UTS: 400 Elon: 17% Proof: 250**
FNG 42.12	Nodular graphite cast iron: Mainly ferritic matrix; designation used by Belgian Standards **DPN: 200 UTS: 420 Elon: 12% Proof: 250**
FNG 50.7	Nodular graphite cast iron: Pearlitic/ferritic structure; designation used by Belgian Standards **DPN: 205 UTS: 500 Elon: 7% Proof: 310**
FNG 60.2	Nodular graphite cast iron: Pearlitic/ferritic structure; designation used by Belgian Standards **DPN: 230 UTS: 600 Elon: 3% Proof: 350**
FNG 70.2	Nodular graphite cast iron: Pearlitic structure; designation used by Belgian Standards **DPN: 265 UTS: 700 Elon: 2% Proof: 400**
FNG 80.2	Nodular graphite cast iron: Pearlitic structure; designation used by Belgian Standards **DPN: 300 UTS: 800 Elon: 2% Proof: 460**
G 10	Fine grained flake graphite cast iron: Designation used by Italian Standards **DPN: 170 UTS: 150**
G 15	Fine grained flake graphite cast iron: Designation used by Italian Standards **DPN: 170 UTS: 150**
G 20	Fine grained flake graphite cast iron: Designation used by Italian Standards **DPN: 195 UTS: 225**
G 25	Fine grained flake graphite cast iron: Designation used by Italian Standards **DPN: 200 UTS: 250**
G 30	Fine grained flake graphite cast iron: Designation used by Italian Standards **DPN: 210 UTS: 300**
G 35	Fine grained flake graphite cast iron: Designation used by Italian Standards **DPN: 250 UTS: 400**
G 2000	Grey cast iron: SAE designation; as cast; formerly SAE 110 **DPN: 187 UTS: 120**
G 3000	3.4% total C 2% Si grey cast iron: SAE designation; as cast; formerly SAE 111 **DPN: 200 UTS: 200**
G 3500	3.5% total C 1.5% Si grey cast iron: SAE designation; as cast; formerly SAE 120 **DPN: 210 UTS: 220**
G 4000	3.4% total C 1.5% Si grey cast iron: SAE designation; as cast; formerly SAE 121 **DPN: 230 UTS: 270**
G 4500	Grey cast iron: SAE designation; as cast; formerly SAE 122 **DPN: 235 UTS: 300**
GA	Cast iron: Can be hardened & tempered; Meehanite; as cast **DPN: 220 UTS: 330**
GA 350	Fine grained flake graphite cast iron: Meehanite; previously grade GA **DPN: 220 UTS: 350**
GB	Cast iron: Meehanite; as cast **DPN: 210 UTS: 300**
GB 300	Fine grained flake graphite cast iron: Meehanite; previously grade GB **DPN: 210 UTS: 300**
GC	Cast iron: Meehanite; as cast **DPN: 195 UTS: 270**
GC 275	Fine grained flake graphite cast iron: Previously grade GC **DPN: 205 UTS: 275**

Note. The following abbreviations and units are used in the tables:

DPN	Hardness, diamond pyramid number
UTS	Ultimate tensile strength, N/mm^2
Elon	Elongation, %
Proof	0.1 % proof strength, N/mm^2

1 N/mm^2=0.1 hbar=0.102 kgf/mm^2=0.06475 tonf/in^2=145.04 lbf/in^2

Symbol	Nominal analysis, supplier, condition and remarks.
GD	Cast iron: Meehanite; as cast **DPN: 185 UTS: 240**
GD 250	Fine grained flake graphite cast iron: Previously grade GD **DPN: 200 UTS: 250**
GE	Cast iron: Meehanite; as cast **DPN: 170 UTS: 200**
GE 200	Fine grained flake graphite cast iron: Meehanite; previously grade GE **DPN: 190 UTS: 200**
GE 225	Fine grained flake graphite cast iron: Meehanite; previously grade GE **DPN: 195 UTS: 225**
GF	Whiteheart cast iron: Britannia; spheroidized **UTS: 480 Elon: 10% Proof: 300**
GF 150	Fine grained flake graphite cast iron: Meehanite **DPN: 170 UTS: 150**
GG 10	Fine grained flake graphite cast iron: Designation used by Austrian Standards
GG 10	Fine grained flake graphite cast iron: Designation used by Danish Standards
GG 10	Fine grained flake graphite cast iron: Designation used by German Standards
GG 15	Fine grained flake graphite cast iron: Designation used by Austrian Standards **DPN: 170 UTS: 150**
GG 15	Fine grained flake graphite cast iron: Designation used by Danish Standards **DPN: 170 UTS: 150**
GG 15	Fine grained flake graphite cast iron: Designation used by German Standards **DPN: 170 UTS: 150**
GG 15	Fine grained flake graphite cast iron: Designation used by Dutch Standards **DPN: 170 UTS: 150**
GG 20	Fine grained flake graphite cast iron: Designation used by Danish Standards **DPN: 190 UTS: 200**
GG 20	Fine grained flake graphite cast iron: Designation used by German Standards **DPN: 190 UTS: 200**
GG 20	Fine grained flake graphite cast iron: Designation used by Austrian Standards **DPN: 195 UTS: 225**
GG 20	Fine grained flake graphite cast iron: Designation used by Danish Standards **DPN: 195 UTS: 225**
GG 20	Fine grained flake graphite cast iron: Designation used by German Standards **DPN: 195 UTS: 225**
GG 20	Fine grained flake graphite cast iron: Designation used by Dutch Standards **DPN: 195 UTS: 225**
GG 25	Fine grained flake graphite cast iron: Designation used by Austrian Standards **DPN: 205 UTS: 275**
GG 25	Fine grained flake graphite cast iron: Designation used by Danish Standards **DPN: 205 UTS: 275**
GG 25	Fine grained flake graphite cast iron: Designation used by Dutch Standards **DPN: 205 UTS: 275**
GG 25	Fine grained flake graphite cast iron: Designation used by German Standards **DPN: 205 UTS: 275**
GG 30	Fine grained flake graphite cast iron: Designation used by Dutch Standards **DPN: 210 UTS: 300**
GG 30	Fine grained flake graphite cast iron: Designation used by German Standards **DPN: 210 UTS: 300**
GG 30	Fine grained flake graphite cast iron: Designation used by Austrian Standards **DPN: 210 UTS: 300**
GG 30	Fine grained flake graphite cast iron: Designation used by Danish Standards **DPN: 210 UTS: 300**
GG 35	Fine grained flake graphite cast iron: Designation used by Dutch Standard **DPN: 220 UTS: 350**
GG 35	Fine grained flake graphite cast iron: Designation used by Austrian Standards **DPN: 220 UTS: 350**
GG 35	Fine grained flake graphite cast iron: Designation used by Danish Standard **DPN: 220 UTS: 350**
GG 35	Fine grained flake graphite cast iron: Designation used by German Standards **DPN: 220 UTS: 350**
GG 40	Fine grained flake graphite cast iron: Danish Standard **DPN: 250 UTS: 400**
GGG 35.3	Nodular graphite cast iron: Mainly ferritic; designation used by German Standards **DPN: 145 UTS: 350 Elon: 24% Proof: 220**
GGG 38	Spheroidal graphite cast iron: Ferritic matrix; German Standard designation
GGG 40	Nodular graphite cast iron: Pearlitic/ferritic graphite; designation used by German Standards **DPN: 165 UTS: 400 Elon: 17% Proof: 250**
GGG 40.3	Nodular graphite cast iron: Mainly ferritic; designation used by German Standards **DPN: 150 UTS: 400 Elon: 20% Proof: 250**
GGG 42	Spheroidal graphite cast iron: Mainly ferritic matrix; designation used by German Standards
GGG 45	Spheroidal graphite cast iron: Ferritic/pearlitic matrix; designation used by German Standards
GGG 50	Nodular graphite cast iron: Pearlitic/ferritic structure; designation used by German Standards **DPN: 205 UTS: 500 Elon: 7% Proof: 310**
GGG 60	Nodular graphite cast iron: Pearlitic/ferritic structure; designation used by German Standards **DPN: 230 UTS: 600 Elon: 3% Proof: 310**
GGG 70	Nodular graphite cast iron: Pearlitic structure; designation German Standards **DPN: 265 UTS: 700 Elon: 2% Proof: 400**
GGG 80	Nodular graphite cast iron: Pearlitic structure; designation used by German Standard **DPN: 300 UTS: 800 Elon: 2% Proof: 460**
GM	Cast iron: Can be hardened & tempered; Meehanite; as cast **DPN: 230 UTS: 360**
GM 400	Fine grained grey cast iron: Meehanite; previously grade GM **DPN: 250 UTS: 400**
GN 38	Nodular graphite cast iron: Pearlitic/ferritic matrix; designation used by Dutch Standards **DPN: 165 UTS: 400 Elon: 17% Proof: 250**
GN 42	Nodular graphite cast iron: Mainly ferritic matrix; designation used by Dutch Standard **DPN: 200 UTS: 420 Elon: 12% Proof: 250**
GN 50	Nodular graphite cast iron: Pearlitic/ferritic structure; designation used by Dutch Standards **DPN: 205 UTS: 500 Elon: 7% Proof: 310**
GN 60	Nodular graphite cast iron: Pearlitic/ferritic structure; designation used by Dutch Standards **DPN: 230 UTS: 600 Elon: 3% Proof: 310**

Symbol	Nominal analysis, supplier, condition and remarks.
GN 70	Nodular graphite cast iron: Pearlitic structure; designation used by Dutch Standards DPN: 265 UTS: 700 Elon: 2% Proof: 400
GOST 1412	General standard for grey cast iron: Russian standard; see designation for details
GOST 7293/54	Nodular graphite cast iron: Specification for Russian Standard; details under designation
GRAIN 21.80	0.08% C 1.3% Mn 0.4% Si Fe: Powder for welding; Esab; for submerged arc; properties of deposit varies with flux used
GRAIN 21.81	0.03% C 0.8% Mn 25% Ni Fe: Powder for welders; Esab; for submerged arc; properties of deposit varies with flux
GRP 38	Nodular graphite cast iron: Pearlitic/ferritic matrix; designation used by Finnish Standards DPN: 165 UTS: 400 Elon: 17% Proof: 250
GRP 40	Nodular graphite cast iron: Mainly ferritic matrix; designation used to Finnish Standard DPN: 200 UTS: 420 Elon: 12% Proof: 250
GRP 50	Nodular graphite cast iron: Pearlitic/ferritic structure; designation used by Finnish Standards DPN: 205 UTS: 500 Elon: 7% Proof: 310
GRP 60	Nodular graphite cast iron: Pearlitic/ferritic structure; designation used by Finnish Standard DPN: 230 UTS: 600 Elon: 3% Proof: 350
GRP 70	Nodular graphite cast iron: Pearlitic structure; designation used by Finnish Standards DPN: 265 UTS: 700 Elon: 2% Proof: 400
GRS 10	Fine grained flake graphite cast iron: Designation used by Finnish Standards
GRS 15	Fine grained flake graphite cast iron: Designation used by Finnish Standards DPN: 170 UTS: 150
GRS 20	Fine grained flake graphite cast iron: Designation used by Finnish Standards DPN: 195 UTS: 225
GRS 30	Fine grained flake graphite cast iron: Designation used by Finnish Standards DPN: 210 UTS: 300
GRS 35	Fine grained flake graphite cast iron: Designation used by Finnish Standards DPN: 220 UTS: 350
GS 370.17	Nodular graphite cast iron: Mainly ferritic structure; designation used by Italian Standards DPN: 180 UTS: 370 Elon: 17% Proof: 230
GS 400.12	Nodular graphite cast iron: Mainly ferritic structure; designation used by Italian Standards DPN: 200 UTS: 400 Elon: 12% Proof: 250
GS 500.7	Nodular graphite cast iron: Pearlitic/ferritic structure; designation used by Italian Standards DPN: 205 UTS: 500 Elon: 7% Proof: 310
GS 600.2	Nodular graphite cast iron: Pearlitic/ferritic structure; designation used by Italian Standards DPN: 230 UTS: 600 Elon: 3% Proof: 350
GS 700.2	Nodular graphite cast iron: Pearlitic structure; designation used by Italian Standards DPN: 265 UTS: 700 Elon: 2% Proof: 400

Note. The following abbreviations and units are used in the tables:

DPN	Hardness, diamond pyramid number
UTS	Ultimate tensile strength, N/mm^2
Elon	Elongation, %
Proof	0.1 % proof strength, N/mm^2

1 N/mm^2=0.1 hbar=0.102 kgf/mm^2=0.06475 tonf/in^2=145.04 lbf/in^2

Symbol	Nominal analysis, supplier, condition and remarks.
GS 800	Nodular graphite cast iron: Pearlitic structure; designation used by Italian Standards DPN: 300 UTS: 800 Elon: 2% Proof: 460
H 11-52	General standard for flake cast iron: Grey; Finnish Standard; see designation for details
H 11-52 GRS40	Fine grained flake graphite cast iron: Finnish Standard DPN: 250 UTS: 400
HA	Cast iron: For use up to 650 °C for burner parts etc.; Meehanite; as cast DPN: 220 UTS: 330
HAYES COMPACTED IRON	Modified grey cast iron: Hayes UTS: 420 Elon: 3%
HAYES HYDRAULIC IRON	Grey cast iron to BS 1452/20 Hayes UTS: 312 Elon: 0%
HAYES HYDRAULIC NODULAR	Spheroidal graphite cast iron to BS 2789 UTS: 500 Elon: 7%
HB	Cast iron: For use up to 700 °C; abrasion resistance for furnace equipment trays etc.; Meehanite; as cast DPN: 300 UTS: 240
HC	Cast iron: For use up to 700 °C; abrasion resistance for oil refinery trays, valves etc.; Meehanite; as cast DPN: 300 UTS: 240
HD	Cast iron: For use up to 620 °C for oil refinery equipment; Meehanite; as cast DPN: 200 UTS: 220
HE	Cast iron: For intermittent temperature service for ingot moulds etc.; Meehanite; as cast DPN: 170 UTS: 210
HR	Cast iron: For use up to 850 °C abrasion resisting for furnace & burner equipment; Meehanite; as cast DPN: 450 UTS: 240
HR 1	Cast iron: For use up to 790 °C; abrasion resisting for oil refinery parts; Meehanite; as cast; similar to HR DPN: 450 UTS: 240
HS	Nodular graphite cast iron: Heat resistant for use up to 900 °C; Meehanite DPN: 200 UTS: 400
INCANITE 1	Cast iron: Incanite DPN: 190 UTS: 210
INCANITE 2	Cast iron: Incanite DPN: 210 UTS: 300
INCANITE 3	Cast iron: Incanite for BS alloys 20 & 23 (BS 1452) DPN: 225 UTS: 330
INCANITE 4	Cast iron: Close grained; Incanite DPN: 225 UTS: 300
INCANITE 5	Cast iron: Incanite DPN: 200 UTS: 340
INCANITE 6	Cast iron: Incanite DPN: 225 UTS: 390
INCANITE HRA	Hematite type cast iron: Incanite; for Al & Zn melting pots, etc. DPN: 190 UTS: 210
INCANITE HRB	Heat resistant cast iron: Superior to grade HRA; Incanite DPN: 210 UTS: 330
JIS G5501	General standard for grey cast iron: Japanese Standard; see designation for details
KC	Cast iron: For use with strong alkalis; Meehanite; as cast
M 3191	General specification for flake cast iron: Grey; Austrian Standard; see designation for details

Symbol	Nominal analysis, supplier, condition and remarks.
M 3191 GG35	Fine grained cast iron: Flake graphite; Austrian Standard **DPN: 250 UTS: 400**
M 3193	General standard for nodular graphite cast iron: Austrian Standard; see designation for details
MNC 706	Swedish Standard; summary of spheroidal graphite cast irons
MNC 707 E	Specification covering several grades of Blackheart malleable cast iron
MNC 1205	General standard for flake graphite cast iron: Grey; Swedish Standard; see SIS for details
N 3	3% C (max) 0.85% Mn 2.2% Si 2.1% Cr 20% Ni cast iron: For furnace fittings; Wellman
NBN 830.01	General specification for flake cast iron: Grey; Belgian Standard; see designation for details
NBN 830.01 FGG40	Fine grained flake graphite cast iron: Belgian Standard **DPN: 250 UTS: 400**
NBN 830.02	General standard for nodular graphite cast iron: Belgian Standard; see designation for details
NEN 6002 A	General standard used for flake cast iron: Grey; Dutch Standard; see designation for details
NEN 6002 D	General standard for nodular graphite cast iron: Dutch Standard; see designation for details
Ni HARD	3.3% C 0.7% Si 2.5% Cr 4.1% Ni 0.4% P white cast iron: Impact 30 J; International Nickel **UTS: 550**
Ni-HARD 4	3% C 6% Ni 8% Cr martensitic white iron: High impact strength; International Nickel **DPN: 560**
Ni-RESIST 1	2.0% C 1.7% Si 15% Ni 6% Cu 2% Cr austenitic cast iron: High corrosion resistance; International Nickel **DPN: 150 UTS: 190 Elon: 2%**
Ni-RESIST 1B	2.0% C 1.9% Si 15% Ni 6% Cu 3% Cr austenitic cast iron: High corrosion resistance; International Nickel **DPN: 180 UTS: 220**
Ni-RESIST 2	2.0% C 1.9% Si 20% Ni 2% Cr austenitic cast iron: For steam service; International Nickel **DPN: 145 UTS: 190**
Ni-RESIST 2B	2.0% C 1.9% Si 20% Ni 5% Cr austenitic cast iron: For high temperature use; International Nickel **DPN: 210 UTS: 220**
Ni-RESIST 3	2.0% C 1.5% Si 30% Ni 3% Cr austenitic cast iron: Low expansion; International Nickel **DPN: 140 UTS: 22**
Ni-RESIST 4	2.0% C 5.0% Si 31% Ni 5% Cr austenitic cast iron: Use with foodstuff; International Nickel **DPN: 180 UTS: 220**
Ni-RESIST 5	2.0% C 1% Si 35% Ni austenitic cast iron: High shock resistance; International Nickel **DPN: 115 UTS: 160**
Ni-RESIST D 2	2.0% C 20% Ni 2% Cr austenitic cast iron: Spheroidal graphite; high expansion; International Nickel **DPN: 170 UTS: 450 Elon: 14% Proof: 240**
Ni-RESIST D 2B	2.0% C 20% Ni 3% Cr austenitic cast iron: Spheroidal graphite; high corrosion resistance; International Nickel **DPN: 180 UTS: 470 Elon: 11% Proof: 240**
Ni-RESIST D 2C	2.0% C 22% Ni austenitic cast iron: Spheroidal graphite; high ductility **DPN: 150 UTS: 440 Elon: 30% Proof: 230**
Ni-RESIST D 2M	2.0% C 4% Mn 22% Ni austenitic cast iron: Spheroidal graphite; low temperature ductility; International Nickel **DPN: 145 UTS: 460 Elon: 40% Proof: 210**

Symbol	Nominal analysis, supplier, condition and remarks.
Ni-RESIST D 3	2.0% C 30% Ni 3% Cr austenitic cast iron: Spheroidal graphite; shock resistant; International Nickel **DPN: 170 UTS: 420 Elon: 11% Proof: 240**
Ni-RESIST D 3A	2.0% C 30% Ni 5% Cr austenitic cast iron: Spheroidal graphite; wear resistant; International Nickel **DPN: 160 UTS: 410 Elon: 15% Proof: 240**
Ni-RESIST D 4	2.0% C 5% Si 30% Ni 5% Cr austenitic cast iron: Spheroidal graphite; high corrosion resistance; International Nickel **DPN: 205 UTS: 470 Elon: 3% Proof: 280**
Ni-RESIST D 5	2.0% C 35% Ni austenitic cast iron: Spheroidal graphite; low expansion; International Nickel **DPN: 155 UTS: 400 Elon: 30% Proof: 230**
Ni-RESIST D 5B	2.0% C 35% Ni 3% Cr austenitic cast iron: Spheroidal graphite; low expansion; International Nickel **DPN: 165 UTS: 410 Elon: 8% Proof: 270**
NITRON	3.2% C 1.5% Si 0.12% S 0.03% P (max) cast iron: Origin unknown
NS 722	General standard for flake graphite cast iron: Grey; Norwegian Standard; see designation for details
NS 11301	Norwegian Standard for nodular graphite cast iron: Details given under designation
NS 11338	Nodular graphite cast iron: Mainly ferritic structure; designation used by Norwegian Standards **DPN: 180 UTS: 370 Elon: 17% Proof: 230**
NS 11342	Nodular graphite cast iron: Mainly ferritic structure; designation used by Norwegian Standards **DPN: 180 UTS: 380 Elon: 15% Proof: 230**
NS 11350	Nodular graphite cast iron: Pearlitic/ferritic struucture; designation used by Norwegian Standards **DPN: 205 UTS: 500 Elon: 7% Proof: 310**
NS 11360	Nodular graphite cast iron: Pearlitic/ferritic structure; designation used by Norwegian Standards **DPN: 235 UTS: 600 Elon: 3% Proof: 350**
NS 11370	Nodular graphite cast iron: Pearlitic structure; designation used by Norwegian Standards **DPN: 265 UTS: 700 Elon: 2% Proof: 400**
R 001	General standard for flake graphite cast iron: Grey; Portugese Standard; see designation for details
SAE 110	3.5% C 0.7% Mn 2.5% Si 0.15% S 0.25% P grey cast iron: As cast **DPN: 187 UTS: 120**
SAE 111	3.3% C 0.8% Mn 2.2% Si 0.15% S 0.2% P grey cast iron: as cast **DPN: 200 UTS: 180**
SAE 113	3.4% C 0.8% Mn 1.5% Si 0.14% S 0.2% P grey cast iron: Alloying elements permitted **DPN: 200 UTS: 180**
SAE 114	3.4% C 0.8% Mn 1.5% Si 0.14% S 0.2% P grey cast iron: Alloying elements permitted **DPN: 230 UTS: 180**
SAE 120	3.3% C 0.8% Mn 2% Si 0.15% S 0.15% P grey cast iron: As cast **DPN: 210 UTS: 240**
SAE 121	3.2% C 0.8% Mn 2% Si 0.15% S 0.12% P grey cast iron: As cast **DPN: 220 UTS: 270**
SAE 122	3.1% C 0.9% Mn 2% Si 0.15% S 0.1% P grey iron: As cast **DPN: 230 UTS: 300**

Symbol	Nominal analysis, supplier, condition and remarks.
SAE 32510	Primary graphite free malleable cast iron: Annealed UTS: 340 Elon: 10% Proof: 210
SAE 35013	Primary graphite free malleable cast iron: Annealed UTS: 370 Elon: 18% Proof: 220
SAE 35018	Malleable cast iron UTS: 370 Elon: 18% Proof: 220
SAE 43010	Pearlitic malleable cast iron: Hardened & tempered DPN: 180 UTS: 420 Elon: 10% Proof: 290
SAE 48005	Pearlitic malleable cast iron: Hardened & tempered DPN: 200 UTS: 500 Elon: 5% Proof: 320
SAE 53004	Pearlitic malleable iron: Casting; hardened & tempered DPN: 200 UTS: 580 Elon: 4% Proof: 370
SAE 60003	Pearlitic malleable iron: Casting; hardened & tempered DPN: 220 UTS: 580 Elon: 3% Proof: 420
SAE 70002	Pearlitic malleable iron: Casting; hardened & tempered DPN: 260 UTS: 600 Elon: 2% Proof: 500
SAE D4018	Nodular graphite cast iron: Ferritic structure DPN: 170 UTS: 415 Elon: 18% Proof: 275
SAE D4512	Nodular graphite cast iron: Pearlitic/ferritic structure DPN: 185 UTS: 450 Elon: 12% Proof: 310
SAE D5506	Nodular graphite cast iron: Pearlitic/ferritic structure DPN: 225 UTS: 550 Elon: 6% Proof: 380
SAE D7003	Nodular graphite cast iron: Pearlitic structure DPN: 270 UTS: 690 Elon: 3% Proof: 480
SAE G 2000	Grey cast iron: As cast DPN: 187 UTS: 120
SAE G 3000	3.4% C (total) 2.0% Si 0.12% S 0.15% P grey cast iron: As cast DPN: 200 UTS: 200
SAE G 3500	3.5% C (total) 1.6% Si 0.12% S 0.15% P grey cast iron: As cast DPN: 220 UTS: 220
SAE G 4000	3.4% C (total) 1.6% Si 0.12% S 0.15% P grey cast iron: As cast DPN: 220 UTS: 270
SAE G 4500	Grey cast iron: As cast DPN: 250 UTS: 290
SAE J434 B	Nodular graphite cast iron: Covers various grades
SANDHOLME GA	Cast iron: Hardened and tempered after machining; Sandholme
SC	Cast iron: For use up to 900 °C; Meehanite; as cast; furnace equipment DPN: 300 UTS: 180
SC 12-28	Fine grained grey cast iron: Designation used by Russian Standards UTS: 150
SC 15-32	Fine grained grey cast iron: Designation used by Russian Standards UTS: 150

Symbol	Nominal analysis, supplier, condition and remarks.
SC 21-40	Fine grained grey cast iron: Designation used by Russian Standards UTS: 225
SC 28-48	Fine grained grey cast iron: Designation used by Russian Standards UTS: 275
SC 32-52	Fine grained grey cast iron: Designation used by Russian Standards UTS: 300
SC 36-56	Fine grained grey cast iron: Designation used by Russian Standards UTS: 350
SC 40-60	Fine grained grey cast iron: Designation used by Russian Standards UTS: 400
SD 51	14.5% Si 4.5% Cr Fe casting: Duriron Co; symbol for Superchlor
SF	Cast iron: Ferritic matrix structure; Meehanite; as cast for BS alloy 27/12 (BS 2789) DPN: 170 UTS: 390 Elon: 15% Proof: 290
SF 400	Nodular graphite cast iron: Pearlitic/ferritic matrix; Meehanite DPN: 165 UTS: 400 Elon: 17% Proof: 250
SFA 420-12	Nodular graphite cast iron: Mainly ferritic matrix; Meehanite; previously grade SF DPN: 200 UTS: 420 Elon: 12% Proof: 250
SFF	Cast iron: Ferritic matrix structure; Meehanite; as cast for BS alloy 24/17 (BS 2789) DPN: 160 UTS: 370 Elon: 20% Proof: 240
SFF 350	Nodular graphite cast iron: Mainly ferritic; Meehanite; previously SFF DPN: 145 UTS: 350 Elon: 24% Proof: 220
SFF 400	Nodular graphite cast iron: Mainly ferritic; Meehanite; previously SFF DPN: 150 UTS: 400 Elon: 20% Proof: 250
SFP 500	Nodular graphite cast iron: Pearlitic/ferritic structure; Meehanite; previously grade SPF DPN: 205 UTS: 500 Elon: 7% Proof: 310
SFS 2113	General standard for nodular graphite cast iron: Finnish Standard; see designation for details
SFT	Cast iron: Ferritic; nodular graphite structure; Meehanite; as cast; for use at temperatures up to 900 °C DPN: 235 UTS: 370 Elon: 2% Proof: 300
SG 38	Nodular graphite cast iron: Pearlitic/ferritic matrix; designation used by Austrian Standards DPN: 165 UTS: 400 Elon: 17% Proof: 250
SG 42	Nodular graphite cast iron: Mainly ferritic matrix; designation used by Austrian Standards DPN: 200 UTS: 420 Elon: 12% Proof: 350
SG 50	Nodular graphite cast iron: Pearlitic/ferritic structure; designation used by Austrian Standards DPN: 205 UTS: 500 Elon: 7% Proof: 310
SG 60	Nodular graphite cast iron: Pearlitic/ferritic structure; designation used by Austrian Standards DPN: 230 UTS: 600 Elon: 3% Proof: 350
SG 70	Nodular graphite cast iron: Pearlitic structure; designation used by Austrian Standards DPN: 265 UTS: 700 Elon: 2% Proof: 400
SH	Cast iron: Nodular graphite type; Meehanite; hardened & tempered for rotating parts, etc. DPN: 350 UTS: 910 Elon: 2% Proof: 600
SH 800	Nodular graphite cast iron: Pearlitic structure; Meehanite; previously grade SM DPN: 300 UTS: 800 Elon: 2% Proof: 460
SH 1000	Nodular graphite cast iron: Pearlitic structure; Meehanite UTS: 1000 Elon: 1%

Note. The following abbreviations and units are used in the tables:

DPN	Hardness, diamond pyramid number
UTS	Ultimate tensile strength, N/mm^2
Elon	Elongation, %
Proof	0.1 % proof strength, N/mm^2

$1 N/mm^2 = 0.1 \text{ hbar} = 0.102 \text{ kgf}/mm^2 = 0.06475 \text{ tonf}/in^2 = 145.04 \text{ lbf}/in^2$

Symbol	Nominal analysis, supplier, condition and remarks.
SHH	Nodular graphite cast iron: For abrasive resistance; hardened & tempered; Meehanite **DPN: 320 UTS: 1000**
SH N	Nodular graphite cast iron: For crankshafts, etc.; normalized; Meehanite **DPN: 270 UTS: 700 Elon: 3% Proof: 500**
SP	Pearlitic nodular cast iron: Meehanite; as cast for BS alloy 37/2 (BS 2789) **DPN: 220 UTS: 580 Elon: 2% Proof: 390**
SPF	Pearlitic nodular cast iron: Meehanite; as cast for BS alloy 32/7 (BS 2789) **DPN: 200 UTS: 500 Elon: 7% Proof: 330**
SIS 08-1002	Blackheart malleable cast iron: Swedish Standard; replaced by 08-1500
SIS 08-1400	Blackheart malleable cast iron: Swedish Standard **DPN: 149 UTS: 300 Elon: 6%**
SIS 08-1500	Blackheart malleable cast iron: Swedish Standard **DPN: 130 UTS: 320 Elon: 12% Proof: 190**
SIS 08-5200	Blackheart malleable cast iron: Swedish Standard **DPN: 160 UTS: 400 Elon: 7% Proof: 240**
SIS 08-5400	Blackheart malleable cast iron: Swedish Standard **DPN: 190 UTS: 500 Elon: 5% Proof: 300**
SIS 08-5403	Blackheart malleable cast iron: Swedish Standard; replaced by SIS 08-5400
SIS 08-5600	Blackheart malleable cast iron: Swedish Standard **DPN: 220 UTS: 600 Elon: 4% Proof: 380**
SIS 08-5603	Blackheart malleable cast iron: Swedish Standard; replaced by SIS 08 5600
SIS 08-6203	Blackheart malleable cast iron: Swedish Standard **DPN: 260 UTS: 700 Elon: 3% Proof: 530**
SIS 08-6403	Blackheart malleable cast iron: Swedish Standard **DPN: 280 UTS: 800 Elon: 2% Proof: 600**
SIS 14-0814	Blackheart malleable iron: Swedish Standard **DPN: 149 UTS: 300 Elon: 5%**
SIS 14-0815	Blackheart malleable iron: Swedish Standard **DPN: 127 UTS: 320 Elon: 12% Proof: 190**
SIS 14-0852	Pearlitic malleable iron: Swedish Standard **DPN: 155 UTS: 400 Elon: 7% Proof: 240**
SIS 14-0854	Pearlitic malleable iron: Swedish Standard **DPN: 190 UTS: 500 Elon: 5% Proof: 300**
SIS 14-0856	Pearlitic malleable iron: Swedish Standard **DPN: 220 UTS: 600 Elon: 4% Proof: 380**
SIS 14-0862	Pearlitic malleable iron: Swedish Standard **DPN: 260 UTS: 700 Elon: 3% Proof: 530**
SIS 14-0864	Pearlitic malleable iron: Swedish Standard **DPN: 290 UTS: 800 Elon: 2% Proof: 600**
SIS 0100	Grey cast iron: As cast; Swedish Standard
SIS 0110	Grey cast iron: As cast; Swedish Standard **DPN: 180**
SIS 0115	Grey cast iron: As cast; Swedish Standard **DPN: 190**
SIS 0120	Grey cast iron: As cast; Swedish Standard **DPN: 210**
SIS 0125	Grey cast iron: As cast; Swedish Standard **DPN: 230**
SIS 0130	Grey cast iron: As cast; Swedish Standard **DPN: 250**
SIS 0135	Grey cast iron: As cast; Swedish Standard **DPN: 260**
SIS 0140	Grey cast iron: As cast; Swedish Standard
SIS 0707	Spheroidal graphite cast iron: As cast; Swedish Standard **DPN: 260**
SIS 0717	Spheroidal graphite cast iron: Annealed; Swedish Standard **DPN: 160**
SIS 0717	Spheroidal graphite cast iron: Swedish Standard **DPN: 150 UTS: 400 Elon: 18% Proof: 250**
SIS 0727	Spheroidal graphite cast iron: Annealed; Swedish Standard **DPN: 200**
SIS 0727	Spheroidal graphite cast iron: Swedish Standard **DPN: 200 UTS: 500 Elon: 7% Proof: 350**
SIS 0732	Cast iron: Spheroidal graphite iron; Swedish Standard **DPN: 230 UTS: 600 Elon: 5% Proof: 400**
SIS 0732	Spheroidal graphite cast iron: Swedish Standard **DPN: 230 UTS: 600 Elon: 5% Proof: 400**
SIS 0737	Sperhoidal graphite cast iron: Normalized; Swedish Standard **DPN: 280**
SIS 0737	Sperhoidal graphite cast iron: Swedish Standard **DPN: 245 UTS: 700 Elon: 3% Proof: 450**
SIS 0810	Blackheart malleable cast iron: Annealed; Swedish Standard **DPN: 140**
SIS 0854	Cast iron: Pearlitic malleable iron; Swedish Standard **DPN: 135 UTS: 500 Elon: 5% Proof: 300**
SIS 0856	Cast iron: Pearlitic malleable iron; Swedish Standard **DPN: 220 UTS: 600 Elon: 4% Proof: 360**
SJ G15	Fine grained flake graphite cast iron: Designation used by Norwegian Standards **DPN: 170 UTS: 150**
SJ G20	Fine grained flake graphite cast iron: Designation used by Norwegian Standards **DPN: 195 UTS: 225**
SJ G25	Fine grained flake graphite cast iron: Designation used by Norwegian Standards **DPN: 195 UTS: 225**
SJ G30	Fine grained flake graphite cast iron: Designation used by Norwegian Standards **DPN: 210 UTS: 300**
SJ G35	Fine grained flake graphite cast iron: Designation used by Norwegian Standards **DPN: 220 UTS: 350**
SJ G40	Fine grained flake graphite cast iron: Designation used by Norwegian Standards **DPN: 250 UTS: 400**
SP 700	Nodular graphite cast iron: Pearlitic structure; Meehanite **DPN: 265 UTS: 700 Elon: 2% Proof: 400**
SPF 600	Nodular graphite cast iron: Pearlitic/ferritic structure; Meehanite; previously SP grade **DPN: 230 UTS: 600 Elon: 3% Proof: 350**
STA 8	Grey iron casting: Replaced by BS 1452/17
STA 9	Grey iron casting: Replaced by BS 1452/20
STA 25	0.8% C 15% Si cast iron: Replaced by BS 1591
SUPERCHLOR	0.9% C 1.5% Mn 14.5% Si 4.5% Cr iron alloy: Duriron Co; excellent corrosion resistance
TENSOLA I	Pearlitic cast iron: Gloucester Foundry **DPN: 180 UTS: 450 Elon: 8% Proof: 300**
TENSOLA II	Pearlitic cast iron: Gloucester Foundry **DPN: 200 UTS: 530 Elon: 6% Proof: 330**
TENSOLA III	Pearlitic cast iron: Gloucester Foundry **DPN: 245 UTS: 690 Elon: 3% Proof: 420**
UNE 36-111	General standard for flake graphite cast irons: Grey; Spanish Standard; see designation for details
UNE 36-118-73	Nodular graphite cast iron: Spanish Standard; details given under designation
UNI 5007	General standard for flake graphite cast iron: Grey; Italian Standard; see designation for details
UNI 14544	Nodular graphite cast iron: Italian Standard; details given under designation
UNS F00001–79999	Cast iron: Grey; malleable; nodular pearlitic malleable; ductile; American Standards System; unified number

Symbol	Nominal analysis, supplier, condition and remarks.
VCR 40-10	Nodular graphite cast iron: Pearlitic/ferritic structure; designation used by Russian Standards DPN: 180 UTS: 390 Elon: 10% Proof: 290
VCh 45-0	Nodular graphite cast iron: Pearlitic/ferritic structure; designation used by Russian Standards DPN: 225 UTS: 440 Proof: 350
VCh 45-5	Nodular graphite cast iron: Pearlitic/ferritic structure; designation used by Russian Standards DPN: 185 UTS: 440 Elon: 5% Proof: 320
VCh 50-1.5	Nodular graphite cast iron: Pearlitic/ferritic structure; designation used by Russian Standards DPN: 225 UTS: 490 Elon: 1.5% Proof: 372
VCH 60.2	Nodular graphite cast iron: Pearlitic/ferritic structure; designation used by Russian Standards DPN: 235 UTS: 585 Elon: 2% Proof: 410
W 22/4	Whiteheart malleable cast iron: Designation used by BS
W 24/8	Whiteheart malleable cast iron: Designation used by BS
WA	Cast iron with good wear resistance: Meehanite; as cast; for gears, brake drums, etc. DPN: 300 UTS: 340
WB	Cast iron: Wear resisting; Meehanite; as cast; for mill liners, sand blast nozzles, etc. DPN: 450 UTS: 300

Symbol	Nominal analysis, supplier, condition and remarks.
WBC	Cast iron: Suitable for chill casting; Meehanite; as cast; for crusher jaws, mill liners, etc. DPN: 550 UTS: 300
WBC	Cast iron: Suitable for chill casting; Meehanite; stress relieved; for rope & wire guildes; truck wheels, etc. DPN: 550 UTS: 210
WH	Cast iron: Wear resisting; Meehanite; as cast; for gravel pipes, sand blast tables, etc. DPN: 600 UTS: 270
WSH	Nodular graphite cast iron: For abrasive resistance; hardened & tempered to give martensitic structure; Meehanite DPN: 600 UTS: 1200

Note. The following abbreviations and units are used in the tables:

DPN	Hardness, diamond pyramid number
UTS	Ultimate tensile strength, N/mm^2
Elon	Elongation, %
Proof	0.1 % proof strength, N/mm^2

$1\ N/mm^2 = 0.1\ hbar = 0.102\ kgf/mm^2 = 0.06475\ tonf/in^2 = 145.04\ lbf/in^2$

20C Iron–nickel alloys – magnetic
With chromium, titanium or cobalt

Specific gravity	8.1	
Density	8100 kg/m³	(0.291 lb/in.³)
Solidus/liquidus	1450 °C	
Thermal conductivity	– W/m °C	
Coefficient of linear expansion (20–100° C)	1–10 × 10⁻⁶/ °C (These are low, and controlled expansion alloys)	
Electrical conductivity	2.1–3.5% IACS (copper 100%)	
Specific resistance	500–800 microhm mm	
Young's modulus of elasticity	145 × 10⁹ N/m²	(21 × 10⁶ lbf/in.²)
Impact	– J	
Fatigue strength	– N/mm²	
Hot strength	– N/mm²	

The above properties have been chosen to show typical values for the specifications listed and may not apply exactly to any one specification. It is possible that with some specifications the values may not be applicable.

General metallurgical characteristics

Iron of a very high purity has been shown to have magnetic properties considerably better than commercially pure iron. This is seldom economically desirable and is briefly discussed in Section 20A.

Iron–silicon alloys with a low carbon content also have useful magnetic properties which are treated as steels and discussed in Section 44B1.

With additions of nickel the magnetic properties of iron first fall, until at about 30% nickel the alloys are completely non-magnetic at room temperature. With further additions of nickel, a series of magnetically useful alloys is available at 50% iron/nickel. With 85% nickel a third series of useful alloys exist which are discussed briefly in Section 27E.

Apart from their magnetic and low expansion properties, the alloys listed have no significance. It is of interest that this property is sometimes useful in the negative sense, being absent, and some of the alloys listed can be made either magnetic or non-magnetic by a relatively small temperature alteration.

The magnetic properties are also considerably influenced by the direction of rolling or cold working – this being known as 'grain orienting', and by the addition of chromium up to about 6%.

Thermal treatment

None of these alloys listed can have their mechanical properties improved by any form of thermal treatment as there is never sufficient carbon present to cause appreciable hardening. The magnetic properties are generally enhanced by full annealing from about 900–1000 °C and with certain of the materials listed, improved or specific magnetic properties can be obtained by varying the rate of cooling. This is generally to within very narrow margins and specialized advice should be sought from the suppliers or the Permanent Magnet Association.

The annealing operation should be carried out after all forming and blanking operations and will generally require some form of inert atmosphere to prevent oxide and scale formation. The most popular atmosphere is hydrogen but recent advances in vacuum technology indicate that this method may soon become economically attractive.

Welding and brazing

Pre-treatment. All oxide, grease, dirt and scale must be removed before welding these materials. They are commonly supplied with a very adherent insulant which must be completely removed in the weld area. The material should be in the annealed condition.

Welding and brazing. This will seldom be required but when necessary, normal electric arc or gas welding techniques using flux or inert gas shielded arc welding give satisfactory results. The most common application will be electric resistance welding which should present no difficulty with the correct equipment and routine control. Brazing using copper based fillers and flux is readily carried out. Welding may affect the magnetic properties and should not be carried out before obtaining technical advice.

Post-treatment. No thermal treatment is necessary. All flux and oxide or scale should be removed to prevent corrosion.

Flaw detection methods

Crack test. This will seldom be required. Magnetic crack test or penetrant oil chalk test can be used, the latter being more flexible.

X-ray. Not required.

Ultra-sonic test. Not required.

Chemical etch. A laboratory type etch for grain orientation is often necessary to prove the direction of cold working. This scraps the test piece.

Corrosion protection

Temporary. The low nickel materials will require to be kept free from moisture and oiled at all times to prevent staining and rust formation. With increasing nickel content the corrosion resistance rises until at 50% nickel no corrosion will occur under normal indoor conditions.

Permanent. Normally these materials will be used under conditions where conventional painting ot plating would not be compatible with the electrical or magnetic requirements.

They are commonly coated with an insulant which should be chosen to act also as the means of corrosion protection.

By virtue of the nature of their application in chokes, transformers and other electrical and electronic equipment the material listed will very seldom be subjected to corrosive atmospheres.

Machinability

These alloys are usually too soft to exhibit good machining characteristics. In general, they are subjected to presswork operations rather than metal removing. When machining is required, light cuts at a high speed gives best results.

Uses

These are confined to magnetic and electronic for chokes, inductances, transformer cores and as screening materials to prevent high frequency interference. Many of the alloys are highly specialized and it is recommended that further information is obtained from the supplier or the Permanent Magnet Association.

Symbol	Nominal analysis, supplier, condition and remarks.
1.3921	0.8% Cr 48% Ni Fe alloy: German Standard
1.3926	46% Ni Fe alloy: German Standard
1.3927	46% Ni Fe alloy: German Standard
2.4472	45% Fe Ni alloy: German Standard
2.4480	0.8% Cr 47% Fe Ni alloy: German Standard
2.4486	6.0% Cr 46% Fe Ni alloy: German Standard

Note. The following abbreviations and units are used in the tables:

DPN	Hardness, diamond pyramid number
UTS	Ultimate tensile strength, N/mm^2
Elon	Elongation, %
Proof	0.1 % proof strength, N/mm^2

$1 N/mm^2 = 0.1 hbar = 0.102 kgf/mm^2 = 0.06475 tonf/in^2 = 145.04 lbf/in^2$

Symbol	Nominal analysis, supplier, condition and remarks.
142 ALLOY	41% Ni Fe alloy: Driver Harris; expansion coefficient 5.3×10^{-6}; electrical conductivity 2.5% IACS **UTS: 800**
146 ALLOY	46% Ni Fe alloy: Driver Harris **UTS: 800**
152 ALLOY	49% Fe Ni alloy: Driver Harris; glass metal seals **UTS: 480**
193 ALLOY	2.7% Cr 32% Ni Fe alloy: Driver Harris; heavy duty rheostats; withdrawn
426 ALLOY	0.08% C (max) 6% Cr 42% Ni Fe alloy: Darwin
ADR	40% Ni Fe alloy: Low expansion; Imphy
ALCHROME D	15% Cr 5.5% Al Fe alloy: For heating elements; Driver
ALNI (high coercivity)	30.0% Ni 12.5% Al 4.0% Cu 0.5% Ti Fe magnetic alloy: Coercivity 680 oersteds; information from PMA

Symbol	Nominal analysis, supplier, condition and remarks.
ALNICO (high remanence)	16.4% Ni 9.2% Al 5.0% Cu 12.3% Co Fe alloy: For magnets; remanence 8000 gauss; information from PMA
ALNICO 3	26% Ni 12% Al Fe alloy: For magnets; Simonds
AMS 5223 A	5.2% Cr 42% Ni 2.3% Ti 0.5% Al Fe alloy: Strip; 10% cold reduced
AMS 5225 A	5.2% Cr 42% Ni 2.3% Ti 0.5% Al Fe alloy: Strip; 50% cold reduced
AMS 5392 G	2.1% Cr 15% Ni 6% Cu Fe alloy: Sand casting
AMS 5393 B	2% Cr 20% Ni Fe alloy: Sand casting
AMS 5394	2% Cr 20% Ni Fe alloy: Sand casting
AMS 5395	22% Ni Fe alloy: Nodular casting
AMS 7701	Ni Fe alloy: Sheet & strip; annealed
AMS 7702	Ni Fe alloy: Sheet & strip; ½-hard
AMS 7705	Ni Fe alloy: Bar & forging
AMS 7717	Ni Fe alloy: Sheet & strip; for magnetic purposes; forming quality
AMS 7718	Ni Fe alloy: Bar, forging, etc.; magnetic alloy
AMS 7719	Ni Fe alloy: Sheet & strip; for magnetic purposes; stamping quality
AMS 7726 A	29% Ni 17% Co Fe alloy: Wire; low expansion alloy; for glass sealing
AMS 7727	29% Ni 17% Co Fe alloy: Bar & forging; low expansion; for glass sealing
AMS 7728 A	29% Ni 17% Co Fe alloy: Sheet & strip; low expansion; for glass sealing
ASC	35% Ni 10% Co Fe alloy: Low expansion; Imphy
ASC A	28% Ni 18% Co Fe alloy: Low expansion; Imphy
AUDIOLLOY	48% Ni Fe alloy: Origin unknown
BS 933/4	Ni Fe alloy: Sheet for magnetic purposes; composition not part of specification
BS 2857	36% Ni Fe alloy: Strip for laminations
BS 2857 B	50% Fe Ni alloy: Strip for laminations
BS 2857 C	35% Ni Fe alloy: For transformer laminations
BS 2857 D	36% Ni Fe alloy: For transformer laminations
BS 3127/9	0.06% C 5.2% Cr 42% Ni 0.5% Al 2.3% Ti Fe alloy: Tubing; annealed; for bourdon tubes **DPN: 160**
BS 3127/9	0.06% C 5.2% Cr 42% Ni 0.5% Al 2.3% Ti Fe alloy: Tubing; lightly cold worked; for bourdon tubes **DPN: 200**
CERAMIC SEALING ALLOY	25% Co 27% Ni Fe alloy: For ceramic metal seal
COLUMNAR HYCOMAX 11	14.5% Ni 7.0% Al 28.5% Co 3.0% Cu 4.0% Ti 2.0% Nb 0.2% S Fe alloy: For magnets; Swift, Lerwick
COMET	2.7% Cr 32% Ni Fe alloy: Driver Harris; heavy duty rheostats; withdrawn **UTS: 600**
DILVER P0	29% Ni 218% Co Fe alloy: Low expansion; Imphy
DILVER P1	29% Ni 18% Co Fe alloy: Low expansion; Imphy
DIN 17745 Ni 48	46% Ni Fe alloy
DIN 17745 Ni 49	0.8% Cr 48% Ni Fe alloy
DIN 17745 Ni Fe 45	45% Fe Ni alloy
DIN 17745 Ni Fe 48 Cr	0.8% Cr 47% Fe Ni alloy
E 10	0.06% C 0.2% Si 0.2% Mn 3.6% Ni Fe alloy: Jessop; controlled expansion alloy
EHC ALLOY	14.0% Ni 7.3% Al 34.0% Co 4.5% Cu 5.5% Ti Fe alloy: For magnets; origin unknown
ELINVAR	9% Cr 35% Ni Fe alloy: Telcon; annealed; expansion 6×10^{-6} **UTS: 690**
ELINVAR	9% Cr 35% Ni Fe alloy: Telcon; cold drawn **DPN: 1540**
GILGRID	45% Fe 10% Mo Ni alloy: Wire for electronic use; Gilbey-Brunton
HCR	50% Fe Ni alloy: Strip; Telcon; annealed; has rectangular hysteresis loop **DPN: 110 UTS: 370**
HIPERNIK	50% Ni Fe alloy: Strip; Westinghouse Electric; for magnetic laminations
HYREM RADIOMETAL	50% Ni Fe alloy: For magnetic purposes; Telcon
HYRHO RADIOMETAL	Ni Fe alloy: High resistivity with high saturation flux density; Telcon; cold rolled **DPN: 250 UTS: 700**
HYRHO RADIOMETAL	50% Ni Fe alloy: With higher resistivity than Radiometal; Telcon
IN 568	2.5% Ni 2.5% Si Fe alloy: International Nickel Co
INCOLOY 805	0.12% C 7.5% Cr 0.5% Mo 36% Ni Fe alloy: For springs; Huntington; low temperature coefficient of modulus of elasticity
INVAR	0.1% C 0.2% Si 0.5% Mn 36% Ni Fe alloy: Latrobe Steel Co; annealed; low expansion; also Telcon **DPN: 131 UTS: 450 Elon: 40% Proof: 220**
INVAR	0.1% C 0.2% Si 0.5% Mn 36% Ni Fe alloy: Latrobe Steel Co; cold worked; low expansion; also Telcon **DPN: 217 UTS: 750 Elon: 8% Proof: 610**
INVAR 36	36% Ni Fe alloy: Low expansion; Telcon; annealed; expansion 2.0×15^{-6}/ °C up to 100 °C **DPN: 140 UTS: 480 Elon: 30% Proof: 260**
INVAR 42	42% Ni Fe alloy: Low expansion; Telcon; annealed; expansion 4.5×10^{-6}/ °C up to 300 °C **DPN: 145 UTS: 520 Elon: 30% Proof: 240**
IRWAR 36	36% Ni Fe alloy: Low expansion; Telcon; metals
IRWAR FM	0.1% C 36% Ni 0.15% Se Fe alloy: Latrobe; low expansion alloy; free machining
IRWAR STANDARD	36% Ni Fe alloy: Low expansion; Imphy
KOERZIT 120K	26.0% Ni 13.0% 4.0% Co 3.0% Cu 1.0% Ti Fe alloy: For magnets; German alloy
KOERZIT 120R	22.0% Ni 12.0% Al 3.0% Co 3.0% Cu 0.5% Ti Fe alloy: For magnets; German alloy
KOERZIT 160	21.0% Ni 10.0% Al 17.0% Co 3.0% Cu 1.0% Ti Fe alloy: For magnets; German alloy
KOERZIT 190	21.0% Ni 10.0% Al 17.0% Co 1.0% Ti 1.0% Ti Fe alloy: For magnets; German alloy
KOERZIT 220	15.0% Ni 7.0% Al 28.0% Co 5.0% Cu 8.0% Ti Fe alloy: For magnets; German alloy
KOERZIT 350	15.0% Ni 7.0% Al 30.0% Co 4.0% Cu 5.0% Ti Fe alloy: For magnets; German alloy
KOERZIT 400K	15.0% Ni 9.0% Al 24.0% Co 3.0% Cu 1.0% Ti Fe alloy: For magnets; German alloy
KOERZIT 450	15.0% Ni 7.0% Al 30.0% Co 4.0% Cu 5.0% Ti Fe alloy: For magnets; German alloy
KOERZIT 500	15.0% Ni 9.0% Al 24.0% Co 3.0% Cu Fe alloy: For magnets; German alloy
KOERZIT 600	15.0% Ni 9.0% Al 24.0% Co 3.0% Cu Fe alloy: For magnets; German alloy
LEGA 42	42% Ni Fe alloy: Low expansion; Driver Harris
LEGA 46	46% Ni Fe alloy: Low expansion; Driver Harris
MAGLOY 1	14.0% Ni 8.0% Al 24.0% Co 3.0% Cu Fe alloy: For magnets; origin unknown
MAGLOY 2	14.0% Ni 8.0% Al 24.0% Co 3.0% Cu 1.2% Ti Fe alloys: For magnets; origin unknown

Note. The following abbreviations and units are used in the tables:

DPN	Hardness, diamond pyramid number
UTS	Ultimate tensile strength, N/mm^2
Elon	Elongation, %
Proof	0.1 % proof strength, N/mm^2

$1 \text{ N/mm}^2 = 0.1 \text{ hbar} = 0.102 \text{ kgf/mm}^2 = 0.06475 \text{ tonf/in}^2 = 145.04 \text{ lbf/in}^2$

Symbol	Nominal analysis, supplier, condition and remarks.
MAGLOY 5	18.0% Ni 10.0% Al 13.0% Co 6.0% Cu Fe alloy: For magnets; origin unknown
MAGLOY 7	21.0% Ni 9.0% Al 20.0% Co Fe alloy: For magnets; origin unknown
MAGLOY 8	17.0% Ni 7.0% Al 30.0% Co 5.0% Ti Fe alloy: For magnets; origin unknown
MAGLOY 10	14.0% Ni 8.0% Al 24.0% Co 3.0% Cu Fe alloy: For magnets; origin unknown
MAGLOY 100	14.0% Ni 8.0% Al 24.0% Co 3.0% Cu Fe alloy: For magnets; origin unknown
N 42	42% Ni Fe alloy: Low expansion; Imphy
N 501	50% Ni 1% Cr Fe alloy: Low expansion; Imphy
NIALCO 1	25.0% Ni 12.0% Al 2.0% Co 4.0% Cu Fe alloy: For magnets; French alloy
NIALCO 1F	25.0% Ni 12.0% Al 2.0% Co 4.0% Cu Fe alloy: For magnets; French alloy
NIACLO II	25.0% Ni 10.0% Al 4.0% Co 4.0% Cu Fe alloy: For magnets; French alloy
NIALCO III	20.0% Ni 10.0% Al 12.0% Co 4.0% Cu Fe alloy: For magnets; French alloy
NIALCO IIIF	20.0% Ni 10.0% Al 12.0% Co 4.0% Cu Fe alloy: For magnets; French alloy
NIALCO IV	20.0% Ni 8.0% Al 10.0% Co 6.0% Cu Fe alloy: For magnets; French alloy
NIALCO IVF	20.0% Ni 8.0% Al 10.0% Co 6.0% Cu Fe alloy: For magnets; French alloy
NIALCO V	21.0% Ni 10.0% Al 17.0% Co 3.0% Cu 1.0% Ti Fe alloy: For magnets; French alloy
NICOSEAL	29% Ni 17% Co Fe alloy: Low expansion; Carpenter
NICOSEL	29% Ni 17% Co Fe alloy: Low expansion; Firth Brown
NIKO	Ni Co Fe alloy: Swift Levick; low expansion **DPN: 170 UTS: 510 Elon: 41% Proof: 360**
NILGRO 42	42% Ni Fe alloy: Low expansion; Darwins; coefficient of expansion $6 \times 10^{-6}/$ °C up to 350 °C **UTS: 550 Elon: 42% Proof: 390**
NILO 36	36% Ni Fe alloy: Wrought; expansion 1.5×10^{-6}; H Wiggin; electrical conductivity 2% IACS
NILO 40	40% Ni Fe alloy: Wrought; expansion 4.1×10^{-6}; H Wiggin; specification Res. 680 microhm/mm; electrical conductivity 2.5% IACS
NILO 42	42% Ni Fe alloy: Wrought; expansion 5.3×10^{-6}; H Wiggin; electrical conducivity 2.75% IACS
NILO 48	48% Ni Fe alloy: Wrought; expansion 8.5×10^{-6}; H Wiggin; electrical conductivity 3.5% IACS
NILO 50	50% Ni Fe alloy: Wrought; expansion 9.3×10^{-6}; H Wiggin; electrical conductivity 3.5% IACS
NILO 475	47% Ni 5% Cr Fe alloy: Wrought; expansion 8.2 $\times 10^{-6}$; H Wiggin; electrical conductivity 2% IACS
NILO K	29% Ni 17% Co Fe alloy: Wrought; expansion 6 $\times 10^{-6}$; H Wiggin; electrical conductivity 3.5% IACS
NILO K45	32.5% Ni 13.0% Co Fe alloy: For glass seals; International Nickel Co
NILO P50	50% Ni Fe alloy: For glass seals; International Nickel Co
NILOMAG 471	47.0% Ni 3.0% Mo Fe alloy: Strip made by powder metallurgy; International Nickel Co
NILVAR	36% Ni Fe alloy: Expansion 1×10^{-6}; Driver Harris; electrical conductivity 2% IACS **UTS: 530**
NIPERMAG	12.0% Al 30.0% Ni 0.4% Ti Fe alloy: For magnets; origin unknown **DPN: 450**
NI-ROD 55	1.5% C 45% Fe Ni alloy: Welding rod; Huntingdon; for welding high phosphorus cast irons
NIROMET 42	42% Ni Fe alloy: For glass/metal seal; Driver
NIROMET 46	46% Ni Fe alloy: For glass/metal seal; Driver
NI-SPAN-C	0.02% C 5.4% Cr 2.4% Ti 42% Ni Fe alloy: Bar & sheet; Huntington; now NI-SPAN-C 902
NI-SPAN-C 902	0.02% C 5.4% Cr 2.4% Ti 42% Ni Fe alloy: Bar & sheet; Huntington; age hardening
No. 4 ALLOY	5.5% Cr 42% Ni Fe alloy: British Driver Harris; glass, metal seals for soft glasses
NS 36	36% Ni Fe alloy: Swift Levick; low expansion **DPN: 135 UTS: 450 Elon: 42% Proof: 240**
NS 42	42% Ni Fe alloy: Swift Levick; low expansion **DPN: 143 UTS: 520 Elon: 46% Proof: 210**
NS 49	49% Ni Fe alloy: Swift Levick; low expansion **DPN: 143 UTS: 530 Elon: 45% Proof: 270**
OERSTIT 120R	22.0% Ni 12.0% Al 3.0% Co 3.0% Cu 0.5% Ti Fe alloys: For magnets; German alloy
ORTHOMUMETAL	Soft magnetic material: Telcon
PERMALLOY 'B'	46% Ni Fe alloy: Sheet, bar, forging, etc.; Standard Telephones; high values of flux density
PERMALLOY 'D'	36% Ni Fe alloy: Standard Telephones; electrical conductivity 1.8% IACS
R 2800	Ni Fe alloy: Telcon; hard rolled; curie point 150 °C **DPN: 170**
RADIO METAL 36	36% Ni Fe alloy: Induction melted castings; Telcon; annealed; similar to Mumetal, but cheaper, see section 27C **DPN: 110 UTS: 520**
RADIO METAL 36	36% Ni Fe alloy: Induction melted castings; Telcon; hard rolled **DPN: 290 UTS: 970**
RADIO METAL 50	50% Fe Ni alloy: Sheet & bar; Telcon; annealed; high initial permeability **DPN: 100 UTS: 400**
RADIO METAL 50	50% Fe Ni alloy: Sheet & bar; Telcon; hard rolled **DPN: 250 UTS: 750**
RECO 100	24.0% Ni 14.0% Al Fe alloy: For magnets; Dutch alloy
RECO 120	26.0% Ni 13.0% Al 4.0% Co 3.0% Cu 1.0% Ti Fe alloy: For magnets; Dutch alloy
RECO 140	24.0% Ni 10.0% Al 5.0% Co 7.0% Cu 0.8% Ti Fe alloy: For magnets; Dutch alloy
RECO 160	18.5% Ni 10.0% Al 13.0% Co 7.5% Cu 1.9% Ti Fe alloy: For magnets; Dutch alloy
RECO 170	24.0% Ni 9.5% Al 10.0% Co 6.0% Cu 5.0% Ti Fe alloy: For magnets; Dutch alloy
RECO 220	15.0% Ni 7.0% Al 26.0% Co 5.0% Cu 7.0% Ti Fe alloy: For magnets; Dutch alloy
RODAR	29% Ni 17% Co Fe alloy: For glass/metal seal; Driver
SANBOLD NA 35	35% Ni Fe alloy: Strip; Sanderson; electrical conductivity 1.8% IACS
SANBOLD NA 47	47% Ni Fe alloy: Strip for transformers, etc.; Sanderson; electrical conductivity 3% IACS
SATMUMETAL	Soft magnetic alloy: 0.025 oersteds coercivity; Telcon
SUPER RADIO METAL 50	50% Fe Ni alloy: Similar to Radiometal 50 Ni alloy; Telcon; annealed; initial permeability 3 times Radio metal 50 **DPN: 110 UTS: 420**
SUPER RADIO METAL 50	50% Fe Ni alloy: Similar to Radiometal 50 Ni alloy; Telcon; hard rolled **DPN: 250 UTS: 750**
SUPER UGIMAX 600	14.0% Ni 8.0% Al 24.0% Co 3.0% Cu Fe alloy: For magnets; French alloy
SUPER UGIMAX 800	14.0% Ni 8.0% Al 24.0% Co 3.0% Cu Fe alloy: For magnets; French alloy
TELCOSEAL I	29% Ni 17% Co Fe alloy: For use with borosilicate glass; Telcon; expansion 5.3×10^{-6}; electrical conductivity 3.75% IACS; annealed **DPN: 150 UTS: 480 Elon: 30%**

Symbol	Nominal analysis, supplier, condition and remarks.
TELCOSEAL II	42% Ni Fe alloy: For use with lead borosilicate glass; Telcon; expansion 5.8×10^{-6}; electrical conductivity 3% IACS; annealed **DPN: 125 UTS: 530 Elon: 27%**
TELCOSEAL III	6% Cr 42% Ni Fe alloy: For use with lead glass; Telcon; expansion 8.7×10^{-6}; electrical conductivity 1.8% IACS; annealed **DPN: 120 UTS: 550 Elon: 35%**
TELCOSEAL V	25% Co Fe alloy: For use with soft lead glass; Telcon; expansion 10.6×10^{-6}; electrical conductivity 2.5% IACS; annealed **DPN: 150**
TELCOSEAL VI	49% Ni Fe alloy: For use with soft lead/lime glass; Telcon; expansion 9.1×10^{-6}; electrical conductivity 4.2% IACS; annealed **DPN: 120 UTS: 480 Elon: 30%**
THERLO	29% Ni 17% Co Fe alloy: Driver Harris; glass/metal seals **UTS: 540**
TICONAL 190	21.0% Ni 12.0% Al 14.0% Co 3.0% Cu Fe alloy: For magnets; Dutch alloy
TICONAL 360	15.0% Ni 8.5% Al 24.0% Co 3.0% Cu 1.5% Ti Fe alloy: For magnets; Dutch alloy
TICONAL 400	14.0% Ni 8.5% Al 24.0% Co 3.0% Cu Fe alloy: For magnets; Dutch alloy
TICONAL 450	14.5% Ni 7.5% Al 34.0% Co 4.5% Cu 5.0% Ti Fe alloy: For magnets; Dutch alloy
TICONAL 500	14.0% Ni 8.5% Al 24.0% Co 3.0% Cu Fe alloy: For magnets; Dutch alloy
TICONAL 600	14.0% Ni 8.5% Al 24.0% Co 3.0% Cu Fe alloy: For magnets; Dutch alloy
TICONAL 600	14.0% Ni 8.0% Al 24.0% Co 3.0% Cu Fe alloy: For magnets; French alloy
TICONAL 700	14.0% Ni 8.0% Al 24.0% Co 3.0% Cu Fe alloy: For magnets; French alloy
TICONAL 750	14.0% Ni 8.0% Al 24.0% Co 3.0% Cu Fe alloy: For magnets; French alloy
TICONAL 800	14.0% Ni 8.0% Al 24.0% Co 3.0% Cu Fe alloy: For magnets; French alloy
TICONAL 650	14.0% Ni 8.5% Al 24.0% Co 3.0% Cu Fe alloy: For magnets; Dutch alloy
TICONAL 750	14.0% Ni 8.5% Al 24.0% Co 3.0% Cu Fe alloy: For magnets; Dutch alloy
TICONAL 1500	15.0% Ni 7.0% Al 35.0% Co 4.0% Cu 5.0% Ti Fe alloy: For magnets; French alloy
TICONAL 1500F	15.0% Ni 7.0% Al 35.0% Co 4.0% Cu 5.0% Ti Fe alloy: For magnets; French alloy

Symbol	Nominal analysis, supplier, condition and remarks.
TICONAL 1800	15.0% Ni 7.0% Al 35.0% Co 4.0% Cu 5.0% Ti Fe alloy: For magnets; French alloy
TICONAL C	13.9% Ni 8.0% Al 24.0% Co 3.0% Cu 0.8% Nb Fe alloy: For magnets; origin unknown
TICONAL FRITTE	14.0% Ni 8.0% Al 24.0% Co 3.0% Cu Fe alloy: For magnets; French alloy
TOPHET D	20% Cr 36% Ni Fe alloy: For heating elements; Driver
UGIMAX 600	14.0% Ni 8.0% Al 24.0% Co 3.0% Cu Fe alloy: For magnets; French alloy
UGIMAX 800	14.0% Ni 8.0% Al 24.0% Co 3.0% Cu Fe alloy: For magnets; French alloy
UNS L05001–L05999	Lead based alloys: American Standard system; unified numbers
VACODIL 36	36% Ni Fe alloy: Low expansion; German proprietary alloy
VACODIL 42	42% Ni Fe alloy: Low expansion; German proprietary alloy
VACODIL 46	46% Ni Fe alloy: Low expansion; German proprietary alloy
VACON 10	28% No 18% Co Fe alloy: Low expansion; German proprietary alloy
VACON 12	28% Ni 18% Co Fe alloy: Low expansion; German proprietary alloy
VACON 20	28% Ni 21% Co Fe alloy: Low expansion; German proprietary alloy
VACON 70	28% Ni 23% Co Fe alloy: Low expansion; German proprietary alloy
VACOVIT 485	48% Ni 5% Cr Fe alloy: Low expansion; German proprietary alloy
VACOVIT 501	49% Ni 1% Cr Fe alloy: Low expansion; German proprietary alloy
ZERALOY	31% Ni 5% Co Fe alloy: Low expansion; Swift Levick

Note. The following abbreviations and units are used in the tables:

DPN	Hardness, diamond pyramid number
UTS	Ultimate tensile strength, N/mm^2
Elon	Elongation, %
Proof	0.1 % proof strength, N/mm^2

1 N/mm^2=0.1 hbar=0.102 kgf/mm^2=0.06475 tonf/in^2=145.04 lbf/in^2

21. Lead Pb

Physical properties

Atomic number	82	
Atomic weight	207.21	
Crystal structure	Face-centred cubic	
Colour	Blue-grey	
Specific gravity	11.34	
Density	11340 kg/m^3	(0.42 lb/in.^{3})
Melting point	327 °C	
Boiling point	1750 °C	
Specific heat	0.129 J/g °C	(0.0308 cal/g °C)
Thermal conductivity	17.6 W/m °C	(8.2 calories/m s °C)
Coefficient of linear expansion (20–100° C)	29.1 × 10^{-6}/ °C	
Latent heat of fusion	24.1 J/g	(5.75 cal/g)
Latent heat of vaporization	850 J/g	(203 cal/g)
Thermal neutron absorption cross-section	0.17 barns/atom	
Electrical conductivity	9% IACS (copper 100%)	
Specific resistance	208 microhm mm	
Temperature coefficient of electrical resistance	0.0043/ °C	
Electrochemical equivalent	3.858 g/A/h	
Electrode potential	-0.126 V	
Magnetic susceptibility	-0.12 × 10^{-6}	
Young's modulus of elasticity	14 × 10^9 N/m^2	(2 × 10^6 lbf/in.^{2})
Tensile strength	18 N/mm^2	(1.2 tonf/in.^{2})
Hardness	5 DPN	

21.1 General notes on lead

Lead occurs in deposits widespread throughout the world and it is of note that this metal, which has been used for plumbing since Roman times at least, makes up only 0.002% of the earth's crust. Lithium, which is looked on as a rare metal is present as 0.004%. Lead ores however, tend to be found in concentrated pockets which makes their mining more economical than lithium, which is evenly distributed.

The excellent ductility of lead which has a complete absence of cold working properties has made this a favourite material which is further enhanced by its excellent corrosion resistance and ease with which it can be cold forge welded.

The Romans had lead water pipes and used lead as a roofing material. Lead pipes are now being replaced by copper, aluminium and plastic tubing, as these have thinner wall sections for the same strength and do not suffer creep failure at normal temperatures. The use of lead for sheathing electric cables is also largely being replaced with aluminium and plastic. Lead will be attacked corrosively by soft or neutral water which contains carbon dioxide. This can cause a health hazard when lead pipe is used.

The largest use of lead and its alloys is in the lead/sulphuric acid electrical storage battery or 'accumulator'. No other material can supply the necessary electrical, chemical and corrosion resistant properties as economically as lead.

The chemical industry still uses large quantities of lead in contact with acids and chromium electroplating equipment is lead lined and uses lead anodes.

Some of the newer plastics, stainless steel and titanium, are now threatening these uses of lead, which suffers from the disadvantage of being the heaviest metal in common use. This added to its low strength and very low creep properties means that it almost always has to be supported by other metals. Additions of antimony to lead increase the strength without affecting its chemical resistance to any great extent, but make it more difficult to weld or solder.

A range of bearing metals exists with lead as the main constituent. Here the lead is used as a soft matrix supporting the hard intermetallic compounds of antimony, copper, tin and cadmium. These alloys are used for high bearing loads, and can stand reasonably high temperatures. Lead bearings do not however have good fatigue strength and must not be used where reciprocating loads are involved unless special precautions are taken, A number of different alloys have been developed which cater for conditions varying from lightly loaded slow moving, up to such purposes as bearings on stone crushing equipment.

It is strongly advised that use is made of the detailed information available from the sources listed for these very special purpose alloys.

Lead is also used as a base for soldering and some fusible alloys.

As an alloying element lead has many uses. It is more economical than tin and has many similar properties being present in many tin based bearing alloys and almost all tin based solders and fusible alloys.

Lead is added to steel, aluminium and copper alloys up to 0.3% to improve machinability. It does not dissolve in any of these materials thus has little effect on their properties, but does form a discontinuity which prevents long curling cuttings. The soft lead also acts to some extent as a built-in solid lubricant, but there remains doubt as to the extent of the practical advantage of this property.

Lead sheeting has been the standard protection until recent years against X-rays and radioactivity.

Lead and its alloys have no natural resonant frequencies, and it is thus a 'silent' metal. This property is being studied at present, as much of the world's considerable noise is caused by the vibration of metals.

The corrosion resistant and bearing properties of lead when applied as an electroplated deposit are now being exploited.

Lead oxide as a base for paints has been used for many years, and there are modern 'paints' chemically producing an adherent deposit of metallic lead on steel which show considerable promise. Metallic lead and most of its compounds are toxic, thus cannot be used in contact with foodstuffs, or on such articles as children's toys.

Although many of the newer materials are joining copper as a competitor to lead and its alloys, there is little doubt that lead will remain an economical material for many purposes for years to come.

Symbol	Nominal analysis, supplier, condition and remarks.
2.3010	99.99% Pb: German Standard
2.3020	99.985% Pb: German Standard
2.3021	0.06% Cu 99.9% Pb: Chemical grade; German Standard
2.3030	99.94% Pb: German Standard
2.3040	99.9% Pb: For chemical use; German Standard
2.3075	0.1% Cu 99.75% Pb: German Standard
2.3085	98.5% Pb: German Standard
2.3131	0.005% Sn & Sb Pb: German Standard
2.3132	0.12% Sb 0.005% Sn Pb: German Standard
2.3137	0.8% Sb 0.005% Sn Pb: German Standard
2.3138	2.5% Sn Pb alloy: German Standard
2.3139	0.04% Te 0.005% Sn & Sb Pb: German Standard
2.3201	0.7% Sb 1.2% As Pb alloy: German Standard
2.3202	0.25% Sb Pb alloy: German Standard
2.3203	3% Sb 1.5% As Pb alloy: German Standard
2.3205	6% Sb Pb alloy: German Standard
2.3208	8% Sb Pb alloy: German Standard
2.3212	12% Sb Pb alloy: German Standard
2.3229	8.8% Sb Pb alloy: German Standard
2.3299	8.8% Sb Pb alloy: German Standard
A 1 BABBITT	Pb base bearing metal: Medium duty; Magnolia Anti-Friction Metal Co **DPN: 26 UTS: 66**
A 5	Ag Pb base soft solder: Mallory Metallurgical; electrical conductivity 7.2% IACS; melting range 304–370 °C **UTS: 37 Elon: 35%**
A 25	Ag Pb base soft solder: Mallory Metallurgical; electrical conductivity 7.2% IACS; melting point 304 °C **UTS: 37 Elon: 35%**
A 36	Lead base white metal: For railway waggons; Anti-Attrition **DPN: 25**
A 105	Lead base white metal: For gland packing; soft; Anti-Attrition
A 110	Lead base white metal: For heavy duty and eccentric bearings; Anti-Attrition

Symbol	Nominal analysis, supplier, condition and remarks.
A 122	Lead base white metal: Capping metal; Anti-Attrition
A 205	Lead base white metal: For gland packing; hard; Anti-Attrition
A 210	Lead base white metal: For railway waggons; Anti-Attrition **DPN: 32**
ALLULOY No. 1	Lead base white metal: For heavy duty & eccentric bearings; Anti-Attrition; as A 110
ALLULOY No. 2	Lead base white metal: Capping metal; Anti-Attrition; as A 122
AMS 4750A	45% Sn Pb alloy: Solder
AMS 4755A	5.5% Ag Pb alloy: Solder
AMS 4756	1.5% Ag 1% Sn Pb alloy: Solder
AMS 7720	6.5% Sb 0.5% Sn Pb alloy: Casting
AMS 7721	6.5% Sb 0.5% Sn Pb alloy: Sheet & extrusion
ANTIMONIAL LEAD	30% Sb Pb: Used in chemical plants battery plates; common name
ARSENIC-LEAD	25% As Pb alloy: Blackwells; primary metal
ASTM B23/7	10% Sn 15% Sb 0.5% Cu Pb alloy: Bearing metal; melting point 240 °C; compressive strength 100 N/mm^2 **DPN: 22**
ASTM B23/8	5% Sn 15% Sb 0.5% Cu Pb alloy: Bearing metal; melting point 237 °C; compressive strength 100 N/mm^2 **DPN: 20**
ASTM B23/13	6% Sn 10% Sb 0.5% Cu Pb alloy: Bearing metal
ASTM B23/15	1% Sn 16% Sb 0.6% Cu Pb alloy: Bearing metal; melting point 248 °C **DPN: 21**
ASTM B29	99.85 – 99.94% Pb pig: Specification covers 4 grades
ASTM B32/1/5 S	1% Sn 1.5% Ag Pb solder: Melting point 309 °C
ASTM B32/2/5 S	2.5% Ag Pb solder: Melting point 304 °C
ASTM B32/2 A	2% Sn Pb solder
ASTM B32/2 B	2% Sn 0.3% Sb Pb solder
ASTM B32/5 A	5% Sn Pb solder: Melting range 270–312 °C
ASTM B32/5 B	5% Sn 0.3% Sb Pb: Solder; melting range 270–312 °C
ASTM B32/10 B	10% Sn 0.3% Sb Pb: Solder; melting range 268–299 °C
ASTM B32/15 B	15% Sn 0.3% Sb Pb: Solder; melting range 227–288 °C
ASTM B32/20 B	20% Sn 0.3% Sb Pb: Solder; melting range 183–277 °C
ASTM B32/20 C	20% Sn 1% Sb Pb: Solder; melting range 183–277 °C
ASTM B32/25 A	25% Sn Pb: Solder; melting range 183–266 °C

Note. The following abbreviations and units are used in the tables:

DPN	Hardness, diamond pyramid number
UTS	Ultimate tensile strength, N/mm^2
Elon	Elongation, %
Proof	0.1 % proof strength, N/mm^2

1 N/mm^2=0.1 hbar=0.102 kgf/mm^2=0.06475 tonf/in^2=145.04 lbf/in^2

Symbol	Nominal analysis, supplier, condition and remarks.
ASTM B32/25 B	25% Sn 0.3% Sb Pb: Solder; melting range 183–266 °C
ASTM B32/25 C	25% Sn 1.3% Sb Pb: Solder; melting range 183–266 °C
ASTM B32/30 A	30% Sn Pb: Solder; melting range 183–255 °C
ASTM B32/30 B	30% Sn 0.3% Sb Pb: Solder; melting range 183–255 °C
ASTM B32/30 C	30% Sn 1.6% Sb Pb: Solder; melting range 183–255 °C
ASTM B32/35 A	35% Sn Pb: Solder; melting range 183–247 °C
ASTM B32/35 B	35% Sn 0.3% Sb Pb: Solder; melting range 183–247 °C
ASTM B32/35 C	35% Sn 1.8% Sb Pb: Solder; melting range 183–247 °C
ASTM B32/40 A	40% Sn Pb: Solder; melting range 183–238 °C
ASTM B32/40 B	40% Sn 0.3% Sb Pb: Solder; melting range 183–238 °C
ASTM B32/40 C	40% Sn 2% Sb Pb: Solder; melting range 183–238 °C
ASTM B32/45 A	45% Sn Pb: Solder; melting range 183–227 °C
ASTM B32/45 B	45% Sn 0.3% Sb Pb: Solder; melting range 183–227 °C
ASTM B102 Y10A	10% Sb Pb alloy: Bearing metal
ASTM B102 YT155A	5% Sn 15% Sb Pb alloy: Bearing metal
ASTM B325	For refined secondary PB; two listed; soft; Cu free and Cu bearing with 0.04% Cu
AUTO A	16% Sb 1% Cu Pb alloy: Bearing metal, Stone Manganese
BABBITT	Sn Cu Sb Pb white metal bearing alloys: Further information from Lead Development Association
BABBITT No. 6	Pb base bearing metal: For general engineering; Eyre Smelting; liquidus 272 °C DPN: 20 UTS: 45 Elon: 3%
BAHN METAL	0.6% Na 0.7% Ca 0.05% Ni Pb: Bearing metal; German origin
BS 219/1S	1.2% Sn 0.1% Sb (max) 1.5% Ag Pb alloy solder: Melting range 309–310 °C
BS 219/5S	5% Sn 0.1% Sb (max) 1.5% Ag Pb alloy solder: Melting range 296–301 °C
BS 219 C	40% Sn 2.2% Pb alloy: Solder; melting range 185–227 °C
BS 219 D	30% Sn 1.5% Sb Pb alloy: Solder; melting range 185–248 °C
BS 219 G	40% Sn 0.4% Sb (max) Pb alloy: Solder; melting range 183–234 °C
BS 219 H	35% Sn 0.3% Sb (max) Pb alloy: Solder; melting range 183–255 °C
BS 219 J	30% Sb 0.3% Sb (max) Pb alloy: Solder; melting range 183–255 °C
BS 219 L	32% Sn 1.8% Sb Pb alloy: Solder, melting range 185–243 °C
BS 219 M	45% Sn 2.5% Sb Pb alloy: Solder; melting range 185–215 °C
BS 219 N	18% Sn 1% Sb Pb alloy: Solder; melting range 185–275 °C
BS 219 R	45% Sn 0.4% Sb (max) Pb alloy: Solder, melting range 183–224 °C
BS 219 V	20% Sn 0.2% Sb (max) Pb alloy: Solder; melting range 183–276 °C
BS 334 A	99.99% Pb chemical lead: Primary material
BS 334 B	Chemical Pb containing protective elements: Primary material 0.005% Bi (max)
BS 335/2	7% Sb Pb alloy: Casting; corrosion resistant for lining tanks; 'Regulus Metal' DPN: 15 UTS: 40
BS 335/3	9% Sb Pb alloy: Casting; 'Regulus Metal'; corrosion resistant; wear resistant DPN: 16 UTS: 45

Symbol	Nominal analysis, supplier, condition and remarks.
BS 335/4	11% Sb Pb alloy: Casting; 'Regulus metal'; corrosion resistant; machinable DPN: 17 UTS: 57
BS 335/5	12% Sb (min) Pb alloy: Casting; 'Regulus Metal'; corrosion resistant; machinable DPN: 19 UTS: 60
BS 602/1	99.8% Pb: Pipe
BS 602/2	99.5% Pb: Pipe
BS 602/3	0.05% Te Pb: Pipe
BS 643	5% Sn 15% Sb Pb alloy: Used for capping wire ropes
BS 801	99.8% Pb: For cable sheathing
BS 801 B	0.9% Sb Pb alloy: For sheathing
BS 801 D	0.25% Cd 0.5% Sb Pb alloy: For sheathing
BS 801 E	0.4% Sn 0.2% Sb Pb alloy: For sheathing
BS 1085	0.004% Ag 0.004% Cu Pb: Pipe
BS 1178	99.9% Pb alloy: Sheet & strip for building
BS 3332/7	12% Sn 13% Sb 1% Cu Pb alloy: Bearing metal
BS 3332/8	5% Sn 15% Sb Pb alloy: Bearing metal
CAPSULE METAL	8% Sn Pb alloy: Origin unknown
CERROBASE	Bi Pb alloy: Casting for pattern metal; Mining & Chemical Ltd; can be cast into paper moulds etc.; melting point 124 °C
COMSOL	Ag Sn Pb base soft solder: Mallory Metallurgical; electrical conductivity 8% IACS; melting point 296 °C UTS: 37 Elon: 40%
CRUSHER	Pb base hardened bearing metal for dusty conditions: Magnolia Anti-Friction Metal Co UTS: 66
DIN 1707 LPB 98.5	98.5% Pb alloy
DIN 1707 LSn 8	8% Sn 0.5% Sb Pb alloy
DIN 1707 LSn 25	25% Sn 1.7% Sb Pb alloy
DIN 1707 LSn 30	30% Sn 2.0% Sb Pb alloy
DIN 1707 LSn 33	33% Sn 2.2% Sb Pb alloy
DIN 1707 LSn 35	35% Sn 2.3% Sb Pb alloy
DIN 1707 LSn 40	40% Sn 2.7% Sb Pb alloy
DIN 1719	Pb specification covers various grades pure lead
DIN 1741 Sb Pb 46	2% Cu 40% Sn 12% Sb Pb alloy DPN: 17 UTS: 60 Elon: 4%
DIN 1741 Sb Pb 59	3% Cu 25% Sn 13% Sb Pb alloy DPN: 18 UTS: 60 Elon: 3%
DIN 1741 Sb Pb 85	5% Sn 10% Sb Pb alloy DPN: 18 UTS: 60 Elon: 8%
DIN 1741 Sb Pb 87	13% Sb Pb alloy DPN: 14 UTS: 45 Elon: 10%
DIN 1741 Sb Pb 97	3% Sb Pb alloy DPN: 9 UTS: 45 Elon: 20%
DIN 16512 Pb Sn 3 Sb 4	3.0% Sn 4% Sb Pb alloy
DIN 16512 Pb Sn 3 Sb 12	3% Sn 12% Sb Pb alloy: Typemetal
DIN 16512 Pb Sn 4 Sb 15	4% Sn 15% Sb Pb alloy
DIN 16512 Pb Sn 5 Sb 12	5% Sn 12% Sb Pb alloy
DIN 16512 Pb Sn 5 Sb 28	5.5% Sn 29% Sb Pb alloy
DIN 16512 Pb Sn 9 Sb 17	9% Sn 17% Sb Pb alloy
DIN 16512 Pb Sn 15 Sb 4	15% Sn 4.5% Sb Pb alloy
DIN 16512 V Pb Sn 5 Sb 28	5% Sn 28% Sb Pb alloy
DIN 16512 V Pb Sn 30 Sb 6	30% Sn 6% Sb Pb alloy
DIN 17640	Specification covering grades of lead
DIN 17641	Pb and Pb alloys: Specification covers several analysis

Symbol	Nominal analysis, supplier, condition and remarks.
DTD 685	1.5% Ag 1.2% Sn Pb alloy: Solder; melting range 309–313 °C
DURASTIC A 1	Cu Pb Sn base bearing for reciprocating engines: Eyre Smelting; melting range 186–350 °C
	DPN: 31 UTS: 90 Elon: 3%
ELECTRO TYPE	3% Sn 3% Sb Pb alloy: White metal; further information from Lead Development Association
ESD 32	10.0% Co 18.0% Fe Pb alloy: Origin unknown
ESD 42	10.0% Co 18.0% Fe Pb alloy: Origin unknown
E QQ S 571/R	2.5% Ag Pb alloy solder: US Federal; melting point 304 °C
FLOWER BRAND	Pb base graphite impregnated bearing metal: Magnolia Anti-Friction Metal Co; constant load & speed
	DPN: 25.5 UTS: 66
HOYT 3 M	Pb base bearing metal: Hoyt
HOYT 4 A	Pb base bearing metal: Hoyt
HOYT 30	30% Sn Pb solder: Contains up to 1.7% Sb; melting range 183–255 °C; Hoyt
HOYT 32	32% Sn 1.8% Sb Pb solder: Melting range 185–243 °C; Hoyt
HOYT 35	35% Sn Pb solder: Melting range 183–255 °C; Hoyt
HOYT 40	40% Sn Pb solder: Slow setting; may contain Sb; melting range 183–234 °C; Hoyt
HOYT 142	Sn Pb base bearing metal: Hoyt
HOYT 155	Sn Pb base bearing metal: Hoyt
HOYT STAR	Pb base bearing metal: Hoyt
h Pb 1	99.9999% Pb: Rod; Light Ltd; high purity metal
h Pb 2	99.9999% Pb: Shot; Light Ltd; high purity metal
h Pb 4b	99.9999% Pb: Sheet 1 mm thick; Light Ltd; high purity metal
h Pb 73a	99.99% Pb single crystal 6.35 mm diameter × 25 mm: Light Ltd
IBIS	Sn Pb base bearing metal for general purpose: Phosphor Bronze Ltd; available in various grades
INTERTYPE METAL	3% Sn 12% Pb alloy: White metal; further information from Lead Development Association
Kb Pb	0.005% Sb & Sn Pb: Designation used by German Standard
Kb Pb Sb	0.12% Sn 0.005% Sn Pb: Designation used by German Standard
Kb Pb Sn 2.5	2.5% Sn Pb alloy: Designation used by German Standard
Kb Pb Te 0.04	0.04% Te 0.005% Sn & Sb Pb: Designation used by German Standard
LEAD	Pb: Powder, stick, sheet, wire, foil, etc.; Blackwell; pure metal
LEAD 6 A	6% Sb Pb alloy: Wire for spraying; Metco Ltd
LINOTYPE METAL	3% Sn 12% Sb Pb alloy: White metal; further information from Lead Development Association
LOCO	Pb base hardened bearing metal for dusty conditions: Magnolia Anti-Friction Metal Co
	DPN: 36
MAGNOLIA METAL	Pb base bearing metal: Further information from the Lead Development Association
MALTEX X No 1	8.7% Sn 16.7% Sb Pb alloy: Bearing metal; Stone Manganese

Symbol	Nominal analysis, supplier, condition and remarks.
MALTEX X No 1 A	4% Sn 14% Sb 1% Cu Pb alloy: Bearing metal; Stone Manganese; as BS 3332/8
MALTEX X No 7	40% Sn 16% Sb 1% Cu Pb alloy: Bearing metal; Stone Manganese
MARINE 1	4% Sn 14% Sb 1% Cu Pb alloy: Bearing metal; Stone Manganese; as BS 3332/8
MARINE II	22.5% Sn 7.7% Sb Pb alloy: Bearing metal; Stone Manganese
MASCOT	Sn Pb base melting metal for general use: Eyre Smelting; melting range 240–300 °C
	DPN: 23 UTS: 66 Elon: 2%
MONOTYPE METAL	9% Sn 17% Sb Pb alloy white metal: Further information from Lead Development Association
Pb Sb 5	6% Sb Pb alloy: Designation used by German Standard
Pb Sb 8	8% Sb Pb alloy: Designation used by German Standard
Pb Sb 9	8.8% Sb Pb alloy: Designation used by German Standard
Pb Sb 9X	8.8% Sb Pb alloy: Designation used by German Standard
Pb Sb 12	12.5% Sb Pb alloy: Designation used by German Standard
PLASTIC METAL	Pb base bearing metal: Further information from Lead Development Association
PLUMBERS SOLDER	33.3% Sn Pb alloy: Further information from Lead Development Association
P M	99.97% Pb: Refined lead; Platt Metals Ltd
QQ L 201 A	99.9% Pb: Sheet; US Federal
QQ L 201 B	99.5% Pb: Sheet; US Federal
QQ S 571 B	40% Sn Pb alloy: Solder; US Federal
QQ S 571 D	0.45% Sb 40% Sn Pb: Solder; US Federal; melting range 182–238 °C
QQ S 571 D Ag 1.5	1.5% Ag 1.0% Sn Pb: Solder; US Federal; melting point 309 °C
QQ S 571 D Ag 2.5	2.5% Ag Pb: Solder; US Federal; melting point 304 °C
QQ S 571 D Ag 5.5	5.5% Ag Pb: Solder; US Federal; melting range 304–366 °C
QQ S 571 D Sn 5	5.0% Sn Pb: Solder; US Federal; melting range 270–313 °C
QQ S 571 D Sn 10	10% Sn Pb; Solder; US Federal; melting range 269–299 °C
QQ S 571 D Sn 20	1.0% Sb 20.5% Sn Pb; Solder; US Federal; melting range 182–277 °C
QQ S 571 D Sn 30	1.6% Sb 30.5% Sn Pb: Solder; US Federal; melting range 182–254 °C
QQ S 571 D Sn 35	1.8% Sb 35.5% Sn Pb: Solder; US Federal; melting range 182–246 °C
QQ S 571 D Pb 65	0.45% Sb 37% Sn Pb: Solder; US Federal; melting range 182–246 °C
QQ S 571 D Pb 80	0.45% Sb 20% Sn Pb: Solder; US Federal; melting range 182–277 °C
RAILWAY A	40% Sn 16% Sb 1% Cu Pb alloy: Bearing metal; Stone Manganese
REGULUS METAL	6–12% Sb Pb alloy: Casting; see BS 335
RM 1 METAL	Pb base hardened bearing metal for heavy dusty conditions: Magnolia Anti-Friction Metal Co
R Pb	0.7% Sb 1.2% As Pb alloy: Designation used by German Standard
SAE 1 A	45% Sn 0.4% Sb (max) Pb alloy: Solder; melting range 182–227 °C
SAE 1 B	43% Sn 1.7% Sb Pb alloy: Solder; melting range 185–223 °C
SAE 2 A	40% Sn 0.4% Sb (max) Pb alloy: Solder; melting range 182–223 °C
SAE 2 B	38% Sn 1.7% Sb Pb alloy: Solder; melting range 185–232 °C
SAE 3 A	30% Sn 0.5% Sb (max) Pb alloy: Solder; melting range 182–254 °C

Note. The following abbreviations and units are used in the tables:

DPN	Hardness, diamond pyramid number
UTS	Ultimate tensile strength, N/mm^2
Elon	Elongation, %
Proof	0.1 % proof strength, N/mm^2

$1\ N/mm^2 = 0.1\ hbar = 0.102\ kgf/mm^2 = 0.06475\ tonf/in^2 = 145.04\ lbf/in^2$

Symbol	Nominal analysis, supplier, condition and remarks.
SAE 3 B	30% Sn 1.0% Sb Pb alloy: Solder; melting range 182–250 °C
SAE 4 A	25% Sn 0.4% Sb (max) Pb alloy; Solder; melting range 182–266 °C
SAE 4 B	25% Sn 1.5% Sb Pb alloy: Solder; melting range 185–260 °C
SAE 5 A	20% Sn 0.4% Sb (max) Pb alloy; Solder; melting range 182–279 °C
SAE 5 B	20% Sn 1.5% Sb Pb alloy: Solder; melting range 185–266 °C
SAE 6 A	15% Sn 0.4% Sb (max) Pb alloy: Solder; melting range 224–290 °C
SAE 6 B	15% Sn 2.7% Sb (max) Pb alloy: Solder; melting range 224–290 °C
SAE 7 A	50% Sn 0.4% Sb (max) Pb alloy: Solder; melting range 182–216 °C
SAE 8 A	35% Sn 0.4% Sb (max) Pb alloy: Solder; melting range 182–245 °C
SAE 9 B	2.7% Sn 5.2% Sb Pb alloy: Solder; melting range 240–290 °C
SAE 13	6% Sn 10% Sb Pb alloy: For bearings
SAE 14	10% Sn 15% Sb Pb alloy: For bearings
SAE 15	1% Sn 15% Sb Pb alloy: For bearings
SAE 16	4.5% Sn 3.5% Sb Pb alloy: For bearings; cast on a sintered matrix
SAE 19	10% Sn Pb alloy: For bearings; this alloy is electroplated onto Cu or Ag based bearings
SAE 190	7% Sn Pb alloy: For bearings; this alloy is electroplated onto Cu or Ag based bearings
SAE 485	3% Sn 46% Cu Pb alloy: For bearings; cast on to steel or sintered backing
SPRABABBITT L	0.25% Cu 10% Sn 13% Sb Pb alloy: Wire for spraying; Metco Ltd
STEREOTYPE	6% Sn 16% Sb Pb alloy: White metal; further information from Lead Development Association

Symbol	Nominal analysis, supplier, condition and remarks.
STA 7 LB1	Pb base bearing metal: Obsolete
STA 7 LB2	Pb base bearing metal: Obsolete; 'Magnolia Metal'
STA 7 TB2A	Pb base bearing metal: Obsolete; 'Babbit'
SX 25	0.25% Cu 2.5% Ag Pb alloy: Soft solder; Sheffield Smelting; melting range 302–305 °C **DPN: 8.0 UTS: 37 Elon: 45% Proof: 23**
TANDEM HDL	Pb base bearing metal: For shock loads; Eyre Smelting; liquidus 350 °C **DPN: 34 UTS: 90 Elon: 1%**
TANDEM RM	Pb base bearing metal: For shock loads; Eyre Smelting; liquidus 350 °C **DPN: 31 UTS: 85 Elon: 3%**
TANDEM SC	Pb base bearing metal: For low speed high duty; Eyre Smelting; liquidus 300 °C **DPN: 27 UTS: 75 Elon: 2%**
UNS L05001–L05999	Lead based alloys: American Standard System
V Pb Sn 5 Sb 28	5% Sn 28% Sb Pb alloy: Designation used by German Standard
V Pb Sn 30 Sb 6	30% Sn 6% Sb Pb alloy: Designation used by German Standard
W E WATSON BRAND	Sn Pb base bearing metal: For heavy duty purposes; Eyre Smelting; melting range 240–370 °C **DPN: 30 UTS: 67 Elon: 1%**

Note. The following abbreviations and units are used in the tables:

DPN	Hardness, diamond pyramid number
UTS	Ultimate tensile strength, N/mm^2
Elon	Elongation, %
Proof	0.1 % proof strength, N/mm^2

$1\ N/mm^2 = 0.1\ hbar = 0.102\ kgf/mm^2 = 0.06475\ tonf/in^2 = 145.04\ lbf/in^2$

22. Lithium Li

Physical properties

Atomic number	3	
Atomic weight	6.94	
Crystal structure	Body-centred cubic	
Colour	Silver-white	
Specific gravity	0.53	
Density	530 kg/m³	(0.019 lb/in.³)
Melting point	186 °C	
Boiling point	1371 °C	
Specific heat	4.02 J/g °C	(0.96 cal/g °C)
Thermal conductivity	71.2 W/m °C	(17 cal/m s °C)
Coefficient of linear expansion (20–100° C)	56 × 10⁻⁶/ °C	
Latent heat of fusion	137 J/g	(32.8 cal/g)
Latent heat of vaporization	19460 J/g	(4648 cal/g)
Thermal neutron absorption cross-section	70 barns/atom	
Electrical conductivity	20% IACS (copper 100%)	
Specific resistance	84 microhm mm	
Temperature coefficient of electrical resistance	0.0045/ °C	
Electrochemical equivalent	0.262 g/A/h	
Electrode potential	-3.02 V	
Magnetic susceptibility	0.5 × 10⁻⁶	
Young's modulus of elasticity	– N/m²	
Tensile strength	– N/mm²	
Hardness	< 5 DPN	

22.1 General notes on lithium

Lithium occurs generally in combination with aluminium as the complex double oxide or fluoride. Approximately 0.004% of the earth's crust is composed of lithium, which is double that of lead, but being more evenly distributed mining is less economical.

The metal is softer than lead, can be cut with a knife and tarnishes rapidly in moist air. It reacts vigorously with water, liberating hydrogen, thus must be contained out of contact with air.

Lithium is the lightest known metal, having a specific gravity one fifth that of aluminium and half that of water.

The low melting point and high specific heat – almost that of water – and high boiling point make lithium an obvious choice as a heat exchange medium. Some use is being made of lithium in preference to sodium and potassium in nuclear submarine reactors, but the corrosive nature of the liquid metal necessitates expensive containers, such as tantalum. Liquid lithium at 200° C attacks most common metals and glass or porcelain.

Alloys of lithium with silver or copper are used as deoxidizers for copper and silver and their alloys, while the alloy with calcium is a powerful scavenging agent for oxygen. The affinity of lithium for oxygen has also been used in heat treatment atmospheres, where the metal is added to the burnt town gas. Any free oxygen is coverted to lithium oxide, but even more important is the ability of lithium to combine with oxygen in water vapour. This is one of the more troublesome components of atmospheric

gas at heat treatment temperatures and the reaction has the secondary effect of releasing hydrogen to enrich the reducing properties of the resultant gas. The more efficient use of hydrogen and cracked ammonia furnace atmospheres, has to a lesser extent reduced the need for lithium enriched atmospheres.

Lithium metal is used in chemical engineering as a catalyst in the production of some synthetic rubbers and forms the base for a series of metallic greases which have exceptional viscosity stability over a range of temperatures.

There are no commercial alloys based on lithium but in small amounts up to 1% lithium has a wide field of application. When added to lead based bearing metals and certain zinc base sheet metal alloys there is a considerable increase in hardness. The aluminium–zinc cast and wrought alloys are strengthened by 0.5% lithium and magnesium alloys have their corrosion resistance increased with no drop in strength.

When added to solder and braze materials lithium improves the wetting properties with in some instances improvement in shear strength. This increase in the ability to wet means enhanced brazing on materials such as chromium and tungsten.

The metal lithium has many useful and interesting properties, although its lack of stability in air precludes its use as the main constituent in normal engineering alloys.

Symbol	Nominal analysis, supplier, condition and remarks.
ASTM B357	99.9% Li metal in ingot form
LITHIUM	Li metal: Blackwells; pure metal
W Li 16	99.98% Li: Ingot; Light Ltd; high purity metal
W Li 20	99.9% Li: Wire 3 mm diameter; Light Ltd
W Li 96	99.8% Li 6 isotope: Light Ltd
W Li 97	99.99% Li 7 isotope: Light Ltd

Note. The following abbreviations and units are used in the tables:

DPN	Hardness, diamond pyramid number
UTS	Ultimate tensile strength, N/mm^2
Elon	Elongation, %
Proof	0.1 % proof strength, N/mm^2

$1\ N/mm^2 = 0.1\ hbar = 0.102\ kgf/mm^2 = 0.06475\ tonf/in^2 = 145.04\ lbf/in^2$

23. Magnesium Mg

Physical properties

Atomic number	12	
Atomic weight	24.32	
Crystal structure	Close-packed hexagonal	
Colour	Silver-white	
Specific gravity	1.74	
Density	1740 kg/m^3	(0.063 lb/in.^{3})
Melting point	650 °C	
Boiling point	1103 °C	
Specific heat	1.03 J/g °C	(0.246 cal/g °C)
Thermal conductivity	159 W/m °C	(38 cal/m s °C)
Coefficient of linear expansion (20–100° C)	26.1 × 10^{-6}/ °C	
Latent heat of fusion	371.8 J/g	(88.8 cal/g)
Latent heat of vaporization	4760 J/g	(1137 cal/g)
Thermal neutron absorption cross-section	0.06 barns/atom	
Electrical conductivity	37% IACS (copper 100%)	
Specific resistance	46 microhm mm	
Temperature coefficient of electrical resistance	0.0040/ °C	
Electrochemical equivalent	0.454 g/A/h	
Electrode potential	-2.37 V	
Magnetic susceptibility	0.55 × 10^{-6}	
Young's modulus of elasticity	64.5 × 10^9 N/m^2	(6.35 × 10^6 lbf/in.^{2})
Tensile strength	annealed – N/mm^2	(10 tonf/in.^{2})
Hardness	35 DPN	

23.1 General notes on magnesium

Magnesium is found generally as the carbonate, the double carbonate with calcium, the sulphate, chloride and silicate, and is the third most common metal. These are widespread throughout the world and are all more or less soluble in water. There is therefore a tendency for them to be dissolved by rivers and streams, and concentrate in the oceans where they form the salt in sea water.

Magnesium is a modern metal owing its place in the table of usefulness almost exclusively to the aircraft industry. The use of higher temperatures and stresses in this industry is to some extent lessening its popularity. The high tonnage requirements however, caused modern plants to be built and reduced the price of magnesium to that of a competitive material where weight-saving is of prime importance.

The affinity of magnesium for oxygen is a considerable disadvantage from the corrosion point of view, necessitating complex corrosion resisting procedures and first quality paint films. This affinity is used for such purposes as metal de-oxidizing and as an oxygen 'getter' in thermionic valves. The well known property of burning in air to give a very white bright light is still used for flares, fireworks and photographic flashes. It forms the base of most incendiary devices.

Magnesium has gained the reputation that it corrodes dangerously rapidly, which is unfortunate. While there is no doubt that severe corrosion can occur with magnesium and its alloys this is not as bad as with steel but can be more spectacular and it is probably easier to protect magnesium permanently than steel. Alkalis and hydrofluoric acid have no effect on magnesium.

Pure magnesium is not often used commercially but there is a well defined range of alloys, some of which can have their mechanical properties enhanced by thermal treatment. Details of these materials are given in the following sections.

Magnesium is also used as an alloying element, particularly with aluminium where it has the effect of increasing strength when cold worked, without detracting from the corrosion resistance. It has a similar though less marked affect on zinc alloys.

There are also heat treatable aluminium magnesium alloys, particularly with silicon, where the magnesium forms intermetallic compounds which take part in the hardening process.

23.2 Magnesium – note on specifications

An explanation of the different methods of classification and designation used with magnesium base materials may be of assistance in identifying these alloys rapidly. This being a comparatively modern metal, there are as yet relatively few specifications and trade names in existence. There are also very few basic suppliers and these are almost all American or Canadian, with the result that the same alloy appears more or less under the same name or symbol to a much greater extent than with the older metals, or even aluminium.

British Standard Institution (BS)

The specifications issued to date are almost all based on the system of classifying by form – sheet, bar, etc. – with the different alloys given a designation, which identifies the analysis. A letter is used to confirm the form.

These specifications are:

BS 2970   Castings – no identifying letter used
BS 3370   Strip – identifying letter 'S'
BS 3371   Tube – identifying letter 'T'
BS 3372   Forging – identifying letter 'F'
BS 3373   Bar – identifying letter 'E'
BS 3374   Plate – identifying letter 'F'

The alloy designation is prefixed by the letters MAG followed by figures which identify the alloy. Each alloy appears three times in the index of this book, for example:

(a) The full title   BS 3370 MAG S111 –
                      Aluminium–zinc–magnesium alloy
                      strip
(b) also   MAG 111 –
                      Aluminium–zinc–magnesium alloy
(c) and   111 – Aluminium–zinc–magnesium
                      alloy

The two last entries are always referred to the BS range of the appropriate section, where the necessary information will be found under the full title. The official British Standard designation for the alloys uses the letters preceding the figures and this is the correct form which appears in other specifications than those listed above.

One exception is the aircraft material specifications which use the letter 'L' as prefix followed by three digits. Aluminium and its alloys use the same prefix followed by two digits.

American Society for Testing and Materials (ASTM)

This body issues separate specifications covering different forms and materials. Each specification is identified by the letters ASTM followed by 'B' indicating a non-ferrous specification, then a number. The specification is then subdivided by analysis, each of which is identified by a letter and figure code which is almost universally used for magnesium alloys.

This uses two or more letters to indicate the alloys content, for example: 'A' for aluminium, 'Z' for zinc, 'M' for magnesium, followed by a two or three digit number identifying the actual alloy. There is no attempt to make the numbers give the percentage alloying elements or other information.

In this book the alloys are always given their full specification number and the designation is also listed in the index as these alloys tend to be called by this rather than their full title. The index entry refers to the appropriate section where the full title appears.

Examples would be:

ASTM B93 AZ 63 A/616% Al 3% Zn Mg alloy ingot – issued in 1961
AZ 63A   6% Al 3% Zn Mg alloy – this is the commonly used designation.

The full title should always be followed by two figures, giving the year of issue, for example, 61 indicates 1961. The letter 'T' following indicates that the specification was tentative at that time. The issue year and other suffix information is given in the index but not in the tabulated section of this book.

Society of Automotive Engineers (SAE)

This body issues a series of specifications with the prefix SAE followed by two or three digits, the first of which is always 5. There has been no attempt to make these figures describe or code the alloy content.

This body also issues a series of Aerospace Material Specifications prefixed with the letters AMS using four digits. A suffix letter denotes the issue number. Again, no effort has been made to code the analysis or form using these figures.

German Standards (DIN)

This uses two distinct systems, the first of which has a letter and figure code descriptive of the form and analysis of the alloy. These are contained in a master specification generally covering the form – that is sheet, bar, etc. – and prefixed by the letters DIN.

The second and newer system uses numbers only, the first of which is always 3 (in common with aluminium alloys) followed by a period then four other digits, the first two classifying the alloy, the second two indicating the quantity of alloying element. This system was devised to suit modern accounting machines and computors and is included in the DIN specification, along with original descriptive method.

In this book; the DIN specification is given and the new figure code. The descriptive designation is indexed and referred to the DIN range in the appropriate section. Examples are:

DIN 1729 Mg Mn 2 1.7% Mn Mg alloy
Mg Mn 2   1.7% Mn Mg alloy: Designation used by German Standard
3.5200   1.7% Mn Mg alloy: German Standard

Trade names and specifications

These very often use the same code or one based on the same system as that attributed to ASTM.

While every effort has been made to include all possible alternate names, it is suggested that if the material cannot be identified immediately, the manufacturer's name or initials should be added or substituted to or from the symbol requiring identification.

23A Magnesium – commercially pure – wrought and cast

Specific gravity	1.738	
Density	1738 kg/m^3	(0.0645 lb/in.^{3})
Solidus/liquidus	650 °C	
Thermal conductivity	157 W/m °C	(37.6 cal/m s °C)
Coefficient of linear expansion (20–100° C)	26.1 × 10^{-6}/ °C	
Electrical conductivity	38% IACS (copper 100%)	
Specific resistance	44.61 microhm mm	
Young's modulus of elasticity	44 × 10^9 N/m^2	(6.4 × 10^6 lbf/in.^{2})
Impact	wrought 16–24 J	(12–20 ft/lb)
Fatigue strength	± 74 N/mm^2	(± 4.8 tonf/in.^{2})
Hot strength	–	

The above properties have been chosen to show typical values for the specifications listed and may not apply exactly to any one specification. It is possible that with some specification the values may not be applicable.

General metallurgical characteristics

Magnesium is the lightest metal commonly used in the service of mankind, the comparative figures being, magnesium 1: aluminium 1.6: steel 4.4: copper 4.9 and titanium 2.5. Commercially pure magnesium however has few properties which make it an attractive engineering material. It has poor casting characteristics, low strength and does not work harden to any appreciable extent, while the corrosion resisting properties of the pure metal are only slightly superior to that of the alloys.

Thermal treatment

Pure magnesium cannot have its properties improved by any form of thermal treatment. Cold work can be removed by annealing at 200–250 °C for about 15 minutes at temperature. The surface must always be protected by one of the chemical treatments prior to this annealing to prevent too rapid oxidation.

Welding and brazing

Pre-treatment. All trace of oil, grease and dirt must be removed.

Welding and brazing. Successful welds can only be achieved using the inert gas shielded arc welding process, with copious supplies of the inert gas to both sides of the weld.

Electric resistance welding using special equipment and inert gas shielding is possible.

Brazing cannot be accomplished on magnesium.

Post-treatment. None required.

Flaw detection methods

Crack test. Penetrant oil chalk test method. Care must be taken to thoroughly dry the parts if they are water washed.

X-ray. Not normally required.

Ultra-sonic test. Not normally required.

Chemical etch. Not required.

Corrosion protection

Temporary. It is essential that magnesium is protected from contact with the atmosphere at all times. Parts should be received fluoride anodized, or dichromate treated and oiled and must be stored in this condition. During machining they should be kept oiled. Coolants are not normally used as magnesium machines readily in the dry condition. If however parts become damp or wet they must be dried immediately in hot oil or air above 100 °C and oiled.

After machining parts should be dichromate treated, then oiled or lanoline greased, or sent immediately for permanent protection. Precautions are necessary when parts are vapour degreased or subjected to alkaline cleaning.

Permanent. Magnesium has a poor reputation regarding corrosion, which is rather unfair as it has superior atmospheric corrosion resistance to most steels under some conditions. There is no doubt however that given correct conditions of humidity and poor protection magnesium can rapidly corrode. Correct treatment relies on first forming an adherent oxide film which prevents further attack, then protecting this with a paint system. The oxide film is produced chemically or electrochemically by dichromate or fluoride anodize treatments, the latter probably being superior for most purposes but requiring specialized equipment. With both methods there can be serious plating effluent problems. Care must always be taken that air pockets do not either prevent the treatment taking place or correct drying of the treated part.

Parts can be used in this oxidized condition provided they are kept dry or oiled at all times. It is however, recommended that the surface is sealed with a stoved lacquer or paint.

For arduous outdoor marine or industrial conditions it is essential that the correct paint system is applied. This should consist of a stoved chromate primer or clear lacquer to seal the surface, followed by good quality primer and top coats of stoved paint.

Where the paint is damaged, or to remove local corrosion the bare magnesium should be thoroughly cleaned, then treated with selenious acid, dried and paint films applied.

Care must always be taken at design to have no pockets or passages which collect moisture or are not free draining. Provided the above precautions are taken magnesium should present no corrosion problem.

Machinability

One of the selling points of magnesium is its excellent machinability.

These pure magnesiums will seldom be machined, but when necessary require no coolant and produce a good surface finish over a wide range of feeds and speeds with very little tool wear. There is a fire risk with finely divided magnesium, necessitating special polishing equipment, which ensures the dust is kept wet. It cannot be too strongly emphasized that there is a considerable danger of serious explosion if magnesium dust is collceted dry along with ferrous material in normal extraction equipment. French chalk should always be held in correctly designed containers near machines cutting magnesium. The machining area should be kept free of large quantities of cuttings to minimize any outbreak. Water must not be used on magnesium fires as this can lead to side reactions releasing hydrogen. Provided normal good housekeeping is practised there is no reason to treat magnesium as a dangerous metal.

Uses

The materials listed are more or less confined to primary metal.

Symbol	Nominal analysis, supplier, condition and remarks.
3.5002	99.95% Mg: German Standard
3.5003	99.8% Mg: German Standard
ASTM B92/9980 A	99.8% Mg: Ingot
ASTM B92/9990 A	99.9% Mg: Ingot
ASTM B92/9990 B	99.9% Mg: Ingot; low iron
ASTM B92/9995 A	99.95% Mg: Ingot
BS L120	99.5% Mg: Ingot; primary metal
DIN 17800 H Mg 99.8	99.8% Mg
DIN 17800 H Mg 99.95	99.95% Mg
HG 2-9990	99.9% Mg: Primary metal; Domal

Symbol	Nominal analysis, supplier, condition and remarks.
HG 2-9995	99.95% Mg: Primary metal; Domal
HG 2-9997	99.97% Mg: Primary metal; Domal
HG 2-9999	99.99% Mg: High purity metal; Domal
MAGNESIUM	Mg: Ingot, stick, cubes, ribbon, wire, etc.; Blackwells; primary metal
MELPURE	99.9% Mg: Ingot; MEL primary metal
SIS 4602	99.95% Mg: Ingot; Swedish Standard; primary metal
SIS 4604	99.8% Mg: Ingot; Swedish Standard; primary metal
H Mg 11a	99.99% Mg: Ingot; Light Ltd; high purity metal

23B Magnesium alloys
With zinc, aluminium, manganese, zirconium, rare earths or thorium

Specific gravity	1.8	
Density	1800 kg/m^3	(0.064 lb/in.^{3})
Solidus/liquidus	600–630 °C	
Thermal conductivity	63–142 W/m °C	(15–34 cal/m s °C)
Coefficient of linear expansion (20–100° C)	26–27 × 10^{-6}/ °C	
Electrical conductivity	10–30% IACS (copper 100%)	
Specific resistance	70–160 microhm mm	
Young's modulus of elasticity	44.8 × 10^9 N/m^2	(6.5 × 10^6 lbf/in.^{2})
Impact	28 J	(21 ft/lb)
Fatigue strength (50 × 10^6 cycles)	wrought ± 100–120 N/mm^2	(± 7–8 tonf/in.^{2})
	cast ± 60–90 N/mm^2	(± 4–6 tonf/in.^{2})

Hot strength

Temperature °C	Tensile strength N/mm^2
20	210
100	180
200	140
300	75
400	45

The above properties have been chosen to show typical values for the specifications listed and may not apply exactly to any one specification. It is possible that with some specifications the values may not be applicable.

General metallurgical characteristics

Magnesium is the lightest metal in general engineering use, being approximately two-thirds the weight of aluminium. The low strength coupled with its low corrosion resistance means that very few uses are found for the pure metal.

The addition of certain alloying elements has the effect of considerably increasing the strength with some improvement in corrosion resistance. Briefly these elements and their effects are as follows:

Zinc. This was the earliest used alloying element and acts as a hardening agent helping to refine the grain. Magnesium alloys containing more than about 1% zinc are however very prone to weld cracking. Zinc is never the sole alloying element, generally being used with aluminium and zirconium. Up to 6% zinc is commonly used in magnesium alloys.

Aluminium. This hardens the alloy with some grain refinement but gives a long freezing range leading to casting porosity in many cases. There are some intermetallic compounds formed which can give age hardening properties. Up to 10% aluminium is commonly used.

Manganese. This has little effect on the strength of the alloys, but improves the corrosion resistance. There is a tendency to coarsen the grain size thus reducing the fatigue strength. Up to 2% manganese is used alone, with considerably less in conjunction with aluminium and zinc.

Zirconium. This has a profound effect, grain refining the cast structure with consequent increase in strength. Zirconium is only slightly soluble in magnesium and is thus always present in very small quantities.

Aluminium and manganese form compounds with zirconium which are insoluble in magnesium and can thus be used to harden by precipitation treatment. Zirconium improves the workability of wrought products. Up to about 0.7% zirconium is generally used.

Rare earth metals. These strengthen magnesium and improve the high temperature properties. The rare earths also reduce the tendency of cracking in the zirconium–zinc–magnesium alloys.

There is also an improvement in the castability, notably in freedom from porosity, when the rare earth elements, especially cerium, are present. This is sufficient to warrant the use of these expensive alloys in some cases where the hot strength is not required.

Thorium. This has a very similar effect to the rare earths in that it improves the creep strength and other high temperature properties, improves the castability and in addition, enhances the fatigue strength. Less than 2% thorium is used for alloying properties.

Silver. This has recently been used added to rare earth and zirconium alloys of magnesium resulting in considerable age hardening properties of castings. Although to date silver is not used in wrought alloys there are indications that this series of silver–rare earth–zirconium alloys will be the basis of magnesium alloys of comparable strength to many aluminium alloys.

All magnesium alloys have good damping characteristics, but are rather more notch sensitive than other common low strength engineering materials. This, coupled with their high galvanic corrosion characteristics, makes the correct design of magnesium articles of prime importance. This should always include generous radii, an absence of pockets to collect moisture and no intimate contact with other metals. Joins must always be carefully sited away from areas of possible stress concentration.

Thermal treatment

In general magnesium alloys are not cold worked to any extent and thus seldom require annealing. Where it is necessary for any reason a temperature of 250–450 °C is required, depending on the alloy content and the degree of previous cold work.

Some of the alloys listed can be hardened by solution treatment and ageing. This requires that the elements added form intermetallic compounds with each other or magnesium, which are soluble in the magnesium matrix at high temperatures but not at room temperature. Thus the 'solution treatment' at a temperature of 380–500 °C for up to 10 hours is used to dissolve these compounds. This is followed by water or air cooling when the compounds remain in solution. In many cases the casting process is used as the solution treatment.

The ageing treatment requires temperatures in the order of 170–350 °C for 10–20 hours when the dissolved compounds start to precipitate, causing strain in the lattice structure resulting in hardening. Over-ageing allows complete precipitation and agglomeration of the compounds, reducing the hardness.

The metallurgical reactions involved are quite complex and vary depending on the alloy and treatment. Most magnesium alloys are supplied in the finally heat treated condition, and where it is necessary for any reason to carry out further thermal treatment affecting the mechanical properties, it is essential that advice is obtained from the source. The metallurgical reactions involved are often sluggish enough not to be too much affected by welding or stress relieving treatments. Stress relieving of magnesium alloys may be necessary after considerable machining or welding, although these alloys are probably more stable at machining than any other common engineering material. The high thermal expansion however, can cause welding stresses which are relieved by a treatment for about one hour at 300 °C.

Welding and brazing

Pre-treatment. The weld area must be freshly prepared and free of all oil, grease and dirt. Large castings and forgings or where there is a rapid change to section near the weld should be pre-heated to about 300 °C.

Welding and brazing. Welding is possible with all alloys except those containing zirconium, using gas or electric arc equipment and fluxes, but this requires considerable skill and there is a constant danger of flux or oxide trapping. This is absent when inert gas shielded electric arc welding is used. This method is suitable for all magnesium alloys, the techniques being very similar to those required for aluminium but with a lower heat input and greater gas flow. Very considerable care is necessary in the choice of filler rod, which should be the same material as the parent metal in most cases, but may be specially formulated to make up any alloying element lost during welding. Other rods are designed to lower the melting point, thus reducing the danger of weld cracking. Use of the wrong material can lead to serious cracking and corrosion troubles. It is also essential to ensure that only very similar alloys are joined by welding to prevent corrosion failure in service.

Brazing and soldering are not usually applied to magnesium alloys.

Post-treatment. Any flux must be removed by washing and scrubbing immediately after welding, otherwise severe corrosion will occur. All weld areas should be wire brushed or filed to eliminate any oxide formed during welding. In many cases it is possible to improve the properties by hammering the weld area.

It is recommended that a stress relieving treatment at about 300 °C for one hour is carried out if there is any shape complication. Removal of distortion can be very effectively carried out by hot stretching at 250 °C and above with no resultant stress remaining. The sluggish nature of the metallurgical reactions means that only in exceptional instances will the parts require any other form of thermal treatment.

Parts should be sent for corrosion protection as soon as possible after welding.

Advice on welding techniques is generally available from the material suppliers.

Flaw detection methods

Crack test. Penetrant oil chalk test method. Care must be taken to ensure thorough drying after water washing. Also there can be chemical etching on magnesium when certain penetrant fluids have been in use for some time. This can lead to corrosion and is usually cured by filtering the penetrant solution. This only applies when the components are immersed in the fluid.

X-ray. This is necessary on all high duty castings and welds and is recommended on a percentage basis on all welds and castings for control purposes.

Ultra-sonic test. Not often used on these alloys, but techniques have been developed which show welding defects and casting flaws.

Chemical etch. Not used.

Corrosion protection

Temporary. These materials must be protected at all times. The use of lanolin or at least an oil film is essential during storage and machining. Soluble oil coolants are not necessary for machinability and should never be used.

Immediately after machining and during storage periods longer than a few weeks the parts should be chromate treated and oiled.

Permanent. Magnesium has a rather unjust reputation for corrodability which in fact is no worse than, and in many cases is superior to steels under many conditions.

It is however, much more expensive than steel and generally more sensitive to the effect of corrosion notches on fatigue strength. There is also a tendency for corrosion to be local and rapid under some conditions, hence it is more spectacular.

Corrosion prevention and protection require careful thought at design, good housekeeping during production and excellent workmanship at final processing, with the use at all times of first class materials.

Design must ensure that no pockets exist to trap moisture, that magnesium is insulated from contact with other metals and that the correct corrosion protection system for the projected use is specified.

Like most metals, magnesium will only corrode in damp or wet conditions, and as it is more electrically negative than all other commonly used metals, it will sacrifice itself when placed in contact with any other metal. It is therefore essential to remove all trace of dirt, swarf or impregnated metal from the surface and prevent contact of the clean surface with the damp atmosphere.

It has recently been proved that cleaning the surface efficiently is more difficult than generally realized. Tests show that a grit blasted surface is prone to corrosion because of fine metallic and non-metallic dust impregnation. Fine magnesium dust in contact with magnesium castings and forgings can be a source of corrosion. An electrolytic treatment in sodium fluoride solution, known as 'fluoride anodize' has been developed, which is proving very successful in cleaning the magnesium surface thoroughly.

Lanolin and paraffin mixtures applied to all stages of machining effectively prevent atmospheric attack. Paraffin washing should be used to remove this protection when necessary. Trichloroethylene vapour degreasing should only be used prior to final permanent protection and then only after thorough paraffin washing. Apart from removing all trace of oil leaving the surface ready to corrode there can be a violent reaction under certain circumstances between finely divided magnesium and dirty trichloroethylene. After machining parts can be fluoride anodized then chemically sealed and painted, or more commonly the parts are dichromate treated, painted with zinc chromate primer then sealed with two or more coats of stoving top coat.

Summarizing, the corrosion protection of magnesium alloys should start by having the part fluoride anodized, be lanolin covered or oiled at all times during machining, should be dichromate treated after final machining then be painted with a zinc chromate primer sealed by two or more stoved top coats. These top coats are not required when parts are to operate in oil.

It is of course, important that the paint film is carefully maintained. If chipped or damaged, the area should be thoroughly wire brushed then treated locally with selenious acid, washed, dried and zinc chromate primed under several coats of sealer top coat.

Provided these reasonable precautions are taken magnesium gives complete satisfaction under all atmospheric conditions, including industrial and marine. Lesser precautions are required under less arduous conditions, for example, no corrosion protection whatever would be required for use in the Sahara desert.

Machinability

Magnesium alloys all have excellent machining characteristics. Tool life is long and more flexibility in feeds and speeds is possible than with almost any other engineering materials.

Generally no coolant whatever is required and soluble oil must not be used as it can cause corrosion. For operations such as drilling and tapping neat lard oil should be used.

There is a fire risk when finely divided magnesium is

heated. Thus blunt tools or any source of ignition such as cigarette ends can start a fire. This is not dangerous as long as good housekeeping is practised in keeping machines and adjacent areas clean and free from large quantities of swarf. The fire is extinguished using chalk, cast iron fillings, dry sand, or any other dry powder which does not support combustion. Water must not be used as magnesium can react releasing hydrogen which then feeds the flames.

This is another reason for not using soluble oil coolants. Normal neat oil coolants correctly used prevent any risk of fire.

Polishing and grinding magnesium must use separate dust extracting equipment from all other metals with provision for water washing the collected dust. It must be noted that fine magnesium dust can cause an explosion if ignited by a spark and this can readily occur inside most dry dust collectors when steel grindings are present.

Uses

Aircraft structural parts such as wheels, landing gear and many items of airframe parts. Aircraft engine carcase parts, pump castings and other non-rotating parts.

The motor industry now uses considerable quantities on such items as brakes, wheels and other unsprung items. Some makers use a large number of magnesium castings which require considerable machining, as the castability and excellent machinability offset the extra cost. Magnesium is used as the anode in cathodically protected structures. Magnesium has excellent vibration damping properties and is thus used to absorb noise.

Symbol	Nominal analysis, supplier, condition and remarks.
3.5161	5.5% Zn 0.5% Zr Mg alloy: Wrought; German Standard
	DPN: 60 **UTS: 250** **Elon: 5%** **Proof: 180**
3.5200	1.8% Mn Mg alloy: Wrought; German Standard
	DPN: 40 **UTS: 180** **Elon: 1.5%** **Proof: 140**
3.5312	3% Al 1% Zn 0.2% Mn Mg alloy: Wrought; German Standard
	DPN: 45 **UTS: 240** **Elon: 10%** **Proof: 150**
3.5612	6% Al 1% Zn 0.2% Mn Mg alloy: Wrought; German Standard
	DPN: 55 **UTS: 240** **Elon: 6%** **Proof: 170**
3.5632	6% Al 3% Zn 0.2% Mn Mg alloy: Casting; German Standard
	DPN: 60 **UTS: 150** **Elon: 5%** **Proof: 90**
3.5812	8% Al 0.8% Zn 0.2% Mn Mg alloy: Casting; German Standard
	DPN: 60 **UTS: 210** **Elon: 5%** **Proof: 120**
3.5912	9% Al 0.8% Zn 0.2% Mn Mg alloy: Casting; German Standard
	DPN: 65 **UTS: 210** **Elon: 3%** **Proof: 140**
3.5922	8.5% Al 1.5% Zn 0.2% Mn Mg alloy: Casting; German Standard
	DPN: 65 **UTS: 200** **Elon: 2%** **Proof: 120**
A 8	8% Al 0.5% Zn 0.3% Mn Mg alloy: Casting; MEL; solution treated; ductile; shock resistant
	UTS: 220 **Elon: 8%** **Proof: 75**
A 2855	8% Al 0.4% Zn 0.3% Mn Mg alloy: Forging; MEL
	DPN: 70 **UTS: 300** **Elon: 11%** **Proof: 180**
AM 503	1.5% Mn Mg alloy: Sheet, bar & tube; MEL; weldable; good corrosion resistance
	UTS: 240 **Elon: 7%** **Proof: 120**
AM 503	1.5% Mn Mg alloy: Sheet & tube; Birmabright for BS alloy 101
AM 503 S	0.7% Mn Mg alloy: MEL for nuclear engineering
AMS 4350 F	6% Al 1% Zn Mg alloy: Extrusions; as extruded; AMS for ASTM alloy AZ 61 A
AMS 4352 A	5.5% Zn Zr Mg alloy: Extrusion; aged; AMS for ASTM alloy ZK 60 A

Note. The following abbreviations and units are used in the tables:

DPN	Hardness, diamond pyramid number
UTS	Ultimate tensile strength, N/mm^2
Elon	Elongation, %
Proof	0.1 % proof strength, N/mm^2

$1\ N/mm^2 = 0.1\ hbar = 0.102\ kgf/mm^2 = 0.06475\ tonf/in^2 = 145.04\ lbf/in^2$

Symbol	Nominal analysis, supplier, condition and remarks.
AMS 4358 A	6% Al 1% Zn Mg alloy: Forging; as forged; AMS for ASTM alloy AZ 61 X
AMS 4360 C	8.5% Al 0.5% Zn Mg alloy: Forging; aged; AMS for ASTM alloy AZ 80 A
AMS 4362	5% Zn Mg alloy: Forging; aged; AMS for ASTM alloy ZK 60 A
AMS 4375 D	3% Al 1% Zn Mg alloy: Sheet; annealed; AMS for ASTM alloy AZ 31 B
AMS 4376 A	3% Al 1% Zn Mg alloy; Plate; AMS for ASTM alloy AZ 31 B
AMS 4377 B	3% Al 1% Zn Mg alloy: Sheet; AMS specification for ASTM alloy AZ 31 B
AMS 4384 B	3.2% Th 0.7% Zr Mg alloy: Sheet; annealed; AMS for ASTM alloy HK 31 A
AMS 4385 C	3.2% Th 0.7% Zr Mg alloy: Sheet; AMS for ASTM alloy HK 31 A
AMS 4387	2.3% Zn 0.6% Zr Mg alloy: Extrusion; as extruded; AMS for ASTM alloy ZK 21 A
AMS 4388	3% Th 1.5% Mn Mg alloy: Extrusion; as extruded; AMS for ASTM alloy HM 31 A
AMS 4389 A	3% Th 1.5% Mn Mg alloy: Extrusion; aged; AMS for ASTM alloy HM 31 A
AMS 4390 B	2% Th 0.8% Mn Mg alloy: Sheet; cold rolled; AMS for ASTM alloy HM 21 A
AMS 4395 A	9% Al 2% Zn Mg alloy: Welding wire; AMS for ASTM alloy AZ 92 A
AMS 4396	2.5% Zn 0.7% Zr 3.3% Ce Mg alloy: Welding wire; AMS for ASTM alloy EZ 33 A
AMS 4418 A	2.5% Ag 2% Dy 1% Zr Mg alloy: Sand casting; solution treated & aged; AMS for ASTM alloy QE 222 A
AMS 4420 H	6% Al 3% Zn Mg alloy: Casting; sand cast; AMS for ASTM alloy AZ 63 A
AMS 4422 J	6% Al 3% Zn Mg alloy: Sand casting; solution treated; AMS for ASTM alloy AZ 63 A
AMS 4424 F	6% Al 3% Zn Mg alloy: Sand casting; solution treated & aged; AMS for ASTM alloy AZ 63 A
AMS 4434 F	9% Al 2% Zn Mg alloy: Sand casting; solution treated & aged; AMS for ASTM alloy AZ 92 A
AMS 4437	8.7% Al 0.1% Zn Mg alloy: Sand casting; solution treated & aged; AMS for ASTM alloy AZ 91 C
AMS 4438	5.7% Zn 1.8% Th Mg alloy: Sand casting; aged; AMS for ASTM alloy ZH 62 A
AMS 4440 A	0.7% Zr 3.5% Ce Mg alloy: Sand casting; aged; AMS for ASTM alloy EK 41 A
AMS 4441 A	0.7% Zr 3.5% Ce Mg alloy: Sand casting; solution treated & aged; AMS for ASTM alloy EK 41 A
AMS 4442 A	2.5% Zn 0.7% Zr 3.3% Ce Mg alloy: Sand casting; aged; AMS for ASTM alloy EZ 33 A

Symbol	Nominal analysis, supplier, condition and remarks.
AMS 4443 A	4.5% Zn 0.7% Zr Mg alloy: Sand casting; aged; AMS for ASTM alloy ZK 51 A
AMS 4444	6% Zn 1% Zr Mg alloy: Sand casting; aged; AMS for ASTM alloy ZK 61 A
AMS 4445 A	3.3% Th 0.8% Zr Mg alloy: Sand casting; solution treated & aged; AMS for ASTM alloy HK 31 A
AMS 4447	2.1% Zn 3.3% Th 0.8% Zr Mg alloy: Sand casting; aged; AMS for ASTM alloy HZ 32 A
AMS 4453	9% Al 2% Zn Mg alloy: Investment casting; solution treated & aged; AMS for ASTM alloy AZ 92 A
AMS 4455	10% Al Mg alloy: Investment casting; solution treated & aged; AMS for ASTM alloy AM 100 A
AMS 4483	10% Al Mg alloy: Permanent mould castings; solution treated & aged; AMS for ASTM alloy AM 100 A
ASM 4484 E	9% Al 2% Zn Mg alloy: Permanent mould casting; solution treated & aged; AMS for ASTM alloy AZ 92 A
AMS 4490 D	9% Al 0.7% Zn Mg alloy: Die casting; as cast; AMS for ASTM alloy AZ 91
AN M 16	9% Al 1.0% Zn Mg alloy: Casting; US Service
AN QQ M 56/1A	9.0% Al 2.0% Zn Mg alloy: Casting; US Service; as cast
	UTS: 140 **Elon: 4%** **Proof: 35**
AN QQ M 56/1C	9.0% Al 2.0% Zn Mg alloy: Casting; US Service; as cast
	UTS: 120 **Proof: 35**
ASTM B80 AM 100 A	10% Al Mg alloy: Sand casting; solution treated & aged
	UTS: 240 **Proof: 100**
ASTM B80 AZ 63 A	6% Al 3% Zn Mg alloy: Sand casting; solution treated & aged
	UTS: 220 **Elon: 3%** **Proof: 90**
ASTM B80 AZ 81 A	7.5% Al 0.7% Zn Mg alloy: Sand casting; solution treated
	UTS: 220 **Elon: 7%** **Proof: 60**
ASTM B80 AZ 91 C	9% Al 0.6% Zn Mg alloy: Sand casting; solution treated & aged
	UTS: 220 **Elon: 3%** **Proof: 90**
ASTM B80 AZ 92 A	9% Al 2% Zn Mg alloy: Sand casting; solution treated & aged
	UTS: 220 **Elon: 1%** **Proof: 100**
ASTM B80 EK 30 A	3% rare earth 0.2% Zr Mg alloy: Sand casting; solution treated & aged
	UTS: 120 **Elon: 2%** **Proof: 75**
ASTM B80 EK 41 A	4% rare earth 0.7% Zr Mg alloy: Sand casting; solution treated & aged
	UTS: 140 **Elon: 1%** **Proof: 90**
ASTM B80 EZ 33 A	2.6% Zn 3.2% rare earth 0.7% Zr Mg alloy: Sand casting; aged
	UTS: 120 **Elon: 2%** **Proof: 75**
ASTM B80 HK 31 A	3.2% Th 0.6% Zr Mg alloy: Sand casting; solution treated & aged
	UTS: 180 **Elon: 4%** **Proof: 75**
ASTM B80 HZ 32 A	2% Zn 3.2% Th 0.7% Zr Mg alloy: Sand casting; aged
	UTS: 180 **Elon: 4%** **Proof: 75**
ASTM B80 K 1 A	0.6% Zr Mg alloy: Sand casting; as cast
	UTS: 150 **Elon: 14%**
ASTM B80 QE 22 A	2.2% rare earth 0.6% Zr Mg alloy: Sand casting; solution treated & aged
	UTS: 240 **Elon: 2%** **Proof: 170**
ASTM B80 ZE 41 A	4.2% Zn 1.2% rare earth 0.6% Zr Mg alloy: Sand casting; aged
	UTS: 180 **Elon: 2.5%** **Proof: 120**
ASTM B80 ZH 62 A	5.8% Zn 2% Th 0.7% Zr Mg alloy: Sand casting; aged
	UTS: 240 **Elon: 5%** **Proof: 150**

Symbol	Nominal analysis, supplier, condition and remarks.
ASTM B80 ZK 51 A	5% Zn 0.7% Zr Mg alloy: Sand casting; aged
	UTS: 220 **Elon: 5%** **Proof: 120**
ASTM B80 ZK 61 A	6% Zn 0.8% Zr Mg alloy: Sand casting; solution treated & aged
	UTS: 270 **Elon: 5%** **Proof: 170**
ASTM B90 AZ 31 B	3% Al 1% Zn Mg alloy: Sheet; annealed
	UTS: 200 **Elon: 10%**
ASTM B90 AZ 31 B	3% Al 1% Zn Mg alloy: Sheet; cold rolled
	UTS: 240 **Elon: 6%** **Proof: 140**
ASTM B90 HK 31 A	3% Al 1% Zn Mg alloy: Sheet
	3.2% Th 0.6% Zr Mg alloy: Sheet; annealed
ASTM B90 HK 31 A	**UTS: 200** **Elon: 12%** **Proof: 90**
	3.2% Th 0.6% Zr Mg alloy: Sheet; cold rolled
ASTM B90 HM 21 A	**UTS: 220** **Elon: 4%** **Proof: 170**
	0.8% Al 2% Th Mg alloy: Sheet; solution treated; cold worked & aged
	UTS: 200 **Elon: 6%** **Proof: 100**
ASTM B90 ZE 10 A	0.2% rare earth 1% Zn Mg alloy: Sheet; annealed
	UTS: 200 **Elon: 14%**
ASTM B90 ZE 10 A	0.2% rare earth 1% Zn Mg alloy: Sheet; cold rolled
	UTS: 220 **Elon: 4%** **Proof: 140**
ASTM B91 AZ 31 B	3% Al 1% Zn Mg alloy: Forging; as forged
	UTS: 220 **Elon: 6%** **Proof: 120**
ASTM B91 AZ 61 A	6.5% Al 1% Zn Mg alloy: Forging; as forged
	UTS: 250 **Elon: 6%** **Proof: 140**
ASTM B91 AZ 80 A	8.5% Al 0.6% Zn Mg alloy: Forging; aged
	UTS: 290 **Elon: 2%** **Proof: 180**
ASTM B91 TA 54 A	3.5% Al Mg alloy: Forging; as forged
	UTS: 280 **Elon: 7%** **Proof: 140**
ASTM B91 ZK 60 A	5.5% Zn Mg alloy: Forging; aged
	UTS: 290 **Elon: 7%** **Proof: 170**
ASTM B93 AM 100 A	10% Al Mg alloy: Ingot
ASTM B93 AZ 63 A	6% Al 3% Zn Mg alloy: Ingot
ASTM B93 AZ 81 A	7.6% Al 0.7% Zn Mg alloy: Ingot
ASTM B93 AZ 91 A	9% Al 0.6% Zn Mg alloy: Ingot
ASTM B93 AZ 91 B	9% Al 6% Zn Mg alloy: Ingot
ASTM B93 AZ 91 C	9% Al 0.6% Zn Mg alloy: Ingot
ASTM B93 AZ 92 A	9% Al 2% Zn Mg alloy: Ingot
ASTM B94 AZ 91 A	9% Al 0.6% Zn 0.1% Cu Mg alloy: Die casting; as cast
	DPN: 63 **UTS: 240** **Elon: 3%** **Proof: 170**
ASTM B94 AZ 91 B	9% Al 0.6% Zn 0.3% Cu Mg alloy: Die casting; as cast
	DPN: 63 **UTS: 240** **Elon: 3%** **Proof: 170**
ASTM B107 AZ 31 B	3% Al 1% Zn Mg alloy: Bar; as drawn
	UTS: 220 **Elon: 7%** **Proof: 150**
ASTM B107 AZ 31 C	3% Al 1% Zn Mg alloy: Bar; as drawn
	UTS: 220 **Elon: 7%** **Proof: 150**
ASTM B107 AZ 61 A	6.5% Al 1% Zn Mg alloy: Bar; as drawn
	UTS: 280 **Elon: 8%** **Proof: 150**
ASTM B107 AZ 80 A	8.5% Al 0.6% Zn Mg alloy: Bar; solution treated
	UTS: 330 **Elon: 4%** **Proof: 220**
ASTM B107 M 1 A	1.2% Mn Mg alloy: Bar; as drawn
	UTS: 210 **Elon: 2%**
ASTM B107 ZK 60 A	5.5% Zn 0.45% Zr Mg alloy: Bar; solution treated
	UTS: 320 **Elon: 4%** **Proof: 250**
ASTM B199 AM 100 A	10% Al Mg alloy: Casting; solution treated & aged
	UTS: 220 **Elon: 2%** **Proof: 100**
ASTM B199 AZ 91 C	9% Al 0.6% Zn Mg alloy: Casting; solution treated & aged
	UTS: 220 **Elon: 3%** **Proof: 90**
ASTM B199 AZ 92 A	9% Al 2% Zn Mg alloy: Casting; solution treated & aged
	UTS: 220 **Proof: 100**
ASTM B199 EK 41 A	4% rare earth 0.6% Zr Mg alloy: Casting; solution treated & aged
	UTS: 150 **Elon: 1%** **Proof: 90**

Symbol	Nominal analysis, supplier, condition and remarks.
ASTM B199 EZ 33 A	2.5% Zn 3% rare earth 0.7% Zr Mg alloy: Casting; aged **UTS: 140 Elon: 2% Proof: 90**
ASTM B217 AZ 31 B	3% Al 1% Zn Mg alloy: Tube; as drawn **UTS: 210 Elon: 8% Proof: 90**
ASTM B217 AZ 31 C	3% Al 1% Zn Mg alloy: Tube; as drawn **UTS: 210 Elon: 8% Proof: 90**
ASTM B217 AZ 61 A	6.5% Al 1% Zn Mg alloy: Tube; as drawn **UTS: 240 Elon: 7% Proof: 90**
ASTM B217 M 1 A	1.2% Mn Mg alloy: Tube; as drawn **UTS: 180 Elon: 2%**
ASTM B217 ZK 60 A	5% Zn 0.4% Zr Mg alloy: Tube; solution treated & aged **UTS: 320 Elon: 4% Proof: 250**
ASTM B260 B Mg 1	9% Al 2% Zn Mg alloy: Braze filler
ASTM B260 B Mg 2	12% Al 5% Zn Mg: Brazing filler metal; braze temperature range 582–610 °C
ASTM B260 B Mg 2a	12% Al 5% Zn 0.005% Be Mg: Brazing filler metal; braze temperature range 582–610 °C
ASTM B403 AM 100 A	10% Al 0.10% (min) Mn Mg alloy: Investment casting
ASTM B403 AZ 81 A	7.5% Al 0.13% (min) Mn 0.7% Zn Mg alloy: Investment casting
ASTM B430 AZ 91 C	8.8% Al 0.13% (min) Mn 0.7% Zn Mg alloy: Investment casting
ASTM B403 AZ 92 A	8.5% Al 0.10% (min) Mn 2.0% Zn Mg alloy: Investment casting
ASTM B403 EZ 33 A	2.5% Zn 0.75% Zr 3.5% rare earth Mg alloy: Investment casting
ASTM B403 HK 31 A	0.7% Zr 3.5% Th Mg alloy: Investment casting
ASTM B403 K 1 A	0.7% Zr Mg alloy: Investment casting
ASTM B403 QE 22 A	0.7% Zr 2.5% Ag 2.2% rare earth Mg alloy: Investment casting
ASTM B403 ZK 61 A	6.0% Zn 0.8% Zr Mg alloy: Investment casting
AZ 21 X	1.7% Al 0.6% Zn Mg alloy: Rod, bar & forging; Domal **UTS: 210 Elon: 5% Proof: 90**
AZ 31	3% Al 1% Zn 0.2% Mn Mg alloy: Wrought; previous German designation
AZ 31	3% Al 1% Zn 0.3% Mn Mg alloy: Sheet & tube; MEL and Birmetal for BS alloy 111
AZ 31 X	3% Al 1% Zn Mg alloy: Rod, tube & forging; Domal **UTS: 220 Elon: 8% Proof: 120**
AZ 61	6% Al 1% Zn 0.3% Mn Mg alloy: Wrought; previous German designation
AZ 61 X	6.2% Al 1.1% Zn Mg alloy: Rod, tube & forging; Domal **UTS: 250 Elon: 7% Proof: 120**
AZ 63	6% Al 3% Zn 0.2% Mn Mg alloy: Casting; previous German designation
AZ 80 X	9% Al 6% Zn Mg alloy: Casting; Domal; as cast **UTS: 150 Elon: 3%**
AZ 80 X	9% Al 0.6% Zn Mg alloy: Casting; Domal; solution treated, room aged **UTS: 220 Elon: 7%**

Symbol	Nominal analysis, supplier, condition and remarks.
AZ 80 X	9% Al 0.6% Zn Mg alloy: Bar & forging; Domal; as forged **UTS: 300 Elon: 6% Proof: 180**
AZ 80 X	9% Al 0.6% Zn Mg alloy: Bar & forging; Domal; aged **UTS: 330 Elon: 3% Proof: 220**
AZ 81	8% Al 0.8% Zn 0.2% Mn Mg alloy: Casting; previous German designation
AZ 91	9% Al 0.8% Zn 0.2% Mn Mg alloy: Casting; previous German designation
AZ 91	9.5% Al 0.5% Zn 0.3% Mn Mg alloy: Casting; MEL solution treated & aged; for pressure die casting **UTS: 220 Elon: 2% Proof: 100**
AZ 91 X	9.5% Al 0.4% Zn 0.3% Mn 0.0015% Be Mg alloy: Casting; MEL; as cast; for die casting **UTS: 210 Elon: 2% Proof: 100**
AZ 91 X	8.7% Al 0.8% Zn 0.3% Mn Mg alloy: Casting; Domal; as cast; die cast **UTS: 140**
AZ 91 X	8.7% Al 0.8% Zn 0.3% Mn Mg alloy: Casting; Domal; solution treated & aged; die cast **UTS: 220 Elon: 3% Proof: 90**
AZ 92	8.5% Al 1.5% Zn 0.2% Mn Mg alloy: Casting; previous German designation
AZ 855	8% Al 0.4% Zn 0.3% Mn Mg alloy: Forging; MEL; for highly stressed forgings **UTS: 300 Elon: 10% Proof: 180**
AZM	6% Al 1% Zn 0.3% Mn Mg alloy: Bar & forging; MEL & Birmetal for BS alloy 121
BIRMAG	Free machining magnesium alloy with properties similar to Birmetal alloy AZM
BS 1272/80	Mg alloy: Ingot & casting; replaced BS 2970
BS 1453 D1	10% Al 0.2% Mn Mg alloy: Weld filler rod for welding Mg alloys
BS 1453 D2	1.5% Mn Mg alloy: Weld filler rod for welding Mg alloys
BS 2901 D1	9.5% Al Mg alloy: Welding rod for inert gas shielded arc welding
BS 2901 D2	1.2% Mn Mg alloy: Rod for inert gas shielded arc welding
BS 2901 D3	7% Al 0.25% Mn Mg alloy: Rod for inert gas shielded arc welding
BS 2901 D4	3% Al 1% Zn 0.4% Mn Mg alloy: Rod for inert gas shielded arc welding
BS 2901 D5	1% Zn 0.8% Zr Mg alloy: Rod for inert gas shielded arc welding
BS 2901 D6	2.0% Zn 2.0% Cd 0.8% Zr Mg alloy: Rod for inert gas shielded arc welding
BS 2901 D7	4.0% Zn 1.2% rare earth 0.6% Zr Mg alloy: Rod for inert gas shielded arc welding
BS 2901 D8	3.2% rare earth 2% Zn 0.8% Zr Mg alloy: Rod for inert gas shielded arc welding
BS 2970 MAG 1 M	8% Al 0.7% Zn 0.3% Mn Mg alloy: Casting; chill cast; as cast **UTS: 180 Elon: 4% Proof: 75**
BS 2970 MAG 1 W	8% Al 0.7% Zn 0.3% Mn Mg alloy: Casting; chill cast; solution treated **UTS: 220 Elon: 10% Proof: 70**
BS 2970 MAG 2 M	8% Al 0.7% Zn 0.4% Mn Mg alloy: Casting; chill cast; high purity **UTS: 180 Elon: 4% Proof: 75**
BS 2970 MAG 2 W	8% Al 0.7% Zn 0.4% Mn Mg alloy: Casting; chill cast; solution treated; high purity **UTS: 220 Elon: 10% Proof: 70**
BS 2970 MAG 3 M	10% Al 0.7% Zn 0.3% Mn Mg alloy: Casting; chill cast; as cast **UTS: 170 Elon: 2% Proof: 90**

Note. The following abbreviations and units are used in the tables:

DPN	Hardness, diamond pyramid number
UTS	Ultimate tensile strength, N/mm^2
Elon	Elongation, %
Proof	0.1 % proof strength, N/mm^2

$1\ N/mm^2 = 0.1\ hbar = 0.102\ kgf/mm^2 = 0.06475\ tonf/in^2 = 145.04\ lbf/in^2$

Symbol	Nominal analysis, supplier, condition and remarks.
BS 2970 MAG 3 W	10% Al 0.7% Zn 0.3% Mn Mg alloy: Casting; chill cast; solution treated **UTS: 210 Elon: 5% Proof: 75**
BS 2970 MAG 3 WP	10% Al 0.7% Zn 0.3% Mn Mg alloy: Casting; chill cast; solution treated & aged **UTS: 210 Elon: 2% Proof: 110**
BS 2970 MAG 4 M	4.5% Zn 0.7% Zr Mg alloy: Casting; chill cast; as cast **UTS: 220 Elon: 10% Proof: 100**
BS 2970 MAG 4 P	4.5% Zn 0.7% Zr Mg alloy: Casting; chill cast; aged **UTS: 240 Elon: 7% Proof: 130**
BS 2970 MAG 5 M	4.5% Zn 0.7% Zr 1.2% rare earth Mg alloy: Casting; chill cast; as cast **UTS: 180 Elon: 4% Proof: 100**
BS 2970 MAG 5 P	4.5% Zn 0.7% Zr 1.2% rare earth Mg alloy: Casting; chill cast; aged **UTS: 210 Elon: 4% Proof: 120**
BS 2970 MAG 6	2.3% Zn 0.6% Zr 3% rare earth Mg alloy: Casting; chill cast; as cast or aged **UTS: 150 Elon: 3% Proof: 90**
BS 2970 MAG 7 M	8% Al 0.9% Zn 0.3% Mn Mg alloy: Casting; chill chill cast; as cast **UTS: 170 Elon: 2% Proof: 75**
BS 2970 MAG 7 W	8% Al 0.9% Zn 0.3% Mn Mg alloy: Casting; chill cast; solution treated **UTS: 210 Elon: 5% Proof: 70**
BS 2970 MAG 7 WP	8% Al 0.9% Zn 0.3% Mn Mg alloy: Casting; chill cast; solution treated & aged **UTS: 210 Elon: 2% Proof: 90**
BS 3370 MAG S101	1.5% Mn Mg alloy; Strip; as rolled **UTS: 200 Elon: 5% Proof: 60**
BS 3370 MAG S111	3% Al 1% Zn 0.5% Mn Mg alloy: Strip; annealed **UTS: 220 Elon: 12% Proof: 100**
BS 3370 MAG S141	1% Zn 0.5% Zr Mg alloy: Strip; as rolled **UTS: 240 Elon: 8% Proof: 150**
BS 3370 MAG S151	3% Zn 0.5% Zr Mg alloy: Strip; as rolled **UTS: 250 Elon: 8% Proof: 170**
BS 3371 MAG T101	1.5% Mn Mg alloy: Tube; as drawn **UTS: 220 Elon: 5% Proof: 100**
BS 3371 MAG T111	3% Al 1% Zn 0.5% Mn Mg alloy: Tube; as drawn **UTS: 220 Elon: 10% Proof: 140**
BS 3371 MAG T121	6% Al 1% Zn 0.3% Mn Mg alloy: Tube; as drawn **UTS: 250 Elon: 7% Proof: 150**
BS 3371 MAG T141	1% Zn 0.5% Zr Mg alloy: Tube; as drawn **UTS: 240 Elon: 4% Proof: 150**
BS 3372 MAG F101	1.5% Mn Mg alloy: Forgings; as forged **UTS: 200 Elon: 5% Proof: 90**
BS 3372 MAG F121	6% Al 1.2% Zn 0.2% Mn Mg alloy: Forging; as forged **UTS: 280 Elon: 8% Proof: 150**
BS 3372 MAG F151	3.25% Zn 0.6% Zr Mg alloy: Forging; as forged **UTS: 250 Elon: 8% Proof: 170**
BS 3373 MAG E101	1.5% Mn Mg alloy: Bar; as extruded **UTS: 210 Elon: 3% Proof: 100**
BS 3373 MAG E111	3% Al 1% Zn 0.3% Mn Mg alloy: Bar; as extruded **UTS: 220 Elon: 10% Proof: 140**
BS 3373 MAG E121	6% Al 1% Zn 0.25% Mn Mg alloy: Bar; as extruded **UTS: 240 Elon: 9% Proof: 150**
BS 3373 MAG E141	1% Zn 0.6% Zr Mg alloy: Bar; as extruded **UTS: 240 Elon: 10% Proof: 150**
BS 3373 MAG E151	3% Zn 0.6% Zr Mg alloy: Bar; as extruded **UTS: 290 Elon: 10% Proof: 200**
BS 3373 MAG E161	5% Zn 0.6% Zr Mg alloy: Bar; as extruded, straightened & heat treated **UTS: 300 Elon: 8% Proof: 210**
BS 3374 MAG P101	1.5% Mn Mg alloy: Plate; as rolled & flattened **UTS: 180 Elon: 5% Proof: 60**

Symbol	Nominal analysis, supplier, condition and remarks.
BS 3374 MAG P111	3% Al 1% Zn 0.3% Mn Mg alloy: Plate; as rolled & flattened **UTS: 210 Elon: 10% Proof: 100**
BS 3374 MAG P141	1% Zn 0.6% Zr Mg alloy: Plate; as rolled & flattened **UTS: 210 Elon: 8% Proof: 100**
BS 3374 MAG P151	3% Zn 0.6% Zr Mg alloy: Plate; as rolled & flattened **UTS: 220 Elon: 8% Proof: 120**
BS L121	8% Al 0.7% Zn 0.3% Mn Mg alloy: Casting; chill cast; as cast **UTS: 180 Elon: 4% Proof: 7.5**
BS L122	8% Al 0.7% Zn 0.3% Mn Mg alloy: Casting; chill cast; solution treated **UTS: 220 Elon: 10% Proof: 60**
BS L123	10% Al 0.7% Zn 0.3% Mn Mg alloy: Casting; chill cast; as cast **UTS: 170 Elon: 2% Proof: 90**
BS L124	10% Al 0.7% Zn 0.3% Mn Mg alloy: Casting; chill cast; solution treated **UTS: 210 Elon: 5% Proof: 75**
BS L125	10% Al 0.7% Zn 0.3% Mn Mg alloy: Casting; chill cast; solution treated & aged **UTS: 210 Elon: 2% Proof: 100**
C	8% Al 1% Zn 0.15% Mn Mg alloy: Casting; MEL; solution treated & aged **UTS: 210 Elon: 2% Proof: 90**
DIN 1729G Mg Al 6 Zn 3	6% Al 3% Zn 0.2% Mn Mg alloy: Casting **DPN: 60 UTS: 150 Elon: 5% Proof: 90**
DIN 1729 Mg Al 3 Zn	3% Al 1% Zn 0.2% Mn Mg alloy: Wrought **DPN: 45 UTS: 240 Elon: 10% Proof: 150**
DIN 1729 Mg Al 6 Zn	6% Al 1% Zn 0.2% Mn Mg alloy: Wrought **DPN: 55 UTS: 240 Elon: 6% Proof: 170**
DIN 1729 Mg Al 8 Zn	8.5% Al 0.6% Zn 0.2% Mn Mg alloy: Wrought **DPN: 60 UTS: 250 Elon: 6% Proof: 180**
DIN 1729 Mg Mn 2	1.8% Mn Mg alloy: Wrought **DPN: 40 UTS: 180 Elon: 1.5% Proof: 140**
DIN 1729 Mg Zn 6 Zr	5.5% Zn 0.5% Zr Mg alloy: Wrought **DPN: 60 UTS: 250 Elon: 5% Proof: 180**
DIN 1729G Mg Al 8 Zn 1	8% Al 0.8% Zn 0.2% Mn Mg alloy: Casting **DPN: 60 UTS: 210 Elon: 5% Proof: 120**
DIN 1729G Mg Al 9 Zn 1	9% Al 0.8% Zn 0.2% Mn Mg alloy: Casting **DPN: 65 UTS: 210 Elon: 3% Proof: 140**
DIN 1729G Mg Al 9 Zn 2	8.5% Al 1.2% Zn 0.2% Mn Mg alloy: Casting **DPN: 65 UTS: 200 Elon: 2% Proof: 120**
DOMAL AZ 21X	1.7% Al 0.6% Zn Mg alloy: Rod, bar & forging; Domal **UTS: 210 Elon: 5% Proof: 90**
DOMAL AZ 31X	3% Al 1% Zn Mg alloy: Rod, tube & forging; Domal **UTS: 220 Elon: 8% Proof: 120**
DOMAL AZ 61X	6.2% Al 1.1% Zn Mg alloy: Rod, tube & forging; Domal **UTS: 250 Elon: 7% Proof: 120**
DOMAL AZ 80X	9% Al 0.6% Zn Mg alloy: Bar & forging; Domal; as forged **UTS: 300 Elon: 6% Proof: 180**
DOMAL AZ 80X	9% Al 0.6% Zn Mg alloy: Bar & forging; Domal; aged **UTS: 330 Elon: 3% Proof: 220**
DOMAL K 1	0.8% Zr Mg alloy: Casting; Domal; as cast; high damping capacity **UTS: 140 Elon: 15% Proof: 450**
DTD 88 C	6% Al 1% Zn 0.2% Mn Mg alloy: Forgings **UTS: 270 Elon: 8% Proof: 150**
DTD 118 B	1.5% Mn Mg alloy: Sheet & strip; annealed **UTS: 200 Elon: 5% Proof: 60**
DTD 140 C	1.5% Mn Mg alloy: Casting; as cast **UTS: 90 Elon: 3%**

Symbol	Nominal analysis, supplier, condition and remarks.
DTD 142 A	1.5% Mn Mg alloy: Bar; as rolled **UTS: 220** **Elon: 4%** **Proof: 120**
DTD 259 A	7% Al 1.5% Zn 0.3% Mn Mg alloy: Bar **UTS: 250** **Elon: 10%** **Proof: 170**
DTD 348 A	6% Al 1% Zn 0.3% Mn Mg alloy: Tube **UTS: 250** **Elon: 8%**
DTD 619	3% Zn 0.7% Zr Mg alloy: Forging; as forged **UTS: 290** **Elon: 8%** **Proof: 200**
DTD 622 A	3% Zn 0.7% Zr Mg alloy: Bar; as rolled & stress released **UTS: 290** **Elon: 10%** **Proof: 200**
DTD 626 B	3% Zn 0.5% Zr Mg alloy: Sheet; as rolled **UTS: 250** **Elon: 8%** **Proof: 170**
DTD 684	8% Al 0.8% Zn 0.4% Mn Mg alloy: Casting; annealed for corrosion resistance purposes **UTS: 180** **Elon: 4%** **Proof: 75**
DTD 690	8% Al 0.7% Zn 0.4% Mn Mg alloy: Casting; solution treated; chill cast **UTS: 220** **Elon: 10%** **Proof: 70**
DTD 708	2.5% Zn 3% rare earth 1% Zr Mg alloy: asting; chill cast; aged **UTS: 150** **Elon: 3%** **Proof: 100**
DTD 711 A	4.5% Zn 0.7% Zr Mg alloy: Casting; chill cast **UTS: 220** **Elon: 7%**
DTD 718	0.5% Zn 3% rare earth 0.7% Zr Mg alloy: Casting; chill cast & aged **UTS: 150** **Elon: 3%** **Proof: 100**
DTD 721 A	4% Zn 0.7% Zr Mg alloy: Casting; chill cast; stress released **UTS: 240** **Elon: 7%** **Proof: 200**
DTD 728	0.7% Zr 3% rare earth Mg alloy: Casting; chill cast & aged **UTS: 150** **Elon: 3%** **Proof: 100**
DTD 729	3% Zn 0.7% Zr Mg alloy: Forging **UTS: 250** **Elon: 8%** **Proof: 170**
DTD 732 A	3% Al 1% Zn 0.3% Mn Mg alloy: Sheet; annealed **UTS: 220** **Elon: 12%** **Proof: 100**
DTD 737	1.5% Mn Mg alloy: Tube **UTS: 220** **Elon: 4%**
DTD 738	4.2% Zn 1.5% rare earth 0.6% Zr Mg alloy: Casting; chill cast **UTS: 180** **Elon: 4%** **Proof: 100**
DTD 742 A	3% Al 1% Zn 0.5% Mn Mg alloy: Sheet; cold rolled; ½-hard **UTS: 240** **Elon: 8%** **Proof: 150**
DTD 748	4.2% Zn 0.6% Zr 1.5% rare earth Mg alloy: Casting; chill cast & stress released **UTS: 210** **Elon: 4%** **Proof: 120**
DTD 749	7% Al 1.5% Zn 0.3% Mn Mg alloy: Bar **UTS: 210** **Elon: 8%** **Proof: 140**
DTD 5001 A	1.2% Zn 0.6% Zr Mg alloy: Sheet; as rolled **UTS: 240** **Elon: 8%** **Proof: 150**
DTD 5005	2.2% Zn 3% Th 0.7% Zr Mg alloy: Casting; chill cast & stress released **UTS: 180** **Elon: 5%** **Proof: 75**
DTD 5011	1.3% Zn 0.7% Zr Mg alloy: Bar; stress released **UTS: 250** **Elon: 10%** **Proof: 170**
DTD 5015	5.5% Zn 1.8% Th 0.7% Zr Mg alloy: Casting; chill cast; stress released **UTS: 250** **Elon: 5%** **Proof: 140**

Symbol	Nominal analysis, supplier, condition and remarks.
DTD 5021	1.3% Zn 0.7% Zr Mg alloy: Tube; stress released **UTS: 240** **Elon: 5%** **Proof: 170**
DTD 5025	2.5% Ag 1.6% rare earth 0.6% Zr Mg alloy: Casting
DTD 5031	5.5% Zn 0.7% Zr Mg alloy: Bar; stress released **UTS: 290** **Elon: 10%** **Proof: 200**
DTD 5035	2.5% Ag 2.5% rare earth 0.6% Zr Mg alloy: Casting; chill cast; solution treated & aged **UTS: 240** **Elon: 2%** **Proof: 170**
DTD 5041	5.5% Zn 0.7% Zr Mg alloy: Bar; stress released **UTS: 300** **Elon: 8%** **Proof: 210**
DTD 5051	1.5% Mn Mg alloy: Plate; annealed **UTS: 200** **Elon: 5%** **Proof: 60**
DTD 5061	3% Al 1% Zn 0.3% Mn Mg alloy: Plate; as rolled **UTS: 220** **Elon: 10%** **Proof: 120**
DTD 5071	1.3% Zn 0.6% Zr Mg alloy: Plate; as rolled; weldable **UTS: 220** **Elon: 12%** **Proof: 120**
DTD 5081	3% Zn 0.6% Zr Mg alloy: Plate; as rolled **UTS: 250** **Elon: 10%** **Proof: 170**
DTD 5091	2.0% Mg 1.0% Mn Mg alloy: Sheet & strip; annealed
DTD 5101	0.2% Zn 1.0% Mn Mg alloy: Sheet & strip; ½-hard
ELEKTRON MCZ	0.7% Zr 3% rare earth Mg alloy; Origin unknown
ELEKTRON Z5Z	0.7% Zr 4.5% Zn Mg alloy: Origin unknown
ELEKTRON ZZ	0.7% Zr 3.0% Zn Mg alloy: Origin unknown
G Mg Al 6 Zn 3	6% Al 3% Zn 0.2% Mn Mg alloy: Casting; designation used by German Standard
GD Mg Al Zn 1	8% Al 0.8% Zn 0.2% Mn Mg alloy: Casting; designation used by German Standard
GD Mg Al 9 Zn 1	9% Al 0.8% Zn 0.2% Mn Mg alloy: Casting; designation used by German Standard
GD Mg Al 9 Zn 2	8.5% Al 1.5% Zn 0.2% Mn Mg alloy: Casting; designation used by German Standard
GK Mg Al 8 Zn 1	8% Al 0.8% Zn 0.2% Mn Mg alloy: Casting; designation used by German Standard
GK Mg Al 9 Zn 1	9% Al 0.8% Zn 0.2% Mn Mg alloy: Casting; designation used by German Standard
GK Mg Al 9 Zn 2	8.5% Al 1.5% Zn 0.2% Mn Mg alloy: Casting; designation used by German Standard
HK 31	3% Th 0.7% Zr Mg alloy: Casting; Domal; solution treated & aged **UTS: 18** **Elon: 4%** **Proof: 75**
HZ 32	2% Zn 3.2% Th 0.7% Zr Mg alloy: Casting; Domal; aged **UTS: 180** **Elon: 4%** **Proof: 75**
K 1	0.8% Zr Mg alloy: Casting; Domal; as cast; high damping capacity **UTS: 140** **Elon: 15%** **Proof: 45**
M 2	1.8% Mn Mg alloy: Wrought; previous German designation
MAGNOX A 12	1% Al 0.01% Be Mg alloy: MEL; for nuclear purposes
MAGNUMINIUM 133	1.5% Mn Mg alloy: Tube, forgings & extrusions; High Duty Alloys for BS alloy MAG 101 **UTS: 210** **Elon: 4%** **Proof: 120**
MAGNUMINIUM 133 X	0.65% Mn Mg alloy: Bar & forging; High Duty Alloys; as drawn or forged **UTS: 240** **Elon: 5%** **Proof: 120**
MAGNUMINIUM 266	6% Al 1% Zn 0.3% Mn Mg alloy: Bar & forging; High Duty Alloys for BS alloy MAG 121 **UTS: 270** **Elon: 10%** **Proof: 140**
MNC 46E	Swedish Standard; summary of magnesium and magnesium alloy ingots and castings
MSR A	0.6% Zr 2.5% Ag 1.7% rare earth Mg alloy: Casting; MEL; solution treated & aged **UTS: 220** **Elon: 4%** **Proof: 150**

Note. The following abbreviations and units are used in the tables:

DPN	Hardness, diamond pyramid number
UTS	Ultimate tensile strength, N/mm^2
Elon	Elongation, %
Proof	0.1 % proof strength, N/mm^2

1 N/mm^2=0.1 hbar=0.102 kgf/mm^2=0.06475 tonf/in^2=145.04 lbf/in^2

Symbol	Nominal analysis, supplier, condition and remarks.
MSR B	0.6% Zr 2.5% Ag 2.5% rare earth Mg alloy: Casting; MEL; solution treated & aged; weldable; used up to 250 °C **UTS: 220 Elon: 2% Proof: 170**
MTZ	0.7% Zr 3% Th Mg alloy: Casting; MEL; solution treated & aged; weldable; used up to 350 °C **UTS: 200 Elon: 5% Proof: 75**
RZ 5	4% Zn 0.7% Zr 1.2% rare earth Mg alloy: Casting; MEL; weldable **UTS: 200 Elon: 4% Proof: 120**
SAE 50	6% Al 3% Zn Mg alloy: Casting; ASTM alloy AZ 63 A; solution treated & aged **UTS: 220 Elon: 3% Proof: 90**
SAE 51	1.2% Mn Mg alloy: Sheet; annealed; ASTM alloy MIA **UTS: 250 Elon: 12%**
SAE 52	3% Al 1% Zn Mg alloy: Extrusion; ASTM alloy AZ 31B; as extruded **UTS: 210 Elon: 7% Proof: 120**
SAE 53	3.5% Al 5% Sn Mg alloy: As forged; ASTM alloy TA 54A **UTS: 240 Elon: 7% Proof: 140**
SAE 500	9% Al 2% Zn Mg alloy: Casting; ASTM alloy AZ 92A; solution treated & aged **UTS: 220 Elon: 1% Proof: 100**
SAE 501	9% Al 0.7% Zn Mg alloy: Die casting; low Cu content; as cast; ASTM alloy AZ 91A **UTS: 220 Elon: 3% Proof: 140**
SAE 501A	9% Al 0.7% Zn 0.3% Cu Mg alloy: Die casting; as cast; ASTM alloy AZ 91B **UTS: 220 Elon: 3% Proof: 140**
SAE 502	10% Al Mg alloy: Casting; solution treated & aged; ASTM alloy AM 100A **UTS: 220 Proof: 100**
SAE 503	9% Al 2% Zn Mg alloy: Casting; solution treated & aged; ASTM alloy AZ 92A **UTS: 220 Proof: 100**
SAE 504	8.7% Al 0.8% Zn Mg alloy: Sand casting; solution treated & aged; ASTM alloy AZ 91C **UTS: 220 Elon: 3% Proof: 90**
SAE 505	7.5% Al 0.7% Zn Mg alloy: Casting; solution treated **UTS: 220 Elon: 7% Proof: 45**
SAE 506	2.5% Zn 3% rare earth 0.7% Zr Mg alloy: Casting; aged **UTS: 120 Elon: 2% Proof: 75**
SAE 507	3.2% Th 0.7% Zr Mg alloy: Casting; solution treated & aged **UTS: 180 Elon: 4% Proof: 75**
SAE 507	3.2% Th 0.7% Zr Mg alloy: Wrought; annealed **UTS: 200 Elon: 12% Proof: 90**
SAE 508	5.7% Zn 2% Th 0.7% Zr Mg alloy: Casting; aged **UTS: 220 Elon: 4% Proof: 140**
SAE 509	4.5% Zn 0.7% Zr Mg alloy: Casting; aged **UTS: 220 Elon: 5% Proof: 120**
SAE 510	3% Al 1% Zn Mg alloy: Sheet; ASTM alloy AZ 31A; annealed **UTS: 270 Elon: 12%**
SAE 510	3% Al 1% Zn Mg alloy: Sheet; cold rolled **UTS: 270 Elon: 4% Proof: 200**
SAE 513	6% Zn 0.8% Zr Mg alloy: Casting; solution treated & aged **UTS: 290 Elon: 5% Proof: 170**
SAE 520	7% Al 1% Zn Mg alloy: Extrusion; ASTM alloy AZ 61A; as extruded **UTS: 250 Elon: 8% Proof: 170**
SAE 522	1.2% Mn Mg alloy: Extrusion; ASTM alloy M1A; as extruded **UTS: 200 Elon: 2%**
SAE 523	9% Al 0.6% Zn Mg alloy: Extrusion; ASTM alloy AZ 80A; aged **UTS: 320 Elon: 4% Proof: 200**
SAE 524	5% Zn Mg alloy: Extrusion; ASTM alloy ZK 60A; as extruded **UTS: 320 Elon: 4% Proof: 250**
SAE 531	6.5% Al 1% Zn Mg alloy: As forged; ASTM alloy AZ 61A **UTS: 250 Elon: 6% Proof: 140**
SAE 532	8.5% Al 0.6% Zn Mg alloy: As forged; ASTM alloy AZ 80A **UTS: 290 Elon: 5% Proof: 170**
SAE 533	1.2% Mn Mg alloy: Forging; as forged; ASTM alloy M1A **UTS: 200 Elon: 3% Proof: 90**
SAE 534	1.2% Zn 0.19% rare earth Mg alloy: Wrought; annealed **UTS: 210 Elon: 5%**
SAE 534	1.2% Zn 0.19% rare earth Mg alloy: Wrought; ½-hard **UTS: 220 Elon: 4%**
SIS 4635	9% Al 0.4% Mn 0.5% Zn 0.001% Be Mg alloy: Casting; Swedish Standard; as cast
SIS 4637	8% Al 0.4% Mn 0.5% Zn Mg alloy: Casting; Swedish Standard
SIS 4640	8% Al 0.2% Mn 1.2% Zn Mg alloy: Casting; Swedish Standard
TZ 6	5.5% Zn 0.7% Zr 1.3% Th Mg alloy: Casting; MEL; aged; weldable **UTS: 270 Elon: 8% Proof: 150**
Z 52	4.5% Zn 0.7% Zr Mg alloy: Casting; MEL; aged **UTS: 240 Elon: 7% Proof: 140**
ZA	0.55% Zr Mg alloy: MEL for nuclear engineering
ZH 62	5.8% Zn 1.8% Th 0.7% Zr Mg alloy: Casting; Domal; aged **UTS: 240 Elon: 4% Proof: 150**
ZK 60	5.5% Zn 0.5% Zr Mg alloy: Bar & forging; Domal; aged **UTS: 320 Elon: 4% Proof: 240**
ZK 60	5.5% Zn 0.5% Zr Mg alloy: Casting; Domal; solution treated & aged **UTS: 270 Elon: 5% Proof: 170**
ZK 60	5.5% Zn 0.5% Zr Mg alloy: Previous German designation
ZK 61	5.5% Zn 0.5% Zr Mg alloy: Bar & forging; Domal; aged **UTS: 320 Elon: 4% Proof: 240**
ZK 61	5.5% Zn 0.5% Zr Mg alloy: Casting; Domal; solution treated & aged **UTS: 270 Elon: 5% Proof: 170**
ZRE 1	2.2% Zn 0.6% Zr 2.7% rare earth Mg alloy: Casting; MEL; annealed; weldable; creep resistant to 250 °C **UTS: 150 Elon: 4% Proof: 75**
ZT 1	2.2% Zn 0.7% Zr 3% Th Mg alloy: Casting; MEL; aged; weldable; used up to 350 °C **UTS: 200 Elon: 7% Proof: 70**
ZTY	0.5% Zn 0.6% Zr 0.75% Th Mg alloy: Sheet & forgings; MEL; as wrought; weldable **UTS: 200 Elon: 9% Proof: 100**
ZW 1	1.2% Zn 0.6% Zr Mg alloy: Sheet, bar & tube; Birmabright for BS alloy 141
ZW 1	1.3% Zn 0.6% Zr Mg alloy: Sheet, plate & tube; MEL; as wrought; weldable **UTS: 220 Elon: 8% Proof: 140**
ZW 3	3% Zn 0.75% Zr Mg alloy: Bar & forging; High Duty Alloys for BS alloy MAG 151
ZW 3	3% Zn 0.6% Zr Mg alloy: Sheet, bar & forging; Birmabright for BS alloy 151

Symbol	Nominal analysis, supplier, condition and remarks.
ZW 3	3% Zn 0.6% Zr Mg alloy: Sheet, plate & forging; MEL; as wrought; weldable UTS: 270 Elon: 8% Proof: 180
ZW 6	5.5% Zn 0.6% Zr Mg alloy: Bar; Birmabright for BS alloy 161
ZW 6	5.5% Zn 0.6% Zr Mg alloy: Extrusion; MEL; aged; not weldable UTS: 300 Elon: 8% Proof: 210

Note. The following abbreviations and units are used in the tables:

DPN	Hardness, diamond pyramid number
UTS	Ultimate tensile strength, N/mm^2
Elon	Elongation, %
Proof	0.1 % proof strength, N/mm^2

$1 N/mm^2 = 0.1$ hbar $= 0.102$ kgf/mm$^2 = 0.06475$ tonf/in$^2 = 145.04$ lbf/in^2

24. Manganese Mn

Physical properties

Atomic number	25	
Atomic weight	54.93	
Crystal structure	Cubic – complex – other structures also known	
Colour	White-grey	
Specific gravity	7.44	
Density	7440 kg/m³	(0.268 lb/in.³)
Melting point	1244 °C	
Boiling point	2150 °C	
Specific heat	0.448 J/g °C	(0.107 cal/g °C)
Thermal conductivity	– W/m °C	
Coefficient of linear expansion (20–100° C)	22.8×10^{-6}/ °C	
Latent heat of fusion	271 J/g	(64.8 cal/g)
Latent heat of vaporization	4091 J/g	(977 cal/g)
Thermal neutron absorption cross-section	12.6 barns/atom	
Electrical conductivity	5.8% IACS (copper 100%)	
Specific resistance	280 microhm mm	
Temperature coefficient of electrical resistance	0.0039/ °C	
Electrochemical equivalent	1.025 g/A/h – divalent	
	0.684 g/A/h – trivalent	
Electrode potential	-1.05 V	
Magnetic susceptibility	11.8×10^{-6}	
Young's modulus of elasticity	159×10^{9} N/m²	(23×10^{6} lbf/in.²)
Tensile strength	480 N/mm²	(32 tonf/in.²)
Hardness	500 DPN	

24.1 General notes on manganese

Native manganese does not exist, but several oxides and hydrated oxides are found in various parts of the world, particularly in the USSR.

High purity manganese metal is hard and brittle, and under certain circumstances has a yellowish tinge. The freshly fractured or cut surfaces tarnish slowly when left in air.

Manganese has no known use, nor does it form the base of many useful alloys. Two which have been produced are a braze metal with cobalt and nickel which has been used to braze stainless steel under certain circumstances and an alloy with copper, nickel and aluminium produced because of its high damping characteristics.

Neither of these alloys is in common use and very little information is available.

Manganese however, is widely used as an alloying element with a considerable number of uses. The most common is a de-oxidizer and de-sulphurizer in the manufacture of steel. Iron oxide and sulphide are brittle refractory materials which tear the plastic steel during rolling or forging. With manganese present these are converted to manganese oxide and sulphide, both of which are plastic at the working temperature of steel. Normally ferro-manganese or spiegal is added to the ladle when tapping the steel furnace, but it has been reported that better physical properties are obtained in steel if high purity manganese is used.

Up to approximately 1% manganese can be present in steel without affecting its properties, except as described above. With austenitic steels there is generally 3–5% manganese present. Additional manganese acts first as a pearlite former in carbon steels and with small amounts of molybdenum present gives air hardening steels.

When above about 12% manganese is present the steel becomes austenitic, but is unstable and readily converted to martensite by cold working.

Manganese also de-sulphurizes copper alloys and when added to brasses imparts a pleasant colour and improves the corrosion resistance with some increase in strength. Higher manganese copper alloys can have a very low coefficient of expansion.

Additions of manganese to nickel gives alloys with special temperature coefficient of resistance.

Aluminium with approximately 2% manganese has a corrosion resistance almost equal to the pure metal, with an increase in strength which can be considerably increased by cold work. Manganese also adds to the strength of the more complex age hardening alloys.

Magnesium has its corrosion resistance improved with additions of small quantities of manganese.

Manganese oxide dust can be a health hazard under certain conditions. There is no evidence that any of the alloys with manganese has ever been dangerous, even when being ground or emery dressed.

Manganese is one of the more useful metallurgical elements, which makes itself felt over a wide range of alloys without forming any useful alloys in its own right. Following are listed the specifications and trade names of manganese and its alloys.

Symbol	Nominal analysis, supplier, condition and remarks.
AMS 4780	16% Ni 16% Co 0.8% B Mn alloy: Braze metal
ASTM A601 A	99.9% Mn metal: Regular grade; electrolytic
ASTM A601 B	99.9% Mn metal: Intermediate hydrogen; electrolytic
ASTM A601 C	99.9% Mn metal: Low hydrogen; electrolytic
ASTM A601 D	94.5% Mn metal 4.5% N: Electrolytic
ASTM A601 E	93.5% Mn metal 6.0% N: Electrolytic
ASTM A601 F	99.9% Mn metal: Powder
ASTM A701	0.08% C (max) 30.0% Si 0.05% P (max) 5% Fe Mn: Ferro-manganese silicon
FERRO-MANGANE SE	0.1% C 3% Si 0.4% P 20% Fe Mn: Metal Alloys; primary metal
FERRO-MANGANE SE	0.08% C 2% Si 0.08% P 20% Fe Mn low P: Metal Alloys; primary metal
h Mn 11	99.99% Mn flake: Light Ltd; high purity metal
MANGANESE	Mn metal: Carbon free; Blackwell; primary metal
MANGANESE 94-95	0.1% C 2.8% Fe 1% Si 1% Al 0.1% Cu Mn metal: Metal Alloys; primary metal
MANGANESE 96-97	0.1% C 1.2% Fe 0.8% Si 0.8% Al 0.1% Cu Mn metal: Metal Alloys; primary metal

Symbol	Nominal analysis, supplier, condition and remarks.
SONOSTON	4% Al 2.5% Ni 30% Cu Mn alloy: Casting; Stone Manganese Marine; high damping capacity UTS: 530 Elon: 20% Proof: 220
UNS R03001–R03999	Mn metal and alloys: American Standard system; unified numbers

Note. The following abbreviations and units are used in the tables:

DPN	Hardness, diamond pyramid number
UTS	Ultimate tensile strength, N/mm^2
Elon	Elongation, %
Proof	0.1 % proof strength, N/mm^2

$1 \ N/mm^2 = 0.1 \ hbar = 0.102 \ kgf/mm^2 = 0.06475 \ tonf/in^2 = 145.04 \ lbf/in^2$

25. Mercury Hg

Physical properties

Atomic number	80	
Atomic weight	200.61	
Crystal structure	Rhombohedral	
Colour	White	
Specific gravity	13.55	
Density	13550 kg/m³	(0.485 lb/in.³)
Melting point	-38.87 °C	
Boiling point	356.58 C	
Specific heat	0.1386 J/g °C	(0.0331 cal/g °C)
Thermal conductivity	8.8 W/m °C	(2.1 cal/m s °C)
Coefficient of linear expansion (20–100° C)	$61 \times 10^{-6}/$ °C	
Latent heat of fusion	11.56 J/g	(2.76 cal/g)
Latent heat of vaporization	292 J/g	(69.7 cal/g)
Thermal neutron absorption cross-section	380 barns/atom	
Electrical conductivity	1.9% IACS (copper 100%)	
Specific resistance	940 microhm mm	
Temperature coefficient of electrical resistance	0.0010/ °C	
Electrochemical equivalent	7.45 g/A/h – monovalent	
	3.73 g/A/h – divalent	
Electrode potential	+ 0.789 V	
Magnetic susceptibility	-0.168×10^{-6}	
Young's modulus of elasticity	This is not applicable to mercury	

25.1 General notes on mercury

A limited amount of native mercury occurs trapped in the ore, which are sulphides called cinnabar, found in areas of extinct volcanoes. The most famous deposit is in Spain where mercury has been mined and refined since 400 BC. Deposits are found all down the western spine of the American continent, but none are as rich or extensive as in the Spanish mines.

It is the only metal remaining liquid at normal temperatures, solidifying at -39 °C when it is ductile, malleable and soft enough to cut with a knife.

The pure metal is used in laboratory type apparatus such as thermometers, and barometers, and has some industrial use as a gas sealing medium. It has always had a certain fascination for its 'quicksilver' qualities. Note should be taken that the metal itself and some of its compounds can be a serious health hazard.

Mercury can be used as a solvent for many metals including all the precious metal elements, nickel, cobalt and iron being the only common metals which are not dis-

solved. At one time use was made of this as a standard method of purification for gold, but it has been almost completely replaced by the cyanide process.

Tin amalgam has been used for silvering mirrors, and amalgams of gold, copper and zinc are still used in dentistry.

Zinc amalgam is used in dry batteries. The action only takes place when a circuit is completed, thus they find use in devices which operate at infrequent intervals where normal batteries would require changing to prevent corrosion.

Mercury is also used in vacuum rectifier equipment, and some use is now being made of the metal in lieu of steam to drive power turbines.

Most mercury is still used in chemical compounds for drugs, although a considerable amount is made into mercury fulminate for explosive detonators and railway fog warning devices.

Some mercury is used as the electrolyte for the production of chlorine and caustic soda.

Note. The following abbreviations and units are used in the tables:

DPN	Hardness, diamond pyramid number
UTS	Ultimate tensile strength, N/mm²
Elon	Elongation, %
Proof	0.1 % proof strength, N/mm²

1 N/mm²=0.1 hbar=0.102 kgf/mm²=0.06475 tonf/in²=145.04 lbf/in²

Symbol	Nominal analysis, supplier, condition and remarks.
AMALGAM	Hg & other metal: Name given to solution of metal in mercury
ARISTALOY	Sn & other metals Hg amalgam: Engelhard Industries; for dental fillings
p Hg 1	99.9999% Hg: Light Ltd; high purity metal
QUICKSILVER	Hg metal: Common name for mercury

221

26. Molybdenum Mo

Physical properties

Atomic number	42	
Atomic weight	95.95	
Crystal structure	Body-centred cubic	
Colour	Dull silver	
Specific gravity	10.3	
Density	10 300 kg/m^3	(0.369 lb/in.^{3})
Melting point	2620 °C	
Boiling point	5560 °C	
Specific heat	0.255 J/g °C	(0.061 cal/g °C)
Thermal conductivity	145 W/m °C	(34.6 cal/m s °C)
Coefficient of linear expansion (20–100° C)	4.9 × 10^{-6}/ °C	
Latent heat of fusion	293 J/g	(70 cal/g)
Latent heat of vaporization	5610 J/g	(1340 cal/g)
Thermal neutron absorption cross-section	2.5 barns/atom	
Electrical conductivity	34% IACS (copper 100%)	
Specific resistance	517 microhm mm	
Temperature coefficient of electrical resistance	0.0039/ °C	
Electrochemical equivalent	1.790 g/A/h	
Electrode potential	-0.2 V	
Magnetic susceptibility	0.93 × 10^{-6} cm/g	
Young's modulus of elasticity	324 × 10^9 N/m^2	(47 × 10^6 lbf/in.^{2})
Tensile strength	annealed 324 × 10^9 N/m^2	(50 tonf/in.^{2})
Hardness	annealed 230 DPN	

26.1 General notes on molybdenum

Molybdenum is generally found as the sulphide ore molybdenite which occurs in local concentrations but is one of the rarer metal elements in the earth's crust. The largest and best known deposit is in Colorado, USA, where a mountain is gradually being removed to recover molybdenum ore containing approximately ten pounds metallic material per ton of mountainside.

Until recently molybdenum was available only as powder and powder compacts. It is now possible to cast the metal using the electric arc principle under high vacuum or inert gas. Ingots up to one ton are produced capable of being forged and rolled.

When heated above 500 °C it is necessary to exclude air from molybdenum to prevent formation of the oxide which does not form a protective coating. The rate of oxide formation is proportional to temperature and is very rapid above 1000 °C. Hot working can be carried out in air after heating in vacuum or inert atmosphere at about 1000 °C. Nitrogen rich atmospheres cause a brittle nitrided layer to form, but salt baths have been successfully used. used.

At present the largest use of molybdenum is as an alloying element in steel. The hardenability of plain carbon and low alloy steel is increased with as little as 0.2% of the metal. This means that larger sections can be hardened or the same size of section can be hardened with a more gentle quenching medium, for example oil in place of water or air instead of oil.

Molybdenum also reduces the danger of 'temper' or 'blue' brittleness which is a condition affecting plain carbon or nickel and chromium low alloy steels tempered in the range 300–500 °C.

Molybdenum in higher quantities up to 5% is present in almost all high speed tool steels, where hot hardness is necessary. The fact that molybdenum forms stable hard carbides with available carbon accounts for these qualities and the increased creep strength of nickel chromium molybdenum steel. Molybdenum has a similar affect to tungsten and vanadium but is generally more economic.

When added to austenitic stainless steel molybdenum considerably enhances the resistance of the metal to corrosive acids and marine atmospheres. The martensitic stainless steels have their hot strength and corrosion resistance improved.

Molybdenum is added to cast iron to increase the strength and heat resistance.

Molybdenum itself has a very low coefficient of friction which can be a problem at forging but is useful for preventing hot frettage where this is a problem. This property is also present in the disulphide compound, and considerable use is made now of molybdenum disulphide as an additive to lubricating oils and greases.

26A Molybdenum – all alloys

Specific gravity	10	
Density	10 000 kg/m³	(0.36 lb/in.³)
Solidus/liquidus	2500–2600 °C	
Thermal conductivity	– W/m °C	
Coefficient of linear expansion	– / °C	
Electrical conductivity	30% IACS (copper 100%)	
Specific resistance	53 microhm mm	
Young's modulus of elasticity	– N/m²	
Impact	– J	
Fatigue strength	tensile ± 300 N/mm²	(± 20 tonf/in.²)
Hot strength	300 N/mm² at 1200 °C in inert atmosphere	

The above properties are typical of the following group, and may not apply exactly to any one specification. It is possible that with certain specifications some of the values may not be applicable.

General metallurgical characteristics

Molybdenum and its alloys are characterized by the ease of working and fabrication at room temperature with excellent mechanical properties at elevated temperatures, but very poor resistance to hot oxidation. The material is now available in most forms and in several alloys although the pure metal is still the most common.

Additions of small percentages of titanium and zirconium increase the creep strength, and raise the temperature required to remove cold work.

Molybdenum metal has a very low coefficient of friction which causes considerable difficulty when hand forging. The property however, can be useful to prevent frettage, or excessive wear.

Thermal treatment

It is not possible to improve the properties of molybdenum or its alloys by thermal treatment, but interstage annealing after warm or cold working is necessary. Temperatures in the order of 1000 °C are required for about one hour. The alloy content and amount of work carried out have an effect on the temperature at which recrystallization takes place. Considerable grain growth can occur with a reduction in mechanical properties if the temperature is too high and as this is critical it is recommended that trials are carried out to ascertain the correct time and temperature. Test pieces cold worked and annealed with the components should always be used.

Stress relieving after cold work is necessary to improve ductility, at a temperature of 850 °C for about 1 hour or lower temperatures for longer periods.

Molybdenum and its alloys can be carburized or nitrided. Carburizing can be by the pack method using activiated charcoal or gas diffusion enriched with carbon monoxide. Both treatments require a temperature of 1000 °C and give a case hardness in excess of 1200 DPN. Undesirable grain growth may occur if cold work has been carried out prior to carburizing. Nitriding requires a cracked ammonia atmosphere at 800 °C to give a hardness equal to the carburized method. No subsequent quenching or heat treatment is necessary after either carburizing or nitriding.

It is also possible to sulphurize molybdenum, forming an adherent layer of low friction molybdenum disulphide. This is carried out in hydrogen sulphide at 570 °C for 20 minutes.

Welding and brazing

Pre-weld treatment. The material must be in the annealed or stress relieved condition.

Absolute and scrupulous cleanliness of the joint surfaces is essential as the welds are very susceptible to contamination. Great care must be taken that the tools and materials used to prepare the joint surfaces are free from contaminating matter.

Welding and brazing. Oxygen and nitrogen pick up cause severe embrittlement thus normal electric arc and gas welding cannot be used. Inert gas shielded arc welding is possible provided there is adequate gas coverage before and after the weld on both sides. Electron beam welding under high vacuum can be used if the parts are small enough and the configuration correct, and this method is suitable for welding difficult metals like molybdenum.

Brazing requires very high vacuum and time at temperature should be kept to a minimum.

Post-weld treatment. Annealing after welding is only necessary if further cold work is to be carried out.

Flaw detection methods

Crack test. Penetrant oil chalk test, using fluorescent fluids, is recommended.

X-ray. Welds generally require examination, at least on a percentage basis.

Castings should be also X-rayed on a percentage quality check basis at least.

Ultra-sonic test. May be used to supplement X-ray and crack testing of large components at an early stage of machining to prove that no large subcutaneous defects are present.

Chemical etch. Not required. Grain size can only be assessed by metallographic techniques which are destructive.

Corrosion protection

Temporary. None required.

Permanent. None required for use up to temperatures of 400 °C. Above this protection against oxidation is essential and considerable research is being carried out on this subject. Existing methods include cladding with nickel, metal spraying with nickel or aluminium, metal diffusion processes with aluminium or chromium, or vapour phase diffusion with silicon.

Painting. No known application.

Plating. No known application.

Machinability.

No great difficulty should be experienced in machining wrought and cast molybdenum and its alloys. More care is required for sintered powder compacts. Where the material has been cold worked different surface textures after machining may be encountered, depending on the direction of grain flow. Stress relieving after heavy machining is recommended. In general these alloys can be compared to austenitic stainless steels.

High speeds or carbide tipped tools are required.

Uses

As grids, anodes, cathodes, supports and leads in electronic vacuum equipments. In electric lamps to support the tungsten filament. It is also used as a mandrel to coil the filament during manufacture.

Electric furnace elements for use above 1500 °C when an inert, hydrogen or vacuum atmosphere is essential. Radiation shields, supports, general furniture and fixtures for vacuum furnaces. Molybdenum alloys are now used for high temperature parts in jet engines and space rocket motors.

The use of molybdenum and its alloys will increase considerably if a method of preventing oxidation can be perfected.

Symbol	Nominal analysis, supplier, condition and remarks.
AMS 78 CD	0.5% Ti 0.08% Zr Mo arc cast alloy: Forging; stress relieved; interim AMS
AMS 5662 B	19% Cr 5% Nb + Ta 3% Mo 1% Ti 19% Fe Mo alloy **DPN: 330 UTS: 1450 Proof: 1110**
AMS 5663 B	19% Cr 5% Nb + Ta 3% Mo 1% Ti 19% Fe Mo alloy **DPN: 330 UTS: 1450 Proof: 1110**
AMS 5664 A	19% Cr 5% Nb + Ta 3% Mo 1% Ti 19% Fe Mo alloy **DPN: 330 UTS: 1450 Proof: 1110**
AMS 7800	Mo-sintered alloy: Sheet & strip; stress relieved
AMS 7801	Mo arc cast alloy: Sheet & strip; stress relieved
AMS 7805	Mo arc cast alloy: Rods; stress relieved
AMS 7806	Mo-sintered alloy: Powder & forgings; stress relieved
AMS 7807	Mo arc cast alloy: Forgings; stress relieved
AMS 7811 A	0.5% Ti Mo arc cast alloy: Sheet & strip; stress relieved
AMS 7813	0.5% Ti Mo arc cast alloy: Bar; stress relieved
AMS 7819	0.5% Ti 0.09% Zr Mo arc cast alloy: Bar; stress relieved
CMX FB 30W 1	0.02% C 30% W Mo billets for forging: Climax **DPN: 200**
CMX FB TZM 1	0.03% C 0.5% Ti 0.08% Zr Mo alloy: Forging; Climax; annealed **DPN: 200**
CMX S 1	0.03% C Mo alloy: Sheet; produced by vacuum arc casting; Climax; stress relieved **UTS: 740 Elon: 11% Proof: 640**
CMX S TZM 1	0.02% C 0.5% Ti 0.1% Zr 99.25% Mo metal; Sheet; Climax **UTS: 1000 Elon: 7% Proof: 830**
CMX WB LC 1	0.005% C Mo alloy: Bar; produced by vacuum arc casting; Climax; stress relieved **DPN: 230 UTS: 420 Elon: 30% Proof: 210**

Symbol	Nominal analysis, supplier, condition and remarks.
CMX WB T 1	0.03% C 0.5% Ti Mo alloy: Bar; wrought; Climax stress relieved **DPN: 260 UTS: 480 Elon: 12% Proof: 240**
CMX WB TZM 1	0.06% C 0.5% Ti 0.08% Zr Mo alloy: Bar; Climax stress relieved **DPN: 270 UTS: 560 Elon: 15% Proof: 320**
CMX WB TZM 2	0.02% C 0.5% Ti 0.1% Zr Mo alloy: Bar; Climax; as rolled & stress relieved **DPN: 270 UTS: 760 Elon: 10% Proof: 600**
ELKONITE 100 M	Mo for rivetting and welding electrodes: Mallory Metallurgical; electrical conductivity 30% IACS **DPN: 190 UTS: 530**
L MOLLA	99.99% Mo rod: 5 mm diameter; Light Ltd; high purity metal
MOLYBDENUM	Mo powder: Wire, rod, sheet, etc.; Blackwells; pure metal
MTC	0.5% Ti Mo alloy: Climax
S Mo 1000	Mo metal: Fansteel Metallurgical
SPRABOND	99% Mo (min): Wire for spraying; Metco; as sprayed **DPN: 382 UTS: 45**
TZM 1	0.02% C 0.5% Ti 0.1% Zr 99.25% Mo metal: Climax; designation used for this specific analysis
UNS R03001–R03999	Mo metal and alloys: American Standards system

Note. The following abbreviations and units are used in the tables:

DPN	Hardness, diamond pyramid number
UTS	Ultimate tensile strength, N/mm^2
Elon	Elongation, %
Proof	0.1 % proof strength, N/mm^2

$1 N/mm^2 = 0.1 hbar = 0.102 kgf/mm^2 = 0.06475 tonf/in^2 = 145.04 lbf/in^2$

27. Nickel Ni

Physical properties

Atomic number	28	
Atomic weight	58.69	
Crystal structure	Face-centred cubic	
Colour	White	
Specific gravity	8.88	
Density	8880 kg/m³	(0.321 lb/in.³)
Melting point	1455 °C	
Boiling point	3380 °C	
Specific heat	0.46 J/g °C	(0.11 cal/g °C)
Thermal conductivity	60.7 W/m °C	(14.5 cal/m s °C)
Coefficient of linear expansion (20–100° C)	13.1×10^{-6}/ °C	
Latent heat of fusion	305.6 J/g	(73 cal/g)
Latent heat of vaporization	5862 J/g	(1400 cal/g)
Thermal neutron absorption cross-section	4.8 barns/atom	
Electrical conductivity	26% IACS (copper 100%)	
Specific resistance	64 microhm mm	
Temperature coefficient of electrical resistance	0.0066/ °C	
Electrochemical equivalent	1.0945 g/A/h	
Electrode potential	-0.25 V	
Magnetic susceptibility	Ferromagnetic	
Young's modulus of elasticity	207×10^9 N/m²	$(30 \times 10^6$ lbf/in.²)
Tensile strength	annealed 45	(30 tonf/in.²)
Hardness	annealed 75 DPN	

27.1 General notes on nickel

Nickel is found as the sulphide, often associated with arsenic and copper and always with cobalt. Meteorites almost always contain some nickel, with iron and cobalt, but apart from this it is not found as the native metal.

The largest deposit occurs in Sudbury, Canada, and this supplies the bulk of the world's wants at present, with recent discoveries in Australia also contributing.

Nickel of 99.9% purity, almost free from cobalt and sulphur is obtained by the Mond process, and this is the material used for the nickel base alloys.

The product however requires careful melting and deoxidizing before it can be rolled, forged or subjected to any degree of hot or cold work.

Information on commercially pure nickel and nickel alloys is given in the following sections.

Nickel is also used as an alloying element in the following fields.

Low alloy steels. Nickel is a ferrite strengthener and thus imparts toughness to steels, increasing the ductility without decreasing the tensile properties.

Engineering steels with 4% nickel were fairly commonplace until nickel scarcity and cost forced metallurgists to look for alternative alloying materials. It was found that almost comparable results could be obtained more economically with chromium, molybdenum and 1–2% nickel.

A range of steels with 9–10% nickel has now been developed which maintain ductility at very low temperatures, and a further series of high nickel, low carbon steels, known as maraging alloys exhibit very high tensile properties with excellent ductility, weldability and ease of thermal treatment.

Austenitic stainless steels. It was found that by adding 8–10% nickel to the 12–18% chromium steels a completely stainless, austenitic steel was formed.

This now accounts for the largest single use of nickel. Details of these steels are given in Section 44N.

Cupro-nickels. Nickel and copper form a series of solid solutions over the complete range of analysis.

Details of these alloys will be found in Sections 14E and 27D.

Nickel also strengthens the alpha phase in copper–zinc alloys and almost all high strength, corrosion resistant brasses and bronzes contain nickel.

The series of 'copper–nickel–zinc' alloys and 'nickel silver' are discussed in Section 14F.

Aluminium alloys. Nickel enhances the high temperature properties of aluminium alloys, particularly the aluminium–copper series.

Cobalt alloys. Many of these complex heat and wear resistant alloys have nickel present, sometimes as the ductile matrix, in others to add malleability without detracting too much from other properties.

Cast iron. Nickel strengthens the ferrite in cast iron and considerably improves the heat resistance. Many special purpose nickel cast irons are now produced.

In general nickel can be said to be one of the most useful and versatile of all alloying elements, quite apart from the range of nickel based high temperature alloys and the many and varied applications found for the pure metal.

27.2 Nickel – notes on specifications

The following information may be of use in helping to find unknown specifications. Although it is only in comparatively modern times that nickel alloys have been used in large quantities in everyday engineering, many nickel based materials have had specialized use for many years. This may be the cause of the greater tendency to use the trade names of nickel alloys rather than national specifications, probably more than with any of the other modern materials. This has been helped by the relatively few producers using a small number of trade names which have the added advantage of being to some extent descriptive.

British Standard Institution (BS)

This body originally issued a series of individual specifications covering separate analysis and form. These have been largely withdrawn in favour of the system where a master specification is used to cover one form – bar, sheet, etc. – within which each different alloy is identified by its own designation. The master specifications are:

BS 3071 Nickel and nickel alloy castings
BS 3072 Nickel and nickel alloy sheet
BS 3073 Nickel and nickel alloy strip
BS 3074 Nickel and nickel alloy tube
BS 3075 Nickel and nickel alloy wire
BS 3076 Nickel and nickel alloy rod and section

To conserve space only the full title is listed in the classified section. The designation is included in the index, and referred to the BS range in the appropriate section.

American Society for Testing and Materials (ASTM)

This body issues a series of separate specifications, each prefixed by the letters ASTM, followed by the letter B signifying a non-ferrous specification, the actual alloy being identified by a number sometimes qualified by letters. The final two numbers, for example, 61, are for the year of issue, while the letter T shows the specification to be tentative. In the interests of clarity the year of issue is not shown in the classified section but is given in the index.

Society of Automotive Engineering (SAE)

The specifications issued by this body use the prefix SAE followed by two or three figures, sometimes having letters in addition to the numers. There is no obvious system or code to addition to the numbers. There is no obvious system or code to indicate alloy content form.

This society also issues Aerospace Material Specifications which carry the prefix AMS followed by four figures. The suffix letter which sometimes appears is used to indicate the issue number. No attempt has been made to code or classify the alloys.

German Standard Specification (DIN)

These are issued under master specifications using the prefix DIN with a number, generally covering the form – bar, sheet, etc. The alloy analysis is then identified using two systems, the earliest having letters and figures to describe and code the analysis and form. The latest method uses the numbers 2.4000 to 2.4999. The first two figures after the point signify the alloy content and form, while the last two figures attempt to give the quantity of the principal alloying elements.

This system was developed for use with modern accounting machines and computers.

Trade names

These very often attempt to describe either the content or the use of the alloy. The trade names of the principal producers, Nimonic and Inconel, cover specific types of alloys, the number being used to identify the analysis of form, the higher numbers indicating more complex alloys.

27A Nickel – commercially pure – wrought and cast

Specific gravity	8.88	
Density	8880 kg/m³	(0.32 lb/in.³)
Solidus/liquidus	1435–1445 °C	
Thermal conductivity	60.7 W/m °C	(14.5 cal/m s °C)
Coefficient of linear expansion	13.3×10^{-6}/ °C	
Electrical conductivity	19% IACS (copper 100%)	
Specific resistance	91 microhm mm	
Young's modulus of elasticity	207×10^9 N/m²	(30×10^6 lbf/in.²)
Impact	cold rolled 160 J	(120 ft/lb) Izod
Fatigue strength	cold rolled ± 290 N/mm²	(± 19 tonf/in.²)
Hot strength		

Temperature °C	Tensile strength N/mm²	Elongation %
20	450	47
150	440	44.5
250	440	45
370	360	61.5
480	220	66

The above properties have been chosen to show typical values for the specification listed and may not apply exactly to any one specification. It is possible that with some specifications the values may not be applicable.

General metallurgical characteristics

The largest single use for pure nickel is as anodes for electroplating. These are generally cast with a controlled amount of oxide to produce a fine grain and are known as 'de-polarized anodes'. These dissolve more evenly, thus more economically than the high purity wrought nickel. Some use is now made of electrolytic nickel which has a sulphur content giving the same effect at lower cost. These can only be used where the plating solution is not poisoned by sulphur contamination.

Some of the materials listed are primary metals used in many industries as their source of nickel. Depending on the method of manufacture the nickels listed will have no contaminants or will have oxygen, carbon, sulphur, or cobalt present in small amounts.

With oxygen and carbon the cold working properties and weldability of nickel are adversely affected, while sulphur affects the corrosion resistance and weldability. Cobalt in small amounts has no serious affect on the properties of nickel.

With oxygen and carbon absent nickel can accept a considerable amount of cold work, which increases the hardness without it becoming too brittle. Thin sheets can be produced with interstage annealing.

Nickel has excellent corrosion resistance and is not attacked by many acids even at elevated temperatures. An adherent oxide film is formed on the surface which successfully protects the metal from further oxidation even at very high temperatures.

Nickel has special electrical and magnetic properties caused by the peculiar configuration of its outer shell of electrons and resembles iron and cobalt in this respect.

Thermal treatment

There are no metallurgical change points with nickel, thus no form of thermal hardening is possible. Where cold work must be removed the parts are heated to above 1020 °C for 20 minutes per inch of section. The rate of cooling is unimportant.

Welding and brazing

Pre-treatment. All traces of oil grease and dirt must be removed. Severe cold work should not be present. Careful edge preparation immediately prior to welding is recommended.

Welding and brazing. The adherent oxide prevents the successful use of gas or electric welding without very active fluxes and considerable skill. Inert gas shielded arc welding with adequate gas flow on both sides of the weld gives very good results. Electric resistance welding is possible with the correct equipment and necessary precautions. Brazing is difficult as nickel does not wet owing to the presence of the oxide film, thus very active fluxes are necessary. Because of the high temperature and corrosion resistance of nickel normal copper alloy braze metals will seldom be used.

High temperature nickel and palladium alloy braze fillers have been developed, but these normally require self bath, hydrogen atmosphere or vacuum furnace equipment. Hydrogen with its active reducing properties probably gives the most economical results, but considerable advances are being made at present in vacuum techniques.

Post-treatment. All fluxes must be completely removed as they are generally hygroscopic and corrosive to other metals as well as the nickel.

Corrosion protection

Temporary. None required.

Permanent. Nickel has excellent corrosion resistance to atmospheric conditions, many acids and all alkalis over a wide range of temperatures. For all normal purposes no further protection is necessary, but where appearance is important and a bright finish necessary, a thin covering of chromium electroplate prevents the nickel tarnishing.

Machinability

Annealed nickel is very soft and tears rather than cuts, giving a poor finish. Cold worked metal can be machined to a better finish. Electrodeposited nickel is generally polished to a bright finish but can be machined, the deposit often having occluded oxygen to give an increased hardness. The harder the plated nickel the poorer will be the adhesion in general, but the easier to machine.

Uses

As an electrodeposited coating under chrome to prevent steel parts corroding. A minimum deposit of 0.025 mm thick is essential. As a cladding on steel to improve the corrosion resistance. Laboratory equipment spatulas, crucibles, balance weights and many other items.

Electronic uses include vacuum tube electrodes. Many countries now use pure nickel for coinage, the magnetic properties making counterfeiting more difficult.

Food manufacturing equipment is often either nickel, nickel clad or nickel plated. Magnestriction is a property of nickel bars where the length can be altered by means of electrical eddy currents. With a.c. current this property can be used in ultrasonic equipment, as the bar dimension varies with the frequency.

The magnetic properties are used for shielding electronic equipment.

Electro-less nickel deposits are being used increasingly. These are chemical deposits of a nickel phosphide alloy which can be hardened to above 500 DPN and has good corrosion resistance.

Symbol	Nominal analysis, supplier, condition and remarks.
1.4876	0.4% Ti Ni: German Werkstoff
2.4050	99.8% Ni: German Standard
2.4051	0.02% Mg 99.8% Ni anode: German Standard
2.4052	0.05% Mg 99.7% Ni anode: German Standard
2.4053	0.07% Mg 99.7% Ni anode: German Standard
2.4056	0.2% Si 99.6% Ni anode: German Standard
2.4060	99.6% Ni: German Standard
2.4062	0.4% Fe 99.4% Ni: German Standard
2.4066	99.2% Ni: German Standard
2.4068	0.02% C 99.0% Ni: German Standard
2.4106	0.8% Mn 98% Ni: German Standard
99 ALLOY	99.8% Ni: Driver-Harris **UTS: 530**
ACTIVE Ni	0.2% Si Mg Ni: For thermionic valves; British Driver-Harris
AMPCO 501	Nickel: Cast; Ampco **DPN: 130**
AMPCOLAY 502	Nickel: Wrought; Ampco for A Nickel
A Nickel	0.06% C 99.5% Ni: Bar, sheet, etc; Huntington; now Nickel 200
A Nickel	99.0% Ni (min): For electrical leads; H Wiggin; annealed; electrical conductivity 20% IACS **DPN: 95 UTS: 370 Elon: 45% Proof: 100**
ASTM A636/75	75% Ni 1.3% Co (max) 0.9% Cu (max) 0.5% Fe (max): Sinter for alloying
ASTM B39 A Shot	0.75% C 0.9% Fe 0.07% S 97.7% Ni: Shot; virgin metal
ASTM B39 Electrolytic	0.1% C 0.6% Fe 0.02% S 99.5% Ni: Virgin metal
ASTM B39 Ingot	98.5% Ni: Virgin metal
ASTM B39 X Shot	0.25% C 0.6% Fe 0.05% S 98.9% Ni: Shot; virgin metal

Symbol	Nominal analysis, supplier, condition and remarks.
ASTM B160	0.15% C 0.3% Mn 0.4% Fe 0.2% Cu Ni: Bar; annealed **UTS: 370 Elon: 40% Proof: 75**
ASTM B161	0.15% C 0.3% Mn 0.4% Fe 0.25% Cu Ni: Tube; annealed **UTS: 370 Elon: 40% Proof: 90**
ASTM B162	0.15% C 0.3% Mn 0.4% Fe 0.25% Cu Ni: Sheet & strip; hard **UTS: 670 Elon: 2% Proof: 480**
ASTM B304 ER Ni 3	0.15% C 93.0% Ni (min): Welding electrode; previous designation ERN 61
ASTM B304 R Ni 2	0.15% C 97.0% Ni (min): Welding electrode; previous designation RN 41
AT NICKEL	0.15% C 0.01% S 0.4% Fe 0.25% Cu 99% Ni: Wrought; H Wiggin; annealed; sheet, bar & plate **DPN: 100 UTS: 370 Elon: 40% Proof: 20**
AT NICKEL	0.15% C 0.01% S 0.4% Fe 0.25% Cu 99% Ni: Wrought; H Wiggin; cold rolled; sheet, bar & plate **DPN: 200 Elon: 10% Proof: 30**
BALLAST NICKEL	99.8% Ni (min): For current limiting controls; Driver-Harris
BS 375 A	0.1% C 99.5% Ni: Refined metal; primary metal
BS 558	98.5% Ni: For anodes; anodes for nickel plating
BS 1525	Nickel and nickel alloy: Sheet; replaced by BS 3072 & 3
BS 1526	Nickel and nickel alloy: Sheet; replaced by BS 3072 & 3
BS 1527	Nickel and nickel alloy: Sheet; replaced by BS 3072 & 3
BS 1528	Nickel and nickel alloy: Bar; replaced by BS 3076
BS 1529	Nickel and nickel alloy: Bar; replaced by BS 3076
BS 1530	Nickel and nickel alloy: Bar; replaced by BS 3076
BS 1531	Nickel and nickel alloy: Tube; replaced by BS 3074
BS 1532	Nickel and nickel alloy: Tube; replaced by BS 3074
BS 1533	Nickel and nickel alloy: Tube; replaced by BS 3074
BS 1534	Nickel and nickel alloy: Wire; replaced by BS 3075
BS 1535	Nickel and nickel alloy: Wire; replaced by BS 3075

Note. The following abbreviations and units are used in the tables:

DPN	Hardness, diamond pyramid number
UTS	Ultimate tensile strength, N/mm^2
Elon	Elongation, %
Proof	0.1 % proof strength, N/mm^2

$1 \text{ N/mm}^2 = 0.1 \text{ hbar} = 0.102 \text{ kgf/mm}^2 = 0.06475 \text{ tonf/in}^2 = 145.04 \text{ lbf/in}^2$

Symbol	Nominal analysis, supplier, condition and remarks.
BS 1536	Nickel and nickel alloy: Wire; replaced by BS 3075
BS 2901 NA32	3% Ti Ni alloy: Rod for all forms of welding
BS 3072 NA11	0.15% C (max) 99% Ni (min): Sheet; annealed
	UTS: 390 Elon: 35% Proof: 90
BS 3072 NA12	0.02% C (max) 99% Ni (min): Sheet; annealed
	UTS: 360 Elon: 35% Proof: 90
BS 3073 NA11	0.15% C (max) 99% Ni (min): Strip; annealed
	DPN: 100 UTS: 390 Elon: 35% Proof: 90
BS 3074 NA11	0.15% C (max) 99% Ni (min): Tube; annealed
	UTS: 360 Elon: 40% Proof: 90
BS 3074 NA11	0.15% C (max) 99% Ni (min): Tube; cold drawn
	UTS: 390 Elon: 15% Proof: 270
BS 3074 NA12	0.02% C (max) 99% Ni: Tube; annealed
	UTS: 340 Elon: 40% Proof: 75
BS 3074 NA12	0.02% C (max) 99% Ni: Tube; cold drawn
	UTS: 400 Elon: 15% Proof: 210
BS 3075 NA11	0.15% C (max) 99% Ni (min): Wire; annealed
	UTS: 340 Elon: 25%
BS 3975 NA11	0.15% C (max) 99% Ni (min): Wire; cold drawn
	UTS: 480
BS 3076 NA11	0.15% C (max) 99% Ni (min): Rod & section; annealed
	UTS: 390 Elon: 35% Proof: 90
BS 3076 NA11	0.15% C (max) 99% Ni (min): Rod & section; cold drawn
	UTS: 530 Elon: 15% Proof: 340
BS 3076 NA12	0.02% C (max) 99% Ni: Bar & section; annealed
	UTS: 330 Elon: 35% Proof: 60
BS 3076 NA12	0.02% C (max) 99% Ni: Bar & section; cold drawn
	UTS: 450 Elon: 15% Proof: 300
CARBONIZED NICKEL	99% Ni (min): For anode plates; Driver-Harris
CATHODE NICKEL	99.5% Ni (min): For electronic valve cathodes; Driver-Harris
CI-NI	1.0% C 97.0% Ni: Electrode; Metrode; for general repair welding
DIN 1701 C Ni 98.5	0.5% C 0.01% S 98.5% Ni
DIN 1701 C Ni 99.5	0.13% C 0.001% S 99.5% Ni
DIN 1701 C Ni 99.8	0.08% C 0.0005% S 99.8% Ni
DIN 1701 E Ni 99.5	0.1% C 0.02% S 99.5% Ni
DIN 1701 E Ni 99.8	0.1% C 0.01% S 99.8% Ni
DIN 1701 M Ni 99.5	0.1% C 0.02% S 99.5% Ni
DIN 1701 W Ni 99	0.1% C 0.015% S 99.0% Ni
DIN 1702	High purity Ni for anodes
DIN 17740 L C Ni99	0.02% C 99.0% Ni
DIN 17740 Ni 99.8	99.8% Ni
DIN 17741 Ni Mn 1	0.8% Mn 98% Ni
DURANICKEL	4.5% Al 0.5% Ti Ni alloy: High carbon bar; H Wiggin for springs & dies; acid resistant
	DPN: 295 Elon: 28%
DURANICKEL	Ni with additions: Driver-Harris; melting point 1435 °C
GFA NICKEL	Low C 99.0% Ni (min): For anodes; H Wiggin
GRADE A	High purity nickel: For thermionic valves; British Driver-Harris
HG NICKEL	0.06% C 99.0% Ni (min): Wire; H. Wiggin; for electronic use

Symbol	Nominal analysis, supplier, condition and remarks.
h Ni 8	99.999% Ni: Sponge; Light Ltd; high purity metal
h Ni 15	99.995% Ni: Wire 1 mm diameter; Light Ltd
HPM	0.2% C 99.97% Ni: Bar & sheet; H Wiggin; now Nickel 270
MNS NICKEL	High C 2.0% Mn Ni alloy: Rod & wire; H Wiggin; improved machinability
Ni 99.8 Mg	0.02% Mg 99.8% Ni: Anode; designation used by German Standard
NICKEL	High purity metal: Impurities less than 15 p.p.m.; Mallory Metallurgical; supplied as sponge, rod, wire & sheet
NICKEL	Ni: Cubes, sticks, powder & wire; Blackwells; pure metal
NICKEL 2	0.05% C 97% Ni: Welding electrode; Metrode
NICKEL 41	0.06% C 99.5% Ni: Filler rod; Huntington; for oxyacetylene welding of nickel
NICKEL 61	0.06% C 3% Ti 96% Ni: Filler rod; Huntington; inert gas shield welding of nickel
NICKEL 131	0.25% C 0.5% Al 3% Ti Ni alloy: Filler rod; Huntington; metal arc welding of nickel
NICKEL 141	0.05% C 0.25% Al 2.2% Ti Ni alloy: Filler rod; Huntington; for welding nickel & steel
NICKEL 200	0.06% C 99.5% Ni: Bar, sheet, forgings, etc; Huntington; annealed
	UTS: 450 Elon: 47% Proof: 120
NICKEL 201	0.01% C 99.5% Ni: Bar, sheet, forging etc; Huntington
NICKEL 205	0.06% C 0.02% Ti 0.04% Mg 99.5% Ni: Bar, sheet, etc; Huntington
NICKEL 213	High C 2.0% Mn Ni alloy: Bar & wire; H Wiggin; improved machinability
NICKEL 220	0.06% Ni 0.02% Ti 0.04% Mg 99.5% Ni: Bar & tube; Huntington
NICKEL 222	0.07% Mg 99.5% Ni (min): Tube & strip; H Wiggin; for electronic use
NICKEL 225	0.06% C 0.02% Ti 0.04% Mg 99.5% Ni: Bar & tube; Huntington
NICKEL 230	0.09% C 0.003% Ti 0.06% Mg 99.5% Ni: Sheet; Huntington
NICKEL 233	0.09% C 0.003% Ti 0.07% Mg 99.5% Ni: Sheet & tube; Huntington
NICKEL 270	0.02% C 99.97% Ni: Bar, sheet & tube; Huntington
NI-ROD	1% C 3% Fe Ni alloy welding electrode: Huntington; welding cast iron etc
NIVAC P	0.007% C Ni; high purity metal: Crucible Steel; annealed; for electrical purposes
	UTS: 270 Elon: 40% Proof: 210
NIVAC P	0.007% C Ni; high purity metal: Crucible Steel; cold rolled
	UTS: 600 Elon: 5% Proof: 580
NORMAL Ni	High purity Ni for thermionic valves: British Driver-Harris; purity between that of passive and active Ni
PA 22 NICKEL	0.07% Mg 99.5% Ni (min): Tube & strip; H. Wiggin; electronic use; now nickel 222
PA 23 NICKEL	0.05% Mg 99.5% Ni (min): Tube & strip; H. Wiggin; electronic use
PASSIVE Ni	High purity Ni for thermionic valves: British Driver Harris
PERMANICKEL	Ni with additions: Driver-Harris
PERMANICKEL 300	0.25% C 0.5% Ti 0.35% Mg 98.6% Ni strip: Huntington; annealed
	UTS: 800 Elon: 41% Proof: 300
X 10 Ni Cr	Al Ti 32/20 0.35% Ti Ni: German Standard
Z-NICKEL	0.02% Cu 0.2% Fe 0.2% Si 0.2% Mn age hardening Ni alloy: Rod or strip; electrical conductivity 12% IACS; origin unknown
	UTS: 690

Note. The following abbreviations and units are used in the tables:

DPN	Hardness, diamond pyramid number
UTS	Ultimate tensile strength, N/mm^2
Elon	Elongation, %
Proof	0.1 % proof strength, N/mm^2

$1\ N/mm^2 = 0.1\ hbar = 0.102\ kgf/mm^2 = 0.06475\ tonf/in^2 = 145.04\ lbf/in^2$

27B Nickel–chromium wrought and cast non-ageing alloys
With iron and other non-ageing elements present

Specific gravity	8.37	
Density	8370 kg/m^3	(0.301 lb/in.^{3})
Solidus/liquidus	1380–1430 °C	
Thermal conductivity	20.9 W/m °C	(5 cal/m s °C)
Coefficient of linear expansion (20–100° C)	12.2 × 10^{-6}/ °C	
Electrical conductivity	1.6% IACS (copper 100%)	
Specific resistance	1100 microhm mm	
Young's modulus of elasticity	186 × 10^9 N/m^2	(27 × 10^6 lbf/in.^{2})
Impact	108 J	(80 ft/lb) Izod
Fatigue strength (10 × 10^6 cycles)	± 180 N/mm^2	(± 12 tonf/in.^{2})
Hot strength		

Temperature °C	Tensile strength N/mm^2	Elongation %
20	780	44
200	740	40
300	740	40
500	720	37
600	520	29
700	340	29
800	200	85
900	100	80
1000	75	100

The above properties are included as a guide to the group of specifications as a whole. There is considerable variation within the group and some of these properties may not be applicable to all the specifications.

General metallurgical characteristics

Chromium strengthens the nickel matrix and by forming a stable, very adherent oxide considerably increases the hot corrosion resistance.

The chromium dissolves in the nickel forming a solid solution stable at room, as well as elevated temperatures. There are no phase changes and no intermetallic compounds to cause precipitation hardening to any measurable extent in the alloys listed.

Iron reduces the corrosion resistance and hot strength, but has some desirable electrical and magnetic properties, and is of course, much cheaper than nickel.

The presence of carbon increases the strength but can cause trouble, as some of the chromium carbide precipitates under certain conditions, resulting in local brittleness. Additions of many elements cauuse formation of intermetallic compounds which can be precipitation age hardened. These alloys are described and listed in the groups following this section. The elements boron and silicon are often present as fluxing agents in those alloys which are applied by metal spraying or used as brazing materials. Tungsten carbide may also be present to strengthen and harden the alloy. Like all nickel alloys those listed are readily hardened by cold work and this is accompanied by a reduction in ductility.

The alloys listed all have excellent corrosion and acid resistance, are relatively easily welded, and maintain their strength and corrosion resistance at high temperatures.

The creep strength of these materials is slightly superior to the austenitic steels but does not compare with the age hardening nickel chromium materials.

Thermal treatment

These alloys cannot be hardened by any form of thermal treatment, but are very easily hardened by cold work. The effects of this are removed by annealing at a temperature of 1020–1060 °C for 15 minutes per cm of section. The rate of cooling is not important. The oxide formed can be removed by acid pickling, sodium hydride treatment or light grit blasting. There are now available liquids which can be sprayed or brushed on prior to annealing which prevent scale formation. The use of these special media is justified if components are too flimsy to blast or pickling is not metallurgically desirable. Also the acid required for pickling is hydrofluoric which is one of the more unpleasant liquids to handle.

Welding and brazing

Pre-treatment. No thermal treatment is required unless cold work is present. This must be removed by annealing, and applies to sheared edges or drilled holes. The surface must be completely free of oxide scale, dirt or grease.

Welding and brazing. Welding using gas or electric arc techniques and corrosive fluxes is possible with these alloys, but it is recommended that the inert gas shielded electric arc welding process is first choice whenever possible. Excellent welds with no danger of flux or oxide trapping are produced and require less skill than the conventional methods.

Resistance welding is readily carried out provided the correct equipment is used under technically controlled conditions and the contact faces are properly prepared. The relatively high electrical resistance of these materials necessitates careful design of resistance welds. Brazing generally requires a flux bath, or hydrogen or vacuum furnace equipment and uses either pure copper or nickel alloy braze materials. Fluxes are often necessary.

Post-treatment. No thermal treatment is necessary although stress relieving of complex structures may be advisable. Fluxes if used must be removed.

Flaw detection

Crack test. Penetrant oil chalk test method.

X-ray. High duty welds often require 100% X-ray. Quality checking on a percentage basis is recommended for all welds, but wrought products seldom require this form of inspection.

Ultra-sonic test. This is used on forgings which require expensive machining operations. Ultra-sonic testing indicates defects below the surface which would become dangerous in the finished part. This method can now be used to identify certain types of welding defects and can be more economical than X-ray.

Chemical etch. Not required.

Corrosion protection

Temporary. None required.

Permanent. None required.

Machinability

These are all rather tough materials which have a propensity to harden by cold work during machining. This is minimized by the use of tipped tools of high feeds and low speeds always ensuring that the tool is cutting and not rubbing. Interrupted cuts and tool dwelling should be avoided.

Where considerable metal has to be removed it may be necessary to anneal after rough machining prior to finish machining to prevent dimensional instability in service.

Grinding is not usually necessary and there is a tendency for the wheels to become loaded, resulting in glazing.

In general these materials are comparable to austenitic stainless steels for machinability.

Uses

Furnace heating elements, furnace furniture, heat treatment jigs, fixtures and trays. Internal combustion engine valves. Combustion equipment on all types of engineering and furnace appliances where high temperature properties apart from high creep strength are required. Springs for high temperature use.

Some of these materials are sprayed or welded on to mechanically stronger base metals to prevent corrosion or high temperature oxidation.

Symbol	Nominal analysis, supplier, condition and remarks.
2.4640	15% Cr 8% Fe Ni alloy: German Standard
2.4867	15% Cr 21% Fe Ni alloy: German Standard
2.4869	1.0% Si 20% Cr Ni alloy: German Standard
2.4870	10% Cr Ni alloy: German Standard
14.75 Mo Nb	0.06% C 15.0% Cr 2.0% Mo 2.0% Nb 8.0% Fe Ni alloy: Electrode; Metrode; for welding Incoloy 800
14.75 Mn Nb	0.06% C 15.0% Cr 6.0% Mn 2.0% Nb 8.0% Fe Ni alloy: Electrode; Metrode; for welding austenitic to ferritic steel
14.75 Nb	0.06% C 15.0% Cr 2% Nb 8.0% Fe Ni alloy: Welding electrode; Metrode
15.60 Mn WNb	0.06% C 6.0% Mn 15.0% Cr 5.0% W 2.0% Nb Ni alloy: Electrode; Metrode
15.60 Nb	0.06% C 15.0% Cr 30% Fe Ni alloy: Welding electrode; Metrode
20.55 4Si Cu W	0.06% C 20.0% Cr 3.5% Mo 4.0% Si 4.0% Cu 2.0% W Ni alloy: Electrode; Metrode

Note. The following abbreviations and units are used in the tables:

DPN	Hardness, diamond pyramid number
UTS	Ultimate tensile strength, N/mm^2
Elon	Elongation, %
Proof	0.1 % proof strength, N/mm^2

1 N/mm^2=0.1 hbar=0.102 kgf/mm^2=0.06475 tonf/in^2=145.04 lbf/in^2

Symbol	Nominal analysis, supplier, condition and remarks.
22H	0.5% C 26% Cr 20% Fe 5% W Ni alloy: Casting; Thompson L'Hospied; annealed **DPN: 180 UTS: 460 Elon: 3%**
25.55.6 Cu W	0.06% C 25.0% Cr 6.0% Mo 6.0% Cu 1.5% W Ni alloy: Electrode; Metrode
50.50	0.08% C 50% Cr Ni alloy: Electrode; Metrode
50.50 Nb	0.08% C 50% Cr 1.5% Nb Ni alloy: Electrode; Metrode
55HSW	0.6% C 1.35% W 17% Cr 30% Fe Ni alloy: For furnace furniture, etc; Wellman
ACCOLAY	15% Cr Ni alloy: Source unknown
ALL-STATE 8.60	57% Ni alloy: Welding electrode for repairing cast iron; All State
ALUMEL	5% Cr Ni alloy: Wire for thermocouples; source unknown
AMPCOLAY 531	0.15% C 15% Cr 7% Fe Ni alloy: Ampco
AMS 4775 A	16.5% Cr 4% Fe 3.8% B 4% Si Ni alloy: Bronze metal
AMS 4776	Low C 16.5% Cr 4% Fe 3.8% B 4% Si Ni alloy: Braze metal
AMS 4777	7% Cr 3% Fe 3% B 4% Si Ni alloy: Braze metal
AMS 5540 F	15.5% Cr 8% Fe Ni alloy: Sheet & strip
AMS 5580 C	15.5% Cr 8% Fe Ni alloy: Seamless tube
AMS 5665 F	15.5% Cr 8% Fe Ni alloy: Bar & forging
AMS 5676	20% Cr Ni alloy: Welding wire
AMS 5677	20% Cr 1.6% Nb + Ta Ni alloy: Electrode; coated wire
AMS 5679 B	15.5% Cr 8% Fe 2% Nb + Ta Ni alloy: Welding wire; cold drawn

Symbol	Nominal analysis, supplier, condition and remarks.
AMS 5682 A	20% Cr Ni alloy: Rod or wire coated electrode
AMS 5683 B	15.5% Cr 8% Fe Ni alloy: Welding wire
AMS 5684 B	15% Cr 9% Fe 2% Nb + Ta Ni alloy: Coated electrode
AMS 5687 C	15.5% Cr 8% Fe Ni alloy: Wire; annealed
ANC 5C	15% Cr 25% Fe Ni alloy: Investment casting; designation used by BS
ANC 8	0.1% C 20% Cr 0.4% Ti 0.8% Al Ni alloy: Investment casting; designation used by BS
AN N4	0.15% C 14% Cr 9% Fe Ni alloy: Welding rod; US Service
AN QQ N 268	13% Cr 9% Fe Ni alloy: Bar & forging; US Service
AN QQ N 271	9% Fe 13% Cr Ni alloy: Sheet; US Service
AN WW T 831	13% Cr 9% Fe Ni alloy: Seamless tube; US Service
AN WW T 833	13% Cr 9% Fe Ni alloy: Welded tube; US Service
	UTS: 720 Elon: 35% Proof: 300
ASTM A494 N12M2B	0.07% C (max) 1.0% Cr (max) 31.5% Mo Ni alloy: Casting
ASTM A494 CW12M2B	0.07% C (max) 3.0% Fe (max) 19% Cr 19% Mo Ni alloy: Casting
ASTM A608 HW50	0.5% C 12.0% Cr 28.0% Fe Ni alloy: Tube; centrifugally cast
ASTM A608 HX50	0.5% C 17.0% Cr 17.0% Fe Ni alloy: Tube; centrifugally cast
ASTM B166	0.15% C 15% Cr 7% Fe Ni alloy: Rod & bar; annealed
	UTS: 580 Elon: 30% Proof: 240
ASTM B166	0.15% C 15% Cr 7% Fe Ni alloy: Rod & bar; cold drawn
	UTS: 780 Elon: 8% Proof: 650
ASTM B167	0.15% C 16% Cr 7% Fe Ni alloy: Tube; annealed
	UTS: 580 Elon: 35% Proof: 210
ASTM B168	0.15% C 16% Cr 7% Fe Ni alloy: Sheet & strip; annealed
	UTS: 580 Elon: 30% Proof: 220
ASTM B168	0.15% C 16% Cr 7% Fe Ni alloy: Sheet & strip; hard
	UTS: 910 Elon: 2% Proof: 650
ASTM B260 B Ni 1	14% Cr 3.3% B 4% Si 4.5% Fe 0.75% C Ni alloy: Braze metal; temperature range 1066–1204 °C
ASTM B260 B Ni 2	7% Cr 3.15% B 4.5% Si 3% Fe Ni alloy: Braze metal; temperature range 1010–1177 °C
ASTM B260 B Ni 5	19% Cr 10.2% Si Ni alloy: Braze metal; temperature range 1149–1204 °C
ASTM B260 B Ni 6	11% P Ni alloy: Braze metal; temperature range 927–1024 °C
ASTM B260 B Ni 7	13% Cr 10% P Ni alloy: Braze metal; temperature range 927–1038 °C
ASTM B260 B Ni Cr	17% Cr 3% B Ni alloy: Braze filler
ASTM B295 E 3 N 12	0.1% C 11% Fe 15% Cr 3% Nb Ni alloy: Welding electrode
	UTS: 580
ASTM B295 E 4 N 12	0.15% C 1% Mn 4% Fe 17% Cr 3% Nb Ni alloy: Welding electrode
ASTM B295 E Ni Cr 1	17.5% Cr (min) 3% Nb Ni alloy: Welding rod; previous designation E 4 N 12

Symbol	Nominal analysis, supplier, condition and remarks.
ASTM B295 E Ni Cr Fe 1	15% Cr 10% Fe 3% Nb Ni alloy: Welding rod; previous designation E 3 N 12
ASTM B295 E Ni Cr Fe 2	15% Cr 2% Nb 2% Mn 9% Fe 1.7% Mo Ni alloy: Welding rod
ASTM B295 E Ni Cr Fe 3	15% Cr 1.5% Nb 7% Mn 8% Fe Ni alloy: Welding rod
ASTM A296	0.12% C 17.5% Cr 18% Mo 5% W 2.5% Co 7% Fe Ni alloy: Casting
CW12M	
ASTM A296 CY40	0.4% C (max) 15.5% Cr 11% Fe Ni alloy: Casting
ASTM A297 HW	0.35% C 12% Cr 38% Fe Ni alloy: Casting; as cast
	UTS: 340
ASTM A297 HX	0.35% C 17% Cr 17% Fe Ni alloy: Casting; as cast
	UTS: 400
ASTM B304 ERN 6N	0.1% C 2% Fe 0.3% Ti 20% Cr Ni alloy: Welding electrode
ASTM B304 ERN 62	0.1% C 8% Fe 16% Cr 2% Nb Ni alloy: Welding electrode
ASTM B304 ER Ni Cr 3	3.0% Mn 20% Cr 1.5% Nb Ni alloy: Welding rod
ASTM B304 R Ni Cr Fe 4	8% Fe 17% Cr Ni alloy: Welding rod; previous designation RN 42
ASTM B304 ER Ni Cr Fe 5	8% Fe 2.5% Nb 17% Cr Ni alloy: Welding rod; previous designation ERN 62
ASTM B304 ER Ni Cr Fe 6	2.3% Mn 8% Fe 16% Cr 3% Ti Ni alloy: Welding rod
ASTM B304 RN 42	0.1% C 8% Fe 16% Cr Ni alloy: Welding electrode
ASTM B344	19% Cr 1% Mn 45% Fe Ni alloy: For resistance wire
ASTM B344/1	20% Cr 2.5% Mn 1% Fe Ni alloy: For resistance wire
ASTM B344/2	16% Cr 1% Mn 23% Fe Ni alloy: For resistance wire
ATGF	15.0% Cr 1.0% Nb 2.5% Ti 0.8% Al 7.0% Fe Ni alloy: Origin unknown
AURIGA XXXII	0.5% C 20% Cr 35% Fe Ni alloy: Casting; D Brown
BLAW KNOX 22H	0.47% C 5% W 26.5% Cr 22% Fe Ni alloy: Casting; for furnace fittings, etc; Wellman
BRIGHTRAY 35	0.1% C (max) 20.0% Cr 0.3% Al 0.1% Tr 35.0% Ni Fe alloy: H. Wiggin; electrical resistance alloy
	UTS: 600
BRIGHTRAY B	0.1% C 0.2% Cu 16% Cr 25% Fe Ni alloy: Wrought; H. Wiggin; annealed; for use up to 950 °C
	DPN: 186 UTS: 680
BRIGHTRAY C	0.1% C 1.6% Si 19% Cr Ni alloy: Wrought; H Wiggin; annealed; for use up to 1150 °C
	DPN: 196 UTS: 740
BRIGHTRAY H	0.1% C 1% Si 0.5% Fe 19% Cr 3.5% Al Ni alloy: Wrought; H. Wiggin; annealed; for use up to 1200 °C
	DPN: 200 UTS: 760
BRIGHTRAY S	0.1% C 1% Fe 20% Cr 0.4% Ti Ni alloy: Wrought; H Wiggin; annealed; for use up to 1150 °C
	DPN: 196 UTS: 730
BS 1648 H	0.5% C 3.0% Si 20% Cr 35% Fe Ni alloy: Casting; as cast; contained in BS 3100
BS 1648 K	0.75% C 3.0% Si 15% Cr 20% Fe Ni alloy: Casting; as cast; contained in BS 3100
BS 1845 N13	7% Cr 4% Si 3% B 3% Fe Ni alloy: For brazing; melting range 955–1000 °C
BS 1845 N16	0.75% C 15% Cr 4% Si 3.2% B 4% Fe Ni alloy: For braze metal; melting range 1000–1060 °C
BS 1845 N17	0.06% C (max) 15% Cr 4% Si 3.2% B 4% Fe Ni alloy: For braze metal; melting range 1010–1070 °C

Note. The following abbreviations and units are used in the tables:

DPN	Hardness, diamond pyramid number
UTS	Ultimate tensile strength, N/mm²
Elon	Elongation, %
Proof	0.1 % proof strength, N/mm²

1 N/mm² = 0.1 hbar = 0.102 kgf/mm² = 0.06475 tonf/in² = 145.04 lbf/in²

Symbol	Nominal analysis, supplier, condition and remarks.
BS 1845 N18	0.15% C (max) 19% Cr 10.2% Si Ni alloy: For braze metal; melting range 1080–1135 °C
BS 2901 NA34	20% Cr Ni alloy: Rod for all forms of welding
BS 3072 NA14	15% Cr 8% Fe Ni alloy: Sheet; annealed
	DPN: 180 UTS: 530 Elon: 30% Proof: 170
BS 3073 NA14	15% Cr 8% Fe Ni alloy: Strip; annealed
	DPN: 180 UTS: 530 Elon: 30% Proof: 180
BS 3074 NA14	15% Cr 8% Fe Ni alloy: Tube; annealed
	UTS: 530 Elon: 35% Proof: 200
BS 3075 NA14	15% Cr 8% Ni alloy: Wire; annealed
	UTS: 530 Elon: 22%
BS 3075 NA14	15% Cr 8% Fe Ni alloy: Wire; cold drawn
	UTS: 770
BS 3076 NA14	15% Cr 8% Fe Ni alloy: Bar & section; annealed
	UTS: 530 Elon: 32% Proof: 150
BS 3076 NA14	15% Cr 8% Fe Ni alloy: Bar & section; cold drawn
	UTS: 700 Elon: 7% Proof: 530
BS 3076 NA17	18% Cr 2.2% Si 1.2% Mn 40% Fe Ni alloy: Rod
BS 3146 ANC5C	15% Cr 25% Fe Ni alloy: Investment casting; as cast
BS 3146 ANC8	0.1% C 20% Cr 0.4% Ti 0.8% Al Ni alloy: Investment casting; as cast
	UTS: 400 Elon: 5%
CALOMIC	15% Cr 20% Fe Ni alloy: Casting; electrical resistance & heating purposes; Telcon Metals
	UTS: 760
CALOMIC	15% Cr 20% Fe Ni alloy: Wrought; electrical resistance up to 1000 °C; Telcon Metals; annealed; specific resistance 1100 microhm mm
	UTS: 760
CALORITE	8% Mn 12% Cr 15% Fe Ni alloy: Heat resistant series; origin unknown
CHLORIMET 2	0.07% C 3.0% Fe 31.5% Mo Ni alloy: Casting; Duriron Co
	UTS: 506 Elon: 20% Proof: 323
CHLORIMET 3	0.07% C (max) 19% Cr 19% Mo Ni alloy: Casting; Duriron Co
	UTS: 500 Elon: 25% Proof: 320
CHLORIMET 3	0.7% C 13% Cr 5% Fe 3% B WC Ni alloy: Origin unknown
CHROMEL A	20% Cr Ni alloy: Wire for thermocouples; source unknown
CHROMEL P	10% Cr Ni alloy: Wire for thermocouples; source unknown
CHRONITE	0.4% Si 1% Al 1% Mn 13.5% Cr 10% Fe Ni alloy: Heat & corrosion resistant used for burners and certain types of valves; origin unknown
CORRONEL 230	35% Cr Ni alloy: Bar & sheet; H Wiggin; pickling and chemical plant
CM 53	7% Cr 4.5% Si 3% B 3% Fe Ni alloy: For hard facing; Dewrance; gas welding
CM 55	11% Cr 4% Si 2% B 4% Fe Ni alloy: For hard facing; Dewrance; gas welding
CM 56	16% Cr 4.5% Si 3.5% B 4% Fe Ni alloy: For hard facing; Dewrance; gas welding
COLMONOY 4	0.4% C 10% Cr 2.5% Fe 2% B Ni alloy: Rod for spraying; Wall Colmonoy; melting point 1110 °C; low friction
	DPN: 370
COLMONOY 4	0.4% C 10% Cr 2.5% Fe 2% B Ni alloy: Casting; Wall Colmonoy; melting point 1110 °C; low friction
	DPN: 370
COLMONOY 5	0.6% C 11.5% Cr 4.2% Fe 2.5% B Ni alloy: Casting; Wall Colmonoy; melting point 1070 °C; low friction
	DPN: 480

Symbol	Nominal analysis, supplier, condition and remarks.
COLMONOY 5	0.6% C 11.5% Cr 4.2% Fe 2.5% B Ni alloy: Rod for hard facing; Wall Colmonoy; melting point 1070 °C; low friction
	DPN: 480
COLMONOY 6	0.7% C 13% Cr 5% Fe 3% B Ni alloy: Rod for hard facing; Wall Colmonoy; melting point 1040 °C; low friction
	DPN: 670
COLMONOY 6	0.7% C 13% Cr 5% Fe 3% B Ni alloy: Castings; Wall Colmonoy; melting point 1040 °C; low friction
	DPN: 670
COLMONOY 8	0.9% C 26% Cr 1% Fe 3.5% B Ni alloy: Powder for spraying; Wall Colmonoy; melting point 1080 °C; low friction
	DPN: 640
COLMONOY 20	0.25% C 5% Cr 3.5% Fe 1% B Ni alloy: Castings; Wall Colmonoy; melting point 1220 °C; good wear resistance
	DPN: 220
COLMONOY 21	0.25% C 5% Cr 2% Fe 1.2% B Ni alloy: Powder for spraying; Wall Colmonoy for repairing moulds
	DPN: 285
COLMONOY 41	0.45% C 2.7% Si 2.7% Fe 10.0% Cr 2.0% B Ni alloy: Bronze powder; Wall Colmonoy; melting point 1105 °C
COLMONOY 45	0.5% C 10% Cr 3.5% Fe 2.2% B Ni alloy: Powder for spraying; Wall Colmonoy; melting point 1085 °C
	DPN: 420
COLMONOY 56	0.7% C 13% Cr 4.5% Fe 2.7% B Ni alloy: Rods for metal spraying; Wall Colmonoy; melting point 1050 °C; low friction corrosion resistance
	DPN: 580
COLMONOY 56	0.7% C 13% Cr 1.5% Fe 2.7% B Ni alloy: Casting; Wall Colmonoy; melting point 1050 °C; low friction
	DPN: 580
COLMONOY 60	0.7% C 3.71% Si 4.3% Fe 14.0% Cr 1.7% Cu 3.0% B 1.75% Mo Ni alloy: Bronze powder; Wall Colmonoy; melting point 1040 °C
COLMONOY 61	0.7% C 3.7% Si 4.6% Fe 14.0% Cr 3.0% B Ni alloy: Bronze powder; Wall Colmonoy; melting point 1025 °C
COLMONY 62	0.7% C 3.7% Si 4.6% Fe 14.0% Cr 3.0% B Ni alloy: Bronze powder; Wall Colmonoy; melting point 1025 °C
COLMONOY 72	0.7% C 3.5% Si 3.5% Fe 13.0% Cr 12.0% W Ni alloy: Bronze metal; Wall Colmonoy; melting point 1060 °C
COLMONOY 75	0.7% C 13% Cr 5% Fe 3% B Wc Ni alloy: Powder for spraying; Wall Colmonoy; melting point 1040 °C; wear resistant
	DPN: 660
DELORO SF 40	7.5% Cr 1.5% Fe 4% Si 1.5% B Ni alloy: Casting; Deloro
	DPN: 380
DELORO SF 50	10% Cr 4% Fe 4% Si 1.5% B Ni alloy: Casting; Deloro
	DPN: 550
DELORO SF 60	15% Cr 4.5% Fe 4.5% Si 3% B Ni alloy: Casting; Deloro
	DPN: 750
DIN 17470 Ni Cr 80/20	20% Cr Ni alloy
DIN 17742 Ni Cr 10	10% Cr Ni alloy
DIN 17742 Ni Cr 15 Fe Mo	15% Fe 16% Cr 7% Mo Ni alloy

Symbol	Nominal analysis, supplier, condition and remarks.
DTD 328	0.2% C (max) 15% Cr 10% Fe 1% Mn Ni alloy: Sheet **UTS: 450**
DTD 703B	0.1% C 20% Cr 0.5% Ti Ni alloy: Sheet; annealed **UTS: 750 Elon: 30% Proof: 300**
DTD 5047	16.5% Cr 35% Fe 3.2% Mo Ni alloy: Sheet & strip; weldable; heat resistant
DTD 5057	18% Cr 14% Co 7.0% Mo Ni alloy: Sheet & strip; weldable; heat resistant
DURCO CY40	0.4% C (max) 11.0% Fe (max) 15.5% Cr Ni alloy: Casting; Duriron Co **UTS: 490 Elon: 30% Proof: 195**
ERA HR 5	20% Cr 20% Fe Ni alloy: Casting; Hadfields specification for BS 1648 K **UTS: 740 Elon: 12% Proof: 330**
FOREMOST HR3	0.2% C 22% Cr 15% Fe Ni alloy: Casting; Swift Levick; annealed **DPN: 350 UTS: 810 Elon: 20% Proof: 610**
FOREMOST HR4	0.1% C 20% Cr 3% Ti 1.0% Al Ni alloy: Casting; Swift Levick; annealed **DPN: 300 UTS: 780 Elon: 44% Proof: 330**
G 63	0.1% C 15.5% Cr Ni alloy: Jessop; scale or creep resistant
HASTELLOY D	0.1% C 10% Si 1% Fe 4% Cu Ni alloy: Casting; Haynes Stellite
HEATING ELEMENTS	0.5% Zr 14% Cr 0.2% Cu 23% Fe Ni alloy: Origin unknown
HS 28W	0.45% C 26% Cr 2% Mo 1% Al 3% W 20% Fe Ni alloy: Thomson L'Hospied **DPN: 180 UTS: 570 Elon: 5% Proof: 370**
HW	0.55% C 12% Cr 23% Fe Ni alloy: Casting; Huntington; aged **UTS: 580 Elon: 4% Proof: 340**
HX	0.55% C 17% Cr 20% Fe Ni alloy: Casting; Thomson L'Hospied **DPN: 172 UTS: 450 Elon: 10% Proof: 250**
INCOLOY 65	0.03% C 21.0% Cr 1.7% Cu 30% Fe Ni alloy: H. Wiggin; for gas shielded arc welding of Incoloy 825
INCOLOY 135	0.05% C 26.0% Fe 1.8% Cu 29.0% Cr 3.2% Mo Ni alloy: H. Wiggin; electrode for metal arc welding of Incoloy 825
INCONEL	0.04% C 15% Cr 7% Fe Ni alloy: International Nickel
INCONEL	15% Cr 10% Fe Ni alloy: H Wiggin; furnace equipment; acid resistant
INCONEL	0.04% C 15.8% Cr 7.2% Fe Ni alloy: Huntington; now Inconel 600; Inconel used as trade name
INCONEL 42	0.04% C 7.2% Fe 16% Cr Ni alloy: Filler; Huntington; for gas welding Inconel 600 & Incoloy 800
INCONEL 62	0.04% C 7.5% Fe 16% Cr 2.3% Nb Ni alloy: Filler; Huntington; electric welding of Inconel 600 & Incoloy 800
INCONEL 112	0.05% C 4.0% Fe 21.5% Cr 9.0% Mo 3.6% Nb + Td Ni alloy: H. Wiggin; electrode for metal arc welding of Inconel 625
INCONEL 132	0.05% C 8.5% Fe 14% Cr 2% Nb 0.1% Co Ni alloy: Filler; Huntington; electric welding of Inconel 600

Symbol	Nominal analysis, supplier, condition and remarks.
INCONEL 182	0.05% C 7.5% Mn 7.5% Fe 14% Cr 2% Nb 0.1% Co Ni alloy: Filler; Huntington; welding Inconel 600 to steel
INCONEL 600	0.2% C 0.7% Cu 8% Fe 16% Cr Ni alloy: Wrought; H Wiggin; annealed **DPN: 180 UTS: 620**
INCONEL 600	15% Cr 8% Fe Ni alloy: Wrought; H Wiggin; annealed; tensile strength at 800 °C 10 tons/in.[2] **DPN: 180 UTS: 600 Elon: 42% Proof: 240**
INCONEL 600	15% Cr 8% Fe Ni alloy: Wrought; H Wiggin; cold drawn **DPN: 280 UTS: 810 Elon: 12% Proof: 760**
INCONEL 600	0.04% C 15.8% Cr 7.2% Fe Ni alloy: Bar, sheet, etc.; Huntington; annealed; Inconel 602 was Inconel 600 **DPN: 150 UTS: 660 Elon: 45% Proof: 350**
INCONEL 600	0.04% C 15.8% Cr 7.2% Fe Ni alloy: Bar, sheet, etc.; Huntington; cold drawn; Inconel 604 was Inconel 600 **DPN: 250 UTS: 910 Elon: 25% Proof: 680**
INCONEL 601	0.1% C (max) 1.5% Al 22.0% Cr 15% Fe Ni alloy: H. Wiggin; heat and corrosion resistance **DPN: 70 UTS: 690 Elon: 50 Proof: 560**
INCONEL 601	23.0% Cr 17.0% Fe 1.2% Al Ni alloy: For hot oxidation resistance; Huntington **DPN: 120 UTS: 600 Elon: 50% Proof: 250**
INCONEL 602	See INCONEL 600
INCONEL 604	0.04% C 15.8% Cr 7.2% Fe 2% Nb Ni alloy: Bar, sheet, etc.; Huntington; previously Inconel 600
INCONEL 610	0.2% C 15.5% Cr 1.0% Nb 9.0% Fe 0.5% Cu Ni alloy: Huntington
INCONEL 705	0.3% C 5.5% Si 15.5% Cr 8.0% Fe 0.5% Cu Ni alloy: Huntington
INCONEL FILLER 82	20.0% Cr 3.0% Mn 3.0% Fe (max) 2.5% Nb Ni alloy: Filler rod; Inco
INCONEL FILLER 92	0.08% C (max) 2.2% Mn 8.0% Fe 15.5% Cr Ni alloy: Filler rod; Inco
KH 70	0.07% C 29.5% Cr 5.0% Fe Ni alloy: Russian Standard designation
KH 701U	0.1% C (max) 29.5% Cr 1.0% Fe (max) Ni alloy: Russian Standard designation
KROMORE	15% Cr Ni alloy: Driver Harris; as Nichrome III; withdrawn
LESCALLOY 600	0.07% C 15% Cr 7% Fe Ni alloy: Latrobe; for nuclear reactor parts
MANAURITE 50W	28% Cr 5% W Ni alloy: Pompey **UTS: 550 Proof: 260**
MANAURITE 60	15% Cr 25% Fe Ni alloy: Pompey **UTS: 450 Elon: 22% Proof: 240**
MAXHETE 5	0.2% C 14.5% Cr 20% Fe Ni alloy: Edgar Allen; heat treatment equipment
MAXHETE 8	0.1% C 19.5% Cr Ni alloy: Edgar Allen; electrical resistance elements
METCO 12C	0.15% C 2.5% Fe 10% Cr 2.5% Si 2.5% B Ni alloy: Powder for spraying; Metco Ltd **DPN: 310**
METCO 15C	1% C 4% Fe 4% Si 17% Cr 3.5% B Ni alloy: Powder for spraying; Metco Ltd **DPN: 802**
METCO 31C	0.5% C 11% Cr 2.5% Si 2.5% Fe 35% WC 2.5% B Ni alloy: Powder for spraying; Metco Ltd; tungsten carbide in Ni alloy matrix **DPN: 865**
METCO 43C	20% Cr Ni alloy: Powder for spraying; Metco Ltd **DPN: 189**
METCOLOY 33	15% Cr 25% Fe Ni alloy: Wire for spraying; Metco Ltd
METCO XP 1150	0.5% C 14.5% Cr 3.5% Si 3.0% B, 3.5% Fe Ni alloy: For metal spray; Metco Ltd

Note. The following abbreviations and units are used in the tables:

DPN	Hardness, diamond pyramid number
UTS	Ultimate tensile strength, N/mm^2
Elon	Elongation, %
Proof	0.1 % proof strength, N/mm^2

1 N/mm^2=0.1 hbar=0.102 kgf/mm^2=0.06475 tonf/in^2=145.04 lbf/in^2

Symbol	Nominal analysis, supplier, condition and remarks.
N 75	20% Cr 0.4% Ti 5% Fe Ni alloy: H Wiggin
Na 22 H	0.5% C 27% Cr 17% Fe Ni alloy: American proprietary alloy listed in SAE year book
NCFI	0.15% C (max) 15.5% Cr 0.5% Cu (max) 8.0% Fe Ni alloy: Japanese Standard designation; Inconel 600.
NH	17% Cr 1% Si 1% Mn 35% Fe Ni: Ingot; primary metal; H Wiggin for addition to cast iron
Ni Cr 15 Fe	15% Cr 8% Fe Ni alloy: Designation used by German Standard
Ni Cr 60/15	15% Cr 21% Fe Ni alloy: Designation used by German Standard
NICHROME	Low C 15% Cr 20% Fe Ni alloy: For resistance elements; British Driver Harris; annealed **UTS: 170**
NICHROME II	18% Cr 14% Fe Ni alloy: British Driver Harris; withdrawn
NICHROME III	15% Cr Ni alloy: British Driver Harris; as Kromore; withdrawn
NICHROME IV	20% Cr Ni alloy: British Driver Harris; withdrawn
NICHROME V	20% Cr Ni alloy: British Driver Harris; use up to 1150 °C **UTS: 870**
NICHROME V	20% Cr Ni alloy: For electrical resistance; British Driver Harris; annealed **UTS: 170**
NICKEL FILLER 61	0.15% C (max) 1.0% Mn 1.5% Al 2.7% Ti 0.03% P (max) Ni: Filler wire for overlaying steel; H. Wiggin
NICKELVAC N	0.07% C 15.5% Cr 8% Fe Ni alloy: Vanadium Alloys Ltd; double vacuum cast
NICRAL Z	16% Cr 2% Fe Ni alloy: Imphy **DPN: 145 UTS: 630 Elon: 35% Proof: 270**
NICREX 3	0.07% C 15% Cr 25% Fe 0.8% Nb nickel alloy: Welding electrode; Murex **UTS: 600 Elon: 35%**
NICREX 4	0.2% C 18% Cr 1.8% Nb Ni alloy: Welding electrode, Murex **UTS: 650 Elon: 30%**
NICROBRAZ 30	0.1% C 19% Cr 10% Si Ni alloy: Braze metal; Wall Colmonoy; melting range 1080–1130 °C
NICROBRAZ 35	19.5% Cr 9.5% Si 9.5% Mn Ni alloy: For brazing; Wall Colmonoy; melting range 1135–1160 °C
NICROBRAZ 50	0.1% C 13% Cr 10% P Ni alloy: Braze metal; Wall Colmonoy; melting point 890 °C
NICROBRAZ 51	25% Cr 10% P Ni alloy: Bronze metal; Wall Colmonoy; melting range 980–1095 °C
NICROBRAZ 120	0.8% C 4.5% Si 13.5% Cr 3.5% B 4.5% Fe Ni alloy: Braze metal; Wall Colmonoy; as 'Standard' but coarser mesh powder
NICROBRAZ 125	0.7% C 14.0% Cr 3.0% B 4.5% Fe Ni alloy: For brazing; Wall Colmonoy; melting range 970–1040 °C
NICROBRAZ 150	0.1% C 15% Cr 2.5% B Ni alloy: Braze metal; Wall Colmonoy; melting point 1050 °C
NICROBRAZ 160	0.45% C 2.5% Si 10% Cr 2% B 2.5% Fe Ni alloy: Braze metal; Wall Colmonoy; melting range 970–1160 °C
NICROBRAZ 170	0.55% C 3.25% Si 11.5% Cr 2.5% B 16% W 3.7% Fe Ni alloy: Braze metal; Wall Colmonoy; melting range 980–1100 °C
NICROBRAZ 180	0.2% C 3% Si 5% Cr 1% B 3.5% Fe Ni alloy: Braze metal; Wall Colmonoy; melting range 970–1180 °C
NICROBRAZ 200	7% Cr 4.5% Si 3.2% B 6% W 3% Fe Ni alloy: Braze metal; Wall Colmonoy; melting range 975–1040 °C
NICROBRAZ 3001	11.5% Cr 6.0% Si Ni alloy: For brazing; Wall Colmonoy; melting range 1220–1235 °C
NICROBRAZ 3002	15.0% Cr 8.0% Si Ni alloy: For brazing; Wall Colmonoy; melting range 1180–1200 °C
NICROBRAZ 3003	17% Cr 9.5% Si 0.1% B Ni alloy: For brazing; Wall Colmonoy; melting range 1150–1180 °C
NICROBRAZ 3004	11.5% Cr 7.0% Si 0.4% B Ni alloy: For brazing; Wall Colmonoy; melting range 1160–1180 °C
NICROBRAZ 3005	13.0% Cr 8.0% Si 0.35% B Ni alloy: For brazing; Wall Colmonoy; melting range 1160–1180 °C
NICROBRAZ LC	0.06% C 13% Cr 4.5% Fe 3.5% B 4.5% Si Ni alloy; Braze metal; Wall Colmonoy; melting range 970–1080 °C
NICROBRAZ LM	0.06% C 6.5% Cr 4.5% Si 3% B 2.5% Fe Ni alloy: Braze metal; Wall Colmonoy; melting range 970–1000 °C
NICROBRAZ LMO1	3.0% Cr 3.0% Si 1.9% B 1.3% Fe Ni alloy: For brazing; Wall Colmonoy; melting range 1090–1155 °C
NICROBRAZ LMO2	2.25% Cr 2.75% Si 1.7% B 1.0% Fe Ni alloy: For brazing; Wall Colmonoy; melting range 1090–1155 °C
NICROBRAZ LMO3	2.5% Cr 3.0% Si 1.8% B 1.1% Fe Ni alloy: For brazing; Wall Colmonoy; melting range 1090–1155 °C
NICROBRAZ LMO4	2.3% Cr 3.0% Si 1.8% B 1.2% Fe Ni alloy: For brazing; Wall Colmonoy; melting range 1090–1160 °C
NICROBRAZ LMO5	2.7% Cr 3.0% Si 1.8% B 1.2% Fe Ni alloy: For brazing; Wall Colmonoy; melting range 1090–1160 °C
NICROBRAZ LMO6	2.3% Cr 2.4% Si 1.5% B 1.0% Fe Ni alloy: For brazing; Wall Colmonoy; melting range 1090–1160 °C
NICROBRAZ LMO7	5.6% Cr 3.5% Si 2.5% B 2.4% Fe Ni alloy: For brazing; Wall Colmonoy; melting range 1120–1160 °C
NICROBRAZ LMO8	3.8% Cr 3.5% Si 2.2% B 1.7% Fe Ni alloy: For brazing; Wall Colmonoy; melting range 1090–1155 °C
NICROBRAZ (STANDARD)	0.8% C 13.5% Cr 4.5% Si 3.5% B 4.5% Fe Ni alloy; Braze metal; Wall Colmonoy; melting range 980–1040 °C
NICROBRAZ WG	0.06% C 11.5% Cr 3.5% Si 3% B 3.5% Fe Ni alloy: Braze metal; Wall Colmonoy; melting range 970–1090 °C; obsolete
NICROTUNG	0.1% C 12.0% Cr 10.0% Co 8.0% W 4.0% Ti 4.0% Al Ni alloy: Origin unknown
NIMOCAST 75	0.1% C 20.0% Cr 0.4% Ti 50% Fe Ni alloy: Casting; H. Wiggin
NIMOCAST 257	0.08% C 20.0% Cr 16.0% Co 1.6% Ti 5.0% Fe Ni alloy: Casting; H. Wiggin
NIMOCAST 258	0.2% C 10.0% Cr 20.0% Co 5.0% Mo 3.7% Ti 4.8% Al 2.0% Fe Ni alloy: Casting; H. Wiggin
NIMOCAST 258	0.2% C 10.0% Cr 20.0% Co 5.0% Mo 3.7% Ti 4.8% Al 2.0% Fe Ni alloy: Casting; H. Wiggin
NIMOCAST PD16	0.13% C 6.0% Cr 2.0% Mo 6.0% Al Ni alloy: H. Wiggin; for turbine blades
NIMOCAST PE10	0.02% C 20.0% Cr 1.0% Co 6.0% Mo 2.5% W 6.5% Nb Ni alloy: Casting; H. Wiggin
NIMONIC 75	20% Cr 0.4% Ti 5% Fe Ni alloy: H. Wiggin; annealed; tensile strength at 750 °C 220 N/mm² **DPN: 200 UTS: 730 Elon: 44% Proof: 340**
NIMONIC 75	20% Cr 0.4% Ti 5% Fe Ni alloy: H. Wiggin; cold rolled **DPN: 300**
PER 1	20.0% Cr 5.0% Co 0.4% Ti Ni alloy: Origin unknown
PER 2	22.0% Cr 5.0% Co 4.0% Ti Ni alloy: Origin unknown
PER 2B	20.0% Cr 20.0% Co 2.0% Ti Ni alloy: Origin unknown

Symbol	Nominal analysis, supplier, condition and remarks.
PER 2U	20.0% Cr 20.0% Co 2.0% Ti 1.0% Al 5.0% Fe, Ni alloy: Origin unknown
PER 2V	19.0% Cr 2.0% Co 2.0% Ti 1.5% Al Ni alloy: Origin unknown
PER 13	13.0% Cr 4.0% Mo 6.0% Al Ni alloy: Origin unknown
PIREKS 60	20% Cr 20% Fe Ni alloy: Casting; Darwins heat treatment equipment
PIREKS 60/13	20% Cr 20% Fe Ni alloy: Rod or wire; Darwins electrical resistors; heating elements
PYROMET 600	0.1% C (max) 15.5% Cr 7% Fe Ni alloy: Carpenter
PYROMIC	20% Cr Ni alloy: For electrical resistance up to 1150 °C; Telcon Metals; annealed
	UTS: 770
QQ C.551	23% Cu 3% Fe 3% Mn Ni alloy: Casting; 2.0% C + Si; US Federal
	UTS: 450 Elon: 25% Proof: 180
R 870G	0.2% C 20.0% Cr 15% Fe Ni alloy: Casting; scaling temperature 1050 °C; Fagerston
RA 333	0.08% C 25% Cr 3.0% Mo 3.5% W 3.5% Co 20% Fe Ni alloy: Simonds
RNC 30	25% Cr 30% Fe Ni alloy: For electrical resistance; Imphy
RNC-CARBIMPHY	20% Cr 35% Fe Ni alloy: For electrical resistance; Imphy; for use in carburizing atmospheres
RNC-SUPERIMPHY	20% Cr Ni alloy: For electrical resistance; Imphy
SAE 70334	0.5% C 12% Cr 25% Fe 0.5% Mo Ni alloy: Casting
SAE 70335	0.5% C 17% Cr 22% Fe 0.5% Mo Ni alloy: Casting
SAE VF 1	0.2% C 20% Cr 1% Fe Ni alloy: For hard facing
SANICRO 71	0.04% C 16% Cr 9% Fe Ni alloy: For tube; Sandvik
	DPN: 160 UTS: 560 Elon: 30% Proof: 230
SF 40	7.5% Cr 1.5% Fe 4% Si 1.5% B Ni alloy: Deloro
SF 50	10% Cr 4% Fe 4% Si 1.5% B Ni alloy: Deloro
SF 60	15% Cr 4.5% Fe 4.5% Si 3% B Ni alloy: Deloro
SIMALLOY 600	0.15% C 16% Cr 8% Fe Ni alloy: Simonds
T 1	10% Cr Ni alloy: Thermocouple; British Driver Harris
	UTS: 650

Symbol	Nominal analysis, supplier, condition and remarks.
TOLOY 50	0.45% C 26% Cr 2% Mo 1% Al 3.0% W 20% Fe Ni alloy: Thomson L'Hospied
	DPN: 180 UTS: 570 Elon: 5% Proof: 370
TOLOY 50/50	0.1% C 50% Cr Ni alloy: Casting; Wellman; superheater tube supports, etc.
TOLOY 65	0.55% C 17% Cr 20% Fe Ni alloy: Casting; Thomson L'Hospied
	DPN: 172 UTS: 450 Elon: 10% Proof: 250
TOLOY 65 HX	0.55% C 17% Cr 18% Fe Ni alloy: Casting for furnace fittings, etc; Wellman
TOPHET 30	30% Cr Ni alloy: For heating elements; Driver Harris
TOPHET A	20% Cr Ni alloy: For heating elements and resistance wire; Driver Harris
TOPHET A	20% Cr Ni alloy: Strip for furnace elements; Gilbey-Brunton
TOPHET C	16% Cr 24% Fe Ni alloy: Strip for furnace elements; Gilbey-Brunton
TOPHET C	15% Cr Ni alloy: For heating elements and resistance wire; Driver Harris
UCAR 75	0.1% C 19.5% Cr Ni alloy: Union Carbide
UCAR 600	0.1% C (max) 15.5% Cr 2.2% Nb Ni alloy: Union Carbide
UCAR 601	0.05% C 23.0% Cr 1.35% Al Ni alloy: Union Carbide
VALRAY 1	20% Cr Ni alloy: With C Mn & Si for facing valves; H. Wiggin
VF 1	0.2% C 20% Cr 1% Fe Ni alloy: For hard facing; designation used by SAE
VIKRO 1	0.7% C 13.5% Cr 24.5% Fe Ni alloy: Casting; Firth Vickers
VIKRO 11	0.4% C 20.0% Cr 35% Fe Ni alloy: Casting; Firth Vickers

Note. The following abbreviations and units are used in the tables:

DPN	Hardness, diamond pyramid number
UTS	Ultimate tensile strength, N/mm^2
Elon	Elongation, %
Proof	0.1 % proof strength, N/mm^2

$1 \ N/mm^2 = 0.1 \ hbar = 0.102 \ kgf/mm^2 = 0.06475 \ tonf/in^2 = 145.04 \ lbf/in^2$

27C Nickel–chromium wrought and cast ageing alloys
With titanium, molybdenum, aluminium and cobalt

Specific gravity	8.1	
Density	8100 kg/m³	(0.291 lb/in.³)
Solidus/liquidus	1340–1390 °C	
Thermal conductivity	113 W/m °C	(27 cal/m s °C)
Coefficient of linear expansion (20–100° C)	11.7×10^{-6}/ °C	
Electrical conductivity	1.4% IACS (copper 100%)	
Specific resistance	1200 microhm mm	
Young's modulus of elasticity	$221 \times 10^9 \ N/m^2$	$(32 \times 10^6 \ lb/in.^2)$
Impact	42–100 J	(30 × 75 ft lb) Izod
Fatigue Strength (15×10^6 cycles)	$\pm 370 \ N/mm^2$ at 750 °C	$(\pm 25 \ tonf/in.^2)$

Hot strength

Temperature °C	Tensile strength N/mm^2	Elongation %
20	1210	33
200	1170	32
400	1080	32
500	1050	32
600	1030	18
700	810	10
800	550	15
900	220	31
1000	60	87

* The hot strength figures quoted are average. Tensile strengths of 170 N/mm^2 at 1000 °C are possible with some of these alloys.

All the above properties are included as a guide to the group of specifications as a whole. There is considerable variation within the group and some of the properties may not be applicable to all the specifications.

General metallurgical characteristics

These alloys are based on the 80/20 nickel/chromium materials described and listed in Section 27B with the addition of elements which form intermetallic compounds with each other and chromium. These take part in the metallurgical reactions necessary to impart age hardening. The principal effect is to increase tensile properties at high temperatures, the most important of which is the creep strength. Most of the alloys listed have excellent high temperature oxidation and corrosion resistance, although some of the more highly alloyed materials with the better creep strength have poorer hot corrosion resistance.

They are all expensive, on the whole difficult to work and are only used when their high temperature, high strength properties are essential.

Thermal treatment

All these alloys harden considerably with very little cold work. They are seldom used in this condition, although for certain purposes some of the alloys are solution treated, cold worked a measured amount then aged.

The solution treatment operation will remove all trace of previous cold work and put the material into its softest possible condition. This requires about four hours at a temperature of 1080–1200 °C followed by water, oil or air cooling. If the parts then receive further cold work it is recommended that a short term solution treat, sometimes called 'skin' anneal, is carried out. This is at the same temperature as the solution treat for about 10–15 minutes followed by water or oil quench. It is advised that this treatment is always applied after machining before ageing, otherwise cracking can occur. Ageing requires a temperature of 750–850 °C for 15–30 hours followed by air cooling. Very little actual scale is formed during these operations, but an adherent dark oxide is produced. This can be readily removed when necessary by grit blasting, hydrofluoric acid pickling, or sodium hydride treatment. This last method is the best but requires very special equipment. The acid pickle requires considerable control and is unpleasant to handle while blasting can be troublesome on thin sheet components and requires some control.

Briefly the metallurgy of these reactions is that the solution treatment dissolves all the intermetallic compounds in the nickel chromium matrix and they remain in solution after quenching. The fact that the treatment requires as long as four hours indicates the sluggish nature of the reaction, although some ageing or precipitation can take place on air cooling. The short time solution treat or 'skin' anneal has no appreciable effect on these intermetallic compounds but removes surface cold work. Any lower temperature, while probably removing some or all the cold work, would also cause some ageing to take place. Ageing (or the precipitation treatment) allows the intermetallic compounds to start coming out of solution and as this strains the lattic structure the hardness is increased. Over-ageing, either by too high a temperature or too long a time, releases some of this strain as the intermetallic compounds regain equilibrium and this reduces the hardness.

Electrical furnaces are recommended for all nickel alloy thermal treatments owing to the danger of sulphur contamination from oil or gas furnaces. Salt baths have been used for some treatments, but require careful choice of the salt being used.

Welding and brazing

Pre-treatment. The material should be in the solution treated, water or oil quenched condition, then grit blasted, acid pickled, or sodium hydride treated to remove scale. No work hardening of the joint area is permissible between the above treatment and the welding operation. Advice should be obtained prior to welding material in the aged condition.

Welding and brazing. Only by using the inert gas shielded arc process can really successful welds be achieved on these materials. Active fluxes and high skill with gas or electric welding can produce joins of a reasonable standard but there is always the danger of flux or oxide trapping.

Brazing generally requires high temperature filler metals necessitating the use of flux bath heating medium, or hydrogen or vacuum furnaces.

Post-treatment. Any fluxes must be removed immediately. Parts must be given the full ageing treatment after welding or brazing, and if there is the slightest chance of the joint area being stressed either at welding or by subsequent work a 'skin' anneal is necessary before ageing to prevent cracking.

Flaw detection methods

Crack test. Penetrant oil chalk test, using fluorescent fluids gives the highest sensitivity.

X-ray. Not used on wrought products. High duty castings should be X-rayed at least on a percentage basis and all welding should be controlled by this method.

Ultra-sonic test. This can be used to detect subcutaneous defects in castings and forgings at an early stage of machining. Certain weld defects can be identified using this method.

Chemical etch. Electropolishing in sulphuric/phosphoric acid is commonly used to identify material defects. Ferric chloride in strong hydrochloric acid shows major defects, grain size and orientation.

Corrosion protection

Temporary. None required.

Permanent. None required for most uses up to high temperatures, provided no sulphur is present.

Aluminizing shows promise for sulphur protection and against very high temperature oxidation. This is a process analogous to carburizing, which requires very careful control, is expensive, and results in a surface which is easily damaged.

Machinability

All the alloys listed harden considerably when cold worked. It is therefore essential that tools are cutting at all times and are not rubbing or subjected to interrupted cuts. High feed rates with slow speeds give best results, but require considerable power and very rigid machines.

Provided sufficient power is available and sharp, tipped tools are used, most of the materials can be machined in the aged condition, but in general best results are obtained with the solution treated, water quenched condition. This can accept considerably more cold work than the aged condition before becoming unmachinable. If there is heavy machining one or more intermediate 'skin' anneals are recommended to remove the machining cold work. A skin anneal is always advisable immediately before final ageing if there is any appreciable machining between the solution treat and age, otherwise cracking at ageing can occur. Grinding is not always successful with these materials which are tough rather than hard, and tend to load the stone, causing glazing. Spark erosion and electrochemical machining techniques are finding considerable use in forming and shaping many of these rather intractable materials.

Uses

Turbine blades in gas and steam turbine engines, combustion equipment requiring high strength at temperature. Internal combustion engine exhaust valves, chemical plant components at the hot end of refineries.

Symbol	Nominal analysis, supplier, condition and remarks.
1/360	0.1% C 10% Cr 5% Mo 2% Cb 6% Al 4.5% Fe 0.3% B Ni alloy: Casting; American proprietary alloy listed in SAE year book
1C 901	0.07% C 13% Cr 6% Mo 3% Ti 40% Fe alloy: Bofors; age hardened **DPN: 340 UTS: 1020 Elon: 12% Proof: 670**
2.4537	0.02% C (max) 15.5% Cr 16.0% Mo 3.7% W 6.0% Fe 2.5% Co 0.35% V Ni alloy
2.4602	6% Fe 16% Cr 17% Mo 4% W Ni alloy: German Standard
2.4605	25% Fe 22% Cr 7% Mo 2.0% Nb Ni alloy: German Standard
2.4889	15% Fe 15% Cr 7% Mo Ni alloy: German Standard
69 INCONEL X	0.04% C 6.8% Fe 15% Cr 0.8% Al 2.4% Ti 0.8% Nb Ni alloy: Filler; Huntington; now Inconel 69

Note. The following abbreviations and units are used in the tables:

DPN	Hardness, diamond pyramid number
UTS	Ultimate tensile strength, N/mm^2
Elon	Elongation, %
Proof	0.1 % proof strength, N/mm^2

1 N/mm^2=0.1 hbar=0.102 kgf/mm^2=0.06475 tonf/in^2=145.04 lbf/in^2

Symbol	Nominal analysis, supplier, condition and remarks.
134 K MONEL	0.25% C 27% Cu 2.5% Mn 2% Al 0.3% Ti Ni alloy: Filler; Huntington; now Monel 134
713C	0.12% C 12.5% Cr 4.2% Mo 2.2% Nb + Ta 6.1% Al 0.8% Ti Ni alloy: International Nickel Co; vacuum melted **UTS: 860 Elon: 8% Proof: 750**
713C	0.12% C 12.5% Cr 4.2% Mo 2.2% Nb 6.1% Al 0.8% Ti 0.12% B 0.1% Zr Ni alloy: Union Carbide
713LC	0.05% C 12% Cr 4.5% Mo 2.3% Nb 6.0% Al 0.7% Ti 0.01% B 0.1% Zr Ni alloy: Union Carbide
713LC	0.05% C 12% Cr 4.5% Mo 2.0% Nb + Ta 5.9% Al 0.6% Ti Ni alloy: International Nickel Co; as cast **UTS: 910 Elon: 15% Proof: 760**
718	0.08% C 19% Cr 3% Mo 18% Fe 1% Ti 0.6% Al 5% Nb + Ta Ni alloy: Crucible Steel Co; solution treated & aged **UTS: 1380 Proof: 1100**
738	0.17% C 8.5% Co 16.0% Cr 1.75% Mo 2.6% W 0.9% Nb 1.75% Ta 3.4% Al 3.4% Ti 0.01% B 0.1% Zr Ni alloy: Union Carbide
AF 1753	0.25% C 16.2% Cr 7.2% Co 1.6% Mo 8.4% W 3.2% Ti 1.9% Al 9.5% Fe 0.06% Zr Ni alloy: Information from SAE handbook
AISI 664	0.06% C 15% Cr 30% Fe 4% Mo 3.6% W 3% Ti 1% Al 0.01% B Ni alloy: Solution treated and aged **UTS: 1470 Elon: 15% Proof: 1020**

Symbol	Nominal analysis, supplier, condition and remarks.
AISI 680	0.1% C 21.5% Cr 9% Mo 0.6% W 18.5% Fe 1.5% Co Ni alloy: Solution treated; has slight ageing properties **UTS: 830 Elon: 52% Proof: 270**
AISI 681	0.05% C 12.5% Cr 37% Fe 6% Mo 2.5% Ti 0.2% Al 0.015% B Ni alloy: Solution treated, double aged **UTS: 1220 Elon: 17% Proof: 740**
AISI 682	0.05% C 12.5% Cr 5.7% Mo 37% Fe 2.8% Ti 0.2% Al 0.015% B Ni alloy: Solution treated, double aged **UTS: 1220 Elon: 17% Proof: 740**
AISI 683	0.09% C 19% Cr 10% Mo 3.1% Ti 1.5% Al 1.8% Fe 11% Co Ni alloy: Solution treated and aged **UTS: 1470 Elon: 14% Proof: 1020**
AISI 684	0.1% C 17.5% Cr 4.2% Mo 0.5% Fe 3% Ti 3% Al 18.5% Co 0.005% B Ni alloy: Solution treated, double aged **UTS: 1250 Elon: 16% Proof: 800**
AISI 685	0.07% C 19.7% Cr 4.45% Mo 3% Ti 1.4% Al 0.7 Fe 13.5% Co 0.005% B Ni alloy: Solution treated, double aged **UTS: 1380 Elon: 25% Proof: 800**
AISI 686	0.12% C 15% Cr 5% Mo 2.5% Ti 2% Al 10% Fe Ni alloy: Solution treated & aged **UTS: 1220 Elon: 21% Proof: 830**
AISI 687	0.07% C 15% Cr 5.25% Mo 3.5% Ti 4.2% Al 0.5% Fe 18.5% Co 0.03% B Ni alloy: Solution treated, double aged **UTS: 1470 Elon: 17% Proof: 760**
AISI 688	15% Cr 2.5% Ti 0.8% Cb 0.8% Al 6.7% Fe Ni alloy: Solution treated, double aged **UTS: 1130 Elon: 24% Proof: 650**
AISI 689	0.15% C 20% Cr 10% Mo 2.6% Ti 1% Al 10% Co 0.005% B Ni alloy: Solution treated & aged **UTS: 1220 Elon: 16% Proof: 880**
AISI 690	0.03% C 18% Cr 3.2% Mo 2.7% Ti 0.2% Al 20% Cr 18% Fe Ni alloy: Solution treated, double aged **UTS: 1020 Elon: 19% Proof: 650**
ALLOY 49	2.5% C 30% Cr 5% Fe 14% W Ni alloy: Casting; Dewrance **DPN: 395 UTS: 770**
ALLOY 63	2% C 26% Cr 20% W 1% V Co alloy: Casting; Dewrance **DPN: 500 UTS: 540**
ALLOY 80	0.15% C 16% Cr 17% Mo 5% W 5% Fe Ni alloy: Casting; Dewrance **DPN: 265 UTS: 540**
ALLOY 81	0.12% C 1% Cr 28% Mo 5% Fe Ni alloy: Casting; Dewrance **DPN: 235 UTS: 540**
ALLOY 713 C	0.12% C 12.5% Cr 4.5% Mo 2.0% Nb 0.6% Ti 6.0% Al 0.012% B 1.0% Fe 0.1% Zr Ni alloy: Information from SAE handbook
ALLVAC 500ZB	0.08% C 19% Cr 4% Mo 17% Co 3% Ti 3% Al 0.006% B 0.06% Zr Ni alloy: Vanadium Alloys; double vacuum cast

Note. The following abbreviations and units are used in the tables:

DPN	Hardness, diamond pyramid number
UTS	Ultimate tensile strength, N/mm^2
Elon	Elongation, %
Proof	0.1 % proof strength, N/mm^2

1 N/mm^2=0.1 hbar=0.102 kgf/mm^2=0.06475 tonf/in^2=145.04 lbf/in^2

Symbol	Nominal analysis, supplier, condition and remarks.
ALLVAC 718	0.05% C 19% Cr 3% Mo 1% Ti 0.5% Al 0.004% B 5.1% Nb + Ta 20% Fe Ni alloy: Vanadium Alloys; double vacuum cast
ALLVAC GMR 235	0.12% C 15.5% Cr 5.5% Mo 2.5% Ti 2% Al 10% Fe Ni alloy: Vanadium Alloys; double vacuum cast
ALLVAC R 41	0.08% C 19% Cr 9.7% Mo 11% Co 3.2% Ti 1.5% Al 0.006% B Ni alloy: Vanadium Alloys; double vacuum cast
ALLVAC M 252	0.14% C 19% Cr 9.7% Mo 10.2% Co 2.6% Ti 1.1% Al 0.06% B Ni alloy: Vanadium Alloys; double vacuum cast
ALLVAC WASPALOY	0.08% C 19.5% Cr 4.25% Mo 14% Co 3% Ti 1.4% Al 0.006% B 0.06% Zr Ni alloy: Vanadium Alloys; double vacuum cast
AMD 57 BX	16.5% Cr 7.5% Co 1.5% Mo 8.5% W 3% Ti 2% Al 9% Fe Ni alloy: Bar & forging; interim AMS
AMPCOLAY 561	30% Fe 20% Cr 3% Mo Ni alloy: Ampco for Ni O NEL
AMS 5384	18% Cr 18% Co 3% Ti 3% Al 4% Mo 2% Fe Ni alloy: Casting; solution treated & aged; vacuum melted
AMS 5388 B	16% Cr 17% Mo 4.5% W 6% Fe Ni alloy: Investment casting
AMS 5389 A	17% Mo 15% Cr 6% Fe 5% W Ni alloy: Sand casting
AMS 5390	22% Cr 1.5% Co 9% Mo 0.6% W 18% Fe Ni alloy: Investment casting
AMS 5391	13% Cr 4.5% Mo 0.7% Ti 6% Al 2.5% Nb + Ta NI alloy: Investment casting; vacuum melted
AMS 5396	28% Mo 5% Fe 0.4% V Ni alloy: Investment casting
AMS 5509	15% Cr 4% W 4% Mo 3% Ti 1% Al 28% Fe Ni alloy: Sheet & strip; vacuum melted
AMS 5536 C	22% Cr 9% Mo 1.5% Co 0.6% W 18% Fe Ni alloy: Sheet
AMS 5541 A	15.5% Cr 2.5% Ti 0.7% Al Fe Ni alloy: Sheet
AMS 5542 C	15.5% Cr 2.5% Ti 1% Nb + Ta 0.7% Al Fe Ni alloy: Sheet
AMS 5544	19.5% Cr 13.5% Co 4.3% Mo 3% Ti 1.4% Al Ni alloy: Sheet; annealed; vacuum melted
AMS 5545	19% Cr 11% Co 10% Mo 3% Ti 1.5% Al Ni alloy: Sheet; solution treated; vacuum melted
AMS 5550A	15.5% Cr 0.7% Ti 3.25% Al Ni alloy: Sheet
AMS 5551	19% Cr 10% Co 10% Mo 1% Al 2.5% Ti Ni alloy: Sheet; solution treated; vacuum melted
AMS 5582	15.5% Cr 7% Fe 2.5% Ti 1% Nb + Ta 0.7% Al Ni alloy: Seamless tube
AMS 5593	25% Cr 20% Fe 3.2% Co 3.2% Mo 3.2% W Ni alloy: Sheet
AMS 5596	19% Cr 3% Mo 5% Nb + Tn 1% Ti 0.6% Al Ni alloy: Sheet; annealed; vacuum melted
AMS 5660 A	12.5% Cr 6% Mo 2.5% Ti 34% Fe Ni alloy: Bar & forging; vacuum melted
AMS 5661	12.5% Cr 5.8% Mo 3% Ti 34% Fe Ni alloy: Bar & forging; vacuum melted
AMS 5667 F	15.5% Cr 7% Fe 2.5% Ti 1% Nb + Ta 0.7% Al Ni alloy: Bar
AMS 5668 D	15.5% Cr 7% Fe 2.5% Ti 1% Nb + Ta 0.7% Al Ni alloy: Bar
AMS 5675	15.5% Cr 7% Fe 3% Ti 2.3% Mn Ni alloy: Welding wire; cold drawn
AMS 5698 B	15.5% Cr 7% Fe 2.5% Ti 1% Nb + Ta 0.7% Al Ni alloy: Wire
AMS 5699 B	15.5% Cr 7% Fe 2.5% Ti 1% Nb + Ta 0.7% Al Ni alloy: Wire; spring temper
AMS 5706	19.5% Cr 13.5% Co 4.5% Mo 3% Ti 1.5% Al Ni alloy: Bar & forging; solution treated; vacuum melted

Symbol	Nominal analysis, supplier, condition and remarks.
AMS 5707 A	19.5% Cr 13.5% Co 4.5% Mo 3% Ti 1.5% Al Ni alloy: Bar & forging; stabilized and aged
AMS 5708	19.5% Cr 13.5% Co 4.5% Mo 3% Ti 1.5% Al Ni alloy: Bar & forging; solution treated
AMS 5709	19.5% Cr 13.5% Co 4.5% Mo 3% Ti 1.5% Al Ni alloy: Bar & forging; solution treated, stabilized and aged
AMS 5712	19% Cr 11% Co 10% Mo 3% Ti 1.5% Al Ni alloy: Bar & forging; solution treated; vacuum melted
AMS 5713	19% Cr 11% Co 10% Mo 3% Ti 1.5% Al Ni alloy: Bar & forging; solution treated and aged; vacuum melted
AMS 5717	25% Cr 3.2% Co 3.2% Mo 3.2% W 30% Fe Ni alloy: Bar & forging
AMS 5746	15% Cr 30% Fe 4% W 4% Mo 1% Al Ni alloy: Bar & forging; vacuum melted
AMS 5750	16% Mo 15.5% Cr 6% Fe 4% W Ni alloy: Bar & forging
AMS 5751	18% Cr 17% Co 3% Ti 3% Al 4% Mo 4% Fe Ni alloy: Bar & forging; solution treated; stabilized and aged
AMS 5753	18% Cr 17% Co 4% Mo 3% Ti 3% Al Ni alloy: Bar & forging; solution treated; vacuum melted
AMS 5754D	22% Cr 1.5% Co 9% Mo 0.6% W 18% Fe Ni alloy: Bar & forging
AMS 5755	5% Cr 24% Mo 5.5% Fe Ni alloy: Bar & forging
AMS 5756	19% Cr 10% Co 10% Mo 1% Al 2.5% Ti Ni alloy: Bar & forging; solution treated; vacuum melted
AMS 5757	19% Cr 10% Co 10% Mo 1% Al 2.5% Ti Ni alloy: Bar & forging; solution treated and aged; vacuum melted
AMS 5778	15.5% Cr 2.5% Ti 1% Nb + Ta 0.7% Al 7% Fe Ni alloy: Electrode
AMS 5779	15% Cr 2% Ti 0.6% Al Nb + Ta Ni alloy: Coated electrode
AMS 5786	5% Cr 24.5% Mo 5.5% Fe Ni alloy: Wire
AMS 5787	4.5% Cr 24% Mo 5.5% Fe Ni alloy: Coated electrode
AMS 5798	22% Cr 1.5% Co 9% Mo 0.6% W 18.5% Fe Ni alloy: Wire
AMS 5799	22% Cr 1.5% Co 9% Mo 0.6% W 18.5% Fe Ni alloy: Coated electrode
AMS 5800 A	19% Cr 11% Co 10% Mo 3.2% Ti 1.5% Al Ni alloy: Welding wire; vacuum melted
AMS 5828	19.5% Cr 13.5% Co 4.3% Mo 3% Ti 4% Al Ni alloy: Welding wire; vacuum melted
ASTM A461/684	0.15% C (max) 17.5% Cr 16.5% Co 4.0% Mo 2.7% Ti 2.8% Al 0.008% B 4.0% Fe Ni alloy: Bar
ASTM A461/685	0.06% C 19.5% Cr 13.5% Co 4.2% Mo 3.0% Ti 1.4% Al 0.008% B 2% Fe Ni alloy: Bar
ASTM A461/688	0.08% C (max) 15.5% Cr 1.0% Co (max) 1.0% Nb + Ta 2.5% Ti 0.7% Al 7.0% Fe Ni alloy: Bar
ASTM A461/689	0.15% C 19% Cr 10% Co 9.7% Mo 2.5% Ti 1.0% Al 0.008% B 5% Fe Ni alloy: Bar
ASTM A567/4	0.12% C (max) 16.5% Cr 2.0% Co 17% W 4.5% V 0.3% Nb + Ta 5.7% Fe Ni alloy: Casting
ASTM A567/5	0.2% C (max) 21.7% Cr 1.7% Co 9% Mo 0.7% W 18.5% Fe Ni alloy: Casting; as AISI 680
ASTM A567/6 V	0.1% C (max) 19% Cr 18% Co 4.0% Mo 2% Fe 3% Al 3% Ti Ni alloy: Casting; as AISI 684

Symbol	Nominal analysis, supplier, condition and remarks.
ASTM A567/7 V	0.14% C 13% Cr 4.5% Mo 2.5% Nb + Ta 6% Al 0.7% Ti 0.1% Zr Ni alloy: Casting
ASTM A567/8	0.15% C 15.5% Cr 5.2% Mo 10% Fe 3% Al 2.0% Ti Ni alloy: Casting
ASTM A567/9 V	0.15% C 15.5% Cr 5.2% Mo 10% Fe 3.0% Al 2.0% Ti Ni alloy: Casting
ASTM A567/10 V	0.15% C 15.5% Cr 5.2% Mo 4.2% Fe 3.6% Al 2.5% Ti Ni alloy: Casting
ASTM A637/80 A	0.1% C (max) 19.5% Cr 2.0% Ti 1.2% Al 3.0% Fe (max) Ni alloy
ASTM A637/684	0.15% C (max) 17.5% Cr 17.5% Co 4.0% Mo 2.8% Ti 3.0% Al 0.008% B 4.0% Fe Ni alloy
ASTM A637/685	0.07% C 19.0% Cr 13.0% Co 4.2% Mo 3.0% Ti 1.4% Al 0.09% Zn 0.008% B 2.0% Fe Ni alloy
ASTM A637/688	0.08% C (max) 15.5% Cr 1.0% Co (max) 1.0% Ta 2.5% Ti 0.8% Al 7.0% Fe Ni alloy
ASTM A637/689	0.15% C 0.5% Mn (max) 19.0% Cr 10.0% Co 9.7% Mo 2.5% Ti 1.0% Al 5.0% Fe Ni alloy
ASTM A637/718	0.08% C (max) 19.0% Cr 1.0% Co (max) 3.0% Mo 5.1% Nb + Ta 0.8% Ti 0.5% Al 20% Fe Ni alloy **UTS: 965 Elon: 30% Proof: 550**
ASTM B260 B Ni 7	13% Cr 10% P Ni alloy: Brazing filler metal; braze temperature range 927–1038 °C
ASTM B295 E3N1B	0.1% C 5% Fe 2% Co 1% Cr 28% Mo Ni alloy: Welding rod; as welded; as E4N1B **UTS: 830**
ASTM B295 E3N1C	0.12% C 5% Fe 2% Co 15% Cr 17% Mo 4% W Ni alloy: Welding electrode; as welded; as E4N1C **UTS: 650**
ASTM B295 E3N19	0.2% C 11% Fe 15% Cr 3% Nb 2% Ti Ni alloy: Welding electrode; age hardened **UTS: 830**
ASTM B295 E4N1B	0.1% C 5% Fe 2% Co 1% Cr 28% Mo Ni alloy: Welding rod; as welded; as E3N1B **UTS: 830**
ASTM B295 E4N1C	0.12% C 5% Fe 2% Co 15% Cr 17% Mo 4% W Ni alloy: Welding electrode; as welded; as E3N1C **UTS: 650**
ASTM B304 ERN7B	0.08% C 6% Fe 2.5% Co 1% Cr 28% Mo 0.4% V Ni alloy: Welding electrode
ASTM B304 ERN7C	0.08% C 6% Fe 2.5% Co 15% Cr 16% Mo 3.5% W Ni alloy: Welding electrode
ASTM B304 ERN7W	0.12% C 6% Fe 2.5% Co 5% Cr 24% Mo 0.6% V Ni alloy: Welding electrode
ASTM B304 ERN69	0.08% C 7% Fe 1% Al 2.5% Ti 16% Cr 1% Nb Ni alloy: Welding electrode
ASTM B322 Ni Mo	1% Cr 5% Fe 2.5% Co 0.4% V 28% Mo Ni alloy: Casting; annealed **UTS: 530 Elon: 6% Proof: 320**
ASTM B322 Ni Mo Cr	16% Cr 4.5% W 6% Fe 2.5% Co 0.3% V 17% Mo Ni alloy: Casting; annealed **UTS: 530 Elon: 4% Proof: 320**
ASTM B333	1% Cr 5% Fe 2.5% Co 0.3% V 28% Mo Ni alloy: Sheet; hot rolled and annealed **UTS: 810 Elon: 45% Proof: 340**
ASTM B334	15% Cr 4% W 6% Fe 2.5% Co 0.3% V 16% Mo Ni alloy: Sheet; hot rolled and annealed **UTS: 810 Elon: 40% Proof: 340**
ASTM B335	1% Cr 5% Fe 2.5% Co 0.3% V 28% Mo Ni alloy: Rod; annealed **UTS: 810 Elon: 32% Proof: 320**
ASTM B336	15% Cr 3.5% W 6% Fe 2.5% Co 16% Mo 0.3% V Ni alloy: Rod; annealed **UTS: 790 Elon: 22% Proof: 320**
ASTROLOY	0.06% C 15% Cr 15% Co 5.2% Mo 3.5% Ti 4.4% Al 0.03% B Ni alloy: American proprietary alloy listed in SAE yearbook
ATG R	0.11% C 20% Cr 0.4% Ti 4% Fe 4% Co Ni alloy: Imphy **UTS: 770 Elon: 40% Proof: 330**

Note. The following abbreviations and units are used in the tables:

DPN	Hardness, diamond pyramid number
UTS	Ultimate tensile strength, N/mm²
Elon	Elongation, %
Proof	0.1 % proof strength, N/mm²

$1 \text{ N/mm}^2 = 0.1 \text{ hbar} = 0.102 \text{ kgf/mm}^2 = 0.06475 \text{ tonf/in}^2 = 145.04 \text{ lbf/in}^2$

Symbol	Nominal analysis, supplier, condition and remarks.
ATG S3	0.1% C 20% Cr 2.0% Ti 1.2% Al 5.0% Fe Ni alloy: Imphy; solution treated & aged **UTS: 1000 Elon: 15% Proof: 590**
ATG S4	0.12% C 20% Cr 2.2% Ti 1.2% Al 20% Co 2.0% Fe Ni alloy: Imphy; solution treated & aged **UTS: 1060 Elon: 15% Proof: 700**
BP 87	0.02% C 3% Mn 20% Cr 0.5% Ti 2.5% Nb Ni alloy: Filler; Huntington; now Inconel 82
BS 3076 NA16	21% Cr 32% Fe 2.2% Cu 3.0% Mo 0.9% Ti Ni alloy: Rod
BS 3976 NA19	18% Co 19% Cr 1.3% Al 2.2% Ti Ni alloy: Rod
BS 3146 ANC9	0.8% C 20% Cr 2.8% Ti 1.4% Al Ni alloy: Investment casting; solution treated and aged **UTS: 570 Elon: 5% Proof: 400**
BS 3146 ANC10	0.1% C 20% Cr 1.2% Al 17% Co Ni alloy: Investment casting; solution treated and aged **UTS: 570 Elon: 5% Proof: 400**
BS 3146 ANC11	0.32% C 21% Cr 10% Mo 10% Co Ni alloy: Investment casting; as cast **UTS: 400 Elon: 5%**
BS 3146 ANC12	0.1% C (max) 21% Cr 10% Mo 2.5% Ti 0.8% Al 1.0% Co Ni alloy: Investment casting; as cast **UTS: 500 Elon: 5%**
BS 3146 ANC16	0.1% C 17% Cr 17% Mo 4.5% W 6% Fe Ni alloy: Investment casting; as cast **UTS: 420 Elon: 5% Proof: 220**
C 242	21.5% Cr 10.5% Mo 10% Co Ni alloy: Casting; H. Wiggin; see Nimonic 242
C 263	20% Cr 5.9% Mo 2.2% Ti 0.5% Al 20% Co Ni alloy: Forging; H. Wiggin; for use up to 850 °C
C 276	15.0% Cr 5.5% Fe 16.0% Mo 4.0% W 2.0% Co Ni alloy: Climax Moly
C 1023	0.15% C 9.75% Co 15.5% Cr 8.3% Mo 4.1% Al 3.6% Ti 0.006% B Ni alloy: Union Carbide
CM 40	2.5% C 29% Cr 10% Co 14% W Ni alloy: For hard facing; Dewrance; gas welding
COLMONOY 70	0.5% C 11.5% Cr 3.7% Fe 2.5% B 16% W Ni alloy: Casting; Wall Colmonoy; melting point 1120 °C; resists high temperature frettage and corrosion **DPN: 580**
COLMONOY 70	0.5% C 11.5% Cr 3.7% Fe 2.5% B 16% W Ni alloy: Rod for hard facing; Wall Colmonoy; melting point 1120 °C; resists high temperature frettage and corrosion **DPN: 580**
CORRONEL 220	0.05% C 0.8% Mn 29% Mo 2% V 3% Fe 0.2% Cu Ni alloy: H. Wiggin; acid resistant **UTS: 1060 Elon: 37% Proof: 550**
CORRONEL 230	0.08% C 1% Cu 5% Fe 36% Cr 1% Ti 0.5% Al Ni alloy: Wrought; H. Wiggin; annealed **DPN: 250 UTS: 800**
CRONITE	0.7% C 17% Cr 0.3% Mo 1.0% W 0.2% Co 25% Fe Ni alloy: Casting; Cronite; as cast **DPN: 195 UTS: 580 Proof: 300**
CRUCIBLE 718	0.08% C 19% Cr 3% Mo 18% Fe 1% Ti 0.06% Al 5% Nb + Ta Ni alloy: Crucible Steel Co; solution treated and aged **UTS: 1300 Proof: 1120**
CRUCIBLE M252	0.15% C 19% Cr 10% Mo 2.5% Ti 10% Co 1% Al Ni alloy: Crucible Steel Co; solution treated and aged; AISI 689 **UTS: 1150 Elon: 16% Proof: 740**
CRUCIBLE X750	0.04% C 15% Cr 2.5% Ti 7% Fe 0.8% Al 0.8% Nb + Ta Ni alloy: Crucible Steel Co; double solution aged & aged; AISI 688 **UTS: 1150 Elon: 23% Proof: 800**
D 979	0.05% C 15% Cr 4% Mo 4% W 3% Ti 1% Al 0.01% B 27% Fe Ni alloy: American proprietary alloy listed in SAE yearbook

Symbol	Nominal analysis, supplier, condition and remarks.
DARWIN 55	0.2% C 23% Cr 4% Mo 2% W 10% Fe 5% Cu Ni alloy: Casting; Darwin; acid resistant
DARWIN 654A	0.1% C (max) 20% Fe 20% Mo Ni alloy: Casting; Darwin; acid resistant; not oxidizing acids
DARWIN 655B	Low C 8% Fe 27% Mo Ni alloy: Casting; Darwin; acid resistant; not oxidizing acids
DARWIN 656C	Low C 14% Cr 17% Mo 5% W 6% Fe Ni alloy: Casting; Darwin; acid resistant
DELORO B	0.1% C 29% Mo 5% Fe Ni alloy: Casting; Deloro **DPN: 270 UTS: 600 Elon: 10% Proof: 360**
DELORO C	0.1% C 17% Cr 5% W 17% Mo 6% Fe Ni alloy: Casting; Deloro **DPN: 275 UTS: 600 Elon: 10% Proof: 370**
DCM	0.08% C 15% Cr 5.2% Mo 3.5% Ti 4.6% Al 5% Fe 0.08% B Ni alloy: Casting; American proprietary alloy listed in SAE yearbook
DIN 17742 Ni Cr 22 Mo	25% Fe 22% Cr 7% Mo 2.0% Nb Ni alloy
DIN 17742 Ni Mo 16 Cr	6% Fe 16% Cr 16% Mo 4% W Ni alloy
DTD 725	0.1% C (max) 19% Cr 1% Al 2% Ti 2% Co (max) Ni alloy: Forging; solution treated and aged; N80 material **DPN: 250**
DTD 736	0.1% C 19% Cr 2.2% Ti 1% Al 5% Fe 2% Co (max) Ni alloy: Forging; solution treated and aged; N80A material **DPN: 250**
DTD 747A	0.1% C 20% Cr 2.2% Ti 1.3% Al 17% Co 5% Fe Ni alloy: Solution treated and aged; N90 **DPN: 250**
DTD 5007	0.2% C 14.5% Cr 5.0% Mo 1.2% Ti 4.7% Al 20% Co Ni alloy: Tensile strength 110 N/mm² at 940 °C; N115 **DPN: 300**
DTD 5017	0.2% C 15% Cr 4.0% Mo 4.0% Ti 5.0% Al 15% Co Ni alloy: Tensile strength 110 N/mm² at 980 °C; N115 **DPN: 330**
DTD 5027	0.13% C 20% Cr 2.2% Ti 1.3% Al 17% Co Ni alloy sheet: Tensile strength 140 N/mm² at 870 °C; solution treated & aged; N90 **DPN: 230 UTS: 1070 Elon: 20% Proof: 630**
DTD 5067	15% Cr 15% Co 4.8% Al 3.8% Ti 3.5% Mo Ni alloy: Vacuum melted
DTD 5077	19.5% Cr 2.25% Ti 1.4% Al Ni alloy: Wrought; for studs, nuts, etc.
EPK 24	See Nimocast PK24
EVANOHM	20% Cr 2.7% Al 2.7% Cu Ni alloy: Wire; Brunton; specific resistance 1340 microhm mm; for use up to 300 °C
FS 718	0.05% C 18.5% Cr 3.0% Mo 0.9% Ti 0.6% Al 5.0% Nb Ta 20% Fe Ni alloy: Firth Sterling; solution treated and aged **UTS: 1500 Elon: 21% Proof: 1350**
FS 901	0.05% C 12.5% Cr 5.7% Mo 2.9% Ti 0.2% Al 37% Fe Ni alloy: Firth Sterling; solution treated and aged **UTS: 1180 Elon: 21% Proof: 850**
FS X 750	0.04% C 15.25% Cr 2.5% Ti 0.9% Nb Ta 0.8% Al 6.7% Fe Ni alloy: Firth Sterling; solution treated and aged **UTS: 1200 Elon: 28% Proof: 770**
G 39	0.5% C 19.5% Cr 3% Mo 3% W 1.5% Nb 5% Fe Ni alloy: Casting; Jessops; used as cast **UTS: 480 Elon: 5%**
G 44	0.08% C 20% Cr 6% Mo 1.6% Ti 8% Co 1% Al Ni alloy: Casting; Jessops; solution treated & aged **UTS: 780 Elon: 40% Proof: 400**

Symbol	Nominal analysis, supplier, condition and remarks.
G 54	0.5% C 19% Cr 3% Mo 3% W 1.5% Nb 1.5% Ta Ni alloy: Casting; Jessops
G 55	0.15% C 15% Cr 4% Mo 2.5% Ti 2% W 2.5% Al Ni alloy: Casting; Jessops; as cast
	UTS: 760 Elon: 8.5% Proof: 580
G 62	0.25% C 15% Cr 6% Mo 1.2% Ti 19% Co 5% Al Ni alloy: Bar & forgings; Jessops; solution treated and aged; vacuum melted
	UTS: 1260 Elon: 15% Proof: 800
G 64	0.1% C 11% Cr 3% Mo 4% W 2% Nb 6% Al 4% Fe Ni alloy: Casting; Jessops; solution treated; vacuum cast
	UTS: 810 Elon: 3.5% Proof: 650
G 66	0.05% C 13% Cr 1.7% Mo 2.2% Ti Ni alloy: Bar & forging; Jessop
	UTS: 910 Elon: 12% Proof: 620
G 67	0.1% C 16% Cr 3% Mo 1% Ti 4% W 6% Al 4% Fe Ni alloy: Casting; Jessops; solution treated; vacuum cast
	UTS: 900 Elon: 1.6% Proof: 760
G 69	0.3% C 20% Cr 3% Mo 3% W 3% Nb 10% Co Ni alloy: Casting; Jessops; as cast
	UTS: 630 Elon: 20% Proof: 310
G 70	0.15% C 15% Cr 18% Co 3% Ti 4% Al 6% Mo Ni alloy: Jessops; solution treated & aged
	UTS: 1270 Elon: 14% Proof: 940
G 73	0.08% C 15% Cr 4.5% Mo 26% Co 2.3% Ti 4.3% Al Ni alloy: Casting; Jessops
G 74	Cr Co W Ti Al Ni alloy: Casting; Jessops; solution treated; vacuum cast
	UTS: 970 Elon: 3% Proof: 830
G 76	0.06% C 21% Cr 10% Mo 2.5% Ti 0.7% Al Ni alloy: Casting; Jessops
G 77	0.04% C 20% Cr 6% Mo 2.5% W 6.5% Nb Ni alloy: Casting; Jessops
G 79	0.14% C 14% Cr 4% Mo 2% Nb 0.8% Ti 6% Al Ni alloy: Casting; Jessops
G 80	0.05% C 20% Cr 2.5% Ti 1.5% Al Ni alloy: Jessops; solution treated & aged
	UTS: 1100 Elon: 28% Proof: 630
G 81	0.05% C 20% Cr 18% Co 2.3% Ti 1.3% Al Ni alloy: Wrought; Jessops
G 82	0.10% C 16.0% Cr 3.0% Ti 2.0% Al Ni alloy: Jessops; scale or creep resistant
G 83	0.07% C 10% Cr 10% Mo 10% Co 3% Ti 2% Al Ni alloy: Wrought; Jessops
G 84	0.17% C 10% Cr 2% W 5% Ti 15% Co 3% Mo 5% Al Ni alloy: Casting; Jessops; vacuum cast
	UTS: 990 Elon: 4% Proof: 760
G 94	0.06% C 9% Cr 3.5% Mo 4.5% W 4% Nb 6% Al Ni alloy: Casting; Jessops
G 95	0.15% C 15.0% Cr 5.0% Mo 2.25% Ti 2.75% Al Ni alloy: Jessops; scale or creep resistant
G 100	0.18% C 10% Cr 3% Mo 0.75% V 15% Co 5% Ti 5.5% Al Ni alloy: Casting; Jessops
G 101	0.06% C 13% Cr 5.5% Mo 35% Fe 2.7% Ti 0.2% Al Ni alloy: Wrought; Jessops
G 103	0.10% C 5.0% Cr 8.0% W 4.0% Mo 15.0% Co Ni alloy: Jessops; scale or creep resistant

Symbol	Nominal analysis, supplier, condition and remarks.
G 157	0.06% C 27% Cr 1.5% Mo 1.5% W 2% Ti 0.7% Al 6% Fe Ni alloy: American proprietary alloy listed in SAE yearbook
G 267	0.05% C 13% Cr 5.5% Mo 40% Fe with Ti and Al Ni alloy: For shear blades; Jessops
	DPN: 320
GMR 235	0.15% C 15% Cr 5.2% Mo 2% Ti 3% Al 10% Fe 0.06% B Ni alloy: Casting; American proprietary alloy listed in SAE yearbook
GMR 235D	0.15% C 15% Cr 5% Mo 2.5% Ti 3.5% Al 4.5% Fe 0.05% B Ni alloy: Casting; American proprietary alloy listed in SAE yearbook
HASTELLOY A	0.1% C 20% Mo 20% Fe Ni alloy: Haynes Stellite
HASTELLOY B	0.05% C 1% Cr 28% Mo 5% Fe 0.5% V 2.5% Co Ni alloy: Wrought; Osborn; resists hydrochloric and phosphoric acid; solution treated & aged
	UTS: 830 Elon: 50% Proof: 370
HASTELLOY B	0.1% C 1% Cr 28% Mo 5% Fe Ni alloy: Casting; Haynes Stellite
HASTELLOY B	0.12% C 1% Cr 28% Mo 5% Fe 0.5% V 2.5% Co Ni alloy: Casting; Osborn; resists hydrochloric and phosphoric acid; solution treated & aged
	UTS: 590 Elon: 12% Proof: 330
HASTELLOY B282	0.02% C 0.6% Cr 28.0% Mo 5.0% Fe 2.0% V Ni alloy: Langley Alloys
HASTELLOY C	0.08% C 14% Cr 15% Mo 3% W 5% Fe 2.5% Co Ni alloy: Wrought; Osborn; solution treated; acid & thermal shock resistant
	UTS: 800 Elon: 50% Proof: 370
HASTELLOY C	0.1% C 16% Cr 17% Mo 4% W 5% Fe Ni alloy: Casting; Haynes Stellite
HASTELLOY C	0.12% C 16% Cr 17% Mo 4% W 5% Fe 2.5% Co Ni alloy: Casting; Osborn; solution treated; acid & thermal shock resistant
	UTS: 550 Elon: 8% Proof: 330
HASTELLOY C276	Low C 15.5% Cr 16.0% Mo 3.7% W 5.5% Fe Ni alloy: Langley Alloys
HASTELLOY F	0.02% C 22% Cr 1.2% Co 6.5% Mo 2.1% Nb 21% Fe Ni alloy: Haynes Stellite
HASTELLOY N	0.06% C 7% Cr 0.5% Co 16% Mo 5% Fe 0.01% B Ni alloy: Haynes Stellite
HASTELLOY R235	0.15% C 15% Cr 2.5% Co 5.5% Mo 2.5% Ti 2% Al 10% Fe Ni alloy: Haynes Stellite
HASTELLOY W	0.12% C 5% Cr 2.5% Co 24% Mo 5% Fe 0.5% V Ni alloy: Haynes Stellite
HASTELLOY X	0.15% C 22% Cr 24% Fe 9% Mo Ni alloy: Haynes Stellite
HASTELLOY X	0.2% C 21% Cr 9% Mo 0.7% W 18% Fe 2% Co Ni alloy: Casting; Osborn; as cast; resists oxidation up to 1200 °C
	UTS: 440 Elon: 11% Proof: 290
HASTELLOY X	0.1% C 21% Cr 9% Mo 0.7% W 18% Fe 2% Co Ni alloy: Wrought; Osborn; solution treated; resists oxidation to 1200 °C
	UTS: 760 Elon: 41% Proof: 360
HEV 2	0.05% C 16% Cr 0.5% Co 3% Ti Ni alloy: For exhaust valves; designation used by SAE; annealed
	UTS: 1100 Elon: 33% Proof: 630
HEV 3	0.05% C 15% Cr 0.7% Co 2.5% Ti 0.7% Al Ni alloy: For valves; designation used by SAE; annealed
	UTS: 1110 Elon: 25% Proof: 630
HEV 5	0.05% C 20% Cr 2% Co 5% Fe 2.5% Ti 1.1% Al Ni alloy: For valves; designation used by SAE; annealed
	UTS: 1140 Elon: 39% Proof: 630

Note. The following abbreviations and units are used in the tables:

DPN	Hardness, diamond pyramid number
UTS	Ultimate tensile strength, N/mm^2
Elon	Elongation, %
Proof	0.1 % proof strength, N/mm^2

$1 \ N/mm^2 = 0.1 \ hbar = 0.102 \ kgf/mm^2 = 0.06475 \ tonf/in^2 = 145.04 \ lbf/in^2$

Symbol	Nominal analysis, supplier, condition and remarks.
HEV 6	0.05% C 20% Cr 18% Co 2.5% Ti 1.1% Al Ni alloy: For valves; designation used by SAE; annealed **UTS: 1220 Elon: 40% Proof: 740**
HYC	0.1% C 0.8% Mn 15% Cr 16% Mo 5% Fe 4% W Ni alloy: Electrode; Murex; for hard facing **DPN: 300**
HX	0.55% C 17% Cr 0.5% Mo 11% Fe Ni alloy: Casting; Huntington; aged **UTS: 510 Elon: 9% Proof: 290**
IC 901	0.05% C 13% Cr 6% Mo 3.0% Ti 37% Fe Ni alloy: Bofors; solution treated and aged **DPN: 320 UTS: 1050 Elon: 12% Proof: 700**
ILLIUM 98	0.05% C 28% Cr 8.5% Mo 1% Fe 5.5% Cu Ni alloy: American proprietary alloy listed in SAE yearbook
ILLIUM B	0.05% C 4.5% Si 28% Cr 8.5% Mo 0.2% B 5% Cu Ni alloy: American proprietary alloy listed in SAE yearbook
ILLIUM G	0.2% C 22.5% Cr 6.5% Mo 6.5% Fe 6.5% Cu Ni alloy: American proprietary alloy listed in SAE yearbook
ILLIUM R	0.07% C 22% Cr 4% Mo 1% Fe 3% Cu Ni alloy: American proprietary alloy listed in SAE yearbook
IN 100	0.18% C 10% Cr 15% Co 3% Mo 5% Ti 5.5% Al 1% Fe 0.015% B 0.05% Zr Ni alloy: Casting; American proprietary alloy listed in SAE yearbook
IN 102	0.06% C 15% Cr 2.9% Nb + Ta 2.9% Mo 3.0% W 7.0% Fe 0.5% Al 0.5% Ti Ni alloy: International Nickel Co; hot worked **UTS: 910 Elon: 50% Proof: 460**
IN 162	0.12% C 10.0% Cr 4.0% Mo 2.0% W 1.0% Nb 2.0% Ta 6.4% Al 0.9% Ti Ni alloy: International Nickel Co; vacuum cast **UTS: 950 Elon: 6% Proof: 190**
IN 722	0.04% C 15% Cr 0.6% Al 2.4% Ti 6.5% Fe Ni alloy: Union Carbide
IN 738	0.17% C 8.5% Co 16.0% Cr 1.7% Mo 2.6% W 1.7% Ta 0.9% Nb 3.4% Al 3.4% Ti Ni alloy: International Nickel Co; vacuum cast **UTS: 1120 Elon: 5% Proof: 970**
INCO 700	0.1% C 2% Mn 15% Cr 29% Co 3% Mo 2.2% Ti 3% Al 4% Fe Ni alloy: Inco
INCO 739	0.07% C 15% Cr 1.7% Ti 2.7% Al 0.5% Fe Ni alloy: Inco
INCO 901	0.05% C 12.8% Cr 5.6% Mo 2.5% Ti 35% Fe Ni alloy: Inco
INCOLOY 804	0.06% C 29.3% Cr 0.25% Al 0.4% Ti 25% Fe Ni alloy: Wrought; Huntington; resists carburization and sulphur penetration
INCOLOY 807	0.08% C 20% Cr 8.0% Co 0.3% Al 0.4% Ti 30% Fe Ni alloy: H. Wiggin; for high temperature use **DPN: 225 UTS: 660**
INCOLOY 825	0.03% C 21.5% Cr 0.9% Ti 3% Mo 30% Fe Ni alloy: Bar & sheet; Huntington; annealed **UTS: 650 Elon: 12% Proof: 220**
INCOLOY 825	0.05% C 22% Cr 35% Fe 3.0% Mo 2.0% Cu Ni alloy: Wrought; H. Wiggin; annealed **DPN: 150 UTS: 640**
INCOLOY 901	0.1% C 13% Cr 2.9% Ti 5.5% Mo 37% Fe Ni alloy: H. Wiggin; solution treated & aged **DPN: 302 UTS: 1120 Elon: 12% Proof: 740**
INCOLOY 901	0.05% C 13.5% Cr 2.5% Ti 6.2% Mo 34% Fe Ni alloy: Bar & sheet; Huntington; age hardened
INCOLOY 901Mod	0.05% C 12.5% Cr 5.8% Mo 2.9% Ti 0.015% B 34.0% Fe Ni alloy
INCONEL 69	0.04% C 6.8% Fe 15% Cr 0.8% Al 2.4% Ti 0.8% Nb Ni alloy: Filler; Huntington; inert gas arc welding of Inconel X750

Symbol	Nominal analysis, supplier, condition and remarks.
INCONEL 82	0.02% C 1% Fe 20% Cr 3% Mn 0.5% Ti 2.5% Nb Ni alloy: Filler; Huntington; inert gas welding of Inconel and steel
INCONEL 92	0.03% C 6.7% Fe 16.5% Cr 2.2% Mn 3.1% Ti Ni alloy: Filler; Huntington; inert gas arc welding of austenitic steels
INCONEL 625	0.05% C 22% Cr 3% Fe 4% Nb 9% Mo Ni alloy: Bar, sheet, etc.; Huntington; annealed **UTS: 1000 Elon: 42% Proof: 600**
INCONEL 700	0.12% C 15% Cr 3% Al 2.2% Ti 28% Co 3.7% Mo Ni alloy: Bar; Huntington; age hardened **UTS: 1210 Elon: 25% Proof: 760**
INCONEL 702	0.04% C 15.6% Cr 3.4% Al 0.7% Ti Ni alloy: Bar, sheet, etc., Huntington
INCONEL 718	0.04% C 19% Cr 18% Fe 0.6% Al 0.8% Ti 5% Nb 3% Mo Ni alloy: Bar, sheet, etc.; H. Wiggin; age hardened **UTS: 1520 Elon: 23% Proof: 1300**
INCONEL 721	0.04% C 16% Cr 2.2% Mn 3% Ti 7.2% Fe Ni alloy: Bar; Huntington
INCONEL 722	0.04% C 15% Cr 0.6% Al 2.4% Ti 6.5% Fe Ni alloy: Bar, sheet, etc.; Huntington; age hardened **UTS: 1210 Elon: 31% Proof: 690**
INCONEL 751	0.04% C 15% Cr 1.2% Al 2.5% Ti 1% Nb 6.7% Fe Ni alloy: Bar; Huntington
INCONEL M	0.04% C 16% Cr 2.2% Mn 3% Ti 7.2% Fe Ni alloy: Huntington; now Inconel 721
INCONEL M	0.04% C 15% Cr 0.6% Al 2.4% Ti 6.5% Fe Ni alloy: Bar, sheet, etc.; Huntington; now Inconel 722
INCONEL W	0.04% C 15% Cr 2.5% Ti 0.6% Al 7% Fe Ni alloy: Inco
INCONEL X	0.04% C 15% Cr 2.5% Ti 1% Al 7% Fe 1% Nb Ni alloy: Inco
INCONEL X	0.04% C 15% Cr 2.5% Ti 0.8% Al 0.8% Nb 6.7% Fe Ni alloy: Bar, sheet, etc.; Huntington; now Inconel X750
INCONEL X550	0.04% C 15% Cr 2.5% Ti 1% Al 7% Fe 1% Nb Ni alloy: Inco
INCONEL X550	0.04% C 15% Cr 1.2% Al 2.5% Ti 1% Nb 6.7% Fe Ni alloy: Bar; Huntington; now Inconel 751
INCONEL X750	0.04% C 15% Cr 2.5% Ti 0.8% Al 0.8% Nb 6.7% Fe Ni alloy: Bar, sheet, etc.; Huntington; age hardened **UTS: 1170 Elon: 22% Proof: 650**
INCO WELD A	0.03% C 2.2% Mn 6.7% Fe 16% Cr 3% Ti Ni alloy: Filler; Huntington; now Inconel 92
J 1500	0.15% C 20% Cr 10% Co 10% Mo 3% Ti 1% Al Ni alloy: American proprietary alloy listed in SAE yearbook
K 42B	0.05% C 18% Cr 22% Co 2.5% Ti 0.2% Al 13% Fe Ni alloy: Westinghouse
KARMA	20% Cr 5% Fe & Al Ni alloy: For electrical resistors; British Driver Harris
KH 70VMTIU	0.08% C (max) 18.5% Cr 4.5% Mo 2.6% Ti 1.2% Al 4.5% W 4.0% Fe (max) Ni alloy: Russian Standard designation
KH 70VMTIU	0.13% C 15.0% Cr 4.0% Mo 1.2% Ti 2.0% Al 3.0% Fe (max) Ni alloy: Russian Standard designation
KH 70VMTIU	0.12% C (max) 14.5% Cr 3.0% Mo 2.1% Ti 2.1% Al 6.0% W 0.3% V 5.0% (max) Ni alloy: Russian Standard designation
KH 70MVTIUB	0.12% C (max) 17.5% Cr 5.0% Mo 2.5% Ti 1.3% Al 2.7% W 0.9% Nb 5.0% Fe (max) Ni alloy: Russian Standard designation
KH 77TIUR	0.06% C (max) 20.5% Cr 2.5% Ti 0.7% Al 4.0% Fe (max) Ni alloy: Russian Standard designation

Symbol	Nominal analysis, supplier, condition and remarks.
KH 80TBIU	0.08% C (max) 16.5% Cr 2.1% Ti 0.7% Al 1.2% Nb 3.0% Fe (max) Ni alloy: Russian Standard designation
KHN 60V	0.1% C (max) 25.0% Cr 0.45% Ti 0.5% Al (max) 4.0% Fe (max) Ni alloy: Russian Standard designation
LANGALLOY 4R	0.12% C 1% Cr 28% Mo 5% Fe 0.5% V 2.5% Co Ni alloy: Casting; Langley; Hastelloy B type **DPN: 220 UTS: 550 Elon: 8% Proof: 360**
LANGALLOY 5R	0.12% C 16% Cr 17% Mo 4% W 5% Fe 2.5% Co Ni alloy: Casting; Langley; Hastelloy C type **DPN: 200 UTS: 480 Elon: 9% Proof: 340**
LANGALLOY 7R	0.08% C 23% Cr 6% Cu 5% Fe 6% Mo 2% W 3% Si Ni alloy: Langley **DPN: 200 UTS: 440 Elon: 8% Proof: 330**
LESCALLOY 718 VAC-ARC	19% Cr 3.0% Mo 0.9% Ti 5.2% Nb + Ta 0.8% Al Ni alloy: Latrobe Steel; solution treated & aged **DPN: 370 UTS: 800 Elon: 54% Proof: 450**
LESCALLOY D979	0.04% C 15% Cr 30% Fe 4% Mo 4% W 3% Ti 1.0% Al Ni alloy: Latrobe; for turbine discs
LESCALLOY X750	0.04% C 15.5% Cr 0.9% Nb + Ta 2.5% Ti 7% Fe Ni alloy: Latrobe
M 21	0.13% C 5.7% Cr 6.0% Al 2.0% Mo 11.0% W 1.5% Nb 0.1% Zr Ni alloy: International Nickel Co; vacuum cast **UTS: 800 Elon: 5.5% Proof: 760**
M 22	0.13% C 5.7% Cr 6.3% Al 2.0% Mo 11.0% W 3.0% Ta 0.6% Zr Ni alloy: International Nickel Co; vacuum cast **UTS: 740 Elon: 5.5% Proof: 690**
M 22B	0.13% C 5.7% Cr 6.3% Al 2.0% Mo 11.0% W 3.0% Ta 0.5% Zr Ni alloy: International Nickel Co; vacuum cast **UTS: 800 Elon: 4% Proof: 720**
M 252	0.15% C 19% Cr 10% Mo 2.5% Ti 10% Co 1% Al Ni alloy: Crucible Steel Co; solution treated & aged - AISI 689 **UTS: 1140 Elon: 16% Proof: 740**
M 313	0.06% C (max) 30% Cr 1.7% Ti 0.9% Al 0.05% Zr Ni alloy: International Nickel Co; solution treated & aged **DPN: 245 UTS: 1130 Elon: 30% Proof: 420**
M 600	0.08% C 19% Cr 7% Mo 2.3% Ti 1.1% Al 13% Fe Ni alloy: American proprietary alloy listed in SAE year book
MACHINETRODE 92.78	0.6% C 0.8% Mn 2.4% Mn 30.0% Cu Ni alloy: Electrodes; Esab **DPN: 150**
MAR M 200	0.15% C 9% Cr 12.5% W 10% Co 1% Nb 5% Al 2% Ti Ni alloy: Casting; Martin Metals Ltd **UTS: 1000 Elon: 6% Proof: 830**
MAR M 246	0.15% C 9% Cr 10% Co 10% W 2.5% Mo 1.5% Ta 1.5% Ti 5.5% Al 0.02% B Ni alloy: Martin Metals Ltd **UTS: 900 Elon: 4% Proof: 780**
MAR M 421	Cr Ni alloy: For creep resistance; Martin Metals Ltd; for turbine blades

Symbol	Nominal analysis, supplier, condition and remarks.
MC 102	0.04% C 20% Cr 6.0% Mo 6.6% Nb + Ta 2.5% W Ni alloy: International Nickel Co; solution treated & aged **UTS: 690 Elon: 5% Proof: 620**
MC 102	0.04% C 20.0% Cr 6.0% Mo 2.5% W 6.6% Nb 3.0% Fe 0.3% Mn 0.25% Si Ni alloy: Union Carbide
METCO 16C	0.5% C 4% Si 16% Cr 3% Cu 3% Mo 2.5% Fe 4% B Ni alloy: Powder for spraying; Metco Ltd **DPN: 765**
METCO 17F	0.5% C 4% Si 17% Cr 3% Fe 2% Cu 2% Mo 3% B Ni alloy: Powder for spraying; Metco Ltd; as sprayed **DPN: 425**
MP 35N	20% Cr 35% Co 10% Mo Ni alloy: Climax Molybdenum
MP 35N	20% Cr 35% Co 10% Mo Ni alloy: Climax Molybdenum
MUMETAL	14% Fe 5% Cu 4% Mo Ni alloy: Sheet & rod; Telcon; annealed; high permeability low hysteresis **DPN: 110 UTS: 530**
MUMETAL	14% Fe 5% Cu 4% Mo Ni alloy: Sheet & rod; Telcon; hard rolled; magnetic properties are reduced **DPN: 290 UTS: 900**
MUMETAL 40	14% Fe 5% Cu 4% Mo Ni alloy: Sheet & rod, Telcon; annealed; high permeability **DPN: 110 UTS: 540**
MUMETAL 40	14% Fe 5% Cu 4% Mo Ni alloy: Sheet & rod; Telcon; hard rolled; permeability reduced **DPN: 290 UTS: 900**
MUMETAL 40CT	Similar to Mumetal 40; vacuum melted material; Telcon
MUMETAL 60	77% Ni alloy with minimum initial permeability of 60 000; Telcon
MUMETAL 60	77% Ni alloy with minimum initial permeability of 60 000; Telcon
NA 22H (c)	0.5% C 27% Cr 16% Fe 6% W Ni alloy: Blaw-Knox
NB	0.04% C 28% Mo 5% Fe Ni alloy: Wrought; Japanese alloy
NC	0.05% C 15% Cr 13% Mo 5% Fe 4% W Ni alloy: Wrought; Japanese alloy
NICKELVAC 700	0.13% C 15% Cr 29% Co 3.7% Mo 2.5% Ti 3.2% Al Ni alloy: Vanadium Alloys; double vacuum melted
NICKELVAC 901	0.06% C 12.5% Cr 6.1% Mo 3% Ti 0.15% B 35% Fe Ni alloy: Vanadium Alloys; double vacuum melted
NICKELVAC W	0.06% C 15.5% Cr 2.5% Ti 0.75% Al 8% Fe Ni alloy: Vanadium Alloys; double vacuum cast
NICKELVAC X	0.06% C 15.5% Cr 2.5% Ti 0.7% Al 6.7% Fe 1% Nb Ni alloy: Vanadium Alloys; double vacuum cast
NICROTUNG	0.1% C 12% Cr 10% Co 8% W 4% Ti 4% Al 0.05% B 0.05% Zr Ni alloy: Casting; American proprietary alloy listed in SAE yearbook
NIMOCAST 80	0.1% C 20% Cr 2.0% Ti 5.0% Fe 1.0% Al 2.0% Co Ni alloy: Casting; H. Wiggin; cast variety of N80A
NIMOCAST 90	0.13% C 20% Cr 2.5% Ti 5.0% Fe 1.5% Al 18% Co Ni alloy: Casting; H. Wiggin; cast variety of N90
NIMOCAST 258	0.2% C 10.0% Cr 20.0% Co 5.0% Mo 3.5% Ti 4.8% Al Ni alloy: Casting; H. Wiggin
NIMOCAST 713	13.4% Cr 4.5% Mo 1.0% Ti 6.2% Al 2.3% Nb Ni alloy: Casting; H. Wiggin; for cast blading used up to 1000 °C
NIMOCAST PE10	20% Cr 6.0% Mo 2.5% W 6.5% Nb Ni alloy: Casting; H. Wiggin; for use up to 850 °C

Note. The following abbreviations and units are used in the tables:

DPN	Hardness, diamond pyramid number
UTS	Ultimate tensile strength, N/mm^2
Elon	Elongation, %
Proof	0.1 % proof strength, N/mm^2

1 N/mm^2=0.1 hbar=0.102 kgf/mm^2=0.06475 tonf/in^2=145.04 lbf/in^2

Symbol	Nominal analysis, supplier, condition and remarks.
NIMOCAST PK24	10% Cr 3.0% Mo 5.2% Ti 5.5% Al 15% Co Ni alloy: Casting; H. Wiggin; vacuum melted; for use up to 1000 °C
NIMOCAST PK24	0.17% C 9.5% Cr 15.0% Co 3.0% Mo 5.5% Ap 4.7% Ti 1.0% V Ni alloy: H. Wiggin; for use up to 1000 °C
NIMOLOY PK37	0.1% C 4.0% Fe 18.5% Cr 18.5% Co 1.0% Al 2.5% Ti Ni alloy: H. Wiggin; abrasion & thermal shock resistance
NIMOCAST T13LC	As T13 with 0.05% C **DPN: 370 UTS: 820**
NIMONIC 80 A	0.1% C 20% Cr 2% Ti 5% Fe 1% Al 2% Co Ni alloy: H. Wiggin; solution treated & aged; tensile strength 600 N/mm^2 at 750 °C **DPN: 300 UTS: 1070 Elon: 40% Proof: 600**
NIMONIC 81	0.05% C (max) 30% Cr 1.8% Ti 0.9% Al 1.0% Co Ni alloy: H. Wiggin; for turbine components **DPN: 230 UTS: 1050**
NIMONIC 90	0.13% C 20% Cr 2.5% Ti 5% Fe 1.5% Al 18% Co Ni alloy: H. Wiggin; solution treated & aged; tensile strength 700 N/mm^2 at 750 °C **DPN: 300 UTS: 1210 Elon: 33% Proof: 750**
NIMONIC 93	19.5% Cr 17% Co 2% Ti 1.5% Al 0.3% Mo 2% Fe Ni alloy: For turbine blades; H. Wiggin
NIMONIC 95	0.15% C 19.5% Cr 18.0% Co 5.0% Fe Ni alloy: H. Wiggin
NIMONIC 100	0.3% C 11.0% Cr 20.5% Co 5.0% Mo 1.5% Ti 5.0% Al 2.0% Fe Ni alloy: H. Wiggin
NIMONIC 100	0.3% C 19.0% Cr 20% Co 5.0% Mo 1.5% Ti 5.0% Al Ni alloy: H. Wiggin
NIMONIC 105	0.2% C 14% Cr 1.2% Ti 1% Fe 4.5% Al 20% Co Ni alloy: H. Wiggin; double solution treated & aged **DPN: 370 UTS: 990 Elon: 7% Proof: 760**
NIMONIC 115	0.15% C 15% Cr 4% Ti 5% Al 3.5% Mo 15% Co Ni alloy: H. Wiggin; double solution treated; tensile strength 1000 N/mm^2 at 800 °C **DPN: 400 UTS: 1210 Elon: 27% Proof: 800**
NIMONIC 118	0.15% C 15% Cr 4% Ti 5% Al 3.5% Mo 15% Co Ni alloy: Wrought; vacuum cast; H. Wiggin; improved high temperature strength over N115 **DPN: 400 UTS: 1210 Elon: 27% Proof: 800**
NIMONIC 120	0.04% C 12.5% Cr 2.5% Ti 4.5% Ap 10.0% Co 5.7% Mo Ni alloy: H. Wiggin; for gas turbine blades **UTS: 1150**
NIMONIC 242	21.5% Cr 10.5% Mo 10% Co Ni alloy: Casting; H. Wiggin; for cast turbine engine blades
NIMONIC 263	20% Cr 20% Co 6% Mo 2.2% Ti 0.5% Al Ni alloy: H. Wiggin **DPN: 320 UTS: 650 Elon: 45% Proof: 400**
NIMONIC 901	12.5% Cr 2.7% Ti 42% Fe Ni alloy: H. Wiggin; for turbine discs & blades **DPN: 360 UTS: 780 Elon: 20% Proof: 580**
NIMONIC 942	0.03% C 12.5% Cr 3.7% Ti 0.6% Al 1.0% Co 6.0% Mo Ni alloy: H. Wiggin; for compressor blades etc.
NIMONIC PE 11	18% Cr 5% Mo 2.5% Ti 0.8% Al 35% Fe Ni alloy: Forging; H. Wiggin; vacuum cast material for use up to 600 °C
NIMONIC PE 13	22% Cr 9.0% Mo 18% Fe 1.5% Co 0.6% W Ni alloy: Sheet; H. Wiggin; for furnace parts
NIMONIC PE 16	16.5% Cr 3.2% Mo 1.2% Ti 1.2% Al 35% Fe Ni alloy: Forging; H. Wiggin; vacuum cast material for use up to 750 °C
NIMONIC PK 25	0.07% C 18.0% Cr 2.9% Ti 2.7% Cl 17.5% Co 4.0% Mo Ni alloy: H. Wiggin; creep resistant **UTS: 1360**
NIMONIC PK 31	20% Cr 4.5% Mo 2.3% Ti 0.4% Al 5.0% Nb 14% Co Ni alloy: Forging; H. Wiggin; gas turbine discs

Symbol	Nominal analysis, supplier, condition and remarks.
NIMONIC PK 33	19% Cr 7.0% Mo 2.0% Ti 2.0% Al 14% Co Ni alloy: Forging; H. Wiggin; vacuum processed for use up to 950 °C
NI O NEL 65	0.03% C 30% Fe 21% Cr 1.7% Cu 1% Ti 3% Mo Ni alloy: Filler; Huntington; inert gas arc welding of Incoloy 825
NI O NEL 825	2% Cu 34% Fe 21% Cr 3% Mo Ni alloy: Wrought; H. Wiggin **UTS: 640 Elon: 30% Proof: 220**
NI O NEL 825	0.03% C 21.5% Cr 0.9% Ti 3% Mo 30% Fe Ni alloy: Bar & sheet; Huntington; now Incoloy 825
NI O NEL 135	0.05% C 31% Fe 19% Cr 1.8% Cu 1% Nb 5.5% Mo Ni alloy: Filler; Huntington; electric welding of Incoloy 825
N 07718	UNS designation for ASTM A637/718
N 07750	UNS designation for ASTM A637/688
N 07500	UNS designation for ASTM A637/684
N 07001	UNS designation for ASTM A637/685
N 07080	UNS designation for ASTM A637/80A
N 07252	UNS designation for ASTM A637/689
PE 10	20% Cr 6.0% Mo 2.5% W 6.5% Nb Ni alloy: Casting; H. Wiggin; see Nimocast PE 10
PERMALLOY C	13% Fe 4% Mo 5% Cu Ni alloy: Standard Telephones; where high initial permeability required
PE 11	18% Cr 5.0% Mo 2.5% Ti 0.8% Al 35% Fe Ni alloy: Forging; H. Wiggin; see Nimonic PE 11
PE 13	22% Cr 9.0% Mo 18% Fe 0.6% W 1.5% Co Ni alloy: Sheet; H. Wiggin; for furnace parts
PE 16	0.06% C 16.5% Cr 3.25% Mo 1.25% Al 1.25% Ti 0.003% B 0.04% Zr 34.15% Fe Ni alloy: Union Carbide
PK 24	10% Cr 3.0% Mo 5.2% Ti 5.5% Al 15% Co Ni alloy: Casting; H. Wiggin; see Nimocast PK 24
PK 31	20% Cr 4.5% Mo 2.3% Ti 0.4% Al 5.0% Nb 14% Co Ni alloy: Forging; H. Wiggin; see Nimonic PK 31
PK 33	19% Cr 7.0% Mo 2.0% Ti 2.0% Al 14% Co Ni alloy: Forging; H. Wiggin; see Nimonic PK 33
PK 42	0.15% C 15% Cr 6% Mo 4% Al 19% Co 3.2% Ti Ni alloy: Forging; H. Wiggin
R 235	0.12% C 15% Cr 1.1% Co 5.5% Mo 2.5% Ti 2.0% Al 10.0% Fe Ni alloy: Information from SAE handbook
REFRACTALOY 26	0.05% C 18% Cr 20% Co 3% Mo 2.8% Ti 0.2% Al 18% Fe Ni alloy: Westinghouse
RENE 41	0.09% C 19% Cr 11% Co 10% Mo 3% Ti 1.5% Al 0.01% B Ni alloy: American proprietary alloy listed in SAE yearbook
REXWELD 66	0.15% C 16% Cr 17% Mo 4.5% W 6% Fe Ni alloy: For hard facing; Crucible Steel Co; work hardened & aged
SAE HEV 2	0.05% C 14% Cr 0.5% Co 3% Ti Ni alloy: For exhaust valves; annealed **UTS: 1110 Elon: 25% Proof: 640**
SAE HEV 3	0.05% C 15% Cr 0.7% Co 2.5% Ti 0.7% Al Ni alloy: For valves; annealed **UTS: 1180 Elon: 25% Proof: 640**
SAE HEV 5	0.05% C 20% Cr 2% Co 5% Fe 2.5% Ti 1.1% Al Ni alloy: For valves; annealed **UTS: 1140 Elon: 39% Proof: 640**
SAE HEV 6	0.06% C 20% Cr 18% Co 2.5% Ti 1.1% Al Ni alloy: For valves; annealed **UTS: 1210 Elon: 40% Proof: 740**
SAE VF 3	2.4% C 29% Cr 10% Co 15% W 6% Fe Ni alloy: For hard facing
SAE VF 4	2.0% C 26% Cr 0.5% Co 8.5% W 4% Fe Ni alloy: For hard facing
SEL 1	0.08% C 22.0% Co 15.0% Cr 4.4% Mo 4.5% Al 2.4% Ti 0.015% B Ni alloy: Union Carbide

Symbol	Nominal analysis, supplier, condition and remarks.
SIMALLOY 750	0.08% C 16% Cr 7.0% Fe 0.7% Al 2.5% Ti 1.0% Nb Ni alloy: Simonds
SUPER MUMETAL 50	14% Fe 4% Cu 4% Mo Ni alloy: Telcon; annealed; better permeability than Mumetal **DPN: 110 UTS: 540**
SUPER MUMETAL 50	14% Fe 4% Cu 4% Mo Ni alloy: Telcon; hard rolled **DPN: 290 UTS: 910**
SUPER MUMETAL 100	14% Fe 5% Co 4% Mo Ni alloy: Telcon; annealed; better permeability than Mumetal **DPN: 110 UTS: 540**
SUPER MUMETAL 100	14% Fe 5% Co 4% Mo Ni alloy: Telcon; hard rolled **DPN: 290 UTS: 910**
SUPER MUMETAL Co	Similar to Mumetal; now referred to as Mumetal 60; Telcon
TICONIUM	0.01% C 23% Cr 31% Co 6% Mo 6% Fe Ni alloy: Westinghouse
TRW 1800	0.09% C 13% Cr 9% W 0.6% Ti 6% Al 0.07% B 0.07% Zr 1% V Ni alloy: Casting; American proprietary alloy listed in SAE yearbook
UCAR 16	0.05% C 16.5% Cr 3.2% Mo 1.2% Ti 1.2% Al 3.0% Fe Ni alloy: Union Carbide
UCAR 80	0.08% C 19% Cr 0.4% Mo 2% Co 2.5% Ti 1.2% Al Ni alloy: Union Carbide; solution treated & aged **UTS: 760 Elon: 15% Proof: 540**
UCAR 90	0.08% C 20% Cr 0.4% Mo 17% Co 2.8% Ti 1.2% Al Ni alloy: Union Carbide; solution treated & aged **UTS: 740 Proof: 540**
UCAR 625	0.05% C 22.0% Cr 9.0% Mo 4.0% Nb 0.2% Ti 0.2% Al 3.0% Fe Ni alloy: Union Carbide
UCAR 700	0.12% C 15.0% Cr 28.5% Co 3.7% Mo 2.1% Ti 3.0% Al 0.7% Fe Ni alloy: Union Carbide
UCAR 702	0.04% C 15.5% Cr 0.7% Ti 3.5% Al 0.4% Fe Ni alloy: Union Carbide
UCAR 713C	0.12% C 13% Cr 4.7% Mo 1% Co 0.7% Ti 2% Nb + Ta 6% Al 0.1% Zr Ni alloy: Union Carbide; as cast; AMS 5391 **UTS: 910 Elon: 5% Proof: 770**
UCAR 713LC	0.05% C 12.0% Cr 4.5% Mo 2.0% Nb 0.6% Ti 6.0% Al Ni alloy: Union Carbide
UCAR 718	0.08% C 19% Cr 3% Mo 1% Co 1% Ti 5.2% Nb + Ta 0.6% Al Ni alloy: Union Carbide; wrought; solution treated & aged **UTS: 1300 Elon: 15% Proof: 1070**
UCAR 718	0.08% C 19% Cr 3% Mo 1% Co 1% Ti 5.2% Nb + Ta 0.6% Al Ni alloy: Union Carbide; vacuum cast; solution treated & aged **UTS: 1070 Elon: 25% Proof: 770**
UCAR 901	0.05% C 13.5% Cr 42.5% Ni 6.2% Mo 2.5% Ti 0.2% Al Fe alloy: Union Carbide
UCAR C130	0.05% C 21.5% Cr 6.0% Mo 2.1% Ti 20.0% Co 0.5% Fe Ni alloy: Union Carbide
UCAR C1023	0.15% C 15.5% Cr 8.3% Mo 9.7% Co 3.6% Ti 4.1% Al Ni alloy: Union Carbide
UCAR GMR 235	0.15% C 15% Cr 5.2% Mo 2% Ti 3% Al 0.07% B Ni alloy: Union Carbide; as cast **UTS: 760 Elon: 4% Proof: 660**
UCAR GMR 235D	0.15% C 16% Cr 5% Mo 2.5% Ti 3.7% Al 0.07% B Ni alloy: Union Carbide; vacuum cast **UTS: 760 Elon: 3% Proof: 720**
UCAR HASTELLOY R 235	0.16% C 15% Cr 5% Mo 2.5% Co 2.5% Ti 2% Al Ni alloy: Sheet; Union Carbide; solution treated & aged **UTS: 1210 Elon: 21% Proof: 820**
UCAR IN100	0.17% C 10% Cr 3% Mo 15% Co 5% Ti 5.5% Al Ni alloy: Union Carbide; vacuum cast; as cast **UTS: 1070 Elon: 9% Proof: 870**

Symbol	Nominal analysis, supplier, condition and remarks.
UCAR IN731	0.18% C 9.5% Cr 10.0% Co 2.5% Mo 4.6% Ti 5.5% Al Ni alloy: Union Carbide
UCAR IN792	0.2% C 12.7% Cr 9.0% Co 2.0% Mo 4.0% W 4.2% Ti 3.2% Al Ni alloy: Union Carbide
UCAR M21	0.21% C 5.7% Cr 6.0% Al 2.0% Mo 11.0% W 1.5% Nb 0.12% Zr 0.02% B Ni alloy: Vacuum cast; Union Carbide
UCAR R 41	0.1% C 19% Cr 10% Mo 11% Co 3.1% Ti 1.5% Al 0.008% B Ni alloy: Sheet & bar; Union Carbide; solution treated & aged **UTS: 1540 Elon: 14% Proof: 1110**
UCAR U500	0.08% C 19.0% Cr 18.0% Co 4.0% Mo 3.0% Ti 3.0% Al 4.0% Fe Ni alloy: Union Carbide
UCAR U520	0.05% C 19.0% Cr 12.0% Co 6.0% Mo 1.0% W 3.0% Ti 2.0% Al Ni alloy: Union Carbide
UCAR U700	0.15% C 15.0% Cr 18.5% Co 5.2% Mo 3.5% Ti 4.2% Al Ni alloy: Union Carbide
UCAR U710	0.07% C 18.0% Cr 3.2% Mo 15.0% Co 1.5% W 5.0% Ti 2.5% Al Ni alloy: Union Carbide
UCAR U722	0.08% C (max) 15.5% Cr 2.4% Ti 0.7% Al Ni alloy: Union Carbide
UCAR WASPALLOY	0.08% C 19% Cr 4.2% Mo 14% Co 3% Ti 1.2% Al 0.1% Zr Ni alloy: Sheet; Union Carbide; solution treated **UTS: 1070 Elon: 25% Proof: 580**
UCAR X	0.1% C 22.0% Cr 9.0% Mo 1.5% Co 0.6% W Ni alloy: Union Carbide
UDIMET 200	12.5% Cr 2.7% Ti 42% Fe Ni alloy: For turbine discs and blades; American proprietary alloy **DPN: 360 UTS: 780 Elon: 20% Proof: 580**
UDIMET 500	0.08% C 19% Cr 19% Co 4% Mo 3% Ti 3% Al 4% Fe 0.01% B Ni alloy: American proprietary alloy listed in SAE yearbook
UDIMET 600	0.1% C 17.5% Cr 16.5% Co 4% Mo 2.8% Ti 4.2% Al 4% Fe 0.04% B Ni alloy: American proprietary alloy listed in SAE yearbook
UDIMET 700	0.15% C 15% Cr 18% Co 5% Mo 3.5% Ti 4.2% Al 1% Fe 0.1% B Ni alloy: American proprietary alloy listed in SAE yearbook
UNITEMP AF 1753	0.2% C 16% Cr 7% Co 1.6% Mo 8.4% W 3.2% Ti 2% Al 10% Fe 0.008% B 0.06% Zr Ni alloy: American proprietary alloy listed in SAE yearbook
UNS N07718	UNS designation for ASTM A670/718
VACUMELTROL 41	0.1% C 19% Cr 9.7% Mo 11% Co 3.15% Ti 1.5% Al 0.007% B 5% Fe Ni alloy: Carpenter
VF 3	2.4% C 29% Cr 10% Co 15% W 6% Fe Ni alloy: For hard facing; designation used by SAE
VF 4	2% C 26% Cr 0.5% Co 8.5% W 4% Fe Ni alloy: For hard facing; designation used by SAE
WASPALOY	0.7% C 19% Cr 14% Co 4.3% Mo 3% Ti 1.3% Al 1% Fe 0.005% B Ni alloy: American proprietary alloy listed in SAE yearbook; contact H. Wiggin
WASPALOY MOD	0.05% C 19% Cr 11.5% Co 7% Mo 2.5% Ti 1.2% Al 1% Fe Ni alloy: American proprietary alloy listed in SAE yearbook; contact H. Wiggin
X 750	0.04% C 15% Cr 2.5% Ti 7% Fe 0.8% Al 0.8% Nb + Ta Ni alloy: Crucible Steel Co; double solution treated & aged; AISI 688 **UTS: 1140 Elon: 23% Proof: 890**

Note. The following abbreviations and units are used in the tables:

DPN	Hardness, diamond pyramid number
UTS	Ultimate tensile strength, N/mm^2
Elon	Elongation, %
Proof	0.1 % proof strength, N/mm^2

1 N/mm^2=0.1 hbar=0.102 kgf/mm^2=0.06475 tonf/in^2=145.04 lbf/in^2

27D Nickel–copper wrought and cast alloys

Specific gravity	8.83	
Density	8830 kg/m^3	(0.319 lb/in.^{3})
Solidus/liquidus	1300–1350 °C	
Thermal conductivity	26 W/m °C	(6.2 cal/m s °C)
Coefficient of linear expansion (20–100° C)	13.6 × 10^{-6}/ °C	
Electrical conductivity	3.6% IACS (copper 100%)	
Specific resistance	480 microhm mm	
Young's modulus of elasticity	165.5 × 10^9 N/m^2	(24 × 10^6 lbf/in.^{2})
Impact	136 J	(100 ft/lb) Izod
Fatigue strength (10 × 10^7 cycles) annealed	± 240 N/mm^2	(± 16 tonf/in.^{2})
cold worked	± 330 N/mm^2	(± 22 tonf/in.^{2})

Hot strength

Temperature °C	Tensile strength N/mm^2	Elongation %
20	530	48
100	530	48
300	520	49
400	450	50
500	340	30
600	240	30
700	140	40
800	75	50
900	50	54
980	34	60

The above properties have been chosen to show typical values for the specifications listed and may not apply exactly to any one specification. It is possible that with some specifications the values may not be applicable.

General metallurgical characteristics

Nickel and copper form a complete series of solid solutions. In the alloys listed the copper is dissolved in the nickel matrix, and this solution persists at room temperature. Thus one phase only is present, resulting in excellent corrosion resistance over a wide range of temperatures. The copper improves the electrical and thermal conductivity over other nickel alloys and also increases the ease of brazing.

Some of these alloys have been in existence for many years, as there are naturally occurring nickel/copper ores which on reduction give a 70/30 nickel/copper alloy. The fact that this is readily worked, easily machined and has good heat and corrosion resistance ensured it a ready market. This alloy was probably the first material which allowed designers to start using the power available when high temperatures were harnessed. Iron may also be present in many of the alloys listed, either as an impurity with adverse affects on corrosion resistance, or as an additive along with manganese and silicon to improve mechanical properties.

With titanium and aluminium additions intermetallic compounds are formed which impart some age hardening properties at the expense of corrosion resistance.

Thermal treatment

Like all nickel alloys these materials are considerably hardened by cold work. This property is used in many cases to strengthen the otherwise unhardenable nickel/copper alloys. The age-hardening alloys can also be cold worked, but on the whole they accept less before becoming brittle.

Annealing to remove cold work requires a temperature of 1000–1040 °C for 10 minutes per 25 mm of section. The rate of cooling is not important with the non-ageing alloys, but all other materials should be water quenched for maximum softening.

By judicious use of furnace temperatures and time the amount of cold work remaining can be accurately controlled to give a product of known tensile strength and ductility.

Alloys with aluminium and titanium present can be age hardened. This requires first that the intermetallic compounds are taken into solution in the nickel/copper matrix and held there. This is achieved by heating to 1000–1040 °C for about 4 hours, then water quenching, and is known as solution treatment. The material is then the softest possible, and any presswork or drawing must be carried out in this condition. The solution treatment must be followed by ageing for 16 hours at 400–450 °C, when the intermetallic compounds agglomerate and start to precipitate, causing strain of the lattice structure and an increase in hardness. This can be additive to that obtained by cold working the solution-treated compounds, but will not always be the case.

Stress relieving of parts which must be dimensionally stable after fabrication should be carried out prior to final machining at either the ageing temperature or the annealing temperature, depending whether the alloy is age hardening or not. The lowest temperature used for stress relieving must always be higher than the operating temperature in service.

All the thermal treatments should be carried out in electrical furnaces, as gas or oil furnaces can have sulphur-contaminated atmospheres which generally corrode these alloys.

Welding and brazing

Pre-treatment. Parts must be in the annealed condition, free from all oil, grease or dirt, and should have the joint faces freshly prepared.

Welding and brazing. Using active fluxes, electric arc and gas welding have been successfully applied to these materials for many years. The heat conductivity is not too high to cause trouble. Inert gas-shielded arc welding techniques give superior results, as there is no flux covering the weld, which can hinder the operator and often become trapped in the weld.

Resistance welding using the correct equipment and with the necessary technical control is commonly carried out.

Brazing presents no difficulties provided the correct fluxes are used.

Post-treatment. Ageing alloys should be solution treated and aged if possible after welding, but must be aged at the very least. Brazed parts should be aged.

Non-ageing alloys require no treatment. All trace of flux must be removed and this should be before any heat treatment operation.

Flaw detection methods

Crack-test. Penetrant oil chalk test. Some of the alloys are magnetic but this does not justify magnetic crack test being specified.

X-ray. Not normally required.

Ultra-sonic test. This can be used on rough-machined forgings which require expensive machining, to prove that no subcutaneous defects are present which would be shown up after machining. It can also be used to identify certain types of welding defect.

Chemical etch. Not required.

Corrosion protection

Temporary. None required.

Permanent. None required. These materials should not however, be used in high sulphur atmospheres.

Machinability

In the annealed condition these materials have a tendency to tear rather than cut. This is reduced if the parts are cold worked and as all the alloys listed harden considerably when cold worked, care must be taken not to allow tool rub or dwell at any time. High feeds, low speeds give best results.

The aged alloys should be machined after ageing with stress relieving at 400 °C before final machining.

Uses

Marine fittings of all types, upper deck, structural, wires, bolts, engine-pump parts and all items in contact with sea water or spray.

General engineering applications where first-class corrosion resistance, economy and medium stress applications are required.

Some low duty combustion equipment uses, such as fireboxes, direct-heated boilers.

Symbol	Nominal analysis, supplier, condition and remarks.
2.4360	2.0% Fe 31% Cu Ni alloy: German Standard
2.4374	3.0% Al 1.5% Fe 30% Cu 0.8% Ti Ni alloy: German Standard
44 K MONEL	0.15% C 29.5% Cu 2.8% Al 0.5% Ti Ni alloy: Filler; Huntington; now Monel 44
64 K MONEL	0.15% C 29.5% Cu 2.8% Al 0.5% Ti Ni alloy: Filler; Huntington; now Monel 64
400	31% Cu 2% Fe 2.0% Mn Ni alloy: International Nickel Co; cold drawn rod **DPN: 220 UTS: 790 Elon: 20% Proof: 570**

Note. The following abbreviations and units are used in the tables:

DPN	Hardness, diamond pyramid number
UTS	Ultimate tensile strength, N/mm^2
Elon	Elongation, %
Proof	0.1 % proof strength, N/mm^2

1 N/mm^2=0.1 hbar=0.102 kgf/mm^2=0.06475 tonf/in^2=145.04 lbf/in^2

Symbol	Nominal analysis, supplier, condition and remarks.
ALLOY 82	0.12% C 9% Si 3% Cu Ni alloy: Casting; Dewrance **DPN: 270 UTS: 680**
AMPCOLAY 551	0.15% C 2.8% Al 0.5% Ti 29.5% Cu 1% Fe Ni alloy: Wrought; Ampco for K Monel
AMPCOLAY 552	30% Cu Ni alloy: Wrought; Ampco for Monel
AMPCOLAY 553	0.12% C 32% Cu 3% Fe 1% Mn 3.7% Si Ni alloy: Casting; Ampco for S Monel
AMPCOLAY 557	0.12% C 32% Cu 3% Fe 1% Mn 2.7% Si Ni alloy: Casting; Ampco for H Monel
AMPCOLAY 558	30% Cu Ni alloy: Weldable; Ampco for Monel
AMPCOLAY 559	30% Cu Ni alloy: Cast; Ampco for Monel
AMS 4544 B	30% Cu Ni alloy: Sheet; annealed
AMS 4574 B	30% Cu Ni alloy: Seamless tube; annealed
AMS 4675 B	30% Cu Ni alloy: Brazed tube; annealed
AMS 4674 C	30% Cu 0.04% S Ni alloy: Bar & forging; free machining
AMS 4675	30% Cu Ni alloy: Bar & forging
AMS 4676	30% Cu 3% Al 0.6% Ti Ni alloy: Bar & forging; K Monel

Symbol	Nominal analysis, supplier, condition and remarks.
AMS 4730C	30% Cu Ni alloy: Wire; annealed; AMS for 400 Monel
AMS 4892	30% Cu 4% Si Ni alloy: Casting; as cast
AMS 4893	30% Cu 4% Si Ni alloy: Casting; heat treated
ASTM B127	30% Cu 2% Fe Ni alloy: Plate, sheet & strip; annealed **UTS: 580** **Elon: 35%** **Proof: 180**
ASTM B127	30% Cu 2% Fe Ni alloy: Plate, sheet & strip; hard **UTS: 740** **Elon: 2%** **Proof: 640**
ASTM B164 A	33% Cu 2.5% Fe 2% Mn Ni alloy: Rod & bar; annealed; low sulphur **UTS: 480** **Elon: 35%** **Proof: 150**
ASTM B164 A	33% Cu 2.5% Fe 2% Mn Ni alloy: Rod & bar; cold drawn; low sulphur **UTS: 760** **Elon: 8%** **Proof: 610**
ASTM B164 B	33% Cu 2.5% Fe 2% Mn Ni alloy: Bar & rod; annealed (0.02–0.06% S) **UTS: 480** **Elon: 35%** **Proof: 150**
ASTM B164 B	33% Cu 2.5% Fe 2% Mn Ni alloy: Bar & rod; cold drawn (0.02–0.06% S) **UTS: 680** **Elon: 8%** **Proof: 340**
ASTM B165	33% Cu 2.5% Fe 2% Mn 0.024% S (max) Ni alloy: Pipe; annealed **UTS: 540** **Elon: 35%** **Proof: 180**
ASTM B295 E3N10	0.4% C 4% Mn 30% Cu 1.5% Al 1% Ti Ni alloy: Welding electrode **UTS: 540**
ASTM B295 E3N14	0.4% C 4% Mn 2% Fe 35% Cu 3% Al Ni alloy: Welding electrode; age hardened **UTS: 740**
ASTM B295 E4N10	0.15% C 2% Mn 2% Fe 30% Cu 3% Nb 1% Ti Ni alloy: Welding electrode
ASTM B295 E Ni Cu 1	0.15% C (max) 3% Nb (max) 4% Mn 30% Cu Ni alloy: Welding rod; previous designation E4N10
ASTM B295 E Ni Cu 2	0.15% C (max) 2.5% Nb (max) 6% Mn 30% Cu Ni alloy: Welding rod
ASTM B295 E Ni Cu 4	0.4% C (max) 4.0% Mn 30% Cu Ni alloy: Welding rod; previous designation E3N10
ASTM B304 ERN60	0.15% C 1% Mn 2% Fe 1% Si 30% Cu 2% Ti Ni alloy: Welding electrode
ASTM B304 ERN64	0.25% C 2% Mn 1% Si 30% Cu 3% Al 0.7% Ti Ni alloy: Welding electrode
ASTM B304 RN40	0.3% C 2% Mn 2% Fe 30% Cu Ni alloy: Welding electrode
ASTM B304 RN43	0.3% C 1% Si 30% Cu Ni alloy: Welding electrode
ASTM B304 R Ni Cu 5	30% Cu Ni alloy: Welding electrode; previous designation RN40
ASTM B304 ER Ni Cu 7	28% Cu Ni alloy: Welding electrode; previous designation ERN60
BS 1537	Cu Ni alloy: Casting; replaced by BS 3071
BS 2901 NA33	2.2% Ti 1.2% Al 2% Fe 25% Cu Ni alloy: Rod for all forms of welding
BS 3071 NA1	30% Cu 1% Mn 1% Si Ni alloy: Casting; as cast **UTS: 420** **Elon: 16%** **Proof: 150**
BS 3071 NA2	30% Cu 1% Mn 2.7% Si Ni alloy: Casting; as cast **UTS: 540** **Elon: 10%** **Proof: 220**
BS 3071 NA3	30% Cu 1% Mn 4% Si Ni alloy: Casting; as cast **DPN: 250** **UTS: 610**

Note. The following abbreviations and units are used in the tables:

DPN	Hardness, diamond pyramid number
UTS	Ultimate tensile strength, N/mm^2
Elon	Elongation, %
Proof	0.1 % proof strength, N/mm^2

1 N/mm^2=0.1 hbar=0.102 kgf/mm^2=0.06475 tonf/in^2=145.04 lbf/in^2

Symbol	Nominal analysis, supplier, condition and remarks.
BS 3072 NA13	32% Cu Ni alloy: Annealed **UTS: 450** **Elon: 33%** **Proof: 150**
BS 3073 NA13	32% Cu Ni alloy: Strip; annealed **DPN: 130** **UTS: 450** **Elon: 33%** **Proof: 150**
BS 3074 NA13	32% Cu Ni alloy: Annealed **UTS: 460** **Elon: 35%** **Proof: 180**
BS 3074 NA13	32% Cu Ni alloy: Tube; cold drawn **UTS: 580** **Elon: 10%** **Proof: 370**
BS 3075 NA13	32% Cu Ni alloy: Wire; annealed **UTS: 450** **Elon: 22%**
BS 3075 NA13	32% Cu Ni alloy: Wire; cold drawn **UTS: 730**
BS 3076 NA13	32% Cu Ni alloy: Bar & section; annealed **UTS: 450** **Elon: 35%** **Proof: 100**
BS 3076 NA13	32% Cu Ni alloy: Bar & section; cold drawn **UTS: 600** **Elon: 14%** **Proof: 450**
BS 3076 NA18	30% Cu 3% Al 0.7% Ti Ni alloy: Rod
BS 3127/4	3% Al 1% Ti 30% Cu Ni alloy: Tube; solution treated for bourdon tubes **DPN: 190**
BS 3127/4	3% Al 1% Ti 30% Cu Ni alloy: Tube; solution treated and cold worked for bourdon tubes **DPN: 250**
BS 3146 ANC18A	0.2% C 30% Cu 1.0% Si Ni alloy: Investment casting; as cast **UTS: 300** **Elon: 15%** **Proof: 100**
BS 3146 ANC18B	0.1% C 30% Cu 2.7% Si Ni alloy: Investment casting; as cast **UTS: 420** **Elon: 10%** **Proof: 220**
BS 3146 ANC18C	0.1% C 30% Cu 4% Si Ni alloy: Investment casting; as cast **UTS: 580** **Proof: 480**
CALMALLOY	2% Fe 29% Cu Ni alloy: Used for temperature compensation shunts; origin unknown
COPEL	45% Cu Ni alloy: Origin unknown
CORRONEL	4% Mn 26% Cu Ni alloy: Origin unknown
DIN 17743 Ni Cu 30 Al	3% Al 1.5% Fe 30% Cu 0.8% Ti Ni alloy
DIN 17743 Ni Cu 30 Fe	2.0% Fe 31% Cu Ni alloy
DTD 10 B	35% Cu 2.5% Fe Ni alloy: Sheet **UTS: 450** **Proof: 100**
DTD 192	35% Cu 2.5% Fe Ni alloy: Bar & forging; annealed **UTS: 540** **Elon: 30%**
DTD 196	32% Cu 1.5% Mn Ni alloy: Bar; annealed **UTS: 450** **Elon: 35%**
DTD 200 A	35% Cu 2.5% Fe Ni alloy: Bar & strip; cold rolled & tempered **UTS: 680** **Elon: 14%** **Proof: 450**
DTD 204 A	31% Cu 2% Fe Ni alloy: Bar & tube; annealed **UTS: 560**
DTD 232	28% Cu 23% Zn 1.5% Fe 1.5% Mn Ni alloy: Sheet; cold rolled **UTS: 870** **Elon: 12%** **Proof: 690**
DTD 237	23% Zn 30% Cu Ni alloy: Sheet & strip; softened **UTS: 540** **Elon: 25%** **Proof: 220**
DTD 268	30% Cu 23% Zn 2% Mn 2% Fe Ni alloy: Bar & wire; annealed **UTS: 540**
DTD 477	30% Cu 1.5% Mn 2% Fe Ni alloy: Tube; annealed; solid drawn **UTS: 490**
DTD 487	30% Cu 3% Al 2% Mn 2% Fe Ni alloy: Bar; annealed & cold drawn to size; for cold headed bolts **UTS: 810** **Elon: 20%** **Proof: 690**
H MONEL	0.12% C 32% Cu 3% Fe 1% Mn 2.7% Si Ni alloy: Casting; H. Wiggin; now Monel 506

Symbol	Nominal analysis, supplier, condition and remarks.
JAE	30% Cu Ni alloy with high temperature coefficient of magnetism: H. Wiggin; curie point 70–75 °C **DPN: 110 UTS: 420 Elon: 52% Proof: 140**
K 500	0.25% C 1% Si 30% Cu 2% Fe 1.5% Mn 1% Ti 3% Al Ni alloy: Wrought; H. Wiggin; annealed; can be age hardened **DPN: 170 UTS: 680**
K MONEL	0.15% C 2.8% Al 0.5% Ti 29.5% Cu 1% Fe Ni alloy: Bar, sheet, etc.; Huntington; now Monel K500
KR MONEL	0.23% C 2.8% Al 0.5% Ti 29.5% Cu Ni alloy: Bar; Huntington; now Monel 501
LANGALLOY 1N	28.5% Cu 1.5% Fe 1.8% Si 1% Mn Ni alloy: Langley **DPN: 120 UTS: 420 Elon: 25% Proof: 180**
LANGALLOY 2N	30% Cu 3% Si 3% Fe Ni alloy: Casting; Langley for BS alloy NA2
LANGALLOY 5N	30% Cu 4% Si 3% Fe Ni alloy: Casting; Langley for BS alloy NA3
LC MONEL	0.12% C 13% Cu 1.3% Fe Ni alloy: Sheet, bar etc.; Huntington; now Monel 406
MONEL 40	0.1% C 31% Cu 1.3% Fe Ni alloy: Filler rod; Huntington; oxyacetylene welding of Monel 400
MONEL 43	0.1% C 40% Cu Ni alloy: Filler rod; Huntington; oxyacetylene welding of Monel 402 and 403
MONEL 44	0.15% C 29.5% Cu 2.8% Al 0.5% Ti Ni alloy: Filler rod; Huntington; oxyacetylene welding of Monel K 500
MONEL 60	0.03% C 30.5% Cu 2.2% Ti Ni alloy: Filler; Huntington; inert gas shielded welding of Monel 400
MONEL 64	0.15% C 29.5% Cu 2.8% Al 0.5% Ti Ni alloy: Filler; Huntington; inert gas shielded welding of Monel K 500
MONEL 130	0.15% C 27% Cu 2.5% Mn 1% Al 0.3% Ti Ni alloy: Filler; Huntington; for welding Monel 400
MONEL 134	0.25% C 27% Cu 2.5% Mn 2% Al 0.3% Ti Ni alloy: Filler; Huntington; for welding Monel K 500
MONEL 140	0.05% C 26% Cu 1.2% Mn 0.3% Al 0.7% Ti 1.5% Cb Ni alloy: Filler; Huntington; for welding Monel 400 to steel
MONEL 180	0.03% C 28% Cu 5% Mn 0.3% Al 0.7% Ti 1.5% Cb Ni alloy: Filler; Huntington; for welding Monel alloys
MONEL 400	0.12% C 31.5% Cu 1.3% Fe Ni alloy: Bar, sheet, tube, etc.; Huntington; annealed **UTS: 550 Elon: 48% Proof: 200**
MONEL 400	0.3% C 32% Cu 3% Fe 1% Mn 1% Si Ni alloy: Casting; H. Wiggin; as cast **DPN: 120 UTS: 420 Elon: 25% Proof: 200**
MONEL 400	0.3% C 32% Cu 2.5% Fe 2% Mn Ni alloy: Wrought; H. Wiggin; annealed **DPN: 140 UTS: 550 Elon: 40% Proof: 220**
MONEL 400	0.3% C 32% Cu 2.5% Fe 2% Mn Ni alloy: Wrought; H. Wiggin; cold rolled **DPN: 200 UTS: 760 Elon: 7% Proof: 690**

Symbol	Nominal analysis, supplier, condition and remarks.
MONEL 401	0.03% C 53% Cu 0.5% Co Ni alloy: Strip; Huntington
MONEL 402	0.12% C 40% Cu Ni alloy: Sheet, bar, etc.; Huntington
MONEL 403	0.12% C 40% Cu 1.8% Mn Ni alloy: Bar, sheet, etc.; Huntington
MONEL 404	0.06% C 0.02% Al 44% Cu Ni alloy: Bar, sheet, etc.; Huntington
MONEL 406	0.12% C 13% Cu 1.3% Fe Ni alloy: Sheet, bar, etc.; Huntington
MONEL 414	0.4% C 32% Cu 2.5% Fe 1.0% Mn Ni alloy: Rod & wire; H. Wiggin; free machining variety of Monel 400
MONEL 501	0.23% C 2.8% Al 0.5% Ti 29.5% Cu Ni alloy: Bar; Huntington
MONEL 505	0.12% C 32% Cu 3% Fe 1% Mn 3.7% Si Ni alloy: Casting; H. Wiggin; as cast **DPN: 240 UTS: 630 Proof: 630**
MONEL 506	0.12% C 32% Cu 3% Fe 1% Mn 2.7% Si Ni alloy: Casting; H. Wiggin; as cast **DPN: 200 UTS: 56 Elon: 12% Proof: 32**
MONEL C & C	0.4% C 32% Cu 2.5% Fe 1.0% Mn Ni alloy: For machining; H. Wiggin; as Monel 400; higher carbon
MONEL K405	0.18% C 31.5% Cu 1.3% Fe Ni alloy: Bar; Huntington
MONEL K500	0.25% C 1% Si 30% Cu 2% Fe 1.5% Mn 1% Ti 3% Al Ni alloy: Wrought; H. Wiggin; annealed; can be age hardened **DPN: 170 UTS: 69**
MONEL K500	0.15% C 2.8% Al 0.5% Ti 19.5% Cu 1% Fe Ni alloy: Bar, sheet, etc.; Huntington; age hardened **DPN: 275 UTS: 107 Elon: 25% Proof: 76**
MONEL K500	0.15% C 2.8% Al 0.5% Ti 29.5% Cu 1% Fe Ni alloy: Huntington; cold drawn & aged **DPN: 300 UTS: 121 Elon: 20%**
NCC PIG	25% Cu 9% Cr 0.5% Si 0.5% Mn 8% Fe Ni alloy: Ingot; H. Wiggin; for addition to cast iron
PLATNAM	33% Cu 13% Sn 0.5% Fe 0.3% Al Ni alloy: Hopkinsons; for valve discs and seats up to 600 °C
R MONEL	0.18% C 31.5% Cu 1.3% Fe Ni alloy: Bar; Huntington; Monel 405
S MONEL	0.12% C 32% Cu 3% Fe 1% Mn 3.7% Si Ni alloy: Casting; H. Wiggin; now Monel 505

Note. The following abbreviations and units are used in the tables:

DPN	Hardness, diamond pyramid number
UTS	Ultimate tensile strength, N/mm^2
Elon	Elongation, %
Proof	0.1 % proof strength, N/mm^2

1 N/mm^2=0.1 hbar=0.102 kgf/mm^2=0.06475 tonf/in^2=145.04 lbf/in^2

27E Nickel–iron alloys
With chromium, copper or molybdenum

Specific gravity	8.5	
Density	8500 kg/m^3	(0.30 lb/in.^{3})
Solidus/liquidus	1425 °C	
Thermal conductivity	– / °C	
Coefficient of linear expansion (20–100° C)	1.5×10^{-6}	
Electrical conductivity	8.5% IACS (copper 100%)	
Specific resistance	200 microhm mm	
Impact	–	
Impact	-	
Fatigue strength	± 210 N/mm^2	(± 14 tonf/in.^{2})
Hot strength		

Temperature °C	Tensile strength N/mm^2
95	580
205	550
310	550
485	520
540	500

The above properties have been chosen to show typical values for the specifications listed and may not apply exactly to any specification. It is possible that with some specifications the values may not be applicable.

General metallurgical characteristics

The alloys of nickel and iron have special magnetic properties and on the whole, this is the sole reason for their existence. As nickel is added to iron the magnetic properties first increase then decrease, the iron/nickel alloys and properties being discussed in Section 20C. Included in Section 20C are the series of alloys varying between 45 and 55% nickel. For the sake of clarity no attempt has been made to include any of these alloys in the present group although all those above 50% nickel are technically members of Section 27E.

Most of the alloys listed here have between 70 and 80% nickel, the remainder being principally iron.

Chromium improves the hot corrosion resistance and strength and also increases the electrical resistance. The copper and molybdenum additions are usually to stabilize some magnetic property.

Thermal treatment

None of these alloys can be hardened by thermal treatment, but in common with all nickel alloys they are considerably hardened by cold work. When necessary, this can be removed by annealing at about 1050 °C for 10 minutes at temperature. The rate of cooling has no effect on the mechanical properties but with some alloys magnetic properties can be altered by the rate of cooling.

These alloys will normally be heated only in electric furnaces to obviate any danger of sulphur or carbon pickup from gas or oil atmospheres. De-scaling will normally be by hydrofluoric acid pickling when necessary. Sodium hydride is satisfactory if the special equipment is available but very often the parts can be used with the slight oxide scale remaining. Grit blasting is not normally possible as this tends to leave an undesirable cold worked surface. Special atmospheres such as hydrogen, nitrogen hydrogen mixture are commonly used, and vacuum treatment is becoming popular.

Welding and brazing

Pre-treatment. Parts should be in the annealed condition if possible, with the joint area thoroughly clean and free from oxide, grease and dirt.

Welding and brazing. Gas and electric arc welding are not recommended on these materials. Inert gas shielded electric arc gives good results and normal electric resistance welding is generally the only method of joining applied.

Brazing using copper alloy filler and the correct flux presents no problems.

Post-treatment. All trace of brazing flux must be removed, but apart from this, no treatment is necessary.

Flaw detection methods

Crack test. This will seldom be required, but when necessary will be by penetrant oil chalk testing.

X-ray. Not required.

Ultra-sonic test. Not normally required.

Chemical etch. Not required.

Corrosion protection

Temporary. None required.

Permanent. None required. These alloys have excellent corrosion resistance and are used for their electrical and magnetic properties, thus are unlikely to be subjected to corrosive atmospheres.

Machinability

Considerable difficulty is liable to be experienced if these alloys have to be machined to fine limits and a good surface finish. Like all soft high nickel alloys the tendency is to tear and gall.

Best results are achieved with the high speed or tipped tools using light cuts at high speeds, ensuring at all times that the tool is cutting and not rubbing or dwelling. Chlorinated liquid coolants are reputed to give good results but require special health precautions such as fume extraction.

Uses

Magnetic shields, inductances, electrical resistors and other electronic, magnetic and telecommunication purposes.

Symbol	Nominal analysis, supplier, condition and remarks.
2.4500–2.4519	2.0% Cr 5% Cu 16% Fe Ni alloy: German Standard numbers
2.4540–2.4559	4% Mo 16% Fe Ni alloy: German Standard numbers
2.4867	15% Cr 20% Fe Ni alloy: German Standard
141 ALLOY	30% Fe Ni alloy: British Driver-Harris; ballast resistance; withdrawn
BALCO	Ni Fe alloy: Gilbey Brunton; high temperature coefficient of resistance
BS 2857 A	24% Fe Ni alloy: Strip for laminations
CI Ni Fe	1.0% C 45% Fe Ni alloy: Electrode; Metrode; for high strength welding
DEF 5192	14% Fe Cu Cr Mn Mo Ni alloy: For high permeability magnetic alloys
DIN 17470 Ni Cr 60/15	15% Cr 20% Fe Ni alloy
DIN 17745 Ni Fe 15 Mo	4% Mo 16% Fe Ni alloy
DIN 17745 Ni Fe 16 Cu Cr	2.0% Cr 5% Cu 16% Fe Ni alloy
DIN 17745 Ni Fe 16 Cu Mo	4% Mo 5% Cu 16% Fe Ni alloy
DIN 17745 Ni Fe 46 Cr	6% Cr 46% Fe Ni alloy
FIXAMPER	28% Fe Ni alloy: Imphy
HIPERNOM	0.05% C 4.5% Mo 16% Fe Ni alloy: Westinghouse; for magnetic shielding
HYTEMCO	30% Fe Ni alloy: British Driver-Harris; ballast resistance
	UTS: 520
KONEL	17% Co 10% Fe Ti Ni alloy: Tensile strength 450 N/mm^2 at 600 °C; source unknown
LEGA 52	49% Fe Ni alloy: Low expansion; Driver Harris
N 52	48% Fe Ni alloy: Low expansion; Imphy
N 54	46% Fe Ni alloy: Low expansion; Imphy
N 58	42% Fe Ni alloy: Low expansion; Imphy

Symbol	Nominal analysis, supplier, condition and remarks.
N 501	49% Fe 1.0% Cr Ni alloy: Low expansion; Imphy
NILGRO 36	36% Fe Ni alloy: Low expansion; Darwins coefficient of expansion 1.5 × 10^{-6} up to 150 °C
	UTS: 480 Elon: 31% Proof: 260
NILO 51	0.15% (max) 49% Fe Ni alloy: H. Wiggin; for sealing to soft glasses
	DPN: 140 UTS: 520
NILO 51	49% Fe Ni alloy: For glass seals; International Nickel Co
NILO 55	0.15% C (max) 42% Fe (max) Ni alloy: H. Wiggin; for cast iron welding electrodes
NILOMAG 77	13.5% Fe 5.0% Cu 4.0% Mo Ni alloy: H. Wiggin; laminations for transducers and switches
NIRON 52	49% Fe Ni alloy: For glass/metal seal; Driver
PERMALLOY F	35% Fe Ni alloy: Standard Telephones; low coercive force
PLATINITE	48% Fe Ni alloy: Low expansion; Imphy
R 2799	Fe Ni alloy: Telcon; hard rolled; Curie point approximately 60 °C
	DPN: 190
SANBOLD NA76	24% Fe Ni alloy: Strip for magnetic shielding, etc.; Sanderson; electrical conductivity 2.8% IACS
VACOVIT 511	49% Fe 1% Cr Ni alloy: Low expansion; German proprietary alloy
VACOVIT 540	46% Fe Ni alloy: Low expansion; German proprietary alloy

Note. The following abbreviations and units are used in the tables:

DPN	Hardness, diamond pyramid number
UTS	Ultimate tensile strength, N/mm^2
Elon	Elongation, %
Proof	0.1 % proof strength, N/mm^2

1 N/mm^2=0.1 hbar=0.102 kgf/mm^2=0.06475 tonf/in^2=145.04 lbf/in^2

27F Nickel – miscellaneous alloys

General metallurgical characteristics

The alloys listed have all a specialized use, based on one or other of the peculiar properties of nickel, modified to a greater or lesser extent by one or more alloying elements. The alloys do not fit into any of the preceding groups and have not sufficient variation in properties to warrant separate groups. The principal effects of the commonest alloying elements are given below, but it should be noted that the effect of any two elements together may result in an alloy of completely different properties.

It must be emphasised that the information available on many of these proprietary materials is rather meagre and the reader is strongly advised to obtain further details from the source if necessary. It was not possible to list the general physical and mechanical properties, as the variations within the group are too great for this to be meaningful.

Aluminium. This can form intermetallic compounds with other alloying elements which may be used to strengthen the nickel matrix. It is also used as a de-oxidizing metal, particularly with nickel alloy welding rods.

Beryllium. This forms intermetallic compounds in the nickel matrix in a manner analogous to the copper/beryllium series – Section 14D. These can be solution treated and aged but to date have not found the same number of applications as the copper/beryllium alloys.

Boron. This is used as a fluxing agent in metal spray alloys and some brazing materials. It lowers the melting point and acts as a vehicle for refractory particles. The boron also helps the material to wet the base metal, thus considerably aiding adhesion.

Copper. This forms a solid solution with nickel and is never the major constituent in the alloys listed. The nickel/copper alloys are discussed in Section 27D.

Iron. This may be added for some electrical purpose, but in general is an economical alloying element which has the minimum effect on other properties.

Manganese. This improves the hot corrosion resistance but the principal use of nickel manganese alloys is for their special electrical properties.

Molybdenum. This improves the resistance of nickel alloys to acids and also increases the hot strength properties. The alloys with a high molybdenum content – above 20% – have very special uses.

Palladium. This is added to certain alloys to give desirable brazing properties such as wetting, without detracting from the excellent corrosion resistance and hot strength of nickel alloys. There is a series of alloys designed for joining nickel/chromium and nickel/copper alloys by brazing.

Silicon. This is a de-oxidizing element which also has considerable fluxing properties and improves the corrosion resistance under certain circumstances.

Titanium. This forms intermetallic compounds which harden the nickel matrix. It is generally used along with aluminium, but can also be used alone as a grain refiner.

Thermal treatment

Certain of the alloys can be solution treated and aged. This will require temperatures above 1000 °C for the solution treatment and 600 °C for ageing.

The majority of the materials listed cannot have their mechanical properties improved by thermal treatment but may require an anneal to remove the affects of cold work. The temperature and time will depend on whether all the cold work is to be removed, in which case 10 minutes at about 1060 °C will be required. If the parts are only to be stress relieved they should be heated to 50 °C above the temperature at which they will be required to operate, for a time of 2 hours at least if below 500 °C or 10 minutes if at 1060 °C.

As with all nickel alloys care must be taken not to contaminate the parts with sulphur at high temperatures. This generally rules out oil heating and care must be exercised using gas heating. Electrical heating gives best results.

De-scaling can be by grit blasting provided the parts are not prone to distort. Acid pickling using hot hydrofluoric acid with nitric or ferric chloride additions can be used, but this is unpleasant to handle. Sodium hydride gives excellent results but requires special equipment. Many of the alloys are used without any form of scale removal as this is generally only a very thin adherent coating.

Further information should always be obtained from the source of the specification before attempting any form of thermal treatment on the alloys listed as many of them have a controlled amount of cold work carried out and this will be removed if the parts are heated above a certain figure.

Welding and brazing

Pre-treatment. Joints must be carefully prepared and free from all grease, oil, dirt and oxide. Grit blasting or sodium hydride pickling give the best results.

The part should be annealed or solution treated if possible, but this will not be the case when the material has been cold worked for specific mechanical properties.

Welding and brazing. Welding by electric arc or gas flame is possible with some of these alloys using an active flux and highly skilled welders. Inert gas shielded electric arc welding is however recommended whenever these materials are to be joined.

Some of the alloys listed are welding rods or braze filler materials and these will be subjected to the conditions laid down for joining the base materials, most of which will be other nickel alloys discussed in the Sections 27.

In general, specialist advice should be sought before welding any of the materials listed.

Post-treatment. Ageing alloys will require to be aged after welding. No post-weld thermal treatment is necessary for any of the other alloys.

Any flux used must be removed immediately after welding.

Flaw detection methods

Crack test. Penetrant oil chalk test, using fluorescent dyes when extra sensitivity is necessary.

X-ray. This may be necessary on a specialized part but will not be a normal requirement.

Ultra-sonic test. This is being used more commonly after rough machining to find subcutaneous defects, thus saving the cost of expensive machining operations. It cannot identify surface flaws if these are very shallow.

Chemical etch. Where grain size or orientation is important a ferric chloride/hydrochloric acid etch will be necessary. This is highly corrosive and the resultant surface will have a reduced fatigue strength, thus should be polished to remove about 0.1 mm per surface.

Corrosion protection

Temporary. None required.

Permanent. None required under most conditions. Many of the alloys listed have been developed for their corrosion resistance to specific media or for their high temperature oxidation resistance.

Machinability

On the whole all the alloys listed will work harden, thus should not be subjected to light rubbing or interrupted cuts. Whenever possible heavy feeds at low speeds using ample power and rigid machines are recommended.

Uses

These alloys cover a wide range of specialist uses, many of which are listed alongside the specifications. These include electrical resistors, low expansion materials and material with controlled electrical or magnetic properties or special expansion characteristics.

Symbol	Nominal analysis, supplier, condition and remarks.
2.4108	0.2% C 1.0% Mn Ni alloy: German Standard
2.4110	2.0% Mn Ni alloy: German Standard
2.4116	5% Mn Ni alloy: German Standard
2.4122	1.5% Al 3% Mn 1.0% Si Ni alloy: German Standard
2.4128	4.5% Al 0.5% Ti Ni alloy: German Standard
2.4132	2.0% Be Ni alloy: German Standard
2.4400–2.4419	3.0% Mo 14% Cu 11% Fe Ni alloy: German Standard
2.4520–2.4539	4.0% Mo 5.0% Cu 16% Fe Ni alloy: German Standard
2.4600	5% Fe 28% Mo Ni alloy: German Standard
26.35 COW Nb	0.5% C 26.0% Cr 5.0% W 15.0% Co 1% Nb 12.0% Fe Ni alloy: Electrode; Metrode
25.35.4 CW Co	0.5% C 25.0% Cr 5.0% W 15.0% Co 15.0% Fe Ni alloy: Electrode; Metrode
25.50 COW Nb Zr	0.5% C 25.0% Cr 9.0% W 12.0% Co 2.0% Nb 3.0% Fe 0.1% Zr Ni alloy: Electrode; Metrode
28.37.4 Cu B	0.05% C 27.0% Cr 4.0% Mo 2.0% Cu 31.0% Fe Ni alloy: Electrode; Metrode; for welding Incoloy 825
133 ALLOY	3.0% Si Ni alloy: Driver Harris; withdrawn **UTS: 910**
200C	2.0% Be Ni alloy: Casting; Brush Beryllium Co
220C	2.2% Be Ni alloy: Casting; Brush Beryllium Co
260C	2.6% Be Ni alloy: Casting; Brush Beryllium Co
288	Pd Ni brazing alloy: Melting point 1237 °C; Engelhard Industries **DPN: 224 UTS: 760 Elon: 30%**
318	Pd Mn Ni brazing alloy: Melting point 1120 °C; Engelhard Industries; electrical conductivity 9% IACS **DPN: 330 UTS: 800 Elon: 31%**
420CR	1.8% Be 1.7% Ti Ni alloy: Casting; Beryllium Corporation
ALLOY 1	15% Mg Ni alloy: Ingot; H. Wiggin; for cast iron additions
ALLOY 1M	15% Mg 0.8% Mischmetal Ni alloy: Ingot; H. Wiggin; for cast iron additions
ALLOY 2	18% Mg 27% Si Ni alloy: Ingot; H. Wiggin; for cast iron additions
AMS 4778 A	4.5% Si 3% B Ni alloy: Braze metal
AMS 4779	3.5% Si 1.8% B Ni alloy: Braze metal
ARC 164	22% Fe 18% Mo Ni alloy: Imphy
ARSENIC-NICKEL	40% As Ni alloy: Blackwells; primary metal

Symbol	Nominal analysis, supplier, condition and remarks.
ASTM A494 Ni Mo	0.12% C (max) 5.0% Fe 2.5% Co 0.5% V 28% Mo Ni alloy: Casting
ASTM A494 Ni Mo Cr	0.12% C (max) 16% Cr 4.5% W 6.0% Fe 2.5% Co 17% Mo Ni alloy: Casting
ASTM B260 B Ni 3	3.15% B 4.5% Si Ni alloy: Brazing metal; temperature range 1010–1177 °C
ASTM B260 B Ni 4	1.6% B 3.5% Si Ni alloy: Brazing metal; temperature range 1010–1177 °C
ASTM B260 B Ni 6	11% P Ni alloy: Brazing metal; braze temperature range 927–1030 °C
ASTM B295 E Ni 1	3% Ti 0.1% C (max) Ni alloy: Welding rod; previous designation E4N11
ASTM B295 E Ni Mo 1	5.5% Fe 2% Co 28% Mo Ni alloy: Welding rod; previous designation E3N1B and E4N1B
ASTM B295 E Ni Mo 2	3.5% Fe 2% Co 15.5% Cr 16.5% Mo 3.7% W Ni alloy: Welding rod; previous designation E3N1C and E4N1C
ASTM B295 E Ni Mo 3	5.5% Fe 4% Cr 2% Co 25% Mo Ni alloy: Welding rod
ASTM B295 E3N11	0.7% C 1.0% Si 1.0% Al 3.0% Ti Ni alloy: Welding electrode **UTS: 370**
ASTM B295 E4N11	0.1% C 1.0% Si 3.0% Ti Ni alloy: Welding electrode
ASTM B304 ERN61	0.15% C 1.0% Mn 1.0% Fe 1.5% Al 2.5% Ti Ni alloy: Welding electrode
ASTM B304 ER Ni Mo 4	5% Fe 2% Co 0.4% V 28% V Ni alloy: Welding electrode; previous designation ERN7B
ASTM B304 ER Ni Mo 5	5% Fe 2% Co 3.5% W 0.2% V 16% Mo Ni alloy: Welding electrode; previous designation ERN7C
ASTM B304 RN41	0.15% C 1.0% Si 0.2% Cu Ni alloy: Welding electrode
BS 1845 N11	11% P Ni alloy: For brazing; melting point 875 °C
BS 1845 N12	13% Cr 10% P Ni alloy: For brazing; melting point 890 °C
BS 1845 N14	4.5% Si 3% B Ni alloy: For braze metal; melting range 960–1040 °C
BS 1845 N15	3.5% Si 1.7% B Ni alloy: For braze metal; melting range 980–1070 °C
BS 1845 PD11	21% Pd 31% Mn Ni alloy: For braze metal; melting point 1120 °C
BS 3146 ANC 15	0.1% C 27% Mo 5% Fe Ni alloy: Investment casting; as cast **UTS: 460 Elon: 10% Proof: 220**
BS 3146 ANC 17	0.1% C 3.0% Cu 9.5% Si Ni alloy: Investment casting; as cast; sulphuric acid resistant
CA NICKEL	4.0% Co Ni alloy: Strip; H. Wiggin; for magnetostriction
CATHODE ALLOY	2% and 4% W Ni alloy: For electronic valve cathodes; Driver-Harris
CINEX	0.7% C 0.1% Mn 2.4% Fe Ni alloy: Welding rod; Murex; for cast iron
CM 52	4.5% Si 3.0% B Ni alloy: For hard facing; Dewrance; gas welding
COBANIC	45% Co Ni alloy: For valve filaments; Driver-Harris

Note. The following abbreviations and units are used in the tables:

DPN	Hardness, diamond pyramid number
UTS	Ultimate tensile strength, N/mm^2
Elon	Elongation, %
Proof	0.1 % proof strength, N/mm^2

$1\ N/mm^2 = 0.1\ hbar = 0.102\ kgf/mm^2 = 0.06475\ tonf/in^2 = 145.04\ lbf/in^2$

Symbol	Nominal analysis, supplier, condition and remarks.
CR1	0.5% C 2.75% Be 0.6% Cr Ni alloy: Casting; Beryllium Corporation
COLMONOY 22	0.15% C 3.5% Si 1.0% Fe 1.35% B Ni alloy: Powder; Wall Colmonoy; for metal spraying; melting point 105 °C
COLMONOY 23	0.1% C 2.3% Si 1.0% Fe 1.25% B Ni alloy: Powder; Wall Colmonoy; for metal spraying; melting point 1065 °C
COLMONOY 23A	0.1% C 2.3% Si 1.0% Fe 1.2% B Ni alloy: Bronze metal; Wall Colmonoy; melting point 1065 °C
COLMONOY 24	0.1% C 2.3% Si 1.0% Fe 1.2% B Ni alloy: Bronze metal; Wall Colmonoy; melting point 1065 °C
COLMONOY 25	0.25% C 4.0% Si 5.0% Cr 2.0% Fe 1.25% B Ni alloy: Powder; Wall Colmonoy; for metal spraying; melting point 1220 °C
COLMONOY 26	0.12% C 3.5% Si 1.0% Fe 1.35% B Ni alloy: Powder; Wall Colmonoy; for metal spraying; melting point 1050 °C
COLMONOY 27	0.1% C 2.3% Si 1.0% Fe 1.25% B Ni alloy: Powder; Wall Colmonoy; for metal spraying; melting point 1065 °C
COLMONOY 28	0.08% C 2.3% Si 1.25% B Ni alloy: Powder; Wall Colmonoy; for metal spraying; melting point 1065 °C
CR 2	0.9% C 2.7% Be 0.6% Cr Ni alloy: Casting; Beryllium Corporation
D NICKEL	0.1% C 4.5% Mn Ni alloy: Bar & sheet; Huntington; now Nickel 211
DELORO 40G	7.5% Cr 4.0% Si 1.2% B 5.0% Fe Ni alloy; Melting range 985–1180 °C; Deloro Stellite UTS: 400 Elon: 1%
DELORO 45	7.5% Cr 4.0% Si 1.5% B 1.5% Fe Ni alloy: Melting range 995–1150 °C; Deloro Stellite DPN: 380
DELORO 50	10% Cr 4.0% Si 1.5% B 4.0% Fe Ni alloy: Melting range 975–1065 °C; Deloro Stellite DPN: 550
DELORO 60	15% Cr 4.5% Si 3.0% B 4.5% Fe Ni alloy: Melting range 965–1005 °C; Deloro Stellite DPN: 750
DELORO PW22	1.2% Cr 2.5% Si 1.3% B Ni alloy: Melting range 940–1260 °C; Deloro Stellite DPN: 225
DIN 17741 Ni Al 4 Ti	4.5% Al 0.5% Ti Ni alloy
DIN 17741 Ni Mn 3 Al	1.5% Al 3.0% Mn 1.0% Si Ni alloy
DIN 17745 Ni Cu 14 Fe Mo	3.0% Mo 14% Cu 11% Fe Ni alloy
DURANICKEL	4.5% Al 0.5% Ti Ni alloy: Bar (high carbon); H. Wiggin; for springs, dies; acid resistant DPN: 295 UTS: 1250 Elon: 28%
DURANICKEL 301	0.15% C 4.5% Al 0.5% Ti Ni alloy: Bar & sheet; Huntington; age hardened UTS: 1350 Elon: 28% Proof: 920
E NICKEL	2.0% Mn Ni alloy: Driver-Harris; International Nickel Co alloy UTS: 610

Note. The following abbreviations and units are used in the tables:

DPN	Hardness, diamond pyramid number
UTS	Ultimate tensile strength, N/mm^2
Elon	Elongation, %
Proof	0.1 % proof strength, N/mm^2

1 N/mm^2=0.1 hbar=0.102 kgf/mm^2=0.06475 tonf/in^2=145.04 lbf/in^2

Symbol	Nominal analysis, supplier, condition and remarks.
E NICKEL	0.1% N 2% Mn Ni alloy: Tube & sheet; Huntington; now Nickel 212
F NICKEL	0.5% C 5.5% Si 1.5% Fe 0.2% Cu 0.06% S Ni; Primary metal; H. Wiggin; supplied as shot or ingot
HASB	0.04% C 28.0% Mo 0.4% V 4.5% Fe Ni alloy: Electrode; Metrode; for welding Hastelloy B type alloys
HASC	0.09% C 15.0% Cr 16.0% Mo 4.0% W 4.5% Fe 0.1% V Ni alloy: Electrode; Metrode; for welding Hastelloy C type alloys
HASF	0.07% C 22.0% Cr 6.0% Mo 2.0% Nb 22.0% Fe Ni alloy: Electrode; Metrode; for welding Hastelloy F type alloys
HASG	0.06% C 22.0% Cr 6.0% Mo 2.0% Cu 0.5% W 19.0% Fe Ni alloy: Electrode; Metrode; for welding Hastelloy E type alloys
HASN	0.04% C 7.0% Cr 16.0% Mo 4.0% Fe Ni alloy: Electrode; Metrode; for welding Hastelloy N type alloys
HASS	0.04% C 15.5% Cr 15.5% Mo Ni alloy: Electrode; Metrode; for welding Hastelloy S type alloys
HASTELLOY C276	0.02% C 15.5% Cr 5.5% Fe 16.0% Mo 3.8% W Ni alloy: Casting; Inco
HASW	0.06% C 5.0% Cr 25.0% Mo 0.4% V 5.0% Fe Ni alloy: Electrode; Metrode; for welding Hastelloy W type alloys
HPW NICKEL	3.5% W Ni alloy: Made by powder metallurgy
IN 504	10% Si 5.7% Ti 3.0% Mo 2.5% Cu Ni alloy: International Nickel Co
INCOCAL 10	0.3% C 0.6% Si 5.5% Ca Ni alloy: Primary metal
INCOLOY 904	0.02% C 14.5% Co 1.6% Ti 50.0% Fe Ni alloy: H. Wiggin; low coefficient of expansion UTS: 920
INCOMAG 1	15% Mg Ni alloy: For cast iron additions; International Nickel Co
INCOMAG 1LC	16% Mg low carbon Ni alloy: For de-oxidation purposes; International Nickel Co
INCOMAG 1M	15% Mg 0.8% Mischmetall Ni alloy: For cast iron additions; International Nickel Co (formerly alloy IM)
INCOMAG 2M	15% Mg 30% Se 0.8% Mischmetall Ni alloy: For cast iron additions; International Nickel Co
INCOMAG 3	4.5% Mg Ni alloy: For cast iron additions; fumeless; International Nickel Co
INCOMAG 4	4.2% Mg 3.4% Fe Ni alloy: For cast iron additions; reduced fume; International Nickel Co
INCOMAG Z	15% Mg 30% Se Ni alloy: For cast iron additions; International Nickel Co (formerly Alloy Z)
INOR 8	0.06% C 7% Cr 0.5% Co 16% Mo 5.0% Fe 0.01% B Ni alloy: Alternative name for Hastelloy N
KH 601U	0.1% C max 16.5% Cr 3.2% Al 26% Fe Ni alloy: Russian Standard designation
KH 75MBTIU	0.1% C (max) 20.5% Cr 19.5% Mo 0.5% Ti 0.5% Al 1.1% Nb 8.0% Fe (max) Ni alloy: Russian Standard designation
KH 78T	0.12% C (max) 20.5% Cr ∩ 2% Ti 6.0% Fe (max) Ni alloy: Russian Standard designation
LANGALLOY 2R	0.7% C 1.0% Fe 1.0% Si 1.0% Mn Ni alloy: Casting; Langley Alloys DPN: 110 UTS: 390 Elon: 25% Proof: 180
M21	0.13% C 5.7% Cr 2.0% Mo 11% W 1.5% Nb 6.0% Al 0.02% B 0.12% Zr Ni alloy: Union Carbide
M22	0.13% C 5.7% Cr 2.0% Mo 11.0% W 3.0% Ta 6.3% Al 0.6% Zr 0.75% Fe 0.6% Si Ni alloy: Union Carbide
M 1628	10% Fe 25% Mo Ni alloy: Imphy

Symbol	Nominal analysis, supplier, condition and remarks.
MAGNO	4.5% Mn Ni alloy: Driver-Harris **UTS: 670**
MANGNOL	5% Mn Ni alloy: For valve grid wire; Driver-Harris
MANGONIC 2	2.0% Mn Ni alloy: H. Wiggin; for filament wires
MANGONIC 3	3.0% Mn Ni alloy: For electrical leads; magnetic; H. Wiggin; annealed; electrical conductivity 12% IACS **DPN: 140 UTS: 540 Elon: 56% Proof: 150**
MANGONIC 5	5.0% Mn Ni alloy: H. Wiggin; filament wire
MAR M432	Ni alloy: Casting for creep property; Martin
MODIFIED HILO	20% Ni alloy: For valve filaments; Driver-Harris
MP 35N MULTIPHASE	35% Co 20% Cr 10% Mo Ni alloy: Latrobe
NICKEL 204	0.06% C 4.5% Co Ni alloy: Bar, sheet, etc.; Huntington
NICKEL 211	0.1% C 4.5% Mn Ni alloy: Bar, sheet, etc.; Huntington
NICKEL 211	5.0% Mn Ni alloy: H. Wiggin; previously Mangonic 5
NICKEL 212	0.1% C 2.0% Mn Ni alloy: Tube & strip; Huntington
NICKEL 212	2.0% Mn Ni alloy: H. Wiggin; previously Mangonic 2
NICKEL 215	5.0% Mn Ni alloy: Wire; International Nickel Co
NICKEL 229	2.0% W 0.05% Mg 0.03% Al Ni: Made by powder metallurgy; International Nickel Co
NICROBRAZ 10	0.1% C 1% P Ni alloy: Braze metal; Wall Colmonoy; melting point 880 °C
NICROBRAZ 60	0.1% C 8% Si 17% Mn Ni alloy: Braze metal; Wall Colmonoy; melting range 1010–1030 °C
NICROBRAZ 65	23% Mn 7.0% Si 4.5% Cu Ni alloy: For brazing; Wall Colmonoy; melting range 980–1010 °C
NICROBRAZ 130	0.06% C 4.5% Si 3.0% B Ni alloy: Braze metal; Wall Colmonoy; melting range 980–1040 °C
NICROBRAZ 135	3.5% Si 1.8% B Ni alloy: Braze metal; Wall Colmonoy; melting range 990–1057 °C
NICROBRAZ 171	0.4% C 10.0% Cr 2.5% B 3.5% Si 12.0% W 3.5% Fe Ni alloy: For brazing; Wall Colmonoy; melting range 970–1095 °C
NICROBRAZ 220	4% Cr 1.0% B 45% Mn Ni alloy: Braze metal; Wall Colmonoy; melting range 995–1080 °C
NICROBRAZ 230	3.5% Cr 2.5% Si 10% B 35% Mn Ni alloy: Braze metal; Wall Colmonoy; melting range 980–1065 °C
NICROBRAZ 1351	1.5% B 3.0% Si Ni alloy: For brazing; Wall Colmonoy; melting range 1090–1180 °C
NICROBRAZ 5040	5.0% Cr 4.0% P 2.1% Si 1.2% N 0.6% Fe Ni alloy: For brazing; Wall Colmonoy; melting range 1010–1150 °C
NICROBRAZ 5060	8.0% Cr 6.0% P 1.4% Si 0.8% B 0.4% Fe Ni alloy: For brazing; Wall Colmonoy; melting range 1010–1120 °C
NICROBRAZ 5075	10.0% Cr 7.5% P 0.9% Si 0.5% B 0.25% Fe Ni alloy: For brazing; Wall Colmonoy; melting range 980–1090 °C
NILOMAG 771	14.0% Fe 5.0% Cu 4.0% Mo Ni alloy: Wire or strip made by powder metallurgy; International Nickel Co

Symbol	Nominal analysis, supplier, condition and remarks.
Ni Mo 30	5% Fe 28% Mo Ni alloy: Designation used by German Standard
NIMOCAST 242	0.35% C 22.0% Cr 10.0% Co 10.5% Mo Ni alloy: H. Wiggin; high thermal shock resistance
NIMOCAST 263	0.06% C 20.0% Cr 20.0% Co 5.8% Mo 2.2% Ti Ni alloy **DPN: 195 UTS: 1000**
Ni–Zr MASTER ALLOY	27.5% Zr 10% Al 5% Fe 7.5% Si Ni alloy: Origin unknown
NMP	Pd Mn Ni brazing alloy: Melting point 1120 °C; Engelhard Industries; electrical conductivity 9% IACS **DPN: 330 UTS: 780 Elon: 7%**
PA 10	4.0% W 0.05% Mg Ni alloy: Tube & strip; H. Wiggin; obsolete
PERMAGRID	4% Ti 0.3% Mg Ni alloy: For valve grid wire; Driver
R 63	1.0% Si 4.0% Mn Ni alloy: Driver-Harris; spark plug electrodes **UTS: 490**
REXWELD 64	0.6% C 3.0% Si 13% Cr 3% B Ni alloy: For hard facing; Crucible Steel Co **DPN: 620**
SYLVALOY	3% Si Ni alloy: For valve filament; Driver-Harris
T 2	5% Al + Mn + Si Ni alloy: Thermocouple; Driver-Harris **UTS: 560**
TRIBALOY 700	32% Mo 15% Cr 3% Si Ni alloy: Intermetallic material for wear resistant purposes; suitable for hard facing; Du pont
UNS P02001–P02999	Os metal and alloys: American standards system; unified numbers
UNS R4001–R49999	Nb metal and alloys: American Standard system; unified numbers
W 5	0.3% Mn 4.0% Si Ni alloy: Wire & strip; H. Wiggin; for spark plug electrodes **DPN: 130 UTS: 550 Elon: 56% Proof: 150**
W 6	0.5% Mn 2.0% Si Ni alloy: Wire & strip; H. Wiggin; for spark plug electrodes **DPN: 120 UTS: 530 Elon: 57% Proof: 140**
W 7	2.75% Mn 1.0% Si Ni alloy: Wire & strip; H. Wiggin; for spark plug electrodes **DPN: 120 UTS: 530 Elon: 57% Proof: 140**
W 9	4.5% Mn 1.0% Si Ni alloy: Wire & strip; H. Wiggin; for spark plug electrodes
W 9	4.5% Mn 1.0% Si with Zr additions Ni alloy: For spark plug electrodes; H. Wiggin

Note. The following abbreviations and units are used in the tables:

DPN	Hardness, diamond pyramid number
UTS	Ultimate tensile strength, N/mm^2
Elon	Elongation, %
Proof	0.1 % proof strength, N/mm^2

$1\ N/mm^2 = 0.1\ hbar = 0.102\ kgf/mm^2 = 0.06475\ tonf/in^2 = 145.04\ lbf/in^2$

28. Niobium Nb (also called columbium Cb)

Physical properties

Atomic number	41	
Atomic weight	92.91	
Crystal structure	Body-centred cubic	
Colour	Steel grey	
Specific gravity	8.6	
Density	8600 kg/m³	
Melting point	2468 °C	
Boiling point	4927 °C	
Specific heat	0.272 J/g °C	(0.065 cal/g °C)
Thermal conductivity	52.3 W/m °C	(12.5 cal/m s °C)
Coefficient of linear expansion (20–100° C)	7.1 × 10⁻⁶/ °C	
Latent heat of fusion	–	(– cal/g)
Latent heat of vaporization	7704 J/g	(1840 cal/g)
Thermal neutron absorption cross-section	1.1 barns/atom	
Electrical conductivity	13.3% IACS (copper 100%)	
Specific resistance	151 microhm mm	
Temperature coefficient of electrical resistance	0.0039/ °C	
Electrochemical equivalent	0.45 g/A/h	
Electrode potential	-1.1 V	
Magnetic susceptibility	2.28 × 10⁻⁶ cgs	
Young's modulus of elasticity	103 × 10⁹ N/m²	(15 × 10⁶ lbf/in.²)
Tensile strength annealed	300 N/mm²	(20 tonf/in.²)
cold worked	600 N/mm²	(40 tonf/in.²)
Hardness annealed	80 DPN	
cold worked	130 DPN	

28.1 General notes on niobium (also called columbium, Cb)

Niobium occurs in nature combined with iron as a complex oxide, generally associated with the similar tantalum–iron oxide. The ore is very scarce and as niobium is generally used as a low percentage alloy very little scrap is recovered.

The metal is quite ductile, capable of being rolled to sheet and bar. It can be welded using the inert gas shielded technique with copious gas flow before and after the weld on both sides.

Niobium resists attack from almost all chemicals, but is in no way superior to tantalum which is slightly easier to produce and more plentiful at present. The ability to withstand oxidation, especially at elevated temperatures, makes niobium potentially useful for gas turbine and rocket motor applications. It has comparable hot strength properties to molybdenum with much superior oxidation resistance. The present scarcity and the high cost prevent its widespread use. It has been used as a canning material in atomic reactors, but has few advantages over tantalum.

By far the best known use for niobium is as a carbide former in austenitic steels and nickel alloys. The carbide is insoluble and being very stable and evenly distributed prevents the formation of other soluble carbides which would be dissolved during welding, and precipitated from solution in the weld area, giving rise to the condition known as 'weld decay'. It is now added in small amounts – 0.01% – to weldable mild steel, as an optional element to aid grain refinement, thus preventing grain growth during welding. This enhances the impact properties.

In the nickel alloys niobium can be present up to 4% when it enhances the creep strength and hot oxidation resistance.

When added to low alloy chrome steels there is some improvement in impact strength and an increase in the rate of nitride penetration.

Some aluminium age-hardening alloys have been produced with up to 0.2% niobium added to refine the grain and strengthen the matrix.

The more complex magnet alloys have niobium added, principally to ensure the presence of very stable carbides.

There are a few alloys based on niobium which have potentially very many high temperature applications. Until a more plentiful supply of the metal is available these applications will continue to be filled by molybdenum and tantalum alloys.

This is the only element on which there has been no agreement regarding the chemical symbol – in Europe it is Nb for niobium – in the Americas it is Cb for columbium. Throughout this book it is referred to as Nb.

Symbol	Nominal analysis, supplier, condition and remarks.
AMS 7850	Nb: Sheet, strip & foil; Columbium
COLUMBIUM	Ferro columbium, free of carbon: Blackwell; primary metal; also called niobium
n Nb 16	99.9% Nb: Rod; 500 p.p.m. Ta; Light Ltd; high purity metal
n Nb 73	99.99% Nb metal: Single crystal; Light Ltd
NIOBIUM	99.6% Nb: Powder, granules & electrodes; Kennametal; columbium
PLATIUM SUBSTITUTE	70% Zr 39.5% Ta Nb alloy: Origin unknown
SU 16	0.08% C 11% W 3% Mo 2% Hf Nb alloy: IMI; solution treated; cold worked & aged
	UTS: 530 **Elon: 22%** **Proof: 440**
SU 31	0.12% C 17% W 3.5% Hf Nb alloy: IMI; heat treated
	UTS: 460 **Elon: 30%** **Proof: 340**

Symbol	Nominal analysis, supplier, condition and remarks.
UNS R4001–R4999	Nb metal and alloys: American Standards system
W Nb 18	99.9% Nb: Powder; Light Ltd

Note. The following abbreviations and units are used in the tables:

DPN	Hardness, diamond pyramid number
UTS	Ultimate tensile strength, N/mm^2
Elon	Elongation, %
Proof	0.1 % proof strength, N/mm^2

$1\ N/mm^2 = 0.1\ hbar = 0.102\ KGF/mm^2 = 0.06475\ tonf/in^2 = 145.04\ LBF/in^2$

29. Osmium Os

Physical properties

Atomic number	76	
Atomic weight	190.2	
Crystal structure	Close-packed hexagonal	
Colour	Blue/white	
Specific gravity	22.5	
Density	22500 kg/m^3	(0.812 lb/in.^{3})
Melting point	3050 °C	
Boiling point	4600 °C	
Specific heat	1.31 J/g °C	(0.312 cal/g °C)
Thermal conductivity	91.67 W/m °C	(22 cal/m s °C)
Coefficient of linear expansion (20–100° C)	6.6 × 10^{-6}/ °C	
Latent heat of fusion	26.8 J/g	(6.4 cal/g)
Latent heat of vaporization	678 J/g	(162 cal/g)
Thermal neutron absorption cross-section	15 barns/atom	
Electrical conductivity	19% IACS (copper 100%)	
Specific resistance	95 microhm mm	
Temperature coefficient of electrical resistance	0.0066/ °C	
Electrochemical equivalent	– g/A/h	
Electrode potential	+ 0.7 V	
Magnetic susceptibility	0.04 × 10^{-6}	
Young's modulus of clasticity	– N/mm^2	
Tensile strength	– N/mm^2	
Hardness	annealed 300 DPN	
	cold worked 670 DPN	

29.1 General notes on osmium

This is one of the platinum group of metals, always being found in association with its fellow members. Much of the metal occurs native in Africa, Alaska and Canada as an alloy with iridium, platinum, rhodium and ruthenium, known as osmiridium. This is often found alongside ores containing gold. Previously this native alloy was used to tip pen nibs and is now used as a source for the platinum group of metals. Pens are now tipped with a powder compact of ruthenium.

The metal, with a density of 22.5, is the heaviest known, and like all members of the platinum group has a high melting point, bluish white colour and excellent resistance to chemical attack. The pure metal is extremely brittle and crumbles to a powder when subjected to cold work, with only a very slight increase in ductility when hot worked. The metal forms an oxide in air at 1000 °C which is very poisonous and once formed is volatile above 200 °C.

Osmium has no properties which make it an attractive alternative to iridium or other more ductile members of its group, and its scarcity, brittleness and readily formed poisonous oxide make it an undesirable first choice.

A limited amount is used as sintered powder wear resistant points for delicate instruments and electrical contacts subject to severe hammering.

The chemical engineering industry has found a limited use for finely divided osmium as a catalyst.

Of all the metals in the platinum group (platium, iridium, osmium, rhodium, ruthenium and palladium) osmium has by far the least attractive properties and the lowest potential usefulness.

Note. The following abbreviations and units are used in the tables:

DPN	Hardness, diamond pyramid number
UTS	Ultimate tensile strength, N/mm^2
Elon	Elongation, %
Proof	0.1 % proof strength, N/mm^2

1 N/mm^2=0.1 hbar=0.102 kgf/mm^2=0.06475 tonf/in^2=145.04 lbf/in^2

Symbol	Nominal analysis, supplier, condition and remarks.
h Os 18	99.97% Os: Powder; Light Ltd; high purity metal
OSMIUM	High purity metal: Impurities 10 p.p.m.; Mallory Metallurgical; supplied as sponge; platinum group metal
UNS P02001–P02999	Os metal and alloys: American Standards system

30. Palladium Pd

Physical properties

Atomic number	46	
Atomic weight	106.4	
Crystal structure	Face-centred cubic	
Colour	White	
Specific gravity	12.02	
Density	12020 kg/m^3	(0.433 lb/in.^{3})
Melting point	1552 °C	
Boiling point	2900 °C	
Specific heat	0.247 J/g °C	(0.059 cal/g °C)
Thermal conductivity	71.2 W/m °C	(17 cal/m s °C)
Coefficient of linear expansion (20–100° C)	11.1 × 10^{-6}/ °C	
Latent heat of fusion	17.6 J/g	(4.12 cal/g)
Latent heat of vaporization	373 J/g	(89 cal/g)
Thermal neutron absorption cross-section	8 barns/atom	
Electrical conductivity	16% IACS (copper 100%)	
Specific resistance	100 microhm mm	
Temperature coefficient of electrical resistance	0.038/ °C	
Electrochemical equivalent	– g/A/h	
Electrode potential	-0.57 V	
Magnetic susceptibility	5.8 × 10^{-6}	
Young's modulus of elasticity	110 × 10^9 N/m^2	(16 × 10^6 lbf/in.^{2})
Tensile strength	annealed 180 N/mm^2	(12 tonf/in.^{2})
Hardness	annealed 37 DPN	

30.1 General notes on palladium

Palladium is a member of the platinum group of metals, the others being iridium, osmium, rhodium and ruthenium.

It is always found associated with these, but being the least chemically resistant member of the group is generally combined with a non-metal rather than as native alloy. The largest deposits are found in Africa, Canada and Russia.

Being the most plentiful and most easily purified member of the group, palladium is the cheapest of the platinum metals.

Like all members of the group it has a high melting point, a bluish colour, and resists chemical attack, although in this last respect it has the least resistance of the six platinum metals.

Pure palladium is very ductile and can be rolled into thin sheets, or drawn to fine wire. It does work harden to some extent and must be annealed between stages of severe cold work. This should be at 1000 °C at least, in an atmosphere free from oxygen, hydrogen, or carbon containing gases, all of which can be absorbed to give a hard brittle product. Suitable atmospheres are nitrogen, argon, or under vacuum.

The fact that palladium oxidizes at temperatures above about 500 °C, coupled with its relatively low chemical resistance, precludes its use as a laboratory tool.

Palladium and its alloys are used in the main as substitutes for platinum. It is much cheaper and more plentiful than platinum and most of its properties are similar to, but not quite as good as, the same platinum property. It is the lightest of the metals in the group, density of 12 compared to 21 for platinum and 22.5 for osmium, and this makes it attractive in the jewellery field.

Most palladium metal is used for this purpose, either as the pure metal or alloyed with silver, molybdenum and other members of its own group.

The same type of alloys are used for electrical purposes, such as potentiometer rubbing contacts and light duty impact contactors. It is not as sensitive or wear resistant as platinum for these purposes, but much cheaper.

Alloyed with gold, palladium is used to protect furnaces. These alloys have good resistance to oxidation up to 1200 °C and melting points from 1000 up to 1200 °C. Having the added advantage of short melting ranges, they are used as fuses inside the furnace melting at the pre-determined temperature, thus cutting off the electrical heating. The oxidation resistance of these alloys at these temperatures is sufficient to ensure stability of electrical resistance and melting point.

The same alloys are used as thermocouple materials in conjunction with other platinum group metals and alloys, as they generate a high e.m.f. over the range 500–1000 °C.

Alloys of palladium with copper, silver and gold are used in the manufacture of false teeth where to good corrosion resistance and silver colour are an added advantage to the good casting characteristics.

A series of palladium alloys has recently been produced for use as high temperature brazing materials. These are used for joining nickel alloys and other high temperature, corrosion resistant alloys which for some reason cannot be welded.

The largest single use for palladium is as a catalyst, either in the finely divided state, or sometimes as a fine wire gauze.

Symbol	Nominal analysis, supplier, condition and remarks.
267	40% Ag Pd alloy: Engelhard; annealed; electrical conductivity 4.1% IACS **DPN: 95**
278	40% Cu Pd alloy: Engelhard; annealed; electrical conductivity 4.9% IACS **DPN: 145**
BS 1845 PD14	40% Ni Pd alloy: For brazing metal; melting point 1235 °C
h Pd 15	99.9% Pd: Wire; 0.5 mm diameter; Light Ltd
h Pd 73	99.99% Pd single crystals 12 mm × 25 mm: Light Ltd
JMC 77	40% Cu Pd alloy: For wiping contacts; Mallory Metallurgical; age hardened **DPN: 350**
JMM 77	Cu Pd alloy: For sliding contacts; 4.6% IACS conductivity; Johnson Matthey **DPN: 210**
PALAURAL	34% Au + Pt Pd alloy: For dental purposes; Mallory Metallurgical; melting range 945–990 °C; hardened **DPN: 258**
PALCO	35% Co Pd alloy: For brazing; melting range 1230–1235 °C; Wesgo

Symbol	Nominal analysis, supplier, condition and remarks.
PALLACAST	Palladium alloy: Used in dentistry; with Au and Ag; Engelhard Industries; annealed; melting range 920–1025 °C **DPN: 172 UTS: 460 Elon: 15% Proof: 330**
PALLACAST	Palladium alloy: Used in dentistry; with Au and Ag; Engelhard Industries; age hardened; melting range 920–1025 °C **DPN: 249 UTS: 850 Elon: 2% Proof: 770**
PALLADIUM	Chemically pure Pd: Engelhard Industries
PALLADIUM	High purity metal: Impurities 10 p.p.m.; Mallory Metallurgical; supplied as sponge & wire
PN 1	40% Ni Pd brazing alloy: Melting point 1237 °C; Engelhard Industries
UNS P03001–P03999	Pd metal and alloys: American Standards system; unified numbers

Note. The following abbreviations and units are used in the tables:

DPN	Hardness, diamond pyramid number
UTS	Ultimate tensile strength, N/mm^2
Elon	Elongation, %
Proof	0 1 % proof strength, N/mm^2

$1\ N/mm^2 = 0.1\ hbar = 0.102\ kgf/mm^2 = 0.06475\ tonf/in^2 = 145.04\ lbf/in^2$

31. Platinum Pt

Physical properties

Atomic number	78	
Atomic weight	195.09	
Crystal structure	Face-centred cubic	
Colour	White	
Specific gravity	21.45	
Density	21450 kg/m^3	(0.774 lb/in.^{3})
Melting point	1769 °C	
Boiling point	3800 °C	
Specific heat	0.134 J/g °C	(0.032 cal/g °C)
Thermal conductivity	69.1 W/m °C	(16.5 cal/m s °C)
Coefficient of linear expansion (20–100° C)	9×10^{-6}/ °C	
Latent heat of fusion	113 J/g	(27 cal/g)
Latent heat of vaporization	2407 J/g	(575 cal/g)
Thermal neutron absorption cross-section	9 barns/atom	
Electrical conductivity	18% IACS (copper 100%)	
Specific resistance	98 microhm mm	
Temperature coefficient of electrical resistance	0.0039/ °C	
Electrochemical equivalent	1.8160 g/A/h	
Electrode potential	+ 1.2 V	
Magnetic susceptibility	1.1×10^{-6}	
Young's modulus of elasticity	152×10^9 N/m^2	(22×10^6 lbf/in.^{2})
Tensile strength	annealed 140 N/mm^2	(9 tonf/in.^{2})
Hardness	annealed 37 DPN	

31.1 General notes on platinum

This is the most important metal in the group of six to which it gives its name, the remaining members being iridium, osmium, palladium, rhodium and ruthenium. The metals are always found in association with each other, are all resistant to chemical attack and have high melting points.

Platinum is outstanding in the group in that it has corrosion resistance equal to the best, coupled with excellent workability. Wire finer than 0.01 mm in diameter is available, capable of withstanding nitric and sulphuric acids at high temperatures.

Platinum and its fellow group members occur in the native state, generally alloyed with each other. The known alluvial deposits are rapidly becoming exhausted, but native alloys have been discovered in Africa and sulphide ores in Canada.

The African deposit is in the form of nuggets, associated with gold, while the Canadian sulphide ores form a valuable byproduct in nickel extraction.

Platinum has been found to be a valuable crucible material for laboratory and chemical engineering use. It withstands repeated rapid heating and cooling, is not readily attacked chemically, thus does not contaminate the material being treated and being soft and ductile withstands the abuse normal to industrial process equipment. Examples of the industries using these crucibles are glass and metal manufacturing when high purity products are necessary.

Platinum is also used for general laboratory ware, where its constant weight under the most severe conditions makes it an invaluable material for items such as electrodes and crucibles.

Platinum and its alloys are the favourite setting materials for diamonds in jewellery. The brilliant permanent lustre and ability to be formed into intricate shapes is responsible for the ever increasing use of platinum alloys as personal jewellery and ornaments.

The dimensional stability of platinum alloys under all normal conditions is responsible for their use as the material for standards, and master instruments. Being capable of forming very thin sheets which can be shaped and readily forge welded, platinum is often the most economical chemical resistant material for sheathing instruments and equipment operating in corrosive, high temperatures. In this field competition from rhodium electroplating is now being felt.

The expansion coefficient of platinum and platinum–rhodium alloys is comparable to that of good quality glass and the corrosion resistant qualities of these materials makes them useful as glass metal seals, or on apparatus where good temperature resistance and ductility are required in conjunction with glass.

Platinum and its alloys are a favourable material for thermocouples, as the e.m.f. generated between, for example, a platinum/platinum–rhodium couple is comparatively high and the couple is robust and withstands hot corrosion and oxidation.

Platinum–rhodium alloys are used as the electrical resistance heating medium in small laboratory type furnaces where high temperatures are required and vacuum or gas atmospheres to protect the elements cannot be used.

Platinum is used as an electrical contact material where a constant resistance over a range of operating conditions and temperatures and under intermittent use, is important. It would not be used where high conductivity is important as plantinum has a reasonably high resistance.

A large amount of platinum is now used as a catalyst in the chemical industry and oil refining. Almost all high octane modern petrols owe their economical production to platinum catalysts.

Some platinum alloys are used as high temperature brazing materials for joining nickel alloy and other high temperature metals where for some reason welding is undesirable.

Platinum, with its excellent corrosion resistantce, high melting point and temperature resistance coupled with good ductility and weldability, makes this one of the more useful metals in the service of mankind. Only its scarcity, difficulty of purification and resultant high price keep this material from general use.

Symbol	Nominal analysis, supplier, condition and remarks.
51	30% Ir Pt alloy: Engelhard; annealed; electrical conductivity 5.3% IACS DPN: 285
52	25% Ir Pt alloy: Engelhard; annealed; electrical conductivity 5.4% IACS DPN: 240
53	20% Ir Pt alloy: Engelhard; annealed; electrical conductivity 5.7% IACS DPN: 200
55	10% Ir Pt alloy: Engelhard; annealed; electrical conductivity 7.0% IACS DPN: 120
102	10% Ru Pt alloy: Englehard; annealed; electrical conductivity 4.1% IACS DPN: 200
103	5% Ru Pt alloy. Engelhard; annealed; electrical conductivity 5.5% IACS DPN: 125
115	Platinum: Commercially pure Pt; Engelhard; electrical conducitivty 15% IACS DPN: 60
450	Au Ag Pt alloy: Engelhard; annealed; electrical conductivity 10.2% IACS DPN: 60
h Pt 10	99.999% Pt: Wire; 0.5 mm diameter; Light Ltd; high purity metal
h Pt 71	99.9999% Pt single crystals 3 mm × 25 mm: Light Ltd
IRRU	Ir Ru Pt alloy: For sliding contacts; 4.4% IACS conductivity; Johnson Matthey DPN: 310
OERSTIT 900 CP	22% Co Pt alloy: For magnets; German alloy listed by PMA
OSMIRIDIUM	30% Ir 35% Os Rh Pt alloy: Natively occurring material

Symbol	Nominal analysis, supplier, condition and remarks.
PALLABRAZE 1237	40% Ni 60% Pd alloy: Brazing metal; Johnson Matthey; melting point 1237 °C
PERMANIT 900 CP	22% Co Pt alloy: For magnets; Austrian alloy; listed by PMA
PGS	Au Ag Pt alloy: Engelhard
PLACO	23% Co Pt alloy: For magnets, German alloy; listed by PMA
PLACOVAR	23% Co Pt alloy: For magnets; American proprietary alloy; consult SAE
PLATINAX 11	Co Pt alloy: For extremely powerful magnets; Johnson Matthey; coercive force 4800 oersted
PLATINUM	High purity metal: Impurities less than 5 p.p.m.; Johnson Matthey; supplied as sponge, wire & sheet
UNS P04001–P04999	Pt metal and alloys: American Standards system, unified number
UNS P05001–P05999	Rh metal and alloys: American Standards system; unified numbers
UNS P06001–P06999	Ru metal and alloys: American Standards system; unified number
UNS R04001–R04999	Re metal and alloys: American Standards system; unified number

Note. The following abbreviations and units are used in the tables:

DPN	Hardness, diamond pyramid number
UTS	Ultimate tensile strength, N/mm^2
Elon	Elongation, %
Proof	0.1 % proof strength, N/mm^2

1 N/mm^2=0.1 hbar=0.102 kgf/mm^2=0.06475 tonf/in^2=145.04 lbf/in^2

32. Plutonium Pu

Physical properties

Atomic number	94	
Atomic weight	239	
Crystal structure	Monoclinic	(generally)
Colour	Yellow/white	
Specific gravity	19	
Density	19000 kg/m³	(0.685 lb/in.³)
Melting point	640 °C	
Boiling point	3235 °C	
Specific heat	Varies with structure	
Thermal conductivity	8.4 W/m °C	(2 cal/m s °C)
Coefficient of linear expansion (20–100° C)	55 × 10⁻⁶/ °C	
Latent heat of fusion	– J/g	(– cal/g)
Latent heat of vaporization	– J/g	(– cal/g)
Thermal neutron absorption cross-section	– barns/atom	
Electrical conductivity	1.2% IACS (copper 100%)	
Specific resistance	1500 microhm mm	
Temperature coefficient of electrical resistance	– / °C	
Electrochemical equivalent	2.8 g/A/h	
Electrode potential	– V	
Magnetic susceptibility	2.2 × 10⁻⁶	
Young's modulus of elasticity	cast 96.5 × 10⁹ N/m²	(14 × 10⁶ lbf/in.²)
Tensile strength	cast 400 N/mm²	(26 tonf/in.²)
Hardness	250 DPN	

32.1 General notes on plutonium

This is an artificial element formed from uranium by the capture of a neutron when nuclear energy is harnessed in an 'atomic pile'. It is, in fact, a member of the second group of 'rare earth elements'.

Plutonium can be separated from the uranium used as the source material by chemical means. The purified plutonium can then be used as the primary material in another type of nuclear reactor.

Although it does not release as much energy as uranium 235 – the higher reactive isotope contained in uranium – it is much more reactive than normal uranium.

Plutonium therefore, is used as a means of producing nuclear energy.

There are no commercially available materials containing plutonium.

Plutonium occurs in minute quantities in nature, having been found in pitchblende in a concentration of 1 part in 10¹².

33. Potassium K

Physical properties

Atomic number	19	
Atomic weight	39.096	
Crystal structure	Body-centred cubic	
Colour	Silver white	
Specific gravity	0.86	
Density	860 kg/m³	(0.03 lb/in.³)
Melting point	63.7 °C	
Boiling point	760 °C	
Specific heat	0.724 J/g °C	(0.173 cal/g °C)
Thermal conductivity	99.2 W/m °C	(23.7 cal/m s °C)
Coefficient of linear expansion (20–100° C)	83 × 10⁻⁶/ °C	
Latent heat of fusion	61.55 J/g	(14.7 cal/g)
Latent heat of vaporization	2077 J/g	(496 cal/g)
Thermal neutron absorption cross-section	2.5 barns/atom	
Electrical conductivity	25% IACS (copper 100%)	
Specific resistance	66.4 microhm mm	
Temperature coefficient of electrical resistance	0.0052/ °C	
Electrochemical equivalent	1.46 g/A/h	
Electrode potential	-2.92 V	
Magnetic susceptibility	0.52 × 10⁻⁶	
Young's modulus of elasticity	– N/m²	(– lb/in.²)
Tensile strength	– N/mm²	
Hardness	– DPN	

33.1 General notes on potassium

Potassium is too reactive an element ever to be found as native metal, but is common in the combined state. It is an essential constituent of all plant life but in quantities far too small to be a source for the metal. The compounds of potassium are all water soluble, yet seawater contains very little of this metal. This would appear to be caused by the greater affinity of soil and plant life for potassium than sodium and magnesium, both of which can be recovered from seawater in economic quantities.

Potassium is a silvery white metal, soft enough to cut with a knife, but tarnishing immediately. If left for any time in air potassium will combine with the moisture in the air to form potassium hydroxide – caustic potash. Potassium reacts violently if brought into contact with water and the metal must always be stored under liquid paraffin in an air tight container.

Potassium is very feebly radioactive but not enough to be either dangerous or useful. This radioactivity is due to the presence of the isotope potassium 40.

An alloy of sodium and potassium is liquid from room temperature to about 400 °C and finds use in thermometers where a wider temperature range than mercury is required.

There are also some lead alloys with small amounts of potassium added as a hardener.

No alloys exist based on potassium and as most of the properties are almost identical to those of sodium this metal is the useful first choice. Sodium is more readily obtained than potassium and thus is more economical.

Symbol	Nominal analysis, supplier, condition and remarks.
WK 16	99.97% K metal: Light Ltd; high purity metal

Note. The following abbreviations and units are used in the tables:

DPN	Hardness, diamond pyramid number
UTS	Ultimate tensile strength, N/mm²
Elon	Elongation, %
Proof	0.1 % proof strength, N/mm²

1 N/mm²=0.1 hbar=0.102 kgf/mm²=0.06475 tonf/in²=145.04 lbf/in²

34. Radium Ra

Physical properties

Atomic number	88	
Atomic weight	226.05	
Crystal structure	–	
Colour	Brilliant white	
Specific gravity	5.0	
Density	5000 kg/m^3	(0.18 lb/in.^{3})
Melting point	700 °C	
Boiling point	1140 °C	
Specific heat	(– cal/g °C)	
Thermal conductivity	(– cal/m s °C)	
Coefficient of linear expansion (20–100° C)	– / °C	
Latent heat of fusion	– cal/g	
Latent heat of vaporization	– cal/g	
Thermal neutron absorption cross-section	– barns/atom	
Electrical conductivity	– % IACS (copper 100%)	
Specific resistance	– microhm mm	
Temperature coefficient of electrical resistance	– / °C	
Electrochemical equivalent	– g/A/h	
Electrode potential	– V	
Magnetic susceptibility	–	
Young's modulus of elasticity	– lbf/in.^{2}	
Tensile strength	– N/mm^2	
Hardness	– DPN	

34.1 General notes on radium

Radium is one of the very rare metal elements which would probably still be undiscovered but for the fact that it is very radioactive. It is almost always found in association with uranium, deposits of which occur in Canada, the Congo and USA. The principal source is the Canadian pitchblende deposits which also contain silver and copper with some nickel. The extraction and purification of the metal is a long drawn out complex process resulting in very little metallic radium from many hundred tons of ore.

Radium is a brilliant white, which rapidly blackens owing to the formation of the nitride. It reacts vigorously with water forming the hydroxide. The best known property of radium is the radioactivity, it being one of the most active, earliest investigated of the radioactive elements. Considerable remance is associated with the early work on these elements and the name of Madame Curie will always be linked with radium.

The metal itself is very chemically active and attacks glass and many types of metal containers. There is still some use made of the metal for radiography, industrially and medicinally, and much of the available metal is used for radiotherapy. The most popular use at present for the metal is in the manufacture of luminous paints, but as these present considerable and serious health hazards the uses are very limited.

The modern radioactive isotopes of cobalt, tantalum and iridium are more easily handled than radium and in many cases give superior X-ray radiographs.

Some of the radiotherapy uses of radium are also being replaced by these materials. There are no industrial uses for any alloys of radium.

The fact that radium is radioactive constitutes a health hazard and users of the metal must comply with stringent regulations regarding hygiene and methods of handling.

Radium metal is an interesting rarity with no important industrial uses and a considerable health hazard.

35. The rare earth elements

Cerium Ce, Dysprosium Dy, Erbium Er, Europium Eu, Gadolium Cd, Holmium Ho, Lutetium Lu, Neodymium Nd, Praseodymium Pr, Promethium Pm, Samarium Sm, Terbium Tb, Thulium Tm, Ytterbium Yb.

Americium Am, Berkelium Bk, Californium Cf, Curium Cm, Einsteinium Es, Fermium Fm, Mendevium Md, Neptunium Np, Plutonium Pu, Protactinium Pa, Thorium Th, Uranium U.

Lanthanium La, Scandium Sc, Yttrium Y.

They fall into two well-defined groups, the first occuring naturally in reasonably large quantities, the second being radioactive and very largely man-made short life elements. A third group resemble the rare earths to such an extent that they are always classified together.

Four of the metals listed as rare earths are discussed separately as some use has been found for them in industry; these are, cerium in the first group and thorium, plutonium and uranium in the second group.

The metals in the first group all occur in monazite sand and are so much alike that it is only in recent years that it has been appreciated that more than two or three elements were present. Monazite sand also contains thorium compounds and is a principal source of this metal, the rare earths generally being a byproduct in its extraction. The extraction and reduction is a complex process, resulting in an alloy 'Mishchmetall' which is about 50% cerium, the remainder being other rare earth metals of the first group with metallic impurities.

The pure metals can only be separated from each other by very complicated and delicate chemical and physical means and apart from cerium there appears very little requirement for them in the pure state.

They are very useful laboratory research materials in that they can be used to illustrate the relationships of elements to each other within the atomic table of elements.

The uses of cerium have been discussed elsewhere, the uses of the remaining elements being few.

Probably the most important use is as grain refiners and strengtheners of magnesium alloys, where no attempt is made to separate the metals and they are always referred to as 'rare earths'.

Other uses are as cores for carbon arcs in the preparation of special optical glass, and some uses are now being found in nuclear engineering as control rod elements.

The metals in the second group are all radio-active and three, plutonium, thorium and uranium are the subject of separate descriptive sections in this book. The remaining elements in this group are mostly man-made, or discovered during the recent researches into nuclear fission. They are all unstable elements as the result of the radio activity, and apart from nuclear engineering, they serve no useful purpose.

The three elements in the last group are generally found with the first group of rare earth metals in Monazite sand. They have no practical significance.

Symbol	Nominal analysis, supplier, condition and remarks.
h Dy 16	99.9% Dy: Ingot; Light Ltd; high purity metal dysprosium
h Dy 20a	99.9% Dy: Wire 1 mm diameter; Light Ltd; high purity metal dysprosium
h Dy 74	99.9% Dy: Single crystals 5 mm × 5 mm × 5 mm; Light Ltd; dysprosium
h Er 16	99.9% Er: Ingot; Light Ltd; high purity erbium
h Er 20a	99.9% Er: Wire 1 mm diameter; Light Ltd; high purity metal erbium
h Eu 16	99.9% Eu: Ingot; Light Ltd; high purity europium
h Eu 20b	99.9% Eu: Wire 1 mm thick; Light Ltd; high purity europium
h Gd 16	99.9% Gd: Ingot; Light Ltd; high purity gadolinium
h Gd 20a	99.9% Gd: Wire 1 mm diameter; Light Ltd; gadolinium
h Gd 74	99.9% Gd: Single crystals 5 mm × 5 mm × 5 mm each; Light Ltd; gadolinium
h Ge 6	99.999% Ge: Ingot; Light Ltd; high purity germanium
h Ge 8	99.999% Ge: Powder; Light Ltd; high purity germanium
h Lu 16	99.9% Lu: Ingot; Light Ltd; lutecium
h Nd 16	99.9% Nd in lumps: Light Ltd; high purity metal
h Nd 20	99.9% Nd: Wire; 1 mm diameter; Light Ltd
h Pr 16	99.9% Pr: Ingot; Light Ltd; high purity metal
h Pr 20	99.9% Pr: Wire; 1 mm diameter; Light Ltd; high purity metal

Symbol	Nominal analysis, supplier, condition and remarks.
h Sc 16	99.95% Sc: Ingot; Light Ltd; high purity metal
h Sc 20	99.9% Sc: Wire; 1 mm diameter; Light Ltd
h Sm 16	99.7% Sm: Ingot; Light Ltd
h Sm 19	99.9% Sm: Foil; Light Ltd
h Tb 16	99.6% Tb: Ingot; Light Ltd; high purity metal
h Tb 74	99.9% Tb: Single crystal 5 mm × 5 mm × 5 mm; Light Ltd
h Tm 16	99.9% Tm: Ingot; Light Ltd; high purity metal
h Tm 20	99.9% Tm: Wire 1 mm diameter; Light Ltd
h Y 16	99.9% Y: Ingot; Light Ltd; high purity metal
h Y 20	99.9% Y: Wire; 1 mm diameter; Light Ltd
h Yb 16	99.9% Yb: Ingot; Light Ltd; high purity metal
h Yb 20	99.9% Yb: Wire; 1 mm diameter; Light Ltd; high purity metal
NEODYMIUM	Nd metal: Blackwells; pure metal

Note. The following abbreviations and units are used in the tables:

DPN	Hardness, diamond pyramid number
UTS	Ultimate tensile strength, N/mm^2
Elon	Elongation, %
Proof	0.1 % proof strength, N/mm^2

$1 \ N/mm^2 = 0.1 \ hbar = 0.102 \ kgf/mm^2 = 0.06475 \ tonf/in^2 = 145.04 \ lbf/in^2$

36. Rhenium Re

Physical properties

Atomic number	75	
Atomic weight	186.22	
Crystal structure	Close-packed hexagonal	
Colour	Silver-white	
Specific gravity	21.2	
Density	21200 kg/m³	(0.723 lb/in.³)
Melting point	3167 °C	
Boiling point	5900 °C	
Specific heat	0.138 J/g °C	(0.033 cal/g °C)
Thermal conductivity	71.2 W/m °C	(17 cal/m s °C)
Coefficient of linear expansion (20–100° C)	6.7×10^{-6}/ °C	
Latent heat of fusion	– J/g	(– cal/g)
Latent heat of vaporization	– J/g	(– cal/g)
Thermal neutron absorption cross-section	86 barns/atom	
Electrical conductivity	8.5% IACS (copper 100%)	
Specific resistance	200 microhm mm	
Temperature coefficient of electrical resistance	0.0031/ °C	
Electrochemical equivalent	– g/A/h	
Electrode potential	– V	
Magnetic susceptibility	–	
Young's modulus of elasticity	460×10^9 N/m²	$(66.7 \times 10^6$ lbf/in.²)
Tensile strength annealed	114 N/mm²	(75 tonf/in.²)
Hardness annealed	170 DPN	
15% cold worked	550 DPN	

36.1 General notes on rhenium

Rhenium was one of the unknown elements when the periodic table was first produced. Not until 1925 was the metal prepared and its properties shown to be in general agreement with those predicted about 1870. It is now known that rhenium is widely distributed, generally as the sulphide associated with molybdenum ores. The first deposits to be worked commercially were in the Rhine area of Germany and this river has given its name to the metal.

Rhenium is silvery white, and can stand very little cold work without intermediate annealing. If the anneal which must be above 1500 °C is carried out in air an oxide forms which diffuses in from the surface causing brittleness. The formation of this oxide also precludes all normal hot working methods. Some success has been achieved by hot forging under hydrogen. The metal has good corrosion resistant properties for all except oxidizing liquids or atmospheres. No serious atmospheric oxidation occurs below red heat and the metal retains its tensile properties at these temperatures.

The high hardness achieved on cold worked rhenium has been utilized on pen nibs and as pivot bearing points in instruments.

Rhenium is readily applied by electroplating and has already been used to coat the inside of tanks for transporting and storing acids. It also has excellent resistance to molten metals such as tin, lead, zinc and silver. Heater elements of rhenium can be used to evaporate these molten metals.

For some purposes rhenium metal is preferred to tungsten for electronic tube electrodes. Rhenium or rhenium-plated electrical contacts have been shown to give excellent results in marine engine magnetos. When used to interrupt high current d.c. circuits rhenium contacts regularly give better results with less arcing than other contact materials, although it does not have particularly good electrical conductivity.

Rhenium when added to molybdenum considerably increases the ductility of the cast metal. Molybdenum–rhenium alloys have therefore been developed as weld filler rods for molybdenum alloys. Sintered alloys with between 50–90% rhenium, tungsten and molybdenum are successfully used for all ball point pens and other purposes where a smooth action is necessary over a protracted period.

Rhenium is one of the very modern metals which is scarce and expensive at present, but which has potential high temperature and corrosion resistant properties, particularly when alloyed with other refractory metals. Research is being carried out at present to reduce the production cost of the metal and prove that the promising potential can be applied in industry.

Symbol	Nominal analysis, supplier, condition and remarks.
RHENIUM	High purity Rh metal: Impurities 10 p.p.m.; Johnson Matthey; supplied as powder & button
RHENIUM	Rh pure metal: Blackwells
UNS R04001–R04999	Re metal and alloys: American Standards system

Note. The following abbreviations and units are used in the tables:

DPN	Hardness, diamond pyramid number
UTS	Ultimate tensile strength, N/mm^2
Elon	Elongation, %
Proof	0.1 % proof strength, N/mm^2

$1\ N/mm^2 = 0.1\ hbar = 0.102\ kgf/mm^2 = 0.06475\ tonf/in^2 = 145.04\ lbf/in^2$

37. Rhodium Rh

Physical properties

Atomic number	45	
Atomic weight	102.91	
Crystal structure	Face-centred cubic	
Colour	Bluish white	
Specific gravity	12.4	
Density	12 400 kg/m³	(0.449 lb/in.³)
Melting point	1960 °C	
Boiling point	4500 °C	
Specific heat	0.243 J/g °C	(0.058 cal/g °C)
Thermal conductivity	151 W/m °C	(36.1 cal/m s °C)
Coefficient of linear expansion (20–100° C)	8.5×10^{-6}/ °C	
Latent heat of fusion	21.8 J/g	(5.2 cal/g)
Latent heat of vaporization	532 J/g	(127 cal/g)
Thermal neutron absorption cross-section	2.6 barns/atom	
Electrical conductivity	40% IACS (copper 100%)	
Specific resistance	43 microhm mm	
Temperature coefficient of electrical resistance	0.0046/ °C	
Electrochemical equivalent	– g/A/h	
Electrode potential	+ 0.8 V	
Magnetic susceptibility	1.14×10^{-6}	
Young's modulus of elasticity	359×10^{9} N/m²	(52×10^{6} lbf/in.²)
Tensile strength	700 N/mm²	(45 tonf/in.²)
Hardness	annealed 100 DPN	
	as plated 800 DPN	

37.1 General notes on rhodium

Rhodium is one of the platinum metals, always being found in association with the remaining members – osmium, iridium, palladium and ruthenium. It is present in small proportions in the natively occurring alloys iridosmium and osmiridium, but also occurs in the palladium rich sulphide ores.

The resultant high purity metal is relatively soft in the annealed condition but work hardens very readily, accepting very little work before crumbling to a powder. Slight impurities also increase the hardness and make cold working impossible. As hot work has to be carried out at high temperatures under reducing atmospheres or vacuum, industry has found very few uses for this rather intractable material in the pure form.

Recent years have shown an increase in the electroplating techniques possible with rhodium. The as plated metal has a high hardness, excellent light reflectivity, resistance to oxidation up to and above red heat, is unaffected by all known acids, and has stable electrical resistance. The metal can only be deposited from acid solutions, thus under coatings of silver or nickel are generally required, to prevent attack and faulty adhesion to base metal of copper or steel.

Rhodium plating is now used on jewellery where its high lustre and hardness make it an attractive economical alternative to platinum. Many silver or silver plated trophies and 'plate' are now flash rhodium plated to prevent, or at least considerably reduce tarnishing. The very thin coat allows the warmer silver colour to shine through, yet prevents scratching of the silver when it is necessary to re-polish.

The good light and heat reflectivity find considerable use as searchlight reflectors and to concentrate heat in infra-red ovens. For best results light reflectors should be polished silver, rhodium plated.

Much chemical equipment and apparatus is now rhodium plated, as this has comparable chemical resistance to platinum and is much harder.

Electrical contact points with both surfaces rhodium plated have a zero electrochemical potential and will stand severe hammering under high corrosive conditions, up to red heat. The high hardness is also being used for 'plug-in' type sockets where softer deposits would be removed by mechanical abrasion.

Molybdenum and its alloys, rhodium plated, are finding some use at the intermediate temperatures where molybdenum is oxidized but rhodium is not. At present this is rather limited, as rhodium starts to diffuse rapidly into the molybdenum matrix about 900 °C. There is however, a sufficient increase in the electrical contact conductivity and surface hardness for some use to be found for rhodium plated molybdenum.

Rhodium used as an alloy hardens platinum, the alloys being workable up to about 40% rhodium with no effect on the resistance to chemical attack. The lower rhodium–platinum alloys are used in the manmade fibre field for such purposes as extrusion nozzles. The same alloys are used for high temperature furnace windings, potentiometers and the gauze and powders which find applications as catalysts in chemical engineering plants.

Although rhodium is for all practical purposes unworkable, the fact that it can be readily electroplated allows the attractive hard wearing and chemical resistant properties to be fully utilized. The use of rhodium electroplating will increase as more economical methods of producing the metal are developed.

Symbol	Nominal analysis, supplier, condition and remarks.
RHODIUM	High purity metal: Impurities 10 p.p.m.; Johnson Matthey; supplied as sponge
UNS P05001–P05999	Rh metal and alloys: American Standards system

Note. The following abbreviations and units are used in the tables:

DPN	Hardness, diamond pyramid number
UTS	Ultimate tensile strength, N/mm^2
Elon	Elongation, %
Proof	0.1 % proof strength, N/mm^2

$1\ N/mm^2 = 0.1\ hbar = 0.102\ kgf/mm^2 = 0.06475\ tonf/in^2 = 145.04\ lbf/in^2$

38. Rubidium Rb

Physical properties

Atomic number	37	
Atomic weight	85.48	
Crystal structure	Body-centred cubic	
Colour	White	
Specific gravity	1.525	
Density	1525 kg/m^3	(0.055 lb/in.^{3})
Melting point	39.0 °C	
Boiling point	685 °C	
Specific heat	0.336 J/g °C	(0.0802 cal/g °C)
Thermal conductivity	– W/m °C	
Coefficient of linear expansion (20–100° C)	90×10^{-6}/ °C	
Latent heat of fusion	25.5 J/g	(6.1 cal/g)
Latent heat of vaporization	888 J/g	(212 cal/g)
Thermal neutron absorption cross-section	0.7 barns/atom	
Electrical conductivity	15% IACS (copper 100%)	
Specific resistance	125 microhm mm	
Temperature coefficient of electrical resistance	0.0052/ °C	
Electrochemical equivalent	– g/A/h	
Electrode potential	-2.9 V	
Magnetic susceptibility	0.2×10^{-6}	
Young's modulus of elasticity	– N/m^2	
Tensile strength	– N/mm^2	
Hardness	– DPN	

38.1 General notes on rubidium

The compounds of rubidium occur sparingly but widespread throughout the world, generally associated with potassium and always with caesium.

The pure metal is soft, silvery white, tarnishing in air and reacting vigorously in water liberating hydrogen. The metal is a laboratory curiosity at present with no industrial use although it has weak radioactivity and has replaced caesium in photo-electric cells for certain purposes.

Rubidium is one of the alkali metals, in the same group as sodium, lithium, potassium, magnesium and caesium.

Although it is comaratively plentiful in nature, the fact that rubidium has no commercial value ensures that it remains a rare metal. No alloys based on the metal, or uses for the metal as an alloying element, have been found to date.

Symbol	Nominal analysis, supplier, condition and remarks.
f Rb 16	99.9% Rb metal: In ampoules; Light Ltd
RUBIDIUM	Pure metal: Blackwells

Note. The following abbreviations and units are used in the tables:

DPN	Hardness, diamond pyramid number
UTS	Ultimate tensile strength, N/mm^2
Elon	Elongation, %
Proof	0.1 % proof strength, N/mm^2

1 N/mm^2=0.1 hbar=0.102 kgf/mm^2=0.06475 tonf/in^2=145.04 lbf/in^2

39. Ruthenium Ru

Physical properties

Atomic number	44	
Atomic weight	101.7	
Crystal structure	Close-packed hexagonal	
Colour	White	
Specific gravity	12.3	
Density	12300 kg/m^3	(0.44 lb/in.^{3})
Melting point	2250 °C	
Boiling point	4110 °C	
Specific heat	0.243 J/g °C	(0.058 cal/g °C)
Thermal conductivity	– W/m °C	
Coefficient of linear expansion (20–100° C)	9.6 × 10^{-6}/ °C	
Latent heat of fusion	193 J/g	(46 cal/g)
Latent heat of vaporization	6196 J/g	(1480 cal/g)
Thermal neutron absorption cross-section	2.6 barns/atom	
Electrical conductivity	24% IACS (copper 100%)	
Specific resistance	72 microhm mm	
Temperature coefficient of electrical resistance	0.0010/ °C	
Electrochemical equivalent	– g/A/h	
Electrode potential	+ 0.45 V	
Magnetic susceptibility	0.56 × 10^{-6}	
Young's modulus of elasticity	414 × 10^9 N/m^2	(60 × 10^6 lbf/in.^{2})
Tensile strength	annealed 370 N/mm^2	(25 tonf/in.^{2})
Hardness	annealed 220 DPN	

39.1 General notes on ruthenium

Ruthenium is a member of the platinum group of metals and is always found associated with iridium, osmium, palladium and rhodium, as well as platinum. Like them it has a high melting point, bluish white colour and excellent resistance to chemical attack. Another similarity is the difficulty with which it is separated from its fellow members and obtained in a pure state.

Ruthenium metal itself has no practical use as it is unworkable, and other members of the platinum group have similar or better properties. To date ruthenium has not found much use as an electroplated deposit. There have been suggestions that it could be used as a very high temperature braze metal for joining refractory metals but there is no evidence of any applications to date.

Many fountain pen nibs are now tipped with a sintered ruthenium alloy containing 13% tungsten 10% molybdenum 3% platinum and cobalt. This is replacing the naturally occurring osmiridium alloy previously exclusively used.

The majority of ruthenium is used alloyed with platinum or palladium both of which are hardened with no reduction in chemical resistance. Above 15% ruthenium these alloys are unworkable.

The platinum ruthenium alloys are used for electrical contacts where they have similar properties to platinum–iridium. They are also generally used as substitutes for platinum jewellery, the alloy being harder and more wear resistant than palladium or platinum but cannot be so readily formed into fine or intricate designs.

Some work has been carried out with ruthenium–molybdenum alloys which show an increase in ductility over pure ruthenium. The resultant alloys have some advantage over molybdenum for high temperature oxidation resistance, but not sufficient to overcome the price advantage of molybdenum. Alloys of ruthenium with molybdenum have also been developed which can be used for brazing refractory materials at temperatures of above 200 °C.

Ruthenium is a close second to osmium in having the least practical use of the platinum metals and for the same reason, in that the metal cannot be readily fabricated.

There are some uses for ruthenium in nuclear reactors but very little information is available.

Symbol	Nominal analysis, supplier, condition and remarks.
Fs	26% Mo 23% Pd Rh Zr 40% Ru alloy: Symbol used for Fissium
FISSIUM	26% Mo 23% Pd Rh Zr 40% Ru alloy: Alloy of varying content in spent nuclear fuel elements
RUTHENIUM	High purity metal: Impurities 10 p.p.m.; Johnson Matthey; supplied as sponge
UNS P06001–P06999	Ru metal and alloys: American Standards system

Note. The following abbreviations and units are used in the tables:

DPN	Hardness, diamond pyramid number
UTS	Ultimate tensile strength, N/mm^2
Elon	Elongation, %
Proof	0.1 % proof strength, N/mm^2

1 N/mm^2=0.1 hbar=0.102 kgf/mm^2=0.06475 tonf/in^2=145.04 lbf/in^2

40. Selenium Se

Physical properties

Atomic number	34	
Atomic weight	78.96	
Crystal structure	Two allotropic forms	
Colour	Steel grey	
Specific gravity	4.81	
Density	4810 kg/m³	(0.174 lb/in.³)
Melting point	220 °C	
Boiling point	685 °C	
Specific heat	0.352 J/g °C	(0.084 cal/g °C)
Thermal conductivity	2.9 W/m °C	(0.7cal/m s °C)
Coefficient of linear expansion (20–100° C)	37 × 10⁻⁶/ °C	
Latent heat of fusion	84 J/g	(20 cal/g)
Latent heat of vaporization	3140 J/g	(750 cal/g)
Thermal neutron absorption cross-section	11.8 barns/atom	
Electrical conductivity	15% IACS (copper 100%)	
Specific resistance	120 microhm mm	
Temperature coefficient of electrical resistance	– / °C	
Electrochemical equivalent	0.5 g/A/h	
Electrode potential	– V	
Magnetic susceptibility	-0.32 × 10⁻⁶	
Young's modulus of elasticity	57.9 × 10⁹ N/m²	(8.4 × 10⁶ lbf/in.²)
Tensile strength	– N/mm²	
Hardness	– DPN	

40.1 General notes on selenium

Selenium has the characteristics of non-metals and metals evenly divided, being a metalloid or semi-metal. The non-metallic characteristics strongly resemble those of sulphur, the range of compounds known as selenides being similar in many respects to the sulphides.

It occurs in nature as metallic selenides generally associated with pyrites – metallic sulphides.

Like the non-metal sulphur, selenium exists in several allotropies the stable form being silvery grey, produced when any form of selenium is heated at 200–230 °C, and allowed to cool. A yellow allotropic form known as 'flowers of selenium' is produced when metallic selenium is allowed to boil.

Most high purity selenium is used as a material for rectifying electric current. Selenium metal rectifiers are now used almost exclusively when large current rectification is necessary. Rotary convertors were previous first choice for this purpose. Germanium and silicon join selenium as rectification metals for electronic and low current use.

Almost all commercial photo-electric devices have selenium as the operating mechanism. This uses a remarkable property in that light falling on the metal generates a small electric current proportional to the amount of light. The exact reason for this is as yet unknown but is believed to be associated with the fact that light causes emission of electrons from the high purity selenium. Photo-electric devices are now used for many purposes including trip mechanisms which operate when a light beam is interrupted, or to switch on artificial light when sunlight is reduced. The same devices convert light into sound in modern films, the sound track running alongside the video film. They also take part in the modern methods of colorimetric analysis whereby accurate quantitative estimation is rapidly carried out.

There have been recent advances in the use of selenium as a corrosive preventative coating on steel. Selenium additions of up to 0.5% are now made to copper alloys and austenitic stainless steels to improve machinability when there is no reduction in the hot working properties and very little in ductility.

The largest single application of selenium makes use of low grade selenium compounds for decolourizing glass. Small quantities added to molten glass eliminate the blue/green tinge caused by traces of iron common to most glass making sands.

Selenious acid is used in the local corrosion protection of magnesium. Selenium and the majority of its compounds are toxic and care must always be taken to prevent ingestion particularly of small quantities over a long period.

Of all the semi-metals or metalloids, selenium is probably the most useful from the purely engineering and metallurgical point of view.

Symbol	Nominal analysis, supplier, condition and remarks.
AR Q	99.99% Se: In shot form; common name selenium for rectifiers
DD Q	99.95% Se: In shot form; common name for selenium for rectifiers
h Se 2	99.999% Se pellets: Light Ltd; high purity metal
SELENIUM	Se metal: Blackwells; pure metal

Note. The following abbreviations and units are used in the tables:

DPN	Hardness, diamond pyramid number
UTS	Ultimate tensile strength, N/mm^2
Elon	Elongation, %
Proof	0.1 % proof strength, N/mm^2

$1 \ N/mm^2 = 0.1 \ hbar = 0.102 \ kgf/mm^2 = 0.06475 \ tonf/in^2 = 145.04 \ lbf/in^2$

41. Silicon Si

Physical properties

Atomic number	14	
Atomic weight	28.06	
Crystal structure	Tetrahedral cubic–amorphous	
Colour	–	
Specific gravity	2.49	
Density	2490 kg/m^3	(0.084 lb/in.^{3})
Melting point	1414 °C	
Boiling point	2600 °C	
Specific heat	0.670 J/g °C	(0.160 cal/g °C)
Thermal conductivity	84 W/m °C	(20 cal/m s °C)
Coefficient of linear expansion (20–100° C)	2.5 × 10^{-6}/ °C	
Latent heat of fusion	1809 J/g	(432 calories/g)
Latent heat of vaporization	10614 J/g	(2535 calories/g)
Thermal neutron absorption cross-section	0.13 barns/atom	
Electrical conductivity	– % IACS (copper 100%)	
Specific resistance	100 000 microhm mm	
Temperature coefficient of electrical resistance	– / °C	
Electrochemical equivalent	– g/A/h	
Electrode potential	-0.45 V	
Magnetic susceptibility	0.13 × 10^{-6}	
Young's modulus of clasticity	112.4 × 10^9 N/m^2	(16.3 × 10^6 lbf/in.^{2})
Tensile strength	– N/mm^2	
Hardness	– DPN	

41.1 General notes on silicon

Silicon is the second most common element occuring on earth, the first being oxygen. Combined with oxygen, silicon forms sand, quartz and flint, all of which are forms of the silicon oxide, silica. It is also the main constituent of granite and many of the metals occur as single or complex silicates.

Silicon is not a true metal and has even less metallic characteristics than most metalloids or semi-metals. It occurs in two forms – amorphous silicon – a brown powder, and crystalline silicon – a black lustrous solid.

The only known industrial use of silicon is to rectify electrical current. For this purpose silicon is superior to selenium for higher temperatures and voltages and thus finds some special purpose applications in the electronic industry. Pure silicon is more difficult to produce, thus being more expensive than either selenium or germanium.

It has no other property resembling that of the true metals and no other industrial or commercial use.

The element has however, found considerable use in the metallurgy of several commonly used metals. These sometimes combine directly with the element, or make use of the abundant oxide – silica, but this being refractory is not usual.

There are now being developed a host of compounds with very special properties, all based on the element silicon, in a manner not unlike the organic compounds based on carbon. Silicone greases, rubbers, cements and so on have all been manufactured, many of which have special low and high temperature applications.

The metallurgical applications of silicon and silica are as follows:

Copper alloys. Silicon acts as a de-oxidizer, but is not generally added for this purpose but to strengthen the matrix and improve the corrosion resistance.

The copper–silicon alloys with up to 3.5% silicon are sometimes used as substitutes for tin bronze where strength and corrosion resistance are of prime importance.

When added to the copper–zinc series silicon imparts some corrosion resistance and strength to the matrix, and are sometimes known as 'silicon brass'.

Aluminium alloys. Silicon acts as a powerful de-oxidizer but is not generally used for this purpose. It lowers the melting point, thus increasing the fluidity of castings and allowing more intricate patterns to be successfully cast. With about 12–13% silicon additions, the lowest melting point of any aluminium alloy is achieved and as these still make sound castings a series of brazing alloys have been developed allowing other aluminium alloys to be brazed.

Silicon also forms intermetallic compounds, particularly with magnesium, which dissolve in the aluminium matrix with correct thermal treatment, then precipitate out, causing 'age hardening'.

Iron. Silicon added to cast iron helps to reduce the carbon present as carbides and increases that present as graphite and is therefore termed a graphitizer. As only carbides harden this has the effect of giving softer, more ductile cast irons. As a secondary effect it increases the fluidity of molten cast iron, thus giving castings of better definition.

High silicon irons have excellent acid resistance, particularly to sulphuric acid.

Steel. Silicon is present in most steels up to 0.35% as it is one of the most commonly used de-oxidizing agents. In these proportions silicon has no affect on the physical or mechanical properties of steel. Between 1–2% silicon steels have excellent hardenability and fatigue strength, thus find use as leaf and coil springs, very often alloyed with manganese and chromium.

Very low carbon steels with 2–4% silicon have special hysteresis properties which make them suitable for transformer cores and other electrical and electronic uses.

Silicon in higher quantities imparts considerable acid resistance, particularly to the oxidizing acids. A process analogous to carburizing has been developed – known as siliconizing or 'Ihrigizing'. With this the parts are packed in silicon carbide and chlorine gas passed over them at 1000 °C. This gives a silicon 'case' of up to 0.075% deep if desired, which has good acid resistance with reasonably good wear. The process is best suited to plain carbon steel or malleable cast iron.

Although silicon cannot truly by classed as a metal it has enough use to metallurgists to warrant the brief note given. There are no alloys based on silicon.

Symbol	Nominal analysis, supplier, condition and remarks.
CMSZ	5% Zr 30% Fe 32.5% Cr 32.5% Si alloy: Origin unknown
L Si 2	99.9999% Si granules: Light Ltd; high purity
L Si 71	Single crystal Si: Light Ltd
SILVAISE	6% Zr 10% Ti 10% V 6% Al 0.5% B Si alloy: Origin unknown

Note. The following abbreviations and units are used in the tables:

DPN	Hardness, diamond pyramid number
UTS	Ultimate tensile strength, N/mm^2
Elon	Elongation, %
Proof	0.1 % proof strength, N/mm^2

$1 \ N/mm^2 = 0.1 \ hbar = 0.102 \ kgf/mm^2 = 0.06475 \ tonf/in^2 = 145.04 \ lbf/in^2$

42. Silver Ag

Physical properties

Atomic number	47	
Atomic weight	107.88	
Crystal structure	Face-centred cubic	
Colour	White	
Specific gravity	10.5	
Density	10500 kg/m³	(0.38 lb/in.³)
Melting point	960.8 °C	
Boiling point	2193 °C	
Specific heat	0.234 J/g °C	(0.056 cal/g °C)
Thermal conductivity	419 W/m °C	(100 cal/m s °C)
Coefficient of linear expansion (20–100° C)	19.6×10^{-6}/ °C	
Latent heat of fusion	105 J/g	(25 cal/g)
Latent heat of vaporization	2332 J/g	(557 cal/g)
Thermal neutron absorption cross-section	63 barns/atom	
Electrical conductivity	105% IACS (copper 100%)	
Specific resistance	15.5 microhm mm	
Temperature coefficient of electrical resistance	0.0041/ °C	
Electrochemical equivalent	4.025 g/A/h	
Electrode potential	+ 0.799 V	
Magnetic susceptibility	0.2×10^{-6}	
Young's modulus of elasticity	76×10^9 N/m²	(11×10^6 lbf/in.²)
Tensile strength	annealed 140 N/mm²	(9 tonf/in.²)
Hardness	annealed 25 DPN	

42.1 General notes on silver

Native silver is still found on occasions generally associated with native copper. The most common ore is the sulphide 'galena', found is South America, Western USA, Australia and Norway.

Silver ores are very often found with lead and copper, the lead mines in this country producing silver as a byproduct as long as enough lead was present to justify extraction.

Some silver has been mined in its own right in Scotland and Cornwall.

Metallic silver is pure white in colour, relatively soft and capable of taking a high polish. In the pure state it is the best conductor of heat and electricity and when polished one of the best reflectors of light.

The metal has excellent resistance to oxidation even at comparatively high temperatures, but tarnishes readily in sulphur-bearing atmospheres.

Silver has resistance to most acids, except nitric acid and concentrated hydrochloric acid, having particularly good resistance to organic materials. This accounts for the long use of silver for cutlery and plate and the modern use for silver plating for the same purpose, and for many utensils and equipment in food manufacture. The high polish which still retains a degree of warmth, not apparent with the platinum metals or chromium, coupled with the ease of manipulation and fabrication makes silver a favourite jewellery material.

Silver is now being used as a ladle material in the pro-

duction of the alkali metals. A considerable quantity of pure silver in the form of very thin sheet is used to line or coat copper alloy pipes and vessels which handle chemicals during manufacture and storage. Silver plating is also used for these applications, some of which are being replaced by the modern acid resistant stainless steels or titanium.

After copper, silver is the most commonly used electrical contact material, having an advantage in that the contact resistance is not increased by oxidation at normal temperatures or by arcing.

Silver should not be used in direct contact with steel as severe corrosion can take place in the steel under certain conditions. Steel parts must be copper - or nickel-plated prior to silver. Stainless steel and nickel alloy parts operating at high temperatures are often silver-plated to prevent seizure of nuts and bolts, thus ensuring easy removal at overhaul.

Nickel silvers contain no silver, the name being applied to the range copper–zinc–nickel alloys which resemble it when polished. This class also applies to German silver.

Several alloys are based on silver, one reason for alloying being to increase the hardness. Coins now have less than 50% silver, the remainder being copper, nickel and zinc.

High quality trophies and ornamental household articles are made from similar types of alloys, sterling silver being 7.5% copper–silver alloy. Modern alloys often have nickel added.

Industrially there is little use for these alloys as the chemical and oxidation resistance is reduced. Silver 'solders' or brazing alloys make up the bulk of the alloys based on silver. These have low melting points, thus reducing the risk of surface oxidation and distortion of the parts being brazed. The resultant join is stronger than copper–zinc brazing as the silver solders have superior wetting and penetrating qualities in addition to their higher intrinsic strength.

These alloys have cadmium, copper, zinc and phosphorus additions to vary the melting range, increase fluidity and for economy reasons. The presence of cadmium is now known to constitute a health hazard at brazing.

Silver is used as an alloying element to pure copper where in small quantities it raises the temperature at which softening occurs after cold work.

Some copper bearing alloys have silver additions, the resultant material standing higher temperatures than the copper–lead series at the expense of frictional properties.

Considerable quantities of silver are used in photographic film and most modern mirrors are silver backed. Recently it has been found that silver can be used catalytically for several important chemical processes.

Silver is probably the most industrially useful of the so-called precious metals. Some of these applications are now finding stainless steels to be a suitable alternative but the ease of manipulation, fabrication and deposition ensure that silver will continue to be used commercially and industrially for as long as it is economically possible.

Symbol	Nominal analysis, supplier, condition and remarks.
120	20% Ni Ag alloy: Engelhard; annealed; electrical conductivity 72% IACS **DPN: 50**
179	28% Cu Ag alloy: Melting point 779 °C; Engelhard annealed; electrical conductivity 80% IACS **DPN: 88**
718	Cu Pd Ag brazing alloy: Melting range 970–1010 °C; Engelhard; electrical conductivity 36% IACS **DPN: 50 UTS: 210 Elon: 28%**
719	Cu Pd Ag brazing alloy: Melting range 1080–1090 °C; Engelhard; electrical conductivity 11% IACS **DPN: 85 UTS: 330 Elon: 30%**
722	Cu Pd Ag brazing alloy: Melting range 807–810 °C; Engelhard; electrical conductivity 45% IACS **DPN: 90 UTS: 440 Elon: 23%**
723	Cu Pd Ag brazing alloy: Melting range 824–852 °C; Engelhard; electrical conductivity 30% IACS **DPN: 75 UTS: 460 Elon: 22%**
724	Cu Pd Ag brazing alloy: Melting range 850–900 °C; Engelhard; electrical conductivity 18% IACS **DPN: 120 UTS: 450 Elon: 20%**
725	Cu Pd Ag brazing alloy: Melting range 901–950 °C; Engelhard; electrical conductivity 12% IACS **DPN: 185 UTS: 500 Elon: 15%**
748	Cu Pd Ag brazing alloy: Melting range 876–898 °C; Engelhard; electrical conductivity 15% IACS **DPN: 140 UTS: 490 Elon: 25%**
1103	Cd Ag alloy: Engelhard; annealed; electrical conductivity 35% IACS **DPN: 40**
1144	Fe Ag alloy: Engelhard; annealed; electrical conductivity 90% IACS **DPN: 65**
1322	5% Pd Ag alloy: Engelhard; annealed; electrical conductivity 45% IACS **DPN: 33**
1323	10% Pd Ag alloy: Engelhard; annealed; electrical conductivity 30% IACS **DPN: 40**

Symbol	Nominal analysis, supplier, condition and remarks.
1325	20% Pd Ag alloy: Engelhard; annealed; electrical conductivity 17% IACS **DPN: 55**
1340	10% Au Ag alloy: Engelhard; annealed; electrical conductivity 48% IACS **DPN: 29**
1379	Cu Ag alloy: Standard silver; Engelhard; annealed; electrical conductivity 90% IACS **DPN: 56**
1380	Cu Ag alloy: Coin silver; Engelhard; annealed; electrical conductivity 86% IACS **DPN: 62**
1386	20% Cu Ag alloy: Engelhard; annealed; electrical conductivity 82% IACS **DPN: 85**
1396	50% Cu Ag alloy: Engelhard; annealed; electrical conductivity 82% IACS **DPN: 95**
1438	30% Ni Ag alloy: Engelhard; annealed; electrical conductivity 62% IACS **DPN: 55**
1439	40% Ni Ag alloy: Engelhard; annealed; electrical conductivity 55% IACS **DPN: 60**
1464	Cd Ag alloy: Engelhard; annealed; electrical conductivity 82% IACS **DPN: 60**
1465	Cd Ag alloy: Engelhard; annealed; electrical conductivity 72.0% IACS **DPN: 70**
1845/4	Cu Zn 61% Ag alloy: Braze metal; Johnson Matthey; BS 18454 melting range 690–737 °C
1845/5	Cu Zn 43% Ag alloy: Braze metal; Johnson Matthhey for BS 1845-5; melting range 698–788 °C
2601	Pd Mn Ag brazing alloy: Melting range 1000–1120 °C; Engelhard; electrical conductivity 19% IACS **DPN: 95 UTS: 270 Elon: 11%**
2602	Pd Mn Ag brazing alloy: Melting range 1180–1200 °C; Engelhard; electrical conductivity 10% IACS **DPN: 130 UTS: 480 Elon: 25%**
2949	Cd Ag alloy: Engelhard; annealed; electrical conductivity 69.0% IACS **DPN: 70**
Ag/Cu EUTECTIC	29% Cu Ag alloy: Braze metal; Johnson Matthey; melting point 778 °C
AMS 4766 A	15% Mn Ag alloy: Braze metal
AMS 4767	7.2% Cu 0.2% Li Ag alloy: Braze metal
AMS 4768	26% Cu 21% Zn 18% Cd Ag alloy: Braze metal
AMS 4769	15% Cu 16% Zn 24% Cd Ag alloy: Braze metal

Note. The following abbreviations and units are used in the tables:

DPN	Hardness, diamond pyramid number
UTS	Ultimate tensile strength, N/mm^2
Elon	Elongation, %
Proof	0.1 % proof strength, N/mm^2

$1 N/mm^2 = 0.1 hbar = 0.102 kgf/mm^2 = 0.06475 tonf/in^2 = 145.04 lbf/in^2$

Symbol	Nominal analysis, supplier, condition and remarks.
AMS 4770 B	15.5% Cu 16.5% Zn 18% Cd Ag alloy: Braze metal
AMS 4771	15.5% Cu 15.5% Zn 3% Ni 16% Cd Ag alloy: Braze metal
AMS 4772 A	40% Cu 5% Zn 1% Ni Ag alloy: Braze metal
AMS 4773	30% Cu 10% Sn Ag alloy: Braze metal
AMS 4774	28.5% Cu 6% Sn 2.5% Ni Ag alloy: Braze metal
ARGO-BRAZE 50	Cu Zn Cd Ni Mn 50% Ag alloy: For brazing; Johnson Matthey; melting range 639–668 °C
ARGO-BRAZE 56	Cn Zn Ni 56% Ag alloy: For brazing; Johnson Matthey; melting range 600–711 °C
ARGO-FLO	Cn Zn Cd 39% Ag alloy: Brazing metal; Johnson Matthey; melting range 605–651 °C
ASTM B260 B Ag 1	15% Cu 17% Zn 24% Cd Ag alloy: Braze filler; melting range 604–617 °C
ASTM B260 B Ag 1 A	15% Cu 16% Zn 18% Cd Ag alloy: Braze filler; melting range 626–635 °C
ASTM B260 B Ag 2	26% Cu 20% Zn 18% Cd Ag alloy: Braze filler; melting range 605–700 °C
ASTM B260 B Ag 3	15% Cu 15% Zn 16% Cd 3% Ni Ag alloy: Braze filler; melting range 631–686 °C
ASTM B260 B Ag 4	30% Cu 29% Zn 2% Ni Ag alloy: Braze filler; melting range 670–776 °C
ASTM B260 B Ag 5	30% Cu 25% Zn Ag alloy: Braze filler; melting range 675–741 °C
ASTM B260 B Ag 6	32% Cu 15% Zn Ag alloy: Braze filler; melting range 686–775 °C
ASTM B260 Ag 7	22% Cu 16% Zn 5% Sn Ag alloy: Braze filler; melting range 617–650 °C
ASTM B260 B Ag 8	28% Cu Ag alloy: Braze filler; melting point 780 °C
ASTM B260 B Ag 8 A	27.7% Cu 0.22% Li Ag alloy: Braze metal; melting range 760–870 °C
ASTM B260 B Ag 9	20% Cu 14% Zn Ag alloy: Braze filler; melting range 692–717 °C
ASTM B260 B Ag 10	20% Cu 10% Zn Ag alloy: Braze filler; melting range 720–750 °C
ASTM B260 B Ag 11	22% Cu 3% Zn Ag alloy: Braze filler; melting range 737–787 °C
ASTM B260 B Ag 13	40% Cu 5% Zn 1% Ni Ag alloy: Braze metal; melting range 855–970 °C
ASTM B260 Ag 18	30% Cu 10% Sn Ag alloy: Braze metal; melting range 718–843 °C
ASTM B260 Ag 19	8.7% Cu 0.22% Li Ag alloy: Braze metal; melting range 877–982 °C
ASTM B260 B Ag Mn	15% Mn Ag alloy: Braze filler; melting range 960–970 °C
BAZAR METAL	9% Ni Ag alloy: Sheet & wire; origin unknown
BS 1561	99.99% Ag: For anodes; cast or rolled; annealed
BS 1845/3	15% Cu 16% Zn 19% Cd Ag alloy: Braze filler; melting range 620–640 °C
BS 1845/4	28% Cu 10% Zn Ag alloy: Braze filler; melting range 690–735 °C
BS 1845/5	37% Cu 19% Zn Ag alloy: Braze filler; melting range 700–775 °C
BS 1845 Ag 1	15% Cu 16% Zn 19% Cd Ag alloy: Braze metal; melting range 620–640 °C
BS 1845 Ag 2	17% Cu 16% Zn 25% Cd Ag alloy: Braze metal; melting range 610–620 °C

Symbol	Nominal analysis, supplier, condition and remarks.
BS 1845 Ag 3	20% Cu 21% Zn 20% Cd Ag alloy: Braze metal; melting range 605–650 °C
BS 1845 Ag 4	28.5% Cu 10% Zn Ag alloy: Braze metal; melting range 690–735 °C
BS 1845 Ag 5	37% Cu 19.5% Zn Ag alloy: Braze metal; melting range 700–775 °C
BS 1845 Ag 6	30% Cu Ag alloy: Braze metal; melting range 600–720 °C
BS 1845 Ag 7	28% Cu Ag alloy: Braze metal; melting point 780 °C
BS 1845 Ag 8	99.99% Ag: For brazing; melting point 960 °C
BS 1845 Pd 1	26.5% Cu 5.0% Pd Ag alloy: For braze metal; melting range 805–810 °C
BS 1845 Pd 2	31.5% Cu 10% Pd Ag alloy: For braze metal; melting range 825–850 °C
BS 1845 Pd 3	22.5% Cu 10% Pd Ag alloy: For braze metal; melting range 830–860 °C
BS 1845 Pd 4	20% Cu 15% Pd Ag alloy: For braze metal; melting range 875–900 °C
BS 1845 Pd 5	28% Cu 20% Pd Ag alloy: For braze metal; melting range 875–900 °C
BS 1845 Pd 6	21% Cu 25% Pd Ag alloy: For braze metal; melting range 900–950 °C
BS 1845 Pd 7	5% Pd Ag alloy: For braze metal; melting range 970–1010 °C
BS 1845 Pd 9	20% Pd 5% Mn Ag alloy: For brazing metal; melting range 1000–1120 °C
BS 1845 Pd 10	33% Pd 3% Mn Ag alloy: For brazing metal; melting range 1180–1200 °C
COIN SILVER	10% Cu Ag alloy: American origin
CUSIL	28% Cu Ag alloy: For brazing; melting point 780 °C Wesgo
DIMPALLOY	Silver solder sheet with flux: Sheffield Smelting; available as MX 18, MX 12, and MX 8
DIN 1734 L Ag 44	25% Zn 32% Cu Ag alloy: Braze metal; working temperature 730 °C
DIN 1734 L Ag 45	15% Zn 20% Cd 19% Cu Ag alloy: Bronze metal; working temperature 620 °C
DIN 1734 L Ag 49	25% Zn 18% Cu 5% Mn 3% Ni Ag alloy: Braze metal; working temperature 690 °C
DIN 1734 L Ag 50	15% Zn 5% Cd 32% Cu Ag alloy: Braze metal; working temperature 700 °C
DIN 1735 L Ag 50 Cd	15% Zu 20% Cu 15% Cd Ag alloy: Braze metal; working temperature 650 °C
DIN 1735 L Ag 60	12% Zn 28% Cu Ag alloy: Braze metal; working temperature 710 °C
DIN 1735 L Ag 60 Cd	10% Zn 25% Cu 2% Sn 2% Cd Ag alloy: Braze metal; working temperature 680 °C
DIN 1735 L Ag 67	8% Zn 25% Cu Ag alloy: Braze metal; working temperature 730 °C
DIN 1735 L Ag 67 Cd	10% Zn 12% Cu 10% Cd Ag alloy: Braze metal; working temperature 710 °C
DIN 1735 L Ag 75	2% Zn 24% Cu Ag alloy: Braze metal; working temperature 770 °C
DIN 1735 L Ag 83	2% Zn 15% Cu Ag alloy: Braze metal; working temperature 830 °C
EASY-FLO 1	Zn Cd Cu 50% Ag alloy: Brazing metal; Johnson Matthey; melting range 620–630 °C; electrical conductivity 22% IACS DPN: 131 UTS: 450 Elon: 35%
EASY FLO 2	Zn Cd Cu 42% Ag alloy: Brazing metal; Johnson Matthey; melting range 608–617 °C; electrical conductivity 20% IACS DPN: 135 UTS: 450 Elon: 30%
EASY-FLO 3	Zn Cd 3% Ni Cu 50% Ag alloy: Braze metal; Johnson Matthey; melting range 634–656 °C; for brazing carbide tips DPN: 135 UTS: 470

Note. The following abbreviations and units are used in the tables:

DPN	Hardness, diamond pyramid number
UTS	Ultimate tensile strength, N/mm^2
Elon	Elongation, %
Proof	0.1 % proof strength, N/mm^2

$1\ N/mm^2 = 0.1\ hbar = 0.102\ kgf/mm^2 = 0.06475\ tonf/in^2 = 145.04\ lbf/in^2$

Symbol	Nominal analysis, supplier, condition and remarks.
ELKONITE 20S	W Ag alloy: For arc resistant contacts; Johnson Matthey; conductivity 43% IACS DPN: 220
ELKONITE 35S	W Ag alloy: For arc resistant contacts; Johnson Matthey; conductivity 52% IACS DPN: 140
ELKONITE 50S	W Ag alloy: For arc resistant contact; Johnson Matthey; conductivity 61% IACS DPN: 115
ELKONITE D54	Cd Ag alloy: For contacts; Johnson Matthey; conductivity 82% IACS
ELKONITE D54L	Cd Ag alloy: For contacts; Johnson Matthey; conductivity 82% IACS
ELKONITE D54X	Cd Ag alloy: For contacts; Johnson Matthey; conductivity 82% IACS
ELKONITE D55X	Cd Ag alloy: For contacts; Johnson Matthey; conductivity 75% IACS
ELKONITE D56	Ni Ag alloy: For contacts; Johnson Matthey; conductivity 72% IACS
ELKONITE D58/1	1% graphite Ag alloy: For sliding contacts; Johnson Matthey; conductivity 96% IACS
ELKONITE D58/2	2% graphite Ag alloy: For sliding contacts; Johnson Matthey; conductivity 87% IACS
ELKONITE D510	Ni Ag alloy: For contacts; Johnson Matthey; conductivity 87% IACS
ELKONITE D520	Ni Ag alloy: For contacts; Johnson Matthey; conductivity 57% IACS
ELKONITE G13	W Ag alloy: For contacts; Johnson Matthey; conductivity 57% IACS DPN: 110
ELKONITE G14	W Ag alloy: For contacts; Johnson Matthey; conductivity 48% IACS DPN: 200
ELKONITE G17	Mo Ag alloy: For contacts; Johnson Matthey; conductivity 48% IACS DPN: 180
ELKONITE G18	Mo Ag alloy: For contacts; Johnson Matthey; conductivity 52% IACS DPN: 150
EUTECTIC ALLOY	29% Cu Ag alloy: Braze metal; Johnson Matthey; melting point 778 °C
G4	W Ag contact material: Sheffield Smelting; electrical conductivity 43% IACS DPN: 225
G 6 Cu Zn	67% Ag alloy: Braze metal for silver; Johnson Matthey; melting range 705–723 °C white colour
G7	W Ag contact material: Sheffield Smelting; electrical conductivity 50% IACS DPN: 165
G9	W Ag contact material: Sheffield Smelting; electrical conductivity 55% IACS DPN: 140
GC15	Graphite Ag contact material: Sheffield Smelting; electrical conductivity 85% IACS DPN: 64
GD10	Cadmium oxide Ag sintered contact material: Sheffield Smelting; electrical conductivity 82% IACS DPN: 60

Note. The following abbreviations and units are used in the tables:

DPN	Hardness, diamond pyramid number
UTS	Ultimate tensile strength, N/mm^2
Elon	Elongation, %
Proof	0.1 % proof strength, N/mm^2

$1\ N/mm^2 = 0.1\ hbar = 0.102\ kgf/mm^2 = 0.06475\ tonf/in^2 = 145.04\ lbf/in^2$

Symbol	Nominal analysis, supplier, condition and remarks.
GD25	Cadmium oxide Ag oxidized contact material: Sheffield Smelting; electrical conductivity 82% IACS DPN: 60
GD35	Cadmium oxide Ag contact material: Sheffield Smelting; electrical conductivity 75% IACS DPN: 65
GN1	Ni Ag contact material: Sheffield Smelting; electrical conductivity 66% IACS DPN: 90
H 12	28% Cu Ag alloy: Braze metal; Sheffield Smelting Co; melting point 778 °C DPN: 106 UTS: 390 Elon: 30% Proof: 290
h Ag 1	99.9999% Ag rod: 12 mm diameter; Light Ltd; high purity metal
h Ag 96	99.0% Ag: Isotope 109; Light Ltd
HARDSILVER	3% Cu Ag alloy: Used for contacts; common name
HW 10C	WC Ag contact material: Sheffield Smelting; electrical conductivity 60% IACS DPN: 130
IRCUSIL 10	10% In 27% Cu Ag alloy: For brazing; melting range 685–730 °C; Wesgo
IRCUSIL 15	14.5% In 24% Cu Ag alloy: For brazing; melting range 630–705 °C; Wesgo
JMC 1715 Mg Ni	Ag alloy: For electrical spring contact; Johnson Matthey; electrical conductivity 55% IACS DPN: 145 UTS: 400 Elon: 14% Proof: 340
JMM 77	Pd Pt Au Ag alloy: For contacts; Johnson Matthey; age hardened DPN: 320 UTS: 1100 Elon: 7% Proof: 1000
JMM 1715	Ni Mg Ag alloy: For contacts; 60% IACS conductivity; Johnson Matthey DPN: 145
MATTHEY 50S	50% W Ag alloy: Sinter for contacts; 61% IACS conductivity; Johnson Matthey DPN: 115
MATTHEY 55X	15% Cd O Ag alloy: For contacts; 75% IACS conductivity; Johnson Matthey DPN: 60
MATTHEY 3045	45% WC Ag alloy: Sinter for contacts; 62% IACS conductivity; Johnson Matthey DPN: 95
MATTHEY D54	10% Cd O Ag alloy: For contacts; 82% IACS conductivity; Johnson Matthey DPN: 50
MATTHEY D54X	10% Cd O Ag alloy: For contacts; 82% IACS conductivity; Johnson Matthey DPN: 58
MATTHEY D55	15% Cd O Ag alloy: For contacts; 72% IACS conductivity; Johnson Matthey DPN: 55
MATTHEY D56	30% Ni Ag alloy: For contacts; 72% IACS conductivity; Johnson Matthey DPN: 68
MATTHEY D58	1% C (graphite) Ag: For contacts; 96% IACS conductivity; Johnson Matthey DPN: 40
MATTHEY D58	2% C (graphite) Ag: For contacts; 86% IACS conductivity; Johnson Matthey DPN: 40
MATTHEY D510	10% Ni Ag alloy: For contacts; 87% IACS conductivity; Johnson Matthey DPN: 40
MATTHEY D520	20% Ni Ag alloy: For contacts; 82% IACS; Johnson Matthey DPN: 48
MATTHEY G13	50% WC Ag alloy: Sinter for contacts; 57% IACS conductivity; Johnson Matthey DPN: 110

Symbol	Nominal analysis, supplier, condition and remarks.
MATTIBRAZE 34	Cu Zn Col 34% Ag alloy: For brazing; Johnson Matthey; melting range 612–668 °C
MELT-ESI MX 12	16.5% Cu 18% Zn 23.5% Cd Ag alloy: Braze metal; Sheffield Smelting Co; melting range 608–621 °C; electrical conductivity 26% IACS; now Thessco
	DPN: 135 UTS: 420 Elon: 27% Proof: 250
MELT-ESI MX 18	15% Cu 15% Zn 22% Cd Ag alloy: Braze filler; Sheffield Smelting Co; melting range 630–640 °C; electrical conductivity 23.8% IACS; now Thessco
	DPN: 115 UTS: 420 Elon: 43% Proof: 300
MELT-ESI MX 18 PLUS	15.5% Cu 15.5% Zn 18% Cd 3% Ni Ag alloy: Braze metal; Sheffield Smelting Co; melting range 640–660 °C; electrical conductivity 17.2% IACS; now Thessco
	DPN: 145 UTS: 450 Elon: 35% Proof: 300
MELT-ESI No 2	16.5% Cu 18% Zn 23.5% Cd Ag alloy: Filler metal; Sheffield Smelting Co; melting range 608–621 °C
	DPN: 135 UTS: 420 Elon: 27% Proof: 250
MX 0	25% Cu 25% Zn 20% Cd Ag alloy: Braze metal; Sheffield Smelting Co; melting range 605–680 °C; electrical conductivity 27.2% IACS
	DPN: 114 UTS: 340 Elon: 10% Proof: 220
MX 5	26% Cu 21% Zn 18% Cd Ag alloy: Braze metal; Sheffield Smelting Co; melting range 608–665 °C
	DPN: 103 UTS: 360 Elon: 15% Proof: 220
MX 8	20% Cu 19% Zn 23% Cd Ag alloy: Braze metal; Sheffield Smelting Co; melting range 606–648 °C; electrical conductivity 25.9% IACS
	DPN: 121 UTS: 420 Elon: 24% Proof: 240
MX 12	16.5% Cu 18% Zn 23.5% Cd Ag alloy: Braze metal; Sheffield Smelting Co; see Melt Esi Mx 12; electrical conductivity 26.0% IACS
MX 18	15% Cu 15% Zn 22% Cd Ag alloy: Braze metal; Sheffield Smelting Co; see Melt Esi MX 18
MX 18 PLUS	15.5% Cu 15.5% Zn 18% Cd 3% Ni Ag alloy: Braze metal; Sheffield Smelting Co; see Melt Esi MX 18 PLUS
NICUSIL 3	0.7% Ni 28% Cu Ag alloy: For brazing; melting range 780–795 °C; Wesgo
PALLABRAZE 1010	5% Pd Ag alloy: Brazing metal; Johnson Matthey; melting range 970–1010 °C
PALLABRAZE 1225	30% Pd Ag alloy: For brazing; Johnson Matthey; melting range 1150–1225 °C
PALCUSIL 5	5.0% Pd 27% Cu Ag alloy: For brazing; melting range 807–810 °C; Wesgo
PALCUSIL 10	10% Pd 32% Cu Ag alloy: For brazing; melting range 824–852 °C; Wesgo
PALCUSIL 15	15% Pd 20% Cu Ag alloy: For brazing; melting range 850–900 °C; Wesgo
PALCUSIL 25	25% Pd 21% Cu Ag alloy: For brazing; melting range 900–950 °C; Wesgo
QQ S 561/4	15% Cu 18% Cd 15% Zn Ag alloy: Solder; US Federal; melting range 600–634 °C
SCP 1	Cu Pd Ag brazing alloy: Melting range 807–810 °C; Engelhard; electrical conductivity 45% IACS
	DPN: 90 UTS: 440 Elon: 23%
SCP 2	Cu Pd Ag brazing alloy: Melting range 824–852 °C; Engelhard; electrical conductivity 30% IACS
	DPN: 75 UTS: 460 Elon: 22%
SCP 3	Cu Pd Ag brazing alloy: Melting range 850–900 °C; Engelhard; electrical conductivity 18% IACS
	DPN: 120 UTS: 450 Elon: 20%
SCP 4	Cu Pd Ag brazing alloy: Melting range 901–950° C; Engelhard; electrical conductivity 12% IACS
	DPN: 185 UTS: 440 Elon: 15%
SCP 5	Cu Pd Ag brazing alloy: Melting range 970–1010 °C; Engelhard; electrical conductivity 36% IACS
	DPN: 50 UTS: 210 Elon: 28%

Symbol	Nominal analysis, supplier, condition and remarks.
SCP 6	Cu Pd Ag brazing alloy: Melting range 1080–1090° C; Engelhard; electrical conductivity 11% IACS
	DPN: 85 UTS: 330 Elon: 30%
SCP 7	Cu Pd Ag brazing alloy: Melting range 876–890° C; Engelhard; electrical conductivity 15% IACS
	DPN: 140 UTS: 480 Elon: 25%
SILBERLOT 44	25% Zn 32% Cu Ag alloy: Braze metal; designation used by German Standard
SILBERLOT 45	15% Zn 20% Cd 19% Cu Ag alloy: Braze metal; designation used by German Standard
SILBERLOT 49	25% Zn 18% Cu 5% Mn 3% Ni Ag alloy: Braze metal; designation used by German Standard
SILBERLOT 50	15% Zn 5% Cd 32% Cu Ag alloy: Braze metal; designation used by German Standard
SILBERLOT 50 Cd	15% Zn 20% Cu 15% Cd Ag alloy: Braze metal; designation used by German Standard
SILBERLOT 60	12% Zn 28% Cu Ag alloy: Braze metal; designation used by German Standard
SILBERLOT 60 Cd	10% Zn 25% Cu 2% Sn 2% Cd Ag alloy: Braze metal
SILBERLOT 67	8% Zn 25% Cu Ag alloy: Braze metal; designation used by German Standard
SILBERLOT 67 Cd	10% Zn 12% Cu 10% Cd Ag alloy: Braze metal; designation used by German Standard
SILBERLOT 75	2% Zn 24% Cu Ag alloy: Braze metal; designation used by German Standard
SILBERLOT 83	2% Zn 15% Cu Ag alloy: Braze metal; designation used by German Standard
SILMANAL	9.0% Mn 4.0% Al Ag alloy: For magnets; American alloy listed by PMA; cold rolled
SILVER	Ag powder: Blackwells; pure metal
SILVER	High purity metal: Impurities 10 p.p.m.; Johnson Matthey; supplied as crystals
SILVER SOLDER	Cu Zn Ag brazing alloys: Covered by BS 1845/3/4 and 5; common name
SIX EIGHTY ALLOY	65% Ag alloy: For dental amalgams; Johnson Matthey
SPM 1	Pd Mn Ag brazing alloy: Melting range 1000–1120 °C; Engelhard; electrical conductivity 19% IACS
	DPN: 95 UTS: 270 Elon: 11%
SPM 1	33% Pd 5% Mn Ag alloy: For brazing; International Nickel; melting range 1000–1115 °C
SPM 2	33% Pd 3% Mn Ag alloy: For brazing; International Nickel; melting range 1180–1200 °C
SPM 2	Pd Mn Ag brazing alloy: Melting range 1180–1200 °C; Engelhard; electrical conductivity 10% IACS
	DPN: 130 UTS: 480 Elon: 25%
STANDARD SILVER	Alternative name for Sterling Ag
STERLING	7.5% Cu Ag alloy: Hall marked alloy of British Commonwealth & USA; common name
THESSCO D	Sn 70% Ag: Dental analgum alloy; Sheffield Smelting Co
THESSCO M	Sn 68% Ag: Dental amalgam alloy; Sheffield Smelting Co
THESSCO MX12	16.5% Cu 18% Zn 23.5% Cd Ag alloy: Braze metal; Sheffield Smelting Co; melting range 608–621 °C; previously Melt-Esi
	DPN: 135 UTS: 430 Elon: 27% Proof: 240
THESSCO MX18	15% Cu 15% Zn 22% Cd Ag alloy: Braze filler; Sheffield Smelting Co; melting range 630–640 °C; previously Melt-Esi
	DPN: 115 UTS: 430 Elon: 43% Proof: 300
THESSCO MX18 PLUS	15.5% Cu 15.5% Zn 18% Cd 3% Ni Ag alloy: Braze metal; Sheffield Smelting Co; melting range 640–660 °C; previously Melt-Esi
	DPN: 145 UTS: 450 Elon: 35% Proof: 300
UNS P07001–P07999	Ag metal and alloys: American Standard system; unified numbers

43. Sodium Na

Physical properties

Atomic number	11	
Atomic weight	22.997	
Crystal structure	Body-centred cubic	
Colour	Silver-white	
Specific gravity	0.971	
Density	971 kg/m³	(0.035 lb/in.³)
Melting point	97.8 °C	
Boiling point	883 °C	
Specific heat	1.235 J/g °C	(0.295 cal/g °C)
Thermal conductivity	135 W/m °C	(32.2 cal/m s °C)
Coefficient of linear expansion (20–100° C)	71×10^{-6}/ °C	
Latent heat of fusion	114 J/g	(27.2 cal/g)
Latent heat of vaporization	4208 J/g	(1005 cal/g)
Thermal neutron absorption cross-section	0.54 barns/atom	
Electrical conductivity	42% IACS (copper 100%)	
Specific resistance	41 microhm mm	
Temperature coefficient of electrical resistance	0.0055 solidus °C	
Electrochemical equivalent	0.86 g/A/h	
Electrode potential	-2.71 V	
Magnetic susceptibility	0.51×10^{-6}	
Young's modulus of elasticity	– N/m²	
Tensile strength	– N/mm²	
Hardness	0.1 DPN	

43.1 General notes on sodium

As would be expected of this very chemically active metal, it is never found in its native state. Combined with chlorine and as the carbonate, vast quantities exist spread over the earth's surface. Sodium chloride – common salt – is found in the sea, and there are enormous deposits of rock salt found all over the world as the result of the evaporation of prehistoric seas. One of these deposits in Poland is said to be 800 km long, 32 km wide, 370 m thick and has been operated as a salt mine for 600 years.

Sodium is a silvery white, soft metal which can be cut with a knife and can be moulded easily. It tarnishes rapidly in air, converting in time to sodium hydroxide by combining with the moisture in the atmosphere. It reacts violently in water with evolution of hydrogen. Sodium must therefore be stored at all times out of contact with air, generally under paraffin in air tight containers. This property reduces the usefulness of the metal which has a low specific gravity, with good heat and electrical conductivity and light reflection.

One use is as a heat transfer medium in high duty internal combustion engine exhaust valves. Here the stem is hollow and filled with sodium which melts at engine running temperature and transfers the heat from the valve head through the stem and valve guides.

The relatively high specific heat and low melting point, coupled with the comparative non-corrosive nature with metals make sodium a potentially attractive heat transfer medium in other fields, and alloyed with potassium it is used for this purpose in certain types of nuclear power generators, and in thermometers where the liquid range from 400 °C to room temperature is wider than for most liquids.

Large quantities of the pure metal are used in the chemical industry in the preparation of such items as sodium cyanide, tetraethyl lead and sodium peroxide.

Some success has been achieved with sodium filled thin wall copper tube for electrical conductors. These can be used where weight is a vital factor, as the final conductor can be almost one-third the weight of the equivalent pure copper wire.

Sodium is used in arc lamps, giving the well-known intense orange light.

The affinity for oxygen finds use as a de-oxidizer in the purification of some of the refractory metals such as titanium and zirconium. Small quantities added to lead increase the hardness.

Some aluminium–silicon alloys have up to 0.1% sodium added as a grain refiner, giving a considerable improvement in mechanical properties.

Sodium has some photo-electric effects, but not sufficient to make it competitive with selenium.

Symbol	Nominal analysis, supplier, condition and remarks.
SODIUM	Na metal: Blackwells; pure metal
w Na 11	99.95% Na: In ampoules; Light Ltd; high purity metal
w Na 16	99.98% Na: In lump form; Light Ltd; high purity metal

Note. The following abbreviations and units are used in the tables:

DPN	Hardness, diamond pyramid number
UTS	Ultimate tensile strength, N/mm^2
Elon	Elongation, %
Proof	0.1 % proof strength, N/mm^2

$$1\ N/mm^2 = 0.1\ hbar = 0.102\ kgf/mm^2 = 0.06475\ tonf/in^2 = 145.04\ lbf/in^2$$

44. Steel

Steel is an alloy of carbon and iron with between 0.05% and 2.0% carbon content; some additions are made to all types of steel, for example manganese. Alloy steels contain one or more additional elements to give properties not obtainable with plain carbon steels.

The old Open Hearth and Bessemer practices of steel making are now obsolete and the bulk of world steel production is made by the basic oxygen steel method, BOS, or variations on this technique. These processes are extremely rapid and use for their feed liquid iron, steel scrap and are blown with oxygen. The liquid steel made by this process can be further treated in various ways to produce high quality steels, for example by vacuum treatments.

Modern aircraft and tool steels are generally manufactured by the electric arc process, or under vacuum. Some of these start with pig iron and scrap, but generally scrap alloy steel with conventional steel ingot is used as the raw material.

Electric melting is confined to the production of high quality alloy steels with the alloying elements added on completion of the fluxing cycle, either to the melt itself during tapping, or to the ladle after tapping. Electric melted steels are used for quality engineering and aircraft steels, tool steels and most of the stainless varieties of steel.

A refinement of this method is to carry out melting, fluxing and casting under vacuum. Some electric melted steels, double vacuum re-melted, are now produced which have tensile properties after forging with the transverse strength almost equal to the longitudinal strength. The use of these steels is confined at present to high duty aircraft parts and ball races.

Of all the metals, indeed of all materials used by man, steel is by far the most versatile. Few objects or articles exist, useful, essential or ornamental, which have not at one time been made of steel in some of the myriad forms in which it exists, from wire a millionth of an inch thick to castings weighing hundreds of tons, from massive ships to delicate jewellery, from tea spoons to tanks.

44.1 The thermal treatment of steel

Several excellent books have been written with this as their sole subject and the information in this section should be taken as no more than brief notes to clear up some common misunderstandings. It has been written in the form of extended definitions, using the same terms as appear in the text. It should however, be noted that there are considerable variations in the use of these terms, dependent on the industry involved and very often the geographical location.

Probably the most abused term is 'heat treated', which tends to be taken as synonymous with the most common thermal treatment carried out in any specific location. It can mean hardened and tempered in one factory and normalized in a second. This book therefore, does not make reference to any part being 'heat treated', but prefers the term 'thermal treatment' when making reference to the process in general.

The term covers the processes where the steel is heated by any means whatever for the purpose of removing mechanical stresses or causing a metallurgical change to the structure, including the method of cooling.

The whole basis of the thermal treatment of steel relies on the fact that iron – also called ferrite – can dissolve iron carbide – also called cementite – and other metallic carbides while remaining in the solid state.

This requires that the steel is heated to above the 'upper critical' point when there is a change in the lattice structure. Below the 'lower critical' point this structure consists of cubes with iron atoms at each corner and in the centre, being known as 'body-centred cubic'. Also present is the compound of iron and carbon, called cementite, or iron carbide. This is insoluble in the iron or ferrite and appears as a separate phase. Thus, below the 'lower critical' point, steel consists of body-centred cubic iron and iron carbide. Above the 'upper critical' point the structure is still cubic with the iron atoms at each corner but the centre atoms have migrated to the faces of the cubes – this being known as a 'face-centred cubic' structure – and this can dissolve the iron carbide, resulting in a single phase structure. The structure is termed austenite and has the appearance under the microscope of a pure metal. For steels which can be hardened by thermal treatment the 'upper critical' point varies from 900 to 700 °C depending on the carbon content and alloying elements, while the 'lower critical' point is at 700 to 720 °C, varied only by alloying elements.

The thermal treatments used to effect the metallurgical condition all make use of the change in the position of the carbon atoms at these critical points which, incidentally, results in a volume change of about 3%.

It must be emphasized that the steel remains solid throughout all the changes and the solution (dissolving) and precipitation of the iron carbides is the result of the migration of the atoms within the iron molecule.

Thermal treatments are applied to steel for the purpose of increasing the hardness or for reducing the hardness and increasing the ductility. The treatments commonly applied for each of these are:

Increase in hardness

Age, carburize, carbo-nitride, cyanide, flame harden, harden, induction harden, nitride, precipitation treat.

Reduce hardness – increase ductility

Anneal, cyclic anneal, normalize, solution treat, spheroidize, stress relief, temper.

As stated above some of these terms can have more than one definition, but following are the meanings ascribed in the text of this book, listed in alphabetical order.

Age. This applies to a limited number of complex ferrous alloys where the upper critical point has been depressed to below room temperature by the addition of alloying elements. These steels are thus austenitic at ambient temperatures. This austenite can be made to dissolve certain other elements by a 'solution treatment' and the steel hardened by precipitating these intermetallic compounds from solution. The ageing process always requires a previous solution treatment and is often called 'precipitation treatment'. This is one of the more complex forms of thermal treatment.

The term 'age' has been used to describe the long term treatment sometimes given to castings by leaving them in the open air. This is in fact a form of 'stress relieving'.

Anneal. This term is often used to denote any method of thermal treatment involving softening but should be applied only when recrystallization has occurred and the material is fully softened. This can only happen when the steel has been taken above the 'upper critical' point by about 30–50 °C for long enough to dissolve all carbides; then cooled very slowly.

The classical anneal requires that the parts remain in the furnace during cooling, resulting in complete equilibrium and the softest possible condition. This is seldom carried out as it is not an economical treatment, the cooling generally being in an insulated pot, in a box packed in sand or in some other refractory material.

The ferritic structure cannot dissolve carbides, whereas these are taken into solution in austenite and at the same time there is a variation in volume in the change from ferrite to austenite. This 'recrystallization', as it is called, completely eliminates all trace of previous thermal treatment, giving a fresh start. The slow cool from the critical temperature maintains equilibrium conditions at all times, allowing the iron carbide to return to its natural position, resulting in a completely stress-free structure.

For many purposes the subcritical anneal described later results in a satisfactory softness, but it must be noted does not cause recrystallization of the grains which is generally necessary if further heavy cold work is to be carried out. The rate of cooling is less critical for low carbon low alloy steels but must be very slow for higher carbon and alloyed steels.

Carburize. This process makes use of the fact that steel held at a temperature in an atmosphere containing carbon will absorb this at the surface until 0.8% carbon is attained. The carbon also diffuses in from the surface giving in effect a higher carbon steel around a lower carbon core

As the percentage of carbon virtually decides the hardness and ductility of steel it is possible using this process to combine high surface hardness – and low ductility, with low core hardness – and high ductility. The process uses two basic methods.

First 'pack carburizing' where the parts are surrounded by activated charcoal inside closed boxes and heated to about 920–950 °C. This requires little special equipment but it is slow and dirty.

The second method is 'gas carburizing' where a carbon-rich atmosphere is used inside special furnaces. The temperature used is in the region of 920–950 °C, the time required being much shorter than with the pack method, as there is no box and large mass of charcoal to be heated. Special furnaces and equipment are essential, but this process is more economical, rapid and cleaner than 'pack carburizing'. For thin or shallow case depths carbo-nitride is commonly used.

Carbo-nitride. This is an alternative method of surface hardening to 'carburize' or to cyanide hardening. It makes use of a blended gas which is basically very similar to gas carburizing, with ammonia being added to give the necessary active nitrogen.

Carbo-nitriding can be carried out over a range of temperatures which coincide with the hardening temperatures of low alloy, low carbon steels. This can therefore be used for direct hardening without the danger of grain growth which occurs with carburizing.

Carbo-nitriding is generally found to be more economical than carburizing for shallow case depths and is probably the most economical of the diffusion-type surface hardening techniques.

Cyanide. This is an alternative method of surface hardening to 'carburize' and makes use of molten sodium cyanide salt as the heating medium and source of carbon and nitrogen. Cyanides are compounds of carbon and nitrogen and depending on the temperatures used, can carburize or nitride. With temperatures in the region of 880–920 °C a carburized case results, whereas temperatures of 500–550 °C give a case which is mostly nitride. Intermediate temperatures result in a true cyanide case partly carburized, partly nitrided.

The cyanide salt bath method is probably the cheapest means of surface hardening, but suffers from the disadvantage that all soluble cyanides are extremely poisonous. Apart from the obvious health hazard for operators there can be considerable trouble and expense with effluent disposal, and this generally makes the 'carbo-nitride' process more economical.

Cyclic anneal. This is a combination of full anneal and subcritical anneal, in that the parts are heated to above the 'upper critical' temperature to dissolve the carbides, then cooled to just above the 'lower critical' temperature, when the carbides are precipitated in the form of spheres. With many of the highly alloyed steels more than one cycle is required to ensure that all carbides are first dissolved, then precipitated as small spheres. This process is used because the resultant structure is the best for machining high carbon steels and is sometimes called spheroidize or spheroidal anneal.

Flame hardening. This makes use of the high temperature of the oxy-acetylene flame. This, when played on the surface of steel, will rapidly raise the surface to above the critical temperature and thus can be used as a means of locally hardening the surface. In all respects, this technique is identical to induction hardening.

It suffers from the disadvantage in the difficulty of controlling accurately the temperature of the oxy-acetylene flame.

Harden. This requires that parts are taken to just above the 'upper critical' point for a sufficient length of time to dissolve all the carbides. The changes in structure at the 'critical points' take the carbon into solution and result in a volume change with the atoms migrating from the centre of each iron molecule to the faces of the cube.

In the hardening process the steel is 'quenched' using brine, water, oil or air as the quenching medium dependent, on the 'hardenability' of the steel. As the steel cools the structure attempts to revert to that of ferrite, but the iron carbide is held in a form of supersaturated solution. This results in an unstable, strained structure called 'martensite' which is the hardest structure obtainable with steel. The hardness is directly proportional to the carbon content, while alloying elements decide the rate of cooling required to give full hardness in addition to having a considerable effect on tempering. Full hardening can only be accomplished when all the carbides have been taken into solution and the cooling rate is sufficient to give the fully martensitic structure. Too rapid a quench gives 100% martensite but results in distortion and cracking. As a general rule martensite is too brittle for practical purposes and must be 'tempered'.

There are means of obtaining the desired tempered structure other than hardening and tempering such as 'martempering', 'austempering' or 'delayed quench hardening', all of which make use of the same principle as outlined above, but require very critical control of times and temperatures and are related to the mass of the part being treated. Technical advice should always be obtained before attempting any of these treatments.

Induction hardening. This makes use of the induction heating technique to raise rapidly the surface of steel to above the critical temperature necessary for hardening. This heating is so rapid that the surface will be at the required temperature while the core or centre remains at ambient. On removing the source of heat, quenching can take place and this will be extremely efficient as the heat will dissipate towards the core as well as from the outside.

Induction hardening is very similar to flame hardening and is generally confined to medium carbon steels such as 0.4% carbon. With alloy steels of high hardenability, there can be problems in the demarcation between case and core being too rapid for good results.

The advantage of induction hardening over flame hardening is that the temperature and the area to be heated can be controlled accurately, with reproducibility of results.

This is the most economical method of producing surface hardened components.

Nitride. The compound of iron and nitrogen, iron nitride, is very hard and retains its hardness to relatively high temperatures, although its formation takes place about 500

°C, with an actual increase in volume.

The process requires special furnaces and equipment using ammonia, which is a compound of hydrogen and nitrogen, as the source of nitrogen. It is a slow, process requiring special steels and is the most expensive of the commonly applied surface hardening treatments. It has the considerable advantage however, that no quenching is involved and the process temperature causes no distortion. There is a growth of about 0.01 mm during a 60-hour nitriding cycle which gives a case depth of about 0.3 mm.

Normalize. This is similar in many respects to the anneal in that the temperature used is about 30–50 °C above the 'upper critical' for sufficient time to take all the carbon into solution and allow recrystallization to take place. The parts are then cooled in air, which can result in a degree of hardness proportional to the carbon content and alloying elements, as alloying has the effect of slowing the speed with which the iron carbide can return to equilibrium.

Normalizing results in complete recrystallization, thus eliminating any heterogeneous grain structure from previous hot or cold work. All unequal stresses are removed although the result is seldom completely strain free for there will always be some air hardening except in low carbon mild steels.

Precipitation treat. See age.

Solution treat. With ferrous metallurgy this term applies only to a limited number of materials which are so highly alloyed that they remain austenitic at room temperature, but have intermetallic compounds present which are soluble in the austenitic matrix.

The solution treatment generally requires a temperature in excess of 1000 °C which dissolves the intermetallic compounds, these being kept in solution by quenching when the steel is in the softest condition.

The steel is then hardened by ageing, when some precipitation of the intermetallic compounds takes place, causing strain.

Spheroidize – or Spheroidal anneal. See cyclic anneal.

Stress relief. This is another term which can mean anything from a full anneal to a 200 °C short term de-embrittlement treatment after electroplating, depending on the circumstances.

Stress relieving treatment should not alter the metallurgical condition of the material and ideally should be at as high a temperature as possible without causing any metallurgical change. Because of the range involved the actual temperature and time required should always be specified, or the reason for stress relieving clearly stated.

The degree of stress relieving achieved is directly proportional to the temperature used and can often result in distortion; thus it should be carried out before machining if fine limits are involved. This distortion is caused by the locked-up stresses becoming greater than the yield strength of the steel, as this drops with rise in temperature. It can sometimes be minimized by reducing the rate of heating.

Subcritical anneal. For this process the steel is heated to just below the 'lower critical' point. At this temperature none of the carbides are taken into solution and no change from ferrite to austenite takes place. Most cold working stresses are released at the same time, but as the steel has not been recrystallized it is not in a suitable condition to accept further heavy cold work.

This subcritical anneal is very often referred to as 'anneal', 'soften', 'stress relief' or 'fully temper'. It is used after normalizing air hardening steels and for softening all steels after carburizing or cold working when they have to be machined prior to further thermal treatment.

Temper. After hardening, steel should be 100% martensite and depending on the carbon content will be hard and brittle. By tempering at 120 °C some increase in ductility with no reduction in hardness is obtained. No hardened steel should be used without being tempered at 120 °C minimum.

As the tempering temperature is increased from this figure so the hardness drops and the ductility increases up to the 'lower critical' point when maximum ductility with almost full softening is achieved. This is caused by the iron carbide, which was trapped in solution in iron and thus held in an unnatural state, migrating to its natural position as a separate phase as the increasing temperature increases the mobility of the iron atoms and releases the iron carbide.

The steel changes in structure from martensite to a ferrite which is soft and ductile at all times. Microscopic examination clearly shows the presence of these constituents and tells the metallurgist whether or not the steel has been correctly treated.

Only steels which are transformed to 100% martensite, then tempered back to the desired structure, give the full mechanical properties. 'Slack' quenching followed by a low temperature temper may give the same hardness figures as a correctly quenched and tempered steel, but not the same proof strength and ductility. The same comments apply to steel which has not been taken above the 'upper critical' temperature prior to quenching.

44.2 Notes on specifications and trade names

As would be expected, there have been a large number of attempts to evolve a standard method of steel classification. These fall into three categories, (a) by analysis, (b) by mechanical properties, (c) by use. In general, the methods based on properties and use are related to specific industries and as such are difficult to describe, but there are many based on analysis. The following short notes may be of assistance in identifying more rapidly specifications which fall into this category.

British Standards Institution (BS)

This body has produced a large number of different specifications, ranging from the single analysis special purpose, to complex methods of classification involving analysis and use. Many of the older single purpose specifications have now been replaced by these classified specifications.

For engineering purposes the designation used was En followed by a number which made no attempt to be descriptive, although in general the lower numbers denoted low carbon mild steels. The number may be followed by a letter indicating a slight difference in analysis. Each En designation detailed analysis range, mechanical properties, condition and other necessary information. BS 971 has been issued as a commentary on the steels in BS 970 and is recommended to those who require further details on why specific steels are used. The En series have now been replaced with a code and BS 970 has been re-issued. Readers are advised to consult British Standard PD 6431 'New designation system for alloy steel' for further information.

In this book the En number is shown with its replacing steel and the BS 970 new reference is given in full with the steel it replaces. The code is also shown by itself as it is assumed that in time engineers will accept the code in the same manner as they use En at present.

There is also a three number designation system used by BS for steels employed in the chemical industry. These are classified according to the form within one specification and include mechanical properties and condition.

For example:

BS 1501 Steel plate section and bar for the chemical, petroleum and allied industries
BS 1503 Steel forgings for these industries
BS 1504 Steel castings for these industries
BS 1506 Steel bolting materials for these industries

Within these specifications covering a form, appears the analysis designated by the three digits. A few examples of these are:

620   0.15% carbon 0.6% chromium 0.6% molybdenum steel
621   0.15% carbon 1.0% chromium 0.5% molybdenum steel
845   18% chromium 10% nickel 2.5% molybdenum steel

Full details of this number system are given in the appendix to BS 1501.

These designations are used within other specifications covering materials for other industries or different material forms.

The examples listed are not intended to be a complete list, the reader is advised to contact the British Standards Institution for further details if necessary

The British Standards Institution have also issued designations headed CLA which are attempts to classify the different materials according to analysis for specific purposes. These are not always material specifications as such, but have all been included in this book for reference purposes.

There is a series of aircraft steel specifications also issued, which should correctly be called British Standards S... but are very commonly referred to as BS S... and have therefore been included in this book under both titles – BS S... and S..., the latter always being referred to the former in the interests of clarity.

American Iron and Steel Institute (AISI) Society of Automotive Engineers (SAE)

These bodies have devised a number code which classifies and describes the steel alloys. In general the same code is used by both bodies sometimes with minor variations, for example, in the prefix letter indicating the method of manufacture used by AISI but not SAE.

The code is based on four digits for plain carbon and low alloy steels, the first two describing the steel, the second two indicating the percentage of carbon present.

Briefly the code is:

10xx	Plain carbon steel – thus 1050 would be a 0.5% carbon steel. B 1025 would be a 0.2% plain carbon steel made by the acid Bessemer process.
11xx	Free machining plain carbon steel – sulphur additions
12xx	Free machining plain carbon steel – sulphur and phosphorus additions
12Lxx	Leaded steels with 0.3% sulphur – free cutting
13xx	Manganese 1.75%
40xx	Molybdenum 0.20 or 0.25%
41xx	Chromium 0.50, 0.80 or 0.95%; molybdenum 0.12, 0.20 or 0.30%
43xx	Nickel 1.83%, chromium 0.50 or 0.80%; molybdenum 0.25%
44xx	Molybdenum 0.53%
46xx	Nickel 0.85 or 1.83%, molybdenum 0.20 or 0.25%
47xx	Nickel 1.05%, chromium 0.45%, molybdenum 0.20 or 0.35%
48xx	Nickel 3.50%, molybdenum 0.25%
50xx	Chromium 0.40%
51xx	Chromium 0.80, 0.88, 0.93, 0.95 or 1.00%
52xx	Carbon 1.04%, chromium 1.03 or 1.45%
61xx	Chromium 0.60 or 0.95%, vanadium 0.13 or 0.15% (min)
86xx	Nickel 0.55%, chromium 0.50%, molybdenum 0.20%
87xx	Nickel 0.55%, chromium 0.50%, molybdenum 0.25%
88xx	Nickel 0.55%, chromium 0.50%, molybdenum 0.35%
92xx	Silicon 2.00%
50Bxx	Chromium 0.28 or 0.50% with boron
51Bxx	Chromium 0.80%
81Bxx	Nickel 0.30%, chromium 0.45%, molybdenum 0.12% with boron
94Bxx	Nickel 0.45%, chromium 0.40%, molybdenum 0.12% with boron

With the stainless varieties of steel the AISI use a three digit code which is also used by SAE but in their case preceded by the figures 50.

The three digit code is:

2xx	Chromium nickel manganese – non-hardenable, austenitic
3xx	Chromium nickel – non-hardenable, austenitic
4xx	Chromium – hardenable and non-hardenable martensitic and ferritic, depending on carbon content
5xx	Chromium – low chromium heat-resisting. This last group are not stainless and appear with the low chromium steels in this book.

There is also an AISI code for tool steels, which has a letter prefix followed by two digits. The letter identifies the type of steel, examples being:

M	Molybdenum steel
T	Tungsten steel
H	Hot work steel
D	Cold work steel – high carbon and chrome
O	Cold work steel – oil hardening

These codes are very often included in the trade name of American proprietary tool steel.

In this book the steels are indexed and classified under the numerical code, also with the AISI and SAE prefix, as it has been found that the steels are equally well known by all these forms. These codes are becoming accepted in the UK. Many UK tool steel manufacturers now use the tool steel classification system.

American Society for Testing and Materials (ASTM)

This body issues individual specifications which always have ASTM as the prefix, followed by the letter A, denoting a ferrous material, then a number identifying the actual specification, which may be followed by letters or numbers subdividing the material by analysis. The AISI code is sometimes used for this purpose. Finally the year of origin is given. A letter T after this denotes a tentative specification. In general each specification covers a steel in a specific form or for a special purpose rather than by analysis.

For clarity, the full specification only appears in the index, the classified lists omitting the year of origin and the letter T.

German Standards (DIN)

There are two systems used at present by the German Specifications. Both systems have a general specification, usually by form carrying a DIN number which is then subdivided. The original method was descriptive, using the chemical symbols and numbers in an attempt to describe the alloy concerned. The latest method consists of numbers only and was devised to suit modern machinery, particularly computors. There is one significant figure which classifies the metal and after the decimal point there are four figures, the first two of which are used to identify the alloy, the last two the quantity. These are called Werkstoff numbers. Most steels are covered by the significant figure 1, but some have no significant figure before the decimal point. Examples are:

DIN 17221-67SiCr5 0.65% carbon, 1.3% silicon,
0.5% chromium spring steel
DIN 17221-7103 0.65% carbon, 1.3% silicon,
0.5% chromium spring steel
DIN 17155-15Mo3 0.18% carbon,
0.3% molybdenum steel for boiler plate
DIN 17155-1.5415 0.18% carbon,
0.3% molybdenum steel for boiler plate

In this book the full DIN specification is given once, using the descriptive form, then the descriptive designation and the Werkstorff number are indexed and classified separately. Examples of the entries would be:

DIN 17221-67SiCr5 0.65% carbon, 1.3% silicon,
0.5% chromium spring steel
67 Si Cr 5 0.65% carbon, 1.3% silicon,
0.5% chromium steel –
designation used by German Standards
0.7103 0.65% carbon, 1.3% silicon,
0.5% chromium steel –
German standard

Thus three separate entries appear for each steel.

Classification system used in this book

All steels appear in Section 44 divided into groups by a system based on analysis, with each basic material given a letter of the alphabet. It is obvious that all carbon contents cannot legitimately be included in one group, and this has been sub-divided and identified by a number. Thus plain carbon steels are divided into four sections 44A1, 44A2, 44A3 and 44A4, according to the carbon content. A slight overlap has been allowed to cope with variations in tolerance, which might otherwise tend to separate materials which should be grouped together.

The decision to include steels with a single alloying element as separate groups was taken because it is believed that in general an alloying element alone has a more marked effect on a steel than when a number of alloying elements are present. It also helps to indicate clearly what each alloy element does to steel.

The problem of residual elements has not been resolved, principally because there is often no clear indication given by the supplier whether or not certain elements are desirable, or will merely be accepted up to a certain limit. Every effort has been made to be consistent, but examples will no doubt be found of low alloy steels wrongly grouped.

44A1 Steel – plain carbon 0.05–0.2 per cent

Specific gravity	7.86	
Density	7860 kg/m^3	(0.28 lb/in.^{3})
Solidus/liquidus	1500 – 1520 °C	
Thermal conductivity	63 W/m °C	(15 cal/m s °C)
Coefficient of linear expansion	11 x 10^{-6}/ °C	
Electrical conductivity	8.4% IACS (copper 100%)	
Specific resistance	20 microhm mm	
Young's modulus of elasticity	92.4 N/m^2	(13.4 x 10^6 lbf/in.^{2})
Impact	55 J	(40 ft/lb) (Izod)
Fatigue strength	-	
Hot strength		

Temperature °C	Tensile strength N/mm^2	Elongation %
100	400	34
200	450	27
300	480	23
400	370	37
500	300	38
600	190	48
700	100	56
800	60	65
900	35	75

The above properties are typical of the following group and may not apply exactly to any one specification. It is possible that with certain specifications some of the values may not be applicable.

General metallurgical characteristics

These steels have too little carbon for them to give any great increase in hardness by thermal treatment. Depending on the method of manufacture and care with which undesirable impurities have been removed, they will exhibit a high degree of ductility with little work hardening at the lower carbon range. The cheaper steels, with varying carbon and higher phosphorus, sulphur and oxygen content, can become relatively brittle and unsuitable for cold working.

The steels are seldom used in the hardened and tempered condition but are commonly normalized. They can however, be carburized to give a high carbon surface which is then hardenable. Without alloying elements this case hardened layer can be rather brittle and there may be a tendency for it to separate from the soft ductile core and to fail by spalling or exfoliation. This can be caused by bending movement, or by heat stresses during hardening or subsequent grinding if the carburizing is not correctly controlled.

Some of the materials listed have elements added to improve machinability. The most common are sulphur, lead and tellurium, all of which act as chip breakers and are reputed to supply some lubricating properties.

This group includes the steels which are the first choice for weldability because of the low hardenability.

Thermal treatment

Many specifications in the following list are used for deep drawing sheet. These are low carbon low impurity steels and only require annealing after considerable deformation.

Other steels with higher carbon and impurity values cannot accept as much cold work and require an anneal after comparatively light deformation to prevent cracking.

The annealing temperature is in the 900–950 °C range for 20 minutes per 25 mm of section, followed by slow cooling. This will result in considerable scale unless carried out under a controlled atmosphere which should be of the reducing rather than neutral type, for example, burnt town gas or cracked ammonia. Where it is not essential to remove all the cold work, subcritical annealing at 650–700 °C is possible. This results in considerably less scale formation and is a satisfactory anneal for a large number of applications.

Normalizing of these materials may be required after operations such as forging and is recommended if they are to be subsequently carburized. This requires 900–950 °C for 20 minutes per 25 mm of section, followed by still air cooling.

Hardening and tempering of these steels is not normally practised as the increase in mechanical properties does not warrant the expense. Where necessary the material must be heated to 870–920 °C and water quenched, followed as soon as possible by tempering at 300–500 °C.

Carburizing can be by the gas, pack or salt bath process, when a temperature of 920–950 °C is generally used. Gas carburizing requires special furnaces, whereas pack carburizing can use normal heat treatment furnaces, but is very much slower. Salt bath carburizing using molten cyanide salts or carbo-nitride are the most common methods of case hardening these steels. With each method the result is similar in that carbon is absorbed into the surface giving a steel of approximately 0.8% carbon. The depth of this is dependent on time, the method used and to some extent the steel being treated. With cyanide or carbo-nitride some nitriding can take place when a lower temperature of 750–800 °C is used, giving a more brittle case.

It is not recommended that case depths less than 0.25 mm or more than 1 mm are used on these particular materials.

After carburizing the parts should be water quenched from 760–780 °C. To prevent excessive grain growth the parts should be cooled after carburizing, then re-heated, but parts are commonly removed from the carburizing atmosphere and either quenched direct or allowed to cool to 760 °C, then quenched. Neither method is recommended for best results. Tempering at 120–150 °C is recommended and is essential if the parts are to be ground.

The temperatures quoted above are for guidance only. Individual specifications or the steel manufacturer should be consulted for the exact times and temperatures necessary for optimum results.

Scale removal after thermal treatment can be grit blasting or acid pickling. This latter is not recommended on case hardened parts owing to the damage of embrittlement.

Welding and brazing

generally required. The weld area No thermal treatment generally required. The weld area should be free from scale and dirt, and correct edge preparation is essential for good results. With some high carbon steels and thicker sections some pre-treatment will be necessary.

Welding and brazing. The low carbon low impurity steels can be readily welded and brazed using any of the conventional electric or gas methods with fluxes. Inert gas shielded arc welding is not normally justified, but there are various processes developed to give very high welding speeds.

All forms of resistance welding can be accomplished with the correct equipment, provided care is taken to remove all traces of grease and oxide from the contact faces.

The best joins will only be obtained with the low carbon, low impurity steels in this group, and as the carbon and impurity level is raised the difficulty of welding increases. With the lower grades and free machining steels listed, welding should only be carried out when no stresses will be present on the finished assembly. The quality of the finished weld will depend on the quality of the steel, the welding process and the skill of the welder.

Brazing using fluxes and copper alloy filler rods can be carried out without trouble.

Post-treatment. Stress relieving at 600 °C after welding may be necessary on complicated parts and is recommended where there is any doubt regarding the quality of the steel or the stresses to be encountered on finished parts. Flux, oxide or scale should be chipped or scraped as soon after the welding as possible and definitely before stress relieving.

Flaw detection methods

Crack test. Magnetic crack test is possible but will seldom be rquired, except on case hardened parts after grinding.

Penetrant oil chalk testing is generally more convenient

and with proper care can be almost as sensitive as magnetic crack testing.

X-ray. This will never be required on normal wrought products, but is a very common requirement after welding. There are radioactive devices developed specifically for this purpose, many of which are portable.

Ultra-sonic test. Not normally required on these steels, but can be a method of weld inspection.

Chemical etch. This may be required to show directional grain flow if severe bending is necessary on a part subject to stress. This requires laboratory etching on sample pieces. A percentage electrolytic sulphuric acid etch is recommended on parts which are case hardened and ground. This will show grinding abuse, and can be used to control grinding quality and reduce the incidence of grinding cracks. This technique will only be used on high integrity parts.

Corrosion protection

Temporary. All the materials listed are very prone to corrosion. This is caused by formation of loosely adhering iron oxides and is accelerated by dampness and acid conditions. Parts should therefore be kept dry and if surface finish is at all important the material should have a film of oil or grease at all times. Soluble oil in good condition prevents corrosion, but when contaminated can be a serious source of corrosion on machined parts.

Permanent. This is one of the most difficult groups of materials to protect permanently from normal atmospheric corrosion. Vast sums of money have been spent on investigation and research into the mechanics of corrosion and methods of prevention and protection, but it remains with the operator to prepare the surface carefully, ensuring that all scale dirt and grease are absent with no trace of moisture evident.

Painting. All traces of corrosion, moisture and dirt must be removed before attempting to apply paint. The most efficient removal method is by grit blasting, but this process must be carefully controlled and the surface brushed clean of all dust, carefully examined for evidence of moisture or corrosion and the primer applied immediately.

It must be emphasized that given the correct conditions of humidity a freshly blasted surface can rust within minutes. The primer can be of the metal base type, zinc, chromate or silicate, and for best results should have the top coats applied as soon as possible after the primer and stoved.

There are various forms of pre-paint phosphate coatings which require dipping or spraying, sometimes with subsequent stoving. These are anti-corrosive coatings in themselves and give a good key for paint adhesion, particularly after grit blasting.

Modern paints based on thermosetting plastics are giving excellent results regarding their ability to withstand hard wear and tear. These are stoving paints which generally require a stoving primer.

In brief there is no easy method of permanently protecting these low carbon steels by painting. Best results rely on blast cleaning, chemical etching, oven drying, priming with metallic oxide or chromate paint, spraying or dipping top coats and stoving. The standard achieved will be related to cost and the standard of quality assurance involved.

Plating. These materials can be electroplated for corrosion protection. Best decorative results appear to be obtained with copper under nickel, the corrosion protection being proportional to the thickness of the plating, the ideal being of the order of 0.05 mm copper, 0.025 mm nickel. Chromium has little effect on the corrosion resistance but does prevent the nickel from dulling. Great care must be taken during preparation and plating, when good results can be obtained.

Zinc, cadmium and tin are commonly applied with good results when appearance is not of prime importance, if sufficient thickness is plated and the correct cleaning and etching techniques used. Zinc is also dip coated to give the well-known galvanized steel. Zinc and cadmium must not be used in contact with foodstuffs. Tin plate is sometimes dip coated but is now more generally electroplated. A chemical seal after plating prevents propagation of microporosity sometimes present in cadmium and zinc.

Metal sprayed coatings are now used, the most popular being zinc and aluminium. For best results the steel must first be dry blasted and the coating should be sealed by wire brushing and chemical treatment immediately after spraying. Final sealing with paint is necessary for use in industrial atmospheres.

Cathodic protection of the materials listed is now used with considerable success in ships, large steel structures and underground pipelines. The part being protected is made the cathode in an electric circuit, the sacrificial, replaceable anode then corrodes, protecting the cathode in the process.

An artificial oxide, produced by immersing the parts in a strong hot sodium hydroxide solution, with oxidizing agents present, prevents further oxidation taking place. This is normally termed 'black oxide' and is a reasonably good corrosion protection provided an oil film is maintained but will not withstand continuous atmospheric attack.

The above remarks must be of a general nature and it is strongly recommended that expert advice is sought on specific problems.

Machinability

It is difficult to obtain a good surface finish with the majority of the steels listed. Those with additives to aid machinability are of course exceptions. The poor finish can be minimized by machining in the hardened condition but this is seldom economical. High surface speeds with constant tool pressures using high speed tools give best results.

Uses

The uses of these materials include 'tin' cans, motor bodies, ship and boiler plate, pipe lines, reinforcer bar steel for structures, corrugated sheeting, fencing, and the host of other uses where a low strength, low unit cost material is required.

Symbol	Nominal analysis, supplier, condition and remarks.
0 STEEL	0.1% C (max) 0.15% Si 0.15% Mn steel: Castings; Edgar Allen; Code No 602 Steel
0.0040	0.12% C 0.3% Mn 0.06% S 0.08% P steel: German Standard
0.0041	0.12% C (max) 0.1% Si 0.3% Mn 0.06% S 0.07% P steel: German Standard
0.0042	0.1% C (max) 0.1% Si 0.3% Mn 0.04% S & P steel: German Standard
0.0043	0.1% C (max) 0.1% Si 0.3% Mn 0.03% S & P steel: German Standard
0.0044	0.1% C (max) 0.08% Si 0.4% Mn 0.05% S & P steel: German Standard
0.0105	0.1% C (max) 0.3% Mn 0.06% S 0.07% P steel: German Standard
0.0115	0.12% C (max) 0.3% Mn 0.06% S 0.08% P steel: German Standard
0.0133	0.12% C (max) 0.3% Mn 0.05% S & P steel: German Standard
0.0402	0.1% C (max) 0.15% Si 0.3% Mn 0.05% S & P steel: German Standard
0.0403	0.1% C (max) 0.15% Si 0.35% Mn 0.05% S & P steel: German Standard
0.0404	0.1% C (max) 0.15% Si 0.35% Mn 0.05% S & P steel: German Standard
0.0415	0.12% C (max) 0.35% Mn 0.06% S 0.08% P steel: German Standard
0.0433	0.12% C (max) 0.15% Si 0.35% Mn 0.04% S & P steel: German Standard
0.0510	0.1% C (max) 0.1% Si 0.35% Mn 0.04% S & P steel: German Standard
0.0511	0.1% C (max) 0.1% Si 0.4% Mn 0.03% S & P steel: German Standard
0.0515	0.1% C (max) 0.1% Si 0.3% Mn 0.03% S & P steel: German Standard
0.0516	0.1% C (max) 0.08% Si 0.3% Mn 0.04% S & P steel: German Standard
0.0517	0.1% C (max) 0.3% Mn 0.04% S & P steel: German Standard
0.0518	0.1% C (max) 0.1% Si 0.3% Mn 0.03% S & P steel: German Standard
0.0524	0.1% C (max) 0.08% Si 0.3% Mn 0.03% S & P steel: German Standard
0.0525	0.1% C (max) 0.08% Si 0.3% Mn 0.04% S & P steel: German Standard
0.0572	0.16% C 0.2% Si 0.3% Mn 0.04% S & P steel: German Standard
0.0612	0.21% C 0.2% Si 0.45% Mn 0.04% S & P steel: German Standard
1.0022	0.13% C 0.3% Mn 0.06% S & P steel: German Standard
1.0061	0.17% C (max) 0.3% Si 0.4% Mn 0.05% S & P steel: German Standard
1.0100	0.17% C (max) 0.05% S 0.08% P steel: German Standard
1.0102	0.17% C (max) 0.05% S & P steel: German Standard
1.0103	0.17% C 0.05% S & P steel: German Standard
1.0104	0.17% C (max) 0.05% S & P steel: German Standard

Symbol	Nominal analysis, supplier, condition and remarks.
1.0106	0.17% C (max) 0.05% S & P steel: German Standard
1.0110	0.2% C (max) 0.05% S 0.08% P steel: German Standard
1.0112	0.2% C (max) 0.05% S 0.06% P steel: German Standard
1.0116	0.2% C (max) 0.05% S & P steel: German Standard
1.0122	0.2% C (max) 0.05% S 0.06% P steel: German Standard
1.0123	0.2% C (max) 0.05% S 0.08% P steel: German Standard
1.0301	0.09% C 0.2% Si 0.35% Mn 0.045% S & P steel: German Standard
1.0305	0.17% C (max) 0.3% Si 0.4% Mn 0.05% S & P steel: German Standard
1.0308	0.18% C (max) 0.05% S & P steel: German Standard
1.0309	0.17% C (max) 0.2% Si 0.4% Mn 0.05% S & P steel: German Standard
1.0318	0.1% C (max) 0.3% Mn 0.04% S & P steel: German Standard
1.0322	0.07% C 0.45% Mn 0.03% S & P steel: German Standard
1.0323	0.09% C 0.45% Mn 0.025% S & P steel: German Standard
1.0325	0.1% C 0.03% S & P steel: German Standard
1.0328	0.1% C 0.6% Mn 0.03% S & P steel: German Standard
1.0329	0.1% C 0.05% Si 0.6% Mn 0.025% S & P steel: German Standard
1.0330	0.1% C (max) 0.3% Mn 0.05% S & P steel: German Standard
1.0331	0.1% C (max) 0.1% Si 0.45% Mn 0.05% S & P steel: German Standard
1.0333	0.1% C (max) 0.1% Si 0.35% Mn 0.04% S & P steel: German Standard
1.0334	0.1% C (max) 0.1% Si 0.45% Mn 0.04% S & P steel: German Standard
1.0336	0.1% C (max) 0.08% Si 0.3% Mn 0.03% S & P steel: German Standard
1.0337	0.1% C (max) 0.08% Si 0.45% Mn 0.03% S & P steel: German Standard
1.0338	0.1% C (max) 0.1% Si 0.3% Mn 0.03% S & P steel: German Standard
1.0356	0.16% C (max) 0.2% Si 0.5% Mn 0.045% S & P steel: German Standard
1.0401	0.16% C 0.2% Si 0.4% Mn 0.045% S & P steel: German Standard
1.0402	0.2% C 0.25% Si 0.45% Mn 0.045% S & P steel: German Standard
1.0405	0.22% C (max) 0.2% Si 0.45% Mn 0.05% S & P steel: German Standard
1.0418	0.22% C (max) 0.2% Si 0.4% Mn 0.05% S & P steel: German Standard
1.0437	0.2% C (max) 0.2% Si 0.6% Mn 0.05% S & P steel: German Standard
1.0439	0.13% C 0.05% Si 0.6% Mn 0.025% S & P steel: German Standard
1.0448	0.13% C 0.05% Si 0.6% Mn 0.03% S & P steel: German Standard
1.0456	0.22% C (max) 0.2% Si 0.5% Mn 0.05% S & P steel: German Standard
1.0611	0.21% C 0.2% Si 0.45% Mn 0.045% S & P steel: German Standard
1.0711	0.12% C (max) 0.7% Mn 0.25% S 0.05% P steel: German Standard; free machining
1.0713	0.13% C (max) 1.1% Mn 0.23% S 0.07% P steel: German Standard; free machining

Note. The following abbreviations and units are used in the tables:

DPN	Hardness, diamond pyramid number
UTS	Ultimate tensile strength, N/mm^2
Elon	Elongation, %
Proof	0.1 % proof strength, N/mm^2

1 N/mm^2=0.1 hbar=0.102 kgf/mm^2=0.06475 tonf/in^2=145.04 lbf/in^2

Symbol	Nominal analysis, supplier, condition and remarks.
1.0719	0.13% C (max) 1.1% Mn 0.23% S 0.03% P steel: German Standard; free machining
1.0716	0.12% C (max) 0.7% Mn 0.25% S 0.05% P 0.2% Pb steel: German Standard; free machining
1.0721	0.09% C 0.2% Si 0.7% Mn 0.22% S 0.07% P steel: German Standard; free machining
1.0723	0.16% C 0.3% Si 0.7% Mn 0.22% S 0.07% P steel: German Standard; free machining
1.0724	0.2% C 0.2% Si 0.7% Mn 0.2% S 0.07% P steel: German Standard; free machining
1.0831	0.2% C 0.5% Si 1.5% Mn 0.05% S & P steel: German Standard
1.0832	0.2% C (max) 0.3% Si 1.5% Mn 0.05% S & P steel: German Standard
1.0833	0.2% C 0.05% S & P steel: German Standard
1.0841	0.2% C (max) 0.05% S & P steel: German Standard
1.1121	0.09% C 0.25% Si 0.4% Mn 0.035% S & P steel: German Standard
1.1141	0.16% C 0.2% Si 0.4% Mn 0.03% S & P steel: German Standard
1.1144	0.16% C 0.2% Si 0.4% Mn 0.035% S & P steel: German Standard
1.1151	0.21% C 0.25% Si 0.5% Mn 0.035% S & P steel: German Standard
1.5034	0.1% C 1.0% Mn 0.03% S & P steel: Welding rod; German Standard
1.5035	0.1% C 1.0% Mn 0.03% S & P steel: Welding rod; German Standard
1.5063	0.13% C 1.5% Mn 0.03% S & P steel: welding rod; German Standard
1.5064	0.13% C 1.6% Mn 0.03% S & P steel: Welding rod; German Standard
1.5074	0.16% C 2.0% Mn 0.02% S & P steel: Sheet & strip; German Standard
1.5083	0.21% C 1.7% Mn 0.03% S & P steel: Welding rod; German Standard
1.5086	0.13% C 2.0% Mn 0.03% S & P steel: Welding rod; German Standard
1.5098	0.13% C 3.0% Mn 0.03% S & P steel: Welding rod; German Standard
1.5331	0.17% C 1.4% Mn 0.2% Ti 0.1% Al steel: Welding rod; German Standard
1.5340	0.17% C 1.1% Mn 0.25% Zr 0.1% Al steel: Welding rod; German Standard
1.5417	0.2% C 1.1% Mn steel: Seamless tube; German Standard
1B	0.08% C 0.25% Mn steel: Sandvik for SIS 1160
1DT	0.06% C 0.04% Si 0.15% Mn steel: Sandvik for SIS 1150
1DTR	0.04% C 0.04% Si 0.15% Mn steel: Sandvik
2LS	0.1% C 0.1% Si 0.4% Mn steel: Sandvik for SIS 1265, 1350
2S	0.1% C 0.15% Si 0.5% Mn steel: Sandvik for SIS 1233/34
2SE	0.1% C 0.15% Si 0.4% Mn 0.03% Cu (max) steel: Sandvik
3BS	0.1% C 0.15% Si 0.45% Mn steel: Sandvik for SIS 1300
3BSX	0.1% C 0.15% Si 0.4% Mn steel (S & P lower than 3BS): Sandvik
3L7	0.1% C 0.2% Si 0.7% Mn steel: Sandvik
3LS	0.15% C 0.25% Si 0.45% Mn steel: Sandvik for SIS 1345
4L7	0.2% C 0.2% Si 0.6% Mn steel: Sandvik for SIS 1434/35
4LM	0.2% C 0.2% Si 1.4% Mn steel: Sandvik
4LS	0.2% C 0.2% Si 0.4% Mn steel: Sandvik for SIS 1357, 1410

Symbol	Nominal analysis, supplier, condition and remarks.
9 S 20	0.12% C (max) 0.7% Mn 0.2% S 0.1% P steel: Designation used by German Standard; free machining
9 S 27	0.12% C (max) 0.8% Mn 0.25% S 0.1% P steel: Designation used by German Standard; free machining
9 S Mn 23	0.13% C (max) 1.1% Mn 0.23% S steel: Designation used by German Standard; free machining
9 S Mn 28	0.14% C (max) 1.1% Mn 0.28% S steel: Designation used by German Standard; free machining
9 S Mn 36	0.15% C (max) 1.2% Mn 0.38% S steel: Designation used by German Standard; free machining
9 S Mn Pb 23	0.13% C (max) 1.1% Mn 0.23% S 0.2% Pb steel: Designation used by German Standard; free machining
9 S Mn Pb 28	0.14% C (max) 1.1% Mn 0.28% S 0.2% Pb steel: Designation used by German Standard; free machining
9 S Mn Pb 36	0.15% C (max) 1.2% Mn 0.38% S 0.2% Pb steel: Designation used by German Standard; free machining
9 S Pb 23	0.12% C (max) 0.7% Mn 0.22% S 0.05% P 0.2% Pb steel: German Standard; free machining
10 M1	0.1% C 0.9% Mn steel: Pompey **UTS: 390 Elon: 25% Proof: 210**
10 S 20	0.1% C 0.3% Si 0.7% Mn 0.22% S 0.07% P steel: German Standard; free machining
11 Mn 4	0.11% C 0.2% Si 1.1% Mn steel: Designation used by German Standard
11 Mn 4 Al	0.11% C 0.1% Si 1.0% Mn steel: Designation used by German Standard
12 Mn 6	0.12% C 1.5% Mn steel: Designation used by German Standard
12 Mn 6 Al	0.12% C 0.18% Si 1.6% Mn steel: Designation used by German Standard
12 Mn 8 Al	0.12% C 0.1% Si 2.0% Mn steel: Designation used by German Standard
12.40	0.1% C 2.0% Mn steel: Wire; copper coated; Esab; for use with submerged arc welding; as welded **DPN: 170 UTS: 600 Elon: 24% Proof: 450**
13 Mn 12	0.13% C 0.22% Si 3.0% Mn steel: Designation used by German Standard
14 Mn 4	0.14% C 1.0% Mn 0.05% S & P steel: Designation used by German Standard
15 S 20	0.15% C 0.7% Mn 0.2% S 0.07% P steel: Designation used by German Standard; free machining; carburizing
17 Mn Ti 6	0.17% C 0.25% Si 1.4% Mn 0.22% Ti 0.1% Al steel: Designation used by German Standard
17 Mn Zr 4	0.17% C 0.5% Si 1.1% Mn 0.25% Zr 0.1% Al steel: Designation used by German Standard
21 Mn 6	0.22% C 0.2% Si 1.7% Mn steel: Designation used by German Standard
22 S 20	0.21% C 0.5% Mn 0.2% S steel: Designation used by German Standard; free machining; carburizing
25CB	0.25% C 0.75% Mn 0.6% Si steel: Electrode; Metrode; weld deposit hardenable **DPN: 375**
30 DAK	0.22% C (max) Mn + Cr 1.6% (max) steel: High yield for construction; Redheugh for BS 4360-55C **UTS: 600 Elon: 18% Proof: 430**
45FG	0.18% C (max) 1.1% Mn 0.04% S & P (max) steel: HOAG; Al killed
48.04	0.07% C 1.2% Mn 0.4% Si steel: For electrodes; Esab **UTS: 550 Elon: 27% Proof: 460**

Symbol	Nominal analysis, supplier, condition and remarks.
48.14	0.07% C 0.85% Mn 0.4% Si steel: For electrodes; Esab
	UTS: 520 **Elon: 28%** **Proof: 420**
50FK	0.2% C (max) 1.3% Mn 0.04% S & P (max) steel: HOAG; Al killed
53.05	0.08% C 0.8% Mn 0.4% Si steel: For electrodes; Esab
	UTS: 570 **Elon: 27%** **Proof: 470**
53.35	0.08% C 1.0% Mn 0.5% Si steel: For electrodes; Esab
	UTS: 540 **Elon: 30%** **Proof: 432**
55FK	0.2% C (max) 1.35% Mn 0.04% S & P (max) steel: HOAG; Al killed
170M	0.2% C 0.2% Si 1.2% Mn steel: Pompey
338	0.05% C 0.1% Mn steel: For carburizing; S Osborn
	DPN: 130
1309	0.07% C (max) 0.25% Mn 0.035% P (max) 0.025% S (max) steel: French Standard designation
A 0	0.12% C 0.5% Mn steel: Cogne
A 1	0.15% C (max) 0.5% Mn (max) steel: Cogne
A 1	0.19% C 0.25% Si 0.8% Mn steel: Jessop
A 2	0.21% C 0.8% Mn steel: Cogne
A 7	0.12% C 0.25% Si 0.3% Mn steel: Jessop
A 7	C not specified 0.04% S & P steel: Plate; Armco; for structural use
	UTS: 500 **Elon: 24%** **Proof: 220**
A 15	0.1% C 0.6% Si 1.2% Mn steel: Welding rod; designation used by BS
A 17	0.15% C 0.5% Si 1.1% Mn 0.03% S & P steel: Welding rod; designation used by BS
A 36	0.26% C 0.2% Si 1.0% Mn 0.04% S & P steel: Plate; Armco; for structural use
	UTS: 540 **Elon: 23%** **Proof: 220**
A 50/1	C not specified 0.07% P (max) 0.05% S (max) steel: Plate; French Standard designation
	UTS: 550 **Proof: 290**
A 50/2	C not quoted 0.045% P & S (max) steel: Plate; French Standard designation
A 52	0.2% C 1.5% Mn 0.05% P & S (max) steel: Belgian specification
A 52 HS	0.2% C 1.5% Mn 0.05% P & S (max) steel: Belgian specification
A 113	C not specified 0.04% S & P steel: Plate; Armco; covers three grades A B & C
	UTS: 450 **Elon: 27%** **Proof: 180**
A 131	0.22% C 0.2% Si 1.0% Mn 0.04% S & P steel: Plate; Armco; cold flanging quality
	UTS: 500 **Elon: 24%** **Proof: 210**
A 283	C not specified 0.04% S & P steel: Plate; Armco; in four grades A B C & D
	UTS: 500 **Elon: 26·** **Proof: 210**
A 299	0.3% C 0.2% Si 1.2% Mn 0.035% S & P steel: Plate; Armco; boilers & pressure vessels
	UTS: 600 **Elon: 20%** **Proof: 270**

Symbol	Nominal analysis, supplier, condition and remarks
A 373	0.26% C 0.2% Si 0.8% Mn 0.04% S & P steel: Plate; Armco; for welded structures
	UTS: 500 **Elon: 24%** **Proof: 210**
A 442	0.23% C 0.2% Si 1.0% Mn 0.04% S & P steel: Plate; Armco; boilers & pressure vessels
	UTS: 490 **Elon: 26%** **Proof: 210**
ABS	0.22% C 0.2% Si 1.0% Mn 0.04% S & P steel: Plate; Armco; cold flanging quality
	UTS: 500 **Elon: 24%** **Proof: 210**
ABS/A	0.25% C (max) 0.6% Mn (min) 0.04% Sn P (max) steel: For ships' hulls; American Bureau of Shipping
	UTS: 450 **Elon: 21%** **Proof: 240**
ABS/AH32	0.18% C (max) 1.3% Mn 0.25% Cr (max) 0.4% Ni (max) 0.08% Mo (max) 0.1% V (max) 0.35% Cu (max) 0.05% Nb (max) steel: For ships' hulls; American Bureau of Shipping
	UTS: 530 **Elon: 19%** **Proof: 320**
ABS/AH36	0.18% C (max) 1.3% Mn 0.25% Cr 0.4% Ni (max) 0.08% Mo (max) 0.1% V (max) 0.35% Cu (max) 0.05% Nb (max) steel: For ships' hulls; American Bureau of Shipping
	UTS: 560 **Elon: 20%** **Proof: 360**
ABS/B	0.21% C (max) 0.85% Mn 0.04% S & P (max) steel: For ships' hulls; American Bureau of Shipping
	UTS: 450 **Elon: 21%** **Proof: 240**
ABS/CS	0.16% C (max) 1.2% Mn 0.04% S & P (max) steel: For ships' hulls; America Bureau of Shipping
	UTS: 450 **Elon: 21%** **Proof: 250**
ABS/D	0.23% C (max) 0.9% Mn (max) steel: Plate; American Bureau of Shipping
	UTS: 440 **Elon: 23%** **Proof: 210**
ABS/D	0.21% C (max) 1.05% Mn 0.04% S & P (max) steel: For ships' hulls; American Bureau of Shipping; impact 27 J at -20 °C
	UTS: 450 **Elon: 21%** **Proof: 250**
ABS/D	0.12% C 0.42% Mn steel: Tube; American Bureau of Shipping
ABS/DH32	0.18% C (max) 1.4% Mn 0.25% Cr (max) 0.4% Ni (max) 0.08% Mo (max) 0.1% V (max) 0.35% Cu (max) 0.05% Nb (max) steel: For ships' hulls; American Bureau of Shipping; impact 34 J at -20 °C
	UTS: 530 **Elon: 20%** **Proof: 320**
ABS/DH36	0.18% C (max) 1.3% Mn 0.25% Cr (max) 0.4% Ni (max) 0.08% Mo (max) 0.1% V (max) 0.35% Cu (max) 0.05% Nb (max) steel: For ships' hulls; American Bureau of Shipping; impact 34 J at -20 °C
	UTS: 560 **Elon: 20%** **Proof: 360**
ABS/DS	0.16% C 1.15% Mn 0.04% S & P (max) steel: For ships' hulls; American Bureau of Shipping
	UTS: 450 **Elon: 21%4240**
ABS/E	0.18% C (max) 1.05% Mn 0.04% S & P (max) steel: For ships' hulls; American Bureau of Shipping; impact 27 J at -40 °C
	UTS: 450 **Elon: 21%** **Proof: 240**
ABS/E	0.25% C (max) 0.9% Mn (max) steel: Plate; American Bureau of Shipping
	UTS: 460 **Elon: 21%** **Proof: 225**
ABS/EH32	0.18% C (max) 1.3% Mn 0.25% Cr (max) 0.4% Ni (max) 0.08% Mo (max) 0.1% V (max) 0.35% Cu (max) 0.05% Nb (max) steel: For ships' hulls; American Bureau of Shipping; impact 34 J at -40 °C
	UTS: 550 **Elon: 20%** **Proof: 320**

Note. The following abbreviations and units are used in the tables:

DPN	Hardness, diamond pyramid number
UTS	Ultimate tensile strength, N/mm^2
Elon	Elongation, %
Proof	0.1 % proof strength, N/mm^2

1 N/mm^2=0.1 hbar=0.102 kgf/mm^2=0.06475 tonf/in^2=145.04 lbf/in^2

Symbol	Nominal analysis, supplier, condition and remarks.
ABS/EH36	0.18% C (max) 1.3% Mn 0.25% Cr (max) 0.4% Ni (max) 0.08% Mo (max) 0.1% V (max) 0.35% Cu (max) 0.05% Nb (max) steel: For ships' hulls; American Bureau of Shipping; impact 34 J at -40 °C
	UTS: 560 Elon: 20% Proof: 360
ABS/G	0.12% C 0.42% Mn steel: Tube; American Bureau of Shipping
ABS/G2A	Flash butt welded or drop forged steel chain: American Bureau of Shipping
	UTS: 570 Elon: 22%
ABS/G2B	Cast steel chain: American Bureau of Shipping
	UTS: 500 Elon: 22%
ABS/G3A	Flash butt welded or drop forged steel chain: American Bureau of Shipping; impact 8 J at 0 °C
	UTS: 700 Elon: 17%
ABS/G3B	Cast steel chain: American Bureau of Shipping; impact 8 J at 0 °C
	UTS: 700 Elon: 17%
ABS/GB	0.22% C (max) 0.9% Mn (max) 0.27% Cu steel: Plate; for intermediate temperature service
	UTS: 380 Elon: 25% Proof: 190
ABS/GA	0.17% C (max) 0.9% Mn (max) 0.25% Cu steel: Plate; for intermediate temperature
	UTS: 330 Elon: 27% Proof: 170
ABS/GC	0.28% C (max) 0.9% Mn (max) 0.27% Cu steel: Plate; for intermediate temperature service
	UTS: 430 Elon: 23% Proof: 210
ABS/GI	Flash butt welded steel chain: American Bureau of Shipping
	UTS: 360 Elon: 30%
ABS/H	0.12% C 0.42% Mn steel: Tube; American Bureau of Shipping
ABS/J	0.27% C (max) 0.93% Mn (max) steel: Tube; American Bureau of Shipping
ABS/K	0.18% C (max) 0.7% Mn steel: Plate; American Bureau of Shipping
	UTS: 430 Elon: 23% Proof: 210
ABS/L	0.23% C (max) 0.75% Mn steel: Plate; American Bureau of Shipping
	UTS: 460 Elon: 24% Proof: 225
ABS/M	0.27% C (max) 1.0% Mn steel: Plate; American Bureau of Shipping
	UTS: 500 Elon: 20% Proof: 245
ABS/N	0.29% C (max) 1.0% Mn steel: Plate; American Bureau of Shipping
	UTS: 550 Elon: 18% Proof: 270
AC1 BAR	0.1% C 0.06% Si 1% Mn 0.04% P 0.3% S steel: Workington Iron & Steel; free machining; hot rolled
	UTS: 420 Elon: 40%
AC1 BAR	0.1% C 0.06% Si 1% Mn 0.04% P 0.3% S steel: Workington Iron & Steel; free machining; bright drawn
	UTS: 600 Elon: 16%
AC1 BEL	0.1% C 0.06% Si 1% Mn 0.04% P 0.3% S 0.2% Pb steel: Workington Iron & Steel; free machining; hot rolled
	UTS: 420 Elon: 40%
AC1 BEL	0.1% C 0.06% Si 1% Mn 0.04% P 0.03% S 0.2% Pb steel: Workington Iron & Steel; free machining; bright drawn
	UTS: 600 Elon: 16%
ACIBOND	0.16% C 1.5% Mn 0.25% Cu 0.05% Al 0.016% B steel: Workington
AGP V 9	0.12% C 0.85% Mn 0.25% S steel: Pompey; free machining
AISI 12 L14	0.15% C (max) 1.0% Mn 0.07% P 0.3% S 0.3% Pb steel: Free machining

Symbol	Nominal analysis, supplier, condition and remarks.
AISI 1110	0.1% C 0.4% Mn 0.04% P 0.1% S steel: Free machining
AISI 1116	0.16% C 1.2% Mn 0.04% P 0.2% S steel: Free machining
AISI 1211	0.13% C (max) 0.8% Mn 0.1% P 0.12% S steel: Free machining
AISI 1212	0.13% C (max) 0.8% Mn 0.1% P 0.2% S steel: Free machining
AISI 1213	0.13% C (max) 0.8% Mn 0.1% P 0.3% S steel: Free machining
AISI 1215	0.09% C (max) 0.9% Mn 0.07% P 0.3% S steel: Free machining
AISI A1320	0.2% C 1.7% Mn 0.04% S & P steel: Obsolete
AISI B1109	0.1% C 0.8% Mn 0.1% S 0.04% P steel: Free cutting
AISI B1111	0.13% C (max) 0.8% Mn 0.1% P 0.12% S steel: Acid Bessemer
AISI B1112	0.13% C (max) 0.9% Mn 0.1% P 0.2% S steel: Free machining
AISI B1113	0.13% C 0.9% Mn 0.2% S 0.1% P steel: Free cutting
AISI C12 L14	0.15% C (max) 1% Mn 0.3% S 0.05% P steel: Free cutting
AISI C1006	0.08% C 0.3% Mn 0.04% S & P steel
AISI C1008	0.1% C 0.4% Mn 0.04% S & P steel
AISI C1009	0.15% C 0.6% Mn 0.04% S & P steel
AISI C1010	0.1% C 0.4% Mn 0.04% S & P steel
AISI C1012	0.12% C 0.4% Mn 0.04% S & P steel
AISI C1015	0.15% C 0.4% Mn 0.04% S & P steel
AISI C1016	0.15% C 0.8% Mn 0.04% S & P steel
AISI C1017	0.18% C 0.4% Mn 0.04% S & P steel
AISI C1018	0.18% C 0.8% Mn 0.04% S & P steel
AISI C1019	0.18% C 0.9% Mn 0.04% S & P steel
AISI C1020	0.2% C 0.4% Mn 0.04% S & P steel
AISI C1021	0.2% C 0.8% Mn 0.04% S & P steel
AISI C1022	0.2% C 0.9% Mn 0.04% S & P steel
AISI C1108	0.1% C 0.7% Mn 0.1% S 0.04% P steel: Free cutting
AISI C1115	0.15% C 0.8% Mn 0.1% S 0.04% P steel: Free cutting
AISI C1117	0.17% C 1.2% Mn 0.1% S 0.04% P steel: Free cutting
AISI C1118	0.17% C 1.5% Mn 0.1% S 0.04% P steel: Free cutting
AISI C1119	0.17% C 1.2% Mn 0.3% S 0.04% P steel: Free cutting
AISI C1120	0.2% C 0.9% Mn 0.1% S 0.04% P steel: Free cutting
ALDUR 45/60	0.21% C 1.3% Mn 0.04% S & P (max) steel: Voest
ALDUR 50	0.21% C 1.3% Mn 0.04% S & P (max) steel: Voest
ALDUR 50/65	0.23% C 1.5% Mn 0.04% S & P (max) steel: Voest
ALDUR 55	0.22% C 1.4% Mn 0.04% S & P (max) steel: Voest
ALDUR 55/63	0.23% C 1.7% Mn 0.04% S & P (max) steel: Voest
ALDUR 58	0.23% C 1.5% Mn 0.04% S & P (max) 0.02‰ Cu steel: Voest
ALDUR 58/72	0.23% C 1.7% Mn 0.04% S & P (max) steel: Voest
ALFORT	0.22% C 1.3% Mn 0.04% S & P steel: Austrian specification
ALFREEZE 1	0.17% C (max) 1.5% Mn 0.05% S & P (max) steel: Appleby-Frodingham
ALFREEZE 11	0.17% C (max) 1.5% Mn 0.04% S & P (max) with N additions steel: Appleby-Frodingham
ALFRIG 5UV	0.18% C 1.1% Mn 0.04% S & P steel: Austrian specification

Symbol	Nominal analysis, supplier, condition and remarks.
ALGO-LOY 1315	0.15% C 1.3% Mn 0.4% Cu steel: Algoma
ALUMOWELD	Aluminium covered steel wire: Copperweld steel; used for overhead lines
AMS 5010 D	0.13% C 0.1% P 0.2% S steel: Bar; AMS for SAE 1112
AMS 5022 F	0.17% C 1.2% Mn steel: Bar & forging; free cutting; AMS spec for SAE 1117
AMS 5030 A	0.06% C (max) steel: Wire for welding
AMS 5031	0.12% C steel-coated welding electrode
AMS 5032 A	0.2% C steel: Wire; annealed; AMS for SAE 1020
AMS 5036 D	0.1% C (max) steel: Sheet & strip; Al coated
AMS 5040 E	0.15% C (max) steel: Sheet & strip for deep forming; cold rolled; AMS for SAE 1010
AMS 5041	0.08% C (max) steel: Sheet & strip; for deep drawing; cold rolled; AMS for SAE 1006
AMS 5042 E	0.15% C (max) steel: For deep forming; cold rolled; AMS for SAE 1010
AMS 5044 C	0.15% C (max) steel: ½-hard; AMS for SAE 1010
AMS 5045 B	0.20% C steel: ½-hard; AMS for SAE 1020
AMS 5047	Low C steel Al killed: For forming; AMS for SAE 1010
AMS 5050 E	0.15% C (max) steel: Annealed; AMS for SAE 1010
AMS 5053 B	0.13% C (max) steel: Tube; welded; annealed; AMS for SAE 1010
AMS 5060 B	0.15% C steel: Bar, forgings & tube; AMS for SAE 1015
AMS 5061 A	0.18% C steel: For bars & wire
AMS 5069	0.18% C steel: Bars, forging & tube; AMS for SAE 1018
AN QQ S 646 1	0.2% C 0.4% Mn 0.05% S & P steel: US Service; as rolled **UTS: 300 Elon: 22% Proof: 220**
AN QQ W 435/3	0.1% C 0.4% Mn 0.05% S & P steel: Wire; US Service; zinc coated; annealed **UTS: 370**
AN S 11	0.2% C 0.4% Mn 0.05% S & P steel: Wire; US Service **UTS: 300 Elon: 22% Proof: 220**
ANSCOL 50	0.18% C (max) 0.7% Mn 0.015% Nb steel: Interlake Steel Co.
API 5A H40	C not specified 0.04% P (max) 0.06% (max) S steel: For casing, tube & drill pipe **UTS: 420 Elon: 20% Proof: 280**
API 5A J55	C not specified 0.04% P (max) 0.06% (max) S steel: For casing, tube & drill pipe **UTS: 520 Elon: 20% Proof: 380**
API 5A K55	C not specified 0.04% P (max) 0.06% (max) S steel: For casing, tube & drill pipe **UTS: 650 Elon: 20% Proof: 380**
API 5A N80	C not specified 0.04% P (max) 0.06% (max) S steel: For casing, tube & drill pipe **UTS: 700 Elon: 20% Proof: 550**
API 5L A	0.22% C (max) 0.9% Mn (max) 0.04% P (max) 0.05% S (max) steel: Line pipe; seamless or welded **UTS: 335 Elon: 25% Proof: 245**

Symbol	Nominal analysis, supplier, condition and remarks.
API 5L A25/1	0.21% C (max) 0.45% Mn 0.045% P (max) 0.06% S (max) steel: Line pipe; seamless or welded **UTS: 310 Elon: 25*MB Proof: 175**
API 5L A25/2	0.21% C (max) 0.45% Mn 0.08% P (max) 0.06% S (max) steel: Line pipe; seamles or welded **UTS: 310 Elon: 25% Proof: 175**
API 5L B	0.27% C (max) 1.15% Mn (max) 0.04% P (max) 0.05% S (max) steel: Line pipe; seamless or welded **UTS: 410 Elon: 25% Proof: 245**
API 5L SA	0.21% C (max) 0.9% Mn (max) 0.04% P (max) 0.05% S (max) steel: Spiral weld line pipe **UTS: 330 Elon: 20% Proof: 210**
API 5L SB	0.26% C (max) 1.15% Mn (max) 0.04% P (max) 0.05% S (max) steel: Spiral weld line pipe **UTS: 330 Elon: 20% Proof: 242**
API 5L X70	0.23% C (max) 1.6% Mn (max) steel: High test line pipe; seamless or welded **UTS: 565 Elon: 20% Proof: 483**
AQUACIDOX	0.11% C 0.3% Si (max) 0.4% Mn 0.3% Cu 0.05% S & P steel: Tube; STD Services (see Resistco)
ARCTIC A	0.15% C 1.5% Mn 0.05% P & S steel: South Durham
ARCTIC B	0.22% C 1.6% Mn 0.05% P & S (max) steel: South Durham
ARCTIC C	0.22% C (max) 1.6% Mn 0.04% P & S (max) steel: South Durham
ARCTIC D	0.14% C (max) 1.5% Mn 0.035% P (max) 0.045% S (max) 0.1% Nb steel: South Durham
ARMCO HS No 4R	0.25% C (max) 1.4% Mn 0.2% Cu steel: Armco
ARMCO HS No 4S	0.25% C (max) 1.4% Mn 0.2% Cu steel: Armco
ARMCO HS No 7	0.13% C 0.7% Mn 0.02% Nb steel: Armco
A St 2	0.1% C (max) 0.15% Si 0.3% Mn 0.05% S & P steel: German Standard
A St 3	0.1% C (max) 0.1% Si 0.4% Mn 0.04% S & P steel: German Standard
A St 4	0.1% C (max) 0.08% Si 0.35% Mn 0.035% S & P steel: German Standard
A St 35/8	0.17% C (max) 0.05% S & P steel: German Standard
A St 45/8	0.22% C (max) 0.05% S & P steel: Designation used by German Standard
ASTM A53 E	Mild steel pipe; 0.08% P: Furnace butt welded; galvanized **UTS: 330 Elon: 18% Proof: 200**
ASTM A 53 F	Mild steel pipe 0.05% P; electric resistance welded; galvanized **UTS: 310 Proof: 190**
ASTM A 53 S	Mild steel pipe; seamless; galvanized **UTS: 330 Proof: 200**
ASTM A67	0.08% (min) C 0.11% (max) P 0.20% (min) Cu: When specified steel for tie plates
ASTM A109/4	0.15% C 0.6% Mn 0.2% Cu (optional) cold rolled steel: Strip; annealed & final light rolled **UTS: 330 Elon: 32%**
ASTM A109/5	0.15% C 0.6% Mn 0.2% Cu (optional) cold rolled steel: Strip; annealed **UTS: 300 Elon: 39%**
ASTM A113	C not specified 0.07% P 0.05% S steel: For locomotives **UTS: 370 Elon: 26% Proof: 180**
ASTM A135	C not specified 0.05% P 0.06% S electric resistance welded steel: Pipe
ASTM A161 (low carbon)	0.15% C 0.5% Mn 0.5% S & P steel: Still tubes
ASTM A178 A	0.12% C 0.4% Mn 0.05% P 0.06% S electric resistance welded steel: Boiler tubes **UTS: 420 Elon: 30% Proof: 260**

Note. The following abbreviations and units are used in the tables:

DPN	Hardness, diamond pyramid number
UTS	Ultimate tensile strength, N/mm^2
Elon	Elongation, %
Proof	0.1 % proof strength, N/mm^2

1 N/mm^2=0.1 hbar=0.102 kgf/mm^2=0.06475 tonf/in^2=145.04 lbf/in^2

Symbol	Nominal analysis, supplier, condition and remarks.
ASTM A179	0.12% C 0.35% Mn 0.05% S & P steel: Cold drawn tube for condenser
ASTM A192	0.12% C 0.45% Mn steel: For boiler tubes
ASTM A194/1	0.15% C (min) steel: For bolts
ASTM A214	0.18% C (max) 0.45% Mn steel: Welded tubes
ASTM A226	0.12% C 0.45% Mn 0.2% Si steel: Tube; electric resistance welded
ASTM A233	Mild steel welding electrodes classified by coating
ASTM A235	C not specified 0.9% Mn 0.05% S & P steel: Forgings; normalized & tempered **UTS: 580 Elon: 21% Proof: 270**
ASTM A237	C not specified 0.05% S & P steel: Forgings; normalized & tempered **UTS: 630 Elon: 21% Proof: 420**
ASTM A251	Mild steel welding rods; classified by strength of deposit
ASTM A254	0.1% C 0.45% Mn 0.05% S & P copper brazed steel: Tube
ASTM A263	Stainless steel clad mild steel sheet
ASTM A264	Austenitic stainless clad mild steel sheet
ASTM A265	Nickel and nickel alloy clad mild steel sheet
ASTM A283 A	Low C 0.05% S & P steel: Plate for structures; as rolled **UTS: 340 Elon: 27% Proof: 170**
ASTM A283 B	Low C 0.05% S & P steel: Plate for structures; as rolled **UTS: 370 Elon: 25% Proof: 180**
ASTM A283 C	Low C 0.05% S & P steel: Plate for structures; as rolled **UTS: 420 Elon: 23% Proof: 200**
ASTM A283 D	Low C 0.05% S & P steel: Plate for structures; as rolled **UTS: 450 Elon: 21% Proof: 220**
ASTM A285 A	0.15% C (max) 0.8% Mn 0.05% S & P steel: Plate
ASTM A285 B	0.2% C (max) 0.8% Mn 0.04% S & P steel: Plate
ASTM A285 C	0.25% C (max) 0.8% Mn 0.04% S & P steel: Plate
ASTM A366	0.15% C (max) 0.5% Mn 0.04% S & P steel: Sheet; cold rolled
ASTM A415	0.15% C 0.4% Mn steel: Sheet; may contain Cu; hot rolled
ASTM A424	0.4% C 0.2% Mn 0.015% P 0.04% S steel: Sheet for porcelain enamelling
ASTM A425	0.15% C (max) 0.45% Mn 0.04% S & P steel: Strip; Cu may be present; commercial quality
ASTM A441	0.25% C (max) 1.3% Mn 0.2% Cu 0.02% V steel: For structural purposes
ASTM A442	0.23% C 0.85% Mn 0.04% S & P 0.2% Si steel: Plate for pressure vessels
ASTM A489	C not specified 0.05% S & P 0.13% Si steel: For eyebolts **UTS: 450 Elon: 30% Proof: 210**
ASTM A502/1	0.19% C 0.6% Mn 0.2% Cu (optional) steel: For structural rivets
ASTM A502/2	0.24% C 1.4% Mn 0.2% Cu (optional) steel: For structural rivets
ASTM A512	Cold drawn plain carbon steel: Tubing; butt welded; graded according to AISI system
ASTM A513	Resistance welded steel tubing; plain carbon and low alloy; graded according to AISI system
ASTM A523 A	0.22% C (max) 0.9% Mn 0.04% S & P steel: Pipe for high pressure; seamless or electric resistance welded
ASTM A524	0.21% C (max) 1.0% Mn 0.05% S & P steel: Seamless pipe for process piping; in grades I and II
ASTM A529	0.27% C 1.2% Mn 0.04% S & P 0.2% Cu (optional) structural steel **UTS: 410 Elon: 19% Proof: 300**
ASTM A537	0.24% C (max) 1.0% Mn 0.04% Si steel: Plate

Symbol	Nominal analysis, supplier, condition and remarks.
ASTM A539	0.15% C (max) 0.6% Mn 0.05% S & P steel: Electric resistance welded coiled tube
ASTM A548	Mild steel wire for screws; graded by AISI system
ASTM A549	Mild steel wire for wood screws; graded by AISI system
ASTM A556 A2	0.18% C (max) 0.5% Mn 0.05% S & P steel: Tube; seamless; cold drawn **UTS: 330 Elon: 35% Proof: 180**
ASTM A558 EN14	0.14% C 2.0% Mn 0.05% Si steel: Electrode
ASTM A558 EL8	0.1% C 0.4% Mn 0.05% Si steel: Electrode
ASTM A558 EL8K	0.1% C 0.4% Mn 0.15% Si steel: Electrode
ASTM A558 EL12	0.12% C 0.5% Mn 0.05% Si steel: Electrode
ASTM A558 EM5K	0.06% C 1.2% Mn 0.6% Si steel: Electrode
ASTM A558 EM12	0.11% C 1.0% Mn 0.05% Si steel: Electrode
ASTM A558 EM12K	0.11% C 1.0% Mn 0.2% Si steel: Electrode
ASTM A558 EM13K	0.13% C 1.2% Mn 0.55% Si steel: Electrode
ASTM A558 EM15K	0.15% C 1.0% Mn 0.2% Si steel: Electrode
ASTM A559 E60S1	0.13% C 0.35% Si steel: Electrode
ASTM A559 E60S2	0.06% C 0.55% Si 0.2% Ti 0.1% Zr steel: Electrode
ASTM A559 E60S3	0.12% C 1.2% Mn 0.55% Si steel: Electrode
ASTM A559 E70S4	0.11% C 0.75% Si steel: Electrode
ASTM A559 E70S5	0.13% C 0.45% Si steel: Electrode
ASTM A559 E70S6	0.11% C 1.6% Mn 1.0% Si steel: Electrode
ASTM A559 E70S6	0.11% C 1.6% Mn 1.0% Si steel: Electrode
ASTM A559 E70T	Low carbon welding electrodes; composite; graded by composition
ASTM A562	0.12% C (max) 1.2% Mn 0.3% Si 0.15% Cu (max) 0.45% Ti steel: Plate for glass or metallic coating
ASTM A569	0.15% C (max) 0.4% Mn 0.04% S & P steel: Strip; Cu may be present
ASTM A572	0.21% C (max) steel: For structural use; may contain V or Nb
ASTM A572	0.22% C (max) steel: For structural use; may contain V or Nb
ASTM A572	0.23% C (max) steel: For structural use; may contain V or Nb
ASTM A572	0.25% C (max) steel: For structural use; may contain V or Nb
ASTM A572	0.26% C (max) steel: For structural use; may contain V or Nb
ASTM A572	0.26% C (max) steel: For structural use; may contain V or Nb
ASTM A573	0.24% C (max) 1.0% Mn 0.2% Si steel: Plate of improved toughness
ASTM A575	Mild and medium carbon hot rolled steel bar; graded by AISI system
ASTM A576	Plain carbon hot rolled steel bars; graded by AISI system (see also 44A2 and 44A3)
ASTM A587	0.15% C (max) 0.45% Mn 0.05% S & P (max) 0.02% Al (min) steel: Pipe; electric welded **UTS: 331 Elon: 40% Proof: 210**
ASTM A589	Low C 0.05% P (max) 0.06% S (max) steel: Pipe; seamless for water pipes **UTS: 330 Elon: 35% Proof: 210**
ASTM A589 B	Low C 0.05% P (max) 0.06% S (max) steel: Pipe seamless for water pipes **UTS: 414 Elon: 30% Proof: 241**
ASTM A589 B	Low C 0.05% P (max) 0.06% S (max) steel: Pipe; **UTS: 310 Elon: 30% Proof: 170**
ASTM A591	Mild steel zinc coated sheet; cold rolled
ASTM A594, 594/1	0.15% C (max) 0.5% Mn (max) 0.02% Al (max) steel: For magnetic purposes
ASTM A594/2	0.1% C (max) 0.6% Mn (max) 0.02% Al (max) steel: For magnetic purposes
ASTM A594/3	0.08% C (max) 0.4% Mn (max) 0.02% Al (max) steel: For magnetic purposes
ASTM A594/4	0.06% C (max) 0.4% Mn (max) 0.015% Al (max) steel: For magnetic purposes

Symbol	Nominal analysis, supplier, condition and remarks.
ASTM A595 A	0.2% C 0.5% Mn 0.04% S & P (max) steel: Tapered tube
	UTS: 450 **Proof: 380**
ASTM A595 A	0.2% C 0.4% Mn steel: Tube tapered for structural purposes
	UTS: 450 **Proof: 380**
ASTM A595 B	0.2% C 0.8% Mn steel: Tube; tapered for structural purposes
	UTS: 480 **Proof: 410**
ASTM A595 B	0.2% C 1.0% Mn 0.05% S & P (max) steel: Tapered tube
	UTS: 480 **Proof: 410**
ASTM A599	Mild steel electro tin plated sheet; cold rolled
ASTM A603	Mild steel wire rope; zinc coated
ASTM A606/2	0.2% C (max) 1.3% Mn (max) 0.06% S (max) alloy steel: Sheet with twice corrosion resistance of plain carbon steel; normalized
	UTS: 450 Elon: 22% Proof: 310
ASTM A606/4	0.26% C (max) 1.3% Mn (max) 0.06% S (max) alloy steel: Sheet with four times corrosion resistance of plain carbon steel; normalized
	UTS: 450 Elon: 22% Proof: 310
ASTM A607/45	0.26% C (max) 1.4% Mn (max) 0.004% Cb or 0.005% V steel: Sheet
	UTS: 410 Elon: 23% Proof: 320
ASTM A607/50	0.27% C (max) 1.4% Mn (max) 0.004% Cb or 0.005% V steel: Sheet
	UTS: 450 Elon: 21% Proof: 345
ASTM A607/55	0.29% C (max) 1.4% Mn (max) 0.004% Cb or 0.005% V steel: Sheet
	UTS: 480 Elon: 19% Proof: 380
ASTM A607/60	0.3% C (max) 1.55% Mn (max) 0.004% Cb or 0.005% V steel: Sheet
	UTS: 520 Elon: 17% Proof: 415
ASTM A607/65	0.3% C (max) 1.55% Mn (max) 0.004% Cb or 0.005% V steel: Sheet
	UTS: 550 Elon: 15% Proof: 450
ASTM A607/70	0.3% C (max) 1.7% Mn (max) 0.004% Cb or 0.005% V steel: Sheet
	UTS: 590 Elon: 14% Proof: 485
ASTM A611 A	0.2% C (max) 0.6% Mn (max) steel: Cold rolled sheet
	UTS: 290 Elon: 26% Proof: 170
ASTM A611 B	0.2% C (max) 0.6% Mn (max) steel: Cold rolled sheet
	UTS: 310 Elon: 24% Proof: 205
ASTM A611 C	0.2% C (max) 0.6% Mn (max) steel: Cold rolled sheet
	UTS: 330 Elon: 22% Proof: 230
ASTM A611 D	0.2% C (max) 0.9% Mn (max) steel: Cold rolled sheet
	UTS: 360 Elon: 20% Proof: 275
ASTM A611 D	0.2% C (max) 0.9% Mn (max) 0.04% S & P (max) 0.2% Cu (max) steel: Cold rolled sheet
ASTM A611 E	0.2% C (max) 0.6% Mn (max) steel: Cold rolled sheet
	UTS: 570 **Proof: 550**
ASTM A611 A,B,C & E	0.2% C (max) 0.04% S & P (max) 0.2% B Cu (max) steel: Cold rolled sheet

Note. The following abbreviations and units are used in the tables:

DPN	Hardness, diamond pyramid number
UTS	Ultimate tensile strength, N/mm^2
Elon	Elongation, %
Proof	0.1 % proof strength, N/mm^2

$1 \ N/mm^2 = 0.1 \ hbar = 0.102 \ kgf/mm^2 = 0.06475 \ tonf/in^2 = 145.04 \ lbf/in^2$

Symbol	Nominal analysis, supplier, condition and remarks.
ASTM A615/40	Plain carbon 0.05% P (max) steel: For concrete reinforcement; deformed
	UTS: 483 Elon: 10% Proof: 276
ASTM A615/60	Plain carbon 0.05% P (max) steel: For concrete reinforcement; deformed
	UTS: 621 Elon: 8% Proof: 414
ASTM A619	0.1% C 0.5% Mn (max) 0.025% P (max) 0.035% S (max) steel: Sheet for deep drawing
ASTM A620	0.1% C (max) 0.5% Mn (max) 0.025% P (max) 0.035% S (max) steel: Sheet for deep drawing
ASTM A621	0.1% C (max) 0.5% Mn (max) 0.025% P (max) 0.035% S (max) steel: Sheet for deep drawing
ASTM A622	0.1% C (max) 0.5% Mn (max) 0.025% P (max) 0.035% S (max) steel: Sheet for deep drawing
ASTM A623 D	0.12% C (max) 0.6% Mn (max) steel: Sheet; tin plated
ASTM A623 L	0.12% C (max) 0.6% Mn (max) 0.12% N Cr + Mo (max) steel: Sheet; tin plated
ASTM A623 MC	0.12% C (max) 0.7% Mn (max) steel: Sheet; tin plated
ASTM A623 MR	0.13% C (max) 0.6% Mn (max) steel: Sheet; tin plated
ASTM A633 A	0.18% C (max) 1.2% Mn steel: For structure; normalized
	UTS: 500 Elon: 19% Proof: 290
ASTM A633 B	0.18% C (max) 1.15% Mn 0.1% V (max) steel: For structure; normalized
	UTS: 500 Elon: 20% Proof: 290
ASTM A633 C	0.2% C (max) 1.2% Mn 0.03% Nb steel: For structures; normalized
	UTS: 520 Elon: 20% Proof: 320
ASTM A633 D	0.2% C (max) 1.0% Mn 0.2% Cr (max) 0.25% Ni (max) steel: For structures; normalized
	UTS: 520 Elon: 20% Proof: 320
ASTM A635	0.15% C (max) 0.5% Mn 0.18% Cu (optional) steel: Sheet
ASTM A640	Steel wire zinc coated for support cables
ASTM A641	Steel wire hot clip zinc coated; annealed
	UTS: 500
ASTM A641	Steel wire hot clip zinc coated; medium drawn
	UTS: 580
ASTM A641	Steel wire hot clip zinc coated; hard drawn
	UTS: 650
ASTM A643 A	0.25% C (max) 1.2% Mn steel: Castings for pressure vessels
	UTS: 485 Elon: 22% Proof: 275
ASTM A650	0.1% C (max) 0.5% (max) 0.025% P (max) 0.035% S (max) steel: Sheet; galvanized for deep drawing
ASTM A650	Low carbon steel plate; double reduced
ASTM A656/2	0.15% C (max) 0.9% Mn (max) 0.2% Ti 0.01% Al (min) steel: For structures
	UTS: 720 Elon: 12% Proof: 550
ASTM A657	Steel plate or strip; cold rolled and chromium plated
ASTM A659	0.2% C steel: Strip & sheet; hot rolled; graded by AISI number
ASTM A660 WCA	0.25% C (max) 0.7% Mn steel: Pipe; centrifugally cast; high temperature use
	UTS: 414 Elon: 24% Proof: 207
ASTM A660 WCB	0.3% C (max) 1.0% Mn steel: Pipe; centrifugally cast; high temperature use
	UTS: 483 Elon: 22% Proof: 207
ASTM A660 WCC	0.25% C (max) 1.2% Mn steel: Pipe; centrifugally cast; high temperature use
	UTS: 483 Elon: 22% Proof: 276
ASTM A662 A	0.17% C (max) 1.1% Mn steel: Plate for low temperature use; 27 J impact at -75 °C
	UTS: 460 Elon: 20% Proof: 275

Symbol	Nominal analysis, supplier, condition and remarks.
ASTM A662 B	0.22% C (max) 1.2% Mn steel: Plate for low temperature use; 20 J impact at -50 °C **UTS: 520 Elon: 20% Proof: 275**
ASTM A663	C not specified 0.04% S & P (max) steel: Hot rolled bars; graded by tensile strength
ASTM A668 A–F	C not specified 1.1% Mn (max) 0.05% S & P (max) steel: For general use
ASTM A668 AH–FH	C not specified 1.1% Mn (max) 0.05% S & P (max) steel: For general use
ASTM A668 G–N	C and Mn not specified 0.04% S & P (max) steel: For general use
ASTM A668 GH–NH	C and Mn not specified 0.04% S & P (max) steel: For general use
ASTM A675	C not specified 0.04% S & P (max) steel: Bars; graded by tensile strength
ASTM A678 A	0.16% C (max) 1.2% Mn steel: Plate for structures; hardened and tempered **UTS: 520 Elon: 22% Proof: 345**
ASTM A678 B	0.2% C (max) 1.0% Mn steel: Plate for structures; hardened and tempered **UTS: 600 Elon: 22% Proof: 414**
ASTM A678 C	0.22% C 1.3% Mn steel: Plate for structures; hardened and tempered **UTS: 700 Elon: 19% Proof: 480**
ASTM A707 L1/1	0.23% C (max) 1.5% Mn steel: Flanges for low temperature use; impact 41 J at -30 °C **UTS: 290 Elon: 22%**
ASTM A707 L1/2	0.23% C (max) 1.5% Mn (max) steel: Flanges for low temperature use; impact 54 J at -30 °C **UTS: 360 Elon: 25%**
ASTM A707 L1/3	0.23% C (max) 1.5% Mn (max) steel: Flanges for low temperature use; impact 68 J at -30 °C **UTS: 415 Elon: 23%**
ASTM A707 L1/4	0.23% C (max) 1.5% Mn (max) steel: Flanges for low temperature use; impact 68 J at -30 °C **UTS: 515 Elon: 20%**
ASTM A709/50 W	0.2% C (max) 1.35% Mn (max) with grain refining elements steel: For bridges **UTS: 485 Elon: 20% Proof: 345**
ASTM A709/100	0.15% C 1.0% Mn with grain refining elements steel: For bridges **UTS: 800 Elon: 17% Proof: 635**
ASTM A709/100 W	0.15% C 1.0% Mn with grain refining elements steel: For bridges **UTS: 800 Elon: 17% Proof: 635**
ASTM A714/1	0.26% C (max) 1.3% Mn (max) 0.18% Cu (min) steel: Pipe; welded or seamless **UTS: 463 Elon: 22% Proof: 345**
ASTM A715	0.15% C (max) 1.65% Mn (max) steel: Sheet; low alloy; high formability; fine grained; graded by alloy content and tensile properties
ASTM A724 A	0.22% C (max) 1.2% Mn steel: For welded pressure vessels **UTS: 700 Elon: 19% Proof: 485**
ASTM A727	0.28% C (max) 1.1% Mn steel: Forging for pipe fittings for notch toughness **UTS: 500 Elon: 22% Proof: 250**
ASTM A730 A	0.15% C (max) 0.45% Mn steel: Forging for railway use
ASTM A730 B	0.2% C 0.45% Mn steel: Forging for railway use
ASTM A732/1 A	0.2% C 0.4% Mn steel: Investment casting; annealed **UTS: 410 Elon: 35% Proof: 280**
ASTM A737 B	0.22% C (max) 1.3% Mn 0.05% Nb (max) steel: Plate for pressure vessels
ASTM A741	Mild steel wire rope; zinc coated for guard rails
ASTM B227	Hard drawn copper clad steel wire for electrical purposes
ASTM B228	Copper clad steel conductor wire; concentric lag

Symbol	Nominal analysis, supplier, condition and remarks.
AURIGA 1A	0.12% C 0.4% Si 0.5% Mn 0.05% S & P steel: Casting; D Brown; high magnetic permeability **UTS: 39 Elon: 22%**
AURIGA 1B	0.17% C 0.4% Si 0.5% Mn 0.05% S & P steel: Casting; D Brown; high magnetic permeability **UTS: 44 Elon: 22%**
AURIGA V1A	0.22% C 1.2% Mn 0.05% S & P steel: Casting; D Brown; normalized & tempered **UTS: 60 Elon: 18%**
AUTROD 12.10	0.1% C 1.0% Mn steel: Weld metal; Esab; copper coated for submerged arc welding; flux can be used to modify analysis of deposit
AUTROD 12.30	0.1% C 1.0% Mn steel: Copper coated welding rod; Esab; for submerged arc welding; properties of the weld metal can be modified with different fluxes
AUTROD 12.51	0.1% C 1.0% Mn 0.6% Si steel: Copper coated wire; Esab; for carbon dioxide or gas shielded welding; as welded **UTS: 580 Elon: 25% Proof: 430**
AUTROD 13.42	0.12% C 1.4% Mn 0.3% Si 0.01% S & P (max) steel: Esab; welding rod; weld deposit; 50 J impact at -20 °C **UTS: 550 Elon: 27% Proof: 440**
AW 10	0.18% C 0.75% Mn 0.02% Nb 0.25% Cu steel: Alan Wood
AW 55	0.16% C 0.55% Mn 0.02% Nb steel: Alan Wood
AW AC	Mild steel Al coated conductivity wire: Copperweld Co
AWX 45	0.12% C 0.45% Mn 0.02% Nb steel: Alan Wood
AWX 50	0.5% Mn 0.02% Nb steel: Alan Wood
AWX 55	0.16% C 0.55% Mn 0.02% Nb steel: Alan Wood
B 0	0.1% C (max) steel: Breda
B 1	0.15% C (max) steel: Breda
B 2	0.2% C (max) steel: Breda
B 4	0.2% C steel: As rolled; Bofors **DPN: 130 UTS: 45 Elon: 24% Proof: 24**
B 4 V	0.15% C 0.7% Mn steel: For case hardening; Bofors; as rolled **DPN: 160**
B 303	0.15% C 0.7% Mn steel: Fagersta
BH 36	0.2% C (max) 1.2% Mn 0.03% P (max) 0.04% S (max) steel: Rheinstahl
BS 14	Low C 0.05% S (max) 0.05% P steel: Structural; pressure parts for marine boilers **UTS: 42 Elon: 24%**
BS 24/1/1	Low C 0.05% S & P steel: For loco axles; Hardened & tempered **UTS: 50 Elon: 25% Proof: 24**
BS 24/1/2	Low C 0.05% S & P steel: For loco axles; hardened and tempered **UTS: 56 Proof: 27**
BS 24/1/3	Low C 0.06% S & P steel: For wagon axles; hardened and tempered **UTS: 56 Proof: 27**
BS 24/3 A 9	Low C 0.06% S & P steel: For railway spring buckles; as rolled **UTS: 39 Elon: 22%**
BS 24/3 A 10	Low C 0.06% S & P steel: For railway spring buckles; as rolled **UTS: 42 Elon: 22%B**
BS 24/3 A 11	Low C 0.06% S & P steel: For railway spring buckles; as rolled **UTS: 53 Elon: 20%**
BS 24/4 A	0.15% C 0.6% Mn 0.05% S & P steel: Forging for railway use; carburized parts **UTS: 39 Elon: 28%**
BS 24/4 B	0.2% C 0.6% Mn 0.05% S & P steel: Forging for railway use; boiler forgings **UTS: 46 Elon: 26%**

302 · 44A1 STEEL – PLAIN CARBON 0.05–0.2 PER CENT

Symbol	Nominal analysis, supplier, condition and remarks.
BS 24/4 E	0.16% C 0.6% Mn 0.06% S & P steel: Forging for railway use **UTS: 40 Elon: 28%**
BS 24/4 F	0.21% C 0.6% Mn 0.06% S & P steel: Forging for railway use **UTS: 45 Elon: 26%**
BS 24/5 611	0.16% C (max) 0.04% S & P steel: Plate; as BS alloy 611; for railway use
BS 24/5/612	Low C 0.05% S & P steel: Plate; as BS alloy 612; for railway use
BS 24/5/613	Low C 0.05% S & P steel: Bar; as BS alloy 613; for railway use
BS 29/22/26	Low C 0.05% S 0.05% P steel: Forging; marine forgings; welding quality **UTS: 36 Elon: 33%**
BS 29/28/32	Low C 0.05% S 0.05% P steel: Forging; marine forgings **UTS: 45 Elon: 28%**
BS 29/32/36	Low C 0.05% S 0.05% P steel: Forging; marine forgings **UTS: 51 Elon: 25%**
BS 29/36/40	Low C 0.05% S 0.05% P steel: Forging; marine forgings **UTS: 58 Elon: 22%**
BS 32/3	Low C free machining steel; withdrawn owing to low resistance to shock
BS 32/4	0.1% C 1% Mn 0.25% S 0.07% P steel: Free machining bar; replaced by BS 970 En 1A
BS 32/5	0.1% C 1.2% Mn 0.45% S 0.06% P steel: Free machining bar; replaced by BS 970 En 1B
BS 75	Wrought steels for automobiles; replaced by BS 970
BS 400	0.2% C 0.3% Si 0.6% Mn 0.045% S & P steel: For gas cylinders
BS 401	0.2% C 0.3% Si 0.7% Mn 0.045% S & P steel: For gas cylinders
BS 640/2	0.12% C 0.04% Si 0.05% Mn 0.05% S & P steel: Electrode
BS 640/3	0.1% C 0.04% Si 0.04% Mn 0.04% S & P steel: Electrode
BS 725	Hot rolled mild steel strip; replaced by BS 1449
BS 782	Electrodes for metal arc welding; replaced by BS 639 & BS 2549
BS 806 A	Low C 0.05% S & P steel: Seamless pipe; cold drawn & annealed **UTS: 390**
BS 806 B	Low C 0.05% S & P steel: Seamless pipe; as drawn **UTS: 390**
BS 806 C	Low C 0.06% S & P steel: Seamless pipe; as drawn **UTS: 370**
BS 806 D	Low C 0.06% S & P steel: Hydraulic lap welded; as welded **UTS: 370**
BS 806 E	Low C 0.06% S & P steel: Roll lap welded pipe; as welded **UTS: 370**
BS 806 F	Low C steel: Welded pipe

Note. The following abbreviations and units are used in the tables:

DPN	Hardness, diamond pyramid number
UTS	Ultimate tensile strength, N/mm^2
Elon	Elongation, %
Proof	0.1 % proof strength, N/mm^2

1 N/mm^2=0.1 hbar=0.102 kgf/mm^2=0.06475 tonf/in^2=145.04 lbf/in^2

Symbol	Nominal analysis, supplier, condition and remarks.
BS 847	Cold rolled mild strip; replaced by BS 1449
BS 970 En 1A	0.1% C 1% Mn 0.25% S 0.07% P steel: Free cutting; cold rolled **UTS: 420 Elon: 14%**
BS 970 En 1B	0.1% C 1.2% Mn 0.4% S 0.06% P steel: Free cutting; cold rolled **UTS: 370 Elon: 12%**
BS 970 En 2	0.2% C (max) 0.8% Mn 0.06% S 0.06% P steel: Rolled cold forming steel **UTS: 300 Elon: 28%**
BS 970 En 2A	0.12% C (max) 0.5% Mn 0.05% S 0.05% P steel: As rolled cold forming steel **UTS: 300 Elon: 28%**
BS 970 En 2A/1	0.1% C (max) 0.5% Mn 0.04% S 0.04% P steel: As rolled cold forming steel **UTS: 300 Elon: 28%**
BS 970 En 2B	0.15% C (max) 0.5% Mn 0.05% S 0.05% P steel: Cold forming steel; as rolled **UTS: 300 Elon: 28%**
BS 970 En 2C	0.2% C 0.5% Mn 0.05% S 0.05% P steel: As rolled cold forming steel **UTS: 300 Elon: 28%**
BS 970 En 2D	0.22% C 0.6% Mn 0.05% S 0.05% P steel: As rolled cold forming steel **UTS: 300 Elon: 28%**
BS 970 En 7	0.2% C 1% Mn 0.15% S 0.06% P steel: Semi-free machining bar; cold drawn **UTS: 600 Elon: 15%**
BS 970 En 32A	0.15% C (max) 0.5% Mn 0.05% S 0.05% P steel: For case hardening; core properties; blank carburized **UTS: 480 Elon: 20%**
BS 970 En 32B	0.15% C 0.9% Mn 0.07% S 0.05% P steel: For case hardening; core properties; blank carburized **UTS: 480 Elon: 20%**
BS 970 En 32C	0.14% C 0.3% Si 0.8% Mn 0.05% S & P steel: For carburizing
BS 970 En 32M	0.15% C 1% Mn 0.12% S 0.05% P steel: For case hardening; core properties; blank carburized; semi free cutting steel **UTS: 480 Elon: 20%**
BS 970 En 201	0.18% C (max) 1.2% Mn 0.05% S 0.05% P steel: For case hardening; blank carburized and double hardened **UTS: 600 Elon: 20%**
BS 970 En 202	0.18% C (max) 1.25% Mn 0.15% S 0.05% P steel: For case hardening; blank carburized and double hardened; free machining **UTS: 580 Elon: 20%**
BS 971	Commentary on steels contained in BS 970 (En series)
BS 980 CDS1	0.2% C 0.05% S 0.05% P steel: Cold drawn tube; annealed **UTS: 370 Proof: 170**
BS 980 CDS2	0.2% C 0.05% S 0.05% P steel: Cold drawn tube; as drawn and tempered **UTS: 420 Proof: 360**
BS 980 CDS3	0.18% C 0.6% Mn steel: Cold drawn tube; as drawn and tempered **UTS: 420 Proof: 360**
BS 980 CDS4	0.15% C 0.8% Mn 0.07% S steel: Cold drawn tube; as drawn and tempered **UTS: 420 Proof: 360**
BS 1052	Low C 0.06% S 0.06% P steel: Wire; annealed **UTS: 390**
BS 1052	Low C 0.06% S 0.06% P steel: Wire; hard drawn **UTS: 690**
BS 1387	Low C 0.06% S & P steel: For steel tubes & sockets **UTS: 390**

Symbol	Nominal analysis, supplier, condition and remarks.
BS 1449 CS1	0.07% C (max) 0.4% Mn 0.03% S & P steel: Strip; extra deep drawing; quality as BS alloy En 2A/1
BS 1449 CS2	0.08% C (max) 0.4% Mn 0.03% S & P steel: Strip; extra deep drawing; quality as BS alloy En 2A/1
BS 1449 CS3	0.1% C (max) 0.5% Mn 0.04% S & P steel: Strip; deep drawing; quality as BS alloy En 2A
BS 1449 CS4	0.12% C (max) 0.5% Mn 0.05% S & P steel: Strip; as BS alloy En 2
BS 1449 CS12	0.12% C 0.5% Mn 0.05% S & P steel: Strip; as BS alloy En 2B
BS 1449 CS17	0.17% C 0.5% Mn 0.05% S & P steel: Strip; as BS alloy En 2C
BS 1449 CS22	0.22% C 0.05% Mn 0.05% S & P steel: Strip; as BS alloy En 2C
BS 1449 HR1	0.09% C 0.45% Mn 0.03% S & P steel: Extra deep drawing (killed) sheet; as BS 970 En 2A/1 **UTS: 320 Elon: 34%**
BS 1449 HR2	0.09% C 0.45% Mn 0.03% S & P steel: Extra deep drawing sheet; as BS 970 En 2A/1 **UTS: 320 Elon: 34%**
BS 1449 HR3	0.1% C 0.5% Mn 0.04% S & P steel: Deep drawing sheet; as BS 970 En 2A **UTS: 330 Elon: 31%**
BS 1449 HR4	0.15% C 0.6% Mn 0.05% S & P steel: Drawing quality sheet; as BS 970 En 2 **UTS: 290**
BS 1449 HR11	0.08% C 0.4% Mn 0.03% S & P steel: Extra deep drawing (killed) plate; as BS 970 En 2A/1 **UTS: 320 Elon: 34%**
BS 1449 HR12	0.09% C 0.4% Mn 0.03% S & P steel: Extra deep drawing plate; as BS 970 En 2A/1 **UTS: 320 Elon: 34%**
BS 1449 HR13	0.1% C 0.5% Mn 0.04% S & P steel: Deep drawing quality plate; as BS 970 En 2A **UTS: 320 Elon: 31%**
BS 1449 HR14	0.15% C 0.6% Mn 0.05% S & P steel: Flanging or drawing quality plate; as BS 970 En 2A **UTS: 270**
BS 1449 HR15	0.2% C 0.9% Mn 0.06% S & P steel: Commercial quality plate; as BS 970 En2 **UTS: 270**
BS 1449 HS1	0.07% C 0.4% Mn 0.03% S & P steel: Plate & sheet **UTS: 270 Elon: 34%**
BS 1449 HS2	0.08% C 0.4% Mn 0.03% S & P steel: Sheet & plate **UTS: 270 Elon: 34%**
BS 1449 HS3	0.1% C 0.5% Mn 0.04% S & P steel: Sheet & plate **UTS: 270 Elon: 30%**
BS 1449 HS4A	0.12% C 0.5% Mn 0.05% S & P steel: Sheet & plate **UTS: 270 Elon: 25%**
BS 1449 HS4B	0.13% C 0.6% Mn 0.06% S & P steel: Sheet & plate **UTS: 270**
BS 1449 HS12	0.12% C 0.5% Mn 0.05% S & P steel: Sheet & plate **UTS: 300 Elon: 25% Proof: 170**
BS 1449 HS17	0.17% C 0.5% Mn 0.05% S & P steel: Sheet & plate **UTS: 340 Elon: 25% Proof: 200**
BS 1449 HS20	0.2% C 0.2% Si 1.5% Mn 0.06% S & P steel: Sheet **UTS: 370 Elon: 18% Proof: 340**
BS 1449 HS22	0.22% C 0.5% Mn 0.05% S & P steel: Sheet & plate **UTS: 390 Elon: 20% Proof: 220**

Symbol	Nominal analysis, supplier, condition and remarks.
BS 1449 HS23	0.22% C 0.6% Mn 0.05% S & P steel: Sheet & plate **UTS: 450 Elon: 20% Proof: 240**
BS 1449 NHR12	0.1% C 0.5% Mn 0.04% S & P steel: Extra deep drawing quality plate **UTS: 290 Elon: 26%**
BS 1449 NHR13	0.12% C 0.5% Mn 0.05% S & P steel: Deep drawing quality plate **UTS: 300 Elon: 23%**
BS 1449 NHR14	0.15% C 0.5% Mn 0.05% S & P steel: Flanging quality plate **UTS: 300 Elon: 22%**
BS 1449 NHR15	0.2% C 0.9% Mn 0.06% S & P steel: Commercial quality plate
BS 1449 NHR21	0.18% C 0.6% Mn 0.05% S & P steel: Plate **UTS: 390 Elon: 20% Proof: 210**
BS 1449 NHR22	0.18% C 0.6% Mn 0.05% S & P steel: Plate **UTS: 440 Elon: 18% Proof: 210**
BS 1449 NHR23	0.18% C 0.6% Mn 0.05% S & P steel: Plate **UTS: 450 Elon: 25% Proof: 240**
BS 1449 NHR25	0.2% C 1.2% Mn 0.06% S & P steel: Plate **UTS: 530 Elon: 14% Proof: 340**
BS 1453 A1	0.1% C 0.6% Mn 0.25% Ni steel: For welding filler rod to give butt weld of 330 N/mm² tensile
BS 1453 A2	0.15% C 0.2% Si 1.2% Mn steel: For welding filler rod to give butt weld of 420 N/mm² tensile
BS 1456 A	0.21% C 0.3% Si 1.5% Mn 0.05% S & P steel: Castings; hardened and tempered; contained in BS 3100 **UTS: 610 Elon: 18%**
BS 1501/101	C not specified 0.06% S & P steel: Plate & bar **UTS: 420 Elon: 20%**
BS 1501/151	0.2% C 0.05% S & P steel: Plate & bar; normalized **UTS: 420 Elon: 20% Proof: 220**
BS 1501/151 A	0.2% C 0.05% S & P steel: Plate & bar; normalized **UTS: 390 Elon: 25% Proof: 180**
BS 1501/151 B	0.2% C 0.05% S & P steel: Plate & bar; normalized **UTS: 390 Elon: 23% Proof: 210**
BS 1501/161 A	0.2% C 0.25% Si 0.50% Mn 0.05% S & P steel (silicon killed): Plate & bar; normalized **UTS: 360 Elon: 25% Proof: 180**
BS 1501/161 B	0.2% C 0.25% Si 0.50% Mn 0.05% S & P steel (silicon killed): Plate & bar; normalized **UTS: 390 Elon: 23% Proof: 200**
BS 1501/161 C	0.2% C 0.25% Si 0.50% Mn 0.05% S & P steel (silicon killed): Plate & bar; normalized **UTS: 450 Elon: 25% Proof: 220**
BS 1501/221	0.2% C 0.25% Si 1.50% Mn 0.05% S & P steel: Plate & bar; normalized **UTS: 500 Elon: 18% Proof: 240**
BS 1506/111	0.2% C 0.3% Si 0.7% Mn 0.06% S & P steel: Bar; as rolled **UTS: 450 Elon: 25% Proof: 220**
BS 1507/101	C not specified 0.06% S & P steel: Seamless pipes; as rolled **UTS: 360 Elon: 30%**
BS 1507/131	C not specified 0.06% S & P steel: Lap welded pipes **UTS: 370 Elon: 30%**
BS 1507/151	C not specified 0.05% S & P steel: Seamless pipes; as rolled **UTS: 370 Elon: 30%**
BS 1507/171	C not specified 0.05% S & P steel: Cold drawn seamless pipes; annealed after drawing **UTS: 370 Elon: 31%**

Symbol	Nominal analysis, supplier, condition and remarks.
BS 1507/181	C not specified 0.05% S & P steel: Lap welded pipes **UTS: 360**
BS 1508/151	C not specified 0.05% S & P steel: Seamless tube; as rolled **UTS: 360 Elon: 29%**
BS 1508/171	C not specified 0.05% S & P steel: Cold drawn tube; annealed after drawing **UTS: 370 Elon: 30%**
BS 1617 A	0.15% C (max) 0.6% Si 0.5% Mn 0.06% S & P steel: Casting; annealed; high magnetic permeability **UTS: 420 Elon: 22%**
BS 1617 B	0.25% C (max) 0.6% Si 0.5% Mn 0.06% S & P steel: Casting; annealed; high magnetic permeability **UTS: 480 Elon: 20% Proof: 200**
BS 1627	0.12% C 0.5% Mn 0.05% S & P steel: Tube
BS 1627	0.15% C 0.3% Si 0.50% Mn 0.05% S & P steel: Cold drawn tubes; annealed **UTS: 300 Elon: 32%**
BS 1633 L	0.18% C (max) 0.2% Si 0.05% S & P steel: For land boilers **UTS: 580 Elon: 21%**
BS 1654	Mild steel tubes; replaced by BS 3059
BS 1678	0.15% C (max) 0.4% Mn 0.05% S & P steel: Tube
BS 1730	0.2% C (max) 0.5% Mn 0.05% S & P steel: Tube
BS 1775 CDS11	Low C 0.06% S & P steel: Seamless tube; cold drawn & annealed **UTS: 300 Proof: 170**
BS 1775 CDS13	Low C 0.06% S & P seamless tube; cold drawn & annealed **UTS: 330 Proof: 200**
BS 1775 CDS16	Low C 0.06% S & P steel: Seamless tube; cold drawn & annealed **UTS: 400 Proof: 240**
BS 1775 CDS20	Low C 0.06% S & P steel: Seamless tube; cold drawn & annealed **UTS: 530 Proof: 300**
BS 1775 CDS 23	Low C 0.06% S & P steel: Seamless tube; cold drawn & annealed **UTS: 480 Proof: 340**
BS 1775 CDS24	Low C 0.06% S & P steel: Seamless tube; cold drawn or cold drawn & tempered **UTS: 420 Proof: 360**
BS 1775 CDS28	Low C 0.06% S & P steel: Seamless tube; cold drawn or cold drawn & tempered **UTS: 530 Proof: 420**
BS 1775 CDS35	Low C 0.06% S & P steel: Seamless tube; cold drawn or cold drawn & tempered **UTS: 630 Proof: 530**
BS 1775 CEW11	Low C 0.06% S & P steel: Electrically welded tube; cold drawn after welding; then annealed **UTS: 300 Proof: 170**
BS 1775 CEW16	Low C 0.06% S & P steel: Electrically welded tube; cold drawn after welding; then annealed **UTS: 370 Proof: 240**

Symbol	Nominal analysis, supplier, condition and remarks.
BS 1775 CEW23	Low C 0.06% S & P steel: Electrically welded tube; cold drawn after welding; then annealed **UTS: 480 Proof: 340**
BS 1775 CEW24	Low C 0.06% S & P steel: Electrically welded tube; cold drawn after welding **UTS: 420 Proof: 360**
BS 1775 CEW28	Low C 0.06% S & P steel: Electrically welded tube; cold drawn after welding **UTS: 530 Proof: 420**
BS 1775 EFW16	Low C 0.06% S & P steel: Electric fusion welded tube; as welded **UTS: 390 Proof: 240**
BS 1775 ERW11	Low C 0.06% S & P steel: Electrically welded tube; as welded **UTS: 300 Proof: 170**
BS 1775 ERW16	Low C 0.06% S & P steel: Electrically welded tube; as welded **UTS: 370 Proof: 240**
BS 1775 ERW20	Low C 0.06% S & P steel: Electrically welded tube; as welded **UTS: 450 Proof: 300**
BS 1775 ERW23	Low C 0.06% S & P steel: Electrically welded tube; as welded **UTS: 480 Proof: 340**
BS 1775 HFS11	Low C 0.06% S & P steel: Seamless tube; as drawn **UTS: 300 Proof: 170**
BS 1775 HFS13	Low C 0.06% S & P steel: Seamless tube; as drawn **UTS: 330 Proof: 200**
BS 1775 HFS16	Low C 0.06% S & P steel: Seamless tube; as drawn **UTS: 400 Proof: 240**
BS 1775 HFS20	Low C 0.06% S & P steel: Seamless tube; as drawn **UTS: 550 Proof: 300**
BS 1775 HFS23	Low C 0.06% S & P steel: Seamless tube; as drawn **UTS: 480 Proof: 340**
BS 1775 HFW13	Low C 0.06% S & P steel: Welded tube; as welded **UTS: 330 Proof: 200**
BS 1775 HFW16	Low C 0.06% S & P steel: Welded tube; as welded **UTS: 400 Proof: 240**
BS 1775 HFW23	Low C 0.06% S & P steel: Welded tube; as welded **UTS: 480 Proof: 340**
BS 1775 HLW16	Low C 0.06% S & P steel: Hydraulically lap welded tube; as welded **UTS: 390 Proof: 240**
BS 1775 OAW11	Low C 0.06% S & P steel: Oxy-acetylene welded tube; as welded **UTS: 300 Proof: 170**
BS 1882	99.0% Ni mild steel to BS alloy 151-154 or 157 sheet
BS 2762 NDI	0.2% C (max) 1.5% Mn 0.05% P 0.06% S steel: Plate; notch ductile for bridges etc. **UTS: 450 Elon: 22% Proof: 220**
BS 2762 NDII	0.2% C 1.5% Mn 0.06% S 0.05% P steel: Plate; notch ductile for bridges etc. **UTS: 450 Elon: 22% Proof: 220**
BS 2762 NDIII	0.17% C (max) 0.3% Si 1.5% Mn 0.05% S & P steel: Plate; notch ductile for bridges, etc. **UTS: 450 Elon: 22% Proof: 210**
BS 2762 NDIV	0.17% C (max) 0.3% Si 1.5% Mn 0.05% S & P steel: Plate; notch ductile for bridges, etc. **UTS: 450 Elon: 22% Proof: 210**

Note. The following abbreviations and units are used in the tables:

DPN	Hardness, diamond pyramid number
UTS	Ultimate tensile strength, N/mm^2
Elon	Elongation, %
Proof	0.1 % proof strength, N/mm^2

1 N/mm^2=0.1 hbar=0.102 kgf/mm^2=0.06475 tonf/in^2=145.04 lbf/in^2

Symbol	Nominal analysis, supplier, condition and remarks.
BS 2772/3	0.2% C 1.5% Mn 0.05% S & P steel: Castings for colliery winding gear; hardened and tempered **DPN: 150 UTS: 480 Elon: 25% Proof: 300**
BS 2858	0.15% C (max) 0.05% Si 0.5% Mn 0.05% S & P steel: Plate for galvanizing pots
BS 2901 A15	0.1% C 1.2% Mn 0.6% Si steel: Rod for gas & electric welding
BS 2901 A17	0.15% C 0.5% Si 1.1% Mn 0.03% S & P steel: Rod for gas & electric welding
BS 3059/1	Low C 0.05% S & P steel: Tube for boilers; seamless; hot finished **UTS: 360**
BS 3059/2	Low C 0.05% S & P steel: Tube for boilers; cold drawn; seamless **UTS: 360**
BS 3059/3	Low C 0.05% S & P steel: Tube for boilers; electrically welded & normalized **UTS: 360**
BS 3059/4	Low C 0.05% S & P steel: Tube for boilers; electrically welded from strip, then cold drawn **UTS: 360**
BS 3100 A4	0.2% C 1.4% 0.05% S & P (max) steel: Casting **UTS: 600 Elon: 16% Proof: 320**
BS 3100 AL1	0.2% C (max) 1.1% Mn 0.04% S & P (max) steel: Casting for low temperature use; 20 J impact at -40 °C **UTS: 430 Elon: 22% Proof: 230**
BS 3100 AM1	0.15% C (max) 0.5% Mn (max) Cr + Mo + Ni + Cu 0.8% (max) steel: Casting with high magnetic permeability **UTS: 400 Elon: 22% Proof: 185**
DS 3100 AM2	0.25% C (max) 0.5% Mn (max) Cr + Mo + Ni + Cu 0.8% (max) steel: Casting with high magnetic permeability **UTS: 450 Elon: 22% Proof: 215**
BS 3100 AW1	0.14% C 0.8% Mn Cr + Mo + Ni + Cu 0.8% (max) steel: Casting for case hardening **UTS: 460 Elon: 12%**
BS 3141 C1A1	0.1% C 0.5% Mn 0.04% S & P steel: Classification system used by BS; discontinued
BS 3141 C2A1	0.12% C (max) 0.5% Mn 0.05% S & P steel: Classification system used by BS; discontinued
BS 3141 C2B1	0.15% C 0.4% Mn 0.05% S & P steel: Classification system used by BS; discontinued
BS 3141 C2E1	0.12% C 0.5% Mn 0.05% S & P steel: Classification system used by BS; discontinued
BS 3141 C2F1	0.12% C 0.8% Mn 0.05% S & P steel: Classification system used by BS; discontinued
BS 3141 C2G1	0.12% C 1.3% Mn 0.05% S & P steel: Classification system used by BS; discontinued
BS 3141 C2K1	0.15% C 0.4% Mn 0.05% S & P steel: Classification system used by BS; discontinued
BS 3141 C2L1	0.15% C 1.1% Mn 0.05% S & P steel: Classification system used by BS; discontinued
BS 3141 C3A1	0.2% C 0.5% Mn 0.05% S & P steel: Classification system used by BS; discontinued
BS 3141 C3B1	0.2% C 0.7% Mn 0.05% S & P steel: Classification system used by BS; discontinued
BS 3141 CF2C1	0.1% C 1% Mn 0.07% P 0.25% S steel: Free cutting; classification system used by BS; discontinued
BS 3141 CF2D1	0.1% C 1.2% Mn 0.06% P 0.5% S steel: Free cutting; classification system used by BS; discontinued
BS 3141 CF2H1	0.14% C 1.1% Mn 0.05% P 0.12% S steel: Free cutting; classification system used by BS; discontinued
BS 3141 CF2J1	0.16% C 1.2% Mn 0.06% P 0.14% S steel: Free cutting; classification system used by BS; discontinued

Symbol	Nominal analysis, supplier, condition and remarks.
BS 3141 CF3D1	0.2% C 1.1% Mn 0.06% P 0.14% S steel: Free cutting; classification system used by BS; discontinued
BS 3141 CF3E1	0.2% C 1.1% Mn 0.05% P 0.14% S steel: Free cutting; classification system used by BS; discontinued
BS 3458	0.2% C 0.5% Cr 0.5% Ni 0.2% Mo steel: For chain slings; hardened and tempered **DPN: 250**
BS 3601 BW	C not specified 0.06% S & P steel: Butt welded pipe; available in grade 22 **UTS: 390**
BS 3601 CDS	C not specified 0.06% S & P steel: Seamless pipe; cold drawn; available in grades 22, 27 and 35 **UTS: 400**
BS 3601 EFW	C not specified 0.06% S & P steel: Electric fusion welded pipe; available in grade 26 **UTS: 420**
BS 3601 ERW	C not specified 0.06% S & P steel: Electrical resistance welded pipes; available in grades 22 and 27 **UTS: 370**
BS 3601 HFS	C not specified 0.06% S & P steel: Seamless pipe; hot finished; available in grades 22, 27 and 35 **UTS: 400**
BS 3601 HLW	C not specified 0.06% S & P steel: Hydraulic welded pipes; available in grade 26 **UTS: 420**
BS 3601 SFW	C not specified 0.06% S & P steel: Spiral seam fusion welded pipe; available in grade 26 **UTS: 420**
BS 3602 CDS	0.22% C 0.2% Si 0.5% Mn 0.05% S & P steel: Seamless pipe; cold drawn; available in grades 23, 27 and 35 **UTS: 420 Proof: 240**
BS 3602 EFW	Analysis not quoted; electric fusion welded steel pipes
BS 3602 ERW	0.22% C 0.05% S & P steel: Electric resistance welded pipes; available in grades 23 and 27 **UTS: 420 Proof: 210**
BS 3602 HFS	0.22% C 0.2% Si 0.5% Mn 0.05% S & P steel: Seamless tube; hot finished; available in grades 23, 27 and 35 **UTS: 420 Proof: 240**
BS 4360/40 A	0.27% C 0.06% S & P steel: As rolled; weldable **UTS: 280 Elon: 22%**
BS 4360/40 B	0.25% C 1.6% Mn 0.06% S & P steel: As rolled; weldable **UTS: 280 Elon: 22%**
BS 4360/40 C	0.22% C 1.6% Mn 0.06% S & P steel: As rolled; weldable **UTS: 280 Elon: 22%**
BS 4360/40 D	0.19% C 1.6% Mn 0.06% S & P steel: Normalized; weldable; with Nb **UTS: 280 Elon: 22%**
BS 4360/40 E	0.19% C 1.6% Mn 0.05% S & P 0.3% Si steel: Normalized; weldable **UTS: 280 Elon: 22%**
BS 4360/43 D	0.19% C 1.6% Mn 0.05% S & P steel: Normalized; weldable; with Nb **UTS: 450 Elon: 22%**
BS 4360/43 E	0.19% C 1.6% Mn 0.3% Si 0.05% S & P steel: Normalized; weldable **UTS: 450 Elon: 22%**
BS 4449	0.25% C (max) 0.05% S & P steel: For concrete reinforcement; as rolled; high yield **UTS: 410 Elon: 14%**
BS 4449	0.25% C (max) 0.06% S & P steel: For concrete reinforcement; as rolled; mild steel **UTS: 250 Elon: 22%**

Symbol	Nominal analysis, supplier, condition and remarks.
BS 4461	0.25% C (max) 0.06% S & P (max) steel: For concrete reinforcement; cold rolled **UTS: 440 Elon: 13%**
BS S3	Low carbon steel: Sheet; 28 ton range suitable for welding; replaced by BS S510
BS S13	0.1% C steel: For carburizing; replaced
BS S14	0.14% C 0.2% Si 0.7% Mn 0.05% S & P steel: Bar & forging; blank carburized; hardened & tempered **UTS: 480 Elon: 20%**
BS S32	Low carbon mild steel: Wire for welding; obsolete
BS S84	Low carbon steel: Sheet; for welding; obsolete (see BS S511)
BS S91	0.15% C (max) 0.2% Si 0.5% Mn 0.04% S & P steel: Normalized; for bearing shells **DPN: 140**
BS S92	0.22% C 0.2% Si 1.5% Mn 0.04% S & P steel: Hardened & tempered; suitable for welding **DPN: 210 UTS: 760 Elon: 17% Proof: 460**
BS S112	0.2% C 0.3% Si 1.0% Mn 0.14% S 0.2% Pb 0.05% P steel: Cold drawn; free machining **UTS: 690 Elon: 15%**
BS S510	0.21% C 0.2% Si 0.6% Mn 0.05% S & P steel: Sheet; normalized; suitable for welding **UTS: 420 Elon: 20% Proof: 240**
BS S511	0.12% C (max) 0.2% Si 0.4% Mn 0.05% S & P steel; Sheet; annealed; suitable for welding **UTS: 330 Elon: 25%**
BS S512	0.12% C (max) 0.2% Si 0.5% Mn 0.05% S & P steel: Tinned sheet; tin to be 99.5% pure and not less than 0.015 mm thick
BS S514	0.2% C 0.2% Si 1.5% Mn 0.05% S & P steel: Sheet; hardened & tempered **UTS: 840 Elon: 12% Proof: 600**
BS S515	0.2% C 0.2% Si 1.5% Mn 0.05% S & P steel: Sheet; annealed; suitable for welding **DPN: 170 UTS: 450 Elon: 20%**
BST 4KP2	0.22% C 0.6% Mn 0.04% P (max) 0.05% S (max) 0.3% Cr Ni Cu (max) steel: Plate; Russian standard
BS T6	0.18% C (max) 0.2% Si 1.0% Mn 0.05% S & P steel: Tube; as drawn & tempered; weldable **UTS: 450 Proof: 420**
BS T26	0.2% C 0.05% S & P steel: Tube; ½-hard
C 1	0.15% C (max) 0.5% Mn steel: Siau
C 2	0.2% C (max) 0.8% Mn steel: Siau
C 8 UNI 4365	0.08% C 0.4% Mn 0.04% S & P steel: Italian Standard
C 9/15	0.15% C 0.4% Mn 0.04% S & P steel: Designation used by German Standard
C 9/22	0.21% C 0.4% Mn 0.04% S & P steel: Designation used by German Standard
C 10	0.1% C 0.3% Mn 0.045% S & P steel: Designation used by German Standard
C10 UNI 4365	0.09% C 0.3% Si 0.5% Mn 0.035% S & P steel: Italian Standard
C 15	0.15% C 0.4% Mn 0.045% S & P steel: For carburizing; designation used by German Standard

Note. The following abbreviations and units are used in the tables:

DPN	Hardness, diamond pyramid number
UTS	Ultimate tensile strength, N/mm^2
Elon	Elongation, %
Proof	0.1 % proof strength, N/mm^2

1 N/mm^2=0.1 hbar=0.102 kgf/mm^2=0.06475 tonf/in^2=145.04 lbf/in^2

Symbol	Nominal analysis, supplier, condition and remarks.
C 15 UNI 4365	0.15% C 0.3% Si 0.5% Mn 0.035% S & P steel: Italian Standard; for bolts
C 20	0.2% C 0.6% Mn steel: Italian Standard designation; normalized **DPN: 180 UTS: 520 Elon: 24% Proof: 240**
C 20 UNI 4365	0.2% C 0.4% Si 0.8% Mn 0.035% S & P steel: Italian Standard; for bolts
C 21 UNI 4365	0.2% C 0.1% Si 0.45% Mn 0.04% S & P steel: Italian Standard; for bolts
C 22	0.21% C 0.4% Mn 0.045% S & P steel: Designation used by German Standard
CASONA	0.1% C (max) 0.25% Si 0.25% Mn steel: For carburizing; Hall & Pickles obsolete
CD 28	0.12% C 0.2% Si 0.5% Mn steel: Fagersta **UTS: 390 Elon: 30%**
CD 32	0.14% C 0.3% Si 0.7% Mn steel: Fagersta **UTS: 440 Elon: 28%**
CD 40	0.17% C 0.3% Si 0.8% Mn steel: Fagersta **UTS: 460 Elon: 25%**
CHAR-PAC	0.2% C (max) 1.0% Mn 0.04% P & S (max) steel: US Steel Corp
CHMS	0.14% C 0.7% Mn steel: For case hardening; ESC for BS alloy En 32
CODE 602	0.1% C (max) 0.15% Si 0.15% Mn steel: Casting; Edgar Allen; o steel
Cq 15	0.15% C 0.2% Si 0.4% Mn 0.04% S & P steel: German Standard
Cq 22	0.22% C 0.25% Si 0.5% Mn 0.04% S & P steel: German Standard
CODE 605	0.15% C 0.25% Si 0.45% Mn steel: Casting; Edgar Allen; T steel
COILEX	0.08% C 0.8% Mn 0.2% Si steel: Welding electrode; Murex **UTS: 420 Elon: 38% Proof: 330**
COILEX F	0.08% C 0.4% Mn 0.15% Si steel: Welding electrode; Murex **UTS: 500 Elon: 25% Proof: 440**
COLCLAD 12 Cr	0.08% C (max) 12% Cr 0.5% Ni stainless steel: Colvilles Stainless to BS alloy En56A
COLCLAD 13/Cr/Al	0.08% C (max) 12% Cr 0.5% Ni 0.2% Al stainless steel on mild steel: Colvilles stainless BS alloy 713
COLCLAD 18/8	0.08% C (max) 19% Cr 10% Ni stainless steel on mild steel: Colvilles stainless to BS alloy En58A
COLCLAD 18/8 ELC	0.03% C (max) 19% Cr 10% Ni stainless steel on mild steel: Colvilles stainless to BS alloy 801C
COLCLAD 18/8 Nb	0.08% C (max) 18% Cr 9% Ni 0.9% Nb stainless steel on mild steel: Colvilles stainless on BS alloy En58
COLCLAD 18.8 Ti	0.08% C (max) 18% Cr 9% Ni 0.5% Ti stainless steel on mild steel: Colvilles stainless to BS alloy En58B
COLCLAD 18/10/2 ELC	0.03% C (max) 18% Cr 12% Ni 2.5% Mo stainless steel on mild steel: Colvilles
COLCLAD 18/10/2 Nb	0.08% C (max) 18% Cr 12% Ni 2.5% Mo 0.9% Nb stainless steel on mild steel: Colvilles stainless to BS alloy 845
COLCLAD 18/10/2 Ti	0.08% C (max) 18% Cr 12% Ni 2.5% Mo 0.5% Ti stainless steel on mild steel: Colvilles stainless to BS alloy En58H
COLCLAD 18/10/3 ELC	0.03% C (max) 18% Cr 12% Ni 3% Mo stainless steel on mild steel: Colvilles
COLCLAD 18/10/3 Nb	0.08% C 18% Cr 12% Ni 3% Mo 0.9% Nb stainless steel on mild steel: Colvilles stainless to BS alloy En58J
COLCLAD INCONEL	0.2% C 16% Cr 8% Fe 0.07% Cu Ni alloy on mild steel: Colvilles; Inconel on mild steel
COLCLAD NICKEL	0.02% C (max) 0.1% Ti (max) 0.2% Cu Ni on mild steel: Colvilles; Ni to BS alloy 1872 - low carbon
COLCLAD	0.15% C (max) 0.1% Ti (max) 0.2% Cu Ni on

Symbol	Nominal analysis, supplier, condition and remarks.
NICKEL	mild steel: Colvilles; Ni to BS alloy 1872
COLORCOAT	Strip steel (also stainless, etc.) with a range of plastic or paint coatings; British Steel
COLTUF 26	0.15% C (max) 0.2% Si 1.2% Mn 0.05% S & P steel: Plate; Colvilles
	UTS: 390 Elon: 26% Proof: 240
COLTUF 28	0.16% C (max) 0.2% Si 1.5% Mn 0.05% S & P steel: Plate; Colvilles
	UTS: 420 Elon: 25% Proof: 240
COLTUF 32	0.22% C (max) 0.3% Si 1.6% Mn 0.5% Ni steel: Plate; Colvilles
	UTS: 540 Elon: 22% Proof: 330
CONLO 1	0.16% C 1.5% Mn 0.2% Si 0.05% S & P steel: Plate; Consett; normalized; suitable for welding; Al killed
	UTS: 440 Elon: 25% Proof: 240
CONLO 1	0.16% C (max) 1.2% Mn 0.05% P & S (max) steel: Consett
CONLO 11	0.16% (max) 1.2% Mn 0.05% P & S (max) steel: Consett
CONLO 11	0.16% C 1.2% Mn 0.2% Si 0.05% S & P steel: Plate; Consett: normalized; suitable for welding; Si killed
	UTS: 440 Elon: 25% Proof: 240
CON-PAC	0.2% C (max) 1.25% Mn 0.25% Cu steel: US Steel Corp.
COPPERWELD	Copper covered steel wire: Copperweld Steel Co; for electronic uses
CORROSTITE	0.11% C 0.3% Si (max) 0.4% Mn 0.3% Cu 0.05% S & P steel: Tube; STD Service (see Resistco)
CREUSELSO 38	0.18% C (max) 1.2% Mn (max) Al killed steel: Creusot
CREUSELSO 42	0.24% C (max) 1.4% Mn (max) 0.03% P & S (max) steel: Creusot
CRO 42	0.2% C 0.6% Ni 0.2% Mo steel: For case hardening; Bofors
	DPN: 220
CRO 53	0.2% C 1.0% Cr 1.2% Ni 0.2% Mo steel: For case hardening; Bofors; annealed
	DPN: 220
Cu 1	0.1% C (max) 0.5% Mn steel: Siau
CUPREX	0.07% C 0.35% Mn 0.5% Cu steel: Welding electrode; Murex
	UTS: 460 Elon: 30% Proof: 440
D 1	0.12% C 0.2% Si 0.5% Mn steel: Pompey
D 1S	0.1% C 0.2% Si 0.7% Mn 0.2% S steel: Pompey; free machining
D 1SS	0.1% C 0.2% Si 1.0% Mn 0.15% S steel: Pompey; free machining
D 2	0.2% C 0.6% Mn steel: Pompey
D 2K	0.2% C 0.3% Si 0.9% Mn 0.35% S steel: Pompey; free machining
D 2Mn	0.2% C 0.2% Si 1.3% Mn steel: Pompey
D 2S	0.2% C 0.25% Si 0.75% Mn 0.15% S steel: Pompey; free machining
D 2SS	0.2% C 0.25% Si 1.0% Mn steel: Pompey
DC 1	0.1% C 0.3% Si 0.5% Mn steel: Pompey
DEEPEX	0.15% C 0.35% Mn steel: Welding electrode; Murex
DEF 13 B1	Mild steel: No properties or specifications listed
DEF 13 B2	Mild steel: Covers BS 970 En 3A 3C 4 etc
	UTS: 420 Elon: 20% Proof: 220
DEF 13 B6	Carbon case hardening steel: Covers BS 970 En 32 etc
	UTS: 480 Elon: 20%
DEF 13 B6 B	Carbon manganese case hardening steel: Covers BS 970 En 201
	UTS: 600 Elon: 20%
DILLINAL 54	0.2% C (max) 1.3% Mn 0.04% P & S (max) steel: German proprietary specification; Al killed

Symbol	Nominal analysis, supplier, condition and remarks.
DILLINAL 54T	0.2% C (max) 1.3% Mn 0.04% P & S (max) steel: German proprietary specification; Al killed
DILLINAL 56	0.2% C (max) 1.3% Mn 0.04% P & S (max) 0.5% Ni steel: German proprietary specification; Al killed
DILLINAL 56T	0.2% C (max) 1.3% Mn 0.5% Ni 0.12% V steel: German proprietary specification; Al killed
DIN 1613 St 35 13K	0.09% C 0.4% Mn steel: For chains
DIN 1624 St 0	0.12% C (max) 0.4% Mn 0.06% S 0.08% P steel: Cold rolled strip
DIN 1624 St 1	0.12% C (max) 0.4% Mn 0.06% S 0.07% P steel: Cold rolled strip
DIN 1624 St 2	0.1% C 0.3% Mn 0.05% S & P steel: Cold rolled strip
DIN 1624 St 3	0.1% C 0.4% Mn 0.04% S & P steel: Cold rolled strip
DIN 1624 St 4	0.1% C (max) 0.4% Mn 0.035% S 0.03% P steel: Cold rolled strip
DIN 1626 42/2	0.25% C (max) 0.06% P (max) 0.05% S (max) steel: Pipe; German standard
	UTS: 460 Proof: 260
DIN 1626 52/3	0.22% C (max) 1.5% Mn (max) 0.05% P & S (max) steel: Pipe; German standard
	UTS: 560 Proof: 360
DIN 1626 St 33	Analysis not specified; steel pipe; German standard
	UTS: 410
DIN 1626 St 34/2	0.17% C (max) 0.05% P & S (max) steel: Pipe; German standard
	UTS: 400 Proof: 240
DIN 1626 St 37	0.2% C (max) 0.08% P (max) 0.05% S (max) steel: Pipe; German standard
	UTS: 400 Proof: 240
DIN 1626 St 37/2	0.2% C (max) 0.06% P (max) 0.05% S (max) steel: Pipe; German Standard
	UTS: 410 Proof: 260
DIN 1626 St 42	0.25% C (max) 0.08% P (max) 0.05% (max) steel: Pipe; German Standard
	UTS: 460 Proof: 260
DIN 1629 St 00	Analysis not specified: Steel tube; German Standard
DIN 1629 St 35	0.18% C (max) 0.05% P & S (max) steel: Pipe; German Standard
	UTS: 400 Proof: 240
DIN 1629 St 35/4	0.17% C (max) 0.4% Mn 0.05% P & S (max) steel: Pipe; German Standard
	UTS: 400 Proof: 240
DIN 1629 St 45	0.25% (max) 0.05% P & S (max) steel: Pipe; German Standard
	UTS: 500 Proof: 260
DIN 1629 St 45/4	0.22% C (max) 0.4% Mn (max) 0.04% P & S (max) steel: Pipe; German Standard
	UTS: 500 Proof: 260
DIN 1629 St 52	0.2% C (max) 1.5% Mn (max) 0.05% P & S (max) steel: Pipe; German Standard
	UTS: 570 Proof: 360
DIN 1629 St 52/4	0.2% C (max) 1.5% Mn (max) 0.05% P & S (max) steel: Pipe; German Standard
	UTS: 570 Proof: 360
DIN 1651 9 S 20	0.12% C (max) 0.7% Mn 0.2% S 0.1% P steel: Free machining
DIN 1651 9 S 27	0.12% C (max) 0.8% Mn 0.25% S 0.1% P steel: Free machining
DIN 1651 9 S Mn 23	0.13% C (max) 1.1% Mn 0.23% S steel: Free machining
DIN 1651 9 S Mn 28	0.14% C (max) 1.1% Mn 0.28% S steel: Free machining
	DPN: 160 UTS: 460

Symbol	Nominal analysis, supplier, condition and remarks.
DIN 1651 9 S Mn 36	0.15% C (max) 1.2% Mn 0.38% S steel: Free machining; German Standard **DPN: 170 UTS: 480**
DIN 1651 9 S Mn Pb 23	0.13% C (max) 1.1%B Mn 0.23% S 0.2% Pb steel: Free machining
DIN 1651 9 S Mn Pb 28	0.14% C (max) 1.1% Mn 0.28% S 0.2% Pb steel: Free machining **DPN: 160 UTS: 460**
DIN 1651 9 S Mn Pb 36	0.15% C (max) 1.2% Mn 0.38% S 0.2% Pb steel: Free machining **DPN: 170 UTS: 480**
DIN 1651 10 S 20	0.09% C 0.7% Mn 0.2% S 0.07% P steel: Free cutting; carburizing
DIN 1651 15 S 20	0.15% C 0.7% Mn 0.2% S 0.07% P steel: Free machining; carburizing
DIN 1651 22 S 20	0.21% C 0.5% Mn 0.2% S steel: Free machining; for carburizing
DIN 1654 C 9 15	0.15% C 0.4% Mn 0.04% S & P steel: For screws
DIN 1654 C 9 22	0.21% C 0.4% Mn 0.04% S & P steel: For screws
DIN 2440	Plain carbon steel: Tubes for screwing; German Standard
DIN 2441	Plain carbon steel: Tubes for screwing; German Standard
DIN 2442	Plain carbon steel: Tubes for screwing; German Standard
DIN 17100 RST 34/1	0.17% C (max) 10.08% P (max) 0.08% P (max) 0.05% S (max) steel: Plate; German specification **UTS: 380 Proof: 200**
DIN 17100 RST 37/1	0.2% C (max) 0.07% P (max) 0.05% S (max) steel: Plate; German specification **UTS: 410 Proof: 230**
DIN 17100 RST 34/2	0.15% C (max) 0.05% S & P (max) 0.007% P (max) steel: Plate; German specification **UTS: 380 Proof: 200**
DIN 17100 RST 37/2	0.17% C (max) 0.05% P & S (max) 0.007% N (max) steel: Plate; German specification **UTS: 410 Proof: 230**
DIN 17100 RST 46/7	0.2% C (max) 0.05% P (max) 0.05% S (max) steel: Plate; German specification **UTS: 490 Proof: 280**
DIN 17100 St 33/1	Chemical analysis not specified: Steel for plate; German specification **UTS: 410 Proof: 190**
DIN 17100 St 33/2	C not specified 0.06% P (max) 0.05% S (max) 0.007% N (max) steel: Plate; German specification **UTS: 410 Proof: 190**
DIN 17100 St 37/3	0.17% C (max) 0.045% P & S (max) 0.009% N (max) steel: Plate; German specification **UTS: 410 Proof: 230**
DIN 17100 St 46/3	2% C (max) 0.045% (max) 0.045% S (max) steel: Plate; German specification **UTS: 490 Proof: 280**
DIN 17100 St 50/1	0.2% C 0.08% P (max) 0.05% S (max) steel: Plate; German standard **UTS: 550 Proof: 290**
DIN 17100 St 52/3	0.21% C (max) 1.5% Mn (max) 0.045% P (max) 0.045% (max) steel: Plate; German specification **UTS: 570 Proof: 350**

Symbol	Nominal analysis, supplier, condition and remarks.
DIN 17100 USt 34/1	0.17% C (max) 0.08% P (max) 0.05% S (max) steel: Plate; German specification **UTS: 380 Proof: 200**
DIN 17100 USt 34/2	0.15% C (max) 0.05% S & P (max) 0.007% N (max) steel: Plate; German specification **UTS: 380 Proof: 200**
DIN 17100 USt 37/1	0.2% C (max) 0.07% P (max) 0.05% S (max) steel: Plate; German specification **UTS: 410 Proof: 230**
DIN 17100 USt 37/2	0.19% C (max) 0.05% P & S (max) 0.007% N (max) steel: Plate; German specification **UTS: 410 Proof: 230**
DIN 17100 USt 42/1	0.25% C (max) 0.08% P (max) 0.05% S (max) steel: Plate; German specification **UTS: 460 Proof: 250**
DIN 17110 MR St 34	0.1% C 0.3% Mn 0.05% S & P steel: For rivets
DIN 17710 MR St 44	0.15% C 0.6% Mn 0.06% S & P steel: For rivets
DIN 17110 MU St 34	0.1% C (max) 0.25% Mn 0.05% S & P steel: For rivets
DIN 17110 Q St 34	0.1% C (max) 0.3% Mn 0.04% S & P steel: For rivets
DIN 17110 TU St 34	0.08% C (max) 0.06% S 0.08% P steel: For rivets
DIN 17155 Gd 1	0.17% C (max) 0.05% S & P steel: Plate; for boiler
DIN 17155 Gd 11	0.23% C (max) 0.05% S & P steel: For boiler plate
DIN 17155 H1	0.16% C (max) 0.4% Mn (max) 0.05% P & S (max) 0.3% Cr (max) steel: Plate; German standard **UTS: 400 Proof: 220**
DIN 17155 H11	0.2% C (max) 0.5% Mn (max) 0.05% P & S (max) 0.3% Cr (max): German Standard **UTS: 440 Proof: 250**
DIN 17155 HM	0.22% C (max) 0.55% Mn (max) 0.3% Cr (max) 0.05% P & S (max) steel: Plate; German Standard
DIN 17155 17 Mn 4	0.17% C 0.1% Mn 0.3% Cr (max) 0.05% P & S (max) steel: Plate; German Standard **UTS: 520 Proof: 280**
DIN 17155 19 Mn 5	0.2% C 1.1% Mn 0.3% Cr (max) 0.05% P & S (max) steel: Plate; German Standard **UTS: 570 Proof: 320**
DIN 17172 RRSt 34/7	0.17% C 0.35% Mn (max) 0.045% P & S steel: Pipe; German Standard **UTS: 390 Proof: 210**
DIN 17172 RRSt 38/7	0.22% C (max) 0.4% Mn (max) 0.04% P & S (max) steel: Pipe; German Standard **UTS: 420 Proof: 250**
DIN 17172 RSt 34/7	0.17% C 0.35% Mn (max) 0.04% P & S steel: Pipe; German Standard **UTS: 390 Proof: 210**
DIN 17172 RSt 38/7	0.22% C (max) 0.4% Mn (max) 0.04% P & S (max) steel: Pipe; German Standard **UTS: 430 Proof: 250**
DIN 17172 St 43/7	0.22% C (max) 0.8% Mn 0.04% P & S (max) steel: Pipe; German Standard **UTS: 490 Proof: 300**
DIN 17172 St 47/7	0.22% C (max) 1.0% Mn 0.04% P & S (max) steel: Pipe; German Standard **UTS: 490 Proof: 300**
DIN 17172 St 53/7	0.22% C (max) 1.2% Mn 0.04% P & S (max) steel: Pipe; German Standard **UTS: 580 Proof: 370**
DIN 17172 USt 34/7	0.17% C 0.35% Mn (max) 0.04% P & S steel: Pipe; German Standard **UTS: 390 Proof: 210**

Note. The following abbreviations and units are used in the tables:

DPN	Hardness, diamond pyramid number
UTS	Ultimate tensile strength, N/mm^2
Elon	Elongation, %
Proof	0.1 % proof strength, N/mm^2

1 N/mm^2=0.1 hbar=0.102 kgf/mm^2=0.06475 tonf/in^2=145.04 lbf/in^2

Symbol	Nominal analysis, supplier, condition and remarks.
DIN 17172 USt 38/7	0.2% C (max) 0.4% Mn (max) 0.04% P & S (max) steel: Pipe; German Standard **UTS: 430** **Proof: 250**
DIN 17175 14 Mn 4	0.14% C 1.0% Mn 0.05% S & P steel: Tube
DIN 17175 St 35/8	0.17% C (max) 0.05% S & P steel: Tube
DIN 17175 St 35/8	0.17% C 0.4% Mn (max) 0.05% P & S (max) steel: Pipe; Japanese Standard; for high temperature **UTS: 400** **Proof: 240**
DIN 17175 St 45/8	0.22% C (max) 0.05% S & P steel: For seamless tube
DIN 17200 C10	0.1% C 0.45% Mn 0.045% P & S (max) steel: For structure; German Standard
DIN 17200 C15	0.15% C 0.45% Mn 0.045% P & S (max) steel: For structure; German Standard
DIN 17200 C22	0.21% C 0.45% Mn 0.045% P & S (max) steel: For structure; German Standard
DIN 17200 C 22	0.21% C 0.4% Mn 0.045% S & P steel
DIN 17200 CK 10	0.1% C 0.45% Mn 0.035% P & S (max) steel: For structures; German Standard
DIN 17200 CK 15	0.15% C 0.45% Mn 0.035% P & S (max) steel: For structures; German Standard
DIN 17200 CK 22	0.22% C 0.45% Mn 0.035% P & S (max) steel: For structures; German Standard
DIN 17200 CK 22	0.21% C 0.4% Mn 0.035% S & P steel
DIN 17200 CM 15	0.15% C 0.45% Mn 0.035% P & S (max) steel: For structures; German Standard
DIN 17210 C 10	0.1% C 0.3% Mn 0.045% S & P steel: For carburizing
DIN 17210 C 15	0.15% C 0.4% Mn 0.045% S & P steel: Bar; for carburizing
DIN 17210 CK 10	0.1% C 0.3% Mn 0.035% S & P steel: For carburizing
DIN 17210 CK 15	0.15% C 0.3% Mn 0.035% S & P steel: Bar; for carburizing
DISCUS	Mild steel corrugated sheet: Galvanized; RTB for BS 3083
DOFASCOLOY M	0.25% C 1.5% Mn 0.2% Cu steel: Dominion
DOFASCOLOY 45 W	0.25% C 1.2% Mn 0.01% Nb steel: Dominion
DOFASCOLOY 50 W	0.25% C 1.25% Mn 0.01% Nb steel: Dominion
DOFASCOLOY 55 W	0.25% C 1.25% Mn 0.01% Nb steel: Dominion
DOFASCOLOY 60 W	0.25% C 1.25% Mn 0.01% Nb steel: Dominion
DOFASCOLOY 100 W	0.25% C 1.25% Mn 0.01% Nb steel: Dominion
DRAGONITE	Low carbon electric-galvanized sheet & plate: Steel co of Wales for BS alloy En 2
DTD 720A	0.15% C 0.7% Mn 0.04% S 0.04% P steel: Softened & cold drawn; for rivets **UTS: 540** **Elon: 27%**
DTD 5199	C steel: Casting; investment cast **UTS: 500**
DTD 5209	C Mn steel: Casting; investment cast **UTS: 620**
DUCOL QT 122	0.17% C (max) 1.4% Mn 0.05% P & S steel: Colvilles
DUNELT 3	0.1% C 0.06% S & P steel: Dunford; normalized **UTS: 340** **Elon: 30%**
DUNELT 4	0.3% C 1% Mn 0.15% S 0.06% P steel: Free machining; Dunford; hardened & tempered **UTS: 600** **Elon: 15%**
DUNELT 5	0.15% C 1.2% Mn 0.15% S 0.06% P steel: Free machining; Dunford; hardened & tempered **UTS: 560** **Elon: 15%**
DUNELT 7	0.15% C 1.2% Mn 0.3% S 0.04% P steel: Free machining; Dunford; cold drawn **UTS: 600** **Elon: 15%**

Symbol	Nominal analysis, supplier, condition and remarks.
DUNELT 14	0.14% C 0.06% S & P steel: For case hardening; Dunford for BS alloy En32
DUNELT 21	0.2% C 1.0% Mn 0.06% S & P steel: Dunford; cold drawn **UTS: 420** **Elon: 17%**
DUNELT 39	0.22% C 1.5% Mn 0.06% S & P steel: Dunford for BS alloy En 14A
EF 206	0.2% C 0.7% Mn 0.035% P & S (max) steel: German proprietory specification; Al killed
ERA 60	0.18% C 1.2% Mn 0.05% S & P carburizing steel: Hadfields for BS alloy En 201
ERA 60M	0.18% C 1.2% Mn 0.05% P 0.15% S steel: Free machining; carburizing; Hadfields for BS alloy En 202
EX-TEN 42	0.21% C 0.9% Mn 0.01% Nb 0.02% V optional steel: US Steel Corp.
FAGESTA 1150	0.07% C (max) 0.25% Mn steel: For deep drawing; Fagesta
FAGESTA 1160	0.08% C (max) 0.25% Mn steel: For deep drawing; Fagesta
FAGESTA 1265	0.1% C 0.35% Mn steel: For carburizing; Fagesta; weldable
FAGESTA 1357	0.2% C 0.6% Mn steel: For carburizing; Fagesta; Weldable
FASTEX 5	0.06% C 0.6% Mn steel: Welding electrode; Murex **UTS: 540** **Elon: 29%** **Proof: 460**
FASTEX 7	0.08% C 0.4% Si 0.6% Mn steel: Welding electrode; Murex **UTS: 540** **Elon: 25%** **Proof: 480**
FASTEX 100	0.07% C 0.4% Mn steel: Welding electrode; Murex **UTS: 420** **Elon: 35%** **Proof: 330**
FB 50	0.2% C (max) 1.2% Mn 0.04% P & S (max) steel: German proprietory specification; Al killed
Fe 33	C not specified 0.07% P (max) 0.07% S (max) steel: Designation used by Italian Standard
Fe 34A	C not specified 0.065% S & P steel: Designation used by Italian Standards
Fe 34B	0.2% C (max) 0.045% S & P (max) steel: Designation used by Italian Standards
Fe 34C	0.2% C (max) 0.05% S & P (max) steel: Designation used by Italian Standards
Fe 37A	0.25% C (max) 0.065% P (max) 0.075% S (max) steel: Designation used by Italian Standards
Fe 37B	0.25% C (max) 0.055% P (max) 0.055% S (max) steel: Designation used by Italian Standards
Fe 37C	0.23% C 0.05% P (max) 0.055% S (max) steel: Designation used by Italian Standards
Fe 37D	0.22% C 0.045% P (max) 0.045% S (max) steel: Designation used by Italian Standards
Fe 42A	0.28% C (max) 0.065% P (max) 0.075% S (max) steel: Designation used by Italian Standard **UTS: 460** **Elon: 23%** **Proof: 245**
Fe 42B	0.24% C (max) 0.055% P (max) 0.055% S (max) 0.011% N (max) steel: Designation used by Italian Standards **UTS: 460** **Elon: 23%** **Proof: 245**
Fe 42C	0.22% C (max) 0.05% P (max) 0.055% S (max) 0.01% N (max) steel: Designation used by Italian Standards **UTS: 460** **Elon: 23%** **Proof: 245**
Fe 42D	0.22% C (max) 0.045% S & P (max) steel: Designation used by Italian Standards **UTS: 460** **Elon: 23%** **Proof: 245**
Fe 44A	0.28% C (max) 0.065% S & P (max) steel: Designation used by Italian Standards
Fe 44B	0.24% C (max) 0.055% S & P (max) steel: Designation used by Italian Standards
Fe 44C	0.24% C (max) 0.045% S & P (max) steel: Designation used by Italian Standards

Symbol	Nominal analysis, supplier, condition and remarks.
Fe 44D	0.22% C (max) 0.045% S & P (max) steel: Designation used by Italian Standards
Fe 50	C not specified 0.045% P 0.06% S steel: Designation used by Italian Standards **UTS: 550 Elon: 18% Proof: 295**
Fe 52B	0.26% C (max) 0.055% S & P steel: Designation used used by Italian Standards
Fe 52C	0.24% C (max) 0.04% S & P (max) steel: Designation used by Italian Standards
Fe 52D	0.23% C (max) 0.045% S & P (max) steel: Designation used by Italian Standards
Fe 60	C not specified 0.045% P (max) 0.06% S (max) steel: Designation used by Italian Standards **UTS: 640 Elon: 15% Proof: 335**
Fe 70	C not soecified 0.055% P (max) 0.06% S (max) steel: Designation used by Italian Standards **UTS: 750 Elon: 10% Proof: 365**
FCCH	0.14% C 0.25% Si 0.6% Mn steel: For carburizing; Firth Brown; for BS970 En32A
FEMAX 33.80	0.080% C 0.50% Sc 0.80% Mn steel: For electrodes; Esab **UTS: 556 Elon: 26% Proof: 479**
FERALSIN 52	0.2% C 1.3% Mn 0.04% P (max) 0.03% S (max) steel: Belgian specification; Al killed
FERALSIN 58	0.25% C 1.5% Mn 0.04% P (max) 0.03% S (max) steel: Belgian specification; Al killed
FILLETRODE 43.33	0.060% C 0.7% Mn 0.40% Sc steel: For electrodes; Esab **UTS: 556 Elon: 27% Proof: 479**
FLT 2A	0.11% C 1.06% Mn 0.12% Cu steel: Fuji; Al killed
FLT 2B	0.11% C 1.16% Mn 0.01% P & S 0.12% Cu steel: Fuji; Al killed
FNB 36A	0.23% C (max) 1.4% Mn 0.035% P & S (max) 0.1% Nb steel: Fuji
FNB 36B	0.23% C (max) 1.4% Mn 0.035% P & S (max) 0.1% Nb steel: Fuji
FNB 40A	0.26% C (max) 1.4Mn 0.035% P & S (max) 0.1% Nb steel: Fuji
FNB 40B	0.26% C (max) 1.4% Mn 0.035% P & S (max) 0.1% Nb steel: Fuji
FORTREX 30	0.07% C 0.4% Si 0.7% Mn steel: Welding electrode; Murex **UTS: 500 Elon: 35% Proof: 440**
FORTREX 35	0.08% C 0.5% Si 0.7% Mn steel: Welding electrode; Murex **UTS: 510 Elon: 34% Proof: 460**
FORTREX 35A	0.075% C 0.4% Si 0.7% Mn steel: Welding electrode; Murex **UTS: 480 Elon: 35% Proof: 440**
FTW 52	0.18% C (max) 1.5% Mn 0.03% P & S (max) steel: Fuji
FTW 55	0.18% C (max) 1.5% Mn 0.03% P & S (max) steel: Fuji
FTW 58	0.18% C (max) 1.5% Mn 0.03% P & S (max) steel: Fuji
G 3101/73/5534	C not specified 0.05% S & P (max) steel: For plate; Japanese specification **UTS: 390 Proof: 200**
GALVAMATT	0.2% C steel: Baldwins Ltd

Symbol	Nominal analysis, supplier, condition and remarks.
GENEX	0.02% C 0.1% Mn steel: Welding electrode; Murex
GLX 45W	0.16% C 0.68% Mn 0.025% Nb steel: National Steel Co.
GLX 50W	0.16% C 0.68% Mn 0.025% Nb steel: National Steel Co.
GLX 55	0.16% C 0.68% Mn 0.025% Nb steel: National Steel Co.
GLX 55	0.22% C (max) 1.35% Mn 0.01% Nb steel: National Steel Co.
GLX 60W	0.16% C 0.68% Mn 0.025% Nb steel: National Steel Co.
GLX 65	0.24% C 1.35% Mn 0.01% Nb steel: National Steel Co.
GLX 70	0.26% C 1.35% Mn 0.01% Nb steel: National Steel Co.
GOST 380 BST2 PS2	0.12% C 0.35% Mn 0.04% P (max) 0.05% S (max) 0.3% Cu Cr Ni (max) steel: Plate; Russian National specification
GOST 380 BST2 PS	0.12% C 0.35% Mn 0.04% P (max) 0.05% S (max) steel: Plate; Russian National specification
GOST 380 BST 25 P	0.18% C (max) 0.45% Mn 0.04% P (max) 0.05% S (max) steel: Plate; Russian National specification
GOST 380 2SP2	0.12% C 0.35% Mn 0.04% P (max) 0.05% S (max) 0.3% Cu Cr Ni (max) steel: Plate; Russian National specification
GOST 380 BST3 GPS	0.22% C (max) 0.9% Mn 0.04% P (max) 0.05% S (max) steel: Plate; Russian National specification **UTS: 440**
GOST 380 BST3 GPS2	0.22% C (max) 0.9% Mn 0.3% G Ni Cu (max) 0.04% P (max) 0.05% S (max) steel: Plate; Russian National specification **UTS: 420 Proof: 240**
GOST 380 BST3 KP2	0.18% C 0.45% Mn 0.04% P (max) 0.05% S (max) 0.3% Cu Cr Ni (max) steel: Plate; Russian National specification
GOST 380 BST3 PS	0.18% C 0.35% Mn 0.04% P (max) 0.05% S (max) steel: Russian National specification
GOST 380 BST3 SP	0.18% C 0.6% Mn 0.04% P (max) 0.05% S (max) steel: Plate; Russian National specification
GOST 380 BST3 SP2	0.18% C 0.5% Mn 0.3% G Ni Cu (max) 0.05% S (max) steel: Plate; Russian National specification
GOST 380 BST4 KP	0.22% C 0.6% Mn 0.4% P (max) 0.05% S (max) steel: Plate; Russian Standard
GOST 380 BST4 PS	0.22% C 0.5% Mn 0.04% P (max) 0.05% S (max) steel: Plate; Russian Standard
GOST 380 BST4 PS2	0.22% C 0.55% Mn 0.04% P (max) 0.05% S (max) 0.3% Cr Cu (max) steel: Plate; Russian specification
GOST 380 BST4 SP2	0.22% C 0.5% Mn 0.3% Cr Ni Cu (max) 0.04% P (max) 0.05% S (max) steel: Plate; Russian Standard
GOST 380 BST5 GPS	0.26% C 1.0% Mn 0.4% P (max) 0.05% S steel: Plate; Russian standard
GOST 380 STO	0.23% C (max) 0.07% P (max) 0.06% S (max) steel: Plate; Russian National specification **UTS: 310**
GOST 380 ST1 KP	0.09% C 0.35% Mn 0.04% P 0.05% S steel: Plate; Russian National specification **UTS: 370**
GOST 380 ST1 KP2	0.09% C 0.35% P (max) 0.3% Cu Cr Ni (max) steel: Plate; Russian National specification **UTS: 370**
GOST 380 ST1 PS	0.09% C 0.35% Mn 0.04% P 0.05% S steel: Plate; Russian National specification **UTS: 370**
GOST 380 ST1 SP	0.09% C 0.35% Mn 0.04% P (max) 0.05% S (max) steel: Plate; Russian National specification **UTS: 370**

Note. The following abbreviations and units are used in the tables:

DPN	Hardness, diamond pyramid number
UTS	Ultimate tensile strength, N/mm^2
Elon	Elongation, %
Proof	0.1 % proof strength, N/mm^2

1 N/mm^2=0.1 hbar=0.102 kgf/mm^2=0.06475 tonf/in^2=145.04 lbf/in^2

Symbol	Nominal analysis, supplier, condition and remarks.
GOST 380 ST1 PS2	0.09% C 0.35% Mn 0.04% P 0.05% S 0.3% Cu Cr Ni steel: Plate; Russian National specification **UTS: 370**
GOST 380 ST2 KP	0.12% C 0.35% Mn 0.04% P (max) 0.05% S (max) steel: Plate
GOST 380 ST2 KP2	0.12% C 0.35% Mn 0.04% P (max) 0.05% S (max) 0.03% Cu Cr Ni (max) steel: Plate; Russian National specification **UTS: 370**
GOST 380 VST2 KP	0.15% C (max) 0.4% Mn 0.04% P (max) 0.05% S (max) steel: Plate; Russian National specification **UTS: 350**
GOST 380 VST2 PS	0.15% C (max) 0.45% Mn 0.04% P (max) 0.05% S steel: Plate; Russian National specification **UTS: 390**
GOST 380 VST2 PS2	0.15% C (max) 0.45% Mn 0.04% P (max) 0.05% S (max) 0.3% Cu Cr Ni (max) steel: Plate; Russian National specification **UTS: 390** **Proof: 220**
GOST 380 VST25 P	0.15% C (max) 0.35% Mn 0.04% P (max) 0.05% S (max) steel: Plate; Russian National specification **UTS: 390**
GOST 380 VST2 SP2	0.15% C (max) 0.35% Mn 0.04% P (max) 0.05% S (max) 0.3% Cu Cr Ni (max) steel: Plate; Russian National specification **UTS: 390** **Proof: 220**
GOST 380 VST3 CPS3	0.22% C (max) 0.9% Mn 0.3% Cr Ni Cu (max) 0.04% P (max) 0.05% S (max) steel: Plate; Russian National specification **UTS: 440** **Proof: 240**
GOST 380 VST3 EPS4	0.22% C (max) 0.9% Mn 0.3% Cr Ni Cu (max) 0.04% P (max) 0.05% S (max) steel: Plate; Russian National specification **UTS: 420** **Proof: 240**
GOST 380 VST3 EPS5	0.22% C (max) 0.9% Mn 0.3% Cr Ni Cu (max) 0.04% P (max) 0.05% S (max) steel: Plate; Russian National specification **UTS: 420** **Proof: 240**
GOST 380 VST3 EPS6	0.22% C (max) 0.9% Mn 0.3% Cr Ni Cu (max) 0.04% P (max) 0.05% S (max) steel: Plate; Russian National specification **UTS: 420** **Proof: 240**
GOST 380 VST3 PS	0.22% C (max) 0.6% Mn 0.04% (max) 0.05% S (max) steel: Plate; Russian National specification **UTS: 430**
GOST 380 VST3 PS2	0.22% C (max) 0.6% Mn 0.03% Cr Ni Cu (max) 0.04% P 0.05% S steel: Plate; Russian National specification **UTS: 430** **Proof: 240**
GOST 380 VST3 PS3	0.22% C (max) 0.6% Mn 0.03% Cr Ni Cu (max) 0.04% P (max) 0.05% S (max) steel: Plate; Russian National specification **UTS: 430** **Proof: 250**
GOST 380 VST3 PS4	0.22% C (max) 0.65% Mn 0.3% Cr Ni Cu (max) 0.04% P 0.05% S steel: Plate; Russian National specification **UTS: 420** **Proof: 240**
GOST 380 VST3 PS5	0.22% C (max) 0.65% Mn 0.3% Cr Ni Cu (max) 0.04% P (max) 0.05% S (max) steel: Plate; Russian National specification **UTS: 420** **Proof: 250**
GOST 380 VST3 PS6	0.22% C (max) 0.6% Mn 0.3% Cr Ni Cu (max) 0.04% P (max) 0.05% S (max) steel: Plate; Russian National specification **UTS: 430** **Proof: 250**
GOST 380 VST3 SP	0.22% C (max) 0.55% Mn 0.04% P (max) 0.05% S (max) steel: Plate; Russian National specification **UTS: 420**

Symbol	Nominal analysis, supplier, condition and remarks.
GOST 380 VST3 SP2	0.22% C (max) 0.55% Mn 0.3% Cr Ni Cu (max) 0.04% P (max) 0.05% S (max) steel: Plate; Russian National specification **UTS: 420** **Proof: 240**
GOST 380 VST3 SP3	0.22% C (max) 0.55% Mn 0.3% Cr Ni Cu (max) 0.04% P (max) 0.05% S (max) steel: Plate; Russian National specification
GOST 380 VST3 SP4	0.22% C (max) 0.55% Mn 0.3% Cr Ni Cu (max) 0.04% P (max) 0.05% S (max) steel: Plate; Russian National specification **UTS: 420** **Proof: 240**
GOST 380 VST3 SP5	0.22% C (max) 0.55% Mn 0.3% G Ni Cu (max) 0.04% P (max) 0.05% S (max) steel: Plate; Russian National specification **UTS: 420** **Proof: 240**
GOST VST4 PS	0.27% C (max) 0.6% Mn 0.04% P (max) 0.05% S (max) steel: Plate; Russian Standard **UTS: 480**
GOST 380 VST4 PS2	0.27% C (max) 0.6% Mn 0.3% Cr Ni Cu (max) 0.04% P (max) 0.05% S (max) steel: Russian Standard **UTS: 480** **Proof: 260**
GOST 380 VST4 PS3	0.27% C (max) 0.6% Mn 0.3% Cr Ni Cu (max) 0.04% P (max) 0.05% S (max) steel: Plate; Russian Standard **UTS: 480** **Proof: 270**
GOST 380 VST4 SP	0.27% (max) 0.5% Mn 0.04% P (max) 0.05% S (max) steel: Plate; Russian Standard **UTS: 480** **Proof: 270**
GOST VST4 SP2	0.27% C (max) 0.5% Mn 0.3% Cr Ni Cu (max) 0.04% P (max) 0.05% S (max) steel: Plate; Russian Standard **UTS: 460** **Proof: 260**
GOST 380 VST4 SP3	0.27% C (max) 0.5% Mn 0.3% Cr Ni Cu (max) 0.04% P (max) 0.05% S (max) steel: Plate; Russian specification **UTS: 480** **Proof: 270**
GOST 380 VST3 SP6	0.22% C (max) 0.55% Mn 0.3% Cr Ni Cu (max) 0.04% P (max) 0.05% S (max) steel: Plate; Russian National specification **UTS: 420** **Proof: 250**
GOST 550 10	0.1% C 0.5% Mn 0.035% P & S (max) 0.25% Cr Ni (max) steel: Pipe; Russian Standard **UTS: 360** **Proof: 220**
GOST 550 10G2	0.12% C 1.4% Mn 0.035% P & S (max) 0.025% Cr Ni (max) steel: Pipe; Russian Standard **UTS: 480** **Proof: 270**
GOST 550 20	0.2% C 0.5% Mn 0.04% P & S (max) 0.25% Cr Ni (max) steel: Pipe; Russian Standard **UTS: 440** **Proof: 260**
GOST 631 D	C not specified 0.045% P & S (max) steel: Pipes Russian Standard **UTS: 650** **Proof: 380**
GOST 613 E	C not specified 0.045% P & S (max) steel: Pipe; Russian Standard **UTS: 750** **Proof: 550**
GOST 631 K	C not specified 0.045% P & S (max) steel: Pipe; Russian Standard **UTS: 700** **Proof: 500**
GOST 631 L	C not specified 0.045% P & S (max) steel: Pipe; Russian Standard **UTS: 800** **Proof: 650**
GOST 631 M	C not specified 0.045% P & S (max) steel: Pipe; Russian Standard **UTS: 900** **Proof: 750**
GOST 632 S	C not specified 0.045% P & S (max) steel: Pipe; Russian Standard **UTS: 550** **Proof: 320**

Symbol	Nominal analysis, supplier, condition and remarks.
GOST 633 D	C not specified 0.045% P & S steel: Pipe; Russian Standard
	UTS: 650 **Proof: 380**
GOST 1050/05 KP	0.06% C (max) 0.4% Mn (max) 0.035% P & S (max) 0.1% Cr 0.25% Ni steel: For structures; Russian Standard
GOST 1050/08	0.08% C 0.5% Mn 0.035% P & S (max) 0.0% Cr (max) 0.2% Ni (max) steel: For structures; Russian Standard
GOST 1050/08 KP	0.08% C 0.4% Mn 0.04% P & S (max) 0.1% Cr (max) 0.25% Ni (max) steel: For structures; Russian Standard
GOST 1050/10	0.1% C 0.5% Mn 0.035% P & S (max) 0.15% Cr (max) 0.2% Ni (max) steel: For structures; Russian Standard
GOST 1050/10 KP	0.1% C 0.4% Mn 0.04% P & S (max) 0.15% Cr (max) 0.2% Ni (max) steel: For structures; Russian Standard
GOST 1050/15	0.15% C 0.5% Mn 0.04% P & S (max) 0.25% Cr & Ni (max) steel: For structures; Russian Standard
GOST 1050/15 G	0.15% C 0.85% Mn 0.04% P & S (max) 0.25% Cr & Ni (max) steel: For structures; Russian Standard
GOST 1050/15 KP	0.15% C 0.4% Mn 0.04% P & S (max) 0.025% Cr & Ni (max) steel: For structures; Russian Standard
GOST 1050/20	0.2% C 0.45% Mn 0.04% P & S (max) 0.25% Cr & Ni (max) steel: For structures; Russian Standard
GOST 1050/20 G	0.2% C 0.85% Mn 0.04% P & S (max) 0.25% Cr & Ni (max) steel: For structures; Russian Standard
GOST 1050/20 KP	0.2% C 0.35% Mn 0.04% P & S (max) 0.25% Cr & Ni (max) steel: For structures; Russian Standard
GOST 1060/10	0.1% C 0.5% Mn 0.035% P & S (max) 0.15% Cr (max) 0.2% Ni (max) steel: Tube; Russian Standard
	UTS: 350
GOST 3262	Steel pipe for water and gas: Russian Standard
GOST 4543/09 C2S	0.12% C (max) 1.5% Mn 0.3% Cr & Ni (max) steel: Russian Standard
GOST 4543/09 G2	0.12% C (max) 1.6% Mn 0.3% Cr & Ni (max) steel: Russian Standard
GOST 4543/10 G2S1	0.12% C (max) 1.5% Mn 0.3% Cr & Ni (max) steel: Russian Standard
GOST 4543/12 GS	0.12% C 1.0% Mn 0.3% Cr & Ni (max) steel: Russian Standard
GOST 4543/14 G	0.14% C 0.85% Mn 0.3% Cr & Ni (max) steel: Russian Standard
GOST 4543/14 G2	0.14% C 1.4% Mn 0.3% Cr & Ni (max) steel: Russian Standard
GOST 4543/15 GF	0.15% C 1.05% Mn 0.3% Cr & Ni (max) steel: Russian Standard
GOST 4543/16 ES	0.16% C 1.05% Mn 0.3% Cr & Ni (max) steel: Russian Standard
GOST 4543/17 GS	0.17% C 1.2% Mn 0.3% Cr & Ni (max) steel: Russian Standard

Symbol	Nominal analysis, supplier, condition and remarks.
GOST 4543/18 G2	0.18% C 1.4% Mn 0.3% Cr & Ni (max) steel: Russian Standard
GOST 4543/18 G2S	0.18% C 1.9% Mn 0.3% P & S (max) steel: Russian Standard
GOST 4543/19 G	0.19% C 0.95% Mn 0.3% Cr & Ni (max) steel: Russian Standard
GOST 4543/27 SG	0.27% C 1.2% Mn 0.25% Cr (max) steel: Russian Standard
GOST 5005/15	0.15% C 0.5% Mn 0.04% P & S (max) 0.25% Cr & Ni (max) steel: Pipe; Russian Standard; welded
	UTS: 480 **Proof: 380**
GOST 5005/20	0.2% C 0.5% Mn 0.04% P & S (max) 0.25% Cr & Ni (max) steel: Pipe; Russian Standard; welded
	UTS: 500 **Proof: 400**
GOST 5058/09 G2	0.12% C (max) 1.6% Mn 0.3% Cr Ni Cu (max) 0.035% P (max) 0.04% S (max) steel: Plate; Russian Standard
GOST 5058/09 G2S	0.12% C 1.5% Mn 0.3% Cr Ni Cu (max) 0.035% P (max) 0.05% S (max) steel: Plate; Russian Standard
	UTS: 450 **Proof: 290**
GOST 5058/10 G2S1	0.12% C (max) 1.45% Mn 0.3% Cr Ni Cu (max) 0.035% P (max) 0.05% S (max) steel: Plate; Russian Standard
	UTS: 500 **Proof: 350**
GOST 5058/12 GS	0.12% C 1.0% Mn 0.3% Cr Ni Cu (max) 0.035% P (max) 0.04% S (max) steel: Plate; Russian Standard
GOST 5058/14 G	0.15% C 0.8% Mn 0.03% Cr Ni Cu (max) 0.0035% P (max) 0.004% P (max) steel: Plate; Russian Standard
	UTS: 400 **Proof: 290**
GOST 5058/14 G2	0.16% C 1.4% Mn 0.3% Cr Ni Cu (max) 0.035% P (max) 0.04% S (max) steel: Plate; Russian Standard
GOST 5058/14 KHGS	0.13% C 1.1% Mn 0.3% Cr Ni Cu (max) 0.035% P (max) 0.04% S (max) steel: Plate; Russian Standard
	UTS: 500 **Proof: 350**
GOST 5058/16 GS	0.15% C 1.1% Mn 0.3% Cr Ni Cu (max) 0.035% P (max) 0.05% S (max) steel: Plate; Russian Standard
	UTS: 440 **Proof: 270**
GOST 5058/17 GS	0.17% C 1.2% Mn 0.3% Cr Ni Cu (max) 0.035% P (max) 0.04% S (max) steel: Plate; Russian Standard
	UTS: 500 **Proof: 340**
GOST 5058/18 G2	0.17% C 1.4% Mn 0.3% Cr Ni Cu (max) 0.035% P 0.04% S steel: Plate; Russian Standard
	UTS: 520 **Proof: 360**
GOST 5058/19 G	0.19% C 0.95% Mn 0.3% Cr Ni Cu (max) 0.55% P (max) 0.04% S (max) steel: Plate; Russian Standard
	UTS: 480 **Proof: 320**
GOST 5520/12 K	0.12% C 0.55% Mn 0.3% Cr Ni Cu (max) 0.04% P & S steel: Plate; Russian Standard
	UTS: 410 **Proof: 220**
GOST 5520/15 K	0.16% C 0.5% Mn 0.3% Cr Ni Cu (max) 0.04% P & S (max) steel: Plate; Russian Standard
	UTS: 440 **Proof: 220**
GOST 5520/18 K	0.18% C 0.65% Mn 0.3% Cr Ni Cu (max) 0.04% P & S steel: Plate; Russian Standard
	UTS: 500 **Proof: 270**
GOST 5520/69/16 K	0.16% C 0.6% Mn 0.3% Cr Ni Cu (max) 0.3% P & S (max) steel: Plate; Russian Standard
	UTS: 460 **Proof: 250**
GOST 5520/69/20 K	0.2% C 0.6% Mn 0.3% Cu Ni Cu (max) 0.04% P & S (max) steel: Plate; Russian Standard
	UTS: 470 **Proof: 240**

Note. The following abbreviations and units are used in the tables:

DPN	Hardness, diamond pyramid number
UTS	Ultimate tensile strength, N/mm^2
Elon	Elongation, %
Proof	0.1 % proof strength, N/mm^2

$1 \text{ N/mm}^2 = 0.1 \text{ hbar} = 0.102 \text{ kgf/mm}^2 = 0.06475 \text{ tonf/in}^2 = 145.04 \text{ lbf/in}^2$

Symbol	Nominal analysis, supplier, condition and remarks.

GOST 5654/10 0.1% C 0.5% Mn 0.035% P & S (max) 0.15% Cr (max) 0.2% Ni (max) steel: Plate; Russian Standard; seamless
UTS: 360 **Proof: 220**

GOST 5654/20 0.12% C 0.5% Mn 0.04% P & S (max) 0.25% Cr (max) 0.2% Ni (max) steel: Plate; Russian Standard; seamless
UTS: 440 **Proof: 260**

GOST 8467 D C not specified 0.045% P & S steel: Pipe; Russian Standard
UTS: 650 **Proof: 380**

GOST 8696/10/G2SD 0.12% C (max) 1.5% Mn 0.04% P & S (max) 0.3% Cr & Ni (max) steel: Pipe; Russian Standard; welded
UTS: 500 **Proof: 350**

GOST 8696 MST2KP 0.11% C 0.35% Mn 0.045% P (max) 0.005% S (max) steel: Pipe; Russian Standard; welded
UTS: 340 **Proof: 220**

GOST 8696 MST3 0.12% C (max) 0.4% Mn 0.08% P (max) 0.08% P (max) 0.05% S (max) steel: Pipe; Russian Standard; welded
UTS: 380 **Proof: 240**

GOST 8696 MST3 0.18% C 0.5% Mn 0.045% P (max) 0.055% S (max) steel: Pipe; Russian Standard; welded
UTS: 380 **Proof: 240**

GOST 8696 MST3KP 0.18% C 0.45% Mn 0.045% P (max) 0.055% S (max) steel: Pipe; Russian Standard; welded
UTS: 380 **Proof: 240**

GOST 8731/10 0.1% C 0.5% Mn 0.035% P & S (max) 0.15% Cr (max) 0.2% Ni (max) steel: Pipe; Russian Standard; seamless
UTS: 340 **Proof: 210**

GOST 8731/10/G2 0.13% C 1.4% Mn 0.035% P & S (max) 0.25% Cr (max) 0.2% Ni (max) steel: Pipe; Russian Standard; seamless
UTS: 480 **Proof: 270**

GOST 8731/20 0.12% C 0.5% Mn 0.04% P & S (max) 0.25% Cr (max) 0.2% Ni (max) steel: Pipe; Russian Standard; seamless
UTS: 420 **Proof: 250**

GOST 8731 BMST4 SP 0.22% C 0.55% Mn 0.045% P (max) 0.055% S (max) steel: Pipe; Russian Standard; seamless
UTS: 420 **Proof: 250**

GOST 8733/10 0.12% C 0.55% Mn 0.035% P & S (max) 0.15% Cr (max) 0.2% Ni (max) steel: Pipe; Russian Standard; seamless
UTS: 340 **Proof: 210**

GOST 8733/10 G2 0.12% C 1.4% Mn 0.035% P & S (max) 0.15% Cr (max) 0.2% Ni (max) steel: Pipe; Russian Standard; seamless
UTS: 430 **Proof: 250**

GOST 8733/20 0.20% C 0.55% Mn 0.035% P & S (max) 0.15% Cr (max) 0.2% Ni (max) steel: Pipe; Russian Standard; seamless
UTS: 420 **Proof: 250**

GOST 10705/08 0.08% C 0.5% Mn 0.035% P & S (max) 0.1% Cr (max) 0.2% Ni (max) steel: Pipe; Russian Standard; welded
UTS: 340 **Proof: 320**

GOST 10705/10 0.1% C 0.5% Mn 0.04% P & S (max) 0.15% Cr (max) 0.2% Ni (max) steel: Pipe; Russian Standard; welded
UTS: 340

GOST 10705/15 0.15% C 0.5% Mn 0.04% P & S (max) 0.25% Cr (max) 0.2% Ni (max) steel: Pipe; Russian Standard; welded
UTS: 380

Symbol	Nominal analysis, supplier, condition and remarks.

GOST 10705/20 0.2% C 0.5% Mn 0.04% P & S (max) 0.25% Cr (max) 0.2% Ni (max) steel: Pipe; Russian Standard; welded
UTS: 400

GOST 10705 BST3 0.12% C (max) 0.4% Mn 0.08% P (max) 0.06% S (max) steel: Pipe; Russian Standard; welded
UTS: 380

GOST 10705 BST4 0.14% C 0.45% Mn 0.08% P (max) 0.06% S (max) steel: Pipe; Russian Standard; welded
UTS: 400

GOST 10705 BKST3 0.18% C 0.55% Mn 0.045% P (max) 0.055% S (max) steel: Pipe; Russian Standard; welded
UTS: 380

GOST 10705 BKST4 0.22% C 0.5% Mn 0.045% P (max) 0.055% S (max) steel: Pipe; Russian Standard; welded
UTS: 400

GOST 10705 MST2 0.13% C 0.4% Mn 0.045% P (max) 0.055% S (max) steel: Pipe; Russian Standard; welded

GOST 10705 MST3 0.18% C 0.45% Mn 0.45% P (max) 0.055% S (max) steel: Pipe; Russian Standard; welded

GOST 10705 MST4 0.22% C 0.55% Mn 0.045% P (max) 0.055% S (max) steel: Pipe; Russian Standard; welded

GOST 10706/09 G2 0.12% C (max) 1.6% Mn 0.04% P & S (max) 0.3% Cr Ni & Cu (max) steel: Pipe; Russian Standard; welded
UTS: 460 **Proof: 310**

GOST 10706/102 SR 0.12% C (max) 1.45% Mn 0.04% P & S (max) 0.3% Cr & Ni (max) 0.22% Cu steel: Pipe; Russian Standard; welded
UTS: 500 **Proof: 350**

GOST 10/06/14 G 0.14% C 0.85% Mn 0.04% P & S (max) 0.3% Cr Ni & Cu (max) steel: Pipe; Russian Standard; welded
UTS: 460 **Proof: 290**

GOST 10706/14/G2 0.16% C 1.4% Mn 0.04% P & S (max) 0.3% Cr Ni & Cu (max) steel: Pipe; Russian Standard
UTS: 480 **Proof: 340**

GOST 10706/15 G2S 0.15% C 1.1% Mn 0.04% P & S (max) 0.3% Cr Ni & Cu (max) steel: Pipe; Russian Standard; welded
UTS: 500 **Proof: 350**

GOST 10706/18 G2S 0.18% C 1.4% Mn 0.04% P & S (max) 0.3% Cr Ni & Cu (max) steel: Pipe; Russian Standard; welded
UTS: 600 **Proof: 400**

GOST 10706/19 G 0.19% C 0.9% Mn 0.04% P & S (max) 0.3% Cr Ni & Cu (max) steel: Pipe; Russian Standard; welded
UTS: 470 **Proof: 300**

GOST 10802/15 GS 0.15% C 1.1% Mn 0.04% P & S (max) 0.3% Cr & Ni (max) steel: Pipe; Russian Standard
UTS: 500 **Proof: 300**

GOST 10802/20 0.13% C 0.5% Mn 0.04% P & S (max) 0.15% Cr (max) 0.25% Ni (max) steel: Pipe; Russian Standard
UTS: 410 **Proof: 250**

GOST 11017/20 0.2% C 0.5% Mn 0.25% Cr (max) 0.2% Ni (max) steel: Plate; Russian Standard; seamless
UTS: 400

GOST 12132/08 0.08% C 0.5% Mn 0.035% P & S (max) 0.1% Cr (max) 0.25% Ni (max) steel: Pipe; Russian Standard; welded
UTS: 320

GOST 12132/10 0.1% C 0.5% Mn 0.035% P & S (max) 0.15% Cr (max) 0.25% Ni (max) steel: Pipe; Russian Standard; welded
UTS: 340

GOST 12132/20 0.2% C 0.5% Mn 0.04% P & S (max) 0.25% Cr & Ni (max) steel: Pipe; Russian Standard; welded
UTS: 420

Symbol	Nominal analysis, supplier, condition and remarks.
GOST 14162/08	0.08% C 0.5% Mn 0.035% P & S (max) 0.1% Cr (max) 0.25% Ni (max) steel: Capillary tube; Russian Standard
	UTS: 320
GOST 14162/10	0.1% C 0.5% Mn 0.035% P & S (max) 0.15% Cr (max) 0.25% Ni (max) steel: Capillary tube; Russian Standard
	UTS: 340
H 1	0.16% C (max) 0.4% Mn 0.05% S & P 0.3% Cr (max) steel: Designation used by German Standard
H1 V	0.26% C (max) 0.3% Cr steel: German Standard
HECLA 26	0.15% C steel: Casting; high permeability; Hadfields for BS 1617 A & B
	UTS: 420 Elon: 22% Proof: 200
HECLA 35	0.2% C 0.5% Mn steel: Hadfields for BS alloys En 2C, En 2D, En 3A, En 3C, En 4
HECLA 46	0.15% C 0.6% Mn 0.5% S & P: Carburizing steel; Hadfields for BS alloy En32A
HECLA CH 31	0.15% C 0.8% Mn 0.06% S & P steel: For carburizing; Hadfields for BS alloy En32B
HEDEX O	0.2% C 1% Mn 0.06% S 0.04% P steel: Wire for forging; Kiveton Park; annealed
	UTS: 530
HIBIL 36	0.2% C (max) 1.5% Mn 0.04% P & S (max) 0.1% Nb steel: Nippon Kokan
HIBIL TYPE A	0.13± C (max) 1.05% Mn 0.35% P & S (max) 0.05% Nb steel: Nippon Kokan
HIBIL TYPE B	0.15% C (max) 1.18% Mn 0.05% Nb steel: Nippon Kokan
HIBIL TYPE C	0.15% C (max) 1.8% Mn 0.05% Nb steel: Nippon Kokan
HICON	0.16% C (max) 1.45% Mn 0.04% P & S (max) 0.05% Nb steel: Nippon Kokan
HICON 36	0.2% C (max) 1.5% Mn 0.04% P & S (max) 0.1% Nb steel: Nippon Kokan
HICON 40	0.2% C (max) 1.5% Mn 0.035% P & S (max) 0.1% Nb steel: Nippon Kokan
HI-F	0.12% C 0.75% Mn 0.2% Cu 0.015% Nb steel: McLouth Steel Corp.
HI-KILLED	0.25% C 1.3% Mn 0.2% Cu steel: Inland Steel Co.
HI-YIELD 45	0.22% C 1.25% Mn 0.025% Nb steel: Granite City Steel Co.
HI-YIELD 50	0.22% C 1.25% Mn 0.02% Nb steel: Granite City Steel Co.
HI-YIELD 55	0.25% C 1.35% Mn 0.02% Nb steel: Granite City Steel Co.
HI-YIELD 60	0.25% C 1.35% Mn 0.02% Nb steel: Granite City Steel Co.
HMS 35	0.1% C 0.25% Si 0.4% Mn 0.03% S & P steel: Krupp
HMS 35/Z	0.12% C 0.25% Si 0.5% Mn 0.05% S & P steel: Krupp
HMS 40	0.15% C 0.25% Si 0.4% Mn 0.03% S & P steel: Krupp
HMS 41/Z	0.15% C 0.25% Si 0.6% Mn 0.05% S & P steel: Krupp
HMS 45	0.22% C 0.25% Si 0.45% Mn 0.035% S & P steel: Krupp
HMS 45/Z	0.17% C 0.25% Si 0.55% Mn 0.05% S & P steel: Krupp

Symbol	Nominal analysis, supplier, condition and remarks.
HSA 1	0.17% C 1.3% Mn 0.2% Cu steel: Toto Steel
HSB	0.23% C 1.3% Mn 0.045% P & S (max) 0.2% Cu steel: Yawata
HSB 40	0.18% C (max) 0.65% Mn 0.04% P & S (max) 0.02% Al steel: Phoenix Rheinrohr
HSB 40S	0.18% C (max) 0.65% Mn 0.04% P & S (max) 0.02% Al (max) steel: Phoenix Rheinrohr
HSB 45	0.18% C (max) 0.85% Mn 0.04% P & S (max) 0.02% Al steel: Phoenix Rheinrohr
HSB 45S	0.18% C (max) 0.85% Mn 0.04% P & S (max) 0.02% Al (max) steel: Phoenix Rheinrohr
HSB 50	0.2% C (max) 1.2% Mn 0.04% P & S (max) 0.02% Al (max) steel: Phoenix Rheinrohr
HSB 50S	0.2% C (max) 1.2% Mn 0.04% P & S (max) 0.02% Al steel: Phoenix Rheinrohr
HTP 52W	0.18% C (max) 1.1% Mn 0.3% Cu steel: Kawasaki Iron & Steel Co.
HTP 57W	0.2% C (max) 1.3% Mn 0.1% Cr 0.03% Cu steel: Kawasaki Iron & Steel Co.
HYPHUS	0.2% C (max) 1% Mn 1.6% Cr + Mn (max): Appleby Frodingham; weldable
	UTS: 540 Elon: 21% Proof: 330
HYPHUS 23	0.2% C (max) 1.5% Mn 0.5% Si 0.05% S & P 0.1% Nb steel: Appleby Frodingham; weldable
	UTS: 550 Elon: 20% Proof: 350
HYPHUS 29	0.22% C (max) 1.6% Mn 0.3% Si 0.05% S & P 0.2% V steel: Appleby Frodingham; weldable
	UTS: 1000 Elon: 20% Proof: 700
HYPLUS 111	0.22% C (max) 1.5% Mn 0.05% Nb steel: Appleby Frodingham
IH 50	0.22% C 1.5% Mn 0.2% Cu steel: IH Wisconsin Steel
IN 1	0.17% C (max) 0.8% Mn 0.01% P & S steel: Yawata
IN 4	0.08% C (max) 0.68% Mn 0.01% P 0.005% S steel: Yawata
IN 5	0.09% C (max) 1.5% Mn 0.015% P 0.008% S steel: Yawata
IN 6	0.11% C 0.78% Mn 0.015% P 0.006% S steel: Yawata
IN 10	0.09% C (max) 0.8% Mn 0.07% Al steel: Yawata
IN A	0.1% C (max) 1.5% Mn 0.01% P & S steel: Yawata
IN B	0.13% C 1.4% Mn 0.015% P & S steel: Yawata
IRONEX 2	0.06% C 0.5% Mn steel: Welding electrode; Murex
	UTS: 460 Elon: 30% Proof: 440
IRONEX 5	0.06% C 0.5% Mn steel: Welding electrode; Murex
	UTS: 480 Elon: 30% Proof: 450
IRONEX 7	0.065% C 0.7% Mn 0.4% Si steel: Welding electrode; Murex
	UTS: 510 Elon: 28% Proof: 480
JALTEN 2	0.15% C (max) 1.4% Mn 0.3% Cu steel: Jones & Laughlin
JALTEN 3	0.25% C (max) 1.6% Mn 0.2% Cu steel: Jones & Laughlin
JALTEN 3R	0.25% C (max) 1.6% Mn 0.2% Cu steel: Jones & Laughlin
JALTEN 3S	0.25% C (max) 1.6% Mn 0.2% Cu steel: Jones & Laughlin
JIS B2351 STPS1	0.2% C (max) 0.45% Mn 0.04% P & S steel: Pipe; Japanese Standard
	UTS: 450 Proof: 180
JIS B2351 STPS2	0.12% C 0.45% Mn 0.035% P & S (max) 0.2% Cu (max) steel: Pipe; Japanese Standard
	UTS: 450 Proof: 180
JIS C8305	0.1% C (max) 0.35% Mn 0.04% P & S (max) steel: Conduit tube; Japanese Standard
	UTS: 280

Note. The following abbreviations and units are used in the tables:

DPN	Hardness, diamond pyramid number
UTS	Ultimate tensile strength, N/mm^2
Elon	Elongation, %
Proof	0.1 % proof strength, N/mm^2

1 N/mm^2=0.1 hbar=0.102 kgf/mm^2=0.06475 tonf/in^2=145.04 lbf/in^2

Symbol	Nominal analysis, supplier, condition and remarks.
JIS G3101 SS41	C not specified 0.05% S & P (max) steel: Plate; National specification; Japanese **UTS: 455** **Proof: 240**
JIS G3101 SS50	C not specified 0.05% P & S (max) steel: Plate; Japanese Standard **UTS: 560** **Proof: 280**
JIS G3103 SB35	0.21% C 0.8% Mn (max) 0.035% P (max) 0.04% S (max) steel: Plate; Japanese Standard **UTS: 380** **Proof: 190**
JIS G3103 SB42	0.25% C 0.8% Mn (max) 0.035% P (max) 0.04% S (max) steel: Plate; Japanese Standard **UTS: 460** **Proof: 230**
JIS G3103 SB46	0.3% (max) 0.9% Mn (max) 0.035% P (max) 0.04% S (max) steel: Plate; Japanese Standard **UTS: 500** **Proof: 250**
JIS G3103 SG30	0.2% C (max) 1.0% Mn (max) 0.04% P & S steel: Plate; Japanese Standard **UTS: 450**
JIS G3103 SP49	0.35% C (max) 0.9% Mn (max) 0.035% P (max) 0.04% S (max) steel: Plate; Japanese Standard **UTS: 540** **Proof: 270**
JIS G3106 B	0.19% C (max) 1.5% Mn (max) 0.04% P & S steel: Plate; Japanese Standard **UTS: 560** **Proof: 320**
JIS G3106 SM41A	0.24% C Mn 2.5% C 0.04% P & S (max) steel: Plate; Japanese National Standard **UTS: 460** **Proof: 250**
JIS G3106 SM41B	0.21% C 0.9% Mn 0.04% P & S (max) steel: Plate; Japanese National Standard **UTS: 460** **Proof: 240**
JIS G3106 SM41C	0.18% C (max) 1.4% Mn (max) 0.04% P & S (max) steel: Plate **UTS: 460** **Proof: 220**
JIS G3106 SM50C	0.18% C (max) 1.5% Mn (max) 0.04% P & S (max) steel: Plate; Japanese Standard **UTS: 560** **Proof: 320**
JIS G3106 SM50Y A & B	8.2% C (max) 1.5% Mn (max) 0.04% P & S (max) steel: Plate; Japanese Standard **UTS: 560** **Proof: 360**
JIS 3106 SM53 B & C	0.2% C (max) 1.5% Mn (max) 0.04% P & S (max) steel: Plate; Japanese Standard **UTS: 490** **Proof: 360**
JIS G3106 SM58	0.18% C (max) 1.5% Mn (max) 0.04% S & P steel: Plate; Japanese Standard; hardened; tempered **UTS: 650** **Proof: 460**
JIS G3114 SMA41 A, B & C	0.2% C (max) 1.4% Mn (max) 0.25% Cr 0.3% Cu 0.04% P & S steel: Plate; Japanese Standard **UTS: 460** **Proof: 240**
JIS G3115 SPV24	0.19% C 1.4% Mn (max) 0.035% P (max) 0.04% S (max) steel: Plate; Japanese Standard **UTS: 450** **Proof: 230**
JIS G3115 SPV32	0.18% C (max) 1.5% Mn (max) 0.035% P (max) 0.04% S (max) steel: Plate; Japanese Standard **UTS: 560** **Proof: 310**
JIS G3115 SPV36	0.2% C (max) 1.6% Mn 0.035% P 0.04% S (max) steel: Plate; Japanese Standard **UTS: 590** **Proof: 340**
JIS G3115 SPV46	0.18% C (max) 1.6% Mn (max) 0.035% P (max) 0.04% S (max) steel: Plate; Japanese Standard; hardened & tempered **UTS: 630** **Proof: 440**
JIS G3115 SPV50	0.18% (max) 0.025% P (max) 0.04% S (max) steel: Plate; Japanese Standard; hardened & tempered **UTS: 670** **Proof: 420**
JIS G3116 SG26	0.2% C (max) 0.3% Mn (max) 0.04% P & S (max) steel: Plate; Japanese Standard **UTS: 410**
JIS G3116 SG30	0.2% C (max) 1.0% Mn (max) 0.04% P & S (max) steel: Plate **UTS: 450** **Proof: 300**
JIS G3116 SG33	0.2% C (max) 1.5% Mn (max) 0.04% P & S (max) steel: Plate; Japanese Standard **UTS: 500** **Proof: 270**
JIS G3116 SG37	0.2% C (max) 1.5% Mn (max) 0.04% P & S (max) steel: Plate; Japanese Standard **UTS: 550** **Proof: 370**
JIS G3118 SGV42	0.25% C 1.0% Mn 0.035% P (max) 0.04% S (max) steel: Plate; Japanese Standard **UTS: 460** **Proof: 230**
JIS G3118 SGV46	0.29% C (max) 1.0% Mn 0.035% P (max) 0.04% S (max) steel: Plate; Japanese Standard **UTS: 500** **Proof: 250**
JIS G3118 SGV49	0.31% C (max) 1.0% Mn 0.035% P (max) 0.04% S (max) steel: Plate; Japanese Standard **UTS: 540** **Proof: 270**
JIS G3126 SLA24A	0.15% C (max) 1.1% Mn 0.035% P & S (max) steel: Plate; Japanese Standard; normalized **UTS: 460** **Proof: 220**
JIS G3126 SLA24B	0.15% C (max) 1.1% Mn 0.035% P & S (max) steel: Plate; Japanese Standard; normalized **UTS: 450** **Proof: 220**
JIS G3126 SLA33A	0.16% C (max) 1.2% Mn 0.035% P & S (max) steel: Plate; Japanese Standard; normalized **UTS: 510** **Proof: 330**
JIS G3126 SLA33B	0.16% C (max) 1.2% Mn 0.035% P & S (max) steel: Plate; Japanese Standard; hardened & tempered **UTS: 510** **Proof: 330**
JIS G3126 SLA37	0.18% C (max) 1.2% Mn 0.035% P & S (max) steel: Plate; Japanese Standard; hardened & tempered **UTS: 560** **Proof: 370**
JIS G3444 STK30	C not specified 0.05% P & S steel: Tube; Japanese Standard **UTS: 300**
JIS G3444 STK41	0.25% C 0.04% P & S steel: Pipe; Japanese Standard **UTS: 410** **Proof: 240**
JIS G3444 STK50	0.18% C (max) 1.5% Mn (max) 0.04% P & S (max) steel: Pipe; Japanese Standard **UTS: 500** **Proof: 400**
JIS G3444 STK51	0.3% C (max) 0.7% Mn 0.04% S & P steel: Pipe; Japanese Standard **UTS: 510** **Proof: 360**
JIS G3444 STK55	0.23% C (max) 1.5% Mn (max) 0.04% P & S steel: Pipe; Japanese Standard **UTS: 550** **Proof: 400**
JIS G3445 STKM11A	0.12% C (max) 0.4% Mn 0.04% P & S (max) steel: Pipe; Japanese Standard **UTS: 300**
JIS G3445 STKM12A	0.2% C (max) 0.4% Mn 0.04% P & S (max) steel: Pipe; Japanese Standard **UTS: 400** **Proof: 280**
JIS G3445 STKM12B	0.2% C (max) 0.5% Mn 0.04% P & S (max) steel: Pipe; Japanese Standard **UTS: 400** **Proof: 280**
JIS G3445 STKM12C	0.2% C (max) 0.45% Mn 0.04% P & S steel: Pipe; Japanese Standard **UTS: 480** **Proof: 360**
JIS G3445 STKM13A	0.25% C (max) 0.6% Mn 0.04% P & S steel: Pipe; Japanese Standard **UTS: 380** **Proof: 220**
JIS G3445 STKM13B	0.25% C (max) 0.6% Mn 0.04% P & S (max) steel: Pipe; Japanese Standard **UTS: 450** **Proof: 310**

Symbol	Nominal analysis, supplier, condition and remarks.

JIS G3445 STKM13C — 0.25% C (max) 0.6% Mn 0.04% P & S (max) steel: Pipe; Japanese Standard
UTS: 520 **Proof: 390**

JIS G3445 STKM18A — 0.18% C (max) 1.5% Mn (max) 0.04% P & S (max) steel: Pipe; Japanese Standard
UTS: 450 **Proof: 280**

JIS G3445 STKM18B — 0.18% C (max) 1.5% Mn (max) 0.04% P & S (max) steel: Pipe; Japanese Standard
UTS: 500 **Proof: 320**

JIS G3445 STKM18C — 0.18% C (max) 1.5% Mn (max) 0.04% P & S (max) steel: Pipe; Japanese Standard
UTS: 520 **Proof: 390**

JIS G3452 SGP — C not quoted 0.05% P & S (max) steel: Tube; Japanese Standard
UTS: 300

JIS G3454 STPG38 — 0.25% C (max) 0.5% Mn 0.04% P & S (max) steel: Pipe; Japanese Standard
UTS: 380 **Proof: 220**

JIS G3455 STS35 — 0.11% C 0.45% Mn 0.035% P & S (max) 0.2% Cu (max) steel: Pipe; Japanese Standard
UTS: 350 **Proof: 180**

JIS G3455 STS38 — 0.25% C (max) 0.5% Mn steel: Pipe; Japanese Standard
UTS: 380 **Proof: 220**

JIS G3456 STPT38 — 0.25% C (max) 0.6% Mn 0.035% P & S (max) 0.2% Cu (max) steel: Pipe; Japanese Standard
UTS: 380 **Proof: 220**

JIS G3457 STPY41 — C not quoted 0.05% P & S (max) steel: Pipe; Japanese Standard
UTS: 410 **Proof: 230**

JIS G3460 STPL39 — 0.25% C (max) 1.35% Mn (max) 0.035% P & S 0.02% Cu (max) steel: Pipe; Japanese Standard
UTS: 390 **Proof: 210**

JIS G3461 STB30 — 0.2% C (max) 0.45% Mn 0.04% P & S steel: Pipe; Japanese Standard
UTS: 300

JIS G3461 STB33 — 0.12% C 0.45% Mn 0.035% P & S 0.2% Cu (max) steel: Pipe; Japanese Standard
UTS: 330 **Proof: 180**

JIS G3461 STB35 — 0.12% C 0.45% Mn 0.035% P & S 0.2% Cu steel: Pipe; Japanese Standard
UTS: 330 **Proof: 180**

JIS G3461 STB42 — 0.32% C (max) 0.55% Mn 0.035% P & S (max) 0.2% Cu (max) steel: Pipe; Japanese Standard
UTS: 420 **Proof: 260**

JIS G3464 STBL39 — 0.25% C (max) 1.35% Mn (max) 0.035% P & S (max) Cu 0.2% (max) steel: Pipe; Japanese Standard
UTS: 390 **Proof: 210**

JIS G3465 STMC55 — C not specified 0.04% P & S (max) steel: Tube for drilling; Japanese Standard
UTS: 550

JIS G3465 STMC65 — C not specified 0.04% P & S (max) steel: Tube for drilling; Japanese Standard
UTS: 650

JIS G3465 STMR60 — C not specified 0.04% P & S (max) steel: Tube for drilling; Japanese Standard
UTS: 600 **Proof: 380**

JIS G3465 STMR70 — C not specified 0.04% P & S (max) steel: Tube for drilling; Japanese Standard
UTS: 700 **Proof: 450**

JIS G3465 STMR80 — C not specified 0.04% P & S (max) steel: Tube for drilling; Japanese Standard
UTS: 800 **Proof: 530**

JIS G3466 STKR41 — 0.25% C (max) 0.04% P & S (max) steel: Pipe; Japanese Standard
UTS: 410 **Proof: 250**

JIS G3466 STKR50 — 0.18% C (max) 1.5% Mn (max) 0.04% P & S (max) steel: Pipe; Japanese Standard
UTS: 500 **Proof: 330**

JIS G4051 S9CK — 0.1% C 0.45% Mn 0.025% P & S (max) steel: For structures; Japanese Standard

JIS G4051 S10C — 0.1% C 0.45% Mn 0.03% P & S (max) steel: For structures; Japanese Standard

JIS G4051 S12C — 0.12% C 0.45% Mn 0.03% P & S steel: For structures; Japanese Standard

JIS G4051 S15C — 0.15% C 0.45% Mn 0.03% P & S (max) steel: For structures; Japanese Standard

JIS G4051 S17C — 0.17% C 0.45% Mn 0.03% P & S (max) steel: For structures; Japanese Standard

JLX W 45 — 0.2% C (max) 0.75% Mn 0.01% Nb steel: Jones & Loughlin

JLX W 50 — 0.21% C (max) 0.75% Mn 0.01% Nb steel: Jones & Loughlin

JLX W 55 — Jones & Loughlin

JLX W 60 — 0.2% C (max) 0.75% Mn 0.01% Nb steel: Jones & Loughlin

JOUVENCEL — Low carbon decarburized steel sheet; non-ageing; Esperance Longdoz
UTS: 310 **Elon: 41%** **Proof: 200**

JS 2 — 0.22% C 0.5% Si 1.5% Mn steel: Casting; Jopling
DPN: 180 **UTS: 620** **Elon: 18%**

J SG4051 S20C — 0.2% C 0.45% Mn 0.03% P & S (max) steel: For structure; Japanese Standard

J SG4106 S Mn 21 — 0.21% C 1.35% Mn 0.03% P & S (max) steel: Japanese Standard

KAISERALOY 45CB — 0.2% C (max) 1.0% Mn 0.01% Nb steel: Kaiser Steel Corp.

KAISERALOY 50MM — 0.27% C (max) 1.35% Mn 0.2% Cu steel: Kaiser Steel Corp.

KAISERALOY 55CB — 0.24% C (max) 1.4% Mn 0.01% Nb steel: Kaiser Steel Corp.

KAISERALOY 60CB — 0.26% C (max) 1.6% Mn 0.01% Nb steel: Kaiser Steel Corp.

KEA 581 — Low C free cutting steel: Kayser Ellison

KF — 0.18% C (max) 1.3% Mn 0.04% S & P (max): Alloy content controlled; DNV

KF 1 — 0.12% C 0.5% Mn 0.05% S & P case hardening steel: Kirkstall for BS alloy En32A

KF 1A — 0.16% C 1.4% Mn 0.05% S & P case hardening steel: Kirkstall for BS alloy EN201

KF 3 — 0.15% C 0.8% Mn 0.05% S & P case hardening steel: Kirkstall for BS alloy En32B

KF 4 — 0.2% C 0.8% Mn 0.06% S & P steel: Kirkstall for BS alloy En 3

KF 23 — 0.1% C 0.1% Si 0.8% Mn 0.25% S 0.07% P steel: Free machining; Kirkstall for BS alloy En 1A

KF 39 — 0.1% C 1% Mn 0.2% S 0.06% P case hardening steel: Free machining; Kirkstall; case hardened & tempered
UTS: 530 **Elon: 22%** **Proof: 370**

KF 39C — 0.15% C 1% Mn 0.12% S 0.05% P case hardening steel: Free machining; Kirkstall for BS alloy En 32M

KF 39E — 0.15% C 1.4% Mn 0.23% S 0.05% P case hardening steel: Free machining; Kirkstall; case hardened & tempered
UTS: 580 **Elon: 20‰**

Note. The following abbreviations and units are used in the tables:

DPN	Hardness, diamond pyramid number
UTS	Ultimate tensile strength, N/mm^2
Elon	Elongation, %
Proof	0.1 % proof strength, N/mm^2

1 N/mm^2=0.1 hbar=0.102 kgf/mm^2=0.06475 tonf/in^2=145.04 lbf/in^2

Symbol	Nominal analysis, supplier, condition and remarks.
KF 39E/1	0.14% C 1.4% Mn 0.14% S 0.05% P case hardening steel: Free machining; Kirkstall for BS alloy En 202
KF 60	0.1% C 0.1% Si 1.2% Mn 0.5% S 0.06% P steel: Free machining; Kirkstall for BS alloy En 1B
KHB 36	0.22% C (max) 1.55% Mn 0.05% P & S (max) steel: German proprietors specification
KIRKASE	0.1% C 1% Mn 0.2% S 0.06% P case hardening steel: Free machining; Kirkstall for KF 39
KORA 2	0.16% C 0.12% Si 0.8% Mn 0.05% S & P steel: Sanderson for BS alloy En32B
KP 2	0.1% C 1% Mn 0.25% S 0.07% P free cutting steel: Kiveton Park for BS alloy En1
KP 3	0.15% C 0.8% Mn steel: For carburizing; Kiveton Park for BS alloy En 32
KP 19	0.2% C 1% Mn 0.15% S 0.06% P steel: For case hardening; Kiveton Park for BS alloy En 7
KUTTWELL	0.14% C 0.9% Mn 0.04% S & P steel: For case hardening; Swift Levick for BS alloy En 32
L 1	0.22% C 0.6% Si 1% Mn steel: Casting; Jessop
LANTERN	0.15% C 0.8% Mn 0.05% S & P steel: For case hardening; Toledo for BS alloys En 32A and En 32B
L St 45/8	0.22% C (max) 0.05% S & P steel: Designation used by German Standard
LT 75N	0.2% C (max) 1.0% Mn steel: Lukens
LT 75QT	0.2% C (max) 1.0% Mn steel: Lukens
LUKENS 45	0.2% C 1.2% Mn 0.05% P & S (max) steel: Lukens
LUKENS 50	0.2% C (max) 1.35% Mn 0.05% P & S (max) steel: Lukens
LUKENS 55	0.22% C (max) 1.35% Mn 0.05% P & S (max) steel: Lukens
LUKENS 60	0.22% C (max) 1.6% Mn 0.05% P & S (max) steel: Lukens
LUKENS 80	0.2% C (max) 1.4% Mn 0.05% P & S (max) 0.3% Cu steel: Lukens
LUKENS 80	0.21% C 1.5% Mn 0.035% S & P 0.3% Cu steel: American proprietary steel; consult SAE for further information **Proof: 580**
LUKENS 440	0.28% C (max) 1.4% Mn 0.05% P & S (max) 0.2% Cu steel: Lukens
M 4	0.21% C 0.25% Si 1.5% Mn steel: Jessop
M 113	C not specified 0.04% S & P steel: Plate; Armco as A113 **UTS: 450 Elon: 27% Proof: 180**
MANGEAR	0.12% C 0.2% Si 1.6% Mn 0.04% S 0.04% P: Steel Peech and Tozer; also known as Phoenix Mangear **DPN: 160 UTS: 500 Elon: 37% Proof: 480**
MAN-TEN	0.25% C (max) 1.4% Mn 0.2% Cu steel: US Steel Corp.
MAN-TEN A440	0.28% C (max) 1.4% Mn 0.2% Cu steel: US Steel Corp.
MARINER	0.28% C 1.1% Mn 0.2% Cu steel: US Steel Corp.
MAX FINE GRAIN	0.2% C (max) 1.0% Mn 0.2% Cu 0.1% Zr steel: National Steel
MAXELOY	0.15% C (max) 1.2% Mn 0.2% Cu steel: Crucible Steel Co.
MAX HIGH MARG	0.22% C 1.25% Mn 0.22% Cu steel: National Steel
MBK 8	0.07% C 0.1% Si 0.4% Mn 0.03% S & P steel: German Standard
MBK 8A1	0.07% C 0.1% Si 0.4% Mn 0.03% S & P Al steel: German Standard
MBK 10	0.1% C 0.05% Si 0.6% Mn 0.025% S & P steel: German Standard
MBK 13	0.13% C 0.05% Si 0.6% Mn 0.025% S & P steel: German Standard

Symbol	Nominal analysis, supplier, condition and remarks.
MED Mn A440	0.28% C (max) 1.5% Mn 0.3% Cu steel: Bethlehem Steel Co.
MIL S 780.9	0.2% C (max) 1.25% Mn steel: Bethlehem Steel Co.
MIL S 12505.3	0.14% C (max) 0.95% Mn with Ni, Mo and Cu steel: Bethlehem Steel Co.
MIL S 12505.6	0.15% C (max) 1.0% Mn 0.3% Cu steel: Bethlehem Steel Co.
MILDTRODE 46.00	0.080% C 0.45% Mn 0.20% Si steel: For electrodes; Esab **UTS: 494 Elon: 25% Proof: 401**
M1-F	0.12% C (max) 0.75% Mn 0.22% Cu 0.015% Nb steel: McLouth
MNC 810	0.19% C 0.17% Mn 0.25% Cr 0.35% Cu: Structural steels; Swedish Standard **UTS: 400 Elon: 24% Proof: 230**
MNC 811	Summary of structural steel tubing; Swedish Standard
MNC 815	List of reinforcing steels: Swedish Standard; individual specification listed
MNC 815E	Specification covering steels for reinforcing: Swedish Standards
MNC 831	Summary of pressure vessel steel tubing; Swedish Standard
MNC 832	Summary of pressure vessel steel for gas cylinders; Swedish Standard
MNC 840	Summary of steels for screws, nuts and rivets; Swedish Standard
MNC 845	Summary of free cutting steels; Swedish Standard
MNC 851	Summary of case hardening steels; Swedish Standard
MNC 915	General specification for cold rolled sheet
MNC 916	General specification for structural steels: Swedish Standard
MNC 920	Summary of cold rolled steel strip; Swedish Standard
MONEMEL I	Low carbon decarburized steel: Sheet; non-ageing; Esperance Longdox; for direct one coat enamelling
MONEMEL 2	Low carbon decarburized steel: Sheet; non ageing; Esperance Longdoz; for direct one coat enamelling and extra deep drawing
MR St 1	0.12% C (max) 0.15% Si 0.3% Mn 0.06% S & P steel: German Standard
MR St 2	0.1% C (max) 0.1% Si 0.3% Mn 0.05% S & P steel: German Standard
MR St 3	0.1% C (max) 0.1% Si 0.3% Mn 0.04% S & P steel: German Standard
MR St 4	0.1% C (max) 0.08% Si 0.35% Mn 0.035% S & P steel: German Standard
MR St 3	0.1% C 0.3% Mn 0.05% S & P steel: Designation used by German Standard
MR St 34/2	0.17% C (max) 0.05% S & P steel: German Standard
MR St 37/2	0.2% C 0.05% S & P steel: German Standard
MR St 42/2	0.25% C (max) 0.05% S & P steel: German Standard
MR St 44	0.15% C 0.6% Mn 0.06% S & P steel: Designation used by German Standard
M St 0	0.12% C (max) 0.3% Mn 0.06% S & P steel: German Standard
M St 34/2	0.17% C (max) 0.05% S & P steel: German Standard
M St 34/3	0.17% C 0.05% S steel: German Standard
M St 37/2	0.2% C (max) 0.05% S & P steel: German Standard
M St 37/3	0.2% C (max) 0.05% S & P steel: German Standard
M St 42/2	0.25% C (max) 0.05% S & P steel: German Standard

Symbol	Nominal analysis, supplier, condition and remarks.
M St 42/3	0.25% C (max) 0.05% S & P steel: German Standard
M St 52/3	0.2% C (max) 0.05% S & P steel: German Standard
MSV	0.1% C 0.5% Mn steel: Pompey
MSVB/W	0.2% C 0.5% Si 1.1% Mn 0.035% S & P steel: Krupp; up to 0.3% Cr may be present
MSVD	0.2% C 0.5% Mn steel: Pompey
MUK 8	0.07% C 0.4% Mn 0.03% S & P steel: German Standard
MUK 10	0.1% C 0.6% Mn 0.03% S & P steel: German Standard
MURAWIRE W1	0.06% C 1.5% Mn 0.06% P 0.6% Si steel: Welding electrode; Murex; copper coated **UTS: 480 Elon: 34% Proof: 390**
MURAWIRE W2	0.06% C 0.7% Mn 0.4% Si steel: Welding electrode; Murex **UTS: 440 Elon: 37% Proof: 330**
MU St 1	0.12% C (max) 0.3% Mn 0.06% S & P steel: German Standard
MU St 2	0.1% C 0.3% Mn 0.05% S & P steel: German Standard
MU St 3	0.1% C (max) 0.3% Mn 0.04% S & P steel: German Standard
MU St 4	0.1% C (max) 0.3% Mn 0.03% S & P steel: German Standard
MU St 34	0.1% C (max) 0.2% Mn 0.05% S & P steel: Designation used by German Standard
MU St 34/2	0.17% C (max) 0.05% S & P steel: German Standard
MU St 37/2	0.2% C (max) 0.05% S & P steel: German Standard
MU St 42/2	0.25% C (max) 0.05% S & P steel: German Standard
N54	0.17% C 1.5% Mn steel **DPN: 157 UTS: 550 Elon: 21%**
NF A35 501/A33	C not specified steel: Plate; French national specification **UTS: 370 Proof: 180**
NF A35 501/A33/2	C not specified 0.05% P & S (max) steel: Plate; French national specification **UTS: 370 Proof: 170**
NF A35 501/A34/1	C not specified 0.06% P (max) 0.05% S (max) steel: Plate; French national specification **UTS: 370 Proof: 160**
NF A35 501/A34/2	C not specified 0.05% P & S (max) 0.007% N (max) steel: Plate; French national specification **UTS: 380 Proof: 150**
NF A35 501/A37/1	0.18% C (max) 0.06% P (max) 0.05% S (max) steel: Plate; French national specification **UTS: 410 Proof: 240**
NF A35 501/A37/2	0.18% C (max) 0.05% P & S 0.007% N (max) steel: Plate; French national specification **UTS: 410 Proof: 210**
NF A35 501/A37/3	0.16% C (max) 0.045% P & S (max) steel: Plate; French national specification **UTS: 410 Proof: 210**
NF A35 501/A37/4	0.16% C (max) 0.04% S & P 0.02% Al (max) steel: Plate; French national specification **UTS: 410 Proof: 210**

Symbol	Nominal analysis, supplier, condition and remarks.
NF A36 205/A37 C1	0.16% C (max) 0.4% Mn (max) 0.04% P & S (max) steel: Plate; French Standard **UTS: 410 Proof: 220**
NF A36 205/A37C2	0.15% C (max) 0.4% Mn (max) 0.035% P & S (max) steel: Plate; French Standard **UTS: 410 Proof: 220**
NF A36 205/A37P1	0.16% C (max) 0.4% Mn (max) 0.04% P & S (max) steel: Plate; French Standard **UTS: 410 Proof: 220**
NF A36 205/A37P2	0.15% C (max) 0.4% Mn 0.035% P & S (max) steel: Plate; French Standard **UTS: 410 Proof: 220**
NF A35 501/A42/1	0.2% C (max) 0.06% P 0.05% S (max) steel: Plate; French Standard **UTS: 460 Proof: 250**
NF A35 501/A42/2	0.2% C (max) 0.05% Mn 0.05% P & S (max) 0.007% N (max) steel: Plate; French Standard **UTS: 460 Proof: 240**
NF A35 501/A42/3	0.18% C (max) 0.045% P & S steel: Plate; French Standard **UTS: 460 Proof: 230**
NF A35 501/A42/4	0.18% C (max) 0.04% P & S (max) 0.02% Al steel: Plate; French Standard **UTS: 460 Proof: 230**
NF A35 501/A47/2	0.24% C (max) 1.3% Mn (max) 0.05% P & S (max) steel: Plate; French Standard **UTS: 520 Proof: 250**
NF A35 501/A47/3	0.2% C (max) 1.3% Mn (max) 0.04% P & S (max) steel: Plate; French Standard **UTS: 520 Proof: 270**
NF A35 501/A47/4	0.2% C (max) 1.3% Mn (max) 0.04% P & S (max) 0.02% Al (max) steel: Plate; French Standard **UTS: 520 Proof: 270**
NF A35 501/A52/2	0.24% C (max) 1.5% Mn (max) 0.05% P & S (max) steel: Plate; French Standard **UTS: 570 Proof: 330**
NF A35 501/A52/3	0.2% C (max) 1.5% Mn (max) 0.045% P & S (max) steel: Plate; French specification **UTS: 570 Proof: 340**
NF A35 501/A52/4	0.2% C (max) 1.5% Mn (max) 0.04% P & S (max) steel: Plate; French specification **UTS: 570 Proof: 340**
NF A35 501/A60/1	C not specified 0.07% P (max) 0.05% S (max) steel: Plate; French Standard **UTS: 660 Proof: 340**
NF A35 501/A60/2	C not specified 0.045% P & S (max) steel: Plate; French Standard **UTS: 660 Proof: 320**
NF A35 501/E24/1	0.18% C (max) 0.06% P (max) 0.05% (max) steel: Plate; French national specification **UTS: 410 Proof: 240**
NF A35 501/E24/2	0.18% C (max) 0.05% P & S (max) 0.007% N (max) steel: Plate; French national specification **UTS: 410 Proof: 220**
NF A35 501/E24/3	0.16% C (max) 0.045% P & S (max) steel: Plate; French national specification **UTS: 410 Proof: 210**
NF A35 501/E24/4	0.16% C (max) 0.04% P & S (max) 0.02% Al (max) steel: Plate; French national specification **UTS: 410 Proof: 220**
NF A35 501/E26/1	0.2% C (max) 0.06% P (max) 0.05% S & P (max) steel: Plate; French Standard **UTS: 460 Proof: 250**
NF A35 501/E26/2	0.2% C (max) 0.05% S & P (max) 0.007% N (max) steel: Plate; French Standard **UTS: 460 Proof: 240**
NF A35 501/E26/3	0.18% C (max) 0.045% S & P (max) steel: Plate; French Standard **UTS: 460 Proof: 230**

Note. The following abbreviations and units are used in the tables:

DPN	Hardness, diamond pyramid number
UTS	Ultimate tensile strength, N/mm^2
Elon	Elongation, %
Proof	0.1 % proof strength, N/mm^2

1 N/mm^2=0.1 hbar=0.102 kgf/mm^2=0.06475 tonf/in^2=145.04 lbf/in^2

Symbol	Nominal analysis, supplier, condition and remarks.

NF A35 501/E26/4 0.18% C (max) 0.04% S & P (max) 0.02% Al steel: Plate; French Standard
UTS: 460 **Proof: 240**

NF A35 501/E30/2 0.24% C (max) 1.3% Mn (max) 0.05% P & S (max) steel: Plate; French Standard
UTS: 520 **Proof: 250**

NF A35 501/E30/3 0.2% C (max) 1.3% Mn (max) 0.045% P & S (max) steel: Plate; French Standard
UTS: 520 **Proof: 270**

NF A35 501/E30/4 0.2% C (max) 1.30% Mn (max) 0.04% P & S (max) 0.02% Al (max) steel: Plate; French Standard
UTS: 520 **Proof: 270**

NF A35 501/E36/2 0.24% C (max) 1.5% Mn (max) 0.05% P & S (max) steel: Plate; French specification
UTS: 570 **Proof: 330**

NF A35 501/E36/3 0.2% C (max) 1.5% Mn (max) 0.045% P & S (max) steel: Plate; French specification
UTS: 570 **Proof: 340**

NF A35 501/E36/4 0.2% C (max) 1.5% Mn (max) 0.04% P & S (max) steel: Plate; French specification
UTS: 570 **Proof: 340**

NF A35 501/E37/2 0.18% C (max) 0.05% P & S (max) 0.007% N (max) steel: Plate; French specification
UTS: 410 **Proof: 220**

NF A35 551/20M5 0.19% C 1.35% Mn steel: Plate; French Standard

NF A35 551/X12 0.13% C 0.45% Mn steel: For structures; French Standard

NF A35 551/XC10 0.09% C 0.045% Mn steel: For structures; French Standard

NF A35 551/XC18 0.18% C 0.5% Mn steel: For structures; French Standard

NF A35 561/S250 0.14% C (max) 1.2% Mn 0.28% S steel: Free machining; French Standard
UTS: 400 **Elon: 23%**

NF A35 561/S250 Pb 0.14% C (max) 1.2% Mn 0.28% S 0.25% Pb steel: Free machining; French Standard
UTS: 400 **Elon: 23%**

NF A35 561/S300 0.15% (max) 0.3% Mn 0.35% S steel: Free machining; French Standard
UTS: 420 **Elon: 21%**

NF A35 561/S300 Pb 0.15% C (max) 1.3% Mn 0.35% S 0.25% Pb steel: Free machining; French Standard
UTS: 420 **Elon: 21%**

NF A35 562/10F2 0.11% C (max) 0.6% Mn 0.2% S steel: Free machining; hardened & tempered; French Standard
UTS: 660 **Elon: 14%** **Proof: 370**

NF A35 562/10 Pb F2 0.11% C 0.6% Mn 0.2% S 0.25% Pb steel: Free machining; hardened & tempered; French Standard
UTS: 660 **Elon: 14%** **Proof: 390**

NF A35 562/12MF4 0.12% C (max) 1% Mn 0.2% S steel: Free machining; hardened & tempered; French Standard
UTS: 880 **Elon: 11%** **Proof: 570**

NF A35 562/20F2 0.18% C (max) 0.7% Mn 0.2% S steel: Free machining; hardened & tempered; French Standard
UTS: 620 **Elon: 17%** **Proof: 390**

NF A35 562/35MF4 0.35% C (max) 1.1% Mn 0.2% S steel: Free machining; hardened & tempered; French Standard
UTS: 890 **Elon: 10%** **Proof: 690**

NF A36 205/A42C1 0.18% C (max) 0.5% Mn (max) 0.04% P & S (max) steel: Plate; French Standard
UTS: 460 **Proof: 240**

NF A36 205/A42C2 0.18% C (max) 0.6% Mn (max) 0.035% P & S (max) steel: Plate; French Standard
UTS: 460 **Proof: 250**

NF A36 205/A42P1 0.18% C (max) 0.5% Mn (max) 0.04% P & S (max) steel: Plate; French Standard
UTS: 460 **Proof: 240**

NF A36 205/A42P2 0.18% C (max) 0.6% Mn (max) 0.035% P & S (max) steel: Plate; French Standard
UTS: 460 **Proof: 250**

NF A36 205/A48C1 0.22% C (max) 0.6% Mn (max) 0.04% P & S (max) steel: Plate; French Standard
UTS: 520 **Proof: 280**

NF A36 205/A48C2 0.2% C (max) 0.8% Mn (max) 0.3% Cr Ni (max) 0.035% P & S steel: Plate; French Standard
UTS: 520 **Proof: 280**

NF A36 205/A48P1 0.22% C (max) 0.6% Mn (max) 0.04% S & P (max) steel: Plate; French Standard
UTS: 520 **Proof: 280**

NF A36 205/A48P2 0.2% C (max) 0.8% Mn (max) 0.2% Cr (max) 0.3% Ni (max) 0.035% P & S (max) steel: Plate; French Standard
UTS: 520 **Proof: 280**

NF A36 205/A52C1 0.22% C (max) 0.9% Mn 0.04% P & S steel: Plate; French Standard
UTS: 570 **Proof: 340**

NF A36 205/A52C2 0.2% C (max) 1.0% Mn (max) 0.2% Cr (max) 0.3% Ni (max) 0.035% P & S (max) steel: Plate; French Standard
UTS: 570 **Proof: 340**

NF A36 205/A52P1 0.22% C (max) 0.9% Mn (max) 0.04% P & S steel: Plate; French Standard
UTS: 570 **Proof: 340**

NF A36 205/A52P2 0.2% C (max) 1.0% Mn (max) 0.2% Cr (max) 0.3% Ni (max) 0.035% P & S (max) steel: Plate; French Standard
UTS: 570 **Proof: 340**

NF A36 208/A42FP1 0.18% C (max) 0.5% Mn (max) 0.04% P & S (max) steel: Plate; French Standard; normalized
UTS: 440 **Proof: 240**

NF A36 208/A42FP2 0.18% C (max) 0.6% Mn (max) 0.035% P & S (max) steel: Plate; French Standard; normalized
UTS: 460 **Proof: 240**

NF A36 208/A48FP1 0.22% C (max) 0.6% Mn (max) 0.04% P & S (max) steel: Plate; French Standard; normalized
UTS: 530 **Proof: 280**

NF A36 208/A48FP2 0.2% C (max) 0.8% Mn (max) 0.2% Cr (max) 0.3% Ni (max) 0.1% Nu (max) 0.05% V (max) steel: Plate; French Standard; normalized
UTS: 520 **Proof: 280**

NF A36 208/A52FP1 0.22% C (max) 0.9% Mn (max) 0.04% P & S (max) steel: Plate; French Standard; normalized
UTS: 570 **Proof: 340**

NK HITEN 50 0.18% C (max) 1.4% C (max) 1.4% Mn 0.035% P & S (max) steel: Nippon Kokan

NK Hiten 55 0.2% C (max) 1.5% Mn 0.035% P & S (max) steel: Nippon Kokan

NV 1-0 0.2% C (max) 0.4% Mn 0.05% (max) S & P 0.009% N (max) steel: Designation used by DNV
UTS: 410 **Elon: 20%** **Proof: 240**

NV 1-1 0.16% C (max) 0.4% Mn 0.05% S & P (max) 0.009% N (max) steel: Designation used by DNV
UTS: 410 **Elon: 20%** **Proof: 240**

NV 1-2 0.14% C (max) 0.4% Mn 0.04% S & P 0.009% N (max) steel: Designation used by DNV
UTS: 410 **Elon: 20%** **Proof: 240**

NV 2-0 0.2% C (max) 0.4% Mn 0.05% S & P (max) 0.009% N (max) steel: Designation used by DNV
UTS: 450 **Elon: 20%** **Proof: 260**

NV 2-1 0.16% C (max) 0.4% Mn 0.05% S & P (max) 0.009% N (max) steel: Designation used by DNV
UTS: 450 **Elon: 20%** **Proof: 260**

NV 2-2 0.16% C (max) 0.4% Mn 0.04% S & P (max) 0.009% N (max) steel: Designation used by DNV
UTS: 450 **Elon: 20%** **Proof: 260**

Symbol	Nominal analysis, supplier, condition and remarks.
NV 2-2	0.14% C (max) 1.1% Mn 0.04% S & P (max) 0.009% N (max) steel: Designation used by DNV **UTS: 450 Elon: 20% Proof: 260**
NV 2-4	0.14% C (max) 0.11% Mn 0.035% S & P (max) 0.009% N (max) steel: Designation used by DNV **UTS: 450 Elon: 20% Proof: 260**
NV 3-0	0.2% C (max) 0.4% Mn 0.05% S & P (max) 0.009% N (max) steel: Designation used by DNV **UTS: 480 Elon: 20% Proof: 260**
NV 3-1	0.16% C (max) 0.4% Mn 0.05% S & P (max) 0.009% N (max) steel: Designation used by DNV **UTS: 480 Elon: 20% Proof: 260**
NV 3-2	0.16% C (max) 0.4% Mn 0.04% S & P (max) 0.009% N (max) steel: Designation used by DNV **UTS: 480 Elon: 20% Proof: 260**
NV 4-0	0.2% C (max) 1.6% Mn (max) 0.05% S & P (max) 0.009% N (max) steel: Designation used by DNV **UTS: 540 Elon: 18% Proof: 330**
NV 4-1	0.16% C (max) 1.6% Mn (max) 0.04% S & P (max) 0.009% N (max) steel: Designation used by DNV **UTS: 540 Elon: 18% Proof: 330**
NV 4-2	0.16% C (max) 1.6% Mn (max) 0.04% S & P (max) 0.009% N (max) steel: Designation used by DNV **UTS: 540 Elon: 18% Proof: 330**
NV 4-3	0.15% C (max) 1.7% Mn (max) 0.04% S & P (max) 0.009% N (max) steel: Designation used by DNV **UTS: 540 Elon: 18% Proof: 330**
NV 4-4	0.15% C (max) 1.7% Mn (max) 0.035% S & P (max) 0.009% N (max) steel: Designation used by DNV **UTS: 540 Elon: 18% Proof: 330**
NV 5-0	0.2% C (max) 1.0% Mn 0.05% S & P (max) 0.009% N (max) steel: Designation used by DNV **UTS: 510 Elon: 19% Proof: 280**
NV 5-1	0.18% C (max) 1.1% Mn 0.04% S & P (max) 0.009% N (max) steel: Designation used by DNV **UTS: 510 Elon: 19% Proof: 280**
NV 5-2	0.16% C (max) 1.2% Mn 0.04% S & P (max) 0.009% N (max) steel: Designation used by DNV **UTS: 510 Elon: 19% Proof: 280**
NV 6-0	0.23% C (max) 1.15% Mn 0.05% S & P (max) 0.009% N (max) steel: Designation used by DNV **UTS: 570 Elon: 18% Proof: 320**
NV 6-1	0.2% C (max) 1.2% Mn 0.04% S & P (max) 0.009% N (max) steel: Designation used by DNV **UTS: 570 Elon: 18% Proof: 320**
NV 6-2	0.18% C (max) 1.25% Mn 0.04% S & P (max) 0.009% N (max) steel: Designation used by DNV **UTS: 570 Elon: 18% Proof: 320**
NVA	C not specified 2.5% Mn (min) 0.05% P & S (max) steel: Plate for ships' hulls; DNV
NVA 27	Analysis as NVA; designation used by DNV **UTS: 470 Elon: 22% Proof: 270**
NVA 32	Analysis as NVA; designation used by DNV **UTS: 520 Elon: 22% Proof: 320**
NVA 36	Analysis as NVA; designation used by DNV **UTS: 550 Elon: 21% Proof: 360**
NVA 40	Analysis as NVA; designation used by DNV **UTS: 600 Elon: 20% Proof: 400**
NVC	0.21% C (max) 1.0% Mn 0.05% P & S (max) steel: Plate for ships' hulls; fine grained steel with Al; normalized; DNV
NVD	0.21% C (max) 1.0% Mn 0.05% P & S (max) steel: Plate for ships' hulls; DNV
NVD 27	Analysis as NVD; designation used by DNV **UTS: 470 Elon: 22% Proof: 270**
NVD 32	Analysis as NVD; designation used by DNV **UTS: 520 Elon: 22% Proof: 320**
NVD 36	Analysis as NVD; designation used by DNV **UTS: 550 Elon: 21% Proof: 360**
NVD 40	Analysis as NVD; designation used by DNV **UTS: 600 Elon: 20% Proof: 400**
NVE	0.18% C (max) 1.1% Mn 0.05% P & S (max) steel: Plate for ships' hulls; normalized; DNV
NVE 27	Analysis as NVE; designation used by DNV **UTS: 470 Elon: 22% Proof: 270**
NVE 32	Analysis as NVE; designation used by DNV **UTS: 520 Elon: 22% Proof: 320**
NVE 36	Analysis as NVE; designation used by DNV **UTS: 550 Elon: 21% Proof: 360**
NVE 40	Analysis as NVE; designation used by DNV **UTS: 600 Elon: 20% Proof: 400**
NVK1	C not specified 0.04% S & P (max) steel: For chains; normalized; designation used by DNV **UTS: 320 Elon: 30%**
NVK2	C not specified 0.04% S & P (max) steel: For chains; normalized; designation used by DNV **UTS: 550 Elon: 22% Proof: 300**
NVK3	C not specified 0.04% S & P (max) steel: For chains; hardened & tempered; designation used by DNV **UTS: 700 Elon: 17% Proof: 400**
NVR 1-1	0.17% C (max) 0.05% S & P (max) steel: Designation used by DNV
NVR 1-2	0.17% C (max) 0.4% Mn 0.05% S & P (max) 0.2% Cr (max) steel: Designation used by DNV
NVR 1-3	0.22% C (max) 0.6% Mn 0.05% S & P (max) 0.2% Cr (max) steel: Designation used by DNV
NVR 1-4	0.2% C (max) 1.1% Mn 0.05% S & P (max) 0.2% Cr (max) steel: Designation used by DNV
NVW	0.2% C (max) 0.5% Mn 0.05% P & S (max) steel: Plate for ships' hulls; DNV
ORELLOY 440	0.19% C 1.3% Mn 0.35% Cu steel: Oregon
OX 520A	0.18% C (max) 1.4% Mn 0.025% Nb 0.009% N steel: Oxelsund
OX 522A	0.18% C (max) 0.3% Mn 0.025% Nb 0.009% N steel: Oxelsund
OX 525B	0.18% C (max) 1.4% Mn 0.025% Nb 0.009% N steel: Oxelsund
P 151	0.15% C 0.8% Mn steel: For case hardening; Carrs
P 282	0.1% C 0.25% Si 0.25% Mn steel: For case hardening; Carrs
PAC	0.09% C 0.4% Mn steel: Pompey
PAC 1	0.12% C 0.4% Mn steel: Pompey
PAC 2	0.15% C 0.6% Mn steel: Pompey
PAR-TEN	0.12% C (max) 0.75% Mn 0.07% Cu steel: US Steel Corp.
PENETRODE 29.10	0.080% C 0.650% Mn 0.06% Si steel: For electrodes; Esab **UTS: 479 Elon: 28% Proof: 355**
PENETRODE 29.10	0.08% C 0.65% Mn 0.06% Si steel: For electrodes; Esab; as welded **UTS: 480 Elon: 28% Proof: 355**

Note. The following abbreviations and units are used in the tables:

DPN	Hardness, diamond pyramid number
UTS	Ultimate tensile strength, N/mm^2
Elon	Elongation, %
Proof	0.1 % proof strength, N/mm^2

1 N/mm^2=0.1 hbar=0.102 kgf/mm^2=0.06475 tonf/in^2=145.04 lbf/in^2

Symbol	Nominal analysis, supplier, condition and remarks.
PG 605	0.15% C 1.25% Mn 0.12% S 0.05% P steel: Free cutting; Park Gate; as rolled **UTS: 450 Elon: 30%**
PGF 1	0.1% C 1.2% Mn 0.35% S 0.09% P steel: Free cutting; Park Gate; as rolled **UTS: 390 Elon: 30%**
PGF 4	0.12% C 1% Mn 0.25% S 0.07% P steel: Free cutting; Park Gate; as rolled **UTS: 340 Elon: 26%**
PGF 5	0.12% C 1.2% Mn 0.5% S 0.06% P steel: Free machining; Park Gate; as rolled **UTS: 340 Elon: 24%**
PHOENIX 1	0.2% C 0.7% Mn 0.06% S 0.06% P steel: Steel Peach & Tozer **UTS: 530 Elon: 20%**
PHOENIX 2	0.2% C 0.7% Mn 0.06% S 0.06% P steel: Steel Peach & Tozer **UTS: 420 Elon: 17%**
PHOENIX 4	0.2% C (max) 0.9% Mn 0.25% S 0.1% P: Free machining steel; Steel Peach & Tozer **UTS: 420 Elon: 14%**
PHOENIX MANGEAR	0.12% C 0.2% Si 1.6% Mn 0.04% S 0.04% P steel: Steel Peach & Tozer; for lifting chains **DPN: 160 UTS: 500 Elon: 37% Proof: 380**
PITT-TEN 2	0.15% C (max) 0.75% Mn 0.035% Cu steel: Pittsburgh Steel
PITT-TEN X45W	0.15% C 0.75% Mn 0.03% Nb steel: Pittsburgh Steel
PITT-TEN X50W	0.15% C 0.75% Mn 0.03% Nb steel: Pittsburgh Steel
PITT-TEN X55W	0.15% C 0.75% Mn 0.03% Nb steel: Pittsburgh Steel
PITT-TEN X60W	0.15% C 0.75% Mn 0.03% Nb steel: Pittsburgh Steel
POSITRODE 46.28	0.080% C 0.45% Mn 0.10% Sl steel: For electrodes; Esab **UTS: 479 Elon: 28% Proof: 386**
PLANEMEL	Low carbon decarburized steel: Sheet; non ageing; Experance Longdoz; for vitreous enamel
PLASTIC HOBBING	0.1% C (max) 1% Si 0.4% Mn steel: Edgar Allen
PM 5	0.18% C 0.14% Si 1.0% Mn steel: Pompey
PM 6	0.2% C 0.15% Si 1.5% Mn steel: Pompey
PRIMEX	0.04% C 0.5% Mn steel: Welding electrode; Murex **UTS: 440 Elon: 33% Proof: 420**
Q St 34	0.1% C 0.3% Mn 0.04% S & P steel: Designation used by German Standard
Q St 34/2	0.17% C (max) 0.05% S & P steel: German Standard
Q St 34/3	0.17% C (max) 0.05% S & P steel: German Standard
Q St 37/2	0.2% C (max) 0.05% S & P steel: German Standard
Q St 37/3	0.2% C (max) 0.05% S & P steel: German Standard
Q St 42/2	0.25% C (max) 0.05% S & P steel: German Standard
Q St 42/3	0.25% C (max) 0.05% S & P steel: German Standard
Q St 52/3	0.2% C (max) 0.05% S & P steel: German Standard
REPUBLIC M1	0.25% C (max) 1.4% Mn 0.2% Cu steel: Republic Steel Co.
REPUBLIC M2	0.25% C (max) 1.4% Mn 0.2% Cu steel: Republic Steel Co.
RESISTCO	0.11% C 0.3% Si (max) 0.4% Mn 0.3% Cu 0.05% S & P steel: Tube; STD Service; for boiler feed tubes **UTS: 390 Elon: 25% Proof: 180**

Symbol	Nominal analysis, supplier, condition and remarks.
RIO 214	0.08% C 26% Cr 5% Ni 1.5% Mo steel: Bofors **DPN: 260 UTS: 600 Elon: 20% Proof: 420**
RIVER ACE 60	0.18% C (max) 1.5% Mn 0.035% P & S (max) steel: Kawasaki Iron & Steel Co.
R St 13	0.1% C (max) 0.1% Si 0.3% Mn 0.04% S & P steel: German Standard
R St 13/03	0.1% C (max) 0.1% Si 0.3% Mn 0.04% S & P steel: German Standard
R St 13/04	0.1% C (max) 0.1% Si 0.3% Mn 0.04% S & P steel: German Standard
R St 13/05	0.1% C (max) 0.1% Si 0.3% Mn 0.04% S & P steel: German Standard
R St 14	0.1% C (max) 0.1% Si 0.3% Mn 0.03% S & P steel: German Standard
R St 14/04	0.1% C (max) 0.1% Si 0.3% Mn 0.03% S & P steel: German Standard
R St 14/05	0.1% C (max) 0.1% Si 0.3% Mn 0.03% S & P steel: German Standard
R St 34	0.17% C (max) 0.05% S & P steel: German Standard
R St 34/2	0.17% C (max) 0.05% S & P steel: German Standard
R St 37	0.2% C (max) 0.05% S 0.08% P steel: German Standard
R St 37/2	0.2% C (max) 0.05% S & P steel: German Standard
R St 37/202	0.2% C (max) 0.05% S & P steel: German Standard
R St 37/203	0.2% C (max) 0.05% S & P steel: German Standard
R St 37/204	0.2% C (max) 0.05% S & P steel: German Standard
R St 37/205	0.2% C (max) 0.05% S & P steel: German Standard
R St 42	0.25% C (max) 0.05% S 0.08% P steel: German Standard
R St 42/2	0.25% C (max) 0.05% S & P steel: German Standard
R St 42/202	0.25% C (max) 0.05% S & P steel: German Standard
R St 42/203	0.25% C (max) 0.05% S & P steel: German Standard
R St 42/204	0.25% C (max) 0.05% S & P steel: German Standard
R St 42/205	0.25% C (max) 0.05% S & P steel: German Standard
S 70	Iron–carbon sintered material: Sintered Products; 23–28% porosity **UTS: 120 Elon: 1%**
SAE 12L14	0.15% C (max) 1% Mn 0.05% P 0.3% S steel: Free cutting
SAE 0022	0.2% C 0.7% Mn 0.6% Si 0.05% S & P cast steel: Suitable for carburizing
SAE 080	0.05% C (max) 0.05% S & P cast steel: Annealed **DPN: 163 UTS: 560 Elon: 18% Proof: 270**
SAE 850	0.2% C Fe: Sinter for bearings; density 5900 kg/m³
SAE 853	0.2% C Fe: Sinter for mechanical parts; density 6300 kg/m³
SAE 870	0.2% C 20% Cu Fe alloy: Sinter; density 7100 kg/m³
SAE 950	0.2% C (max) 1.25% Mn 0.05% S 0.15% P steel: General specification covering weldable steels; as rolled **UTS: 480 Elon: 20% Proof: 320**
SAE 1005	0.06% C (max) 0.35% Mn 0.04% P 0.05% S steel: Bar, etc.
SAE 1006	0.08% C 0.3% Mn 0.04% P 0.05% S steel: Hot rolled as AISI C1006 **DPN: 86 UTS: 300 Elon: 30% Proof: 150**

Symbol	Nominal analysis, supplier, condition and remarks.
SAE 1006	0.08% C 0.3% Mn 0.04% P 0.05% S steel: Cold drawn
	DPN: 95 UTS: 330 Elon: 20% Proof: 440
SAE 1008	0.1% C 0.4% Mn 0.04% P 0.05% S steel: Hot rolled as AISI C1008
	DPN: 86 UTS: 300 Elon: 30% Proof: 150
SAE 1008	0.1% C 0.4% Mn 0.04% P 0.05% S steel: Cold drawn
	DPN: 95 UTS: 340 Elon: 20% Proof: 440
SAE 1009	0.15% C 0.6% Mn 0.04% P 0.05% S steel: Hot rolled as AISI C1009
	DPN: 86 UTS: 300 Elon: 30% Proof: 150
SAE 1009	0.15% C 0.6% Mn 0.04% P 0.05% S steel: Cold drawn
	DPN: 95 UTS: 330 Elon: 20% Proof: 440
SAE 1010	0.1% C 0.4% Mn 0.04% P 0.05% S steel: Hot rolled as AISI C1010
	DPN: 95 UTS: 330 Elon: 28% Proof: 170
SAE 1010	0.1% C 0.4% Mn 0.04% P 0.05% S steel: Cold drawn
	DPN: 105 UTS: 370 Elon: 20% Proof: 300
SAE 1012	0.12% C 0.4% Mn 0.04% P 0.05% S steel: Hot rolled as AISI C1012
	DPN: 95 UTS: 330 Elon: 28% Proof: 180
SAE 1012	0.12% C 0.4% Mn 0.04% P 0.05% S steel: Cold drawn
	DPN: 105 UTS: 370 Elon: 19% Proof: 320
SAE 1013	0.13% C 0.7% Mn 0.04% P 0.05% S steel: Bar, etc.
SAE 1015	0.15% C 0.4% Mn 0.04% P 0.05% S steel: Hot rolled as AISI C1015
	DPN: 101 UTS: 340 Elon: 28% Proof: 180
SAE 1015	0.15% C 0.4% Mn 0.04% P 0.05% S steel: Cold drawn
	DPN: 111 UTS: 390 Elon: 18% Proof: 330
SAE 1016	0.15% C 0.8% Mn 0.04% P 0.05% S steel: Bar; hot rolled as AISI C1016
	DPN: 111 UTS: 390 Elon: 25% Proof: 200
SAE 1016	0.15% C 0.8% Mn 0.04% P 0.05% S steel: Bar; cold drawn
	DPN: 121 UTS: 440 Elon: 18% Proof: 360
SAE 1017	0.18% C 0.4% Mn 0.04% P 0.05% S steel: Bar; hot rolled as AISI C1017
	DPN: 105 UTS: 370 Elon: 26% Proof: 200
SAE 1017	0.18% C 0.4% Mn 0.04% P 0.05% S steel: Bar; cold drawn
	DPN: 116 UTS: 420 Elon: 18% Proof: 340
SAE 1018	0.18% C 0.8% Mn 0.04% P 0.05% S steel: Bar; hot rolled as AISI C1018
	DPN: 116 UTS: 400 Elon: 25% Proof: 210
SAE 1018	0.18% C 0.8% Mn 0.04% P 0.05% S steel: Bar; cold drawn
	DPN: 126 UTS: 450 Elon: 15% Proof: 370
SAE 1019	0.18% C 0.9% Mn 0.04% P 0.05% S steel: Bar; hot rolled as AISI C1019
	DPN: 116 UTS: 420 Elon: 25% Proof: 210
SAE 1019	0.18% C 0.9% Mn 0.04% P 0.05% S steel: Bar; cold drawn
	DPN: 131 UTS: 460 Elon: 15% Proof: 390

Note. The following abbreviations and units are used in the tables:

DPN	Hardness, diamond pyramid number
UTS	Ultimate tensile strength, N/mm^2
Elon	Elongation, %
Proof	0.1 % proof strength, N/mm^2

1 N/mm^2=0.1 hbar=0.102 kgf/mm^2=0.06475 tonf/in^2=145.04 lbf/in^2

Symbol	Nominal analysis, supplier, condition and remarks.
SAE 1020	0.2% C 0.4% Mn 0.04% P 0.05% S steel: Bar; hot rolled as AISI C1020
	DPN: 111 UTS: 390 Elon: 25% Proof: 200
SAE 1020	0.2% C 0.4% Mn 0.04% P 0.05% S steel: Bar; cold drawn
	DPN: 121 UTS: 440 Elon: 15% Proof: 360
SAE 1021	0.2% C 0.8% Mn 0.04% P 0.05% S steel: Bar
SAE 1022	0.2% C 0.9% Mn 0.04% P 0.05% S steel: Bar; hot rolled as AISI C1022
	DPN: 121 UTS: 440 Elon: 23% Proof: 220
SAE 1022	0.2% C 0.9% Mn 0.04% P 0.05% S steel: Bar; cold drawn
	DPN: 137 UTS: 500 Elon: 15% Proof: 400
SAE 1023	0.22% C 0.4% Mn 0.04% P 0.05% S steel: Bar; hot rolled as AISI C1023
	DPN: 111 UTS: 390 Elon: 25% Proof: 210
SAE 1023	0.22% C 0.4% Mn 0.04% P 0.05% S steel: Bar; cold drawn
	DPN: 121 UTS: 440 Elon: 15% Proof: 370
SAE 1108	0.1% C 0.7% Mn 0.04% P 0.1% S steel: Free cutting; hot rolled as AISI C1108
	DPN: 101 UTS: 340 Elon: 30% Proof: 180
SAE 1108	0.1% C 0.7% Mn 0.04% P 0.1% S steel: Free cutting; cold drawn
	DPN: 121 UTS: 390 Elon: 20% Proof: 330
SAE 1109	0.1% C 0.8% Mn 0.04% P 0.1% S steel: Free cutting; hot rolled as AISI B1109
	DPN: 101 UTS: 340 Elon: 30% Proof: 180
SAE 1109	0.1% C 0.8% Mn 0.04% P 0.1% S steel: Free cutting; cold drawn
	DPN: 121 UTS: 390 Elon: 20% Proof: 330
SAE 1110	0.1% C 0.4% Mn 0.04% P 0.1% S steel: Free machining
SAE 1111	0.13% C 0.8% Mn 0.1% P 0.12% S steel: Free cutting; hot rolled as AISI B1112
	DPN: 121 UTS: 420 Elon: 25% Proof: 220
SAE 1111	0.13% C 0.8% Mn 0.1% P 0.12% S steel: Free cutting; cold drawn
	DPN: 131 UTS: 480 Elon: 10% Proof: 390
SAE 1112	0.13% C 0.9% Mn 0.1% P 0.2% S steel: Free cutting; hot rolled as AISI B1111
	DPN: 121 UTS: 440 Elon: 25% Proof: 220
SAE 1112	0.13% C 0.9% Mn 0.1% P 0.2% S steel: Free cutting; cold drawn
	DPN: 137 UTS: 480 Elon: 10% Proof: 400
SAE 1113	0.13% C 0.9% Mn 0.1% P 0.3% S steel: Free cutting; hot rolled as AISI B1113
	DPN: 121 UTS: 440 Elon: 25% Proof: 220
SAE 1113	0.13% C 0.9% Mn 0.3% S 0.1% P steel: Free cutting; cold drawn
	DPN: 137 UTS: 480 Elon: 10% Proof: 400
SAE 1115	0.15% C 0.8% Mn 0.04% P 0.1% S steel: Free cutting; hot rolled as AISI C1115
	DPN: 111 UTS: 390 Elon: 25% Proof: 200
SAE 1115	0.15% C 0.8% Mn 0.04% P 0.1% S steel: Free cutting; cold drawn
	DPN: 121 UTS: 440 Elon: 20% Proof: 360
SAE 1116	0.16% C 1.2% Mn 0.04% P 0.2% S steel: Free machining
SAE 1117	0.17% C 1.2% Mn 0.04% P 0.1% S steel: Free cutting; hot rolled as AISI C1117
	DPN: 121 UTS: 440 Elon: 23% Proof: 220
SAE 1117	0.17% C 1.2% Mn 0.04% P 0.1% S steel: Free cutting; cold drawn
	DPN: 137 UTS: 520 Elon: 15% Proof: 400
SAE 1118	0.17% C 1.5% Mn 0.04% P 0.1% S steel: Free cutting; hot rolled as AISI C1118
	DPN: 131 UTS: 460 Elon: 23% Proof: 240

Symbol	Nominal analysis, supplier, condition and remarks.
SAE 1118	0.17% C 1.5% Mn 0.04% P 0.1% S steel: Free cutting; cold drawn
	DPN: 143 UTS: 510 Elon: 15% Proof: 440
SAE 1119	0.17% C 1.2% Mn 0.04% P 0.3% S steel: Free cutting; hot rolled as AISI C1119
	DPN: 121 UTS: 440 Elon: 23% Proof: 220
SAE 1119	0.17% C 1.2% Mn 0.04% P 0.3% S steel: Free cutting; cold drawn
	DPN: 137 UTS: 500 Elon: 15% Proof: 400
SAE 1120	0.2% C 0.9% Mn 0.04% P 0.1% S steel: Free cutting; hot rolled as AISI C1120
	DPN: 121 UTS: 440 Elon: 23% Proof: 220
SAE 1120	0.2% C 0.9% Mn 0.04% P 0.1% S steel: Free cutting; cold drawn
	DPN: 137 UTS: 500 Elon: 15% Proof: 400
SAE 1215	0.09% C 0.9% Mn 0.06% P 0.3% S steel: Free machining
SAE 1320	0.2% C 1.7% Mn 0.04% S & P steel: Obsolete
SAE 1513	0.13% C 1.2% Mn 0.04% P 0.05% S steel: For bar
SAE 1518	0.18% C 1.2% Mn 0.04% P 0.05% S steel: Bar, etc.
SAE 1522	0.22% C 1.2% Mn 0.04% P 0.05% S steel: Bar, etc.
SAE 12413	0.13% C 0.9% Mn 0.1% P 0.3% S steel: Free machining
SAN BOLD 3	0.2% C 0.3% Si 0.6% Mn steel: Sanderson; for BS alloy En3A
SB 36F	0.15% C 1.2% Mn 0.04% P & S (max) steel: German proprietors specification; Al killed
SB 36FT	0.15% C 1.25% Mn 0.04% P & S (max) steel: German proprietors specification; Al killed
SB 38F	0.2% C (max) 1.2% Mn 0.04% P & S (max) steel: German proprietors specification; Al killed
SB 38FT	0.2% C (max) 1.2% Mn 0.04% P & S (max) steel: German proprietors specification
SB 39FT	0.16% C 1.3% Mn 0.03% V 0.03% Tl 0.02% Al (max) steel: German proprietors specification
SB 40F	0.2% C (max) 1.4% Mn 0.04% P & S (max) steel: German proprietors specification; Al killed
SB 40FT	0.2% C (max) 1.3% Mn 0.04% P & S (max) steel: German proprietors specification; Al killed
SB 42FT	0.16% C 1.3% Mn 0.035% P & S (max) 0.1% V steel: German proprietors specification; Al killed
SCH 1	0.14% C 0.2% Si 1.0% Mn 0.06% S & P steel: Spencer; for BS alloy En32
SE 90	Iron carbon sintered materials: Sintered Products; 4–8% porosity
	UTS: 390 Elon: 15%
SELCO 53	0.22% C (max) 1.5% Mn 0.04% P & S (max) steel: Italian proprietors specification
SELCO 53V	0.2% C (max) 1.5% Mn 0.04% P & S (max) steel
SH 13-54	0.15% C 1.0% Mn 0.02% P 0.015% S 0.4% Cu steel: Fuji
SH 13-60	0.18% C 1.05% Mn steel: Fuji
SHEF-LO TEMP	0.2% C (max) 1.0% Mn steel: Armco
SHEF-SUPER LO TEMP	0.2% C (max) 1.1% Mn steel: Armco
SIS 14-11-11	0.08% C 0.4% Mn 0.04% P & S steel: For drawn wire; Swedish Standard
	UTS: 590
SIS 14-11-42	0.1% C 0.5% Mn 0.04% P & S steel: For cold rolled sheet; drawing quality; Swedish Standard
	UTS: 345 Elon: 25%
SIS 14-12-11	0.12% C 0.4% Mn 0.06% P & S steel: For drawn wire; Swedish Standard
	UTS: 500
SIS 14-1232	0.13% C 0.2% Si 0.4% Mn 0.05% S & P 0.009% N steel: Swedish Standard
	UTS: 360 Elon: 25%

Symbol	Nominal analysis, supplier, condition and remarks.
SIS 14-13-05	0.18% C 0.6% Mn 0.05% P 0.04% S steel: For castings
	DPN: 135 UTS: 410 Elon: 25% Proof: 230
SIS 14-13-12	0.2% C 0.5% Mn structural steel: Swedish Standard
	DPN: 125 UTS: 400 Elon: 19% Proof: 215
SIS 14-13-70	0.15% C 0.75% Mn 0.035% P & S steel: For case hardening; Swedish Standard
	DPN: 300 UTS: 100 Elon: 12% Proof: 48
SIS 14-14-50	0.22% C 0.7% Mn steel: For general engineering; Swedish Standard
	DPN: 135 UTS: 480 Elon: 24% Proof: 230
SIS 14-21-06	0.2% C 1.6Mn 0.035% P & S steel: For pressure vessels; Swedish Standard
	UTS: 510 Elon: 22% Proof: 350
SIS 14-21-07	0.2% C 1.6% Mn 0.035% P & S steel: For pressure vessels; Swedish Standard
	UTS: 510 Elon: 22% Proof: 350
SIS 14-21-16	0.2% C 1.8% Mn 0.035% P & S steel: For pressure vessels; Swedish Standard
	UTS: 530 Elon: 20% Proof: 390
SIS 14-21-17	0.2% C 1.8% Mn 0.035% P & S steel: For pressure vessels; Swedish Standard
	UTS: 530 Elon: 20% Proof: 390
SIS 14-21-32	0.2% C 1.6% Mn 0.035% P & S steel: Structural steel; Swedish Standard
SIS 14-21-33	0.2% C 1.6% Mn 0.035% P & S structural steel: Swedish Standard
SIS 14-21-34	0.2% C 1.6% Mn 0.036% P & S structural steel: Swedish Standard
	UTS: 510 Elon: 22% Proof: 350
SIS 14-21-35	0.2% C 1.6% Mn 0.035% P & S structural steel: Swedish Standard
	UTS: 510 Elon: 22% Proof: 350
SIS 14-21-42	0.2% C 1.8% Mn 0.035% P & S structural steel: Swedish Standard
	UTS: 530 Elon: 20% Proof: 390
SIS 14-21-43	0.2% C 1.8% Mn 0.035% P & S structural steel: Swedish Standard
	UTS: 530 Elon: 20% Proof: 390
SIS 14-21-44	0.2% C 1.8% Mn 0.035% P & S structural steel: Swedish Standard
	UTS: 530 Elon: 20% Proof: 390
SIS 14-21-45	0.2% C 1.8% Mn 0.035% P & S structural steel: Swedish Standard
	UTS: 530 Elon: 20% Proof: 390
SIS 14-11-60	0.08% C 0.35% Mn 0.03% P 0.04% S steel: For cold rolled strips; Swedish Standard
	DPN: 130 UTS: 400 Elon: 30% Proof: 200
SIS 14-12-65	0.1% C 0.35% Mn 0.03% P 0.04% S steel: For cold rolled strips; Swedish Standard
	DPN: 150 UTS: 550 Elon: 30% Proof: 350
SIS 14-21-68	0.28% C 1.6% Mn 0.06% P 0.05% S steel: For reinforcing; Swedish Standard
	UTS: 600
SIS 1140	0.15% C 0.5% Mn 0.06% S & P steel: Sheet; commercial grade; Swedish Standard
SIS 1142	0.1% C 0.5% Mn 0.05% S & P steel: Sheet for pressing; Swedish Standard
SIS 1142	0.1% C (max) 0.5% Mn (max) 0.04% S & P steel: Sheet; cold rolled
SIS 1145	0.1% C 0.5% Mn 0.04% S & P steel: Sheet; for deep drawing; Swedish Standard
SIS 1146	0.1% C 0.5% Mn 0.04% S & P steel: Sheet; for deep drawing; Swedish Standard
SIS 1146	0.1% C (max) 0.5% Mn (max) 0.04% S & P steel: Sheet; cold rolled; Swedish Standard
SIS 1147	0.08% C (max) 0.45% Mn (max) 0.03% S & P steel: Sheet; cold rolled; Swedish Standard

Symbol	Nominal analysis, supplier, condition and remarks.
SIS 1147	0.1% C 0.5% Mn 0.03% S & P steel: Sheet; for extra deep drawing; Swedish Standard
SIS 1148	0.1% C 0.5% Mn 0.03% S & P steel: Cold rolled; for extra deep drawing; Swedish Standard
SIS 1150	0.07% C 0.06% Si 0.3% Mn 0.03% S & P 0.1% Cu steel: Strip; Swedish Standard; annealed **DPN: 95 UTS: 370**
SIS 1150	0.07% C 0.06% Si 0.3% Mn 0.03% S & P 0.1% Cu steel: Strip; Swedish Standard; cold rolled **DPN: 130 UTS: 370**
SIS 1151	0.12% C (max) 0.6% Mn (max) 0.04% S & P steel: Sheet for hot dip galvanizing; Swedish Standard **UTS: 500**
SIS 1151	0.12% C (max) 0.6% Mn 0.04% S & P steel: For cold rolled sheet; Swedish Standard **UTS: 500**
SIS 1152	0.12% C (max) 0.5% Mn (max) 0.04% S & P steel: Sheet for galvanizing; Swedish Standard
SIS 1157	0.08% C (max) 0.45% Mn (max) 0.03% S & P steel: Sheet for galvanizing; Swedish Standard
SIS 1160	0.08% C (max) 0.3% Mn 0.04% (max) 0.03% P (max) steel: Swedish Standard **DPN: 100 UTS: 300 Elon: 35%**
SIS 1160	0.08% C (max) 0.35% Mn (max) 0.03% P 0.04% S steel: For cold rolled strip; Swedish Standard
SIS 1232	0.13% C 0.2% Si 0.4% Mn 0.05% S & P 0.009% N steel: Swedish Standard **UTS: 360 Elon: 25%**
SIS 1250	0.12% C 0.2% Si 0.4% Mn 0.05% S & P steel: Strip; Swedish Standard; for cold rolling
SIS 1265	0.1% C 0.35% Mn 0.03% S & P steel: Strip; Swedish Standard; annealed **DPN: 115 UTS: 340**
SIS 1265	0.1% C steel: Case hardening; possible Swedish Standard
SIS 1265	0.1% C 0.35% Mn 0.03% S & P steel: Strip; Swedish Standard; cold rolled **DPN: 210 UTS: 640**
SIS 1270	0.15% C (max) 1.0% Mn (max) 0.05% S (max) 0.05% P (max) steel: Plate; Swedish Standard **UTS: 290 Elon: 18%**
SIS 1270	0.15% C (max) 1.0% Mn 0.05% S & P steel: Sheet; Swedish Standard **UTS: 330 Elon: 18% Proof: 250**
SIS 1305	0.18% C (max) 0.6% Mn 0.04% S & P steel: Swedish Standard; annealed **UTS: 390 Elon: 25% Proof: 210**
SIS 1311	0.14% C 0.5% Mn 0.06% S & P structural steel: Swedish Standard; as rolled **UTS: 390 Elon: 24% Proof: 220**
SIS 1311	0.13% C 0.5% Mn 0.06% S 0.08% P steel: For reinforcing; Swedish Standard **Elon: 20% Proof: 220**
SIS 1312	0.12% C 0.6% Mn 0.05% S & P 0.09% N structural steel: Swedish Standard; as rolled **UTS: 390 Elon: 24% Proof: 220**
SIS 1313	0.12% C 0.7% Mn 0.05% S & P 0.009% N structural steel: Swedish Standard; as rolled **UTS: 390 Elon: 24% Proof: 220**

Note. The following abbreviations and units are used in the tables:

DPN	Hardness, diamond pyramid number
UTS	Ultimate tensile strength, N/mm^2
Elon	Elongation, %
Proof	0.1 % proof strength, N/mm^2

$1 N/mm^2 = 0.1 hbar = 0.102 kgf/mm^2 = 0.06475 tonf/in^2 = 145.04 lbf/in^2$

Symbol	Nominal analysis, supplier, condition and remarks.
SIS 1316	0.15% (max) 0.7% Mn (max) 0.05% S & P steel: Plate; for cold rolling; Swedish Standard
SIS 1350	0.15% C 0.6% Mn 0.05% S & P steel: Swedish Steel; normalized **DPN: 120 UTS: 400 Elon: 27% Proof: 210**
SIS 1350-6	0.15% steel: Bar; Swedish Standard; work hardened **DPN: 135 UTS: 420 Elon: 10% Proof: 360**
SIS 1357	0.2% C 0.35% Mn 0.03% S & P steel: Strip; Swedish Standard; annealed **DPN: 135 UTS: 460**
SIS 1357	0.2% C 0.35% Mn 0.03% S & P steel: Strip; Swedish Standard; cold rolled 225 740
SIS 1360	0.15% C (max) 1.2% Mn 0.05% S & P steel: Sheet; Swedish Standard **UTS: 360 Elon: 18% Proof: 280**
SIS 1360	0.15% C (max) 1.2% Mn 0.05% P & S steel: For construction; Swedish Standard **UTS: 360 Elon: 18% Proof: 280**
SIS 1370	0.15% C 0.7% Mn 0.03% S & P steel: Swedish Standard
SIS 1386	Steel: Bar for reinforcing; Swedish Standard; included in MNC 815
SIS 1387	Steel: For reinforcing; Swedish Standard included in MNC 815
SIS 1411	0.17% C 0.7% Mn 0.06% S 0.08% P steel: For reinforcing; Swedish Standard **Elon: 20% Proof: 260**
SIS 1412	0.2% C 0.4% Cu 0.3% Cr (max) steel: Structural; Swedish Standard
SIS 1413	0.18% C 0.4% Cu 0.3% Cr (max) steel: Structural; Swedish Standard
SIS 1414	0.18% C 0.4% Cu 0.3% Cr (max) steel: Structural; Swedish Standard
SIS 1426	0.2% C (max) 1.0% Mn (max) 0.05% S & P steel: Sheet for cold rolling; Swedish Standard
SIS 1674	0.5% C 0.3% Si 0.8% Mn 0.035% S (max) 0.035% P (max) steel: Hardened & tempered; Swedish Standard **UTS: 700 Elon: 15% Proof: 440**
SIS 1912	0.09% C 0.6% Mn 0.08% P 0.25% S steel: Free machining; Swedish Standard; cold rolled **UTS: 440**
SIS 1914	0.11% C 1.1% Mn 0.3% S 0.25% Pb steel: Free machining; Swedish Standard
SIS 1922	0.16% C 1% Mn 0.05% P 0.15% S steel: Free machining; Swedish Standard; cold rolled **UTS: 760**
SIS 1926	0.15% C 1.0% Mn 0.2% S 0.25% Pb steel: Free machining; Swedish Standard
SIS 2032	Steel: For reinforcing; Swedish Standard; included in MNC 815
SIS 2106	0.18% C 1.4% Mn 0.035% S (max) 0.035% P (max): For pressure vessels; Swedish Standard **UTS: 400 Elon: 21%**
SIS 2106	0.2% C (max) 1.6%-Mn (max) 0.02% N (max) steel: For pressure vessels; Swedish Standard; normalized **UTS: 510 Elon: 22% Proof: 350**
SIS 2107	0.2% C (max) 1.4% Mn 0.035% S (max) 0.035% P (max): For pressure vessels; Swedish Standard **UTS: 400 Elon: 22%**
SIS 2107	0.2% C (max) 1.6% Mn (max) 0.02% N (max) steel: For pressure vessels; Swedish Standard; normalized **UTS: 510 Elon: 22% Proof: 350**
SIS 2116	0.2% C (max) 1.8% Mn 0.03% P & S (max) 0.02% N (max) steel: For pressure vessels; Swedish Standard

Symbol	Nominal analysis, supplier, condition and remarks.
SIS 2117	0.2% C (max) 1.8% Mn (max) 0.035% P & S (max) 0.02% N (max) steel: For pressure vessels; Swedish Standard
SIS 2121	0.2% C (max) 1.5% Mn 0.05% P & S (max) steel: Sheet; Swedish Standard; galvanized **UTS: 400 Elon: 16% Proof: 320**
SIS 2122	0.2% C (max) 1.0% Mn 0.05% S 0.05% P steel: Plate; Swedish Standard **UTS: 390 Elon: 14%**
SIS 2122	0.2% C (max) 1.5% Mn 0.05% P & S (max) steel: Plate; Swedish Standard; galvanized **UTS: 430 Elon: 14% Proof: 350**
SIS 2122	0.2% C (max) 1.5% Mn 0.05% S & P steel: Sheet; Swedish Standard **UTS: 430 Elon: 14% Proof: 350**
SIS 2132	0.2% C (max) 1.6% Mn 0.035% P & S (max) 0.02% N (max) steel: For structures; Swedish Standard
SIS 2133	0.2% C (max) 1.6% Mn 0.035% P & S (max) 0.02% N steel: For structures; Swedish Standard
SIS 2134	0.2% C (max) 1.6% Mn 0.035% P & S (max) 0.02% N steel: For structures; Swedish Standard
SIS 2135	0.2% C (max) 1.6% Mn 0.035% P & S (max) 0.02% N steel: For structures; Swedish Standard
SIS 2136	0.2% C (max) 1.5% Mn (max) 0.05% S & P steel: Sheet for cold rolling; Swedish Standard
SIS 2136	0.2% C (max) 1.5% Mn 0.05% S & P steel: Sheet; Swedish Standard **UTS: 400 Elon: 16% Proof: 320**
SIS 2137	Plain carbon steel: For reinforcing **UTS: 1000 Elon: 3% Proof: 800**
SIS 2142	0.2% C (max) 1.8% Mn 0.035% P & S (max) 0.02% N steel: For structures; Swedish Standard
SIS 2143	0.2% C (max) 1.8% Mn 0.035% P & S 0.02% N steel: For structures; Swedish Standard
SIS 2144	0.2% C (max) 1.8% Mn 0.035% P & S (max) 0.02% N steel: For structures; Swedish Standard
SIS 2145	0.2% C (max) 1.8% Mn 0.035% P & S (max) 0.02% N steel: For structures; Swedish Standard
SIS 2164	Steel: For reinforcing; Swedish Standard; included in MNC 815
SIS 2165	0.24% C (max) 1.6% Mn (max) 0.05% S 0.06% P steel: For reinforcing; Swedish Standard **Elon: 15% Proof: 370**
SIS 1265	0.1% C 0.35% Mn 0.03% S & P steel: Strip; cold rolled
SIS 2167	Steel: For reinforcing; Swedish Standard; included in MNC 815
SIS 2168	0.28% C (max) 1.6% Mn (max) 0.05% S 0.06% P steel: For reinforcing; Swedish Standard **Elon: 12% Proof: 590**
SIS 1270	0.15% C (max) 1.0% Mn 0.05% P & S steel: Sheet; Swedish Standard; galvanized **UTS: 330 Elon: 18% Proof: 250**
SIS 2172-04	0.2% C (max) steel: For case hardening; Swedish Standard; normalized **UTS: 500 Elon: 21% Proof: 320**
SIS 2172-21	0.2% C (max) 1.5% Mn (max) 0.5% Si steel: Casting; Swedish Standard; normalized **UTS: 500 Elon: 18% Proof: 300**
SIS 8001	0.2% C sintered iron: Swedish Standard
SIS 8020	0.2% C (max) 0.35% P sintered steel: Swedish Standard
SKF 21T2	0.18% C 1.4% Mn hollow steel: Bar **DPN: 170 UTS: 570 Elon: 23% Proof: 360**
SKF CEAX	0.5% C 0.7% Mn 0.05% S 0.05% P steel: For springs; SKF
SKF TERMEK	Steel: Free machining; bar; SKF

Symbol	Nominal analysis, supplier, condition and remarks.
SMOOTHTRODE 43.23	0.07% C 0.68% Mn 0.4% Si steel: For electrodes; Esab **UTS: 556 Elon: 26% Proof: 279**
SMR St 2	0.1% C (max) 0.1% Si 0.45% Mn 0.05% S & P steel: German Standard
SMR St 3	0.1% C (max) 0.1% Si 0.45% Mn 0.04% S & P steel: German Standard
SMR St 4	0.1% C (max) 0.07% Si 0.45% Mn 0.03% S & P steel: German Standard
SMU St 2	0.1% C 0.45% Mn 0.05% S & P steel: German Standard
SMU St 3	0.1% C (max) 0.45% Mn 0.04% S & P steel: German Standard
SMU St 4	0.1% C (max) 0.45% Mn 0.03% S & P steel: German Standard
SOUNDOTENAX 52	0.2% C 1.3% Mn 0.32% Cr steel: Cockerill-Ougree
SOUNDOTENAX 56	0.2% C 1.3% Mn 0.32% Cr 0.42% Ni 0.22% Mo steel: Cockerill-Ougree
SOUNDOTENAX 66	0.2% C 1.3% Mn 0.3% Cr 0.9% Ni 0.4% Mo steel: Cockerill-Ougree
SP 25	Sintered Ni alloy steel similar to SP 24; Durasint **DPN: 210 UTS: 410 Elon: 2%**
SP 26	Sintered Ni alloy steel similar to SP 20; Durasint **DPN: 210 UTS: 550 Elon: 2%**
SPEAR B8	0.06% C 0.1% Si 0.4% Mn steel: Spear & Jackson; obsolete
SPELTAFAST	Mild steel: Sheet; galvanized; RTB for BS 2989
SPRASTEEL 10	0.1% C 0.5% Mn 0.04% S & P steel: Wire for spraying; Metco; as sprayed **DPN: 189 UTS: 200**
SPRASTEEL 25	0.23% C 0.6% Mn 0.04% S & P steel: Wire for spraying; Metco; as sprayed **DPN: 189 UTS: 220**
S St 2	0.1% C 0.15% Si 0.35% Mn 0.05% S & P steel: German Standard
S St 3	0.1% C (max) 0.1% Si 0.4% Mn 0.04% S & P steel: German Standard
S St 4	0.1% C (max) 0.08% Si 0.4% Mn 0.035% S & P steel: German Standard
St 10	0.15% C (max) 0.35% Mn 0.06% S & P steel: German Standard
St 10/01	0.15% C (max) 0.35% Mn 0.06% S & P steel: German Standard
St 10/02	0.15% C (max) 0.35% Mn 0.06% S & P steel: German Standard
St 10/03	0.15% C (max) 0.35% Mn 0.06% S & P steel: German Standard
St 34/2	0.17% C (max) 0.05% S & P steel: German Standard
St 34/3	0.17% C 0.05% S & P 0.01% N steel: German Standard
St 35	0.18% C (max) 0.05% S & P steel: German Standard
St 35/4	0.17% C (max) 0.3% Si 0.4% Mn 0.05% S & P steel: German Standard
St 35/13 K	0.09% C 0.4% Mn steel: Designation used by German Standard
St 37	0.2% C (max) 0.05% S 0.08% P steel: German Standard
St 37/2	0.2% C 0.05% S 0.06% P steel: German Standard
St 37/3	0.2% C (max) 0.05% S & P steel: German Standard
St 42	0.25% C (max) 0.05% S 0.08% P steel: German Standard
St 42/2	0.25% C (max) 0.05% S & P steel: German Standard
St 42/3	0.25% C (max) 0.05% S & P steel: German Standard

Symbol	Nominal analysis, supplier, condition and remarks.
St 45	0.25% C (max) 0.3% Cr 0.05% S & P steel: German Standard
St 45/4	0.22% C (max) 0.2% Si 0.4% Mn 0.05% S & P steel: German Standard
St 52	0.2% C (max) 0.5% Si 1.5% Mn 0.05% S & P steel: German Standard
St 52/3	0.2% C (max) 0.05% S & P steel: German Standard
St 52/302	0.2% C (max) 0.05% S & P steel: German Standard
St 52/4	0.2% C (max) 0.3% Si 1.5% Mn 0.05% S & P steel: German Standard
STA 5 V1A	0.1% C steel: Free machining; replaced by BS alloy En1A
STA 5 V1B	0.2% C steel: Semi-free cutting; replaced by BS alloy En 7A
STA 5 V2	0.2% C steel: Cold drawn; replaced by BS alloy En 2
STA 5 V2A	0.12% C steel: Replaced by BS alloy En 2A
STA 5 V2A/1	0.1% C steel: Replaced by BS alloy En 2A/1
STA 5 V2B	0.15% C steel: Replaced by BS alloy En 2B
STA 5 V2C	0.2% C steel: Replaced by BS alloy En 2C
STA 5 V2D	0.22% C steel: Replaced by BS alloy En 2D
STA 5 V3	0.2% C steel: Replaced by BS alloy En 3A
STA 5 V15	0.15% C (max) steel: Replaced by BS alloy En 32C
STA 5 V15/1	0.15% C (max) steel: Replaced by BS alloy En 32C
STA 5 V15/A	0.18% C (max) steel: Replaced by BS alloy En 201
STA 5 V15A/1	0.18% C steel: Obsolete
STA 5 V15AM	0.18% C (max) steel: Free cutting; replaced by BS alloy En 202
STA 38 B	Low carbon killed steel: For drawn shell bodies
STC 1	0.2% C 0.2% Si 0.8% Mn 0.05% S & P steel: Spencer; normalized DPN: 150 UTS: 450 Elon: 25%
STELCO C845	0.2% C 1.2% Mn 0.005% Nb 0.2% Cu steel: Steel Co. of Canada
STELCO C850	0.2% C 1.2% Mn 0.005% Nb 0.2% Cu steel: Steel Co. of Canada
STELCO C855	0.2% C 1.2% Mn 0.005% Nb 0.2% Cu steel: Steel Co. of Canada
STELCO C860	0.2% C 1.2% Mn 0.005% Nb 0.2% Cu steel: Steel Co. of Canada
STELCOLOY G	0.12% C 0.75% Mn 0.45% Cr 0.45% Ni 0.45% Cu steel: Steel Co. of Canada
STELCOLOY S	0.15% C 1.35% Mn 0.32% Cr 0.3% Ni 0.02% V 0.3% Cu steel: Steel Co. of Canada
STELVETITE G	PVC on one side of hot dip galvanized steel: British Steel
STELVETITE R	PVC on both sides of zinc-plated steel: British Steel
STELVETITE Z	Coloured PVC on one side of zinc-plated steel: British Steel
SUPAVIT	Mild steel: Sheet decarburized for vitreous enamelling; Richard Thomas & Baldwins
SUPERELSO 60	0.2% C (max) 1.2% Mn 0.03% P & S steel: Crusot

Symbol	Nominal analysis, supplier, condition and remarks.
SUPERTOUGH B20	0.2% C 1.5% Mn steel: ESC for BS alloy En 14A
SUPER TUFCOR	0.15% C 0.9% Mn steel: Jonas; core properties hardened and tempered DPN: 170 UTS: 580 Elon: 20%
T	0.15% C 0.3% Si 0.45% Mn steel: Casting; Edgar Allen; Code No 605 steel
T 708	0.09% C 1.0% Mn 0.05% P 0.25% S steel: Free machining; Fagersta UTS: 390 Elon: 28%
T 709	0.09% C 1.0% Mn 0.05% P 0.25% S steel: Free machining; Fagersta UTS: 390 Elon: 28%
T 715	0.15% C 0.25% Si 0.9% Mn 0.05% P 0.15% S steel: Free machining; Fagersta UTS: 460 Elon: 25%
T 725	0.25% C 0.25% Si 0.9% Mn 0.05% P 0.15% S steel: Free machining; Fagersta UTS: 530 Elon: 22%
TENAPSO	0.07% C 0.2% Si 0.35% Mn 0.35% Cu steel: Pompey
TENSITRODE 48.30	0.08% C 1% Mn 0.6% Si steel: For electrodes; Esab UTS: 589 Elon: 28% Proof: 494
TENSITRODE 55.00	0.08% C 1.5% Mn 0.60% Si steel: For electrodes; Esab UTS: 602 Elon: 28% Proof: 570
TOLEDO 15	0.2% C 0.8% Mn 0.06% S & P steel: Toledo for BS alloy En 2
TOLEDO 20	0.2% C 1% Mn 0.06% S & P steel: Toledo for BS alloys En3 En 3A & En 3B
TONCAN	0.03% C 0.45% Cu 0.07% Mo steel: Corrosion resistant; no further information
T P W	0.06% C 0.5% Mn steel: Welding electrode; Murex
T St 0	0.12% C (max) 0.3% Mn 0.06% S 0.08% P steel: German Standard
T St 10	0.15% C (max) 0.35% Mn 0.06% S 0.08% P steel: German Standard
T St 10/01	0.15% C (max) 0.35% Mn 0.06% S 0.08% P steel: German Standard
T St 10/02	0.15% C (max) 0.35% Mn 0.06% S 0.08% P steel: German Standard
T St 10/03	0.15% C (max) 0.35% Mn 0.06% S 0.08% P steel: German Standard
TT St 35	0.16% C (max) 0.2% Si 0.5% Mn 0.05% S & P steel: German Standard
TT St 41	0.2% C (max) 0.2% Si 0.6% Mn 0.05% S & P steel: German Standard
TT St 45	0.22% C (max) 0.2% Si 0.5% Mn 0.05% S & P steel: German Standard
TUBROD 15.10	0.06% C 1.7% Mn 0.9% Si steel: Tubular welding rod; Esab; as welded UTS: 675 Elon: 24% Proof: 575
TUFCOR	0.15% C 0.7% Mn steel: Jonas; core properties; hardened & tempered DPN: 148 UTS: 480 Elon: 20%
TU St 1	0.12% C (max) 0.3% Mn 0.06% S & P steel: German Standard
TU St 2	0.1% C (max) 0.3% Mn 0.05% S & P steel: German Standard
TU St 34	0.08% C 0.25% Mn 0.06% S 0.08% P steel: Designation used by German Standard
TU St 37	0.2% C (max) 0.05% S 0.08% P steel: German Standard
TU St 37/02	0.2% C (max) 0.05% S 0.08% P steel: German Standard
TU St 37/03	0.2% C (max) 0.05% S 0.08% P steel: German Standard
TU St 37/04	0.2% C (max) 0.05% S 0.08% P steel: German Standard

Note. The following abbreviations and units are used in the tables:

DPN	Hardness, diamond pyramid number
UTS	Ultimate tensile strength, N/mm^2
Elon	Elongation, %
Proof	0.1 % proof strength, N/mm^2

1 N/mm^2 = 0.1 hbar = 0.102 kgf/mm^2 = 0.06475 tonf/in^2 = 145.04 lbf/in^2

Symbol	Nominal analysis, supplier, condition and remarks.
TU St 37/05	0.2% C (max) 0.05% S 0.08% P steel: German Standard
TYPE M	Mild steel: Multi wire Al-coated wire; Copperweld Co.
TYPE XX	0.02% C 0.1% Mn steel: Welding electrode; Murex
	UTS: 400 Elon: 30% Proof: 370
UHB 3	0.15% C (max) 0.15% Si 0.5% Mn 0.03% S & P steel: Tube; Uddelholm
	UTS: 370 Elon: 30% Proof: 210
UHB 4M10	0.22% C 0.2% Si 0.8% Mn 0.03% S & P steel: Tube; Uddelholm
	UTS: 450 Elon: 30% Proof: 240
UMS 45	0.22% C 0.25% Si 0.45% Mn 0.045% S & P steel: Krupp
UNBREAKABLE	0.14% C 0.9% Mn 0.04% S & P steel: For case hardening; Swift Levick for BS alloy En 32
UNI 743	Replaced by UNI 5334
UNI 815	Replaced by UNI 5335
UNI 2952	Replaced by UNI 5331
UNI 2953	Replaced by UNI 5331
UNI 4365	Italian Standard covering: Plain carbon and low alloy steels for bolts
UNI 5331 C 10	0.1% C 0.5% Mn steel: For carburizing
UNI 5331 C 16	0.15% C 0.5% Mn steel: For carburizing
UNI 5334 Fe 33	C not specified: Hot rolled steel section
	UTS: 40 Elon: 17%
UNI 5334 Fe 34	0.17% C hot rolled steel: Section; graded by S & P content
	UTS: 380 Elon: 28% Proof: 180
UNI 5334 Fe 37	0.2% C hot rolled steel: Section; graded by S & P content
	UTS: 400 Elon: 26% Proof: 210
UNI 5334 Fe 42	0.2% C hot rolled steel: Section; graded by S & P content
	UTS: 460 Elon: 23% Proof: 220
UNI 5334 Fe 50	C not specified hot rolled steel: Section; graded by S & P content
	UTS: 550 Elon: 19% Proof: 300
UNI 5335 Fe 33	0.08% P 0.06% S steel: Plate; hot rolled
	UTS: 350 Elon: 17%
UNI 5335 Fe 34	0.17% C steel: Plate; hot rolled; graded by S & P content
	UTS: 360 Elon: 27% Proof: 180
UNI 5335 Fe 37	0.2% C steel: Plate; hot rolled; graded by S & P content
	UTS: 400 Elon: 25% Proof: 210
UNI 5335 Fe 42	0.21% C steel: Plate; hot rolled; graded by S & P content
	UTS: 440 Elon: 22% Proof: 230
UNI 5335 Fe 50	C not specified steel: Plate; hot rolled; graded by S & P content
	UTS: 540 Elon: 19% Proof: 280
UNION 32	0.2% C 1.2% Mn 0.15% Cr 0.25% Cu 0.05% Al steel: Dortmund-Horder
UNION 36	0.22% C (max) 1.4% Mn 0.04% P & S steel: Dortmund-Horder; Al killed
UNION 40	0.24% C (max) 1.5% Mn 0.04% P & S (max) steel: Dortmund-Horder; Al killed
UNION 45	0.24% C (max) 1.6% Mn 0.04% P & S (max) 0.15% Ti steel: Dortmund-Horder; Al killed
UNION HB36	0.22% C 1.4% Mn 0.05% P & S (max) steel: Dortmund-Horder; Al killed
UNION Q38	0.2% C (max) 1.1% Mn 0.04% P & S (max) steel: Dortmund-Horder
UNION Q42	0.22% C (max) 1.4% Mn 0.04% P & S (max) steel: Dortmund-Horder
UNITRODE 48.00	0.080% C 0.70% Mn 0.60% Si steel: For electrodes; Esab
	UTS: 525 Elon: 30% Proof: 417

Symbol	Nominal analysis, supplier, condition and remarks.
U St 12	0.1% C (max) 0.3% Mn 0.05% S & P steel: German Standard
U St 12/03	0.1% C (max) 0.3% Mn 0.05% S & P steel: German Standard
U St 12/04	0.1% C (max) 0.3% Mn 0.05% S & P steel: German Standard
U St 12/05	0.1% C (max) 0.3% Mn 0.05% S & P steel: German Standard
U St 13	0.1% C (max) 0.3% Mn 0.04% S & P steel: German Standard
U St 13/03	0.1% C (max) 0.3% Mn 0.04% S & P steel: German Standard
U St 13/04	0.1% C 0.3% Mn 0.04% S & P steel: German Standard
U St 13/05	0.1% C (max) 0.3% Mn 0.04% S & P steel: German Standard
U St 14	0.1% C (max) 0.3% Mn 0.03% S & P steel: German Standard
U St 14/04	0.1% C (max) 0.3% Mn 0.03% S & P steel: German Standard
U St 14/05	0.1% C (max) 0.3% Mn 0.03% S & P steel: German Standard
U St 34/2	0.17% C (max) 0.05% S & P steel: German Standard
U St 37	0.2% C (max) 0.05% S 0.08% P steel: German Standard
U St 37/2	0.2% C (max) 0.05% S & P steel: German Standard
U St 37/202	0.2% C (max) 0.05% S & P steel: German Standard
U St 37/203	0.2% C (max) 0.05% S & P steel: German Standard
U St 37/204	0.2% C (max) 0.05% S & P steel: German Standard
U St 37/205	0.2% C (max) 0.05% S & P steel: German Standard
U St 42	0.25% C (max) 0.05% S & P steel: German Standard
U St 42/2	0.25% C (max) 0.05% S & P steel: German Standard
U St 42/202	0.25% C (max) 0.05% S & P steel: German Standard
U St 42/203	0.25% C 0.05% S & P steel: German Standard
U St 42/204	0.25% C (max) 0.05% S & P steel: German Standard
U St 42/205	0.25% C 0.05% S & P steel: German Standard
VERSITRODE 46.58	0.08% C 0.045% Mn 0.15% Si steel: For electrodes; Esab
	UTS: 510 Elon: 28% Proof: 401
VODEX	0.06% C 0.4% Mn steel: Welding electrode; Murex
	UTS: 500 Elon: 33% Proof: 450
W 6	0.11% C steel: Welding electrode; Cu covered; Armco; as welded
	UTS: 530 Elon: 27% Proof: 450
W 61	0.11% C 0.2% Si 1.0% Mn steel: Welding electrode; Cu covered; Armco; as welded
	UTS: 620 Elon: 25% Proof: 540
WELCON 2H	0.18% C (max) 1.35% Mn 0.04% P & S (max) steel: Japan Steel Works
WELCON 50	0.18% C (max) 1.35% Mn 0.035% P & S (max) steel: Japan Steel Works
WEL-TEN 50	0.18% C 1.3% Mn 0.035% P & S (max) steel: Yawata
WEL-TEN 55	0.18% C 1.35% Mn 0.035% P & S (max) steel: Yawata
WU St 12	0.1% C (max) 0.3% Mn 0.05% S & P steel: German Standard
WU St 12/03	0.1% C 0.3% Mn 0.05% S & P steel: German Standard

Symbol	Nominal analysis, supplier, condition and remarks.
WU St 12/04	0.1% C (max) 0.3% Mn 0.05% S & P steel: German Standard
WU St 12/05	0.1% C (max) 0.3% Mn 0.05% S & P steel: German Standard
WU St 37/2	0.2% C 0.05% S 0.06% P steel: German Standard
WU St 37/202	0.2% C (max) 0.05% S 0.06% P steel: German Standard
WU St 37/203	0.2% C (max) 0.05% S 0.06% P steel: German Standard
WU St 37/204	0.2% C (max) 0.05% C 0.06% P steel: German Standard
WU St 37/205	0.2% C (max) 0.05% S 0.06% P steel: German Standard
XL CUT	0.09% C (max) 0.9% Mn 0.3% S steel: British Steel; free machining
XL CUT	0.09% C (max) 0.97% Mn 0.05% P (max) 0.3% S steel: British Steel Co.; free cutting
XL CUT Pb	0.09% C (max) 0.9% Mn 0.3% S 0.15% Pb (min) steel: British Steel; free machining
XL CUT Pb	0.09% C (max) 0.97% Mn 0.06% P (max) 0.3% S 0.15% Pb (min) steel: British Steel Co.; free machining
XL CUT SPb	0.08% C (max) 0.97% Mn 0.09% P (max) 0.3% S 0.2% Pb (min) 0.008% N steel: British Steel Co.; free machining
XL CUT SPb	0.08% C (max) 0.9% Mn 0.3% S 0.2% Pb (min) steel: British Steel Co.; free machining
Y 12	0.17% C (max) 0.25% Mn 0.035% P (max) 0.025% S (max): French Standard

Symbol	Nominal analysis, supplier, condition and remarks.
YAW-TEN 50	0.16% C (max) 1.1% Mn 0.3% Ti 0.25% Cu steel: Yawata
YB TEN	0.12% C (max) 0.75% Mn 0.04% P & S steel: Youngstown Steel Co.
YEO 40	0.22% C 1.3% Mn 0.1% Ti steel: Yamata
YND 33	0.1% Cr 1.05% Mn 0.035% P & S (max) steel: Yamata
YOLOY 50W	0.17% C 0.7% Mn 0.05% P & S (max) steel: Youngstown Steel Co.
YOLOY M, G & A	0.25% C (max) 1.6% Mn 0.2% Cu steel: Youngstown Steel Co.
YO-MAN B	0.25% C (max) 1.6% Mn 0.2% Cu steel: Youngstown Steel Co.
ZINCGRIP	Mild steel: Sheet; Zn coated; Armco
ZINCROMETAL	Mild steel sheet coated with zinc rich paint; weldable; Dacral

Note. The following abbreviations and units are used in the tables:

DPN	Hardness, diamond pyramid number
UTS	Ultimate tensile strength, N/mm^2
Elon	Elongation, %
Proof	0.1 % proof strength, N/mm^2

$1 N/mm^2 = 0.1 \text{ hbar} = 0.102 \text{ kgf/mm}^2 = 0.06475 \text{ tonf/in}^2 = 145.04 \text{ lbf/in}^2$

44A2 Steel – plain carbon 0.25–0.45 per cent

Specific gravity	7.86	
Density	7860 kg/m^3	(0.28 lb/in.^{3})
Solidus/liquidus	1470–1500 °C	
Thermal conductivity	41.9 W/m °C	(10 cal/m s °C)
Coefficient of linear expansion (20–100° C)	$11 \times 10^{-6}/$ °C	
Electrical conductivity	8–10% IACS (copper 100%)	
Specific resistance	180–250 microhm mm	
Young's modulus of elasticity	– kg/m^2	
Impact	88–115 J	(65–85 ft/lb)(Izod)
Fatigue strength		
Hot strength		

Temperature °C	Tensile strength N/mm^2	Elongation %
200	450	37
300	450	39
400	420	38
500	200	38
650	210	45

The above properties are typical of the following group and may not apply to any one specification. It is possible that with certain specifications some of the values may not be applicable.

General metallurgical characteristics

The steels in this group are all hardenable by thermal treatment and cold work. The degree of hardenability is directly proportional to the carbon content, modified to some extent by the presence of impurities which tend to increase the hardness, while decreasing the ductility by a disproportional amount. All the steels listed have some silicon and manganese, these being added as de-oxidizing and de-sulphurizing elements during manufacture. The quantities present have no significant effect on the properties.

The quality of the steels can be roughly assessed from the phosphorus and sulphur contents, as both these elements should be as low as possible, although some steels have high sulphur, lead or tellurium additions to aid machinability, but this is not normally required on these medium carbon materials.

Some of the lower carbon steels are used in the carburized condition.

Small additions of copper have been made to certain of the alloys listed to improve the atmospheric corrosion resistance and the tensile properties. Many of the steels will have residual elements present which could have an adverse effect on welding as they tend to increase the hardenability.

Thermal treatment

None of the steels listed can be cold worked to any great extent without intermediate annealing. Ideally this should be at 880–900 °C for 20 minutes per 24 mm of section, followed by slow cooling. In practice a sub-critical anneal at 650–670 °C will suffice except where severe cold work has to be followed by further cold work.

After forging or other hot work or welding, the parts should be normalized. This carried out at 880–900 °C for 10 minutes per 25 mm of section, followed by cooling in still air. Many of the steels listed are used in this condition, which homogenizes the structure, removes all forging stresses and ensures maximum ductility.

Hardening requires the parts to be taken to 840–880 °C for 10 minutes per 25 mm of section and water quenched. If distortion is a problem, oil quenching can be used, but will not result in maximum hardness. The parts should be tempered at 300–600 °C as soon as possible after hardening. All of the lower carbon steels listed can be case hardened, when the same remarks apply as in Section 44A1.

It must be emphasized that the temperatures quoted are for guidance only and may not apply to the whole group of materials listed. Readers are strongly advised to refer to the full specification or to consult the supplier on treatments of individual steels, as only this will ensure the optimum properties from the chosen material.

Normal electric, gas or oil fired furnaces can be used but some care is necessary at the high temperatures to ensure that local decarburization does not occur. These materials seldom warrant the use of controlled atmospheres although these may be necessary with high duty components which cannot be machined after hardening.

Scale removal can be by blasting or acid pickling.

Welding and brazing

Pre-treatment. The weld area should be free from scale, grease and oil. Joint preparation and design are of paramount importance for high quality welds.

For welding the parts should be of similar type steel and in the normalized or annealed condition. Generally some pre-heating will be necessary with these materials.

Welding and brazing. These materials are all hardenable, thus care should be taken not to quench the weld area, particularly near corners or other stress raisers. A large cold mass of steel adjacent to the weld can be an efficient quench, giving very high hardness figures near the weld with consequent very low ductility, hence the importance of adequate pre-heating.

Gas or electric arc welding with fluxes, in the hands of a trained operator, gives satisfactory joins. Inert gas shielded arc welding is sometimes justified where scale formation could be troublesome, but is generally not economical.

Resistance welding with its intense local heat can cause cracking adjacent to the weld, unless careful thought is given to the weld design.

Brazing, using copper alloys and fluxes, is relatively trouble free and can be carried out on hardened and tempered or normalized parts, without affecting the properties, if the correct braze metal and techniques are chosen.

Post-treatment. All the material listed may have a hard, brittle zone adjacent to the weld and this must be tempered if the parts are to be subject to any form of stress whatever. With correct pre-heating this may not be necessary.

Ideally they should be designed to be used in the normalized condition with this carried out after welding. With the careful design, eliminating all sharp corners and stress raisers, welded assemblies can be hardened and tempered. At the very least the weld area must be stress relieved by playing a gas flame on the weld and the weld zone to temper or stress release the part locally. There are now available electrically heated 'blankets' which allow stress relieving of the largest welded structures. The same equipment can be used for pre-heating.

All fluxes and weld scale should be removed by chipping or blasting before any heat treatment.

Flaw detection methods

Crack Test. Magnetic crack test can be applied to all the steels listed. Penetrant oil chalk testing can be as sensitive as the magnetic test and is more easily applied on local areas, or large structures.

X-ray. Not required on wrought products, but very commonly applied to production welds. Portable equipment using radioactive metal isotopes has been specially developed for this purpose.

Ultra-sonic test. Not normally required, but can be used to identify certain types of welding defects.

Chemical etch. Not normally required, except as a laboratory check on the quality of bars or forgings. This usually scraps the part being examined.

Corrosion protection

Temporary. Parts should be kept dry and out of contact with a moist atmosphere if corrosion is to be prevented.

Wherever possible, oil or grease should be applied and this is essential on finished machined areas. Soluble cutting oils in good condition prevent corrosion, but can cause rusting when dirty and contaminated with metal dust.

Permanent. It is extremely difficult to protect the materials listed. Iron forms oxides in the presence of oxygen with moisture, and as these are non-adherent they are readily removed, leaving the surface free for further attack. As the action can be very rapid, it is essential that the protection is applied immediately after surface preparation.

The success of corrosion protection rests largely on the pre-treatment surface preparation. All traces of scale and rust must be removed and oil and loose particles of dirt or dust must be absent as this will affect the adhesion of the protective coating.

Painting. The best method of preparation is shot blasting. This process is open to abuse and must be carefully controlled to secure optimum results. All traces of dust should be removed by brushing after blasting and it must be realized that a freshly blasted surface can corrode within minutes in humid conditions. Wire brushing and emery dressing are also possible but are not generally as effective as blasting. Pickling chemicals must be used with caution as rapid corrosion can occur if the correct control is not used.

Phosphating the prepared surfaces acts as a corrosion protection in its own right and is an excellent key for the primer. The parts must then be oven dried before painting to drive off moisture.

The prepared surface should have a primer applied and this would ideally be of the zinc silicate or chromate type, followed as soon after drying as possible with a top coat.

Plastic paints are now available which resist abrasion and damage, and are non-porous. The best of these generally require stoving and still require the metal oxide or chromate type of primer.

Plating. Many of the steels listed require decorative chromium plating. They must have an adequate undercoating, generally of copper and nickel. It is recommended that 0.05 mm copper and 0.025 mm nickel, plated on correctly cleaned and prepared surfaces should be the minimum for outdoor corrosion resistance. The chromium plate is very thin and serves only to prevent the nickel tarnishing and maintain the bright finish.

Cadmium, tin and zinc, can all be applied, the coating in this case being up to 0.025 mm thick, which is adequate provided careful pre-plating treatment is carried out. Tin and zinc are also applied as dip coatings. Zinc and cadmium must not be used in contact with any foodstuffs.

Many of these steels will require stress releasing after plating to prevent hydrogen embrittlement.

Metal spraying, particularly of zinc and aluminium, can be successful, provided the surface is blasted to clean and give a key prior to spraying, and wire brushed, then chemically sealed or painted after spraying.

Cathodic protection makes use of the electroplating principles by completing an electric circuit with the part to be protected and a sacrificial anode, which corrodes and is replaced. This method should be supplemented by other forms of surface coating and is finding considerable application on pipe lines, marine structures and ship as well as building foundations and steel structures.

Black oxide treatment uses a hot concentrated caustic soda solution with additions to give an adherent dark oxide which is successful in preventing further oxidation provided an oil film is maintained.

The permanent protection of the steels listed is neither cheap nor easy, but the initial cost of correct preparation, followed by good quality materials properly applied, is very much more economical than the cost of removing a failed coating with the consequent rust, then re-protecting. Experience also indicates that it is more difficult to apply a corrosion protection at the second attempt than the first. Success relies on very careful and complete preparation of the surface followed by the correct application of good quality materials to the proper specification.

Machinability

These materials can all be readily machined in the hardened and tempered condition, the optimum results being obtained with a hardness of 250–300 diamond number – 600–750 N/mm^2 tensile range.

Of all the ferrous materials available, those in this group are probably the best from the point of view of a good surface finish coupled with long tool life, apart from the specially developed free machining materials.

In the annealed condition and to some extent with normalizing, there is a tendency for tearing resulting in a poor finish.

Uses

These steels are the general purpose engineering and constructional first choice for low price with medium strength.

Ship plate, boiler plate, sections for building, roof trusses, lifting equipment gears, springs, low stressed static and moving parts in all forms of engines, pipe lines, pylons, rails and sheeting where this forms part of the stressed structure.

Symbol	Nominal analysis, supplier, condition and remarks.
0.0652	0.36% C 0.2% Si 0.6% Mn 0.04% S & P steel: German Standard
0.0967	0.38% C 1.5% Si 0.7% Mn steel: For springs; German Standard
0.0968	0.46% C 1.7% Si 0.7% Mn steel: For springs; German Standard
1.0130	0.25% C (max) 0.05% S 0.08% P steel: German Standard
1.0132	0.25% C (max) 0.05% S 0.06% P steel: German Standard
1.0136	0.25% C (max) 0.05% S & P steel: German Standard
1.0142	0.25% C (max) 0.05% S 0.06% P steel: German Standard
1.0143	0.25% C (max) 0.05% S & P steel: German Standard
1.0507	0.36% C 0.05% S & P steel: German Standard
1.0509	0.36% C 0.2% Si 0.4% Mn 0.05% S & P steel: German Standard
1.0530	0.3% C 0.05% S 0.08% P steel: German Standard
1.0532	0.3% C 0.05% S 0.06% P steel: German Standard
1.0540	0.4% C 0.05% S 0.08% P steel: German Standard
1.0542	0.4% C 0.05% S 0.06% P steel: German Standard
1.0651	0.36% C 0.2% Si 0.6% Mn 0.045% S & P steel: German Standard
1.0726	0.38% C 0.2% Si 0.7% Mn 0.2% S 0.07% P steel: Free machining; German Standard
1.0727	0.46% C 0.2% Si 0.7% Mn 0.2% S 0.07% P steel: Free machining; German Standard
1.0903	0.51% C 0.65% Mn 0.045% S & P (max) steel: For springs; German Standard
1.0970	0.38% C 0.65% Mn 0.045% S & P (max) steel: For springs; German Standard
1.1174	0.36% C 0.2% Si 0.5% Mn 0.035% S & P steel: German Standard
1.1181	0.36% C 0.2% Si 0.5% Mn 0.035% S & P steel: German Standard
1.1183	0.36% C 0.25% Si 0.5% Mn 0.035% S & P steel: German Standard
1.1191	0.46% C 0.25% Si 0.6% Mn 0.035% S & P steel: German Standard
1.1193	0.46% C 0.25% Si 0.7% Mn 0.035% S & P steel: German Standard
1.1194	0.46% C 0.25% Si 0.7% Mn 0.035% S & P steel: German Standard
1.1730	0.45% C 0.7% Mn 0.035% P & S (max) steel: German Standard designation
1.5038	0.4% C 1.0% Mn 0.035% S & P steel: Forging; German Standard
1.5066	0.3% C 1.3% Mn 0.3% Cr 0.03% S & P steel: Forging; German Standard
1.5067	0.36% C 1.3% Mn 0.035% S & P steel: Forging; German Standard
1.5120	0.38% C 0.8% Si 1.1% Mn steel: Forging; German Standard
1.5121	0.46% C 0.8% Si 1.1% Mn steel: Forging; German Standard
1.5122	0.37% C 1.2% Si 1.2% Mn steel: For springs; German Standard

Note. The following abbreviations and units are used in the tables:

DPN	Hardness, diamond pyramid number
UTS	Ultimate tensile strength, N/mm^2
Elon	Elongation, %
Proof	0.1 % proof strength, N/mm^2

1 N/mm^2=0.1 hbar=0.102 kgf/mm^2=0.06475 tonf/in^2=145.04 lbf/in^2

Symbol	Nominal analysis, supplier, condition and remarks.
6 LM	0.3% C 0.2% Si 1.4% Mn steel: Sandvik
7 L	0.35% C 0.25% Si 0.5% Mn steel: Sandvik for SIS 1550
8 LM	0.4% C 0.2% Si 1.3% Mn steel: Sandvik for SIS 2120
9 L	0.45% C 0.2% Si 0.4% Mn steel: Sandvik for SIS 1650, 1660
9 L 7	0.45% C 0.25% Si 0.75% Mn steel: Sandvik
9 LM	0.45% C 0.2% Si 1.4% Mn steel: Sandvik
30 NCD 12 UNI 4365	0.3% C 0.8% Cr 2.9% Ni 0.45% Mo steel: Italian Standard; for bolts
35 S 20	0.35% C 0.6% Mn 0.2% S 0.05% P steel: Free machining; designation used by German Standard
36 Mn 5	0.36% C 1.4% Mn 0.25% Si steel: Designation used by German Standard
37CB	0.37% C 0.8% Mn 0.4% Si steel: Electrode; Metrode
38 Mn Si 4	0.38% C 1.1% Mn 0.8% Si steel: Designation used by German Standard
38 NCD 4 UNI 4365	0.38% C 0.7% Cr 0.8% Ni 0.2% Mo steel: Italian Standard; for bolts
38 Si 6	0.38% C 1.5% Si 0.6% Mn steel: Designation used by German Standard
38 Si 8	0.38% C 0.7% Mn 1.5% Si steel: Designation used by German Standard
40 NCD 7 UNI 4365	0.4% C 0.7% Cr 1.8% Ni 0.25% Mo steel: Italian Standard; for bolts
42 Mn V 7	0.42% C 1.8% Mn 0.25% Si 0.1% V steel: Designation used by German Standard
45 S 20	0.46% C 0.7% Mn 0.2% S 0.05% P steel: Free machining; designation used by German Standard
46 Si 7	0.46% C 1.7% Si 0.7% Mn steel: Designation used by German Standard
50 Si 7	0.49% C 0.6% Mn 1.75% Si steel: Italian Standard designation
	DPN: 250 UTS: 1450 Elon: 7% Proof: 1150
55 Si 8	0.55% C 0.85% Mn 2.0% Si steel; Italian Standard designation
	DPN: 250 UTS: 1550 Elon: 5% Proof: 1250
1022	0.22% C 0.9% Mn 0.04% S & P steel: Designation used by AISI
1029	0.29% C 0.8% Mn 0.05% S & P steel: Designation used by AISI
1044	0.47% C 0.4% Mn 0.04% S & P steel: Designation used by AISI
1105	0.65% C 0.2% Mn 0.02% P & S (max) steel: French Standard designation
1139	0.39% C 1.5% Mn 0.04% P 0.17% S steel: Free machining; designation used by AISI
1306	0.55% C 0.25% Mn 0.035% P (max) 0.025% S (max) steel: French Standard designation
1307	0.45% C 0.25% Mn 0.035% P (max) 0.025% S (max) steel: French Standard designation
1308	0.38% C 0.25% Mn 0.035% P (max) 0.025% S (max) steel: French Standard designation
2321	0.45% C steel: French Standard designation
A 2	0.23% C 0.25% Si 0.8% Mn steel: Jessop
A 3	0.32% C 0.25% Si 0.8% Mn steel: Jessop
A 3	0.32% C (max) 0.8% Mn steel: Cogne
A 4	0.42% C (max) 0.9% Mn steel: Cogne
A 4	0.35% C 0.25% Si 0.8% Mn steel: Jessop
A 5	0.4% C 0.25% Si 0.8% Mn steel: Jessop
A 5	0.52% C (max) 0.9% Mn steel: Cogne
A 6	0.45% C 0.25% Si 0.8% Mn steel: Jessop
A 16	0.27% C 0.4% Si 1.4% Mn 0.03% S & P steel: Welding rod; designation used by BS
A 201 A	0.25% C 0.2% Si 0.8% Mn 0.35% S & P steel: Plate; Armco; flange & firebox grades
	UTS: 450 Elon: 28% Proof: 200

Symbol	Nominal analysis, supplier, condition and remarks.
A 201 B	0.3% C 0.2% Si 0.8% Mn 0.035% S & P steel: Plate; Armco; flange & firebox quality **UTS: 530 Elon: 25% Proof: 210**
A 212 A	0.3% C 0.2% Si 0.9% Mn 0.35% S & P steel: Plate; Armco; flange & firebox quality **UTS: 540 Elon: 24% Proof: 220**
A 212 B	0.32% C 0.2% Si 0.9% Mn 0.35% S & P steel: Plate; Armco; flange & firebox quality **UTS: 600 Elon: 22% Proof: 240**
A 285	0.24% C 0.8% Mn 0.035% S & P steel: Plate; Armco; flange & firebox quality in three grades A, B & C **UTS: 340 Elon: 30% Proof: 180**
A 455	0.33% C 0.1% Si 1.0% Mn 0.04% S & P steel: Plate; Armco **UTS: 580 Elon: 16% Proof: 220**
ABS/F	0.3% C (max) 0.9% Mn (max) steel: Plate; American Bureau of Shipping **UTS: 500 Elon: 19% Proof: 245**
ABS/F	0.35% C (max) 0.8% Mn (max) steel: Tube; American Bureau of Shipping
ABS/G	0.33% C (max) 0.9% Mn (max) steel: Plate; American Bureau of Shipping **UTS: 550 Elon: 17% Proof: 270**
AISI 1029	0.29% C 0.8% Mn 0.05% S & P steel
AISI 1139	0.39% C 1.5% Mn 0.04% P 0.17% S steel: Free machining
AISI 1330	0.3% C 0.3% Si 1.8% Mn steel
AISI 1335	0.35% C 0.3% Si 1.8% Mn steel
AISI 1340	0.4% C 0.3% Si 1.8% Mn steel
AISI 1345	0.4% C 0.3% Si 1.8% Mn steel
AISI C1023	0.22% C 0.4% Mn 0.04% S & P steel
AISI C1024	0.23% C 1.5% Mn 0.04% S & P steel
AISI C1025	0.25% C 0.4% Mn 0.04% P 0.05% S steel
AISI C1026	0.26% C 0.7% Mn 0.04% S & P steel
AISI C1027	0.25% C 1.3% Mn 0.04% P 0.05% S steel
AISI C1030	0.3% C 0.8% Mn 0.04% P 0.05% S steel
AISI C1033	0.33% C 0.9% Mn 0.04% P 0.05% S steel
AISI C 1035	0.35% C 0.8% Mn 0.04% P 0.05% S steel
AISI C1036	0.34% C 1.2% Mn 0.04% P 0.05% S steel
AISI C1037	0.35% C 0.9% Mn 0.04% P 0.05% S steel
AISI C1038	0.38% C 0.8% Mn 0.04% P 0.05% S steel
AISI C1039	0.4% C 0.9% Mn 0.04% P 0.05% S steel
AISI C1040	0.4% C 0.8% Mn 0.04% P 0.05% S steel
AISI C1041	0.4% C 1.5% Mn 0.04% P 0.05% S steel
AISI C1042	0.43% C 0.8% Mn 0.04% P 0.05% S steel
AISI C1043	0.43% C 0.9% Mn 0.04% P 0.05% S steel
AISI C1045	0.46% C 0.8% Mn 0.04% P 0.05% S steel
AISI C1046	0.46% C 0.9% Mn 0.04% P 0.05% S steel
AISI C1126	0.26% C 0.9% Mn 0.1% S 0.04% P steel: Free cutting
AISI C1132	0.31% C 1.5% Mn 0.1% S 0.04% P steel: Free cutting
AISI C1137	0.36% C 1.5% Mn 0.1% S 0.04% P steel: Free cutting
AISI C1138	0.37% C 0.9% Mn 0.1% S 0.04% P steel: Free cutting
AISI C1140	0.4% C 0.9% Mn 0.1% S 0.04% P steel: Free cutting

Symbol	Nominal analysis, supplier, condition and remarks.
AISI C1141	0.41% C 1.5% Mn 0.1% S 0.04% P steel: Free cutting
AISI C1144	0.44% C 1.5% Mn 0.3% S 0.04% P steel: Free cutting
AISI C1145	0.45% C 0.9% Mn 0.04% P 0.05% S steel: Free cutting
AISI C1146	0.46% C 0.9% Mn 0.04% P 0.1% S steel: Free cutting
AM 4	0.35% C 0.25% Si 1.5% Mn steel: Jessop for BS alloy En 15
A Mn	0.25% C Mn steel: Casting; Firth Brown; normalized; tempered **DPN: 170 UTS: 550 Elon: 25% Proof: 340**
AMS 5024 C	0.37% C 1.5% Mn steel: Bar & forging; free cutting; AMS for SAE 1137
AMS 5062 A	0.25% C (max) steel: For bar, forging, tube, etc.
AMS 5070 B	0.22% C steel: Bar & forging; AMS for SAE 1022 **UTS: 370**
AMS 5075 A	0.25% C steel: Seamless tube; AMS for SAE 1025 **UTS: 370**
AMS 5077A	0.25% C steel: Welded tube; AMS for SAE 1025 **UTS: 370**
AMS 5080A	0.35% C steel: Bar, forging & tube; AMS for SAE 1035
AMS 5082	0.35% C steel: Seamless tube; AMS for SAE 1035 **UTS: 650**
AN QQ W 429/1	Plain carbon steel: Wire; US Service **UTS: 1830**
AN S 4	0.35% C 0.7% Mn 0.05% S & P steel: US Service
AN T 4/1	0.25% C 0.4% Mn 0.05% S & P steel: Tube; welded; US Service **UTS: 300 Elon: 22% Proof: 210**
AN WW T 846/1	0.25% C 0.4% Mn 0.05% S & P steel: Seamless tube; US Service **UTS: 300 Elon: 22% Proof: 210**
API 5A C75/1	0.5% C (max) 1.9% Mn (max) 0.2% Mo 0.5% Cr + Ni + Cu (max) steel: Casing & tubing; normalized & tempered **UTS: 655 Elon: 20% Proof: 515**
API 5A C75/2	0.4% C (max) 1.5% Mn steel: Casing & tubing; hardened & tempered **UTS: 655 Elon: 20% Proof: 515**
API 5A C75/3	0.42% C 0.85% Mn 1.0% Cr 0.2% Mo steel: Casing & tubing; normalized & tempered **UTS: 655 Elon: 20% Proof: 515**
API 5A C95	0.45% C 1.9% Mn steel: Casing & tubing; hardened & tempered **UTS: 730 Elon: 20% Proof: 655**
API 5L S	Spiral weld line pipe; See API designation for grades
API 5LU X80	0.26% C (max) 1.4% Mn (max) with other elements as agreed steel: Ultra-high test line pipe **UTS: 690 Elon: 20% Proof: 550**
API 5LU X100	0.26% C (max) 1.4% Mn (max) with other elements as agreed steel: Ultra-high test line pipe **UTS: 830 Elon: 20% Proof: 690**
API 5L X42	0.29% C 1.25% Mn (max) 0.04% P (max) 0.05% S (max) steel: High test line pipe; seamless & welded **UTS: 415 Elon: 20% Proof: 290**
API 5L X46	0.30% C (max) 1.35% Mn (max) 0.04% P (max) 0.05% S (max) steel: High test line pipe; seamless & welded **UTS: 435 Elon: 20% Proof: 315**
API 5L X52	0.30% C (max) 1.35% Mn (max) 0.04% P (max) 0.05% S (max) steel: High test line pipe; seamless & welded **UTS: 470 Elon: 20% Proof: 360**

Note. The following abbreviations and units are used in the tables:

DPN	Hardness, diamond pyramid number
UTS	Ultimate tensile strength, N/mm^2
Elon	Elongation, %
Proof	0.1 % proof strength, N/mm^2

1 N/mm^2=0.1 hbar=0.102 kgf/mm^2=0.06475 tonf/in^2=145.04 lbf/in^2

Symbol	Nominal analysis, supplier, condition and remarks.
API 5L X56	0.26% C (max) 0.26% C (max) 1.35% Mn (max) with Nb V or Ti steel: High test line pipe; seamless or welded
	UTS: 500 Elon: 20% Proof: 385
API 5L X60	0.26% C (max) 1.35% Mn (max) with Nb, V or Ti steel: High test line pipe; seamless or welded
	UTS: 530 Elon: 20% Proof: 415
API 5L X65	0.26% C (max) 1.4% Mn with Nb and V steel: High test line pipe; seamless or welded
	UTS: 540 Elon: 20% Proof: 445
ASM 122	0.25% C (max) steel: French specification
ASTM A27/60/30	0.3% C (max) 0.6% Mn 0.05% S & P steel: Casting
ASTM A27/65/35	0.3% C (max) 0.7% Mn 0.05% S & P steel: Casting
ASTM A27/70/36	0.35% C (max) 0.7% Mn 0.05% S & P steel: Casting
ASTM A27/70/40	0.25% C (max) 1.0% Mn 0.05% S & P steel: Casting
ASTM A27 N1	0.25% C (max) 0.7% Mn 0.05% S & P steel: Casting
ASTM A27 N2	0.35% C (max) 0.6% Mn 0.05% S & P steel: Casting
ASTM A27 U60/30	0.25% C (max) 0.7% Mn 0.05% S & P steel: Casting
ASTM A31 A	C not specified 0.4% Mn 0.05% P (max) 0.06% S (max) steel: For rivets
	UTS: 350 Elon: 27% Proof: 160
ASTM A31 B	0.3% C 0.5% Mn 0.05% P (max) 0.06% S (max) steel: For rivets
	UTS: 440 Elon: 22% Proof: 200
ASTM A36	0.30% C 1.0% Mn 0.10% Si steel: Plate for structural use
ASTM A36	0.30% C 0.75% Mn steel: Bar for structural use
ASTM A105/1	0.35% C 0.9% Mn 0.05% S 0.05% P steel: For flanges; normalized
ASTM A105/11	0.35% C 0.9% Mn 0.3% Si 0.05% S 0.05% P steel: For flanges; normalized
ASTM A106 A	0.25% C (max) 0.5% Mn 0.05% S & P steel: Pipe
ASTM A106 B	0.3% C (max) 0.6% Mn 0.05% S & P pipe
ASTM A106 C	0.35% C (max) 0.6% Mn 0.05% S & P pipe
ASTM A108	Carbon steel: Bars & shafting; cold finished; graded to AISI code
ASTM A109/1	0.25% C 0.6% Mn 0.2% Cu (optional) cold rolled steel: Strip; hard
	UTS: 640 Elon: 3%
ASTM A109/2	0.25% C 0.6% Mn 0.2% Cu (optional) cold rolled steel: Strip; ½-hard
	UTS: 450 Elon: 10%
ASTM A109/3	0.25% C 0.6% Mn 0.2% Cu (optional) cold rolled steel: Strip; ¼-hard
	UTS: 370 Elon: 20%
ASTM A131 Grade A	C not specified 0.04% (max) P 0.05% (max) S steel: For ships
ASTM A131 Grade B	0.21% C 0.95% Mn 0.04% (max) P 0.05% (max) S steel: For ships
ASTM A131 Grade C	0.23% C 0.75% Mn 0.04% (max) P 0.05% (max) S steel: For ships
ASTM A139 A	C not specified 0.7% Mn 0.04% S & P steel: Arc welded; pipe
	UTS: 330 Elon: 35% Proof: 200
ASTM A139 B	0.3% C (max) 0.7% Mn 0.4% S & P steel: Arc welded; pipe
	UTS: 400 Elon: 30% Proof: 240
ASTM A148	Steel: Casting; high strength; graded by mechanical properties
ASTM A178 C	0.35% C 0.8% Mn 0.05% P 0.06% S electric resistance welded boiler tubes
ASTM A181/1	0.35% C (max) 0.9% Mn steel: For pipe flanges
	UTS: 440 Elon: 22% Proof: 220

Symbol	Nominal analysis, supplier, condition and remarks.
ASTM A181/11	0.35% C 0.9% Mn steel: For pipe flanges
	UTS: 480 Elon: 18% Proof: 250
ASTM A183	0.30% C (min) steel: For bolts
ASTM A194/2	0.4% C (min) steel: For bolts
ASTM A210 A1	0.27% C (max) 0.9% Mn 0.1% Si steel: Seamless tube
ASTM A210 C	0.35% C (max) 0.7% Mn 0.1% Si steel: Seamless tube
ASTM A216 WCA	0.25% C (max) 0.7% Mn 0.6% Si steel: Casting for high temperature service
ASTM A216 WCB	0.3% C (max) 1.0% Mn 0.6% Si steel: Casting for high temperature service
ASTM A216 WCC	0.25% C (max) 1.2% Mn 0.6% Si steel: Casting for high temperature service
ASTM A217 WC1	0.25% C (max) 0.65% Mn 0.5% Mo steel: Casting
ASTM A242	0.25% C 1.25% Mn 0.05% S steel: For structural use; as rolled
	UTS: 480 Elon: 19% Proof: 340
ASTM A266/1 & 2	0.35% C (max) 0.6% Mn steel: For drawn forging
ASTM A266/3	0.5% C (max) 0.65% Mn steel: For drawn forging
ASTM A284A	0.22% C 0.2% Si 0.9% Mn 0.04% S & P steel: Plate; as rolled
	UTS: 330 Elon: 28% Proof: 150
ASTM A284 B	0.3% C 0.2% Si 0.9% Mn 0.04% S & P steel: Plate; as rolled
	UTS: 370 Elon: 27% Proof: 180
ASTM A284 C	0.33% C 0.2% Si 0.9% Mn 0.04% S & P steel: Plate; as rolled
	UTS: 420 Elon: 25% Proof: 200
ASTM A284 D	0.35% C 0.2% Si 0.9% Mn 0.04% S & P steel: Plate; as rolled
	UTS: 420 Elon: 24% Proof: 220
ASTM A290 A	0.3% C 0.8% Mn 0.05% S & P steel: Forging; hardened & tempered
	DPN: 160 UTS: 580 Elon: 20% Proof: 270
ASTM A290 B	0.35% C 0.8% Mn 0.05% S & P steel: Forging; hardened & tempered
	DPN: 160 UTS: 580 Elon: 20% Proof: 270
ASTM A290 C1	0.45% C 0.8% Mn 0.05% S & P steel: Forging; hardened & tempered
	DPN: 170 UTS: 610 Elon: 20% Proof: 300
ASTM A291/1	0.4% C 0.5% Mn 0.05% S & P steel: Forging; hardened & tempered
	UTS: 610 Elon: 19% Proof: 340
ASTM A291/2	0.42% C 0.05% S & P steel: Forging; hardened & tempered
	UTS: 680 Elon: 20% Proof: 480
ASTM A299	0.3% C 1.2% Mn 0.2% Si 0.04% S & P steel: Plate for pressure vessels
ASTM A306/45	Plain carbon steel: Bar; as rolled
	UTS: 330 Elon: 33% Proof: 150
ASTM A306/50	Plain carbon steel: Bar; as rolled
	UTS: 370 Elon: 30% Proof: 170
ASTM A306/55	Plain carbon steel: As rolled
	UTS: 370 Elon: 26% Proof: 180
ASTM A306/60	Plain carbon steel: As rolled
	UTS: 440 Elon: 22% Proof: 200
ASTM A306/65	Plain carbon steel: As rolled
	UTS: 460 Elon: 20% Proof: 210
ASTM A325	0.3% C 0.5% Mn steel: Hardened & tempered
	UTS: 710 Elon: 14% Proof: 530
ASTM A333/1	0.3% C 0.7% Mn steel: Pipe for low temperature use
ASTM A333/6	0.3% C 0.7% Mn steel: Pipe for low temperature use
ASTM A350 LF1	0.3% C (max) 1% Mn 0.04% S & P steel: For low temperature use
	UTS: 400 Elon: 25% Proof: 200

Symbol	Nominal analysis, supplier, condition and remarks.
ASTM A350 LF2	0.3% C (max) 1.3% Mn 0.2% Si 0.04% S & P steel: For low temperature use **UTS: 460 Elon: 22% Proof: 240**
ASTM A352 LCB	0.3% C 0.6% Si 1.0% Mn steel: Normalized & tempered for low temperature service **UTS: 440 Elon: 24% Proof: 220**
ASTM A356/1	0.35% C (max) steel: Casting for turbines
ASTM A372 Class I	0.30% (max) 1.0% (max) Mn 0.27% Si steel: Forging for pressure vessels
ASTM A372 Class II	0.40% C (max) 1.29% (max) Mn 0.27% Si steel: Forging for pressure vessels
ASTM A372 Class III	0.48% C (max) 1.65% (max) Mn 0.27% Si steel: Forging for pressure vessels
ASTM A372 Class VII	0.4% C 0.7% Mn 0.3% Si 1.8% Ni 0.8% Cr 0.2% Mo steel: For pressure vessels
ASTM A374	0.22% C (max) 1.2% Mn steel: Sheet & strip; cold rolled
ASTM A375	0.22% C (max) 1.25% Mn steel: Sheet & strip; hot rolled
ASTM A381	0.32% C (max) 1.3% Mn 0.05% S & P steel: For high pressure pipe; mechanical properties vary with grades
ASTM A413	0.33% C (max) 1% Mn 0.05% S & P steel: For chains
ASTM A414	0.25% C steel: Sheet of firebox quality; various grades according to C content
ASTM A433	0.2% C 0.25% Pb free machining steel: For pressure vessels; graded according to C content
ASTM A440	0.3% C 0.06% S & P 0.3% Si 0.2% Cu steel: For structural purposes
ASTM A449	0.42% C 0.6% Mn 0.04% S & P steel: For studs; hardened & tempered **DPN: 230**
ASTM A454	0.3% C 1% Mn steel: For conveyor chain; classified by tensile properties
ASTM A455	0.3% C 1.0% Mn 0.04% S & P 0.2% Si steel: Plates for pressure vessels; graded by carbon content
ASTM A465 LI	0.3% C (max) 0.9% Mn 0.05% S & P 0.2% Pb steel: For pipe flanges; as forged **UTS: 400 Elon: 22% Proof: 200**
ASTM A465 LII	0.3% C (max) 0.9% Mn 0.05% S & P 0.2% P steel: For pipe flanges; normalized **UTS: 400 Elon: 25% Proof: 200**
ASTM A465 LIII	0.35% C (max) 0.9% Mn 0.05% S & P 0.2% Pb steel: For pipe flanges; as forged **UTS: 460 Elon: 28% Proof: 240**
ASTM A465 LIV	0.35% C (max) 0.9% Mn 0.05% S & P 0.2% Pb steel: For pipe flanges; normalized **UTS: 460 Elon: 22% Proof: 240**
ASTM A466	0.25% C 0.05% S & P steel: Chain; weldless
ASTM A467	0.33% C (max) 1% Mn 0.05% S & P steel: Machine & coil chain
ASTM A470/1	0.45% C (max) 0.03% V steel: For rotors; vacuum treated
ASTM A486	Steel: Casting for bridges; classified by tensile properties
ASTM A487/3 N	0.35% C (max) 1.5% Mn 0.4% Mo steel: Casting for pressure; normalized; 3Q grade is hardened and tempered

Note. The following abbreviations and units are used in the tables:

DPN	Hardness, diamond pyramid number
UTS	Ultimate tensile strength, N/mm^2
Elon	Elongation, %
Proof	0.1 % proof strength, N/mm^2

1 N/mm^2=0.1 hbar=0.102 kgf/mm^2=0.06475 tonf/in^2=145.04 lbf/in^2

Symbol	Nominal analysis, supplier, condition and remarks.
ASTM A490	0.45% C 0.04% S & P steel: For bolts
ASTM A500	0.26% C (max) 0.04% S & P 0.2% Cu steel: For structural tubing; cold formed
ASTM A501	0.26% C (max) 0.04% S & P 0.2% Cu steel: For structural tubing; cold formed
ASTM A515	0.25% C steel: Plate for fusion welded pressure vessels; graded by tensile strength
ASTM A516	0.24% C steel: Plate for fusion welded pressure vessels; graded by tensile strength
ASTM A519	Seamless steel mechanical tubing: Low and medium carbon or alloy steel: graded according to AISI system
ASTM A523 B	0.26% C (max) 1.1% Mn 0.04% S & P steel: Pipe for high pressure; seamless or electric resistance welded
ASTM A541/1	0.35% C (max) 0.06% V steel: Forging
ASTM A556 B2	0.27% C (max) 0.6% Mn 0.05% S & P steel: Seamless tube; cold drawn **UTS: 420 Elon: 30% Proof: 260**
ASTM A556 C2	0.3% C (max) 0.65% Mn 0.05% S & P steel: Seamless tube; cold drawn **UTS: 490 Elon: 30% Proof: 280**
ASTM A557	Electric welded steel tubes; see ASTM A556 for analysis etc. of grades
ASTM A563	0.35% C steel: For nuts; different grades indicate type of steel and heat treatment
ASTM A570	0.25% C steel: Sheet & strip; graded by Mn content
ASTM A573/70	0.27% C (max) 1.0% Mn 0.2% Si steel of improved toughness
ASTM A612	0.3% C (max) 1.2% Mn 0.04% S & P (max) steel: Plate for pressure vessels **UTS: 600 Elon: 20% Proof: 345**
ASTM A618/1	0.26% C (max) 1.3% Mn (max) steel: Tube; seamless or welded for structure **UTS: 20% Proof: 345**
ASTM A649/2	0.5% C (max) 0.7% Mn steel: For rolls; corrugated
ASTM A649/4	0.35% C (max) 0.8% Mn steel: For rolls; corrugated
ASTM A680	Cold rolled carbon steel: Strip; see ASTM A682 for analysis; see also 44A3 and 44A4
ASTM A694	0.3% C (max) 1.5% Mn (max) steel: For pipe flanges and fittings; graded by yield strength
ASTM A695 A	C not specified 0.04% S & P (max) steel: Bar; graded by yield strength
ASTM A695 B	0.35% C 1.1% Mn 0.04% S & P (max) steel: Bar; graded by yield strength
ASTM A695 C	C not specified 0.04% S & P (max) steel: Bar; graded by yield strength
ASTM A695 D	C not specified 0.04% S & P (max) 0.2% Pb steel: Bar; graded by yield strength
ASTM A696 B	0.32% C (max) 1.04% Mn (max) 0.04% S & P (max) steel: Bar for pressure use **UTS: 415 Elon: 20% Proof: 240**
ASTM A696 B	0.32% C (max) 1.0% Mn (max) steel: For pressure use **UTS: 415 Elon: 20% Proof: 240**
ASTM A696 C	0.32% C (max) 1.0% Mn (max) steel: For pressure use **UTS: 485 Elon: 18% Proof: 275**
ASTM A696 C	0.32% C (max) 1.04% Mn 0.04% S & P (max) steel: Bar for pressure use **UTS: 485 Elon: 18% Proof: 275**
ASTM A707 L2/1	0.33% C (max) 1.4% Mn steel: Flanges for low temperature use; impact 41 J at -46 °C **UTS: 290 Elon: 22%**
ASTM A707 L2/2	0.33% C (max) 1.4% Mn steel: Flanges for low temperature use; impact 54 J at -46 °C **UTS: 360 Elon: 25%**

Symbol	Nominal analysis, supplier, condition and remarks.
ASTM A707 L2/3	0.33% C (max) 1.4% Mn steel: Flanges for low temperature use; impact 68 J at -46 °C **UTS: 415 Elon: 23%**
ASTM A707 L2/4	0.33% C (max) 1.4% Mn steel: Flanges for low temperature use; impact 68 J at -46 °C **UTS: 515 Elon: 20%**
ASTM A709/36	0.26% C 1.0% Mn with grain refining elements steel: For bridges **UTS: 400 Elon: 20% Proof: 250**
ASTM A709/50	0.23% C (max) 1.35% Mn (max) with grain refining elements steel: For bridges **UTS: 450 Elon: 20% Proof: 345**
ASTM A711	Steel for forging; graded by AISI designation
ASTM A722	C not specified 0.04% S & P (max) steel: Bar; for prestressed concrete **UTS: 4%35 Proof: 810**
ASTM A732/2 A	0.3% C 0.8% Mn steel: Investment casting; annealed **UTS: 450 Elon: 25% Proof: 310**
ASTM A732/2 Q	0.3% C 0.8% Mn steel: Investment casting; hardened & tempered **UTS: 590 Elon: 10% Proof: 415**
ASTM A732/3 A	0.4% C 0.8% Mn steel: Investment casting; annealed **UTS: 515 Elon: 25% Proof: 330**
ASTM A732/3 Q	0.4% C 0.8% Mn steel: Investment casting; hardened & tempered **UTS: 690 Elon: 10% Proof: 620**
ASTM A738	0.24% C (max) 1.5% Mn (max) steel: Plate for pressure use; 27 J impact at -45 °C; hardened & tempered **UTS: 600 Elon: 20% Proof: 310**
AURIGA I	0.25% C 0.4% Si 0.7% Mn 0.05% S & P steel: Casting; D Brown **UTS: 460 Elon: 22%**
AURIGA III	0.42% C 0.3% Si 0.8% Mn 0.05% S & P steel: Casting; D Brown **UTS: 630 Elon: 12%**
AURIGA IIIA	0.35% C 0.4% Si 0.7% Mn 0.05% S & P steel: Casting; D Brown **UTS: 580 Elon: 15%**
AURIGA VIB	0.29% C 1.5% Mn 0.04% S & P steel: Casting; D Brown; hardened & tempered **UTS: 760 Elon: 15%**
B 3	0.3% C (max) steel: Breda
B 3 Mn	0.32% C 0.3% Si 1.2% Mn steel: Pompey
B 4	0.4% C (max) steel: Breda
B 5	0.5% C (max) steel: Breda
B 7	0.35% C steel: As rolled; Bofors **DPN: 160 UTS: 540 Elon: 22% Proof: 250**
B 9 S	0.45% C steel: Bofors **DPN: 190 UTS: 660 Elon: 16%**
B 103	0.46% C 0.8% Mn steel: Fagersta; normalized **DPN: 200 UTS: 630 Elon: 18%**
B 805	0.4% C 1.3% Mn steel: Fagersta
B METAL	0.25% C 0.3% Si 0.7% Mn steel: Casting; Edgar Allen; code 603 steel
BS 15/1	0.25% C 0.06% S 0.06% P steel: General purpose structural steel **UTS: 450 Elon: 25% Proof: 240**
BS 15/2	0.25% C 0.06% S 0.06% P 0.3% Cu steel: General purpose structural steel **UTS: 450 Elon: 25% Proof: 240**
BS 15/3	0.25% C 0.06% S 0.06% P 0.4% Cu steel: General purpose structural steel **UTS: 450 Elon: 25% Proof: 240**
BS 24/2/4	Medium C 0.05% S & P steel: For loco tyres; hardened & tempered **UTS: 1120 Elon: 9%**

Symbol	Nominal analysis, supplier, condition and remarks.
BS 24/2/5	Medium C 0.06% S & P steel: For wagon tyres; hardened & tempered **UTS: 1120 Elon: 8%**
BS 24/3A/3	0.4% C 1.7% Si 0.8% Mn steel: For wagon springs; as BS alloy En 46
BS 24/3A/4	0.4% C 1.7% Si 0.8% Mn steel: For wagon springs; as BS alloy En 46
BS 24/4C	0.32% C 0.1% Si 0.7% Mn 0.05% S & P steel: Forging; for railway use in couplings etc. **UTS: 540 Elon: 24%**
BS 24/4D	0.42% C 0.1% Si 0.7% Mn 0.05% S & P steel: Forging; for railway use in couplings etc. **UTS: 630 Elon: 18%**
BS 32/1	0.4% C 0.7% Mn 0.06% S 0.06% P steel: Bar; replaced by BS 970 En 6A
BS 32/2	0.25% C 1% Mn 0.06% S 0.06% P steel: Bar; replaced by BS 970 En 3B
BS 47A	0.25% C 0.15% Si 0.8% Mn 0.075% S & P steel: For railway fishplates
BS 47B	0.36% C 0.15% Si 0.8% Mn 0.075% S & P steel: For railway fishplates
BS 53	Cold drawn weldless steel tubes for boilers; replaced by BS 3059
BS 165	Hard drawn steel: Wire for re-inforcement; replaced by BS 785
BS 399	0.44% C 0.3% Si 0.7% Mn 0.05% S & P steel: For gas cylinders
BS 494	Cold drawn mild steel: Tube for boilers; replaced by BS 3059
BS 548	0.3% C (max) 0.05% S 0.05% P structural steel: As rolled **UTS: 610 Elon: 18% Proof: 320**
BS 592 A	0.25% C (max) 0.6% Si 1.0% Mn 0.06% S & P steel: Casting; contained in BS 3100 **UTS: 420 Elon: 22% Proof: 210**
BS 592 B	0.35% C (max) 0.6% Si 1.0% Mn 0.06% S & P steel: Casting; contained in BS 3100 **UTS: 480 Elon: 20% Proof: 240**
BS 592 C	0.45% C (max) 0.6% Si 1.0% Mn 0.06% S & P steel: Casting; contained in BS 3100 **UTS: 530 Elon: 15% Proof: 240**
BS 640/1	0.37% C 0.1% Si 0.9% Mn 0.05% S & P steel: Electrodes
BS 785	0.3% C (max) 0.05% S 0.05% P steel: For concrete reinforcement; high tensile form (C content near top limit) **UTS: 610 Elon: 16% Proof: 330**
BS 785	0.3% C (max) 0.05% S 0.05% P steel: For Concrete reinforcement; medium tensile form (higher C than mild steel) **UTS: 540 Elon: 16% Proof: 240**
BS 785	0.3% C (max) 0.05% S 0.05% P steel: For concrete reinforcement; low carbon form (mild steel) **UTS: 450 Elon: 18%**
BS 970 En 3	0.25% C (max) 1% Mn 0.06% S 0.06% P steel: Bar & forging; as rolled **UTS: 450 Elon: 25%**
BS 970 En 3A	0.2% C 0.7% Mn 0.06% P steel: Bar & forging; normalized **UTS: 420 Elon: 25%**
BS 970 En 3B	0.25% C (max) 1% Mn 0.06% S 0.06% P steel: Bar; cold drawn **UTS: 420 Elon: 17%**
BS 970 En 4	0.3% C (max) 1% Mn 0.06% S 0.06% P steel: Bar & forging; normalized **DPN: 160 UTS: 480 Elon: 25%**
BS 970 En 4A	0.3% C (max) 1% Mn 0.06% S 0.06% P steel: Bar; cold drawn **UTS: 550 Elon: 12%**

Symbol	Nominal analysis, supplier, condition and remarks.
BS 970 En 5	0.3% C 0.8% Mn 0.06% S 0.06% P steel: Bar & forging; hardened & tempered
	DPN: 200 UTS: 610 Elon: 18% Proof: 370
BS 970 En 5A	0.27% C 0.8% Mn 0.06% S 0.06% P steel: Bar & forging; supplied to composition only
BS 970 En 5B	0.3% C 0.8% Mn 0.06% S 0.06% P steel: Bar & forging; supplied to composition only
BS 970 En 5C	0.32% C 0.8% Mn 0.06% S 0.06% P steel: Bar & forging; supplied to composition only
BS 970 En 5D	0.3% C 0.9% Mn 0.06% S 0.06% P steel: Bar; hardened, tempered & bright drawn
	DPN: 230 UTS: 610 Elon: 18% Proof: 370
BS 970 En 5K	0.3% C 0.3% Si 0.8% Mn 0.05% S & P steel: Bar & forging
BS 970 En 6	0.4% C (max) 0.7% Mn 0.06% S 0.06% P steel: Bar; cold drawn; Izod not less than 20 ft/lb
	UTS: 610 Elon: 15%
BS 970 En 6A	0.4% C (max) 0.3% Si 0.7% Mn 0.06% S & P steel: Bar
BS 970 En 6B	0.4% C (max) 0.7% Mn 0.06% S & P steel: Bar; cold drawn
	UTS: 610 Elon: 15%
BS 970 En 6K	0.35% C (max) 0.3% Si 0.7% Mn 0.05% S & P steel: Bar; cold drawn
BS 970 En 8	0.4% C 0.9% Mn 0.06% S 0.06% P steel: Bar & forging; hardened & tempered
	DPN: 200 UTS: 610 Elon: 22% Proof: 420
BS 970 En 8A	0.36% C 0.8% Mn 0.06% S 0.06% P steel: Bar & forging; hardened & tempered; supplied to composition only
BS 970 En 8B	0.38% C 0.8% Mn 0.06% S 0.06% P steel: Bar & forging; hardened & tempered; supplied to composition only
BS 970 En 8C	0.41% C 0.8% Mn 0.06% S 0.06% P steel: Bar & forging; hardened & tempered; supplied to composition only
BS 970 En 8D	0.43% C 0.8% Mn 0.06% S 0.06% P steel: Bar & forging; hardened & tempered; supplied to composition only
BS 970 En 8E	0.37% C 1% Mn 0.06% S 0.06% P steel: Bar & forging; hardened & tempered; supplied to composition only
BS 970 En 8K	0.4% C 0.3% Si 0.8% Mn 0.05% S & P steel: Bar & forging
BS 970 En 8M	0.4% C 1% Mn 0.16% S 0.07% P free machining steel: Bar; hardened & tempered
	DPN: 200 UTS: 610 Elon: 22% Proof: 420
BS 970 En 15	0.35% C 1.5% Mn steel: Bar & forging; hardened & tempered
	DPN: 250 UTS: 760 Elon: 20% Proof: 580
BS 970 En 15A	0.35% C 1.5% Mn steel: Bar & forging; hardened & tempered
	DPN: 250 UTS: 760 Elon: 20% Proof: 540
BS 970 En 15B	0.37% C 1.2% Mn steel: Bar & forging; hardened & tempered
	DPN: 250 UTS: 760 Elon: 20% Proof: 540
BS 970 En 16A	0.27% C 1.5% Mn steel: Bar & forging; hardened & tempered; mechanical properties not quoted; replaced by 605 A 27

Note. The following abbreviations and units are used in the tables:

DPN	Hardness, diamond pyramid number
UTS	Ultimate tensile strength, N/mm^2
Elon	Elongation, %
Proof	0.1 % proof strength, N/mm^2

1 N/mm^2=0.1 hbar=0.102 kgf/mm^2=0.06475 tonf/in^2=145.04 lbf/in^2

Symbol	Nominal analysis, supplier, condition and remarks.
BS 970 En 16B	0.32% C 1.5% Mn steel: Bar & forging; hardened & tempered; mechanical properties not quoted; replaced by 605 A 32
BS 970 En 16C	0.37% C 1.5% Mn steel: Bar & forging; hardened & tempered; mechanical properties not quoted; replaced by 605 A 37
BS 970 En 46	0.4% C 1.75% Si 0.8% Mn steel: For springs; water hardened & tempered
BS 980 CDS5	0.3% C 0.7% Mn steel: Tube; cold drawn; annealed
	UTS: 390 Proof: 240
BS 980 CDS6	0.3% C 0.6% Mn steel: Tube; cold drawn; as drawn & tempered
	UTS: 530 Proof: 420
BS 980 CDS7	0.45% C 0.6% Mn steel: Tube; cold drawn; annealed
	UTS: 500 Proof: 330
BS 980 CDS9	0.25% C 1.5% Mn steel: Tube; cold drawn; annealed
	UTS: 450 Proof: 270
BS 980 CDS10	0.25% C 1.5% Mn steel: Tube; cold drawn; as drawn & tempered
	UTS: 610 Proof: 450
BS 980 CEW3	0.3% C 0.06% S & P steel: Tube; cold drawn; electric welded
BS 980 CEW4	0.3% C 0.06% S & P steel: Tube; cold drawn; electric welded
BS 980 ERW2	0.3% C (max) 0.06% S & P steel: Tube
BS 980 ERW3	0.4% C (max) 0.06% S & P steel: Tube
BS 1045	0.4% C (max) 0.3% Si 1.5% Mn 0.05% S & P steel: For gas cylinders; normalized
	UTS: 670 Elon: 15% Proof: 390
BS 1144	Cold rolled and twisted mild steel: Bar; for concrete reinforcing
	UTS: 450 Elon: 13% Proof: 370
BS 1287	.44% C 0.7% Mn 0.045% S & P steel: For gas cylinders
BS 1414B	0.35% C (max) 0.9% Mn 0.05% S & P steel: For valves
BS 1449 HR5A	0.25% C 0.8% Mn 0.05% S & P rimming steel: Sheet; as BS 970 En 2C/A
	UTS: 400 Elon: 25% Proof: 210
BS 1449 HR5B	0.25% C 0.8% Mn 0.05% S & P balanced steel: Sheet; as BS 970 En 2C/B
	UTS: 400 Elon: 25% Proof: 210
BS 1449 HR5C	0.25% C 0.8% Mn 0.05% S & P steel: Sheet; killed
	UTS: 400 Elon: 25% Proof: 210
BS 1449 HR6A	0.27% C 0.8% Mn 0.05% S & P rimming steel: Sheet
	UTS: 450 Elon: 20% Proof: 240
BS 1449 HR6B	0.27% C 0.8% Mn 0.05% S & P balanced steel: Sheet
	UTS: 450 Elon: 20% Proof: 240
BS 1449 HR6C	0.27% C 0.8% Mn 0.05% S & P steel: Sheet; killed
	UTS: 450 Elon: 20% Proof: 240
BS 1449 HR16A	0.25% C 0.8% Mn 0.05% S & P rimming steel: Plate
	UTS: 400 Elon: 28% Proof: 220
BS 1449 HR16B	0.25% C 0.8% Mn 0.05% S & P balanced steel: Plate
	UTS: 400 Elon: 28% Proof: 220
BS 1449 HR16C	0.25% C 0.8% Mn 0.05% S & P steel: Plate; killed
	UTS: 400 Elon: 28% Proof: 220
BS 1449 HR17A	0.27% C 0.8% Mn 0.05% S & P rimming steel: Plate
BS 1449 HR17B	0.27% C 0.8% Mn 0.05% S & P balanced steel: Plate
BS 1449 HR17C	0.27% C 0.8% Mn 0.05% S & P steel: Plate; killed

Symbol	Nominal analysis, supplier, condition and remarks.
BS 1449 HS30	0.3% C 0.2% Si 0.8% Mn 0.06% S & P steel: Sheet
	UTS: 480 Elon: 18% Proof: 270
BS 1449 HS40	0.4% C 0.2% Si 0.8% Mn 0.06% S & P steel: Sheet
	UTS: 530 Elon: 16% Proof: 280
BS 1449 NHR24	0.3% C 0.8% Mn 0.06% S & P steel: Sheet
	UTS: 520 Elon: 15% Proof: 270
BS 1453 A3	0.27% C 0.4% Si 1.5% Mn steel: For welding filler rod; to give butt weld strength of 480 N/mm^2
BS 1456 B	0.3% C 0.4% Si 1.5% Mn 0.05% S & P steel: Casting; normalized or hardened & tempered; contained in BS 3100
	UTS: 610 Elon: 15%
BS 1458	Composition to suit mechanical properties 0.05% S & P steel: Casting; normalized or hardened & tempered; contained in BS 3100
	DPN: 230 UTS: 610 Elon: 15% Proof: 480
BS 1459	Composition to suit mechanical properties 0.05% S & P steel: Casting; normalized or hardened & tempered; contained in BS 3100
	DPN: 280 UTS: 610 Elon: 12% Proof: 580
BS 1503/161 A & B	0.25% C 0.2% Si 0.05% S & P steel: Forging; normalized
	UTS: 420 Proof: 200
BS 1503/161 C	0.3% C 0.2% Si 1.0% Mn 0.05% S & P steel: Forging; normalized
	UTS: 530 Proof: 240
BS 1503/221	0.23% C 0.25% Si 1.5% Mn 0.05% S & P steel: Forging; normalized
	UTS: 540 Proof: 240
BS 1504/101 A	C not specified 0.06% S & P steel: Casting; normalized
	UTS: 420 Elon: 20% Proof: 200
BS 1504/101 B	C not specified 0.06% S & P steel: Casting; normalized
	UTS: 460 Elon: 20% Proof: 210
BS 1504/101 C	C not specified 0.06% S & P steel: Casting; normalized
	UTS: 560 Elon: 15% Proof: 240
BS 1504/161 A	0.25% C (max) 0.5% Si 0.9% Mn 0.05% S & P steel: Casting; normalized
	UTS: 420 Elon: 22% Proof: 210
BS 1504/161 B	0.3% C (max) 0.5% Si 0.9% Mn 0.05% S & P steel: Casting; normalized
	UTS: 460 Elon: 22% Proof: 240
BS 1560 B	0.35% C (max) 0.05% S & P steel: For pipe flanges
BS 1570 B	0.35% C (max) 0.05% S & P steel: For valves
BS 1633 A	0.28% C (max) 0.2% Si 0.6% Mn 0.05% S & P steel: For land boilers; normalized
	UTS: 390 Elon: 30%
BS 1633 B	0.28% C (max) 0.2% Si 0.6% Mn 0.05% S & P steel: For land boilers; normalized
	UTS: 420 Elon: 28%
BS 1633 C	0.28% C (max) 0.2% Si 0.6% Mn 0.05% S & P steel: For land boilers; normalized
	UTS: 450 Elon: 26%
BS 1633 D	0.28% C (max) 0.2% Si 0.6% Mn 0.05% S & P steel: For land boilers; normalized
	UTS: 480 Elon: 23%
BS 1633 E	0.32% C (max) 0.2% Si 0.6% Mn 0.05% S & P steel: For land boilers; normalized
	UTS: 550 Elon: 22%
BS 1663	0.3% C (max) 0.05% S & P steel: For electrically welded chain
BS 1717 CDS 101	0.2% C 0.05% S & P steel: Seamless tube; annealed
	UTS: 300 Proof: 160

Symbol	Nominal analysis, supplier, condition and remarks.
BS 1717 CDS101	0.2% C 0.05% S & P steel: Seamless tube; normalized
	UTS: 360 Proof: 210
BS 1717 CDS102	0.2% C 0.05% S & P steel: Seamless tube; as drawn
	UTS: 420 Proof: 360
BS 1717 CDS102	0.2% C 0.05% S & P steel: Seamless tube; after welding, brazing or annealing
	UTS: 300 Proof: 170
BS 1717 CDS103	0.25% C 0.6% Mn 0.05% S & P steel: Seamless tube; annealed
	UTS: 360 Proof: 210
BS 1717 CDS103	0.25% C 0.6% Mn 0.05% S & P steel: Seamless tube; normalized
	UTS: 420 Proof: 270
BS 1717 CDS104	0.25% C 0.6% Mn 0.05% S & P steel: Seamless tube; as drawn
	UTS: 480 Proof: 390
BS 1717 CDS104	0.25% C 0.6% Mn 0.05% S & P steel: Seamless tube; after welding brazing or annealing
	UTS: 360 Proof: 210
BS 1717 CDS105	0.35% C 0.6% Mn 0.05% S & P steel: Seamless tube; annealed
	UTS: 420 Proof: 270
BS 1717 CDS105	0.35% C 0.6% Mn 0.05% S & P steel: Seamless tube; normalized
	UTS: 480 Proof: 330
BS 1717 CDS106	0.35% C 0.6% Mn 0.05% S & P steel: Seamless tube; as drawn
	UTS: 530 Proof: 420
BS 1717 CDS106	0.35% C 0.6% Mn 0.05% S & P steel: Seamless tube; after annealing or brazing
	UTS: 420 Proof: 270
BS 1717 CDS107	0.5% C 0.6% Mn 0.05% S & P steel: Seamless tube; annealed
	UTS: 500 Proof: 330
BS 1717 CDS107	0.5% C 0.6% Mn 0.05% S & P steel: Seamless tube; normalized
	UTS: 580 Proof: 420
BS 1717 CDS108	0.5% C 0.6% Mn 0.05% S & P steel: Seamless tube; as drawn
	UTS: 690 Proof: 580
BS 1717 CDS108	0.5% C 0.6% Mn 0.05% S & P steel: Seamless tube; after welding, brazing or annealing
	UTS: 500 Proof: 330
BS 1717 CEW101	0.2% C 0.6% Mn 0.06% S & P steel: Electrically welded tube; cold drawn strip; annealed after welding
	UTS: 300 Proof: 170
BS 1717 CEW102	0.2% C 0.6% Mn 0.06% S & P steel: Electrically welded tube; cold drawn strip prior to welding
	UTS: 420 Proof: 360
BS 1717 CEW102	0.2% C 0.6% Mn 0.06% S & P steel; Electrically welded tube; after annealing or brazing
	UTS: 300 Proof: 170
BS 1717 CEW103	0.25% C 0.6% Mn 0.06% S & P steel; Electrically welded tube; cold drawn strip, annealed after welding
	UTS: 360 Proof: 210
BS 1717 CEW104	0.25% C 0.6% Mn 0.06% S & P steel: Electrically welded tube; cold drawn strip prior to welding
	UTS: 480 Proof: 390
BS 1717 CEW104	0.25% C 0.6% Mn 0.06% S & P steel: Electrically welded tube; after annealing or brazing
	UTS: 360 Proof: 210
BS 1717 ERW101	0.1% C 0.6% Mn 0.06% S & P steel: Electrically welded tube; as welded
	UTS: 300 Proof: 170

Symbol	Nominal analysis, supplier, condition and remarks.
BS 1717 ERW102	0.2% C 0.6% Mn 0.06% S & P steel: Electrically welded tube; as welded **UTS: 370** **Proof: 220**
BS 1717 ERW102	0.2% C 0.6% Mn 0.06% S & P steel: Electrically welded tube; annealed or brazed **UTS: 300** **Proof: 170**
BS 1717 ERW103	0.3% C 0.6% Mn 0.06% S & P steel: Electrically welded tube; as welded **UTS: 450** **Proof: 300**
BS 1717 ERW103	0.3% C 0.6% Mn 0.06% S & P steel: Electrically welded tube; annealed or brazed **UTS: 370** **Proof: 220**
BS 2763	C not specified 0.050% S & P steel: Wire for rope; specification covers bright and galvanized; tensile varies with diameter
BS 2772/2	0.27% C 0.2% Si 0.05% S & P steel: For colliery winding equipment; hardened & tempered
BS 2901 A 16	0.27% C 0.4% Si 1.4% Mn 0.03% S & P steel: Rod for gas & electric welding
BS 3059/5	0.25% C (max) 0.05% S & P steel: Tube for boilers; seamless; hot finished **UTS: 480**
BS 3059/6	0.25% C 0.05% S & P steel: Tube for boilers; seamless; cold drawn **UTS: 480**
BS 3083	Mild steel: Sheet; galvanized & corrugated
BS 3100	Steel: Casting for general engineering purposes, covering a range of BS specifications
BS 3100 A1	0.25% C (max) 0.9% Mn (max) 0.06% S & P (max) 0.8% Cr + Mo + Ni + Cu (max) steel: Casting **UTS: 430 Elon: 22% Proof: 230**
BS 3100 A2	0.35% C (max) 1.0% Mn (max) 0.06% S & P (max) steel: Casting **UTS: 490 Elon: 18% Proof: 260**
BS 3100 A3	0.45% C (max) 1.0% Mn (max) 0.06% S & P (max) steel: Casting **UTS: 540 Elon: 14% Proof: 295**
BS 3100 A5	0.3% C 1.4% Mn 0.015% S & P (max) steel: Casting **UTS: 700 Elon: 13% Proof: 370**
BS 3100 A6	0.3% C 1.4% Mn 0.05% S & P (max) steel: Casting **UTS: 800 Elon: 13% Proof: 495**
BS 3100 AW2	0.45% C 1.0% Mn (max) 0.8% Cr + Mo + Ni + Cu (max) steel: Casting for surface hardening **UTS: 620 Elon: 12% Proof: 325**
BS 3100 AW3	0.55% C 1.0% Mn (max) 0.8% Cr + Mo + Ni + Cu (max) steel: Casting for surface hardening **UTS: 690 Elon: 8% Proof: 370**
BS 3100 BT1	C not specified 0.05% S & P (max) steel: Casting for high tensile use **UTS: 750 Elon: 11% Proof: 495**
BS 3100 BT2	C not specified 0.04% S & P (max) steel: Casting **UTS: 925 Elon: 8% Proof: 585**
BS 3100 BT3	C not specified 0.03% S & P (max) steel: Casting **UTS: 1080 Elon: 6% Proof: 695**
BS 3127/8	0.35% C 1.2% Mn 0.05% S & P steel: Tube; lightly cold worked; for bourdon tubes **DPN: 170**

Note. The following abbreviations and units are used in the tables:

DPN	Hardness, diamond pyramid number
UTS	Ultimate tensile strength, N/mm^2
Elon	Elongation, %
Proof	0.1 % proof strength, N/mm^2

1 N/mm^2=0.1 hbar=0.102 kgf/mm^2=0.06475 $tonf/in^2$=145.04 lbf/in^2

Symbol	Nominal analysis, supplier, condition and remarks.
BS 3141 C3C1	0.28% C (max) 0.7% Mn 0.05% S & P steel: Classification system used by BS; discontinued
BS 3141 C3E1	0.25% C 0.5% Mn 0.06% S & P steel: Classification system used by BS; discontinued
BS 3141 C3G1	0.25% C 0.7% Mn 0.05% S & P steel: Classification system used by BS; discontinued
BS 3141 C3H1	0.28% C 0.8% Mn 0.06% S & P steel: Classification system used by BS; discontinued
BS 3141 C4A1	0.32% C (max) 0.7% Mn 0.05% S & P steel: Classification system used by BS; discontinued
BS 3141 C4B1	0.3% C 0.8% Mn 0.06% S & P steel: Classification system used by BS; discontinued
BS 3141 C4C1	0.3% C 0.5% Mn 0.06% S & P steel: Classification system used by BS; discontinued
BS 3141 C4D1	0.3% C 0.8% Mn 0.05% S & P steel: Classification system used by BS; discontinued
BS 3141 C4E1	0.32% C 0.8% Mn 0.06% S & P steel: Classification system used by BS; discontinued
BS 3141 C4F1	0.32% C 1% Mn 0.05% S & P steel: Classification used by BS; discontinued
BS 3141 C4G1	0.36% C 0.8% Mn 0.06% S & P steel: Classification system used by BS; discontinued
BS 3141 C4J1	0.35% C 0.8% Mn 0.05% S & P steel: Classification system used by BS; discontinued
BS 3141 C4K1	0.4% C 0.8% Mn 0.06% S & P steel: Classification system used by BS; discontinued
BS 3141 C4L1	0.37% C 1% Mn 0.05% S & P steel: Classification system used by BS; discontinued
BS 3141 C5A1	0.41% C 0.8% Mn 0.06% S & P steel: Classification system used by BS; discontinued
BS 3141 C5C1	0.4% C 0.8% Mn 0.05% S & P steel: Classification system used by BS; discontinued
BS 3141 C5D1	0.4% C 1% Mn 0.06% S & P steel: Classification system used by BS; discontinued
BS 3141 C5F1	0.42% C 0.8% Mn 0.05% S & P steel: Classification system used by BS; discontinued
BS 3141 C5G1	0.42% C 1% Mn 0.05% S & P steel: Classification system used by BS; discontinued
BS 3141 C5J1	0.45% C 0.6% Mn 0.05% S & P steel: Classification system used by BS; discontinued
BS 3141 C5K1	0.45% C 1.1% Mn 0.06% S & P steel: Classification system used by BS; discontinued
BS 3141 C5L1	0.47% C 0.8% Mn 0.06% S & P steel: Classification system used by BS; discontinued
BS 3141 CF4H1	0.35% C 1.1% Mn 0.06% P 0.18% S steel: Free cutting; classification system used by BS; discontinued
BS 3141 CF4M1	0.37% C 1.1% Mn 0.06% P 0.18% S steel: Free cutting; classification system used by BS; discontinued
BS 3141 CF5B1	0.4% C 1.1% Mn 0.06% P 0.18% S steel: Free cutting; classification system used by BS; discontinued
BS 3141 CF5E1	0.4% C 1.1% Mn 0.06% P 0.18% S steel: Free cutting; classification system used by BS; discontinued
BS 3141 CF5H1	0.42% C 1.2% Mn 0.06% P 0.18% S steel: Free cutting; classification system used by BS; discontinued
BS 3146 CLA3	C not specified 0.05% S & P steel: Investment casting; hardened & tempered **DPN: 220 UTS: 690 Elon: 15% Proof: 480**
BS 3146 CLA4	C not specified 0.05% S & P steel: Investment casting; hardened & tempered **DPN: 275 UTS: 920 Elon: 12% Proof: 580**
BS 3146 CLA5A	C not specified 0.02% S 0.025% P steel: Investment casting; hardened & tempered **DPN: 300 UTS: 920 Elon: 12% Proof: 670**

Symbol	Nominal analysis, supplier, condition and remarks.
BS 3146 CLA5B	C not specified 0.02% S 0.025% P steel: Investment casting; hardened & tempered **DPN: 360 UTS: 1140 Elon: 7% Proof: 910**
BS 3706/1	0.25% C 0.06% S & P steel: As rolled **UTS: 420 Elon: 20%**
BS 3706/2	0.25% C 0.06% S & P 0.3% Cu steel: As rolled **UTS: 420 Elon: 20%**
BS 3706/3	0.25% C 0.06% S & P 0.45% Cu steel: As rolled **UTS: 420 Elon: 20%**
BS 3740	Mild steel plate clad with stainless may be 13% Cr or one of the 18/8 Cr/Ni type
BS 4360/43 A1	0.3% C 0.06% S & P steel: As rolled; weldable **UTS: 300 Elon: 22%**
BS 4360/43 A	0.3% C 0.06% S & P steel: As rolled; weldable **UTS: 300 Elon: 22%**
BS 4360/43 B	0.26% C 1.6% Mn 0.06% S & P steel: As rolled; weldable **UTS: 450 Elon: 22%**
BS 4360/43 C	0.22% C 1.6% Mn 0.06% S & P steel: As rolled; weldable **UTS: 450 Elon: 22%**
BS 4360/50 A	0.27% C 1.7% Mn 0.06% S & P steel: Sheet; as rolled; weldable **UTS: 530 Elon: 18%**
BS 4360/50 B	0.24% C 1.6% Mn 0.06% S & P steel: Sheet; as rolled; weldable; with Nb **UTS: 530 Elon: 18%**
BS 4360/50 C	0.24% C 1.6% Mn 0.06% S & P steel: Weldable; with Nb **UTS: 530 Elon: 18%**
BS 4360/50 D	0.22% C 1.6% Mn 0.3% Si 0.05% S & P steel: Normalized; weldable; with Nb **UTS: 530 Elon: 18%**
BS 4360/55 C	0.26% C 1.7% Mn 0.05% S & P steel: Plate; as rolled; weldable; with Nb **UTS: 610 Elon: 17% Proof: 440**
BS 4360/55 E	0.26% C 1.7% Mn 0.05% S & P steel: Plate; normalized; with Nb **UTS: 610 Elon: 17% Proof: 440**
BS 5006	Cold worked steel: Bar & strip for automobiles; withdrawn; replaced by BS 1449 and BS 2453
BS 5009	Steel: Tube for automobiles; withdrawn; replaced by BS 980
BS 5010	Steel: For laminated springs for automobiles; withdrawn; replaced with BS 970
BS 5028	Steel: Castings (No 1 & 2 grades) for automobiles; withdrawn
BS S1 A	0.2% C 0.2% Si 0.7% Mn 0.05% S & P steel: Bar; cold drawn **UTS: 530 Elon: 12%**
BS S1 B	0.3% C 0.2% Si 0.7% Mn 0.05% S & P steel: Bar; cold drawn **UTS: 530 Elon: 12%**
BS S1 C	0.35% C 0.2% Si 0.7% Mn 0.05% S & P steel: Bar; cold drawn **UTS: 530 Elon: 12%**
BS S6	0.4% C steel: Normalized; replaced by BS/S93
BS S20	Steel: Sheet; tinned; replaced by BS/S512
BS S21	0.25% C (max) 0.2% Si 0.8% Mn 0.05% S & P steel: Cold drawn **UTS: 420 Elon: 17%**
BS S23	Medium carbon steel: Replaced
BS S24	Bright steel: Bar for keys; replaced
BS S26	Medium carbon steel: Forgings; 600 N/mm²; replaced
BS S27	Medium carbon steel: Bar; 600 N/mm²; replaced
BS S31	Steel: Wire for forging bolts; replaced
BS S71	0.3% C steel: Normalized; obsolete
BS S76	0.4% C steel: Hardened & tempered; obsolete
BS S77	0.3% C steel: Hardened & tempered; obsolete

Symbol	Nominal analysis, supplier, condition and remarks.
BS S93	0.4% C 0.2% Si 0.7% Mn 0.04% S & P steel: Normalized **DPN: 180 UTS: 610 Elon: 20% Proof: 300**
BS S105	0.4% C 0.2% Si 0.9% Mn 0.04% S & P steel: Wire for forged bolts; hardened & tempered after forging **UTS: 610 Elon: 18% Proof: 660**
BS S113	0.4% C 0.2% Si 0.8% Mn 0.2% Pb 0.04% S & P steel: Hardened & tempered; Pb content optional **DPN: 230 UTS: 760 Elon: 15%**
BS S116	0.4% C 0.2% Si 0.9% Mn 0.04% S & P steel: Bar; hardened & tempered **DPN: 280 UTS: 920 Elon: 18% Proof: 660**
BS S516	0.45% C 0.2% Si 1.5% Mn 0.05% S & P steel: Sheet; hardened & tempered **UTS: 920 Elon: 8% Proof: 760**
BS T1	0.4% C (max) 0.3% Si 1.5% Mn 0.05% S & P steel: Tube; hardened & tempered **UTS: 530**
BS T5	Medium carbon steel: Tube; obsolete
BS T14	Medium carbon steel: Tube for axles; tempered; obsolete
BS T21	Carbon steel: Tube; annealed; obsolete
BS T35	0.25% C (max) 0.2% Si 1.5% Mn 0.05% S & P steel: Tube; hardened & tempered; weldable **UTS: 530 Proof: 450**
BS T45	0.25% C (max) 0.2% Si 1.5% Mn 0.05% S & P steel: Tube; hardened & tempered; weldable **UTS: 670 Proof: 610**
BS T54	0.25% C 0.2% Si 0.6% Mn 0.05% S & P steel: Tube; as drawn & tempered; weldable **UTS: 530 Proof: 450**
BS W1	C not specified 0.04% S & P galvanized steel: Wire; high tensile wire
BS W2	C not specified 0.04% S & P galvanized steel: Wire
BS W3	0.45% C 0.3% Si 1.2% Mn 0.05% S & P steel: Wire **UTS: 840 Elon: 10%**
BS W8	0.45% C 0.2% Si 1.2% Mn 0.05% S & P steel: Tie rods **UTS: 840 Elon: 10%**
BS W9	C not specified 0.04% S & P galvanized steel: Wire
C 3	0.32% C (max) 0.8% Mn steel: Siau
C 4	0.42% C (max) 0.8% Mn steel: Siau
C 5	0.5% C (max) 0.8% Mn steel: Siau
C 30	0.3% C 0.7% Mn steel: Italian Standard designation; normalized **DPN: 200 UTS: 620 Elon: 21% Proof: 280**
C 30 UNI 4365	0.3% C 0.4% Si 0.8% Mn 0.035% S & P steel: Italian Standard; for bolts
C 35	0.35% C 0.7% Mn steel: Italian Standard designation; normalized **DPN: 210 UTS: 620 Elon: 19% Proof: 300**
C 35 UNI 4365	0.35% C 0.35% Si 0.7% Mn 0.04% S & P steel: Italian Standard; for bolts
C 40	0.4% C 0.7% Mn steel: Italian Standard designation; normalized **DPN: 240 UTS: 670 Elon: 17% Proof: 340**
C 40 UNI 4365	0.4% C 0.4% Si 0.8% Mn 0.035% S & P steel: Italian Standard; for bolts
C 45	0.45% C 8.7% Mn steel: Italian Standard designation; normalized **DPN: 235 UTS: 720 Elon: 16% Proof: 370**
C 45	0.46% C 0.7% Mn steel: Italian Standard **DPN: 200 UTS: 1300 Elon: 8% Proof: 1000**
C 45W3	0.45% C 0.7% Mn 0.035% P & S (max) steel: German Standard designation

Symbol	Nominal analysis, supplier, condition and remarks.
C 50	0.5% C 0.75% Mn steel: Italian Standard designation **DPN: 245 UTS: 750 Elon: 15% Proof: 400**
C 140	0.4% C 1.5% Mn 0.25% Si 0.02% S & P steel: SHD drills, etc.
CASE 1280	0.28% C 0.2% Si 1.1% Mn 0.035% S & P steel: Plate; Armco; pressure vessels **UTS: 560 Elon: 22% Proof: 240**
CCU	0.32% C 0.3% Si 0.7% Mn 0.9% Cu steel: Casting; Edgar Allen; code 642
CEAX	0.45% C 0.6% Mn 0.035% P & S steel: Bars; SKF Ltd.
Cf 35	0.36% C 0.2% Si 0.6% Mn 0.035% S & P steel: German Standard
Cf 43	0.46% C 0.2% Si 0.7% Mn 0.035% S & P steel: German Standard
C METAL	0.32% C 0.3% Si 0.7% Mn steel: Casting; Edgar Allen; code 604
CMN 1	0.25% C 0.25% Si 1.5% Mn steel: Firth Brown
CMN 2	0.35% C 0.25% Si 1.5% Mn steel: Firth Brown for BS alloy En 15 & 15A
CODE 603	0.25% C 0.3% Si 0.7% Mn steel: Casting; Edgar Allen; B metal
CODE 604	0.32% C 0.22% Si 0.7% Mn steel: Casting; Edgar Allen; C metal
CODE 606	0.45% C 0.3% Si 0.8% Mn steel: Casting; Edgar Allen; D metal
CODE 609	0.32% C 0.6% Si 0.7% Mn steel: Casting; Edgar Allen; 'Retort'
CODE 610	0.25% C 1.4% Mn steel: Casting; Edgar Allen; MT 'P' steel
CODE 611	0.37% C 0.3% Si 0.7% Mn steel: Casting; Edgar Allen; 'Matte Ladle'
CODE 626	0.3% C 1.4% Mn steel: Casting; Edgar Allen; MT 'R'
CODE 642	0.32% C 0.3% Si 0.7% Mn 0.9% Cu steel: Casting; Edgar Allen; code 642
CPV	0.2% V steel: Balfour Darwin; three ranges of carbon
Cq 35	0.35% C 0.25% Si 0.6% Mn 0.04% S & P steel: German Standard
Cq 45	0.45% C 0.25% Si 0.65% Mn 0.04% S & P steel: German Standard
D 3	0.32% C 0.3% Si 0.7% Mn steel: Pompey
D 3S	0.35% C 0.25% Si 0.8% Mn steel: Pompey
D 4	0.4% C 0.3% Si 0.6% Mn steel: Pompey
D 4Mn	0.4% C 0.3% Si 1.3% Mn steel: Pompey
D 48	0.48% C 0.3% Si 0.6% Mn steel: Pompey
DEF 13 B3	Carbon and carbon manganese steel: Covers BS alloys En 5, 6, 8 and includes En 14 **UTS: 550 Elon: 15% Proof: 240**
DEF 13 B3A	Carbon and low alloy hardened and tempered steel: Covers BS alloy En 5, 8, 15 and includes En 12 and 14 **UTS: 610 Elon: 20% Proof: 370**
DIN 1629 St 55	0.36% C (max) 0.05% P & S (max) steel: Pipe; German Standard **UTS: 600 Proof: 300**

Symbol	Nominal analysis, supplier, condition and remarks.
DIN 1629 St 55/4	0.36% C (max) 0.4% Mn (max) 0.05% P & S (max) steel: Pipe; German Standard **UTS: 600 Proof: 300**
DEF 13 B3B	Carbon and low alloy hardened and tempered steel: Covers BS alloys En 5 and 8 and includes En 15, 17, 29 and 100 **UTS: 670 Elon: 20% Proof: 450**
DIN 1651 35 S 20	0.35% C 0.6% Mn 0.2% S 0.05% P steel: Free machining
DIN 1651 45 S 20	0.46% C 0.7% Mn 0.2% S 0.05% P steel: Free machining
DIN 1654 C9/35	0.36% C 0.6% Mn 0.04% S & P steel: For screws
DIN 1654 C9/45	0.46% C 0.7% Mn 0.04% S & P steel: For screws
DIN 17100 RSt 42/1	0.25% C (max) 0.08% P (max) 0.05% S (max) steel: Plate; German specification **UTS: 460 Proof: 250**
DIN 17100 RSt 42/2	0.23% C (max) 0.05% P (max) 0.05% S (max) steel: Plate; German specification **UTS: 460 Proof: 250**
DIN 17100 St 50/2	0.3% C 0.05% P & S (max) steel: Plate **UTS: 550 Proof: 290**
DIN 17100 St 60/1	0.35% C 0.08% P (max) 0.05% S (max) steel: Plate; German Standard **UTS: 660 Proof: 330**
DIN 17100 St 60/2	0.4% C 0.05% P & S steel: Plate; German Standard **UTS: 660 Proof: 330**
DIN 17100 UST 42/2	0.25% C (max) 0.05% P (max) 0.05% S (max) steel: Plate; German specification **UTS: 460 Proof: 250**
DIN 17155 H1V	0.26% C (max) 0.05% S & P steel: For boiler plate
DIN 17155 HIV	0.26% (max) 0.6% Mn (max) 0.3% Cr (max) 0.05% P & S (max) steel: Plate; German Standard **UTS: 520 Proof: 280**
DIN 17155 H1VA	0.26% C (max) 0.05% S & P steel: For boiler plate
DIN 17155 H1VL	0.26% C (max) 0.05% S & P steel: For boiler plate
DIN 17200 30 Mn 5	0.3% C 1.4% Mn 0.035% S & P steel: Forging
DIN 17200 37 Mn Si 5	0.37% C 1.2% Si 1.2% Mn steel: Forging
DIN 17200 40 Mn 4	0.4% C 0.9% Mn 0.035% S & P steel
DIN 17200 C 35	0.35% C 0.65% Mn 0.045% P & S (max) steel: For structures; German Standard
DIN 17200 C 45	0.45% C 0.65% Mn 0.045% P & S (max) steel: For structures; German Standard
DIN 17220 C45	0.46% C 0.7% Mn 0.045% S & P steel
DIN 17200 CK35	0.36% C 0.6% Mn 0.04% S & P steel
DIN 17200 CK 35	0.35% C 0.65% Mn 0.035% P & S (max) steel: For structures; German Standard
DIN 17200 CK 45	0.45% C 0.65% Mn 0.035% P & S (max) steel: For structures; German Standard
DIN 17200 CK 45	0.46% C 0.7% Mn 0.035% S & P steel
DIN 17200 CM 35	0.35% C 0.65% Mn 0.035% P & S (max) steel: For structures; German Standard
DIN 17200 CM 45	0.45% C 0.65% Mn 0.035% P & S (max) steel: For structure; German Standard
DIN 17200 28 Mn 6	0.28% C 1.5% Mn 0.035% P & S (max) steel: German Standard
DIN 17200 40 Mn 4	0.4% C 0.95% Mn 0.035% P & S (max) steel: German Standard
DIN 17221 38 Si 6	0.38% C 1.5% Si 0.7% Mn steel: For springs
DIN 17221 38 Si 7	0.38% C 0.65% Mn 0.045% P & S (max) steel: For springs; German Standard; heat treated
DIN 17221 46 Si 7	0.46% C 1.7% Si 0.6% Mn steel: For springs
DISQUE E	0.45% C 1.75% Si 0.7% Mn steel: Pompey
DISQUE H	0.55% C 1.75% Si 0.7% Mn steel: Pompey
D METAL	0.45% C 0.3% Si 0.85% Mn steel: Casting; Edgar Allen; code 606

Note. The following abbreviations and units are used in the tables:

DPN	Hardness, diamond pyramid number
UTS	Ultimate tensile strength, N/mm^2
Elon	Elongation, %
Proof	0.1 % proof strength, N/mm^2

1 N/mm^2=0.1 hbar=0.102 kgf/mm^2=0.06475 tonf/in^2=145.04 lbf/in^2

Symbol	Nominal analysis, supplier, condition and remarks.
DMS	0.35% C 1.2% Si 1.2% Mn steel: Pompey
DOUBLE EXTRA	1.0% C tool steel: Balfour Darwin
DTD 503 A	0.25% C 1.5% Mn 0.05% S 0.05% P steel: Tube; annealed
	UTS: 370
DTD 666	C not specified 0.02% S 0.025% P steel: Casting; casting to be X-rayed
	DPN: 269 UTS: 910 Elon: 12% Proof: 630
DTD 705	C not specified 0.02% S 0.025% P steel: Casting; casting to be X-rayed
	DPN: 340 UTS: 1210 Elon: 7% Proof: 910
DTD 5032	0.4% C 0.8% Mn 0.05% S 0.05% P steel: Swaged; blued if necessary at 350 °C
	UTS: 910 Elon: 10%
DTD 5072	C not specified 0.02% S 0.025% P steel: Casting; hardened, tempered & X-rayed
	DPN: 370 UTS: 1210 Elon: 7% Proof: 910
DUCOL HT	0.3% C (max) 1.6% Mn 0.05% S & P steel: Plate & section; Colvilles alloy for BS 548
	UTS: 560 Elon: 18% Proof: 330
DUNELT	0.4% C 0.06% S & P steel: Dunford for BS alloy En 8
DUNELT 40	0.35% C 1.6% Mn 0.06% S & P steel: Dunford for BS alloy En 15A
DUNELT 41	0.38% C 1.2% Mn 0.15% S 0.06% P steel: Free machining; Dunford; hardened & tempered
	DPN: 250 UTS: 720 Elon: 15%
E 5M	0.5% C 0.3% Si 0.6% Mn steel: Pompey
ERA 51	0.35% C 1.5% Mn steel: Hadfield for BS alloy En 15 & 15A
ERA 53	0.37% C 1.2% Mn steel: Hadfield for BS alloy En 15B
FAGERSTA 1660	0.45% C 0.6% Mn steel: Fagersta
FAGERSTA B805	0.4% C 1.3% Mn steel: Fagersta
F F 3	0.35% C 0.25% Si 0.5% Mn steel: Pompey
	UTS: 780 Elon: 12% Proof: 530
FF 3H	0.37% C 0.3% Si 0.7% Mn steel: Pompey
	UTS: 740 Elon: 10% Proof: 450
F F 4	0.42% C 0.3% Si 0.7% Mn steel: Pompey
	UTS: 920 Elon: 9% Proof: 660
FF 38	0.38% C 0.3% Si 0.7% Mn steel: Pompey
FUJI CON SUPER	0.33% C 1.2% Mn 0.015% P & S steel: Fuji
G 10300	UNS designation for type 1030 steel
G 10350	UNS designation for type 1035 steel
G 10400	UNS designation for type 1040 steel
GOST 380 BST5 SP	0.33% C 0.65% Mn 0.04% P (max) 0.05% S (max) steel: Plate; Russian Standard
GOST 380 BST5 SP2	0.33% C 0.65% Mn 0.3% Cr Ni Cu (max) 0.04% P (max) 0.05% S (max) steel: Plate; Russian Standard
GOST 380 BST5 PS	0.33% C 0.65% Mn 0.04% P (max) 0.05% S (max) steel: Plate; Russian Standard
GOST 380 BST6 PS	0.42% C 0.65% Mn 0.04% P (max) 0.05% S (max) steel: Plate; Russian Standard
GOST 380 BST6 PS2	0.42% C 0.65% Mn 0.3% Cr Ni Cu (max) 0.04% P (max) 0.05% S (max) steel: Plate; Russian Standard
GOST 380 BST6 PS2	0.33% C 0.65% Mn 0.3% Cr Ni Cu (max) 0.04% P (max) 0.05% S (max) steel: Plate; Russian Standard
GOST 380 BST6 SP	0.42% C 0.65% Mn 0.04% P (max) 0.05% S (max) steel: Plate; Russian Standard
GOST 380 BST6 SP2	0.42% C 0.65% Mn 0.3% Cr Ni Cu (max) 0.04% P (max) 0.05% S (max) steel: Plate; Russian Standard
GOST 380 VST5 GPS	0.3% C (max) 1.0% Mn 0.04% P 0.05% S (max) steel: Plate; Russian Standard
	UTS: 480
GOST 380 VST5 GPS2	0.3% C (max) 1.0% Mn 0.3% Cr Ni Cu (max) 0.04% (max) 0.05% S (max) steel: Plate; Russian Standard
GOST 380 VST5 PS	0.37% C (max) 0.75% Mn 0.04% P 0.05% S (max) steel: Plate; Russian Standard
	UTS: 570
GOST 380 VST5 PS2	0.37% C (max) 0.75% Mn 0.3% Cr Ni Cu (max) 0.04% P 0.05% S steel: Plate; Russian Standard
	UTS: 570 Proof: 280
GOST 380 VST5 SP	0.37% C (max) 0.75% Mn 0.04% S 0.05% P steel: Plate; Russian Standard
	UTS: 570
GOST 380 VST5 SP2	0.37% C (max) 0.75% Mn 0.3% Cr Ni Cu (max) 0.04% P (max) 0.05% S (max) steel: Plate; Russian Standard
	UTS: 570 Proof: 280
GOST 1050/25	0.26% C 0.65% Mn 0.04% P & S (max) 0.25% Cr & Ni (max) steel: For structures; Russian Standard
GOST 1050/25 G	0.26% C 0.85% Mn 0.04% P & S (max) 0.25% Cr & Ni (max) steel: For structures; Russian Standard
GOST 1050/30	0.3% C 0.65% Mn 0.04% P & S (max) 0.25% Cr & Ni (max) steel: For structures; Russian Standard
GOST 1050/30 G	0.3% C 0.85% Mn 0.04% P & S (max) 0.25% Cr & Ni (max) steel: For structures; Russian Standard
GOST 1050/35	0.35% C 0.65% Mn 0.04% P & S (max) 0.25% Cr Ni (max) steel: For structures; Russian Standard
GOST 1050/35 G	0.35% C 0.85% Mn 0.04% P & S (max) 0.25% Cr & Ni (max) steel: For structures; Russian Standard
GOST 1050/40	0.4% C 0.65% Mn 0.04% P & S (max) 0.25% Cr & Ni (max) steel: For structures; Russian Standard
GOST 1050/40 G	0.4% C 0.85% Mn 0.04% P & S (max) 0.25% Cr & Ni (max) steel: For structures; Russian Standard
GOST 1050/45	0.45% C 0.65% Mn 0.04% P & S (max) 0.25% Cr & Ni (max) steel: For structures; Russian Standard
GOST 1050/45 G	0.45% C 0.85% Mn 0.04% P & S (max) 0.25% Cr & Ni (max) steel: For structures; Russian Standard
GOST 2052/50 S2	0.51% C 0.75% Mn 0.04% P & S (max) 0.3% Cr & Ni (max) steel: Bar; Russian Standard
GOST 2052/55 SG	0.55% C 0.9% Mn 0.04% P & S (max) 0.3% Cr & Ni (max) steel: Bar; Russian Standard
GOST 2052/55 SG	0.55% C 0.9% Mn 0.04% P & S (max) 0.3% Cr (max) 0.4% Ni (max) 0.25% Cu (max) steel: For springs; Russian Standard
GOST 4543/25 G2S	0.25% C 1.4% Mn 0.3% Cr & Ni (max) steel: Russian Standard
GOST 4543/35 G2	0.35% C 1.6% Mn 0.25% Cr (max) steel: Russian Standard
GOST 4543/35 GS	0.34% C 1.0% Mn 0.3% Cr & Ni (max) steel: Russian Standard
GOST 4543/35 SG	0.35% C 1.25% Mn 0.25% Cr (max) steel: Russian Standard
GOST 4543/36 G2S	0.36% C 1.65% Mn 0.25% Cr (max) steel: Russian Standard
GOST 4543/40 G2	0.4% C 1.6% Mn 0.25% Cr (max) steel: Russian Standard
GOST 4543/45 G2	0.45% C 1.6% Mn 0.25% Cr (max) steel: Russian Standard
GOST 7909/36 Cr 2S	0.36% C 0.04% P (max) 0.045% S (max) 0.3% Cr (max) 0.4% Ni (max) steel: Pipe; Russian Standard
	UTS: 700 Proof: 500

Symbol	Nominal analysis, supplier, condition and remarks.
GOST 8467/36 G2S	0.36% C 1.65% Mn 0.045% P & S (max) 0.3% Cr (max) 0.4% Ni (max) steel: Pipe; Russian Standard **UTS: 700** **Proof: 500**
GOST 8731 BMST5 SP	0.32% C 0.65% Mn 0.045% P (max) 0.055% S (max) steel: Pipe; Russian Standard; seamless **UTS: 500** **Proof: 270**
GOST 8731/35	0.35% C 0.65% Mn 0.04% P & S (max) 0.25% Cr (max) 0.2% Ni (max) steel: Pipe; Russian Standard seamless **UTS: 520** **Proof: 300**
GOST 8731/45	0.45% C 0.65% Mn 0.04% P & S (max) 0.25% Cr (max) 0.2% Ni (max) steel: Pipe; Russian Standard; seamless **UTS: 600** **Proof: 330**
GOST 8733/35	0.35% C 0.65% Mn 0.04% P & S (max) 0.15% Cr (max) 0.2% Ni (max) steel: Pipe; Russian Standard; seamless **UTS: 520** **Proof: 300**
GOST 8733/45	0.45% C 0.65% Mn 0.04% P & S (max) 0.15% Cr (max) 0.2% Ni (max) steel: Pipe; Russian Standard; seamless **UTS: 600** **Proof: 330**
GOST 10706/24 G	0.24% C 0.9% Mn 0.04% P & S (max) 0.3% Cr Ni & Cu (max) steel: Pipe; Russian Standard; welded **UTS: 490** **Proof: 330**
GOST 10706/25 G2S	0.24% C 1.4% Mn 0.04% P & S (max) 0.3% Cr Ni & Cu (max) steel: Pipe; Russian Standard; welded **UTS: 600** **Proof: 400**
GOST 12132/35	0.35% C 0.65% Mn 0.04% P & S (max) 0.25% Cr & Ni (max) steel: Pipe; Russian Standard; welded **UTS: 520**
GOST 12132/45	0.45% C 0.05% Mn 0.04% P & S (max) 0.25% Cr & Ni (max) steel: Pipe; Russian Standard; welded **UTS: 600**
HARDFLEX 9	0.45% C steel: Strip; Sandvik; supplied hardened & tempered **UTS: 900** **Proof: 750**
HARDFLEX II	0.55% C steel: Strip; Sandvik; supplied hardened & tempered **UTS: 1000** **Proof: 850**
HECLA 37	0.4% C 0.8% Mn steel: Hadfield for BS alloys En 8 & 8A, B & C
HECLA 37M	0.4% C 1.1% Mn 0.15% S steel: Free machining; Hadfields for BS alloy En 8M
HECLA 40	0.3% C 0.8% Mn steel: Hadfields for BS alloys En 5 & 5A, B, C & K
HEDEX 1	0.27% C 0.9% Mn 0.05% S 0.04% P steel: Wire for forging; Kiveton Park; annealed **UTS: 530**
HEDEX 1A	0.32% C 0.8% Mn 0.05% S 0.04% P steel: Wire for forging; Kiveton Park; annealed **UTS: 530**
HEDEX 1B	0.37% C 0.8% Mn 0.05% S 0.04% P steel: Wire for forging; Kiveton Park; annealed **UTS: 560**
HEDEX 2	0.42% C 0.8% Mn 0.05% S 0.04% P steel: Wire for forging; Kiveton Park; annealed **UTS: 610**

Symbol	Nominal analysis, supplier, condition and remarks.
HEDEX 3	0.38% C 0.8% Mn 0.05% S 0.04% P steel: Wire for forging; Kiveton Park; annealed **UTS: 560**
HEDEX 12	0.22% C 1.5% Mn 0.05% S 0.04% P steel: Wire for forging; Kiveton Park; annealed **UTS: 550**
HEDEX 13	0.37% C 0.8% Mn 0.05% S 0.04% P steel: Wire for forging; Kiveton Park; annealed **UTS: 560**
HI-MAN HI-STRESS	0.3% C 1.3% Mn 0.2% Cu steel: Inland Steel Co. Medium carbon hot rolled deformed steel: Section; Park Gate; for concrete reinforcing **Proof: 280**
HI STRENGTH C	0.24% C 1.3% Mn 0.05% Nb steel: Armco; weldable; in 4 grades **UTS: 530** **Elon: 15%** **Proof: 420**
HITENSPEED 45	0.38% C 1.3% Mn 0.04% P (max) 0.26% S steel: Hardened & tempered **UTS: 700** **Elon: 12%** **Proof: 500**
HITENSPEED 45A	0.42% C 1.3% Mn 0.04% P (max) 0.26% S steel: Hardened & tempered **UTS: 700** **Elon: 12%** **Proof: 540**
HMS 55	0.35% C 0.25% Si 0.55% Mn 0.035% S & P steel: Krupp
HMS 65	0.45% C 0.25% Si 0.65% Mn 0.035% S & P steel: Krupp
HTC	0.4% C 0.8% Mn steel: ESC for BS alloy En 8
JIS G3101 SS55	0.3% C 1.6% Mn (max) 0.04% P & S (max) steel: Plate; Japanese Standard **UTS: 550** **Proof: 410**
JIS G3106 SM50A	0.21% C 1.5% Mn (max) 0.04% P & S (max) steel: Plate; Japanese Standard **UTS: 560** **Proof: 330**
JIS G3445 STKM14A	0.3% C (max) 0.7% Mn 0.04% P & S (max) steel: Pipe; Japanese Standard **UTS: 420** **Proof: 250**
JIS G3445 STKM14B	0.3% C (max) 0.7% Mn 0.04% P & S (max) steel: Pipe; Japanese Standard **UTS: 510** **Proof: 360**
JIS G3445 STKM14C	0.3% C (max) 0.7% Mn 0.04% P & S (max) steel: Pipe; Japanese Standard **Proof: 420**
JIS G3445 STKM15A	0.3% C 0.7% Mn 0.04% S & P steel: Pipe; Japanese Standard **UTS: 480** **Proof: 280**
JIS G3445 STKM15C	0.3% C 0.7% Mn 0.04% P & S (max) steel: Pipe; Japanese Standard **UTS: 590** **Proof: 440**
JIS G3445 STKM16A	0.4% C 0.7% Mn 0.04% P & S steel: Pipe; Japanese Standard **UTS: 520** **Proof: 330**
JIS G3445 STKM16C	0.4% C 0.7% Mn 0.04% P & S (max) steel: Pipe; Japanese Standard **UTS: 630** **Proof: 470**
JIS G3454 STPG42	0.3% C (max) 0.5% Mn 0.04% P & S steel: Pipe; Japanese Standard **UTS: 420** **Proof: 250**
JIS G3455 STS42	0.3% C (max) 0.6% Mn 0.035% P & S 0.2% Cu (max) steel: Pipe; Japanese Standard
JIS G3455 STS49	0.33% C (max) 0.7% Mn 0.035% P & S 0.2% Cu steel: Pipe; Japanese Standard **UTS: 490** **Proof: 280**
JIS G3456 STPT49	0.33% C (max) 0.65% Mn 0.2% Cu (max) 0.035% P & S (max) steel: Pipe; Japanese Standard **UTS: 490** **Proof: 280**
JIS G3456 STPT42	0.3% C (max) 0.65% Mn 0.2% Cu (max) 0.035% P & S (max) steel: Pipe; Japanese Standard **UTS: 420** **Proof: 250**

Note. The following abbreviations and units are used in the tables:

DPN	Hardness, diamond pyramid number
UTS	Ultimate tensile strength, N/mm^2
Elon	Elongation, %
Proof	0.1 % proof strength, N/mm^2

1 N/mm^2=0.1 hbar=0.102 kgf/mm^2=0.06475 tonf/in^2=145.04 lbf/in^2

Symbol	Nominal analysis, supplier, condition and remarks.
JIS 4051/28 C	0.28% C 0.75% Mn 0.03% P & S (max) steel: For structures; Japanese Standard
JIS G4051 S33C	0.33% C 0.75% Mn 0.03% P & S (max) steel: For structures; Japanese Standard
JIS G4051 S35C	0.35% C 0.75% Mn 0.03% P & S (max) steel: For structures; Japanese Standard
JIS G4051 S38C	0.38% C 0.75% Mn 0.03% P & S (max) steel: For structures; Japanese Standard
JIS G4051 S45C	0.45% C 0.75% Mn 0.03% P & S (max) steel: For structures; Japanese Standard
JIS G4051 S43C	0.43% C 0.75% Mn 0.03% P & S (max) steel: For structures; Japanese Standard
JIS G4051 S22C	0.22% C 0.45% Mn 0.03% P & S (max) steel: For structures; Japanese Standard
JIS G4051 S25C	0.25% C 0.45% Mn 0.03% P & S (max) steel: For structures; Japanese Standard
JIS G4051 S30C	0.3% C 0.75% Mn 0.03% P & S (max) steel: For structures; Japanese Standard
JIS G4051 S40C	0.4% C 0.75% Mn 0.03% P & S (max) steel: For structures; Japanese Standard
JIS G4106 S Mn 1	0.33% C 0.03% P & S (max) steel: Japanese Standard
JIS G4106 S Mn 2	0.38% C 1.5% Mn 0.03% P & S (max) steel: Japanese Standard
JIS G4106 S Mn 3	0.43% C 1.5% Mn 0.03% P & S (max) steel: Japanese Standard
JS 3	0.45% C (max) 0.6% Si 1.0% Mn steel: Casting; Jopling **DPN: 170 UTS: 530 Elon: 15% Proof: 240**
JS 12	0.3% C 1.5% Mn steel: Casting; Jopling **DPN: 230 UTS: 760 Elon: 15%**
JS 16	0.35% C (max) 0.6% Si 1.0% Mn steel: Casting; Jopling **DPN: 160 UTS: 480 Elon: 20% Proof: 240**
KF 4A	0.22% C 1.6% Mn 0.06% S P steel: Kirkstall; hardened & tempered **UTS: 640 Elon: 20% Proof: 450**
KF 5	0.22% C 0.8% Mn 0.06% S & P steel: Kirkstall for BS alloy En 4
KF 8	0.38% C 0.8% Mn 0.06% S & P steel: Kirkstall for BS alloy En 8
KF 9	0.42% C 1.2% Mn 0.06% S & P steel: Kirkstall for BS alloy En 8K
KF 10	0.44% C 0.7% Mn 0.06% S & P steel: Kirkstall for BS alloy En 8K
KF 10A	0.48% C 0.8% Mn 0.06% S & P steel: Kirkstall for BS alloy En 43A
KF 40	0.36% C 1.5% Mn 0.05% S & P steel: Kirkstall for BS alloy En 15
KF 40A	0.28% C 1.5% Mn 0.06% S & P steel: Kirkstall for BS alloy En 14B
KF 51	0.4% C 0.2% Si 1.2% Mn 0.17% S 0.06% P steel: Free machining; Kirkstall for BS alloy En 8M
KF 59	0.25% C 0.25% Si 1.2% Mn 0.14% S 0.06% P steel: Free machining; Kirkstall for BS alloy En 7
KF 63	0.38% C 0.2% Si 1.5% Mn 0.2% S 0.06% P steel: Free machining; Kirkstall for BS alloy En 15 AM
KF 63A	0.44% C 0.3% Si 1.5% Mn 0.3% S 0.05% P steel: Free machining; Kirkstall
KF 64	0.3% C 0.3% Si 1.4% Mn 0.16% S 0.05% P steel: Free machining; Kirkstall; hardened & tempered **UTS: 610 Elon: 17% Proof: 530**
KP 4	0.3% C 1% Mn 0.06% S & P steel: Kiveton Park for BS alloy En 4
KP 5	0.3% C 0.9% Mn 0.06% S & P steel: Kiveton Park for BS alloy En 5
KP 6	0.4% C 0.8% Mn 0.06% S & P steel: Kiveton Park for BS alloy En 8
KP 20	0.35% C 1.5% Mn 0.06% S & P steel: Kiveton Park for BS alloy En 15

Symbol	Nominal analysis, supplier, condition and remarks.
KP 21	0.25% C 1.5% Mn 0.06% S & P steel: Kiveton Park for BS alloy En 14
L 2	0.28% C 0.6% Si 1% Mn steel: Casting; Jessop; grade B
L 3	0.37% C 0.6% Si 1% Mn steel: Casting; Jessop; grade C
M 3	0.46% C 0.22% Si 1.6% Mn steel: Jessop
MACHINERY 30	0.3% C 0.2% Si 0.8% Mn 0.04% S & P steel: Atlas; as AISI 1030
MACHINERY 40	0.4% C 0.2% Si 0.8% Mn 0.04% S & P steel: Atlas; as AISI 1040
MATTE LADLE	0.37% C 0.3% Si 0.7% Mn steel: Casting; Edgar Allen; code 611
MBV	0.4% C 0.4% Si 0.9% Mn 0.035% S & P steel: Krupp
MBV/m	0.36% C 0.25% Si 1.35% Mn 0.035% S & P steel: Krupp; up to 0.3% Cr may be present
MBV/w	0.3% C 0.25% Si 1.35% Mn 0.035% S & P steel: Krupp; up to 0.3% Cr may be present
MEL-TROL TGS	0.2% C 1.3% Mn 0.2% Si steel: Carpenter
MNC 720	Swedish Standard; summary of steel castings
MNC 852	General standard for SIS range of steels for hardening and tempering; Swedish Standard
MNC 854	Swedish Standard; summary of steel for induction and flame hardening
MNC 870	Covers spring steels of various types; Swedish Standard; includes detailed specifications
MS 34	0.38% C 0.8% Si 1.05% Mn 0.035% S & P steel: Krupp
MS 43	0.46% C 0.8% Si 1.05% Mn 0.035% S & P steel: Krupp
MS BV	0.37% C 1.25% Si 1.25% Mn 0.035% S & P steel: Krupp
M St 50/2	0.3% C 0.05% S 0.06% P steel: German Standard
M St 60/2	0.4% C 0.05% S 0.06% P steel: German Standard
MS V3	0.3% C 0.75% Mn steel: Pompey
MT 'P'	0.25% C 1.4% Mn steel: Casting; Edgar Allen; code 610 steel
MT 'R'	0.3% C 1.4% Mn steel: Casting; Edgar Allen; code 626
N 82	0.4% C 1.3% Mn steel: Bofors; hardened & tempered **DPN: 230 UTS: 730 Elon: 17% Proof: 480**
N 91	0.45% C 0.8% Mn steel: For surface hardening; Bofors; normalized; hardened by flame or induction heating **DPN: 200 UTS: 670 Elon: 18% Proof: 300**
NF A35 551/35 M5	0.36% C 1.2% Mn steel: French Standard
NF A35 551/XC25	0.26% C 0.55% Mn steel: For structures; French Standard
NF A35 551/XC32	0.32% C 0.65% Mn steel: For structures; French Standard
NF A35 551/XC38	0.38% C 0.65% Mn steel: For structures; French Standard
NF A35 551/XC42	0.42% C 0.65% Mn steel: For structures; French Standard
NF A35 571/41S7	0.41% C 0.65% Mn 0.035% P & S (max) steel: Bar; French Standard
NF A35 571/46S7	0.46% C 0.65% Mn 0.035% P & S (max) steel: Bar; French Standard
NF A35 571/50S7	0.5% C 0.65% Mn 0.05% P & S (max) steel: Bar; French Standard
NF A35 571/51S7	0.51% C 0.65% Mn 0.035% P & S (max) 0.3% Cr (max) steel: Bar; French Standard
OCM 6	0.35% C 1.5% Mn steel: W Marrison for BS alloy En 15
Q BRAND	0.4% C 0.8% Mn steel: For cars; die inserts, wedges, etc. **UTS: 650**

Symbol	Nominal analysis, supplier, condition and remarks.
QQ W 409a	Low & medium carbon steel: Wire for cold heating; US Federal specification covering a range of AISI alloys
QQ W 414a	Carbon steel: Wire; for book binding; US Federal specification
QQ W 418a	Carbon steel: Wire; for concrete reinforcing; US Federal specification; cold drawn
RETORT	0.32% C 0.6% Si 0.7% Mn steel: Casting; Edgar Allen; code 609
RS	0.45% C 1.7% Si 0.6% Mn steel: Pompey
RS 1	0.45% C 1.8% Si 0.7% Mn steel: Pompey
S 3F	0.35% C 0.3% Si 0.6% Mn steel: Pompey
S 4F	0.4% C 0.3% Si 0.6% Mn steel: Pompey
S 4M	0.45% C 0.2% Si 0.7% Mn steel: Pompey
S 4S	0.4% C 0.3% Si 0.6% Mn steel: Pompey
SAE 0025	0.25% C 0.7% Mn 0.8% Si steel: Casting; as cast; suitable for welding DPN: 187 UTS: 420 Elon: 22% Proof: 200
SAE 003 C	0.3% C (max) 0.6% Si 0.7% Mn 0.05% S & P cast steel: Hardened & tempered; weldable DPN: 131 UTS: 450 Elon: 24% Proof: 220
SAE 0035	0.3% C 0.6% Si 0.8% Mn 0.05% S & P investment cast steel: Hardened & tempered UTS: 480 Elon: 22% Proof: 210
SAE 0050	0.45% C 0.4% Si 0.7% Mn 0.05% S & P cast steel: Hardened & tempered DPN: 200 UTS: 730 Elon: 10% Proof: 480
SAE 090	C not specified 0.05% S & P cast steel: Hardened & tempered DPN: 190 UTS: 670 Elon: 20% Proof: 420
SAE 0105	C not specified 0.05% S & P cast steel: Hardened & tempered DPN: 220 UTS: 760 Elon: 17% Proof: 610
SAE 0120	C not specified 0.05% S & P cast steel: Hardened & tempered DPN: 250 UTS: 890 Elon: 14% Proof: 740
SAE 0150	C not specified 0.05% S & P cast steel: Hardened & tempered DPN: 310 UTS: 1070 Elon: 9% Proof: 910
SAE 0175	C not specified 0.05% S & P cast steel: Hardened & tempered DPN: 360 UTS: 1300 Elon: 6% Proof: 1070
SAE 851	0.4% Fe sinter: For bearings; density 5900 kg/m³
SAE 861	0.3% C 4.0% Cu Fe sintered metal: Specific density 5.8–6.2 UTS: 150
SAE 862	0.25% C 9.0% Cu Fe alloy sinter: For bearings; density 6000 kg/m³
SAE 863	0.25% C 22% Cu Fe alloy sinter: For bearings; density 6000 kg/m³
SAE 1024	0.23% C 1.5% Mn 0.04% P 0.05% S steel: Bar; hot rolled as AISI C1024 DPN: 149 UTS: 530 Elon: 20% Proof: 300
SAE 1024	0.23% C 1.5% Mn 0.04% P 0.05% S steel: Bar; cold drawn DPN: 163 UTS: 600 Elon: 12% Proof: 580
SAE 1025	0.25% C 0.4% Mn 0.04% P 0.05% S steel: Bar; hot rolled as AISI C1025 DPN: 116 UTS: 500 Elon: 25% Proof: 210

Note. The following abbreviations and units are used in the tables:

DPN	Hardness, diamond pyramid number
UTS	Ultimate tensile strength, N/mm²
Elon	Elongation, %
Proof	0.1 % proof strength, N/mm²

1 N/mm²=0.1 hbar=0.102 kgf/mm²=0.06475 tonf/in²=145.04 lbf/in²

Symbol	Nominal analysis, supplier, condition and remarks.
SAE 1025	0.25% C 0.4% Mn 0.04% P 0.05% S steel: Bar; cold drawn DPN: 126 UTS: 450 Elon: 15% Proof: 370
SAE 1026	0.26% C 0.7% Mn 0.04% S & P steel: As AISI C1026
SAE 1027	0.25% C 1.3% Mn 0.04% P 0.05% S steel: Bar; hot rolled DPN: 149 UTS: 540 Elon: 18% Proof: 300
SAE 1027	0.25% C 1.3% Mn 0.04% P 0.05% S steel: Bar; cold drawn DPN: 163 UTS: 610 Elon: 12% Proof: 500
SAE 1029	0.29% C 0.75% Mn 0.04% P 0.05% S steel: Bar, etc.
SAE 1030	0.3% C 0.8% Mn 0.04% P 0.05% S steel: Bar; hot rolled DPN: 137 UTS: 480 Elon: 20% Proof: 240
SAE 1030	0.3% C 0.8% Mn 0.04% P 0.05% S steel: Bar; cold drawn DPN: 149 UTS: 550 Elon: 12% Proof: 450
SAE 1033	0.33% C 0.9% Mn 0.04% P 0.05% S steel: Bar; hot rolled DPN: 143 UTS: 530 Elon: 18% Proof: 240
SAE 1033	0.33% C 0.9% Mn 0.04% P 0.05% S steel: Bar; cold drawn DPN: 163 UTS: 580 Elon: 12% Proof: 480
SAE 1035	0.35% C 0.8% Mn 0.04% P 0.05% S steel: Bar; hot rolled DPN: 143 UTS: 530 Elon: 18% Proof: 240
SAE 1035	0.35% C 0.8% Mn 0.04% P 0.05% S steel: Bar; cold drawn DPN: 163 UTS: 580 Elon: 12% Proof: 480
SAE 1036	0.34% C 1.2% Mn 0.04% P 0.05% S steel: Bar; hot rolled DPN: 163 UTS: 610 Elon: 16% Proof: 320
SAE 1036	0.34% C 1.2% Mn 0.04% P 0.05% S steel: Bar; cold drawn DPN: 187 UTS: 660 Elon: 12% Proof: 540
SAE 1037	0.35% C 0.9% Mn 0.04% P 0.05% S steel: Bar; hot rolled DPN: 143 UTS: 540 Elon: 18% Proof: 240
SAE 1037	0.35% C 0.9% Mn 0.04% P 0.05% S steel: Bar; cold drawn DPN: 167 UTS: 600 Elon: 12% Proof: 480
SAE 1038	0.38% C 0.8% Mn 0.04% P 0.05% S steel: Bar; hot rolled DPN: 149 UTS: 550 Elon: 18% Proof: 300
SAE 1038	0.38% C 0.8% Mn 0.04% P 0.05% S steel: Bar; cold drawn DPN: 163 UTS: 610 Elon: 12% Proof: 500
SAE 1039	0.4% C 0.9% Mn 0.04% P 0.05% S steel: Bar; hot rolled DPN: 156 UTS: 580 Elon: 16% Proof: 300
SAE 1039	0.4% C 0.9% Mn 0.04% P 0.05% S steel: Bar; cold drawn DPN: 179 UTS: 630 Elon: 12% Proof: 530
SAE 1040	0.4% C 0.8% Mn 0.04% P 0.05% S steel: Bar; hot rolled DPN: 149 UTS: 550 Elon: 18% Proof: 300
SAE 1040	0.4% C 0.8% Mn 0.04% P 0.05% S steel: Bar; cold drawn DPN: 170 UTS: 630 Elon: 12% Proof: 550
SAE 1041	0.4% C 1.5% Mn 0.04% P 0.05% S steel: Bar; hot rolled DPN: 187 UTS: 660 Elon: 15% Proof: 360
SAE 1041	0.4% C 1.5% Mn 0.04% P 0.05% S steel: Bar; cold drawn DPN: 207 UTS: 750 Elon: 10% Proof: 630
SAE 1042	0.43% C 0.8% Mn 0.04% P 0.05% S steel: Bar; hot rolled DPN: 163 UTS: 580 Elon: 16% Proof: 300

Symbol	Nominal analysis, supplier, condition and remarks.
SAE 1042	0.43% C 0.8% Mn 0.04% P 0.05% S steel: Bar; cold drawn **DPN: 179 UTS: 650 Elon: 12% Proof: 550**
SAE 1043	0.43% C 0.9% Mn 0.04% P 0.05% S steel: Bar; hot rolled **DPN: 163 UTS: 600 Elon: 16% Proof: 320**
SAE 1043	0.43% C 0.9% Mn 0.04% P 0.05% S steel: Bar; cold drawn **DPN: 179 UTS: 660 Elon: 12% Proof: 560**
SAE 1045	0.46% C 0.8% Mn 0.04% P 0.05% S steel: Bar; hot rolled **DPN: 163 UTS: 600 Elon: 16% Proof: 320**
SAE 1045	0.46% C 0.8% Mn 0.04% P 0.05% S steel: Bar; cold drawn **Free cutting; cold drawn 0.46% C 0.9% Mn** 0.04% P 0.05% S steel: Bar; hot rolled 0.39% C 1.5% Mn 0.04% P 0.18% S steel: Free machining 0.46% C 0.9% Mn 0.04% P 0.05% S steel: Bar; 0.4% C 0.9% Mn 0.04% P 0.1% S steel: Free cutting; hot rolled
SAE 1126	0.26% C 0.9% Mn 0.04% P 0.1% S steel: 0.4% C 0.9% Mn 0.04% P 0.1% S steel: Free cutting; cold drawn
SAE 1126	0.26% C 0.9% Mn 0.04% P 0.1% S steel: 0.41% C 1.5% Mn 0.04% P 0.1% S steel: Free cutting; hot rolled
SAE 1132	0.31% C 1.5% Mn 0.04% P 0.1% S steel: 0.41% C 1.5% Mn 0.04% P 0.1% S steel: Free cutting; cold drawn
SAE 1132	0.31% C 1.5% Mn 0.04% P 0.1% S steel: 0.44% C 1.5% Mn 0.4% P 0.3% S steel: Free cutting, hot rolled
SAE 1137	0.36% C 1.5% Mn 0.04% P 0.1% S steel: 0.44% C 1.5% Mn 0.4% P 0.3% S steel: Free cutting; cold drawn
SAE 1137	0.36% C 1.5% Mn 0.04% P 0.1% S steel: 0.45% C 0.9% Mn 0.04% P 0.05% S steel: Free cutting; hot rolled
SAE 1138	0.37% C 0.9% Mn 0.04% P 0.1% S steel: 0.45% C 0.9% Mn 0.04% P 0.05% S steel: Free cutting; cold drawn
SAE 1138	0.37% C 0.9% Mn 0.04% P 0.1% S steel: 0.46% C 0.9% Mn 0.04% P 0.1% S steel: Free cutting; hot rolled
SAE 1139	0.39% C 1.5% Mn 0.04% P 0.18% S free machining 0.46% C 0.9% Mn 0.04% P 0.1% S steel: Free cutting; cold drawn 0.4% C 0.9% Mn 0.04% P 0.1% S free cutting steel: Hot rolled **DPN: 156 UTS: 580 Elon: 16% Proof: 300**
SAE 1140	0.4% C 0.9% Mn 0.04% P 0.1% S free cutting steel: Cold drawn **DPN: 170 UTS: 640 Elon: 12% Proof: 530**
SAE 1141	0.41% C 1.5% Mn 0.04% P 0.1% S free cutting steel: Hot rolled **DPN: 187 UTS: 680 Elon: 15% Proof: 360**
SAE 1141	0.41% C 1.5% Mn 0.04% P 0.1% S free cutting steel: Cold drawn **DPN: 212 UTS: 770 Elon: 10% Proof: 630**
SAE 1144	0.44% C 1.5% Mn 0.4% P 0.3% S free cutting steel: Hot rolled **DPN: 197 UTS: 700 Elon: 15% Proof: 370**
SAE 1144	0.44% C 1.5% Mn 0.4% P 0.3% Free cutting steel: Cold drawn **0.36% C 1.3% Mn 0.04% P 0.05% S steel: Bar,** etc. 0.45% C 0.9% Mn 0.04% P 0.05% S free cutting steel: Hot rolled **DPN: 170 UTS: 630 Elon: 15% Proof: 320**
SAE 1145	0.45% C 0.9% Mn 0.04% P 0.05% S free cutting steel: Cold drawn **DPN: 187 UTS: 680 Elon: 12% Proof: 580**
SAE 1146	0.46% C 0.9% Mn 0.04% P 0.1% S free cutting steel: Hot rolled **DPN: 170 UTS: 630 Elon: 15% Proof: 320**
SAE 1146	0.46% C 0.9% Mn 0.04% P 0.1% S free cutting steel: Cold drawn **DPN: 187 UTS: 680 Elon: 12% Proof: 580**
SAE 1330	0.3% C 0.3% Si 1.8% Mn steel: Mechanical properties not specified
SAE 1335	0.35% C 0.3% Si 1.8% Mn steel: Mechanical properties not specified
SAE 1340	0.4% C 0.3% Si 1.8% Mn steel: Mechanical properties not specified
SAE 1345	0.45% C 0.3% Si 1.8% Mn steel: Mechanical properties not specified
SAE 1524	0.24% C 1.5% Mn 0.04% P 0.05% S steel: Bar, etc.
SAE 1525	0.25% C 1.0% Mn 0.04% P 0.05% S steel: Bar, etc. **0.26% C 1.2% Mn 0.04% P 0.05% S steel: Bar,** etc.
SAE 1527	0.27% C 1.3% Mn 0.04% P 0.05% S steel: Bar, etc. **0.34% C 1.3% Mn 0.04% P 0.05% S steel: Bar,** etc.
SAE 1541	0.41% C 1.5% Mn 0.04% P 0.05% S steel: Bar, etc. **0.47% C 1.5% Mn 0.04% P 0.05% S steel: Bar,** etc.
SAE NV1	hardening & tempering and for flame & induction hardening; as SAE 1041 **0.47% C 0.2% Si 1.5% Mn steel: For inlet valves;** as SAE 1047
SANBOLD 8	0.4% C 0.2% Si 0.8% Mn steel: Sanderson for **DPN: 200 UTS: 750 Elon: 14% Proof: 400**
SANBOLD 15	0.35% C 0.2% Si 1.5% Mn steel: Sanderson for BS alloy En 15
SANBOLD 52	0.35% C 0.2% Si 1.5% Mn steel: Sanderson for **replaced by Sanbold 15**
SANBOLD 3890	0.4% C 0.2% Si 0.8% Mn steel: Sanderson; replaced by Sanbold 8 **0.65% C 1.5% Mn 1.0% Si steel: Origin unknown**
SIMANCON	0.23% C 0.6% Si 1.6% Mn 0.045% S & P 0.04% **welding** **UTS: 600 Elon: 22% Proof: 390**
SIS 14-1606	0.23% C 0.6% Si 1.6% Mn 0.045% S & P 0.04% **casting; Swedish Standard** **DPN: 205 UTS: 650 Elon: 10 Proof: 300**
SIS 14-1650	0.42% C 0.7% Mn steel: For general engineering; **DPN: 200 UTS: 600 Elon: 20% Proof: 300** **DPN: 195 UTS: 660 Elon: 12 Proof: 290**
SIS 14-1655	0.52% C 0.7% Mn steel: For general engineering; Swedish Standard **DPN: 230 UTS: 760 Elon: 12 Proof: 350**
SIS 14-1672	0.5% C 0.7% Mn 0.035% P & S steel: For 0.25% C 0.02% Si 0.9% Mn steel: Structural purposes; Swedish Standard **DPN: 225 UTS: 750 Elon: 16 Proof: 350**
SIS 14-1674	0.5% C 0.75% Mn 0.035% P & S steel: For hardening & tempering and for flame & induction hardening; Swedish Standard **DPN: 200 UTS: 750 Elon: 14 Proof: 400**
SIS 14-2120	0.42% C 1.25% Mn 0.035% P & S steel: For hardening & tempering; Swedish Standard; hardened and tempered **DPN: 220 UTS: 425 Elon: 16 Proof: 750**
SIS 14-2128	0.43% C 1.5% Mn 0.05% P & S steel: For gas cylinders; Swedish Standard **DPN: 225 UTS: 750 Elon: 15 Proof: 420**

Symbol	Nominal analysis, supplier, condition and remarks.
SIS 14-1505	0.3% C 0.6% Mn 0.05% P 0.04% S steel: For castings; Swedish Standard **DPN: 160 UTS: 500 Elon: 18 Proof: 250**
SIS 14-1550	0.46% C 0.7% Mn steel: For general engineering; Swedish Standard **DPN: 160 UTS: 550 Elon: 15 Proof: 310**
SIS 14-1572	0.35% C 0.7% Mn 0.035% P & S steel: For hardening & tempering and for flame & induction hardening; Swedish Standard **DPN: 200 UTS: 600 Elon: 20 Proof: 300**
SIS 14-2165	0.25% C 1.6% Mn 0.05% S & P steel: Swedish Standard
SIS 14-2168	0.25% C 1.6% Mn 0.05% S & P steel: Swedish Standard; for reinforcing
SIS 1410	0.25% C 0.02% Si 0.9% Mn structural steel: Swedish Standard **UTS: 460 Elon: 20% Proof: 240**
SIS 1412	0.2% C 0.3% Cr (max) 0.4% Cu (max) steel: Structural purposes; Swedish Standard
SIS 1450	0.25% C 0.6% Mn 0.05% S & P steel: Bar, forging, etc.; Swedish Standard; normalized **DPN: 130 UTS: 450 Elon: 24% Proof: 220**
SIS 1450-0	0.25% C steel: Swedish Standard; as rolled; suffix indicates condition **DPN: 135 UTS: 420 Elon: 24% Proof: 220**
SIS 1550-06	0.35% C steel: Bar; Swedish Standard; work hardened **DPN: 165 UTS: 530 Elon: 8% Proof: 450**
SIS 1505	0.3% C 0.6% Mn 0.04% S & P steel: Swedish Standard; annealed **UTS: 460 Elon: 18% Proof: 240**
SIS 1510	0.3% C 1.2% Mn structural steel: Swedish Standard; as rolled **UTS: 580 Elon: 20% Proof: 330**
SIS 1550	0.35% C 0.6% Mn 0.05% S & P steel: Bar, forging, etc.; Swedish Standard; normalized **DPN: 160 UTS: 530 Elon: 22% Proof: 240**
SIS 1572	0.35% C steel: For flame & induction hardening; mechanical properties not specified; Swedish Standard
SIS 1572	0.35% C steel: Swedish Standard; hardened & tempered **UTS: 600 Elon: 19% Proof: 370**
SIS 1572-05	0.35% C steel: for flame & induction hardening; Swedish Standard; hardened & tempered; suffix indicates condition **DPN: 220 UTS: 720 Elon: 17% Proof: 400**
SIS 1650	0.45% C 20.6% Mn 0.04% S & P steel: Plate, bar & forging; Swedish Standard; normalized **DPN: 190 UTS: 630 Elon: 18% Proof: 300**
SIS 1650-06	0.45% C steel: Bar; Swedish Standard; work hardened; suffix indicates condition **DPN: 195 UTS: 620 Elon: 7% Proof: 530**
SIS 1660	0.44% C 0.4% Mn 0.03% S & P steel: Strip; Swedish Standard; annealed **DPN: 155 UTS: 650**
SIS 1660	0.44% C 0.4% Mn 0.03% S & P steel: Strip; Swedish Standard; cold rolled **DPN: 255 UTS: 760**

Symbol	Nominal analysis, supplier, condition and remarks.
SIS 1672	0.45% C 0.8% Mn 0.03% S & P steel: For flame hardening; Swedish Standard; annealed **DPN: 200**
SIS 1672	0.46% C steel: Swedish Standard; hardened & tempered **UTS: 700 Elon: 16% Proof: 420**
SIS 1674	0.52% C steel: Swedish Standard **UTS: 740 Elon: 13% Proof: 510**
SIS 1674	0.52% C steel: Swedish Standard; hardened & tempered **UTS: 740 Elon: 13% Proof: 510**
SIS 1757	Reinforcing steel: Cold rolled **UTS: 1000 Elon: 2.5%**
SIS 1940	0.26% C 1% Mn 0.05% P 0.15% S free machining steel: Swedish Standard; normalized **UTS: 480**
SIS 1957	0.38% C 1% Mn 0.05% P 0.15% S free machining steel: Swedish Standard; normalized **UTS: 530**
SIS 2120	0.41% C 1.2% Mn 0.03% P & S steel: Bar & forging; Swedish Standard; hardened & tempered **DPN: 290 UTS: 1030 Elon: 8% Proof: 680**
SIS 2120	0.41% C 1.2% Mn steel: Swedish Standard; hardened & tempered **UTS: 700 Elon: 12% Proof: 450**
SIS 2128	0.4% C 1.6% Mn 0.05% S & P steel: For gas cylinders; Swedish Standard; as forged **DPN: 220 UTS: 750 Elon: 15% Proof: 420**
SIS 2165	0.25% C 1.6% Mn 0.05% S & P steel: Swedish Standard; included in MNC 815
SIS 2168	0.25% C 1.6% Mn 0.05% S & P steel: Swedish Standard; reinforced steel; included in MNC 815
SIS 2168	0.28% C (max) 1.6% Mn (max) 0.05% S & P steel: For reinforcing **Elon: 12% Proof: 590**
SIS 8010	0.4% C sintered iron: Swedish Standard
SIS 8051	0.4% C 2.0% Cu sintered steel
SK 42	0.27% C (max) 1.2% Mn 0.04% P & S (max) steel: Phoenix
SPEAR 412	0.35% C 1.3% Mn 0.5% Ni steel: Spear & Jackson for BS alloy En 12
St 50	0.3% C 0.05% S 0.08% P steel: German Standard
St 50/2	0.3% C 0.05% S 0.06% P steel: German Standard
St 50/202	0.3% C 0.05% S 0.06% P steel: German Standard
St 50/203	0.3% C 0.05% S 0.06% P steel: German Standard
St 55	0.36% C 0.05% S & P steel: German Standard
St 55/4	0.35% C 0.2% Si 0.3% Mn 0.05% S & P steel: German Standard
St 60	0.4% C 0.05% S 0.08% P steel: German Standard
St 60/2	0.4% C 0.05% S 0.06% P steel: German Standard
St 60/202	0.4% C 0.05% S 0.06% P steel: German Standard
St 60/203	0.4% C 0.05% S 0.06% P steel: German Standard
STA 5 V4	0.3% C steel: Replaced by BS alloy En 5
STA 5 V4/1	0.3% C steel: Replaced by BS alloy En 5A
STA 5 V4/2	0.4% C steel: Replaced by BS alloy En 8A
STA 5 V4/3	0.3% C steel: Replaced by BS alloy En 5A
STA 5 V4/4	0.3% C steel: Replaced by BS alloy En 5C
STA 5 V4A	0.4% C steel: Replaced by BS alloy En 8K
STA 5 V4AM	0.35% C 0.25% Si 1.0% Mn 0.05% P 0.15% S En 8M
STA 5 V4A/1	0.4% C steel: Replaced by BS alloy En 8B
STA 5 V4A/2	0.4% C steel: Replaced by BS alloy En 8D
STA 5 V4A/3	0.4% C steel: Replaced by BS alloy En 8C
STA 5 V7	0.35% C steel: Replaced by BS alloy En 15
STA 28	Medium carbon steel: Strip; mainspring quality; hardened & tempered
STA 30	Medium carbon steel: For rifle barrels, etc.
STA 36 A	Medium carbon steel: Forging for shell bodies; hardened & tempered

Note. The following abbreviations and units are used in the tables:

DPN	Hardness, diamond pyramid number
UTS	Ultimate tensile strength, N/mm^2
Elon	Elongation, %
Proof	0.1 % proof strength, N/mm^2

1 N/mm^2=0.1 hbar=0.102 kgf/mm^2=0.06475 tonf/in^2=145.04 lbf/in^2

Symbol	Nominal analysis, supplier, condition and remarks.
STC 2	0.3% C 0.2% Si 1.0% Mn 0.06% S & P steel: Spencer; normalized
	DPN: 150 UTS: 650 Elon: 25%
STC 3	0.35% C 0.2% Si 1.0% Mn 0.05% S & P steel: Spencer; normalized
	DPN: 150 UTS: 530 Elon: 25%
STC 4	0.4% C 0.2% Si 1.0% Mn 0.05% S & P steel: Spencer; normalized
	DPN: 170 UTS: 620 Elon: 20%
STRENLITE 50	0.33% C 1.65% Mn steel: Steel Co. of Canada
STRENLITE 440	0.27% C 1.4% Mn 0.2% Cu steel: Steel Co. of Canada
SUPERTOUGH B 35	0.35% C 1.5% Mn steel: ESC for BS alloy En 15
T 735	0.35% C 0.25% Si 1.0% Mn 0.05% P 0.15% S steel: Fagersta; free machining
	UTS: 580 Elon: 20%
TOLEDO 30	0.3% C 1% Mn 0.06% S & P steel: Toledo for BS alloy En 5
TOLEDO 40	0.4% C 0.8% Mn 0.06% S & P steel: Toledo for BS alloy En 6A
TOWER 50	0.33% C 1.2% Mn 0.04% P & S (max) steel: Algoma
TOWER 55	0.33% C 1.65% Mn 0.04% P & S (max) steel: Algoma
UMS 55	0.35% C 0.25% Si 0.55% Mn 0.045% S & P steel: Krupp

Symbol	Nominal analysis, supplier, condition and remarks.
UMS 65	0.45% C 0.25% Si 0.65% Mn 0.045% S & P steel: Krupp
VIGILANT	0.55% C 0.7% Mn steel: Origin unknown
WHC	0.58% C 0.5% Mn steel: Origin unknown
WLM WARRANTED	Plain carbon steel 0.15–0.55% C: W Marrison
Y 35	0.35% C 0.25% Mn 0.035% P (max) 0.025% S (max) steel: French Standard designation
Y 45	0.45% C 0.25% Mn 0.035% P (max) 0.025% S (max) steel: French Standard designation
Y 55	0.55% C 0.25% Mn 0.035% P (max) 0.025% S (max) steel: French Standard designation
Y 55S7	0.55% C 0.8% Mn 0.4% Cr (max) steel: French Standard designation
Y 45S7	0.45% C 0.6% Mn steel: French Standard

Note. The following abbreviations and units are used in the tables:

DPN	Hardness, diamond pyramid number
UTS	Ultimate tensile strength, N/mm^2
Elon	Elongation, %
Proof	0.1 % proof strength, N/mm^2

$1\ N/mm^2 = 0.1\ hbar = 0.102\ kgf/mm^2 = 0.06475\ tonf/in^2 = 145.04\ lbf/in^2$

44A3 Steel – plain carbon 0.5–0.8 per cent

Specific gravity	7.86	
Density	7860 kg/m^3	(0.28 $lb/in.^3$)
Solidus/liquidus	1460–1490 °C	
Thermal conductivity	46.1 W/m °C	(11 cal/m s °C)
Coefficient of linear expansion	$11 \times 10^{-6}/$ °C	
Electrical conductivity	8–10% IACS (copper 100%)	
Specific resistance	180–250 microhm mm	
Young's modulus of elasticity	– N/m^2	
Impact	61–115 J	(45–85 ft/lb)(Izod)
Fatigue strength (no. of cycles)		
Hot strength		

Temperature °C	Tensile strength N/mm^2	Elongation %
200	800	21
300	840	17.5
400	660	27
500	460	35
600	300	41

The above properties are typical of the following group, and may not apply exactly to any one specification. It is possible that with certain specifications some of the values may not be applicable.

General metallurgical characteristics

This group is the medium high carbon range capable of 1500 N/mm^2 tensile, provided a low ductility can be accepted. They are available in all the normal forms of sheet, bar, tube, forgings and castings, but thin section sheet and tubing is not normally available as the materials cold work and require frequent anneals to prevent embrittlement. They are reasonably hardenable but not under still air conditions. Almost without exception the materials listed are used in the normalized, hardened and tempered, or cold drawn condition.

Thermal treatment

While these steels can be readily cold worked, as the carbon content rises the need for interstage annealing increases. This should be at 820–850 °C for 20 minutes per 25 mm of section, followed by slow cooling. In addition to the formation of scale there will be some de-carburization at the surface if this is carried out in air and the use of a controlled atmosphere is therefore recommended. For many purposes a sub-critical anneal at 650–700 °C will be sufficient. This causes some scale but no appreciable de-carburization. Normalizing requires 820–850 °C for 10 minutes at temperature, followed by cooling in still air, which removes all forging and welding stresses leaving a homogeneous structure very commonly used as the final condition for structural steel.

In order to harden the steels must be taken to 800–840 °C, held for 10 minutes per 25 mm of section, then water quenched. Oil quenching is recommended to reduce distortion but will not give full hardening.

All of these steels will be glass hard and brittle in the water quenched condition and should be tempered as soon after hardening as possible. Care must be exercised when handling parts before tempering and they should not have any rapid changes of section or very small radii.

Tempering will be at 200–500 °C for 2–4 hours with a tendency to brittleness in the 350–500 °C range. This can be prevented by quenching after tempering but it is advisable to use a molybdenum bearing steel if the mechanical properties given by this temper range are essential.

Normal electric, gas or oil fired furnace equipment can be used, but when high duty parts are hardened with no subsequent machining a controlled atmosphere will be necessary. Scale removal should be by blasting. Acid pickling after hardening may result in hydrogen embrittlement but can be used after normalizing or annealing. Sodium hydride treatment probably gives the best results but requires very special equipment.

These steels must not be carburized, as the core will be brittle when the case is hard, thus obviating the main purpose of the process. Many of the steels listed can however, be selectively hardened by the induction heating process. By the use of correctly designed coils a small area can be heated above 840 °C in seconds, then quenched with water. Parts must be carefully designed and the correct material chosen for this process to be successful.

Flame hardening can be used in a similar but slightly less selective manner and the same remarks apply. It is also very difficult to achieve reproduceable results using flame hardening.

Welding and brazing

Pre-treatment. The parts should be in the annealed or normalized condition and the weld area must be free from scale, grease and dirt.

Welding and brazing. Welding is not recommended with these materials as it will almost invariably result in a brittle zone adjacent to the weld. The effect of this can be reduced but not eliminated by subsequent normalizing or correct pre-heating.

In the hands of a skilled operator reasonable joins can be produced using gas or electric arc welding, but it is not recommended that these materials are used as the first design choice when welding is involved.

Brazing will soften parts which have been hardened and tempered but it is satisfactory for normalized components. Copper alloy braze fillers with suitable fluxes give the most economical results.

Post-treatment. All scale and flux must be removed. The parts must be stress relieved or tempered at 550–700 °C as a very minimum. If possible the assembly should be normalized. Electrically heated blankets are now available capable of stress relieving the largest of welded structures. The same equipment can be used for pre-heating.

With careful weld design to reduce stress raisers to a minimum, parts can be re-hardened and tempered, but it is again emphasized that these materials should not be welded by choice.

Brazed parts in the normalized condition need no subsequent treatment. If in doubt they sould be tempered at 650 °C minimum for two or four hours.

Flaw detection methods

Crack test. Magnetic crack testing is the most common. Penetrant oil chalk test with modern fluids and in skilled hands gives comparable results and is more portable. Selected areas can be searchingly examined by chalk testing.

X-ray. This will not normally be required on the wrought products, and welding should not be specified on stressed parts which might require X-ray.

Ultra-sonic test. Not normally required.

Chemical etch. Not normally required. Laboratory control etching to prove the quality of bars or forgings generally scraps the part.

Corrosion protection

Temporary. These steels must always be kept dry and oiled to prevent rust appearing. Good quality soluble cutting oils have considerable corrosion inhibition, but contaminated or incorrectly mixed coolants can give rise to severe and rapid rusting. Fine limits must be oiled at all times.

Permanent. There is no cheap, easy method of permanently protecting any of the materials listed. The standard of protection relies firstly on the preparation, then on the care with which the protection is applied and finally on the quality of the materials used.

The preparation for any method of protection is to remove all trace of existing corrosion, dirt and grease. It is not always realized that a minute trace of corrosion left at the bottom of a rust pit will propagate under almost any of the standard finishes, causing first a blister, then failure of the protecting medium.

Grit blasting correctly carried out is the most satisfactory method of ensuring a good surface. This fresh, clean material rusts very readily and should have the subsequent operation carried out as soon as possible. For best results the blasting operation must be carefully controlled and the surface brushed free of dust.

Painting. After the initial preparation the parts can be phosphate treated either by dipping or spraying. This gives the paint better key and has some corrosion preventative properties but requires oven drying before painting. A primer of the zinc or lead oxide or chromate type should always be used, whether or not the parts have been phosphate treated. Top coats of stoving paint give the best results. Modern plastic type paints are now available which are non-porous with excellent abrasive and shock resistance but these still require metallic type primers for best results.

Plating. Electroplating for purely decorative purposes generally uses chrome plating on nickel. Best results are obtained with a coating of 0.05 mm thick copper, followed by 0.025 mm thick nickel. The chromium layer is very thin, and acts only to prevent the nickel tarnishing.

Cadmium, tin and zinc are commonly plated, a thickness of less than 0.025 mm being sufficient to prevent corrosion under normal atmospheric conditions. A chemical seal immediately after plating is recommended as this prevents any micro-porosity propagating. Cadmium and zinc should not be used for parts in contact with food.

All forms of electroplating are subject to a phenomenan known as hydrogen embrittlement. This gives rise to brittle failures under certain conditions at less than the calculated tensile strength of the material. The effect is proportional to the tensile strength of the plated steel, being non-existent below 500–600 N/mm² 300 DPN, and very serious on steels above 1500 N/mm² – 450 DPN. High tensile steels which have been subjected to machining stresses should be stress relieved before plating and all steels above 300 DPN must be stress relieved after plating. The temperature of both stress relieving treatments should be 150–200 °C minimum for 2 hours.

Cathodic protection is now being practised on structures, marine installations and craft, pipe lines and pylons. With this the part is made negative in a low voltage electric circuit, the positive or anode of which will then corrode at the expense of the cathode, and is then replaced. The system should be additive to paint where corrosive conditions are severe or where full protection cannot be guaranteed by conventional means.

Metal spraying of zinc or aluminium on a correctly pretreated surface can give excellent results, particularly if the coating is wire brushed and chemically sealed or painted.

Chemical oxidation, commonly called 'black oxide' is often applied to certain of these steels. This has a dark blue or brown colour and is a very fine adherent oxide which prevents further oxidation. There are various systems all based on hot strong sodium hydroxide with powerful oxidizing agents added. The resultant surface must be kept oiled and will not withstand long exposure to atmospheric conditions. It has long been used as the standard finish for guns.

Machinability

In the fully hardened condition, these materials can be more than 500 DPN and can only be machined with difficulty and a very short tool life.

Normal practice is to rough and semi-finish machine in the normalized condition, harden and temper if required and complete the machining of fine limits by grinding. As these materials require water quenching for their full mechanical properties some distortion should be expected and an allowance must be made.

Cold drawn wire and bar can only be ground as it has similar properties to the hardened and tempered condition.

Uses

Springs, collets, fixtures, mandrels, shear blades of all kind. General engineering purposes where medium high hardness is required and little or no shock resistance.

Symbol	Nominal analysis, supplier, condition and remarks.
0.0722	0.48% C 0.2% Si 0.7% Mn 0.045% S & P steel: German Standard
0.0740	0.53% C 0.35% Si 0.5% Mn 0.045% S & P steel: German Standard
0.0761	0.68% C 0.35% Si 0.7% Mn 0.045% S & P steel: German Standard
0.0763	0.74% C 0.2% Si 0.45% Mn 0.045% S & P steel: German Standard
0.0073	0.75% C 0.35% Si 0.7% Mn 0.045% S & P steel: German Standard

Note. The following abbreviations and units are used in the tables:

DPN	Hardness, diamond pyramid number
UTS	Ultimate tensile strength, N/mm²
Elon	Elongation, %
Proof	0.1 % proof strength, N/mm²

1 N/mm²=0.1 hbar=0.102 kgf/mm²=0.06475 tonf/in²=145.04 lbf/in²

Symbol	Nominal analysis, supplier, condition and remarks.
0.0931	0.6% C 1.2% Si 1.0% Mn steel: For springs; German Standard
0.0969	0.51% C 1.7% Si 0.3% Mn steel: For springs; German Standard
0.0970	0.56% C 1.7% Si 0.8% Mn steel: For springs; German Standard
0.0971	0.64% C 1.7% Si 0.8% Mn steel: For springs; German Standard
0.1210	0.53% C 0.35% Si 0.5% Mn 0.035% S & P steel: German Standard
0.1221	0.61% C 0.35% Si 0.7% Mn 0.035% S & P steel: German Standard
0.1231	0.68% C 0.35% Si 0.7% Mn 0.035% S & P steel: German Standard
0.1248	0.75% C 0.2% Si 0.5% Mn 0.035% S & P steel: German Standard
0.5028	0.65% C 1.7% Si 0.8% Mn steel: For springs; German Standard
0.5029	0.71% C 1.7% Si 0.7% Mn steel: For springs; German Standard

Symbol	Nominal analysis, supplier, condition and remarks.
0.70 Li Pja	0.7% C 0.25% Si 0.5% Mn steel: Fagersta
1.0505	0.53% C 0.4% Si 0.55% Mn 0.05% S & P steel: German Standard
1.0601	0.61% C 0.35% Si 0.7% Mn 0.045% S & P steel: German Standard
1.0603	0.68% C 0.4% Si 0.7% Mn 0.045% S & P steel: German Standard
1.0605	0.75% C 0.4% Si 0.7% Mn 0.045% S & P steel: German Standard
1.0632	0.5% C 0.05% S & P steel: German Standard
1.0728	0.62% C 0.3% Si 0.7% Mn 0.2% S 0.07% P steel: German Standard
1.0751	0.61% C 0.25% Si 0.7% Mn 0.045% S & P steel: German Standard
1.1210	0.53% C 0.35% Si 0.5% Mn 0.035% S & P steel: German Standard
1.1213	0.53% C 0.25% Si 0.6% Mn 0.035% S & P steel: German Standard
1.1221	0.61% C 0.35% Si 0.7% Mn 0.035% S & P steel: German Standard
1.1231	0.68% C 0.35% Si 0.7% Mn 0.035% S & P steel: German Standard
1.1248	0.75% C 0.2% Si 0.5% Mn 0.035% S & P steel: German Standard
1.1249	0.72% C 0.2% Si 0.3% Mn 0.03% S & P steel: German Standard
1.1740	0.6% C 0.7% Mn 0.035% S & P (max) steel: German Standard designation
46 Mn Si 4	0.48% C 0.8% Si 1.1% Mn: Designation used by German Standard
50 Mn Si 4	0.5% C 0.8% Si 1.1% Mn steel: Designation used by German Standard
53 Mn Si 4	0.54% C 0.9% Si 1.0% Mn steel: Designation used by German Standard
1046	0.46% C 0.9% Mn 0.04% P 0.05% S steel: Designation used by AISI
1047	0.47% C 1.5% Mn 0.04% S & P steel: Designation used by AISI
1104	0.75% C 0.02% P & S (max) steel: French Standard designation
1164	0.75% C 0.2% Mn 0.02% P & S (max) steel: French Standard designation
1165	0.65% C 0.2% Mn 0.02% P & S (max) steel: French Standard designation
1204	0.75% C 0.25% Mn 0.025% P & S (max) steel: French Standard designation
1304	0.75% C 0.25% Mn 0.035% P (max) 0.025% S (max) steel: French Standard designation
1305	0.65% C 0.25% Mn 0.035% P (max) 0.025% S (max) steel: French Standard designation
A 6	0.62% C (max) 0.9% Mn steel: Cogne
A 32 STEEL	0.55% C 0.8% Mn steel: ESC for BS alloy En 9
ABRADUR 2	0.7% C 0.35% Si 1.0% Mn steel: Pompey
ACIBRADE	0.68% C 0.05% Si 1.5% Mn 0.06% S & P steel: Workington Iron & Steel Co; hot rolled
AISI 1051	0.51% C 1.0% Mn 0.05% S & P steel
AISI 1052	0.52% C 1.3% Mn 0.05% S & P steel
AISI 1053	0.53% C 0.8% Mn 0.05% S & P steel
AISI C1048	0.48% C 1.3% Mn 0.04% S & P steel

Symbol	Nominal analysis, supplier, condition and remarks.
AISI C1049	0.49% C 0.8% Mn 0.04% P 0.05% S steel
AISI C1050	0.51% C 0.8% Mn 0.04% P 0.05% S steel
AISI C1052	0.51% C 1.2% Mn 0.04% P 0.05% S steel
AISI C1055	0.55% C 0.8% Mn 0.04% P 0.05% S steel
AISI C1060	0.6% C 0.8% Mn 0.04% P 0.05% S steel
AISI C1064	0.63% C 0.6% Mn 0.04% P 0.05% S steel
AISI C1065	0.65% C 0.8% Mn 0.04% P 0.05% S steel
AISI C1070	0.7% C 0.8% Mn 0.04% P 0.05% S steel
AISI C1074	0.75% C 0.6% Mn 0.04% P 0.05% S steel
AISI C1078	0.78% C 0.5% Mn 0.04% P 0.05% S steel
AISI C1080	0.82% C 0.8% Mn 0.04% P 0.05% S steel
AISI C1151	0.51% C 0.9% Mn 0.04% P 0.1% S steel: Free cutting
AQUATOUGH 70	0.7% C 0.3% Mn steel: Clyde Alloy **DPN: 750**
AMS 5085	0.50% C steel: Sheet & strip; AMS for SAE alloy 1050
AMS 5110 A	0.8% C steel: Music wire; commercial quality; AMS for SAE alloy 1080
AMS 5115 B	0.7% C steel: Spring wire; AMS for SAE alloy 1070
AMS 5120 D	0.75% C steel: Spring & strip; annealed; AMS for SAE alloy 1074
A STEEL	0.5% C 0.8% Mn steel: ECS for BS alloy En 43A
ASTM A1	0.68% C 0.8% Mn 0.15% Si steel: Rails
ASTM A21	0.48% C 0.75% Mn steel: Axles for railway use
ASTM A25 Class A	0.57% C (max) 0.67% Mn steel: For steel wheels for electrical service
ASTM A25 Class B	0.62% C 0.67% Mn steel: For steel wheels for electrical service
ASTM A25 Class C	0.72% C 0.67% Mn steel: For steel wheels for electrical service
ASTM A25 Class V	0.72% C 0.67% Mn steel: for steel wheels for electrical service
ASTM A227	0.55% C 0.2% Si 0.9% Mn 0.045% S & P steel: Spring wire; hard drawn; actual tensile varies with diameter **UTS: 1550**
ASTM A228	0.85% C 0.2% Si 0.4% Mn 0.03% S & P steel: Spring wire; cold drawn; piano wire **UTS: 3400**
ASTM A229 A	0.6% C 0.2% Si 1.0% Mn 0.045% S & P steel: Spring wire; hardened & tempered **UTS: 1840**
ASTM A229 B	0.6% C 0.2% Si 0.8% Mn 0.045% S & P steel: Spring wire; hardened & tempered **UTS: 1840**
ASTM A230	0.65% C 0.2% Si 0.7% Mn 0.03% S & P steel: Spring wire; hardened & tempered for valve springs **UTS: 1550**
ASTM A241	0.58% C steel: For tie plates
ASTM A290 C2	0.55% C 0.8% Mn 0.05% S & P steel: Forging; hardened & tempered **DPN: 170 UTS: 610 Elon: 20% Proof: 300**
ASTM A306/70	Plain carbon steel: As rolled **UTS: 540 Elon: 18% Proof: 220**
ASTM A306/75	Plain carbon steel: As rolled **UTS: 550 Elon: 18% Proof: 240**
ASTM A306/80	Plain carbon steel: As rolled **UTS: 550 Elon: 17% Proof: 270**
ASTM A321	0.55% C 0.8% Mn steel: Hardened & tempered **UTS: 630 Elon: 18% Proof: 450**
ASTM A407	0.57% C 0.9% Mn steel: Wire for coil springs; upholstery springs
ASTM A417	0.62% C 0.9% Mn 0.04% S & P steel: Wire for upholstery springs
ASTM A421	0.85% C 0.8% Mn 0.04% S & P 0.2% Si steel: Wire for pre-stressed concrete
ASTM A504	0.6% C 0.7% Mn 0.05% S & P steel: For wheels

Note. The following abbreviations and units are used in the tables:

DPN	Hardness, diamond pyramid number
UTS	Ultimate tensile strength, N/mm^2
Elon	Elongation, %
Proof	0.1 % proof strength, N/mm^2

1 N/mm^2=0.1 hbar=0.102 kgf/mm^2=0.06475 tonf/in^2=145.04 lbf/in^2

Symbol	Nominal analysis, supplier, condition and remarks.
ASTM A551	0.7% C 0.7% Mn 0.2% Si 0.05% S & P steel: Tyres graded by C content
ASTM A583	0.7% C steel: Castings for railway wheels; classified by use
ASTM A631	0.7% C 0.7% Mn: Cast steel wheels for railway service; classified by use
ASTM A648/1	0.55% C 0.8% Mn steel: Wire; hard drawn for reinforcing **UTS: 1300**
ASTM A648/11	0.65% C 0.8% Mn steel: Wire; hard drawn for reinforcing **UTS: 1500**
ASTM A648/111	0.7% C 0.8% Mn steel: Wire; hard drawn for reinforcing
ASTM A679	0.7% C 0.8% Mn steel: Wire; hard drawn **UTS: 2000**
ASTM A680	Medium high carbon steel: Strip; cold rolled for springs
ASTM A682	Cold rolled carbon steel: Strip; coded by AISI type; see also 44A1 44A2 and 44A4
ASTM A682	Medium high carbon steel: Strip; cold rolled graded by AISI system
ASTM A684	Medium high carbon steel: Strip for springs
ASTM A713	High carbon steel: For springs; graded by AISI designation
ASTM A713	Medium high carbon steel: Wire for springs; graded to AISI system
ASTM A729	0.6% C (max) 1.5% Mn steel: For railway axles; hardened & tempered **UTS: 690 Elon: 20% Proof: 450**
ASTM A730 C, D & E	0.47% C 0.75% Mn steel: Forgings for railway use
ASTM A730 F	0.52% C 0.75% Mn steel: Forgings for railway use
ASTM A730 G & H	C not specified 0.75% Mn steel: Forgings for railway use
ASTM A732/4 A	0.5% C 0.8% Mn steel: Investment casting; annealed **UTS: 620 Elon: 20% Proof: 345**
ASTM A732/4 Q	0.5% C 0.8% Mn steel: Investment casting; hardened & tempered **UTS: 860 Elon: 5% Proof: 680**
ASTM B606	0.65% C 1.0% Mn steel: Wire; galvanized; for core wire on A1 electrical conductors
ATERCLIFFE	Plain carbon tool steels 0.6%–1.5% C: Sanderson; carbon content specified by a temper no.
AURIGA V	0.55% C 0.4% Si 0.7% Mn 0.05% S & P steel: Casting; D Brown **UTS: 670 Elon: 10%**
AW 5	0.55% C 0.6% Mn steel: ESC; spanners, pliers, etc. **DPN: 690**
AW 8	0.8% C 0.8% Mn steel: ESC; picks, hammers, etc. **DPN: 690**
B 1	0.55% C 0.25% Si 0.6% Mn steel: Jessop for BS alloy En 9
B 6	0.6% C (max) steel: Breda
B 10	0.5% C steel: Normalized; Bofors **DPN: 190 UTS: 610 Elon: 18% Proof: 300**
B 12	0.6% C steel: Normalized; Bofors; obsolete **DPN: 220 UTS: 720 Elon: 15% Proof: 330**
B 14	0.7% C steel: Normalized; Bofors **DPN: 230 UTS: 770 Elon: 13% Proof: 370**
BS 9	0.55% C 1.0% Mn 0.06% P & S steel: For bull head rails
BS 11	0.55% C 1.0% Mn 0.06% P & S steel: For flat bottom rails
BS 24/3 A1	0.55% C 1.7% Si 0.8% Mn steel: For wagon springs; as BS alloy En 45
BS 24/3 A2	0.55% C 1.7% Si 0.8% Mn steel: For wagon springs; as BS alloy En 45

Symbol	Nominal analysis, supplier, condition and remarks.
BS 24/3 A5	0.77% C 0.3% Si 0.6% Mn steel: For wagon springs; as BS alloy En 42
BS 24/3 A6	0.77% C 0.3% Si 0.6% Mn steel: For wagon springs; as BS alloy En 42
BS 24/3 A7	0.5% C 0.3% Si 0.7% Mn steel: For wagon springs; as BS alloy En 43
BS 24/3 A8	0.55% C 0.3% Si 0.7% Mn steel: For wagon springs; as BS alloy En 43
BS 24/3 BB	0.6% C 1.8% Si 0.8% Mn 0.05% S & P steel: For springs; as BS alloy En 45
BS 24/3 BC	0.6% C 1.9% Si 0.6% Mn 0.05% S & P steel: For springs; as BS alloy En 45A
BS 224/1	0.6% C 0.7% Mn 0.05% P 0.05% S steel: For die blocks; hardened & tempered **DPN: 250**
BS 970 En 9	0.55% C 0.7% Mn 0.06% S 0.06% P steel: Bar & forging; hardened & tempered **DPN: 250 UTS: 760 Elon: 18% Proof: 450**
BS 970 En 9K	0.55% C 0.3% Si 0.7% Mn 0.05% S & P steel: Bar & forging
BS 970 En 42	0.8% C 0.6% Mn 0.05% S 0.05% P steel: For springs; oil hardened & tempered for springs
BS 970 En 42A	0.85% C 0.6% Mn 0.05% S 0.05% P steel: For springs; oil hardened & tempered for springs
BS 970 En 42B	0.65% C 0.3% Si 0.7% Mn 0.05% S & P steel: Bar for springs
BS 970 En 43	0.5% C 0.7% Mn 0.05% S 0.05% P steel: For springs; water hardened & tempered
BS 970 En 43A	0.5% C 0.9% Mn 0.06% S 0.06% P steel: For bar & forgings; hardened & tempered; can be cold drawn **DPN: 50 UTS: 760 Elon: 18% Proof: 450**
BS 970 En 43B	0.48% C 0.9% Mn 0.06% S 0.06% P steel: Bar & forgings; mechanical properties not part of specification
BS 970 En 43C	0.52% C 0.9% Mn 0.06% S 0.06% P steel: Bar & forgings; mechanical properties not part of specification
BS 970 En 43D	0.62% C 0.5% Mn 0.06% S 0.06% P steel: Bar & forgings; mechanical properties not part of specification
BS 970 En 43E	0.67% C 0.8% Mn 0.06% S 0.06% P steel: Bar & forgings; mechanical properties not part of specification
BS 970 En 43F	0.5% C 0.6% Mn 0.05% S 0.05% P steel: For springs; water hardened & tempered
BS 970 En 45	0.6% C 0.9% Mn 1.75% Si steel: For springs; oil hardened & tempered
BS 970 En 45A	0.6% C 0.8% Mn 1.8% Si steel: For springs; oil hardened & tempered; C held to tighter limits than En 45
BS 970 En 49	0.6% C 1% Mn 0.05% S 0.05% P steel: For spring wire; hard drawn
BS 970 En 49A	0.6% C 1% Mn 0.05% S 0.05% P steel: For spring wire; hard drawn
BS 970 En 49B	0.65% C 1% Mn 0.05% S 0.05% P steel: For springs; hard drawn
BS 970 En 49C	0.7% C 0.75% Mn 0.04% S 0.04% P steel: For springs; hard drawn
BS 970 En 49D	0.75% C 0.6% Mn 0.04% S 0.04% P steel: For springs; hard drawn
BS 980 CDS 8	0.5% C 0.6% Mn steel: Cold drawn tube; as drawn & tempered **UTS: 690 Proof: 580**
BS 1408	0.7% C 1% Mn 0.3% Si steel: Cold drawn; composition complies with BS 970 En 49A BC & D **UTS: 1550**

Symbol	Nominal analysis, supplier, condition and remarks.
BS 1429	0.6% - 1.0% C steel: Wire for springs; annealed; covered by BS alloys En 42 44 En 45 En 47 & En 50
BS 1429 En 42C	0.75% C 0.7% Mn 0.05% S & P steel: For springs
BS 1449 HS50	0.5% C 0.2% Si 0.8% Mn 0.05% S & P steel: Sheet
BS 1449 HS60	0.6% C 0.2% Si 0.7% Mn 0.05% S & P steel: Sheet
BS 1449 HS70	0.7% C 0.2% Si 0.7% Mn 0.05% S & P steel: Sheet
BS 1449 HS80	0.8% C 0.2% Si 0.7% Mn 0.05% S & P steel: Sheet
BS 1506/162	0.6% C (max) 0.2% Si 0.8% Mn 0.05% S & P steel: Bar; normalized **UTS: 760 Elon: 18% Proof: 330**
BS 1760 A	0.45% C 0.6% Si 1.0% Mn 0.06% S & P steel: Casting for surface hardening; contained in BS 3100 **UTS: 610 Elon: 12% Proof: 300**
BS 1760 B	0.6% C 0.6% Si 1.0% Mn 0.06% S & P steel: Casting for surface hardening; contained in BS 3100 **UTS: 670 Elon: 10% Proof: 330**
BS 2453	0.4% C 0.25% Si 0.8% Mn 0.05% S & P steel: For spokes; may be zinc coated **UTS: 1070**
BS 2691	0.8% C 0.2% Si 0.7% Mn 0.05% S & P steel: Wire for concrete; cold drawn & stress released **UTS: 1680 Proof: 1380**
BS 2803	0.6% C 0.8% Mn 0.3% Si steel: Wire; ground; different grades define slight difference **UTS: 1540**
BS 3111/1	0.4% C 0.2% Si 0.8% Mn 0.05% S & P steel: Wire; annealed & drawn; for cold forged bolts **UTS: 610**
BS 3141 C6A1	0.5% C 0.6% Mn 0.05% S & P steel: Classification system used by BS; discontinued
BS 3141 C6B1	0.5% C 0.8% Mn 0.06% S & P steel: Classification system used by BS; discontinued
BS 3141 C6C1	0.5% C 1.1% Mn 0.06% S & P steel: Classification system used by BS; discontinued
BS 3141 C6G1	0.52% C 0.9% Mn 0.06% S & P steel: Classification system used by BS; discontinued
BS 3141 C6E1	0.5% C 0.4% Mn 0.05% S & P steel: Classification system used by BS; discontinued
BS 3141 C6F1	0.5% C 0.7% Mn 0.05% S & P steel: Classification system used by BS; discontinued
BS 3141 C6G1	0.55% C 0.8% Mn 0.06% S & P steel: Classification system used by BS; discontinued
BS 3141 C6H1	0.55% C 1.1% Mn 0.06% S & P steel: Classification system used by BS; discontinued
BS 3141 C7A1	0.6% C 0.8% Mn 0.05% S & P steel: Classification system used by BS; discontinued
BS 3141 C7B1	0.62% C 0.5% Mn 0.05% S & P steel: Classification system used by BS; discontinued
BS 3141 C7C1	0.65% C 0.8% Mn 0.05% S & P steel: Wire; classification system used by BS; discontinued
BS S25	High carbon steel: Bar; replaced

Symbol	Nominal analysis, supplier, condition and remarks.
BS S70	0.55% 0.2% Si 0.7% Mn 0.04% S & P steel: Normalized **DPN: 230 UTS: 760 Elon: 15%**
BS S79	0.55% 0.2% Si 0.7% Mn 0.04% S & P steel: Hardened & tempered **DPN: 280 UTS: 920 Elon: 13%**
BS S513	0.8% C 0.3% Si 0.7% Mn 0.05% S & P steel: For springs; hardened & tempered **DPN: 450**
BS S517	0.47% C 0.2% Si 1.5% Mn 0.05% S & P steel: Sheet; hardened & tempered **UTS: 1210 Elon: 5% Proof: 1000**
BS W6	C not specified 0.04% S & P steel: Galvanized wire for balloon cables **UTS: 2310**
BULL HEAD RAIL	0.55% C 1% Mn 0.06% S & P steel: See BS 9
BULLS HEAD	0.6% C (min) steel: Marsh Brothers; applies to a range of plain C steels
C 6	0.6% C (max) 0.8% Mn steel: Siau
C 53	0.53% C 0.35% Si 0.5% Mn 0.045% S & P steel: German Standard
C 60	0.6% C 0.75% Mn steel: Italian Standard designation; normalized **DPN: 260 UTS: 820 Elon: 12% Proof: 450**
C 60	0.61% C 0.7% Mn steel: Italian Standard designation **DPN: 230 UTS: 1450 Elon: 6% Proof: 1050**
C 70	0.69% C 0.7% Mn steel: Italian Standard designation **DPN: 250 UTS: 1450 Elon: 5% Proof: 1050**
C 75	0.75% C 0.55% Mn steel: Italian Standard designation **DPN: 265 UTS: 1450 Elon: 5% Proof: 1050**
C 90	0.9% C 0.55% Mn steel: Italian Standard designation **DPN: 270 UTS: 1500 Elon: 4% Proof: 1100**
C 60 W3	0.6% C 0.7% Mn 0.035% P & S (max) steel: German Standard designation
C 175	0.75% C 0.25% Si 0.3% Mn 0.02% S & P steel: Drills, etc.; SHD
C 255	0.55% C 0.25% Si 0.7% Mn 0.04% S & P steel: Drills, etc.; SHD
CD 60	0.5% C 0.4% Si 0.8% Mn steel: Fagersta **UTS: 760 Elon: 18%**
Cf 70	0.71% C 0.2% Si 0.3% Mn 0.03% S & P steel: German Standard
CK 53	0.53% C 0.35% Si 0.5% Mn 0.035% S & P steel: German Standard
CK 60	0.61% C 0.7% Mn 0.04% S & P steel: Designation used by German Standard
CK 67	0.67% C 0.35% Si 0.7% Mn 0.035% S & P steel: German Standard
COLORADO	Plain carbon tool steels 0.6% 1.5% C: Sanderson; carbon content specified by temper no.
COMMON HARDENING	0.55% C 0.7% Mn steel: 0rigin unknown
CRESCENT 6	0.65% C steel: Spencer
D 5	0.55% C 0.3% Si 0.7% Mn steel: Pompey
D 6	0.6% C 0.3% Si 0.8% Mn steel: Pompey
D 7	0.7% C 0.3% Si 0.7% Mn steel: Pompey
D 8	0.8% C 0.3% Si 0.6% Mn steel: Pompey
DIN 1651 60 S 20	0.6% C 0.7% Mn 0.2% S 0.05% P steel: Free machining
DIN 17100 St 70/2	0.5% C 0.05% P & S (max) 0.007% N (max) steel: Plate; German Standard **UTS: 770 Proof: 360**
DIN 17200 C 55	0.55% C 0.75% Mn 0.045% P & S steel: For structures; German Standard
DIN 17200 C 60	0.6% C 0.75% Mn 0.045% P & S steel: For structures; German Standard

Note. The following abbreviations and units are used in the tables:

DPN	Hardness, diamond pyramid number
UTS	Ultimate tensile strength, N/mm^2
Elon	Elongation, %
Proof	0.1 % proof strength, N/mm^2

1 N/mm^2=0.1 hbar=0.102 kgf/mm^2=0.06475 tonf/in^2=145.04 lbf/in^2

Symbol	Nominal analysis, supplier, condition and remarks.
DIN 17200 C60	0.61% C 0.7% Mn 0.045% S & P steel
DIN 17200 CK 55	0.55% C 0.75% Mn 0.035% S & P (max) steel: For structures; German Standard
DIN 17200 CK 60	0.6% C 0.75% Mn 0.035% P & S (max) steel: For structures; German Standard
DIN 17200 CK60	0.61% C 0.7% Mn 0.4% S & P steel
DIN 17200 CM 55	0.55% C 0.75% Mn 0.035% P & S (max) steel: For structures; German Standard
DIN 17200 CM 60	0.6% C 0.75% Mn 0.035% P & S (max) steel: For structures; German Standard
DIN 17221 51 S 7	0.51% C 1.7% Si 0.6% Mn steel: For springs
DIN 17221 55 S 7	0.55% C 1.7% Si 0.9% Mn steel: For springs
DIN 17221 60 Si Mn 5	0.6% C 1.2% Si 1.0% Mn steel: For springs
DIN 17221 65 Si 7	0.65% C 1.7% Si 0.8% Mn steel: For springs
DIN 17221 66 Si 7	0.66% C 1.7% Si 0.8% Mn steel: For springs
DIN 17222 60 Si Mn 5	0.6% C 1.2% Si 1.0% Mn steel: For springs
DIN 17222 65 Si 7	0.65% C 1.7% Si 0.8% Mn steel: For springs
DIN 17222 66 Si 7	0.66% C 1.7% SI 0.8% Mn steel: For springs
DIN 17222 71 Si 7	0.71% C 1.7% Si 0.7% Mn steel: For springs
DIN 17222 C67	0.68% C 0.7% Mn 0.035% S & P steel: For springs
DIN 17222 C75	0.75% C 0.7% Mn 0.045% S & P steel: For springs
DIN 17222 M75	0.75% C 0.4% Mn 0.045% S & P steel: For springs
DIN 17222 MK75	0.75% C 0.5% Mn 0.035% S & P steel: For springs
DTD 5 A	0.75% C 1% Mn 0.04% S 0.04% P steel: For springs; hard drawn & tempered **UTS: 1680**
DTD 215 A	0.7% C 0.5% Mn 0.03% S 0.03% P steel: Wire; cold drawn & tempered; not for valve springs **UTS: 1550**
DTD 239 A	0.8% C 1% Mn 0.045% S 0.04% P steel: For springs; hardened & tempered; not for valve springs **UTS: 1380**
DUNELT 53	0.55% C 0.7% Mn 0.06% S & P steel: Dunford for BS alloy En 9
E 7	0.7% C 0.25% Si 0.3% Mn steel: Pompey
E 8 FX	0.8% C 0.12% Si 0.3% Mn steel: Pompey
EXTRA	C as specified 0.25% Mn steel: For tools; Braeburn
FLAT BOTTOM RAILS	0.55% C 1% Mn 0.06% P & S steel: See BS 11
FABIS	Plain carbon tool steels 0.6%–1.5% C: Sanderson; carbon content specified by temper no.
FAGESTA 1680	0.6% C 0.3% Mn steel: Fagesta
FAGESTA 1773	0.75% C 0.3% Mn steel: Fagesta
FAGESTA 1778	0.7% C 0.6% Mn steel: For springs; Fagesta
FAGESTA 1779	0.85% C 0.5% Mn steel: For springs; Fagesta
G 10500	UNS designation for type 1050 steel
G 10550	UNS designation for type 1055 steel
G 10600	UNS designation for type 1060 steel
G 10640	UNS designation for type 1064 steel
G 10650	UNS designation for type 1065 steel
G 10700	UNS designation for type 1070 steel
G 10740	UNS designation for type 1074 steel
G 10800	UNS designation for type 1080 steel
G 10850	UNS designation for type 1085 steel
G 10860	UNS designation for type 1086 steel
GOST 1050/50	0.5% C 0.75% Mn 0.03% P & S steel: For structures; Russian Standard
GOST 1050/50 G	0.5% C 0.85% Mn 0.04% P & S (max) 0.25% Cr & Ni (max) steel: For structures; Russian Standard

Symbol	Nominal analysis, supplier, condition and remarks.
GOST 1050/55	0.55% C 0.65% Mn 0.04% P & S (max) 0.25% Cr & Ni (max) steel: For structures; Russian Standard
GOST 1050/60	0.6% C 0.75% Mn 0.04% P & S (max) 0.25% Cr & Ni (max) steel: For structures; Russian Standard
GOST 1050/60 G	0.6% C 0.85% Mn 0.04% P & S (max) 0.25% Cr & Ni (max) steel: For structures; Russian Standard
GOST 1050/65	0.65% C 0.75% Mn 0.04% P & S (max) 0.25% Cr & Ni (max) steel: For structures; Russian Standard
GOST 1050/65 G	0.65% C 1.05% Mn 0.04% P & S (max) 0.25% Cr & Ni (max) steel: For structures; Russian Standard
GOST 1050/70	0.7% C 0.65% Mn 0.04% P & S (max) 0.25% Cr & Ni (max) steel: For structures; Russian Standard
GOST 1050/70 G	0.7% C 1.05% Mn 0.04% P & S (max) 0.25% Cr & Ni (max) steel: For structures; Russian Standard
GOST 1050/75	0.75% C 0.65% Mn 0.04% P & S (max) 0.25% Cr & Ni (max) steel: For structures; Russian Standard
GOST 1050/80	0.8% C 0.75% Mn 0.04% P & S steel: For structures; Russian Standard
GOST 2052/55 S2	0.55% C 0.75% Mn 0.04% P & S (max) 0.3% Cr & Ni (max) steel: Bar; Russian Standard
GOST 2052/60 S2A	0.6% C 0.85% Mn 0.03% P & S (max) 0.3% Cr (max) 0.4% Ni (max) 0.25% Cu (max) steel: For springs; Russian Standard
GOST 2052/60 SG	0.6% C 0.9% Mn 0.04% P & S (max) 0.4% Cr & Ni (max) steel: Bar; Russian Standard
GOST 2052/60 SG	0.6% C 0.9% Mn 0.04% P & S (max) 0.3% Cr 0.4% Ni 0.25% Cu steel: For springs; Russian Standard
GOST 2052/60 SGA	0.6% C 0.9% Mn 0.03% P & S (max) 0.3% Cr & Ni (max) steel: Bar; Russian Standard
GOST 2052/60 SGA	0.6% C 0.9% Mn 0.03% P & S (max) 0.3% Cr (max) 0.4% Ni (max) 0.25% Cu (max) steel: For springs; Russian Standard
GOST 2052/62 S2A	0.62% C 0.85% Mn 0.03% P & S (max) 0.3% Cr (max) 0.4% Ni (max) 0.25% Cu (max) steel: For springs; Russian Standard
GOST 2052/65	0.65% C 0.04% P & S (max) 0.25% Cr & Ni (max) steel: Bar; Russian Standard
GOST 2052/65 G	0.65% C 1.05% Mn 0.04% P & S (max) 0.4% Cr & Ni (max) steel: Bar; Russian Standard
GOST 2052/70	0.7% C 0.65% Mn 0.04% P & S 0.25% Cr & Ni (max) steel: Bar; Russian Standard
GOST 2052/70 S3A	0.7% C 0.75% Mn 0.03% P & S (max) 0.3% Cr & Ni (max) steel: Bar; Russian Standard
GOST 2052/75	0.75% C 0.65% Mn 0.045% P & S (max) 0.3% Cr & Ni (max) 0.25% Cu (max) steel: For springs; Russian Standard
GOST 4543/50 G2	0.5% C 1.6% Mn 0.25% Cr (max) steel: Russian Standard
HARDFLEX 13M	0.75% C steel: Strip; Sandvik; supplied hardened & tempered **UTS: 1300 Proof: 1100**
HDB 1	0.6% C 0.3% Si 0.7% Mn steel: Huntsman; large press dies
HECLA 18	0.65% C 0.4% Mn steel: Hadfields for BS alloy En 42E
HECLA 36	0.85% C 0.7% Mn steel: Hadfields for BS alloy En 42D
HECLA 41	0.55% C 0.7% Mn steel: Hadfields for BS alloy En 9
HECLA 42	0.8% C 0.65% Mn steel: Hadfields for BS alloy En 42

Symbol	Nominal analysis, supplier, condition and remarks.
HECLA D17	0.55% C 0.7% Mn steel: Hadfields for BS alloy En 9
HECLA S55	0.55% C 0.85% Mn steel: Hadfields for BS alloys En 45, 45A & 46
HEDEX 11	0.55% C 0.7% Mn 0.05% S 0.04% P steel: Wire for forging; Kiveton Park for BS alloy En 9
HEDEX 26	0.62% C 1% Mn steel: Wire for forging; Kiveton Park; annealed **UTS: 670**
HMS 75	0.53% C 0.25% Si 0.65% Mn 0.035% S & P steel: Krupp
HMS 80	0.6% C 0.25% Si 0.65% Mn 0.035% S & P steel: Krupp
ICSW 1	0.8% C steel: For castings; Inman
JEM No.5	0.8% C steel: Jonas
JEM No.6	0.65% C steel: Jonas
JIS G3445 STKM17C	0.5% C 0.7% Mn 0.04% P & S (max) steel: Pipe; Japanese Standard **UTS: 660** **Proof: 490**
JIS G3445 STKM17A	0.5% C 0.7% Mn 0.04% P & S (max) steel: Pipe; Japanese Standard **UTS: 560** **Proof: 350**
JIS G4801 SUP7	0.6% C 0.85% Mn 0.035% P & S (max) steel: For springs; Japanese Standard
JIS G405 S48C	0.48% C 0.75% Mn 0.03% P & S (max) steel: For structures; Japanese Standard
JIS G4051 S50C	0.5% C 0.75% Mn 0.03% P & S (max) steel: For structures; Japanese Standard
JIS G4051 S53G	0.53% C 0.75% Mn 0.03% P & S (max) steel: For structures; Japanese Standard
JIS G4051 S55C	0.55% C 0.75% Mn 0.03% P & S (max) steel: For structures; Japanese Standard
JIS G4051 S58C	0.58% C 0.75% Mn 0.03% P & S (max) steel: For structures; Japanese Standard
KF 10B	0.52% C 0.8% Mn 0.06% S & P steel: Kirkstall for BS alloy En 9
KF 10C	0.58% C 0.7% Mn 0.06% S & P steel: Kirkstall for BS alloy En 9
KF 10D	0.63% C 0.7% Mn 0.06% S & P steel: Kirkstall; hardened & tempered **UTS: 910** **Elon: 12%** **Proof: 760**
KP 7	0.55% C 0.7% Mn 0.06% S & P steel: Kiveton Park for BS alloy En 9
KP 8	0.75% C 0.6% Mn 0.05% S & P steel: Kiveton Park for BS alloy En 42
KP 25	0.5% C 1.7% Si 0.8% Mn steel: For springs; Kiveton Park for BS alloy En 45
L 7	0.7% C steel: Pompey
LC 60	0.65% C steel: Water quenched; Low Moor; obsolete **DPN: 790**
LC 80	0.8% C steel: Water quenched; Low Moor; obsolete **DPN: 790**
LRT	0.5% C 0.2% Si 0.5% Mn steel: Pompey
M 75	0.75% C 0.4% Mn 0.045% S & P steel: Designation used by German Standard
MC 7	0.75% C 0.25% Si 0.3% Mn steel: Pompey

Symbol	Nominal analysis, supplier, condition and remarks.
MK 75	0.75% C 0.5% Mn 0.035% S & P steel: Designation used by German Standard
MR 6	0.6% C 0.3% Si 0.7% Mn steel: Pompey
MS 45	0.5% C 0.8% Si 1.05% Mn 0.035% S & P steel: Krupp
MS 50	0.53% C 0.9% Si 1.05% Mn 0.035% S & P steel: Krupp
NF A35 351/XC55	0.55% C 0.65% Mn steel: For structures; French Standard
NF A35 501/A70/2	C not specified 0.045% P & S (max) steel: Plate; French Standard
NF A35 551/XC48	0.48% C 0.65% Mn steel: For structures; French Standard
NF A35 551/XC65	0.65% C 0.75% Mn steel: For structures; French Standard
NF A35 551/XC70	0.7% C 0.75% Mn steel: For structures; French Standard
NF A35 551/XC80	0.8% C 0.65% Mn steel: For structures; French Standard
NF A35 571/56S7	0.56% C 0.75% Mn 0.035% P & S (max) 0.45% Cr (max) steel: For springs; French Standard
NF A35 571/60S7	0.6% C 0.85% Mn 0.05% P & S (max) steel: Bar; French Standard
NF A35 571/61S7	0.6% C 0.85% Mn 0.035% P & S (max) 0.45% Cr (max) steel: For springs; French Standard
No.7 TEMPER	0.7% C 0.2% Mn steel: Picks, screw-drivers, etc.; ESC **DPN: 630**
No.8 TEMPER	0.8% C 0.2% Mn steel: Rock drills, etc.; ESC **DPN: 630**
P 256	0.5% C 0.7% Mn steel: Chisels drifts, etc.; Carr's **DPN: 580**
PEGASE S	0.5% C 0.08% Si 0.5% Mn steel: Pompey
QQ W 461f	Carbon steel: Wire; round; US Federal specification
RS 2	0.5% C 1.7% Si 0.6% Mn steel: Pompey
RS 3	0.55% C 1.7% Si 0.8% Mn steel: Pompey
RS 4	0.6% C 1.85% Si 0.85% Mn steel: Pompey
SSM	0.54% C 0.3% Si 0.65% Mn steel: Pompey
S 6F	0.6% C 0.35% Si 0.7% Mn steel: Pompey
S 7F	0.65% C 0.3% Si 0.7% Mn steel: Pompey
S 8	0.8% C 0.3% Si 0.7% Mn steel: Pompey
S 65	0.65% C 0.3% Si 0.7% Mn steel: Pompey
S 75	0.75% C 0.3% Si 0.5% Mn steel: Pompey
S 145	0.55% C 1.7% Si 0.7% Mn steel: For springs; Bofors; hardened & tempered **DPN: 450** **UTS: 1550** **Elon: 7%** **Proof: 1150**
S 310	0.55% C 21.7% Si 30.8% Mn steel: Fagersta **DPN: 610**
SAE 864 A	10.8% C 21.5% Cu Fe: Sintered material; porosity 18%; specific gravity 5.7–6.1 **UTS: 140**
SAE 864 B	0.8% C 1.5% Cu Fe: Sintered material; specific gravity 6.1–6.5 **UTS: 180**
SAE 865 A	0.8% C 4.0% Cu Fe: Sintered material; porosity 18% - specific gravity 5.7–6.1 **UTS: 160**
SAE 865 B	0.8% C 4.0% Cu Fe: Sintered material; specific gravity 6.1–6.5 **UTS: 210**
SAE 866 A	0.8% C 8.5% Cu Fe: Sintered material; porosity 18% - specific gravity 5.7–6.1 **UTS: 160**
SAE 866 B	0.8% C 8.5% Cu Fe: Sintered material; specific gravity 6.1–6.5 **UTS: 230**
SAE 867 A	0.8% C 20% Cu Fe: Sintered material; porisity 18% - specific gravity 5.7–6.1 **UTS: 330**

Note. The following abbreviations and units are used in the tables:

DPN	Hardness, diamond pyramid number
UTS	Ultimate tensile strength, N/mm^2
Elon	Elongation, %
Proof	0.1 % proof strength, N/mm^2

1 N/mm^2=0.1 hbar=0.102 kgf/mm^2=0.06475 tonf/in^2=145.04 lbf/in^2

Symbol	Nominal analysis, supplier, condition and remarks.
SAE 967 B	0.8% C 20% Cu Fe: Sintered material; specific gravity 6.1–6.5 **UTS: 210**
SAE 1044	0.47% C 0.4% Mn 0.04% S & P steel: Mechanical properties not quoted
SAE 1047	0.47% C 1.5% Mn 0.04% S & P steel: Mechanical properties not quoted
SAE 1048	0.48% C 1.3% Mn 0.04% S & P steel: Mechanical properties not quoted
SAE 1049	0.49% C 0.8% Mn 0.04% P 0.05% S steel: Bar; hot rolled **DPN: 179 UTS: 630 Elon: 15% Proof: 330**
SAE 1049	0.49% C 0.8% Mn 0.04% P 0.05% S steel: Bar; cold drawn **DPN: 197 UTS: 740 Elon: 10% Proof: 450**
SAE 1050	0.51% C 0.8% Mn 0.04% P 0.05% S steel: Bar; hot rolled **DPN: 179 UTS: 660 Elon: 15% Proof: 340**
SAE 1050	0.51% C 0.8% Mn 0.04% P 0.05% S steel: Bar; cold drawn **DPN: 197 UTS: 740 Elon: 10% Proof: 610**
SAE 1051	0.51% C 1.0% Mn 0.04% P 0.05% S steel: Bar, etc.
SAE 1052	0.51% C 1.2% Mn 0.04% P 0.05% S steel: Bar; hot rolled **DPN: 217 UTS: 770 Elon: 12% Proof: 420**
SAE 1053	0.53% C 0.8% Mn 0.04% P 0.05% S steel: Bar, etc.
SAE 1055	0.55% C 0.3% Mn 0.04% P 0.05% S steel: Bar; hot rolled **DPN: 192 UTS: 530 Elon: 12% Proof: 360**
SAE 1060	0.6% C 0.8% Mn 0.04% P 0.05% S steel: Bar; hot rolled **DPN: 201 UTS: 560 Elon: 12% Proof: 370**
SAE 1061	0.61% C 0.9% Mn 0.04% P 0.05% S steel: Bar, etc.
SAE 1064	0.63% C 0.6% Mn 0.04% P 0.05% S steel: Bar; hot rolled **DPN: 201 UTS: 560 Elon: 12% Proof: 370**
SAE 1065	0.65% C 0.8% Mn 0.04% P 0.05% S steel: Bar; hot rolled **DPN: 207 UTS: 730 Elon: 12% Proof: 390**
SAE 1066	0.66% C 1.0% Mn 0.04% P 0.05% S steel: Bar, etc.
SAE 1069	0.69% C 0.6% Mn 0.04% P 0.05% S steel: Bar, etc.
SAE 1070	0.7% C 0.8% Mn 0.04% P 0.05% S steel: Bar; hot rolled **DPN: 212 UTS: 750 Elon: 12% Proof: 390**
SAE 1072	0.72% C 1.2% Mn 0.04% P 0.05% S steel: Bar
SAE 1074	0.75% C 0.6% Mn 0.04% P 0.05% S steel: Bar; hot rolled **DPN: 217 UTS: 770 Elon: 12% Proof: 400**
SAE 1078	0.78% C 0.5% Mn 0.04% P 0.05% S steel: Bar; hot rolled **DPN: 207 UTS: 730 Elon: 12% Proof: 390**
SAE 1080	0.8% C 0.8% Mn 0.04% P 0.05% S steel: Bar; hot rolled **DPN: 229 UTS: 820 Elon: 10% Proof: 440**
SAE 1151	0.51% C 0.9% Mn 0.04% P 0.1% S steel: Free cutting; hot rolled **DPN: 187 UTS: 670 Elon: 15% Proof: 340**
SAE 1151	0.51% C 0.9% Mn 0.04% P 0.1% S steel: Free cutting; cold drawn **DPN: 207 UTS: 750 Elon: 10% Proof: 630**
SAE 1548	0.48% C 1.2% Mn 0.04% P 0.05% S steel: Bar, etc.
SAE 1551	0.51% C 1.0% Mn 0.04% P 0.05% S steel: Bar, etc.

Symbol	Nominal analysis, supplier, condition and remarks.
SAE 1552	0.52% C 1.35% Mn 0.04% P 0.05% S steel: Bar, etc.
SAE 1561	0.61% C 0.9% Mn 0.04% P 0.05% S steel: Bar, etc.
SAE 1566	0.66% C 1.0% Mn 0.04% P 0.05% S steel: Bar, etc.
SAE 1572	0.72% C 1.15% Mn 0.04% P 0.05% S steel: Bar, etc.
SANBOLD 9	0.53% C 0.3% Si 0.6% Mn steel: Sandersons for BS alloy En 9
SANBOLD 42	0.8% C 0.2% Si 0.6% Mn steel: Sanderson for BS alloy En 42
SANBOLD 43	0.5% C 0.3% Si 0.7% Mn steel: Sanderson for BS alloy En 43
SANBOLD 45	0.55% C 1.7% Si 0.9% Mn steel: Sanderson for BS alloy En 45
SANBOLD 50	0.55% C 0.3% Si 0.6% Mn steel: Sanderson; replaced by Sanbold 9
SANBOLD 59	0.5% C 0.3% Si 0.7% Mn steel: Sanderson; replaced by Sanbold 43
SANBOLD 60	0.8% C 0.2% Si 0.6% Mn steel: Sanderson; replaced by Sanbold 42
SANBOLD 8058	0.55% C 1.7% Si 0.9% Mn steel: Sanderson; replaced by Sanbold 45
SH 4	0.48% C 0.35% Si 0.7% Mn steel: Pompey
SH 5	0.55% C 0.35% Si 0.8% Mn steel: Pompey
Si Mn	0.57% C 1.8% Si 0.8% Mn steel: For springs; ESC for BS alloy En 45 & 45A
SIS 14-17-74	0.8% C 0.5% Mn 0.035% P & S steel: Spring; Swedish Standard **UTS: 2500 Proof: 1500**
SIS 14-17-78	0.7% C 0.65% Mn 0.03% P & S (max) steel: For cold rolled strips; Swedish Standard **DPN: 500 UTS: 1500**
SIS 1606	0.5% C 0.6% Mn 0.04% S & P steel: Annealed; Swedish Standard **UTS: 630 Elon: 10% Proof: 300**
SIS 1655	0.55% C 0.65% Mn 0.05% S & P steel: Swedish Standard
SIS 1655/00	0.55% C steel: Swedish Standard; as rolled or forged; suffix indicates condition **DPN: 225 UTS: 700 Elon: 12% Proof: 340**
SIS 1665	0.55% C 0.4% Mn 0.03% S & P steel: Strip; annealed; Swedish Standard **DPN: 165 UTS: 400**
SIS 1665	0.55% C 0.4% Mn 0.3% S & P steel: Strip; cold rolled; Swedish Standard **DPN: 260 UTS: 780**
SIS 1672/08	0.47% C steel: For flame & induction hardening; Swedish Standard; as rolled **DPN: 200 UTS: 650 Elon: 16% Proof: 300**
SIS 1674	0.52% C steel: Sheet & plate for hardening & tempering; Swedish Standard **UTS: 700 Elon: 15% Proof: 440**
SIS 1674	0.52% C steel: For induction or flame hardening; Swedish Standard
SIS 1678	0.61% C steel: Swedish Standard; hardened & tempered **UTS: 850 Elon: 11% Proof: 580**
SIS 1678	0.61% C 0.2% Si 0.75% Mn 0.35% S & P steel: Swedish Standard
SIS 1770	0.7% C 0.6% Mn 0.035% S & P steel: Hardened & tempered; Swedish Standard **UTS: 1320**
SIS 1774	0.65% C 0.25% Si 0.5% Mn 0.035% S & P spring steel
SIS 1778	0.7% C 0.5% Mn 0.03% S & P steel: Strip; annealed; Swedish Standard **DPN: 170 UTS: 600**

Symbol	Nominal analysis, supplier, condition and remarks.
SIS 1778	0.7% C 0.5% Mn 0.03% S & P steel: Cold rolled; Swedish Standard **DPN: 270 UTS: 820**
SIS 1778	0.75% C 0.65% Mn 0.03% S & P steel: Strip; cold rolled **DPN: 210 UTS: 650 Elon: 20% Proof: 440**
SIS 1778	0.75% C 0.65% Mn 0.03% S & P steel: Strip; cold rolled
SIS 1973	0.5% C 1% Mn 0.05% P 0.15% S steel: Free machining; normalized; Swedish Standard **UTS: 630**
SIS 2090	0.56% C 1.75% Si 0.8% Mn 0.3% Cr (max) steel: For springs **DPN: 400 UTS: 1300**
SIS 8011	0.75% C sintered steel: Swedish Standard
SK 6	0.75% C 0.5% Mn (max) 0.2% Cr & Ni (max) steel: Japanese Standard designation
SK 7	0.65% C 0.5% Mn 0.2% Cr & Ni (max) steel: Japanese Standard designation
SKC 3	0.77% C 0.5% Mn (max) 0.2% Cr (max) 0.25% V (max) steel: For hollow drills; Japanese Standard designation
SMN	0.55% C 1.8% Si 0.8% Mn steel: Punches, chisels, etc.; Huntsman **DPN: 610**
SPECIAL	C not specified 0.25% Mn steel: For tools; Braeburn
SPEAR 50	0.52% C steel: Spear & Jackson; hardened & tempered **DPN: 240 UTS: 780**
SPEAR 75	0.7% C steel: 9For press tools; Spear & Jackson
St 70/2	0.5% C 0.05% S & P steel: German Standard
St 70/202	0.5% C 0.05% S & P steel: German Standard
St 70/203	0.5% C 0.05% S & P steel: German Standard
STA 3	0.75% C steel: Wire; hard drawn; replaced by BS alloy En 49D
STA 4	0.7% C steel: Wire; hard drawn; replaced by BS alloy En 49C
STA 5 V4B	0.52% C steel: Replaced by BS alloy En 43A
STA 5 V4B/1	0.52% C steel: Replaced by BS alloy En 43B
STA 5 V4B/2	0.52% C steel: Replaced by BS alloy En 43C
STA 5 V5	0.55% C steel: Replaced by BS alloy En 9K
STA 5 V5A	0.5% C steel: Replaced by BS alloy En 43D
STA 5 V5A/1	0.5% C steel: Replaced by BS alloy En 43D
STA 5 V5A/2	0.5% C steel: Replaced by BS alloy En 43E
STA 5 V22B	0.5% C steel: Replaced by BS alloy En 43
STA 5 V22B/1	0.5% C steel: Replaced by BS alloy En 43
STA 5 V22B/2	0.5% C steel: Replaced by BS alloy En 43
STA 5 V23	0.6% C 1.8% Si 0.8% Mn steel: Replaced by BS alloy En 45
STA 5 V23A	0.6% C 1.8% Si 0.8% Mn steel: Replaced by BS alloy En 45
STA 41	0.5% C steel: Strip; hardened & tempered; for springs
STANDARD	C not specified 0.25% Mn steel: For tools; Braeburn
STC 5	0.5% C 0.2% Si 0.6% Mn 0.06% S & P steel: Spencer; normalized **DPN: 240 UTS: 720 Elon: 18%**
STC 6	0.55% C 0.2% Si 0.5% Mn 0.05% S & P steel: Spencer; normalized **DPN: 240 UTS: 690 Elon: 18%**

Symbol	Nominal analysis, supplier, condition and remarks.
T 11	0.72% C 0.2% Si 0.2% Mn steel: Jessop
T 12	0.81% C 0.2% Si 0.2% Mn steel: Jessop
T 21	0.72% C 0.25% Si 0.25% Mn steel: Jessop
T 22	0.8% C 0.25% Si 0.25% Mn steel: Jessop
T 31	0.72% C 0.27% Si 0.27% Mn steel: Jessop
T 32	0.8% C 0.27% Si 0.27% Mn steel: Jessop
T 750	0.5% C 0.25% Si 1.0% Mn 0.05% P 0.15% S steel: Fagersta; free machining
TEMPER No.2	0.6% C steel: Scissors, knives, auger bits, etc.; Sanderson
TEMPER No.2½	0.7% C steel: Crowbars, hammers, stone bits, etc.; Sanderson
TEMPER No.3	0.8% C steel: Punches, shear blades, stone bits, etc,; Sanderson
TOLEDO 55	0.55% C 0.6% Mn 0.06% S & P steel: Toledo for BS alloy En 9
U 7	0.7% C 0.3% Mn 0.2% Cr (max) 0.25% Ni (max) steel: Russian Standard designation
U 7A	0.7% C 0.25% Mn 0.15% Cr (max) 0.2% Ni (max) steel: Russian Standard designation
U 8A	0.8% C 0.25% Mn 0.15% Cr (max) 0.2% Ni (max) steel: Russian Standard designation
UHB 11	0.5% C 0.2% Si 0.5% Mn 0.035% S steel: Uddelholm
UMS 75	0.53% C 0.25% Si 0.65% Mn 0.045% S & P steel: Krupp
UMS 80	0.6% C 0.25% Si 0.65% Mn 0.045% S & P steel: Krupp
VSMD	0.65% C 0.2% Si 0.3% Mn steel: Hammers, rock drills, etc.; Vulcan
VSS	0.66% C 0.2% Si 0.5% Mn steel: Springs; Vulcan
VSSM	0.57% C 1.5% Si 1% Mn steel: Heavy duty vehicle springs; Vulcan **DPN: 460**
W 6	C not specified 0.4% S & P steel: Galvanized wire; designation used by BS
W 21	0.6% C steel: Welding electrode; Cu covered; Armco; for hard surfacing
WLM BEST	0.7% C 0.35% S & P steel: Temper grade varies with carbon content; W Marrison
WTS 0.75%	0.75% C 0.15% Si 0.35% Mn steel: Centres vice jaws etc.; Marsh Brothers
Y 65	0.65% C 0.25% Mn 0.035% P (max) 0.025% S (max) steel: French Standard designation
Y 75	0.75% C 0.25% Mn 0.035% P (max) 0.025% S (max) steel: French Standard designation
Y 65V	0.65% C 0.02% P & S (max) steel: French Standard designation
Y 75V	0.75% C 0.25% Mn 0.02% P & S (max) steel: French Standard designation
XBP	0.6% C 0.3% Si 1% Mn 0.04% S & P steel: Toledo for BS alloy En 49

Note. The following abbreviations and units are used in the tables:

DPN	Hardness, diamond pyramid number
UTS	Ultimate tensile strength, N/mm²
Elon	Elongation, %
Proof	0.1 % proof strength, N/mm²

1 N/mm²=0.1 hbar=0.102 kgf/mm²=0.06475 tonf/in²=145.04 lbf/in²

44A4 steel – plain carbon 0.8 per cent minimum

Specific gravity	7.86	
Density	7860 kg/m³	(0.28 lb/in.³)
Solidus/liquidus	1430–1480 °C	
Thermal conductivity	41.9 W/m °C	(10 cal/m s °C)
Coefficient of linear expansion	11 × 10⁻⁶/ °C	
Electrical conductivity	7–10% IACS (copper 100%)	
Specific resistance	180–280 microhm mm	
Young's modulus of elasticity	– N/m²	
Impact	– J	
Fatigue strength (no. of cycles)	–	
Hot strength	–	

The above properties are typical of the following group, and may not apply exactly to any one specification. It is possible that with certain specifications some of the values may not be applicable.

General metallurgical characteristics

The steels in this group are all hyper-eutectic, that is, there is no free ferrite present at any time. This means that very little cold work can be accomplished without cracking or fracturing the parts and this limits the forms available to forging and relatively heavy section sheet and bar. Cold drawn wire is available when the carbon is at the lower end of the range.

The steels are relatively easily hardened and only in very special cases are they used in anything except the hardened and tempered or cold drawn condition.

There may be residual elements present, very small quantities of which can have considerable effect on the hardenabilty. Where air hardening is undesirable it is advisable that the supplier be informed of this fact when ordering.

Thermal treatment

When it is necessary to fully anneal these steels they must be heated to 760–780 °C for 30 minutes per 25 mm of section, then very slowly cooled. This puts them in their softest possible condition, but will cause scale formation and decarburization. For most purposes a sub-critical anneal at 650–670 °C will suffice and for machining purposes is probably superior to the full anneal. This does not cause decarburization and produces very little scale. Normalizing of these materials is carried out at 760–780 °C for 20 minutes per 25 mm of section followed by still air cooling. This eliminates all forging and cold working stresses, homogenizes the grain structure and makes it machinable. Relatively small amounts of chromium, molybdenum, or vanadium impurities can make these steels air harden at the normalize and if this occurs normalizing must be followed by a sub-critical anneal.

Hardening requires a temperature 730–760 °C for 10 minutes per 25 mm of section followed by water quenching. The steel is glass hard and brittle in this condition and must be immediately tempered, care being taken in handling. Considerable thought must be given to the design of parts at the thermal treatment stage. There should be no rapid change of section, sharp re-entrant corners, or other forms of stress raiser from which cracking could propagate at quenching. Tempering should be at 150–350 °C and as higher temperatures can cause brittleness unless quenched after temper, the use of alloy steels to obtain these desired mechanical properties is recommended. These materials cannot be carburized, the carbon content of the steel being at least equivalent to a carburized case.

Induction heating can be used for local hardening. The part should be in the normalized or annealed condition and must be carefully designed to make the best use of the process as there is a danger of cracking in the zone adjacent to the hard area. Heating is by means of an induction coil which rapidly heats the chosen area to 750 °C followed by water quench. Flame heating can also be used but is not as readily controlled. The parts must be tempered after these forms of hardening.

It is important to note that if hardened and tempered parts are too rapidly heated cracking can take place. This does not apply to the normalized or annealed condition to the same extent, but heating of these materials should always be carried out slowly.

All normal types of furnace equipment, electric, gas or oil fired can be used on these steels. Being high carbon there is always the danger of surface decarburization which can be reduced or prevented by the use of protective atmospheres for thermal treatments above 700 °C. These atmospheres can be the normal treated gas, or cracked ammonia. With gas furnaces some atmosphere control can be obtained by judicious variation of the air/gas ratio.

Scale removal should be by grit blasting. Acid pickling must not be applied to any of these materials after hardening, but they can all be pickled in the normalized or annealed condition. Sodium hydride treatment gives excellent results but requires very specialized equipment.

Welding and brazing

Pre-treatment. The parts should be normalized or annealed and must be free of scale, dirt and grease. Preheating is essential on all these steels before welding.

Welding and brazing. It is not recommended that these materials are welded or brazed. Cracking in the weld zone, especially with the local heat of electric welding, is always extremely likely. With skilled welding, using the gas flame to pre-heat and post-heat the weld area, welding can be accomplished but will always have a brittle area. This can be minimized by subsequent hardening or normalizing but cannot be eliminated. Resistance welding can seldom be carried out without causing cracking. Brazing can only be carried out in the normalized or annealed condition, and unless special braze fillers are used cannot be subsequently hardened and tempered.

Neither welding or brazing should be the first choice for joining these materials. If either is necessary for any reason, considerable care must be taken to minimize the serious drop in properties of the as welded condition.

Post-treatment. With these materials it is essential that some form of thermal treatment immediately follows welding and brazing. As a minimum the area should be stress relieved locally using a gas flame or preferably an electric blanket. If possible the parts should be hardened and tempered but this is not generally practicable without causing cracking.

Normalizing removes most of the welding stresses but does not give the mechanical properties of hardening and tempering.

Flaw detection methods

Crack test. Magnetic crack test, or alternatively penetrant oil chalk test. With modern techniques and fluids this can be as sensitive as the magnetic method and is more flexible.

X-ray. This will seldom be required on wrought products and welding is only occasionally carried out.

Ultra-sonic test. This can be used on expensive forgings to prove the material at an early stage of machining.

Chemical etch. In the hardened and tempered condition these materials are very susceptible to grinding abuse, which can cause softening and in extreme cases cracking. A 50% sulphuric acid electrolytic etch can be used as a visual check for this defect without scrapping any pieces. Nitric acid etches can also be used for this purpose. Other laboratory etches to prove the material generally scrap the part.

Corrosion protection

Temporary. Parts must always be kept dry and should be oiled or greased at all times. High quality soluble oil coolant in good condition is a satisfactory temporary corrosion inhibitor, but when dirty or contaminated can be a serious cause of corrosion. Finished diameters must be oiled immediately and parts should be stored in the oily condition inside plastic containers.

Permanent. Corrosion protection of these materials is expensive and relatively difficult. Also as they are generally used in the hardened and tempered condition, corrosion pitting can be a serious stress raiser problem, causing early failure sometimes with disastrous suddenness.

With all forms of corrosion protection it is essential that the last trace of existing corrosion, moisture and dirt of any kind is removed. As this is generally easier on new parts, it is one reason why it is much more economical to apply the correct protection first time rather than removing a faulty protection with corrosion, and then re-coat. It is very difficult to ensure the removal of the last trace of active rust from corrosion pits and unless this is done corrosion will propagate under any coating from this nucleus.

Preparation for any method of protection can include shot blasting. With these hard materials very little key is obtained from the blasting but it is an excellent method of removing dirt and rust. Alkaline rust removers in good condition remove all traces of rust but acid must not be used as it can cause brittleness.

Painting. Preparation should be by blasting or wire brushing followed by phosphate treatment or a proprietary etch primer. Some phosphate treatments can cause embrittlement and if the top coats are not to be stoved, this should be followed by stress relieving. The primer should be of the zinc chromate type. Top coats should preferably be stoved and applied as soon as possible after the primer. Modern plastic paints are very abrasion resistant and nonporous, but should have the metallic type of primer.

Plating. These materials will seldom be decoratively electroplated. It is advised that all of these materials are stress relieved at 150–200 °C before electroplating if the part has had any form of machining. Where decorative chromium plating is necessary best results are obtained with 0.05 mm copper followed by 0.025 mm nickel under the very thin layer of chromium plating. Cadmium, tin or zinc plating can be applied less than 0.025 mm thick with considerable long term corrosion resistance to atmospheric attack. The success of all electro-plating depends on careful preplating treatment. Post-plating stress relieving is essential on all of the listed materials when they have been hardened and tempered. Without this treatment inexplicable brittle failure below the stipulated tensile strength can take place owing to hydrogen embrittlement. As there is no visual evidence that the stress relieving treatment has been carried out, it is recommended that plating is only applied to these materials when parts are not to be used for high duty purposes.

Metal spraying using zinc or aluminium is possible, but adhesion will be poor as this generally relies on the key from the pre-treatment blast pattern, which is very shallow on these hard steels. Modern techniques of spraying may overcome this defect.

Cathodic protection makes the part negative in an electric circuit, thus protecting it at the expense of the positive or anode. These steels are not normally used in circumstances where this would be possible.

There are various forms of artificial oxidation which can be applied which result in a dark blue or brown coloration, the process being known as 'black oxide'. There are various proprietary processes most of which are based on a strong, hot sodium hydroxide solution containing powerful oxidizing agents. Apart from the inherent corrosion resistance of the film it has oil retention properties but will not withstand outdoor atmospheric conditions. It has been the standard finish for guns for many years, and is an adequate protection for hand tools and fixtures used and stored indoors.

Machinability

In the hardened and tempered condition these steels must be ground, as they are too hard for conventional machining. It is advised that rough machining is carried out in the normalized condition which has excellent machinability; fine limits can then be ground after hardening and tempering.

Sharp corners and stress raisers should not be present before hardening as the quench can cause cracking.

Uses

Cutting tools, press tools and fixtures, hard wearing parts not subject to sudden shock. Shear blades, scissors, springs and knives.

Symbol	Nominal analysis, supplier, condition and remarks.
0.0783	0.84% C 0.2% Si 0.4% Mn 0.05% S & P steel: German Standard
0.1274	1.0% C 0.2% Si 0.4% Mn 0.035% S & P steel: German Standard
1.1264	0.8% C 0.2% Si 0.3% Mn 0.025% S & P steel: German Standard
1.1274	1.0% C 0.2% Si 0.4% Mn 0.035% S & P steel: German Standard
1.1525	0.08% C 0.17% Mn 0.02% P & S (max) steel: German Standard designation
1.1545	1.05% C 0.17% Mn 0.02% P & S (max) steel: German Standard designation
1.1830	0.85% C 0.6% Mn 0.025% P & S (max) steel: German Standard designation
1.2002	1.0% C 0.5% Mn steel: German Standard
02	0.9% C 1.6% Mn steel: Designation used by AISI
16	0.8% C steel: Fagersta
17	1% C 0.25% Si 0.45% Mn steel: Sandvik for SIS 1870 1880
17 AP	1% C 0.4% Mn 0.05% S 0.2% Pb steel: Sandvik; annealed; free machining **UTS: 660**
20	1.0% C steel: Fagersta
20	1.15% C 0.25% Si 0.35% Mn steel: Sandvik for SIS 1885
20 P	1.15% C 0.35% Mn 0.2% Pb steel: Sandvik; free machining
24	1.2% C steel: Fagersta
40 D 35	1.0% C 0.25% Si 0.45% Mn steel: Fagersta
44 D 35	1.1% C 0.3% Si 0.5% Mn steel: Fagersta
48 D 35	1.2% C 0.3% Si 0.5% Mn steel: Fagersta **DPN: 400**
1101	1.2% C 0.12% Mn 0.02% P & S steel: French Standard designation
1102	1.05% C 0.02% P & S steel: French Standard designation
1103	0.9% C 0.02% P & S (max) steel: French Standard designation
1161	1.2% C 0.2% Mn 0.02% P & S (max) steel: French Standard designation
1162	1.05% C 0.2% Mn 0.02% P & S (max) steel: French Standard designation
1163	0.9% C 0.2% Mn 0.02% P & S (max) steel: French Standard designation
1200	1.35% C 0.25% Mn 0.025% P & S (max) steel: French Standard designation

Note. The following abbreviations and units are used in the tables:

DPN	Hardness, diamond pyramid number
UTS	Ultimate tensile strength, N/mm^2
Elon	Elongation, %
Proof	0.1 % proof strength, N/mm^2

1 N/mm^2=0.1 hbar=0.102 kgf/mm^2=0.06475 tonf/in^2=145.04 lbf/in^2

Symbol	Nominal analysis, supplier, condition and remarks.
1201	1.2% C 0.25% Mn 0.025% P & S (max) steel: French Standard designation
1202	1.05% C 0.25% Mn 0.025% P & S (max) steel: French Standard designation
1203	0.9% C 0.25% Mn 0.025% P & S (max) steel: French Standard designation
1303	0.9% C 0.25% Mn 0.3% P 0.2% S steel: French Standard designation
AISI C1084	0.86% C 0.8% Mn 0.04% P 0.05% S steel
AISI C1086	0.86% C 0.4% Mn 0.04% P 0.05% S steel
AISI C1090	0.91% C 0.8% Mn 0.04% P 0.05% S steel
AISI C1095	0.95% C 0.4% Mn 0.04% P 0.05% S steel
AQUATOUGH 100	0.95% C 0.3% Mn steel: Clyde Alloy **DPN: 770**
AMS 5112E	0.9% C steel: Music & spring wire; best quality; AMS for SAE 1090
AMS 5121B	0.95% C steel: Spring & strip; annealed; AMS for SAE 1095
AMS 5122B	0.95% C steel: Spring & strip; rolled; AMS for SAE 1095
AMS 5132C	0.95% C steel: Bar; AMS for SAE 1095
AN QQ A/666/1	1.0% C 0.4% Mn 0.04% S & P steel: Strip; US Service
AN QQ W 441/3	0.85% C 0.2% Si 0.4% Mn 0.03% S & P steel: Wire; US Service; music wire **UTS: 2000**
AN S5	1.0% C 0.4% Mn 0.05% S & P steel: US Service
ASTM A68	1.0% C 0.2% Si 0.4% Mn steel: Bars for springs
ASTM A679	0.8% C 0.7% Mn steel: Wire; hard drawn for springs **UTS: 2000**
ATLAS ALPHA 8	0.8% C 0.2% Si 0.25% Mn steel: Atlas; as AISI type W1
ATLAS X10	1.05% C 0.2% Si 0.2% Mn steel: Atlas; as AISI type W1
ATLAS X12	1.2% C 0.20% Si 0.25% Mn steel: Atlas; as AISI type W1
ATLAS XX-95	0.95% C 0.3% Si 0.3% Mn steel: Atlas; as AISI type W1
B 15T	0.75% C steel: Bofors; obsolete **DPN: 820**
B 15V	0.8% C steel: Bofors; obsolete **DPN: 820**
B 20V	1.0% C steel: Bofors **DPN: 820**
B 24V	1.2% C steel: Bofors **DPN: 850**
B 28V	1.4% C steel: Bofors **DPN: 850**
BC	1.0% C tool steel: Origin unknown
BELL B & P	0.6–1.2% C steel: John Vessey; a range of steels based on C content
BEST WARRANTED	0.9% C tool sheet: Balfour Darwin
BLACK LABEL	1.0% C tool steel: Jessop
BLUE LABEL	1.0% C steel: Origin unknown

Symbol	Nominal analysis, supplier, condition and remarks.
BLUE LABEL	0.98% C 0.25% Si 0.25% Mn tool sheet: Huntsman **DPN: 630**
BS 24/3Ba	1.0% C 0.6% Mn 0.05% S & P steel: For springs; as BS alloy En 44
BS 970 En 42D	0.85% C 0.3% Si 0.7% Mn 0.05% S & P steel: For springs
BS 970 En 44	1% C 0.6% Mn 0.05% S 0.05% P steel: For springs; oil hardened & tempered
BS 970 En 44A	1.1% C 0.7% Mn 0.05% S 0.05% P steel: Strip for springs; strip may be cold rolled or hardened & tempered
BS 970 En 44C	1.1% C 0.6% Mn 0.05% S & P steel: For springs
BS 1407	1.1% C 0.3% Mn 0.3% Si 0.5% Cr (optional) steel: Supplied ground; 'Silver Steel'
BS 1423	1.0% C 0.3% Si 0.6% Mn 0.5% Cr (optional) steel: For springs; hardened & tempered **DPN: 780**
BS 1429 En 44B	0.95% C 0.6% Mn 0.05% S & P steel: Bar
BS 1449 HS90	0.9% C 0.2% Si 0.5% Mn 0.04% S & P steel: Sheet
BS 1449 HS100	1.0% C 0.2% Si 0.4% Mn 0.04% S & P steel: Sheet
BS 3141 C9A1	0.8% C 0.5% Mn 0.05% S & P steel: Wire; Classification system used by BS; discontinued
BS 3141 C9B1	0.8% C 0.8% Mn 0.04% S & P steel: Wire; classification system used by BS; discontinued
BS 3141 C9C1	0.85% C 0.4% Mn 0.05% S & P steel: Classification system used by BS; discontinued
BS 3141 C9D1	0.85% C 0.7% Mn 0.05% S & P steel: Wire; classification system used by BS; discontinued
BS 3141 C11A1	1% C 0.5% Mn 0.05% S & P steel: Wire; classification system used by BS; discontinued
BS 3141 C11B1	1.0% C 0.5% Mn 0.04% S & P steel: Wire; classification system used by BS; discontinued
BS 3141 C12A1	1.1% C 0.5% Mn 0.05% S & P steel: Wire; classification system used by BS; discontinued
BS 3141 C13A1	1.1% C 0.3% Mn 0.04% S & P steel: Wire; classification system used by BS; discontinued
C 80W1	0.8% C 0.17% Mn 0.02% P & S (max) steel: German Standard designation
C 85WS	0.85% C 0.6% Mn 0.025% P & S (max) steel: German Standard designation
C 100	1.0% C 0.42% Mn steel: Italian Standard designation **DPN: 270 UTS: 1450 Elon: 4% Proof: 110**
C 105W1	1.05% C 0.17% Mn 0.02% P & S (max) steel: German Standard designation
C 1105	1.05% C 0.4% Mn 0.25% Si 0.02% S & P steel: Drills, etc.; SHD
CARBON	1.05% C 0.2% Si 0.2% Mn 0.08% Cr steel: Latrobe
CELFOR	1.0% C steel: Shear blades, dies, etc.; Sanderson **DPN: 950**
CHISEL STEEL	0.9% C steel: For cold chisels & punches; origin unknown
CHQ	Plain C die steel: Analysis vanes; Firth Sterling
COH	0.9% C 0.25% Si 1.7% Mn steel: Spencer; collets, gauges, etc. **DPN: 750**

Symbol	Nominal analysis, supplier, condition and remarks.
COLUMBIA ELECTREX	1.06% C steel: Columbia; AISI type W1
COLUMBIA EXTRA	1.06% C steel: Columbia; AISI type W1
COLUMBIA EXTRA HEADERDIE	0.95% C steel: Columbia; AISI type W1
COLUMBIA STANDARD	1.06% C steel: Columbia; AISI type W1
CONQUEROR TEMP 1–6	1.0% C steel: C content and hardness varies; origin unknown
CRESCENT	1.1% C steel: Spencer
CRESCENT 1	1.3% C steel: Spencer
CRESCENT 3	1.0% C steel: Spencer
CRESCENT 4	0.85% C steel: Spencer
CRESCENT 5	0.75% C steel: Spencer
CRUSCA	0.8% C 0.15% Si 0.3% Mn steel: Crucible Steel Co; for drills
DC	0.8% C 0.3% Si 0.35% Mn steel: Pompey
DIN 17222 M85	0.85% C 0.4% Mn 0.045% S & P steel: For springs
DIN 17222 MK101	1.0% C 0.4% Mn 0.045% S & P steel: For springs
DOH	0.9% C 0.25% Si 1.7% Mn steel: Spencer; collets, gauges, etc. **DPN: 750**
DTD 5 B	0.8% C 0.2% Si 0.5% Mn steel: Wire; cold drawn & tempered **UTS: 1550**
DTD 488	1.0% C 1.0% Mn 0.05% S 0.05% P steel: Strip; hardened & tempered **UTS: 1840 Proof: 1620**
E 11C	1.2% C 0.45% Mn steel: Pompey
ECLIPSE CHD	1.0% C steel: Origin unkown
EXTRA	1.0% C tool steel: Origin unknown
FAGERSTA 16	0.8% C 0.2% Si 0.3% Mn steel: Fagersta
FAGERSTA 20	1.0% C 0.2% Si 0.3% Mn steel: Fagersta **DPN: 730**
FAGERSTA 24	1.2% C 0.2% Si 0.3% Mn steel: Fagersta **DPN: 800**
FAGERSTA 1870	1.0% C 0.5% Mn steel: For springs; Fagersta
FAGERSTA 1884	1.2% C 0.3% Mn steel: For springs & cutting tools; Fagersta
FILE	1.2% C 0.18% Si 0.3% Mn 0.05% S & P steel: Also produced with up to 3% Cr additions; Toledo
G 10950	UNS designation for type 1095 steel
GAUGE STEEL	1.0% C steel: Hardened & ground; general trade term
GENUINE STUBS	1.1% C 0.35% Mn 0.2% Si 0.045% S & P steel: For BS 1407; Peter Stubs
GOST 1050/85	0.85% C 0.65% Mn 0.04% P & S (max) steel: For structures; Russian Standard
GOST 2052/60 S2	0.61% C 0.85% Mn 0.04% P & S (max) 0.3% Cr (max) 0.4% Ni (max) 0.25% Cu (max) steel: For springs; Russian Standard
GOST 2052/85	0.85% C 0.65% Mn 0.04% P & S (max) 0.13% Cr & Ni 0.25% Cu (max) steel: For springs; Russian Standard
GRAPH A1	1.5% C 0.2% Si 0.3% Mn 0.15% A1 steel: Free graphite, Timken; mandrels, dies, etc. **DPN: 850**
GREEN LABEL	1.2% C 0.2% Mn 0.2% Si steel: Carpenter; for drill rod
GREEN LABEL	0.8% C 0.25% C 0.25% Mn tool steel: Huntsman **DPN: 650**
GS 6	0.9% C 2% Mn steel: Cutting tools; Osborn; obsolete
H9 DOUBLE HEADER	0.9% C 0.4% Mn 0.4% Si steel: Carpenter for AISI type W1
HECLA 34	1.0% C 0.6% Mn steel: Hadfield for BS alloys En 44B C D & E

Note. The following abbreviations and units are used in the tables:

DPN	Hardness, diamond pyramid number
UTS	Ultimate tensile strength, N/mm^2
Elon	Elongation, %
Proof	0.1 % proof strength, N/mm^2

1 N/mm^2=0.1 hbar=0.102 kgf/mm^2=0.06475 tonf/in^2=145.04 lbf/in^2

Symbol	Nominal analysis, supplier, condition and remarks.
HM 1	0.9% C 1.6% Mn tool steel: Origin unknown
HYFORM	1.0% C steel: Origin unknown
ICS	1.0% C steel: Origin unknown
JEM No.1	1.4% C steel: Jonas
JEM No.2	1.2% C steel: Jonas
JEM No.3	1.1% C steel: Jonas
JEM No.4	0.9% C steel: Jonas
JIS G4801 54P3	0.82% C 0.45% Mn 0.035% P & S (max) steel: For springs; Japanese Standard
JIS G4801 84P4	1.0% C 0.45% Mn 0.035% P & S (max) steel: Japanese Standard
KE 1006	1.0% C 0.2% Si 0.2% Mn steel: Coining dies, etc.; Kayser Ellison **DPN: 910**
KEA 108	1.0% C 0.5% Mn steel: Sanderson Kayser
KEA 108	1.0% C 0.5% Mn 0.2% Si steel: Free cutting; Kayser Ellison
KP 9	0.86% C (min) steel: Kiveton Park
L 11	1.1% C steel: Pompey
L 12	1.2% C steel: Pompey
L 13	1.3% C steel: Pompey
LC 100	0.9% C steel: Hardened & tempered; water quenched; Low Moor; obsolete **DPN: 800**
LC 110	1.1% C steel: Hardened & tempered; water quenched; Low Moor; obsolete **DPN: 800**
LC 120	1.2% C steel: Hardened & tempered; water quenched; Low Moor; obsolete **DPN: 800**
LC 140	1.4% C steel: Hardened & tempered; water quenched; Low Moor; obsolete **DPN: 800**
LT 20	0.9% C 2.0% Mn steel: Non shrinking tool steel; Low Moor **DPN: 800**
MOTOR BRAND	1.0% C steel: Supplied as ground flat stock, Carrs
MT	1.7% C 8.0% Al steel: Casting or forging for magnets; Japanese alloy listed by PMA
No. 1 HARDENITE	0.9% C tool steel: Origin unknown
No.1 HARDENITE	1.0% C steel: For tools; origin unknown
No.2 HARDENITE	0.9% C tool steel: Origin unknown
No.3 HARDENITE	0.9% C tool steel: Origin unknown
No.9 TEMPER	0.9% C 0.2% Mn steel: Reamers & dies; ESC **DPN: 750**
No.10 TEMPER	1.0% C 0.2% Mn steel: Hand chisels, axes etc.; ESC **DPN: 750**
No.11 TEMPER	1.1% 0.2% Mn steel: Knives, reamers, etc.; ESC **DPN: 750**
No.12 TEMPER	1.2% C 0.2% Mn steel: Woodmaking tools, razors, etc.; ESC **DPN: 750**
No.14 TEMPER	1.4% C 0.2% Mn steel: Razors & surgical instruments; ESC **DPN: 750**
NOLL SPECIAL	1.05% C 0.2% Mn 0.2% Si steel: As AISI W1; Carpenter
NONVAR	0.92% C 1.75% Mn steel: Firth Brown
NSS 3	1% C 2% Mn steel: Press tools, jig brushes, etc.; Balfour **DPN: 660**
OTTOWA	0.8% C 0.15% Si 0.2% Mn 0.08% S & P steel: Tube; Atlas; for hollow rock drills
P 1	10.1% C steel: For moulds; designation used by AISI
QQ W 428	High carbon steel: Wire for springs; US Federal specification
QQ W 432	Carbon steel: Wire; US Federal specification

Symbol	Nominal analysis, supplier, condition and remarks.
QQ W 470	0.85% C 0.2% Si 0.4% Mn 0.03% S & P steel: Wire; US Federal specification; cold drawn; music wire **UTS: 2280**
RCS	1.0% C tool steel: Origin unknown
RED LABEL	0.9% C 0.25% Si 0.25% Mn tool steel: Huntsman **DPN: 750**
REMOUNT	1.0% C tool steel: Origin unknown
S 90	0.9% C 0.3% Si 0.5% Mn steel: Pompey
SAE 852	0.8% C Fe: Sinter for mechanical parts; density 5900 Kg/m³
SAE 855	0.8% C Fe: Sinter for mechanical parts; density 6300 Kg/m³
SAE 872	0.8% C 20% Cu Fe alloy: Sinter; density 7100 kg/m³
SAE 1084	0.86% C 0.8% Mn 0.04% P 0.05% S steel: Bar; hot rolled **DPN: 241 UTS: 900 Elon: 10% Proof: 460**
SAE 1085	0.86% C 0.9% Mn 0.04% P 0.05% S steel: Bar; hot rolled **DPN: 248 UTS: 910 Elon: 10% Proof: 480**
SAE 1086	0.86% C 0.4% Mn 0.04% P 0.05% S steel: Bar; hot rolled **DPN: 229 UTS: 830 Elon: 10% Proof: 440**
SAE 1090	0.91% C 0.8% Mn 0.04% P 0.05% S steel: Bar; hot rolled **DPN: 248 UTS: 910 Elon: 10% Proof: 480**
SAE 1095	0.95% C 0.4% Mn 0.04% P 0.05% S steel: Bar; hot rolled **DPN: 248 UTS: 900 Elon: 10% Proof: 460**
SAE W108	0.8% C tool steel
SAE W109	0.9% C tool steel
SAE W110	1.0% C tool steel
SAE W112	1.2% C tool steel
SANBOLD 44	1.0% C 0.3% Si 0.6% Mn steel: For BS alloy En 44; Sanderson
SILVER STEEL	1.1% C 0.35% Mn 0.2% Si 0.04% S & P steel: Common name; as rolled; supplied centreless ground; BS 1407
SIS 14-18	70 1.0% C 0.45% Mn 0.03% P & S steel: For cold rolled strips; Swedish Standard **DPN: 550 UTS: 1800**
SIS 14-18-80	1.0% C 0.3% Mn 0.03% P 0.02% S steel: Cold work; for tools; Swedish Standard **DPN: 95**
SIS 1774	0.8% C 0.6% Mn 0.2% Si steel: Swedish Standard; included in MNC 870
SIS 1870	1.0% C 0.45% Mn 0.03% S & P steel: Strip; cold rolled **DPN: 620 UTS: 2000**
SIS 1870	1.0% C 0.45% Mn 0.03% S & P steel: Strip; cold rolled; hardened **DPN: 600 UTS: 2000**
SIS 1870	0.95% C 0.5% Mn 0.03% S & P steel: Strip; annealed; Swedish Standard **DPN: 185 UTS: 630**
SIS 1870	0.95% C 0.5% Mn 0.03% S & P steel: Strip; cold rolled; Swedish Standard **DPN: 280 UTS: 850**
SIS 1880	1.0% C 0.3% Mn 0.025% S & P steel: Bar & forging; annealed; Swedish Standard **DPN: 195**
SIS 1885	1.2% C 0.3% Mn steel: Annealed; Swedish Standard **DPN: 195**
SK 1	1.45% C 0.5% Mn (max) 0.2% Cr & Ni (max) steel: Japanese Standard designation
SK 2	1.2% C 0.5% Mn (max) 0.2% Cr & Ni (max) steel: Japanese Standard designation

Symbol	Nominal analysis, supplier, condition and remarks.
SK 3	1.05% C 0.5% Mn (max) 0.2% Cr & Ni (max) steel: Japanese Standard designation
SK 4	0.95% C 0.5% Mn (max) 0.2% Cr & Ni (max) steel: Japanese Standard designation
SK 5	0.85% C 0.5% Mn (max) 0.2% Cr & Ni (max) steel: Japanese Standard designation
SOD	0.9% C 1.6% Mn steel: Braeburn; AISI type 02
SPEAR MD4	1.0% C 0.2% Si 0.2% Mn 0.03% S & P steel: Spear & Jackson; for coin dies
SPEAR No.2	1.0% C steel: Spear & Jackson
SPEAR No.3	0.85% C steel: Spear & Jackson
SPRASTEEL 80	0.8% C 0.7% Mn 0.04% S & P steel: Wire for spraying; Metco; as sprayed **DPN: 352 UTS: 180**
STA 5/V22A	0.8% C steel: Replaced by BS alloy En 42
STA 5/V22A/1	0.8% C steel: Replaced by BS alloy En 42
STA 5/V22A/2	0.8% C steel: Replaced by BS alloy En 42
STA 27	1.0% C steel: Free cutting; for pinions, etc.
STENTOR	0.9% C 1.6% Mn 0.2% Si steel: Carpenter for AISI 02
STERLING	1.0% C steel: Firth Sterling
STUBS	1.1% C 0.2% Si 0.35% Mn 0.45% S & P steel: As rolled; supplied centreless ground; Peter Stubs **DPN: 220 UTS: 760 Elon: 32% Proof: 580**
T 13	0.9% C 0.2% Si 0.2% Mn steel: Jessop
T 14	1% C 0.2% Si 0.2% Mn steel: Jessop
T 15	1.15% C 0.2% Si 0.2% Mn steel: Jessop
T 16	1.3% C 0.2% Si 0.2% Mn steel: Jessop
T 23	0.9% C 0.25% Si 0.25% Mn steel: Jessop
T 24	1% C 0.25% Si 0.25% Mn steel: Jessop
T 25	1.15% C 0.25% Si 0.25% Mn steel: Jessop
T 26	1.3% C 0.25% Si 0.25% Mn steel: Jessop
T 27	1.4% C 0.25% Si 0.25% Mn steel: Jessop
T 33	0.9% C 0.27% Si 0.27% Mn steel: Jessop
T 34	1.0% C 0.27% Si 0.27% Mn steel: Jessop
T 35	1.15% C 0.27% Si 0.27% Mn steel: Jessop
T 36	1.3% C 0.27% Si 0.27% Mn steel: Jessop
T 37	1.4% C 0.27% Si 0.27% Mn steel: Jessop
TD	1.0% C tool steel: Origin unknown
TEMPER No.4	1.0% C steel: Knives, wood tools, chisels, etc.; Sanderson
TEMPER No.3½	0.9% C steel: Shear blades, knives, chisels, etc.; Sanderson
TEMPER No.4	1.1% C steel: Knives, chisels, wood tools, etc.; Sanderson
TEMPER No.5	1.2% C steel: Chisels, knives, wood tools, etc.; Sanderson
TEMPER No.5½	1.3% C steel: Knives, granite drills, etc.; Sanderson
TEMPER No.6	1.4% C steel: Scrapers, surgical tools, etc.; Sanderson
TM	1.0% C tool steel: Origin unknown
TMS	0.92% C 1.75% Mn steel: Firth Brown
T QUALITY	1.0% C tool steel: Origin unknown
TREBLE EXTRA	1.0% C steel: Origin unknown
TREBLE EXTRA	0.7–1.4% C range of steels: Firth Brown
U 8	0.8% C 0.25% Mn 0.25% Ni (max) 0.2% Cr (max) steel: Russian Standard designation

Symbol	Nominal analysis, supplier, condition and remarks.
U 85	0.85% C 0.5% Mn 0.25% Ni (max) 0.2% Cr (max) steel: Russian Standard designation
U 9	0.9% C 0.25% Mn 0.2% Cr (max) 0.25% Ni (max) steel: Russian Standard designation
U 9A	0.9% C 0.22% Mn 0.15% Cr (max) 0.2% Ni (max) steel: Russian Standard designation
U 10	.0% C 0.25% Mn 0.2% Cr (max) 0.25% Ni (max) steel: Russian Standard designation
U 10A	1.0% C 0.22% Mn 0.15% Cr (max) 0.2% Ni (max) steel: Russian Standard designation
U 11	1.1% C 0.25% Mn 0.2% C (max) 0.25% Ni (max) steel: Russian Standard designation
U 11A	1.1% C 0.22% Mn 0.15% Cr (max) 0.2% Ni (max) steel: Russian Standard designation
U 12	1.2% C 0.25% Mn 0.2% Cr (max) 0.25% Ni (max) steel: Russian Standard designation
U 12A	1.2% C 0.22% Mn 0.15% Cr (max) 0.2% Ni (max) steel: Russian Standard designation
U 13	1.3% C 0.25% Mn 0.2% (max) 0.25% Ni (max) steel: Russian Standard designation
U 13A	1.3% C 0.22% Mn 0.15% Cr (max) 0.2% Ni (max) steel: Russian Standard designation
U BRAND	0.7–1.7% C range of steels: Firth Brown
UGA	0.85% C 0.52% Mn 0.15% Cr (max) 0.2% Ni (max) steel: Russian Standard designation
UHB	1.0% C steel: For springs, Uddelholm
UHB 20	1.04% C 0.2% Si 0.25% Mn steel: Uddelholm; for type W1
VEDAS	0.8–1.2% C steel: Available in 4 grades depending on C content; Hall & Pickles; obsolete
VEDAS	1.0% C steel: Origin unknown
VHRD	1.0% C tool steel: Origin unknown
VIZOR	0.8–1.2% C steel: Available in 4 grades depending on C content; Hall & Pickles; obsolete
VIZOR	1.0% C steel: Origin unknown
W 1	1.0% C steel: C varies iver wide limits; designation used by AISI
W 1	1.0% C steel: Osborn
W 110	1.0% C steel: For tools; designation used by SAE
W 112	1.2% C steel: For tools; designation used by SAE
WLM 9015	0.9% C 1.8% Mn steel: Press tools & shear blades; W Marrison **DPN: 720**
WLM SPECIAL	0.7–1.3% C 0.03% S & P steel: Temper grade varies with C content; W Marrison
WORTLE PLATE	2.5% C steel: Used in the manufacture of wire drawing dies; origin unknown
WTS 1%	1.0% C 0.15% Si 0.35% Mn steel: Blanking tools & shear blades; Marsh Brothers
Y 90	0.9% C 0.25% Mn 0.03% P 0.02% S steel: French Standard designation
Y 90V	0.9% C 0.25% Mn 0.025% P & S (max) steel: French Standard designation
Y 105	1.05% C 0.02% P & S (max) steel: French Standard designation
Y 105	1.05% C 0.25% Mn 0.025% P & S (max) steel: French Standard designation
Y 105	1.05% C 0.25% Mn 0.025% P & S (max) steel: French Standard designation
Y 120	1.2% C 0.02% P & S (max) steel: French Standard designation
Y 120	1.20% C 0.02% P & S (max) steel: French Standard designation
Y 120	1.2% C 0.25% Mn 0.025% P & S (max) steel: French Standard designation
Y 135	1.35% C 0.25% Mn 0.025% P & S (max) steel: French Standard designation
YELLOW LABEL	0.9% C 0.03% S & P steel: T Turton

Note. The following abbreviations and units are used in the tables:

DPN	Hardness, diamond pyramid number
UTS	Ultimate tensile strength, N/mm^2
Elon	Elongation, %
Proof	0.1 % proof strength, N/mm^2

1 N/mm^2=0.1 hbar=0.102 kgf/mm^2=0.06475 tonf/in^2=145.04 lbf/in^2

44B1 Steel – low carbon with silicon
0.05% carbon – 3.5% silicon

Specific gravity	7.5–7.7	
Density	7500–7700 kg/m^3	(0.27–0.28 lb/in.^{3})
Solidus/liquidus	1470–1500 °C	
Thermal conductivity	50.2 W/m °C	(12 cal/m s °C)
Coefficient of linear expansion	10 × 10^{-6}/ °C	
Electrical conductivity	– IACS (copper 100%)	
Specific resistance	– microhm mm	
Young's modulus of elasticity	– N/m^2	
Impact	– J	
Fatigue strength (no. of cycles)	–	
Hot strength	–	

The above properties are typcial of the following group, and may not apply exactly to any one specification. It is possible that with certain specifications some of the values may not be applicable.

General metallurical characteristics

This is a completely ferritic material sometimes known as silicon iron and would be more correctly shown with the pure irons. The materials however, are generally known as 'silicon steels' and are conveniently listed here.

There is insufficient carbon present for any thermal hardening, but the silicon gives some increase in strength with a tendency to brittleness and causes the material to work harden. Silicon also improves the corrosion resistance. Very often Swedish iron is used to manufacture the steels listed as this has a minimum of residual elements and is inherently low in sulphur and phosphorus.

The materials have peculiar electrical and magnetic properties which are the reason for their existence, the only useful form being thin sheet. This is generally so that the grains are preferentially oriented, thus improving the magnetic properties in one direction.

Thermal treatment

These materials are always used annealed and this must be carried out in the finished condition to remove the cold work of blanking. Annealing requires a temperature of 920–950 °C and as it is always on thin sheet for electrical use some form of controlled atmosphere is essential, hydrogen or cracked ammonia being the most common. If the parts are closely packed the time at temperature must be increased to give at least 10 minutes per 25 mm of section of the closely packed mass, not the individual sheets.

None of the materials listed is hardenable, thus the rate of cooling after annealing is not critical, but a controlled cooling rate is sometimes necessary to give the best electrical or magnetic properties.

Welding and brazing

Pre-treatment. Parts should be annealed and free from rust. All traces of oil or grease must be removed to prevent carbon pick up local to the weld.

Welding and brazing. All normal forms of welding give satisfactory results. Resistance welding is the one most commonly required. Brazing is readily carried out using copper alloy filler rod and flux.

Post-treatment. No thermal treatment necessary. All traces of flux must be removed.

Flaw detection methods

Crack test. Magnetic crack test may be required, but is not generally necessary.

X-ray. Not required.

Ultra-sonic test. Not required.

Chemical etch. Not required on production parts. A laboratory etch to show the grain flow for parts with preferred orientation may be necessary. This must be carried out on a test piece which is then scrap.

Corrosion protection

Temporary. The sheet and strip must be oiled or greased during storage. Parts can be lightly oiled during the manufacturing cycle and must be kept free of moisture at all times.

Permanent. The fact that there is very little carbon present, added to the presence of silicon, gives these materials considerable built-in corrosion resistance.

They are seldom used under arduous corrosion conditions and thus require very little additional protection. This protection must be carefully chosen not to interfere with the electrical or magnetic properties required in the finished part.

Painting, using zinc or lead oxide or chromate primers with normal air drying top coat, generally gives sufficient protection. Care must always be taken to ensure freedom from dirt and corrosion before applying the primer. This is best achieved by blasting.

Phosphate treatment, by itself or as a paint base, can be used.

In general these materials are not used in the open air and have sufficient natural corrosion résistance for most purposes. They are often used with an electrical insulator which is chosen to prevent corrosion. They may also have surface oxidation treatments applied to produce an adherent oxide which acts as an insulator and corrosion inhibitor.

Machinability

These materials are seldom machined. The presence of silicon means there is considerable abrasive present which reduces tool life. This applies also to press tools.

Uses

These steels are used for laminations on all types of transformers, including those on loud speakers.

Symbol	Nominal analysis, supplier, condition and remarks.
IS 15	0.05% C (max) 1.4% Si 0.1% Cu steel: Sandvik
IS 35	0.05% C (max) 3.5% Mn steel: Sandvik
AMS 7711	1.5% Si steel: Sheet & strip for electrical purposes
AMS 7712	2.5% Si steel: Sheet & strip for electrical purposes
AMS 7714	3% Si steel: Sheet & strip for electrical purposes
AMS 7715	4.25% Si steel: Sheet & strip for electrical purposes
BS 933/3	Si steel: Sheet for magnetic purposes; composition not part of specification
LOSIL 17	2.3% Si hot rolled dynamo steel: Electrical resistivity 400 microhm mm BS 601/1; Steel Co of Wales **DPN: 175 UTS: 460 Elon: 19% Proof: 320**
LOSIL 17C	2.3% Si hot rolled & annealed dynamo steel: Electrical resistivity 400 microhm mm; Steel Co of Wales **DPN: 175 UTS: 460 Elon: 22% Proof: 320**
LOSIL 19	1.7% Si hot rolled dynamo steel: Electrical resistivity 330 microhm mm; Steel Co of Wales **DPN: 150 UTS: 420 Elon: 21% Proof: 270**
LOSIL 19C	1.7% Si hot rolled & annealed dynamo steel: Electrical resistivity 330 microhm mm; Steel Co of Wales **DPN: 150 UTS: 420 Elon: 25% Proof: 270**
LOSIL 22	1.3% Si hot rolled dynamo steel: Electrical resistivity 280 microhm mm; Steel Co of Wales; BS 601/1 **DPN: 130 UTS: 390 Elon: 22% Proof: 240**
LOSIL 22C	1.3% Si hot rolled & annealed dynamo steel: Electrical resistivity 220 microhm mm; Steel Co of Wales **DPN: 105 UTS: 340 Elon: 28% Proof: 200**
LOSIL 25	0.4% Si hot rolled dynamo steel: Electrical resistivity 170 microhm mm; Steel Co of Wales; BS 601/1 **DPN: 90 UTS: 320 Elon: 25% Proof: 170**
LOSIL 25C	0.4% Si hot rolled & annealed dynamo steel: Electrical resistivity 170 microhm mm; Steel Co of Wales **DPN: 90 UTS: 320 Elon: 31% Proof: 170**
M 6W	Low C Si steel: Strip; grain oriented; Armco
M 7W	Low C Si steel: Strip; grain oriented; Armco
M 7X	Low C Si steel: Strip; grain oriented; Armco
M 8X	Low C Si steel: Strip; grain oriented; Armco
TRANSIL 74	4.2% Si hot rolled steel: For transformers; electrical resistivity 620 microhm mm; Steel Co of Wales; BS 601/1 **DPN: 265 UTS: 560 Elon: 4% Proof: 460**

Symbol	Nominal analysis, supplier, condition and remarks.
TRANSIL 80	4.1% Si hot rolled steel: For transformers; electrical resistivity 610 microhm mm; Steel Co of Wales; BS 601/1 **DPN: 260 UTS: 560 Elon: 6% Proof: 460**
TRANSIL 86	4% Si hot rolled steel: For transformers; electrical resistivity 590 microhm mm; Steel Co of Wales; BS 601/1 **DPN: 255 UTS: 560 Elon: 9% Proof: 450**
TRANSIL 92	3.9% Si hot rolled steel: For transformers; electrical resistivity 580 microhm mm; Steel Co of Wales; BS 601/1 **DPN: 250 UTS: 560 Elon: 11% Proof: 450**
TRANSIL 100	3.6% Si hot rolled transformer steel: Electrical resistivity 550 microhm mm; Steel Co of Wales; BS 601/1 **DPN: 235 UTS: 560 Elon: 13% Proof: 420**
TRANSIL 107	3.4% Si hot rolled transformer steel: Electrical resistivity 520 microhm mm; Steel Co of Wales; BS 601/1 **DPN: 225 UTS: 540 Elon: 14% Proof: 400**
TRANSIL 115	3.2% Si hot rolled transformer steel: Electrical resistivity 500 microhm mm; Steel Co of Wales; BS 601/1 **DPN: 215 UTS: 530 Elon: 15% Proof: 390**
UNISIL 46	3.1% Si cold reduced steel: For transformers; electrical resistivity 480 microhm mm; Steel Co of Wales; BS 601/11 **DPN: 165 UTS: 360 Elon: 13% Proof: 300**
UNISIL 51	3.1% Si cold reduced steel: For transformers; electrical resistivity 480 microhm mm; Steel Co of Wales; BS 601/11 **DPN: 165 UTS: 360 Elon: 13% Proof: 300**
UNISIL 56	3.1% Si cold reduced steel: For transformers; electrical resistivity 480 microhm mm; Steel Co of Wales; BS 601/11 **DPN: 165 UTS: 360 Elon: 13% Proof: 300**
UNISIL 62	3.1% Si cold reduced steel: For transformers; electrical resistivity 480 microhm mm; Steel Co of Wales; BS 601/11 **DPN: 165 UTS: 360 Elon: 13% Proof: 300**

Note. The following abbreviations and units are used in the tables:

DPN	Hardness, diamond pyramid number
UTS	Ultimate tensile strength, N/mm^2
Elon	Elongation, %
Proof	0.1 % proof strength, N/mm^2

$1\ N/mm^2 = 0.1\ hbar = 0.102\ kgf/mm^2 = 0.06475\ tonf/in^2 = 145.04\ lbf/in^2$

44B2 Steel – medium carbon with silicon

Specific gravity	7.86	
Density	7860 kg/m³	(0.28 lb/in.³)
Solidus/liquidus	1460–1500 °C	
Thermal conductivity	41.9 W/m °C	(10 cal/m s °C)
Coefficient of linear expansion	$10 \times 10^{-6}/$ °C	
Electrical conductivity	– IACS (copper 100%)	
Specific resistance	– microhm mm	
Young's modulus of elasticity	– N/m²	
Impact	– J	
Fatigue strength	–	
Hot strength	–	

The above properties are typical of the following group, and may not apply exactly to any one specification. It is possible that with certain specifications some of the values may not be applicable.

General metallurgical characteristics

Silicon raises the heat treatment temperatures required for all thermal treatments and is a graphite former, thus reducing the carbon available for hardening and aiding decarburization. Practically without exception all steels have some silicon added as it is a very efficient and cheap de-oxidizer, the product of which has little effect on ductility if trapped in the steel.

When less than 0.5% silicon is present it is not classified as an alloying element as it has no appreciable effect on the mechanical properties. Above this figure there is some increase in corrosion resistance and fatigue strength although not to such a marked degree as with other alloying elements.

There is a process analogous to carburizing which absorbs silicon into the steel surface, converting all the carbon to graphite and giving a high silicon layer which has excellent resistance to many acids.

The silicon steels all have fairly good fatigue strength with the very great advantage that notches for any reason have a minimum effect on reducing this fatigue strength. It is seldom used as the sole alloying element, generally being present with chromium and very often vanadium.

Thermal treatment

These steels require annealing after relatively little cold work and this requires a temperature of 760–860 °C for 20 minutes per 25 mm of section followed by slow cooling. The tendency of silicon to accentuate decarburization must always be remembered and unless parts are to be machined after heat treatment efforts must be taken to exclude air. At annealing use can be made of cast iron turnings, sand or other refractory material in the form of small particles.

Sub-critical annealing at 550–600 °C for two hours will be found suitable for most purposes and can be carried out using an air atmosphere.

Normalizing requires 880–920 °C for 20 minutes per 25 mm of ruling section followed by still air cooling. Best results are obtained by heating in a box with cuttings or sand and removing the parts for cooling, there should be no appreciable air hardening with these steels. Hardening requires a temperature of 840–900 °C for 10 minutes per 25 mm section followed by water or oil quenching. Parts can be heated in enclosed boxes along with a handful of charcoal, the lid being removed to allow the parts to be efficiently quenched.

Controlled atmosphere furnaces or salt baths can be used for heating, but in general these materials are used for the less sophisticated parts manufactured without the aid of such expensive equipment. They are often stabilized at 520–600 °C for three hours to prevent dimensional movement during service.

None of these materials can be surface hardened by any known method.

Welding and brazing

Pre-treatment. The weld area must be clean and free from oxide and scale.

Welding and brazing. When used in springs it is strongly recommended that parts are not welded or brazed without first obtaining metallurgical advice. For parts other than springs, gas or electric arc welding with flux in skilled hands, followed by the correct thermal treatment, can minimize but cannot eliminate the cast structure which always results in some weakening.

Brazing is possible but will generally cause softening. Use of high melting point fillers allows the parts to be rehardened after brazing.

None of the materials listed should have welding or brazing specified by design.

Post-treatment. All trace of flux must be removed. The parts must be at least stress relieved at 500–600 °C and should be hardened and tempered if at all possible.

Flaw detection

Crack test. Magnetic crack test. The penetrant oil chalk test method with modern sensitive fluids can give comparable results and is often more flexible.

X-ray. Not required.

Ultra-sonic test. Not normally required.

Chemical etch. Not required.

Corrosion protection

Temporary. The materials should be kept free from moisture, and oiled or greased at all times. After machining parts must be dried, then thoroughly greased until ready for use.

Permanent. These materials are generally used for purposes such as road and rail vehicle springs and similar duties where corrosion protection of a permanent nature is not economically possible.

The parts should be blasted, then treated with one of the common surface treatments such as phosphate, chromate, or chemical oxidation. Proprietary solutions are available for all of these treatments. A thin coat of primer paint is sometimes used to prevent corrosion until the parts are put into use.

During operation the parts should be kept well oiled or greased.

Machinability

In the hardened and tempered condition machining produces a good finish but results in considerable tool wear. Apart from the hardness, the presence of silicon increases the abrasive nature of the steel.

In the annealed condition there is a tendency for tearing to take place with little or no reduction in tool wear.

Uses

The use of these steels is confined to low duty leaf type springs and tool steels.

Symbol	Nominal analysis, supplier, condition and remarks.
AISI 9250	0.5% C 2% Si steel: Obsolete
AISI 9255	0.55% C 2.0% Si steel
AISI 9260	0.6% C 2.0% Si steel
ASTM A59	0.6% C 0.8% Mn 2% Si steel: Bars for springs; obsolete; see A552
CRD SILICO	0.55% C 2% Si 0.8% Mn steel: Shear blades; C Denton **DPN: 540**
S 4	0.55% C 0.8% Mn 2.0% Si steel: Designation used by AISI
SAE 9250	0.5% C 2% Si steel: Obsolete
SAE 9255	0.55% C 0.8% Mn 2% Si steel: Mechanical properties not specified

Symbol	Nominal analysis, supplier, condition and remarks.
SAE 9260	0.6% C 0.8% Mn 2% Si steel: Mechanical properties not specified

Note. The following abbreviations and units are used in the tables:

DPN	Hardness, diamond pyramid number
UTS	Ultimate tensile strength, N/mm^2
Elon	Elongation, %
Proof	0.1 % proof strength, N/mm^2

$1 N/mm^2 = 0.1 hbar = 0.102 kgf/mm^2 = 0.06475 tonf/in^2 = 145.04 lbf/in^2$

44C Steel – low and medium carbon with boron

Specific gravity	7.86	
Density	7860 kg/m³	
Solidus/liquidus	1490–1520 °C	
Thermal conductivity	50.2 W/m °C	(12 cal/m s °C)
Coefficient of linear expansion	$11 \times 10^{-6}/$ °C	
Electrical conductivity	8–10% IACS (copper 100%)	
Specific resistance	180–250 microhm mm	
Young's modulus of elasticity	$90 \times 10^9 N/m^2$	($13 \times 10^6 lbf/in.^2$)
Impact	54 J	(40 ft/lb)
Fatigue strength (no. of cycles)	–	

Hot strength

Temperature °C	Tensile strength N/mm²	Elongation %
100	420	34
200	450	27
300	470	23
400	370	37
500	300	38
600	200	48
700	100	56
800	60	65
900	35	75

The properties above are typical of the following group and may not apply exactly to any one specification. It is possible that with certain specifications some of the values may not be applicable.

General metallurgical characteristics

Boron in very small amounts has a startling affect on the hardenability of steel. For example low carbon sheet material which would not air harden in plain carbon steel will be fully air hardenable with the addition of 0.0015% boron. Similar differences are found with thicker sections which can be successfully oil hardened where previously water quenching only achieved a degree of surface hardening. Boron also enhances the hardenability of other alloying elements and in the United States of America at least, the element is being used as a very economical substitute for some of the more expensive elements.

Larger quantities of boron result in brittle, unworkable steels.

The beneficial effects of boron are only apparent with low and medium carbon steels, there being no real increase in hardenability above 0.6% carbon. The number of plain boron steels in use is limited but it is commonly found in conjunction with vanadium, molybdenum and chromium. There is no improvement in corrosion resistance but the reduction of alloying elements often improves the machinability.

The weldability of boron steels is one of the principal reasons for their use in this country at present.

The presence of this alloying element is often not noticed or ignored because of the very small amounts present in steel.

Thermal treatment

Boron has no appreciable affect on the critical change points, thus boron steels require the same temperatures as plain carbon steels. Boron steels have the same cold working properties as plain carbon steels, thus the need for interstage annealing will depend on the carbon content. Where necessary this should be at 880–980 °C for 30 minutes per 25 mm of section and slowly cooled. Heating to 650–670 °C for two hours removes most cold work and is called 'sub-critical annealing' which is not suitable for interstage annealing when further deep drawing is to be carried out. Normalizing also requires that the part is heated to 880–980 °C for 15 minutes per 25 mm of section followed by cooling in still air. This removes all forging stresses, homogenizes the structure and results in some air hardening which usually necessitates a temper before parts can be machined. This is the most usual condition in which the steels are used.

Hardening requires about 10 minutes per 25 mm of section at a temperature of 840–940 °C followed by oil quenching then tempering. Boron unlike many alloying elements narrows the range of temperatures for correct thermal treatment, and it is therefore most important that the origin of the specification being treated is consulted for exact times and temperatures.

Welding and brazing

Pre-treatment. Parts must be in the normalized, or hardened and tempered condition and it is essential that all oil, grease, dirt scale and rust are removed. Edge preparation must be carefully carried out to ensure that the best results are obtained from these weldable steels. Pre-heating will usually be required.

Welding and brazing. These steels have generally been developed specifically for their weldability, having a controlled hardenability with a fine grained structure.

Gas or electric arc welding, using flux, or inert gas shielded arc welding, can all be used with excellent results.

Resistance welding of all types is successful provided the correct equipment is used with the necessary technical control. Automatic equipment is often used to give very rapid weld speeds with these materials.

Brazing presents no difficulty provided care is taken not to heat the part above the hardening temperature as this will result in local softening. The use of high temperature braze metal allows the parts to be hardened and tempered after brazing and is recommended wherever possible.

Post-treatment. Welds must be tempered. The advantage of these steels is that the mechanical properties of the welded assembly can be almost fully recovered by the simple temper treatment. Where necessary this can be carried out locally using gas torches or portable electric blankets.

The same treatment can be applied to parts which have been high temperature brazed. All trace of flux and oxide must be removed after welding or brazing.

Flaw detection

Crack test. Magnetic crack test or penetrant oil chalk test.

X-ray. Wrought products are not normally inspected by this method, but mass production welds and high duty welds and brazes are commonly subjected to X-ray. Portable equipment using radioactive metal isotopes is available which gives excellent results under all conditions.

Ultra-sonic test. This must be required to prove that forgings with expensive machining operations had no massive defects which would become dangerous after machining. There are also techniques now available which show certain types of weld defects.

Chemical etch. Not required.

Corrosion protection

Temporary. These steels must be kept free from moisture at all times and should be oiled or greased during storage.

During machining and welding there may be operations when the parts become wet. If possible the coolant should be inhibited and if this is not possible the part should be dried immediately and oiled to prevent rust forming.

Permanent. These materials have superior hardenability to plain carbon steel, but have no better corrosion resistance and the result of corrosion pitting can be more serious with the increase in hardness. In order to protect against corrosion it is essential to remove all trace of rust, then prevent any contact between the atmosphere and the steel. Rusting is the formation of complex oxides which are only loosely adherent and leave the surface free for further corrosion when it is removed. This leads to pitting which can have a serious effect on the fatique strength. Rust always has moisture and air occluded within itself and this can continue to act even when covered by a paint film or plating deposit. This is the most common reason for failure of corrosion protection films.

The best method for cleaning and removing existing corrosion is by grit blasting. This must be carefully controlled and the resultant light covering of dust removed by brushing. Blasted surfaces can corrode in a very short time in humid conditions and thus should not be treated until immediately before the blasted surface is to be protected.

Painting. Ideally the parts should be phosphate or chromate treated and oven dried prior to painting. A primer of the zinc chromate type should then be applied followed by stoved top coats to seal and supply the impact and abrasion resistance.

The phosphate and chromate coatings have certain corrosion preventative qualities and are excellent oil retentive surfaces. Provided this oil film is maintained, adequate indoor corrosion protective is generally afforded.

Plating. These steels will seldom be electroplated but present no difficulties provided normal pre-plating, cleaning and etching are carried out. Chromium plating should have at least 0.05 mm copper followed by 0.025 mm nickel deposited under the thin chromium deposit.

Cadmium and zinc require less than 0.025 mm thick deposits for use under normal atmospheric conditions. Parts with a hardness above 350 DPN (900 N/mm^2 tensile) should be stress relieved at 200 °C after plating, and if considerable machining is involved a pre-plating stress relieve is recommended. These are necessary to remove the danger of hydrogen embrittlement which can cause brittle failure below the calculated tensile strength. Welds are particularly prone to this defect which is entirely removed by the stress release. Unfortunately there is no visual method of ensuring that stress relieving has been carried out and it is recommended that an alternative to electroplating is used where high duty welds are involved. Metal spraying, particulary with zinc or aluminium gives excellent results provided a good grit blast key is obtained before spraying, and the coating is sealed by wire brushing and chemical treatment or painting after spraying.

Cathodic protection uses the electroplating principles of a sacrificial anode electrically connected to the part being protected which is then negative or cathodic. This system is used on ships and marine structures as well as pipe lines, buildings and pylons.

The corrosion protection of these steels relies on careful preparation and good quality materials applied under controlled conditions. The degree of protection obtained is proportional to this care and the quality of the materials used.

Machinability

In the soft condition there is a tendency to tear rather than cut. When normalized and tempered these steels exhibit machinable characteristics which are superior to their plain carbon equivalent.

Uses

Welded structures, pipes, high duty sheet metal engineering components. Bosses, brackets and studs which require welding. At the time of going to press no steels containing boron without other alloying elements have been found. This section has however, been retained to supply the necessary information on the affect of boron on steels. Boron steels will be found in Sections 44G1 and 44K1.

44D Steel–manganese alloys

Specific gravity	7.7	
Density	7700 kg/m³	(0.28 lb/in.³)
Solidus/liquidus	1430–1520 °C	
Thermal conductivity	50.2 W/m °C	(12 cal/m s °C)
Coefficient of linear expansion	Variable	
Electrical conductivity	– IACS (copper 100%)	
Specific resistance	– microhm mm	
Young's modulus of elasticity	– N/m²	
Impact	– J	
Fatigue strength	–	
Hot strength	–	

The above properties are typical of the following group, and may not apply exactly to any one specification. It is possible that with certain specifications some of the values may not be applicable.

General metallurgical characteristics

Manganese is present in almost all steels but is not acknowledged as an alloying element unless present above 1.65%. It is essentially a de-oxidizer and desulphurizer which is popular because any manganese oxide or sulphide trapped in the steel instead of the slag is more ductile than the steel itself. This means that all hot and cold working operations can take place without refractory oxides or sulphides tearing cavities in the more ductile steel.

Manganese also lowers the critical heat treatment temperatures. With more than about 15% manganese the critical heat treatment temperature is below room temperature thus forming an austenitic, non-heat treatable steel.

Manganese also strengthens steel and this is particularly the case with higher carbon steels where 2–5% manganese gives a steel which is toughened and hardened by cold work. The presence of silicon aids this toughening process.

Thermal treatment

These steels all exhibit considerable cold working properties and require full annealing after very limited deformation. This requires about 30 minutes per 25 mm of section at 760–780 °C followed by slow cooling. Sub-critical annealing at 650–670 °C for two hours generally removes sufficient cold work to allow machining but not further heavy cold work. Normalizing requires 20 minutes per 25 mm of section at 760–780 °C followed by air cooling. This removes any forging or cold work stress and homogenizes the grain structure, generally with a little air hardening, which can be considerable if residual traces of nickel, chromium or molybdenum are present.

Hardening requires 700–820 °C for about 10 minutes per 25 mm of section followed by oil or water quenching. This results in less distortion or dimensional changes than equivalent plain carbon steels with no reduction in hardness. Tempering at 180–300 °C reduces the hardness and improves the ductility of the steels.

These steels cannot be carburized as the carbon content is always greater than that achieved by this process.

Welding and brazing

Pre-treatment. All traces of grease and dirt must be removed.

Welding and brazing. Welding is seldom necessary or desirable on these materials as the local cast area will always have a tendency to be brittle. With skilled operation and correct post-weld treatment this can be minimized but never eliminated. Resistance butt welding with the cast area squeezed out gives good results, but requires special equipment with considerable technical control.

Brazing presents no difficulties, except that local softening around the braze will be encountered. This can be prevented by the use of braze fillers which melt above the hardening temperature, and re-hardening after brazing.

Post-treatment. Welds should be normalized, then hardened and tempered if possible but at least should be stress relieved at the tempering temperature.

As stated above, brazing using high temperature filler materials should also be hardened and tempered.

All fluxes and slag must be removed before heat treatment to prevent serious corrosion.

Flaw detection

Crack test. Magnetic crack test. Penetrant oil chalk test with modern high sensitive fluids can give comparable results to the magnetic method and is more flexible.

X-ray. Not normally required.

Ultra-sonic test. Not normally required.

Chemical etch. Not required.

Corrosion protection

Temporary. These steels must be oiled or greased at all times and stored under cover. During machining and subsequent storage parts should be kept free from moisture and oiled or greased.

Permanent. These steels are used under conditions which make permanent corrosion protection difficult if not impossible. They should be kept free from moisture and when not in use oiled at all times. The retention of this oil film on non-working surfaces is helped by phosphate, chromate or black oxide treatments. These are all easily controlled proprietary chemical dip processes which kill active corrosion to some extent and supply a measure of protection. If a more permanent treatment is required these steels should be treated in a similar manner to the plain high carbon steels described in Section 44A4.

Even in the annealed condition there will be a considerable increase in hardness with certain machining operations particularly if blunt tools or interrupted cuts are used. Heavy cuts at low speeds probably give best results and at no time should the tool be allowed to rub without cutting. This applies particularly to drilling operations.

Machinability

All these steels harden considerably with cold work and as this is additive to the hardness achieved by heat treatment they are seldom machinable after heat treatment.

Uses

Chisels and picks, low duty low cost drills, taps, drill bushes, shear blades and other duties calling for economical, low life, wear resistance.

Symbol	Nominal analysis, supplier, condition and remarks.
1.3402	1.1% C 14.0% Mn 0.4% Si steel: German Standard
1.3405	0.9% C 17.5% Mn 0.8% Si steel: German Standard
1.3802	1.2% C 12.5% Mn steel: German Standard
13 Mn R	0.9% C 13.0% Mn steel: Electrode metrode; for hard facing; work hardens from 200 to 500 DPN
17 Mn Mo R	0.9% C 17.0% Mn 1.0% Mo steel: Electrode metrode; for hard facing; work hardens from 200 to 500 DPN
ABRADUR M14	1.2% C 0.35% Si 13% Mn steel: Pompey; wear resistant; annealed DPN: 200 UTS: 1070 Elon: 55% Proof: 360
ABRADUR M14N	1.2% C 0.7% Si 13.5% Mn 2.7% Ni steel: Pompey; wear resistant
AM	1.25% Mn austenitic steel: Casting; Wokingham DPN: 229
ASTM A128 A	1.2% C 11.0% Mn steel: Casting
ASTM A128 B1	1.0% C 12.5% Mn steel: Casting
ASTM A128 B2	1.1% C 12.5% Mn steel: Casting
ASTM A128 B3	1.2% C 12.5% Mn steel: Casting
ASTM A128 B4	1.28% C 12.5% Mn steel: Casting
ASTM A128 C	1.2% C 12.5% Mn 2.0% Cr steel: Casting
ASTM A128 D	1.0% C 12.5% Mn 1.0% Mo steel: Casting
ASTM A128 E1	1.0% C 12.5% Mn 1.0% Mo steel: Casting
ASTM A128 E2	1.2% C 12.5% Mn 2.0% Mo steel: Casting
AURIGA X	1.2% C 12% Mn steel: Casting; D Brown; austenitic; cold working
AURIGA XI	0.8% C 12% Mn 3.5% Ni steel: Casting; D Brown; normalized; austenitic; cold working
AURIGA XII	1.2% C 12% Mn 1.7% Cr steel: Casting; D Brown; austenitic; cold working
BS 1457	1.2% C 11.0% Mn steel: Casting; annealed; contained in BS 3100
CODE 630	1.3% C 14% Mn 1.6% Cr steel: Casting; Edgar Allen; 'Duraloy'
CODE 631	1.1% C 12% Mn steel: Casting; Edgar Allen; 'Ord Mn 5'
CODE 633	1.5% C 17.5% Mn steel: Casting; Edgar Allen; 'Mang 18'
CODE 634	1% C 12% Mn steel: Casting; Edgar Allen; 'Tumblers'
CRUCIBLE 13% Mn	1.2% C 0.2% Si 13.0% Mn steel: Annealed; Crucible Steel DPN: 180 UTS: 940 Elon: 55% Proof: 360
DURALOY	1.3% C 14% Mn 1.6% Cr steel: Casting; Edgar Allen; Code 630
EPOK	13% Mn steel: Casting; annealed; Firth Brown DPN: 220

Symbol	Nominal analysis, supplier, condition and remarks.
HARDTRODE 86.08	1.10% C 13.0% Mn steel: For electrodes; Esab
IMPERIAL MANGANESE	1.15% C 12.5% Mn steel: Chisels; Edgar Allen
JS 14	1.2% C 11.0% Mn steel: Casting; Jopling; for wear resistance DPN: 230
MANG 18	1.5% C 17.5% Mn steel: Casting; Edgar Allen; Code 633
NN 3	1.1% C 13% Mn steel: Bofors
ORD Mn5	1.1% C 12% Mn steel: Casting; Edgar Allen; Code 631
RED DIAMOND	0.45% C 1.75% Mn steel: Plate for wear; Redheugh DPN: 270 UTS: 940 Elon: 15% Proof: 550
SIS 2183	1.2% C 12.5% Mn steel: Casting; Swedish Standard
TITAN	1.2% C 13% Mn steel: Chisel steel; Osborn DPN: 210 UTS: 940 Elon: 50% Proof: 340
TITAN MANGANESE	1.2% C 13.0% Mn steel: Cold work; Osborn
TRM 45	0.4% C 15% Mn 5% Ni steel: Colt Ind; hot rolled DPN: 171 UTS: 830 Elon: 70% Proof: 260
TUBROD 15.60	1.0% C 13% Mn steel: Tubular weld rod; Esab; for wear resistance; cold worked DPN: 520
TUMBLERS	1% C 12% Mn steel: Casting; Edgar Allen; Code 634
VS MANG	1.15% C 0.3% Si 12% Mn steel: Cold working steel; Vulcan DPN: 200
X 90 Mn 18	0.9% C 17.5% Mn 0.8% Si steel: Designation used by German Standard
X 110 Mn 14	1.1% C 14.0% Mn 0.45% Si steel: Designation used by German Standard
X 120 Mn 12	1.2% C 12.5% Mn steel: Designation used by German Standard

Note. The following abbreviations and units are used in the tables:

DPN	Hardness, diamond pyramid number
UTS	Ultimate tensile strength, N/mm^2
Elon	Elongation, %
Proof	0.1 % proof strength, N/mm^2

1 N/mm^2=0.1 hbar=0.102 kgf/mm^2=0.06475 tonf/in^2=145.04 lbf/in^2

44E1 Steel – low carbon with chromium
0.1–0.35% carbon, up to 9% chromium

Specific gravity	7.84	
Density	7840 kg/m³	(0.28 lb/in.³)
Solidus/liquidus	1470–1510 °C	
Thermal conductivity	50.2 W/m °C	(12 cal/m s °C)
Coefficient of linear expansion	11 × 10⁻⁶/ °C	
Electrical conductivity	8–10% IACS (copper 100%)	
Specific resistance	180–250 Microhm mm	
Young's modulus of elasticity	– N/m²	
Impact	– J	
Fatigue Strength (no. of cycles)	–	
Hot strength		

Temperature °C	Tensile strength N/mm²	Elongation %
200	480	38
425	400	36
480	340	38
625	210	55

The above properties are typical of the following group, and may not apply exactly to any one specification. It is possible that with certain specifications some of the values may not be applicable.

General metallurical characteristics

Chromium forms stable carbides with available carbon, the carbon combining with chromium in preference to iron. These are harder than iron carbide and more sluggish in their metallurgical reactions, thus a longer time at temperature is necessary to allow solution to take place and the quenching rate can be slower.

As these carbides are inherently harder the unhardened steel is harder than an equivalent plain carbon steel. Chromium in fact is the most efficient of the common hardening elements.

Chromium also imparts corrosion resistance as any chromium at the steel surface becomes oxidized and this improves the resistance to rusting. Unlike rust it is a stable very adherent oxide which inhibits further oxidation. Corrosion resistance is proportional to the chromium content but none of the steels listed can be classed as stainless or corrosion resistant although they do not rust as readily as plain carbon steel.

Being a carbide former chromium aids carburization and some of the lower carbon steels listed were developed for carburizing steels. The case formed tends to be brittle, and as the core is relatively soft these steels should not be used for high duty purposes.

Many chrome steels become brittle and notch sensitive at sub-zero temperatures.

Some of the steels have small amounts of copper additions. This improves the corrosion resistance and it is reported helps the adhesion of paint.

Larger quantities of chromium are present in all stainless steels which are described in Sections 44M and 44N.

Thermal treatment

These steels require annealing after relatively small amounts of cold work. This will be at 850–880 °C for 1 hour per 25 mm of section followed by slow cooling. For many purposes a sub-critical anneal at 650–700 °C will drop the hardness to the required extent with less distortion and scale formation. Normalizing is recommended after forging, especially if the parts are to be carburized, or if final dimensional stability is important. This requires 850–880 °C for 30 minutes per 25 mm of section then cooled in still air. As the chromium content imparts a degree of air hardening the properties will be improved to some extent and some of the materials listed use this normalize as the final condition. Cooling in a draught free situation is important to prevent unequal stresses.

Hardening requires the part to be at 820–850 °C for 20 minutes per 25 mm of section followed by oil or water quenching. The higher carbon chrome steels should not be water quenched as there is some danger of quench cracking. Tempering at 150–300 °C should follow immediately after hardening. Some of these steels can become brittle if tempered in the 350–500 °C range, this being known as 'temper brittleness' and can be reduced by quenching after tempering, but it is recommended that an alternative molybdenum bearing steel is chosen if the desired hardness can only be obtained by tempering in this range.

As chromium carbides are only taken into solution slowly, it is important that the correct time and temperature is used. The figures given above are for guidance only, reference must always be made to the specification or material supplier for detailed information.

These materials can be carburized by any of the normal methods. It is not recommended that case depths of less than 0.25 mm or more than 1 mm be used on these steels. The surface of the case will be 0.8% carbon which will be very hard and brittle, supported by a relatively soft core. This can lead to separation of case from core with subsequent exfoliation or spalling.

Carburizing is at 920–950 °C, using either the gas or pack method, or by cyanide molten salt. Gas carburizing is recommended as it is faster and more controllable than the pack method, but requires a special furnace.

The parts should be re-heated to 750–780 °C and oil quenched, followed immediately by a 120–200 °C temper.

The practice of quenching direct from the carburizing temperature, or allowing to cool to the estimated temperature and quenching is not recommended with these steels which have a tendency to grain growth, and thus to be brittle.

Welding and brazing

Pre-treatment. No thermal treatment required. The area should be clear, free from oil, rust and scale, and careful joint preparation is essential.

Welding and brazing. The lower carbon and chrome steels can be welded or brazed using conventional electric arc or gas heating with fluxes. Inert shielded arc welding gives similar results but is seldom economically justified. There will always be a danger of weld cracking with the higher chrome and carbon steels because of their air hardening properties, but this can be reduced by pre-heating and controlling the rate of cooling, but is seldom economical as weldable steels of equal strength are readily available.

Resistance welding is possible only with the low alloy steels, the other steels cracking on cooling.

Brazing using fluxes and copper alloys is trouble free.

Post-treatment. Welding slag, oxide and fluxes should be completely removed. It is strongly recommended that the welded parts are normalized at least, and hardened and tempered if possible after welding. At the very least the parts should be stress released at 600–650 °C, this being carried out with welding torches or electric blankets. The only exception to this are those low carbon low chrome steels specially developed for their weldability.

Flaw detection methods

Crack test. Small parts can be magnetically crack tested. Penetrant oil and chalk test is now almost as sensitive as the magnetic method and can be locally applied more readily.

X-ray. This will not be required on wrought products but will be required on high duty welds.

Ultra-sonic test. Not normally required.

Chemical etch. This is recommended on carburized parts which are hardened and ground. An electrolytic sulphuric etch can be used to show grinding abuse and thus control the process, preventing scrap caused by grinding cracks.

Corrosion protection

Temporary. Parts should be oiled or greased and free from moisture at all times. Correctly maintained soluble oil coolants inhibit corrosion, but dirty coolant encourages staining and rust. Good housekeeping during manufacture can prevent formation of rust and considerably ease the problem of permanent protection.

Permanent. Although the chrome content of these steels increases the corrosion resistance to a limited extent, none can be classed as corrosion resistant. It is important that all evidence of staining or corrosion is removed before any attempt is made to protect these steels. They are rather notch sensitive which makes rust pitting more dangerous than with less brittle materials.

Painting. Preparation by dry blasting is recommended. Unless carefully controlled this process is open to abuse and it is worth noting that a freshly blasted surface can start to corrode within minutes under humid conditions. Also there is generally a fine dust left on the surface which should be removed by brushing.

Phosphate coatings applied by dipping or spraying after blasting give some measure of corrosion protection and act as a key for the paint primer. They should be oven dried before painting.

A primer is desirable and should be of the zinc silicate or chromate type, applied by spray or brush. The top coats should be applied as soon as possible, the stoving type of modern plastic paint giving the best result.

The degree of protection obtained can usually be measured by the amount of careful preparation carried out and quality of the paints applied. It is recommended that advice is obtained from a competent authority for specific problems.

Plating. Electroplating using normal cleaning and etching techniques is possible on all the steels listed.

Chromium plating for appearance purposes should be on top of 0.05 mm copper and 0.025 mm nickel for guaranteed results, the degree of corrosion protection being proportional to the thickness of this plating. The actual chrome plate is always very thin. Cadmium, zinc and tin plating all give adequate corrosion protection with deposits of under 0.025 mm if correctly applied. These steels are not usually dip tinned or galvanized (zinc dipped). Cadmium and zinc coatings must not be used in contact with foodstuffs.

Metal spraying of zinc or aluminium after grit blasting is proving to be an efficient method of corrosion protection, provided the sprayed deposit is sealed immediately by wire brushing and chemical treatment.

Cathodic protection of large structures is now common practice. With this, sacrificial electrodes are electrically connected to the structure and low voltage electric current passed, making the structure cathodic. This protects it from corrosion at the expense of the anode which is replaceable. Ships, pipe lines, pylons and building foundations have all been protected in this way. The process can be additive to a paint system.

Machinability

Chromium increases the machinability of these materials over their plain carbon equivalents. Even in the annealed condition they give a reasonably good finish. In the normalized or hardened and tempered condition the finish should be good, but tool wear will increase. High speed steel should satisfactorily machine all the materials listed.

Uses

Pressure vessels, structural steel, and for some low duty engineering purposes. The higher chrome steels are used for dies, plastic moulds and some fixtures.

Symbol	Nominal analysis, supplier, condition and remarks.
0.7033	0.33% C 1.0% Cr steel: German Standard
1.0345	0.16% C (max) 0.3% Cr steel: German Standard
1.0408	0.25% C (max) 0.3% Cr steel: German Standard
1.0425	0.2% C (max) 0.3% Cr steel: German Standard
1.0435	0.22% C (max) 0.3% Cr steel: German Standard
1.0445	0.26% C (max) 0.3% Cr steel: German Standard
1.0501	0.36% C 0.5% Cr steel: German Standard
1.0844	0.17% C 1.0% Mn 0.3% Cr steel: For boiler plate; German Standard
1.0845	0.2% C 1.2% Mn 0.3% Cr steel: Boiler plate; German Standard
1.0916	0.17% C 1.0% Mn 0.3% Cr steel: For boiler plate; German Standard
1.0935	0.2% C 1.2% Mn 0.3% Cr steel: Boiler plate; German Standard
1.2162	0.21% C 1.2% Cr tool steel: German Standard designation
1.4712	0.12% C (max) 2.2% Si 6.0% Cr steel: Heat resistant; German Standard
1.4713	0.12% C (max) 6.5% Cr 0.8% Al steel: Heat resistant; German Standard
1.4716	0.1% C (max) 9% Cr steel: Heat resistant; German Standard
1.4903	0.1% C (max) 8% Cr 5.5% Al steel: High conductivity of heat; German Standard
1.5053	0.2% C 1.2% Mn 0.3% Cr steel: Forging; German Standard
1.7015	0.15% C 0.7% Cr steel: For carburizing; German Standard
1.7033	0.35% C 1.1% Cr steel: Cold drawn for bolts; German Standard
1.7034	0.38% C 1.1% Cr steel: German Standard
1.7071	0.18% C 2.4% Cr steel: For high pressure; German Standard
1.7083	0.1% C 2.8% Cr steel: For high pressure; German Standard
1.7147	0.2% C 1.3% Mn 1.2% Cr steel: For carburizing; German Standard
1.8401	0.3% C 1.1% Mn 0.9% Cr 0.25% Ti 0.1% Al steel: Welding rod; German Standard
1.8504	0.33% C 1.4% Cr 1.0% Al steel: For nitriding; German Standard
1.8506	0.33% C 1.2% Cr 1.0% Al steel: For nitriding; German Standard
3 LC 2	0.15% C 0.5% Mn 0.65% Cr steel: Sandvik
3 MC 2	0.15% C 1.2% Mn 1% Cr steel: Sandvik
6 LMC	0.3% C 1.4% Mn 0.035% Cr steel: Sandvik

Note. The following abbreviations and units are used in the tables:

DPN	Hardness, diamond pyramid number
UTS	Ultimate tensile strength, N/mm^2
Elon	Elongation, %
Proof	0.1 % proof strength, N/mm^2

1 N/mm^2=0.1 hbar=0.102 kgf/mm^2=0.06475 tonf/in^2=145.04 lbf/in^2

Symbol	Nominal analysis, supplier, condition and remarks.
10 Cr 11	0.1% C 0.4% Mn 2.9% Cr steel: Designation used by DIN
16 Cr 9	0.16% C 0.4% Mn 2.4% Cr steel: Designation used by DIN
19 Mn 5	0.2% C 1.2% Mn 0.3% Cr steel: Designation used by DIN
20 Mn 5	0.2% C 1.1% Mn 0.3% Cr steel: Designation used by DIN
21 Mn Cr 5	0.2% C 1.2% Cr steel: For tool; German Standard designation
30 Mn Cr Ti 4	0.3% C 1.1% Mn 0.9% Cr 0.25% Ti steel: Designation used by DIN
34 Cr Al 6	0.33% C 0.8% Mn 1.4% Cr 1.0% Al steel: Designation used by DIN
0708 Si Al	0.07% C 6.5% Cr 0.8% Si 0.8% Al steel: NYBY annealed DPN: 185 UTS: 280 Elon: 20% Proof: 160
5115	0.15% C 0.8% Mn 0.8% Cr steel: Designation used by AISI
A 202	0.2% C 1.2% Mn 0.45% Cr steel: Plate; Armco; for pressure tanks UTS: 630 Elon: 20% Proof: 300
ACM	Cr steel: For magnets; Holzer; further information from Permanent Magnet Association
AISI 5015	0.15% C 0.4% Cr steel
AISI 5117	0.17% C 0.8% Cr steel: Obsolete
AISI 5120	0.20% C 0.8% Cr steel
AISI 5130	0.3% C 0.8% Mn 0.9% Cr steel
AISI 5132	0.32% C 0.7% Mn 0.9% Cr steel
AISI 5135	0.35% C 0.7% Mn 0.9% Cr steel
APS 10	0.12% C 2.0% Cr 0.9% Al steel: Pompey UTS: 460 Elon: 20% Proof: 300
APS 10C	0.12% C 2.5% Cr 0.4% Al steel: Pompey UTS: 460 Elon: 20% Proof: 300
APS 20	0.12% C 4.0% Cr 1.0% Al steel: Pompey UTS: 460 Elon: 20% Proof: 300
APS 20C	0.12% C 4.0% Cr 0.4% Al steel: Pompey UTS: 460 Elon: 20% Proof: 300
APS 30	0.1% C 6.5% Cr 1.3% Al steel: Pompey UTS: 460 Elon: 20% Proof: 300
ASN 151	0.22% C 1.5% Mn 0.5% Cr 0.5% Cu steel: Australian Iron & Steel Co.; normalized
ASTM A202 A	0.17% C (max) 1.2% Mn 0.8% Si 45% Cr steel: For pressure vessels
ASTM A202 B	0.25% C (max) 1.2% Mn 0.8% Si 0.5% Cr steel: For pressure vessels
ASTM A588 D	0.15% C 0.7% Cr 0.1% Zl steel: For structures UTS: 440 Elon: 21% Proof: 320
ASTM A588 Cr	0.2% C (max) 1.2% Mn (max) 0.7% Cr 0.4% Cu steel: For structures UTS: 440 Elon: 21% Proof: 320
ATLANTES HT	0.2% C 0.9% Mn 0.65% Cr 0.5% Cu steel: South Durham
BS 681	Carbon chromium steel: Replaced by BS 970
BS 968	0.2% C 1.5% Mn 0.5% Cr steel: Normalized UTS: 530 Elon: 23% Proof: 340
BS 970/523 A14	0.15% C 0.4% Mn 0.4% Cr steel: For carburizing; replaces BS 970 En 206

Symbol	Nominal analysis, supplier, condition and remarks.
BS 970/523 M15	0.15% C 0.4% Mn 0.4% Cr steel: For carburizing supplied to mechanical properties; replaces BS 970 En 351
BS 970/527 A19	0.2% C 0.8% Mn 0.8% Cr steel: For carburizing; replaces BS 970 En 207
BS 970/527 M20	0.2% C 0.8Mn 0.8% Cr steel: For carburizing; supplied to mechanical properties
BS 970/530 A30	0.30% C 1% Cr steel: Supplied to chemical composition
BS 970/530 A32	0.32% C 1% Cr steel: Supplied to chemical composition
BS 970/530 A36	0.36% C 1% Cr steel: Supplied to chemical composition
BS 970/530 H30	0.30% C 1.0% Cr steel: Supplied to hardenability requirement
BS 970/530 H32	0.32% C 0.7% Mn 1.0% Cr steel: Supplied to hardenability requirement
BS 970 En 18A	0.3% C 0.7% Mn 1% Cr steel: Bar & forgings; hardened & tempered; properties not part of specification
BS 970 En 18B	0.32% C 0.7% Mn 1% Cr steel: Bar & forgings; hardened & tempered; properties not part of specification
BS 970 En 18C	0.36% C 0.7% Mn 1% Cr steel: Bar & forgings; hardened & tempered; properties not part of specification
BS 970 En 206	0.15% C 0.4% Cr steel: For case hardening; mechanical properties not part of specification
BS 970 En 207	0.18% C 0.7% Cr steel: For case hardening; mechanical properties not part of specification
BS 1956 A	0.5% C 0.7% Mn 1.0% Cr steel: Casting; abrasion resisting; normalized or hardened & tempered; contained in BS 3100 **DPN: 207 UTS: 690 Elon: 10%**
BS 1956 B	0.5% C 0.7% Mn 1.0% Cr steel: Casting; abrasion resisting; normalized or hardened & tempered; contained in BS 3100 **DPN: 293**
BS 3551/A	0.35% C 1.5% Mn 0.25% Cr steel: For shackles; hardened & tempered **DPN: 275**
BS S19	Chromium steel: Forgings for valves; replaced
BS/S115	0.4% C 1.0% Cr steel: Wire for forged bolts; hardened & tempered after forging **UTS: 910 Elon: 18% Proof: 660**
BS S117	0.4% C 1.0% Cr steel: Bar & forging; hardened and tempered **DPN: 280 UTS: 910 Elon: 18% Proof: 660**
C 1K	0.12% C 0.9% Cr steel: Pompey
C 1MK	0.16% C 1.2% Mn 0.9% Cr steel: Pompey **UTS: 115037% Proof: 770**
C 2K	0.16% C 0.9% Cr steel: Pompey **UTS: 980 Elon: 8% Proof: 770**
C 2MK	0.2% C 1.3% Cr steel: Pompey
C 35	0.36% C 0.5% Cr steel: German Standard
CA 30	0.34% C 1.35% Cr 0.95% Al steel: For nitriding; Krupp
CA 30 A	0.34% C 1.15% Cr 0.95% Al steel: For nitriding; Krupp

Note. The following abbreviations and units are used in the tables:

DPN	Hardness, diamond pyramid number
UTS	Ultimate tensile strength, N/mm^2
Elon	Elongation, %
Proof	0.1 % proof strength, N/mm^2

1 N/mm^2=0.1 hbar=0.102 kgf/mm^2=0.06475 tonf/in^2=145.04 lbf/in^2

Symbol	Nominal analysis, supplier, condition and remarks.
CBV	0.15% C 0.25% Si 0.65% Cr steel: Krupp
CBV 1	0.34% C 1.05% Cr steel: Krupp
CHROMADOR	0.3% C 0.8% Mn 0.8% Cr 0.3% Cu steel: Dorman Long Ltd; as rolled; resists corrosion **DPN: 180 UTS: 660 Elon: 17% Proof: 360**
CMBV/h	0.2% C 0.25% Si 1.15% Cr steel: Krupp
CONICRO	0.23% C (max) 0.95% Mn 0.6% Cr 0.4% Cu steel: Consett
CMBV/w	0.16% C 0.25% Si 0.95% Cr steel: Krupp
COR-TEN A	0.12% C (max) 1% Cr 0.49% Cu steel: Appleby; Frodingham; weldable **UTS: 570 Elon: 22% Proof: 330**
COR-TEN B	0.14% C 0.5% Cr 0.3% Cu 0.06% V steel: Appleby; Frodingham **UTS: 300 Elon: 21% Proof: 210**
D 3K	0.32% C 0.4% Cr steel: Pompey
DIN 1654 34 Cr 4	0.34% C 1.1% Cr steel: For cold forged bolts
DIN 17110 17 Mn 4	0.17% C 1.0% Mn 0.3% Cr steel: For boiler plate
DIN 17155 17 Mn 4	0.17% C 1.0% Mn 0.3% Cr steel: For boiler plate
DIN 17155 19 Mn 4	0.2% C 1.2% Mn 0.3% Cr steel: Boiler plate
DIN 17155 H	0.2% C (max) 0.3% Cr steel: For boiler plate
DIN 17155 H 1	0.16% C (max) 0.4% Mn 0.3% Cr steel: Plate
DIN 17155 H 11a	0.2% C (max) 0.3% Cr steel: For boiler plate
DIN 17155 H 111	0.22% C 0.3% Cr steel: Seamless tube
DIN 17200 15 Cr 3	0.15% C 0.55% Mn 0.55% Cr steel: German Standard
DIN 17200 16 Mn Cr 5	0.16% C 1.15% Mn 0.9% Cr steel: German Standard
DIN 17200 16 Mn Cr S 5	0.16% C 1.15% Mn 0.95% Cr steel: German Standard
DIN 17200 20 Mn Cr 5	0.2% C 1.2% Mn 1.15% Cr steel: German Standard
DIN 17200 20 Mn Cr S 5	0.2% C 1.2% Mn 1.15% Cr steel: German Standard
DIN 17200 34 Cr 4	0.39% C 0.75% Mn 1.05% Cr steel: German Standard
DIN 17200 34 Cr 4	0.34% C 1.1% Cr steel: Forging
DIN 17200 34 Cr S 4	0.34% C 0.75% Mn 1.05% Cr steel: German Standard
DIN 17210 15 Cr 3	0.15% C 0.6% Cr steel: For carburizing
DIN 17210 16 Mn Cr 5	0.16% C 1.1% Mn 1% Cr steel: For carburizing
DIN 17210 20 Mn Cr 5	0.2% C 1.2% Mn 1.1% Cr steel: For carburizing
DIN 17470 Cr Al 8/5	0.1% C (max) 8.0% Cr 5.5% Al steel: For conducting heat
DK 3	0.32% C 1.0% Cr steel: Pompey **UTS: 980 Elon: 9% Proof: 820**
DK 3E	0.32% C 1.0% Cr steel: Pompey
DK 35E	0.35% C 1.0% Cr steel: Pompey
DUK TEN	0.15% C (max) 0.5% Mn 3.5% Ni steel: German proprietors specification
DURAPSO	0.19% C 0.4% Cr 0.4% Cu steel: Pompey
DURAPSO 3	0.2% C 0.4% Cr 0.5% Cu steel: Pompey
FDK 3	0.33% C 0.9% Cr steel: Pompey **UTS: 950 Elon: 11% Proof: 740**
GOST 550 KH5	0.15% C (max) 0.5% Mn (max) 5.0% Cr steel: Pipe; Russian Standard **UTS: 400 Proof: 220**
GOST 4543/14 KHCr	0.14% C 1.1% Mn 0.65% Cr 0.3% Ni (max) 0.3% Cu (max) steel: Russian Standard
GOST 4543/15 KH	0.15% C 0.55% Mn 0.85% Cr steel: Russian Standard
GOST 4543/15 KHA	0.15% C 0.55% Mn 0.85% Cr steel: Russian Standard
GOST 4543/15 KHR	0.15% C 0.55% Mn 0.85% Cr 0.0035% B steel: Russian Standard
GOST 4543/15 KHRA	0.15% C 0.55% Mn 0.85% Cr 0.0035% B: Russian Standard

Symbol	Nominal analysis, supplier, condition and remarks.
GOST 4543/18 KHG	0.18% C 1.1% Mn 1.05% Cr steel: Russian Standard
GOST 4543/18 KHGT	0.2% C 0.9% Mn 1.15% Cr 0.09% Ti steel: Russian Standard
GOST 4543/20 KH	0.2% C 0.65% Mn 0.85% Cr steel: Russian Standard
GOST 4543/20 KHG2C	0.22% C 1.7% Mn 1.1% Cr 0.3% Ni (max) 0.3% Cu (max) 0.1% Zr steel: Russian Standard
GOST 4543/20 KHGR	0.2% C 0.9% Mn 0.95% Cr steel: Russian Standard
GOST 4543/20 KHGSA	0.2% C 0.95% Mn 0.95% Cr steel: Russian Standard
GOST 4543/25 KHGSA	0.25% C 0.95% Mn 0.95% Cr steel: Russian Standard
GOST 4543/30 KH	0.3% C 0.65% Mn 0.95% Cr steel: Russian Standard
GOST 4543/30 KHGT	0.28% C 0.95% Mn 1.15% Cr steel: Russian Standard
GOST 4543/30 KHGS	0.3% C 0.95% Mn 0.95% Cr steel: Russian Standard
GOST 4543/33 KHS	0.33% C 0.45% Mn 1.45% Cr steel: Russian Standard
GOST 4543/35 KH	0.35% C 0.65% Mn 0.95% Cr steel: Russian Standard
GOST 4543/35 KHRA	0.35% C 0.65% Mn 0.95% Cr steel: Russian Standard
GOST 5632 KH5	0.15% C (max) 0.5% Mn (max) 5.25% Cr steel: Plate; Russian Standard; annealed **UTS: 400** **Proof: 170**
GOST 5632 KH5	0.15% C (max) 5.2% Cr steel: Russian Standard
GOST 5632 KH6S14	0.15% C (max) 6.2% Cr 0.9% Al steel: Russian Standard
GOST 8733/15 KH	0.16% C 0.95% Mn 0.85% Cr steel: Pipe; Russian Standard; seamless **UTS: 420**
GOST 8733/20 KH	0.2% C 0.65% Mn 0.85% Cr 0.2% Ni (max) steel: Pipe; Russian Standard; seamless **UTS: 440**
GOST 8731/20 KH	0.2% C 0.65% Mn 0.035% Mn 0.85% Cr 0.2% Ni (max) steel: Pipe; Russian Standard; seamless **UTS: 440**
GOST 12132/15 KH	0.15% C 0.55% Mn 0.85% Cr steel: Pipe; Russian Standard; welded **UTS: 420**
GOST 12132/30 KHGSA	0.3% C 0.95% Mn 0.95% Cr 0.25% Ni (max) steel: Pipe; Russian Standard; welded **UTS: 500**
HARDTRODE 83.28	0.10% C 0.750% Mn 3.50% Cr steel: For electrodes; Esab **DPN: 320**
HECLA 260	0.15% C 0.4% Mn 0.4% Cr steel: Case hardening; Hadfields for BS alloy En 206
HECLA 207	0.18% C 0.7% Mn 0.7% Cr steel: Case hardening; Hadfields for BS alloy En 207
HEDEX 25	0.18% C 0.7% Cr steel: Wire for forging; Kiveton Park for BS alloy En 207
HSCR	0.12% C (max) 0.35% Mn 0.8% Cr 0.4% Cu steel: Patent Shop Steel Co.
HT2	0.09% C 2% Si 0.35% Mn 5% Cr steel: Sandvik
HTS	0.2% C (max) 1.5% Mn 0.5% Cr 0.5% Cu steel: Patent Shop Steel Co.
HYPLUS 1	0.2% C (max) 1.5% Mn 0.5% Cr 1.6% Mn + Cr (max) steel: Appleby; Fredingham
IDROTUB 52	0.18% C (max) 1.2% Mn 0.3% Cr steel: Italian proprietor's specification
JIS G4106 S Cr 2	0.3% C 0.7% Mn 1.05% Cr steel: Japanese Standard
JIS G4106 S Cr 3	0.35% C 0.7% Mn 1.05% Cr steel: Japanese Standard

Symbol	Nominal analysis, supplier, condition and remarks.
JIS G4106 S Cr 21	0.15% C 0.7% Mn 1.05% Cr steel: Japanese Standard
JIS G4106 S Cr 22	0.22% C 0.7% Mn 1.05% Cr steel: Japanese Standard
JIS G4106 S Mn C 21	0.2% C 1.3% Mn 0.5% Cr steel: Japanese Standard
KF 18	0.3% C 1% Cr steel: Kirkstall for BS alloy En 18
KF 18 A	0.32% C 1% Cr steel: Kirkstall for BS alloy En 18
KF 18 B	0.37% C 1% Cr steel: Kirkstall for BS alloy En 18
KF 30 A	0.18% C 0.7% Cr steel: Case hardening; Kirkstall for BS alloy En 207
LS 55	0.12% C 6% Cr 0.55% Mo: Low Moor; obsolete **DPN: 220** **UTS: 480** **Elon: 23%** **Proof: 400**
MAFERITE 6	C not quoted 6% Cr steel: Pompey **DPN: 200**
MAFERITE 8	C not quoted 8% Cr steel: Pompey **DPN: 200**
MANOIR APS10	0.12% C 2.0% Cr 0.65% Al steel: Pompey; normalized **UTS: 550** **Elon: 14%** **Proof: 340**
MANOIR APS20	0.12% C 4.0% Cr 0.65% Al steel: Pompey; normalized **UTS: 550** **Elon: 14%** **Proof: 340**
MANOIR APS25	0.12% C 4.0% Cr 0.9% Al 0.9% Ni steel: Pompey; normalized **UTS: 870** **Elon: 8%** **Proof: 670**
MANOIR APS30	0.12% C 7.0% Cr 1.5% Al 1.0% Si steel: Pompey; normalized **UTS: 600** **Elon: 10%** **Proof: 300**
MIL S12505.4	0.14% C (max) 0.7% Mn 0.5% Cr 0.1% Zr steel· Bethlehem
Mn Cr	0.17% C 0.3% Cr steel: Pompey
NF A35 551/16MC5	0.16% C 1.15% Mn 0.95% Cr steel: French Standard
NF A35 551/20MCS	0.2% C 1.2% Mn 1.15% Cr steel: French Standard
NF A35 551/32C	0.32% C 0.75% Mn 1.0% Cr steel: French Standard
P 5	0.1% C 2.25% Cr steel: Designation used by AISI
PLMA/2	0.15% C 0.5% Mn 1% Cr steel: For case hardening; ESC for plastic moulds
PM 12	0.18% C 0.15% Mn 0.15% Cr steel: Pompey
SAE 5015	0.15% C 0.4% Cr steel
SAE 5115	0.15% C 0.8% Mn 0.8% Cr steel: Mechanical properties not specified
SAE 5117	0.17% C 0.8% Mn 0.8% Cr steel: Mechanical properties not specified
SAE 5120	0.2% C 0.8% Mn 0.8% Cr steel: Mechanical properties not specified
SAE 5130	0.3% C 0.8% Mn 0.9% Cr steel: Mechanical properties not specified
SAE 5132	0.32% C 0.7% Mn 0.9% Cr steel: Mechanical properties not specified
SAE 5135	0.35% C 0.7% Mn 0.9% Cr steel: Mechanical properties not specified
SAMSON EXTRA	0.1% C 2.3% Cr steel: Carpenter; for AISI type P5
SIS 14-12-32	0.13% C 0.5% Mn 0.25% Cr 0.3% Cu steel: For pressure vessels; Swedish Standard **UTS: 370** **Elon: 25%** **Proof: 200**
SIS 14-12-32E	0.13% C 0.5% Mn 0.25% Cr 0.3% Cu (max) steel: For pressure vessels; Swedish Standard **UTS: 390** **Elon: 18%** **Proof: 290**
SIS 14-13-06	0.18% C 1.1% Mn 0.3% Cr Cu steel: For casting; Swedish Standard **DPN: 135** **UTS: 410** **Elon: 25%** **Proof: 23**
SIS 14-13-30	0.17% C 0.9% Mn 0.25% Cr 0.3% Cu steel: For pressure vessels; Swedish Standard **UTS: 410** **Elon: 26%** **Proof: 215**

Symbol	Nominal analysis, supplier, condition and remarks.
SIS 14-13-30E	0.15% C 0.6% Mn 0.25% Cr 0.3% Cu steel: For pressure vessels; Swedish Standard **UTS: 420 Elon: 26% Proof: 210**
SIS 14-14-30	0.15% C 0.7% Mn 0.25% Cr 0.3% Cu steel: For pressure vessels; Swedish Standard **UTS: 480 Elon: 24% Proof: 285**
SIS 14-14-30E	0.18% C 0.7% Mn 0.25% Cr 0.30% Cu (max) steel: For pressure vessels; Swedish Standard **UTS: 470 Elon: 24% Proof: 260**
SIS 14-14-32	0.12% C 1.1% Mn 0.25% Cr 0.3% Cu steel: For pressure vessels; Swedish Standard **UTS: 480 Elon: 24% Proof: 255**
SIS 14-14-32E	0.14% C 1.1% Mn 0.25% Cr (max) 0.3% Cu steel: For pressure vessels; Swedish Standard **UTS: 470 Elon: 24% Proof: 250**
SIS 14-14-34	0.22% C 0.6% Mn 0.25% Cr 0.3% Cu steel: For pressure vessels; Swedish Standard **DPN: 150 UTS: 800 Elon: 23% Proof: 260**
SIS 14-14-35	0.22% C 0.8% Mn 0.25% Cr 0.3% Cu steel: For pressure vessels; Swedish Standard **UTS: 500 Elon: 21% Proof: 260**
SIS 14-14-35E	0.22% C 0.8% Mn 0.25% Cr (max) 0.30% Cu (max) steel: For pressure vessels; Swedish Standard **UTS: 500 Elon: 21% Proof: 260**
SIS 14-21-01	0.18% C 1.2% Mn 0.25% Cr 0.3% Cu steel: For pressure vessels; Swedish Standard **UTS: 600 Elon: 21% Proof: 300**
SIS 14-21-01E	0.2% C 1.2% Mn 0.25% Cr (max) 0.3% Cu (max) steel: For pressure vessels; Swedish Standard **UTS: 550 Elon: 21% Proof: 300**
SIS 14-21-03	0.15% C 1.2% Mn 0.25% Cr 0.3% Cu steel: For pressure vessels; Swedish Standard **UTS: 550 Elon: 21% Proof: 300**
SIS 14-21-03E	0.15% C 1.3% Mn 0.25% Cr 0.3% Cu steel: For pressure vessels; Swedish Standard; normalized **UTS: 550 Elon: 21% Proof: 300**
SIS 1225	0.08% C 0.1% Cr 0.2% Cu 0.009% N steel: For chains; Swedish Standard; normalized **UTS: 340 Elon: 30%**
SIS 1233	0.17% C 0.2% Cr 0.3% Cu 0.009% N steel: Bar; Swedish Standard; as rolled **UTS: 370 Elon: 25%**
SIS 1234	0.17% C 0.2% Cr 0.3% Cu 0.009% N steel: Bar; Swedish Standard; as rolled **UTS: 370 Elon: 25%**
SIS 1306	0.18% C (max) 1% Mn 0.3% Cr 0.3% Cu steel: Swedish Standard; annealed **UTS: 390 Elon: 25% Proof: 210**
SIS 1330	0.12% C 0.5% Mn 0.25% Cr 0.4% Cu 0.009% N steel: For pressure vessels; Swedish Standard; normalized **UTS: 390 Elon: 26% Proof: 220**
SIS 1331	0.11% C 0.7% Mn 0.25% Cr 0.4% Cu 0.009% N steel: For pressure vessels; Swedish Standard; normalized **UTS: 390 Elon: 26% Proof: 220**
SIS 1332	0.1% C 0.7% Mn 0.25% Cr 0.4% Cu 0.009% N steel: For pressure vessels; Swedish Standard; normalized **UTS: 390 Elon: 26% Proof: 220**
SIS 1411	0.2% C 0.9% Mn 0.3% Cr 0.4% Cu structural steel: Swedish Standard; as rolled **UTS: 450 Elon: 20% Proof: 240**
SIS 1430	0.15% C 0.9% Mn 0.25% Cr 0.4% Cu 0.009% N steel: For pressure vessels; Swedish Standard; normalized **UTS: 450 Elon: 24% Proof: 240**

Symbol	Nominal analysis, supplier, condition and remarks.
SIS 1431	0.12% C 1% Mn 0.25% Cr 0.4% Cu 0.009% N steel: For pressure vessels; Swedish Standard; normalized **UTS: 450 Elon: 24% Proof: 240**
SIS 1432	0.12% C 1% Mn 0.25% Cr 0.4% Cu 0.009% N steel: For pressure vessels; Swedish Standard; normalized **UTS: 450 Elon: 24% Proof: 240**
SIS 1434	0.22% C (max) 0.2% Cr 0.3% Cu 0.009% N steel: For pressure vessels; Swedish Standard; hot rolled **DPN: 150 UTS: 450 Elon: 26% Proof: 240**
SIS 1435	0.22% C (max) 0.2% Cr 0.3% Cu 0.009% N steel: For pressure vessels; Swedish Standard; hot rolled **UTS: 450 Elon: 21% Proof: 240**
SIS 2101	0.18% C 1% Mn 0.25% Cr 0.4% Cu 0.009% N steel: For pressure vessels; Swedish Standard; normalized **UTS: 510 Elon: 21% Proof: 320**
SIS 2102	0.14% C 1.6% Mn 0.25% Cr 0.4% Cu 0.009% N steel: For pressure vessels; Swedish Standard; normalized **UTS: 510 Elon: 18% Proof: 320**
SIS 2103	0.14% C 1.6% Mn 0.25% Cr 0.4% Cu 0.009% N steel: For pressure vessels; Swedish Standard; normalized **UTS: 510 Elon: 18% Proof: 320**
SIS 2108	0.2% C 1.4% Mn 0.1% Cr 0.2% Cu 0.009% N steel: For chains; Swedish Standard; hardened & tempered **DPN: 180 UTS: 660 Elon: 18%**
SIS 2172	0.2% C 1.3% Mn 0.3% Cr 0.4% Cu 0.009% N structural steel: Swedish Standard; as rolled **UTS: 530 Elon: 21% Proof: 320**
SIS 2173	0.14% C 1.2% Mn 0.3% Cr 0.4% Cu 0.009% N structural steel: Swedish Standard; normalized **UTS: 530 Elon: 18% Proof: 320**
SIS 2174	0.14% C 1.2% Mn 0.2% Cr 0.3% Cu 0.009% N structural steel: Swedish Standard; normalized **UTS: 530 Elon: 18% Proof: 320**
THIRTY-OAK	0.22% C 0.2% Si 1.0% Mn 0.5% Cr 1.6% Mn + Cr (max) steel: Round Oak Steel; weldable; may have 0.5% Cu **UTS: 630 Elon: 17% Proof: 450**
THIRTY OAK 1	0.22% C (max) 1.5% Mn 0.5% Cr 0.08% Nb steel: Round Oak Steel Co.
THIRTY OAK 2	0.22% C (max) 1.5% Mn 0.5% Cr 0.08% Nb steel: Round Oak Steel Co.
TUBROD 15.40	0.09% C 2.0% Mn 3.8% Cr steel: Tubular welding rod; Esab; for use with submerged arc; properties vary with flux chased; as welded **DPN: 380**
WLM 4060	0.3% C 1% Cr steel: 600–900 N/m² tensile; W Marrison; analysis varies over wide range
X 8 Cr 9	0.1% C (max) 1.5% Si 1.5% Mn 9.0% Cr steel: Designation used by German Standard
X 10 Cr Al 7	0.12% (max) C 6.5% Cr 0.7% Al steel: Designation used by German Standard
X 10 Cr Si 6	0.12% (max) C 2.2% Si 6.0% Cr steel: Designation used by German Standard

Note. The following abbreviations and units are used in the tables:

DPN	Hardness, diamond pyramid number
UTS	Ultimate tensile strength, N/mm²
Elon	Elongation, %
Proof	0.1 % proof strength, N/mm²

1 N/mm²=0.1 hbar=0.102 kgf/mm²=0.06475 tonf/in²=145.04 lbf/in²

44E2 Steel – medium carbon with chromium
0.4–0.65% carbon, up to 9% chromium with silicon

Specific gravity	7.84	
Density	7840 kg/m^3	(0.28 lb/in^3.)
Solidus/liquidus	1400–1470 °C	
Thermal conductivity	– W/m °C	
Coefficient of linear expansion	11 x 10^{-6}/ °C	
Electrical conductivity	8–10% IACS (copper 100%)	
Specific resistance	180–230 microhm mm	
Young's modulus of elasticity	– N/m^2	
Impact	41–54 J	(30–40 ft/lb)
Fatigue strength (no. of cycles)	–	
Hot strength	–	

The above properties are typical of the following group, and may not apply exactly to any one specification. It is possible that with certain specifications some of the values may not be applicable.

General metallurgical characteristics

Chromium has a twin effect on steel, the principal one, with the steels listed, being to act as a carbide former. This improves the hardenability, thus allowing deeper hardening with less drastic quenching media. The carbides slow down the metallurgical reactions and retain their hardness at higher temperatures, the steels also being harder even in the annealed condition as the chromium carbides are harder than iron carbide.

The second effect of chromium on steel is to improve the corrosion resistance. This is achieved by an adherent stable oxide which forms on the surface. With the steels listed, there is never sufficient chromium present to make the steels stainless or even corrosion resistant, but they have better resistance than plain carbon steels, this being in proportion to the chromium present.

These steels are capable of a very high hardness, but like many plain carbon single alloy steels with a carbide former they are notch sensitive with a tendency to brittleness, particularly at low temperatures.

Some of the listed steels have up to 3.0% silicon additions which improve the fatigue and corrosion resistance, thus giving economical spring steels.

Thermal treatment

A pre-heat at about 600 °C is recommended before any thermal treatment to prevent cracking.

These steels can accept very little cold work without becoming brittle and unworkable. Annealing requires a temperature of 800–850 °C for about 1 hour per 25 mm section followed by slow cooling. Care must be taken not to keep the parts too long at this high temperature, otherwise the carbides agglomerate into large spheres which have a deleterious effect on the mechanical properties.

Normalizing requires the same temperature as annealing, for about ¾ hour per 25 mm of section, followed by air cooling. This results in considerable air hardening and must always be followed by a tempering or sub-critical annealing operation.

Probably the best method of softening these steels for machining is to use the cyclic or spheroidizing anneal. This consists of heating to the normalizing temperature, then transferring to the subcritical annealing temperature. This cycle must be repeated at least once and results in the hard chromium carbides being formed into small spheres dispersed in a soft iron matrix.

Hardening requires that parts are heated to at least 780–820 °C for 30 minutes per 25 mm of section, then oil or air quenched. The time can be reduced by increasing the temperature with a consequent danger of some grain growth. Parts must be tempered as soon as possible after quenching, as they are very brittle in the hardened condition.

With the sluggish metallurgical reactions of these steels, it is often necessary to balance a long time at the correct temperature, resulting in considerable scale with less grain growth, against the shorter time at higher temperature, giving more grain growth with less scale and a reduction in cost.

It is essential that the full specification is consulted or detailed information obtained from the material supplier before attempting any thermal treatment of these steels.

None of the steels should be carburized as the core will be almost as hard as the case, giving no ductility to support the brittle hard surface.

Scale removal after thermal treatment is generally by blasting. Sodium hydride treatment is satisfactory if the correct equipment is available, but acid pickling must not be used as these steels are susceptible to hydrogen embrittlement.

No special furnace equipment is necessary, oil, gas, electric or salt bath heating, all giving satisfactory results.

Welding and brazing

Pre-treatment. Absolute cleanliness of the joint area is essential and pre-heating will always be required.

Welding and brazing. Welding is not recommended with these steels, as there will always be a danger of cracking in the weld zone. None of the steels listed should be chosen by designers when welding is involved during manufacture.

With careful control, particularly regarding the rate of cooling after welding, satisfactory joints can be obtained using gas welding and a skilled operator. This control is not generally possible using electric arc welding.

Brazing is often possible on fully hardened and tempered parts, provided the heat input is kept to a minimum, low melting point filler materials are used and there is no dwell of the brazing torch to cause softening. This is only possible because of the sluggish nature of the metallurgical reactions.

Post-treatment. All traces of flux and oxide must be removed immediately after welding or brazing.

Welded parts should be normalized, hardened and tempered if possible. As an absolute minimum, they must be tempered even if this is only by local heating of the weld zone, using gas torches or electric blankets. Unless this is carried out there will be a very hard, brittle area adjacent to the weld. Brazing requires no further treatment.

Flaw detection methods

Crack test. Magnetic crack detection is the most common. With modern sensitive fluids, penetrant oil chalk testing gives excellent results and is more flexible than the magnetic method.

X-ray. Not normally required.

Ultra-sonic test. This may be applied to large forgings with expensive machining operations, when carried out at an early machining stage. This shows subcutaneous defects which would be dangerous on the finished part.

Chemical etch. Not normally required.

Corrosion protection

Temporary. The chromium content is never sufficient to give any of the listed steels any appreciable corrosion protection. They should therefore always be stored under cover and kept free from moisture, oiled or greased at all times.

During machining the normal coolants used generally act as temporary rust inhibitors, but it should be noted that contaminated or dirty soluble oil can be a very potent source of corrosion.

Finished parts must be cleaned and dried then oiled or greased and stored in plastic containers out of contact with air.

Permanent. Corrosion of steel is caused by the formation of iron oxide in the presence of moisture. This is loosely adherent thus is readily brushed or washed away, leaving a surface ready for further attack, resulting in pitting and local scaling. This can be particularly dangerous with steels of this type which are notch sensitive, as even one small corrosion pit can act as the notch to cause failure.

It is essential that all trace of corrosion is removed and that parts are thoroughly cleaned before applying any form of corrosion protection. Corrosion products always have moisture and oxygen occluded and even when sealed off from the atmosphere can propagate. This results in an increase in volume which stresses and eventually causes failure of the protecting medium.

Grit blasting is probably the best means of ensuring removal of all adherent dirt, oxide and rust, but must be carefully controlled and the resultant fine dust deposit removed by brushing.

Alkaline rust removers are technically good, but rather expensive. Acid rust removers or pickles must not be used, as they cause hydrogen embrittlement on these steels.

Phosphate treatment after blasting gives a surface capable of retaining oil or acting as an excellent key to hold paint. These steels should be stress relieved at about 200 °C after.phosphating, this treatment also drying the part prior to painting. Providing the oil film is maintained and parts are stored and used under cover, phosphate treated tools can give many years of corrosion free service.

Painting. These steels are generally used because of their high surface hardness and abrasion resistance, thus will not often be painted. Where non-working areas require protection or for appearance purposes, the part should have a zinc silicate or chromate primer applied after phosphating and this sealed with one or more coats of stoving top coat. Modern plastic paints have good adhesion and are shock and abrasion resisting, but should be applied on top of the metallic type primers.

Plating. These materials are very notch sensitive and unless a stress relieving treatment can be guaranteed before and after electroplating, there will be considerable danger from hydrogen embrittlement. This defect causes brittle failure below the calculated tensile strength and is particularly dangerous with these high tensile notch-sensitive materials. A stress relieving treatment at 200 °C before and after plating completely eliminates the danger, but there is no visual method of inspecting to ensure that the stress relieve treatment has in fact been carried out.

Decorative chromium plating will seldom be required. When necessary it must be preceded by a nickel or copper/nickel plating deposit at least 0.075 mm thick. The chromium plating is extremely thin and porous, thus without the undercoat corrosion failure would be rapid. Failure of chromium plating can almost always be attributed to faulty preparation or insufficient undercoat and not to the chromium layer.

Cadmium or zinc plating, less than 0.025 mm thick, gives excellent corrosion protection. Metal spraying for corrosion protection is not commonly applied as these steels are generally too hard to be roughened by blasting and this is the only means at present of obtaining satisfactory adhesion. Modern high velocity metal spray guns give good results without the need for this surface roughening, but the process is expensive.

An artificial oxide – called black oxide – can be produced in hot sodium hydroxide solutions with oxidizing agents added. This gives adequate indoor corrosion protection for many years provided an oil film is maintained.

Machinability

In the hardened and tempered condition these steels can only be ground, but where intricate shapes are involved spark erosion or electrochemical machining may be justified.

The spheroidized or cyclic annealed condition is the best for normal machining operations, a final grinding allowance or 0.1–0.2 mm being allowed after hardening. Distortion at hardening may occur on intricate shapes or where heavy machining has been performed, but this will never be as much as with plain carbon steels.

Uses

Fixtures, hand tools, cutting tools where high hardness is required and a degree of brittleness and notch sensitivity can be accepted. Applications where medium to high abrasion resistance is required.

Symbol	Nominal analysis, supplier, condition and remarks.
0.7035	0.41% C 1.0% Cr steel: German Standard
0.7103	0.67% C 1.3% Si 0.5% Cr steel: For springs; German Standard
1.0503	0.46% C 0.5% Cr steel: German Standard
1.0961	0.6% C 0.85% Mn 0.3% Cr steel: For springs; German Standard
1.0961	0.6% C 0.85% Mn 0.3% Cr steel: For springs; German Standard
1.2721	0.5% C 1.1% Cr tool steel: German Standard designation
1.4704	0.45% C 4.0% Si 2.7% Cr steel: For valves; German Standard
1.4718	0.45% C 3.0% Si 9% Cr steel: For valves; German Standard
1.7035	0.41% C 1.1% Cr steel: For cold forged bolts; German Standard
1.8403	0.5% C 1.1% Mn 0.9% Cr 0.2% Ti 0.1% Al steel: Welding rod; German Standard
1.8404	0.6% C 0.9% Mn 0.9% Cr 0.2% Ti 0.1% Al steel: Welding rod; German Standard
4KHS	0.4% C 1.45% Cr steel: Russian Standard designation
6KHS	0.65% C 1.15% Cr steel: Russian Standard designation
5 LC	0.5% C 1.7% Si 0.12% Cr steel: Pompey
No. 6 HARDENITE	1.0% C 1.0% Mn 1.35% Cr steel: For tools; origin unknown
6 LC	0.55% C 1.8% Si 0.25% Cr steel: Pompey
6 LCK	0.6% C 1.85% Si 0.35% Cr steel: Pompey
7KH3	0.67% C 3.5% Cr steel: Russian Standard designation
12 C	0.65% C 0.15% Cr steel: Sandvik
34 Cr Al S5	0.33% C 1.2% Cr 1.0% Al steel: Designation used by German Standard
35 Cr Mn 5	0.35% C 0.95% Mn 1.15% Cr steel: Italian Standard designation; as rolled. DPN: 230
37 Cr 4	0.38% C 1.1% Cr steel: Designation used by German Standard
40 Cr 4	0.4% C 0.65% Mn 1.05% Cr steel: Italian Standard designation; as rolled. DPN: 230
41 Cr 4	0.41% C 1.1% Cr steel: Designation used by German Standard
50 Mn Cr Ti 4	0.5% C 1.1% Mn 0.9% Cr 0.25% Ti steel: Designation used by German Standard
50 Ni Cr 13	0.5% C 1.1% Cr steel: For tools; German Standard designation
60 Mn Cr Ti 4	0.6% C 0.9% Cr 0.25% Ti steel: Designation used by German Standard

Note. The following abbreviations and units are used in the tables:

DPN	Hardness, diamond pyramid number
UTS	Ultimate tensile strength, N/mm²
Elon	Elongation, %
Proof	0.1 % proof strength, N/mm²

$1 \text{ N/mm}^2 = 0.1 \text{ hbar} = 0.102 \text{ kgf/mm}^2 = 0.06475 \text{ tonf/in}^2 = 145.04 \text{ lbf/in}^2$

Symbol	Nominal analysis, supplier, condition and remarks.
60 Si Cr 8	0.6% C 0.85% Mn 0.32% Cr steel: Italian Standard designation. **DPN: 250 UTS: 1600 Elon: 5% Proof: 1350**
67 Si Cr 5	0.67% C 1.3% Si 0.5% Cr steel: Designation used by German Standard
1234	0.75% C 0.25% Mn 0.32% Cr steel: French Standard designation
2323	0.6% C 0.6% Cr steel: French Standard designation
9254	0.55% C 1.5% Si 0.7% Cr steel: Designation used by German Standard
AB 75	0.6% C 2.0% Si 0.25% Cr tool steel: Balfour; chisels, dies, etc. **DPN: 540**
ABRADUR SK	0.64% C 1.7% Si 0.8% Cr steel: Pompey **DPN: 660**
ADAMANT	0.52% C Cr steel: Casting; Firth Brown; normalized & tempered **DPN: 210**
AISI 50 B44	0.44% C 0.5% Cr 0.0005% B steel
AISI 50 B46	0.46% C 0.3% Cr 0.0005% B steel
AISI 50 B50	0.5% C 0.5% Cr 0.0005% B steel
AISI 50 B60	0.6% C 0.5% Cr 0.0005% B steel
AISI 51 B60	0.6% C 0.8% Cr 0.0005% B steel
AISI 5045	0.45% C 0.6% Cr steel: Obsolete
AISI 5046	0.46% C 0.3% Cr steel
AISI 5140	0.4% C 0.8% Cr steel
AISI 5145	0.45% C 0.8% Cr steel
AISI 5147	0.47% C 1% Cr steel
AISI 5150	0.5% C 0.8% Cr steel
AISI 5152	0.52% C 1% Cr steel
AISI 5155	0.55% C 0.8% Cr steel
AISI 5160	0.6% C 0.8% Cr steel
AISI 9261	0.6% C 2% Si 0.2% Cr steel
AISI 9262	0.6% C 2.0% Si 0.3% Cr steel
ANTICHOC 2	0.48% C 1.7% Si 0.12% Cr steel: Pompey
ASTM A401	0.55% C 1.5% Si 0.6% Cr steel: For springs; cold drawn & tempered
AURIGA VII	0.5% C 0.7% Mn 1% Cr 0.4% S & P steel: Casting; D Brown; normalized & tempered **UTS: 680 Elon: 10% Proof: 370**
BCTA	0.6% C 0.65% Cr steel: ESC; hot stamping dies **DPN: 460**
BCTB	0.6% C 0.6% Cr steel: ESC; BS 970 En 11
BS 970/530 A40	0.4% C 1% Cr steel: Supplied to chemical composition
BS 970/530 H36	0.37% C 0.7% Mn 1.0% Cr steel: Supplied to hardenability requirement
BS 970/530 H40	0.40% C 0.7% Mn 1.0% Cr steel: Supplied to hardenability requirement
BS 970 En 11	0.6% C 0.7% Cr steel: Bar & forging; hardened & tempered **DPN: 300 UTS: 770 Elon: 15% Proof: 610**
BS 970 En 18	0.4% C 1% Cr steel: Bar & forging; hardened & tempered **DPN: 260 UTS: 840 Elon: 18% Proof: 640**
BS 970 En 18D	0.40% C 1% Cr steel: Bar & forging; hardened & tempered; properties not specified
BS 970 En 48	0.5% C 1.2% Cr steel: For springs

Symbol	Nominal analysis, supplier, condition and remarks.
BS 970 En 52	0.45% C 3.5% Si 8% Cr steel: Bar & forging; hardened & tempered; for valves **DPN: 275**
BS 970 En 53	0.6% C 1.5% Si 6.2% Cr steel: Bar & forging; hardened & tempered; for valves **DPN: 260**
BS 1453 A5	0.4% C 0.5% Si 1% Mn 1% Cr: Weld filler rod for the repair of wearing surfaces; railway points
BS 3100 BW2	0.5% C 0.7% Mn 1.0% Cr steel: Casting for abrasion resistance
BS 3100 BW3	0.5% C 0.7% Mn 1.0% Cr steel: Casting for abrasion resistance
BS 3111/3	0.4% C 1.0% Cr steel: Wire; annealed & drawn; for cold forged bolts **UTS: 680**
BS 3146 CLA12A	0.5% C 1% Cr steel: Investment casting; hardened & tempered; for abrasion resistance **DPN: 207 UTS: 680 Elon: 10%**
BS 3146 CLA12B	0.5% C 1% Cr steel: Investment casting; hardened & tempered; for abrasion resistance **DPN: 293**
C 35	0.36% C 0.5% Cr steel: German Standard
CA 30	0.34% C 1.35% Cr 0.95% Al steel: For nitriding; Krupp
CA 30 A	0.34% C 1.15% Cr 0.95% Al steel: For nitriding; Krupp
CBV 2	0.37% C 1.05% Cr steel: Krupp
CBV 2h	0.36% C 1.5% Cr steel: Krupp
CBV/Z	0.41% C 1.05% Cr steel: Krupp
CCR 2	0.6% C 0.6% Cr steel: Huntsman; picks, axes, chisels, etc. **DPN: 600**
CCR 350	0.7% C 3.5% Cr steel: Origin unknown
CHISEL STEEL	0.35% C 1.5% Cr steel: Origin unknown
CL 60	0.6% C 0.7% Mn 0.6% Cr steel: Origin unknown
CMN	0.65% C 12% Mn 2.5% Cr 0.4% Ni steel: American proprietary alloy listed in SAE year book
CODE 620	0.5% C 0.8% Cr steel: Casting; Edgar Allen; as Cromax F
CP1	0.43% C 1% Si 1.4% Cr tool steel: Osborn; chisel; shock resisting; obsolete **DPN: 600**
CPG	0.5% C 1% Si 1.25% Cr steel: Jonas; chisels, hammers, etc. **DPN: 600**
CR	0.4% C 1% Cr steel: ESC for BS alloy En 18
CROMANSIL	0.2% C 1.2% Mn 0.7% Si 0.5% Cr steel: Origin unknown
CROMAX F	0.5% C 0.9% Cr steel: Casting; Edgar Allen; for wear resistance; Code 620 **DPN: 250**
DEF 13 B 5A	Alloy steel: Hardened & tempered; covers En 18; also includes En 15, 21, 22, 100, etc. **UTS: 760 Elon: 20% Proof: 550**
DIN 1654 41 Cr 4	0.41% C 1.1% Cr steel: For cold forged bolts
DIN 17200 34 Cr Al S 5	0.34% C 0.75% Mn 1.15% Cr 1.0% Al steel: German Standard
DIN 17200 37 Cr 4	0.37% C 0.75% Mn 1.05% Cr steel: German Standard

Symbol	Nominal analysis, supplier, condition and remarks.
DIN 17200 37 Cr S 4	0.37% C 0.75% Mn 1.05% Cr steel: German Standard
DIN 17200 38 Cr 2	0.38% C 0.6% Mn 0.5% Cr steel: German Standard
DIN 17200 41 Cr 4	0.41% C 1.1% Cr steel: Forging; German Standard
DIN 17200 46 Cr 2	0.46% C 0.65% Mn 0.27% Cr 0.0005 B (max) steel: German Standard
DIN 17221 55 Cr 3	0.55% C 0.85% Mn 0.75% Cr 0.3% Cu (max) steel: For springs; German Standard
DIN 17221 55 Cr 3	0.55% C 0.85% Mn 0.85% Cr steel: For springs; German Standard
DIN 17221 60 Si Cr 7	0.6% C 0.85% Mn 0.3% Cr steel: For springs; German Standard
DIN 17221 60 Si Cr 7	0.6% C 0.85% Mn 0.3% Cr steel: For springs; German Standard
DIN 17221 67 Si Cr 5	0.67% C 1.3% Si 0.5% Cr steel: For springs; German Standard
DIN 17222 67 Si Cr 5	0.67% C 1.3% Si 0.5% Cr steel: For springs; German Standard
DIN 17225 67 Si Cr 5	0.67% C 1.3% Si 0.5% Cr steel: For springs; German Standard
DK 4	0.38% C 1.0% Cr steel: Pompey **UTS: 1070 Elon: 8% Proof: 870**
DK 6	0.6% C 0.75% Cr steel: Pompey
DUNELT 33	0.4% C 1.0% Cr steel: Dunford for BS alloy En 18
DUNELT 83	0.45% C 3.5% Si 8% Cr steel: For valves; Dunford for BS alloy En 52
Fagersta B126	0.6% C 1.7% Mn 0.2% Cr steel: For diaphragms; Fagersta
FAGERSTA C182	0.9% C 0.4% Mn 0.5% Cr steel: For wood saws; Fagersta
FAGERSTA C261	0.4% C 0.5% Mn 0.5% Cr steel: Bar; Fagersta
FAGERSTA C525	0.75% C 0.6% Mn 0.3% Cr steel: For springs; Fagersta
FDK 4	0.38% C 0.9% Cr steel: Pompey **UTS: 1070 Elon: 10% Proof: 760**
FF 4H	0.48% C 0.17% Cr steel: Pompey **UTS: 870 Elon: 8% Proof: 610**
FIRTHAG	0.45% C 0.95% Cr steel: Firth Brown for BS alloy En 18
GOST 2052/60 S2KHA	0.6% C 0.55% Mn 0.85% Cr steel: Bar; Russian Standard
GOST 2052/70 S2KHA	0.7% C 0.5% Mn 0.3% Cr steel: Bar; Russian Standard
GOST 4543/30 KHGSA	0.3% C 0.95% Mn 0.95% Cr steel: Russian Standard
GOST 4543/35 KHG2	0.36% C 1.7% Mn 0.55% Cr steel: Russian Standard
GOST 4543/35 KHGSA	0.35% C 0.95% Mn 1.25% Cr steel: Russian Standard
GOST 4543/38 KHA	0.38% C 0.65% Mn 0.95% Cr steel: Russian Standard
GOST 4543/38 KHIU	0.38% C 0.35% Mn 1.65% Cr 0.65% Al steel: Russian Standard
GOST 4543/38 KHS	0.38% C 0.45% Mn 1.45% Cr steel: Russian Standard
GOST 4543/40 KH	0.4% C 0.65% Mn 0.95% Cr steel: Russian Standard
GOST 4543/40 KHG	0.4> C 1.1> Mn 1.0% Cr steel: Russian Standard
GOST 4543/40 KHGR	0.4% C 0.85% Mn 0.95% Cr 0.0035% steel: Russian Standard
GOST 4543/40 KHR	0.4% C 0.65% Mn 0.95% Cr 0.0035% B steel: Russian Standard
GOST 4543/40 KHS	0.4% C 0.45% Mn 1.45% Cr steel: Russian Standard
GOST 4543/45 KH	0.45% C 0.65% Mn 0.95% Cr steel: Russian Standard
GOST 4543/45 KHC	0.46% C 0.65% Mn 0.95% Cr 0.2% Zr 0.5Al (max) steel: Russian Standard

Note. The following abbreviations and units are used in the tables:

DPN	Hardness, diamond pyramid number
UTS	Ultimate tensile strength, N/mm^2
Elon	Elongation, %
Proof	0.1 % proof strength, N/mm^2

1 N/mm^2=0.1 hbar=0.102 kgf/mm^2=0.06475 tonf/in^2=145.04 lbf/in^2

Symbol	Nominal analysis, supplier, condition and remarks.
GOST 4543/50 KH	0.5% C 0.65% Mn 0.95% Cr steel: Russian Standard
GOST 5632/4 KH9S2	0.4% C 9.0% Cr steel: Russian Standard
GOST 5632/4 KH952	0.4% C 9.0% Cr steel: Plate; Russian Standard
GOST 7909/30 KHGS	0.31% C 0.95% Cr 0.25% Ni (max) steel: Pipe; Russian Standard **UTS: 1100 Proof: 850**
GOST 7909/40 KH	0.4% C 0.65% Mn 0.95% Cr steel: Pipe; Russian Standard **UTS: 1000 Proof: 800**
GOST 8467/30 KHGS	0.31% C 0.95% Mn 0.95% Cr 0.25% Ni (max) steel: Pipe; Russian Standard **UTS: 1100 Proof: 850**
GOST 8467/40 KH	0.4% C 0.65% Mn 0.95% Cr 0.25% Ni (max) steel: Pipe; Russian Standard **UTS: 1000 Proof: 800**
GOST 8731/30 KHGSA	0.31% C 0.9% Mn 0.9% Cr 0.2% Ni (max) steel: Pipe; Russian Standard; seamless **DPN: 700**
GOST 8731/40 KH	0.4% C 0.65% Mn 0.9% Cr 0.2% Ni (max) steel: Pipe; Russian Standard; seamless **UTS: 670**
GOST 8733/30 KHGSA	0.31% C 0.95% Mn 0.95% Cr steel: Pipe; Russian Standard; seamless **UTS: 500**
GOST 8733/40 KH	0.4% C 0.05% Mn 0.95% Cr steel: Pipe; Russian Standard; seamless **UTS: 630**
HIV	0.26% C (max) 0.3% Cr steel: German Standard
H 3	0.6% C 1.5% Si 2% Cr steel: Jessop for BS alloy En 53
H 18	0.45% C 3.3% Si 8.5% Cr steel: Jessop specification for BS alloy En 52
H 23	0.48% C 1.3% Cr steel: Jessop
H 28	0.4% C 1% Cr steel: Jessop for BS alloy En 18
H 30	0.6% C 0.65% Cr steel: Jessop
HECLA 104	0.6% C 0.65% Cr steel: Hadfields BS alloy En 11
HECLA 105	0.5% C 1.2% Cr steel: Hadfields for BS alloy En 48
HECLA 120	0.45% C 3.5% Si 8% Cr steel: Hadfields for BS alloy En 52
HECLA 148	0.65% C 3.5% Cr steel: Origin unknown
HECLA 166	0.6% C 1.5% Si 6.2% Cr steel: Hadfields for BS 970 En 53
HECLA 198	0.55% C 1.5% Si 0.75% Cr steel: Hadfields for BS alloy En 48A
HEDEX 4	0.4% C 1% Cr steel: Wire for forging; Kiveton Park for BS alloy En 18
HEDEX 29	0.4% C 0.8% Cr steel: Wire for forging; Kiveton Park; annealed **UTS: 610**
HNV 2	0.4% C 3.9% Si 2.2% Cr steel: For inlet valves; designation used by SAE
HNV 3	0.45% C 3.3% Si 8.5% Cr steel: For valves; designation used by SAE
JIS G4106 S Cr 4	0.4% C 0.7% Mn 1.02% Cr steel: Japanese Standard
JIS G4106 S Cr 5	0.45% C 0.7% Mn 1.05% Cr steel: Japanese Standard
JIS G4106 S Mn C 3	0.43% C 1.5% Mn 0.5% Cr steel: Japanese Standard
JIS G4801 SUP9	0.55% C 0.8% Mn 0.8% Cr steel: For springs; Japanese Standard
JIS G4106 SUP10	0.5% C 0.85% Mn 0.95% Cr 0.15% V (max) steel: Japanese Standard
JIS G4801 SUP11	0.55% C 0.8% Mn 0.8% Cr steel: For springs; Japanese Standard
KF 18C	0.4% C 1% Cr steel: Kirkstall for BS alloy En 18
KF 18D	0.45% C 1% Cr steel: Kirkstall for BS alloy En 18

Symbol	Nominal analysis, supplier, condition and remarks.
KP 32	0.4% C 1% Cr steel: Kiveton Park for BS alloy En 18
L 5	0.42% C 1.45% Mn 0.85% Cr steel: Casting; Jessop
LT 36	0.45% C 1.5% Cr steel: Low Moor; obsolete **DPN: 540**
LV 52	0.45% C 3.5% Si 8% Cr steel: For valves; Low Moor; obsolete **DPN: 320**
MONARCH GENERAL UTILITY	0.6% C 0.7% Mn 0.7% Cr steel: For tools; origin unknown
MSC 40	0.43% C 0.5% Cr steel: Krupp
MSC 50	0.5% C 0.7% Cr steel: Krupp
MY 0	0.55% C 0.95% Cr steel: Origin unknown
NFA35 551/38C2	0.38% C 0.75% Mn 0.45% Cr steel: French Standard
NF A35 551/38C4	0.38% C 0.75% Mn 1.0% Cr steel: French Standard
NF A35 551/42C2	0.43% C 0.75% Mn 0.45% Cr steel: French Standard
NF A35 551/42C4	0.42% C 0.75% Mn 1.0% Cr steel: French Standard
NF A35 551/10006	0.98% C 0.3% Mn 1.5% Cr 0.3% Ni (max) 0.1% Mo (max) steel: French Standard
NF A35 571/45C4	0.45% C 0.75% Mn 1.0% Cr steel: Bar; French Standard
NF A35 571/60SC7	0.6% C 0.75% Mn 0.52% Cr steel: For bars; French Standard
NITA	0.53% C 1.5% Cr 0.2% Al 1.1% Ac steel: Origin unknown
NVG	0.5% C 0.8% Cr steel: For inlet valves; designation used by SAE
OCM 9	0.4% C 1% Cr steel: W Marrison for BS alloy En 18
P 609	0.4% C 0.8% Mn 1% Cr steel: Carrs; pressure dies for casting Zn Pb Sn base alloys **DPN: 400**
PN	0.42% C 0.45% Mn 0.45% Cr steel: Origin unknown
RMK	0.45% C 1.0% Cr steel: Pompey
RMKS	0.45% C 1.5% Cr steel: Pompey
RS 3K	0.6% C 1.8% Si 0.4% Cr steel: Pompey
RS 4K	0.65% C 1.7% Si 0.8% Cr steel: Pompey
RSMK	0.6% C 0.5% Cr steel: Pompey
S 4K	0.45% C 0.5% Cr steel: Pompey
SAE 50 B40	0.4% C 0.5% Cr steel with B
SAE 50 B44	0.44% C 0.5% Cr 0.0005% B steel
SAE 50 B46	0.46% C 0.3% Cr 0.0005% B steel
SAE 50 B50	0.5% C 0.5% Cr 0.0005% B steel
SAE 50 B60	0.6% C 0.5% Cr 0.0005% B steel
SAE 51 B60	0.6% C 0.8% Cr 0.0005% B steel
SAE 5045	0.45% C 0.7% Cr steel: Obsolete
SAE 5046	0.46% C 0.3% Cr steel: Mechanical properties not specified
SAE 5060	0.6% C 0.5% Cr steel
SAE 5140	0.4% C 0.8% Cr steel: Mechanical properties not specified
SAE 5145	0.45% C 0.8% Cr steel: Mechanical properties not specified
SAE 5147	0.47% C 1% Cr steel: Mechanical properties not specified
SAE 5150	0.5% C 0.8% Cr steel: Mechanical properties not specified
SAE 5152	0.52% C 1% Cr steel: Mechanical properties not specified
SAE 5155	0.55% C 0.8% Cr steel: Mechanical properties not specified
SAE 5160	0.6% C 0.8% Cr steel: Mechanical properties not specified

Symbol	Nominal analysis, supplier, condition and remarks.
SAE 9254	0.55% C 1.5% Si 0.7% Cr steel: Mechanical properties not specified
SAE 9261	0.6% C 2.0% Si 0.2% Cr steel: Mechanical properties not specified
SAE 9262	0.6% C 2.0% Si 0.3% Cr steel: Mechanical properties not specified
SAE HNV 2	0.4% C 3.9% Si 2.2% Cr steel: For inlet valves
SAE HNV 3	0.45% C 3.3% Si 8.5% Cr steel: For valves **UTS: 980 Elon: 22% Proof: 740**
SAE NVG	0.5% C 0.8% Cr steel: For inlet valves as SAE 5150
SANBOLD 11	0.6% C 0.65% Cr steel: Sanderson for BS alloy En 11
SANBOLD 18	0.4% C 1% Cr steel: Sanderson for BS alloy En 18
SANBOLD 54	0.4% C 1% Cr steel: Sanderson; replaced by Sanbold 18
SCRI	0.47% C 1% Si 1.4% Cr steel: Huntsman; shear blades; punches, etc. **DPN: 600**
SIS 14-29-12E	0.6% C 0.65% Mn 0.3% Mo 0.3% Cr (max) 0.3% Cu steel: For pressure vessels; Swedish Standard; normalized **UTS: 490 Elon: 23% Proof: 270**
SIS 14-2090	0.5% C 0.7% Mn 0.3% Cr spring steel: Swedish Standard **DPN: 480 UTS: 1400 Elon: 7%**
SIS 2230	0.51% C 1.0% C 0.15% V steel: For springs; Swedish Standard
SIS 2085	0.55% C 1.7% Si 0.3% Cr steel: Bar & forging; Swedish Standard; hardened & tempered **DPN: 480 UTS: 1380 Elon: 5% Proof: 1260**
SIS 2090	0.55% C 1.7% Si 0.3% Cr steel: Bar & forging; Swedish Standard; hardened & tempered **DPN: 480 UTS: 1380 Elon: 5% Proof: 1260**
SIS 2230	0.51% C 1.0% C 0.15% V steel: For springs; Swedish Standard
SIS 2245	0.41% C 1.0% Cr steel: Swedish Standard **DPN: 255 UTS: 920**
SLV	0.45% C 3.4% Si 8.3% Cr steel: Firth Brown for BS alloy En 52 **DPN: 285 UTS: 1070 Elon: 15%**
SPKS	0.45% C 9.0% Cr steel: Pompey
STA 5 V9C	0.4% C 1.0% Cr steel: Replaced by BS alloy En 18
STA 5 V9C/a	0.4% C 1.0% Cr steel: Replaced by BS alloy En 18B

Symbol	Nominal analysis, supplier, condition and remarks.
STA 5 V9C/2	0.4% C 1.0% Cr steel: Replaced by BS alloy En 18C
STA 5 V9C/3	0.4% C 1.0% Cr steel: Replaced by BS alloy En 18D
STA 5 V9C/4	0.4% C 1.0% Cr steel: Replaced by BS alloy En 18A
STA 5 V 24	0.45% C 3.5% Si 8.5% Cr steel: Replaced by BS alloy En 52
SUH 1	0.45% C 32% Si 8.5% Cr steel: Heat resisting; Japanese Standard designation
TR 3KD	0.38% C 0.9% Cr steel: Pompey
TR 4	0.45% C 0.15% Cr steel: Pompey
TR 5	0.52% C 0.15% Cr steel: Pompey
TRK	0.42% C 0.15% Cr steel: Pompey
TRIPLE SIX	0.6% C 0.6% Cr steel: John Vessay **DPN: 700**
VALVEX 518D	0.45% C 3.5% Si 8% Cr steel: For valves; Swift Levick for BS alloy En 52
VIPER 666	0.6% C 0.6% Cr steel: Hall & Pickles; lathe centres, leather cutters, etc.; obsolete **DPN: 600**
VS 50	0.45% C 3.5% Si 8.5% Cr steel: Darwins for BS alloy En 52
VSCS	0.55% C 0.8% Cr steel: Vulcan; heavy duty vehicle springs **DPN: 440**
VS 1C	0.4% C 1% Cr steel: Vulcan **UTS: 1070**
Y 60SC7	0.6% C 0.8% Mn 0.6% Cr steel: French Standard designation
Y 75V	0.75% C 0.25% Mn 0.32% Cr steel: French Standard designation
Z 45CS9	0.45% C 0.8% Mn (max) 8.0% Cr steel: French Standard designation

Note. The following abbreviations and units are used in the tables:

DPN	Hardness, diamond pyramid number
UTS	Ultimate tensile strength, N/mm²
Elon	Elongation, %
Proof	0.1 % proof strength, N/mm²

$1 \text{ N/mm}^2 = 0.1 \text{ hbar} = 0.102 \text{ kgf/mm}^2 = 0.06475 \text{ tonf/in}^2 = 145.04 \text{ lbf/in}^2$

44E3 Steel – high carbon with chromium
0.7% carbon minimum – 5% chromium maximum

Specific gravity	7.8	
Density	7800 kg/m³	(0.23 lb/in³.)
Solidus/liquidus	1400–1450 °C	
Thermal conductivity	– W/m °C	
Coefficient of linear expansion	$11 \times 10^{-6}/$ °C	
Electrical conductivity	8–10% IACS (copper 100%)	
Specific resistance	170–230 microhm mm	
Young's modulus of elasticity	200×10^9 N/m²	$(29 \times 10^6$ lbf/in.²)
Impact	– J	
Fatigue strength	± 570 N/mm²	(± 37.5 tonf/in².)
Hot strength	–	

The above properties are typical of the following group, and may not apply exactly to any one specification. It is possible that with certain specifications some of the values may not be applicable.

General metallurgical characteristics

Chromium forms very hard stable carbides with the carbon present in steel. These are harder than iron carbides and retain their hardness at higher temperatures. Being stable the carbides are very sluggish in their metallurgical reactions and care must always be taken that the correct times and temperatures are used. As these steels are all high carbon there is considerable carbon available to form chromium carbides and even in the annealed condition they have considerable hardness. Chromium imparts some corrosion resistance by virtue of the tenacious uncorrodable oxide, which it forms, but by no stretch of the imagination can any of the steels listed be classed as corrosion resistant. These steels are all air hardening, being the most readily hardened of any of the low alloy carbon steels.

In the hardened or tempered condition the steels have very little ductility and are notoriously notch sensitive. They will always be used in the hardened and tempered condition. Larger quantities of chromium (over 12%) are present in the stainless steels, which are described in sections 44M and 44N. Small quantities of titanium or aluminium are sometimes added to control grain growth.

Thermal treatment

Pre-heating at about 600 °C is advisable before any thermal treatment of these steels to prevent cracking. They cannot accept any degree of cold work without becoming embrittled. Full annealing requires a temperature of 850–930 °C for one hour per 25 mm of section, followed by a very slow cool. For many purposes a sub-critical anneal at 650–670 °C for two hours will suffice. Cyclic annealing is necessary sometimes and consists of varying the temperature between 900 and 660 °C several times. This gives the best structure for machining.

Normalizing requires a temperature of 850–930 °C for 30 minutes per 25 mm of section followed by still air cooling. This will remove any forging stresses, refine and homogenize the grain but cause considerable hardening. The parts are softened for machining by a two hour subcritical anneal at 650–670 °C or by cyclic annealing.

To harden, the parts must be taken slowly up to 820–900 °C and held for 20 minutes per 25 mm of section, oil quenched or air cooled and then immediately tempered at 150–350 °C depending on the required hardness. As previously stated, the chromium carbides make the metallurgical reactions sluggish and it is important that the parts are held long enough at temperature for the reactions to take place. The use of higher temperatures speeds up these reactions but leads to undesirable grain growth. Reference should be made to the source of the specification or material supplier for detailed information on the treatment necessary.

These materials cannot be case hardened, but they can be induction or flame hardened locally. The desired area is heated by means of a high frequency coil or oxyacetylene flame. Removal of the heat source is generally sufficient to cause full hardening by air quenching.

The sluggish nature of the steels do not make them the economical or technical first choice for this method of hardening, as the long time at temperature necessary for solution of the carbides is not compatible with these very rapid sources of heating.

Normal electric, gas or oil fired furnaces, or salt baths can be used, some care being necessary to prevent severe local decarburization when a slightly oxidizing atmosphere is present. Unless an inert or reducing atmosphere can be guaranteed, it is necessary to have the free scaling conditions obtained from an oxidizing atmosphere.

Scale removal should be by grit blasting or sodium hydride. Acid pickling must not be used with these steels in the hardened and tempered condition as it causes brittleness, but is a suitable method of scale removal when soft.

Welding and brazing

Pre-treatment. The steels should be in the fully annealed condition. The weld area must be clean and freshly prepared, and careful pre-heating is essential.

Welding and brazing. None of these materials should be welded or brazed without taking considerable care to prevent cracking in the weld zone, caused by hardening stresses. This cracking can be prevented by careful pre- and post-weld treatment which should aim at a slow rate of heating and cooling.

None of the materials listed should be first choice for welding, although a skilled operator can produce good welds. Gas welding generally gives better results than electric as there is less heat shock.

Brazing can be carried out but will cause local softening. There are high melting point braze materials available which allow parts to be re-hardened and tempered after brazing. Parts should always be annealed if possible before brazing, or at least pre-heated using the gas torch to prevent cracking.

Post-treatment. Whenever possible parts should be normalized then hardened and tempered. As an absolute minimum the weld area must be stress relieved at 250–350 °C to temper the fully hardened areas.

Fluxes and scale must be removed before any thermal treatment, otherwise severe corrosion can take place.

Flaw detection methods

Crack test. Magnetic crack test. Penetrant oil chalk test with skilled operators is now almost as sensitive as the magnetic test and is more flexible.

X-ray. Not normally required.

Ultra-sonic test. This is commonly applied to find subcutaneous defects on large expensive forgings at an early stage in machining. This cannot detect surface defects.

Chemical etch. An electrolytic sulphuric acid etch to show grinding abuse is recommended, at least as a quality check on the standard of grinding. This will highlight defective grinding, which can cause surface softening before it is serious enough to cause cracking.

Corrosion protection

Temporary. These steels have a limited amount of corrosion resistance proportional to the chromium content but this is not sufficient to justify disregarding temporary corrosion protection. The material must always be oiled or greased and never stored or allowed to lie in moist conditions.

Soluble oil coolant in good condition will not cause corrosion but when it becomes contaminated, especially with metallic particles it can become a source of rusting.

It is recommended that finished parts are cleaned, oiled and stored in air-tight plastic wherever possible.

Permanent. These steels are almost always used under cover. As their high surface hardness is the principal reason for their existance it is very seldom that ideal permanent protection can be carried out. Because of the high hardness it is not generally possible to roughen the surface by blasting, but this process can still be used to clean and remove traces of corrosion. Alkaline rust removers give good results if properly controlled, but acid must not be used as it will result in embrittlement.

Phosphating after blasting ensures that any last trace of rust has been eliminated. A subsequent stress release at 200 °C is generally necessary to prevent embrittlement. The resultant surface is a good key for oil or paint and many tools are phosphate treated and oiled as their permanent protection, this being sufficient provided the parts are kept indoors with a film of oil at all times.

Zinc silicate or chromate primer paints should be applied if painting is essential, followed by stoved top coats for best results.

Plating. Electroplating of these high tensile steels will very seldom be for decorative purposes. Cadmium, tin and zinc can be applied, the deposit never being more than 0.025 mm thick. Plating of these steels is not generally recommended as there is a grave danger of hydrogen embrittlement which can cause brittle failure below the calculated tensile strength. This is caused by the preparatory cleaning and etching as well as the plating, failure sometimes taking place in the plating vat.

Stress relieving immediately after plating at 200 °C eliminates this tendency and a similar treatment before plating obviates risk of failure during plating. It is however, difficult to guarantee that these operations are in fact carried out.

Metal spraying is possible but it is very difficult to roughen the surface enough to obtain a proper key, thus adhesion of the sprayed coating is seldom satisfactory. With modern methods of spraying using high particle velocities this is being overcome to some extent.

Artificial oxidation processes using hot strong sodium hydroxide solutions with oxidizing agents added give an oil-retaining corrosion resistant surface. These 'black oxide' treatments are in reality dark brown or blue, the finely divided iron oxide being firmly adherent to the base metal preventing further oxidation. The protection afforded is not sufficient to withstand constant exposure to the atmosphere but is sufficient for normal indoor damp conditions provided it is kept oiled. This is the finish which has been applied to guns for centuries.

Machinability

These steels must be machined in the normalized and sub-critical annealed condition. Best results are obtained with the carbides spheroidized, this being achieved by cyclic annealing, varying the temperature between that of the full anneal and the sub-critical anneal for several cycles.

Any fine limits and areas which must not be de-carburized should be machined after final hardening. Care must be taken to ensure that no stress raisers are present which would cause cracking at heat treatment. A minimum of 0.2 mm must be ground off to ensure removal of de-carburization after heat treatment by conventional methods.

Grinding must be very carefully carried out to prevent softening or cracking of the glass hard material. A stress release at about 150 °C between rough and final grinding can be helpful in reducing grinding trouble.

Uses

Cutting tools, press dies and fixtures, pivot points, wear resistant slides, ball and roller races. Springs are made from the lower carbon steels.

Symbol	Nominal analysis, supplier, condition and remarks.
1.2067	1.15% C 1.5% Cr steel: German Standard designation
1.2067	10.5% C 1.4% Cr steel: German Standard
1.3501	1.0% C 0.5% Cr steel: For bearings; German Standard
1.3503	1.02% C 0.35% Mn 1.5% Cr steel: For bearings; German Standard

Note. The following abbreviations and units are used in the tables:

DPN	Hardness, diamond pyramid number
UTS	Ultimate tensile strength, N/mm^2
Elon	Elongation, %
Proof	0.1 % proof strength, N/mm^2

1 N/mm^2=0.1 hbar=0.102 kgf/mm^2=0.06475 tonf/in^2=145.04 lbf/in^2

Symbol	Nominal analysis, supplier, condition and remarks.
1.3503	1.0% C 1.0% Cr steel; For ball bearings; German Standard
1.3505	1.0% C 1.5% Cr steel: For ball bearings; German Standard
1.3520	1.0% C 1.1% Mn 1.5% Cr steel: For ball races; German Standard
1.7131	1.6% C 1.1% Mn 1.0% Cr steel: German Standard
1.7176	0.55% C 0.85% Mn 0.85% Cr 0.3% Cu (max) steel: For springs; German Standard
1.7176	0.55% C 0.85% Mn 0.75% Cr steel: For springs; German Standard
1.7305	0.41% C 0.7% Mn 1.05% Cr steel: For bearings; German Standard
1.8405	0.7% C 2.0% Mn 1% Cr 0.2% Ti 0.1% Al steel: Welding rod; German Standard
1.8425	1.1% C 2.0% Mn 1.8% Cr 0.2% Ti 0.1% Al steel: Welding rod; German Standard

Symbol	Nominal analysis, supplier, condition and remarks.
8KH3	0.8% C 3.5% Cr steel: Russian Standard designation
9KHS	0.9% C 1.1% Cr steel: Russian Standard designation
10 K 2	1.0% C 1.45% Cr steel: Pompey
11KH	1.1% C 0.55% Cr steel: Russian Standard designation
13KH	1.32% C 0.45% Mn 0.55% Cr steel: Russian Standard designation
14 Si C1	0.75% C 0.35% Cr steel: Sandvik
17C	1.0% C 0.15% Cr steel: Sandvik; annealed **UTS: 660**
17 C1	1.0% C 0.5% Cr steel: Sandvik
21C	1.25% C 0.15% Cr steel: Sandvik; annealed **UTS: 690**
22C	1.3% C 0.15% Cr steel: Sandvik
22C1	1.25% C 0.45% Cr steel: Sandvik; annealed **UTS: 720**
73.08	0.08% C 0.8% Mn 0.6% Ni 0.4% Cu steel: For electrodes; Esab standard **UTS: 590 Elon: 28% Proof: 490**
73.68	0.06% C 0.7% Mn 2.4% Ni 0.3% Si 0.01% S 0.01% P steel: For electrodes; Esab **UTS: 600 Elon: 28% Proof: 500**
41 Cr 4	0.41% C 0.7% Mn 1.05% Cr steel: For bearings; German Standard
67 Si Cr 5	0.67% C 1.3% Si 0.5% Cr steel: Designation used by German Standard
70 Mn Cr Ti 8	0.7% C 1.1% Cr 0.2% Ti steel: Designation used by German Standard
73/3.5% Cr	1.0% C 3.5% Cr steel: For magnets; Simonds
100 Cr 6	1.15% C 1.5% Cr steel: German Standard designation
100 Cr 6 (W3)	1.0% C 1.55% Cr steel: Designation used by German Standard
100 Cr Mn 6 (W4)	1.0% C 1.5% Cr steel: Designation used by German Standard
105 Cr 2 (W1)	1.05% C 0.5% Cr steel: Designation used by German Standard
105 Cr 4 (W2)	1.05% C 1.1% Cr steel: Designation used by German Standard
105 Cr 6	1.02% Cr 0.3% Mn 1.5% Cr steel: For bearings; German Standard
110 Mn Cr Ti 8	1.1% C 1.8% Cr 0.25% Ti steel: Designation used by German Standard
150	1.0% C 1.45% Cr steel: F Parkin; ball races, collets, etc. **DPN: 870**
180	1.0% C 0.5% Cr steel: F Parkin; reamers, taps, etc. **DPN: 820**
1230	1.35% C 0.3% Mn 0.35% Cr steel: French Standard designation
1230	1.35% C 0.2% Mn 0.32% Cr steel: French Standard designation
1231	1.2% C 0.2% Mn 0.32% Cr steel: French Standard designation
1232	1.05% C 0.2% Mn 0.32% Cr steel: French Standard designation

Note. The following abbreviations and units are used in the tables:

DPN	Hardness, diamond pyramid number
UTS	Ultimate tensile strength, N/mm^2
Elon	Elongation, %
Proof	0.1 % proof strength, N/mm^2

1 N/mm^2=0.1 hbar=0.102 kgf/mm^2=0.06475 tonf/in^2=145.04 lbf/in^2

Symbol	Nominal analysis, supplier, condition and remarks.
1233	0.9% C 0.25% Mn 0.32% Cr steel: French Standard designation
ADAMANT GR	0.77% C Cr steel: Casting; Firth Brown; normalized **DPN: 300**
AFNOR 100 Cb	1.0% C 1.5% Cr steel: French Standard
AGS	1.05% C 1.3% Cr steel: Sheet; Balfour; templates, bushes, etc. **DPN: 770**
AISI E50100	1.0% C 0.5% Cr steel
AISI E51100	1.0% C 1.00% Cr steel
AISI E52100	1.0% C 1.5% Cr steel
AMS 6440D	1.0% C 1.5% Cr steel: Bar & forging; AMS for SAE 52100
AMS 6441B	1.0% C 1.5% Cr steel: Seamless tube; AMS for SAE 52100
AMS 6442B	1.0% C 0.5% Cr steel: Bar & forging; AMS for SAE 50100
AMS 6443B	1.0% C 1.0% Cr steel: For ball bearings; vacuum melted; AMS for SAE 51100
AMS 6444B	1.0% C 1.5% Cr steel: For ball bearings; vacuum melted; AMS for SAE 52100
AMS 6445A	1.0% C 1.0% Cr steel: For ball bearings; vacuum melted; AMS for SAE 51100
AMS 6446	1.0% C 1.0% Cr steel: Bar & forging; AMS for SAE 51100
AMS 6447	1.0% C 1.5% Cr steel: Bar & forging; AMS for SAE 52100
ASTM A485/1	1.0% C 1.0% Mn 0.5% Si 1.0% Cr steel: High hardenability
ASTM A485/2	0.97% C 1.3% Mn 0.7% Si 1.6% Cr steel: High hardenability
ASTM A732/15 A	1.0% C 1.45% Cr steel: Investment casting; annealed **DPN: 240**
ATLAS Q	1.2% C 0.5% Cr steel: Atlas; as AISI type W5
B 4 CR	1.0% C 1.4% Cr steel: ESC; bushes, taps, etc. **DPN: 850**
BRI	1.0% C 1.0% Cr steel: Jonas; press tools **DPN: 720**
BRUNSWICK V 981	1.0% C 1.4% Cr steel: John Vessey **DPN: 800**
BS 970/520 M60	0.6% C 0.65% Mn 0.65% Cr steel: Supplied to mechanical properties; replaces En 11
BS 970/534 A99	0.99% C 0.3% Mn 1.4% Cr steel: Supplied to chemical composition
BS 970/535 A99	0.99% C 0.65% Mn 1.8% Cr steel: Supplied to chemical composition
BS 970 En 31	1.0% C 1.3% Cr steel: Bar & forgings; ball races
C 46	1.3% C 0.5% Cr steel: Fagersta **DPN: 910**
C 71	0.9% C 1.5% Si 1.0% Cr steel: Fagersta **DPN: 750**
C 182	0.9% C 0.5% Cr steel: Fagersta **DPN: 800**
C 525	0.75% C 1.3% Si 0.3% Cr steel: Fagersta
CH 4	1.25% C 0.7% Cr steel: Forez
CH 5	1.0% C 0.5% Cr steel: Forez
CNS NON SHRINK	1.0% C 1.0% Cr steel: T Turton
CODE 623	0.72% C 2.0% Cr steel: Casting; Edgar Allen; Cromax H
Cr 030	0.9% C 3.3% Cr steel: Forging; for magnets; German Standard included in DIN 17410
Cr 035	0.9% C 1.0% Mn 5.0% Cr steel: Forging; for magnets; German Standard included in DIN 17410
CRC	1.0% C 1.45% Cr steel: ESC for BS alloy En 31
CROMAX N	0.73% C 2.0% Cr steel: Casting; Edgar Allen; for wear resistant parts **DPN: 250**

Symbol	Nominal analysis, supplier, condition and remarks.
DCCM	1.05% C 1.3% Cr tool steel: For cold work; Balfour Darwin
DTD 5022	1.0% C 1.5% Cr steel **DPN: 750**
DUNELT 90	1.0% C 1.2% Cr steel: Dunford for BS alloy En 31
E 10	1.0% C 0.13% Cr steel: Pompey
EH 4	1.0% C 1.0% Si 4.0% Cr steel: Forging; for magnets; Polish alloy listed by PMA
EKh	1.0% C 3.3% Cr steel: For magnets; Russian alloy listed by PMA
FAGERSTA C250	1.2% C 0.35% Mn 0.17% Cr steel: Bar; Fagersta
FAGERSTA N164	0.8% C 0.4% Mn 2.0% Cr steel: For wood saws; Fagersta
GOST 801 SHKH6	1.1% C 0.3% Mn 0.55% Cr 0.3% Ni (max) steel: For bearings; Russian Standard
GOST 801 SHKH9	1.05% C 0.3% Mn 1.05% Cr 0.3% Ni (max) steel: For bearings; Russian Standard
GOST 801 SHKH9	1.05% C 0.3% Mn 1.1% Cr steel: For bearings; Russian Standard
GOST 801 SHKH15	1.0% C 0.3% Mn 1.5% Cr steel: For bearings; Russian Standard
GOST 801 SHKH15	1.0% C 0.3% Mn 1.5% Cr 0.3% Ni (max) steel: For bearings; Russian Standard
GOST 801 SHKH15SG	1.0% C 1.05% Mn 1.5% Cr 0.3% Ni (max) steel: For bearings; Russian Standard
GOST 4543	0.3% C 0.65% Mn 1.15% Cr 0.0035% B steel: Russian Standard
H 2	1.0% C 1.5% Cr steel: Jessops **DPN: 800**
H 17	1.0% C 6.0% Cr steel: For magnets; Jessops
H 44	1.0% C 1.5% Cr steel: Jessops **DPN: 850**
H 63	1.0% C 0.5% Cr steel: Jessops
H 64	0.97% C 1.0% Cr steel: Jessops
HAO	1.0% C 6.0% Cr steel: For magnets; Belgian alloy listed by PMA
HECLA 108	1.0% C 1.3% Cr steel: Hadfields for BS alloy En 31
HEDEX 18	1.0% C 1.3% Cr steel: Wire for forgings; Kiveton Park for BS alloy En 31
HEDEX 41	1.0% C 0.32% C steel: Wire for forgings; Kiveton Park; annealed **UTS: 630**
HRS	1.0% C 1.3% Cr steel: Jonas; press tools **DPN: 680**
HRS	1.0% C 1.4% Cr steel: Osborn; dies, balls, rollers, races; obsolete **DPN: 720**
ID 2.02	0.9% C 1.7% Mn 0.25% Cr steel: Inman
JCNS	0.9% C 0.4% Cr 1.3% Mn steel: Jonas
JIS G4805 SUJ1	1.0% C 0.3% Mn 1.05% Cr 0.08% Mo (max) steel: For bearings; Japanese Standard
JIS G4805 SUJ2	1.05% C 0.5% Mn (max) 1.45% Cr 0.08% Mo (max) steel: For bearings; Japanese Standard
JIS G4805 SUJ3	1.02% C 1.0% Mn 1.1% Cr 0.08% Mo (max) steel: For bearings; Japanese Standard
KE 4L	1.0% C 0.2% Cr steel: Kayser Ellison; for ball bearings; needles, etc.

Symbol	Nominal analysis, supplier, condition and remarks.
KE 160	1.2% C 0.4% Mn 0.5% Cr steel: Kayser Ellison; taps, punches, etc.
KE 839	1.0% C 1.5% Cr steel: Kayser Ellison; taps, dies, thread rolls, etc. **DPN: 870**
KH	1.02% C 1.45% Cr steel: Russian Standard designation
KP 24	1.0% C 1.5% Cr steel: Kiveton Park for BS alloy En 31
KR 6	1.0% C 1.5% Cr steel: Pompey
L 1	1.0% C 1.25% Cr steel: Designation used by AISI
LESCALLOY 52100	1.0% C 1.5% Cr steel: For bearings; Latrobe; as AISI 52100
LK	1.4% C 0.7% Cr steel: Pompey
LT 2	1.0% C 0.5% Cr steel: Low Moor; obsolete **DPN: 870**
LT 22	1.0% C 1.5% Cr steel: Low Moor; obsolete **DPN: 770**
MARSH MN	0.9% C 0.25% Cr steel: Marsh Brothers; gauges, taps, etc.
M CHROME REGENT	1.0% C 1.45% Cr 0.1% Mo 0.15% Ni 0.15% Cu steel: Latrobe
MIC	0.95% C 0.35% Cr steel: ESC; symbol for 'Microlim' **DPN: 770**
MIC 2	0.95% C 0.35% Cr steel: Origin unknown
MICROLIM	0.95% C 0.3% Cr steel: ESC; punches, rivet, dies, etc. **DPN: 770**
MN	0.9% C 0.25% Cr steel: Marsh Brothers; gauges, taps, etc.
Mn Cr	0.6% C 0.8% Cr steel: ESC; springs
NF A35-565/100C2	1.02% C 0.3% Mn 0.5% Cr steel: For bearings; French Standard
NF A35-565/100C6	1.02% C 0.3% Mn 1.5% Cr 0.1% Mo (max) steel: For bearings; French Standard
NF A35-565/100C6	1.05% C 0.3% Mn 1.5% Cr 0.08% Mo (max) steel: For bearings; French Standard
NSC	1.0% C 1.4% Cr steel: Huntsman; centres, small dies, reamers, etc. **DPN: 800**
NSCD	0.9% C 0.3% Cr steel: Huntsman; slitters, gauges, reamers, etc. **DPN: 720**
NSCT	1.0% C 1.4% Cr steel: Huntsman; ball races, press tools, taps, etc. **DPN: 800**
P 280	0.6% C 0.6% Cr steel: Carrs; hot press, dies, etc. **DPN: 450**
P 704	0.95% C 0.5% Cr steel: Carrs; vice jaws, punches, etc. **DPN: 630**
P 720	1.0% C 1.3% Cr steel: Carrs; centres, ball bearings, etc. **DPN: 730**
PRESTO	1.0% C 1.4% Cr steel: Carpenter; for AISI type L1
RED DIAMOND II	1.1% C 0.6% Mn 1.5% Cr steel: For wear; Redheugh **DPN: 295 UTS: 1020 Elon: 10% Proof: 870**
REGENT VAC-ARC	1.0% C 1.5% Cr steel: Latrobe Steel; for SAE 52100 **DPN: 800**
SAE 50100	1.0% C 0.5% Cr steel: Mechanical properties not specified
SAE 51100	1.0% C 1.0% Cr steel: Mechanical properties not specified
SAE 52100	1.0% C 1.5% Cr steel: Mechanical properties not specified

Note. The following abbreviations and units are used in the tables:

DPN	Hardness, diamond pyramid number
UTS	Ultimate tensile strength, N/mm^2
Elon	Elongation, %
Proof	0.1 % proof strength, N/mm^2

1 N/mm^2=0.1 hbar=0.102 kgf/mm^2=0.06475 tonf/in^2=145.04 lbf/in^2

Symbol	Nominal analysis, supplier, condition and remarks.
SANBOLD 31	1.1% C 1.3% Cr steel: Sanderson for BS alloy En 31
SANBOLD 57	1.1% C 1.3% Cr steel: Sanderson; replaced by Sanbold 31
SIS 14-2258	1.0% C 0.3% Mn 1.5% Cr steel: Alloyed; for structures; Swedish Standard **DPN: 1250**
SIS 2092	1.0% C 1.0% Cr steel: Swedish Standard; annealed **DPN: 240**
SIS 2258	1.0% C 1.5% Cr steel: Swedish Standard; annealed **DPN: 210**
SKC 11	0.97% C 0.5% Mn (max) 1.2% Cr 0.4% Mo (max) 0.25% V (max) steel: For roller drills; Japanese Standard designation
SKF 3	1.0% C 1.5% Cr steel: Bar for bearings; Japanese Standard designation
SKS 8	1.45% C 0.35% Cr steel: Japanese Standard designation
SKS 93	1.05% C 0.95% Mn 0.4% Cr steel: Japanese Standard designation
SKS 94	0.95% C 0.9% Mn 0.4% Cr steel: Japanese Standard designation
SKS 95	0.85% C 0.9% Mn 0.4% Cr steel: Japanese Standard designation
SK SILVER STEEL	1.2% C 0.4% Mn 0.4% Cr steel: Sanderson Kayser
SL 934	1.1% C 1.25% Cr steel: Swift Levick; press, dies, punches, etc.
SPEAR CRH	0.75% C 0.4% Cr steel: Origin unknown
SPEAR GFS	0.95% C 0.5% Cr steel: Origin unknown
SR 1855	1.0% C 1.5% Si 1.0% Cr steel: Bofors **DPN: 850**
STA 5 V14	1.1% C 1.3% Cr steel: Replaced by BS alloy En 31
SUJ 2	1.0% C 1.45% Cr steel: Japanese Standard designation
T 1040	1.0% C 0.25% Cr steel: Origin unknown
TERFLING 2	0.6% C 0.7% Cr steel: Uddeholm; for type S4
TOLEDO BR	1.1% C 1.3% Cr steel: Toledo for BS alloy En 31
VAMPIRE	0.9% C 0.3% Cr steel: Hall & Pickles; press tools, gauges, etc.; obsolete **DPN: 800**
VANSTAR	1.0% C 5.0% Cr steel: Pompey
VIPER	1.0% C 1.0% Cr steel: Hall & Pickles; gauges, ball bearings etc.; obsolete **DPN: 800**

Symbol	Nominal analysis, supplier, condition and remarks.
VSCM	0.9% C 1.0% Cr steel: Vulcan; press tools; milling cutters, etc. **DPN: 720**
W 3	1.0% C 0.5% Cr steel: Designation used by AISI; obsolete
W 4	1.0% C 0.25% Cr steel: Carbon varies over wide limits; designation used by AISI
W 5	1.1% C 0.5% Cr steel: Designation used by AISI
WATERDIE STANDARD	1.0% C 0.5% Cr steel: Columbia; AISI type W5
WATERDIE EXTRA	1.0% C 0.5% Cr steel: Columbia; AISI type W5
WLM 1015	1.0% C 1.4% Cr steel: W Marrison; for ball races
X 45 CrSi9	0.45% C 3.0% Si 9.0% Cr steel: Designation used by German Standard
X 45 CrSi9/3	0.45% C 3.0% Si 9.0% Cr steel: Designation used by German Standard
X 45 SiCr4	0.45% C 4.0% Si 2.7% Cr steel: Designation used by German Standard
XX 9CR	0.9% C 0.5% Cr steel: ESC; mandrels, punches, etc. **DPN: 750**
Y 90V	0.9% C 0.25% Mn 0.32% Cr steel: French Standard designation
Y 100 C6	1.05% C 1.4% Cr steel: French Standard
Y 105V	1.05% C 0.2% Mn 0.32% Cr steel: French Standard designation
Y 120V	1.20% C 0.2% Mn 0.32% Cr steel: French Standard designation
Y 135C	1.35% C 0.3% Mn 0.35% Cr steel: French Standard designation
Y 135V	1.35% C 0.2% Mn 0.32% Cr steel: French Standard designation
Y 2120 C	1.2% C 0.4% Cr steel: French Standard

Note. The following abbreviations and units are used in the tables:

DPN	Hardness, diamond pyramid number
UTS	Ultimate tensile strength, N/mm^2
Elon	Elongation, %
Proof	0.1 % proof strength, N/mm^2

$1 N/mm^2 = 0.1 \text{ hbar} = 0.102 \text{ kgf}/mm^2 = 0.06475 \text{ tonf}/in^2 = 145.04 \text{ lbf}/in^2$

44F1 Steel – low carbon with nickel
0.35% carbon maximum

Specific gravity	7.8	
Density	7800 kg/m^3	(0.28 $lb/in.^3$)
Solidus/liquidus	1470–1510 °C	
Thermal conductivity	– W/m °C	
Coefficient of linear expansion	11 × 10^{-6}/ °C	
Electrical conductivity	6–9% IACS (copper 100%)	
Specific resistance	230–270 microhm mm	
Young's modulus of elasticity	– N/m^2	
Impact	54–105 J	(40–80 ft/lb)
Fatigue strength	–	
Hot strength	–	

The above properties are typical of the following group, and may not apply exactly to any one specification. It is possible that with certain specifications some of the values may not be applicable.

General metallurgical characteristics

Nickel is a ferrite strengthener and toughener which also reduces the critical temperatures and widens the range of temperatures at which optimum hardening is obtained. The increased strength and toughness are apparent in the normalized or softened condition, even without the slight air hardening properties of nickel steels.

One important property of some nickel steels is to retain their strength and ductility at very low temperatures when other materials become embrittled. The ductility and notch resistance is roughly proportional to the nickel and inversely proportional to the carbon content.

With the higher nickel steels there is some corrosion resistance, but under 4% nickel they are little better than carbon steels.

The good ductility, toughness and flexible heat treatment of these low carbon nickel steels make them good case hardening materials.

The relatively large amounts of nickel required, the high cost and scarcity resulted in research on alternative materials, which has been largely successful. Similar or superior properties are now available with low nickel, chromium and molybdenum steels to those obtainable from 3–5% straight nickel, with the exception of low temperature impact values.

Thermal treatment

Although exhibiting relatively good ductility none of these steels can accept a high percentage of cold work without intermediate anneals, which require 840–900 °C for 30 minutes per 25 mm of section, followed by slow cooling. A sub-critical anneal at 650–670 °C for two hours will remove the majority of cold work and soften the steel for most purposes.

Normalizing also requires a temperature of 840–900 °C for 20 minutes per 25 mm section, followed by cooling in still air. This refines and homogenizes the grain, removing all forging or cold working stresses and results in a degree of hardening and toughening from the air quench which is never sufficient to require a subsequent anneal prior to machining.

In order to harden, these steels must be heated to 820–880 °C followed by oil quench. The parts should then be tempered at 200–500 °C to the required hardness.

Some of the steels listed have been specially formulated for case hardening and most of the remainder are suitable with the exception of those with higher carbon, which would tend to have too high a core hardness for most purposes. Gas, pack of salt bath carburizing are all possible. Gas carburizing is the most rapid, and probably gives a superior case, but requires special furnace equipment. Pack carburizing is slow and dirty, giving a slightly harder, more brittle surface, but can use normal furnace equipment. Cyanide salt bath is the cheapest method but the case has some nitrogen which can make it more brittle. Carburizing should be at 900–930 °C for the required time to give the necessary case depth, which should not be less than 0.1 mm except for special purposes.

The parts should be cooled after carburizing, then reheated to 740–780 °C oil quenched and tempered at 150–250 °C. For optimum results parts should be refined between carburizing and hardening. This requires quenching from 840–860 °C, but is not necessary with modern good quality steels, which can have additions to ensure they are fine grained. The practice of quenching direct from the carburizing temperature is not recommended for best results.

Where machining to remove local areas of case is necessary, the part should be sub-critical annealed after carburizing or refining if this has been carried out. The subsequent hardening after machining is as described above.

The times and temperatures quoted are for guidance only and full details of the treatments must always be obtained from the actual supplier or origin of the specification.

Normal electric, gas, oil fired or salt bath heating can be used with these steels and there is seldom any need to have special furnace atmospheres.

Scale removal can be by grit blasting, acid pickling or sodium hydride. Acid pickling must not be used on case hardened parts after hardening, as there is a danger of embrittlement.

Welding and brazing

Pre-treatment. Although not essential it is advisable to weld these materials in the normalized or annealed condition. The weld area must be free from scale and oil, and correct edge preparation is important. Case hardened parts should never be welded or brazed. Many of these steels will require pre-heating.

Welding and brazing. Normal electric and gas welding with fluxes gives good results. Inert gas shielded arc welding requires no flux but is seldom economical.

Special equipment for very fast automatic electric welding is available for use on these steels.

Brazing is readily accomplished using fluxes and copper alloy filler but will cause local softening on hardened parts.

Post-treatment. Stress relieving at 650–670 °C or local tempering above 350 °C is an essential minimum requirement.

Ideally the parts should be normalized, then hardened and tempered and this is often possible with correct design to minimize stress raisers.

Electric heated blankets or gas heating are available for local tempering or stress relieving. All traces of flux and slag must be removed before thermal treatment, otherwise corrosion will result.

Flaw detection methods

Crack test. Magnetic crack testing. Modern penetrants with skilled operation make the oil chalk test method almost as sensitive and more flexible than the magnetic method.

X-ray. Not used on wrought products but commonly applied to mass production welds. Radioactive isotope sources now give very portable efficient means of X-ray examination.

Ultra-sonic testing. This is used to show defects in large expensive forgings. It cannot identify surface faults. Techniques have now been developed which identify certain types of weld defect.

Chemical etch. An electrolytic sulphuric acid etch is recommended as a control on the grinding of carburized parts. This will clearly show the extent of grinding abuse before visible grinding cracking makes itself known.

Corrosion protection

Temporary. The nickel content of these steels has no significant effect on their corrosion resistance, thus normal precautions regarding storing in moisture free conditions, and oiling or greasing at all times apply.

Good quality soluble oil coolant in good condition inhibits corrosion but dirty coolant, particularly that loaded with metallic particles, is a serious hazard.

Finished carburized parts should be cleaned, oiled and sealed in plastic before storing.

Permanent. There is no cheap simple method of protecting these materials against atmospheric corrosion. As they are relatively expensive, they warrant the correct treatment. This implies that all traces of corrosion, moisture and dirt are removed, and the surface immediately sealed off from all contact with air. It is not always appreciated that corrosion products remain active even when completely covered and as their volume expands as the reaction proceeds, this ruptures the protective coating.

Grit blasting correctly carried out removes rust and dirt and leaves a roughened surface which acts as a key for the subsequent treatment. Control of the blast process is essential, as is the removal of the surface dust before the next process.

Carburized hardened parts are not visibly affected by blasting, but corrosion and dirt will have been removed. Alkaline de-rusters are highly efficient when correctly controlled. Acid de-rusters must not be used on carburized hardened parts as they can cause embrittlement, but all other steels in this group can be acid pickled.

Painting. After blasting the parts should ideally be phosphate treated by dipping or spraying, stove dried, then a zinc silicate or chromate primer applied. As soon as possible after drying the parts should have the top coats laid on and these should be of the stoving type. Modern plastic paints have good abrasion and shock resistance and are non-porous.

Plating. Electroplating for decorative purposes relies on the thickness and quality of the copper and nickel substrates for corrosion resistance. This should be 0.05 mm of copper and 0.025 mm of nickel for best results. The chromium top coat is very thin, supplying only the bright lustre and preventing the nickel from tarnishing.

Cadmium, tin and zinc are commonly electrodeposited less than 0.025 mm thick. Provided pre-plating cleaning and etching are adequate these metals efficiently protect against normal atmospheric corrosion. Cadmium and zinc must not be used in contact with foodstuffs.

Parts hardened to above 300 DPN and all carburized parts should be stress relieved at 200 °C after electroplating to remove the effects of hydrogen embrittlement. This can cause failure in a brittle manner below the calculated tensile strength and is especially dangerous on case hardened items.

Metal spraying using zinc or aluminium gives very good results, provided adequate blasting is carried out before spraying and the parts are wire-brushed and chemically sealed or painted immediately after metal spraying. Normal metal spray does not adhere well to case hardened areas as it relies on the blast key which is absent in these instances.

Cathodic protection uses sacrificial anodes electrically connected to the structure being protected. This is made electrically negative when the positive or anode will corrode, protecting the negative or cathode in the process. Cathodic protection is often additive to painting or plating. It is applied to ships, marine structures, pipe lines, steel foundations and pylons.

Black oxide is a form of artificial oxide formed when the parts are immersed in strong hot sodium hydroxide solution, with oxidizing agents added. It produces a good surface for oil retention and will withstand normal indoor usage. If kept dry and oiled there is no atmospheric attack. This finish has been applied to guns for many centuries.

Machinability

These materials tend to tear in the normalized condition. For best results parts should be hardened and tempered to about 280–300 DPN which will reduce the tool life somewhat but give a superior finish.

Case hardened areas must of course, be ground as they are too hard for conventional machining.

Uses

General structural steel where high strength to weight is required. Case hardened part where core ductility is important, such as gear shafts which must flex.

Symbol	Nominal analysis, supplier, condition and remarks.
1.5613	0.24% C 1.2% Ni steel: Forging: German Standard
1.5622	0.14% C 1.4% Ni steel: For low temperature use; German Standard
1.5633	0.24% C 2.1% Ni steel: Forging; German Standard
1.5680	0.2% C (max) 4.7% Ni steel: German Standard
1.6215	0.09% C 1.1% Mn 0.4% Ni steel: Welding rod; German Standard
1.6216	0.17% C 0.8% Ni steel: Welding rod; German Standard
2NiB	0.07% C 2.5% Ni steel: Electrode: Metrode; for low temperature use
2.2D8	0.2% C 0.75% Mn 2.2% Ni steel: Greusot-Loire
3NiB	0.05% C 2.3% Ni steel: Electrode: Metrode; for low temperature use
3.5D8	0.19% C 0.75% Mn 3.5% Ni steel: Creusot-Loire
3NS	0.3% C 3% Ni steel: ESC for BS alloy En 21
5NSR	0.15% C 5% Ni steel: For case hardening; Jonas; core properties
	DPN: 321 UTS: 1000 Elon: 13%
8CN2	0.08% C 2.0% Ni steel: Pompey
9 Mn Ni 4	0.09% C 1.1% Mn 0.45% Ni steel: Designation used by German Standard
10 N9	0.1% C 9.0% Ni steel: Pompey
	UTS: 610 Elon: 22% Proof: 390
12 Ni19	0.2% C (max) 0.45% Ni steel: Designation used by German Standard
14 N3	0.15% C 3.45% Ni steel: Pompey
	UTS: 460 Elon: 25% Proof: 270
14 Ni6	0.14% C 1.4% Ni steel: Designation used by German Standard
17 Mn Ni4	0.17% C 0.7% Ni steel: Designation used by German Standard
A 203A	0.2% C 2.2% Ni steel: Plate; Armco; for low temperature use
	UTS: 670 Elon: 24% Proof: 220
A 203B	0.24% C 2.2% Ni steel: Plate; Armco; for low temperature use
	UTS: 610 Elon: 22% Proof: 270
A 203D	0.2% C 3.5% Ni steel: Plate, Armco; for low temperature use
	UTS: 500 Elon: 24% Proof: 220
A 203E	0.23% C 3.5% Ni steel: Plate; Armco; for low temperature use
	UTS: 580 Elon: 23% Proof: 270
A 440	0.28% C 1.4% Mn 0.4% Ni 0.2% Cu steel: Armco
	UTS: 480 Elon: 20% Proof: 300
AISI 2317	0.17% C 3% Ni steel
AISI 2515	0.15% C 5% Ni steel
AISI A2330	0.3% C 3.5% Ni steel: Obsolete
AISI E2512	0.12% C 5% Ni steel: Obsolete
AISI E2517	0.17% C 5% Ni steel
ALFRIG Ni3	0.15% C 1.5% Mn 3.5% Ni steel: Austrian specification
AMS 6242C	0.17% C 5% Ni steel: Bar & forging; AMS for SAE 2517
AN QQ S 676/1	0.25% C 1.75% Mn 0.2% Ni steel: US Service
	UTS: 460 Elon: 15% Proof: 370
AN QQ S 689/1	0.3% C 3.5% Ni steel: US Service; normalized & tempered
	DPN: 230 UTS: 840 Elon: 17% Proof: 740

Symbol	Nominal analysis, supplier, condition and remarks.
ARMCO HS No. 1	0.15% C (max) 0.5% Mn 0.75% Ni 0.6% Cu steel: Armco
ARMCO HS No. 2	0.15% C (max) 0.6% Mn 0.75% Ni 0.6% Cu steel: Armco
ASTM A203 A	0.2% C 0.75% Mn 2.25% Ni steel: Plate for pressure vessels
ASTM A203 B	0.25% C 0.75% Mn 2.25% Ni steel: Plate for pressure vessels
ASTM A 203 D	0.2% C 0.75% Mn 3.5% Ni steel: Plate for pressure vessels
ASTM A203 E	0.21% C 0.75% Mn 3.5% Ni steel: Plate for pressure vessels
ASTM A316 E8016C	0.12% C Ni steel: Welding electrodes; code varies with analysis
ASTM A316 E8016C1	0.12% C 2.4% Ni steel: Electrodes for arc welding
ASTM A316 E8016C2	0.12% C 3.4% Ni steel: Electrodes for arc welding
ASTM A316 E8018C1	0.12% C 2.4% Ni steel: Electrodes for arc welding
ASTM A316 E8018C2	0.12% C 3.4% Ni steel: Electrodes for arc welding
ASTM A320 L10	0.18% C 3.5% Ni steel: Normalized
	UTS: 460 Elon: 25% Proof: 270
ASTM A333/3	0.18% C 3.5% Ni steel: Pipe for low temperature use
ASTM A333/7	0.19% C 2.2% Ni steel: Pipe for low temperature use
ASTM A333/8	0.13% C 9.0% Ni steel: Pipe for low temperature use
ASTM A350 LF3	0.2% C (max) 3.5% Ni steel: For low temperature use
	UTS: 460 Elon: 25% Proof: 270
ASTM A352 IC3	0.15% C 3.5% Ni steel: Normalized & tempered for low temperature service
	UTS: 440 Elon: 24% Proof: 270
ASTM A352 LC2	0.25% C 2.5% Ni steel: Normalized & tempered for low temperature service
	UTS: 440 Elon: 24% Proof: 270
ASTM A352 LC3	0.15% C 0.65% Mn 3.5% Ni steel: Casting for low temperature use
ASTM A 353 A & B	0.13% C 9.0% Ni steel: Normalized & tempered
	UTS: 630 Elon: 21% Proof: 420
ASTM A522	0.13% C (max) 9.0% Ni steel: For valve fitting; low temperature service
ASTM A553	0.13% C (max) 9.0% Ni steel: Plate for pressure vessels
	UTS: 750 Elon: 20% Proof: 580
ASTM A588	0.2% C (max) 0.8% Mn 0.6% Ni steel: For structures
	UTS: 440 Elon: 21% Proof: 320
ASTM A658	0.1% C (max) 0.5% Mn (max) 0.5% Cr (max) 36.0% Ni steel: Plate for pressure vessels
	UTS: 500 Elon: 30% Proof: 240
ASTM A690	0.22% C (max) 0.7% Mn 0.52% Ni steel: For marine piles
	UTS: 485 Elon: 18% Proof: 345
ASTM A707 L7/1	0.2% C (max) 1.0% Mn (max) 3.4% Ni steel: Flanges; for low temperature use; impact 41 J at -73 °C
	UTS: 290 Elon: 22%
ASTM A707 L7/2	0.2% C (max) 1.0% Mn (max) 3.4% Ni steel: Flanges for low temperature use; impact 54 J at -73 °C
	UTS: 360 Elon: 25%
ASTM A707 L7/3	0.2% C (max) 1.0% Mn (max) 3.4% Ni steel: Flanges for low temperature use; impact 68 J at -73 °C
	UTS: 415 Elon: 23%

Note. The following abbreviations and units are used in the tables:

DPN	Hardness, diamond pyramid number
UTS	Ultimate tensile strength, N/mm^2
Elon	Elongation, %
Proof	0.1 % proof strength, N/mm^2

1 N/mm^2=0.1 hbar=0.102 kgf/mm^2=0.06475 tonf/in^2=145.04 lbf/in^2

Symbol	Nominal analysis, supplier, condition and remarks.
ASTM A707 L7/4	0.2% C (max) 1.0% Mn (max) 3.4% Ni steel: Flanges for low temperature use; impact 68 J at -73 °C **UTS: 515 Elon: 20%**
ASTM A714/5 F	0.2% C (max) 1.4% Mn 1.0% Cu 1.9% Ni steel: Pipe; welded or seamless **UTS: 379 Elon: 27%**
ASTM A714/5 E or S	0.2% C (max) 1.4% Mn 1.0% Cu 1.9% Ni steel: Pipe; welded or seamless **UTS: 448 Proof: 317**
ASTM A732/11 Q	0.2% C 1.8% Ni 0.25% Mo steel: Investment casting; hardened & tempered **UTS: 830 Elon: 10% Proof: 690**
AURIGA XVII	0.15% C 3.5% Ni steel: Casting; D Brown; normalized & tempered for sub-zero use **UTS: 440 Elon: 25% Proof: 270**
AUSTEN 56	0.23% C 1.6% Mn 0.6% Ni steel: Australian Iron & Steel Co.
BS 970 En 14B	0.25% C 1.5% Mn 0.4% Ni steel: Bar & forging; hardened & tempered **DPN: 215 UTS: 670 Elon: 20% Proof: 450**
BS 1501/503	0.15% C 3.50% Ni steel: Plate & bar; normalized **UTS: 440 Elon: 20% Proof: 260**
BS 1501/509	0.1% C (max) 9.0% Ni 0.2% Al (min) steel: Plate; hardened & tempered for pressure vessels; Charpy impact 47 J at 160 °C **UTS: 700 Elon: 18% Proof: 530**
BS 1501/510	0.1% C (max) 9.2% Ni 0.02% Al (min) steel: Plate; hardened & tempered for pressure vessels; Charpy impact 47 J at 160 °C **UTS: 700 Elon: 18% Proof: 600**
BS 1503/503	0.15% C 3.5% Ni steel: Forging; normalized & tempered **UTS: 460 Elon: 20% Proof: 270**
BS 1504/503	0.15% C (max) 3.5% Ni steel: Casting; normalized & tempered **UTS: 440 Elon: 25% Proof: 240**
BS 3100 BL2	0.12% C (max) 3.5% Ni steel: Casting for low temperature use; impact -20 J at -60 °C **UTS: 460 Elon: 20% Proof: 280**
BS 3146 CIA10	0.15% C 3% Ni steel: Investment casting; for case hardening; refined & hardened core properties **UTS: 670 Elon: 18%**
BSN 3	0.3% C 3% Ni steel: Toledo for BS alloy En 21
BS S4	Nickel steel: Sheet; obsolete
BS S8	3.0% Ni steel: Hardened & tempered; replaced
BS S9	3.0% Ni steel: Forging; hardened & tempered; replaced
BS S10	3.0% Ni steel: Bar & forging; hardened & tempered; replaced
BS S15	0.12% C 3.0% Ni steel: For carburizing; blank carburized hardened & tempered **UTS: 670 Elon: 18%**
BS S17	5% Ni steel: For carburizing; replaced
BS S46	Nickel alloy steel: Strip; obsolete
BS S67	Low Carbon 5% Ni steel: For carburizing; obsolete
BS S83	Low carbon 5% Ni steel: For carburizing; obsolete
B & W Ni 57	0.2% C 3.5% Ni steel: Tube; Babcock & Wilcox (USA)
CASALLOY	0.1% C 1.0% Ni steel: For case hardening; Clyde Alloy; annealed **DPN: 180**
CH 3N	0.13% C 3.1% Ni steel: For case hardening; ESC for BS alloy En 33
CMN 1	0.25% C 0.25% Si 1.5% Mn steel: Firth Brown for BS alloy En 14A & B
CRYONIC 5	0.13% C 0.27% Mo 5% Ni 0.1% V steel: Armco Steel; for low temperature use **UTS: 69 Elon: 20% Proof: 45**

Symbol	Nominal analysis, supplier, condition and remarks.
DOFASCOLOY	0.18% C 1.0% Mn 0.9% Ni 0.6% Cu steel: Dominion
DOFASCOLOY 2	0.15% C 1.0% Mn 0.9% Ni 0.6% Cu steel: Dominion
DUCOL QT131	0.2% C 1.6% Mn 0.5% Ni steel: Colvilles; Al killed
E 0	0.12% C 3% Ni steel: Jessop for BS alloy En 33
E 1A	0.3% C 3% Ni steel: Jessop for BS alloy En 21
E 4	0.11% C 4.8% Ni steel: Jessop for BS alloy En 37
FLT 2C	0.1% C 0.56% Mn 2.5% Ni 0.1% Cu steel: Fuji
FLT 2D	0.07% C 0.5% Mn 3.5% Ni 0.14% Cu steel: Fuji
FN 30	0.3% C 3% Ni steel: Firth Brown **UTS: 840 Elon: 20%**
HARDY NICKEL IRON	0.08% C 0.2% Si 2.0% Ni steel: Atlas; for low duty sub-zero parts **DPN: 120 UTS: 340 Elon: 30% Proof: 210**
IN 787B	0.06% C (max) 1.3% Ni 1.15% Cu 0.02% Nb (min) steel: International Nickel Co.; solution treated & aged impact 170 J ° at -60 °C **UTS: 630 Elon: 26% Proof: 550**
JIS G3460	0.18% C (max) 0.45% Mn 3.6% Ni 0.02% Cu (max) steel: Pipe; Japanese Standard **UTS: 460 Proof: 250**
JIS G3464 STBL46	0.18% C (max) 0.5% Mn 3.6% Ni steel: Pipe; Japanese Standard **UTS: 460 Proof: 250**
KF 13	0.32% C 3% Ni steel: Kirkstall for BS alloy En 21
KF 15/3	0.12% C 3.2% Ni case hardening steel: Kirkstall for BS alloy En 33
KOERZIT 120	27% Ni 13% Al steel: For magnets; German alloy listed by PMA
KP 11	Low C 2% Ni steel: For case hardening; Kiveton Park
LT 3N	0.12% C 3.5% Ni steel: For low temperature use; ESC; normalized & tempered **UTS: 480 Elon: 24% Proof: 270**
LT 9N	0.1% C 9% Ni steel: For low temperature use; ESC; normalized & tempered **UTS: 670 Elon: 22% Proof: 450**
MAXELOY 4	0.15% C (max) 1.2% Mn 0.5% Ni 0.2% Cu steel: Crucible Steel Co.
MIL S 12505.1	0.23% C (max) 1.25% Mn 0.75% Ni 0.45% Cu steel: Bethlehem Steel Co.
N 1/57	0.2% C 3.5% Ni steel: Tube; Babcock & Wilcox (USA)
N 3CH	0.14% C 3.1% Ni steel: For carburizing; Firth Brown for BS 970 En 33
N 5CH	0.1% C 5.0% Ni steel: For carburizing; Firth Brown for BS 970 En 37
NBV 1	0.24% C 1.15% Ni steel: Krupp
NBV 1/h	0.34% C 1.35% Ni steel: Krupp
NBV 2/w	0.14% C 1.5% Ni steel: Krupp
NBV 5/w	0.12% C 5.0% Ni steel: Krupp
NF A36 208/2.25Ni	0.15% C (max) 0.8% Mn (max) 2.0% Ni steel: Plate; French Standard; normalized **UTS: 500 Proof: 270**
NF A36 208/3.5Ni	0.15% C (max) 0.8% Mn (max) 3.5% Ni steel: Plate; French Standard **UTS: 500 Proof: 270**
NF A36 208/9Ni	0.1% C (max) 0.8% Mn (max) 9.0% Ni steel: Plate; French Standard **UTS: 700 Proof: 600**
NICLOY	0.2% C 3.5% Ni steel: Tube; Babcock & Wilcox (USA)
NICLOY 5	0.14% C 5.0% Ni steel: Tube; Babcock & Wilcox (USA)
NICLOY 9	0.12% C 9% Ni 0.3% Cu steel: Tube; Babcock & Wilcox (USA)

Symbol	Nominal analysis, supplier, condition and remarks.
NICUAGE TYLER	0.06% C (max) 0.5% Mn 0.85% Ni 1.2% Cu 0.02% Nb (min) steel: Information from Inco; age hardened Impact 50 J at -20 °C **DPN: 230 UTS: 690 Elon: 23% Proof: 620**
Ni Cu Cb	0.06% C (max) 0.5% Mn 0.85% Ni 1.15% Cu 0.02% Nb (min) steel: International Nickel Co; age hardened **Elon: 29%**
Ni Cu Ti	0.15% C (max) 1.0% Mn 0.7% Ni 0.05% Ti 0.2% Cu steel: Jones & Laughlin
NICUAGE	0.06% C 0.5% Mn 0.85% Ni 1.15% Cu 0.02% Nb steel: Strip, bar, section & tube; International Nickel Co.; normalized; aged **DPN: 215 UTS: 640 Elon: 22% Proof: 590**
NICUAGE 1	0.06% C 0.85% Ni 1.1% Cu 0.02% Nb steel: Imco; age hardened **DPN: 235 UTS: 690 Elon: 23% Proof: 640**
NKR	0.3% C 3% Ni steel: Jonas **UTS: 920 Elon: 20% Proof: 760**
NMS	0.23% C 1.6% Mn 0.6% Ni steel: Australian Iron & Steel Co
NSR	0.15% C 3% Ni steel: For case hardening; Jonas; core properties **DPN: 207 UTS: 690 Elon: 18%**
NV 20.0	0.1% C (max) 0.55% Mn 3.25% Ni (min) steel: Designation used by DNV **UTS: 450 Elon: 21% Proof: 350**
NV 20.00	0.1% C (max) 0.55% Mn 2.25% Ni (min) steel: Designation used by DNV **UTS: 400 Elon: 21% Proof: 320**
NV 20.1	0.1% C (max) 0.55% Mn 4.75% (min) steel: Designation used by DNV **UTS: 500 Elon: 25% Proof: 400**
NV 20.2	0.08% C (max) 0.55% Mn 9.0% Ni (min) steel: Designation used by DNV **UTS: 650 Elon: 20% Proof: 450**
NVR 20.0	0.15% C (max) 0.45% Mn 3.25% Ni (min) steel: Designation used by DNV
NVR 20.00	0.15% C (max) 0.45% Mn 2.25% Ni (min) steel: Designation used by DNV
NVR 20.1	0.1% C (max) 0.55% Mn 4.75% Ni (min) steel: Designation used by DNV
NVR 20.2	0.08% C (max) 0.6% Mn 8.6% Ni (min) steel: Designation used by DNV
P 153	0.12% C 3% Ni steel: For case hardening; Carrs
PITT-TEN 1	0.12% C (max) 0.8% Mn 0.75% Ni 0.8% Cu steel: Pittsburgh Steel
PPT	0.2% C 4.0% Ni 1.2% Al steel: Origin unknown
SAE 2317	0.17% C 3.5% Ni steel: Obsolete
SAE 2330	0.3% C 3.5% Ni steel: Obsolete
SAE 2512	0.12% C 5% Ni steel: Obsolete
SAE 2515	0.15% C 5% Ni steel: Mechanical properties not specified
SAE 2517	0.17% C 0.5% Mn 5% Ni steel: Mechanical properties not specified
SANBOLD 14	0.25% C 1.5% Mn 0.4% Ni steel: Sanderson for En 14B
SANBOLD 51	0.25% C 1.5% Mn 0.4% Ni steel: Sanderson; replaced by Sanbold 14
SHEF-TEN	0.28% C 1.4% Mn 0.4% Ni 0.2% Cu steel: Armco **UTS: 480 Elon: 20% Proof: 300**

Symbol	Nominal analysis, supplier, condition and remarks.
SIS 8110	0.2% C (max) 1.0% Cu 3.0% Ni steel: Sintered
SL 2N26	0.17% C (max) 0.7% Mn (max) 2.2% Ni steel: Plate; Japanese Standard; draft specification **UTS: 500 Proof: 260**
SL 3N26	0.15% C (max) 0.7% Mn (max) 3.5% Ni steel: Plate; Japanese Standard; draft specification **UTS: 500 Proof: 260**
SL 3N45	0.15% C (max) 0.7% Mn 3.5% Ni steel: Plate; French Standard; draft; hardened & tempered **UTS: 620 Proof: 450**
SL 9N53	0.12% C (max) 9.0% Ni steel: Plate; Japanese Standard; draft; normalized; tempered **UTS: 770 Proof: 530**
SL 9N60	0.12% C (max) 0.9% Mn (max) 9.0% Ni steel: Plate; Japanese Standard; draft; hardened & tempered **UTS: 770 Proof: 600**
SN 5	Nickel steel sintered material: Sintered Products Ltd; 10–15% porosity **UTS: 400 Elon: 5%**
SN 10	Nickel steel sintered material: Sintered Products Ltd; 10–15% porosity **UTS: 630 Elon: 3%**
SN 15	Nickel steel sintered material: Sintered Products Ltd; 10–15% porosity **UTS: 840 Elon: 2%**
SPEAR 433	0.12% C 3.0% Ni steel: Spear & Jackson as BS alloy En 33
SPEAR 521	0.3% C 3.2% Ni steel: Spear & Jackson as BS alloy En 21
SUPER TOUGH B25	0.25% C 1.5% Mn 0.4% Ni steel: ESC for BS alloy En 14
TUBROD 14.18	0.1% C 0.8% Mn 1.5% Ni 0.6% Al steel: Welding wire; Esab; obsolete
TUBROD 15.18	0.08% C 1.2% Mn 0.5% Ni 0.4% Si steel: Tubular weld rod; Esab; for hard facing; as welded **UTS: 625 Elon: 25% Proof: 470**
VS 3N	0.1% C 3% Ni steel: For carburizing; Vulcan **UTS: 690**
VS 5N	0.1% C 5% Ni steel: For carburizing; Vulcan **UTS: 1070**
VS 15M	0.25% C 1.5% Mn 0.4% Ni steel: Vulcan
VS 35N	0.35% C 3.5% Ni steel: Vulcan
YND 37	0.14% C (max) 1.5% Mn 0.7% Ni steel: Yamata
YOLOY S	0.2% C (max) 1.0% Mn 1.9% Ni 1.0% Cu steel: Youngstown Steel Co.

Note. The following abbreviations and units are used in the tables:

DPN	Hardness, diamond pyramid number
UTS	Ultimate tensile strength, N/mm^2
Elon	Elongation, %
Proof	0.1 % proof strength, N/mm^2

1 N/mm^2=0.1 hbar=0.102 kgf/mm^2=0.06475 tonf/in^2=145.04 lbf/in^2

44F2 Steel – medium carbon with nickel
0.4% carbon minimum

Specific gravity	7.8	
Density	7800 kg/m³	(0.28 lb/in.³)
Solidus/liquidus	1450–1500 °C	
Thermal conductivity	– W/m °C	
Coefficient of linear expansion	11 × 10⁻⁶/ °C	
Electrical conductivity	6–9% IACS (copper 100%)	
Specific resistance	230–270 microhm mm	
Young's modulus of elasticity	– N/m²	
Impact	- 102 J	(75 ft/lb) (Izod)
Fatigue strength	± 539 N/mm²	(± 33.5 tonf/in.²)
Hot strength		

Temperature °C	Tensile strength N/mm²	Elongation %
200	750	25.5
400	730	21.5
600	270	40
800	100	71
1000	40	72.5

The above properties are typical of the following group, and may not apply exactly to any one specification. It is possible that with certain specifications some of the values may not be applicable.

General metallurgical characteristics

Nickel additions in steel lower the critical change points, in proportion to the nickel content. This means that parts do not require to be taken to as high a temperature as with plain carbon steels. Nickel also toughens the ferrite, resulting in better mechanical properties even in the normalized or annealed conditions. There is some increase in corrosion resistance, again proportional to the nickel content, but this is never sufficient in the steels listed to be noticeable.

The one time scarcity and high cost of nickel resulted in research, culminating in the use of low nickel, chromium and molybdenum steels in lieu of the 4% nickel series and this use has remained long after the scarcity disappeared.

There is still considerable use of the low nickel steels which show a degree of toughness in the normalized condition, and also nickel bearing steels for good impact properties at low temperatures.

Thermal treatment

These steels cannot accept much cold work without interstage annealing. Full annealing requires a temperature of 760–820 °C for 30 minutes per 25 mm of section, followed by slow cooling. A subcritical anneal for two hours at 640–660 °C removes most of the cold work and is sufficient for many purposes, including machinability. Normalizing requires a temperature of 760–820 °C for 20 minutes per 25 mm of section, followed by still air cooling.

This treatment removes all forging and cold work stresses, refines and homogenizes the structure, and re-sults in very little air hardening. It is the treatment applied very often to large structural components.

Hardening requires 740–800 °C for 10 minutes per 25 mm of section, then oil quenching followed by tempering at 300–500 °C for two hours. These steels will not be carburized, as the core will be too hard and brittle to support the case.

The origin of the specification should always be consulted for more exact information on the temperatures required.

No special furnace equipment is necessary for treating any of the steels listed. Electric, gas or oil fired furnaces, or salt baths, can all be used for heating, some care always being desirable to prevent local intergranular decarburization, which can be prevented by ensuring free scaling conditions. Where no machining after treatment is possible and no decarburization is permissible, a controlled atmosphere of burnt town gas or cracked ammonia will be necessary.

Scale removal is generally by grit blasting, but acid pickling can be used except when the parts are harder than 350 DPN. Sodium hydride de-scaling is very suitable but requires special equipment.

Welding and brazing

Pre-treatment. Parts should be in the normalized condition and not hardened and tempered. All traces of scale and rust should be removed and the weld area cleaned. Careful joint preparation is essential and pre-heating will be required.

Welding and brazing. Normal electric or gas welding using flux gives good results, as does inert gas shielded arc, but this is seldom economical. These materials are sometimes welded using the modern high speed automatic welding techniques but more care and control are necessary than with the low carbon steels.

Electrical resistance welding using special equipment is also possible. Brazing presents no problems, but cannot be subsequently hardened unless special high melting point filler rods are used.

Post-treatment. The weld area will have a hardened zone, the degree of hardness being decided by the severity of the quench caused by the surrounding mass of cold metal, and can be largely eliminated by careful pre- and post-weld heating. As an absoluted minimum, the weld zone must be tempered to soften this hard area and improve the difficulty. Electric blankets or local gas heating can be used.

Ideall the parts should be hardened and tempered and for this care must be taken not to have stress raisers. The nickel content considerably reduces the problem of quench cracking from stress raiser, but does not result in its elimination.

Most welded assemblies in these materials are normalized or sub-critically annealed to remove welding stresses and soften off the hardened zone. All traces of flux and slag must be removed prior to heat treatment, otherwise severe corrosion can occur.

Flaw detection methods

Crack test. Magnetic crack test. Oil penetrant chalk testing fluids and techniques are now available, almost as sensitive as the magnetic method and much more flexible.

X-rays. This is seldom applied to wrought items, but is a standard requirement on high duty or mass production welds. Special radioactive metal isotope equipment is highly mobile, giving well-defined radiographs and has been developed for this purpose.

Ultra-sonic test. This is often required on important forgings to prove that there are no subcutaneous defects. It will not detect surface defects and is generally called for at an early stage of machining.

Chemical etch. Not required.

Corrosion protection

Temporary. The nickel content has no appreciable effect on the corrosion resistance of these materials. It is therefore important that they are stored in moisture-free conditions and oiled or greased at all times. Normal good quality soluble oil coolants act as reasonable corrosion inhibitors, but when dirty or contaminated with metal dust those can be a cause of staining and corrosion.

Finished machined parts should be oiled or greased and sealed in plastic.

Permanent. Steel corrodes when in contact with moist air. The corrosion products are loosely adherent, readily washed or rubbed off, leaving a surface free for fresh corrosion. This continuous metal removal, resulting in pitting, weakens the structure and makes rusting a serious problem.

It is necessary first to remove all trace of existing corrosion, then to cover and seal the surface completely from the atmosphere. It should be appreciated that rust has moisture and air occluded which can allow it to propagate even when sealed by paint. There is then an increase in volume which can fail the film, allowing access of air for further corrosion to take place.

Grit blasting is a good method of cleaning and also roughens the surface, giving the paint some key. The blast process must be carefully controlled. After blasting, ideally the parts should be phosphate treated as this further improves the paint key and has some measure of corrosion resistance.

Zinc silicate or chromate type primers should be applied next with the top coats, laid on as soon as possible and stoved for best results. Modern plastic type paints have good wear resistance but should have the metal oxide primers as a base.

Plating. Electroplating for appearance only will not normally be applied to these materials. Cadmium, tin and zinc plating present no difficulties using normal cleaning and etching techniques but parts above 350 DPN (1000 N/mm^{-2} tensile) should be stress released immediately after plating to remove the danger of hydrogen embrittlement. Zinc and cadmium plating must not be used in contact with foodstuffs.

Metal spraying using zinc or aluminium is very commonly applied now to large structures as well as individual components. Best results require grit blasting before spraying, and wire brushing and chemical sealing after spraying. In some instances painting is carried out to ensure complete sealing.

Cathodic protection uses an electric circuit with the component or structure made the negative or cathode when the anode or positive corrodes and can be replaced at regular intervals. This system is often used in conjunction with painting. Pipe lines, ships, marine structures, pylons and buildings are all now protected in this manner.

An artificial oxide can be produced by immersing the parts in strong hot sodium hydroxide with oxidizing agents added. The result is a brown or blue adherent oxide film commonly called 'black oxide'. Provided it is kept dry and oiled, this can be an adequate corrosion protection and has been used for centuries by gunsmiths.

Machinability

In the normalized, annealed or fully tempered condition, these materials have a tendency to tear.

Best results for finish are achieved when hardened and tempered about 300 DPN, thus resulting in higher tool wear but a better surface finish with less galling.

Uses

Structural steel (low nickel steels). Medium strength shafts and stressed items requiring good ductility and fatigue strength, chisels, hammers and pneumatic tools subjected to continuous shock loading.

Symbol	Nominal analysis, supplier, condition and remarks,
14N 3	0.75% C 2.6% Ni steel: Sandvik
13728	12% Al 27% Ni 2% Co 3% Cu Steel: For magnets; German Standard alloy for Alni 120 in DIN 17410
AISI A2340	0.4% C 3.5% Ni steel: Obsolete
AISI A2345	0.45% C 3.5% Ni steel: Obsolete
ALNI	12% Al 28% Ni 0.4% Cu 0.5% Ti steel: Casting for magnets; PMA
Al Ni	12% Al 22% Ni steel: Casting, German Standard
Alni 1	24.5% Ni 6% Cu 12.5% Al steel: For magnets; casting; Jessops
Al Ni 120	13% Al 27.5% Ni 2.5% Cu steel: Casting for magnets; German Standard
ALNICO 3	12% Al 26% Ni 2.0% Cu steel: Casting for magnets; American alloy; information from PMA
AM 4 Ni	0.4% C 1.5% Mn 0.8% Ni steel: Jessop for BS alloy En 12
ASTM A320 L9	0.4% C 3.5% Ni steel: Hardened & tempered **UTS: 880 Elon: 16% Proof: 720**
ASTM A352 LC2	0.25% C (max) 0.65% Mn 2.5% Ni steel: Casting for low temperature use
BS 970/503 A37	0.37% C 0.85% Ni steel: Supplied to chemical composition
BS 970/503 A42	0.42% C 0.85% Ni steel: Supplied to chemical composition
BS 970/503 H37	0.38% C 0.85% Mn 0.85% Ni steel: Supplied to hardenability requirement
BS 970/503 H42	0.43% C 0.85% Mn 0.85% Ni steel: Supplied to hardenability requirement
BS 970/503 M40	0.4% C 0.85% Mn 0.85% Ni steel: Supplied to mechanical properties; replaces En 12
BS 970/530 M40	0.4% C 0.75% Mn 1.1% Ni steel: Supplied to mechanical properties; replaces En 18
BS 970 En 10	0.55% C 0.7% Ni steel: Bar & forging; mechanical properties not part of specification
BS 970 En 12	0.4% C 1.5% Mn 0.8% Ni steel: Bar & forging; hardened & tempered **DPN: 200 UTS: 610 Elon: 22% Proof: 450**
BS 980 CDS14	0.4% C 3% Ni steel: Cold drawn tube; hardened & tempered **UTS: 690 Proof: 580**
BSN 1	0.4% C 1.5% Mn 0.8% Ni steel: Toledo for BS alloy En 12
BSN 35	0.4% C 3.5% Ni steel: Toledo
CHISEL STEEL	.4% C 3% Ni steel: Origin unknown
CL 15	0.4% C 0.5% Mn 3.25% Ni steel: Origin unknown
DIN 17200 41 Cr 4	0.41% Cr 0.65% Mn 1.05% Cr steel: German Standard
DIN 17200 41 Cr S 4	0.41% C 0.65% Mn 0.05% Cr steel: German Standard
DMA 3	26% Ni 12% Al 2.0% Cu steel: For magnets; Japanese alloy listed by PMA
DUNELT 26	0.4% C 0.7% Ni steel: Dunford for BS alloy En 12
E 1	0.4% C 3.5% Ni steel: Jessop
EXPANDAL	0.6% C 5.7% Mn 10% Ni steel: Latrobe; high expansion steel
GOST 2052/60 S2N2A	0.6% C 0.55% Mn 0.3% Cr (max) 1.55% Ni steel: Russian Standard
HDB 2	0.6% C 1.25% Ni steel: Huntsman; large press dies
HECLA 10	0.55% C 0.65% Ni steel: Hadfields for BS alloy En 10

Symbol	Nominal analysis, supplier, condition and remarks.
HECLA 115	0.4% C 1.5% Mn 0.8% Ni steel: Hadfields for BS alloy En 12
HEDEX 24	0.4% C 0.8% Ni steel: Wire for forgings; Kiveton Park
HTCN	0.4% C 0.75% Ni steel: ESC for BS alloy En 12
KF 12A	0.4% C 1.5% Mn 0.8% Ni steel: Kirkstall for BS alloy En 12
KOERZIT 90	21% Ni 12% Al steel: For magnets; German alloy listed by PMA
KP 10	0.4% C 1.5% Mn 0.8% Ni steel: Kiveton Park for BS alloy En 12
LE	0.4% C 3.20% Ni tool steel: Osborn; chisel steel; obsolete **DPN: 600**
LT 226	0.4% C 3.2% Ni steel: Low Moor; obsolete **DPN: 600**
MANOIR 15N4	0.15% C 3.8% Ni steel: Pompey **UTS: 550 Elon: 18% Proof: 320**
MARSH N3	0.4% C 3.2% Ni steel: Marsh Brothers, chisels
MAX MITH	0.4% C 3.25% Ni steel: Edgar Allen; blacksmiths' tools, chisels, etc.
N 3	0.4% C 3.2% Ni steel: Marsh Brothers; chisels
N 10	Ni steel sinter: Firth Cleveland; specific garvity 6.6–7.0; hardened & tempered; file hard **UTS: 700 Elon: 0.8%**
N 20	Ni steel sinter: Firth Cleveland; specific gravity 6.8–7.2; hardened & tempered; file hard **UTS: 1010 Elon: 2%**
N 30	Ni steel sinter: Firth Cleveland; specific gravity 6.7–7.1; hardened & tempered; file hard **UTS: 890 Elon: 1%**
Ni Al 0	22% Ni 10% Al steel: Casting for magnets; Belgian alloy listed by PMA
Ni Al 1	30% Ni 12% Al Fe alloy: Casting for magnets; Belgian alloy listed PMA
NIPERMAG	12.0% Al 30.0% Ni 0.4% Ti Fe alloy: For magnets; origin unknown
ON 1	0.4% C 1.5% Mn 0.8% Ni steel: W Marrison for BS alloy En 12
S 4	Nickel steel: Sheet; obsolete; see BS/S4
S 50	Ni Mn steel sinter: Firth Cleveland; specific gravity 6.4–6.8; hardened & tempered; file hard **UTS: 700 Elon: 0.8%**
SAE 2340	0.4% C 3.5% Ni steel: Obsolete
SAE 2345	0.45% C 3.5% Ni steel: Obsolete
SANBOLD 12	0.4% C 1.5% Mn 1% Ni steel: Sanderson for BS alloy En 12
SANBOLD 39	0.4% C 1.5% Mn 1.0% Ni steel: Sanderson; replaced by Sanbold 12
SPEAR D3	0.4% C 0.5% Mn 3.25% Ni steel: Origin unknown
STA 5/V5/1	0.55% C 0.65% Ni steel: Replaced by BS alloy En 10
T 3N	0.4% C 3.2% Ni steel: ESC; punches, etc.

Note. The following abbreviations and units are used in the tables:

DPN	Hardness, diamond pyramid number
UTS	Ultimate tensile strength, N/mm^2
Elon	Elongation, %
Proof	0.1 % proof strength, N/mm^2

$1 N/mm^2 = 0.1 hbar = 0.102 kgf/mm^2 = 0.06475 tonf/in^2 = 145.04 lbf/in^2$

44F3 Steel – low carbon, high nickel – maraging with molybdenum, cobalt, aluminium and titanium

Specific gravity	8.05	
Density	8050 kg/m³	(0.289 lb/in.³)
Solidus/liquidus	– °C	
Thermal conductivity	– W/m °C	
Coefficient of linear expansion	10×10^{-6}/ °C	
Electrical conductivity	aged 4% IACS (copper 100%)	
Specific resistance	aged 400 microhm mm	
Young's modulus of elasticity	186×10^9 N/m²	$(27 \times 10^2$ lbf/in.²)
Impact	– J	
Fatigue strength	± 900 N/mm²	(± 58 tonf/in.²)
Hot strength		

Temperature °C	Tensile strength N/mm²	Elongation %
400	1400	18
500	1300	20
550	1000	22

The above properties are typical of the following group, and may not apply exactly to any one specification. It is possible that with certain specifications some of the values may not be applicable.

General metallurgical characteristics

These steels have over 15% nickel, with titanium, generally some cobalt, molybdenum and aluminium. The carbon, phosphorus and sulphur are all held at low levels.

They are modern steels developed by the International Nickel Company and are not yet in general use.

The materials have very high tensile and proof strength, in the order of 2000 N/mm² with good elongation at 10-15% and unique notch resistance for any material of comparable strength.

The metallurgy depends on a complex precipitation reaction caused by intermetallic compounds of nickel, cobalt, titanium or aluminium hardening a fully martensitic structure. This martensite is very ductile owing to the low carbon and high nickel. The steels are martensitic in the annealed condition but can accept considerable cold work which can be used to enhance the final mechanical properties by subsequent ageing. They have no outstanding corrosion resistance, being only slightly better than normal alloy steels. Their excellent notch resistance however, considerably reduces the danger normally associated with rust and high strength steels.

These steels are still at the development stage but the high strength, excellent notch strength and weldability would indicate a large number of potential uses.

Thermal treatment

These steels can accept a considerable amount of cold work without interstage annealing, but when necessary require a temperature of 700–750 °C for 30 minutes per 25 mm section.

The hardening treatment for the steels varies somewhat, depending on the analysis, but as with normal alloy steels, the first step is always to obtain a fully martensitic structure. This may be obtained by quenching followed by deep freeze at -70 °C if necessary, or quench followed by 25–50% cold reduction. Some of the higher nickel steels require an intermediate treatment at about 650 °C for four hours after annealing. This has been called an 'ausage' and must be followed by deep freezing at -70 °C to complete the transformation of austenite to martensite.

All of the steels are quite soft and very ductile in the martensitic condition, being hardened by an age at 400–450 °C for 2–4 hours.

It must be emphasized that the times and temperatures quoted above are general. These are expensive steels designed to give very good mechanical properties which will only be obtained with the correct treatment. Readers are therefore urged to contact their material supplier for exact times and temperatures relating to each steel. The metallurgy of the thermal treatments relies on first, the very low carbon martensite, formed at relatively slow rates of cooling, and second, the precipitation from this tough matrix of very hard intermetallic compounds. This second process is very complex. One great advantage of these steels is that there are no rapid cooling cycles to cause distortion or even undue stress. One disadvantage is that it is not possible to refine the grain by thermal treatment, but only by controlled hot working or forging.

No special furnace equipment or atmospheres are necessary for these steels.

Scale removal can be by grit blasting or sodium hydride, provided the special plant is available for the later method. Acid pickling is not recommended for these steels owing to the danger of embrittlement.

Welding and brazing

Pre-treatment. Careful joint design and scrupulous cleanliness are necessary with these steels.

Welding and brazing. Using inert gas shielded arc welding, very good results are possible with most of these steels. Welding can be carried out with the steels in any condition provided care is taken to keep the heat input to an absolute minimum, using chills behind the weld when necessary. This restricts the sectional thickness that can be welded successfully.

Pre-heating and post-heating during welding are not recommended, and the flow of inert gas should be above that used for austenitic stainless steels. Resistance welding requires techniques similar to those used for austenitic stainless steels.

Brazing presents no difficulty provided the above precautions regarding heat input are observed. This care is necessary to prevent the formation of stable austenite which would at least necessitate subsequent annealing and often requires cold work in addition to transform the austenite to martensite.

Post-treatment. The weld area must at least be re-aged after welding or brazing. It is preferable that the whole assembly is subjected to this treatment wherever possible.

Where the weld area has been subjected to excessive heat, no hardening will occur after ageing. These parts will require to be annealed and re-aged. There is still considerable research and development being carried out on the welding and brazing of these steels and it is advised that contact is made with the origin of the specification for detailed information.

Flaw detection methods

Crack test. Magnetic crack test with the correct equipment and controlled techniques gives the best results. Penetrant oil chalk test using modern highly sensitive fluids and skilled personnel can give comparable results and is often more flexible.

X-ray. Welding and brazing should always be subjected to this inspection technique at least on a control basis, but perferably 100%.

Ultra-sonic test. This will be commonly applied to forgings after rough machining to identify subcutaneous defects, thus saving further expensive machining operations. Certain types of weld defects can be found using the ultra-sonic method which is cheaper than X-raying.

Chemical etch. None required.

Corrosion protection

Temporary. These are very expensive materials with relatively poor corrosion resistance. They should not be stored out-of-doors and must be kept oiled or protected at all times.

Soluble oil coolant used during machining can give some measure of protection, but this is seldom suitable for periods longer than a few weeks. It should be noted that dirty or contaminated coolants, especially those with metallic particles, can be a serious cause of corrosion. Finished machined parts must be cleaned, dried, oiled and sealed in plastic.

Permanent. The steels listed will almost always be used for their good mechanical properties and will thus be seldom required to withstand atmospheric corrosion. The high nickel content ensures a natural corrosion resistance superior to the majority of alloy steels but this is in no way comparable with any of the corrosion resistant or stainless steels. As already stated, the good notch resistance of the maraging alloys acts as a safety factor, considerably reducing the danger normally associated with rust and high tensile steels.

It is however, still essential that all evidence of existing rust and oxide are removed before applying any form of surface protection. Grit blasting is probably the best method provided it is adequately controlled and carried out immediately before the surface protection. It should always be realized that freshly blasted surfaces are very prone to rapid rusting.

For some purposes phosphate treatment after blasting may be sufficient protection, or it can be used under a paint system. A stress relieving treatment is necessary to remove any danger of hydrogen embrittlement.

Painting. This can be on top of the phosphate coating which must have been oven dried before painting. It should be noted however, that most phosphate coatings break down at about 200–250 °C to give off water vapour. Many parts in maraging steels will be designed to operate above these temperatures and thus cannot be phosphate treated and painted.

Zinc silicate or chromate primers below one or more stoved top sealer coats give very good corrosion protection. Modern plastic paints have excellent abrasion resistance and are non-porous, but should be applied on top of the metallic oxide or chromate primers wherever possible for best results.

Plating. These materials can be electroplated using normal cleaning and etching techniques. They will never require chromium plating for decorative purposes, but may be cadmium, tin or zinc plated, usually to a thickness of under 0.025 mm.

It is essential that all parts are stress relieved at 200 °C before and after the plating operations to remove hydrogen embrittlement. Unless it is possible to guarantee that these operations are always carried out, plating should not be requested on these materials and it must be noted that there can be no visual proof that the parts have been at 200 °C. Hydrogen embrittlement can cause failure in a brittle manner below the calculated tensile strength and with these high duty tensile steels this can be particularly dangerous.

Metal spraying using zinc or aluminium gives excellent results when applied to a blasted surface and subsequently wire brushed, chemically sealed or painted.

Machinability

In the annealed or solution treated condition these materials resemble the austenitic stainless steels except that they do not work harden. They are tough with a tendency to tear and gall rather than cut.

After ageing the materials machine to give a superior finish. Use of tipped tools backed by machines with adequate power gives best results.

Uses

These steels are still being developed, but potential uses are associated with high duty rotating parts where the excellent ratio of unnotched to notched strength gives the designer a greater factor of safety. This applies to such items as shafts of all types, compressor and turbine discs in steam and gas turbines. The strength to weight ratio of these steels is the best for any metal.

Symbol	Nominal analysis, supplier, condition and remarks.
1.6351	0.03% C 17.0% Ni 4.7% Mo 10.2% Co 0.3% Ti steel: Casting; maraging; German standard
1.6354	.03% C 18.0% Ni 4.9% Mo 8.7% Co 0.75% Ti steel: Maraging; German Standard
1.6359	0.03% C 18.0% Ni 4.9% Mo 7.7% Co 0.45% Ti steel: Maraging; German Standard
17 Ni 1600	17.0% Ni 4.6% Mo 10.0% Co 0.3% Ti 0.05% Al steel: Castings; maraging; Inco **Proof: 1600**
18 Ni 1400	18.0% Ni 3.0% Mo 8.5% Co 0.2% Ti 0.1% Al steel: Maraging; Inco **Proof: 1400**
18 Ni 1700	18.0% Ni 5.0% Mo 8.0% Co 0.4% Ti 0.1% Al steel: Maraging; Inco **Proof: 1700**
18 Ni 1900	18.0% Ni 5.0% Mo 9.0% Co 0.6% Ti 0.1% Al steel: Maraging; Inco **Proof: 1900**
18 Ni 2400	17.5% Ni 3.75% Mo 12.5% Co 1.8% Ti 0.15% Al steel: Maraging; Inco **Proof: 2400**
18 Ni MA	0.03% C 18.0% Ni 8.0% Co 4.0% Mo 0.25% Ti steel: Electrode; metrode; maraging
18 Ni MA	Low C 18.0% Ni 8.0% Co 4.0% Mo 0.25% Ti steel: Electrode; metrode; maraging; aged **DPN: 570**
250 0.02% C	18.3% Ni 5.0% Mo 8.0% Co 0.4% Ti 0.1% Al steel: Maraging; information from Climax Molybdenum
300 0.02% C	18.3% Ni 5.0% Mo 9.0% Co 0.65% Ti 0.1% Al steel: Maraging; information from Climax Molybdenum
648A	0.02% C 18% Ni 3.2% Mo 8.5% Co 0.2% Ti 0.1% Al steel: Firth brown; maraging **DPN: 138**
648B	0.02% C 18% Ni 4.8% Mo 7.7% Co 0.4% Ti 0.1% Al steel: Firth Brown; maraging **UTS: 1700 Elon: 12% Proof: 1600**
648C	0.02% C 18% Ni 5.0% Mo 9.0% Co 0.6% Ti 0.1% Al steel: Firth Brown; maraging **UTS: 2000 Elon: 12% Proof: 1900**
ABS/H	0.22% C (max) 0.9% Mn (max) 0.5% Mo steel: Plate; American Bureau of Shipping **UTS: 500 Elon: 19% Proof: 260**
ABS/I	0.23% C (max) 0.9% Mn (max) 0.52% Mo steel: Plate; American Bureau of Shipping **UTS: 550 Elon: 17% Proof: 280**
ABS/J	0.25% C (max) 0.9% Mn 0.52% Mo steel: Plate American Bureau of Shipping **UTS: 580 Elon: 16% Proof: 305**
ABS/K	0.15% C 0.55% Mn 0.55% Mo steel: Tube; American Bureau of Shipping

Symbol	Nominal analysis, supplier, condition and remarks.
ABS/L	0.3% C 0.6% Mn 0.55% Mo steel: Tube; American Bureau of Shipping
ABS/M	0.14% C (max) 0.55% Mn 0.55% Mo steel: Tube; American Bureau of Shipping
ABS/N	0.15% C (max) 0.5% Mn 0.55% Mo steel: Tube; American Bureau of Shipping
ABS/O	0.15% C (max) 0.5% Mn 0.55% Mo steel: Tube; American Bureau of Shipping
ABS/P	0.15% C (max) 0.5% Mn 1.0% Mo steel: Tube; American Bureau of Shipping
AL18 NiCoMo (250)	0.02% C 18% Ni 5% Mo 0.4% Ti 7% Co 0.1% Al steel: Allegheny Ludlum; solution treated & aged; maraged **DPN: 500 UTS: 1700 Elon: 12% Proof: 1600**
AL18 NiCoMo (300)	0.02% C 18% Ni 5% Mo 0.6% Ti 9% Co 0.1% Al steel: Allegheny Ludlum; solution treated & aged; maraged **UTS: 2200 Elon: 14% Proof: 2100**
AL20 Ni (250)	0.02% C 20% Ni 1.4% Ti 0.2% Al 0.5% Nb steel: Allegheny Lulum; solution treated & aged; maraged **DPN: 630 UTS: 2300 Elon: 1.5% Proof: 2200**
AL25 Ni (250)	0.02% C 25% Ni 1.4% Ti 0.2% Al 0.5% Nb steel: Allegheny Ludlum; solution treated & aged; maraged **UTS: 2400 Elon: 2% Proof: 2200**
ALMAR 18/200	0.03% C 18% Ni 32% Mo 0.2% Ti 8.5% Co 0.02% Zr steel: Allegheny Ludlum; solution treated & aged; maraged **UTS: 1550 Elon: 15% Proof: 1500**
ALMAR 18/250	0.03% C 18% Ni 5% Mo 0.4% Ti 8% Co 0.1% Al 0.02% Zr steel: Allegheny Ludlum; solution treated & aged; maraged **DPN: 580 UTS: 2000 Elon: 12% Proof: 1850**
ALMAR 18/300	0.03% C 18.5% Ni 5% Mo 0.7% Ti 9% Co 0.1% Al 0.02% Zr steel: Allegheny Ludlum; solution treated & aged; maraged **DPN: 610 UTS: 2200 Elon: 11% Proof: 2100**
ALMAR 20	0.03% C 19.5% Ni 1.5% Ti 0.2% Al 0.02% Zr 0.5% Nb steel: Allegheny Ludlum; solution treated & aged; maraged **UTS: 1900 Elon: 8% Proof: 1700**
ALMAR 25	0.03% C 25.5% Ni 1.5% Ti 0.2% Al 0.02% Zr 0.5% Nb steel: Allegheny Ludlum; solution treated & aged; maraged **UTS: 2000 Elon: 8% Proof: 1850**
ASTM A538 A	0.03% C 18% Ni 7.7% Co 4.2% Mo 0.2% Ti steel: Maraging **UTS: 1440 Elon: 8% Proof: 1380**
ASTM A538 B	0.03% C 18% Ni 7.7% Co 4.8% Mo 0.4% Ti steel: Maraging **UTS: 1650 Elon: 6% Proof: 1600**
ASTM A538 C	0.03% C 18.5% Ni 8.7% Co 4.9% Mo 0.7% Ti steel: maraging **UTS: 1800 Elon: 6% Proof: 1700**
AATM A579/71	0.03% C (max) 18% Ni 3.2% Mo 0.2% Ti 8.5% Co 0.1% Al steel: Forging; maraging
ASTM A579/72	0.03% C (max) 18% Ni 4.9% Mo 0.4% Ti 8.0% Co 0.1% Al steel: Forging; maraging
ASTM A579/73	0.03% C (max) 18.5% Ni 4.9% Mo 0.65% Ti 9.0% Co 0.1% Al steel: Forging; maraging

Note. The following abbreviations and units are used in the tables:

DPN	Hardness, diamond pyramid number
UTS	Ultimate tensile strength, N/mm^2
Elon	Elongation, %
Proof	0.1 % proof strength, N/mm^2

1 N/mm^2=0.1 hbar=0.102 kgf/mm^2=0.06475 tonf/in^2=145.04 lbf/in^2

Symbol	Nominal analysis, supplier, condition and remarks.
ASTM A579/74	0.03% C (max) 5.0% Cr 12% Ni 3.0% Mo 0.1% Ti 0.3% Al steel: Forging; maraging
ASTM A579/75	0.03% C 5.0% Cr 12% Ni 3.0% Mo 0.17% Ti 0.4% Al steel: Forging; maraging 0.03% C (max) 5.0% Cr 12.0% Ni 3.0% Mo 0.27% Ti 0.4% Al (max) steel: Maraging
	UTS: 1300 Elon: 14% Proof: 1250
CFS	1.1% C 0.5% Cr 0.1% Ni steel: Latrobe
DTD 5212	18% Ni Co Mo steel: Maraging; double vacuum melted
	UTS: 1800
DTD 5232	18% Ni Co Mo steel: Maraging; vacuum remelted
	UTS: 1800
FE PA95	0.03% C 18.0% Ni 4.9% Mo 8.0% Co 0.4% Ti steel: Maraging; origin unknown
G 140	0.02% C 18.0% Ni 4.8% Mo 8.0% Co 0.4% Ti steel: Jessop
MAR 100	0.02% C 10.3% Co 4.65% Mo 0.11% Al 0.37% Ti 16.75% Ni Fe alloy; Union Carbide
MARAGE 200	0.03% C 18.0% Ni 3.2% Mo 8.5% Co 0.1% Ti steel
MARAGE 250	0.03% C (max) 18.0% Ni 4.9% Mo 7.7% Mo 0.4% Ti 0.08% Al steel
MARVAC 18	0.03% C 18% Ni 7.5% Co 5.0% Mo 0.4% Ti 0.1% Al steel: Latrobe Steel; maraged
	UTS: 1900 Elon: 11% Proof: 1700
MARVAC 18A	0.03% C 18% Ni 9.0% Co 5.0% Mo 0.6% Ti 0.1% Al steel: Latrobe Steel; maraged
	UTS: 2100 Elon: 10% Proof: 2000
MARVAC 18LT	0.03% C 18% Co 8.0% Co 4.5% Mo 0.2% Ti 0.1% Al steel: Latrobe Steel; maraged
	UTS: 11% Proof: 1650
MARVAC 200	0.02% C 18% Ni 7.5% Co 4.25% Mo 0.2% Ti 0.1% Al steel: Latrobe; maraging
MARVAC 250	0.02% C 18% Ni 8.0% Co 5.0% Mo 0.4% Ti 0.1% Al steel: Latrobe; maraging
MARVAC 300	0.02% C 18% Ni 9.0% Co 5.0% Mo 0.6% Ti 0.1% Al steel: Latrobe; maraging

Symbol	Nominal analysis, supplier, condition and remarks.
Ni MARK 1	0.3% C (max) 18.5% Ni 7.5% Co 4.9% Mo 0.4% Ti 0.1% Al steel: Carpenter; maraging
Ni MARK 300	0.03% C (max) 18.5% Ni 4.9% Mo 8.5% Co 0.7% Ti 0.1% Al steel: Carpenter; maraging
TELMAR 90	0.01% C 18% Ni 8.5% Co 3.2% Mo 0.2% Ti steel: Maraging; Telcon
	DPN: 450 UTS: 1500 Elon: 11% Proof: 1420
TELMAR 110	0.01% C 18% Ni 7.7% Co 4.8% Mo 0.4% Ti steel: Maraging; Telcon
	DPN: 520 UTS: 1800 Elon: 12% Proof: 1700
TELMAR 125	0.01% C 18.5% Ni 9.0% Co 4.9% Mo 0.6% Ti steel: Maraging; Telcon
	DPN: 560 UTS: 1900 Elon: 12% Proof: 1850
UCAR MARAGING 110	0.01% C 18.0% Ni 8.0% Co 5.0% Mo 0.5% Ti 0.1% Al maraging steel: Union Carbide
ULTIMATE 110	0.01% C 17.5% Ni 8.0% V 4.8% Mo with Ti and Al steel: For die casting aluminium; Jessop; maraging
	DPN: 600
VASCOMAX 250	0.03% C (max) 18% Ni 5% Mo 4% Ti 7% Co 1% Al steel: Vanadium Alloys; solution treated & aged; maraging steel
	DPN: 540 UTS: 1900 Elon: 10% Proof: 1700
VASCOMAX 300 CVM	0.03% C 18% Ni 5% Mo 6% Ti 9% Co 1% Al steel: Vanadium Alloys; solution treated & aged; maraged
	DPN: 610 UTS: 2200 Elon: 11% Proof: 2000

Note. The following abbreviations and units are used in the tables:

DPN	Hardness, diamond pyramid number
UTS	Ultimate tensile strength, N/mm^2
Elon	Elongation, %
Proof	0.1 % proof strength, N/mm^2

$1\ N/mm^2 = 0.1\ hbar = 0.102\ kgf/mm^2 = 0.06475\ tonf/in^2 = 145.04\ lbf/in^2$

44G1 Steel – low carbon with molybdenum
0.05–0.35% carbon, 0.7% molybdenum maximum

Specific gravity	7.85	
Density	7850 kg/m^3	(0.28 lb/in.^{3})
Solidus/liquidus	– °C	
Thermal conductivity	– W/m °C	
Coefficient of linear expansion	11 × 10^{-6}/ °C	
Electrical conductivity	8–10% IACS (copper 100‰)	
Specific resistance	180–250 microhm mm	
Young's modulus of elasticity	– N/m^2	
Impact	54–108 J	(40–80 ft/lb)
Fatigue strength (no. of cycles)	–	
Hot strength	–	

The above properties are typical of the following group, and may not apply exactly to any one specification. It is possible that with certain specifications some of the values may not be applicable.

General metallurgical characteristics

Molybdenum forms very stable carbides which are harder than, and formed in preference to, iron carbides. Because of this molybdenum steels have better creep strength, maintain hardness at higher temperature and have increased hardenability over plain carbon steels. It also acts as a grain refiner, preventing excessive grain growth, Molybdenum has many similar effects to tungsten, which is generally cheaper but requires larger quantities to achieve the same purpose.

None of the steels listed has more than 0.7% molybdenum and very few have as much as 0.5%. Traces of boron are sometimes added in addition to the molybdenum. This has a similar effect and is additive to those of molybdenum as far as hardenability is concerned.

These steels are seldom used in the cold worked condition.

Thermal treatment

The lower carbon steels are capable of accepting considerable cold work, but not as much as the plain carbon equivalent. Interstage annealing will be required between stages of cold working and this requires a temperature of 840–900 °C for 30 minutes per 25 mm section followed by slow cooling. Sub-critical annealing at 650–670 °C to remove most of the cold work is satisfactory for many purposes.

Normalizing at 840–900 °C for 20 minutes per 25 mm section, then still air cooling, removes all forging stresses, refines and homogenizes the grain and is a common condition for using many of the steels listed. There will be some air hardening and comparatively small amounts of nickel or chromium which may be present as residual elements can increase this considerably. Tempering or sub-critical annealing after normalizing is advisable.

By heating to 820–870 °C for 15 minutes per 25 mm of section and water or oil quenching the parts will be fully hardened and must be tempered at 300–600 °C for 2 hours, as soon as possible after hardening. One of the important characteristics of molybdenum is to increase the depth of hardening over that obtained with plain carbon steels. This property is proportional to the molybdenum content and is helped by the presence of traces of boron.

These steels can be carburized using the gas, pack or salt bath methods. In all instances the case will be rather brittle with a tendency to spall or exfoliate at grinding or under any bending or twisting stresses. Wherever possible steels with some nickel, or nickel and chromium, in addition to molybdenum, are recommended for case hardening.

After case hardening the parts should be cooled, reheated to 720–740 °C and oil quenched, then immediately tempered at 150–250 °C.

No special furnace equipment is necessary for any of the steels listed which can all be treated in electrically heated, gas or oil fired furnaces, or salt baths. De-carburization is not normally a problem with these relatively low carbon materials.

Scale removal is generally by grit blasting or acid pickling although some of the higher carbon steels may be above the hardness of 350 DPN at which hydrogen embrittlement becomes a serious problem if they are acid pickled.

Welding and brazing

Pre-treatment. The weld area must be clean, free of all traces of oil or grease, which could cause carburizing of the low carbon steels. The steels should always be in the annealed, or normalized and tempered condition, before welding. Pre-heating may be necessary.

Welding and brazing. Many of these steels have been specially developed for welding. Good results are obtained using normal gas or electric welding with flux or inert gas shielded arc welding. Resistance welding also results in excellent joins. Modern automatic high speed welding equipment can be used with these materials. The degree of air hardening in the welding steels is controlled but residual elements in other steels can increase this, causing cracking in certain cases. Care must therefore be taken to ensure that the correct steel specification is chosen and supplied.

Brazing presents no problems, but it is advisable not to heat any areas above the hardening temperature of about 750 °C.

Post-treatment. The welding steels are designed to give some air hardening in the weld area and this must be tempered to regain the ductility. Ideally the assembly should be normalized and tempered after welding as this removes all welding stresses and improves the mechanical properties in the weld zone. As a very minimum the weld area must be tempered using gas torches or electric blankets. The use of gas torches is not recommended as it is all too easy locally to overheat, causing hard spots.

Brazed parts cannot generally be normalized without melting the braze, but can be tempered.

Flaw detection methods

Crack test. Magnetic crack test. Penetrate oil chalk testing in skilled hands can be as sensitive as the magnetic method and is much more flexible.

X-ray. Seldom required on wrought products but should be carried out on welds. Portable equipment using radioactive metal isotopes has been developed for inspecting welds on site.

Ultra-sonic test. Seldom required. Can identify subcutaneous defects in large forgings at an early stage, but these are seldom made into high duty parts using any of the listed materials. This inspection technique can be used for certain types of welding defects.

Chemical etch. This would only be required as a control of grinding on case hardened parts, when electrolytic sulphuric acid etching will show grinding abuse, thus identifying this defect before it is serious enough to cause cracking.

Corrosion protection

Temporary. Molybdenum has no beneficial effects on the corrosion resistance and as these steels are often used under conditions of stress where pitting could reduce the fatigue strength it is important that they are protected wherever possible. This means keeping them free from moisture and oiled or greased at all times.

Soluble oil coolants can act as corrosion inhibitors when they are kept clean and fresh, but when dirty and contaminated with metal dust they can encourage rust.

Permanent. Rusting is a complex oxidation process, with the resultant scale only very loosely adherent. As it is washed or brushed off it leaves a fresh surface ready to corrode further, thus encouraging deep pits and local areas of metal loss. Corrosion protection requires the complete elimination of corrosion products and dirt, as any rust left always has a trace of moisture and air which allows corrosion to proceed even under an air tight film, with resultant failure of the protecting medium. The presence of dirt prevents the formation of the necessary air tight film in the first instance.

Grit blasting correctly carried out gives a surface which in addition to being rust free has a roughness which acts as a key for the subsequent coating. The surface must be brushed free of the loose dust always present, and protected as soon as possible. A freshly blasted surface can rust within minutes in humid conditions.

Painting. For best results the freshly blasted surface should be phosphate or chromate treated, oven dried, then painted with zinc silicate or chromate primer. As soon as possible top coats of stoving enamel should be applied. Modern plastic paints have excellent abrasion and shock resistance but should have the metal oxide type of primer as a base.

Plating. These materials will seldom have decorative plating applied. If necessary, best results require copper and nickel undercoats at least 0.075 mm thick in total, below the chromium plating.

Zinc or cadmium electroplating requires less than 0.025 mm but should be chemically sealed immediately after plating and must not be used on parts in contact with foodstuffs. As with all plating processes the pre-treatment must be carefully carried out.

Dipping in molten zinc after cleaning and etching results in a galvanized coating.

Metal spraying, particularly zinc or aluminium, gives excellent results provided the parts are correctly blasted. The coating should be subsequently wire brushed and chemically sealed or painted.

Cathodic protection is given by completing an electric circuit with the structure as the negative or cathode, and an anode or positive which then corrodes and by its sacrifice protects the cathode. This system is applied to ships, marine structures, pipe lines, pylons and steel buildings but should preferably be additive to a paint system.

Artificial oxides can be produced which adhere to machined items by immersing in a strong hot sodium hydroxide solution to which oxidizing agents have been added. The resultant surface is dark blue or brown, commonly called 'black oxide' and has been known to gunsmiths for many years. Provided it is kept dry and oiled the corrosion protection is excellent.

Machinability

The inherent even fine-grained structure of these steels reduces the machinability regarding surface finish. High speed steel tools are generally satisfactory, with tipped tools sometimes being required for the higher carbon steels particularly for long production runs. The hard carbides being abrasive, rapidly wear plain carbon cutting tools.

There will be an increased tendency for tearing rather than cutting when in the soft condition. This is reduced as the hardness is increased with an optimum about 300 DPN.

Uses

Large welded structure, boiler tubes and plate, gas cylinders, tubing for pipe lines.

Sheet products for welded combustion equipment. Low duty welded shafts from forgings or bar.

Symbol	Nominal analysis, supplier, condition and remarks.
1.2 Mo6/7	0.22% C 1.1% Mn 0.5% Mo steel: Creusot-Loire
1.3 Mo6/7	0.22% C 1.25% Mn 0.5% Mo steel: Creusot-Loire
1.5415	0.16% C 0.3% Mo steel: For boiler plate; German Standard
1.5416	0.15% C 0.3% Mo steel: German Standard
1.5424	0.12% C 0.5% Mo 0.2% Cu steel: Welding rod; German Standard
1.5425	0.1% C 0.5% Mo steel: welding rod; German Standard
1.5426	0.15% C 0.5% Mo steel: Welding rod; German Standard
1.5427	0.13% C 2.0% Mn 0.5% Mo steel: Welding rod; German Standard
1.5428	0.15% C 3.0% Mn 0.5% Mo steel: Welding rod; German Standard
3 Mo	0.15% C 0.35% Mo steel: Sandvik
3 Mo 1	0.13% C 0.45% Mo steel: Sandvik

Symbol	Nominal analysis, supplier, condition and remarks.
9 Mn Mo 4/5	0.09% C 1.0% Mn 0.5% Mo 0.22% Cu steel: Designation used by German Standard
13 Mn Mo 3/5	0.12% C 0.5% Mo 0.22% Cu steel: Designation used by German Standard
13 Mn Mo 6/5	0.14% C 1.5% Mn 0.5% Mo steel: Designation used by German Standard
13 Mn Mo 8/5	0.13% C 2.0% Mn 0.5% Mo steel: Designation used by German Standard
13 Mn Mo 12/5	0.14% C 3.0% Mn 0.5% Mo steel: Designation used by German Standard
15 Mo 3	0.15% C 0.3% Mo steel: Designation used by German Standard
20 Mn Mo 4	0.2% C 1.1% Mn 0.25% Mo steel: Designation used by German Standard
20 Mo 3	0.15% C 0.3% Mo steel: Designation used by German Standard
A 204 A	0.23% C 0.5% Mo steel: Plate; Armco; pressure vessels used at high temperature **UTS: 480 Elon: 24% Proof: 240**
A 204 B	0.25% C 0.5% Mo steel: Plate; Armco; pressure vessels used at temperature **UTS: 530 Elon: 23% Proof: 270**
A 204 C	0.26% C 0.5% Mo steel: Plate; Armco; pressure vessels used at temperature **UTS: 640 Elon: 22% Proof: 300**

Note. The following abbreviations and units are used in the tables:

DPN	Hardness, diamond pyramid number
UTS	Ultimate tensile strength, N/mm^2
Elon	Elongation, %
Proof	0.1 % proof strength, N/mm^2

1 N/mm^2=0.1 hbar=0.102 kgf/mm^2=0.06475 tonf/in^2=145.04 lbf/in^2

Symbol	Nominal analysis, supplier, condition and remarks.
ABS/H	0.22% C (max) 0.9% Mn (max) 0.5% Mo steel: Plate; American Bureau of Shipping **UTS: 500 Elon: 19% Proof: 260**
ABS/I	0.23% C (max) 0.9% Mn (max) 0.52% Mo steel: Plate; American Bureau of Shipping **UTS: 550 Elon: 17% Proof: 280**
ABS/J	0.25% C (max) 0.9% Mn 0.52% Mo steel: Plate; American Bureau of Shipping **UTS: 580 Elon: 16% Proof: 305**
ABS/K	0.15% C 0.55% Mn 0.55% Mo steel: Tube; American Bureau of Shipping
ABS/L	0.2% C 0.6% Mn 0.55% Mo steel: Tube; American Bureau of Shipping
ABS/M	0.14% C (max) 0.55% Mn 0.55% Mo steel: Tube; American Bureau of Shipping
ABS/N	0.15% C (max) 0.5% Mn 0.55% Mo steel: Tube; American Bureau of Shipping
ABS/O	0.15% C (max) 0.5% Mn 0.55% Mo steel: Tube; American Bureau of Shipping
ABS/P	0.15% C (max) 0.5% Mn 1.0% Mo steel: Tube; American Bureau of Shipping
AISI 4012	0.12% C 0.2% Mo steel
AISI 4023	0.23% C 0.25% Mo steel
AISI 4024	0.22% C 0.25% Mo steel
AISI 4027	0.27% C 0.25% Mo steel
AISI 4028	0.27% C 0.25% Mo steel
AISI 4028	0.27% C 0.25% Mo steel
AISI 4032	0.32% C 0.8% Mn 0.25% Mo steel
AISI 4037	0.37% C 0.25% Mo steel
AISI 4419	0.2% C 0.5% Mo steel
AISI 4422	0.22% C 0.4% Mo steel
AISI 4427	0.27% C 0.4% Mo steel
AISI 4520	0.2% C 0.5% Mo steel: Obsolete
A Mn Mo	0.25% C Mn Mo steel: Casting; Firth Brown; normalized & tempered **DPN: 200 UTS: 700 Elon: 17% Proof: 450**
AMS 6300	0.37% C 0.25% Mo steel: Bar & forging; AMS for SAE 4037
AN S 9/1	0.37% C 0.25% Mo steel: US Service; hardened & tempered **DPN: 230 UTS: 840 Elon: 17% Proof: 740**
ASTM A161T1	0.15% C 0.5% Mn 0.05% S & P 0.5% Mo steel: Tube
ASTM A204A	0.22% C 0.9% Mn 0.5% Mo steel: Plate for pressure vessels
ASTM A204 B	0.23% C 0.9% Mn 0.5% Mo steel: Plate for pressure vessels
ASTM A204 C	0.24% C 0.9% Mn 0.5% Mo steel: Plate for pressure vessels
ASTM A209 T1	0.15% C 0.5% Mn 0.55% Mo steel: Seamless tube
ASTM A209 T1A	0.2% C 0.5% Mn 0.55% Mo steel: Seamless tube
ASTM A209 T1B	0.14% C (max) 0.5% Mn 0.55% Mo steel: Seamless tube
ASTM A250 T1	0.15% C 0.5% Mn 0.55% Mo steel: Electric welded tube
ASTM A250 T1A	0.2% C 0.5% Mn 0.55% Mo steel: Electric welded tube
ASTM A250 T1B	0.14% C (max) 0.5% Mn 0.55% Mo steel: Electric welded tube
ASTM A295 F1	0.25% C 0.5% Mo steel: Seamless drum forgings; normalized & tempered **UTS: 500 Elon: 20% Proof: 270**
ASTM A316 E7010A1	0.12% C 0.6% Mn 0.4% Si 0.52% Mo steel: Electrodes for arc welding
ASTM A316 E7011A1	0.12% C 0.6% Mn 0.4% Si 0.52% Mo steel: Electrodes for arc welding
ASTM A316 E7015A1	0.12% C 0.9% Mn 0.6% Si 0.52% Mo steel: Electrodes for arc welding
ASTM A316 E7016A1	0.12% C 0.9% Mn 0.6% Si 0.52% Mo steel: Electrodes for arc welding
ASTM A316 E7018A1	0.12% C 0.9% Mn 0.8% Si 0.52% Mo steel: Electrodes for arc welding
ASTM A316 E7020A1	0.12% C 0.6% Mn 0.4% Si 0.52% Mo steel: Electrodes for arc welding
ASTM A316 E7027A1	0.12% C 0.52% Mo steel: Electrodes for arc welding
ASTM A316 E9015/8D1	0.12% C 1.5% Mn 0.3% Mo steel: Welding electrode; code varies with analysis
ASTM A316 E9015D1	0.12% C 0.35% Mo steel: Electrodes for arc welding
ASTM A316 E9018D1	0.12% C 0.35% Mo steel: Electrodes for arc welding
ASTM A316 E10015/8D2	0.15% C 1.8% Mn 0.3% Mo steel: Welding electrodes; code varies with analysis
ASTM A316 E10015D2	0.15% C 0.35% Mo steel: Electrodes for arc welding
ASTM A316 E10016D2	0.15% C 0.35% Mo steel: Electrodes for arc welding
ASTM A316 E10018D2	0.15% C 0.35% Mo steel: Electrodes for arc welding
ASTM A335 P1	0.15% C 0.5% Mo steel: Seamless pipe
ASTM A335 P15	0.15% C (max) 1.4% Si 0.5% Mo steel: Seamless pipe
ASTM A336 F1	0.25% C 0.5% Mo: Seamless drum forging **UTS: 460 Elon: 20% Proof: 270**
ASTM A352 LC1	0.25% C 0.55% Mo steel: Normalized & tempered for low temperature service **UTS: 440 Elon: 24% Proof: 220**
ASTM A356/2	0.25% C 0.5% Mo steel: Casting; normalized & tempered; for turbines **UTS: 440 Elon: 22% Proof: 220**
ASTM A356/3	0.25% C 1.0% Mo steel: Casting; normalized & tempered; for turbines **UTS: 530 Elon: 18% Proof: 330**
ASTM A369 FP1	0.15% C 0.5% Mo steel: For forged & bored pipe
ASTM A426 CP1	0.15% C 0.5% Mo steel: Cast pipe
ASTM A426 CP15	0.15% C (max) 0.5% Mo steel: Cast pipe
ASTM A487/2 N	0.3% C (max) 1.2% Mn 0.2% Mo steel: Casting for pressure; normalized; 2Q grade is quenched & tempered
ASTM A514 J	0.17% C 0.56% Mo 0.003% B steel: Plate
ASTM A514 K	0.15% C 0.50% Mo 0.003% B steel: Plate
ASTM A517 C	0.15% C 0.25% Mo 0.003% B steel: Boiler plate
ASTM A517 J	0.17% C 0.58% Mo 0.003% B steel: Boiler plate
ASTM A517 K	0.15% C 0.5% Mo 0.003% B steel: Boiler plate
ASTM A533 A	0.25% C (max) 1.25% Mo 0.52% Mo steel: Plate
ASTM A588 E	0.15% C (max) 1.2% Mn (max) 0.17% Mo 0.65% Cu steel: For structures **UTS: 440 Elon: 21% Proof: 320**
ASTM A692	0.22% C 0.7% Mn 0.5% Mo steel: Pipe; seamless for pressure use **UTS: 500 Elon: 20% Proof: 290**
ASTM A699/1	0.06% C (max) 1.7% Mn 0.3% Mo 0.06% Nb steel **UTS: 700 Elon: 12% Proof: 485**
ASTM A699/2	0.06% C (max) 1.7% Mn 0.3% Mo 0.06% Nb steel **UTS: 700 Elon: 12% Proof: 515**
ASTM A699/3	0.06% C (max) 1.7% Mn 0.3% Mo 0.06% Nb steel **UTS: 680 Elon: 12% Proof: 485**

Note. The following abbreviations and units are used in the tables:

DPN	Hardness, diamond pyramid number
UTS	Ultimate tensile strength, N/mm^2
Elon	Elongation, %
Proof	0.1 % proof strength, N/mm^2

$1 \ N/mm^2 = 0.1 \ hbar = 0.102 \ kgf/mm^2 = 0.06475 \ tonf/in^2 = 145.04 \ lbf/in^2$

Symbol	Nominal analysis, supplier, condition and remarks.
ASTM A699/4	0.06% C (max) 1.7% Mn 0.3% Mo 0.06% Nb steel **UTS: 680 Elon: 12% Proof: 515**
ASTM A707 L6/1	0.09% C (max) 1.8% Mn 0.3% Mo 0.07% Nb steel: Flanges for low temperature use; impact 41 J at -62 °C **UTS: 290 Elon: 22%**
ASTM A707 L6/2	0.09% C (max) 1.8% Mn 0.3% Mo 0.07% Nb steel: Flanges for low temperature use; impact 54 J at -62 °C **UTS: 360 Elon: 25%**
ASTM A707 L6/2	0.09% C (max) 1.8% Mn 0.3% Mo 0.07% Nb steel: Flanges for low temperature use; impact 68 J at -62 °C **UTS: 415 Elon: 23%**
ASTM A707 L6/3	0.09% C (max) 1.8% Mn 0.3% Mo 0.07% Nb steel: Flanges for low temperature use; impact 68 J at -62 °C **UTS: 515 Elon: 20%**
ASTM A732/6 N	0.35% C (max) 1.5% Mn 0.4% Mo steel: Investment casting; normalized & tempered **UTS: 620 Elon: 20% Proof: 415**
ASTM A735	0.08% C 1.5% Mn 0.35% Mo 0.08% Nb steel: Plate for pressure vessels
ASTM A737 C	0.24% C (max) 1.3% Mn 0.1% V steel: Plate for pressure vessels
AURIGA VIII LT	0.18% C 0.5% Mo steel: Casting; D Brown normalized & tempered for sub-zero use **UTS: 440 Elon: 25% Proof: 240**
AUTROD 12.24	0.1% C 1.0% Mn 0.5% Mo steel: Welding wire; Esab; copper coated; used for submerged arc welding
BEARCOMO	0.16% C 1.5% Mn 0.25% Mo 0.5% Cu steel: Plate; Consett; normalized; suitable for welding; creep resisting **UTS: 460 Elon: 22% Proof: 300**
BS 806M	0.17% C 0.55% Mo steel: Seamless pipe; cold drawn & normalized **UTS: 480**
BS 970/605 A32	0.32% C 1.5% Mn 0.27% Mo steel: Supplied to chemical composition
BS 970/605 A37	0.37% C 1.5% Mn 0.27% Mo steel: Supplied to chemical composition
BS 970/605 H32	0.32% C 1.5% Mn 0.27% Mo steel: Supplied to hardenability requirement
BS 970/605 H37	0.37% C 1.5% Mn 0.27% Mo steel: Supplied to hardenability requirement
BS 970/605 M30	0.3% C 1.5% Mn 0.27% Mo steel: Supplied to mechanical properties; replaces En 16D
BS 970/605 M36	0.36% C 1.5% Mn 0.27% Mo steel: Supplied to mechanical properties; replaces En 16
BS 970/606 M36	0.36% C 1.5% Mn 0.27% Mo 0.2% S steel: Free machining; supplied to mechanical properties; replaces En 16M
BS 970/608 H37	0.38% C 1.5% Mn 0.47% Mo steel: Supplied to hardenability requirement
BS 970/608 M38	0.38% C 1.5% Mn 0.47% Mo steel: Supplied to mechanical properties; replaces En 17
BS 970 En 16	0.3% C 1.5% Mn 0.3% Mo steel: Bar & forging; hardened & tempered **DPN: 320 UTS: 1000 Elon: 16% Proof: 770**
BS 970 En 17	0.35% C 1.5% Mn 0.4% Mo steel: Bar & forging hardened & tempered **DPN: 320 UTS: 1000 Elon: 16% Proof: 770**
BS 980 CDS11	0.25% C 1.5% Mn 0.2% Mo steel: Cold drawn tube; as drawn & tempered; can be welded then heat treated **UTS: 700 Proof: 580**
BS 1398	0.2% C 0.5% Mo steel: Casting; normalized; contained in BS 3100 **UTS: 450 Elon: 20% Proof: 240**

Symbol	Nominal analysis, supplier, condition and remarks.
BS 1453 A6	0.15% C (max) 0.5% Mo: Weld filler rod
BS 1501/240	0.15% C 0.5% Mo steel: Plate & bar; normalized **UTS: 400 Elon: 22% Proof: 210**
BS 1501/261	0.13% C 0.5% Mo 0.003% B steel: Plate; normalized; for pressure vessels **UTS: 600 Elon: 16% Proof: 430**
BS 1503/240 A	0.2% C 0.5% Mo steel: Forging; normalized **UTS: 450 Proof: 210**
BS 1503/240 B	0.2% C 0.5% Mo steel: Forgings; normalized **UTS: 480 Elon: 20% Proof: 270**
BS 1504/240	0.2% C 0.6% Mo steel: Casting; normalized **UTS: 450 Elon: 20% Proof: 240**
BS 1506/240	0.3% C 0.5% Mo steel: Bar; hardened & tempered **UTS: 610 Elon: 22% Proof: 370**
BS 1507/240	0.20% C 0.5% Mo steel: Pipe; normalized **UTS: 460 Elon: 20%**
BS 1507/240 A	0.2% C 0.5% Mo steel: Seamless pipe; normalized **UTS: 450 Elon: 25%**
BS 1507/240 B	0.25% C 0.5% Mo steel: Seamless pipe; normalized **UTS: 460 Elon: 25%**
BS 1508/240	0.17% C 0.6% Mo steel: Tube; normalized **UTS: 450 Elon: 22%**
BS 1652	0.5% Mo steel: Tube; replaced by BS 3059
BS 1717 CDS109	0.26% C 0.2% Mo steel: Seamless tube; as drawn **UTS: 690 Proof: 610**
BS 1717 CDS109	0.26% C 0.2% Mo steel: Seamless tube; after welding, brazing or annealing **UTS: 550 Proof: 370**
BS 2493 A	0.12% C 0.5% Mo steel: Welding electrode
BS 3100 D1	0.2% C (max) 0.7% Mn 0.55% Mn Cr + Ni + Cu 0.8% (max) steel: Casting for high temperature use **UTS: 460 Elon: 18% Proof: 260**
BS 3100 BL1	0.2% C (max) 0.55% Mo steel: Casting for low temperature use; impact 20 J at -50 °C **UTS: 460 Elon: 18% Proof: 260**
BS S102	0.35% C 0.3% Mo steel: Hire for forged bolts; hardened & tempered after forging **UTS: 920 Elon: 18% Proof: 660**
BS S114	0.34% C 0.3% Mo steel: Hardened & tempered **DPN: 280 UTS: 920 Elon: 18% Proof: 660**
C Mo 15	0.15% C 0.6% Mo steel: ESC; normalized **UTS: 400 Elon: 20% Proof: 210**
C Mo 25	0.25% C 0.55% Mo steel: ESC **UTS: 610 Elon: 22% Proof: 370**
CODE 629	0.21% C 0.25% Mo steel: Casting; Edgar Allen; MT Mo 'P'
CODE 641	0.3% C 0.25% Mo steel: Casting; Edgar Allen; MT Mo 'R'
CODE 644	0.25% C 0.6% Mo steel: Casting; Edgar Allen
DIN 17155 15 Mo 3	0.16% C 0.6% Mn 0.3% Mo steel: Plate German Standard **UTS: 500 Proof: 270**
DIN 17155 15 Mo 3	0.15% C 0.3% Mo steel: For boiler plate
DIN 17175 15 Mo 3	0.15% C 0.7% Mn 0.3% Mo steel: For boiler plate
DTD 740	0.13% C 0.5% Mo 0.5% Mo 0.003% B steel: Tube; normalized & tempered; weldable **UTS: 610 Proof: 450**
DTD 5062	0.13% C 0.5% Mo 0.003% B steel: Sheet; normalized; weldable **UTS: 610 Elon: 12% Proof: 420**
DTD 5202	0.5% Mo B steel: Wrought **UTS: 600**
DUNELT 50	0.32% C 1.6% Mn 0.3% Mo steel: Dunford for BS alloy En 16
DUNELT 51	0.35% C 1.5% Mn 0.45% Mo steel: Dunford for BS alloy En 17
ERA 59	0.35% C 1.5% Mn 0.3% Mo steel: Hadfields for BS alloy En 16

Symbol	Nominal analysis, supplier, condition and remarks.
ERA 156	0.35% C 1.5% Mn 0.45% Mo steel: Hadfields for BS alloy En 17
ESSHETE D 4C	0.12% C 0.5% Mo steel: Samuel Fox; weldable **UTS: 450**
FASTEX 100M	0.07% C 0.5% Mo steel: Welding electrode; Murex **UTS: 450 Elon: 30% Proof: 360**
FORTIWELD	0.16% C (max) 0.6% Mo 0.005% B steel: United Steel Co.; normalized; suitable for welding **UTS: 600 Elon: 16% Proof: 440**
GOST 4543 38KHGN	0.38% C 0.95% Mn 0.65% Cr 0.85% Ni steel: Russian Standard
GOST 4543 38KHVF1UA	0.38% C 0.3% Mn 1.65% Cr 0.15% V 0.3% W 0.5% Al steel: Russian Standard
GOST 4543 38KHVFU	0.38% C 0.3% Mn 1.65% Cr 0.15% V 0.3% W 0.5% Al steel: Russian Standard
GOST 4543 38KHM1UA	0.38% C 0.55% Mn 1.5% Cr 0.2% Mo 1.0% Al steel: Russian Standard
JALTEN 1	0.15% C (max) 1.3% Mn 0.05% V 0.3% Cu steel: Jones & Laughlin
JALLOY No. 1	0.16% C 1.15% Mn 0.15% Mo 0.2% Cu steel: Jones & Laughlin
JALLOY No. 3	0.28% C 1.5% Mo steel: Jones & Laughlin
JALLOY AR 280	0.28% C 1.5% Mn 0.15% Mo 0.2% Cu 0.0005% B steel: Jones & Laughlin
JALLOY AR 320	0.28% C 1.5% Mn 0.15% Mo 0.2% Cu 0.0005% B steel: Jones & Laughlin
JALLOY AR 360	0.28% C 1.5% Mn 0.15% Mo 0.2% Cu 0.0005% B steel: Jones & Laughlin
JALLOY AR 400	0.28% C 1.5% Mn 0.15% Mo 0.2% Cu 0.0005% B steel: Jones & Laughlin
JALLOY S 90	0.15% C 1.25% Mn 0.25% Mo 0.003% B steel: Jones & Laughlin
JALLOY S 100	0.15% C 1.25% Mn 0.25% Mo 0.003% B steel: Jones & Laughlin
JALLOY S 110	0.15% C 1.35% Mn 0.25% Mo steel: Jones & Laughlin
JALLOY S 340	0.15% C 1.3% Mn 0.25% Mo steel: Jones & Laughlin
JIS G3103 SB46M	0.25% C (max) 0.9% Mn (max) 0.5% Mo steel: Plate; Japanese Standard **UTS: 500 Proof: 250**
JIS G3103 SB49M	0.27% C (max) 0.9% Mn 0.55% Mo steel: Plate; Japanese Standard **UTS: 540 Proof: 270**
JIS G3103 SB56M	0.27% C (max) 1.2% Mn 0.5% Mo steel: Plate; Japanese Standard **UTS: 630 Proof: 350**
JIS G3458 STPA12	0.15% C 0.55% Mn 0.55% Mo steel: Pipe; Japanese Standard **UTS: 390 Proof: 210**
JIS G3462 STBA12	0.15% C 0.55% Mn 0.55% Mo steel: Pipe; Japanese Standard **UTS: 390 Proof: 210**
JIS G3462 STBA13	0.2% C 0.55% Mn 0.55% Mo steel: Pipe; Japanese Standard **UTS: 420 Proof: 210**
JIS G3120 SQV1A	0.25% C (max) 1.2% Mn 0.5% Mo steel: Plate; Japanese Standard **UTS: 630 Proof: 350**

Symbol	Nominal analysis, supplier, condition and remarks.
JIS G3119 SBV1A	0.25% C (max) 1.1% Mn 0.5% Mo steel: Plate; Japanese Standard **UTS: 600 Proof: 320**
JIS G3119 SBV1B	0.25% C (max) 1.2% Mn 0.5% Mo steel: Plate; Japanese Standard **UTS: 630 Proof: 350**
JIS G3120 SQV1B	0.25% C (max) 1.2% Mn 0.5% Mo steel: Plate; Japanese Standard **UTS: 720 Proof: 490**
JS 6	0.2% C 0.5% Mo steel: Casting; Jopling; for use up to 425 °C **DPN: 180 UTS: 450 Elon: 20% Proof: 240**
JS 81	0.32% C 1.5% Mn 0.3% Mo steel: Casting; Jopling **DPN: 230 UTS: 760 Elon: 15% Proof: 480**
KAISALOY 70MB	0.15% C (max) 0.6% Mn 0.6% Mo 0.001% B steel: Kaiser Steel Corp.
KF 46	0.35% C 1.5% Mn 0.3% Mo steel: Kirkstall for BS alloy En 16
KF 46A	0.34% C 1.5% Mn 0.45% Mo steel: Kirkstall for BS alloy En 17
KP 14	0.35% C 1.5% Mn 0.3% Mo steel: Kiveton Park for BS alloy En 16
M 1	0.35% C 1.5% Mn 0.27% Mo steel: Jessop for BS alloy En 16
M 2	0.35% C 1.5% Mn 0.45% Mo steel: Jessop for BS alloy En 17
MANOIR PFO	0.15% C 0.5% Mo steel: Pompey; hardened & tempered **UTS: 450 Elon: 24% Proof: 260**
MAXEL 1B	0.18% C 0.1% Mo 0.04% P 0.08% S steel: Free machining; Crucible Steel Co.; may be carburized **DPN: 156 UTS: 480 Elon: 30% Proof: 330**
MM 01	0.32% C 1.6% Mn 0.27% Mo steel: Firth Brown for BS alloy En 16
MM 02	0.33% C 1.6% Mn 0.4% Mo steel: Firth Brown for BS alloy En 17
Mo B	0.07% C 0.8% Mn 0.5% Mo steel: Electrode; metrode; for BS 2493
MOLEX	0.06% C 0.5% Mo steel: Welding electrode; Murex **UTS: 460 Elon: 33% Proof: 440**
MOLYTRODE 74.56	0.08% 0.06% Mn 0.6% Mo steel: For electrodes; Esab **UTS: 556 · Elon: 28 Proof: 463**
MT Mo 'P'	0.21% C 1.4% Mn 0.25% Mo steel: Casting; Edgar Allen; Code 629
MT 'R'	0.3% C 1.4% Mn 0.25% Mo steel: Casting; Edgar Allen; Code 641
NV 7-1	0.2% C (max) 0.6% Mn 0.35% Mo steel: Designation used by DNV **UTS: 440 Elon: 22% Proof: 280**
NV 7-2	0.18% C (max) 0.6% Mn 0.55% Mo steel: Designation used by DNV **UTS: 440 Elon: 22% Proof: 300**
NVR 7-1	0.2% C (max) 0.6% Mn 0.32% Mo steel: Designation used by DNV
OCM 7	0.35% C 1.6% Mn 0.3% Mo steel: W Marrison for BS alloy En 16
PFO	0.24% C 0.5% Mo steel: Pompey
PFO A	0.12% C 0.5% Mo steel: Pompey
REYNOLDS 531	0.26% C 1.5% Mn 0.2% Mo steel: Tube; Reynolds; suitable for brazing **UTS: 690 Elon: 10% Proof: 610**
SAE 4012	0.12% C 0.2% Mo steel
SAE 4023	0.23% C 0.25% Mo steel: Mechanical properties not specified
SAE 4024	0.22% C 0.25% Mo steel: Mechanical properties not specified
SAE 4027	properties not specified
SAE 4028	0.28% C 0.25% Mo steel: Mechanical properties not specified

Note. The following abbreviations and units are used in the tables:

DPN	Hardness, diamond pyramid number
UTS	Ultimate tensile strength, N/mm^2
Elon	Elongation, %
Proof	0.1 % proof strength, N/mm^2

1 N/mm^2=0.1 hbar=0.102 kgf/mm^2=0.06475 tonf/in^2=145.04 lbf/in^2

Symbol	Nominal analysis, supplier, condition and remarks.
SAE 4032	0.32% C 0.25% Mo steel: Mechanical properties not specified
SAE 4037	0.37% C 0.25% Mo steel
SAE 4419	0.2% C 0.5% Mo steel: Mechanical properties not quoted 0.22% C 0.4% Mo steel: Mechanical properties not quoted
SAE 4427	0.27% C 0.4% Mo steel: Mechanical properties not quoted
SANBOLD 16	0.35% C 1.5% Mn 0.3% Mo steel: Sanderson for BS alloy En 16
SANBOLD 17	0.35% C 1.5% Mn 0.4% Mo steel: Sanderson for BS alloy En 17
SANBOLD 53	0.35% C 1.5% Mn 0.4% Mo steel: Sanderson; replaced by Sanbold 17
SANBOLD 74	0.35% C 1.5% Mn 0.3% Mo steel: Sanderson; replaced by Sanbold 16
SIS 2912	0.2% C 0.35% Mo steel: For pressure vessel tubes; Swedish Standard; as drawn **UTS: 480 Elon: 22% Proof: 190**
SMM	0.35% C 1.5% Mn 0.5% Mo steel: Spencer for BS alloy En 17
SPEAR 416	0.37% C 1.5% Mn 0.25% Mo steel: Spear & Jackson for BS alloy En 16
STA 5 V9A	0.32% C 1.5% Mn 0.3% Mo steel: Replaced by BS alloy En 16
STA 5 V9A/1	0.32% C 1.5% Mn 0.27% Mo steel: Replaced by BS alloy En 16A

Symbol	Nominal analysis, supplier, condition and remarks.
STA 5 V9B	0.35% C 1.5% Mn 0.45% Mo steel: Replaced by BS alloy En 17
STA 5 V9B/1	0.35% C 1.5% Mn 0.45% Mo steel: Obsolete
SUPERTOUGH C40	0.35% C 1.6% Mn 0.28% Mo steel: ESC for BS alloy En 16
SUPERTOUGH C40M	0.35% C 1.6% Mn 0.45% Mo steel: ESC for BS alloy En 17
UHB STATO 21	0.18% C 0.3% Mo steel: Tube; Uddelholm; normalized **UTS: 480 Elon: 30% Proof: 270**
VSMM	0.35% C 1.5% Mn 0.35% Mo steel: Vulcan **UTS: 840**
W 7	0.11% C 0.6% Mo steel: Welding electrode; Cu covered; Armco; as welded **UTS: 680 Elon: 22% Proof: 580**

Note. The following abbreviations and units are used in the tables:

DPN	Hardness, diamond pyramid number
UTS	Ultimate tensile strength, N/mm^2
Elon	Elongation, %
Proof	0.1 % proof strength, N/mm^2

$1 N/mm^2 = 0.1 hbar = 0.102 kgf/mm^2 = 0.06475 tonf/in^2 = 145.04 lbf/in^2$

44G2 Steel – medium and high carbon with molybdenum
0.4% carbon minimum, 0.7% molybdenum maximum

Specific gravity	7.8	
Density	7800 kg/m^3	(0.28 lb/in.^{3})
Solidus/liquidus	– °C	
Thermal conductivity	– W/m °C	
Coefficient of linear expansion	11 × 10^{-6}/ °C	
Electrical conductivity	8–10% IACS (copper 100%)	
Specific resistance	180–250 microhm mm	
Young's modulus of elasticity	– N/m^2	
Impact	– J	
Fatigue strength (no. of cycles)	–	
Hot strength		

Temperature °C	Tensile strength N/mm^2	Elongation %
200	460	35
425	400	35
480	330	40
650	210	50

The above properties are typical of the following group and may not apply exactly to any one specification. It is possible that with certain specifications some of the values may not be applicable.

General metallurgical characteristics

Carbon will always combine with molybdenum in preference to iron, thus the molybdenum content of these steels is always present as the carbide. This increases the creep strength and helps the steels to retain their hardness at higher temperatures. The stability of the carbides increases the hardenability and gives an inherent fine-grained structure which improves the toughness of the correctly heat treated parts. There is no apparent difference in corrosion resistance between these and plain carbon steels. In many respects moolybdenum has the same effect as tungsten but less is required to achieve the same result.

Molybdenum steels are noted for their lack of notch sensitivity in the hardened condition, being superior to chromium steels in this respect.

These materials are almost invariably used hardened and tempered, seldom or never in the cold worked condition.

Thermal treatment

None of the steels listed is capable of accepting very much cold work without becoming embrittled. This effect is proportional to the carbon content and can be removed by annealing for 30 minutes per 25 mm of section at 750–850 °C and cooling slowly. Most cold work is removed by a sub-critical anneal at 650–670 °C for two hours and this is generally satisfactory unless further heavy cold deformation is necessary.

Because of the stable nature of the carbides it may be necessary to cyclic anneal certain parts to improve their machinability. For this the parts should be heated to above 750–850 °C then transferred to a furnace at 650–700 °C. This should be repeated at least once and with higher carbon steels may require several cycles. The result is a steel with the hard carbides present as spheroids in the soft ferrite matrix, the process being referred to as cyclic annealing or spheroidizing.

Normalizing requires the same temperature as annealing, that is 750–850 °C depending on the steel, for about 20 minutes per 25 mm of section, followed by cooling in still air. This removes all forging or cold work stresses, refines and homogenizes the grain structure and results in some air hardening. This will often be sufficient to require a temper or sub-critical anneal before further work. Any residual elements will considerably increase this air hardenability and can prove troublesome in certain conditions.

In order to harden, the steel must be heated to 700–800 °C depending on the analysis, and oil or water quenched. Time at temperature should be 15 minutes per 25 mm of section. This results in full hardening, necessitating immediate tempering at 200–300 °C to give the required hardness. The higher carbon steels in particular will be extremely brittle in the 'as quenched' condition. At all times these steels should be taken to temperature slowly to prevent cracking caused by the volume changes which take place at the metallurgical change points.

Although in general it can be said that the higher carbon steels require the lower temperatures and the lower carbon materials the higher temperatures, it is essential that the origin of the specification is consulted for exact details of times and temperatures. These molybdenum steels are more critical than most regarding the correct temperature. These materials must not be carburized, the core hardness being equal to that of any case hardness with the majority of the steels listed.

Local flame or induction hardening is possible but not recommended with these steels as comparable results are obtainable with plain carbon steel.

The furnaces used for processing these steels can be of the electric, gas or oil fired type. Some care is necessary, particularly with the higher carbon steels, to ensure that the atmospheric conditions are never slightly oxidizing as this can lead to inter-granular de-carburization. An oxidizing atmosphere giving free scaling should always be aimed for, unless a neutral or reducing atmosphere is available, and this will seldom be necessary with these steels. Scale removal is generally by grit blasting, but sodium hydride gives excellent results if the specialized plant is available. These steels must not be acid pickled in the hardened condition above 350 DPN. Below this hardness acid pickling does not cause appreciable embrittlement and can be used for scale removal.

Welding and brazing

Pre-treatment. Parts must be in the normalized or annealed condition, and the joint area should be cleaned and freshly prepared and must be pre-heated.

Welding and brazing. Care must be taken that residual elements such as chrome or nickel which could increase the hardenability of these materials are absent, otherwise cracking in the weld zone can occur.

Provided the steel analysis is controlled and parts are correctly designed to ensure no rapid chills, gas or electric arc welding with fluxes is possible with skilled operators. Inert gas shielded arc welding with its narrow heated zone can give very good welds provided the join is properly designed.

With resistance welding the same general remarks apply.

Brazing must be very carefully carried out to ensure no areas are heated to above the hardening temperature unless special high temperature braze fillers, melting above the normalizing temperature, are used.

Post-treatment. It is essential that the weld zone is tempered. Ideally the parts should be normalized, or hardened and tempered. Very careful design to eliminate all stress raisers is necessary if the welded parts are to be hardened. It is essential that all welded components are at least locally tempered at the weld zone. This can be by gas heating but is preferably by electrically heated blankets.

Provided care is taken in the choice of materials, with skilled operators and most important, correct pre- and post-weld treatment, good joints are possible with most materials in this group.

Flaw detection methods

Crack test. Magnetic crack test is the general first choice, but good results are possible using penetrant oil chalk testing, with modern highly sensitive fluids and skilled operators. This method is more flexible than the magnetic.

X-ray. Not normally required, although it is recommended where welds are stressed in service.

Ultra-sonic test. This is not normally carried out on parts made from the materials listed, but can be used to identify certain types of weld defects.

Chemical etch. High carbon parts in the fully hardened and tempered condition are very prone to grinding abuse, resulting in softening and in some cases cracking.

An electrolytic sulphuric etch after grinding can clearly show the extent of this abuse, and it is recommended that this is used as a control of grinding. Nitric acid etches give similar results.

Corrosion protection

Temporary. Molybdenum has no appreciable effect on the corrosion resistance of steel, thus parts must be oiled or greased and kept dry at all times. During machining mineral oil and good quality clean soluble coolants act as corrosion inhibitors. It should however, be noted that any coolant loaded with metallic dust can cause severe rusting.

Finished diameters should be cleaned and oiled as soon as produced, and finished parts oiled then sealed in plastic.

Permanent. The materials listed are almost always used in the hardened and tempered condition when they are not very ductile, thus corrosion pitting can be a serious problem. It is recommended that care is taken at all times to prevent formation of rust rather than the use of expensive methods of rust removal. These materials are used for their hardness and wear resistance and this considerably reduces the choice of protective coatings.

In order to prevent corrosion the surface must be free from all existing rust and then sealed from contact with air. Removal of rust can be by blasting which is quite effective on these materials but will not result in a rough surface acting as the key for any subsequent coating. Alkaline rust removers are also quite effective. Acid should not be used on these parts owing to the danger of hydrogen embrittlement.

Because of the uses to which these steels are put the means of preventing damp air acting on the surface is very often an oil or grease film.

Phosphate treatments are to some extent corrosion resisting and as they absorb oil, many tools use this as their permanent protection. A stress release is essential after phosphating to obviate hydrogen embrittlement.

Painting. Painting is seldom applied to these steels, but when necessary should be on top of the phosphate, followed by zinc chromate type primer, then a good quality top coat applied and stoved.

Plating. It will be very seldom that these steels will be electroplated for appearance purposes. Where necessary copper and nickel undercoats with a total thickness of 0.075 mm are recommended beneath the thin film of decorative chromium.

Cadmium or zinc electrodeposits are also uncommon although they can be applied, but must not be used on parts in contact with foodstuffs. Great care must be taken to remove machining stresses before plating and the hydrogen which is occluded during plating. Unless this is successfully carried out failure in a brittle manner below the design tensile strength can occur. Stress relieving at 200 °C before and after plating eliminates this defect, but because of the difficulty in proving that this has been correctly carried out and the disastrous affects of umpredictable failures from this cause, electroplating is not recommended.

Metal spraying is not very successful as this relies on the key effect from blasting for adhesion and this is almost absent with these high hardness steels. Techniques using high velocity metal particles are having some success.

Cathodic protection is not normally feasible as these steels are used for portable items of equipment.

'Black oxide' treatment can be very successful provided the film is properly maintained and kept oiled. The film is produced using a hot strong sodium hydroxide solution with oxidizing agents added and is actually firmly adherent oxides of iron.

Machinability

In the hardened and tempered condition all of these steels are too hard to be machined using conventional tools and can only be ground. Adequate cooling, light cuts and frequent wheel dressing are necessary to prevent grinding abuse. Spark erosion or electrochemical machining techniques can remove metal after hardening but are not normally used on these steels.

Rough machining is carried out in the normalized and sub-critical annealed condition, or with the higher carbon varieties after cyclic annealing.

Where dimensional stability is necessary and heavy cuts are required at rough machining it is recommended that parts are sub-critically annealed after machining before hardening and tempering, then finally ground.

Uses

Cutting tools for free machining non-ferrous materials, shear blades, press tools and fixtures. Machining fixtures which require through hardening and are too large for plain carbon steels.

Taps, dies, drills and cutting tools for short production runs.

Symbol	Nominal analysis, supplier, condition and remarks.
06	1.45% C 1.0% Si 0.25% Mo steel: Designation used by AISI
19F	1.5% C 1.0% Si 0.25% Mo steel: Carrs; press & bending tools, etc. DPN: 730
2121	1.4% C 1.0% Mn 0.3% Mo steel: French Standard designation
AISI 4042	0.42% C 0.25% Mo steel
AISI 4047	0.47% C 0.25% Mo steel
AISI 4053	0.53% C 0.25% Mo steel
AISI 4063	0.63% C 0.25Mo steel
AISI 4068	0.68% C 0.25% Mo steel
AISI A4068	0.68% C 0.25% Mo steel: Obsolete
ASTM A194/4	0.45% C 0.8% Mn 0.25% Mo steel: For bolts
ASTM A194/7	0.42% C 0.9% Mn 0.2% Mo steel: For bolts
ATLAS CM	0.4% C 0.15% Mo steel: Atlas; hardened & tempered DPN: 230 UTS: 760 Elon: 23% Proof: 610
ATLAS SHAFTING	0.4% C 0.15% Mo steel: Bar; Atlas; normalized; centreless ground UTS: 660
BS 3111/2	0.4% C 0.3% Mo steel: Wire; annealed & drawn; UTS: 670
CRD 515	0.5% C 1.0% Si 0.5% Mo steel: C Denton; shear blades; obsolete DPN: 540
E 356	1.5% C 0.45% Mo steel: Fagersta
GRAPH Mo	1.45% C 1.0% Mn 1.25% Si 0.25% Mo steel: Free graphite; Timken; tools for wear resistance DPN: 425 UTS: 1200 Elon: 10% Proof: 1100
GRAY DIAMOND	1.45% C 1.1% Si 0.25% Mo steel: Latrobe; free graphite
HEDEX 6	0.4% C 0.3% Mo steel: Wire for forging; Kiveton Park; annealed UTS: 580
HEDEX 22	0.42% C 0.25% Mo steel: Wire for forging; Kiveton Park; annealed UTS: 640
JALLOY No. 3AR	0.55% C 1.5% Mn 0.15% Mo steel: Jones & Laughlin
JALLOY No. 7	0.55% C 1.5% Mn 0.15% Mo steel: Jones & Laughlin
KF 57	0.38% C 1.6% Mn 0.3% Mo 0.2% S 0.05% P steel: Free machining; Kirkstall UTS: 920 Elon: 16% Proof: 660
LT 38	0.6% C 2.0% Si 0.25% Mo steel: Low Moor; obsolete DPN: 670

Symbol	Nominal analysis, supplier, condition and remarks.
LTG	1.45% C 1.0% Si 0.25% Mo steel: Low Moor; obsolete
M1C 8	1.4% C 1.0% Si 0.25% Mo steel: Taps, punches, rolls, etc., ESC DPN: 600
M11	0.63% C 0.25% Mo steel: Round Oak
M12	0.67% C 0.25% Mo steel: Round Oak
MAXEL 2B	0.4% C 0.1% Mo 0.08% S steel: Free machining; Crucible Steel Co DPN: 220 UTS: 760 Elon: 18% Proof: 390
MONARK 1	0.5% C 0.45% Mo steel: Atlas; as AISI type S2
S 60	Ni Mo steel sinter: Firth Cleveland; specific gravity 6.4–6.8; hardened & tempered; file hard UTS: 780 Elon: 0.5%
SAE 06	1.5% C 0.25% Mo tool steel
SAE 4042	0.42% C 0.25% Mo steel: Mechanical properties not specified
SAE 4047	0.47% C 0.25% Mo steel: Mechanical properties not specified
SAE 4053	0.53% C 0.25% Mo steel: Mechanical properties not specified
SAE 4063	0.63% C 0.25% Mo steel: Mechanical properties not specified
SAE 4068	0.68% C 0.25% Mo steel: Mechanical properties not specified 0.9% C 0.25% Mo steel: Simonds; AISI type 02
SMO	0.55% C 1.7% Si 0.35% Mo steel: Huntsman; chisels, shear blades, collets, etc. DPN: 670
SOLAR	0.5% C 0.5% Mo 0.4% Mn 1.0% Si steel: Carpenter; for AISI type S2
TG	1.5% C 1.0% Si 0.3% Mo graphitic tool steel: Osborn; brushes, gauges, etc.; self-lubricating steel; obsolete DPN: 600
TRITON	0.5% C 0.6% Mo steel: Braeburn; AISI type S2

Note. The following abbreviations and units are used in the tables:

DPN	Hardness, diamond pyramid number
UTS	Ultimate tensile strength, N/mm^2
Elon	Elongation, %
Proof	0.1 % proof strength, N/mm^2

1 N/mm^2=0.1 hbar=0.102 kgf/mm^2=0.06475 tonf/in^2=145.04 lbf/in^2

44H Steel – all carbon contents with tungsten

Specific gravity	7.8	
Density	7800 kg/m³	(0.28 lb/in.³)
Solidus/liquidus	– °C	
Thermal conductivity	– N/m °C	
Coefficient of linear expansion	11 × 10⁻⁶/ °C	
Electrical conductivity	– IACS (copper 100%)	
Specific resistance	– microhm mm	
Young's modulus of elasticity	– N/m²	
Impact	– J	
Fatigue strength	–	
Hot strength	–	

The above properties are typical of the following group and may not apply exactly to any one specification. It is possible that with certain specifications some of the values may not be applicable.

General metallurgical characteristics

Tungsten combines with the carbon in steel to form very hard carbides which retain their hardness above red heat. These tungsten carbides form in preference to iron carbide, thus all available tungsten exists as the carbide. This inhibits grain growth and by slowing down the metallurgical reactions increases the hardenability.

Fine grained steels tend to be less notch sensitive and less brittle than coarse grained, very hard steels. Tungsten is therefore useful in that while it increases the hardenability by an amount almost equal to chromium, it has a controlling effect on the grain size which tends to make them tougher and less notch sensitive. The intrinsically hard carbide particles ensure that these steels are never very soft and it is their hot hardness properties which find the greatest use. There are comparatively few steels containing tungsten only as it is generally present with chromium, molybdenum or vanadium.

Thermal treatment

It is advisable that these steels are pre-heated at about 500 °C before transferring to a higher temperature, otherwise cracking may occur.

These materials are not capable of accepting any cold work without becoming brittle and shattering. Full annealing requires the parts to be heated to 740–840 °C for about an hour per 25 mm of section followed by slow cooling, but this treatment by itself will seldom be required.

Sub-critical annealing at 650–670 °C for two hours will remove machining stresses and most cold work stress, but again is not often used as it is generally necessary to cyclic anneal these materials to soften them for machining. This requires heating to the high temperature for 30 minutes per 25 mm of section, then transferring to a furnace at 650–670 °C for one hour. This cycle should be repeated at least once, the result being that the carbides appear as spheroids in the softer matrix of ferrite. Normalizing will seldom be required on these materials as spheroidal annealing gives the best machining characteristics and successfully removes all forging stresses. To normalize the parts should be heated to 740–840 °C and cooled in still air. This will always result in considerable hardening which is not always tempered by sub-critically annealing at 650–670 °C, hence the use of cyclic annealing.

These steels are hardened by heating to 720–820 °C, those with carbon of 0.7% and upwards requiring the 720 °C temperature, the 0.3–0.5° carbon steels requiring 780–820 °C. The time at temperature should be 30 minutes per 25 mm of section, followed by oil quenching. Water quenching will almost always cause cracking, but air cooling does not always give even hardening unless the component is of regular shape and section. The parts will always be very hard and brittle after quenching and should be tempered as soon as possible at 150 °C minimum. Higher hardening temperatures than those quoted can be used for shorter times, thus reducing scaling at the expense of some grain growth. These steels are never carburized as they can achieve an equal through hardness to that obtained in any carburized case. Induction or flame hardening is possible but much better results are obtained with higher alloyed steels.

It is advised that the above information is used only as a guide and the specification origin consulted for full details of times and temperatures.

No special furnace equipment is necessary but it must be appreciated that high carbon steels will always lose some surface carbon unless protected from oxidizing atmospheres. Salt baths are very often used for this reason.

De-scaling is generally by grit blasting but sodium hydride gives excellent results provided the specialist equipment is available. Acid pickling of the hardened and tempered steels is not possible owing to the danger of embrittlement.

Welding and brazing

Pre-treatment. Parts should be clean, free of grease, oil and any dirt, oxide or scale. Welding should be in the annealed condition and will require careful pre-heating.

Welding and brazing. It is strongly advised that no welding is attempted on these materials as cracking of the weld zone caused by the very local hardening stresses is highly probable. Gas welding in skilled hands using the torch to pre-heat and post-heat the weld zone can result in passable welds.

Brazing using induction heating can be carried out on fully hardened parts because of the rapid means of heating and the sluggish metallurgical nature of these steels, but great care is necessary not to overheat or hold too long at temperature. Normal gas heated brazing generally softens and can cause brittleness but again is possible provided the heat input is carefully controlled.

Post-treatment. It is essential that immediately after welding the weld area is at least stress relieved or tempered. Ideally the parts should be hardened and tempered after welding but great care is required to prevent cracking from sharp corners. Care must also be taken to prevent de-carburization.

Brazed parts do not normally require any treatment but stress relieving at the tempering temperature is always advisable. All fluxes must be removed otherwise severe corrosion can occur.

Flaw detection methods

Crack test. Magnetic crack test. Penetrant oil chalk test using modern sensitive penetrant fluids can give comparable results to the magnetic method and is more flexible.

X-ray. Not normally required.

Ultra-sonic test. Large expensive forgings can be examined after rough machining to ensure the absence of subcutaneous defects. Ultra-sonic testing does not find surface defects and is therefore generally only used to save carrying out expensive machining operations on potentially scrap parts.

Chemical etch. These materials are all susceptible to grinding abuse which causes local softening and cracking. An electrolytic sulphuric acid etch on finished parts can show this, thus enabling control to be established over the grinding.

Corrosion protection

Temporary. These materials must be oiled or greased at all times and stored in dry conditions. During machining the parts must be kept oiled and this is essential once finished areas have been produced. Soluble oil coolant in good condition acts as a temporary rust inhibitor but contaminated or dirty coolants, especially when loaded with metal particles, causes staining which can rapidly change to severe corrosion. Finished parts should be cleaned, oiled and stored in plastic.

Permanent. Tungsten has no appreciable affect on the corrosion resistance properties of steel, but the effect of corrosion pitting on a fully hardened tungsten steel can be much more serious than on a more ductile material.

Steel corrosion is caused by the formation of complex iron oxides which are very loosely adherent, thus easily removed, to allow fresh attack on the uncovered surface. Formation of rust can only be prevented by completely excluding air and moisture. It must also be appreciated that corrosion is self-propagating and even the most effi-cient corrosion protecting film will fail if it is applied on top of rust.

The materials listed are almost always used indoors and the principal reason for their existence is hardness and wear resistance. This means that corrosion protective coatings are rapidly worn off the working areas. Many tools are grit blasted then phosphate treated. This efficiently removes all live corrosion and produces an oil-retaining surface. Provided the parts are kept oiled this can prevent corrosion for many years under indoor conditions.

Artificially forming an adherent 'black oxide' film by immersing in proprietary solutions of strong hot sodium hydroxide with added oxidizing agents also gives an efficient oil-retaining surface.

Painting. Painting on top of phosphating gives good results provided zinc chromate type primers are used. These should then be sealed with a top coat. Alkaline rust removers can be used to kill active rust before painting but these require careful control. Acid should not be used for this purpose as it causes embrittlement which can lead to failure. Tools are commonly painted on their non-working surfaces while the working areas are protected by oil when not in use.

Plating. Electroplating is not recommended on these high hardness materials as they are adversely affected by hydrogen embrittlement, which causes failure in a brittle manner below the calculated tensile strength. The embrittling effect can be eliminated by stress relieving at 200 °C for two hours before and after plating, but it is difficult to organize a system to ensure this is carried out and there is no method of inspection which can prove positively that parts were stress relieved. Apart from this there is no difficulty in electroplating these materials provided the correct pre-plating etch is used. This is more critical to ensure adhesion than with the more ductile steels.

Metal spraying relies on the key from the grit blasted surface and this is absent with these hard materials. Techniques using high velocity metal particles can overcome this deficiency.

Machinability

These materials should be machined in the spheroidize annealed condition when the surface finish will generally be very good, with tool wear necessitating high speed or tipped tools at all times. After hardening only grinding, spark erosion or electrochemical machining techniques can be used. A minimum of 0.1 mm metal per surface must be removed to elimate the de-carburized layer generally formed during hardening. It is advised that sharp corners, rapid change of section or other stress raisers are formed after the thermal treatment, otherwise cracking or distortion during quenching may occur.

Uses

Cutting tools, press tools and fixtures of simple regular shape, shear blades.

Symbol	Nominal analysis, supplier, condition and remarks.
AK	0.7% C 7% W steel: For tools; Osborn obsolete **DPN: 950**
BRUNSWICK V104	1.2% C 2.0% W steel: John Vessey; chisels, punches **DPN: 540**
BRUNSWICK V105	0.3% C 9% W steel: John Vessey; hot work dies **DPN: 620**
BRUNSWICK V414	0.4% C 14% W steel: John Vessey **DPN: 540**
BS S68	High tungsten steel: Obsolete
F 1	1.0% C 1.25% W steel: Designation used by AISI
F 2	1.25% C 3.5% W steel: Designation used by AISI
FCW 5	1.35% C 5% W steel: Firth Brown
HALLAMITE	C not specified 14% W steel: For tools; Hallamshire
HALLAMITIER	C not specified 18% W steel: For tools; Hallamshire
HALLAMITIEST	C not specified 22% W steel: For tools; Hallamshire
INTRA	1.2% C 1.5% W steel: Jonas; taps, reamers etc. **DPN: 670**
JO	1.15% C 1.4% W tool steel: Osborn; taps, reamers, small tools; obsolete **DPN: 850**
KE 621	1.05% C 0.4% W steel: Kayser Ellison; taps, punches, etc.
KE 637	1.0% C 1.7% Mn 0.25% W steel: Kayser Ellison; gauges, etc. **DPN: 850**
KAO	1.3% C 3.0% W steel: For tools; Huntsman; cold drawing dies **DPN: 790**

Symbol	Nominal analysis, supplier, condition and remarks.
KAOX	1.3% C 5% W steel: Huntsman; cold drawing dies **DPN: 790**
LT5	1.1% C 1.5% W steel: For tools; Low Moor; obsolete **DPN: 850**
LT 13	1.7% C 7% W steel: For tools; Low Moor; obsolete **DPN: 950**
MEL-TROL K-W	1.3% C 3.5% F steel: Carpenter for AISI type F3
SAE HNV 7	0.5% C 3.5% Si 14% W steel: For inlet valves
SPEAR No. 1	1.0% C W steel: Spear & Jackson; for hand tools
SUPERDRAW	1.3% C 2.25% W steel: Edgar Allen; cold drawing dies **DPN: 950**
TDB	1.4% C 5% W steel: Jonas; dies **DPN: 670**
T 60601	UNS designation for type F1 tool steel
T 60602	UNS designation for type F2 tool steel
TDC	1.2% C 1.25% W steel: Firth Brown

Note. The following abbreviations and units are used in the tables:

DPN	Hardness, diamond pyramid number
UTS	Ultimate tensile strength, N/mm²
Elon	Elongation, %
Proof	0.1 % proof strength, N/mm²

1 N/mm²=0.1 hbar=0.102 kgf/mm²=0.06475 tonf/in²=145.04 lbf/in²

44J Steel – all carbon contents with vanadium

Specific gravity	7.8	
Density	7800 kg/m³	(0.28 lb/in.³)
Solidus/liquidus	– °C	
Thermal conductivity	– W/m °C	
Coefficient of linear expansion	11 × 10⁻⁶/ °C	
Electrical conductivity	– IACS (copper 100%)	
Specific resistance	– microhm mm	
Young's modulus of elasticity	– N/m²	
Impact	– J	
Fatigue strength	–	
Hot strength	–	

The above properties are typical of the following group and may not apply exactly to any one specification. It is possible that with certain specifications some of the values may not be applicable.

General metallurgical characteristics

Vanadium is an excellent de-oxidizer, carbide former and grain refiner which is very expensive, and relatively scarce. Vanadium is seldom used purely as a de– oxidizer and is not often found as the sole alloying element, being generally present with chromium. Its grain refining properties are associated with the carbide forming. These carbides, being stable, do not readily dissolve in the iron matrix and prevent formation of large carbide masses. Carbon combines with vanadium in preference to iron, thus all the

vanadium exists as the carbide, the remaining carbon forming iron carbide. Vanadium carbide is very hard and stable, consequently even annealed steel is hard.

Vanadium improves the fatigue strength on the one hand, but increases the notch sensitivity on the other and has no appreciable effect on the corrosion resistance.

Most of the steels listed are of the high carbon type (0.8% carbon and upwards) which can be made glass hard by thermal treatment.

Thermal treatment

These steels will seldom be subjected to any degree of cold work as the high carbon steels particularly become very brittle. Where necessary full annealing requires 730–770 °C for the high carbon range and 800–920 °C for the medium and low carbon steels followed in each case by slow cooling. A time of one hour per 25 mm of section is necessary at temperature.

Sub-critical annealing for 2 hours at 650–670 °C removes most evidence of cold work and machining stresses but fails to soften the hard carbides.

Normalizing for 30 minutes per 25 mm of section at the same temperature as the full anneal followed by still air cooling removes all cold work, refines the grain, homogenizes the structure and results in some air hardening, which cannot be readily removed by tempering or sub-critical annealing. The ideal method of softening for machining is cyclic or spheroidize annealing. This requires heating of the part to above the annealing temperature for 20–30 minutes per 25 mm of section, then cooling to the sub-critical temperature for about 1 hour. This is repeated at least once and results in the carbides forming small hard spheroids in the soft iron (or ferrite) matrix.

To harden the high carbon steels should be heated to 740–780 °C for 20 minutes per 25 mm of section and oil quenched, while the low and medium carbon steels require 780–860 °C for the same time, then quenched. This must be immediately followed by tempering at 150–350 °C, depending on the hardness required. These steels are all air hardening but unless the parts are of regular shape and equal section, oil quenching should be employed. If air hardening is necessary to prevent cracking or distortion, a more highly alloyed steel should be used.

Although electric, gas or oiled fired furnaces can be used, most of these materials will be hardened using salt baths. This applies particularly to the higher carbon steels where the danger of de-carburization is greatest.

Scale removal is generally by grit blasting but sodium hydride is satisfactory, provided the specialist equipment is available. Acid pickling must not be employed on any of these steels after hardening and tempering, as it can cause severe embrittlement.

Welding and brazing

Pre-treatment. The area must be free of grease, scale and dirt. If possible the parts should be cyclic annealed, and pre-heating is essential.

Welding and brazing. It is recommended that the high carbon varieties of these steels are not welded as it is very difficult to prevent quench cracking adjacent to the weld. A skilled operator using gas welding with pre-heating and immediate post-heating to delay the cooling rate can produce satisfactory welds. Electric arc welding is more difficult.

Resistance welding requires careful design and is not recommended.

Brazing can be accomplished on fully hardened and tempered parts provided rapid heating is used, but this requires very careful control.

The low carbon steels can be very successfully welded by any standard method and are not prone to weld cracking. The medium carbon steels require more care.

Post-treatment. Welds must be stress relieved or tempered, at least locally, as soon after welding as possible. With correct design it is possible to harden and temper after welding but care is necessary to prevent quench cracking and de-carburization. Brazing requires no thermal treatment apart from stress relieving but complete removal of all flux is essential.

Flaw detection methods

Crack test. Magnetic crack test, with penetrant oil chalk test almost as sensitive and more flexible.

X-ray. Not normally required.

Ultra-sonic test. May be required to prove that large expensive forgings have no subcutaneous defects which would show after machining. This test does not show surface defects and is generally used to save the cost of expensive finish machining operations on parts which could then be shown to be scrap.

Chemical etch. The high carbon steels can be maltreated at grinding unless great care is taken. This results in softening and in severe cases grinding cracking. An electrolytic sulphuric acid etch after grinding shows whether or not these have been abused, allowing the grinding to be controlled without scrapping the part.

Corrosion protection

Temporary. These steels must be stored in the oiled or greased conditions and out of contact with moist air. During machining, finished areas must be cleaned and oiled immediately.

Good quality soluble oil coolants act to some extent as temporary corrosion inhibitors but it must always be noted that if allowed to become contaminated, particularly with metallic particles, these liquids can cause staining and corrosion.

Finished parts must be cleaned, dried and oiled then stored in plastic or other non-porous containers.

Permanent. Vanadium does not enhance the corrosion resisting properties of alloy steels.

Vanadium steels however are generally harder than plain carbon steels and are also more notch sensitive, thus corrosion pitting is potentially more serious.

Corrosion protection relies on a complete covering of non-porous material preventing the atmosphere contacting the part being protected. It must always be remembered that corrosion has air and moisture occluded in the rust and is thus self-propagating. The increase in volume of the rust as it propagates is generally the reason for failure of protecting films. The high hardness of the listed materials means that grit blasting, while removing corrosion products, does not roughen the surface to any great extent.

Alkaline rust removers are probably more efficient provided they are properly controlled. Acid rust removers must not be used as they can cause brittleness in these materials.

Phosphate treatment can kill light corrosion and act as an excellent paint base and oil-retentive surface. With correct maintenance of this oil film many tools have a long corrosion free life.

Painting. Painting will very seldom be carried out on these materials as they are almost invariably used for their hard wear resistance properties. Phosphated tools are however, often painted on their non-working areas. This paint should be of the zinc chromate type primer with a sealing top coat applied as soon as possible afterwards.

Plating. Decorative chromium plating will not often be required. When necessary, undercoats of copper and nickel to a minimum of 0.075 mm thick are essential.

Cadmium and zinc plating less than 0.025 mm thick give adequate protection. With all these materials there is a grave danger of hydrogen embrittlement during plating. This can be eliminated by stress relieving at 200 °C before and after plating, but there is no visual indication that these operations have been carried out and without it inexplicable brittle failures below the calculated tensile strength can and do take place.

Metal spraying is sometimes carried out on the lower carbon varieties of the steels listed where the hardness is low enough to allow blasting to roughen the surface. Without this blast key the adhesion of the sprayed coating is negligible. Modern metal spray techniques using very high velocities of the metal particles are showing promise on these hard metals. The coating should always be sealed by wire brushing, then chemical treatment or painting. Cathodic protection is not normally required on the high carbon steels as they are used to make portable tools, not large outdoor structures. The low carbon variety may require this system, which makes the structure negative or cathodic in an electric circuit the anode or positive then corrodes protecting the structure. The anode can be replaced at regular intervals.

Machinability

The high carbon varieties cannot be machined after hardening except by grinding, spark erosion or electrochemical machining. Rough machining should be in the cyclic or spheroidize annealed condition when good surface finishes are obtained but tool life will be low as the carbides present are very abrasive. High speed or tipped tools are therefore essential.

An allowance must always be made for the removal of at least 0.1 mm per surface after hardening to remove the decarburized layer. Design of the parts at the heat treatment stage is also important to ensure there are no rapid changes of section to act as stress raisers for crack propagation. The low carbon vanadium steels generally exhibit good machinability in the heat treated condition and tend to tear in the soft state.

Uses

Cutting tools, press tools, shear blades. The low carbon steels are used for structural and engineering purposes where an increase in hardenability and fatigue strength over plain carbon steels is necessary.

Symbol	Nominal analysis, supplier, condition and remarks.
1.2833	1.05% C 1.15% V steel: German Standard
1.2838	1.4% C 3.4% V steel: German Standard
1.5213	0.2% C 1.4% Mn 0.1% V steel: For seamless drums; German Standard
1.5223	0.41% C 1.7% Mn 0.1% V steel: Forging; German Standard
02	0.9% C 1.7% Mn 0.15% V steel: For cold work; Osborn
06S	0.95% C 0.25% V steel: Carrs; centres, cold heading dies, etc. **DPN: 750**
12W1	1.1% C 0.2% V steel: Pompey
15VDT	0.8% C 0.1% V steel: Sandvik
17VDT	1.0% C 0.1% V steel: Sandvik
20 Mn V6	0.2% C 1.5% Mn 0.12% V steel: Designation used by German Standard
42 Mn V7	0.41% C 1.7% Mn 0.1% V steel: Designation used by German Standard
90 MV8	0.9% C 2.0% Mn 0.2% V steel: French Standard designation
145 V33	1.45% C 3.3% V steel: German Standard designation
1162	1.05% C 0.15% V steel: French Standard designation

Note. The following abbreviations and units are used in the tables:

DPN	Hardness, diamond pyramid number
UTS	Ultimate tensile strength, N/mm²
Elon	Elongation, %
Proof	0.1 % proof strength, N/mm²

$1 N/mm^2 = 0.1 hbar = 0.102 kgf/mm^2 = 0.06475 tonf/in^2 = 145.04 lbf/in^2$

Symbol	Nominal analysis, supplier, condition and remarks.
1163	0.9% C 0.15% V steel: French Standard designation
2211	0.9% C 2.0% Mn 0.2% V steel: French Standard designation
9257	1.0% C 0.25% V tool steel: Origin unknown
A 441	0.22% C 1.2% Mn 0.2% V 0.2% Cu steel: Armco **UTS: 480 Elon: 22% Proof: 320**
A 540	0.8% C 0.1% V steel: Fagersta **DPN: 630**
A 575	0.73% C 0.18% V steel: SHD; drills
ARMCO HS No. 3	0.1% C 0.6% Mn 0.02% V 0.2% Cu steel: Armco
ARMCO HS No. 5	0.22% C (max) 1.25% Mn 0.02% V 0.2% V 0.2% Cu steel: Armco
ARMCO HS No. 6	0.17% C 0.7% Mn 0.02% Nb steel: Armco
ASTM A225 A	0.18% C 1.45% Mn 0.12% V steel: Normalized **UTS: 510 Elon: 22%**
ASTM A225 B	0.2% C 1.45% Mn 0.12% V steel: Normalized **UTS: 550 Elon: 21%**
ASTM A292 1	0.45% C 0.1% V steel: For generator motors; hardened & tempered; obsolete; see A 469 **UTS: 530 Elon: 24% Proof: 270**
ASTM A293/1	0.45% C 0.1% V steel: For turbine rotors; hardened & tempered **UTS: 530 Elon: 24% Proof: 270**
ASTM A356/1	0.35% C 0.03% V steel: Casting; normalized & tempered **UTS: 460 Elon: 20% Proof: 240**
ASTM A469/1	0.45% C 0.2% V steel: For rotors; vacuum treated
ASTM A487 IN	0.3% C (max) 0.1% V steel: Casting for pressure; normalized; IQ grade is quenched and tempered
ASTM A508/1	0.35% C (max) 0.6% Mn 0.05% V (max) steel: For pressure vessels; vacuum treated

Symbol	Nominal analysis, supplier, condition and remarks.
ASTM A618/11	0.26% C (max) 1.3% Mn (max) 0.18% Cu 0.01% V (min) steel: Tube; heat formed welded or seamless **UTS: 483 Elon: 22% Proof: 345**
ASTM A618/11	0.26% Cr (max) 1.3% Mn (max) 0.01% V (min) steel: Tube; seamless & welded; for structures 483 19% 345
ASTM A618/111	0.27% C (max) 1.4% Mn (max) 0.01% V (min) steel: Seamless & welded tube for structures **UTS: 448 Elon: 19% Proof: 345**
ASTM A618/111	0.27% C 1.4% Mn 0.01% V (min) steel: Tube; hot formed, welded & seamless
ASTM A633 E	0.22% C (max) 1.3% Mn 0.08% V 0.02% N steel: For structures; normalized **UTS: 590 Elon: 20% Proof: 415**
ASTM A656/1	0.18% C (max) 1.6% Mn (max) 0.1% V 0.02% Al (min) 0.01% N steel: For structures **UTS: 720 Elon: 12% Proof: 550**
ASTM L3/1	0.25% C (max) 1.3% Mn 0.1% V 0.18% Cu (min) steel: Flanges for low temperature use; impact 41 J at -46 °C **UTS: 290 Elon: 22%**
ASTM L3/2	0.25% C (max) 1.3% Mn 0.1% V 0.18% Cu (min) steel: Flanges for low temperature use; impact 54 J at -46 °C **UTS: 360 Elon: 25%**
ASTM A707 L3/3	0.25% C (max) 1.3% Mn 0.1% V 0.18% Cu (min) steel: Flanges for low temperature use; impact 68 J at -46 °C **UTS: 415 Elon: 23%**
ASTM A707 L3/4	0.25% C (max) 1.3% Mn 0.1% V 0.18% Cu (min) steel: Flanges for low temperature use; impact 68 J at -46 °C **UTS: 515 Elon: 20%**
ASTM A714/2	0.26% C (max) 1.3% Mn 0.18% Cu (min) 0.01% V steel: Pipe; welded or seamless **UTS: 463 Elon: 22% Proof: 345**
ASTM A714/3	0.27% C (max) 1.4% Mn (max) 0.18% Cu (min) 0.01% V steel: Pipe; welded or seamless **UTS: 448 Elon: 20% Proof: 345**
ASTM A732 SN	0.3% C (max) 0.8% Mn 0.1% V steel: Investment casting; normalized & tempered **UTS: 590 Elon: 22% Proof: 380**
ASTM A737 A	0.22% C (max) 1.2% Mn 0.1% V (max) steel: Plate for pressure vessels
ATLAS SPECIAL 10	1.05% C 0.2% V steel: Atlas; as AISI type W2
AW 441	0.22% C (max) 1.25% Mn 0.02% V 0.2% Cu steel: Alan Wood
AWV 50	0.22% C (max) 1.25% Mn 0.02% V 0.015% B steel: Alan Wood
AWV 55	0.22% C (max) 1.15% Mn 0.02% V 0.015% B steel: Alan Wood
AWV 60	0.22% C (max) 1.15% Mn 0.02% V 0.015% B steel: Alan Wood
AWV 65	0.22% C (max) 1.6% Mn 0.02% V 0.015% B steel: Alan Wood
B 188	0.8% C 2.1% Mn 0.1% V steel: Fagersta **DPN: 790**

Note. The following abbreviations and units are used in the tables:

DPN	Hardness, diamond pyramid number
UTS	Ultimate tensile strength, N/mm^2
Elon	Elongation, %
Proof	0.1 % proof strength, N/mm^2

1 N/mm^2=0.1 hbar=0.102 kgf/mm^2=0.06475 tonf/in^2=145.04 lbf/in^2

Symbol	Nominal analysis, supplier, condition and remarks.
BC 3V	1.4% C 3.5% V steel: Origin unknown
BC 8V	0.82% C 0.18% V steel: ESC; rock drills, punches, etc. **DPN: 730**
BC 10V	1.0% C 0.2% V steel: ESC; twist drills, dental burrs **DPN: 730**
BH 39	0.17% C (max) 1.3% Mn 0.17% V steel: Rheinstahl
BH 36S	0.16% C (max) 1.3% Mn 0.025% P & S (max) 0.1% V steel: Rheinstahl
BH 395	0.16% C (max) 1.3% Mn 0.17% V steel: Rheinstahl
BLUE & WHITE LABEL	0.97% C 0.5% V steel: For tools; Huntsman **DPN: 750**
CB/V45	0.2% C 1.2% Mn 0.01% Nb 0.01% V steel: Algoma
CB/V50	0.2% C 1.2% Mn 0.01% V 0.01% Nb steel: Algoma
CB/V55	0.2% C 1.2% Mn 0.01% V 0.01% Nb steel: Algoma
CN/V60 022/C	0.2% C 1.25% Mn 0.01% V 0.01% Nb steel: Algoma
CLAY-LOY	0.22% C 1.25% Mn 0.02% V 0.2% Nb 0.02% Nb steel: Phoenix Steel Corp
COLDIE	0.87% C 0.25% Mn 0.2% V steel: Braeburn
CRD 90C	0.9% C 0.25% V steel: C Denton; shear blades; obsolete **DPN: 690**
CRD 100C	1.0% C 0.25% V steel: C Denton; shear blades; obsolete **DPN: 690**
CRD 120C	1.15% C 0.25% V steel: C Denton; shear blades **DPN: 690**
DIAMOND 1	1.3% C 0.25% V steel: Spencer
DIAMOND 2	1.1% C 0.25% V steel: Spencer
DIAMOND 3	1.0% C 0.25% V steel: Spencer
DIAMOND 4	0.85% C 0.25% V steel: Spencer
DIAMOND 5	0.75% C 0.25% V steel: Spencer
DIAMOND 6	0.65% C 0.25% V steel: Spencer
DIN 17200 42 Mn V 7	0.41% C 1.8% Mn 0.1% V steel: Forging
DOFASCOLOY MY	0.22% C 1.2% Mn 0.02% V 0.2% Cu steel: Dominion
DOUBLE EXTRA	1.0% C 0.25% V steel: Origin unknown
DOUBLE CONQUEROR VANADIUM	1.0% C 0.25% V tool steel: Origin unknown
DP 5	1.0% C 0.25% V tool steel: Osborn; cutting tools, cold forming dies; obsolete **DPN: 750**
EX-TEN 45	0.22% C 1.25% Mn 0.01% Nb or 0.01% V steel: US Steel Corp.
EX-TEN 50	0.22% C 1.35% Mn 0.01% V or 0.01% Nb steel: US Steel Corp.
EX-TEN 55	0.24% C 1.35% Mn 0.01% V or 0.01% Nb steel: US Steel Corp.
EX-TEN 60	0.25% C 1.35% Mn 0.02% V or 0.01% Nb steel: US Steel Corp.
EX-TEN 65	0.26% C 1.35% Mn 0.02% V or 0.01% Nb steel: US Steel Corp.
EX-TEN 70	0.26% C 1.35% Mn 0.02% V or 0.01% Nb steel: US Steel Corp.
EXTRA TOUGH HARD V	0.9% C 0.15% V steel: Schoeller-Bleckmann **DPN: 730**
F	1.0% C 0.3% V steel: Russian Standard designation
FB 50AK	0.2% C (max) 1.3% Mn 0.03% P & S (max) 0.1% V 0.015% N steel: German proprietors specification; Al killed

Symbol	Nominal analysis, supplier, condition and remarks.
FMP 035	1.05% C 0.35% V steel: F Parkin; as tool steel W2
FTW 60	0.15% C (max) 1.23% Mn 0.01% P 0.005% S 0.08% V steel: Fuji
GLS 441	0.22% C 1.25% Mn 0.02% N 0.2% Cu steel: National Steel Co.
GOST 5058/15 GF	0.15% C 1.0% Mn 0.3% Cr Ni Cu (max) 0.075% V steel: Plate; Russian Standard · **UTS: 520**　　**Proof: 360**
GS 10 Mn 7	0.08% C 1.7% Mn 0.06% V 0.04% Nb cast steel: Thyssen; precipitation hardened
HECLA 28	1.0% C 0.25% V steel: Origin unknown
HI STRENGTH B	0.22% C 1.2% Mn 0.2% V 0.2% Cu steel: Armco **UTS: 480**　**Elon: 22%**　**Proof: 320**
HI YIELD 42	0.21% C 0.9% Mn 0.02% V steel: Granitility Steel Co.
IH 65	0.22% C 1.65% Mn 0.01% V or 0.01% Nb 0.2% Cu steel: I H Wisconsin Steel Co.
IHX 45	0.2% C 1.0% Mn 0.01% V or 0.01% Nb steel: I H Wisconsin Steel
IHX 50	0.22% C 1.1% Mn 0.01% V or 0.01% Nb steel: I H Wisconsin Steel
IHX 55	0.24% C 1.4% Mn 0.01% V or 0.01% Nb steel: I H Wisconsin
IHX 60	0.26% C 1.55% Mn 0.01% V or 0.01% Nb steel: I H Wisconsin
IHX 65	0.26% C 1.6% Mn 0.01% V or 0.01% Nb steel: I H Wisconsin
IHX 70	0.26% C 1.65% Mn 0.01% V or 0.01% Nb steel: I H Wisconsin
INX 42	0.2% C 0.9% Mn 0.01% V or 0.01% Nb steel: Inland Steel Co.
INX 45	0.2% C 1.0% Mn 0.01% V or 0.01% Nb steel: Inland Steel Co.
INX 50	0.22% C 1.25% Mn 0.01% V or 0.01% Nb steel: Inland Steel Co.
INX 55	0.24% C 1.4% Mn 0.01% V or 0.01% Nb steel: Inland Steel Co.
INX 60	0.26% C 1.65% Mn 0.01% V or 0.01% Nb steel: Inland Steel Co.
INX 65	0.26% C 1.65% Mn 0.01% V or 0.01% Nb steel: Inland Steel Co.
INX 70	0.26% C 1.65% Mn 0.01% V or 0.01% Nb steel: Inland Steel Co.
J 4V	0.9% C 0.25% V steel: Jonas; cold swaging dies **DPN: 570**
JLX 42	0.2% C (max) 1.0% Mn 0.01% V or 0.01% Nb steel: Jones & Laughlan
JLX 65	0.26% C (max) 1.5% Mn 0.01% V or 0.01% Nb steel: Jones & Laughlan
JLX 70	0.26% C (max) 1.65% Mn 0.01% V or 0.01% Nb steel: Jones & Laughlin
KAISALOY 42CV	0.2% C (max) 0.9% Mn 0.01% V or 0.01% Nb steel: Kaiser Steel Corp.
KAISALOY 50MV	0.22% C (max) 1.25% Mn 0.02% V 0.2% Cu steel: Kaiser Steel Corp.
KE A205	1.4% C 3.4% V steel: Kayser Ellison; dies, cold heading punches **DPN: 910**
LT 1	1.0% C 0.25% V steel: Low Moor; obsolete **DPN: 850**
MEL-TROL MIRROMOLD	0.1% C 0.2% Mn 0.1% V steel: Carpenter for AISI type P1
MIL S 12505/5	0.23% C (max) 1.25% Mn 0.02% V 0.4% Cu steel: Bethlehem
MLX 45	0.15% C (max) 1.0% Mn 0.02% V or 0.005% Nb steel: McLouth
MLX 50	0.2% C (max) 1.0% Mn 0.02% V or 0.005% Nb steel: McLouth

Symbol	Nominal analysis, supplier, condition and remarks.
MLX 55	0.24% C (max) 1.2% Mn 0.02% V or 0.005% Nb steel: McLouth
MLX 60	0.26% C (max) 1.5% Mn 0.2% V or 0.005% Nb steel: McLouth
MNV	0.22% C (max) 1.25% Mn 0.02% V 0.2% Cu steel: Lukens
MNV A441	0.22% C (max) 1.25% Mn 0.02% V 0.2% Cu steel: Bethlehem
NORESCO EXTRA TOUGH HARD V	0.9% C 0.15% V steel: Schoeller-Bleckmann **DPN: 730**
ORELLOY 441	0.22% C 1.25% Mn 0.02% V 0.2% Cu steel: Oregon
P121	0.6% C 0.1% V steel: Bofors **DPN: 800**
P171	0.85% C 0.1% V steel: Bofors **DPN: 800**
P 181	0.9% C 0.1% V steel: Bofors; obsolete **DPN: 800**
P 211	1.05% C 0.1% V steel: Obsolete; Bofors **DPN: 850**
PINK LABEL	0.9% C 0.2% V steel: T Turton
PM 15	0.14% C 1.5% Mn 0.07% V steel: Pompey
PM 15 E	0.14% C 1.5% Mn 0.07% V steel: Pompey
RENOWN	1.0% C 0.2% V steel: Latrobe
REPUBLIC A441	0.22% C (max) 1.25% Mn 0.02% V 0.2% Cu steel: Republic Steel Co.
REPUBLIC X42W	0.22% C (max) 1.1% Mn 0.01% V or 0.01% Nb steel: Republic Steel Co.
REPUBLIC X45W	0.2% C 0.75% Mn 0.01% V or 0.01% Nb steel: Republic Steel Co.
REPUBLIC X50W	0.2% C 0.75% Mn 0.01% V or 0.01% Nb steel: Republic Steel Co.
REPUBLIC X60W	0.25% C 1.5% Mn 0.01% V or 0.01% Nb steel: Republic Steel Co.
REPUBLIC X65W	0.25% C 1.5% Mn 0.01% V or 0.01?Nb steel: Republic Steel Co.
REPUBLIC X65W	0.26% C 1.5% Mn 0.01% V or 0.01% Nb steel: Republic Steel Co.
REPUBLIC X70W	0.26% C 1.65% Mn 0.01% V or 0.01% Nb steel: Republic Steel Co.
SAE W209	0.9% C 0.2% V steel: For tools
SAE W210	1.0% C 0.2% V steel: For tools
SAE W310	1.0% C 0.4% V steel: For tools
SB 39F	0.16% C 1.3% Mn 0.035% P & S (max) 0.03% V 0.03% Ti steel: German proprietor's specification
SB 42F	0.16% C 1.3% Mn 0.035% P & S (max) 0.1% V steel: German proprietor's specification; Al killed
SHEFFIELD HIGH STRENGTH B	0.22% C (max) 1.20% Mn 0.02% V 0.2% Cu steel: Armco
SILVAN STAR	1.0% C with V steel: Firth Sterling
SIS 2900	0.8% C 0.1% V steel: Swedish Standard; annealed **DPN: 195**
SIX STAR VANADIUM	1.0% C 0.2% Si 0.2% Mn 0.15% V steel: For tools; Jessop
SKA 45	0.22% C (max) 1.25% Mn 0.02% V steel: Phoenix
SKA 50	0.26% C (max) 1.3% Mn 0.02% V steel: Phoenix
SKS 43	1.05% C 0.17% V steel: Japanese Standard designation
SKS 44	0.85% C 0.17% V steel: Japanese Standard designation
SPECIAL V	C as specified 0.25% Mn 0.2% V steel: For tools; Braeburn
SPECIAL CONQUEROR VANADIUM	1.0% C 0.25% V steel: For tools; origin unknown
STELCO VANADIUM	0.22% C 1.2% Mn 0.02% V 0.2% Cu 0.2% Nb steel: Steel Co. of Canada
STERLING V	1.0% C with V steel: Firth Sterling

Symbol	Nominal analysis, supplier, condition and remarks.
TCR	0.8% C 0.3% V steel: Swift Levick; chisels, punches
TOWER 60	0.33% C 1.65% Mn 0.05% V steel: Algoma
TRI-STEEL	0.2% C 1.25% Mn 0.02% Cu steel: Inland Steel
TRI-TEN	0.22% C (max) 1.25% Mn 0.02% V 0.02% Cu steel: US Steel Co.
UHB 19 Va	0.9% C 0.07% V steel: Uddeholm for type W2
V 42	0.22% C (max) 1.25% Mn 0.02% V steel: Bethlehem Steel
V 45	0.22% C (max) 1.25% Mn 0.02% V steel: Bethlehem
V 50	0.22% C (max) 1.25% Mn 0.02% V 0.015% N steel: Bethlehem
V 55	0.22% C (max) 1.25% Mn 0.02% V 0.015% N steel: Bethlehem
V 60	0.22% C (max) 1.25% Mn 0.02% V 0.015% N steel: Bethlehem Steel
V 65	0.22% C (max) 1.25% Mn 0.02% V steel: Bethlehem Steel
V 90V	0.9% C 0.15% V steel: French Standard designation
VANQUISH	0.95% C 0.25% V steel: Hall & Pickles; obsolete DPN: 670
VS 4	1.05% C 0.25% V steel: Edgar Allen; dies, hammers. etc.

Symbol	Nominal analysis, supplier, condition and remarks.
W 2	1.0% C 0.25% V steel: Carbon varies over wide limits; designation used by AISI
W 2	1.0% C 0.2% V steel: Osborn
VANADIUM EXTRA	1.06% C 0.2% V steel: Columbia; AISI type W2
VANADIUM STANDARD	1.06% C 0.2% V steel: Columbia; AISI type W2
Y 1 105V	1.05% C 0.15% V steel: French Standard
Y 105V	1.05% C 0.15% V steel: French Standard designation
YEO 36	0.18% C 1.3% Mn 0.1% V steel: Yamata
YOLOY A 242	0.22% C (max) 1.25% Mn 0.02% V 0.2% Cu steel: Youngstown Steel

Note. The following abbreviations and units are used in the tables:

DPN	Hardness, diamond pyramid number
UTS	Ultimate tensile strength, N/mm^2
Elon	Elongation, %
Proof	0.1 % proof strength, N/mm^2

1 N/mm^2=0.1 hbar=0.102 kgf/mm^2=0.06475 tonf/in^2=145.04 lbf/in^2

44K1 Steel – low carbon alloys
0.05–0.35% carbon with two or more of chromium, nickel, molybdenum, tungsten or vanadium

Specific gravity	7.86	
Density	7860 kg/m^3	(0.283 lb/in.^{3})
Solidus/liquidus	– °C	
Thermal conductivity	37.7 W/m °C	(9 cal/m s °C)
Coefficient of linear expansion	11 × 10^{-6}/ °C	
Electrical conductivity	7–9% IACS (copper 100%)	
Specific resistance	230–270 microhm mm	
Young's modulus of elasticity	206 × 10^9 N/m^2	(30 × 10^6 lbf/in.^{2})
Impact	54–108 J	(40–80 ft/lb)(Izod)
Fatigue strength	–	
Hot strength		

Temperature °C	Tensile strength N/mm^2	Elongation %
100	800	24
200	800	23
300	800	23
400	750	25
500	640	24
600	500	26
650	370	29

The properties above are typical of the following group and may not apply exactly to any one specification. It is possible that with certain specifications some of the values may not be applicable.

General metallurgical characteristics

These steels are tougher, more hardenable, have better fatigue strength and, to some extent, corrosion resistance than the plain carbon or single alloy steels.

At least two of the alloying elements are present in all steels listed, while some steels have all the elements present, the majority having some chromium. Some of the steels are designed specially for carburizing and all of the steels can be carburized. Tungsten and vanadium should not be present however, as there is no technical advantage and with the very hard carbide formed there is a danger of the case being too hard and brittle.

One characteristic of alloy steel is that it is inherently fine grained, particularly with vanadium present. This improves the fatigue strength and allows a greater laxity of temperatures during thermal treatment.

Very briefly the effect of each alloying element is:

Chromium. This is a carbide former which aids carburization and slows down the metallurgical reactions, thus increasing hardenability. Chromium improves the properties of steel at high temperatures, including oxidation resistance. It has a tendency to promote grain growth, and as it raises the critical temperatures, there is a tendency for chromium to produce large grained brittle steels.

Nickel. This strengthens and toughens the ferritic matrix of steel, improving the impact strength, particularly at low temperatures. Some 10% nickel steels have been specially developed for very low temperature application.

Nickel added to carburized steels gives a tough ductile core supporting a hard but not brittle case.

Nickel lowers the temperatures required for heat treatments and widens the range of suitable thermal treatment temperatures, thus giving more flexibility.

Molybdenum. This is a carbide former, the resultant carbides being well distributed, small and very stable. Thus it gives fine grained steels, improves the hot strength and creep strength, increases the depth of hardness achieved and improves the fatigue strength.

In addition molybdenum added to nickel and nickel chromium steels eliminates temper brittleness. This is a brittle condition obtained when some steels are tempered in the range 250–450 °C and is caused by carbide precipitation. The molybdenum carbides inhibit this tendency by stabilizing the carbides.

Tungsten. This forms very hard stable carbides and acts in many ways as does molybdenum. More tungsten is required to give the same effect as very little molybdenum. The use is generally confined to tool steels where large quantities of the very hard carbides are required.

Vanadium. This is a powerful de-oxidizer in addition to being a carbide former almost as powerful as molybdenum. The carbides are always very small and evenly distributed throughout the mass, making vanadium the best grain refining element. This prevents agglomeration of these and other carbides during long term thermal treatments. Vanadium also forms nitrides and is present in most nitriding steels. It is commonly used with chromium where it considerably improves the fatigue strength because of its grain refining property, thus overcoming the tendency of chromium to give large grains.

Copper, boron and aluminium. These are sometimes also present in these steels. The copper enhances the corrosion resistance and there are claims that its presence improves the adhesion of paint films. Boron improves the hardenability to a much greater extent than any other element and also increases this function in other alloying ele-

ments. It reduces the range of temperatures for successful heat treatments. Aluminium is a powerful de-oxidizer, but the refractory nature of the oxide formed reduces the usefulness for this purpose. Small amounts of aluminium added to the de-oxidizer steel increases the ability of these steels to nitride, and has some grain refining properties.

All steels listed will almost always be used in the hardened and tempered, or surface hardened condition.

Thermal treatment

These steels cannot accept any great amount of cold work without becoming too brittle for practical purposes. There are however, some wire specifications which are used in the cold drawn and tempered condition.

Normally after any amount of cold work an anneal would be required. This will be a temperature of 880–920 °C for 30 minutes per 25 mm of section, followed by very slow cooling.

For most purposes a sub-critical anneal at 650–700 °C for 2 hours will give a material soft enough for machining.

Where the carbide forming elements are present in any considerable quantity, it may be necessary to cyclic or spheroidize anneal. This consists of heating to 880–900 °C for 30 minutes, then cooling to 650 °C for one hour and repeating the cycle at least once. This results in small globules or spheres of carbide in the soft ferritic matrix and is the best structure for machining. Normalizing also requires 880–900 °C for about 20 minutes per 25 mm of section, followed by still air cooling. Most of the steels listed will air harden with this treatment, requiring a sub-critical anneal for two hours at 650 °C before machining. Some of the higher alloyed, higher carbon steels will not soften sufficiently and should always be cyclic annealed.

Hardening requires 840–900 °C for about 20 minutes per 25 mm of section, followed by oil quenching and parts should then be tempered as soon as possible for 2–4 hours at 200–600 °C, depending on the hardness required. Care is required with the chromium/nickel steels which can become brittle if slowly cooled through the 250–400 °C range after tempering. This defect, known as 'temper brittleness', is absent in the presence of molybdenum, tungsten or vanadium.

Most of these steels can be carburized by either the pack, gas or cyanide salt bath method the nickel/chromium/molybdenum steels giving the best results. Pack carburizing is the slowest, but requires only standard furnace equipment, although it is a dirty, lengthy process. Gas carburizing gives a superior case, requires special furnaces, is much more rapid and relatively clean. The cyanide salt bath method gives a slightly inferior case, is very rapid, but has certain dangers in operation and usually presents an effluent problem.

The steels should be carburized at 900–930 °C for a time depending on the case depth. Finished depths of less than 0.01 mm should not be specified and depths greater than about one-third of the core thickness are not advisable, as this obviates the main purposes of case hardening - a hard surface supported by a ductile core. The use of a very thin hard skin is sometimes necessary to prevent frettage but will not support a high load or allow flexing.

After carburizing the parts should be cooled, re-heated

to 840–900 °C and oil quenched then reheated to 730–780 °C, oil quenched then tempered at 150–200 °C. For low duty parts it may not be necessary to refine (or harden) the core. This is the 840–900 °C process carried out after carburizing, before hardening and is the identical treatment to hardening the steel in the uncarburized condition. The subsequent hardening of the case fully tempers the core. The refine treatment is not necessary with these steels which are all inherently fine grained when used for low duty components.

The practice of quenching direct from the carburizing temperature or allowing the parts to cool in the air to the estimated hardening temperature is not advised. If desired, the parts can be transferred to a furnace at the hardening temperature for sufficient time for them to be at the correct temperature prior to quenching.

Areas which are not to be case hardened can be stopped off, using copper plating or special lacquers, which are generally copper loaded. If machining is necessary to remove undesired case, the parts must be sub-critical annealed after carburizing and hardened after machining.

Many of these steels can be nitride hardened. This requires special furnace equipment and is a lengthy expensive process, but results in a very hard case. The treatment is at 500 °C which is below the tempering temperature of nitriding steels and thus can be carried out with no risk of distortion after final hardening and tempering, and after finish machining. A stabilize treatment at the tempering temperature is recommended if there is any heavy machining after hardening and final dimensions are held to fine limits. A growth of 0.01 mm per surface takes place during a 60 hour nitriding cycle, using good quality nitriding steels. These should always have some aluminium and vanadium present and will generally have chromium. Without aluminium the rate of penetration is slow, while the vanadium prevents the carbides agglomerating during the long time at the nitriding temperature.

Although the alloying elements give greater laxity at heat treatment than the plain carbon or single alloy steels, it is essential that the correct treatment is given to obtain the best mechanical properties from these complex expensive materials. The information given is a rough guide and it is essential that contact is made with the origin of the specifications used for exact times and temperatures.

There are also more exotic thermal treatments such as 'austempering' and 'martempering' possible with the materials listed, which require detailed instructions for each specific steel but can result in slightly superior mechanical properties with much less distortion.

No special furnace equipment is necessary for these steels, but they are all rather expensive and therefore merit correctly designed furnaces with proper temperature control. As with all steels, slightly oxidizing conditions can lead to serious inter-granular decarburization and it is therefore advisable to ensure that fully oxidizing, free scaling conditions exist, unless a completely neutral or reducing atmosphere can be guaranteed. This may be desirable for hardening carburized components which cannot be subsequently ground, but otherwise is not normally necessary on these steels.

Scale removal is generally by grit blasting, although sodium hydride is satisfactory if the correct equipment is available. Only steels below a hardness of about 350 DPN can be acid pickled to remove scale. The harder steels will all be embrittled by this treatment.

Welding and brazing

Pre-treatment. The steels should be in the annealed or normalized and sub-critical annealed condition for welding. The weld area must be free of grease, dirt and scale, and will generally require pre-heating.

Welding and brazing. Most of the steels can be welded, provided some precaution is taken against too severe quenching in the weld area. Many of the lower alloy steels have been specially developed for their weldability and controlled hardenability. Gas welding probably gives the best results as it is possible to pre-heat and post-heat during the welding operation. Electric welding tends to have a much more local heat which can be dangerous, but there are now special weld processes which use a heavy flux blanket to slow down the rate of cooling. Fluxes are essential for gas and electric welding, but not for inert gas shielded arc welding.

Resistance welding requires careful design to prevent chilling, causing quench cracks.

In general it can be said that welding difficulties will increase with the carbon and alloying content, but all these steels are weldable with proper attention to technique and post-weld treatment. Brazing will generally cause local softening and unless special high temperature filler metals are used the parts cannot be re-hardened.

Post-treatment. All traces of fluxes must be removed, as they cause severe corrosion.

Ideally the parts should be hardened and tempered after welding, care being taken at design and planning to remove stress raisers from which cracking could propagate. Some of the weldable steels are designed with controlled hardenability requiring tempering after welding.

As a minimum, welds must be stress relieved or tempered locally to reduce the high hardness which will be present in the weld zone. This local tempering can be with gas heating or using portable electric blankets.

Flaw detection

Crack test. Magnetic crack test is possible on all of these steels. Penetrant oil chalk test fluids are now available which give comparable results and are more flexible than the magnetic method.

X-ray. Not required on the wrought products, but will be required on mass production welds and high duty welds. Special portable equipment using radioactive metal isotopes has been developed for this purpose.

Ultra-sonic test. This is commonly applied to large forgings which are expensive to machine and shows subcutaneous defects which would be dangerous on the finished part. It thus can save expensive machining operations being carried out on scrap components. Some forms of weld defect can be identified with ultra-sonic testing.

Chemical etch. Case hardened parts after hardening are usually ground. This can give rise to softening and even cracking unless carefully controlled. An electrolytic sulphuric acid etch can identify this abuse and help to control the grinding process.

Corrosion protection

Temporary. These are all expensive materials which have very little built-in corrosion resistance and should never be stored out of doors. The steels should have all machined surfaces cleaned and greased during temporary storage.

During machining, finished areas should be oiled at all times. Soluble oil coolants provide adequate corrosion protection for short periods, provided they are in good condition. Dirty coolant can cause staining and incipient corrosion which can rapidly propagate.

Finished components should be cleaned, dried and oiled, then stored in plastic containers.

Permanent. Corrosion is caused by the action of oxygen and moisture on iron, forming a bulky, loosely adherent mixture of oxides. These are readily removed to allow further attack on the base metal, thus leading to local pitting. The steels listed are generally used under conditions of stress and this pitting can seriously reduce the fatigue strength of finished parts, particularly on case hardened items.

To prevent corrosion it is essential to have an air tight adherent film, strong enough to withstand normal usage. It must always be appreciated that corrosion contains within itself moisture and air, and can thus propagate with an increase in volume which ruptures the protecting film. It is therefore essential that all trace of existing corrosion is removed prior to applying any protecting system. Experience shows that it is much cheaper to apply the correct treatment first time, rather than add the cost of removing a faulty treatment to that of re-application. Grit blasting is probably the best method of ensuring adequate cleanliness and freedom from rust. The surface must be brushed free from dust and the protection applied as soon afterwards as possible. The blast process must be carefully controlled.

Phosphate or chromate treatments coat the steel surface with a film which has corrosion resisting properties and acts as an excellent key for paint adhesion. Alkaline rust removers give good results if correctly controlled and acid rust removers can be used on parts which are below 350 DPN. Steels of a higher hardness can be seriously embrittled by acid treatments.

Painting. Zinc silicate or chromate primers should be applied to the phosphate or chromate treated surface which should ideally have been first blasted, then oven dried after the chemical treatment. As soon after priming as possible the top coats should be laid on and stoved for best results. Modern plastic paints are available which are non-porous and abrasion resisting, but these should still have the preparatory coatings listed above for really permanent protection.

Case hardened areas always have poorer paint adhesion than any soft areas, as there is less mechanical keying. If these parts are chemically treated before priming they should be stress relieved to remove the danger of hydrogen embrittlement. Oven drying at 150 °C will generally be sufficient.

Plating. Plating presents no real problems with any of the steels listed, provided the correct cleaning and etching sequences are carried out. There is however, a danger of hydrogen embrittlement affecting case hardened parts and to a lesser extent any through hardened parts, harder than 350 DPN (900 N/mm² tensile). This defect arises at plating and in extreme cases can cause failure during plating. This is by brittle fracture below the calculated tensile strength and can be eliminated by stress relieving at 200 °C before and after plating, but unfortunately there is no visual evidence that this treatment has been carried out.

Where decorative chromium plating is required, this should have undercoats of copper and nickel to a thickness of 0.075 mm, the top coat of chromium being very thin only prevents the nickel tarnishing and is seldom the cause of chromium plating failure.

Cadmium and zinc plating presents no difficulty, the coat being always less than 0.025 mm thick, but these must not be used in contact with foodstuffs.

Metal spraying using zinc or aluminium gives excellent corrosion protection provided the surface is first cleaned and roughened by blasting and the coating sealed chemically or by painting. Cathodic protection uses electroplating principles, making the part to be protected negative (cathodic) to a sacrificial anode or positive, which is replaceable. This system is applied to marine structures, ships, pipe lines, pylons, and steel structures, and should be additive to a paint system.

Many of the steels listed will be used for tools which can be adequately protected with the artificial oxide formed when they are treated in a hot strong sodium hydroxide solution with oxidizing agents added. There are numbers of these proprietary 'black oxide' treatments available, all of which give good results provided an oil film is maintained on the parts at all times. The oil retaining structure of the surface film considerably helps in keeping parts oiled.

These are all expensive steels which warrant more than reasonable care to ensure that careful design and manufacture are not thwarted by poor corrosion protection. The degree of protection obtained is directly proportional to the care given to the pre-treatment, the application of the protection system and the quality of the materials used.

Machinability

This will vary depending on the carbon content, alloying elements and heat-treatment condition.

With low carbon, low alloy steels in the soft condition, tearing takes place. An improvement will almost always be found by hardening and tempering to about 250–350 DPN when tool life may decrease somewhat, but surface finish will be improved.

With the higher carbon, high alloyed steels tool wear may be too high in any except the spheroidize annealed condition.

Case hardened parts cannot be machined after hardening except by grinding, spark erosion or electrochemical machining.

Where the finished part has a hardness in excess of 350 DPN it is recommended that all rough machining is carried out in the annealed condition and finish ground after hardening. High speed or tipped tools should be used at all times. The hardening process usually results in some decarburization and at least 0.1 mm must be ground off finished parts to remove this and regain the full surface hardness.

Uses

Hot working tools, dies, and casting equipment, punches, shears, drills, cutting tools for nonferrous parts. High duty structural steels.

Gears, shafts, studs, nuts, bolts in automobile and aircraft engineering.

Symbol	Nominal analysis, supplier, condition and remarks.
03 NCMB	0.03% C 0.5% Ni 0.5% Cr 0.16% Mo steel: Electrode; Metrode
0.5% FO	0.2% C (max) 0.7% Mn 0.7% Cr 0.5% Mo steel: Creusot-Loire
07 NCMB	0.07% C 0.5% Ni 0.5% Cr 0.16% Mo steel: Electrode; Metrode
0.5467	0.3% C 2.4% Cr 4.2% W 0.6% V steel: For springs; German Standard
1.5404	0.21% C 0.3% Cr 0.3% Ni 0.5% Mo 0.3% V steel: Forging; German Standard
1.5406	0.18% C 0.3% Cr 0.3% Ni 0.9% Mo 0.35% V steel: Forging; German Standard
1.5419	0.21% C 0.5% Mn 0.3% Cr 0.35% Mo steel: Forging; German Standard
1.5620	0.34% C 0.6% Cr 1.3% Ni steel: Forging; German Standard
1.5919	0.17% C 1.6% Cr 1.6% Ni steel: For carburizing; German Standard
1.5920	0.18% C 2.0% Cr 2.0% Ni steel: For carburizing; German Standard
1.5924	0.15% C 1.5% Cr 1.5% Ni steel: For carburizing; German Standard
1.5934	0.18% C 2.0% Cr 2.0% Ni steel: For carburizing; German Standard
1.6001	0.24% C 2.2% Cr 0.7% Ni steel: For pressure vessels; German Standard
1.6511	0.36% C 1.1% Cr 1.1% Ni 0.20% Mo steel: Forging; German Standard
1.6513	0.28% C 1.2% Cr 1.2% Ni 0.25% Mo steel: Forging; German Standard
1.6582	0.34% C 1.5% Cr 1.5% Ni 0.2% Mo steel: Forging; German Standard
1.6590	0.3% C 2.0% Cr 2.0% Ni 0.3% Mo steel: Forging; German Standard
1.6592	0.28% C 1.2% Cr 2.0% Ni 0.35% Mo 0.1% V steel: Forging; German Standard
1.6604	0.3% C 2.1% Cr 2.1% Ni 0.35% Mo steel: German Standard
1.6761	0.28% C 1.2% Cr 1.2% Ni 0.45% Mo steel: Forging; German Standard
1.7205	0.14% C 0.8% Cr 0.15% Mo steel: Bar; German Standard
1.7207	0.18% C 0.7% Cr 0.15% Mo steel: German Standard

Symbol	Nominal analysis, supplier, condition and remarks.
1.7214	0.25% C 1.1% Cr 0.2% Mo steel: Forging; German Standard
1.7218	0.25% C 1.1% Cr 0.2% Mo steel: Forging; German Standard
1.7220	0.34% C 1.1% Cr 0.6% Ni 0.2% Mo steel: Forging; German Standard
1.7242	0.17% C 1.1% Cr 0.4% Ni 0.25% Mo steel: Forging; German Standard
1.7251	0.23% C 0.8% Cr 0.25% Mo steel: For seamless drums; German Standard
1.7254	0.25% C 1.0% Cr 0.25% Mo steel: German Standard
1.7258	0.24% C 1.1% Cr 0.6% Ni 0.25% Mo steel: For high temperature fasteners; German Standard
1.7273	0.24% C 2.4% Cr 0.8% Ni 0.25% Mo steel: For high pressures; German Standard
1.7281	0.16% C 2.2% Cr 0.35% Mo steel: For high pressures; German Standard
1.7283	0.2% C 2.2% Cr 0.8% Ni 0.3% Mo steel: For high pressures; German Standard
1.7307	0.17% C 4.0% Cr 1.0% Mo 0.1% V steel: Welding rod; German Standard
1.7324	0.2% C 0.4% Cr 0.45% Mo steel: Welding rod; German Standard
1.7334	0.2% C 0.5% Cr 0.4% Mo steel: For carburizing; German Standard
1.7335	0.14% C 0.8% Cr 0.45% Mo steel: For boiler plate; German Standard
1.7337	0.16% C 1.1% Cr 0.4% Ni 0.45% Mo steel: Forging; German Standard
1.7345	0.1% C 1.2% Cr 0.5% Mo steel: Welding rod; German Standard
1.7346	0.12% C 1.1% Cr 0.5% Mo steel: Welding rod; German Standard
1.7350	0.22% C 1.1% Cr 0.6% Ni 0.45% Mo steel: Forging; German Standard
1.7356	0.1% C (max) 1.7% Cr 0.5% Mo steel: Welding rod; German Standard
1.7361	0.32% C 3.0% Cr 0.6% Ni 0.4% Mo steel: German Standard
1.7362	0.15% C 5% Cr 0.55% Mo steel: For high pressure; German Standard
1.7373	0.15% C (max) 6.0% Cr 0.7% Mo steel: Welding rod; German Standard
1.7380	0.15% C (max) 2.2% Cr 1.1% Mo steel: German Standard
1.7384	0.1% C (max) 3.0% Cr 1.0% Mo steel: Welding rod; German Standard
1.7394	0.1% C (max) 9.5% Cr 1.0% Mo steel: Welding rod; German Standard
1.7513	0.22% C 1.1% Cr 0.2% V steel: German Standard
1.7704	0.3% C 2.5% Cr 0.2% Mo 0.15% V steel: Forging; German Standard
1.7707	0.3% C 2.5% Cr 0.6% Ni 0.2% Mo 0.2% V steel: Forging; German Standard

Note. The following abbreviations and units are used in the tables:

DPN	Hardness, diamond pyramid number
UTS	Ultimate tensile strength, N/mm^2
Elon	Elongation, %
Proof	0.1 % proof strength, N/mm^2

1 N/mm^2=0.1 hbar=0.102 kgf/mm^2=0.06475 tonf/in^2=145.04 lbf/in^2

Symbol	Nominal analysis, supplier, condition and remarks.
1.7708	0.16% C 1.1% Cr 0.3% Mo 0.2% V steel: German Standard
1.7709	0.14% C 1.1% Cr 0.25% Mo 0.2% V steel: German Standard
1.7710	0.24% C 1.3% Cr 0.22% Mo 0.25% V steel: German Standard
1.7733	0.24% C 1.4% Cr 0.6% Ni 0.55% Mo 0.2% V steel: For high temperature fasteners; German Standard
1.7766	0.17% C 2.7% Cr 0.25% Mo 0.15% V steel: For high pressures; German Standard
1.7779	0.2% C 3.2% Cr 0.55% Mo 0.5% V steel: For high pressures; German Standard
1.8070	0.21% C 1.3% Cr 0.6% Ni 1.1% Mo 0.3% V steel: For high temperature fasteners; German Standard
1.8212	0.21% C 2.8% Cr 0.4% Mo 0.4% W 0.8% V steel: For pressure vessels; German Standard
1.8507	0.33% C 1.1% Cr 0.2% Mo 1.0% Al steel: For nitriding; German Standard
1.8514	0.3% C 2.5% Cr 0.2% Mo 0.15% V steel: For nitriding; German Standard
1.8519	0.3% C 2.5% Cr 0.2% Mo 0.15% V steel: For nitriding; German Standard
1.8544	0.32% C 1.1% Cr 0.2% Mo 1.1% Al steel: For nitriding; German Standard
1.8550	0.33% C 1.7% Cr 1.0% Ni 0.2% Mo 1.0% Al steel: For nitriding; German Standard
1 CrMoB	0.07% C 1.3% Cr 0.5% Mo steel: Electrode; Metrode for BS 2493
1.1% FO	0.17% C (max) 0.5% Mn 1.0% Cr 0.5% Mo steel: Creusot-Loire
1.2% FO	0.17% C (max) 0.5% Mn 1.25% Cr 0.5% Mo steel: Creusot-Loire
1.2567	0.3% C 2.4% Cr 0.6% V 4.3% W steel: German Standard designation
1.26365	0.32% C 3.0% Cr 2.8% Mo 0.5% Mo steel: German Standard designation
1.2764	0.19% C 1.3% Cr 4.1% Ni 0.2% Mo 0.4% W tool steel: German Standard designation
1.2766	0.32% C 1.2% Cr 4.1% Ni 0.2% Mo steel: German Standard
1.3344	1.2% C 4.1% Cr 5.0% Mo 6.3% W 3.0% V steel: For tools; German Standard designation
2 Cr Mo B	0.07% C 2.2% Cr 1.0% Mo steel: Electrode; Metrode
2 Cr Mo C B	0.1% C 2.2% Cr 1.0% Mo steel: Electrode; Metrode
2 N3C1	0.12% C 0.55% Mn 0.7% Cr 3% Ni steel: Sandvik for SIS 2514
2.2% FO	0.15% C (max) 0.45% Mn 2.25% Cr 1.0% Mo steel: Creusot-Loire
3% FO	0.15% C (max) 0.45% Mn 3.0% Cr 1.0% Mo steel: Creusot-Loire
3 GMoB	0.07% C 3.0% Cr 1.0% Mo steel: Electrode; Metrode
3 KH2V8F	0.35% C 2.35% Cr 0.35% V 8.2% W steel: Russian Standard designation
3 Ni MoLB	0.04% C 3.2% Ni 0.5% Mo steel: Electrode; Metrode; for low temperature use

---

Note. The following abbreviations and units are used in the tables:

DPN	Hardness, diamond pyramid number
UTS	Ultimate tensile strength, N/mm^2
Elon	Elongation, %
Proof	0.1 % proof strength, N/mm^2

1 N/mm^2=0.1 hbar=0.102 kgf/mm^2=0.06475 tonf/in^2=145.04 lbf/in^2

Symbol	Nominal analysis, supplier, condition and remarks.
4 KH2V5FM	0.35% C 2.5% Cr 0.8% Mo 0.8% V 5.0% W steel: Russian Standard designation
4 NHD	0.35% C 1.5% Cr 4% Ni 5.7% W 0.35% V steel: For tools; Huntsman; hot stamping dies **DPN: 530**
4 WHD	0.35% C 1.2% Cr 4% W 0.25% V steel: Huntsman; pressure die casting dies, etc **DPN: 500**
4.5% DO	0.15% C (max) 0.5% Mn (max) 0.5% Cr 4.0% Ni 0.4% Mo steel: Creusot-Loire
5 C 2 Mo	0.25% C 1.1% Cr 0.2% Mo steel: Sandvik for SIS 2225
5 Cr Mo 10	0.1% C 5% Cr 0.55% Mo steel: ESC; annealed for BS 1501/625
	UTS: 400 Elon: 20% Proof: 210
5 Cr Mo 22	0.22% C 0.45% Mn 5% Cr 0.55% Mo steel: ESC for BS 1506/625
	UTS: 670 Elon: 19% Proof: 580
5 Cr Mo B	0.07% C 5.0% Cr 0.5% Mo steel: Metrode
6 N 4 C 2	0.3% C 1.3% Cr 4.3% Ni steel: Sandvik
6 N 3 C2 Mo	0.3% C 1% Cr 3.3% Ni 0.25% Mo steel: Sandvik for SIS 2534
7 C 2 Mo	0.35% C 1.1% Cr 0.2% Mo steel: Sandvik
7 Cr Mo 7/5	0.1% C (max) 1.7% Cr 0.5% Mo steel: Designation used by German Standard
7 Cr Mo 12/10	0.1% C (max) 3.0% Cr 1.0% Mo steel: Designation used by German Standard
7 Cr Mo B	0.07% C 7.0% Cr 0.5% Mo steel: Electrode; Metrode
9 Cr Mo 4/5	0.1% C 1.2% Cr 0.5% Mo steel: Designation used by German Standard
9 Cr Mo B	0.07% C 9.0% Cr 1.0% Mo steel: Electrode; Metrode
9 Cr Mo L B	0.05% C 9.0% Cr 1.0% Mo steel: Electrode; Metrode
10 CNK 1	0.09% C 1.0% Cr 1.4% Ni 1.0% Mo steel: Pompey
10 CNK 3	0.09% C 0.75% Cr 2.7% Ni steel: Pompey
10 CNK 32	0.09% C 0.75% Cr 3.0% Ni steel: Pompey
10 Cr Mo 9/10	0.15% C (max) 2.2% Cr 1.0% Mo steel: Designation used by German Standard
10 NC12	0.1% C 0.8% Cr 3.0% Ni steel: French Standard designation
10 NCMB	0.1% C 0.5% Ni 0.5% Cr 0.16% Mo steel: Electrode; Metrode
11 Cr Mo 4/5	0.12% C 1.1% Cr 0.5% Mo steel: Designation used by German Standard
12 CNK 3	0.14% C 0.75% Cr 2.7% Ni steel: Pompey
12 Cr Mo 19/5	0.15% C (max) 0.5% Cr 0.55% Mo steel: Designation used by German Standard
12 Cr Mo V B	0.07% C 12.0% Cr 0.7% Mo 0.15% V steel: Electrode; Metrode
12 KH1MF	0.11% C 0.55% Mn 1.05% Cr 0.3% MO 0.22% V steel: Russian Standard designation
12 MKH	0.13% C 0.55% Mn 0.5% Cr 0.5% Mo steel: Russian Standard designation
12 NCD 6	0.12% C 0.9% Cr 1.4% Ni 0.2% V steel: Pompey
12 NCMB	0.12% C 0.5% Ni 0.5% Cr 0.16% Mo steel: Metrode
12 S	0.3% C 3% Cr 8.5% W 0.25% V steel: Carrs; hot work, punches, dies, etc. **DPN: 460**
13 Cr Mo 4/4	0.14% C 0.8% Cr 0.45% Mo steel: Designation used by German Standard
13 Cr Mo V 4/2	0.14% C 1% Cr 0.25% Mo 0.2% V steel: Designation used by German Standard
14 CNK 3	0.14% C 0.75% Cr 3.45% Ni steel: Pompey
14 CNK 4	0.14% C 0.65% Cr 4.25% Ni steel: Pompey
14 CNK 32	0.13% C 0.75% Cr 3.0% Ni steel: Pompey

Symbol	Nominal analysis, supplier, condition and remarks.
15 NCMB	0.15% C 1.0% Ni 0.5% Cr 0.16% Mo steel: Electrode; Metrode
14 NKD 2	0.14% C 1.0% Cr 1.4% Ni steel: Pompey
15 Cr Mo 3	0.14% C 0.8% Cr 0.15% Mo steel: Designation used by German Standard
15 Cr Mo 6	0.15% C 1.5% Cr 1.5% Si steel: For carburizing; designation used by German Standard
15 NKD	0.14% C 0.5% Cr 0.55% Ni 0.2% Mo steel: Pompey
16 CNK 1	0.16% C 1.0% Cr 1.4% Ni steel: Pompey
16 Cr Mo 4	0.16% C 1.1% Cr 0.25% Mo 0.4% Ni steel: Designation used by German Standard
16 Cr Mo 4/4	0.17% C 1.1% Cr 0.4% Mo 0.45% Ni steel: Designation used by German Standard
16 Cr Mo 9/3	0.16% C 2.2% Cr 0.35% Mo steel: Designation used by German Standard
16 Cr Mo V 4/2	0.16% C 1.1% Cr 0.3% Mo 0.2% V steel: Designation used by German Standard
16 NCK 3	0.16% C 0.8% Cr 2.75% Ni steel: Pompey
17/22 AS	0.30% C 1.2% Cr 0.5% Mo 0.25% V steel: Timken; hardened & tempered; for high temperature use **DPN: 415 UTS: 1550 Elon: 15% Proof: 1300**
17/22 AV	0.27% C 1.2% Cr 0.5% Mo 0.8% V steel: Timken; hardened & tempered; for high temperature use **DPN: 444 UTS: 1550 Elon: 14% Proof: 1380**
17 CNK	0.17% C 0.6% Cr 0.6% Ni steel: Pompey
17 Cr Mo V 10	0.18% C 2.7% Cr 0.25% Mo 0.15% V steel: Designation used by German Standard
17 Cr Mo V 16/10	0.17% C 4.0% Cr 1.0% Ni 0.12% V steel: Designation used by German Standard
18 Mo V 8/4	0.18% C 0.3% Cr 0.3% Ni 0.9% Mo 0.35% V steel: Designation used by German Standard
17 ND 2	0.17% C 2.0% Ni 0.2% Mo steel: Pompey
17 ND 3	0.17% C 3.5% Ni 0.2% Mo steel: Pompey
18 Cr Mo 3	0.18% C 0.7% Cr 0.15% Mo steel: Designation used by German Standard
18 Cr Ni 8	0.18% C 2.0% Cr 2.0% Ni carburizing steel: Designation used by German Standard
18 KH3MV	0.18% C 0.4% Mn 2.75% Cr 0.6% Mo 0.07% V 0.65% W steel: Russian Standard designation
20 Cr Mo 9	0.2% C 2.3% Cr 0.8% Ni 0.3% Mo steel: Designation used by German Standard
20 Cr Mo V 13/5	0.2% C 3.2% Cr 0.55% Mo 0.5% V steel: Designation used by German Standard
20 DN32.12	0.2% C 3.0% Ni 3.2% Mo steel: French Standard designation
20 KH3MVF	0.2% C 0.4% Mn 3.0% Cr 0.45% Mo 0.7% V 0.45% W steel: Russian Standard designation
20 ND 2K	0.19% C 0.3% Cr 1.7% Ni 0.25% Mo steel: Pompey
20 NK 1	0.2% C 1.0% Cr 1.4% Ni steel: Pompey
20 NK 3	0.2% C 0.75% Cr 2.75% Ni steel: Pompey
20 NXD	0.2% C 0.5% Cr 0.55% Ni 0.2% Mo steel: Pompey
21 Cr Mo 3	0.22% C 0.9% Cr 0.25% Mo steel: Designation used by German Standard

Symbol	Nominal analysis, supplier, condition and remarks.
21 Cr V Mo W 12	0.22% C 2.9% Cr 0.4% Mo 0.8% V 0.4% W steel: Designation used by German Standard
21 Mo V 5/3	0.21% C 0.3% Cr 0.3% Ni 0.5% Mo 0.3% V steel: Designation used by German Standard
21 Cr Mo V 5/11	0.21% C 1.3% Cr 0.6% Ni 1.1% Mo 0.3% V steel: Designation used by German Standard
22 Cr Mo 4/4	0.23% C 1.1% Cr 0.6% Ni 0.45% Mo steel: Designation used by German Standard
22 Cr V 4	0.22% C 1.1% Cr 0.2% V steel: Designation used by German Standard
22 Mo 4	0.22% C 0.3% Cr 0.3% Mo steel: Designation used by German Standard
22 NKD	0.22% C 0.5% Cr 0.55% Ni 0.2% Mo steel: Pompey
24 Cr Mo 5	0.24% C 1.1% Cr 0.6% Ni 0.25% Mo steel: Designation used by German Standard
24 Cr Mo 10	0.24% C 2.5% Cr 0.8% Ni 0.25% Mo steel: Designation used by German Standard
24 Cr Mo V 5/2	0.24% C 1.3% Cr 0.22% Mo 0.25% V steel: Designation used by German Standard
24 Cr Mo V 5/5	0.24% C 1.8% Cr 0.6% Ni 0.55% Mo 0.2% V steel: Designation used by German Standard
24 Cr Ni 9	0.24% C 2.2% Cr 0.7% Ni steel: Designation used by German Standard
24 Ni 4	0.24% C 0.3% Cr 1.2% Ni steel: Designation used by German Standard
24 Ni 8	0.24% C 0.3% Cr 1.2% Ni steel: Designation used by German Standard
24 Ni 8	0.24% C 0.3% Cr 2.1% Ni steel: Designation used by German Standard
24S	0.24% C 3% Cr 2.5% Ni 8.5% W steel: Carrs; hot extrusion & press dies, etc. **DPN: 500**
25 Cr Mo 4	0.25% C 1.1% Cr 0.2% Mo steel: Designation used by German Standard
25 Cr Mo 4	0.24% C 0.65% Mn 0.15% Cr 0.2% Mo steel: Italian Standard designation; as rolled **DPN: 210**
25 KH1MF	0.18% C 0.55% Mn 1.65% Cr 0.3% Mo 0.22% V steel: Russian Standard designation
25 KH2MF	0.25% C 0.55% Mn 2.3% Cr 1.05% Mo 0.32% V steel: Russian Standard designation
25 NCD11	0.26% C 0.5% Mn 0.75% Cr 0.27% Ni steel: Origin unknown
25 NKI	0.25% C 1.0% Cr 1.55% Ni steel: Pompey
25 S	0.25% C 1.5% Cr 4.25% Ni 6.7% W steel: Carrs; hot extrusion & press dies, etc. **DPN: 525**
27 Cr Mo Ni	0.27% C 1.1% Cr 0.3% Mo 0.6% Ni steel: Electrode; Metrode
27 Cr Mo Ni 1	0.27% C 1.1% Cr 0.3% Mo 1.0% Ni steel: Electrode; Metrode
27 Cr Mo Ni 2	0.27% C 1.1% Cr 0.3% Mo 2.0% Ni steel: Electrode; Metrode
28 F	0.1% C 0.3% Mn 5% Cr 1% Mo 0.3% V steel: For case hardening; Carrs
28 Ni Cr Mo 4	0.28% C 1.2% Cr 1.2% Ni 0.25% Mo steel: Designation used by German Standard
28 Ni Cr Mo 4/4	0.28% C 1.2% Cr 1.2% Ni 0.45% Mo steel: Designation used by German Standard
28 Ni Cr Mo 7/4	0.28% C 1.3% Cr 2.0% Ni 0.35% Mo 0.11% V steel: Designation used by German Standard
28 NKD 3	0.3% C 1.55% Cr 3.2% Ni 0.55% Mo steel: Pompey
30 Cr Mo V 9	0.3% C 2.5% Cr 0.6% Ni 0.2% Mo 0.15% V steel: Forging; designation used by German Standard
30 Cr Ni Mo 8	0.3% C 2.0% Cr 2.0% Ni steel: Designation used by German Standard
30 DCV28	0.3% C 2.8% Cr 2.8% Mo 0.5% V steel: French Standard designation

Note. The following abbreviations and units are used in the tables:

DPN	Hardness, diamond pyramid number
UTS	Ultimate tensile strength, N/mm^2
Elon	Elongation, %
Proof	0.1 % proof strength, N/mm^2

1 N/mm^2=0.1 hbar=0.102 kgf/mm^2=0.06475 tonf/in^2=145.04 lbf/in^2

Symbol	Nominal analysis, supplier, condition and remarks.
30 DKCV28	0.3% C 2.8% Cr 2.8% Mo 0.5% V 2.5% Co steel: French Standard designation
30 NCD2	0.3% C 0.8% Mn 0.4% Cr 0.5% Ni steel: Origin unknown
30 NCD16	0.3% C 0.45% Mn 1.8% Cr 4.0% Ni 0.4% Mo steel: For low temperature use; origin unknown UTS: 1200 Elon: 11% Proof: 900
30 NK 3	0.3% C 0.75% Cr 2.7% Ni steel: Pompey
30 NKD 2	0.3% C 2.0% Cr 2.0% Ni 0.4% Mo steel: Pompey
30 NKD 3	0.3% C 0.9% Cr 3.0% Ni 0.35% Mo steel: Pompey
30 NKDE	0.3% C 0.6% Cr 0.6% Ni 0.2% Mo steel: Pompey
30 W Cr V 17/9	0.3% C 2.3% Cr 4.2% W 0.6% V steel: Designation used by German Standard
31 Cr Mo V 9	0.3% C 2.5% Cr 0.2% Mo 0.15% V steel: Designation used by German Standard
32 Cr Mo 12	0.32% C 3.0% Cr 0.6% Ni 0.4% Mo steel: Designation used by German Standard
32 Cr Mo B	0.32% C 1.2% Cr 0.3% Mo steel: Electrode; Metrode
32 NDC18.12	0.32% C 0.5% Cr 4.5% Ni 1.2% Mo steel: French Standard designation
34 Cr Al Mo 5	0.33% C 1.2% Cr 0.2% Mo 1.0% Al steel: Designation used by German Standard
34 Cr Al Ni 7	0.33% C 1.7% Cr 1.0% Ni 0.2% Mo 1.0% Al steel: Designation used by German Standard
34 Cr Mo 4	0.34% C 1.1% Cr 0.6% Ni 0.2% Mo steel: Designation used by German Standard
34 Cr Ni Mo 6	0.34% C 1.5% Cr 1.5% Ni 0.2% Mo steel: Designation used by German Standard
34 Ni 5	0.34% C 0.6% (max) Cr 1.4% Ni steel: Designation used by German Standard
35 CP	0.35% C 1.0% Cr 0.25% Mo steel: Pompey
35 NK 1	0.35% C 1.0% Cr 1.45% Ni steel: Pompey
35 NKD 1	0.35% C 0.55% Cr 1.15% Ni 0.15% Mo steel: Pompey
35 NKD 2	0.35% C 1.0% Cr 1.4% Ni 0.15% Mo steel: Pompey
36 Cr Ni Mo 4	0.36% C 1.1% Cr 1.1% Ni 0.2% Mo steel: Designation used by German Standard
36 Cr Ni Mo 6	0.34% C 1.6% Cr 1.6% Ni 0.2% Mo steel: Designation used by German Standard
40 FK	0.16% C (max) 1.0% Mn 0.04% P & S (max) steel: HOAG; Al killed
43 BV 12	0.12% C 0.5% Cr 1.8% Ni 0.25% Mo 0.03% V steel: Designation used by SAE; obsolete
43 BV 14	0.14% C 0.5% Cr 1.8% Ni 0.1% Mo 0.03% V steel: Designation used by SAE; obsolete
46 B 12	0.12% C 0.6% Mn 0.3% Si 1.8% Ni 0.25% Mo boron steel: Designation used by SAE
58 S	0.35% C 1% Si 5% Cr 2% Mo 1.2% W 0.25% V steel: Carrs; hot extrusion dies, shears, etc. DPN: 550
74 S	0.3% C 2.3% Cr 0.3% Mo 4.25% W 0.6% V steel: Carrs; hot extrusion dies, hot shears, etc. DPN: 560
75.75	0.050% C 1.40% Mn 1.80% Ni 0.350% Cr 0.40% Mo steel: For electrode; Esab UTS: 800 Elon: 18% Proof: 730
80 HLES	0.15% C (max) 0.5% Mn (max) 0.5% Cr 4.0% Ni 0.4% Mo 0.1% V steel: Creusot-Loire
158	0.1% C 1.5% Cr 3.5% Ni steel: Carpenter
227	0.3% C 3.2% Cr 9% W 0.35% V steel: Balfour; hot dies for forging & casting DPN: 520
293	0.36% C 3.25% Cr 9% W 0.35% V steel: For hot work; Balfour; for tools; non-ferrous casting die punches DPN: 570

Symbol	Nominal analysis, supplier, condition and remarks.
505	0.32% C 0.3% Mn 2.25% Cr 0.5% Mo 9% W 0.5% V steel: F Parkin; hot forming; extrusion & forging dies DPN: 540
796	0.3% C 1.5% Cr 6.3% W 0.3% V steel: F Parkin; hot punches & dies DPN: 500
0908 Mo	0.07% C 9.0% Cr 1.0% Mo steel: NYBY annealed DPN: 190 UTS: 280 Elon: 20% Proof: 160
1418	0.3% C 1.25% Cr 4% Ni 0.3% Mo steel: For tools; Osborn; plastic dies DPN: 570 UTS: 155
1433	0.15% C 1.2% Cr 4.2% Ni steel: Case hardening; Osborn DPN: 340
2504	0.15% C (max) 0.5% Mn (max) 0.5% Cr 4.0% Ni 0.4% Mo 0.1% V steel: Creusot-Loire
2831	0.08% C 5.0% Cr 1.0% Mo 0.3% V steel: French Standard designation
2881	0.1% C 1.0% Cr 1.5% Ni steel: French Standard designation
2882	0.1% C 0.8% Cr 3.0% Ni steel: French Standard designation
3115	0.15% C 0.6% Cr 1.2% Ni steel: Designation used by AISI; obsolete
3120	0.2% C 0.6% Cr 1.2% Ni steel: Designation used by AISI
3130	0.3% C 0.6% Cr 1.2% Ni steel: Designation used by AISI
3215	0.15% C 1% Cr 1.7% Ni steel: Designation used by AISI; obsolete
3220	0.2% C 1% Cr 1.7% Ni steel: Designation used by AISI; obsolete
3230	0.30% C 1% Cr 1.7% Ni steel: Designation used by AISI; obsolete
3310	0.1% C 1.5% Cr 3% Ni steel: Designation used by AISI
3312	0.1% C 1.5% Cr 3.5% Ni steel: Designation used by AISI; obsolete
3316	0.16% C 1.5% Cr 3.5% Ni steel: Designation used by AISI
3325	0.25% C 1.5% Cr 3.5% Ni steel: Designation used by AISI; obsolete
3335	0.35% C 1.5% Cr 3.5% Ni steel: Designation used by AISI; obsolete
3382	0.35% C 1.8% Cr 4.0% Ni 0.4% Mo 0.1% V steel: French Standard designation
3383	0.32% C 0.5% Cr 4.5% Ni 1.2% Mo steel: French Standard designation
3415	0.15% C 0.8% Cr 3% Ni steel: Designation used by AISI; obsolete
3436	0.35% C 0.8% Cr 3% Ni steel: Designation used by AISI; obsolete
3451	0.3% C 2.8% Cr 2.8% Mo 0.5% V steel: French Standard designation
3455	0.2% C 3.0% Ni 3.2% Mo steel: French Standard designation
3543	0.3% C 3.0% Cr 0.4% V 9.0% W steel: French Standard designation
4119	0.19% C 0.5% Cr 0.25% Mo steel: Designation used by AISI
4125	0.25% C 0.5% Cr 0.25% Mo steel: Designation used by AISI; obsolete
4330	0.3% C 0.82% Cr 1.8% Ni 0.41% Mo 0.07% V steel
4335	0.35% C 0.72% Cr 1.8% Ni 0.35% Mo 0.2% V steel
4617	0.17% C 1.8% Ni 0.25% Mo steel: Designation used by AISI

Symbol	Nominal analysis, supplier, condition and remarks.
4812	0.12% C 3.5% Ni 0.25% Mo steel: Designation used by AISI
6115	0.15% C 1% Cr 0.15% V steel: Designation used by AISI; obsolete
6125	0.25% C 1% Cr 0.15% V steel: Designation used by AISI; obsolete
6130	0.3% C 1% Cr 0.15% V steel: Designation used by AISI; obsolete
6135	0.35% C 1% Cr 0.15% V steel: Designation used by AISI; obsolete
A 1000	0.32% C 0.65% Cr 2.5% Ni 0.5% Mo steel: Edgar Allen; hardened & tempered **UTS: 2000 Elon: 20% Proof: 1380**
A 242	0.12% C 0.6% Cr 0.7% Ni 0.1% Mo 0.4% Cu steel: Armco **UTS: 500 Elon: 22% Proof: 330**
A 325	0.25% C 3.2% Cr 0.5% Mo steel: SHD; drills, etc.
A 330	0.28% C 2.1% Cr 0.3% Ni 0.3% Mo steel: SHD; drills, etc.
A 387 A	0.2% C 0.7% Cr 0.55% Mo steel: Plate; Armco; high temperature & slightly corrosive use **UTS: 520 Elon: 24% Proof: 270**
A 387 B	0.17% C 1.0% Cr 0.55% Mo steel: Plate; Armco; high temperature & slightly corrosive use **UTS: 500 Elon: 24% Proof: 240**
A 387 C	0.17% C 1.25% Cr 0.55% Mo steel: Plate; Armco; for high temperature & slightly corrosive use **UTS: 500 Elon: 24% Proof: 240**
A 387 D	0.15% C 2.2% Cr 1.0% Mo steel: Plate; Armco; for high temperature & slightly corrosive use **UTS: 520 Elon: 20% Proof: 200**
A 387 E	0.15% C 3.0% Cr 1.0% Mo steel: Plate; Armco; for high temperature & slightly corrosive use **UTS: 520 Elon: 28% Proof: 200**
A 420	0.2% C 0.5% Cr 1.75% Ni 0.25% Mo steel: SHD; drills
AB 213	0.3% C 1% Cr 0.3% Ni 1.75% W steel: Balfour; chisels, punches, etc. **DPN: 570**
ADIC	0.37% C 1.0% Si 5% Cr 1.3% Mo 1.0% V steel: For tools; hot work; Balfour; Al and Mg die casting moulds, etc. **DPN: 570**
ADS	0.35% C 5% Cr 1.5% Mo 0.5% V steel: Inman; for aluminium die casting
AEI/M	0.12% C 1.2% Cr 4.2% Ni 0.3% Mo steel: For case hardening; Swift Levick for BS alloy En 39B
AGT VAC-ARC	0.1% C 1.2% Cr 3.2% Ni 0.12% Mo steel: For carburizing; Latrobe Steel
AHT 28	0.3% C 1.4% Cr 4.0% Ni 0.2% Mo steel: Atlas; hardened & tempered **DPN: 375 UTS: 1380 Elon: 12% Proof: 1180**
AIS	0.2% C 1.5% Mn 0.7% Cr 0.3% Mo steel: Australian Iron & Steel Co
AISI 94 B17	0.17% C 0.4% Cr 0.45% Ni 0.1% Mo 0.0005% B steel
AISI 94 B30	0.3% C 0.4% Cr 0.45% Ni 0.1% Mo 0.0005% B (min) steel

Note. The following abbreviations and units are used in the tables:

DPN	Hardness, diamond pyramid number
UTS	Ultimate tensile strength, N/mm^2
Elon	Elongation, %
Proof	0.1 % proof strength, N/mm^2

1 N/mm^2=0.1 hbar=0.102 kgf/mm^2=0.06475 tonf/in^2=145.04 lbf/in^2

Symbol	Nominal analysis, supplier, condition and remarks.
AISI 501	0.1% C (min) 5% Cr 0.5% Mo steel
AISI 502	0.1% C 5% Cr 0.5% Mo steel
AISI 602	0.3% C 1.2% Cr 0.5% Mo 0.25% V steel: Hardened & tempered **DPN: 350 UTS: 1210 Elon: 18% Proof: 1000**
AISI 603	0.27% C 1.25% Cr 0.5% Mo 0.85% V steel: Hardened & tempered **DPN: 444 UTS: 1550 Elon: 14% Proof: 1300**
AISI 604	0.2% C 1% Cr 1% Mo 0.1% V steel: Normalized & tempered **UTS: 1000 Proof: 840**
AISI 3135	0.35% C 0.6% Cr 1.2% Ni steel
AISI 4118	0.2% C 0.5% Cr 0.12% Mo steel
AISI 4130	0.3% C 1% Cr 0.2% Mo steel
AISI 4137	0.37% C 1% Mn 0.2% Mo steel
AISI 4317	0.17% C 0.5% Cr 1.8% Ni 0.25% Mo steel: Obsolete
AISI 4320	0.2% C 0.5% Cr 1.8% Ni 0.25% Mo steel
AISI 4330	0.3% C 0.8% Cr 1.8% Ni 0.4% Mo 0.07% V steel
AISI 4335	0.35% C 0.72% Cr 1.8% Ni 0.35% Mo 0.2% V steel
AISI 4337	0.37% C 0.8% Cr 1.8% Ni 0.25% Mo steel
AISI 4608	0.08% C 1.5% Cr 0.2% Mo steel: Obsolete
AISI 4615	0.15% C 1.8% Ni 0.25% Mo steel
AISI 4620	0.2% C 0.5% Mn 1.8% Ni 0.25% Mo steel
AISI 4621	0.21% C 0.8% Mn 1.8% Ni 0.25% Mo steel
AISI 4626	0.26% C 0.8% Ni 0.2% Mo steel
AISI 4718	0.18% C 0.45% Cr 1% Ni 0.35% Mo steel
AISI 4720	0.2% C 0.45% Cr 1.0% Ni 0.2% Mo steel
AISI 4815	0.15% C 3.5% Ni 0.25% Mo steel
AISI 4817	0.12% C 3.5% Ni 0.25% Mo steel: Obsolete
AISI 4820	0.2% C 3.5% Ni 0.25% Mo steel
AISI 6117	0.17% C 0.8% Cr 0.1% V steel
AISI 6118	0.18% C 0.6% Cr 0.12% V steel
AISI 6120	0.2% C 0.8% Cr 0.1% V steel
AISI 8615	0.15% C 0.5% Cr 0.5% Ni 0.2% Mo steel
AISI 8617	0.17% C 0.5% Cr 0.6% Ni 0.2% Mo steel
AISI 8620	0.2% C 0.5% Cr 0.6% Ni 0.2% Mo steel
AISI 8622	0.22% C 0.5% Cr 0.6% Ni 0.2% Mo steel
AISI 8625	0.25% C 0.5% Cr 0.6% Ni 0.2% Mo steel
AISI 8627	0.27% C 0.5% Cr 0.6% Ni 0.2% Mo steel
AISI 8630	0.3% C 0.5% Cr 0.6% Ni 0.2% Mo steel
AISI 8632	0.32% C 0.5% Cr 0.5% Ni 0.2% Mo steel: Obsolete
AISI 8635	0.35% C 0.5% Cr 0.6% Ni 0.2% Mo steel
AISI 8637	0.37% C 0.5% Cr 0.6% Ni 0.2% Mo steel
AISI 8715	0.15% C 0.5% Cr 0.6% Ni 0.25% Mo steel: Obsolete
AISI 8717	0.17% C 0.5% Cr 0.6% Ni 0.25% Mo steel: Obsolete
AISI 8719	0.2% C 0.5% Cr 0.6% Ni 0.25% Mo steel: Obsolete
AISI 8720	0.2% C 0.5% Cr 0.6% Ni 0.25% Mo steel
AISI 8735	0.35% C 0.5% Cr 0.6% Ni 0.25% Mo steel: Obsolete
AISI 8822	0.22% C 0.5% Cr 0.6% Ni 0.35% Mo steel
AISI 9437	0.37% C 0.4% Cr 0.4% Ni 0.1% Mo steel: Obsolete
AISI A3115	0.15% C 0.6% Cr 1.2% Ni steel: Obsolete
AISI A3120	0.2% C 0.6% Cr 1.2% Ni steel: Obsolete
AISI A3130	0.3% C 0.6% Cr 1.2% Ni steel: Obsolete
AISI A4119	0.2% C 0.5% Cr 0.25% Mo steel: Obsolete
AISI A4125	0.25% C 0.5% Cr 0.25% Mo steel: Obsolete
AISI E3316	0.16% C 1.5% Cr 3.5% Ni steel: Obsolete
AISI E9310	0.1% C 1.2% Cr 3.2% Ni 0.1% Mo steel
AISI E9315	0.15% C 1.2% Cr 3.2% Ni 0.1% Mo steel: Obsolete
AISI E9317	0.17% C 1.2% Cr 3.2% Ni 0.1% Mo steel: Obsolete
AISI X4620	0.2% C 1.8% Ni 0.25% Mo steel: Obsolete

Symbol	Nominal analysis, supplier, condition and remarks.
ALCODIE	0.36% C 5.0% Cr 1.5% Mo 1.3% W 0.3% V steel: Columbia; AISI type H12
ALDECOR	0.15% C (max) 2% Ni 1% Cr 0.2% Mo 1% Cu steel: Source unknown
ALGO-COR-TEN	0.14% C 1.1% Mn 0.5% Cr 0.15% V 0.3% Cu steel: Algoma
ALGO-LOY 1317	0.22% C 1.6% Mn 0.7% Cr 0.05% Mo 0.2% Cu steel: Algoma
ALGOMA	0.12% C 1.6% Mn 3.4% Ni 0.4% Mo 0.02% V steel: Algoma
ALGOMA 90	0.12% C 1.6% Mn 3.3% Ni 0.4% Mo 0.02% V steel: Algoma
ALGOTUF 50	0.2% C 1.1% Mn 0.5% Ni 0.02% V steel: Algoma
ALLOY MOULD	0.32% C 1.2% Cr 4.1% Ni 0.2% Mo steel: Sanderson Kayser
AM 1.	0.35% C 5.0% Cr 1.5% Mo 0.45% V 1.35% W steel: For hot marking; tools; Edgar Allen
AMI	0.35% C 5% Cr 1.5% Mo 1.35% W 0.5% V steel: Edgar Allen; brass forging dies & inserts **DPN: 550**
AMI SPECIAL	0.35% C 5% Cr 1.5% Mo 1.35% W 0.5% V steel: Edgar Allen; also known as AMI
AMS 5502 A	0.1% C 5% Cr 0.5% Mo steel: Sheet & strip; AMS for SAE 51501 + Mo
AMS 5602 A	0.1% C 5% Cr 0.5% Mo steel: Bar & forging; AMS for SAE 51501
AMS 6250 E	0.1% C 1.5% Cr 3.5% Ni steel: Bar & forging, etc.; AMS for SAE 3310
AMS 6260 F	0.1% C 1.2% Cr 3.25% Ni 0.1% Mo steel: Bar, forging, etc.; AMS for SAE 9310
AMS 6263 C	0.15% C 1.2% Cr 3.25% Ni 0.1% Mo steel: Bar, forging, etc.; AMS for SAE 9315
AMS 6264 C	0.17% C 1.2% Cr 3.25% Ni 0.1% Mo steel: Bar, forging, etc.; AMS for SAE 9317
AMS 6265 B	0.1% C 1.2% Cr 3.2% Ni 0.1% Mo steel: Bar, forging, etc.; AMS for SAE 9310; premium quality; vacuum melted
AMS 6266 B	0.11% C 0.5% Cr 1.8% Ni 0.25% Mo B steel: Bar & forging; AMS for SAE 43BV12
AMS 6267	0.1% C 1.2% Cr 3.2% Ni 0.1% Mo steel: Bar & forging; AMS for SAE 9310; premium quality
AMS 6270 F	0.15% C 0.5% Cr 0.5% Ni 0.2% Mo steel: Bar & forging; AMS for SAE 8615
AMS 6272 D	0.17% C 0.5% Cr 0.55% Ni 0.2% Mo steel: Bar & forging; AMS for SAE 8617
AMS 6274 F	0.2% C 0.5% Cr 0.55% Ni 0.2% Mo steel: Bar & forging; AMS for SAE 8620
AMS 6275 A	0.17% C 0.4% Cr 0.45% Ni 0.1% Mo B steel: Bar & forging; AMS for SAE 94B17
AMS 6276 B	0.2% C 0.5% Cr 0.55% Ni 0.2% Mo steel: Bar & forging; AMS for SAE 8620; premium quality
AMS 6277	0.2% C 0.5% Cr 0.55% Ni 0.2% Mo steel: Bar & forging; AMS for SAE 8620
AMS 6280 D	0.3% C 0.5% Cr 0.55% Ni 0.2% Mo steel: Bar & forging; AMS for SAE 8630
AMS 6231 B	0.3% C 0.5% Cr 0.55% Ni 0.2% Mo steel: Tube; AMS for SAE 8630
AMS 6282 C	0.35% C 0.5% Cr 0.55% Ni 0.25% Mo steel: Tube; AMS for SAE 8735
AMS 6290 C	0.15% C 1.8% Ni 0.25% Mo steel: Bar & forging; AMS for SAE 4615
AMS 6292 C	0.17% C 1.8% Ni 0.25% Mo steel: Bar & forging; AMS for SAE 4617
AMS 6294 C	0.2% C 1.8% Ni 0.25% Mo steel: Bar & forging; AMS for SAE 4620
AMS 6299	0.2% C 0.5% Cr 1.7% Ni 0.25% Mo steel: Bar & forging; AMS for SAE 4320
AMS 6302 A	0.3% C 1.25% Cr 0.5% Mo 0.25% V steel: Bar & forging
AMS 6303	0.27% C 1.25% Cr 0.5% Mo 0.85% V steel: Bar & forging
AMS 6320 E	0.35% C 0.5% Cr 0.5% Ni 0.25% Mo steel: Bar & forging; AMS for SAE 8735
AMS 6330 A	0.35% C 0.6% Cr 1.25% Ni steel: Bar & forging; AMS for SAE 3135
AMS 6350 C	0.3% C 1% Cr 0.2% Mo steel: Sheet & strip; annealed; AMS for SAE 4130
AMS 6352 B	0.35% C 1% Cr 0.2% Mo steel: Sheet & strip; annealed; AMS for SAE 4135
AMS 6354	0.15% C 0.6% Cr 0.2% Mo 0.1% Zr steel: Sheet & strip
AMS 6355 F	0.3% C 0.5% Cr 0.5% Ni 0.2% Mo steel: Sheet & strip; AMS for SAE 8630
AMS 6356	0.32% C 1% Cr 0.2% Mo steel: Sheet & strip
AMS 6357 C	0.35% C 0.5% Cr 0.5% Ni 0.25% Mo steel: Sheet & strip; AMS for SAE 8735
AMS 6360 D	0.3% C 1% Cr 0.2% Mo steel: Seamless tube; normalized; AMS for SAE 4130
AMS 6361	0.3% C 1% Cr 0.2% Mo steel: Seamless tube; hardened & tempered **UTS: 920**
AMS 6362	0.3% C 1% Cr 0.2% Mo steel: Seamless tube; hardened & tempered **UTS: 1100**
AMS 6365 C	0.35% C 1% Cr 0.2% Mo steel: Seamless tube; normalized; AMS for SAE 4135
AMS 6370 E	0.3% C 1% Cr 0.2% Mo steel: Bar & forging; AMS for SAE 4130
AMS 6371 C	0.3% C 1% Cr 0.2% Mo steel: Tube; AMS for SAE 4130
AMS 6372 C	0.35% C 1% Cr 0.2% Mo steel: Tube; AMS for SAE 4135
AMS 6385	0.3% C 1.25% Cr 0.5% Mo 0.25% V steel: Sheet & strip
AMS 6407 A	0.3% C 1.2% Cr 2% Ni 0.45% Mo steel: Bar & forging
AMS 6412 E	0.37% C 0.8% Cr 1.8% Ni 0.25% Mo steel: Bar & forging; AMS for SAE 4337
AMS 6413 C	0.37% C 0.8% Cr 1.8% Ni 0.25% Mo steel: Tube; AMS for SAE 4337
AMS 6418 B	0.25% C 1.5% Si 0.3% Cr 1.8% Ni 0.4% Mo steel: Bar & forging
AMS 6421	0.37% C 0.8% Cr 0.8% Ni 0.2% Mo B steel: Bar & forging; AMS for SAE 98B37
AMS 6427 C	0.3% C 0.8% Cr 1.8% Ni 0.4% Mo 0.07% V steel: Bar & forging
AMS 6428	0.35% C 0.8% Cr 1.8% Ni 0.35% Mo 0.2% V steel: Bar & forging
AMS 6429	0.35% C 0.8% Cr 1.8% Ni 0.35% Mo 0.2% V steel: Bar & forging; vacuum melted
AMS 6430	0.35% C 0.8% Cr 1.8% Ni 0.35% Mo 0.2% V steel: Bar & forging; special grade
AMS 6433	0.35% C 0.8% Cr 1.8% Ni 0.35% Mo 0.2% V steel: Sheet & strip; special grade
AMS 6434	0.35% C 0.8% Cr 1.8% Ni 0.35% Mo 0.2% V steel: Sheet & strip
AMS 6435	0.35% C 0.8% Cr 1.8% Ni 0.35% Mo 0.2% V steel: Sheet & strip; vacuum melted
AMS 6436	0.22% C 1.25% Cr 0.5% Mo 0.8% V steel: Sheet & strip
AMS 6458 B	0.3% C 1.2% Cr 0.5% Mo 0.3% V steel: Welding wire; vacuum melted
AMS 6460	0.15% C 0.7% Si 0.6% Cr 0.2% Mo 0.1% Zr steel: Welding wire
AMS 6461 B	0.3% C 1% Cr 0.2% V steel: Welding wire; AMS for SAE 6130
AMS 6462 B	0.3% C 1% Cr 0.2% V steel: Welding wire; AMS for SAE 6130

Symbol	Nominal analysis, supplier, condition and remarks.
AMS 6464	0.1% C 1% Mo 0.2% V steel: Coated welding electrode
AMS 6466	0.1% C (max) 5% Cr 0.5% Mo steel: Welding wire; cold drawn
AMS 6467	0.1% C (max) 5% Cr 0.5% Mo steel: Coated welding electrode
AMS 6475 B	0.25% C 1.2% Cr 3.5% Ni 0.2% Mo 1.25% Al steel: Bar & forging; nitriding steel
AMS 6530 D	0.3% C 0.5% Cr 0.5% Ni 0.2% Mo steel: Seamless tube; normalized; AMS for SAE 8630
AMS 6535 C	0.35% C 0.5% Cr 0.5% Ni 0.25% Mo steel: Seamless tube; normalized; AMS for SAE 8735
AMS 6550 D	0.3% C 0.5% Cr 0.5% Ni 0.2% Mo steel: Welded tube; normalized; AMS for SAE 8630
AN 1 A	0.3% C 1.3% Cr 4.1% Ni steel: W Marrison for BS alloy En 30A
AN 1 B	0.3% C 1.3% Cr 4.1% Ni 0.3% Mo steel: W Marrison for BS alloy En 30B
ANC 1	0.3% C 1.25% Cr 4.25% Ni steel: Firth Brown for BS alloy En 30A
ANCM	0.3% C 1.25% Cr 4.25% Ni 0.25% Mo steel: Firth Brown for BS alloy En 30B
AN QQ S684a	0.3% C 1.0% Cr 0.2% Mo steel: US Service; normalized & tempered **DPN: 230 UTS: 840 Elon: 18% Proof: 740**
AN QQ S685/4	0.3% C 1.0% Cr 0.2% Mo steel: Strip & plate; US Service **UTS: 690 Elon: 18% Proof: 480**
AN QQ S686/1	0.35% C 1.0% Cr 0.2% Mo steel: Strip & plate; US Service; annealed **UTS: 620**
AN S 12/2	0.3% C 0.5% Cr 0.5% Ni 0.2% Mo steel: Strip; US Service; annealed **UTS: 620**
AN S 13a	0.2% C 0.5% Cr 0.5% Ni 0.2% Mo steel: US Service
AN S 14a	0.3% C 0.5% Cr 0.5% Ni 0.2% Mo steel: US Service; normalized & tempered **DPN: 230 UTS: 840 Elon: 16% Proof: 740**
AN S 15a	0.35% C 0.5% Cr 0.6% Ni 0.25% Mo steel: US Service; hardened & tempered **DPN: 230 UTS: 840 Elon: 16% Proof: 740**
AN S 22/1	0.35% C 0.5% Cr 0.6% Ni 0.25% Mo steel: Strip; US Service; annealed **UTS: 620**
AN T 3	0.3% C 1.0% Cr 0.2% Mo steel: Tube; welded; US Service; annealed **UTS: 640**
AN T 15/2	0.3% C 0.5% Cr 0.6% Ni 0.2% Mo steel: Seamless tube; US Service; normalized **UTS: 630 Elon: 12% Proof: 460**
AN T 22/2	0.35% C 0.5% Cr 0.5% Ni 0.25% Mo steel: Seamless tube; US Service; annealed **UTS: 740**
AN T 33/1	0.3% C 0.5% Cr 0.6% Ni 0.2% Mo steel: Welded tube; US Service; annealed **UTS: 630**
AN WW T 850/1	0.3% C 1.0% Cr 0.2% Mo steel: Seamless tube; US Service; normalized **UTS: 630 Elon: 12% Proof: 460**

Note. The following abbreviations and units are used in the tables:

DPN	Hardness, diamond pyramid number
UTS	Ultimate tensile strength, N/mm^2
Elon	Elongation, %
Proof	0.1 % proof strength, N/mm^2

1 N/mm^2=0.1 hbar=0.102 kgf/mm^2=0.06475 tonf/in^2=145.04 lbf/in^2

Symbol	Nominal analysis, supplier, condition and remarks.
AN WW T 852	0.35% C 1.0% Cr 0.2% Mo steel: Seamless tube; US Service **UTS: 700**
APS 10 M	0.12% C 2.0% Cr 0.1% Mo 1.0% Al steel: Pompey **UTS: 460 Elon: 20% Proof: 300**
APS 10 M4	0.12% C 2.0% Cr 0.35% Mo 0.35% Al steel: Pompey **UTS: 460 Elon: 20% Proof: 300**
APS 20 M	0.12% C 4.0% Cr 0.12% Mo 0.9% Al steel: Pompey **UTS: 460 Elon: 20% Proof: 300**
APS 25	0.15% C 4.0% Cr 0.7% Ni 0.7% Al steel: Pompey **UTS: 820 Elon: 15% Proof: 580**
AR COL 360	0.13% C 1.35% Mn 0.8% Cr 0.4% Mo steel: Colvilles; abrasion resisting **DPN: 320**
ARO 75	0.4% C 1.7% Cr 0.3% Mo 1.1% Al steel: For nitriding; Bofors; hardened & tempered **DPN: 280 UTS: 920 Elon: 15% Proof: 620**
ARMOUR PLATE	0.3% C 2.75% Cr 2.5% Ni 0.25% Mo steel: Plate; common name
ASTM A182 F1	0.3% C 0.6% Mn 0.5% Mo steel: For pipe fittings
ASTM A182 F5	0.15% C (max) 0.45% Mn 5% Cr 0.5% Mo steel: For pipe fittings
ASTM A182 F5a	0.25% C (max) 0.6% Mn 0.5% Cr 0.5% Mo steel: For pipe fittings
ASTM A182 F7	0.15% C (max) 0.45% Mn 0.8% Si 7% Cr 0.5% Mo steel: For pipe fittings
ASTM A182 F9	0.15% C (max) 0.45% Mn 0.7% Si 9% Cr 1.0% Mo steel: For pipe fittings
ASTM A182 F11	0.15% C 0.5% Mn 0.7% Si 1.2% Cr 0.5% Mo steel: For pipe fittings
ASTM A182 F12	0.15% C 0.5% Mn 0.4% Si 1% Cr 0.5% Mo steel: For pipe fittings
ASTM A182 F21	0.15% C (max) 0.4% Mn 3% Cr 0.9% Mo steel: For pipe fittings
ASTM A182 F22	0.15% C (max) 0.45% Mn 2.2% Cr 0.95% Mo steel: For pipe fittings
ASTM A193 B5	0.1% C 1% Mn 5% Cr 0.5% Mo steel: For bolts, etc.; hardened & tempered **UTS: 730 Elon: 16% Proof: 560**
ASTM A194/3	0.1% C (min) 5% Cr 0.5% Mo steel: For bolts
ASTM A199 T3b	0.15% C (max) 2.0% Cr 0.5% Mo steel: Condenser tube
ASTM A199 T4	0.15% C (max) 0.7% Si 2.5% Cr 0.5% steel: Condenser tube
ASTM A199 T5	0.15% C (max) 5.0% Cr 0.5% Cr steel: Condenser tube
ASTM A199 T7	0.15% C (max) 0.7% Si 7.0% Cr 0.5% Mo steel: Condenser tube
ASTM A199 T9	0.15% C (max) 9% Cr 1.0% Mo steel: Condenser tube
ASTM A199 T11	0.15% C (max) 1.2% Cr 0.5% Mo steel: Condenser tube
ASTM A199 T21	0.15% C 3.0% Cr 0.95% Mo steel: Condenser tube
ASTM A199 T22	0.15% C 2.2% Cr 1.0% Mo steel: Condenser tube
ASTM A200	Alloy steel still tube; grades as in ASTM A199
ASTM A217 C5	0.2% C (max) 0.5% Mn 5.2% Cr 0.55% Mo steel: Casting
ASTM A217 C12	0.2% C (max) 0.5% Mn 9% Cr 1.1% Mo steel: Casting
ASTM A217 WC4	0.2% C (max) 0.65% Mn 0.9% Ni 0.65% Cr 0.55% Mo steel: Casting
ASTM A217 WC5	0.2% C (max) 0.55% Mn 0.8% Ni 0.7% Cr 1.1% Mo steel: Castings
ASTM A217 WC6	0.2% C (max) 0.65% Mn 1.2% Cr 0.55% Mo steel: Castings
ASTM A217 WC9	0.18% C (max) 0.5% Mn 2.3% Cr 1.0% Mo steel: Castings

Symbol	Nominal analysis, supplier, condition and remarks.
ASTM A290 D	0.35% C 1% Cr 0.2% Mo steel: Forging; hardened & tempered
	DPN: 223 UTS: 760 Elon: 18% Proof: 530
ASTM A290 D2	0.35% C 1% Cr 0.2% Mo steel: Forging; hardened & tempered
	DPN: 245 UTS: 840 Elon: 15% Proof: 760
ASTM A290 E	0.35% C 0.8% Cr 1.8% Ni 0.25% Mo steel: Forging; hardened & tempered
	DPN: 302 UTS: 980 Elon: 16% Proof: 760
ASTM A290 F	0.35% C 0.8% Cr 1.8% Ni 0.25% Mo steel: Forging; hardened & tempered
	DPN: 302 UTS: 1070 Elon: 13% Proof: 840
ASTM A290 G	0.35% C 0.8% Cr 1.8% Ni 0.25% Mo steel: Forging; hardened & tempered
	DPN: 341 UTS: 1210 Elon: 11% Proof: 1000
ASTM A292/2	0.3% C 0.5% Cr 2% Ni 0.2% Mo 0.1% V steel: For generator rotors; hardened & tempered; obsolete
	UTS: 530 Elon: 22% Proof: 370
ASTM A292/3	0.3% C 0.5% Cr 2% Ni 0.2% Mo 0.1% V steel: For generator rotors; hardened & tempered; obsolete
	UTS: 640 Elon: 20% Proof: 500
ASTM A292/4	0.35% C 0.7% Cr 2.5% Ni 0.2% Mo 0.1% V steel: For generator rotors; hardened & tempered; obsolete
	UTS: 760 Elon: 18% Proof: 580
ASTM A292/5	0.35% C 0.7% Cr 2.5% Ni 0.2% Mo 0.1% V steel: For generator rotors; hardened & tempered; obsolete
	UTS: 840 Elon: 16% Proof: 640
ASTM A293/2	0.3% C 0.7% Cr 2% Ni 0.25% Mo 0.1% V steel: For turbine rotors; hardened & tempered
	UTS: 580 Elon: 22% Proof: 370
ASTM A293/3	0.3% C 0.7% Cr 2% Ni 0.25% Mo 0.1% V steel: For turbine rotors; hardened & tempered
	UTS: 700 Elon: 20% Proof: 500
ASTM A293/4	0.35% C 0.7% Cr 2.5% Ni 0.25% Mo 0.1% V steel: For turbine rotors; hardened & tempered
	UTS: 760 Elon: 18% Proof: 580
ASTM A293/5	0.35% C 1.25% Cr 2.5% Ni 0.25% Mo 0.03% V steel: For turbine rotors; hardened & tempered
	UTS: 880 Elon: 17% Proof: 700
ASTM A293/6	0.37% C 1% Mn 1% Cr 0.05% Ni 1.2% Mo 0.25% V steel: For turbine rotors; hardened & tempered
	UTS: 780 Elon: 16% Proof: 620
ASTM A294 C	0.35% C 0.7% Cr 2.5% Ni 0.2% Mo 0.1% V steel: For turbine wheels
ASTM A295 F2	0.15% C 1% Cr 0.5% Mo steel: Seamless drum forging; normalized & tempered
	UTS: 500 Elon: 18% Proof: 270
ASTM A295 F5	0.15% C 5% Cr 0.5% Mo steel: Seamless drum forging; normalized & tempered
	UTS: 500 Elon: 18% Proof: 270
ASTM A295 F5A	0.25% C 5% Cr 0.5% Mo steel: Seamless drum forging; normalized & tempered
	UTS: 580 Elon: 19% Proof: 340
ASTM A295 F22	0.15% C 2.2% Cr 1% Mo steel: Seamless drum forging; normalized & tempered
	UTS: 500 Elon: 18% Proof: 270
ASTM A295 F31	0.35% C 2.5% Ni 0.3% Mo 0.15% V steel: Seamless drum forging; normalized & tempered
	UTS: 700 Elon: 18% Proof: 500
ASTM A295 F32	0.35% C 3.3% Cr 0.7% Ni 0.4% Mo 0.1% V steel: Seamless drum forging; normalized & tempered
	UTS: 740 Elon: 18% Proof: 420
ASTM A298 E502	0.1% C 5% Cr 0.5% Mo steel: Welding rod annealed after welding, for plain carbon steels
	UTS: 400 Elon: 20%

Symbol	Nominal analysis, supplier, condition and remarks.
ASTM A302 A	0.21% C 1.1% Mn 0.5% Mo steel: Plate for pressure vessels
ASTM A302 B	0.22% C 1.3% Mn 0.5% Mo steel: Plate for pressure vessels
ASTM A302 C	0.22% C 1.3% Mn 0.5% Mo 0.5% Ni steel: Plate for pressure vessels
ASTM A302 D	0.22% C 1.3% Mn 0.5% Mo 0.8% Ni steel: Plate for pressure vessels
ASTM A302 Grade C	0.23% C 1.30% Mn 0.22% Si 0.52% Mo 0.55% Ni steel: Plate for pressure vessels
ASTM A302 Grade D	0.23% C 1.20% Mn 0.22% Si 0.52% Mo 0.85% Ni steel: Plate for pressure vessels
ASTM A316 E7000A1	0.12% C 0.7% Mn 0.05% Mo steel: Welding electrodes; code varies with analysis
ASTM A316 E7020G	1.0% (min) Mn 0.8% (min) Si 0.05% (min) Ni 0.3% (min) Cr 0.2% (min) Mo 0.1% (min) V steel: Electrodes for arc welding
ASTM A316 E8015/18B	0.1% Cr and Mo steel: Electrodes; code varies with analysis
ASTM A316 E8015B4L	0.05% C 2.0% Cr 0.52% Mo steel: Electrodes for arc welding
ASTM A316 E8016B1	0.12% C 0.52% Cr 0.52% Mo steel: Electrodes for arc welding
ASTM A816 E8018B1	0.12% C 0.52% Cr 0.52% Mo steel: Electrodes for arc welding
ASTM A316 E8015B2L	0.05% C 1.25% Cr 0.52% Mo steel: Electrodes for arc welding
ASTM A316 E8016B2	0.12% C 1.25% Cr 0.52% Mo steel: Electrodes for arc welding
ASTM A316 E8018B2	0.12% C 1.25% Cr 0.52% Mo steel: Electrodes for arc welding
ASTM A316 E9015B3L	0.05% C 2.25% Cr 1.05% Mo steel: Electrodes for arc welding
ASTM A316 E8016C3	0.12% C 0.95% Ni 0.15% Cr 0.35% Mo 0.05% V steel: Electrodes for arc welding
ASTM A316 E8018C3	0.12% C 0.95% Ni 0.15% Cr 0.35% Mo 0.05% V steel: Electrodes for arc welding
ASTM A316 E9015B3	0.12% C 2.25% Cr 1.05% Mo steel: Electrodes for arc welding
ASTM A316 E9016B3	0.12% C 2.25% Cr 1.05% Mo steel: Electrodes for arc welding
ASTM A316 E9018B3	0.12% C 0.25% Cr 1.05% Mo steel: Electrodes for arc welding
ASTM A316 E9016/8B3	0.1% C 2.2% Cr Mo steel: Welding electrodes; code varies with analysis
ASTM A316 E9018M	0.10% C 1.6% Ni 0.15% Cr 0.35% Mo 0.05% V steel: Electrodes for arc welding
ASTM A316 E10018M	0.10% C 1.75% Ni 0.35% Cr 0.37% Mo 0.05% V steel: Electrodes for arc welding
ASTM A316 E11018M	0.10% C 1.9% Ni 0.40% Cr 0.42% Mo 0.05% V steel: Electrodes for arc welding
ASTM A316 E9018M	0.1% C 1% Mn 1.5% Ni 0.3% Mo steel: Welding electrodes
ASTM A316 E10018M	0.1% C 1.2% Mn 1.8% Ni 0.3% Mo steel: Welding electrodes
ASTM A316 E11018M	0.1% C 1% Mn 1.8% Ni 0.4% Cr 0.4% Mo steel: Welding electrodes
ASTM A316 E12018M	0.1% C 1.8% Mn 2% Ni 0.9% Ni 0.4% Mo steel: Welding electrodes
ASTM A316 E12018M	0.10% C 2.0% Ni 0.9% Cr 0.42% Mo 0.05% V steel: Electrodes for arc welding
ASTM A316 EXX10G	C not specified 1.0% (min) Mn 0.8% (min) Si 0.5% (min) Ni 0.3% (min) Cr 0.2% (min) Mo 0.1% (min) V steel: Electrodes for arc welding
ASTM A316 EXX11G	C not specified 1.0% (min) Mn 0.8% (min) Si 0.5% (min) Ni 0.3% (min) Cr 0.2% (min) Mo 0.1% (min) V steel: Electrodes for arc welding
ASTM A316 EXX13G	C not specified 1.0% (min) Mn 0.8% (min) Si 0.5% (min) Ni 0.3% (min) Cr 0.2% (min) Mo 0.1% (min) V steel: Electrodes for arc welding

Symbol	Nominal analysis, supplier, condition and remarks.
ASTM A316 EXX15G	C not specified 1.0% (min) Mn 0.8% (min) Si 0.5% (min) Ni 0.3% (min) Cr 0.2% (min) Mo 0.1% (min) V steel: Electrodes for arc welding
ASTM A316 EXX16G	C not specified 1.0% (min) Mn 0.8% (min) Si 0.5% (min) Ni 0.3% (min) Cr 0.2% (min) Mo 0.1% (min) V steel: Electrodes for arc welding
ASTM A316 EXX18G	C not specified 1.0% (min) Mn 0.8% (min) Si 0.5% (min) Ni 0.3% (min) Cr 0.2% (min) Mo 0.1% (min) V steel: Electrodes for arc welding
ASTM A335 P2	0.15% C 0.7% Cr 0.5% Mo steel: Seamless pipe
ASTM A335 P5	0.15% C (max) 5.0% Cr 0.5% Mo steel: Seamless pipe
ASTM A335 P5b	0.15% C (max) 5.0% Cr 0.5% Mo steel: Seamless pipe
ASTM A335 P5c	0.12% C (max) 5.0% Cr 0.5% Mo steel: Seamless pipe
ASTM A335 P7	0.15% C (max) 7.0% Cr 0.5% Mo steel: Seamless pipe
ASTM A335 P9	0.15% C (max) 9.0% Cr 1.0% Mo steel: Seamless pipe
ASTM A335 P11	0.15% C (max) 1.2% Cr 0.5% Mo steel: Seamless pipe
ASTM A335 P12	0.15% C (max) 1.0% Cr 0.5% Mo steel: Seamless pipe
ASTM A335 P21	0.15% C (max) 3.0% Cr 0.95% Mo steel: Seamless pipe
ASTM A335 P22	0.15% C (max) 2.3% Cr 1.0% Mo steel: Seamless pipe
ASTM A336 F2	0.15% C 1.0% Cr 0.5% Mo steel: Seamless drum forging **UTS: 460 Elon: 18% Proof: 270**
ASTM A336 F5	0.15% C (max) 5.0% Cr 0.5% Mo steel: Seamless drum forging **UTS: 400 Elon: 19% Proof: 240**
ASTM A336 F5a	0.25% C (max) 5.0% Cr 0.5% Mo steel: Seamless drum forging **UTS: 540 Elon: 19% Proof: 330**
ASTM A336 F22	0.15% C (max) 2.2% Cr 1.0% Mo steel: Seamless drum forging **UTS: 460 Elon: 18% Proof: 270**
ASTM A336 F31	0.35% C (max) 2.5% Ni 0.4% Mo 0.15% V (max) steel: Seamless drum forging **UTS: 630 Elon: 18% Proof: 330**
ASTM A336 F32	0.35% C (max) 3.3% Ni 3.3% Cr 0.4% Mo 0.1% V steel: Seamless drum forging **UTS: 670 Elon: 18% Proof: 400**
ASTM A350 LF4	0.12% C (max) 0.7% Ni 0.7% Cr 0.5% Cu 0.2% Al steel: For low temperature use **UTS: 400 Elon: 25% Proof: 200**
ASTM A355 C	0.23% C 1.2% Cr 3.5% Ni 0.2% Mo 1.0% Al steel: Hardened & tempered **DPN: 250**
ASTM A356/4	0.2% C 1.0% Mo 0.2% V steel: Casting; normalized & tempered **UTS: 620 Elon: 16% Proof: 400**
ASTM A356/5	0.25% C 0.5% Cr 0.5% Mo steel: Casting; normalized & tempered **UTS: 460 Elon: 22% Proof: 270**

Symbol	Nominal analysis, supplier, condition and remarks.
ASTM A356/6	0.2% C 1.2% Cr 0.5% Mo steel: Casting; normalized & tempered **UTS: 460 Elon: 22% Proof: 300**
ASTM A356/7	0.2% C 1.2% Cr 0.5% Mo 0.2% V steel: Casting; normalized & tempered **UTS: 460 Elon: 22% Proof: 270**
ASTM A356/8	0.25% C 1.2% Cr 1.0% Mo steel: Casting; normalized & tempered **UTS: 530 Elon: 18% Proof: 330**
ASTM A356/9	0.2% C 1.2% Cr 1.0% Mo 0.2% V steel: Casting; normalized & tempered **UTS: 640 Elon: 15% Proof: 400**
ASTM A356/10	0.2% C 2.4% Cr 1.0% Mo steel: Casting; normalized & tempered **UTS: 580 Elon: 20% Proof: 360**
ASTM A357	0.15% C 5.0% Cr 0.5% Mo steel: Plate; annealed **UTS: 460 Elon: 20% Proof: 200**
ASTM A369 FP1	0.15% C 0.7% Cr 0.5% Mo steel: For forged or bored pipe
ASTM A369 FP3b	0.15% C (max) 2.0% Cr 0.5% Mo steel: For forged or bored pipe
ASTM A369 FP5	0.15% C (max) 5.0% Cr 0.5% Mo steel: For forged or bored pipe
ASTM A369 FP7	0.15% C (max) 7.0% Cr 0.5% Mo steel: For forged or bored pipe
ASTM A369 FP9	0.15% C (max) 9.0% Cr 1.0% Mo steel: For forged or bored pipe
ASTM A369 FP11	0.15% C (max) 1.2% Cr 0.5% Mo steel: For forged or bored pipe
ASTM A369 FP12	0.15% C (max) 1.1% Cr 0.5% Mo steel: For forged or bored pipe
ASTM A369 FP21	0.15% C (max) 3.0% Cr 1.0% Mo steel: For forged or bored pipe
ASTM A369 FP22	0.15% C (max) 2.3% Cr 1.0% Mo steel: For forged or bored pipe
ASTM A372 VI	0.18% C (max) 0.30% Mn 0.25% Si 2.7% Ni 1.4% Cr 0.30% Mo steel: Forging for pressure vessels
ASTM A387 A	0.21% C (max) 0.65% Cr 0.52% Mo steel: Plate for pressure vessels; normalized & tempered **UTS: 550 Elon: 20% Proof: 300**
ASTM A387 B	0.17% C (max) 1.0% Cr 0.5% Mo steel: Plate for pressure vessels; normalized & tempered **UTS: 510 Elon: 20% Proof: 270**
ASTM A387 C	0.17% C (max) 1.2% Cr 0.5% Mo steel: Plate for pressure vessels; normalized & tempered
ASTM A387 D	0.15% C (max) 2.2% Cr 1.0% Mo steel: Plate for pressure vessels; normalized & tempered **UTS: 600 Elon: 18% Proof: 300**
ASTM A387 E	0.15% C (max) 3.0% Cr 1.0% Mo steel: Plate for pressure vessels; normalized & tempered
ASTM A389 C23	0.2% C (max) 1.2% Cr 0.5% Mo 0.2% V steel: Casting for high temperature use **UTS: 460 Elon: 18% Proof: 270**
ASTM A389 C24	0.2% C (max) 1.0% Cr 1.0% Mo 0.2% V steel: Casting for high temperature use **UTS: 540 Elon: 15% Proof: 330**
ASTM A333/4	0.1% C 0.7% Mn 0.7% Ni 0.8% Cr 0.6% Cu 0.2% Al steel: Pipe for low temperature use
ASTM A404 F24	0.2% C (max) 1.0% Cr 1.0% Mo 0.2% V steel: For high temperature use; flanges, etc.
ASTM A405 P24	0.15% C (max) 1.0% Cr 1.0% Mo 0.2% V steel: For high temperature use; seamless pipe
ASTM A410	0.12% C (max) 0.8% Cr 0.7% Ni 0.5% Cu 0.2% Al steel: Plate for pressure vessels
ASTM A423/1	0.15% C (max) 0.65% Cr 0.4% Ni steel: Welded tube
ASTM A423/2	0.15% C (max) 0.75% Ni 0.1% Mo steel: Welded tube
ASTM A426 CP2	0.15% C 0.65% Cr 0.5% Mo steel: Cast pipe

Note. The following abbreviations and units are used in the tables:

DPN	Hardness, diamond pyramid number
UTS	Ultimate tensile strength, N/mm^2
Elon	Elongation, %
Proof	0.1 % proof strength, N/mm^2

1 N/mm^2 = 0.1 hbar = 0.102 kgf/mm^2 = 0.06475 tonf/in^2 = 145.04 lbf/in^2

Symbol	Nominal analysis, supplier, condition and remarks.
ASTM A426 CP5	0.15% C (max) 5.0% Cr 0.5% Mo steel: Cast pipe
ASTM A426 CP5b	0.15% C (max) 5.0% Cr 0.5% Mo steel: Cast pipe
ASTM A426 CP7	0.15% C (max) 7.0% Cr 0.5% Mo steel: Cast pipe
ASTM A426 CP9	0.15% C (max) 9.0% Cr 1.0% Mo steel: Cast pipe
ASTM A426 CP11	0.15% C (max) 1.2% Cr 0.5% Mo steel: Cast pipe
ASTM A426 CP12	0.15% C (max) 1.0% Cr 0.5% Mo steel: Cast pipe
ASTM A426 CP21	0.15% C (max) 3.0% Cr 0.95% Mo steel: Cast pipe
ASTM A426 CP22	0.15% C (max) 2.3% Cr 1.0% Mo steel: Cast pipe
ASTM A469/2	0.25% C 2.5% Ni 0.5% Cr (max) 0.3% Mo 0.03% V steel: For rotors; vacuum treated
ASTM A487/4 N	0.3% C (max) 0.6% Ni 0.6% Cr 0.2% Mo steel: Casting for pressure; normalized; 4Q grade is hardened & tempered
ASTM A487/5 N	0.3% C (max) 1.2% Mn 0.6% Ni 0.6% Cr 0.2% Mo steel: Casting for pressure; normalized; 5Q grade is hardened & tempered
ASTM A487/7 Q	0.2% C (max) 0.9% Ni 0.6% Cr 0.5% Mo 0.08% V 0.004% B steel: Casting for pressure; hardened & tempered
ASTM A487/8 N	0.2% C (max) 2.5% Cr 1.0% Mo steel: Casting for pressure; normalized; 8Q grade is hardened & tempered
ASTM A487/9 N	0.33% C (max) 0.9% Cr 0.25% Mo steel: Casting for pressure; normalized; 9Q grade is hardened & tempered
ASTM A487/10 N	0.3% C (max) 1.7% Ni 0.75% Cr 0.3% Mo steel: Casting for pressure; normalized; 10Q grade is hardened & tempered
ASTM A508/2	0.27% C (max) 0.7% Mn 0.7% Ni 0.35% Cr 0.6% Mo 0.05% V steel: For pressure vessels; vacuum treated
ASTM A508/3	0.2% C 1.3% Mn 0.6% Ni 0.52% Mo 0.05% V (max) steel: For pressure vessels; vacuum treated
ASTM A508/4	0.23% C (max) 3.3% Ni 1.7% Cr 0.5% Mo 0.03% V (max) steel: For pressure vessels; vacuum treated
ASTM A508/5	0.23% C (max) 3.3% Ni 1.7% Cr 0.05% Mo 0.1% V steel: For pressure vessels; vacuum treated
ASTM A514 A	0.17% C 0.65% Cr 0.22% Mo 0.1% Zr 0.0025% B (max) steel: For plate
ASTM A514 B	0.14% C 0.52% Cr 0.2% Mo 0.05% V 0.02% Ti 0.001% B steel: For plate
ASTM A514 C	0.15% C 0.25% Mo 0.003% B steel: Plate
ASTM A514 D	0.17% C 1.0% Cr 0.2% Mo 0.08% Ti 0.03% Cu 0.003% B steel: Plate
ASTM A514 E	0.16% C 1.7% Cr 0.5% Mo 0.07% Ti 0.3% Cu 0.003% B steel: Plate
ASTM A514 F	0.15% C 0.85% Ni 0.5% Cr 0.5% Mo 0.005% V 0.3% Cu 0.004% B steel: Plate
ASTM A514 G	0.18% C 0.7% Cr 0.5% Mo 0.1% Zr 0.0025% B (max) steel: Plate
ASTM A514 H	0.17% C 0.5% Ni 0.48% Cr 0.25% Mo 0.05% V 0.001% B steel: Plate
ASTM A517 A	0.18% C 0.65% Cr 0.23% Mo 0.1% Zr 0.0025% B (max) steel: Boiler plate
ASTM A517 B	0.15% C 0.5% Cr 0.2% Mo 0.005% V 0.02% Ti 0.001% B steel: Boiler plate
ASTM A517 D	0.17% C 1.0% Cr 0.2% Mo 0.07% Ti 0.3% Cu 0.003% B steel: Boiler plate
ASTM A517 E	0.16% C 1.7% Cr 0.5% Mo 0.07% Ti 0.3% Cu 0.003% B steel: Boiler plate
ASTM A517 F	0.15% C 0.9% Ni 0.5% Cr 0.5% Mo 0.05% V 0.4% Cu 0.004% B steel: Boiler plate
ASTM A517 G	0.18% C 0.7% Cr 0.5% Mo 0.1% Zr 0.0025% B steel: Boiler plate
ASTM A517 H	0.17% C 0.5% Ni 0.5% Cr 0.25% Mo 0.05% V 0.0005% B steel: Boiler plate

Symbol	Nominal analysis, supplier, condition and remarks.
ASTM A533 B	0.25% C (max) 1.25% Mn 0.52% Mo 0.55% Ni steel: Plate
ASTM A533 C	0.25% C (max) 1.25% Mn 0.52% Mo 0.85% Ni steel: Plate
ASTM A534	carburizing steels; graded by AISI system; see also Section 44A1:
ASTM A535	steels for ball and rolls bearings; graded by AISI system; see Sections 44K2 and 44K3:
ASTM A54/2	0.27% C (max) 0.7% Ni 0.35% Cr 0.62% Mo 0.06% V steel: Forging
ASTM A541/3	0.2% C 1.3% Mn 0.6% Ni 0.52% Mo 0.06% V steel: Forging
ASTM A541/4	0.18% C 0.25% Ni (max) 0.15% Cr (max) 0.1% V steel: Forging
ASTM A541/5	0.15% C 1.25% Cr 0.52% Mo steel: Forging
ASTM A541/6	0.15% C (max) 2.2% Cr 1.0% Mo steel: Forging
ASTM A541/7	0.23% C (max) 3.2% Ni 1.5% Cr 0.5% Mo 0.03% V steel: Forging
ASTM A541/8	0.23% C (max) 3.2% Ni 1.5% Cr 0.5% Mo 0.1% V steel: Forging
ASTM A542	0.15% C (max) 2.25% Cr 1.0% Mo steel: Plate
ASTM A543	0.23% C (max) 3.2% Ni 1.75% Cr 0.5% Mo steel: Plate; Ni content varies with thickness **UTS: 780 Elon: 14% Proof: 650**
ASTM A574	0.33% C (min) 0.04% S & P steel: For cap screws; may contain Cr Ni Mo or V to ensure mechanical properties are obtained
ASTM A579/11	0.25% C 1.5% Cr 3.0% Ni 0.9% Mo 0.05% Nb steel: Forging
ASTM A579/12	0.12% C (max) 0.55% Cr 5.0% Ni 0.5% Mo 0.07% V steel: Forging
ASTM A579/13	0.3% C 0.95% Cr 0.2% Mo 0.07% V steel: Forging
ASTM A579/31	0.25% C 0.3% Cr 1.8% Ni 0.4% Mo steel: Forging
ASTM A588 A	0.14% C 1.1% Mn 0.52% Cr 0.06% V 0.32% Cu steel: For structures **UTS: 440 Elon: 21% Proof: 320**
ASTM A588 B	0.2% C (max) 1.0% Mn 0.55% Cr 0.32% Ni 0.05% V 0.3% Cu steel: For structures **UTS: 440 Elon: 320**
ASTM A588 C	0.15% C (max) 1.1% Mn 0.4% Cr 0.06% V 0.35% Cu steel: For structures **UTS: 440 Elon: 21% Proof: 320**
ASTM A588 F	0.15% C 0.75% Mn 0.7% Ni 0.15% Mo 0.05% V 0.7% Cu steel: For structures **UTS: 440 Elon: 21% Proof: 320**
ASTM A588 H	0.2% C (max) 1.25% Mn (max) 0.16% Cr 0.45% Ni 0.06% V 0.27% Cu steel: For structures **UTS: 440 Elon: 21% Proof: 320**
ASTM A592 A	0.18% C 1.0% Mn 0.65% Cr 0.2% Mo 0.1% Zr steel: For pressure vessels **UTS: 850 Elon: 18% Proof: 650**
ASTMA 592 E	0.17% C 0.5% Mn 1.7% Cr 0.5% Mo 0.07% Ti 0.25% Cu 0.003% B steel: For pressure vessels **UTS: 850 Elon: 18% Proof: 650**
ASTM A592 F	0.15% C 0.6% Cr 0.9% Ni 0.5% Mo 0.05% V steel: For pressure vessels **UTS: 850 Elon: 18% Proof: 650**
ASTM A595 C	0.15% C (max) 0.35% Mn 1.0% Cr 0.7% Ni (max) 0.4% Cu steel: Tapered tube **UTS: 480 Proof: 410**
ASTM A595 C	0.15% C (max) 0.4% Mn 0.3% Cu 0.7% Cr 0.08% Ni (max) steel: Tube; tapered for structural use
ASTM A597	0.35% C 5.2% Cr 1.5% Mo 0.35% V 1.3% W steel: Cast; for tools
ASTM A597 CH13	0.36% C 5.2% Cr 1.5% Mo 1.0% V steel: Cast; for tools

Symbol	Nominal analysis, supplier, condition and remarks.
ASTM A643 B	0.25% C (max) 1.3% Mn 0.7% Ni 0.5% Mo steel: Casting for pressure vessels **UTS: 650 Elon: 22% Proof: 345**
ASTM A643 C1	0.2% C (max) 2.3% Cr 1.0% Mo steel: Casting for pressure vessels **UTS: 690 Elon: 20% Proof: 380**
ASTM A643 C2	0.2% C (max) 2.3% Cr 1.0% Mo steel: Casting for pressure vessels **UTS: 750 Elon: 18% Proof: 515**
ASTM A643 C3	0.2% C (max) 2.3% Cr 1.0% Mo steel: Casting for pressure vessels **UTS: 830 Elon: 15% Proof: 585**
ASTM A643 D1	0.2% C (max) 1.7% Cr 3.2% Ni 0.5% Mo steel: Casting for pressure vessels **UTS: 830 Elon: 15% Proof: 585**
ASTM A643 D2	0.2% C (max) 1.7% Cr 3.2% Ni 0.5% Mo steel: Casting for pressure vessels **UTS: 900 Elon: 13% Proof: 690**
ASTM A645	0.15% C (max) 0.4% Mn 5.0% Ni 0.25% Mo 0.1% Al steel: For pressure vessels; impact value 22 J at -170 °C **UTS: 720 Elon: 20% Proof: 450**
ASTM A646/1111	0.4% C 5.0% Cr 1.3% Mo 0.5% V steel
ASTM A649/3	0.35% C (max) 0.55% Mn 1.0% Cr 0.2% Mo steel: For rolls; corrugated
ASTM A687/1	0.36% C 1.1% Mn 0.3% Cr 0.22% Ni 0.08% Mo steel: For bolts; studs; impact 20 J at -20 °C **UTS: 1034 Elon: 15% Proof: 724**
ASTM A687/11	0.4% C 0.9% Mn 0.9% Cr 0.2% Mo steel: For bolts & studs; impact 20 J at -30 °C **UTS: 1034 Elon: 15% Proof: 724**
ASTM A706	0.3% C (max) 1.5% Mn (max) Cr + Ni + V + Mo at controlled (max) to give C equivalent 0.55% (max) **UTS: 550 Elon: 14% Proof: 500**
ASTM A707 L4/1	0.2% C (max) 0.5% Mn 1.8% Ni 0.21% Mo steel: Flanges for low temperature use; impact 41 J at -62 °C **UTS: 290 Elon: 22%**
ASTM A707 44/2	0.2% C (max) 0.5% Mn 1.8% Ni 0.21% Mo steel: Flanges for low temperature use; impact 41 J at -62 °C **UTS: 360 Elon: 25%**
ASTM A707 L4/3	0.2% C (max) 0.5% Mn 1.8% Ni 0.21% Mo steel: Flanges for low temperature use; impact 68 J at -62 °C **UTS: 415 Elon: 23%**
ASTM A707 L4/4	0.2% C (max) 0.5% Mn 1.8% Ni 0.21% Mo steel: Flanges for low temperature use; impact 68 J at -62 °C **UTS: 515 Elon: 20%**
ASTM A707 L5/1	0.09% C (max) 0.75% Cr 0.8% Ni 0.2% Mo 1.1% Cu 0.02% Nb (min) steel: Flanges for low temperature use; impact at 41 J at -62 °C **UTS: 290 Elon: 22%**
ASTM A707 L5/2	0.09% C (max) 0.75% Cr 0.8% Ni 0.2% Mo 1.1% Cu 0.2% Nb (min) steel: Flanges for low temperature use; impact 64 J at -62 °C **UTS: 360 Elon: 25%**

Symbol	Nominal analysis, supplier, condition and remarks.
ASTM A707 L5/3	0.09% C (max) 0.75% Cr 0.8% Ni 0.2% Mo 1.1% Cu 0.02% Nb (min) steel: Flanges for low temperature use; impact 68 J at -62 °C **UTS: 415 Elon: 23%**
ASTM A707 L5/4	0.09% C (max) 0.75% Cr 0.8% Ni 0.2% Mo 1.1% Cu 0.02% Nb (min) steel: Flanges for low temperature use; impact 68 J at -62 °C **UTS: 515 Elon: 20%**
ASTM A707 L8/1	0.22% C (max) 0.3% Mn 1.8% Cr 3.2% Ni 0.5% Mo 0.04% V steel: Flanges for low temperature use; impact 41 J at -73 °C **UTS: 290 Elon: 22%**
ASTM A707 L8/2	0.22% C (max) 0.3% Mn 1.8% Cr 3.2% Ni 0.5% Mo 0.04% V steel: Flanges for low temperature use; impact 54 J at -73 °C **UTS: 360 Elon: 25%**
ASTM A707 L8/3	0.22% C (max) 0.3% Mn 1.8% Cr 3.2% Ni 0.5% Mo 0.04% V steel: Flanges for low temperature use; impact 68 J at -73 °C **UTS: 415 Elon: 23%**
ASTM A707 L8/4	0.22% C (max) 0.3% Mn 1.8% Cr 3.2% Ni 0.5% Mo 0.04% V steel: Flanges for low temperature use; impact 68 J at -73 °C **UTS: 515 Elon: 20%**
ASTM A710 A	0.07% C (max) 0.75% Cr 0.85% Ni 0.2% Mo 0.02% Nb (max) steel
ASTM A710 B	0.06% C (max) 1.3% Ni 1.1% Cu 0.02% Nb (min) steel
ASTM A714 4	0.13% C (max) 0.3% Cu 1.0% Cr 0.35% Ni steel: Pipe; welded or seamless **UTS: 400 Proof: 248**
ASTM A714/6 E & S	0.18% C (max) 0.8% Mn 0.6% Cu 0.3% Cr (max) 0.8% Ni 0.15% Mo steel: Pipe; welded or seamless **UTS: 448 Proof: 317**
ASTM A723/1	0.35% C (max) 1.4% Cr 0.3% Mo 0.2% V (max) steel: For pressure applications
ASTM A723/2	0.4% C (max) 1.4% Cr 0.4% Mo 0.2% V (max) steel: For pressure applications
ASTM A723/3	0.4% C (max) 1.4% Cr 0.6% Mo 0.2% V (max) steel: For pressure application
ASTM A732/7 Q	0.3% C 0.5% Mn 1.0% Cr 0.2% Mo steel: Investment casting; hardened & tempered **UTS: 1000 Elon: 7% Proof: 800**
ASTM A732/8 Q	0.4% C 0.8% Mn 1.0% Cr 0.2% Mo steel: Investment casting; hardened & tempered **UTS: 1240 Elon: 5% Proof: 1000**
ASTM A732/9 Q	0.3% C 0.8% Cr 1.8% Ni 0.25% Mo steel: Investment casting; hardened & tempered **UTS: 1030 Elon: 7% Proof: 795**
ASTM A732/10 Q	0.4% C 0.8% Cr 1.8% Ni 0.25% Mo steel: Investment casting; hardened & tempered **UTS: 1240 Elon: 5% Proof: 1000**
ASTM A732/13 Q	0.2% C 0.5% Cr 0.5% Ni 0.2% Mo steel: Investment casting; hardened & tempered **UTS: 720 Elon: 10% Proof: 590**
ASTM A732/14 Q	0.3% C 0.5% Cr 0.5% Ni 0.2% Mo steel: Investment casting; hardened & tempered **UTS: 1030 Elon: 7% Proof: 790**
ASTM A734 A	0.14% C (max) 0.6% Mn 1.0% Cr 1.1% Ni 0.3% Mo steel: Plate for pressure vessels; hardened & tempered **UTS: 600 Elon: 20% Proof: 450**
ASTM A734 B	0.19% C (max) 1.6% Mn (max) 0.3% Cr (max) 0.13% V steel: Plate for pressure vessels; hardened & tempered **UTS: 600 Elon: 20% Proof: 450**
ASTM A736	0.09% C (max) 0.55% Mn 0.75% Cr 0.8% Ni 0.2% Mo 1.1% Cu 0.02% Nb (min) steel: Plate for pressure use

Note. The following abbreviations and units are used in the tables:

DPN	Hardness, diamond pyramid number
UTS	Ultimate tensile strength, N/mm²
Elon	Elongation, %
Proof	0.1 % proof strength, N/mm²

1 N/mm²=0.1 hbar=0.102 kgf/mm²=0.06475 tonf/in²=145.04 lbf/in²

Symbol	Nominal analysis, supplier, condition and remarks.
ASTM B432	Mild or low alloy steel plate clad with Cu or copper alloy
ATLANTES MT	0.12% C (max) 0.5% Mn 0.8% Cr 0.5% Ni 0.06% Cu steel: South Durham
AURIGA VIII	0.2% C 0.4% Si 0.25% Cr 0.5% Mo steel: Casting; D Brown; normalized & tempered for high temperature use **UTS: 450 Elon: 20% Proof: 240**
AURIGA VIIIB	0.18% C 0.3% Si 0.25% Cr 0.5% Mo 0.3% V steel: Casting; D Brown; normalized & tempered for high temperature use **UTS: 530 Elon: 20% Proof: 300**
AURIGA VIIIC	0.15% C 0.3% Si 2.25% Cr 1% Mo steel: Casting; D Brown; normalized & tempered for high temperature use **UTS: 460 Elon: 20% Proof: 270**
AURIGA IX	0.32% C 0.8% Mn 1.2% Cr 1% Ni 0.3% Mo steel: Casting; D Brown; hardened & tempered **UTS: 920 Elon: 12% Proof: 700**
AURIGA IXB	0.32% C 0.7% Mn 1.2% Cr 1.2% Ni 0.4% Mo steel: Casting; D Brown; hardened & tempered **UTS: 1110 Elon: 10% Proof: 780**
AURIGA XV	0.25% C 0.4% Si 3% Cr 0.5% Mo steel: Casting; D Brown; hardened & tempered for high temperature use **UTS: 620 Elon: 18% Proof: 360**
AURIGA XVI	0.3% C 0.7% Mn 0.8% Cr 2% Ni 0.4% Mo steel: Casting; D Brown; hardened & tempered **UTS: 1110 Elon: 10% Proof: 780**
AURIGA XXI	0.15% C 0.5% Si 5% Cr 0.5% Mo steel: Casting; D Brown; hardened & tempered for high temperature use **UTS: 620 Elon: 18% Proof: 400**
AURIGA XXII	0.15% C 0.8% Si 9% Cr 1% Mo steel: Casting; D Brown; hardened & tempered for high temperature use **UTS: 620 Elon: 18% Proof: 400**
AURIGA XXXIV	0.3% C 1.5% Mn 0.6% Cr 0.8% Ni 0.4% Mo steel: Casting; D Brown; hardened & tempered **UTS: 1110 Elon: 10% Proof: 780**
AUTROD 13/12	0.1% C 1.3% Mn 0.6% Si 1.0% Cr 0.5% Mo steel: Welding wire; copper coated; Esab; for gas shielded welding; as welded **UTS: 630 Elon: 26% Proof: 530**
A W DYNALLOY 10	0.15% C (max) 0.8% Mn 0.6% Ni 0.1% Mo 0.45% Cu steel: Alan Wood
A W DYNALLOY 50	0.15% C (max) 1.0% Mn 0.55% Ni 0.15% Mo 0.45% Cu steel: Alan Wood
AZ	0.3% C 3.2% Cr 8.5% W 0.25% V steel: For hot dies; S Osborn; tools; extrusion dies, rivet dies; now H31
AZN	0.25% C 8.5% W 3% Cr 0.25% V 2.2% Ni steel: For hot dies; tools; S Osborn; now H21N
B 45CH	0.15% C 0.6% Ni 0.65% Cr 0.12% Mo steel: For carburizing; Firth Brown for BS 970 En 361
B 55CH	0.2% C 0.6% Ni 0.65% Cr 0.12% Mo steel: For carburizing; Firth Brown for BS 970 En 362
B 65CH	0.24% C 0.6% Ni 0.65% Cr 0.12% Mo steel: For carburizing; Firth Brown for BS 970 En 363
BENUM	0.3% C 1.25% Cr 4.1% Ni 0.3% Mo steel: Origin unknown
BCD 37	0.37% C 1.6% Cr 2.0% W 0.3% V steel: Carrs; chisels
BCH	0.18% C (max) 0.8% Cr 3.5% Ni steel: Toledo for BS alloy En 36
BCMF	0.35% C 0.6% Cr 2.5% Ni 0.5% Mo steel: Toledo for BS alloys En 25 & 26
BCMO	0.3% C 0.9% Cr 3.25% Ni 0.5% Mo steel: Toledo for BS alloy En 27

Symbol	Nominal analysis, supplier, condition and remarks.
BCN	0.3% C 0.7% Cr 3.25% Ni steel: Toledo for BS alloy En 23
BCN (Mo)	0.3% C 0.7% Cr 3.25% Ni 0.06% Mo steel: Toledo for BS alloy En 23
BH 43	0.18% C (max) 1.3% Mn 0.55% Ni 0.18% V steel: Rheinstahl
BH 43S	0.17% C (max) 1.3% Mn 0.55% Ni 0.18% V steel: Rheinstahl
BH 47	0.19% C (max) 1.3% Mn 0.55% Ni 0.2% V steel: Rheinstahl
BH 47S	0.18% C (max) 1.3% Mn 0.55% Ni 0.2% V steel: Rheinstahl
BH 51	0.2% C (max) 1.5% Mn 0.55% Ni 0.22% V steel: Rheinstahl
BH 51	0.2% C (max) 1.5% Mn 0.65% Ni 0.16% V steel: German proprietor's steel
BH 51S	0.2% C (max) 1.5% Mn 0.55% Ni 0.22% V steel: Rheinstahl
BH 57V	0.16% C 1.3% Mn 0.55% Ni 0.16% V steel: Rheinstahl; Al killed
BH 57VT	0.16% C 1.3% Mn 0.55% Ni 0.16% V steel: Rheinstahl; Al killed
BH 65V	0.18% C 1.5% Mn 0.55% Ni 0.18% V steel: Rheinstahl; Al killed
BH 65VT	0.18% C 1.5% Mn 0.55% Ni 0.18% V steel: Rheinstahl; Al killed
BL 30	0.25% C 1.5% Cr 0.8% Ni 0.7% Mo 0.2% V steel: Pompey
BRUNSWICK V108	0.35% C 5% Cr + Mo + V steel: John Vessey; hot work tools **DPN: 425**
BRUNSWICK V/110/CNA	0.3% C 1.2% Cr 4% Ni 0.3% Mo steel: John Vessey; stamping dies, shear blades, etc. **DPN: 400**
BS 24/5/614	0.23% C (max) 1.5% Mn Ni Cr steel: Plate & bar for railway use
BS 682	3% Ni Cr case hardening steel: Obsolete
BS 806 P	0.15% C 0.8% Cr 0.5% Mo steel: Seamless pipe; cold drawn & normalized **UTS: 500**
BS 806 Q	0.15% C 0.8% Cr 0.5% Mo steel: Seamless pipe; hot drawn & normalized **UTS: 500**
BS 970/635 A14	0.15% C 0.8% Mn 0.85% Ni 0.65% Cr 0.1% Mo steel: For carburizing
BS 970/635 M15	0.15% C 0.8% Mn 0.9% Ni 0.6% Cr steel: For carburizing; supplied to mechanical properties
BS 970/637 A16	0.16% C 0.8% Mn 1.0% Ni 0.8% Cr 0.1% Mo steel: For carburizing
BS 970/637 H17	0.17% C 0.7% Mn 1.0% Ni 0.8% Cr steel: For carburizing; supplied on hardenability
BS 970/637 M17	0.17% C 0.75% Mn 1.0% Ni 0.8% Cr steel: For carburizing; supplied to mechanical properties; replaces BS En 352
BS 970/640 A35	0.35% C 1.3% Mn 0.65% Cr steel: Supplied to chemical composition
BS 970/640 H35	0.35% C 0.75% Mn 1.25% Ni 0.65% Cr steel: Supplied to hardenable requirements
BS 970/653 M31	0.31% C 0.55% Mn 3.0% Ni 1.1% Cr steel: Supplied to mechanical properties; replaces En 23
BS 970/635 A12	0.13% C 0.5% Mn 3.25% Ni 0.85% Cr steel: For carburizing
BS 970/655 H13	0.13% C 0.5% Mn 3.5% Ni 0.8% Cr steel: For carburizing; supplied to hardenability
BS 970/655 M13	0.13% C 0.45% Mn 3.3% Ni 0.85% Cr steel: For carburizing; supplied to mechanical properties; replaces BS 970 En 36A
BS 970/659 A15	0.16% C 0.4% Mn 4.1% Ni 1.3% Cr steel: For carburizing

Symbol	Nominal analysis, supplier, condition and remarks.
BS 970/659 H15	0.15% C 0.4% Mn 4.1% Ni 1.2% Cr steel: For carburizing; supplied to hardenability
BS 970/659 M15	0.15% C 0.35% Mn 4.1% Ni 1.2% Cr steel: For carburizing; supplied to mechanical properties; replaces BS 970 En 39A
BS 970/665 A17	0.17% C 0.5% Mn 1.8% Ni 0.25% Mo steel: For carburizing
BS 970/665 A19	0.2% C 0.5% Mn 1.8% Ni 0.25% Mo steel: For carburizing
BS 970/665 A22	0.22% C 0.5% Mn 1.8% Ni 0.25% Mo steel: For carburizing; was BS 970 En35A
BS 970/665 A24	0.25% C 0.5% Mn 1.8% Ni 0.25% Mo steel: For carburizing; was BS 970 En 35B
BS 970/665 H17	0.17% C 0.55% Mn 1.75% Ni 0.25% Mo steel: For carburizing; supplied to hardenability
BS 970/665 H20	0.2% C 0.5% Mn 1.75% Ni 0.25% Mo steel: For carburizing; supplied to hardenability
BS 970/665 H23	0.23% C 0.5% Mn 1.75% Ni 0.25% Mo steel: For carburizing; supplied to hardenability
BS 970/655 M17	0.17% C 0.5% Mn 1.75% Ni 0.25% Mo steel: For carburizing; supplied to mechanical properties; replaces BS En 34 UTS: 750 Elon: 12%
BS 970/665 M20	0.3% C 0.5% Mn 1.75% Ni 0.25% Mo steel: For carburizing; supplied to mechanical properties UTS: 820 Elon: 11%
BS 970/665 M23	0.23% C 0.5% Mn 1.75% Ni 0.25% Mo steel: For carburizing; supplied to mechanical properties; replaces BS 970 En 35 UTS: 900 Elon: 10%
BS 970/722 M24	0.24% C 0.6% Mn 3.2% Cr 0.55% Mo steel: Supplied to mechanical properties; replaces En 40B
BS 970/785 M19	0.19% C 1.6% Mn 0.55% Ni 0.25% Mo steel: Supplied to mechanical properties; replaces En 13
BS 970/805 A15	0.16% C 0.8% Mn 0.5% Ni 0.5% Cr 0.2% Mo steel: For carburizing
BS 970/805 A17	0.17% C 0.8% Mn 0.5% Ni 0.5% Cr 0.2% Mo steel: For carburizing
BS 970/805 A20	0.2% C 0.8% Mn 0.5% Ni 0.5% Cr 0.2% Mo steel: For carburizing
BS 970/805 A22	0.22% C 0.8% Mn 0.5% Ni 0.5% Cr 0.2% Mo steel: For carburizing
BS 970/805 A24	0.24% C 0.8% Mn 0.5% Ni 0.5% Cr 0.2% Mo steel: For carburizing
BS 970/805 H17	0.17% C 0.8% Mn 0.5% Ni 0.5% Cr 0.2% Mo steel: For carburizing; supplied to hardenability
BS 970/805 H20	0.2% C 0.8% Mn 0.5% Ni 0.5% Cr 0.2% Mo steel: For carburizing; supplied to hardenability
BS 970/805 H22	0.22% C 0.8% Mn 0.5% Ni 0.5% Cr 0.2% Mo steel: For carburizing; supplied for hardenability
BS 970/805 H25	0.25% C 0.8% Mn 0.5% Ni 0.5% Cr 0.2% Mo properties; replaces BS 970 En 362
BS 970/805 M17	0.17% C 0.75% Mn 0.5% Ni 0.5% Cr 0.2% Mo steel: For carburizing; supplied to mechanical properties; replaces BS 970 En 360 UTS: 750 Elon: 12%
BS 970/805 M20	0.2% C 0.7% Mn 0.5% Ni 0.5% Cr 0.2% Mo steel: For carburizing; supplied to mechanical

Note. The following abbreviations and units are used in the tables:

DPN	Hardness, diamond pyramid number
UTS	Ultimate tensile strength, N/mm^2
Elon	Elongation, %
Proof	0.1 % proof strength, N/mm^2

1 N/mm^2=0.1 hbar=0.102 kgf/mm^2=0.06475 tonf/in^2=145.04 lbf/in^2

Symbol	Nominal analysis, supplier, condition and remarks.
BS 970/805 M22	0.22% C 0.75% Mn 0.5% Ni 0.5% Cr 0.2% Mo steel: For carburizing; supplied to mechanical properties UTS: 900 Elon: 10%
BS 970/805 M25	0.25% C 0.75% Mn 0.5% Ni 0.5% Cr 0.2% Mo steel: For carburizing; supplied to mechanical properties
BS 970/815 A16	0.16% C 0.8% Mn 1.4% Ni 1.0% Cr 0.15% Mo steel: For carburizing
BS 970/815 H17	0.17% C 0.75% Mn 1.5% Ni 1.0% Cr 0.15% Mo steel: For carburizing; supplied to hardenability
BS 970/815 M17	0.17% C 0.75% Mn 1.5% Ni 1.0% Cr 0.15% Mo steel: For carburizing; supplied to mechanical properties; replaces BS 970 En 353 UTS: 1050 Elon: 8%
BS 970/820 A16	0.1% C 0.8% Mn 1.75% Ni 1.0% Cr 0.15% Mo steel: For carburizing
BS 970/820 H17	0.17% C 0.8% Mn 1.75% Ni 1.0% Cr 0.15% Mo steel: For carburizing; supplied to hardenability
BS 970/820 M17	0.17% C 0.75% Mn 1.75% Ni 1.0% Cr 0.15% Mo steel: For carburizing; supplied to mechanical properties; replaces BS 970 En 355 UTS: 1050 Elon: 8%
BS 970/822 A17	0.17% C 0.8% Mn 2.0% Ni 1.5% Cr 0.2% Mo steel: For carburizing
BS 970/822 H17	0.17% C 0.2% Mn 2.0% Ni 1.5% Cr 0.2% Mo steel: For carburizing; supplied to hardenability
BS 970/822 M17	0.17% C 0.55% Mn 2.0% Ni 1.5% Cr 0.2% Mo steel: For carburizing; supplied to mechanical properties UTS: 1200 Elon: 8%
BS 970/823 M30	0.3% C 2.0% Ni 2.0% Cr 0.4% Mo steel: Supplied to mechanical properties
BS 970/826 M31	0.31% C 2.5% Ni 0.65% Cr 0.5% Mo steel: Supplied to mechanical properties; replaces En 25
BS 970/830 M31	0.31% C 3.0% Ni 1.0% Cr 0.3% Mo steel: Supplied to mechanical properties; replaces En 27
BS 970/832 H13	0.13% C 0.5% Mn 3.3% Ni 0.8% Cr 0.2% Mo steel: For carburizing; supplied to hardenability
BS 970/832 M13	0.13% C 0.5% Mn 3.3% Ni 0.85% Cr 0.15% Mo steel: For carburizing; supplied to mechanical properties; replaces BS 970 En 36C UTS: 1050 Elon: 8%
BS 970/835 A15	0.15% C 0.4% Mn 4.1% Ni 1.3% Cr 0.2% Mo steel: For carburizing
BS 970/835 H15	0.15% C 0.4% Mn 4.1% Ni 1.2% Cr 0.2% Mo steel: For carburizing; supplied to hardenability
BS 970/835 M15	0.15% C 0.35% Mn 4.1% Ni 1.2% Cr 0.2% Mo steel: For carburizing; supplied to mechanical properties; replaces BS 970 En 39B UTS: 1020 Elon: 8%
BS 970/835 M30	0.3% C 4.1% Ni 1.2% Cr 0.27% Mo steel: Supplied to mechanical properties; replaces En 30B
BS 970/905 M31	0.31% C 1.6% Cr 0.2% Mo 1.0% Al steel: Supplied to mechanical properties; for nitriding; replaces En 41A
BS 970 En 13	0.2% C 1.5% Mn 0.5% Ni 0.2% Mo steel: Bar & forging; hardened & tempered DPN: 200 UTS: 620 Elon: 22% Proof: 450
BS 970 En 14A	0.2% C 1.5% Mn (max) 0.25% Cr 0.4% Ni (max) steel: Bar & forging; hardened & tempered DPN: 215 UTS: 700 Elon: 20% Proof: 450
BS 970 En 20	0.35% C 1% Cr 0.6% Mo steel: Bar & forging; hardened & tempered DPN: 320 UTS: 1000 Elon: 16% Proof: 780
BS 970 En 21	0.3% C 0.3% Cr 3% Ni steel: Bar & forging; hardened & tempered DPN: 250 UTS: 770 Elon: 20% Proof: 580

Symbol	Nominal analysis, supplier, condition and remarks.
BS 970 En 23	0.3% C 0.75% Cr 3% Ni steel: Bar & forging; hardened & tempered **DPN: 320 UTS: 1000 Elon: 16% Proof: 780**
BS 970 En 25	0.3% C 0.7% Cr 2.5% Ni 0.5% Mo steel: Bar & forging; hardened & tempered **DPN: 320 UTS: 1000 Elon: 16% Proof: 780**
BS 970 En 27	0.3% C 1% Cr 3.5% Ni 0.5% Mo steel: Bar & forging; hardened & tempered **DPN: 320 UTS: 1000 Elon: 16% Proof: 780**
BS En 28	0.3% C 1% Cr 3.5% Ni 0.4% Mo steel: Bar & forging; hardened & tempered **DPN: 320 UTS: 1000 Elon: 16% Proof: 780**
BS 970 En 29	0.25% C 3% Cr 0.4% Ni 0.5% Mo steel: Bar & forging; hardened & tempered **DPN: 320 UTS: 1000 Elon: 16% Proof: 780**
BS 970 En 30A	0.3% C 0.5% Mn 1.2% Cr 4.25% Ni steel: Bar & forging; hardened & tempered **DPN: 444 UTS: 1550 Elon: 10% Proof: 1300**
BS 970 En 30B	0.3% C 1.2% Cr 4.25% Ni 0.3% Mo steel: Bar & forging; hardened & tempered **DPN: 444 UTS: 1550 Elon: 10% Proof: 1300**
BS 970 En 33	0.12% C 0.3% Cr 3% Ni steel: For case hardening; core properties; blank carburized **UTS: 710 Elon: 18%**
BS 970 En 34	0.17% C 1.75% Ni 0.25% Mo steel: For case hardening; core properties; blank carburized **UTS: 710 Elon: 18%**
BS 970 En 35	0.25% C 1.75% Ni 0.25% Mo steel: For case hardening; core properties; blank carburized **UTS: 840 Elon: 15%**
BS 970 En 35A	0.22% C 1.75% Ni 0.25% Mo steel: For case hardening
BS 970 En 35B	0.25% C 1.75% Ni 0.25% Mo steel: For case hardening
BS 970 En 36	0.18% C (max) 0.9% Cr 3.5% Ni steel: For case hardening; core properties; blank carburized **UTS: 920 Elon: 14%**
BS 970 En 37	0.16% C (max) 0.3% Cr 5% Ni steel: For case hardening; core properties; blank carburized **UTS: 760 Elon: 20%**
BS 970 En 38	0.16% C (max) 5% Ni 0.2% Mo steel: For case hardening; core properties; blank carburized **UTS: 1000 Elon: 13%**
BS 970 En 39A	0.15% C 1.2% Cr 4.25% Ni steel: For case hardening; core properties; blank carburized **UTS: 1300 Elon: 12%**
BS 970 En 39B	0.15% C 1.2% Cr 4.25% Ni 0.2% Mo steel: For case hardening; core properties; blank carburized **UTS: 1300 Elon: 12%**
BS 970 En 40A	0.15% C 3% Cr 0.4% Ni 0.6% Mo steel: Hardened & tempered; suitable for nitriding **DPN: 280 UTS: 920 Elon: 17% Proof: 740**
BS 970 En 40B	0.25% C 3% Cr 0.4% Ni 0.6% Mo steel: Hardened & tempered; suitable for nitriding **DPN: 280 UTS: 920 Elon: 17% Proof: 270**
BS 970 En 41	0.25% C 1.5% Cr 0.4% Ni 0.2% Mo 1.0% Al steel: Hardened & tempered; suitable for nitriding **DPN: 230 UTS: 700 Elon: 20% Proof: 480**
BS 970 En 51	0.3% C 0.3% Cr 3% Ni steel: Bar & forging; hardened & tempered for valves **DPN: 229**
BS 970 En 100A	0.27% C 1.3% Mn 0.4% Cr 0.7% Ni 0.2% Mo steel: Bar & forging; mechanical properties not part of specification
BS 970 En 100B	0.32% C 1.3% Mn 0.4% Cr 0.7% Ni 0.2% Mo steel: Bar & forging
BS 970 En 100C	0.37% C 1.3% Mn 0.4% Cr 0.7% Ni 0.2% Mo steel: Bar & forging

Symbol	Nominal analysis, supplier, condition and remarks.
BS 970 En 111	0.35% C 0.6% Cr 1.25% Ni steel: Bar & forging; hardened & tempered **DPN: 280 UTS: 840 Elon: 18% Proof: 680**
BS 970 En 111A	0.35% C 0.6% Cr 1.25% Ni steel: Bar & forging held to tighter limit than En 111
BS 970 En 320	0.17% C 2% Cr 2% Ni 0.2% Mo steel: For case hardening; blank carburized; double hardened **UTS: 1300 Elon: 12%**
BS 970 En 325	0.2% C 0.5% Cr 1.75% Ni 0.25% Mo steel: For case hardening; blank carburized; double hardened **UTS: 840 Elon: 15%**
BS 980 CDS12	0.25% C 1% Cr 0.2% Mo steel: Cold drawn tube; as drawn & tempered; can be welded & heat treated **UTS: 700 Proof: 580**
BS 980 CDS13	0.35% C 1% Cr 0.2% Mo steel: Cold drawn tube; hardened & tempered **UTS: 700 Proof: 580**
BS 980 CDS16	0.25% C 4% Ni 1.2% Cr 0.2% Mo steel: Cold drawn tube; hardened & tempered **UTS: 1210 Elon: 7% Proof: 1070**
BS 980 CDS17	0.3% C 4% Ni 1.2% Cr 0.2% Mo steel: Cold drawn tube; hardened & tempered **UTS: 1380 Elon: 5% Proof: 1140**
BS 1453 A4	0.3% C 0.3% Cr 3% Ni steel: Weld filler rod; for welds to be hardened & tempered
BS 1461	0.25% C (max) 3.0% Cr 0.45% Mo steel: Casting; hardened & tempered; contained in BS 3100 **UTS: 700 Elon: 18% Proof: 400**
BS 1462	0.2% C (max) 5.0% Cr 0.55% Mo steel: Casting; hardened & tempered; contained in BS 3100 **UTS: 610 Elon: 18% Proof: 400**
BS 1463	0.2% C (max) 9.0% Cr 1.1% Mo steel: Casting; hardened & tempered; contained in BS 3100 **UTS: 610 Elon: 18% Proof: 400**
BS 1501/271	0.14% C 1.2% Mn 0.55% Cr 0.23% Mo 0.1% V steel: Plate; normalized & tempered; for pressure vessels **UTS: 650 Elon: 16% Proof: 400**
BS 1501/281	0.13% C 1.1% Mn 0.9% Ni 0.55% Cr 0.24% Mo 0.1% V steel: Plate; normalized & tempered; for pressure vessels **UTS: 650 Elon: 16% Proof: 400**
BS 1501/282	0.14% C 1.1% Mn 1.5% Ni 0.5% Cr 0.35% Mo 0.1% V steel: Plate; normalized & tempered; for pressure vessels **UTS: 660 Elon: 18% Proof: 420**
BS 1501/620/27	0.12% C 1.0% Cr 0.5% Mo steel: Plate; normalized & tempered; for pressure vessels **UTS: 480 Elon: 21% Proof: 300**
BS 1501/620/31	0.15% C 1.0% Cr 0.5% Mo steel: Plate; normalized & tempered; for pressure vessels **UTS: 530 Elon: 18% Proof: 330**
BS 1501/620A	0.15% C 0.65% Cr 0.6% Mo steel: Plate & bar; normalized **UTS: 420 Elon: 20% Proof: 220**
BS 1501/620B	0.15% C 0.9% Cr 0.6% Mo steel: Plate & bar; normalized **UTS: 440 Elon: 20% Proof: 220**
BS 1501/621	0.12% C 1.2% Cr 0.55% Mo steel: Plate **UTS: 520 Elon: 18% Proof: 330**
BS 1501/622/31	0.13% C 2.2% Cr 1.0% Mo steel: Plate; normalized & tempered; for pressure vessels **UTS: 550 Elon: 18% Proof: 290**
BS 1501/622/45	0.15% C 2.2% Cr 1.0% Mo steel: Plate; normalized & tempered; for pressure vessels **UTS: 760 Elon: 15% Proof: 570**
BS 1501/625	0.15% C 0.55% Mo 5.0% Cr steel: Plate & bar; annealed **UTS: 400 Elon: 20% Proof: 210**

Symbol	Nominal analysis, supplier, condition and remarks.
BS 1503/620	0.20% C 1.0% Cr 0.5% Mo steel: Forging; normalized & tempered **UTS: 460 Elon: 20% Proof: 270**
BS 1503/622	0.15% C 2.3% Cr 1.0% Mo steel: Forging; normalized & tempered **UTS: 460 Elon: 20% Proof: 270**
BS 1503/623	0.2% C 3% Cr 0.4% Mo steel: Normalized & tempered **UTS: 670 Elon: 23% Proof: 420**
BS 1503/625	0.25% C 5.0% Cr 0.5% Mo steel: Forgings; hardened & tempered **UTS: 610 Elon: 20% Proof: 440**
BS 1504/621	0.2% C (max) 1.2% Cr 0.5% Mo steel: Casting; normalized & tempered **UTS: 460 Elon: 20% Proof: 270**
BS 1505/622	0.18% C (max) 2.5% Cr 1.0% Mo steel: Casting; normalized & tempered **UTS: 460 Elon: 20% Proof: 270**
BS 1504/623	0.25% C (max) 3.0% Cr 0.5% Mo steel: Casting; hardened & tempered **UTS: 700 Elon: 18% Proof: 360**
BS 1504/625	0.2% C (max) 5% Cr 0.5% Mo steel: Casting; hardened & tempered **UTS: 610 Elon: 18% Proof: 400**
BS 1504/629	0.2% C (max) 9% Cr 1.0% Mo steel: Casting; hardened & tempered **UTS: 610 Elon: 18% Proof: 400**
BS 1506/625	0.3% C (max) 5.0% Cr 0.5% Mo steel: Bar; hardened & tempered **UTS: 700 Elon: 19% Proof: 580**
BS 1506/661	0.3% C 1.0% Cr 0.6% Mo 0.25% V steel: Bar; normalized & tempered **UTS: 840 Elon: 16% Proof: 720**
BS 1507/621	0.15% C 1% Cr 0.5% Mo steel: Seamless pipe; normalized **UTS: 450 Elon: 24%**
BS 1507/623	0.15% C 3% Cr 0.5% Mo steel: Pipe; normalized **UTS: 450 Elon: 22%**
BS 1507/625	0.15% C 0.5% Mo 5% Cr steel: Pipe; normalized **UTS: 450 Elon: 20%**
BS 1508/621	0.15% C 1% Cr 0.5% Mo steel: Tube; normalized **UTS: 450 Elon: 22%**
BS 1508/622	0.15% C 2.5% Cr 1% Mo steel: Tube; normalized **UTS: 400 Elon: 20%**
BS 1508/625	0.15% C 5% Cr 0.5% Mo steel: Tube; normalized **UTS: 450 Elon: 18%**
BS 1628 A	0.15% C 2.5% Cr 1% Mo steel: Cold drawn tube; annealed **UTS: 400 Elon: 27%**
BS 1628 B	0.15% C 5% Cr 0.5% Mo steel: Cold drawn tube; annealed **UTS: 400 Elon: 27%**
BS 1653	Cr Mo steel: Tube; replaced by BS 3059
BS 1717 CDS110	0.26% C 1% Cr 0.25% Mo steel: Seamless tube; as drawn **UTS: 700 Proof: 370**
BS 1717 CDS110	0.26% C 1% Cr 0.25% Mo steel: Seamless tube; after welding, brazing or annealing **UTS: 550 Proof: 370**

Note. The following abbreviations and units are used in the tables:

DPN	Hardness, diamond pyramid number
UTS	Ultimate tensile strength, N/mm²
Elon	Elongation, %
Proof	0.1 % proof strength, N/mm²

1 N/mm²=0.1 hbar=0.102 kgf/mm²=0.06475 tonf/in²=145.04 lbf/in²

Symbol	Nominal analysis, supplier, condition and remarks.
BS 2493 B	0.1% C 1% Cr 0.5% Mo steel: Welding electrode
BS 2493 C	0.1% C 2.2% Cr 1% Mo steel: Welding electrode
BS 2493 D	0.1% C 5% Cr 0.5% Mo steel: Welding electrode
BS 3059/7	0.15% C (max) 0.3% Ni 0.5% Mo 0.3% Cu steel: Tube; seamless; hot finished & normalized **UTS: 440**
BS 3059/8	0.15% C (max) 0.3% Ni 0.5% Mo 0.3% Cu steel: Tube; cold drawn; seamless; normalized **UTS: 440**
BS 3059/9	0.12% C (max) 0.8% Cr 0.3% Ni 0.5% Mo 0.3% Cu steel: Tube; hot finished; seamless; normalized **UTS: 440**
BS 3059/10	0.12% C 0.8% Cr 0.3% Ni 0.5% Mo 0.3% Cu steel: Tube; cold drawn; seamless; normalized **UTS: 440**
BS 3059/11	0.1% C 2.2% Cr 0.3% Ni 1.0% Mo steel: Tube; seamless; hot finished; normalized **UTS: 460**
BS 3059/12	0.1% C 2.2% Cr 0.3% Ni 1.1% Mo steel: Tube; seamless; cold drawn & normalized **UTS: 460**
BS 3100 B2	0.2% C (max) 0.7% Mn 1.2% Cr 0.5% Mo steel: Casting for high temperature use **UTS: 480 Elon: 17% Proof: 280**
BS 3100 B3	0.18% C (max) 0.6% Mn 2.3% Cr 1.0% Mo steel: Casting for high temperature use **UTS: 540 Elon: 17% Proof: 325**
BS 3100 B4	0.25% C (max) 0.5% Mn 3.0% Cr 0.5% Mo steel: Casting for high temperature use **UTS: 620 Elon: 13% Proof: 370**
BS 3100 B5	0.2% C (max) 0.5% Mn 5.0% Cr 0.5% Mo steel: Casting for high temperature use **UTS: 620 Elon: 13% Proof: 420**
BS 3100 B6	0.2% C (max) 0.5% Mn 9.0% Cr 1.0% Mo steel: Casting for high temperature use **UTS: 620 Elon: 13% Proof: 420**
BS 3100 B7	0.12% C 0.5% Mn 0.4% Cr 0.5% Mo 0.25% V steel: Casting for high temperature use **UTS: 510 Elon: 17% Proof: 300**
BS 3100 BW1	0.16% C 0.45% Mn 0.85% Cr 0.2% Mo 3.3% Ni steel: Casting for case hardening **UTS: 1000 Elon: 7%**
BS 3100 BW4	0.6% C 0.7% Mn 1.1% Cr 0.3% Mo steel: Casting for abrasion resistance
BS 3111/4	0.35% C 0.6% Cr 1.2% Ni steel: Wire; annealed & drawn; for cold forged bolts **UTS: 700**
BS 3127/6	0.3% C 1% Cr 0.2% Mo steel: Tube; annealed; for bourdon tubes **DPN: 165**
BS 3146 CLA1A	0.25% C 1% Mn low Ni Cr Mo Cu steel: Investment casting; hardened & tempered **DPN: 150 UTS: 420 Elon: 22% Proof: 210**
BS 3146 CLA1B	0.35% C 1% Mn low Ni Cr Mo Cu steel: Investment casting; hardened & tempered **DPN: 160 UTS: 480 Elon: 20% Proof: 210**
BS 3146 CLA2A	0.21% C 1.5% Mn low Ni Cr Mo Cu steel: Investment casting; hardened & tempered **DPN: 175 UTS: 530 Elon: 18%**
BS 3146 CLA2B	0.29% C 1.5% Mn low Cr Ni Mo Cu steel: Investment casting; hardened & tempered **DPN: 220 UTS: 700 Elon: 15%**
BS 3146 CLA6	0.2% C low Cr Ni 0.6% Mo steel: Investment casting; normalized **DPN: 150 UTS: 450 Elon: 20% Proof: 240**
BS 3146 CLA7	0.25% C 3% Cr 0.5% Mo steel: Investment casting; hardened & tempered **DPN: 200 UTS: 610 Elon: 18% Proof: 360**

Symbol	Nominal analysis, supplier, condition and remarks.
BS 3146 CLA9	0.15% C low Cr Ni Mo Cu steel: Investment casting; for case hardening; refined & hardened core **UTS: 480 Elon: 20%**
BS 3146 CLA11	0.25% C 3.2% Cr 0.6% Mo steel: Investment casting; hardened & tempered; for nitriding **DPN: 260 UTS: 840 Elon: 12% Proof: 580**
BS 3551 C	0.2% C 0.6% Cr 0.6% Ni 0.2% Mo steel: For shackles; hardened & tempered **DPN: 270**
BS S16	Nickel chrome steel: For carburizing; replaced
BS S28	0.3% C 1.2% Cr 4.2% Ni 0.3% Mo steel: Bar & forging; hardened & tempered **DPN: 444 UTS: 1550 Elon: 12% Proof: 1030**
BS S82	0.15% C 1.2% Cr 4.1% Ni 0.2% Mo steel: For carburizing; blank carburized; hardened & tempered **UTS: 1300 Elon: 10%**
BS S90	Low C 5% Ni case hardening steel: Obsolete; see BS/S107
BS S96	0.32% C 0.7% Cr 2.5% Ni 0.55% Mo steel: Hardened & tempered **DPN: 280 UTS: 920 Elon: 18% Proof: 670**
BS S97	0.32% C 0.7% Cr 2.5% Ni 0.55% Mo steel: Hardened & tempered **DPN: 320 UTS: 1070 Elon: 16% Proof: 820**
BS S103	0.35% C 0.7% Cr 1.2% Ni steel: Wire for forged bolts; hardened & tempered after forging **UTS: 920 Elon: 18% Proof: 670**
BS S106	0.24% C 3.2% Cr 0.55% Mo steel: For nitriding; hardened & tempered **DPN: 290 UTS: 1060 Elon: 17% Proof: 700**
BS S107	0.13% C 0.8% Cr 3.3% Ni 0.15% Mo steel: For carburizing; blank carburized; hardened & tempered **UTS: 1070 Elon: 12%**
BS S120	0.32% C 0.6% Cr 2.5% Ni 0.55% Mo steel: Hardened & tempered **DPN: 450 UTS: 1550 Elon: 12% Proof: 1030**
BS S122	0.35% C 0.6% Cr 1.2% Ni steel: Bar; hardened & tempered **UTS: 920 Elon: 18% Proof: 650**
BS S518	0.25% C (max) 1.0% Cr 0.2% Mo steel: Sheet; hardened & tempered; weldable **UTS: 840 Elon: 12% Proof: 610**
BS S519	0.25% C (max) 1.0% Cr 0.2% Mo steel: Sheet; hardened & tempered; weldable **UTS: 1210 Elon: 5% Proof: 1000**
BSS 1462	0.2% C 5% Cr 0.55% Mo steel: Jessop
BS STA 5 V20	0.3% C 1.5% Cr 0.2% Mo 1% Al steel: Replaced by BS 970 En 41
BS T2	0.25% C 1.0% Cr 4.0% Ni 0.2% Mo steel: Tube; hardened & tempered **UTS: 1450 Elon: 7% Proof: 1210**
BS T53	0.25% C (max) 1.0% Cr 0.2% Mo steel: Tube; hardened & tempered; weldable **UTS: 700**
BS T56	0.25% C (max) 1.0% Cr 0.2% Mo steel: Tube; hardened & tempered; weldable **UTS: 530 Proof: 450**
BS T57	0.25% C 1.0% Cr 4.0% Ni 0.2% Mo steel: Tube; hardened & tempered **UTS: 1210 Elon: 10% Proof: 1000**
BS T59	0.25% C 1.0% Cr 0.2% Mo steel: Tube; hardened & tempered; weldable **UTS: 760 Proof: 700**
BS T60	0.25% C (max) 1.0% Cr 0.2% Mo steel: Tube; hardened & tempered; weldable **UTS: 1210 Elon: 10% Proof: 1120**

Symbol	Nominal analysis, supplier, condition and remarks.
BVR 20h/W	0.28% C 1.15% Cr 1.15% Ni 0.25% Mo steel: Krupp
BVR 20T	0.28% C 1.15% Cr 1.15% Ni 0.45% Mo steel: Krupp
B & W WL2	0.15% C 9.0% Cr 0.5% Mo steel: Tube; Babcock & Wilcox (USA)
BYR 1	0.28% C 1.25% Cr 1.95% Ni 0.35% Mo steel: Krupp
C 5	0.35% C 5% Cr 1.5% Mo 1% W steel: Turton Bros
C 642	0.25% C 1.0% Cr 0.2% Mo steel: Fagersta
C 643	0.34% C 1.0% Cr 0.2% Mo steel: Fagersta
CANO 30	0.34% C 1.65% Cr 1.0% Ni 0.2% Mo 0.95% Al steel: For nitriding; Krupp
CAO 30	0.34% C 1.15% Cr 0.2% Mo 0.95% Al steel: For nitriding; Krupp
CARPENTER 345	0.35% C 5.0% Cr 1.25% W 1.5% Mo steel: Carpenter for AISI type H12
CARPENTER 345	0.35% C 5.0% Cr 1.25% W 1.5% Mo tool steel: Carpenter for AISI type H12
CASCADE	0.2% C 0.25% Cr 4.1% Ni 0.2% V 1.2% Al steel: Latrobe; solution treated & aged; type P21 **DPN: 230 UTS: 1300 Elon: 16% Proof: 1180**
CBS 600	0.22% C 1.5% Cr 1.0% Mo steel: For case hardening; Timken; for bearings at high temperature **DPN: 330 UTS: 1210 Elon: 17% Proof: 920**
CBVO	0.34% C 1.05% Cr 0.2% Mo steel: Krupp
CBVO/3R	0.32% C 3.0% Cr 0.4% Mo steel: Krupp
CBVO/M	0.25% C 1.05% Cr 0.2% Mo steel: Krupp
CBVO/TM	0.24% C 1.05% Cr 0.25% Mo steel: Krupp
CBVOV	0.3% C 2.5% Cr 0.2% Mo 0.15% V steel: Krupp
CBVOV/N	0.31% C 2.5% Cr 0.2% Ni 0.15% V steel: For nitriding; Krupp
CH 5	0.37% C 5% Cr 1.4% Mo 0.6% V steel: Marsh Brothers; hot forging dies etc
CH 45D	0.15% C 0.6% Cr 0.8% Ni steel: For case hardening; ESC for BS alloy En 351
CH 55B	0.16% C 0.8% Cr 1% Ni steel: For case hardening; ESC for BS alloy En 352
CH 65B	0.17% C 1% Cr 1.3% Ni steel: For case hardening; ESC for BS alloy En 353
CH 65J	0.16% C 0.85% Cr 1.5% Ni 0.2% Mo steel: For case hardening; ESC **UTS: 1000 Elon: 13%**
CH 75B	0.18% C 1% Cr 1.8% Ni 0.15% Mo steel: For case hardening; ESC for BS alloy En 354
CH 85A	0.18% C 1.5% Cr 2% Ni 0.2% Mo steel: For case hardening; ESC for BS alloy En 355
CH 361	0.15% C 0.7% Cr 0.6% Ni 0.1% Mo steel: For case hardening; ESC for BS alloy En 361
CH 362	0.2% C 0.7% Cr 0.6% Ni 0.1% Mo steel: For case hardening; ESC for BS alloy En 362
CH 363	0.25% C 0.7% Cr 0.6% Ni 0.1% Mo steel: For case hardening; ESC for BS alloy En 363
CHD	0.33% C 5.0% Cr 1.5% W 0.3% V 1.5% Mo tool steel: W Spencer
CHD 3	0.34% C 1% Si 5% Cr 1.6% Mo 1.3% W 0.4% V tool steel: Huntsman; hot punches, shear, etc. **DPN: 500**
CHIPPEWA	0.3% C 0.4% Cr 3.0% Ni 0.25% Mo steel: Tube; Atlas; for detachable rock drill bits
CHNC 55	0.15% C 0.8% Cr 3.2% Ni 0.17% Mo steel: For case hardening; ESC for BS alloy En 36A
CHNC 65	0.15% C 0.8% Cr 3.2% Ni 0.17% Mo steel: For case hardening; ESC for BS alloy En 36 B & C
CHNM 45	0.17% C 1.8% Ni 0.25% Mo steel: For case hardening; ESC for BS alloy En 34
CHNM 55	0.22% C 1.8% Ni 0.25% Mo steel: For case hardening; ESC for BS alloy En 35

Symbol	Nominal analysis, supplier, condition and remarks.
CHROMOLOY	0.2% C 1% Cr 1% Mo 0.1% V steel: American proprietary alloy listed in SAE yearbook
CHROMOTRODE 76/12	0.07% C 0.8% Mn 1.0% Cr 0.5% Mo steel: For electrodes; Esab
	UTS: 710 **Elon: 26%** **Proof: 602**
CHROMOTRODE 76/19	0.07% C 0.9% Mn 1.25% Cr 0.5% Mo steel: For electrodes; Esab
	UTS: 726 **Elon: 21%** **Proof: 618**
CHROMOTRODE 76/22	0.07% C 0.9% Mn 2.25% Cr 1.0% Mo steel: For electrodes; Esab
	UTS: 726 **Elon: 22%** **Proof: 618**
CHROMOTRODE 76/28	0.07% C 0.9% Mn 2.25% Cr 1.0% Mo steel: For electrodes; Esab standard
	UTS: 772 **Elon: 22%** **Proof: 679**
CHRO-MOW	0.3% C 1% Si 1.25% W 5% Cr 1.35% Mo steel: Used for die holders; forging tools & extrusion punches; origin unknown
CHROMVA-W	0.18% C 2.7% Cr 0.5% Mo 0.05% W 0.7% V steel: Firth Brown; for use at temperatures up to 580°C
	UTS: 920 **Elon: 15%** **Proof: 700**
CKD 1	0.12% C 0.25% C 1.0% Ni steel: Pompey
CKD 2	0.18% C 1.0% Cr 0.25% Mo steel: Pompey
	UTS: 1150 **Elon: 7%** **Proof: 860**
CL 222	0.35% C 0.8% Cr 0.5% Mo 0.5% W steel: For tools; origin unknown
CL 444	0.35% C 5.0% Cr 1.5% Mo 0.3% V 1.5% W steel: Origin unknown
CLW	0.3% C 3.3% Cr 9% W 0.5% V steel: Casting; Latrobe; for precision casting of forging dies **DPN: 572**
CM 1	0.07% C 5% Cr 1% Mo 0.3% V steel: Huntsman for case hardened dies for Zn casting
CM 70	0.18% C 1.2% Mn 0.6% Ni + Cr + Mo steel: For carburizing; Steel Peech & Tozer; core properties; hardened & tempered
	UTS: 1070 **Elon: 22%** **Proof: 760**
CM 80F	0.2% C 1.2% Mn 0.8% Cr + Mo + Cu 0.004% B steel: Steel Peech & Tozer; can be carburized
	UTS: 1230 **Elon: 10%** **Proof: 920**
CM 90F	0.22% C 1.5% Mn 0.8% Ni + Cr + Mo 0.004% B steel: Steech Peech & Tozer
CMV	0.35% C 1% Si 5% Cr 1.5% Mo 0.9% V steel: Sanderson; forging dies, etc. **DPN: 570**
CMW	0.35% C 1% Si 0.3% Mn 5% Cr 1.2% W steel: Sanderson; forging dies **DPN: 560**
CN 1	0.15% C 0.3% Cr 3.1% Ni steel: For case hardening; W Marrison for BS alloy En 33
CN 2	0.18% C 0.9% Cr 3.4% Ni steel: For case hardening; W Marrison for BS alloy En 36
CN 3A	0.15% C 1.2% Cr 4.2% Ni steel: For case hardening; W Marrison for BS alloy En 39A
CN 3 B	0.15% C 1.2% Cr 4.2% Ni 0.3% Mo steel: For case hardening; W Marrison for BS alloy En 39B
CN 5	0.18% C 2.0% Cr 2.0% Ni 0.2% Mo steel: For case hardening; W Marrison for BS alloy En 320

Symbol	Nominal analysis, supplier, condition and remarks.
CN 9	0.2% C 0.6% Cr 0.8% Ni 0.1% Mo steel: For case hardening; W Marrison for BS alloy En 351
CN 10	0.22% C 0.8% Cr 1.1% Ni 0.1% Mo steel: For case hardening; W Marrison for BS alloy En 352
CN 11	0.22% C 1% Cr 1.3% Ni 0.15% Mo steel: For case hardening; W Marrison for BS alloy En 353
CN 12	0.22% C 1% Cr 1.8% Ni 0.2% Mo steel: For case hardening; W Marrison for BS alloy En 354
CN 13	0.2% C 1.6% Cr 2.0% Ni 0.2% Mo steel: For case hardening; W Marrison for BS alloy En 355
CNBN 1/W	0.15% C 1.5% Cr 1.5% Ni steel: Krupp
CNBV 2V/W	0.18% C 2.0% Cr 1.8% Ni 0.1% V steel: Krupp
CNBV 2/W	0.18% C 2.0% Cr 2.0% Ni steel: Krupp
CNBV 3/W	0.14% C 0.75% Cr 2.5% Ni steel: Krupp
CNBV 4/W	0.14% C 0.75% Cr 3.5% Ni steel: Krupp
CNBV 5/W	0.14% C 1.1% Cr 4.5% Ni steel: Krupp
CNBVO	0.36% C 1.05% Cr 1.05% Ni 0.2% Mo steel: Krupp
CNBVO 1	0.34% C 1.55% Cr 1.55% Ni 0.25% Mo steel: Krupp
CNBVO 1/1	0.17% C 1.65% Cr 1.55% Ni 0.35% Mo steel: Krupp
CNBVO 2	0.3% C 1.95% Cr 1.95% Ni 0.3% Mo steel: Krupp
CND	0.1% C 1.0% Ni 0.15% Mo steel: Pompey
CNKD1	0.16% C 0.6% Cr 1.2% Ni 0.15% Mo steel: Pompey; hardened & tempered
	UTS: 1030 **Elon: 9%** **Proof: 740**
CNKD2	0.18% C 1.0% Cr 1.3% Ni 0.2% Mo steel: Pompey; hardened & tempered
	UTS: 1360 **Elon: 7%** **Proof: 920**
CNKD3	0.16% C 1.0% Cr 3.2% Ni 0.2% Mo steel: Pompey; hardened & tempered
	UTS: 1360 **Elon: 8%** **Proof: 990**
CODE 107	0.3% C 0.7% Cr 2.5% Ni 0.5% Mo steel: Casting; Edgar Allen; 'A 100'
CODE 617	0.2% C (max) 5% Cr 0.5% Mo steel: Casting; Edgar Allen
CODE 625	0.22% C 3.1% Cr 0.5% Mo steel: Casting; Edgar Allen; 'Cr Mo 329'
COMPAX	0.13% C 1.4% Cr 3.7% Ni steel: Uddeholm for type PG
CORTEN 50	0.12% C (max) 0.35% Mn 0.8% Cr 0.65% Ni 0.5% Cu steel:
CP 30	0.3% C 1.6% Cr 0.7% Ni 0.4% Mo steel: ESC
	UTS: 1000 **Elon: 16%** **Proof: 800**
CRD 9% W	0.3% C 3.5% Cr 9% W 0.1% V steel: C Denton; hot shear blades; obsolete **DPN: 460**
CRD HYKROM	0.35% C 5.0% Cr 1.5% Mo 0.4% V steel: Origin unknown
CRD HYKROM B	0.35% C 3.0% Cr 0.5% Mo 0.1% V steel: Origin unknown
CRD SPECIAL HOT DIE	0.3% C 1% Si 5% Cr 1.5% Mo 1% V steel: C Denton; hot shear blades; obsolete **DPN: 460**
CREUSABRO 32	0.2% C 1.2% Mn 1.3% Cr 0.25% Mo 0.2% Cu steel: Creusot-Loire
CREUSABRO 32	0.2% C 1.2% Mn 1.3% Cr 0.25% Mo 0.2% Cu steel: For abrasion resistance; Creusot-Loire
CREUSELSO 32	0.2% C 1.2% Mn 1.3% Cr 1.2% Mo steel: Creusot-Loire
CREUSELSO 42	0.2% C 1.6% Mn 0.7% Cr + Ni + Mo + Cu steel: Creusot-Loire
CREUSELSO 47	0.18% C (max) 1.7% Mn (max) 0.5% Ni 0.07% V steel: Creusot-Loire
CRM 3	0.3% C 3% Cr 0.5% Mo steel: Firth Bronn for BS alloy En 29B
Cr Mo 18	0.18% C 1.2% Cr 0.4% Ni 0.5% Mo 0.3% Cu steel: Casting; Wolsingham
	UTS: 460 **Elon: 20%** **Proof: 270**

Note. The following abbreviations and units are used in the tables:

DPN	Hardness, diamond pyramid number
UTS	Ultimate tensile strength, N/mm^2
Elon	Elongation, %
Proof	0.1 % proof strength, N/mm^2

$1 \ N/mm^2 = 0.1 \ hbar = 0.102 \ kgf/mm^2 = 0.06475 \ tonf/in^2 = 145.04 \ lbf/in^2$

Symbol	Nominal analysis, supplier, condition and remarks.
Cr Mo 23	0.22% C 3% Cr 0.4% Ni 0.5% Mo 0.3% Cu steel: Casting; Wolsingham
	DPN: 190 UTS: 700 Elon: 18% Proof: 360
Cr Mo 33	0.32% C 1.2% Cr 0.4% Ni 0.5% Mo 0.3% Cu steel: Casting; Wolsingham
	UTS: 610
Cr Mo 329	0.22% C 3.2% Cr 0.5% Mo steel: Casting; Edgar Allen; Code 625
CRO 42	0.2% C 0.5% Cr 0.6% Ni 0.2% Mo steel: For carburizing; Bofors; hardened & tempered; core properties
	DPN: 300 UTS: 1000 Elon: 12% Proof: 500
CRO 53	0.2% C 1.0% Cr 1.2% Ni 0.1% Mo steel: For carburizing; Bofors; hardened & tempered; core properties
	DPN: 360 UTS: 1200 Elon: 11% Proof: 600
CRO 684 BOVAC	0.3% C 1.7% Cr 0.4% Mo 0.6% Ni steel: Bofors; vacuum melted
CRO 861	0.35% C 1.4% Cr 1.4% Ni 0.2% Mo steel: Bofors; hardened & tempered
	DPN: 300 UTS: 1000 Elon: 14% Proof: 700
COR 7146	0.32% C 2.7% Cr 0.8% Ni 0.6% Mo steel: Bofors
CRODI	0.35% C 5.0% Cr 1.4% Mo 1.2% W 0.3% V steel: Atlas as AISI type H12
CROLOY ½	0.15% C 0.7% Cr 0.5% Mo steel: Tube; Babcock & Wilcox (USA)
CROLOY 1	0.15% C 1.0% Cr 0.5% Mo steel: Tube; Babcock & Wilcox (USA)
CROLOY 1¼	0.15% C 1.25% Cr 0.5% Mo steel: Tube; Babcock & Wilcox (USA)
CROLOY 1¾	0.15% C 1.75% Cr 0.7% Mo steel: Tube; Babcock & Wilcox (USA)
CROLOY 2	0.15% C 2.0% Cr 0.5% Mo steel: Tube; Babcock & Wilcox (USA)
CROLOY 2¼	0.15% C 2.2% Cr 1.0% Mo steel: Tube; Babcock & Wilcox (USA)
CROLOY 2½	0.15% C 2.5% Cr 0.5% Mo steel: Tube; Babcock & Wilcox (USA)
CROLOY 3	0.15% C 3.0% Cr 1.0% Mo steel: Tube; Babcock & Wilcox (USA)
CROLOY 5	0.15% C 5.0% Cr 0.5% Mo steel: Tube; Babcock & Wilcox (USA)
CROLOY 5Si	0.15% C 1.5% Si 5.0% Cr 0.5% Mo steel: Tube; Babcock & Wilcox (USA)
CROLOY 5Ti	0.12% C 5.0% Cr 0.5% Mo 0.7% Ti steel: Tube; Babcock & Wilcox (USA)
CROLOY 7	0.15% C 7.0% Cr 0.5% Mo steel: Tube; Babcock & Wilcox (USA)
CROLOY 9½ Mo	0.15% C 9.0% Cr 0.5% Mo steel: Tube; Babcock & Wilcox (USA)
CROLOY 9M	0.15% C 9.0% Cr 1.0% Mo steel: Tube; Babcock & Wilcox (USA)
CROMODIE	0.35% C 1.0% Si 5% Cr 1.5% Mo 1.0% V steel: Firth Brown
CROMODIE W	0.31% C 0.9% Si 5% Cr 1.7% Mo 1.1% W 0.2% V steel: Firth Brown
CROVAN	0.35% C 5.0% Cr 1.4% Mo 0.9% V steel: Atlas; as AISI H13
CTHV	0.3% C 3.4% Cr 8.5% W 0.35% V steel: Marsh Brothers; hot forging dies, etc.
CTM	0.35% C 2.5% Cr 2.5% Ni 0.5% Mo 2.5% W 0.5% V steel: Swift Levick; hot dies
CTU	0.37% C 1.0% Si 5% Cr 1.3% Mo 1.7% W 1.0% V steel: For tools; Marsh Brothers; hot work dies
Cu Ni Mo	0.21% C (max) 0.95% Mn 1.4% Ni 0.25% Mo 0.65% Cu steel: US Steel Corp.
CUP-TEN	0.12% C (max) 0.6% Mn 0.6% Cr 0.22% Mo 0.35% Cu steel: Nippon-Kokan

Symbol	Nominal analysis, supplier, condition and remarks.
CUP-TEN 6	0.12% C (max) 0.6% Mn 0.8% Cr 0.35% Mo 0.1% V 0.35% Cu steel: Nippon-Kokan
CUP-TEN 50	0.18% C (max) 1.2% Mn 0.8% Cr 0.35% Mo 0.1% V 0.35% Cu steel: Nippon-Kokan
CUP-TEN 53	0.18% C (max) 1.2% Mn 0.8% Cr 0.35% Mo 0.1% V 0.35% Cu steel: Nippon-Kokan
CVM	0.36% C 1.0% Si 5% Cr 1.3% Mo 0.3% V steel: ESC; extrusion, forging & die casting dies
	DPN: 500
CVM 2	0.36% C 2.0% Si 5.0% Cr 1.3% Mo 1.4% W 0.3% V steel: ESC; hot working dies
	DPN: 460
CVM5	0.32% C 5.0% Cr 1.6% Mo 0.3% V 1.5% W steel: Origin unknown
D 66	0.3% C 3.0% Cr 9.0% W 0.3% V steel: Fagersta
	DPN: 550
DEF 13 B4	Weldable structural steel covers BS 968; BS alloy En 14A etc
	UTS: 480 Elon: 14% Proof: 290
DEF 13 B6C	Alloy case hardening steel covers BS alloy En 34; also includes En 351, etc.
	UTS: 700 Elon: 18%
DEF 13 B6D	Alloy case hardening steel covers BS alloy En 35; also includes En 36, En 325, etc.
	UTS: 840 Elon: 15%
DEF 13 B6E	Alloy case hardening steel covers BS alloy En 36; also includes En 353, etc.
	UTS: 1000 Elon: 12%
DEF 13 B6F	Alloy case hardening steel covers BS alloy En 354
	UTS: 1140 Elon: 12%
DEF 13 B6G	Alloy case hardening steel covers BS alloy En 39A 39B; also includes En 355, etc.
	UTS: 1300 Elon: 12%
DOUBLE GRIFFIN	0.35% C 1.7% Cr 3.5% Ni steel: Balfour; chisels, wedges, etc.
	DPN: 400
DIN 17155 13 Cr Mo 4/4	0.14% C 0.6% Mn 0.8% Cr 0.45% Mo steel: Plate; German Standard; normalized & tempered
	UTS: 510 Proof: 300
DIN 17155 13 Cr Mo 4/4	0.14% C 0.8% Cr 0.45% Mo steel: For boiler plate; German Standard
DIN 17175 10 Cr Mo 9/10	0.15% C (max) 2.2% Cr 1.0% Mo steel: German Standard
DIN 17175 13 Cr Mo 4/4	0.14% C 0.8% Cr 0.45% Mo steel: For boiler plate; German Standard
DIN 17200 17 Cr Mo 6	0.17% C 0.5% Mn 1.65% Cr 1.55% Ni 0.3% Mo steel: German Standard
DIN 17200 18 Cr Ni 8	0.18% C 0.5% Mn 1.95% Cr 1.95% Ni steel: German Standard
DIN 17200 15 Cr Ni 6	0.15% C 0.5% Mn 1.6% Cr 1.6% Ni steel: German Standard
DIN 17200 20 Mo Cr 4	0.2% C 0.75% Mn 0.4% Cr 0.45% Mo steel: German Standard
DIN 17200 20 Mo Cr S 4	0.2% C 0.75% Mn 0.41% Cr 0.45% Mo steel: German Standard
DIN 17200 25 Cr Mo 4	0.25% C 0.65% Mn 1.05% Cr 0.22% Mo steel: German Standard
DIN 17200 25 Cr Mo 4	0.25% C 1.1% Cr 0.2% Mo steel: Forging; German Standard
DIN 17200 25 Mo Cr 4	0.25% C 0.75% Mn 0.5% Cr 0.45% Mo steel: German Standard
DIN 17200 25 Mo Cr S 4	0.25% C 0.75% Mn 0.5% Cr 0.45% Mo steel: German Standard
DIN 17200 30 Cr Mo V 9	0.3% C 2.5% Cr 0.6% Ni 0.2% Mo 0.15% V steel: Forging; German Standard
DIN 17200 30 Cr Ni Mo 8	0.3% C 2.0% Cr 2.0% Ni 0.3% Mo steel: Forging; German Standard
DIN 17200 30 Cr Ni Mo 8	0.3% C 0.45% Mn 2.0% Cr 2.0% Ni 0.45% Mo steel: German Standard

Symbol	Nominal analysis, supplier, condition and remarks.
DIN 17200 31 Cr Mo 12	0.31% C 0.55% Mn 3.0% Cr 0.45% Mo 0.2% V steel: German Standard
DIN 17200 32 Cr Mo 12	0.32% C 0.55% Mn 3.0% Cr 0.45% Mo steel: German Standard
DIN 17200 34 Cr Al Mo 5	0.34% C 0.6% Mn 1.15% Cr 0.2% Mo 1.0% Al steel: German Standard
DIN 17200 34 Cr Al Ni 7	0.34% C 0.55% Mn 1.65% Cr 1.0% Ni 0.2% Mo 1.0% Al steel: German Standard
DIN 17200 34 Cr Mo 4	0.33% C 0.65% Mn 1.05% Cr 0.22% Mo steel: German Standard
DIN 17200 34 Cr Mo 4	0.34% C 1.1% Cr 0.4% Ni 0.2% Mo steel: Forging; German Standard
DIN 17200 34 Cr Mo S 4	0.33% C 0.65% Mn 1.05% Cr 0.22% Mo steel: German Standard
DIN 17200 34 Cr Ni Mo 6	0.34% C 0.55% Mn 1.5% Cr 1.5% Ni 0.22% Mo steel: German Standard
DIN 17200 34 Cr Ni Mo 6	0.34% C 1.5% Cr 1.5% Ni 0.2% Mo steel: Forging; German Standard
DIN 17200 36 Cr Ni Mo 4	0.36% C 1.1% Cr 1.1% Ni 0.2% Mo steel: German Standard
DIN 17200 39 Cr Mo V	0.39% C 0.55% Mn 3.25% Cr 0.95% Mo 0.2% V steel: German Standard
DIN 17210 15 Cr Ni 6	0.15% C 1.5% Cr 1.5% Ni steel: For carburizing
DIN 17210 18 Cr Ni 8	0.17% C 2.0% Cr 2.0% Ni steel: For carburizing
DIN 17225 30 W Cr V 1/7/9	0.3% C 2.3% Cr 4.2% W 0.6% V steel: For springs
DIN 17240 21 Cr Mo V 5/11	0.21% C 1.3% Cr 0.6% Ni 1.1% Mo 0.3% V steel: For high temperature fasteners
DIN 17240 24 Cr Mo 5	0.24% C 1.1% Cr 0.6% Ni 0.25% Mo steel: For high temperature fasteners
DIN 17240 24 Cr Mo V 5/5	0.24% C 1.3% Cr 0.6% Ni 0.55% Mo 0.2% V steel: For high temperature fasteners
DNV	0.35% C 3.25% Cr 9.5% W 0.4% V steel: Simonds; AISI type H21
DOFASCOLOY P	0.16% C 0.6% Mn 0.6% Cr 0.9% Ni 0.6% Cu steel: Dominion
DR 34	0.15% C 0.8% Cr 1.5% Ni steel: For case hardening; Bofors; annealed DPN: 200
DR 44	0.2% C 0.8% Cr 1.5% Ni steel: For case hardening; Bofors; annealed DPN: 220
DRO 86 BOVAC	0.35% C 1.5% Cr 0.2% Mo 1.5% Ni steel: Bofors; vacuum melted
DTD 713	0.3% C 0.6% Cr 2.5% Ni 0.5% Mo steel: Tube; hardened & tempered UTS: 1140 Elon: 14%
DTD 723	0.3% C 0.7% Cr 2.5% Ni 0.5% Mo steel: Tube; hardened & tempered UTS: 1380 Elon: 12%
DTD 730	0.35% C 3% Cr 1% Mo 0.2% V steel: For nitriding; hardened & tempered DPN: 400 UTS: 1380 Elon: 12% Proof: 1070
DTD 5002	0.15% C 0.6% Cr 0.8% Ni steel: For case hardening; blank carburized UTS: 700 Elon: 18%

Symbol	Nominal analysis, supplier, condition and remarks.
DTD 5082	0.22% C 1% Cr 0.2% Mn steel: Bar & forging; hardened & tempered; weldable DPN: 360 UTS: 1140 Elon: 10% Proof: 840
DTD 5112	0.32% C 0.9% Cr 0.2% Mo steel: Sheet; hardened & tempered UTS: 1210 Elon: 4% Proof: 1040
DTD 5123	0.32% C 1% Cr 0.2% Mo steel: Forging & bar; hardened & tempered UTS: 1140 Elon: 4% Proof: 920
DTD 5132	0.32% C 1% Cr 0.2% Mo steel: Tube; hardened & tempered UTS: 1210 Elon: 4%
DTD 5142	0.32% C 1% Cr 0.2% Mo steel: Tube; hardened & tempered UTS: 1210 Elon: 4%
DTD 5152	0.32% C 1% Cr 0.2% Mo steel: Wire for welding filler; hardened & tempered; supplied annealed & copper coated DPN: 380
DTD 5219	1% Cr Mo steel: Casting; investment cast UTS: 800
DTD 5229	3% Cr Mo steel: Casting; investment cast UTS: 710
DTD 5239	3% Ni steel: Casting for case hardening; investment cast UTS: 700
DTD 5249	3% Cr Mo steel: Casting for nitriding; investment cast UTS: 920
DUCOL QT28	0.17% C (max) 0.3% Cr 0.5% Ni 0.3% Mo steel: Colvilles
DUCOL QT342	0.17% C (max) 1.5% Mn 0.7% Cr 0.3% Ni 0.25% Mo 0.1% V steel: Colvilles
DUCOL QT35	0.15% C (max) 1.2% Mn 1.0% Cr 1.2% Ni 0.5% Mo 0.12% V steel: Colvilles
DUCOL QT445A	0.17% C (max) 1.2% Mn 0.5% Cr 0.25% Mo 1.04% V steel: Colvilles
DUCOL W21	0.23% C (max) 1.6% Mn 0.5% Cr 0.25% Ni steel: Plate & section; Colvilles; Cu may be present up to 0.5% UTS: 510 Elon: 23% Proof: 330
DUCOL W23	0.2% C (max) 1.5% Mn 0.5% Cr 0.25% Ni steel: Plate & section; Colvilles alloy for BS 968 0.5% Cu optional; Mn + Cr 1.6% (max) UTS: 480 Elon: 23% Proof: 340
DUCOL W23D	0.2% C (max) 1.5% Mn 0.5% Cr 0.25% Ni 0.5% Cu: With grain refining elements; Colvilles
DUCOL W24	0.22% C (max) 1.5% Mn 0.5% Cr 0.25% Ni steel: Plate & section; Colvilles alloy for BS 968; 0.5% Cu optional; Cr + Mn 1.6% (max) UTS: 480 Elon: 22% Proof: 330
DUCOL W25	0.2% C (max) 1.5% Mn 0.5% Cr 0.25% Ni steel: Plate; Colvilles; 0.5% Cu optional; Mn + Cr 1.6% (max) UTS: 510 Elon: 23% Proof: 360
DUCOL W25	0.2% C (max) 1.5% Mn 0.5% Cr 0.25% Ni steel: Colvilles
DUCOL W25D	0.2% C (max) 1.5% Mn 0.5% Cr 0.25% Ni: With grain refining elements; Colvilles
DUCOL W30	0.17% C (max) 1.5% Mn 0.7% Cr 0.28% Mo 0.1% V steel: Plate; Colvilles; weldable UTS: 610 Elon: 17% Proof: 450
DUNELT 10	0.32% C 0.3% Cr 3% Ni steel: Dunford for BS alloy En 21
DUNELT 15	0.12% C 0.3% Cr 3.2% Ni steel: For case hardening; Dunford for BS alloy En38
DUNELT 16	0.12% C 1% Cr 3.2% Ni steel: For carburizing; Dunford for BS alloy En 36
DUNELT 17	0.12% C 0.3% Cr 5.0% Ni steel: For case hardening; Dunford for BS alloy En 37

Note. The following abbreviations and units are used in the tables:

DPN	Hardness, diamond pyramid number
UTS	Ultimate tensile strength, N/mm^2
Elon	Elongation, %
Proof	0.1 % proof strength, N/mm^2

$1 \ N/mm^2 = 0.1 \ hbar = 0.102 \ kgf/mm^2 = 0.06475 \ tonf/in^2 = 145.04 \ lbf/in^2$

Symbol	Nominal analysis, supplier, condition and remarks.
DUNELT 18	0.13% C 0.3% Cr 5.0% Ni 0.2% Mo steel: For case hardening; Dunford for BS alloy En 38
DUNELT 20	0.18% C 1.7% Ni 0.25% Mo steel: For case hardening; Dunford for BS alloy En 34
DUNELT 55	0.3% C 0.7% Cr 3.2% Ni steel: Dunford for BS alloy En 23
DUNELT 60	0.34% C 0.7% Cr 3.3% Ni steel: Dunford; hardened & tempered **DPN: 300 UTS: 1040 Elon: 17%**
DUNELT 64A	0.32% C 0.7% Cr 2.5% Ni 0.5% Mo steel: Dunford; hardened & tempered **DPN: 350 UTS: 1000 Elon: 14%**
DUNELT 64B	0.4% C 0.7% Cr 2.5% Ni 0.5% Mo steel: Dunford; hardened & tempered **DPN: 375 UTS: 1070 Elon: 12%**
DUNELT 65	0.27% C 1.2% Cr 3.2% Ni 0.6% Mo 0.2% V steel: Dunford for BS alloy En 27
DUNELT 67	0.25% C 3.2% Cr 0.4% Ni (max) 0.5% Mo steel: For nitriding; Dunford for BS alloy En 40B
DUNELT 82M	0.16% C 1.2% Cr 4.2% Ni 0.25% Mo steel: For case hardening; Dunford for BS alloy S82
DUNELT 128	0.3% C 1.3% Cr 4.2% Ni steel: Dunford for BS alloy En 30A
DUNELT 129	0.3% C 1.2% Cr 4.2% Ni 0.3% Mo steel: Dunford for BS alloy En 30B
DUREHETE 1050	0.2% C 1% Cr 1% Mo 0.7% V steel: For use at high temperature; Samuel Fox; capable of operating up to 500 °C **UTS: 860 Elon: 22% Proof: 770**
DUREHETE 1055	0.2% C 1.1% Cr 0.95% Mo 0.7% V 0.1% Ti steel: Samuel Fox; tempered **UTS: 840 Elon: 13% Proof: 720**
E 4 Mo	0.12% C 5% Ni 0.2% Mo steel: Jessop for BS alloy En 38
E 6	0.17% C 1.7% Ni 0.25% Mo steel: Jessop for BS alloy En 34
E 9	0.24% C 1.7% Ni 0.25% Mo steel: Jessop for BS alloy En 35
E 11	0.19% C 1.8% Ni 0.25% Mo steel: Jessop
ECLIPSE LDC	0.35% C 5.0% Cr 1.5% Mo 0.4% V steel: Origin unknown
ECO/m	0.23% C 1.0% Cr 0.2% Mo steel: Krupp
EEEE	0.28% C 2.15% Cr 0.3% Ni 0.3% Mo steel: Crucible Steel Co for drills
EHW 1	0.25% C 4.0% Cr 15% W 0.5% V steel: Latrobe; type H25
EOC/m	0.25% C 0.45% Cr 0.45% Mo steel: Krupp
EOC/m	0.2% C 0.4% Cr 0.45% Mo steel: Krupp
ER 1	0.32% C 3% Cr 2.8% Mo 0.5% V tool steel: Balfour; die casting brass, etc. **DPN: 550**
ERA 164	0.2% C 1.6% Mn 0.25% Cr 0.4% Mo steel: Hadfields for BS alloy En 14A & B
ERA 165	0.2% C 1.6% Mn 0.6% Ni 0.25% Mo steel: Hadfields for BS alloy En 13
ESSHETE CRM 2	0.15% C (max) 2% Cr 1% Mo steel: Samuel Fox; normalized & tempered; high temperature use **UTS: 630 Elon: 30% Proof: 550**
ESSHETE CRM 5	0.12% C (max) 5% Cr 0.5% Mo steel: Samuel Fox; annealed; not easily welded **UTS: 450 Elon: 41%**
ESSHETE CML	0.12% C 1% Cr 0.5% Mo steel: Creep resisting; Samuel Fox; weldable **UTS: 510**
ESSHETE MV	0.15% C 0.3% V 0.5% Mo 0.3% Cr steel: Creep resisting; Samuel Fox; weldable **UTS: 610**
ETAD 1	0.3% C 1.35% Cr 3.8% Ni 0.5% V steel: Pompey; hardened & tempered **UTS: 1610 Elon: 6% Proof: 1300**

Symbol	Nominal analysis, supplier, condition and remarks.
ETAD 3	0.35% C 1.85% Cr 3.85% Ni 0.5% Mo steel: Pompey; hardened & tempered **UTS: 1630 Elon: 6% Proof: 1300**
ETANO	0.35% C 1.7% Cr 3.75% Ni steel: Pompey; hardened & tempered **UTS: 1680 Elon: 6% Proof: 1300**
EX 22	0.15% C 0.9% Mn 0.5% Cr 0.25% Mo steel: For carburizing; SAE designation replacing 8615
EX 23	0.17% C 0.9% Mn 0.5% Cr 0.25% Mo steel: For carburizing; SAE designation replacing 8617
EX 24	0.2% C 0.9% Mn 0.5% Cr 0.25% Mo steel: For carburizing; SAE designation replacing 8620
EX 25	0.22% C 0.9% Mn 0.5% Cr 0.25% Mo steel: For carburizing; SAE designation replacing 8622
EX 26	0.25% C 0.9% Mn 0.5% Cr 0.25% Mo steel: For carburizing; SAE designation for 8625
EX 27	0.27% C 0.9% Mn 0.5% Cr 0.25% Mo steel: For carburizing; SAE designation for 8627
EX 28	0.18% C 0.9% Mn 0.5% Cr 0.5% Ni 0.35% Mo steel: For carburizing; SAE designation replacing 4718
EX 29	0.2% C 0.9% Mn 0.5% Cr 0.5% Ni 0.35% Mo steel: For carburizing; SAE designation replacing 4320
EX 30	0.15% C 0.8% Mn 0.5% Cr 0.85% Ni 0.52% Mo steel: For carburizing; SAE designation replacing 4865
EX 31	0.17% C 0.8% Mn 0.5% Cr 0.85% Ni 0.52% Mo steel: For carburizing; SAE designation replacing 4817
EX 32	0.2% C 0.8% Mn 0.5% Cr 0.85% Ni 0.52% Mo steel: For carburizing; SAE designation replacing 4820
EX 55	0.17% C 0.9% Mn 0.5% Cr 1.85% Ni 0.72% Mo steel: For carburizing; SAE temporary designation
EX 56	0.1% C 0.9% Mn 0.5% Cr 1.8% Ni 0.72% Mo steel: For carburizing; SAE temporary designation
EXD 1	0.3% C 1.5% Cr 5.5% W 3.7% Ni 5.5% W steel: For tools; Balfour; hot extrusion dies for non-ferrous metal, etc. **DPN: 480**
F 9	0.2% C 2% Cr 2.4% W 0.25% V steel: Jessop
F 45CH	0.15% C 0.8% Ni 0.6% Cr steel: For carburizing; Firth Brown for BS 970 En 351
F 55CH	0.18% C 1.0% Ni 0.8% Cr steel: For carburizing; Firth Brown for BS 970 En 352
F 65CH	0.18% C 1.3% Ni 1.0% Mo steel: For carburizing; Firth Brown for BS 970 En 353
F 75CH	0.18% C 1.75% Ni 1.0% Cr 0.15% Mo steel: For carburizing; Firth Brown for BS 970 En 354
F 85CH	0.18% C 2.0% Ni 1.6% Cr 0.2% Mo steel: For carburizing; Firth Brown for BS 970 En 355
F 543	0.09% C 3.9% Cr 4.8% Ni 3.0% Mo steel: Firth Brown
FAGERSTA K353	0.15% C 0.8% Mn 0.8% Cr 1.5% Ni steel: For case hardening; Fagersta
FAGERSTA N104	0.1% C 0.4% Mn 2.1% Ni steel: For case hardening; Fagersta
FB 60	0.18% C (max) 1.4% Mn 0.7% Ni 0.18% V 0.015% N steel: German proprietor's standard; Al killed
FB 60AK	0.18% C (max) 1.4% Mn 0.7% Ni 0.18% V 0.015% N steel: German proprietor's specification; Al killed
FB 70	0.2% C (max) 1.5% Mn 0.7% Ni 0.18% V 0.015% N steel: Mannesmann; Al killed
FB 70AK	0.2% C (max) 1.5% Mn 0.7% Ni 0.18% V 0.015% N steel: Mannesmann; Al killed
FB 80	0.22% C (max) 1.6% Mn 0.7% Ni 0.18% V 0.02% N steel: Mannesmann; Al killed

Symbol	Nominal analysis, supplier, condition and remarks.
FB 80AK	0.22% C (max) 1.6% Mn 0.7% Ni 0.22% V 0.02% N steel: Mannesmann; Al killed
FIRTH OB	0.09% C 0.5% Cr 1.25% Ni steel: Firth Brown for plastic moulds
FKD 3	0.35% C 1.0% Cr 0.25% Mo steel: Pompey **UTS: 1140 Elon: 10% Proof: 950**
fmp 505	0.32% C 9.2% W 3.2% Cr 0.5% Mo steel: F Parkin; as tool steel H 21
fmp 507	0.3% C 9.0% W 3.0% Cr 0.3% V 0.5% Mo 2.5% Ni steel: F Parkin
fmp 513	0.35% C 1.25% W 5.25% Cr 0.35% V 1.5% Mo steel: F Parkin; as tool steel H 12
FNCR	0.35% C 0.6% Cr 1.25% Ni steel: Firth Brown for BS alloy En 111
FOREMOST A	0.3% C 3.0% Cr 9.0% W 0.4% V steel: Swift Levick; hot dies
FORMITE 2	0.33% C 3.3% Cr 9.2% W 0.5% V steel: Columbia; AISI type H21
FORTREX 45	0.06% C 1.5% Ni 0.4% Mo steel: Welding electrode; Murex **UTS: 680 Elon: 27%**
FORTREX 55	0.07% C 1.1% Mn 0.6% Cr 2% Ni 0.4% Mo steel: Welding electrode; Murex **UTS: 820 Elon: 24% Proof: 670**
GO	0.3% C 0.7% Cr 3% Ni steel: Jessop for BS alloy En 23
GI SPECIAL	0.3% C 1.3% Cr 4% Ni 0.3% Mo steel: Jessop
G 8	0.27% C 1.2% Cr 3% Ni 0.4% Mo steel: Jessop
G 11	0.3% C 0.6% Cr 2.5% Ni 0.5% Mo steel: Jessop for BS alloy En 25
G 15	0.12% C 0.9% Cr 3.2% Ni steel: Jessop for BS alloy En 36A
G 16	0.15% C 1.2% Cr 4.2% Ni 0.25% Mo steel: Jessop for BS alloy En 39B
G 27	0.37% C 3% Cr 0.4% Ni 1% Mo 0.2% V steel: Jessop for BS alloy En 40C
G 27	0.30% C 1.0% Cr 3.3% Ni 0.35% Mo steel: Jessop for BS alloy En 27 and 28
G 28	0.12% C 0.5% Cr 1.3% Ni steel: Jessop
G 36	0.2% C 0.3% Cr 1.7% Ni 0.23% Mo steel: Jessop
G 45	0.35% C 0.6% Cr 1% Ni steel: Jessop for BS alloy En 111
G 46	0.15% C 0.6% Cr 0.8% Ni steel: Jessop for BS alloy En 351
G 47	0.2% C 0.8% Cr 1% Ni steel: Jessop for BS alloy En 352
G 48	0.2% C 1% Cr 1.25% Ni 0.11% Mo steel: Jessop for BS alloy En 353
G 49	0.2% C 1% Cr 1.7% Ni 0.15% Mo steel: Jessop for BS alloy En 354
G 50	0.2% C 1.5% Cr 2% Mo steel: Jessop for BS alloy En 355
G 51	0.15% C 0.7% Cr 0.6% Ni 0.11% Mo steel: Jessop for BS alloy En 361
G 52	0.2% C 0.7% Cr 0.6% Ni 0.11% Mo steel: Jessop for BS alloy En 362
G 53	0.24% C 0.7% Cr 0.6% Ni 0.11% Mo steel: Jessop for BS alloy En 363

Symbol	Nominal analysis, supplier, condition and remarks.
G 86	0.10% C 6.0% Cr 11.0% W 2.0% Mo 1.5% Nb 8.0% Al: Jessop; scale or creep resistant
G 110	0.3% C 1.2% Cr 4.1% Ni 0.3% Mo steel: Toledo for BS alloy En 30A (no Mo) and En 30B
GEARITE	0.31% C 1.2% Cr 4.2% Ni steel: Swift Levick for BS alloy En 30
GEARMAC	0.31% C 0.7% Cr 2.5% Ni 0.6% Mo steel: Swift Levick for BS alloy En 25
GK 3	0.35% C 2.0% Cr 0.25% Mo 0.15% V steel: For carburizing; Firth Brown
GK 5	0.25% C 2.0% Cr 0.25% Mo 0.15% V steel: For carburizing; Firth Brown
GK 7	0.18% C 2.0% Cr 0.25% Mo 0.15% V steel: For carburizing; Firth Brown
GOST 550/12 KHM	0.16% C (max) 0.55% Mn 0.95% Cr 0.47% Mo steel: Pipe; Russian Standard **UTS: 420 Proof: 250**
GOST 550 KH5M	0.15% C (max) 0.5% Mn (max) 5.0% Cr 0.55% Mo steel: Pipe; Russian Standard **UTS: 400 Proof: 220**
GOST 550 KH5YF	0.15% C (max) 0.5% Mn (max) 5.0% Cr 0.55% W 0.5% V steel: Pipe; Russian Standard **UTS: 400 Proof: 220**
GOST 4543/12 KH2NGA	0.12% C 0.45% Mn 1.4% Cr 3.4% Ni steel: Russian standard
GOST 4543/12 KHN2	0.12% C 0.45% Mn 0.75% Cr 1.7% Ni steel: Russian standard
GOST 4543/12 KHN3	0.12% C 0.45% Mn 0.75% Cr 3.0% Ni steel: Russian Standard
GOST 4543/13 N2KHA	0.12% C 0.45% Ni 0.35% Cr 1.9% Ni steel: Russian Standard
GOST 4543/15 KH2GN2T	0.15% C 0.85% Mn 1.6% Cr 1.6% Ni 0.1% W steel: Russian Standard
GOST 4543/15 KH2GN2TA	0.15% C 0.85% Mn 1.6% Cr 1.6% Ni steel: Russian Standard
GOST 4543/15 KH2GN2TRA	0.15% C 0.85% Mn 1.6% Cr 1.6% Ni 0.003% V 0.1% W steel: Russian Standard
GOST 4543/15 KHF	0.15% C 0.5% Mn 0.95% Cr 0.15% V steel: Russian Standard
GOST 4543/15 KHGNT	0.15% C 0.85% Mn 0.85% Cr 1.6% Ni 0.1% W steel: Russian Standard
GOST 4543/15 KHGNTA	0.15% C 0.85% Mn 0.85% Cr 1.6% Ni 0.1% W steel: Russian Standard
GOST 4543/15 KHM	0.15% C 0.55% Mn 0.95% Cr 0.47% Mo steel: Russian Standard
GOST 4543/15 NM	0.15% C 0.55% Mn 0.3% Cr (max) 1.7% Ni 0.25% Mo steel: Russian Standard
GOST 4543/16 KHSN	0.16% C 0.45% Mn 0.95% Cr 0.75% Ni steel: Russian Standard
GOST 4543/18 KH2N4VA	0.17% C 0.4% Mn 1.5% Cr 4.2% Ni 1.0% W steel: Russian Standard
GOST 4543/18 KHGN	0.18% C 0.95% Mn 0.55% Cr 0.55% Ni steel: Russian Standard
GOST 4543/18 SNRA	0.18% C 0.75% Mn 0.95% Cr 0.95% Ni 0.0035% V steel: Russian Standard
GOST 4543/20 KH2NGA	0.19% C 0.45% Mn 1.4% Cr 3.4% Ni steel: Russian Standard
GOST 4543/20 KHF	0.2% C 0.65% Mn 0.95% Cr 0.15% V steel: Russian Standard
GOST 4543/20 KHGNR	0.2% C 0.85% Mn 0.85% Cr 0.95% Ni 0.0035% steel: Russian Standard
GOST 4543/20 KHN	0.2% C 0.6% Mn 0.6% Cr 1.2% Ni steel: Russian Standard
GOST 4543/20 KHN3A	0.2% C 0.45% Mn 0.75% Cr 3.0% Ni steel: Russian Standard
GOST 4543/20 KHN4FA	0.2% C 0.4% Mn 0.95% Cr 4.0% Ni 0.22% B steel: Russian Standard
GOST 4543/20 KHNR	0.2% C 0.75% Mn 0.9% Cr 0.95% Ni 0.0025% B steel: Russian Standard

Note. The following abbreviations and units are used in the tables:

DPN	Hardness, diamond pyramid number
UTS	Ultimate tensile strength, N/mm^2
Elon	Elongation, %
Proof	0.1 % proof strength, N/mm^2

1 N/mm^2=0.1 hbar=0.102 kgf/mm^2=0.06475 tonf/in^2=145.04 lbf/in^2

Symbol	Nominal analysis, supplier, condition and remarks.
GOST 4543/20 NM	0.21% C 0.55% Mn 0.3% Cr (max) 1.7% Ni 0.25% Mo steel: Russian Standard
GOST 4543/25 KH2GNTA	0.25% Cr 0.95% Mn 1.5% Cr 1.1% Ni 0.1% W steel: Russian Standard
GOST 4543/25 KH2N4VA	0.25% C 0.4% Mn 1.5% Cr 4.2% Ni 1.0% W steel: Russian Standard
GOST 4543/30 KH2GN2	0.3% C 0.95% Mn 1.55% Cr 1.6% Ni steel: Russian Standard
GOST 4543/30 KH2NVA	0.30% C 0.45% Mn 1.8% Cr 1.6% Ni 1.4% W steel: Russian Standard
GOST 4543/30 KH2NVFA	0.3% C 0.45% Mn 1.8% Cr 1.6% Ni 1.4% W 0.21% B steel: Russian Standard
GOST 4543/30 KHGNA	0.3% C 0.75% Mn 1.1% Cr 0.45% Ni steel: Russian Standard
GOST 4543/30 KHGSNA	0.3% C 1.15% Mn 1.05% Cr 1.2% Ni steel: Russian Standard
GOST 4543/30 KHM	0.3% C 0.5% Mn 0.95% Cr 0.2% Mo steel: Russian Standard
GOST 4543/30 KHMA	0.3% C 0.5% Mn 0.95% Cr 0.2% Mo steel: Russian Standard
GOST 4543/30 KHN2VFA	0.3% C 0.45% Mn 0.75% Cr 2.2% Ni 0.65% W 0.2% B steel: Russian Standard
GOST 4543/30 KHN3A	0.3% C 0.45% Mn 0.75% Cr 3.0% Ni steel: Russian Standard
GOST 4543/30 KHNVA	0.3% C 0.45% Mn 0.75% Cr 1.45% Ni 0.65% W steel: Russian Standard
GOST 4543/35 KHM	0.36% C 0.55% Mn 0.95% Cr 0.20% Mo steel: Russian Standard
GOST 4543/38 KHNVA	0.38% C 0.45% Mn 1.5% Cr 1.45% Ni 0.65% W steel: Russian Standard
GOST 4543/38 KHVA	0.38% C 0.4% Mn 1.1% Cr 0.65% W steel: Russian Standard
GOST 4543/45 KHN	0.45% C 0.65% Mn 0.6% Cr 1.2% Ni steel: Russian Standard
GOST 5058/10 KHSND	0.12% C (max) 0.65% Mn 0.8% Cr 0.75% Ni 0.5% Cu steel: Plate; Russian Standard; hardened & tempered UTS: 540 Proof: 400
GOST 5058/15 KHSND	0.15% C 0.6% Mn 0.75% Cr 0.75% Ni 0.3% Cu steel: Plate; Russian Standard; hardened & tempered UTS: 600 Proof: 500
GOST 5632 KH5M	0.15% C (max) 5.2% Cr 0.52% Mo steel: Russian Standard
GOST 5632 KH5VF	0.15% C (max) 5.2% Cr 0.5% V 0.5% W steel: Russian Standard
GOST 5632 KH6SIU	0.15% C (max) 6.2% Cr 0.9% Al steel: Plate; Russian Standard
GOST 5632 KH6SM	0.15% C (max) 0.7% Mn (max) 5.75% Cr 0.55% Mo steel: Plate; Russian Standard; annealed UTS: 580 Proof: 310
GOST 5632 KH17 G9AN4	0.14% C (max) 1.7% Cr 4.0% Ni 0.2% N steel: Russian Standard
GOST 5632 KHCSM	0.15% C (max) 6.2% Cr 0.52% Mo steel: Russian Standard
GOST 5632/1 KH8VF	0.12% C 7.7% Cr 0.4% V 0.8% W steel: Russian Standard
GOST 5632/1 KH11 MF	0.15% C 10.7% Cr 0.7% Mo 0.3% V steel: Plate; Russian Standard
GOST 8731/12 KHN2	0.13% C 0.45% Mn 0.75% Cr 1.7% Ni steel: Pipe; Russian Standard; seamless UTS: 500 Proof: 400
GOST 8731/15 KHM	0.14% C 0.5% Mn 0.9% Cr 0.5% Mo steel: Pipe; Russian Standard; seamless UTS: 440 Proof: 230
GOST 8731/30 KHMA	0.3% C 0.5% Mn 0.9% Cr 0.2% Mo steel: Pipe; Russian Standard; seamless UTS: 600 Proof: 400

Symbol	Nominal analysis, supplier, condition and remarks.
GOST 8733/15 KH	0.14% C 0.55% Mn 0.95% Cr 0.5% Mo steel: Pipe; Russian Standard; seamless UTS: 440 Proof: 230
GOST 10802/12 K2 MFB	0.1% C 0.55% Mn 2.3% Cr 0.6% Mo 0.27% V 0.25% Cu (max) steel: Pipe; Russian Standard; seamless UTS: 420 Proof: 210
GOST 10802/12 KH1MF	0.11% C 0.55% Mn 0.95% Cr 0.3% Mo 0.22% V steel: Pipe; Russian Standard; seamless UTS: 500 Proof: 260
GOST 10802/12 KH2 MFSR	0.11% C 0.55% Mn 1.75% Cr 0.61% Mo 0.27% V 0.65% Nb 0.25% Cu (max) steel: Pipe; Russian Standard; seamless UTS: 480 Proof: 260
GOST 10802/15 KH1ME	0.13% C 0.55% Mn 1.25% Cr 1.0% Mo 0.27% V 0.2% Cu (max) steel: Pipe; Russian Standard; seamless UTS: 620 Proof: 350
GOST 10802/15 KHM	0.14% C 0.55% Mn 0.95% Cr 0.47% Mo steel: Pipe; Russian Standard; seamless UTS: 450 Proof: 240
GOST 12132/30 KHMA	0.3% C 0.5% Mn 0.95% Cr 0.2% Mo 0.25% Ni (max) steel: Pipe; Russian Standard; welded UTS: 600
GPS 1	0.15% C 1.5% Cr 4% Ni 0.25% Mo steel: For case hardening; Jonas; core properties DPN: 415 UTS: 1300 Elon: 12%
GPS 2	0.15% C 1% Cr 3% Ni steel: For case hardening; Jonas; core properties DPN: 341 UTS: 1070 Elon: 13%
GPS 3	0.15% C 0.6% Cr 2.75% Ni steel: For case hardening; Jonas; core properties DPN: 269 UTS: 840 Elon: 18%
GS 10 Mn Mo7/4	0.08% C 1.7% Mn 0.4% Mo 0.06% V 0.04% Nb steel: Cast; Thyssen; precipitation hardened DPN: 155 UTS: 530 Elon: 13% Proof: 405
H 5DM	0.15% C 0.5% Mn 0.5% Cr 4.0% Ni 0.4% Mo 0.1% V steel: Creusot-Loire
H 11	0.35% C 5% Cr 1.5% Mo 0.4% V steel: For tools; designation used by AISI
H 12	0.35% C 5% Cr 1.5% Mo 1.5% W 0.3% V steel: For tools; designation used by AISI
H 12	0.35% C 5.0% Cr 1.4% W 1.6% Mo 0.25% V steel: For hot work; Osborn
H 13	0.35% C 5% Cr 1.5% Mo 1% V steel: For tools; designation used by AISI
H 14	0.28% C 1% Cr 0.2% Mo steel: Jessop
H 19	0.2% C 5% Cr 0.55% Mo steel: Jessop
H 20	0.35% C 2.0% Cr 9.0% W steel: For tools; designation used by AISI
H 21	0.35% C 3.5% Cr 9% W 0.4% V steel: For tools; designation used by AISI
H 21	0.3% C 3.0% Cr 8.8% W 0.3% V steel: For hot work; Osborn
H 21N	0.25% C 3.0% Cr 8.8% W 0.3% V 2.3% Ni steel: For hot work; Osborn
H 22	0.35% C 2.0% Cr 11.0% W steel: For tools; For tools; designation used by AISI
H 22	0.2% C 1.5% Cr 0.5% Mo steel: Jessop
H 25	0.25% C 4.0% Cr 15.0% W steel: For tools; designation used by AISI
H32	0.24% C 3.2% Cr 0.3% Ni 0.5% Mo steel: Jessop for BS alloy En 40B
H 40	0.20% C 2.5% Cr 0.5% Mo 0.3% Ni 0.75% V steel: Jessop; creep resisting DPN: 300 UTS: 990 Elon: 20% Proof: 840
H 50	0.37% C 1.1% Si 5% Cr 1.3% Mo 1.1% V steel: Jessop DPN: 450

Symbol	Nominal analysis, supplier, condition and remarks.
H 51	0.2% C 0.6% Mo 0.25% V steel: Jessop
H 57	0.15% C 2.2% Cr 1% Mo steel: Jessop
H 65	0.12% C 1.02% Cr 0.55% Mo steel: Jessop
H 600/2	0.3% C 1.3% Si 4.5% Cr 1.0% Mo 0.3% V steel: Electrode; Metrode; hardenable deposit **DPN: 600**
HA	0.2% C 9% Cr 1% Mo steel: Casting; Huntingdon; aged **UTS: 770 Elon: 21% Proof: 570**
HARDTRADE 85/58	0.35% C 10% Mn 1.5% Cr 8.0% W 0.8% W 2.0% Co steel: For electrodes; Esab **DPN: 480**
HCM5	0.3% C 3.0% Cr 0.4% Mo steel: For nitriding; Firth Brown for BS 970 En 40B
HCM 7	0.2% C 3.0% Cr 0.4% Mo steel: For nitriding; Firth Brown for BS 970 En 40A
HCRS	0.25% C 0.75% Cr 0.6% Mo steel: ESC for BS alloy En 20A
HCRS 3	0.13% C 0.9% Cr 0.5% Mo steel: ESC; normalized & tempered BS alloy 620 **UTS: 440 Elon: 19% Proof: 220**
HCRS 4	0.13% C 2.25% Cr 1% Mo steel: ESC; good creep resistance
HD 3	0.32% C 3.5% Cr 9% W 0.3% V steel: T Turton
HD 3M	0.3% C 3.0% Cr 2.8% Mo 0.5% V steel: Origin unknown
HD 10	0.35% C 3.0% Cr 0.5% Mo 4.2% W steel: Spencer; forging dies, casting dies, etc. **DPN: 400**
HD 12	0.35% C 2.3% Cr 2% Ni 10% W 0.25% V steel: Turton Bros
HDS	0.28% C 3.2% Cr 9.5% W 0.3% V steel: Jessop; dies for casting Cu **DPN: 400**
HDZ	0.3% C 3.4% Cr 0.35% V 8.4% W steel: Origin unknown
HECLA 66	0.12% C 0.3% Cr 3.2% Ni steel: Case hardening; Hadfields for BS alloy En 33
HECLA 67	0.3% C 1.2% Cr 4.1% Ni steel: Hadfields for BS alloy En 30A
HECLA 67B	0.3% C 1.2% Cr 4.1% Ni 0.3% Mo steel: Hadfields for BS alloy En 30B
HECLA 73	0.3% C 0.3% Cr 3% Ni steel: Hadfields for BS alloy En 21
HECLA 78	0.16% C 0.3% Cr 5% Ni case hardening steel: Hadfields for BS alloy En 37
HECLA 98	0.35% C 0.6% Cr 1.2% Ni steel: Hadfields for BS alloy En 111 & En 111A
HECLA 116	0.3% C 0.7% Cr 3.2% Ni 0.6% Mo steel: Hadfields for BS alloy En 23 & 27
HECLA 138	0.3% C 0.6% Cr 2.5% Ni 0.5% Mo steel: Hadfields for BS alloy En 25
HECLA 142	0.3% C 1.2% Cr 3.5% Ni 0.45% Mo steel: Hadfields for BS alloy En 28
HECLA 143	0.15% C 0.8% Cr 3.5% Ni steel: Hadfields for BS alloy En 36A
HECLA 143B	0.15% C 0.8% Cr 3.5% Ni 0.2% Mo steel: Case hardening; Hadfields for BS alloy En 36C
HECLA 146	0.16% C 1.2% Cr 4% Ni steel: Case hardening; Hadfields for BS alloy En 39A

Note. The following abbreviations and units are used in the tables:

DPN	Hardness, diamond pyramid number
UTS	Ultimate tensile strength, N/mm^2
Elon	Elongation, %
Proof	0.1 % proof strength, N/mm^2

1 N/mm^2=0.1 hbar=0.102 kgf/mm^2=0.06475 tonf/in^2=145.04 lbf/in^2

Symbol	Nominal analysis, supplier, condition and remarks.
HECLA 146B	0.16% C 1.2% Cr 4% Ni 0.25% Mo steel: Case hardening; Hadfields
HECLA 149C	0.35% C 3.5% Cr 0.4% V 9.0% W steel: For tools; origin unknown
HECLA 151	0.16% C 1.75% Ni 0.25% Mo steel: Case hardening; Hadfields for BS alloy En 34
HECLA 151B	0.24% C 1.75% Ni 0.25% Mo steel: Case hardening; Hadfields for BS alloy En 35
HECLA 153	0.25% C 0.7% Cr 0.7% Mo steel: Hadfields for BS alloy En 20A
HECLA 163	0.16% C 0.3% Cr 5% Ni 0.25% Mo steel: Case hardening; Hadfields for BS alloy En 38
HECLA 174	0.35% C 5.0% Mo 0.4% V steel: Origin unknown
HECLA 177	0.35% C 5.0% Cr 1.5% Mo 0.3% V steel: Origin unknown
HECLA 181	0.2% C 0.6% Cr 0.8% Ni 0.1% Mo steel: Case hardening; Hadfields for BS alloy En 351
HECLA 182	0.2% C 0.8% Cr 1% Ni 0.1% Mo steel: Case hardening; Hadfields for BS alloy En 352
HECLA 183	0.2% C 1% Cr 1.25% Ni 0.12% Mo steel: Case hardening; Hadfields for BS alloy En 353
HECLA 184	0.2% C 1% Cr 1.75% Ni 0.15% Mo steel: Case hardening; Hadfields for BS alloy En 354
HECLA 185	0.2% C 1.5% Cr 2% Ni 0.2% Mo steel: Case hardening; Hadfields for BS alloy En 355
HECLA 191	0.15% C 0.7% Cr 0.6% Ni 0.1% Mo steel: Case hardening; Hadfields
HECLA 192	0.2% C 0.7% Cr 0.6% Ni 0.1% Mo steel: Case hardening; Hadfields for BS alloy En 362
HECLA 193	0.24% C 0.7% Cr 0.6% Ni 0.1% Mo steel: Case hardening; Hadfields for BS alloy En 363
HECLA 196	0.3% C 1.6% Cr 0.4% Ni 0.2% Mo 1.1% Al steel: Nitriding; Hadfields for BS alloy En 41 A & B
HECLA 197	0.22% C 0.5% Cr 1.7% Ni 0.25% Mo steel: Case hardening; Hadfields for BS alloy En 325
HECLA 317	0.15% C 3.2% Cr 0.4% Ni 0.5% Mo steel: Nitriding; Hadfields for BS alloys En 40 A & B, En 29 A & B
HEDEX 7	0.3% C 0.55% Cr 1.2% Ni steel: Wire for forging; Kiveton Park for BS alloy En 111
HEDEX 16	0.12% C 0.3% Cr(max) 3% Ni steel: Wire for forging; Kiveton Park for BS alloy En 33
HEDEX 17	0.2% C 0.6% Cr 0.8% Ni 0.1% Mo steel: Wire for forging; Kiveton Park for BS alloy En 351
HEDEX 19	0.2% C 0.5% Cr 0.55% Ni 9.25% Mo steel: Wire for forging; Kiveton Park; annealed **UTS: 580**
HEDEX 20	0.3% C 1.2% Cr 4.2% Ni steel: Wire for forging; Kiveton Park for BS alloy En 30A
HEDEX 21	0.35% C 0.5% Cr 0.6% Ni 0.25% Mo steel: Wire forging; Kiveton Park; annealed **UTS: 610**
HEDEX 23	0.27% C 0.5% Cr 0.5% Ni 0.25% Mo steel: Wire for forging; Kiveton Park; annealed **UTS: 610**
HEDEX 27	0.2% C 0.5% Cr 0.55% Ni 0.2% Mo steel: Wire for forging; Kiveton Park; annealed **UTS: 580**
HEDEX 28	0.17% C 0.5% Cr 0.55% Ni 0.2% Mo steel: Wire for forging; Kiveton Park; annealed **UTS: 540**
HEDEX 30	0.17% C 0.3% Cr (max) 1.7% Ni 0.25% Mo steel: Wire for forging; Kiveton Park for BS alloy En 34
HEDEX 31	0.12% C 0.8% Cr 3.5% Ni steel: Wire for forging; Kiveton Park for BS alloy En 36B
HEDEX 32	0.3% C 0.9% Cr 0.2% Mo steel: Wire for forging; Kiveton Park; annealed **UTS: 580**
HEDEX 34	0.31% C 0.7% Cr 2.5% Ni 0.6% Mo steel: Wire for forging; Kiveton Park for BS alloy En 25

Symbol	Nominal analysis, supplier, condition and remarks.
HEDEX 38	0.16% C 0.2% Cr 4.2% Ni 0.2% Mo steel: Wire for forging; Kiveton Park for BS alloy En 39B
HEDEX 39	0.18% C (max) 0.8% Cr 3.5% Ni steel: Wire for forging; Kiveton Park for BS alloy En 36
HEDEX 40	0.27% C 0.5% Cr 0.55% Ni 0.2% Mo steel: Wire for forging; Kiveton Park; annealed **UTS: 580**
HEDEX 43	0.15% C 0.5% Cr 0.55% Ni 0.2% Mo steel: Wire for forging; Kiveton Park; annealed **UTS: 540**
HEDEX 46	0.37% C 0.9% Cr 0.2% Mo steel: Wire for forging; Kiveton Park; annealed **UTS: 580**
HI 440	0.12% C 0.7% Mn 0.5% Ni 0.18% Mo 0.9% Cu steel: Inland Steel Co
HI-STEEL	0.12% 0.75% Mn 0.5% Ni 0.18% Mo 0.9% Cu steel: Inland Steel Co
HI STRENGTH A	0.12% C 0.6% Cr 0.7% Ni 0.1% Mo 0.4% Cu steel: Armco; weldable **UTS: 500 Elon: 22% Proof: 330**
HI TUFMOLD	0.35% C 1.2% Cr 0.4% Mo steel: Jessop
HI Z80	0.18% C (max) 0.9% Mn 0.6% Cr 0.85% Ni 0.5% Mo 0.07% V 0.2% Cu 0.004% B steel: Fuji
HK 5	0.25% C 2.0% Cr 1.0% Mo 0.5% V 0.6% Al steel: Firth Brown for nitriding
HOT DIE No 5	0.32% C 3.25% Cr 9.5% W 0.35% V steel: Edgar Allen; hot extrusion dies, etc. **DPN: 500**
HOTFORM No 1	0.35% C 5% Cr 1.4% Mo 1.4% W 0.45% V steel: Vanadium Alloys Ltd **DPN: 600**
HOTFORM No 2	0.35% C 5% Cr 1.3% Mo 0.45% V steel: Vanadium Alloys Ltd **DPN: 600**
HOV	0.3% C 1% Si 5% Cr 1.5% Mo 1% W steel: Jonas; gravity dies for Al casting **DPN: 550**
HOWARD A	0.35% C 5.0% Cr 1.5% Mo 0.4% V steel: Simmonds; AISI type H 11
HOWARD B	0.35% C 5.0% Cr 1.5% Mo 1.5% W 0.4% V steel: Simmonds; AISI type H 12
HOWARD C	0.35% C 5.0% Cr 1.5% Mo 1.0% V steel: Simmonds; AISI type H 13
HP	0.32% C 3.2% Cr 9% W 0.45% V steel: For tools; Huntsman; for hot work **DPN: 560**
HR 33	0.12% C 0.7% Cr 3.0% Ni steel: For case hardening; Bofors; annealed **DPN: 200**
HSB 51	0.18% C (max) 1 2% Mn 0.45% Ni 0.1% V 0.35% Cu steel: Phoenix Rheinrohr
HSB 51S	0.18% C (max) 1.2% Mn 0.45% Ni 0.1% V 0.35% Cu steel
HSB 52	0.18% C (max) 1.2% Mn 0.55% Ni 0.2% V 0.45% Cu steel: Phoenix Rheinrohr
HSB 52S	0.18% C (max) 1.2% Mn 0.55% Ni 0.12% V 0.45% Cu steel
HSB 55C	0.12% C 1.2% Mn 0.65% Ni 0.15% V 0.6% Cu steel: Phoenix Rheinrohr
HSB 77V	0.15% C (max) 1.2% Mn 0.5% Cr 1.2% Ni 0.3% Mo 0.1% V steel: Phoenix Rheinrohr
HSM/W9A	0.3% C 2.8% Cr 9.5% W 0.35% V die steel: ESC; for hot work **DPN: 460**
HT 1	0.09% C 1.5% Si 5% Cr 0.5% Mo steel: Sandvik
HT 3	0.1% C 5% Cr 0.5% Mo steel: Sandvik
HT 5	0.13% C 1% Cr 0.45% Mo steel: Sandvik for SIS 2216
HT 7	0.1% C 9% Cr 1% Mo steel: Sandvik

Symbol	Nominal analysis, supplier, condition and remarks.
HT 8	0.1% C 2.3% Cr 1% Mo steel: Sandvik for SIS 2218
HT 51	0.12% C 1.2% Cr 0.5% Mo steel: Sandvik
HW 1	0.3% C 3.5% Cr 9% W 0.5% V steel: For tools; Darwin; dies, die inserts **DPN: 480**
HW 2N	0.28% C 0.85% Cr 2.25% Ni 0.5% Mo 0.3% V 2.1% W steel: Origin unknown
HW 4	0.3% C 2.8% Cr 3% Ni 9% W 0.25% V steel: For tools; Darwin; dies punches **DPN: 480**
HW 5	0.3% C 5% Cr 1% W 2% Mo 0.25% V 1% Si steel: For tools; Darwin; dies **DPN: 480**
HW 6 NV	0.33% C 1.5% Cr 3.7% Ni 5.5% W 0.27% V steel: ESC; hot forging dies, punches, etc. **DPN: 470**
HWD 1	0.35% C 5.0% Cr 1.5% Mo 1.4% H 0.3% V steel: Firth Sterling
HWT 7	0.3% C 1.5% Cr 3.7% Ni 5.5% W steel: For tools; origin unknown
HY 80 ARMCO	0.18% C 0.25% Mn 1.4% Cr 2.7% Ni 0.6% Mo 0.03% V 0.25% Cu steel: Armco Steel Co
HY 100	0.2% C 0.25% Mn 1.4% Cr 2.9% Ni 0.6% Mo steel: Armco
HY 140	0.12% C (max) 0.55% Cr 5.0% Ni 0.5% Mo 0.07% V 0.15% Cu (max) steel: Origin unknown
HY 150	0.18% C 0.5% Mn 1.5% Cr 3.75% Ni 0.4% Mo 0.1% V steel: Armco
HYDRA E	0.3% C 3.5% Cr 8.5% W 0.35% V steel: Hall & Pickles; hot extrusion of copper alloys; obsolete **DPN: 550**
HYDRA M	0.35% C 1.5% Cr 4% Ni 0.5% Mo 4% W steel: Hall & Pickles; die casting dies for Al; etc.; obsolete **DPN: 500**
HYDRA Z	0.26% C 3% Cr 2.5% Ni 0.5% Mo 8.5% W 0.25% V steel: Hall & Pickles; forging dies; obsolete **DPN: 500**
HYKRO	0.28% C 3.2% Cr 0.5% Mo steel: ESC; nitriding steel; BS alloy En 40 A & B **UTS: 1430 Elon: 17% Proof: 1140**
HYKROM A	0.35% C 5% Cr 1.5% Mo 0.3% V steel: C Denton; extrusion dies, casting dies, etc.; obsolete **DPN: 610**
HYKROM B	0.35% C 3% Cr 0.5% Mo 0.1% V steel: C Denton; drop forging dies, etc.; obsolete **DPN: 550**
HY-TUF	0.25% C 1.3% Mn 1.5% Si 0.3% Cr 1.8% Ni 0.4% Mo steel: Crucible Steel Co **UTS: 1680 Elon: 14% Proof: 1380**
IBD	0.32% C 1.2% Cr 4.2% Ni 0.35% Mo steel: Inman; can be nitrided
IDROTUB 56	0.2% C (max) 1.4% Mn 0.3% Cr 0.3% Ni steel: Italian proprietory specification
IDROTUB 58T	0.18% C (max) 1.3% Mn 0.3% Cr 0.35% Ni 0.22% Mo steel: Italian proprietory specification
IDROTUB 62T	0.16% C (max) 1.3% Mn 0.3% Cr 0.25% Ni 0.22% Mo 0.11% V steel: Italian proprietory specification
IMPACTO	0.16% C 1.7% Ni 0.25% Mo steel: For carburizing; Atlas
IN 12	0.14% C (max) 0.8% Mn 0.55% Cr 1.0% Ni 0.35% Mo 0.07% V steel: Yawata
IN 787	0.04% C 0.5% Mn 0.6% Cr 1.2% Ni 0.2% Mo 0.05% Nb 1.2% Cu steel: Solution treated & aged; information from Climax Machy **UTS: 615 Elon: 30% Proof: 540**

444 · 44K1 STEEL – LOW CARBON ALLOYS

Symbol	Nominal analysis, supplier, condition and remarks.
IN 787A	=10.07% C (max) 0.75% Cr 0.85% Ni 0.2% Mo 1.2% Cu 0.02% Nb (min) steel: International Nickel Co; solution treated & aged; impact 176 J at -60 °C **UTS: 630 Elon: 26% Proof: 550**
INKOMO	0.3% C 0.7% Cr 3% Ni 0.5% Mo steel: Sanderson; hardened & tempered **DPN: 320 UTS: 1000 Elon: 14% Proof: 780**
IR 34	0.13% C 0.7% Cr 3.5% Ni steel: For case hardening; Bofors; annealed; obsolete **DPN: 220**
IR 74	0.3% C 0.8% Cr 3.5% Ni steel: Bofors; hardened & tempered; obsolete **DPN: 300 UTS: 1000 Elon: 15% Proof: 530**
IRO 743	0.35% C 1.1% Cr 3.4% Ni 0.3% Mo steel: Bofors **DPN: 475 UTS: 1500 Elon: 7%**
JO	0.37% C 3.2% Cr 9.5% W 0.3% V steel: Jessop
JO Sp1	0.27% C 3.2% Cr 9.5% W 0.3% V steel: Jessop
J 7	0.35% C 3.5% Cr 2.5% W 0.25% V steel: Jessop
J 12	0.26% C 2.7% Cr 3% Ni 0.5% Mo 10% W 0.3% V steel: Jessop; dies for casting Cu **DPN: 400**
J 21	0.21% C 3.2% Cr 4.2% Mo 1.1% W steel: Jessop
J 23	0.35% C 2.5% Cr 4.2% Mo 1.8% W 1% V steel: Jessop
JAVELIN	0.26% C 3.2% Cr 0.15% Ni 0.55% Mo steel: Tube; Atlas; for detachable rock drill bits
JC 20	0.3% C 2.75% Cr 10% W 0.4% V steel: Jonas; extrusion dies for nonferrous metals **DPN: 500**
JC 20N	0.3% C 2.5% Cr 2.25% Ni 10% W 0.25% V steel: Jonas; extrusion dies for non-ferrous metals **DPN: 560**
JH 2	0.3% C 3% Cr 0.55% Mo nitriding steel: S Osborn; obsolete **UTS: 920**
JIS G3114 SMA50A, B & C	0.19% C (max) 1.4% Mn (max) 0.8% Cr 0.5% Cu with Mo Nb Ni or V steel: Plate; Japanese Standard **UTS: 560 Proof: 360**
JIS G3114 SMA58	0.19% C (max) 1.4% Mn (max) 0.75% Cr 0.5% Cu uith Mo Ni V etc. steel: Plate; Japanese Standard; hardened & tempered **UTS: 650 Proof: 460**
JIS G3119 SBV3	0.25% C (max) 1.2% Mn 0.9% Ni 0.5% Mo steel: Plate; Japanese Standard; normalized **UTS: 630 Proof: 350**
JIS G3120 SQV2A	0.25% C (max) 1.25% Mn 0.6% Ni 0.55% Mo steel: Plate; Japanese Standard; hardened & tempered **UTS: 630 Proof: 350**
JIS G3120 SQV2B	0.25% C (max) 1.2% Mn 0.6% Ni 0.5% Mo steel: Plate; Japanese Standard; hardened & tempered **UTS: 720 Proof: 490**
JIS G3120 SQV3A	0.25% C (max) 1.25% Mn 0.85% Ni 0.5% Mo steel: Plate; Japanese Standard; hardened & tempered **UTS: 630 Proof: 350**

Symbol	Nominal analysis, supplier, condition and remarks.
JIS G3120 SQV3B	0.25% C (max) 1.2% Mn 0.9% Ni 0.5% Mo steel: Plate; Japanese Standard; hardened & tempered **UTS: 730 Proof: 490**
JIS G3125 SPAH	0.12% C (max) 0.35% Mn 0.8% Cr 0.5% Cu steel: Plate; Japanese Standard **UTS: 500 Proof: 360**
JIS G3179 SBV2	0.25% C (max) 1.2% Mn 0.6% Ni 0.5% Mo steel: Plate; Japanese Standard; normalized **UTS: 630 Proof: 350**
JIS G3441 STKS1A	0.3% C 0.6% Mn 1.0% Cr 0.2% Mo steel: Pipe; Japanese Standard **UTS: 670**
JIS G3441 STKS1B	0.3% C 0.6% Mn 1.0% Cr 0.2% Mo steel: Pipe; Japanese Standard **UTS: 560 Proof: 400**
JIS G3441 STKS1C	0.3% C 0.6% Mn 1.0% Cr 0.2% Mo steel: Pipe; Japanese Standard **UTS: 630 Proof: 500**
JIS G3441 STKS1D	0.3% C 0.6% Mn 1.0% Cr 0.2% Mo steel: Pipe; Japanese Standard **UTS: 880 Proof: 700**
JIS G3441 STKS1E	0.3% C 0.6% Mn 1.0% Cr 0.2% Mo steel: Pipe; Japanese Standard **UTS: 1050 Proof: 950**
JIS G3441 STKS2A	0.3% C 0.75% Mn 0.5% Cr 0.55% Ni 0.2% Mo steel: Pipe; Japanese Standard **UTS: 670**
JIS G3441 STKS2B	0.3% C 0.75% Mn 0.5% Cr 0.55% Ni 0.2% Mo steel: Pipe; Japanese Standard **UTS: 560 Proof: 400**
JIS G3441 STKS2C	0.3% C 0.75% Mn 0.5% Cr 0.55% Ni 0.2% Mo steel: Pipe; Japanese Standard **UTS: 630 Proof: 500**
JIS G3441 STKS2D	0.3% C 0.75% Mn 0.5% Cr 0.55% Ni 0.2% Mo steel: Pipe; Japanese Standard **UTS: 880 Proof: 700**
JIS G3441 STKS2E	0.3% C 0.75% Mn 0.5% Cr 0.55% Ni 0.2% Mo steel: Pipe; Japanese Standard **UTS: 1050 Proof: 950**
JIS G3441 STK3A	0.35% C 0.6% Mn 1.0% Cr 0.2% Mo steel: Pipe; Japanese Standard **UTS: 700**
JIS G3441 STK3B	0.35% C 0.6% Mn 1.0% Cr 0.2% Mo steel: Pipe; Japanese Standard **UTS: 560 Proof: 400**
JIS G3441 STK3C	0.35% C 0.6% Mn 1.0% Cr 0.2% Mo steel: Pipe; Japanese Standard **UTS: 670 Proof: 500**
JIS G3441 STK3D	0.35% C 0.6% Mn 1.0% Cr 0.2% Mo steel: Pipe; Japanese Standard **UTS: 880 Proof: 700**
JIS G3441 STK3E	0.35% C 0.6% Mn 1.0% Cr 0.2% Mo steel: Pipe; Japanese Standard **UTS: 1050 Proof: 950**
JIS G3441 STKS4A	0.34% C 0.85% Mn 0.55% Cr 0.5% Ni 0.25% Mo steel: Pipe; Japanese Standard **UTS: 700**
JIS G3441 STKS4B	0.34% C 0.85% Mn 0.55% Cr 0.5% Ni 0.25% Mo steel: Pipe; Japanese Standard **UTS: 560 Proof: 400**
JIS G3441 STKS4C	0.34% C 0.85% Mn 0.55% Cr 0.5% Ni 0.25% Mo steel: Pipe; Japanese Standard **UTS: 670 Proof: 500**
JIS G3441 STKS4D	0.34% C 0.85% Mn 0.55% Cr 0.5% Ni 0.25% Mo steel: Pipe; Japanese Standard **UTS: 880 Proof: 700**
JIS G3441 STKS4E	0.34% C 0.85% Mn 0.55% Cr 0.5% Ni 0.25% Mo steel: Pipe; Japanese Standard **UTS: 1050 Proof: 95**

Note. The following abbreviations and units are used in the tables:

DPN	Hardness, diamond pyramid number
UTS	Ultimate tensile strength, N/mm^2
Elon	Elongation, %
Proof	0.1 % proof strength, N/mm^2

1 N/mm^2=0.1 hbar=0.102 kgf/mm^2=0.06475 tonf/in^2=145.04 lbf/in^2

Symbol	Nominal analysis, supplier, condition and remarks.
JIS G3458 STPA22	0.15% C (max) 0.45% Mn 1.0% Cr 0.55% Cu 0.035% S P (max) steel: Plate; Japanese Standard **UTS: 420** **Proof: 210**
JIS G3458 STPA23	0.15% C (max) 0.45% Mn 1.25% Cr 0.55% Mo 0.035% S P (max) steel: Pipe; Japanese Standard **UTS: 420** **Proof: 210**
JIS G3458 STPA24	0.15% C (max) 0.45% Mn 2.25% Cr 1.0% Mo 0.03% S P (max) steel: Pipe; Japanese Standard **UTS: 420** **Proof: 210**
JIS G3458 STPA25	0.15% C (max) 0.45% Mn 5% Cr 0.55% Mo steel: Pipe; Japanese Standard **UTS: 420** **Proof: 210**
JIS G3458 STPA26	0.15% C (max) 0.45% Mn 9% Cr 1% Mo steel: Pipe; Japanese Standard **UTS: 420** **Proof: 210**
JIS G3462 STBA20	0.15% C 0.45% Mn 0.65% Cr 0.52% Mo steel: Pipe; Japanese Standard **UTS: 420** **Proof: 210**
JIS G3462 STBA22	0.15% C (max) 0.45% Mn 1.0% Cr 0.55% Mo steel: Pipe; Japanese Standard **UTS: 420** **Proof: 210**
JIS G3462 STBA23	0.15% C (max) 0.45% Mn 1.25% Cr 0.55% Mo steel: Pipe; Japanese Standard **UTS: 420** **Proof: 210**
JIS G3462 STBA24	0.15% C (max) 0.45% Mn 2.3% Cr 1.0% Mo steel: Pipe; Japanese Standard **UTS: 420** **Proof: 210**
JIS G3462 STBA25	0.15% C (max) 0.45% Mn 5.0% Cr 0.55% Mo steel: Pipe; Japanese Standard **UTS: 420** **Proof: 210**
JIS G3462 STBA26	0.15% C (max) 9.0% Cr 1.0% Mo steel: Pipe; Japanese Standard **UTS: 420** **Proof: 210**
JIS G4106 SCM1	0.32% C 0.45% Mn 1.2% Cr 0.22% Mo steel: Japanese Standard
JIS G4106 SCM2	0.31% C 0.7% Mn 1.05% Cr 0.22% Mo steel: Japanese Standard
JIS G4106 SCM3	0.36% C 0.7% Mn 1.05% Cr 0.22% Mo steel: Japanese Standard
JIS G4106 SCM21	0.16% C 0.7% Mn 1.05% Cr 0.22% Mo steel: Japanese Standard
JIS G4106 SCM22	0.21% C 0.7% Mn 1.05% Cr 0.22% Mo steel: Japanese Standard
JIS G4106 SCM23	0.2% C 0.85% Mn 1.05% Cr 0.22% Mo steel: Japanese Standard
JIS G4106 SCM24	0.23% C 0.7% Mn 1.05% Cr 0.4% Mo steel: Japanese Standard
JIS G4106 SNC2	0.31% C 0.5% Mn 0.8% Cr 2.75% Ni steel: Japanese Standard
JIS G4106 SNC 21	0.15% C 0.5% Mn 0.35% Cr 2.25% Ni steel: Japanese Standard
JIS G4106 SNC22	0.15% C 0.5% Mn 1.85% Cr 3.25% Ni steel: Japanese Standard
JIS G4106 SNCM1	0.31% C 0.75% Mn 0.8% Cr 1.8% Ni 0.22% Mo steel: Japanese Standard
JIS G4106 SNCM2	0.25% C 0.5% Mn 1.25% Cr 3.25% Ni 0.22% Mo steel: Japanese Standard
JIS G4106 SNCM5	0.3% C 0.5% Mn 3.0% Cr 3.0% Ni 0.6% Mo steel: Japanese Standard
JIS G4106 SNCM21	0.21% C 0.85% Mn 0.5% Cr 0.5% Ni 0.22% Mo steel: Japanese Standard
JIS G4106 SNCM22	0.15% C 0.55% Mn 0.5% Cr 1.8% Ni 0.22% Mo steel: Japanese Standard
JIS G4106 SNCM25	0.15% C 0.45% Mn 0.85% Cr 4.25% Ni 0.22% Mo steel: Japanese Standard
JIS G4106 SNCM26	0.17% C 1.0% Mn 1.6% Cr 3.0% Ni 0.5% Mo steel: Japanese Standard

Symbol	Nominal analysis, supplier, condition and remarks.
JIS G4109 SCMV1	0.21% C (max) 0.65% Mn 0.65% Cr 0.5% Mo steel: Plate; Japanese Standard; normalized and tempered **UTS: 560** **Proof: 320**
JIS G4109 SCMV2	0.17% C (max) 0.5% Mn 1.0% Ni 0.5% Mo steel: Plate; Japanese Standard; normalized & tempered **UTS: 530** **Proof: 280**
JIS G4109 SCMV3	0.17% C (max) 0.5% Mn 1.25% Cr 0.55% Mo steel: Plate; Japanese Standard; normalized & tempered **UTS: 600** **Proof: 320**
JIS G4109 SCMV4	0.15% C (max) 0.45% Mn 2.25% Cr 1.0% Mo steel: Plate; Japanese Standard; normalized & tempered **UTS: 610** **Proof: 320**
JIS G4109 SCMV5	0.15% C (max) 3.0% Cr 1.0% Mo steel: Plate; Japanese Standard; normalized & tempered **UTS: 610** **Proof: 320**
JIS G4109 SCMV6	0.15% C (max) 0.45% Mn 5.0% Cr 0.5% Mo steel: Plate; Japanese Standard; normalized & tempered **UTS: 610** **Proof: 320**
JS 1	0.25% C (max) 0.6% Si 1.0% Mn Ni Cr Mo and Cu at 1% (max) steel: Casting; Jopling; weldable **DPN: 130** **UTS: 420** **Elon: 22%** **Proof: 210**
JS 1B	0.25% C (max) 0.4% Ni 0.25% Cr 0.1% Mo 0.4% Cu steel: Casting; Jopling; for magnetic properties **DPN: 140** **UTS: 440** **Elon: 20%** **Proof: 200**
JS 4	0.15% C (max) 0.4% Ni 0.2% Cr 0.1% Mo 0.4% Cu steel: Casting; Jopling; for magnetic properties **DPN: 110** **UTS: 420** **Elon: 22%**
JS 9	0.35% C (max) 1.0% Ni (max) 1.0% Cr (max) 0.6% Mo (max) steel: Casting; Jopling **DPN: 230** **UTS: 760** **Elon: 15%** **Proof: 480**
JS 10	0.35% C (max) 1.0% Ni (max) 1.0% Cr (max) 0.6% Mo (max) steel: Casting; Jopling **DPN: 275** **UTS: 920** **Elon: 12%** **Proof: 580**
JS 40	0.2% C (max) 0.4% Ni 1.2% Cr 0.5% Mo 0.4% Cu (max) steel: Casting; Jopling; for use up to 54 °C **DPN: 160** **UTS: 460** **Elon: 20%** **Proof: 270**
K 18	0.32% C 5% Cr 1.7% Mo 1.4% W steel: Jessop **DPN: 440**
K 291	0.13% C 0.7% Cr 3.0% Ni steel: Fagersta
K 336	0.18% C 0.7% Cr 3.0% Ni steel: Fagersta
K 353	0.15% C 0.8% Cr 1.5% Ni steel: Fagersta
K 354	0.2% C 0.8% Cr 1.5% Ni steel: Fagersta
K 825	0.36% C 1.4% Cr 1.4% Ni 0.2% Mo steel: Fagersta
KAISALOY No. 1	0.2% C (max) 1.25% Mn 0.25% Cr 0.6% Ni 0.15% Mo 0.02% V 0.35% Cu 0.25% Al steel: Kaiser Steel Corp.
KAISALOY No. 2	0.12% C (max) 0.6% Mn 0.25% Cr 0.6% Ni 0.1% Mo 0.005% Ti 0.02% V 0.3% Cu steel: Kaiser Steel Co.
KAISALOY 45FG	0.12% C (max) 0.6% Mn 0.25% Cr 0.6% Ni 0.1% Mo 0.005% Ti 0.02% V 0.3% Cu steel: Kaiser Steel Corp.
KAISALOY 50CR	0.2% C (max) 1.25% Mn 0.25% Cr 0.6% Ni 0.15% Mo 0.02% V 0.005% Ti 0.35% Cu steel: Kaiser Steel Corp.
KD 2	0.25% C 1.0% Cr 0.25% Mo steel: Pompey. **UTS: 1070** **Elon: 10%** **Proof: 870**
KD 2SE	0.25% C 1.0% Cr 0.25% Mo steel: Pompey **UTS: 990** **Elon: 11%** **Proof: 760**
KD 3	0.35% C 1.0% Cr 0.25% Mo steel: Pompey **UTS: 1150** **Elon: 9%** **Proof: 990**
KE 41	0.12% C 1.35% Cr 4.2% Ni 0.2% Mo steel: For case hardening; Kayser Ellison; hardened & tempered; core properties **DPN: 375** **UTS: 1300** **Elon: 15%**

Symbol	Nominal analysis, supplier, condition and remarks.
KE 169	0.12% C 0.9% Cr 3.2% Ni 0.2% Mo steel: For case hardening; Kayser Ellison; hardened & tempered; core properties **DPN: 302 UTS: 1000 Elon: 16%**
KE 287	0.12% C 0.3% Cr 3.2% Ni 0.2% Mo steel: For case hardening; Kayser Ellison; hardened & tempered; core properties **DPN: 235 UTS: 790 Elon: 22%**
KE 339	0.25% C 0.7% Si 3% Cr 2.2% Ni 0.4% Mo 0.4% Mo 9.5% W 0.5% V steel: Kayser Ellison; for hot working copper **DPN: 470**
KE 660	0.12% C 0.3% Cr 4.7% Ni 0.2% Mo steel: For case hardening; Kayser Ellison; hardened & tempered; core properties **DPN: 311 UTS: 1020 Elon: 16%**
KE 805	0.35% C 1% Cr 1.5% Ni 0.2% Mo steel: Kayser Ellison; hardened & tempered **DPN: 350 UTS: 1150 Elon: 17% Proof: 1030**
KE 897	0.3% C 1.2% Cr 4.2% Ni 0.2% Mo steel: Kayser Ellison; hardened & tempered **DPN: 470 UTS: 1680 Elon: 12%**
KE A145	0.35% C 1% Si 5.5% Cr 1.3% Mo 1% V steel: Kayser Ellison; for plastic extrusion dies **DPN: 402**
KF 15A	0.14% C 1% Cr 3.4% Ni steel: Case hardening; Kirkstall for BS alloy En 36A
KF 15B	0.16% C 1% Cr 3.5% Ni 0.25% Mo steel: Case hardening; Kirkstall for BS alloy En 36C
KF 15C	0.16% C 1.2% Cr 4.2% Ni 0.3% Mo steel: Case hardening; Kirkstall for BS alloy En 39B
KF 16	0.32% C 0.9% Cr 3.2% Ni steel: Kirkstall for BS alloy En 23
KF 19	0.3% C 1.3% Cr 4.1% Ni 0.3% Mo steel: Kirkstall for BS alloy En 30B
KF 28	0.32% C 0.8% Cr 3.2% Ni 0.6% Mo steel: Kirkstall for BS alloy En 27
KF 28B	0.32% C 0.6% Cr 2.5% Ni 0.6% Mo steel: Kirkstall for BS alloy En 25
KF 31B	0.19% C 0.5% Cr 1.7% Ni 0.25% Mo steel: Case hardening; Kirkstall for BS alloy En 325
KF 38A	0.35% C 0.6% Cr 1.2% Ni steel: Kirkstall for BS alloy En 111
KF 42	0.17% C 1.7% Ni 0.25% Mo steel: Case hardening; Kirkstall for BS alloy En 34
KF 42A	0.22% C 1.7% Ni 0.25% Mo steel: Case hardening; Kirkstall for BS alloy En 35
KF 50A	0.2% C 1.6% Mn 0.6% Ni 0.25% Mo steel: Kirkstall for BS alloy En 13
KF 70	0.2% C (max) 0.6% Cr 0.8% Ni 0.1% Mo steel: Case hardening; Kirkstall for BS alloy En 351
KF 71	0.2% C 0.8% Cr 1% Ni 0.1% Mo steel: Case hardening; Kirkstall for BS alloy En 352
KF 72	0.2% C 1% Cr 1.2% Ni 0.11% Mo steel: Case hardening; Kirkstall for BS alloy En 353
KF 73	0.2% C (max) 1% Cr 1.7% Ni 0.15% Mo steel: Case hardening; Kirkstall for BS alloy En 354
KF 74	0.2% C (max) 1.5% Cr 2% Ni 0.2% Mo steel: Case hardening; Kirkstall for BS alloy En 355

Symbol	Nominal analysis, supplier, condition and remarks.
KF 75	0.15% C 0.7% Cr 0.6% Ni 0.11% Mo steel: Case hardening; Kirkstall for BS alloy En 361
KF 76	0.2% C 0.7% Cr 0.5% Ni 0.11% Mo steel: Case hardening; Kirkstall for BS alloy En 362
KF 77	0.24% C 0.7% Cr 0.5% Ni 0.11% Mo steel: Case hardening; Kirkstall for BS alloy En 363
KH 5M	0.15% C (max) 5.2% Cr 0.52% Mo steel: Russian Standard designation
KH 5VF	0.15% C (max) 5.2% Cr 0.5% V 0.5% W steel: Russian Standard designation
KO	0.15% C 0.8% Mn 0.6% Cr 0.85% Ni 0.5% Mo 0.07% V 0.3% Cu steel: Kawasaki Iron & Steel Co.
KP 12	0.12% C 0.3% Cr 3% Ni steel: Case hardening; Kiveton Park for BS alloy En 33
KP 13	0.3% C 0.5% Mn 0.3% Cr 3% Ni steel: Kiveton Park for BS alloy En 21
KP 17	0.3% C 0.8% Cr 3.25% Ni 0.6% Mo steel: Kiveton Park for BS alloy En 23
KP 23	0.3% C 1% Cr 3.5% Ni 0.4% Mo steel: Kiveton Park for BS alloy En 27
KP 31	0.18% C (max) 0.8% Cr 3.5% Ni steel: For case hardening; Kiveton Park for BS alloy En 36
KP 35	0.16% C 1.2% Cr 4% Ni steel: For case hardening; Kiveton Park for BS alloy En 39
KP 36	0.3% C 1.2% Cr 4.2% Ni 0.3% Mo steel: Kiveton Park for BS alloy En 30B
KP 37	0.2% C 1.7% Ni 2.25% Mo steel: For case hardening; Kiveton Park for BS alloy En 34
KR 35	0.13% C 1.2% Cr 4.2% Ni steel: For case hardening; Bofors; annealed **DPN: 240**
KR 75	0.35% C 1.2% Cr 4.2% Ni steel: Bofors; hardened & tempered **DPN: 340 UTS: 1160 Elon: 12% Proof: 870**
L 2N	0.17% C 1.8% Ni 0.25% Mo steel: For case hardening; Swift Levick for BS alloy En 34
L 197	0.36% C 1.4% Cr 1.4% Ni 0.2% Mo steel: Fagersta; hardened & tempered **DPN: 400 UTS: 1360 Elon: 5% Proof: 1160**
L 536	0.3% C 1.0% Cr 3.2% Ni 0.25% Mo steel: Fagersta
LARPORT	0.3% C 1.25% Cr 4.25% Ni 0.3% Mo steel: Hall & Pickles; plastic dies & moulds; hardened & tempered; obsolete **UTS: 1140 Elon: 20% Proof: 1020**
LCHD	0.35% C 3.5% Cr 0.4% C 9.0% W steel: Origin unknown
LESCALLOY 4330 +V	0.3% C 0.85% Cr 1.8% Ni 0.4% Mo 0.07% V steel: Latrobe
LESCALLOY 4335 +V	0.35% C 0.8% Cr 1.85% Ni 0.35% Mo 0.2% V steel: Latrobe
LESCALLOY 9310	0.1% C 3.2% Ni 1.2% Cr 0.12% Mo steel: Latrobe; for carburizing
LESCALLOY BG31	0.12% C 1.5% Cr 5.0% Mo 2.8% Ni steel: Latrobe; for carburizing
LESCALLOY UT19	0.16% C 1.2% Cr 4.25% Ni 0.25% Mo steel: Latrobe; for carburizing
LESCO HW114	0.3% C 9.75% W 2.75% Cr 1.65% Ni steel: Latrobe; for hot working
LHG	0.12% C 0.8% Cr 3.2% Ni steel: For case hardening; Swift Levick for BS alloy En 36
LK 5	0.3% C 1.6% Cr 0.2% Mo 1.1% Al steel: For nitriding; Firth Brown for En 41A
LK 7	0.2% C 1.6% Cr 0.2% Mo 1.1% Al steel: For nitriding; Firth Brown
LKP	0.24% C 1.3% Cr 3.5% Ni 0.3% Mo 1.0% Al steel: Firth Brown; for nitriding

Note. The following abbreviations and units are used in the tables:

DPN	Hardness, diamond pyramid number
UTS	Ultimate tensile strength, N/mm^2
Elon	Elongation, %
Proof	0.1 % proof strength, N/mm^2

1 N/mm^2=0.1 hbar=0.102 kgf/mm^2=0.06475 tonf/in^2=145.04 lbf/in^2

Symbol	Nominal analysis, supplier, condition and remarks.
LOYCON N	0.15% C (max) 1.2% Mn 0.6% Cr 1.6% Ni 0.3% Mo 0.12% V steel: Consett; normalized **UTS: 610 Elon: 16% Proof: 360**
LOYCON QT	0.15% C (max) 1.2% Mn 0.6% Cr 1.6% Ni 0.3% Mo 0.12% V steel: Consett; hardened & tempered **UTS: 770 Elon: 18% Proof: 710**
LPD	0.35% C 1% Si 5% Cr 1.6% Mo 1.3% W 0.3% V steel: Casting; Latrobe; steel for precision casting of forging dies **DPN: 550**
LT 39	0.35% C 5.0% Cr 1.3% Mo 1.0% V steel: For tools; Low Moor
LT 41W	0.35% C 1.0% Si 5% Cr 1.6% Mo 1.5% W 0.25% V steel: Low Moor; for hot work tools; obsolete **DPN: 550**
LT 43	0.3% C 3.2% Cr 9% W 0.2% V steel: Low Moor; for hot work tools; obsolete **DPN: 620**
LT 45	0.25% C 3% Cr 2.5% Ni 8.5% W 0.25% V steel: Low Moor; for hot work tools; obsolete **DPN: 630**
LT 52	0.35% C 1.25% Cr 4.25% Ni 0.3% Mo steel: Low Moor; die steel; obsolete **DPN: 580**
LT 53	0.12% C 1.25% Cr 4% Ni steel: Low Moor; carburizing die steel; obsolete
LT FORGING	0.33% C 3.5% Cr 9.5% W 0.5% V steel: Firth Stirling
M Mo 30	0.3% C 1.5% Mn 0.25% Cr 0.4% Ni 0.3% Mo steel: Casting; Wolsingham **DPN: 225 UTS: 760 Elon: 15% Proof: 480**
MANOIR ABRADIER 600	Cr W V steel: For carburizing; Pompey
MANOIR PF1	0.15% C 0.5% Cr 0.5% Mo steel: Pompey **UTS: 500 Elon: 18% Proof: 280**
MANOIR PF2	0.15% C 1.0% Cr 0.5% Mo steel: Pompey **UTS: 550 Elon: 16% Proof: 320**
MANOIR PF5	0.15% C 5.0% Cr 0.5% Mo steel: Pompey **UTS: 600 Elon: 18% Proof: 260**
MANOIR PF6	0.15% C 2.2% Cr 1.0% Mo steel: Pompey **UTS: 500 Elon: 15% Proof: 240**
MANOIR PF15	0.2% C 0.8% Cr 0.4% Mo steel: Pompey **UTS: 650 Elon: 16% Proof: 400**
MANOIR PFV55	0.15% C 2.0% Cr 0.35% Mo + V + Al steel: Pompey **UTS: 550 Elon: 15% Proof: 280**
MANOIR PM17	0.12% C with Mo and V steel: Pompey **UTS: 600 Elon: 16% Proof: 420**
MANOIR PM35	0.17% C with Cr Mo and V steel: Pompey **UTS: 1100 Elon: 8% Proof: 800**
MARSH CH 5	0.37% C 5% Cr 1.4% Mo 0.6% V steel: Marsh Brothers; hot forging dies, etc.
MARSH CTHY	0.3% C 3.4% Cr 8.5% W 0.35% V steel: Marsh Brothers; hot forging dies, etc.
MAX HIGH TENSILE	0.15% C (max) 0.75% Mn 0.55% Cr 0.2% Mo 0.2% Cu 0.07% Zr steel: National Steel
MAXINIUM	0.32% C 5.25% Cr 0.5% Mo 5% W steel: Edgar Allen; die casting moulds **DPN: 570**
MAXNAP	0.3% C 1.05% Cr 0.2% V steel: Edgar Allen; rivet hammers, lathe centres, etc. **DPN: 540**
MAXTRESS	0.35% C 1% Cr 3.5% Ni 0.5% Mo steel: Jonas **UTS: 1380 Elon: 13% Proof: 1300**
MAYARI R	0.12% C (max) 0.75% Mn 0.7% Cr 1.0% Ni 0.1% Zr steel: Bethlehem Steel Co.
MCHT	0.25% C 0.4% Cr 1.7% Ni 0.4% Mo steel: Pompey

Symbol	Nominal analysis, supplier, condition and remarks.
MCL	0.3% C 3.7% Cr 6.25% Mo 1.0% W 0.75% V steel: Latrobe; hot work **DPN: 450**
MCN	0.33% C 0.7% Cr 3% Ni 0.3% Mo steel: For hot work tools; Osborn; stamping dies; casting dies, etc., obsolete **DPN: 350**
MCV 24	0.27% C 0.75% Cr 0.1% V steel: Krupp
MEL-TROL SUPER SAMSON	0.1% C 5.0% Cr 0.9% Mo 0.25% V steel: Carpenter; for AISI type P4
MEL-TROL TK	0.35% C 3.5% Cr 9.0% W 0.4% V steel: Carpenter
MIL S 12505/2	0.12% C (max) 1.0% Mn 0.75% Cr 0.7% Ni 0.35% Cu steel: Bethlehem Steel Co.
MK 2	0.3% C 3.0% Cr 0.4% V 8.8% W steel: Origin unknown
ML 50	0.13% C 0.48% Mn 0.4% Cr 0.61% Ni 0.01% Mo 0.2% Cu steel: McLouth Steel Co
ML 60	0.13% C 0.9% Mn 0.51% Cr 0.68% Ni 0.01% Mo 0.3% Cu steel: McLouth
ML 70	0.16% C 0.95% Mn 0.55% Cr 0.73% Ni 0.01% Mo 0.3% Cu steel: McLouth
MNC 831	General standard for SIS range of pressure vessel steel tubes; Swedish Standard
MNC 832	General standard for SIS range of seamless gas pressure vessels; Swedish Standard
MNC 853	General standard for SIS range of alloyed steels intended for heat treatment; Swedish Standard
Mo 18	0.18% C 0.25% Cr 0.4% Ni 0.55% Mo 0.3% Cu steel: Casting; Wolsingham **UTS: 450 Elon: 20% Proof: 270**
MODAL 5	0.3% C 2.1% Cr 2.1% Ni 0.4% Mo steel: Pompey
MOHD	0.36% C 3.5% Cr 6% Mo 1% W 0.75% V tool steel: Huntsman; hot work dies **DPN: 600**
MOHD	0.36% C 3.5% Cr 6.0% Mo 0.75% V 1.0% V steel: Origin unknown
MONARCH DW8	0.3% C 2.75% Cr 0.45% V 7.7% W tool steel: Origin unknown
MONARCH NCG	0.35% C 1.5% Cr 4.0% Ni 0.3% Mo steel: Origin unknown
MONARCH SPECIAL TAN	0.3% C 0.75% Cr 2.5% Ni 0.6% Mo tool steel: Origin unknown
MOSTAR	0.32% C 5.0% Cr 1.15% Mo 1.2% V 0.35% V steel: Pompey
MOV	0.18% C 0.6% Mo 0.3% V steel: ESC; creep resisting; for steam turbines
MOVA	0.2% C 0.8% Cr 0.3% Ni 0.7% Mo 0.25% V steel: Firth Brown; for use at temperatures up to 550 °C **UTS: 840 Elon: 15% Proof: 610**
MTA 2	0.35% C 1.7% Cr 3.75% Ni steel: Pompey
MTA 3	0.35% C 1.7% Cr 4.0% Ni 0.5% Mo steel: Pompey
MTR	0.13% C 0.8% Mn 0.55% Cr 1.4% Ni 0.5% Mo 0.08% V 0.25% Cu steel: Imphy
MVW	0.35% C 1.75% Cr 1.25% Ni 0.5% Mo 0.5% W 0.5% V steel: Swift Levick; for hot dies
N 1	0.2% C 0.5% Cr 0.5% Ni 0.2% Mo steel: Jessop
N 2	0.22% C 0.5% Cr 0.55% Ni 0.2% Mo steel: Jessop
N 2MCH	0.17% C 1.8% Ni 0.25% Mo steel: For carburizing; Firth Brown for BS 970 En 34
N 3	0.17% C 0.5% Cr 0.55% Ni 0.2% Mo steel: Jessop
N 3CCH	0.12% C 3.3% Ni 0.8% Cr steel: For carburizing; Firth Brown for BS 970 En 36A & B
N 3CMCH	0.12% C 3.3% Ni 0.8% Cr 0.2% Mo steel: For carburizing; Firth Brown for BS 970 En 36
N 3MCH	0.24% C 1.8% Ni 0.25% Mo steel: For carburizing; Firth Brown for BS 970 En 35
N 4CCH	0.14% C 4.25% Ni 1.25% Cr steel: For carburizing; Firth Brown for BS 970 En 39A

Symbol	Nominal analysis, supplier, condition and remarks.
N 4CMCH	0.14% C 4.25% Ni 1.25% Cr 0.25% Mo steel: For carburizing; Firth Brown for BS 970 En 39B
N 5MCH	0.12% C 5.1% Ni 0.2% Mo steel: For carburizing; Firth Brown for BS 970 En 38
N 12	0.12% C 0.2% Cr 3.5% Ni 0.15% Mo 0.3% Cu steel: Casting; Wolsingham
	UTS: 440 Elon: 25% Proof: 240
N A XTRA	0.15% C 0.85% Mn 0.6% Cr 0.3% Mo 0.0025% B steel: National Steel
N A XTRA 55	0.2% C (max) 1.0% Mn 0.7% Cr 0.25% Mo 0.07% Zr steel: Hoag
N A XTRA 60	0.2% C (max) 1.0% Mn 0.7% Cr 0.25% Mo 0.07% Zr steel: Hoag
N A XTRA 65	0.2% C (max) 1.0% Mn 0.7% Cr 0.25% Mo 0.07% Zr steel: Hoag
N A XTRA 70	0.2% C (max) 1.0% Mn 0.7% Cr 0.25% Mo 0.07% Zr steel: Hoag
N A XTRA 75	0.2 °C (max) 1.0% Mn 0.7% Cr 0.25% Mo 0.07% Zr steel: Hoag
N A XTRA 80	0.2% C (max) 0.85% Mn 0.6% Cr 0.3% Mo 0.05% Zr steel: National Steel
N A XTRA 93	0.15% C 0.8% Mn 0.6% Cr 0.3% Mo steel: National Steel
N A XTRA 100	0.15% C 0.85% Mn 0.6% Cr 0.3% Mo steel: National Steel
NATIONALLOY 14	0.35% C 1.0% Cr 3.5% Ni 0.5% Mo 0.12% V steel: Source unknown
NBVOC 1/W	0.17% C 0.5% Cr 1.6% Ni 0.45% Mo steel: Krupp
NCM 1	0.34% C 1.1% Cr 1.5% Ni 0.2% Mo steel: Huntsman
	DPN: 600
NCM 2	0.35% C 0.6% Cr 2.5% Ni 0.6% Mo steel: Huntsman; punches, collets, moulds, etc.
	DPN: 560
NCM 3	0.35% C 1.2% Cr 3.5% Ni 0.4% Mo steel: Huntsman; punches, collets, moulds, etc.
	DPN: 550
NCM 3	0.3% C 0.6% Cr 2.4% Ni 0.45% Mo steel: Firth Brown for BS alloy En 25
NCM 4	0.35% C 1.3% Cr 4% Ni 0.3% Mo steel: Huntsman; dies, punches, moulds, etc.
	DPN: 550
NCM 6	0.3% C 0.8% Cr 3.25% Ni 0.3% Mo steel: Firth Brown for BS alloy En 27
NCM 30	0.3% C 1% Cr 1.7% Ni 0.35% Mo 0.3% Cu steel: Casting; Wolsingham
	DPN: 230 UTS: 760 Elon: 15% Proof: 480
NCM 40	0.4% C 1% Cr 1.7% Ni 0.35% Mo 0.3% Cu steel: Casting; Wolsingham
NCMo	0.25% C Cr Mo steel: Casting; Firth Brown; hardened & tempered
	DPN: 250 UTS: 840 Elon: 17% Proof: 610
NCM SPECIAL	0.32% C 1.3% Cr 4.2% Ni 0.3% Mo steel: Swift Levick for BS alloy En 30B
NCR 2	0.3% C 0.75% Cr 3.25% Ni steel: Firth Brown for BS alloy En 23
NF A35 551/16N06	0.15% C 0.75% Mn 1.0% Cr 1.4% Ni steel: French Standard
NF A35 551/ 18NCDG	0.18% C 0.75% Mn 1.0% Cr 1.4% Ni 0.2% Mo steel: French Standard

Note. The following abbreviations and units are used in the tables:

DPN	Hardness, diamond pyramid number
UTS	Ultimate tensile strength, N/mm^2
Elon	Elongation, %
Proof	0.1 % proof strength, N/mm^2

1 N/mm^2=0.1 hbar=0.102 kgf/mm^2=0.06475 tonf/in^2=145.04 lbf/in^2

Symbol	Nominal analysis, supplier, condition and remarks.
NF A35 551/20NC6	0.18% C 0.7% Mn 1.0% Cr 1.4% Ni steel: French Standard
NF A35 551/20NCD2	0.2% C 0.75% Mn 0.5% Cr 0.55% Ni 0.22% Mo steel: French Standard
NF A35 551/25CD4	0.26% C 0.75% Mn 1.0% Cr 0.22% Mo steel: French Standard
NF A35 551/ 30CAD6/12	0.3% C 0.65% Mn 1.65% Cr 0.32% Mo 1.15% Al steel: French Standard
NF A35 551/30CD4	0.31% C 0.75% Mn 0.95% Cr 0.22% Mn steel: French Standard
NF A35 551/30CD12	0.31% C 0.55% Mn 3.0% Cr 0.45% Mo steel: French Standard
NF A35 551/30CND8	0.3% C 0.45% Mn 2.0% Cr 2.0% Ni 0.45% Mo steel: French Standard
NF A35 551/35CD4	0.36% C 0.75% Mn 0.95% Cr 0.22% Mo steel: French Standard
NF A35 551/35NC6	0.35% C 0.75% Mn 1.0% Cr 1.4% Ni steel: French Standard
NF A35 551/35NCD6	0.34% C 0.75% Mn 1.0% Cr 1.4% Ni 0.22% Mo steel: French Standard
NF A35 565/ 16NCD13	0.15% C 0.5% Mn (max) 1.0% Cr 3.25% Ni 0.22% Mo steel: For bearings; French Standard
NF A35 565/18NCD4	0.101% C 0.65% Mn 0.42% Cr 1.1% Ni 0.22% Mo steel: For bearings; French Standard
NF A35 565/20NCD2	0.2% C 0.8% Mn 0.5% Cr 0.55% Ni 0.22% Mo steel: For bearings; French Standard
NF A35 565/20NCD7	0.19% C 0.55% Mn 0.4% Cr 1.8% Ni 0.25% Mo steel: For bearings; French Standard
NF A36 206/ 10CD910	0.15% C (max) 0.6% Mn 2.25% Cr 1.0% Mo 0.04% V (max) steel: Plate; French Standard; normalized & tempered
	UTS: 570 Proof: 300
NF A36 206/ 15CD205	0.18% C (max) 0.7% Mn 0.5% Cr 0.5% Mo 0.04% V (max) steel: Plate; French Standard; normalized & tempered
	UTS: 510 Proof: 270
NF A36 206/ 15CD405	0.18% C (max) 0.6% Mn 1.0% Cr 0.5% Mo 0.04% V (max) steel: Plate; French Standard; normalized & tempered
	UTS: 510 Proof: 300
NF A36 206/15D3	0.18% C (max) 0.65% Mn 0.3% Cr (max) 0.3% Mo 0.04% V (max) steel: Plate; French Standard
	UTS: 490 Proof: 260
NF A36 206/ 15MDV405	0.18% C (max) 1.2% Mn 0.3% Cr (max) 0.5% Mo 0.08% V steel: Plate; French Standard; normalized & tempered
	UTS: 570 Proof: 350
NF A36 206/ 18MD405	0.2% C (max) 1.2% Mn 0.3% Cr (max) 0.5% Mo 0.04% V (max) steel: Plate; French Standard; normalized & tempered
	UTS: 570 Proof: 350
NF A36 208/05NiA	0.14% C (max) 0.2% Cr (max) 0.5% Ni 0.1% Mo (max) 0.05% V (max) steel: Plate; French Standard
	UTS: 460 Proof: 270
NF A36 208/1.5Ni	0.18% C (max) 1.5% Mn (max) 0.2% Cr (max) 1.5% Ni 0.1% Mo (max) 0.05% V (max) steel: Plate; French Standard
	UTS: 550 Proof: 340
NF A36 208/5Ni	0.1% C (max) 0.8% Mn (max) 0.2% Cr (max) 5.0% Ni 0.1% Mo (max) 0.05% V (max) steel: Plate; French Standard
	UTS: 550 Proof: 380
NF A36 208/A52FP2	0.2% C (max) 0.1% Mn (max) 0.2% Cr (max) 0.3% Ni (max) 0.1% (max) 0.05% V (max) steel: Plate; French Standard; normalized
	UTS: 570 Proof: 340
NH	0.3% C 4.1% Ni 1.3% Cr 0.3% Mo steel: Obsolete; S Osborn; replaced by RABI

Symbol	Nominal analysis, supplier, condition and remarks.
NHP	0.26% C 3% Cr 2.5% Ni 0.5% Mo 9% W 0.3% V steel: For tools; Huntsman; forging dies; for nimonic alloys **DPN: 460**
NI MO	0.26% C (max) 0.3% Mn 3.0% Ni 0.6% Mo steel: Isaacson
NiCr 322	0.3% C 3.2% Ni 0.75% Cr 0.3% Mo steel: Wrought; Belgian National specification
NiCr 342	0.3% C 4.2% Ni 1.2% Cr steel: Wrought; Belgian national specification
NiCrMo 335	0.3% C 3.3% Ni 0.9% Cr 0.4% Mo steel: Wrought; Belgian national specification
NiCrMo 342	0.3% C 4.2% Ni 1.3% Cr 0.3% Mo steel: Wrought; Belgian national specification
NiCrMo 415	0.3% C 1.5% Ni 1.2% Cr 0.28% Mo steel: Wrought; Belgian national specification
NICUAGE 1	0.06% C 0.85% Ni 1.15% Cu 0.02% Nb steel: International Nickel Co.; aged weldable **DPN: 230 UTS: 700 Elon: 23% Proof: 640**
NICUAGE 1	0.04% C 0.8% Ni 1.1% Cu 0.02% (min) Nb steel: Age hardening; for use in high strength structural work; International Nickel Co **DPN: 230 UTS: 680 Elon: 23% Proof: 650**
NIT-KD	0.25% C 3.2% Cr 0.3% Mo steel: Pompey
NITRAL 4	0.34% C 1.65% Cr 0.2% Mo 0.9% Al steel: Pompey
NITRALLOY 5	0.3% C 1.5% Cr 0.2% Mo 1.1% Al steel: ESC; BS alloy En 41A; nitriding steel
NITRALLOY LK7	0.2% C 1.6% Cr 0.2% Mo 1.1% Al steel: For nitriding; Firth Brown
NK HITEN 60A	0.18% C (max) 1.5% Mn 0.3% Mo 0.1% V steel: Nippon Kokan
NK HITEN 60B	0.16% C (max) 1.35% Mn 0.4% Cr 0.6% Ni 0.3% Mo 0.15% Cu steel: Nippon Kokan
NK HITEN 70	0.18% C (max) 1.2% Mn 0.8% Cr 1.0% Ni 0.6% Mo 0.22% Cu steel: Nippon Kokan
NK HITEN 80	0.18% C (max) 1.0% Ni 0.8% Cr 1.0% Ni 0.6% Mo 0.1% V 0.3% Cu 0.006% B steel: Nippon Kokan
NK HITEN 100 HW90	0.18% C (max) 0.9% Mn 0.6% Cr 0.9% Ni 0.5% Mo 0.15% V 0.3% Cu steel: Nippon Kokan
NO. 158	0.1% C 1.5% Cr 3.5% Ni steel: Carpenter for AISI type P6
NORESCO SG	0.26% C 2.75% Cr 9% W 0.15% V steel: Schoeller-Bleckmann **UTS: 1500**
NORESCO NSW	0.3% C 1.3% Cr 0.35% Mo 4.25% W steel: Schoeller-Bleckmann **UTS: 1530**
NORESCO PARFORCE SPECIAL 10	0.15% C 0.75% Cr 3.6% Ni steel: For case hardening; Schoeller-Bleckmann; for plastic moulds
NORESCO SGW	0.23% C 2.5% Cr 1.5% Ni 0.15% Mo 9.5% W 0.15% V steel: Schoeller-Bleckmann **UTS: 1500**
NORESCO WCD	0.35% C 1% Si 5% Cr 1.2% Mo 1.2% W 0.35% V steel: Schoeller-Bleckmann **UTS: 1550**
NORESCO WCD 2	0.33% C 1% Si 5% Cr 1.1% Mo 0.5% V steel: Schoeller-Bleckmann **UTS: 1500**
NSW	0.3% C 1.3% Cr 0.35% Mo 4.25% W steel: Schoeller-Bleckmann **UTS: 1500**
N-TUP CR196	0.13% C 0.2% Mo 5.5% Ni steel: Nippon Steel Corp.; information from Climax Molybden **UTS: 760 Elon: 22% Proof: 600**

Symbol	Nominal analysis, supplier, condition and remarks.
NV 2	0.35% C 1% Si 1.5% W 5% Cr 1.6% Mo 0.2% V tool steel: For hot working; S Osborn; dies, shears, blades, punches, etc.; obsolete **DPN: 460**
NV 7-3	0.18% C (max) 0.45% Mn 2.25% Cr 1.0% Mo steel: Designation used by DNV **UTS: 450 Elon: 20% Proof: 280**
NVR 7-2	0.15% C (max) 0.6% Mn 0.9% Cr 0.55% Mo steel: Designation used by DNV
NVR 7-3	0.15% C (max) 0.45% Mn 2.25% Cr 1.0% Mo steel: Designation used by DNV
OCM 5	0.2% C 1.5% Mn 0.25% Cr 0.4% Ni steel: W Marrison for BS alloy En 14
ON 3	0.3% C 0.3% Cr 3.1% Ni steel: W Marrison for BS alloy En 21
ON 5	0.3% C 0.8% Cr 3.1% Ni 0.6% Mo steel: W Marrison for BS alloy En 23
ON 7	0.3% C 0.7% Cr 2.6% Ni 0.6% Mo steel: W Marrison for BS alloy En 25
ON 9	0.3% C 0.9% Cr 3.4% Ni 0.5% Mo steel: W Marrison for BS alloy En 27
ON 10	0.35% C 1.1% Cr 4% Ni 0.5% Mo steel: W Marrison for BS alloy En 28
ON 12	0.35% C 0.6% Cr 1.3% Ni steel: W Marrison for BS alloy En 111
ORELLOY 242	0.22% C 0.5% Mn 0.86% Cr 0.45% Ni 0.4% Cu steel: Oregon
OSBORN 501	0.2% C 5% Cr 0.5% Mo steel: Osborn; previously PY grade
P 2	0.07% C 2.0% Cr 0.2% Mo 0.5% Co steel: Designation used by AISI
P 3	0.11% C (max) 0.55% Cr 1.25% Ni tool steel: Designation used by AISI
P 4	0.07% C 5.0% Cr 0.75% Mo steel: Designation used by AISI
P 20	0.35% C 1.25% Cr 0.4% Mo steel: Designation used by AISI
P 21	0.2% C 0.25% Cr 4.1% Ni 0.2% V steel: For tools; listed in ASTM A681 1977
P 21	0.2% C 0.25% Cr 0.2% V 4.1% Ni steel: For tools; designation used by AISI
P 155	0.15% C 1% Cr 3.5% Ni steel: For case hardening; Carrs
P 157	0.15% C 1.2% Cr 4.25% Ni steel: For case hardening; Carrs
P 552	0.3% C 1.25% Cr 4.25% Ni 0.3% Mo steel: Carrs; gears, shafts, bolts, etc.; can be case hardened **UTS: 1300**
P 558	0.3% C 0.7% Cr 2.5% Ni 0.5% Mo steel: Carrs; crankshafts, gears, spindles, etc. **DPN: 350 UTS: 1070**
P 564	0.3% C 0.75% Cr 3% Ni steel: Carrs; crankshafts, bolts, spindles, etc. **DPN: 350 UTS: 1000**
P 612	0.3% C 1.6% Cr 0.2% Mo 1% Al steel: For nitriding; Carrs **UTS: 610**
P 615	0.3% C 3% Cr 0.3% Ni 0.4% Mo steel: For nitriding; Carrs **UTS: 1000**
PARFORCE SPECIAL 10	0.15% C 0.75% Cr 3.6% Ni steel: For case hardening; Schoeller-Bleckmann; for plastic moulds
PCS	0.35% C 1% Si 5% Cr 1.5% Mo 1.4% W 0.25% V steel: T Turton
PDX	0.28% C 0.85% Cr 2.25% H 0.3% V 0.5% Mo 2.25% Ni steel: For tools; Balfour Darwin
PF 1	0.15% C 0.5% Cr 0.5% Mo steel: Pompey
PF 2	0.12% C 1.1% Cr 0.5% Mo steel: Pompey
PF 2M	0.1% C 1.1% Cr 0.5% Mo steel: Pompey

Symbol	Nominal analysis, supplier, condition and remarks.
PF 5	0.12% C 4.4% Cr 0.5% Mo steel: Pompey
PF 5K	0.08% C 5.0% Cr 0.6% Mo steel: Pompey
PF 6	0.12% C 2.25% Cr 1.0% Mo steel: Pompey
PF 15	0.15% C 0.8% Cr 0.4% Mo steel: Pompey
PFDV	0.13% C 1.35% Cr 0.8% Mo 0.25% V steel: Pompey
PFV 55	0.12% C 2.0% Cr 0.35% Mo 0.08% V 0.4% Al steel: Pompey
PLASMOLD	0.35% C 1.3% Cr 4.3% Ni 0.3% Mo steel: Firth Brown; for plastic dies
PLASTIFORM	0.3% C 1.3% Cr 4% Ni 0.25% Mo steel: Clyde Alloy **DPN: 550**
PLMA/5	0.1% C 5% Cr 0.7% Mo 0.2% V steel: ESC for plastic moulds; case hardened
PLMC/3	0.16% C 1.1% Cr 4.2% Ni 0.25% Mo steel: ESC for plastic moulds; case hardened
PM 17	0.14% C 1.55% Mn 0.12% Mo 0.07% V steel: Pompey
PM 18	0.18% C 1.5% Mn 0.25% Cr 0.4% Ni 0.15% Mo 0.3% Cu steel: Casting; Wolsingham **UTS: 480 Elon: 20%**
PM 23	0.23% C 1.4% Mn 0.25% Cr 0.4% Ni 0.15% Mo 0.3% Cu steel: Casting; Wolsingham **UTS: 610 Elon: 20%**
PM 24	0.19% C 0.4% Mo 0.1% V steel: Pompey
PM 30	0.3% C 1.5% Mn 0.25% Cr 0.4% Ni 0.15% Mo 0.3% Cu steel: Casting; Wolsingham **UTS: 760 Elon: 17%**
PM 35	0.35% C 1.5% Mn 0.25% Cr 0.4% Ni 0.15% Mo 0.3% Cu steel: Casting; Wolsingham **UTS: 610 Elon: 17%**
PM 35	0.18% C 1.5% Cr 0.2% Mo 0.07% V steel: Pompey
PM 38	0.28% C 1.7% Cr 0.8% Ni 0.5% Mo 0.1% V steel: Pompey
PNUSNAP WH	0.35% C 1% Cr 1.8% W steel: Firth Brown; punches, shears
PREMO	0.05% C 3.9% Cr 0.5% Mo 0.1% V steel: Uddelholm for type P4
PRESSURDIE/2	0.35% C 5.0% Cr 1.45% Mo 1.2% W 0.3% V steel: Braeburn; AISI type H12
PREXI	0.15% C 1.1% Cr 0.25% Mo steel: Uddelholm
PX 80 PLUS 0.2%	0.9% Mn 0.7% Cr 0.3% Mo 0.0025% B steel: Phoenix
PX 90 PHIS	0.2% C 0.9% Mn 0.7% Cr 0.3% Mo 0.0025% B steel: Phoenix
PX 100 PHIS	0.2% C 0.9% Mn 0.7% Cr 0.3% Mo 0.0025% B steel: Phoenix
PX 110 PHIS	0.2% C 0.9% Mn 0.7% Cr 0.3% Mo 0.0025% B steel: Phoenix
PX 360 PHIS	0.2% C 0.9% Mn 0.7% Cr 0.3% Mo 0.0025% B steel: Phoenix
PY 8720	0.2% C 0.5% Cr 0.5% Ni 0.2% Mo steel: Pompey
PYROTOUGH	0.35% C 2.75% Cr 0.5% W 0.25% V steel: Clyde Alloy **DPN: 460**

Note. The following abbreviations and units are used in the tables:

DPN	Hardness, diamond pyramid number
UTS	Ultimate tensile strength, N/mm^2
Elon	Elongation, %
Proof	0.1 % proof strength, N/mm^2

1 N/mm^2=0.1 hbar=0.102 kgf/mm^2=0.06475 tonf/in^2=145.04 lbf/in^2

Symbol	Nominal analysis, supplier, condition and remarks.
R 9030	0.35% C 0.7% Cr 2.7% Ni 0.5% Mo steel: For hot work tools; Balfour; moulding dies; piercing punches **DPN: 550**
RABI	0.3% C 4.1% Ni 1.3% Cr 0.3% Mo steel: S Osborn; mould steel; hardened & tempered **DPN: 500**
RBD	0.3% C 3% Cr 10% W 0.5% V steel: Firth Brown
RBD-E	0.22% C 2.2% Cr 2.2% Ni 0.45% Mo 10% W 0.2% V steel: Firth Brown
RED DIAMOND 20H	0.25% C 1.0% Cr 0.2% Mo steel: For wear; Redheugh **DPN: 260 UTS: 820 Elon: 18% Proof: 630**
RED DIAMOND 20S	0.25% C 1.0% Cr 0.2% Mo steel: For wear; Redheugh **DPN: 220 UTS: 710 Elon: 25% Proof: 600**
RED INDIAN	0.35% C 5.0% Cr 0.3% Mo 4.5% W 0.3% V 0.5% Co steel: Atlas; as AISI type H14
REPUBLIC 50	0.15% C (max) 0.7% Mn 0.3% Cr 0.7% Ni 0.1% Mo 0.7% Cu steel: Republic Steel
REPUBLIC 65	0.15% C 1.0% Mn 1.2% Ni 0.25% Mo 1.1% Cu steel: Republic Steel
REPUBLIC 70	0.2% C (max) 1.0% Mn 1.5% Ni 0.25% Mo 1.2% Cu steel: Republic Steel Co
RHB 65	0.2% C (max) 1.4% Mn 0.7% Ni 0.18% V steel: German proprietor's specification
RIVER ACE 70	0.18% C (max) 1.2% Mn 0.4% Cr 1.0% Ni 0.4% Mo 0.08% V 0.4% Cu 0.005% B steel: Kawasaki Iron & Steel Co
RIVER ACE KO	0.18% C (max) 1.0% Mn 0.8% Cr 1.5% Ni 0.6% Mo 0.08% V 0.5% Cu 0.006% B steel: Kawasaki Iron & Steel Co
RO 211	0.12% C 2.5% Cr 1.1% Mo steel: Bofors; normalized & tempered **DPN: 190 UTS: 610 Elon: 26% Proof: 460**
RO 346	0.16% C 1.0% Cr 0.6% Mo steel: Bofors; hardened & tempered **DPN: 240 UTS: 770 Elon: 20% Proof: 680**
RO 651	0.3% C 1.0% Cr 0.2% Mo steel: Bofors **DPN: 325 UTS: 1100 Elon: 11%**
RO 653	0.25% C 1.1% Cr 0.2% Mo steel: Bofors; hardened & tempered **DPN: 300 UTS: 990 Elon: 13% Proof: 700**
RO 663	0.28% C 1.2% Cr 0.3% Mo steel: Bofors
RO 752	0.35% C 1.1% Cr 0.2% Mo steel: Bofors; hardened & tempered **DPN: 280 UTS: 820 Elon: 14% Proof: 580**
RO 4154	0.25% C 3.0% Cr 0.5% Mo steel: Bofors; hardened & tempered **DPN: 260 UTS: 840 Elon: 20% Proof: 710**
RO 7155	0.35% C 3.0% Cr 0.5% Mo steel: Bofors; hardened & tempered **DPN: 350 UTS: 1140 Elon: 12% Proof: 890**
ROP 10	0.08% C 3.0% Cr 0.8% Mo 0.2% V steel: For carburizing; Bofors
ROP 63	0.3% C 2.5% Cr 0.5% Ni 0.3% Mo 0.3% V steel: For nitriding; Bofors; hardened & tempered; obsolete **DPN: 320 UTS: 1030 Elon: 12% Proof: 820**
ROP 5462	0.24% C 1.3% Cr 0.5% Mo steel: Bofors; hardened & tempered **DPN: 260 UTS: 820 Elon: 20% Proof: 700**
ROPT	0.2% C 3.0% Cr 0.6% Mo 0.5% W 0.8% V steel: Bofors; obsolete
RS 2	0.17% C 1.8% Ni 0.25% Mo steel: Toledo for BS alloy En 34 & En 35
RS 3	0.12% C 0.3% Cr 3% Ni steel: Toledo for BS alloy En 33

Symbol	Nominal analysis, supplier, condition and remarks.
RT 27	0.35% C 1.0% Cr 3.5% Ni 5.5% W 0.2% V steel: Bofors; obsolete **DPN: 350**
RT 45	0.3% C 3.0% Cr 10% W 0.3% V steel: Bofors; obsolete **DPN: 450**
SABEX	0.3% C 2.2% Cr 1.7% Ni 0.1% V steel: Sanderson; hot shear blades, casting dies **DPN: 610**
SAE 43 BV12	0.12% C 0.5% Cr 1.8% Ni 0.25% Mo 0.03% V steel: Obsolete
SAE 43 BV14	0.14% C 0.5% Cr 1.8% Ni 0.1% Mo 0.03% V steel: Obsolete
SAE 46 B12	0.12% C 1.85% Ni 0.25% Mo steel
SAE 94 B15	0.15% C 0.5% Ni 0.35% Cr 0.11% Mo steel with B
SAE 94 B17	0.17% C 0.4% Cr 0.45% Ni 0.1% Mo 0.0005% B steel
SAE 94 B30	0.3% C 0.4% Cr 0.45% Ni 0.1% Mo 0.0005% B (min) steel
SAE 3115	0.15% C 0.6% Cr 1.2% Ni steel: Obsolete
SAE 3120	0.2% C 0.6% Cr 1.2% Ni steel: Obsolete
SAE 3130	0.3% C 0.6% Cr 1.2% Ni steel
SAE 3135	0.35% C 0.6% Cr 1.2% Ni steel
SAE 3215	0.15% C 1% Cr 1.7% Ni steel: Obsolete
SAE 3220	0.2% C 1% Cr 1.7% Ni steel: Obsolete
SAE 3230	0.3% C 1% Cr 1.7% Ni steel: Obsolete
SAE 3240	0.4% C 1% Cr 1.7% Ni steel: Obsolete
SAE 3310	0.1% C 3% Ni 1.5% Cr steel: As AISI 3310
SAE 3312	0.1% C 1.5% Cr 3.5% Ni steel: Obsolete
SAE 3316	0.16% C 1.6% Cr 3.5% Ni steel: Obsolete
SAE 3325	0.25% C 1.5% Cr 3.5% Ni steel: Obsolete
SAE 3335	10.35% C 1.5% Cr 3.5% Ni steel: Obsolete
SAE 3415	0.15% C 0.8% Cr 3% Ni steel: Obsolete
SAE 3435	0.35% C 0.8% Cr 3.0% Ni steel
SAE 3436	0.35% C 0.8% Cr 3% Ni steel: Obsolete
SAE 4118	0.2% C 0.5% Cr 0.12% Mo steel
SAE 4119	0.19% C 0.5% Cr 0.25% Mo steel
SAE 4125	0.25% C 0.5% Cr 0.25% Mo steel: Obsolete
SAE 4130	0.3% C 1% Cr 0.2% Mo steel
SAE 4135	0.35% C 1.0% Cr 0.2% Mo steel
SAE 4137	0.37% C 1% Cr 0.2% Mo steel
SAE 4317	0.17% C 0.5% Cr 1.8% Ni 0.25% Mo steel: Obsolete
SAE 4320	0.2% C 1.8% Ni 0.5% Cr 0.25% Mo steel
SAE 4337	0.37% C 0.8% Cr 1.8% Ni 0.25% Mo steel
SAE 4608	0.08% C 1.5% Ni 0.2% Mo steel
SAE 4615	0.15% C 1.8% Ni 0.25% Mo steel
SAE 4617	0.17% C 1.8% Ni 0.25% Mo steel
SAE 4620	0.2% C 1.8% Ni 0.25% Mo steel
SAE 4621	0.21% C 1.8% Ni 0.25% Mo steel
SAE 4626	0.26% C 0.8% Ni 0.2% Mo steel
SAE 4718	0.18% C 0.45% Cr 1% Ni 0.35% Mo steel
SAE 4720	0.2% C 0.45% Cr 0.1% Ni 0.2% Mo steel
SAE 4812	0.12% C 3.5% Ni 0.25% Mo steel
SAE 4815	0.15% C 3.5% Ni 0.25% Mo steel
SAE 4817	0.17% C 3.5% Ni 0.25% Mo steel
SAE 4820	0.2% C 3.5% Ni 0.25% Mo steel
SAE 6115	0.15% C 1% Cr 0.15% V steel: Obsolete
SAE 6117	0.17% C 0.8% Cr 0.1% V steel
SAE 6118	0.18% C 0.6% Cr 0.12% V steel
SAE 6120	0.2% C 0.8% Cr 0.1% V steel
SAE 6125	0.25% C 1% Cr 0.15% V steel: Obsolete
SAE 6130	0.3% C 1% Cr 0.15% V steel: Obsolete
SAE 6135	0.35% C 1% Cr 0.15% V steel: Obsolete
SAE 8115	0.15% C 0.3% Ni 0.4% Cr 0.11% Mo steel
SAE 8615	0.15% C 0.5% Cr 0.5% Ni 0.2% Mo steel
SAE 8617	0.17% C 0.5% Cr 0.6% Ni 0.2% Mo steel
SAE 8620	0.2% C 0.5% Cr 0.6% Ni 0.2% Mo steel
SAE 8622	0.22% C 0.5% Cr 0.6% Ni 0.2% Mo steel

Symbol	Nominal analysis, supplier, condition and remarks.
SAE 8625	0.25% C 0.5% Cr 0.6% Ni 0.2% Mo steel
SAE 8627	0.27% C 0.5% Cr 0.6% Ni 0.2% Mo steel
SAE 8630	0.3% C 0.5% Cr 0.6% Ni 0.2% Mo steel
SAE 8632	0.32% C 0.5% Cr 0.6% Ni 0.2% Mo steel: Obsolete
SAE 8635	0.35% C 0.5% Cr 0.6% Ni 0.2% Mo steel
SAE 8637	0.37% C 0.5% Cr 0.6% Ni 0.2% Mo steel
SAE 8715	0.15% C 0.5% Cr 0.6% Ni 0.25% Mo steel
SAE 8717	0.17% C 0.5% Cr 0.6% Ni 0.25% Mo steel: Obsolete
SAE 8719	0.2% C 0.5% Cr 0.6% Ni 0.25% Mo steel: Obsolete
SAE 8720	0.2% C 0.5% Cr 0.6% Ni 0.25% Mo steel
SAE 8735	0.35% C 0.5% Cr 0.6% Ni 0.25% Mo steel: Obsolete
SAE 8822	0.22% C 0.5% Cr 0.6% Ni 0.35% Mo steel
SAE 9310	0.1% C 1.2% Cr 3.2% Ni 0.1% V steel
SAE 9315	0.15% C 1.2% Cr 3.2% Ni 0.1% Mo steel
SAE 9317	0.17% C 1.2% Cr 3.2% Ni 0.1% Mo steel
SAE 9437	0.37% C 0.4% Cr 0.4% Ni 0.1% Mo steel: Obsolete
SAE 51501	0.1% C (min) 5% Cr 0.5% Mo steel
SAE 51502	0.1% C 5% Cr 0.5% Mo steel
SAE 60502	0.25% C 5% Cr 0.5% Mo steel: Casting
SAE 70502	0.25% C (max) 5.5% Cr 0.5% Mo steel: Casting
SAE H11	0.35% C 5% Cr 1.5% Mo 0.4% V steel: For tools
SAE H12	0.35% C 5% Cr 1.5% Mo 1.5% H 0.3% V steel: For tools
SAE H13	0.35% C 5% Cr 1.5% Mo 1% V steel: For tools
SAE H21	0.35% C 3.5% Cr 9% W 0.4% V steel: For tools
SAE X4620	0.2% C 1.8% Ni 0.25% Mo steel
SANBOLD 21	0.3% C 0.3% Cr 3% Ni steel: Sanderson for BS alloy En 21
SANBOLD 23	0.3% C 0.8% Cr 3% Ni 0.5% Mo steel: Sanderson for BS alloy En 23; Mo optional
SANBOLD 25	0.3% C 0.7% Cr 2.5% Ni 0.5% Mo steel: Sanderson for BS alloy En 25
SANBOLD 30	0.32% C 1.3% Cr 4.2% Ni 0.3% Mo steel: Sanderson for BS alloy En 30
SANBOLD 33	0.12% C 0.3% Cr 3.25% Ni steel: Sanderson for BS alloy En 33
SANBOLD 34	0.18% C 1.7% Ni 0.25% Mo steel: Sanderson for BS alloy En 34
SANBOLD 36	0.16% C 0.3% Cr 3.5% Ni steel: Sanderson for BS alloy En 36A, B or C; C varied to suit
SANBOLD 37	0.16% C 0.3% Cr 5% Ni steel: Sanderson for BS alloy En 37
SANBOLD 38	0.16% C 0.3% Cr 5% Ni 0.2% Mo steel: Sanderson for BS alloy En 38
SANBOLD 39	0.16% C 1.2% Cr 4.2% Ni 0.2% Mo steel: Sanderson for BS alloy En 39
SANBOLD 56	0.3% C 0.7% Cr 2.5% Ni 0.5% Mo steel: Sanderson; replaced by Sanbold 25
SANBOLD 58	0.18% C 1.7% Ni 0.25% Mo steel: Sanderson; replaced by Sanbold 34
SANBOLD 68	0.16% C 1.2% Cr 4.2% Ni 0.2% Mo steel: Sanderson; replaced by Sanbold 39
SANBOLD 69	0.16% C 0.3% Cr 5% Ni 0.2% Mo steel: Sanderson; replaced by Sanbold 38
SANBOLD 77	0.32% C 1.3% Cr 4.2% Ni 0.3% Mo steel: Sanderson; replaced by Sanbold 30
SANBOLD 101	0.16% C 0.8% Cr 3.5% Ni steel: Sanderson; replaced by Sanbold 36
SANBOLD N 2832	0.12% C 0.3% Cr 3.25% Ni steel: Sanderson; replaced by Sanbold 33
SANBOLD N 3603	0.16% C 0.3% Cr 5% Ni steel: Sanderson; replaced by Sanbold 37
SANBOLD N 5366	0.3% C 0.3% Cr 3% Ni steel: Sanderson; replaced by Sanbold 21

Symbol	Nominal analysis, supplier, condition and remarks.
SANBOLD NK 8055	0.3% C 0.8% Cr 3% Ni 0.5% Mo steel: Sanderson; replaced by Sanbold 23
SB 46F	0.15% C 1.2% Mn 0.45% Ni 0.2% V 0.35% Cu steel: German proprietor's specification
SB 46FT	0.15% C 1.2% Mn 0.45% Ni 0.2% V 0.35% Cu steel: German proprietor's specification
SCORPION	0.3% C 1.3% Cr 3.4% Ni steel: T Turton
SD 20	0.35% C 1.2% Cr 1.7% Ni steel: Balfour; collets, gears, etc. **DPN: 550**
SENECA	0.35% C 3.2% Cr 9.5% W 0.4% V steel: Atlas; as AISI type H 21
SG	0.26% C 2.75% Cr 9% W 0.15% V steel: Schoeller-Bleckmann **UTS: 1420**
SGW	0.23% C 2.5% Cr 1.5% Ni 0.15% Mo 9.5% W 0.15% V steel: Schoeller-Bleckmann **UTS: 1420**
SGW 1	0.35% C 1% Cr 1.5% Ni 0.25% Mo steel: Jonas **UTS: 1060 Elon: 16% Proof: 1000**
SGW 2	0.3% C 0.75% Cr 3% Ni steel: Jonas; hardened & tempered **UTS: 1300 Elon: 13% Proof: 1140**
SGW 3	0.3% C 0.6% Cr 3% Ni steel: Jonas; hardened & tempered **UTS: 1210 Elon: 13% Proof: 1070**
SHEFFIELD HIGH STRENGTH A	0.12% C 0.4% Mn 0.6% Cr 0.7% Ni 0.1% Mo 0.03% Ti 0.4% Cu steel: Armco Steel
SHEF-LO-TEMP	0.18% C 0.2% Cr 0.2% Ni 0.07% Mo 0.2% Cu steel: Armco; for low temperature use; hardened & tempered **UTS: 580 Elon: 23% Proof: 340**
SHEF-SUPER-LO-T EMP	0.16% C 0.18% Cr 0.23% Ni 0.07% Mo 0.2% Cu steel: Armco; for low temperature use; hardened & tempered **UTS: 670 Elon: 23% Proof: 390**
SHNC	0.3% C 1.25% Cr 4.25% Ni steel: ESC; for BS alloy En 30A
SIS 14-34E	0.22% C 0.8% Mn 0.25% Cr 0.30% Cu 0.045% P S steel: For pressure vessels; Swedish Standard **UTS: 500 Elon: 23% Proof: 260**
SIS 14-2203	0.15% C 0.45% Mn 9% Cr 1% Mn 0.25% Cu steel: For pressure vessels; Swedish Standard **UTS: 680 Elon: 18% Proof: 390**
SIS 14-2203E	0.15% C 0.45% Mn 0.85% Cr 1.0% Mo 0.25% Cu steel: For pressure vessels; Swedish Standard; normalized & annealed **UTS: 650 Elon: 18% Proof: 390**
SIS 14-2216	0.14% C 0.6% Mn 1.0% Cr 0.5% Mo 0.25% Cu steel: For pressure vessels; Swedish Standard **UTS: 510 Elon: 20% Proof: 300**
SIS 14-2218	0.1% C 0.5% Mn 2.3% Cr 1.0% Mo 0.25% Cu steel: For pressure vessels; Swedish Standard **UTS: 600 Elon: 17% Proof: 265**
SIS 14-2218E	0.11% C 0.5% Mn 2.2% Cr 1.0% Mo 0.25% Cu steel: For pressure vessels; normalized & annealed; Swedish Standard **UTS: 550 Elon: 18% Proof: 265**

Symbol	Nominal analysis, supplier, condition and remarks.
SIS 14-2223	0.18% C 0.7% Mn 0.9% Cr 0.6% Mo 0.4% Ni 0.3% Cr 0.04% P S steel: For casting; Swedish Standard **DPN: 160 UTS: 500 Elon: 28% Proof: 280**
SIS 14-2224	0.18% C 0.7% Mn 2.0% Cr 1.0% Mo 0.4% Ni 0.3% Cu 0.04% P S steel: For castings; Swedish Standard **DPN: 180 UTS: 500 Elon: 28% Proof: 280**
SIS 14-2225	0.25% C 0.65% Mn 1.1% Cr 0.3% Ni 0.2% Mo steel: For hardening & tempering; Swedish Standard **DPN: 300 UTS: 900 Elon: 13% Proof: 690**
SIS 14-2233	0.3% C 0.6% Mn 1.0% Cr 0.2% Mo steel: For gas cylinders; Swedish Standard **DPN: 330 UTS: 930 Elon: 13% Proof: 710**
SIS 14-2234	0.3% C 0.6% Mn 1.0% Cr 0.2% Mo steel: For hardening & tempering; Swedish Standard **DPN: 300 UTS: 1030 Elon: 12% Proof: 750**
SIS 14-2317	0.26% C 0.5% Mn 11.0% Cr 0.5% Ni 1.2% Mo 0.3% V steel: High strength high tempered martensitic; Swedish Standard **DPN: 260 UTS: 800 Elon: 50% Proof: 640**
SIS 14-2506	0.2% C 0.75% Mn 0.5% Cr 0.55% Ni 0.2% Mo 0.035% 0.04% P S steel: For case hardening; Swedish Standard **DPN: 275 UTS: 80 Elon: 10% Proof: 50**
SIS 14-2511	0.15% C 0.9% Mn 0.8% Cr 41.0% Ni 0.1% Mo 0.035% P S steel: For case hardening; Swedish Standard **DPN: 300 UTS: 100 Elon: 10% Proof: 50**
SIS 14-2512	0.2% C 0.9% Mn 0.8% Cr 1% Ni 0.1% Mo steel: For case hardening; Swedish Standard **DPN: 350 UTS: 125 Elon: 9% Proof: 60**
SIS 14-2523	0.2% C 0.9% Mn 1.0% Cr 1.2% Ni 0.12% Mo steel: For case hardening; Swedish Standard **DPN: 350 UTS: 1000 Elon: 8% Proof: 750**
SIS 2203	0.1% C 9% Cr 1.0% Mo steel: Fagersta
SIS 2216	0.15% C 0.9% Cr 0.6% Mo steel: For pressure vessels; Swedish Standard; as forged **UTS: 500 Elon: 22% Proof: 300**
SIS 2218	0.15% C 2.2% Cr 1% Mo steel: For pressure vessels; Swedish Standard; as forged **UTS: 500 Elon: 20% Proof: 250**
SIS 2223	0.18% C 1% Cr 0.6% Mo steel: Casting; Swedish Standard; normalized & annealed **DPN: 160 UTS: 480 Elon: 20% Proof: 270**
SIS 2224	0.18% C 2.2% Cr 1% Mo steel: Casting; Swedish Standard; normalized & annealed **DPN: 180 UTS: 480 Elon: 20% Proof: 270**
SIS 2225	0.25% C 1.1% Cr 0.25% Ni 0.2% Mo steel: Bar, plate & forging; Swedish Standard; hardened & tempered **DPN: 350 UTS: 1070 Elon: 8% Proof: 860**
SIS 2225	0.25% C 1.05% Cr 0.2% Mo steel: Swedish Standard; hardened & tempered **UTS: 780 Elon: 14% Proof: 600**
SIS 2233	0.3% C 1% Cr 0.25% Mo steel: For gas cylinders; Swedish Standard; hardened & tempered **DPN: 280 UTS: 930 Elon: 13% Proof: 720**
SIS 2234	0.33% C 1.1% Cr 0.2% Mo steel: Bar & forging; Swedish Standard; hardened & tempered **DPN: 280 UTS: 840 Elon: 12% Proof: 700**
SIS 2240	0.3% C 3% Cr 0.3% Ni 0.5% Mo steel: For nitriding; Swedish Standard; hardened & tempered **DPN: 370 UTS: 1210 Elon: 12% Proof: 880**
SIS 2317	0.23% C 11.7% Cr 0.5% Ni 1.2% Mo 0.3% Mo 0.045% B steel: Swedish Standard **DPN: 260 UTS: 910 Elon: 16% Proof: 640**

Note. The following abbreviations and units are used in the tables:

DPN	Hardness, diamond pyramid number
UTS	Ultimate tensile strength, N/mm^2
Elon	Elongation, %
Proof	0.1 % proof strength, N/mm^2

1 N/mm^2=0.1 hbar=0.102 kgf/mm^2=0.06475 tonf/in^2=145.04 lbf/in^2

Symbol	Nominal analysis, supplier, condition and remarks.
SIS 2506	0.2% C 0.5% Cr 0.55% Ni 0.2% Mo steel: For construction; Swedish Standard **DPN: 320 UTS: 100 Elon: 10% Proof: 550**
SIS 2511	0.15% C 0.9% Cr 1.5% Ni steel: Bar & forging; Swedish Standard
SIS 2512	0.2% C 0.9% Cr 1.5% Ni steel: Bar & forging; Swedish Standard
SIS 2514	0.12% C 0.8% Cr 3% Ni steel: Bar & forging; Swedish Standard
SIS 2515	0.17% C 0.8% Cr 3% Ni steel: Bar & forging; Swedish Standard
SIS 2523	0.2% C 1.0% Cr 1.2% Ni 0.12% Mo steel: Swedish Standard; annealed **DPN: 220 UTS: 110 Elon: 8% Proof: 75**
SIS 2534	0.31% C 1% Cr 3.2% Ni 0.25% Mo steel: Bar & forging; Swedish Standard
SIS 2730	0.3% C 3% Cr 9% W 0.3% V steel: Swedish Standard; annealed **DPN: 250**
SIS 8100	0.2% C (max) 1.5% Cu 1.75% Ni 0.5% Mo steel: Sintered
SKC 31	0.18% C 0.9% Mn 1.6% Cr 3.0% Ni 0.6% Mo steel: For hollow drills; Japanese Standard designation
SKD 2	0.18% C 1.0% Cr 0.2% Mo steel: Pompey
SKD 4	0.3% C 2.5% Cr 0.4% V 5.5% W steel: Japanese Standard designation
SKD 5	0.3% C 2.5% Cr 0.4% V 9.5% W steel: Japanese Standard designation
SL 503 M	0.25% C 2.7% Cr 2.5% Ni 0.5% Mo 9.0% W 0.3% V steel: Swift Levick; hot dies
SL 598	0.25% C 1.5% Cr 4.0% Ni 6.0% W 0.3% V steel: Swift Levick; hot dies
SL 853	0.35% C 5.25% Cr 4.5% W 0.5% Co steel: Swift Levick; for hot dies
SL 942	0.35% C 2.5% Cr 4.3% Mo 2.0% W 1.0% V steel: Swift Levick; for hot dies
SMA 58	See JIS G3114 SMA58
SN 3	0.3% C 0.2% Cr 3.2% Ni steel: Spencer for BS alloy En 21
SN 5/½	sintered Ni alloy steel: High impact strength; can be hardened; Durasint; specific gravity 6.6; 16% porosity **DPN: 160 UTS: 470 Elon: 6%**
SNC 1	0.3% C 0.8% Cr 0.5% Mo steel: Spencer for BS alloy En 23
SNC 3G	0.3% C 1.2% Cr 4.2% Ni 0.3% Mo (optional) steel: Spencer for BS alloy En 30
SNCH 6	0.16% C 0.9% Cr 3.5% Ni steel: For case hardening; Spencer for BS alloy En 36
SNCH 9	0.16% C 1.2% Cr 4.2% Ni 0.2% Mo (optional) steel: Spencer for BS alloy En 39
SNCM	0.3% C 0.7% Cr 2.5% Ni 0.5% Mo steel: Spencer for BS alloy En 25
SNCM 26	0.32% C 1.2% Cr 3.7% Ni 0.5% Mo steel: Spencer for BS alloys En 27 & 28
SNH 3	0.12% C 0.2% Cr 3.2% Ni steel: For case hardening; Spencer for BS alloy En 33
SNH 5	0.14% C 0.2% Cr 4.8% Ni steel: For carburizing; Spencer for BS alloy En 37
SP 10	Sintered alloy steel: Good wear resistance; specific gravity 6.0; 22% porosity; Durasint **DPN: 190 UTS: 440 Elon: 1%**
SP 20	Sintered Ni alloy steel: Machinable; specific gravity 6.7; porosity 15%; Durasint **DPN: 260 UTS: 520 Elon: 1%**
SP 22	Sintered Ni Mo steel: Specific gravity 6.9; Durasint **DPN: 300 UTS: 720 Elon: 1%**
SP 24	Sintered Ni alloy steel: Good wear resistance; can be carbo-nitrided; specific gravity 6.8; 13% porosity **DPN: 125 UTS: 390 Elon: 4%**
SPEAR 451	0.14% C 0.5% Cr 0.75% Ni steel: Spear & Jackson; as BS alloy En 351
SPEAR 524	0.32% C 1.1% Cr 1.5% Ni 0.2% Mo steel: Spear & Jackson for BS alloy En 24
SPEAR 525	0.3% C 0.6% Cr 2.5% Ni 0.6% Mo steel: Spear & Jackson; as BS alloy En 25
SPEAR 536	0.12% C 1.0% Cr 3.0% Ni steel: Spear & Jackson; as BS alloy En 36
SPEAR 552	0.17% C 0.8% Cr 1.0% Ni steel: Spear & Jackson; as BS alloy En 352
SPEAR 653	0.18% C 0.9% Cr 1.25% Ni 0.11% Mo steel: Spear & Jackson; as BS alloy En 353
SPEAR 855	0.18% C 1.6% Cr 2.0% Ni 0.2% Mo steel: Spear & Jackson; as BS alloy En 355
SPEAR 1030	0.32% C 1.2% Cr 4.1% Ni 0.25% Mo steel: Spear & Jackson; as BS alloy 30B
SPEAR B1	0.32% C 1.2% Cr 4.2% Ni 0.3% Mo steel: Spear & Jackson
SPEAR B4	0.32% C 0.6% Mn 2.1% Cr 0.2% Mo 0.15% V steel: Origin unknown
SPEAR D5	0.35% C 5.7% Cr 1.5% Mo 1.0% V steel: Spear & Jackson; for aluminium die casting moulds
SPEAR D7	0.32% C 5.0% Cr 1.5% Mo 1.5% W 0.4% V steel: Spear & Jackson; for extrusion dies
SPEAR D9	0.3% C 3.2% Cr 9.5% W 0.3% V steel: Spear & Jackson; for hot work
SPEAR DX	0.28% C 0.85% Cr 2.2% Ni 0.5% Mo 0.25% V 2.2% W steel: Origin unknown
SPECIAL BB	0.35% C 3.5% Cr 9.0% W 0.4% V steel: For tools; origin unknown
SPECIAL BB	0.28% C 3.2% Cr 9.5% W 0.3% V steel: Jessop; dies for casting Cu **DPN: 400**
SPECIAL HDS	0.35% C 3.5% Cr 9.0% W 0.4% V steel: For tools; origin unknown
SPENARD	0.32% C 1.25% Cr 4.1% Ni 0.3% Mo steel: Spencer
SPRASTEEL LS	0.04% C 2% Mn 1.5% Cr 4% Ni 1.3% Mo steel: Wire for spraying; Metco; as sprayed **DPN: 261 UTS: 220**
SSG	0.3% C 1.5% Cr 4% Ni steel: Jonas; hardened & tempered **UTS: 1430 Elon: 15% Proof: 1300**
SS 100	0.16% C 1.7% Cr 0.5% Mo 0.07% Ti 0.003% B 0.3% Cu steel: Armco; hardened & tempered **UTS: 840 Elon: 18% Proof: 740**
SSS 100	0.16% C 0.55% Mn 1.7% Cr 0.5% Mo 0.07% Ti 0.06% V 0.3% Cu steel: Armco
SSS 100A	0.17% C 1.0% Cr 0.2% Mo 0.07% Ti 0.003% B 0.3% Cu steel: Armco; hardened & tempered **UTS: 840 Elon: 18% Proof: 740**
SSS 100A	0.17% C 0.55% Mn 1.0% Cr 0.2% Mo 0.06% Ti 0.3% Cu 0.06% V steel: Armco
SSS 110	0.16% C 0.55% Mn 1.8% Cr 0.5% Mo 0.06% Ti 0.06% V 0.3% Cu steel: Armco
SSS 321	0.19% C 0.55% Mn 1.4% Cr 0.4% Mo 0.1% Ti 0.4% Cu 0.0015% B steel: Armco
SSS 360	0.19% C 0.5% Mn 1.4% Cr 0.4% Mo 0.1% Ti 0.3% Cu 0.0015% B steel: Armco
STA 5 V6	0.2% C 0.2% Cr 0.3% Ni steel: Replaced by BS alloy En 14
STA 5 V6A	0.2% C 0.2% Cr 0.3% Ni steel: Replaced by BS alloy En 14A
STA 5 V6B	0.2% C 0.2% Cr 0.3% Ni steel: Obsolete

Symbol	Nominal analysis, supplier, condition and remarks.
STA 5 V8	0.20% C 0.6% Ni 0.25% Mo steel: Replaced by BS alloy En 13
STA 5 V9E	0.35% C 0.6% Cr 1.25% Ni steel: Replaced by BS alloy En 111
STA 5 V9E/1	0.35% C 0.6% Cr 1.25% Ni steel: Replaced by BS alloy En 111
STA 5 V11	0.31% C 0.7% Cr 2.5% Ni 0.6% Mo steel: Replaced by BS alloy En 25
STA 5 V13	0.3% C 1.2% Cr 4.2% Ni steel: Replaced by BS alloy En 30A
STA 5 V13A	0.3% C 1.2% Cr 4.2% Ni steel: Replaced by BS alloy En 30B
STA 5 V16A	0.17% C 1.75% Ni 0.25% Mo steel: Replaced by BS alloy En 34
STA 5 V16A/1	0.17% C 1.7% Ni 0.2% Mo steel: Obsolete
STA 5 V16B	0.24% C 1.75% Ni 0.25% Mo steel: Replaced by BS alloy En 35
STA 5 V16B/1	0.24% C 1.75% Ni 0.25% Mo steel: Replaced by BS alloy En 35A
STA 5 V16B/2	0.24% C 1.75% Ni 0.25% Mo steel: Replaced by BS alloy En 35B
STA 5 V17	0.12% C 0.25% Cr 3.2% Ni steel: Replaced by BS alloy En 33
STA 5 17/1	0.12% C 0.25% Cr 3.2% Ni steel: Obsolete
STA 5 V18	0.16% C 0.8% Cr 3.5% Ni steel: Replaced by BS alloy En 36A
STA V18/1	0.16% C 0.8% Cr 3.5% Ni steel: Obsolete
STA 5 V18A	0.16% C 0.8% Cr 3.5% Ni steel: Replaced by BS alloy En 36B
STA 5 V19	0.15% C 1.2% Cr 4.2% Ni steel: Replaced by BS alloy En 39A
STA 5 V19/1	0.15% C 1.2% Cr 4.2% Ni steel: Obsolete
STA 5 V20A	0.3% C 1.6% Cr 0.3% Ni 0.2% Mo 1.0% Al steel: Replaced by BS alloy En 41
STA 5 V20B	0.3% C 1.6% Cr 0.3% Ni 0.2% Mo 1.0% Al steel: Replaced by BS alloy En 41
STA 5 V20C	0.3% C 1.6% Cr 0.3% Ni 0.2% Mo 1.0% Al steel: Replaced by BS alloy En 41
STA 5 V20D	0.3% C 1.6% Cr 0.3% Ni 0.2% Mo 1.0% Al steel: Obsolete
STA 39	0.25% C 0.4% Ni 0.2% Mo steel: Suitable for welding
STA 40	0.2% C 0.6% Ni 0.25% Mo steel: Weldable
STAGMOLD	0.32% C 1.3% Cr 4.1% Ni 0.3% Mo steel: For tools; for cold marking; Edgar Allen
SUPER AW 23	0.3% C 1.2% Cr 4.1% Ni 0.2% Mo steel: ESC; punches, etc. **DPN: 500**
SUPER CHNC	0.15% C 1.2% Cr 4.2% Ni 0.25% Mo steel: For case hardening; ESC for BS alloy En 39B
SUPERELSO	0.18% C (max) 0.9% Mn 1.1% Cr 1.5% Mn 0.25% Mo steel: Creusot-Loire
SUPERELSO 70	0.18% C (max) 1.1% Mn (max) 1.0% Cr 1.7% Ni 0.4% Mo 0.01% V steel: Creusot-Loire
SUPERELSO 600	0.09% C 1.1% Mn 1.7% Cr 0.8% Ni 0.2% Mo steel: Creusot-Loire
SUPERMOLD	0.3% C 1.7% Cr 0.4% Mo steel: Latrobe; dies & mould; type P20 **DPN: 340**
SUPER NITRALLOY	0.23% C 0.5% Cr 2.0% Al 0.25% Mo 0.1% V steel: Latrobe; for nitriding
SUPER NOVEX	0.3% C 0.8% Cr 3% Ni 0.25% Mo steel: Jonas **DPN: 320 UTS: 1050 Elon: 18% Proof: 920**
SUPER SAMSON	0.1% C 5.0% Cr 0.9% Mo 0.25% V steel: Carpenter for AISI type P4
SUPER SHNC	0.3% C 1.25% Cr 4.25% Ni 0.25% Mo steel: ESC for BS alloy En 30B
SUPERTOUGH C20	0.18% C 1.6% Mn 0.6% Ni 0.25% Mo steel: ESC for BS alloy En 13
SUPERVITAC	Free machining grade of steel by Creusot-Loire
SUPER VNCA	0.3% C 0.75% Cr 3% Ni 0.25% Mo steel: ESC for BS alloy En 23
SUPREX A	0.08% C 0.9% Mn 1.25% Cr 0.5% Mo steel: Welding electrode; Murex **UTS: 750 Elon: 24% Proof: 640**
SUPREX B	0.08% C 2.2% Cr 1% Mo steel: Welding electrode; Murex **UTS: 750 Elon: 22% Proof: 660**
SUPREX C	0.08% C 4.5% Cr 0.6% Mo steel: Welding electrode; Murex **UTS: 750 Elon: 23% Proof: 660**
SUPREX M	0.07% C 0.05% Cr 0.05% Mo steel: Welding electrode; Murex **UTS: 640 Elon: 30% Proof: 560**
SUPRIMPACTO	0.12% C 1.5% Cr 3.7% Ni steel: For carburizing; Atlas
SV 10KH5M	0.12% C (max) 5.0% Cr 0.5% Mo steel: For welding wire; Russian Standard designation
SVL	0.3% C 1.3% Cr 4% Ni 0.3% Mo steel: Jessop
T 20821	UNS designation for H21 type tool steel
T 20822	UNS designation for H22 type tool steel
T 20823	UNS designation for H23 type tool steel
T 51602	UNS designation for P2 type tool steel
T 51603	UNS designation for P3 type tool steel
T 51604	UNS designation for P4 type tool steel
T 51605	UNS designation for P5 type tool steel
T 51606	UNS designation for P6 type tool steel
T 51620	UNS designation for P20 type tool steel
T 51621	UNS designation for P21 type tool steel
T ALLOY A	0.33% C 3.5% Cr 9.2% W 0.5% V steel: Braeburn; AISI type H21
T ALLOY C	0.25% C 4.0% Cr 0.5% V 15% W steel: Braeburn
TENSITRODE 75.55	0.07% C 0.6% Mn 1.6% Ni 0.5% Mo 0.1% V steel: For electrodes; Esab standard **UTS: 726 Elon: 23% Proof: 618**
TENSITRODE 75.65	0.07% C 0.9% Mn 1.6% Ni 0.75% Mo 0.25% V steel: For electrodes; Esab **UTS: 726 Elon: 23% Proof: 618**
THERMENOL	0.05% C 3.3% Mo 16% Al 0.3% V steel: American proprietary alloy listed in SAE yearbook
TI	0.15% C 0.8% Mn 0.6% Cr 0.8% Ni 0.5% Mo 0.06% V 0.35% Cu 0.004% B steel: Rheinstahl
TI	0.15% C 0.8% Mn 0.5% Cr 0.85% Ni 0.5% Mo 0.07% V 0.3% Cu steel: Lukens Steel Co
TI TYPE A	0.17% C 0.8% Mn 0.6% Cr 0.2% Mo 0.06% V 0.02% Ti 0.003% B steel: Rheinstahl
TI TYPE A	0.15% C 0.85% Mn 0.5% Cr 0.3% Mo 0.02% Ti 0.05% V steel: Lukens Steel Co
TI TYPE B	0.16% C 1.1% Mn 0.5% Cr 0.55% Ni 0.25% Mo 0.05% V 0.3% Cu steel: Lukens Steel Co
TOBA	0.37% C 1% Cr 0.2% V steel: T Turton
TUBROD 15.42	0.15% C 2.0% Mn 4.5% Cr 0.5% Ni 0.5% Mo steel: Tubular weld rod; Esab for hard facing; as welded **DPN: 430**
TUBROD 15.50	0.3% C 2.5% Mn 5.0% Cr 0.8% Mo steel: Tubular weld rod; Esab; for hard facing; as welded **DPN: 520**

Note. The following abbreviations and units are used in the tables:

DPN	Hardness, diamond pyramid number
UTS	Ultimate tensile strength, N/mm^2
Elon	Elongation, %
Proof	0.1 % proof strength, N/mm^2

$1 \text{ N/mm}^2 = 0.1 \text{ hbar} = 0.102 \text{ kgf/mm}^2 = 0.06475 \text{ tonf/in}^2 = 145.04 \text{ lbf/in}^2$

Symbol	Nominal analysis, supplier, condition and remarks.
TUREX 9	0.3% C 2.4% Cr 0.1% V steel: Pompey
TY	0.28% C 1.5% Cr 4% Ni 5% W 0.25% V steel: For tools; S Osborn; for hot extrusions dies, punches, etc. **DPN: 500**
UHB STATO 23	0.15% C 0.9% Cr 0.45% Mo steel: Tube; Uddelholm; normalized **UTS: 500 Elon: 28% Proof: 290**
UHB STATO 28	0.15% C 2.25% Cr 1.0% Mo steel: Tube; Uddelholm; normalized **UTS: 480 Elon: 28% Proof: 250**
UHB UDDCO 2	0.13% C 5% Cr 0.5% Mo steel: Uddelholm; annealed **DPN: 150 UTS: 480 Elon: 20% Proof: 240**
UHB UDDCO 3	0.06% C 3% Cr 0.5% Mo steel: Uddelholm; annealed **DPN: 150 UTS: 440 Elon: 20% Proof: 240**
UHB UDDCO 9	0.1% C 9.0% Cr 1.0% Mo steel: Tube; Uddelholm; normalized **UTS: 610 Elon: 25% Proof: 370**
UNI 5331 16 Cr Ni 4	0.15% C 1.0% Cr 1.0% Ni steel: For carburizing
UNI 5331 20 Cr Ni 4	0.2% C 1.0% Cr 1.0% Ni steel: For carburizing
UNI 5331 12 Ni Cr 3	0.12% C 0.5% Cr 0.6% Ni steel: For carburizing
UNI 5331 16 Ni Cr 11	0.14% C 0.7% Cr 2.7% Ni steel: For carburizing
UNI 5331 16 Ni Cr Mo 2	0.15% C 0.5% Cr 0.5% Ni 0.2% Mo steel: For carburizing
UNI 5331 16 Ni Cr Mo 12	0.16% C 1.0% Cr 3.0% Ni 0.35% Mo steel: For carburizing
UNI 5331 18 Ni Cr Mo 5	0.18% C 0.8% Cr 1.3% Ni 0.2% Mo steel: For carburizing
UNI 5331 18 Ni Cr Mo 7	0.18% C 0.6% Cr 1.6% Ni 0.25% Mo steel: For carburizing
UNI 5331 20 Ni Cr Mo 2	0.2% C 0.5% Cr 0.5% Ni 0.2% Mo steel: For carburizing
VALAND 1	0.28% C 3.0% Cr 9.4% W 0.3% V steel: Uddelholm for type H21
VANACHROM	0.35% C 5.5% Cr 1.35% Mo 1.0% V steel: Swift Levick for hot dies
VCN 15	0.35% C 1.25% Ni 0.5% Cr steel: Wrought; Austrian National specification
VCN 25w(g)	0.3% C 3.2% Ni 0.7% Cr 0.3% Mo steel: Wrought; Austrian national specification
VCN 35h	0.3% C 3.2% Ni 0.7% Cr 0.25% Mo steel: Wrought; Austrian national specification
VCN 35w(e)	0.3% C 3.2% Ni 0.7% Cr 0.3% Mo steel: Wrought; Austrian national specification
VCN 45(d)	0.3% C 4.2% Ni 1.2% Cr steel: Wrought; Austrian national specification
VIBRAC V30	0.3% C 0.6% Cr 2.5% Ni 0.5% Mo steel: ESC for BS alloy En 25
VMC	0.35% C 5.0% Cr 1.5% Mo 1.0% V steel: Origin unknown
VMC(H)	0.33% C 3.0% Cr 0.5% Mo 0.15% V steel: Origin unknown
VNCA	0.23% C 0.75% Cr 3% Ni steel: ESC for BS alloy En 23
VS 351	0.1% C 1% Cr 3.5% Ni steel: For carburizing; Vulcan; hardened & tempered; core properties **UTS: 1140**
VSHD	0.3% C 3% Cr 10% W 0.5% V steel: Vulcan; hot dies **DPN: 440**
VS INC	0.35% C 1% Cr 1.25% Ni 0.3% Mo steel: Vulcan **UTS: 920**
VS 3 NC	0.35% C 1% Cr 3% Ni 0.35% Mo steel: Vulcan **UTS: 920**

Symbol	Nominal analysis, supplier, condition and remarks.
VS 4 NC	0.3% C 1.3% Cr 4.1% Ni 0.35% Mo steel: Vulcan **UTS: 920**
VWMC	0.35% C 5.0% Cr 1.5% Mo 1.5% V steel: Origin unknown
W 9B	0.22% C 2.5% Cr 2.7% Ni 0.45% Mo 0.3% V 9.5% W steel: Origin unknown
WC 10	0.1% C 0.25% Cr 0.4% Ni 0.15% Mo 0.3% Cu (all max) steel: Casting; Wolsingham **UTS: 390 Elon: 22% Proof: 180**
WC 18	0.18% C 0.25% Cr 0.4% Ni 0.15% Mo 0.3% Cu (all max) steel: Casting; Wolsingham **UTS: 450 Elon: 22% Proof: 210**
WC 23	0.23% C 0.25% Cr 0.4% Ni 0.15% Mo 0.3% Cu (all max) steel: Casting; Wolsingham **UTS: 480 Elon: 20% Proof: 220**
WC 28	0.28% C 0.25% Cr 0.4% Ni 0.15% Mo 0.3% Cu (all max) steel: Casting; Wolsingham **UTS: 520 Elon: 20% Proof: 240**
WC 33	0.34% C 0.25% Cr 0.4% Ni 0.15% Mo 0.3% Cu (all max) steel: Casting; Wolsingham **UTS: 550 Elon: 20% Proof: 250**
WC 181	0.18% C 0.25% Cr 0.4% Ni 0.15% Mo 0.3% Cu all max steel: Casting; Wolsingham **UTS: 450 Elon: 30% Proof: 200**
WC 182	0.18% C 0.25% Cr 0.4% Ni 0.15% Mo 0.3% Cu (all max) steel: Casting; Wolsingham **UTS: 440 Elon: 24% Proof: 240**
WCD	0.35% C 1% Si 5% Cr 1.2% Mo 1.2% W 0.35% V steel: Schoeller-Bleckmann **UTS: 1550**
WCD 2	0.33% C 1% Si 5% Cr 1.1% Mo 0.5% V steel: Schoeller-Bleckmann **UTS: 1420**
WELCON 2H.S	0.12% C 0.9% Mn 0.5% Cr 1.0% Ni 0.5% Mo steel: Japan Steel Works
WELCON 2H.U	0.12% C 0.9% Mn 0.8% Cr 0.5% Ni 0.7% Mo 0.32% Cu steel: Japan Steel Works
WEL-MONIX	0.15% C (max) 1.4% Mn 0.3% Mo 0.14% V steel: German proprietary specification
WEL-TEN 60	0.16% C (max) 1.2% Mn 0.4% Cr 0.6% Ni 0.15% V steel
WEL-TEN 60H	0.18% C (max) 1.2% Mn 1.0% Ni 0.15% Ti with V steel: Yawata
WEL-TEN 62	0.18% C (max) 1.2% Mn 0.3% Cr 0.6% Ni 0.12% V steel: Yawata
WEL-TEN 80	0.18% C (max) 0.9% Mn 0.6% Cr 1.5% Ni 0.6% Mo 0.1% V 0.006% B 0.3% Cu steel: Yawata
WEL-TEN 100W	0.18% C (max) 0.9% Mn 0.6% Cr 1.5% Ni 0.6% Mo 0.1% V 0.3% Cu steel: Yawata
WL 2	0.15% C 9.0% Cr 0.5% Mo steel: Tube; Babcock & Wilcox (USA)
WLM 77	0.35% C 0.8% Cr 0.7% Mo 0.2% V steel: W Marrison; chisels, shears
WLM 1030	0.3% C 3.5% Cr 10% W steel: W Marrison; hot dies **DPN: 550**
WLM 1530	0.18% C 1% Cr 3% Ni steel: For case hardening; W Marrison; analysis varies over wide limits
WLM 3033	0.3% C 0.9% Cr 3.4% Ni 0.5% Mo steel: W Marrison; hardened & tempered **DPN: 320 UTS: 1000 Elon: 16% Proof: 800**
WLM 3042	0.3% C 1.3% Cr 4.1% Ni steel: Mo optional; W Marrison; hardened & tempered **DPN: 444 UTS: 1550 Elon: 10% Proof: 1300**
WLM 4053	0.35% C 0.5% Cr 0.5% W steel: W Marrison; chisels, punches
X 7 Cr Mo 6/1	0.15% C (max) 6.0% Cr 0.7% Mo steel: Designation used by German Standard
X 7 Cr Mo 10/1	0.1% (max) 9.5% Cr 1.0% Mo steel: Designation used by German Standard

Symbol	Nominal analysis, supplier, condition and remarks.
X 19 Ni Cr Mo4	0.19% C 1.3% Cr 4.1% Ni 0.2% Mo 0.4% W steel: For tools; German Standard designation
X 30 W Cr V 53	0.3% C 2.4% Cr 0.6% V 4.3% W steel: German Standard designation
X 32 Cr Mo V 33	0.32% C 3.0% Cr 2.8% Mo 0.5% V steel: German Standard designation
X 4620	0.2% C 1.8% Ni 0.25% Mo steel: Designation used by SAE; obsolete
XAR 15	0.15% C 0.9% Mn 0.6% Cr 0.22% Mo 0.1% Zr steel: Republic Steel
XAR 30	0.3% C (max) 0.8% Mn 0.6% Cr 0.21% Mo 0.1% Zr steel: Republic Steel
Y 10 N66	0.1% C 1.0% Cr 1.5% Ni steel: French Standard designation
Y 35 NCD16	0.35% C 1.8% Cr 4.0% Ni 0.4% Mo 0.1% V steel: French Standard designation
Y 35 NCD16 (3)	0.32% C 1.2% Cr 4.1% Ni 0.2% Mo steel: French Standard
YD1	0.25% C 2.25% Cr 9.5% W steel: Jonas; hot extrusion dies **DPN: 400**
YDC	0.3% C 2.25% Cr 8.5% W 0.35% V steel: Jonas; hot extrusion dies **DPN: 520**
YND 58	0.14% C (max) 0.7% Mn 0.5% Cr 2.3% Ni 0.5% Mo steel: Yamata

Symbol	Nominal analysis, supplier, condition and remarks.
YOLOY	0.15% C (max) 0.75% Mn 1.7% Ni 0.1% V steel: Youngstoun Steel Co
YOLOY EHS	0.18% C (max) 1.0% Mn 0.4% Cr 0.7% Ni 0.4% Mo 0.4% Cu steel: Youngstoun Steel Co
YOLOY HS	0.15% C (max) 1.0% Mn 1.0% Ni 0.2% Mo 0.75% Cu steel: Youngstoun Steel Co
YWA	0.28% C 1.3% Cr 3.4% Ni 5.8% W 0.25% V steel: Edgar Allen; extrusion dies, mandrels, etc. **DPN: 450**
Z 30WCV9	0.3% C 3.0% Cr 0.4% V 9.0% W steel: French Standard designation
Z 8CDV5	0.08% C 5.0% Cr 1.0% Mo 0.3% V steel: French Standard designation

Note. The following abbreviations and units are used in the tables:

DPN	Hardness, diamond pyramid number
UTS	Ultimate tensile strength, N/mm^2
Elon	Elongation, %
Proof	0.1 % proof strength, N/mm^2

$1\ N/mm^2 = 0.1\ hbar = 0.102\ kgf/mm^2 = 0.06475\ tonf/in^2 = 145.04\ lbf/in^2$

44K2 Steel – medium carbon alloys
0.4–0.65% carbon with two or more of chromium, nickel, molybdenum, tungsten or vanadium

Specific gravity	7.84	
Density	7840 kg/m³	(0.282 lb/in.²)
Solidus/liquidus	– °C	
Thermal conductivity	41.9 W/m °C	(10 cal/m s °C)
Coefficient of linear expansion	11×10^{-6}/ °C	
Electrical conductivity	7–10% IACS (copper 100%)	
Specific resistance	190–280 microhm mm	
Young's modulus of elasticity	$32.07 \times 10^9\ N/m^2$	$(30 \times 10^6\ lbf/in.^2)$
Impact	40–80 J	(30–60 ft/lb) (Izod)
Fatigue strength	–	
Hot strength		

Temperature °C	Tensile strength N/mm²	Elongation %
100	930	25
200	920	21
300	920	21
400	840	29
500	710	27
550	610	32
600	520	39

The above properties are typical of the following group and may not apply exactly to any one specification. It is possible that with certain specifications some of the values may not be applicable.

General metallurgical characteristics
These steels have a high enough carbon content to ensure a minimum strength of about 9000 N/mm² when fully tempered.

The top limit of the carbon present is around the eutectoid point, thus even in the fully hardened condition there should never be sufficient free cementite present to make the material too brittle. At least two of the alloying elements are present in all the steels with chromium present in the majority, while a few steels have all five alloying elements present.

Some of the lower carbon steels listed can be carburized but in general the core hardness is too high for general engineering applications. Many of these steels can be nitride hardened, but generally this requires special formulation with some aluminium which speeds up the process.

The effect of each alloying element is described under the section where the single alloy steels are listed. When two or more of these elements are present the combined effect is the sum total of the individuals.

Briefly the elements affect the steels as follows:

Chromium. This is a carbide former, with the carbides being harder and more stable than iron carbides. Carbon will combine with chromium in preference to iron, thus all the chromium is generally present as carbide.

Chromium increases the hardenability of steel in that larger sections can be through hardened, or less drastic oil quenching can be used in place of water quenching, compared to plain carbon steels. There is however, a tendency for chromium steels to be notch sensitive. Chromium also increases the corrosion resistance, although none of the steels listed can be classed as stainless. The higher chromium stainless steels are listed and described in Sections 44M and 44N. Chromium raises the critical temperatures for steel heat treatment but slows down the speed at which the metallurgical reations take place.

Nickel. This strengthens the ferritic matrix of the steel, imparting toughness rather than hardness and increasing the fatigue strength. Nickel lowers the critical heat treatment temperatures and widens the range required for successful treatment. It also improves the ductility of steel at low temperatures. A range of 10% nickel steels have been developed with good ductility at the temperatures required for liquid air and other low temperature uses.

The range of 4% nickel steels has largely been replaced with the more economical chromium, molybdenum, 1% nickel steels which have equal fatigue strength and ductility, with better tensile properties.

Molybdenum. This is a carbide former, the resultant molybdenum carbide being inherently small grained and evenly distributed. This ensures that the steels are fine grained, which improves the fatigue strength. These stable carbides slow down the metallurgical reactions, ensuring deeper hardening, greater hardenability and the elimination of temper brittleness. This is a defect caused by the undesirable precipitation of carbides during tempering between 250 and 500 °C which occurs with chromium/nickel steels. As carbon always combines with the molybdenum in preference to other carbide formers including chromium and stays in solution, this obviates carbide precipitation. Vanadium and tungsten have similar effects.

Tungsten. This is a carbide former which results in very hard, rather massive tungsten carbide particles. Generally tungsten is used with chromium and vanadium which aid the distribution and prevent agglomeration of the carbides. Tungsten has similar properties to molybdenum but larger quantities are usually required to achieve the same effect.

Vanadium. This forms very small carbides and also influences other carbide formers to produce tiny particles which in turn ensures a fine grain steel. Thus vanadium has a considerable effect on improving the fatigue strength and is very often used with chromium to achive this purpose. The presence of vanadium aids the nitriding process by preventing carbide particles agglomerating. Vanadium has little effect on the temperatures at which the metallurgical changes take place, but slows down these reactions, thus improving the hardenability.

None of these elements in the quantities used has any appreciable effect on the corrosion resistance although these alloy steels will always have slightly better resistance than plain carbon steels.

The alloying elements actually used depend on whether maximum hardness or maximum ductility is required, with the economy of the final product the deciding factor.

Thermal treatment

None of these steels can be cold worked to any great extent without becoming too brittle for practical use. There are some wire specifications which are cold drawn and tempered, and some bar alloys have final cold sizing passes, but in general full annealing between cold working operations will rarely be required as these materials are almost always hot worked.

It is important that the steels are taken slowly to temperature for all thermal treatments, otherwise there is a danger of distortion and in extreme cases cracking. The use of pre-heat at 500–600 °C before all high temperature processes is recommended when any change of section is involved.

Sub-critical annealing at 650–670 °C for two hours removes most cold work and is suitable for many of the lower alloyed materials prior to machining.

When chromium, vanadium and tungsten are present in any quantity cyclic or spheroidal annealing is recommended. This requires that the parts are heated to 800–920 °C for about 30 minutes then transferred to a furnace at 670–700 °C for two hours, this cycle being repeated at least once, with some of the higher alloyed steels requiring three or more such treatments. This results in the hard carbides being formed into small balls or spheroids distributed evenly throughout the soft iron or ferritic matrix and is the best structure for machining these steels. The value of molybdenum and vanadium as grain refiners and in preventing carbides agglomerating is appreciated with these prolonged treatments. Only those steels which it is not necessary to cyclic anneal should be normalized. This requires 800–920 °C for 30 minutes per 25 mm of section, followed by still air cooling and will almost invariably result in considerable air hardening. This can be removed by a sub-critical anneal at 650–670 °C for two hours, provided the steel is not too highly alloyed, when cyclic annealing in place of the normalize and anneal is recommended.

The steels are hardened by heating to 780–900 °C for 20 minutes per 25 mm of section and oil or air quenching.

With small parts of even section almost full hardening may be obtained in still air, but generally an air blast at least is necessary. As soon as possible after hardening the parts should be tempered. Because of the very sluggish metallurgical reactions of many of these steels multiple tempering may be necessary to obtain the best properties. A deep freeze at a temperature of minus 80–100 °C, either after hardening or between tempers, can be beneficial in dimensionally stabilizing the parts and obtaining the maximum hardness. These steels are not generally carburized as the core is too hard to supply the necessary ductile support for the hard case, although some of the lower carbon steels can be carburized when they should be treated as described previously in section 44K1.

Scale removal after thermal treatments is usually by blasting. Acid pickling is not recommended owing to the danger of hydrogen embrittlement, but sodium hydride can be used if available. Some of these steels will be treated in atmosphere controlled furnaces which prevent or reduce scale formation.

Many of the listed steels are specially designed nitriding steels. This process requires special furnace equipment and is very slow, with the presence of aluminium and vanadium essential if any appreciable nitride depth is required. It is carried out at 490–520 °C for periods of up to 100 hours' furnace duration, resulting in a very hard brittle case. The principal advantage of this over a carburized case is that no subsequent hardening operations are necessary, thus there is no scale or distortion problem.

Parts to be nitrided should be normalized and sub-critical annealed, rough machined, hardened and tempered to the final requirement, machined to within about 0.1 mm of finished vital dimensions, then stabilized at about 30 °C below the tempering temperature for about 4 hours. If parts are straightened at this stage they should be re-stabilized. After finish grinding the parts are nitrided when there will be a growth of approximately 0.01 mm per surface. This should be chemically removed by etching, or lapped off, as the top surface of nitriding is glass hard, very brittle and notch sensitive. The fatigue strength is thus improved considerably when this layer is removed. It is essential to remove chemically this brittle layer if any grinding is to be carried out, otherwise there will be a danger of grinding cracking.

The above information can only be of a very general nature and reference should be made to the origin of the specification for full details.

No mention has been made of the more specialized methods of thermally treating these materials such as austempering and martempering. These require very tight control of times, temperatures and weights and details must be obtained from the source or a competent laboratory.

Welding and brazing

Pre-treatment. All scale, oil, grease and dirt must be removed and parts should be in the soft condition. Pre-heating will be essential with all these steels.

Welding and brazing. A small number of the steels listed have been developed with controlled hardenability and are suitable for welding, but in the main the weld area or adjacent zone will always have some case or overheated structure which cannot be completely removed by subsequent treatment. There is also the danger of quench cracking during welding unless the components are correctly designed. Skilled welders using gas, electric or inert gas shielded arc welding with correct post-weld treatment can minimize these defects, but can never eliminate them completely. These materials should therefore not be the first design choice when welding is essential, unless specialist advice is obtained.

Brazing hardened parts with low melting point filler metals and very careful control of the heating medium to ensure that the tempering temperature is never exceeded is possible but difficult and only applies when tempering is above 550 °C.

Post-treatment. All fluxes and slag must be removed. The parts should be normalized, then hardened and tempered for best results, but this will seldom be possible unless very careful thought is given to design.

As an absolute minimum the weld area must be stress relieved or tempered to prevent brittle failure adjacent to the weld.

Flaw detection methods

Crack test. Magnetic crack test. Very often penetrant oil chalk testing can give comparable results and is more flexible.

X-ray. Not normally required.

Ultra-sonic test. Large forgings which will have expensive machining can be ultra-sonic inspected at an early stage. Any subcutaneous defects are identified, thus saving the cost of further machining.

Chemical etch. Not required.

Corrosion protection

Temporary. These are all expensive materials, thus it is worth taking care at all times to ensure that they are protected. They must be oiled, or preferably greased, always kept dry and stored under cover. During machining the parts should be oiled at all times. The use of soluble oil coolants if they are kept clean helps to inhibit corrosion but dirty coolant, especially when contaminated with metallic particles, can be a serious source of corrosion. Finished parts should be cleaned, dried, oiled and packed in plastic. Many cardboards and papers contain corrosive substances which can cause staining and rust even uhen the parts are oiled.

Permanent. Rusting or steel corrosion is the formation of complex oxides of iron, which being only loosely adherent are readily removed to allow further oxidation to take place. This causes formation of pits and roughness, which being surface stress raisers can seriously reduce the fatigue properties of these high strength steels.

For permanent protection it is essential first that all previous corrosion is removed and as soon as possible a non-porous film is applied to prevent ingress of moisture and air. This film must be strong enough to withstand the usage for which the part is designed. Rust covered by a non-porous film can propagate, as it contains air and moisture; there is then an increase in volume causing failure of the film.

Grit blasting when properly controlled is the best method of cleaning and preparing the surface as it leaves a roughened surface to act as a key for the subsequent film. The surface must be well brushed to remove the dust caused by blasting, and the subsequent treatment must be carried out as soon after blasting as possible as the freshly prepared surface can corrode very rapidly.

Phosphate treatment acts as an excellent paint base and is in itself a corrosion resistant film and oil retentive surface. Many of the steels listed are used in this condition and provided the oil film is maintained, long corrosion free life can be achieved under indoor conditions. Steels above 350 DPN (1000 N/mm² tensile) should be stress relieved to eliminate hydrogen embrittlement.

Painting. This should be on the blasted, phosphated surface which must be oven dried before application of zinc silicate or chromate primer. The top coats of paint should be stoved for best results and applied as soon after the primer as possible.

Plating. Decorative chromium plating is not often required on these steels. When necessary at least 0.05 mm of nickel, or copper and nickel undercoat should be deposited below the very thin coat of chromium. Cadmium and zinc plating 0.025 mm thick gives excellent corrosion protection under all normal conditions, but must not be used for parts in contact with foodstuffs. The majority of these steels will be above 350 DPN (1000 N/mm² tensile) and these require stress relieving at 200 °C after plating. If considerable machining stress exists a stress relieve before plating is recommended. This is required to eliminate hydrogen embrittlement which can cause brittle failure below the calculated tensile strength. This defect becomes more serious as the hardness increases. Although entirely removed by the 200 °C stress relief treatment there can never be any visual proof that this is always carried out on every part. As brittle failure of high tensile steel parts can be disastrous it is recommended that parts above 1200–1400 N/mm² tensile do not specify electroplating as the protective finish.

Metal spraying of zinc or aluminium is not recommended except where the steel is soft enough to be roughened by grit blasting as this supplies the key for adhesion of the sprayed metal. Post-treatment should be wire brush, then chemical seal and paint. New techniques of metal spraying where the particles are given very high velocities are being developed, which rely less on the surface roughness. Chemical oxidation, forming an adherent coating, can be produced with proprietary 'black oxide' solutions which are basically hot strong sodium hydroxide with oxidizing additions. If used indoors with the oil film maintained these can give excellent corrosion resistance.

In general it can be said that the standard of corrosion protection obtained relies on the care taken at preparation, the quality of the protection media and standard to which it is applied. All the steels listed are expensive materials which warrant considerable care. Experience also clearly shows that it is much cheaper to apply correctly a good quality protective coating first time rather than have the expense of removing the failed coating and re-applying the protecting medium. Also it is more difficult to obtain successful results at the second attempt.

Many of the components made from these steels will be used without protection as they operate under oily conditions. Considerable care must always be taken that these parts do not rust before being assembled, or while lying around at overhaul.

Machinability

Many of the materials listed will be too hard to machine after hardening and tempering as they will be above 400 DPN. The best condition for machinability is hardened and tempered at 250–300 DPN and for some purposes it may be worth calling for this condition and re-hardening and tempering prior to final grinding. For normal purposes the steels should be rough machined in the normalized and annealed or cyclic annealed condition, which generally results in some tearing, then finally ground after hardening and tempering. Where parts are machined from the rough in the finally hardened and tempered condition it may be advisable to stress relieve at the tempering temperature before finish machining, to remove any stress. High speed or tipped tools are essential for cutting the steels listed.

Spark erosion and electrochemical machining techniques can be used to shape fully hardened parts, but will seldom be economical.

Uses

Gears, shafts, couplings, springs, studs, bolts, nuts and all forms of stressed engineering compounds.

Cutting tools of all kinds, press tools, jigs and fixtures, hand tools of all types.

Wear resisting components.

Note. The following abbreviations and units are used in the tables:

DPN	Hardness, diamond pyramid number
UTS	Ultimate tensile strength, N/mm²
Elon	Elongation, %
Proof	0.1 % proof strength, N/mm²

1 N/mm²=0.1 hbar=0.102 kgf/mm²=0.06475 tonf/in²=145.04 lbf/in²

Symbol	Nominal analysis, supplier, condition and remarks.
00	0.6% C 1.1% Cr 1.9% W 0.3% V tool steel: Balfour; punches, chisels, etc. **DPN: 710**
0.7225	0.42% C 1.1% Cr 0.6% Ni 0.2% Mo steel: Forging; German Standard
0.7561	0.42% C 1.5% Cr 0.1% V steel: For bolts; German Standard
0.7727	0.45% C 1.4% Cr 0.7% Mo 0.3% V steel: For springs; German Standard
0.8159	0.5% C 1.1% Cr 0.1% V steel: German Standard

Symbol	Nominal analysis, supplier, condition and remarks.
0.8161	0.58% C 1.0% Cr 0.1% V steel: For springs; German Standard
0.8239	0.65% C 3.7% Cr 0.85% Mo 8.5% W 0.7% V steel: For springs; German Standard
1.1232	0.38% C 5.3% Cr 1.1% Mo 0.4% V steel: German Standard designation
1.2241	0.5% C 1.1% Cr 0.2% V steel: German Standard
1.2344	0.38% C 5.2% Cr 1.35% Mo 1.0% V steel: German Standard
1.2344	0.4% C 5.3% Cr 1.4% Mo 1.0% V steel: German Standard designation
1.2541	0.35% C 1.1% Cr 0.2% V 2.0% W steel: German Standard designation
1.2542	0.45% C 1.1% Cr 0.2% V 2.0% W steel: German Standard designation
1.2547	0.5% C 1.5% Cr 0.2% C 2.2% W steel: German Standard
1.2550	0.6% C 1.1% Cr 0.2% V 2.0% W steel: German Standard designation
1.2711	0.54% C 0.7% Cr 1.7% Ni 0.3% Mo 0.1% V steel: For tools; German Standard designation
1.2713	0.55% C 0.7% Cr 1.7% Ni 0.3% Mo 0.1% V steel: Russian Standard designation
1.2714	0.55% C 1.0% Cr 1.7% Ni 0.5% Mo 0.1% V steel: German Standard designation
1.2714	0.63% C 1.0% Cr 1.45% Ni 0.2% Mo steel: German Standard
1.2721	0.50% C 1.1% Cr 3.3% Ni steel: For tools; German Standard designation
1.2767	0.45% C 1.4% Cr 4.1% Ni 0.3% Mo 0.5% W steel: For tools; German Standard designation
1.7701	0.52% C 0.9% Mn 1.05% Cr 0.2% Mo 0.1% V steel: For springs; German Standard
1.7225	0.42% C 1.1% Cr 0.6% Ni (max) 0.2% Mo steel: Forging; German Standard
1.7228	0.5% C 1.1% Cr 0.2% Mo steel: Forging; German Standard
1.7561	0.42% C 1.6% Cr 0.1% V steel: German Standard
1.8154	0.52% C 1.1% Cr 0.1% V steel: For springs; German Standard
1.8159	0.5% C 0.85% Mn 1.05% Cr 0.15% V steel: For springs; German Standard
1.8159	0.51% C 0.85% 1.05% Cr 0.15% V steel: For springs; German Standard
1.8159	0.51% C 1.1% Cr 0.1% V steel: German Standard
2 CV	0.45% C 2% Si 2% Cr 0.3% V steel: T Turton
4KH8V2	0.4% C 7.5% Cr 2.5% W steel: Russian Standard designation
4KHV2S	0.4% C 1.15% Cr 2.25% W steel: Russian Standard designation
5/317	0.5% C 1.0% Cr 1.75% Ni steel: Carpenter
5KHNGM	0.55% C 0.75% Cr 0.22% Mo steel: Russian Standard designation
5KHNM	0.55% C 0.65% Cr 1.6% Ni 0.22% Mo steel: Russian Standard designation
5KHNV	0.55% C 0.65% Cr 1.6% Ni 0.55% W steel: Russian Standard designation
5KHV2S	0.5% C 1.15% Cr 2.25% W steel: Russian Standard designation

Symbol	Nominal analysis, supplier, condition and remarks.
7KHF	0.68% C 0.55% Cr 0.22% V steel: Russian Standard designation
8 MC 2 Mo	0.4% C 1% Cr 0.2% Mo steel: Sandvik for SIS 2244
8 NIC 2	0.4% C 0.85% Cr 1.3% Ni steel: Sandvik for SIS 2530
11.11	0.4% C 5.0% Cr 1.3% Mo 0.5% V steel: Origin unknown
14HD	0.45% C 3.0% Cr 15.0% W steel: For tools; origin unknown
14 HD	0.4% C 3.5% Cr 14% W 0.5% V steel: For tools; Huntsman; cutting tools; extrusion dies DPN: 630
17/22A	0.45% C 1.25% Cr 0.5% Mo 0.25% V steel: Timken; hardened & tempered for high temperature use DPN: 415 UTS: 1550 Elon: 13% Proof: 1380
18 D	0.5% C 4.0% Cr 1.0% V 18.0% W steel: For tools; origin unknown
18 HD	0.5% C 4% Cr 18% W 1% V steel: For tools; Huntsman; cutting tools; Cu extrusion dies DPN: 700
30 Ni Cr Mo 12	0.3% C 0.65% Mn 0.8% Cr 2.9% Ni 0.45% Mo steel: Italian Standard designation; as rolled DPN: 260
34 Cr Ni Mo 8(e.f.)	0.4% C 1.5% Ni 1.2% Cr 0.28% Mo steel: Wrought; Austrian national specification
35 NC5 UNI 4365	0.36% C 0.9% Cr 1.2% Ni steel: Italian Standard; for bolts
35 NC9 UNI 4365	0.36% C 0.8% Cr 2.3% Ni steel: Italian Standard; for bolts
35 Ni Cr9	0.35% C 0.65% Mn 0.75% Cr 2.2% Ni steel: Italian Standard designation DPN: 240
35 Ni Cr Mo 15	0.34% C 0.5% Mn 0.65% Cr 3.7% Ni 0.3% Mo steel: Italian Standard designation; as rolled DPN: 275
35 W Cr V7	0.35% C 1.1% Cr 0.2% V 2.0% W steel: German Standard designation
38 CD4 UNI 4365	0.38% C 1.0% Cr 0.2% Mo steel: Italian Standard; for bolts
38 Ni Cr Mo 4	0.38% C 0.85% Mn 0.85% Cr 0.85% Ni 0.2% Mo steel: Italian Standard designation; as rolled DPN: 240
40 Cr Mo 4	0.4% C 0.85% Mn 1.05% Cr 0.2% Mo steel: Italian Standard designation; as rolled DPN: 245
40 Ni Cr Mo 7	0.4% C 0.65% Mn 0.75% Cr 1.75% Ni 0.25% Mo steel: Italian Standard designation; as rolled DPN: 250
40 NKD	0.4% C 0.5% Cr 0.55% Ni 0.2% Mo steel: Pompey
42 Cr Mo 4	0.42% C 1.1% Cr 0.6% Ni (max) 0.2% Mo steel: Designation used by German Standards
42 Cr V 6	0.42% C 1.5% Cr 0.1% V steel: Designation used by German Standards
45 CP	0.42% C 0.95% Cr 0.25% Mo steel: Pompey
45 Cr 5 Ni 5	0.45% C 5.0% Cr 4.5% Ni steel: Electrode; Metrode; hardenable deposit DPN: 400
45 Cr Mo V 6-7	0.45% C 1.4% Cr 0.7% Mo 0.3% V steel: Designation used by German Standards
45 W Cr V 7	0.45% C 1.1% Cr 0.2% V 2.0% W steel: German Standard designation
50 Cr Mo 4	0.5% C 1.1% Cr 0.2% Mo steel: Designation used by German Standards
50 Cr V4	0.5% C 0.8% Mn 1.0% Cr 0.15% V steel: Italian Standard designation DPN: 250 UTS: 1550 Elon: 6% Proof: 150

Note. The following abbreviations and units are used in the tables:

DPN	Hardness, diamond pyramid number
UTS	Ultimate tensile strength, N/mm^2
Elon	Elongation, %
Proof	0.1 % proof strength, N/mm^2

$1 \text{ N/mm}^2 = 0.1 \text{ hbar} = 0.102 \text{ kgf/mm}^2 = 0.06475 \text{ tonf/in}^2 = 145.04 \text{ lbf/in}^2$

Symbol	Nominal analysis, supplier, condition and remarks.
50 Cr V 4	0.5% C 1.1% Cr 0.1% V steel: Designation used by German Standards
50 Ni Cr 13	0.5% C 1.1% Cr 3.3% Ni steel: For tools; German Standard designation
50 WCV20	0.5% C 1.5% Cr 0.2% C 2.2% W steel: French Standard
52 Si Cr Ni5	0.53% C 0.8% Mn 1.4% Si 0.05% Cr 0.6% Ni steel: Italian Standard designation **DPN: 260 UTS: 700 Elon: 5% Proof: 1350**
53 S	0.4% C 1% Si 5% Cr 1.4% Mo 1% V steel: Can be nitrided; Carrs; hot extrusion dies, mandrels, punches, etc. **DPN: 530**
54 Ni Cr Mo V6	0.54% C 0.7% Cr 1.7% Ni 0.3% Mo 0.1% V tool steel: German Standard designation
55 Ni Cr Mo V6	0.55% C 0.7% Cr 1.7% Ni 0.3% Mo 0.1% V steel: German Standard designation
55 Ni NCDV7	0.55% C 0.8% Cr 1.75% Ni 0.3% Mo 0.2% V steel: French Standard designation
55 WC20	0.55% C 1.0% Cr 2.0% W steel: French Standard designation
56 Ni Cr Mo V7	0.55% C 1.0% Cr 1.7% Ni 0.5% Mo 0.1% V steel: German Standard designation
58 Cr V 4	0.58% C 1.1% Cr 0.1% V steel: Designation used by German Standards
60 W Cr V7	0.6% C 1.1% Cr 0.2% V 2.0% W steel: German Standard designation
65 W Mo 34-8	0.65% C 3.7% Cr 0.85% Mo 8.5% W 0.7% V steel: Designation used by German Standards
81 B40	0.4% C 0.35% Ni 0.4% Cr 0.1% Mo B steel: Designation used by AISI
86 B45	0.45% C 0.6% Ni 0.5% Cr 0.2% Mo 0.0005% B steel: Designation used by AISI
300M	0.4% C 0.8% Cr 1.8% Ni 0.4% Mo 0.08% V steel: American proprietary alloy listed in SAE yearbook
328	0.4% C 1% Si 5.5% Cr 0.8% Mo 0.5% V steel: F Parkin; Al die cast moulds, etc. **DPN: 440**
348	0.4% C 1.6% Cr 4% Ni 0.3% Mo steel: F Parkin; shear blades, press dies **DPN: 700**
361	0.52% C 1.1% Cr 1.9% W 0.3% V steel: For hot work tools; Balfour; forging dies, punches, etc. **DPN: 460**
399	0.5% C 1.3% Cr 2.3% W 0.2% V steel: F Parkin; trimming dies, chisels, etc. **DPN: 600**
528	0.4% C 4.2% Cr 4.2% W 2.2% V 4.2% Co tool steel: Jessop; forging dies for copper alloys
1213	0.45% C 0.6% Cr 0.2% Mo steel: French Standard designation
1413	0.38% C 1.2% Cr 1.6% Ni 0.25% Mo tool steel: S Osborn; plastic dies, zinc die cast dies; obsolete **DPN: 580**
1850	0.57% C 0.35% Mn 4% Cr 18% W 0.7% V steel: F Parkin; hot dies; can be used at higher hardness **DPN: 500**
2321	0.38% C 5.0% C 1.25% Mo 0.5% V steel: French Standard designation
2331	0.42% C 1.0% Cr 0.2% Mo steel: French Standard designation
2332	0.5% C 1.0% C 0.15% V steel: French Standard designation
2341	0.55% C 1.0% Cr 2.0% W steel: French Standard designation
3145	0.40% C 0.8% Cr 1.2% Ni steel: Designation used by AISI; obsolete
3150	0.5% C 0.8% Cr 1.2% Ni steel: Designation used by AISI; obsolete

Symbol	Nominal analysis, supplier, condition and remarks.
3240	0.4% C 1% Cr 1.7% Ni steel: Designation used by AISI; obsolete
3245	0.45% C 1% Cr 1.7% Ni steel: Designation used by AISI; obsolete
3250	0.5% C 1% Cr 1.7% Ni steel: Designation used by AISI; obsolete
3340	0.4% C 1.5% Cr 3.5% Ni steel: Designation used by AISI; obsolete
3381	0.55% C 0.8% Cr 1.75% Ni 0.3% Mo 0.2% V steel: French Standard designation
3432	0.38% C 5.0% Cr 1.25% Mo 0.5% V 1.25% W steel: French Standard designation
3450	0.5% C 0.8% Cr 3% Ni steel: Designation used by AISI; obsolete
3541	0.4% C 4.0% Cr 0.5% Mo 0.5% V 5.0% W steel: French Standard designation
3548	0.65% C 4.0% Cr 5.0% Mo 2.0% V 6.0% W steel: French Standard designation
4337	0.37% C 0.8% Cr 1.8% Ni 0.25% Mo steel: Designation used by AISI
4620	0.4% C 1.8% Ni 0.25% Mo steel: Designation used by AISI; obsolete
4640	0.4% C 1.8% Ni 0.25% Mo steel: Designation used by AISI
6140	0.4% C 1% Cr 0.15% V steel: Designation used by AISI; obsolete
7260	0.6% C 0.8% Cr 1.8% W steel: Designation used by AISI; obsolete
9763	0.63% C 0.2% Cr 0.5% Ni 0.2% Mo steel: Designation used by AISI; obsolete
71360	0.6% C 3.5% Cr 14% W steel: Designation used by AISI; obsolete
71660	0.6% C 3.5% Cr 16% W steel: Designation used by AISI; obsolete
A 6	0.7% C 2.0% Mn 1.0% Cr 1.0% Mo steel: For tools; designation used by AISI
A 8	0.55% C 5.0% Cr 1.25% Mo 1.25% W steel: For tools; designation used by AISI
A 9	0.5% C 5.0% Cr 1.5% Ni 1.4% Mo 1.0% V steel: For tools; designation used by AISI
A 13	0.4% C 1.15% Cr 1.5% Ni 0.3% Mo steel: Edgar Allen; hardened & tempered **DPN: 500 UTS: 1620 Elon: 11% Proof: 1500**
A 442	0.42% C 0.35% Cr 3% Ni 0.27% Mo steel: SHD; drills, etc.
ABRAZO	0.47% C 1.5% Mn 0.5% Cr 0.2% Mo steel: Plate; Colvilles; abrasion resisting **DPN: 230**
ACP	0.5% C 1% Cr 0.15% V steel: Toledo for BS alloy En 47
ADS HB	0.35% C 5.0% Cr 1.0% V 1.5% Mo tool steel: For hot working; Inman
AIRMO	0.7% C 1.0% Cr 1.35% Mo steel: Firth Sterling
AISI 81 B45	0.45% C 0.45% Cr 0.3% Ni 0.1% Mo 0.0005% B steel
AISI 601	0.46% C 1% Cr 0.5% Mo 0.3% V steel: Hardened & tempered **UTS: 920 Elon: 29% Proof: 760**
AISI 610	0.4% C 5% Cr 1.3% Mo 0.5% V steel: Hardened & tempered **UTS: 2150 Elon: 7% Proof: 1920**
AISI 3140	0.4% C 0.6% Cr 1.2% Ni steel
AISI 4140	0.4% C 1% Cr 0.2% Mo steel
AISI 4142	0.42% C 1% Cr 0.2% Mo steel
AISI 4145	0.45% C 1% Cr 0.2% Mo steel
AISI 4147	0.47% C 1% Cr 0.2% Mo steel
AISI 4150	0.5% C 1% Cr 0.2% Mo steel
AISI 4161	0.61% C 0.8% Cr 0.3% Mo steel
AISI 4340	0.4% C 0.8% Cr 1.8% Ni 0.25% Mo steel
AISI 6145	0.45% C 1% Cr 0.15% V steel

Symbol	Nominal analysis, supplier, condition and remarks.
AISI 6150	0.5% C 1% Cr 0.15% V steel
AISI 8640	0.4% C 0.5% Cr 0.6% Ni 0.2% Mo steel
AISI 8641	0.4% C 0.5% Cr 0.6% Ni 0.2% Mo steel
AISI 8642	0.42% C 0.5% Cr 0.6% Ni 0.2% Mo steel
AISI 8645	0.45% C 0.5% Cr 0.6% Ni 0.2% Mo steel
AISI 8647	0.47% C 0.5% Cr 0.6% Ni 0.2% Mo steel: Obsolete
AISI 8650	0.5% C 0.5% Cr 0.6% Ni 0.2% Mo steel
AISI 8653	0.53% C 0.7% Cr 0.6% Ni 0.2% Mo steel
AISI 8655	0.55% C 0.5% Cr 0.6% Ni 0.2% Mo steel
AISI 8660	0.6% C 0.5% Cr 0.6% Ni 0.2% Mo steel
AISI 8740	0.4% C 0.5% Cr 0.6% Ni 0.25% Mo steel
AISI 8742	0.42% C 0.5% Cr 0.6% Ni 0.25% Mo steel
AISI 8745	0.45% C 0.5% Cr 0.6% Ni 0.25% Mo steel: Obsolete
AISI 8750	0.5% C 0.5% Cr 0.6% Ni 0.25% Mo steel
AISI 9440	0.4% C 0.4% Cr 0.4% Ni 0.1% Mo steel: Obsolete
AISI 9442	0.42% C 0.5% Cr 0.4% Ni 0.1% Mo steel: Obsolete
AISI 9445	0.45% C 0.4% Cr 0.4% Ni 0.1% Mo steel: Obsolete
AISI 9447	0.47% C 0.4% Cr 0.4% Ni 0.1% Mo steel: Obsolete
AISI 9747	0.47% C 0.2% Cr 0.5% Ni 0.2% Mo steel: Obsolete
AISI 9763	0.63% C 0.2% Cr 0.5% Ni 0.2% Mo steel: Obsolete
AISI 9840	0.4% C 0.8% Cr 1% Ni 0.25% Mo steel
AISI 9845	0.45% C 0.8% Cr 1% Ni 0.25% Mo steel: Obsolete
AISI 9850	0.5% C 0.8% Cr 1% Ni 0.25% Mo steel
AISI A3141	0.4% C 0.8% Cr 1.2% Ni steel: Obsolete
AISI A3145	0.45% C 0.8% Cr 1.2% Ni steel: Obsolete
AISI A3150	0.5% C 0.8% Cr 1.2% Ni steel: Obsolete
AISI A3240	0.4% C 1% Cr 1.7% Ni steel: Obsolete
AISI A4640	0.40% C 1.8% Ni 0.25% Mo steel: Obsolete
AISI E4340	0.40% C 0.8% Cr 1.8% Ni 0.25% Mo steel
AISI H11	0.4% C 5% Cr 1.3% Mo 0.5% V steel: Similar to AISI 610
ALHD	0.4% C 1% Si 5% Cr 1.3% Mo 1% V steel: Jonas; gravity dies for Al casting **DPN: 620**
ALLOY 10	0.6% C 1.8% Si 0.25% V 0.4% Mo steel: Braeburn
ALZ	0.4% C 5.0% Cr 1.4% Mo 1.0% V steel: Origin unknown
AM 3	0.4% C 1% Si 5% Cr 1.35% Mo 1.1% V steel: Edgar Allen for Al die casting moulds **DPN: 600**
AMS 6304 B	0.45% C 1% Cr 0.5% Mo 0.3% V steel: Bar & forging; AMS for alloy 17/22A
AMS 6312 A	0.4% C 1.8% Ni 0.25% Mo steel: Bar & forging; AMS for SAE 4640
AMS 6317 B	0.4% C 1.8% Ni 0.25% Mo steel: Bar & forging; hardened & tempered; AMS for SAE 4640 **UTS: 920**
AMS 6321	0.4% C 0.4% Cr 0.3% Ni 0.1% Mo B steel: Bar & forging; AMS for SAE 81 B40
AMS 6322 E	0.4% C 0.5% Cr 0.5% Ni 0.25% Mo steel: Bar & forging; AMS for SAE 8740

Symbol	Nominal analysis, supplier, condition and remarks.
AMS 6323 C	0.4% C 0.5% Cr 0.25% Mo steel: Tube; AMS for SAE 8740
AMS 6324 B	0.4% C 0.6% Cr 0.7% Ni 0.25% Mo steel: Bar & forging; AMS for SAE 8740
AMS 6325 C	0.4% C 0.5% Cr 0.5% Ni 0.25% Mo steel: Bar & forging; hardened & tempered; AMS for SAE 8740 **UTS: 760**
AMS 6327 C	0.4% C 0.5% Cr 0.5% Ni 0.25% Mo steel: Bar & forging; hardened & tempered; AMS for SAE 8740 **UTS: 920**
AMS 6328 D	0.5% C 0.5% Cr 0.5% Ni 0.25% Mo steel: Bar & forging; AMS for SAE 8750
AMS 6342 C	0.4% C 0.8% Cr 1% Ni 0.25% Mo steel: Bar & forging; AMS for SAE 9840
AMS 6351	0.4% C 1% Cr 0.2% Mo steel: Sheet & strip; spheroidized; AMS for SAE 4130
AMS 6358 A	0.4% C 0.5% Cr 0.5% Ni 0.25% Mo steel: Sheet & strip; AMS for SAE 8740
AMS 6359 A	0.4% C 0.8% Cr 1.8% Ni 0.25% Mo steel: Sheet & strip; AMS for SAE 4340
AMS 6378	0.4% C 1% Cr 0.2% Mo steel: Bar; drawn & tempered; AMS for SAE 4140 **Proof: 920**
AMS 6379	0.5% C 1% Cr 0.2% Mo steel: Bar; drawn & tempered; AMS for SAE 4150 **Proof: 1210**
AMS 6381 A	0.4% C 1% Cr 0.2% Mo steel: Tube; AMS for SAE 4140
AMS 6382 E	0.4% C 1% Cr 0.2% Mo steel: Bar & forging; AMS for SAE 4140
AMS 6390	0.4% C 1% Cr 0.2% Mo steel: Seamless tube; AMS for SAE 4140; special quality
AMS 6406	0.44% C 1.6% Si 2% Cr 5% Mo 0.05% V steel: Sheet & strip
AMS 6414	0.4% C 0.8% Cr 1.8% Ni 0.25% Mo steel: Bar & forging; AMS for SAE 4340
AMS 6415 F	0.4% C 0.8% Cr 1.8% Ni 0.25% Mo steel: Bar & forging; AMS for SAE 4340
AMS 6416	0.45% C 1.6% Si 0.8% Cr 1.8% Ni 0.4% Mo 0.07% V steel: Bar & forging
AMS 6422 B	0.4% C 0.8% Cr 0.8% Ni 0.2% Mo B steel: Bar & forging; AMS for SAE 98 B40
AMS 6423	0.42% C 0.9% Cr 0.7% Ni 0.5% Mo B steel: Bar & forging
AMS 6437 A	0.4% C 5% Cr 1.3% Mo 0.5% V steel: Sheet & strip
AMS 6448 B	0.5% C 1% Cr 0.15% V steel: Bar & forging; AMS for SAE 6150
AMS 6450 B	0.5% C 1% Cr 0.2% V steel: Wire for springs; annealed; AMS for SAE 6150
AMS 6445 B	0.5% C 1% Cr 0.15% V steel: Sheet & strip for springs; annealed; AMS for SAE 6150
AMS 6470 E	0.4% C 1.6% Cr 0.35% Mo 1.1% Al steel: Bar & forging; nitriding steel
AMS 6485 A	0.4% C 5% Cr 1.3% Mo steel: Bar & forging
AMS 6487 B	0.4% C 5% Cr 1.3% Mo 0.5% V steel: Bar & forging; vacuum melted; premium quality
AMS 6488	0.4% C 5% Cr 1.3% Mo 0.5% V steel: Bar & forging; premium quality
AN QQ S690/1	0.4% C 0.6% Cr 1.2% Ni steel: US Service; normalized & tempered **DPN: 230**
AN QQ S752a	0.38% C 1.0% Cr 0.2% Mo steel: US Service; hardened & tempered **DPN: 230 UTS: 840 Elon: 18% Proof: 740**
AN QQ S756a	0.4% C 0.75% Cr 1.8% Ni 0.25% Mo steel: US Service; hardened & tempered **UTS: 1070 Elon: 16% Proof: 870**

Note. The following abbreviations and units are used in the tables:

DPN	Hardness, diamond pyramid number
UTS	Ultimate tensile strength, N/mm^2
Elon	Elongation, %
Proof	0.1 % proof strength, N/mm^2

$1 \ N/mm^2 = 0.1 \ hbar = 0.102 \ kgf/mm^2 = 0.06475 \ tonf/in^2 = 145.04 \ lbf/in^2$

Symbol	Nominal analysis, supplier, condition and remarks.

AN S 16a — 0.41% C 0.5% Cr 0.6% Ni 0.25% Mo steel: US Service; hardened & tempered
DPN: 230   UTS: 840   Elon: 16%   Proof: 740

ANTICHOC IX — 0.6% C 1.6% Si 0.37% Cr 0.1% V steel: Pompey

ANTICHOC H — 0.45% C 1.1% Cr 1.9% W 0.15% V steel: Pompey

ARDHO 2 — 0.45% C 1.3% Cr 2.3% W 0.3% V steel: Spencer

ASTM A60 — 0.5% C 1% Cr 0.1% V steel: Bar for springs; obsolete

ASTM A193 B7 — 0.4% C 1% Cr 0.2% Mo steel: For bolts, etc.; hardened & tempered
UTS: 890   Elon: 16%   Proof: 700

ASTM A193 B14 — 0.45% C 1% Cr 0.35% Mo 0.25% V steel: For bolts; hardened & tempered
UTS: 920   Elon: 16%   Proof: 760

ASTM A193 B16 — 0.4% C 1% Cr 0.5% Mo 0.3% V steel: For bolts; hardened & tempered
UTS: 840   Elon: 17%   Proof: 700

ASTM A231 — 0.5% C 1% Cr 0.15% V steel: Spring wire; supplied annealed or cold drawn

ASTM A232 — 0.5% C 1% Cr 0.2% V steel: Spring wire; cold drawn & tempered
UTS: 1900

ASTM A291/3 — 0.45% C 1.25% Cr 3% Ni 0.15% Mo 0.1% V steel: Forging; hardened & tempered
UTS: 760   Elon: 19%   Proof: 580

ASTM A291/4 — 0.4% C 0.6% Cr 1.6% Ni 0.2% Mo 0.1% V steel: Forging; hardened & tempered
UTS: 840   Elon: 14%   Proof: 700

ASTM A291/5 — 0.4% C 0.6% Cr 1.6% Ni 0.2% Mo 0.1% V steel: Forging; hardened & tempered
UTS: 1000   Elon: 14%   Proof: 760

ASTM A291/6 — 0.4% C 0.6% Cr 1.6% Ni 0.2% Mo 0.1% V steel: Forging; hardened & tempered
UTS: 1070   Elon: 13%   Proof: 840

ASTM A291/7 — 0.45% C 0.6% Cr 1.6% Ni 0.2% Mo 0.1% V steel: Forging; hardened & tempered
DPN: 80   UTS: 1000   Elon: 12%

ASTM A294 A — 0.4% C 1.2% Cr 0.15% Mo steel: For turbine wheels

ASTM A294 B — 0.45% C 1.2% Cr 2.5% Ni 0.2% Mo steel: For turbine wheels

ASTM A295 F30 — 0.45% C 0.5% Mo 0.2% V steel: Seamless drum forging; normalized & tempered
UTS: 580   Elon: 21%   Proof: 340

ASTM A320 L7 — 0.43% C 0.9% Cr 0.2% Mo steel: Hardened & tempered
UTS: 860   Elon: 16%   Proof: 730

ASTM A320 L43 — 0.4% C 0.8% Cr 1.8% Ni 0.2% Mo steel: Hardened & tempered
UTS: 860   Elon: 16%   Proof: 730

ASTM A331 — Specification covering alloy steel bars; AISI designations are used to define the alloys

ASTM A332/8650 — 0.5% C 0.5% Ni 0.2% Mo steel: For springs; obsolete

ASTM A332/8655 — 0.55% C 0.5% Cr 0.6% Ni 0.2% Mo steel: For springs; obsolete

ASTM A332/8660 — 0.6% C 0.5% Cr 0.5% Ni 0.2% Mo steel: For springs; obsolete

ASTM A336 F30 — 0.45% C (max) 0.45% Cr 0.27% V steel: Seamless drum forgings
UTS: 540   Elon: 21%   Proof: 330

ASTM A354 — Alloy steel with 0.04% S & P steel: For bolts and nuts; analysis not part of specification; grades vary with mechanical properties

ASTM A355 A — 0.41% C 1.6% Cr 0.4% Mo 1.0% Al steel: Hardened & tempered
DPN: 250

ASTM A355 B — 0.35% C 1.2% Cr 0.2% Mo 0.2% Si 1.0% Al steel: Hardened & tempered
DPN: 250

Symbol	Nominal analysis, supplier, condition and remarks.

ASTM A355 D — 0.35% C 1.1% Cr 0.2% Mo 1.0% Al steel: Hardened & tempered
DPN: 250

ASTM A372 Class IV — 0.45% C 1.60% Mn 0.27% Si 0.22% Mo steel: Forging for pressure vessels

ASTM A372 Class V Type A — 0.30% C 0.55% Mn 0.27% Si 1.0% Cr 0.2% Mo steel: Forging for pressure vessels

ASTM A372 Class V Type B — 0.35% C 0.85% Mn 0.27% Si 1.0% Cr 0.2% Mo steel: Forging for pressure vessels

ASTM A372 Class V Type C — 0.30% C 0.52% Cr 0.55% Ni 0.20% Mo steel: For pressure vessels

ASTM A372 Class V Type D — 0.35% C 0.52% Cr 0.55% Ni 0.20% Mo steel: Forging for pressure vessels

ASTM A372 Class V Type E — 0.42% C 1.0% Cr 0.20% Mo steel: Forging for pressure vessels

ASTM A391 — 0.3% C (max) 0.04% S & P (max) steel: For chain; alloying elements will be present

ASTM A469/3 — 0.27% C 2.5% Ni 0.5% Cr (max) 0.35% Mo steel: For rotors, vacuum treated

ASTM A469/4 — 0.27% C 3.0% Ni 0.5% Cr (max) 0.3% Mo steel: For rotors; vacuum treated

ASTM A469/5 — 0.3% C 3.0% Ni 0.5% Cr (max) 0.45% Mo steel: For rotors, vacuum treated

ASTM A469/6 — 0.28% C 3.5% Ni 1.5% Cr 0.45% Mo 0.1% V steel: For rotors; vacuum treated

ASTM A470/2 — 0.25% C (max) 2.5% Ni 0.75% Cr (max) 0.25% Mo (min) steel: For turbine rotors; vacuum treated

ASTM A470/3 & 4 — 0.28% C (max) 2.5% Ni 0.75% Cr (max) 0.25% Mo (min) steel: For turbine rotors; vacuum treated

ASTM A470 5, 6 & 7 — 0.28% C 3.5% Ni 1.5% Cr 0.45% Mo 0.1% V steel: For turbine rotors; vacuum treated

ASTM A470/8 — 0.3% C 0.7% Ni 1.2% Cr 1.2% Mo 0.25% V steel: For turbine rotors; vacuum treated

ASTM A471 — 0.4% C (max) 3.0% Ni 2.0% Cr (max) 0.5% Mo 0.05% V steel: For turbine discs; vacuum treated; grades 1, 2, 3, 4, 5 and 6

ASTM A471/7 — 0.31% C 0.5% Ni (max) 1.1% Cr 1.2% Mo 0.25% V steel: For turbine discs; vacuum treated

ASTM A487 GN — 0.38% C (max) 1.5% Mn 0.6% Cr 0.6% Ni 0.35% Mo steel: Castings for pressure; normalized; 6Q grade is hardened and tempered

ASTM A540 B21 — 0.4% C 0.55% Mo 1.0% Cr 0.3% V steel: For bolts

ASTM A540 B22 — 0.42% C 1.0% Cr 0.2% Mo steel: For bolts

ASTM A540 B23 — 0.4% C 0.8% Cr 1.7% Ni 0.25% Mo steel: For bolts

ASTM A540 B24 — 0.4% C 0.8% Cr 1.8% Ni 0.35% Mo steel: For bolts

ASTM A579/21 — 0.35% C 0.8% Cr 1.85% Ni 0.5% Mo 0.2% V steel: Forging

ASTM A579/22 — 0.4% C 0.8% Cr 1.8% Ni 0.5% Mo 0.07% V steel: Forging

ASTM A579/23 — 0.47% C 1.0% Cr 0.6% Ni 1.0% Mo 0.11% V steel: Forging

ASTM A579/32 — 0.42% C 1.6% Si 0.8% Cr 1.85% Ni 0.42% Mo steel: Forging

ASTM A579/38 — 0.43% C 1.5% Si 2.1% Cr 0.5% Mo 0.05% V steel: Forging

ASTM A579/41 — 0.4% C 5.0% Cr 1.3% Mo 0.5% V steel: Forging

ASTM A597 CS5 — 0.6% C 2.0% Si .035% Cr (max) 0.5% Mo 0.35% V (max) steel: Cast; for tools

ASTM A628 — Steel plate for security purposes to resist drilling composition by agreement

ASTM A646/98 BV40 — 0.43% C 0.9% Cr 0.75% Ni 0.5% Mo 0.03% V steel

ASTM A646 — See Nit 135

ASTM A649/1 A — 0.42% C 1.0% Cr 0.3% Mo steel: For rolls; corrugated

Symbol	Nominal analysis, supplier, condition and remarks.
ASTM A649/1 B	0.4% C 0.8% Cr 1.8% Ni 0.25% Mo steel: For rolls; corrugated
ASTM A732/12 Q	0.5% C 0.95% Cr 0.15% V (min) steel: Investment casting; hardened & tempered **UTS: 1300 Elon: 4% Proof: 1170**
ASTRA	0.55% C 1.0% Cr 2.0% W steel: Origin unknown
AT	0.45% C 0.5% Cr 3.0% Ni steel: For cold work; Osborn
AUSTENITE	0.6% C 4% Cr 14% W 0.5% V steel: Inman; cutting tools
AWCW	0.63% C 4% Cr 0.7% V 0.5% Mo 14% W steel: For tools; ESC; shear blades, punches, etc. **DPN: 790**
BCC	0.5% C 1% Si 1.2% Cr 2.25% W 0.25% V steel: Carrs; snaps, drifts, etc. **DPN: 580**
BCD 48	0.5% C 1.6% Cr 2% W 0.3% V steel: Carrs; chisels, snaps, etc. **DPN: 600**
BESTEM	C not specified 3.0% Cr Ni Mo steel: Origin unknown
BKM	0.4% C 1.5% Cr 4.25% Ni steel: Jonas; bakelite moulds, etc. **DPN: 460**
BRAKE DIE	0.5% C 0.6% Cr 0.2% Mo steel: Atlas; hardened & tempered **DPN: 330 UTS: 1140 Elon: 13% Proof: 1070**
BRUNSWICK V 109	Medium C 0.6% Cr + Mo 1.5% Ni steel: John Vessey; stamping dies **DPN: 440**
BRUNSWICK V 114	5% Cr 1.5% Mo 1% W steel: John Vessey **DPN: 424**
BRUNSWICK V 130	Medium C 3% Cr 0.8% W steel: John Vessey **DPN: 820**
BS 224/2	0.6% C 0.3% Cr 1.2% Ni steel: For die blocks; hardened & tempered **DPN: 250**
BS 224/5	0.55% C 0.6% Cr 1.5% Ni 0.3% Mo steel: For die blocks; hardened & tempered **DPN: 300**
BS 970/640 M40	0.4% C 0.75% Mn 1.3% Ni 0.65% Cr steel: Supplied to mechanical properties; replaces En III
BS 970/708 A37	0.37% C 1% Cr 0.2% Mo steel: Supplied to chemical composition
BS 970/708 A42	0.42% C 1% Cr 0.2% Mo steel: Supplied to chemical composition
BS 970/708 H37	0.37% C 0.85% Mn 1.0% Cr 0.2% Mo steel: Supplied to hardenability requirements
BS 970/708 H42	0.42% C 0.85% Mn 1.0% Cr 0.2% Mo steel: Supplied to hardenability requirements
BS 970/708 M40	0.4% C 0.85% Mn 1.0% Cr 0.21% Mo steel: Supplied to mechanical properties; replaces En 19A
BS 970/709 M40	0.4% C 0.85% Mn 1.0% Cr 0.3% Mo steel: Supplied to mechanical properties; replaces En 19
BS 970/816 M40	0.4% C 0.5% Mn 1.5% Ni 1.2% Cr 0.15% Mo steel: Supplied to mechanical properties; replaces En 110
BS 970/817 M40	0.4% C 1.5% Ni 1.2% Cr 0.27% Mo steel: Supplied to mechanical properties; replaces En 24

Symbol	Nominal analysis, supplier, condition and remarks.
BS 970/826 M40	0.4% C 2.5% Ni 0.65% Cr 0.5% Mo steel: Supplied to mechanical properties; replaces En 26
BS 970/897 M39	0.39% C 3.2% Cr 1.0% Mo 0.2% V steel: Supplied to mechanical properties; replaces En 40C
BS 970/905 M39	0.39% C 1.6% Cr 0.2% Mo 1.1% Al steel: Supplied to mechanical properties; for nitriding; replaces En 41B
BS 970/945 A4D	0.4% Cr 0.8% Ni 0.5% Cr 0.2% Mo steel: Supplied to chemical composition
BS 970/945 M38	0.38% C 0.75% Ni 0.5% Cr 0.2% Mo steel: Supplied to mechanical properties; replaces En 100
BS 970 En 19	0.4% C 1.2% Cr 0.3% Mo steel: Bar & forging; hardened & tempered; suitable for nitriding **DPN: 350 UTS: 1070 Elon: 15% Proof: 880**
BS 970 En 19A	0.4% C 1.1% Cr 0.25% Mo steel: Bar & forging; hardened & tempered; suitable for nitriding **DPN: 320 UTS: 1000 Elon: 16% Proof: 770**
BS 970 En 19B	0.37% C 1.1% Cr 0.25% Mo steel: Bar & forging; hardened & tempered; for nitriding; properties not part of specification
BS 970 En 19C	0.42% C 1.1% Cr 0.25% Mo steel: Bar & forging; hardened & tempered; for nitriding; properties not part of specification
BS 970 En 22	0.4% C 0.3% Cr 3.5% Ni steel: Bar & forging; hardened & tempered **DPN: 270 UTS: 840 Elon: 18% Proof: 690**
BS 970 En 24	0.4% C 1% Cr 1.5% Ni 0.3% Mo steel: Bar & forging; hardened & tempered **DPN: 320 UTS: 1000 Elon: 17% Proof: 770**
BS 970 En 26	0.4% C 0.6% Cr 2.5% Ni 0.6% Mo steel: Bar & forging; hardened & tempered **DPN: 320 UTS: 1000 Elon: 16% Proof: 770**
BS 970 En 40C	0.4% C 3% Cr 0.4% Ni 1.0% Mo 0.2% V steel: Suitable for nitriding; properties at least En 40A
BS 970 En 47	0.5% C 1% Cr 0.15% V steel: For springs
BS 970 En 50	0.45% C 1.25% Cr 0.15% V steel: For wire; hardened & tempered; for valve springs **UTS: 1550**
BS 970 En 100	0.4% C 1.3% Mn 0.4% Cr 0.7% Ni 0.2% Mo steel: Bar & forging; hardened & tempered **DPN: 300 UTS: 920 Elon: 17% Proof: 730**
BS 970 En 100D	0.42% C 1.3% Mn 0.4% Cr 0.7% Ni 0.2% Mo steel: Bar & forging; mechanical properties not part of specification
BS 970 En 110	0.4% C 1.2% Cr 1.5% Ni 0.15% Mo steel: Bar & forging; hardened & tempered **DPN: 300 UTS: 920 Elon: 17% Proof: 730**
BS 970 En 160	0.4% C 1.75% Ni 0.3% Mo steel: Bar & forging; hardened & tempered **DPN: 280 UTS: 840 Elon: 18% Proof: 690**
BS 970 En 160A	0.4% C 1.75% Ni 0.3% Mo steel: Bar & forging; mechanical properties not part of specification; C held to tighter limits than En 160
BS 980 CDS15	0.55% C 1% Cr 0.2% Mo steel: Cold drawn tube; hardened & tempered **UTS: 1070 Elon: 10% Proof: 760**
BS 1506/621 A	0.4% C 1.1% Cr 0.3% Mo steel: Bar; hardened & tempered **UTS: 840 Elon: 18% Proof: 700**
BS 1506/621 B	0.4% C 1.3% Cr 0.7% Mo steel: Bar; hardened & tempered **UTS: 920 Elon: 17% Proof: 740**
BS 9156/C	0.6% C 1.2% Cr 0.3% Mo steel: Casting; normalized or hardened & tempered; in BS 3100 **DPN: 341**
BS 3111/5	0.4% C 1.2% Cr 0.3% Mo steel: Wire; annealed & drawn; for cold forged bolts **UTS: 730**

Note. The following abbreviations and units are used in the tables:

DPN	Hardness, diamond pyramid number
UTS	Ultimate tensile strength, N/mm^2
Elon	Elongation, %
Proof	0.1 % proof strength, N/mm^2

1 N/mm^2=0.1 hbar=0.102 kgf/mm^2=0.06475 tonf/in^2=145.04 lbf/in^2

Symbol	Nominal analysis, supplier, condition and remarks.
BS 3111/6	0.4% C 1.2% Cr 1.5% Ni 0.15% Mo steel: Wire; annealed & drawn; for cold forged bolts **UTS: 730**
BS 3146 CLA1C	0.45% C 1% Mn low Ni Cr Mo Cu steel: Investment casting; hardened & tempered **DPN: 180 UTS: 530 Elon: 15% Proof: 250**
BS 3146 CLA8	0.4% C low Cr Mo Cu steel: Investment casting; normalized & tempered **0.6% C 1.2% Cr 0.3% Mo steel: Investment** casting; hardened & tempered; for abrasion resistance **DPN: 341**
BS 3551 B	0.4% C 1.3% Mn 0.5% Cr 0.7% Ni 0.2% Mo steel: For shackles; hardened & tempered **DPN: 270**
BS 5005	Wrought steels: For automobiles; withdrawn; replaced by BS 970
BS 5007	Sheet steels: For automobiles; withdrawn; replaced by BS 1449
BS 5008	Valve steels: For automobiles; withdrawn; replaced by BS 970
BS N35	0.4% C 0.3% Cr 3.5% Ni steel: Toledo for BS alloy En 22
BS S2	Medium carbon alloy steel: 850 N/mm² range; replaced by BS S94, S95 & S96
BS S11	Nickel chromium steel: 900 N/mm² range; replaced by BS S94, S95 & S96
BS S12	Alloy steel: Bar & forging; high tensile; replaced
BS S18	Alloy steel: 1300 N/mm² tensile range; replaced
BS S33	Medium carbon alloy steel: 900 N/mm² tensile; obsolete
BS S34	Medium carbon alloy steel: Bar; hardened & tempered; obsolete
BS S40	Medium carbon alloy steel: Strip; high tensile; obsolete
BS S43	Medium carbon alloy steel: Strip; obsolete
BS S65	Medium carbon nickel chrome steel: Replaced by BS/S97
BS S69	Medium carbon 3.5% Ni steel: Obsolete; see BS S94, S95 & S96
BS S81	Medium carbon chrome nickel steel: 1070 N/mm² tensile; obsolete; replaced by BS/S97
BS S86	Medium carbon chrome nickel steel: Sheet; obsolete; see BS S514
BS S87	Medium carbon chrome nickel steel: Sheet; obsolete
BS S88	Medium carbon chrome nickel steel: Sheet; obsolete; see BS S517
BS S94	0.4% C 0.5% Cr 0.7% Ni 0.2% Mo steel: Hardened & tempered **DPN: 280 UTS: 920 Elon: 18% Proof: 670**
BS S95	0.4% C 1.2% Cr 1.5% Ni 0.3% Mo steel: Hardened & tempered **DPN: 280 UTS: 920 Elon: 18% Proof: 670**
BS S98	0.4% C 0.7% Cr 2.5% Ni 0.55% Mo steel: Hardened & tempered **DPN: 360 UTS: 1210 Elon: 14% Proof: 1000**
BS S99	0.4% C 0.7% Cr 2.5% Ni 0.55% Mo steel: Hardened & tempered **DPN: 390 UTS: 1300 Elon: 14% Proof: 1070**
BS S118	0.4% C 1.2% Cr 1.5% Ni 0.15% Mo steel: Hardened & tempered **DPN: 280 UTS: 920 Elon: 18% Proof: 670**
BS S119	0.4% C 1.2% Cr 1.5% Ni 0.3% Mo steel: Hardened & tempered **DPN: 320 UTS: 1070 Elon: 16% Proof: 820**
BS T50	0.5% C (max) 1.0% Cr 0.2% Mo steel: Tube; hardened & tempered **UTS: 760 Proof: 700**

Symbol	Nominal analysis, supplier, condition and remarks.
BUSTER ALLOY	0.52% C 1.35% Cr 2.25% W 0.25% V steel: Columbia; AISI type S1
B & W – WL 1	0.4% C 0.8% Cr 1.5% Ni 0.2% Mo steel: Tube; Babcock & Wilcox (USA)
C 5	5.0% Cr 1.5% Mo steel: For tools; hot forging; Turton B100; can be used cast
C 108	0.45% C 3.0% Cr 1.5% Mo steel: Fagersta **DPN: 620**
C 320	0.45% C 3.1% Cr 0.4% Mo 0.1% V steel: Fagersta **0.39% C 1.0% Si 5.3% Cr 1.4% Mo 1.0% V steel:** Fagersta **DPN: 570**
C 345	0.42% C 1.1% Cr 0.2% Mo steel: Fagersta
C 424	0.5% C 1.0% Cr 0.15% V steel: Fagersta
CARPENTER 5/317	0.5% C 1.0% Cr 1.75% Ni steel: Carpenter
CARPENTER 481	0.55% C 1.9% Si 0.8% Mn 0.25% Cr 0.4% Mo steel: For tools; Carpenter for AISI S5
CARPENTER 883	0.37% C 1.0% Si 5.25% Cr 1.0% V 1.3% Mo steel: For tools; Carpenter for AISI H13
CBVO/h	0.5% C 1.05% Cr 0.2% Mo steel: Krupp
CBVO/z	0.42% C 1.05% Cr 0.2% Mo steel: Krupp
CCS	0.4% C 1.15% Si 4.25% W 5.25% Cr steel: For tools; origin unknown
CCV	0.4% C 1.25% Cr 0.15% V steel: Origin unknown
CCW	0.52% C 1.2% Cr 2.3% W steel: Firth Brown; chisels, punches
CDS	0.4% C 1% Si 5% Cr 1.35% Mo 0.7% V steel: T Turton
CDS 2	0.4% C 1% Si 5% Cr 1.35% Mo 1.1% V steel: T Turton
CDV	0.39% C 0.5% Cr 1.0% V 1.5% Mo steel: For tools
CEC SMOOTHCUT	0.58% C 0.3% Cr 0.25% V steel: Free cutting; Columbia; AISI type S4S
CHD 1	0.38% C 1% Si 5% Cr 1.3% Mo 1% V steel: Huntsman; for nitrided dies for Al & Mg casting **DPN: 550**
CHD 2	0.4% C 1% Si 5% Cr 0.3% Mo 5% W 0.3% V steel: Huntsman; dies for Al & brass casting **DPN: 540**
CHIMO	0.55% C 2.0% Si 0.2% Cr 0.5% Mo 0.2% V steel: Firth Sterling
CHISEL STEEL	Low alloy W Cr steel: Origin unknown
CHROMETOUGH	0.55% C 0.5% Cr 0.2% V steel: Clyde Alloy **DPN: 600**
CHROME TUNGSTEN	0.55% C 7.4% W 7.5% Cr 0.16% Ni steel: Latrobe; for hot working
CHW	0.5% C 2.8% Cr 15.1% W 0.5% V steel: Latrobe; casting & forging dies for brass; type H24 **DPN: 570**
CL 40X	0.55% C 0.3% Cr 0.4% Mo 0.2% V steel: Origin unknown
CL 99	0.4% C 1.2% Cr 4.2% Ni 0.25% Mo steel: Origin unknown
CL 224	0.55% C 0.75% Cr 1.5% Ni 0.3% Mo steel: Origin unknown
CL 225	0.5% C 0.5% Cr 1.25% Ni 0.25% Mo steel: Origin unknown
CLARITE EW 50	0.53% C 4.0% Cr 18% W 1.0% V steel: Columbia; AISI type H26
CLOSELOY	Medium C Ni Cr Mo steel: Armstrong Whitworth; range of steels with varying C
CMN	0.52% C 0.6% Cr 1.5% Ni 0.22% Mo steel: Marsh Brothers; forging dies
CMVM	0.5% C 0.9% Cr 0.2% V steel: Huntsman; chisels, screwdrivers; dies; etc. **DPN: 500**
CODE 628	0.45% C 0.8% Cr 0.3% Mo steel: Casting; Edgar Allen; Cr Mo 'P' steel

Symbol	Nominal analysis, supplier, condition and remarks.
CODE 638	0.48% C 1.7% Si 0.6% Cr 4.2% Ni 0.25% Mo steel: Casting; Edgar Allen; 'YCW STEM'
CODE 639	0.5% C 0.8% Cr 0.3% Mo steel: Casting; Edgar Allen; Cr Mo/F Mo steel
COLLET STEEL	0.6% C 0.8% Cr 0.4% Ni 0.2% Mo steel: Crucible Steel Co **DPN: 570**
COLMONOY 155	0.5% C 0.1% Si 1.5% Mn 2.5% Cr 0.4% Mo iron alloy: Bronze metal; Wall Colomony; melting point 1450 °C
COMMANDO	0.5% C 1.4% Cr 2.1% W 0.2% V steel: Simonds; AISI type S1
CONQUEROR LC	0.45% C 3.0% Cr 15.0% W steel: For tools; origin unknown
CONSUMET	0.5% C 4.0% Cr 0.1% Ni 1.0% V 4.5% Mo steel: Carpenter for bearings
CP 2	0.53% C 1.4% Cr 2% W 0.2% V steel: For tools; S Osborn; shock & Fatigue resistant tools; obsolete **DPN: 570**
CP 40	0.4% C 1.6% Cr 0.7% Ni 0.5% Mo steel: ESC **UTS: 1070 Elon: 15% Proof: 880**
CR 43	0.43% C 1% Cr 0.4% Ni 0.15% Mo 0.3% Cu steel: Casting; Wolsingham **DPN: 200 UTS: 610 Elon: 12%**
CR 50	0.5% C 1% Cr 0.4% Ni 0.15% Mo 0.3% Cu steel: Casting; Wolsingham **DPN: 207 UTS: 700 Elon: 10%**
CR 83	0.4% C 0.8% Cr 1.3% Ni steel: Bofors; obsolete **DPN: 260 UTS: 840 Elon: 15% Proof: 580**
CR 501	0.5% C 1% Cr 0.4% Ni 0.15% Mo 0.3% Cu steel: Casting; Wolsingham **DPN: 293**
CRD 23	0.4% C 1.2% Cr 1.5% Ni 1.2% Cr steel: Origin unknown
CRD 24	0.4% C 1.2% Cr 1.5% Ni 0.3% Mo steel: C Denton; shear blades; obsolete **DPN: 550**
CRD NICHROMA	0.45% C 1% Si 0.65% Cr 3% Ni steel: C Denton; shear blades; obsolete **DPN: 550**
CRD NTCS	0.5% C 1% Cr 0.25% V steel: C Denton; shear blades; obsolete **DPN: 550**
CRM 1	0.4% C 1.2% Cr 0.3% Mo steel: Firth Brown for BS alloy En 19 & 19A
CRM 2	0.4% C 1.3% Cr 0.8% Mo steel: Firth Brown for BS alloy En 20B
CrMo60	0.6% C 1.1% Cr 0.4% Ni 0.3% Mo 0.3% Cu steel: Casting; Wolsingham **DPN: 341**
CRO 684 BOVAC	0.3% C 1.7% Cr 0.6% Ni 0.4% Mo steel: Bofors; vacuum de-gassed for plastic moulds; hardened & tempered **DPN: 300 UTS: 900 Elon: 18% Proof: 750**
Cr Mo/F Mo	0.5% C 0.8% Cr 0.3% Mo steel: Casting; Edgar Allen; Code 639 steel
Cr Mo 'P'	0.45% C 0.8% Cr 0.3% Mo steel: Casting; Edgar Allen; Code 628 steel

Note. The following abbreviations and units are used in the tables:

DPN	Hardness, diamond pyramid number
UTS	Ultimate tensile strength, N/mm^2
Elon	Elongation, %
Proof	0.1 % proof strength, N/mm^2

1 N/mm^2=0.1 hbar=0.102 kgf/mm^2=0.06475 tonf/in^2=145.04 lbf/in^2

Symbol	Nominal analysis, supplier, condition and remarks.
CROMOL	0.4% C 1.1% Cr 0.3% Mo steel: ESC for BS alloy En 19
CROMAX Cr Mo F Mo	0.53% C 0.8% Cr 0.3% Mo steel: Casting; Edgar Allen for wear resistance **DPN: 250**
CROWN	0.4% C 1.3% Cr 2.3% W 0.15% V steel: Jessop **DPN: 620**
CRV	0.5% C 1% Cr 0.2% V spring steel: ESC for BS alloy En 47
CTC 1	0.45% C 1.25% Cr 2% W steel: Marsh Brothers; chisels, caulking irons, etc.
CTH	0.43% C 0.25% Cr 0.5% W steel: Marsh Brothers; hot forging dies, etc.
CV 4	0.45% C 2.3% Cr 0.3% V steel: Origin unknown
CV 40h	0.42% C 1.5% Cr 0.1% V steel: Krupp
cV 50	0.5% C 1.05% Cr 0.1% V steel: Krupp
CV 58	0.58% C 1.05% Cr 0.1% V steel: Krupp
CVHD	0.6% C 0.7% Cr 0.2% V steel: Huntsman; dies for zinc, casting, extrusion dies, etc. **DPN: 335**
CV Punch & Die	0.45% C 1% Cr 0.25% V steel: T Turton
CVM 3	0.4% C 5% Cr 1.0% Si 1.3% Mo 1.1% V die steel: ESC; forging dies, blanks, etc. **DPN: 550**
CVM 4	0.58% C 5.0% Cr 1.35% Mo 1.1% V steel: Origin unknown
CVM 6	0.38% C 5.0% Cr 1.75% Mo 0.95% V 2.0% W steel: Origin unknown
D 6A	0.46% C 1% Cr 0.5% Ni 1% Mo steel: American proprietary alloy listed in SAE yearbook
D 6AC	0.46% C 1.0% Cr 0.5% Ni 1.0% Mo steel: Origin unknown
D 6AC	0.5% C 1.05% Cr 0.6% Ni 1.01% Mo 0.1% V steel
D 249	0.48% C 1.2% Cr 0.25% Mo 2.2% W 0.15% V Fagersta **DPN: 550**
D 421	0.45% C 1.3% Cr 2.3% W 0.3% V steel: Spencer
D 444	0.4% C 4.2% Cr 0.5% Ni 4.2% Mo 2.2% V 4.2% C steel: Fagersta RC51
DAMASCUS	0.55% C 2.0% Si 0.25% Cr 0.3% V steel: Latrobe
DART	0.41% C 3.3% Cr 0.35% V 2.4% Mo: Latrobe; for hot working
DEF 13B5B	Alloy steel: Hardened & tempered; covers En 19 and includes En 16, 18, 22, 100, etc. **UTS: 840 Elon: 18% Proof: 630**
DEF 13B5C	Alloy steel: Hardened & tempered; covers En 19 and includes En 16, 100, etc. **UTS: 920 Elon: 17% Proof: 710**
DEF 13B5D	Alloy steel: Hardened & tempered; covers En 19 and includes En 16, 17, 24, etc. **UTS: 1000 Elon: 16% Proof: 760**
DEF 13B5E	Alloy steel: Hardened & tempered; covers En 24, 110 and includes En 19, etc. **UTS: 1070 Elon: 14% Proof: 840**
DEF 13B5F	Alloy steel: Hardened & tempered; covers En 24, 25, etc. **UTS: 1140 Elon: 14% Proof: 900**
DEF 13B5G	Alloy steel: Hardened & tempered; covers En 24, 26, etc. **UTS: 121 Elon: 14% Proof: 99**
DIN 50CR V4	0.5% C 0.85% Mn 1.05% Cr 0.15% V steel: For springs; German Standard
DIN 1654/ 42 Cr Mo 4	0.42% C 1.1% Cr 0.6% Ni 0.2% Mo steel: Forging
DIN 1654/42 Cr V 6	0.42% C 1.5% Cr 0.1% V steel: For bolts
DIN 17200/ 36 Cr Ni Mo 4	0.36% C 0.65% Mn 1.05% Cr 1.05% Ni 0.22% Mo steel: German Standard
DIN 17200/ 42 Cr Mo 4	0.42% C 1.1% Cr 0.6% Ni 0.2% Mo steel: Forging

Symbol	Nominal analysis, supplier, condition and remarks.
DIN 17200/ 42 Cr Mo 4	0.41% C 0.65% Mn 1.05% Cr 0.22% Mo steel: German Standard
DIN 17200/ 42 Cr Mo S 4	0.42% C 0.65% Mn 1.05% Cr 0.22% Mo steel: German Standard
DIN 17200/ 50 Cr Mo 4	0.5% C 1.1% Cr 0.2% Mo steel: Forging
DIN 17200/ 50 Cr Mo 4	0.5% C 0.65% Mn 1.05% Cr 0.22% Mo steel: German Standard
DIN 17200/ 50 Cr V 4	0.51% C 0.9% Mn 0.95% Cr 0.15% V steel: German Standard
DIN 17221/ 50 Cr V 4	0.5% C 1.1% Cr 0.1% V steel: For springs
DIN 17221/ 51 Cr Mo V 4	0.52% C 0.9% Mn 1.05% Cr 0.2% Mo 0.1% V steel: For springs; German Standard
DIN 17221/ 58 Cr V 4	0.58% C 1.1% Cr 0.1% V steel: For springs: German Standard
DIN 17222/ 50 Cr V 4	0.5% C 1.1% Cr 0.1% V steel: For springs: German Standard
DIN 17222/ 58 Cr V 4	0.58% C 1.1% Cr 0.1% V steel: For springs: German Standard
DIN 17225/ 45 Cr Mo V 6/7	0.45% C 1.4% Cr 0.7% Mo 0.3% V steel: For springs: German Standard
DIN 17225/ 50 Cr V 4	0.5% C 1.1% Cr 0.1% V steel: For springs: German Standard
DIN 17225/ 65 W Mo 34/8	0.65% C 3.7% Cr 0.85% Mo 8.5% W 0.7% V steel: For springs: German Standard
DM 3	0.42% C 1.5% Cr 3.5% W 0.35% V steel: T Turton
DOUBLE FLYGO	High C 18% W steel: For tools; high speed air hardening; Turton Bros
DOUBLE ZEBRA	0.4% C 0.3% Cr 3.5% Ni steel: Sanderson; chisels, drifts, etc. DPN: 610
DRO 86 BOVAC	0.35% C 1.5% Cr 1.5% Ni 0.2% Mo steel: Bofors; vacuum de-gassed; for plastic moulds; hardened & tempered DPN: 300 UTS: 900 Elon: 18% Proof: 750
DRO 1133	0.55% C 0.6% Cr 1.6% Ni 0.3% Mo steel: Bofors; dies DPN: 400
DTD 4 C	0.45% C 1.25% Cr 0.2% V steel: Springs for engine valves UTS: 1550
DTD 87 B	0.4% C 1.6% Cr 0.4% Ni 0.2% Mo 1% Al steel: For bar & forging; suitable for nitriding DPN: 286 UTS: 920 Elon: 17% Proof: 640
DTD 167 A	0.4% C 1.0% Cr 0.2% Mo steel: Tube; hardened & tempered DPN: 200 UTS: 700 Proof: 610
DTD 5012	0.4% C 3.2% Cr 1% Mo 0.2% V steel: Air hardened & tempered DPN: 444 UTS: 1550 Elon: 12% Proof: 1040
DTD 5042	0.42% C 1.25% Cr 1.75% Ni 1% Mo 0.2% V steel: Bar; hardened & tempered DPN: 385 UTS: 1300 Elon: 12% Proof: 1070
DTD 5052	0.42% C 1.25% Cr 1.75% Ni 1% Mo 0.2% V steel: Plate; hardened & tempered DPN: 380 UTS: 1300 Elon: 12% Proof: 1070
DTD 5162	High tensile steel: For bolts; high metallurgical quality UTS: 900
DTD 5172	Low alloy steel: Castings; hardened & tempered DPN: 265 UTS: 920 Elon: 12% Proof: 670
DTD 5192	Ni Cr Mo V steel: Wrought; vacuum remelted UTS: 1900
DTD 5222	5% Cr Mo V steel: Vacuum melted; steel for bolts UTS: 1800
DUNELT 34	0.38% C 1.3% Cr 0.3% Mo steel: Dunford for BS alloy En 19
DUNELT 36	0.42% C 1.2% Cr 0.2% V steel: Dunford for BS alloy En 50
DUNELT 69	0.4% C 0.3% Cr (max) 3.5% Ni steel: Dunford for BS alloy En 22
DUNELT 100	0.4% C 1.2% Cr 1.5% Ni steel: Dunford; hardened & tempered DPN: 280 UTS: 920 Elon: 18%
DUNELT 101	0.4% C 1.2% Cr 1.5% Ni 0.25% Mo steel: Dunford for BS alloy En 24
DUREHETE 900	0.4% C 1.25% Cr 0.7% Mo steel: Samuel Fox; for use up to 500 °C UTS: 1010 Elon: 22% Proof: 930
DUREHETE 950	0.4% C 1.25% Cr 0.7% Ni 0.25% V steel: Samuel Fox; for use up to 500 °C UTS: 980 Elon: 25% Proof: 920
DUX 4	0.5% C 1.1% Cr 2.25% W steel: ESC; shear blades, punches, etc. DPN: 600
DYCAST 1	0.4% C 5.4% Cr 0.8% Mo 0.5% V steel: Latrobe; hot punches, etc.; type H 11 DPN: 430
DYNAFLEX VAC-ARC	0.4% C 0.9% Si 5.0% Cr 1.3% Mo 0.5% V steel: Latrobe steel; hardened & tempered UTS: 2000 Elon: 8% Proof: 1550
E	0.65% C 3.75% Cr 14% W 0.5% V steel: S osborn; obsolete DPN: 790
E 50	0.5% C 3% Cr 14% W 0.5% V tool steel: For hot working; S Osborn; for dies, inserts, covers, etc.; obsolete DPN: 550
E 4340	0.40% C 0.8% Cr 1.8% Ni 0.25% Mo steel: Designation used by AISI
ECHO	0.65% C 4% Cr 14.5% W 0.6% V steel: John Vessey
ECONO	0.4% C 3.75% Cr 5.7% Mo 1.0% W 0.7% V steel: Braeburn
ELECTERN	0.55% C 0.75% Cr 1.5% Ni 0.28% Mo steel: Origin unknown
ELECTRITE 5	0.6% C 4.1% Cr 18% W 1.1% V steel: Latrobe; hot work; type H26
ELECTRITE 7	0.68% C 4.2% Cr 5.0% Mo 6.4% W 2.0% V steel: Latrobe; hot work; type H42
ERA 171	0.4% C 3% Cr 0.4% Ni 1% Mo 0.2% V nitriding steel: Hadfields for BS alloy En 40C
ETAD	0.4% C 2.5% Cr 4.6% Ni 0.5% Mo steel: Pompey
ETAD 2	0.35% C 1.3% Cr 4.0% Ni 0.5% Mo steel: Pompey
F 1	0.44% C 1.25% Cr 0.2% V steel: Jessop
F 2	0.6% C 0.7% Cr 0.2% V steel: Jessop DPN: 340
F 5	0.44% C 1.3% Cr 2.3% W 0.12% V steel: Jessop
F 7	0.5% C 1.5% Cr 0.7% Ni 1.2% W 0.12% V steel: Jessop DPN: 620
F 10	0.55% C 0.35% Cr 0.3% Mo 0.2% V steel: Shock resistant; Jessop DPN: 640
FAGERSTA D234	1.1% C 0.4% 0.6% Cr 1.9% W steel: For saw blades; Fagersta
FAGERSTA K825	0.36% C 0.7% Mn 1.4% Cr 1.4% Ni 0.2% Mo steel: Fagersta for valves
FALCON 4	0.45% C 1.5% Cr 0.25% V steel: Atlas; as AISI type S1
FALCON 6	0.55% C 1.5% Cr 2.0% W 0.25% V steel: Atlas; as AISI type S1
F C V	0.4% C 2.5% Cr 0.2% V steel: Jonas; dies for casting Al DPN: 441

Symbol	Nominal analysis, supplier, condition and remarks.
FIREDIE	0.4% C 5.0% Cr 1.4% Mo 0.5% V steel: Columbia; AISI type H11
FIREDIE 13	0.4% C 5.2% Cr 1.4% Mo 1.0% V steel: Columbia; AISI type H13
FIREDIE 13 SMOOTHCUT	0.4% C 5.2% Cr 1.4% Mo 1.0% V steel: Columbia; AISI type H13 S; free cutting
FIREX	0.4% C 1.2% Cr 4.25% Ni tool steel: Darwin; chisels, shear blades, etc. DPN: 620
FKD 4	0.42% C 1.0% Cr 0.25% Mo steel: Pompey UTS: 1250 Elon: 9% Proof: 1040
fmp 328	0.4% C 5.2% Cr 0.4% V 1.35% Mo steel: F Parkin; as tool steel H11
fmp 329	0.4% C 5.25% Cr 1.0% V 1.35% Mo steel: F Parkin; as tool steel H13
fmp 348	0.43% C 1.5% Cr 0.3% Mo 4.0% Ni steel: F Parkin
fmp 399	0.5% C 2.2% W 1.45% r 0.2% V steel: F Parkin; as tool steel S1
fmp 682	0.6% C 4.0% Cr 2.0% V 8.25% Mo steel: F Parkin; as tool steel H43
fmp 1850	0.55% C 18% W 4.1% Cr 0.7% V 1% Mo steel: F Parkin; as tool steel H26
FN 35	0.4% C 0.25% Cr 3.5% Ni steel: Firth Brown for BS alloy En 22
FOREMOST 257	Medium C Cr Ni Mo steel: Swift Levick for hot dies DPN: 300
FOREMOST PNEUMATIC	0.45% C 1.7% Cr 2% W 0.25% V steel: Swift Levick; punches, shears
FORMDIE SMOOTHCUT	0.51% C 5.2% Cr 1.5% Ni 1.4% Mo 1.05% V steel: Free cutting; Columbia; AISI type A9S
FORMITE 3	0.51% C 3.0% Cr 15% W 0.5% V steel: Columbia; AISI type H24
G 5 Sp1	0.4% C 1.1% Cr 1.5% Ni 0.27% Mo steel: Jessop for BS alloy En 24
G 7	0.5% C 0.6% Cr 1.7% Ni 0.3% Mo steel: Jessop
G 12	0.4% C 0.6% Cr 2.5% Ni 0.55% Mo steel: Jessop for BS alloy En 26
G 14 A	0.42% C 1% Cr 3.7% Ni steel: Jessop
G 14 C	0.4% C 1.5% Cr 3.5% Ni 0.5% Mo steel: Jessop
G 20	0.4% C 1.3% Mn 0.5% Cr 0.75% Ni 0.2% Mo steel: Jessop for BS alloy En 100
G 26	0.4% C 1.1% Cr 1.4% Ni 0.15% Mo steel: Jessop for BS alloy En 110
G 33	0.4% C 0.7% Cr 1.37% Ni 0.22% Mo steel: Jessop
G 105	0.4% C 1.2% Cr 1.5% Ni 0.25% Mo steel: Toledo for BS alloy En 24; also produced with lower Mo to alloy En 110
GEAROL	0.4% C 1.2% Cr 1.5% Ni 0.3% Mo steel: Swift Levick for BS alloy En 24
GJ 2	0.5% C 1% Cr 0.2% V steel: For tools; S Osborn; punches, spanners, tool holders; obsolete DPN: 620
GOST 2052/50 KHFA	0.5% C 0.75% Mn 0.95% Cr 0.15% V steel: Bar; Russian Standard
GOST 2052/50 KHFA	0.5% C 0.75% Mn 0.95% Cr 0.15% V 0.4% Cu (max): Russian Standard
GOST 2052/50 KHGFA	0.51% C 0.9% Mn 1.05% Cr 0.2% V steel: Bar; Russian Standard

Note. The following abbreviations and units are used in the tables:

DPN	Hardness, diamond pyramid number
UTS	Ultimate tensile strength, N/mm^2
Elon	Elongation, %
Proof	0.1 % proof strength, N/mm^2

1 N/mm^2=0.1 hbar=0.102 kgf/mm^2=0.06475 tonf/in^2=145.04 lbf/in^2

Symbol	Nominal analysis, supplier, condition and remarks.
GOST 2052/50 KHGFA	0.52% C 0.9% Mn 1.1% Cr 0.2% V 0.25% Cu (max) steel: For springs; Russian Standard
GOST 2052/60 S2KHFA	0.6% C 0.55% Mn 1.05% Cr 0.15% V steel: Bar; Russian Standard
GOST 2052/65 S2VA	0.65% C 0.85% Mn 0.3% Cr (max) 1.0% W steel: Bar; Russian Standard
GOST 4543/38 KHN3MFA	0.38% C 0.4% Mn 1.35% Cr 3.2% Ni 0.4% Mo 0.15% B steel: Russian Standard
GOST 4543/38 KHN3VA	0.38% C 0.35% Mn 1.0% Cr 3.0% Ni 0.65% W steel: Russian Standard
GOST 4543/38 KHN3VFA	0.38% C 0.4% Mn 1.2% Cr 3.2% Ni 0.65% W 0.15% B steel: Russian Standard
GOST 4543/40 KHFA	0.4% C 0.65% Mn 0.95% Cr 0.15% V steel: Russian Standard
GOST 4543/40 KHN	0.4% C 0.65% Mn 0.6% Cr 1.2% Ni steel: Russian Standard
GOST 4543/40 KHNMA	0.4% C 0.65% Mn 0.75% Cr 1.5% Ni 0.2% Mo steel: Russian Standard
GOST 4543/40 KHNVA	0.4% C 0.65% Mn 0.75% Cr 1.45% Ni 1.0% W steel: Russian Standard
GOST 4543/50 KHN	0.5% C 0.65% Mn 0.6% Cr 1.2% Ni steel: Russian Standard
GOST 5632/4 KH10S2M	0.4% C 9.7% Cr 0.8% Mo steel: Plate; Russian Standard
GOST 5632/4 KH10S2M	0.4% C 9.7% Cr 0.8% Mo steel: Russian Standard
GRANE 2	0.55% C 1.5% Cr 3.0% Ni steel: Uddelholm for type L6
GV	0.5% C 1% Cr 3.0% Ni 0.3% V 0.2% Mo steel: For tools; S Osborn; shears, punches, dies; obsolete DPN: 630
H 10	0.41% C 3.3% Cr 0.35% V 2.4% Mo steel: Designation used by AISI
H 13	0.39% C 5.0% Cr 1.4% Mo 1.0% V steel: For hot work; Osborn
H 14	0.4% C 5.0% Cr 5.0% W tool steel: Designation used by AISI
H 15	0.4% C 5.0% Cr 5.0% Mo steel: Designation used by AISI; obsolete
H 16	0.55% C 7.0% Cr 7.0% W steel: For tools; designation used by AISI
H 24	0.45% C 3.0% Cr 15.0% W steel: For tools; designation used by AISI
H 24	0.4% C 1.2% Cr 0.3% Mo steel: Jessop for BS alloy En 19
H 24	0.52% C 4.1% Cr 14.5% W 0.45% V steel: For hot work; Osborn
H 26	0.5% C 4.0% Cr 18.0% W 1.0% V steel: For tools; designation used by AISI
H 31	0.4% C 1.2% Cr 0.7% Mo steel: Jessop for BS alloy En 20B
H 41	0.65% C 4.0% Cr 8.0% Mo 1.5% W 1.0% V steel: For tools; designation used by AISI
H 42	0.6% C 4.0% Cr 5.0% Mo 6.0% W 2.0% V steel: For tools; designation used by AISI
H 43	0.55% C 4.0% Cr 8.0% Mo 2.0% V steel: For tools; designation used by AISI
H 55	0.55% C 1% Cr 0.5% Mo 0.07% V steel: Jessop
H 56	0.55% C 1% Cr 0.45% Mo 0.05% V with S steel: Jessop
H 600.1	0.4% C 1.5% Mn 3.0% Cr 3.0% Mo steel: Electrode; Metrode; hardenable deposit DPN: 600
H 750	0.6% C 1.5% Mn 5.0% Cr 5.0% Mo steel: Electrode; Metrode; hardenable deposit DPN: 700
HAVOC	0.5% C 0.5% Mo 0.2% V steel: Simonds; AISI type S2

Symbol	Nominal analysis, supplier, condition and remarks.
HCM 3	0.4% C 3.0% Cr 1.0% Mo 0.25% V steel: Firth Brown for BS 970 En 40C
HCRS 2	0.4% C 1.25% Cr 0.6% Mo steel: ESC for BS alloy En 20B
HCRS 5	0.4% C 1.25% Cr 0.7% Mo 0.25% V steel: ESC; hardened & tempered **UTS: 840 Elon: 16% Proof: 720**
HD 2	0.4% C 3.35% Cr 2.4% W 0.25% V steel: Spencer; forging dies and inserts **DPN: 400**
HD 3	W steel: For tools; hot forging; Turton Bros
HD 12	10% W steel: For tools; hot forging; Turton Bros
HDB 3	0.5% C 0.6% Cr 1.5% Ni steel: Huntsman; long run press dies
HDB 5	0.5% C 0.6% Cr 1.6% Ni 0.2% Mo steel: Huntsman; long run press dies
HECLA 70	0.4% C 0.3% Cr 3.2% Ni steel: Hadfields for BS alloy En 22
HECLA 76	0.5% C 1% Cr 0.15% V steel: Hadfields for BS alloy En 47, En 50
HECLA 100	0.4% C 1.2% Mn 0.4% Cr 0.7% Ni 0.2% Mo steel: Hadfields for BS alloy En 100, En 100 A, B, C & D
HECLA 110	0.4% C 1.2% Cr 1.5% Ni 0.15% Mo steel: Hadfields for BS alloy En 19 & 19A
HECLA 135	0.6% C 2.0% Cr 2.0% Ni 0.5% Mo steel: Origin unknown
HECLA 139	0.55% C 0.7% Cr 1.75% Ni 0.25% Mo steel: Origin unknown
HECLA 157	0.4% C 1.2% Cr 0.3% Mo steel: Hadfields for BS alloy En 19 & 19A
HECLA 160C	0.38% C 1.5% Cr 0.5% Mo 0.2% V 3.6% W steel: Origin unknown
HEDEX 8	0.4% C 1.2% Cr 1.5% Ni 0.15% Mo steel: Wire for forging; Kiveton Park for BS alloy En 110
HEDEX 9	0.4% C 1.2% Cr 0.3% Mo steel: Wire for forging; Kiveton Park for BS alloy En 19
HEDEX 10	0.4% C 0.5% Cr 0.55% Ni 0.25% Mo steel: Wire for forging; Kiveton Park; annealed **UTS: 640**
HEDEX 15	0.4% C 0.5% Cr 0.55% Ni 0.2% Mo steel: Wire for forging; Kiveton Park; annealed **UTS: 650**
HEDEX 33	0.45% C 1.2% Cr 0.15% V steel: Wire for forging; Kiveton Park for BS alloy En 50
HEDEX 47	0.4% C 1.2% Cr 1.5% Ni 0.3% Mo steel: Wire for forging; Kiveton Park for BS alloy En 24
HNV 1	0.55% C 1.5% Si 0.8% Cr 0.75% Mo steel: For inlet valves; designation used by SAE
HNV 4	0.45% C 3.3% Si 7% Cr 1% Ni steel: For inlet valves; designation used by SAE
HR 119	0.55% C 1.5% Cr 3.0% Ni steel: Bofors **DPN: 400**
HRO 1243	0.55% C 1.0% Cr 3.0% Ni 0.3% Mo steel: Bofors; dies **DPN: 440**
HW 7	0.45% C 5.0% Cr 1.0% Mo 3.7% W 0.3% V 0.5% Co steel: Atlas; for hot working **DPN: 400**
HWD 2	0.38% C 5.25% Cr 1.35% Mo 0.5% V steel: Firth Sterling
HWD 3	0.4% C 5.25% Cr 1.25% Mo 1.05% V steel: Firth Sterling
HWD (Mod)	0.55% C 5.0% Cr 1.5% Mo 1.4% W 0.3% V steel: Firth Sterling
HYBLADE	0.55% C 0.3% Cr 0.4% Mo 0.2% V steel: Origin unknown
HYDRA HD	0.4% C 4% Cr 14% W 0.7% V steel: Hall & Pickles; forging dies; obsolete **DPN: 630**
HYDRA VK	0.4% C 4.75% Cr 1.2% Mo 1% V steel: Hall & Pickles; dies for casting Al; extrusion dies, etc.; obsolete **DPN: 550**
HYKRO V	0.4% C 3% Cr 0.9% Mo 0.2% V steel: ESC; nitriding steel **UTS: 1380 Elon: 10% Proof: 1070**
HY-TEN	0.5% C 1% Mn 0.7% Cr 0.2% Mo 0.08% S steel: Wheelock Lovejoy & Co; for machinability
IDEOR	0.4% C 1.0% Si 1% Cr 2% W 0.15% V tool steel: Shock resisting; Darwin; chisels, shear blades, etc. **DPN: 650**
IMPAX	0.36% C 1.4% Cr 0.2% Mo 1.4% Ni steel: Uddelholm
IVORESCO WM EXTRA	0.38% C 1.05% Si 2.1% Cr 4.25% Mo 0.32% V steel: Schoeller-Bleckman **UTS: 1550**
J 27	0.42% C 3.2% Cr 13% W 0.25% V steel: Jessop
JAYSEE	0.45% C 0.5% Cr 0.5% W steel: Jonas; chisels, hammers, etc. **DPN: 530**
JG 3	0.5% C 1% Mn 1% Cr 0.3% Mo steel: For tools; S Osborn; chisels; obsolete
JIS G4106 SCM4	0.4% C 0.75% Mn 0.95% Cr 0.22% Mo steel: Japanese Standard
JIS G4106 SCM5	0.45% C 0.72% Mn 1.05% Cr 0.22% Mo steel: Japanese Standard
JIS G4106 SNC1	0.36% C 0.65% Mn 0.7% Cr 1.25% Ni steel: Japanese Standard
JIS G4106 SNC3	0.36% C 0.5% Mn 0.8% Cr 3.25% Ni steel: Japanese Standard
JIS G4106 SNCM1	0.45% C 0.6% Mn (max) 1.5% Cr 0.3% Ni (max) 0.22% Mo 1.0% Al steel: Japanese Standard
JIS G4106 SNCM6	0.4% C 0.85% Mn 0.52% Cr 0.55% Ni 0.22% Mo steel: Japanese Standard
JIS G4106 SNCM7	0.45% C 0.85% Mn 0.52% Cr 0.55% Ni 0.22% Mo steel: Japanese Standard
JIS G4106 SNCM8	0.4% C 0.75% Mn 0.8% Cr 1.8% Ni 0.22% Mo steel: Japanese Standard
JIS G4801 SUP10	0.5% C 0.8% Mn 0.95% Cr 0.2% V steel: For springs; Japanese Standard
JL	0.4% C 1.0% Si 5% Cr 1.4% Mo 0.4% V tool steel: For tools; hot working; S Osborn; hot extrusion dies, etc.; obsolete **DPN: 460**
JS	0.5% C 1.4% Cr 2.5% W 0.25% V steel: Firth Sterling
JS 5	0.42% C 0.4% Ni 0.2% Cr 0.1% Mo 0.4% Cu steel casting: Jopling - for induction hardening **DPN: 200 UTS: 620 Elon: 12% Proof: 300**
JS 13	0.52% C 0.4% Ni 0.2% Cr 0.1% Mo 0.4% Cu steel casting: Jopling **DPN: 230 UTS: 700 Elon: 10% Proof: 330**
JS 26	0.5% C 0.4% Ni 0.2% Cr 0.1% Mo 0.4% Cu steel casting: Jopling **DPN: 200 UTS: 620 Elon: 12% Proof: 300**
K 19	0.4% C 1.5% Cr 0.55% Mo 3.5% W 0.3% V steel: Jessop
KAISALOY No.3	0.35% C 1.5% Mn 0.25% Cr 0.4% Ni 0.1% Mo 0.005% Ti 0.05% V 0.35% Cu steel: Kaiser Steel Co
KD 4	0.4% C 1.0% Cr 0.25% Mo steel: Pompey **UTS: 1250 Elon: 8% Proof: 1070**
KE 241	0.55% C 3% Si 8.5% Cr 2% W steel: Kayser Ellison; valves
KE 275	0.4% C 3% Cr 0.4% Mo 9.5% W 0.5% V steel: Kayser Ellison for hot working copper **DPN: 530**

Symbol	Nominal analysis, supplier, condition and remarks.
KE 355	0.43% C 1.5% Cr 4.2% Ni 0.2% Mo steel: Kayser Ellison; hardened & tempered for plastic dies **DPN: 460 UTS: 1680**
KE 396	0.6% C 1% Cr 1.5% Ni 0.2% Mo steel: Kayser Ellison; for punches, shear blades **DPN: 700**
KE 896	0.5% C 1% Cr 0.2% V steel: Kayser Ellison; hardened & tempered **DPN: 430 UTS: 1500 Elon: 10% Proof: 133**
KE 960	0.5% C 1.5% Cr 2.2% W 0.2% V steel: Kayser Ellison; shear blades, screwdriver bits **DPN: 670**
KE 1029	0.55% C 1.4% Si 6.2% Cr 0.25% W steel: Kayser Ellison; for valves
KEA 227	0.5% C 1.2% Mn 0.65% Cr 0.2% Mo steel: Sanderson Kayser
KF 26	0.4% C 1.2% Cr 1.5% Ni 0.25% Mo steel: Kirkstall for BS alloy En 24
KF 26A	0.4% C 1.2% Cr 1.4% Ni 0.15% Mo steel: Kirkstall for BS alloy En 110
KF 28C	0.4% C 0.6% Cr 2.5% Ni 0.6% Mo steel: Kirkstall for BS alloy En 26
KF 48	0.4% C 1.2% Cr 0.3% Mo steel: Kirkstall for BS alloy En 19
KF 53	0.42% C 1.2% Cr 0.2% Ni 0.7% Mo steel: Kirkstall for BS alloy En 20B
KF 56	0.42% C 1.4% Mn 0.45% Cr 0.7% Ni 0.2% Mo steel: Kirkstall for BS alloy En 100
KHNSV	0.55% C 1.5% Cr 1.0% Ni 0.55% W steel: Russian Standard designation
KLAH	0.55% C 1% Si 1.75% Cr 1.75% W steel: Jonas; chisels, punches, etc. **DPN: 700**
KP 16	0.4% C 0.3% Cr 3.5% Ni steel: Kiveton Park for BS alloy En 22
KP 22	0.4% C 1.2% Cr 1.5% Ni 0.3% Mo steel: Kiveton Park for BS alloy En 24
KP 26	0.5% C 1% Cr 0.15% V steel: Kiveton Park for BS alloy En 47
KP 33	0.4% C 1% Cr 0.3% Mo steel: Kiveton Park for BS alloy En 19
KP 34	0.4% C 1.4% Mn 0.5% Cr 0.8% Ni 0.2% Mo steel: Kiveton Park for BS alloy En 100
L 94	0.6% C 0.6% Cr 1.5% Ni 0.25% Mo steel: Fagersta **DPN: 730**
L 97	0.55% C 1.0% Cr 3.0% Ni 0.3% Mo steel: Fagersta **DPN: 600**
LANARK	0.55% C 1.9% Si 0.16% Cr 0.28% V 1.3% Mo steel: Latrobe; for coldwork
LCKD	0.6% C 0.12% Cr 0.25% Mo steel: Pompey
LESCALLOY 300M	0.42% C 1.8% Ni 0.8% Cr 0.4% Mo 0.07% V steel: Latrobe
LESCALLOY 4340	0.4% C 0.8% Cr 1.8% Ni 0.25% Mo steel: Latrobe
LESCALLOY HP 9-4-30	0.3% C 8.5% Ni 1.0% Cr 1.0% Mo 0.1% V 4.0% Co steel: Latrobe
LESCALLOY	0.45% C 1.1% Cr 0.6% Ni 1.0% Mo 0.12% V

Symbol	Nominal analysis, supplier, condition and remarks.
LADISH D6AC VAC-ARC	steel: Latrobe steel; hardened & tempered **UTS: 1680 Elon: 7% Proof: 1300**
LESCALLOY UT18	0.4% C 3.2% Cr 0.95% Mo 0.2% V steel: Latrobe
LK 1	0.5% C 1.6% Cr 0.2% Mo 1.1% Al steel: For nitriding; Firth Brown
LK 3	0.4% C 1.5% Cr 0.2% Mo 1.1% Al steel: For nitriding; Firth Brown for En 41B
LNCM	0.4% C 1.2% Cr 1.3% Ni 0.15% Mo steel: ESC for BS alloy En 110
LOMINIUM	0.47% C 1.75% Cr 0.2% V steel: Edgar Allen for Zn die casting moulds **DPN: 600**
LOW CARBON TATMO	0.6% C 1.7% W 3.75% Cr 1.0% V 8.7% Mo steel: Latrobe; for hot working
LOW CARBON TNW	0.6% C 4.0% Cr 2.0% V 8.0% Mo steel: Latrobe; for hot working
LS	0.5% C 3.2% Cr 1.7% Mo steel: Braeburn
LT 8	0.45% C 3.0% Ni 0.5% Cr steel: Low Moor
LT 9	0.5% C 1.0% Mn 1% Cr 0.3% Mo steel: Low Moor **DPN: 600**
LT 31	0.5% C 1.0% Si 1.5% Cr 2% W 0.2% V steel: Low Moor; for hot work tools; obsolete **DPN: 790**
LT 41	0.4% C 1.0% Si 4.7% Cr 1.4% Mo 0.4% V steel: Low Moor; for hot work tools; obsolete **DPN: 630**
LT 42	0.4% C 1.0% Si 5% Cr 0.4% Mo 4% W steel: Low Moor; for hot work & tools; obsolete **DPN: 550**
LT 47	0.5% C 3% Cr 14% W 0.5% V steel: Low Moor; for hot work tools; obsolete **DPN: 670**
LT 51	0.4% C 1.2% Cr 1.7% Ni 0.2% Mo steel: Low Moor; die steel; obsolete **DPN: 550**
LUNDIE	0.4% C 4.75% W 5.25% Cr steel: Latrobe; for hot working
LV 52 Mo	0.5% C 1.5% Si 7.5% Cr 0.5% Mo steel: Low Moor; valve steel; obsolete **DPN: 320**
M Mo 38	0.38% C 1.5% Mn 0.25% Cr 0.4% Ni 0.3% Mo 0.3% Cu steel: Casting; Wolsingham **DPN: 375 UTS: 920 Elon: 12% Proof: 580**
MALLOY	0.6% C 1.1% Cr 0.25% Mo steel: Origin unknown
MANOFORT 160	0.3% C 2% Ni 1.0% Cr 0.4% Mo steel: Pompey **UTS: 1600 Elon: 8% Proof: 1250**
MANOFORT 180	0.4% C 3.5% Ni 1.4% Cr 0.4% Mo steel: Pompey **UTS: 1700 Elon: 5% Proof: 1400**
MANOIR ABRADUR 240	0.5% C 1.0% Mn 1.0% Cr steel: Pompey **UTS: 1200 Elon: 5% Proof: 100**
MANOIR ABRADUR 300	0.6% C 1.6% Si 0.8% Cr steel: Pompey **UTS: 1500 Elon: 3% Proof: 1300**
MANOIR ABRADUR 400	0.3% C 1.8% Ni 1.0% Cr 35% Mo steel: Pompey **UTS: 1500 Elon: 5% Proof: 1150**
MANOIR ABRADUR 500	0.4% C 3.5% Ni 1.4% Cr 0.3% Mo steel: Pompey **UTS: 1600 Elon: 5% Proof: 1300**
MANOIR ABRADUR PM38	0.5% C 1.5% Mn 1.5% Cr + Mo + V + Ni steel: Pompey **UTS: 1650 Elon: 4% Proof: 1300**
MANOIR RS1	0.45% C 1.8% Si 0.7% Mn steel: Pompey **UTS: 1320 Elon: 4% Proof: 1150**
MARSH CMN	0.52% C 0.6% Cr 1.5% Ni 0.22% Mo steel: Marsh Brothers; forging dies
MARSH CTC 1	0.45% C 1.25% Cr 2% W steel: Marsh Brothers; chisels, caulking irons, etc.
MARSH CTH	0.43% C 0.25% Cr 0.5% W steel: Marsh Brothers; hot forging dies, etc.
MARSH NC	0.45% C 1.1% Cr 3.7% Ni steel: Marsh Brothers; chisels

Note. The following abbreviations and units are used in the tables:

DPN	Hardness, diamond pyramid number
UTS	Ultimate tensile strength, N/mm²
Elon	Elongation, %
Proof	0.1 % proof strength, N/mm²

1 N/mm²=0.1 hbar=0.102 kgf/mm²=0.06475 tonf/in²=145.04 lbf/in²

Symbol	Nominal analysis, supplier, condition and remarks.
MARSH NC 1	0.4% C 1% Cr 1.5% Ni steel: Marsh Brothers; collets
MAS	0.52% C 1.0% Cr 2.0% Ni 0.4% Mo 0.2% V steel: Pompey
MAX CHIP	0.42% C 1.7% Si 1.05% Cr 4.1% Ni steel: Edgar Allen
MAXEL 3½	0.5% C 1.25% Mn 0.65% Cr 0.18% Mo 0.08% S steel: Free machining; Crucible Steel Co **DPN: 250 UTS: 880 Elon: 22% Proof: 370**
MC	0.4% C 0.6% Cr 1% Ni steel: Jonas **UTS: 1000 Elon: 17% Proof: 840**
MCH	0.5% C 3.7% Cr 6.2% Mo 1.0% W 0.75% V steel: Latrobe; hot work **DPN: 550**
MCV	0.5% C 1.2% Mn 0.6% Cr 0.5% Ni 0.25% Mo steel: Sanderson; has additions of B, Ti, Zr & V **DPN: 255 UTS: 800 Elon: 23% Proof: 770**
MGR	0.55% C 5.0% Cr 1.2% Mo 1.2% W steel: Latrobe; punch & dies; type A8 **DPN: 630**
MINERVA HC	0.53% C 1.8% Cr 1.9% W 0.2% V steel: Edgar Allen; punches, shear blades
MINERVA LC	0.43% C 1.8% Cr 1.9% W 0.2% V steel: Edgar Allen; chisels, punches, shear blades
MNC 851	General standard for SIS range of case hardening steels; Swedish Standard
MOHD	0.36% C 3.5% Cr 6.0% Mo 1.0% W 0.75% V steel: For tools; Huntsman; extrusion dies **DPN: 600**
MOHICAN 6	0.62% C 3.7% Cr 8.7% Mo 1.7% W 1.0% V steel: Atlas; as AISI type II51
MOLITE HW60	0.63% C 4.0% Cr 8.2% Mo 1.9% V steel: Columbia; AISI type H43
MONARCH BLA	0.4% C 5.0% Cr 1.4% Mo 1.0% V steel: For tools; origin unknown
MONARCH NCV	0.4% C 1.5% Cr 4.2% Ni 0.2% Mo steel: Origin unknown
MONARK 2	0.6% C 0.3% Cr 0.2% Mo steel: Atlas; as AISI S5
MUSHET	0.65% C 3.7% Cr 0.5% V 14.0% W steel: For tools; origin unknown
N 4	0.42% C 1.2% Cr 1.7% Ni 1% Mo 0.2% V steel: Jessop
NA	0.6% C 2% Si 0.3% Mo 0.2% V steel: For tools; S Osborn; punches, picks, shear blades; obsolete **DPN: 620**
NARVE	0.65% C 1.1% Cr 0.57% Mo steel: Uddelholm
NAT	0.32% C 1.0% Si 5.0% Cr 0.4% V 1.4% Mo tool steel: Balfour Darwin
NC	0.45% C 1.1% Cr 3.7% Ni steel: Marsh Brothers; chisels
NC 1	0.4% C 1% Cr 1.5% Ni steel: Marsh Brothers; collets, etc.
NCM 1	0.4% C 1.1% Cr 1.6% Ni 0.3% Mo steel: Firth Brown for BS alloy En 24
NCM 4	0.4% C 0.65% Cr 2.4% Ni 0.45% Mo steel: Firth Brown for BS alloy En 26
NC MO	0.38% C 1.2% Cr 1.4% Ni 0.15% Mo steel: Firth Brown for BS alloy En 110
NCMS	0.6% C 0.6% Cr 0.6% Ni 0.15% Mo steel: For springs; ESC
NCMV	0.42% C 1.2% Cr 1.75% Ni 1% Mo 0.2% V steel: ESC; nitriding, fatigue resistant **UTS: 1840 Elon: 8% Proof: 1550**
NF	0.38% C 1.6% Ni 1.2% Cr 0.25% Mo steel: S Osborn; for moulds; hardened & tempered **DPN: 500**
NF A35 551/42CD4	0.42% C 0.75% Mn 0.95% Cr 0.22% Mo steel: French Standard

Symbol	Nominal analysis, supplier, condition and remarks.
NF A35 551/50CV4	0.5% C 0.85% Mn 1.0% Cr 0.15% V steel: French Standard
NF A35 571/45SCD6	0.46% C 0.65% Mn 0.62% Cr 0.22% Mo steel: Bar; French Standard
NF A35 571/50CV4	0.5% C 0.85% Mn 1.0% Cr 0.15% V steel: For springs; French Standard
Ni Cr Si YCW STEM	0.48% C 1.7% Si 0.6% Cr 4.2% Ni 0.25% Mo steel: Casting; Edgar Allen; code 638 steel
NIT 135	0.41% C 1.55% Cr 0.35% Mo 1.1% Al steel
NITRAL	0.42% C 1.65% Cr 0.3% Mo steel: Pompey
NITRALLOY 3	0.4% C 1.5% Cr 0.2% Mo 1.1% Al steel: ESC for BS alloy En 41 B nitriding steel
NITRALLOY LK1	0.5% C 1.6% Cr 0.2% Mo 1.1% Al steel: For nitriding; Firth Brown
NITRALLOY LK3	0.4% C 1.6% Cr 0.2% Mo 1.1% Al steel: For nitriding; Firth Brown
NITRALLOY LK5	0.3% C 1.6% Cr 0.2% Mo 1.1% Al steel: For nitriding; Firth Brown
NITRALLOY N	0.25% C 1.25% Cr 1.1% Al 0.25% Mo 3.5% Ni steel: Latrobe; nitriding steel
NITRALLOY N135 Mod	0.4% C 1.6% Cr 1.1% Al 0.35% Mo steel: Latrobe; for nitriding
NITRIDING STEEL	0.4% C 1.6% Cr 0.35% Mo 1.1% Al steel: For nitriding; Crucible Steel Co **DPN: 320 UTS: 1140 Elon: 17% Proof: 1000**
NMCM	0.4% C 1.3% Mn 0.4% Cr 0.75% Ni 0.2% Mo steel: Firth Brown for BS alloy En 100
No.1 MONARCH	0.65% C 4.0% Cr 0.6% V 14.0% W steel: For tools; origin unknown
NO 481	0.55% C 0.25% Cr 0.4% Mo steel: Carpenter; for AISI type S5
NORESCO PARFORCE SPECIAL 2	0.4% C 1.4% Cr 4.25% Ni 1% W steel: Schoeller-Bleckmann; for large plastic dies **UTS: 1000**
NORESCO SPK	0.45% C 2.35% Cr 0.3% V steel: Schoeller-Bleckmann **UTS: 1210**
NORESCO TYRANT EXTRA	0.45% C 1% Si 1.05% Cr 2% W steel: Schoeller-Bleckmann **DPN: 550**
NORESCO WAW	0.45% C 1.5% Cr 0.5% Mo 0.5% W 0.9% V steel: Schoeller-Bleckmann **UTS: 1380**
NTC	0.4% C 0.4% Cr 0.5% W steel: For tools; Huntsman; punches, chisels, etc. **DPN: 550**
NUSHANK	0.43% C 0.4% Cr 3.0% Ni 0.25% Mo steel: Tube; Atlas; for detachable rock drill bits
NV 3	0.5% C 0.4% Cr 0.3% Ni 0.15% Mo steel: For inlet valves; designation used by SAE
NV 4	0.4% C 0.6% Cr 1.2% Ni steel: For inlet valves; designation used by SAE
NV 5	0.45% C 0.5% Cr 0.5% Ni 0.2% Mo steel: For inlet valves; designation used by SAE
NYBLADE	0.48% C 1.1% Si 0.65% Cr 3% Ni steel: Sanderson; chisels, picks, shear blades, etc. **DPN: 600**
NZ 2	0.55% C 0.7% Cr 1.5% Ni 0.2% Mo steel: For hot work tools; S Osborn; forging dies, die holders, etc.; obsolete **DPN: 302**
OCM 10	0.4% C 1.2% Cr 0.3% Mo steel: W Marrison for BS alloy En 19
OK	0.4% C 1.3% Cr 2.3% W 0.15% V steel: Jessop **DPN: 620**
OK CROWN	0.5% C 1.5% Cr 0.4% Mo 0.2% V 2.0% W steel: Origin unknown
ON 2	0.4% C 0.5% Cr 0.8% Ni 0.2% Mo steel: W Marrison for BS alloy En 100

Symbol	Nominal analysis, supplier, condition and remarks.
ON 4	0.4% C 0.3% Cr 3.5% Ni steel: W Marrison for BS alloy En 22
ON 6	0.4% C 1.2% Cr 1.6% Ni 0.3% Mo steel: W Marrison for BS alloy En 24
ON 8	0.4% C 0.7% Cr 2.6% Ni 0.6% Mo steel: W Marrison for BS alloy En 26
ON 11	0.4% C 1.2% Cr 1.4% Ni 0.2% Mo steel: W Marrison for BS alloy En 110
OOC	0.6% C 1.1% Cr 0.3% V 1.9% W steel: For tools; origin unknown
ORLEANS	0.55% C 0.25% Cr 0.35% Mo 0.2% V steel: Simonds; AISI type S5
ORVAR 2	0.37% C 5.3% Cr 1.4% Mo 1.0% V steel: Uddelholm; for type H13
P 553	0.4% C 1.1% Cr 1.5% Ni 0.3% Mo steel: Carrs; crankshafts, gears, etc. **DPN: 370 UTS: 920**
P 602	0.4% C 1.1% Cr 0.3% Mo steel: Carrs; pressure dies for Zn Pb & Sn base castings **DPN: 460**
P 614	0.4% C 1.6% Cr 0.2% Mo 1% Al steel: For nitriding; Carrs **UTS: 760**
P 618	0.4% C 3% Cr 0.3% Ni 1% Mo 0.25% V steel: For nitriding; Carrs **UTS: 1380**
PARFORCE SPECIAL 2	0.4% C 1.4% Cr 4.25% Ni 1% W steel: Schoeller-Bleckmann; for large plastic dies **UTS: 1000**
PAX 2	0.5% C 1.5% Cr 2.2% W 0.2% V steel: Sanderson; shear blades, punches **DPN: 700**
PAX NON-BREAK	0.4% C 1.5% Cr 2.2% W 0.2% V steel: Sanderson; chisels, dies, punches, etc. **DPN: 570**
PCSK	0.52% C 1.5% Cr .04% Mo 0.2% V 2.0% W steel: Origin unknown
PENUMO	0.45% C 1.25% Cr 2.0% W steel: For tools; origin unknown
PLM B/1	0.55% C 0.75% Cr 1.6% Ni 0.3% Mo steel: ESC; plastic moulds; short runs **UTS: 950**
PLM B/2	0.42% C 0.6% Cr 2.6% Ni 0.6% Mo steel: ESC plastic moulds; short runs **UTS: 1100**
PLM C/1	0.42% C 0.6% C 2.6% Ni 0.6% Mo steel: ESC; plastic moulds **DPN: 500**
PLM C/2	0.4% C 1.4% Cr 4.2% Ni 0.25% Mo steel: ESC; plastic moulds **DPN: 550**
PN 1	0.5% C 1.6% Si 0.4% Mo 0.23% V steel: ESC; rivet snaps, air hammer tools, etc. **DPN: 630**
PNUSNAP OH	0.43% C 1% Si 1% Cr 1.8% W steel: Firth Brown; shear blades
PNEUTOUGH	0.5% C 1.5% Cr 2.25% W 0.25% V steel: Clyde Alloy **DPN: 570**

Symbol	Nominal analysis, supplier, condition and remarks.
PREGA	0.45% C 3.0% Cr 0.45% Mo steel: Uddelholm
PRESSURDIE	0.5% C 5.0% Cr 2.0% V 1.3% Mo steel: Braeburn
PRESSURDIE/1	0.38% C 5.0% Cr 0.25% Mo 5.0% W 0.2% V steel: Braeburn; AISI type H 14
PRESSURDIE/3	0.39% C 5.5% Cr 1.2% Mo 1.0% V steel: Braeburn; AISI type H 13
PRESSURDIE/3L	0.39% C 5.5% Cr 1.2% Mo 0.4% V steel: Braeburn; AISI type H 11
PRESSURDIE/5	0.38% C 3.5% Cr 1.0% Mo 1.2% W 1.0% V steel: Braeburn
PRESSURDIE/16	0.5% C 5.0% Cr 1.25% Mo 1.25% W 0.2% V steel: Braeburn; AISI type A8
PYRODIE	0.4% C 1% Si 5% Cr 1.3% Mo 1% V steel: Clyde Alloy **DPN: 500**
PYROMET 882	0.4% C 5.0% Cr 1.5% Mo 0.4% V steel: Carpenter for AISI type H 11
PW	0.4% C 1% Si 5% Cr 0.4% Mo 4.0% W steel: For hot working tools; S Osborn; hot dies, punches, mandrels, etc.; obsolete **DPN: 460**
QQ W 405a	Alloy steel: Wire; US Federal Specification; covers a range of steels prefixed FS with AISI code following
QQ W 412	Alloy steel: Wire for springs; US Federal Specification; covers several analyses
RABI	0.3% C 4.1% Ni 1.3% Cr 0.3% Mo steel: Osborn
RED DIAMOND 21	0.4% C 1.1% Cr 0.25% Mo steel: For wear; Redheugh **DPN: 375 UTS: 1300 Elon: 10% Proof: 1100**
RED DIAMOND 22	0.4% C 1.5% Ni 1.25% Cr 0.3% Mo steel: Redheugh **DPN: 325 UTS: 1180 Elon: 15% Proof: 1070**
REGIN 3	0.5% C 1.2% Cr 2.5% W 0.2% V steel: Uddelholm for type S1
RKV	0.5% C 1.0% Cr 0.15% V steel: Pompey
RO 952	0.4% C 1.1% Cr 0.2% Mo steel: Bofors; hardened & tempered **DPN: 290 UTS: 990 Elon: 12% Proof: 700**
RO 8155	0.4% C 3.0% Cr 0.5% Mo steel: For dies; Bofors; annealed **DPN: 230**
ROP 19	0.4% C 5.3% Cr 1.4% Mo 1.0% V steel: Bofors **DPN: 450**
ROP 9653	0.45% C 1.5% Cr 0.7% Mo 0.3% V steel: Bofors **DPN: 500**
RP 1152	0.5% C 1.1% Cr 0.15% V steel: For springs; Bofors; hardened & tempered **DPN: 460 UTS: 1620 Elon: 7% Proof: 1210**
RSKD	0.45% C 0.8% Cr 0.25% Mo steel: Pompey
RTO 712	0.4% C 1.2% Cr 0.3% Mo 2.5% W 0.2% V steel: Bofors; obsolete **DPN: 640**
RTO 912	0.5% C 1.2% Cr 0.3% Mo 2.5% W 0.2% V steel: Bofors **DPN: 500**
S 1	0.52% C 1.4% Cr 2.2% W 0.2% V steel: For cold work; Osborn
S 1	0.5% C 1.5% Cr 0.4% Mo 2% W 0.2% V steel: Designation used by SAE
S 2	0.5% C 1.0% Si 0.5% Mo 0.25% V steel: For tools; designation used by SAE
S3	0.5% C 0.75% Cr 1.0% W steel: Designation used by SAE; obsolete
S 4	0.55% C 0.25% Cr 0.2% V steel: Designation used by AISI
S 5	0.55% C 2.0% Si 0.3% Cr 0.4% Mo 0.2% V steel: For tools; designation used by SAE
S 6	0.45% C 1.4% Mn 2.2% Si 1.5% Cr 0.4% Mo steel: Designation used by SAE

Note. The following abbreviations and units are used in the tables:

DPN	Hardness, diamond pyramid number
UTS	Ultimate tensile strength, N/mm^2
Elon	Elongation, %
Proof	0.1 % proof strength, N/mm^2

$1 \text{ N/mm}^2 = 0.1 \text{ hbar} = 0.102 \text{ kgf/mm}^2 = 0.06475 \text{ tonf/in}^2 = 145.04 \text{ lbf/in}^2$

Symbol	Nominal analysis, supplier, condition and remarks.
S 7	0.5% C 3.2% Cr 1.4% Mo steel: Designation used by SAE
SAE 81 B45	0.45% C 0.45% Cr 0.3% Ni 0.1% Mo 0.0005% B steel
SAE 86 B45	0.45% C 0.5% Cr 0.6% Ni 0.2% Mo 0.0005% B steel
SAE 94 B40	0.40% C 0.4% Cr 0.45% Ni 0.11% Mo steel
SAE 3140	0.4% C 0.6% Cr 1.2% Ni steel
SAE 3145	0.45% C 0.8% Cr 1.2% Ni steel: Obsolete
SAE 3150	0.5% C 0.8% Cr 1.2% Ni steel: Obsolete
SAE 3245	0.45% C 1% Cr 1.7% Ni steel: Obsolete
SAE 3250	0.5% C 1% Cr 1.7% Ni steel: Obsolete
SAE 3340	0.40% C 1.5% Cr 3.5% Ni steel: Obsolete
SAE 3450	0.5% C 0.8% Cr 3% Ni steel: Obsolete
SAE 4140	0.4% C 1% Cr 0.2% Mo steel
SAE 4142	0.42% C 1% Cr 0.2% Mo steel
SAE 4145	0.45% C 1% Cr 0.2% Mo steel
SAE 4147	0.47% C 1% Cr 0.2% Mo steel
SAE 4150	0.5% C 1% Cr 0.2% Mo steel
SAE 4161	0.61% C 0.8% Cr 0.3% Mo steel
SAE 4340	0.4% C 0.8% Cr 1.8% Ni 0.25% Mo steel
SAE 4620	0.4% C 1.8% Ni 0.25% Mo steel: Obsolete
SAE 4640	0.4% C 1.8% Ni 0.25% Mo steel
SAE 6140	0.4% C 1% Cr 0.15% V steel: Obsolete
SAE 6145	0.45% C 1% Cr 0.15% V steel: Obsolete
SAE 6145	0.45% C 1% Cr 0.15% V steel
SAE 6150	0.5% C 1% Cr 0.15% V steel
SAE 7260	0.6% C 0.8% Cr 1.8% W steel: Obsolete
SAE 8640	0.4% C 0.5% Cr 0.6% Ni 0.2% Mo steel
SAE 8641	0.40% C 0.5% Cr 0.6% Ni 0.2% Mo steel (S higher than 8640)
SAE 8642	0.42% C 0.5% Cr 0.6% Ni 0.2% Mo steel
SAE 8645	0.45% C 0.5% Cr 0.6% Ni 0.2% Mo steel
SAE 8647	0.47% C 0.5% Cr 0.6% Ni 0.2% Mo steel: Obsolete
SAE 8650	0.5% C 0.5% Cr 0.6% Ni 0.2% Mo steel
SAE 8653	0.53% C 0.7% Cr 0.6% Ni 0.2% Mo steel
SAE 8655	0.55% C 0.5% Cr 0.6% Ni 0.2% Mo steel
SAE 8740	0.4% C 0.5% Cr 0.6% Ni 0.25% Mo steel
SAE 8742	0.42% C 0.5% Cr 0.6% Ni 0.25% Mo steel
SAE 8745	0.45% C 0.5% Cr 0.6% Ni 0.25% Mo steel: Obsolete
SAE 8750	0.5% C 0.5% Cr 0.6% Ni 0.25% Mo steel
SAE 9440	0.4% C 0.4% Cr 0.4% Ni 0.1% Mo steel: Obsolete
SAE 9442	0.42% C 0.4% Cr 0.4% Ni 0.1% Mo steel: Obsolete
SAE 9445	0.45% C 0.4% Cr 0.4% Ni 0.1% Mo steel: Obsolete
SAE 9447	0.47% C 0.4% Cr 0.5% Ni 0.1% Mo steel: Obsolete
SAE 9747	0.47% C 0.2% Cr 0.5% Ni 0.2% Mo steel: Obsolete
SAE 9763	0.63% C 0.2% Cr 0.5% Ni 0.2% Mo stccl: Obsolete
SAE 9840	0.4% C 0.8% Cr 1% Ni 0.25% Mo steel
SAE 9845	0.45% C 0.3% Cr 1% Ni 0.25% Mo steel: Obsolete
SAE 9850	0.5% C 0.8% Cr 1% Ni 25% Mo steel
SAE 71360	0.6% C 3.5% Cr 14% W steel: Obsolete
SAE 71660	0.6% C 3.5% Cr 16% W steel: Obsolete
SAE E 4340	0.4% C 0.8% Cr 1.75% Ni 0.25% Mo steel
SAE HNV 1	0.55% C 1.5% Si 0.8% Cr 0.75% Mo steel: For inlet valves
SAE HNV 4	0.45% C 3.3% Si 7% Cr 1% Ni steel: For inlet valves
SAE NV 3	0.5% C 0.4% Cr 0.3% Ni 0.15% Mo steel: For inlet valves
SAE NV 4	0.4% C 0.6% Cr 1.2% Ni steel: For inlet valves; as SAE 3140

Symbol	Nominal analysis, supplier, condition and remarks.
SAE NV 5	0.45% C 0.5% Cr 0.5% Ni 0.2% Mo steel: For inlet valves; as SAE 8645
SAE S 1	0.5% C 1.5% Cr 0.4% Mo 2% W 0.2% V steel
SAE S 2	0.5% C 0.5% Mo 0.25% V steel: For tools
SAE S 5	0.55% C 0.3% Cr 0.4% Mo 0.2% V steel: For tools
SAE X 3140	0.4% C 0.8% Cr 1.2% Ni steel: Obsolete
SANBOLD 19	0.4% C 1.2% Cr 0.3% Mo steel: Sanderson for BS alloy En 19
SANBOLD 22	0.4% C 0.3% Cr 3.5% Ni steel: Sanderson for BS alloy En 22
SANBOLD 24	0.4% C 1.2% Cr 1.5% Ni 0.3% Mo steel: Sanderson for BS alloy En 24
SANBOLD 66	0.4% C 0.3% Cr 3.5% Ni steel: Sanderson; replaced by Sanbold 22
SANBOLD 155	0.4% C 1.2% Cr 0.3% Mo steel: Sanderson; replaced by Sanbold 19
SANBOLD CVS	0.5% C 1% Cr 0.15% V steel: Sanderson for BS alloy En 47
SANBOLD NK 2833	0.4% C 1.2% Cr 1.5% Ni 0.3% Mo steel: Sanderson; replaced by Sanbold 24
SBM	0.58% C 1.0% Cr 0.25% V steel: Origin unknown
SEARCHER	0.45% C 1% Cr 2% W steel: Hall & Pickles; chisels, wedges, shear blades, etc.; obsolete **DPN: 620**
SEARCHER A1	0.63% C 0.4% Cr 1.6% W steel: Origin unknown
SHEARTOUGH	0.45% C 1.15% C 2.25% Ni 0.2% V steel: Clyde Alloy **DPN: 550**
SHC 1	0.44% C 1% Si 0.65% Cr 3% Ni steel: Firth Brown; for chisels & punches
SILICO ALLOY	0.58% C 1.95% Si 0.3% Cr 0.5% Mo steel: Columbia; AISI type S5
SILMO	0.55% C 2.0% Si 0.35% Cr 0.3% Mo 0.2% V steel: Jessop; shock resistant
SIS 14-2216 E	0.4% C 0.6% Mn 1.0% Cr 0.5% Mo 0.25% Cu steel: For pressure vessels; Swedish Standard; normalized & annealed **UTS: 500 Elon: 19% Proof: 290**
SIS 14-2230	0.5% C 0.8% Mn 1.0% Cr 0.15% V steel: For springs; Swedish Standard **DPN: 450 UTS: 1400 Elon: 7% Proof: 1150**
SIS 14-2242	0.4% C 0.45% Mn 5.0% Cr 1.4% Mo 1.0% V steel: For tools & springs; hot work; Swedish Standard **DPN: 350 UTS: 140 Elon: 8% Proof: 10**
SIS 14-2244	0.4% C 0.75% Mn 1.0% Cr 0.2% Mo steel: For hardening & tempering; Swedish Standard **DPN: 330 UTS: 1030 Elon: 12% Proof: 750**
SIS 14-2550	0.55% C 0.4% Mn 1.0% Cr 3.0% Ni 0.3% Mo steel: For tools; hot work steel; Swedish Standard **DPN: 260**
SIS 14-2710	0.5% C 0.3% Mn 1.0% Cr 0.25% Mo 2.5% W 0.2% V steel: For tools and cold end; hot work steel; Swedish Standard **DPN: 245**
SIS 2230	0.52% C 0.22% Si 0.9% Mn 1.1% Cr 0.15% V steel: For springs **DPN: 400 UTS: 1300**
SIS 2234	0.33% C 0.65% Mn 0.05% Cr 0.22% Mo steel: Swedish Standard; hardened & tempered **UTS: 1080 Elon: 10% Proof: 880**
SIS 2242	0.38% C 1% Si 5.5% Cr 1.4% Mo 1% V steel: Swedish Standard; hardened & tempered **DPN: 410 UTS: 1380 Elon: 8% Proof: 1070**
SIS 2244	0.42% C 1.1% Cr 0.2% Mo steel: Bar & forging; Swedish Standard; hardened & tempered **DPN: 350 UTS: 1140 Elon: 10% Proof: 880**

Symbol	Nominal analysis, supplier, condition and remarks.
SIS 2244	0.42% C 0.75% Mn 1.05% Cr 0.22% Mo steel: Swedish Standard; hardened & tempered **UTS: 1080 Elon: 10%14880**
SIS 2541	0.35% C 1.4% Cr 1.4% Ni 0.2% Mo steel: Swedish Standard
SIS 2550	0.55% C 1% Cr 3% Ni 0.3% Mo steel: Bar & forging; Swedish Standard; annealed **DPN: 260**
SIS 2710	0.5% C 1.2% Cr 0.25% Mo 2.5% W 0.15% V steel: Swedish Standard; annealed **DPN: 245**
SIS 2940	0.41% C 1.7% Cr 0.25% Mo 1.1% Al steel: For nitriding; Swedish Standard; hardened & tempered **DPN: 270 UTS: 920 Elon: 15% Proof: 630**
SIS 8101	0.4% C 1.5% Cu 1.7% Ni 0.5% Mo steel: Sintered
SIS 8102	0.6% C 1.5% Cu 1.75% Ni 0.5% Mo steel: Sintered
SKC 24	0.38% C 0.65% Mn 0.5% Cr 3.0% Ni 0.22% Mo steel: For hollow drills; Japanese Standard designation
SKD 6	0.37% C 5.2% Cr 1.25% Mo 0.4% V steel: Japanese Standard designation
SKD 61	0.37% C 5.0% Cr 1.25% Mo 1.0% V steel: Japanese Standard designation
SKD 62	0.38% C 5.0% Cr 1.25% Mo 0.4% V 1.25% W steel: Japanese Standard designation
SKF 327A	0.4% C 1.2% Cr 0.3% Mo steel: Hardened & tempered **UTS: 930 Proof: 720**
SKS 4	0.5% C 0.75% Cr 0.75% W steel: Japanese Standard designation
SKS 41	0.4% C 1.25% Cr 3.0% W steel: Japanese Standard designation
SKT 2	0.55% C 1.0% Mn 1.0% Cr 0.2% V steel: Japanese Standard designation
SKT 3	0.55% C 1.05% Cr 0.42% Ni 0.4% Mo 0.2% V (max) steel: Japanese Standard designation
SKT 4	0.55% C 0.8% Mn 0.85% Cr 1.7% Ni 0.35% Mo 0.2% V (max) steel: Japanese Standard designation
SKT 5	0.55% C 0.8% Mn 1.25% Cr 0.35% Mo 0.22% V steel: Japanese Standard designation
SKT 6	0.75% C 0.8% Mn 0.95% Cr 2.75% Ni 0.42% Mo 0.2% V (max) steel: Japanese Standard designation
SL 823	0.4% C 5.0% Cr 1.0% Mo 0.4% V steel: Swift Levick; for hot dies
SL 866	0.4% C 5.0% Cr 1.25% Mo 1.0% W 0.3% V steel: Swift Levick for hot dies
SL 962	0.4% C 1.5% Cr 0.5% Mo 3.5% Mo 3.5% W 0.35% V steel: Swift Levick for hot dies
SL 983	Medium C 2% Cr Mo steel: Swift Levick **DPN: 400**
SM	0.58% C 0.25% Cr 0.35% Mo 0.2% V steel: Forez
SN 3½	0.4% C 0.2% Cr 3.5% Ni steel: Spencer for BS alloy En 22
SNC4G	0.4% C 1.2% Cr 1.5% Ni 0.3% Mo steel: Spencer for BS alloy En 24
SOMDIE	0.48% C 1.2% Cr 1.0% Ni 0.5% Mo 0.07% V steel: rigin unknown
SOMDIE SPECIAL	0.5% C 3.2% Cr 1.4% Mo steel: Origin unknown

Note. The following abbreviations and units are used in the tables:

DPN	Hardness, diamond pyramid number
UTS	Ultimate tensile strength, N/mm^2
Elon	Elongation, %
Proof	0.1 % proof strength, N/mm^2

$1 \text{ N/mm}^2 = 0.1 \text{ hbar} = 0.102 \text{ kgf/mm}^2 = 0.06475 \text{ tonf/in}^2 = 145.04 \text{ lbf/in}^2$

Symbol	Nominal analysis, supplier, condition and remarks.
SPARTAN 5	0.5% C 4.0% Cr 18% W 1.0% V steel: Atlas; as AISI type H26
SPEAR 419	0.4% C 1.2% Cr 0.3% Mo steel: Spear & Jackson for BS alloy En 19
SPEAR 500	0.4% C 0.45% Cr 0.2% Mo steel: Spear & Jackson for BS alloy En 100
SPEAR 510	0.4% C 1.2% Cr 1.5% Ni 0.15% Mo steel: Spear & Jackson for BS alloy En 110
SPEAR 626	0.4% C 0.6% Cr 2.5% Ni 0.6% Mo steel: Spear & Jackson; as BS alloy En 26
SPEAR AHC	0.5% C 1.5% Cr 2.0% W steel: Spear & Jackson
SPEAR CV	0.78% C 0.45% Cr 0.15% V steel: Origin unknown
SPEAR D1	0.5% C 0.8% Cr 0.25% V steel: For zinc die casting mould; Spear & Jackson; hardened & tempered **UTS: 870**
SPEAR D2	0.5% C 2.5% Cr 0.2% V steel: Origin unknown
SPEAR EXM	0.73% C 0.2% Cr 1.5% Ni steel: Origin unknown
SPEAR LEAPFROG	0.65% C 3.75% Cr 0.6% V 14.0% W steel: For tools; origin unknown
SPECIAL HW No.3	0.37% C 3.1% Cr 0.45% Mo 0.25% V 11.25% W steel: Origin unknown
SPECIAL ARDHO	0.4% C 0.8% Cr 3.0% Ni steel: Spencer; chisels, punches, etc.
SPECIAL WOLFRAM	0.5% C 1.2% Cr 2.2% W steel: Swift Levick; hot dies
SPECIFICATION 55	0.68% C 0.7% Cr 0.65% Ni 0.2% Mo steel: Crucible Steel Co; for saw blades, etc.
SPEEDICUT 14	0.6% C 3.5% Cr 14% W 0.6% V steel: Firth Brown
SPK	0.45% C 2.35% Cr 0.3% V steel: Schoeller-Bleckmann **UTS: 1210**
SPKD	0.4% C 10.5% Cr 1.0% Mo steel: Pompey
SPS	0.5% C 0.9% Si 0.9% Cr 2% W 0.2% V steel: For tools; Huntsman; chisels, shear blades, punches, etc. **DPN: 600**
SPS 245	0.4% C 0.6% Cr 1.25% Ni 0.15% Mo steel: Atlas; hardened & tempered **DPN: 340 UTS: 1140 Elon: 16% Proof: 1070**
SRO	0.45% C 2.8% Si 9% Cr 0.3% Mo steel: Bofors; obsolete
SSC	0.58% C 1.05% Cr 0.3% V steel: Origin unknown
STA 5/V9D	0.4% C 1.2% Cr 0.3% Mo steel: Replaced by BS alloy En 19
STA 5/V9D/1	0.4% C 1.2% Cr 0.3% Mo steel: Replaced by BS alloy En 19B
STA 5/V9D/2	0.4% C 1.2% Cr 0.3% Mo steel: Replaced by BS alloy En 19C
STA 5/V10	0.4% C 1.2% Cr 1.5% Ni 0.3% Mo steel: Replaced by BS alloy En 24
STA 5/V12	0.4% C 0.7% Cr 2.5% Ni 0.6% Mo steel: Replaced by BS alloy En 26
STA 5/V21A	0.4% C 3.0% Cr 0.3% Ni 1.0% Mo 0.2% V steel: Replaced by BS alloy En 40
STA 5/V21B	0.4% C 3.0% Cr 0.3% Ni 1.0% Mo 0.2% V steel: Replaced by BS alloy En 40
STAG 14	0.67% C 3.5% Cr 0.37% V 15.0% W steel: For tools; origin unknown
STAG C	0.45% C 2.5% Cr 0.35% V steel: Edgar Allen
STAMINAL	0.55% C 0.4% Cr 2.7% Ni 0.45% Mo 0.13% V steel: Latrobe; dies, punches & stamps **DPN: 600**
SUPER Si Mn	0.57% C 0.8% Mn 1.8% Si 0.8% Cr 0.25% Mo steel: For springs; ESC
SUPER TOUGH D	0.4% C 1.3% Mn 0.5% Cr 0.75% Ni 0.2% Mo steel: ESC for BS alloy En 100
T 20810	UNS designation for H10 type tool steel
T 20811	UNS designation for H11 type tool steel

Symbol	Nominal analysis, supplier, condition and remarks.
T 20812	UNS designation for H12 type tool steel
T 20813	UNS designation for H13 type tool steel
T 20814	UNS designation for H14 type tool steel
T 20824	UNS designation for H24 type tool steel
T 20826	UNS designation for H26 type tool steel
T 20841	UNS designation for H41 type tool steel
T 20842	UNS designation for H42 type tool steel
T 20843	UNS designation for H43 type tool steel
T 41901	UNS designation for S1 type tool steel
T 41902	UNS designation for S2 type tool steel
T 41904	UNS designation for S4 type tool steel
T 41905	UNS designation for S5 type tool steel
T 41906	UNS designation for S6 type tool steel
T 41907	UNS designation for S7 type tool steel
T-ALLOY	0.38% C 3.5% Cr 10.5% W 0.4% V steel: Braeburn; AISI type H22
T-ALLOY-B	0.5% C 3.0% Cr 15% W 0.5% V steel: Braeburn; AISI type H 24
TH	0.38% C 2.1% Cr 0.15% V steel: For hot work tools; S Osborn; casting dies; obsolete **DPN: 400**
THERMODIE	0.55% C 0.9% Cr 2.1% Ni 0.8% Mo steel: Origin unknown
TITANIC 141	0.6% C 2% Si 0.3% Mo 0.2% V steel: S Osborn; name for NA; obsolete
TM	0.4% C 5.0% Cr 1.2% Mo 0.5% V steel: Forez
TPC	0.45% C 1.2% Cr 0.15% V steel: Toledo for BS alloy En 50
TRKD	0.42% C 1.0% Cr 0.25% Mo steel: Pompey
TRNKD	0.4% C 0.5% Cr 0.5% Ni 0.2% Mo steel: Pompey
TYRANT EXTRA	0.45% C 1% Si 1.05% Cr 2% W steel· Schoeller-Bleckmann **DPN: 550**
UCX 2	0.4% C 1.1% Cr 0.25% Mo 0.15% V 1.0% Co steel: American proprietary alloy listed in SAE yearbook
ULTIMO 4	0.4% C 0.7% Cr 1.7% Ni 0.4% Mo steel: Atlas; hardened & tempered **DPN: 375 UTS: 1300 Elon: 15% Proof: 1250**
ULTIMO 6	0.55% C 1.0% Cr 1.6% Ni 0.75% Mo steel: Atlas; forging die inserts, etc. **DPN: 400**
VA	0.63% C 2% Cr 2% Ni 0.5% Mo steel: For hot work tools; S Osborn; obsolete **DPN: 400**
VAGS	0.4% C 1.1% Cr 1.5% Ni 0.3% Mo steel: ESC for BS alloy En 24
VDC	0.4% C 1% Si 5% Cr 1% V 1.2% Mo steel: Casting; Latrobe; for precision casting of forging dies **DPN: 550**
VERSATILE	0.4% C 1% Cr 1.5% Ni steel: Hall & Pickles; swaging tools, etc; obsolete **UTS: 1150 Elon: 17% Proof: 1080**
VIADUCT 15	0.4% C 1.1% Cr 1.9% W 0.3% V steel: Balfour; chisels, caulking tools, etc. **DPN: 550**
VIBRAC V 45	0.4% C 0.6% Cr 2.5% Ni 0.5% Mo steel: ESC for BS alloy En 26
VIBRO	0.5% C 1.4% Cr 2.1% W 0.25% V steel: Braeburn; AISI type S1
VINCO HOT WORK	0.5% C 4.0% Cr 18% W 1.0% V steel: Braeburn; AISI H 26
VISCOUNT 20	0.4% C 5.0% Cr 1.2% Mo 1.0% V steel: With sulphides; Latrobe; free machining form of VDC; type H13 **DPN: 650**

Symbol	Nominal analysis, supplier, condition and remarks.
VISCOUNT 44	0.4% C 5.0% Cr 1.2% Mo 1.0% V steel: With sulphides; Latrobe; free machining form of VDC; type H13 **DPN: 650**
VS 35 V	0.45% C 1.5% Cr 0.35% V steel: Vulcan **UTS: 1070**
VSCV	0.55% C 1% Cr 0.25% V steel: Vulcan; vehicle springs **DPN: 460**
VSNT	0.45% C 0.4% Cr 0.6% W steel: Vulcan; chisels, punches, etc. **DPN: 560**
WAW	0.45% C 1.5% Cr 0.5% Mo 0.5% W 0.9% V steel: Schoeller-Bleckmann **UTS: 1380**
WC 38	0.38% C 0.25% Cr 0.4% Ni 0.15% Mo 0.3% Cu (all max) steel: Casting; Wolsingham **DPN: 163 UTS: 580 Elon: 15% Proof: 270**
WC 43	0.42% C 0.25% Cr 0.4% Ni 0.15% Mo 0.3% Cu (all max) steel: Casting; Wolsingham **DPN: 175 UTS: 610 Elon: 12% Proof: 300**
WC 48	0.48% C 0.25% Cr 0.4% Ni 0.15% V 0.3% Cu (all max) steel: Casting; Wolsingham **DPN: 185 UTS: 630 Elon: 12% Proof: 300**
WC 53	0.53% C 0.25% Cr 0.4% Ni 0.15% V 0.3% Cu (all max) steel: Casting; Wolsingham **DPN: 190 UTS: 700 Elon: 10% Proof: 330**
WC 58	0.58% C 0.25% Cr 0.4% Ni 0.15% Mo 0.3% Cu (all max) steel: Casting; Wolsingham **DPN: 200 UTS: 700 Elon: 10% Proof: 330**
WC 63	0.63% C 0.25% Cr 0.4% Ni 0.15% Mo 0.3% Cu (all max) steel: Casting; Wolsingham **DPN: 210 UTS: 700 Elon: 10% Proof: 330**
WCPS	0.48% C 1.5% Cr 0.7% Ni 0.2% Mo 1.2% W 0.15% V tool steel: Huntsman; chisels, shear blades, punches, etc. **DPN: 670**
WL 1	0.4% C 1.0% Cr 1.8% Ni 0.25% Mo steel: Tube; Babcock & Wilcox (USA)
WLM 1214	0.4% C 1.3% Si 1% Cr 2% W steel: W Marrison; chisels, punches
WM EXTRA	0.38% C 1.05% Si 2.1% Cr 4.25% Mo 0.32% V steel: Schoeller-Bleckmann **UTS: 1550**
WZ	0.4% C 3.0% Cr 0.4% Mo 0.25% V 11.0% W steel: Origin unknown
X 38 Cr Mo V 51	0.38% C 5.3% Cr 1.1% Mo 0.4% V steel: German Standard designation
X 40 Cr Mo V 51	0.4% C 5.3% Cr 1.4% Mo 1.0% V steel: German Standard designation
X 45 Ni Cr Mo 4	0.45% C 1.4% Cr 4.1% Ni 0.3% Mo 0.5% W steel: For tools; German Standard designation
X 3140	0.4% C 0.8% Cr 1.2% Ni steel: Designation used by AISI; obsolete
XDH	0.55% C 4.0% Cr 18% W 1.0% V steel: Firth Sterling
XDL	0.48% C 3.5% Cr 14% W 0.75% V steel: Firth Sterling
XL CHISEL	0.55% C 2.0% W 1.35% Cr 0.25% V 0.1% Mo steel: Latrobe; tool steel
XP 180	0.5% C 5.0% Cr 2.0% V 1.3% Mo steel: Braeburn; as Pressurdie 7
Y 42CD4	0.42% C 1.0% Cr 0.2% Mo steel: French Standard designation
Y 45SCD6	0.45% C 0.6% Cr 0.2% Mo steel: French Standard designation
Y 50CV4	0.5% C 1.1% Cr 0.2% V steel: French Standard
Y 50CV4	0.5% C 1.0% Cr 0.15% V steel: French Standard designation

Symbol	Nominal analysis, supplier, condition and remarks.
YCW STEM	0.48% C 1.7% Si 0.6% Cr 4.2% Ni 0.25% Mo steel: Casting; Edgar Allen; code 638 steel
Z 38CDV5	0.38% C 5.0% Cr 1.25% Mo 0.5% V steel: French Standard designation
Z 38CDV5	0.38% C 5.2% Cr 1.3% Mo 1.0% V steel: French Standard
Z 38CDWV5	0.38% C 5.0% Cr 1.25% Mo 0.5% V 1.25% W steel: French Standard designation
Z 40CSD10	0.4% C 0.8% Mn (max) 10.5% Cr 1.0% Mo steel: French Standard designation
Z 40WCV5	0.4% C 4.0% Cr 0.5% Mo 0.5% V 5.0% W steel: French Standard designation

Symbol	Nominal analysis, supplier, condition and remarks.
Z 65WDCV605	0.65% C 4.0% Cr 5.0% Mo 2.0% V 6.0% W steel: French Standard designation

Note. The following abbreviations and units are used in the tables:

DPN	Hardness, diamond pyramid number
UTS	Ultimate tensile strength, N/mm^2
Elon	Elongation, %
Proof	0.1 % proof strength, N/mm^2

$1 \ N/mm^2 = 0.1 \ hbar = 0.102 \ kgf/mm^2 = 0.06475 \ tonf/in^2 = 145.04 \ lbf/in^2$

44K3 Steel – high carbon alloys
nickel, molybdenum, tungsten or vanadium

Specific gravity	7.87	
Density	7870 kg/m^3	(0.29 lb/in.^{3})
Solidus/liquidus	– °C	
Thermal conductivity	– W/m °C	
Coefficient of linear expansion	11 × 10^{-6}/ °C	
Electrical conductivity	– IACS (copper 100%)	
Specific resistance	– microhm mm	
Young's modulus of elasticity	203.4 × 10^9 N/m^2	(29.5 × 10^6 lbf/in.^{2})
Impact	– J	
Fatigue strength	–	
Hot strength	–	

The above properties are typical of the following group and may not apply exactly to any one specification. It is possible that with certain specifications some of the values may not be applicable.

General metallurgical characteristics

These are all very hard brittle materials which have been developed for their high hardness which is only achieved after hardening and tempering and are thus only used in this condition. This hardness persists at fairly high temperatures dependent of the amount of alloying carried out.

None of the alloying elements enhances the corrosion resistance to any great extent whereas the high hardness and low ductility reduce the notch sensitivity, thus making the presence of corrosion pitting particularly dangerous.

All the alloys have sufficient carbon present to ensure the presence at all times of free cementite which is inherently hard even in the annealed state.

Each steel has at least two of the listed alloying elements and many have all five present. The effect of each element is as described in the single alloy steel bearing that name, and when two or more elements are present the effect is additive.

Briefly the effects of the alloying elements are:

Chromium. This is a carbide former, with the carbides being harder and more stable than iron carbides. Chromium increases the hardenability of steel in that larger sections can be through hardened, or less drastic oil or air quenching used in place of water, compared to plain or lower alloy carbon steels. Chromium also increases the corrosion resistance although none of the steels listed can be classed as stainless. The higher chromium stainless steels are listed and described in separate sections. Chromium raises the critical temperatures for steel heat treatment and slows down the speed at which the metallurgical reactions take place and has a tendency to promote grain growth.

Nickel. This strengthens the ferrite matrix of the steel, imparting toughness rather than hardness and increases the fatigue strength. Nickel lowers the critical heat treatment temperatures and widens the range required for successful treatment. It also improves the ductility of steel at low temperatures. Nickel is seldom an important constituent in the steels listed as hardness and abrasion resistance are their prime purpose, and nickel has less effect on these high carbon steels than the lower carbon varieties.

Molybdenum. This is a carbide former, the resultant molybdenum carbide being inherently small grained and evenly distributed. This ensures that the steels are fine grained, thus improving the fatigue strength. These stable carbides slow down the metallurgical reactions, ensuring deeper hardening, great hardenability and the elimination of temper brittleness. They also help to prevent agglomeration of the carbides and grain growth with prolonged high temperature treatments. As carbon always combines with the molybdenum in preference to other carbide formers, including iron, and stays in solution, this obviates carbide precipitation. Vanadium and tungsten have similar effects.

Tungsten. This is a carbide former which results in very hard, rather massive tungsten carbide particles. Generally tungsten is used with chromium and vanadium which aid the distribution and prevent agglomeration of the carbides. Molybdenum has similar properties to tungsten with generally less required to give equal results. Tungsten was the first element to be used in quantity in high carbon steels to improve the hot hardness.

Vanadium. This forms very small carbides and also influences other carbide formers to produce tiny particles which in turn ensure a fine grain steel. Thus vanadium has a considerable effect on improving the fatigue strength and is very often associated with chromium for this purpose. Vanadium has little effect on the temperatures at which the metallurgical changes take place, but slows down these reactions, thus improving the hardenability.

The steels listed here may have slightly better corrosion resistance than plain carbon steels but cannot be classified as stainless. Also the effect of corrosion pitting on the rather notch sensitive materials is much more serious than on more ductile steels.

The choice of elements in the steels is a complex process, balancing hardness, hot hardness, ductility and workability with the overriding factor that the final product must be economically saleable.

Thermal treatment

These steels cannot be cold worked to any appreciable extent as they become brittle and will actually shatter in many instances.

The use of pre-heat at 600–800 °C is recommended before taking these materials to a higher temperature, otherwise, cracking or distortion is a possibility.

The presence of the alloying elements ensures very sluggish metallurgical reactions and this applies particularly to softening. Practically without exception these steels will require to be cyclic or spheroidize annealed to reduce the hardness for machining. This consists of heating for about one hour at 850–950 °C then transferring to a furnace at 670–800 °C for two hours and repeating at least once. Higher carbon, higher alloyed steels, particularly those with molybdenum and vanadium, may require three or four cycles to achieve the desired softness and there are considerable variations in the recommended times and temperatures.

Normalizing will always cause some hardening and must be followed by a tempering operation, often referred to as 'anneal'.

These steels can be readily cracked by rapid heating or cooling through the change points, hence the recommendation always to pre-heat by heating to about 650–850 °C then transferring to the hardening temperatures. This will vary from 900–1200 °C depending on the analysis and time at temperature. The carbides of chromium, tungsten and vanadium are all very slow in being taken into sol-

ution, and this can be speeded up by increasing the temperature which in turn tends to cause grain growth. This grain growth is inhibited to some extent by vanadium, molybdenum and tungsten, thus the heat treatment of these high carbon alloy steels is a matter of balancing time against temperature. Too short a time or too low a temperature means that full hardening is not achieved, while the use of too long a time leads to surface de-carburization, overheating and grain growth which causes brittleness and may result in quench cracking from local intergranular de-carburization.

Most of these steel formulations have considerable flexibility regarding times and temperatures, but for optimum results laboratory advice and trials should be obtained and information sought from the origin of the material chosen. The result of faulty treatment can be spectacular – as with quench cracking – but is more commonly much more insidious, being evident in shorter tool life which is very difficult to assess.

Many of these steels require a double temper with a deep freeze operation between, at minus 20–60 °C. This ensures that the correct metallurgical reactions take place to give a homogeneous through hardened and tempered structure which is dimensionally stable.

Tempering is usually in the 300–650 °C range.

It is also possible to harden these steels by martempering or austempering. These processes require tight control of time, temperature and weight, thus should only be attempted under strict laboratory control, after carrying out trials.

Salt bath heating is most often used for these materials to prevent the excessive oxidation and de-carburization which would take place in air furnaces. Some de-carburization will always occur even with the best quality salts correctly controlled. Vacuum furnace heat treatment is showing promise for many of these steels but is expensive with existing equipment.

Scale removal is by grit blasting or sodium hydride treatment when the specialist equipment is available. Acid pickling must never be attempted on these steels after hardening owing to the danger of embrittlement.

Of all metallurgical thermal treatments carried out on commonly used steels, these are the most complicated and difficult to control. The materials involved are all of high cost and the advantages to be gained in tool life from the correct treatment are considerable. There is little doubt that very many good quality expensive tool steels are maltreated by faulty thermal treatments to end up no better than simpler, cheaper steels correctly treated.

This situation can be remedied by co-operation between the user, heat treater and metallurgist.

Welding and brazing

Pre-treatment. Welding should be in the annealed condition, with all trace of dirt, grease and oxide removed. Pre-heating is essential for all these steels.

Welding and brazing. Welding is not advised for these materials as the resultant cast zone will always be weak and brittle. If it is possible to upset forge weld the material, then successful welds can be produced as this squeezes out the cast structure. This requires special equipment and very careful control. Normal welding with electric or gas equipment is not recommended and should not be carried out. All the above strictures apply when a homogeneous stressed part is required. It is often possible to design parts such that the weld is unstressed, therefore acceptable. For example these expensive materials are often welded to plain carbon steels for economical reasons, the weld area being sited remote from any stress concentration.

Brazing is often used on the materials, which have high tempering temperatures, using low melting point filler metals. Because of the very slow metallurgical reactions of these steels there is little danger of softening provided reasonably fast brazing is carried out, using for example induction heating.

Post-treatment. Welded parts should be hardened and tempered, care being taken that if two materials are involved special techniques may be required. Parts which have been welded in the hardened and tempered condition must be tempered and should be closely examined for cracking.

Flaw detection methods

Crack test. Magnetic crack test, or penetrant oil chalk test.

X-ray. Not normally required.

Ultra-sonic test. Not normally required, but may be used on parts which have extensive machining. The ultra-sonic test shows up subcutaneous defects, thus saving the cost of machining.

Chemical etch. These materials are very prone to grinding abuse after hardening. An electrolytic sulphuric etch can be used to identify this grinding softening at a stage long before it is serious enough to cause grinding cracks and is recommended as a means of controlling grinding techniques. A nitric acid etch can be used to give similar results.

Corrosion protection

Temporary. These are expensive materials which can corrode very easily and must therefore be kept in moisture free conditions and oiled or greased at all times.

During manufacture parts should be kept oiled. Soluble oil coolant in good condition is an excellent temporary corrosion inhibitor but when dirty or contaminated with metallic dust is conducive to rusting.

Finished parts should be cleaned, dried, oiled and packed in plastic. The use of paper, cardboard or even wood should be carefully examined as these may contain compounds which can cause staining and rusting even when the parts are oiled.

Permanent. The alloy content of these steels has a slight beneficial effect on their corrosion resistance, but there is never sufficient to prevent rusting even under normal indoor conditions. The high hardness and wear resistance of the steels listed is the sole reason for their existence and this means that it is seldom possible to protect permanently the finished parts.

Steel corrosion or rust is caused by moisture and air, and can only be prevented by their complete exclusion. The corrosion products are complex iron oxides which are readily removed to allow further rusting to take place with resultant pitting, particularly dangerous with these hard notch sensitive steels.

Removal of existing corrosion prior to protection can be alkaline de-rusting which requires careful control but does not cause hydrogen embrittlement. Acid de-rusters must not be used for this reason. Grit blasting can remove corrosion but the high hardness prevents any surface roughening or key. Phosphate treatment successfully converts light rust and staining, leaving a surface capable of retaining an oil film, and provided this is maintained many years of corrosion-free life can be obtained under indoor conditions. A stress relieve at 200 °C is recommended after the phosphating to obviate any danger of hydrogen embrittlement.

Painting. The phosphate treatment is an excellent key for a paint system. This is often purely for identification purposes, but may be to protect non-working areas.

Zinc silicate or chromate primers are advised with a top sealer coat which should be stoved when additional protection is desired.

Plating. It is essential that these steels are stress relieved at 200 °C before and after any plating operations to prevent hydrogen embrittlement causing failure. This occurs in a brittle manner below the calculated tensile stress and can be particularly dangerous with these high strength steels.

Because of the fact that there is no known means of inspecting to ensure that the stress relieving operations have been carried out and the catastrophic nature of failure on these parts, electroplating is not advised.

Plating itself presents no great difficulty provided correct etching techniques are used.

Black oxide treatment, whereby an artificial adherent oxide film is formed using proprietary hot alkaline solutions, gives satisfactory results provided the surface is kept dry and oiled.

These materials are not normally used out of doors and can seldom be treated with the normal permanent protection media because the components are almost always put to arduous use. It must be remembered that they are expensive materials with very little built-in corrosion resistance. The liberal use of oil and grease coupled with the exclusion of moisture is strongly recommended, but this will do little to prevent corrosion propagating once it has started. Routine examination and early removal of staining or incipient corrosion should be practised wherever this is feasible.

Machinability

These materials are never easy to machine, as even in the softest or cyclic annealed condition they have a high hardness with abrasive carbides.

After hardening and tempering only grinding, spark erosion or electrochemical machining is possible.

Very great care must be taken when grinding these materials to prevent softening or in extreme cases cracking. There is a mistaken impression that because many of these materials have a good hot hardness they can be dry ground with heavy cuts. Even very careful dry grinding will always cause some surface softening and at no time should sparking be allowed if the best results are to be obtained.

Spark erosion and electrochemical machining require rather exotic machines but give excellent results.

Uses

Cutting tools of all types, milling, drilling, tapping, broaching, where tough or hard materials are to be machined. Good quality hand tools of all types, such as spanners. Hot work dies for foundry and forge. Certain engineering applications where wear and temperature use coincide, such as high duty ball and roller bearings.

Certain arduous shock and abrasion resisting uses such as pneumatic chisels, rock cutting drills, crusher jaws, excavator bucket teeth.

Many of these materials are used in conjunction with cheaper more ductile steels as inserts or tips, brazed, bolted, welded or glued. The design of the joint must be very carefully considered to ensure it is relatively stress free, or in pure compression.

Symbol	Nominal analysis, supplier, condition and remarks.
01	0.9% C 0.5% Cr 0.5% W steel: Designation used by AISI
02	0.9% C 1.4% Mn 0.3% Cr 0.3% Mo 0.2% V steel: For tools; designation used by SAE
06/05/04/02	0.85% C 4.0% Cr 5.0% Mo 6.0% W 2.0% V steel: For tools; French Standard designation
06/05/04/04	1.3% C 4.0% Cr 4.5% Mo 5.5% W 4.0% V steel: For tools; French Standard designation
07	1.2% C 0.75% Cr 1.75% W steel: Designation used by AISI
08/04/02/02	0.85% C 4.2% Cr 5.0% Mo 6.2% W 1.8% V steel: For tools; French Standard designation
09B	1% C 1% Mn 0.6% Cr 1% W steel: Carrs Ltd; shear blades, rivet dies, etc. **DPN: 750**
1.2210	1.15% C 0.7% Cr 0.1% V steel: German Standard designation
1.2363	1.0% C 5.25% Cr 1.0% Mo 0.2% V steel: German Standard
1.2510	0.95% C 0.5% Cr 0.2% V 0.5% W steel: German Standard
1.2552	0.8% C 1.1% Cr 0.3% V 1.2% W steel: German Standard designation
1.2842	0.9% C 2.0% Mn 0.3% Cr 0.1% V steel: German Standard designation
1.3333	0.99% C 4.0% Cr 2.65% Mo 2.85% W 2.35% V steel: For tools; German Standard designation
1.3342	1.0% C 4.1% Cr 5.0% Mo 6.3% W 2.0% V steel: For tools; German Standard designation
1.3343	0.88% C 4.1% Cr 5.0% Mo 6.3% W 1.85% V steel: For tools; German Standard designation
1.3343	0.83% C 4.1% Cr 5.0% Mo 1.9% V 6.4% W steel: For tools; German Standard
1.3346	0.8% C 3.9% Cr 8.5% Mo 1.1% V 1.5% W steel: For tools; German Standard
1.3355	0.75% C 4.1% Cr 1.1% V 18.0% W steel: For tools; German Standard
2AS	0.82% C 4.2% Cr 5.0% Mo 6.2% W steel: Forez
3AS	0.76% C 4.2% Cr 18% W 1.0% V steel: Forez
4AS	0.8% C 4.2% Cr 0.7% Mo 18% W 1.4% V steel: Forez
5/6/2 quality	0.83% C 4.2% Cr 5% Mo 6.4% W 1.9% V steel: T Turton

Symbol	Nominal analysis, supplier, condition and remarks.
5CC	1.0% C 0.6% Mn 5.2% Cr 1% Mo 0.5% V steel: ESC; shear blades, punches, etc. **DPN: 750**
6/5/2	0.85% C 4% Cr 6% W 2% V steel: For tools; Darwins; reamers, punches, etc.
8KH4V4F1(P4)	0.8% C 4.5% Cr 1.2% V 4.5% W steel: Russian Standard designation
8KHF	0.75% C 0.55% Cr 0.22% V steel: Russian Standard designation
8N2	4% Cr 8% Mo 1.5% W 1% V steel: For tools; Vanadium Alloys Ltd; M1 type; AISI code M1A
0 KX	0.9% C 0.8% Cr 0.12% V steel: Pompey
9 KH	0.87% C 1.52% Cr 0.5% V 1.25% W steel: Russian Standard designation
9KHF	0.85% C 0.55% Cr 0.22% V steel: Russian Standard designation
9KH5F	0.92% C 5.0% Cr 0.22% V 1.0% W steel: Russian Standard designation
9KH5VF	0.9% C 0.25% Mn 5.0% Cr 0.22% V 1.0% W steel: Russian Standard designation
9KHVG	0.9% C 1.1% Mn 0.65% Cr 0.1% V 0.65% W steel: Russian Standard designation
10 KD	1.0% C 1.0% Cr 0.2% Mo steel: Pompey
12 KV	1.15% C 0.9% Cr 0.12% V steel: Pompey
12 W 12 C1	0.75% C 0.25% Mn 0.4% Cr 5.8% W steel: Sandvik
14 NC	0.8% C 0.35% Mn 0.15% Cr 0.5% Ni steel: Sandvik; annealed **UTS: 660**
16 C1 V	0.9% C 0.55% Mn 0.6% Cr 0.1% V steel: Sandvik
20 W 1V	1.15% C 0.25% Cr 0.6% W 0.1% V steel: Sandvik
21 T 10	1.25% C 0.35% Mn 0.3% Cr 1.7% W 0.1% V steel: Sandvik
21 T 10 P	1.25% C 0.3% Mn 0.25% Cr 1.7% W 0.1% V 0.2% Pb steel: Sandvik
32 B	1.0% C 0.7% Mn 5.25% Cr 1.15% Mo 0.3% V steel: Carrs; coining tools, swaging rolls, etc. **DPN: 680**
66 HS	4% Cr 5% Mo 2% V 6% W tool steel: Bethlehem Steel Co; M2 type; AISI code M2E
67 S	1.4% C 0.5% Cr 3.25% W steel: Carrs; extrusion dies, etc. **DPN: 800**
70 H	0.9% C 0.5% Cr 0.5% W steel: For tools; origin unknown
80 DCV 42.16	0.8% C 4.0% Cr 4.25% Mo 1.0% V steel: French Standard designation
80 W Cr V 8	0.8% C 1.1% Cr 0.3% V 1.2% W steel: German Standard designation
90MCW5	0.9% C 1.25% Mn 0.5% Cr 0.5% W steel: French Standard designation
90 MCW V5	0.95% C 0.5% Cr 0.2% V 0.5% W steel: French Standard

Note. The following abbreviations and units are used in the tables:

DPN	Hardness, diamond pyramid number
UTS	Ultimate tensile strength, N/mm^2
Elon	Elongation, %
Proof	0.1 % proof strength, N/mm^2

1 N/mm^2=0.1 hbar=0.102 kgf/mm^2=0.06475 tonf/in^2=145.04 lbf/in^2

Symbol	Nominal analysis, supplier, condition and remarks.
90 Mn V 8	0.9% C 2.0% Mn 0.3% Cr 0.1% V steel: German Standard designation
100C3	1.0% C 0.75% Cr 0.15% V steel: French Standard designation
100WC10	1.0% C 0.5% Cr 0.2% V 1.0% W steel: French Standard designation
115 Cr V 3	1.15% C 0.7% Cr 0.1% V steel: German Standard designation
110WC20	1.1% C 0.3% Mn 0.75% Cr 2.0% W steel: French Standard designation
140C3	1.4% C 0.75% Cr 0.15% V steel: French Standard designation
140SMD4	1.4% C 1.0% Mn 0.3% Mo steel: French Standard designation
175	0.85% C 1.9% Mn 0.15% Cr 0.3% V steel: F Parkin; reamers, gauges, etc. **DPN: 830**
200	0.9% C 1.3% Mn 0.5% Cr 0.5% W steel: F Parkin; chisels, gauges, etc. **DPN: 850**
212	1.3% C 0.75% Mn 1.5% Cr 1% W steel: F Parkin; forming dies, gauges, etc. **DPN: 850**
300	1.5% C 0.3% Mn 0.8% Cr 5.5% W steel: F Parkin; cold draw dies, marble cutting tools, etc. **DPN: 800**
321	.25% C 0.5% Mn 0.75% Cr 0.4% Mo 2.3% W 0.2% V steel: F Parkin; taps & dies, etc. **DPN: 870**
562	0.8% C 4.25% Cr 6.5% W 2% V steel: F Parkin; cutting tools, form tools **DPN: 800**
562	0.85% C 4.75% Cr 5.0% Mo 6.0% W 2.0% V steel: Spear & Jackson
599	0.68% C 4.2% Cr 14% W 0.7% V steel: F Parkin; taps, dies, etc. **DPN: 800**
622	0.75% C 4% Cr 18% W 1% V steel: F Parkin; cutting tools **DPN: 800**
2131	1.0% C 0.75% Cr 0.15% V steel: French Standard designation
2132	1.4% C 0.75% Cr 0.15% V steel: French Standard designation
2133	0.8% C 0.35% Cr 1.0% Ni steel: French Standard designation
2141	1.0% C 0.5% Cr 0.2% V 1.0% W steel: French Standard designation
2142	1.1% C 0.3% Mn 0.75% Cr 2.0% W steel: French Standard designation
2212	0.9% C 0.5% Cr 0.5% W steel: French Standard designation
2231	1.0% C 5.0% Cr 1.0% Mo 0.3% V steel: German Standard designation
3551	0.8% C 4.0% Cr 4.25% Mo 1.0% V steel: French Standard designation
4151	0.8% C 4.0% Cr 2.0% Mo 12.0% W 2.0% V steel: For tools; French Standard designation
4161	0.8% C 4.0% Cr 12.0% W 3.5% V steel: For tools; French Standard designation

Note. The following abbreviations and units are used in the tables:

DPN	Hardness, diamond pyramid number
UTS	Ultimate tensile strength, N/mm^2
Elon	Elongation, %
Proof	0.1 % proof strength, N/mm^2

1 N/mm^2=0.1 hbar=0.102 kgf/mm^2=0.06475 tonf/in^2=145.04 lbf/in^2

Symbol	Nominal analysis, supplier, condition and remarks.
4201	0.8% C 4.0% Cr 18.0% W 1.0% V steel: For tools; French Standard designation
4203	0.85% C 4.0% Cr 18.0% Mo 2.0% V steel: For tools; French Standard designation
4301	0.85% C 4.0% Cr 5.0% Mo 6.0% W 2.0% V steel: For tools; French Standard designation
4361	1.3% C 4.2% Cr 4.5% Mo 6.2% W 4.0% V steel: For tools; French Standard designation
4441	0.85% C 4.2% Cr 5.0% Mo 6.2% W 1.9% W steel: For tools; French Standard designation
6542 MOLY	0.85% C 4.0% Cr 5.0% Mo 2.0% V 6.0% W steel: For tools; origin unknown
A 2	1.0% C 5% Cr 1.2% Mo 0.4% V steel: For tools; designation used by AISI
A 2	1.0% C 5.0% Cr 1.05% Mo 1.3% V steel: For cold work; Osborn
A 3	1.25% C 5.0% Cr 1.0% Mo 1.0% V steel: For tools; designation used by AISI
A 4	1.0% C 2.0% Mn 1.0% Cr 1.0% Mo steel: For tools; designation used by AISI
A 5	1.0% C 3.0% Mn 1.0% Cr 1.0% Mo steel: For tools; designation used by AISI
A 7	2.25% C 5.2% Cr 1.0% Mo 1.0% W 4.7% V steel: For tools; designation used by AISI
A 7	2.3% C 5.3% Cr 1.0% W 1.0% Mo 4.7% V steel: For cold work; Osborn
A 7W	2.25% C 5.25% Cr 1.0% Mo 1.0% W 4.7% V steel: Simonds; AISI type A7
A 10	1.35% C 1.8% Mn 1.2% Si 1.8% Ni 1.5% Mo steel: For tools; designation used by AISI
A 395	0.95% C 1.1% Cr 0.25% Mo steel: SHD; drills
ADAMANT HP	0.85% C Cr Mo steel: Casting; Firth Brown; hardened & tempered **DPN: 370**
ADAMANT XL	1.25% C Cr Mo steel: Casting; Firth Brown; hardened & tempered **DPN: 400**
ADAMANT XTRA	0.77% C Cr Mo steel: Casting; Firth Brown: hardened & tempered **DPN: 330**
AH CHROME DIE	1.0% C 5.0% Cr 1.2% Mo 0.4% V steel: Origin unknown
AIRQUE	1.0% C 5.2% Cr 1.1% Mo 0.25% V steel: Braeburn; AISI type A2
AIRQUE 4	1.0% C 2.5% Mn 1.1% Cr 1.1% Mo steel: Braeburn
AIRQUE SPECIAL	0.8% C 5.25% Cr 1.1% Mo 0.25% V steel: Braeburn; AISI type A2
AIRQUE V	1.25% C 5.25% Cr 1.1% Mo 1.0% V steel: Braeburn; AISI type A3
AIRTRUE	1.0% C 5.0% Cr 1.0% Mo 0.25% V steel: Simonds; AISI type A2
AIRVAN	1.0% C 5.25% Cr 1.15% Mo 0.25% V steel: Firth Sterling
AISI 611	0.84% C 4.2% Cr 5% Mo 6.3% W 1.9% V steel: Hardened & tempered; for cutting tools **UTS: 780 Elon: 21% Proof: 370**
AISI 612	0.87% C 4% Cr 8.2% Mo 1.9% V steel: Cutting tools **DPN: 910**
AISI 613	0.8% C 4% Cr 4.25% Mo 1% V steel: Hardened & tempered **DPN: 870 UTS: 2760 Elon: 2% Proof: 2420**
ALLOY B	1.25% C 0.3% Si 0.3% Mn 1.1% Cr 4.5% W 0.25% V steel: Jessop
AMS 5626A	4% Cr 18% W 1% V steel: Bar & forging
AMS 6490A	0.8% C 4% Cr 4.2% Mo 1% V steel: Bar & forging; vacuum melted; premium quality M50

Symbol	Nominal analysis, supplier, condition and remarks.
AN 2	0.78% C 4.25% Cr 22% W 1% V steel: Osborn; obsolete **DPN: 872**
ARDENT	1.5% C 0.5% Cr 4% W steel: Hall & Pickles; cold drawing dies; obsolete **DPN: 750**
ARK SUPERIOR	0.8% C 4.25% Cr 18% W 1.3% V steel: Jessop **DPN: 750**
ARK SUPERIOR EXTRA	0.73% C 4.2% Cr 1.4% V 22.0% W steel: For tools; origin unknown
ARK TRIUMPH	0.67% C 0.3% Si 0.3% Mn 3.7% Cr 13% W 0.6% V steel: Jessop
ARK TRIUMPHANT	1.23% C 4.5% Cr 13% W 3.7% V steel: Jessop
ARLEY	1.4% C 3% Cr + W steel: John Vessey; extrusion dies **DPN: 870**
ARNE	0.9% C 0.5% Cr 0.5% W 0.1% V steel: Uddelholm; for type 01
ASTM A485/3	1.02% C 0.8% Mn 0.25% Si 1.3% Cr 0.25% Mo steel: High hardenability
ASTM A485/4	1.02% C 1.2% Mn 0.3% Si 1.3% Cr 0.5% Mo steel: High hardenability
ASTM A597	0.92% C 1.2% Mn 0.7% Cr 0.3% V (max) 0.5% W steel: For cast tools
ASTM A597 CA2	1.0% C 5.2% Cr 1.2% Mo 0.3% V steel: For cast tools
ASTM A600	High speed tool steel specification analysis given under grade AISI system
ATLAS M3/2	4% Cr 6% Mo 6% W 3% V steel: For tools; Atlas Steels Ltd; AISI code M3/2
ATLAS M4	4% Cr 4.5% Mo 5.5% W 4% V steel: For tools; Atlas Steels Ltd; AISI code M4Z
ATLAS XXX	1.3% C 0.35% Cr 3.7% W steel: Atlas; as AISI type F2
AW	0.7% C 4% Cr 14% W 1% V steel: ESC; cutting tools **DPN: 910**
AW 4 V	1.2% C 4.2% Cr 13.8% W 4% V steel: ESC; cutting, punches **DPN: 910**
AXL	0.7% C 4.5% Cr 18% W 1.2% V steel: Fagersta **DPN: 750**
B 1	1.12% C 0.27% Cr 0.2% V 1.0% W steel: Russian Standard designation
B 2/2 (3)	2.8% C 5.25% Cr 4.5% V 1.1% Mo steel: Latrobe
B 6	4% Cr 18% W 1% V steel: For tools; Universal Cuclops; AISI code T1D
B 9	4% Cr 18% W 2% V steel: For tools; Universal Cyclops; AISI code T2D
BADGER	0.94% C 0.5% Cr 0.5% W steel: Latrobe; punches & dies; type θ1 **DPN: 740**
BESTEM	High C 3% Ni Mo steel: For dies; Somers
BFD	1.2% C 0.6% Cr 1.65% W 0.2% V steel: Simonds; AISI type 07
BLUE CHIP	0.73% C 4.0% Cr 18% W 1.15% V steel: Firth Sterling
BOLSTER STEEL	0.5% C 1.2% Mn 0.65% Cr 0.2% Mo steel: Sanderson Kayser
BP 4	1.25% C 3% W 0.15% V steel: Osborn; cold draw dies; burnishing tools; obsolete **DPN: 950**
BR 3	2.8% C 5.25% Cr 4.5% V 1.1% Mo steel: Latrobe
BR 3 DIE STEEL	2.8% C 5.25% Cr 1.1% Mo 4.5% V steel: Latrobe; abrasion resistant **DPN: 700**
BRAEBURN M7	4% Cr 8.7% Mo 1.7% W 2% V steel: For tools; Braeburn Alloy Steel; AISI code M71

Symbol	Nominal analysis, supplier, condition and remarks.
BRAEFOUR	1.27% C 4.5% Cr 4.5% Mo 5.5% W 4.0% V steel: Braeburn; AISI type M4
BRAEMOW	0.84% C 4.2% Cr 5.0% Mo 6.5% W 2.0% V steel: Braeburn; AISI type M2
BRAEMOW 2	4% Cr 5% Mo 6% W 2% V steel: For tools; Braeburn steel; AISI code M21
BRAEMOW SPECIAL	0.66% C 4.0% Cr 1.7% V 6.5% W 5.0% Mo steel: Braeburn
BRAEVAN	4% Cr 6% Mo 6% W 2.4% V steel: For tools; Braeburn Alloy Steel; AISI code M31
BRAEVAN/1	1.02% C 4.0% Cr 5.7% Mo 6.2% W 2.5% V steel: Braeburn; AISI type M3
BRAEVAN/2	1.15% C 4.0% Cr 5.5% Mo 5.7% W 3.3% V steel: Braeburn; AISI type M3
BRUNSWICK V150	1.1% C 0.8% W + Cr steel: John Vessey **DPN: 770**
BRUNSWICK V888	1.0% C 5% Cr 1.0% Mo steel: John Vessey **DPN: 750**
BSS	0.8% C 4.2% Cr 9% Mo 2% W 1.3% V steel: Huntsman; thread rolling dies; cutting tools **DPN: 870**
C 253	1.0% C 1.0% Cr 0.25% Mo steel: Fagersta **DPN: 750**
C 254	1.0% C 1.2% Cr 0.35% Mo steel: Fagersta **DPN: 750**
C 550	1.0% C 5.2% Cr 1.1% Mo 0.2% V steel: Fagersta **DPN: 680**
CAPITAL 305	0.8% C 4.0% Cr 1.7% W 1.1% V 8.7% Mo steel: For tools; Balfour Darwin
CAPITAL 395	1.2% C 4.5% Cr 5.5% W 4.0% V 4.5% Mo steel: For tools; Balfour Darwin
CAPITAL 562	0.83% C 4.25% Cr 5% Mo 6.4% W 1.9% V steel: For tools; Balfour; reamers, twist drills, etc. **DPN: 850**
CARBON COLD HEADER	0.93% C 0.05% Cr 1.18% V steel: Latrobe; for cold work
CARPENTER 10 STAR	4% Cr 8% Mo 2% V steel: For tools; Carpenter Steel Co; AISI code M10K
CARPENTER SUPER SPEED STAR	4% Cr 6% Mo 6% W 2.4% V steel: For tools; Carpenter Steel Co; AISI code M3K
CARVITE	4% Cr 18% W 4% V steel: For tools; Columbia Tool Steel Co; AISI code T9H
CASTLE 6/6/2	0.85% C 4.0% Cr 5.2% Mo 2.0% V 6.0% W steel: For tools; origin unknown
CASTLE SUPERIOR	0.75% C 4.0% Cr 1.0% V 18.0% W steel: For tools; origin unknown
CDD	1.9% C 0.6% Cr 6% W steel: Firth Brown
CENTURY	1.0% C 4.0% Cr 2.0% V 0.8% W 8.0% Mo steel: Braeburn
CHD 4	0.6% C 0.3% Si 0.3% Mn 3.7% Cr 0.6% Mo 0.4% V steel: Huntsman; hot blanking, shears, etc. **DPN: 530**
CL 45	0.8% C 1.0% Cr 0.2% V steel: Origin unknown
CL 652	0.85% C 4.0% Cr 5.0% Mo 2.0% V 6.0% W steel: For tools; origin unknown
CLARITE	4% Cr 18% W 1% V steel: For tools; Columbia Tool Steel Co; AISI code T1H
CLYDALL 18	0.8% C 4% Cr 18% W 1% V steel: Clyde Alloy **DPN: 800**
CLYDMO	0.83% C 4% Cr 5% Mo 6.2% W 1.8% V steel: Clyde Alloy **DPN: 800**
CMC	0.9% C 0.3% Cr 0.3% Mo 0.2% V steel: Origin unknown
COL-GRAPH	1.45% C 0.2% Cr 0.25% Mo steel: Columbia; AISI type 06
COLUMBIA SPECIAL	1.06% C 0.2% Cr 0.05% V steel: Columbia; AISI type W1

Symbol	Nominal analysis, supplier, condition and remarks.
CONQUEROR 14	0.75% C 4.0% Cr 2.0% V 14.0% W steel: For tools; origin unknown
CORONA 2	0.8% C 0.25% Mn 4% Cr 5% Mo 6.5% W 2% V steel: Sandvik
CRD 18	0.7% C 4.2% Cr 18% W 0.75% V steel: C Denton; shear blades; obsolete **DPN: 580**
CRD 22	0.78% C 4.5% Cr 22% W 1.7% V steel: C Denton; drills, taps, etc.; obsolete
CRD 652	0.85% C 4.5% Cr 5.3% Mo 6% W 2% V steel: C Denton; punches, saws, etc.; obsolete
CRD DIAMOND YELLOW	0.9% C 1.2% Mn 0.5% Cr 0.5% W 0.1% V steel: C Denton; blanking dies, gauges, etc.; obsolete **DPN: 540**
CRM 1	1.0% C 0.2% Si 0.6% Mn 5% Cr 1% Mo 0.25% V steel: For tools; Huntsman; cold punches, shear blades, etc. **DPN: 770**
CRM 2	0.6% C 0.25% Si 0.6% Mn 0.6% Cr 0.3% Mo steel: Huntsman; rivet tools, arbors, wrenches, etc. **DPN: 650**
CROMODIE HC	1.0% C 5.25% Cr 1.1% Mo 0.25% V steel: Firth Brown
CROMODIE HCV	2.4% C 5.25% Cr 1.15% Mo 4.1% V steel: Firth Brown
CRO-MO-LOY	1.0% C 5.0% Cr 1.0% Mo 0.2% V steel: Atlas; as AISI type A2
CROWN(4)SUPERB	0.75% C 0.95% Cr 0.18% V steel: Latrobe
CRP	0.93% C 1.0% Cr 0.15% V steel: Origin unknown
CRV 14	0.68% C 3.75% Cr 14% W 0.65% V steel: Huntsman; cutting tools **DPN: 870**
CRV 1444	1.2% C 4.4% Cr 13.7% W 3.7% V steel: Huntsman; cutting tools **DPN: 850**
CV 18	0.7% C 4.2% Cr 18% W 1.3% V steel: Huntsman; cutting tools **DPN: 870**
CV 22	0.8% C 4.5% Cr 22% W 1.4% V steel: Huntsman; cutting tools **DPN: 910**
CV 1842	0.8% C 4.2% Cr 18% W 2% V steel: Huntsman; cutting tools **DPN: 870**
CY	0.78% C 4.5% Cr 0.7% Mo 18% W 1.25% V steel: For tools; Osborn; obsolete **DPN: 910**
CYC 18/4/1	0.75% C 4% Cr 18% W 1% V steel: Marsh Brothers; cutting tools
CYCLONE 4V	1.3% C 4.0% Cr 4.5% Mo 4.0% V 5.5% W steel: For tools; origin unknown
CYCLONE 56	0.8% C 4.2% Cr 15% Mo 6.5% W 2% V steel: For tools; ESC; cutting tools & drills **DPN: 870**
CYLONE 56CW	0.78% C 4.2% Cr 5% Mo 6.5% W 2% V steel: For tools; ESC; blanking dies, thread rolling dies **DPN: 870**
CYCLONE 92	0.8% C 4.0% Cr 8.0% Mo 1.0% V 1.5% W steel: For tools; origin unknown

Symbol	Nominal analysis, supplier, condition and remarks.
CYCLONE 92CW	0.8% C 4.0% Cr 8.0% Mo 1.0% V 1.5% W steel: For tools; origin unknown
D 61	0.9% C 1.2% Mn 0.5% Cr 0.5% W 0.1% V steel: Fagersta **DPN: 750**
D 416	1.4% C 1.6% Cr 5.5% W 0.5% V steel: Fagersta **DPN: 800**
D 941	0.82% C 4.0% Cr 5.0% Mo 6.5% W 1.9% V steel: Fagersta **DPN: 870**
D 943	0.8% C 3.8% Cr 9.0% Mo 1.7% W 1.2% V steel: Fagersta **DPN: 750**
D 953	0.95% C 4.0% Cr 2.6% Mo 2.8% W 2.3% V steel: Fagersta **DPN: 680**
D 954	1.0% C 4.1% Cr 8.8% Mo 2.0% W 2.0% V steel: Fagersta **DPN: 750**
DBL 2	4% Cr 5% Mo 6% W 2% V steel: For tools; Allegheny-Ludlum; AISI code M2G
DBL 2½	4% Cr 6% Mo 6% W 2.4% V steel: For tools; Allegheny-Ludlum; AISI code M3G
DBL 3	4% Cr 6% Mo 6% W 3% V steel: For tools; Allegheny-Ludlum; AISI code M3G
DCM	1.0% C 0.6% Mn 5% Cr 1% Mo 0.25% V steel: Jonas, shear, blades, etc. **DPN: 600**
DOUBLE DIAMOND	0.9% C 0.95% Cr 0.3% Mo steel: Crucible Steel Co; for drills
DOUBLE FLYGO	0.73% C 4% Cr 18% W 1% V steel: Turton Bros
DOUBLE MUSKET ND	0.7% C 4.0% Cr 1.0% V 18.0% W steel: Origin unknown
DOUBLE SPECIAL	1.4% C 0.8% Cr 5% W steel: Sanderson; turning tools for cast iron copper, etc. **DPN: 720**
DOUBLE MUSKET	0.75% C 4.25% Cr 1.25% V 18.0% W steel: For tools; origin unknown
DOUBLE MUSKET	0.75% C 18% W 4.25% Cr 1.25% V steel: For tools; Osborn; obsolete
DUREX	0.75% C 4.5% Cr 18.5% W 1% V tool steel: Darwin; cold heading dies, etc.
DUX 12	1.2% C 0.75% Cr 1.4% W steel: ESC; drills, taps, dental burrs, etc. **DPN: 750**
ECHO MOLYBDENUM	0.82% C 4.1% Cr 5% Mo 6.4% W 1.9% V steel: John Vessey
ECLIPSE 103	0.7% C 4.0% Cr 1.0% V 18.0% W steel: Origin unknown
EF 10	1.0% C 0.1% Cr 0.03% Mo 0.03% V steel: Pompey
EGALIT	0.85% C 2.1% Mn 0.07% Cr 0.16% V steel: Pompey
EL	0.78% C 4.2% Cr 18% W 2% V steel: For tools; Osborn; obsolete **DPN: 850**
ELECTEM	1.5% Ni Cr Mo steel: For dies, Somers
ELECTRITE 1	4% Cr 18% W 1% V steel: For tools; Latrobe Steel Co; AISI code T1F
ELECTRITE 19	4% Cr 18% W 2% V steel: For tools; Latrobe Steel Co; AISI code T2F
ELECTRITE CORSAIR	4% Cr 6% Mo 6% W 2.4% V steel: For tools; Latrobe Steel Co; AISI code M3F
ELECTRITE CORSAIR XL	1.02% C 4.0% Cr 6.0% Mo 6.1% W 2.4% V steel: Latrobe cutting tools; M3 Type 1 **DPN: 850**
ELECTRITE CRUSADER	4% Cr 6% Mo 6% W 3% V steel: For tools; Latrobe Steel Co; AISI code M3F

Note. The following abbreviations and units are used in the tables:

DPN	Hardness, diamond pyramid number
UTS	Ultimate tensile strength, N/mm^2
Elon	Elongation, %
Proof	0.1 % proof strength, N/mm^2

1 N/mm^2=0.1 hbar=0.102 kgf/mm^2=0.06475 tonf/in^2=145.04 lbf/in^2

Symbol	Nominal analysis, supplier, condition and remarks.
ELECTRIDE CRUSADER XL	1.2% C 4.1% Cr 6.0% Mo 6.0% W 3.2% V steel: Latrobe; cutting tools; M3 type 2 **DPN: 850**
ELECTRITE DOUBLE 6	4% Cr 5% Mo 6% W 2% V steel: For tools; Latrobe Steel Co; AISI code M2F
ELECTRITE DOUBLE 6 M2XL	0.85% C 4.1% Cr 5.0% Mo 6.3% W 1.8% V steel: Latrobe; cutting tools **DPN: 800**
ELECTRITE MV-1	0.8% C 0.15% W 4.1% Cr 1.8% V 4.5% Mo steel: Latrobe; for tools
ELECTRITE MV2	0.88% C 1.1% W 4.1% Cr 1.8% V 4.5% Mo steel: Latrobe; for tools
ELECTRITE No. 1 XL	0.75% C 4.1% Cr 0.7% Mo 18% W 1.1% V steel: Latrobe; cutting tools; type T1 with sulphides **DPN: 800**
ELECTRITE No. 7	0.68% C 4.25% Cr 5.0% Mo 6.4% W 1.9% V steel: Latrobe; cutting tools; type H42
ELECTRITE No. 19	0.85% C 4.1% Cr 0.6% Mo 18% W 2.1% V steel: Latrobe; cutting tools; type T2
ELECTRITE STARK	1.28% C 4.5% Cr 4.5% Mo 5.5% W 4.0% V steel: Latrobe; cutting tools; type M4 **DPN: 850**
ELECTRITE TATMO	4% Cr 8% Mo 1.5% W 1% V steel: For tools; Latrobe Steel Co; AISI code M1F
ELECTRITE TATMO V	1.0% C 3.7% Cr 8.7% Mo 1.7% W 2.0% V steel: Latrobe; cutting tools; type M7
ELECTRITE TATMO XL	0.8% C 4.0% Cr 8.5% Mo 1.5% W 1.0% V steel: Latrobe; cutting tools; type M1
ELECTRITE TNW	4% Cr 8% Mo 2% V steel: For tools; Latrobe Steel Co; AISI code M10F
ELECTRITE TNW XL	0.87% C 4.0% Cr 8.0% Mo 1.9% V steel: Latrobe; cutting tools; type M10
ELECTRITE U	4% Cr 14% W 2% V steel: For tools; Latrobe Steel Co; AISI code T7B
ELECTRITE VANADIUM	4% Cr 18% W 3% V steel: For tools; Latrobe Steel Co; AISI code T3F
ESA	1.45% C 4.0% W 0.5% Cr 0.25% V steel: Latrobe
ETA	0.9% C 0.15% V 0.5% W steel: Origin unknown
EV 5	0.75% C 0.5% Cr 6.0% W steel: For magnets; Russian alloy listed by PMA
EW 6	0.7% C 0.5% Cr 6.0% W steel: For magnets; Polish alloy listed by PMA
EVM	4% Cr 18% W 2% V steel: For tools; Vanadium Alloys; AISI code T2A
EXL-DIE	0.95% C 0.5% Cr 0.1% V steel: Columbia; AISI type 01
˜XTRA DOUBLE MUSHET	0.76% C 4.2% Cr 0.3% Mo 0.8% V 18.2% W 2.2% steel: For tools; origin unknown
EXTRA SUPER	0.79% C 4.5% Cr 1.4% V 22.0% W steel: For tools; origin unknown
EXTRA TRIPLE CONQUEROR	1.55% C 0.55% Cr 5.5% W steel: For tools; origin unknown
EXTRA TRIPLE GRIFFIN	1.7% C 0.2% Cr 6% W steel: For tools; Balfour; tube drawing, dies, etc. **DPN: 850**
E-Z-DIE SMOOTHCUT	1.0% C 5.25% Cr 1.1% Mo 0.2% V steel: Columbia; AISI type A2S; free cutting
E-Z-DIE V	2.23% C 5.25% Cr 1.1% Mo 1.1% W 4.8% V steel: Columbia; AISI type A7
EXTRA VANADIUM	0.7% C 4% Cr 18% W 1.25% V steel: For tools; Vulcan; cutting tools **DPN: 850**
F 1	1.23% C 1.35% W 0.45% Cr 0.2% V 0.25% Mo steel: Designation used by AISI
F 3	1.25% C 0.75% Cr 3.5% W steel: Designation used by AISI
F 4	0.95% C 0.2% Si 1% Mn 0.5% Cr 0.2% V steel: Jessop
FAGERSTA D941	0.82% C 0.3% Mn 4.0% Cr 5.0% Mo 1.9% V 6.5% W steel: For saw blades; Fagersta
FAGERSTA D953	0.95% C 0.3% Mn 4.0% Cr 2.6% Mo 2.3% V 2.8% W steel: Fagersta
FAVORIT	0.9% C 0.2% Cr 0.15% V steel: Schoeller-Bleckmann **DPN: 540**
FHS 18	0.75% C 4.0% Cr 1.0% V 18.0% W steel: For tools; origin unknown
FLASHKUT	0.95% C 3.7% Cr 0.25% Mo 0.18% V steel: Origin unknown
FMP	0.95% C 1.25% Mn 0.55% Cr 0.5% W 0.15% V steel: F Parkin **DPN: 800**
fmp 200	0.95% C 0.5% W 0.5% Cr 0.17% V steel: F Parkin; as tool steel 01
fmp 379	1.0% C 5.0% Cr 0.3% V 1.1% Mo steel: F Parkin; as tool steel A2
fmp 470	0.8% C 22.0% W 4.5% Cr 1.5% V 1.0% Mo steel: F Parkin
fmp 501	0.8% C 2.0% W 3.9% Cr 1.2% V 9.0% Mo steel: F Parkin; as tool steel M1
fmp 504	1.3% C 5.7% W 4.4% Cr 4.0% V 4.6% Mo steel: F Parkin; as tool steel M4
fmp 562	0.82% C 6.2% W 4.1% Cr 2.0% V 5.0% Mo steel: F Parkin; as tool steel M2
fmp 563	1.0% C 6.0% W 4.0% Cr 2.3% V 5.0% Mo steel: F Parkin; as tool steel M3/1
fmp 599	0.7% C 14.0% W 4.2% Cr 0.8% V steel: F Parkin
fmp 842	0.84% C 18.5% W 4.2% Cr 2.2% V 1.0% Mo steel: F Parkin; as T2 tool steel
fmp 922	1.0% C 1.7% W 4.0% Cr 2.0% V 8.7% Mo steel: F Parkin
fmp 948	0.88% C 3.0% Cr 2.0% V 8.25% Mo steel: F Parkin; as tool steel M10
FRAMDIE	1.1% C 1.35% Cr 0.4% Mo steel: Columbia; AISI type L7
FS M2½	1.02% C 4.0% Cr 6.1% Mo 6.1% W 2.5% V steel: Firth Sterling
FS M10	0.87% C 4.0% Cr 8.2% Mo 2.0% V steel: Firth Sterling
FURIOUS	1.25% C 1.2% Cr 4.5% W 0.25% V steel: Hall & Pickles; broaches, taps, punches, etc.; obsolete **DPN: 750**
G 23	0.62% C 0.2% Si 0.6% Mn 2% Cr 2% Ni 0.45% Mo steel: Jessop
G 24	3.0% C 4.5% Cr 13% Ni 4.5% Cu steel: Casting; Jessop
G 25	0.7% C 0.2% Si 0.3% Mn 0.2% Cr 0.45% Ni steel: Jessop
GS 8	0.9% C 1.2% Mn 0.6% Cr 0.5% W steel: For tools; Osborn; taps, dies, broaches, etc.; obsolete **DPN: 772**
GFS	0.9% C 0.5% Cr 0.5% W steel: Origin unknown
GOAHED	0.7% C 4.3% Cr 18% W 1.3% V steel: John Vessey
GRAPH-AIR	1.35% C 1.8% Ni 1.5% Mo steel: Free graphite; Timken; tools to withstand wear **UTS: 1950 Proof: 1620**
GRAPH-TUNG	1.5% C 0.5% Mo 2.8% W steel: Free graphite; Timken; dies & tools to withstand wear **DPN: 800**
H 4	0.9% C 1.8% Mn 0.5% Cr 0.5% W steel: Jessop **DPN: 800**
H 4 SPECIAL	0.9% C 1.4% Mn 0.3% Cr 0.3% Mo 0.2% V steel: Origin unknown
H 4 Sp1	0.9% C 0.2% Si 2% Mn 0.35% Cr 0.35% W steel: Jessop
H 7	2.3% C 5.2% Cr 1.1% Mo 4.5% V steel: Jessop Refractory Moulds **DPN: 730**

Symbol	Nominal analysis, supplier, condition and remarks.
H 15	0.8% C 0.2% Si 0.5% Mn 0.6% Cr 0.2% Mo steel: Jessop
H 33	1.0% C 0.4% Si 0.3% Mn 6.2% Cr 1% Mo steel: Jessop
	DPN: 750
H 62	1.0% C 0.2% Si 1.1% Mn 0.4% Cr 0.3% Ni 0.35% V steel: Jessop
HARDTRODE 85.65	0.9% C 1.3% Mn 4.5% Cr 7.5% Mo 2.0% W 1.6% V steel: For electrode; Esab
	DPN: 700
HEDERVAN	1.4% C 0.15% Cr 0.1% Mo 3.5% V steel: Latrobe; punches, dies, etc.
	DPN: 820
HECLA 15	0.9% C 0.3% Cr 0.3% Mo 0.2% V steel: Origin unknown
HECLA 114	0.7% C 0.8% Cr 1.5% Ni 0.2% Mo steel: Origin unknown
HECLA 175	1.0% C 5.0% Cr 1.2% Mo 0.4% V steel: For tools; origin unknown
HI WEAR 64	1.5% C 1.0% Cr 1.0% Mo 4.0% W steel: Carpenter
	DPN: 870
HI-Mo	4% Cr 8% Mo 1.5% W 1% V steel: For tools; Firth Sterling Inc; AISI code M1B
H M HIGH SPEED	4% Cr 8% Mo 1.5% W 1% V steel: For tools; Bethlehem Steel Co; AISI code M1E
HORSEMAN SUPER	0.75% C 4.0% Cr 1.0% W 18.0% W steel: For tools; origin unknown
H RV	1.0% C 1.5% Cr 0.25% V steel: Osborn; cutting tools
	DPN: 850
HRW	1.0% C 0.6% Mn 1.5% Cr 0.5% W steel: Jonas; press tools
	DPN: 720
HS 22	0.75% C 5% Cr 22% W 2% V steel: For tools; Darwin; form cutters, etc.
HSP 41	0.7% C 3.7% Cr 0.75% V 14.5% W steel: For tools; origin unknown
HURCO EXCELLENT	0.75% C 4.0% Cr 1.0% W 18.0% W steel: For tools; origin unknown
HUSKY	0.8% C 4.5% Cr 22% W 1.5% V steel: Hall & Pickles; cutting tools; obsolete
	DPN: 800
HV BLUE CHIP	4% Cr 18% W 2% V steel: For tools; Firth Sterling Inc; AISI code T2B
HV Hi Mo	4% Cr 8.7% Mo 1.7% W 2% V steel: For tools; Firth Sterling Inc; AISI code M7B
HYDRA	0.65% C 4% Cr 15% W 0.5% V steel: Hall & Pickles; drills, taps, reamers, etc.; obsolete
HYDRA HUSKY	0.8% C 4.5% Cr 22% W 1.5% V steel: Hall & Pickles; cutting tools; obsolete
	DPN: 800
HYDRA VANTAGE	1.25% C 4.5% Cr 13.5% W 4% V steel: Hall & Pickles; taps, reamers, etc.; obsolete
	DPN: 800
HYPEAK 14	0.8% C 4% Cr 14% W 1% V steel: Swift Levick; cutting tools
HYPEAK 101	0.8% C 4% Cr 5% Mo 6.5% W 2% V steel: Swift Levick; cutting tools

Symbol	Nominal analysis, supplier, condition and remarks.
HYPEAK 202	0.8% C 4.2% Cr 0.5% Mo 18% W 1.2% V steel: Swift Levick; cutting tools
IAS	1.2% C 0.75% Cr 1.75% W steel: Origin unknown
IDI	1.3% C 1.3% Mn 1.1% Cr 0.5% W steel: Inman
IMPERIAL	1.4% C 0.75% Cr 4.5% W steel: Edgar Allen; cutting tools
INMOLYTE M2	0.85% C 4.0% Cr 2.0% V 6.0% W 5.0% Mo steel: For tools; Inman
INVARD	0.9% C 0.5% Cr 0.5% W 0.2% V steel: Firth Sterling
INVINCIBLE 18	0.7% C 4.0% Cr 1.0% V 18.0% W steel: For tools; origin unknown
INVINCIBLE 22	0.75% C 4.0% Cr 1.25% V 22.0% W steel: For tools; origin unknown
J 1	0.67% C 0.3% Si 0.3% Mn 3.7% Cr 13.2% W 0.6% V steel: Jessop
J 13	1.23% C 4.5% Cr 13% W 3.7% V steel: Jessop
J 24	0.8% C 4% Cr 3.4% W 5% Mo 1.9% V with S steel: High speed; Jessop; cutting tools
	DPN: 850
J 30	0.75% C 0.3% Si 0.3% Mn 4.2% Cr 18% W 1.1% V steel: Jessop
J 34	0.8% C 4% Cr 5% Mo 6.5% W 1.9% V steel: Jessop
	DPN: 750
J 35	0.81% C 4% Cr 8.8% Mo 1.6% W 1.1% V steel: Jessop
J 38	0.8% C 4% Cr 0.75% Mo 18% W 2% V steel: Jessop
J 39	0.77% C 0.3% Si 0.3% Mn 4.2% Cr 21% W 1.5% V steel: Jessop
J 40	0.78% C 4.2% Cr 0.6% Mo 9.0% W 2.0% V steel: Jessop; high speed
J 41	0.85% C 4.5% Cr 0.8% Mo 11.5% W 3.0% V steel: Jessop; high speed
JA 3	0.97% C 5% Cr 1% Mo 0.3% V steel: Osborn; dies, cold forming tools, etc.; obsolete
	DPN: 770
JIS G4805 SUJ4	1.02% C 0.5% Mn (max) 1.45% Cr 0.17% Mo steel: For bearings; Japanese Standard
JIS G4805 SUJ5	1.02% C 1.0% Mn 1.05% Cr 0.17% Mo steel: For bearings; Japanese Standard
K 4	0.9% C 1.2% Mn 0.5% Cr 0.5% W steel: Jessop
	DPN: 800
K 4 SPECIAL	0.9% C 0.5% Cr 0.5% W steel: Origin unknown
K 4 Spl	0.92% C 0.5% Cr 0.5% W steel: Jessop; carbon and alloy tools
K 6	1.25% C 0.2% Si 0.3% Mn 0.3% Cr 1.3% W steel: Jessop
K 6 Spl	1.4% C 0.2% Si 0.3% Mn 0.3% Cr 1.5% W steel: Jessop
K 8 HP	0.9% C 0.2% Si 0.9% Mn 1.1% Cr 1.5% W steel: Jessop
K 9	1.0% C 0.8% Mn 0.75% Cr 0.4% W steel: Edgar Allen; dies, gauges, etc.
	DPN: 750
K 10 Spl	1.6% C 0.3% Si 0.3% Mn 0.6% Cr 5.5% W steel: Jessop
K 15	1.25% C 0.3% Si 0.3% Mn 1.1% Cr 4.5% W 0.25% V steel: Jessop
K 16	0.9% C 0.2% Si 1.2% Mn 0.5% Cr 0.5% W 0.2% V steel: Jessop
K 17	0.7% C 0.15% Si 0.2% Mn 0.4% Cr 6% W 0.5% Co steel: For magnets; Jessop
K 20	1.5% C 0.2% Si 1.1% Mn 0.6% Cr 0.7% W steel: Jessop
K 21	1.05% C 0.2% Si 0.5% Mn 1% Cr 0.6% W steel: Jessop

Note. The following abbreviations and units are used in the tables:

DPN	Hardness, diamond pyramid number
UTS	Ultimate tensile strength, N/mm^2
Elon	Elongation, %
Proof	0.1 % proof strength, N/mm^2

1 N/mm^2=0.1 hbar=0.102 kgf/mm^2=0.06475 tonf/in^2=145.04 lbf/in^2

Symbol	Nominal analysis, supplier, condition and remarks.
KAOC	1.3% C 0.3% Si 0.5% Mn 1% Cr 3% W steel: For tools; Huntsman; taps, reamers, etc. **DPN: 750**
KE 127	0.95% C 0.4% Mn 0.4% Cr 0.15% V steel: Kayser Ellison; ball & roller bearings
KE 226	1.0% C 0.5% Cr 2.1% W steel: Kayser Ellison; piercing punches, etc. **DPN: 950**
KE 595	1.25% C 0.8% Mn 1.2% Cr 1.2% W steel: Kayser Ellison; taps, dies, etc. **DPN: 910**
KE 672	1.1% C 1.5% Cr 0.5% W steel: Kayser Ellison; press tools **DPN: 870**
KE 708	1.4% C 0.35% Mn 0.6% Cr 1.5% W steel: Kayser Ellison; punches, taps, etc.
KEA 162	1.0% C 0.6% Mn 5.2% Cr 1% Mo 0.2% V steel: Kayser Ellison; cutting tools **DPN: 850**
KE DIAMOND 10	1.5% C 0.4% Mn 0.6% Cr 5.7% W steel: Kayser Ellison; cutting tools **DPN: 960**
KEEWATIN	0.9% C 0.5% Cr 0.5% W 0.2% V steel: Atlas; as AISI type 01
KELOCK 237	0.75% C 4.1% Cr 1.1% V 18.0% W steel: For tools; Sanderson Kayser
KELOCK 795	0.7% C 4% Cr 14% W 0.6% V steel: Kayser Ellison; cutting tools, shear blades
KELOCK 1014	0.75% C 6% Cr 0.75% Mo 14% W 1.4% V steel: Kayser Ellison; cutting tools
KELOCK A157	0.8% C 4% Cr 5% Mo 6% W 2% V steel: Kayser Ellison; cutting tools
KELOCK A182	0.8% C 3.9% Cr 8.5% Mo 1.15% V 1.5% W steel: For tools; Sanderson Kayser
KELOCK A229	1.05% C 3.7% Cr 9.5% Mo 1.1% V 1.5% W steel: For tools; Sanderson Kayser
KH 6VF	1.1% C 6.2% Cr 0.55% V 13.0% W steel: Russian Standard designation
KHV 5	1.35% C 0.55% Cr 0.22% V 0.45% W steel: Russian Standard designation
KHVG	1.0% C 1.0% Mn 1.05% Cr 1.4% W steel: Russian Standard designation
KHVSG	1.0% C 0.85% Cr 0.1% V 0.85% W steel: Russian Standard designation
KISKI	0.9% C 0.5% Cr 0.6% W 0.2% V steel: Braeburn; AISI type 01
KK	1.1% C 1.4% Cr 0.4% Mo steel: Atlas **DPN: 520**
KK 3	0.83% C 4.2% Cr 5% Mo 6.5% W 2% V steel: For tools; Osborn; obsolete **DPN: 850**
KONCOR	1.1% C 1% Si 5.2% Cr 1.1% Mo 4% V steel: Latrobe; precision casting of forging dies **DPN: 610**
KSA	0.9% C 1.8% Mn 0.5% Cr 0.5% W steel: Jessop **DPN: 800**
KSA	0.9% C 1.4% Mn 0.3% Cr 0.3% Mo 0.2% V steel: Origin unknown
L 2	0.8% C 1.0% Cr 0.2% V steel: Carbon varies over wide limits; designation used by AISI
L 3	1.0% C 1.5% Cr 0.2% V steel: Designation used by AISI
L 3	1.0% C 1.4% Cr 0.25% V steel: For cold work; Osborn
L 4	1.0% C 0.6% Mn 1.5% Cr 0.25% V steel: Designation used by AISI; obsolete
L 5	1.0% C 1.0% Mn 1.0% Cr 0.25% Mo steel: Designation used by AISI; obsolete
L 6	0.7% C 0.8% Cr 0.2% Mo 1.5% Ni steel: Designation used by AISI

Symbol	Nominal analysis, supplier, condition and remarks.
L 7	1.0% C 1.5% Cr 0.4% Mo steel: Designation used by AISI
LESCO HS29XL	0.08% C 6.3% W 4.1% Cr 1.8% V 5.05% Mo steel: Latrobe; tools; contains sulphides
LF	1.25% C 4.5% W 1.15% Cr 0.25% V steel: For tools; Osborn; broaches, reamers, taps, obsolete **DPN: 850**
LHS 14	0.7% C 3.5% Cr 14% W 0.5% V steel: For tools; Low Moor; obsolete **DPN: 850**
LHS 18	0.75% C 4% Cr 18% W 1.2% V steel: For tools; Low Moor; obsolete **DPN: 870**
LHS 22	0.75% C 4.5% Cr 22% W 1.2% V steel: For tools; Low Moor; obsolete **DPN: 910**
LHS 652	0.8% C 4% Cr 5% Mo 6% W 1.9% V steel: For tools; Low Moor; obsolete **DPN: 870**
LMW	High C 4% Cr 8% Mo 1.5% W 1% V steel: For tools; Allegheny Ludlum; AISI code M1G
LMW-V	High C 4% Cr 8.7% Mo 1.7% W 2% V steel: For tools; Allegheny Ludlum; AISI code M7G
LOCKPORT SPECIAL	High C 4% Cr 18% W 2% V tool steel: Simonds; AISI code T2N
LT 12	1.25% C 3% W 0.15% V steel: Low Moor; for tools; obsolete **DPN: 950**
LT 22V	1.0% C 1.5% Cr 0.2% V steel: Low Moor; non-shrinking; tools; obsolete **DPNɪ 800**
LT 23	0.9% C 1.25% Mn 0.5% Cr 5% W steel: Low Moor; non-shrinking; tools; obsolete **DPN: 800**
LT 24	1.0% C 1% Mn 0.8% Cr 0.7% W steel: Low Moor; non-shrinking; obsolete **DPN: 850**
LT 40	1.0% C 5% Cr 1% Mo 0.3% V steel: Low Moor; obsolete **DPN: 770**
LTAH	0.7% C 2% Mn 0.35% Si 1% Cr 1.5% Mo steel: Sanderson; forming dies, shear blades, etc. **DPN: 680**
LTTS	1.15% C 0.3% Si 0.3% Mn 0.7% Cr 1.3% W 0.25% V steel: For tools; Huntsman; centres, hacksaw blades, etc. **DPN: 720**
LXX	4% Cr 18% W 1% V steel: For tools; Allegheny Ludlum; AISI code T1G
M 1	0.8% C 4% Cr 8% Mo 1.5% W 1% V steel: For tools; designation used by AISI; the supplier is coded by a suffix letter
M 1	0.8% C 4.0% Cr 1.8% W 8.75% Mo 1.1% V steel: Osborn
M 2	0.85% C 4% Cr 5% Mo 6% W 2% V steel: For tools; designation used by AISI; the supplier is coded by a suffix letter
M 2	0.85% C 4.0% Cr 6.3% W 5.0% Mo 1.9% V steel: Osborn
M 3/1	1.05% C 4% Cr 6% Mo 6% W 2.4% V steel: For tools; designation used by AISI; the supplier is coded by a suffix letter
M 3/2	1.2% C 4% Cr 6% Mo 6% W 3% V steel: For tools; designation used by AISI; the supplier is coded by a suffix letter
M 4	1.3% C 4% Cr 4.5% Mo 5.5% W 4% V steel: For tools; designation used by AISI; the supplier is coded by a suffix letter

Symbol	Nominal analysis, supplier, condition and remarks.
M 7	1.0% C 4% Cr 8.7% Mo 1.7% W 2% V steel: For tools; designation used by AISI; the supplier is coded by a suffix letter; (higher C than M1)
M 8	0.8% C 5.0% Cr 5.0% Mo 5.0% W 1.5% V 1.2% Nb steel: Designation used by AISI; obsolete
M 10	0.9% C 4% Cr 8% Mo 2% V steel: For tools; designation used by AISI; the supplier is coded by a suffix letter
M 50	0.8% C 4.1% Cr 4.25% Mo 1.1% V steel: Latrobe
M 52	0.88% C 1.1% W 4.1% Cr 1.8% V 4.5% Mo steel: Designation used by AISI; for tools
MANGDIE	0.95% C 1.25% Mn 0.5% Cr 0.5% W 0.2% V steel: Clyde Alloy **DPN: 750**
MANOIR ABRADUR 220	0.7% C 1.0% Mn 0.04% S & P steel: Pompey **UTS: 770** **Elon: 7%** **Proof: 500**
MARSH CYC1841	0.75% C 4% Cr 18% W 1% V steel: Marsh Brothers; cutting tools
MAXIMUM 18	0.75% C 4.2% Cr 18% W 1.1% V steel: Firth Brown
MC 2	0.8% C 3.75% Cr 8.7% Mo 1.7% W 1.1% V steel: For tools; Osborn; obsolete **DPN: 850**
MCMO	0.7% C 2% Mn 1% Cr 1.3% Mo steel: For tools; Huntsman; blanking dies, punches, etc. **DPN: 720**
MCT	0.95% C 1.1% Mn 0.5% Cr 0.5% W steel: Firth Brown
MEL-TROL 484FM	1.0% C 5.2% Cr 1.1% Mo 0.2% V steel: Carpenter for AISI type A2; contains sulphides
MEL-TROL HI SHOCK 60	0.68% C 1.0% Cr 0.5% Ni 1.0% Mo 2.5% Cu 0.15% V steel: For tools; Carpenter
MEL-TROL HI SHOCK 60	0.7% C 1.0% Cr 0.5% Ni 1.0% Mo 2.5% Cu 0.15% V steel: Carpenter
MEL-TROL HI WEAR 64	1.5% C 0.9% Cr 1.0% Mo 4.0% W steel: Carpenter **DPN: 750**
MEL-TROL SPEED STAR	0.8% C 4.2% Cr 5.0% Mo 6.2% W 1.9% V steel: Carpenter for AISI type M2
MEL-TROL STAR ZENITH	0.7% C 4.0% Cr 18.2% W 1.1% V steel: Carpenter for AISI type T1
MEL-TROL VEGA	0.7% C 0.3% Si 2.0% Mn 1.0% Cr 1.35% Mo steel: For tools; Carpenter for AISI A6
MEL-TROL VEGA-FM	0.7% C 2.2% Mn 1.0% Cr 1.35% Mo steel: For tools; Carpenter; for AISI A6
MERMAID	0.75% C 4.0% Cr 18% W 1.4% V steel: Spear & Jackson
MIC 3	0.95% C 1.5% Mn 0.5% Cr 0.5% W steel: ESC; taps, punches, reamers, etc. **DPN: 750**
MIC 4	0.95% C 1.25% Mn 0.5% Cr 0.5% W 0.2% V steel: ESC; taps, dies, reamers, etc. **DPN: 800**
MIC 9	1.0% C 0.8% Mn 0.7% Cr 0.4% W steel: ESC; small punches, form tools, etc. **DPN: 750**
ML	4% Cr 18% W 2% V steel: For tools; Allegheny Ludlum; AISI code T2G
MOCARB	1.0% C 4.0% Cr 1.9% V 6.5% W 5.0% Mo steel: Braeburn

Note. The following abbreviations and units are used in the tables:

DPN	Hardness, diamond pyramid number
UTS	Ultimate tensile strength, N/mm^2
Elon	Elongation, %
Proof	0.1 % proof strength, N/mm^2

1 N/mm^2 = 0.1 hbar = 0.102 kgf/mm^2 = 0.06475 tonf/in^2 = 145.04 lbf/in^2

Symbol	Nominal analysis, supplier, condition and remarks.
MO CUT	4% Cr 8% Mo 1.5% W 1% V steel: For tools; Braeburn Alloy Steel; AISI code M11
MOGUL	4% Cr 8% Mo 1.5% W 1% V steel: For tools; Jessop; AISI code M1W
MOHICAN 8	4% Cr 8% Mo 1.5% W 1% V steel: For tools; Atlas Steel Ltd; AISI code M1Z
MOLITE	4% Cr 5% Mo 6% W 2% V steel: For tools; Columbia Tool Steel Co; AISI code M2H
MOLITE 1	0.82% C 4.0% Cr 8.5% Mo 1.6% W 1.1% V steel: Columbia; AISI type M1
MOLITE 3	4% Cr 6% Mo 6% W 2.4% V steel: For tools; Columbia Tool Steel Co; AISI code M3H
MOLITE 3 SMOOTHCUT	1.03% C 4.0% Cr 6.2% Mo 6.2% W 2.5% V steel: Columbia; AISI type M3S/1; free machining
MOLITE 4	1.28% C 4.5% Cr 4.5% Mo 5.5% W 4.0% V steel: Columbia
MOLITE SMOOTHCUT	0.85% C 4.1% Cr 5.0% Mo 6.2% W 1.9% V steel: Columbia; AISI type M2S; free machining
MOLVA	0.82% C 4.0% Cr 8.0% Mo 1.9% V steel: Simonds; AISI type M10
MOLVA T	0.82% C 4.0% Cr 5.0% Mo 6.0% W 1.9% V steel: Simonds; AISI type M2
MOLVA TC	1.2% C 4.0% Cr 6.25% Mo 6.25% W 2.8% V steel: Simonds; AISI type M3
MOLYCUT 562	0.8% C 4.1% Cr 5% Mo 6.4% W 1.9% V steel: Firth Brown
MONARCH	0.75% C 7.5% Cr 1.3% V 22.0% W steel: For tools; origin unknown
MONARCH 652	0.85% C 4.0% Cr 5.0% Mo 2.0% V 6.0% W steel: For tools; origin unknown
MONARCH OHB	0.9% C 0.5% Cr 0.5% W steel: For tools; origin unknown
MONARCH PCS	0.8% C 1.25% Cr 0.5% Ni 1.9% W steel: For tools; origin unknown
MONARCH TAN	0.7% C 0.8% Cr 0.2% Mo 1.5% Ni steel: For tools; origin unknown
MOTEMP	0.88% C 4.0% Cr 8.0% Mo 2.0% V steel: Braeburn; AISI type M10
MO-TEMP M10	4% Cr 8% Mo 2% V steel: For tools; Braeburn Alloy Steel M10 type; AISI code M10
MOTOR MAGNUS	0.8% C 4.25% Cr 5% Mo 6.5% W 2% V steel: Carrs; form tools, gear cutters, etc. **DPN: 910**
MOTOR MAXIMUM	0.75% C 4.25% Cr 18% W 1.2% V steel: Carrs; boring tools, chisels **DPN: 910**
MOTOR SPECIAL	0.7% C 4% Cr 14% W 0.25% V steel: Carrs; twist drills, reamers, etc. **DPN: 850**
MOTUF	1.0% C 3.7% Cr 8.7% Mo 1.7% W 2.1% V steel: Braeburn; AISI type M7
MOTUNG	4% Cr 8% Mo 1.5% W 1% V steel: For tools; Universal Cyclops; AISI code M1D
MOTUNG 652	4% Cr 5% Mo 6% W 2% V steel: For tools; Universal Cyclops; AISI code M2D
MOTUNG CV	4% Cr 8.7% Mo 1.7% W 2% V steel: For tools; Universal Cyclops; AISI code M7D
MOVAN	4% Cr 8% Mo 2% V steel: For tools; Universal Cyclops; AISI code M10D
MOW 562	0.83% C 4% Cr 5% Mo 6% W 2% V steel: Huntsman; cutting tools Jessop; AISI code M2W
MUSHET HIGH VANADIUM	1.25% C 4.6% Cr 3.7% V 13.0% W steel: For tools; origin unknown
MUSHET M	0.83% C 4.2% Cr 5.0% Mo 1.9% V 6.5% W steel: For tools; origin unknown
MUSHET MOLYB.	0.8% C 3.7% Cr 8.7% Mo 1.1% V 1.7% W steel: For tools; origin unknown
MUSKET MKK	0.85% C 4.0% Cr 5.0% Mo 2.0% V 6.0% W steel: For tools; origin unknown

Symbol	Nominal analysis, supplier, condition and remarks.
MUSTANG	4% Cr 5% Mo 6% W 2% V steel: For tools; AISI code M2W
MUSTANG M3/1	4% Cr 6% Mo 6% W 2.4% V steel: For tools; Jessop; AISI code M3W
MUSTANG M3/2	4% Cr 6% Mo 6% W 3% V steel: For tools; Jessop; AISI code M3W
MV-1-VAC-ARC	0.8% C 4.1% Cr 4.25% Mo 1.1% V steel: Latrobe; ball races **DPN: 750**
NATRONA	0.88% C 4.0% Cr 1.75% V 0.5% W 4.0% Mo steel: Braeburn
NCM IHC	0.6% C 0.25% Si 0.45% Mn 0.9% Cr 0.5% Ni 0.3% Mo steel: Huntsman; thread rolling dies, forming tools, etc. **DPN: 720**
ND	0.75% C 18% W 4.25% Cr 1.25% V steel: For tools; Osborn 'Double Mushet'; obsolete **DPN: 850**
NDS	0.75% C 1.0% Cr 1.7% Ni steel: Latrobe
NEATRO	4% Cr 4.5% Mo 5.5% W 4% V steel: For tools; Vanadium Alloys; AISI code M14A
NEW CAPITAL	0.7% C 3.7% Cr 14.5% W 0.75% V steel: For tools; Balfour; punches, dies, etc. **DPN: 850**
NEWHALL	0.9% C 1.2% Mn 0.5% Cr 0.7% W 0.15% V steel: Sanderson; centres, collets, etc. **DPN: 750**
NF A35 565/100CD7	1.02% C 0.3% Mn 1.8% Cr 0.2% Mo steel: For bearings; French Standard
No.4 HARDENITE VAN	1.4% C 0.4% Cr 0.4% Mo 3.6% V steel: For tools; origin unknown
No.7 HARDENITE	0.9% C 0.5% Cr 0.5% W steel: For tools; origin unknown
No.8 HARDENITE	0.85% C 1.35% Cr 0.2% V 1.0% W steel: For tools; origin unknown
No.10 HARDENITE	1.2% C 3.0% Cr 1.0% W steel: For tools; origin unknown
NORESCO 88 HP	0.74% C 4.25% Cr 18% W 1% V steel: Schoeller-Bleckmann **DPN: 870**
NORESCO 99 HP	0.74% C 4.25% Cr 0.9% Mo 18.5% W 1.15% V steel: Schoeller-Bleckmann **DPN: 910**
NORESCO DHS	0.84% C 4.25% Cr 5% Mo 6.25% W 1.9% V steel: Schoeller-Bleckmann **DPN: 870**
NORESCO FAVORIT	0.9% C 0.2% Cr 0.15% V steel: Schoeller-Bleckmann **DPN: 540**
NORVAR	0.9% C 1.4% Mn 0.3% Cr 0.3% Mo 0.2% V steel: Origin unknown
NOVO	0.7% C 4% Cr 14% W 0.5% V steel: Jonas; cutting tools **DPN: 850**
NOVO 6/6/2	0.85% C 4.0% Cr 5.0% Mo 2.0% V 6.0% W steel: For tools; origin unknown
NOVO 9/2	0.75% C 1.5% Cr 3.75% Cr 1.0% V 9.0% Mo tool steel: Jonas; type M1
NOVO C	0.75% C 6.2% Cr 0.7% Mo 13.5% W 1.25% V steel: Jonas; cutting or parting off tools **DPN: 870**
NOVO SUPERIOR	0.75% C 4% Cr 18% W 1% V steel: Jonas; cutting tools **DPN: 850**
NOVO SUPERIOR 6/6/2	0.85% C 4% Cr 5.25% Mo 6% W 2% V steel: Jonas; cutting tools **DPN: 910**
NOVO SUPERIOR SS	0.75% C 4.5% Cr 21% W 1.25% V steel: Jonas; cutting form tools **DPN: 910**
NOVO TCV	1.5% C 11% W 5.0% Cr 5.0% V tool steel: Jonas Jonas
NRO 146	0.7% C 2.1% Mn 1.0% Cr 1.3% Mo 0.1% S steel: Bofors; annealed; for tools **DPN: 240**
NUTHERM	0.7% C 1.0% Cr 1.3% Mo steel: Atlas; AISI type A6
OHD	1.0% C 1.6% Cr 0.5% W steel: For tools; origin unknown
OI	0.93% C 0.5% Cr 0.5% W 0.2% V steel: For cold work; Osborn
OILDIE SMOOTHCUT	1.05% C 1.6% Cr 0.5% W steel: Free cutting; Columbia; AISI type 03S
P 9E5	1.45% C 4.1% Cr 9.7% W 4.6% V steel: For tools; Russian Standard designation
P 10	0.75% C 4.0% Cr 18% W 1.2% V steel: Bofors; cutting tools; obsolete **DPN: 850**
P 14F4	1.25% C 4.3% Cr 1.37% W 3.7% V steel: For tools; Russian Standard designation
P 86 OH	1.0% C 1% Mn 0.5% Cr 1% W 0.2% V steel: T Turton
PGT	1.15% C 0.8% Mn 0.75% Cr 1.5% W steel: Jonas; taps, dies, etc. **DPN: 630**
PILGER ROLL STEEL	1.1% C Cr Mo steel: Casting; Firth Brown; hardened & tempered **DPN: 340**
PITHO	0.9% C 1.2% Mn 0.5% Cr 0.7% W 0.15% V steel: Sanderson; alternative name to Newhall **DPN: 750**
PLUTOCRAT	0.8% C 4.25% Cr 0.5% Mo 22% W 1% V steel: Carrs; reamers, heavy duty drills, etc. **DPN: 910**
PO 2	0.85% C 4.0% Cr 5.0% Mo 6.5% W 2.0% V steel: Bofors; obsolete **DPN: 850**
PO 3	0.85% C 4.0% Cr 3.0% Mo 6.5% W 2.0% V steel: Bofors; obsolete **DPN: 850**
PO 7	1.0% C 3.8% Cr 9.0% Mo 1.8% W 2.0% V steel: Bofors; high speed steel as type M7
PP	1.0% C 1% Mn 0.8% Cr 0.8% W steel: Osborn; cutting tools; obsolete **DPN: 800**
PRN 2	1.05% C 1.3% Mn 1.3% Cr steel: Balfour; blanking tools, gauges, etc. **DPN: 660**
PYRO-VAN	0.75% C 5.2% Cr 1.1% Mo 2.5% V steel: Latrobe; hot work **DPN: 470**
QV	0.7% C 4.0% Cr 14.5% W 1.0% V steel: S Osborn; high speed steel
QV/E	0.7% C 4.0% Cr 14.5% W 1.0% V steel: Obsolete; S Osborn; replaced by QV
R 9	0.9% C 4.1% Cr 9.2% W 2.3% V steel: For tools; Russian Standard designation
R 12	0.85% C 3.4% Cr 12.5% W 1.7% V steel: For tools; Russian Standard designation
R 18	0.75% C 4.1% Cr 17.7% W 1.2% V steel: For tools; Russian Standard designation
R 18F2	0.9% C 4.1% Cr 17.7% W 2.1% V steel: For tools; Russian Standard designation
RDS	0.7% C 1.0% Cr 1.75% Ni steel: Carpenter; for AISI type L6
RED CUT SUPERIOR	4% Cr 18% W 1% V steel: For tools; Vanadium Alloys; AISI code T1A
RED STREAK	4% Cr 18% W 1% V steel: For tools; Simonds Steel Mills; AISI code T1N

Symbol	Nominal analysis, supplier, condition and remarks.
REX 4V	4% Cr 18% W 4% steel: For tools; Crucible Steel Co; AISI code T9X
REX 939	4% Cr 18% W 3% V steel: For tools; Crucible Steel Co; AISI code T3X
REX AA	4% Cr 18% W 1% V steel: For tools; Crucible Steel Co; AISI code T1X
REX CHAMPION	4% Cr 14% W 2% V steel: For tools; Crucible Steel Co; AISI code T7X
REX M2	4% Cr 5% Mo 6% W 2% V steel: For tools; Crucible Steel Co; AISI code M2X
REX M3/1	4% Cr 6% Mo 6% W 2.4% V steel: For tools; Crucible Steel Co; AISI code M3X
REX M3/2	4% Cr 6% Mo 6% W 3% V steel: For tools; Crucible Steel Co; AISI code M3X
REX M7	4% Cr 8.7% Mo 1.7% W 2% V steel: For tools; Crucible Steel Co; AISI code M7X
REX SUPER VAN	4% Cr 18% W 2% V steel: For tools; Crucible Steel Co; AISI code T2X
REX TMO	4% Cr 8% Mo 1.5% W 1% V steel: For tools; Crucible Steel Co; AISI code M1X
REX VM	4% Cr 8% Mo 2% V steel: For tools; Crucible Steel Co; AISI code M10X
RIGOR	1.0% C 5.3% Cr 1.1% Mo 0.2% V steel: Uddelholm for type A2
ROLLO	1.25% C 2.9% Cr 6.0% W steel: Origin unknown
ROP 21	1.0% C 5.5% Cr 1.1% Mo 0.2% V steel: Bofors DPN: 800
RSD	1.0% C 5.2% Cr 0.2% V 1.1% Mo steel: For tools; Balfour & Darwin
RT 1733	0.9% C 0.6% Cr 0.6% W 0.1% V steel: Bofors DPN: 800
RV	1.0% C 1.4% Cr 0.25% V steel: Obsolete; S Osborn; replaced by L3
RW 2	1.25% C 4.5% Cr 13% W 3.75% V steel: For tools; S Osborn; obsolete DPN: 870
S 3-3.2	1.0% C 4.2% Cr 2.65% Mo 2.85% W 2.3% V steel: For tools; German Standard designation
S 6-5.2	0.88% C 4.2% Cr 5.0% Mo 6.3% W 1.85% V steel: For tools; German Standard designation
S 6-5.3	1.22% C 4.2% Cr 5.0% Mo 6.3% W 2.0% V steel: For tools; German Standard designation
SABEN 652	0.81% C 4.1% Cr 5% Mo 6.5% W 2% V steel: Sanderson; cutting tools DPN: 770
SABEN EXTRA	0.75% C 4% Cr 18% W 1% V steel: Sanderson; cutting tools DPN: 770
SABEN HC	1.25% C 4.5% Cr 0.3% Mo 13.5% W 3.7% V steel: Sanderson; cutting tools DPN: 800
SABEN KERAU	0.75% C 4.25% Cr 0.25% Mo 22% W 1.5% V steel: Sanderson; cutting tools DPN: 800
SAE 01	0.9% C 1.2% Mn 0.5% Cr 0.5% W 0.2% V steel: For tools
SAE 02	0.9% C 1.4% Mn 0.3% Cr 0.3% Mo 0.2% V steel: For tools
SAE 6195	0.95% C 1% Cr 0.15% V steel: Obsolete

Symbol	Nominal analysis, supplier, condition and remarks.
SAE 8660	0.6% C 0.5% Cr 0.6% Ni 0.2% Mo steel
SAE A2	1.0% C 5% Cr 1.2% Mo 0.4% V steel: For tools
SAE I6	0.7% C 0.8% Cr 0.2% Mo 0.2% V steel
SAE I7	1.0% C 1.5% Cr 0.4% Mo steel
SAE M1	0.8% C 4% Cr 9% Mo 1.7% W 1.1% V steel: For tools
SAE M2	0.85% C 4.2% Cr 5% Mo 6.2% W 2% V steel: For tools
SAE M3	1.1% C 4.2% Cr 6% Mo 6% W 2.7% V steel: For tools
SAE M4	1.2% C 4.5% Cr 5% Mo 6% W 4.2% V steel: For tools
SAE T1	0.7% C 4.2% Cr 18% W 1% V steel: For tools
SAE T2	0.8% C 4.1% Cr 0.8% Mo 18% W 2% V steel: For tools
SC 6-5-2	1.0% C 4.2% Cr 5.0% Mo 6.3% W 1.85% V steel: For tools; German Standard designation
SCV	1.0% C 1.0% Cr 0.2% V steel: Sanderson; cutting dies, blanking tools, etc. DPN: 540
SCV 3	0.95% C 1.1% Cr 0.25% V steel: Spencer; ball races, gauges, etc.
SCV 5	1.0% C 5.0% Cr 0.25% V 0.95% Mo steel: For tools; W Spencer
SE4AS	0.82% C 4.5% Cr 0.8% Mo 18.2% W 1.7% V steel: Forez
SEARCHER A1	0.63% C 1.12% Si 0.4% Cr 1.6% W steel: Hall & Pickles; wedges, shear blades, obsolete DPN: 750
SELECT BFM	1.0% C 5.25% Cr 1.1% Mo 0.25% V steel: Latrobe; contains sulphides; AISI type A2 die steel DPN: 600
SHOCKTOUGH	High C 1.2% Cr + Ni 2% W steel: John Vessey; press tools, punches DPN: 580
SILVANITE	1.3% C 4.0% Cr 8.0% W 0.5% V steel: Columbia; AISI type F8
SIS 14-2092	1.0% C 0.75% Mn 1.0% Cr steel: Cold water tool steel; Swedish Standard DPN: 240
SIS 14-2140	1.0% C 1.2% Mn 0.5% Cr 0.5% W 0.1% V steel: Cold water tool steel
SIS 14-2260	1.0% C 0.6% Mn 5.0% Cr 1.1% Mo 0.2% V steel: Cold water tool steel; Swedish Standard DPN: 240
SIS 14-2310	1.5% C 0.45% Mn 12% Cr 0.8% Mo 0.85% V steel: Cold water tool steel; Swedish Standard DPN: 260
SIS 14-2722	0.85% C 0.3% Mn 4.0% Cr 5% Mo 6% W 2% V 0.03% P & S steel: High speed tool steel; Swedish Standard DPN: 260
SIS 14-2723	0.85% C 0.3% Mn 4.0% Cr 5.0% Mo 6.5% W 5.0% C 2.0% V 0.03% P & S steel: High speed tool steel; Swedish Standard DPN: 300
SIS 14-2724	0.85% C 0.3% Mn 4.0% Cr 3.2% Mo 6.5% W 2% V 0.03% P & S steel: High speed tool steel; Swedish Standard DPN: 260
SIS 14-2724	0.85% C 0.3% Mn 4.0% Cr 3.0% Mo 6.5% W 2.0% V steel: High speed tool steel; Swedish Standard
SIS 14-2782	1% C 0.3% Mn 4.0% Cr 8.5% Mo 2.7% V 2% V 0.03% P & S steel: High speed tool steel; Swedish Standard DPN: 260

Note. The following abbreviations and units are used in the tables:

DPN	Hardness, diamond pyramid number
UTS	Ultimate tensile strength, N/mm^2
Elon	Elongation, %
Proof	0.1 % proof strength, N/mm^2

$1\ N/mm^2 = 0.1\ hbar = 0.102\ kgf/mm^2 = 0.06475\ tonf/in^2 = 145.04\ lbf/in^2$

Symbol	Nominal analysis, supplier, condition and remarks.
SIS 2140	0.9% C 1.2% Mn 0.5% Cr 0.5% W 0.1% V steel: Swedish Standard; annealed **DPN: 220**
SIS 2260	1.0% C 5.2% Cr 1.1% Mo 0.2% V steel: Swedish Standard; annealed **DPN: 240**
SIS 2722	0.82% C 4% Cr 5% Mo 6.5% W 2% V steel: Swedish Standard; annealed **DPN: 260**
SIS 2724	0.88% C 4% Cr 3.2% Mo 6.5% W 2.1% V steel: Swedish Standard; annealed **DPN: 260**
SIW	0.67% C 0.6% Cr 1.3% W 1.15% Si steel: Balfour; chisels, coal cutter blades **DPN: 580**
SIXIX	4% Cr 5% Mo 6% W 2% V steel: For tools; Atlas Steels Ltd; AISI code M2Z
SIXIX-fm	0.84% C 4.0% Cr 5.0% Mo 6.5% W 1.9% V 0.12% S steel: Atlas; as AISI type M2; free machining
SKD 12	1.0% C 5.0% Cr 0.5% Ni (max) 1.0% Mo 0.35% V steel: Japanese Standard designation
SKF 3	1.0% C 0.3% Mn 1.5% Mn 0.35% Mo steel: For bearings; SKF Ltd.
SKF 24	1.0% C 1.8% Cr 0.2% Mo steel: Tube; SKF Ltd
SKF 25	1.0% C 1.8% Cr 0.35% Mo steel: Tube; SKF Ltd
SKH 2	0.77% C 4.1% Cr 18.0% W 1.0% V steel: For tools; Japanese Standard designation
SKH 9	0.85% C 4.2% Cr 5.0% Mo 6.1% W 1.9% V steel: For tools; Japanese Standard designation
SKH 52	1.05% C 4.2% Cr 5.5% Mo 6.2% W 2.4% V steel: For tools; Japanese Standard designation
SKH 53	1.2% C 4.3% Cr 5.5% Mo 6.2% W 3.1% V steel: For tools; Japanese Standard designation
SKH 54	1.32% C 4.2% Cr 5.0% Mo 5.9% W 4.2% V steel: For tools; Japanese Standard designation
SKS 1	1.35% C 0.75% Cr 4.5% W 0.2% V steel: For tools; Japanese Standard designation
SKS 2	1.05% C 0.8% Cr 1.25% W 0.2% V (max) steel: For tools; Japanese Standard designation
SKS 2	1.05% C 0.8% Cr 1.25% W 0.2% V (max) steel: Japanese Standard designation
SKS 3	0.95% C 0.75% Cr 0.75% W steel: Japanese Standard designation
SKS 5	0.8% C 0.32% Cr 1.0% Ni steel: Japanese Standard designation
SKS 7	1.15% C 0.35% Cr 0.2% V (max) 2.25% W steel: Japanese Standard designation
SKS 11	1.25% C 0.35% Cr 3.5% W steel: For tools; Japanese Standard designation
SKS 21	1.05% C 0.35% Cr 0.8% W steel: For tools; Japanese Standard designation
SKS 31	1.0% C 1.0% Mn 1.0% Cr 1.25% W steel: Japanese Standard designation
SKS 42	0.8% C 0.38% Cr 0.22% V 2.0% W steel: Japanese Standard designation
SKS 51	0.8% C 0.35% Cr 1.6% Ni steel: For tools; Japanese Standard designation
SL 173	1.0% C 1.2% Mn 0.5% Cr 0.5% W steel: Swift Levick; punches
SL 874	1.4% C 0.5% Cr 4.0% W steel: Swift Levick; shears, punches
SL 954	1.2% C 1.5% Cr 1.25% W steel: Swift Levick; press dies, punches
SNSC	0.9% C 0.28% Si 1% Mn 0.5% Cr 0.5% W 0.2% V steel: For tools; Huntsman; taps, shear blades, etc. **DPN: 720**
SOMDIE	1% Cr 1% Ni 0.5% Mo steel: For dies; Somers

Symbol	Nominal analysis, supplier, condition and remarks.
SPARTAN 7	4% Cr 18% W 1% V steel: For tools; Atlas Steels Ltd; AISI code T1Z
SPEAR 35	0.8% C 0.2% Cr 2.5% Ni steel: Origin unknown
SPEAR 562	0.85% C 4.0% Cr 5.0% Mo 2.0% V 6.0% W steel: For tools; origin unknown
SPEAR MERMAID	0.7% C 4.0% Cr 1.0% V 18.0% W steel: Origin unknown
SPEAR PS	1.05% C 1.5% Cr 0.25% Mo steel: Spear & Jackson
SPEED STAR	4% Cr 5% Mo 6% W 2% V steel: For tools; Carpenter Steel Co; AISI code M2K
SPECIAL ECHO	0.73% C 4.3% Cr 18.2% W 1.3% V steel: John Vessey
SPECIAL HS	4% Cr 18% W 1% V steel: For tools; Bethlehem Steel Co; AISI code T1E
SPEEDICUT MAXIMUM 18	0.75% C 4.2% Cr 18% W 1.1% V steel: Firth Brown
SPEEDICUT MAX 22	0.73% C 4.2% Cr 1.4% V 22.0% W steel: For tools; origin unknown
SPM	0.9% C 1.4% Mn 0.5% Cr 0.5% W steel: Jonas; punches; blanking & forming tools **DPN: 610**
STAG MO	0.85% C 4.75% Cr 5% Mo 6% W 2% V steel: Edgar Allen; cutting tools
STAG MO 562	0.85% C 4.75% Cr 5% Mo 6% W 2% V steel: Edgar Allen; also known as Stag Mo
STAG SPECIAL	0.7% C 5% Cr 19% W 1.5% V steel: Edgar Allen
STAR BLUE CHIP	4% Cr 14% W 2% V steel: For tools; Firth Sterling Inc; AISI code T7X
STAR COLUMBRIUM	4% Cr 5% Mo 5% W 1.5% V 1.2% Nb steel: For tools; Carpenter Steel Co, AISI code M8K
STAR MAX	4% Cr 8% Mo 1.5% W 1.0% V steel: For tools; Carpenter Steel Co; AISI code M1K
STAR MO 68	1.02% C 4.0% Cr 5.0% Mo 6.4% W 2.0% V steel: Firth Sterling
STAR MO M2	4% Cr 5% Mo 6% W 2% V steel: For tools; Firth Sterling Inc; AISI code M2B
STAR ZENITH	4% Cr 18% W 1.0% V steel: For tools; Carpenter Steel Co; AISI code T1K
STM	4% Cr 8% Mo 1.5% W 1.0% V steel: For tools; Simonds Steel Mills; AISI code M1N
STYR	0.78% C 4.5% Cr 0.5% Mo 21.5% W 1.4% V steel: For tools; ESC; symbol for SUPER TYR **DPN: 910**
SUPER AUSTENITE	0.7% C 4.5% Cr 0.5% Mo 18% W 1.0% V steel: Inman; cutting tools
SUPER AUSTENITE TI	0.7% C 4.0% Cr 1.0% V 18.0% W steel: For tools; Inman
SUPER DOUBLE MUSHET	0.78% C 4.25% Cr 1.0% V 22.5% W steel: For tools; origin unknown
SUPER ECLIPSE	1.0% C 1.0% Cr 0.25% Mo steel: Origin unknown
SUPER HS	4% Cr 18% W 2% V steel: For tools; Vulcan Crucible Steel Co; AISI code T2P
SUPER HYDRA	0.75% C 4.25% Cr 18% W 1.1% V steel: For tools; Hall & Pickles **DPN: 770**
SUPER MONARCH	0.7% C 4.0% Cr 18.0% W 1.0% V steel: For tools; origin unknown
SUPER RAPID	0.8% C 4.0% Cr 1.2% V 19.0% W steel: For tools; origin unknown
SUPER RAPID	0.7% C 4.0% Cr 1.0% V 18.0% W steel: Origin unknown
SUPER Si Mn	0.57% C 0.8% Mn 1.8% Si 0.3% Cr 0.25% Mo steel: For springs; ESC
SUPER SPECIAL ECHO	0.7% C 4.4% Cr 21% W 1.4% V steel: John Vessey
SUPER TYR	0.78% C 4.5% Cr 0.5% Mo 21.5% W 1.4% V steel: ESC; cutting tools **DPN: 910**

Symbol	Nominal analysis, supplier, condition and remarks.
SUPERIOR 5	1.35% C 6.2% Cr 6.2% V 1.1% Mo steel: Braeburn
SUPERIOR SPUR	0.72% C 4% Cr 0.3% Mo 18% W 1.0% V steel: T Turton
SUPREMUS	4% Cr 18% W 1.0% V steel: For tools; Jessop; AISI code T1W
SUPREMUS EXTRA	4% Cr 18% W 2% V steel: For tools; Jessop; AISI code T2W
T 1	0.7% C 4% Cr 18% W 1.0% V steel: For tools; designation used by AISI; the supplier is coded by a suffix letter
T 1	0.75% C 4.0% Cr 18.0% W 1.1% V steel: Osborn
T 2	0.8% C 4% Cr 18% W 2% V steel: For tools; designation used by AISI; the supplier is coded by a suffix letter
T 3	1.05% C 4.0% Cr 18.0% W 3.0% V steel: Designation used by AISI; obsolete
T 7	0.75% C 4% Cr 14% W 2% V steel: For tools; designation used by AISI; the supplier is coded by a suffix letter
T 9	1.2% C 4% Cr 18% W 4% V steel: For tools; designation used by AISI; the supplier is coded by a suffix letter
T 10	0.9% C 2% W 0.5% V steel: Turton Bros
T 11301	UNS designation for M1 type tool steel
T 11302	UNS designation for M2 type tool steel
T 11304	UNS designation for M4 type tool steel
T 11307	UNS designation for M7 type tool steel
T 11310	UNS designation for type M 10 tool steel
T 11313	UNS designation for M3/1 type tool steel
T 11323	UNS designation for M3/2 type tool steel
T 12001	UNS designation for T1 type tool steel
T 12002	UNS designation for T2 type tool steel
T 30102	UNS designation for A2 type tool steel
T 30103	UNS designation for A3 type tool steel
T 30104	UNS designation for A4 type tool steel
T 30105	UNS designation for A5 type tool steel
T 30106	UNS designation for A6 type tool steel
T 30107	UNS designation for A7 type tool steel
T 30108	UNS designation for A8 type tool steel
T 30109	UNS designation for A9 type tool steel
T 30110	UNS designation for A10 type tool steel
T 31501	UNS designation for 01 type tool steel
T 31502	UNS designation for 02 type tool steel
T 31506	UNS designation for 06 type tool steel
T 31507	UNS designation for 07 type tool steel
T 61202	UNS designation for L2 type tool steel
T 61203	UNS designation for L3 type tool steel
T 61206	UNS designation for L6 type tool steel TBS 600 0.98% C 1.5% Cr 0.3% Mo steel: Timken; for bearings at high temperature DPN: 800
TBS 1000	0.8% C 1.1% Cr 5.0% Mo 1.1% V steel: Timken; bearings for high temperature use DPN: 700
TCA 2	0.65% C 1.5% Cr 2.5% W 0.25% V steel: T Turton
TDBC	1.4% C 0.3% Mn 0.7% Cr 5% W steel: Jonas; dies DPN: 720

Note. The following abbreviations and units are used in the tables:

DPN	Hardness, diamond pyramid number
UTS	Ultimate tensile strength, N/mm^2
Elon	Elongation, %
Proof	0.1 % proof strength, N/mm^2

1 N/mm^2=0.1 hbar=0.102 kgf/mm^2=0.06475 tonf/in^2=145.04 lbf/in^2

Symbol	Nominal analysis, supplier, condition and remarks.
TEENAX	0.9% C 0.5% Cr 0.5% W 0.15% V steel: Simonds; AISI type 01
TM 6	4% Cr 5% Mo 6% W 2% V steel: For tools; Vulcan Crucible Steel Co; AISI code M2P
TMS	0.9% C 1.4% Mn 0.3% Cr 0.3% Mo 0.2% V steel: Origin unknown
TOH	0.9% C 1.3% Mn 0.4% Cr 0.5% W steel: Balfour; blanking dies, etc. DPN: 650
TPM	0.9% C 1.4% Mn 0.5% Cr 0.5% W steel: Jonas; fine grained grade of SPM DPN: 660
TRIPLE CONQUEROR	1.25% C 0.75% Cr 3.5% W steel: For tools; origin unknown
TRIPLE CRESCENT	1.25% C 1.15% Cr 4.25% W 0.3% V steel: Spencer; cutting tools, broaches, etc. DPN: 780
TRIPLE GRIFFIN	1.3% C 0.3% Cr 2.7% W steel: For tools; Balfour; tube drawing dies etc DPN: 850
TRIUMPH SUPER	0.7% C 4.0% Cr 1.0% V 18.0% W steel: Origin unknown
TRIUMPH SUPERB	0.8% C 4.25% Cr 18% W 1.3% V steel: Jessop DPN: 750
TROJAN	4% Cr 18% W 2% V steel: For tools; Atlas Steels Ltd; AISI code T2Z
TRS	1.2% C 1.15% Cr 4.5% W 0.25% V steel: Jonas; taps; reamers, etc. DPN: 650
TUNGSTEN DIAMOND	1.25% C 0.75% Cr 3.5% W steel: For tools; origin unknown
TUNGSTEN SPUR	0.67% C 3.8% Cr 14% W 0.7% V steel: T Turton
TWIN VAN	4% Cr 18% W 2% V steel: For tools; Braeburn Alloy Steel: AISI code T21
ULTRA CAPITAL	0.76% C 4.25% Cr 18.5% W 1.2% V steel: For tools; Balfour; reamers, twist drills, etc. DPN: 870
ULTRA CAPITAL 22	0.76% C 4.2% Cr 22% W 1.25% V steel: For tools; Balfour; reamers, twist drills, etc. DPN: 850
UNICUT	4% Cr 6% Mo 6% W 2.4% V steel: For tools; Universal Cyclops; AISI code M3D
UNI-DIE SMOOTHCUT	0.71% C 1.0% Cr 1.3% Mo steel: Columbia; AISI type A6S; free cutting
V 175	0.8% C 4.2% Cr 0.5% Mo 18% W 1.7% V steel: Spencer; cutting tools DPN: 780
VALIANT	1.5% C 0.5% Cr 4% W steel: Hall & Pickles; obsolete DPN: 750
VAN CHIP	4% Cr 6% Mo 6% W 3% V steel: For tools; Firth Sterling Inc; AISI code M3B
VAN CUT 1	4% Cr 6% Mo 6% W 2.4% V steel: For tools; Vanadium Alloys; AISI code M3A
VAN CUT 2	4% Cr 6% Mo 6% W 3% V steel: For tools; Vanadium Alloys; AISI code M3A
VANGUARD	1.3% C 0.7% Mn 1.5% Cr 1% W steel: Hall & Pickles; taps, dies, gauges, etc.; obsolete DPN: 850
VANITE	4% Cr 18% W 2% V steel: For tools; Columbia Tool Steel Co; AISI code T2H
VAN-LOM	4% Cr 8% Mo 2% W steel: For tools; Vanadium Alloys; AISI code M10A
VANTAGE	1.25% C 4.5% Cr 13.5% W 4% V steel: Hall & Pickles; taps, reamers, etc.; obsolete DPN: 800
VAP	0.75% C 4.1% Cr 18.2% W 1.2% V steel: For tools; ESC; cutting tools, taps & blades DPN: 872

Symbol	Nominal analysis, supplier, condition and remarks.
VAP 2V	0.82% C 4.2% Cr 2.0% V 18.0% W steel: For tools; origin unknown
VAPCW	0.7% C 4% Cr 0.8% Mo 18% W 1.2% V steel: For tools; ESC; shear blades, chisels, etc. **DPN: 800**
VASCO M2	4% Cr 5% Mo 6% W 2% V steel: For tools; Vanadium Alloys; AISI code M2A
VASCO M7	4% Cr 8.7% Mo 1.7% W 2% V steel: For tools; Vanadium Alloys; AISI code M7A
VELOS	0.85% C 4.2% Cr 6.4% W 1.9% V 5.2% Mo steel: For tools; W Spencer
VELOS UR	0.75% C 4.2% Cr 0.5% Mo 18% W 1.7% V steel: Spencer; cutting tools **DPN: 780**
VIBRESIST	0.9% C 1.0% Cr 0.25% Mo steel: Tube; Atlas; for hollow rockdrills
VICTOR	1.0% C 5.25% Cr 1.1% Mo 0.25% V steel: Hall & Pickles; press tools, gauges, etc.; obsolete **DPN: 680**
VIKING	1.1% C 1.35% Cr 0.45% Mo steel: Braeburn
VINCO	4% Cr 18% W 1% V steel: For tools; Braeburn Alloy Steel; AISI code T11
VLM	4% Cr 8% Mo 2% V steel: For tools; Allegheny Ludlum; AISI code M10G
VMC	5% Cr 1.5% Mo 1.0% V steel: For dies; Somers
VS 1/2	1.2% C 0.5% Cr 2% W steel: Vulcan; twist drills, hacksaw blades, etc. **DPN: 720**
VS 6/6	0.8% C 4% Cr 6% Mo 6% W 2% V steel: Vulcan; cutting tools **DPN: 850**
VS 14	0.65% C 3.5% Cr 14% W 0.75% V steel: Vulcan; cutting tools **DPN: 770**
VS 16/15	0.75% C 4.5% Cr 1.5% Mo 16% W 1.0% V steel: Vulcan; cutting tools **DPN: 850**
VS 22	0.8% C 4% Cr 0.35% Mo 22% W 1.25% V steel: Vulcan; cutting tools **DPN: 870**
VSCW	1.2% C 0.75% Cr 2.25% W 0.15% V steel: Vulcan; dies, reamers **DPN: 750**
VSD	1.0% C 5% Cr 1.0% Mo 0.15% V steel: Vulcan; punching dies, shear blades **DPN: 720**
VSM	0.7% C 3.0% Cr 5.2% Mo steel: Carpenter; for high temperature use
VSMNC	0.8% C 4% Cr 2% Mo 7.5% W 2.4% V steel: Vulcan; cutting tools **DPN: 850**
VSR	1.2% C 4% W 0.75% Cr steel: Vulcan; reamers, etc. **DPN: 770**
VWMC	5% Cr 1.5% Mo 1.5% W 0.5% V steel: For dies, Somers
W7	1.0% C 0.5% Cr 0.2% V steel: Designation used by AISI; obsolete
WEAREX	0.7% C 4% Cr 14.5% W 1% V steel: Darwin; shaping tools, hacksaw blades, etc.
WG	1.0% C 0.25% Cr 0.25% V steel: Designation used by AISI; obsolete

Symbol	Nominal analysis, supplier, condition and remarks.
WKE	0.8% C 4.5% Cr 1.2% Mo 18.5% W 1.4% V 5.5% Co steel: Fagersta **DPN: 770**
WLM 1175	0.7% C 1% Si 0.5% Cr 1.5% W steel: W Marrison; shear blades
WLM 1261	1.25% C 2% Cr 6% W steel: W Marrison; cutting tools
WLM 1841	0.75% C 4% Cr 18% W 1% V steel: W Marrison; cutting tools
WLM 2241	0.75% C 4% Cr 22% W 1% V steel: W Marrison; cutting tools
WM	0.86% C 4.0% Cr 3.2% Mo 6.5% W 2.1% V steel: Fagersta **DPN: 850**
WOLFRAM	4% Cr 18% W 1% V steel: For tools; Vulcan Crucible Steel Co; AISI code T1P
WS	1.07% C 0.3% Si 1.2% Mn 1% Cr 1.7% W steel: Huntsman; turning tools, taps, etc. **DPN: 750**
WS 44	0.9% C 1.2% Mn 0.5% Cr 0.5% W 0.15% V steel: John Vessey; press tools, collets, etc. **DPN: 750**
W TAP	1.23% C 1.35% W 0.45% Cr 0.2% V 0.25% Mo steel: Latrobe
WULCRO	1.3% C 1% Cr 4.5% W 0.15% V steel: Vulcan; extruding dies **DPN: 870**
XP 550	1.3% C 6.2% Cr 6.2% V 1.1% Mo steel: Braeburn; as Superior 5
Y 100C6	1.0% C 1.5% Cr 0.15% V steel: French Standard designation
Z 13.0WDCV	1.3% C 4.0% Cr 4.5% Mo 5.5% W 4.0% V steel: For tools; French Standard designation
Z80WCDV 12/0402/02	0.8% C 4.0% Cr 2.0% Mo 12.0% W 2.0% V steel: For tools; French Standard designation
Z 80WCV 18/04/01	0.8% C 4.0% Cr 18% W 1.0% V steel: For tools; French Standard designation
Z 85DCWV	0.85% C 4.2% Cr 5.0% Mo 6.2% W 1.8% V steel: For tools; French Standard designation
Z 85WDCV	0.85% C 4.0% Cr 5.0% Mo 6.0% W 2.0% V steel: For tools; French Standard designation
Z 85WCV 18/04/2	0.85% C 4.0% Cr 18.0% W 2.0% V steel: For tools; French Standard designation
Z 100CDV 5	1.0% C 5.0% Cr 1.0% Mo 0.3% V steel: German Standard designation
Z 100CDV 5	1.0% C 5.3% Cr 1.0% Mo 0.2% V steel: French Standard
Z 130WCV 12/04/04	0.8% C 4.0% Cr 12.0% W 3.5% V steel: For tools; French Standard designation

Note. The following abbreviations and units are used in the tables:

DPN	Hardness, diamond pyramid number
UTS	Ultimate tensile strength, N/mm^2
Elon	Elongation, %
Proof	0.1 % proof strength, N/mm^2

$1 N/mm^2 = 0.1 hbar = 0.102 kgf/mm^2 = 0.06475 tonf/in^2 = 145.04 lbf/in^2$

44L Steel–cobalt alloys
High carbon with chromium, nickel, molybdenum tungsten or vanadium

Specific gravity	7.9	
Density	7900 kg/m³	(0.296 lb/in.³)
Solidus/liquidus	1288–1354 °C	
Thermal conductivity	61 W/m °C	(14.6 cal/m s °C)
Coefficient of linear expansion	17.5×10^{-6}/ °C	
Electrical conductivity	2% IACS (copper 100%)	
Specific resistance	930 microhm mm	
Young's modulus of elasticity	– N/m²	
Impact	– J	
Fatigue strength	–	
Hot strength	–	

The above properties are typical of the following group, and may not apply exactly to any one specification. It is possible that with certain specifications some of the values may not be applicable.

General metallurgical characteristics

Cobalt is probably unique in that it is never present alone but always as an addition to already highly alloyed steels. Like nickel and silicon, it does not form carbides but dissolves in the ferritic matrix.

Nickel, iron and cobalt are the only common ferromagnetic elements and considerable advances have been made in recent years with complex cobalt alloys which have excellent magnetic, high temperature and permeability characteristics. These alloys are briefly discussed in Section 13A.

Additions of up to 30% cobalt to ferrous alloys also considerably improve certain magnetic properties and many of the materials listed have been developed for this reason. Among the valuable properties given by cobalt are the stability of magnets for use at high temperatures and under more arduous conditions than normal.

Apart from magnets, cobalt is added in quantities of up to 10% to alloy steels when it imbues them with unique hot strength properties and results in less galling and pick up under friction conditions. These are all comparatively new materials and much of the metallurgy remains to be explained.

Cobalt is similar in many respects to nickel, but as well as strengthening the ferrite, appears to stabilize the carbides and maintains their properties to much higher temperatures, while at the same time giving these high carbon steels a degree of toughness not obtainable by other means.

The other alloying elements found in these steels have the same effect as that described in Section 44K3.

Thermal treatment

These materials will never be cold worked as they become brittle and unusable after comparatively little cold deformation. They thus never require interstage annealing, always being subjected to hot work during fabrication.

It is much more important that these steels are protected from oxidation than it is for other steels. If the steels are heated in air for the times and temperatures necessary to allow the metallurgical reactions to take place, the surface will be seriously denuded of cobalt.

Annealing will require a temperature of 800–1000 °C for at least an hour per 25 mm of section and must always be carried out under reducing atmosphere or packed in cast iron turnings. The parts require to be very slowly cooled from the annealing temperature.

Cyclic or spheroidal annealing gives superior results regarding machinability, but as it relies on temperature cycling between about 900 and 700 °C this is difficult unless very flexible atmosphere controlled furnaces or expensive salt baths are used.

Hardening requires a temperature of 900–1250 °C depending on the alloy, for about 30 minutes per 25 mm of section and the parts must always be pre-heated to 650–800 °C. Air quenching is often applied to these alloys as long as the part is of a uniform sectional thickness under 50 mm. Oil quenching is the most common method but it is recommended that the parts are withdrawn from the oil while still fairly hot and allowed to cool in still air or immediately transferred to the temper at 400–650 °C, depending on the hardness required. Double tempering with a deep freeze between at minus 800–100 °C is commonly applied to these steels to ensure complete transformation, as the metallurgical reactions can be extremely sluggish, requiring considerable time or extremes of temperature to ensure completion.

Many of the materials listed will not be subjected to any of the conventional metallurgical treatments but will have special requirements allied to their magnetic properties. These usually require controlled cooling rates through certain temperature ranges, and information on the treatments must be obtained from the origin of the specification or the Permanent Magnet Association.

These cobalt materials are all very expensive, and more than any other group of steels require accurate thermal treatment under carefully controlled conditions. It cannot be too strongly emphasized that the information given above is for guidance only and must be supplemented by detailed information.

These materials only give of their best, whether it be the magnetic or abrasive resistant property, if they are correctly treated, when they become the outstanding materials in their own field. Without this correct treatment they are no better, if not worse, than less highly alloyed and cheaper steels.

These cobalt bearing steels should only be heated in neutral salt baths or under controlled atmospheric conditions. Modern vacuum furnace techniques are already showing that for certain applications this rather expensive method is economical on these materials.

Under no circumstances should high cobalt steels be subjected to long term high temperature treatments under oxidizing conditions.

Scale removal should never present any problem with these steels provided the above points are adhered to, but grit blasting is recommended to clean the surface of all trace of salt or other adhering matter.

Welding and brazing

Pre-treatment. All oxide, scale, grease, and dirt must be absent and the steel should be in the soft condition for welding. Pre-heating is essential.

Welding and brazing. These materials should never be welded or brazed when the join is to be subjected to any degree of stress.

As a rule the only welding applied is to join these expensive materials to a cheaper plain carbon steel. This very often uses the flash butt welding technique which requires very special equipment but results in a forge weld with the cast structure rejected from the joint area. Provided the correct thermal treatment can be subsequently applied this gives excellent welds.

Inert gas shielded electric arc welding can give adequate joins provided some method of controlled heating and cooling of the weld area can be devised. Normal electric arc or gas welding should not be used as this tends to be oxidizing, thus reducing the cobalt content at the weld area.

Brazing can be successfully applied provided a very rapid means of heating is applied as these steels have very slow metallurgical reaction, thus do not soften unless held at temperature for an appreciable time. Induction heating is often used for this purpose but gas heating can be applied as long as the flame is used to melt the filler and not allowed to dwell on one spot.

Post-treatment. Welded assemblies should be hardened and tempered if possible, and as an absolute minimum parts must be stress relieved for several hours at as near the tempering temperature as possible. All trace of flux must be removed before any thermal treatment is carried out otherwise corrosion can occur.

Flaw detection methods

Crack test. Magnetic crack test is possible with all of these materials but it is recommended that the oil penetrant method is used on the magnetic materials at least, to prevent peculiar results and possible bad effects. The oil penetrant method using modern fluids and techniques is as sensitive as magnetic crack testing and more flexible.

X-ray. The wrought materials will seldom or never require X-ray but castings should be subjected to this form of inspection at least as a quality check.

Ultra-sonic test. Parts with expensive machining operations may be ultrasonically tested at an early machining stage. This identifies subcutaneous defects, thus saving further machining operations on faulty parts.

Chemical etch. These materials in the hardened and tempered condition are prone to grinding abuse which can cause softening or even cracking. An electrolytic sulphuric acid etch successfully identifies this defect, making possible control of the grinding operation without causing scrap.

Corrosion protection

Temporary. Apart from the high cobalt materials these steels have little or no built-in corrosion protection, although they are superior to mild steels. They must be oiled or greased at all times and protected against damp or humid conditions.

During machining, soluble oil coolants can be a source of corrosion and staining if allowed to become contaminated, but are excellent temporary corrosion inhibitors if kept in good condition.

After machining the parts should be cleaned, dried and oiled, then stored in plastic. Many types of paper, cardboard or even wood contain compounds which can cause staining or corrosion even on oiled components.

Permanent. Steel corrosion or rusting is a complex oxidation process which results in a loosely adherent bulky deposit, easily removed to allow further corrosion. This often leads to local metal loss and pitting which can be exceedingly dangerous with these rather notch sensitive steels if used under conditions of high stress.

The fact that they are often used for their hardness and abrasion resistance means that normal application of protective coating would have little effect.

Before any method of protection can be applied it is essential that all evidence of existing corrosion is removed. Unless this is efficiently carried out, the remaining rust which has air and moisture occluded will propagate under any protective media applied.

Grit blasting efficiently removes all trace of rust if carried out under controlled conditions but the surface must be brushed clear of dust and protected as soon as possible, otherwise fresh rusting will occur.

Alkaline rust removers can also be used to eliminate corrosion and again require considerable control to be effective and economical. Acids must not be used as they can cause severe embrittlement.

Phosphating is an efficient means of killing slight corrosion and leaves a surface which has excellent oil or paint retention properties. Many tools with long corrosion-free lives were originally phosphated and oiled, with the oil film maintained at regular intervals. Parts should be stress relieved at 200° C after phosphating to remove any tendency to hydrogen embrittlement.

Painting. This should always be applied after phosphating. A zinc chromate or silicate primer is advised, sealed with a good quality top coat. Many tools are painted on their non-working surfaces to corrosion protect and identify.

Plating. These steels can all be plated provided care is taken at pre-plating cleaning and etching. It is however, essential that a stress release at 200 °C is carried out after plating, and if much machining has been carried out, parts must also be stress relieved before plating. This is necessary to eliminate the danger of hydrogen embrittlement which causes failure in a brittle manner below the calculated tensile strength. As there is no visual method of inspection to ensure that stress relieving has in fact been carried out it is recommended that plating is not specified if the parts are to be used where they will be subjected to high stresses.

Metal spraying of zinc or aluminium is seldom practicable on these hard materials as this relies on the mechanical bond supplied from the roughened surface produced by grit blasting, which is not possible on very hard steels. Metal spray techniques using high velocities have been developed which may overcome this defect.

Chemical oxidation, using hot sodium hydroxide solutions with oxidizing agents added, gives an artificial, very adherent film which has oil retaining properties. This is known as 'black oxide' and has been used successfully by gunsmiths for many centuries. Provided the oil film is carefully maintained parts can give many years of corrosion-free service.

Machinability

The presence of cobalt enhances the machinability of these steels in the annealed condition as cobalt reduces the tendency for galling and pick up. After hardening it is only possible for parts to be ground, or for the new techniques of spark erosion and chemical machining to be used for metal removal.

Grinding must be carefully controlled to obviate any danger of softening or cracking, and dry grinding or heavy cuts must be avoided.

Uses

Magnets, high speed cutting tools for arduous conditions, uses where hot abrasion strength is essential. Some indication of the use is given where possible in the individual alloy description but this can never be explicit or complete.

Symbol	Nominal analysis, supplier, condition and remarks.
06/05/05/04/01	0.85% C 4.0% Cr 5.0% Mo 6.0% W 2.0% V 5.0% Co steel: For tools; French Standard designation
07/05/05/05/04	1.5% C 4.0% Cr 5.0% Mo 6.5% W 5.0% V 5.0% Co steel: For tools; French Standard designation
09/08/04/02/01	1.1% C 4.0% Cr 9.0% Mo 1.5% W 1.0% V 8.0% Co steel: For tools; French Standard designation
1.3202	1.42% C 4.2% Cr 0.85% Mo 12.0% W 3.7% V 5.0% Co steel: For tools; German Standard designation
1.3207	1.27% C 4.1% Cr 4.0% Mo 10.2% W 3.25% V 10.5% Co steel: For tools; German Standard designation
1.3243	0.92% C 4.1% Cr 5.0% Mo 6.3% W 1.85% V 5.0% Co steel: For tools; German Standard designation
1.3246	1.1% C 4.1% Cr 3.8% Mo 6.8% W 1.85% V 5.0% Co steel: For tools; German Standard designation
1.3247	1.05% C 3.7% Cr 9.5% Mo 1.1% V 1.5% W 8.0% Co steel: German Standard
1.3255	0.75% C 4.2% Cr 0.6% Mo 1.1% V 18.5% W 5.5% Co steel: For tools; German Standard
1.3255	0.8% C 4.2% Cr 0.65% Mo 17.7% W 1.61% V 4.7% Co steel: For tools; German Standard designation
1.3265	0.75% C 5.0% Cr 0.6% Mo 1.7% V 18.5% W 9.5% Co steel: For tools; German Standard
2 Co 4 Cr	1.0% C 4.0% Cr 0.5% W 2.0% Co steel: For magnets; PMA
3% Co	1.0% C 0.1% Si 0.5% Mn 9% Cr 1.2% Mo 3% Co steel: For magnets; Jessop

Symbol	Nominal analysis, supplier, condition and remarks.
6% Co	1.0% C 0.1% Si 0.5% Mn 9% Cr 1.2% Mo 6% Co steel: For magnets; Jessop
8-N-2 COBALT	4% Cr 8% Mo 2% W 1.2% V 5% Co steel: For tools; Vanadium Alloys; AISI code M30A
08S	1.4% C 14% Cr 0.8% Mo 3% Co steel: Carrs; punches, dies, etc. **DPN: 710**
9% Co	1.0% C 0.1% Si 0.5% Mn 9% Cr 1.2% Mo 9% Co steel: For magnets; Jessop
10/07/05/05/04	1.75% C 4.0% Cr 5.0% Mo 6.5% W 5.0% V 10.0% Co steel: For tools; French Standard designation
15% Co	1.0% C 0.1% Si 0.5% Mn 9% Cr 1.2% Mo 15% Co steel: For magnets; Jessop
17 Co	0.7% C 2.5% Cr 8.0% W 17% Co steel: For magnets; American alloy listed by PMA
30 DCKY/28	0.32% C 3.0% Cr 2.8% Mo 0.35% V 3.0% Co steel: French Standard
35% Co	0.9% C 0.1% Si 0.5% Mn 6% Cr 4% W 35% Co steel: For magnets; Jessop
81/18.5% Co	0.75% C 3.7% Cr 5.0% W 18.5% Co steel: For magnets; Simonds
83/3% Co	1.0% C 4.0% Cr 3.2% Co steel: For magnets; Simonds
178 Co	0.95% C 0.6% Mn 5% Cr 4% W 35.5% Co steel: For magnets; Sandvik
333	1.55% C 4.7% Cr 12.5% W 5.0% V 5.0% Co steel: Forez
444	1.55% C 4.75% Cr 19.5% W 5.0% V 14.5% Co steel: Forez
455	0.78% C 4.2% Cr 0.4% Mo 18.5% W 1.3% V 5.5% Co steel: F Parkin; cutting tools **DPN: 820**
808	0.8% C 4.2% Cr 1% Mo 19% W 1.6% V 10.5% Co steel: F Parkin; cutting tools **DPN: 850**
3500/37% Co	0.75% C 3.8% Cr 5.0% W 38% Co steel: For magnets; Simonds
4171	1.3% C 4.0% Cr 12.0% W 5.0% V 5.0% Co steel: For tools; French Standard designation
4475	1.1% C 4.0% Cr 9.0% Mo 1.5% W 1.0% V 8.0% Co steel: For tools; French Standard designation

Note. The following abbreviations and units are used in the tables:

DPN	Hardness, diamond pyramid number
UTS	Ultimate tensile strength, N/mm^2
Elon	Elongation, %
Proof	0.1 % proof strength, N/mm^2

1 N/mm^2=0.1 hbar=0.102 kgf/mm^2=0.06475 tonf/in^2=145.04 lbf/in^2

Symbol	Nominal analysis, supplier, condition and remarks.
4175	1.65% C 4.0% Cr 12.0% W 5.0% V 10.0% Co steel: For tools; French Standard designation
4271	0.8% C 4.0% Cr 18.0% W 1.0% V 5.0% Co steel: For tools; French Standard designation
4275	0.8% C 4.0% Cr 18.0% W 1.6% V 10.0% Co steel: For tools; French Standard designation
4371	0.85% C 4.0% Cr 5.0% Mo 6.0% W 2.0% V 5.0% Co steel: For tools; French Standard designation
4373	1.5% C 4.0% Cr 5.0% Mo 6.5% W 5.0% V 5.0% Co steel: For tools; French Standard designation
4375	1.75% C 4.0% Cr 5.0% Mo 6.5% W 5.0% V 10.0% Co steel: For tools; French Standard designation
13743	11.5% Al 24.5% Ni 11% Co 3% Cu steel: For magnets; German Standard alloy for Alnico 160 listed in DIN 17410
13745	11% Al 20% Ni 16% Co 3% Cu steel: For magnets; German Standard alloy for Alnico 190 listed in DIN 17410
13756	7% Al 15% Ni 27% Co 4% Cu 7% Ti steel: For magnets; German Standard alloy for Alnico 220 listed in DIN 17410
13758	7% Al 15% Ni 30% Co 4% Cu 5% Ti steel: For magnets; German Standard alloy for Alnico 350 listed in DIN 17410
13760	9% Al 15.5% Ni 4% Cu 24% Co steel: For magnets; German Standard for Alnico 400 listed in DIN 17410
13761	8.5% Al 15% Ni 24% Co 3% Cu steel: For magnets; German Standard for Alnico 500 listed in DIN 17410
ACMITE	4% Cr 18% W 1% V 5% Co steel: For tools; Columbia Tool Steel Co; AISI code T4H
AISI 661	0.12% C 20.7% Cr 19.8% Ni 19% Co 2.9% Mo 2.3% W 1.1% Cb Fe alloy: Solution treated & aged **UTS: 840 Elon: 43% Proof: 370**
AJ 2	0.76% C 4% Cr 0.3% Mo 18% W 0.8% V 2.25% Co steel: For tools; Osborn; obsolete **DPN: 820**
ALCOMAX I	7.5% Al 11% Ni 3.0% Cu 1.5% Ti 25% Co steel: Casting; PMA; anistropic
ALCOMAX II	8.0% Al 11.5% Ni 4.5% Cu 24% Co steel: Casting; PMA; anistropic
ALCOMAX II	11% Ni 21% Co 4% Cu 9% Al steel: For magnets (casting); Jessop; remanence 13 000 gauss; coercivity 580 oersted
ALCOMAX II Sc	8.0% Al 11.5% Ni 4.5% Cu 24% Co steel: Casting; PMA; anistropic; semi-columnar
ALCOMAX II Sc	11% Ni 21% Co 4% Cu 9% Al steel: Casting; crystal oriented magnet; Jessops; remanence 13 700 gauss; coercivity 600 oersted
ALCOMAX III	8.0% Al 13.5% Ni 3.0% Cu 0.8% Nb 24% Co steel: Casting; PMA; anistropic
ALCOMAX III	13.5% Ni 24% Co 3% Cu 8% Al 0.7% Nb steel: Cast magnets; Jessops; remanance 12 600 gauss; coercivity 650 oersted
ALCOMAX III Sc	8.0% Al 13.5% Ni 3.0% Cu 0.8% Nb 24% Co steel: Casting; PMA; anistropic; semi-columnar
ALCOMAX III Sc	13.5% Ni 24% Co 3% Cu 8% Al 0.7% Nb steel: Cast crystal oriented magnet; Jessop; remanence 13 200 gauss; coercivity 700 oersted
ALCOMAX IV	8.0% Al 1.35% Ni 3.0% Cu 2.5% Nb 24% Co steel: Casting; PMA; amistropic
ALCOMAX IV	13% Ni 24% Co 3% Cu 8% Al 2.5% Nb steel: Cast magnets; Jessops; remanence 11 500 gauss; coercivity 750 oersted
ALCOMAX IV Sc	13% Ni 24% Co 3% Cu 8% Al 2.5% Nb steel: Crystal oriented cast magnet; Jessops; remanence 12 200 gauss; coercivity 780 oersted
ALCONIT-AN-O	Al Ni Cu Ti Co steel: Casting for magnets; Von Roll; further information from PMA
ALCONIT AN 16	Al Ni Cu Ti Co steel: Casting for magnets; Von Roll; further information from PMA
ALLOY 8	2.3% C 16% Cr 6% Ni 20% Co Fe alloy: Casting; Dewrance **DPN: 501 UTS: 800**
ALLOY 15	3.75% C 16% Cr 4% Ni 6.5% Mo 1.5% Si 20% Co Fe alloy: Casting; Dewrance **DPN: 680 UTS: 610**
ALLOY 18	2.5% C 25% Cr 15% Ni 8% Mo 25% Co Fe alloy: Casting; Dewrance **DPN: 400 UTS: 680**
ALNICO	10% Al 18% Ni 5.0% Cu 0.5% Ti 12% Co steel: Casting for magnets; PMA
ALNICO I	22% Ni 0.35% Ti 12% Al 5.0% Co steel: For magnets; Simonds
ALNICO I	18% Ni 12.5% Co 6.5% Cu 0.7% Ti 10% Al steel: Casting for magnets; Jessop; remanence 7250 gauss; coercivity 500 oersted
ALNICO 1A	22.5% Ni 5.0% Co 12.0% Al Fe steel: For magnets; origin unknown
ALNICO 1B	21.0% Ni 5.0% Co 3.0% Cu 12.0% Al Fe alloy: For magnets; origin unknown
ALNICO 1C	20.0% Ni 5.0% Co 2.0% Cu 12.0% Al Fe alloy: For magnets; origin unknown
ALNICO II	16% Ni 12% Co 6% Cu 0.7% Ti 9% Al steel: Casting for magnets, Jessops; remanence 6200 gauss; coercivity 480 oersted
ALNICO 2	17% Ni 0.45% Ti 10% Al 6.0% Cu 12.5% Co steel: For magnets; Simonds
ALNICO 2A	18.0% Ni 13.0% Co 5.0% Cu 10.0% Al Fe alloy: For magnets; origin unknown
ALNICO 2B	19.0% Ni 13.0% Co 3.0% Cu 10.0% Al Fe alloy: For magnets; origin unknown
ALNICO 2C	21% Ni 0.3% Ti 10% Al 6% Cu 12.5% Co steel: For magnets; Simonds
ALNICO 2H	19.0% Ni 14.5% Co 3.0% Cu 10.0% Al Fe alloy: For magnets; origin unknown
ALNICO III	20% Ni 12% Co 6% Cu 0.7% Ti 10% Al steel: Casting for magnets; Jessops; remanence 6500 gauss; coercivity 620 oersted
ALNICO 3A	26.0% Ni 3.0% Cu 12.0% Al Fe alloy: For magnets; origin unknown
ALNICO 3B	25.0% Ni 3.0% Cu 12.0% Al Fe alloy: For magnets; origin unknown
ALNICO 3C	24.0% Ni 3.0% Cu 12.0% Al Fe alloy: For magnets; origin unknown
ALNICO 4	25% Ni 0.4% Ti 12% Al 5.0% Co steel: For magnets; Simonds
ALNICO 4A	28.0% Ni 5.0% Co 12.0% Al Fe alloy: For magnets; origin unknown
ALNICO 4B	27.0% Ni 5.0% Co 12.0% Al Fe alloy: For magnets; origin unknown
ALNICO 5	14.5% Ni 8% Al 3.0% Cu 24% Co steel: For magnets; Simonds
ALNICO 5/7	1.45% Ni 8% Al 3.0% Cu 24% Co steel: For magnets; Simonds
ALNICO 5A	15.0% Ni 24.0% Co 3.0% Cu 8.0% Al Fe alloy: For magnets; origin unknown
ALNICO 5AB	14.5% Ni 24.0% Co 3.0% Cu 8.0% Al Fe alloy: For magnets; origin unknown
ALNICO 5A,B,D & G	14.5% Ni 24.0% Co 3.0% Cu 8.0% Al Fe alloy: For magnets; origin unknown
ALNICO 5B	14.0% Ni 24.0% Co 3.0% Cu 8.0% Al Fe alloy: For magnets; origin unknown

Symbol	Nominal analysis, supplier, condition and remarks.
ALNICO 5B,D & G	14.0% Ni 24.0% Co 3.0% Cu 8.0% Al Fe alloy: For magnets; origin unknown
ALNICO 5DG	14% Ni 8% Al 24% Co 3.0% Cu steel: For magnets; columnar form; anistropic; American Alloy; consult PMA
ALNICO 5E	15% Ni 0.45% Ti 8% Al 3.6% Cu 22.5% Co steel: For magnets; Simonds
ALNICO 6	15% Ni 1.25% Ti 8% Al 3.4% Cu 24% Co steel: For magnets; Simonds
ALNICO 6A	17.0% Ni 23.0% Co 3.0% Cu 8.0% Al Fe alloy: For magnets; origin unknown
ALNICO 6B	16.0% Ni 24.0% Co 3.0% Cu 8.0% Al 1.0% Ti Fe alloy: For magnets; origin unknown
ALNICO 7	20% Ni 5.0% Ti 8.5% Al 3.0% Cu 23.5% Co steel: For magnets; Simonds
ALNICO 7	18.0% Ni 24.0% Co 3.25% Al 8.5% Al 5.0% Ti Fe alloy: For magnets; origin unknown
ALNICO 8	15% Ni 5.0% Ti 7.5% Al 4.0% Cu 35% Co steel: For magnets; Simonds
ALNICO 8A	14.0% Ni 38.0% Co 3.0% Cu 7.5% Al 8.0% Ti Fe alloy: For magnets; origin unknown
ALNICO 8B	7% Al 15% Ni 35% Co 4% Cu 5% Ti steel: For magnets; American alloy listed by PMA
ALNICO 9	15.0% Ni 35.0% Co 4.0% Cu 7.0% Al 5.0% Ti Fe alloy: For magnets; origin unknown
ALNICO 12	18% Ni 6.0% Al 35% Co 8.0% Ti steel: For magnets; American alloy; consult PMA; anistropic
ALNICO 160	11.5% Al 24.5% Ni 11% Co 3.0% Cu steel: For magnets; German Standard
ALNICO 190	11% Al 20% Ni 16% Co 3% Cu steel: For magnets; German alloy listed by PMA
ALNICO 220	7% Al 15% Ni 27% Co 4% Cu 7% Ti steel: For magnets; German alloy listed by PMA
ALNICO 300	10% Al 18% Ni 20% Co 4% Cu steel: For magnets; German alloy listed by PMA
ALNICO 350	7% Al 15% Ni 30% Co 4% Cu 5% Ti steel: For magnets; German alloy listed by PMA
ALNICO 400	9.0% Al 15.5% Ni 4.0% Cu 24% Co steel: For magnets; German Standard; anistropic
ALNICO 450	7% Al 15% Ni 34% Co 4% Cu 5% Ti steel: For magnets; German alloy listed by PMA
ALNICO 500	8.5% Al 15% Ni 24% Co 3% Cu steel: For magnets; German alloy listed by PMA
ALNICO 580	8.5% Al 15% Ni 24% Co 3% Cu steel: For magnets; German alloy listed by PMA
ALNICO 700	8.5% Al 15% Ni 24% Co 3% Cu steel: For magnets; German alloy listed by PMA
AM 367	0.02% C 15.2% Cr 3.4% Ni 2.0% Mo 0.3% Ti 15.5% Co steel: Allegheny Ludlum
AMS 5376 B	21% Cr 20% Co 3% Mo 2.5% W 1% Nb + Ti Fe alloy: Casting
AMS 5531	20% Cr 20% Ni 20% Co 3% Mo 2% W 0.7% Nb + Ti Fe alloy: Sheet
AMS 5532 B	20% Cr 20% Ni 20% Co 3% Mo 2% W 1% Nb Fe alloy: Sheet
AMS 5533 A	20% Cr 20% Ni 20% Co 4% Mo 4% W 4% Nb Fe alloy: Sheet
AMS 5585	20% Cr 20% Ni 20% Cu 3% Mo 2% W 1% Nb + Ti Fe alloy: Welded tube

Symbol	Nominal analysis, supplier, condition and remarks.
AMS 5768 E	20% Cr 20% Ni 20% Co 3% Mo 2% W 1% Nb + Ti Fe alloy: Bar & forging; solution treated & aged
AMS 5769	20% Cr 20% Ni 20% Co 3% Mo 2% W 1% Nb + Ti Fe alloy: Bar; solution treated
AMS 5770 B	20% Cr 20% Ni 20% Co 4% Mo 4% W 4% Nb + Ti Fe alloy: Bar; solution treated & aged
AMS 5794 A	20% Cr 20% Ni 20% Co 3% Mo 2% W 1% Nb + Ti Fe alloy: Welding wire
AMS 5795 B	20% Cr 20% Ni 20% Co 3% Mo 2% W 1% Nb + Ti Fe alloy: Coated electrode
ANSEMAX	8% Al 14% Ni 24% Co 3% Cu steel: For magnets; Italian alloy; listed by PMA
ARK SUPERLATIVE	0.75% C 4.0% Cr 1.0% V 18.0% W 5.0% Co steel: Origin unknown
ASTM A579/81	0.27% C 0.5% Cr 8.0% Ni 0.45% Mo 0.09% V 4.0% Co steel: Forging
ASTM A579/82	0.31% C 1.0% Cr 7.7% Ni 1.0% Mo 0.09% V 4.5% Co steel: Forging
ASTM A579/83	0.45% C 0.27% Cr 7.7% Ni 0.27% Mo 0.09% V 4.0% Co steel: Forging
ASTM A605	0.2% C 0.3% Mn 0.72% Cr 9.0% Ni 1.0% Mo 4.8% Co 0.08% V steel: For pressure vessels **UTS: 1430 Elon: 14% Proof: 1200**
ASTM A646 HP9/4/20	0.2% C 0.75% Cr 9.0% Ni 1.0% Mo 0.09% V 4.5% Co steel
ASTM A646 HP9/4/30	0.31% C 1.0% Cr 7.5% Ni 1.0% Mo 0.1% V 4.5% Co steel
B 7	4% Cr 18% W 1% V 5% Co steel: For tools; Universal Cyclops; AISI code T4D
B 8	4% Cr 14% W 2% V 5% Co steel: For tools; Universal Cyclops; AISI code T8D
B 10	4% Cr 18% W 2% V 8% Co steel: For tools; Universal Cyclops; AISI code T5D
BCC	1.4% C 0.35% Si 0.35% Mn 12% Cr 0.8% Mo 0.3% V 2.7% Co steel: For tools; Huntsman; shear blades; intricate punches, etc. **DPN: 750**
BONDED CARBIDE	4.5% Cr 20% W 1.5% V 12% Co steel: For tools; Braeburn Alloy Steel; AISI code T6I
BONDED CARBIDE Jr	4% Cr 18% W 2% V 8% Co steel: For tools; Braeburn Alloy Steel; AISI code T5I
BONDED CARBIDE Sr	0.75% C 4.25% Cr 2.0% V 18.5% W 0.7% Mo 7.6% Co steel: Braeburn Alloy Steel
BRAEBURN M33	0.9% C 3.75% Cr 1.15% V 1.75% W 9.5% Mo 8.25% Co steel: Braeburn Alloy Steel
BRAEBURN T15	1.5% C 4.5% Cr 4.7% V 13.0% W 0.5% Mo 4.5% Co steel: Braeburn Alloy Steel
BRAECUT	1.5% C 4.25% Cr 6.25% Mo 5.2% W 2.2% V 12% Co steel: Braeburn Alloy Steel
BRAEMAX	1.1% C 3.7% Cr 1.1% V 1.5% W 9.5% Mo 8.0% Co steel: Braeburn Alloy Steel
BRAETUF	1.1% C 4.0% Cr 2.0% V 5.0% W 6.7% Mo 7.25% Co steel: Braeburn Alloy Steel
BU 8	0.78% C 4.5% Cr 0.75% Mo 18% W 1.25% V 8.25% Co steel: For tools; Osborn; obsolete **DPN: 850**
BV 10	0.78% C 4.5% Cr 0.75% Mo 18% W 1.25% V 10% Co steel: For tools; Osborn; obsolete **DPN: 850**
C 445	0.3% C 2.8% Cr 2.8% Mo 0.5% V 2.8% Co steel: For tools; Fagersta **DPN: 460**
CALDUR	0.27% C 10.5% Cr 7.0% W 4.0% Co steel: Uddelholm
CALMAX	0.28% C 11.5% Cr 7.5% W 0.5% V 9.5% Co steel: Uddelholm

Note. The following abbreviations and units are used in the tables:

DPN	Hardness, diamond pyramid number
UTS	Ultimate tensile strength, N/mm^2
Elon	Elongation, %
Proof	0.1 % proof strength, N/mm^2

1 N/mm^2=0.1 hbar=0.102 kgf/mm^2=0.06475 tonf/in^2=145.04 lbf/in^2

Symbol	Nominal analysis, supplier, condition and remarks.
CAPITAL 398	0.82% C 4.15% Cr 6.4% W 1.9% V 5% Co 5% Mo steel: For tools; Balfour; reamers, toolbits, etc. **DPN: 950**
CCM 10	0.8% C 4.5% Cr 20% W 1.6% V 10.5% Co steel: Huntsman; cutting tools (used as tips) **DPN: 850**
CCMP 15	0.8% C 4.7% Cr 22% W 1.6% V 15% Co steel: Huntsman; cutting tools (used as tips) **DPN: 800**
CIRCLE C	0.77% C 4.5% Cr 1.0% Mo 18.5% W 2.0% V 9.0% Co steel: Firth Sterling
CIRCLE H	1.28% C 4.2% Cr 5.1% Mo 8.25% W 1.6% V 5.4% Co steel: Firth Sterling
CIRCLE M	4% Cr 5% Mo 6% W 8% Co tool steel: Firth Sterling; AISI code M36B
CIRCLE T 15	1.5% C 4.7% Cr 12.5% W 5.0% V 5.0% Co steel: Firth Sterling
CLYDALL 5 SPECIAL	0.78% C 4% Cr 18.5% W 1.3% V 5.5% Co steel: For tools; Clyde Alloy **DPN: 790**
CLYDALL 12 SPECIAL	0.78% C 4% Cr 22% W 1.35% V 12% Co steel: For tools; Clyde Alloy **DPN: 790**
CM 5	0.8% C 4.6% Cr 18% W 1.5% V 5.7% Co steel: Huntsman; cutting tools **DPN: 910**
CM 18	2.5% C 25% Cr 15% Ni 8% Mo 25% Co Fe alloy: For hard facing gas welding; Dewrance
CM 32	1.25% C 23% Cr 8% Mo 10% Co Fe alloy: For hard facing; Dewrance; gas welding
CM 118	2.5% C 25% Cr 15% Ni 8% Mo 25% Co Fe alloy: For hard facing; Dewrance; electric welding
CM 132	1.25% C 23% Cr 8% Mo 10% Co Fe alloy: For hard facing; Dewrance; electric welding
CM 1255	1.5% C 4.7% Cr 12% W 5% V 5% Co steel: Huntsman; cutting tools **DPN: 910**
Co 040	1.0% C 4.0% Cr 0.7% W 2.0% Co steel: For magnets; German Standard; included in DIN 17410
Co 050	1.0% C 8.5% Cr 1.3% Mo 6.5% Co steel: For magnets; German Standard; included in DIN 17410
Co 060	1.0% C 8.5% Cr 1.6% Mo 11% Co steel: For magnets; German Standard; included in DIN 17410
Co 070	1.0% C 9% Cr 1.6% Mo 16% Co steel: For magnets; German Standard; included in DIN 17410
Co 090	0.9% C 5.0% Cr 0.4% Mo 5.0% W 31% Co steel: For magnets; German Standard; included in DIN 17410
COALNI	11.5% Al 25.5% Ni 4% Co 4% Cu steel: For magnets; Italian alloy listed by PMA
COALNIMAX	8% Al 14% Ni 24% Co 3% Cu steel: For magnets; Italian alloy listed by PMA
COBALT	0.74% C 4.0% Cr 0.8% Mo 18% W 1.0% V 5.0% Co steel: Braeburn Alloy Steel; AISI type T4
COBITE	0.81% C 4.2% Cr 0.6% Mo 18.5% W 2.0% V 8.7% Co steel: Columbia; AISI type T5
CODE P	0.12% C 10.5% Cr 4.7% Mo 1.25% Cu 6.0% Co steel: American proprietary alloy
COLUMAX	13.5% Ni 8% Al 3.0% Cu 2.0% Nb 24% Co steel: For magnets; PMA; columnar crystal form of Alcomax III
COMO	0.82% C 4.0% Cr 8.5% Mo 1.5% W 1.2% V 5.0% Co steel: Braeburn Alloy Steel; AISI type M30
COMO	17% Mo 12% Co steel (probably): Japanese alloy listed by PMA
COMOKUT	4% Cr 18% W 1% V 5% Co steel: For tools; Bethlehem Steel Co; AISI code T4E
COMOL	17% Mo 12% Co steel: For magnets; American alloy listed by PMA
CONGO	0.78% C 4.0% Cr 5.0% Mo 4.0% W 1.4% V 12% Co steel: Braeburn Alloy Steel; AISI type M6
CONGO HOT-WORK	0.1% C 4.0% Cr 0.75% V 4.0% W 5.0% Mo 23.5% Co steel: Braeburn Alloy Steel
COVAN	1.5% C 5% Cr 12.5% W 5% V 5% Co steel: Hall & Pickles; cutting tools; obsolete **DPN: 820**
CRD 339	1.25% C 4.0% Cr 3.5% V 8.5% W 8.5% Co steel: For tools; origin unknown
CRD 206	0.78% C 4.5% Cr 0.5% Mo 19% W 2% V 5% Co steel: C Denton; drills & cutters; obsolete
CRD 2212	0.8% C 5.0% Cr 0.5% Mo 21% W 2.0% V 10% Co steel: C Denton; cutting tools; obsolete
CRD TVC	1.5% C 4.0% Cr 5.0% V 12.0% W 5.0% Co steel: For tools; origin unknown
CX	0.8% C 4.4% Cr 1.0% Mo 18.5% W 1.6% V 10% Co steel: For tools; ESC symbol for Cyclone Extra **DPN: 910**
CY	0.7% C 4.5% Cr 0.7% Mo 18% W 1.25% V steel: For tools; Osborn; obsolete **DPN: 910**
CYC 5% Co	0.7% C 4.5% Cr 1% Mo 1.5% V 18% W 5% Co steel: Marsh Brothers; cutting tools
CYC 275	0.8% C 4.5% Cr 1.2% Mo 1.2% V 20% W 10% Co steel: Marsh Brothers; cutting tools
CYCLONE EXTRA	0.8% C 4.4% Cr 1% Mo 18.5% W 1.6% V 10% Co steel: For tools; ESC; cutting tools **DPN: 910**
CYCLONE MC6	4.0% Cr 5.0% Mo 2.0% V 6.0% W 5.0% Co steel: For tools; origin unknown
CYCLONE MC 6V	1.25% C 4.3% Cr 5.5% Mo 7% W 2.8% V 6% Co tool steel: ESC; cutting tools **DPN: 910**
CYCLONE MC 33	0.9% C 4.0% Cr 9.5% Mo 1.15% V 1.5% W 8.0% Co steel: For tools; origin unknown
CYCLONE MC42	1.05% C 4.0% Cr 9.8% Mo 1.1% V 1.5% W 8.3% Co steel: For tools; origin unknown
CYCLONE MC 98	0.9% C 3.7% Cr 8.7% Mo 1.6% W 2% V 8.2% Co steel: ESC; cutting tools **DPN: 910**
CYCLONE SUPERCUT	1.5% C 4.7% Cr 12.5% W 5% V 5% Co steel: ESC; cutting tools **DPN: 910**
D 444	0.4% C 4.2% Cr 0.5% Mo 4.2% W 2.2% V 4.2% Co steel: Fagersta
DARWIN 505	0.7% C 4.5% Cr 18.5% W 1% V 5% Co steel: For tools; Darwin; milling cutters, etc.
DARWIN 1366	0.73% C 4.5% Cr 20% W 2% V 12% Co steel: For tools; Darwin; hobs, borers, etc.
DGC	0.8% C 4.5% Cr 18.5% W 1.4% V 5.7% Co steel: For tools; ESC symbol for DG cyclone **DPN: 910**
DG CYCLONE	0.8% C 4.5% Cr 18.5% W 1.4% V 5.7% Co steel: ESC; cutting tools, drills, etc. **DPN: 910**
DMA 1	14% Ni 8.0% Al 3.0% Cu 24% Co steel: For magnets; Japanese alloy listed by PMA
DMA 2	14% Ni 8.0% Al 3.0% Cu 2.0% Nb 24% Co steel: For magnets; Japanese alloy listed by PMA
DMA 4	22% Ni 12% Al 5.0% Co steel: For magnets; Japanese alloy listed by PMA
DOUBLE ECHO COBALT	0.83% C 4.4% Cr 0.5% Mo 18% W 1.6% V 10% Co steel: John Vessey
DOUBLE RAPID	0.75% C 4.0% Cr 1.0% W 18.0% W 5.0% Co steel: Origin unknown
DOUBLE RAPID	0.8% C 4.75% Cr 0.5% Mo 1.45% V 19.0% W 6.25% Co steel: For tools; origin unknown

Symbol	Nominal analysis, supplier, condition and remarks.
DOUBLE SEVEN	1.25% C 0.3% Mn 12.5% Cr 1.5% Mo 2.75% Co steel: Edgar Allen; punches, dies, etc. **DPN: 730**
DOUBLE SUPER HYDRA	0.8% C 4.5% Cr 18% W 1.7% V 5% Co steel: Hall & Pickles; cutting tools; obsolete **DPN: 820**
ECHO COBALT	0.83% C 4.7% Cr 0.5% Mo 18% W 1.4% V 5.2% Co steel: John Vessey
ECHO SPECIAL VANADIUM	1.5% C 5% Cr 13% W 5% Co steel: John Vessey
ECHO SUPERCUT	1.08% C 3.7% Cr 9.5% Mo 1.5% W 1.15% V 8% Co steel: John Vessey
ECLIPSE H5	0.8% C 4.0% Cr 2.0% V 18.0% W 8.0% Co steel: Origin unknown
EHK 3	1.0% C 6.0% Cr 2.0% Co steel: Forging for magnets; Polish alloy listed by PMA
EK 5	1.0% C 6.0% Cr 6.0% Co steel: For magnets; Russian alloy listed by PMA
EK 6	1.0% C 6.0% Cr 6.0% Co steel: Forging for magnets; Polish alloy listed by PMA
EK 15	1.0% C 8.0% Cr 1.5% Mo 15% Co steel: For magnets; Russian alloy listed by PMA
ELECTRITE Co 9	0.9% C 4.1% Cr 5.2% Mo 5.7% W 1.9% V 9.0% Co steel: Latrobe Steel Co; cutting tools; type M36
ELECTRITE Co	4% Cr 5% Mo 4% W 1.5% V 12% Co steel: For tools; Latrobe Steel Co; AISI code M6F
ELECTRITE COBALT	0.73% C 4.1% Cr 0.7% Mo 18% W 1.1% V 5.0% Co steel: Latrobe Steel Co; cutting tools; type T4
ELECTRITE DYNACUT	1.26% C 3.7% Cr 8.7% Mo 1.8% W 2.0% V 8.2% Co steel: Latrobe Steel Co; cutting tools; Type M43
ELECTRITE DYNAVAN	4% Cr 12% W 5% V 5% Co steel: For tools; Latrobe Steel Co; AISI code T15F
ELECTRITE DYNAVAN XL	1.57% O 4.7% Cr 0.5% Mo 12.5% W 4.7% V 5.0% Co steel: Latrobe Steel Co; cutting tools; type T15
ELECTRITE KELVAN	0.87% C 1.7% W 3.7% Cr 1.1% V 9.5% Mo 5.0% Co steel: Latrobe Steel Co
ELECTRITE LACOMO	4% Cr 8% Mo 2% W 1.2% V 5% Co steel: For tools; Latrobe Steel Co; AISI code M30F
ELECTRITE SUPER	0.85% C 4.1% Cr 0.8% Mo 18.7% W 1.9% V 9.0% Co steel: Latrobe Steel Co; cutting tools; type T5
ELECTRITE SUPER COBALT	4% Cr 18% W 2% V 8% Co steel: For tools; Latrobe Steel Co; T5 type; AISI code TSF
ELECTRITE TATMO	0.9% C 4.0% Cr 8.7% Mo 1.5% W 2.0% V 8.5% Co steel: Latrobe Steel Co; cutting tools; type M34
ELECTRITE TATMO COBALT	4% Cr 8% Mo 2% W 2% V 8% Co steel: For tools; Latrobe Steel Co; M34; AISI code M34F
ELECTRITE UB	4% Cr 14% W 2% V 5% Co steel: For tools; Latrobe Steel Co; AISI code T8F
ELECTRITE ULTRA	4.5% Cr 20% W 1.5% V 12% Co steel: For tools; Latrobe Steel Co; AISI code T6F
ELECTRITE ULTRAVAN	1.5% C 4.2% Cr 5.0% Mo 6.3% W 4.7% V 5.0% Co steel: Latrobe Steel Co; cutting tools; type M15
EMK 10	1.0% C 6.0% Cr 1.5% Mo 10% Co steel: For magnets; Polish alloy listed by PMA
EMK 15	1.0% C 9% Cr 1.5% Mo 15% Co steel: Polish alloy listed by PMA
ER Co 3	0.32% C 3% Cr 0.5% V 3.0% Co 2.8% Mo steel: For tools; Balfour Darwin

Symbol	Nominal analysis, supplier, condition and remarks.
EWK 30	0.75% C 6.0% Cr 6.0% W 30% Co steel: For magnets; Polish alloy listed by PMA
EXTRA NAP SUPERIOR	0.8% C 4.5% Cr 0.5% Mo 20% W 1.25% V 10% Co steel: For tools; Vulcan; cutting tools **DPN: 910**
EXTRA SUPERIOR SPUR	0.72% C 4.5% Cr 18.5% W 1.5% V 3.7% Co steel: For tools; T Turton
EXTRA TRIPLE MUSHET	0.78% C 4.5% Cr 0.75% Mo 1.25% V 18.25% W 8.25% Co steel: For tools; origin unknown
F ALLOY	0.06% C 0.4% Mn 28% Ni 17.5% Co steel: Darwins
FHS 20/5	0.8% C 4.5% Cr 1.5% V 20.0% W 5.0% Co steel: For tools; origin unknown
FHS 20/10	0.83% C 4.5% Cr 1.5% V 20.0% W 10.0% Co steel: Origin unknown
fmp 455	0.8% C 18.5% W 4.2% Cr 1.3% V 5.0% Co 0.7% Mo steel: F Parkin; as tool steel T4
fmp 526	1.1% C 6.7% W 4.2% Cr 2.0% V 5.0% Co 3.7% Mo steel: F Parkin; as tool steel M41
fmp 530	0.8% C 2.0% W 4.0% Cr 1.25% V 5.0% Co 8.2% Mo steel: F Parkin; as tool steel M30
fmp 536	1.5% C 6.5% W 4.0% Cr 5.0% V 5.0% Co 3.5% Mo steel: F Parkin; as tool steel M15
fmp 542	1.05% C 1.5% W 3.7% Cr 1.1% V 8.0% Co 9.5% Mo steel: F Parkin; as tool steel M42
fmp 555C	1.5% C 12.7% W 4.7% Cr 5.0% V 5.0% Co 1.0% Mo steel: F Parkin; as T15 tool steel
fmp 808	0.8% C 18.5% W 4.2% Cr 1.6% V 10% Co 1.0% Mo steel: F Parkin; as tool steel T6
fmp 828	0.8% C 18.5% W 4.5% Cr 2.0% V 8.0% Co 1.0% Mo steel: F Parkin; as T5 tool steel
fmp 928	0.9% C 2.0% W 4.0% Cr 2.0% V 8.0% Co 8.5% Mo steel: F Parkin; as tool steel M34
fmp 929	1.25% C 1.8% W 3.7% Cr 2.0% V 8.2% Co 8.7% Mo steel: F Parkin; as tool steel M43
fmp 933	1.3% C 9.2% W 4.2% Cr 3.5% V 3.7% Mo 10.0% Co steel: F Parkin
FS 2.5	4% Cr 14% W 2% V 5% Co steel: For tools; Firth Sterling Inc; AISI code T8
FS M34	0.9% C 4.0% Cr 8.0% Mo 2.0% W 2.0% V 8.0% Co steel: Firth Sterling
GECALLOY	General name for magnetic powders; Salford Electric
GOLD STAR	4% Cr 14% W 2% V 5% Co steel: For tools; Carpenter Steel Co; AISI code T8K
GREY CUT COBALT	4.5% Cr 20% W 1.5% V 12% Co steel: For tools; Vanadium Alloys; AISI code T6A
GZ 6	0.8% C 4.8% Cr 0.75% Mo 18% W 1.25% V 5% Co steel: For tools; Osborn **DPN: 850**
H 10A	0.37% C 3.0% Cr 2.75% Mo 0.45% V 3.0% Co steel: Osborn
H 19	0.4% C 4.25% Cr 4.2% W 2.0% V 4.25% Co steel: For tools; designation used by AISI
H 19	0.39% C 4.3% Cr 4.25% W 0.4% Mo 2.2% V 4.3% Co steel: For hot work; Osborn
H 43	0.9% C 0.15% Si 0.5% Mn 4% Cr 0.5% W 2% Co steel: For magnets; Jessop
H 61	0.32% C 0.35% Si 0.37% Mn 2.75% Cr 2.75% Mo 0.2% V 2.7% Co steel: Jessop
HA 1	1.0% C 6.0% Cr 3.0% Co steel: For magnets; Belgian alloy listed by PMA
HA 2	1.0% C 8.0% Cr 12% Co steel: For magnets; Belgian alloy listed by PMA
HA 3	1.0% C 10% Cr 15% Co steel: For magnets; Belgian alloy listed by PMA
HA 4	0.7% C 6.0% Cr 35% Co steel: For magnets; Belgian alloy listed by PMA

Note. The following abbreviations and units are used in the tables:

DPN	Hardness, diamond pyramid number
UTS	Ultimate tensile strength, N/mm²
Elon	Elongation, %
Proof	0.1 % proof strength, N/mm²

$1 \text{ N/mm}^2 = 0.1 \text{ hbar} = 0.102 \text{ kgf/mm}^2 = 0.06475 \text{ tonf/in}^2 = 145.04 \text{ lbf/in}^2$

Symbol	Nominal analysis, supplier, condition and remarks.
HAYNES 56	0.25% C 20% Cr 13% Ni 11% Co 4.5% Mo 1% W 1% Nb Fe alloy: American proprietary alloy listed in SAE year book
HAYNES 93	3% C 17% Cr 16% Mo 6% Co steel: Osborn; abrasion & corrosion resistant **DPN: 850**
HEV 1	0.1% C 21.3% Cr 20% Ni 20% Co 2.5% W 3% Mo steel: Designation used by SAE; annealed **UTS: 800 Elon: 63% Proof: 370**
HIPERCO 27	27% Co 0.7% Cr 0.8% Mn 0.7% Ni steel: For magnets; Westinghouse
HIPERCO 50	49% Co 2% V iron alloy: For magnets; Westinghouse
HD 3MX	0.33% C 3.0% Cr 2.8% Mo 0.9% V 1.0% W 3.0% Co steel: For tools; origin unknown
HP 9-4-30	See ASTM A646
HS 95	0.15% C 21% Cr 20% Ni 20% Co 3% Mo 2.5% W 1% Nb Fe alloy: Union Carbide
HSP 15/17	1.3% C 4.2% Cr 3.0% Mo 3.5% V 9.0% W 8.0% Co steel: For tools; origin unknown
HV 5	1.5% C 4.7% Cr 12.5% W 5.0% V 5.0% Co 0.5% Mo steel: Spencer
HYCOMAX I	21% Ni 9% Al 20% Co 1.6% Cu Fe alloy: Casting; PMA; anistropic
HYCOMAX II	7.0% Al 15% Ni 29% Co 4% Cu 2% Nb 4% Ti Fe alloy: PMA; cast for magnets
HYCOMAX III	7% Al 15% Ni 34% Co 4% Cu 5% Ti Fe alloy: For magnets; PMA; cast & sintered
HYCOMAX IV	7.5% Al 14% Ni 38% Co 3% Cu 8% Ti steel: For magnets; cast PMA
HYDRA COVAN	1.5% C 5.0% Cr 12.5% W 5.0% Co steel: Hall & Pickles; cutting tools **DPN: 820**
HYDRA MULTICO	0.85% C 4.5% Cr 0.75% Mo 22% W 1.5% V 11% Co steel: Hall & Pickles; cutting tools **DPN: 790**
HYNICO II	20% Ni 8% Al 20% Co 4.0% Cu 4.0% Ti 0.8% Nb Fe alloy: Casting; PMA
HYPEAK 5 V 5	1.45% C 5% Cr 13% W 5% V 5% Co steel: Swift Levick; cutting tools
HYPEAK 303	0.75% C 4% Cr 0.25% Mo 18% W 1.25% V 5.5% Co steel: Swift Levick; cutting tools
INDALLOY	17% Mo 12% Co Fe alloy: Material for magnets; American material listed by PMA
INMANITE	0.8% C 4.5% Cr 1% Mo 20% W 1.5% V 5% Co steel: Inman; cutting tools
INMARITE T4	0.75% C 4.0% Cr 1.0% V 18.0% W 5.0% Co steel: For tools; Inman
J 5	0.82% C 0.3% Si 0.3% Mn 4.2% Cr 18.5% W 1.6% V 5% Co steel: Jessop; cutting tools
J 26	0.81% C 4.2% Cr 20% Mo 1.6% V 10% Co steel: For tools; Jessop
J 36	1.5% C 4.7% Cr 12.5% W 5% Bv 5% Co steel: For tools; Jessop **DPN: 720**
J 37	1.3% C 4.5% Cr 4% Mo 9.5% W 3.5% V 10% Co steel: For tools; Jessop **DPN: 720**
J 42	1.0% C 3.8% Cr 1.5% W 9.5% Mo 1.1% V 8% Co steel: Jessop; high speed cutting tools **DPN: 975**
KEA 218	0.32% C 3.0% Cr 2.8% Mo 0.35% V 3.0% Co steel: For tools; Sanderson Kayser
KELOCK 873	0.7% C 4.2% Cr 0.7% Mo 18.2% W 1% V 5% Co steel: Kayser Ellison; cutting tools
KELOCK 1021	0.7% C 4.2% Cr 0.7% Mo 18.2% W 1.4% V 10.5% Co steel: Kayser Ellison; cutting tools
KELOCK A 72	0.87% C 4.2% Cr 0.5% Mo 18.7% W 1% V 5.5% Co steel: Kayser Ellison; cutting tools

Symbol	Nominal analysis, supplier, condition and remarks.
KERAU WUMDA	0.75% C 4% Ör 0.5% Mo 18% W 1.25% V 5.5% Co steel: Sanderson; now Saben Wunda **DPN: 790**
KING COBALT	4.5% Cr 20% W 1.5% V 12% Co steel: For tools; Jessop; AISI code T6W
KOERZIT 130	25% Ni 12% Al 5.0% Co 4.0% Cu steel: Casting for magnets; German alloy listed by PMA
KOERZIT 160	21% Ni 13% Al 14% Co 3% Cu steel: For magnets; German alloy listed by PMA
KOERZIT 190	20% Ni 9% Al 16% Co 3.0% Cu steel: For magnets; German alloy listed by PMA
KOERZIT 250	17% Ni 8.0% Al 26% Co 4.0% Cu 7.0% Ti Fe alloy: Casting; German alloy listed by PMA
KOERZIT 300	17% Ni 9% Al 20% Co 3% Cu steel: For magnets; German alloy listed by PMA
KOERZIT 400	14% Ni 8% Al 24% Co 3% Cu steel: For magnets; German alloy listed by PMA
KOERZIT 580	14% Ni 8% Al 24% Co 3% Cu steel: For magnets; German alloy listed by PMA
KOERZIT 700	14% Ni 8% Al 24% Co 3% Cu steel: For magnets; German alloy listed by PMA
KOERZIT VS 55	7.5% Al 14% Ni 38% Co 3% Cu 8% Ti steel: For magnets; German alloy listed by PMA
KOVA 57	1.3% C 4.2% Cr 3.0% Mo 9.0% W 3.5% V 8.5% Co steel: Origin unknown
KOVAR	29% Ni 17% Co 0.4% Mn iron alloy: Westinghouse; low expansion alloy
LEDA	0.75% C 4% Cr 0.7% Mo 18% W 1.4% V 5.2% Co steel: For tools; Firth Brown
LHS 2C	0.75% C 4% Cr 0.3% Mo 18.25% W 1.8% V 2% Co steel. For tools; Low Moor; obsolete **DPN: 850**
LHS 5C	0.8% C 4.5% Cr 0.75% Mo 18.25% W 1.25% V 5% Co steel: For tools; Low Moor; obsolete **DPN: 910**
LHS 8C	0.75% C 4.25% Cr 0.75% Mo 18.25% W 1.25% Co steel: For tools; Low Moor; obsolete **DPN: 910**
LHS 10C	0.75% C 4.5% Cr 0.7% Mo 19% W 1.25% V 10% Co steel: For tools; Low Moor; obsolete **DPN: 950**
LHS 227	1.5% C 4.7% Cr 3% Mo 6.5% W 5% V 5% Co steel: For tools; Low Moor; obsolete **DPN: 975**
LHS 235	1.3% C 4% Cr 3.2% Mo 9% W 3% V 9.5% Co steel: For tools; Low Moor; obsolete **DPN: 975**
LT 49	0.4% C 4.2% Cr 0.4% Mo 4.2% W 2.2% V 4.2% Co steel: Low Moor; for hot work tools; obsolete **DPN: 572**
LU	0.75% C 4% Cr 0.7% Mo 20% W 1% V 10% Co steel: Osborn; obsolete **DPN: 850**
M 6	4% Cr 5% Mo 4% W 1.5% V 12% Co steel: For tools; designation used by AISI; the supplier is coded by a suffix letter
M 15	4% Cr 3.5% Mo 6.5% W 5% V 5% Co steel: For tools; designation used by AISI; the supplier is coded a suffix letter
M 15	1.55% C 4.7% Cr 6.5% W 3.0% Mo 5.0% V 5.0% Co steel: Osborn
M 30	4% Cr 8% Mo 2% W 1.2% V 5% Co steel: For tools; designation used by AISI; the supplier is coded by a suffix letter
M 33	0.9% C 4.0% Cr 9.5% Mo 1.5% W 1.15% V 8.0% Co steel: For tools; designation used by AISI
M 34	4% Cr 8% Mo 2% W 2% V 8% Co steel: For tools; designation used by AISI; the supplier is coded by a suffix letter

Symbol	Nominal analysis, supplier, condition and remarks.
M 35	4% Cr 5% Mo 6% W 2% V 5% Co steel: For tools; designation used by AISI; the supplier is coded by a suffix letter
M 36	4% Cr 5% Mo 6% W 2% V 8% Co steel: For tools; designation used by AISI; the supplier is coded by a suffix letter
M 40	4% Cr 8% Mo 1.5% V 8% Co steel with boron: For tools; designation used by AISI; the supplier is coded by a suffix letter
M 41	4.2% Cr 3.8% Mo 6.2% W 2.0% W 5.0% Co steel: Designation used by AISI
M 41	1.1% C 4.2% Cr 3.8% Mo 2.0% V 6.8% W 5.2% Co steel: For tools; AISI designation
M 42	1.07% C 4.1% Cr 1.5% W 9.8% Mo 1.1% V 8.3% Co steel: Osborn
M 43	1.2% C 2.7% W 3.7% Cr 1.6% V 8.0% Mo 8.25% Co steel: For tools; designation used by AISI
M 44	1.15% C 4.2% Cr 6.5% Mo 2.0% V 5.3% W 11.8% Co steel: For tools; AISI designation
M 44	1.1% C 4.2% Cr 5% W 3.2% Mo 2.2% V 12% Co steel: For tools; designation used by AISI
M 46	1.28% C 4.0% Cr 8.2% Mo 3.2% V 2.0% W 8.2% Co steel: For tools; AISI designation
M 46	1.2% C 4% Cr 2% W 8.2% Mo 3.2% V 8.2% Co steel: For tools; designation used by AISI
M 47	1.1% C 3.7% Cr 9.7% Mo 1.2% V 1.5% W 5.0% Co steel: For tools; AISI designation
M 47	1.1% C 3.7% Cr 1.5% W 9.5% Mo 1.2% V 5% Co steel: For tools; designation used by AISI
MAGLOY 1	Co Ni Al Fe alloy: Casting for magnets; remanence 12 700 gauss; Preformations Ltd
MAGLOY 5	Co Ni Al Fe alloy: Casting for magnets; remanence 7700 gauss; Preformations Ltd
MAGLOY 6	Co Ni Al Fe alloy: Casting for magnets; remanence 6200 gauss; Preformations Ltd
MAGLOY 7	Co Ni Al Fe alloy: Casting for magnets; remanence 9500 gauss; Preformations Ltd
MAGLOY 8A	Co Ni Al Fe alloy: Casting for magnets; remanence 9000 gauss; Preformations Ltd
MAGLOY 8B	Co Ni Al Fe alloy: Casting for magnets; remanence 8500 gauss; Preformations Ltd
MAGLOY 10	Co Ni Al Fe alloy: Casting for magnets; remanence 12 900 gauss; Preformations Ltd
MAGLOY 15	Co Ni Al Fe alloy: Casting for magnets; remanence 11 600 gauss; Preformations Ltd
MAGLOY 100	Co Ni Al Fe alloy: Casting for magnets; remanence 13 400 gauss; Preformations Ltd
MAGNACUT	0.75% C 4.5% Cr 1.25% V 18.0% W 10.0% Co steel: For tools; origin unknown
MAGNICO	14% Ni 8.0% Al 3.0% Cu 24% Co Fe alloy: Casting; Russian alloy listed by PMA
MARAGE 300	0.03% C (max) 18.5% Ni 5.0% Mo 9.0% Co 0.6% Ti 0.08% Ti steel
MARSH CYC 5% Co	0.7% C 4.5% Cr 1% Mo 18% W 1.5% V 5% Co steel: Marsh Brothers; cutting tools
MARSH CYC 275	0.8% C 4.5% Cr 1.2% Mo 20% W 1.2% V 10% Co steel: Marsh Brothers; cutting tools

Symbol	Nominal analysis, supplier, condition and remarks.
MAXITE	4% Cr 14% W 2% V 5% Co steel: For tools; Columbia Tool Steel Co; AISI code T8H
MS 6	6.0% Co steel: For magnets; Swiss alloy listed by PMA
MS 15	15% Co steel: For magnets; Swiss alloy listed by PMA
MS 35	35% Co steel: For magnets; Swiss alloy listed by PMA
MULTICO	0.85% C 4.5% Cr 0.75% Mo 22% W 1.5% V 11% Co steel: Hall & Pickles; cutting tools DPN: 790
MULTIMET	0.1% C 21% Cr 20% Ni 20% Co 3% Mo 2% W 1.0% Nb iron alloy: Wrought; Osborn; solution treated; heat resistant alloy UTS: 800 Elon: 49% Proof: 370
MUSKET HIGH V Co	1.5% C 4.7% Cr 5.0% V 12.5% W 5.0% Co steel: For tools; origin unknown
MUSKET SPECIAL	1.5% C 4.7% Cr 3.0% Mo 5.0% V 6.5% W 5.0% Co steel: For tools; origin unknown
MUSKET SPECIAL VG	1.5% C 4.7% Cr 3.0% Mo 5.0% V 6.5% W 5.0% Co steel: For tools; origin unknown
N 153	0.3% C 17% Cr 15% Ni 12% Co 2% W 1% Nb Fe alloy: Union Carbide
N 155	0.15% C 21% Cr 20% Ni 20% Co 3% Mo 2.5% W 1% Nb Fe alloy: Union Carbide
N 156	0.3% C 17% Cr 33% Ni 24% Co 3% Mo 2% W 1% Nb Fe alloy: Union Carbide
NAP SUPERIOR	0.8% C 4.5% Cr 0.5% Mo 20% W 1.25% V 5% Co steel: For tools; Vulcan; cutting tools DPN: 910
NEW KS	17.7% Ni 27.2% Co 3.7% Al 6.7% Ti Fe alloy: For magnets; origin unknown
Ni Al Co 2	28% Ni 10% Al 6.0% Co Fe alloy: Casting for magnets; Belgian alloy listed by PMA
Ni Al Co 2B	26% Ni 12% Al 8.0% Co Fe alloy: Casting for magnets; Belgian alloy listed by PMA
Ni Al Co 6	20% Ni 10% Al 12% Co 6.0% Cu Fe alloy: Casting for magnets; Belgian alloy listed by PMA
NICOLLOY 15	0.4% C 1.25% Mn 19% Cr 10% Ni 1.5% Mo 1.5% W 1.5% V 15% Co steel: Swift Levick; for hot dies
NIPIGON	4% Cr 18% W 2% V 8% Co steel: For tools; Atlas Steels Ltd; AISI code T5Z
NORESCO K5	0.74% C 4.25% Cr 0.65% Mo 18.5% W 1.4% V 4.5% Co steel: For tools; Schoeller-Bleckmann DPN: 850
NORESCO K10	0.68% C 4.25% Cr 0.6% Mo 18.5% W 10% Co steel: For tools; Schoeller-Bleckmann DPN: 830
NORESCO K12M	1.25% C 4.25% Cr 5.5% Mo 7% W 2.8% V 12.5% Co steel: For tools; Schoeller-Bleckmann DPN: 910
NORESCO WU	0.3% C 2.7% Cr 0.15% Mo 10% W 0.25% V 2.2% Co steel: For tools; Schoeller-Bleckmann UTS: 1000
NORESCO WVC	0.4% C 4.2% Cr 0.4% Mo 4.2% W 2.2% V 4.2% Co steel: For tools; Schoeller-Bleckmann UTS: 1000
NOVO ENORMOUS	0.75% C 4.5% Cr 18% W 1.25% V 10% Co steel: For tools; Jonas DPN: 850
NOVA MAX	0.75% C 4.5% Cr 18% W 1.25% V 5% Co steel: Jonas; cutting tools DPN: 850
NOVO SUPERB	1.25% C 4.25% Cr 3.75% Mo 10.5% W 3.2% V 10% Co steel: Jonas; cutting tools DPN: 950
NOVA V	1.45% C 14.0% W 4.75% Cr 5.0% V 5.0% Co steel: For tools; Jonas; type T15

Note. The following abbreviations and units are used in the tables:

DPN	Hardness, diamond pyramid number
UTS	Ultimate tensile strength, N/mm^2
Elon	Elongation, %
Proof	0.1 % proof strength, N/mm^2

1 N/mm^2=0.1 hbar=0.102 kgf/mm^2=0.06475 tonf/in^2=145.04 lbf/in^2

Symbol	Nominal analysis, supplier, condition and remarks.
NOVA VHC	1.05% C 1.75% W 3.75% Cr 1.2% Cr 1.2% V 9.5% Mo 8.0% Co steel: For tools; Jonas; type M42
NOX 1S4	12.0% Cr 15.0% Ni 10.0% Co 3.0% Mo 2.0% W Fe alloy: Origin unknown
OERSTIT	14% Ni 8% Al 3% Cu 24% Co Fe alloy: International Nickel Co; magnetic alloy
OERSTIT 120R	12% Al 22% Ni 3% Co 3% Cu 0.5% Ti steel: For magnets; German alloy listed by PMA
OERSTIT 160	10% Al 21% Ni 17% Co 3% Cu 1% Ti steel: For magnets; German alloy listed by PMA
OERSTIT 190	10% Al 21% Ni 17% Co 3% Cu 0.5% Ti steel: For magnets; German alloy listed by PMA
OERSTIT 220	7% Al 15% Ni 28% Co 5% Cu 8% Ti steel: For magnets; German alloy listed by PMA
OERSTIT 300	9% Al 17% Ni 22% Co 3% Cu 2% Ti steel: For magnets; German alloy listed by PMA
OERSTIT 350	7% Al 15% Ni 30% Co 4% Cu 5% Ti steel: For magnets; German alloy listed by PMA
OERSTIT 400K	9% Al 15% Ni 24% Co 3% Cu 1% Ti steel: For magnets; German alloy listed by PMA
OERSTIT 400R	9% Al 14% Ni 24% Co 3% Cu steel: For magnets; German alloy listed by PMA
OERSTIT 450	7% Al 15% Ni 32% Co 4% Cu 5% Ti steel: For magnets; German alloy listed by PMA
OERSTIT 500	8% Al 15% Ni 24% Co 3% Cu steel: For magnets; German alloy listed by PMA
OERSTIT 600	8% Al 15% Ni 24% Co 3% Cu steel: For magnets; German alloy listed by PMA
OERSTIT 700	8% Al 15% Ni 24% Co 3% Cu steel: For magnets; German alloy listed by PMA
P 3	0.1% C 0.6% Cr 1.25% Co steel: For tools; designation used by AISI
P 6	0.1% C 1.5% Cr 3.5% Co steel: For tools; designation used by AISI
P 15	0.8% C 4.0% Cr 1.0% Mo 18% W 1.5% V 2.0% Co steel: Bofors; cutting tools **DPN: 840**
P 21	0.2% C 1.2% Ni 4.0% Co steel: For tools; designation used by AISI
PANTHER 5	4% Cr 12% W 5% V 5% Co steel: For tools; Allegheny Ludlum; AISI code T15G
PANTHER SPECIAL	4% Cr 18% W 1% V 5% Co steel: For tools; Allegheny Ludlum; AISI code T4G
PB 2	1.5% C 4.75% Cr 12% W 5% V 5% Co steel: For tools; Osborn; obsolete **DPN: 910**
PBD	0.35% C 1% Cr 0.5% Mo 5% W 4% Co steel: Jonas; dies for casting Cu **DPN: 570**
PDCO	0.3% C 2.5% Cr 10% W 2% Co steel: Jonas; dies for casting Cu & also hot press dies **DPN: 547**
PERMET PF2	3.0% Co Fe alloy: For magnets; origin unknown
PERMANIT	Prefix for Austrian alloys using same grade numbers as Oerstit; listed by PMA
PLUTO PARAMOUNT	0.8% C 4.5% Cr 0.7% Mo 18% W 1.25% V 10% Co steel: Carrs; turning tools **DPN: 950**
PLUTO PERFECTUM	0.8% C 4.8% Cr 0.8% Mo 18% W 1.25% V 5% Co steel: For tools; Carrs; milling cutters, reamers, etc. **DPN: 820**
PLUTO PREMIER	1.5% C 4.5% Cr 3% Mo 6.3% W 5% V 5% Co steel: Carrs; cutting tools; broaches **DPN: 950**
POWHATAN	4% Cr 18% W 1% V 5% Co steel: For tools; Atlas Steels Ltd; AISI code T4Z
PO 41	1.1% C 4.3% Cr 3.8% Mo 6.8% W 2.0% V 5.0% Co steel: Bofors

Symbol	Nominal analysis, supplier, condition and remarks.
PRESSURDIE C	0.38% C 4.2% Cr 0.5% Mo 4.2% W 2.2% V 4.2% Co steel: Braeburn; AISI type H19
PURPLE LABEL	4% Cr 1% V 18% W 1% V 5% Co steel: For tools; Jessop; AISI code T4W
PURPLE LABEL EXTRA	4% Cr 18% W 2% V 8% Co steel: For tools; Jessop; T5 type; AISI code T5W
PYROMET X12	0.12% C 10.5% Cr 4.7% Mo 6.0% Co 1.2% Cu 0.08% N steel: Carpenter
Q 5	0.85% C 4.0% Cr 1.0% Mo 18% W 1.7% V 5.0% Co steel: Bofors; obsolete **DPN: 850**
Q 10	0.85% C 4.0% Cr 1.0% Mo 18% W 1.7% V 10% Co steel: Bofors; obsolete **DPN: 850**
QRO 45	0.3% C 2.8% Cr 2.8% Mo 0.5% V 2.8% Co steel: Bofors **DPN: 400**
QX 1	1.1% C 4.3% Cr 5.0% Mo 6.5% W 2.6% V 5.3% Co steel: Bofors; high speed steel similar to type M35
QX 2	1.1% C 4.0% Cr 5.0% Mo 6.5% W 2.5% V 10% Co steel: Bofors; cutting tools; obsolete **DPN: 850**
R 9K5	0.95% C 4.1% Cr 9.7% W 2.3% V 5.5% Co steel: For tools; Russian Standard designation
R 9K10	0.95% C 4.1% Cr 9.7% W 2.3% V 10.0% Co steel: For tools; Russian Standard designation
R 10K5F5	1.5% C 4.5% Cr 10.7% W 4.6% V 5.5% Co steel: For tools; Russian Standard designation
R 18K5F2	0.9% C 4.1% Cr 17.7% W 2.1% V 5.5% Co steel: For tools; Russian Standard designation
RC 32	1.25% C 23% Cr 8.0% Mo 10% Co Fe alloy: Casting; Dewrance **DPN: 460 UTS: 810**
RECO 2A	20% Ni 7.0% Al 20% Co 7.0% Cu 6.5% Ti Fe alloy: Casting for magnets; Belgian alloy listed by PMA
RECO 3A	20% Ni 9.5% Al 14% Co 6.0% Cu 0.5% Ti Fe alloy: Casting for magnets; Belgian alloy listed by PMA
RED CHIP	4% Cr 18% W 1% V 5% Co steel: For tools; Firth Sterling; AISI code T4B
RED CUT COBALT	4% Cr 18% W 1% V 5% Co steel: For tools; Vanadium Alloys; AISI code T4A
RED CUT COBALT B	4% Cr 18% W 2% V 8% Co steel: For tools; Vanadium Alloys; AISI code T5A
RED SABRE	4% Cr 12% W 5% V 5% Co tool steel: Bethlehem Steel; AISI code T15E
REMALLOY	17% Mo 12% Co Fe alloy: For magnets; American alloy listed by PMA
REMALLOY 17	17% Mo 12% Co steel: For magnets; Simonds
REMALLOY 20	20% Mo 12% Co steel: For magnets; Simonds
REX 95	4% Cr 14% W 2% V 5% Co steel: For tools; Crucible Steel: AISI code T8X
REX 440	4.5% Cr 20% W 1.5% V 12% Co steel: For tools; Crucible Steel; AISI code T6X
REX AAA	4% Cr 18% W 1% V 5% Co steel: For tools; Crucible Steel; AISI code T4X
REX M25	4% Cr 5% Mo 6% W 2% V 5% Co steel: For tools; Crucible Steel; AISI code M35X
REX SUPER CUT	4% Cr 18% W 2% V 8% Co steel: For tools; Crucible Steel; AISI code T5X
REX T Mo 5	4% Cr 8% Mo 2% W 1.2% V 5% Co steel: For tools; Crucible Steel; AISI code M30X
S 6-5-2-5	0.92% C 4.2% Cr 5.0% Mo 6.3% W 1.85% V 4.7% Co steel: For tools; German Standard designation
S 7-4-2-5	1.10% C 4.2% Cr 3.8% Mo 6.8% W 1.85% V 5.0% Co steel: For tools; German Standard designation

Symbol	Nominal analysis, supplier, condition and remarks.
S 10-4-3-10	1.28% C 4.2% Cr 4.0% Mo 10.2% W 3.2% V 10.5% Co steel: For tools; German Standard designation
S 12-1-4-5	1.4% C 4.2% Cr 0.85% Mo 12.0% W 3.7% V 5.0% Co steel: For tools; German Standard designation
S 18-1-2-5	0.8% C 4.2% Cr 0.65% Mo 17.8% W 1.6% V 5.0% Co steel: For tools; German Standard designation
S 497	0.4% C 14% Cr 20% Ni 19% Co 4% Mo 4% W 4% Nb Fe alloy: Allegheny Ludlum
S 590	0.4% C 20% Cr 20% Ni 20% Co 4% W 4% Nb Fe alloy: Allegheny Ludlum
SABEN TENCO	0.75% C 5% Cr 0.5% Mo 18% W 1.7% V 9.5% Co steel: Sanderson **DPN: 790**
SABEN WUNDA	0.75% C 4% Cr 0.5% Mo 18% W 1.25% V 5.5% Co steel: Sanderson; cutting tools **DPN: 790**
SABRE	4% Cr 12% W 5% V 5% Co steel: For tools; Atlas Steel Ltd; AISI code T15Z
SAE HEV 1	0.1% C 21.3% Cr 20% Ni 3.0% Mo 2.5% W 20% Co steel: Annealed
SAE T4	0.75% C 4% Cr 0.8% Mo 18% W 1% V 5% Co steel: For tools
SAE T5	0.8% C 4% Cr 0.8% Mo 18.5% W 2.1% V 8% Co steel: For tools
SAE T8	0.8% C 4.2% Cr 0.8% Mo 13% W 2% V 5% Co steel: For tools
SC	0.8% C 4.5% Cr 21.5% W 1.5% V 12% Co tool steel: ESC; code for super cyclone **DPN: 910**
SERMALLOY A1	7.5% Al 14% Ni 3.8% Co 3% Cu 8% Ti steel: For magnets; French alloy listed by PMA
SIS 2733	0.88% C 0.3% Mn 4.0% Cr 5.0% Mo 6.5% W 1.9% V 5.0% Co steel: For cutting tools; Swedish Standard
SERMALLOY A2	7.5% Al 12% Ni 24% Co 2% Cu 0.2% Si steel: For magnets; French alloy listed by PMA
SIS 2750	0.7% C 4.5% Cr 0.5% Mo 18% W 1.2% V 0.6% Co steel: Swedish Standard; annealed **DPN: 265**
SIS 2754	0.8% C 4.5% Cr 1.4% Mo 18% W 1.6% V 5.5% Co steel: Swedish Standard; annealed **DPN: 290**
SIS 2756	0.8% C 4.5% Cr 1% Mo 18% W 1.6% V 10% Co steel: Swedish Standard; annealed **DPN: 310**
SKH 3	0.77% C 4.2% Cr 18.0% W 1.0% V 5.0% Co steel: For tools; Japanese Standard designation
SKH 4A	0.77% C 4.1% Cr 18.0% W 1.25% V 9.5% Co steel: For tools; Japanese Standard designation
SKH 4B	0.77% C 4.2% Cr 19.0% W 1.25% V 15.0% Co steel: For tools; Japanese Standard designation
SKH 5	0.3% C 4.2% Cr 18.5% W 1.25% V 16.5% Co steel: For tools; Japanese Standard designation
SKH 10	1.52% C 4.2% Cr 12.5% Cr 12.5% W 5.0% V 4.7% Co steel: For tools; Japanese Standard designation

Symbol	Nominal analysis, supplier, condition and remarks.
SKH 55	0.85% C 4.2% Cr 5.2% Mo 6.2% W 2.0% V 5.0% Co steel: For tools; Japanese Standard designation
SKH 56	0.85% C 4.1% Cr 5.5% Mo 6.1% W 2.0% V 8.0% Co steel: For tools; Japanese Standard designation
SKH 57	1.2% C 4.2% Cr 3.5% Mo 10.0% W 3.3% V 10.0% Co steel: For tools; Japanese Standard designation
SL 972	0.1% C 3.5% Cr 5.0% Mo 4.0% W 0.5% V 25% Co steel: Swift Levick; for hot dies
SL 977	0.25% C 12% Cr 8.0% W 0.5% V 10% Co steel: Swift Levick; for hot dies
SO 12/21	0.75% C 4.0% Cr 0.75% Mo 1.0% V 20.0% W 10.0% Co steel: Origin unknown
SOVB	0.78% C 4.5% Cr 0.75% Mo 1.25% V 18.75% W 10.25% Co steel: For tools; origin unknown
SOBV	0.78% C 4.5% Cr 0.75% Mo 18% W 1.25% V 10% Co steel: For tools; Osborn; obsolete
SPEAR DOUBLE CENTURY	1.5% C 4.5% Cr 4.0% Mo 4.8% V 9.0% W 7.5% Co steel: Origin unknown
SPEAR SUPERIOR	0.8% C 4.5% Cr 1.5% V 20.0% W 12.0% Co steel: Origin unknown
SPEAR TRIPLE MERMAID	0.75% C 4.0% Cr 1.0% V 18.0% W 5.0% Co steel: Origin unknown
SPEEDICUT LEDA	0.75% C 4% Cr 0.7% Mo 18% W 1.4% V 5.2% Co steel: Firth Brown
SPEEDICUT SUPERLEDA	0.78% C 4.8% Cr 1% Mo 18.8% W 2% V 9.3% Co steel: Firth Brown
SPEEDICUT VANLEDA	1.5% C 4.5% Cr 12.2% W 5% V 4.7% Co steel: Firth Brown
STAG EXTRA SPECIAL	0.8% C 5% Cr 0.5% Mo 19% W 1.5% V 6% Co steel: Edgar Allen
STAG MAJOR	0.8% C 5% Cr 0.5% Mo 21% W 1.5% V 11.5% Co steel: Edgar Allen
STAR Mo M 2/5	4% Cr 5% Mo 6% W 2% V 5% Co steel: For tools; Firth Sterling; N 135 type AISI code M35B
STAG VANCO	1.5% C 5% Cr 0.5% Mo 17.5% W 5% V 9% Co steel: Edgar Allen
STAR BORON	4% Cr 8% Mo 1.5% V 8% Co B steel: For tools; Carpenter; M40 type; AISI code M40K
SUPER CAPITAL	1.3% C 4.25% Cr 3% Mo 9% W 3.5% V 8.5% Co steel: For tools; Balfour, reamers, hobs **DPN: 975**
SUPER COBALT	4% Cr 18% W 2% V 8% Co steel: For tools; Simonds; T5 type AISI code T5N
SUPER CYCLONE	0.8% C 4.5% Cr 21.5% W 1.5% V 12% Co steel: For tools; ESC; cutting tools **DPN: 910**
SUPER DBL	4% Cr 5% Mo 6% W 2% V 8% Co steel: For tools; Allegheny Ludlum; M36 type AISI code M36G
SUPER HALLAMITE	18% W 5% Co steel (C not specified): For tools; Hallams
SUPER Hi-Mo	4% Cr 8% Mo 2% W 1.2% V 5% Co steel: For tools; Firth Sterling; M30 type AISI code M30B
SUPER INMANITE	0.8% C 4.5% Cr 1% Mo 22% W 1.5% V 10% Co steel: Inman; cutting tools
SUPER INVINCIBLE ADVANCE 10	0.8% C 4.0% Cr 2.0% V 18.0% W 8.0% Co steel: For tools; origin unknown
SUPER INVINCIBLE 5	0.75% C 4.0% Cr 1.0% V 18.0% W 5.0% Co steel: For tools; origin unknown
SUPER INVINCIBLE TB	0.75% C 4.0% Cr 2.0% V 14.0% W 5.0% Co steel: For tools; origin unknown
SUPER INVINCIBLE ADVANCE 10	0.8% C 4.0% Cr 2.0% V 18.0% W 8.0% Co steel: For tools; origin unknown
SUPERLEDA	0.78% C 4.8% Cr 1% Mo 18.8% W 2% V 9.3% Co steel: Firth Brown

---

Note. The following abbreviations and units are used in the tables:

DPN	Hardness, diamond pyramid number
UTS	Ultimate tensile strength, N/mm^2
Elon	Elongation, %
Proof	0.1 % proof strength, N/mm^2

1 N/mm^2=0.1 hbar=0.102 kgf/mm^2=0.06475 tonf/in^2=145.04 lbf/in^2

Symbol	Nominal analysis, supplier, condition and remarks.
SUPER LMW	4% Cr 8% Mo 2% W 1.2% V 5% Co steel: For tools; Allegheny Ludlum; M30 type AISI code M30G
SUPER Mo CHIP	4% Cr 8% Mo 1.5% V 8% Co B steel: For tools; Firth Sterling; M40 type AISI code M40B
SUPER MOTUNG	4% Cr 8% Mo 2% W 1.2% V 5% Co steel: For tools; Universal Cyclops; M30 type AISI code M30D
SUPER MOTUNG SPECIAL	4% Cr 8% Mo 2% W 2% V 8% Co steel: For tools; Universal Cyclops; M34 type AISI code M34D
SUPER PANTHER	4% Cr 18% W 2% V 8% Co steel: For tools; Allegheny Ludlum; T5 type AISI code T5G
SUPER-SUPER MONARCH	0.73% C 4.0% Cr 18.0% W 0.8% Mo 1.1% V 4.7% Co steel: For tools; origin unknown
SUPER UNICUT	4% Cr 3.5% Mo 6.5% W 5% V 5% Co steel: For tools; Universal Cyclops; AISI code M15D
T 4	0.75% C 4% Cr 18% W 1% V 5% Co steel: For tools; designation used by AISI; the supplier is coded by a suffix letter
T 4	0.73% C 4.0% Cr 18.0% W 0.8% Mo 1.1% V 4.7% Co steel: Osborn
T 5	0.8% C 4% Cr 18% W 2% V 8% Co steel: For tools; designation used by AISI; the supplier is coded by a suffix letter
T 6	0.8% C 4.5% Cr 20% W 1.5% V 12% Co steel: For tools; designation used by AISI; the supplier is coded by a suffix letter
T 6	0.8% C 4.0% Cr 20.5% W 0.8% Mo 1.5% V 11.5% Co steel: Osborn
T 8	0.75% C 4% Cr 14% W 2% V 5% Co steel: For tools; designation used by AISI; the supplier is coded by a suffix letter
T 15	1.5% C 4% Cr 12% W 5% V 5% Co steel: For tools; designation used by AISI; the supplier is coded by a suffix letter
T 11306	UNS designation for M6 type tool steel
T 11330	UNS designation for M30 type tool steel
T 11333	UNS designation for M33 type tool steel
T 11334	UNS designation for M34 type tool steel
T 11336	UNS designation for M36 type tool steel
T 11341	UNS designation for M41 type tool steel
T 11342	UNS designation for M42 type tool steel
T 11343	UNS designation for M43 type tool steel
T 11344	UNS designation for M44 type tool steel
T 11346	UNS designation for M46 type tool steel
T 11347	UNS designation for M47 type tool steel
T 12004	UNS designation for T4 type tool steel
T 12005	UNS designation for T5 type tool steel
T 12006	UNS designation for T6 type tool steel
T 12008	UNS designation for T8 type tool steel
T 12015	UNS designation for T15 type tool steel
T 20819	UNS designation for H19 type tool steel
THREE CASTLES	0.75% C 4.5% Cr 1.25% V 18.0% W 5.0% Co steel: For tools; origin unknown
TICONAL C	14% Ni 8.0% Al 24% Co 3.0% Cu 1.0% Nb steel: Casting; Mullard alloy listed by PMA
TICONAL D	14% Ni 8.0% Al 24% Co 3.0% Cu 1.0% Ti steel: Casting; Mullard alloy listed by PMA
TICONAL E	14% Ni 8.0% Al 24% Co 3.0% Cu 1.5% Ti steel: Casting; Mullard alloy listed by PMA
TICONAL F	14% Ni 8.0% Al 24% Co 3.0% Cu 0.5% Ti steel: Casting; Mullard alloy listed by PMA
TICONAL G	14% Ni 8.0% Al 24% Co 3.0% Cu steel: Casting; Mullard alloy listed by PMA
TICONAL GX	14% Ni 8.0% Al 24% Co 3.0% Cu steel: Casting; Mullard alloy listed by PMA; columnar crystal form
TICONAL H	14% Ni 8.0% Al 24% Co 2.5% Nb steel: Casting; Mullard alloy listed by PMA
TICONAL L	14.5% Ni 7.0% Al 19.5% Co 1.5% Cu 0.8% Si steel: Casting; Mullard alloy listed by PMA
TICONAL S	14% Ni 8.0% Al 24% Co 3.0% Cu steel: Sintered; Mullard alloy listed by PMA
TICONAL X	15% Ni 7.0% Al 34% Co 5.0% Ti steel: Casting; Mullard alloy listed by PMA
TREBLE SUPER MONARCH	0.8% C 4.5% Cr 20.0% W 1.5% V 12.0% Co steel: For tools; origin unknown
TRIPLE 5C	1.5% C 5% Cr 13% W 5% V 5% 5% Co steel: For tools; F Parkin DPN: 910
TRIPLE 5 MONARCH	1.5% C 4.0% Cr 5.0% V 12.0% W 5.0% Co steel: For tools; origin unknown
TRIPLE FLYGO	0.73% C 4.7% Cr 1% Mo 18% Cr 1% V 5% Co steel: For tools; Turton Bros
TRIPLE FLYGO	High C tool steel with 5.0% Co high speed: Turton Bros
TRIPLE MERMAID	0.8% C 4.5% Cr 18% W 1.5% V 5.5% Co steel: For tools; Spear & Jackson
TRIPLE MUSHET	0.8% C 4.8% Cr 0.75% Mo 1.25% V 18.2% W 5.0% Co steel: For tools; origin unknown
TRIPLE MUSKET GZ	0.75% C 4.0% Cr 1.0% V 18.0% W 5.0% Co steel: Origin unknown
TRIPLE SPUR	0.82% C 4.5% Cr 1% Mo 20% W 1.7% V 11% Co steel: For tools; T Turton
TRIPLE VELOS	0.8% C 4.2% Cr 0.5% Mo 18% W 1.15% V 5.0% Co steel: Spencer; cutting tools
TWO SPUR	0.78% C 4.5% Cr 0.7% Mo 21% W 1.7% V 6% Co steel: For tools; T Turton
TRIUMPH SUPERB 1000	0.8% C 4.7% Cr 18.5% W 1.6% V 5 7% Co steel: For tools; Jessop DPN: 730
TRIUMPH SUPERB DOUBLE	0.8% C 4.5% Cr 1.5% V 20.0% W 12.0% Co steel: Origin unknown
TRIUMPH SUPERB DOUBLE 1000	0.8% C 4.7% Cr 20% W 1.6% V 10% Co steel: For tools; Jessop DPN: 730
TTQ	0.8% C 4.7% Cr 20% W 1.6% V 10% Co steel: For tools; Jessop DPN: 730
TTQ	0.8% C 4.5% Cr 1.5% V 20.0% W 12.0% Co steel: Origin unknown
TUNCO	4% Cr 18% W 1% V 5% Co steel: For tools; Simonds Steel Mills; AISI code T4N
ULTRA CAPITAL +1	0.85% C 4.25% Cr 18.5% W 1.2% V 5% Co steel: For tools; Balfour; reamers, screwing dies, etc. DPN: 910
ULTRA CAPITAL +2	0.76% C 4.25% Cr 20% W 1.5% V 10% Co steel: For tools; Balfour; cutting tools for hard material DPN: 830
VANCO 5	1.5% C 5% Cr 12% W 5% V 5% Co steel: For tools; Darwin; broaches
VANLEDA	1.5% C 4.5% Cr 12.2% W 5.0% V 4.7% Co steel: For tools; Firth Brown
VASCO SUPREME	4% Cr 12% W 5% V 5% Co steel: For tools; Vanadium Alloys; AISI code T15A
VASCO SUPREME A	4% Cr 3.5% Mo 6.5% W 5% V 5% Co steel: For tools; Vanadium Alloys; AISI code M15A
VC12	0.8% C 4.2% Cr 0.5% Mo 20% W 1.8% V 12% Co steel: Spencer; cutting tools DPN: 800
VELOS 42	1.05% C 4.0% Cr 1.6% W 1.2% V 9.5% Mo 8.5% Co steel: W Spencer
VF	0.4% C 4.2% Cr 0.4% Mo 4.2% W 2.2% V 4.2% Co steel: For tools; Osborn; for hot working; obsolete DPN: 550
VG	1.5% C 4.7% Cr 3% Mo 6.5% W 5% V 5% Co steel: For tools; Osborn; obsolete DPN: 950

Symbol	Nominal analysis, supplier, condition and remarks.
VS 10	High C tool steel with 10.0% Co high speed: Turton Bros
VS 35	0.6% C 2% Cr 6% W 35% Co steel: For magnets; Vulcan
WE	1.35% C 4% Cr 3% Mo 9% W 3% V 9% Co steel: For tools; Osborn; obsolete **DPN: 1000**
WKE	0.8% C 4.5% Cr 1.2% Mo 18.5% W 1.4% V 5.5% Co steel: Fagersta; milling cutters
WKE 4	1.25% C 4.1% Cr 3.1% Mo 9.0% W 3.1% V 9.0% Co steel: For tools; Fagersta **DPN: 670**
WKE 44	1.2% C 4.8% Cr 3.2% Mo 7.3% W 3.3% V 5.3% Co steel: Fagersta; parting off tools, milling cutters
WKE 45	1.4% C 4.2% Cr 3.5% Mo 9.0% W 3.9% V 12.5% Co steel: For tools; Fagersta **DPN: 850**
WKE 46	0.82% C 4.0% Si 5.0% Mn 6.5% W 1.9% V 5.0% Co steel: Fagersta **DPN: 790**
WKE Extra	0.8% C 4.5% Cr 1.0% Mo 18.5% W 1.4% V 10.0% Co steel: For tools; Fagersta **DPN: 671**
WLM 1020	0.8% C 4% Cr 20% W 1.25% V 10% Co steel: W Marrison; cutting tools
WLM 5018	0.8% C 4% Cr 18% W 1.25% V 5% Co steel: W Marrison; cutting tools
WOLFRAM COBALT	4% Cr 18% W 1% V 5% Co steel: For tools; Vulcan; AISI code T4P
WU	0.3% C 2.7% Cr 0.15% Mo 10% W 0.25% V 2.2% Co steel: For tools; Schoeller-Bleckmann **UTS: 1550**
WVC	0.4% C 4.2% Cr 0.4% Mo 4.2% W 2.2% V 4.2% Co steel: For tools; Schoeller-Bleckmann **UTS: 1550**
WV Co	1.5% C 4.7% Cr 12.5% W 5% Co steel: For tools; ESC symbol for cyclone supercut **DPN: 910**

Symbol	Nominal analysis, supplier, condition and remarks.
YCM 2B	15% Ni 1.25% Ti 8% Al 3.4% Cu 24% Co steel: For magnets; Japanese alloy listed by PMA
YMDC	0.35% C 1% Si 1.5% Cr 3% Ni 5% W 0.5% Co steel: For tools; Jonas; hot extrusion dies **DPN: 580**
Z80WKCV 18-05/04-01	0.8% C 4.0% Cr 18.0% W 1.0% V 5.0% Co steel: For tools; French Standard designation
Z80WKCV 18-10/04-02	0.8% C 4.0% Cr 18.0% W 1.6% V 10.0% Co steel: For tools; French Standard designation
Z85WDKCV	0.85% C 4.0% Cr 5.0% Mo 6.0% W 2.0% V 5.0% Co steel: For tools; French Standard designation
Z 110DKCWV	1.1% C 3.8% Cr 9.5% Mo 1.5% W 1.15% V 8.0% Co steel: For tools; French Standard designation
Z 150WDKCV	1.5% C 4.0% Cr 5.0% Mo 6.5% W 5.0% V 5.0% Co steel: For tools; French Standard designation
Z 150WKCV 12-05/05-04	1.3% C 4.0% Cr 12.0% W 5.0% V 5.0% Co steel: For tools; French Standard designation
Z 165WKCV	1.65% C 4.0% Cr 12.0% W 5.0% V 10.0% Co steel: For tools; French Standard designation
Z 175KWDCV	1.75% C 4.0% Cr 5.0% Mo 6.5% W 5.0% V 10.0% Co steel: For tools; French Standard designation

Note. The following abbreviations and units are used in the tables:

DPN	Hardness, diamond pyramid number
UTS	Ultimate tensile strength, N/mm^2
Elon	Elongation, %
Proof	0.1 % proof strength, N/mm^2

1 N/mm^2=0.1 hbar=0.102 kgf/mm^2=0.06475 tonf/in^2=145.04 lbf/in^2

44M1 Steel – low and medium carbon, high chromium
0.8% carbon, 10–20% chromium maximum with titanium; molybdenum and nickel

Specific gravity	7.7	
Density	7700 kg/m³	(0.276 lb/in.³)
Solidus/liquidus	1470 °C	
Thermal conductivity	18.8 W/m °C	(4.5 cal/ms °C)
Coefficient of linear expansion	12.3×10^{-6}/ °C	
Electrical conductivity	2.4% IACS (copper 100%)	
Specific resistance	720 microhm mm	
Young's modulus of elasticity	207×10^9 N/m²	$(30 \times 10^6$ lbf/in.²)
Impact	7–9.5 J	(5–70 ft/lb)
Fatigue strength endurance limit	±300 N/mm²	(± 20 tonf/in.²)
Hot strength		

Temperature °C	Tensile strength N/mm²	Elongation %
450	760	60
500	700	68
600	450	81
650	370	85

The above properties are typical of the following group and may not apply exactly to any one specification. It is possible that with certain specifications some of the values may not be applicable.

General metallurgical characteristics

These steels have a minimum of about 11% chromium, sufficient to ensure that the surface is always protected with a film of chromium oxide, which is firmly adherent and successfully prevents the formation of the loose iron oxides which normally form. None of the steels are truly stainless as they will corrode if used outdoors in damp climates.

As the carbon content is decreased the corrosion resistance improves, until with the carbon content held below that which allows hardening the corrosion resistance is excellent.

There are actually two distinct groups of steel under this heading as the low carbon or ferritic high chrome steels are different in many respects from the medium carbon martensitic high chrome steels, It was found impossible however, to make the division on analysis alone as the same steel can be ferritic or martensitic depending on the thermal treatment.

The very low carbon steels are always ferritic, being non-heat treatable, but from 0.1% carbon the steels are almost always used in the hardened and tempered martensitic condition.

The lower carbon steels very seldom have any alloy content apart from chromium. The general effect of the alloying elements in the higher carbon steels is to improve the creep and other hot strength properties. The alloy elements are usually carbide formers which help to prevent grain growth, stabilize the structure and generally slow down all the metallurgical reactions. Nickel up to about 2% is sometimes present to improve the fatigue strength and low temperature properties which are poor in general with these steels.

Thermal treatment

Where corrosion resistance is more important than strength the steels are used fully annealed and thus become ferritic, but this seldom applies to those with a carbon content above 0.2%.

Annealing requires a temperature of 900–1200 °C for about one hour per 25 mm of section, followed by slow cooling. Similar results are obtained by normalizing where the only difference is cooling in still air, which produces some hardening and decreases the corrosion resistance if the carbon content is above about 0.15%. With the higher carbon steels above 0.25% it will be necessary to follow the normalize with a sub-critical anneal at 650–670 °C. These steels will not normally be fully annealed. Hardening requires a temperature of 860–1150 °C for 30 minutes per 25 mm of section, followed by oil quenching. Forced air cooling is sometimes used where distortion must be held to a minimum. Tempering must be carried out as soon as possible for 2–4 hours at 300–750 °C. The very low carbon steels – below 0.1% carbon – can never be hardened as there is insufficient carbon present to form any appreciable martensite. These steels require quenching after tempering to prevent brittleness if tempered above 350 °C.

Pre-heating at 500–600 °C is recommended on the medium and higher carbon steels, especially when other alloying elements are present, otherwise cracking or distortion can take place.

It is essential that reference is made to the specification origin for full details of these thermal treatments.

These steels are prone to harden when worked or machined. When further severe work is to be carried out a full anneal is necessary, otherwise the sub-critical anneal will suffice. Where corrosion resistance is of paramount importance the parts should be fully annealed as local cold work reduces corrosion resistance.

No surface hardening by carburizing or nitriding is carried out with the steels listed, although most of the lower carbon steels could be carburized if this was ever necessary. Many of the steels are used in the cold worked condition when some corrosion resistance is sacrificed for strength.

Normal furnace equipment is satisfactory for all the steels listed, but if gas or oil fired heating is used, care must be taken to prevent carburizing conditions as this may lead to local carbon penetration which can be undesirable.

Controlled atmospheres are seldom necessary as the high chromium content ensures an adherent oxide film preventing excessive scale formation. When required, they must be of the inert gas type or hydrogen, for example cracked ammonia, or vacuum. Atmospheres containing carbon are not suitable owing to the danger of carbon pick up.

There are now available fusible grit paints which can be applied before thermal treatments. These require careful application but can result in virtual scale-free processing and are economical when grit blasting is not possible and sodium hydride not available.

Scale removal can be by grit blasting provided the parts are robust enough to withstand this treatment which requires high pressures and large particles to remove the very adherent oxide which forms above 900 °C.

Sodium hydride probably gives the best results but requires specialist expensive equipment and constant control.

Pickling requires hot hydrofluoric acid, ferric chloride mixtures which are unpleasant liquids to handle and can cause embrittlement of parts above about 350 DPN.

Welding and brazing

Pre-treatment. The steels should be fully annealed prior to welding. Absolute cleanliness is essential, particularly from all oil or grease which could result in carbon pick up at the weld. The higher carbon, higher alloyed materials will require pre-heating to reduce distortion.

Welding and brazing. The ferritic steels can be readily welded, preferably using the inert gas shielded electric arc method. With the higher carbon steels there is the danger of quench cracking which can be overcome by controlled cooling and pre-heating. Resistance welding should be restricted to the ferritic steels as it is difficult to prevent quench cracking adjacent to the electrode.

Special steels have been developed for their welding characteristics, but advice should always be sought before welding the higher carbon and higher alloyed steels in this group.

Brazing is readily carried out and provided the heating is rapidly applied will not cause softening. Fluxes are necessary.

Post-treatment. Ferritic steels should be fully annealed. Martensitic steels should be hardened and tempered. As an absolute minimum the weld area must be locally stress relieved or tempered.

All fluxes must be completely removed as they are corrosive.

Flaw detection methods

Crack test. Magnetic crack test is possible but unless the correct equipment and techniques are available, it is advised that the penetrant oil chalk test method is employed. This is more flexible and with modern fluids can be as sensitive as magnetic crack test.

X-ray. Certain high duty welds and brazes may require to be X-ray examined either 100% or as quality checks. Normal wrought products seldom require this method of inspection, but high duty castings should have their quality controlled by routine X-ray.

Ultra-sonic test. Forgings or castings may be ultra-sonic inspected at an early machined stage when subcutaneous defects will be detected which could be dangerous on the finished part. Also certain types of weld defect can be identified.

Chemical etch. Electropolish using chrome phosphoric acid can be used to identify material defects and improve corrosion resistance, but this will only be required on very high duty parts. Both hot and cold chemical etches have been devised to highlight surface defects. Some of these improve the surface finish.

Corrosion protection

Temporary. None required. These steels should not be stored out of doors for long periods.

Permanent. None normally required. Chemical or electropolishing is sometimes used to improve the surface finish and passivate the surface. These steels are not truly stainless, thus should not be used under damp or corrosive conditions or in contact with other metals. Corrosion once started can propagate rapidly, thus parts should always be kept clean and designed to be free of moisture traps.

Machinability The ferritic steels have a tendency to tear rather than cut, thus giving a poor finish. High surface speeds, with low tool pressure and single point cutting action, give the best results.

The martensitic steels give best results when in the hardened and tempered condition around 250 DPN. Tipped tools are always necessary to combat the abrasive carbides.

Uses

Cutlery, golf clubs, steam turbine blading, compressor blades in gas turbines, motor car trim, household ornaments, steel tubing and sheet for medium corrosive condition, but not for marine or food use. Architectural uses for inside trim or outside use, but not in marine or industrial atmosphere without special treatment.

Symbol	Nominal analysis, supplier, condition and remarks.
0.4021	0.19% C 13% Cr steel: German Standard
1 C 27	0.08% C 13.5% Cr 1.0% Mo steel: Tube
	DPN: 160 UTS: 520 Elon: 13% Proof: 350
1C 36	0.12% C 17% Cr steel: Sandvik
1C 342	0.08% C 16.7% Cr 0.5% Ti steel: Sandvik
1CN	0.2% C 14% Cr 1% Ni steel: Inman; for plastic extrusion dies
1KH11MF	0.15% C 10.7% Cr 0.7% Mo 0.32% V steel: Russian Standard designation
1KH12V2MF	0.13% C 12.0% Cr 0.75% Mo 0.22% V 2.0% W steel: Russian Standard designation
1KH12N2VMF	0.13% C 11.5% Cr 1.65% Ni 0.4% Mo 0.22% V 1.8% W steel: Russian Standard designation
1KH12VNMF	0.15% C 12.0% Cr 0.6% Ni 0.6% Mo 0.2% V 0.9% W steel: Russian Standard designation
1.2625	0.38% C 12.0% Cr 1.0% V 12.0% W steel: For tools; German Standard
1.3811	0.15% C (max) 25% Cr steel: Bar; German Standard
1.3827	0.1% C (max) 17% Cr 0.8% Ti steel: Forging; German Standard
1.400	0.08% C (max) 13.0% Cr steel: German Standard
1.4000	0.08% C (max) 13% Cr steel: German Standard
1.4001	0.08% C (max) 14% Cr steel: German Standard
1.4002	0.08% C (max) 13.0% Cr 0.2% Al steel: German Standard
1.4006	0.1% C 13% Cr steel: Forging; German Standard
1.4007	0.35% C 14% Cr steel: Welding rod; German Standard
1.4009	0.1% C (max) 14.5% Cr steel: Welding rod; German Standard
1.4015	0.1% C (max) 17.5% Cr steel: Forging; German Standard
1.4016	0.1% C (max) 16.5% Cr steel: Forging; German Standard
1.4021	0.19% C 13% Cr steel: For springs; German Standard
1.4024	0.15% C 13% Cr steel: Forging; German Standard
1.4034	0.41% C 13% Cr steel: For bearings; German Standard
1.4044	0.2% C 16% Cr 2% Ni steel: German Standard
1.4057	0.2% C 17% Cr 2% Ni steel: Forging; German Standard
1.4084	0.1% C (max) 28% Cr steel: German Standard
1.4104	0.13% C 16% Cr 0.25% Mo 0.2% S steel: German Standard; free machining
1.4113	0.07% C (max) 17.0% Cr 1.05% Mo steel: German Standard
1.4116	0.46% C 14.4% Cr 0.12% V steel: German Standard
1.4119	0.15% C 13% Cr 1.1% Mo steel: German Standard
1.4120	0.2% C 13% Cr 1.0% Ni 1.2% Mo steel: German Standard
1.4122	0.38% C 16% Cr 1.0% Ni 1.1% Mo steel: German Standard
1.4137	0.1% C (max) 30% Cr 2.0% Ni 2.1% Mo steel: German Standard
1.4502	0.1% C (max) 17.5% Cr 0.5% Ti steel: Forging; German Standard

Symbol	Nominal analysis, supplier, condition and remarks.
1.4510	0.1% C (max) 17% Cr 0.8% Ti steel: Forging; German Standard
1.4511	0.1% C (max) 17.5% Cr 1.2% Nb steel: Forging; German Standard
1.4523	0.1% C (max) 17% Cr 1.8% Mo 0.8% Ti steel: German Standard
1.4525	0.1% C (max) 17.5% Cr 1.0% Ni 1.1% Mo 0.6% Ti steel: German Standard
1.4722	0.12% C (max) 13% Cr steel: Forging; German Standard
1.4723	0.12% C (max) 14.5% Cr 1.0% Ni steel: Forging; German Standard
1.4724	0.12% C (max) 13% Cr 1.0% Al steel: Forging; German Standard
1.4741	0.12% C (max) 18% Cr steel: Forging; German Standard
1.4742	0.12% C (max) 18% Cr 1.0% Al steel: Forging; German Standard
1.4762	0.12% C (max) 24% Cr 1.5% Al steel: Forging; German Standard
1.4765	25% Cr 5% Al steel: German Standard
1.4767	20% Cr 5% Al steel: German Standard
1.4772	0.12% C (max) 29% Cr steel: Forging; German Standard
1.4773	0.1% C (max) 30% Cr steel: Forging; German Standard
1.4774	0.1% C (max) 29% Cr 4.5% Al steel: Heat conductor; German Standard
1.4820	0.15% C (max) 26% Cr 4.5% Ni steel: Forging; German Standard
1 4821	0.2% C 25% Cr 4.5% Ni steel: Forging; German Standard
1.4905	0.1% C (max) 20% Cr 4.5% Al steel: Heat conductor; German Standard
1.4920	0.17% C 11.5% Cr 1.1% Mo 0.3% V steel: German Standard
1.4922	0.20% C 11.5% Cr 1.1% Mo 0.3% V steel: German Standard
1.4924	0.22% C 11% Cr 1.1% Mo 0.3% V steel: German Standard
1.4934	0.22% C 11% Cr 0.5% Ni 1.0% Mo 0.5% W 0.3% V steel: German Standard
1.4935	0.2% C 11.5% Cr 0.5% Ni 1.0% Mo 0.5% W 1.0% V steel: German Standard
1.4994	0.45% C 23.0% Cr 5.0% Ni 2.7% Mo steel: German Standard
2C 27	0.08% C 13.5% Cr steel: Sandvik
2C 34	0.12% C 17.2% Cr steel: Sandvik
2KH12VMBFR	0.18% C 12.0% Cr 0.5% Mo 0.2% V 0.5% W 0.3% Nb steel: Russian Standard designation
2R 27	0.12% C 13% Cr steel: Bofors; obsolete
2R 29	0.09% C 17.5% Cr steel: Bofors; obsolete
2R 47	0.21% C 13% Cr steel: Bofors
2R 57	0.3% C 13% Cr 0.5% Ni steel: Bofors; obsolete
2R 77	0.27% C 13% Cr steel: Bofors
2R 107	0.55% C 14% Cr steel: Bofors; hardened & tempered obsolete
	DPN: 325 UTS: 1070 Elon: 10% Proof: 870
2RL 2	0.1% C 13.5% Cr 0.2% S steel: Free machining Bofors; hardened & tempered; obsolete
	DPN: 210 UTS: 700 Elon: 25% Proof: 530
2RL 3	0.12% C 16.5% Cr 0.3% Mo 0.2% S steel: Free machining; Bofors; annealed
	DPN: 185 UTS: 620 Elon: 20% Proof: 300
2RL 14	0.2% C 13.5% Cr 1.0% Mn 0.2% S steel: Free machining; Bofors; hardened & tempered
	DPN: 315 UTS: 750 Elon: 14% Proof: 680
2RMO	0.05% C 13% Cr 6% Ni 1.5% Mo steel: Weldable; Bofors; hardened & tempered
	DPN: 290 UTS: 810 Elon: 15% Proof: 600

Note. The following abbreviations and units are used in the tables:

DPN	Hardness, diamond pyramid number
UTS	Ultimate tensile strength, N/mm^2
Elon	Elongation, %
Proof	0.1 % proof strength, N/mm^2

1 N/mm^2=0.1 hbar=0.102 kgf/mm^2=0.06475 tonf/in^2=145.04 lbf/in^2

Symbol	Nominal analysis, supplier, condition and remarks.
2RM 2	0.05% C 12.5% Cr 5.5% Ni steel: Bofors; hardened & tempered; suitable for welding **DPN: 265 UTS: 800 Elon: 15% Proof: 600**
2RO 26	0.1% C 12% Cr 0.5% Ni 0.5% Mo steel: Bofors; hardened & tempered **DPN: 220 UTS: 720 Elon: 22% Proof: 580**
2RO 27	0.1% C 13.5% Cr 1.2% Mo steel: Bofors
2RO 46	0.2% C 12% Cr 0.5% Ni 1.2% Mo steel: Bofors; obsolete
3RE 60	0.03% C (max) 18.5% Cr 4.7% Ni 2.7% Mo steel: Tube; Sandvik; weldable **UTS: 836 Elon: 44% Proof: 420**
3RE 60	0.03% C (max) 18.5% Cr 4.7% Ni 2.7% Mo steel: Sandvik; ferritic; austenitic; weldable **DPN: 260 UTS: 820 Elon: 22% Proof: 500**
4C 27	0.2% C 12.5% Cr steel: Sandvik
4C 27A	0.22% C 13% Cr 1% Ni 1.3% Mo 0.2% S steel: Sandvik
4C 54	0.18% C 27% Cr steel: Sandvik
4N 2C 36	0.25% C 17% Cr 2% Ni steel: Sandvik
6C 27	0.32% C 14% Cr steel: Sandvik
6C 283	0.28% C 14% Cr 0.5% Mo 0.25% Cu steel: Sandvik
7C 27	0.27% C 13% Cr steel: Sandvik
7C 27 Mo2	0.35% C 13.5% Cr 1.3% Mo steel: Sandvik
10C 27	0.5% C 14% Cr steel: Sandvik
10 RE 20	0.08% C 26.3% Cr 5% Ni steel: Sandvik
10 RE 21	0.08% C (max) 26.5% Cr 5% Ni 1.5% Mo steel: Sandvik
11 Cr 9 Mn B	0.4% C 9.0% Mn 11.0% Cr steel: Electrode; Metrode; work hardens from 200 to 400 DPN
12 Cr MOB	0.07% C 12.0% Cr 1.0% Mo steel: Electrode; Metrode
13BMP	0.07% C 13.0% Cr steel: Electrode; Metrode
13RMP	0.07% C 13.0% Cr steel: Electrode; Metrode
13.2R	0.06% C 13.0% Cr 2.0% Ni steel: Electrode; Metrode
13.4LB	0.04% C 13.0% Cr 3.5% Ni steel: Electrode; Metrode
13.4 Mo LR	0.04% C 13.0% Cr 3.5% Ni 0.4% Mo steel: Electrode; Metrode
17 BMP	0.07% C 17.0% Cr steel: Electrode; Metrode
17RMP	0.07% C 17.0% Cr steel: Electrode; Metrode
18.2	18.0% Cr 2.0% Mo steel: Resists chlorides; has good low temperature impacts; Climax Molybdenum
20RMP	0.07% C 20.0% Cr steel: Electrode; Metrode; resists scaling up to 950 °C
23C 27	0.12% C 13% Cr steel: Sandvik
28RMP	0.07% C 28.0% Cr steel: Electrode; Metrode; resists scaling up to 1000 °C
45 Cr 13	0.45% C 13.0% Cr steel: Electrode; Metrode; hardenable corrosion resist deposit **DPN: 550**
45 Cr 25	0.45% C 25.0% Cr steel: Electrode; Metrode; for hardenable corrosion resistant deposit **DPN: 300**
45 Cr 25	0.45% C 1.0% Mn 25.0% Cr 1.0% Ni steel: Electrode; Metrode; for repair welding of castings

Symbol	Nominal analysis, supplier, condition and remarks.
249H	0.25% C 17% Cr 2% Ni steel: Avesta
249T	0.12% C 17% Cr steel: Avesta
300 Cr 15 Mn 6	3.0% C 6.0% Mn 2.0% Si 15.0% Cr steel: Electrode; Metrode; for hard facing **DPN: 550**
393	0.12% C 13% Cr steel: Avesta
419	0.25% C 11.5% Cr 0.5% Mo 2.5% W steel: American proprietary alloy listed in SAE year book
422	0.2% C 12% Cr 1.0% Mo 1.0% W 0.3% V steel: Hardened & tempered; designation used by AISI **DPN: 320 UTS: 1050 Elon: 18% Proof: 920**
422M	0.85% C 12% Cr 2.25% Mo 1.7% W 0.5% V steel: American proprietary alloy listed in SAE year book
510	0.12% C 13% Cr steel: Soderfors
511	0.21% C 13% Cr steel: Soderfors
512	0.27% C 13% Cr steel: Soderfors
522	0.12% C 17% Cr steel: Soderfors
525	0.25% C 17% Cr 2% Ni steel: Soderfors
636	0.22% C 13% Cr 0.7% Ni 1.0% Mo 1.0% W 0.35% V steel: Carpenter for AISI type 422
739	0.21% C 13% Cr steel: Avesta
739H	0.27% C 13% Cr steel: Avesta
831	0.12% C 21% Cr steel: Avesta
1408	0.07% C 13.0% Cr 0.2% Al steel: Nyby; annealed **DPN: 190 UTS: 420 Elon: 20% Proof: 290**
1410	0.12% C 13.0% Cr steel: Nyby
1410 Mo	0.08% C 14.0% Cr 1.0% Mo steel: Nyby; annealed **DPN: 200 UTS: 480 Elon: 20% Proof: 290**
1415	0.15% C 13.0% Cr steel: Nyby
1415 Mo	0.14% C 13.0% Cr 1.0% Mo steel: Nyby; annealed **DPN: 190 UTS: 480 Elon: 20% Proof: 290**
1435	0.27% C 13% Cr steel: Nyby
1706	0.05% C 17.0% Cr steel: Nyby; annealed **DPN: 180 UTS: 420 Elon: 20% Proof: 290**
1708	0.07% C 17.0% Cr steel: Nyby; annealed **DPN: 200 UTS: 420 Elon: 20% Proof: 290**
1708 Mo	0.07% C 17.0% Cr 1.0% Mo steel: Nyby; annealed **DPN: 200 UTS: 420 Elon: 20% Proof: 290**
1708 Mo Nb	0.07% C 17.0% Cr 1.0% Mo 0.9% Nb steel: Nyby; annealed **DPN: 200 UTS: 480 Elon: 20% Proof: 290**
1708 Mo Ti	0.07% C 17.0% Cr 1.0% Mo 0.5% Ti steel: Nyby; annealed **DPN: 200 UTS: 480 Elon: 20% Proof: 290**
1708 Nb	0.07% C 17.0% Cr 0.9% Nb steel: Nyby; annealed **DPN: 180 UTS: 420 Elon: 20% Proof: 290**
1708 Ti	0.07% C 17.0% Cr 0.5% Ti steel: Nyby; annealed **DPN: 180 UTS: 420 Elon: 20% Proof: 290**
1710 Si	0.10% C 2.2% Si 17.0% Cr steel: Nyby; annealed **DPN: 230 UTS: 530 Elon: 15% Proof: 330**
2013	0.2% C 13.0% Cr steel: F Parkin; plastic moulds
2520	0.20% C 25.0% Cr 0.15% Ni steel: Nyby; annealed **DPN: 230 UTS: 480 Elon: 20% Proof: 330**
2731	0.7% C 15.0% Cr steel: For tools; French Standard designation
2732	0.4% C 14.0% Cr steel: For tools; French Standard designation
2733	0.3% C 13.0% Cr steel: For tools; French Standard designation
4379	0.3% C 13.0% Cr steel: Sanderson; 'Non-Stain' **DPN: 277 UTS: 940 Elon: 20% Proof: 800**
AH	0.2% C 19% Cr steel: Annealed; furnace equipment. Osborn **UTS: 530 Elon: 20% Proof: 450**
AISI 403	0.15% C 12% Cr steel: Hardened & tempered **DPN: 400 UTS: 1500 Elon: 2% Proof: 1350**

Note. The following abbreviations and units are used in the tables:

DPN	Hardness, diamond pyramid number
UTS	Ultimate tensile strength, N/mm^2
Elon	Elongation, %
Proof	0.1 % proof strength, N/mm^2

1 N/mm^2=0.1 hbar=0.102 kgf/mm^2=0.06475 tonf/in^2=145.04 lbf/in^2

Symbol	Nominal analysis, supplier, condition and remarks.
AISI 405	0.08% C 12.5% Cr 0.2% Al steel: Annealed **DPN: 180 UTS: 420 Elon: 20% Proof: 210**
AISI 410	0.15% C (max) 12% Cr steel: Hardened & tempered **DPN: 400 UTS: 1500 Elon: 2% Proof: 1350**
AISI 414	0.1% C 12.5% Cr 2% Ni steel: Hardened & tempered **DPN: 440 UTS: 1550 Elon: 2% Proof: 1350**
AISI 416	0.15% C (max) 13% Cr steel: Hardened & tempered **DPN: 280 UTS: 1030 Elon: 15% Proof: 840**
AISI 416 Se	0.15% C 13% Cr 0.15% Se steel: Free cutting
AISI 418	0.15% C 13% Cr 3% W steel: Hardened & tempered **DPN: 400 UTS: 1500 Elon: 10% Proof: 1350**
AISI 420	0.35% C 13% Cr steel **DPN: 550 UTS: 1840 Elon: 2% Proof: 1610**
AISI 420F	0.3% C 13% Cr steel: Hardened & tempered **DPN: 500 UTS: 1620 Elon: 5% Proof: 1350**
AISI 430	0.12% C (max) 16% Cr steel: Hardened & tempered **DPN: 500 UTS: 1100 Elon: 3% Proof: 880**
AISI 430 F	0.12% C (max) 16% Cr 0.15% S steel: Hardened & tempered **DPN: 300 UTS: 1100 Elon: 2% Proof: 880**
AISI 430 F Se	0.12% C 16% Cr 0.15% Se steel: Free machining
AISI 431	0.2% C (max) 16% Cr 2% Ni steel: Hardened & tempered **DPN: 440 UTS: 1580 Elon: 10% Proof: 1380**
AISI 434	0.12% C 17.0% Cr 1% Mo steel: Sheet, strip & wire **DPN: 164 UTS: 530 Elon: 23% Proof: 370**
AISI 440 A	0.7% C 17% Cr 0.7% Mo steel: Hardened & tempered **DPN: 550 UTS: 1920 Elon: 2% Proof: 1700**
AISI 440 B	0.8% C 17% Cr 0.7% Mo steel
AISI 442	0.2% C (max) 20% Cr steel: Annealed **DPN: 200 UTS: 580 Elon: 20% Proof: 300**
AISI 446	0.25% C (max) 26% Cr steel: Annealed **DPN: 200 UTS: 450 Elon: 20% Proof: 300**
AISI 614	0.12% C 12.2% Cr steel: Annealed **UTS: 620 Elon: 30% Proof: 200**
AISI 614	0.12% C 12.2% Cr steel: Hardened & tempered **DPN: 255 UTS: 880 Elon: 21% Proof: 760**
AISI 615	0.17% C 13% Cr 2% Ni 0.2% Mo 2.9% W steel: Hardened & tempered **UTS: 1140 Elon: 16% Proof: 1000**
AISI 616	0.23% C 12% Cr 1% Mo 1% W 0.25% V steel: Hardened & tempered **UTS: 1070 Elon: 20% Proof: 880**
AISI 619	0.3% C 11.4% Cr 2.7% Mo 0.25% V steel: Hardened & tempered **UTS: 1000 Elon: 18% Proof: 700**
AL 419	0.2% C 1.2% Cr 1.2% Ni 3% W 0.5% Mo 0.5% V steel: Hardened & tempered; designation used by ASTM **DPN: 330 UTS: 1210 Elon: 15% Proof: 1000**
AMS 5340	14% Cr 4.2% Ni 2.5% Mo 3.2% Cu 0.25% Nb + Ti steel: Investment casting
AMS 5342	16% Cr 4% Ni 3.1% Cu steel: Investment casting; solution treated & aged **UTS: 950**
AMS 5343	16% Cr 4% Ni 3% Cu steel: Investment casting; solution treated & aged **UTS: 1110**
AMS 5344	16% Cr 4% Ni 3% Cu steel: Investment casting; solution treated & aged **UTS: 1350**
AMS 5349	13% Cr 0.25% S steel: Investment casting
AMS 5350 D	12.5% Cr steel: Investment casting

Symbol	Nominal analysis, supplier, condition and remarks.
AMS 5351 B	0.15% C 12.5% Cr steel: Sand casting; AMS for SAE 60410
AMS 5353	16% Cr 1.8% Ni steel: Investment casting
AMS 5354 B	13% Cr 2% Ni 3% W steel: Investment casting
AMS 5355 A	16% Cr 4% Ni 3% Cu steel: Investment casting
AMS 5359	15% Cr 4% Ni 2.3% Mo 0.1% N steel: Sand casting
AMS 5368	15% Cr 4% Ni 2.3% Mo 1.0% N steel: Investment casting
AMS 5372	15.8% Cr 2% Ni steel: Sand casting
AMS 5398 A	16% Cr 4% Ni 3% Cu Fe steel: Sand casting
AMS 5503	0.2% C 17% Cr steel: Sheet & strip; AMS for SAE 51431
AMS 5504 D	0.15% C 12.5% Cr steel: Sheet & strip; AMS for SAE 51410
AMS 5505	0.15% C 12.5% Cr steel: Sheet & strip; ferrite controlled; AMS for SAE 51410
AMS 5506 A	0.35% C 13% Cr steel: Sheet & strip; AMS for SAE 51420
AMS 5508 A	13% Cr 2% Ni 3% W steel: Sheet & strip
AMS 5591 D	0.15% C 12.5% Cr steel: Seamless tube; AMS for SAE 51410
AMS 5609	0.15% C 12.5% Cr 0.12% Nb steel: Tubing; ferrite controlled; AMS for SAE 51410
AMS 5610 E	0.35% C 13% Cr steel: Bar & forgings; free machining; AMS for SAE 51416F
AMS 5612	0.15% C 12.5% Cr steel: Bar or tube; ferrite controlled; AMS for SAE 51410
AMS 5613 E	0.15% C 12.5% Cr steel: Bar or tube; AMS for SAE 51410
AMS 5614	0.15% C 12% Cr 0.5% Mo steel: Bar & forging; AMS for SAE 51410 + Mo
AMS 5615 B	0.1% C 12.5% Cr 2% Ni steel: Bar & forging; AMS for SAE 51414
AMS 5616 D	13% C 2% Ni 3% W steel: Bar & forging
AMS 5620 D	0.35% C 13% Cr steel: Bar & forging; free machining; AMS for SAE 51420F
AMS 5621	0.35% C 13% Cr steel: Bar & forging; AMS for SAE 51420
AMS 5627	0.2% C 17% Cr steel: Bar & forging
AMS 5628 B	0.2% C 16% Cr 2% Ni steel: Bar & forging; AMS for SAE 51431
AMS 5631	0.7% C 17% Cr 0.75% Mo steel: Bar & forging; AMS for SAE 51440A
AMS 5655	12.5% Cr 0.75% Ni 1% Mo 1% W steel: Bar & forging
AMS 5710 B	0.8% C 20% Cr 1.3% Ni 2.3% Si steel: Bar & forging
AMS 5776	12.5% Cr steel: Welding wire; AMS for SAE 51410
AMS 5777	12.5% Cr steel: Coated electrode; AMS for SAE 51410
AMS 5817 A	13% Cr 2% Ni 3% W steel: Welding wire
AMS 5821 A	12.5% Cr steel: Welding wire; special grade; AMS for SAE 51410
ANORINOX 35 AM	0.25% C 16% Cr 1.5% Mo steel: Anor
AN QQ S770/2/1	0.17% C 16.5% Cr 2.0% Ni steel: US Service; hardened & tempered **UTS: 1210 Elon: 13% Proof: 940**
AN QQ S770/2/11	0.17% C 16.5% Cr 2.0% Ni steel: US Service; hardened & tempered **UTS: 780 Elon: 15% Proof: 610**
ANTINIT KW 10M	0.1% C 13% Cr 1.0% Mo steel: Gebruder Bohler
ANTINIT KW 15M	0.15% C 13% Cr 1.0% Mo steel: Gebruder Bohler
ANTINIT KW 20M	0.2% C 13% Cr 1.0% Mo steel: Gebruder Bohler
ARD 3	0.38% C 17% Cr 0.6% Ni 1.2% Mo steel: Hardened & tempered; Schoeller-Bleckmann **DPN: 260 UTS: 870 Elon: 14% Proof: 580**

Symbol	Nominal analysis, supplier, condition and remarks.
ARH	0.40% C 14.5% Cr steel: Hardened & tempered; Schoeller-Bleckmann
	DPN: 400 UTS: 1280 Elon: 6% Proof: 1110
ARH TOUGH	0.22% C 14% Cr 0.4% Ni steel: Hardened & tempered Schoeller-Bleckmann
	DPN: 240 UTS: 800 Elon: 15% Proof: 460
ARKS	0.13% C 17% Cr 0.5% Ni S steel: Free machining; hardened & tempered; Schoeller-Bleckmann
	DPN: 230 UTS: 760 Elon: 12% Proof: 450
ARL	0.22% C 17% Cr 1.4% Ni steel: Hardened & tempered; Schoeller-Bleckmann
	DPN: 260 UTS: 840 Elon: 14% Proof: 580
ARMCO 12	0.15% C 12% Cr steel: Armco
ARMCO 12/2	0.15% C 12% Cr 2.0% Ni steel: Armco
ARMCO 12 Al	0.03% C 12% Cr 0.2% Al steel: Armco
ARMCO 12 FM	0.15% C 13% Cr 0.07% S or Se or Mo steel: Armco; free machining
ARMCO 12T	0.15% C 12% Cr steel: Armco
ARMCO 13-C-35	0.15% C 13% Cr 0.2% S steel: Armco; free machining
ARMCO 16/2	0.2% C 16% Cr 2.0% Ni steel: Armco
ARMCO 17	0.12% C 16% Cr steel: Armco
ARMCO 17-C-60	0.65% C 17% Cr 0.7% Mo steel: Armco
ARMCO 27	0.35% C 25% Cr 0.25% N steel: Armco
ARW	0.1% C 13.5% Cr steel: Bar & sheet; hardened & tempered; Schoeller-Bleckmann
	DPN: 200 UTS: 640 Elon: 18% Proof: 440
ARWA	0.08% C (max) 13% Cr Al steel: Annealed; Schoeller-Bleckmann
	DPN: 150 UTS: 530 Elon: 20% Proof: 290
ARWB	0.1% C 14% Cr steel: Annealed; Schoeller-Bleckmann
	DPN: 160 UTS: 530 Elon: 20% Proof: 290
ARWD	0.08% C (max) 17% Cr 1.7% Mo steel: For welding; annealed; Schoeller-Bleckmann
	DPN: 160 UTS: 550 Elon: 20% Proof: 290
ARWDN	0.08% C (max) 17% Cr 1.7% Mo Nb steel: Annealed; Schoeller-Bleckmann
	DPN: 160 UTS: 550 Elon: 20% Proof: 290
ARWDT	0.08% C (max) 17% Cr 1.7% Mo Ti steel: Annealed; Schoeller-Bleckmann
	DPN: 160 UTS: 550 Elon: 20% Proof: 290
ARWF	0.08% Cr (max) 12% Cr steel: Hardened & tempered; Schoeller-Bleckmann
	DPN: 200 UTS: 650 Elon: 18% Proof: 440
ARWK	0.08% C 16% Cr steel: Schoeller-Bleckmann
	DPN: 150 UTS: 510 Elon: 20% Proof: 290
ARWKIN	0.08% C (max) 18% Cr Nb steel: For welding; annealed; Schoeller-Bleckmann
	DPN: 150 UTS: 510 Elon: 20% Proof: 290
ARWKT	0.08% C (max) 18% Cr Ti steel: For welding; annealed; Schoeller-Bleckmann
	DPN: 150 UTS: 510 Elon: 20% Proof: 290
ARZ	0.2% C 13.5% Cr steel: Hardened & tempered; Schoeller-Bleckmann
	DPN: 260 UTS: 840 Elon: 14% Proof: 530
ASM 123	0.3% C 13% Cr 1.25% Mo steel: Sambre & Meuse
ASTM A176/403	0.15% C 12% Cr steel: Plate
	DPN: 202 UTS: 480 Elon: 25% Proof: 200

Note. The following abbreviations and units are used in the tables:

DPN	Hardness, diamond pyramid number
UTS	Ultimate tensile strength, N/mm^2
Elon	Elongation, %
Proof	0.1 % proof strength, N/mm^2

1 N/mm^2=0.1 hbar=0.102 kgf/mm^2=0.06475 tonf/in^2=145.04 lbf/in^2

Symbol	Nominal analysis, supplier, condition and remarks.
ASTM A176/405	0.08% C 1% Si 13% Cr steel: Plate
	DPN: 202 UTS: 450 Elon: 22% Proof: 150
ASTM A176/410	0.15% C 12% Cr 0.7% Ni steel: Plate
	DPN: 202 UTS: 450 Elon: 22% Proof: 200
ASTM A176/410 S	0.08% C 12% Cr 0.6% Ni steel: Plate
	DPN: 202 UTS: 420 Elon: 22% Proof: 200
ASTM A176/430	0.12% C 16% Cr 0.7% Ni steel: Plate
	DPN: 202 UTS: 450 Elon: 22% Proof: 200
ASTM A176/442	0.35% C 20% Cr steel: Plate
	DPN: 202 UTS: 530 Elon: 20% Proof: 270
ASTM A176/446	0.2% C 25% Cr 0.25% N steel: Plate
	DPN: 217 UTS: 530 Elon: 20% Proof: 270
ASTM A182 F6	0.12% C (max) 1.0% Mn 12% Cr steel: For pipe fittings
ASTM A193 B6	0.15% C 12% Cr steel: For bolts, etc.; hardened & tempered as AISI 410-416
	UTS: 800 Elon: 15% Proof: 610
ASTM A194/6	0.15% C (max) 13% Cr steel: For bolts
ASTM A240/405	0.08% C 13% Cr 0.2% Al steel: For pressure vessels
ASTM A240/410	0.15% C 12.5% Cr steel: For pressure vessels
ASTM A240/410 S	0.08% C 12.5% Cr steel: For pressure vessels
ASTM A240/430 A	0.12% C 15% Cr steel: For pressure vessels
ASTM A240/430 B	0.12% C 17% Cr steel: For pressure vessels
ASTM A268 TP329	0.2% C 26% Cr 1.5% Mo steel: Tube
ASTM A268 TP405	0.08% C 12% Cr 0.2% Al steel: Tube
ASTM A268 TP410	0.15% C 12% Cr steel: Tube
ASTM A268 TP430	0.12% C 16% Cr steel: Tube
ASTM A268 TP433	0.2% C 21% Cr 1% Cu steel: Tube
ASTM A268 TP466	0.2% C 16% Cr 0.2% N steel: Tube
ASTM A276/403	0.15% C 12% Cr steel: Bar
ASTM A276/405	0.08% C 13% Cr 0.2% Al steel: Bar
ASTM A276/410	0.15% C 12% Cr steel: Bar
ASTM A276/414	0.15% C 12% Cr 2% Ni steel: Bar
ASTM A276/416	0.15% C 13% Cr 0.6% Mo steel: Bar
ASTM A276/416 Se	0.15% C 13% Cr 0.15% Se steel: Bar; free machining
ASTM A276/420	0.15% C 13% Cr steel: Bar
ASTM A276/430	0.12% C 16% Cr steel: Bar
ASTM A276/430 F	0.12% C 16% Cr 0.6% Mo steel: Bar
ASTM A276/430 F Se	0.12% C 16% Cr 0.15% Se steel: Bar
ASTM A276/431	0.2% C 16% Cr 2% Ni steel: Bar
ASTM A276/440 A	0.7% C 7% Cr 0.7% Mo steel: Bar
ASTM A276/440 B	0.8% C 7% Cr 0.7% Mo steel: Bar
ASTM A276/446	0.2% C 25% Cr 0.2% Ni steel: Bar
ASTM A295 F6	0.12% C 1% Mn 12% Cr steel: Seamless drum forgings; normalized & tempered
	UTS: 530 Elon: 18% Proof: 300
ASTM A296 CA15	0.3% C 12% Cr steel: Casting; annealed
	UTS: 610 Elon: 18% Proof: 440
ASTM A296 CA40	0.2% C 12% Cr steel: Casting; annealed
	UTS: 610 Elon: 18% Proof: 440
ASTM A296 CB30	0.3% C 20% Cr steel: Casting; annealed
	UTS: 440 Proof: 200
ASTM A296 CC50	0.5% C 28% Cr steel: Casting; annealed
	UTS: 360
ASTM A297 HC	0.5% C 28% Cr steel: Casting; as cast
	UTS: 360
ASTM A297 HD	0.5% C 28% Cr 5% Ni steel: Casting; as cast
	UTS: 500 Elon: 8% Proof: 240
ASTM A298 E410	0.12% C 13% Cr steel: Welding rod; annealed after welding chromium stainless steels
	UTS: 460 Elon: 20%
ASTM A298 E430	0.1% C 17% Cr steel: Welding rod; annealed after welding chromium stainless steels
	UTS: 460 Elon: 20%
ASTM A336 F6	0.12% C (max) 12.0% Cr steel: Seamless drum forgings
	UTS: 5710 Elon: 8% Proof: 360

Symbol	Nominal analysis, supplier, condition and remarks.
ASTM A351 CA15	0.15% C 13% Cr 1.0% Ni 0.5% Mo steel
ASTM A437	0.22% C 0.6% Ni 11.7% Cr 1.1% Mo 0.25% V 1.0% W steel: For high temperature bolts
ASTM A565	Martensitic steel: Forging; grades 615, 616 and 619 available
ASTM A579/51	0.15% C (max) 12% Cr steel: Forging
ASTM A579/52	0.2% C 12.2% Cr 1.0% Ni 1.0% Mo 0.25% V 1.0% W steel: Forging
ASTM A579/53	0.2% C (max) 16% Cr 2.0% Ni steel: Forging
ASTM A579/61	0.07% C (max) 16.5% Cr 4.0% Ni 4.0% Ti 0.3% V steel: Forging
ASTM A581/416	0.15% C (max) 13% Cr 0.15% S (min) steel: Wire; free machining
ASTM A581/416 Se	0.15% C (max) 13% Cr 0.15% Se (min) steel: Wire; free machining
ASTM A581/430 F	0.12% C (max) 16% Cr 0.15% S (min) steel: Wire; free machining
ASTM A581/430 F Se	0.12% C (max) 16% Cr 0.15% Se (min) steel: Wire; free machining
ASTM A581/XM6	0.15% C (max) 13% Cr 0.15% S (min) 0.6% Mo steel: Wire; free machining
ASTM A651 TP430 Ti	0.1% C 17.7% Cr 0.75% Ti (max) 0.15% Al (max) steel: Tube for water
ASTM A651 TP434	0.12% C 17.0% Cr 1.0% Mo steel: Tube for water
ASTM A651 TP409	0.08% C (max) 11.0% Cr 0.75% Ti (max) steel: Tube for water
ASTM A651 TP430	0.12% C 17.0% Cr steel: Tube
ASTM A651 TPXM8	0.07% C 18.0% Cr 1.1% Ti 0.15% Al (max) steel: Tube for water
ASTM A731 TPXM33	0.06% C (max) 26.0% Cr 1.2% Mo 0.6% Ti steel: Pipe; welded & seamless **UTS: 450 Elon: 20% Proof: 275**
ASTM A731 TPXM27	0.01% C (max) 26.2% Cr 1.2% Mo steel: Pipe; welded & seamless **UTS: 450 Elon: 20% Proof: 275**
AURIGA XXIII FCG	0.15% C 1% Si 12% Cr 1% Ni 0.3% Mo steel: Casting; hardened & tempered; D Brown **UTS: 700 Elon: 12% Proof: 440**
AUSTINOX F	0.1% C 18% Cr 0.5% Ni 0.2% Mo 0.2% S steel: Pompey **UTS: 580 Elon: 50% Proof: 240**
AVESTA 393 S	0.08% C 14.5% Cr 1.0% Mo steel: Avesta
AVESTA 739 C	0.14% C 14.5% Cr 1.0% Mo steel: Avesta
AVESTA 739 SG	0.12% C 13.0% Cr 1.1% Ni 1.1% Mo steel: Avesta casting
AVESTA 739 SH	0.20% C 14.0% Cr 1.0% Mo steel: Avesta
B 3422C	0.10% C 2.2% Si 17.0% Cr steel: French Standard **DPN: 230 UTS: 530 Elon: 15% Proof: 330**
B 3423C	0.20% C 25.0% Cr 0.15% Ni steel: French Standard **DPN: 230 UTS: 480 Elon: 20% Proof: 330**
BREARLEY A	0.15% C (max) 12% Cr steel: Bar, billet, etc.; Brown Bayley for BS alloy En 56A
BREARLEY C	0.12% C (max) 17% Cr steel: Bar, billets, etc.; Brown Bayley for BS alloy En 60
BS 970/403 S17	0.08% C (max) 13% Cr steel
BS 970/410 S21	0.11% C 12% Cr steel: Hardened & tempered **UTS: 360 Elon: 20% Proof: 330**
BS 970/416 S21	0.13% C 12.5% Cr 0.6% Mo (max) 0.2% S steel: Free machining; hardened & tempered **UTS: 360 Elon: 15% Proof: 330**
BS 970/416 S29	0.17% C 12.5% Cr 0.6% Mo (max) 0.2% S steel: Free machining; hardened & tempered **UTS: 510 Elon: 11% Proof: 480**
BS 970/416 S41	0.13% C 12.5% Cr 0.6% Mo (max) steel: Hardened & tempered **UTS: 360 Elon: 15% Proof: 330**
BS 970/416 S37	0.24% C 13% Cr 0.6% Mo (max) 0.2% S steel: Free machining; hardened & tempered **UTS: 580 Elon: 10% Proof: 540**

Symbol	Nominal analysis, supplier, condition and remarks.
BS 970/420 S29	0.17% C 12.5% Cr steel: Hardened & tempered **UTS: 510 Elon: 15% Proof: 480**
BS 970/420 S37	0.24% C 13% Cr steel: Hardened & tempered **UTS: 510 Elon: 15% Proof: 480**
BS 970/420 S45	0.42% C 13% Cr steel: Hardened & tempered **UTS: 510 Elon: 15% Proof: 480**
BS 970/430 S15	0.1% C (max) 17% Cr steel
BS 970/431 S29	0.16% C 2.5% Ni 16.5% Cr steel: Hardened & tempered **UTS: 670 Elon: 11% Proof: 620**
BS 970/441 S29	0.16% C 2.5% Ni 16.5% Cr 0.6% Mo (max) 0.2% S steel: Free machining; hardened & tempered **UTS: 670 Elon: 8% Proof: 600**
BS 970/441S49	0.16% C 2.5% Ni 16.5% Cr 0.6% Mo (max) steel: Hardened & tempered **UTS: 670 Elon: 8% Proof: 600**
BS 970 En 56A	0.12% C 13% Cr steel: Hardened & tempered **DPN: 220 UTS: 700 Elon: 20%**
BS 970 En 56B	0.15% C 13% Cr steel: Hardened & tempered **DPN: 220 UTS: 700 Elon: 20%**
BS 970 En 56C	0.21% C 13% Cr steel: Hardened & tempered **DPN: 220 UTS: 700 Elon: 20%**
BS 970 En 56D	0.27% C 13% Cr steel: Hardened & tempered **DPN: 220 UTS: 700 Elon: 20%**
BS 970 En 56M	0.15% C 13% Cr 0.75% S steel: Hardened & tempered; free machining; C varies with En 56 steels
BS 970 En 57	0.25% C 17% Cr 2% Ni steel: Hardened & tempered **DPN: 248 UTS: 890 Elon: 15%**
BS 970 En 59	0.8% C 2.0% Si 20% Cr 1.5% Ni steel: Bar & forging; hardened & tempered; for valves **DPN: 480**
BS 970 En 60	0.12% C 17% Cr steel
BS 970 En 61	0.12% C 21% Cr steel
BS 980 CDS 18	0.2% C 13% Cr steel: Tube; annealed; cold drawn **UTS: 420 Proof: 270**
BS 1449/403 S17	0.08% C (max) 13% Cr steel: Plate & sheet; annealed **DPN: 170 UTS: 430 Elon: 20% Proof: 250**
BS 1449/405 S17	0.08% C (max) 13% Cr 0.2% Al steel: Plate & sheet; annealed **DPN: 170 UTS: 430 Elon: 20% Proof: 250**
BS 1449/410 S21	0.1% C 12% Cr steel: Plate & sheet; hardened & tempered **DPN: 183 UTS: 620 Elon: 18% Proof: 380**
BS 1501/713	0.08% C 12% Cr 0.2% Al steel: Plate; annealed **UTS: 400 Elon: 21% Proof: 180**
BS 1502/713	0.08% C 12.5% Cr 0.2% Al steel: Forging; hardened & tempered **DPN: 260 UTS: 610 Elon: 20% Proof: 420**
BS 1504/713	0.18% C 12% Cr steel: Casting; hardened & tempered **UTS: 610 Elon: 18% Proof: 440**
BS 1506/713	0.12% C (max) 12% Cr steel: Bar; hardened & tempered **UTS: 580 Elon: 25% Proof: 370**
BS 1554 En 56B	0.15% C 13% Cr steel: Wire; hardened & tempered **UTS: 610**
BS 1554 En 57B	0.25% C 17.5% Cr steel: Wire; hardened & tempered **UTS: 920**
BS 1630 A	0.15% C (max) 12.5% Cr steel: Casting; hardened & tempered; contained in BS 3100 **DPN: 180 UTS: 530 Elon: 20% Proof: 360**
BS 1630 B	0.16% C 12.5% Cr steel: Casting; hardened & tempered; contained in BS 3100 **DPN: 200 UTS: 610 Elon: 18% Proof: 440**

Symbol	Nominal analysis, supplier, condition and remarks.
BS 1630 C	0.25% C 12.5% Cr steel: Casting; hardened & tempered; contained in BS 3100
	DPN: 225 UTS: 700 Elon: 15% Proof: 450
BS 1648 A	0.25% C (max) 14% Cr steel: Casting; as cast; contained in BS 3100
BS 2056 En 56A	0.12% C (max) 13% Cr steel: Wire for springs; as drawn
	UTS: 700
BS 2056 En 56B	0.16% C 13% Cr steel: Wire for springs; as drawn
	UTS: 700
BS 2056 En 56C	0.2% C 13% Cr steel: Wire for springs; as drawn
	UTS: 760
BS 2056 En 56D	0.3% C 13% Cr steel: Wire for springs; as drawn
	UTS: 840
BS 2056 En 57	0.25% C (max) 18% Cr 2% Ni steel: Wire for springs; as drawn
	UTS: 840
BS 2926/13	0.08% C (max) 12% Cr steel: For welding electrodes
BS 2926/17	0.1% C (max) 16% Cr steel: For welding electrodes
BS 3100/410 C21	0.15% C (max) 1.25% Cr steel: Casting
BS 3100/420 C29	0.2% C (max) 12.5% Cr steel: Casting
BS 3100/420 C24	0.25% C (max) 14.0% Cr steel: Casting
BS 3100/425 C11	0.1% C 12.5% Cr 3.8% Ni steel: Casting
BS 3146 ANC1A	0.15% C 12% Cr steel: Investment casting; hardened & tempered
	DPN: 180 UTS: 530 Elon: 20% Proof: 360
BS 3146 ANC1B	0.18% C 12% Cr steel: Investment casting; hardened & tempered
	DPN: 200 UTS: 610 Elon: 18% Proof: 440
BS 3146 ANC1C	0.25% C 12% Cr steel: Investment casting; hardened & tempered
	DPN: 220 UTS: 700 Elon: 15% Proof: 450
BS 3146 ANC2	0.2% C 17% Cr 2% Ni steel: Investment casting; hardened & tempered
	DPN: 280 UTS: 840 Elon: 10% Proof: 700
BS S61	0.12% C (max) 12% Cr steel: Hardened & tempered
	DPN: 180 UTS: 610 Elon: 22% Proof: 330
BS S62	0.22% C 13% Cr steel: Hardened & tempered
	DPN: 220 UTS: 760 Elon: 17% Proof: 480
BS S80	0.17% C 16.5% Cr 2% Ni steel: Hardened & tempered
	DPN: 280 UTS: 920 Elon: 12% Proof: 620
BS S124	0.2% C 13% Cr 0.3% S steel: Hardened & tempered; free machining
	DPN: 230 UTS: 770 Elon: 15% Proof: 420
BVT 130	0.2% C 12% Cr 1.2% Mo steel: Bochumer Verein
CA 6NM	0.06% C 13.0% Cr 4% Ni 0.7% Mo steel: Casting; hardened & tempered; information from Climax Molybdenum
	UTS: 84 Elon: 24% Proof: 70
CA 15	0.15% C (max) 12.2% Cr 1.0% Ni (max) 0.5% Mo (max) steel: Casting; information from Department of Mines, Canada
CA 15	0.3% C 12% Cr steel: Casting; designation used by ASTM

Symbol	Nominal analysis, supplier, condition and remarks.
CA 15	0.15% C 1.5% Si 12.5% Cr 1% Ni 0.5% Mo steel: Casting; aged; Huntington
	DPN: 390 UTS: 1420 Elon: 7% Proof: 1070
CA 40	0.2% C 12% Cr steel: Casting; designation used by ASTM
CA 40	0.3% C 13% Cr 1% Ni 0.5% Mo steel: Casting; aged; Huntington
	DPN: 470 UTS: 1630 Elon: 1% Proof: 1210
CB 30	0.3% C 20% Cr 2% Ni steel: Casting; aged; Huntington
	DPN: 195 UTS: 700 Elon: 15% Proof: 370
CB 30	0.3% C 20% Cr steel: Casting; designation used by ASTM
CC 50	0.5% C 28% Cr steel: Casting; designation used by ASTM
CC 50	0.5% C 28% Cr 4% Ni steel: Casting; as cast; Huntington
	DPN: 210 UTS: 720 Elon: 18% Proof: 420
CEKAS M151	0.15% C 13% Cr 1.0% Mo steel: Kuhbler & Son
CEKAS M152	0.2% C 13% Cr 1.0% Mo steel: Kuhbler & Son
CHROMEX 1	0.07% C 13.5% Cr steel: Welding electrode; Murex
	UTS: 640 Elon: 32%
CHROMEX 2	0.07% C 18% Cr steel: Welding electrode; Murex
	UTS: 570 Elon: 25%
CHROMEX 3	0.1% C 30% Cr steel: Welding electrode; Murex
	UTS: 570 Elon: 25%
CHROMODUR II	0.18% C 12% Cr 0.3% Ni 1.0% Mo steel: Krupp
COBALT ASCOLOY	0.2% C 12.2% Cr 3% W 5% Co 0.2% V steel: American proprietary alloy listed in SAE year book
CONSUMET 355	0.12% C 15.5% Cr 4.5% Ni 2.9% Mo 0.1% N steel: Carpenter
CORNIX 1	0.2% C 13% Cr 15% Mo steel: Rochling
CORRESIST 13 HMo	0.2% C 13% Cr 1.0% Mo steel: Pose-Marre
CR Al 20/5	0.1% C 20% Cr 4.5% Al steel: Designation used by German Standard
Cr Al 25/5	25% Cr 5% Al steel: Designation used by German Standard
Cr Al 30/5	0.1% C (max) 30% Cr 4.5% Al steel: Designation used by German Standard
CRH 4/17	0.1% C 13% C 0.6% Ni 0.4% Mo steel: Tokushu Seiko Co
CRO 13 Mo	0.15% C 13% Cr 0.6% Ni steel: Officine Metallurgiche
CROLOY 12	0.15% C 12% Cr 0.5% Ni steel: Babcock & Wilcox (USA)
CROLOY 12 Al	0.08% C 12% Cr 0.5% Ni 0.2% Al steel: Babcock & Wilcox (USA)
CROLOY 18	0.12% C 16% Cr 0.5% Ni steel: Babcock & Wilcox (USA)
CROLOY 22	0.2% C 21% Cr 0.5% Ni 1.0% Cu steel: Babcock & Wilcox (USA)
CROLOY 27	0.2% C 27% Cr 0.5% Ni steel: Babcock & Wilcox (USA)
CROLOY 27/4/1	0.2% C 25% Cr 4% Ni 1.5% Mo steel: Babcock & Wilcox (USA)
CROMIMPHY 1 bis Mo	0.8% C 13% Cr 0.5% Ni 0.5% Mo steel: Imphy
CROMIMPHY 33	0.3% C 12% Cr 1.0% Ni 0.3% Mo steel: Imphy
CROMIMPHY A8Mo	0.08% C 13.0% Cr 0.5% Mo steel: Imphy
CROMIMPHY A10Mo	0.15% C (max) 13.0% Cr 0.5% Mo steel: Imphy
CROMIMPHY A15Mo	0.18% C 13.0% Cr 0.5% Mo steel: Imphy
CROMIMPHY A18Mo	0.25% C (max) 13.0% Cr 0.5% Mo steel: Imphy
CROMIMPHY A20Mo	0.22% C 12.5% Cr 1.0% Mo steel: Imphy

Note. The following abbreviations and units are used in the tables:

DPN	Hardness, diamond pyramid number
UTS	Ultimate tensile strength, N/mm^2
Elon	Elongation, %
Proof	0.1 % proof strength, N/mm^2

$1\ N/mm^2 = 0.1\ hbar = 0.102\ kgf/mm^2 = 0.06475\ tonf/in^2 = 145.04\ lbf/in^2$

Symbol	Nominal analysis, supplier, condition and remarks.
CROMIMPHY A1007Mo	0.12% C (max) 13% Cr 0.7% Mo steel: Imphy
CRUCIBLE 26-1	0.06% C (max) 26.0% Cr 0.5% Ni 1.2% Mo 0.6% Ti 0.2% Cu (max) steel: Colt Ind **UTS: 530 Elon: 30% Proof: 375**
CRUCIBLE 401	0.15% C 12% Cr steel: Hardened & tempered; Crucible Steel Co **DPN: 250 UTS: 1070 Elon: 20% Proof: 920**
CRUCIBLE 403	0.15% C 12.5% Cr steel: Hardened & tempered; Crucible Steel Co **DPN: 200 UTS: 1000 Elon: 20% Proof: 760**
CRUCIBLE 416	0.15% C 13% Cr 0.6% Mo or Zr steel: Hardened & tempered; Crucible Steel Co **DPN: 300 UTS: 1070 Elon: 20% Proof: 920**
CRUCIBLE 420	0.15% C (min) 13% Cr steel: Hardened & tempered; Crucible Steel Co **DPN: 350 UTS: 1170 Elon: 15% Proof: 1000**
CRUCIBLE 430	0.12% C 16% Cr steel: Annealed; Crucible Steel Co **DPN: 160 UTS: 530 Elon: 30% Proof: 290**
CRUCIBLE 431	0.2% C 16% Cr 2% Ni steel: Hardened & tempered; Crucible Steel Co **DPN: 400 UTS: 1380 Elon: 20% Proof: 1070**
CRUCIBLE 440A	0.7% C 17% Cr 0.7% Mo steel: Crucible Steel Co **UTS: 610**
CRUCIBLE 440B	0.8% C 17% Cr 0.7% Mo steel: Crucible Steel Co **DPN: 640**
CRUCIBLE 442	0.2% C 21% Cr steel: Annealed; Crucible Steel Co **DPN: 175 UTS: 570 Elon: 22% Proof: 370**
CRUCIBLE 446	0.2% C 25% Cr 0.25% N steel: Annealed; Crucible Steel Co **DPN: 180 UTS: 580 Elon: 20% Proof: 370**
CUTLERY	0.15% C (min) 13% Cr steel: Bar; billet; Brown Bayley for BS alloy En 56D
D 12	Symbol of Durco CA15; Durion Co
D 8512	0.2% C 12% Cr 1.0% Mo steel: Grusstahlwerk
DARWIN 168	0.75% C 17% Cr steel: Surgical instruments, dies, rollers, etc.; Darwin **DPN: 640**
DAUPHINOX TP Mo	0.25% C 13% Cr 0.8% Mo steel: Bonpertius **UTS: 275 Proof: 535**
DIEHARD LC	0.8% C 12% Cr 0.5% Mo 0.5% V steel: Firth Brown
DILVER O	25% Cr Fe alloy: Low expansion; Imphy
DILVER T	20% Cr Fe alloy: Low expansion; Imphy
DIN 17224 X 20 Cr 13	0.19% C 13% Cr steel: For springs; German Standard
DIN 17440 X 6 Cr Mo 17	0.07% C (max) 1.0% Mn (max) 17% Cr 1.0% Mo 0.045% P (max) 0.05% S (max) steel: Plate; German Standard; annealed
DIN 17440 X 7 Cr 13	0.08% C (max) 1.30% Cr steel: Plate; German Standard; annealed **UTS: 610 Proof: 255**
DIN 17440 X 7 Cr Al 13	0.08% C 1.0% Mn (max) 13% Cr 0.2% Al 0.045% P (max) 0.03% S (max) steel: Plate; German Standard; annealed **UTS: 560 Proof: 255**
DIN 17440 X 8 Cr 17	0.10% C (max) 16.5% Cr steel: Plate; German Standard; annealed **UTS: 600 Proof: 275**
DIN 17440 X 8 Cr Nb 17	0.1% C (max) 17.0% Cr 1.2% Nb steel: Plate; German Standard; annealed **UTS: 510 Proof: 275**
DIN 17440 X 8 Cr Ti 17	0.1% C (max) 17.0% Cr 0.7% Ti steel: Plate; German Standard; annealed **UTS: 485 Proof: 275**
DIN 17440 X 10 Cr 13	0.1% C 13.0% Cr steel: Plate; German Standard; annealed **UTS: 610 Proof: 300**
DIN 17440 X 12 Cr Mo S 17	0.13% C 16.5% Cr 0.25% Mo steel: Plate; German Standard; annealed **UTS: 700 Proof: 310**
DIN 17440 X 15 Cr 13	0.14% C 13.0% Cr steel: Plate; German Standard; annealed **UTS: 760**
DIN 17440 X 20 Cr 13	0.2% C 13.0% Cr steel: Plate; German Standard; annealed **UTS: 760**
DIN 17440 X 22 Cr Ni 17	0.19% C 1.0% Mn (max) 19% Cr 2% Ni 0.045% P (max) 0.03% S (max) steel: Plate; German Standard; annealed **DPN: 969**
DIN 17440 X 40 Cr 13	0.45% C 13.0% Cr steel: Plate; German Standard; annealed **UTS: 810**
DIN 17440 X 48 Cr Mo V 15	0.46% C 14.4% Cr 0.5% Mo 0.12% V steel: Plate; German Standard; annealed **UTS: 920**
DIN 17470 Cr Al 20/5	0.1% C (max) 20% Cr 4.5% Al steel: Heat conductor; German Standard
DIN 17470 Cr Al 25/5	25% Cr 5% Al steel: Heat resisting; German Standard
DIN 17470 Cr Al 30/5	0.1% C (max) 29% Cr 4.5% Al steel: Heat conductor; German Standard
DN	0.18% C 17% Cr 1.5% Ni steel: Osborn; obsolete **UTS: 840 Elon: 20% Proof: 610**
DN 5	0.17% C 17% Cr 1.5% Ni steel: Osborn **UTS: 840 Elon: 20% Proof: 610**
DNH	0.22% C 17% Cr 1.5% Ni steel: Valve seats, etc.; Osborn; obsolete **UTS: 840 Elon: 20% Proof: 610**
DOMINIAL RM 13 Mo	0.2% C 13% Cr 1.0% Mo steel: Kind & Co
DOUBLE TWELVE	0.32% C 12% Cr 1% V 12% W steel: Pressure die casting moulds; Edgar Allen **DPN: 450**
DSC	0.35% C 13.5% Cr steel: Darwin **DPN: 610**
DTD 97B	0.15% C (max) 13% Cr steel: Tube; annealed **UTS: 420 Proof: 270**
DTD 161A	0.12% C (max) 13% Cr steel: Annealed **UTS: 530**
DTD 203B	0.15% C 13% Cr steel: Tube; hardened & tempered **UTS: 920 Proof: 760**
DTD 271	0.3% C 13% Cr steel: Strip; hardened & tempered **UTS: 1550 Proof: 1210**
DTD 326A	0.3% C 13% Cr steel: Wire for springs; hardened & tempered; not engine valve springs **UTS: 1420**
DTD 525	0.3% C 12% Cr 0.7% S 0.5% P steel: Hardened & tempered; free machining **DPN: 240 UTS: 760 Elon: 15% Proof: 420**
DTD 715	0.25% C 18% Cr 2% Ni 0.3% S steel: Hardened & tempered; free cutting **DPN: 270 UTS: 920 Elon: 15%**
DTD 5046	12% Cr Mo V steel: Sheet & strip **UTS: 1020**
DTD 5065	12% Cr steel: For bolts, studs, etc.; heat resisting **UTS: 1100**
DUNELT 61	0.15% C 12% Cr steel: Dunford for BS alloy En 56A
DUNELT 62	0.22% C 12% Cr steel: Dunford for BS alloy En 56B
DUNELT 63	0.25% C 18% Cr 1% Ni steel: Dunford for BS alloy En 57
DURCO CA15	0.15% C (max) 1.0% Ni (max) 12.5% Cr 0.5% Mo (max) steel: Casting; Duriron Co **UTS: 630 Elon: 18% Proof: 460**

Symbol	Nominal analysis, supplier, condition and remarks.
EK	0.3% C 31% Cr steel: Annealed; static furnace equipment; Osborn
	UTS: 500 Elon: 20% Proof: 300
ENDURO FC	0.13% C 12.5% Cr 0.4% Ni 0.5% Mo steel: Fiat
ENGINEERING	0.15% C (min) 13% Cr steel: Bar, billet, etc.; Brown Bayley for BS alloy En56C
ERA HR4	29% Cr steel: Casting; Hadfields for BS 1648B
ERA HR4 HARD	29% C high C steel: Casting; Hadfields for BS 1648C
EV 1	0.45% C 23% Cr 4.8% Ni 2.8% Mo steel: For exhaust valves; designation used by SAE
	UTS: 920 Elon: 3%
EV 2	0.4% C 24% Cr 3.8% Ni 1.4% Mo steel: For exhaust valves; designation used by SAE
	UTS: 1070 Elon: 3%
EXD 5	0.3% C 12% Cr 12% W 1.1% V steel: Extrusion dies for brass, etc.; Balfour
	DPN: 395
F 1	0.12% C 13% Cr steel: Ugine
F 15	0.12% C 13% Cr 0.75% S steel: Free machining; Ugine
F 17	0.12% C 17% Cr steel: Ugine
FAGERSTA R100	0.05% C 14.5% Cr steel: Fagersta
FAGERSTA R290	0.05% C 17.5% Cr steel: For oil burners; Fagersta
FAGERSTA R700	0.2% C 13.5% Cr steel: For cutting tools; Fagersta
FAGERSTA R710	0.3% C 13.5% Cr steel: For cutting tools; Fagersta
FAGERSTA R720	0.4% C 13.5% Cr steel: For cutting tools; Fagersta
FAGERSTA R730	0.5% C 13.5% Cr steel: For cutting tools; Fagersta
FAGERSTA R740	0.3% C 14.0% Cr 0.4% Ni 0.6% Mo steel: For cutting tools; Fagersta
FAL	0.1% C 13% Cr 4.2% Al steel: High electrical resistance; Firth Vickers
	UTS: 610 Elon: 25% Proof: 450
FAS	0.43% C 11.5% Cr steel: Firth Vickers
	DPN: 240 UTS: 760 Elon: 20% Proof: 530
FC1	0.1% C 13% Cr 0.4% Mo steel: Free machining; Firth Vickers
	DPN: 180 UTS: 630 Elon: 28% Proof: 420
FCS	0.2% C 13% Cr 0.4% Mn 0.23% S steel: Free machining; Firth Vickers
	DPN: 220 UTS: 680 Elon: 25% Proof: 440
FECRALLOY	0.03% C (max) 18.5% Cr 4.6% Al 0.2% Y steel: Resistalloy Ltd; oxidation resistant; annealed
	UTS: 510 Elon: 23% Proof: 375
FG	0.25% C 13% Cr steel: Hardened & tempered; Firth Vickers
	DPN: 450 UTS: 1500 Elon: 18% Proof: 1350
FG (L)	0.15% C 13% Cr steel: Hardened & tempered; Firth Vickers
	DPN: 450 UTS: 1500 Elon: 18% Proof: 1350
FH	0.25% C 13% Cr steel: Firth Vickers
FHM	0.8% C 16.5% Cr 0.5% Mo steel: Firth Vickers
	DPN: 600
FI	0.12% C 12% Cr 0.7% Ni steel: Hardened & tempered; Firth Vickers
	DPN: 172 UTS: 570 Elon: 33% Proof: 360
FI 17	0.08% C 17% Cr 2.5% Ni steel: Hardened & tempered; Firth Vickers; sheet
	DPN: 175 UTS: 840 Elon: 28% Proof: 360

Symbol	Nominal analysis, supplier, condition and remarks.
FI 20	0.07% C 21% Cr steel: Firth Vickers
	UTS: 480 Elon: 30%
FI Ti	0.06% C (max) 11.5% Cr 1.0% Ni (max) 0.5% Ti steel: Firth Vickers
FLUGINOX 51	0.05% C 12.5% Cr 0.5% Mo 0.2% Al steel: Ugine
FLUGINOX 60	0.25% C 13% Cr 0.5% Mo steel: Ugine
FNZ	0.25% C 17% Cr 2% Ni steel: Firth Vickers
	DPN: 380 UTS: 1300 Elon: 15% Proof: 1140
FOREMOST 14 Cr	0.12-0.35% C 13% Cr range of steels: Swift Levick for BS alloys En 56 A, B & C
FOREMOST 467	0.25% C 18% Cr 1% Ni steel: Swift Levick for BS alloy En 57
FOREMOST HR 2	0.2% C 17% Cr 2% Ni steel: Casting; annealed; Swift Levick
	DPN: 150 UTS: 530 Elon: 25% Proof: 340
FOREMOST IRON	0.12% C 13% Cr 1% Ni steel: Hardened & tempered; Swift Levick
	DPN: 170 UTS: 600 Elon: 30%
FV 403	0.12% C 12% Cr steel: Hardened & tempered; Firth Vickers
	DPN: 172 UTS: 570 Elon: 33% Proof: 360
FV 406	0.1% C 13% Cr 4.25% Al steel: High electrical resistance; 1350 microhm mm at 600 °C; Firth Vickers
	UTS: 610 Elon: 25% Proof: 450
FV 410	0.1% C 13% Cr steel: Hardened & tempered; Firth Vickers
	DPN: 172 UTS: 570 Elon: 33% Proof: 360
FV 420	0.25% C 13% Cr steel: Hardened & tempered; Firth Vickers
	DPN: 450 UTS: 1500 Elon: 18% Proof: 1350
FV 430	0.06% C 16.5% Cr steel: Firth Vickers
	DPN: 170 UTS: 530 Elon: 32% Proof: 330
FV 431	0.16% C 16% Cr 2.5% Ni steel: Firth Vickers
	DPN: 380 UTS: 1300 Elon: 15% Proof: 1140
FV 440B	0.8% C 16.5% Cr 0.5% Mo steel: Firth Vickers
	DPN: 600
FV 446	0.09% C 29% Cr 1.8% Ni steel: Firth Vickers
	DPN: 180 UTS: 530 Elon: 25%
FV 448	0.1% C 11% Cr 0.75% Ni 0.7% Mo steel: Firth Vickers
	UTS: 940 Elon: 17% Proof: 800
FV 507	0.12% C 10.5% Cr 0.8% Mo 0.15% V 0.4% Nb steel: Casting; Firth Vickers; ferritic stainless steel; hardened & tempered
	DPN: 300 UTS: 970 Elon: 12% Proof: 820
FV 535	0.07% C 10.5% Cr 0.3% Ni 6% Co 0.75% Mo steel: Creep resistant; Firth Vickers
	UTS: 1020 Elon: 18% Proof: 880
FV 566	0.12% C 11.5% Cr 2.3% Ni 1.4% Mo 0.15% V 0.3% Nb steel: Firth Vickers
FV 607	0.15% C 11% Cr 0.6% Ni 0.8% Mo 0.25% V steel: Hardened & tempered; Firth Vickers
	UTS: 980 Elon: 23% Proof: 820
FV 702	0.03% C 15.7% Cr 2.5% Ni 1.0% Mo 0.5% Nb steel: Firth Vickers
FV Fl	0.08% C 13% Cr steel: For turbine blades; Firth Vickers
	UTS: 610 Elon: 33% Proof: 390
FX	0.1% C 21% Cr steel: Annealed; furnace & domestic equipment; Osborn
	UTS: 450 Elon: 30% Proof: 300
G 85	0.06% C 20.0% Cr 13.5% W 4.5% Mo 3.0% Ti 1.5% Al steel: Jessop; scale or creep resistant
G 4301 Sec/1	0.15% C (max) 12% Cr steel: Japanese Standard
G 4301 Sec/4	0.15% C (min) 13% Cr steel: Japanese Standard
G 4301 Sec/5	0.12% C 16% Cr steel: Japanese Standard
GALA	0.1% C 13% Cr steel: Hadfields for BS alloy BS/S61

Note. The following abbreviations and units are used in the tables:

DPN	Hardness, diamond pyramid number
UTS	Ultimate tensile strength, N/mm^2
Elon	Elongation, %
Proof	0.1 % proof strength, N/mm^2

1 N/mm^2=0.1 hbar=0.102 kgf/mm^2=0.06475 tonf/in^2=145.04 lbf/in^2

Symbol	Nominal analysis, supplier, condition and remarks.
GALAHAD A	0.1% C 13% Cr steel: Hadfields for BS 1630A **UTS: 530 Elon: 20% Proof: 360**
GALAHAD A	0.12% C 13% Cr 1% Ni steel: Wrought; Hadfields for BS alloy En 56A
GALAHAD AFC	0.12% C 13% Cr 1% Ni 0.6% Mo 0.75% S steel: Free machining; Hadfields for BS alloy En 56AM
GALAHAD B	0.18% C 13% Cr steel: Casting; Hadfields for BS 1630B **UTS: 610 Elon: 18% Proof: 440**
GALAHAD B	0.16% C 13% Cr 1% Ni steel: Wrought; Hadfields for BS alloy En 56B
GALAHAD BFC	0.15% C 13% Cr 0.6% Mo 0.75% S steel: Free machining; Hadfields for BS alloy En 56BM
GALAHAD C	0.25% C 13% Cr steel: Casting; Hadfields for BS 1630C **UTS: 700 Elon: 15% Proof: 450**
GALAHAD C	0.21% C 13% Cr 1% Ni steel: Wrought; Hadfields for BS alloy En 56C
GALAHAD CFC	0.21% C 13% Cr 1% Ni 0.6% Mo 0.75% S steel: Free machining; Hadfields for BS alloy En 56CM
GALAHAD D	0.3% C 13% Cr 1% Ni steel: Hadfields for BS alloy En 56D
GALAHAD DFC	0.3% C 13% Cr 1% Ni 0.8% Mo 0.75% S steel: Free machining; Hadfields for BS alloy En 56DM
GALAHAD E	0.25% C 17% Cr 2% Ni steel: Hadfields for BS alloy En 57
GALAHAD F	0.12% C 17% Cr 0.5% Ni steel: Hadfields for BS alloy En 60
GASC	0.5% C 14% Cr steel: Jonas; plastic moulds **DPN: 480**
GASD	0.35% C 14% Cr steel: Jonas; plastic moulds **DPN: 460**
GILBRUN 10	0.12% C (max) 13.0% Cr steel: Wire; Gilly Brunton; annealed **UTS: 500 Elon: 33% Proof: 310**
GILBRUN Cr 17	0.07% C 17.0% Cr steel: Wire; Gilly Brunton **UTS: 540 Elon: 33% Proof: 340**
GILBRUN Cr 20	0.07% C (max) 20.0% Cr steel: Wire; Gilly Brunton **UTS: 540 Elon: 33% Proof: 340**
GOST 5632/0 KH13	0.08% C (max) 0.6% Mn (max) 12.0% Cr steel: Russian Standard
GOST 5632/0 KH13	0.08% C (max) 12.0% Cr steel: Russian Standard
GOST 5632/0 KH13	0.08% C (max) 0.6% Mn (max) 12% Cr 0.03% P (max) 0.025% S (max): Russian Standard; annealed steel plate **UTS: 400**
GOST 5632/0 KH17T	0.08% C (max) 17.0% Cr 0.8% Ti steel: Russian Standard
GOST 5632/0 KH2 N5T	0.08% C (max) 21.0% Cr 5.2% Ni 0.45% Ti steel: Plate; Russian Standard
GOST 5632/0 KH17T	0.08% C (max) 17.0% Cr 0.8% Ti steel: Plate; Russian Standard; annealed
GOST 5632/1 KH11 MF	0.16% C 10.7% Cr 0.7% Mo 0.32% W steel: Russian Standard
GOST 5632/1 KH12N2VFM	0.13% C 11.2% Cr 1.8% Ni 0.4% Mo 0.2% V 1.8% W steel: Plate; Russian Standard
GOST 5632/1 KH12N2VMF	0.13% C 11.2% Cr 1.7% Ni 0.42% Mo 0.2% V 1.8% W steel: Russian Standard
GOST 5632/1 KH12S1U	0.1% C 13.0% Cr 1.4% Al steel: Russian Standard
GOST 5632/1 KH12S1U	0.9% C 13.0% Cr 1.4% Al steel: Plate; Russian Standard
GOST 5632/1 KH12V2MF	0.13% C 12.0% Cr 0.75% Mo 0.2% V 2.0% W steel: Plate; Russian Standard
GOST 5632/1 KH12V2MF	0.13% C 12.0% Cr 0.75% Mo 0.2% V 2.0% W steel: Russian Standard
GOST 5632/1 KH12VNMF	0.15% C 12.0% Cr 0.6% Ni 0.6% Mo 0.2% V 0.9% W steel: Russian Standard
GOST 5632/1 KH12VNMF	0.15% C 12.0% Cr 0.6% Ni 0.6% Mo 0.25% V 0.9% W steel: Plate; Russian Standard
GOST 5632/1 KH13	0.12% C 13.0% Cr steel: Russian Standard
GOST 5632/1 KH13	0.12% C 0.6% Cr 0.6% Mn (max) 13% Cr 0.03% P (max) 0.025% S (max) steel: Plate; Russian Standard; annealed **UTS: 550 Proof: 280**
GOST 5632/1 KH13N3	0.12% C 13.5% Cr 2.7% Ni steel: Russian Standard
GOST 5632/1 KH13N3	0.1% C 13.5% Cr 2.6% Ni steel: Russian Standard
GOST 5632/1 KH17N2	0.14% C 17.0% Cr 2.0% Ni steel: Russian Standard
GOST 5632/1 KH21N5T	0.12% C 21.0% Cr 5.2% Ni 0.7% Ti steel: Plate; Russian Standard
GOST 5632/2 KH12VMBFR	0.18% C 12.0% Cr 0.5% Mo 0.22% V 0.5% W 0.3% Nb 0.003% B steel: Russian Standard
GOST 5632/2 KH12VMBMR	0.18% C 12.0% Cr 0.5% Mo 0.2% V 0.6% W 0.3% Nb 0.003% B steel: Plate; Russian Standard
GOST 5632/2 KH13	0.2% C 13.0% Cr steel: Plate; Russian Standard; annealed **UTS: 500 Proof: 250**
GOST 5632/2 KH13	0.2% C 13.0% Cr steel: Russian Standard
GOST 5632/2 KH13N4G9	0.22% C 13.0% Cr 4.2% Ni steel: Plate; Russian Standard
GOST 5632/2 KH17N2	0.25% C 17.0% Cr 2.0% Ni steel: Russian Standard
GOST 5632/2 KH17N2	0.24% C 17.0% Cr 2.0% Ni steel: Russian Standard
GOST 5632/2 KH17N2	0.25% C 0.8% Mn 17.0% Cr 2.0% Ni steel: Russian Standard
GOST 5632/2 KH17N2	0.26% C 17.0% Cr 2.0% Ni steel: Plate; Russian Standard
GOST 5632/3 KH13	0.3% C 13.0% Cr steel: Russian Standard
GOST 5632/3 KH13	0.3% C 13.0% Cr steel: Plate; Russian Standard; annealed **UTS: 550**
GOST 5632/3 KH13N7S2	0.3% C 13.0% Cr 7.2% Ni steel: Russian Standard
GOST 5632/4 KH13	0.4% C 13.0% Cr steel: Plate; Russian Standard; annealed **UTS: 520**
GOST 5632/4 KH13	0.4% C 13.0% Cr steel: Russian Standard
GOST 5632/15 KH12VMF	0.15% C 12.0% Cr 0.6% Ni 0.6% Mo 0.22% V 0.9% W steel: Plate; Russian Standard
GOST 5632 KH13N3	0.11% C 13.5% Cr 2.6% Ni steel: Plate; Russian Standard
GOST 5632 KH14	0.15% C (max) 0.7% Mn (max) 14.0% Cr steel: Russian Standard
GOST 5632 KH14	0.15% C (max) 14.0% Cr steel: Plate; Russian Standard
GOST 5632 KH14	0.15% C (max) 14.0% Cr steel: Russian Standard
GOST 5632 KH14	0.7% C 14.0% Cr steel: Russian Standard
GOST 5632 KH14G14N	0.12% C (max) 14.0% Cr 1.25% Ni steel: Russian Standard
GOST 5632 KH14G14N	0.12% C (max) 14.0% Cr 1.2% Ni Steel: Plate; Russian Standard
GOST 5632 KH14G14N3T	0.1% C (max) 14.0% Cr 3.0% Ni 0.5% Ti steel: Plate; Russian Standard
GOST 5632 KH14G14N13T	0.1% C (max) 14.0% Cr 3.0% Ni 0.6% Ti steel: Russian Standard
GOST 5632 KH17	0.12% C (max) 17.0% Cr steel: Russian Standard
GOST 5632 KH17	0.12% C 17.0% Cr steel: Plate; Russian Standard; annealed **UTS: 450 Proof: 280**
GOST 5632 KH17AG14	0.15% C (max) 17.0% Cr 0.6% Ni (max) 0.45% N steel: Russian Standard
GOST 5632 KH17AG14	0.15% C (max) 17.0% Cr 0.35% N steel: Plate; Russian Standard

Symbol	Nominal analysis, supplier, condition and remarks.
GOST 5632 KH17G9AN4	0.12% C (max) 17.0% Cr 4.0% Ni 0.2% N steel: Russian Standard
GOST 5632 KH17H2	0.14% C 0.8% Mn (max) 19% Cr 2% Ni 0.03% P (max) 0.025% S (max) steel: Plate; Russian Standard solution treated
	UTS: 1100 **Proof: 800**
GOST 5632 KH18S1U	0.15% C 0.5% Mn (max) 18.5% Cr 0.85% Al steel: Russian Standard
GOST 5632 KH18S1U	0.15% C (max) 18.5% Cr 1.0% Al steel: Plate; Russian Standard; Annealed
	UTS: 500 **Proof: 300**
GOST 5632 KH25T	0.15% C (max) 0.8% Mn (max) 2.5% Cr 0.8% Ti 0.035% P 0.025% S (max) steel: Plate; Russian Standard
	UTS: 450 **Proof: 300**
GOST 5632 KH25T	0.15% C (max) 25.5% Cr 0.8% Ti steel: Russian Standard
GOST 5632 KH28	0.15% C (max) 28.5% Cr steel: Russian Standard
GOST 5632 KH28	0.15% C (max) 29.0% Cr steel: Plate; Russian Standard
GOST 5632 KH28AN	0.15% C 26.5% Cr 1.3% Ni 0.2% N steel: Plate; Russian Standard
GOST 5632 KH28AN	0.15% C (max) 1.5% Mn (max) 26.5% Cr 1.3% Ni 0.2% N steel: Russian Standard
GOST 5632/0 KH17T	0.08% C (max) 17.0% Cr 0.8% Ti steel: Russian Standard
GOST 9940/0 KH13	0.08% C (max) 12.0% Cr steel: Pipe; Russian Standard; seamless
	UTS: 380
GOST 9940/0 KH17T	0.08% C (max) 17.0% Cr 0.6% Ti steel: Pipe; Russian Standard; seamless
GOST 9940/1 KH13	0.12% C 0.6% Mn (max) 13.0% Cr steel: Pipe; Russian Standard; seamless
	UTS: 400
GOST 9940/1 KH17	0.12% C (max) 17.0% Cr steel: Pipe; Russian Standard; seamless
	UTS: 450
GOST 9940 KH28 T	0.15% C (max) 25.5% Cr 0.8% Ti steel: Pipe; Russian Standard; seamless
	UTS: 450
GOST 9940 KH28	0.15% C (max) 28.5% Cr steel: Pipe; Russian Standard; seamless
	UTS: 450
GOST 9941 KH17	0.12% C (max) 17.0% Cr steel: Pipe; Russian Standard; seamless
	UTS: 450
GOST 9941/0 KH13	0.08% C (max) 12.0% Cr steel: Pipe; Russian Standard; seamless
	UTS: 380
GOST 9941/0 KH17T	0.08% C (max) 17.0% Cr 0.8% Ti steel: Pipe; Russian Standard; seamless
GOST 9941/1 KH13	0.12% C 13.0% Cr steel: Pipe; Russian Standard; seamless
	UTS: 408
GOST 9941 KH25 T	0.15% C (max) 25.5% Cr 0.8% Ti steel: Pipe; Russian Standard; seamless
	UTS: 450
GOST 10498/1 KH13S2M2	0.12% C 13% Cr 0.6% Mo steel: Pipe; Russian Standard; seamless
	UTS: 550

Note. The following abbreviations and units are used in the tables:

DPN	Hardness, diamond pyramid number
UTS	Ultimate tensile strength, N/mm²
Elon	Elongation, %
Proof	0.1 % proof strength, N/mm²

$1 \text{ N/mm}^2 = 0.1 \text{ hbar} = 0.102 \text{ kgf/mm}^2 = 0.06475 \text{ tonf/in}^2 = 145.04 \text{ lbf/in}^2$

Symbol	Nominal analysis, supplier, condition and remarks.
GOST 10802/1 KH11B2MF	0.13% C 0.65% Mn 11.0% Cr 0.75% Mo 0.2% V 2.0% W 0.32% Cu (max) steel: Pipe; Russian Standard; seamless
	UTS: 600 **Proof: 400**
GOST 14162/1 KH13	0.12% C 0.6% Mn (max) 13.0% Cr steel: Capillary tube; Russian Standard
GOST 14162/2 KH13	0.2% C 13.0% Cr steel: Capillary tube; Russian Standard
GQ	0.85% C 17% Cr steel: Osborn; ball races, gears, etc.; obsolete
	UTS: 760 **Elon: 20%** **Proof: 370**
GREEK ASCOLOY	0.17% C 13% Cr 2.0% Ni 3.0% W steel: Firth Sterling; hardened & tempered
	UTS: 1000 **Elon: 20%** **Proof: 760**
GX 5 Cr Ni 13.4	0.07% C 13.0% Cr 3.7% Ni 0.5% Mo steel: Casting; designation used by Swiss Standards
GX 5 Cr Ni 13.4	0.07% C 12.7% Cr 4.2% Ni 0.7% Mo (max) steel: Casting; designation used by German Standards
	UTS: 850 **Elon: 10%** **Proof: 580**
H 23	0.3% C 12.0% Cr 12.0% W steel: Designation used by AISI
H 29	0.8% C 2% Si 0.4% Mn 20% Cr 1.4% Ni steel: Jessop for BS alloy En 59
H 46	0.16% C 11.5% Cr 0.6% Ni 0.6% Mo 0.3% V steel: Jessop
H 53	0.08% C 10% Cr 0.7% Ni 0.8% Mo 0.5% W 0.4% V 6.5% Co 0.4% Nb steel: Jessop; high creep strength
H 59	0.07% C 11% Cr 3% Ni 1.5% Mo 0.4% V steel: Jessop
	UTS: 1070 **Elon: 20%** **Proof: 1000**
HARDINOX 3	0.3% C 13% Cr steel: Pompey
HARDINOX 4	0.4% C 13% Cr steel: Pompey
HARDTRODE 84.42	0.12% C 0.3% Mn 13.0% Cr steel: For electrodes; Esab
	DPN: 460
HARDTRODE 84.52	0.25% C 0.3% Mn 13.0% Cr steel: For electrodes; Esab
	DPN: 500
HC	0.5% C 28% Cr 4% Ni 0.5% Mo steel: Casting; Huntington; aged
	UTS: 820 **Elon: 18%** **Proof: 530**
HCA	0.3% C 12% Cr 12% W 0.7% V steel: Braeburn; AISI type H23
HD	0.5% C 28% Cr 5.5% Ni 0.5% Mo steel: Casting; Huntington; as cast
	UTS: 610 **Elon: 16%** **Proof: 330**
HECLA SNS	0.8% C 2% Si 19% Cr 1.5% Ni steel: Hadfields for BS alloy En 59
HEDEX 5	0.12% C (max) 17% Cr steel: Wire for forging; Kiveton Park; annealed
	UTS: 630
HNV 6	0.8% C 2.3% Si 20% Cr 1.3% Ni steel: For valves; designation used by SAE
HRM 2M1	0.15% C 13.0% Cr 1.15% Mo steel: Henricot
HRM 2M2	0.20% C 13.0% Cr 1.15% Mo steel: Henricot
HT 9	0.2% C 12.0% Cr 0.55% Ni 0.1% Mo 0.5% W 0.3% V steel: Sandvik
HWX	0.36% C 13.5% Cr 5.8% Ni 0.65% V 2.8% W steel: For hot working; origin unknown
ICN	0.2% C 1.0% Mn 14.0% Cr 1.0% Ni steel: Origin unknown
IMMUNIT R10/13 Mo Ni	0.1% C 12.5% Cr 0.45% Ni 0.4% Mo steel: Bochum
IMMUNIT R12/13 Mo A	0.15% C 13.0% Cr 0.30% Mo steel: Bochum
IMMUNIT R15/13 Mo	0.15% C 13.0% Cr 1.0% Mo steel: Bochum
IMMUNIT R20/13 Mo	0.2% C 13.0% Cr 1.0% Mo steel: Bochum

Symbol	Nominal analysis, supplier, condition and remarks.
IMPERIAL CT	0.3% C 13% Cr steel: Edgar Allen for BS alloy En 56D
IMPERIAL EQ	0.2% C 13% Cr steel: Edgar Allen for BS alloy En 56C
IMPERIAL R1	0.12% C (max) 13% Cr steel: Edgar Allen
IMPERIAL R10	0.06% C (max) 13% Cr steel: Edgar Allen
IMPERIAL S80	0.25% C 18% Cr 1.5% Ni steel: Edgar Allen for BS alloy En 57
INOFO 1	0.12% C 13% Cr steel: Forez
INOFO 2	0.3% C 13% Cr steel: Forez
INOFO 4	0.1% C 17% Cr steel: Forez
INOFO 4S	0.12% C 17% Cr steel: Forez
INOFO 5	0.18% C 16% Cr 1.5% Ni steel: Forez
INOX 1	0.1% C 13% Cr steel: Pompey
INOX 1F	0.12% C 13% Cr 0.2% Mo 0.2% S steel: Pompey; free machining UTS: 760 Elon: 15% Proof: 580
INOX 2	0.2% C 13% Cr 0.4% Ni steel: Pompey
INOX 3	0.3% C 13% Cr 0.4% Ni steel: Pompey
INOX 16	0.08% C 17% Cr 0.4% Ni steel: Pompey UTS: 480 Elon: 17% Proof: 240
INOX 17T	0.1% C 16% Cr steel: Pompey
INOX 430F	0.1% C 17% Cr 0.4% Ni steel: Pompey UTS: 530 Elon: 21% Proof: 330
INOX 431	0.15% C 16% Cr 2.25% Ni steel: Pompey; hardened & tempered UTS: 870 Elon: 14% Proof: 730
INVENTOR	0.15% C (min) 13% Cr steel: Bar, billet, etc.; Brown Hayley for BS alloy En 56B
IOX KM	0.11% C 12.75% Cr 0.50% Mo steel: Nazionale Cogne
IRRUBIGO 1 Mo	0.15% C 13.0% Cr 1.2% Mo steel: Schmidt & Clemens
IRRUBIGO 2 Mo	0.18% C 13.0% Cr 1.2% Mo steel: Schmidt & Clemens
JETHETE	0.14% C 12.0% Cr 2.4% Ni 1.8% Mo 0.35% V 0.05% N steel: Origin unknown
JETHETE M140	0.08% C 13.0% Cr 1.0% Mo steel: Samuel Fox
JETHETE M151	0.1% C 12% Cr 1.5% Ni 0.6% Mo 0.3% V steel: Samuel Fox; weldable UTS: 930 Elon: 10% Proof: 780
JETHETE M152	0.1% C 12% Cr 2.5% Ni 1.8% Mo 0.3% V steel: Samuel Fox; creep resisting; weldable UTS: 1040 Elon: 20% Proof: 780
JETHETE M153	0.15% C 12% Cr 2% Ni 1.5% Mo steel: Samuel Fox; creep resistant; weldable UTS: 840 Elon: 24% Proof: 610
JETHETE M154	0.1% C 12.0% Cr 2.5% Ni 17.5% Mo 0.37% V steel: Samuel Fox; hardened & tempered UTS: 650 Elon: 11% Proof: 570
JETHETE M160	0.2% C 12% Cr 1.25% Ni 0.7% Mo 0.4% V 0.7% Nb steel: Samuel Fox; creep resistant UTS: 1110 Elon: 20% Proof: 870
JIS G3446 SUS410	0.15% C (max) 12.5% Cr steel: Tube for structural Japanese Standard UTS: 420 Proof: 210
JIS G3446 SUS430	0.15% C (max) 17.0% Cr steel: Tube for structural; Japanese Standard UTS: 420 Proof: 250
JIS G3463 SUS410TB	0.15% C (max) 12.5% Cr steel: Pipe; Japanese Standard UTS: 420 Proof: 210
JIS G3463 SUS430TB	0.12% C (max) 17.0% Cr steel: Pipe; Japanese Standard UTS: 420 Proof: 210
JIS G4304 SUS403	0.15% C (max) 1.0% Mn (max) 12% Cr 0.6% Ni steel: Plate; Japanese Standard; annealed UTS: 450 Proof: 210

Symbol	Nominal analysis, supplier, condition and remarks.
JIS G4304 SUS405	0.08% C (max) 1.0% Mn (max) 13% Cr 0.6% Ni (max) 0.2% Al 0.04% P (max) 0.03% S steel: Plate; Japanese Standard; annealed UTS: 420 Proof: 180
JIS G4304 SUS410	0.15% C 12.5% Cr steel: Plate; Japanese Standard; annealed UTS: 450 Proof: 210
JIS 4304 SUS429	0.12% C (max) 15.00% Cr steel: Plate; Japanese Standard; annealed UTS: 460 Proof: 210
JIS 4304 SUS430	0.12% C (max) 17.0% Cr steel: Plate; Japanese Standard; annealed UTS: 460 Proof: 210
JIS G4304 SUS434	0.12% C (max) 1.0% Me (max) 17% Cr 0.6% Ni (max) 1.0% Mo 0.04% P (max) 0.03% S (max) steel: Plate; Japanese Standard; annealed UTS: 460 Proof: 210
JIS G4312 SUH446	0.2% C 1.5% Mn 2.5% Cr steel: Plate; Japanese Standard; solution treated UTS: 570 Proof: 210
K20L	0.2% C 13% Cr 1.0% Mo steel: Zapp
KALKOS	0.3% C 12% Cr 12% W 1.0% V steel: Latrobe; extrusion & forging dies for brass; type H23 DPN: 390
KANTHAL Al	0.06% C 23% Cr 2.0% Co 5.5% Al Fe alloy: Kanthal AB; for furnace elements DPN: 230 UTS: 770 Elon: 20% Proof: 550
KANTHAL D	0.1% C 22% Cr 2.0% Co 5.7% Al Fe alloy: Kanthal AB; for furnace elements DPN: 200 UTS: 700 Elon: 20% Proof: 550
KARONI 40/13 Mo	0.2% C 13% Cr 1.2% Mo steel: Kabel
KE 15	0.1% C 13% Cr steel: Kayser Ellison; hardened & tempered for plastic dies UTS: 760
KE 25	0.25% C 13% Cr steel: Kayser Ellison for BS alloy En 56C
KE 35	0.35% C 13% Cr steel: Kayser Ellison
KE 40A	0.1% C 13% Cr steel: For free cutting agent; Kayser Ellison for BS alloy En 56AM
KE 43	0.17% C 17% Cr 1.75% Ni steel: Kayser Ellison for BS alloy En 57
KEA 28	0.45% C 13.75% Cr 1% Ni steel: Kayser Ellison; hardened & tempered for plastic dies DPN: 560
KEA 138	0.35% C 12% Cr 12% W 1.05% V steel: Kayser Ellison for hot hobbing
KEA 203	0.85% C 17.5% Cr 0.55% Mo steel: Kayser Ellison
KEA 505	0.17% C 17.0% Cr 1.75% Ni with Se steel: Free machining; Kayser Ellison
KEA 505	0.17% C 16.0% Cr 2.2% Ni with Se steel: Sanderson Kayser
KEA 508	0.25% C 13.0% Cr with Se: Free machining; Kayser Ellison
KK BREARLEY	0.15% C 13% Cr 0.15% Se steel: Bar, billet, etc.; Brown Bayley for BS alloy En 56AM
KK ENGINEERING	0.21% C 13% Cr with Se Zr or Mo steel: Bar, billet, etc., Brown Bayley for BS En 56CM
KK TWOSCORE	0.25% C 17% Cr 2% Ni steel: Brown Bayley; free cutting form of BS alloy En 57
KORROSIL 2M	0.2% C 13% Cr 1.2% Mo steel: Klockner-Werke
KP 1	0.13% C 12.2% Cr 0.7% Mo steel: USSR
KP 27	0.25% C (max) 17% Cr 2% Ni steel: Kiveton Park for BS alloy En 57
KP 28	0.25% C 13% Cr steel: Kiveton Park for BS alloy En 56C
KP 29	0.12% C (max) 13% Cr steel: Kiveton Park for BS alloy En 56A
LANGALLOY 6V	0.08% C 12% Cr 0.6% Ni 0.2% Al steel: Langley Alloys for BS alloy 713

Symbol	Nominal analysis, supplier, condition and remarks.
LANGALLOY 80V	0.2% C 17% Cr 2% Ni steel: Langley Alloys **UTS: 870 Elon: 16% Proof: 650**
LAPELLOY	0.3% C 11.5% Cr 0.3% Ni 2.7% Mo 0.25% V steel: Latrobe Steel; hardened & tempered **DPN: 231 UTS: 1070 Elon: 15% Proof: 840**
LAPELLOY C	0.22% C 11.5% Cr 2.75% Mo 2% Cu steel: Latrobe Steel
LESCALLOY 5616	0.18% C 13% Cr 2.0% Ni 3.0% W steel: Latrobe Steel; hardened & tempered **DPN: 321 UTS: 1030 Elon: 17% Proof: 880**
LESCALLOY S616	0.18% C 13% Cr 3.0% W 2.0% Ni steel: Latrobe
LINCO	0.3% C 11.5% Cr 0.3% Ni 2.7% Mo 0.25% V steel: Latrobe Steel; hardened & tempered **DPN: 331 UTS: 1070 Elon: 15% Proof: 840**
LOWSCORE	0.2% C 20% Cr steel: Bar, billets, etc.; Brown Bayley for BS alloy En 61
LS 1	0.08% C 13% Cr steel: Low Moor for BS/S61; annealed; obsolete
LS 16	0.1% C 17% Cr steel: Low Moor for BS alloy En 60; obsolete
LS 22	0.1% C 21% Cr steel: Low Moor for BS alloy En 61; obsolete
LS 61	0.12% C 13% Cr steel: Low Moor for BS/S61; obsolete
LS 62	0.2% C 13% Cr steel: Low Moor for BS/S62; obsolete
LS 80	0.18% C 17% Cr 2% Ni steel: Low Moor for BS S80; obsolete
LS CUT	0.3% C 13% Cr steel: Low Moor; for cutlery; obsolete **DPN: 480 UTS: 1550 Elon: 3% Proof: 1330**
LS CUT/HC	0.4% C 13% Cr steel: For cutlery; Low Moor; high carbon LS Cut; obsolete **UTS: 650**
LT 54	0.1% C (max) 13% Cr steel: Low Moor; can be carburized for die steel; obsolete
LT 55	0.25% C 13% Cr steel: Low Moor; die steel; obsolete **DPN: 480**
LV 19	0.85% C 19% Cr 2.5% Mo 0.5% V steel: Low Moor; valve steel; obsolete **DPN: 340**
LV 20	0.8% C 2.0% Si 20% Cr 1.5% Ni steel: Low Moor; valve steel; obsolete **DPN: 340**
LV 24	0.45% C 24% Cr 4.7% Ni 2.7% Mo steel: Low Moor; valve steel; obsolete
M10 FIRTH CLEVELAND	Martensitic stainless steel sinter: Firth Cleveland; specific gravity 6.4–6.8 hardened and tempered **UTS: 590 Elon: 1%**
MAFERITE 14	14% Cr steel: Pompey **UTS: 2000**
MAFERITE 17	17% Cr steel: Pompey **UTS: 2400**
MAFERITE 20	20% Cr 1.0% Mo steel: Pompey **UTS: 2500**
MAFERITE 25A	25% Cr 1.5% Al 1.5% Si steel: Pompey
MAFERITE 28N	28% Cr 4% Ni steel: Pompey
MAFERITE 25	25% Cr 1.0% Mo steel: Pompey **UTS: 2500**

Symbol	Nominal analysis, supplier, condition and remarks.
MAFERITE 30	30% Cr steel: Pompey **UTS: 2500**
MAFERITE 30P	30% Cr + Mo steel: Pompey
MARKER S14M	0.18% C 13% Cr 1.15% Mo steel: Schmidt & Clemens
MAXHETE 3	0.2% C (max) 28% Cr steel: Edgar Allen; furnace equipment
MB	0.08% C 15% Cr steel: Osborn; annealed for cutlery; obsolete **UTS: 530 Elon: 30% Proof: 450**
MB 2	0.1% C 17% Cr steel: Osborn; annealed; cutlery, motor trim; obsolete **UTS: 600 Elon: 32% Proof: 370**
MEL-TROL STAINLESS 2	0.35% C 13.0% Cr steel: Carpenter
METCO 42C	0.2% C 16% Cr 2% Ni steel: Wire for spraying; coarser grade than 42F; Metco **DPN: 360**
METCO 42F	0.2% C 16% Cr 2% Ni steel: Powder for spraying; Metco; finer grade than 42C **DPN: 420**
METCOLOY 2	0.3% C 13% Cr steel: Wire for spraying; Metco; as sprayed **DPN: 284 UTS: 270**
MINOX 10	0.1% C 13% Cr steel: Pompey; hardened & tempered **DPN: 380 UTS: 1200 Elon: 6% Proof: 1000**
MINOX 15	0.2% C 13% Cr 1.0% Mo steel: Pompey; castings
MINOX 20	0.2% C 14% Cr steel: Pompey; hardened & tempered **UTS: 1500 Elon: 4% Proof: 1200**
MINOX 30	0.3% C 14% Cr steel: Pompey; hardened & tempered **UTS: 1800 Elon: 4% Proof: 1400**
MINOX 1610	0.1% C 16% Cr steel: Pompey **DPN: 170 UTS: 520 Proof: 280**
MINOX 1620	0.2% C 16% Cr 2% Ni steel: Pompey; hardened & tempered **UTS: 1300 Elon: 5% Proof: 1000**
MINOX 1820	0.15% C 19% Cr 1.0% Ni steel: Pompey; annealed **DPN: 190 UTS: 550 Proof: 300**
MNC 900	General standard for SIS range of stainless steels; Swedish Standard
MOLY ASCOLOY	0.08% C 13% Cr 2% Mo steel: American proprietary alloy listed in SAE year book
MTS 1	0.18% C 11.5% Cr 1.0% Mo steel: Deutsche
N 23	0.32% C 14% Cr steel: Vulcan; hardened & tempered **UTS: 1140 Elon: 33%**
NC 2 (Mo)	0.85% C 17% Cr 0.7% Mo steel: Bar & forging; Brown Bayley
NEUTROTHERM KW20M	0.2% C 13.0% Cr 1.0% Mo steel: Bohler
NF A35 572/Z6C13	0.08% C (max) 12.5% Cr 0.5% Ni (max) steel: French Standard
NF A35 572/Z6CA13	0.08% C (max) 12.5% Cr 0.2% Al steel: French Standard
NF A35 572/Z8C17	0.10% C (max) 17.0% Cr 0.5% Ni (max) steel: French Standard
NF A35 572/ Z8CD17/01	0.1% C (max) 17.0% Cr 0.5% Ni (max) 1.1% Mo steel: French Standard
NF A35 572/ Z10CF17	0.12% C (max) 17.0% Cr 0.5% Ni (max) 0.6% Mo (max) steel: French Standard
NF A35 572/Z12C13	0.11% C 12.5% Cr 0.5% Ni (max) steel: French Standard
NF A35 572/ Z12CF13	0.15% C (max) 13.0% Cr 0.6% Mo (max) steel: French Standard
NF A35 572/Z20C13	0.20% C 13.0% Cr 1.0% Ni (max) steel: French Standard
NF A35 572/Z30C13	0.3% C 13.0% Cr 1.0% Ni (max) steel: French Standard

Note. The following abbreviations and units are used in the tables:

DPN	Hardness, diamond pyramid number
UTS	Ultimate tensile strength, N/mm^2
Elon	Elongation, %
Proof	0.1 % proof strength, N/mm^2

1 N/mm^2=0.1 hbar=0.102 kgf/mm^2=0.06475 tonf/in^2=145.04 lbf/in^2

Symbol	Nominal analysis, supplier, condition and remarks.
NF A35 572/Z40C14	0.4% C 13.5% Cr 1.0% Ni (max) steel: French Standard
NF A35 572/Z100CD17	1.05% C 17.0% Cr 0.5% Ni (max) 0.5% Mo (max) steel: French Standard
NF A35 573/Z6C13	0.08% C (max) 12.5% Cr steel: Plate; French Standard **UTS: 480** **Proof: 270**
NF A35 573/Z6CA13	0.08% C 12.5% Cr 0.25% Al steel: Plate; French Standard
NF A35 573/Z8C17	0.1% C (max) 17.0% Cr steel: Plate; French Standard **UTS: 500** **Proof: 250**
NF A35 573/Z8CD17/01	0.1% C 1.0% Mn 17% Cr 0.5% Ni (max) 1.0% Mo 0.04% P (max) 0.05% S steel: Plate; French Standard **UTS: 600** **Proof: 290**
NF A35 573/Z12C13	0.12% C 12.5% Cr steel: Plate; French Standard **UTS: 700** **Proof: 430**
NICRAL S	25% C 2.0% Si steel: Imphy **DPN: 190** **UTS: 600** **Elon: 20%** **Proof: 400**
NIROSTA VK5M	0.20% C 13.0% Cr 1.2% Mo steel: Krupp
NIROSTA VK7M	0.15% C 13.0% Cr 1.2% Mo steel: Krupp
NON STAIN	0.3% C 1.0% Mn 13.0% Cr 0.5% Ni steel: Sanderson
NOVO 18/2	0.15% C 17% Cr 2% Ni steel: Jonas **DPN: 255** **UTS: 840** **Elon: 25%** **Proof: 630**
NOVO AS 1	0.1% C 14% Cr steel: Jonas; hardened & tempered **DPN: 200** **UTS: 610** **Elon: 30%** **Proof: 480**
NOVO AS 20	0.1% C 19% Cr steel: Jonas; annealed **DPN: 160** **UTS: 570**
NOVO ASFC	0.1% C 14% Cr 0.35% S steel: Jonas; free cutting
NOVO GAS	0.3% C 14% C steel: Jonas **DPN: 207** **UTS: 700** **Elon: 23%** **Proof: 530**
NOVO GBS	0.2% C 14% Cr steel: Jonas **DPN: 217** **UTS: 720** **Elon: 26%** **Proof: 530**
NOVO NOX T131	0.2% C 13% Cr 1.2% Mo steel: Sudwestfalen
NOVO NOX TT131	0.15% C 13% Cr 1.2% Mo steel: Sudwestfalen
NOVO THERM	0.2% C 30% Cr 0.75% Mo steel: Jonas
NST M1	0.1% C 13.0% Cr 0.50% Ni 0.5% Mo steel: Japan Metal Industries
NST M2	0.15% C 12.3% Cr 0.50% Ni (max) 1.0% Mo steel: Japan Metal Industries
NVR 7-4	0.12% C 0.45% Mn 9.0% Cr 1.0% Mo steel: Designation used by DNV
NVR 7-5	0.22% C (max) 0.6% Mn 11.7% Cr 0.5% Ni 1.0% Mo steel: Designation used by DNV
NYBY 1410 Mo	0.08% C 14.0% Cr 1.0% Mo steel: Nyby
NYBY 1415 Mo	0.14% C 13.0% Cr 1.0% Mo steel: Nyby
OSBORN 405	0.08% C 14% Cr steel: Osborn; previously RPA grade
OSBORN 410	0.1% C 13% Cr steel: Osborn; previously RP grade
OSBORN 420C	0.18% C 13% Cr steel: Osborn; previously SA grade
OSBORN 420D	0.35% C 13% Cr steel: Osborn; previous PH grade
OSBORN 430AL	0.12% C (max) 18% Cr 1.0% Al steel: Osborn; previously MBA grade
OSBORN 431	0.18% C (max) 17% Cr 1.7% Ni steel: Osborn; previously DN grade
OSBORN 440B	0.6% C 17% Cr 0.5% Mo steel: Osborn; previously GQ grade
P 4	0.18% C 27.0% Cr steel: Sandvik
P 12	0.27% C 13% Cr steel: Ugine
P 1000	0.1% C 13% Cr steel: Carrs for BS alloy En 56
P 1001	0.16% C 13% Cr steel: Carrs for BS alloy En 56
P 1002	0.21% C 13% Cr steel: Carrs for BS alloy En 56
P 1003	0.3% C 13% Cr steel: Carrs for BS alloy En 56
P 1008	0.35% C 13% Cr 1% Ni steel: Plastic moulds & dies; Carrs

Symbol	Nominal analysis, supplier, condition and remarks.
P 1009	0.18% C 17% Cr 2.1% Ni steel: Carrs for BS alloy En 56
PANTEG 430	0.06% C 17% Cr steel: Sheet **UTS: 540** **Elon: 25%**
PARALLOY M2	0.2% C (max) 1.3% Cr steel: APV Paramount
PARALLOY M3	0.3% C (max) 13% Cr steel: APV Paramount
PARALLOY M15	0.15% C (max) 13% Cr steel: APV Paramount
PARALLOY MPH	0.08% C (max) 17% Cr 4% Ni 2.5% Cu steel: APV Paramount
PARALLOY MPH2	0.08% C (max) 1.4% Cr 4% Ni 2% Mo 2.5% Cu steel: APV Paramount
PARALLOY ML	17% Cr 2% Ni steel: APV Paramount
PHENIX HD 32	0.2% C 12.75% Cr 0.4% Ni 0.15% Mo steel: Schoeller-Bleckmann
PHENIX HD 301	0.15% C 12.75% Cr 0.5% Ni 1.15% Mo steel: Schoeller-Bleckmann
PKH	0.3% C 13% Cr steel: Bar; RTB for BS alloy En 56D
PKI	0.06% C 17% Cr steel: Sheet; RTD for BS alloy En 60
PKL	0.1% C 13% Cr steel: Bar; RTB for BS alloy En 56A
PKM	0.2% C 13% Cr steel: Bar; RTB for BS alloy En 56C
PKN	0.15% C 17% Cr 2% Ni steel: Bar; RTB for BS alloy En 57
PURO M Mo	0.2% C 13% Cr 1.0% Mo steel: Hagener
PURO M Mo2	0.15% C 13% Cr 1.0% Mo steel: Hagener
PYRISTA	0.09% C 29% Cr 1.8% Ni steel: Firth Vickers **DPN: 180** **UTS: 530** **Elon: 25%**
QQ S 763/6	0.15% C 13% Cr 0.5% Ni 0.6% Mo S and Se steel: US Service specification **DPN: 200** **UTS: 450** **Elon: 25%** **Proof: 200**
R 1	0.1% C 13% Cr steel: Jessop **UTS: 720** **Elon: 29%** **Proof: 510**
R 2	0.2% C 13% Cr 0.75% Ni steel: Jessop **UTS: 670** **Elon: 21%** **Proof: 690**
R 4	0.07% C 25% Cr 4% Ni 1.5% Mo steel: Japanese steel
R 4	0.16% C 16.25% Cr 2.2% Ni steel: Jessop for BS alloy En 57; austenitic
R 12	0.1% C 27% Cr 1.25% Ni steel: Jessop
R 15	0.15% C 13% Cr 0.23% S steel: Free machining; Jessop for BS alloy En 56 BM
R 19	0.3% C 13% Cr 0.2% S steel: Free machining; Jessop
R 26	0.1% C 13% Cr 0.22% S steel: Free cutting; Jessop for BS alloy En 56AM
R 29	0.1% C 21% Cr steel: Jessop
R 34	0.3% C 13% Cr steel: Jessop for BS alloy En 56D
R 50	0.08% C 17.0% Cr 4.25% Ni 2.75% Mo steel: Jessop; corrosion resistant steel
R 55	0.10% C 15% Cr 4.25% Ni 2.5% Mo steel: Jessop; corrosion resistant steel
R 100	0.06% C 14.5% Cr steel: Annealed; Fagersta **UTS: 620** **Elon: 25%** **Proof: 390**
R 110	0.1% C 13.5% Cr steel: Hardened & tempered; Fagersta **DPN: 220** **UTS: 740** **Elon: 14%** **Proof: 530**
R 120G	0.12% C 13.0% Cr steel: Casting; Fagersta
R 140	0.1% C 13.5% Cr 1.1% Ni steel: Hardened & tempered; Fagersta **DPN: 220** **UTS: 740** **Elon: 14%** **Proof: 530**
R 140G	0.12% C 13.0% Cr 1.0% Mo steel: Casting; Fagersta
R 200	0.09% C 17.5% Cr steel: Annealed; Fagersta **DPN: 200** **UTS: 450** **Elon: 23%** **Proof: 240**
R 220	0.17% C 17.5% Cr 2% Ni steel: Hardened & tempered; Fagersta **DPN: 260** **UTS: 820** **Elon: 14%** **Proof: 630**

Symbol	Nominal analysis, supplier, condition and remarks.
R 290	0.06% C 17.5% Cr steel: Annealed; Fagersta
	DPN: 200 UTS: 420 Elon: 23% Proof: 240
R 600G	0.25% C (max) 24% Cr steel: Casting; scaling temperature 1070 °C; Fagersta
R 620	0.1% C 28% Cr 4.3% Ni steel: Annealed; Fagersta
	DPN: 260 UTS: 620 Elon: 20% Proof: 370
R 640	0.1% C 26% Cr 5.3% Ni 1.6% Mo steel: Annealed; Fagersta
	DPN: 260 UTS: 620 Elon: 20% Proof: 400
R 700	0.2% C 13.5% Cr steel: Hardened & tempered; Fagersta
	DPN: 270 UTS: 870 Elon: 15% Proof: 690
R 710	0.3% C 13.5% Cr steel: Hardened & tempered; Fagersta
	DPN: 300 UTS: 920 Elon: 9% Proof: 740
R 720	0.4% C 13.5% Cr steel: Hardened & tempered; Fagersta
	DPN: 300 UTS: 920 Elon: 9% Proof: 740
R 730	0.58% C 13.5% Cr steel: Fagersta
	DPN: 630
R 740	0.3% C 14% Cr 0.4% Ni 0.6% Mo steel: Hardened & tempered; Fagersta
	DPN: 300 UTS: 920 Elon: 9% Proof: 740
RCA 33	30% Cr 5.0% Al Fe alloy: Imphy
RCA 44	20% Cr 5.0% Al Fe alloy: Imphy
RE 39	0.2% C 17% Cr 2% Ni steel: Bofors; hardened & tempered
	DPN: 290 UTS: 820 Elon: 14% Proof: 630
REMANIT 1520 Mo	0.2% C 12.5% Cr 1.2% Mo steel: Deutsche
RESOURCE	0.3% C 13% Cr steel: Plastic moulds; Hall & Pickles; obsolete
	DPN: 610
RFK	0.20% C 13.0% Cr 1.0% Mo steel: Friedr Lohmann
RHM 5	0.30% C 13.0% Cr 0.4% Mo steel: Tokushu Seiko
RHM 7	0.17% C 0.6% Mn 12.0% Cr 0.7% Ni 1.10% Mo steel: Tokushu Seiko
RHM 8	0.14% C 0.5% Mn 12.0% Cr 0.6% Ni 1.1% Mo steel: Tokushu Seiko
RHM 9	0.2% C (max) 13.0% Cr 1.0% Ni 1.0% Mo steel: Tokushu Seiko
RHM 10	0.13% C 13.0% Cr 0.6% Ni (max) 0.45% Mo steel: Tokushu Seiko
RHM 11	0.15% C 12.5% Cr 0.5% Ni (max) 1.0% Mo steel: Tokushu Seiko
RHM 37	0.13% C 12.25% Cr 0.6% Ni (max) 0.45% Mo steel: Tokushu Seiko
RNOM	3.20% C 13.0% Cr 0.3% Ni (max) 1.2% Mo steel: Rochling-Buderus
RNO Mo	0.15% C 13.0% Cr 1.0% Ni (max) 1.2% Mo steel: Rochling-Buderus
ROSTODUR R3	0.20% C 13.5% Cr 1.0% Mo steel: Hoffmann
ROSTODUR R7	0.15% C 13.0% Cr 1.0% Mo steel: Hoffmann
ROP 43	0.2% C 12% Cr 0.6% Ni 1.2% Mo 0.3% V steel: Bofors; hardened & tempered
	DPN: 280 UTS: 750 Elon: 15% Proof: 590
ROP 46	0.2% C 12% Cr 0.5% Mo 0.4% V 0.2% Nb steel: Bofors; hardened & tempered
	DPN: 290 UTS: 950 Elon: 16% Proof: 800

Symbol	Nominal analysis, supplier, condition and remarks.
RP	0.1% C 13% Cr steel: Plastic dies; Osborn
RP 1	0.12% C 13% Cr steel: Annealed; turbine blades etc.; Osborn
	UTS: 530 Elon: 30% Proof: 450
RPA	0.08% C 13% Cr steel: Annealed; golf clubs; rivets; Osborn
	UTS: 530 Elon: 30% Proof: 450
RRJ 11	0.12% C 13% Cr steel: Fagersta
RRM 20	0.12% C 17% Cr steel: Fagersta
RRM 22	0.25% C 17% Cr 2.0% Ni steel: Fagersta
RRS 70	0.21% C 13% Cr steel: Fagersta
RRS 71	0.27% C 13% Cr steel: Fagersta
RS (4)	0.42% C 13% Cr steel: Hard wearing parts with medium corrosion resistance; Osborn
	UTS: 700 Elon: 30% Proof: 300
RS (A)	0.17% C 13% Cr steel: Scissors & knives; Osborn
	UTS: 700 Elon: 30% Proof: 580
RS (B)	0.22% C 13% Cr steel: Osborn
	UTS: 740 Elon: 25% Proof: 610
RS (C)	0.32% C 13% Cr steel: Osborn
	UTS: 700 Elon: 30% Proof: 300
RT 46	0.3% C 12% Cr 12% W 0.5% V steel: Bofors; obsolete
	DPN: 380
R THERM 2012 Mo	0.2% C 12% Cr 1.2% Mo steel: Bochum
S 1R	0.1% C 25% Cr steel: Bofors; annealed
S 410	Sintered martensitic stainless steel SG 6.9; Durasint
	DPN: 170 UTS: 500 Elon: 7%
S 40300	UNS designation for type 403 steel
S 40500	UNS designation for type 405 steel
S 41000	UNS designation for type 410 steel
S 41040	UNS designation for type XM30 steel
S 43000	UNS designation for type 430 steel
S 43035	UNS designation for type XM8 steel
S 44625	UNS designation for type XM27 steel
SAE 51403	0.15% C 12% Cr steel: Mechanical properties not quoted
SAE 51405	0.08% C 12.5% Cr 0.2% Al steel: Mechanical properties not quoted
SAE 51409	0.08% C 11.25% Cr 0.75% Ti (max) steel: Non-hardenable
SAE 51410	0.15% C (max) 12% Cr steel: Mechanical properties not specified
SAE 51414	0.1% C 12.5% Cr 2% Ni steel: Mechanical properties not specified
SAE 51416	0.15% C (max) 13% Cr 0.6% Zr or Mo steel
SAE 51416 F	0.15% C (max) 13% Cr 0.07% S or Se steel: Free machining; mechanical properties not specified
SAE 51416 Se	0.15% C 13% Cr 0.15% Se steel: Free cutting; mechanical properties not specified
SAE 51420	0.35% C 13% Cr steel: Mechanical properties not specified
SAE 51420 F	0.35% C 13% Cr 0.07% S or Se steel: Mechanical properties not specified
SAE 51420 F Se	0.35% C 13% Cr 0.15% Se steel: Mechanical properties not quoted
SAE 51430	0.12% C (max) 16% Cr steel: Mechanical properties not specified
SAE 51430 F	0.12% C (max) 16% Cr 0.07% S or Se steel: Free machining
SAE 51430 F Se	0.12% C 16% Cr 0.15% Se steel: Free machining; mechanical properties not quoted
SAE 51431	0.2% C (max) 16% Cr 2% Ni steel: Mechanical properties not specified
SAE 51434	0.12% C 16% Cr 1% Mo steel: Mechanical properties not quoted
SAE 51436	0.12% C 16% Cr 1% Mo 0.5% Nb steel: Mechanical properties not quoted

Note. The following abbreviations and units are used in the tables:

DPN	Hardness, diamond pyramid number
UTS	Ultimate tensile strength, N/mm^2
Elon	Elongation, %
Proof	0.1 % proof strength, N/mm^2

1 N/mm^2=0.1 hbar=0.102 kgf/mm^2=0.06475 tonf/in^2=145.04 lbf/in^2

Symbol	Nominal analysis, supplier, condition and remarks.
SAE 51440 A	0.7% C 17% Cr 0.7% Mo steel: Mechanical properties not specified
SAE 51440 B	0.85% C 17% Cr 0.7% Mo steel: Mechanical properties not specified
SAE 51442	0.2% C (max) 20% Cr steel: Mechanical properties not specified
SAE 51446	0.25% C (max) 26% Cr 1% Ni steel: Mechanical properties not specified
SAE 51499	0.08% C 11% Cr 0.5% Ni 0.7% Nb steel: Mechanical properties not quoted
SAE 60326	0.07% C 16% Cr 4.2% Ni 2.8% Cu steel: Casting
SAE 60328	0.04% C 26% Cr 5.2% Ni 2% Mo 3% Cu steel: Casting
SAE 60410	0.15% C 13% Cr 1% Ni 0.5% Mo steel: Casting; ASTM alloy CA 15
SAE 60420	0.3% C 13% Cr 1% Ni 0.5% Mo steel: Casting; ASTM alloy CA 40
SAE 60442	0.3% C 20% Cr 2% Ni steel: Casting; ASTM alloy CB 30
SAE 60446	0.5% C 28% Cr 4% Ni steel: Casting; ASTM alloy CC 50
SAE 70446	0.5% C (max) 28% Cr 4% Ni 0.5% Mo steel: Casting; ASTM alloy HC
SAE EV 1	0.45% C 23% Cr 4.8% Ni 2.8% Mo steel: For exhaust valves **UTS: 920 Elon: 3%**
SAE EV 2	0.4% C 24% Cr 3.8% Ni 1.4% Mo steel: For exhaust valves **UTS: 1070 Elon: 3%**
SAE HNV 6	0.8% C 2.3% Si 20% Cr 1.3% Ni steel: For valves **UTS: 1140 Elon: 15% Proof: 800**
SANBRON BB	0.25% C 13% Cr steel: Sanderson for BS alloy En 56C or D
SANBRON DD	0.25% C 18% Cr 1.7% Ni steel: Sanderson for BS alloy En 57
SC 40	0.4% C 12% Cr steel: Valves, pump spindles; Balfour **DPN: 630**
SC 45	0.75% C 17% Cr steel: Plastic & rubber dies, etc.; Balfour **DPN: 660**
SC 140 Mo	0.14% C 14.5% Cr 1.0% Mo steel: Stavange
SEC 1	0.15% C (max) 12% Cr steel: Designation used by Japanese Standard
SEC 4	0.15% C (min) 13% Cr steel: Designation used by Japanese Standard
SEC 5	0.12% C 16% Cr steel: Designation used by Japanese Standard
SF 10	0.12% C 13% Cr 1% Ni steel: Annealed; Samuel Fox **UTS: 550 Elon: 25%**
SF 10 ELC	0.03% C (max) 13% Cr 1% Ni steel: Weldable; Samuel Fox
SF 11	0.15% C 13% Cr 1% Ni steel: Hardened & tempered; Samuel Fox **UTS: 720 Elon: 30%**
SF 12	0.20% C 13% Cr 1% Ni steel: Hardened & tempered; Samuel Fox **UTS: 740 Elon: 28%**
SF 13	0.3% C 13% Cr 1% Ni steel: Annealed; Samuel Fox **UTS: 26%**
SF 14	0.4% C 14% Cr 1% Ni steel: Hardened & tempered; Samuel Fox **UTS: 750 Elon: 26%**
SF 15	0.7% C 14% Cr 1% Ni steel: Hardened & tempered; Samuel Fox **UTS: 780 Elon: 25%**
SF 17	0.1% C 17% Cr steel: Annealed; Samuel Fox **UTS: 510 Elon: 27%**
SF 18	0.2% C 17.5% Cr 2% Ni steel: Hardened & tempered; Samuel Fox **UTS: 900 Elon: 24%**
SF 610	0.12% C 13% Cr 1% Ni 0.2% Zr steel: Free machining; Samuel Fox **UTS: 540 Elon: 25% Proof: 360**
SF 611	0.15% C 13% Cr 1% Ni 0.2% Zr steel: Free machining; Samuel Fox **UTS: 730 Elon: 30% Proof: 580**
SF 612	0.21% C 13% Cr 1% Ni 0.2% Zr steel: Free machining; Samuel Fox **UTS: 740 Elon: 28% Proof: 600**
SF 613	0.27% C 13% Cr 1% Ni 0.2% Zr steel: Free machining; Samuel Fox **UTS: 540 Elon: 26% Proof: 270**
SFG	0.08% C 14% Cr steel: Samuel Fox **UTS: 530 Elon: 31%**
SIC HMo	0.12% C 13% Cr 1.1% Mo steel: Stavanger
SIS 14-2301	0.08% C 14% Cr 0.2% Ni steel: Annealed; Swedish Standard **DPN: 210 UTS: 420 Elon: 16% Proof: 210**
SIS 14-2302	0.12% C 13% Cr steel; Swedish Standard **DPN: 200**
SIS 14-2303	0.2% C 12% Cr steel: Annealed; Swedish Standard **DPN: 230 UTS: 770**
SIS 14-2303	0.2% C 12% Cr steel: Hardened & tempered; Swedish Standard **DPN: 300 UTS: 990 Elon: 15% Proof: 700**
SIS 14-2304	0.35% C 13% Cr steel: Annealed; Swedish Standard **DPN: 240 UTS: 820**
SIS 14-2304	0.35% C 13% Cr steel: Hardened & tempered; Swedish Standard **DPN: 550 UTS: 1500 Proof: 1070**
SIS 14-2320	0.1% C (max) 17% Cr steel: Annealed; Swedish Standard **DPN: 200**
SIS 14-2320	0.1% C 1% Mn 17% Cr 0.5% Ni; Swedish Standard **DPN: 210 UTS: 530 Elon: 18% Proof: 250**
SIS 14-2321	0.21% C 1% Mn 17% Cr 1.8% Ni stainless steel: Swedish Standard **DPN: 300 UTS: 1010 Elon: 12% Proof: 640**
SIS 14-2321	0.25% C 16% Cr 2.0% Ni steel: Annealed; Swedish Standard **DPN: 270**
SIS 14-2321	0.25% C 16% Cr 2.0% Ni steel: Hardened & tempered; Swedish Standard **DPN: 280 UTS: 920 Elon: 14% Proof: 640**
SIS 14-2322	0.25% C 24% Cr steel: Annealed; Swedish Standard **DPN: 210 UTS: 820**
SIS 14-2322	0.25% C 1.5% Mn 2.6% Cr stainless steel: Swedish Standard **DPN: 235 UTS: 740**
SIS 14-2323	0.12% C 25% Cr 4.0% Ni steel: Annealed; Swedish Standard **DPN: 260 UTS: 920**
SIS 14-2324	0.1% C 25% Cr 4.5% Ni 1.3% Mo steel: Annealed; Swedish Standard **DPN: 260 UTS: 820 Elon: 20% Proof: 400**
SIS 14-2325	0.08% C 17.5% Cr 0.2% Ni 1.8% Mo steel: Annealed; Swedish Standard **DPN: 180 UTS: 500 Elon: 25% Proof: 330**
SIS 14-2325	0.08% C 1% Mn 17% Cr 1.7% Mo 0.5% Ni stainless steel: Swedish Standard **DPN: 190 UTS: 500 Elon: 25% Proof: 330**
SIS 14-2380	0.11% C 1.5% Mn 13% Cr 1.0% Ni 0.6% Mo steel: Stainless; free cutting; Swedish Standard **DPN: 223 UTS: 650 Elon: 13% Proof: 340**

Symbol	Nominal analysis, supplier, condition and remarks.
SIS 14-2383	0.14% C 1.5% Mn 17% Cr 0.5% Ni steel: Stainless; free cutting; Swedish Standard **DPN: 240 UTS: 735 Elon: 12% Proof: 440**
SIS 2102	0.21% C 17.0% Cr 1.9% Ni steel: Swedish Standard **DPN: 270 UTS: 1000 Elon: 12% Proof: 640**
SIS 2301	0.08% C 14% Cr 0.2% Ni steel: Wrought; annealed; Swedish Standard **DPN: 210 UTS: 420 Elon: 16% Proof: 210**
SIS 2302	0.12% C 13% Cr steel: Swedish Standard; included in MNC 900 **DPN: 200**
SIS 2302	0.12% C 13.0% Cr 1.0% Ni (max) steel: For construction; Swedish Standard; hardened & tempered **UTS: 700 Elon: 16% Proof: 410**
SIS 2302	0.12% C 13.0% Cr 1.0% Ni (max) steel: Swedish Standard; hardened & tempered **UTS: 670 Elon: 16% Proof: 390**
SIS 2303	0.2% C 12% Cr steel: Annealed; Swedish Standard **DPN: 230 UTS: 770**
SIS 2303	0.2% C 12% Cr steel: Hardened & tempered; Swedish Standard **DPN: 300**
SIS 2303	0.22% C 13.0% Cr 1.0% Ni (max) steel: Swedish Standard; hardened & tempered **UTS: 980 Elon: 14% Proof: 690**
SIS 2303	0.22% C 13.0% Cr 1.0% Ni (max) steel: Swedish Standard **UTS: 1000 Elon: 14% Proof: 690**
SIS 2304	0.35% C 13% Cr steel: Plate, bar, forging, etc.; annealed; Swedish Standard **DPN: 240 UTS: 820**
SIS 2304	0.35% C 13% Cr steel: Plate, bar, forging, etc.; hardened & tempered; Swedish Standard **DPN: 550 UTS: 1500 Proof: 1070**
SIS 2317	0.23% C 11.7% Cr .05% Ni 1.2% Mo 0.3% V 0.05% B steel: Bar or forging; Swedish Standard; hardened & tempered **UTS: 870 Elon: 16% Proof: 640**
SIS 2317	0.23% C 0.55% Mn 12.2% Cr 0.5% Ni 1.2% Mo 0.3% V 0.045% N steel: Heat resisting; Swedish Standard **DPN: 260 UTS: 875 Elon:16%14640**
SIS 2320	0.1% C (max) 17% Cr steel: Swedish Standard included in MNC 900 **DPN: 200**
SIS 2321	0.25% C 16% Cr 2% Ni steel: Annealed; Swedish Standard **DPN: 270**
SIS 2321	0.21% C 17.0% Cr 1.9% Ni steel: For construction; Swedish Standard; hardened & tempered **UTS: 1000 Elon: 12% Proof: 640**
SIS 2321	0.25% C 16% Cr 2% Ni steel: Hardened & tempered; Swedish Standard **DPN: 280 UTS: 920 Elon: 14% Proof: 640**
SIS 2322	0.25% C 24% Cr steel: Annealed; Swedish Standard **DPN: 210 UTS: 820**

Note. The following abbreviations and units are used in the tables:

DPN	Hardness, diamond pyramid number
UTS	Ultimate tensile strength, N/mm^2
Elon	Elongation, %
Proof	0.1 % proof strength, N/mm^2

$1\ N/mm^2 = 0.1\ hbar = 0.102\ kgf/mm^2 = 0.06475\ tonf/in^2 = 145.04\ lbf/in^2$

Symbol	Nominal analysis, supplier, condition and remarks.
SIS 2323	0.12% C 25% Cr 4% Ni steel: Plate, bar & forging; annealed; Swedish Standard **DPN: 260 UTS: 920**
SIS 2324	0.1% C 25% Cr 4.5% Ni 1.3% Mo steel: Plate, bar, etc.; annealed; Swedish Standard **DPN: 260 UTS: 820 Elon: 20% Proof: 400**
SIS 2325	0.08% C 17.5% Cr 0.2% Ni 1.8% Mo steel: Annealed **DPN: 180 UTS: 500 Elon: 25% Proof: 330**
SIS 2326	0.025% C (max) 18.0% Cr 0.5% Ni (max) 2.2% Mo 0.6% Ti steel: Swedish Standard **DPN: 210 UTS: 550 Elon: 25% Proof: 340**
SIS 2380	0.12% C 13.0% Cr 0.22% S steel: Free machining; Swedish Standard; hardened & tempered **DPN: 223 UTS: 590 Elon: 12% Proof: 410**
SIS 2380	0.12% C 0.2% S 13% Cr 0.6% Mo (max) steel: Swedish Standard
SIS 2383	0.14% C 17% Cr 0.2% S steel: Free machining; Swedish Standard
SL 440A	0.7% C 17% Cr 0.7% Mo steel: Hardened & tempered; Swift Levick **DPN: 600**
SL 440B	0.85% C 17% Cr 0.7% Mo steel: Hardened & tempered; Swift Levick **DPN: 600**
SOLEIL 4	0.14% C 14.5% Cr 1.0% Mo steel: Firminy
SOLEIL B2	0.08% C 12.5% Cr 0.6% Ni steel: Creusot-Loire
SOLEIL B3	0.08% C 13.0% Cr 0.6% Ni steel: Creusot-Loire
SOLEIL B4	0.12% C 17.2% Cr 0.75% Ni steel: Creusot-Loire
SPEAR B2	0.1% C 13.5% Cr steel: Spear & Jackson
SPW	0.25% C 13.0% Cr steel: Origin unknown
SRO 2	0.45% C 2.8% Si 13% Cr 1.0% Mo steel: Bofors; obsolete
S S1	0.12% C 13% Cr steel: Uddelholm
S S2	0.12% C 17% Cr steel: Uddelholm
S S6	0.27% C 13% Cr steel: Uddelholm
S S22	0.25% C 17% Cr Ni steel: Uddelholm
S S31	0.15% C 13% Cr steel: Uddelholm
STA 5 V25	0.2% C 13% Cr steel: Replaced by BS alloy En 56C
STA 5V25A	0.3% C 13% Cr steel: Replaced by BS alloy En 56D
STA 5 V25B	0.12% C 13% Cr steel: Replaced by BS alloy En 56A
STA 5 V25B/1	0.12% C 13% Cr steel: Replaced by BS alloy En 56A
STA 5 V25B/2	0.16% C 13% Cr steel: Replaced by BS alloy En 56B
STA 5 V25M	0.25% C (average) 13% Cr steel: Free machining grade; replaced by BS alloy En 56 appropriate M
STAINLESS 16	0.27% C 13.5% Cr 1.5% Mo steel: Uddelholm
STAINLESS 51	0.1% C 14% Cr 1.0% Ni steel: Uddelholm
STRAINTRODE 6815	0.05% C 0.5% Mn 13.5% Cr steel: For electrodes; Esab standard **UTS: 880**
SUH 3	0.4% C 11.0% Cr 1.0% Mo steel: Heat resisting; Japanese Standard designation
SUH 600	0.17% C 11.5% Cr 0.6% Mo steel: Heat resisting; Japanese Standard
SUH 616	0.22% C 12.0% Cr 1.0% Mo 1.0% W 0.2% W 0.25% V steel: Heat resisting; Japanese Standard designation
SUPERIOR 4	1.0% C 12% Cr 0.8% V 0.7% V steel: Braeburn
SV 06KH 14	0.08% C (max) 0.5% Mn 14.0% Cr steel: For welding; Russian Standard designation
SV 08KH 14GT	0.1% C (max) 1.1% Mn 14.0% Cr 0.8% Ti steel: For welding wire; Russian Standard designation
SV 10KH 11MFN	0.12% C 11.2% Cr 0.75% Ni 0.75% Mo 0.32% V steel: For welding wire; Russian Standard designation

Symbol	Nominal analysis, supplier, condition and remarks.
SV 10KH 11VMFN	0.1% C 0.5% Mn 11.2% Cr 1.0% Ni 1.1% Mo 0.3% V 1.2% W steel: For welding wire; Russian Standard designation
SV 10KH 13	0.12% C 0.5% Mn 13.0% Cr steel: For welding wire; Russian Standard designation
SV 10 KH 17T	0.12% C (max) 0.7% Mn (max) 17.0% Cr 0.5% Ti (max) steel: For welding wire; Russian Standard designation
SV 13KH 25T	0.15% C (max) 0:8% Mn (max) 24.5% Cr 0.5% Ti (max) steel: For welding wire; Russian Standard designation
TEMPEST	0.16% C 17% Cr 1.8% Ni steel: Glass moulds & inserts; Hall & Pickles **UTS: 1130 Elon: 20% Proof: 920**
TRUBRITE B41	Low C 17% Cr steel: Arthur Lee
TRUBRITE B42	Low C 20% Cr steel: Arthur Lee
TRUBRITE C72	Low C 12% Cr steel: Hardenable; Arthur Lee
TRUBRITE C74	Low C 12% Cr steel: For cutlery; Arthur Lee
TRUBRITE C76	12% Cr steel: For cutlery; Arthur Lee
TUBROD 15.23	0.3% C 0.5% Mn 13.0% Cr steel: Tubular welding rod; Esab; for hard facing; as welded **DPN: 525**
TUBROD 15.70	0.06% C 0.5% Mn 13.0% Cr steel: Tubular weld rod for submerged arc welding; corrosion resistant; Esab **DPN: 330 UTS: 900**
TUBROD 15.73	0.3% C 0.4% Mn 13.0% Cr steel: Tubular welding rod for submerged arc welding; Esab; corrosion resistant; as welded **DPN: 520**
TWOSCORE	0.2% C 16% Cr 2% Ni steel: Bar, billet, etc.; Brown Bayley for BS alloy En 57
U 12	0.17% C 13% Cr steel: Ugine
UHB STAINLESS 1	0.1% C 13.5% Cr steel: Uddelholm; annealed **DPN: 160 UTS: 480 Elon: 18% Proof: 290**
UHB STAINLESS 2	0.09% C 17.5% Cr steel: Uddelholm; annealed **DPN: 150 UTS: 490 Elon: 23% Proof: 330**
UHB STAINLESS 5	0.25% C 25% Cr steel: Uddelholm; annealed **DPN: 170 UTS: 530 Elon: 18% Proof: 330**
UHB STAINLESS 6	0.33% C 13.5% Cr steel: Uddelholm; hardened & tempered **DPN: 250 UTS: 820 Elon: 11% Proof: 630**
UHB STAINLESS 21	0.06% C 13.5% Cr steel: Uddelholm; annealed **DPN: 140 UTS: 440 Elon: 25% Proof: 250**
UHB STAINLESS 22	0.2% C 17% Cr 1.5% Ni steel: Uddelholm; hardened & tempered **DPN: 260 UTS: 820 Elon: 11% Proof: 580**
UHB STAINLESS 31	0.22% C 13.5% Cr 0.7% Ni steel: Uddelholm; hardened & tempered **DPN: 220 UTS: 710 Elon: 15% Proof: 540**
UHB STAINLESS 41	0.13% C 13.5% Cr steel: Free cutting; Uddelholm; hardened & tempered **DPN: 200 UTS: 630 Elon: 15% Proof: 480**
UHB STAINLESS 51	0.1% C 13.5% Cr 1.0% Mo steel: Uddelholm; hardened & tempered **DPN: 190 UTS: 650 Elon: 15% Proof: 500**
UHB STAINLESS 72	0.12% C 17% Cr 0.25% Mo steel: Free cutting; Uddelholm; annealed **DPN: 150 UTS: 530 Elon: 20% Proof: 290**
UHB STAINLESS 716	0.35% C 13.5% Cr 1.0% Mo steel: Uddelholm; hardened & tempered **DPN: 250 UTS: 820 Elon: 11% Proof: 630**
UHB STAINLESS 721	0.06% C 12% Cr steel: Uddelholm
UHB UDD Co 12	0.23% C 11.7% Cr 0.5% Ni 1.0% Mo 0.5% W 0.3% V steel: Tube; Uddelholm; hardened & tempered **UTS: 760 Elon: 20% Proof: 480**
UNI 3992 X43 CS8	0.42% C 8.7% Cr steel: For valves
UNI 4047	0.16% C (max) 19% Cr 2.0% Ni steel
UNI 4047 X8 CA13	0.08% C (max) 12.5% Cr 0.2% Al steel
UNI 4047 X12 C17	0.12% C (max) 17% Cr steel
UNI 4047 X12 CA12	0.12% C (max) 12% Cr 1.2% Al steel
UNI 4047 X15 C13	0.15% C (max) 12.5% Cr steel
UNI 4047 X15 CF13	0.15% C (max) 13% Cr 0.6% Mo (max) steel
UNI 4047 X20 C13	0.2% C 13% Cr steel
UNI 4047 X20 CN16	0.2% C (max) 16% Cr 2.0% Ni steel
UNI 4047 X12 CA23	0.12% C (max) 23% Cr 1.6% Al steel
UNI 4047 X25 C26	0.25% C (max) 25.5% Cr steel
UNI 4047 X32 C13	0.3% C 13% Cr steel
UNI 4047 X40 C14	0.4% C .14% Cr steel
UNI S115	Replaced by UNI 3992 and UNI 4047
URANUS 50	0.06% C (max) 21.0% Cr 2.3% Mo 1.2% Cu steel: Creusot-Loire
V 3M	0.27% C 13% Cr steel: Krupps
V 5M	0.15% C 13% Cr steel: Krupps
V 13F	0.12% C 13% Cr steel: Krupps
V 13F Al	0.08% C 13% Cr 0.2% Al steel: Krupps
V 13 FB	0.08% C 14% Cr steel: Krupps
V 13F Supra	0.08% C 13% Cr steel: Krupps
V 17F	0.1% C 16.5% Cr steel: Krupps
V 17F Extra	0.1% C 17% Cr 0.7% Ti steel: Krupps
V 17F X Extra	0.1% C 17% Cr 1.0% Nb steel: Krupps
VALVIC 857	0.8% C 2.1% Si 20% Cr 1.5% Ni steel: For valves; Swift Levick **DPN: 500 UTS: 1620**
VH 217	0.21% C 13% Cr steel: Wykmanshyttan
VH 273	0.12% C 13% Cr steel: Wykmanshyttan
VH 417	0.27% C 13% Cr steel: Wykmanshyttan
VH 620	0.12% C 17% Cr steel: Wykmanshyttan
VH 626	0.25% C 17% Cr 2.0% Ni steel: Wykmanshyttan
VK 17F	0.07% C 17% Cr 1.0% Mo steel: Krupps
VS 17	0.1% C 17% Cr steel: Annealed; Vulcan **UTS: 840 Elon: 33%**
VS S61	0.1% C 13% Cr steel: Hardened & tempered; Vulcan **UTS: 950 Elon: 30%**
VS S80	0.2% C 16% Cr 1.5% Ni steel: Hardened & tempered; Vulcan **UTS: 1280 Elon: 24%**
W 12	0.12% C 13% Cr steel: Welding electrode; Cu coated; Armco
WCR 100	0.7% C 14% Cr 20% W 3.5% B steel: Rod for facing cutting tools, etc.; melting point 1200 °C; Wall Colmonoy **DPN: 950**
WEARTRODE 84.58	0.7% C 0.7% Mn 10.0% Cr steel: For electrodes; Esab **DPN: 635**
X 6 Cr 13	0.08% C (max) 12.5% Cr 0.6% Ni steel: Italian Standard
X 6 Cr Mo 17	0.07% C (max) 17.0% Cr 1.05% Mo steel: German Standard
X 6 Cr Mo 17	0.07% C (max) 17.0% Cr 1.0% Mo steel: Designation used by German Standard
X 7 Cr 14	0.08% C (max) 14.0% Cr steel: Designation used by German Standard
X 7 Cr Al 13	0.08% C (max) 13.0% Cr 0.2% Al steel: German Standard
X 7 Cr Al 13	0.08% C (max) 13.0% Cr 0.2% Al steel: Designation used by German Standard
X 8 Cr 13	0.07% C 13% Cr 0.2% Al steel **DPN: 190 UTS: 420 Elon: 20% Proof: 290**
X 8 Cr 14	0.1% C (max) 14.5% Cr steel: Designation used by German Standard
X 8 Cr 15	0.12% C (max) 14.5% Cr steel: Designation used by German Standard
X 8 Cr 17	0.1% C (max) 16.5% Cr steel: Designation used by German Standard

Symbol	Nominal analysis, supplier, condition and remarks.
X Cr 18	0.1% C (max) 17.5% Cr steel: Designation used by German Standard
X 8 Cr 28	0.1% C (max) 28.0% Cr steel: Designation used by German Standard
X 8 Cr 30	0.1% C (max) 30% Cr 2.0% Ni steel: Designation used by German Standard
X 8 Cr Mo 302	0.1% C (max) 2.0% Si 30.0% Cr 2.25% Mo 2.0% Ni 12% N steel: Designation used by erman Standard
X 3 Cr Mo Ti 17	0.1% C 17.0% Cr 1.75% Mo 0.8% Ti steel: Designation used by German Standard
X 8 Cr Mo Ti 18	0.1% C (max) 17.5% Cr 1.2% Mo 0.6% Ti steel: Designation used by German Standard
X 8 Cr Nb 17	0.1% C (max) 17.5% Cr 1.3% Nb steel: Designation used by German Standard
X 8 Cr Ti 17	0.1% C (max) 17.0% Cr 1.0% Ti steel: Designation used by German Standard
X 8 Cr Ti 18	0.1% C (max) 17.5% Cr 0.6% Ti steel: Designation used by German Standard
X 10 C18	0.07% C 17.0% Cr steel: Italian Standard **DPN: 200 UTS: 180 Elon: 10% Proof: 290**
X 10 Cr 13	0.1% C 13.0% Cr steel: Designation used by German Standard
X 10 Cr 25	0.15% C (max) 25.0% Cr steel: Designation used by German Standard
X 10 Cr Al 13	0.12% C (max) 13.0% Cr 1.0% Al steel: Designation used by German Standard
X 10 Cr Al 18	0.12% C (max) 18.0% Cr 1.0% Al steel: Designation used by German Standard
X 10 Cr Al 24	0.12% C (max) 24.0% Cr 1.5% Al steel: Designation used by German Standard
X 10 Cr Si 13	0.12% C (max) 2.1% Si 13.0% Cr steel: Designation used by German Standard
X 10 Cr Si 18	0.12% C (max) 2.1% Si 18.0% Cr steel: Designation used by German Standard
X 10 Cr Si 29	0.12% C (max) 29.0% Cr steel: Designation used by German Standard
X 12 C17	0.07% C 17.0% Cr steel: Italian Standard **DPN: 200 UTS: 180 Elon: 20% Proof: 290**
X 12 Cr Mo S 17	0.15% C 16.5% Cr 0.25% Mo 0.2% S steel: Designation used by German Standard
X 15 Cr 13	0.15% C 13.0% Cr steel: Designation used by German Standard
X 15 Cr Mo 12/1	0.15% C 11.5% Cr 1.1% Mo steel: Designation used by German Standard
X 15 Cr Mo 13	0.15% C 13.0% Cr 1.1% Mo steel: Designation used by German Standard
X 20 C 28	0.2% C 25.0% Cr 0.15% N steel: Italian Standard **DPN: 230 UTS: 480 Elon: 20% Proof: 330**
X 20 Cr 13	0.19% C 13% Cr steel: Designation used by German Standard
X 20 Cr Mo 13	0.2% C 13.0% Cr 1.1% Mo steel: Designation used by German Standard
X 20 Cr Mo W V 12/1	0.2% C 12% Cr 0.5% Ni 1.0% Mo 0.5% W 0.3% V steel: German Standard
X 20 Cr Ni Si 25/4	0.2% C 25.0% Cr 4.5% Ni steel: Designation used by German Standard
X 22 Cr Mo V 12/1	0.22% C 11.5% Cr 0.6% Ni 1.1% Mo 0.3% V steel: Designation used by German Standard

Symbol	Nominal analysis, supplier, condition and remarks.
X 22 Cr Mo W V 12/1	0.2% C 12.0% Cr 0.5% Ni 1.0% Mo 0.3% V 0.5% W steel: Designation used by German Standard
X 22 Cr Ni 17	0.2% C 17.0% Cr 2.0% Ni steel: Designation used by German Standard
X 25 C26	0.2% C 25.0% Cr 0.15% N steel: Italian Standard **DPN: 230 UTS: 480 Elon: 20% Proof: 330**
X 35 Cr 14	0.35% C 14.0% Cr steel: Designation used by German Standard
X 35 Cr Mo 17	0.38% C 16.5% Cr 1.1% Mo steel: Designation used by German Standard
X 40 Cr 13	0.4% C 13.0% Cr steel: Designation used by German Standard
X 45 Cr Mo V 15	0.44% C 14.2% Cr 0.52% Mo 0.12% V steel: Designation used by German Standard
X 45 Cr Mo V 15	0.45% C 14.4% Cr 0.13% V steel: German Standard
X 45 Cr Ni Mo 23/5	0.45% C 23.0% Cr 5.0% Ni 2.7% Mo steel: Designation used by German Standard
X 80 Cr Ni Si 20	0.8% C 2.25% Si 20.0% Cr 1.5% Ni steel: Designation used by German Standard
X B	0.78% C 2% Si 19% Cr 1.5% Ni steel: Firth Vickers **DPN: 300 UTS: 970 Elon: 15%**
X B	0.8% C 17% Cr steel: Bar & forging; Brown Bayley for BS alloy En 59
XM 8	0.07% C 18.0% Cr 0.5% Ni (max) 0.6% Nb (min) 0.15% Al (max) steel: Designation used by ASTM
XM 27	0.01% C (max) 26.0% Cr 0.1% Mo steel: Designation used by ASTM
XM 30	0.18% C 12.5% Cr 0.2% Nb steel: Designation used by ASTM
Z 6C13	0.08% C (max) 12.5% Cr 0.6% Ni steel: French National Standard
Z 6 C13	0.15% C 12% Cr steel: French National Standard
Z 6 C13 Al	0.07% C 13% Cr 0.2% Al steel: French National Standard **DPN: 190 UTS: 420 Elon: 20% Proof: 290**
Z 6CA13	0.08% C (max) 13.0% Cr 0.6% Ni steel: French National Standard
Z 6CND13/04M	0.08% C 12.5% Cr 3.0% Ni 1.0% Mo steel: Casting; designation used by French Standard
Z 6CNDNb17/12	0.08% C (max) 17.5% Cr 12.0% Ni 2.5% Mo steel: French National Standard
Z 8C17	0.12% C (max) 17.2% Cr 0.75% Ni steel: French National Standard
Z 8 C17	0.12% C 16% Cr 2% Ni steel: French National Standard **DPN: 200 UTS: 1130 Elon: 20% Proof: 290**
Z 10 C13	0.12% C 13% Cr steel: French National Standard
Z 10 C17	0.07% C 17.0% Cr steel: French National Standard **DPN: 200 UTS: 180 Elon: 20% Proof: 290**
Z 10 C18	0.12% C 17% Cr steel: French National Standard **DPN: 200 UTS: 180 Elon: 20% Proof: 290**
Z 12 C13	0.15% C 12% Cr steel: French National Standard
Z 12 C18	0.12% C 16% Cr 2% Ni steel: French National Standard
Z 15 C27	0.2% C 25% Cr 0.2% N steel: French National Standard **DPN: 230 UTS: 480 Elon: 20% Proof: 330**
Z 15 CN16/2	0.2% C 16% Cr 2% Ni steel: French National Standard
Z 20 C13	0.21% C 13% Cr steel: French National Standard
Z 20 C25	0.2% C 25% Cr 0.15% N steel: French National Standard **DPN: 230 UTS: 480 Elon: 20% Proof: 330**
Z 30 C13	0.15% C 13% Cr steel: French National Standard
Z 30C13	0.3% C 13.0% Cr steel: French Standard designation

Note. The following abbreviations and units are used in the tables:

DPN	Hardness, diamond pyramid number
UTS	Ultimate tensile strength, N/mm^2
Elon	Elongation, %
Proof	0.1 % proof strength, N/mm^2

1 N/mm^2=0.1 hbar=0.102 kgf/mm^2=0.06475 tonf/in^2=145.04 lbf/in^2

Symbol	Nominal analysis, supplier, condition and remarks.
Z 70C14	0.4% C 14.0% Cr steel: French Standard designation
Z 70C15	0.7% C 15.0% Cr steel: French Standard designation
Z 80CSN20/02	0.8% C 0.8% Mn (max) 20.0% Cr 1.3% Ni steel: French Standard designation

Note. The following abbreviations and units are used in the tables:

DPN	Hardness, diamond pyramid number
UTS	Ultimate tensile strength, N/mm^2
Elon	Elongation, %
Proof	0.1 % proof strength, N/mm^2

$1 N/mm^2 = 0.1 hbar = 0.102 kgf/mm^2 = 0.06475 tonf/in^2 = 145.04 lbf/in^2$

44M2 Steel – high carbon, high chromium
0.8% carbon minimum, 13% chromium, with titanium and molybdenum

Specific gravity	7.77	
Density	7770 kg/m³	(0.28 lb/in.³)
Solidus/liquidus	1450 °C	
Thermal conductivity	18.8 W/m °C	(4.5 cal/m s °C)
Coefficient of linear expansion	12 × 10⁻⁶/ °C	
Electrical conductivity	2% IACS (copper 100%)	
Specific resistance	700 microhm mm	
Young's modulus of elasticity	– N/m²	
Impact	– J	
Fatigue strength	–	
Hot strength	–	

The above properties are typical of the following group, and may not apply exactly to any one specification. It is possible that with certain specifications some of the values may not be applicable.

General metallurgical characteristics

These steels are very hard with a tendency to brittleness even in the annealed condition. The high chromium content ensures a degree of corrosion resistance but this does not equal that of the lower carbon high chrome steels owing to the presence of the carbides. Additions of titanium and molybdenum ensure grain refinement, improve the hot strength characteristics and raise the tempering temperature. The corrosion resistance of these steels is improved by hardening and tempering and they are almost always used in this condition.

Thermal treatment

As these steels cannot accept cold work without becoming hard, brittle and unworkable, they are never used in the cold drawn condition. Annealing between machining operations may be necessary and certainly is prior to machining after forging or casting. This will almost always be a cyclic or spheroidal anneal, where the part is first heated to 1000 °C for about two hours, then cooled to 670–700 °C for two hours and this cycle repeated at least once, preferably twice. The resultant structure consists of small balls or spheres of hard carbide evenly distributed throughout the ductile matrix. The presence of grain refining elements such as titanium or molybdenum ensures that the spheres are small, whereas the simple chromium steels will have massive globules which are more difficult to machine and take longer to dissolve during thermal treatments. For annealing between machining operations to remove local work hardening, the sub-critical anneal at 670–700 °C for 2–4 hours will suffice in some instances. Only under exceptional circumstances will any of these steels be normalized.

Hardening requires temperatures in the region of 1000–1250° C, the parts being held at temperature for about 30 minutes per 25 mm of section as the metallurgical reactions with these steels are extremely sluggish. These can be speeded up by increasing the temperatures with consequent danger of grain growth, but this is inhibited to some extent by the presence of titanium, molybdenum or vanadium. Grain growth increases hardenability, improves machinability, but reduces considerably the fatigue strength and impact resistance, and it is generally undesirable.

These steels will all air harden provided the mass is not too great and can also be oil quenched provided no rapid changes of section are present. As hardened these steels are extremely brittle and must be carefully handled, being tempered as soon as possible at 400–700 °C for 2–4 hours.

Normal furnace equipment using electric, gas or oil heating, or salt baths, can be used on all these alloys. Being high carbon, they can be de-carburized but the high chromium content considerably reduces the hazard. Where no machining can be carried out after hardening and tempering and the part is to be subjected to fatigue stress the use of a controlled atmosphere or salt baths is advised. In both cases it is essential that slight oxidizing conditions are avoided as this can lead to serious local intergranular de-carburization.

Scale removal should be by grit blasting or sodium hydride if a suitable plant for the latter process is available. Acid pickling must never be carried out on these steels after hardening and tempering as they are prone to hydrogen embrittlement.

Welding and brazing

Pre-treatment. The joint area must be clean and free from scale and oxide. Parts should be in the annealed condition if possible. Pre-heating is essential.

Welding and brazing. Welding must not be carried out on any of the steels listed when the weld area is to be subjected to stress. With careful skilled welding, the correct equipment and the specified post-treatment, welds of a reasonable standard can be achieved. The cast structure however, will always constitute a brittle zone which cannot be eliminated.

Gas, electric arc or inert gas shielded electric arc welding can be used; the last method, requiring no flux, probably gives the best results. Superior welds are obtained if the parts can be welded with the weld area held above 300–500 °C and allowed to cool slowly.

Resistance heating is sometimes used, followed by forge welding. This requires special equipment with careful control, when excellent welds free of any cast structure can be obtained. Brazing can usually be accomplished on these steels without causing softening, provided rapid local heating is used.

Post-treatment. Fluxes must be removed as they can cause corrosion.

After welding the parts should preferably be hardened and tempered, care being taken to remove stress raisers particularly from the weld area. This gives a homogeneous structure with some loss of strength at the weld zone. As an absolute minimum, the weld area must be locally stress released using electrically heated blankets or gas welding torches.

Flaw detection methods

Crack-test. Magnetic crack test can be used but penetrant oil chalk test probably gives as good results, requires less equipment and is more flexible.

X-ray. Not required.

Ultra-sonic test. Large forgings which require costly machining can be examined by ultra-sonics at a very early machining stage. This will show up subcutaneous defects, thus saving the expense of machining scrap parts.

Chemical etch. These parts are subject to softening and cracking caused by faulty grinding. Electrolytic sulphuric etching of finished parts can identify the effects of this grinding abuse, thus giving a measure of control over grinding. Dilute nitric acid etching can be used if deserved.

Corrosion protection

Temporary. These steels have a fair measure of corrosion protection, but should always be stored in the oiled or greased condition. No special precautions are required during machining, but it should be noted that dirty coolants can cause corrosion on these steels, particularly when contaminated with metal dust.

After machining the parts should be cleaned, dried, oiled and stored out of contact with air.

Permanent. Painting is seldom required, but is not difficult provided zinc silicate or chromate primer is used with a top sealer coat and care is taken to ensure absolute metal cleanliness and a degree of roughness to supply some mechanical bond.

Plating. Cadmium or zinc is sometimes used, but these steels are very prone to hydrogen embrittlement which causes sudden failure in a brittle manner below the design stress.

The danger can be eliminated by stress relieving at 200 °C before and after plating, but this treatment produces no visual evidence to indicate it has been carried out. Because of the disastrous results of failure from hydrogen embrittlement, plating is not recommended unless adequate control of the stress release can be guaranteed.

These steels will not be metal sprayed or cathodically protected under normal conditions.

Machinability

In the hardened and tempered condition these materials can only be ground, spark eroded or electrochemically machined. Rough machining can be carried out in the spheroidize annealed condition with sub-critical anneal after very heavy machining if dimensional accuracy is important. These steels are hard even in the annealed condition, requiring the use of tipped tools and rigid powerful machines with low speed and feeds. It is important not to allow tools to dwell or rub, and interrupted cuts should be avoided.

Uses

Plastic dies and moulds, forging dies, surgical equipment, cutlery, hot and cold rolls for non-ferrous materials, ball and roller bearings.

Symbol	Nominal analysis, supplier, condition and remarks
1.2080	2.0% C 12.0% Cr steel: German Standard designation
1.2080	2.1% C 12.5% Cr 0.2% V steel: German Standard designation
1.2419	1.05% C 1.0% Cr 1.2% W steel: German Standard designation
1.2436	2.0% C 12.0% Cr 0.7% W steel: German Standard designation
1.2601	1.65% C 12.0% Cr 0.6% Mo 0.1% V 0.5% W steel: German Standard designation

Note. The following abbreviations and units are used in the tables:

DPN	Hardness, diamond pyramid number
UTS	Ultimate tensile strength, N/mm^2
Elon	Elongation, %
Proof	0.1 % proof strength, N/mm^2

1 N/mm^2=0.1 hbar=0.102 kgf/mm^2=0.06475 tonf/in^2=145.04 lbf/in^2

Symbol	Nominal analysis, supplier, condition and remarks.
1.2601	1.5% C 12.0% Cr 0.85% Mo 0.28% V steel: German Standard
1.4112	0.9% C 18% Cr 1.2% Mo 0.1% V steel: German Standard
1.4721	2.1% C 12% Cr steel: For valves; German Standard
1.4747	0.8% C 20% Cr 1.5% Ni steel: For valves; German Standard
1 CW	2.0% C 13% Cr steel: Inman
1 R3	2.3% C 11.75% Cr 0.25% V steel: Thread roll dies, punches, etc; Osborn
	DPN: 540
2 RO 189	1.0% C 17% Cr 0.6% Mo steel: Bofors; hardened & tempered
	DPN: 350 UTS: 1070 Elon: 8% Proof: 820
14 Cr 4 Mo	1% C 14.5% Cr 4.0% Mo 0.12% V steel: Proprietary alloy; information from SAE handbook
14S	2% C 14% Cr 0.5% Ni 0.3% Mo 0.3% V steel: Thread rolls, press tools, etc.; Carrs
	DPN: 660
18 C 283	1.05% C 14% Cr 0.5% Mo 0.5% Co 0.2% Cu steel: Sandvik
23S	2% C 13% Cr 0.3% Mo steel: Plastic moulds, shear blades, etc.; Carrs
	DPN: 700
25.51C	1.0% C 25.0% Cr 5.0% Ni steel: Electrode; for cavitation resistance
69S	1.5% C 12% Cr 0.7% Mo 0.25% V steel: Press tools, thread rolls, etc.; Carrs
	DPN: 660
100 Cr 29	1.0% C 29.0% Cr steel: Electrode; Metrode; hard corrosion resistant deposit
100 Cr 29	1.0% C 1.0% Mn 29.0% Cr steel: Electrode; Metrode; for repair welding or casting
105 W Cr 6	1.06% C 1.0% Cr 1.2% W steel: German Standard designation
120 Cr 25	1.2% C 25.0% Cr steel: Electrode; Metrode; for hard surfacing
	DPN: 630
120 Cr 25	1.2% C 1.0% Mn 25.0% Cr 1.0% Ni steel: Electrode; Metrode; for repair welding of castings
140 Cr 29	1.4% C 29.0% Cr steel: Electrode; Metrode; for hard facing
	DPN: 640
150 Cr 25	1.5% C 25.0% Cr steel: Electrode; Metrode; for hard facing
	DPN: 640
300 Cr 12 Mo W	3.0% C 12.0% Cr 2.0% Mo 2.0% W steel: Electrode; Metrode; for hard facing
	DPN: 550
336	1.5% C 0.3% Mn 12% Cr 0.6% Mo 0.6% V steel: F Parkin; thread roll dies, gauges, etc.
	DPN: 630
338	2.0% C 13% Cr steel: Plastic moulds, press tools, etc.; F Parkin
	DPN: 800
350 Cr 20	3.5% C 1.0% Mn 20.0% Cr steel: Electrode; Metrode; for hard facing
	DPN: 620

Note. The following abbreviations and units are used in the tables:

DPN	Hardness, diamond pyramid number
UTS	Ultimate tensile strength, N/mm^2
Elon	Elongation, %
Proof	0.1 % proof strength, N/mm^2

1 N/mm^2=0.1 hbar=0.102 kgf/mm^2=0.06475 tonf/in^2=145.04 lbf/in^2

Symbol	Nominal analysis, supplier, condition and remarks.
476	1.5% C 12% Cr 1% Mo 0.2% V steel: Blanking dies, plastic moulds; Sanderson
476 SPECIAL	2.2% C 12% Cr 1% Mo 0.2% V steel: Blanking dies & punches etc.; Sanderson
	DPN: 750
2233	2.0% C 12.0% Cr steel: French Standard designation
2234	2.0% C 12.0% Cr 0.8% Mo 0.2% V steel: French Standard designation
2235	1.6% C 12.0% Cr 0.8% Mo 0.4% V steel: French Standard designation
2236	1.8% C 12.0% Cr 0.8% Mo 0.5% V 3.0% Co steel: French Standard designation
2237	2.3% C 12.0% Cr 1.0% Mo 4.0% V steel: French Standard designation
A 6	1.5% C 11.7% Cr 0.5% Mo 1.3% V steel: T Turton
A 11	2.0% C 13% Cr 0.2% Mo 0.2% V steel: T Turton
A 22	1.5% C 14% Cr 0.6% Ni 1.5% Mo 0.2% V steel: T Turton
AISI 440 C	1.1% C 17% Cr 0.7% Mo steel
	DPN: 600 UTS: 2000 Elon: 1% Proof: 1840
AISI 440 F	1.0% C 17% Cr 0.07% Se steel: Free machining
	DPN: 600 UTS: 2000 Elon: 1% Proof: 1840
AISI 617	1.1% C 17.5% Cr 0.5% Mo steel: Annealed
	DPN: 230 UTS: 760 Elon: 14% Proof: 450
AISI 617	1.1% C 17.5% Cr 0.5% Mo steel: Hardened & tempered
	DPN: 580 UTS: 2000 Elon: 2% Proof: 1920
AISI 618	1.05% C 14.5% Cr 4% Mo steel: Hardened & tempered for ball bearings, etc.
	DPN: 820
ALLOY 1	4% C 16% Cr 6% Ni 0.5% V Fe alloy: Casting; Dewrance
	DPN: 500 UTS: 530
ALLOY 4	4% C 16% Cr 6% Ni 4.5% Si 0.5% V Fe alloy: Casting; Dewrance
	DPN: 750 UTS: 500
ALLOY 10 Mod	4.2% C 16% Cr 2% Ni 8% Mo 1.5% Si 1% V Fe alloy: Casting; Dewrance
	DPN: 870 UTS: 360
ALLOY 19	1.25% C 32% Cr 2% Cu Fe alloy: Casting; Dewrance
	DPN: 370 UTS: 800
ALLOY C	2.3% C 13% Cr steel: Jessop
	DPN: 870
ALLOY CWPS	1.5% C 12.0% Cr 1.0% Mo steel: Origin unknown
AMS 5352 A	1.0% C 17% Cr 0.5% Mo steel: Investment casting
AMS 5630 C	1.1% C 17% Cr 0.5% Mo steel: Bar & forging; AMS for SAE 51440 C
AMS 5632 B	1.0% C 17% Cr 0.5% Mo steel: Bar & forging; free machining; AMS for SAE 51440 F
ARH 8	1.05% C 16% Cr 0.8% Mo Co steel: Hardened & tempered; Schoeller-Bleckmann
	DPN: 770
ARMCO 17 C 80	0.85% C 17% Cr 0.7% Mo steel: Armco
ARMCO 17 C 100	1.0% C 17% Cr 0.7% Mo steel: Armco
ARMCO 17 C 100 FM	1.0% C 17% Cr 30.2% Se steel: Armco
ARS	2.35% C 12% Cr 1.0% Mo 4.0% V steel: Simonds; AISI type B7
ASTM A457/650	0.08% C (max) 16.2% Cr 25.5% Ni 6.2% Mo steel: Plate
ASTM A276/440 C	1.1% C 17% Cr 0.7% Mo steel: Bar
ASTM A597 CD2	1.5% C 12.0% Cr 1.0% Mo 0.7% V 0.9% Co cast steel: For tools
ASTM A597 CD5	1.5% C 12.0% Cr 1.0% Mo 0.4% V 3.0% Co cast steel: For tools
AT 2	2.2% C 12% Cr 0.9% Mo 0.25% V steel: Braeburn; AISI type D4

Symbol	Nominal analysis, supplier, condition and remarks.
ATMODIE	1.55% C 11.9% Cr 0.8% Mo 0.9% V steel: Columbia; AISI type D2
ATMODIE SMOOTHCUT	1.55% C 11.9% Cr 0.8% Mo 0.9% V steel: Columbia; AISI type D2S; free machining
BCD	2% C 0.3% Si 0.3% Mn 12% Cr 0.3% V steel: For tools; Huntsman; plastic mould dies, shear blades, etc. **DPN: 770**
BCHV	1.8% C 12% Cr 0.8% Mo 0.25% V steel: For tools; light stamping dies, shear blades, etc.; Huntsman **DPN: 750**
BCRS	1.0% C 13% Cr 0.8% Mo 0.3% V steel: For tools; dies, shear blades, etc.; Huntsman **DPN: 800**
BCW	2.0% C 12% Cr 1% W steel: For tools; plastic mould dies; Huntsman **DPN: 770**
BKV	2.2% C 12.5% Cr 0.25% V steel: Origin unknown
BR 4 FM	2.3% C 12.5% Cr 1.1% Mo 4.0% V steel: Latrobe; die steel; type D7 **DPN: 830**
BRUNSWICK V106	2.0% C 13% Cr + Mo steel: Plastic moulds, shear blades; John Vessey **DPN: 720**
BS 648/B	1.0% C (max) 27% Cr steel: Casting; as cast; contained in BS 3100
BS 1648/C	1.5% C 27% Cr steel: Casting; as cast; contained in BS 3100
BS 3100/452 C11	1.0% C (max) 27.0% Cr 4.0% Ni (max) 1.5% Mo (max) steel: Casting
BS 3100/452 C12	1.5% C (max) 27.0% Cr 4.0% Ni (max) 1.5% Mo (max) steel: Casting
C 12	1.8% C 12.5% Cr 0.5% V steel: Origin unknown
C 265	1.6% C 12% Cr 0.8% Mo 0.8% V steel: Fagersta **DPN: 680**
CCM	1.75% C 13.5% Cr 0.7% Mo steel: For dies; Swift Levick
CCM	1.5% C 12% Cr 0.85% Mo 0.9% V steel: Simonds; AISI type D2
CD	2.1% C 12.5% Cr steel: Plastic moulding, dies, etc.; ESC **DPN: 870**
CDV 1	2.25% C 12.0% Cr 1.0% Mo steel: Origin unknown
CDV 2	1.5% C 12.0% Cr 1.0% Mo steel: Origin unknown
CDV 4	1.5% C 12.0% Cr 1.0% Mo 0.8% V steel: For tools; origin unknown
CDW	1.5% C 12.0% Cr 1.0% Mo steel: Origin unknown
CHIPPER KNIFE	0.81% C 11.5% Cr 0.4% Mo steel: Latrobe
CMD	2.0% C 12% Cr 0.25% V steel: Turton Bros
CMD	High C High Cr steel: For cold forging; Turton Bros
COBALT CHROME FM	1.5% C 12.25% Cr 0.85% Mo 3.1% Co steel: Latrobe; die steel; type D5 **DPN: 770**
COBALTCROM	1.4% C 13% Cr 3% Co 0.75% Mo steel: Blanking dies, shear blades, etc.; Darwin **DPN: 680**

Note. The following abbreviations and units are used in the tables:

DPN	Hardness, diamond pyramid number
UTS	Ultimate tensile strength, N/mm^2
Elon	Elongation, %
Proof	0.1 % proof strength, N/mm^2

1 N/mm^2=0.1 hbar=0.102 kgf/mm^2=0.06475 tonf/in^2=145.04 lbf/in^2

Symbol	Nominal analysis, supplier, condition and remarks.
COLMONOY 1	0.9% C 11% Cr 2.5% B steel: Rods for facing; impact & abrasion resistant; melting point 1320 °C; Wall Colmonoy **DPN: 730**
COLMONOY 1 SPECIAL	1.0% C 13% Cr 3% B steel: Rod for facing; abrasion resistant; melting point 1350 °C; Wall Colmonoy **DPN: 800**
CRD DIAMOND	High carbon 2.2% C 12.5% Cr steel: Origin unknown
CRD DIAMOND RED	1.5% C 12% Cr 0.8% Mo 0.2% V steel: Blanking tools, gauges; C Denton **DPN: 680**
CROMOVAN	1.5% C 12% Cr 1.0% Mo 1.0% V steel: Firth Sterling
CROSS PIPES	1.5% C 11% Cr 2% Co steel: Press tools & forging dies; John Vessey **DPN: 720**
CROSTAR	2.0% C 12.5% Cr 0.1% V steel: Pompey
CRUCIBLE 440 BM	1.0% C 18% Cr 0.6% Mo 0.15% V steel: Crucible Steel Co **DPN: 660**
CRUCIBLE 440 C	1.1% C 17% Cr 0.7% Mo steel: Crucible Steel Co **DPN: 660**
D 1	1.0% C 12.0% Cr 1.0% Mo steel: Designation used by AISI
D 2	1.5% C 12.0% Cr 1.0% Mo steel: Designation used by AISI
D 2	1.5% C 12.0% Cr 0.9% Mo 0.9% V steel: For cold work; Osborn
D 3	2.25% C 12.0% Cr steel: Designation used by AISI
D 3	2.2% C 12.5% Cr 0.25% V steel: S Osborn; cold work steel; hardened & tempered **DPN: 700**
D 4	2.25% C 12.0% Cr 1.0% Mo steel: Designation used by AISI
D 5	1.5% C 12% Cr 1% Mo 0.8% V 3% Co steel: For tools; designation used by AISI
D 6	1.5% C 12% Cr 1.0% Mo steel: Designation used by AISI; obsolete
D 7	2.35% C 120% Cr 1.0% Mo 4.0% V steel: Designation used by AISI
D 20	1.5% C 12% Cr 1.0% Mo steel: John Vessey **DPN: 770**
D 65	2.0% C 13% Cr 1.2% W steel: Fagersta **DPN: 800**
D 165	2.1% C 13% Cr 1.4% W 4.0% V steel: Fagersta **DPN: 740**
DELCROME 50V	2.75% C 27% Cr 0.75% V Fe alloy: Melting range 1375–1400 °C; Delcro Stellite **DPN: 580**
DELCROME 450	0.2% C 12.0% Cr Fe alloy: Delcro Stellite **DPN: 450**
DELCROME C	3.7% C 21% Cr Fe alloy: Casting; Deloro **DPN: 610**
DELCROME R	3.0% C 30% Cr 6.0% Mn Fe alloy: Delcro Stellite **DPN: 580**
DELFOR	2.5% Fe 13.5% Cr 5.5% W 16.5% Co 9% Mo Fe alloy: Casting; Deloro **DPN: 610 UTS: 440 Elon: 1%**
DELFER B	3.2% C 18% Cr 6.0% Co 16% Mo 2.0% V Fe alloy: Melting point 1178 °C; Delcro Stellite **DPN: 820**
DIEHARD HCD	2.1% C 13% Cr 0.5% Mo steel: Firth Brown
DIEHARD STANDARD	1.5% C 12% Cr 0.9% Mo 0.9% V steel: Firth Brown
DOMINATOR	1.9% C 12.5% Cr steel: Schoeller-Bleckmann **DPN: 640**

Symbol	Nominal analysis, supplier, condition and remarks.
DOMINATOR VM	1.55% C 11.5% Cr 0.8% Mo 1% V steel: Schoeller-Bleckmann **DPN: 740**
DOMINATOR Z	1.9% C 11.5% Cr 2% W steel: Schoeller-Bleckmann **DPN: 740**
DOUBLE SIX	1.9% C 0.3% Mn 12.5% Cr 0.8% Mo 0.25% V steel: Punches, dies, etc.; Edgar Allen **DPN: 750**
DRS 16	0.8% C 21% Cr 1.5% Ni steel: Bofors; obsolete
DT	2.2% C 13% Cr steel: Forez
DTva	1.55% C 12% Cr 0.8% Mo 0.8% V 0.4% Co steel: Forez
ECLIPSE 91A	1.5% C 12.0% Cr 1.0% Mo steel: Origin unknown
FASTWORK	1.5% C 12.5% Cr 0.25% V 0.75% Mo tool steel: Balfour Darwin
fmp 336	1.5% C 12.0% Cr 0.9% V 0.8% Mo steel: F Parkin; as tool steel D2
fmp 338	2.05% C 13.0% Cr steel: F Parkin; as tool steel D3
FNS	1.5% C 12% Cr 0.8% Mo 0.85% V steel: Atlas; as AISI type D2
FNS fm	1.5% C 12% Cr 0.8% Mo 0.85% V 0.12% S steel: Atlas; as AISI type D2; free machining
GOST 5632/9 KH18	9.95% C 18.0% Cr steel: Russian Standard
GOST 5632/9 KH18	0.95% C 0.7% Mn (max) 18% Ci 0.03% P (max) 0.025% S (max) steel: Plate; Russian Standard; annealed **UTS: 700**
GSN	1.5% C 11.5% Cr 0.8% Mo 0.25% V steel: Broaches, ring gauges, etc.; S Osborn **DPN: 740**
GSN	2.1% C 13.0% Cr steel: Latrobe; die steel; type D3 **DPN: 830**
GSN + Mo	2.2% C 12.0% Cr 0.5% V 0.8% Mo steel: Latrobe
H 5	1.5% C 12% Cr 1% Mo 1% V steel: Jessop
H 9	2.1% C 13% Cr steel: Jessop
H 9 (alloy C)	2.1% C 12.7% Cr steel: Jessop; carbon & alloy tool steels
H 42	1.6% C 13% Cr 0.7% Mo 0.3% V steel: Jessop **DPN: 770**
H 142	1.6% C 13% Cr 0.7% Mo 0.3% V with S steel: For tools; Jessop **DPN: 730**
HAYNES 90	2.7% C 27% Cr steel: Abrasion & corrosion resistant; S. Osborn **DPN: 740**
HECLA 125	2.25% C 12.0% Cr steel: Origin unknown
HECLA 159	1.5% C 12.0% Cr 1.0% Mo steel: Origin unknown
HFC	2.1% C 12% Cr steel: Press tools, extruding dies, etc.; Marsh Bros
ICS	2.0% C 13.0% Cr steel: Inman
ICW	2.25% C 12.0% Cr steel: Origin unknown
ICW PLUS D3	1.5% C 12.0% Cr 1.0% Mo steel: For dies; Inman
IR	2.2% C 12.5% V steel: Obsolete; S Osborn; replaced by D3
IU	1.9% C 12.5% Cr 0.75% Mo 0.3% V steel: S Osborn; cold work steel; hardened & tempered **DPN: 700**
JJ	1.4% C 14% Cr 3.25% Co 0.5% Ni 0.7% Mo steel: Shear blades, etc.; S. Osborn **DPN: 740**
KE 200	1.4% C 13% Cr 0.6% Mo 3.5% Co steel: Press tools; Kayser Ellison
KE 954	1.4% C 13% Cr 0.6% Mo 3.5% Co steel: For valves; Kayser Ellison
KE 961	1.5% C 12.5% Cr 0.5% W steel: For plastic dies; Kayser Ellison **DPN: 660**

Symbol	Nominal analysis, supplier, condition and remarks.
KE 970	2.0% C 13.5% Cr 0.5% W steel: Press tools & dies; Kayser Ellison **DPN: 872**
KE 1036	0.9% C 13% Cr steel: Kayser Ellison
KEA 180	1.55% C 12% Cr 0.9% Mo 0.2% V steel: Hot shear blades; Kayser Ellison **DPN: 750**
KEA 180/476	1.55% C 12.0% Cr 0.85% Mo 0.3% V steel: Sanderson Kayser
KEA 207	1.05% C 17.0% Cr 0.6% Mo steel: Kayser Ellison
KH 12	12.2% C 12.2% Cr steel: French Standard designation
KH 12F1	11.7% C 0.8% V steel: Russian Standard designation
KH 12F1	1.32% C 11.7% Cr 0.8% V steel: Russian Standard designation
KH 12M	1.55% C 11.7% Cr 0.5% Mo 0.22% V steel: Russian Standard designation
KKK	2.0% C 12.5% Cr 0.8% Ni 0.6% Mo steel: Spencer; press tools; drawing dies **DPN: 640**
KMV	1.5% C 12.0% Cr 1.0% Mo steel: Origin unknown
LESCALLOY BG 42	1.15% C 14.5% Cr 4.0% Mo 1.2% V steel: Latrobe; for bearings
LESCALLOY BG 66	0.85% C 4.15% Cr 6.3% W 1.85% V 5.0% Mo steel: Latrobe
LESCO BC42 VAC ARC	1.15% C 14.5% Cr 4.0% Mo 1.2% V steel: Latrobe Steel; for bearings **DPN: 400**
LONGWAIR	2.2% C 13% Cr steel: For tools; BSC (Clyde Alloy), previously Nonwear **DPN: 550**
LT 25	2.0% C 12% Cr steel: Low Moor **DPN: 640**
LT 27	2.3% C 11.75% Cr 0.25% V steel: Low Moor
LT 33	1.5% C 12% Cr 0.75% Mo 0.25% V steel: Low Moor **DPN: 720**
LT 35	1.9% C 12.5% Cr 0.75% Mo 0.25% V steel: Low Moor
MAFERITE 30C	1.2% C 30% Cr steel: Pompey **DPN: 350**
MARSH HFC	2.1% C 12% Cr steel: Press tools, extruding dies, etc.; Marsh Bros
MCHD	2.0% C 13% Cr steel: Plastic moulds; John Vessey
MEL-TROL 610FM	1.5% C 12.0% Cr 0.8% Mo 0.9% V steel with sulphides: Carpenter for AISI type D2
MEL-TROL HAMPDEN	2.1% C 12.5% Cr 0.5% Ni steel: Carpenter for type D3
MONARCH HCR	1.5% C 12.0% Cr 1.0% Mo steel: For tools; origin unknown
NEOR	2.0% C 13% Cr steel: Press tools, lathe centres, etc.; Darwin **DPN: 700**
NN	2.25% C 12% Cr 0.8% Mo 0.25% V steel: Atlas; as AISI type D4
NONWAIR	2.2% C 13% Cr steel: Clyde Alloy **DPN: 770**
NORESCO DOMINATOR	1.9% C 12.5% Cr steel: Schoeller-Bleckmann **DPN: 640**
NORESCO DOMINATOR VM	1.55% C 11.5% Cr 0.8% Mo 1% V steel: Schoeller-Bleckmann **DPN: 720**
NORESCO DOMINATOR Z	1.9% C 11.5% Cr 2% W steel: Schoeller-Bleckmann **DPN: 740**
NRM	1.7% C 11.5% Cr 0.8% Mo 0.25% V steel: Press tools; Jonas **DPN: 740**

Symbol	Nominal analysis, supplier, condition and remarks.
NRW	2.2% C 12.5% Cr steel: Press tools; Jonas **DPN: 770**
OHIO DIE	1.5% C 12% Cr 0.8% Mo 0.8% V steel: Vanadium Alloys **DPN: 750**
OLYMPIC FM	1.5% C 12.0% Cr 0.75% Mo 1.0% V steel: Latrobe; die steel; type D2 **DPN: 800**
QQ S 763/10	1.0% C 16% Cr 0.5% Ni 0.5% Mo 1.5% Cu steel: US Federal; annealed **DPN: 240**
R 5	1.05% C 17% Cr steel: Jessop
R 680G	2.5% C 27.0% Cr steel: Casting; scaling temperature 1070 °C; Fagersta
R 770	0.95% C 13.5% Cr steel: Fagersta **DPN: 660**
REXWELD 54	4.0% C 30% Cr 4% Ni Fe alloy: For hard facing; Crucible Steel Co **DPN: 540**
ROP 57	1.55% C 12% Cr 0.8% Mo 0.9% V steel: Bofors **DPN: 760**
RT 60	2.0% C 13% Cr 1.2% W steel: Bofors **DPN: 760**
RVM	1.55% C 11.5% Cr 0.7% Mo 0.25% V steel: Clyde Alloy **DPN: 750**
SAE D2	1.5% C 12% Cr 1% Mo 0.8% V 0.6% Co steel
SAE D3	2.2% C 12% Cr 0.8% Mo 0.7% W 0.8% V steel
SAE D5	1.5% C 12% Cr 1% Mo 0.8% V 3% Co steel
SAE D7	2.2% C 12% Cr 1% Mo 4% V steel
SAE 51440 C	1.1% C 17% Cr 0.7% Mo steel: Mechanical properties not specified
SAE 51440 F	1.1% C 17% Cr 0.07% S or Se 0.7% Mo steel: Mechanical properties not specified
SAE 51440 F Se	1.1% C 17% Cr 0.15% Se steel: Mechanical properties not specified; free machining
SC 13	2.0% C 13% Cr 0.2% V 0.3% Si steel: Blanking dies, hobs, etc.; Balfour **DPN: 740**
SC 25	1.6% C 13% Cr 0.2% V 0.7% Mo steel: Blanking dies, etc.; Balfour **DPN: 620**
SC 26	1.8% C 12% Cr 0.3% Si steel: Mandrels, forming punches, etc.; Balfour **DPN: 630**
SC 38	1.5% C 12% Cr 1.1% V 0.7% Mo steel: Blanking dies, shear blades, etc.; Balfour **DPN: 660**
SIMONDS 168	0.9% C 12% Cr 0.75% Mo steel: Simonds; AISI type D1
SIMONDS 12225	2.1% C 13% Cr steel: Simonds; AISI type D3
SIS 14-2312	2.0% C 0.75% Mn 13.0% Cr 12.5% W steel: For tools; cold water steel; Swedish Standard **DPN: 280**
SIS 2183	1.2% C 12.5% Cr steel: Swedish Standard
SIS 2310	1.5% C 12% Cr 0.8% Mo 0.9% V steel: Swedish Standard; annealed **DPN: 260**

Symbol	Nominal analysis, supplier, condition and remarks.
SIS 2312	2.0% C 12.5% Cr 1.2% W steel: Swedish Standard; annealed **DPN: 280**
SKD 1	2.1% C 13.5% Cr 0.3% V (max) steel: Japanese Standard designation
SKD 2	2.0% C 13.5% Cr 0.5% Ni (max) 3.0% W steel: Japanese Standard designation
SKD 11	1.55% C 12.0% Cr 1.0% Mo 0.35% V steel: Japanese Standard designation
SL 440 B	0.85% C 17% Cr 0.7% Mo steel: Swift Levick; hardened & tempered **DPN: 600**
SL 440 C	1.1% C 17% Cr 0.7% Mo steel: Swift Levick; hardened & tempered **DPN: 600**
SPEAR D12	2.0% C 13% Cr steel: Spear & Jackson
SL 951	High C High Cr Ni W Co steel: Casting; Swift Levick **DPN: 400**
SL 958	High C High Cr Ni W Co steel: Casting; Swift Levick **DPN: 400**
SOLAR ECLIPSE	2.25% C 12.0% Cr steel: Origin unknown
SONCOLD	1.5% C 12.0% Cr 1.0% Mo steel: Designation unknown
SPEAR D 13	1.9% C 13.5% Cr 1.3% Mo steel: Spear & Jackson
SPEAR D 14	1.3% C 13% Cr 1.5% Mo 2.7% Co steel: Spear & Jackson
SPEAR D16	1.5% C 12.0% Cr 1.0% Mo steel: Origin unknown
SPECIAL K	2.25% C 12.0% Cr steel: Origin unknown
STA 5 V24A	0.8% C 19.5% Cr 1.4% Ni steel: Replaced by BS alloy En 59
SUPERDIE	2.13% C 11.6% Cr 0.8% W steel: Columbia; AISI type D3
SUH 4	0.8% C 19.7% Cr 1.4% Ni steel: Heat resisting; Japanese Standard designation
SUPER C12	1.5% C 12.0% Cr 1.0% Mo steel: Origin unknown
SUPER HICRO D5	1.5% C 12.0% Cr 1.0% Mo 3.0% Co steel: For dies; Inman
SUPERIOR 1	2.15% C 12% Cr 0.6% V steel: Braeburn; AISI type D3
SUPERIOR 2	1.4% C 12.5% Cr 0.5% Ni 0.8% Mo 3.5% Co steel: Braeburn; AISI type D5
SUPERIOR 3	1.5% C 12% Cr 0.8% Mo 0.8% V steel: Braeburn; AISI type D2
SUPERWEAR	2.0% C 13% Cr 0.5% Co steel: John Vessey **DPN: 750**
SVERKER 3	2.05% C 12.7% Cr 1.25% W steel: Uddelholm for type D6
SVERKER 21	1.55% C 11.8% Cr 0.8% Mo 0.9% V steel: Uddelholm for type D2
T 30402	UNS designation for D2 type tool steel
T 30403	UNS designation for D3 type tool steel
T 30404	UNS designation for D4 type tool steel
T 30405	UNS designation for D5 type tool steel
T 30407	UNS designation for D7 type tool steel
TRIPLE DIE	2.2% C 12% Cr 1.0% Mo 1.0% V steel: Firth Sterling
UNI 3992 X80 CSN20	0.8% C 19.5% Cr steel: For valves
UHB STAINLESS AEB	0.95% C 13.5% Cr steel: Uddelholm; hardened & tempered **DPN: 390 UTS: 1250 Elon: 3% Proof: 950**
VASCO 7152	1.7% C 17.4% Cr steel: Vanadium Alloys **DPN: 660**
VESUVIUS	1.0% C 29.0% Cr 0.3% Ni steel: Firth Vickers
VITAL	2.0% C 13% Cr steel: Blanking dies, thread rolls; Hall & Pickles **DPN: 740**

Note. The following abbreviations and units are used in the tables:

DPN	Hardness, diamond pyramid number
UTS	Ultimate tensile strength, N/mm^2
Elon	Elongation, %
Proof	0.1 % proof strength, N/mm^2

1 N/mm^2=0.1 hbar=0.102 kgf/mm^2=0.06475 tonf/in^2=145.04 lbf/in^2

Symbol	Nominal analysis, supplier, condition and remarks.
VITAL X	1.5% C 12% Cr 0.8% Mo 0.25% V steel: Press tools, punches, etc.; Hall & Pickles **DPN: 680**
VS 13 C	2.0% C 13% Cr 0.25% V steel: Rolling dies, gauges, etc.; Vulcan **DPN: 720**
WALLEX	2.5% C 12% Cr 1.7% Mn 6% Mo steel: Rod for facing; abrasion, impact, corrosion resistant; melting point 1180 °C; Wall Colmonoy **DPN: 720**
WCD	1.3% C 13% Cr 0.5% W steel: John Vessey **DPN: 800**
WCR 100	0.75% C 14% Cr 20% W 3.5% B Fe alloy: For hard facing; melting point 1200 °C; Wall Colmonoy
WLM 2014	2.0% C 14% Cr steel: Press tools; W Marrison **DPN: 740**
WLM 2014 SPECIAL	2.0% C 14% Cr 1.5% W steel: Press tools; W Marrison **DPN: 740**
WLM 3141	1.4% C 14% Cr 0.7% Mo 3% Co steel: Plastic moulds & dies; W Marrison **DPN: 680**
WPS	2.3% C 0.4% Mn 13% Cr steel: Jessop **DPN: 870**
X 90 CrMoV18	0.9% C 18.0% Cr 1.1% Mo 0.10% V steel: Designation used by German Standard
X 165 Cr Mo V 12	1.65% C 12.0% Cr 0.6% Mo 0.1% V 0.5% W steel: German Standard designation

Symbol	Nominal analysis, supplier, condition and remarks.
X 210 Cr 12	2.1% C 11.5% Cr steel: Designation used by German Standard
X 210 Cr 12	2.0% C 12.0% Cr steel: German Standard designation
X 210 Cr W 12	2.0% C 12.0% Cr 0.7% W steel: German Standard designation
Z 160CDV12	1.55% C 12.0% Cr 0.8% Mo 0.3% V steel: French Standard
Z 200C12	2.1% C 12.5% Cr 0.2% V steel: French Standard
Z 160CDV12	1.6% C 12.0% Cr 0.8% Mo 0.4% V steel: French Standard designation
Z 180CKD1203	1.8% C 12.0% Cr 0.8% Mo 0.5% V 3.0% Co steel: French Standard designation
Z 200C12	2.0% C 12.0% Cr steel: French Standard designation
Z 200CD12	2.0% C 12.0% Cr 0.8% Mo 0.2% V steel: French Standard designation
Z 230CDV12-04	2.3% C 12.0% Cr 1.0% Mo 4.0% V steel: French Standard designation

Note. The following abbreviations and units are used in the tables:

DPN	Hardness, diamond pyramid number
UTS	Ultimate tensile strength, N/mm^2
Elon	Elongation, %
Proof	0.1 % proof strength, N/mm^2

$1 \ N/mm^2 = 0.1 \ hbar = 0.102 \ kgf/mm^2 = 0.06475 \ tonf/in^2 = 145.04 \ lbf/in^2$

44N1 Steel – chromium–nickel – austenitic wrought and cast
With niobium, titanium, molybdenum and tungsten

Specific gravity	8.05	
Density	8050 kg/m³	(0.29 lb/in.³)
Solidus/liquidus	1230–1280 °C	
Thermal conductivity	16.3 W/m °C	(3.9 cal/m s °C)
Coefficient of linear expansion	$18 \times 10^{-6}/$ °C	
Electrical conductivity	3% IACS (copper 100%)	
Specific resistance	700 microhm mm	
Young's modulus of elasticity	$207 \times 10^9 \ N/m^2$	$(30 \times 10^6 \ lbf/in.^2)$
Impact	cold worked 20 J	(15 ft/lb) (Izod)
	annealed 108–136 J	(80–100 ft/lb)
Fatigue strength	–	
Hot strength		

Temperature °C	Tensile strength N/mm^2	Elongation %
650	580	40
700	450	45
800	330	45
900	170	58

The above properties are typical of the following group, and may not apply exactly to any one specification. It is possible that with certain specifications some of the values may not be applicable.

General metallurgical characteristics

When 18% chromium and 8% nickel are added to steel the effect is so to depress the critical change points that the austenitic or gamma phase is present at room temperature. This is a single phase solid solution and like all single phase materials has excellent corrosion resistance.

In this group are all the truly stainless as distinct from corrosion resistant steels. The fact that they are non-magnetic is a rapid but not very precise means of identification, as cold work can cause the steels to become slightly magnetic.

These materials being single phase are incapable of being hardened by thermal treatment, but can have their strength improved by cold working, when they can become magnetic as some austenite is transformed to martensite. The carbon content of these steels is important in that the higher the carbon, the harder they can be made by cold work.

The presence of carbon however, can cause the appearance of a second phase of martensite or chromium carbide, either of which can considerably reduce the corrosion resistance. The presence of martensite after cold working can generally be accepted and design allowance made for the reduction in corrosion resistance. The chromium carbides however, contribute little to the mechanical properties and as the corrosion resistance is drastically reduced at grain boundaries, it can have a serious effect on the fatigue strength as well as corrosion resistance. This is known as 'weld decay'.

By the addition of carbide stabilizing elements such as niobium or titanium which combine with carbon in preference to chromium and remain in solution, the problem of carbide precipitation at grain boundaries is eliminated. These steels are said to be 'weld stabilized' and are also suitable for use at high temperatures.

The low carbon steels generally have better ductility and corrosion resistance with lower tensile properties, but are all weldable as there are very few carbides present to be precipitated.

Many of the steels listed have considerably more chromium and nickel than the 18/8 steels. This generally improves their hot oxidation resistance and ability to withstand temperature shock without very much improving the creep strength.

Other elements are added to these alloys for specific purposes as follows:

Manganese. This is almost always present up to about 2.5% as a de-oxidizing and de-sulphurizing agent. No attempt is made to list this element in the alloy descriptions of this group, and when present in greater amounts the alloys are listed and discussed in Section 44P.

Molybdenum. This considerably enhances the acid resistance without affecting the weldability. It also confers some strength and can actually be used to prevent carbide precipitation, thus making the higher carbon versions weldable. Because of the high cost and scarcity of molybdenum its use is confined to improving the acid resistance at present.

Tungsten. This stabilizes the carbides, preventing their precipitation during welding or heat cycling. Tungsten also increases the strength and acid resistance. It is seldom found by itself, being generally associated with molybdenum.

Titanium. This forms very stable carbides which are taken into solution in the austenitic matrix and are not readily precipitated thus preventing the formation of a second phase. Carbon combines with titanium in preference to iron or chromium thus preventing the formation of these carbides.

Sufficient titanium must always be present to combine with all the carbon and most specifications are written calling for a minimum of eight times as much titanium as carbon. These are known as titanium stabilized weldable austenitic steels. There is some indication that these can present some difficulty when brazing or melt spraying is involved.

Niobium (columbium). This is the only element with two names and different symbols. Niobium is the common European form, and throughout this book use of this and the symbol Nb has been standardized. Columbium (Cb) is the American version.

The element acts in an analogous manner to titanium, forming stable carbides which are not readily precipitated from solid solutions. Tantalum and niobium are always found together and are difficult to separate. As tantalum acts in an identical manner to niobium the sum total of these elements is generally accepted. This should be ten times the carbon content to prevent carbide precipitation.

Tellurium, lead and sulphur are sometimes present to improve the machinability but there is some reduction in the corrosion resistance and mechanical properties.

Thermal treatment

These steels can all be cold worked, the ability to be hardened being in proportion to the carbon and other alloy content. With very severe cold work the alloys become magnetic owing to the formation of some martensite. It is not possible to harden any of the steels listed by any form of thermal treatment, but if subjected to very low temperatures it is possible to convert some or all the austenite to martensite. This is unstable and has no practical significance except to note that care should be taken if using the steels listed for cryogenic purposes.

Unless required hardened by cold work these steels should be annealed. This necessitates heating to a temperature of 1040–1100 °C for about one hour per 25 mm of section. The rate of cooling is not generally important but the unstabilized steels should be water quenched to keep the carbides in solution.

These steels can be adversely affected by carbon and sulphur pick-up, thus some care is necessary if gas or oil heating is used to ensure an oxidizing atmosphere at all times and sulphur free fuel. Electric furnaces probably give the best results.

Controlled atmospheres are seldom necessary as all the steels listed have excellent hot oxidation resistance. When required for such purposes as annealing bright drawn strip, carbon bearing atmospheres must not be used. Normally cracked ammonia, giving a hydrogen and nitrogen atmosphere, is used, but dry hydrogen or an inert gas gives equally good results. Vacuum furnaces also give excellent results but are expensive.

There are now available fusible frit paints which are applied prior to processing and readily removed afterwards. Provided great care is taken to ensure complete and proper coverage, these paints give excellent results and are economical when oxide free parts which are too delicate to blast are essential and acid pickling or sodium hydride is not available.

Scale removal when required, will be by grit blasting, acid pickling or sodium hydride treatment.

Grit blasting can cause distortion on sheet metal parts and acid pickling uses unpleasant hydrofluoric acid solutions. Sodium hydride gives the best results but requires very special equipment and techniques.

Welding and brazing

Pre-treatment. It is essential that there is no trace of oil, grease or other organic matter to prevent carbon pick-up in the weld. Parts must be in the annealed condition and no cropping or machining of the joint area should be permitted after annealing. This should be cleaned by de-scaling, wire brushing or machining with sharp tools to prevent work hardening, which can cause cracking of the weld.

Welding and brazing. Welding should always be with the inert gas shielded electric arc process for best results as no fluxes are involved, thus reducing the danger of flux or oxide trapping. Gas or electric arc welding is possible provided fluxes are used but requires more care and skill.

Resistance welding is readily accomplished provided the contact area is clean and free of oxide, and as the electrical resistance is relatively high, good quality welds can be easily produced.

Brazing requires special active fluxes in order to wet the surface. High temperature brazing is often necessary using salt bath, hydrogen or vacuum brazing techniques, but for most purposes torch or induction heating is satisfactory.

Post-treatment. Provided the carbon content is below 0.08%, or stabilizing elements such as niobium or titanium are present, no post-weld thermal treatment is required. With the unstabilized materials it is essential that an anneal at 1040–1100 °C is carried out to re-dissolve the precipitated carbides. Without this, the defect known as 'weld decay' will occur. This takes the form of severe local corrosion around the weld area where the carbides have precipitated. Where fluxes have been used these must be completely removed as they are corrosive and generally hygroscopic.

Flaw detection methods

Crack test. Penetrant oil chalk test methods.

X-ray. High duty welds are generally 100% X-ray examined. All welding and brazing should have their quality controlled by routine percentage X-ray checking.

Castings may require X-ray when used for pressure vessels or other high duty purposes.

Ultra-sonic test. Forgings or castings with costly machining operations should be ultra-sonic inspected at an early stage. This shows the presence of defects below the surface which can be dangerous on the finished part, thus saving the cost of machining. It can also identify certain types of welding defects.

Chemical etch. Not required. There is available a laboratory etch which indicates the susceptibility of the steel to weld decay, but this scraps the sample.

Corrosion protection

Temporary. None required.

Permanent. None required. These are expensive materials developed purely for their excellent corrosion resistance which is superior to all paint and electroplated coatings.

Machinability

These materials are all relatively soft except when cold worked and have a tendency to tear, thus giving a poor surface finish. They also harden readily by cold work, thus blunt tools, very low feeds causing rubbing, and interrupted cuts should be avoided.

High feeds, low speeds using tipped tools and robust powerful machine tools are recommended. Grinding is difficult as the stones become loaded and glaze the surface.

Interstage annealing 1040–1100 °C may be necessary where considerable machining or forming is carried out, and this is recommended after heavy machining if corrosion resistance is of paramount importance.

Uses

Chemical plant, marine use, food and drink manufacturing and distribution, including kitchenware for canteen and domestic use, transport of chemicals, laboratory equipment, household ornaments.

General engineering low strength, very high atmospheric corrosion resistance purposes. The acid resistance varieties are used for chemical engineering equipment of all types.

The higher alloyed materials are used where high temperature oxidation resistance is necessary.

Note. The following abbreviations and units are used in the tables:

DPN	Hardness, diamond pyramid number
UTS	Ultimate tensile strength, N/mm^2
Elon	Elongation, %
Proof	0.1 % proof strength, N/mm^2

$1 N/mm^2 = 0.1 hbar = 0.102 kgf/mm^2 = 0.06475 tonf/in^2 = 145.04 lbf/in^2$

Symbol	Nominal analysis, supplier, condition and remarks.
0.4310	0.15% C (max) 17% Cr 7.5% Ni steel: German Standard
0.4401	0.07% C (max) 17% Cr 11% Ni 2.2% Mo steel: German Standard
1.4300	0.12% C (max) 18% Cr 9% Ni steel: German Standard
1.4301	0.07% C (max) 18% Cr 10% Ni steel: German Standard

Symbol	Nominal analysis, supplier, condition and remarks.
1.4302	0.06% C (max) 19% Cr 9.5% Ni steel: Forging; German Standard
1.4303	0.07% C (max) 18.5% Cr 11.0% Ni steel: German Standard
1.4304	0.15% C (max) 18% Cr 9% Ni steel: German Standard
1.4305	0.15% C (max) 18% Cr 9% Ni 0.15% S steel: German Standard; free machining
1.4306	0.03% C 18.5% Cr 11.2% Ni steel: German Standard
1.4306	0.03% C (max) 19.0% Cr 9.5% Ni steel: German Standard
1.4307	0.1% C (max) 12% Cr 13% Ni steel: German Standard
1.4310	0.15% C (max) 17% Cr 7.5% Ni steel: German Standard
1.4311	0.03% C (max) 18.0% Cr 10.2% Ni 0.18% N steel: German Standard
1.4314	0.07% C (max) 18% Cr 10% Ni steel: German Standard
1.4324	0.15% C (max) 17% Cr 7.5% Ni steel: German Standard
1.4401	0.07% C (max) 17.5% Cr 11.5% Ni 2.3% Mo steel: German Standard
1.4402	0.06% C (max) 19% Cr 10.0% Ni 2.2% Mo steel: German Standard
1.4404	0.03% C (max) 17.0% Cr 12.0% Ni 2.2% Mo steel: German Standard
1.4404	0.03% C (max) 17.5% Cr 12.5% Ni 2.2% Mo steel: German Standard
1.4406	0.03% C (max) 17.5% Cr 12.0% Ni 2.2% Mo 0.18% N steel: German Standard
1.4429	0.03% C (max) 17.5% Cr 13.0% Ni 2.7% Mo 0.18% N steel: German Standard
1.4435	0.03% C (max) 17.5% Cr 12.0% Ni 2.7% Mo steel: German Werkstoff number
1.4436	0.07% C (max) 17.5% Cr 13.0% Ni 2.7% Mo steel: German Standard
1.4438	0.03% C (max) 19.0% Cr 13.0% Ni 3.5% Mo steel: German Werkstoff number
1.4438	0.03% C (max) 18.0% Cr 16.0% Ni 3.5% Mo steel: German Standard
1.4447	0.06% C 18.0% Cr 13.5% Ni 4.7% Mo steel: German Standard
1.4449	0.07% C (max) 17.0% Cr 13.5% Ni 4.5% Mo steel: German Standard
1.4504	0.09% C (max) 17% Cr 7% Ni 1.0% Al steel: German Standard
1.4505	0.07% C (max) 17.5% Cr 20% Ni 2.2% Mo 2.0% Cu 0.7% Nb steel: German Standard
1.4507	0.08% C (max) 18.5% Cr 21.0% Ni 2.2% Mo 2.0% Cu 0.9% Nb steel: German Standard
1.4514	0.9% C 15% Cr 7% Ni 2.5% Mo 1.0% Al steel: German Standard
1.4541	0.1% C (max) 18% Cr 10% Ni 0.5% Ti steel: Forging; German Standard
1.4550	0.1% C (max) 18% Cr 10% Ni 1.0% Nb steel: Forging; German Standard

Symbol	Nominal analysis, supplier, condition and remarks.
1.4551	0.1% C (max) 19% Cr 9% Ni 1.5% Nb steel: Forging; German Standard
1.4554	0.1% C (max) 19% Cr 9% Ni 1.5% Nb steel: Welding rod; German Standard
1.4568	0.09% C (max) 17.0% Cr 7.2% Ni 1.1% Al steel: German Standard
1.4570	0.1% C (max) 17.5% Cr 11.5% Ni 1.4% Mo steel: German Standard
1.4571	0.1% C (max) 17.5% Cr 11.5% Ni 2.2% Mo 0.6% Ti steel: German Standard
1.4573	0.1% C (max) 17.5% Cr 13.0% Ni 2.7% Mo 0.6% Ti steel: German Standard
1.4574	0.09% C (max) 15% Cr 7.0% Ni 2.5% Mo 1.0% Al steel: German Standard
1.4577	0.06% C (max) 25.0% Cr 25.0% Ni 2.2% Mo 0.4% Ti steel: German Standard
1.4580	0.1% C (max) 17.5% Cr 11.5% Ni 2.3% Mo 1.0% Nb steel: German Standard
1.4581	0.1% C (max) 19.0% Cr 10.5% Ni 2.2% Mo 1.3% Nb steel: German Standard
1.4583	0.1% C (max) 17.5% Cr 13.0% Ni 2.7% Mo 1.0% Nb steel: German Standard
1.4584	0.1% C (max) 19.0% Cr 11.5% Ni 2.7% Mo 1.3% Nb steel: German Standard
1.4587	0.1% C (max) 26.0% Cr 25.0% Ni 2.2% Mo 1.3% Nb steel: German Standard
1.4589	0.07% C (max) 18.0% Cr 13.5% Ni 4.7% Mo 1.0% Nb steel: German Standard
1.4828	0.2% C (max) 20% Cr 12% Ni steel: Forging; German Standard
1.4829	0.15% C (max) 22.0% Cr 11.0% Ni steel: German Standard
1.4833	0.08% C (max) 23.0% Cr 13.5% Ni steel: German Werkstoff number
1.4841	0.2% C (max) 2.0% Si 25.0% Cr 20.0% Ni steel: German Standard
1.4842	0.15% C (max) 25.0% Cr 20% Ni steel: German Standard
1.4843	0.2% C (max) 24.0% Cr 19.0% Ni steel: German Standard
1.4844	0.2% C (max) 2.0% Si 25.0% Cr 20.0% Ni steel: German Standard
1.4845	0.15% C (max) 25.0% Cr 21.0% Ni steel: German Standard
1.4854	0.15% C (max) 25.0% Cr 20% Ni steel: German Standard
1.4860	0.2% C (max) 21.0% Cr 30.0% Ni Fe steel: German Standard
1.4878	0.15% C (max) 18% Cr 10% Ni 0.6% Ti steel: Forging; German Standard
1.4943	0.45% C 2.5% Si 18.0% Cr 9.0% Ni 1.0% W steel: German Standard
1.4944	0.08% C (max) 15% Cr 25% Ni 1.2% Mo 0.3% V 2.1% Ti steel: German Standard
1.4954	0.08% C (max) 14% Cr 25% Ni 0.4% V 2.0% Ti steel: German Standard
1.4984	0.1% C (max) 16.5% Cr 16.5% Ni 1.7% Mo 1.0% Nb + Ta steel: German Standard
1.4986	0.07% C 16.8% Cr 16.7% Ni 1.8% Mo 0.8% Nb 0.07% B steel: German Standard
1.6903	0.1% C (max) 18% Cr 10% Ni 0.5% Ti steel: For low temperature use; German Standard
1.6905	0.1% C (max) 18% Cr 10% Ni 1.0% Nb steel: For low temperature use; German Standard
1KH25N25TR	0.1% C 24.5% Cr 25.5% Ni 1.3% Ti 0.01% B (max) steel: Russian Standard designation
1R2	0.16% C (max) 18% Cr 9.0% Ni steel: Sandvik
2R	0.2% C 18% Cr 10% Ni 3.0% Mo steel: Sandvik
2R1	0.16% C (max) 13% Cr 13% Ni steel: Sandvik

Note. The following abbreviations and units are used in the tables:

DPN	Hardness, diamond pyramid number
UTS	Ultimate tensile strength, N/mm^2
Elon	Elongation, %
Proof	0.1 % proof strength, N/mm^2

$1 N/mm^2 = 0.1 hbar = 0.102 kgf/mm^2 = 0.06475 tonf/in^2 = 145.04 lbf/in^2$

Symbol	Nominal analysis, supplier, condition and remarks.
2R2A	0.16% C 18% Cr 9% Ni 0.2% steel: Sandvik; free machining
2R16	0.02% C (max) 19.4% Cr 10.5% Ni steel: Sandvik
3KH19N9MVBT	0.32% C 19.0% Cr 9.0% Ni 1.25% Mo 0.35% Ti 0.35% Nb 1.25% W steel: Russian Standard designation
3R12	0.03% C (max) 18.5% Cr 11% Ni steel: Sandvik
3R16	0.03% C (max) 19.4% Cr 10.5% Ni steel: Sandvik; annealed
	UTS: 700
3R17	0.03% C (max) 20.6% Cr 9.7% Ni steel: Sandvik
3R54	0.03% C (max) 18.5% Cr 10.5% Ni 1.5% Mo steel: Sandvik
3R60	0.03% C (max) 17% Cr 13.6% Ni 2.8% Mo steel: Sandvik
3R61	0.03% C (max) 18.5% Cr 10.5% Ni 2.5% Mo steel: Sandvik
3R62	0.03% C 19.5% Cr 13% Ni 2.3% Mo steel: Sandvik
3R63	0.03% C (max) 17.8% Cr 12.2% Ni 2.8% Mo steel: Sandvik
3R64	0.03% C 18.5% Cr 14.5% Ni 3.5% Mo steel: Sandvik
3R65	0.03% C (max) 18% Cr 14.5% Ni 4.5% Mo steel: Sandvik
6R10	0.06% C (max) 18.5% Cr 9% Ni steel: Sandvik for SIS 2333
6R55	0.06% C (max) 17.5% Cr 11% Ni 1.5% Mo steel: Sandvik for SIS 2341
6R60	0.06% C (max) 17.5% Cr 13.5% Ni 2.8% Mo steel: Sandvik for SIS 2343
6R64	0.06% C (max) 18.5% Cr 14% Ni 3.5% Mo steel: Sandvik
7RE12	0.07% C (max) 25.5% Cr 21% Ni steel: Sandvik
8R30	0.08% C 17.5% Cr 11.5% Ni 0.4% Ti steel: Sandvik for SIS 2337
8R40	0.08% C (max) 17.5% Cr 11.5% Ni 0.8% Nb steel: Sandvik for SIS 2338
8R41	0.08% C 16.5% Cr 13% Ni 0.8% Nb steel: Sandvik
8R42	0.08% C 18.5% Cr 9.5% Ni 0.8% Nb steel: Sandvik
8R70	0.08% C 17.5% Cr 12.4% Ni 2.5% Mo 0.4% Ti steel: Sandvik for SIS 2344
8R80	0.08% C 17.5% Cr 13.5% Ni 2.5% Mo 0.8% Nb steel: Sandvik for SIS 2365
8R81	0.08% C (max) 18.5% Cr 11% Ni 2.3% Mo 0.8% Nb steel: Sandvik
10R13	0.08% C 18% Cr 11% Ni steel: Sandvik
10R52	0.08% C 17.5% Cr 9% Ni 1.5% Mo steel: Sandvik for SIS 2340
10R53	0.08% C 17% Cr 11% Ni 1.5% Mo steel: Sandvik
11R50	0.09% C 17% Cr 8% Ni 0.7% Mo steel: Sandvik
11R51	0.09% C 17% Cr 8% Ni 0.7% Mo steel: Sandvik
12/12 EL	0.05% C 12% Cr 13% Ni steel: Nyby; annealed
	DPN: 160 UTS: 480 Elon: 55% Proof: 180
12R10	0.1% C 18% Cr 9% Ni steel: Sandvik for SIS 2331
12R10 HV	0.1% C 18% Cr 9.0% Ni steel: Wire for springs; Sandvik; vacuum melted
	UTS: 2400
12R11	0.1% C 18% Cr 7.7% Ni steel: Sandvik for SIS 2331
15/35	0.1% C 15% Cr 35% Ni steel: Eastern Stainless for AISI 330
15.35.2 CB	0.2% C 15.0% Cr 35.0% Ni steel: Electrode; Metrode
15RA10	0.12% C 18% Cr 10% Ni 0.5% Mo 0.2% S steel: Free cutting; Sandvik for SIS 2346

Symbol	Nominal analysis, supplier, condition and remarks.
15RE10	0.12% C 23% Cr 22% Ni steel: Sandvik for SIS 2361
15RE11	0.12% C 25.5% Cr 21% Ni steel: Sandvik
15RE12	0.12% C 25.5% Cr 21% Ni steel: Sandvik
16/13 L Nb	0.07% C 16% Cr 13% Ni 0.7% Nb steel: Nyby; annealed
	DPN: 190 UTS: 530 Elon: 35% Proof: 210
16/14 L Mo Nb V	0.07% C 16% Cr 16% Ni 1.8% Mo 0.7% V 0.7% Nb steel: Nyby; annealed
	DPN: 190 UTS: 530 Elon: 30% Proof: 220
16/16 L Mo Nb	0.07% C 16% Cr 16% Ni 1.8% Mn 0.7% Nb steel: Nyby; annealed
	DPN: 190 UTS: 530 Elon: 35% Proof: 220
16/25/6	0.08% C 16% Cr 25% Ni 6% Mo 0.15% N steel: American proprietary alloy listed in SAE year book
17/7	0.15% C 17% Cr 7% Ni steel: Eastern Stainless; annealed
	UTS: 700 Elon: 55% Proof: 270
17/7	0.15% C 17% Cr 7% Ni steel: Eastern Stainless; ½-hard
	UTS: 920 Elon: 25% Proof: 530
17/7	0.15% C 17% Cr 7% Ni steel: Eastern Stainless; hard
	UTS: 1380 Elon: 8% Proof: 1000
17/7	0.10% C 17.0% Cr 7.0% Ni steel: Nyby; annealed
	DPN: 180 UTS: 480 Elon: 45% Proof: 180
17.8.2 RCF	0.08% C 17.0% Cr 9.0% Ni 2.0% Mo steel: Electrode; Metrode; creep resisting; ferritic content 3–8%
17.12.2.4 MN	0.06% C 17.0% Cr 12.0% Ni 2.0% Mo 4.0% Mn steel: Electrode; Metrode
17/12 UL	0.03% C (max) 17.0% Cr 12.0% Ni steel: Nyby; annealed
	DPN: 160 UTS: 450 Elon: 45% Proof: 180
17/12 ULR	0.03% C (max) 17.0% Cr 12.0% Ni 0.05% Co 0.05% Ni 0.25% Cu steel: Nyby; annealed nuclear reactor quality
	DPN: 160 UTS: 450 Elon: 45% Proof: 180
17/14 Cu Mo	0.12% C 16% Cr 14% Ni 2.5% Mo 0.25% Ti 0.5% Nb 3% Cu steel: Armco
17.17.4 LR	0.04% C 17.0% Cr 17.0% Ni 4.0% Mo steel: Electrode; Metrode
17.38.4 CB	0.4% C 17.0% Cr 38.0% Ni steel: Electrode; Metrode
17.38.4 CR	0.4% C 17.0% Cr 38.0% Ni steel: Electrode; Metrode
18/8	0.15% C 18% Cr 9% Ni steel: Eastern Stainless; annealed
	UTS: 700 Elon: 50% Proof: 300
18/8	0.10% C 18.0% Cr 8.0% Ni steel: Nyby; annealed
	DPN: 180 UTS: 480 Elon: 45% Proof: 180
18/8 EL	0.05% C (max) 18.0% Cr 10.0% Ni steel: Nyby; annealed
	DPN: 180 UTS: 480 Elon: 45% Proof: 180
18/8 EMo	0.05% C 17.0% Cr 10.0% Ni 1.6% Mo steel: Nyby; annealed
	DPN: 180 UTS: 480 Elon: 45% Proof: 200
18/8 FM	0.15% C 18% Cr 9% Ni 0.15% S or Se steel: Eastern Stainless; annealed; free machining
	UTS: 630 Elon: 40% Proof: 270
18/8 L	0.07% C 18.0% Cr 9.0% Ni steel: Nyby; annealed
	DPN: 180 UTS: 480 Elon: 45% Proof: 180
18/8 LNb	0.03% C (max) 18.0% Cr 10.0% Ni 0.3% Nb (max) steel: Nyby; annealed
	DPN: 190 UTS: 480 Elon: 40% Proof: 210
18/8 LNbR	0.03% C (max) 18.0% Cr 10.0% Ni 0.3% Nb (max) steel: Nyby; annealed; nuclear reactor quality
	DPN: 190 UTS: 480 Elon: 40% Proof: 210

Symbol	Nominal analysis, supplier, condition and remarks.
18/8 LT	0.07% C 18.0% Cr 10.0% Ni 0.35% Ti steel: Nyby; annealed
	DPN: 190 UTS: 480 Elon: 40% Proof: 180
18/8 Si	0.15% C 2.5% Si 18% Cr 9% Ni steel: Eastern Stainless; annealed; for AISI 302B
	UTS: 700 Elon: 45% Proof: 330
18/8 UL	0.03% C (max) 19.0% Cr 11.0% Ni steel: Nyby; annealed
	DPN: 180 UTS: 450 Elon: 45% Proof: 180
18/8 ULR	0.03% C (max) 19.0% Cr 11.0% Ni steel: Nyby; annealed; nuclear reactor quality
	DPN: 180 UTS: 450 Elon: 45% Proof: 180
18/8 UMo	0.03% C 17.0% Cr 11.0% Ni 1.6% Mo steel: Nyby; annealed
	DPN: 180 UTS: 450 Elon: 45% Proof: 180
18/10 Cb	0.08% C 18% Cr 11% Ni 1% Nb + Ta steel: Eastern Stainless; annealed
	UTS: 630 Elon: 45% Proof: 320
18/10 EMo	0.05% C 17.0% Cr 11.0% Ni 2.3% Mo steel: Nyby; annealed
	DPN: 180 UTS: 480 Elon: 45% Proof: 210
18/10 LMoNb	0.07% C 17.0% Cr 12.0% Ni 2.3% Mo 0.7% (max) Nb steel: Nyby; annealed
	DPN: 190 UTS: 480 Elon: 40% Proof: 220
18/10 LMoT	0.07% C 17.0% Cr 12.0% Ni 2.3% Mo 0.35% (max) Ti steel: Nyby; annealed
	DPN: 190 UTS: 480 Elon: 40% Proof: 210
18/10 Ti	0.08% C 18% Cr 10.5% Ni 0.5% Ti steel: Eastern Stainless; annealed
	UTS: 610 Elon: 48% Proof: 240
18/10 UMo	0.03% C 17.0% Cr 11.0% Ni 2.3% Mo steel: Nyby; annealed
	DPN: 180 UTS: 480 Elon: 45% Proof: 180
18/11 EL	0.05% C 18% Cr 11% Ni steel: Nyby; annealed
	DPN: 180 UTS: 480 Elon: 45% Proof: 180
18/12	0.12% C 18% Cr 12% Ni steel: Eastern Stainless
	UTS: 570 Elon: 47% Proof: 240
18/12 EMo	0.05% C 17.0% Cr 12.0% Ni 2.8% Mo steel: Nyby; annealed
	DPN: 180 UTS: 480 Elon: 45% Proof: 210
18/12 LMo	0.07% C 17.0% Cr 11.0% Ni 2.8% Mo steel: Nyby; annealed
	DPN: 180 UTS: 480 Elon: 45% Proof: 210
18/12 LMoNB	0.07% C 17.0% Cr 13.0% Ni 2.8% Mo 0.7% (max) Nb steel: Nyby; annealed
	DPN: 190 UTS: 480 Elon: 40% Proof: 220
18/12 LMoT	0.07% C 17.0% Cr 13.0% Ni 2.8% Mo 0.35% (max) Ti steel: Nyby; annealed
	DPN: 190 UTS: 480 Elon: 40% Proof: 210
18/12 Mo	0.08% C 17% Cr 12% Ni 2.5% Mo steel: Eastern stainless; annealed
	UTS: 600 Elon: 49% Proof: 240
18/12 MoL	0.03% C 17% Cr 12% Ni 2.5% Mo steel: Eastern Stainless; annealed
	UTS: 570 Elon: 46% Proof: 240
18/12 UMo	0.03% C (max) 17.0% Cr 12.0% Ni 2.8% Mo steel: Nyby; annealed
	DPN: 180 UTS: 480 Elon: 45% Proof: 200

Note. The following abbreviations and units are used in the tables:

DPN	Hardness, diamond pyramid number
UTS	Ultimate tensile strength, N/mm^2
Elon	Elongation, %
Proof	0.1% proof strength, N/mm^2

1 N/mm^2=0.1 hbar=0.102 kgf/mm^2=0.06475 tonf/in^2=145.04 lbf/in^2

Symbol	Nominal analysis, supplier, condition and remarks.
16/12 UMoR	0.03% C (max) 17.0% Cr 12.0% Ni 2.8% Mo steel: Nyby; annealed; nuclear reactor quality
	DPN: 180 UTS: 480 Elon: 45% Proof: 200
18/13 EMo	0.05% C 17.0% Cr 14.0% Ni 2.8% Mo steel: Nyby; annealed
	DPN: 190 UTS: 480 Elon: 45% Proof: 200
18/14 EMo	0.05% C 19.0% Cr 14.0% Ni 3.5% Mo steel: Nyby; annealed
	DPN: 190 UTS: 480 Elon: 45% Proof: 210
18/14 UMo	0.03% C (max) 19.0% Cr 14.0% Ni 3.5% Mo steel: Nyby; annealed
	DPN: 190 UTS: 480 Elon: 45% Proof: 200
18/15 EMo	0.05% C 17.0% Cr 15.0% Ni 4.5% Mo steel: Nyby; annealed
	DPN: 190 UTS: 480 Elon: 45% Proof: 210
18/15 UMo	0.03% C (max) 17.0% Cr 15.0% Ni 4.5% Mo steel: Nyby; annealed
	DPN: 190 UTS: 480 Elon: 45% Proof: 200
18.18.3 Cu LR	0.04% C 18.0% Cr 18.0% Ni 3.0% Mo 1.0% Cu steel: Electrode; Metrode
18.20.2 Cu Nb	0.09% C 18.0% Cr 20.0% Ni 2.0% Mo 2.0% Cu 1.0% Nb steel: Electrode; Metrode
19/9	0.08% C 19% Cr 10% Ni steel: Eastern Stainless; annealed
	UTS: 630 Elon: 50% Proof: 330
19/9 DL	0.3% C 19% Cr 10% Ni 1.5% Mo 1.5% W 0.2% Ti 0.4% Nb + Ta 0.5% Cu steel: Eastern Stainless; annealed
	UTS: 700 Elon: 30% Proof: 330
19/9 DL	0.3% C 19% Cr 9% Ni 1.25% Mo 1.2% W 0.3% Ti 0.4% Nb steel: Universal Cyclops
19/9 DX	0.3% C 19.2% Cr 9% Ni 1.5% Mo 1.2% W 0.5% Ti steel: Universal Cyclops
19/9 L	0.03% C (max) 19% Cr 10% Ni steel: Eastern Stainless; annealed
	UTS: 570 Elon: 52% Proof: 240
19.9 LB	0.03% C 19.0% Cr 9.5% Ni steel: Electrode; Metrode
19.9 LRMP	0.03% C 19.0% Cr 9.5% Ni steel: Electrode; Metrode
19.9 NbB	0.06% C 19.0% Cr 9.5% Ni 0.8% Nb steel: Electrode; Metrode
19.9 NbLR	0.04% C (max) 19.0% Cr 9.5% Ni 0.7% Nb steel: Electrode; Metrode
19.9 NbR	0.06% C 19.0% Cr 9.5% Ni 0.8% Nb steel: Electrode; Metrode
19.9%R	0.06% C 19.0% Cr 9.5% Ni steel: Electrode; Metrode
19/9 WMo	0.1% C 19% Cr 9% Ni 0.4% Mo 1.3% W 0.4% Ti 0.4% Nb steel: Universal Cyclops
19/9 WX	0.1% C 20% Cr 8.5% Ni 0.5% Mo 1.5% W 1.3% Nb 0.2% Ti steel: American proprietary alloy listed in SAE year book
19.9.1 R	0.03% C 19.0% Cr 9.5% Ni steel: Electrode; Metrode
19.12.3 B	0.06% C 19.0% Cr 11.0% Ni 3.0% Mo steel: Electrode; Metrode
19.12.3 ELRCF	0.03% C (max) 18.0% Cr 14.0% Ni 3.0% Mo steel: Electrode; Metrode; ferrite controlled
19.12.3 LR	0.03% C (max) 19.0% Cr 11.0% Ni 3.0% Mo steel: Electrode; Metrode; ferrite content 7–10%
19.12.3 LRCF	0.03% C (max) 18.0% Cr 14.0% Ni 3.0% Mo steel: Electrode; Metrode; ferrite content 1.5% (max)
19.12.3 LRCF2	0.03% C (max) 18.0% Cr 14.0% Ni 3.0% Mo steel: Electrode; Metrode; ferrite content 2% (max)
19.12.3 LRNF	0.03% C (max) 17.0% Cr 15.0% Ni 3.0% Mo steel: Electrode; Metrode; ferrite content 0.5% (max)
19.12.3 NbR	0.06% C 18.0% Cr 10.5% Ni 3.0% Mo 0.8% Nb steel: Electrode; Metrode

Symbol	Nominal analysis, supplier, condition and remarks.
19.12.3 NLB	0.04% C 17.0% Cr 12.5% Ni 3.0% Mo 0.2% N steel: Electrode; Metrode
19.12.3 R	0.06% C 19.0% Cr 11.0% Ni 3.0% Mo steel: Electrode; Metrode
19.12.3 RMP	0.06% C 19.0% Cr 11.0% Ni 3.0% Mo steel: Electrode; Metrode
19.12.3 ULRCF	0.025% C (max) 18.0% Cr 14.0% Ni 3.0% Mo steel: Electrode; Metrode
19/12 Mo	0.08% C 19% Cr 12% Ni 3.5% Mo steel: Eastern Stainless; annealed
	UTS: 630 Elon: 45% Proof: 330
19/12 MoL	0.03% C 19% Cr 12% Ni 3.5% Mo steel: Eastern Stainless; annealed
	UTS: 610 Elon: 48% Proof: 290
19.13.4 ELRCF	0.03% C (max) 18.0% Cr 15.0% Ni 4.0% Mo steel: Electrode; Metrode; ferrite content 1–5%
19.13.4 LRCF	0.03% C 18.0% Cr 15.0% Ni 4.0% Mo steel: Electrode; Metrode; ferrite content 1–5%
19.13.4 R	0.06% C 19.0% Cr 12.0% Ni 4.0% Mo steel: Electrode; Metrode
19.15.3 MnBNF	0.03% C 3.0% Mn 19.0% Cr 15.0% Ni 3.0% Mo steel: Electrode; Metrode; nil ferrite content
19.15.3 MnRNF	0.03% C 3.0% Mn 19.0% Cr 15.0% Ni 3.0% Mo steel: Electrode; Metrode; nil ferrite content
20.9.3 B	0.08% C 19.0% Cr 9.0% Ni 3.5% Mo steel: Electrode; Metrode; ferrite content 20%
20.9.3 R	0.08% C 19.0% Cr 9.0% Ni 3.5% Mo steel: Electrode; Metrode; ferrite content 20%
20.9.3 RMP	0.08% C 19.0% Cr 9.0% Ni 3.5% Mo steel: Electrode; Metrode; ferrite content 20%
20/10	0.08% C 20% Cr 11% Ni steel: Eastern Stainless for AISI 30B
20.10.3 CB	0.3% C 20.0% Cr 10.0% Ni steel: Electrode; Metrode; for welding stainless steel castings
20/12 L	0.07% C 20.0% Cm 11.0% Ni steel: Nyby; annealed
	DPN: 180 UTS: 530 Elon: 50% Proof: 210
20/12 Si	0.12% C 20.0% Cr 12.0% Ni 2.0% Si steel: Nyby; annealed
	DPN: 190 UTS: 580 Elon: 40% Proof: 290
20.15.3 CB	0.3% C 20.0% Cr 15.0% Ni steel: Electrode; Metrode
20.15.3 CR	0.3% C 20.0% Cr 15.0% Ni steel: Electrode; Metrode
20.23.3 CuNb	0.07% C 20.0% Cr 23.0% Ni 3.0% Mo 3.0% Mo 3.0% Cu 1.0% Nb steel: Electrode; Metrode
20/25 N6	0.03% C 20.0% Cr 0.3% Ni 0.7% Mn 0.6% Si 25% Ni Fe alloy: Union Carbide
20/25 Ti	0.03% C 20.0% Cr 0.3% Ti 0.7% Mn 0.6% Si 25% Ni Fe alloy: Union Carbide
20.25.4 NbR	0.04% C 20.0% Cr 25.0% Ni 4.5% Mo 1.0% Nb steel: Electrode; Metrode
20.29.3 CuNbB	0.06% C 20.0% Cr 25.0% Ni 2.0% Mo 3.0% Cu 0.8% Nb steel: Electrode; Metrode
20.32 NbB	0.1% C 20.0% Cr 32.0% Ni 1.1% Nb steel: Electrode; Metrode; for welding Incoloy 800
20.34.3 CuNbB	0.06% C 20.0% Cr 34.0% Ni 3.0% Mo 3.0% Cu 0.8% Nb steel: Electrode; Metrode
21.26.5 CuNbR	0.04% C 21.0% Cr 26.0% Ni 4.5% Mo 1.5% Cu 0.8% Nb steel: Electrode; Metrode
22A	0.04% C 20% Cr 22% Ni 2% Mo 1.5% Cu steel: Japanese alloy; Japanese Metal Industry
23.12.2 LR	0.03% C 24.0% Cr 12.0% Ni 2.5% Mo steel: Electrode; Metrode
23.12.2 NbR	0.06% C 24.0% Cr 12.0% Ni 2.5% Mo 0.8% Nb steel: Electrode; Metrode
23.12.2 R	0.06% C 24.0% Cr 12.0% Ni 2.5% Mo steel: Electrode; Metrode
23.12 LR	0.03% C 24.0% Cr 12.0% Ni steel: Electrode; Metrode

Symbol	Nominal analysis, supplier, condition and remarks.
23.12 NbLR	0.03% C 24.0% Cr 12.0% Ni 0.8% Nb steel: Electrode; Metrode
23.12 NbR	0.06% C 24.0% Cr 12.0% Ni 0.8% Nb steel: Electrode; Metrode
23.12 R	0.06% C 24.0% Cr 12.0% Ni steel: Electrode; Metrode; ferrite content 15%
23.12 WR	0.1% C 24.0% Cr 12.0% Ni 3.0% W steel: Electrode; Metrode
23/14 L	0.07% C 23.0% Cr 13.0% Ni steel: Nyby; annealed
	DPN: 190 UTS: 530 Elon: 40% Proof: 240
23/14 LNb	0.07% C 23.0% Cr 14.0% Ni 0.7% (max) Nb steel: Nyby; annealed
	DPN: 200 UTS: 530 Elon: 40% Proof: 240
24.14.2 CWR	0.25% C 24.0% Cr 14.0% Ni 3.0% W steel: Electrode; Metrode
24.24.3 CNB	0.3% C 24.0% Cr 24.0% Ni 1.7% Nb steel: Electrode; Metrode
25.5.1 C	1.0% C 25.0% Cr 5.0% Ni steel: Electrode; Metrode; for resistance to cavitation
	DPN: 380
25.6.2 CuLR	0.03% C 25.0% Cr 6.0% Ni 2.0% Mo 3.0% Cu steel: Electrode; Metrode; for wear resistance
25.6.2 CuR	0.06% C 25.0% Cr 6.0% Ni 2.0% Mo 3.0% Cu steel: Electrode; Metrode; for corrosion & wear resistance
	DPN: 250
25.6.3 CuR	0.06% C 25.0% Cr 6.0% Ni 3.0% Mo 2.0% Cu steel: Electrode; Metrode
25/12	0.08% C 23% Cr 13% Ni steel: Eastern Stainless; annealed
	UTS: 620 Elon: 45% Proof: 330
25.12.4 CR	0.4% C 25.0% Cr 12.0% Ni steel: Electrode; Metrode
25/20	0.08% C 25% Cr 21% Ni steel: Eastern Stainless; annealed
	UTS: 620 Elon: 43% Proof: 330
25.20 B	0.17% C 25.0% Cr 20.0% Ni 5.0% Mn steel: Electrode; Metrode
25.20 HR	0.4% C 1.5% Mn 25.0% Cr 20.0% Ni steel: Electrode; Metrode
25.20 HR	0.4% C 1.5% Mn 25.0% Cr 20.0% Ni steel: Electrode; Metrode
25.20 NbB	0.1% C 25.0% Cr 20.0% Ni 1.0% Nb 5.0% Mn steel: Metrode
25.20 MoB	0.1% C 25.0% Cr 20.0% Ni 2.5% Mo steel: Electrode; Metrode
25/20 Si	0.08% C 2.5% Si 25% Cr 21% Ni steel: Eastern Stainless; for AISI 314
25.20.2 CB	0.2% C 1.5% Mn 25.0% Cr 20.0% Ni steel: Electrode; Metrode
25.20.3 CB	0.3% C 1.5% Mn 25.0% Cr 20.0% Ni steel: Electrode; Metrode
25/21 L	0.07% C 25.0% Cr 20.0% Ni steel: Nyby; annealed
	DPN: 190 UTS: 530 Elon: 40% Proof: 240
25/21 Si	0.12% C 25.0% Cr 21.0% Ni 2.0% Si steel: Nyby; annealed
	DPN: 200 UTS: 580 Elon: 40% Proof: 240
25/24 EMo	0.05% C 25.0% Cr 25.0% Ni 2.3% Mo steel: Nyby; annealed
	DPN: 180 UTS: 530 Elon: 40% Proof: 240
25.25.2 Nb	0.07% C 25.0% Cr 25.0% Ni 2.1% Mo 1.0% Nb steel: Electrode; Metrode
23.35.4 CR	0.4% C 25.0% Cr 35.0% Ni steel: Electrode; Metrode
26.33.4 CNbR	0.4% C 26.0% Cr 33.0% Ni 1.5% Nb steel: Electrode; Metrode
27/4 L	0.07% C 27.0% Cr 4.0% Ni steel: Nyby; annealed
	DPN: 260 UTS: 580 Elon: 20% Proof: 370

Symbol	Nominal analysis, supplier, condition and remarks.
27/5 LMo	0.07% C 27.0% Cr 5.0% Ni 1.6% Mo steel: Nyby; annealed **DPN: 260 UTS: 630 Elon: 20% Proof: 400**
28.6.4 CB	0.4% C 28.0% Cr 6.0% Ni steel: Electrode; Metrode
29.9 R	0.09% C 29.0% Cr 9.0% Ni steel: Electrode; Metrode; ferrite content 30–35%
30A	0.04% C 20% Cr 30% Ni 2.5% Mo 3.5% Cu steel: Japanese alloy
248 SV	0.03% C 16% Cr 5.0% Ni 1.0% Mo steel: Weldable; Avesta
254 EM	0.08% C 25% Cr 21% Ni steel: Avesta **UTS: 650 Elon: 40% Proof: 300**
254 SLX	0.02% C 20% Cr 25% Ni 4.5% Mo 1.5% Cu steel: Avesta **UTS: 590 Elon: 40% Proof: 270**
302 NILSTAIN	0.15% C 18% Cr 8% Ni steel for springs: Driver
304 NILSTAIN	0.08% C (max) 19% Cr 10% Ni steel: Driver
304 ELC NILSTAIN	0.03% C (max) 19% Cr 10% Ni steel: Driver
305 NILSTAIN	0.12% C (max) 18% Cr 11.5% Ni steel: Driver
309 NILSTAIN	0.2% C (max) 23% Cr 13.5% Ni steel: Driver
310 NILSTAIN	0.25% C (max) 25% Cr 20.5% Ni steel: Driver
314 NILSTAIN	0.25% C (max) 24.5% Cr 20.5% Ni steel: Driver
316 NILSTAIN	0.08% C (max) 17% Cr 12% Ni 2.5% Mo steel: Driver
316 ELC NILSTAIN	0.03% C (max) 17% Cr 12% Ni 2.5% Mo steel: Driver
317 NILSTAIN	0.08% C 19% Cr 13% Ni 3.5% Mo steel: Driver
321 NILSTAIN	0.08% C (max),18% Cr 10.5% Ni 0.4% Ti steel: Driver
330 NILSTAIN	15% Cr 34% Ni steel: Driver
347 NILSTAIN	0.08% C (max) 18% Cr 11% Ni 0.8% Nb steel: Driver
430 NILSTAIN	0.12% C (max) 16% Cr steel: Driver
453S	0.1% C 26% Cr 5.0% Ni 1.5% Mo steel: Avesta **UTS: 650 Elon: 20% Proof: 500**
553	0.08% C 18% Cr 9% Ni steel: Soderfors
554	0.16% C (max) 18% Cr 9% Ni steel: Soderfors
556	0.16% C 18% Cr 9% Ni 0.2% S steel: Free machining; Soderfors
559	0.16% C (max) 13% Cr 13% Ni steel: Soderfors
564	0.12% C 18% Cr 10% Ni 3.0% Mo steel: Soderfors
832C	0.16% C 18% Cr 8% Ni 0.2% S steel: Free machining; Avesta
832K	0.12% C 18% Cr 10% Ni 3.0% Mo steel: Avesta
832M	0.1% C 18% Cr 0.9% Ni steel: Avesta
832MM	0.06% C 18.5% Cr 10.5% Ni steel: Avesta **UTS: 600 Elon: 70% Proof: 270**
832MP	0.07% C 17.5% Cr 8.5% Ni steel: Avesta **UTS: 660 Elon: 70% Proof: 300**
832MV	0.15% C 18% Cr 8% Ni steel: Avesta
832MVN	0.05% C 18.5% Cr 8.5% Ni 0.2% N steel: Avesta
832MVR	0.03% C 18.5% Cr 10.5% Ni steel: Avesta **UTS: 540 Elon: 60% Proof: 250**
832MVRN	0.03% C 18.5% Cr 10.5% Ni 0.2% N steel: Avesta
832MVT	0.08% C 17.5% Cr 10.5% Ni 0.4% Ti steel: Avesta **UTS: 600 Elon: 50% Proof: 280**
832P	0.16% C 13% Cr 13% Ni steel: Avesta

Symbol	Nominal analysis, supplier, condition and remarks.
832SF	0.05% C 17% Cr 11% Ni 2.3% Mo steel: Avesta **UTS: 600 Elon: 55% Proof: 290**
832SFR	0.03% C 17% Cr 11.5% Ni 2.3% Mo steel: Avesta **UTS: 580 Elon: 55% Proof: 290**
832SFT	0.08% C 17% Cr 12% Ni 2.2% Mo 0.4% Ti steel: Avesta **UTS: 600 Elon: 50% Proof: 290**
832SK	0.05% C 17% Cr 11.5% Ni 2.7% Mo steel: Avesta **UTS: 600 Elon: 55% Proof: 290**
832SKR	0.03% C 17.5% Cr 13% Ni 2.7% Mo steel: Avesta **UTS: 580 Elon: 55% Proof: 290**
832SKRN	0.03% C 17.5% Cr 13% Ni 2.7% Mo 0.2% N steel: Avesta
832SN	0.05% C 18.5% Cr 14.5% Ni 3.5% Mo steel: Avesta **UTS: 600 Elon: 50% Proof: 300**
832SV	0.05% C 17% Cr 10% Ni 1.5% Mo steel: Avesta **UTS: 590 Elon: 60% Proof: 280**
832T	0.15% C 18% Cr 8% Ni 0.6% T steel: Avesta
832V	0.12% C 18% Cr 10% Ni 2.0% Mo steel: Avesta
3632	0.12% C 25.0% Cr 20.0% Ni steel: French Standard designation
3636	0.12% C 18.0% Cr 37.0% Ni steel: French Standard designation
3682	0.06% C 15.0% Cr 25.0% Ni 1.25% Mo 0.3% V 2.0% Ti steel: French Standard designation
A 1/57	0.08% C 2.5% Si 18% Cr 12% Ni steel: Babcock & Wilcox (USA)
A 10	Austenitic stainless steel sinter: Firth Cleveland; specific gravity 6.4–6.8; as sintered **DPN: 100 UTS: 360 Elon: 6%**
A 16X	0.03% C (max) 20.0% Cr 24.0% Ni 6.5% Mo stainless steel: Allegheny Ludlum **UTS: 650 Elon: 45% Proof: 310**
AGT	0.16% C 13% Cr 13% Ni steel: Ugine
AISI 301	0.1% C 17% Cr 7% Ni steel **DPN: 180 UTS: 740 Elon: 50% Proof: 220**
AISI 302	0.1% C 18% Cr 8% Ni steel **DPN: 180 UTS: 580 Elon: 50% Proof: 200**
AISI 302B	0.15% C 18% Cr 9% Ni steel **DPN: 180 UTS: 580 Elon: 40% Proof: 180**
AISI 303	0.15% C (max) 18% Cr 9% Ni with S or Se steel **DPN: 160 UTS: 530 Elon: 40% Proof: 200**
AISI 303 Se	0.15% C 18% Cr 9% Ni 0.15% Se steel: Free cutting
AISI 304	0.08% C (max) 19% Cr 9% Ni steel **DPN: 180 UTS: 580 Elon: 50% Proof: 200**
AISI 304L	0.03% C 19% Cr 10% Ni steel **DPN: 180 UTS: 480 Elon: 40% Proof: 150**
AISI 305	0.12% C (max) 18% Cr 12% Ni steel **DPN: 180 UTS: 480 Elon: 50% Proof: 150**
AISI 308	0.08% C 20% Cr 11% Ni steel **UTS: 530 Elon: 50% Proof: 200**
AISI 309	0.2% C 23% Cr 13% Ni steel **DPN: 200 UTS: 580 Elon: 40% Proof: 200**
AISI 309S	0.08% C 23% Cr 14% Ni steel **DPN: 200 UTS: 530 Elon: 40% Proof: 200**
AISI 310	0.25% C (max) 25% Cr 21% Ni steel: Annealed **DPN: 180 UTS: 530 Elon: 40% Proof: 200**
AISI 310S	0.08% C 25% Cr 21% Ni steel
AISI 314	0.29% C 25% Cr 21% Ni steel: Annealed **DPN: 180 UTS: 530 Elon: 40% Proof: 200**
AISI 316	0.08% C (max) 17% Cr 12% Ni 2.5% Mo steel: Annealed **DPN: 200 UTS: 480 Elon: 40% Proof: 200**
AISI 316 L	0.03% C 17% Cr 12% Ni 2.5% Mo steel: Annealed **DPN: 180 UTS: 480 Elon: 40% Proof: 180**
AISI 316N	0.08% C 17.0% Cr 12.0% Ni 2.5% Mo 0.13% N steel: Stainless

Note. The following abbreviations and units are used in the tables:

DPN	Hardness, diamond pyramid number
UTS	Ultimate tensile strength, N/mm^2
Elon	Elongation, %
Proof	0.1 % proof strength, N/mm^2

1 N/mm^2=0.1 hbar=0.102 kgf/mm^2=0.06475 tonf/in^2=145.04 lbf/in^2

Symbol	Nominal analysis, supplier, condition and remarks.	Symbol	Nominal analysis, supplier, condition and remarks.
AISI 317	0.08% C (max) 19% Cr 13% Ni 3.5% Mo steel: Annealed	AMS 5521 B	0.25% C 25% Cr 20% Ni steel: Sheet & strip for deep drawing; AMS for SAE 30310
	DPN: 200 Elon: 40% Proof: 200	AMS 5522 B	2% Si 25% Cr 20% Ni steel: Sheet & strip
AISI 317 L	0.03% C 20% Cr 14% Ni 3.5% Mo steel	AMS 5523	0.2% C 23% Cr 13% Ni steel: Sheet & strip; AMS for SAE 30309
AISI 318	0.08% C 20% Cr 14% Ni 3.5% Mo steel: Annealed	AMS 5524 B	0.08% C 18% Cr 13% Ni 2.5% Mo steel: Sheet & strip; AMS for SAE 30316
	DPN: 200 UTS: 530 Elon: 40% Proof: 180	AMS 5552 A	20% Cr 32% Ti 1% Ti Fe alloy: Sheet
AISI 321	0.08% C (max) 18% Cr 10% Ni 0.4% Ti steel: Annealed	AMS 5556 A	0.08% C 18% Cr 11% Ni Nb + Ta steel: Tube; AMS for SAE 30347
	DPN: 200 UTS: 530 Elon: 40% Proof: 180	AMS 5557 A	0.08% C 18% Cr 11% Ni Ti steel: Tube; AMS for SAE 30321
AISI 321H	0.06% C 18.0% Cr 10.5% Ni 0.7% Ti (max) steel: Stainless	AMS 5558	0.08% C 18% Cr 11% Ni Nb + Ta steel: Welded tube; AMS for SAE 30347
AISI 325	0.25% C (max) 9% Cr 21% Ni 1.2% Cu steel	AMS 5559 A	0.08% C 18% Cr 10% Ni Ti steel: Welded tube; AMS for SAE 30321
AISI 327	0.1% C 25.5% Cr 4.8% Ni steel	AMS 5560 D	0.08% C 19% Cr 10% Ni steel: Seamless tube; AMS for SAE 30304
AISI 329	0.09% C 26% Cr 5% Ni 1.5% Mo steel: Obsolete	AMS 5565 D	0.08% C 19% Cr 9% Ni steel: Welded tube; AMS for SAE 30304
AISI 347	0.08% C (max) 18% Cr 11% Ni 0.8% Nb steel: Annealed	AMS 5566 D	0.08% C 19% Cr 10% Ni steel: Seamless or welded tube; AMS for SAE 30304
	DPN: 200 UTS: 580 Elon: 40% Proof: 200	AMS 5567	0.08% C 19% Cr 10% Ni steel: Tube; AMS for SAE 30304
AISI 347 F	0.08% C 18% Cr 12% Ni 0.8% Nb 0.2% Se steel: Annealed	AMS 5570 G	0.08% C 18% Cr 11% Ni Ti steel: Seamless tube; AMS for SAE 30321
	DPN: 200 UTS: 580 Elon: 40% Proof: 200	AMS 5571 B	0.08% C 18% Cr 11% Ni Nb + Ta steel: Seamless tube; AMS for SAE 30347
AISI 347H	0.6% C 18.0% Cr 11.0% Ni 1.0% Nb (max) steel: Stainless	AMS 5572 B	0.25% C 25% Cr 20% Ni steel: Seamless tube; AMS for SAE 30310
AISI 348	0.08% C 18.0% Cr 11.0% Ni 0.8% Nb + Ta (min) steel	AMS 5573 C	0.08% C 17% Cr 12% Ni 2.5% Mo steel: Seamless tube; AMS for SAE 30316
AISI 348	0.08% C 18% Cr 11% Ni 0.1% Ta 0.8% Nb steel	AMS 5574	0.2% C 23% Cr 13% Ni steel: Seamless tube; AMS for SAE 30309
AISI 348H	0.06% C 18.0% Cr 11.0% Ni 0.6% Nb (min) steel	AMS 5575 F	0.08% C 18% Cr 11% Ni Nb + Ta steel: Welded tube; AMS for SAE 30347
ALCRESS	7.5% Al 20% Cr Fe alloy: Source unknown	AMS 5576 C	0.08% C 18% Cr 10% Ni Ti steel: Welded tube; AMS for SAE 30321
AMS 5358	18% Cr 9% Ni steel: Investment casting	AMS 5577 A	0.25% C 25% Cr 20% Ni steel: Welded tube; AMS for SAE 30310
AMS 5360 B	0.08% C 17% Cr 13% Ni 2% Mo steel: Investment casting; AMS for SAE 60316	AMS 5579	20% Cr 9% Ni 1.5% Mo 1.5% W 0.4% Nb + Ta 0.2% Ti steel: Welded tube
AMS 5361 B	0.08% C 18% Cr 13% Ni 2% Mo steel: Sand casting; AMS for SAE 60316	AMS 5592	18% Cr 35% Ni 1.15% Si Fe alloy: Sheet
AMS 5362 D	0.08% C 19% Cr 12% Ni Nb + Ta steel: Investment casting; AMS for SAE 60347	AMS 5635	0.15% C 18% Cr 9% Ni steel: Bar & forging; free machining; AMS for SAE 30303 F
AMS 5363 B	0.08% C 18% Cr 10.5% Ni Nb + Ta steel: Sand casting; AMS for SAE 60347	AMS 5636 A	0.1% C 18% Cr 8% Ni steel: Bar; cold drawn; AMS for SAE 30302
AMS 5365 A	0.2% C 25% Cr 20% Ni steel: Sand casting; AMS for SAE 60310		**UTS: 740**
AMS 5366 A	0.2% C 25% Cr 20% Ni steel: Investment casting; AMS for SAE 60310	AMS 5637 A	0.1% C 18% Cr 8% Ni steel: Bar; cold drawn; AMS for SAE 30302
AMS 5369 A	20% Cr 9% Ni 1.4% Mo 1.4% Nb Ti steel: Sand casting		**UTS: 920**
AMS 5370	Low C 19.5% Cr 10% Ni steel: Investment casting	AMS 5639 A	0.08% C 19% Cr 9% Ni steel: Bar, forgings, etc; AMS for SAE 30304
AMS 5371	Low C 19.5% Cr 10% Ni steel: Sand casting	AMS 5640 F	0.15% C 18% Cr 9% Ni steel: Bar & forging; free machining; AMS for SAE 30303 F
AMS 5507	0.08% C 17% Cr 13% Ni 2.5% Mo steel: Sheet & strip; AMS for SAE 30316	AMS 5641 A	0.15% C 18.5% Cr 10% Ni steel: Bar & forging; free machining; for swaging; AMS for SAE 30303 F
AMS 5510 H	0.08% C 18% Cr 10% Ni Ti steel: Sheet & strip; AMS for SAE 30321	AMS 5624 C	0.15% C 18% Cr 11% Ni Cb + Ta steel: Bar & forging; free machining; AMS for SAE 30303 F
AMS 5511 A	Low C 18% Cr 8% Ni steel: Sheet & strip; AMS for SAE 30304	AMS 5645 G	0.08% C 18% Cr 10% Ni Ti steel: Bar, forging, etc.; AMS for SAE 30321
AMS 5512 B	0.08% C 18% Cr 11% Ni Nb + Ta steel: Sheet & strip; AMS for SAE 30347	AMS 5646 E	0.08% C 18% Cr 11% Ni Nb + Ta steel: Bar, forging; AMS for SAE 30347
AMS 5513	0.08% C 19% Cr 9% Ni steel: Sheet & strip; AMS for SAE 30304	AMS 5647 A	0.08% C 18% Cr 8% Ni steel: Bar & forging; AMS for SAE 30304
AMS 5514 A	0.1% C 18% Cr 11% Ni steel: Sheet & strip for deep drawing; AMS for SAE 30305	AMS 5648 C	0.08% C 18% Cr 13% Ni 2.5% Mo steel: Bar & forging; AMS for SAE 30316
AMS 5515 D	0.1% C 18% Cr 8% Ni steel: Sheet & strip for deep drawing; AMS for SAE 30302	AMS 5649	0.08% C 18% Cr 13% Ni 2% Mo steel: Bar & forging; free machining; AMS for SAE 30316 F
AMS 5516 E	0.1% C 18% Cr 8% Ni steel: Sheet & strip; cold rolled; AMS for SAE 30302		
	UTS:530		
AMS 5517 D	0.1% C 18% Cr 8% Ni steel: Sheet & strip; cold rolled; AMS for SAE 30301		
	UTS: 920		
AMS 5518 C	0.1% C 18% Cr 8% Ni steel: Sheet & strip cold rolled; AMS for SAE 30301		
	UTS: 1120		
AMS 5519 E	0.1% C 18% Cr 8% Ni steel: Sheet & strip; cold rolled; AMS for SAE 30301		
	UTS: 1380		

Symbol	Nominal analysis, supplier, condition and remarks.
AMS 5650 A	0.2% C 23% Cr 13.5% Ni steel: Bar & forging; AMS for SAE 30309
AMS 5651 D	0.25% C 25% Cr 20% Ni steel: Bar & forging; AMS for SAE 30310
AMS 5652 B	2% Si 25% Cr 20% Ni steel: Bar & forging
AMS 5653	0.08% C 7% Cr 13% Ni 2.5% Mo steel: Bar & forging; AMS for SAE 30316
AMS 5680 B	0.08% C 18% Cr 11% Ni Nb + Ta steel: Welding wire; AMS for SAE 30347
AMS 5681 A	0.08% C 19% Cr 10% Ni Nb + Ta steel: Coated electrode; AMS for SAE 30347
AMS 5685 C	0.12% C 18% Cr 11% Ni steel: Wire; AMS for SAE 30305
AMS 5686 A	0.12% C 18% Cr 11% Ni steel: Wire for rivetting; AMS for SAE 30305
AMS 5688 D	0.1% C 18% Cr 8% Ni steel: Wire; spring temper; AMS for SAE 30302
AMS 5689	0.08% C 18% Cr 9.5% Ni Ti steel: Screen wire; AMS for SAE 30321
AMS 5690 E	0.08% C 17% Cr 12% Ni 2.5% Mo steel: Screen wire; AMS for SAE 30316
AMS 5691 B	0.08% C 18% Cr 13% Ni 2% Mo steel: Coated electrode; AMS for SAE 30316
AMS 5694 B	0.25% C 25% Cr 20% Ni steel: Welding wire; cold drawn; AMS for SAE 30310
AMS 5695 A	0.25% C 25% Cr 20% Ni steel: Coated electrode; AMS for SAE 30310
AMS 5697	0.08% C 19% Cr 9% Ni steel: Wire; AMS for SAE 30304
AMS 5700 B	14% Cr 14% Ni 0.4% Mo 2.5% W steel: Bar & forging
AMS 5705 A	2.5% Si 12.8% Cr 8% Ni steel: Bar & forging
AMS 5725 A	16% Cr 25% Ni 6% Mo Fe alloy: Bar; up to 40 mm in diameter
AMS 5727 B	16% Cr 25% Ni 6% Mo Fe alloy: Forging
AMS 5728 B	16% Cr 25% Ni 6% Mo Fe alloy: Forging; special casting process
AMS 5733 C	13.5% Cr 26% Ni 3% Mo 1.8% Ti Fe alloy: Bar & forging
AMS 5738	0.15% C 18% Cr 9% Ni steel: Bar; free machining; AMS for SAE 30303 F
AMS 5741 C	13.5% Cr 26% Ni 1.7% Mo 3% Ti Fe alloy: Bar & forging; vacuum melted
AMS 5742 A	20% Cr 32% Ni 1% Ti Fe alloy: Bar & forging
AMS 5784	29% Cr 9% Ni steel: Wire
AMS 5785 B	29% Cr 9% Ni steel: Coated electrode
ANKA	0.15% C (max) 18% Cr 9% Ni steel: Bar, billets, etc.; Brown Bayley for BS En 58A
ANKA E	0.08% C (max) 19% Cr 10% Ni steel: Bar, billet, etc.; Brown Bayley for BS alloy En 58E
ANKA M	0.16% C 12% Cr 12% Ni steel: Bar, billet, etc.; Brown Bayley for BS alloy En 58D
AN QQ S 757/3	0.1% C 17% Cr 7% Ni 1.0% Nb 0.5% Ti steel: US Service; annealed
	UTS: 700 Elon: 40%
AN QQ S 771/4/FM	0.12% C 17% Cr 7% Ni 0.7% Mo 0.18% S + P steel: US Service; annealed
	UTS: 690 Elon: 35%

Symbol	Nominal analysis, supplier, condition and remarks.
AN QQ S 771/4/G	0.12% C 17% Cr 7% Ni 0.5% Cu steel: US Service; cold drawn
	UTS: 840 Elon: 12% Proof: 700
AN QQ S 771/4/MCR	0.1% C 17% Cr 7% Ni 2.0% Mo 0.5% Cu steel: US Service; cold drawn
	UTS: 840 Elon: 12% Proof: 700
AN QQ S 772/G	0.15% C 17% Cr 6.5% Ni 0.5% Cu Ti or Nb steel: US Service; annealed
	UTS: 540 Elon: 40% Proof: 200
AN QQ S 772/MCR	0.1% C 17% Cr 10% Ni 2.0% Mo steel: US Service
	UTS: 760 Elon: 40% Proof: 200
AN QQ W 423/1	0.2% C 12% Cr 14% Ni steel: US Service; annealed
	UTS: 760
AN T 14/1	0.1% C 17% Cr 7% Ni 0.5% Cu Nb or Ti stabilized steel: Flexible tube; US Service
AN T 43	0.12% C 17% Cr 6.5% Ni 0.5% Cu steel: Tube; US Service; annealed
	UTS: 740 Elon: 30% Proof: 200
AN WW T 855/1	0.12% C 17% Cr 7% Ni 0.5% Cu steel: US Service; annealed
	UTS: 700 Elon: 30% Proof: 200
AN WW T 858/1	0.1% C 17% Cr 7% Ni Ti or Nb stabilized steel: Seamless tube; US Service; annealed
	UTS: 740 Elon: 35%
AN WW T 861/2	0.1% C 17% Cr 7% Ni 0.5% Cu Ti or Nb stabilized steel: Welded tube; US Service
	UTS: 740
ARC/098	25% Cr 14% Ni 1.25% Mo steel: Imphy
ARC 2233	18% Cr 10% Ni 2.4% Mo steel: Imphy
ARC 2266	18% Cr 10% Ni 3.0% Mo steel: Imphy
ARMCO/17/7	0.12% C 17% Cr 7% Ni steel: Armco
ARMCO 18/8	0.12% C 18% Cr 9% Ni steel: Armco
ARMCO 18/8 Si	0.11% C 2.5% Si 18% Cr 9% Ni steel: Armco
ARMCO 18/10 Cb Ta	0.08% C 18% Cr 10% Ni 1.7% Nb + Ta steel: Armco
ARMCO 18/10 Ti	0.08% C 18% Cr 10% Ni 0.5% Ti steel: Armco
ARMCO 18/12 Mo	0.1% C 17% Cr 12% Ni 2.0% Mo steel: Armco
ARMCO 20/10	0.08% C 20% Cr 11% Ni steel: Armco
ARMCO 25/12	0.2% C 23% Cr 13% Ni steel: Armco
ARMCO 25/20	0.25% C 25% Cr 21% Ni steel: Armco
ARMEX 2	0.08% C 20% Cr 8.5% Ni 3.5% Mo steel: Welding electrode; Murex
	UTS: 760 Elon: 35%
ARMEX 3	0.08% C 19.5% Cr 8.5% Ni 2.5% Mo steel: Welding electrode; Murex
	UTS: 680 Elon: 37%
ARMEX GT	0.07% C 17% Cr 8% Ni 2.0% Mo steel: Welding electrode; Murex
	UTS: 610 Elon: 35%
ASTM A167/301	0.15% C 17% Cr 7% Ni steel: Sheet & plate
	DPN: 202 UTS: 530 Elon: 40% Proof: 200
ASTM A167/302	0.15% C 18% Cr 9% Ni steel: Sheet & plate
	DPN: 202 UTS: 530 Elon: 40% Proof: 200
ASTM A167/302 B	0.15% C 2.5% Si 18% Cr 9% Ni steel: Plate & sheet
	DPN: 217 UTS: 530 Elon: 40% Proof: 200
ASTM A167/304	0.08% C 19% Cr 10% Ni steel: Sheet & plate
	DPN: 202 UTS: 530 Elon: 40% Proof: 200
ASTM A167/304 L	0.03% C 19% Cr 10% Ni steel: Sheet & plate
	DPN: 202 UTS: 480 Elon: 40% Proof: 170
ASTM A167/305	0.12% C 18% Cr 11% Ni steel: Sheet & plate
	DPN: 202 UTS: 500 Elon: 40% Proof: 170
ASTM A167/308	0.08% C 20% Cr 11% Ni steel: Sheet & plate
	DPN: 202 UTS: 530 Elon: 40% Proof: 200
ASTM A167/309	0.2% C 23% Cr 13% Ni steel: Sheet & plate
	DPN: 217 UTS: 530 Elon: 40% Proof: 200
ASTM A167/309 S	0.08% C 23% Cr 13% Ni steel: Sheet & plate
	DPN: 217 UTS: 530 Elon: 40% Proof: 200

Note. The following abbreviations and units are used in the tables:

DPN	Hardness, diamond pyramid number
UTS	Ultimate tensile strength, N/mm^2
Elon	Elongation, %
Proof	0.1 % proof strength, N/mm^2

1 N/mm^2=0.1 hbar=0.102 kgf/mm^2=0.06475 tonf/in^2=145.04 lbf/in^2

Symbol	Nominal analysis, supplier, condition and remarks.
ASTM A167/310	0.25% C 25% Cr 21% Ni steel: Sheet & plate **DPN: 217 UTS: 530 Elon: 40% Proof: 200**
ASTM A167/310 S	0.08% C 25% Cr 21% Ni steel: Sheet & plate **DPN: 217 UTS: 530 Elon: 40% Proof: 200**
ASTM A167/316	0.08% C 17% Cr 12% Ni 2.5% Mo steel: Sheet & plate **DPN: 217 UTS: 530 Elon: 40% Proof: 200**
ASTM A167/316 L	0.03% C 17% Cr 12% Ni 2.5% Mo steel: Sheet & plate **DPN: 217 UTS: 480 Elon: 40% Proof: 170**
ASTM A167/317	0.08% C 19% Cr 13% Ni 3.5% Mo steel: Sheet & plate **DPN: 217 UTS: 530 Elon: 35% Proof: 200**
ASTM A167/317 L	0.03% C 19% Cr 14% Ni 3.5% Mo steel: Sheet & plate **DPN: 217 UTS: 530 Elon: 35% Proof: 200**
ASTM A167/321	0.08% C 18% Cr 10% Ni 0.4% Ti steel: Sheet & plate **DPN: 202 UTS: 530 Elon: 40% Proof: 200**
ASTM A167/347	0.08% C 18% Cr 11% Ni 1% Nb steel: Sheet & plate **DPN: 202 UTS: 530 Elon: 40% Proof: 200**
ASTM A167/348	0.08% C 18% Cr 11% Ni 1% Nb steel: Sheet & plate **DPN: 202 UTS: 530 Elon: 40% Proof: 200**
ASTM A177	0.12% C 17% Cr 7% Ni steel: Sheet & strip; ½-hard **UTS: 920 Elon: 25% Proof: 530**
ASTM A177	0.12% C 17% Cr 7% Ni steel: Sheet & strip; hard **UTS: 1380 Elon: 4% Proof: 1030**
ASTM A182 F10	0.14% C 20.5% Ni 8% Cr steel: For pipe fittings
ASTM A182 F304	0.08% C (max) 9.5% Ni 19% Cr steel: For pipe fittings
ASTM A182 F304H	0.07% C 9.5% Ni 19% Cr steel: For pipe fittings
ASTM A182 F304L	0.035% C (max) 9.5% Ni 19% Cr steel: For pipe fittings
ASTM A182 F310	0.15% C (max) 20.5% Ni 25% Cr steel: For pipe fittings
ASTM A182 F316	0.08% C (max) 12% Ni 17% Cr 2.5% Mo steel: For pipe fittings
ASTM A182 F316H	0.07% C 12% Ni 17% Cr 2.5% Mo steel: For pipe fittings
ASTM A182 F316L	0.03% C (max) 12.5% Ni 17% Cr 2.5% Mo steel: For pipe fittings
ASTM A182 F321	0.08% C (max) 9% Ni (min) 17% Cr (min) steel: For pipe fittings
ASTM A182 F321H	0.07% C 9% Ni (min) 17% Cr (min) steel: For pipe fittings
ASTM A182 F347	0.08% C (max) 11% Ni 18.5% Cr steel: For pipe fittings
ASTM A182 F347H	0.07% C 11% Ni 18.5% Cr steel: For pipe fittings
ASTM A182 F348	0.08% C (max) 11% Ni 18.5% Cr steel: For pipe fittings
ASTM A182 F348H	0.07% C 11% Ni 18.5% Cr steel: For pipe fittings
ASTM A193 B8	0.08% C 19% Cr 10% Ni steel: Bar **UTS: 530 Elon: 35% Proof: 200**
ASTM A193 B8C	0.08% C 18% Cr 11% Ni 1% Nb steel: Bar **UTS: 530 Elon: 35% Proof: 200**
ASTM A193 B8M	0.08% C 17% Cr 12% Ni 2.5% Mo steel: Bar **UTS: 530 Elon: 35% Proof: 200**
ASTM A193 B8T	0.08% C 18% Cr 10% Ni 0.5% Ti steel: Bar **UTS: 530 Elon: 35% Proof: 200**
ASTM A194/8	0.08% C (max) 10% Ni 19% Cr steel: For bolts
ASTM A194/8 C	0.08% C (max) 11% Ni 18% V 0.08% Nb Cr steel: For bolts
ASTM A194/8 M	0.08% C (max) 12% Ni 17% Cr 2.5% Mo steel: For bolts
ASTM A194/8 T	0.08% C (max) 11% Ni 18% Cr 0.4% Ti steel: For bolts

Symbol	Nominal analysis, supplier, condition and remarks.
ASTM A240/302	0.15% C 18% Cr 9% Ni steel: For pressure vessels
ASTM A240/304	0.08% C 19% Cr 10% Ni steel: For pressure vessels
ASTM A240/304 L	0.03% C 19% Cr 10% Ni steel: For pressure vessels
ASTM A240/305	0.12% C 18% Cr 12% Ni steel: For pressure vessels
ASTM A240/309S	0.08% C 23% Cr 13% Ni steel: For pressure vessels
ASTM A240/310S	0.08% C 25% Cr 21% Ni steel: For pressure vessels
ASTM A240/316	0.08% C 17% Cr 12% Ni 2.5% Mo steel: For pressure vessels
ASTM A240/316 L	0.03% C 17% Cr 12% Ni 2.5% Mo steel: For pressure vessels
ASTM A240/317	0.08% C 19% Cr 13% Ni 3.5% Mo steel: For pressure vessels
ASTM A240/317 L	0.03% C 19% Cr 13% Ni 3.5% Mo steel: For pressure vessels
ASTM A240/321	0.08% C 18% Cr 11% Ni 0.5% Ti steel: For pressure vessels
ASTM A240/347	0.08% C 18% Cr 11% Ni 1.0% Nb + Ta steel: For pressure vessels
ASTM A240/348	0.08% C 18% Cr 11% Ni 1.0% Nb + Ta steel: For pressure vessels
ASTM A249 TP304	0.08% C (max) 9.5% Ni 19% Cr steel: For boiler tube
ASTM A249 TP304H	0.07% C 9.5% Ni 19% Cr steel: For boiler tube
ASTM A249 TP304L	0.035% C (max) 10% Ni 19% Cr steel: For boiler tube
ASTM A249 TP309	0.15% C (max) 13.5% Ni 23% Cr steel: For boiler tube
ASTM A249 TP310	0.15% C (max) 21% Ni 25% Cr steel: For boiler tube
ASTM A249 TP316	0.08% C (max) 12.5% Ni 17% Cr 2.5% Mo steel: For boiler tube
ASTM A249 TP316H	0.07% C 12.5% Ni 17% Cr 2.5% Mo steel: For boiler tube
ASTM A249 TP316L	0.035% C (max) 12.5% Ni 17% Cr 2.5% Mo steel: For boiler tube
ASTM A249 TP317	0.08% C (max) 12.5% Ni 19% Cr 3.5% Mo steel: For boiler tube
ASTM A249 TP321	0.08% C (max) 11% Ni 18.5% Cr steel: For boiler tube
ASTM A249 TP321H	0.07% C 11% Ni 18.5% Cr steel: For boiler tube
ASTM A249 TP347	0.08% C (max) 11% Ni 18.5% Cr steel: For boiler tube
ASTM A249 TP347H	0.07% C (max) 11% Ni 18.5% Cr steel: For boiler tube
ASTM A249 TP348	0.08% C (max) 11% Ni 18.5% Cr 0.1% Ta steel: For boiler tube
ASTM A249 TP348H	0.07% C 11% Ni 18.5% Cr 0.1% Ta steel: For boiler tube
ASTM A269 TP304	0.08% C 19% Cr 9% Ni steel: Tube
ASTM A269 TP304L	0.03% C 19% Cr 10% Ni steel: Tube
ASTM A269 TP316	0.08% C 17% Cr 12% Ni steel: Tube
ASTM A269 TP316L	0.03% C 17% Cr 12% Ni 2.5% Mo steel: Tube
ASTM A269 TP317	0.08% C 19% Cr 12% Ni 3.5% Mo steel: Tube
ASTM A269 TP321	0.08% C 18% Cr 11% Ni 0.4% Ti steel: Tube
ASTM A269 TP347	0.08% C 18% Cr 12% Ni 1.0% Nb + Ta steel: Tube
ASTM A269 TP348	0.08% C 18% Cr 11% Ni 1.0% Nb + Ta steel: Tube
ASTM A271	Stainless steel: Still tubes; see AISI range for grades
ASTM A276/302	0.15% C 18% Cr 9% Ni steel: Bar
ASTM A276/302 B	0.15% C 2.5% Si 18% Cr 9% Ni steel: Bar
ASTM A276/303	0.15% C 18% Cr 9% Ni 0.6% Mo steel: Bar

Symbol	Nominal analysis, supplier, condition and remarks.
ASTM A276/303 Se	0.15% C 18% Cr 9% Ni 0.15% Se steel: Bar
ASTM A276/304	0.08% C 19% Cr 10% Ni steel: Bar
ASTM A276/304 L	0.03% C 19% Cr 10% Ni steel: Bar
ASTM A276/308	0.08% C 20% Cr 11% Ni steel: Bar
ASTM A276/309	0.2% C 23% Cr 13% Ni steel: Bar
ASTM A276/309 S	0.08% C 23% Cr 13% Ni steel: Bar
ASTM A276/310	0.25% C 25% Cr 21% Ni steel: Bar
ASTM A276/310 S	0.08% C 25% Cr 21% Ni steel: Bar
ASTM A276/314	0.25% C 25% Cr 21% Ni steel: Bar
ASTM A276/316	0.08% C 17% Cr 12% Ni 2.5% Mo steel: Bar
ASTM A276/316 L	0.03% C 17% Cr 12% Ni 2.5% Mo steel: Bar
ASTM A276/321	0.08% C 18% Cr 11% Ni 0.4% Ti steel: Bar
ASTM A276/347	0.08% C 18% Cr 11% Ni 1.0% Nb + Ta steel: Bar
ASTM A276/348	0.08% C 18% Cr 11% Ni 1.0% Nb + Ta steel: Bar
ASTM A289 C	0.05% C 5% Cr 25% Ni steel: For non-magnetic retaining rings
	UTS: 1140 **Elon: 12%** **Proof: 840**
ASTM A295 F8	0.08% C 19% Cr 10% Ni steel: Seamless drum forging; annealed
	UTS: 500 **Elon: 30%** **Proof: 200**
ASTM A295 F8C	0.08% C 18% Cr 10% Ni 1.0% Nb steel: Seamless drum forging; annealed
	UTS: 500 **Elon: 30%** **Proof: 200**
ASTM A295 F8M	0.08% C 17% Cr 12% Ni 2.5% Mo steel: Seamless drum forging; annealed
	UTS: 500 **Elon: 30%** **Proof: 200**
ASTM A295 F8T	0.08% C 17% Cr 9% Ni 0.4% Ti steel: Seamless drum forging; annealed
	UTS: 500 **Elon: 30%** **Proof: 200**
ASTM A295 F10	0.15% C 8% Cr 21% Ni steel: Seamless drum forging; annealed
	UTS: 580 **Elon: 25%** **Proof: 200**
ASTM A295 F25	0.15% C 25% Cr 21% Ni steel: Seamless drum forging; annealed
	UTS: 530 **Elon: 30%** **Proof: 200**
ASTM A296 CE30	0.3% C 29% Cr 9% Ni steel: Casting; annealed
	UTS: 540 **Elon: 10%** **Proof: 240**
ASTM A296 CF3	0.03% C 19% Cr 9% Ni steel: Casting; annealed
	UTS: 440 **Elon: 35%** **Proof: 180**
ASTM A296 CF3M	0.03% C 19% Cr 10% Ni 3.0% Mo steel: Casting; annealed
	UTS: 460 **Elon: 30%** **Proof: 200**
ASTM A296 CF8	0.08% C 20% Cr 10% Ni steel: Casting; annealed
	UTS: 450 **Elon: 35%** **Proof: 180**
ASTM A296 CF8C	0.08% C 19% Cr 10% Ni 0.8% Nb steel: Casting; annealed
	UTS: 460 **Elon: 30%** **Proof: 200**
ASTM A296 CF8M	0.08% C 19% Cr 10% Ni Mo steel: Casting; annealed
	UTS: 460 **Elon: 30%** **Proof: 200**
ASTM A296 CF16F	0.16% C 19% Cr 9% Ni 0.3% Se 1.5% Mo cast steel: Annealed
	UTS: 460 **Elon: 25%** **Proof: 200**
ASTM A296 CF20	0.2% C 19% Cr 9% Ni steel: Casting; annealed
	UTS: 460 **Elon: 30%** **Proof: 200**
ASTM A290 CG8M	0.08% C 19% Cr 11% Ni 4.0% Mo steel: Casting; annealed
	UTS: 500 **Elon: 25%** **Proof: 240**

Symbol	Nominal analysis, supplier, condition and remarks.
ASTM A296 CG12	0.12% C 22% Cr 12% Ni steel: Casting; annealed
	UTS: 460 **Elon: 35%** **Proof: 180**
ASTM A296 CH20	0.2% C 25% Cr 12% Ni steel: Casting; annealed
	UTS: 460 **Elon: 30%** **Proof: 200**
ASTM A296 CK20	0.2% C 25% Cr 20% Ni steel: Casting; annealed
	UTS: 440 **Elon: 30%** **Proof: 180**
ASTM A296 CN7M	0.07% C (max) 20.5% Cr 29.5% Ni 2.5% Mo 3.5% Cu steel: Casting
ASTM A297 HE	0.2% C 29% Cr 9% Ni steel: Casting; as cast
	UTS: 580 **Elon: 9%** **Proof: 270**
ASTM A297 HF	0.20% C 19% Cr 9% Ni steel: Casting; as cast
	UTS: 460 **Elon: 25%** **Proof: 240**
ASTM A297 HH	0.2% C 25% Cr 12% Ni steel: Castings; as cast
	UTS: 500 **Elon: 10%** **Proof: 240**
ASTM A297 HI	0.2% C 28% Cr 15% Ni steel: Casting; as cast
	UTS: 460 **Elon: 10%** **Proof: 240**
ASTM A297 HK	0.2% C 25% Cr 20% Ni steel: Casting; as cast
	UTS: 440 **Elon: 10%** **Proof: 240**
ASTM A297 HL	0.2% C 29% Cr 20% Ni steel: Casting; as cast
	UTS: 440 **Elon: 10%** **Proof: 240**
ASTM A297 HN	0.2% C 20% Cr 25% Ni steel: Casting; as cast
	UTS: 420 **Elon: 8%**
ASTM A297 HT	0.35% C 15% Cr 35% Ni steel: Casting; as cast
	UTS: 440 **Elon: 4%**
ASTM A297 HU	0.55% C 18% Cr 40% Ni Fe alloy: Casting
ASTM A297 HV	0.35% C 19% Cr 39% Ni steel: Casting; as cast
	UTS: 440 **Elon: 4%**
ASTM A298 E308	0.08% C 19% Cr 10% Ni steel: Welding rod; as welded
	UTS: 540 **Elon: 35%**
ASTM A298 E308ELC	0.04% C 19% Cr 10% Ni steel: Welding rod; as welded
	UTS: 500 **Elon: 35%**
ASTM A298 E309	0.15% C 24% Cr 13% Ni steel: Welding rod; as welded
	UTS: 540 **Elon: 35%**
ASTM A298 E309 Cb	0.12% C 24% Cr 13% Ni 0.8% Nb steel: Welding rod; as welded
	UTS: 540 **Elon: 30%**
ASTM A298 E309 Mo	0.12% C 24% Cr 13% Ni 2.0% Mo steel: Welding rod; as welded
	UTS: 540 **Elon: 35%**
ASTM A298 E310	0.2% C 27% Cr 21% Ni steel: Welding rod; as welded
	UTS: 540 **Elon: 30%**
ASTM A298 E310 Cb	0.12% C 27% Cr 21% Ni 0.8% Nb steel: Welding rod; as welded
	UTS: 540 **Elon: 25%**
ASTM A298 E310 Mo	0.12% C 27% Cr 21% Ni 2.0% Mo steel: Welding rod; as welded
	UTS: 540 **Elon: 30%**
ASTM A298 E312	0.15% C 29% Cr 9% Ni steel: Welding rod; as welded
	UTS: 580 **Elon: 22%**
ASTM A298 E316	0.08% C 19% Cr 13% Ni 2.0% Mo steel: Welding rod; as welded
	UTS: 540 **Elon: 30%**
ASTM A298 E316ELC	0.04% C 19% Cr 13% Ni 2.0% Mo steel: Welding rod; as welded
	UTS: 500 **Elon: 30%**
ASTM A298 E317	0.08% C 20% Cr 13% Ni 3.0% Mo steel: Welding rod; as welded
	UTS: 540 **Elon: 30%**
ASTM A298 E318	0.08% C 19% Cr 13% Ni 2% Mo 1% Nb steel: Welding rod; as welded
	UTS: 540 **Elon: 25%**
ASTM A298 E330	0.25% C 16% Cr 33% Ni Fe alloy: Welding rod; as welded
	UTS: 500 **Elon: 30%**

Note. The following abbreviations and units are used in the tables:

DPN	Hardness, diamond pyramid number
UTS	Ultimate tensile strength, N/mm^2
Elon	Elongation, %
Proof	0.1 % proof strength, N/mm^2

1 N/mm^2=0.1 hbar=0.102 kgf/mm^2=0.06475 tonf/in^2=145.04 lbf/in^2

Symbol	Nominal analysis, supplier, condition and remarks.
ASTM A298 E347	0.08% C 20% Cr 10% Ni 1% Nb steel: Welding rod; as welded
	UTS: 540 **Elon: 30%**
ASTM A312	Austenitic steel: Pipe; graded as ASTM A249
ASTM A313	0.15% C 19% Cr 8.5% Ni steel: Spring wire; cold drawn
	UTS: 2280
ASTM A320 B8	0.08% C 19% Cr 10% Ni steel: Annealed
	UTS: 500 **Elon: 35%** **Proof: 200**
ASTM A320 B8C	0.08% C 18% Cr 11% Ni 0.8% Nb steel: Annealed
	UTS: 500 **Elon: 35%** **Proof: 200**
ASTM A320 B8D	0.08% C 18% Cr 11% Ni 0.8% Nb 0.1% Ta steel: Annealed
	UTS: 500 **Elon: 35%** **Proof: 200**
ASTM A320 B8F	0.15% C 18% Cr 9% Ni 0.6% Mo or 0.15% Se steel: Annealed
	UTS: 500 **Elon: 35%** **Proof: 200**
ASTM A320 B8T	0.08% C 18% Cr 10% Ni 0.4% Ti steel: Annealed
	UTS: 500 **Elon: 35%** **Proof: 200**
ASTM A336 F8	0.08% C 9.5% Ni 19% Cr steel: Seamless drum forging
	UTS: 460 **Elon: 30%** **Proof: 200**
ASTM A336 F8M	0.08% C 12% Ni 17% Cr 2.5% Mo steel: Seamless drum forging
	UTS: 460 **Elon: 30%** **Proof: 200**
ASTM A336 F8C	0.08% C 10.5% Ni 18% Cr steel: Seamless drum forging
	UTS: 460 **Elon: 30%** **Proof: 200**
ASTM A336 F8T	0.08% C (max) 9.0% Ni 17.0% Cr steel: Seamless drum forging
	UTS: 460 **Elon: 30%** **Proof: 200**
ASTM A336 F10	0.15% C 20% Ni 8% Cr steel: Seamless drum forging
	UTS: 540 **Elon: 25%** **Proof: 200**
ASTM A336 F25	0.15% C (max) 21% Ni 25% Cr steel: Seamless drum forging
	UTS: 500 **Elon: 30%** **Proof: 200**
ASTM A351 CF3	0.03% C 19.0% Cr 10% Ni steel: Annealed
	UTS: 460 **Elon: 35%** **Proof: 200**
ASTM A351 CF3M	0.03% C 19.0% Cr 11.0% Ni 2.0% Mo steel: Annealed
	UTS: 460 **Elon: 30%** **Proof: 200**
ASTM A351 CF8	0.08% C 19.0% Cr 9.0% Ni steel: Annealed
	UTS: 460 **Elon: 35%** **Proof: 200**
ASTM A351 CF8C	0.08% C 19.0% Cr 10.0% Ni 1.0% Nb steel: Annealed
	UTS: 460 **Elon: 30%** **Proof: 200**
ASTM A351 CF8M	0.08% C 19.0% Cr 10.0% Ni 2.0% Mo steel: Annealed
	UTS: 460 **Elon: 30%** **Proof: 200**
ASTM A351 CF10MC	0.1% C 16.0% Cr 14.0% Ni 2.0% Mo steel: Annealed
	UTS: 460 **Elon: 20%** **Proof: 200**
ASTM A351 CH8	0.08% C 24.0% Cr 13.0% Ni steel: Annealed
	UTS: 440 **Elon: 30%** **Proof: 180**
ASTM A351 CH10	0.10% C 24.0% Cr 13.0% Ni steel: Annealed
	UTS: 460 **Elon: 30%** **Proof: 200**
ASTM A351 CH20	0.20% C 24.0% Cr 13.0% Ni steel: Annealed
	UTS: 460 **Elon: 30%** **Proof: 200**
ASTM A351 CK45	0.35% C 24.0% Cr 20% Ni steel: Annealed
	UTS: 440 **Elon: 10%** **Proof: 220**
ASTM A351 CT35	0.35% C 14.0% Cr 34.0% Ni 0.5% Mo steel: Annealed
	UTS: 440 **Elon: 15%** **Proof: 180**
ASTM A362	Heat resistant tubular castings; analysis as ASTM A296 and A297
ASTM A371 ER308	0.08% C 20.25% Cr 10% Ni steel: Corrosion resistant for welding rods & electrodes
ASTM A371 ER308L	0.03% C 20.25% Cr 10% Ni steel: Corrosion resistant for welding rods & electrodes

Symbol	Nominal analysis, supplier, condition and remarks.
ASTM A371 ER309	0.12% C 24% Cr 13.0% Ni steel: Corrosion resistant for welding rods & electrodes
ASTM A371 ER310	0.09% C 26.5% Cr 21.25% Ni steel: Corrosion resistant for welding rods & electrodes
ASTM A371 ER312	0.09% C 30.0% Cr 9.25% Ni steel: Corrosion resistant for welding rods & electrodes
ASTM A371 ER316	0.08% C 19.0% Cr 12.5% Ni 2.5% Mo steel: Corrosion resistant for welding rods & electrodes
ASTM A371-ER316L	0.03% C 19.0% Cr 12.5% Ni 2.5% Mo steel: Corrosion resistant for welding rods & electrodes
ASTM A371-ER317	0.08% C 19.5% Cr 14.0% Ni 3.5% Mo steel: Corrosion resistant for welding rods & electrodes
ASTM A371-ER318	0.08% C 19.0% Cr 12.5% Ni 2.5% Mo 0.8% Ta + Nb steel: Corrosion resistant for welding rods & electrodes
ASTM A376	Stainless steel: Pipe for high temperature service; see AISI range for grades
ASTM A403	Stainless steel: For welders fittings; see AISI range for grades
ASTM A409	Stainless steel: For pipes; see AISI for grades
ASTM A430	Stainless steel: Forged or bored pipe; see AISI range for grades
ASTM A451 CPF8	0.08% C (max) 9.5% Ni 19.5% Cr cast steel: Pipe
ASTM A451 CPF8C	0.08% C (max) 10.5% Ni 19.5% Cr 0.9% Nb cast steel: Pipe
ASTM A451 CPF8M	0.08% C (max) 9.5% Ni 19.5% Cr 2.5% Mo cast steel: Pipe
ASTM A451 CPF10MC	0.1% C (max) 14.5% Ni 16.5% Cr 2.0% Mo 1.0% Nb cast steel: Pipe
ASTM A451 CPH8	0.08% C 13.5% Ni 24% Cr cast steel: Pipe
ASTM A451 CPH10	0.2% C (max) 13.5% Ni 24% Cr cast steel: Pipe; alternate name for CPH 20
ASTM A451 CPH20	0.2% C (max) 13.5% Ni 24% Cr cast steel: Pipe; alternate name for CPH 10
ASTM A451 CPK20	0.2% C (max) 20.5% Ni 25% Cr cast steel: Pipe
ASTM A452	Austenitic steel: Cast pipe; subsequently cold worked and annealed; three grades as AISI standards
ASTM A473	Stainless steel: For forgings; graded by AISI system; includes steel in Section 44M1
ASTM A478	Stainless steel: Wire for wearing; graded by AISI system
ASTM A479	Stainless steel: Bar & shapes for boilers, etc.; graded by AISI system
ASTM A492	Stainless steel: For wire rope; graded by AISI system
ASTM A493	Stainless steel: Wire for cold heading; graded by AISI system
ASTM A511	Austenitic seamless steel: Mechanical tubing; graded according to AISI system
ASTM A554	Stainless steel: Welded tubing; AISI symbols used for different grades
ASTM A567/3	0.2% C (max) 1.5% Mn 21.2% Cr 20% Ni 19.7% Co 3.0% Mo 2.5% W 1.0% Nb + Ta iron: Casting as AISI 661
ASTM A567 HK40	0.4% C 26% Cr 20% Ni 0.1% Ni Fe alloy: Casting
ASTM A567 HK50	0.5% C 26% Cr 20% Ni 0.1% Ni Fe alloy: Casting
ASTM A579	0.09% C (max) 17.0% Cr 7.0% Ni 1.2% Al steel: Forging
ASTM A579	0.09% C (max) 14.75% Cr 7.2% Ni 2.3% Mo 1.0% Al steel: Forging
ASTM A580	Stainless steel: Wire classified by AISI system; see also Sections 44M1 & 44M2
ASTM A581/303	0.15% C (max) 18% Cr 9% Ni 0.15% S (min) steel: Wire; stainless; free machining
ASTM A581/303 Se	0.15% C (max) 18% Cr 9% Ni 0.15% Se (min) steel: Wire; stainless; free machining
ASTM A581 XM2	0.15% C (max) 18% Cr 9% Ni 0.13% S 0.5% Mo 0.8% Al steel: Wire; free machining

Symbol	Nominal analysis, supplier, condition and remarks.
ASTM A581 XM3	0.15% C (max) 18% Cr 9% Ni 0.2% S 0.6% Mo 0.2% Pb steel: Wire; free machining
ASTM A582	Stainless steel: Wire; graded in same manner as ASTM A581; see also Sections 44M1 and 44P
ASTM A608 H135	0.35% C 28.0% Cr 16.0% Ni 0.5% Mo (max) steel: Tube; centrifugally cast
ASTM A608 HC30	0.7% Mn 4.0% Ni 28.0% Cr 0.5% Mo (max) steel: Tube; centrifugally cast; properties at 760 °C **UTS: 365 Elon: 40%**
ASTM A608 HD50	0.5% C 28.0% Cr 5.5% Ni 0.5% Mo (max) steel: Tube; centrifugally cast; properties at 760 °C **UTS: 514**
ASTM A608 HE35	0.3% C 21.0% Cr 11.0% Ni 0.5% Mo (max) steel: Tube; centrifugally cast
ASTM A608 HF30	0.3% C 21.0% Cr 10.5% Ni 0.5% Mo (max) steel: Tube; centrifugally cast; properties at 760 °C **UTS: 179 Elon: 7%**
ASTM A608 HH30	0.3% C 26.0% Cr 12.5% Ni 0.5% Mo (max) steel: Tube; centrifugally cast
ASTM A608 HH33	0.32% C 25% Cr 13% Ni 0.5% Mo (max) steel: Tube; centrifugally cast
ASTM A608 HK30	0.3% C 25.0% Cr 20.5% Ni 0.5% Mo (max) steel: Tube; centrifugally cast; properties at 760 °C **UTS: 179**
ASTM A608 HK40	0.4% C 25.0% Cr 20.5% Ni 0.5% Mo (max) steel: Tube; centrifugally cast; properties at 760 °C **UTS: 200 Elon: 7%**
ASTM A608 HL30	0.3% C 30.0% Cr 20.0% Ni 0.5% Mo (max) steel: Tube; centrifugally cast
ASTM A608 HL40	0.4% C 30.0% Cr 20.0% Ni 0.5% Mo (max) steel: Tube; centrifugally cast
ASTM A608 HN40	0.4% C 21.0% Cr 25.0% Ni 0.5% Mo (max) steel: Tube; centrifugally cast
ASTM A608 HT50	0.5% C 17.0% Cr 35% Ni 0.5% Mo (max) steel: Tube; centrifugally cast
ASTM A608 HU50	0.5% C 19.0% Cr 39% Ni 0.5% Mo (max) steel: Tube; centrifugally cast
ASTM A632 TP304	0.08% C (max) 19.0% Cr 9.5% Ni steel: Tube
ASTM A632 TP304L	0.04% C (max) 19.0% Cr 10.5% Ni steel: Tube
ASTM A632 TP310	0.15% C 25.0% Cr 20.5% Ni steel: Tube
ASTM A632 TP316	0.08% C 17.0% Cr 12.5% Ni 2.5% Mo steel: Tube
ASTM A632 TP316L	0.04% C 17.0% Cr 15.5% Ni 2.5% Mo steel: Tube
ASTM A632 TP317	0.08% C 19.0% Cr 12.5% Ni 3.5% Mo steel: Tube
ASTM A632 TP321	0.08% C 18.5% Cr 11.0% Ni steel: Tube
ASTM A632 TP347	0.08% C 18.5% Cr 11.0% Ni steel: Tube
ASTM A632 TP348	0.08% C 18.5% Cr 11.0% Ni steel: Tube
ASTM A651 TP304	0.08% C 19.0% Cr 9.5% Ni steel: Tube for water
ASTM A651 TP316	0.08% C 17.0% Cr 12.5% Ni 2.5% Mo steel: Tube for water
ASTM A666	Austenitic stainless steel: For structures; graded by AISI system
ASTM A666	Stainless steel: Sheet, plate, bar, etc.; for structural purposes; graded by AISI system (see also Section 44P)
ASTM A669	0.03% C (max) 1.6% Mn 18.5% Cr 4.75% Ni 2.7% Mo steel: Tube; ferritic & austenitic structure **DPN: 290 UTS: 630 Elon: 30% Proof: 440**
ASTM A688 TP304	0.08% C 19.0% Cr 9.5% Ni steel: Welded tube

Note. The following abbreviations and units are used in the tables:

DPN	Hardness, diamond pyramid number
UTS	Ultimate tensile strength, N/mm^2
Elon	Elongation, %
Proof	0.1 % proof strength, N/mm^2

1 N/mm^2=0.1 hbar=0.102 kgf/mm^2=0.06475 tonf/in^2=145.04 lbf/in^2

Symbol	Nominal analysis, supplier, condition and remarks.
ASTM A688 TP304L	0.035% C 19.0% Cr 10.5% Ni steel: Welded tube
ASTM A688 TP316	0.08% C 17.0% Cr 12.5% Ni 2.5% Mo steel: Welded tube
ASTM A688 TP316L	0.035% C 17.0% Cr 12.5% Ni 2.5% Mo steel: Welded tube
ASTRANIT N/Z	0.08% C 18% Cr 9.0% Ni steel: Krupp
ASTRANIT SST/Z	0.08% C 18% Cr 10% Ni steel: Krupp
ASV	6% Cr 42% Ni Fe alloy: Low expansion; Imphy
AURIGA XXIVB	0.12% C 18% Cr 9% Ni 1.0% Nb steel: Casting; D Brown; annealed **UTS: 450 Elon: 20% Proof: 200**
AURIGA XXIVB FCG	0.12% C 18% Cr 9% Ni 0.3% Mo 1% Nb steel: Casting; D Brown; annealed **UTS: 450 Elon: 15% Proof: 200**
AURIGA XXIVC	0.12% C 17% Cr 10% Ni 2.5% Mo 1% Nb steel: Casting; D Brown; annealed **UTS: 450 Elon: 15% Proof: 200**
AURIGA XXVII	0.3% C 24% Cr 13% Ni 3% W steel: Casting; D Brown; annealed **UTS: 530 Elon: 10% Proof: 220**
AURIGA XXVIIA	0.3% C 25% Cr 12% Ni steel: Casting; D Brown; annealed **UTS: 530 Elon: 10% Proof: 220**
AURIGA XXVIII	0.25% C 29% Cr 3% Ni steel: Casting; D Brown **UTS: 370**
AURIGA XXXIII	0.5% C 18% Cr 27% Ni Fe alloy: Casting; D Brown **UTS: 420**
AUSTINOX	0.06% C 18% Cr 10% Ni steel: Pompey; annealed **UTS: 540 Elon: 48% Proof: 200**
AUSTINOX 3	0.02% C 18% Cr 10% Ni steel: Pompey **UTS: 460 Elon: 45% Proof: 180**
AUSTINOX B	0.06% C 18% Cr 12% Ni steel: Pompey **UTS: 580 Elon: 45% Proof: 210**
AUSTINOX BB	0.08% C 18% Cr 12% Ni 2.7% Mo steel: Pompey
AUSTINOX S	0.06% C 18% Cr 10% Ni 0.3% Ti steel: Pompey **UTS: 600 Elon: 45% Proof: 210**
AUSTINOX SB	0.06% C 18% Cr 12% Ni 2.2% Mo 0.3% Ti steel: Pompey **UTS: 570 Elon: 40% Proof: 210**
AUSTINOX SN	0.06% C 18% Cr 10% Ni 0.6% Nb steel: Pompey
AUTAAS 101	0.04% C (max) 19.2% Cr 9.5% Ni steel: Origin unknown
AUTOROD 16.10	0.025% C 20.0% Cr 9.0% Ni steel: Wire for welding; Esab; as welded **UTS: 580 Elon: 45% Proof: 390**
AUTOROD 16.12	0.025% C 20% Cr 10.0% Ni steel: For welding wire; Esab; as welded **UTS: 600 Elon: 45% Proof: 380**
AUTOROD 16.30	0.025% C 20% Cr 12.0% Ni 2.8% Mo steel: Welding wire; Esab; as welded **UTS: 620 Elon: 35% Proof: 340**
AUTOROD 16.32	0.025% C 20.0% Cr 12.0% Ni 2.8% Mo steel: Welding wire; Esab; as welded **UTS: 600 Elon: 40% Proof: 380**
B 3312a	0.05% C 17% Cr 10% Ni 1.6% Mo steel: French National Standard **DPN: 180 UTS: 480 Elon: 45% Proof: 180**
BB 0K	0.12% C 19% Cr 10% Ni 2% Mo steel: Bar, billets, etc.; Brown Bayley for BS alloy En 58H
BB 2K	0.08% C (max) 17% Cr 12% Ni 2.5% Mo steel: Bar, billets, etc.; Brown Bayley for BS alloy En 58H & J
BB 4K	0.08% C (max) 19% Cr 14% Ni 3.5% Mo steel: Bar, billets, etc.; Brown Bayley for BS alloy En 58J
BBMK	18% Cr 18% Ni 2% Mo 2% Cu steel: Bar, billets, etc.; Brown Bayley
BLAW KNOX Mo-RE1	0.5% C (max) 1.3% W 22% Cr 30% Ni Fe alloy: Casting; for furnace fittings, etc.; Wellman

Symbol	Nominal analysis, supplier, condition and remarks.
BRIGHTRAY F	0.15% C 2% Si 18% Cr 37% Ni Fe alloy: Wrought; H Wiggin; annealed; for use up to 1000 °C when carburizing **DPN: 186 UTS: 720**
BS	0.12% C 18% Cr 8% Ni steel: Osborn **UTS: 700 Elon: 50% Proof: 270**
BS 2	0.08% C 14% Cr 13% Ni steel: Osborn; for deep drawing **UTS: 610 Elon: 60% Proof: 220**
BS 970/302 S25	0.12% C (max) 9.5% Ni 18% Cr steel: Annealed; replaces En 58A **DPN: 183 UTS: 500 Elon: 40% Proof: 200**
BS 970/303 S21	0.12% C (max) 9.5% Ni 18% Cr steel: Annealed; replaces En 58M **DPN: 183 UTS: 500 Elon: 40% Proof: 200**
BS 970/303 S41	0.12% C (max) 9.5% Ni 18% Cr steel: Annealed; replaces En 58M **DPN: 183 UTS: 500 Elon: 40% Proof: 200**
BS 970/304 S12	0.03% C (max) 10.5% Ni 18.2% Cr steel: Annealed **DPN: 183 UTS: 450 Elon: 40% Proof: 170**
BS 970/304 S15	0.06% C (max) 9.5% Ni 18.2% Cr steel: Annealed; replaces En 58E **DPN: 183 UTS: 450 Elon: 40% Proof: 170**
BS 970/310 S24	0.15% C (max) 20.5% Ni 24.5% Cr steel: Annealed **DPN: 207 UTS: 750 Elon: 40% Proof: 210**
BS 970/315 S16	0.07% C (max) 10% Ni 17.5% Cr 1.5% Mo steel: Annealed; replaces En 58H **DPN: 183 UTS: 450 Elon: 40% Proof: 170**
BS 970/316 S12	0.03% C (max) 12.5% Ni 17.5% Cr 2.7% Mo steel: Annealed **DPN: 183 UTS: 450 Elon: 40% Proof: 170**
BS 970/316 S16	0.07% C 11.5% Ni 17.5% Cr 2.7% Mo steel: Annealed **DPN: 183 UTS: 450 Elon: 40% Proof: 170**
BS 970/317 S12	0.03% C (max) 15.5% Ni 18.5% Cr 3.5% Mo steel: Annealed **DPN: 183 UTS: 450 Elon: 40% Proof: 170**
BS 970/317 S16	0.06% C (max) 13.5% Ni 18.5% Cr 3.5% Mo steel: Annealed **DPN: 183 UTS: 450 Elon: 40% Proof: 170**
BS 970/320 S17	0.08% C (max) 12.5% Ni 17.5% Cr 2.7% Mo 0.5% Ti steel: Annealed; replaces En 58J **DPN: 183 UTS: 470 Elon: 40% Proof: 190**
BS 970/321 S12	0.08% C (max) 10.5% Ni 18% Cr 0.5% Ti steel: Annealed; replaces En 58B and En 58C **DPN: 183 UTS: 470 Elon: 40% Proof: 200**
BS 970/321 S20	0.12% C (max) 9.5% Ni 18% Cr 0.5% Ti steel: Annealed; replaces En 58B and En 58C **DPN: 183 UTS: 500 Elon: 40% Proof: 210**
BS 970/325 S21	0.12% C (max) 9.5% Ni 18% Cr 0.6% Ti steel: Annealed; replaces En 58M **DPN: 183 UTS: 500 Elon: 40% Proof: 200**
BS 970/326 S36	0.12% C (max) 11.5% Ni 17.5% Cr 2.7% Mo steel: Annealed **DPN: 183 UTS: 500 Elon: 40% Proof: 200**
BS 970/347 S17	0.08% C (max) 10.5% Ni 18% Cr 0.8% Nb steel: Annealed; replaces En 58F and En 58G **DPN: 183 UTS: 500 Elon: 40% Proof: 210**
BS 970 En 54	0.42% C 13% Cr 10% Ni 3% W steel: Bar & forging; annealed; for valves **DPN: 302**
BS 970 En 55	0.3% C 17% Cr 9% Ni 3% W steel: Bar & forging; annealed; for valves **DPN: 302**
BS 970 En 58A	0.16% C (max) 18% Cr 9% Ni steel: Annealed **UTS: 530 Elon: 30% Proof: 180**
BS 970 En 58B	0.15% C (max) 18% Cr 9% Ni 0.6% Ti steel: Annealed; weldable **UTS: 530 Elon: 30% Proof: 180**
BS 970 En 58C	0.15% C (max) 19% Cr 11% Ni 0.6% Ti steel: Annealed; weldable **UTS: 530 Elon: 30% Proof: 180**
BS 970 En 58D	0.16% C (max) 18% Cr 13% Ni steel: Sheet; annealed for deep drawing
BS 970 En 58E	0.08% C (max) 18% Cr 9% Ni steel: Annealed **UTS: 530 Elon: 30% Proof: 180**
BS 970 En 58F	0.15% C (max) 18% Cr 8% Ni Nb steel: Annealed; weldable **UTS: 530 Elon: 30% Proof: 180**
BS 970 En 58G	0.15% C (max) 18% Cr 10% Ni Nb steel: Annealed; weldable **UTS: 530 Elon: 30% Proof: 180**
BS 970 En 58H	0.12% C (max) 18% Cr 9% Ni 2.0% Mo steel: Annealed; weldable; Ti or Nb optional; acid resistant **UTS: 530 Elon: 30% Proof: 180**
BS 970 En 58J	0.12% C (max) 18% Cr 9% Ni 3% Mo steel: Annealed; weldable; Ti or Nb optional; acid resistant **UTS: 530 Elon: 30% Proof: 180**
BS 970 En 58M	Free machining grade of En 58; specification to be agreed by suppliers
BS 980 CDS 19	0.16% C (max) 18% Cr 8% Ni steel: Tube; cold drawn & annealed **UTS: 530 Proof: 220**
BS 980 CDS 20	0.16% C (max) 18% Cr 8% Ni Ti steel: Tube; cold drawn & annealed; weldable **UTS: 530 Proof: 220**
BS 1449/304 S12	0.03% C 10.5% Ni 18% Cr steel: Plate & sheet; weldable **DPN: 192 UTS: 500 Elon: 40% Proof: 200**
BS 1449/304 S15	0.06% C 9% Ni 18% Cr steel: Plate & sheet; annealed **DPN: 192 UTS: 520 Elon: 40% Proof: 210**
BS 1449/309 S24	0.15% C (max) 14.5% Ni 23.5% Cr steel: Plate & sheet; annealed **DPN: 207 UTS: 550 Elon: 40% Proof: 220**
BS 1449/310 S24	0.15% C (max) 20.5% Ni 24.5% Cr steel: Plate & sheet; annealed **DPN: 207 UTS: 550 Elon: 40% Proof: 220**
BS 1449/312 S24	0.15% C (max) 17.5% Ni 24.5% Cr steel: Plate & sheet; annealed **DPN: 207 UTS: 550 Elon: 40% Proof: 220**
BS 1449/316 S12	0.03% C (max) 12.5% Ni 17.5% Cr 2.7% Mo steel: Plate & sheet; annealed **DPN: 197 UTS: 500 Elon: 40% Proof: 200**
BS 1449/316 S16	0.07% C 11.5% Ni 17.5% Cr 2.7% Mo steel: Annealed **DPN: 207 UTS: 550 Elon: 40% Proof: 210**
BS 1449/317 S16	0.06% C (max) 13.5% Ni 18.5% Cr 3.5% Mo steel: Plate & sheet; annealed **DPN: 207 UTS: 550 Elon: 35% Proof: 210**
BS 1449/320 S17	0.08% C 12.5% Ni 17.5% Cr 2.7% Mo 0.5% Ti steel: Plate & sheet; annealed **DPN: 207 UTS: 550 Elon: 40% Proof: 210**
BS 1449/321 S12	0.08% C 10.5% Ni 18% Cr 0.6% Ti steel: Plate & sheet; weldable; annealed **DPN: 202 UTS: 520 Elon: 40% Proof: 210**
BS 1449/347 S17	0.08% C (max) 10.5% Ni 18% Cr 0.9% Nb steel: Plate & sheet; weldable; annealed **DPN: 202 UTS: 520 Elon: 40% Proof: 210**
BS 1453 A8 Nb	0.1% C (max) 18% Cr 9% Ni 1% Nb steel: Weld filler rod; for welding austenitic steel
BS 1453 A10	0.15% C (max) 25% Cr 12% Ni steel: Weld filler rod; for welding heat resistant stainless steel

Symbol	Nominal analysis, supplier, condition and remarks.
BS 1453 A10 Nb	0.15% C (max) 25% Cr 12% Ni 1.2% Nb steel: Weld filler rod; for welding heat resistant stainless steel
BS 1453 A11	0.15% C (max) 1.5% Si 25% Cr 21% Ni steel: Weld filler rod; for welding heat resistant stainless steels
BS 1453 A11 Nb	0.15% C (max) 1.5% Si 25% Cr 21% Ni 1.2% Nb steel: Weld filler rod; for welding heat resistant stainless steels
BS 1453 A12	0.08% C (max) 17% Cr 11% Ni 3% Mo steel: Weld filler rod; for welding Mo bearing stainless steels
BS 1453 A12 Nb	0.08% C (max) 17% Cr 11% Ni 3% Mo 1.2% Nb steel: Weld filler rod; for welding Mo bearing stainless steels
BS 1501/801 B	0.08% C 19% Cr 10% Ni steel: Plate & bar; annealed **UTS: 530 Elon: 25% Proof: 210**
BS 1501/801 C	0.03% C 20% Cr 10.0% Ni steel: Plate & bar; annealed **UTS: 500 Elon: 25% Proof: 200**
BS 1501/821 Nb	0.08% C 20% Cr 9.0% Ni 0.8% Nb steel: Plate & bar; annealed **UTS: 530 Elon: 25% Proof: 200**
BS 1501/821 Ti	0.12% C 20% Cr 10% Ni 0.5% Ti steel: Plate & bar; annealed **UTS: 530 Elon: 25% Proof: 210**
BS 1501/845 B	0.08% C 18% Cr 10% Ni 2.5% Mo steel: Plate & bar; annealed **UTS: 530 Elon: 25% Proof: 210**
BS 1501/845 Ti	0.08% C 18% Cr 10% Ni 2.5% Mo 0.35% Ti steel: Plate & bar; annealed **UTS: 530 Elon: 25% Proof: 210**
BS 1501/846	0.08% C 20% Cr 12% Ni 3.5% Mo steel: Plate & bar; annealed **UTS: 530 Elon: 25% Proof: 210**
BS 1503/801	0.08% C 18.5% Cr 10.0% Ni steel: Forging; annealed **UTS: 530 Elon: 30% Proof: 210**
BS 1503/821 Nb	0.08% C 18% Cr 9.0% Ni 1.0% Nb steel: Forging; annealed **UTS: 530 Elon: 30% Proof: 210**
BS 1503/821 Ti	0.12% C 18% Cr 8% Ni 0.5% Ti steel: Forging; annealed **UTS: 530 Elon: 30% Proof: 210**
BS 1503/845 B	0.08% C 17.5% Cr 10% Ni 2.5% Mo steel: Forging; annealed **UTS: 530 Elon: 30% Proof: 200**
BS 1503/845 Ti	0.08% C 17.5% Cr 10% Ni 2.5% Mo 0.5% Ti steel: Annealed **UTS: 530 Elon: 30% Proof: 200**
BS 1503/846	0.08% C 19% Cr 13% Ni 3.5% Mo steel: Forging; annealed **UTS: 530 Elon: 30% Proof: 200**
BS 1504/801	0.12% C (max) 18% Cr 8% Ni steel: Casting; annealed **UTS: 450 Elon: 20% Proof: 200**
BS 1504/821 Nb	0.12% C (max) 18% Cr 9% Ni 1.0% Nb steel: Casting; annealed **UTS: 450 Elon: 20% Proof: 200**

Symbol	Nominal analysis, supplier, condition and remarks.
BS 1504/821 Ti	0.12% C (max) 18% Cr 8% Ni 0.5% Ti steel: Casting; annealed **UTS: 450 Elon: 20% Proof: 200**
BS 1504/845 B	0.08% C (max) 17% Cr 11% Ni 2.5% Mo steel: Casting; annealed **UTS: 450 Elon: 15% Proof: 200**
BS 1504/845 Nb	0.12% C (max) 17% Cr 11% Ni 2.5% Mo steel: Casting; annealed **UTS: 450 Elon: 15% Proof: 200**
BS 1504/846	0.08% C 19% Cr 12% Ni 3.5% Mo steel: Casting; annealed **UTS: 450 Elon: 15% Proof: 200**
BS 1506/801 A	0.16% C (max) 18% Cr 9% Ni steel: Bar; annealed **UTS: 530 Elon: 30% Proof: 200**
BS 1506/801 AM	0.16% C (max) 18% Cr 9% Ni 0.3% S steel: Bar; free machining; annealed **UTS: 530 Elon: 30% Proof: 200**
BS 1506/801 B	0.08% C (max) 18% Cr 9% Ni steel: Bar; annealed **UTS: 530 Elon: 30% Proof: 200**
BS 1506/801 C	0.03% C 18% Cr 11% Ni steel: Bar; annealed **UTS: 530 Elon: 30% Proof: 200**
BS 1506/821 Nb	0.08% C (max) 18% Cr 9% Ni 1.0% Nb steel: Bar; annealed **UTS: 530 Elon: 30% Proof: 200**
BS 1506/821 Ti	0.12% C 18% Cr 8% Ni 0.6% Ti steel: Bar; annealed **UTS: 530 Elon: 30% Proof: 200**
BS 1506/821 Ti M	0.12% C (max) 18% Cr 8% Ni 0.5% Ti 0.5% Zr steel: Bar; free machining; annealed **UTS: 530 Elon: 30% Proof: 200**
BS 1506/845	0.08% C (max) 17% Cr 10% Ni 2.7% Mo steel: Bar; annealed **UTS: 530 Elon: 30% Proof: 200**
BS 1507/821	0.15% C 2% Mn 10% Ni 18% Cr Ti or Nb stabilized seamless steel: Pipe; annealed **UTS: 530 Elon: 30%**
BS 1507/825	0.15% C 24% Cr 21% Ni 1% Ti seamless steel: Pipe; annealed **UTS: 530 Elon: 30%**
BS 1507/845	0.08% C 17% Cr 10% Ni 3% Mo seamless steel: Pipe; annealed; weldable **UTS: 530 Elon: 30%**
BS 1508/801	0.16% C 18% Cr 10% Ni steel: Tube; annealed **UTS: 530 Elon: 30%**
BS 1508/821	0.15% C 18% Cr 10% Ni 0.7% Ti steel: Tube; annealed **UTS: 530 Elon: 30%**
BS 1508/825	0.15% C 24% Cr 22% Ni 1.0% Ti steel: Tube; annealed **UTS: 530 Elon: 30%**
BS 1508/845	0.08% C 17% Cr 10% Ni 3.0% Mo steel: Tube; annealed **UTS: 530 Elon: 30%**
BS 1554 En 58A	0.16% C 18% Cr 9% Ni steel: Wire
BS 1554 En 58B	0.15% C 2% Mn 18% Cr 8% Ni steel: Wire
BS 1554 En 58C	0.15% C 18% Cr 11% Ni steel: Wire
BS 1554 En 58D	0.16% C 13% Cr 13% Ni steel: Wire
BS 1554 En 58E	0.08% C 18% Cr 10% Ni steel: Wire
BS 1554 En 58F	0.15% C 18% Cr 8% Ni steel: Wire
BS 1554 En 58G	0.15% C 18% Cr 11% Ni steel: Wire
BS 1554 En 58H	0.12% C 18% Cr 10% Ni steel: Wire
BS 1554 En 58J	0.12% C 18% Cr 10% Ni steel: Wire
BS 1631 A	0.12% C (max) 17% Cr 8% Ni steel: Casting; annealed; contained in BS 3100 **UTS: 450 Elon: 20% Proof: 200**
BS 1631 B Nb	0.12% C (max) 18% Cr 9% Ni 1.0% Nb steel: Casting; annealed; contained in BS 3100 **UTS: 450 Elon: 20% Proof: 200**

Note. The following abbreviations and units are used in the tables:

DPN	Hardness, diamond pyramid number
UTS	Ultimate tensile strength, N/mm^2
Elon	Elongation, %
Proof	0.1 % proof strength, N/mm^2

1 N/mm^2=0.1 hbar=0.102 kgf/mm^2=0.06475 tonf/in^2=145.04 lbf/in^2

Symbol	Nominal analysis, supplier, condition and remarks.
BS 1631 B Ti	0.12% C (max) 18% Cr 8% Ni 0.6% Ti steel: Casting; annealed; contained in BS 3100 **UTS: 450 Elon: 20% Proof: 200**
BS 1632 A	0.08% C (max) 19% Cr 12% Ni 3.5% Mo steel: Casting; annealed; contained in BS 3100 **UTS: 450 Elon: 15% Proof: 200**
BS 1632 B	0.08% C (max) 17% Cr 16% Ni 2.5% Mo steel: Casting; annealed; contained in BS 3100 **UTS: 450 Elon: 15% Proof: 200**
BS 1632 C Nb	0.12% C (max) 17% Cr 10% Ni 2.5% Mo 1.0% Nb steel: Casting; annealed; contained in BS 3100 **UTS: 450 Elon: 15% Proof: 200**
BS 1632 C Ti	0.12% C (max) 17% Cr 10% Ni 2.5% Mo 0.6% Ti steel: Casting; annealed; contained in BS 3100 **UTS: 450 Elon: 15% Proof: 200**
BS 1632 D	0.08% C (max) 17% Cr 8% Ni 2.5% Mo steel: Casting; annealed; contained in BS 3100 **UTS: 450 Elon: 15% Proof: 200**
BS 1648 D	0.4% C (max) 19% Cr 6% Ni steel: Casting; as cast; contained in BS 3100
BS 1648 E	0.5% C (max) 25% Cr 12% Ni steel: Casting; as cast; contained in BS 3100
BS 1648 F	0.5% C (max) 25% Cr 20% Ni steel: As cast; contained in BS 3100
BS 1648 G	0.5% C (max) 17% Cr 25% Ni steel: Casting; as cast; contained in BS 3100
BS 2901 A8 Nb	0.1% C (max) 18% Cr 9% Ni 1.0% Nb steel: Rod for all forms of welding
BS 2901 A8 Ti	0.1% C (max) 18% Cr 9% Ni 0.7% Ti steel: Rod for all forms of welding
BS 2901 A10	0.12% C 25% Cr 12% Ni steel: Rod for all forms of welding
BS 2901 A10 Nb	0.12% C 25% Cr 12% Ni 1.0% Nb steel: Rod for gas and electric welding
BS 2901 A10 Ti	0.12% C 25% Cr 12% Ni 0.7% Ti steel: Rod for all forms of welding
BS 2901 A11	0.12% C 25% Cr 21% Ni steel: Rod for all forms of welding
BS 2901 A11 Ti	0.12% C 25% Cr 21% Ni 0.7% Ti steel: For all forms of welding
BS 2901 A12	0.08% C 17% Cr 11% Ni 3% Mo steel: For all forms of welding
BS 2901 A12 Nb	0.08% C 17% Cr 11% Ni 3% Mo 1% Nb steel: Rod for gas & electric welding
BS 2901 A12 Ti	0.08% C 17% Cr 11% Ni 3% Mo 0.5% Ti steel: Rod for all forms of welding
BS 2926/19/9	0.08% C (max) 10% Ni 20% Cr steel: For welding electrodes
BS 2926/19/9L	0.04% C (max) 10% Ni 20% Cr steel: For welding electrodes
BS 2926/19/9 Nb	0.1% C 10% Ni 20% Cr 1.0% Nb steel: For welding electrodes
BS 2926/19/12/3	0.08% C (max) 12% Ni 19% Cr 3.0% Mo steel: For welding electrodes
BS 2926/19/12/3 L	0.04% C (max) 12% Ni 19% Cr 3.0% Mo steel: For welding electrodes
BS 2926/19/12/3 Nb	0.1% C (max) 12% Ni 19% Cr 3.0% Mo 1.0% Nb steel: For welding electrodes
BS 2926/19/13/4	0.08% C (max) 13% Ni 19% Cr 4.5% Mo steel: For welding electrodes
BS 2926/19/13/4 L	0.04% C 13% Ni 19% Cr 4.5% Mo steel: For welding electrodes
BS 2926/19/13/4 Nb	0.1% C (max) 13% Ni 19% Cr 4.5% Mo 1.0% Nb steel: For welding electrodes
BS 2926/23/12	0.15% C (max) 12.5% Ni 23.5% Cr steel: For welding electrodes
BS 2926/23/12 Nb	0.1% C (max) 12.5% Ni 23.5% Cr 1.0% Nb steel: For welding electrodes
BS 2926/23/12/2	0.1% C (max) 12.5% Ni 23.5% Cr 2.5% Mo steel: For welding electrodes
BS 2926/23/12 W	0.2% C (max) 12.5% Ni 23.5% Cr 3.0% W steel: For welding electrodes
BS 2926/25/20	0.2% C (max) 20% Ni 26% Cr steel: For welding electrodes
BS 2926/25/20 H	0.4% C 20% Ni 26% Cr steel: For welding electrodes
BS 2926/25/20 Nb	0.12% C (max) 20% Ni 26% Cr 1.0% Nb steel: For welding electrodes
BS 2926 A	0.08% C 19% Cr 10% Ni steel: Welding electrode; as welded **UTS: 540 Elon: 35%**
BS 2926 B	0.1% C 19% Cr 9% Ni 1.0% Nb steel: Welding electrode; as welded **UTS: 540 Elon: 30%**
BS 2926 C	0.08% C 19% Cr 12% Ni 3.5% Mo 1.0% Nb steel: Welding electrode; as welded **UTS: 540 Elon: 30%**
BS 2926 D	0.5% C 17% Cr 11% Ni 3% Mo 1.0% Nb steel: Welding electrode; as welded **UTS: 540 Elon: 25%**
BS 2926 E	0.6% C 19% Cr 9% Ni 3% Mo steel: Welding electrode; as welded **UTS: 540 Elon: 25%**
BS 2926 F	0.6% C 24% Cr 12% Ni steel: Welding electrode; as welded **UTS: 540 Elon: 35%**
BS 2926 G	1% C 25% Cr 12% Ni 3% W steel: Welding electrode; as welded **UTS: 540 Elon: 25%**
BS 3014/1	0.08% C 18% Cr 9% Ni steel: Welded tube; annealed **UTS: 530 Elon: 45% Proof: 180**
BS 2014/2	0.15% C 18% Cr 8% Ni steel: Welded tube; annealed **UTS: 530 Elon: 45% Proof: 180**
BS 3014/3	0.08% C 18% Cr 9% Ni 0.5% Ti steel: Welded tube; annealed **UTS: 530 Elon: 45% Proof: 180**
BS 3014/4	0.15% C 18% Cr 9% Ni 0.6% Ti steel: Welded tube; annealed **UTS: 530 Elon: 45% Proof: 180**
BS 3014/5	0.08% C 19% Cr 11% Ni 0.8% Nb: Welded tube; annealed **UTS: 530 Elon: 45% Proof: 180**
BS 3014/6	0.08% C 17.5% Cr 11% Ni 3.0% Mo steel: Welded tube; annealed **UTS: 530 Elon: 45% Proof: 180**
BS 3076 NA 15	21% Cr 32% Ni 0.4% Al 0.4% Ti Fe alloy: Rod
BS 3100/302 C25	0.12% C 19.0% Cr 8.0% Ni steel: Casting
BS 3100/304 C12	0.03% C 19.0% Cr 8.0% Ni steel: Casting
BS 3100/304 C15	0.08% C 19.0% Cr 8.0% Ni steel: Casting
BS 3100/347 C17	0.08% C 19.0% Cr 8.5% Ni 1.0% Nb steel: Casting
BS 3100/302 C35	0.3% C (max) 19.5% Cr 8.0% Ni steel: Casting
BS 3100/309 C30	0.5% C (max) 23.5% Cr 12.0% Ni 1.0% Mo steel: Casting
BS 3100/309 C40	0.5% C (max) 27.5% Cr 10.0% Ni 1.0% Mo steel: Casting
BS 3100/310 C45	0.5% C (max) 24.5% Cr 19.5% Ni 1.0% Mo steel: Casting
BS 3100/315 C16	0.08% C 19.0% Cr 8.0% Ni 1.3% Mo steel: Casting
BS 3100/316 C12	0.03% C 19.0% Cr 10.0% Ni 2.5% Mo steel: Casting
BS 3100/316 C16	0.08% C 19.0% Cr 10.0% Ni 2.5% Mo steel: Casting
BS 3127/7	0.08% C 17% Cr 12% Ni 2.5% Mo steel: Tube; lightly cold worked; for bourdon tubes **DPN: 220**

Symbol	Nominal analysis, supplier, condition and remarks.
BS 3146 ANC 3A	0.1% C 17% Cr 8% Ni steel: Investment casting; annealed **UTS: 450 Elon: 25%**
BS 3146 ANC 3BNb	0.1% C 19% Cr 9% Ni 1.0% Nb steel: Investment casting; annealed **UTS: 450 Elon: 25% Proof: 210**
BS 3146 ANC 4A	0.08% C (max) 19% Cr 13% Ni 3.5% Mo steel: Investment casting; annealed **UTS: 450 Elon: 15% Proof: 210**
BS 3146 ANC 4B	0.08% C (max) 17% Cr 10% Ni 2.5% Mo steel: Investment casting; annealed **UTS: 450 Elon: 15% Proof: 210**
BS 3146 ANC 4CNb	0.12% C (max) 17% Cr 10% Ni 2.5% Mo 1.0% Nb steel: Investment casting; annealed **UTS: 450 Elon: 15% Proof: 210**
BS 3146 ANC 5A	0.5% C 25% Cr 20% Ni steel: Investment casting; as cast
BS 3146 ANC 5B	0.5% C 20% Cr 40% Ni steel: Investment casting; as cast
BS 3146 ANC 6A	0.2% C 22% Cr 12% Ni steel: Investment casting; as cast **DPN: 250 UTS: 450 Elon: 20%**
BS 3146 ANC 6B	0.22% C 22% Cr 12% Ni 3% W steel: Investment casting; as cast **DPN: 250 UTS: 450 Elon: 20%**
BS 3146 ANC 7	0.35% C 19% Cr 13% Ni 1.7% Mo 2.5% W 3% Nb 10% Co steel: Casting; as cast **UTS: 530 Proof: 300**
BS 3532/2	Austenitic stainless steel for implants and surgery tools classified into types 316 S16, 316 S12, 317 S10
BS 4534/1	0.08% C 12% Ni 18% Cr 0.5% Mo (max) cast steel: Tube weldable; as cast **UTS: 450 Elon: 26%**
BS 4534/1 F	0.06% C 8% Ni 18.5% Cr cast steel: Tube; weldable; as cast **UTS: 480 Elon: 26%**
BS 4534/2	0.06% C 12.5% Ni 18% Cr 1.0% Nb 0.5% Mo (max) cast steel: Tube; weldable; as cast **UTS: 450 Elon: 22%**
BS 4534/2 F	0.06% C 8% Ni 18.5% Cr 1.0% Nb cast steel: Tube; weldable; as cast **UTS: 480 Elon: 22%**
BS 4534/3	0.06% C 13.5% Ni 17% Cr 2.5% Mo cast steel: Tube **UTS: 450 Elon: 26%**
BS 4534/3 F	0.06% C 8% Ni 18.5% Cr 2.5% Mo cast steel: Tube; weldable; as cast **UTS: 480 Elon: 26%**
BS 4534/4	0.3% C 10% Ni 19.5% Cr 0.5% Mo (max) cast steel: Tube **UTS: 480 Elon: 22%**
BS 4534/5	0.3% C 12% Ni 25% Cr 0.5% Mo (max) cast steel: Tube weldable; as cast; can be aged **UTS: 510 Elon: 9%**
BS 4534/6	0.4% C 20% Ni 25% Cr 0.5% Mo (max) cast steel: Tube **UTS: 480 Proof: 9%**

Symbol	Nominal analysis, supplier, condition and remarks.
BS 4534/7	0.15% C (max) 21% Ni 25% Cr 0.5% Mo (max) cast steel: Weldable tube; as cast **UTS: 450 Elon: 22%**
BS 4534/8	0.45% C 35% Ni 18% Cr 0.5% Mo (max) cast steel: Tube **UTS: 450 Elon: 7%**
BS 4534/9	0.5% C 35% Ni 18% Cr 0.5% Mo (max) cast steel: weldable tube; as cast **UTS: 450 Elon: 7%**
BS 4534/10	0.15% C (max) 35% Ni 18% Cr 0.5% Mo (max) cast steel: Weldable tube; as cast **UTS: 420 Elon: 22%**
BS 4534/11	0.1% C (max) 32% Ni 20.5% Cr 0.5% Mo (max) cast steel: Weldable tube; as cast **UTS: 400 Elon: 26%**
BS S85	Austenitic steel: Sheet; obsolete
BS S108	0.2% C (max) 23% Cr 14% Ni 1.5% Nb steel: Annealed **UTS: 530 Elon: 30% Proof: 200**
BS S109	0.2% C (max) 23% Cr 18% Ni 1.5% Nb steel: Annealed **UTS: 530 Elon: 30% Proof: 200**
BS S110	0.16% C (max) 18% Cr 9% Ni Ti or Nb steel: Annealed; stabilized **UTS: 530 Elon: 30% Proof: 200**
BS S111	0.42% C 14% Cr 14% Ni 0.4% Mo 2.8% W steel: Annealed **DPN: 269**
BS S125	0.15% C (max) 23% Cr 15% Ni 0.7% Ti steel: Annealed **UTS: 530 Elon: 30% Proof: 200**
BS S126	0.15% C (max) 23% Cr 14% Ni 1.1% Nb steel: Annealed **UTS: 530 Elon: 30% Proof: 200**
BS S127	0.15% C (max) 25% Cr 17% Ni 0.7% Ti steel: Annealed **UTS: 530 Elon: 30% Proof: 200**
BS S128	0.15% C (max) 25% Cr 17% Ni 1.1% Nb steel: Annealed **UTS: 530 Elon: 30% Proof: 200**
BS S129	0.12% C (max) 18% Cr 10% Ni 0.7% Ti steel: Annealed **UTS: 530 Elon: 30% Proof: 200**
BS S130	0.08% C 18% Cr 10% Ni 1.0% Nb steel: Annealed **UTS: 530 Elon: 30% Proof: 200**
BS S520	0.16% C (max) 18% Cr 10% Ni Ti or Nb steel: Sheet; stabilized; cold rolled & tempered **UTS: 920 Elon: 15% Proof: 620**
BS S521	0.16% C (max) 18% Cr 10% Ni Ti or Nb steel: Sheet; stabilized; annealed **UTS: 530 Elon: 30% Proof: 200**
BS S522	0.16% C (max) 23% Cr 14% Ni Ti or Nb steel: Sheet; stabilized; annealed **UTS: 530 Elon: 30%**
BS S523	0.16% C (max) 23% Cr 18% Ni Ti or Nb steel: Sheet; stabilized; annealed **UTS: 530 Elon: 30%**
BS M	0.06% C 18% Cr 8% Ni steel: Osborn; weldable **UTS: 700 Elon: 60% Proof: 270**
BS Cb	0.08% C 18% Cr 8% Ni steel: Osborn; weld stabilized **UTS: 700 Elon: 55% Proof: 270**
BS T	0.08% C 18% Cr 8% Ni steel: Osborn; weld stabilized **UTS: 700 Elon: 55% Proof: 270**
BS T55	0.16% C 18% Cr 10% Ni Ti or Nb stabilized steel: Tube; annealed **UTS: 530 Proof: 230**

Note. The following abbreviations and units are used in the tables:

DPN	Hardness, diamond pyramid number
UTS	Ultimate tensile strength, N/mm^2
Elon	Elongation, %
Proof	0.1 % proof strength, N/mm^2

$1 \ N/mm^2 = 0.1 \ hbar = 0.102 \ kgf/mm^2 = 0.06475 \ tonf/in^2 = 145.04 \ lbf/in^2$

Symbol	Nominal analysis, supplier, condition and remarks.
BS T58	0.16% C (max) 18% Cr 10% Ni Ti or Nb steel: Tube; stabilized; cold drawn & tempered **UTS: 760** **Proof: 700**
BS T61	0.16% C 23% Cr 17% Ni Ti or Nb steel: Tube; stabilized; annealed **UTS: 530** **Proof: 220**
BS W10	0.2% C 12% Cr 13% Ni steel: Wire
BS W11	0.2% C (max) 12% Cr 14% Ni steel: Wire
B & W A1	0.15% C 25% Cr 20% Ni steel: Tube; Babock & Wilcox (USA)
B & W A1/57	0.08% C 2.5% Si 18% Cr 12% Ni steel: Babcock & Wilcox (USA)
CALITE	4.5% Al 5.5% Cr 40% Ni Fe alloy: Heat resistant; origin unknown
CALMET	25% Cr 12% Ni steel: Casting; Calorizing Corp; heat resistant properties **DPN: 160** **UTS: 540** **Proof: 330**
CAMLOY	15% Cr 30% Ni Fe alloy: For marine valves; Cameron & Son; obsolete **UTS: 840** **Elon: 35%**
CARPENTER 20	0.05% C 20% Cr 28.5% Ni 2.25% Mo 3.25% Cu steel: Jessops
CD 4M	Symbol for Durcornet 100; Duriron Co
CE 30	0.3% C 28% Cr 9% Ni steel: Casting; Huntington; as cast **DPN: 170** **UTS: 700** **Elon: 15%** **Proof: 370**
CF 3	0.03% C 19% Cr 10% Ni steel: Casting; Huntington
CF 3M	0.03% C 19% Cr 11% Ni 2.5% Mo steel: Casting; Huntington
CF 8	0.08% C 19% Cr 9% Ni steel: Casting; Huntington; as cast **UTS: 540** **Elon: 55%** **Proof: 230**
CF 8C	0.08% C 19% Cr 11% Ni 1.0% Nb steel: Casting; Huntington; as cast **UTS: 540** **Elon: 9%** **Proof: 220**
CF 8 M	0.08% C 19% Cr 11% Ni 2.5% Mo steel: Casting; Huntington; as cast **UTS: 530** **Elon: 50%** **Proof: 240**
CF 12 M	0.12% C 19% Cr 11% Ni 2.5% Mo steel: Casting; Huntington; as cast **UTS: 530** **Elon: 50%** **Proof: 270**
CF 16 F	0.16% C 19% Cr 11% Ni 1.5% Mo 0.3% Se steel: Casting; Huntington; as cast; free machining **UTS: 540** **Elon: 52%** **Proof: 240**
CF 20	0.2% C 19% Cr 9% Ni steel: Casting; Huntington; as cast **UTS: 540** **Elon: 50%** **Proof: 210**
CG 8 M	0.08% C 19% Cr 9% Ni 3% Mo steel: Casting; Huntington
CH 20	0.2% C 24% Cr 13% Ni steel: Casting; Huntington; as cast **UTS: 620** **Elon: 38%** **Proof: 300**
CHROMAX	18% Cr 37% Ni Fe alloy: British Driver Harris; furnace elements **UTS: 700**
CK 20	0.2% C 25% Cr 21% Ni steel: Casting; Huntington; as cast **UTS: 520** **Elon: 37%** **Proof: 220**
CN 7 M	0.07% C 20% Cr 29% Ni 2.5% Mo 3.5% Cu steel: Casting; Huntington; as cast **UTS: 450** **Elon: 48%** **Proof: 180**
COLMONOY C290	0.45% C 2.5% Si 13.2% Cr 37.0% Ni 1.5% B iron alloy: For brazing; Wall Colmonoy
COLMONOY C395	0.5% C 2.5% Si 1.5% B 13.2% Cr 37.0% Ni Fe alloy bronze metal: Wall Colmonoy
CONTRACID	Cr Ni 10% Mo Fe alloy: Corrosion resistant alloy; no further information

Symbol	Nominal analysis, supplier, condition and remarks.
CRM 6D	1.0% C 5.0% Mn 20% Cr 5.0% Ni 1.0% Mo 1.0% W Fe alloy: Information from SAE handbook
Cr Ni 25-20	0.2% C (max) 24.0% Cr 19.0% Ni steel: Designation used by German Standard
CROLOY 15/15N	0.15% C 2% Mn 16% Cr 15% Ni 1.5% Mo 1.4% W 1% Nb steel: Babcock & Wilcox (USA)
CROLOY 16/13/3	0.08% C 17% Cr 12% Ni 2.5% Mo steel: Tube; Babcock & Wilcox (USA)
CROLOY 16/13/3 ELC	0.035% C 17% Cr 14% Ni 2.5% Mo steel: Tube; Babcock & Wilcox (USA)
CROLOY 16/13/3 H	0.07% C 17% Cr 12% Ni 2.5% Mo steel: Tube; Babcock & Wilcox (USA)
CROLOY 18/8 Cb	0.08% C 18% Cr 10% Ni 1.0% Nb steel: Tube; Babcock & Wilcox (USA)
CROLOY 18/8 Cb H	0.6% C 18% Cr 10% Ni 1.0% Nb steel: Babcock & Wilcox (USA)
CROLOY 18/8 Cb Ta	0.08% C 18% Cr 10% Ni 1.0% Nb + Ta steel: Tube; Babcock & Wilcox (USA)
CROLOY 18/8 Cb Ta H	0.6% C 18% Cr 10% Ni 1.0% Nb + Ta steel: Babcock & Wilcox (USA)
CROLOY 18/8 ELC	0.035% C 19% Cr 12% Ni steel: Tube; Babcock & Wilcox (USA)
CROLOY 18/8 EM	0.15% C 18% Cr 10% Ni 0.1% Se steel: Tube; Babcock & Wilcox (USA); free machining
CROLOY 18/8 H	0.07% C 19% Cr 9% Ni steel: Tube; Babcock & Wilcox (USA)
CROLOY 18/8 HC	0.12% C 18% Cr 9% Ni steel: Tube; Babcock & Wilcox (USA)
CROLOY 18/8 S	0.08% C 19% Cr 9% Ni steel: Tube; Babcock & Wilcox (USA)
CROLOY 18/8 Si	0.08% C 2.5% Si 18% Cr 12% Ni steel: Tube; Babcock & Wilcox (USA)
CROLOY 18/8 S Ti H	0.7% C 18% Cr 10% Ni 0.6% Ti steel: Tube; Babcock & Wilcox (USA)
CROLOY 18/8 Ti	0.08% C 19% Cr 11% Ni 0.6% Ti steel: Tube; Babcock & Wilcox (USA)
CROLOY 18/12	0.12% C 18% Cr 12% Ni steel: Tube; Babcock & Wilcox (USA)
CROLOY 18/13/3	0.08% C 19% Cr 12% Ni 3.5% Mo steel: Tube; Babcock & Wilcox (USA)
CROLOY 25/12	0.15% C 23% Cr 13% Ni steel: Tube; Babcock & Wilcox (USA)
CROLOY 25/20	0.15% C 25% Cr 20% Ni steel: Tube; Babcock & Wilcox (USA)
CRONITE 25/12	0.5% C (max) 1.0% Mn 12% Ni 25% Cr Fe alloy: Casting; heat resistant **UTS: 550** **Elon: 10%** **Proof: 100**
CRONITE 25/20	0.5% C 24% Cr 20% Ni 0.7% W steel: Casting; Cronite; as cast **DPN: 170** **UTS: 530** **Elon: 18%** **Proof: 300**
CRONITE 37/18	0.3% C 36% Ni 19% Cr Fe alloy: Casting; heat resistant **UTS: 520** **Elon: 8%** **Proof: 290**
CRONITE 37/18/Nb	0.3% C (max) 36% Ni 19% Cr 2.0% Nb Fe alloy: Casting; heat resistant **UTS: 570** **Elon: 8%** **Proof: 330**
CRONITE 428	0.5% C 23% Cr 15% Ni 0.25% Mo 2.5% W steel: Casting; Cronite; as cast **DPN: 165** **UTS: 570** **Elon: 20%** **Proof: 300**
CROWN MAX	0.2% C 23% Cr 11.5% Ni 2.8% W steel: Firth Vickers for BS alloy En 55
CRUCIBLE 301	0.15% C 17% Cr 7% Ni steel: Crucible Steel Co; annealed **UTS: 800** **Elon: 60%** **Proof: 270**
CRUCIBLE 301	0.15% C 17% Cr 7% Ni steel: Crucible Steel Co; ½-hard **UTS: 920** **Elon: 25%** **Proof: 530**

Symbol	Nominal analysis, supplier, condition and remarks.
CRUCIBLE 301	0.15% C 17% Cr 7% Ni steel: Crucible Steel Co; hard
	UTS: 1380 **Elon: 8%** **Proof: 1000**
CRUCIBLE 303	0.15% C 18% Cr 9% Ni 0.6% Mo or Zr steel: Crucible Steel Co; annealed
	UTS: 650 **Elon: 50%** **Proof: 220**
CRUCIBLE 304	0.08% C 19% Cr 10% Ni steel: Crucible Steel Co; annealed
	DPN: 150 **UTS: 620** **Elon: 55%** **Proof: 240**
CRUCIBLE 309	0.2% C 23% Cr 13% Ni steel: Crucible Steel Co; annealed
	DPN: 170 **UTS: 700** **Elon: 50%** **Proof: 270**
CRUCIBLE 316	0.08% C 17% Cr 12% Ni 2.5% Mo steel: Crucible Steel Co; annealed
	DPN: 160 **UTS: 630** **Elon: 50%** **Proof: 270**
CRUCIBLE 321	0.08% C 18% Cr 11% Ni 0.5% Ti steel: Crucible Steel Co; annealed
	DPN: 160 **UTS: 630** **Elon: 50%** **Proof: 250**
CRUCIBLE 347	0.08% C 18% Cr 11% Ni 1% Nb + Ta steel: Crucible Steel Co; annealed
	DPN: 160 **UTS: 690** **Elon: 50%** **Proof: 170**
CRUCIBLE 348	0.08% C 18% Cr 11% Ni 1% Nb + Ta (Ta 0.1% (max) steel: Crucible Steel Co; annealed
	DPN: 160 **UTS: 690** **Elon: 50%** **Proof: 290**
D 2	Symbol for Durco CF8; Duriron Co
D 2L	Symbol for Durco CF 3; Duriron Co
D 4	Symbol for Durco CF 8M; Duriron Co
D 4L	Symbol for Durco FC 3M; Duriron Co
D 20	Symbol for Durimet 20; Duriron Co
DARWIN K	0.15% C 23% Cr 2% Mo 23% Ni 0.25% Cu Fe alloy: Casting; Darwin
DDQ	0.08% C 12.5% Cr 12.5% Ni steel: Firth Vickers
	DPN: 160 **UTS: 570** **Elon: 50%**
DIN 17440 X 2 Cr Ni 18/9	0.03% C (max) 2.0% Mn (max) 18.5% Cr 11% Ni steel: Plate; German Standard
	UTS: 586 **Proof: 178**
DIN 17440 X 2 Cr Ni 18/10	0.03% Cr 2.0% Mn (max) 18% Cr 10.3% Ni steel: Plate; German Standard
	UTS: 660 **Proof: 275**
DIN 17440 X 2 Cr Ni Mo 18/10	0.03% C (max) 17.5% Cr 13.75% Ni 2.75% Mo steel: Plate; German Standard; solution treated
	UTS: 590 **Proof: 240**
DIN 17440 X 2 Cr Ni Mo 18/12	0.03% C (max) 17.5% Cr 13.75% Ni 2.75% Mo steel: Plate; German Standard; solution treated
	UTS: 590 **Proof: 240**
DIN 17440 X 2 Cr Ni Mo N18/2	0.03% C (max) 17.5% Cr 12.0% Ni 2.25% Mo 0.18% N steel: Plate; German Standard; solution treated
	UTS: 700 **Proof: 325**
DIN 17440 X 2 Cr Ni Mo 18/16	0.03% C (max) 2.0% Mn (max) 18% Cr 16% Ni 3.5% Mo 0.045% P (max) 0.03% Ni (max) steel: Plate; German Standard; solution treated
	UTS: 612 **Proof: 199**
DIN 17440 X 2 Cr Ni Mo N 18/13	0.03% C (max) 17.5% Cr 13.25% Ni 2.75% Mo 0.18% N steel: Plate; German Standard; solution treated
	UTS: 700 **Proof: 350**
DIN 17440 X 5 Cr Ni 18/9	0.07% C (max) 2.0% Mo (max) 18% Cr 9% Ni steel: Plate; German Standard
	UTS: 201 **Proof: 189**

Symbol	Nominal analysis, supplier, condition and remarks.
DIN 17440 X 5 Cr Ni 19/11	0.07% C (max) 18.5% Cr 11.2% Ni steel: Plate; German Standard; solution treated
	UTS: 230 **Proof: 190**
DIN 17440 X 5 Cr Ni Mo 18/10	0.07% C (max) 17.5% Cr 12.0% Ni 2.25% Mo steel: Plate; German Standard; solution treated
	UTS: 600 **Proof: 250**
DIN 17440 X 5 Cr Ni Mo 18/12	0.07% C (max) 17.5% Cr 12.75% Ni 2.75% Mo steel: Plate; German Standard; solution treated
	UTS: 600 **Proof: 250**
DIN 17440 X 10 Cr Ni Mo Nb 18/10	0.1% C (max) 17.5% Cr 12.0% Ni 2.25% Mo 0.8% Nb steel: Plate; German Standard; solution treated
	UTS: 610 **Proof: 230**
DIN 17440 X 10 Cr Ni Mo Ti 18/10	0.1% C (max) 17.5% Cr 12.2% Ni 2.25% Mo 0.5% Ti steel: Plate; German Standard; solution treated
	UTS: 610 **Proof: 270**
DIN 17440 X 10 Cr Ni Nb 18/9	0.1% C (max) 2.0% Mn (max) 18% Cr 10% Ni 0.8% Nb 0.04% P (max) 0.03% S (max) steel: Plate; German Standard; solution treated
	UTS: 640 **Proof: 209**
DIN 17440 X 10 Cr Ni Ti 18/9	0.1% C (max) 2.0% Mn (max) 18% Cr 10% Ni 0.5% Ti (min) 0.045% P (max) 0.03% S (max) steel: Plate; German Standard; solution treated
	UTS: 640 **Proof: 209**
DIN 17440 X 12 Cr Ni 5188	0.15% C (max) 2% Mn (max) steel: Plate; German specification (18.0% Cr 9.0% Ni)
	UTS: 610 **Proof: 219**
DIN 17470 Cr Ni 25/20	25% Cr 20% Ni Fe alloy
DIN 17470 Ni Cr 30/20	20% Cr 30% Ni Fe alloy
DISCALOY	26% Ni 13.5% Cr 2.8% Mo 1.8% Ti steel: Westinghouse; for high temperature use
DTD 189 A	0.15% C 18% Cr 8% Ni Ti or Nb steel: Wire; weldable
	UTS: 530
DTD 712 A	0.06% C (max) 18% Cr 10% Ni Ti or Nb steel: Sheet annealed; weldable
	UTS: 450 **Elon: 40%** **Proof: 150**
DTD 734	0.08% C 18% Cr 9% Ni: Obsolete
DTD 5006	0.15% C 18% Cr 8% Ni 0.6% Ti or 1.0% Nb steel: Cold drawn
	UTS: 1550
DTD 5016	0.1% C (max) 18% Cr 11% Ni Nb steel: Tube; annealed; weldable
	UTS: 620 **Elon: 40%** **Proof: 180**
DTD 5036	Low C Cr Ni steel: For rivets & split pins; weldable
DTD 5056	Ni Cr steel: For bolts, etc.; suitable for cold forming
DTD 5269	Ni/Cr steel: Casting; investment cast; Nb stabilized
	UTS: 470
DTD 5259	Ni/Cr steel: Casting; investment cast; not stabilized; for use up to 350 °C
	UTS: 470
DTD 5279	Ni/Cr 2.5% Mo steel: Casting; investment cast
	UTS: 500
DTD 5289	Ni/Cr 3.5% Mo steel: Casting; investment cast
	UTS: 500
DUNELT 84	0.43% C 14% Cr 10% Ni 0.5% Mo 3% W steel: Dunford for BS alloy En 54
DUNELT 88	0.2% C 18% Cr 9% Ni steel: Dunford; annealed; can be stabilized for welding
	UTS: 530 **Elon: 30%**
DURCO CF 3	0.03% C (max) 10.0% Ni 19.0% Cr steel: Casting; Duriron Co
	UTS: 455 **Elon: 35%** **Proof: 195**

Note. The following abbreviations and units are used in the tables:

DPN	Hardness, diamond pyramid number
UTS	Ultimate tensile strength, N/mm^2
Elon	Elongation, %
Proof	0.1 % proof strength, N/mm^2

1 N/mm^2 = 0.1 hbar = 0.102 kgf/mm^2 = 0.06475 tonf/in^2 = 145.04 lbf/in^2

Symbol	Nominal analysis, supplier, condition and remarks.
DURCO CF 3M	0.03% C (max) 11.0% Ni 19.0% Cr 2.5% Mo steel: Casting; Duriron Co **UTS: 490 Elon: 30% Proof: 210**
DURCO CF 8	0.08% C (max) 9.0% Ni 19.5% Cr steel: Casting; Duriron Co **UTS: 460 Elon: 35% Proof: 195**
DURCO CF 8M	0.08% C (max) 11.0% Ni 19.5% Cr 2.5% Mo steel: Casting; Duriron Co **UTS: 490 Elon: 30% Proof: 211**
DURCORNET 100	0.04% C (max) 5.2% Ni 25.5% Cr 3.0% Cu steel: Casting; Duriron Co **UTS: 700 Elon: 20% Proof: 490**
DURIMET 20	0.07% C (max) 29.0% Ni 20.5% Cr 2.5% Mo steel: Casting; Duriron Co **UTS: 500 Elon: 20% Proof: 323**
EMS	0.06% C 18.5% Cr 10% Ni steel: Sheet; Firth Vickers; weldable **DPN: 160 UTS: 620 Elon: 60%**
ER 308	0.08% C 20.5% Cr 10.0% Ni steel: Designation used by ASTM
ERA 1414	0.4% C 14% Cr 10% Ni 3% W 0.2% Nb steel: Hadfields for BS alloy En 54 & 54A
ERA CR 1	0.1% C 18% Cr 8% Ni steel: Casting; Hadfields **DPN: 170 UTS: 580 Elon: 20% Proof: 270**
ERA CR 1	0.16% C 18% Cr 9% Ni steel: Wrought; Hadfields for BS alloy En 58A
ERA CR 2	0.3% C 21% Cr 9% Ni 2% Cu 4% W steel: Forging; Hadfield; not stabilized for welding **DPN: 235 UTS: 800 Elon: 31% Proof: 450**
ERA CR 2	0.3% C 1.5% Si 21% Cr 9% Ni 2% Cu 4% W steel: Casting; Hadfields **DPN: 229 UTS: 700 Elon: 15% Proof: 370**
ERA CR 4	18% Cr 12% Ni 3% Mo steel: Casting; Hadfields for BS 1632 A **UTS: 450 Elon: 15% Proof: 200**
ERA CR 4	0.1% C 19% Cr 9% Ni 3% Mo steel: Hadfields **DPN: 187 UTS: 710 Elon: 50% Proof: 330**
ERA CR 4FC	18% Cr 10% Ni 2% Mo S steel: Hadfields for BS alloy En 58 M; free machining
ERA CR 4SFC	18% Cr 10% Ni 2% Mo steel: Hadfields for BS alloy En 58 M
ERA CR 4S	0.1% C 19% Cr 9% Ni 3% Mo 0.6% Ti steel: Forging; Hadfields; stabilized for welding **DPN: 187 UTS: 710 Elon: 50% Proof: 300**
ERA CR 4S	0.12% C 19% Cr 9% Ni 3% Mo 0.6% Ti casting: Hadfields **DPN: 212 UTS: 690 Elon: 25% Proof: 300**
ERA CR 4S (Cb)	18% Cr 12% Ni 2.5% Nb steel: Casting; Hadfields for BS 1632 C **UTS: 450 Elon: 15% Proof: 200**
ERA CRI FC	18% Cr 9% Ni S steel: Hadfields for BS alloy En 58 M; free machining
ERA CRIS	0.1% C 18% Cr 9% Ni 0.6% Ti steel: Casting; Hadfields **DPN: 187 UTS: 620 Elon: 18% Proof: 300**
ERA CRIS	0.1% C 18% Cr 9% Ni 0.6% Ti steel: Forging; Hadfields; stabilized for welding **DPN: 170 UTS: 580 Elon: 60% Proof: 300**
ERA CRIS (Cb)	18% Cr 8% Ni Nb stabilized steel: Casting; Hadfields for BS 1631B **UTS: 450 Elon: 20% Proof: 200**
ERA CRIS (Cb)	0.15% C 18% Cr 9% Ni 1.0% Nb steel: Wrought; Hadfields for BS alloy En 58 F & 58 G
ERA CRI SFC	18% Cr 9% Ni 0.6% Ti S steel: Hadfields for BS alloy En 58 M; free machining
ERA HR 1	0.3% C 17% Cr 9% Ni 3% W steel: Wrought; Hadfields for BS alloy En 55
ERA HR 1	21% Cr 7% Ni 4% W steel: Casting; Hadfields for BS 1648 D **UTS: 740 Elon: 15% Proof: 450**
ERA HR 1S	21% Cr 7% Ni 3.5% Si steel: Casting; Hadfields for BS 1648 D
ERA HR 2	21% Cr 7% Ni steel: Casting; Hadfields for BS 1648 D **UTS: 700 Elon: 15% Proof: 300**
ERA HR 2W	21% Cr 7% Ni 2% W steel: Casting; Hadfields for BS 1648 D
ERA HR 3	25% Cr 20% Ni steel: Casting; Hadfields for BS 1648 F **UTS: 600 Elon: 16%**
ERA HR 6	25% Cr 12% Ni steel: Casting; Hadfields for BS 1648 E **UTS: 450 Elon: 20%**
ERA HR 6W	25% Cr 12% Ni 3.5% W steel: Casting; Hadfields for BS 1648 E **UTS: 450 Elon: 20%**
ERA HR 7	15% Cr 15% Ni steel: Casting; Hadfields for BS 1648 H
ERA HR 8	25% Cr 25% Ni W steel: Casting; Hadfields for BS 1648 H
ESCALLOY 20	0.7% C 20% Cr 28% Ni 2.5% Mo 1% Nb 3.2% Cu steel: Eastern Stainless
ESSHETE 316	0.07% C 17.0% Cr 12.5% Ni 2.3% Mo steel: Samuel Fox; solution treated **UTS: 500 Elon: 30% Proof: 210**
ESSHETE 321	0.07% C 18.0% Cr 10.0% Ni 0.7% (max) Ti steel: Samuel Fox; solution treated **UTS: 500 Elon: 30% Proof: 210**
ESSHETE 347	0.07% C 18.0% Cr 11.0% Ni 0.7% Nb (min) steel: Samuel Fox; solution treated **UTS: 500 Elon: 30% Proof: 220**
EV 3	0.2% C 21% Cr 11.5% Ni steel: For exhaust valves; designation used by SAE **UTS: 840 Elon: 26% Proof: 450**
EV 10	1% C 14% Cr 14% Ni steel: Casting for exhaust valves; designation used by SAE
EVERSHYNE	Osborn Trade name for austenitic steel; not weld stabilized
EW 2	0.08% C 18% Cr 10% Ni 1.5% Mo steel: Osborn; weldable; acid resistant **UTS: 740 Elon: 45% Proof: 300**
EW 3	0.08% C 18% Cr 10% Ni 2.5% Mo steel: Osborn; weldable; acid resistant **UTS: 760 Elon: 45% Proof: 330**
EW 4	0.06% C 18% Cr 10% Ni 4% Mo steel: Osborn; austenitic; weldable; acid resistant **UTS: 760 Elon: 45% Proof: 370**
EZEFORM 35	18% Cr 8% Ni steel: Atlas
FAGERSTA R310	0.07% C 18.0% Cr 10.0% Ni steel: For deep drawing; Fagersta
FAGERSTA R320	0.06% C 18.0% Cr 9.0% Ni steel: For deep drawing; Fagersta
FAGERSTA R350	0.04% C 18.5% Cr 9.0% Ni steel: For deep drawing; Fagersta
FAGERSTA R358	0.04% C 18.0% Cr 9.5% Ni Nb: Stabilized steel; Fagersta; for heat resistant
FAGERSTA R359	0.04% C (max) 18.0% Cr 9.5% Ni Ti: Stabilized steel; Fagersta; for heat resistance
FAGERSTA R360	0.03% C 18.5% Cr 10.0% Ni steel: For tubes; Fagersta
FAGERSTA R390	0.05% C 18.0% Cr 11.5% Ni steel: For deep drawing & spinning; Fagersta
FAGERSTA R428	0.04% C 17.5% Cr 11.0% Ni 2.3% Mo with Nb steel: Fagersta
FAGERSTA R440	0.05% C 17.0% Cr 12.0% Ni 2.8% Mo steel: Acid resistant; Fagersta
FAGERSTA R450	0.04% C 18.0% Cr 10.5% Ni 1.5% Mo steel: For acid resistance; Fagersta
FAGERSTA R460	0.03% C 17.0% Cr 12.5% Ni 2.8% Mo steel: For acid resistance; Fagersta

Symbol	Nominal analysis, supplier, condition and remarks.
FAGERSTA R590	0.06% C 13.5% Cr 12.0% Ni steel: For deep drawing; Fagersta
FAGERSTA R800	0.08% C 20.5% Cr 19.0% Ni steel: Tube for heat resistance; Fagersta
FAGERSTA R820	0.05% C 25.0% Cr 20.5% Ni steel: Tube for heat resistance; Fagersta
FCB	0.06% C 18% Cr 10% Ni 0.7% Nb steel: Firth Vickers
	DPN: 175 UTS: 630 Elon: 58% Proof: 250
FCB (T)	0.08% C 17.5% Cr 12% Ni 0.8% Nb steel: Firth Vickers
	UTS: 580 Elon: 45%
FDP	0.05% C 18% Cr 9% Ni 0.5% Ti steel: Firth Vickers
	DPN: 160 UTS: 620 Elon: 50% Proof: 240
FDP (L)	0.05% C 18% Cr 9% Ni 0.5% Ti steel: Firth Vickers
FMB	0.06% C 18% Cr 11% Ni 2.8% Mo steel: Firth Vickers
FMB (L)	0.03% C 17.5% Cr 12% Ni 2.8% Mo steel: Firth Vickers
FMB Ti	0.06% C 18% Cr 12% Ni 2.8% Mo 0.3% Ti steel: Firth Vickers; weldable
	DPN: 180 UTS: 620 Elon: 50% Proof: 270
FML	0.07% C 18% Cr 9.5% Ni 1.2% Mo steel: Firth Vickers; for acid resistance
	DPN: 170 UTS: 620 Elon: 50% Proof: 250
FNB	0.4% C 14% Cr 13.5% Ni 2.5% W steel: Darwins for BS alloy En 54
FOREMOST 18/8	0.15% C 18.5% Cr 9% Ni steel: Swift Levick for BS alloy En 58A
FOREMOST 18/8 Mo	0.12% C 18% Cr 9% Ni 2.7% Mo steel: Swift Levick for BS alloy En 58J
FOREMOST 18/8 Nb	0.1% C 18.5% Cr 9% Ni 1% Nb steel: Swift Levick for BS alloy En 58F
FOREMOST 18/8 Ti	0.1% C 18.5% Cr 9% Ni 0.5% Ti steel: Swift Levick for BS alloy En 58B
FOREMOST HR 1	0.3% C 25% Cr 20% Ni 3% W steel: Casting; Swift Levick
	DPN: 220 UTS: 760 Elon: 27% Proof: 420
FOREMOST HR 5	0.3% C 20% Cr 8% Ni 4% W steel: Casting; Swift Levick; annealed
	DPN: 220 UTS: 830 Elon: 49% Proof: 330
FOREMOST HR 6	0.1% C 18% Cr 18% Ni 3.5% Mo 0.6% Ti 2.5% Cu steel: Casting; Swift Levick; annealed
	DPN: 163 UTS: 390 Elon: 12% Proof: 180
FOREMOST HR 7	0.2% C 25% Cr 12% Ni 3% W steel: Casting; Swift Levick; annealed
	DPN: 220 UTS: 710 Elon: 32% Proof: 270
FOREMOST HR 8	0.2% C 25% Cr 20% Ni steel: Casting; Swift Levick; annealed
	DPN: 200 UTS: 630 Elon: 48%
FORMWELL	0.11% C 17% Cr 9.5% Ni steel: Atlas; annealed
	DPN: 130 UTS: 620 Elon: 63% Proof: 220
FS 16/25/6	0.05% C 16% Cr 25% Ni 6.5% Mo steel: Firth Sterling
	DPN: 201 UTS: 840 Elon: 35% Proof: 390
FSL	0.06% C 18.5% Cr 10% Ni steel: Firth Vickers

Note. The following abbreviations and units are used in the tables:

DPN	Hardness, diamond pyramid number
UTS	Ultimate tensile strength, N/mm^2
Elon	Elongation, %
Proof	0.1 % proof strength, N/mm^2

$1 \ N/mm^2 = 0.1 \ hbar = 0.102 \ kgf/mm^2 = 0.06475 \ tonf/in^2 = 145.04 \ lbf/in^2$

Symbol	Nominal analysis, supplier, condition and remarks.
FSL (L)	0.05% C 18.5% Cr 10% Ni steel: Sheet; Firth Vickers
	UTS: 610 Elon: 60% Proof: 220
FSQ	0.1% C 18% Cr 11% Ni steel: Firth Vickers
FST	0.08% C 18% Cr 9% Ni steel: Firth Vickers
	DPN: 170 UTS: 610 Elon: 50%
FST (L)	0.06% C 18.5% Cr 10% Ni steel: Firth Vickers
FV 301	0.1% C 17.5% Cr 7.5% Ni steel: Firth Vickers; annealed
	UTS: 740 Elon: 55%
FV 301	0.1% C 17.5% Cr 7.5% Ni steel: Firth Vickers; ½-hard
	UTS: 880 Elon: 50%
FV 301	0.1% C 17.5% Cr 7.5% Ni steel: Firth Vickers; ½-hard
	UTS: 980 Elon: 35%
FV 301	0.1% C 17.5% Cr 7.5% Ni steel: Firth Vickers; hard
	UTS: 1050 Elon: 25%
FV 301	0.1% C 17.5% Cr 7.5% Ni steel: Firth Vickers; extra hard
	UTS: 1210 Elon: 10%
FV 302	0.08% C 18% Cr 9% Ni steel: Firth Vickers
	DPN: 170 UTS: 610 Elon: 50%
FV 303	0.1% C 18% Cr 9% Ni 0.6% Ti 0.25% S 0.3% Mo steel: Firth Vickers; free cutting; weldable
	UTS: 600 Elon: 53% Proof: 240
FV 304	0.06% C 18.5% Cr 10% Ni steel: Sheet; Firth Vickers; weldable
	DPN: 160 UTS: 610 Elon: 60%
FV 304	0.05% C 18.5% Cr 10% Ni steel: Sheet; Firth Vickers; weldable
	UTS: 610 Elon: 60% Proof: 220
FV 304 (L)	0.03% C 18% Cr 11% Ni steel: Firth Vickers; weldable
	DPN: 160 UTS: 580 Elon: 60% Proof: 220
FV 305	0.1% C 18% Cr 11% Ni steel: For deep drawing; Firth Vickers; annealed
	UTS: 580 Elon: 60% Proof: 240
FV 309	0.20% C 23% Cr 11.5% Ni 2.8% W steel: Firth Vickers
	DPN: 220 UTS: 740 Elon: 28%
FV 310	0.1% C 23% Cr 21% Ni steel: Firth Vickers
	DPN: 190 UTS: 700 Elon: 35%
FV 316	0.06% C 18% Cr 11% Ni 2.8% Mo steel: Firth Vickers; for acid resistance; sulphuric, chlorides, etc.
	DPN: 180 UTS: 610 Elon: 50% Proof: 270
FV 316 (L)	0.03% C 17.5% Cr 12% Ni 2.8% Mo steel: Firth Vickers; weldable
	DPN: 160 UTS: 580 Elon: 50% Proof: 240
FV 321	0.05% C 18% Cr 9% Ni 0.5% Ti steel: Firth Vickers; weldable
	DPN: 160 UTS: 610 Elon: 50% Proof: 240
FV 347	0.06% C 18% Cr 10% Ni 0.7% Nb steel: Firth Vickers; weldable
	DPN: 175 UTS: 620 Elon: 58% Proof: 250
FV 555	0.05% C 16.5% Cr 10.5% Ni 2.4% Mo steel: Firth Vickers; this steel is developed for its weldability
	UTS: 540 Elon: 67% Proof: 200
FVS	0.4% C 13.6% Cr 14% Ni 2.6% W steel: Firth Vickers
	DPN: 230 UTS: 760 Elon: 35%
G 2	0.42% C 14% Cr 14% Ni 2.6% W steel: Wrought; Jessop; for BS alloy En 54
G 4	0.12% C 16% Cr 13% Ni 3% W 0.5% Ti steel: Jessop
G 9	0.12% C 16% Cr 11% Ni 1% W 0.4% Ti steel: Jessop

Symbol	Nominal analysis, supplier, condition and remarks.
G 19	0.4% C 19% Cr 13% Ni 1.8% Mo 2.6% W 3.1% Nb 10% Co steel: Casting; Jessop; as cast; creep resisting **UTS: 580 Elon: 4% Proof: 300**
G 19	0.4% C 19% Cr 13% Ni 1.7% Mo 2.5% W 10% Co 3% Nb steel: Jessop; wrought
G 21	0.4% C 13% Cr 13% Ni 2.5% W 1.0% Nb steel: Jessop; creep resisting **UTS: 730 Elon: 47% Proof: 250**
G 38	0.17% C 16% Cr 12% Ni 2.5% Mo 1.3% W 1.25% V steel: Jessop
G 40	0.23% C 20% Cr 25% Ni 3% Mo 1% Ti 1% W steel: Jessop **UTS: 840 Elon: 18% Proof: 630**
G 71	0.10% C 18.0% Cr 37.0% Ni steel: Jessop; as cast; scale or creep resistant
G 72	0.10% C 21.0% Cr 32.0% Ni steel: Jessop; as cast; scale or creep resistant
G 125	0.02% C 18.5% Ni 5.0% Mo 9.0% Cr 0.65% Ti 0.9% Al steel: Jessop
G 3426 STC 52D	0.08% C 17% Cr 12% Ni 2.5% Mo steel: Japanese Standard
G 4301 SEC 7	0.15% C (max) 18% Cr 9% Ni steel: Japanese Standard
G 4301 SEC 8	0.08% C (max) 19% Cr 10% Ni steel: Japanese Standard
G 4301 SEC 9	0.03% C (max) 19% Cr 10% Ni steel: Japanese Standard
G 4301 SEC 10	0.08% C 18% Cr 10.5% Ni 0.5% Ti steel: Japanese Standard
G 4301 SEC 13	0.03% C 17% Cr 12% Ni 2.5% Mo steel: Japanese Standard
G 4302 SEH 5	0.25% C 25% Cr 20% Ni steel: Japanese Standard
GAMMA Cb	0.4% C 15.2% Cr 24.5% Ni 4.1% Mo 2.2% Nb Fe alloy: Origin unknown
GOST 5632/1 KH 13N3	0.12% C 13.5% Cr 22.5% Ni steel: Russian Standard
GOST 5632/1 KH14N16B	0.1% C 14.0% Cr 15.5% Ni 1.1% Nb steel: Plate; Russian Standard
GOST 5632/1 KH14N16B	0.1% C 14.0% Cr 15.0% Ni 1.1% Nb steel: Russian Standard
GOST 5632/1 KH14N16BR	0.1% C (max) 14.0% Cr 15.5% Ni 1.1% Nb 0.005% B (max) 0.02% Zi (max) steel: Plate; Russian Standard
GOST 5632/1 KH14N16BR	0.1% C 15.0% Cr 14.0% Ni 1.1% Nb 0.005% B 0.02% Ce steel: Russian Standard
GOST 5632/1 KH14N18V2B	0.1% C 14.5% Cr 19.0% Ni 1.0% Nb 2.2% W steel: Plate; Russian Standard
GOST 5632/1 KH14N18V2B	0.1% C 14.0% Cr 19.0% Ni 2.3% W 1.5% Nb steel: Russian Standard
GOST 5632/1 KH14N18V2BR	0.1% C 14.0% Cr 19.0% Ni 1.0% Nb 2.2% W 0.005% B (max) 0.02% Zi (max) steel: Plate; Russian Standard
GOST 5632/1 KH14N18V2BR	0.1% C 14.0% Cr 19.0% Ni 2.3% W 1.1% Nb 0.005% B 0.02% Ce steel: Russian Standard
GOST 5632/1 KH14N18V2BR1	0.1% C 14.0% Cr 19.0% Ni 1.0% Nb 2.2% W 0.025% B (max) 0.02% Zi (max) steel: Plate; Russian Standard
GOST 5632/1 KH14Ni18V2BR1	0.1% C 14.0% Cr 19.0% Ni 2.3% W 1.1% Nb 0.025% B 0.02% Ce steel: Russian Standard
GOST 5632/1 KH16N13M2B	0.09% C 16.0% Cr 13.5% Ni 2.2% Mo 1% Nb steel: Russian Standard
GOST 5632/1 KH16N13M2B	0.09% C 16.0% Cr 13.5% Ni 2.25% Mo 1.1% Nb steel: Russian Standard
GOST 5632/1 KH16N13M2B	0.09% C 16.0% Cr 13.5% Ni 2.25% Mo 1.1% Nb steel: Plate; Russian Standard
GOST 5632/1 KH16N13M2B	0.09% C 16.0% Cr 13.0% Ni 2.2% Mo 1.1% Nb steel: Russian Standard
GOST 5632/1 KH21N5T	0.12% C 0.8% Mn (max) 21.0% Cr 5.2% Ni 0.8% Ti steel: Russian Standard
GOST 5632/1 KH25N25TR	0.1% C 24.5% Cr 25.5% Ni 1.3% Ti 0.01% B steel: Russian Standard
GOST 5632/2 KH18N9	0.17% C 18.0% Cr 9.0% Ni steel: Russian Standard
GOST 5632/2 KH18N9	0.17% C 1.6% Mn 18% Cr 9% Nc steel: Plate; Russian Standard **UTS: 675 Proof: 280**
GOST 5632/2 KH18N9	0.17% C 18.0% Cr 9.0% Ni steel: Russian Standard
GOST 5632/3 KH13N7S2	0.3% C 13.0% Cr 6.5% Ni steel: Plate; Russian Standard
GOST 5632/3 KH19N9MVBT	0.31% C 19.0% Cr 9.0% Ni 1.2% Mo 1.2% W 0.35% Ti 0.35% Nb steel
GOST 5632/4 KH12N8G8MFB	0.38% C 12.5% Cr 8.0% Ni 1.2% Mo 1.3% V 0.3% Ti steel: Plate; Russian Standard
GOST 5632/4 KH12N8G8MFB	0.38% C 12.5% Cr 8.0% Ni 1.3% Mo 1.35% V 0.3% Nb steel: Russian Standard
GOST 5632/4 KH14N14V2M	0.45% C 14.0% Cr 14.0% Ni 0.32% Mo 2.2% W steel: Plate; Russian Standard
GOST 5632/4 KH14N14V2M	0.45% C 14.0% Cr 14.0% C 0.32% Mo 2.3% W steel: Russian Standard
GOST 5632/4 KH14N7G7F2MS	0.42% C 15.0% Cr 7.0% Ni 0.8% Mo 1.7% V steel: Plate; Russian Standard
GOST 5632/4 KH15N17G7F2MS	0.41% C 15.0% Cr 7.0% Ni 0.8% Mo 1.7% V steel: Russian Standard
GOST 5632/4 KH18N2582	0.36% C 18.0% Cr 24.5% Ni steel: Russian Standard
GOST 5632/4 KH18N2552	0.36% C 18.0% Cr 24.5% Ni steel: Russian Standard
GOST 5632/4 KH18N2552	0.36% C 18.0% Cr 24.5% Ni steel: Russian Standard
GOST 5632/4 KH18N2552	0.36% C 18.0% Cr 24.5% Ni steel: Plate; Russian Standard
GOST 5632/610 KH18N10	0.08% C (max) 1.5% Mn 18% Cr 10% Ni 0.035% P (max) 0.02% S (max) steel: Plate; Russian Standard **UTS: 630 Proof: 220**
GOST 5632 KH12N20T3R	0.1% C (max) 11.2% Cr 19.5% Ni 2.9% Ti 0.8% Al 0.0014% B steel: Russian Standard
GOST 5632 KH12N22T3MR	0.1% C (max) 11.2% Cr 22.5% Ni 1.3% Mo 2.9% Ti 0.8% Al 0.02% B steel: Russian Standard
GOST 5632 KH15N9IU	0.09% C (max) 0.8% Mn (max) 15.0% Cr 7.7% Ni 1.0% Al steel: Russian Standard
GOST 5632 KH16N15M3B	0.09% C (max) 16.0% Cr 15.0% Ni 2.7% Mo 0.7% Nb steel: Russian Standard
GOST 5632 KH16N15M3B	0.09% C (max) 16.0% Cr 15.0% Ni 2.75% Mo 0.7% Nb steel: Plate
GOST 5632 KH16N15M3B	0.09% C (max) 16.0% Cr 15.0% Ni 2.75% Mo 0.75% Nb steel: Russian Standard
GOST 5632 KH16N15M3B	0.09% C (max) 16.0% Cr 15.0% Ni 2.7% Mo 0.75% Nb steel: Russian Standard
GOST 5632 KH17N71U	0.09% C (max) 17% Cr 7.0% Ni 1.0% Al steel: Plate; Russian Standard
GOST 5632 KH17N7IU	0.09% C (max) 0.8% Mn (max) 0.8% Mn 17.0% Cr 7.0% Ni 1.05% Al steel: Russian Standard
GOST 5632 KH17N13M2T	0.1% C (max) 17.0% Cr 15.0% Ni 2.1% Mo 0.45% Ti steel: Russian Standard
GOST 5632 KH17N13M2T	0.1% C (max) 17.0% Cr 13.0% Ni 2.1% Mo 0.45% Ti steel: Plate; Russian Standard **UTS: 520 Proof: 220**
GOST 5632 KH17N13M2T	0.1% C (max) 17.0% Cr 13.0% Ni 2.2% Mo 0.45% Ti steel: Plate
GOST 5632 KH17N13M3T	0.1% C (max) 17.0% Cr 13.0% Ni 3.5% Mo 0.45% Ti steel: Plate; Russian Standard
GOST 5632 KH17N13M3T	0.1% C (max) 17.0% Cr 13.0% Ni 3.5% Mo 0.5% Ti steel: Russian Standard
GOST 5632 KH17N13M3T	0.1% C (max) 17.0% Cr 13.0% Ni 3.5% Mo 0.45% Ti steel: Russian Standard
GOST 5632 KH17N13M3T	0.1% C (max) 17.0% Cr 13.0% Ni 3.5% Mo 0.45% Ti steel: Russian Standard

Symbol	Nominal analysis, supplier, condition and remarks.
GOST 5632 KH18N9	0.12% C (max) 18.0% Cr 9.0% Ni steel: Russian Standard
GOST 5632 KH18N9	0.12% C (max) 18.0% Cr 9.0% Ni steel: Russian Standard
GOST 5632 KH18N9	0.12% C 1.5% Mn 18% Cr 9.0% Ni steel: Plate; Russian Standard
GOST 5632 KH18N9	0.12% C (max) 18.0% Cr 9.0% Ni steel: Russian Standard
GOST 5632 KH18N9T	0.12% C (max) 1.5% Mn 19% Cr 8.5% Ni 0.7% Ti steel: Plate; Russian Standard; solution treated **UTS: 550 Proof: 200**
GOST 5632 KH18N9T	0.12% C (max) 18.0% Cr 8.7% Ni 0.7% Ti steel: Russian Standard
GOST 5632 KH18N10E	0.12% C (max) 18.0% Cr 10.0% Ni 0.24% Se or Te steel: Russian Standard
GOST 5632 KH18N10E	0.12% C (max) 15% Mn 10% Ni 18% Cr steel: Plate; Russian specification
GOST 5632 KH18N10E	0.12% C (max) 18.0% Cr 10.0% Ni 0.25% Te or Se steel: Russian Standard
GOST 5632 0KH18N10T	0.08% C (max) 18.0% Cr 10.0% Ni 0.6% Ti steel: Russian Standard
GOST 5632 KH18N10T	0.12% C (max) 18.0% Cr 10.0% Ni 0.7% Ti steel: Russian Standard
GOST 5632 KH18N10T	0.12% C (max) 1.5% Mn 18% Cr 10% Ni 0.7% Ti steel: Plate; Russian Standard; solution treated **UTS: 520 Proof: 200**
GOST 5632 KH18N12T	0.12% C (max) 1.5% Mn 18% Cr 12% Ni 0.7% Ti steel: Plate; Russian Standard; solution treated **UTS: 550 Proof: 200**
GOST 5632 KH18N12T	0.12% C (max) 18.0% Cr 12.0% Ni 0.7% Ti steel: Russian Standard
GOST 5632 KH20N1452	0.2% C (max) 20.5% Cr 13.5% Ni steel: Russian Standard
GOST 5632 KH20N1452	0.2% C (max) 20.5% Cr 13.5% Ni steel: Plate; Russian Standard
GOST 5632 KH20N1452	0.2% C (max) 20.5% Cr 13.5% Ni steel: Russian Standard
GOST 5632 KH20N14S2	0.2% C (max) 1.5% Mn (max) 20.5% Cr 13.5% Ni steel: Russian Standard
GOST 5632 KH20N14S2	0.08% C (max) 20.5% Cr 13.5% Ni steel: Russian Standard
GOST 5632 KH23NB	0.2% C (max) 2.0% Mn (max) 23.5% Cr 13.5% Ni steel: Russian Standard
GOST 5632 KH23N13	0.2% C (max) 23.5% Cr 13.5% Ni steel: Russian Standard
GOST 5632 KH23N13	0.2% C (max) 23.5% Cr 13.5% Ni steel: Plate; Russian Standard; solution treated **UTS: 500 Proof: 300**
GOST 5632 KH23N13	0.2% C (max) 23.5% Cr 13.5% Ni steel: Russian Standard
GOST 5632 0KH23N18	0.1% C (max) 23.5% Cr 18.5% Ni steel: Plate; Russian Standard
GOST 5632 KH23N18	0.2% C (max) 23.5% Cr 18.5% Ni steel: Russian Standard
GOST 5632 KH23N18	0.2% C (max) 23.5% Cr 18.5% Ni steel: Plate; Russian Standard
GOST 5632 KH23N18	0.2% C (max) 23.5% Cr 18.5% Ni steel: Russian Standard
GOST 5632 KH23N18	0.2% C (max) 23.5% Cr 18.5% Ni steel: Russian Standard

Note. The following abbreviations and units are used in the tables:

DPN	Hardness, diamond pyramid number
UTS	Ultimate tensile strength, N/mm^2
Elon	Elongation, %
Proof	0.1 % proof strength, N/mm^2

1 N/mm^2=0.1 hbar=0.102 kgf/mm^2=0.06475 tonf/in^2=145.04 lbf/in^2

Symbol	Nominal analysis, supplier, condition and remarks.
GOST 5632 KH25N16G7AR	0.12% C (max) 6.0% Mn 24.5% Cr 17.5% Ni 0.37% N 0.02% B steel: Russian Standard
GOST 5632 KH25N20S2	0.2% C (max) 25.5% Cr 19.5% Ni steel: Russian Standard
GOST 5632 KH25N20S2	0.2% C (max) 25.5% Cr 19.5% Ni steel: Plate; Russian Standard; solution treated **UTS: 600 Proof: 300**
GOST 5632 KH25N20S2	0.2% C (max) 25.5% Cr 19.5% Ni steel: Russian Standard
GOST 5632 KH25H25TR	0.1% C 24.5% Cr 25.5% Ni 1.4% Ti 0.01% B (max) steel: Plate; Russian Standard
GOST 5632 KH10N20T2	0.08% C 11.0% Cr 19.0% Ni 1.0% Al (max) 2.0% Ti steel: Plate; Russian Standard
GOST 5632/0 KH10N20T2	0.08% C 11.0% Cr 19.0% Ni 2.0% Ti 1.0% Al (max) steel: Russian Standard
GOST 5632/0 KH14N28V3T31UR	0.08% C (max) 14.0% Cr 27.5% Ni 3.2% W 2.8% Ti 0.8% Al 0.02% B steel: Russian Standard
GOST 5632 KH15N91U	0.09% C (max) 15.0% Cr 8.1% Ni 1.0% Al steel: Plate; Russian Standard
GOST 5632/0 KH17N16M3T	0.08% C (max) 17.0% Cr 16.0% Ni 2.7% Mo 0.45% Ti steel: Plate; Russian Standard
GOST 5632/0 KH17N16M3T	0.08% C (max) 17.0% Cr 16.0% Ni 2.7% Mo 0.45% Ti steel: Russian Standard
GOST 5632 KH18N9T	0.12% C (max) 18.0% Cr 8.7% Ni 0.7% Ti steel: Russian Standard
GOST 5632/0 KH18N10	0.08% C (max) 18.0% Cr 10.0% Ni steel: Russian Standard
GOST 5632/0 KH18N10	0.04% C (max) 18.0% Cr 10.0% Ni steel: Russian Standard
GOST 5632/0 KH18N10	0.08% C (max) 18.0% Cr 10.0% Ni steel: Russian Standard
GOST 5632/0 KH18N10	0.08% C (max) 18.0% Cr 10.0% Ni steel: Russian Standard
GOST 5632/0 KH18N10E	0.12% C (max) 18.0% Cr 10.0% Ni Te 0.25% Se steel: Russian Standard; free machining
GOST 5632/0 KH18N10T	0.08% C 1.5% Mn 18% Cr 10% Ni 0.5% Ti steel: Plate; Russian Standard; solution treated **UTS: 500 Proof: 200**
GOST 5632 KH18N10T	0.12% C (max) 18.0% Cr 10.0% Ni 0.7% Ti steel: Russian Standard
GOST 5632/0 KH18N10T	0.08% C (max) 18.0% Cr 10.0% Ni 0.6% Ti steel: Russian Standard
GOST 5632/0 KH18N11	0.06% C (max) 18.0% Cr 11.0% Ni steel: Plate; Russian Standard; solution treated **UTS: 520**
GOST 5632/0 KH18N11	0.06% C (max) 18.0% Cr 11.0% Ni steel: Russian Standard
GOST 5632/0 KH18N11	0.06% C (max) 18.0% Cr 11.0% Ni steel: Russian Standard
GOST 5632/0 KH18N12B	0.08% C (max) 1.5% Mn 18% Cr 12% Ni 0.9% Nb 0.035% P (max) 0.02% S (max) steel: Plate; Russian Standard; solution treated **UTS: 550 Proof: 200**
GOST 5632/0 KH18N11	0.06% C (max) 18.0% Cr 11.0% Ni steel: Russian Standard
GOST 5632/0 KH18N12B	0.08% C (max) 18.0% Cr 12.5% Ni 1.2% Nb steel: Russian Standard
GOST 5632/0 KH18N12B	0.08% C (max) 18.0% Cr 12.0% Ni 1.2% Nb steel: Russian Standard
GOST 5632/0 KH18N12T	0.08% C (max) 18.0% Cr 12.5% Ni 0.5% Ti steel: Russian Standard
GOST 5632/0 KH18N12T	0.08% C (max) 1.5% Mn 18% Cr 12% Ni 0.5% Ti steel: Plate; Russian Standard; solution treated
GOST 5632 KH18N12T	0.12% C (max) 18.0% Cr 12.5% Ni 0.7% Ti steel: Russian Standard
GOST 5632/0 KH18N12T	0.08% C (max) 18.0% Cr 12.0% Ni 0.6% Ti steel: Russian Standard
GOST 5632/0 KH20N14S2	0.08% C (max) 20.5% Cr 13.5% Ni steel: Plate; Russian Standard

Symbol	Nominal analysis, supplier, condition and remarks.
GOST 5632/0 KH20N14S2	0.08% C (max) 1.5% Mn (max) 20.5% Cr 13.5% Ni steel: Russian Standard
GOST 5632/0 KH20N14S2	0.08% C (max) 20.5% Cr 13.5% Ni steel: Russian Standard
GOST 5632/0 KH21N5T	0.08% C (max) 0.8% Mn 21.0% Cr 5.2% Ni 0.45% Ti steel: Russian Standard
GOST 5632/0 KH21N6M2T	0.08% C (max) 21.0% Cr 6.0% Ni 2.1% Mo 0.3% Ti steel: Plate; Russian Standard
GOST 5632/0 KH21N6M2T	0.08% C (max) 0.8% Mn (max) 21.0% Cr 6.2% Ni 2.1% Mo 0.3% Ti steel: Russian Standard
GOST 5632/0 KH23N18	0.1% C (max) 23.5% Cr 18.5% Ni steel: Russian Standard
GOST 5632/0 KH23N18	0.1% C (max) 23.5% Cr 18.5% Ni steel: Russian Standard
GOST 5632/0 KH23N18	0.1% C (max) 23.5% Cr 18.5% Ni steel: Russian Standard
GOST 5632/0 KH23N28M2T	0.06% C (max) 23.5% Cr 27.5% Ni 2.2% Mo 0.55% Ti steel: Russian Standard
GOST 5632/0 KH23N28M2T	0.06% C (max) 23.5% Cr 26.5% Ni 2.2% Mo 0.6% Ti steel: Plate; Russian Standard
GOST 5632/0 KH23N28M3DeT	0.06% C (max) 23.5% Cr 27.5% Ni 2.7% Mo 0.55% Ti 3.0% Cu steel: Russian Standard
GOST 5632/0 KH23N28M3D3T	0.06% C (max) 23.5% Cr 27.5% Ni 2.7% Mo 0.6% Ti steel: Plate; Russian Standard
GOST 5632/00 KH18N10	0.04% C (max) 18.0% Cr 10.0% Ni steel: Russian Standard
GOST 5632/00 KH18N10	0.04% C (max) 1.5% Mn 18% Cr 10% Ni steel: Plate; Russian Standard
	UTS: 450 **Proof: 160**
GOST 5632/00 KH18N10	0.04% C (max) 18.0% Cr 10.0% Ni steel: Russian Standard
GOST 9940/1 KH14N18V2BR	0.1% C 14.0% Cr 19.0% 2.3% W 1.1% Nb steel: Plate; Russian Standard; seamless
	UTS: 550
GOST 9940/1 KH21N5T	0.12% C 21.0% Cr 5.2% Ni 0.45% Ti steel: Pipe; Russian Standard; seamless
GOST 9940/0 KH21N5T	0.08% C (max) 21.0% Cr 5.2% Ni 0.45% Ti steel: Pipe; Russian Standard; seamless
GOST 9940/2 KH18N9	0.12% C (max) 18.0% Cr 9.0% Ni steel: Pipe; Russian Standard; seamless
	UTS: 540
GOST 9940 KH17NBM2T	0.1% C (max) 17.0% Cr 13.0% Ni 2.1% Mo 0.45% Ti steel: Pipe; Russian Standard; seamless
	UTS: 540
GOST 9940 KH18N9	0.12% C (max) 18.0% Cr 9.0% Ni steel: Pipe; Russian Standard; seamless
	UTS: 540
GOST 9940 KH18N10T	0.12% C (max) 18.0% Cr 10.0% Ni 0.7% Ti steel: Pipe; Russian Standard; seamless
	UTS: 540
GOST 9940 KH18N12T	0.12% C (max) 18.0% Cr 12.0% Ni 0.7% Ti steel: Pipe; Russian Standard; seamless
	UTS: 540
GOST 9940/0 KH17N16M3T	0.08% C (max) 17.0% Cr 16.0% Ni 2.7% Mo 0.45% Ti steel: Pipe; Russian Standard; seamless
	UTS: 500
GOST 9940/0 KH18N10T	0.08% C (max) 18.0% Cr 10.0% Ni 0.6% Ti steel: Pipe; Russian Standard; seamless
	UTS: 520
GOST 9940/0 KH18N12B	0.08% C (max) 18.0% Cr 12.0% Ni 1.2% Nb steel: Pipe; Russian Standard; seamless
	UTS: 520
GOST 9940/0 KH18N12T	0.08% C (max) 18.0% Cr 12.0% Ni 0.6% Ti steel: Pipe; Russian Standard; seamless
	UTS: 520
GOST 9940/0 KH20N1452	0.08% C (max) 20.5% Cr 13.5% Ni steel: Pipe; Russian Standard; seamless
GOST 9940/0 KH23N18	0.1% C (max) 23.5% Cr 18.5% Ni steel: Pipe; Russian Standard; seamless
	UTS: 500
GOST 9940/00 KH18N10	0.04% C (max) 18.0% Cr 10.0% Ni steel: Pipe; Russian Standard; seamless
	UTS: 450
GOST 9941/1 KH14N18V2BR	0.1% C 21.0% Cr 5.1% Ni 0.8% Ti steel: Pipe; Russian Standard; seamless
GOST 9941/1 KH21N5T	0.12% C 21.0% Cr 5.2% Ni 0.45% Ti steel: Pipe; Russian Standard; seamless
GOST 9941/1 KH18N9	0.17% C 18.0% Cr 9.0% Ni steel: Pipe; Russian Standard; seamless
	UTS: 540
GOST 9941 KH17N13M2T	0.1% C (max) 14.5% Cr 19.0% Ni 2.2% W 1.0% Nb steel: Pipe; Russian Standard; seamless
	UTS: 550
GOST 9941 KH18N9	0.12% C (max) 18.0% Cr 10.0% Ni steel: Pipe; Russian Standard; seamless
	UTS: 520
GOST 9941 KH18N12T	0.12% C (max) 18.0% Cr 12.0% Ni 0.6% Ti steel: Pipe; Russian Standard; seamless
	UTS: 520
GOST 9941 KH21N5T	0.08% C (max) 21.5% Cr 13.5% Ni steel: Pipe; Russian Standard; seamless
GOST 9941/0 KH17N16M3T	0.08% C (max) 17.0% Cr 13.0% Ni 2.1% Mo 0.45% Ti steel: Pipe; Russian Standard; seamless
	UTS: 540
GOST 9941/0 KH18N10	0.08% C (max) 18.0% Cr 10.0% Ni steel: Pipe; Russian Standard; seamless
	UTS: 450
GOST 9940/0 KH18N10	0.08% C (max) 18.0% Cr 10.0% Ni steel: Pipe; Russian Standard; seamless
	UTS: 520
GOST 9941/0 KH18N10T	0.08% C (max) 18.0% Cr 10.0% Ni 0.6% Ti steel: Pipe; Russian Standard; seamless
	UTS: 520
GOST 9941/0 KH18N12B	0.08% C (max) 18.0% Cr 12.0% Ni 0.7% Ti steel: Pipe; Russian Standard; seamless
	UTS: 540
GOST 9941/0 KH18N12B	0.08% C (max) 18.0% Cr 12.0% Ni 0.7% Ti steel: Pipe; Russian Standard; seamless
	UTS: 540
GOST 9941/0 KH18N12T	0.08% C (max) 18.0% Cr 10.0% Ni 0.7% Ti steel: Pipe; Russian Standard; seamless
	UTS: 540
GOST 9941/0 KH20N1452	0.08% C (max) 20.5% Cr 13.5% Ni steel: Pipe; Russian Standard; seamless
GOST 9941/0 KH23N18	0.1% C (max) 23.5% Cr 18.5% Ni steel: Pipe; Russian Standard; seamless
	UTS: 500
GOST 9941/00 KH18N10	0.04% C (max) 17.0% Cr 16.0% Ni 2.7% Mo 0.45% Ti: Russian Standard; seamless
	UTS: 500
GOST 10498/1 KH16N15M3B	0.06% C 16.0% Cr 15.0% Ni 3.0% Mo 0.6% Nb steel: Pipe; Russian Standard; seamless
	UTS: 550
GOST 10498/0 KH16N15M3	0.03% C (max) 15.5% Cr 15.0% Ni 3.0% Mo steel: Pipe; Russian Standard; seamless
	UTS: 560
GOST 10498/0 KH18N10T	0.05% C 18.0% Cr 10.0% Ni 0.6% Ti steel: Pipe; Russian Standard; seamless
	UTS: 540
GOST 10498/0 KH18N10T	0.85% C 18.0% Cr 10.0% Ni 0.8% Ti steel: Pipe; Russian Standard; seamless
	UTS: 560
GOST 10498/00 KH16N15M3B	0.04% C (max) 16.0% Cr 15.0% Ni 2.7% Mo 0.5% Nb steel: Pipe; Russian Standard; seamless
	UTS: 520
GOST 11068/1 KH21N5T	0.11% C 21.0% Cr 5.2% Ni 0.8% Ti steel: Pipe; Russian Standard; seamless
GOST 11068 KH17N13M2T	0.1% C (max) 17.0% Cr 13.0% Ni 2.1% Mo 0.45% Ti steel: Pipe; Russian Standard; seamless

Symbol	Nominal analysis, supplier, condition and remarks.
GOST 11068 KH17N13M3T	0.1% C 17.0% Cr 13.0% Ni 3.5% Mo 0.45% Ti steel: Pipe; Russian Standard; seamless
GOST 11068 KH18N10T	0.12% C (max) 18.0% Cr 10.0% Ni 0.7% Ti steel: Pipe; Russian Standard; seamless **UTS: 560**
GOST 11068/0 KH18N10T	0.08% C (max) 18.0% Cr 10.0% Ni 0.6% Ti steel: Pipe; Russian Standard; seamless **UTS: 540**
GOST 11068 KH18N12T	0.12% C (max) 18.0% Cr 12.0% Ni 0.7% Ti steel: Pipe; Russian Standard; seamless **UTS: 560**
GOST 11068/0 KH18N12T	0.08% C (max) 18.5% Cr 12.0% Ni 0.6% Ti steel: Pipe; Russian Standard; seamless **UTS: 540**
GOST 11068/0 KH21N5T	0.08% C (max) 21.0% Cr 5.2% Ni steel: Pipe; Russian Standard; seamless
GOST 11068/0 KH23N28M3D3T	0.06% C (max) 23.5% Cr 27.5% Ni 3.0% Mo 0.5% Ti steel: Pipe; Russian Standard; seamless
GOST 11068/0 KH23N28M2T	0.06% C (max) 23.5% Cr 27.5% Ni 2.1% Mo 0.5% Ti steel: Pipe; Russian Standard; seamless
GOST 11068/00 KH18N10T	0.04% C (max) 18.0% Cr 10.0% Ni 0.5% Ti steel: Tube; Russian Standard; seamless **UTS: 500**
GOST 10802/1 KH14N18V2BR	0.09% C 14.0% Cr 19.0% Ni 2.4% W 1.1% Nb steel: Pipe; Russian Standard; seamless **UTS: 550** **Proof: 240**
GOST 10802 KH16N14V2BR	0.09% C 16.5% Cr 14.0% Ni 2.4% W 1.1% Nb steel: Pipe; Russian Standard; seamless **UTS: 550** **Proof: 220**
GOST 10802 KH16N16V2MBR	0.08% C 16.0% Cr 16.0% Ni 0.65% Mo 2.5% W 0.8% Nb 0.3% Cu (max) steel: Pipe; Russian Standard; seamless **UTS: 550** **Proof: 240**
GOST 10802 KH18N12T	0.12% C (max) 18.0% Cr 12.0% Ni 0.7% Ti steel: Pipe; Russian Standard; seamless **UTS: 540** **Proof: 220**
GOST 14162 KH18N9	0.12% C (max) 18.0% Cr 9.0% Ni steel: Capillary tube; Russian Standard
GOST 14162 KH18N10T	0.12% C (max) 18.0% Cr 9.5% Ni 0.7% Ti steel: Capillary tube; Russian Standard **UTS: 560**
GOST 14162 KH18N12T	0.12% C (max) 18.0% Cr 12.0% Ni 0.7% Ti steel: Capillary tube; Russian Standard **UTS: 520**
GOST 14162/0 KH18N10T	0.08% C (max) 18.0% Cr 9.5% Ni 0.6% Ti steel: Capillary tube; Russian Standard **UTS: 540**
GOST 14162/0 KH18N12T	0.08% C (max) 18.0% Cr 12.0% Ni 0.6% Ti steel: Capillary tube; Russian Standard **UTS: 520**
GT 45	0.12% C 16% Cr 14% Ni 2.5% Mo 0.25% Ti 0.5% Nb 3% Cu steel: Armco
HE	0.35% C 28% Cr 9% Ni 0.5% Mo steel: Casting; Huntington; aged **UTS: 620** **Elon: 10%** **Proof: 370**
HEADWELL	0.07% C 16% Cr 18% Ni steel: Atlas; annealed **DPN: 130** **UTS: 530** **Elon: 52%** **Proof: 200**
HEDEX 14	0.08% C 18% Cr 10% Ni steel: Wire for forging; Kiveton Park for BS alloy En 58E

Symbol	Nominal analysis, supplier, condition and remarks.
HEDEX 35	0.08% C (max) 18% Cr 11% Ni steel: Wire for forging; Kiveton Park for BS alloy En 58E
HEDEX 36	0.12% C (max) 19% Cr 10% Ni 3% Mo steel: Wire for forging; Kiveton Park for BS alloy En 58J
HEDEX 37	0.05% C (max) 19% Cr 10% Ni (min) steel: Wire for forging; Kiveton Park for BS alloy En 58E
HEDEX 44	0.08% C (max) 18% Cr 14% Ni steel: Wire for forging; Kiveton Park; annealed **UTS: 690**
HEDEX 45	0.08% C (max) 18% Cr 16% Ni steel: Wire for forging; Kiveton Park; annealed **UTS: 690**
HF	0.3% C 20% Cr 11% Ni 0.5% Mo steel: Casting; Huntington; aged **UTS: 710** **Elon: 25%** **Proof: 330**
HH	0.3% C 26% Cr 12% Ni 0.5% Mo 0.2% N steel: Casting; Huntington; aged **UTS: 630** **Elon: 8%** **Proof: 300**
HI	0.35% C 28% Cr 16% Ni 0.5% Mo steel: Casting; Huntington; aged **UTS: 630** **Elon: 6%** **Proof: 450**
Hi Ni 1	3.7% C 16% Cr 8% Ni 0.5% V Fe alloy: Casting; Dewrance **DPN: 540** **UTS: 580**
HI-PROOF 304	0.06% C (max) 19% Cr 9.5% Ni steel: S Fox **UTS: 580** **Elon: 35%** **Proof: 300**
HI-PROOF 304L	0.03% C (max) 19% Cr 10% Ni steel: S Fox **UTS: 610** **Elon: 35%** **Proof: 350**
HI-PROOF 316	0.07% C (max) 17% Cr 12% Ni 2.5% Mo steel: S Fox **UTS: 580** **Elon: 35%** **Proof: 320**
HI-PROOF 316L	0.03% C (max) 17% Cr 12% Ni 2.5% Mo steel: S Fox **UTS: 580** **Elon: 35%** **Proof: 320**
HI-PROOF 347	0.08% C (max) 18% Cr 10% Ni steel: S Fox **UTS: 610** **Elon: 35%** **Proof: 350**
HK	0.4% C 26% Cr 20% Ni 0.5% Mo steel: Casting; Huntington; aged **UTS: 610** **Elon: 10%** **Proof: 330**
HL	0.4% C 30% Cr 20% Ni 0.5% Mo steel: Casting; Huntington; as cast **UTS: 570** **Elon: 19%** **Proof: 330**
HN	0.35% C 21% Cr 25% Ni 0.5% Mo steel: Casting; Huntington; as cast **UTS: 450** **Elon: 17%** **Proof: 220**
HNM	0.3% C 18.5% Cr 9.5% Ni 0.23% P steel: American proprietary alloy listed in SAE year book
HOTSPUR A	0.4% C 14% Cr 10% Ni 3% W steel: Bar, billet, etc.; Brown Bayley for BS alloy En 54
HOTSPUR C	0.2% C 25% Cr 20% Ni steel: Bar, billet, etc.; Brown Bayley
HOTSPUR D	0.35% C 17% Cr 9% Ni 3% W steel: Bar, billet, etc.; Brown Bayley for BS alloy En 55
HOTSPUR F	0.2% C 25% Cr 12% Ni steel: Bar, billet, etc.; Brown Bayley
HR CROWN 1	0.2% C 24.0% Cr 12.0% Ni 0.5% W steel: Casting; Firth Vickers; annealed **DPN: 220** **UTS: 530** **Elon: 32%** **Proof: 270**
HR CROWN MAX	0.2% C 23% Cr 11.5% Ni 2.8% W steel: Firth Vickers for BS alloy En 55 **DPN: 220** **UTS: 740** **Elon: 28%**
HS 88	0.07% C 12.5% Cr 15% Ni 2% Mo 0.6% W 0.6% Ti 0.15% B steel: Haynes Stellite
HT	0.55% C 15% Cr 35% Ni 0.5% Mo steel: Casting; Huntington; aged **UTS: 540** **Elon: 5%** **Proof: 300**

Note. The following abbreviations and units are used in the tables:

DPN	Hardness, diamond pyramid number
UTS	Ultimate tensile strength, N/mm^2
Elon	Elongation, %
Proof	0.1 % proof strength, N/mm^2

1 N/mm^2=0.1 hbar=0.102 kgf/mm^2=0.06475 tonf/in^2=145.04 lbf/in^2

Symbol	Nominal analysis, supplier, condition and remarks.
HYDRA XL	0.25% C 23% Cr 11% Ni 3% W steel: Hall & Pickles; cores; die pins for non-ferrous casting **UTS: 700 Elon: 33% Proof: 390**
HU	0.55% C 19% Cr 39% Ni 0.5% Mo steel: Casting; Huntington; aged **UTS: 530 Elon: 5% Proof: 290**
ICL 164	0.08% C (max) 17.0% Cr 12.0% Ni 2.2% Mo steel: Creusot-Loire
ICL 164 BC	0.03% C (max) 17.0% Cr 12.0% Ni 2.2% Mo steel: Creusot-Loire
ICL 164 Nb	0.08% C (max) 17.5% Cr 12.0% Ni 2.5% Mo 0.4% Ti steel: Creusot-Loire
ICL 164 T	0.08% C (max) 17.5% Cr 12.0% Ni 2.5% Mo 0.9% Nb steel: Creusot-Loire
ICL 166	0.08% C 17.5% Cr 12.0% Ni 2.75% Mo steel: Creusot-Loire
ICL 166 BC	0.03% C 17.5% Cr 12.0% Ni 2.75% Mo steel: Creusot-Loire
ICL 167 CN	0.045% C (max) 17.6% Cr 12.0% Ni 2.4% Mo steel: Creusot-Loire
ICL 168	0.08% C (max) 19.0% Cr 13.0% Ni 3.5% Mo steel: Creusot-Loire
ICL 168 BC	0.03% C (max) 19.0% Cr 13.0% Ni 3.5% Mo steel: Creusot-Loire
ICL 472	0.08% C (max) 19.0% Cr 9.2% Ni steel: Creusot-Loire
ICL 472 BC	0.03% C (max) 19.0% Cr 10.0% Ni steel: Creusot-Loire
ICL 472 Nb	0.08% C (max) 18.0% Cr 10.5% Ni 1.0% Nb + Ta steel: Creusot-Loire
ICL 472 T	0.08% C (max) 18.0% Cr 10.5% Ni 0.4% Ti steel: Creusot-Loire
ICL 473 BC	0.04% C (max) 19.2% Cr 9.5% Ni steel: Creusot-Loire
ICL 473 Nb	0.05% C (max) 19.0% Cr 9.5% Ni steel: Creusot-Loire
IMMACULATE	0.3% C 20% Cr 7.5% Ni 2.0% W steel: Wrought; Firth Vickers **DPN: 220 UTS: 830 Elon: 50% Proof: 380**
IMMACULATE 2W	0.25% C 1.0% Mn 21.0% Cr 9.0% Ni steel: Casting; Firth Vickers; annealed **DPN: 200 UTS: 530 Elon: 30% Proof: 250**
IMMACULATE 5	0.1% C 23% Cr 21% Ni steel: Firth Vickers **DPN: 180 UTS: 700 Elon: 35%**
IMMACULATE 5T	0.1% C 22% Cr 17% Ni 0.7% Ti steel: Firth Vickers **DPN: 180 UTS: 610 Elon: 45%**
INCOLOY	0.1% C 20.5% Cr 32% Ni Fe alloy: Inco
INCOLOY	20% Cr 32% Ni Fe alloy: Heat resistant; H Wiggin; furnace equipment
INCOLOY 800	0.1% C 0.5% Cu 21% Cr 32% Ni Fe alloy: Wrought; H Wiggin; annealed **DPN: 180 UTS: 580**
INCOLOY 800H	20% Cr 32% Ni stainless steel: Wrought; H Wiggin
INCOLOY 801	0.04% C 20.5% Cr 1% Ti 32% Ni Fe alloy: Bar, sheet, etc.; Huntington
INCOLOY 810	0.25% C 21.0% Cr 32% Ni 0.5% Cu Fe alloy: Information from SAE handbook
INCOLOY D S	18% Cr 37% Ni 2% Se Fe alloy: Heat resistant; H Wiggin; furnace equipment; sulphur resistant
INCOLOY T	0.1% C 20.5% Cr 32% Ni 1.0% Ti Fe alloy: Inco
INOFO 10	0.16% C 18% Cr 9% Ni steel: Forez
INOFO 10B	0.07% C 18% Cr 10% Ni steel: Forez
INOFO 10Ti	0.12% C 18% Cr 10% Ni 0.5% Ti steel: Forez
INOFO 11	0.03% C 18% Cr 10% Ni steel: Forez
INOFO 12	0.07% C 18% Cr 11% Ni 2.5% Mo steel: Forez
INOFO 12B	0.04% C 18% Cr 11% Ni 2.5% Mo steel: Forez
INOFO 13S	0.12% C 18% Cr 10% Ni steel: Forez

Symbol	Nominal analysis, supplier, condition and remarks.
IR 4	0.15% C (max) 18% Cr 9% Ni 0.6% Ti steel: Sandvik
IR 41	0.15% C (max) 18% Cr 9% Ni 1.0% Nb steel: Sandvik
JIS G4303 SUS	Suffix in agreement with AISI etc. steels
JIS G4304 SUS302	15% C (max) 2% Mn (max) 18% Cr 9% Ni steel: Plate; Japanese specification **UTS: 530 Proof: 210**
JIS G4304 SUS304	0.3% C (max) 2% Mn (max) 19.0% Cr 10.5% Ni: Japanese specification **UTS: 530 Proof: 210**
JIS G4304 SUS304L	0.08% C (max) 2.0% Mn (max) 19% Cr 9.5% Ni steel: Plate; Japanese Standard **UTS: 530min Proof: 210**
JIS G4304 SUS305	0.12% C (max) 2.0% Mn (max) 18% Cr 11.5% Ni steel: Plate; Japanese specification; solution treated **Proof: 180**
JIS G4304 SUS309S	0.08% C (max) 23% Cr 13.5% Ni steel: Plate; Japanese Standard; solution treated **UTS: 530 Proof: 210**
JIS G4304 SUS310S	0.08% C (max) 25.0% Cr 20.5% Ni steel: Plate; Japanese Standard; solution treated **UTS: 530 Proof: 210**
JIS G4304 SUS316	0.08% C (max) 17.0% Cr 12.0% Ni 2.5% Mo steel: Plate; Japanese Standard; solution treated **UTS: 530 Proof: 210**
JIS G4304 SUS316JI	0.08% C (max) 18.0% Cr 12.0% Ni 2.0% Mo 1.2% Cu steel: Plate; Japanese Standard; solution treated **UTS: 530 Proof: 210**
JIS G4304 SUS316JIL	0.03% C (max) 18.0% Cr 14.0% Ni 2.0% Mo 1.75% Cu steel: Plate; Japanese Standard; solution treated **UTS: 490 Proof: 180**
JIS G4304 SUS316L	0.03% C (max) 17.0% Cr 13.5% Ni 2.5% Mo steel: Plate; Japanese Standard; solution treated **UTS: 490 Proof: 180**
JIS G4304 SUS317	0.08% C (max) 2.0% Mn (max) 19% Cr 13% Ni 3.5% Mo 0.04% P (max) 0.03% S (max) steel: Plate; Japanese Standard; solution treated **UTS: 530 Proof: 210**
JIS G4304 SUS317L	0.03% C (max) 2.0% Mn (max) 19% Cr 13% Ni 3.5% Mo 0.04% P (max) 0.03% S (max) steel: Plate; Japanese Standard; solution treated **UTS: 490 Proof: 180**
JIS G4304 SUS321	0.08% C (max) 2.0% Mn 18% Cr 10.5% Ni 0.4% Ti (min) 0.04% P (max) 0.03% S (max) steel: Plate; Japanese Standard; solution treated **UTS: 530 Proof: 210**
JIS G4304 SUS329J1	0.08% C 25.5% Cr 4.5% Ni 2.0% Mo steel: Plate; Japanese Standard; solution treated **UTS: 600 Proof: 400**
JIS G4304 SUS347	0.08% C (max) 2.0% Mn (max) 18% Cr 10% Ni 0.8% Nb Ta 0.04% P (max) 0.03% S (max) steel: Plate; Japanese Standard; solution treated **UTS: 530 Proof: 210**
JIS G4304 SUS631	0.09% C (max) 1.0% Mn 17% Cr 6.8% Ni 1.5% Al 0.04% P (max) 0.03% S (max) steel: Plate; Japanese Standard
JIS G4312 SUH309	0.2% C (max) 23.0% Cr 13.5% Ni steel: Plate; Japanese Standard; solution treated **UTS: 570 Proof: 210**
JIS G4312 SUH310	0.25% C (max) 25.0% Cr 20.5% Ni steel: Plate; Japanese Standard; solution treated **UTS: 600 Proof: 210**
JIS G4312 SUH330	0.15% C (max) 15.5% Cr 35.0% Ni steel: Plate; Japanese Standard; solution treated **UTS: 570 Proof: 210**

Symbol	Nominal analysis, supplier, condition and remarks.
JIS 3446 SU304	0.08% C (max) 19% Cr 9.5% Ni steel: Tube for structural use; Japanese Standard **UTS: 600** **Proof: 210**
JIS G3447 SUS304LTBS	0.3% C (max) 19.0% Cr 11.0% Ni steel: Pipe; Japanese Standard **UTS: 490**
JIS G3447 SUS304TBS	0.08% C (max) 19.0% Cr 9.5% Ni steel: Pipe; Japanese Standard **UTS: 530**
JIS G3446 SUS316	0.08% C (max) 17% Cr 12% Ni 2.5% Mo steel: Tube for structural use; Japanese Standard **UTS: 600** **Proof: 210**
JIS G3447 SUS316LTBS	0.03% C (max) 17.0% Cr 14.0% Ni 2.5% Mo steel: Pipe; Japanese Standard **UTS: 490**
JIS G3447 SUS316TBS	0.08% C (max) 17.0% Cr 12.0% Ni 2.5% Mo steel: Pipe; Japanese Standard **UTS: 530**
JIS G3446 SUS321	0.08% C (max) 18% Cr 11.0% Ni 0.4% Ti steel: Tube for structural use; Japanese Standard **UTS; 800** **Proof: 530**
JIS G3446 SUS347	0.08% C (max) 18.0% Cr 11.0% Ni 0.8% Nb + Ta steel: Tube for structural use; Japanese Standard **UTS: 800** **Proof: 530**
JIS G3459 SUS304HTP	0.07% C 19.0% Cr 9.5% Ni steel: Pipe; Japanese Standard **UTS: 530** **Proof: 210**
JIS G3459 SUS304LTP	0.03% C (max) 19.5% Cr 11.0% Ni steel: Pipe; Japanese Standard **UTS: 490** **Proof: 180**
JIS G3459 SUS304TP	0.08% C (max) 19.0% Cr 9.5% Ni steel: Pipe; Japanese Standard **UTS: 530** **Proof: 210**
JIS G3459 SUS309TP	0.15% C 23.5% Cr 13.5% Ni steel: Pipe; Japanese Standard **UTS: 530** **Proof: 210**
JIS G3459 SUS310TP	0.15% C (max) 25.0% Cr 20.5% Ni steel: Pipe; Japanese Standard **UTS: 530** **Proof: 210**
JIS G3459 SUS316HTP	0.7% C 17.0% Cr 12.5% Ni 2.5% Mo steel: Pipe; Japanese Standard **UTS: 530** **Proof: 210**
JIS G3459 SUS316LTP	0.03% C (max) 17.0% Cr 14.0% Ni 2.5% Mo steel: Pipe; Japanese Standard **UTS: 490** **Proof: 210**
JIS G3459 SUS316TP	0.08% C (max) 17.0% Cr 12.0% Ni 2.5% Mo steel: Pipe; Japanese Standard **UTS: 530** **Proof: 210**
JIS G3459 SUS321HTP	0.07% C 18.5% Cr 11.0% Ni 0.6% Ti steel: Pipe; Japanese Standard **UTS: 530** **Proof: 210**
JIS G3459 SUS321TP	0.08% C (max) 18.0% Cr 11.0% Ni 0.4% Ti steel: Pipe; Japanese Standard **UTS: 530** **Proof: 210**
JIS G3459 SUS329JITP	0.08% C (max) 25.5% Cr 4.5% Ni 2.0% Mo steel: Pipe; Japanese Standard **UTS: 600** **Proof: 400**
JIS G3459 SUS347HTP	0.07% C 18.5% Cr 11.0% Ni 0.8% Nb + Ta steel: Pipe; Japanese Standard **UTS: 530** **Proof: 210**

Symbol	Nominal analysis, supplier, condition and remarks.
JIS G3459 SUS347TP	0.08% C (max) 18.0% Cr 11.0% Ni 0.8% Nb + Ta steel: Pipe; Japanese Standard **UTS: 530** **Proof: 210**
JIS G3463 SUS304HTB	0.07% C 19.0% Cr 9.0% Ni steel: Pipe; Japanese Standard **UTS: 530** **Proof: 210**
JIS G3463 SUS304LTB	0.03% C (max) 19.0% Cr 11.0% Ni steel: Pipe; Japanese Standard **UTS: 490** **Proof: 180**
JIS G3463 SUS304TB	0.08% C (max) 19.0% Cr 19.0% Ni steel: Pipe; Japanese Standard **UTS: 530** **Proof: 210**
JIS G3463 SUS309STB	0.15% C 23.0% Cr 13.5% Ni steel: Pipe; Japanese Standard **UTS: 530** **Proof: 210**
JIS G3463 SUS310STB	0.15% C 25.0% Cr 20.5% Ni steel: Pipe; Japanese Standard **UTS: 530** **Proof: 210**
JIS G3463 SUS316HTB	0.07% C 17.0% Cr 12.5% Ni 2.5% Mo steel: Pipe; Japanese Standard **UTS: 530** **Proof: 210**
JIS G3463 SUS 316LTB	0.03% C (max) 17.0% Cr 14.0% Ni 2.5% Mo steel: Pipe; Japanese Standard **UTS: 490** **Proof: 180**
JIS G3463 SUS316TB	0.08% C (max) 17.0% Cr 12.0% Ni 2.5% Mo steel: Pipe; Japanese Standard **UTS: 530** **Proof: 210**
JIS G3463 SUS321HTB	0.07% Cu 18.5% Cr 11.0% Ni 0.6% Ti steel: Pipe; Japanese Standard **UTS: 530** **Proof: 210**
JIS G3463 SUS321TB	0.08% C (max) 18.0% Cr 11.0% Ni 0.4% Ti steel: Pipe; Japanese Standard **UTS: 530** **Proof: 210**
JIS G3463 SUS329J1TB	0.08% C (max) 25.5% Cr 5.5% Ni 2.0% Mo steel: Pipe; Japanese Standard **UTS: 600** **Proof: 400**
JIS G3463 SUS347HTB	0.07% C 18.5% Cr 11.0% Ni 1.0% Nb + Ta steel: Pipe; Japanese Standard **UTS: 530** **Proof: 210**
JIS G3463 SUS347TB	0.08% C (max) 18.0% Cr 11.0% Ni 0.8% Nb + Tb steel: Pipe; Japanese Standard
KE 965	0.4% C 13% Cr 13.5% Ni 0.55% Mo 2.75% W steel: Kayser Ellison; for valves
KEA 23	0.1% C 18.5% Cr 9.0% Ni and Ti steel: Weldable; Kayser Ellison
KEA 23	0.07% C 18.0% Cr 9.0% Ni + Ti steel: Sanderson Kayser
KEA 221	0.07% C 18.0% Cr 11.5% Ni 2.6% Mo steel: Sanderson Kayser
KEA 221	0.09% C 18.5% Cr 10.0% Ni 2.7% Mo steel: Kayser Ellison
KEA 507	0.1% C 18.5% Cr 8.0% Ni with Se steel: Free machining; Kayser Ellison
KEA 507	0.09% C 18.0% Cr 8.5% Ni + Se steel: Sanderson Kayser
KEA 521	0.07% C 18.0% Cr 11.5% Ni 2.6% Mo steel: Sanderson Kayser
KEA 521	0.09% C 18.5% Cr 10.0% Ni 2.7% Mo and Se steel: Free machining; Kayser Ellison
KEH 22/A507	0.1% C 20% Cr 7.5% Ni + free cutting agent steel: Kayser Ellison; annealed **UTS: 530 Elon: 30% Proof: 180**
KH 18N10T	0.12% C (max) 18.0% Cr 10.0% Ni 0.7% Ti steel: Russian Standard designation
KH 18N12T	0.12% C (max) 18.0% Cr 12.0% Ni 0.7% Ti steel: Russian Standard designation
KH 25N16G7AR	0.12% C (max) 24.5% Cr 16.5% Ni 0.4% N 0.02% B (max) steel: Russian Standard designation

Note. The following abbreviations and units are used in the tables:

DPN	Hardness, diamond pyramid number
UTS	Ultimate tensile strength, N/mm^2
Elon	Elongation, %
Proof	0.1 % proof strength, N/mm^2

1 N/mm^2=0.1 hbar=0.102 kgf/mm^2=0.06475 tonf/in^2=145.04 lbf/in^2

Symbol	Nominal analysis, supplier, condition and remarks.
KHN 35VMT	0.12% C (max) 15.0% Cr 34.0% Ni 2.5% Mo 1.3% Ti 2.7% W Fe alloy: Russian Standard designation
KH 35VT	0.12% C (max) 15.0% Cr 36.0% Ni 1.3% Ti 3.2% W Fe alloy: Russian Standard designation
KH 35VT1U	0.08% C (max) 15.0% Cr 35.0% Ni 2.8% Ti 1.0% Al 3.2% W Fe alloy: Russian Standard designation
KH 35VTR	0.1% C (max) 15.0% Cr 36.5% Ni 1.3% Ti 4.5% W Fe alloy: Russian Standard designation
KH 38VT	0.09% C 21.5% Cr 37.0% Ni 1.0% Ti 3.2% V Fe alloy: Russian Standard designation
KK WELDANKA	0.15% C (max) 18% Cr 9% Ni 0.15% Se steel: Bar, billets, etc.; Brown Bayley for BS alloy En 58M
LANGALLOY 1V	0.08% C 18% Cr 9% Ni 1.0% Nb steel: Langley for BS alloy 821
LANGALLOY 2V	0.08% C 18% Cr 10% Ni 2.5% Mo steel: Langley for BS alloy 845; free machining
LANGALLOY 3V	0.08% C 18% Cr 10% Ni 2.5% Mo steel: Langley for BS alloy 845
LANGALLOY 4V	25% Cr 12% Ni W steel: Langley DPN: 170 UTS: 580 Elon: 15% Proof: 270
LANGALLOY 5V	0.08% C 18% Cr 10% Ni 2.5% Mo steel: Langley for BS alloy 845 with low magnetic permeability; free machining
LANGALLOY 10V	0.08% C 19% Cr 10% Ni steel: Langley; for low temperature use UTS: 480 Elon: 40% Proof: 240
LANGALLOY 12V	0.08% C 20% Cr 12% Ni 3.5% Mo steel: Langley for BS alloy 846; free machining
LANGALLOY 13V	0.08% C 20% Cr 12% Ni 3.5% Mo steel: Langley for BS alloy 846
LANGALLOY 15V	0.15% C 18% Cr 9% Ni steel: Langley UTS: 480 Elon: 40% Proof: 240
LO	0.07% C 18% Cr 10% Ni 1.0% Mo steel: Osborn; weldable; acid resistant UTS: 710 Elon: 50% Proof: 270
LS 12	0.1% C 13% Cr 13% Ni steel: Low Moor DPN: 160 UTS: 530 Elon: 45% Proof: 220
LS 18	0.1% C 18% Cr 9% Ni steel: Low Moor; not weldable DPN: 150 UTS: 570 Elon: 50% Proof: 240
LS 18/2	0.08% C 18% Cr 10% Ni 2.0% Mo steel: Low Moor; acid resisting; weldable DPN: 160 UTS: 610 Elon: 48% Proof: 250
LS 18/3	0.08% C 18% Cr 10% Ni 2.75% Mo steel: Low Moor; acid resistant; weldable DPN: 165 UTS: 610 Elon: 48% Proof: 270
LS 18/4	0.07% 18% Cr 10% Ni 3.75% Mo steel: Low Moor; acid resistant; weldable DPN: 160 UTS: 610 Elon: 48% Proof: 270
LS 18L	0.06% C 18% Cr 9% Ni steel: Low Moor; corrosion resistance reduced by welding DPN: 150 UTS: 570 Elon: 50% Proof: 240
LS 20/12	0.15% C 20% Cr 13% Ni steel: Low Moor; heat resistant DPN: 160 UTS: 610 Elon: 48% Proof: 250
LS 25/12/3	0.2% C 24% Cr 11% Ni 3% W steel: Low Moor; heat resistant DPN: 230 UTS: 840 Elon: 35% Proof: 440
LS 25/20	0.15% C 25% Cr 18% Ni steel: Low Moor DPN: 200 UTS: 630 Elon: 45% Proof: 300
LS WF	0.08% C 18% Cr 9% Ni Ti or Nb steel: Low Moor DPN: 160 UTS: 610 Elon: 50% Proof: 250
LV 10	0.4% C 2.5% Si 19% Cr 9% Ni steel: Low Moor; valve steel
LV 21/12	0.25% C 21% Cr 12% Ni steel: Low Moor; valve steel

Symbol	Nominal analysis, supplier, condition and remarks.
LV 21/12N	0.2% C 21% Cr 12% Ni 0.2% Nb steel: Low Moor; valve steel
LV 54	0.4% C 13% Cr 11% Ni 3% W steel: Low Moor; valve steel DPN: 260
LV 55	0.45% C 2.5% Si 19% Cr 9% Ni 1.1% W steel: Low Moor; valve steel
M 5	0.03% C 18% Cr 16% Ni 5% Mo 0.2% Cu steel: Japanese specification
M 7	0.06% C 10% Cr 17% Ni 7% Mo steel: Japanese specification
MA 1	0.12% C 18% Cr 9.5% Ni steel: Schoeller-Bleckmann; annealed DPN: 175 UTS: 540 Elon: 50% Proof: 210
MA 3	0.09% C (max) 12.5% Cr 13% Ni steel: Schoeller-Bleckmann; annealed DPN: 170 UTS: 540 Elon: 55% Proof: 200
MA 4	0.08% C (max) 18% Cr 8.5% Ni steel: Schoeller-Bleckmann; annealed DPN: 175 UTS: 540 Elon: 50% Proof: 210
MACDN	0.06% C 18% Cr 20% Ni 2.2% Mo Cu Nb steel: Schoeller-Bleckmann; annealed DPN: 185 UTS: 610 Elon: 40% Proof: 220
MACM	0.07% C 18% Cr 22% Ni 3% Mo Cu Nb steel: Schoeller-Bleckmann; annealed DPN: 185 UTS: 610 Elon: 40% Proof: 220
MAN	0.08% C 18% Cr 9.5% Ni Nb steel: Weldable; Schoeller-Bleckmann; annealed DPN: 185 UTS: 570 Elon: 40% Proof: 240
MANAURITE 8S	18% Cr 8% Ni steel: Pompey UTS: 500 Elon: 20% Proof: 240
MANAURITE 10	22% Cr 10% Ni steel: Pompey UTS: 500 Elon: 18% Proof: 260
MANAURITE 12	25% Cr 12% Ni steel: Pompey UTS: 540 Elon: 18% Proof: 260
MANAURITE 20	25% Cr 20% Ni steel: Pompey UTS: 560 Elon: 20% Proof: 260
MANAURITE 35	15% Cr 35% Ni steel: Pompey UTS: 500 Elon: 20% Proof: 240
MANGANESE P	0.1% C 2.0% Mn 20% Cr 8% Ni steel: Welding electrode; Murex; used as a buffer layer when hard facing DPN: 200
MANIFLEX	0.2% C 21% Cr 11.5% Ni steel: Carpenter
MAO	0.05% C (max) 18% Cr 9.5% Ni steel: Schoeller-Bleckmann; annealed DPN: 175 UTS: 540 Elon: 50% Proof: 200
MAO SUPERIOR	0.03% C (max) 18% Cr 9.5% Ni steel: Schoeller-Bleckmann; annealed DPN: 175 UTS: 540 Elon: 50% Proof: 200
MAOY	0.06% C (max) 18% Cr 9.5% Ni steel: Schoeller-Bleckmann; annealed DPN: 175 UTS: 540 Elon: 50% Proof: 200
MASBw	0.06% C 17% Cr 13% Ni 4.2% Mo steel: Schoeller-Bleckmann; annealed DPN: 185 UTS: 610 Elon: 45% Proof: 220
MASD	0.06% C (max) 17% Cr 9% Ni 1.5% Mo steel: Schoeller-Bleckmann; annealed DPN: 180 UTS: 610 Elon: 45% Proof: 210
MASN	0.08% C 18% Cr 11.5% Ni 2.2% Mo Nb steel: Schoeller-Bleckmann; annealed DPN: 185 UTS: 570 Elon: 40% Proof: 240
MASO	0.05% C 18% Cr 11% Ni 2.2% Mo steel: Schoeller-Bleckmann; annealed DPN: 175 UTS: 540 Elon: 45% Proof: 200
MASO SUPERIOR	0.03% C (max) 18% Cr 11% Ni 2.2% Mo steel: Schoeller-Bleckmann; annealed DPN: 175 UTS: 540 Elon: 45% Proof: 200

Symbol	Nominal analysis, supplier, condition and remarks.
MASOY	0.05% C 18% Cr 12% Ni 2.6% Mo steel: Schoeller-Bleckmann; annealed
	DPN: 175 UTS: 570 Elon: 45% Proof: 200
MASOY SUPERIOR	0.03% C 18% Cr 12% Ni 2.6% Mo steel: Schoeller-Bleckmann; annealed
	DPN: 175 UTS: 570 Elon: 45% Proof: 200
MASW	0.07% C 18% Cr 11.5% Ni 2.2% Mo Ti steel: Schoeller-Bleckmann; annealed
	DPN: 185 UTS: 570 Elon: 40% Proof: 240
MASWY	0.07% C 17% Cr 12% Ni 2.7% Mo Ti steel: Schoeller-Bleckmann; annealed
	DPN: 185 UTS: 610 Elon: 40% Proof: 240
MAT	0.07% C 18% Cr 9.5% Ni Ti steel: Weldable; Schoeller-Bleckmann; annealed
	DPN: 185 UTS: 570 Elon: 40% Proof: 240
MAUSTENOX A	0.12% C (max) 18% Cr 9% Ni steel: Pompey
	UTS: 500 Elon: 30% Proof: 240
MAUSTINOX Abc	0.08% C (max) 18% Cr 9% Ni steel: Pompey
	UTS: 500 Elon: 35% Proof: 240
MAUSTINOX Atbc	0.05% C (max) 18% Cr 10% Ni steel: Pompey
	UTS: 470 Elon: 40% Proof: 220
MAUSTINOX B	0.12% C (max) 18% Cr 10% Ni 2.5% Mo steel: Pompey
	UTS: 540 Elon: 30% Proof: 240
MAUSTINOX Bbc	0.08% C 18% Cr 10% Ni 2.5% Mo steel: Pompey
	UTS: 500 Elon: 35% Proof: 240
MAUSTINOX C	0.12% C 20% Cr 8% Ni 2.5% Mo steel: Pompey
	UTS: 550 Elon: 25% Proof: 310
MAUSTINOX F	0.1% C 24% Cr 13% Ni steel: Pompey
	UTS: 570 Elon: 30% Proof: 240
MAUSTINOX H	0.1% C 25% Cr 20% Ni steel: Pompey
	UTS: 520 Elon: 30% Proof: 240
MAUSTINOX SA	0.08% C (max) 18% Cr 9% Ni + Nb + Ta steel: Pompey
	UTS: 500 Elon: 35% Proof: 240
MAUSTINOX SB	0.08% C 18% Cr 10% Ni 2.5% Mo + Nb + Ta steel: Pompey
MAUSTINOX X	0.1% C 15% Cr 30% Ni 4.5% Mo steel: Pompey
	UTS: 470 Elon: 35% Proof: 230
MAUSTINOX XD	0.1% C 15% Cr 30% Ni 6% Mo steel: Pompey
	UTS: 470 Elon: 35% Proof: 230
MAUSTINOX Y	0.1% C 20% Cr 20% Ni 4.5% Mo steel: Pompey
MAUSTINOX YD	0.07% C 20% Cr 20% Ni 6% Mo steel: Pompey
	UTS: 480 Elon: 32% Proof: 260
MAUSTINOX YN	0.07% C 20% Cr 30% Ni 5% Mo Cu steel: Pompey
	UTS: 480 Elon: 32% Proof: 260
MAZ	0.14% C 18% Cr 8.5% Ni S steel: Free machining; Schoeller-Bleckmann; annealed
	DPN: 175 UTS: 540 Elon: 45% Proof: 210
MAXHETE 1	0.35% C 18.5% Cr 8.5% Ni 2.5% W steel: Edgar Allen
MAXHETE 1A	0.4% C 13% Cr 13% Ni steel: Edgar Allen; internal combustion engine valves
MAXHETE 2	0.25% C (max) 25% Cr 13.5% Ni steel: Edgar Allen; high temperature uses
MAXHETE 4	0.25% C (max) 25% Cr 20% Ni steel: Edgar Allen; furnace equipment
MAXILVRY	0.12% C (max) 18% Cr 8% Ni steel: Edgar Allen for BS alloy En 58A

Symbol	Nominal analysis, supplier, condition and remarks.
MAXILVRY ADS	0.12% C (max) 12.5% Cr 12.5% Ni steel: Edgar Allen for BS alloy En 58D
MAXILVRY AWP	0.1% C (max) 18% Cr 8% Ni steel: Edgar Allen for BS alloy En 58B
MAXILVRY CB	0.1% C (max) 18% Cr 8% Ni steel: Edgar Allen for BS alloy En 58F
MCS 6	0.05% C 17.5% Cr 13.5% Ni 2.5% Mo 1.2% Cu steel: Japanese specification
METCO 41C	0.1% C 1.0% Si 18% Cr 12% Ni 2% Mo steel: Powder for spraying; Metco
	DPN: 189
METCOLOY 1	0.08% C 19% Cr 9% Ni steel: Wire for spraying; Metco; as sprayed
	DPN: 146 UTS: 200
METCOLOY 4	0.08% C 17% Cr 12% Ni 2.5% Mo steel: Wire for spraying; Metco
MNC 900	Covers stainless steel: Swedish Standard; includes detailed specification
NB	0.15% C (max) 18% Cr 9% Ni 1.0% Nb steel: Avesta
NCF 2	0.1% C (max) 21.0% Cr 32.5% Ni 0.35% Al 0.35% Ti Fe alloy: Japanese Standard designation; Inconil 800
NF A35 572/ Z2CN18/10	0.04% C (max) 18.0% Cr 10.0% Ni steel: French Standard
NF A35 572/ Z2CND17/13	0.03% C (max) 17.0% Cr 13.0% Ni 2.7% Mo steel: French Standard
NF A35 572/ Z2CND19/15	0.03% C (max) 19.0% Cr 15.0% Ni 3.5% Mo steel: French Standard
NF A35 572/ Z5CNDU21/08	0.06% C (max) 21.0% Cr 8.0% Ni 2.5% Mo steel: French Standard
NF A35 572/ Z6CN18/09	0.07% C (max) 18.0% Cr 9.0% Ni steel: French Standard
NF A35 572/ Z6CNDNb17/12	0.08% C (max) 17.0% Cr 12.0% Ni 2.2% Mo Nb + Ta 0.8% steel: French Standard
NF A35 572/ Z6CNDNb17/13	0.08% C (max) 17.0% Cr 13.0% Ni 2.7% Mo 0.8% Nb + Ta steel: French Standard
NF A35 572/ Z6CNNb18/11	0.08% C (max) 18.0% Cr 11.0% Ni 1.0% Nb + Ta steel: French Standard
NF A35 572/ Z6CNT18/11	0.08% C (max) 18.0% Cr 11.0% Ni 0.06% Ti steel: French Standard
NF A35 572/ Z8CN13/13	0.1% C (max) 13.0% Cr 13.0% Ni steel: French Standard
NF A35 572/ Z8CN18/12	0.10% C (max) 18.0% Cr 12.0% Ni steel: French Standard
NF A35 572/ Z8CNDT17/12	0.1% C (max) 17.0% Cr 12.0% Ni 2.2% Mo 0.6% Ti steel: French Standard
NF A35 572/ Z10CN18/09	0.12% C 18.0% Cr 9.0% Ni steel: French Standard
NF A35 572/ Z10CNF18/09	0.12% C (max) 18.0% Cr 9.0% Ni steel: French Standard
NF A35 572/ Z10CNT18/11	0.12% C (max) 18.0% Cr 11.0% Ni 0.8% Ti steel: French Standard
NF A35 572/ Z12CN17/08	0.11% C 17.0% Cr 7.5% Ni steel: French Standard
NF A35 572/ Z2CND17/11	0.07% C (max) 17.0% Cr 11.0% Ni 2.25% Mo steel: French Standard
NF A35 572/ Z2CND17/12	0.07% C (max) 17.0% Cr 12.0% Ni 2.7% Mo steel: French Standard
NF A35 573/ Z2CN18/10	0.03% C (max) 2.0% Mo (max) 18% Cr 10% Ni 0.04% (max) 0.03% S (max) steel: Plate; French Standard
	UTS: 185 Proof: 570
NF A35 573/ Z2CND17/12	0.03% C (max) 17.0% Cr 12.0% Ni 2.25% Mo steel: Plate; French Standard
	UTS: 600 Proof: 240
NF A35 573/ Z2CND17/13	0.03% C (max) 17.0% Cr 13.0% Ni 2.75% Mo steel: Plate; French Standard
	UTS: 600 Proof: 240

Note. The following abbreviations and units are used in the tables:

DPN	Hardness, diamond pyramid number
UTS	Ultimate tensile strength, N/mm^2
Elon	Elongation, %
Proof	0.1 % proof strength, N/mm^2

1 N/mm^2=0.1 hbar=0.102 kgf/mm^2=0.06475 tonf/in^2=145.04 lbf/in^2

Symbol	Nominal analysis, supplier, condition and remarks.
NF A35 573/ Z2CND19/15	0.03% C (max) 2.0% Mn (max) 19% Cr 15% Ni 3.5% Mo 0.04% P 0.03% S steel: Plate; French Standard
	UTS: 600 **Proof: 210**
NF A35 573/ Z5CNUD21/08	0.06% C (max) 21.0% Cr 8.0% Ni 2.6% Mo 1.5% Cu steel: Plate; French Standard
	UTS: 600 **Proof: 350**
NF A35 573/ Z6CN18/09	0.07% C (max) 2.0% 78% Cr 9% Ni steel: Plate; French Standard
	UTS: 620 **Proof: 200**
NF A35 573/ Z6CND17/11	0.07% C (max) 17.0% Cr 11.0% Ni 2.25% Mo steel: Plate; French Standard
	UTS: 620 **Proof: 250**
NF A35 573/ Z6CNDNb17/12	0.08% C (max) 17.0% Cr 12.0% Ni 2.25% Mo 1.0% Nb + Ta steel: Plate; French Standard
	UTS: 650 **Proof: 260**
NF A35 573/ Z6CNNb18/11	0.08% C (max) 2.0% Mn (max) 18% Cr 11% Ni 0.9% Nb Ta 0.04% P (max) 0.03% S (max) steel: Plate; French Standard
	UTS: 600 **Proof: 210**
NF A35 573/ Z6CNT18/11	0.08% C (max) 2.0% Mn (max) 18% Cr 11% Ni 0.5% Ti 0.04% P (max) 0.03% S (max) steel: Plate; French Standard
	UTS: 600 **Proof: 210**
NF A35 573/ Z8CN13/13	0.1% C (max) 13.0% Cr 13% Ni 2.0% Cu steel: Plate; French Standard
	UTS: 500 **Proof: 240**
NF A35 573/ Z8CNDT17/12	0.1% C (max) 17.0% Cr 12.0% Ni 2.2% Mo 0.6% Ti steel: Plate; French Standard
	UTS: 650 **Proof: 260**
NF A35 573/ Z8CNDT17/12	0.1% C (max) 17.0% Cr 12.0% Ni 2.2% Mo 0.6% Ti steel: Plate; French Standard
	UTS: 650 **Proof: 260**
NF A35 573/ Z8CN18/12	0.1% C (max) 2.0% Mn (max) 18.0% Cr 12.0% Ni steel: Plate; French Standard
	UTS: 600 **Proof: 220**
NF A35 573/ Z10CNT18/11	0.12% C (max) 2.0% Mn 18% Cr 11% Ni 0.15% Ti 0.04% P (max) 0.03% S (max) steel: Plate; French Standard
	UTS: 650 **Proof: 225**
NF A35 573/ Z10CN18/09	0.12% C (max) 2% Mn (max) 18% Cr 9% Ni steel: Plate; French Specification
	UTS: 650 **Proof: 225**
NF A35 573/ Z12CN17/08	1.1% C (max) 2% Mn (max) 17% Pn 7.5% Ni steel: Plate; French specification
	UTS: 750 **Proof: 250**
NF A36 209/ Z2CN18/10	0.03% C (max) 2.0% Mn (max) 18% Cr 10% Ni 0.04% P (max) 0.03% S (max) steel: Plate; French Standard
	UTS: 185 **Proof: 570**
NF A36 209/ Z6CN18/09	0.01% C (max) 2.0% Mn (max) 18% Cr 9% Ni steel: Plate; French Standard
	UTS: 620 **Proof: 200**
NF A36 209/ Z6CNT18/11	0.08% C (max) 1.5% Mn 18% Cr 12% Ni 0.5% Ti 0.035% P (max) 0.02% S (max) steel: Plate; French Standard
	UTS: 600 **Proof: 210**
Ni Cr 30/20	0.2% C (max) 21.0% Cr 30.0% Ni steel: Designation used by German Standard
NICRAL C	36% Ni 18% Cr steel: Imphy
	DPN: 190 **UTS: 650** **Elon: 30%** **Proof: 350**
NICRAL C	0.08% C (max) 18.5% Cr 35.5% Ni steel: Creusot-Loire
NICRAL D	0.08% C (max) 25.0% Cr 20.5% Ni steel: Creusot-Loire
NICRAL D	25% C 20% Ni steel: Imphy
	DPN: 175 **UTS: 650** **Elon: 35%** **Proof: 350**
NICRAL DM	25% Cr 20% Ni steel: Imphy
	DPN: 175 **UTS: 600** **Elon: 15%** **Proof: 300**

Symbol	Nominal analysis, supplier, condition and remarks.
NICRAL H	25% Cr 12% Ni steel: Imphy
	DPN: 175 **UTS: 650** **Elon: 30%** **Proof: 300**
NICRAL H	0.08% C (max) 23.0% Cr 13.5% Ni steel: Creusot-Loire
NICRAL HR2	0.15% C 19.0% Cr 9.0% Ni steel: Creusot-Loire
NICRAL K	22% Cr 40% Ni Fe alloy: Imphy
NICRALT	0.12% C 16% Cr 13% Ni 3.0% W 0.5% Ti steel: Imphy; annealed
	UTS: 840 **Elon: 35%** **Proof: 220**
NICREX 1	0.1% C 26% Cr 20% Ni 0.4% Mo steel: Welding electrode; Murex
	UTS: 630 **Elon: 40%**
NICREX 1 Cb	0.1% C 26% Cr 20% Ni 0.4% Mo 1% Nb steel: Welding electrode; Murex; Nb also called Cb
	UTS: 740 **Elon: 35%**
NICREX 2	0.1% C 23% Cr 18% Ni 3.2% Mo 1% Nb steel: Welding electrode; Murex
	UTS: 450 **Elon: 27%**
NICREX AC	0.08% C 18% Cr 8% Ni 0.8% Nb steel: Welding electrode; Murex
	UTS: 670 **Elon: 40%**
NICROMAZ B	15% Cr 40% Ni steel: Imphy; resistant to hot concentrated sulphuric acid
NICROMAZ C	20% Cr 25% Ni 5% Mo 1.5% Cu steel: Imphy
NISIMAZ	2.5% Cr 21% Ni 7.0% Si steel: Imphy; resistant to hot acids
NOVO 18/3	0.1% C 18% Cr 8% Ni steel: Jonas; fully softened; not weld stabilized
	DPN: 183 **UTS: 460** **Elon: 65%** **Proof: 320**
NOVO 18/8W	0.06% C 18% Cr 8% Ni steel: Jonas; annealed; stabilized with 0.35% Ti or 0.6% Nb
	DPN: 183 **UTS: 460** **Elon: 65%** **Proof: 320**
NS 20E	0.16% C 19% Cr 9% Ni steel: Ugine
NS 21A	0.08% C (max) 19% Cr 10% Ni steel: Ugine
NS 21S	0.15% C (max) 19% Cr 9% Ni 0.6% Ti steel: Ugine
NS 9S	0.15% C (max) 19% Cr 9% Ni 1.0% Nb steel: Ugine
NSCD	18% Cr 16% Ni 6% Mo 3% Cu steel: Ugine
NSM 21	0.12% C (max) 19% Cr 10% Ni 3.0% Mo steel: Ugine
NSMC	0.12% C (max) 19% Cr 10% Ni 3.0% Mo steel: Ugine
NSU	18% Cr 9% Ni steel: Ugine; free machining
NTK M7	0.06% C 10% Cr 17% Ni 7% Mo steel: For use with hydrochloric acid. Kawasaki; annealed
	UTS: 840 **Elon: 60%** **Proof: 370**
NV 25 1	0.05% C (max) 18.5% Cr 8.0% Ni steel: Designation used by DNV
	UTS: 500 **Elon: 40%** **Proof: 200**
NV 25 2	0.08% C (max) 18.5% Cr 8.5% Ni steel: Designation used by DNV
	UTS: 500 **Elon: 40%** **Proof: 200**
NV 25 3	0.05% C (max) 19.0% Cr 9.5% Ni 2.5% Mo steel: Designation used by DNV
	UTS: 500 **Elon: 40%** **Proof: 200**
NV 25 4	0.08% C (max) 19.0% Cr 10.0% Ni 2.5% Mo steel: Designation used by DNV
	UTS: 500 **Elon: 40%** **Proof: 200**
NVR 25 1	0.06% C (max) 18.5% Cr 10.0% Ni steel: Designation used by DNV
NVR 25 2	0.03% C (max) 18.5% Cr 11.0% Ni steel: Designation used by DNV
NVR 25 3	0.06% C (max) 18.2% Cr 11.0% Ni 1.8% Mo steel: Designation used by DNV
NVR 25 4	0.06% C 18.2% Cr 12.0% Ni 2.8% Mo steel: Designation used by DNV
NVR 25 5	0.03% C 18.2% Cr 13.7% Ni 2.8% Mo steel: Designation used by DNV
OR 2	0.08% C 18% Cr 9% Ni steel: Sandvik
OR 11	0.12% C 19% Cr 10% Ni 3.0% Mo steel: Sandvik

Symbol	Nominal analysis, supplier, condition and remarks.
OSBORN 303	0.07% C 18% Cr 9.0% Ni steel: Osborn; previously BSMF grade
OSBORN 304	0.05% C 18% Cr 9.5% Ni steel: Osborn; previously BSM grade
OSBORN 309	0.08% C 22.5% Cr 14.5% Ni steel: Osborn; previously IK grade
OSBORN 310	0.1% C 24.5% Cr 20.0% Ni steel: Osborn; previously FD grade
OSBORN 316	0.05% C 18% Cr 11% Ni 2.7% Mo steel: Osborn; previously EW grade
OSBORN 317	0.05% C 18% Cr 13.5% Ni 3.7% Mo steel: Osborn; previously EWA grade
OSBORN 321	0.06% C 18% Cr 9.5% Ni steel: Osborn; previously BST grade
OSBORN 330	0.1% C (max) 18% Cr 37.5% Ni 2.0% Si steel: Osborn; previously FL grade
OSBORN 347	0.07% C 18% Cr 9.5% Ni 0.6% Nb steel: Osborn; previously BSC grade
P 1010	0.12% C 18.5% Cr 8.5% Ni steel: Carrs for BS alloy En 58A
P 1011	0.14% C 18% Cr 8% Ni 0.5% Ti steel: Carrs for BS alloy En 58B
P 1012	0.14% C 18% Cr 10% Ni 0.5% Ti steel: Carrs for BS alloy En 58C
P 1013	0.14% C 12% Cr 12% Ni steel: Carrs for BS alloy En 58
P 1014	0.07% C 18% Cr 10% Ni steel: Carrs for BS alloy En 58E
P 1015	0.14% C 18% Cr 8% Ni 1.0% Nb steel: Carrs for BS alloy En 58F
P 1016	0.14% C 18% Cr 10% Ni 1.0% Nb steel: Carrs for BS alloy En 58G
P 1017	0.1% C 18% Cr 10% Ni 2.0% Mo steel: Carrs for BS alloy En 58H
P 1018	0.1% C 18% Cr 10% Ni 3.0% Mo steel: Carrs for BS alloy En 58J
PANTEG 304	0.05% C 18% Cr 9.75% Ni steel: Sheet; Richard Thomas & Baldwin; air cooled **UTS: 600 Elon: 50%**
PANTEG 315	0.06% C 17.5% Cr 11.0% Ni 1.5% Mo steel: Sheet; Richard Thomas & Baldwin; air cooled **UTS: 650 Elon: 45%**
PANTEG 316	0.05% C 17.5% Cr 11.8% Ni 3.0% Mo steel: Sheet; Richard Thomas & Baldwin; air cooled **UTS: 660 Elon: 40%**
PANTEG 321	0.05% C 17.5% Cr 9.6% Ni 0.4% Ti steel: Sheet; Richard Thomas & Baldwin; air cooled **UTS: 640 Elon: 50%**
PARALLOY 0	18% Cr 8% Ni steel: APV Paramount
PARALLOY 0KW	19% Cr 10% Ni Nb: Stabilized APV Paramount
PARALLOY 0L	17% Cr 12% Ni steel (low magnetic permeability): APV Paramount
PARALLOY 0LC	18% Cr 8% Ni 0.03% C (max) steel: APV Paramount
PARALLOY 0LS	17% Cr 1.2% Ni steel: Free machining; low magnetic permeability; APV Paramount
PARALLOY 0S	18% Cr 8% Ni steel: Free machining; APV Paramount

Symbol	Nominal analysis, supplier, condition and remarks.
PARALLOY 0SW	18% Cr 8% Ni Nb steel: Free machining; stabilized; APV Paramount
PARALLOY 0W	18% Cr 8% Ni Nb steel: Stabilized; APV Paramount
PARALLOY 3	18% Cr 8% Ni 3% Mo steel: APV Paramount
PARALLOY 3K	17% Cr 10% Ni 3% Mo steel: APV Paramount
PARALLOY 3KLC	0.03% C (max) 17% Cr 10% Ni 3% Mo steel: APV Paramount
PARALLOY 3L	18% Cr 14% Ni 3% Mo steel: Low magnetic permeability; APV Paramount
PARALLOY 3LS	18% Cr 14% Ni 3% Mo steel: Free machining; low magnetic permeability; APV Paramount
PARALLOY 3S	18% Cr 8% Ni 3% Mo steel: Free machining; APV Paramount
PARALLOY 3W	17% Cr 10% Ni 3% Mo Nb steel: Stabilized; APV Paramount
PARALLOY 3WS	17% Cr 10% Ni 3% Mo Nb steel: Stabilized; free machining; APV Paramount
PARALLOY 4K	19% Cr 12% Ni 4% Mo steel: APV Paramount
PARALLOY 4KLC	0.03% C (max) 19% Cr 12% Ni 4% Mo steel: APV Paramount
PEL	0.06% C 17.5% Cr 8.5% Ni steel: Sheet; RTB for BS alloy En 58A
PET	0.07% C 17.5% Cr 9% Ni 0.4% Ti steel: Sheet; RTB for BS alloy En 58B
PMH	0.06% C 17.5% Cr 10.5% Ni 3.0% Mo steel: Sheet; RTB for BS alloy En 58J
PML	0.06% C 17.5% Cr 9.5% Ni 1.5% Mo steel: Sheet; RTB for BS alloy En 58H
PYROMET 860	0.1% C 14% Cr 42% Ni 6% Mo 4% Co 3% Ti 1.0% Al 0.01% B Fe alloy: Carpenter
QQ S 763/7	0.12% C 17% Cr 8% Ni 0.7% Mo 0.2% Se 0.7% Zr steel: US Federal; annealed **UTS: 630 Elon: 28% Proof: 200**
QQ S 763/8	0.1% C 17% Cr 7% Ni 1.0% Nb or 0.5% Ti steel: US Federal; annealed **UTS: 450 Elon: 35% Proof: 150**
QQ S 766/1	0.08% C 18% Cr 8% Ni steel: US Federal; annealed **UTS: 530 Elon: 40%**
QQ W 423a	Corrosion resisting steel: Wire; US Federal specification
R 3	0.1% C 18% Cr 9% Ni steel: Jessop for BS alloy En 58A
R 9	0.08% C 18.5% Cr 9.5% Ni 2.0% Mo 0.3% Ti steel: Jessop for BS alloy En 58H
R 10	0.08% C 12% Cr 12% Ni steel: Jessop for BS alloy En 58D
R 11	0.25% C 1.2% Si 23% Cr 14% Ni steel: Jessop
R 16	0.08% C 18% Cr 9% Ni 0.3% Ti steel: Jessop for BS 970 En 58B
R 18	0.32% C 18% Cr 7.5% Ni 2.7% W steel: Jessop for BS 970 En 55
R 20	0.1% C 19% Cr 12% Ni 1.25% Nb steel: Jessop **UTS: 580 Elon: 52% Proof: 200**
R 22	0.2% C 22% Cr 11% Ni 2.7% W steel: Casting; Jessop; as cast **UTS: 510 Elon: 28%**
R 22	0.17% C 22% Cr 11.5% Ni 3% W steel: Jessop
R 23	0.1% C 25% Cr 22% Ni steel: Jessop **UTS: 720 Elon: 45% Proof: 270**
R 24	0.08% C 18% Cr 9% Ni 0.22% S steel: Free machining; Jessop for BS alloy En 58AM
R 25	0.08% C 18% Cr 9.5% Ni 2.8% Mo steel: Jessop for BS alloy En 58J
R 27	0.05% C 20% Cr 28.5% Ni 2.25% Mo 3.25% Cu steel: Jessop
R 27 Nb	0.05% C 20% Cr 28.5% Ni 2.25% Mo 3.25% Cu Nb steel: Jessop
R 35	0.03% C 20% Cr 25% Ni 0.3% Nb steel: Jessop

Note. The following abbreviations and units are used in the tables:

DPN	Hardness, diamond pyramid number
UTS	Ultimate tensile strength, N/mm^2
Elon	Elongation, %
Proof	0.1 % proof strength, N/mm^2

1 N/mm^2=0.1 hbar=0.102 kgf/mm^2=0.06475 tonf/in^2=145.04 lbf/in^2

Symbol	Nominal analysis, supplier, condition and remarks.
R 41	0.06% C 17% Cr 11% Ni 2.35% Mo steel: Jessop
R 42	0.06% C 19% Cr 9% Ni steel: Jessop
R 43	0.1% C 17% Cr 8% Ni 1.0% Nb steel: Jessop
R 45	0.12% C 21.5% Cr 17.5% Ni 0.5% Ti steel: Jessop
R 47	0.45% C 2.2% Si 19% Cr 9% Ni 1.4% W steel: Jessop
R 48	0.09% C 18% Cr 9% Ni 0.6% Nb 0.22% S steel: Jessop; free machining; weld stabilized
R 51	0.08% C 23.0% Cr 15.0% Ni 0.32% Ti steel: Jessop; corrosion resistant steel
R 52	0.08% C 23.0% Cr 15.0% Ni 0.04% Nb steel: Jessop; corrosion resistant steel
R 53	0.08% C 24.0% Cr 17.0% Ni 0.32% Ti steel: Jessop; corrosion resistant steel
R 54	0.08% C 24.0% Cr 17.0% Ni 0.64% Nb steel: Jessop; corrosion resistant steel
R 300	0.1% C 18% Cr 8.5% Ni steel: Fagersta; annealed **DPN: 200 UTS: 530 Elon: 45% Proof: 200**
R 310	0.07% C 18% Cr 10% Ni steel: Fagersta; annealed **DPN: 180 UTS: 530 Elon: 45% Proof: 180**
R 320	0.07% C 18% Cr 9% Ni steel: Fagersta; annealed **DPN: 180 UTS: 530 Elon: 55% Proof: 200**
R 350	0.04% C 18.5% Cr 9% Ni steel: Fagersta; annealed **DPN: 180 UTS: 530 Elon: 50% Proof: 200**
R 350G	0.05% C 18.0% Cr 8.5% Ni steel: Casting; Fagersta
R 358	0.05% C 18.0% Cr 9.5% Ni Nb steel: Wrought; stainless; type 347; Fagersta
R 358G	0.03% C 18.0% Cr 9.5% Ni 0.8% Nb steel: Casting; Fagersta
R 359	0.05% C 18.0% Cr 9.5% Ni Ti steel: Wrought; stainless; type 321; Fagersta
R 360	0.03% C 18% Cr 10% Ni steel: Fagersta; annealed **DPN: 180 UTS: 530 Elon: 50% Proof: 200**
R 360G	0.03% C (max) 18.5% Cr 10.0% Ni steel: Casting; Fagersta
R 380	0.1% C 18% Cr 8.5% Ni 0.15% Se steel: Fagersta; free machining **UTS: 520 Elon: 40%**
R 390	0.05% C 18.0% Cr 11.5% Ni steel: Wrought; stainless; type 304; Fagersta
R 400	0.09% C 18% Cr 9% Ni 1.5% Mo steel: Fagersta; annealed **DPN: 200 UTS: 530 Elon: 45% Proof: 200**
R 420	0.05% C 17.5% Cr 11% Ni 2.3% Mo steel: Fagersta; annealed **DPN: 180 UTS: 530 Elon: 45% Proof: 200**
R 428	0.05% C 17.5% Cr 11.0% Ni 2.3% Mo Nb steel: Wrought; stainless; Fagersta
R 429	0.05% C 17.0% Cr 11.0% Ni 2.3% Mo Ti stainless steel: Wrought; Fagersta
R 440	0.05% C 17% Cr 12% Ni 2.8% Mo steel: Fagersta; annealed **DPN: 180 UTS: 530 Elon: 45% Proof: 200**
R 440G	0.05% C 18.0% Cr 10.5% Ni 2.8% Mo steel: Casting; Fagersta
R 450	0.05% C 18% Cr 10.5% Ni 1.5% Mo steel: Fagersta; annealed **DPN: 180 UTS: 530 Elon: 45% Proof: 200**
R 460	0.03% C 17% Cr 12% Ni 2.8% Mo steel: Fagersta; annealed **DPN: 180 UTS: 530 Elon: 45% Proof: 20**
R 460G	0.03% C (max) 17.0% Cr 12.0% Ni 2.8% Mo steel: Casting; Fagersta
R 470	0.05% C 17% Cr 14.5% Ni 4.5% Mo steel: Fagersta; annealed **DPN: 180 UTS: 530 Elon: 45% Proof: 200**
R 470G	0.04% G 17.5% Cr 15.5% Ni 3.7% Mo steel: Casting; Fagersta

Symbol	Nominal analysis, supplier, condition and remarks.
R 500	0.06% C 18% Cr 9.5% Ni 1% Nb steel: Fagersta; annealed **DPN: 200 UTS: 530 Elon: 45% Proof: 240**
R 510	0.06% C 18% Cr 9.5% Ni 0.5% Ti steel: Fagersta; annealed **DPN: 200 UTS: 530 Elon: 45% Proof: 240**
R 520	0.08% C 18% Cr 4.8% Ni 8.0% Mn steel: Fagersta; annealed **DPN: 240 UTS: 580 Elon: 50% Proof: 270**
R 530	0.06% C 17.5% Cr 11% Ni 2.3% Mo + Ti steel: Fagersta; annealed **DPN: 180 UTS: 530 Elon: 40% Proof: 240**
R 590	0.06% C 13.5% Cr 12% Ni steel: Fagersta; annealed **DPN: 170 UTS: 480 Elon: 50% Proof: 180**
R 620G	0.1% C 24.0% Cr 5.0% Ni steel: Casting; scaling temperature 1070 °C; Fagersta
R 640G	0.1% C 24.0% Cr 5.0% Ni 1.5% Mo steel: Casting; scaling temperature 1070 °C; Fagersta
R 690G	0.05% C 13.0% Cr 6.0% Ni steel: Casting; Fagersta
R 800	0.08% C 20.5% Cr 19% Ni steel: Fagersta; annealed **DPN: 220 UTS: 530 Elon: 40% Proof: 200**
R 800G	0.2% C 1.5% Si 20.0% Cr 19.0% Ni steel: Casting; scaling temperature 1050 °C; Fagersta
R 820	0.07% C 25% Cr 20% Ni steel: Fagersta; annealed **DPN: 230 UTS: 580 Elon: 40% Proof: 240**
R 830	0.1% C 22.5% Cr 23.5% Ni steel: Fagersta; annealed **DPN: 230 UTS: 580 Elon: 30% Proof: 240**
R 830G	0.2% C 1.5% Si 23.0% Cr 22.0% Ni steel: Casting; scaling temperature 1125 °C; Fagersta
R 850G	0.2% C 1.5% Si 23.0% Cr 11.5% Ni steel: Casting; scaling temperature 1125 °C; Fagersta
R 860G	0.2% C 20.0% Cr 40% Ni steel: Casting; scaling temperature 1050 °C; Fagersta
R 890Gr	0.05% C 20.0% Cr 24.0% Ni 3.0% Mo 2.5% Cu steel: Casting
RA 330	0.08% C 18% Cr 35% Ni steel: Simonds
RCK 3	0.15% C 25% Cr 13% Ni 0.2% N steel: Bofors; annealed **DPN: 170 UTS: 630 Elon: 54% Proof: 340**
RCK 4	0.2% C 26% Cr 12% Ni 0.2% N steel: Bofors; obsolete
RCT	0.45% C 15% Cr 12% Ni 3.0% W steel: Bofors; obsolete
RCT 3	0.2% C 21% Cr 13% Ni 3.0% W steel: Bofors; obsolete
REB 210	0.18% C 25% Cr 21% Ni steel: Bofors; obsolete
RIM 21	0.08% C 18% Cr 11% Ni 1.5% Mo steel: Bofors; obsolete
RIM 29	0.16% C 18% Cr 8% Ni steel: Bofors
RIM 210	0.05% C 18% Cr 11% Ni 1.5% Mo steel: Bofors; obsolete
RIM 213	0.07% C 18% Cr 11% Ni 2.3% Mo 0.4% Ti steel: Bofors; annealed **DPN: 165 UTS: 580 Elon: 50% Proof: 200**
RIM 215	0.12% C (max) 18% Cr 10% Ni 3.0% Mo steel: Bofors
RIM 217	0.02% C 18% Cr 13% Ni 2.8% Mo steel: Bofors **DPN: 180 UTS: 500 Elon: 45% Proof: 200**
RIM 290	0.02% C 18% Cr 10% Ni steel: Bofors; annealed; obsolete **DPN: 140 UTS: 510 Elon: 60% Proof: 200**
RIM 291	0.05% C 18% Cr 9% Ni steel: Bofors; annealed **DPN: 155 UTS: 570 Elon: 65% Proof: 200**
RIM 294	0.15% C (max) 18% Cr 9% Ni 0.6% Ti steel: Bofors
RIM 295	0.15% C 19% Cr 9% Ni 0.6% Ti steel: Bofors

Symbol	Nominal analysis, supplier, condition and remarks.
RIO 214	0.09% C 26% Cr 5% Ni 1.5% Mo steel: Bofors; annealed
	DPN: 240 UTS: 710 Elon: 28% Proof: 530
RLH 2	18% Cr 9% Ni S steel: Free machining; Bofors
RNC 0	12% C 12% Ni Fe alloy: For electric resistance; Imphy
RNC 1	12% C 35% Ni Fe alloy: For electrical resistance; Imphy
RRN J 30	0.15% C 18% Cr 9% Ni steel: Fagersta
RRN J 31	0.08% C (max) 19% Cr 9% Ni steel: Fagersta
RRN J 38	0.1% C 18% Cr 9% Ni S steel: Free machining; Fagersta
RRN J 42	0.12% C (max) 18% Cr 10% Ni 2.0% Mo steel: Fagersta
RRN J 44	0.12% C (max) 18% Cr 10% Ni 3.0% Mo steel: Fagersta
RRN J 50	0.15% C 18% Cr 9% Ni 1.0% Nb steel: Fagersta
RRN J 51	0.15% C 18% Cr 9% Ni 0.6% Ti steel: Fagersta
RRN J 59	18% Cr 9% Ni steel: For deep drawing; Fagersta
S 18/8	Stainless steel: Sintered material; Sintered Products Ltd; 15–20% porosity
	UTS: 400 Elon: 22%
S 18/8	Sintered 18/8 stainless steel: Specific gravity 6.9; Durasint
	DPN: 130 UTS: 440 Elon: 10%
S 495	0.4% C 14% Cr 20% Ni 4% Mo 4% W 4% Nb steel: Allegheny Ludlum
S 588	0.4% C 18.5% Cr 20% Ni 4% Mo 4% W 4% Nb steel: Allegheny Ludlum
S 20910	UNS designation for type XM19 steel
S 30200	UNS designation for 302 type steel
S 30400	UNS designation for type 304 steel
S 30403	UNS designation for type 304L steel
S 30409	UNS designation for 304H type steel
S 30451	UNS designation for 304N type steel
S 31008	UNS designation for 310S type steel
S 31600	UNS designation for 316 type steel
S 31603	UNS designation for type 316L steel
S 31609	UNS designation for type 316H steel
S 31651	UNS designation for 316N type steel
S 32100	UNS designation for 321 type steel
S 32109	UNS designation for 321H type steel
S 34700	UNS designation for 347 type steel
S 34709	UNS designation for 347H type steel
S 34800	UNS designation for 348 type steel
S 34809	UNS designation for 348H type steel
SAE 30301	0.1% C 17% Cr 7% Ni steel: Mechanical properties not specified; AISI type 301
SAE 30302	0.1% C 18% Cr 9% Ni steel: Mechanical properties not specified; AISI type 302
SAE 30302 B	0.15% C 18% Cr 9% Ni steel: As SAE 30302 with different C range
SAE 30303	0.15% C (max) 18% Cr 9% Ni 0.6% Zi or Mo steel
SAE 30303 F	0.15% C (max) 18% Cr 9% Ni 0.6% Mo steel: Free machining; mechanical properties not specified; P or S or Se 0.07%; AISI type 303F
SAE 30303 Se	0.15% C 18% Cr 9% Ni 0.5% Se (min) steel: Free cutting; mechanical properties not specified

Note. The following abbreviations and units are used in the tables:

DPN	Hardness, diamond pyramid number
UTS	Ultimate tensile strength, N/mm^2
Elon	Elongation, %
Proof	0.1 % proof strength, N/mm^2

1 N/mm^2=0.1 hbar=0.102 kgf/mm^2=0.06475 tonf/in^2=145.04 lbf/in^2

Symbol	Nominal analysis, supplier, condition and remarks.
SAE 30304	0.08% C (max) 19% Cr 9% Ni steel: Mechanical properties not specified; AISI type 304
SAE 30304 L	0.03% C (max) 19% Cr 10% Ni steel: Mechanical properties not specified
SAE 30305	0.12% C (max) 18% Cr 12% Ni steel: Mechanical properties not specified; AISI type 305
SAE 30308	0.08% C 20% Cr 11% Ni steel
SAE 30309	0.2% C (max) 23% Cr 14% Ni steel: Mechanical properties not specified; AISI type 309
SAE 30309 S	0.08% C 23% Cr 14% Ni steel: Mechanical properties not specified
SAE 30310	0.25% C (max) 25% Cr 20% Ni steel: Mechanical properties not specified; AISI type 310
SAE 30310 S	0.08% C 25% Cr 21% Ni steel: Mechanical properties not specified
SAE 30314	0.25% C 24% Cr 21% Ni steel: Mechanical properties not specified
SAE 30316	0.08% C (max) 17% Cr 12% Ni 2.5% Mo steel: Mechanical properties not specified; AISI type 316
SAE 30316 L	0.03% C 17% Cr 12% Ni 2.5% Mo steel: Mechanical properties not specified
SAE 30317	0.08% C (max) 19% Cr 13% Ni 3.5% Mo steel: Mechanical properties not specified; AISI type 317
SAE 30321	0.08% C (max) 18% Cr 10% Ni 0.5% Ti steel: Mechanical properties not specified; AISI type 321
SAE 30325	0.25% C (max) 8% Cr 21% Ni 1.2% Cu steel: Mechanical properties not specified; AISI type 325
SAE 30329	0.09% C 26% Cr 5% Ni 1.5% Mo steel: Obsolete
SAE 30330	0.25% C (max) 15% Cr 35% Ni steel: Mechanical properties not specified
SAE 30330 A	0.45% C 15% Cr 35% Ni steel: Mechanical properties not specified
SAE 30347	0.08% C 18% Cr 10% Ni 1.0% Nb steel: Mechanical properties not specified; AISI type 347
SAE 30348	0.08% C 18% Cr 11% Ni 0.1% Ta 0.8% Nb steel: Mechanical properties not specified
SAE 60303	0.16% C 19% Cr 10% Ni 1.5% Mo 0.3% Se steel: Casting; ASTM alloy CF 16F
SAE 60304	0.08% C 19% Cr 9% Ni steel: Casting; ASTM alloy CF 8
SAE 60304 L	0.03% C 1.5% Mn 19% Cr 10% Ni steel: Casting; as cast; weldable
SAE 60309	0.2% C 24% Cr 13% Ni steel: Casting; ASTM alloy CH 20
SAE 60310	0.2% C 24% Cr 21% Ni steel: Casting; ASTM alloy CK20
SAE 60312	0.3% C 28% Cr 9% Ni steel: Casting; ASTM alloy CE30
SAE 60316	0.08% C 19% Cr 11% Ni 2.5% Mo steel: Casting; ASTM alloy CF 8M
SAE 60316 L	0.03% C 19% Cr 11% Ni 2.5% Mo steel: Casting; as cast; weldable
SAE 60317	0.08% C 19% Cr 11% Ni 3.5% Mo steel: Casting; as cast
SAE 60347	0.08% C 19% Cr 10% Ni 0.7% Nb steel: Casting
SAE 70308	0.3% C 20% Cr 10% Ni 0.5% Mo steel: Casting; ASTM alloy HF
SAE 70309	0.35% C 26% Cr 12% Ni 0.5% Mo 0.2% N steel: Casting; ASTM alloy HH
SAE 70310	0.5% C 26% Cr 20% Ni 0.5% Mo steel: Casting; ASTM alloy HK
SAE 70310 A	0.5% C 30% Cr 20% Ni 0.5% Mo steel: Casting; ASTM alloy HL
SAE 70311	0.45% C 21% Cr 25% Ni 0.5% Mo steel: Casting
SAE 70312	0.35% C 28% Cr 10% Ni 0.5% Mo steel: Casting; ASTM alloy HE
SAE 70327	0.5% C (max) 28% Cr 6% Ni 0.5% Mo steel: Casting; ASTM alloy HD
SAE 70330	0.5% C 15% Cr 35% Ni 0.5% Mo steel: Casting; ASTM alloy HT

Symbol	Nominal analysis, supplier, condition and remarks.
SAE 70331	0.5% C 19% Cr 39% Ni 0.5% Mo steel: Casting; ASTM alloy HU
SAE EV 3	0.2% C 21% Cr 11.5% Ni steel: For exhaust valves
	UTS: 840 Elon: 26% Proof: 450
SAE EV 10	1% C 14% Cr 14% Ni steel: Casting for exhaust valves
SAE HNV 5	0.35% C 2.5% Si 13% Cr 8% Ni 0.5% Mo steel: For inlet valves
SANBRON AA	0.16% C 19% Cr 8% Ni steel: Sanderson for BS alloy En 58A
SANICRO 31	0.04% C 21% Cr 31% Ni Fe alloy: Tube; Sandvik
	DPN: 150 UTS: 490 Elon: 30% Proof: 170
SEC 7	0.15% C (max) 18% Cr 9% Ni steel: Designation used by Japanese Standard
SEC 8	0.08% C (max) 19% Cr 10% Ni steel: Designation used by Japanese Standard
SEC 9	0.03% C (max) 19% Cr 10% Ni steel: Designation used by Japanese Standard
SEC 10	0.08% C 18% Cr 10.5% Ni 0.5% Ti steel: Designation used by Japanese Standard
SEC 13	0.03% C 17% Cr 12% Ni 2.5% Mo steel: Designation used by Japanese Standard
SEH 5	0.25% C 25% Cr 20% Ni steel: Designation used by Japanese Standard
SF 24	0.08% C 13% Cr 13% Ni steel: Annealed; Samuel Fox
	UTS: 540 Elon: 56%
SF 25	0.1% C 18% Cr 9% Ni 2.0% Mo + Ti steel: Weldable; Samuel Fox
	UTS: 690 Elon: 46%
SF 50	0.12% C 17% Cr 10% Ni Mo Ti S steel: Weldable; free machining; Samuel Fox
	UTS: 630 Elon: 56%
SF 301	0.15% C 17% Cr 7% Ni steel: Annealed; Samuel Fox
	UTS: 920 Elon: 55%
SF 302	0.08% C 18% Cr 9% Ni steel: Annealed; Samuel Fox
	UTS: 620 Elon: 57%
SF 304	0.06% C 19% Cr 10% Ni steel: Annealed; Samuel Fox
	UTS: 600 Elon: 57%
SF 304 L	0.03% C 19% Cr 10% Ni steel: Annealed; weldable; Samuel Fox
	UTS: 600 Elon: 57%
SF 305	0.12% C 18% Cr 12% Ni steel: Annealed; Samuel Fox
	UTS: 580 Elon: 54%
SF 316	0.08% C 17% Cr 12% Ni 2.5% Mo steel: Weldable; up to 10 mm; Samuel Fox
	UTS: 610 Elon: 50%
SF 316 ELC	0.03% C (max) 17% Cr 12% Ni 2.5% Mo steel: Annealed; weldable; Samuel Fox
	UTS: 610 Elon: 50% Proof: 32
SF 316 Ti	0.08% C 17% Cr 12% Ni 2.5% Mo + Ti steel: Weldable; Samuel Fox
	UTS: 610 Elon: 50%
SF 317	0.08% C 19% Cr 13% Ni 3.5% Mo steel: Weldable; Samuel Fox
	UTS: 630 Elon: 50%
SF 321	0.08% C 18% Cr 10% Ni Ti steel: Weldable; Samuel Fox
	UTS: 610 Elon: 53%
SF 347	0.08% C 18% Cr 10% Ni Nb steel: Weldable; Samuel Fox
	UTS: 620 Elon: 54%

Symbol	Nominal analysis, supplier, condition and remarks.
SF 835	0.08% C 17% Cr 12% Ni 3.5% Mo + Ti steel: Weldable for use on check plates; annealed; Samuel Fox
	UTS: 610 Elon: 50%
SIRIUS 3	0.08% C (max) 25.0% Cr 20.5% Ni steel: Creusot-Loire
SIRIUS 35	0.08% C (max) 18.5% Cr 35.5% Ni steel: Creusot-Loire
SIRIUS 345	0.08% C (max) 23.0% Cr 13.5% Ni steel: Creusot-Loire
SIRIUS S12	0.15% C (max) 19.0% Cr 9.0% Ni steel: Creusot-Loire
SIS 14-2324	0.11% C 2% Mn 25% Cr 5% Ni 1.5% Mo stainless steel: Swedish Standard
	DPN: 260 UTS: 600 Elon: 19% Proof: 410
SIS 14-2324	0.1% C 26% Cr 5% Ni 1.5% Mo steel: Swedish Standard
	DPN: 260
SIS 14-2330	0.12% C 18% Cr 9% Ni steel: Annealed; Swedish Standard
	DPN: 200
SIS 14-2331	0.12% C 18% Cr 8% Ni steel: For springs; Swedish Standard
SIS 14-2332	0.08% C 18% Cr 10% Ni steel: Annealed; Swedish Standard
	DPN: 180
SIS 14-2332	0.07% C 2% Mn 18% Cr 10% Ni steel: Swedish Standard; stainless
	DPN: 200
SIS 14-2333	0.05% C 2.0% Mn 18% Cr 10% Ni steel: Swedish Standard; stainless
	DPN: 200 UTS: 600 Elon: 45% Proof: 200
SIS 14-2333	0.06% C 18% Cr 10% Ni steel: Annealed; Swedish Standard
	DPN: 180
SIS 14-2337	0.08% C 18% Cr 10% Ni 0.6% Ti steel: Swedish Standard
	DPN: 190
SIS 14-2337	0.08% C 2.0% Mn 18% Cr 10% Ni 0.6% Ti steel: Swedish Standard; stainless
	DPN: 210 UTS: 600 Elon: 40% Proof: 200
SIS 14-2338	0.08% C 2.0% Mn 18% Cr 10% Ni N 0.9% Nb 0.5% Ta steel: Swedish Standard; stainless
	DPN: 210 UTS: 600 Elon: 40% Proof: 210
SIS 14-2338	0.08% C 2.0% Mn 18% Cr 10% Ni 0.9% Nb 0.5% Ta steel: Swedish Standard; stainless
SIS 14-2338	0.08% C 18% Cr 11% Ni 1% Nb + Ta steel: Swedish Standard
	DPN: 190
SIS 14-2341	0.06% C 18% Cr 11% Ni 1.7% Mo steel: Annealed; Swedish Standard
	DPN: 180
SIS 14-2343	0.06% C 17% Cr 12% Ni 2.7% Mo steel: Annealed; Swedish Standard
	DPN: 180
SIS 14-2343	0.05% C 2.0% Mn 19% Cr 18% Cr 12% Ni 2.5% Mo steel: Swedish Standard; stainless
	DPN: 200 UTS: 690 Elon: 45% Proof: 220
SIS 14-2344	0.08% C 18% Cr 12% Ni 2.7% Mo 0.6% Ti steel: Annealed; Swedish Standard
	DPN: 190
SIS 14-2345	0.08% C 18% Cr 12% Ni 2.7% Mo 1% Nb + Ta steel: Swedish Standard
	DPN: 190
SIS 14-2346	0.12% C 2% Mn 18% Cr 0.6% Mo 9% Ni steel: Swedish Standard; stainless free cutting
	DPN: 220 UTS: 650 Elon: 35% Proof: 210
SIS 14-2347	0.05% C 2.0% Mn 17% Cr 12% Ni 2.25% Mo steel: Swedish Standard; stainless
	DPN: 200 UTS: 590 Elon: 45% Proof: 230

Symbol	Nominal analysis, supplier, condition and remarks.
SIS 14-2348	0.03% C 2.0% Mn 17.0% Cr 12.0% Ni 2.0% Mo steel: Swedish Standard; stainless DPN: 200 UTS: 590 Elon: 45% Proof: 220
SIS 14-2350	0.08% C 2.0% Mn 17.0% Cr 12% Ni 2.0% Mo 0.6% Ti steel: Swedish Standard; stainless DPN: 210 UTS: 590 Elon: 40% Proof: 230
SIS 14-2352	0.03% C 2.0% Mn 18% Cr 10% Ni steel: Swedish Standard; stainless DPN: 190 UTS: 590 Elon: 45% Proof: 210
SIS 14-2352	0.03% C 2.0% Mn 18% Cr 10.5% Ni steel: Swedish Standard; stainless DPN: 195 UTS: 590 Elon: 40% Proof: 210
SIS 14-2352	0.03% C 18% Cr 11% Ni steel: Annealed; Swedish Standard DPN: 180
SIS 14-2353	0.03% C 17.5% Cr 13% Ni 2.7% Mo steel: Annealed; Swedish Standard DPN: 180
SIS 14-2353	0.03% C 2.0% Mn 17% Cr 13% Ni 2.8% Mo steel: Swedish Standard; stainless DPN: 200 UTS: 590 Elon: 45% Proof: 230
SIS 14-2361	0.08% C 2.0% Mn 25% Cr 2.0% Ni steel: Swedish Standard; stainless DPN: 220 UTS: 830
SIS 14-2361	0.08% C 25% Cr 21% Ni steel: Annealed; Swedish Standard DPN: 230 UTS: 610
SIS 14-2366	0.05% C 2.0% Mn 18.0% Cr 14.0% Ni 3.5% Mo steel: Swedish Standard; stainless DPN: 200 UTS: 590 Elon: 40% Proof: 220
SIS 14-2367	0.03% C 2.0% Mn 18.5% Cr 15.5% Ni 3.5% Mo steel: Swedish Standard; stainless DPN: 200 UTS: 590 Elon: 40% Proof: 230
SIS 14-2370	0.05% C 2.0% Mn 18.0% Cr 9.5% Ni steel: Swedish Standard; stainless DPN: 220 UTS: 690 Elon: 40% Proof: 300
SIS 14-2374	0.05% C 2.0% Mn 18% Cr 12% Ni 3% Mo steel: Swedish Standard; stainless DPN: 200 UTS: 690 Elon: 40% Proof: 320
SIS 14-2375	0.03% C 2.0% Mn 17% Cr 12% Ni 2.5% Mo steel: Swedish Standard; stainless DPN: 220 UTS: 690 Elon: 40% Proof: 290
SIS 14-2570	0.08% C 2.0% Mn 15.0% Cr 25.0% Ni 1.0% Mo 0.3% V 2.2% Ti 0.35% Al steel: Swedish Standard; high strength; high temperature; austenitic DPN: 300 UTS: 750 Elon: 45% Proof: 500
SIS 14-2570	See SIS 2570, Section 44N1
SIS 2324	0.1% C 20% Cr 5% Ni 1.5% Mo steel: Swedish Standard; included in MNC 900 DPN: 260
SIS 2326	0.025% C (max) 18.0% Cr 0.5% Ni (max) 2.2% Mo 0.3% Ti steel: Swedish Standard
SIS 2330	0.15% C 17% Cr 7% Ni steel: Plate, bar, etc.; Swedish Standard DPN: 200 UTS: 770
SIS 2331	0.15% C 17% Cr 7% Ni steel: Sheet; cold rolled; Swedish Standard UTS: 920 Elon: 3% Proof: 770
SIS 2332	0.08% C 18% Cr 10% Ni steel: Bar, forging, etc.; annealed; Swedish Standard DPN: 180 UTS: 480 Elon: 45% Proof: 200
SIS 2333	0.06% C (max) 18% Cr 10% Ni steel: Bar, forging, etc.; annealed; Swedish Standard DPN: 180 UTS: 480 Elon: 45% Proof: 200
SIS 2334	0.15% C (max) 18% Cr 9.0% Ni Ti or Nb steel: Swedish Standard
SIS 2337	0.08% C 18% Cr 10% Ni 0.5% Ti steel: Bar, forging, etc.; annealed; Swedish Standard DPN: 190 UTS: 480 Elon: 40% Proof: 200
SIS 2338	0.08% C 18% Cr 10% Ni 0.8% Nb + Ta steel: Bar, forging, etc.; annealed; Swedish Standard DPN: 190 UTS: 500 Elon: 40% Proof: 200
SIS 2338	0.08% C (max) 18.0% Cr 10.5% Ni 0.8% Nb + Ta steel: Swedish Standard
SIS 2338	0.b8% C (max) 18.0% Cr 10.5% Ni 0.85% Nb + Ta steel: Swedish Standard
SIS 2338	0.08% C (max) 18.0% Cr 10.5% Ni steel: Swedish Standard DPN: 90 UTS: 590 Elon: 40% Proof: 210
SIS 2338	0.08% C (max) 18.0% Cr 10.5% Ni 0.6% Ti steel: Swedish Standard
SIS 2340	0.1% C (max) 2.0% Mn (max) 9% Ni 17% Cr 1.5% Mo steel: For reinforcing; Swedish Standard Elon: 7% Proof: 780
SIS 2340	0.1% C 17% Cr 8% Ni 1.3% Mo steel: Bar, forging, etc.; annealed; Swedish Standard DPN: 200 UTS: 580 Elon: 40% Proof: 200
SIS 2341	0.06% C 18% Cr 10% Ni 1.8% Mo steel: Bar, forging, etc.; annealed; Swedish Standard DPN: 180 UTS: 500 Elon: 45% Proof: 200
SIS 2342	0.1% C 17% Cr 9% Ni 2.3% Mo steel: Bar, forging, etc.; annealed; Swedish Standard DPN: 200 UTS: 580 Elon: 40% Proof: 200
SIS 2343	0.06% C 17% Cr 12% Ni 2.7% Mo steel: Bar, forging, etc.; annealed; Swedish Standard DPN: 180 UTS: 530 Elon: 45% Proof: 210
SIS 2344	0.08% C 17% Cr 12% Ni 2.7% Mo 0.6% Ti steel: Bar, forging, etc.; Swedish Standard DPN: 190 UTS: 540 Elon: 40% Proof: 200
SIS 2345	0.08% C 17% Cr 12% Ni 2.7% Mo 0.9% Nb + Ta steel: Bar, forging, etc.; Swedish Standard DPN: 190 UTS: 540 Elon: 40% Proof: 200
SIS 2346	0.15% C 17% Cr 10% Ni 0.7% Mo steel: Plate & bar; annealed; Swedish Standard DPN: 200 UTS: 530 Elon: 40% Proof: 200
SIS 2346	0.15% C 17% Cr 10% Ni 0.7% Mo steel: Plate & bar; cold rolled; Swedish Standard DPN: 310 UTS: 680 Elon: 20% Proof: 480
SIS 2347	0.06% C (max) 17.5% Cr 12% Ni 2.2% Mo steel: Swedish Standard DPN: 180
SIS 2347	0.05% C (max) 17.2% Cr 12.2% Ni 2.2% Mo steel: Swedish Standard
SIS 2347	0.05% C (max) 17.2% Cr 12.2% Ni 2.2% Mo steel: Swedish Standard
SIS 2347	0.05% C (max) 17.0% Cr 13.0% Ni 2.0% Mo steel: Swedish Standard DPN: 190 UTS: 600 Elon: 45% Proof: 210
SIS 2348	0.03% C (max) 17.2% Cr 12.5% Ni 2.2% Mo steel: Swedish Standard
SIS 2348	0.03% C 17.2% Cr 12.5% Ni 2.2% Mo steel: Swedish Standard
SIS 2348	0.03% C (max) 17.0% Cr 12.5% Ni 2.2% Mo steel: Swedish Standard DPN: 190 UTS: 603 Elon: 45% Proof: 200
SIS 2350	0.08% C (max) 17.2% Cr 12.2% Ni 2.2% Mo 0.6% Ti steel: Swedish Standard

Note. The following abbreviations and units are used in the tables:

DPN	Hardness, diamond pyramid number
UTS	Ultimate tensile strength, N/mm^2
Elon	Elongation, %
Proof	0.1 % proof strength, N/mm^2

1 N/mm^2=0.1 hbar=0.102 kgf/mm^2=0.06475 tonf/in^2=145.04 lbf/in^2

Symbol	Nominal analysis, supplier, condition and remarks.
SIS 2350	0.08% C 17.0% Cr 12.0% Ni 2.2% Mo steel: Swedish Standard **DPN: 200 UTS: 600 Elon: 40% Proof: 210**
SIS 2352	0.03% C 18.5% Cr 10.5% Ni steel: Swedish Standard; included in MNC 900 **DPN: 180**
SIS 2353	0.03% C 17% Cr 12% Ni 2.7% Mo steel: Swedish Standard; included in MNC 900 **DPN: 180**
SIS 2360	0.12% C 19% Cr 18% Ni steel: Plate, bar, forging; annealed; Swedish Standard **DPN: 220 UTS: 820**
SIS 2361	0.2% C 22% Cr 19% Ni steel: Bar, forging, etc.; annealed; Swedish Standard **DPN: 230 UTS: 770**
SIS 2366	0.06% C (max) 18.5% Cr 15.2% Ni 3.5% Mo steel: Casting; Swedish Standard **UTS: 550 Elon: 35% Proof: 200**
SIS 2366	0.05% C 18.5% Cr 14.5% Ni 3.5% Mo steel: Swedish Standard **DPN: 190 UTS: 600 Elon: 40% Proof: 220**
SIS 2366	0.06% C (max) 18.5% Cr 14.7% Ni 3.5% Mo steel: Casting; Swedish Standard
SIS 2366	0.05% C 18.5% Cr 14.5% Ni 3.5% Mo steel: Swedish Standard
SIS 2366	0.06% C (max) 19.0% Cr 5.7% Ni 3.5% Mo steel: Swedish Standard
SIS 2367	0.03% C (max) 18.5% Cr 15.5% Ni 3.5% Mo steel: Swedish Standard
SIS 2367	0.03% C (max) 18.5% Cr 16.0% Ni 3.5% Mo steel: Swedish Standard **DPN: 190 UTS: 600 Elon: 48% Proof: 210**
SIS 2370	0.05% C (max) 18.0% Cr 9.5% Ni steel: Swedish Standard **DPN: 195 UTS: 700 Elon: 40% Proof: 290**
SIS 2370	0.05% C (max) 18.0% Cr 8.5% Ni 0.2% N steel: Swedish Standard
SIS 2370	0.05% C (max) 18.5% Cr 9.5% Ni 0.18% N steel: Swedish Standard
SIS 2570	0.08% C 15% Cr 25% Ni 1.2% Mo 2.1% Ti 0.2% V steel: Swedish Standard; included in MNC 870 **DPN: 300**
SIS 2371	0.03% C (max) 10.5% Ni 18.5% Cr steel: Swedish Standard
SIS 2371	0.03% C (max) 18.0% Cr 10.5% Ni 0.19% N steel: Swedish Standard
SIS 2374	0.05% C (max) 17.2% Cr 12.2% Ni 2.7% Mo 0.19% N steel: Swedish Standard
SIS 2374	0.05% C (max) 17.5% Cr 12.0% Ni 3.0% Mo steel: Swedish Standard **DPN: 190 UTS: 700 Elon: 40% Proof: 300**
SIS 2374	0.05% C (max) 17.2% Cr 10.7% Ni 2.7% Mo 0.2% N steel: Swedish Standard
SIS 2375	0.03% C (max) 12.5% Ni 17.5% Cr 2.7% Mo steel: Swedish Standard
SIS 2570	0.08% C 15% Cr 25% Ni 1.2% Mo 0.2% V 2.0% Ti B steel: For high temperature use; Swedish Standard **DPN: 200**
SKF 304	As AISI 304 SKF
SKF 316	As AISI 316 SKF
SKF 321	As AISI 321 SKF
SPW	0.28% C 20% Cr 10% Ni 0.2% Mo steel: Pompey
S R ALLOY	0.6% C 20% Cr 37% Ni 0.4% Mo 2.0% W steel: Casting; Cronite; as cast **DPN: 170 UTS: 450 Elon: 9% Proof: 250**
S S 3M	0.16% C 18% Cr 9.0% Ni steel: Uddelholm
S S 3MM	0.08% C 19% Cr 9% Ni steel: Uddelholm

Symbol	Nominal analysis, supplier, condition and remarks.
S S 4	0.12% C 19% Cr 10% Ni 2.0% Mo steel: Uddelholm
S S 24	0.12% C (max) 19% Cr 10% Ni 3.0% Mo steel: Uddelholm
S S 33	0.16% C 12% Cr 12% Ni steel: For deep drawing; Uddelholm
S S 43	18% Cr 9.0% Ni S steel: Free machining; Uddelholm; C content to be agreed
S S 53	0.15% C 18% Cr 9.0% Ni 0.6% Ti steel: Uddelholm
STA 5 V26	0.42% C 13% Cr 10% Ni 3% W steel: Replaced by BS alloy En 56
STA 5 V27	0.16% C (max) 18% Cr 9% Ni steel: Replaced by BS alloy En 58A
STA 5 V27A	0.15% C 18% Cr 10% Ni steel: Weld stabilized; replaced by BS alloy En 58; stabilized grade
STA 5 V27M	0.1% C 18% Cr 10% Ni steel: Free machining; replaced by BS alloy EN 58M
STA 5 V28	0.2% C 18% Cr 2.0% Ni steel: Replaced by BS alloy En 57
STAYBRITE	18% Cr 8% Ni steel: Firth Vickers; general trade name covering stainless steels
STC 52D	0.08% C 17% Cr 12% Ni 2.5% Mo steel: Designation used by Japanese Standard
STRAINTRODE 64.30	0.03% C 0.6% Mn 19.0% Cr 10.0% Ni 0.8% Si steel: For electrodes; Esab; 6% ferrite **UTS: 564 Elon: 46% Proof: 420**
STRAINTRODE 61.81	0.06% C 1.5% Mn 20.0% Cr 9.5% Ni 0.8% Si 0.7% Nb steel: For electrodes; Esab; 8% ferrite **UTS: 700 Elon: 30% Proof: 570**
STRAINTRODE 62.30	0.03% C 0.6% Mn 18.5% Cr 10.0% Ni 1.7% Mo 0.7% Si steel: For electrodes; Esab; 8% ferrite **UTS: 570 Elon: 45% Proof: 460**
STRAINTRODE 63.30	0.03% C 0.6% Mn 18.5% Cr 12.5% Ni 2.8% Mo steel: For electrodes; Esab; 9% ferrite **UTS: 590 Elon: 35% Proof: 490**
STRAINTRODE 63.35	0.05% C 0.7% Mn 18.5% Cr 12.5% Ni 2.8% Mo 0.5% Si steel: For electrodes; Esab; 15% ferrite **UTS: 590 Elon: 35% Proof: 490**
STRAINTRODE 63.80	0.03% C 1.5% Mn 18.5% Cr 12.5% Ni 2.8% Mo 0.6% Nb steel: For electrodes; Esab; 10% ferrite **UTS: 664 Elon: 35% Proof: 455**
STRAINTRODE 64.30	0.03% C 0.7% Mn 19.0% Cr 13.5% Ni 4.0% Mo steel: For electrodes; Esab; 11% ferrite **UTS: 602 Elon: 35% Proof: 455**
STRAINTRODE 67.15	0.1% C 1.7% Mn 26.0% Cr 20.0% Ni steel: For electrodes; Esab Standard **UTS: 610 Elon: 35% Proof: 450**
STRAINTRODE 67.45	0.05% C 6.5% Mn 17.5% Cr 9.0% Ni steel: For electrodes; Esab Standard **UTS: 620 Elon: 42% Proof: 390**
STRAINTRODE 67.70	0.04% C 1.0% Mn 22.0% Cr 12.0% Ni 2.5% Mo steel: For electrodes; Esab Standard; 15% ferrite **UTS: 730 Elon: 30% Proof: 640**
STRAINTRODE 67.75	0.05% C 3.0% Mn 23.0% Cr 12.0% Ni 0.5% Mo steel: For electrodes; Esab; 17% ferrite **UTS: 630 Elon: 30% Proof: 510**
STRAINTRODE 68.60	0.1% C 1.5% Mn 26.0% Cr 5.0% Ni 1.5% Mo steel: For electrodes; Esab Standard **UTS: 650 Elon: 35%**
SUH 31	0.4% C 15.0% Cr 14.0% Ni 2.5% W steel: Heat resisting; Japanese Standard designation
SV 02KH19N9	0.04% C (max) 19.5% Cr 9.0% Ni steel: For welding wire; Russian Standard designation
SV 04KH19N9	0.06% C (max) 19.0% Cr 9.0% Ni steel: For welding wire; Russian Standard designation
SV 04KH19N952	0.06% C (max) 19.0% Cr 9.0% Ni steel: For welding wire; Russian Standard designation
SV 04KH19N11M3	0.06% C (max) 19.0% Cr 11.0% Ni 2.5% Mo steel: For welding wire; Russian Standard designation

Symbol	Nominal analysis, supplier, condition and remarks.
SV 05KH19N9F352	0.07% C (max) 19.0% Cr 9.0% Ni 2.5% V steel: For welding wire; Russian Standard
SU 06KH19N9T	0.08% C (max) 19.0% Cr 9.0% Ni 0.7% Ti steel: For welding wire; Russian Standard designation
SV 07KH18N9T1U	0.09% C (max) 18.0% Cr 9.0% Ni 1.2% Ti 0.7% Al steel: For welding wire; Russian Standard designation
SV 07KH25N13	0.09% C (max) 24.5% Cr 13.0% Ni steel: For welding wire; Russian Standard designation
SV 08KH19N9F252	0.1% C (max) 19.0% Cr 9.0% Ni 2.1% V steel: . For welding wire; Russian Standard designation
SV 08KH19N10B	0.07% C 19.5% Cr 9.7% Ni 1.35% Nb steel: For welding wire; Russian Standard designation
SV 08KH19N12M3	0.08% C 19.5% Cr 12.2% Ni 2.6% Mo steel: For welding wire; Russian Standard designation
SV 08KH20N10G6	0.1% C (max) 6.0% Mn 21.0% Cr 10.0% Ni steel: For welding wire; Russian Standard designation
SV 08KH25N5TMF	0.1% C (max) 25.0% Cr 5.2% Ni 0.09% Mo 0.11% V 0.14% Ti 0.15% N steel: For welding wire; Russian Standard designation
SV 08KH20N9G7T	0.1% C (max) 7.0% Mn 19.5% Cr 9.0% Ni 0.75% Ti steel: For welding wire; Russian Standard designation
SV 10KH16N25M6	0.1% C 16.2% Cr 25.5% Ni 6.2% Mo 0.15% N steel: For welding wire; Russian Standard designation
SV 10KH20N15	0.12% C (max) 20.5% Cr 15.0% Ni steel: For welding wire; Russian Standard designation
SV 13KH25N18	0.15% C (max) 25.5% Cr 18.5% Ni steel: For welding wire; Russian Standard designation
SV 25KH25N16G7	0.24% C 7.0% Mn 25.5% Cr 16.0% Ni steel: For welding wire; Russian Standard designation
SV 30KH15N35V3B 3T	0.08% C (max) 19.0% Cr 10.0% Ni 2.5% Mo 0.7% Ti 2.7% W steel: For welding wire; Russian Standard designation
SV 30KH15N35V3B 3T	0.29% C 15.0% Cr 35.0% Ni 0.8% Ti 3.0% Nb steel: For welding wire; Russian Standard designation
THERMALLOY 40 A2	0.5% C 26% Cr 15% Ni 1% Nb 0.13% N steel: Casting; American proprietary alloy listed in SAE year book
THERMALLOY 40 E	0.9% C 26% Cr 12% Ni steel: Casting; American proprietary alloy listed in SAE year book
THERMALLOY 47	0.4% C 26% Cr 20% Ni 0.1% N steel: Casting; American proprietary alloy listed in SAE year book
THERMALLOY 50 CQ	0.5% C 15% Cr 35% Ni 1% Nb steel: Casting; American proprietary alloy listed in SAE year book
THERMALLOY D47	0.5% C 28% Cr 20% Ni 0.1% N steel: Casting; American proprietary alloy listed in SAE year book
THERMINOX	0.12% C 25% Cr 20% Ni steel: Pompey
THERMINOX 13	0.1% C 25% Cr 13% Ni steel: Pompey
THERMINOX 20	0.07% C 25% Cr 20% Ni steel: Pompey
THERMINOX 821	0.2% C 25% Cr 5.0% Ni steel: Pompey
TIMKEN 16/25/6	0.1% C 16% Cr 25% Ni 6% Mo 0.15% N steel: Timken

Note. The following abbreviations and units are used in the tables:

DPN	Hardness, diamond pyramid number
UTS	Ultimate tensile strength, N/mm^2
Elon	Elongation, %
Proof	0.1 % proof strength, N/mm^2

1 N/mm^2=0.1 hbar=0.102 kgf/mm^2=0.06475 tonf/in^2=145.04 lbf/in^2

Symbol	Nominal analysis, supplier, condition and remarks.
TOLOY 9 TYPE HE	0.35% C 28% Cr 9.5% Ni steel: Casting; resistant to sulphur bearing atmosphere; Wellman
TOLOY 10 TYPE HF	0.3% C 21% Cr 10.5% Ni steel: Casting for use 600–850 °C; Wellman
TOLOY 12	0.3% C 25% Cr 12% Ni steel: Casting; Thompson L'Hospied; annealed **DPN: 180 UTS: 630 Elon: 15% Proof: 340**
TOLOY 12 HH2	0.45% C 24.5% Cr 12% Ni steel: Casting for use between 925–1050 °C; Wellman
TOLOY 16	0.3% C 27% Cr 16% Ni steel: Casting; Thompson L'Hospied; annealed **DPN: 175 UTS: 610 Elon: 8% Proof: 480**
TOLOY 20	0.4% C 26% Cr 20% Ni steel: Casting; Thompson L'Hospied; annealed **DPN: 180 UTS: 530 Elon: 16% Proof: 340**
TOLOY 20 HK40	0.4% C 26% Cr 20% Ni steel: Casting; for use at temperatures above 1050 °C; Wellman
TOLOY 25 HN	0.37% C 21% Cr 25% Ni steel: Casting; good high temperature corrosion; Wellman
TOLOY 37	0.55% C 19% Cr 37% Ni steel: Casting; Thompson L'Hospied; annealed **DPN: 165 UTS: 540 Elon: 12% Proof: 360**
TOLOY 37 HU	0.45% C 18.5% Cr 37.5% Ni Fe casting: Resists sulphur and carbon attack at high temperature; Wellman
TOPHET 12.12	0.16% C (max) 12.0% Cr 13.0% Ni steel: Wire; Getly-Brunton for weaving close mesh
TOPHET 18.8	0.16% C (max) 18.0% Cr 9.0% Ni steel: Getly-Brunton; can be hard drawn
TOPHET 18.8 Low C	0.08% C 18.0% Cr 9.5% Ni steel: Getly-Brunton; hard drawn **DPN: 160**
TOPHET 25.20	0.1% C 25.0% Cr 20.0% Ni steel: Wire; Getly-Brunton; annealed **UTS: 725 Elon: 55% Proof: 360**
TOPHET Cb	0.08% C (max) 18.0% Cr 9.2% Ni 0.6% Nb (min) steel: Getly-Brunton; annealed **UTS: 700 Elon: 55% Proof: 300**
TOPHET Mo	0.08% C (max) 18.0% Cr 10.0% Ni 2.85% Mo steel: Getly-Brunton; annealed **UTS: 720 Elon: 55% Proof: 360**
TOPHET Ti	0.08% C (max) 18.0% Cr 9.2% Ni 0.3% Ti (min) steel: Getly-Brunton; hard drawn **DPN: 165**
TRUBRITE	18% Cr 8% Ni Ti stabilized steel: Arthur Lee
TRUBRITE A1	12% Cr 12% Ni steel: For deep drawing; Arthur Lee
TRUBRITE A3	18% Cr 8% Ni steel: Arthur Lee
TRUBRITE A6	Low C 19% Cr 10% Ni steel: Weldable; Arthur Lee
TURBALOY 13	0.13% C 17.5% Cr 23.5% Ni 2.5% Mo 1% W 1.5% Nb 1.5% Al Fe alloy: General Electric
UCAR 11	0.05% C 18.0% Cr 39.0% Ni 5.2% Mo 0.5% Co 2.5% Ti Fe alloy: Casting; Union Carbide
UCAR 706	0.06% C (max) 16.0% Cr 41.5% Ni 1.0% Co 3.0% Nb 1.7% Mo Fe alloy: Union Carbide
UCAR 800	0.05% C 21.0% Cr 32.5% Ni 0.4% Ti 0.4% Al Fe alloy: Union Carbide
UHB STAINLESS 3	0.1% C 18% Cr 8.5% Ni steel: Uddelholm; annealed **DPN: 160 UTS: 630 Elon: 50% Proof: 220**
UHB STAINLESS 3L	0.03% C 18% Cr 10% Ni steel: Uddelholm; annealed **DPN: 160 UTS: 580 Elon: 50% Proof: 200**
UHB STAINLESS 3M	0.08% C 18% Cr 8.5% Ni steel: Uddelholm; annealed **DPN: 160 UTS: 630 Elon: 50% Proof: 220**
UHB STAINLESS 3MM	0.06% C 18% Cr 9% Ni steel: Uddelholm; annealed **DPN: 160 UTS: 630 Elon: 50% Proof: 220**

Symbol	Nominal analysis, supplier, condition and remarks.
UHB STAINLESS 4	0.09% C 17.5% Cr 9.5% Ni 1.5% Mo steel: Uddelholm; annealed
	DPN: 160 UTS: 630 Elon: 50% Proof: 240
UHB STAINLESS 4MM	0.06% C 17.5% Cr 10.5% Ni 1.5% Mo steel: Uddelholm; annealed
	DPN: 160 UTS: 630 Elon: 50% Proof: 240
UHB STAINLESS 15	0.07% C 18.5% Cr 20.5% Ni steel: Uddelholm; annealed
	DPN: 160 UTS: 580 Elon: 35% Proof: 250
UHB STAINLESS 24	0.06% C 17.5% Cr 12% Ni 2.5% Mo steel: Uddelholm; annealed
	DPN: 160 UTS: 630 Elon: 45% Proof: 240
UHB STAINLESS 24L	0.03% C 17.5% Cr 13% Ni 2.8% Mo steel: Uddelholm; annealed
	DPN: 160 UTS: 580 Elon: 50% Proof: 220
UHB STAINLESS 25	0.07% C 23.5% Cr 21.5% Ni steel: Uddelholm; annealed
	DPN: 160 UTS: 580 Elon: 35% Proof: 240
UHB STAINLESS 33MM	0.06% C 12.5% Cr 12.5% Ni steel: Uddelholm; annealed
	DPN: 130 UTS: 530 Elon: 55% Proof: 210
UHB STAINLESS 34	0.05% C 17.5% Cr 14% Ni 4.3% Mo steel: Uddelholm; annealed
	DPN: 170 UTS: 530 Elon: 40% Proof: 270
UHB STAINLESS 34L	0.03% C 17.5% Cr 14.5% Ni 4.2% Mo steel: Uddelholm
UHB STAINLESS 43	0.08% C 17.5% Cr 9.5% Ni 0.5% Mo steel: Free cutting; Uddelholm; annealed
	DPN: 160 UTS: 630 Elon: 45% Proof: 220
UHB STAINLESS 44	0.1% C 25% Cr 5.3% Ni 1.5% Mo steel: Uddelholm; annealed
	DPN: 230 UTS: 680 Elon: 20% Proof: 500
UHB STAINLESS 45	0.1% C 25.5% Cr 4.8% Ni steel: Uddelholm; annealed
	DPN: 220 UTS: 630 Elon: 25% Proof: 450
UHB STAINLESS 53	0.06% C 18% Cr 9.5% Ni Ti steel: Uddelholm; annealed
	DPN: 160 UTS: 630 Elon: 45% Proof: 220
UHB STAINLESS 55	0.15% C 23% Cr 13% Ni steel: Uddelholm; annealed
	DPN: 160 UTS: 630 Elon: 40% Proof: 290
UHB STAINLESS 63	0.06% C 18% Cr 10% Ni Nb steel: Uddelholm; annealed
	DPN: 160 UTS: 630 Elon: 40% Proof: 220
UHB STAINLESS 63H	0.08% C 16% Cr 13% Ni 1.0% Nb steel: Tube; Uddelholm; annealed
	DPN: 210 UTS: 610 Elon: 35% Proof: 210
UHB STAINLESS 64	0.06% C 18% Cr 18% Ni 2.2% Mo 2.0% Cu + Nb steel: Uddelholm; annealed
	DPN: 160 UTS: 530 Elon: 40% Proof: 240
UHB STAINLESS 524	0.06% C 17.5% Cr 13% Ni 2.8% Mo Ti steel: Uddelholm; annealed
	DPN: 160 UTS: 620 Elon: 50% Proof: 250
UHB STAINLESS 624	0.06% C 17.5% Cr 13% Ni 2.8% Mo Nb steel: Uddelholm; annealed
	DPN: 160 UTS: 620 Elon: 40% Proof: 250
UHB STAINLESS 673	0.1% C 17% Cr 7.5% Ni steel: Uddelholm
UHB STAINLESS 703	0.09% C 17.5% Cr 10.5% Ni steel: Uddelholm; annealed
	DPN: 160 UTS: 620 Elon: 50% Proof: 220
UHB STAINLESS 724	0.06% C 17% Cr 13.5% Ni 2.75% Mo steel: Tube; Uddelholm; annealed
	DPN: 180 UTS: 580 Elon: 45% Proof: 210
UHB STAINLESS 734L	0.03% C 17.5% Cr 13.5% Ni 3.4% Mo steel: Uddelholm
UHB STAINLESS 734 Nb	0.06% C 17.5% Cr 13.5% Ni 3.4% Mo steel: Uddelholm

Symbol	Nominal analysis, supplier, condition and remarks.
UHB STAINLESS 753	0.08% C 18% Cr 10% Ni 0.5% Ti steel: Tube; Uddelholm; annealed
	DPN: 190 UTS: 630 Elon: 40% Proof: 200
UNI 3992 X45 CNW 1909	0.45% C 19% Cr 9% Ni 1.0% W steel: For valves
UNI 3992 X50 CNW 1414	0.5% C 14% Cr 14% Ni 2.5% W steel: For valves
UNI 4047 X3 CN 1911	0.03% C (max) 19% Cr 10.5% Ni steel
UNI 4047 X6 CN 1911	0.06% C (max) 19% Cr 10.5% Ni steel
UNI 4047 X8 CN 1910	0.08% C (max) 19% Cr 10.5% Ni steel
UNI 4047 X15 CN 1808	0.15% C (max) 18% Cr 9% Ni steel
UNI 4047 X15 CNF 1808	0.15% C (max) 18% Cr 9% Ni 0.6% Mo steel
URANUS 50	20% Cr 8% Ni 2.5% Mo 1.5% Cu steel: Holtzer-Loire
URANUS 50	0.06% C (max) 7.5% Cr 21.0% Ni 2.5% Mo 1.5% Cu steel: Creusot-Loire
URANUS 65	0.03% C (max) 24.5% Cr 20.5% Ni with Nb steel: Creusot-Loire
URANUS A5	0.06% C (max) 17.5% Ni 2.25% Mo 2.0% Cu steel: Creusot-Loire
URANUS B6	Low C 20% Cr 25% Ni 4.5% Mo 1.5% Cu steel: Holtzer-Loire
URANUS B6	0.02% C (max) 20.5% Cr 25.5% Ni 4.4% Mo 1.2% Cu steel: Creusot-Loire
URANUS B6	0.02% C 20.5% Cr 25.5% Ni 4.4% Mo 1.5% Cu steel: Creusot-Loire
URANUS S1	0.015% C (max) 1.5% Mn 17.5% Cr 14.2% Ni with Nb steel: Creusot-Loire
V 2A	0.16% C 18% Cr 9.0% Ni steel: Krupps
V 2A EXTRA	0.15% C 18% Cr 9.0% Ni 0.6% Ti steel: Krupps
V 2A FH	0.15% C 17% Cr 7.5% Ni steel: Krupps
V 2A NORMAL	0.12% C 18% Cr 9% Ni steel: Krupps
V 2A SUPRA	0.07% C 18% Cr 10% Ni steel: Krupps
V 2A SUPRA NK	0.03% C 18% Cr 11% Ni steel: Krupps
V 2A XET	0.08% C 18% Cr 10% Cr 0.2% Mo 1.0% Nb steel: Krupps
V 2A X EXTRA	0.1% C 18% Cr 10% Ni 1.0% Nb steel: Krupps
V 4A EXTRA	0.1% C 17.5% Cr 11.5% Ni 2.2% Mo 0.5% Ti steel: Krupps
V 4A SUPRA	0.07% C 17.5% Cr 11.5% Ni 2.2% Mo steel: Krupps
V 4A X EXTRA	0.1% C 17.5% Cr 11.5% Ni 2.2% Mo 1.0% Nb steel: Krupps
V 8A SUPRA	0.12% C 18% Cr 10% Ni 2.0% Mo steel: Krupps
V 12A SUPRA	0.16% C 12% Cr 12% Ni steel: Krupps
V 14A SUPRA	0.07% C 17% Cr 13.5% Ni 4.5% Mo steel: Krupps
V 16A EXTRA	0.07% C 17.5% Cr 20% Ni 2.2% Mo 0.3% Ti 2.0% Cu steel: Krupps
V 16A X EXTRA	0.07% C 17.5% Cr 20% Ni 2.2% Mo 0.5% Nb 2.0% Cu steel: Krupps
V 18A SUPRA NK	0.03% C 18.5% Cr 16.5% Ni 3.5% Mo steel: Krupps
V 24A EXTRA	0.06% C 25% Cr 25% Ni 2.2% Mo 0.3% Ti steel: Krupps
V 25AF X EXTRA	0.06% C 25% Cr 7% Ni 1.5% Mo 0.4% Nb steel: Krupps
V 44A EXTRA	0.1% C 17.5% Cr 13% Ni 2.7% Mo 0.5% Ti steel: Krupps
V 44A SUPRA	0.07% C 17.5% Cr 13% Ni 2.7% Mo steel: Krupps
V 44A SUPRA NK	0.03% C 17.5% Cr 12% Ni 2.2% Mo steel: Krupps
V 44A X EXTRA	0.1% C 17.5% Cr 13% Ni 2.7% Mo 1.0% Nb steel: Krupps
V 57	0.05% C 14.7% Cr 27.2% Ni 1.3% Mo 3.0% Ti 0.2% Al 0.3% V stainless steel: Information from SAE handbook

Symbol	Nominal analysis, supplier, condition and remarks.
VALVO 829	0.42% C 13.5% Cr 13.5% Ni 2.7% W steel: For valves; Swift Levick for BS alloy En 54
VERSALLOY	0.11% C 17% Cr 8% Ni steel: Atlas; annealed **DPN: 150 UTS: 730 Elon: 55% Proof: 240**
VH 274	0.16% C 18% Cr 9.0% Ni steel: Wykmanshyttan
VH 650	0.12% C 18% Cr 10% Ni 3.0% Mo steel: Wykmanshyttan
VIKRO 10	0.35% C 15.0% Cr 25.0% Ni steel: Casting; Firth Vickers; annealed **UTS: 500 Elon: 8% Proof: 280**
VITRESIST	0.06% C (max) 2% Si 23% Ni 23% Cr 5.5% Mo 1.5% Cu 0.4% Nb steel: D Brown; for use with hot sulphuric acid
VS 12/12	0.08% C 13% Cr 12.5% Ni steel: Vulcan; annealed **UTS: 920 Elon: 65%**
VS 18/8	0.09% C 18% Cr 8% Ni steel: Vulcan; annealed **UTS: 1070 Elon: 60%**
VS 18/8 Ti	0.06% C 18% Cr 8.5% Ni Ti steel: Vulcan; annealed **UTS: 1110 Elon: 56%**
VS 18/8/2	0.08% C 18.5% Cr 8.5% Ni 2.5% Mo steel: Vulcan; annealed **UTS: 1140 Elon: 51%**
W 18/8	18% Cr 8% Ni Nb steel: Welding electrode; Cu coated; Armco
W 19/9	0.03% C 19% Cr 9% Ni steel: Welding electrode; Cu coated; Armco
WELDANKA A	0.15% C (max) 18% Cr 9% Ni steel: Castings; Brown Bayley
WELDANKA B	0.08% C 18% Cr 10% Ni 0.4% Ti steel: Bar, billet, etc.; Brown Bayley for BS alloy En 58B and C
WELDANKA CB	0.08% C 18% Cr 12% Ni 0.8% Nb + Ta steel: Bar, billet, etc.; Brown Bayley for BS alloy En 58F and G
X 2 Cr Ni 18/9	0.03% C (max) 19.0% Cr 10% Ni steel: German Standard
X 2 Cr Ni 18/9	0.03% C (max) 18.5% Cr 11.2% Ni steel: Designation used by German Standard
X 2 Cr Ni 18/10	0.03% C (max) 18.0% Cr 10.2% Ni 0.17% N steel: German Standard
X 2 Cr Ni 18/11	0.03% C (max) 19.0% Cr 10.0% Ni steel: German Standard
X 2 Cr Ni 18/19	0.03% C (max) 18.5% Cr 11.2% Ni steel: German Standard
X 2 Cr Ni Mo 17/12	0.03% C (max) 17.0% Cr 12.0% Ni 2.2% Mo steel: German Standard
X 2 Cr Ni Mo 17/12	0.03% C (max) 17.5% Cr 12.0% Ni 2.7% Mo steel: German Standard
X 2 Cr Ni Mo 18/10	0.03% C (max) 17.0% Cr 12.0% Ni 2.2% Mo steel: German Standard
X 2 Cr Ni Mo 18/10	0.03% C (max) 17.5% Cr 12.5% Ni steel: Designation used by German Standard
X 2 Cr Ni Mo 18/10	0.03% C (max) 18.0% Cr 16.0% Ni 35% Mo steel: German Standard
X 2 Cr Ni Mo 18/10	0.03% C (max) 17.5% Cr 12.5% Ni 2.2% Mo steel: German Standard
X 2 Cr Ni Mo 18/12	0.03% C (max) 17.5% Cr 13.5% Ni 2.7% Mo steel: German Standard

Note. The following abbreviations and units are used in the tables:

DPN	Hardness, diamond pyramid number
UTS	Ultimate tensile strength, N/mm^2
Elon	Elongation, %
Proof	0.1 % proof strength, N/mm^2

$1\ N/mm^2 = 0.1\ hbar = 0.102\ kgf/mm^2 = 0.06475\ tonf/in^2 = 145.04\ lbf/in^2$

Symbol	Nominal analysis, supplier, condition and remarks.
X 2 Cr Ni Mo 18/12	0.03% C (max) 17.5% Cr 12.0% Ni 2.7% Mo steel: German Standard
X 2 Cr Ni Mo 18/12	0.03% C (max) 17.5% Cr 13.7% Ni steel: Designation used by German Standard
X 2 Cr Ni Mo 18/16	0.03% C (max) 19.0% Cr 13.0% Ni 3.5% Mo steel: German Standard
X 2 Cr Ni Mo 18/16	0.03% C 18.0% Cr 16.0% Ni 3.5% Mo steel: Designation used by German Standard
X 2 Cr Ni N 18/10	0.03% C (max) 18.0% Cr 10.2% Ni 0.18% N steel: Designation used by German Standard
X 2 Cr Ni Mo N 18/12	0.03% C (max) 17.5% Cr 12.0% Ni 2.2% Mo 0.16% N steel: Designation used by German Standard
X 2 Cr Ni Mo Ni 18/12	0.03% C (max) 17.5% Cr 12.0% Ni 2.2% Mo 0.18% N steel: German Standard
X 2 Cr Ni Mo N 18/13	0.03% C (max) 17.5% Cr 13.2% Ni 2.7% Mo 0.16% N steel: Designation used by German Standard
X 3 CN 19 11	0.03% C 19% Cr 11% Ni steel: Italian Standard; annealed **DPN: 180 UTS: 450 Elon: 45% Proof: 170**
X 5 Cr Ni 18/9	0.07% C (max) 18.0% Cr 10.0% Ni steel: Designation used by German Standard
X 5 Cr Ni 18/10	0.08% C (max) 19.0% Cr 9.2% Ni steel: German Standard
X 5 Cr Ni 19/9	0.06% C (max) 19.0% Cr 9.5% Ni steel: Designation used by German Standard
X 5 Cr Ni 19/11	0.07% C (max) 18.5% Cr 11.2% Ni steel: Designation used by German Standard
X 5 Cr Ni Mo 17/12	0.08% C (max) 17.0% Cr 12.0% Ni 2.2% Mo steel: German Standard
X 5 Cr Ni Mo 17/13	0.07% C (max) 17.0% Cr 13.5% Ni 4.5% Mo steel: Designation used by German Standard
X 5 Cr Ni Mo 18/10	0.07% C (max) 17.5% Cr 11.5% Ni 2.2% Mo steel: Designation used by German Standard
X 5 Cr Ni Mo 18/12	0.07% C 17.5% Cr 13.0% Ni 2.7% Mo steel: Designation used by German Standard
X 5 Cr Ni Mo 18/13	0.06% C (max) 18.0% Cr 13.5% Ni 4.7% Mo steel: Designation used by German Standard
X 5 Cr Ni Mo 19/10	0.06% C (max) 19.0% Cr 10.0% Ni 2.2% Mo steel: Designation used by German Standard
X 5 Cr Ni Mo Cu Nb 18/18	0.07% C (max) 17.5% Cr 20.0% Ni 2.2% Mo 2.0% Cr 1.0% Nb steel: Designation used by German Standard
X 5 Cr Ni Mo Cu Nb 20/18	0.06% C (max) 17.5% Cr 20.5% Ni 2.2% Mo 2.0% Cu 0.5% Nb steel: German Standard
X 5 Cr Ni Mo Ti 25/25	0.06% C (max) 25% Cr 25% Ni 2.2% Mo 0.5% Ti steel: Designation used by German Standard
X 6 CN 1911	0.05% C 18.0% Cr 10.0% Ni steel: Italian Standard; annealed **DPN: 180 UTS: 480 Elon: 45% Proof: 180**
X 6 Cr Ni 24/18	0.08% C (max) 23.0% Cr 13.5% Ni steel: Italian Standard
X 6 Cr Ni 25/20	0.08% C (max) 25.0% Cr 20.5% Ni steel: Italian Standard
X 6 Cr Ni Mo Nb 17/12	0.08% C (max) 17.5% Cr 12.0% Ni 2.5% Mo steel: Italian Standard
X 6 Cr Ni Mo Ti 17/12	0.08% C (max) 17.5% Cr 12.0% Ni 2.5% Mo steel: Italian Standard
X 6 Cr Ni Nb 18/11	0.08% C (max) 18.0% Cr 10.5% Ni Nb steel: Italian Standard
X 6 Cr Ni Ti 18/11	0.08% C (max) 18.0% Cr 10.5% Ni Ti steel: Italian Standard
X 7 Cr Ni 23/14	0.08% C (max) 23.0% Cr 13.5% Ni steel: German Standard
X 7 Cr Ni Al 17/7	0.09% C (max) 17.0% Cr 7.2% Ni 1.1% Al steel: German Standard
X 8 CN 1910	0.07% C 18.0% Cr 9.0% Ni steel: Italian Standard; annealed **DPN: 180 UTS: 480 Elon: 45% Proof: 180**

Symbol	Nominal analysis, supplier, condition and remarks.
X 8 CN 2520	0.07% C 25.0% Cr 20.0% Ni steel: Italian Standard; annealed **DPN: 190 UTS: 530 Elon: 40% Proof: 240**
X 8 CND 1712	0.05% C 17.0% Cr 11.0% Ni 2.3% Mo steel: Italian Standard; annealed **DPN: 180 UTS: 480 Elon: 45% Proof: 210**
X 8 CND 1712	0.05% C 17.0% Cr 12.0% Ni 2.8% Mo steel: Italian Standard; annealed **DPN: 190 UTS: 480 Elon: 45% Proof: 200**
X 8 CN Nb 1811	0.07% C 18.0% Cr 10.0% Ni 0.7% (max) Nb steel: Italian Standard; annealed **DPN: 190 UTS: 480 Elon: 40% Proof: 210**
X 8 CNT 1808	0.07% C 18.0% Cr 10.0% Ni 0.35% (max) Ti steel: Italian Standard; annealed **DPN: 190 UTS: 480 Elon: 40% Proof: 180**
X 8 CNT 1810	0.07% C 18.0% Cr 10.0% Ni 0.35% (max) Ti steel: Italian Standard; annealed **DPN: 190 UTS: 480 Elon: 40% Proof: 180**
X 8 Cr Ni 12/12	0.1% C (max) 13.0% Cr 12.5% Ni steel: Designation used by German Standard
X 8 Cr Ni Mo B Nb 16/16K	0.08% C 16.8% Cr 16.7% Ni 1.8% Mo 1.0% Nb 0.08% B steel: Designation used by German Standard
X 8 Cr Ni Mo Cu Nb 20/18	0.08% C (max) 18.5% Cr 21.0% Ni 2.2% Mo 2.0% Cu 0.8% Nb steel: Designation used by German Standard
X 8 Cr Ni Mo Nb 17/13	0.07% C (max) 18.0% Cr 13.5% Ni 4.7% Mo 1.0% Nb steel: Designation used by German Standard
X 8 Cr Ni Mo Nb 19/10	0.1% C (max) 19.0% Cr 10.5% Ni 2.2% Mo 1.3% Nb steel: Designation used by German Standard
X 8 Cr Ni Mo Nb 19/12	0.1% C (max) 19.0% Cr 11.5% Ni 2.7% Mo 1.2% Nb steel: Designation used by German Standard
X 8 Cr Ni Mo Nb 25/25	0.1% C (max) 26.0% Cr 25.0% Ni 2.2% Mo 1.3% Nb steel: Designation used by German Standard
X 8 Cr Ni Mo V Nb 16/13	0.1% C (max) 23% Cr 13% Ni 20% Mo 0.7% V 1.0% Nb steel: German Standard
X 8 Cr Ni Nb 19/9	0.1% C (max) 19.0% Cr 9.0% Ni 1.3% Nb steel: Designation used by German Standard
X 8 Ni Cr Mo Cu Nb	0.08% C 18.5% Cr 21% Ni 2.2% Mo 2.0% Cu 1.0% Nb steel: German Standard
X 10 CND 1808	0.05% C 17% Cr 10% Ni 1.6% Mo steel: Italian Standard; annealed **DPN: 180 UTS: 480 Elon: 45% Proof: 180**
X 10 CNDT 1808	0.07% C 17% Cr 12% Ni 2.3% Mo 0.3% Ti steel: Italian Standard; annealed **DPN: 120 UTS: 480 Elon: 40% Proof: 210**
X 10 Cr Ni Mo Nb 18/10	0.1% C (max) 17.5% Cr 11.5% Ni 2.2% Mo 0.8% Nb steel: Designation used by German Standard
X 10 Cr Ni Mo Nb 18/12	0.1% C (max) 17.5% Cr 13.0% Ni 2.7% Mo 0.8% Nb steel: Designation used by German Standard
X 10 Cr Ni Mo Ti 18/10	0.1% C (max) 17.5% Cr 11.5% Ni 2.2% Mo 0.6% Ti steel: Designation used by German Standard
X 10 Cr Ni Mo Ti 18/10/1	0.1% C (max) 17.5% Cr 11.5% Ni 1.4% Mo 0.5% Ti steel: Designation used by German Standard
X 10 Cr Ni Mo Ti 18/12	0.1% C (max) 17.5% Cr 13.0% Ni 2.7% Mo 0.6% Ti steel: Designation used by German Standard
X 10 Cr Ni Nb 18/9	0.1% C (max) 18.0% Cr 10.0% Ni 0.8% Nb steel: Designation used by German Standard
X 10 Cr Ni Nb 18/10	0.1% C (max) 18.0% Cr 10.0% Ni 0.8% Nb steel: Designation used by German Standard
X 10 Cr Ni Ti 18/9	0.1% C (max) 18.0% Cr 10.0% Ni 0.6% Ti steel: Designation used by German Standard
X 10 Cr Ni Ti 18/10	0.1% C (max) 18.0% Cr 10.0% Ni 0.6% Ti steel: Designation used by German Standard
X 12 Cr Ni 17/7	0.15% C (max) 17.0% Cr 7.5% Ni steel: Designation used by German Standard
X 12 Cr Ni 18/8	0.12% C (max) 18.0% Cr 9.0% Ni steel: Designation used by German Standard
X 12 Cr Ni 18/9	0.12% C 18.0% Cr 9.0% Ni steel: Designation used by German Standard
X 12 Cr Ni 22/12	0.15% C (max) 22.0% Cr 11.0% Ni steel: Designation used by German Standard
X 12 Cr Ni 25/4	0.15% C (max) 26.0% Cr 4.5% Ni steel: Designation used by German Standard
X 12 Cr Ni 25/20	0.15% C (max) 25.0% Cr 20% Ni steel: Designation used by German Standard
X 12 Cr Ni 25/21	0.15% C (max) 25.0% Cr 20.0% Ni steel: Designation used by German Standard
X 12 Cr Ni S 18/8	0.15% C (max) 18.0% Cr 9.0% Ni 0.15% S steel: Designation used by German Standard
X 12 Cr Ni Si 36/16	0.08% C (max) 18.5% Cr 35.5% Ni steel: German Standard
X 12 Cr Ni Ti 18/9	0.15% C (max) 18.0% Cr 10.0% Ni 0.7% Ti steel: Designation used by German Standard
X 12 Ni Cr 36/18	0.2% C 18% Cr 38% Ni steel: German Standard
X 12 Ni Cr Si 36/16	0.15% C 16% Cr 36% Ni steel: German Standard
X 15 CN 1707	0.10% C 17.0% Cr 7.0% Ni steel: Italian Standard; annealed **DPN: 180 UTS: 480 Elon: 45% Proof: 180**
X 15 CN 1808	0.10% C 18.0% Cr 8.0% Ni steel: Italian Standard; annealed **DPN: 180 UTS: 480 Elon: 45% Proof: 180**
X 15 CN 2412	0.07% C 23.0% Cr 13.0% Ni steel: Italian Standard; annealed **DPN: 190 UTS: 530 Elon: 40% Proof: 240**
X 15 CN 2420	0.07% C 25.0% Cr 20.0% Ni steel: Italian Standard; annealed **DPN: 190 UTS: 530 Elon: 40% Proof: 240**
X 15 Cr Ni 25/20	0.15% C 25% Cr 20% Ni steel: German Standard
X 15 Cr Ni Si 20/12	0.2% C (max) 20.0% Cr 12.0% Ni steel: Designation used by German Standard
X 15 Cr Ni Si 20/12	0.15% C (max) 19.0% Cr 9.0% Ni steel: German Standard
X 15 Cr Ni Si 25/20	0.2% C (max) 25.0% Cr 20.0% Ni steel: Designation used by German Standard
X 20 CN 245	0.07% C 27.0% Cr 4.0% Ni steel: Italian Standard; annealed **DPN: 260 UTS: 630 Elon: 20% Proof: 370**
X 20 CN 2412	0.07% C 23.0% Cr 13.0% Ni steel: Italian Standard; annealed **DPN: 190 UTS: 530 Elon: 40% Proof: 240**
X 20 T2	23.0% Cr 12.0% Ni 4.0% W Fe alloy: Origin unknown
X 25 CN 2520	0.12% C 25.0% Cr 21.0% Ni 2.0% Si steel: Italian Standard; annealed **DPN: 200 UTS: 580 Elon: 40% Proof: 290**
X 45 Cr Ni W 18/9	0.45% C 18.0% Cr 9.0% Ni 1.0% W steel: Designation used by German Standard
XM 19	0.06% C 21.5% Cr 12.5% Ni 2.5% Mo 0.2% Nb 0.2% V steel: Designation used by ASTM
Z 1 NCDU25/20	0.02% C (max) 20.5% Cr 25.5% Ni 4.4% Mo 1.5% Cu steel: French National Standard
X 2 CN18/10	0.03% C 19.0% Cr 11.0% Ni steel: French Standard; annealed **DPN: 180 UTS: 450 Elon: 45% Proof: 170**
Z 2 CND17/12	0.03% C (max) 17.0% Cr 12.0% Ni 2.2% Mo steel: French National Standard
Z 2 CND17/13	0.03% C (max) 17.5% Cr 12.0% Ni 2.7% Mo 1.0% Nb steel: French National Standard
Z 2 CND 18/10	0.03% C 17.0% Cr 12.0% Ni 2.8% Mo steel: French Standard; annealed **DPN: 180 UTS: 480 Elon: 45% Proof: 180**
Z 2 CND19/15	0.03% C (max) 19.0% Cr 13.0% Ni 3.5% Mo steel: French National Standard

Symbol	Nominal analysis, supplier, condition and remarks.
Z 3 CN18/08	0.03% C (max) 19.0% Cr 9.0% Ni steel: French Standard; annealed
	DPN: 180 UTS: 450 Elon: 45% Proof: 170
Z 3 CN18/10	0.03% C 19% Cr 10% Ni steel: French National Standard; annealed
	DPN: 180 UTS: 480 Elon: 45% Proof: 180
Z 3 CND18/12	0.03% C 17% Cr 12% Ni 2.5% Mo steel: French National Standard; annealed
	DPN: 180 UTS: 480 Elon: 45% Proof: 180
Z 5 CN18/08	0.07% C 18% Cr 9% Ni steel: French National Standard; annealed
	DPN: 180 UTS: 480 Elon: 45% Proof: 180
Z 5 CND18/10	0.12% C 18% Cr 10% Ni 3.0% Mo steel: French National Standard; annealed
	DPN: 180 UTS: 480 Elon: 45% Proof: 180
Z 5 CNDU21/08	0.06% C (max) 7.5% Cr 21.0% Ni 2.5% Mo 1.5% Mo steel: French National Standard
Z 6 CN18/09	0.08% C (max) 19.0% Cr 9.2% Ni steel: French National Standard
Z 6 CN18/10	0.08% C 19% Cr 10% Ni steel: French National Standard; annealed
	DPN: 180 UTS: 480 Elon: 45% Proof: 180
Z 6 CND17/11	0.08% C (max) 17.0% Cr 12.0% Ni 2.2% Mo steel: French National Standard
Z 6 CND17/12	0.08% C (max) 17.5% Cr 12.0% Ni 2.7% Mo 0.4% Ti steel: French National Standard
Z 6 CND18/12	0.08% C 17% Cr 12% Ni 2.5% Mo steel: French National Standard; annealed
	DPN: 180 UTS: 480 Elon: 45% Proof: 210
Z 6 CNDT17/12	0.08% C (max) 17.5% Cr 12.0% Ni 2.5% Mo steel: French National Standard
Z 6 CNT18/10	0.08% C (max) 18.0% Cr 10.5% Ni 0.9% Ti steel: French National Standard
Z 6 NCT25/15	0.06% C 15.0% Cr 25.0% Ni 1.25% Mo 0.3% V 2.0% Ti steel: French Standard designation
Z 7 CN18/10	0.07% C 18.0% Cr 9.0% Ni steel: French National Standard; annealed
	DPN: 180 UTS: 480 Elon: 45% Proof: 180
Z 8 CNDNb18/12	0.07% C 17.0% Cr 13.0% Ni 2.8% Mo 0.7% (max) Nb steel: French National Standard; annealed
	DPN: 190 UTS: 480 Elon: 40% Proof: 220
Z 8 CNDT18/12	0.07% C 17.0% Cr 12.0% Ni 2.3% Mo 0.35 (max) Ti steel: French National Standard; annealed
	DPN: 120 UTS: 480 Elon: 40% Proof: 210
Z 10 CN11/12	0.16% C 12% Cr 12% Ni steel: For deep drawing; French National Standard
Z 10 CN12/12	0.05% C 12% Cr 13% Ni steel: French National Standard; annealed
	DPN: 160 UTS: 480 Elon: 55% Proof: 200
Z 10 CNNb18/10	0.08% C 18% Cr 12% Ni 1.0% Nb steel: French National Standard; annealed
	DPN: 190 UTS: 480 Elon: 40% Proof: 210
Z 10 CNDNb18/10	0.07% C 17% Cr 12% Ni 2.3% Mo 0.7% Nb steel: French National Standard; annealed
	DPN: 190 UTS: 480 Elon: 40% Proof: 220
Z 10 CNDT18/12	0.07% C 17% Cr 12% Ni 2.3% Mo 0.3% Ti steel: French National Standard; annealed
	DPN: 120 UTS: 480 Elon: 40% Proof: 210
Z 10 CNS25/13	0.2% C 23% Cr 14% Ni steel: French National Standard; annealed
	DPN: 190 UTS: 530 Elon: 40% Proof: 240
Z 10 CNS25/20	0.25% C 25% Cr 21% Ni steel: French National Standard; annealed
	DPN: 200 UTS: 580 Elon: 40% Proof: 290
Z 10 CNT18/08	0.07% C 18% Cr 10% Ni 0.35% Ti steel: French National Standard; annealed
	DPN: 190 UTS: 480 Elon: 40% Proof: 180
Z 10 CNT18/10	0.08% C 18% Cr 11% Ni 0.4% Ti steel: French National Standard; annealed
	DPN: 190 UTS: 480 Elon: 40% Proof: 180
Z 10 NNb18/10	0.15% C (max) 18% Cr 9.0% Ni 1.2% Nb steel: French National Standard; annealed
Z 12 CN	0.15% C 18% Cr 9% Ni steel: French National Standard
Z 12 CN18/08	0.08% C 18% Cr 9.0% Ni steel: French National Standard; annealed
	DPN: 180 UTS: 480 Elon: 45% Proof: 180
Z 12 CN18/10	0.1% C 18% Cr 8% Ni steel: French National Standard; annealed
	DPN: 180 UTS: 480 Elon: 45% Proof: 180
Z 12 CNS25/20	0.08% C (max) 25.0% Cr 20.5% Ni steel: French National Standard
Z 12 CNS25/20	0.12% C 25.0% Cr 20.0% Ni steel: French National Standard
Z 12 NCS37/18	0.08% C (max) 18.5% Cr 35.5% Ni steel: French National Standard
Z 12 NCS37/18	0.12% C 18.0% Cr 37.0% Ni steel: French National Standard
Z 15 CN25/13	0.08% C (max) 23.0% Cr 13.5% Ni steel: French National Standard
Z 15 CNS20/10	0.12% C 20% Cr 12% Ni 2.0% Si steel: French National Standard; annealed
	DPN: 190 UTS: 580 Elon: 40% Proof: 290
Z 15 CNS20/12	0.15% C (max) 19.0% Cr 9.0% Ni steel: French National Standard
Z 15 CNS25/13	0.12% C 20.0% Cr 12.0% Ni 2.0% Si steel: French National Standard; annealed
	DPN: 190 UTS: 580 Elon: 40% Proof: 290
Z 15 CNS25/20	0.25% C 25% Cr 20% Ni steel: French National Standard
Z 20 C25	0.07% C 27.0% Cr 4.0% Ni steel: French National Standard; annealed
	DPN: 260 UTS: 630 Elon: 20% Proof: 370
Z 20 CNS24/13	0.07% C 23.0% Cr 13.0% Ni steel: French National Standard; annealed
	DPN: 190 UTS: 530 Elon: 40% Proof: 240
Z 20 CNS24/20	0.12% C 25.0% Cr 21.0% Ni 2.0% Si steel: French National Standard; annealed
	DPN: 200 UTS: 580 Elon: 40% Proof: 290
Z 20 CN25/05	0.07% C 27.0% Cr 4.0% Ni steel: French National Standard; annealed
	DPN: 260 UTS: 630 Elon: 20% Proof: 370
Z 25 CNWS20/09	0.25% C 20.0% Cr 9.0% Ni 2.0% W steel: French National Standard
Z 35 CNS14/14	0.35% C 14.0% Cr 14.0% Ni 2.5% W steel: French National Standard
Z 52 CMN21/09	0.52% C 9.0% Mn 21.0% Cr 3.7% Ni 0.45% N steel: French National Standard

Note. The following abbreviations and units are used in the tables:

DPN	Hardness, diamond pyramid number
UTS	Ultimate tensile strength, N/mm^2
Elon	Elongation, %
Proof	0.1 % proof strength, N/mm^2

1 N/mm^2=0.1 hbar=0.102 kgf/mm^2=0.06475 tonf/in^2=145.04 lbf/in^2

44N2 Steel – chromium–nickel – wrought and cast, age hardening
With molybdenum, niobium, tungsten, titanium, aluminium or nitrogen

Specific gravity	7.83	
Density	7830 kg/m^3	(0.226 lb/in.^{3})
Solidus/liquidus	1435 °C	
Thermal conductivity	16.7 W/m °C	(4 cal/m s °C)
Coefficient of linear expansion	11.5 × 10^{-6}/ °C	
Electrical conductivity	20% IACS (copper 100%)	
Specific resistance	850 microhm mm	
Young's modulus of elasticity	214.4 × 10^9 N/m^2	(31.1 × 10^6 lbf/in.^{2})
Impact	aged 61 J	(45 ft/lb) Izod
	over aged 115 J	(85 ft/lb)
Fatigue strength (endurance limit)	aged ± 480 N/mm^2	(± 32 tonf/in.^{2})
	over aged ± 600 N/mm^2	(± 40 tonf/in.^{2})

Hot strength

Temperature °C	Tensile strength N/mm^2	Elongation %
420	1150	22
530	1130	21
650	1000	20
700	880	12

The above properties are typical of the following group, and may not apply exactly to any one specification. It is possible that with certain specifications some of the values may not be applicable.

General metallurgical characteristics

These steels are based on the 18/8 chromium nickel series discussed in Section 44N1. The alloys listed can all be solution treated and aged to increase their mechanical properties. This relies on the alloying elements forming intermetallic compounds and carbides which are dissolved in the austenitic matrix then precipitated, causing some strain in the lattice structure.

A few of the steels have the alloying elements so balanced that the austenitic structure becomes unstable and can be hardened by controlled ageing. Both processes rely on complex metallurgical reactions requiring strict control during thermal treatments and deformation.

The increased strength is obtained at the expense of some corrosion resistance but many of the alloys have been developed for their high temperature tensile properties associated with a high degree of hot oxidation resistance.

In the annealed or solution treated condition, many of these steels are almost identical to those listed in Section 44N1 and it is possible that some of those listed should in fact be in Section 44N2.

Thermal treatment

All these steels can be considerably hardened by cold work when some of the austenite is transformed to martensite, the change being identified by the steel becoming magnetic. It is not advised that any of these materials should be used in the work hardened condition without subsequent ageing, as there are more economical materials which can achieve the mechanical properties of this condition with equal or better corrosion resistance. Some of these steels are designed to be work hardened between solution treatment and ageing for maximum tensile strength.

As stated above, there are two types of these steels, each requiring approximately the same temperatures but with different metallurgical reactions. The first type is hardened by changing the austenite to martensite. This change can take place during cooling from the solution treatment temperature and is followed by ageing, which is a complex process depending on the steel analysis for exact times and temperatures, causing an increase in the martensite hardness. Alternatively the steel can be quenched to austenite from the solution treatment, then transformed to martensite by a sub-zero treatment at about minus 70 °C. This martensite is then further hardened by the ageing treatment.

With the second type carbides are dissolved at the solution treatment, then precipitated at ageing.

The solution treatment for all the steels is between 1020–1150 °C, the ageing temperature being 450–850 °C depending on the steel and the hardness required. In many cases a double age at different temperatures is necessary.

These are all very expensive materials with rather complex metallurgical thermal treatments necessary to ensure the best mechanical properties. It is essential to refer to the origin of the specification being heat treated for details of times and temperatures. The above information can be no more than a guide to the range of temperatures and equipment required.

As with all stainless steels, care must be taken to eliminate sulphur and carbon bearing atmospheres. For this reason electric furnaces are preferred, and only hydrogen, inert gas or vacuum atmospheres should be used, although for most purposes air is suitable, followed by scale removal.

This can be by grit blasting, acid pickle or sodium hydride. Grit blasting can cause distortion of sheet metal parts, while the pickle requires hydrofluoric acid which is unpleasant and dangerous to handle, requiring special tanks. Sodium hydride requires very specialized equipment but gives the best results.

There are now available coatings which are applied by spray, dip or brush before heat treatment and prevent oxides forming. The use of these materials can only be justified in special cases where conventional de-scaling is impossible or difficult.

Welding and brazing

Pre-treatment. The steels should be in the solution treated condition for welding. The joint area must be carefully prepared and free of all traces of dirt, grease, oil and cold work. Pre-heating is not required or advised on these materials.

Welding and brazing. Inert gas shielded electric arc welding is the only method of welding recommended with these steels. They are very similar to the 18/8 chromium/nickel stainless steels regarding weldability, provided some care is taken during cooling to ensure this is rapid enough to prevent local ageing. This can usually be achieved by the use of copper chills adjacent to the weld on parts other than sheet metal components, where the heat build up would cause ageing.

Normal gas or electric arc welding requires active fluxes and much greater manipulative skill. Resistance welding gives good results provided the contact points are scrupulously clean. The high electrical resistance ensures rapid heat input, thus there is seldom any ageing problem during welding.

Brazing requires careful consideration at the design stage to ensure that the filler rod melting temperature is remote from the ageing temperature. This should either be high enough to allow re-solution treatment, or low enough not to affect the ageing. For the former, furnace brazing under hydrogen or vacuum is essential. Provided very rapid heating is employed, for example induction heating, some of the alloys can be brazed without affecting the solution treatment or even in the finally aged condition. Before this is attempted, advice should be sought from the supplier or other competent authority.

Post-treatment. All fluxes if used must be removed before any thermal treatment. Welded parts require to be aged.

As stated above, brazing requires consideration at the design stage and the post-braze treatment will depend on the type of brazing applied.

Flaw detection methods

Crack test. Oil penetrant chalk test, which should be carried out after final ageing.

X-ray. This is commonly applied to castings as a means of quality control and would be required on high quality welds. It is recommended that all welding is controlled by routine X-ray examination on a percentage basis.

Ultra-sonic test. These expensive materials are often inspected ultra-sonically at an early stage in machining. This identifies internal defects, thus saving the cost of expensive machining on faulty components. Techniques are now available whereby ultrasonic testing can identify certain types of weld defect.

Chemical etch. Not required.

Corrosion protection

Temporary. None required.

Permanent. None required under most circumstances. These steels, although not as corrosion resistant as those in Section 44N1, all have excellent resistance to atmospheric corrosion. Coatings are being developed which protect them against attack from hot atmospheres, particularly sulphur and oxidation. These include diffusion coatings of chromium and aluminium, sprayed coatings of ceramics, metal oxides or corrosion resistant metals. Most of the processes are proprietary and require considerable care and constant control to achieve repetitive results.

Machinability

These materials may be machined in the annealed condition when they have the same characteristics as the austenitic steels - Section 44N1, that is, they tear and gall, resulting in poor surface finish.

In the aged condition the materials are harder, thus tool wear may be greater, but surface finish is much superior with less tendency for galling.

All the materials harden with cold work, thus interrupted cuts, tool dwell or light rubbing cuts must be avoided. Robust machines using low speeds and high feeds with tipped tooling give best results.

Because of the tendency to cold work it may be technically necessary to rough machine then re-solution treat and age before final machining. In these instances a part-age before rough machining should be considered as this can often result in much superior machinability over the solution treated condition.

Uses

Steam and gas turbine high duty rotating parts which operate at fairly high temperatures – 200–400 °C – such as turbine discs, compressor wheels, compressor blading and steam generator turbine blading. Chemical plant components. Internal combustion engine exhaust valves.

Symbol	Nominal analysis, supplier, condition and remarks.
1.4974	0.12% C 21% Cr 20% Ni 3.0% Mo 2.5% W 1.0% Nb & Ta 20% Co steel: German Standard
1KH14N16B	0.1% C 14.0% Cr 15.5% Ni 1.1% Nb steel: Russian Standard
1KH14N18V2B	0.1% C 14.0% Cr 19.0% Ni 2.3% W 1.1% Nb steel: Russian Standard
1KH14N18V2BR	0.1% C 14.0% Cr 19.0% Ni 2.3% W 1.1% Nb 0.005% B (max) 0.02% Ce (max) steel: Russian Standard
1KH14N18V2BR1	0.1% C 14.0% Cr 19.0% Ni 2.3% W 1.1% Nb 0.005% B (max) 0.02% Ce (max) steel: Russian Standard
3RE60	0.03% C (max) 18.5% Cr 4.7% Ni 2.7% Mo steel: Tube, strip or bar; Sandvik; solution treated **DPN: 260 UTS: 800 Elon: 22% Proof: 500**
4KH14N14V2M	0.45% C 14.0% Cr 14.0% Ni 0.32% Mo 2.3% W steel: Russian Standard designation
13.4.2Cu R	0.06% C 13.0% Cr 4.0% Ni 2.0% Mo 2.0% Cu steel: Electrodes; Metrode; precipitation hardening
14/4PH	0.03% C 14.1% Cr 4.2% Ni 3.3% Mo 0.25% Nb 3.2% Cu steel: Age hardenable; information from SAE handbook
14.6.2 R	0.05% C 13.5% Cr 5.8% Ni 1.7% Mo steel: Electrode; Metrode
15.4 Mn Si R	0.05% C 15.0% Cr 4.0% Ni 3.0% Mn 3.5% Si steel: Electrode; Metrode
16/25/6	0.5% C 16% Cr 25% Ni 6.0% Mo 0.15% N steel: Information from SAE handbook
17.4 Cu R	0.06% Cu 16.0% Cr 4.0% Ni 2.0% Cu steel: Electrodes; Metrode; precipitation hardening
17/4PH	0.04% C 16.5% Cr 4% Ni 0.3% Cb 3.5% Cu steel: Age hardening, American proprietary alloy listed in SAE year book
17.4.1 R	0.06% C 16.0% Cr 4.0% Ni 1.0% Mo steel: Electrode; Metrode; for sea water resistance
17/7PH	0.07% C 17% Cr 7% Ni 1.1% Al steel: Age hardening; American proprietary alloy listed in SAE year book
17/14 Cu Mo	0.12% C 16% Cr 14% Ni 2.5% Mo 0.4% Nb 0.2% Ti 3% Cu steel: Age hardening; American proprietary alloy listed in SAE year book
19/9 DL	0.3% C 19% Cr 9% Ni 1.25% Mo 1.2% W 0.4% Nb 0.3% Al steel: Age hardening; American proprietary alloy listed in SAE year book
19/9W Mo	0.1% C 19% Cr 9% Ni 0.4% Mo 1.3% W 0.4% Nb 0.4% Al steel: Age hardening; American proprietary alloy listed in SAE year book
22/4/9	0.5% C 8.5% Mn 20% Cr 3.5% Ni 0.4% N steel: Age hardening; American proprietary alloy listed in SAE year book
901	0.1% C 12.5% Cr 42% Ni 6% Mo 2.7% Ti 0.015% B steel: Fe alloy; Carpenter
A 286	0.06% C 15% Cr 25% Ni 0.3% V 1.3% Mo 2.0% Ti steel: Bofors; solution treated & aged **DPN: 290 UTS: 1030 Elon: 24% Proof: 720**
A 286	0.08% C 15% Cr 26% Ni 1.25% Mo 2% Ti steel: Allegheny Ludlum; solution treated & aged **UTS: 1000 Elon: 25%**

Symbol	Nominal analysis, supplier, condition and remarks.
AISI 630	0.04% C 16% Cr 4.25% Ni 0.27% Nb 3.3% Cu steel: Solution treated & aged (aged 1 hour at 430° C) **DPN: 420 UTS: 1420 Elon: 14% Proof: 1340**
AISI 631	0.07% C 17% Cr 7.1% Ni 1.17% Al steel: Solution treated, deep frozen & aged **DPN: 430 UTS: 1420 Elon: 10% Proof: 1300**
AISI 632	0.07% C 15.1% Cr 7.1% Ni 2.25% Mo 1.17% Al steel: Solution treated; deep frozen & aged **UTS: 1750 Elon: 6% Proof: 1620**
AISI 632	0.07% C 15.1% Cr 7.1% Ni 2.25% Mo 1.17% Al steel: Annealed **UTS: 920 Elon: 30% Proof: 370**
AISI 633	0.1% C 16.5% Cr 4.25% Ni 2.75% Mo 0.09% N steel: Solution treated, deep frozen & aged **UTS: 1550 Elon: 12% Proof: 1300**
AISI 634	0.13% C 15.5% Cr 4.25% Ni 2.7% Mo 0.1% N steel: Solution treated, deep frozen & aged **UTS: 1580 Elon: 19% Proof: 1300**
AISI 635	0.06% C 17% Cr 7% Ni 0.2% Al 0.8% Ti steel: Solution treated & aged **DPN: 460**
AISI 650	0.05% C 16% Cr 25% Ni 6% Mo 0.15% N steel: Warm worked **UTS: 630 Elon: 23% Proof: 500**
AISI 651	0.32% C 18.5% Cr 9% Ni 1.4% Mo 1.4% W 0.2% Ti 0.4% Nb steel: Solution treated & aged **DPN: 189 UTS: 450 Elon: 36% Proof: 150**
AISI 652	0.3% C 18.5% Cr 9% Ni 1.6% Mo 1.35% W 0.55% Ti steel: Solution treated & aged **DPN: 189 UTS: 450 Elon: 36% Proof: 150**
AISI 653	0.12% C 15.9% Cr 14.1% Ni 2.5% Mo 0.45% Nb 0.25% Ti steel: Solution treated & aged **UTS: 610 Elon: 45% Proof: 220**
AISI 660	0.05% C 14.7% Cr 25.2% Ni 1.3% Mo 2.15% Ti 0.2% Al steel: Solution treated & aged **UTS: 1000 Elon: 25% Proof: 610**
AISI 662	0.04% C 13.5% Cr 26% Ni 2.7% Mo 1.7% Ti 0.07% Al 0.005% B Fe alloy: Solution treated & aged **UTS: 1040 Elon: 24% Proof: 650**
AISI 663	0.05% C 14.7% Cr 27.2% Ni 1.3% Mo 3% Ti 0.2% Al 0.3% V 0.01% B Fe alloy: Solution treated & aged **DPN: 341**
AISI 665	0.03% C 13.5% Cr 26% Ni 1.7% Mo 3% Ti 0.15% Al 0.02% B Fe alloy: Solution treated & aged **DPN: 350 UTS: 1210 Elon: 13% Proof: 970**
AL 350	0.1% C 0.5% Mn 17% Cr 4.2% Ni 2.7% Mo steel: Solution treated; double aged or freeze & temper; listed in SAE year book **DPN: 400 UTS: 1550 Elon: 9% Proof: 1350**
AM 350	0.08% C 16.5% Cr 4.3% Ni 2.7% Mo 0.1% N steel: Allegheny Ludlum; solution treated & double aged **DPN: 400 UTS: 1380 Elon: 14% Proof: 1100**
AM 355	0.13% C 15.5% Cr 4.3% Ni 2.7% Mo 0.1% N steel: Allegheny Ludlum; solution treated & double aged **UTS: 1380 Elon: 13% Proof: 1070**
AMS 5520 A	15% Cr 7% Ni 2.5% Mo 1% Al steel: Sheet & strip
AMS 5221 A	5.2% Cr 42% Ni 2.3% Ti 0.5% Al Fe alloy: Strip; solution treated; AMS for SAE 902
AMS 5525 B	15% Cr 26% Ni 1.3% Mo 2% Ti 0.3% V steel: Sheet & strip
AMS 5528 A	17% Cr 7% Ni 1% Al steel: Sheet & strip

Note. The following abbreviations and units are used in the tables:

DPN	Hardness, diamond pyramid number
UTS	Ultimate tensile strength, N/mm²
Elon	Elongation, %
Proof	0.1 % proof strength, N/mm²

1 N/mm²=0.1 hbar=0.102 kgf/mm²=0.06475 tonf/in²=145.04 lbf/in²

Symbol	Nominal analysis, supplier, condition and remarks.
AMS 5529 A	17% Cr 7% Ni 1% Al steel: Sheet & strip; age hardened
	UTS: 1500
AMS 5538	20% Cr 9% Ni 1.6% Mo 1.4% W Ti steel: Sheet & strip
AMS 5539	20% Cr 9% Ni 1.6% Mo 1.4% W Ti steel: Sheet & strip; hot rolled & stress relieved
	UTS: 920
AMS 5541 A	15.5% Cr 4.5% Ni 3% Mo 0.1% N steel: Sheet & strip
AMS 5543	13.5% Cr 26% Ni 1.75% Mo 3% Ti steel: Sheet; vacuum melted; solution treated
AMS 5546	16.5% Cr 4.5% Ni 3% Mo 0.1% N steel: Sheet & strip; cold rolled & tempered
AMS 5548 A	16.5% Cr 4.5% Ni 3% Mo 0.1% N steel: Sheet; high temperature annealed
AMS 5549 A	15.5% Cr 4.5% Ni 3% Mo 0.1% N steel: Plate; solution treated
AMS 5554	16.5% Cr 4.5% Ni 3% Mo 0.1% N steel: Seamless tube
AMS 5657	15% Cr 7% Ni 2.5% Mo 1% Al steel: Bar & forging
AMS 5568	17% Cr 7% Ni 1% Al steel: Welded tube; aged
AMS 5594	15.5% Cr 4.5% Ni 3% Mo 0.1% N steel: Plate
AMS 5638	18% Cr 9% Ni 0.5% Mo 0.8% Al steel: Free machining
AMS 5643 F	17% Cr 4% Ni 4% Cu steel: Bar & forging
AMS 5644 B	17% Cr 7% Ni 1% Al steel: Bar & forging
AMS 5673 A	17% Cr 7% Ni 1% Al steel: Wire; aged
AMS 5720 A	20% Cr 9% Ni 1.5% Mo 1.5% W Nb + Ta Ti steel: Bar; up to 40 mm diameter
AMS 5721 B	20% Cr 9% Ni 1.5% Mo 1.5% W Nb + Ta Ti steel: Bar; up to 25 mm in diameter
AMS 5722 A	20% Cr 9% Ni 1.5% Mo 1.5% W Nb Ti steel: Bar & forging
AMS 5723 A	20% Cr 9% Ni 1.5% Mo 1.5% W 0.6% Ti steel: Bar & forging etc
AMS 5724	20% Cr 9% Ni 1.5% Mo 1.5% W 0.6% Ti steel: Bar; up to 25 mm in diameter
AMS 5526 C	20% Cr 9% Ni 1.5% Mo 1.5% W Nb Ti steel: Sheet & strip
AMS 5527 A	20% Cr 9% Ni 1.5% Mo 1.5% W 0.4% Nb + Ta 0.2% Ti steel: Sheet & strip; hot rolled & stress relieved
	UTS: 92
AMS 5729	20% Cr 9% Ni 1.5% Mo 1.5% W 0.6% Ti steel: Bar; up to 40 mm in diameter
AMS 5731 A	15% Cr 26% Ni 1.3% Mo 2% Ti 0.3% V steel: Bar, forging, etc.; solution treated
AMS 5732 A	15% Cr 26% Ni 1.3% Mo 2% Ti 0.3% V steel: Bar, forging, etc.; solution treated & aged
AMS 5734	15% Cr 26% Ni 1.3% Mo 2% Ti 0.3% V steel: Bar, forging, etc.; annealed
AMS 5735 E	15% Cr 26% Ni 1.5% Mo 2% Ti 0.3% V steel: Bar, forging, etc.
AMS 5736 B	15% Cr 26% Ni 1.5% Mo 2% Ti 0.3% V steel: Bar & forging; solution treated
AMS 5737 B	15% Cr 26% Ni 1.5% Mo 2% Ti 0.3% V steel: Bar, forging, etc.; annealed & aged

Note. The following abbreviations and units are used in the tables:

DPN	Hardness, diamond pyramid number
UTS	Ultimate tensile strength, N/mm^2
Elon	Elongation, %
Proof	0.1 % proof strength, N/mm^2

1 N/mm^2=0.1 hbar=0.102 kgf/mm^2=0.06475 tonf/in^2=145.04 lbf/in^2

Symbol	Nominal analysis, supplier, condition and remarks.
AMS 5743 B	15.5% Cr 4.5% Ni 3% Mo 0.1% N steel: Bar & forging
AMS 5745	16.5% Cr 4.5% Ni 3% Mo 0.1% N steel: Bar & forging
AMS 5774	16.5% Cr 4.5% Ni 3% Mo 0.1% N steel: Wire for welding
AMS 5775	16.5% Cr 4.5% Ni 3% Mo 0.1% N steel: Wire coated electrode
AMS 5780	15.5% Cr 4.5% Ni 3% Mo 0.1% N steel: Welding wire
AMS 5781	15.5% Cr 4.5% Ni 3% Mo 0.1% N steel: Coated electrode
AMS 5782 A	19% Cr 9% Ni 0.5% Mo 1.5% W 1% Nb + Ta 0.2% Ti steel: Wire for welding
AMS 5783 B	19% Cr 9% Ni 0.5% Mo 1.5% W 1% Nb + Ta steel: Coated electrode
AMS 5804 A	15% Cr 26% Ni 1.3% Mo 2.2% Ti 0.3% V steel: Welding wire
AMS 5805 A	15% Cr 26% Ni 1.3% Mo 2.2% Ti 0.3% V steel: Welding wire; vacuum melted
AMS 5812 B	15% Cr 7% Ni 2.5% Mo 1% Al steel: Welding wire; vacuum melted
AMS 5813 A	15% Cr 7% Ni 2% Mo 1.0% Al steel: Welding wire
AMS 5825 A	17% Cr 5% Ni 0.2% Nb + Ta 3.5% Cu steel: Welding wire
AMS 5827	17% Cr 5% Ni 0.2% Nb + Ta 3.5% Cu steel: Welding electrode
ASTM A453/651	0.32% C 9.5% Ni 19.5% Cr 1.5% W 0.2% Ti 0.4% Nb steel: For high expansion bolts
ASTM A453/660	0.08% C (max) 25.5% Ni 1.45% Cr 1.2% Mo 2.1% Ti 0.3% Al steel: For bolts
ASTM A453/662	0.08% C (max) 26% Ni 13.5% Cr 2.8% Mo 2.0% Ti 0.3% Al 0.005% B steel: For high expansion bolts
ASTM A453/665	0.08% C (max) 26% Ni 14% Cr 1.8% Mo 3.0% Ti 0.2% Al 0.04% B steel: For high expansion bolts
ASTM A457/651	0.32% C 19% Cr 9.5% Ni 1.3% Mo 1.3% W 0.45% Nb 0.2% Ti steel: Plate
ASTM A457/652	0.32% C 19% Cr 9.5% Ni 1.6% Mo 1.3% W 0.5% Ti steel: Plate
ASTM A458	Stainless steel: Bar; graded as ASTM A457
ASTM A461/630	0.07% C 16.5% Cr 0.3% Nb & Ta 4.0% Cu 4.0% Ni steel: Bar
ASTM A461/631	0.09% C (max) 17% Cr 1.2% Al 7.2% Ni steel: Bar
ASTM A461/634	0.12% C 15.5% Cr 2.8% Mo 4.5% Ni 0.1% N steel: Bar
ASTM A461/632	0.09% C (max) 15% Cr 2.5% Mo 1.2% Al 7.2% Ni steel: Bar
ASTM A461/660	0.08% C (max) 14.7% Cr 25.5% Ni 1.2% Mo 2.1% Ti 0.005% B steel: Bar
ASTM A461/661	0.12% C 21.5% Cr 20% Ni 3.0% Mo 2.5% W 1.0% Nb + Ta 20% Co Fe alloy: Bar
ASTM A461/662	0.08% C 13.5% Cr 26% Ni 3.2% Mo 1.8% Ti 0.005% B steel: Bar
ASTM A477	Stainless steel: For forgings used at high temperature; graded by AISI system
ASTM A564	Stainless steel: Bars & shapes; hot and cold rolled; graded by AISI system
ASTM A579	0.12% C 15.5% Cr 4.5% Ni 2.8% Mo 0.1% N steel: Forging
ASTM A638/660	0.08% C (max) 15.0% Cr 26.0% Ni 1.2% Mo 2.1% Ti 0.25% V 0.005% B Fe alloy
ASTM A638/662	0.08% C (max) 13.5% Cr 26.0% Ni 3.0% Mo 1.8% Ti 0.005% B Fe alloy
ASTM A639/661	0.12% C 21.2% Cr 20.0% Ni 3.0% Mo 2.5% W 1.0% Nb + Ta 19.7% Co Fe alloy

Symbol	Nominal analysis, supplier, condition and remarks.
ASTM A693/630	0.07% C 16.5% Cr 4.0% Ni 4.0% Cu steel: Precipitation hardened **UTS: 1200 Elon: 5% Proof: 1050**
ASTM A693/631	0.09% C 17.0% Cr 7.2% Ni 1.2% Al steel: Precipitation hardened **UTS: 1300 Elon: 5% Proof: 1200**
ASTM A693/632	0.09% C 15.0% Cr 7.2% Ni 1.2% Al 25% Mo steel: Precipitation hardened **UTS: 1480 Elon: 3% Proof: 1250**
ASTM A693/634	0.2% C 15.5% Cr 4.5% Ni 2.9% Mo steel: Precipitation hardened **UTS: 1250 Elon: 12% Proof: 1080**
ASTM A693/635	0.08% C 16.8% Cr 6.8% Ni 0.4% Al 0.8% Ti steel: Precipitation hardened **UTS: 1280 Elon: 7% Proof: 1050**
ASTM A693 XM9	0.05% C 14.2% Cr 6.5% Ni 0.1% Al 0.3% Mo 0.75% Mo steel: Precipitation hardened **UTS: 1200 Elon: 8% Proof: 1050**
ASTM A693 XM12	0.07% C 14.7% Cr 4.0% Ni 3.5% Cu steel: Precipitation hardened **UTS: 1200 Elon: 8% Proof: 1050**
ASTM A693 XM13	0.05% C 12.7% Cr 8.0% Ni 1.1% Al 2.2% Mo steel: Precipitation hardened **UTS: 1400 Elon: 8% Proof: 1350**
ASTM A693 XM16	0.06% C 11.7% Cr 8.5% Ni 0.5% Mo 1.1% Ti 1.0% Cu steel: Precipitation hardened **UTS: 1520 Elon: 3% Proof: 1400**
ASTM A693 XM25	0.05% C 15.0% Cr 6.0% Ni 0.7% Mo 1.5% Cu steel: Precipitation hardened **UTS: 1100 Elon: 6% Proof: 1000**
ASTM A705	Austenitic age hardening stainless steel: Forgings; graded by AISI designation
ATGB	0.4% C 13.0% Cr 13.0% Ni 10.0% Co 2.0% Mo 2.5% W 3.0% Nb steel: Origin unknown
ATGB	0.4% C 13% Cr 13% Ni 2.0% Mo 2.5% W 3.0% Nb 10% Co steel: Imphy; annealed **UTS: 760 Elon: 15% Proof: 500**
ATV 57	0.1% C 18.5% Cr 30.0% Ni 20.0% Co 2.0% Ti 1.0% Al Fe alloy: Origin unknown
ATV S7	0.1% C 19% Cr 30% Ni 2.0% Ti 1.0% Al 20% Co Fe alloy: Imphy **DPN: 270 UTS: 890 Elon: 15% Proof: 480**
CA 6NM	0.06% C (max) 12.2% Cr 4.0% Ni 0.7% Mo steel: Casting; information from Department of Mines, Canada **UTS: 770 Elon: 15% Proof: 560**
CG 27	0.05% C 13% Cr 38% Ni 5.7% Mo 1.6% Al 2.5% Ti 0.7% Nb steel: Crucible Steel Co; solution treated & aged **UTS: 1420 Elon: 20% Proof: 1000**
CRM 15 D	1.0% C 5.0% Mn 20.0% Cr 5.0% Ni 2.0% Mo 2.0% W 2.0% Nb 0.2% N Fe alloy: Information from SAE handbook
CRUCIBLE 901	0.05% C 12.5% Cr 42% Ni 5.7% Mo 3% Ti 0.2% Al steel: Crucible Steel Co; solution treated & aged **UTS: 1210 Elon: 15% Proof: 920**
CRUCIBLE A286	0.05% C 15% Cr 25% Ni 1.25% Mo 2.2% Ti 0.3% V 0.2% Al steel: Crucible Steel Co; solution treated & aged **UTS: 1130 Elon: 25% Proof: 800**
CRUCIBLE CG27	0.05% C 13% Cr 38% Ni 5.7% Mo 1.6% Al 2.5% Ti 0.7% Nb steel: Crucible Steel Co; solution treated & aged **UTS: 1420 Elon: 20% Proof: 1000**
CUSTOM 455	0.03% C 11.75% Cr 9.0% Ni 1.2% Ti 2.2% Cu 0.3% Nb + Ta steel: Carpenter; age hardening
D 51	0.24% C 6.0% Ni 12% Cr 0.6% Mo 1.0% V steel: ESC; solution treated & aged **DPN: 550 UTS: 1500 Elon: 17%**

Symbol	Nominal analysis, supplier, condition and remarks.
D 70	0.03% C 4.2% Ni 12% Cr 4.2% Mo 14.5% Co 0.6% Ti steel: ESC; solution treated & aged **UTS: 1600**
DISCOLOY 24	0.04% C 13% Cr 26% Ni 2.7% Mo 1.7% Ti 0.1% Al steel: Westinghouse
DTD 5086	17% Cr 7% Ni steel with additions: Age hardening for rod, wire & springs
DTD 5299	Ni/Cr with Cu & Mo steel: Casting; investment cast; age hardenable **UTS: 950**
DTD 5309	Ni/Cr with Cu & Mo steel: Casting; investment cast; age hardenable **UTS: 1250**
EME	0.1% C 19% Cr 12% Ni 3.2% W 1.2% Nb 0.15% N steel: American proprietary alloy listed in SAE year book
ESSHETE 1250	0.15% C 15% Cr 10% Ni 1% Mo 0.4% V 1.0% Nb steel: Samuel Fox; solution treated; weldable **UTS: 450 Elon: 67% Proof: 210**
EV 4	0.2% C 21% Cr 11.5% Ni 0.2% N steel: For exhaust valves; designation used by SAE
EV 5	0.38% C 19% Cr 8% Ni steel: For exhaust valves; designation used by SAE
EV 6	0.38% C 19% Cr 8% Ni 0.2% N steel: For exhaust valves; designation used by SAE
EV 7	0.2% C 21% Cr 4.5% Ni 0.3% N steel: For exhaust valves; designation used by SAE
EV 8	0.53% C 9% Mn 21% Cr 3.7% Ni 0.42% N steel: For exhaust valves; designation used by SAE
EV 9	0.45% C 14% Cr 14% Ni 2.4% W 0.3% Mo steel: For exhaust valves; designation used by SAE
EV 11	0.7% C 6% Mn 21% Cr 2% Ni 0.2% N steel: For exhaust valves
FERRALIUM	25% Cr 5% Ni Mo Cu steel: Transformable; Langley Alloys; solution treated & aged **DPN: 320 UTS: 650 Elon: 20% Proof: 550**
FS A 286	0.05% C 15% Cr 26% Ni 1.25% Mo 2.1% Ti 0.2% Al steel: Firth Sterling; solution treated & aged **UTS: 1030 Elon: 25% Proof: 760**
FV 326	0.25% C 17% Cr 18.5% Ni 2.5% Mo 7% Co 1.75% Nb steel: Firth Vickers; age hardened **UTS: 730 Elon: 44% Proof: 300**
FV 467	0.18% C 14% Cr 10% Ni 2% Mo 2.5% Cu 0.8% Ti steel: Firth Vickers; age hardened **UTS: 650 Elon: 52% Proof: 250**
FV 520(B)	0.07% C 14% Cr 5.4% Ni 1.8% Mo 1.8% Cu 0.7% Nb steel: Forging; Firth Vickers; aged **DPN: 430 UTS: 1210 Elon: 12% Proof: 1020**
FV 520(S)	0.06% C 15.8% Cr 5.4% Ni 1.8% Mo 1.9% Cu 0.2% Ti steel: Sheet; Firth Vickers; aged **DPN: 410 UTS: 1250 Elon: 10% Proof: 1070**
FV 548	0.08% C 16.5% Cr 11.5% Ni 1.5% Mo 1.0% Nb stainless steel: Firth Vickers; age hardened **UTS: 630 Elon: 52% Proof: 250**
G 18B	0.4% C 13% Cr 13% Ni 1.7% Mo 2.5% W 10% Co 3% Nb steel: Jessop; solution treated & aged **DPN: 216 UTS: 710 Elon: 35% Proof: 270**
G 19	0.4% C 19% Cr 13% Ni 1.8% Mo 2.6% W 3.1% Nb 10% Co steel: Casting; Jessop; as cast; creep resistant **UTS: 580 Elon: 4% Proof: 300**
C 38	0.17% C 16% Cr 12% Ni 1.2% V 2.5% Mo 1.3% W steel: Jessop; solution treated & warm worked **UTS: 820 Elon: 34% Proof: 710**
G 40	0.23% C 20% Cr 25% Ni 3% Mo 1% Ti 1% W steel: Jessop **UTS: 840 Elon: 18% Proof: 630**

Symbol	Nominal analysis, supplier, condition and remarks.
G 41	0.17% C 16% Cr 12% Ni 1.3% W 1.3% Nb 1.3% V steel: Jessop; annealed & warm worked **UTS: 730 Elon: 26% Proof: 530**
G 68	0.08% C 15% Cr 26% Ni 1.25% Mo 2% Ti steel: Jessop; solution treated & aged **UTS: 1000 Elon: 25%**
GOST 5632 KH12H22T3MP	0.1% C (max) 15% Cr 23.5% Ni 1.3% Mo 3.0% Ti 0.8% Al (max) 0.02% B (max) steel: Plate; Russian Standard
GOST 5632 KH12N20T3R	0.1% C (max) 11.5% Cr 19.5% Ni 0.8% Al 3.0% Ti 0.012% B steel: Plate; Russian Standard
GOST 5632 KH25N16G7AR	0.12% C (max) 24.5% Cr 16.5% Ni 0.4% N 0.02% B (max) steel: Plate; Russian Standard
GOST 5632/ 0KH14N28V3T31UR	0.08% C (max) 14.0% Cr 27.5% Ni 2.8% Ti 3.1% W 0.9% Al 0.02% B (max) steel: Plate; Russian Standard
GOST 5632/ 3KH19N9MVBT	0.32% C 19.0% Cr 9.0% Ni 1.25% Mo 0.25% Nb 1.25% W 0.35% Ti steel: Plate; Russian Standard
H 1	0.35% C 28% Cr 16% Ni 0.5% Mo steel: Casting; Huntington; age hardened **UTS: 410 Elon: 6% Proof: 300**
J 1300	0.08% C 14% Cr 33% Ni 4% Mo 6.5% W 2% Ti 0.2% Al 0.2% Fe alloy: American proprietary alloy listed in SAE year book
JIS G4312 SUH 661	0.12% C 21.5% Cr 20% Ni 3.0% Mo 2.7% W 20% Co 1.0% Nb + Ta steel: Plate; Japanese Standard; solution treated **UTS: 700 Proof: 320**
K 66220	UNS designation for ASTM A638/662
K 66286	UNS designation for ASTM A638/660
KH 12N20T3R	0.1% C (max) 11.2% Cr 19.0% Ni 2.9% Ti 0.8% Al (max) 0.01% B steel: Russian Standard
KH 12N22T3MR	0.1% C (max) 11.2% Cr 23.5% Ni 1.3% Mo 2.9% Ti 0.8% Al (max) 0.2% B (max) steel: Russian Standard
LANGALLOY 20V	0.05% C 20% Cr 29% Ni 3.5% Cu 3% Mo 0.5% Nb Fe alloy: Langley **DPN: 140 UTS: 420 Elon: 25% Proof: 210**
LANGALLOY 40V	25% Cr 5% Ni transformable steel: Langley; solution treated & aged **DPN: 320 UTS: 1020 Elon: 15% Proof: 760**
LANGALLOY 40V	25% Cr 5% Ni transformable steel: Langley; solution treated **DPN: 260 UTS: 730 Elon: 20% Proof: 630**
LESCALLOY 901 VAC-ARC	0.1% C 12% Cr 42% Ni 1.0% Co 6.0% Mo 3.0% Ti 0.3% Al 0.015% B steel: Latrobe Steel; solution treated & aged **UTS: 1300 Elon: 18% Proof: 920**
LESCALLOY V57	0.05% C 14.7% Cr 0.5% V 27% Ni 1.25% Mo 3% Ti 0.25% Al steel: Latrobe; age hardening
LESCALLOY V57 VAC-ARC	0.08% C 15% Cr 26% Ni 1.2% Mo 3.0% Ti 0.5% V 2% Al 0.01% B steel: Latrobe Steel; solution treated & aged **UTS: 880 Elon: 12% Proof: 700**
LT 21/42	0.5% C 9.0% Mn 4.5% Ni 21% Cr 2.0% Nb 0.4% N steel: Low Moor
LV 21/69	0.08% C 9% Mn 6.0% Ni 21% Cr 0.4% N steel: Sheet & bar; Low Moor

Symbol	Nominal analysis, supplier, condition and remarks.
M 30B	0.08% C 14% Cr 33% Ni 4% Mo 6.5% W 2% Ti 0.2% Al 0.2% Fe alloy: American proprietary alloy listed in SAE year book
M 813	0.08% C 18% Cr 35% Ni 4% Mo 2.2% Ti 1.4% Al Fe alloy: American proprietary alloy listed in SAE year book
NF A35 572/ Z6CNUD15/04	0.07% C (max) 14.5% Cr 4.0% Ni 1.3% Mo steel: French Standard
NF A35 572/ Z6CNU17/04	16.5% Cr 4.0% Ni steel: French Standard
NICKELVAC A286	0.06% C 14.7% Cr 1.25% Mo 25% Ni 2.1% Ti 0.2% Al 1.5% Mn Fe alloy: Vanadium Alloys; double vacuum cast
PH 15-7 Mo	0.07% C 15% Cr 7% Ni 2.2% Mo 1.1% Al steel: Age hardening; American proprietary alloy listed in SAE year book
R 30155	UNS designation for ASTM A639/661
REX 559	0.05% C 15.5% Cr 25.5% Ni 1.3% Mo 2.1% Ti 0.3% V 0.2% Al 0.006% B steel: Firth Vickers; solution treated **UTS: 1090 Elon: 33% Proof: 800**
S 13800	UNS designation for type XM 13 steel
S 15500	UNS designation for type XM 12 steel
S 15700	UNS designation for 632 type steel
S 17400	UNS designation for 630 type steel
S 17700	UNS designation for 631 type steel
S 21600	UNS designation for XM 17 type steel
S 21603	UNS designation for XM 18 type steel
S 24000	UNS designation for XM 29 type steel
S 35500	UNS designation for 634 type steel
S 36200	UNS designation for XM 9 type steel
S 45000	UNS designation for XM 25 type steel
S 45500	UNS designation for XM 16 type steel
SAE EV 4	0.2% C 21% Cr 11.5% Ni 0.2% N steel: For exhaust valves; solution treated & aged **UTS: 840 Elon: 26% Proof: 450**
SAE EV 5	0.38% C 19% Cr 8% Ni steel: For exhaust valves; solution treated & aged **UTS: 1140 Elon: 34% Proof: 450**
SAE EV 6	0.38% C 19% Cr 8% Ni 0.2% N steel: For exhaust valves; solution treated & aged **UTS: 1140 Elon: 34% Proof: 450**
SAE EV 7	0.2% C 21% Cr 4.5% Ni 0.3% N steel: For exhaust valves; solution treated & aged
SAE EV 8	0.53% C 9% Mn 21% Cr 3.7% Ni 0.42% N steel: For exhaust valves; solution treated & aged **UTS: 1210 Elon: 9% Proof: 760**
SAE EV 9	0.45% C 14% Cr 14% Ni 2.4% W 0.3% Mo steel: For exhaust valves; solution treated & aged
SAE EV 11	0.7% C 6% Mn 21% Cr 2% Ni 0.2% N steel: For exhaust valves; solution treated & aged **UTS: 1110 Elon: 20% Proof: 530**
SIS 2570	0.08% C 14% Cr 25% Ni 1.2% Mo 0.4% V 2.1% Ti steel: For springs; Swedish Standard; solution treated cold worked & aged **DPN: 350 UTS: 1070 Elon: 8% Proof: 1000**
STAINLESS W	0.12% C 17% Cr 7% Ni 1% Ti 1% Al 0.2% N steel: Age hardening; American proprietary alloy listed in SAE year book
UNITEMP 212	0.8% C 16% Cr 20% Ni 0.5% Cb 4% Ti 0.05% Al 0.07% B 0.05% Zr steel: American proprietary alloy listed in SAE year book
UNS S13800	0.05% C 12.75% Cr 8.0% Ni 1.1% Al 2.2% Mo steel: American Standard XM 13 type steel
UNS S15500	0.07% C 14.7% Cr 4.5% Ni 3.5% Cu steel: American Standard for XM 12 type steel
UNS S15700	0.09% C 15.0% Cr 7.2% Ni 1.2% Al 2.5% Mo steel: American Standard for 632 type steel
UNS S17400	0.07% C 16.5% Cr 4.0% Ni steel: American Standard for 630 type steel

Note. The following abbreviations and units are used in the tables:

DPN	Hardness, diamond pyramid number
UTS	Ultimate tensile strength, N/mm^2
Elon	Elongation, %
Proof	0.1 % proof strength, N/mm^2

1 N/mm^2=0.1 hbar=0.102 kgf/mm^2=0.06475 tonf/in^2=145.04 lbf/in^2

Symbol	Nominal analysis, supplier, condition and remarks.
UNS S17700	0.09% C 17.0% Cr 7.2% Ni 1.2% Al steel: American Standard for 631 type steel
UNS S35500	0.12% C 15.5% Cr 4.5% Ni 3.0% Mo steel: American Standard for 634 type steel
UNS S36200	0.05% C 14.2% Cr 6.8% Ni 0.1% Al 0.3% Mo 0.8% Ti steel: American Standard for XM 9 type steel
UNS S45000	0.05% C 15.0% Cr 6.0% Ni 0.7% Mo 1.5% Cu steel: American Standard for XM 25 type steel
UNS S45500	0.05% C 11.7% Cr 8.5% Ni 0.5% Mo 1.1% Ti 2.0% Cu steel: American Standard for XM 16 type steel
V 57	0.08% C (max) 14.7% Cr 27.5% Ni 1.2% Mo 3% Ti 0.1% Al steel: Carpenter
VK	0.07% C 17% Cr 7% Ni 1% Al steel: Osborn; age hardened **DPN: 380 UTS: 1210 Elon: 15% Proof: 760**
W 545	0.08% C 13.5% Cr 26% Ni 1.5% Mo 2.8% Ti 0.2% Al 0.08% B steel: American proprietary alloy listed in SAE year book
XM 9	0.05% C (max) 14.2% Cr 6.7% Ni 0.1% Al 0.3% Mo 0.7% Ti steel: Precipitation hardened **UTS: 1240 Elon: 10% Proof: 1205**
XM 12	0.07% C (max) 14.2% Cr 4.5% Ni 3.5% Cu steel: Precipitation hardened **UTS: 1070 Elon: 12% Proof: 1000**

Symbol	Nominal analysis, supplier, condition and remarks.
XM 13	0.05% C (max) 12.7% Cr 8.0% Ni 1.1% Al 2.2% Mo steel: Precipitation hardened **UTS: 1415 Elon: 10% Proof: 1310**
XM 16	0.05% C (max) 11.7% Cr 8.5% Ni 0.5% Mo 1.1% Ti 2.0% Cu steel: Precipitation hardened **UTS: 1380 Elon: 10% Proof: 1275**
XM 17	0.08% C 20.0% Cr 6.0% Ni 8.0% Mn 2.5% Mo 0.35% N steel: Designation used by ASTM
XM 18	0.03% C 20.0% Cr 6.0% Ni 2.5% Mo 0.35% N steel: Designation used by ASTM
XM 25	0.05% C (max) 15.0% Cr 6.0% Ni 0.7% Mo 1.5% Cu steel: Precipitation hardened **UTS: 1100 Elon: 12% Proof: 1035**
XM 29	0.08% C 18.0% Cr 3.0% Ni 12.5% Mn 0.3% N steel: Designation used by ASTM

Note. The following abbreviations and units are used in the tables:

DPN	Hardness, diamond pyramid number
UTS	Ultimate tensile strength, N/mm^2
Elon	Elongation, %
Proof	0.1 % proof strength, N/mm^2

1 N/mm^2=0.1 hbar=0.102 kgf/mm^2=0.06475 tonf/in^2=145.04 lbf/in^2

44P Steel – high manganese austenitic

Specific gravity	8.0	
Density	8000 kg/m^3	(0.28 lb/in.^{3})
Solidus/liquidus	– °C	
Thermal conductivity	– W/m °C	
Coefficient of linear expansion	– / °C	
Electrical conductivity	– IACS (copper 100%)	
Specific resistance	– microhm mm	
Young's modulus of elasticity	– N/m^2	
Impact	– J	
Fatigue strength	–	
Hot strength	–	

The above properties are typical of the following group, and may not apply exactly to any one specification. It is possible that with certain specifications some of the values may not be applicable.

General metallurgical characteristics

Manganese alone has the effect of depressing the critical points of steel and with more than 18% manganese, steel becomes austenitic.

These high manganese steels are discussed in Section 44D. The steels in the following list are all austenitic and contain above 4 per cent manganese with varying amounts of chromium and nickel to ensure a stable austenitic structure. Manganese has a considerable influence on the cold working properties of steel and this increases in proportion to the manganese content which may be as high as 20%.

These high manganese steels are relatively unworkable and are generally used as castings or sprayed coatings for abrasion resistant purposes.

Chromium improves the corrosion resistant properties, particularly the hot oxidation resistance and when nickel is also present the hot strength of the alloys.

There are certain specially formulated 4–6% manganese chromium nickel alloys which have a high coefficient of linear expansion which can be useful when steel is associated with aluminium.

Thermal treatment

These steels are all hardened by cold work, but cannot have their mechanical properties improved by any form of thermal treatment. Very little cold work considerably hardens the material and reduces the ductility, very often forming martensite. The steels can be annealed by heating to 1000–1040 °C for 20 minutes per 25 mm of section and cooling by any convenient method. This removes all the work hardening and is not often required in practice.

Controlled annealing at temperatures in the order of 500–850 °C can be used to remove excessive work hardening without eliminating that required to give the desired mechanical properties. It is recommended that trials are carried out on sample batches as it is impossible to re-harden the parts except by further cold work if too high a temperature is used.

Welding and brazing

Pre-treatment. The parts should be annealed for welding, and the joint area thoroughly cleaned.

Welding and brazing. Welding will very seldom be required as it locally removes the work hardening. When necessary it should be by the inert gas shielded electric arc method. Some brittleness in addition to softening in the weld zone can be expected from the carbide precipitation. Welding in stressed areas should therefore always be avoided. Some of the alloys listed are in the form of rods used for hard facing. These are applied with a reducing oxyacetylene flame.

Brazing can be accomplished using fluxes and copper alloy fillers, provided some local softening is permissible. This can be minimized by using rapid heating methods.

Post-treatment. No thermal treatment is necessary, but all fluxes must be removed.

Flaw detection methods

Crack test. Oil penetrant chalk test.

X-ray. May be required on high duty castings and should be used as a means of controlling casting techniques. X-ray will not normally be required on wrought products.

Ultra-sonic test. These are expensive materials which are exceptionally difficult to machine. It is therefore recommended that ultra sonic examination is carried out at an early stage of machining to obviate the possibility of finishing machining parts which would then be shown to be scrap.

Chemical etch. Not required.

Corrosion protection

Temporary. None required.

Permanent. None required. These are fully austenitic stainless steels with a corrosion resistance almost equal to the 18/8 type steels described in Section 44N1. If severely work hardened some of the steels may transform to martensite with a reduction in corrosion resistance.

Machinability

These steels are extremely difficult to machine owing to their work hardening properties which cause rapid increases in hardness, even with the use of sharp, correctly designed cutting tools. It is essential that the tools always cut without any suggestion of rubbing. High feed rates at slow surface speeds with adequate power and well maintained tools give best results. Constant feed with no interrupted cutting, or 'nibbling' action is essential.

It may be necessary to carry out one or more controlled anneals to remove local excess hardening, which under certain conditions can prevent all further machining.

Grinding is not often successful as the stones become glazed, requiring very frequent dressing. These materials lend themselves to the new metal removal processes, such as spark erosion, and chemical machining where no work hardening occurs. Special equipment however, is essential.

Uses

Studs. Inserts and similar items to be used with aluminium alloys at high temperatures where normally the high expension of aluminium would cause loosening. If possible alternative methods of design should be sought to obviate the use of these expensive intractable steels.

Other uses include wear resistant items in corrosive atmospheres.

Symbol	Nominal analysis, supplier, condition and remarks.
1.3817	0.4% C 18.0% Mn 4.0% Cr steel: German Standard
1.3960	0.45% C 13.0% Mn 5.5% Cr 6.5% Ni 0.9% V steel: German Standard
1.3962	0.12% C 6.0% Mn 11.5% Cr 10.0% Ni steel: German Standard
1.4370	0.15% C (max) 6.5% Mn 18.5% Cr 8.5% Ni steel: German Standard
1.4371	0.1% C (max) 8.5% Mn 18.0% Cr 5.5% Ni 0.15% N steel: German Standard

Note. The following abbreviations and units are used in the tables:

DPN	Hardness, diamond pyramid number
UTS	Ultimate tensile strength, N/mm²
Elon	Elongation, %
Proof	0.1 % proof strength, N/mm²

$1 \text{ N/mm}^2 = 0.1 \text{ hbar} = 0.102 \text{ kgf/mm}^2 = 0.06475 \text{ tonf/in}^2 = 145.04 \text{ lbf/in}^2$

Symbol	Nominal analysis, supplier, condition and remarks.
1.4451	0.15% C (max) 18.0% Mn 12.0% Cr 2.0% Ni 0.5% Mo steel: German Standard
1.5151	0.4% C 18.0% Mn 4.0% Cr 0.1% Ni steel: German Standard
1.5152	0.4% C 22.5% Mn 4.0% Cr steel: German Standard
1.5161	0.15% C (max) 18.0% Mn 12.0% Cr 2.0% Ni 0.5% Mo steel: German Standard
10.17.5.9 Mn	0.15% C 10.0% Cr 17.0% Ni 5.0% Mo 9.0% Mn steel: Electrode; Metrode
14 Cr 14 Mn R	0.35% C 14.0% Mn 14.0% Cr 0.7% Mo 2.0% Ni 0.3% V steel: Electrode; Metrode; for hard facing; work hardens from 250 to 450 DPN
17.12.1.4 Mn Nb	0.08% C 17.0% Cr 12.0% Ni 1.5% Mo 4.0% Mn 1.0% Nb steel: Electrode; Metrode
18/5/9	0.08% C 18% Mn 5.0% Cr 9.0% Ni steel: Nyby, annealed DPN: 230 UTS: 580 Elon: 45% Proof: 270
18.6.9 Mn Si Nb	0.06% C 18.0% Cr 6.0% Ni 2.0% Si 1.0% Nb 9.0% Mn steel: Electrode: Metrode

Symbol	Nominal analysis, supplier, condition and remarks.
18.8.3 Mn Si Nb N	0.06% C 18.0% Cr 8.0% Ni 3.0% Mo 2.0% Si 0.5% Nb 7.0% Mn 0.2% N steel: Electrode; Metrode
21/4N	0.55% C 9% Mn 21% Cr 4% Ni 0.4% N steel: Jessop
65 Mn 12 Cr 5 Ni 3	0.65% C 12.0% Mn 4.5% Cr 3.0% Ni steel: Electrode; Metrode; work hardening deposit
664 M	0.06% C 6.5% Mn 17% Cr 5.5% Ni 0.1% N steel: Avesta
664 MV	0.05% C 8.0% Mn 18% Cr 5.5% Ni 0.2% N steel: Avesta
1031 ALLOY	0.3% C 6% Mn 13% Ni 2% Cu Fe alloy: Darwin
AF 71	0.3% C 18% Mn 12.5% Cr 3% Mo 0.2% B 0.2% N 8% V steel: American proprietary alloy listed in SAE year book
AF 183	0.3% C 18% Mn 12.5% Cr 3% Mo 0.8% V 0.2% N steel: American proprietary alloy listed in SAE year book
AISI 201	0.15% C 6% Mn 17% Cr 4.5% Ni 0.25% N steel
AISI 202	0.15% C (max) 8.7% Mn 18.0% Cr 5.0% Ni steel
AKMC/Z	0.4% C 18% Mn 4.0% Cr 0.1% N steel: Krupps
AM 88	0.2% C 0.3% Si 8.0% Mn 8.0% Cr 5.7% Ni steel: Pompey
AMS 5623	0.6% C 5.5% Mn 9.5% Ni steel: Bar & forging; high expansion steel; annealed
AMS 5624 A	0.55% C 4.5% Mn 4% Cr 12.5% Ni steel: Bar; high expansion
AMS 5625 A	0.6% C 5.5% Mn 9.5% Ni steel: Bar; cold drawn; high expansion
ASTM	A666 stainless steel; Sheet, plate, bar etc.; for structural purposes; graded by AISI system (see also Section 44N1)
ASTM A289 A	0.55% C 8% Mn 5% Cr 8% Ni steel: For non-magnetic retaining rings **UTS: 1070 Elon: 20% Proof: 920**
ASTM A289 B	0.5% C 18% Mn 5% Cr 2% Ni steel: For non-magnetic retaining rings **UTS: 1070 Elon: 20% Proof: 920**
ASTM A412/201	0.15% C (max) 6.5% Mn 27% Cr 4.5% Ni steel: Sheet, etc.
ASTM A412/202	0.15% C (max) 8.7% Mn 18% Cr 5% Ni steel: Sheet, etc.
ASTM A429/201	0.15% C (max) 6.5% Mn 17% Cr 4.5% Ni steel: Bar
ASTM A429/202	0.15% C (max) 8.7% Mn 18% Cr 5% Ni steel: Bar
ASTM A581 XM1	0.08% C 5.7% Mn 17% Cr 5.7% Ni 0.2% S steel: Wire; free machining
ASTM A688 TPXM29	0.06% C 13.0% Mn 18% Cr 3.0% Ni 0.3% N steel: Tube; welded **UTS: 690 Elon: 35% Proof: 380**
ASTRANIT M/Z	0.12% C 18% Mn 12% Cr 2.0% Ni 0.5% Mo steel: Krupps
AURIGA XXXI	0.27% C 8% Mn 8% Cr 8% Ni steel: Casting; D Brown; annealed **UTS: 540 Elon: 24% Proof: 240**
BS 2926 H	0.6% C 6% Mn 26% Cr 19% Ni 1% Mo 2% Nb steel: Welding electrode; as welded **UTS: 540 Elon: 25%**

Note. The following abbreviations and units are used in the tables:

DPN	Hardness, diamond pyramid number
UTS	Ultimate tensile strength, N/mm^2
Elon	Elongation, %
Proof	0.1 % proof strength, N/mm^2

1 N/mm^2=0.1 hbar=0.102 kgf/mm^2=0.06475 tonf/in^2=145.04 lbf/in^2

Symbol	Nominal analysis, supplier, condition and remarks.
CRICIBLE 201	0.15% C 6% Mn 17% Cr 4.5% Ni 0.25% N steel: Crucible Steel Co; annealed **UTS: 800 Elon: 55% Proof: 370**
CRUCIBLE 202	0.15% C 8.5% Mn 18% Cr 5% Ni 0.25% N steel: Crucible Steel Co; annealed **UTS: 760 Elon: 55% Proof: 370**
CSA	0.25% C 4% Mn 18% Cr 5% Ni 1.3% Mo 1.3% W 1% Nb steel: Crucible Steel Co
DTD 247	0.7% C (max) 4.5% Mn 3% Cr 13% Ni steel: Bar & forging; as rolled or softened; high expansion steel **UTS: 620 Elon: 25%**
G 192	0.6% C 8.5% Mn 22% Cr steel: American proprietary alloy listed in SAE year book
GOST 5632 ICHITE9AN4	12% C (max) 9% Mn 17% Cr 4% Ni 2% N steel: Russian specification **UTS: 700 Proof: 350**
GOST 5632/2 KH13N4G9	0.22% C 9.0% Mn 13.0% Cr 4.2% Ni steel: Russian Standard
GOST 5632 KH17G9AN4	0.12% C 9.2% Mn 17.0% Cr 4.0% Ni 0.2% N steel: Russian Standard
H 35	0.08% C 3.25% Mn 11.7% Cr 4.2% Ni 0.6% Nb steel: Jessop
HTX	0.04% C 8.5% Mn 21% Cr 8% Ni 1.5% Mo 0.2% N 0.23% P steel: American proprietary alloy listed in SAE year book
JIS G4303 SUS201	0.15% C (max) 6.5% Mn 17.0% Cr 4.5% Ni steel: Japanese Standard
JIS G4303 SUS202	0.15% C 8.2% Mn 18.0% Cr 5.0% Ni steel: Japanese Standard
JIS G4 SUS202	0.15% C (max) 8.5% Mn (max) 18% Cr 5% Ni 0.75% N steel: Plate; Japanese specification
LV 21/2 N	0.5% C 20% Cr 2.0% Ni 8.5% Mn steel: For valves; S Osborn
LV 21/4 N	0.5% C 21% Cr 4% Ni 9% Mn 0.4% N steel: Low Moor; for valves
LV 21/42	0.5% C 21.0% Cr 4.0% Ni 9.0% Mn 2.0% Nb steel: For valves; S Osborn
LV 21/43	0.5% C 21.0% Cr 4.0% Ni 9.0% Mn 1.5% Nb steel: For valves; S Oshorn
MANGANESE Ni	0.8% C 13% Mn 3.5% Ni steel: Welding electrode; Murex; for hard facing; work hardens **DPN: 300**
METCOLOY 5	0.15% C 8% Mn 18% Cr 5% Ni steel: Wire for spraying; Metco
NICOLLOY 4	0.52% C 9.0% Mn 21% Cr 4% Ni 0.4% N steel: Swift Levick for hot dies
R 32	0.55% C 4.5% Mn 3.7% Cr 12.5% Ni steel: Jessop
R 46	0.55% C 9% Mn 21% Cr 4% Ni 0.4% N steel: Jessop
RED DIAMOND 14H	1.1% C 12.5% Mn steel: For wear; Redheugh **DPN: 360 UTS: 520 Elon: 8%**
RED DIAMOND 14S	1.1% C 12.5% Mn steel: For wear; Redheugh **DPN: 280 UTS: 630 Elon: 12%**
RNK 29	0.09% C 8.5% Mn 18% Cr 5% Ni 0.2% N steel: Bofors
SAE 30201	0.15% C 6% Mn 17% Cr 4.5% Ni 0.25% N steel: Mechanical properties not quoted
SAE 30202	0.15% C 8% Mn 18% Cr 5% Ni 0.25% N (max) steel: Mechanical properties not quoted
SIS 2357	0.12% C 8% Mn 18% Cr 5% Ni 0.2% N steel: Bar, forging etc.; Swedish Standard; annealed **DPN: 220 UTS: 580 Elon: 45% Proof: 270**
UNS R05001–R05999	Ta metal and alloys: American Standards system
UNS R06001–R06999	Ti metal and alloys: American Standards system
UNS L13001–L13999	Tin based alloys: American Standards system
VALMAG 984	High Mn Cr Ni austenitic steel: For valves; Swift Levick; solution treated & aged **DPN: 350 UTS: 1110 Elon: 7%**

Symbol	Nominal analysis, supplier, condition and remarks.
WALMANG 3	0.7% C 14% Mn 3.5% Ni steel: Rod for facing; Wall Colmonoy; for repairing high Mn alloys; melting point 1430 °C **DPN: 630**
X 8 Cr Mn Ni 18/9	0.1% C (max) 8.5% Mn 18.0% Cr 5.5% Ni 0.15% N steel: Designation used by German Standard
X 12 Mn Cr 18/10	0.15% C (max) 18.0% Mn 12.0% Cr 2.0% Ni 0.6% Mo steel: Designation used by German Standard
X 12 Mn Cr 18/11	0.15% C (max) 18.0% Mn 12.0% Cr 2.0% Ni 0.5% Mo steel: Designation used by German Standard
X 15 Cr Ni Mn 18/8	0.15% C (max) 6.5% Mn 19.0% Cr 8.5% Ni steel: Designation used by German Standard
X 20 Cr Ni Mn 12/9	0.12% C 6.0% Mn 11.5% Cr 10.0% Ni steel: Designation used by German Standard
X 40 Mn Cr 18	0.4% C 18.0% Mn 4.0% Cr steel: Designation used by German Standard
X 40 Mn Cr 22	0.4% C 23.0% Mn 4.0% Cr steel: Designation used by German Standard
X 40 Mn Cr N 18	0.4% C 18.0% Mn 4.0% Cr 0.1% N steel: Designation used by German Standard

Symbol	Nominal analysis, supplier, condition and remarks.
X 55 Mn Ni Cr 14	0.45% C 13.0% Mn 5.5% Cr 6.5% Ni 0.9% V steel: Designation used by German Standard
Z 10 CMN 18/7	0.08% C 18% Cr 5.0% Ni 9.0% Mn steel: French National Standard; annealed **DPN: 230 UTS: 580 Elon: 45% Proof: 270**
Z 10 CMN 19/9	0.08% C 18% Cr 5.0% Ni 9.0% Mn steel: French National Standard; annealed **DPN: 230 UTS: 580 Elon: 45% Proof: 270**
Z 12 CMN 18/7	0.08% C 18% Cr 5.0% Ni 9.0% Mn steel: French National Standard; annealed **DPN: 230 UTS: 580 Elon: 45% Proof: 270**

Note. The following abbreviations and units are used in the tables:

DPN Hardness, diamond pyramid number
UTS Ultimate tensile strength, N/mm^2
Elon Elongation, %
Proof 0.1 % proof strength, N/mm^2

1 N/mm^2=0.1 hbar=0.102 kgf/mm^2=0.06475 $tonf/in^2$=145.04 lbf/in^2

45. Strontium Sr

Physical properties

Atomic number	38	
Atomic weight	87.63	
Crystal structure	Face-centred cubic	
Colour	Yellowish white	
Specific gravity	2.6	
Density	2600 kg/m³	(0.094 lb/in.³)
Melting point	770 °C	
Boiling point	1366 °C	
Specific heat	0.3077 J/g °C	(0.0735 cal/g °C)
Thermal conductivity	– W/m °C	
Coefficient of linear expansion (20–100° C)	– / °C	
Latent heat of fusion	105 J/g	(25 cal/g)
Latent heat of vaporization	1717 J/g	(410 cal/g)
Thermal neutron absorption cross-section	1.16 barns/atom	
Electrical conductivity	7.5% IACS (copper 100%)	
Specific resistance	250 microhm mm	
Temperature coefficient of electrical resistance	0.0038/ °C	
Electrochemical equivalent	– g/A/h	
Electrode potential	-2.89 V	
Magnetic susceptibility	-0.2 × 10⁻⁶	
Young's modulus of elasticity	– N/m²	
Tensile strength	– N/mm²	
Hardness	– DPN	

45.1 General notes on strontium

Strontium is always found as a compound, the sulphate or the carbonate being the most common generally associated with lead ores. It was first isolated in 1790 in the lead mines then worked near the village of Strontian in the North-West of Scotland.

The metal is yellowish white when freshly cut, but tarnishes rapidly and decomposes water, releasing hydrogen. Strontium resembles calcium and as it has no unique properties justifying extraction on a commercial scale, there are therefore no alloys based on the metal.

It exhibits photo-electrical properties, but these cannot compete economically with selenium. A limited amount of strontium has been used to de-oxidize copper and some of its alloys, but the high cost of the metal makes this an uneconomical method. Lead is hardened when traces of strontium are added along with small amounts of tin.

Strontium compounds are used in small quantities in pharmacy and the electronic industry. They have been added to open hearth furnace brick work when it is claimed lower sulphur steels are obtained. The strong red colour given off when compounds are ignited is used in fireworks and distress or identification flares. Strontium sulphate is a stable powder which can be ground to a very fine flour and finds some use as a filler material for plastics and rubber.

Strontium is an alkaline earth metal which has no real value to mankind, but has a radioactive isotope, which is a decay product of materials produced by nuclear fission and is in the fallout. As this particular isotope takes a long time to decompose (has a long half-life) and is absorbed into the bone and stores there, it constitutes a danger to health.

Symbol	Nominal analysis, supplier, condition and remarks.
STRONTIUM	Pure metal: Blackwells

Note. The following abbreviations and units are used in the tables:

DPN	Hardness, diamond pyramid number
UTS	Ultimate tensile strength, N/mm²
Elon	Elongation, %
Proof	0.1 % proof strength, N/mm²

1 N/mm²=0.1 hbar=0.102 kgf/mm²=0.06475 tonf/in²=145.04 lbf/in²

46. Tantalum Ta

Physical properties

Atomic number	73	
Atomic weight	181.0	
Crystal structure	Body-centred cubic	
Colour	Iron grey	
Specific gravity	16.65	
Density	16650 kg/m^3	(0.6 lb/in.^{3})
Melting point	2950 °C	
Boiling point	5300 °C	
Specific heat	0.153 J/g °C	(0.0365 cal/g °C)
Thermal conductivity	54.4 W/m °C	(13 cal/m s °C)
Coefficient of linear expansion (20–100° C)	6.5 x 10^{-6}/ °C	
Latent heat of fusion	159 J/g	(38 cal/g)
Latent heat of vaporization	– J/g	
Thermal neutron absorption cross-section	21.3 barns/atom	
Electrical conductivity	13.9% IACS (copper 100%)	
Specific resistance	125 microhm mm	
Temperature coefficient of electrical resistance	0.0031 °C	
Electrochemical equivalent	– g/A/h	
Electrode potential	4.1 V	
Magnetic susceptibility	0.90 × 10^{-6}	

Young's modulus of elasticity	cold worked	186 × 10^9 N/m^2	(27 × 10^6 lbf/in.^{2})
Tensile strength	annealed	450 N/mm^2	(30 tonf/in.^{2})
	cold worked	900 N/mm^2	(60 tonf/in.^{2})
Hardness	annealed	100 DPN	
	cold worked	200 DPN	

46.1 General notes on tantalum

This occurs as an oxide in combination with iron, always associated with the niobium–iron oxide.

Metal tantalum of 99.9% purity is now commercially available but it is still an expensive material and will remain so as long as the present scarcity of workable ores exists.

The basic product is a powder which must be sintered and forged under vacuum. With modern techniques of vacuum casting and electron beam melting, a product capable of further hot reduction is available which may prove more economical than vacuum sintering.

Tantalum is a white colour as polished, but in air immediately forms an adherent oxide which gives the metal an iron grey coloration.

It is capable of accepting considerable cold work without hardening to any great extent, thus does not require repeated anneals. As tantalum combines with most gases in contact with it from 500 °C upwards, all heating must be in vacuum. None of the inert gases combine with the metal but it is always necessary to produce a high vacuum to remove atmospheric gases which would contaminate the inert gases. Tantalum can be readily welded provided adequate flow of an inert gas (generally argon or helium) is available on both sides of the weld, as well as fore and aft the weld pool. Best results are obtained by vacuum welding or electron beam welding if this is possible. Tantalum castings can be produced, provided melting, casting and cooling take place under vacuum.

Tantalum has excellent resistance to chemical attack from all acids except hydrofluoric and hot concentrated sulphuric acids. It has good electrical properties and maintains these at a high temperature. Oxidation resistance is very good up to 450 °C when it falls off rapidly, but above 500 °C tantalum must be used under vacuum or one of the inert gases.

This chemical resistance together with an excellent heat transfer and its workability makes tantalum a favourite metal in the chemical industry, there being no doubt that only the high cost of the metal restricts its use at present. One method of reducing the initial capital outlay is to use a thin coating of tantalum on a strong cheaper base metal. Austenitic stainless steel is often used and electroplating techniques are being developed for this purpose.

Tantalum has various uses in the electronic industry, partly for its high strength at elevated temperatures where it is used as anode and grid supports in valves. The ability to combine with atmospheric gases makes it a popular, if expensive, 'getter' to remove the last traces of these gases from the vacuum glass envelope.

Tantalum also has peculiar properties connected with its oxide film which make it a useful material in the manufacture of capacitors and rectifiers.

Vacuum furnace equipment sometimes uses tantalum for any shields or supports which require considerable shaping and welding during manufaxture and high strength at the elevated working temperatures.

The good corrosion characteristics and proven compatibility to body fluids allows the use of tantalum implants in surgery.

Tantalum is similar in many respects to niobium, and like this metal has its use restricted by the high price and scarcity. Alloyed with stainless steels and nickel alloys it

prevents carbide precipitation (weld decay) but is generally second choice to niobium. A series of sintered carbide tool materials includes tantalum, tungsten, and titanium carbides, generally bound in a nickel or cobalt matrix. Tantalum has been used as an alloying element in high speed steels, but has little to offer over tungsten and is more expensive.

Some tantalum compounds are used as catalysts in chemical engineering, particularly in the manufacture of certain synthetic rubbers.

Symbol	Nominal analysis, supplier, condition and remarks.
AMS 7848	10% W Ta alloy: Bar
AMS 7849	Ta: Sheet, strip or foil
ASTM B364	High purity Ta: Ingot & sheet; cold worked UTS: 530 Elon: 2%
ASTM B364	High purity Ta: Ingot & sheet; stress relieved UTS: 370 Elon: 8%
ASTM B364	High purity Ta: Ingot & sheet; annealed UTS: 220 Elon: 27%
ASTM B365	High purity Ta: Rod & wire; cold worked UTS: 500 Elon: 1%
ASTM B365	High purity Ta: Rod & wire; annealed UTS: 270 Elon: 10%
h Ta 16	99.98% Ta: Ingot; Light Ltd; high purity metal
h Ta 73	99.95% Ta: Single crystal 3 mm diameter; Light Ltd
T 222	0.01% C 9.6% W 2.2% Ha Ta alloy: Westinghouse Electrical Corporation

Symbol	Nominal analysis, supplier, condition and remarks.
TANTALUM	99.6% Ta: Powder; Kennametal
TANTALUM	Ta metal: Blackwells; pure metal
TANTALUM	Ta: Ingot, rod, sheet & wire; Murex
TANTALUM	Ta: Rod, sheet & wire; Tungsten Manufacturing Co

Note. The following abbreviations and units are used in the tables:

DPN	Hardness, diamond pyramid number
UTS	Ultimate tensile strength, N/mm^2
Elon	Elongation, %
Proof	0.1 % proof strength, N/mm^2

$1 \ N/mm^2 = 0.1 \ hbar = 0.102 \ kgf/mm^2 = 0.06475 \ tonf/in^2 = 145.04 \ lbf/in^2$

47. Tellurium Te

Physical properties

Atomic number	52	
Atomic weight	127.61	
Crystal structure	–	
Colour	Shining white	
Specific gravity	6.24	
Density	6240 kg/m³	(0.225 lb/in.³)
Melting point	452 °C	
Boiling point	1007 °C	
Specific heat	0.201 J/g °C	(0.048 cal/g °C)
Thermal conductivity	1.3 W/m °C	(0.3 cal/m s °C)
Coefficient of linear expansion (20–100° C)	16.8×10^{-6}/ °C	
Latent heat of fusion	30.6 J/g	(7.3 cal/g)
Latent heat of vaporization	666 J/g	(159 cal/g)
Thermal neutron absorption cross-section	4.7 barns/atom	
Electrical conductivity	1.0% IACS (copper 100%)	
Specific resistance	1500 microhm mm	
Temperature coefficient of electrical resistance	–	
Electrochemical equivalent	0.99 g/A/h	
Electrode potential	– V	
Magnetic susceptibility	-0.31×10^{-6}	
Young's modulus of elasticity	– N/m²	
Tensile strength	– N/mm²	
Hardness	– DPN	

47.1 General notes on tellurium

Natural tellurium occurs in small amounts and is fairly widespread in complex gold, silver, lead, antimony, bismuth and copper compounds. Tellurium is a metalloid very similar to selenium and like all metalloids has non-metallic characteristics, including poor heat and electrical conductivity. It has few uses in the pure state and there is no alloy system based on tellurium. Some use has been made of the material in the vapour form in lamps when it is claimed artificial sunlight can be obtained. In the finely divided form tellurium is used as an additive to rubber to improve the oxidation resistance.

When electroplated on steel, tellurium resists the attack of most mild acids and alkalis and as the plating technique appears to be relatively simple there is considerable potential use in this field.

In small quantities of less than 1% it considerably improves the machinability of copper and can be used for the same purpose in steel where the effect is identical to that of selenium.

Tellurium improves the mechanical properties and corrosion resistance of lead and has been added to lead and tin base bearing metals for the same reasons. In cast irons minute quantities of about 0.05% (maximum) tellurium inhibit graphite formation, thus increasing the hardness. It is often used in mould washes to cause chill hardening of the cast surfaces, as the surface carbon is then present as high carbides instead of graphite.

Light has no effect on the electrical resistance of tellurium, thus unlike selenium it has no use in light meters. Some of its compounds however, particularly bismuth telluride, are finding use in the electronic industry as semiconductors.

Tellurium and most of its compounds are to a degree toxic, but to a lesser extent than selenium compounds. There is never sufficient tellurium present in metallic alloys in common use to present any health hazard whatever.

There are no alloy systems based on the metal.

Note. The following abbreviations and units are used in the tables:

DPN	Hardness, diamond pyramid number
UTS	Ultimate tensile strength, N/mm²
Elon	Elongation, %
Proof	0.1 % proof strength, N/mm²

1 N/mm²=0.1 hbar=0.102 kgf/mm²=0.06475 tonf/in²=145.04 lbf/in²

Symbol	Nominal analysis, supplier, condition and remarks
h Te 6	99.999% Te metal: Lump form; Light Ltd; high purity metal
h Te 73	99.995% Te single crystals, 6 mm diameter: Light Ltd
TELLURIUM	High purity metal: impurities 5 p.p.m.; Johnson Matthey; supplied as cast bar

48. Thallium Tl

Physical properties

Atomic number	81	
Atomic weight	204.39	
Crystal structure	Close-packed hexagonal to 262 °C	
	Body-centred cubic above 262 °C	
Colour	White	
Specific gravity	11.85	
Density	11850 kg/m^3	(0.46 lb/in.^{2})
Melting point	304 °C	
Boiling point	1450 °C	
Specific heat	0.136 J/g °C	(0.0326 cal/g °C)
Thermal conductivity	39.4 W/m °C	(9.4 cal/m s °C)
Coefficient of linear expansion (20–100° C)	28 × 10^{-6}/ °C	
Latent heat of fusion	20.9 J/g	(5.0 cal/g)
Latent heat of vaporization	1231 J/g	(194 cal/g)
Thermal neutron absorption cross-section	3.4 barns/atom	
Electrical conductivity	10% IACS (copper 100%)	
Specific resistance	150 microhm mm	
Temperature coefficient of electrical resistance	0.0051/ °C	
Electrochemical equivalent	2.5 g/A/h	
Electrode potential	-0.336 V	
Magnetic susceptibility	-0.24 × 10^{-6}	
Young's modulus of elasticity	– N/m^2	
Tensile strength	7.5 N/mm^2	(0.5 tonf/in.^{2})
Hardness	2 DPN	

48.1 General notes on thallium

Thallium has a high specific gravity and resembles lead in colour and hardness, being readily cut by a knife. It oxidizes if left in air and slowly decomposes water, thus must be stored under paraffin or alcohol.

The use of thallium and many of its compounds is severely restricted by its toxic nature, added to the ease with which it oxidizes.

Alloys with silver and lead up to about 20% thallium have been developed which appear to have excellent corrosion resistance under certain conditions. Silver with 10–20% thallium does not blacken in contact with sulphide compounds as do most other silver alloys.

The lead–thallium alloys have higher melting points than other lead alloys and have been used for heat fuses,
while thallium added to lead/tin bearing alloys considerably improves the corrosion resistance and increases the hardness. Claims have been made that a tin/thallium/lead based anode has considerably less corrosion attack than the antimony/lead anodes normally used for chromium plating.

Some thallium compounds are sensitive to infra-red light and find some uses in equipment analogous to photoelectric cells using selenium in normal light.

In general thallium is a little used metal whose potential, if any, is severely restricted by the toxic nature of the metal and its compounds.

There are no alloy systems based on thallium.

Note. The following abbreviations and units are used in the tables:

DPN	Hardness, diamond pyramid number
UTS	Ultimate tensile strength, N/mm^2
Elon	Elongation, %
Proof	0.1 % proof strength, N/mm^2

1 N/mm^2=0.1 hbar=0.102 kgf/mm^2=0.06475 tonf/in^2=145.04 lbf/in^2

Symbol	Nominal analysis, supplier, condition and remarks.
THALLIUM	High purity metal: Impurities 5 p.p.m.; Johnson Matthey; supplied as wire in sealed containers
THALLIUM w Th Tl 18	Pure metal: Blackwells 99.9% Th Tl powder: Light Ltd

49. Thorium Th

Physical properties

Atomic number	90	
Atomic weight	232.12	
Crystal structure	Face-centred cubic	
Colour	–	
Specific gravity	11.3	
Density	11300 kg/m^3	(0.406 lb/in.^{3})
Melting point	1800 °C	
Boiling point	3500 °C	
Specific heat	0.116 J/g °C	(0.0276 cal/g °C)
Thermal conductivity	37.7 W/m °C	(9 cal/m s °C)
Coefficient of linear expansion (20–100° C)	11.1 × 10^{-6}/ °C	
Latent heat of fusion	82.9 J/g	(19.8 cal/g)
Latent heat of vaporization	2617 J/g	(625 cal/g)
Thermal neutron absorption cross-section	– barns/atom	
Electrical conductivity	14% IACS (copper 100%)	
Specific resistance	130 microhm mm	
Temperature coefficient of electrical resistance	0.0038/ °C	
Electrochemical equivalent	– g/A/h	
Electrode potential	-2.1 V	
Magnetic susceptibility	0.11 × 10^{-6}	
Young's modulus of elasticity	73.1 × 10^9 N/m^2	(10.6 × 10^6 lbf/in.^{2})
Tensile strength	200 N/mm^2	(14 tonf/in.^{2})
Hardness	35 DPN	

49.1 General notes on thorium

Thorium does not exist in the native state but the oxide is fairly widespread and deposits of the silicate are also known. The sand monazite contains thorium oxide along with the rare earth metals and this is the principal source. Thoria (thorium oxide) crucibles are used in the extraction and working of many of the refractory metals as it withstands temperatures up to 2500 °C.

High purity thorium can now be produced and there are indications that this will result in cheaper metal of a purity capable of accepting cold work without becoming too brittle.

Thorium is greyish white and quite soft. Its freshly cut surfaces rapidly oxidize but this prevents further oxide formation. It does not ignite spontaneously but will continue burning in air, forming the very white powder thoria – thorium oxide. The fact that thorium is radioactive restricts its widespread use and its low mechanical strength reduces its value in nuclear engineering where most of the metal is used. It finds some use as a substitute for radium as it is considerably cheaper to produce, although not radioactive. Thorium is also used in X-ray equipment and as a source of radioactivity for non-destructive flaw detection. Certain types of photo-electric cells use thorium, but the demand for the metal at present is very low.

There are no alloys based on thorium but a range of magnesium alloys exists with 2% thorium which have good ductility and improved high temperature properties while retaining the tensile strength of other magnesium alloys.

As previously stated the oxide thoria is a valuable refractory and considerable quantities at one time were used to make gas mantles. This was the highly purified form as the white crystalline structure gave the desired incandescent effect.

There are only limited uses for the metal thorium and these are not likely to increase much even with a reduction in the present high cost of the metal. The low mechanical properties coupled with the radioactivity account for this unpopularity.

Note. The following abbreviations and units are used in the tables:

DPN	Hardness, diamond pyramid number
UTS	Ultimate tensile strength, N/mm^2
Elon	Elongation, %
Proof	0.1 % proof strength, N/mm^2

1 N/mm^2=0.1 hbar=0.102 kgf/mm^2=0.06475 tonf/in^2=145.04 lbf/in^2

Symbol	Nominal analysis, supplier, condition and remarks.
THORIUM	Pure metal: Blackwells

50. Tin Sn

Physical properties

Atomic number		50	
Atomic weight		118.7	
Crystal structure	alpha	Tetrahedral cubic – 18 °C	
	beta	Body-centred tetragonal – 18–161 °C	
	gamma	Close-packed hexagonal – 161 °C	
Colour		Bright white	
Specific gravity	alpha	5.81	
	beta	7.29	
	gamma	6.56	
Density	alpha	5810 kg/m^3	(0.21 lb/in.^{3})
	beta	7290 kg/m^3	(0.26 lb/in.^{3})
	gamma	6560 kg/m^3	0.236 lb/in.^{3})
Melting point		231.84 °C	
Boiling point		2270 °C	
Specific heat		0.234 J/g °C	(0.056 cal/g °C)
Thermal conductivity		63.2 W/m °C	(15.1 cal/m s °C)
Coefficient of linear expansion (20–100° C)		23.8 × 10^{-6}/ °C	
Latent heat of fusion		59.5 J/g	(14.2 cal/g)
Latent heat of vaporization		2399 J/g	(573 cal/g)
Thermal neutron absorption cross-section		0.65 barns/atom	
Electrical conductivity		14% IACS (copper 100%)	
Specific resistance		1150 microhm mm	
Temperature coefficient of electrical resistance		0.0044/ °C	
Electrochemical equivalent		– g/A/h	
Electrode potential		-0.14 V	
Magnetic susceptibility		-0.25 × 10^{-6}	
Young's modulus of elasticity		41.4 × 10^9 N/m^2	(6 × 10^6 lbf/in.^{2})
Tensile strength		220 N/mm^2	(15 tonf/in.^{2})
Hardness		– DPN	

50.1 General notes on tin

Tin is never found in the native state but deposits of the oxide ore – called tinstone – are widespread throughout the world. These deposits are seldom large and it is becoming progressively harder to win the ore.

Metal of 99.75–99.99% purity is commonly available, the most undesirable elements being lead, arsenic and bismuth, all of which embrittle tin.

The metal is silver-white and although it rapidly loses its original lustre it does not readily tarnish or discolour in air. It is a soft, ductile metal, being readily cut by a knife, and in the pure state is not subject to cold work. When block tin or tin pipes are bent a characteristic noise known as 'tin creak' is heard.

Tin is capable of being rolled into very thin sheets, but owing to its low strength cannot be drawn into thin wire.

The present high cost of tin prevents it being used for many purposes for which it is eminently suitable. The complete freedom from toxicity and its extreme ease of manipulation make it ideal as food and drink containers and at one time water pipes and 'silver' paper were made in pure tin.

The largest single use of tin is to coat steel sheet which is subsequently used to manufacture the well-known 'tins' used to store all types of food and drink. Although plastics and, to a greater extent, aluminium are now entering this field, tin plate remains the most economical and one of the commonest means of permanently storing foods of all kinds.

Tin can be applied to sheet steel, cast iron, aluminium, copper and its alloys generally by dip tinning after cleaning and fluxing, but also by electroplating and metal spraying techniques.

The low melting point, high fluidity and good wetting properties ensure excellent coverage by dip tinning, while the position of tin in the electrochemical series ensures that it sacrifices itself to protect the base metal if there is any break in the coating. Tin has comparable corrosion resistance in most atmospheres to zinc and cadmium without any danger from toxicity.

Apart from use in food containers, tin is used to coat copper conductor wires to prevent corrosion and ease the soldering of connections. Pure tin is also used to coat bearing shells and moulds prior to white metalling, and electrical connections of all types before soldering.

A limited quantity of pure tin piping and foil is still used for piping beer and wrapping foods, such as certain cheeses which corrode aluminium foil, but in general stainless steel and plastics have taken over in these fields.

About 50% of all tin mined is used for these purposes, while the remainder goes to the many alloys which use tin. The tin based alloys are described in the section following, but in addition it is a valuable alloying element, particularly in conjunction with copper. The copper–tin alloys take the generic name 'bronze' and are used where bearing properties, strength and corrosion resistance are required. The alloys are discussed in more detail under Section 14, particularly Section 14K.

Tin is also commonly found in lead alloys where it is used to strengthen. Small quantities increase the creep strength of lead while the lead–tin solders cover a wide range of analysis for a variety of purposes relating to melting temperatures, corrosion resistance and strength. In general the higher the tin content, the higher duty the alloys.

Tin is also added to lead based bearing alloys where it increases the load bearing properties and reduces the frictional coefficient.

It is of interest that tin added to lead almost always has a beneficial effect, whereas lead added to tin results in deleterious embrittlement and reduction in other properties.

Some zinc/tin and nickel/tin alloys are available which find most use as electrodeposits where an attractive silver-type lustre is desired. The nickel–tin deposit can have the analysis adjusted to give a bright plate which requires very little polishing. These deposits are ralatively expensive.

There is no doubt that tin would find considerably more use if the price could be reduced. This is unlikely as tin is readily and economically obtained from its ores which are scarce and difficult to win. Tin has been used by mankind for centuries and the discovery of new easily obtained ore deposits is therefore highly improbable.

Following are listed the brand names and available information on pure tin.

Symbol	Nominal analysis, supplier, condition and remarks.
ASTM B339	99–99.8% Sn: Pig; primary metal; covers 5 grades
BANKA	0.03% Pb 0.006% Cu 0.014% As 99.935% Sn: Brand name; further information from Tin Research Institute
BS 3252 T1	99.9% Sn: Ingot; high purity tin
BS 3252 T2	99.75% Sn: Ingot; refined tin
BS 3252 T3	99.0% Sn: Ingot; common tin
CORNISH REFINED	0.065% Pb 0.34% Cu 0.03% As 99.82% Sn: Brand name; further information from Tin Research Institute
E S COY Ltd	0.03% Pb 0.012% Cu 0.031% As 99.899% Sn: Brand name; further information from Tin Research Institute
HAWTHORNE	0.028% Pb 0.028% Cu 0.029% As 99.891% Sn: Brand name; further information from Tin Research Institute
HERBERTON	0.18% Pb 0.03% Cu 0.037% As 99.715% Sn: Brand name; further information from Tin Research Institute
LAMB & FLAG	0.38% Pb 0.025% Cu 0.02% As 99.56% Sn: Brand name; further information from Tin Research Institute
LAMB & FLAG COMMER	0.763% Pb 0.03% Cu 0.029% As 99.128% Sn: Brand name; further information from Tin Research Institute
LONGHORN 3 STAR	0.04% Pb 0.03% Cu 0.024% As 99.86% Sn: Brand name; further information from Tin Research Institute
MELLANEAR GUAR 99.9	0.037% Pb 0.009% Cu 0.008% As 99.915% Sn: Brand name; further information from Tin Research Institute
MELLANEAR REFINED	0.065% Pb 0.022% Cu 0.025% As 99.827% Sn: Brand name; further information from Tin Research Institute
M & T ELECTROLYTIC	0.008% Pb 0.001% Cu 99.98% Sn: Brand name; further information from Tin Research Association
M & T No 1	0.01% Pb 0.001% Cu 0.016% As 99.96% Sn: Brand name; further information from Tin Research Institute

Symbol	Nominal analysis, supplier, condition and remarks.
PASS No 1	0.001% Pb 0.0002% Cu 0.003% As 99.975% Sn: Brand name; further information from Tin Research Institute
PYRMANT	0.048% Pb 0.022% Cu 0.043% As 99.85% Sn: Brand name; further information from Tin Research Institute
REGIS	0.011% Pb 0.001% Cu 0.004% As 99.972% Sn: Brand name; further information from Tin Research Association
ROSE	0.002% Pb 0.012% Cu 99.955% Sn: Brand name; further information from Tin Research Association
TIN	Sn: Ingot, sheet, sponge, etc.; Blackwells; primary metal
TOYO	0.04% Pb 0.03% Cu 0.03% As 99.8% Sn: Brand name; further information from Tin Research Institute
TULIP	0.086% Pb 0.01% Cu 0.008% As 99.87% Sn: Brand name; further information from Tin Research Institute
UMHK	0.12% Pb 0.013% Cu 0.006% As 99.966% Sn: Brand name; further information from Tin Research Institute
ZTM	0.09% Pb 0.04% Cu 0.063% As 99.78% Sn: Brand name; further information from Tin Research Institute

Note. The following abbreviations and units are used in the tables:

DPN	Hardness, diamond pyramid number
UTS	Ultimate tensile strength, N/mm^2
Elon	Elongation, %
Proof	0.1 % proof strength, N/mm^2

$1\ N/mm^2 = 0.1\ hbar = 0.102\ kgf/mm^2 = 0.06475\ tonf/in^2 = 145.04\ lbf/in^2$

50A Tin alloys

Specific gravity	7.35–7.95	
Density	7350–7950 kg/m³	(0.266–0.287 lb/in.³)
Solidus/liquidus	185–370 °C	
Thermal conductivity	– W/m °C	
Coefficient of linear expansion	– / °C	
Electrical conductivity	–	
Specific resistance	– microhm mm	
Young's modulus of elasticity	– N/m²	
Impact	–	
Hot strength		

Temperature °C	Tensile strength N/mm²
20	31
100	16
150	9

The above properties are typical of the following group, and may not apply exactly to any one specification. It is possible that with certain specifications some of the values may not be applicable.

General metallurgical characteristics

Tin based alloys are expensive and thus are only used when their properties cannot be reproduced using cheaper metals or other methods. Both the principal uses of the tin alloys also employ lead based alloys which are much cheaper. The first and greatest use of tin in the alloy form is as solder. This always has some lead present and may have other elements added for specific purposes.

In general the high tin solders have better corrosion resistance and are stronger than the high lead varieties. Silver increases the strength at temperature. The temperature range over which solders melt is of considerable importance as this decides the workability of the metal. Where high strength, mass produced articles are joined by soldering, a short solidification range is generally desirable to prevent movement and give speedy joins. Where it is necessary to work the solder during solidification, as with plumbers' or electricians' wiped joints, a long slow solidification range is essential. The alloy content decides the liquidus/solidus points and hence the choice of alloy for any particular use.

The second largest use for tin alloys is as a bearing metal where it again has lead as a second best competitor. Tin bearing alloys must be low in lead to prevent embrittlement, generally have under 0.5% lead with up to 10% antimony and 5% copper.

Tin has an excellent frictional coefficient, and even when alloyed is soft enough to accept grit particles, which become embedded and covered with tin, causing little or no scoring. There appears to be some affinity between tin and oil which considerably aids the tin bearing metals and allows them to run for longer periods in oil-starved conditions than most other bearing alloys.

There is also a considerable increase in temperature characteristics over the lead bearing alloys. Except for low load high speed duties, the tin bearing alloys must always be supported by a stronger metal shell on to which they are cast. The bond of tin alloy to steel or bronze must be excellent to prevent exfoliation and failure by plucking. Tin alloy bearings are now being replaced with new design bearings and alternative materials.

Thermal treatment

None of the tin alloys can have any of their characteristics altered by any form of thermal treatment. Because of their relatively low melting point, it is possible to cast tin bearing alloys on to heat treated steel shells. Provided care is taken it is even possible to cast on to steel which has been case hardened. It is not possible to carry out any useful form of heat treatment on the steel backing after casting the tin alloy without the danger of re-melting the tin.

Welding and brazing

These alloys melt far below the temperatures generally associated with the terms 'welding and brazing'. They can however, be very readily joined by the heat fusion process, and large bearings may be repaired by this method.

As stated above many of the alloys are solders used to join other metals where temperature of use and joint strength are low. The techniques for casting the metals, repairing by melting and soldering are generally similar.

All traces of oxide, oil grease and dirt must first be removed, particular care being taken when the tin alloy has to be bonded to steel. The final cleaning is often carried out by means of a flux, but this must not be relied upon to remove excessive dirt or oxide. The cleaned surface is then 'tinned' using pure tin or a high tin solder, either by dipping in the molten metal or rubbing the heated part with a stick of solder. It is the quality of this tinned layer which largely decides the standard of the final bond. Good quality fluxes considerably help this operation but may be acid in nature, resulting in subsequent corrosion unless carefully removed.

The final operation is the casting or soldering, depending on the function. This will vary from each part but in general it can be said that best results are obtained when a temperature just above the melting point of the tin alloy is used. The base metal should be at almost the same temperature, with rapid quenching as soon as solidification commences.

Flaw detection methods

Crack test. Oil penetrant chalk test. By carefully selecting the method of machining it is generally possible to prove the quality of the tin alloy casting by chalk testing at a stage where it can be easily re-cast if porosity or faulty bonding is present.

X-ray. Not required, except on special purpose high duty bearings.

Ultra-sonic test. This can be used on some bearings after rough machining to prove that the bond is satisfactory. This will also show porosity.

Symbol	Nominal analysis, supplier, condition and remarks.
2.3820	11% Zn Sn alloy: Working temperature 210 °C; German Standard
2.3830	40% Zn Sn alloy: Working temperature 260 °C; German Standard
2.3852	45% Pb 10% Zn + Cd Sn alloy: Working temperature 220 °C; German Standard
A 85	Tin base white metal: For diesel engine bearings; Anti-Attrition Ltd
A 89	Tin base white metal: For marine petrol engine bearings; Anti-Attrition Ltd
A 139	Tin base white metal: For railway carriage bearings; Anti-Attrition Ltd
A 127	Tin base white metal: For electric motor bearings; Anti-Attrition Ltd
A 129	Tin base white metal: For electric motor bearings; Anti-Attrition Ltd
ADASTRAL 'A'	3% Cu 4.5% Sb Sn alloy: Bearing metal; Stone Manganese for BS 3332/1
AERO No 2	Sn base Pb free bearing metal: High load; high speed; Magnolia Anti-Friction Metal Co

Note. The following abbreviations and units are used in the tables:

DPN	Hardness, diamond pyramid number
UTS	Ultimate tensile strength, N/mm^2
Elon	Elongation, %
Proof	0.1 % proof strength, N/mm^2

1 N/mm^2=0.1 hbar=0.102 kgf/mm^2=0.06475 tonf/in^2=145.04 lbf/in^2

Chemical etch. Not required.

Corrosion protection

Temporary. None required for tin alloys. The base metal will require to be treated in the same manner as that described in the appropriate section. Particular care must be taken if fluxes have been used as these are often of a very corrosive nature.

Permanent. No protection required for tin alloys which have excellent corrosion resistance to all except certain acids. The low melting point however, restricts the use of stove enamels in many cases and it must be remembered they are very often used in conjunction with mild steels which have very poor corrosion resistance.

Machinability

Tin alloys are subject to the same difficulties as all soft metals in that they tend to tear rather than cut.

Tools must be sharp and only single point cutting should be attempted. High speeds with low tool pressures and copious coolant as a lubricant give best results.

Uses

Solders for high duty, high strength, corrosion resistant purposes. For tinning before casting lead or tin based bearing metals.

Bearing metals for high duty purposes where the temperatures and pressures do not warrant the use of copper base alloys. The tin alloys have superior self-lubricating properties to all other common metal bearing alloys.

Symbol	Nominal analysis, supplier, condition and remarks.
ALGER METAL	Gn Sb alloy containing 10% Sb 0.3% Cu Sn alloy: Source unknown
ALGIERS METAL	10% Sb 0.3% Cu Sn alloy: Source unknown
ALICIA	Sn base bearing metal: Medium duty; Magnolia Anti-Friction Metal Co
AMS 4800 A	4.5% Cu 4.5% Sb Sn alloy: Bearing metal; 'Babbit'
ARGENTINE METAL	14.5% Sb Sn alloy: For toys; source unknown
ARGENTINE METAL	15% Sb Sn alloy
ARMA	Sb Cu Sn base bearing metal: For high duty loads; Phosphor Bronze Co
ARSENIC-TIN	5% As Sn alloy: Blackwells; primary metal
ASHBERRY	16% Sb 3% Cb 2% Zn Sn alloy: White metal; information from Tin Research Institute
ASTM B23/1	4.5% Sb 4.5% Cu 0.3% Pb Sn alloy: Bearing metal; melting point 223 °C; compressive strength 90 N/mm^2 **DPN: 17**
ASTM B23/2	7.5% Sb 0.3% Pb 3.5% Cu Sn alloy: Bearing metal; melting point 241 °C; compressive strength 100 N/mm^2 **DPN: 24**
ASTM B23/3	8% Sb 0.3% Pb 8% Cu Sn alloy: Bearing metal; melting point 240 °C; compressive strength 120 N/mm^2 **DPN: 27**
ASTM B32/50 A	50% Pb Sn: Solder; melting range 183–216 °C

Symbol	Nominal analysis, supplier, condition and remarks.
ASTM B32/50 B	50% Pb 0.3% Sb Sn: Solder; melting range 183–216 °C
ASTM B32/60 A	40% Pb Sn: Solder; melting range 183–190 °C
ASTM B32/60 B	40% Pb 0.3% Sb Sn: Solder; melting range 183–190 °C
ASTM B32/63 A	37% Pb Sn: Solder (eutectic solder); melting point 183 °C
ASTM B32/63 B	37% Pb 0.3% Sb Sn:. Solder (eutectic solder); melting point 183 °C
ASTM B32/70 A	30% Pb Sn: Solder; melting range 183–192 °C
ASTM B32/70 B	30% Pb 0.3% Sb Sn: Solder; melting range 183–192 °C
ASTM B32/95 TA	5% Sb Sn: Solder; melting range 234–240 °C
ASTM B102 CY44A	4.5% Sb 4.5% Cu Sn alloy: Bearing metal
ASTM B102 PY1815A	15% Sb 18% Pb 2% Cu Sn alloy: Bearing metal
ASTM B102 YC135A	13% Sb 5% Cu Sn alloy: Bearing metal
ATLAS ADMIRALTY 'A'	Zn Sn base bearing alloy: For stem tube bearings; Eyre Smelting Co.; melting range 200–425 °C; compressive strength 60 N/mm² **DPN: 30 UTS: 75**
ATLAS ADMIRALTY 'B'	Sn base Pb free bearing metal: For steam turbines; Eyre Smelting Co; melting range 240–325 °C; compressive strength 70 N/mm² **DPN: 31 UTS: 90**
ATLAS AMACOL	Pb Sn base bearing alloy: For reciprocating engines; Eyre Smelting Co; melting range 190–370 °C; compressive strength 70 N/mm² **DPN: 33 UTS: 90**
ATLAS D8	Sn base bearing metal: For diesels; Eyre Smelting Co; melting range 240–405 °C; compressive strength 70 N/mm² **DPN: 35 UTS: 90**
ATLAS DD	Cn Sn base Pb free bearing metal: For diesels; Eyre Smelting Co; melting range 240–370 °C; compressive strength 70 N/mm² **DPN: 32 UTS: 80**
ATLAS DG	Sn base bearing metal: For diesels; Eyre Smelting Co; melting range 230–376 °C; compressive strength 70 N/mm² **DPN: 33 UTS: 90**
ATLAS FDS	Sn base Pb free bearing metal: For steam turbines; Eyre Smelting Co; melting range 240–330 °C; compressive strength 70 N/mm² **DPN: 32 UTS: 90**
ATLAS INFRANGA	Sn base Pb free bearing metal: For high speed diesels, Eyre Smelting Co; melting range 240–330 °C; compressive strength 70 N/mm² **DPN: 32 UTS: 90**
ATLAS TENAXAS	Pb Sn base bearing alloy: For reciprocating engines; Eyre Smelting Co; melting range 186–340 °C; compressive strength 70 N/mm² **DPN: 30 UTS: 80**
AUTO 'C'	4% Pb 11% Sb 4% Cu Sn alloy: Bearing metal; Stone Manganese for BS 3332/3
BABBITT	Sb Cu Pb Sn white metal bearing alloys: Information from Tin Research Institute

Note. The following abbreviations and units are used in the tables:

DPN	Hardness, diamond pyramid number
UTS	Ultimate tensile strength, N/mm²
Elon	Elongation, %
Proof	0.1 % proof strength, N/mm²

1 N/mm²=0.1 hbar=0.102 kgf/mm²=0.06475 tonf/in²=145.04 lbf/in²

Symbol	Nominal analysis, supplier, condition and remarks.
BABBITT No 1	Sn base bearing metal: For high speed use; Eyre Smelting Co; melting point 376 °C **DPN: 33 UTS: 90 Elon: 5%**
BRITTANIA	7% Sb 2% Cu Sn alloy: White metal; information from Tin Research Institute
BS 2B 21	4% Cu 9% Sb Sn alloy: White metal; bearing metal
BS 2B 22	4% Cu 4% Sb 0.3% Ni Sn alloy: White metal; bearing metal
BS 219/95 A	5% Sb 0.07% Pb Sn alloy: Solder; melting range 236–243 °C
BS 219A	35% Pb 0.6% Sb Sn alloy: Solder; melting range 183–185 °C
BS 219B	47% Pb 2.7% Sb Sn alloy: Solder; melting range 185–204 °C
BS 219F	50% Pb 0.5% Sb (max) Sn alloy: Solder; melting range 183–212 °C
BS 219K	40% Pb 0.5% Sb (max) Sn alloy: Solder; melting range 183–188 °C
BS 3332/1	7% Sb 3% Cu Sn alloy: Bearing metal
BS 3332/2	9% Sb 4% Cu Sn alloy: Bearing metal
BS 3332/3	10% Sb 5% Cu 4% Pb Sn alloy: Bearing metal
BS 3332/4	12% Sb 3% Cu 10% Pb Sn alloy: Bearing metal
BS 3332/5	7% Sb 3% Cu 15% Pb Sn alloy: Bearing metal
BS 3332/6	10% Sb 3% Cu 27% Pb Sn alloy: Bearing metal
BS 3332/9	1.5% Cu 30% Zn Sn alloy: Bearing metal
CERROBEND	Bi Cd Pb Sn alloy: Casting filler metal for tube bending; Mining & Chemical Products Ltd; melting point 70 °C; expands on solidification
CERROMATRIX	Bi Sb Pb Sn alloy: Casting for short run dies; Mining & Chemical Products Ltd; melting range 102–227 °C; expands on solidification
CERROSEAL - 35	50% Tn Sn alloy: Mining & Chemical Products Ltd; melting range 117–127 °C; for fusible links
CERROTRU	Bi Sn alloy: Casting for pattern metal; Mining & Chemical Products Ltd; melting point 137.5 °C; no volume changes on solidification
CONDENSER FOIL	15% Pb 2% Sb Sn alloy: Used as the dielectric in electronic industry; origin unknown
CUSIL	20% Cu Ag alloy: Free of trapped carbon; melting point 780 °C; Brazing alloy; Wesgo
DECARBONIZED	
CUSIL VT	20% Cu Ag alloy: For brazing; melting point 780 °C; Wesgo
CY 44A	4.5% Sb 4.5% Cu Sn alloy: Designation used by ASTM
DE	Sn base Pb free bearing metal: High load; high speed; Magnolia Anti-Friction Metal Co
DIN 1704	99.0% Sn: Supplied in four grades
DIN 1707 L Sn 50	46% Pb 3.3% Sb Sn alloy
DIN 1707 L Sn 60	35% Pb 3.5% Sb Sn alloy
DIN 1707 L Sn 90	8% Pb 1.3% Sb Sn alloy
DIN 1742 Sg Sn 50	4% Cu 33% Pb 13% Sb Sn alloy **DPN: 26 UTS: 45 Elon: 2%**
DIN 1742 Sg Sn 60	4% Cu 23% Pb 13% Sb Sn alloy **DPN: 28 UTS: 45 Elon: 1.5%**
DIN 1742 Sg Sn 70	5% Cu 10% Pb 15% Sb Sn alloy **DPN: 30 UTS: 100 Elon: 1%**
DIN 1742 Sg Sn 75	5% Cu 3% Pb 17% Sb Sn alloy **DPN: 30 UTS: 100 Elon: 2%**
DIN 1742 Sg Sn 78	4% Cu 1% Pb 17% Sb Sn alloy **DPN: 30 UTS: 180 Elon: 3%**
DIN 8512 L Sn Pb Zn	45% Pb 10% Zn + Cd Sn alloy: Working temperature 220 °C
DIN 8512 L Sn Zn 10	11% Zn Sn alloy: Working temperature 210 °C
DIN 8512 L Sn Zn 40	40% Zn Sn alloy: Working temperature 260 °C
DSYL	7% Cu 9% Sb 0.3% Pb Sn alloy: Bearing metal; Stone Manganese for BS 3332/2

Symbol	Nominal analysis, supplier, condition and remarks.
DTD 214	3.5% Cu 7% Sb 5% Cb Sn alloy: White metal; suitable for bearings
DTD 244	6% Cu 6.5% Sb 0.5% Ni Sn alloy: White metal; suitable for bearings
h Sn 1	99.9999% Sn: Bar; Light Ltd; high purity metal
h Sn 73	99.99% Sn: Single crystal; 25 mm × 25 mm; Light Ltd
HOYT 11D	Sb Cu Sn base bearing metal: Hoyt Ltd
HOYT 11R	Sn base Pb free bearing metal: Hoyt Ltd
HOYT 11Z3	Sn base Pb free bearing metal: Hoyt Ltd
HOYT 38	Sb Cu Sn base bearing metal: Hoyt Ltd
HOYT 71	Sb Cu Sn base bearing metal: Hoyt Ltd
HOYT 133C	Sn base bearing metal: Hoyt Ltd
HOYT 156B	Sn base bearing metal: Hoyt Ltd
HOYT 175	Sn base bearing metal: Hoyt Ltd
	DPN: 37 UTS: 80 Elon: 1.5%
HOYT BLOWPIPE SOLDER	35.5% Pb Sn: Solder; quick setting; melting range 183–185 °C; Hoyt
HOYT FIFTY	50% Pb Sn: Solder; may contain Sb; melting range 183–212 °C; Hoyt
HOYT MARINE A	Sn base bearing metal: Hoyt Ltd
HOYT SIXTY	40% Pb Sn: Solder; melting range 183–188 °C; Hoyt
HOYTICE	Sn base bearing metal: Hoyt Ltd
	DPN: 27 UTS: 65 Elon: 6%
L Sn 50	46% Pb 3.3% Sb Sn alloy: Designation used by German Standard
L Sn 60	45% Pb 3.2% Sb Sn alloy: Designation used by German Standard
L Sn 90	8% Pb 1.3% Sb Sn alloy: Designation used by German Standard
L Sn Pb Zn	45% Pb 10% Zn + Cd Sn alloy: Designation used by German Standard
L Sn Zn 10	11% Zn Sn alloy: Designation used by German Standard
L Sn Zn 40	40% Zn Sn alloy: Designation used by German Standard
LION BRAND	7% Cu 5% Sb Sn alloy: White metal; Blackwells; marine; main bearings
LM 10A	Ag Cu Sn alloy: Soft Solder; Johnson Methey; electrical conductivity 13% IACS; melting range 214–275 °C
	UTS: 70 Elon: 16%
NAVY	4% Cu 5% Sb 15% Pb Sn alloy: Bearing metal; Stone Manganese for BS 3332/5
No 75	Sn base bearing metal: High speed; medium load; Magnolia Anti-Friction Metal Co
	DPN: 28 UTS: 80
OE 1 METAL	Sn base bearing metal: High speed; medium load; Magnolia Anti-Friction Metal Co
OE 2 METAL	Sn base bearing metal: Medium duty; Magnolia Anti-Friction Metal Co
PEWTER	12% Pb Sn alloy: White metal; information from Tin Research Institute
PLUMBSOL	Ag Sn base soft solder: Johnson Matthey; electrical conductivity 13.3% IACS; melting range 221–225 °C
	UTS: 2.5 Elon: 60%
PY 1815A	15% Sb 18% Pb 2% Cu Sn alloy: Designation used by ASTM range

Note. The following abbreviations and units are used in the tables:

DPN	Hardness, diamond pyramid number
UTS	Ultimate tensile strength, N/mm^2
Elon	Elongation, %
Proof	0.1 % proof strength, N/mm^2

1 N/mm^2=0.1 hbar=0.102 kgf/mm^2=0.06475 tonf/in^2=145.04 lbf/in^2

Symbol	Nominal analysis, supplier, condition and remarks.
QQ M 161/1	4.0% Sb 4.0% Cu Sn alloy: For bearings; US Federal
QQ M 161/2	7.5% Sb 4.0% Cu Sn alloy: For bearings; US Federal
QQ M 161/3	8.0% Sb 8.0% Cu 0.3% Pb Sn alloy: For bearings; US Federal
QQ M 161/4	13% Sb 5.5% Cu 0.2% Pb Sn alloy: For bearings; US Federal
QQ M 161/5	10% Sb 3.0% Cu 25% Pb Sn alloy: For bearings; US Federal
QQ S 571 A	50% Pb Sn alloy: Solder; US Federal
QQ S 571b Sn 60	0.45% Sb 40% Pb Sn: Solder; US Federal; melting range 182–189 °C
QQ S 571d Pb 70	0.45% Sb 31% Sn: Solder; US Federal; melting range 182–254 °C
QQ S 571d Sb 5	5.0% Sb Sn: Solder; US Federal; melting range 232–240 °C
QQ S 571d Sn 50	0.45% Sb 50% Pb Sn: Solder; US Federal; melting range 182–216 °C
QQ S 571d Sn 62	0.45% Sb 38% Pb Sn: Solder; US Federal; melting range 177–189 °C
QQ S 571d Sn 63	0.45% Sb 37% Pb Sn: Solder; US Federal; melting point 182 °C
QQ S 571d Sn 70	0.45% Sb 29% Pb Sn: Solder; US Federal; melting range 182–193 °C
QQ S 571d Sn 96	0.1% Pb (max) 4.0% Ag Sn: Solder; US Federal; melting point 221 °C
RAILWAY C	5% Pb 11% Sb 4% Cu Sn alloy: Bearing metal; Stone Manganese for BS 3332/3
SAE 10	4.5% Cu 4.5% Sb Sn alloy: For bearings
SAE 11	5.7% Cu 6.7% Sb Sn alloy: For bearings
SAE 12	3.5% Cu 7.5% Sb Sn alloy: For bearings
Sg Sn 50	4% Cu 33% Pb 13% Sb Sn alloy: Designation used by German Standard
Sg Sn 60	4% Cu 23% Pb 13% Sb Sn alloy: Designation used by German Standard
Sg Sn 70	5% Cu 10% Pb 15% Sb Sn alloy: Designation used by German Standard
Sg Sn 75	5% Cu 3% Pb 17% Sb Sn alloy: Designation used by German Standard
Sg Sn 78	4% Cu 1% Pb 17% Sb Sn alloy: Designation used by German Standard
SILVER BABBITT	Ag Cd Ni Sn base bearing metal: Highest duty; Magnolia Anti-Friction Metal Co
	DPN: 33.6
SOVEREIGN 8TS	3.0% Cu 8.9% Sb 1.1% Cd Sn alloy: Bearing metal; Glacier Metals Co
	DPN: 30 UTS: 98 Elon: 15% Proof: 84
SOVEREIGN 8TS	8.8% Sb 3.0% Cu 1% Cd Sn: Bearing metal; Glacier Metals
	DPN: 30 UTS: 100 Elon: 16% Proof: 80
SPRA BABBITT	3.5% Cu 0.25% Pb 7.5% Sb Sn alloy: Wire for spraying; Metco
STA 7 TB1B	8% Sb 8% Cu 0.3% Pb Sn alloy: Bearing metal; obsolete
STA 7 TB2	Sn base bearing metal: Obsolete
STERN TUBE METAL	30% Zn 1.5% Cu Sn alloy: White metal; information from Tin Research Institute
SX 4	4% Ag Sn alloy: For soldering; Sheffield Smelting Co; melting range 221–224 °C
	DPN: 14 UTS: 50 Elon: 48% Proof: 45
TANDEM DE	Sn base bearing metal: For reciprocating engines; Eyre Smelting Co; melts at 315 °C
	DPN: 27 UTS: 80 Elon: 24%
TANDEM HP	Sn base bearing metal: For reciprocating engines; Eyre Smelting Co; melts at 330 °C
	DPN: 32 UTS: 80 Elon: 6%
TANDEM ME	Sn base bearing metal: For high speed use; Eyre Smelting Co; melts at 370 °C
	DPN: 32 UTS: 90 Elon: 4%

Symbol	Nominal analysis, supplier, condition and remarks.
TANDEM ML	Sn base bearing metal: For high speed use; Eyre Smelting Co; melts at 340 °C **DPN: 30 UTS: 80 Elon: 3%**
TANDEM PLUS	Sn base bearing metal: For reciprocating engines; Eyre Smelting Co; melts at 330 °C **DPN: 33 UTS: 90 Elon: 16%**
TC	Ag Cd Ni Sn base bearing metal: Highest duty; Magnolia Anti-Friction Metal Co **UTS: 95**
TCS	Ag Cd Ni Sn base bearing metal: Highest duty; Magnolia Anti-Friction Metal Co; silver 'Babbitt' **DPN: 33.6**
TENAXAS Al	Pb Sn base bearing metal; For reciprocating engines; Eyre Smelting Co; melting range 186–340 °C **DPN: 30 UTS: 80 Elon: 3.5%**
TINMANS SOLDER	33.3% Pb Sn alloy: White metal: Information from Tin Research Institute
TURBEX	Sn base Pb free bearing metal: High speed; high load; Magnolia Anti-Friction Metal Co **DPN: 25.5 UTS: 80**
UNDER-WATER	30% Zn 1.5% Cu 0.5% Pb 0.5% Sb Sn alloy: Bearing metal; Stone Manganese for BS 3332/9
VULCAN	Sn base bearing metal: For high duty loads; Phosphor Bronze Ltd

Symbol	Nominal analysis, supplier, condition and remarks.
WAGNER'S ALLOY	10% Sb 1% Cu 3% Zn 0.8% Bi Sn alloy: Origin unknown
WARNE'S METAL	26% Ni 26% Bi 11% Co Sn alloy: Used for jewellery; origin unknown
WELCH'S ALLOY	48% Ag Sn alloy: Used for dental purposes; origin unknown
WAMATO METAL	4.5% Cu 4.5% Sb 1% Pb 1% Ni 1% Co (max) Sn alloy: Used for main bearings in internal combustion engines; origin unknown
YC 135A	13% Sb 5% Cb Sn alloy: Designation used by ASTM
ZINN	0.7% Pb 0.3% Cu Sn alloy: Used for hard service bearings; origin unknown

Note. The following abbreviations and units are used in the tables:

DPN	Hardness, diamond pyramid number
UTS	Ultimate tensile strength, N/mm^2
Elon	Elongation, %
Proof	0.1 % proof strength, N/mm^2

1 N/mm^2=0.1 hbar=0.102 kgf/mm^2=0.06475 tonf/in^2=145.04 lbf/in^2

51. Titanium Ti

Physical properties

Atomic number	22	
Atomic weight	47.9	
Crystal structure	alpha Close-packed hexagonal up to 880 °C	
	beta Body-centred cubic above 880 °C	
Colour	Dark grey	
Specific gravity	4.5	
Density	4500 kg/m^3	(0.162 lb/in.^{3})
Melting point	1680 °C	
Boiling point	2800 °C	
Specific heat	0.528 J/g °C	(0.126 cal/g °C)
Thermal conductivity	17 W/m °C	(4.07 cal/m s °C)
Coefficient of linear expansion (20–100° C)	8.9×10^{-6}/ °C	
Latent heat of fusion	435.4 J/g	(104 cal/g)
Latent heat of vaporization	– J/g	
Thermal neutron absorption cross-section	5.6 barns/atom	
Electrical conductivity	3% IACS (copper 100%)	
Specific resistance	554 microhm mm	
Temperature coefficient of electrical resistance	0.0026/ °C	
Electrochemical equivalent	– g/A/h	
Electrode potential	+ 0.2 V	
Magnetic susceptibility	1.25×10^{-6}	
Young's modulus of elasticity	116×10^9 N/m^2	$(16.8 \times 10^6$ lbf/in.$^2)$
Tensile strength	220 N/mm^2	(15 tonf/in.^{2})
Hardness	60 DPN	

51.1 General notes on titanium

Titanium is one of the more common metals in the earth and is found widespread in abundant quantities, but never as the native metal. Titanium oxide made from 'ilmenite' has been used for many years as a white pigment for paints which is not subject to yellowing, as is zinc oxide when in contact with sulphur or sulphide.

It is only in recent years that it has been possible to produce titanium in commercial quantities and it is still relatively expensive. Titanium is now available in all the normal forms such as castings, forging, bar, tube and extruded sections. There is also a well-defined series of alloys based on titanium which are discussed and listed in the following sections.

The metal is steel grey in appearance, having a density twice that of aluminium and two-thirds that of steel. The mechanical properties at room temperature are comparable with many steels, and at high temperature titanium and its alloys are superior to all except the special creep resistant steels.

Titanium metal resists the attack of almost all acids over a very wide range of temperatures and this with its high temperature properties makes it a favourite material with chemical, nuclear and aerospace engineers.

As the usage increases so the price drops and further uses for titanium and its alloys become an economical proposition.

Titanium was a laboratory curiosity until very recent years, but already the quantity used annually can be measured in thousands of tons. The fact that it is available in commercial quantities of a purity to allow working with conventional equipment at easily obtainable temperatures accounts for this rapid increase.

In this book titanium and its alloys have been divided into three groups, Section 51A commercially pure, Section 51B alloys with no apparent increase in mechanical properties by thermal treatment and Section 51C alloys with some mechanical property improved by thermal treatment. It should be appreciated that there is some overlap with Sections 51B and C particularly with alloys which appear in Section 51C which are used in the annealed condition in service. For the purposes of this book if the alloy is heat treatable it has been included in Section 51C.

51A Titanium – commercially pure

Specific gravity	4.51	
Density	–	(0.163 lb/in.^{3})
Solidus/liquidus	1725 °C	
Thermal conductivity	–	(400 cal/m s °C
Coefficient of linear expansion	9.1×10^{-6}	
Electrical conductivity	3.0% IACS (copper 100%)	
Specific resistance	550 microhm mm	
Young's modulus of elasticity	$103–124 \times 10^9$ N/m^2	($15–18 \times 10^6$ lbf/in.^{2})
Impact	– J	
Fatigue strength – endurance limit	± 220–300 N/mm^2	(± 15–20 tonf/in.^{2})
Hot strength		

Temperature °C	Tensile strength N/mm^2	Elongation %
100	450	25
200	340	28
300	270	28
400	220	28
500	180	28
600	140	150

The above properties are typical of the following group, and may not apply exactly to any one specification. It is possible that with certain specifications some of the values may not be applicable.

General metallurgical characteristics

The specifications listed do not contain any alloying metallic elements. Controlled amounts of oxygen however, up to 0.2%, affect the mechanical properties. It is important that carbon, iron and hydrogen are held to low limits, otherwise there will be a considerable drop in ductility. The materials listed have excellent corrosion resistance, can be formed and welded, and have reasonable mechanical strength.

Titanium is approximately two-thirds the weight of steel and twice the weight of aluminium, and depending on the purity and cold work can be the same strength as medium alloy steel. The higher the strength, the more difficult to weld in general. There is some evidence that titanium is more notch sensitive than other materials of comparative strength, thus stress raisers must be absent.

Thermal treatment

None of these materials can have their properties improved by heat treatment. If desired cold work can be removed by annealing at 500–550 °C in normal electric furnaces, with air atmosphere, but the time at temperature should be held to a minimum to prevent oxide formation. This treatment will effectively remove all traces of cold work, thus should not be used where the added strength obtained from work hardening is desired. Titanium must not be heated above 700 °C as the oxide penetration then becomes rapid, embrittling the surface layer to a considerable depth.

In many cases an interstage anneal will be required between forming operations, or even after heavy machining cuts where dimensional stability is essential.

Peculiar effects have been noted regarding hardness and brittleness after annealing which have been attributed to oxygen diffusing throughout the part, particularly thin sections. This can be reduced by ensuring that the parts are held at temperature for as short a time as possible and that too high a temperature is never employed.

Normal controlled atmospheres containing hydrogen, nitrogen or any carbonaceous gas must not be used as they all have some embrittling effects, generally worse than oxygen. Where necessary, inert gases such as argon or helium can be used, but it is difficult to purge furnace chambers to ensure a complete absence of air unless vacuum techniques are used. With modern equipment it is possible to carry out annealing cycles under vacuum and this is possibly the only technically satisfactory method of heating above 700 °C.

Scale removal can be by grit blasting but as the oxide produced by annealing in air is hard and adherent, it is often necessary to use high pressures and large sized grit particles which can cause distortion on flimsy parts.

Salt baths have been developed to de-scale titanium, but this can cause undesirable local oxide penetration or metal attack unless very carefully controlled. Sodium hydride baths must not be used as hydrogen will be absorbed with resultant brittleness.

Acid pickling using hot strong phosphoric acid or hydrofluoric/nitric acid mixtures can be used provided the necessary equipment is available. The solutions must be carefully controlled as used solutions can cause rapid local attack.

Welding and brazing

Pre-treatment. Where any cold work, caused by machining, press work operations or shearing is present, this should be removed by annealing prior to welding.

The weld area must be thoroughly cleaned, then emery dressed or blasted. Care must be taken that iron contamination of the weld area is not caused by using wire brushes or metal blasting grit.

Welding and brazing. This must be inert gas shielded arc welding, with the inert gas also applied to the underside of the weld. There must be a sufficient volume of the gas at all times to allow the weld to cool out of contact with air. Only by ensuring absence of carbon (from organic materials, oils, etc.), hydrogen (from moisture), iron (from tools, wire brushes, blasting media) and oxygen (from air), can ductile porous free welds be obtained.

It is seldom necessary to use elaborate evacuated or gas filled chambers to obtain these conditions. Care in obtaining absolute cleanliness and use of adequate inert gas flow both sides of the weld are however, essential; smooth flowing movements are necessary to ensure a minimum disturbance to the gas curtain.

Resistance and flash butt welding are possible using special equipment, provided the above precautions are carried out.

Electric arc and gas welding are not possible even with the highest skill and best possible conditions.

Brazing has no known application at present. If required a vacuum furnace would be necessary.

Post-treatment. None required.

Flaw detection methods

Crack test. Penetrant oil and chalk test is the usual method. Where maximum sensitivity is required fluorescent oil penetrant examined under ultra-violet light can be used, with the surface previously etched in hydrofluoric/nitric acid.

X-ray. Not normally required except as a means of controlling the quality of welding.

Ultra-sonic test. This is commonly carried out at an early stage of manufacture to prove the absence of large defects and subcutaneous flaws which would become dangerous on the finished part. Titanium is an expensive material to machine and considerable economies can be effected by ensuring that all defects are identified at as early a stage as possible.

Acid etch. A hydrofluoric acid/ferric sulphate etch has been found satisfactory as a grain size and directional indicator. It also highlights forging defects and is an excellent preparation for crack testing. This etch has an adverse effect on the fatigue properties and the parts should therefore have the surface layer removed by polishing or blasting, preferably vapour blasting after examination.

Corrosion protection

Temporary. None required.

Permanent. These specifications stand up to the most corrosive media without any further treatment. Their resistance to hot oxidation is excellent, being superior to most austenitic stainless steels. Above about 750 °C this drops off and advice should be obtained before using these materials at these temperatures.

Machinability

These materials readily cold work, thus tools must be kept sharp. Better results are obtained taking heavy cuts than light skims. The specifications listed behave in many respects like annealed 18/8 type stainless steels. Interrupted cuts and tool dwell should be avoided at all times.

Some care is necessary when polishing or dry grinding titanium as the dust can be explosive. As long as it is not allowed to accumulate in closed containers there is no danger. Keeping the dust wet removes the danger, but water must not be used on titanium fires as this can cause an explosion. Dry chalk or fine sand should be available and used to confine rather than extinguish the fire.

Uses

The chemical industry uses these materials as they stand up to most chemicals, except fuming nitric, hot concentrated phosphoric and sulphuric acids.

The aircraft industry also uses them where weight saving is vital and aluminium cannot be used because of temperature limitations.

The present high cost of titanium inhibits its use in many other applications where its strength/weight ratio make it desirable material, for example reciprocating components in engineering. As wider use is made and manufacturing techniques improve this price disadvantage will be reduced.

Titanium has the disadvantage that it is subject to greater frictional wear damage than other materials of comparable hardness. It is also difficult at present to electroplate successfully or metal spray to improve the surface.

Symbol	Nominal analysis, supplier, condition and remarks.
3.7024	Commercially pure Ti: German Standard
3.7025	Commercially pure Ti: German Werkstoff number
3.7034	Commercially pure Ti: German Standard
3.7064	Commercially pure Ti: German Standard
12	Commercially pure Ti: Krupp
15	Commercially pure Ti: Krupp
18S	Commercially pure Ti: Krupp
30	Commercially pure Ti: Atlas

Note. The following abbreviations and units are used in the tables:

DPN	Hardness, diamond pyramid number
UTS	Ultimate tensile strength, N/mm^2
Elon	Elongation, %
Proof	0.1 % proof strength, N/mm^2

$1\ N/mm^2 = 0.1\ hbar = 0.102\ kgf/mm^2 = 0.06475\ tonf/in^2 = 145.04\ lbf/in^2$

Symbol	Nominal analysis, supplier, condition and remarks.
35	Commercially pure Ti: Ugine
40	Commercially pure Ti: Ugine
50	Commercially pure Ti: Ugine
A 40	Commercially pure Ti: Crucible Steel Co **Proof: 270**
A 55	Commercially pure Ti: Crucible Steel Co **Proof: 370**
A 70	Commercially pure Ti: Crucible Steel Co **Proof: 480**
AIR 9/182 T35	0.1% C (max) 0.2% Fe Ti: Commercially pure; origin unknown
AMS 4900 B	Commercially pure Ti: Sheet & strip; annealed; AMS for alloy A 55 **Proof: 370**
AMS 4901 C	Commercially pure Ti: Sheet & strip; annealed; AMS for commercial alloy A 70 **Proof: 500**
AMS 4902 A	Commercially pure Ti: Sheet & strip; annealed; AMS for commercial alloy A 40
AMS 4921 A	Commercially pure Ti: Bar & forging; annealed; AMS for commercial alloy A 70 **Proof: 500**
AMS 4941	Commercially pure Ti: Welded tube; annealed; AMS for commercial alloy A 40 **Proof: 270**
AMS 4942	Commercially pure Ti: Seamless tube; annealed **Proof: 270**
AMS 4951 A	Commercially pure Ti: Wire; annealed
ASTM B265/1	0.1% C 0.01% H 0.2% Fe Ti: Sheet; commercially pure titanium **UTS: 270 Elon: 22% Proof: 200**
ASTM B265/2	0.1% C 0.01% H 0.2% Fe Ti: Sheet; commercially pure titanium **UTS: 340 Elon: 20% Proof: 270**
ASTM B265/3	0.1% C 0.01% H 0.2% Fe Ti: Sheet; commercially pure titanium **UTS: 420 Elon: 18% Proof: 340**
ASTM B265/4	0.1% C 0.01% H 0.2% Fe Ti: Sheet; commercially pure titanium **UTS: 580 Elon: 15% Proof: 500**
ASTM B299 MD120	0.02% C 0.005% H 99.3% Ti: Sponge; primary metal **DPN: 120**
ASTM B299 MD160	0.05% C 0.005% H 99.1% Ti: Sponge; primary metal **DPN: 160**
ASTM B299 ML120	0.025% C 0.03% H 99.1% Ti: Sponge; primary metal **DPN: 120**
ASTM B299 ML140	0.06% C 0.05% H 99.1% Ti: Sponge; primary metal **DPN: 140**
ASTM B299 ML160	0.05% C 0.05% H 99.1% Ti: Sponge; primary metal **DPN: 160**
ASTM B299 SL120	0.025% C 0.0125% H 99.1% Ti: Sponge; primary metal **DPN: 120**

Note. The following abbreviations and units are used in the tables:

DPN	Hardness, diamond pyramid number
UTS	Ultimate tensile strength, N/mm^2
Elon	Elongation, %
Proof	0.1 % proof strength, N/mm^2

1 N/mm^2=0.1 hbar=0.102 kgf/mm^2=0.06475 tonf/in^2=145.04 lbf/in^2

Symbol	Nominal analysis, supplier, condition and remarks.
ASTM B299 SL140	0.06% C 0.015% H 99.1% Ti: Sponge; primary metal **DPN: 140**
ASTM B299 SL160	0.03% C 0.015% H 99.1% Ti: Sponge; primary metal **DPN: 160**
ASTM B337/1	0.1% C 0.015% H 0.05% N Ti: Tube; commercially pure titanium **UTS: 270 Elon: 22% Proof: 200**
ASTM B337/2	0.1% C 0.015% H 0.05% N Ti: Tube; commercially pure titanium **UTS: 340 Elon: 20% Proof: 270**
ASTM B337/3	0.1% C 0.015% H 0.07% Ni Ti: Tube; commercially pure titanium **UTS: 420 Elon: 18% Proof: 340**
ASTM B337/4	0.15% C 0.015% H 0.07% N Ti: Tube; commercially pure titanium **UTS: 580 Elon: 15% Proof: 500**
ASTM B338/1	0.1% C 0.015% H 0.25% O Ti: Tube; commercially pure; annealed **UTS: 270 Elon: 22%**
ASTM B338/2	0.1% C 0.015% H 0.25% O Ti: Tube; commercially pure; annealed **UTS: 340 Elon: 20%**
ASTM B338/3	0.1% C 0.015% H 0.35% O Ti: Tube; commercially pure; annealed **UTS: 420 Elon: 18%**
ASTM B338/4	0.15% C 0.015% H 0.45% O Ti: Tube; commercially pure; annealed **UTS: 580 Elon: 15%**
ASTM B348/1	0.1% C 0.012% H 0.2% O Ti: Bar; commercially pure; annealed **UTS: 270 Elon: 22% Proof: 200**
ASTM B348/2	0.1% C 0.001% H 0.2% O Ti: Bar; commercially pure; annealed **UTS: 270 Elon: 22% Proof: 200**
ASTM B348/3	0.1% C 0.012% H 0.3% O Ti: Bar; commercially pure; annealed **UTS: 420 Elon: 18% Proof: 340**
ASTM B348/4	0.15% C 0.012% H 0.4% O Ti: Bar; commercially pure; annealed **UTS: 580 Elon: 15% Proof: 500**
ASTM B367 C1	0.1% C 0.01% H 0.04% O Ti: Casting; annealed; commercially pure **UTS: 450 Elon: 12% Proof: 370**
ASTM B381 F1	0.1% C 0.01% H 0.2% O Ti: Forging; annealed; commercially pure **UTS: 270 Elon: 22% Proof: 200**
ASTM B381 F2	0.1% C 0.01% H 0.2% O Ti: Forging; annealed **UTS: 330 Elon: 20% Proof: 270**
ASTM B381 F3	0.1% C 0.012% H 0.3% O Ti: Forging; annealed; commercially pure **UTS: 420 Elon: 18% Proof: 330**
ASTM B381 F4	0.15% C 0.012% H 0.4% O Ti: Forging; annealed; commercially pure **UTS: 580 Elon: 15% Proof: 480**
ASTM B382 ER Ti	Commercially pure Ti: Welding rod
A Ti 24	Ti commercially pure: For deep drawing; Avesta **UTS: 350 Elon: 25% Proof: 220**
A Ti 30	Ti commercially pure: For lining vessels etc.; Avesta **UTS: 480 Elon: 16% Proof: 270**
A Ti 35	Ti commercially pure: Avesta **UTS: 530 Elon: 13% Proof: 330**
A Ti Pd	0.2% Pd Ti alloy: Avesta; for use with dilute acids **UTS: 360 Elon: 25% Proof: 220**
BS 3531/1.5	99.0% Ti: Wrought; for implants and surgery tools
CRUCIBLE A 40	Commercially pure Ti: Crucible Steel Co **Proof: 270**

Symbol	Nominal analysis, supplier, condition and remarks.
CRUCIBLE A 55	Commercially pure Ti: Crucible Steel Co **Proof: 370**
CRUCIBLE A 70	Commercially pure Ti: Crucible Steel Co **Proof: 480**
DTD 5003 B	99.7% Ti: Commercially pure bar; annealed **UTS: 530 Elon: 20% Proof: 300**
DTD 5013 B	0.013% H Ti: Commercially pure bar; annealed **UTS: 450 Elon: 25% Proof: 200**
DTD 5023 B	99.7% Ti: Commercially pure sheet; annealed; suitable for welding **UTS: 530 Elon: 20% Proof: 330**
DTD 5033 B	99.7% Ti: Commercially pure sheet; annealed **UTS: 450 Elon: 25% Proof: 200**
DTD 5063 A	99.7% Ti: Commercially pure sheet; annealed **UTS: 700 Elon: 15% Proof: 450**
DTD 5073	0.1% C 0.2% Fe Ti: Commercially pure Ti tube; annealed **UTS: 370 Proof: 270**
DTD 5183	Commercially pure Ti: Sheet & strip **UTS: 450 Elon: 22% Proof: 270**
DTD 5193	Commercially pure Ti: Sheet & strip **UTS: 610 Elon: 18% Proof: 370**
h Ti 18	99.9% Ti: Sponge; Koch-Light Ltd
h Ti 24	99.6% Ti: Sheet; Koch-Light Ltd
h Ti 72	99.999% Ti single crystals 3 mm × 6 mm: Koch-Light Ltd
HYLITE 1	High purity Ti: Chemical grade; Jessops; obsolete
HYLITE 10	0.013% H (max) commercially pure Ti: Bar, forgings, sheet, etc.; Jessops; annealed; obsolete **UTS: 400 Elon: 30% Proof: 240**
HYLITE 15	0.013% H (max) commercially pure Ti: Bar & forging; Jessop; annealed; obsolete **UTS: 500 Elon: 25% Proof: 300**
HYLITE 15	0.013% H (max) commercially pure Ti: Sheet; Jessop; cold rolled & annealed; obsolete **UTS: 520 Elon: 18% Proof: 370**
HYLITE 15 H	Commercially pure Ti: Jessop; obsolete
ICI 115	Commercially pure Ti: ICI; now IMI see Ti 115
ICI 125	Commercially pure Ti: ICI; now IMI see Ti 125
ICI 130	Commercially pure Ti: ICI; now IMI see Ti 130
IMI 260	0.2% Pd Ti alloy: IMI
MST 0.2 Pd	0.1% C 0.01% H 0.2% Fe Ti: Wrought; Reactive Metal Products Ltd
MST 30	Commercially pure Ti: Reactive Metal Products Ltd; annealed **UTS: 250 Elon: 22% Proof: 200**
MST 40	Commercially pure Ti: Reactive Metal Products Ltd; annealed **UTS: 330 Elon: 20% Proof: 250**
MST 55	Commercially pure Ti: Reactive Metal Products Ltd; annealed **UTS: 450 Elon: 18% Proof: 370**
MST 70	Commercially pure Ti: Reactive Metal Products Ltd; annealed **UTS: 580 Elon: 15% Proof: 480**
RM I 0.2% Pd	Commercially pure Ti: Reactive Metal Products Ltd; annealed **UTS: 340 Elon: 22% Proof: 250**
RM I 30	Commercially pure Ti: Reactive Metal Products Ltd; annealed **UTS: 250 Elon: 25% Proof: 200**
RM I 40	Commercially pure Ti: Reactive Metal Products Ltd; annealed **UTS: 340 Elon: 22% Proof: 250**

Symbol	Nominal analysis, supplier, condition and remarks.
RM I 55	Commercially pure Ti: Reactive Metal Products Ltd; annealed **UTS: 530 Elon: 20% Proof: 370**
RM I 70	Commercially pure Ti: Reactive Metal Products Ltd; annealed **UTS: 570 Elon: 15% Proof: 480**
SB 265	See ASTM B265/1
T 35	Commercially pure Ti: French National Standard
T 40	Commercially pure Ti: French National Standard
T 50	Commercially pure Ti: French National Standard
T 60	Commercially pure Ti: French National Standard
Ti 0.15 Pd	0.08% C 0.05% N 0.015% H 0.25% Fe 0.15% Pd Ti: Sheet, bar, etc.; Titanium Metals Ltd; annealed for chemical industry **DPN: 200 UTS: 370 Elon: 22% Proof: 300**
Ti 35 A	0.08% C 0.05% N 0.015% H 0.25% Fe Ti: Sheet, bar, etc.; Titanium Metals Ltd; annealed **DPN: 120 UTS: 240 Elon: 25% Proof: 170**
Ti 55 A	0.08% C 0.05% N 0.015% H 0.25% Fe Ti: Sheet, bar, etc.; Titanium Metals Ltd; annealed **DPN: 200 UTS: 360 Elon: 22% Proof: 300**
Ti 65 A	0.08% C 0.05% N 0.015% H 0.25% Fe Ti: Sheet, bar, etc.; Titanium Metals Ltd; annealed **DPN: 220 UTS: 440 Elon: 20% Proof: 360**
Ti 75 A	0.08% C 0.05% N 0.015% H 0.25% Fe Ti: Sheet, bar, etc.; Titanium Metals Ltd; annealed **DPN: 265 UTS: 530 Elon: 15% Proof: 480**
Ti 100 A	0.08% C 0.05% N 0.01% H 0.25% Fe Ti: Bar & wire; Titanium Metals Ltd; annealed **DPN: 295 UTS: 540 Elon: 15% Proof: 460**
Ti 115	Commercially pure Ti: IMI; annealed **UTS: 390 Elon: 30% Proof: 200**
Ti 120	Commercially pure Ti: Rod, sheet & tube; IMI; withdrawn
Ti 125	Commercially pure Ti: IMI; annealed **UTS: 450 Elon: 22% Proof: 270**
Ti 130	Commercially pure Ti: IMI; annealed **UTS: 530 Elon: 20% Proof: 330**
Ti 150	Commercially pure Ti: IMI; annealed **UTS: 610 Elon: 18% Proof: 370**
Ti 155	Commercially pure Ti: Sheet; IMI; annealed **UTS: 650 Elon: 24% Proof: 550**
Ti 160	Commercially pure Ti: IMI; annealed **UTS: 700 Elon: 15% Proof: 450**
Ti 260	Commercially pure Ti with addition of Pd: IMI; for use with non-oxidizing acids **UTS: 360 Elon: 25% Proof: 200**
Ti Po 1	Commercially pure Ti: AICMA
Ti Po 2	Commercially pure Ti: AICMA
Ti Po 2	Commercially pure Ti: AICMA
TITANIUM	Ti 99–100% purity: Blackwells
U 99/100	99.95% Ti 0.03% Ag: Commercially pure Ti; origin unknown

Note. The following abbreviations and units are used in the tables:

DPN	Hardness, diamond pyramid number
UTS	Ultimate tensile strength, N/mm^2
Elon	Elongation, %
Proof	0.1 % proof strength, N/mm^2

$1 \ N/mm^2 = 0.1 \ hbar = 0.102 \ kgf/mm^2 = 0.06475 \ tonf/in^2 = 145.04 \ lbf/in^2$

51B Titanium alloys – not solution treated

Specific gravity	4.5	
Density	4314 kg/m^3	(0.16 lb/in.^{3})
Solidus/liquidus	1510–1660 °C	
Thermal conductivity	14.7 W/m °C	(3.5 cal/m s °C)
Coefficient of linear expansion	8.7 × 10^{-6}/ °C	
Electrical conductivity	1.0% IACS (copper 100%)	
Specific resistance	1500 microhm mm	
Young's modulus of elasticity	103–117 × 10^9 N/m^2	(15–17 × 10^6 lbf/in.^{2})
Impact	15 J	(11 ft/lb) (Izod)
Fatigue strength – endurance limit	± 450 N/mm^2	(± 30 tonf/in.^{2})
Hot strength		

Temperature °C	Tensile strength N/mm^2	Elongation %
100	840	20
200	760	20
300	720	20
400	700	18
500	580	18
600	270	150

The above properties are typical of the following group, and may not apply exactly to any one specification. It is possible that with certain specifications some of the values may not be applicable.

General metallurgical characteristics

These materials have alloying additions of aluminium, manganese, tin or vanadium which strengthen the titanium matrix, particularly when cold worked. With the alloys listed, the alloying elements are dissolved in the titanium and remain in solution at all temperatures, thus these alloys cannot be strengthened by any form of thermal treatment.

Titanium alloys have a fatigue endurance limit above which failure by fatigue does not occur and thus differ from other non-ferrous metals. They are however, rather notch sensitive in that the fatigue strength falls by a disproportionate amount when stress raisers are present. Care must therefore be exercised not to design changes of section, joins or other stress raisers where there are stress situations.

Also listed in this section are several titanium carbide specifications. These are hard, unworkable, special purpose materials.

Thermal treatment

These alloys can all be hardened by cold work and most of them will accept a considerable amount of deformation before becoming too brittle for normal use.

It is recommended that parts are stress relieved after cold working to ensure dimensional stability. This should be for two hours at a temperature 20–50 °C above that at which the part will be used in service or 500 °C, whichever is the lower.

When all trace of cold working must be removed, these materials must be annealed for about 30 minutes at 750–800 °C. Time at temperature must be as short as possible to prevent undesirable grain growth. This anneal will cause surface hardening by pick up of hydrogen, oxygen and nitrogen, which must be removed by pickling as it is very abrasive in nature and being hard and brittle could have a serious effect on the fatigue strength.

Stress relieving at 450–500 °C will remove most of the cold work and results in little or no surface embrittlement. This treatment is satisfactory for most purposes.

Some de-scaling can be carried out by grit blasting but this must be carefully controlled to prevent surface contamination from the grit used. As the surface layer is relatively thick, hard and very adherent, the pressures required are high and large sized grit particles are necessary. This will result in distortion of all but the sturdiest components and often mitigates against the use of blasting for de-scaling.

Acid pickles using warm hydrofluoric and nitric acid mixtures remove the scale but require considerable control to prevent local attack on the base metal. Oxidizing molten salt baths can also be used but again require control to prevent oxygen pick-up.

Sodium hydride baths must not be used on titanium owing to the danger of hydrogen contamination and embrittlement.

Welding and brazing

Pre-treatment. It is essential that all traces of grease, oil and dirt are removed prior to welding. It is more important that titanium alloys are scrupulously clean for welding than for most metals.

Edge preparation should be carried out immediately prior to joining and should be designed to produce the minimum of cold work. If necessary, the parts should be annealed after machining the edge, then pickled or carefully blasted immediately before welding.

The design of the joint must be carefully considered to

ensure that adequate supplies of inert gas will be at both sides of the weld.

Welding and brazing. As titanium allous are embrittled by oxygen, nitrogen and hydrogen, none of the normal electric arc or gas welding techniques can be used.

Some techniques have been developed using vacuum chambers or inert gas filled plastic containers but these tend to be cumbersome and make manipulation awkward. For most purposes inert gas shielded electric arc welding gives satisfactory results provided certain precautions are followed. Careful design of the joint and edge to ensure complete gas coverage of both sides of the weld at all times is essential when this is above about 600 °C. This generally necessitates backing strips and implies full access to both sides of the weld. Welders must use smooth movements to prevent disturbances of the gas layer and the equipment must be capable of separate control of gas flow and electric arc, so that a gas curtain is available before the arc is struck and is maintained while the weld pool is cooling. Provided these precautions are taken, the alloys listed can all be welded to give excellent results.

Resistance welding can often be carried out without the use of inert gas, provided some slight reduction in ductility can be accepted. This is achieved by the relatively high electrical resistance of titanium alloys which allows the use of high currents with rapid welding speeds of the order of 20–50 cycles of a.c. current. A high standard of surface cleanliness is essential and constant control of the welding technique.

Brazing of titanium can only be carried out under vacuum or with an inert gas atmosphere. Hydrogen brazing is not possible.

Post-treatment. No thermal treatment is necessary after welding or brazing, except on complicated structures to remove thermal stresses.

Flaw detection methods

Crack test. Penetrant oil chalk test. Where high sensitivity is essential fluorescent fluids are recommended with the surface prepared by acid etching when the highest sensitivity is required.

X-ray. Not normally required on wrought products but should always be called for on welds. High duty joins should be examined 100% and all welds controlled by percentage quality checking.

Ultra-sonic test. This is applied after rough machining to prove the absence of sub-cutaneous defects which would otherwise only be found after further expensive machining. Techniques are available for identifying porosity and lack of fusion in welds.

Chemical etch. Hot phosphoric or hydrofluoric acid etch can be used to show undesirable grain growth. It is also useful in preparing the surface for crack detection when very high sensitivity is essential.

Corrosion protection

Temporary. None required.

Permanent. None required for most uses and up to 500 °C. Above this temperature, diffusion of oxygen and nitrogen from the surface increases with a resultant drop in corrosion resistance and fatigue strength.

The titanium alloys listed all have excellent corrosion resistance obtained from the tenacious oxide coating. These alloys resist attack from all normal acids and alkalis at room temperature and often at high temperatures. One exception is fuming nitric acid which can react violently – even explosively – under certain conditions at room temperature. Unlike other metals, alloying elements have little affect on the corrosion properties of titanium.

Machinability

Titanium alloys have a tendency to gall and weld themselves to cutting tools, resulting in poor finish and low tool life.

This can be minimized using special coolants of the chlorinated carbon type.

It is important to keep the tool cutting at all times, obviated interrupted cuts and tool rubbing. This generally requires robust machines of high rigidity to overcome the inherent high strength of the alloys. Care must be taken to ensure complete removal of the scale formed at annealing as this is very abrasive.

In many respects the alloys resemble the tougher austenitic stainless steels.

Titanium dust can be made to ignite readily and given the correct conditions this can result in an explosion. Provided good housekeeping is practised at all times and accumulations of the dust forbidden or kept wet there is no danger. Titanium fires must not be tackled with water as this causes an explosion. Polishing and grinding areas handling titanium should have containers of dry chalk or fine sand and this should be used to confine rather than extinguish the blaze.

Uses

Aircraft structural parts where reasonable strength at temperature or corrosion resistance are essential, such as fireproof engine bulkheads. Gas turbine engine components such as jet pipes, exhaust units, etc.

In chemical engineering these titanium alloys are used for similar application to commercially pure titanium where greater strength is required.

Symbol	Nominal analysis, supplier, condition and remarks.
2 Cr 2 Fe 2 Mo	0.05% C 2.2% Cr 2.2% Mo 2.2% Fe Ti alloy: Alpha–beta grade; information from SAE handbook
4 Al 4 Mn	0.05% C 4.0% Mn 4.0% Al Ti alloy: Alpha–beta alloy; information from SAE handbook
5 Al 1.25 Fe 2.75 Cr	0.08% C 2.7% Cr 5.0% Al 1.2% Fe Ti alloy: Alpha–beta grade; information from SAE handbook
5 Al 2.5 Sn	0.15% C 2.5% Sn 5.0% Al Ti alloy: Alpha grade; information from SAE handbook
5 Al 5 Wn 5 Zr	0.02% C 4.8% Sn 5.0% Al 5.2% Zr Ti alloy: Alpha grade; information from SAE handbook
7 Al 2 Cb 1 Ta	0.04% C 1.0% W 2.0% Nb 7.0% Al Ti alloy: Alpha grade; information from SAE handbook
7 Al 4 Mo	0.05% C 4.0% Mo 6.9% Al Ti alloy: Alpha–beta grade; information from SAE handbook
7 Al 12 Zr	0.02% C 7.0% Al 12.0% Zr Ti alloy: Alpha grade; information from SAE handbook
8 Al 1 Mo 1 V	0.04% C 1.0% Mo 8.0% Al 1.0% V Ti alloy: Alpha grade; information from SAE handbook
8 Mn	0.1% C 8.0% Mn Ti alloy: Alpha–beta grade; information from SAE handbook
A 110 AT	5% Al 2.5% Sn alloy: Crucible Steel Co; annealed UTS: 920 Elon: 15% Proof: 440
AMS 4908 A	8% Mn Ti alloy: Sheet & strip; annealed Proof: 760
AMS 4910	5% Al 2.5% Sn Ti alloy: Sheet & strip; annealed; AMS for commercial alloy A 110AT Proof: 760
AMS 4923	2% Cr 2% Fe 2% Mo Ti alloy: Bar & forging; annealed Proof: 880
AMS 4925 A	4% Al 4% Mn Ti alloy: Bar & forging; annealed; AMS for commercial alloy C130 AM Proof: 930
AMS 4926	5% Al 2.5% Sn Ti alloy: Bar; annealed; AMS for commercial alloy A 110AT Proof: 810
AMS 4927	5% Cr 3% Al Ti alloy: Bar & forging
AMS 4953	5% Al 2.5% Sn Ti alloy: Welding wire; annealed; AMS for commercial alloy A 110AT
AMS 4966	5% Al 2.5% Sn Ti alloy: Forging; annealed; AMS for commercial alloy A 110AT Proof: 800
AMS 4967 A	11% Sn 5% Zr 2.25% Al 1% Mo 0.2% Si Ti alloy
AMS 4974	11% Sn 4% Mo 2.2% Al 40.2% Si Ti alloy
ASTM B265/6	0.1% C 0.01% H 5% Al 2.5% Sn Ti alloy: Sheet UTS: 840 Elon: 10% Proof: 800
ASTM B265/7	0.15% C 0.01% H 7% Mn Ti alloy: Sheet UTS: 880 Elon: 10% Proof: 780
ASTM B348/6	0.1% C 0.02% H 5% Al 2.5% Sn Ti alloy: Bar; annealed UTS: 840 Elon: 10% Proof: 780
ASTM B348/7	0.1% C 0.01% H 4% Al 4% Mn Ti alloy: Bar; annealed UTS: 1050 Elon: 10% Proof: 980
ASTM B367 C3	0.1% C 0.01% H 5% Al 2.5% Sn Ti alloy: Casting; annealed UTS: 840 Elon: 8% Proof: 760

Note. The following abbreviations and units are used in the tables:

DPN	Hardness, diamond pyramid number
UTS	Ultimate tensile strength, N/mm²
Elon	Elongation, %
Proof	0.1 % proof strength, N/mm²

1 N/mm²=0.1 hbar=0.102 kgf/mm²=0.06475 tonf/in²=145.04 lbf/in²

Symbol	Nominal analysis, supplier, condition and remarks.
ASTM B381 F6	0.1% C 0.02% H 5% Al 2.5% Sn Ti alloy: forging; annealed UTS: 820 Elon: 10% Proof: 760
ASTM B381 F7	0.1% C 0.012% H 4% Al 4% Mn Ti alloy: Forging; annealed UTS: 1030 Elon: 10% Proof: 920
ASTM B382 ERTI 2.5 Al 16 V	2.5% Al 16% V Ti alloy: Welding rod
ASTM B382 ERTI 3Al	3% Al Ti alloy: Welding rod
ASTM B382 ERTI 5 Al 2.5 Sn	5% Al 2.5% Sn Ti alloy: Welding rod
ASTM B382 ERTI 8 Al 2 Cb 1 Ta	8% Al 2% Nb 1% Ta Ti alloy: Welding rod
C130 AM	4% Al 4% Mn Ti alloy: Crucible Steel Co; annealed DPN: 1070 Elon: 12% Proof: 1000
CRUCIBLE A110AT	5% Al 2.5% Sn Ti alloy: Crucible Steel Co; annealed UTS: 920 Elon: 15% Proof: 760
CRUCIBLE C130AM	4% Al 4% Mn Ti alloy: Crucible Steel Co; annealed UTS: 1070 Elon: 12% Proof: 1000
DTD 5043 B	1.5% Al 1.5% Mn Ti alloy: Bar; annealed UTS: 730 Elon: 20% Proof: 450
DTD 5053	4% Al 4% Mn Ti alloy: Bar; stress relieved UTS: 950 Elon: 12% Proof: 870
DTD 5083	5% Al 22.5% Sn Ti alloy: Bar; annealed UTS: 760 Elon: 12% Proof: 700
DTD 5093	5% Al 2.5% Sn Ti alloy: Sheet; annealed UTS: 760 Elon: 12% Proof: 700
DTD 5123	2.5% Cu Ti alloy: Bar; annealed UTS: 530 Elon: 20% Proof: 370
DTD 5133	2.5% Cu Ti alloy: Sheet; annealed UTS: 580 Elon: 20% Proof: 440
DTD 5143	4.0% Al 4.0% Mn Ti alloy: Forging; annealed UTS: 930 Elon: 12% Proof: 870
DTD 5213	11% Sn Al Mo Si Ti alloy: Bar & billet; annealed UTS: 1210 Elon: 10% Proof: 1080
DTD 5203	Al Mo Sn Si Ti alloy: Bar & billet UTS: 1150
DTD 5223	Al Mo Sn Si Ti alloy: Forging UTS: 1150
DTD 5233	2.5% Cu Ti alloy UTS: 800
DTD 5243	2.5% Cu Ti alloy UTS: 800
DTD 5253	2.5% Cu Ti alloy UTS: 770
DTD 5263	2.5% Cu Ti alloy UTS: 770
DTD M159	11% Sn Al Mo Si Ti alloy: Draft specification UTS: 1210 Elon: 10% Proof: 1050
DTD M160	11% Sn Al Mo Si Ti alloy: Draft specification UTS: 1210 Elon: 10% Proof: 1050
FERRO-TITANIUM	0.1% C 8% Al 1% Si 0.1% Cu 40% Fe Ti alloy: Metal Alloys Ltd; primary metal
HYLITE 20	5% Al 2.5% Sn 0.013% H (max) Ti alloy: Annealed; Jessop; obsolete UTS: 860 Elon: 11% Proof: 720
HYLITE 25	2.5% Cu Ti alloy: Jessop; obsolete
HYLITE 55	3% Al 6% Sn 5% Zr 0.5% Si Ti alloy: Jessop; obsolete
ICI 230	2% Cu Ti alloy: ICI; now IMI; see Ti 230
ICI 314 A	4% Al 4% Mn Ti alloy: ICI; now IMI; see Ti 314A
ICI 314 C	2% Al 2% Mn Ti alloy: ICI; now IMI; see Ti 314C

Symbol	Nominal analysis, supplier, condition and remarks.
ICI 317	5% Al 2.5% Sn Ti alloy: ICI; now IMI; see Ti 317
K 138 A	Ti C with Co binder: Kennametal; seal rings, bearings, etc.; in contact with Co bound WC
K 151 A	Ti C with Ni bonder: Kennametal; difficult to wet; used for brazing fixtures, etc.
K 162 B	Ti C with No Mo binder: Kennametal; seal rings, valve parts
K 163 B1	Ti C with No Mo binder: Kennametal; stronger than K 162B
K 164 B	Ti C with Ni Mo binder: Kennametal; stands thermal & mechanical shock
K 165	Ti C with Ni Mo binder: Kennametal; wear resistant
KENTANIUM	Titanium carbide: Available in various grades; Kennametal
MST 3 Al 2.5 V	0.05% C 0.3% Fe 3% Al 2.5% V 0.015% H Ti alloy: Wrought; Reactive Metals Ltd
MST 4 Al 4 Mn	0.08% C 0.4% Fe 4% Al 4% Mn 0.01% H Ti alloy: Wrought; Reactive Metals Ltd
MST 5 Al 2.5 Sn	0.08% C 5% Al 2.5% Sn low N & H Ti alloy: Wrought; Reactive Metals Ltd; for low temperature use
MST 7 Al 2 Cb 1 Ta	7% Al 2% Nb 1% Ta 0.08% O Ti alloy: Wrought; Reactive Metals Ltd
MST 7 Al 12 Zr	0.04% C 7% Al 12% Zr Ti alloy: Wrought; Reactive Metals Ltd
MST 8 Mn	0.2% C 8% Mn low N & H Ti alloy: Wrought; Reactive Metals Ltd
RMI 4 Al 4 Mn	0.08% C 0.4% Fe 4% Al 4% Mn 0.01% H Ti alloy: Wrought; Reactive Metals Ltd; annealed **UTS: 1000 Elon: 10% Proof: 920**
RMI 5 Al 2.5 Sn	0.08% C 5% Al 2.5% Sn 0.015% H Ti alloy: Reactive Metals Ltd; annealed **UTS: 870 Elon: 10% Proof: 780**
RMI 5 Al 2.5 Sn EL1	0.05% C 5% Al 2.5% Sn 0.015% H Ti alloy: Reactive Metals Ltd; annealed; low impurity grade **UTS: 730 Elon: 10% Proof: 630**
RMI 5 Al 6 Sn 2 Zr 1 Mo Si	0.05% C 5% Al 6% Sn 2% Zr 1.0% Mo 0.2% Si Ti alloy: Reactive Metals Ltd; alpha alloy **UTS: 650 Elon: 12% Proof: 560**
RMI 6 Al 2 Cb 1 Ta 1 Mo	0.05% C 6% Al 2% Nb 1.0% Ta 1.0% Mo Ti alloy: Bar, etc.; Reactive Metals Ltd
RMI 6 Al 2 Cb 1 Ta 1 Mo	0.05% C 6% Al 2% Nb 1.0% Ta 1.0% Mo Ti alloy: Reactive Metals Ltd; alpha alloy **UTS: 540 Elon: 12% Proof: 490**
RMI 7 Al 2 Cb 1 Ta	0.04% C 7% Al 2% Nb 1% Ta Ti alloy: Reactive Metals Ltd; annealed **UTS: 820 Elon: 10%**
RMI 7 Al 12 Zr	0.04% C 7% Al 12% Zr Ti alloy: Reactive Metals Ltd; annealed **UTS: 940 Elon: 10% Proof: 870**
RMI 8 Al 1 Mo 1 V	0.08% C 8% Al 1% Mo 1% V 0.015% H Ti alloy: Wrought; Reactive Metals Ltd; annealed **UTS: 990 Elon: 10% Proof: 880**
RMI 8Mn	0.2% C 8% Mn 0.12% H Ti alloy: Reactive Metals Ltd; annealed **UTS: 950 Elon: 10% Proof: 780**
T 283	2.5% Al 15% V Ti alloy: Jessop; obsolete
TA 2M	2% Al 2% Mn Ti alloy: French National Standard
TA 4DE	4% Al 4% Mo 2% Sn 0.5% Si Ti alloy: French National Standard
TA 4M	4% Al 4% Mn Ti alloy: French National Standard **UTS: 950 Elon: 15% Proof: 870**

Symbol	Nominal analysis, supplier, condition and remarks.
TA 5E	5.0% Al 2.5% Sn Ti alloy: French National Standard **UTS: 760 Elon: 12% Proof: 700**
TA 6Z-5D	Al Zr Mo Si Ti alloy: For creep resistance; French National Standard
TA 6ZW	5% Al 5% Zr 1% W 0.3% Si Ti alloy: For creep strength; French National Standard
TE11DA	11% Sn 4% Mo 2½% Al 0.2% Si Ti alloy: French National Standard
Ti 5 Al 2.5 Sn	0.08% C 4.6% Al 2.5% Sn low N & H Ti alloy: Sheet, bar, etc.; Titanium Metals Ltd; annealed **DPN: 352 UTS: 820 Elon: 10% Proof: 770**
Ti 5 Al 2.5 Sn EL1	0.08% C 5% Al 2.5% Sn low N & H Ti alloy: Sheet, bar, etc.; Titanium Metals Ltd; annealed; for low temperature use **DPN: 340 UTS: 690 Elon: 10% Proof: 620**
Ti 5 Al 6 Sn 5 Zr	0.04% C 5% Al 5% Sn 5% Zr Ti alloy: Sheet, bar, etc.; Titanium Metals Ltd; annealed; high creep strength **UTS: 800 Elon: 10% Proof: 750**
Ti 7 Al 12 Zr	0.04% C 7% Al 12% Zr Ti alloy: Sheet, bar, etc.; Titanium Metals Ltd; annealed; high creep strength **UTS: 900 Elon: 10% Proof: 820**
Ti 8 Mn	0.2% C 8% Mn low N & H Ti alloy: Sheet, bar, etc.; Titanium Metals Ltd; annealed **UTS: 820 Elon: 10% Proof: 750**
Ti 314 A	4% Al 4% Mn Ti alloy: Plate; IMI **UTS: 930 Elon: 15% Proof: 870**
Ti 314 C	2% Al 2% Mn Ti alloy: Plate; IMI **UTS: 740 Elon: 20% Proof: 450**
Ti 315	2.0% Al 2.0% Mn Ti alloy: IMI **UTS: 730 Elon: 20% Proof: 450**
Ti 317	5% Al 2.5% Sn Ti alloy: Bar & plate; IMI **UTS: 760 Elon: 12% Proof: 700**
Ti 550	4% Al 4% Mo 2% Sn 0.5% Si Ti alloy: Bar for creep use up to 400 °C; IMI **UTS: 1140 Elon: 9% Proof: 1000**
Ti 551	4% Al 4% Mo 4% Sn 0.5% Si Ti alloy: Bar; IMI **UTS: 1240 Elon: 8% Proof: 1100**
Ti 680	11% Sn Al Mo Si Ti alloy: IMI **UTS: 1210 Elon: 10% Proof: 1050**
Ti 684	6% Al 5% Zr 1% W 0.3% Si Ti alloy: Bar; IMI; for creep use up to 520 °C **UTS: 1000 Elon: 6% Proof: 880**
Ti 685	Al Zr Mo Si Ti: Bar; IMI; for creep use up to 550 °C **UTS: 1000 Elon: 6% Proof: 880**
Ti P62	4.0% Al 4.0% Mn Ti alloy: AICMA **UTS: 930 Elon: 15% Proof: 870**
UNS R50001–R59999	Ti metal and alloys: American Standards System; unified numbers
VR 65	Mo Ni Ti: Sinter material; VR/Wesson specific gravity 5.9; hardness 1870 DPN

Note. The following abbreviations and units are used in the tables:

DPN	Hardness, diamond pyramid number
UTS	Ultimate tensile strength, N/mm²
Elon	Elongation, %
Proof	0.1 % proof strength, N/mm²

$1 \text{ N/mm}^2 = 0.1 \text{ hbar} = 0.102 \text{ kgf/mm}^2 = 0.06475 \text{ tonf/in}^2 = 145.04 \text{ lbf/in}^2$

51C Titanium alloys – solution treated

Specific gravity	4.42–4.96	
Density	4420–4960 kg/m³	(0.16–0.10 lb/in.³)
Solidus/liquidus	1510–1650 °C	
Thermal conductivity	6.3 W/m °C	(1.5 cal/m s °C)
Coefficient of linear expansion	$8 \times 10^{-6}/$ °C	
Electrical conductivity	1.0% IACS (copper 100%)	
Specific resistance	1700 microhm mm	
Young's modulus of elasticity	$103–117 \times 10^9$ N/m²	$(15–17 \times 10^6$ lbf/in.²)
Impact	16 J	(12 ft/lb) Izod
Fatigue strength – endurance limit	$\pm$ 370–450 N/mm²	($\pm$ 25–30 tonf/in.²)
Hot strength		

Temperature °C	Tensile strength N/mm²
100	910
200	780
300	760
400	690
500	610

The above properties are typical of the following group, and may not apply exactly to any one specification. It is possible that with certain specifications some of the values may not be applicable.

General metallurgical characteristics

With these alloys use is made of the phase change which occurs in pure titanium at about 880 °C, from alpha to beta structure. This is caused by changing from a hexagonal close-packed crystal structure to a body-centred cubic crystal structure. This change is analogous to that which occurs in iron and is the basis for all thermal treatments of steel where one phase dissolves iron carbide while the other does not. To date no element analogous to carbon in steel has been found to give a similar response with titanium but all the alloys listed can have their mechanical properties improved by some form of thermal treatment.

The alloys form two groups, the more common being hardened by a complex metallurgical process relying on the instability of the beta phase.

The second series of alloys are much more recent and are to some extent in an early stage of development. With these the beta phase has been stabilized at room temperature instead of 880 °C with pure titanium; this is achieved by alloying and is analogous to the 18/8 chromium/nickel stainless steels. Hardening is then achieved by precipitating intermetallic compounds from solution in a manner similar to the complex stainless steel alloys discussed in section 44N2.

All the alloys have good workability in the annealed or solution treated condition. The corrosion resistance of titanium is seldom markedly decreased by the addition of alloying elements but little benefit is achieved from these elements, particularly regarding hot oxidation and hydrogen contamination, thus use is limited to below 500 °C for most purposes.

Unlike other non-ferrous metals, titanium alloys have an endurance limit below which failure does not occur by fatigue. They have however, rather poor notch sensitivity characteristics, thus considerable care must be taken to remove potential stress raisers from areas of stress concentration. These include joints, rapid changes in section, damage marks or corrosion pits.

Thermal treatment

The alloys listed are all capable of accepting considerable cold work in the annealed or solution treated condition but very little after final aging. With some alloys the best possible mechanical properties are obtained by a measured amount of cold work after solution treatment before ageing.

Solution treating and annealing are virtually identical in that the only way to remove all cold work and prevent some age hardening is to heat to 700–850 °C, depending on the alloy and water quenching. This homogenizes the grain and removes all trace of previous hot or cold work, while the quench suppresses any tendency to age. Any intermetallic compounds are dissolved and retained in solution.

Where it is only necessary to remove the cold work, stress relieving or annealing can be carried out at 600–650 °C when the cooling rate is relatively unimportant. No appreciable scale is produced by this treatment whereas the solution treatment causes formation of a brittle hard oxide which must be removed. Lower stress relieving temperatures remove some degree of cold work in proportion to the temperature used.

These materials are age hardened at a temperature between 350 and 450 °C for a time of up to 16 hours. The actual time at temperature can be varied for each alloy but in general best results are obtained by the lowest temperature for the longest time. The age allows some of the dissolved intermetallic to start precipitating from solution and this results in some lattice strain and hardening.

The age operation must always be preceded by solution treatment but some cold work can be carried out prior to ageing with beneficial results. Excessive cold work can result in cracking occurring at the age, and any stress relieving carried out after solution treatment will reduce the efficiency of the age.

Temperatures and times are critical for all these alloys and it is essential that advice is obtained from the supplier regarding the correct information before attempting any form of thermal treatment. Too high a temperature at solution treatment results in undesirable grain growth and excessive scale, too low a temperature in incomplete solution, thus improper hardening. The wrong time or temperature at ageing results in a drop in hardness.

Heating of these alloys above about 700 °C in any atmosphere containing hydrogen, nitrogen or oxygen results in a hard brittle surface with this brittleness penetrating into the section when hydrogen is part of the atmosphere. Titanium alloys should therefore not be heated above 700 °C in a reducing or moist atmosphere. Provided scale removal facilities are available an oxidizing atmosphere is acceptable. Inert gas atmospheres or vacuum should be used when scale removal is not possible.

Electric furnaces, using dry air if possible, give satisfactory results. If gas or oil fired furnaces are used, it is essential that the burners are adjusted to give an oxygen rich atmosphere. Neutral salt baths can be used provided very careful control is maintained to ensure the salt remains neutral or slightly oxidizing.

Scale removal is essential if the alloys have been heated above about 700 °C, unless under vacuum or an inert gas. The oxide formed is hard, abrasive and brittle with resultant deleterious effects on cutting tools or fatigue strength if not removed.

Grit blasting, using a coarse abrasive and high pressures, is reasonably efficient provided the parts are substantial enough not to be distorted. Care must also be taken that the surface is not contaminated with metallic particles as this would seriously affect the corrosion resistance. Molten salt baths are available which attack the surface layer. These require considerable control to prevent local attack on the metal itself and generally a subsequent acid pickle. Acid pickling with nitric/hydrofluoric acid used warm gives satisfactory results provided routine control is carried out, as the rate of attack increases and becomes selective with older acid solutions.

Sodium hydride must not be used as this causes considerable embrittlement.

Welding and brazing

Pre-treatment. Absolute cleanliness of the joint area is essential. Freedom from moisture, grease, oil and dirt is more important when titanium alloys are to be joined than with any other metal.

Design of the joint must be carefully considered to ensure adequate inert gas coverage on both sides of the weld and give the welder freedom of movement.

The parts should be annealed or solution treated, the actual weld area being carefully blasted or acid pickled immediately prior to welding.

Welding and brazing. Normal electric arc or gas welding is not possible with any of these alloys as the oxygen, nitrogen and hydrogen picked up cause a very brittle useless weld.

The use of vacuum chambers and pliable, inert gas filled plastic bags has resulted in successful joints but these are not suitable methods for production. Inert gas shielded electric arc welding is now giving successful welds provided an adequate flow of gas in maintained at both sides of the weld at all times, when the metal is above about 600 °C. This requires separate controls for gas and the electric arc to allow the gas to flow before striking the arc and after the weld is completed. Special fixtures are required to maintain the gas curtain at the underside of the weld. Careful design of the joint can considerably aid this problem. The welder must ensure the minimum disturbance to the gas curtain by the use of downhand welding and slow deliberate movements.

Provided these precautions are followed, excellent fusion welds can be accomplished.

Resistance welding can be successful without the use of gas curtain, provided the correct equipment and technical control is available. This is possible because the high electrical resistance of titanium allows high currents to be used for very short periods of no more than 10–20 cycles of alternating current.

Brazing must be carried out under vacuum and is seldom economically justified. Hydrogen atmospheres must not be used as this leads to serious embrittlement.

Post-treatment. Parts must be aged to obtain the maximum properties. Generally the welding will be carried out after solution treatment and ageing is all that is subsequently necessary. If the part was not solution treated this must be carried out before ageing.

Flaw detection methods

Crack test. Penetrant oil chalk test. Where the highest sensitivity is required, fluorescent dye penetrant oils should be used preceded by an acid etch.

X-ray. All high duty welds should be X-rayed and it is recommended that all welding is controlled by a percentage X-ray. Forgings will not normally be subjected to this form of inspection.

Ultra-sonic test. This should be applied at an early stage of manufacture to parts with expensive machine operations. It can find sub-cutaneous defects which would scrap the finished part.

Techniques have now been developed which can identify certain types of welding defects.

Chemical etch. A hot phosphoric or nitric/hydrofluoric acid etch can be used to show large grain or other surface defects. It is also useful in preparing the surface for crack test when extra sensitivity is necessary.

Corrosion protection

Temporary. None required.

Permanent. None required under most circumstances. These alloys have approximately the same corrosion resistance as the pure grades of titanium. With the exception of fuming nitric acid there is no appreciable attack of any acid or alkali at room temperature and considerable resistance to many acids at high temperatures.

Up to about 600 °C there is no appreciable attack in normal atmospheres but above this figure a hard brittle oxide forms and this becomes excessive, at higher temperatures. There is also considerable attack from hydrogen which rapidly penetrates normal sections. Hydrogen attack can take place if moisture is present above about 600 °C. Considerable work is in progress on suitable means of protecting titanium alloys from these high temperature effects.

Machinability

These alloys have a high strength in the aged condition and this added to the poor machinability makes it difficult to achieve a good finish. There is a tendency for titanium alloys to gall and weld to the cutting edge resulting in tearing and plucking.

The alloys listed are probably best machined in the annealed or solution treated condition, then aged. It may be necessary to re-solution treat to remove excessive cold work prior to ageing if very heavy machining is involved. Where distortion must be minimized, parts can be finally machined after ageing.

Best results will be obtained using rigid machines with ample power, taking heavy cuts at low speeds. It is important to prevent the tool rubbing and interrupted cuts should be avoided if at all possible.

It is also important to ensure that all oxide scale is removed as this is very hard and abrasive.

Titanium dust can ignite with explosive violence under certain conditions thus some precautions are necessary where these alloys are ground or polished. Provided normal good housekeeping is practised and the dust not allowed to accumulate there is no danger. Containers of dry chalk or fine sand should always be available as water must not be used on titanium fires. The chalk or sand should be used to confine rather than extinguish the blaze. Polishing or grinding dust should be kept wet at all times.

Uses

The expense of these materials restricts their use to the aircraft industry at present. They are used for high duty rotating parts such as compressor blading, shafts and wheels.

A limited number of chemical engineering uses have been found for these alloys where their light weight, corrosion resistance and strength are valuable.

Symbol	Nominal analysis, supplier, condition and remarks.
1 Al 8 V 5 Fe	0.05% C 1.0% Al 5.0% Fe 8.0% V Ti alloy: Beta grade; information from SAE handbook
2.5 Al 16 V	0.04% C 2.5% Al 16% V Ti alloy: Alpha–beta grade; information from SAE handbook
3.7164	6.0% Al 4.0% V Ti alloy: German Standard **UTS: 95 Elon: 10% Proof: 700**
3 Al 12.5 V	0.02% C 3.0% Al 2.5% V Ti alloy: Alpha–beta grade; information from SAE handbook
3 Al 13 V 11 Cr	0.02% C 11.0% Cr 3.0% Al 13.5% V Ti alloy: Beta grade; information from SAE handbook
4 Al 3 Mo 1 V	0.04% C 3.0% Mo 4.25% Al 1.0% V Ti alloy: Alpha–beta alloy; information from SAE handbook
6 Al 4 V	0.02% C 6.1% Al 3.8% V Ti alloy: Alpha–beta grade; information from SAE handbook
6 Al 6 V 2 Sn	0.02% C 2.0% Sn 5.5% Al 5.5% V Ti alloy: Alpha–beta grade; information from SAE handbook
AMS 4911 A	6% Al 4% V Ti alloy: Sheet & strip; annealed; AMS for commercial alloy C 120 AV **Proof: 880**
AMS 4912	4% Al 3% Mo 1% V Ti alloy: Sheet & strip; solution treated
AMS 4913	4% Al 3% Mo 1% V Ti alloy: Sheet & strip; solution treated & aged

Symbol	Nominal analysis, supplier, condition and remarks.
AMS 4917	13.5% V 11% Cr 3% Al Ti alloy: Sheet & strip; solution treated; AMS for commercial alloy B 120 VCA
AMS 4928 A	6% Al 4% V Ti alloy: Bar & forgings; annealed; AMS for commercial alloy C 120 AV **Proof: 880**
AMS 4929	5.4% Al 1.5% Cr 1.3% Fe 1.25% Mo Ti alloy: Bar; annealed; AMS for commercial alloy Ti 155A **Proof: 1000**
AMS 4935	6% Al 4% V Ti alloy: Bar; multiple vacuum melted; annealed; AMS for commercial alloy C 120 AV
AMS 4954 A	6% Al 4% V Ti alloy: Welding wire; AMS for commercial alloy C 120 AV
AMS 4969	5.5% Al 1.5% Cr 1.3% Fe 1.25% Mo Ti alloy: Forging; annealed; AMS for commercial alloy Ti 155 A
ASTM B265/5	0.1% C 0.01% H 6.2% Al 4% V Ti alloy: Sheet **UTS: 950 Elon: 10% Proof: 88**
ASTM B348/5	0.1% C 0.012% H 6% Al 4% V Ti alloy: Bar; annealed **UTS: 960 Elon: 10% Proof: 880**
ASTM B367 C2	0.1% C 0.01% H 6% Al 4% V Ti alloy: Casting; annealed **UTS: 950 Elon: 6% Proof: 870**
ASTM B381 F5	0.1% C 0.012% H 6% Al 4% V Ti alloy: Forging; annealed **UTS: 920 Elon: 10% Proof: 840**
ASTM B381 F8	0.1% C 0.012% H 6% Al 5.5% V 2% Sn 0.7% Cu Ti alloy: Forging; annealed **UTS: 1070 Elon: 12% Proof: 1000**
ASTM B381 F9	0.1% C 0.012% H 6.5% Al 4% Mo Ti alloy: Forging; annealed **UTS: 1070 Elon: 10% Proof: 1000**

Note. The following abbreviations and units are used in the tables:

DPN	Hardness, diamond pyramid number
UTS	Ultimate tensile strength, N/mm²
Elon	Elongation, %
Proof	0.1 % proof strength, N/mm²

$1 \text{ N/mm}^2 = 0.1 \text{ hbar} = 0.102 \text{ kgf/mm}^2 = 0.06475 \text{ tonf/in}^2 = 145.04 \text{ lbf/in}^2$

Symbol	Nominal analysis, supplier, condition and remarks.
ASTM B382 ER Ti 4 Al 3 Mo 1 V	4% Al 1% V 3% Mo Ti alloy: Welding rod
ASTM B382 ER Ti 4 Al 4 V	4% Al 4% V Ti alloy: Welding rod
ASTM B382 ER Ti 5 Al 4 Fe Cr	5% Al 2.8% Cr 1% Fe Ti alloy: Welding rod
ASTM B382 ER Ti 6 Al 4 V	6% Al 4% V Ti alloy: Welding rod
ASTM B382 ER Ti 13 V 11 Cr 3 Al	3% Al 13% V 11% Cr Ti alloy: Welding rod
B 120 VCA	3% Al 13% V 11% Cr Ti alloy: Crucible Steel Co; solution treated & aged UTS: 1520 Elon: 6% Proof: 1250
C 120 AV	6% Al 4% V Ti alloy: Crucible Steel Co; solution treated & aged UTS: 1110 Elon: 7% Proof: 1050
C 135 AMO	7% Al 4% Mo Ti alloy: Crucible Steel Co; solution treated & aged UTS: 6%10 Proof: 1140
CRUCIBLE B120 VCA	3% Al 13% V 11% Cr Ti alloy: Crucible Steel Co; solution treated & aged UTS: 1410 Elon: 6% Proof: 1250
CRUCIBLE C 120 AV	6% Al 4% V Ti alloy: Crucible Steel Co; solution treated & aged UTS: 1110 Elon: 7% Proof: 1050
CRUCIBLE C 135 AMo	7% Al 4% Mo Ti alloy: Crucible Steel Co; solution treated & aged UTS: 1410 Elon: 6% Proof: 1140
DTD 5103	4% Al 4% Mo 2% Sn 0.5% Si Ti alloy: Bar; solution treated & aged UTS: 1070 Elon: 10% Proof: 880
DTD 5113	11% Sn 5% Zr 2.2% Al 1% Mo 0.4% Si Ti alloy: Bar; solution treated & aged UTS: 1030 Elon: 10% Proof: 850
DTD 5153	4.0% Al 4.0% Mo 2.0% Sn 0.5% Si Ti alloy: Forging; solution treated & aged UTS: 1070 Elon: 10% Proof: 880
DTD 5163	6% Al 4% V Ti alloy: Sheet; solution treated & aged UTS: 950 Elon: 8% Proof: 870
DTD 5173	6% Al 4% V Ti alloy: Bar; annealed UTS: 950 Elon: 10% Proof: 870
DTD 5203	4% Al 4% Mo 0.5% Si 2% Sn Ti alloy
DTD 5223	4% Al 4% Mo 0.5% Si 4% Sn Ti alloy
DTD M200	6% Al 0.3% Si 1.0% W 5.0% Zr Ti alloy
DTD M201	6% Al 1.0% Cu 4.0% Mo 0.2% Si 5.0% Zr Ti alloy
EX 684	Al Zr Si Ti alloy: Bar & forging; IMI; weldable; beta stabilized UTS: 1030
HA 7146	0.08% C 7% Al 4% Mo Ti alloy: Harvey Aluminium; solution treated & aged UTS: 920 Elon: 6% Proof: 750
HA 8116	0.08% C 8.0% Al 1.0% Mo 1.0% V Ti alloy: Harvey Aluminium; annealed DPN: 350 UTS: 580 Elon: 10% Proof: 530
HYLITE 30	1.5% Al 1.5% Mn Ti alloy: Jessop; solution treated; obsolete DPN: 230 UTS: 690 Elon: 20% Proof: 530

Symbol	Nominal analysis, supplier, condition and remarks.
HYLITE 40	4% Al 4% Mn Ti alloy: Jessop; solution treated; obsolete DPN: 310 UTS: 950 Elon: 18% Proof: 880
HYLITE 45	6% Al 4% V Ti alloy: Jessop; solution treated & aged; obsolete UTS: 940 Elon: 22% Proof: 880
HYLITE 50	4% Al 2% Sn 0.5% Si 4% Mo Ti alloy: Jessop; solution treated & aged; obsolete DPN: 350 UTS: 1210 Elon: 15% Proof: 1080
HYLITE 51	4% Al 4% Sn 0.5% Si 4% Mo Ti alloy: Jessop; solution treated & aged; obsolete UTS: 1300 Elon: 14% Proof: 1160
HYLITE 60	3% Al 6% Sn 5% Zr 0.5% Si 2% Mo Ti alloy: Jessop; solution treated & aged; obsolete UTS: 1070 Elon: 14% Proof: 1050
HYLITE 65	Al Sn Zr Si Mo alloy: Forging; Jessop; obsolete UTS: 1070
ICI 318A	6% Al 4% V Ti alloy: ICI; now IMI; see Ti 318A
MST 8 Al 1 Mo 1 V	0.08% C 8% Al 1% Mo 1% V 0.01% H Ti alloy: Wrought; Reactive Metals Ltd
RMI 1 Al 8 V 5 Fe	0.05% C 5% Fe 1.2% Al 8% V 0.012% H Ti alloy: Wrought; Reactive Metals Ltd; solution treated & aged UTS: 1550 Elon: 8% Proof: 1410
RMI 3 Al 2.5 V	0.05% C 0.3% Fe 3% Al 2.5% V 0.015% H Ti alloy: Wrought; Reactive Metals Ltd; annealed UTS: 610 Elon: 16% Proof: 530
RMI 4 Al 3 Mo 1 V	0.08% C 0.2% Fe 4% Al 3% Mo 1% V 0.015% H Ti alloy: Wrought; Reactive Metals Ltd; annealed UTS: 880 Elon: 10% Proof: 820
RMI 4 Al 3 Mo 1 V	0.08% C 0.2% Fe 4% Al 3% Mo 1% V 0.015% H Ti alloy: Wrought; Reactive Metals Ltd; solution treated & aged UTS: 1300 Elon: 4% Proof: 1120
RMI 6 Al 2 Sn 4 Zr 2 Mo	6% Al 2% Sn 4% Zr 2% Mo Ti alloy: Wrought; Reactive Metals Ltd; alpha–beta alloy UTS: 580 Elon: 10% Proof: 540
RMI 6 Al 4 V	0.08% C 6.2% Al 4% V 0.012% H Ti alloy: Reactive Metals Ltd; solution treated & aged UTS: 1070 Elon: 7% Proof: 1000
RMI 6 Al 4 V EL 1	0.09% C 6% Al 4% V 0.012% H Ti alloy: Reactive Metals Ltd; solution treated & aged; low impurities UTS: 1070 Elon: 7% Proof: 1000
RMI 6 Al 6 V 2 Sn	0.05% C 5.5% Al 5.5% V 2% Sn 0.6% Cu Ti alloy: Reactive Metals Ltd; solution treated & aged UTS: 1250 Elon: 8% Proof: 1170
RMI 7 Al 4 Mo	0.08% C 7% Al 4% Mo 0.012% H Ti alloy: Reactive Metals Ltd; solution treated & aged UTS: 1160 Elon: 8% Proof: 1070
RMI 13 V 11 Cr 3 Al	0.05% C 3% Al 10.5% Cr 13% V 0.02% H Ti alloy: Wrought; Reactive Metals Ltd; annealed UTS: 920 Elon: 10% Proof: 780
RMI 13 V 11 Cr 3 Al	0.05% C 3% Al 10.5% Cr 13% V 0.02% H Ti alloy: Wrought; Reactive Metals Ltd; solution treated & aged UTS: 1280 Elon: 3% Proof: 1180
RMI 16 V 2.5 Al	0.06% C 0.3% Fe 2.5% Al 17% V 0.15% H Ti alloy: Wrought; Reactive Metals Ltd; solution treated & aged UTS: 1180 Elon: 4% Proof: 1070
RS 115	0.08% C 4% Al 3% Mo 1% V low N & H Ti alloy: Republic Steel Co; aged
RS 135	0.08% C 7% Al 4% Mo Ti alloy: Republic Steel Co; aged UTS: 920 Elon: 6% Proof: 750

Note. The following abbreviations and units are used in the tables:

DPN	Hardness, diamond pyramid number
UTS	Ultimate tensile strength, N/mm^2
Elon	Elongation, %
Proof	0.1 % proof strength, N/mm^2

1 N/mm^2=0.1 hbar=0.102 kgf/mm^2=0.06475 tonf/in^2=145.04 lbf/in^2

Symbol	Nominal analysis, supplier, condition and remarks.
RS 811 X	0.08% C 8.0% Al 1.0% Mo 1.0% V Ti alloy: Republic Steel Co; annealed **DPN: 350 UTS: 580 Elon: 10% Proof: 530**
T 443	3% Al 11% Cr 13% V Ti alloy: Jessop; obsolete
T 713	2.2% Al 11% Sn 0.3% Si 4% Mo Ti alloy: Jessop; obsolete
TA 6V	6.0% Al 4.0% V Ti alloy: French National Standard **UTS: 950 Elon: 10% Proof: 1000**
Ti 4 Al 3 Mo 1 V	0.08% C 4% Al 3% Mo 1% V low N & H Ti alloy: Sheet; Titanium Metals Ltd; annealed **DPN: 340 UTS: 840 Elon: 10% Proof: 770**
Ti 4 Al 3 Mo 1 V	0.08% C 4% Al 3% Mo 1% V low N & H Ti alloy: Sheet; Titanium Metals Ltd; solution treated & aged **DPN: 426 UTS: 1210 Elon: 5% Proof: 1050**
Ti 5 Al 4 Fe Cr	0.1% C 5.25% Al 1.5% Fe 2.7% Cr low N & H Ti alloy: Sheet, bar, etc.; Titanium Metals Ltd; annealed **DPN: 340 UTS: 1050 Elon: 10% Proof: 970**
Ti 5 Al 4 Fe Cr	0.1% Cr 5.25% Al 1.5% Fe 2.7% Cr low N & H Ti alloy: Sheet, bar, etc; Titanium Metals Ltd; solution treated & aged **DPN: 390 UTS: 1210 Elon: 4% Proof: 1050**
Ti 6 Al 4 V	0.8% C 6% Al 4% V low N & H Ti alloy: Sheet, bar, etc.; Titanium Metals Ltd; annealed **DPN: 352 UTS: 880 Elon: 10% Proof: 830**
Ti 6 Al 4 V	0.08% C 6% Al 4% V low N & H Ti alloy: Sheet, bar, etc.; Titanium Metals Ltd; solution treated & aged **UTS: 1080 Elon: 10% Proof: 1050**
Ti 6 Al 4 V EL 1	0.08% C 6% Al 4% V low N & H Ti alloy: Sheet, bar, etc.; Titanium Metals Ltd; annealed; for low temperature use **DPN: 340 UTS: 880 Elon: 10% Proof: 820**
Ti 6 Al 6 V 2 Sn	0.05% C 5.5% Al 5.5% V 2% Sn 0.7% Fe 0.7% Cu Ti alloy: Bar; Titanium Metals Ltd; annealed **UTS: 1050 Elon: 9% Proof: 950**
Ti 6 Al 6 V 2 Sn	0.05% C 5.5% Al 5.5% V 2% Sn 0.7% Fe 0.7% Cu Ti alloy: Bar; Titanium Metals Ltd; solution treated & aged **UTS: 1210 Elon: 7% Proof: 1080**
Ti 7 Al 4 Mo	0.08% C 7% Al 4% Mo Ti alloy: Bar & wire; Titanium Metals Ltd; annealed **DPN: 380 UTS: 990 Elon: 10% Proof: 920**
Ti 7 Al 4 Mo	0.08% C 7% Al 4% Mo Ti alloy: Bar & wire; Titanium Metals Ltd; solution treated and aged **UTS: 1160 Elon: 7% Proof: 1080**
Ti 8 Al 1 Mo 1 V	0.08% C 8% Al 1% Mo 1% V low N & H Ti alloy: Sheet, bar, etc.; Titanium Metals Ltd; annealed **DPN: 350 UTS: 880 Elon: 10% Proof: 820**
Ti 8 Al 1 Mo 1 V	0.08% C 8% Al 1% Mo 1% V low N & H Ti alloy: Sheet, bar, etc.; Titanium Metals Ltd; solution treated & aged **UTS: 920 Elon: 10% Proof: 840**
Ti 13 V 11 Cr 3 Al	0.05% C 3% Al 10.5% Cr 13% V low N & H Ti alloy: Sheet, bar, etc.; Titanium Metals Ltd; annealed **DPN: 340 UTS: 860 Elon: 10% Proof: 810**
Ti 13 V 11 Cr 3 Al	0.05% C 3% Al 10.5% Cr 13% V low N & H Ti alloy: Sheet, bar, etc.; Titanium Metals Ltd; solution treated & aged **UTS: 1160 Elon: 4% Proof: 1080**
Ti 140 A	2.0% Fe 2.0% Cr 2.0% Mo Ti alloy: Titanium Metals Ltd
Ti 155 A	0.08% C 5.5% Al 1.5% Fe 1.5% Cr 1.2% Mo low N & H Ti alloy: Bar; Titanium Metals Ltd; annealed **DPN: 385 UTS: 1000 Elon: 9% Proof: 920**

Symbol	Nominal analysis, supplier, condition and remarks.
Ti 155 A	0.08% C 5.5% Al 1.5% Fe 1.5% Cr 1.2% Mo low N & H Ti alloy: Bar; Titanium Metals Ltd; solution treated & aged **UTS: 1180 Elon: 9% Proof: 1080**
Ti 205	15% Mo Ti alloy: IMI; solution treated & aged **UTS: 1060 Elon: 4% Proof: 900**
Ti 230	2% Cu Ti alloy: IMI; solution treated & aged **UTS: 760 Elon: 23% Proof: 580**
Ti 318 A	6% Al 4% V Ti alloy: IMI **UTS: 950 Elon: 10% Proof: 700**
Ti 550	4% Al 4% Mo 0.5% Si 2% Sn Ti alloy: Rod; IMI; solution treated & aged **UTS: 1230 Elon: 12% Proof: 1120**
Ti 551	4% Al 4% Mo 0.5% Si 4% Sn Ti alloy: Rod; IMI; air cooled & aged **UTS: 1350 Elon: 10% Proof: 1250**
Ti 679	11% Sn 5% Zr Al Mo Si Ti alloy: IMI; solution treated & aged **UTS: 800 Elon: 10% Proof: 580**
Ti 684	6% Al 0.3% Si 1% W 5% Zr Ti alloy: Rod; IMI; solution treated & aged **UTS: 1030 Elon: 10% Proof: 930**
Ti 685	6% Al 0.5% Si 0.5% Mo 5% Zr Ti alloy: Rod; IMI; soluiton treated & aged **UTS: 1030 Elon: 10% Proof: 930**
Ti 700	6% Al 5% Zr 4% Mo 1% Cu Ti: Bar; IMI; for creep use up to 400 °C; solution treated & aged **UTS: 1350 Elon: 6% Proof: 1240**
Ti 700	6.0% Al 1.0% Cu 4.0% Mo 0.2% Si 5.0% Zr Ti alloy: Rod; IMI; oil quenched & aged **UTS: 1400 Elon: 10% Proof: 1300**
UTA 8 DV	0.08% C 8% Al 1.0% Mo 1.0% V Ti alloy: French National Standard; annealed **DPN: 350 UTS: 880 Elon: 10% Proof: 810**
SAE 11	5.7% Cu 6.7% Sb Sn alloy: For bearings
SAE 12	3.5% Cu 7.5% Sb Sn alloy: For bearings
Sg Sn 50	4% Cu 33% Pb 13% Sb Sn alloy: Designation used by German Standard
Sg Sn 60	4% Cu 23% Pb 13% Sb Sn alloy: Designation used by German Standard
Sg Sn 70	5% Cu 10% Pb 15% Sb Sn alloy: Designation used by German Standard
Sg Sn 75	5% Cu 3% Pb 17% Sb-Sn alloy: Designation used by German Standard
Sg Sn 78	4% Cu 1% Pb 17% Sb Sn alloy: Designation used by German Standard
SILVER BABBITT	Ag Cd Ni Sn base bearing metal: Highest duty; Magnolia Anti-Friction Metal Co **DPN: 33.6**
SOVEREIGN 8TS	3.0% Cu 8.9% Sb 1.1% Cd Sn alloy: Bearing metal; Glacier Metals Co **DPN: 30 UTS: 98 Elon: 15% Proof: 84**
SOVEREIGN 8TS	8.8% Sb 3.0% Cu 1% Cd Sn: Bearing metal; Glacier Metals **DPN: 30 UTS: 100 Elon: 16% Proof: 80**
SPRA BABBITT	3.5% Cu 0.25% Pb 7.5% Sb Sn alloy: Wire for spraying; Metco
STA 7 TB1B	8% Sb 8% Cu 0.3% Pb Sn alloy: Bearing metal; obsolete
STA 7 TB2	Sn base bearing metal: Obsolete
STERN TUBE METAL	30% Zn 1.5% Cu Sn alloy: White metal; information from Tin Research Institute
SX 4	4% Ag Sn alloy: For soldering; Sheffield Smelting Co; melting range 221–224 °C **DPN: 14 UTS: 50 Elon: 48% Proof: 45**
TANDEM DE	Sn base bearing metal: For reciprocating engines; Eyre Smelting Co; melts at 315 °C **DPN: 27 UTS: 80 Elon: 24%**

Symbol	Nominal analysis, supplier, condition and remarks.
TANDEM HP	Sn base bearing metal: For reciprocating engines; Eyre Smelting Co; melts at 330 °C **DPN: 32** **UTS: 80** **Elon: 6%**
TANDEM ME	Sn base bearing metal: For reciprocating engines;

Note. The following abbreviations and units are used in the tables:

DPN	Hardness, diamond pyramid number
UTS	Ultimate tensile strength, N/mm^2
Elon	Elongation, %
Proof	0.1 % proof strength, N/mm^2

$1 \ N/mm^2 = 0.1 \ hbar = 0.102 \ kgf/mm^2 = 0.06475 \ tonf/in^2 = 145.04 \ lbf/in^2$

52. Tungsten W

Physical properties

Atomic number	74	
Atomic weight	184.0	
Crystal structure	Body-centred cubic	
Colour	Steel grey	
Specific gravity	19.3	
Density	19300 kg/m^3	(0.69 lb/in.^{3})
Melting point	3370 °C	
Boiling point	5930 °C	
Specific heat	0.134 J/g °C	(0.032 cal/g °C)
Thermal conductivity	163.3 W/m °C	(39.7 cal/m s °C)
Coefficient of linear expansion (20–100° C)	4.4×10^{-6}/ °C	
Latent heat of fusion	184.2 J/g	(44 cal/g)
Latent heat of vaporization	– J/g	
Thermal neutron absorption cross-section	19.2 barns/atom	
Electrical conductivity	31% IACS (copper 100%)	
Specific resistance	56.5 microhm mm	
Temperature coefficient of electrical resistance	0.0048/ °C	
Electrochemical equivalent	3.430 g/A/h	
Electrode potential	4.5 V	
Magnetic susceptibility	0.33×10^{-6}	
Young's modulus of elasticity	345×10^9 N/m^2	(50×10^6 lbf/in.^{2})
Tensile strength sintered	15	(10 tonf/in.^{2})
drawn wire	450	(300 tonf/in.^{2})
Hardness sintered	255 DPN	
drawn & annealed	480 DPN	

52.1 General notes on tungsten

Tungsten exists in nature as wolframite, a complex iron manganese tungsten oxide, and scheelite, a calcium tungsten oxide. The metal is still commonly called wolfram and it is from this that the chemical symbol is derived. The ores are generally found in association with those of other refractory metals, such as molybdenum, and also with tin ores.

Tungsten of at least 99.7% purity is available, being steel grey in colour and resembling molybdenum in many respects. The tensile strength varies from 150 N/mm^2 in the fully annealed condition to over 4500 N/mm^2 for fine wire finally cold drawn using diamond dies.

Tungsten does not oxidize appreciably below about 500 °C and does not form a volatile oxide as readily as molybdenum. It has excellent resistance to the attack of most of the common acids, even at high temperature.

These properties make it a valuable element in nuclear engineering and there is little doubt that only the difficulty in working the pure metal precludes much greater use of tungsten in this field.

Pure tungsten wire is widely used in electric lamps when it generally has some thoria present to prevent sag at temperature.

The low thermal expansion of tungsten makes it useful when glass metal seals are required and this property is enhanced by the reasonably good electrical conductivity.

Tungsten is generally used as the non-consumable electrode in inert gas shielded arc welding, again because of the reasonably good electrical conductivity, coupled this time with the high melting point.

Because of the excellent refractory nature of the metal, considerable work is in progress to obtain a more ductile material. Using vacuum arc and electron beam melting and re-melting, reasonably ductile tungsten has been obtained, which would appear to indicate that if the undesirable impurities could be identified and eliminated a workable tungsten could be produced.

Probably the best known use of tungsten is in combination with carbon as sintered tungsten carbide for cutting tools and other uses, where abrasion at high temperature has to be overcome.

The largest quantity of tungsten is used as an alloying element in steel, where it combines with the carbon and gives the steel the ability to retain its hardness at temperatures up to red heat. Tungsten is seldom the sole alloying element, generally being present up to 18–20% with chromium and vanadium in lesser amounts. The degree of hot hardness is in direct proportion to the tungsten present. Tungsten is also used as an alloying element with other metals where its refractory qualities are useful. Many of the cobalt and nickel high temperature alloys include tungsten and it is also a valuable addition to some creep resistant steels.

Alloyed with copper, tungsten gives a hard material with good electrical conductivity, useful for resistance welding electrodes.

Some sintered alloys with about 90% tungsten have a specific gravity almost double that of lead. These are used for balance weights where space saving is important, for example high speed rotating parts and aircraft control surfaces.

These alloys are also used in nuclear engineering and medicine as X-ray barriers.

Tungsten is one of the modern metals which have valuable properties regarding temperature and corrosion resistance. It is however, difficult to obtain in pure state, and present knowledge and techniques are such that very great difficulty exists in working the metal.

Following are the various grades of tungsten, including the available trade names for tungsten carbides.

Symbol	Nominal analysis, supplier, condition and remarks.
2A 3	11.0% Co WC: Sinter material; Vr/Wesson; specific gravity 14.2; hardness 1470 DPN
2A 5	6.0% Co WC: Sinter material; VR/Wesson as GI; specific gravity 14.85; hardness 1860 DPN
2A 6	8.0% Co WC: Sinter material; VR/Wesson; specific gravity 147; hardness 1480 DPN
2A 7	4.0% Co WC: Sinter material; VR/Wesson; specific gravity 15.1; hardness 1875 DPN
2A 68	6.0% Co WC: Sinter material; VR/Wesson; specific gravity 14.85; hardness 1700 DPN
26	7.0% Co 10.0% Ta C WC: Sinter material; VR/Wesson; specific gravity 12.4; hardness 1790 DPN
44 A	6.0% WC: For tool tips; General Electric Co **UTS: 4500 Elon: 1.0% Proof: 1420**
55A	13% WC: For tool tips; General Electric Co **UTS: 3700 Elon: 1.9% Proof: 530**
55B	16% Co WC: For tool tips; General Electric Co **UTS: 2700 Elon: 2.7% Proof: 700**
78	8.0% Co 4.0% TaC 12% TiC WC: For tool tips; General Electric Co **UTS: 4600 Elon: 1.0% Proof: 1700**
78B	9.0% Co 8.0% TiC WC: For tool tips; General Electric Co **UTS: 3900 Elon: 2.0% Proof: 610**
90	10% Co WC: For tool tips; General Electric Co **UTS: 4200 Elon: 1.9% Proof: 920**
120	12% Co WC: For tool tips; General Electric Co **UTS: 3700 Elon: 3.5% Proof: 840**
190	25% Co WC: For tool tips; General Electric Co **UTS: 3300 Elon: 3.5% Proof: 370**
330	12% Ni 2.0% Mo 9.5% TaC 30% TiC WC: For tool tips; General Electric Co **UTS: 3700 Elon: 0.5% Proof: 1750**
350	4.5% Co 12.2% TaC 12.5% TiC WC: For tool tips; General Electric Co **UTS: 4600 Elon: 0.9% Proof: 1200**
370	8.5% Co 11.5% TaC 8.0% TiC WC: For tool tips; General Electric Co **UTS: 4600 Elon: 1.0% Proof: 1700**
779	9.0% Co WC: For tool tips; General Electric Co **UTS: 4200 Elon: 1.5% Proof: 920**
860	5.0% Co 4.0% TaC WC: For tool tips; General Electric Co **UTS: 4800 Elon: 1.1% Proof: 2000**

Symbol	Nominal analysis, supplier, condition and remarks.
883	6.0% Co WC: For tool tips; General Electric Co **UTS: 4500 Elon: 0.8% Proof: 2000**
900	3.0% Co 4.0% TaC WC: For tool tips; General Electric Co **UTS: 4500 Elon: 0.7% Proof: 2500**
907	6.0% Co 20% TaC WC: For tool tips; General Electric Co **UTS: 4800 Elon: 1.7% Proof: 1700**
999	3.0% Co WC: For tool tips; General Electric Co **UTS: 4500 Elon: 0.5% Proof: 2400**
3047	WC: Tungsten carbide; Kennametal; high strength; high impact resistance
3109	WC: With Co binder; Kennametal; extrusion punches
3411	WC: Tungsten carbide; Kennametal; high hardness and impact value
AMD 78 BU	W: Sintered forgings; Interim AMS
AMS 7725	W base high density sintered alloy shapes
AMS 7898	W: Sintered sheet, strip & foil
AMS 7899	W: Sheet, strip & foil
ASTM B297 EW P	99.5% W: Electrode for arc welding
ASTM B297 EW Th 1	1.0% Th W: Electrode for arc welding
ASTM B297 EW Th 2	2.0% Th W: Electrode for arc welding
ASTM B297 EW Zr	0.4% Zr W: Electrode for arc welding
ASTM B346	6% Ni 3% Cu W alloy: Powder compact; physical properties vary with weight of part **UTS: 58 Elon: 1% Proof: 450**
BS 4276 BS1	5.0% Co: Hard metal; grain size 0.5–1 μm **DPN: 1750 UTS: 1700**
BS 4276 BS2	5.5% Co: Hard metal; grain size 0.5–1.5 μm **DPN: 1720 UTS: 1700**
BS 4276 BS3	5.5% Co: Hard metal; grain size 0.5–2 μm **DPN: 1650 UTS: 1700**
BS 4276 BS4	6.0% Co: Hard metal; grain size 1–3 μm **DPN: 1550 UTS: 2000**
BS 4276 BS5	8% Co: Hard metal; grain 1–5 μm **DPN: 1450 UTS: 2000**
BS 4276 BS6	10.5% Co: Hard metal; grain size 1–5 μm **DPN: 1350 UTS: 2200**
BS 4276 BS7	12% Co: Hard metal; grain size 1–5 μm **DPN: 1240 UTS: 2300**
BS 4276 BS8	15% Co: Hard metal; grain size 1–5 μm **DPN: 1150 UTS: 2500**
CARBALOY	General Electric name for tungsten carbide grades covered by designation numbers
COLMONOY 705	2.4% C 2.0% Si 6.8% Cr 2.2% Fe 1.5% B 2.0% Co 39% Ni W alloy: Powder for metal spraying; melting point 1040 °C; Colmonoy
COPELMET	Cu + W or WC sintered material: Metro-Cutanit; range of materials
CUTANIT	Co + Ti + Ta carbides with WC sintered materials: Metro-Cutanit; range of materials

Note. The following abbreviations and units are used in the tables:

DPN	Hardness, diamond pyramid number
UTS	Ultimate tensile strength, N/mm^2
Elon	Elongation, %
Proof	0.1 % proof strength, N/mm^2

1 N/mm^2=0.1 hbar=0.102 kgf/mm^2=0.06475 tonf/in^2=145.04 lbf/in^2

Symbol	Nominal analysis, supplier, condition and remarks.
F 26	WC: Tungsten carbide; Kennametal; high hardness and impact valve
GE 125	25% Rh W alloy: Sintered & wrought; billet & tube; General Electric Co
GE HEVIMET	6.0% Ni 4.0% Cu W alloy: General Electric Co; specific gravity 17.0 **UTS: 2000**
GI	6.0% Co WC: Sinter material; VR/Wesson now 2A5
HEAVY METAL	Cu Ni W alloy: Specific gravity 16–18; Tungsten Mfg Co.; sintered **DPN: 250**
HEVIMET	6.0% Ni 4.0% Cu W alloy: General Electric Co; specific gravity 17.0 **UTS: 2000**
HR	9.0% Co 8.0% T C 10% Ta + Nb WC: Sinter material; VR/Wesson now VR R77
HV	3.0% Co 13.0% Ti C 1.0% Mo WC: Sinter material; VR/Wesson; specific gravity 11.9; hardness 1890 DPN
hW 11	99.99% W: Rod; Koch-Light Ltd
hW 72	99.999% W: Single crystal 3mm diameter; Koch-Light Ltd
K 1	WC with Co binder: Kennametal; highest impact valve
K 2S	WC with WTiC2: Kennametal; heavy to medium loads and interrupted cuts
K 3H	WC with high percentage WTiC2: Kennametal; medium heavy cuts on low C steel
K 4H	WC with WTiC2: Kennametal; light load form tools
K 5H	WC with high percentage WTiC2: Kennametal; for semi-finishing cuts
K 6	WC with Co binder: Kennametal; highest compressive strength
K 7H	WC with high percentage WTiC2: Kennametal; high speed low load machining
K 8	WC with Co binder: Kennametal; best wear resistance
K 9	WC: Tungsten carbide; Kennametal; high wear resistance
K 11	WC with Co binder: Kennametal; hardest grade
K 21	WC with WTiC2: Kennametal; medium loads with interrupted cuts
K 81	WC with WTiC & Co binder: Kennametal; for cold punches
K 82	WC with WTiC2 & Co binder: Kennametal; impact extrusion dies
K 84	WC with WTiC2 & Co binder: Kennametal; tube sizing mandrels
K 86	WC with WTiC2 and Co binder: Kennametal; draw dies, ball valves, etc.
K 90	WC with Co binder: Kennametal; blanking dies, thrust bearings
K 90A	WC with Co binder: Kennametal; heavy leading dies
K 91	WC with Co binder: Kennametal; medium blanking dies, crushing hammers
K 92	WC with Co binder: Kennametal; medium blanking dies, etc.

Note. The following abbreviations and units are used in the tables:

DPN	Hardness, diamond pyramid number
UTS	Ultimate tensile strength, N/mm^2
Elon	Elongation, %
Proof	0.1 % proof strength, N/mm^2

1 N/mm^2=0.1 hbar=0.102 kgf/mm^2=0.06475 tonf/in^2=145.04 lbf/in^2

Symbol	Nominal analysis, supplier, condition and remarks.
K 94	WC with Co binder: Kennametal; light blanking dies, etc.
K 95	WC with Co binder: Kennametal; light blanking dies, valve parts, etc.
K 96	WC with Co binder: Kennametal; compacting dies, seal rings, etc.
K 601	WC and TaC: Binder free; tantalum & tungsten carbide; Kennametal; corrosion and wear resistant
K 701	WC with Cr & Co binder: Kennametal; corrosion & wear resistant
K 801	WC with Ni binder: Kennametal; corrosion resistant & high hardness
KENNAMETAL	Tungsten carbide: Available in various grades; Kennametal
KENNERTIUM W2	W alloy: For powder metallurgy; Kennametal; high density; specific gravity 18.5 **UTS: 720 Elon: 3% Proof: 580**
KENNERTIUM W10	W alloy: For powder metallurgy; Kennametal; high density; specific gravity 17.0 **UTS: 920 Elon: 15% Proof: 640**
KM	WC with high percentage WTiC2: Kennametal; for interrupted rough cuts
MATTHEY 1W3	40% Cu W alloy: Sinter for contacts; conductivity 41% IACS; Johnson Matthey **DPN: 140**
MATTHEY 3W3	32% Cu W alloy: Sinter for contacts; conductivity 33% IACS; Johnson Matthey **DPN: 160**
MATTHEY 30W3	22% Cu W: Sinter alloy for contacts; conductivity 28% IACS; Johnson Matthey **DPN: 240**
MATTHEY 205	27% Ag W: Sintered alloy for contacts; conductivity 43% IACS; Johnson Matthey **DPN: 220**
MATTHEY 35S	35% Ag W alloy: Sinter for contacts; conductivity 52% IACS; Johnson Matthey **DPN: 140**
MATTHEY 2355	45% Ag W alloy: Sinter for contacts; conductivity 60% IACS; Johnson Matthey **DPN: 125**
MATTHEY 2265	35% Ag W: Sinter alloy for contacts; conductivity 50% IACS; Johnson Matthey **DPN: 135**
MATTHEY 2373	27% Ag W: Sinter alloy for contacts; conductivity 45% IACS; Johnson Matthey **DPN: 180**
MATTHEY G14	40% Ag WC: Sinter alloy for contacts; conductivity 36% IACS; Johnson Matthey **DPN: 200**
MUTCO 32C	0.1% C 0.8% Si 3.5% Cr 14% Ni 0.8% B WC alloy: Powder for spraying; Metco; tungsten carbine in stainless matrix **DPN: 865**
PROLITE	Tungsten carbide: Murex Ltd
SILVELMET	Ag + W or WC sintered material: Metro-Cutanit; range of materials
TUNGSTEN	W: Powder, sheet, wire & amorphous powder; Blackwells
UNS R07001	R07999 W: Metal & alloys; American Standards system
VOLOMIT	4.5% C 2% Fe W alloy: Origin unknown
VR 13	10.0% Co WC: Sinter material; VR/Wesson; specific gravity 14.4; hardness 1350
VR 14	11.5% Co WC: Sinter material; VR/Wesson; specific gravity 14.35; hardness 1250 DPN
VR 15	13.0% Co WC: Sinter material; VR/Wesson; specific gravity 14.2; hardness 1180 DPN
VR 52	4.0% Co WC: Sinter material for abrasion resistance; VR/Wesson; specific gravity 15.15; hardness 1870 DPN

Symbol	Nominal analysis, supplier, condition and remarks.
VR 54	8.0% Co WC: Sinter material; VR/Wesson; specific gravity 14.6; hardness 1860 DPN
VR 71	6.0% Co 18.0% Ti C 10.0% Ta C + NbC WC: Sinter material; VR/Wesson; specific gravity 10.7; hardness 1875 DPN
VR 73	6.5% Co 12.0% Ti C 10.0% Ta C + NbC WC: Sinter material; VR/Wesson; specific gravity 11.9; hardness 1860 DPN
VR 75	7.5% Co 8.0% Ti C 10.0% Ta C + NbC WC: Sinter material; VR/Wesson; specific gravity 12.65; hardness 1790 DPN
VR 77	9.0% Co 8.0% Ti C 10.0% Ta C + NbC: Sinter material; VR/Wesson as HR; specific gravity 12.55; hardness 1710 DPN
VR 87	17.0% Co 28.0% Ta C WC: Sinter material; VR/Wesson; specific gravity 13.5; hardness 894 DPN
VR 89	11.0% Co 18.0% Ta C WC: Sinter material; VR/Wesson; specific gravity 14.05; hardness 1480 DPN
W 18	99.9% W: Powder; Koch-Light Ltd; high purity metal
W 367	99.9% W Fe + Mo 0.04% (max): Sintered and swaged; Tungsten Mfg Co; suitable for brazing; for electrical contacts
W 408	99.9% W 0.02% Fe + Mo (max): Sintered and swaged; Tungsten Mfg Co; suitable for brazing; for electrical contacts
W 410L	99.9% W: Sintered & swaged; Tungsten Mfg Co; suitable for brazing; for electrical contacts

Symbol	Nominal analysis, supplier, condition and remarks.
W 410M	99.9% W 0.02% Fe + Mo (max): Sintered & swaged; Tungsten Mfg Co; suitable for brazing; for electrical contacts
WALLEX 55	3.5% C 1.4% Si 12.5% Cr 0.5% Fe 1.5% B 5.5% Ni 28% Co W alloy: Powder for metal spraying; melting point 1120 °C; Colmonoy
WALLEX 505	2.4% C 1.4% Si 12.5% Cr 0.5% Fe 1.5% B 5.6% Ni 24.0% Co W alloy: Powder for metal spraying; melting point 1120 °C; Colmonoy
WH	7.0% Co 13.0% Ti C 2.0% Ta C WC: Sinter material; VR/Wesson; specific gravity 11.75; hardness 1860 DPN
WM	10.0% Co 13.0% Ti C 2.0% Ta C + NbC WC: Sinter material; VR/Wesson; specific gravity 11.5; hardness 1710 DPN
WOLFRAM	WO: Tungsten oxide; Blackwells; primary material
WS	13.0% Co 4.0% Ti C 2.0% Ta C + NbC WC: Sinter material; VR/Wesson; specific gravity 13.2; hardness 1400 DPN

Note. The following abbreviations and units are used in the tables:

DPN	Hardness, diamond pyramid number
UTS	Ultimate tensile strength, N/mm^2
Elon	Elongation, %
Proof	0.1 % proof strength, N/mm^2

$1\ N/mm^2 = 0.1\ hbar = 0.102\ kgf/mm^2 = 0.06475\ tonf/in^2 = 145.04\ lbf/in^2$

53. Uranium U

Physical properties

Atomic number	92	
Atomic weight	238.07	
Crystal structure	Body-centred cubic	
Colour	Lustrous silvery white	
Specific gravity	19.07	
Density	19070 kg/m^3	(0.689 lb/in.^{3})
Melting point	1689 °C	
Boiling point	3818 °C	
Specific heat	0.1156 J/g °C	(0.0276 cal/g °C)
Thermal conductivity	26.8 W/m °C	(6.4 cal/m s °C)
Coefficient of linear expansion (20–100° C)	19 × 10^{-6}/ °C	
Latent heat of fusion	– J/g	
Latent heat of vaporization	– J/g	
Thermal neutron absorption cross-section	694 barns/atom	
Electrical conductivity	6.0% IACS (copper 100%)	
Specific resistance	300 microhm mm	
Temperature coefficient of electrical resistance	0.0034/ °C	
Electrochemical equivalent	– g/A/h	
Electrode potential	-1.4 V	
Magnetic susceptibility	–	
Young's modulus of elasticity	190 × 10^9 N/m^2	(27.5 × 10^6 lbf/in.^{2})
Tensile strength cast & annealed	340 N/mm^2	(23 tonf/in.^{2})
Hardness	190 DPN	

53.1 General notes on uranium

This is classed as a rare earth metal, belonging to the radioactive group. It is however, well known as a source of atomic energy and as such is discussed in this section separately from the much lesser known metals classed as 'rare earths'.

Uranium is not widely distributed, being present it is estimated at about four parts per million of the earth's surface. It is thus a rare element in the true sense. Like radium it is found in pitchblende and also as carnotite.

Uranium is a lustrous silvery white metal. It is quite hard and with a specific gravity of almost 20 is one of the heaviest known elements.

Apart from its use in nuclear engineering, uranium has comparatively few applications. These are sufficient however, for the metal to have been known and used before the atomic bomb was exploded. It has photo-electric properties useful in the ultra-violent range, and has been used in glow discharge lamps.

As an alloying element in certain fields it is claimed that uranium establishes the carbides and prevents agglomeration, thus maintaining an even distribution of fine carbide, ensuring good ductility and fatigue properties.

Some of the compounds of uranium are used in the glass and ceramic industries to impart special properties and colours.

With the development of nuclear fission uranium has become an important element, with a literature comparable in quantity to many of the older metals which have served mankind for centuries. The reason for this is that an isotope of uranium – uranium 235 – can have its nucleus broken fairly readily by bombarding it with neutrons and this releases energy. This breaking or splitting or nuclei of elements has been known for some time, but until uranium 235 the energy input always exceeded the energy output. The products of the fission are other elements – generally radioactive – which themselves are unstable and change to other elements through a lengthy process culminating in a stable non-radioactive element such as lead.

The newly discovered elements with atomic numbers 93–102 which are discussed with the rare earths in Section 35, are all products of nuclear bombardment and generally have uranium as the starting element.

Uranium itself is radioactive, and is thus a health hazard along with some of its alloys and compounds. Not all the compounds exhibit this effect and many can be safely handled, but is suggested however, that where there is any doubt, advice is obtained before handling uranium compounds and salts.

Symbol	Nominal analysis, supplier, condition and remarks.
COMMERCIAL U1	U with total impurities 1000 p.p.m.: British Nuclear Fuels
COMMERCIAL U2	U with total impurities 500 p.p.m.: British Nuclear Fuels
COMMERCIAL U3	U with total impurities 200 p.p.m.: British Nuclear Fuels
COMMERCIAL U4	High purity U: British Nuclear Fuels
h U 22	99.7% U: Chips; Koch-Light Ltd
UNS R08001–R089999	V metal and alloys: American Standards system
URANIUM 1	Commercial metal: British Nuclear Fuels
URANIUM 2	Commercial as hexafluoride: British Nuclear Fuels
URANIUM 3	Commercial as dioxide: British Nuclear Fuels

Symbol	Nominal analysis, supplier, condition and remarks.
URANIUM 4	Commercial as tetrafluoride: British Nuclear Fuels
URANIUM U	Commercial & high purity forms; Blackwells

Note. The following abbreviations and units are used in the tables:

DPN	Hardness, diamond pyramid number
UTS	Ultimate tensile strength, N/mm^2
Elon	Elongation, %
Proof	0.1 % proof strength, N/mm^2

$1 \ N/mm^2 = 0.1 \ hbar = 0.102 \ kgf/mm^2 = 0.06475 \ tonf/in^2 = 145.04 \ lbf/in^2$

54. Vanadium V

Physical properties

Atomic number	23	
Atomic weight	50.95	
Crystal structure	Body-centred cubic	
Colour	Brilliant white	
Specific gravity	6.11	
Density	6110 kg/m^3	(0.221 lb/in.^{3})
Melting point	1735 °C	
Boiling point	3000 °C	
Specific heat	0.502 J/g °C	(0.120 cal/g °C)
Thermal conductivity	31.0 W/m °C	(7.4 cal/m s °C)
Coefficient of linear expansion (20–100° C)	8.33 × 10^{-6}/ °C	
Latent heat of fusion	– J/g	
Latent heat of vaporization	– J/g	
Thermal neutron absorption cross-section	4.7 barns/atom	
Electrical conductivity	7.0% IACS (copper 100%)	
Specific resistance	248 microhm mm	
Temperature coefficient of electrical resistance	0.0028/ °C	
Electrochemical equivalent	– g/A/h	
Electrode potential	-1.5 V	
Magnetic susceptibility	1.4 × 10^{-6}	
Young's modulus of elasticity	125.5 × 10^9 N/m^2	(18.2 × 10^6 lbf/in.^{2})
Tensile strength cold rolled	800 N/mm^2	(53 tonf/in.^{2})
Hardness	170 DPN	

54.1 General notes on vanadium

This metal occurs sparingly, very often in conjunction with the rare earth metals, and is never found in the native state.

The concentration, extraction and purification are complex processes, with the added disadvantage that the pentoxide is toxic and this is usually present at one stage of the purification.

The metal is usually obtained as a powder requiring compacting and further purification to give a ductile product. The vast majority of vanadium is used in steel as an alloying element and for this purpose a high purity product is not normally required provided it is free from undesirable impurities such as sulphur.

The pure metal is a brilliant white colour, quite stable at room temperature but combining with oxygen and nitrogen when heated. The metal is also attacked by the strong oxidizing acids.

In the highly purified state, vanadium can be hot and cold worked, showing very little tendency to harden under the influence of excessive cold work – a reduction of over 80% being possible between anneals.

Because of the affinity for oxygen and the fact that nitrogen and hydrogen have deleterious effects, the metal can only be heated under vacuum or using argon.

Vanadium is machined very easily and can be welded using the inert gas shielded arc process.

The difficulty in extraction, together with the scarcity of the ore, prices the metal out of competition with materials of comparable properties.

By far the largest use for vanadium is as an alloy in steel, where it combines with carbon to form hard stable carbides. These are always minute and evenly dispersed and have a considerable influence on the grain size of the steel. It has also been shown that small quantities of vanadium inhibit the tendency of chromium carbides to agglomate. These properties have a considerable effect on the ductility, fatigue strength and notch sensitivity of the high tensile chromium and tungsten–chromium steels which are notoriously lacking in these properties.

Vanadium is also added to cast iron to improve the hardenability and fatigue strength, while small additions to the age hardening aluminium casting alloys refine the grain, increase the response to thermal treatment and improve the fatigue strength.

Vanadium has considerable potential as a general purpose material which can be very easily worked and fabricated but has low resistance to atmospheric attack at higher temperatures, although the mechanical properties are reasonable. The scarcity and resultant high cost precludes the use of the material at present and there are no known alloys.

Symbol	Nominal analysis, supplier, condition and remarks.
h V 22	99.8% V: Granules; Koch-Light Ltd
h V 74	99.9% V: Single crystals 3 mm diameter; Koch-Light Ltd
VANADIUM	V metal: Blackwells; pure metal

Note. The following abbreviations and units are used in the tables:

DPN	Hardness, diamond pyramid number
UTS	Ultimate tensile strength, N/mm^2
Elon	Elongation, %
Proof	0.1 % proof strength, N/mm^2

$1 \ N/mm^2 = 0.1 \ hbar = 0.102 \ kgf/mm^2 = 0.06475 \ tonf/in^2 = 145.04 \ lbf/in^2$

55. Zinc Zn

Physical properties

Atomic number	30	
Atomic weight	65.38	
Crystal structure	Close-packed hexagonal	
Colour	Bluish white	
Specific gravity	7.1	
Density	7100 kg/m^3	(0.26 lb/in.^{3})
Melting point	419.5 °C	
Boiling point	907 °C	
Specific heat	0.3898 J/g °C	(0.0931 cal/g °C)
Thermal conductivity	112.2 W/m °C	(26.8 cal/m s °C)
Coefficient of linear expansion (20–100° C)	31.2 × 10^{-6}/ °C	
Latent heat of fusion	110 J/g	(26.3 cal/g)
Latent heat of vaporization	1754 J/g	(419 cal/g)
Thermal neutron absorption cross-section	1.06 barns/atom	
Electrical conductivity	28% IACS (copper 100%)	
Specific resistance	59 microhm mm	
Temperature coefficient of electrical resistance	0.0041/ °C	
Electrochemical equivalent	1.219 g/A/h	
Electrode potential	-0.76 V	
Magnetic susceptibility	-0.157 × 10^{-6}	
Young's modulus of elasticity	96.5 × 10^9 N/m^3	(14 × 10^6 lbf/in.^{2})
Tensile strength	cast 37 N/mm^2	(2.5 tonf/in.^{2})
Hardness	30 DPN	

55.1 General notes on zinc

Zinc ores occur in large deposits, generally easily mined and widespread throughout the world. Important deposits are worked in Australia and Canada.

Zinc is a bluish white metal which is comparatively brittle at normal temperatures, but soft and ductile at 100 °C. It has been known for many years as an excellent corrosion protective coating for steel. Galvanized corrugated steel sheeting was a favourite building material where economy had first consideration and appearance was not considered. The degree of protection afforded by the zinc coating is proportional to the coating thickness, but varies considerably depending on the atmospheric conditions. A coating of one ounce per square foot (300 g/m^2 per surface should have an indefinite life indoors (at least 60 years) twenty-five years in inland rural conditions, twenty years in rural maritime, and less than ten years in industrial outdoor conditions, these figures being halved for coatings for half an ounce per square foot (150 g/m^2). The method of application has little effect on the life as long as the correct pre-treatment is followed.

Zinc can be applied by hot dipping, by electroplating, by sherardizing, by metal spraying when it must be sealed, or as a zinc pigmented paint.

Considerable development has been carried out in recent years on the continuous production of a zinc coating on steel strip and the coating has ductility and adhesion sufficient to withstand relatively deep deformation on presses.

The advantage of zinc coating is that it is cheap to apply and sacrificial by nature. This means that if slightly damaged to bare the steel substrate, the zinc will corrode itself to protect the steel – this is the galvanic cell action, hence the name 'galvanizing'.

Considerable quantities of zinc are used as the can is 'dry' batteries where it acts as the sacrificial electrode in the production of electric current. This galvanic action is similar to that taking place when steel is protected by zinc and the coating is damaged.

Zinc of commercial purity is used in the form of sheet or strip for building applications such as roofing, flashings and rainwater goods. T-metal (see under Section 55A), an alloy which possesses better mechanical properties, particularly as regards creep strength, is also used for roofing.

Rolled zinc photo-engraving plates of controlled fine grain size are used by the printing industry for letterpress work and reproducing half-tone illustrations.

Rolled and cast zinc anodes are used to protect buried and immersed steelwork cathodically.

Zinc is also used as an alloying element with several distinct purposes as follows:

Copper alloys. Zinc is the main constituent along with copper in the series of alloys known as brass. These range from about 25 to 40% zinc and are the commonest of copper alloys.

In small quantities zinc is used to de-oxidize copper and as an economical addition which has little effect on the attractive properties of copper. These alloys are discussed in Section 14B and 14C.

Zinc is always present in the nickel silvers where it is a much more economical addition than nickel when colour is more important than strength. These alloys are discussed in Section 14F.

Aluminium alloys. Zinc additions with copper, magnesium and chromium form the strongest series of wrought aluminium alloys. Zinc is usually present from 5–7%. These alloys are discussed in Section 1G.

55A Zinc alloys

General metallurgical characteristics

The elements added to zinc all strengthen the alloys without in any way detracting from the excellent casting characteristics, in fact, by increasing the range of temperatures over which solidification occurs, the castability is improved for some purposes.

Some of the alloys have slightly improved tensile properties at higher temperatures while others have been developed to give specific casting shrinkage or growth. Zinc alloys are used for their economy when competing with brass, aluminium and other casting materials, but also in their own right for the ability to reproduce accurately intricate shapes. They can also compete with iron and steel in that they are much more readily melted, thus although not as cheap initially, they are more economical for short run production purposes, where they can be readily remelted and used again. This cannot be so easily carried out with cast iron or steel where more specialized foundry equipment is necessary.

Thermal treatment

None of the alloys can have their mechanical properties improved to any great extent by cold working or thermal treatment. There is a measurable surface hardening with cold work and this property is used in certain alloys to improve the life of press dies, but in the main heavy deformation causes brittleness.

Some of the alloys listed show improved mechanical properties after natural ageing for a long period – up to two years – but this is unusual and of little significance.

Welding and brazing

Die cast alloys can be gas welded using a filler rod of identical composition and it is possible to melt and cast metal to fill up holes. Modern plastic adhesives give better results than welding or brazing.

Flaw detection methods

Crack test. These low duty alloys will seldom require this inspection. When necessary oil penetrant chalk test should be used.

Zinc is also added to some aluminium casting alloys where its main function is economy. These alloys appear in Section 1M.

Magnesium alloys. The addition of zinc to magnesium, along with zirconium gives a range of alloys with very fine grain which have excellent hot strength properties. These are listed in Section 23B.

Zinc also forms a series of alloys which are used for a variety of purposes. These alloys are discussed in Section 55A, following which are listed all the names and specifications of zinc and its alloys.

X-ray. Not required except as a very occasional quality control check.

Ultra-sonic test. Not required.

Chemical etch. Not required.

Corrosion protection

Temporary. None required.

Permanent. These alloys have adequate built-in corrosion resistance for indoor use and require no protection except to be kept dry. Corrosion takes the form of a white powder which, while unsightly, seldom propagates to any extent under normal conditions.

For outdoor use and for appearance purposes, painting and plating are commonly applied.

Painting. This requires a scrupulously clean surface if it is to withstand outdoor conditions. The parts must be chromate treated or an etch primer used to ensure adhesion of the paint system. The use of a primer coat is recommended, followed as soon as possible by a sealing top coat. The best results are obtained with stoving paints which should always be specified when the parts are to be used for outdoor or arduous purposes.

Plating. Special etching techniques are necessary for these alloys, but provided these are used, normal electroplating procedures give satisfactory results. It is of interest that it is seldom possible to electroplate these alloys after stripping off a failed coating. This is particularly difficult if the article has been in use for any time when replating is seldom economical.

Zinc alloys have an intrinsically good corrosion resistance and are generally treated for appearance purposes rather than corrosion protection.

Machinability

These alloys are seldom machined to any great extent as they are generally die cast to a finished form. When necessary, high surface speeds with low feeds and tool pressure, using sharp single point tools give best results.

High speed or tipped tools will only be necessary for long mass production runs.

Uses

General purpose die casting material for light weight, low duty components, door handles, covers, casings for all household and transport articles. This is now first choice material where economy is of prime importance.

Some alloys are also used for press tools, bolsters and dies for short production runs.

Gravity castings are used for this purpose and also other tooling where the life is short and the shape complicated. It can be readily cast to a master component which can be made of plaster of Paris.

Symbol	Nominal analysis, supplier, condition and remarks.
2Z 4	97.75% Zn: Ingot; designation used by French Standard
2Z 5	98.5% Zn: Ingot; designation used by French Standard
2Z 6	98.75% Zn: Ingot; designation used by French Standard
2.2360	40% Cd 2.0% Al Zn alloy: Working temperature 300 °C; German Standard
AC 41A	1.0% Cu 4% Al Zn alloy: Die casting; designation used by ASTM
AG 40A	0.2% Cu 4% Al Zn alloy: Die casting; designation used by ASTM
ALBALOY	Electrodeposited bright Cu Sn Zn alloy: Origin unknown
ALMEN 305	30% Al 5% Cu Zn alloy: Casting as cast; for bearings; Hill Alzen
	DPN: 100 UTS: 200 Elon: 2% Proof: 120
ALZEN 305K	30% Al 5% Cu Zn alloy: Extrusion for bearings Hill Alzen
	DPN: 110 UTS: 250 Elon: 12% Proof: 150
AMS 4803A	4% Al 0.04% Mg Zn alloy: Casting; as cast
ARSENIC-ZINC	40% As Zn alloy: Blackwells; primary metal
ASTM B6	98.5–99.99% Zn: Slab; spelter; primary metal; covers 5 grades
ASTM B69	Rolled zinc: Sheet & plate; gives information on commercially pure Zn
ASTM B86 AC41A	1.0% Cu 4% Al Zn alloy: Die casting
	DPN: 91 UTS: 33 Elon: 7%
ASTM B86 AG40A	0.2% Cu (max) 4% Al Zn alloy: Die casting
	DPN: 82 UTS: 29 Elon: 10%
ASTM B240 AC41A	4.0% Al 0.04% Mg Zn: Ingot for die casting; as cast
ASTM B240 AG40A	4.0% Al 0.04% Mg Zn: Ingot for die casting; as cast
AVONMOUTH	98.5% Zn ingot: Imperial Smelting Co Ltd; brand name for BS 3436/4
BATT/Z	High purity Zn sheet for battery cases: London Zinc; Zincon H
BBS/Z	Zn alloy: London Zinc; Zincon/G
BIDENY METAL	88.5% Zn balance Pb and Cu: Used for domestic utensils; Indian origin
BINDING METAL	3.7% Sb 28% Sn Zn alloy: Used for wire ropes; origin unknown
BIRMINGHAM PLATINA	25% Cu Zn alloy: Origin unknown
BISMUTH-ZINC	40% Bi Zn alloy: Blackwells; primary metal
BS 220	Fine zinc: Replaced by BS 3436
BS 221	Special zinc: Replaced by BS 3436
BS 222	Foundry zinc: Replaced by BS 3436

Note. The following abbreviations and units are used in the tables:

DPN	Hardness, diamond pyramid number
UTS	Ultimate tensile strength, N/mm^2
Elon	Elongation, %
Proof	0.1 % proof strength, N/mm^2

$1\ N/mm^2 = 0.1\ hbar = 0.102\ kgf/mm^2 = 0.06475\ tonf/in^2 = 145.04\ lbf/in^2$

Symbol	Nominal analysis, supplier, condition and remarks.
BS 849	1.2% Pb Zn metal: Sheet; for roofing purposes
BS 1003	Zn: High purity; replaced by BS 3436
BS 1004 A	4.1% Al 0.05% Mg Zn: Ingot & die casting; as cast
	DPN: 83 UTS: 270 Elon: 15%
BS 1004 B	4.0% Al 1.0% Cu 0.05% Mg Zn alloy: For die casting; after 1 year age
	DPN: 74 UTS: 200 Elon: 11%
BS 1141	Secondary zinc alloy: Die castings; withdrawn
BS 2656	0.003% Pb 0.003% Cd 0.004% Hg Zn anodes
BS 3436 Zn 1	99.99% Zn: Ingot; primary metal
BS 3436 Zn 2	99.95% Zn: Ingot; primary metal
BS 3436 Zn 3	99.5% Zn: Ingot; primary metal
BS 3436 Zn 4	98.5% Zn: Ingot; primary metal
DEF 17A	4.1% Al 0.05% Mg Zn: Die castings; defence specification covering BS 1004/A material
DI-METAL	4% Al 3% Cu Zn alloy: Used for die-castings; origin unknown
DICO	Zinc: Casting for press tools; melting point 390 °C; Platt metal Co
DIN 1706	97.5–99.995% Zn (min) pure metal
DIN 1743 G Zn Al 4 CU 1	4.0% Al 0.8% Cu 0.03% Mg Zn alloy: German Standard; for sand castings
	DPN: 70 UTS: 180 Elon: 0.5%
DIN 1743 G Zn Al 6 Cu 1	5.8% Al 1.4% Cu Zn alloy: Casting; German Standard; for sand castings
	DPN: 80 UTS: 180 Elon: 1.0%
DIN 1743 GD Zn Al 4	4.0% Al 0.3% Cu 0.03% Mg Zn alloy: Casting; German Standard
DIN 1743 GD Zn Al 4 Cu 1	4.0% Al 0.8% Cu 0.03% Mg Zn alloy: German National Standard; for pressure castings
	DPN: 80 UTS: 270 Elon: 2%
DIN 1743 GK Zn Al 4 Cu 1	4.0% Al 0.8% Cu 0.03% Mg Zn alloy: German Standard; for die castings
	DPN: 70 UTS: 200 Elon: 1.0%
DIN 1743 GK Zn Al 6 Cu 1	5.8% Al 1.4% Cu Zn alloy: German Standard; for die castings
DIN 1743 Z400	4.0% Al 0.3% Cu 0.03% Mg Zn alloy: Casting; German Standard
	DPN: 70 UTS: 250 Elon: 1.5%
DIN 1743 Z410	4.0% Al 0.8% Cu 0.03% Mg Zn alloy: Casting; German Standard
DIN 1743 Z610	5.8% Al 1.4% Cu Zn alloy: Casting; German Standard
DIN 8512 L Zn Cd 40	40% Cd 2.0% Al Zn alloy: Working temperature 300 °C
DOLER	4% Al 3% Cu Zn alloy: Used for die castings; origin unknown
EHRHARDT	3% Cu 6% Sn 2% Pb Zn alloy: Origin unknown
ELEC/Z	Zn alloy: London Zinc; now Zincon 1
ERAYDE	0.1% Ag 2% Cu Zn alloy: Used for radio shields; origin unknown
EZDA	Cu free Zn alloy: Die casting; Electrolytic Zinc Co; further information from Zinc Development Corporation
FENTON	6% Cu 14% Sn Zn alloy: Used for bearings; origin unknown
FONTAIN MOREAU	6% Cu 1% Pb 1% Fe Zn alloy: Used for ornamental work; origin unknown

Symbol	Nominal analysis, supplier, condition and remarks.
h Zn 1	99.9999% Zn: Rod; Koch-Light Laboratories Ltd; high purity metal
h Zn 73a	99.99% Zn: Single crystals 6.3 mm × 25 mm; Koch-Light Laboratories Ltd
ILZRO 12	12% Al 1% Cu 0.02% Mg Zn alloy: For casting; International Lead Zinc Research Organization; further information from Zinc Development Corporation **DPN: 115 UTS: 230 Elon: 6% Proof: 140**
ILZRO 14	0.02% Al 1.2% Cu 0.27% Ti Zn alloy: For casting; International Lead Zinc Organization; further information from Zinc Development Corporation
KAYEM	Zinc: Rolled sheet; Imperial Smelting Corporation; blanking dies; melting point 390 °C **DPN: 96 UTS: 320 Elon: 12%**
KAYEM	Zinc: Casting; Imperial Smelting Corporation; melting point 390 °C for press tools **DPN: 109 UTS: 220 Elon: 1.6%**
KAYEM 2	Zinc: Casting; Imperial Smelting Corporation; melting point 358 °C; press tools **DPN: 140 UTS: 140**
KB 90	Iron copper zinc: Sintered material; Sintered Products Ltd; 3.8% porosity **UTS: 340 Elon: 12%**
KIRKSITE A	Zn base alloy: For press tools; Hoyt Ltd
L Zn Cd 40	40% Cd 2.0% Al Zn alloy: Designation used by German Standard
L M 15	Ag Cd Zn base: Soft solder; Johnson Matthey; electrical conductivity 15% IACS; melting range 280–320 °C **UTS: 180 Elon: 5%**
MAIN METAL	Al Zn alloy: Casting for bearings; Main Metal **DPN: 110 UTS: 220 Elon: 2% Proof: 150**
MAIN METAL	Al Zn alloy: Extrusion for bearings; Main Metal **DPN: 140 UTS: 310 Elon: 16% Proof: 260**
MAZAK 3	4.0% Al 0.05% Mg Zn alloy: For die casting; Imperial Smelting Co; results after 1 year natural ageing **DPN: 70 UTS: 250 Elon: 25%**
MAZAK 5	4.0% Al 1% Cu 0.05% Mg Zn alloy: For die casting; Imperial Smelting Co; results after 1 year natural ageing **DPN: 74 UTS: 300 Elon: 11%**
MAZAK 7	4% Al low Mg Cu Ni Zn alloy: For die casting; Imperial Smelting Co
MNC 71	General standard for SIS range of zinc die casting alloys: Swedish Standard
NF A55 101/Z5	98.0% Zn: Ingot; French National Standard
NF A55 101/Z6	98.5% Zn: Ingot; French National Standard
NF A55 101/Z7	99.5% Zn: Ingot; intermediate; French National Standard
NF A55 101/Z8	99.95% Zn: Ingot; fine; French National Standard
NF A55 101 Z9	99.993% Zn: Ingot; extra fine; French National Standard
NF A55 102 ZA4G	4.0% Al 0.04% Mg Zn alloy: Ingot; French National Standard
NF A55 102 ZA4 U1G	4.0% Al 1.0% Cu 0.04% Mg Zn alloy: Ingot; French National Standard

Symbol	Nominal analysis, supplier, condition and remarks.
NF A55 102 ZA4 U3G	4.0% Al 3.0% Cu 0.04% Mg Zn alloy: Ingot; French National Standard
NF A55 101/2Z4	99.75% Zn: Ingot; French National Standard
NF A55 101/2Z5	98.5% Zn: Ingot; French National Standard
NF A55 101/2Z6	98.75% Zn: Ingot; French National Standard
OC/Z	Zn alloy: London Zinc; now Zincon F
QQ Z 35/a	Zinc: Slab or spelter; US Federal specification
QQ Z 100/a	Zinc: Alloy, sheet & strip; US Federal specification
QQ Z 285	Zinc: For anodes; US Federal specification
QQ Z 301/c	Zinc: Sheet & strip; US Federal specification
QQ Z 351/a	Zinc: Slab or spelter; US Federal specification
QQ Z 363/a	Zinc base alloys: For die castings; US Federal specification
SAE 903	4% Al Zn alloy: Casting; as cast **DPN: 82 UTS: 270 Elon: 10%**
SAE 925	4% Al 1% Cu Zn alloy: Casting; as cast **DPN: 91 UTS: 300 Elon: 7%**
SALGE	4% Cu 10% Sn 1% Sb 1% Pb Zn alloy: Used for novelties; origin unknown
SCHULZ	3% Al 6% Cu Zn alloy: Type metal; origin unknown
SEVERN	99.5% Zn: Ingot; Imperial Smelting Co Ltd; brand name for BS 3436/3
SIS 7020	4% Al Zn: Die casting; Swedish Standard
SIS 7020	4.0% Al 0.04% Mg Zn alloy: Die casting; Swedish Standard
SIS 7030	4.0% Al 1.0% Cu 0.04% Mg Zn alloy: Die casting; Swedish Standard
SIS 7030	4% Al 1% Cu Zn: Die casting; Swedish Standard
SOREL	10% Cu 10% Fe Zn alloy: Ornamental use; origin unknown
STZ 20	Cu Ti Zn alloy: Made by Stolberger Zinc AG; information from Zinc Development Corporation
STZ 30	Cu Ti Zn alloy: Made by Stolberger Zinc AG; information from Zinc Development Corporation
SWANSEA VALE	98.5% Zn: Ingot; Imperial Smelting Co Ltd; brand name for BS 3436/4
TANDEM PTA	Zinc: Casting for press tools; melting point 390 °C; Fry Metal
T METAL	0.8% Cu 0.1% Ti 0.003% Mn 0.002% Cr Zn alloy: Wrought; London Zinc Mills; heat treated **UTS: 220 Elon: 45% Proof: 120**
UNI 3718 Gp Zn Al 4	4.0% Al 0.04% Mg Zn alloy: Italian Standard
UNI 3718 Gp Zn Al 4 Cu 1	4.1% Al 1.0% Cu 0.04% Mg Zn alloy: Italian Standard
UNI 3718 Gp Zn Al 4 Cu 3	4.1% Al 3.1% Cu 0.04% Mg Zn alloy: Italian Standard
UNS L14001–L14999	Zinc based alloys: American Standards system; unified numbers
VAUCHERS ALLOY	18% Sn 4.5% Pb 2.5% Sb Zn alloy: For bearings; origin unknown
W Zn 3	99.9999% Zn: Powder; Koch-Light Laboratories Ltd
Z 5	98.0% Zn: Ingot; designation used by French Standard
Z 6	99.5% Zn: Ingot; designation used by French Standard
Z 7 INTERMEDIARE	99.5% Zn: Ingot; designation used by French Standard
Z 8 FIN	99.95% Zn: Ingot; designation used by French Standard
Z 9 EXTRA-FIN	99.993% Zn: Ingot; designation used by French Standards
ZA 4	Designation used by Italian Standard for UNI 3718 Gp Zn Al 4
A 4 C1	Designation used by Italian Standard for UNI 3718 Gp Zn Al 4 Cu 1

Note. The following abbreviations and units are used in the tables:

DPN	Hardness, diamond pyramid number
UTS	Ultimate tensile strength, N/mm^2
Elon	Elongation, %
Proof	0.1 % proof strength, N/mm^2

$1 \ N/mm^2 = 0.1 \ hbar = 0.102 \ kgf/mm^2 = 0.06475 \ tonf/in^2 = 145.04 \ lbf/in^2$

Symbol	Nominal analysis, supplier, condition and remarks.
ZA 4 C3	Designation used by Italian Standard for UNI 3718 Gp Zn Al 4 Cu 3
ZA 4 U1G	4.0% Al 1.0% 0.04% Mg Zn alloy: Ingot; designation used by French Standard
ZA 4 U3G	4.0% Al 3.0% Cu 0.04% Mg Zn alloy: Ingot; designation used by French Standard
ZA 4 G	4.0% Al 0.04% Mg Zn alloy: Ingot; designation used by French Standard
ZAMA	4% Al Zn alloy: Die casting; Italian alloy; further information from Zinc Development Corporation
ZAMAC	4% Al Zn alloy: For die castings; French alloy; further information from Zinc Development Corporation
ZAMAK	4% Al Zn alloy: Die casting; American alloy; further information from Zinc Development Corporation
ZAM METAL	Zinc base alloy containing Al and Hg: Used for Zn plating anodes; origin unknown
ZELCO	15% Al 2% Cu Zn: Solder; origin unknown
ZILLAY	1% Cu 0.8% Col 0.1% Mg Zn alloy: Used for roofing; origin unknown
ZIMAL	4.15% Al 3.0% Cu Zn alloy: Origin unknown
ZINC	Zn: Ingot, sticks, powder, dust, etc.; Blackwells
ZINCON F	0.7% Pb Zn alloy: Sheet & strip for roofing; London Zinc Mills; OC/Z

Symbol	Nominal analysis, supplier, condition and remarks.
ZINCON G	0.7% Pb Zn alloy: Sheet; London Zinc Mills; high quality; BBS/Z
ZINCON H	0.6% Pb Zn alloy: Strip & calots; London Zinc Mills; for dry battery cases Batt/Z
ZINCON I	99.99% Zn: Sheet; high purity; London Zinc Mills; elec/Z
ZINCON J	0.7% Cu 0.08% Ti Zn alloy: Sheet & strip; London Zinc Mills; heat treatable; T metal **UTS: 150 Elon: 45% Proof: 80**
ZINC TINSEL	40% Pb Zn alloy: Origin unknown
ZISKON	29% Al Zn alloy: Origin unknown
Zn Al 4	4.0% Al 0.04% Mg Zn alloy: For die casting; Swedish Standard designation
Zn Al Cu 1	4.0% Al 1.0% Cu 0.04% Mg Zn alloy: Die casting; Swedish Standard designation

Note. The following abbreviations and units are used in the tables:

DPN	Hardness, diamond pyramid number
UTS	Ultimate tensile strength, N/mm^2
Elon	Elongation, %
Proof	0.1 % proof strength, N/mm^2

1 N/mm^2=0.1 hbar=0.102 kgf/mm^2=0.06475 tonf/in^2=145.04 lbf/in^2

56. Zirconium Zr

Physical properties

Atomic number	40	
Atomic weight	91.22	
Crystal structure	alpha Close-packed hexagonal up to 860 °C	
	beta Body-centred cubic – 860 °C to melting point	
Colour	Silvery white	
Specific gravity	6.53	
Density	6530 kg/m³	(0.234 lb/in.³)
Melting point	1850 °C	
Boiling point	3580 °C	
Specific heat	0.285 J/g °C	(0.068 cal/g °C)
Thermal conductivity	16.7 W/m °C	(4 cal/m s °C)
Coefficient of linear expansion (20–100° C)	5.8×10^{-6}/ °C	
Latent heat of fusion	251 J/g	(60 cal/g)
Latent heat of vaporization	– J/g	
Thermal neutron absorption cross-section	0.18 barns/atom	
Electrical conductivity	4% IACS (copper 100%)	
Specific resistance	390 microhm mm	
Temperature coefficient of electrical resistance	0.004/ °C	
Electrochemical equivalent	0.844 g/A/h	
Electrode potential	– V	
Magnetic susceptibility	1.32×10^{-6}	
Young's modulus of elasticity	94.5×10^9 N/m²	$(13.7 \times 10^6$ lbf/in.²)
Tensile strength	annealed 330 N/mm²	(22 tonf/in.²)
Hardness	annealed 150 DPN	

56.1 General notes on zirconium

This is too chemically active an element ever to be found in the native state. It occurs in considerable quantities as zircon, which is zirconium silicate, and in various other less common forms. Although it is treated as a rare metal zirconium is in fact the twelfth most common element in order of abundance, the scarcity being solely caused by extreme difficulty in extraction.

The pure metal is soft malleable and ductile, silvery white in colour, but becomes very brittle when impure. Zirconium cannot be heated above about 850 °C in air without becoming embrittled by absorbing the oxide and nitride which starts to form on the surface at about 400 °C but which do not diffuse in until about 850 °C.

Satisfactory hot working can be carried out between 600 and 800 °C.

There are no known means of improving the mechanical properties of zirconium or its alloys by thermal treatment, but cold working can be used to increase the tensile strength without causing too rapid a drop in ductility. Annealing at 600 °C in air removes most of the effects of cold work but full annealing requires 800 °C for at least one hour which necessitates inert atmosphere or vacuum to prevent oxide and nitride diffusion.

The scale formed when heated in air is difficult to remove except by grit blasting and must be eliminated before machining as it is very abrasive.

Machining of zirconium presents no great difficulty as it does not harden rapidly with cold work. In many respects it resembles the austenitic stainless steels but note must be taken of the fire danger that exists when zirconium is in the finely divided state. There is no evidence that the metal or any of its compounds presents any other hazard to health.

Zirconium metal resists most acids and all alkalis and is not appreciably attacked by many molten metals, including the sodium–potassium alloy used in nuclear reactors. The acid resistance is not as good as in tantalum, particularly when oxidizing acids such as nitric are involved.

The pure metal has been found to be very compatible to body fluids, and tissue adheres to metal inserts in a satisfactory manner. Zirconium thus finds use, with tantalum, for repairing bones and in some cases seized joints are replaced using these metals.

The largest use of zirconium is in the nuclear field where the excellent corrosion resistance joins with the low neutron absorption properties to make this the first choice for 'canning' the reactor material in certain types of nuclear power installations.

For this purpose the metal must be free from hafnium but is sometimes alloyed with small amounts of tin, chromium and nickel.

The excellent corrosion resistance of zirconium finds some uses in chemical engineering but these are neither as many, nor as varied as with tantalum. Rayon spinnerets are now made from zirconium and the metal is also used as electric lamp filaments when the power required per candle power produced is lower than with the tungsten filament lamps.

The metal can be welded but requires either high vacuum or copious supplies of inert gas around the weld pool to prevent inclusions of the oxide or nitride embrittling the weld. Wherever possible welding should be in a gas tight chamber under inert atmosphere. Fusion and resistance welding are possible, and zirconium can be welded to other refractory materials such as tungsten and molybdenum using these methods.

Although it resists the attack of most acids zirconium is actually very chemically reactive and this becomes apparent when it is in a finely divided form. This has been put to use as a means of producing a smokeless white flash for photographic and other purposes and is also used as a primer for ammunition. It replaces the normal mercury fulminate or lead azide primers when smokeless ammunition is required.

Zirconium metal is also a popular 'getter' in vacuum apparatus as it can be made into anodes or grids which combine the last traces of gas in the vacuum and then has sufficient strength at high temperatures to prevent distortion and sagging.

This same activity becomes a serious disadvantage when zirconium dust or fine turnings are involved as they are prone to spontaneous combustion under certain conditions. Zirconium fires become explosive if water is added and continue fuming under carbon dioxide and many chlorinated compounds. Dry powder or inert material such as lime, fluorspar or graphite should be used first to confine the blaze then to reduce the ignited material. Provided good housekeeping is practised and fine zirconium dust or turnings are not stored in quantity, machining and fabrication zirconium should present no greater a hazard than magnesium. It is however, necessary to comply with certain Factory Act requirements before handling zirconium.

There are very few alloys with zirconium as the base metal, additions are made in amounts of under 0.5% to about 40% zirconium.

Zirconium is now a popular metal for de-oxidizing high quality steels. Some sulphides and nitrides are removed at the same time, resulting in better cold working properties and increased ductility.

It is also used as an alloying element in certain stainless and tool steels when it is reputed to improve the wear resistance. It is added to heat resistant cast irons when there is an improvement in the machinability with no appreciable reduction in other properties. Higher zirconium contents with silicon cast iron, improve the resistance to hot sulphuric acid. Zirconium added to copper acts as a de-oxidizer and residual amounts are sometimes found in the brasses and bronzes.

A new alloy of copper and zirconium has been developed which has certain age hardening characteristics and is similar in many respects to copper–chromium alloys.

Higher zirconium contents in copper give alloys which have mechanical properties after thermal treatment comparable to copper–beryllium.

Added to magnesium, zirconium acts as a grain refiner and gives improved response to age hardening, particularly in the magnesium–zinc range of alloys.

An alloy of zirconium and lead is used in the manufacture of flints for lighters.

The oxide, and purified silicate, is used as a refractory material particularly the silicate, zircon, which is a popular lining material for electric furnaces.

Other uses for the oxide include abrasive polishing dust, refractory cement, ceramic and porcelain glaze, and as filler material for plastics and rubber.

Symbol	Nominal analysis, supplier, condition and remarks.
ASTM B349 R1	Zr: Sponge; high purity; for nuclear application primary metal **DPN: 150**
ASTM B350 R1	Zr: Ingot; high purity; for nuclear application **DPN: 160**
ASTM B350 RA1	1.5% Sn 0.1% Cr 0.05% Ni Zr: Ingot; for nuclear applications **DPN: 200**
ASTM B351 R1	Zr: Bar; high purity; for nuclear applications; annealed **DPN: 290** **Elon: 18%** **Proof: 120**
ASTM B351 RA1	1.5% Sn 0.1% Cr 0.05% Ni Zr: Bar; for nuclear application; annealed **UTS: 420** **Elon: 14%** **Proof: 220**
ASTM B352 R1	Zr: Sheet & plate; high purity; for nuclear application; annealed **UTS: 300** **Elon: 18%** **Proof: 170**

Note. The following abbreviations and units are used in the tables:

DPN	Hardness, diamond pyramid number
UTS	Ultimate tensile strength, N/mm^2
Elon	Elongation, %
Proof	0.1 % proof strength, N/mm^2

1 N/mm^2=0.1 hbar=0.102 kgf/mm^2=0.06475 tonf/in^2=145.04 lbf/in^2

Symbol	Nominal analysis, supplier, condition and remarks.
ASTM B352 RA1	1.5% Sn 0.1% Cr 0.05% Ni Zr: Sheet & plate; for nuclear application; annealed **UTS: 440** **Elon: 14%** **Proof: 300**
ASTM B353 R1	Zr: Tube; high purity; for nuclear application; annealed **UTS: 320** **Elon: 18%** **Proof: 120**
ASTM B353 R1	Zr: Tube; high purity; for nuclear application; ¼-hard **UTS: 420** **Elon: 8%** **Proof: 300**
ASTM B353 R1	Zr: Tube; high purity; for nuclear application; ½-hard **UTS: 580** **Elon: 5%** **Proof: 450**
ASTM B353 RA1	1.5% Sn 0.1% Cr 0.05% Ni Zr: Tube; for nuclear application; annealed **UTS: 420** **Elon: 14%** **Proof: 220**
ASTM B353 RA1	1.5% Sn 0.1% Cr 0.05% Ni Zr: Tube; for nuclear application; ¼-hard **UTS: 530** **Elon: 8%** **Proof: 450**
ASTM B353 RA1	1.5% Sn 0.1% Cr 0.5% Ni Zr: Tube for nuclear application; ½-hard **UTS: 700** **Elon: 3%** **Proof: 610**
ASTM B356 R1	Zr: Forging; high purity; for nuclear application; annealed **UTS: 300** **Elon: 18%** **Proof: 170**
ASTM B356 RA1	1.5% Sn 0.1% Cr 0.05% Ni Zr: Forging for nuclear application; annealed **UTS: 420** **Elon: 14%** **Proof: 270**

Symbol	Nominal analysis, supplier, condition and remarks.
ATR	0.58% Mo 0.56% Cn Zr alloy: Jessop
h Zr 25	99.5% Zr: Wire 1 mm diameter; Koch-Light Ltd
h Zr 74	99.5% Zr: Single crystals 3 mm × 6 mm; Koch-Light Ltd
R 1	High purity Zr: Designation used by ASTM
RA 1	1.5% Sn 0.1% Cr 0.05% Ni Zr: Designation used by ASTM
UNS R60001–R69999	Zr metal and alloys: American Standards system; unified numbers
Z 1	Commercially pure Zr: Reactor grade; Jessop
Z 2	0.12% Fe 0.1% Cr 0.05% Ni 1.4% Sn Zr alloy: Jessop
Z 3	0.58% Mo 0.56% Cr Zr alloy: Jessop
Z 5	Commercially pure Zr: Non-reactor grade; Jessop
ZIRCALLOY 11	0.12% Fe 0.1% Cr 0.05% Ni 1.4% Sn Zr alloy: Jessop; annealed
	UTS: 500 Elon: 28% Proof: 290
ZIRCONIUM	Zr 99–100% purity: Blackwells
ZIRCONIUM 10	Pure Zr: For reactors; IMI
	UTS: 250 Elon: 25% Proof: 160

Symbol	Nominal analysis, supplier, condition and remarks.
ZIRCONIUM 20	1.5% Sn 0.12% Fe 0.1% Cr 0.05% Ni Zr alloy: IMI
	UTS: 300 Elon: 21% Proof: 220
ZIRCONIUM 30	0.5% Cu 0.55% Mo Zr alloy: IMI
	DPN: 150 UTS: 320 Elon: 21% Proof: 200
ZIRCONIUM 40	1.5% Sn 0.2% Fe 0.1% Cr Zr alloy: IMI

Note. The following abbreviations and units are used in the tables:

DPN	Hardness, diamond pyramid number
UTS	Ultimate tensile strength, N/mm^2
Elon	Elongation, %
Proof	0.1 % proof strength, N/mm^2

$1\ N/mm^2 = 0.1\ hbar = 0.102\ kgf/mm^2 = 0.06475\ tonf/in^2 = 145.04\ lbf/in^2$

Appendix I

The following is a list of names and addresses of the firms and organizations whose materials are listed in this book.

It must be emphasized that although permission to use the information has invariably been given, this does not in any way involve them in the responsibility for the extracts. It must also be reiterated that much of the information is covered by trade marks or patents and nothing in this book can be construed as giving permission to make use of any material in any way that might infringe upon these rights.

Appendix II lists the trade names with the supplier with whom contact should always be made in the first instance.

AA – see Aluminium Association of America.
AALCO, 31 Davies Street, London W1, U.K.
ABS – see American Bureau of Shipping.
AEI Limited, Trafford Park, Manchester 17, U.K.
Aerospace Material Specification – issued by SAE.
AFNOR – see Association Français de Normalisation.
AICMA – see Association des Constructeurs de Material Aerospatial.
AISI – see American Iron and Steel Institute.
Alcan (Germany) GMBH, Postfach 85, Nurnberg, Germany.
Alcan Industries Limited, Bush House, Aldwych, London WC2, U.K.
ALCOA – see Aluminium Company of America.
Allegheny Ludlum Steel Company, Brackenbridge, Pennsylvania, U.S.A.
All-State Welding Alloys Company Incorporated, White Plains, New York, U.S.A.
Aluminium Association of America, 420 Lexington Avenue, New York, U.S.A.
Aluminium Company Ltd – now British Aluminium Company Ltd Norfolk House, St James Square, London SW1, U.K.
Aluminium Company of America (ALCOA), now Imperial Aluminium Company, P.O. Box 216, Witton, Birmingham 6, U.K.
Aluminium Federation, Portland House, Stag Place, London SW1, U.K.
Aluminium Français, 23 rue Balzac, Paris 8, France.
Aluminium Wire and Cable Company Limited, Port Tenant, Swansea, Glamorgan, U.K.
Aluminium Zentral ev, 4 Dusseldorf 10, Postfach 10008, Germany.
American Bureau of Shipping, 45 Broad Street, New York, New York 10004, U.S.A.
American Iron and Steel Institute, 150 East 42nd Street, New York, U.S.A.
American Petroleum Institute, 300 Corrigan Tower Building, Dallas, Texas 75201, U.S.A.
American Society for Testing and Materials (ASTM), 1916 Race Street, Philadelphia, Pennsylvania, U.S.A.
AMPCO Metal Incorporated, Milwaukee, U.S.A.
AMS – see Aerospace Material Specification.

Anglo-Swiss Aluminium Company, Queen's Building, 55 Queen Street, Sheffield, U.K.
Anor (Acieries and Forges d') – information from Climax Molybdenum Company.
Anti-Attrition Company Limited, Gorton Lane, Manchester 18, U.K.
API – see American Petroleum Institute
Appleby-Frodingham Steel Company, Scunthorpe, Lincolnshire, U.K.
Armco Limited, 76 Grosvenor Street, London W1, U.K.
Armstrong Whitworth (Metal Industries) Limited, Western Road Works, Jarrow, County Durham, U.K.
N.C. Ashton, St Andrew's Road, Huddersfield, U.K.
Association des Constructeurs de Material Aerospatial, Paris, France.
Association Français de Normalisation (AFNOR), 23 rue Notre Dame des Victoires, Paris 2, France.
ASTM – see American Society for Testing and Materials.
Atlas Steel Company, 828 Depot Street, Parkersburg, West Virginia, U.S.A.
Atlas Steels Limited, Welland, Ontario, Canada.
Austrian Specification Borsegasse 18, 1010 Vienna – copies held by British Standards Institution.
Avesta Jernwerks, Aktiebolag, Avesta, Sweden.

Babcock and Wilcox (U.S.A.), Baberton, Ohio, U.S.A.
Baird and Scottish Steel Co. Ltd – no longer exists.
Baldwins Limited, Victoria Embankment, London SE1, U.K.
Balfour and Darwins Limited, Capital Steel Works, Sheffield 3, U.K.
Barker and Allan Limited, Nickel Silver Works, Springhill, Birmingham, U.K.
Batterium Metal Limited, Market Harborough, Leicester, U.K.
Belgian Specification, (Institute Belge Normalisation, NBN), 29 avenue de la Brabanconne, Brussels 4, Belgium.
Beryllium Corporation, Reading, Pennsylvania, U.S.A.
Bethlehem Steel Company, Bethlehem, Pennsylvania, U.S.A.

T.M. Birkett, Billington and Newton Limited, Hanley, Stoke-on-Trent, Staffs., U.K.

Birmetals Limited, Clapgate Lane, Quinton, Birmingham 32, U.K.

Birmingham Aluminium Company, Birmid Works, Dartmouth Road, Smethwick,, Birmingham, U.K.

Birmingham Battery and Metal Company Limited, Selly Oak, Birmingham 29, U.K.

BKL Alloys Limited, Factory Centre, Kings Norton, Birmingham, U.K.

Blackwells Metallurgical Limited, Russel Road, Liverpool 19, U.K.

Blaw-Knox Limited, Rochester, Kent, U.K.

BNFL – see British Nuclear Fuels.

Bochumer Verein – information from Climax Molybdenum Company.

Bofors, Aktiebolaget, Bofors, Sweden.

Bofors Steel Limited, Dowsetts Lane, Ramsden Heath, Billericay, Essex, U.K.

Bofors Steels Incorporated, 2 Henderson Drive, West Caldwell, New Jersey, U.S.A.

Gebruder Bohler and Company AG – information from Climax Molybdenum Company.

G. Bohler and Company, Aktiengesellschaft, Vienna, Austria.

Bohler Brothers and Company (London) Limited, 46 Kingsway, London WC2, U.K.

Thomas Bolton and Sons Limited, Froghall, Stoke-on-Trent, Staffs., U.K.

Bonperithis (Forge and Acieries de) – information from Climax Molybdenum Company.

James Booth Aluminium Limited, Kitts Green, Birmingham 33, U.K.

Bradley and Foster Limited, Darlaston Ironworks, Darlaston, Staffs., U.K.

Braeburn Steel Company, Braeburn, Pennsylvania, U.S.A.

Brandhurst Company Limited, Vintry House, Queen Street Place, London EC4, U.K.

Brimabright Limited, Woodgate Works, Quinton, Birmingham 32, U.K.

Britannia Iron and Steel Works Limited, Bedford, U.K.

British Aluminium Company Limited, Norfolk House, St James Square, London SW1, U.K.

British Driver-Harris Company Limited, Cheadle Heath, Cheadle, Cheshire, U.K.

British Metal Corporation Limited, Princes House, 93 Gresham Street, London EC2, U.K.

British Nuclear Fuels Limited, Risley, Warrington, U.K.

British Rollery Mills Limited, Brymill Steel Works, P.O. Box 10, Tipton, Staffs., U.K.

British Rolling Mills Limited, Bloomfield Road, P.O. Box 10, Tipton, Staffs., U.K.

British Standards Institution (BS), 2 Park Street, London W1, U.K.

British Steel Corporation, P.O. Box No. 403, 33 Grosvenor Place, London SW10, U.K.

Brown Bayley Steels Limited, Riverdale, Riverdale Road, Sheffield 10, U.K.

David Brown Industries Limited, Foundries Division, Penistone, nr. Sheffield, U.K.

Brush Beryllium Company, 17876 St Clair Avenue, Cleveland, Ohio 44110, U.S.A.

Cameron & Sons – no longer exists.

Canadian Standards Association, 235 Montreal Road, Ottawa 7, Canada.

Carobronze, School Road, Belmont Road, London W4, U.K.

Carpenter Steel Company, 135 West Barn Street, Reading, Pennsylvania, U.S.A.

R. Carr and Company Limited, Pluto Works, Wadsley Bridge, Sheffield 6, U.K.

CDA – see Copper Development Association.

CDAA – see Copper Development Association of America.

C. Clifford Limited, Dog Pool Mills, Birmingham 30, U.K.

Climax Molybdenum Company of Europe Limited, 2 Cavendish Place, London W1, U.K.

Climax Molybdenum Corporation, 1270 Avenue of Americas, New York 20, U.S.A.

Clyde Alloy Steel Company Limited, Hallside Works, Cambuslang, Scotland, U.K. - now British Steel Corp.

Cobalt Information Centre, Chichester House, 278/282 High Holborn, London WC1, U.K.

Columbia Tool Steel Company, 500 Lincoln Highway, Chicago Heights, Illinois, U.S.A.

Colvilles Limited, 195 West George Street, Glasgow C2, Scotland, U.K. (now Part of B.S.C.).

Comalco (Europe) Limited, P.O. Box 133, 6 St James Square, London SW1, U.K.

Consett Iron Company Limited, Consett, County Durham, U.K.

Consolidated Beryllium Limited, St Andrew's Road, Avonmouth, Bristol, U.K.

Constrictor Ltd – no longer exists.

Copper Development Association (CDA), Orchard House, Mutton Lane, Potters Bar, Hertfordshire, U.K.

Copper Development Association of America (CDAA), 405 Lexington Avenue, New York, U.S.A.

Copperweld Steel Company, Glassport, Pennsylvania 15045, U.S.A.

Phillip Cornes and Company Limited, Claybrook Drive, Washford, Redditch, Worcs., U.K.

Creusot-Loire Steel Company Limited, 27 Broadwick Street, London W1, U.K.

Cronite Foundry Company Limited, Lawrence Road, Tottenham, London N15, U.K.

Crucible Steel Company - now Crucible, The Materials Research Centre P.O. Box 88, Parle Way West, Route 80, Pittsburg PA 15230, U.S.A.

Cutokumfu, Helsinki, Finland.

Dacral, 164 rue Ambroise Croizat, 93204 St Denis, France.

Danish Specification – see Danish Standards Association.

Danish Standards Association (Dansk Standardiserings-rad) Aurehøjvej 12, 2900 Hellerup, Denmark, Copies held by British Standards Institution.

Darwins Limited, Tinsley, Sheffield, U.K.

DEF – Ministry of Aviation, St Giles Court, 1-13 St Giles High Street, London WC2, U.K.

Deloro Stellite Limited, Stratton St Margaret, Swindon, Wilts, U.K.

Delta Metal Company, Dartmouth, Birmingham, U.K.

C. Denton Steel and Tool Company, Westwick Steel Works, Solly Street, Sheffield, U.K.

Det Norske Veritas, Grenseveien 92, Oslo, Norway.

Deutsche Edelstahlwerke AG – information from Climax Molybdenum Company.

Dewrance and Company Limited, Special Alloy Division, Great Dover Street, London SE1,

DIN – see German Standards.

Director of Materials Research and Development (DTD), Ministry of Aviation, St Giles Court, 1–13 St Giles High Street, London WC2, U.K.

DNN – see Det Norske Veritas.

DOMAL – see Dominion Magnesium Limited.

Dominion Magnesium Limited, Haley, Ontario, Canada.

Dorman Long Steel Limited, P.O. Box 4, Lackenly Works, Middlesborough, U.K.

Driver-Harris, 201 Middlesex Street, Harrison, New Jersey, U.S.A.

DTD – see Director of Materials Research and Development.

Dunford and Elliott, Attercliffe Wharf Works, Sheffield 9, U.K.

Dunford and Hadfield Limited, East Hecla Works, Sheffield 9, U.K.

Durasint Products Limited, Hamilton Road, Sutton-in-Ashfield, Notts., U.K.

Duriron Company, c/o Premaberg Limited, 22/24 High Street Halstead, Essex, U.K.

Duriron Company, Dayton, Ohio, U.S.A.

Eastern Stainless Steel Company, Baltimore, U.S.A.

Edgar and Allen and Company Limited, P.O. Box 93, Imperial Steel Works, Sheffield 9, U.K.

Electrolytic Zinc Company, Crown Works, 138 Wednesbury Road, Walsall, Staffs., U.K.

Elm Engineering Limited, Southern Road, Aylesbury, Bucks., U.K.

Enfield Rolling Mills Limited, Brimsdown, Enfield, Mddx., U.K.

Engelhard Industries, 52 High Holborn, London WC1, U.K.

English Steel Corporation (ESC), River Don Works, Sheffield, U.K.

ESAB Limited, Gillingham, Kent, U.K.

ESC – see English Steel Corporation.

Esperance Longder, 60 rue d'Harscamp, Liege, Belgium.

Eyre Smelting Company, Tandem Works, Merton Abbey, London SW19, U.K.

Fagersta Bruk AB, Fagersta, Sweden.

Fagersta Steels Limited, Kinwarton Farm Road, Alcester, War., U.K.

Fansteel Metallurgical Corporation, No. 1 Tantallan Place, North Chicago, U.S.A.

Fiat Spa, Milan, Italy.

Firminy (Forge de) – information from Climax Molybdenum Company.

Firth Brown, Atlas Works, Sheffield, U.K.

Firth Cleveland Sintered Products Limited, Treforest, Pontypridd, Glamorgan. U.K.

Firth Sterling Incorporated, 3113 Forbes Street, Pittsburg, Pennsylvania, U.S.A.

Firth Vickers, Staybrite Works, Weldon Street, Sheffield 9, U.K.

Forez (Acieres du), Saint Etienne, France.

Samuel Fox and Company Limited, Stocksbridge Works, Sheffield, U.K.

French Afnor, Association Français de Normalisation, 23 rue Notre Dame des Victoires, Paris 2, France.

French National Standard, Association Français de Normalisation 23 rue Notre Dame des Victoires, Paris 2, France.

Friedr Lohmann – information from Climax Molybdenum Company.

S. Fry and Company Limited, Christchurch Road, Merton Abbey, London SW19, U.K.

Furnival Steel Company Limited, Furnival Steel Works, Sheffield 4, U.K.

William Gallimore and Sons Limited, Sheffield, U.K.

A.B. Galnol and Company, Row, Birmingham, U.K.

Gebruder Bohler, Kapfenberg, Steiermark, Austria.

Gebruder Bohler and Company AG – information from Climax Molybdenum Company.

General Alloys Company, Boston, Massachusetts, U.S.A.

General Electric Company, Detroit 32, Michigan, U.S.A.

General Electric Company, Metallurgical Products Department, Detroit, U.S.A.

General Steel Industries Incorporated, Eddystone, Pa, U.S.A.

General Tool and Die Company, East Orange, New Jersey, U.S.A.

Georgsmanienwerke Selesiastahl GMBH, Georgsmarienhutte, Germany.

Gerhardi and Company, Ladenscheid, Germany.

German Standards, 175 Uhlandstrasse, 1 Berlin 15, Germany.

Gilby Brunton Limited, Seamill, Musselburgh, Edinburgh, U.K.

Gilby-Foder, Trefileries et Lamenoirsde Precision, Rueil-Malmaison, France.

Gimo-Osterby Bruks AB, Gimo, Sweden.

Giulini Werke AG, Rohrsbach, Germany.

Glacier Metal Company Limited, 368 Ealing Road, Alperton, U.K.

Gloucester Foundry, Emlyn Works, Gloucester, U.K.

T.R. Goldschmidt AG, Abteilung Metalle, Essen, Germany.

Gorham Tool Company, Detroit, Michigan, U.S.A.

GOST – Russian Specification – Copies held at British Institution.

Gottingen Aluminiumwerke GMBH, Gottingen, Germany.

Granite City Steel Company, Granite City, Illinois, U.S.A.

Grant and West Limited, London, U.K.

Graylom Steel Company, New York, U.S.A.

Great Lakes Steel Company, Detroit, Michigan, U.S.A.

Great Western Steel Division, Hayland Steel Company, Los Angeles, California, U.S.A.

Gruss Stahlwerk, Witten AG, Germany.

Guronitwerke Vervoort GMBH, Dusseldorf, Germany.

Gusstahl-Hondels GMBH, Van Dohlen-Stahl, Wetter, Ruhr, Germany.

Hadfields Limited, East Hecla Works, Sheffield 9, U.K.

Hagener Gusstahlwerke, Remy, Germany.

Hall and Pickles, Port Street, Manchester 1, U.K. – now part of Osborne Steels.

Hallamshire Steel Company Limited, Sheffield 3, U.K.

Hallamskie Steel Company – now part of Sheffield Rolling Mills.

Hanricot – information from Climax Molybdenum Company.

Harrison Fischer – no information available.

Harvey Aluminium Limited, Torrance, California, U.S.A.

Haynes Stellite, Samuel Osborn and Company Limited, P.O. Box 1, Clyde Steel Works, Sheffield 3, U.K.

HDA – see High Duty Alloys

A. Heckford Limited, Birmingham Metal Works, Frederick Street Birmingham, U.K.

High Duty Alloys Limited (HDA), Buckingham Avenue, Trading Estate, Slough, Bucks., U.K.

Hill Alzen (Sales) Limited, P.O. Box 22, Stringer Lane, Willenhill, Staffs., U.K.

Hoffmann – information from Climax Molybdenum Company.

I. Holroyd and Company Limited, P.O. Box 24, Holfus Works, Rochdale, U.K.

Holtzer-Loire – no information available.

Hopkinsons, P.O. Box 27, Brittania Works, Huddersfield, Yorks., U.K.

Hoyt Metal Company (Great Britain) Limited, Deodar Road, Putney, London SW15, U.K.

Huntingdon Alloy Products, The International Nickel Company, 67 Wall Street, New York 5, U.S.A.

B. Huntsman, Crucible Steel Works, Attercliffe, Sheffield 9, U.K.

ICI Limited – see Imperial Metal Industries.

IMI – see Imperial Metal Industries.

Imperial Aluminium Company Limited, P.O. Box 216, Witton, Birmingham 6, U.K.

Imperial Metal Industries Limited (IMI; formerly ICI), P.O. Box 216, Kynoch Works, Witton, Birmingham 6, U.K.

Imperial Smelting Corporation Limited, P.O. Box 133, 6 St James Square, London SW1, U.K.

Imphy, Siege Social, 84 rue de Lille, Paris, France.

Incanite Foundries Limited, Cornwall Road, Smethwick 40, U.K.

INCO – see International Nickel Company.

I. Inman and Company Limited, Brittania Steel Works, Furnival Road, Sheffield 4, U.K.

INTAL – see International Alloys Limited.

International Alloys Limited (INTAL), Haydon Hill, Aylesbury, Bucks., U.K.

International Lead Zinc Research Organisation Incorporated 292 Madison Avenue, New York, New York 10017, U.S.A.

International Meehanite Company Limited, Meerion House, Albert Road North, Reigate, Surrey, U.K.

International Nickel Company (INCO), Thames House, Millbank, London SW1, U.K.

Italian Standards, Ente Nazionale Italiano de Unificazione Milano, Italy.

Japan Metal Industry, Kawasaki, Japan.

Jessop-Saville Limited, Brightside Works, Sheffield, U.K.

Johnson Matthey, 73–83 Hatton Garden, London EC1, U.K.

Jones and Colver Limited, Sheffield 9, U.K.

Jones and Rouke, 86–92 Northwood Street, Birmingham, U.K.

Kawasaki Plant, Japan Metal Industries Company Limited, Kawasaki, Japan.

Kawecki Berylco Incorporated, St Andrew's Road, Avonmouth, Bristol, U.K.

E. and E. Kaye Limited, Ponders End, Enfield, Mddx., U.K.

Kayser-Ellison and Company Limited, Carlisle Steel Works, Sheffield 4, U.K.

Kennametal (G.B.) Limited, 82–84 Coleshill Street, Birmingham 4, U.K.

Kind and Company, Edelstahlwerk – information from Climax Molybdenum Company.

Kirkstall Forge Engineering Limited, Leeds 5, Yorks., U.K.

Kiveton Park Steel Company, Kiveton Park, Yorks., U.K.

Klockner-Werke AG – information from Climax Molybdenum Company.

Krupps Steel Works, Fried, Essen, Germany.

C. Kuhbler and Son – information from Climax Molybdenum Company.

L'Aluminium Français, 23 rue Balzac, Paris 8, France.

Langley Alloys Limited, Langley, Slough, Bucks., U.K.

Latrobe Steel Company, Pennsylvania, U.S.A.

Lead Development Association, 34 Berkeley Square, London W1, U.K.

Arthur Lee and Sons Limited, P.O. Box 54, Sheffield, U.K.

Light Alloys Limited, St Leonards Road, Willesden, London NW10, U.K.

Light Laboratories Limited, Poyle, Colnbrook, Slough, Bucks., U.K.

London Zinc Mills Limited, Enfield, Mddx., U.K.

Low Moor Alloy Steelworks Limited, Low Moor, Bradford, U.K. - now Osborne Steel Extensions.

Magnesium Elecktron Limited, P.O. Box 6, Lumms Lane, Clifton Junction, Nr Manchester, U.K.

Magnolia Anti-Friction Metal Company, 34 Victoria Street, London SW1, U.K.

Main Metal (Great Britain) Limited, 69–71 Monmouth Street, London WC2, U.K.

Mallory Metallurgical Products Limited, Exhibition Grounds, Wembley, Mddx., U.K.

Manganese Bronze Limited, Elton Park Works, P.O. Box 19, Hadleigh Road, Ipswich, U.K.

W.L. Marrison Limited, Bower Street, Sheffield 3, U.K.

Marsh Brothers and Company Limited, Ponds Steel Works, Shude Lane, Sheffield, U.K.

Martin Marietta Company, Wheeling, Illinois, U.S.A.

Martin Metals Company, Division Martin Marietta Corporation, Wheeling, Illinois, U.S.A.

McKechnie Metals Limited, Middlemore Lane, Aldridge, Near Walsall, Staffs., U.K.

Meehanite (International Meehanite Company Limited), Murion House, Albert Road North, Reigate, Surrey, U.K.

Meigh Casting Company Limited, Werkinton Foundry, Nr Swindon, Cheltenham, Glos., U.K.

Mel-Magnesium Elecktron Limited, P.O. Box 6, Lumms Lane, Clifton Junction, Nr Manchester, U.K.

Metal Alloys Limited, Treforest Estate, Pontypridd, Wales, U.K.

Metco Limited, Chobham, Woking, Surrey, U.K.

Metro-Cutanit Limited, Grappenhall, Warrington, Lans., U.K.

Metrode Products Limited, Hanworth Lane, Chertsey, Surrey, U.K.

William Mills and Company Limited, Wednesbury, Staffs., U.K.

Mining and Chemical Products Limited, Alperton, Wembley, Mddx., U.K.

Miralite Co Ltd. – no longer exists.

Murex Limited, Rainham, Essex, U.K.

Nazionale Cogne – information from Climax Molybdenum Company.

NBN – see Belgian Specification.

Nelco Metal Corporation, Canaan, Connecticut, U.S.A.

Netherlands Standardisation Institute (Nederlands Normalisatie Institut; Dutch Specification), Polaweg 5, Ruswuk, Netherlands.

Northern Aluminium Company Limited – now Alcan Industries Limited, Bush House, Aldwych, London WC2, U.K.

Nyby Stainless Steels, 41–43 Mincing Lane, London EC3, U.K.

Officine Metallurgiche de Pont, St Martin, France.

Samuel Osborne and Company Limited, P.O. Box 1, Clyde Steel Works, Sheffield 3, U.K.

Park Gate Iron and Steel Company Limited, Park Gate Works, P.O. Box 23, Rotherham, U.K.

F. Parkin (Sheffield) Limited, St Thomas Steel Works, Sheffield 8, U.K.

Patent Shaft Steel, England.

Permanent Magnet Association (PMA), 301 Glossop Road, Sheffield 10, U.K.

Phosphor Bronze Company Limited, Temple Manor Works, Rochester, Kent, U.K.

Platt Metals Limited, Enfield, Mddx., U.K.

PMA – see Permanent Magnet Association.

Pompey Society of Acieres, 61 rue de Monceau, Paris 8, France.

Posse-Marre – information from Climax Molybdenum Company.

Kabel C. Pouplier Stahlwerk – information from Climax Molybdenum Company.

Preformations Limited, Cheney Manor, Swindon, Wilts., U.K.

Quebec Metallurgical Corporation, Quebec, Canada.

Reactive Metals Incorporated, Niles, Ohio, U.S.A.

Redheugh·Iron and Steel Company Limited, Teams, Gateshead, Co. Durham, U.K.

Republic Steel Company, Republic Building, Cleveland 1, Ohio, U.S.A.

Resistalloy Limited, Woodside Works, Rutland Road, Sheffield, U.K.

Reynolds Tubes Company, Hay Hall Works, Tysley, Birmingham 11, U.K.

Richards, Thomas and Baldwins Limited, RTB House, 151 Gower Street, London WC1, U.K. – now B.S.C. Ltd.

Rochling-Budenis – information from Climax Molybdenum Company.

Round Oak Steel, P.O. Box 3, Brierly Hill, Staffs., U.K.

Russian Specifications (GOST) – information held at British Standards Institution.

SAE – see Society of Automotive Engineers.

Salford Electric, Peel Works, Barten Lane, Eccles, Manchester M30, U.K.

Sambre and Meuse (Acieries de) – information from Climax Molybdenum Company.

Sanderson Brothers and Newbould Limited – now Sanderson Kayser Limited.

Sanderson Kayser Limited, Attercliffe Steelworks, Newhall Road, Sheffield S9, U.K.

Sandvik Steel Incorporated, Halesowen, Worcester, U.K.

Schmidt and Clemens Edelstahlwerk – information from Climax Molybdenum Company.

Schoeller-Bleckmann Steels (Great Britain) Limited, P.O. Box 9, Taylor's Lane, Oldbury, Warley, Worcs., U.K.

Sheffield Hollow Drill Steel Company Limited (SHD), Cardbrook Rolling Mills, Sheffield 9, U.K.

Sheffield Smelting Company, Royds Mill Street, Sheffield 4, U.K.

Simonds, 110 Safety First Bank Buildings, Fitchburgh, Massachusetts, U.S.A.

Sintered Products Limited, Sutton-in-Ashfield, Notts., U.K.

SIS – see Swedish Standardizing Commission.

Fred Smith and Company, Anaconda Works, Salford 3, Lancs., U.K.

Society Metallurgigue d'Imphy, Siege Social, 84 rue de Lille, Paris, France.

Society of Automotive Engineers (SAE), 485 Lexington Avenue, New York 17, U.S.A.

Soderfors Steel Works, Soderfors, Sweden.

Walter Somers, Haywood Forge, P.O. Box No. 7, Hales Owen, Worcs., U.K.

Southern Forge Limited – now Imperial Aluminium Company Limited, P.O. Box 246, Witton, Birmingham, U.K.

Spanish National Standardization Institute (Instituto Nationale de Racionalización del Trabajo), Sarrano 150, Madrid, Spain.

Spear and Jackson Industrial Tool Steel, Aetna
Works, Saville Street, Sheffield 4, U.K.
Walter Spenser and Company Limited, Crescent
Steel Works, Warren Street, Sheffield 6, U.K.
STA – The Under Secretary of State, War Office
(DS), First Avenue House, High Holborn, London
WC1, U.K.
Standard Telephones and Cables Limited,
Connaught House, 63 Aldwych, London WC2, U.K.
Standards Association of Australia, Standards
House, 80-86 Arthur St., North Sydney, NSW 2060
Australia.
Stavanger Elektre Stalverk, Stavanger, Norway.
STD Services Limited, TI House, Five Ways,
Birmingham 16, U.K.
Steel Company of Wales, Steel Division, Port
Talbot, Glamorgan – Now British Steel Corp.
Steel Peech and Tozer Limited, P.O. Box 50, The
Ickles, Rotherham, U.K. – now British Steel Corp.
J. Stone Limited, Woolwich Road, London SE7, U.K.
Stone Manganese, J. Stone and Company
(Charlton) Limited, Anchor and Hope Lane, London
SE17, U.K.
Peter Stubs, Scotland Road, Warrington, Lancs.,
U.K.
Sudwestfalen-Stahlwerk AG – information from
Climax Molybdenum Company.
Suffolk Iron Company Limited, Sifbronze Works,
Stowmarket, U.K.
Swedish Standardisering Kommission, Metals
Department, Box 3.295, Stockholm, Sweden.
Swift Levick and Sons Limited, Clarence Steel
Works, Leveson Street, Sheffield 4, U.K.

Telcon Metals Limited, P.O. Box 12, Manor Royal,
Crawley, Sussex, U.K.
Thomson L'Hospied and Company Limited,
Ablecote, Stourbridge, U.K.
Thyssen Glesserei AG, Mulheim, West Germany.
Timken Roller Bearing Company, Canton, Ohio,
U.S.A.
Tin Research Institute, Fraser Road, Perivale,
Greenford, Mddx., U.K.
Titanium Metal Company of America, 233
Broadway, New York, U.S.A.
Tokushu Seiko Company Limited – information
from Isfan Iron and Steel Federation, Tekko
Building, Tokyo.
Toledo Steel Works, Neepsend Lane, Sheffield 3,
U.K.
Toshiba, Tokyo Shibaura Electric Company, Tokyo,
Japan.
The Tungum Company Limited, The White House,
Arle, Cheltenham, Glos., U.K.
T. Turton and Sons Limited, P.O. Box 116, Sheaf

Works, Sheffield 4, U.K.
Turton Brothers and Matthews Limited, Rutland
Road, P.O. Box 40, Sheffield 3, U.K.

Uddelholm Aktiebolag, Uddelholm, Sweden.
Uddelholm Swedish Steels Limited, Crown Works,
Rubery, Birmingham, U.K.
Ugine (Acieries d'), Paris, France.
The Under Secretary of State (STA), War Office
(DS), First Avenue House, High Holborn, London
WC1, U.K.
Union Carbide Limited, Shepley Street, Glossop,
Derbys., U.K.
United Steel Company, Samuel Fox, Stocksbridge
Works, Sheffield, U.K.

Universal Cyclops Steel Corporation, Bridgeville,
U.S.A.
U.S. Federal Specifications – copies held by British
Standards Institution, 2 Park Street, London W1,
U.K.
U.S. Service Specifications, U.S. Government
Printing Office, Washington DC, U.S.A.

Vanadium Alloy Steel Company, Latrobe,
Pennsylvania, U.S.A.
John Vessey and Sons Limited Brunswick Steel
Works, Arley Street Sheffield 2, U.K. – now Hall
and Pickles.
Vickers Armstrong Limited, Vickers House,
Broadway, Westminster, London SW1, U.K.
The Vulcan Steel and Tool Company Limited,
Cornish Steel Works, Green Lane, Sheffield 3, U.K.

Wai-Met Alloys Company, Division of Howe Sound
Company, Dearborn, Michigan, U.S.A.
Wall Colmonoy (Canada) Limited, Pontadaine,
Glamorgan, U.K.
Wellman Alloys Limited, Ambiecote, Stourbridge,
Worcs., U.K.
V.R. Wesson Company, 800 Market Street,
Waukegan, Illinois, U.S.A.
Westinghouse Electric Corporation, East Pittsburgh,
Pa, U.S.A.
Henry Wiggin, Holmer Road, Hertford, U.K.
Wilson Walton Incorporated.
Wolsingham Steel Company Limited, Wolsingham,
Bishop Auckland, Co. Durham, U.K.

Yorkshire Imperial Metals Limited, P.O. Box 166,
Leeds, Yorks., U.K.

Zapp-Robt – information from Climax Molybdenum
Company.
Zinc Development Association, 34 Berkeley Square,
London W1, U.K.

Appendix II

This is a list of trade names which appear in the book, with the names of the company or association which supplied the information. It is important to realize that this company may not be the holder of the trade name, but will be the licensee in the U.K. The full name and address of the company is given in Appendix I.

While every effort has been made to list all the trade names it must be appreciated that some may have been missed. It is again emphasized that the information in this book must not be used in any manner to infringe trade names or practices, and this applies whether or not the material is listed in this Appendix.

ABRADUR	Pompey	CROMAX	Edgar Allen
ADAMANT	Firth Brown	CROMIMPHY	D'Imphy
ADVANCE	British Driver Harris	CROMODIE	Firth Brown
ALCAN	Alcan	CRONITE	Cronite Foundry
ALCOMAX	P.M.A.	CROTORITE	Manganese Bronze
ALDURAL	J. Booth	DOMAL	Dominion Magnesium
ALMAR	Allegheny Ludlum	DOMINATOR	Schoeller Bleckmann
ALNICO	P.M.A.	DUCOL	Colvilles
ALUMAGNESE	E. Kaye	DUNELT	Dunford
ANKA	Brown Bayley	DURAL	J. Booth
AQUATOUGH	Clyde Alloy	DURALUMIN	J. Booth
ARK	Jessop-Saville	DURANICKEL	Driver Harris
ARMCO	Armco	DURCILIUM	E. Kaye
ASTRANIT	Krupps	DUREHETE	S. Fox
ATLAS	Eyre Smelting	ELECTRITE	Latrobe
AURIGA	D. Brown	ELIVAR	Telcon
AUSTINOX	Pompey	ELKONITE	Johnson Matthey
AWCO	Aluminium Wire Co.	ERA	Hadfield
BATNICKON	Birmingham Battery	ESSHETE	S. Fox
BERYLCO	Beryllium Corporation	FORMOST	Swift Levick
BIRMETAL	Birmabright	FORTIWELD	United Steel Co.
BOLTOMET	Thomas Bolton	GALAHAD	Hadfields
BOROFIL	IMI	HASTELLOY	S. Osborn
BREARLEY	Brown Bayley	HAYNES	S. Osborn
BRIGHTRAY	H. Wiggin	HECLA	Hadfields
BRUNSWICK	John Vessey	HEDEX	Kiveton Park
CALMET	Calorizing Corporation	HIDUMINIUM	High Duty Alloys
CALOMIC	Telcon Metals	HIDURAX	Langley Alloys
CAPITAL	Balfour	HIDUREL	Langley Alloys
CERALUMIN	J. Stone	HIDURIT	Langley Alloys
CERRO ALLOYS	Mining & Chemical	HYDRA	Hall & Pickles
CHROMADOR	Dorman Long	HYLITE	Jessop Saville
CHROMAX	Driver Harris	HYPEAK	Swift Levick
CHROMEX	Murex	HYTEMPO	Driver Harris
CLYDALL	Clyde Alloys	IMMACULATE	Firth Vickers
COGWHEEL	Phosphor Bronze	IMMADIUM	Manganese Bronze Co.
COLCLAD	Colvilles	IMPALCO	Imperial Al. Co.
COLMONOY	Wall Colmonoy	INCANITE	Incanite
COLTUF	Colvilles	INCOLOY	H. Wiggin
COMALCO	Comalco	INCONEL	Huntington
COMBARLEY	T. Bolton	JEM	Jonas
COMET	Driver Harris	JETHETE	S. Fox
CONLO	Consett	KANTHAL	Kanthal AB
COPELMET	Metro Cutanit	KARMA	Driver Harris
CORRONEL	H. Wiggin	KAYEM	Imperial Smelting Corporation
COR-TEN	Appleby Frodingham	KENNERTEUM	Kennametal
CRESCENT	Spencer	MONEL	H. Wiggin
CROLOY	Babcock & Wilcox	KUNIFER	IMI

KUTHERM	IMI	NORAL	Northern Al. Co.
KYCUBE	IMI	NORESCO	Schoeller-Bleckmann
KYNAL	IMI	NOVO	Jonas
LANGALLOY	Langley	OROBRAZE	Johnson Matthey
LEDA	Firth Brown	OROCAST	Engelhard Industries
LESCALLOY	Latrobe Steel	PALLABRAZE	Johnson Matthey
LOHM	Driver Harris	PALLACAST	Engelhard Industries
LOSIL	Steel Co. of Wales	PERALUMAN	Anglo Swiss
MAGNO	Driver Harris	PERMALLOY	Standard Telephones
MAGNUMINIUM	High Duty Alloys	PERMANICKEL	Driver Harris
MALLORY	Johnson Matthey	PERMENDUR	Telcon
MALTEX	Stone Manganese	PHOENIX	Steel Peech & Tozer
MANGEAR	Steel Peech & Tozer	PIREKS	Darwins
MANGONIC	H. Wiggin	RADIOMETAL	Telcon
MAXHETE	Edgar Allen	SANBOLD	Sanderson
MAXILVRY	Edgar Allen	SPRABOND	Metco
MAZAK	Imperial Smelting Corporation	STAG	Edgar Allen
MELT-ESI	Sheffield Smelting	STELLITE	Deloro
MULTIMET	Osborn	TANDEM	Eyre Smelting
MUMETAL	Telcon	TAURUS	D. Brown
NICHROME	Driver Harris	TELCALLOY	Telcon
NICKELVAC	Vanadium Alloys	TOPHET	Brunton
NICREX	Murex	TRAN-COR	Armco
NICROBRAZE	Wall Colmonoy	TRANSIL	Steel Co. of Wales
NILO	H. Wiggin	TRUBRITE	Arthur Lee
NILVAR	Driver Harris	TRUCAST	Engelhard Industries
NIMOCAST	H. Wiggin	TURBISTON	Stone Manganese Marine
NIMONIC	H. Wiggin	UCAR	Union Carbide
NI-O-NEL	H. Wiggin	UNISIL	Steel Co. of Wales
NI-SPAN-C	Huntington	VICALLOY	Telcon
NITRALLOY	ESC	WALLEX	Wall Colmonoy
NITROFIL	IMI	WELDANKA	Brown Bayley

Appendix III

List of elements with their symbols

The section of the book under which each metal is discussed is shown alongside the element name.

	Element	Symbol		Element	Symbol
	Actinium	Ac	25	Mercury	Hg
1	Aluminium	Al	26	Molybdenum	Mo
35	Americium	Am	35	Neodymium	Nd
2	Antimony	Sb		Neon	Ne
	Argon	Ar	35	Neptunium	Np
3	Arsenic	As	27	Nickel	Ni
	Astatine	At	28	Niobium	Nb
4	Barium	Ba		Nitrogen	N
35	Berkelium	Bk		Nobelium	No
5	Beryllium	Be	29	Osmium	Os
6	Bismuth	Bi		Oxygen	O
7	Boron	B	30	Palladium	Pd
	Bromine	Br		Phosphorus	P
8	Cadmium	Cd	31	Platinum	Pt
9	Caesium	Cs	32	Plutonium	Pu
10	Calcium	Ca		Polonium	Po
35	Californium	Cf	33	Potassium	K
	Carbon	C	35	Praseodymium	Pr
11	Cerium	Ce	35	Promethium	Pm
	Chlorine	Cl	35	Protectinium	Pa
12	Chromium	Cr	34	Radium	Ra
13	Cobalt	Co		Radon	Rn
14	Copper	Cu	36	Rhenium	Re
35	Curium	Cm	35	Rhodium	Rh
35	Dysprosium	Dy	38	Rubidium	Rb
35	Einsteinium	Es	39	Ruthenium	Ru
35	Erbium	Er	35	Samarium	Sm
35	Europium	Eu	35	Scandium	Sc
35	Fermium	Fm	40	Selenium	Se
	Fluorine	F	41	Silicon	Si
	Francium	Fr	42	Silver	Ag
35	Gadolinium	Gd	43	Sodium	Na
15	Gallium	Ga	45	Strontium	Sr
16	Germanium	Ge		Sulphur	S
17	Gold	Au	46	Tantalum	Ta
	Hafnium	Hf		Technetium	Tc
	Helium	He	47	Tellurium	Te
35	Holmium	Ho	35	Terbium	Tb
	Hydrogen	H	48	Thallium	Tl
18	Indium	In	49	Thorium	Th
	Iodine	I	38	Thulium	Tm
19	Iridium	Ir	50	Tin	Sn
20	Iron	Fe	51	Titanium	Ti
	Krypton	Kr	52	Tungsten	W
35	Lanthanum	La	53	Uranium	U
	Lawrencium	Lw	54	Vanadium	V
21	Lead	Pb		Xenon	Xe
22	Lithium	Li	35	Ytterbium	Yb
35	Lutetium	Lu	35	Yttrium	Y
23	Magnesium	Mg	53	Zinc	Zn
24	Manganese	Mn	56	Zirconium	Zr
35	Mendelevium	Md			

Abbreviations used in this book

A	ampere
°C	degree Celsius (Centigrade)
DPN	diamond pyramid number (identical to VPN), a hardness value
Elon	elongation – there are various methods of calculating this value but the results are considered to be meaningful for the purposes of this book.
g	gram
hbar	hectobar
in.	inch
IACS	International Annealed Conductivity Standard – copper is taken as 100%
J	joule
kg	kilogram (mass) (1000 g)
kgf	kilogram (force)
lb	pound (mass)
lbf	pound (force)
m	metre
mm	millimetre
N	newton
Proof	0.1% proof strength unless otherwise stated
s	second
tonf	ton (force)
UTS	ultimate tensile strength given in hbar
V	volt
W	watt

In some instances firms' names and associations have been abbreviated in the tabulated sections of the book. These are listed in full in Appendix I.

Temperature conversion table

C		F	C		F	C		F	C		F	C		F	C		F
-17.8	0	32	18.9	66	150.8	210	410	770	577	1070	1958	943	1730	3146	1310	2390	4334
-17.2	1	33.8	19.4	67	152.6	216	420	788	582	1080	1976	949	1740	3164	1316	2400	4352
-16.7	2	35.6	20.0	68	154.4	221	430	806	588	1090	1994	954	1750	3182	1321	2410	4370
-16.1	3	37.4	20.6	69	156.2	227	440	824	593	1100	2012	960	1760	3200	1327	2420	4388
-15.6	4	39.2	21.1	70	158.0	232	450	842	599	1110	2030	966	1770	3218	1332	2430	4406
-15.0	5	41.0	21.7	71	159.8	238	460	860	604	1120	2048	971	1780	3236	1338	2440	4424
-14.4	6	42.8	22.2	72	161.6	243	470	878	610	1130	2066	977	1790	3254	1343	2450	4442
-13.9	7	44.6	22.8	73	163.4	249	480	896	616	1140	2084	982	1800	3272	1349	2460	4460
-13.3	8	46.4	23.3	74	165.2	254	490	914	621	1150	2102	988	1810	3290	1354	2470	4478
-12.8	9	48.2	23.9	75	167.0	260	500	932	627	1160	2120	993	1820	3308	1360	2480	4496
-12.2	10	50.0	24.4	76	168.8	266	510	950	632	1170	2138	999	1830	3326	1366	2490	4514
-11.7	11	51.8	25.0	77	170.6	271	520	968	638	1180	2156	1004	1840	3344	1371	2500	4532
-11.1	12	53.6	25.6	78	172.4	277	530	986	643	1190	2174	1010	1850	3362	1377	2510	4550
-10.6	13	55.4	26.1	79	174.2	282	540	1004	649	1200	2192	1016	1860	3380	1382	2520	4568
-10.0	14	57.2	26.7	80	176.0	288	550	1022	654	1210	2210	1021	1870	3398	1388	2530	4586
-9.44	15	59.0	27.2	81	177.8	293	560	1040	660	1220	2228	1027	1880	3416	1393	2540	4604
-8.89	16	60.8	27.8	82	179.6	299	570	1058	666	1230	2246	1032	1890	3434	1399	2550	4622
-8.33	17	62.6	28.3	83	181.4	304	580	1076	671	1240	2264	1038	1900	3452	1404	2560	4640
-7.78	18	64.4	28.9	84	183.2	310	590	1094	677	1250	2282	1043	1910	3470	1410	2570	4658
-7.22	19	66.2	29.4	85	185.0	316	600	1112	682	1260	2300	1049	1920	3488	1416	2580	4676
-6.67	20	68.0	30.0	86	186.8	321	610	1130	688	1270	2318	1054	1930	3506	1421	2590	4694
-6.11	21	69.8	30.6	87	188.6	327	620	1148	693	1280	2336	1060	1940	3524	1427	2600	4712
-5.56	22	71.6	31.1	88	190.4	332	630	1166	699	1290	2354	1066	1950	3542	1432	2610	4730
-5.00	23	73.4	31.7	89	192.2	338	640	1184	704	1300	2372	1071	1960	3560	1438	2620	4748
-4.44	24	75.2	32.2	90	194.0	343	650	1202	710	1310	2390	1077	1970	3578	1443	2630	4766
-3.89	25	77.0	32.8	91	195.8	349	660	1220	716	1320	2408	1082	1980	3596	1449	2640	4784
-3.33	26	78.8	33.3	92	197.6	354	670	1238	721	1330	2426	1088	1990	3614	1454	2650	4802
-2.78	27	80.6	33.9	93	199.4	360	680	1256	727	1340	2444	1093	2000	3632	1460	2660	4820
-2.22	28	82.4	34.4	94	201.2	366	690	1274	732	1350	2462	1099	2010	3650	1466	2670	4838
-1.67	29	84.2	35.0	95	203.0	371	700	1292	738	1360	2480	1104	2020	3668	1471	2680	4856
-1.11	30	86.0	35.6	96	204.8	377	710	1310	743	1370	2498	1110	2030	3686	1477	2690	4874
-0.56	31	87.8	36.1	97	206.6	382	720	1328	749	1380	2516	1116	2040	3704	1482	2700	4892
0	32	89.6	36.7	98	208.4	388	730	1346	754	1390	2534	1121	2050	3722	1488	2710	4910
0.56	33	91.4	37.2	99	210.2	393	740	1364	760	1400	2552	1127	2060	3740	1493	2720	4928
1.11	34	93.2	38	100	212	399	750	1382	766	1410	2570	1132	2070	3758	1499	2730	4946
1.67	35	95.0	43	110	230	404	760	1400	771	1420	2588	1138	2080	3776	1504	2740	4964
2.22	36	96.8	49	120	248	410	770	1418	777	1430	2606	1143	2090	3794	1510	2750	4982
2.78	37	98.6	54	130	266	416	780	1436	782	1440	2624	1149	2100	3812	1516	2760	5000
3.33	38	100.4	60	140	284	421	790	1454	788	1450	2642	1154	2110	3830	1521	2770	5018
3.89	39	102.2	66	150	302	427	800	1472	793	1460	2660	1160	2120	3848	1527	2780	5036
4.44	40	104.0	71	160	320	432	810	1490	799	1470	2678	1166	2130	3866	1532	2790	5054
5.00	41	105.8	77	170	338	438	820	1508	804	1480	2696	1171	2140	3884	1538	2800	5072
5.56	42	107.6	82	180	356	443	830	1526	810	1490	2714	1177	2150	3902	1543	2810	5090
6.11	43	109.4	88	190	374	449	840	1544	816	1500	2732	1182	2160	3920	1549	2820	5108
6.67	44	111.2	93	200	392	454	850	1562	821	1510	2750	1188	2170	3938	1554	2830	5126
7.22	45	113.0	99	210	410	460	860	1580	827	1520	2768	1193	2180	3956	1560	2840	5144
7.78	46	114.8	100	212	414	466	870	1598	832	1530	2786	1199	2190	3974	1566	2850	5162
8.33	47	116.6	104	220	428	471	880	1616	838	1540	2804	1204	2200	3992	1571	2860	5180
8.89	48	118.4	110	230	446	477	890	1634	843	1550	2822	1210	2210	4010	1577	2870	5198
9.44	49	120.2	116	240	464	482	900	1652	849	1560	2840	1216	2220	4028	1582	2880	5216
10.0	50	122.0	121	250	482	488	910	1670	854	1570	2858	1221	2230	4046	1588	2890	5234
10.6	51	123.8	127	260	500	493	920	1688	860	1580	2876	1227	2240	4064	1593	2900	5252
11.1	52	125.6	132	270	518	499	930	1706	866	1590	2894	1232	2250	4082	1599	2910	5270
11.7	53	127.4	138	280	536	504	940	1724	871	1600	2912	1238	2260	4100	1604	2920	5288
12.2	54	129.2	143	290	554	510	950	1742	877	1610	2930	1243	2270	4118	1610	2930	5306
12.8	55	131.0	149	300	572	516	960	1760	882	1620	2948	1249	2280	4136	1616	2940	5324
13.3	56	132.8	154	310	590	521	970	1778	888	1630	2966	1254	2290	4154	1621	2950	5342
13.9	57	134.6	160	320	608	527	980	1796	893	1640	2984	1260	2300	4172	1627	2960	5360
14.4	58	136.4	166	330	626	532	990	1814	899	1650	3002	1266	2310	4190	1632	2970	5378
15.0	59	138.2	171	340	644	538	1000	1832	904	1660	3020	1271	2320	4208	1638	2980	5396
15.6	60	140.0	177	350	662	543	1010	1850	910	1670	3038	1277	2330	4226	1643	2990	5414
16.1	61	141.8	182	360	680	549	1020	1868	916	1680	3056	1282	2340	4244	1649	3000	5432
16.7	62	143.6	188	370	698	554	1030	1886	921	1690	3074	1288	2350	4262	1705	3100	5612
17.2	63	145.4	193	380	716	560	1040	1904	927	1700	3092	1293	2360	4280	1760	3200	5792
17.8	64	147.2	199	390	734	566	1050	1922	932	1710	3110	1299	2370	4298	1816	3300	5972
18.3	65	149.0	204	400	752	571	1060	1940	938	1720	3128	1304	2380	4316	1871	3400	6152

To convert any temperature in Centigrade or Fahrenheit degrees to the other, take the figure in the centre column, and if converting from Centigrade to Fahrenheit, read off to the right-hand column; if converting from Fahrenheit to Centigrade, read off to the left-hand column.

For example: 720° C = 1328° F
                         1200° F = 649° C

Strength conversion table

hbar	N/mm²	tonf/in.²	lbf/in.²	kgf/mm²	hbar	N/mm²	tonf/in.²	lbf/in.²	kgf/mm²	hbar	N/mm²	tonf/in.²	lbf/in.²	kgf/mm²
1	10	0.647	1 450	1.02	81	810	52.45	117 500	82.60	161	1610	104.2	233 500	164.2
2	20	1.295	2 900	2.04	82	820	53.09	118 900	83.62	162	1620	104.9	235 000	165.2
3	30	1.942	4 350	3.06	83	830	53.74	120 400	84.64	163	1630	105.5	236 400	166.2
4	40	2.590	5 800	4.08	84	840	54.39	121 800	85.65	164	1640	106.2	237 900	167.2
5	50	3.237	7 250	5.10	85	850	55.04	123 300	86.67	165	1650	106.8	239 300	168.3
6	60	3.885	8 700	6.12	86	860	55.68	124 700	87.69	166	1660	107.5	240 800	169.3
7	70	4.532	10 150	7.14	87	870	56.33	126 200	88.71	167	1670	108.1	242 200	170.3
8	80	5.180	11 600	8.16	88	880	56.98	127 600	89.73	168	1680	108.8	243 700	171.3
9	90	5.827	13 050	9.18	89	890	57.63	129 100	90.75	169	1690	109.4	245 100	172.3
10	100	6.475	14 500	10.20	90	900	58.27	130 500	91.77	170	1700	110.1	246 600	173.3
11	110	7.122	15 950	11.22	91	910	58.92	132 000	92.79	171	1710	110.7	248 000	174.4
12	120	7.770	17 400	12.24	92	920	59.57	133 400	93.81	172	1720	111.4	249 500	175.4
13	130	8.417	18 850	13.26	93	930	60.22	134 900	94.83	173	1730	112.0	250 900	176.4
14	140	9.065	20 300	14.28	94	940	60.86	136 300	95.85	174	1740	112.7	252 400	177.4
15	150	9.712	21 750	15.30	95	950	61.51	137 800	96.87	175	1750	113.3	253 800	178.4
16	160	10.36	23 200	16.32	96	960	62.16	139 200	97.89	176	1760	114.0	255 300	179.5
17	170	11.01	24 650	17.33	97	970	62.80	140 700	98.91	177	1770	114.6	256 700	180.5
18	180	11.65	26 100	18.35	98	980	63.45	142 100	99.93	178	1780	115.3	258 200	181.5
19	190	12.30	27 550	19.37	99	990	64.10	143 600	101.0	179	1790	115.9	259 600	182.5
20	200	12.95	29 000	20.39	100	1000	64.75	145 000	102.0	180	1800	116.5	261 100	183.5
21	210	13.60	30 450	21.41	101	1010	65.37	146 500	103.0	181	1810	117.2	262 500	184.6
22	220	14.24	31 900	22.43	102	1020	66.04	147 900	104.0	182	1820	117.8	264 000	185.6
23	230	14.89	33 350	23.45	103	1030	66.69	149 400	105.0	183	1830	118.5	265 400	186.6
24	240	15.54	34 800	24.47	104	1040	67.34	150 800	106.0	184	1840	119.1	266 900	187.6
25	250	16.19	36 250	25.49	105	1050	67.99	152 300	107.1	185	1850	119.8	268 300	188.6
26	260	16.83	37 700	26.51	106	1060	68.63	153 700	108.1	186	1860	120.4	269 800	189.7
27	270	17.48	39 150	27.53	107	1070	69.28	155 200	109.1	187	1870	121.1	271 200	190.7
28	280	18.13	40 600	28.55	108	1080	69.93	156 600	110.1	188	1880	121.7	272 700	191.7
29	290	18.78	42 050	29.57	109	1090	70.58	158 100	111.1	189	1890	122.4	274 100	192.7
30	300	19.42	43 500	30.59	110	1100	71.22	159 500	112.2	190	1900	123.0	275 600	193.7
31	310	20.07	44 950	31.61	111	1110	71.87	161 000	113.2	191	1910	123.7	277 000	194.8
32	320	20.72	46 400	32.63	112	1120	72.52	162 400	114.2	192	1920	124.3	278 500	195.8
33	330	21.37	47 850	33.65	113	1130	73.17	163 900	115.2	193	1930	125.0	279 900	196.8
34	340	22.01	49 300	34.67	114	1140	73.81	165 300	116.2	194	1940	125.6	281 400	197.8
35	350	22.66	50 750	35.69	115	1150	74.46	166 800	117.3	195	1950	126.3	282 800	198.8
36	360	23.31	52 200	36.71	116	1160	75.11	168 200	118.3	196	1960	126.9	284 300	199.9
37	370	23.96	53 650	37.73	117	1170	75.76	169 700	119.3	197	1970	127.6	285 700	200.9
38	380	24.60	55 100	38.75	118	1180	76.40	171 100	120.3	198	1980	128.2	287 200	201.9
39	390	25.25	56 550	39.77	119	1190	77.05	172 600	121.3	199	1990	128.9	288 600	202.9
40	400	25.90	58 000	40.79	120	1200	77.70	174 000	122.4	200	2000	129.5	290 100	203.9
41	410	26.55	59 450	41.81	121	1210	78.35	175 500	123.4	201	2010	130.1	291 500	205.0
42	420	27.19	60 900	42.83	122	1220	78.99	176 900	124.4	202	2020	130.8	293 000	206.0
43	430	27.84	62 350	43.85	123	1230	79.64	178 400	125.4	203	2030	131.4	294 400	207.0
44	440	28.49	63 800	44.87	124	1240	80.29	179 800	126.4	204	2040	132.1	295 900	208.0
45	450	29.14	65 250	45.89	125	1250	80.93	181 300	127.5	205	2050	132.7	297 300	209.0
46	460	29.78	66 700	46.91	126	1260	81.58	182 800	128.5	206	2060	133.4	298 800	210.1
47	470	30.43	68 150	47.93	127	1270	82.23	184 200	129.5	207	2070	134.0	300 200	211.1
48	480	31.08	69 600	48.95	128	1280	82.88	185 700	130.5	208	2080	134.7	301 700	212.1
49	490	31.73	71 050	49.97	129	1290	83.53	187 100	131.5	209	2090	135.3	303 100	213.1
50	500	32.37	72 500	50.99	130	1300	84.17	188 600	132.6	210	2100	136.0	304 600	214.1
51	510	33.02	73 950	52.00	131	1310	84.82	190 000	133.6	211	2110	136.6	306 000	215.2
52	520	33.67	75 400	53.02	132	1320	85.47	191 500	134.6	212	2120	137.3	307 500	216.2
53	530	34.32	76 850	54.04	133	1330	86.12	192 900	135.6	213	2130	137.9	308 900	217.2
54	540	34.96	78 300	55.06	134	1340	86.76	194 400	136.6	214	2140	138.6	310 400	218.2
55	550	35.61	79 750	56.08	135	1350	87.41	195 800	137.7	215	2150	139.2	311 800	219.2
56	560	36.26	81 200	57.10	136	1360	88.06	197 300	138.7	216	2160	139.9	313 300	220.3
57	570	36.91	82 650	58.12	137	1370	88.71	198 700	139.7	217	2170	140.5	314 700	221.3
58	580	37.55	84 100	59.14	138	1380	89.35	200 200	140.7	218	2180	141.2	316 200	222.3
59	590	38.20	85 550	60.16	139	1390	90.00	201 600	141.7	219	2190	141.8	317 600	223.3
60	600	38.85	87 000	61.18	140	1400	90.65	203 100	142.8	220	2200	142.4	319 100	224.3
61	610	39.50	88 450	62.20	141	1410	91.30	204 500	143.8	221	2210	143.1	320 500	225.4
62	620	40.14	89 900	63.22	142	1420	91.94	206 000	144.8	222	2220	143.7	322 000	226.4
63	630	40.79	91 350	64.24	143	1430	92.59	207 400	145.8	223	2230	144.4	323 400	227.4
64	640	41.44	92 800	65.26	144	1440	93.24	208 900	146.8	224	2240	145.0	324 900	228.4
65	650	42.09	94 250	66.28	145	1450	93.89	210 300	147.9	225	2250	145.7	326 300	229.4
66	660	42.74	95 700	67.30	146	1460	94.53	211 800	148.9	226	2260	146.3	327 800	230.5
67	670	43.38	97 150	68.32	147	1470	95.18	213 200	149.9	227	2270	147.0	329 200	231.5
68	680	44.02	98 600	69.34	148	1480	95.83	214 700	150.9	228	2280	147.6	330 700	232.5
69	690	44.68	100 050	70.36	149	1490	96.48	216 100	151.9	229	2290	148.3	332 100	233.5
70	700	45.32	101 500	71.38	150	1500	97.12	217 600	153.0	230	2300	148.9	333 600	234.5
71	710	45.97	103 000	72.40	151	1510	97.77	219 000	154.0	231	2310	149.6	335 000	235.6
72	720	46.62	104 400	73.42	152	1520	98.42	220 500	155.0	232	2320	150.2	336 500	236.6
73	730	47.27	105 900	74.44	153	1530	99.07	221 900	156.0	233	2330	150.9	337 900	237.6
74	740	47.91	107 300	75.46	154	1540	99.71	223 400	157.0	234	2340	151.5	339 400	238.6
75	750	48.56	108 800	76.48	155	1550	100.4	224 800	158.1	235	2350	152.2	340 800	239.6
76	760	49.21	110 200	77.50	156	1560	101.0	226 300	159.1	236	2360	152.8	342 300	240.6
77	770	49.86	111 700	78.52	157	1570	101.7	227 700	160.1	237	2370	153.5	343 700	241.7
78	780	50.50	113 100	79.54	158	1580	102.3	229 200	161.1	238	2380	154.1	345 200	242.7
79	790	51.15	114 600	80.56	159	1590	103.0	230 600	162.1	239	2390	154.8	346 600	243.7
80	800	51.80	116 000	81.58	160	1600	103.6	232 100	163.2	240	2400	155.4	348 100	244.7

Hardness and tensile values approximate conversion table

Brinell 10 mm ball, 3000 kg load		Approximate equivalent tensile strength for steel		Vickers hardness no. (DPN or VPN)	Rockwell 120° cone			1/16 in. ball	Scleroscope
Diameter (mm)	Hardness no.	(tonf/in.²)	(hbar)		C scale (150 kg)	D scale (100 kg)	A scale (60 kg)	B scale (100 kg)	
2.00	945			1250	71	80	87	—	—
2.05	898			1150	70	79	87	—	—
2.10	856			1050	69	79	86	—	—
2.15	816			1000	68	78	86	—	—
2.20	781			975	67	78	85	—	—
2.25	745	163	251.7	950	66	77	85	—	106
2.30	712	155	239.4	910	65	76	84	—	100
2.35	683	150	231.7	850	64	75	84	—	95
2.40	653	143	220.9	790	62	73	83	—	91
2.45	627	137	211.6	750	61	72	82	—	87
2.50	601	132	203.9	715	59	71	81	—	84
2.55	578	127	196.1	671	57	69	80	—	81
2.60	555	122	188.4	633	56	68	79	—	78
2.65	534	117	180.7	599	54	67	78	—	75
2.70	514	112	173.0	572	52	65	77	—	72
2.75	495	108	166.8	547	50	64	76	—	70
2.80	477	105	162.2	523	49	63	75	—	67
2.85	461	101	156.0	501	48	62	75	—	65
2.90	444	98	151.4	479	47	61	74	—	63
2.95	429	95	146.7	459	45	60	73	—	61
3.00	415	92	142.1	441	44	59	73	—	59
3.05	401	88	135.9	424	42	58	72	—	57
3.10	388	85	131.3	409	41	57	71	—	55
3.15	375	82	126.6	395	40	56	71	—	54
3.20	363	80	123.6	382	39	55	70	—	52
3.25	352	77	118.9	369	37	53	69	—	51
3.30	341	75	115.8	358	36	52	68	—	49
3.35	331	73	112.7	344	34	51	67	—	48
3.40	321	71	109.7	332	33	50	67	—	46
3.45	311	68	105.0	321	32	50	67	—	45
3.50	302	66	101.9	310	31	49	66	—	44
3.55	293	64	98.8	299	30	49	66	—	43
3.60	285	63	97.3	290	29	48	65	—	42
3.65	277	61	94.2	282	27	46	64	—	41
3.70	269	59	91.1	274	26	45	64	—	40
3.75	262	58	89.6	267	25	45	63	—	39
3.80	255	56	86.5	260	24	44	63	—	38
3.85	248	55	84.9	253	23	43	62	—	37
3.90	241	53	81.9	246	22	42	62	—	36
3.95	235	51	78.8	240	21	41	61	100	35
4.00	229	50	77.2	234	20	41	61	99	34
4.05	223*	49	75.7	228	19	40	60	98	33
4.10	217	48	74.1	222	18	—	60	97	32
4.15	212	46	71.0	217	17	—	59	96	31
4.20	207	45	69.5	212	16	—	58	95	31
4.25	201	44	68.0	206	15	—	57	94	30
4.30	197	43	66.4	202	13	—	57	93	30
4.35	192	42	64.9	197	12	—	56	92	29
4.40	187	41	63.3	192	10	—	56	91	28
4.45	183	40	61.8	188	9	—	55	90	28
4.50	179	39	60.2	184	8	—	55	89	27
4.55	174	38	58.7	179	7	—	54	88	27
4.60	170	38	58.7	175	6	—	54	87	26
4.65	167	38	58.7	172	4	—	53	86	26
4.70	163	37	57.1	168	3	—	52	84	25
4.75	159	36	55.6	164	2	—	51	83	24
4.80	156	36	55.6	161	1	—	51	82	24
4.85	152	35	54.1	157	—	—	50	81	23
4.90	149	34	52.5	154	—	—	50	80	23
4.95	146	33	51.0	151	—	—	49	79	22
5.00	143	33	51.0	148	—	—	49	78	22
5.05	140	32	49.4	145	—	—	48	76	21
5.10	137	31	47.9	142	—	—	47	75	21
5.15	134	31	47.9	139	—	—	47	74	21
5.20	131	30	46.3	136	—	—	46	73	20
5.25	128	30	46.3	133	—	—	45	72	20
5.30	126	29	44.8	131	—	—	45	71	20
5.35	123	28	43.2	128	—	—	44	69	—
5.40	121	28	43.2	126	—	—	44	68	—
5.45	118	27	41.7	123	—	—	43	67	—
5.50	116	27	41.7	121	—	—	43	65	—
5.55	114	26	40.2	119	—	—	42	64	—
5.60	112	25	38.6	117	—	—	41	63	—
5.65	109	25	38.6	115	—	—	40	62	—
5.70	107	24	37.1	113	—	—	40	60	—
5.75	105	24	37.1	111	—	—	39	59	—
5.80	103	23	35.5	109	—	—	39	57	—
5.85	101	23	35.5	107	—	—	38	55	—
5.90	99	22	34.0	104	—	—	38	54	—
5.95	97	22	34.0	102	—	—	37	53	—
6.00	95	21	32.4	100	—	—	35	51	—

Conversion factors

g/A/h	to	mg/C	× 0.278
hbar	to	N/mm^2	× 10
hbar	to	kgf/mm^2	× 1.0197
hbar	to	tonf/in.^{2}	× 0.64749
hbar	to	lbf/in.^{2}	× 1450.4
J	to	ft lb	× 1.35
J/g	to	cal/g	× 0.239
J/g °C	to	cal/g °C	× 0.239
J/g °C	to	British Thermal Unit/lb °F	× 0.239
kg/m^3	to	lb/in.^{2}	× 3.61 × 10^{-5}
microhm mm	to	microhm/cm/cm^2	× 0.1
N/m^2	to	lbf/in.^{2}	× 1.45 × 10^{-4}
W/m °C	to	cal/m s °C	× 0.24
W/m °C	to	cal/cm/cm^2 s °C	× 0.0024
W/m °C	to	British Thermal Unit/ft/h °F	× 0.58

Metallic Materials Specification Handbook: Fourth Edition

Modern computer technology makes it possible for reference books such as *Metallic Materials Specification Handbook* to be made available as on-line services in addition to the conventional book versions.

It is essential that such a system is planned with needs of its potential users firmly in mind. We hope, therefore, that users of this book who might be interested in access to its information in a computer data base will complete the card below and return it to us. In the UK your reply will need no postage stamp; in the USA please reply to our New York Office: Methuen, Inc., 733 Third Avenue, New York, NY 10017.

When our plans for the next edition of the book are under way, we will consult all those who have expressed an interest in a service of this kind.

Metallic Materials Specification Handbook : Fourth Edition

Please send details to :

Name.. Position ....

Company ....

Address ....

....

Index

1.4580	44N1	1.5120	44A2
1.4581	44N1	1.5121	44A2
1.4583	44N1	1.5122	44A2
1.4584	44N1	1.5151	44P
1.4587	44N1	1.5152	44P
1.4589	44N1	1.5161	44P
1.4704	44E2	1.5213	44J
1.4712	44E1	1.5223	44J
1.4713	44E1	1.5331	44A1
1.4716	44E1	1.5340	44A1
1.4718	44E2	1.5404	44K1
1.4721	44M2	1.5406	44K1
1.4722	44M1	1.5415	44G1
1.4723	44M1	1.5416	44G1
1.4724	44M1	1.5417	44A1
1.4741	44M1	1.5419	44K1
1.4742	44M1	1.5424	44G1
1.4747	44M2	1.5425	44G1
1.4762	44M1	1.5426	44G1
1.4765	44M1	1.5427	44G1
1.4767	44M1	1.5428	44G1
1.4772	44M1	1.5613	44F1
1.4773	44M1	1.5620	44K1
1.4774	44M1	1.5622	44F1
1.4820	44M1	1.5633	44F1
1.4821	44M1	1.5680	44F1
1.4828	44N1	1.5919	44K1
1.4829	44N1	1.5920	44K1
1.4833	44N1	1.5924	44K1
1.4841	44N1	1.5934	44K1
1.4842	44N1	1.6001	44K1
1.4843	44N1	1.6215	44F1
1.4844	44N1	1.6216	44F1
1.4845	44N1	1.6351	44F3
1.4854	44N1	1.6354	44F3
1.4860	44N1	1.6359	44F3
1.4876	27A	1.6511	44K1
1.4878	44N1	1.6513	44K1
1.4903	44E1	1.6582	44K1
1.4905	44M1	1.6590	44K1
1.4920	44M1	1.6592	44K1
1.4922	44M1	1.6604	44K1
1.4924	44M1	1.6761	44K1
1.4934	44M1	1.6903	44N1
1.4935	44M1	1.6905	44N1
1.4943	44N1	1.7015	44E1
1.4944	44N1	1.7033	44E1
1.4954	44N1	1.7034	44E1
1.4974	44N2	1.7035	44E2
1.4984	44N1	1.7071	44E1
1.4986	44N1	1.7083	44E1
1.4994	44M1	1.7131	44E3
1.5034	44A1	1.7147	44E1
1.5035	44A1	1.7176	44E3
1.5038	44A2	1.7205	44K1
1.5053	44E1	1.7207	44K1
1.5063	44A1	1.7214	44K1
1.5064	44A1	1.7218	44K1
1.5066	44A2	1.7220	44A2
1.5067	44A2	1.7225	44K2
1.5074	44A1	1.7228	44K2
1.5083	44A1	1.7242	44K1
1.5086	44A1	1.7251	44K1
1.5098	44A1	1.7254	44K1

1.7258	44K1
1.7273	44K1
1.7281	44K1
1.7283	44K1
1.7305	44E3
1.7307	44K1
1.7324	44K1
1.7334	44K1
1.7335	44K1
1.7337	44K1
1.7345	44K1
1.7346	44K1
1.7350	44K1
1.7356	44K1
1.7361	44K1
1.7362	44K1
1.7373	44K1
1.7380	44K1
1.7384	44K1
1.7394	44K1
1.7513	44K1
1.7561	44K2
1.7701	44K2
1.7704	44K1
1.7707	44K1
1.7708	44K1
1.7709	44K1
1.7710	44K1
1.7733	44K1
1.7766	44K1
1.7779	44K1
1.8070	44K1
1.8154	44K2
1.8159	44K2
1.8212	44K1
1.8401	44E1
1.8403	44E2
1.8404	44E2
1.8405	44E3
1.8425	44E3
1.8504	44E1
1.8506	44E1
1.8507	44K1
1.8514	44K1
1.8519	44K1
1.8544	44K1
1.8550	44K1
1.26365	44K1
1/360	27C
1 A	
see BS range	1A
1A	14G
1 Al 8 V 5 Fe	51C
1 B	
see BS range	1A
1B	44A1
1 C	
see BS range	1A
1C	14A
1 C 27	44M1
1C 36	44M1
1C 342	44M1
1C 901	27C
1CN	44M1

AISI 316	44N1	AISI 662	44N2	AISI 4340	44K2
AISI 316 L	44N1	AISI 663	44N2	AISI 4419	44G1
AISI 316N	44N1	AISI 664	27C	AISI 4422	44G1
AISI 317	44N1	AISI 665	44N2	AISI 4427	44G1
AISI 317 L	44N1	AISI 670	13A	AISI 4520	44G1
AISI 318	44N1	AISI 671	13A	AISI 4608	44K1
AISI 321	44N1	AISI 680	27C	AISI 4615	44K1
AISI 321H	44N1	AISI 681	27C	AISI 4620	44K1
AISI 325	44N1	AISI 682	27C	AISI 4621	44K1
AISI 327	44N1	AISI 683	27C	AISI 4626	44K1
AISI 329	44N1	AISI 684	27C	AISI 4718	44K1
AISI 347	44N1	AISI 685	27C	AISI 4720	44K1
AISI 347 F	44N1	AISI 686	27C	AISI 4815	44K1
AISI 347H	44N1	AISI 687	27C	AISI 4817	44K1
AISI 348	44N1	AISI 688	27C	AISI 4820	44K1
AISI 348H	44N1	AISI 689	27C	AISI 5015	44E1
AISI 403	44M1	AISI 690	27C	AISI 5045	44E2
AISI 405	44M1	AISI 1029	44A2	AISI 5046	44E2
AISI 410	44M1	AISI 1051	44A3	AISI 5117	44E1
AISI 414	44M1	AISI 1052	44A3	AISI 5120	44E1
AISI 416	44M1	AISI 1053	44A3	AISI 5130	44E1
AISI 416 Se	44M1	AISI 1110	44A1	AISI 5132	44E1
AISI 418	44M1	AISI 1116	44A1	AISI 5135	44E1
AISI 420	44M1	AISI 1139	44A2	AISI 5140	44E2
AISI 420F	44M1	AISI 1211	44A1	AISI 5145	44E2
AISI 430	44M1	AISI 1212	44A1	AISI 5147	44E2
AISI 430 F	44M1	AISI 1213	44A1	AISI 5150	44E2
AISI 430 F Se	44M1	AISI 1215	44A1	AISI 5152	44E2
AISI 431	44M1	AISI 1330	44A2	AISI 5155	44E2
AISI 434	44M1	AISI 1335	44A2	AISI 5160	44E2
AISI 440 A	44M1	AISI 1340	44A2	AISI 6117	44K1
AISI 440 B	44M1	AISI 1345	44A2	AISI 6118	44K1
AISI 440 C	44M2	AISI 2317	44F1	AISI 6120	44K1
AISI 440 F	44M2	AISI 2515	44F1	AISI 6145	44K2
AISI 442	44M1	AISI 3135	44K1	AISI 6150	44K2
AISI 446	44M1	AISI 3140	44K2	AISI 8615	44K1
AISI 501	44K1	AISI 3310		AISI 8617	44K1
AISI 502	44K1	see 3310		AISI 8620	44K1
AISI 601	44K2	AISI 4012	44G1	AISI 8622	44K1
AISI 602	44K1	AISI 4023	44G1	AISI 8625	44K1
AISI 603	44K1	AISI 4024	44G1	AISI 8627	44K1
AISI 604	44K1	AISI 4027	44G1	AISI 8630	44K1
AISI 610	44K2	AISI 4028	44G1	AISI 8632	44K1
AISI 611	44K3	AISI 4032	44G1	AISI 8635	44K1
AISI 612	44K3	AISI 4037	44G1	AISI 8637	44K1
AISI 613	44K3	AISI 4042	44G2	AISI 8640	44K2
AISI 614	44M1	AISI 4047	44G2	AISI 8641	44K2
AISI 615	44M1	AISI 4053	44G2	AISI 8642	44K2
AISI 616	44M1	AISI 4063	44G2	AISI 8645	44K2
AISI 617	44M2	AISI 4068	44G2	AISI 8647	44K2
AISI 618	44M2	AISI 4118	44K1	AISI 8650	44K2
AISI 619	44M1	AISI 4130	44K1	AISI 8653	44K2
AISI 630	44N2	AISI 4137	44K1	AISI 8655	44K2
AISI 631	44N2	AISI 4140	44K2	AISI 8660	44K2
AISI 632	44N2	AISI 4142	44K2	AISI 8715	44K1
AISI 633	44N2	AISI 4145	44K2	AISI 8717	44K1
AISI 634	44N2	AISI 4147	44K2	AISI 8719	44K1
AISI 635	44N2	AISI 4150	44K2	AISI 8720	44K1
AISI 650	44N2	AISI 4161	44K2	AISI 8735	44K1
AISI 651	44N2	AISI 4317	44K1	AISI 8740	44K2
AISI 652	44N2	AISI 4320	44K1	AISI 8742	44K2
AISI 653	44N2	AISI 4330	44K1	AISI 8745	44K2
AISI 660	44N2	AISI 4335	44K1	AISI 8750	44K2
AISI 661	44L	AISI 4337	44K1	AISI 8822	44K1

Term	Code	Term	Code	Term	Code
SPEAR D9	44K1	SPRABRONZE P	14K	SSC	44K2
SPEAR D12	44M2	SPRABRONZE TM	14C	SSG	44K1
SPEAR D 13	44M2	SPRASTEEL 10	44A1	SSM	44A3
SPEAR D 14	44M2	SPRASTEEL 25	44A1	S Sn Bz 6	
SPEAR D16	44M2	SPRASTEEL 80	44A4	see DIN range	14K
SPEAR DOUBLE		SPRASTEEL LS	44K1	S Sn Bz 12	
CENTURY	44L	SPS	44K2	see DIN range	14K
SPEAR DX	44K1	SPS 245	44K2	S So Ms	14C
SPEAR EXM	44K2	SPV 32		SSS 100	44K1
SPEAR GFS	44E3	see JIS G3115		SSS 100A	44K1
SPEAR LEAPFROG	44K2	SPV 32	44A1	SSS 110	44K1
SPEAR MD4	44A4	SPV 36		SSS 321	44K1
SPEAR MERMAID	44K3	see JIS G3115		SSS 360	44K1
SPEAR No. 1	44H	SPV 36	44A1	S St 2	44A1
SPEAR No.2	44A4	SPV 46		S St 3	44A1
SPEAR No.3	44A4	see JIS G3115		S St 4	44A1
SPEAR PS	44K3	SPV 46	44A1	St 00	
SPEAR SUPERIOR	44L	SPV 50		see DIN 1629	44A1
SPEAR TRIPLE		see JIS G3115		St 0	
MERMAID	44L	SPV 50	44A1	see DIN range	44A1
SPECIAL	44A3	SPW	44N1	St 1	
SPECIAL ADVANCE	14E	SPW	44M1	see DIN range	44A1
SPECIAL ARDHO	44K2	SQV 1A		ST 2	
SPECIAL BB	44K1	see JIS G3120		see GOST 380	44A1
SPECIAL CONQUEROR		SQV 1A	44G1	St 2	
VANADIUM	44J	SQV 1B		see DIN range	44A1
SPECIAL ECHO	44K3	see JIS G3120		ST 3	
SPECIAL HDS	44K1	SQV 1B	44G1	see GOST 380	44A1
SPECIAL HS	44K3	SQV 2A		St 3	
SPECIAL HW No.3	44K2	see JIS G3120		see DIN range	44A1
SPECIAL K	44M2	SQV 2A	44K1	St 4	
SPECIAL V	44J	SQV 2B		see DIN range	44A1
SPECIAL WOLFRAM	44K2	see JIS G3120		St 10	44A1
SPECIFICATION 55	44K2	SQV 2B	44K2	St 10/01	44A1
SPECULUM	14K	SQV 3A		St 10/02	44A1
SPEDEX	14F	see JIS G3120		St 10/03	44A1
SPEEDICUT 14	44K2	SQV 3A	44K1	St 33	
SPEEDICUT LEDA	44L	SQV 3B		see DIN 1626	44A1
SPEEDICUT MAX 22	44K3	see JIS G3120		St 34	
SPEEDICUT MAXIMUM		SQV 3B	44K1	see DIN range	44A1
18	44K3	SR 200	5.1	St 34/2	44A1
SPEEDICUT		SR 1855	44E3	St 34/3	44A1
SUPERLEDA	44L	S R ALLOY	44N1	St 35	44A1
SPEEDICUT VANLEDA	44L	SRO	44K2	St 35/4	44A1
SPEED STAR	44K3	SRO 2	44M1	St 35/8	
SPELTAFAST	44A1	SS	14C	see DIN range	44A1
SPENARD	44K1	S S1	44M1	St 35/13 K	44A1
SPF	20B	SS 1	14C	St 37	
SPF 600	20B	S S2	44M1	see DIN 1626	44A1
SPINNING BRASS	14B	S S 3M	44N1	St 37/2	44A1
SPK	44K2	S S 3MM	44N1	St 37/3	44A1
SPKD	44K2	S S 4	44N1	St 42	44A1
SPKS	44E2	S S6	44M1	St 42/2	44A1
SPM	44K3	SS 15	14F	St 42/3	44A1
SPM 1	42.1	S S22	44M1	St 43.7	
SPM 2	42.1	S S 24	44N1	see DIN range	44A1
SPOOLARC GRADE C	14K	S S31	44M1	St 45	44A1
SPRA BABBITT	51C	S S 33	44N1	St 45/4	44A1
SPRA BABBITT	50A	SS 41		St 45/8	
SPRABABBITT L	21.1	see JIS G3101		see DIN range	44A1
SPRABOND	26A	SS 41	44A1	St 47.7	
SPRABRASSY	14B	S S 43	44N1	see DIN range	44A1
SPRABRONZE AA	14G	S S 53	44N1	St 50	44A2
SPRABRONZE C	14B	SS 100	44K1	St 50/2	44A2

W 1	44A4	W 260	1G	WEL-TEN 60	44K1
W 2		W 310		WEL-TEN 60H	44K1
see BS W 2	44A2	see SAE W 310	44J	WEL-TEN 62	44K1
W 2	44J	W 367	52.1	WEL-TEN 80	44K1
W 3	44E3	W 408	52.1	WEL-TEN 100W	44K1
W 3	1B	W 410L	52.1	WELCH'S ALLOY	50A
W 4	44E3	W 410M	52.1	WELCON 2H	44A1
W 4	1D	W 545	44N2	WELCON 2H.S	44K1
W 5	44E3	WA	20B	WELCON 2H.U	44K1
W 5	27F	WAGNER'S ALLOY	50A	WELCON 50	44A1
W 5	1D	WAI-MET 50	13A	WELDANKA A	44N1
W 6	44A3	W Al 8	1A	WELDANKA B	44N1
W 6	44A1	WALLEX	44M2	WELDANKA CB	44N1
W 6	27F	WALLEX 1	13A	WELSBACH	11.1
W 6	1D	WALLEX 1NE	13A	WESSELS ALLOY	14F
W7	44K3	WALLEX 4	13A	W E WATSON BRAND	21.1
W 7	44G1	WALLEX 6	13A	WF 11	13A
W 7	27F	WALLEX 6NE	13A	WF 31	13A
W 8		WALLEX 7NE	13A	W Fe 18	20A
see BS W 3	44A2	WALLEX 12	13A	WG	44K3
W 9		WALLEX 12NE	13A	WH	52.1
see BS W 9	44A2	WALLEX 50	13A	WH	20B
W 9	27F	WALLEX 55	52.1	WHC	44A2
W 9	1E	WALLEX 505	52.1	WHEEL BRASS	14B
W 9B	44K1	WALMANG 3	44P	WHITE ARSENIC	3.1
W 10		WAMATO METAL	50A	WHITE BENEDICT	
see BS W 10	44N1	WARNE'S METAL	50A	METAL	14F
W 10	1E	WAS	14B	WHITE GOLD SOLDER	17.1
W 010	1D	WASPALOY	27C	WHITE PINE LAKE	
W 11		WASPALOY MOD	27C	COPPER	14A
see BS W 11	44N1	WATERDIE EXTRA	44E3	WI 52	13A
W 11	1F	WATERDIE STANDARD	44E3	WILMIL	1C
W 011	1D	WAW	44K2	WILMIL M	1K
W 12	44M1	WB	20B	WINNS BRONZE	14F
W 12	1F	w Ba 21	4.1	WIPTAM	13A
W 012	1D	WBC	20B	WK 16	33.1
W 16	1G	WC 10	44K1	WKE	44L
W 17	1F	WC 18	44K1	WKE	44K3
W 18	52.1	WC 23	44K1	WKE 4	44L
W 18	1F	WC 28	44K1	WKE 44	44L
W 18/8	44N1	WC 33	44K1	WKE 45	44L
W 19/9	44N1	WC 38	44K2	WKE 46	44L
W 20	1E	WC 43	44K2	WKE Extra	44L
W 21	44A3	WC 48	44K2	WL 1	44K2
W 22/4	20B	WC 53	44K2	WL 2	44K1
W 24/8	20B	WC 58	44K2	W Li 16	22.1
W 26	1G	WC 63	44K2	W Li 20	22.1
W 30	1E	WC 181	44K1	W Li 96	22.1
W 61	44A1	WC 182	44K1	W Li 97	22.1
W 108		W Ca 17	10.1	WLM 77	44K1
see SAE W 108	44A4	WCC	2.1	WLM 1015	44E3
W 109		WCD	44M2	WLM 1020	44L
see SAE W 109	44A4	WCD	44K1	WLM 1030	44K1
W 110		WCD 2	44K1	WLM 1175	44K3
see SAE W 110	44A4	W Co 23	13.1	WLM 1214	44K2
W 112		WCPS	44K2	WLM 1261	44K3
see SAE W 112	44A4	WCR 100	44M2	WLM 1530	44K1
W 150	1F	WCR 100	44M1	WLM 1841	44K3
W 152	1F	WE	44L	WLM 2014	44M2
W 160	1G	WEAREX	44K3	WLM 2014 SPECIAL	44M2
W 209		WEARTRODE 84.58	44M1	WLM 2241	44K3
see SAE W 209	44J	WEL-MONIX	44K1	WLM 3033	44K1
W 210		WEL-TEN 50	44A1	WLM 3042	44K1
see SAE W 210	44J	WEL-TEN 55	44A1	WLM 3141	44M2